D1129906

MOLECULAR CELL BIOLOGY

MOLECULAR CELL BIOLOGY

Revised Printing with Expanded Index

JAMES DARNELL

Vincent Astor Professor
Rockefeller University

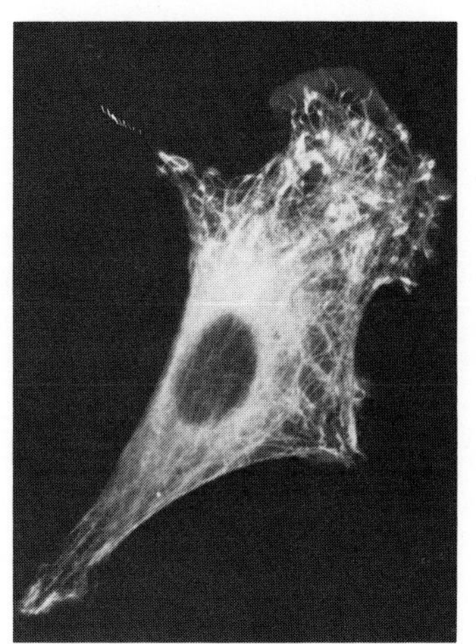

HARVEY LODISH

Member of the Whitehead Institute for Biomedical Research
Professor of Biology, Massachusetts Institute of Technology

DAVID BALTIMORE

Director of the Whitehead Institute for Biomedical Research
Professor of Biology, Massachusetts Institute of Technology

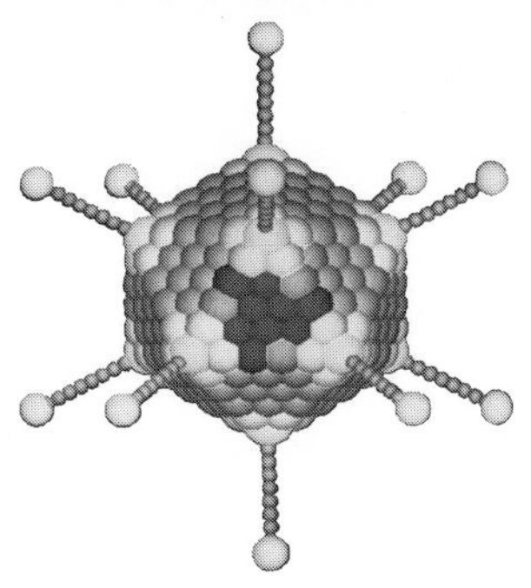

SCIENTIFIC
AMERICAN
BOOKS

Front cover photograph courtesy of J. Victor Small

Back cover photograph courtesy of Richard Feldman

Library of Congress Cataloging-in-Publication Data

Darnell, James E.
Molecular cell biology.

Includes bibliographies and index.
1. Cytology. 2. Molecular biology. I. Lodish,
Harvey F. II. Baltimore, David. III. Title.
[DNLM: 1. Cells. 2. Molecular Biology. QH 581.2 D223m]
QH581.2.D37 1986 574.87′6042 86-1881
ISBN 0-7167-1448-5
ISBN 0-7167-6001-0 (international student ed.)

Printed in the United States of America

Scientific American Books is a subsidiary of Scientific American, Inc.
Distributed by W. H. Freeman and Company, 41 Madison Avenue,
New York, New York 10010

4 5 6 7 8 9 0 KP 4 3 2 1 0 8 9 8 7

To our families:
Jane, Chris, Bobby, and Jon
Pam, Heidi, Martin, and Stephanie
Alice and Teak

Preface

BIOLOGY today is scarcely recognizable as the subject that biologists knew and taught 10 years ago. A decade ago, gene structure and function in the simple cells of bacteria were known in considerable detail. But now we also know that a different set of molecular rules governs gene organization and expression in all eukaryotic cells, including those of humans. We are learning about the genes and regulatory proteins that control not only single metabolic steps but complicated developmental events such as the formation of a limb, a wing, or an eye. In addition to these advances in understanding the genetic machinery and its regulation, great progress has been made in the study of the structure and function of cell organelles and of specialized cell proteins. To comprehend fully what has been learned requires a reformulation of a body of related information formerly classified under the separate headings of genetics, biochemistry, and cell biology. *Molecular Cell Biology* aims to present the essential elements of this new biology.

Traditionally, the sciences of genetics, biochemistry, and cell biology—the three areas in which the greatest progress has been made in the last 25 to 30 years—used different experimental approaches and often different experimental material. Classical geneticists sought mutations in specific genes to begin identifying the gene products and characterizing their physiological functions. Biochemists tried to understand the actions of proteins, especially enzymes, from their sequences and three-dimensional structures. Cell biologists attempted to discover how specific proteins took part in the construction and operation of specialized cell structures. These subjects were taught as three courses, albeit with varying degrees of overlap.

A group of techniques collectively referred to as molecular genetics is mainly responsible for unifying the three disciplines. Not only do these techniques provide a powerful analytic force, but they also serve to unify all experimental biology in its language and concerns. With the tools of molecular genetics, genes for all types of proteins—enzymes, structural proteins, regulatory proteins—can be purified, sequenced, changed at will, reintroduced into individual cells of all kinds (even into the germ lines of organisms) and expressed there as proteins. Most of experimental biology now relies heavily on molecular genetics.

In addition to the outstanding advances in molecular genetics, comparable advances have been made in culturing the cells of vertebrates, invertebrates, and plants, including the cells that produce various individual monoclonal antibodies. The use of cell cultures has greatly unified and simplified experimental designs. Finally, very sophisticated instrumentation has become available. Powerful electron microscopes and advanced techniques in electron microscopy have greatly improved our understanding of cell substructure. Modern computers have arrived in time to store—and then to compare—rapidly accumulating information (such as protein and nucleic acid sequences), as well as to present graphic displays of molecular structures. Of equal importance, computers rapidly complete elaborate calculations so that x-ray crystallographic analysis (or other kinds of image analysis) can be performed in days instead of months or years.

Those who teach biology at the undergraduate or graduate level and in medical schools can convey this comprehensive and integrated experimental approach in the classroom only when they have access to appropriate teaching materials. Our book is intended to fill the need for such materials: We wrote it to solve our own problems as teachers. It was our purpose to teach a one-year course that integrates molecular biology with biochemistry, cell biology, and genetics and that applies this coherent insight to such fascinating problems as development, immunology, and cancer. We hope that the availability of this material in a unified form will stimulate the teaching of molecular cell biology as an integral subject and that such integrated courses will be offered to students as early as possible in their undergraduate education. Only then will students be truly able to grasp the findings of the new biology and its relation to the specialized areas of cell biology, genetics, and biochemistry.

We have aimed to provide a college textbook that is no more difficult than the basic textbooks encountered by undergraduate physics and chemistry students in their respective programs of study. That there will be complaints about the scope and depth of a textbook this large seems inevitable. But in addition to dividing the book into parts, we have clearly identified the parts of the chapters themselves, by means of descriptive subheadings. This organization will enable students and teachers to be selective in their reading. We recognize that some teachers of one-semester courses may wish to continue teaching molecular biology and cell biology as separate courses. The book is organized so that, in such situations, emphasis can easily be given to either the gene or the cell.

Whichever path students and teachers choose to follow, we believe that the focus of teaching an experimental science such as biology should be the experiments themselves. We have devoted much space to presentations of experiments. Scientists make advances by phrasing the unknown as an experimental question, designing an experiment to answer the question, and assembling the experimental results to produce a coherent answer. Students who see how biological progress is intimately connected to experimentation will be more likely to keep pace with the progress of biology in the future and perhaps even to contribute to it.

Most students today begin an undergraduate major in biology with an introductory course in general biology and at least one course in chemistry. Such students need a clear explanation of how basic biochemical and biological principles relate to the central developing areas of molecular cell biology. This focus is provided in Part I of the book. Chapter 1 presents some of the key ideas in the history of modern biology. Chapters 2 and 3 deal with the fundamentals of biochemical structure, function, and energetics. Chapter 4 presents cellular polymer synthesis, including the flow of genetic information from DNA to RNA to protein. Then, Chapter 5 covers the basic principles of cell structure and function. Students who are well prepared probably will not need to restudy most of the material in the first five Chapters. All students should read Chapters 6 and 7, for these present the new (now standard) cell and molecular techniques that are changing the face of experimental biology. These chapters will enable students to follow the experiments described in the remainder of the book.

With the foundations established, we go on to describe research results and conclusions drawn from work in molecular and cell biology.

In Part II (Chapters 8 and 9) we stress RNA biosynthesis and gene expression before we discuss gene structure and chromosomal organization (Chapters 10 and 11). We use this order of presentation because the diversity of organisms, while surely encoded in DNA, is brought to reality by the programmed expression of genes during the differentiations of each organism. Thus, we concentrate on genes as transcription units and deal extensively with how genes are controlled (Chapter 12) as well as how they are replicated and repaired (Chapter 13).

In Part III we turn to the ways in which proteins—the ultimate gene products—work together to make a living cell. Instructing the new biology student in the names and shapes of the parts of a cell can by itself be a satisfying experience because of the wealth of pictorial detail available. But it is the integration of the structural detail with molecular function that gives meaning to cell activity, and many aspects of this integration have been brought to an advanced level within the last few years. We show how many fundamental properties of cells are explained by properties of specific membrane proteins (Chapter 14). These activities include the regulated entry and exit of molecules into and out of cells (Chapter 15), transmission of signals between cells (Chapter 16), the characteristics of nerve cells and electrical properties of membranes (Chapter 17), cellular movements and cell shape (Chapters 18, 19), and the generation and use of ATP (Chapter 20). We consider how a cell is assembled from its component nucleic acids, proteins, carbohydrates, and lipids, and, in particular, how proteins are "targeted" to their appropriate destination in the cell (Chapter 21). Special attention is given to organelles and cells that illustrate phenomena of general significance in many organisms—the contraction of muscle (Chapter 18), beating of flagella (Chapter 19), regulation of carbohydrate metabolism (Chapter 20), and the propagation of nerve impulses (Chapter 17).

Advances in molecular cell biology have led to significant discoveries about such changes in cell function and behavior as the development of higher organisms, the immunologic response, cancer, and cell evolution. These are discussed in Part IV of the book. Chapter 22 discusses the classical and modern experimental work that has led to the view that development is an organized train of cellular changes. Although much remains to be learned, development doubtless depends upon the expression of "the right gene in the right place at the right time." Chapter 23 makes the point that cancer, which is caused by alterations of normal cell growth control and tissue architecture, likely arises as a disorganization of the structure or function of relatively few genes. Chapter 24 is devoted to the cells of the immune system because they employ a very special mechanism of change during their differentiation: The DNA encoding antibodies is rearranged to form a virtually unlimited and constantly changing set of proteins that protects animals from the ever present threat of microbial infection. Finally all the variations that we see and study in cells are part of a continuum of ongoing evolutionary changes that began about 3.5 billion years ago. Molecular cell biology has taught us to look at the earliest stages of evolution in a new way, and so we close the book with Chapter 25, which examines those new insights.

Over the past seven years, from the earliest planning stages to completion, this book has occupied a significant portion of our energies. During this period we have often exchanged views and read each others' work, so that our book is truly a joint responsibility and a joint product.

We have called upon our friends and colleagues very liberally and cannot possibly mention all of those who gave us useful advice at various stages of development of the manuscript. We specifically thank:

Wayne M. Becker, University of Wisconsin, Madison
Stephen Benson, California State University, Hayward
Sherman Beychok, Columbia University
David Bloch, University of Texas, Austin
David Clayton, Rockefeller University
Francis Crick, Salk Institute
Robert Davenport, Massachusetts Institute of Technology
David De Rosier, Brandeis University
Ford Doolittle, Dalhousie University
Ernest DuBrul, University of Toledo
Charles Emerson, University of Virginia
Richard Firtel, University of California, San Diego
Ursula Goodenough, Washington University
Jeffrey Flier, Harvard Medical School
Barry Gumbiner, University of California Medical Center at San Francisco
James Hageman, New Mexico State University
Maija Hinkle, Cornell University
Johns Hopkins, Washington University
Richard Hynes, MIT
Warren Jelinek, New York University Medical School
David Kabat, Oregon Health Sciences University
Flora Katz, University of California Medical Center at San Francisco
Thomas Kreis, European Molecular Biology Laboratory, Heidelberg
Scott Landfear, Harvard School of Public Health
John Lis, Cornell University
Maurice Liss, Boston College
Mary Nijhout, Duke University
Daniel O'Kane, University of Pennsylvania
Nan Orme-Johnson, Tufts University School of Medicine
Leslie Orgel, Salk Institute
Larry Puckett, Immuno Nuclear Corporation
Charles Richardson, Harvard Medical School
Karin Rodland, Reed College
Alan Schwartz, Dana Farber Cancer Center
Richard D. Simoni, Stanford University
Roger Sloboda, Dartmouth College
Frank Solomon, MIT
David R. Soll, University of Iowa
Pamela Sperry, California State Polytechnic Institute at Pomona
Rocky Tuan, University of Pennsylvania
Joseph Viles, Iowa State University
Michael Young, Rockefeller University
Norton Zinder, Rockefeller University

for reading one or more chapters. In addition, many scientists have generously supplied drawings and photo-

acknowledgement in the appropriate places in the book.

Our editorial colleagues at Scientific American Books have provided the very substantial resources necessary to complete such a large project and given us their most thoughtful, careful advice and professional skills in unlimited amounts. When our spirits or energies sagged, Patty Mittelstadt, Andrew Kudlacik, Janet Wagner, Donna McIvor, Sally Immerman, James Funston, Betsy Galbraith, Faye Webern, and in particular, Neil Patterson and Linda Chaput could always be counted on to get us over a rough spot. In many, many places in the book not only have outright mistakes been removed but a clearer sentence, a more orderly line of thinking, a better picture or drawing exists because of their tireless efforts. The errors and infelicities that remain we must acknowledge are our own. We are also thankful for the talented work of the illustrators, Shirley Baty and George Kelvin, and the lively cooperation of the production team: Mike Suh, Margo Dittmer, Melanie Neilson, and Ellen Cash. Finally we wish to give a very special vote of thanks to our secretaries—Miriam Boucher, Lois Cousseau, Audrey English, Ginger Pierce, and Marilyn Smith—for their endless patience in dealing with the many necessary drafts of the book.

To our friends and associates in our laboratories and, most of all, to our families we apologize for the long absences and the vacant stares that frequently came in the wake of long hours of working on the book. If our efforts are successful in helping to unify the teaching of molecular cell biology we will be deeply grateful.

Jim Darnell
Harvey Lodish
David Baltimore

April 1986

Contents in Brief

Contents

1

The History of Molecular Cell Biology

The aim of modern biology is to interpret the properties of the organism by the structure of its constituent molecules.

FRANÇOIS JACOB,
The Logic of Life
Translated by Betty E. Spillman
Copyright 1973 by Pantheon Books, a Division of Random House, Inc.

We have complete confidence that further research of the intensity given to genetics will eventually provide man with the ability to describe with completeness the essential features that constitute life.

JAMES D. WATSON,
The Molecular Biology of the Gene, 1973, 3d ed., The Benjamin/Cummings Publishing Company

MODERN biology aims at no less than a full understanding of cell function in molecular terms. The preceding statements by two of the architects of modern biology presage the determination of today's biologists to carry the spectacular successes of the 1950s and 1960s—the discoveries of the structure of DNA, the roles of RNA in protein synthesis, the genetic code, and the nature of gene regulation in bacteria—into studies of the cells and organs of higher organisms, including human beings. The molecular approach to biology has already affected and will increasingly affect every traditional biological discipline—histology, cytology, anatomy, embryology, genetics, physiology, and evolution. Most of these areas of biology now deal with *eukaryotic* cells (cells with a nucleus) and organisms constructed of eukaryotic cells. However, in the early phases of the molecular approach, biologists relied on studies of *prokaryotic* cells (cells without a nucleus)—particularly studies of a single bacterial type, *Escherichia coli* (*E. coli*). It therefore became the custom in the 1960s to introduce the biology student to the molecular events in cell function mainly through descriptions of these bacterial experiments. *E. coli* and the viruses that infect it became the core subject matter of molecular biology.

The use of bacterial models to illustrate the most elementary and general principles by which all cells function remains logical and sound. But many of the characteristics that make the cells from higher organisms different from bacteria are now being revealed at a molecular level. A molecular cell biology that embraces all types of cells is upon us. Many of the general principles of eukaryotic cell structure and function, and the proteins responsible for this structure and function, have no counterparts in bacterial cells. Not only do the cells of higher organisms contain proteins and cellular structures that are not found in bacteria, but there are many differences in the organization and expression of their genes.

If a molecular description of cell function is to serve as a foundation for the study of cytology, embryology, physiology, and evolution—again, subjects concerned mainly with higher organisms built of eukaryotic cells—then a description of molecular cell biology should feature the current knowledge about eukaryotic cells, which is what this book aims to do.

This introductory chapter briefly traces the historical background for the discoveries that illuminated the chemical nature of the gene—the elementary unit of heredity—and the discovery of the general principles of eukaryotic cell structure. Much of the excitement in today's molecular cell biology has arisen at the interface of these two formerly separate worlds of biology. No longer does the microscopist simply display the wonderful internal structures of cells. Biochemists have purified and identified the protein components of these structures and have begun to work out their functions. Molecular biologists using the exciting new techniques of molecular genetics have started to isolate the genes responsible for many important elements of cell structure and function. The merging of these related studies, ranging from genetics to protein chemistry, has created a new discipline. The more comprehensive approach constitutes molecular cell biology.

Figure 1-1 Charles Darwin (1809–1882), about 5 years before publication of the *Origin of Species* (1859). *Courtesy of G. E. Hutchinson.*

Evolution and the Cell Theory

Three central questions about life and living things have recurred since ancient times: How did life originate? Why does "like beget like"? How does an individual plant or animal develop from a fertilized egg or seed? Until the nineteenth century, these questions were addressed primarily through religion or philosophy, or both. In the last half of the nineteenth century, there arose two theories—the *theory of evolution* and the *cell theory*—that catalyzed the conversion of biology from an observational pastime into an active experimental science.

The Theory of Evolution Arises from Naturalistic Studies

In their theories of evolution, Charles Darwin and Alfred Wallace made the first modern scientific responses to the three central questions listed above. Through charting the geographic distributions of animals and plants and observing the anatomic similarities and differences among closely related groups of organisms, Darwin and Wallace recognized the inconstancy of the biological world; when examined over long periods of time, the extant species changed. They hypothesized that certain forces of the environment—changing land masses, fluctuations in local temperature and rainfall, and long-term climatic trends—have acted as agents of "natural selection" on succeeding generations of organisms, in which heritable changes have occurred. In such a selective environment, new species can continuously arise and old species that are no longer suited to the environment die out. To quote from Darwin's *Origin of Species* (1859):

> As many more individuals of each species are born than can possibly survive, and as consequently there is a frequently recurring struggle for existence, it follows that any being, if it vary in any manner profitable to itself, under the complex and sometimes varying conditions of life, will have a better chance of survival and thus be *naturally selected* (italics ours). From the strong principle of inheritance, any selected variety will tend to propagate its new and modified form.

The Cell Theory Comes to Prominence in the Late Nineteenth Century

In the last half of the seventeenth century, the Dutch inventor Anton van Leeuwenhoek devised the first light microscope. It revealed to him a bewildering array of particles too small to be visible to the naked eye. These structures, which Leeuwenhoek called "animalcules," were later recognized as single-celled organisms. (Some of Leeuwenhoek's original samples, which still contain algal and other cells, have recently been recovered: see Figure 1-2.) At about the same time, Robert Hooke observed microscopic units that make up cork (a dead tissue). He termed the units "cells" (Figure 1-3). However, not until nearly two centuries later was the living cell recognized as the basic unit of life. When improved microscopes, techniques for preserving tissues, and tools for slicing them into thin sections became available early in the nineteenth century, investigators observed not only that tissues were composed of units or cells, but also that the cells could *divide*. It began to appear that each individual cell possessed life. In 1847 Theodor Schwann described the construction of tissues as follows:

> The elementary parts of all tissues are formed of cells in an analogous, though very diversified, manner, so that it may be asserted that there is one universal principle of development for the elementary parts of organisms, however different, and that this principle is the formation of cells.

This formulation was elaborated by Matthias Jakob Schleiden, who studied plants:

> Each cell leads a double life, one independent pertaining to its development, the other an intermediary, since it has become an integrated part of a plant.

A few years later, in 1858, the German pathologist Rudolph Virchow articulated what soon became the accepted form of the cell theory:

> Every animal appears as a sum of vital units, each of which bears in itself the complete characteristics of life. . . .

Up to the time of Schleiden and Schwann, many philosophers and scientists believed in *vitalism*. According to the vitalists, no single part of an organism was alive; instead, properties of living matter were somehow shared by the whole organism. Further, it was thought that a primitive form of protoplasm, a "primitive blastema," was the germinal material of most organisms. The cell theory greatly weakened these beliefs with the proposition that organisms could arise from the growth of constituent cells. One implication of the cell theory was most important: if individual cells could grow and divide they were proper subjects for the study of living organisms.

Before Louis Pasteur, single-celled organisms of the sort Leeuwenhoek had seen were thought to be capable of spontaneous generation and were not considered to have an inheritance of their own. But Pasteur showed that microbes would grow in a culture medium only if a few microbes were seeded into the medium: these were classic experiments that weighed heavily against the concept of the spontaneous generation of life in any form. By the end of the nineteenth century, the cell theory was widely accepted and the foundations of modern biology were in place.

(a)

(b)

Figure 1-2 (a) Examples of the dried specimens prepared by Anton van Leeuwenhoek and sent to the Royal Society of London between 1674 and 1687. The specimens were discovered by Brian J. Ford in 1981. (b) A scanning electron micrograph of one of Leeuwenhoek's specimens reveals some of the items that Leeuwenhoek must have seen in his samples of brackish water: for example, algal cells (Al) and diatom chains (D). [See *Nature*, 1981, **292**:407.] *Courtesy of Brian J. Ford.*

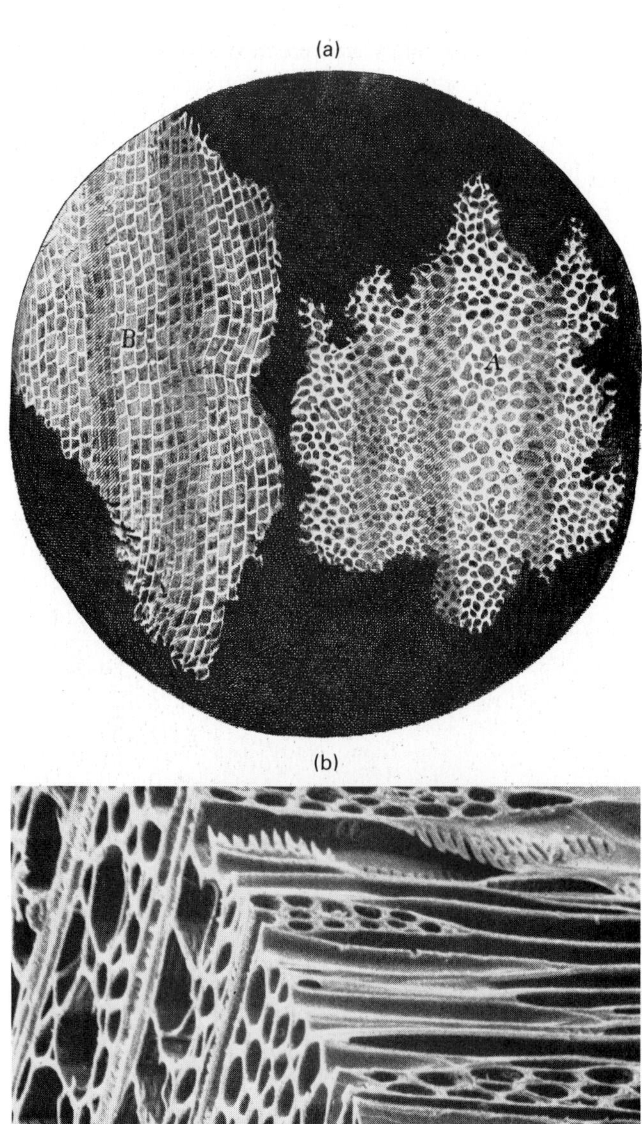

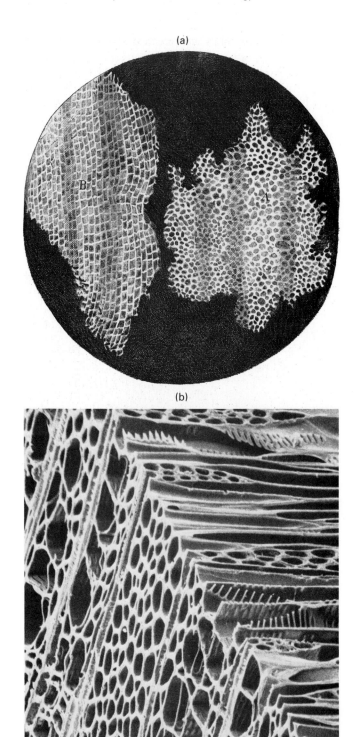

Figure 1-3 (a) In 1665, Robert Hooke drew the cut surface of cork that he observed through a light microscope; he called the spaces in the pattern "cells." (b) A scanning electron micrograph of a cut surface of wood. *Photograph (a) courtesy of Chapin Library, Williams College. Photograph (b) courtesy of B. A. Meylan.*

Classical Biochemistry and Genetics

The theory of evolution combined with the cell theory provided the intellectual framework on which biology developed as an experimental science. Two major branches of the new experimental science grew independently of one another for many years: (1) the isolation and chemical characterization of cell substances, or *biochemistry;* and (2) the study of the inheritance of characteristics by whole animals and plants, or *genetics.*

Biochemistry Begins with the Demonstration That Chemical Transformations Take Place in Cell Extracts

Early in the nineteenth century it was discovered that a major constituent of extracts of both plant and animal cells was a complicated material that made a fibrous precipitate when the extracts were heated or mixed with acid. The precipitate contained carbon, hydrogen, oxygen, and nitrogen in approximately equal amounts. G. J. Mulder concluded in 1838 that this fibrous material was

> without doubt the most important of the known components of living matter, and it would appear that without it life would not be possible. This substance has been named protein.

The question of how protein and other cellular compounds were formed became a controversial one. Did the laws of chemistry apply to cells? One important attack on these problems was to try to synthesize organic compounds in the laboratory. Friedrich Wöhler, a German chemist, was able to form urea (a substance present in the blood and excreted in urine) and oxalic acid (a prominent constituent of spinach and many other vegetables) from simpler organic substances. Marcelin Berthelot succeeded in making acetylene, a simple hydrocarbon, from carbon and hydrogen, which proved that the most widely occurring linkage in organic matter could be made in the laboratory. Finally, in 1897, Eduard Buchner showed that organic chemical transformation could be performed by cell extracts. He ground up yeast cells, carefully filtered out any remaining living cells, and showed that glucose could be converted into ethyl alcohol by the cell extract, just as occurs in the production of wine by yeast cells. The central belief of the enlightened chemists and cell physiologists of the time was expressed by Claude Bernard in 1872:

> Just as physics and chemistry discover the mineral components of compound bodies of experimental investigation, so to comprehend the phenomena of life, that are so complex, it is necessary to go deep into the organism and to analyze the organs and tissues in order to reach the organic components.

The science of biochemistry had at its start the dual goals of chemical analysis of cell constituents, on the one hand, and analysis of the chemical reactions carried out by these components, on the other. An important aspect of early biochemistry was the research on the nature of the chemical units that make up protein. By 1900, 16 of the 20 standard amino acids that serve as the building blocks of proteins were known (although threonine, the last to be discovered, was not isolated until 1935). And by 1900 Emil Fischer had proposed the correct mechanism for the formation of the chemical links in protein—the peptide bonds between adjacent amino acids. Fischer deduced that these bonds are created by the elimination of water when the α-amino group of one amino acid is joined to the carboxyl group of its neighbor.

The other major cell constituents—lipids, carbohydrates, and nucleic acids—were recognized and partially purified from various sources in the last half of the nineteenth century. For example, by 1871 Friedrich Miescher, in his studies of the constituents of cells, had isolated what must have been primarily deoxyribonucleic acid (DNA) from the nuclei of dead white blood cells. But the groundwork had not yet been laid for the linkage of a specific cellular component to heredity. Almost 50 years passed before biochemists resumed making investigations that would play an important role in understanding heredity.

Figure 1-4 Gregor Mendel (1822–1884), the father of genetics. *Courtesy of the Moravian Museum, Brno.*

Classical Genetics Begins with the Controlled Breeding Studies of Gregor Mendel

The discovery of the basic rules of heredity by the Austrian monk Gregor Mendel is an oft-repeated tale. However, it is hard to overemphasize the methodological, as well as the conceptual, leaps that Mendel made. First, he chose a simple organism, the garden pea, which could be easily grown and fertilized. The flowers normally self-pollinate, but they can be opened up and artificially pollinated without damaging them or their ability to form seed. Thus Mendel was able to make experimental genetic crosses by fertilizing the ovules of one strain with the pollen from another. Second, he was careful to select a particular variety of garden pea (*Pisum sativum*) that had been purified by repeated self-pollination so that each pure strain always "bred true"; that is, all plants produced had identical characteristics. Third, the "characters," or traits, that Mendel chose to study—such as the color and texture of seeds and seed coats, the position of flowers on the stems, and the length of the stems—could be determined unambiguously so that no confusion arose in counting the number of offspring of each type after each cross (Figure 1-5). Finally, and most important, he reached his conclusions with the aid of rigorous mathematical analyses of large numbers of descendants of his experimental crosses.

Mendel used his pure strains of peas to demonstrate that if certain traits are expressed in either parent, all offspring will express those traits. Such traits are called *dominant*. Other traits are expressed in offspring only if they are expressed in both parents (each a pure strain). These *recessive* traits are alternate forms of the dominant traits. Of crucial importance was Mendel's discovery that if two pure strains—one having the dominant form of a trait and the other the recessive form—were crossed, the recessive form would disappear in the first filial (F_1) generation; however, if F_1 plants were crossed, the recessive form of the trait would reappear in one-fourth of the plants of the second filial (F_2) generation (Figure 1-6). Because of this masking and reappearance of recessive traits, Mendel concluded that each parental plant had two hereditary units for each trait, but that these units segregated in the formation of gametes (pollen or ovules) so that each gamete had only one unit.

Mendel then went on to follow several dominant/recessive trait pairs simultaneously through two generations of cross-breeding. He showed that different trait pairs could behave independently of one another. This led him to formulate a law of independent assortment of hereditary units for different traits (Figure 1-7).

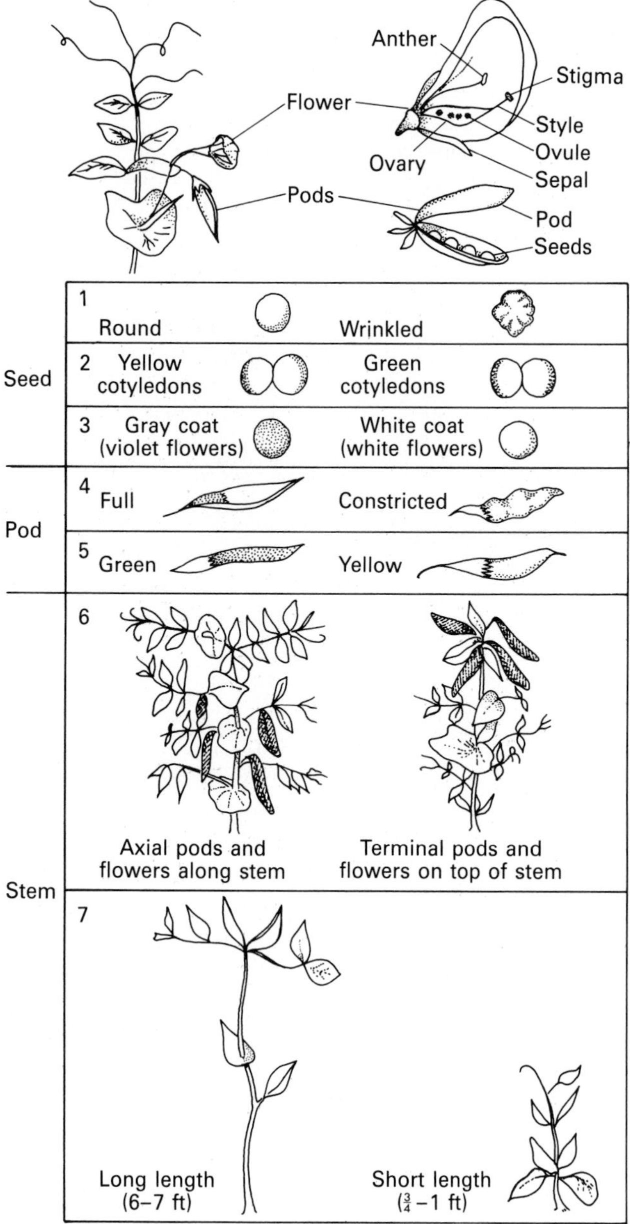

Figure 1-5 The seven "characters" observed and described by Mendel. To cross two strains, the flower of one strain is opened before self-pollination takes place, the anthers are removed, and the ovules are dusted with pollen from the other strain. The offspring of such a cross are called the F_1 generation. *From M. W. Strickberger, 1976,* Genetics, *2d ed., Macmillan, p. 115.*

These remarkably clear experiments were described fully by Mendel in 1865, but his conclusions were intellectually so far ahead of his contemporaries that they were ignored. Not until 1900 were they rediscovered and widely accepted. It was necessary for the cell theory to become more firmly rooted before the early cell biologists could make the connection between Mendel's genetics and cell division.

Chromosomes Are the Carriers of the Mendelian Units of Heredity

Once it became widely accepted that each cell within an organism is the product of the division of another cell, attention turned to the sperm and the egg cells, the union of which represented the first step in all cell division in higher organisms. In animals, the egg cell is comparatively large and consists mostly of cytoplasm. The much smaller sperm cell is composed mainly of a nucleus. It was reasoned that if, when the two unite, both make equal hereditary contributions to the offspring, the nucleus and not the cytoplasm must hold the key to genetic transfer. In 1892 August Weismann stated that

> the essence of heredity is the transmission of a nuclear substance of specific molecular structure.

He further suggested that the sperm and the egg cells contain a substance

> that by its physical and chemical properties and by its molecular nature is able to become a new individual of the same species.

Soon after the nucleus had become an important subject of study, filamentous structures were observed microscopically in the nuclei of both plant and animal cells. These ubiquitous structures, the *chromosomes,* were constant in their sizes and in their number within the somatic cells (general body cells) of a given species, but their sizes and shapes varied from species to species. In each somatic cell there were two copies of each morphologic type of chromosome—that is, the chromosomes of somatic cells were seen to exist in homologous pairs. During every somatic-cell division, the chromosomes somehow doubled in number and separated into two groups so that each daughter cell received the same number and kinds of chromosomes as the mother cell—two of each morphologic type.

In 1903 Walter Sutton linked the chromosomes to mendelian heredity. He observed that in the cell division that produced germ cells (sperm and egg cells, or gametes) the threadlike chromosomes divided so that each gamete received only *one* chromosome of each morphologic type. Since mendelian cross-breeding had shown that each parent contributes one hereditary unit for each trait in the offspring, Sutton reasoned that the chromosomes were the carriers of Mendel's units of heredity and that the parental sperm and egg each contributed one set of chromosomes to every new individual.

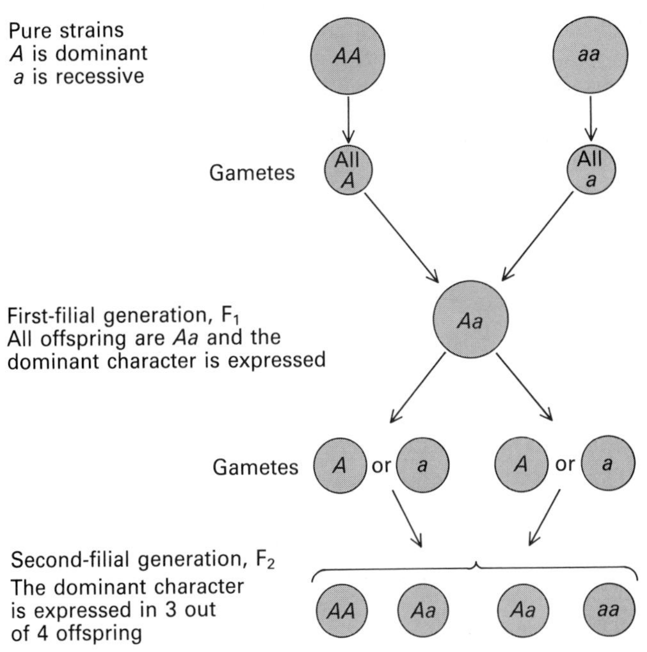

Figure 1-6 Mendel's experiments revealed that an individual organism has two alternate hereditary units for a given trait, represented here by *A* for the dominant unit and *a* for the recessive unit. In crossing pure strains of peas, he discovered that although all F_1 plants expressed the dominant unit, the recessive unit was apparently still present because it emerged in the F_2 generation. Thus he concluded that the two hereditary units do not "blend" in the parent plants but remain discrete and segregate in the formation of pollen and ovules (gametes). Each ovule or grain of pollen receives only one of the two parental units. This conclusion has come to be known as Mendel's first law.

The Reduction in Chromosome Numbers in the Formation of Germ Cells Is Crucial to the Development of the Chromosome Theory of Heredity

The reason for and the meaning of the difference in chromosomal behavior between somatic cells and germ cells was then given close attention. As mentioned earlier, both the sperm cells and the egg cells of a given species have one chromosome of each type. Each gamete thus contributes equally to the fertilized egg, which has twice as many chromosomes as either gamete alone. In calling this contribution of chromosomes "equal" we ignore for the moment the different sex chromosomes contributed by sperm and egg; this topic will be discussed in Chapters 5 and 10.

Just before the first cell division of a fertilized egg and before all divisions thereafter, the individual chromosomes duplicate so that each appears as a pair of threadlike structures (*chromatids*) that are attached at one point, the *centromere*. These "sister" chromatids split longitudinally at division so that each daughter cell receives the same number of chromosomes as the fertilized egg contained. The number of chromosomes in a gamete is called the *haploid* number, n, for the species; the number in a somatic cell is called the *diploid* number, $2n$. This process of the equal partitioning of sister chromatids at cell division is called *mitosis*. It was first described around 1880 by Walter Flemming, an important early cytologist.

As Sutton noted, the formation of sperm or egg cells entails a qualitatively different kind of chromosomal division. This process is called *meiosis*. As in mitosis, the process begins with diploid cells whose chromosomes have duplicated to form pairs of sister chromatids attached at their centromeres. In the first meiotic division the sister chromatids do not separate; each of the two daughter cells receives both copies of either the maternal or the paternal member of a homologous pair. In the second meiotic division, the chromatids do separate longitudinally. Each of the resulting four nuclei contains only one copy of each chromosome type. This produces the haploid number of chromosomes. Although four germ cells can arise from each cell that enters the meiotic cycle, it is often the case that only one mature egg cell results, rather than four. The other three meiotic products degenerate. Sutton realized that meiotic cell division was the basis for the distribution of traits in mendelian heredity. He described the assortment of chromosomes in the course of gamete formation for a hypothetical organism whose somatic cells contained four homologous pairs of chromosomes—A, B, C, and D being the paternal contributions, and a, b, c, and d the maternal ones:

> Each of the ripe germ cells arising from the reduction divisions must receive one member from each of the "synaptic pairs" [the sister chromatids, which are still coupled after the first meiotic division]. And there are 16 possible combinations of originally maternal and paternal chromosomes that will form a complete series: to wit, a,B,C,D; A,b,C,D; A,B,c,D; A,B,C,d; a,b,C,D; a,B,C,d; a,b,c,d; and their conjugates, A,b,c,d; etc.

Further, he noted that

> the number of possible combinations in the germ-products of a single individual of any species is represented by the simple formula 2^n, in which n represents the number of chromosomes in the reduced [gametic] series.

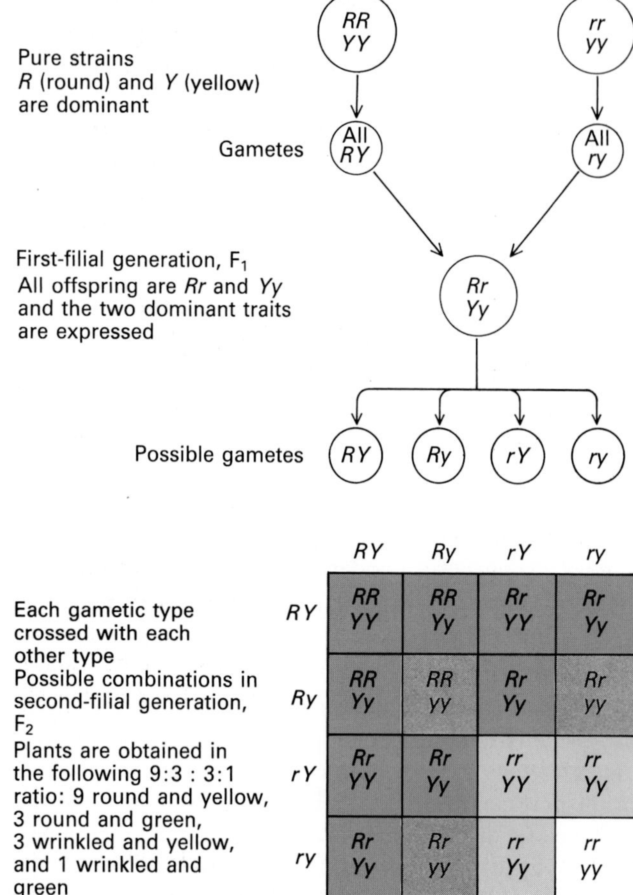

Figure 1-7 When Mendel crossed strains of peas that were pure for each of two traits—say, round *(R)* and yellow *(Y)* peas with wrinkled *(r)* and green *(y)* peas—the plants of the F_2 generation had four different combinations of the two traits. The four combinations always appeared in a certain ratio—9:3:3:1. Mendel accounted for this ratio by proposing that the two traits—form and color—were assorted among the offspring independently of one another. The independent assortment of traits has come to be known as Mendel's second law.

Chromosomes Contain Linear Arrays of Genes

The study of the assortment of mendelian units of heredity, or *genes* (as they were named by the Danish geneticist Wilhelm Johannsen in 1909), and of the relation of this assortment to chromosomal structure was greatly expanded by Thomas Hunt Morgan and his students in the early 1900s. Morgan chose the fruit fly, *Drosophila melanogaster,* for his studies because it could mature through an entire developmental cycle from fertilized egg to adult in about 2 weeks. The early studies showed that one particular chromosome was responsible for sex determination. Morgan's group also discovered, by crossing genetically marked flies, that the genes of *Drosophila* segregated as four sets; these sets were termed *linkage groups.* The presence of four linkage groups in the genetic crosses was correlated with the presence of four visible chromosomes in *Drosophila* cells. It was then found in studies of corn that the number of genetic linkage groups (10) corresponded to the number of chromosomes; similar results were obtained for peas (7 linkage groups and 7 chromosomes). By 1920 the chromosome theory of heredity had become the common currency of the new field of genetics.

Another important accomplishment of the *Drosophila* geneticists was the discovery of how genes are arranged in chromosomes. Both by genetic techniques and by visible observation of chromosomes, it was determined that the *Drosophila* chromosomes are *regular* arrays of genes with consistent positioning of particular genes. Subsequently, it was found that gene order in the chromosomes could change. In 1931 Barbara McClintock correlated visible rearrangements of gene segments within and between chromosomes with the distribution of specific genetic

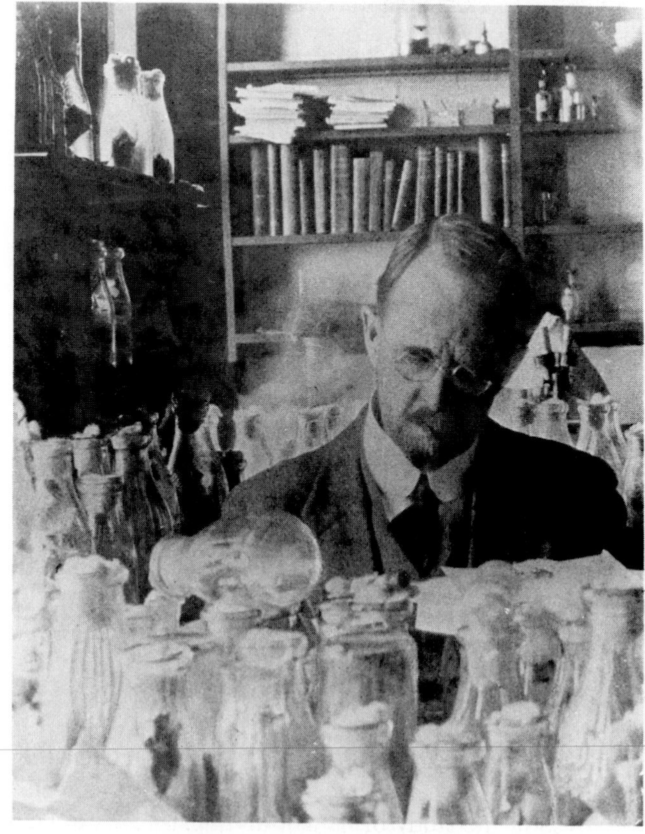

Figure 1-8 Thomas Hunt Morgan (1866–1945), the founder of modern experimental genetics, shown in his laboratory at Columbia University. *Courtesy of Mrs. Curt Stern.*

Figure 1-9 Barbara McClintock (1902–) lecturing on her genetic work with corn, which showed for the first time microscopic evidence of chromosome breakage and reunion during recombination. Later she showed that some chromosome segments are "mobile" in the genome (see Chapter 11), for which she received the Nobel Prize in 1983. *Nik Kleinberg/Picture Group.*

traits in strains of corn (*Zea mays*). These classical genetic studies gave the genes a physical reality, so that by the mid-1930s genes were no longer remote or "theoretical" entities but widely accepted determinants of biological specificity. The discovery of the chemical nature of the gene was a consequence of the merging of genetics and biochemistry between 1930 and 1950.

The Merging of Genetics and Biochemistry

The first instance in which a relation was discovered between a genetic defect and a biochemical abnormality occurred in the study of human disease. Archibald Garrod had realized by 1909 that the disease *alkaptonuria* was caused by a rare recessive mutation that was inherited according to mendelian rules. Patients having this disease eventually suffer severe arthritis, but their condition can be diagnosed prior to the onset of arthritic pain because they excrete a large amount of homogentisic acid in their urine. This phenolic compound, which turns the urine black, had been recognized by the 1890s as a product on the pathway to the breakdown of the amino acids tyrosine and phenylalanine, each of which contains a six-membered aromatic ring. Garrod surmised the nature of the genetic defect in his treatise on "Inborn Errors of Metabolism":

> We may further conceive that the splitting of the benzene ring in normal metabolism is the work of a special enzyme, [and] that in congenital alkaptonuria this enzyme is wanting.

Garrod was correct: the missing enzyme is homogentisic acid oxidase, which is predominantly a liver enzyme that catalyzes breakage of the benzene ring during the degradation of phenolic compounds. But it was too early in the history of genetics or biochemistry for the realization that genes do in fact control the structures of enzymes.

Appropriately enough, an important early step in the unification of genetics and biochemistry was the result of an experiment with fruit flies by George Beadle and Boris Euphrussi.

Genes Encode Enzymes

As already mentioned, experiments with fruit flies yielded much of the early information about genes. Among the most intensively studied genetic characteristics of *Drosophila* was eye color. The normal color is a dark red that is due, it is now known, to the production of two chemically distinct pigments. Morgan and his colleagues had collected many flies bearing changes, or mutations, in the gene(s) that controlled eye color. Flies with either of two mutant genes, *cinnabar* or *vermillion,* have eyes that are a much lighter red. Beadle and Euphrussi grafted eye tissue from larvae of one or the other of these mutant stocks of flies onto normal, or *wild-type,* larvae. Extra eyes then developed in the grafts, and these eyes were normal in color. Beadle and Euphrussi concluded that some metabolic product from the surrounding normal tissue diffused into the mutant graft and allowed the formation of the normal pigments in the grafted tissue. Thus, the correction of a genetic defect by a metabolic product was possible. Because a graft of *cinnabar* larval tissue onto a *vermillion* larva (or vice versa) did not correct the eye color in the grafted tissue, the experimenters reasoned that *cinnabar* and *vermillion* represented mutations in different enzymes that were both necessary for pigment formation. One of the components of the dark red pigment was then shown to be a metabolic product of the amino acid tryptophan. It finally proved true that the mutations *cinnabar* and *vermillion* produce defects in two different enzymes that catalyze two different chemical reactions necessary for transforming tryptophan into the dark red pigment. *Thus, the connection between gene activity and biochemical action was established.*

Beadle, together with Edward Tatum, then sought a genetic system even simpler than *Drosophila*—one in which biochemistry could be brought still closer to genetics. They turned to the bread mold *Neurospora crassa,* a lowly fungus that nevertheless derives its genetic material from two parents, just as higher organisms do. Therefore, like garden peas and *Drosophila, Neurospora* could be used in genetic crosses. But *Neurospora* offered the enormous advantage that it could be cultured in the laboratory from a single spore on a few simple substances: water, a sugar (a carbon source), salts containing ammonium ions (a nitrogen source), and one other supplement, biotin. Each wild-type *Neurospora* spore gives rise to a single colony that can grow equally well on a simple medium or on a medium to which enrichments have been added. However, a rare spore (1 in approximately 1000) requires an enriched medium for growth. Beadle and Tatum found that a number of such "mutants" resumed normal growth on a simple medium to which a single substance had been added—for example, an amino acid or a vitamin. They concluded that each mutant cell that could be restored to growth by the addition of a single compound carried a single gene defect that impaired the production of an enzyme necessary for a single metabolic step: in other words, *one gene was responsible for one enzyme.*

Figure 1-10 Joshua Lederberg (1925–) around 1958, when he shared a Nobel Prize with Edward L. Tatum (1909–1975) and George W. Beadle (1903–). Lederberg's work on genetic exchange between strains of *Escherichia coli* helped to establish this organism as the one most widely used in molecular biology. *Courtesy of J. Lederberg.*

Beadle and Tatum's powerful conclusion—that genes somehow control protein (enzyme) structure—led inevitably to studies of the chemical structure of a gene, as the first step in elucidating the molecular basis for the genetic control of protein synthesis. Successful studies of the structure of genes and the control of protein synthesis by genes form the recent history of molecular cell biology; present-day experiments in this area will be fully discussed in later chapters. Here it suffices to say that once Beadle and Tatum unveiled the power of microbial systems in genetics, the use of bacterial cells soon followed. In the mid-1940s Joshua Lederberg found that the genes of bacteria could be defined not only by nutritional experiments, but also by the observation of gene exchange during the "mating" of two bacterial cells. The use of bacteria as the simplest genetic system became extremely popular. An experiment with bacteria (one that actually predated Lederberg's work) led to a discovery that was essential to our understanding of the chemical nature of the gene.

DNA Is Identified as the Genetic Material, Paving the Way for the Study of the Molecular Basis of Gene Structure and Function

In 1944 Oswald Avery, Colin MacLeod, and MacLyn McCarty were studying *Streptococcus pneumoniae,* a bacterium that is frequently isolated from patients who have pneumonia. The more active disease-causing strains are called *smooth,* or S, because the presence of a gelatinous outer capsule on each cell makes the colonies glisten. The less virulent strains lack the gelatinous capsule and thus form more ragged-looking colonies; these strains are called *rough,* or R. The two strains can also be distinguished by antibacterial antibodies in the serum of convalescing patients. Avery and his colleagues followed the lead of the British physician Fred Griffiths, who showed that the R strain of *S. pneumoniae* could be converted into the S strain during infection of mice. Avery's group extracted the deoxyribonucleic acid (DNA) from a culture of S bacteria and mixed it with cells of the R type in a test tube. When the DNA-treated pneumococci were grown into colonies on bacterial plates, some of the bacteria had smooth, or S, characteristics, and these cells continued thereafter to exhibit the S type. The colonies

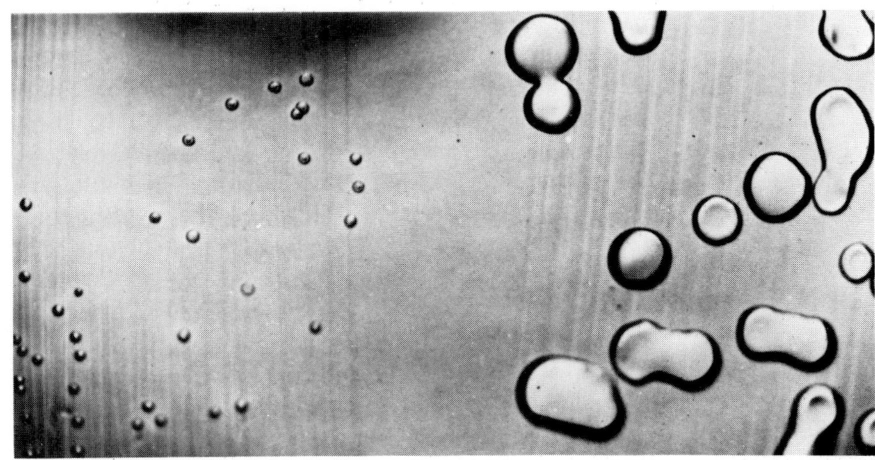

Figure 1-11 Colonies of *Streptococcus pneumoniae*. Small populations that appear rather dull or "rough" are shown at the left, and large, glistening, "smooth" colonies are shown at the right. The DNA of the organisms in smooth colonies can transform rough organisms into the smooth form. *From O. T. Avery, C. M. MacLeod, and M. McCarty, 1944,* J. Exp. Med. *78:137.*

were said to be "transformed." The substance that had caused the transformation—that is, the *transforming principle*—was shown to be DNA (Figure 1-11). The proof was obtained not only by biochemical testing (e.g., for the presence of deoxyribose and the absence of ribose), but also by developing another experimental plan that would be repeated many times in later years: Avery, MacLeod, and McCarty used a biological—in fact, an enzymatic—test to prove a biochemical point. They found that addition of a variety of proteases (enzymes that digest proteins) did not alter the transforming principle. However, tiny amounts of purified deoxyribonuclease (DNase), a then newly recognized enzyme that destroys DNA, immediately inactivated the transforming principle. This proved that DNA was the transforming principle that could genetically alter bacteria. Like Mendel's work 80 years earlier and Garrod's work 40 years earlier, this experiment was not immediately appreciated by the scientific community. Several more years of experience in genetic studies with bacteria were needed to prepare biologists and geneticists for accepting another experiment that also used bacterial systems and that also showed DNA to be the genetic material.

Bacterial cells, like almost all other cells in nature, are susceptible to viral infections. Many viruses contain DNA that has the ability, once inside a host cell, to dictate the reproduction of more viruses; most bacterial viruses, or bacteriophages, have DNA as their only nucleic acid. These viruses exhibit heritable traits—for example, fast or slow growth, growth only on certain cells, or growth in a restricted temperature range; such traits can be used to identify specific strains of virus. Moreover, different strains of bacteriophages can exchange genes when two strains infect the same cell: this indicates that the infecting viruses participate in a genetic process within the host cell. In 1952 Alfred Hershey and Martha Chase showed that only the DNA of the bacterial virus and not its protein portion must enter the host bacterium to initiate infection. Thus, the genetic information that causes new virus production resides in the DNA and not in the protein of the bacterial virus. Hershey and Chase's result, plus the earlier findings of Avery and his colleagues on bacterial transformation, established that DNA is the genetic material. [Some viruses use ribonucleic acid (RNA) as their genetic material. The nature of viruses as cell-dependent genetic systems will be discussed in Chapter 6].

This conclusion spawned some tentative answers to the age-old questions about the origins of diversity in life, about heredity, and about animal and plant development. The knowledge that DNA was the universal genetic material of cells suggested that: (1) early evolution must have depended on the development of a cell carrying sufficient instructions in its DNA to grow; (2) the random variation and selection that, according to the ideas of Darwin and Wallace, led to changes in species must have resulted from random changes in the DNA; (3) the faithful reproduction of DNA from generation to generation is what causes "like to beget like"; and (4) the programmed unfurling of the genetic endowment in the DNA underlies the development of every new plant or animal.

The conceptual advances that occurred between 1859, when Darwin's *Origin of Species* was published, and 1952, when a small group of informed scientists knew that DNA was the controlling molecule of life, were of enormous consequence. Then, no sooner had the biological importance of DNA been recognized than its physical structure was discovered.

The Birth of Molecular Biology

The modern era of molecular cell biology, which has been mainly concerned with how genes govern cell activity, began in 1953 when James D. Watson and Francis H. C. Crick deduced the double-helical structure of DNA. Two

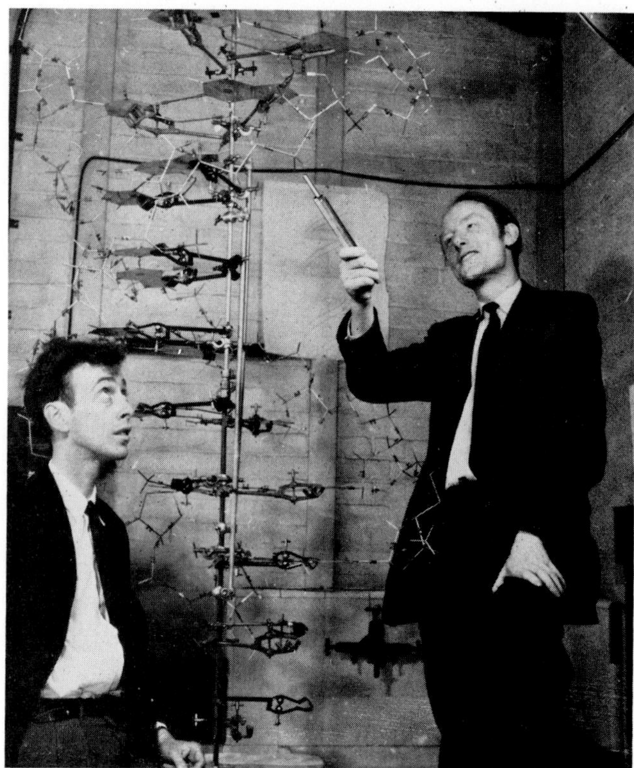

Figure 1-12 James D. Watson (1928–) *(left)* and Francis H. C. Crick (1916–) *(right)* ponder and admire the double-helical model of DNA that they constructed in 1952–1953. *From J. D. Watson,* The Double Helix, *Atheneum, p. 215, copyright 1968 by J. D. Watson. Photograph by A. C. Barrington Brown.*

major clues led Watson and Crick to build a correct model of the DNA molecule. First, Erwin Chargaff had separated and measured the four nucleic acid bases—thymine, cytosine, adenine, and guanine—in DNA from various sources. He found that the amount of guanine always equalled that of cytosine, and the amount of adenine always equalled that of thymine; this is conventionally written as A = T and G = C and is referred to as *base pairing.* [Note that the total amount of adenine plus thymine does not always equal the total amount of guanine plus cytosine; i.e., the (A + T)/(G + C) ratio can vary in DNA from different sources.] Second, x-ray diffraction patterns of DNA fibers had been obtained by Maurice Wilkins and his collaborators, particularly Rosalind Franklin. These images, formed by the refractions of an x-ray beam by the regularly spaced atoms of the DNA molecule, revealed extensive details about the structure: its helical nature; the diameter of the linear molecule [2 nanometers (nm)]; the distance between adjacent bases (0.34 nm); and, because a turn equalled 3.4 nm, the approximate number of bases per turn of the helix (10). By fitting together the base-pairing data and the structural data obtained from x-rays, Watson and Crick built an accurate model of DNA.

The marvelously simple, self-complementary organization of DNA suggested a means by which it could be copied in each generation. Using the principles of base complementarity, scientists also discovered how information in DNA is accurately transferred into protein structure. This subject is discussed in Chapter 4 and in Chapters 8 through 12.

Before it was used in determining the structure of DNA, x-ray crystallography had been applied to crystals of simple substances and to pure proteins. Linus Pauling had suggested a regular helical arrangement for certain parts of protein chains in 1951. Max Perutz and John Kendrew at Cambridge, England, had been analyzing the x-ray patterns produced by crystals of small proteins for more than a decade by 1959, when Kendrew finally determined the position in space of each atom in sperm whale myoglobin; a part of the structure was indeed a helix. The power of x-ray crystallography in molecular biology lies in its contributions to three-dimensional models of complex biological molecules. This technique not only aids in determining the sites on the molecule at which important activities take place, but the x-ray studies also frequently suggest the physicochemical basis for a specific biochemical capacity. For example, from the three-dimensional study of hemoglobin, the oxygen-carrying protein of blood, Perutz and his colleagues determined the exact nature of the oxygen-binding site: a "pocket" in which the iron-containing heme molecule is bound. When oxygen is bound by the heme in the pocket, the protein chains assume a physical arrangement, or conformation, different from that when no oxygen is bound. By now perhaps 100 three-dimensional protein structures, as well as the first RNA crystal structures (those for tRNA), have been determined at a resolution of 0.3 to 0.5 nm.

Another major discipline on which molecular cell biology depended throughout the first half of the twentieth century was traditional biochemistry. Experiments in which cells are broken apart and assays are devised for specific proteins, which are then purified, are the domain of biochemists. Our understanding of the nature of chemical reactions and the discovery of individual enzymes have depended to a great extent on research in which critical biochemical steps have been reconstructed from purified proteins in the test tube. Since Eduard Buchner's early experiments showing that yeast cell extracts could ferment glucose to ethanol, the efforts of hundreds of biochemists have resulted in the identification of virtually all of the enzymes responsible for the most important metabolic pathways. Vitamins have been recognized as enzyme cofactors. In addition, the importance of high-energy phosphates, particularly of adenosine triphosphate (ATP), as the "energy currency of the cell" has been proved, and the mechanics of ATP generation have been

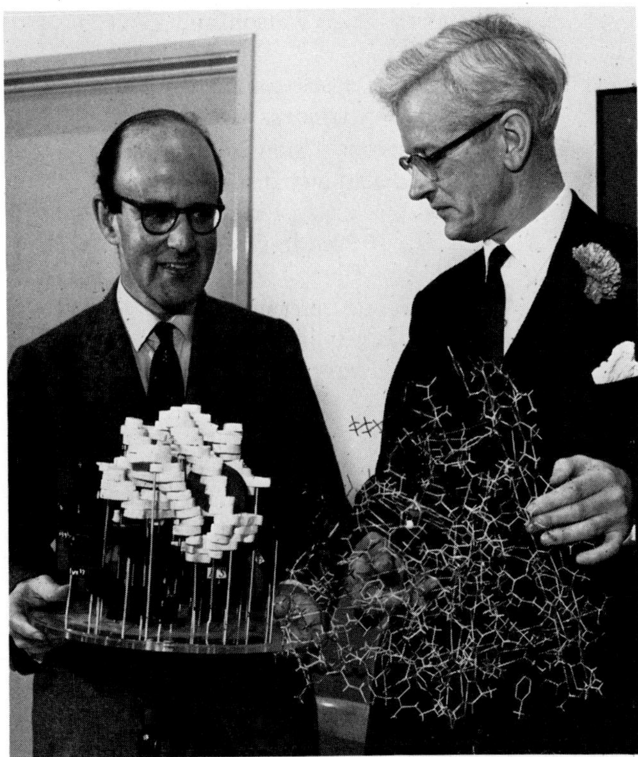

Figure 1-13 (Max Perutz (1914–) *(left)* and John Kendrew (1917–) *(right)* with the three-dimensional models of hemoglobin *(left)* and myoglobin *(right)* that they developed using x-ray crystallographic techniques. *Keystone Press Agency.*

illuminated (Chapters 2, 3, and 20). A knowledge of these topics is essential for developing an understanding of how cell structures carry out biochemical functions.

Electron Microscopy: A Modern View of Cell Structure

Throughout the 1930s and early 1940s, while geneticists were seeking the identity of the gene and biochemists were engaged in finding enzymes and unlocking many secrets of the cell's chemistry, relatively little attention was paid to the organization of the higher structures within the cell. The physical limitations of the light microscope imposed a great limitation on inquiry into cell structure. The resolution of the best light microscopes is dictated by the average wavelength of visible light; thus, objects less than 500 nm in diameter could not be resolved from one another. In that era only the larger intracellular structures—the nuclei, nucleoli, and mitochondria in animal cells, and the plastids, such as chloroplasts, in plant cells—had been identified with the ordinary microscopes. When the electron microscope became available, a veil was lifted from the eyes of biologists. Its practical resolving power even in its early forms was from 1 to 5 nm, and its theoretical resolving power is 10 times higher—from 0.1 to 0.5 nm. Suddenly a world of structure never before observed was displayed before a spellbound audience. And very soon a striking fact emerged: that all eukaryotic cells have certain cell structures in common. This important conclusion, which has been elaborated over the past several years, is the basis for much of Chapter 5 and also of extensive discussion in Chapters 14 through 21.

Biochemical Activities Can Be Assigned to Specific Subcellular Structures

By the end of the 1950s, the fundamental observations that had been made independently by geneticists, biochemists, and structural cell biologists were ready to be stitched together into a coherent pattern that would become the design for molecular cell biology. Two different experimental approaches provided key contributions.

First, a correlation between cell structure and biochemical activity was achieved by breaking open plant and animal cells and separating, or fractionating, the broken cell extracts. Each fraction was studied by using both the electron microscope and biochemical assay. Much of the early electron-microscopic identification of cell structures and biochemical characterization of cell fractions was the work of Albert Claude and three of his associates—Christian de Duve, George Palade, and Keith Porter. The current sophisticated picture of cell structure and function is in a very large way due to their research.

The Activity of Genes Is Highly Regulated

The second experimental approach that had a great unifying effect was genetic analysis. The work of Beadle, Tatum, Lederberg, and others in the 1940s and early 1950s had made it clear that genes encoded proteins, but the means by which gene activity was controlled was still totally unclear. And yet it was widely supposed that the differences between cells having the same genes were due to differences in the activity of those genes. Two French scientists, François Jacob and Jacques Monod (whose work is described in Chapter 8), did the most to reveal how genes are controlled. They proposed that certain genes regulate the activity of other genes. This principle finally unified the diverse approaches to the subject that we now call molecular cell biology. The concept of the cell has evolved a long way from its original characterization as a simple unit of living matter: the cell has become an organism in which the controlled and integrated actions of genes produce specific sets of proteins that build characteristic structures and carry out characteristic enzymatic activities.

(a)

(b)

Figure 1-14 The architects of modern cell biology. (a) *From left to right:* George Palade (1912–), Albert Claude (1899–1983), and Keith Porter (1912–). (b) Christian de Duve (1917–). Dr. Claude's group at Rockefeller University was the first to achieve success in the use of cell fractionation and biochemical analysis of fractions (mainly the work of Drs. Claude, de Duve, and Palade), and in electron microscopy (primarily done by Drs. Porter and Palade). *Photograph (a) courtesy of Columbia University, photograph by Manny Warman. Photograph (b) courtesy of Rockefeller University.*

The Molecular Approach Is Applied to Eukaryotic Cells

Since 1961, the year in which Jacob and Monod presented their findings on regulatory genes, there have been three advances of great significance: (1) proof that messenger RNA carries the information from DNA to the protein-synthesizing machinery; (2) the discovery of the genetic code, by which information is stored in the nucleic acids; and (3) the discovery that proteins are translated by transfer RNA and ribosomes, small nucleoprotein granules that were first discovered by electron microscopy.

Much of the research that contributed to these achievements was done by biochemists and molecular geneticists working with *E. coli*. However, two developments in the 1960s have led an ever-increasing number of workers into the study of mammalian cells: the successful culturing of mammalian cells in vitro, particularly through the efforts of such pioneers as Harry Eagle and Theodore Puck; and the emergence of animal virology, notably through the work of Renato Dulbecco and his students. Although many aspects of protein and nucleic acid synthesis and of cell growth and cell function were elucidated in the 1960s and early 1970s, the lack of detailed genetic information about higher organisms impeded progress. However, since 1975 new techniques have allowed any desired segment of DNA to be isolated and purified in large amounts; other breakthroughs have resulted in the sequencing and the synthesis of DNA and of peptides. It

Figure 1-15 *From left to right:* Harry Eagle (1905–), Renato Dulbecco (1914–), and Theodore T. Puck (1916–) in 1972, when they were honored for their work on the successful in vitro growth of animal cells and the development of quantitative techniques in animal virology. *Courtesy of Harry Eagle.*

is now commonplace for molecular biologists to rearrange DNA sequences at will, and it has become possible for genes to be transferred between cells. These fantastic strides forward, combined with the improved methods now being developed for examining cells and cell components both biochemically and structurally, have given rise to great optimism about the success of future research. Surely few problems in molecular cell biology, even those of finding a cure for cancer and determining the molecular basis of differentiation and development, will remain insoluble for much longer.

References

Classical Biochemistry and Genetics

CREIGHTON, H. S., and B. MCCLINTOCK. 1931. A correlation of cytological and genetical crossing-over in *Zea mays*. *Proc. Nat'l. Acad. Sci. USA* **17**:492–497. Reprinted in J. A. Peters, 1959, *Classic Papers in Genetics*, Prentice-Hall.

GABRIEL, M. L., and S. FOGEL, eds. 1955. *Great Experiments in Biology*. Prentice-Hall.

JACOB, F. 1973. *The Logic of Life: A History of Heredity*. Pantheon.

MENDEL, G. 1865. Versuche über Pflanzen-Hybriden. *Verh. Naturforshung Ver. Brünn* **4**:3–47. Translated by W. A. Bateson as "Experiments in plant hybridization," and reprinted in J. A. Peters, 1959, *Classic Papers in Genetics*, Prentice-Hall.

MORGAN, T. H. 1910. Sex linked inheritance in *Drosophila*. *Science* **32**:120–122. Reprinted in J. A. Peters, 1959, *Classic Papers in Genetics*, Prentice-Hall.

PETERS, J. A. 1959. *Classic Papers in Genetics*. Prentice-Hall.

SUTTON, W. S. 1903. The chromosomes in heredity. *Biol. Bull.* **4**:231–251. Reprinted in J. A. Peters, 1959, *Classic Papers in Genetics*, Prentice-Hall.

The Merging of Genetics and Biochemistry

GARROD, A. 1909. *Inborn Errors of Metabolism*. Oxford University Press.

MORGAN, T. H. 1926. *The Theory of the Gene*. Yale University Press.

WILSON, E. B. 1925. *The Cell in Development and Inheritance*, 3d ed. Macmillan.

Genes Encode Enzymes

BEADLE, G. W., and E. L. TATUM. 1941. Genetic control of biochemical reactions in *Neurospora*. *Proc. Nat'l. Acad. Sci. USA* **27**:499–506. Reprinted in J. A. Peters, 1959, *Classic Papers in Genetics*, Prentice-Hall.

LEDERBERG J., and E. L. TATUM. 1946. Gene recombination in *Escherichia coli*. *Nature* **158**:588. Reprinted in J. A. Peters, 1959, *Classic Papers in Genetics*, Prentice-Hall.

ZINDER, N. D., and J. LEDERBERG. 1952. Genetic exchange in *Salmonella*, *J. Bacteriol.* **64**:679–699. Reprinted in J. A. Peters, 1959, *Classic Papers in Genetics*, Prentice-Hall.

DNA Is Identified as the Genetic Material

AVERY, O. T., C. M. MACLEOD, and M. MCCARTY. 1944. Studies on the chemical nature of the substance inducing transformation of pneumococcal types. *J. Exp. Med.* **79**:137–158.

HERSHEY, A. D., and M. CHASE. 1952. Independent functions of viral protein and nucleic acid in growth of bacteriophage. *J. Gen. Physiol.* **36**:39–56.

MCCARTY, M., and O. T. AVERY. 1946. I. Studies on the chemical nature of the substance inducing transformation of pneumococcal types. II. Effect of desoxyribonuclease on the biological activity of the transforming substance. *J. Exp. Med.* **83**:89–96.

The Birth of Molecular Biology

CAIRNS, J., G. S. STENT, and J. D. WATSON, eds. 1966. *Phage and the Origins of Molecular Biology*. Cold Spring Harbor Laboratory.

CHARGAFF, E. 1950. Chemical specificity of the nucleic acids and mechanisms of their enzymatic degradation. *Experimentia* **6**:201–240.

KENDREW, J. C. 1963. Myoglobin and the structure of proteins. *Science* **139**:1259–1266.

PERUTZ, M. F. 1964. The hemoglobin molecule. *Sci. Am.* **211**(5):64–76.

STENT, G. S., and R. CALENDAR. 1978. *Molecular Genetics: An Introductory Narrative*, 2d ed. W. H. Freeman and Company.

WATSON, J. D. 1968. *The Double Helix*. Atheneum.

WATSON, J. D., and F. H. C. CRICK. 1953. General implications of the structure of deoxyribonucleic acid. *Nature* **171**:964–967.

WATSON, J. D., and F. H. C. CRICK. 1953. A structure for deoxyribose nucleic acid. *Nature* **171**:737–738.

Electron Microscopy: A Modern View of Cell Structure

Biochemical Activities Can Be Assigned to Specific Subcellular Structures

CLAUDE, A. 1946. Fractionation of mammalian liver cells by differential centrifugation, I and II. *J. Exp. Med.* **84**:51–171.

CLAUDE, A. 1975. The coming of age of the cell. *Science* **189**:433–435.

DE DUVE, C. 1975. Exploring cells with a centrifuge. *Science* **189**:186–194.

PALADE, G. 1975. Intracellular aspects of the process of protein synthesis. *Science* **189**:347–358.

PORTER, K. R., A. CLAUDE, and E. FULLAM. 1945. A study of tissue culture cells by electron microscopy: methods and preliminary observations. *J. Exp. Med.* **81**:233–243.

The Activity of Genes Is Highly Regulated

JACOB, F., and J. MONOD. 1961. Genetic regulatory mechanisms in the synthesis of proteins. *J. Mol. Biol.* **3**:318–356.

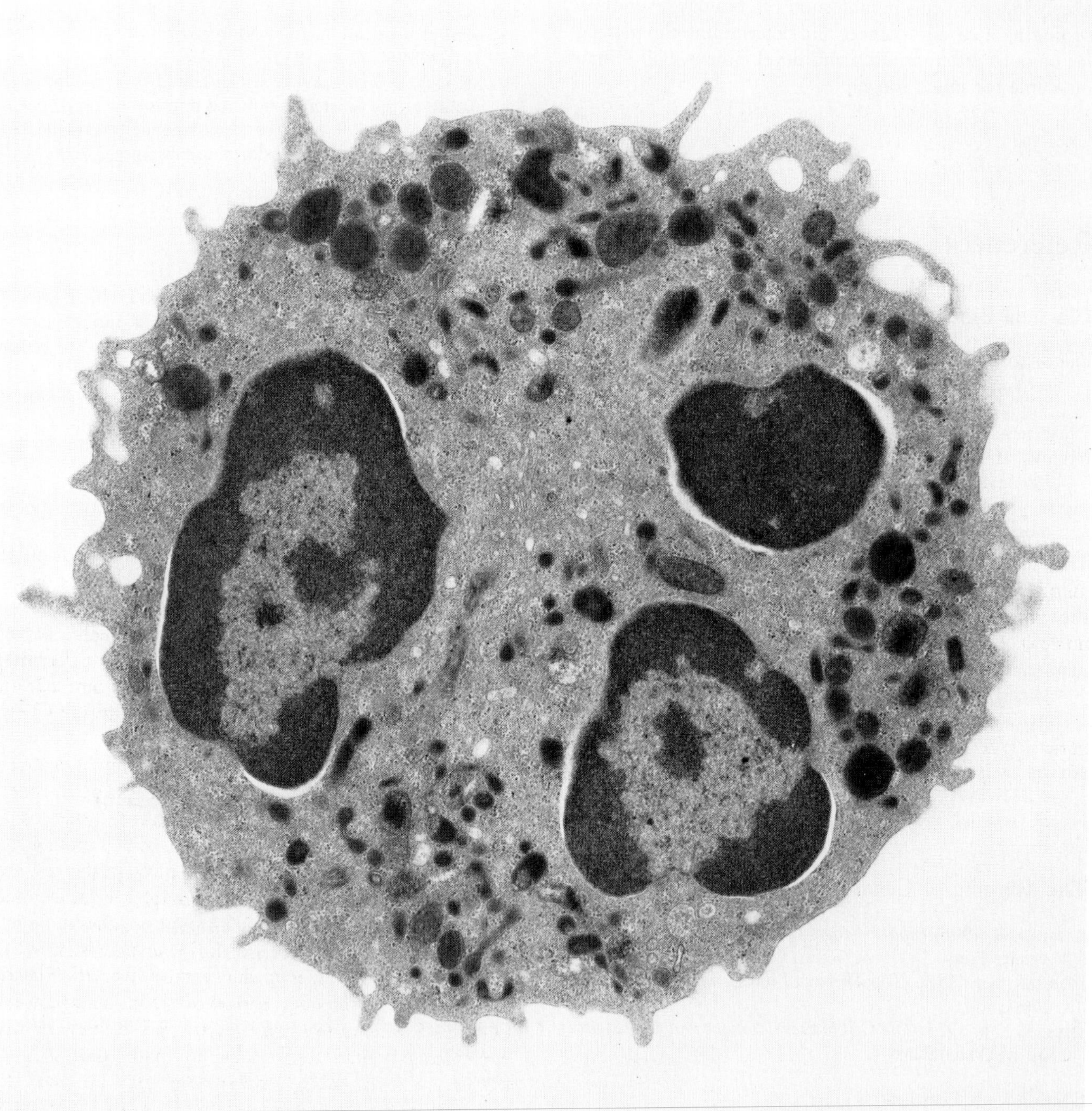

I

Introduction: Molecules, Cells, and Experimental Techniques

Thin-section transmission electron micrograph of human leukocyte, showing lobes of the single nucleus. *Courtesy of M. J. Karnovsky.*

PART I presents the necessary background material for ready comprehension of the topics in the rest of the book. The principles of basic chemistry that especially apply to important biological problems are reviewed in Chapter 2. The major biological molecules, both the basic building blocks and the macromolecules, are described in Chapter 3. A central idea to take from these chapters is that biological reactions obey the laws of chemistry and that the enzymological function of proteins is to speed up reactions that would otherwise occur much too slowly under the conditions present inside cells. Among the most highly organized and surely among the most important set of chemical reactions in the cell are those leading to protein synthesis. Chapter 4 outlines the complicated overall process of protein synthesis, which requires three separate roles of RNA, and the somewhat simpler but crucial steps in the synthesis of nucleic acid. Chapter 5 is a descriptive review of the major structures of cells. Fractionation and microscopic techniques are explained as well. Students who have had a thorough introduction to chemistry and biology may not need to make more than a cursory examination of Chapters 2 through 5.

The subject of Chapters 6 and 7 is popular experimental organisms and basic experimental methods used by molecular cell biologists. Chapter 6 features biological information: how cells are grown, how genetic techniques are applied to cells, the limitations that the study of whole organisms imposes on cellular research, and how such limitations are combated. Chapter 7 deals with basic chemical and physical experimental techniques, including the use of radioisotopes, the indispensable markers used throughout biology. Among the many methods discussed are various procedures for separating proteins and nucleic acids on the basis of their length, which is perhaps the most frequent experimental manipulation in molecular biology. Also described in this chapter is the wealth of clever new molecular biological approaches to the study of gene structure and gene expression.

The material in Chapters 6 and 7 is provided so that in the remainder of the book important experimental conclusions can be presented by discussing the actual experiments that underlie the conclusions. It is a cliche, nevertheless a relevant one, that biology is an active experimental science. Simply to deliver conclusions and not lead the reader through the excitement of experimental discovery not only loses the essence of modern biology but seems basically unfair. Therefore we suggest that students read Chapters 6 and 7 in their entirety before delving into the chapters that follow and then return to Chapters 6 and 7 whenever they need to review information about an experimental organism or technique.

2

Energy and Chemical Reactions

THIS chapter is a review of some concepts that are usually taught in general chemistry courses and are required for a comprehension of the workings of a cell. It is assumed that the users of this book will have had introductory college-level chemistry; some will also have taken more advanced courses. Nevertheless, readers may wish to examine this chapter to refresh their memories, because topics dealt with here are important in understanding the concepts and experiments presented later on.

The chapter begins with a discussion of energy and its transformations. It proceeds with a review of the different types of bonds that can be formed between individual atoms in a molecule, between groups of atoms within larger molecules, and between different molecules. Because water is the major constituent of cells and of the spaces between cells, the next topics treated are the structure of water, the reactions of acids and bases, and the structures and reactions of chemicals dissolved in water. Finally, the factors controlling the rate and the equilibrium of chemical transformations are discussed, to lay the groundwork for later chapters on cellular metabolism. Chapter 3 will consider the chemistry of the major classes of molecules that make up living systems.

Energy

Mastery of any aspect of chemistry requires an understanding of energy—both as it governs the strength of chemical bonds and as it determines the direction and rate of chemical reactions. The production of energy, its storage, and its use are as central to the economy of the cell as they are to the management of the world's resources.

Energy may be defined as the ability to do work. This is a concept that is fairly easy to grasp when it is applied to automobile engines and electric power plants. When we consider the energy associated with chemical bonds and chemical reactions within cells, the concept of work becomes less intuitive. Nonetheless, cells do require energy to do work: when, for example, they synthesize glucose from carbon dioxide and water in the process of photosynthesis, or when they play host to the reactions involved in contracting muscles or in replicating DNA.

Energy Exists in Several Forms

There are two principal forms of energy: kinetic and potential. *Kinetic energy* is the energy of movement—the motion of a car, for example, or the motion of molecules. *Heat* is what we call the energy of the molecular motion of matter. We measure heat indirectly by measuring temperature. Through the use of mechanical systems we can harness heat to do certain kinds of work, such as running cars and generating electricity. For heat to do work, it must flow from a region of higher temperature to one of lower temperature. Differences in temperature often exist between the internal and external environments of cells; however, cells generally cannot harness these heat differentials to do work. Even in higher animals and plants that have evolved a mechanism for thermoregulation, the kinetic energy of molecules is used chiefly to maintain constant organismic temperatures.

The energy that usually concerns us when we study biological or chemical systems is *potential energy,* or stored energy. Consider a ball at the top of a hill; it has potential energy by virtue of its location. The ball will lose potential energy and acquire kinetic energy as it rolls down the hill. Likewise, atoms and molecules have potential energy by virtue of their ability to undergo energy-releasing chemical reactions—their ability to form and break chemical bonds. For example, the sugar glucose is high in potential energy. Cells degrade it continuously; the kinetic energy released when its chemical bonds are broken is harnessed to do many kinds of work. However, cells store only small quantities of free glucose. Liver and muscle cells convert the excess into long polymers called glycogen, the form in which most glucose is stored in cells. Glycogen thus represents accumulated potential energy.

All Forms of Energy Are Interconvertible

Energy exists in many forms, among which are thermal, electric, radiant, and chemical. Thermal energy is, as we have noted, the kinetic energy of molecular motion; radiant energy can be considered the kinetic energy of photons, or waves of light. Electric energy can be either kinetic (when it refers to the energy of moving electrons or other charged particles) or potential (when it refers to separated electrons or charged particles). All forms of energy—both in inanimate objects and in the living world—are interconvertible, in accordance with the first law of thermodynamics, which states that *energy is neither created nor destroyed.* In photosynthesis, for example, the radiant energy of light is transformed into the chemical potential energy of the bonds between the atoms in a glucose molecule. In muscles and nerves chemical potential energy is transformed into mechanical and electric energy, respectively. Most of the biochemical reactions that are described in this book involve the making or breaking of at least one chemical bond.

Another form of potential energy to which we shall refer often is that in concentration gradients. When a substance is present on one side of a barrier, such as a membrane, in a concentration different from its concentration on the other side, the result is a concentration gradient. All cells form concentration gradients by selectively taking up nutrients from their surroundings. The energy in chemical bonds is utilized for this process.

Because all forms of energy can be interconverted, all can be expressed in the same units of measurement—namely, the calorie (cal) or the kilocalorie (1 kcal = 1000 cal).*

Chemical Bonds

Several kinds of chemical bonds between atoms and molecules are important in biological systems. As a rule, the strongest and most stable are the covalent bonds between atoms. Noncovalent bonds such as hydrogen bonds, ionic bonds, and van der Waals interactions are weaker, but they help to stabilize biological structures such as proteins and nucleic acids, as well as complexes composed of several large molecules. Hydrophobic interactions also are weaker, noncovalent bonds, but these, too, are essen-

*A calorie is defined as the amount of thermal energy required to heat 1 cm^3 of water at 14°C by 1°C. Many biochemistry textbooks use the joule (J) instead: 1 J = 4.18 cal, so the two systems can be interconverted quite readily. The energy changes in chemical reactions, such as the making or breaking of chemical bonds, are measured in kilocalories per mole (kcal/mol) in this book. One mole of any substance is the amount that contains 6.02×10^{23} molecules, which is known as Avogadro's number. The weight of a mole of a substance, in grams, is the same as the molecular weight. For example, the molecular weight of water is 18, so 1 mol of water weighs 18 g.

tial for maintaining the architecture of many biological macromolecules and membranes.

The Most Stable Bonds between Atoms Are Covalent

In a *covalent bond,* the nuclei of two atoms are held close together because electrons in their outermost shells move in orbitals that are shared by both atoms. Within the molecule, these bonding electrons no longer belong to one atom or the other but are free to move around both nuclei and in the space between them. Most of the molecules in living systems contain only six different atoms: carbon, hydrogen, oxygen, nitrogen, phosphorus, and sulfur. The outer electron shell of each atom has a characteristic number of electrons:

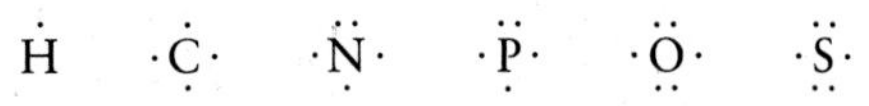

H· ·C· ·N· ·P· ·O· ·S·

These atoms readily form covalent bonds with other atoms and rarely exist as isolated entities. All of them except hydrogen have other electrons closer to the nucleus, but only the outer unpaired electrons participate in covalent bonds. As a rule, each type of atom forms a characteristic number of covalent bonds with other atoms. A hydrogen atom, with its one electron, forms only one bond. A carbon atom, with four electrons in its outer shell, generally forms four bonds, as in methane (CH_4):

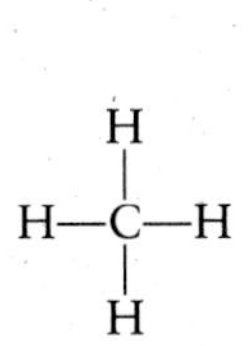

```
    H
    |
H—C—H
    |
    H
```

Although oxygen and sulfur contain six electrons in their outermost shells, an atom of oxygen or sulfur usually forms only two covalent bonds, as in molecular oxygen (O_2) or hydrogen sulfide (H_2S):

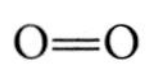

```
           H
           |
O=O    H—S
```

A sulfur atom can, however, form as many as six covalent bonds, as in sulfur trioxide (SO_3) or sulfuric acid (H_2SO_4):

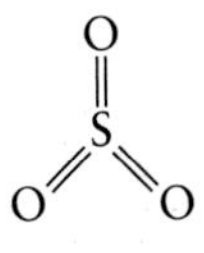

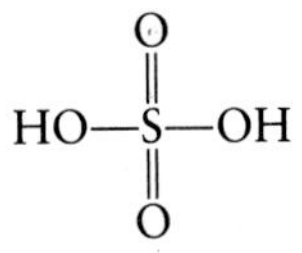

```
   O            O
   ‖            ‖
   S       HO—S—OH
  // \\         ‖
 O     O        O
```

Nitrogen and phosphorus have five electrons in their outer shells, and these atoms can form either three covalent bonds, as in ammonia (NH_3), or five, as in phosphoric acid (H_3PO_4):

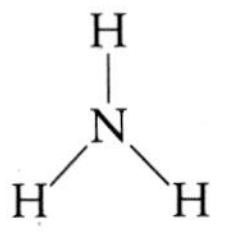

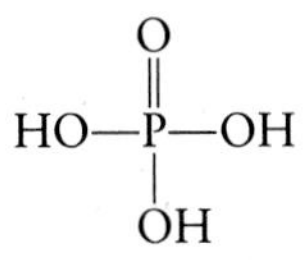

```
    H            O
    |            ‖
    N       HO—P—OH
   / \           |
  H   H          OH
```

The Large Energy Changes Associated with the Making and Breaking of Covalent Bonds Chemical bonds that require a relatively large amount of energy to break them are considered stronger, or more stable, than those that require less energy to disrupt them. The input of energy, ΔE, that is required to break a covalent bond is quite high; for a typical C—C bond in the gas ethane, it is about 83 kcal/mol:

$$H_3C{:}CH_3 \longrightarrow \cdot CH_3 + \cdot CH_3 \qquad \Delta E = +83 \text{ kcal/mol}$$

ΔE represents the difference between the energy of the reactants and that of the products.

Each of the products of the above reaction contains an unpaired electron in its outer orbital. Such molecules, termed free radicals, are extremely reactive and chemically unstable; thus, considerable energy is required to form them. Other products that could result from breaking the C—C bond in ethane are equally unstable: $^+CH_3$ and $^-CH_3$.

According to the first law of thermodynamics, *formation* of the C—C bond in ethane *releases* the same amount of energy as is required to break the bond: 83 kcal/mol. Note that when energy is released, ΔE is negative:

$$\cdot CH_3 + \cdot CH_3 \longrightarrow H_3C{:}CH_3 \qquad \Delta E = -83 \text{ kcal/mol}$$

At room temperature (25°C), well under 1 in 10^{12} ethane molecules exists as a pair of $\cdot CH_3$ radicals. This is because very few ethane molecules have sufficient kinetic energy at room temperature to break the covalent C—C bond.

The most useful measure of the strength of a chemical bond in a biological system is the amount of *free energy* that is released in its formation. Thus, we say that the most stable chemical bonds are those that release the largest amount of free energy when they are formed. A discussion of the concept of free energy is deferred until page 36.

The Precise Orientations of Covalent Bonds When two or more atoms form covalent bonds with another central atom, these bonds are oriented at precise angles to one another. The angles are determined by the mutual repulsion of the electrons of the outer electron orbitals of the central atom. In methane, for example, the central atom (carbon) is bonded to four other atoms (hydrogens) whose positions define the four points of a tetrahedron, so that the angle between any two bonds is 109.5° (Figure 2-1). However, if a carbon atom is linked to only three

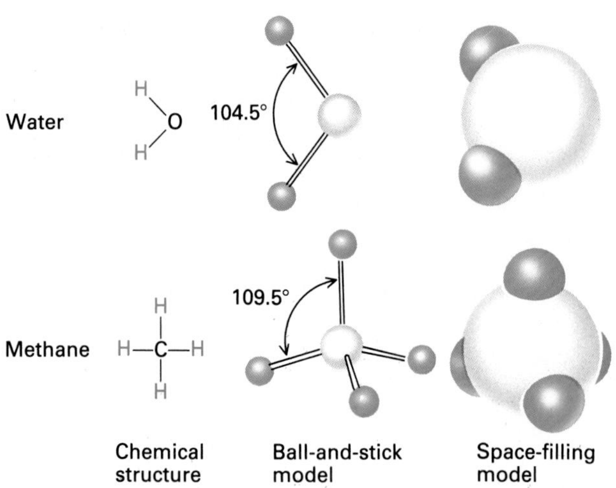

Figure 2-1 Bond orientations in a water molecule and in a methane molecule. Each molecule is represented in three ways. The atoms in the ball-and-stick models are smaller than they really are in relation to bond length, so that the bond angles can be shown clearly. The sizes of the electron clouds in the space-filling models are more accurate.

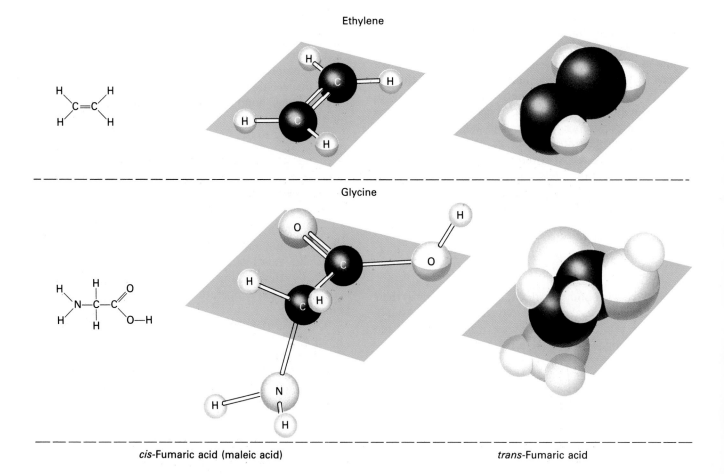

Ethylene

Glycine

cis-Fumaric acid (maleic acid)

trans-Fumaric acid

Figure 2-2 In an ethylene molecule, the carbon atoms are connected by a double bond, which causes all the atoms to lie in the same plane. (Atoms connected by a single bond can rotate freely about the bond axis, but those connected by a double bond cannot.) Similarly, in the amino acid glycine, all of the atoms linked to one of the carbon atoms (here the one bonded to oxygen) lie in the same plane with it, because of the double bond between that carbon and an oxygen. The rigidity of the structure of fumaric acid, with its double-bonded carbons, causes it to exist in two isomers, trans and cis (called maleic acid). Only the trans form is generally found in cells.

other atoms, two of its outer electrons participate in the formation of a double bond:

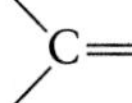

In this case, the carbon atom and all three of the atoms linked to it lie in the same plane, because atoms connected by a double bond cannot rotate freely about the bond axis (Figure 2-2). As a result, many molecules that contain a double bond, such as fumaric acid, exist in two isomeric forms, termed *cis* and *trans,* that are not interconvertible. As will become evident in Chapter 3, the rigid planarity imposed by double bonds has enormous significance for the shape of large biological molecules such as peptides and nucleic acids.

The outer electron orbitals not involved in covalent bond formation also contribute to the configurations of molecules. The outer shell of the oxygen atom has two pairs of electrons that are not involved in covalent bond formation. In the water molecule, the orbitals of oxygen's nonbonding electrons have a specific orientation in space (Figure 2-3). The high electron densities of these orbitals compress the angle between the covalent H—O—H bonds of water so that it is 104.5°, rather than 109.5° as in methane.

Table 2-1 The electronegativities of some atoms that are important in biological systems

Element	Electronegativity
Fluorine (F)	4.0 (most electronegative)
Oxygen (O)	3.5
Chlorine (Cl)	3.0
Nitrogen (N)	3.0
Bromine (Br)	2.8
Sulfur (S)	2.5
Carbon (C)	2.5
Iodine (I)	2.5
Phosphorus (P)	2.1
Hydrogen (H)	2.1
Magnesium (Mg)	1.2
Calcium (Ca)	1.0
Lithium (Li)	1.0
Sodium (Na)	0.9
Potassium (K)	0.8 (least electronegative)

Dipoles: The Result of Unequal Sharing of Electrons in Covalent Bonds In a covalent bond, one or more pairs of electrons are shared between two atoms. In certain cases, the bonded atoms exert differing attractions for the electrons of the bond. This results in unequal sharing of the electrons. The power of an atom in a molecule to attract electrons to itself is called *electronegativity.* Electronegativity is measured on a scale from 4.0, for fluorine (the most electronegative atom), to a hypothetical zero (Table 2-1). In a covalent bond in which the atoms either are identical or have the same electronegativity, the bonding electrons are shared equally. Such a bond is said to be *nonpolar.* This is the case for C—C bonds. However, if two atoms differing in electronegativity are bonded, one will exert a greater attraction on the electrons of the bond. A bond of this type is said to be *dipolar;* one end will be slightly negatively charged (δ^-), and the other end will be slightly positively charged (δ^+). In a water molecule, for example, the oxygen atom, which has an electronegativity of 3.5, attracts the electrons of the bonds more than do the hydrogen atoms, which each have an electronegativity of 2.1; that is, the bonding electrons in their orbital motion spend more time around the oxygen atom than around the hydrogens. Because both hydrogen atoms are on the same side of the oxygen atom, that side of the molecule has a slight positive charge, whereas the other side has a slight negative charge. A molecule that incorporates separated positive and negative charges is called a *dipole* (Figure 2-4).

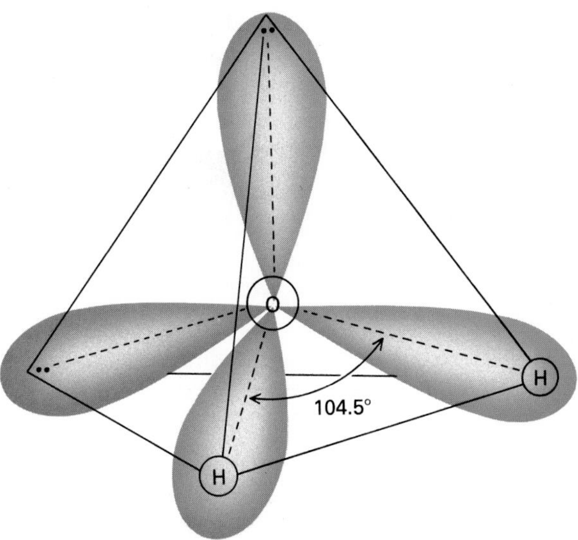

Figure 2-3 An oxygen atom in a water molecule has four outer electron orbitals that point approximately toward the corners of a tetrahedron. Each orbital contains two electrons. The electrons in two of the orbitals are shared with hydrogen atoms. The other two orbitals contain pairs of electrons that belong only to the oxygen atom.

If two strongly electronegative atoms are bonded together, as is the case for the two oxygen atoms in hydrogen peroxide (H—O—O—H), the bond is relatively unstable and reactive because the bonding electrons are pulled toward both electronegative oxygen atoms. Thus,

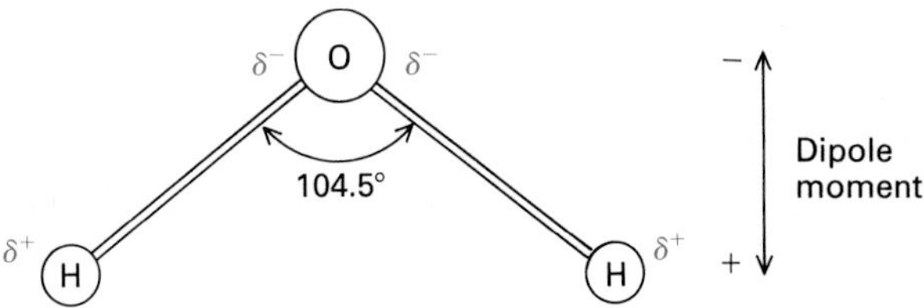

Figure 2-4 The water molecule as an electric dipole. The symbol δ represents a partial charge—i.e., a charge weaker than that on an electron or a proton. The direction of the charge separation is termed the dipole moment.

in the laboratory hydrogen peroxide can participate in a number of reactions in which one of its oxygen atoms is transferred to an acceptor molecule. This occurs, for example, in the conversion of methionine, a normal amino acid, to the abnormal amino acid methionine sulfone:

$$\underset{\substack{\text{Methionine}\\ \text{(normal amino acid)}}}{H_3C{-}S{-}CH_2{-}CH_2{-}CH(NH_2)COOH} + 2H_2O_2 \longrightarrow$$

$$\underset{\substack{\text{Methionine sulfone}\\ \text{(abnormal amino acid)}}}{H_3C{-}\overset{\overset{\displaystyle O}{\|}}{\underset{\underset{\displaystyle O}{\|}}{S}}{-}CH_2{-}CH_2{-}CH(NH_2)COOH} + 2H_2O$$

Table 2-2 The strengths of some bonds that are important in biological systems*

Type of bond	Energy (kcal/mol)	Type of bond	Energy (kcal/mol)
SINGLE BOND		DOUBLE BOND	
O—H	110	C=O	170
H—H	104	C=N	147
P—O	100	C=C	146
C—H	99	P=O	120
N—H	93	TRIPLE BOND	
C—O	84	C≡C	195
C—C	83	HYDROGEN BOND†	
S—H	81	NH···O	4–5
C—N	70	OH···N	4–5
C—S	62	NH···N	4–5
N—O	53	OH···O	4–5
S—S	51		

*The value given for each bond is the amount of energy required to break it. Note that double and triple bonds are stronger than single bonds.

†In aqueous solutions, the strength of a hydrogen bond is only 1 to 2 kcal/mol, as explained on page 26.

Weak Bonds Stabilize the Structures of Biological Molecules

The strongest forces that hold the atoms of a molecule together are covalent bonds. Typical energies released in covalent bond formation are shown in Table 2-2. The values range from 50 to 200 kcal/mol. However, not all of the bonds responsible for maintaining the structures of large molecules are covalent. The forces that stabilize the three-dimensional architecture of individual large molecules and that bind one molecule to another are often much weaker. The energy released in the formation of these weak bonds is only 1 to 5 kcal/mol. Because the average kinetic energy of molecules at room temperature (25°C) is about 0.6 kcal/mol, many molecules will have enough energy to break these weak bonds. At physiological temperatures (25 to 37°C), weak bonds have a transient existence, but when many act together they can produce very stable structures.

To understand the source of this stability, let us consider a simple reaction of enormous importance in biological systems: the noncovalent binding of a small molecule, A, to a protein or other large molecule, P*:

$$A_{free} + P \rightleftharpoons A_{bound}{\cdot}P$$

*This type of interaction occurs, for instance, between a hormone (A) and its receptor (P).

The greater the energy released in the binding of A to P, the more stable the A·P complex is relative to the reactants. To find the numerical relation between stability and binding energy, we use the *Boltzmann distribution*. This equation quantifies the amount of each interconvertible species—here, A_{bound} and A_{free}—with respect to the difference in their energy content:

$$\frac{A_{bound}}{A_{free}} = \exp\left(-\frac{E_{bound} - E_{free}}{RT}\right) = \exp\left(-\frac{\Delta E}{RT}\right)$$

or

$$\ln\frac{A_{bound}}{A_{free}} = -\frac{\Delta E}{RT}$$

where $E_{bound} - E_{free}$, or ΔE, is the difference in energy between the bound and free forms of A; where R is the gas constant, 1.987 cal/(degree · mol); and where T is the absolute temperature (at 25°C, $T = 25 + 273$ K = 298 K).

Suppose that $\Delta E = -1$ kcal/mol, a typical energy for a single weak, noncovalent bond. Expressing this value in calories, we have

$$\ln\frac{A_{bound}}{A_{free}} = -\frac{-1000}{1.987 \times 298} = 1.69$$

$$\frac{A_{bound}}{A_{free}} = 5.42$$

Thus,

$$\frac{A_{\text{bound}}}{A_{\text{total}}} = \frac{A_{\text{bound}}}{A_{\text{bound}} + A_{\text{free}}} = \frac{5.42}{5.42 + 1} = 0.84$$

so 84 percent of A will be bound to the polymer P. Suppose, however, that A is bound to P by two weak bonds, each of which has a strength of 1 kcal/mol. The total binding energy will be 2 kcal per mole of A bound, and we have

$$\ln \frac{A_{\text{bound}}}{A_{\text{free}}} = (-2)\frac{-1000}{1.987 \times 298} = 3.38$$

$$\frac{A_{\text{bound}}}{A_{\text{free}}} = 29.37$$

In this case, $A_{\text{bound}}/A_{\text{total}} = 29.37/30.37 = 0.97$, so 97 percent of A will be bound. Similarly, if A is bound to P by three weak bonds of 1 kcal/mol each, then $A_{\text{bound}}/A_{\text{free}} = 161$, and 99.4 percent of A will be bound. The important conclusion is that the stability of binding, as measured by the fraction of molecules bound to P, increases exponentially with the total bond strength. Three weak bonds result in an interaction that is vastly more stable than that of one weak bond.

In certain cases, the weak bonds that hold molecules together can break independently of one another. Consider, for example, two molecules bound together by four weak bonds. If two of the bonds break, the molecules will still be held together by the other two, and the two broken bonds will probably re-form before the two intact bonds have broken (Figure 2-5). This is why certain molecules can maintain stable associations with other molecules if enough weak bonds work together. At any instant at least a few will be intact.

Because weak bonds play such crucial roles in biological structures, any student of the cell should be well acquainted with them. The four main types are *the hydrogen bond, the ionic bond, the van der Waals interaction,* and the *hydrophobic bond* or *interaction.* We shall consider each of these in turn.

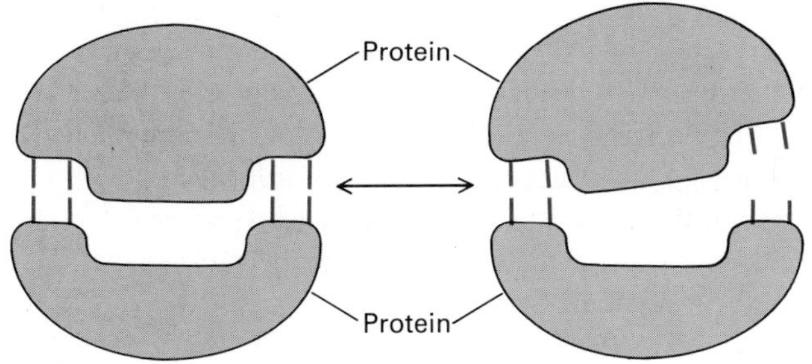

Figure 2-5 The importance of multiple weak bonds in stabilizing an association between two molecules. In the complex on the left, four noncovalent bonds bind the two protein molecules together. Even if two of the bonds are broken, as in the complex on the right, the remaining two bonds will facilitate the re-formation of the broken ones.

Table 2-3 Typical hydrogen bond lengths*

Bond	Length (nm)
OH···O⁻	0.26
OH···O	0.27
OH···N	0.28
⁺NH···O	0.29
NH···O	0.30
NH···N	0.31

*The values listed are the distances between the nuclei of the donor and the acceptor atom.

The Hydrogen Bond A hydrogen atom can normally form a covalent bond with only one other atom at a time. However, a covalently bonded hydrogen atom may form an additional bond: a *hydrogen bond* is a weak association between an electronegative atom (the *acceptor* atom) and a hydrogen atom that is covalently bonded to another atom (the *donor* atom). The hydrogen atom is closer to the donor (D) than to the acceptor (A):

$$\text{D—H} + \text{A} \rightleftharpoons \text{D—H}\underbrace{\cdots}_{\text{Hydrogen bond}}\text{A}$$

The covalent bond between the donor and the hydrogen atom must be dipolar, and the outer shell of the acceptor atom must have nonbonding electrons that attract the δ^+ charge of the hydrogen atom. The hydrogen bond in water is a classic example: a hydrogen atom in one molecule is attracted to a pair of electrons in the outer shell of an adjacent oxygen atom (Figure 2-6). The strength of a hydrogen bond in water is about 5 kcal/mol, which is much weaker than a covalent H—O bond; the strengths of hydrogen bonds between other donor and acceptor atoms are similar (see Table 2-2).

An important feature of all hydrogen bonds, including that of water, is directionality. The strongest hydrogen bonds are those in which all three participating atoms—the donor, the hydrogen atom, and the acceptor—lie in a straight line. The distance between the nuclei of the hydrogen and oxygen atoms of adjacent hydrogen-bonded molecules in water is approximately 0.27 nm, which is about twice the length of covalent H—O bonds in water. This is one reason why hydrogen bonds are much weaker than covalent bonds. Most hydrogen bonds are between 0.26 and 0.31 nm long (Table 2-3). The length of the covalent bond between the donor and the hydrogen atom of a hydrogen bond is a bit longer than it would be if there

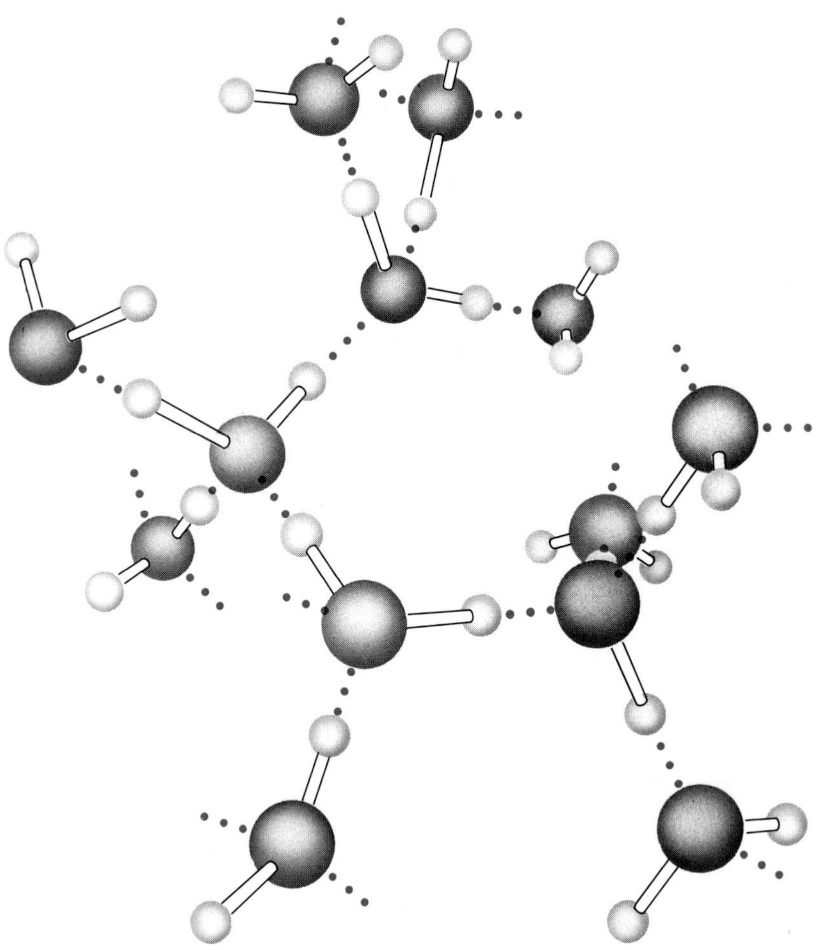

Figure 2-6 In liquid water, each H_2O molecule apparently forms transient hydrogen bonds with several others, which creates a fluid network of hydrogen-bonded molecules. The precise structure of liquid water is still not known with certainty (see page 27).

were no acceptor for the hydrogen, because the acceptor "pulls" the hydrogen away from the donor.

Because all N—H bonds are dipolar, nitrogen can act as a donor in a hydrogen bond; so can oxygen in the O—H bonds of molecules other than water. Nitrogen and oxygen can also be acceptors in the formation of hydrogen bonds with adjacent molecules. The amino (—NH_2) and hydroxyl (—OH) groups are those most frequently involved in hydrogen bond formation in biological systems. Such groups confer solubility in water on many molecules. For instance, methanol (CH_3OH) and methylamine (CH_3NH_2) can form many hydrogen bonds with water, enabling these molecules to dissolve in water at high concentrations (Figure 2-7).

Hydrogen bonds among molecules dissolved in nonpolar solutions are stronger than those among molecules in aqueous solutions. The energy of a hydrogen bond between two molecules dissolved in a nonpolar solvent is approximately 5 to 8 kcal/mol. Formation of hydrogen bonds between two molecules dissolved in water releases much less than 5 to 8 kcal/mol, because the molecules will often be hydrogen-bonded to water when they are not hydrogen-bonded to each other. For example, the formation of a hydrogen bond between molecules containing a donor —OH oxygen and an acceptor —NH_2 nitrogen requires the breaking of hydrogen bonds with water (Figure 2-8). The energy released overall may be only 1 to 2 kcal/mol.

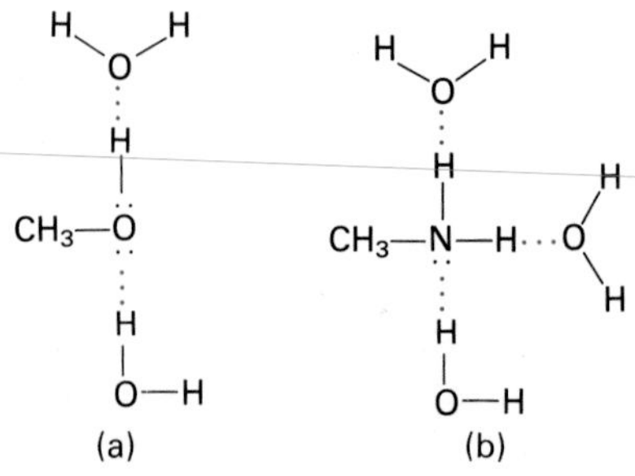

Figure 2-7 Hydrogen bonds between (a) methanol and water, and (b) methylamine and water. Each of the two pairs of nonbonding electrons in the outer shell of oxygen can accept a hydrogen atom in a hydrogen bond. The single pair of unshared electrons in the outer shell of nitrogen is similarly capable of becoming an acceptor in a hydrogen bond. The —OH oxygen and the —NH_2 nitrogen can also be donors for hydrogen bonds to oxygen atoms in H_2O.

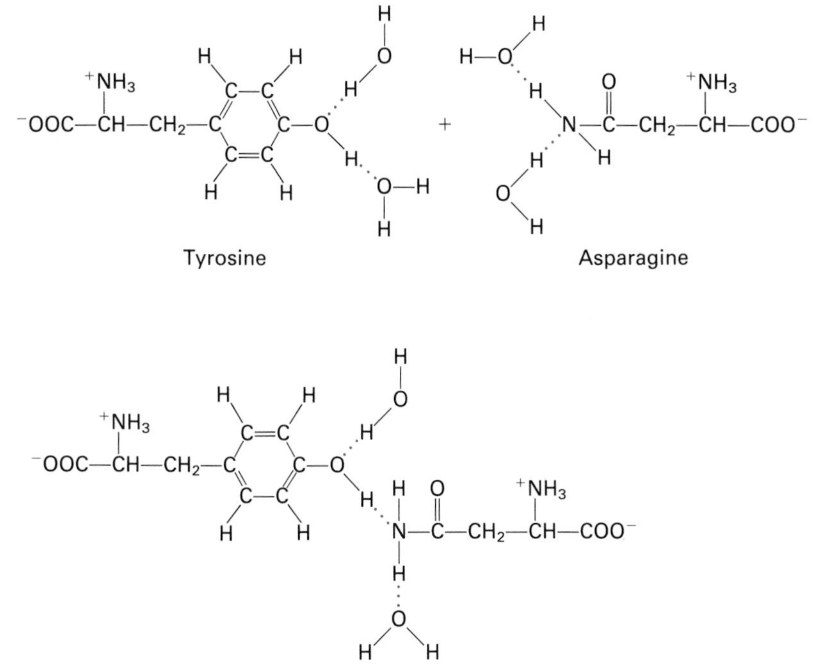

Figure 2-8 Both the —OH group of tyrosine and the $—NH_2$ of the amide ($—CONH_2$) group of asparagine can form multiple hydrogen bonds with water molecules *(top)*. The —OH oxygen can also be the donor in a hydrogen bond with the $—NH_2$ nitrogen *(bottom)*. For this tyrosine-asparagine bond to be formed, hydrogen bonds between the —OH and water and between the $—NH_2$ and water must be broken. Thus, the tyrosine-asparagine hydrogen bond is less stable in an aqueous solution than in a nonpolar solvent, where it is quite strong.

Another reason why a single hydrogen bond is not very stable at 25°C is that, as mentioned earlier, the average kinetic energy of molecules at room temperature is about 0.6 kcal/mol. However, because the aggregate strength of many weak bonds can be enormous, most large biological molecules are rigid in structure, due to the stability conferred by multiple hydrogen bonds between different parts of the molecules.

The Structures of Liquid Water and of Ice Hydrogen bonding among water molecules is of crucial importance for life on earth because all life requires an aqueous environment. The mutual attraction of its molecules causes water to have melting and boiling points at least 100°C higher than they would be if water were nonpolar; in the absence of these intermolecular attractions, water on earth would exist primarily as a gas.

Ordinary ice is a crystal. Each oxygen atom forms two covalent bonds with two hydrogen atoms, and two hydrogen bonds with hydrogen atoms of adjacent molecules. The oxygen atom is at the center of a tetrahedron (Figure 2-9), and the two bonds in which each hydrogen atom participates are collinear:

$$O—H\cdots O$$

The exact structure of liquid water is still unknown. It is believed to contain many icelike, maximally hydrogen-bonded networks, but these are presumably so transient and small in volume that stable crystals do not form. Most likely, water molecules are in rapid motion, so that they constantly make and break hydrogen bonds with adjacent molecules. As the temperature of water increases toward 100°C, the kinetic energy of its molecules becomes greater than the energy of the hydrogen bonds connecting them, and the gaseous form of water appears.

Ionic Bonds Covalent bonds between different atoms are generally asymmetric, because one of the atoms is usually more electronegative than the other. The more electronegative atom attracts the shared electrons more, which makes the bond dipolar (see Table 2-1). When a dipolar bond breaks, the bonding electrons often stay with the more electronegative atom, which then becomes a negatively charged ion, or an *anion;* the other part of the molecule becomes a positively charged ion, or a *cation*. In some compounds, the atoms are so different in electronegativity that the bonding electrons are always found around the more electronegative atom—that is, the electrons are never shared. In sodium chloride (NaCl), for example, the bonding electron contributed by the sodium atom is completely transferred to the chlorine atom. Even in solid crystals of NaCl, the sodium and chlorine atoms are ionized, so it is more accurate to write the formula for the compound as Na^+Cl^-. Because electrons are not shared, the bonds in such compounds cannot be considered covalent. They are called *ionic bonds,* and they result from the attraction of a positive charge for a negative charge. Unlike covalent or hydrogen bonds, ionic bonds do not have fixed geometric orientations. This is because the electrostatic field around an ion—that is, its attraction for an opposite charge—is uniform in all directions.

In aqueous solutions, simple ions of biological significance such as Na^+, K^+, Ca^{2+}, Mg^{2+}, and Cl^-, do not exist as free, isolated entities. Instead, each is surrounded

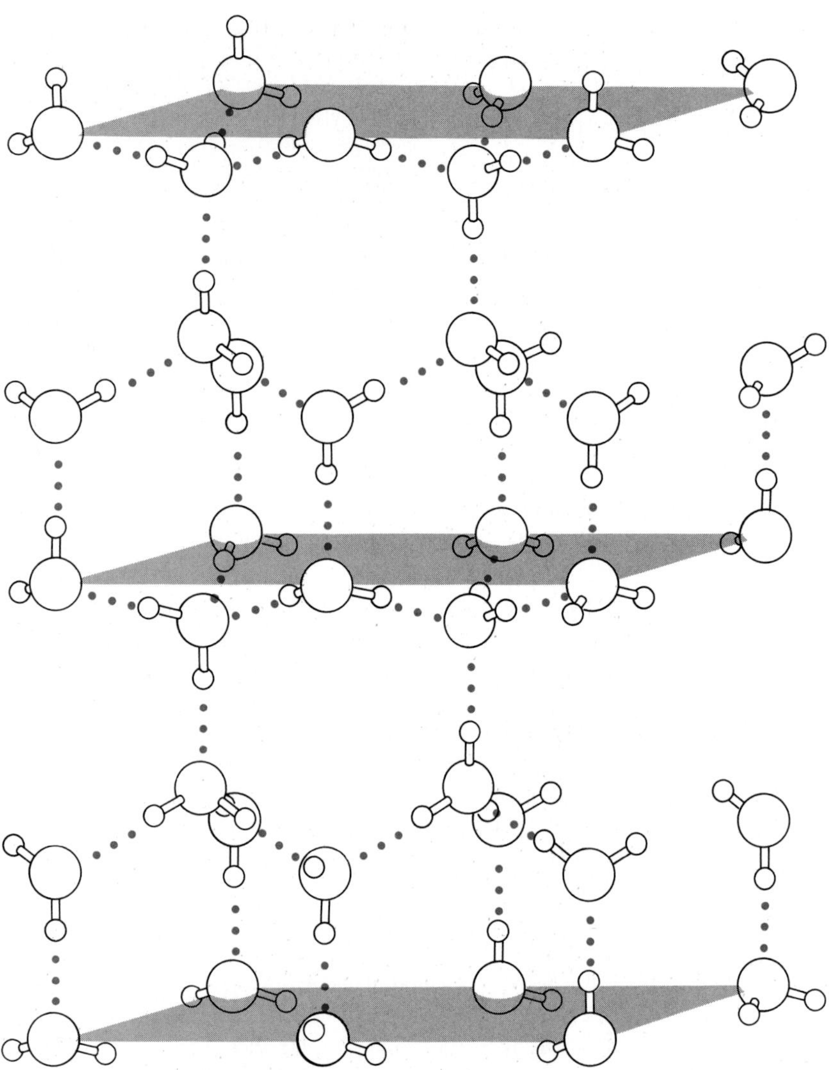

Figure 2-9 In an ice crystal, each oxygen atom is at the center of a tetrahedron, where it forms four bonds, one directed toward each vertex. Two of the bonds are covalent bonds with two hydrogen atoms; the other two are hydrogen bonds in which the oxygen is an acceptor. Thus, ice is a rigid lattice of H_2O molecules.

by a stable, tightly held shell of water molecules. The primary interaction is between the ion and the oppositely charged end of the water dipole; for example:

$$\mathrm{K^{+}}\ \ {}^{\delta-}\mathrm{O}\begin{matrix}\diagup\ \mathrm{H}^{\delta+} \\ \diagdown\ \mathrm{H}^{\delta+}\end{matrix}$$

A diagram of an Mg^{2+} ion and its six tightly bound water molecules is shown in Figure 2-10. As will be discussed in Chapters 15 and 17, ions play an important biological role when they pass through narrow pores, or channels, in membranes. Ionic movements through membranes are essential for the conduction of nerve impulses and for the stimulation of muscle contraction. In estimating the sizes of such ions, it is important to include the shell of water molecules.

The favorable energy of hydration of ions—that is, the large amount of energy released when ions tightly bind water molecules—is why most ionic compounds are quite soluble in water. Ions that are oppositely charged are shielded from one another by the water and thus tend not to recombine. Molecules having dipolar bonds also can attract water molecules, as can those that easily form hydrogen bonds. Such molecules are called *polar molecules*. Polar molecules are hydrophilic (from the Greek for "water-loving"), meaning that they can dissolve in water. Typical chemical groups that interact well with water are the hydroxyl, —OH; the amino, $—NH_2$; the peptide bond,

$$\begin{matrix}\mathrm{O} & & \mathrm{H} \\ \| & & | \\ —\mathrm{C} & — & \mathrm{N}—\end{matrix}$$

and the ester bond,

$$\begin{matrix}\mathrm{O} & & \\ \| & & \\ —\mathrm{C} & — & \mathrm{O}—\end{matrix}$$

Van der Waals Interactions When two atoms approach one another closely, they create a nonspecific, weak attractive force that produces a *van der Waals interaction*, named for Johannes Diderik van der Waals (1837–1923), the Dutch physicist who first described it.

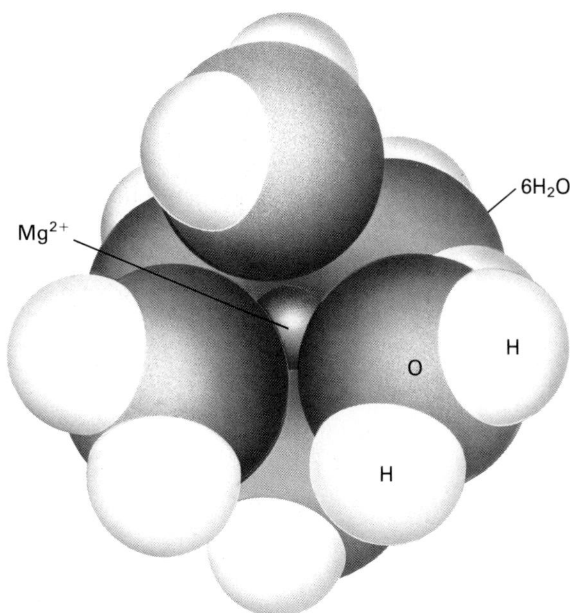

Figure 2-10 A magnesium ion (Mg^{2+}) in water has six H_2O molecules clustered around it. They are held in place by electrostatic interactions between the positive Mg^{2+} and the partially negative oxygen of the water dipole.

In any atom, momentary random fluctuations in the distribution of its electrons give rise to transient dipoles. If two noncovalently bonded atoms are close enough together, the transient dipole in one atom will perturb the electron cloud of the other. This perturbation will generate a transient dipole in the second atom, which will attract the dipole in the first. The interaction between these transiently forming dipoles results in a net weak attraction (Figure 2-11). Van der Waals interactions occur among all types of molecules, both polar and nonpolar. In particular, they are responsible for the cohesion among the molecules of nonpolar liquids and solids, such as heptane,

```
   H  H  H  H  H  H  H
   |  |  |  |  |  |  |
H—C—C—C—C—C—C—C—H
   |  |  |  |  |  |  |
   H  H  H  H  H  H  H
```

which cannot form ionic or hydrogen bonds with other molecules. The attraction decreases rapidly with increasing distance, so it is effective only when atoms are quite close to one another. However, if atoms get too close together, they become repelled by the negative charges in their outer electron shells.

When the van der Waals attraction between two atoms exactly balances the repulsion between their two electron clouds, the atoms are said to be in *van der Waals contact.* Each type of atom has a radius—the *van der Waals radius*—at which it is in van der Waals contact with other atoms. The van der Waals radius is a measure of the size of the atom. Two covalently bonded atoms are closer together than two atoms that are merely in van der Waals contact (Figure 2-12). Table 2-4 lists the van der Waals and covalent radii of some atoms that are important in biological systems.

The energy of the van der Waals interaction amounts to about 1 kcal/mol, a value only slightly higher than the average thermal (kinetic) energy of molecules at 25°C. Thus, the van der Waals interaction is even weaker than the hydrogen bond, which in aqueous solutions typically has an energy of 1 to 2 kcal/mol. The attraction between two large molecules can be appreciable, however, if they have precisely complementary shapes, so that they make many van der Waals contacts when they come into proximity. The van der Waals contacts are among the interactions that take place between an antibody molecule and its specific antigen, and between many enzymes and their specifically bound substrates, as will be discussed in Chapter 3.

Hydrophobic Interactions *Nonpolar molecules* contain neither ions nor dipolar bonds, and thus these molecules do not become hydrated. Because they are insoluble or almost insoluble in water, they are called *hydrophobic* (from the Greek for "water-fearing"). The covalent bonds between two carbon atoms and between carbon and hydrogen atoms are the most common nonpolar bonds in biological systems. *Hydrocarbons*—molecules made only of carbon and hydrogen—are virtually insoluble in water. The large hydrophobic molecule tristearin (Figure 2-13), a component of animal fat, is also insoluble in water even though it has six oxygen atoms that participate in some slightly dipolar bonds between carbon and oxygen. The core or central section of all biological membranes, including the surface membranes of cells, is com-

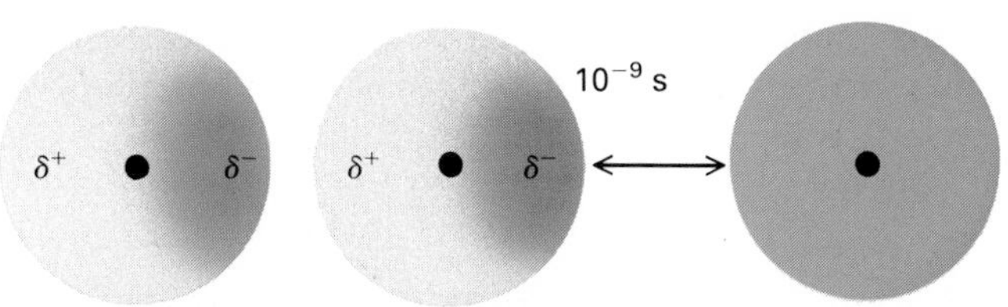

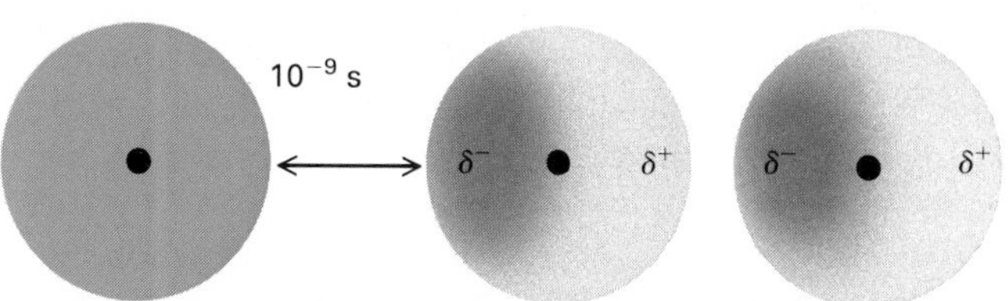

Figure 2-11 Momentary fluctuations in the distribution of the electrons of adjacent atoms produce the nonspecific attractions called van der Waals interactions. A transient dipole in the electron cloud around one atom induces a similar dipole in the other atom. The electrostatic attraction between the two transient dipoles creates the van der Waals interaction.

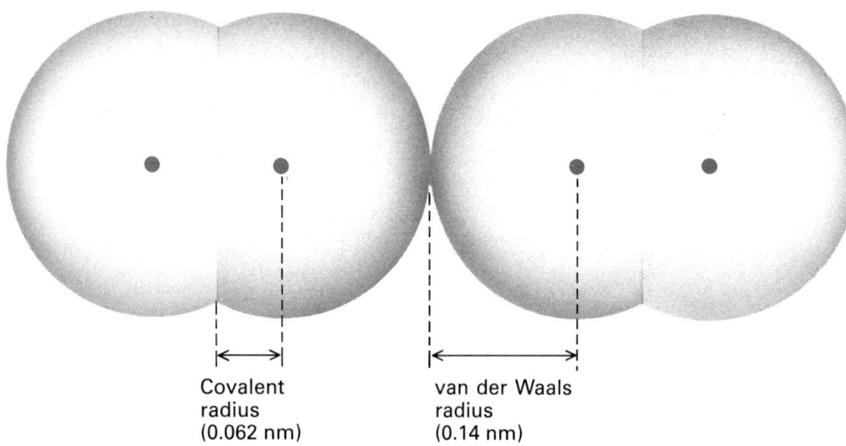

Figure 2-12 Two oxygen molecules in van der Waals contact. Each type of atom has a van der Waals radius, at which van der Waals interactions with other atoms are optimal. Because atoms repel one another if they are close enough together for their outer electron shells to overlap, the van der Waals radius is a measure of the size of the electron cloud surrounding an atom. (Note that the covalent radius indicated here is for the double bond of O=O; the single-bond covalent radius of oxygen is slightly longer, as shown in Table 2-4.)

Table 2-4 Van der Waals radii and covalent (single-bond) radii of some biologically important atoms*

Atom	Van der Waals radius (nm)	Covalent radius for a single bond (nm)
H	0.10	0.030
O	0.14	0.074
F	0.14	0.071
N	0.15	0.073
C	0.17	0.077
S	0.18	0.103
Cl	0.18	0.099
Br	0.20	0.114
I	0.22	0.133

*The internuclear distance for a covalent bond or a van der Waals interaction is approximately the sum of the values for the two participating atoms. Note that the van der Waals radius is about twice as long as the covalent radius.

posed almost exclusively of the hydrocarbon portions of molecules and contains few, if any, water molecules or other polar molecules.

The force that causes hydrophobic molecules or nonpolar portions of molecules to aggregate together in water is called the hydrophobic bond or, more precisely, the *hydrophobic interaction.* The only actual attraction between the molecules is a van der Waals attraction; the "bond" occurs by default, due to the exclusion of the hydrophobic molecules from the hydrogen-bonded water lattice.

To understand the source of the energy of hydrophobic interactions, consider that for nonpolar molecules to dissolve in water, they must insert themselves into the hydrogen-bonded lattice. This requires the breaking of hydrogen bonds among the water molecules, and the formation of a hydrogen-bonded "cage" or lattice around each hydrocarbon molecule. The effect is to increase the structural organization of the water. Because water molecules are normally in constant, random motion, the formation of such cages restricts the freedom of motion of a number of the water molecules. This situation is energetically unfavorable; it is said to decrease the entropy, or randomness, of the population of water molecules (see page 36 for a further discussion of entropy). The opposition of water molecules to having their motion restricted by dissolved portions of hydrophobic compounds is the major reason why molecules such as heptane and tristearin are essentially insoluble in water. The hydrophobic molecules are "corralled" by their mutual exclusion from the hydrogen-bonded water lattice.

In some cases, water does contain dissolved or partially dissolved hydrocarbons, and such mixtures may have more hydrogen bonds than pure water does. However, with the solutes present, the water structure becomes more rigid and organized—lower in entropy—and it is therefore unstable.

Small hydrocarbons such as butane,

$$CH_3—CH_2—CH_2—CH_3$$

are slightly soluble in water [90 g/l at 25°C and 1 atmosphere (atm) of pressure]. Because these molecules are small, the formation of cages of water molecules around them restricts the freedom of relatively few water molecules. Also, the number of hydrogen bonds between the water molecules forming the cages may be equal to or even greater than the number of hydrogen bonds in pure water. Note, however, that 1-butanol,

$$CH_3—CH_2—CH_2—CH_2OH$$

mixes completely with water in all proportions. Thus, the replacement of just one hydrogen with the dipolar —OH group allows the molecule to form hydrogen bonds with water and greatly increases its solubility.

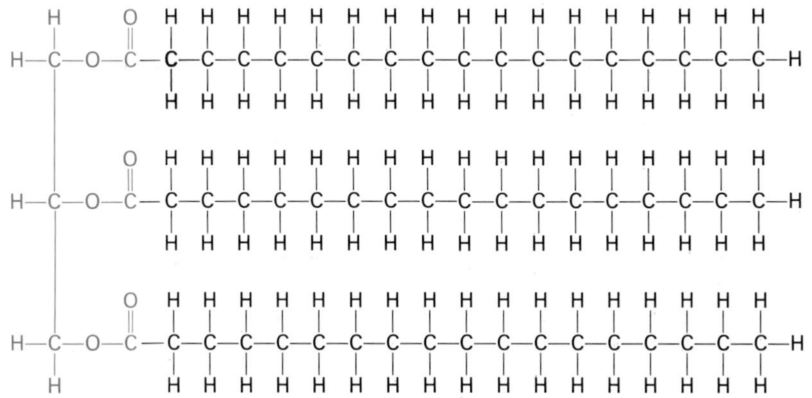

Figure 2-13 The chemical structure of tristearin, or tristearoyl glycerol, a component of natural fats. It contains three molecules of stearic acid,

$$CH_3—(CH_2)_{16}—COOH$$

esterified to one molecule of glycerol,

$$HOCH_2—CH(OH)—CH_2OH$$

The atoms at the polar, or hydrophilic, end of the molecule are colored; the rest of the molecule is highly hydrophobic.

The Binding Specificity Conferred by Weak Interactions We have already noted that multiple weak bonds can contribute to the stability of large biological molecules. These interactions can also confer *specificity* by determining how large molecules will fold or which regions of different molecules will bind together. All weak bonds are effective only over a short range and require close contact between the reacting groups. Thus, for weak bonds to form properly there must be a complementarity between the sites on two interacting surfaces. Figure 2-14 illustrates how several different weak bonds and interactions can bind two protein chains together. Almost any other arrangement of the same groups on the two surfaces would not allow the molecules to bind so tightly.

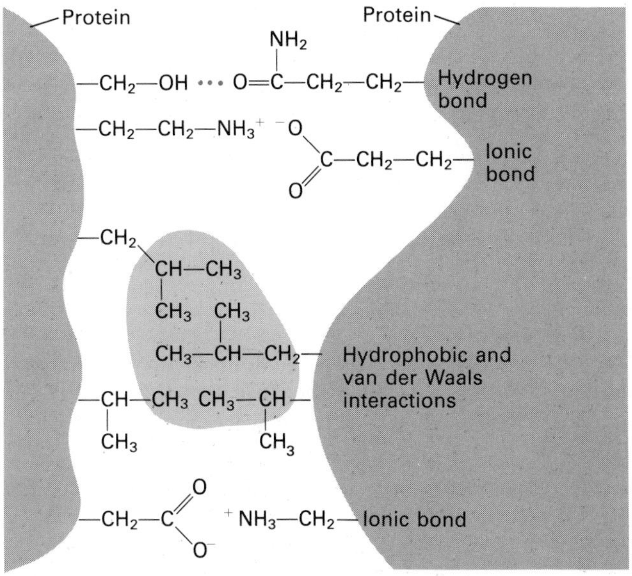

Figure 2-14 The binding of a hypothetical pair of proteins by two ionic bonds, one hydrogen bond, and one large hydrophobic interaction. The structural complementarity of the surfaces of the two molecules gives rise to this particular combination of weak bonds, and hence to the specificity of binding between the molecules.

pH and the Concentration of Hydrogen Ions

The solvent inside cells and in all extracellular fluids is water. A major characteristic of any aqueous solution is the concentration of positively charged hydrogen ions (H^+) and negatively charged hydroxyl ions (OH^-). These ions are important for several reasons. As the dissociation products of H_2O, they are the ubiquitous constituents of all living systems. They are also readily liberated by many reactions that take place between organic molecules.

Water Dissociates into Hydronium and Hydroxyl Ions

When a water molecule dissociates, one of its dipolar H—O bonds breaks; the result is a positively charged hydrogen ion (H^+) and a negatively charged hydroxyl ion (OH^-). The hydrogen ion—a proton—has a short lifetime as a free particle; it combines with a water molecule to form a *hydronium ion* (H_3O^+). This understood, it is still convenient to refer to the concentration of hydrogen ions in a solution, $[H^+]$, even though what is really meant is the concentration of hydronium ions, $[H_3O^+]$.

The dissociation of water is reversible,

$$H_2O \rightleftharpoons H^+ + OH^-$$

and at 25°C,

$$[H^+][OH^-] = 10^{-14}\ M^2$$

where $[H^+]$ and $[OH^-]$ are the concentrations of the respective ions in moles per liter or molarity, symbolized by *M*. Because 1 mole of water weighs 18 g and 1 liter of water weighs 1000 g, pure water is about 55 *M* H_2O. In a pure aqueous solution, $[H^+] = [OH^-] = 10^{-7}\ M$.

The concentration of hydrogen ions in a solution is expressed conventionally as its pH:

$$pH = -\log [H^+] = \log \frac{1}{[H^+]}$$

In pure water at 25°C, $[H^+] = 10^{-7}$ *M*, so pH = $-\log 10^{-7} = 7.0$. On the pH scale from 0 to 14, 7.0 is considered neutral. pH values below 7.0 indicate acidic solutions, whereas values above 7.0 indicate basic or alkaline solutions (Table 2-5). In a 0.1 *M* solution of hydrogen chloride (HCl) in water, $[H^+]$ is 0.1 *M* because all of the HCl has dissociated into H^+ and Cl^- ions. Thus, for this solution,

$$pH = -\log 0.1 = 1.0$$

One of the most important properties of a biological fluid is its pH. The pH of the cytoplasm of cells is normally about 7.2. Cells of higher organisms contain organelles, such as lysosomes, in which the pH is much lower—about 5. This represents a value of $[H^+]$ that is more than 100 times higher than its value in the cytoplasm. Lysosomes contain many degradative enzymes that function optimally in such an acidic environment, whereas their action would be inhibited in the near-neutral environment of the cytoplasm. Thus, maintenance of a specific pH is imperative for the proper functioning of some cellular structures. On the other hand, dramatic shifts in cellular pH may play an important role in controlling cellular activity. For example, the pH of the cytoplasm of an unfertilized sea urchin egg is 6.6. Within 1 min of fertilization, the pH rises to 7.2 (i.e., $[H^+]$ decreases to about one-fourth of its original value). The change in pH somehow triggers the growth and division of the egg.

Table 2-5 The pH scale

	Concentration of hydrogen ions (mol/l)	pH	Example
Increasing acidity ↑	10^{-0}	0	
	10^{-1}	1	Gastric fluids
	10^{-2}	2	Lemon juice
	10^{-3}	3	Vinegar
	10^{-4}	4	Acid soil
	10^{-5}	5	Lysosomes
	10^{-6}	6	Cytoplasm of contracting muscle
Neutral	10^{-7}	7	Pure water and cytoplasm
Increasing alkalinity ↓	10^{-8}	8	Sea water
	10^{-9}	9	Very alkaline natural soil
	10^{-10}	10	Alkaline lakes
	10^{-11}	11	Household ammonia
	10^{-12}	12	Lime (saturated solution)
	10^{-13}	13	
	10^{-14}	14	

Acids Release Hydrogen Ions and Bases Combine with Hydrogen Ions

In general, any molecule or ion that can release a hydrogen ion is called an *acid;* any molecule or ion than can combine with a hydrogen ion is called a *base*. Thus, hydrogen chloride is an acid. Hydroxide ion is a base, as is ammonia (NH_3), which readily picks up a hydrogen ion to become an ammonium ion (NH_4^+). Many organic molecules are acidic because they have a carboxyl group (—COOH), which tends to dissociate to form the negatively charged *carboxylate ion* ($—COO^-$):

$$X—C(=O)OH \rightleftharpoons X—C(=O)O^- + H^+$$

where X represents the rest of the molecule. The amino group ($—NH_2$), which is part of many important biological molecules, is a base because, like ammonia, it can take up a hydrogen ion:

$$X—NH_2 + H^+ \rightleftharpoons X—NH_3^+$$

When acid is added to a solution, $[H^+]$ increases (i.e., the pH goes down). Consequently, $[OH^-]$ decreases because the hydroxyl ions combine with the hydrogen ions. On the other hand, adding a base to a solution decreases $[H^+]$ (i.e., the pH goes up). Because $[H^+][OH^-] = 10^{-14}\ M^2$, any increase in $[H^+]$ is coupled with a reduction in $[OH^-]$, and vice versa. No matter how strongly acidic or alkaline a solution is, it always contains both kinds of ions—neither $[OH^-]$ nor $[H^+]$ is ever zero. For example, if $[H^+] = 0.1\ M$ (i.e., pH = 1.0), then $[OH^-] = 10^{-13}\ M$.

The degree to which a dissolved acid releases hydrogen ions, or to which a base takes them up, depends partly on the pH of the solution. Amino acids have the general formula

$$H—C(NH_2)(R)—COOH$$

(where R stands for the rest of the molecule), but in neutral solutions (pH = 7.0), they exist predominantly in the doubly ionized form

$$H—C(NH_3^+)(R)—COO^-$$

Such a dipolar ion is called a *zwitterion*. In solutions at low pH, carboxylate ions recombine with the abundant hydrogen ions, so the predominant form of amino acid molecules is

$$\begin{array}{c} NH_3^+ \\ | \\ H{-}C{-}COOH \\ | \\ R \end{array}$$

At high pH, the scarcity of hydrogen ions decreases the chances that an amino group or a carboxylate ion will pick up a hydrogen ion; thus, the predominant form of amino acid molecules is

$$\begin{array}{c} NH_2 \\ | \\ H{-}C{-}COO^- \\ | \\ R \end{array}$$

Many Biological Molecules Contain Multiple Acidic or Basic Groups

Many of the chemicals used by cells have multiple acidic or basic groups, and in the laboratory it is often essential to know the precise state of dissociation of each of these groups at various pH values. The dissociation of a simple acid HA, such as acetic acid (CH_3COOH), is described by

$$HA \rightleftharpoons H^+ + A^-$$

The equilibrium constant K_a for this reaction is

$$K_a = \frac{[H^+][A^-]}{[HA]}$$

By taking the logarithm of both sides and rearranging the result, we can derive the very important relation known as the *Henderson-Hasselbalch equation:*

$$\log K_a = \log \frac{[H^+][A^-]}{[HA]}$$

$$\log K_a = \log [H^+] + \log \frac{[A^-]}{[HA]}$$

$$-\log [H^+] = -\log K_a + \log \frac{[A^-]}{[HA]}$$

Substituting pH for $-\log [H^+]$ and pK_a for $-\log K_a$, we have

$$pH = pK_a + \log \frac{[A^-]}{[HA]}$$

The pK_a of any acid is equal to the pH at which half of the molecules are dissociated and half are neutral. This can be derived by observing that if $pK_a = pH$, then $\log([A^-]/[HA]) = 0$, or $[A^-] = [HA]$. The Henderson-Hasselbalch equation allows us to calculate the degree of dissociation of an acid if both the pH of the solution and the pK_a of the acid are known.

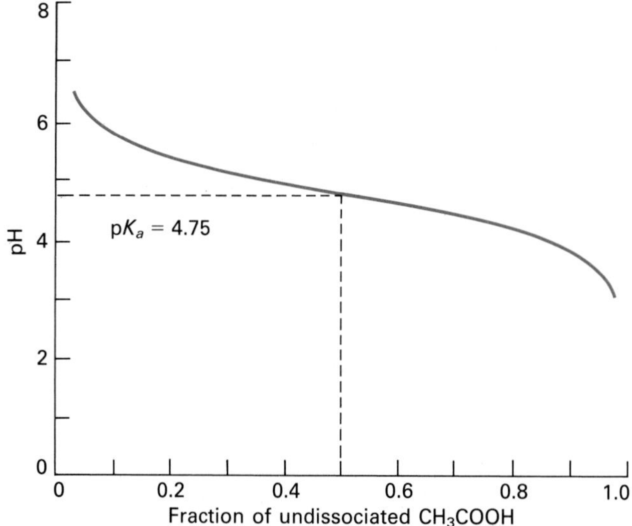

Figure 2-15 The titration curve of acetic acid (CH_3COOH). The pK_a for the dissociation of acetic acid to H^+ and CH_3COO^- is 4.75. Because pH is measured on a logarithmic scale, the solution changes from 91 percent CH_3COOH at pH 3.75 to 9 percent CH_3COOH at pH 5.75. In this pH range the acid has its maximum buffering capacity (see text).

A *titration curve* shows the fraction of molecules in the acid form (HA) as a function of pH (Figure 2-15). At one pH unit below the pK_a of an acid, 91 percent of the molecules are in the HA form; at one pH unit above the pK_a, 91 percent are in the A^- form. Suppose that we have a solution of an acid (or base) at its pK_a value, and we add to it additional acid or base. The pH of the solution changes, but less than would occur if the first acid were not present: protons released by an added acid are taken up by the A^- form of the first acid; or, the hydroxyl ions generated by the added base are neutralized by protons released by the first acid. This property of acids (and bases) is called buffering. An acid's or base's buffering capacity falls off rapidly at more than one pH unit from the pK_a.

Phosphoric acid is an example of a physiologically important acid that has multiple groups that are capable of dissociating:

$$\begin{array}{c} O \\ \| \\ HO{-}P{-}OH \\ | \\ OH \end{array}$$

The three protons do not dissociate simultaneously; loss of each can be described by a discrete pK_a constant:

$$\begin{array}{llll} H_3PO_4 & \rightleftharpoons H_2PO_4^- & + H^+ & pK_a = 2.1 \\ H_2PO_4^- & \rightleftharpoons HPO_4^{2-} & + H^+ & pK_a = 7.2 \\ HPO_4^{2-} & \rightleftharpoons PO_4^{3-} & + H^+ & pK_a = 12.7 \end{array}$$

Figure 2-16 shows the titration curve of phosphoric acid. Phosphate ions are present in cells in considerable quantities, and they play an important role in maintaining, or buffering, the pH of the cytoplasm. The pK_a for the dissociation of the second proton, pH 7.2, is similar to the pH of the cytoplasm. Because pK_a = pH = 7.2, then, according to the middle equation above and the Henderson-Hasselbalch equation, 50 percent of cellular phosphate is in the HPO_4^{2-} form and the other 50 percent is $H_2PO_4^-$.*

In nucleic acids, phosphate is found as a diester. It is linked to two carbon atoms of adjacent ribose sugars:

$$-\overset{|}{\underset{|}{C}}-O-\overset{\overset{O}{\|}}{\underset{\underset{OH}{|}}{P}}-O-CH_2-$$

The pK_a for the dissociation of the —OH proton is about 3, which is similar to the pK_a for the dissociation of the first proton from phosphoric acid; thus, at neutral pH, each phosphate residue in DNA or RNA is negatively charged:

$$-\overset{|}{\underset{|}{C}}-O-\overset{\overset{O}{\|}}{\underset{\underset{O^-}{|}}{P}}-O-CH_2-$$

This is why DNA and RNA are called nucleic *acids*.

Chemical Reactions

Any chemical—whether in a cell or in free solution—can, in principle, undergo many possible chemical reactions. To predict which reactions will in fact take place, we must determine two different parameters of each reaction: the *rate* at which the reaction can occur under the given conditions (such as concentration, temperature, and pressure), and the *extent* to which the reaction can proceed. An enzyme, as we shall see, is a biological catalyst that accelerates the rate of a particular chemical reaction, but that does not affect the extent of the reaction.

Every Chemical Reaction Has a Defined Equilibrium Constant

When reactants first come together—that is, before any products have been formed—they react at a rate determined in part by their initial concentrations:

$$\text{Reactants} \longrightarrow \text{products}$$

As the reaction products accumulate, the concentrations of reactants decrease and so does the reaction rate. Mean-

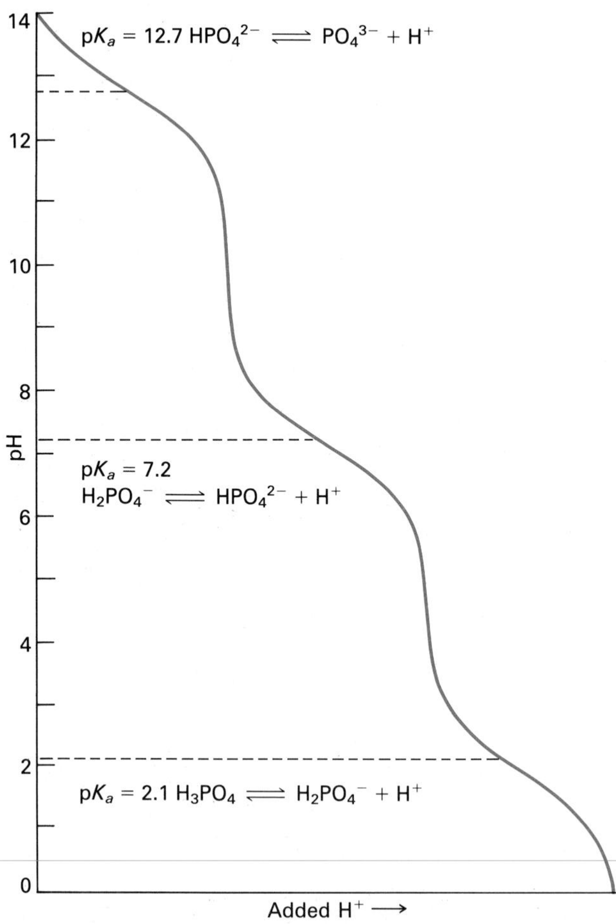

Figure 2-16 The titration curve of phosphoric acid (H_3PO_4). This biologically ubiquitous molecule has three hydrogen atoms that dissociate from the phosphate group at different pH values; thus, phosphoric acid has three pK_a values. In dilute solution, the pK_a's are 2.1, 7.2, and 12.7.

*Cell biologists often need to prepare buffered solutions with defined pH values. Suppose that we need a 0.1 *M* solution of sodium phosphate, pH = 6.8. NaH_2PO_4 dissociates to Na_2HPO_4 via the reaction

$$NaH_2PO_4 + NaOH \longrightarrow Na_2HPO_4 + H_2O$$

or, considering only the phosphate,

$$H_2PO_4^- \longrightarrow HPO_4^{2-} + H^+$$

The pK_a for this reaction is 7.2 (see Figure 2-16). We can use the Henderson-Hasselbalch equation to calculate the relative quantities of Na_2HPO_4 and NaH_2PO_4 in a solution at pH 6.8:

$$6.8 = 7.2 + \log \frac{HPO_4^{2-}}{H_2PO_4^-}$$

$$-0.4 = \log \frac{HPO_4^{2-}}{H_2PO_4^-}$$

$$0.398 = \frac{HPO_4^{2-}}{H_2PO_4^-}$$

Thus, mixing 0.1 *M* solutions of Na_2HPO_4 and NaH_2PO_4 in a ratio of 0.398:1 will yield a solution of pH 6.8.

while, some of the product molecules begin to participate in the reverse reaction, which re-forms the reactants:

$$\text{Reactants} \longleftarrow \text{products}$$

This reaction is slow at first, but it speeds up as the concentrations of products increase. Eventually, the rates of the forward and reverse reactions become equal, so that the concentrations of reactants and products stop changing. The mixture is said to be in *chemical equilibrium.*

The ratio of products to reactants at equilibrium depends on the nature of the compounds and also on temperature and pressure. Under standard physical conditions (i.e., 25°C and 1 atm), the ratio of products to reactants at equilibrium, which is always the same for a given reaction, is termed the *equilibrium constant.* For a simple reaction $A + B \rightleftharpoons X + Y$, the equilibrium constant is given by

$$K_{eq} = \frac{[X][Y]}{[A][B]}$$

In general, for a reaction

$$aA + bB + cC + \ldots \rightleftharpoons zZ + yY + xX + \ldots$$

where the capital letters represent particular molecules or atoms and the lower-case letters represent the number of each in the reaction formula, the *equilibrium constant* is given by

$$K_{eq} = \frac{[Z]^z[Y]^y[X]^x \ldots}{[A]^a[B]^b[C]^c \ldots}$$

where the brackets indicate the equilibrium concentrations.

To illustrate several points concerning equilibrium, we shall use one fairly simple biochemical reaction: the interconversion of the compounds glyceraldehyde 3-phosphate (G3P) and dihydroxyacetone phosphate (DHAP). This reaction occurs in glycolysis, and it is catalyzed by the enzyme triosephosphate isomerase, as shown in Figure 2-17. The equilibrium constant for the reaction under standard conditions is 22.2:

$$K_{eq} = \frac{[\text{DHAP}]}{[\text{G3P}]} = 22.2$$

Thus, when the reaction reaches equilibrium, the ratio of the concentrations of DHAP and G3P is 22.2:1.

In reactions such as this one, which involves only a single reactant and a single product, at equilibrium the ratio of product to reactant is independent of the amounts of product and reactant initially present and is equal to the equilibrium constant K_{eq}. It is also independent of the rate of the reaction. In the presence of an enzyme or other catalyst, the rate of the reaction may increase. However, the final ratio of product to reactant will always be the same. Reactions associated with a large equilibrium constant are able to occur spontaneously, but the magnitude of this constant has no bearing on the rate of the reaction or on whether the reaction will take place at all under normal conditions. For example, despite the large equilibrium constant for the conversion of G3P into DHAP, in an aqueous solution in which the enzyme catalyst is absent, the reaction is so slow that it is undetectable.

Dihydroxyacetone phosphate

triosephosphate isomerase

Glyceraldehyde 3-phosphate

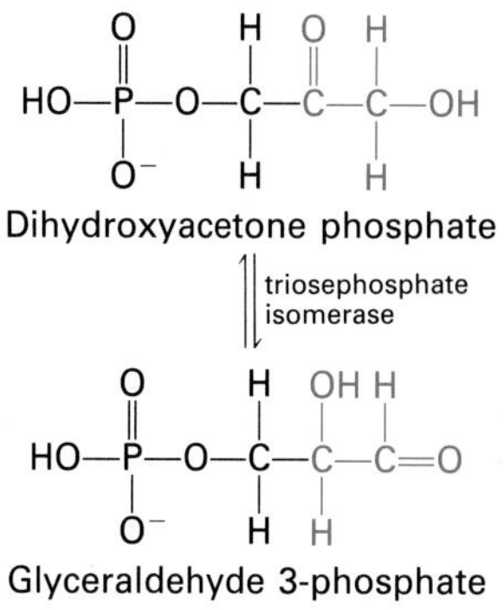

Figure 2-17 The interconversion of glyceraldehyde 3-phosphate and dihydroxyacetone phosphate is catalyzed by the enzyme triosephosphate isomerase. This reaction occurs as part of the metabolism of glucose to pyruvic acid by the glycolytic pathway, which is discussed in Chapter 5. The reacting portions of the molecules are shown in color.

In reactions involving multiple reactants and/or products, the equilibrium concentration of a particular product or particular reactant is a function of the initial amounts of all reactants and products and the equilibrium constant. Consider, for example, hydrolysis (cleavage by addition of water) of the dipeptide glycylalanine (GA) to glycine (G) and alanine (A):

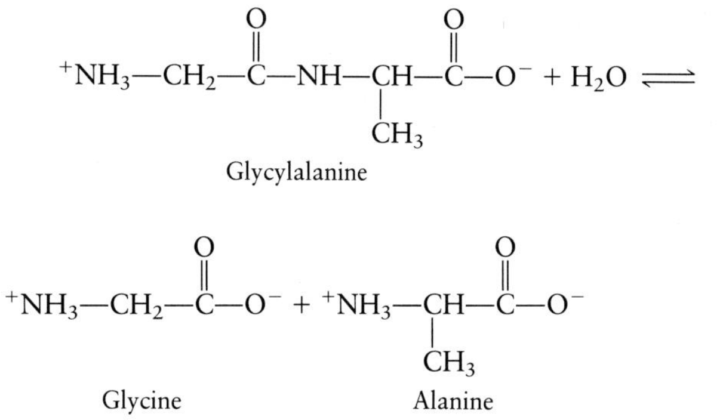

$$^{+}NH_3{-}CH_2{-}C(=O){-}NH{-}CH(CH_3){-}C(=O){-}O^- + H_2O \rightleftharpoons$$

Glycylalanine

$$^{+}NH_3{-}CH_2{-}C(=O){-}O^- + {}^{+}NH_3{-}CH(CH_3){-}C(=O){-}O^-$$

Glycine Alanine

$$K_{eq} = \frac{[G][A]}{[GA]}$$

(The concentration of water is, by convention, not included in the calculation of such equilibrium ratios.) The equilibrium is strongly in the direction of the formation of glycine and alanine. However, suppose that the initial reaction mixture contains a small amount of glycylalanine and a large amount of alanine. As the reaction proceeds, the concentration of alanine will always greatly exceed the concentration of glycine produced by hydrolysis. This excess alanine will drive the reaction in the reverse direction, so that at equilibrium more glycine will be found within the dipeptide than in its free form. Because the opposite is true for other initial concentrations of reactants and products, the equilibrium ratio of [G] to [GA] clearly depends on the initial concentrations of all reactants and products.

The Change in Free Energy, ΔG, Determines the Direction of a Chemical Reaction

The application of thermodynamics allows chemists and biologists to predict the direction that a reaction will take. Given a certain mixture of reactants and products, we can determine how their proportions will change in order to reach equilibrium. Because biological systems are held at constant temperature and pressure, it is possible to use a measure of potential energy to predict the direction of chemical reactions under these conditions. This measure of potential energy is called *free energy*. It is symbolized by G, after the American chemist Josiah Willard Gibbs (1839–1903), a founder of the science of thermodynamics. We are interested in what happens to the free energy when one molecule or configuration is changed into another. Thus, our concern is with relative, rather than absolute, values of free energy—in particular, with the difference between values before and after the change. This difference is written as ΔG. Gibbs showed that at constant temperature and pressure, "all systems change in such a way that free energy is minimized."

Here it is important that we define what is meant by a "system." For our purposes a system is an entity that does not exchange mass with its surroundings, but that can exchange heat. Thus, a test tube containing a chemical reaction is a system if it is surrounded by a water bath at constant temperature; heat can be exchanged through the walls of the test tube, but not matter. A cell can be a system, provided that it is not exchanging molecules with the extracellular solution.

In mathematical terms, Gibbs' law that systems change to minimize free energy is a statement about ΔG: if ΔG is negative for a chemical reaction or mechanical process, the reaction or process will be able to take place spontaneously. If ΔG is positive, it will not. Another way of stating this law is: if any system of particles (e.g., single oxygen atoms) has free energy G_1, and the system can change to have free energy G_2 (e.g., by pairing the oxygen atoms), then the change can occur if and only if G_2 is less than G_1 (this is the case with single oxygen atoms, and they always pair up spontaneously).

Table 2-6 summarizes the thermodynamic terms we shall be using in the remainder of this section.

Table 2-6 Summary of thermodynamic terms

Term	Definition
S	Entropy: a measure of the degree of randomness or disorder of a system
ΔS	Change in entropy: $\Delta S = S_{\text{products}} - S_{\text{reactants}}$
H	Enthalpy, or heat: a measure of the energy in chemical bonds
ΔH	Change in enthalpy: $\Delta H = H_{\text{products}} - H_{\text{reactants}}$
G	Free energy: $G = H - TS$
ΔG	Change in free energy: $\Delta G = G_{\text{products}} - G_{\text{reactants}} = \Delta H - T\,\Delta S$
$\Delta G^{\circ\prime}$	Value of ΔG under standard conditions: 298 K (25°C), 1 atm, pH 7.0, and reactants at 1 M
K_{eq}	Equilibrium constant: $K_{\text{eq}} = 10^{-\Delta G^{\circ\prime}/2.3RT}$
E	Electric potential (volts)
ΔE	Change in electric potential in a reaction: $\Delta G = -n\mathcal{F}\,\Delta E$
E_0'	Value of E under standard conditions: 298 K, 1 atm, and reactants at 1 M

The Dependence of the ΔG of a Reaction on the Changes in Heat and Entropy The ΔG of a reaction (at any constant temperature and pressure) is a composite of two factors: the change in heat content between reactants and products, and the change in the entropy of the system. The heat or *enthalpy, H*, of reactants or products is equal to their total bond energies. Enthalpy is released or absorbed in a chemical reaction when bonds are formed or broken; hence, the overall change in enthalpy, represented by ΔH, is equal to the overall change in bond energies. In an *exothermic* reaction, heat is given off and ΔH is negative (the products contain less energy than the reactants did); in an *endothermic* reaction, heat is absorbed and ΔH is positive. According to the first law of thermodynamics, the total amount of energy in the system and its surroundings cannot change. Thus, in an *exergonic reaction*, the energy lost in the transformation of reactants into products appears as heat—as an increase in the kinetic energy of the molecules. Reactions will tend to proceed if they liberate energy (i.e., if $\Delta H < 0$), but this is only one of the two important parameters of free energy to consider.

Entropy, symbolized by S, is a measure of the degree of randomness or disorder of a system. A change in entropy is denoted by ΔS. Entropy increases as a system becomes more disordered, and decreases as it becomes more structured. According to the second law of thermodynamics, a reaction tends to occur spontaneously when the total entropy of the system and its surroundings increases. Consider, for example, the potential energy stored in a concentration gradient. The diffusion of solutes from one solution to a solution in which their concentration is lower is an example of a biologically important process that is driven only by an increase in entropy; in such a process, ΔH is very near zero. Suppose that a solution of 0.1 *M* NaCl is separated from a 0.01 *M* solution by a membrane through which Na^+ and Cl^- ions can diffuse. Eventually, the concentration of ions on either side of the membrane will be the same. After diffusion, the ions from the 0.1 *M* solution can move over a larger volume than they could before, as can the ions from the more dilute solution, and thus the disorder of the system has increased (Figure 2-18). The maximum entropy is achieved when all ions can diffuse freely in the largest volume possible—that is, when the concentrations of all ions are the same on both sides of the membrane. If the degree of hydration of Na^+ does not change significantly upon dilution, ΔH will be approximately zero and thus the negative free energy of the reaction

$$0.1\ M\ \text{solution} \overset{\text{membrane}}{|} 0.1\ M\ \text{solution} \rightarrow 0.055\ M\ \text{solution}$$

is due solely to a positive value of ΔS.

As mentioned earlier (page 30), the relatively high entropy of pure water promotes the hydrophobic interaction between hydrocarbon molecules in water. In a flask of water, all H_2O molecules are unrestrained in their motion and move randomly, making and breaking hydrogen bonds. If a long hydrocarbon molecule, such as heptane or tristearin, were to dissolve in the water, the presence of the hydrocarbon would restrict the freedom of motion of some of the H_2O molecules by causing them to form a cage around it. This would impose a high degree of order on their arrangement and lower the entropy of the system (i.e., $\Delta S < 0$). Because of the negative entropy change, hydrophobic molecules do not dissolve in aqueous solutions.

Gibbs showed that free energy, G, can be defined as

$$G = H - TS$$

where H is the heat energy or enthalpy of the system, T is its temperature in kelvins, and S is its entropy. If temperature and pressure are not allowed to change, a reaction will proceed spontaneously only if there is a negative change in free energy—a negative ΔG—in the equation

$$\Delta G = \Delta H - T\,\Delta S$$

For example, an exothermic reaction (one in which $\Delta H < 0$) that increases entropy ($\Delta S > 0$) occurs spontaneously ($\Delta G < 0$). An endothermic reaction ($\Delta H > 0$) can still occur spontaneously if ΔS is positive, and if the $T\,\Delta S$ term is large enough to overcome the positive ΔH. If ΔG is zero—that is, if there is no change in free energy in the conversion of reactants into products—the system will be at equilibrium: any conversion of reactants into products will be balanced by an equal conversion of products into reactants.

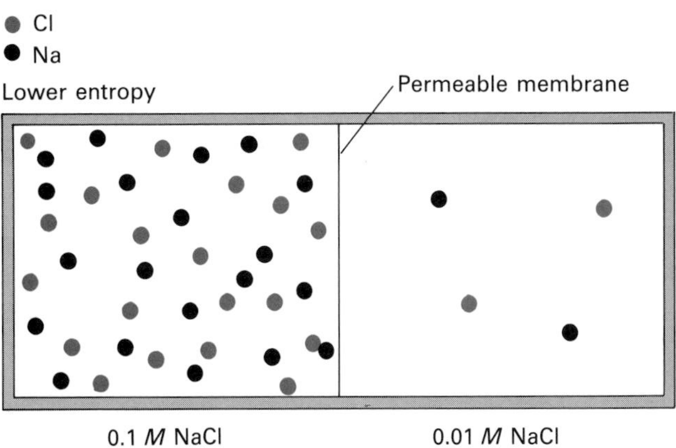

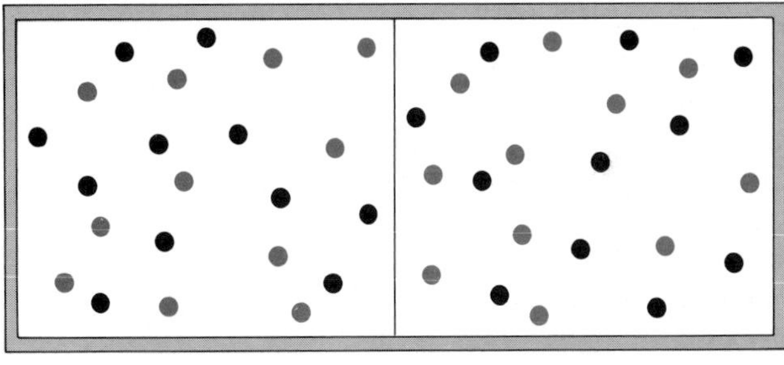

Figure 2-18 The entropy of a concentrated salt solution increases when the solution is allowed to mix with a dilute one. In this example, a membrane permeable to Na^+ and Cl^- ions is placed between the concentrated and the dilute solutions; the Na^+ and Cl^- diffuse through the membrane until their concentration is uniform on both sides of it. This gives the ions from both the concentrated and the dilute solutions more room and hence more freedom of movement, with the result that the randomness of the system is increased.

The thermodynamic principle that is important in predicting whether reactions, under any conditions, will proceed is called the second law of thermodynamics. It states that the change in entropy, ΔS, of a system and its surroundings must be positive for a reaction to proceed; that is, a reaction can take place only if $\Delta S_{total} > 0$, where

$$\Delta S_{total} = \Delta S_{system} + \Delta S_{surroundings}$$

In other words, the *overall* degree of disorder in the system and its surroundings must increase.

It is true that many biological reactions lead to an increase in order rather than an increase in entropy; an obvious example is the reaction that links amino acids together to form a protein. A solution of protein molecules has a lower entropy than does a solution of the same amino acids unlinked, because the freedom of movement of any amino acid in a protein is restricted by its being bound in a long chain (Figure 2-19). For the linking reac-

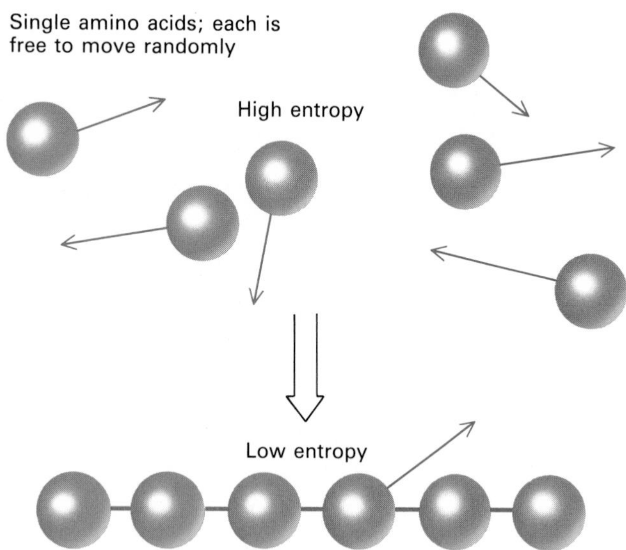

Figure 2-19 Single amino acids are free to move randomly. When they are linked together to form a peptide, however, their motion is restricted; although there is some rotation about the bonds that connect the amino acids together, lateral movements of individual amino acids are inhibited. When a peptide is formed, there is thus an increase in the order of the system, or a decrease in entropy ($\Delta S < 0$). There is also an increase in enthalpy ($\Delta H > 0$), due to the input of heat required to form the linking bonds. Because $\Delta G = \Delta H - T\,\Delta S$, ΔG is positive for the linking reaction, which will not occur spontaneously unless it is coupled to a reaction with a negative ΔG (see page 42).

tion to proceed, there must be a compensatory increase in entropy elsewhere in the system or in its surroundings.

There are two equivalent ways of describing the second law of thermodynamics: $\Delta S_{\text{total}} > 0$, and $\Delta G_{\text{system}} < 0$. To understand this, let us see how Gibbs' definition of ΔG relates to the definition of ΔS in the second law of thermodynamics. ΔG refers only to the system under study, and not to the system plus its surroundings; that is why it is so useful for the study of enzymatic reactions or total cellular metabolism. Although ΔG_{system} is a calculable parameter of the system, and ΔS_{total}, referring as it does to the system plus its surroundings, is less easy to measure, the two parameters are intimately related. A loss of heat from the system ($\Delta H_{\text{system}} < 0$) represents a gain of heat in the surroundings; this is manifested in the increased molecular motion, or entropy, of the surroundings. In fact,

$$\Delta H_{\text{system}} = -T\,\Delta S_{\text{surroundings}}$$

where T is, again, the temperature in kelvins. If we combine this equation with Gibbs' formula for ΔG, we have

$$\begin{aligned}\Delta G_{\text{system}} &= \Delta H_{\text{system}} - T\,\Delta S_{\text{system}}\\ &= -T\,\Delta S_{\text{surroundings}} - T\,\Delta S_{\text{system}} = -T\,\Delta S_{\text{total}}\end{aligned}$$

Thus, ΔG is always negative if ΔS_{total} is positive, which indicates that *any* reaction that increases the entropy of a system plus its surroundings will tend to proceed spontaneously.

Temperature, Concentrations of Reactants, and Other Parameters That Affect the ΔG of a Reaction

We have mentioned that the change in free energy for a reaction is influenced by several factors—temperature, pressure, and the initial concentrations of reactants and products; in addition, most biological reactions, like others that take place in aqueous solutions, are affected by the pH of the solution. Thus, reference books on biochemistry give values for $\Delta G^{\circ\prime}$, which is the free-energy change of a reaction under certain standard conditions: a temperature of 298 K (25°C), a pressure of 1 atm, a pH of 7.0 (as in pure water), and initial concentrations of 1 *M* for all reactants and products (except protons, which are kept at pH 7.0) (Table 2-7). $\Delta G^{\circ\prime}$ is termed the standard free-energy change of the reaction. The sign of $\Delta G^{\circ\prime}$ depends on the direction in which the reaction is written. If the reaction A → B has a $\Delta G^{\circ\prime}$ of $-a$ kcal/mol, then the reverse reaction, B → A, will have a $\Delta G^{\circ\prime}$ value of $+a$ kcal/mol.

The conditions of most biological reactions differ from the standard conditions, particularly in the concentrations of reactants. However, we can estimate free-energy changes for different temperatures and beginning concentrations by using the following equation:

$$\Delta G = \Delta G^{\circ\prime} + RT \ln Q = \Delta G^{\circ\prime} + RT \ln \frac{[\text{products}]}{[\text{reactants}]}$$

where R is the gas constant [1.987 cal/(degree · mol)], T is the temperature (in kelvins), and Q is the initial ratio of products to reactants (constructed like the equilibrium constant, page 35). For example, in the interconversion

$$\text{DHAP} \rightleftharpoons \text{G3P}$$

we have

$$Q = \frac{[\text{G3P}]}{[\text{DHAP}]} \quad \text{and} \quad \Delta G^{\circ\prime} = +1840 \text{ cal/mol}$$

The equation for ΔG becomes

$$\Delta G = 1840 + 1.987T \ln \frac{[\text{G3P}]}{[\text{DHAP}]}$$

If both [DHAP] and [G3P] are 1 *M*, then $\Delta G = \Delta G^{\circ\prime} =$ 1840 cal/mol, because $RT \ln 1 = 0$. The reaction will tend to proceed from right to left, i.e., in the direction of formation of DHAP. Suppose, however, that [DHAP] is 0.1 *M* and that [G3P] is 0.001 *M*, with other conditions being standard. Then $Q = 0.001/0.1 = 0.01$, and

Table 2-7 Values of $\Delta G^{\circ\prime}$, the standard free-energy change, for some important biochemical reactions

Reaction	$\Delta G^{\circ\prime}$ (kcal/mol)
HYDROLYSIS	
Acid anhydrides:	
Acetic anhydride + $H_2O \rightarrow$ 2 acetate	−21.8
$PP_i + H_2O \rightarrow 2P_i$*	−8.0
ATP + $H_2O \rightarrow$ ADP + P_i	−7.3
Esters:	
Ethylacetate + $H_2O \rightarrow$ ethanol + acetate	−4.7
Glucose 6-phosphate + $H_2O \rightarrow$ glucose + P_i	−3.3
Amides:	
Glutamine + $H_2O \rightarrow$ glutamate + NH_4^+	−3.4
Glycylglycine + $H_2O \rightarrow$ 2 glycine (a peptide bond)	−2.2
Glycosides:	
Sucrose + $H_2O \rightarrow$ glucose + fructose	−7.0
Maltose + $H_2O \rightarrow$ 2 glucose	−4.0
ESTERIFICATION	
Glucose + $P_i \rightarrow$ glucose 6-phosphate + H_2O	+3.3
REARRANGEMENT	
Glucose 1-phosphate $\rightarrow$ glucose 6-phosphate	−1.7
Fructose 6-phosphate $\rightarrow$ glucose 6-phosphate	−0.4
Glyceraldehyde 3-phosphate $\rightarrow$ dihydroxyacetone phosphate	−1.8
ELIMINATION	
Malate $\rightarrow$ fumarate + H_2O	+0.75
OXIDATION	
Glucose + $6O_2 \rightarrow 6CO_2 + 6H_2O$	−686
Palmitic acid + $23O_2 \rightarrow 16CO_2 + 16H_2O$	−2338
PHOTOSYNTHESIS	
$6CO_2 + 6H_2O \rightarrow$ glucose + $6O_2$	+686

*PP_i = pyrophosphate; P_i = phosphate.

SOURCE: A. L. Lehninger, 1975, *Biochemistry*, 2d ed., Worth, p. 397.

$$\begin{aligned}\Delta G &= 1840 + 1.987(298) \ln 0.01 \\ &= 1840 + 1.987(298)(-4.605) \\ &= 1840 - 2727 = -887 \text{ cal/mol}\end{aligned}$$

Clearly, the reaction will now proceed in the direction of formation of G3P.

In a reaction A + B $\rightleftharpoons$ C, in which two molecules combine to form a third, the equation for ΔG becomes

$$\Delta G = \Delta G^{\circ\prime} + RT \ln \frac{[C]}{[A][B]}$$

The direction of the reaction will be shifted more toward the right (toward formation of C) if *either* [A] or [B] is increased.

The Relation between the Standard Free-Energy Change, $\Delta G^{\circ\prime}$, and the Equilibrium Constant, K_{eq} A chemical mixture at equilibrium is already in a state of minimal free energy: no free energy is generated or released. Thus, for a system at equilibrium we can write

$$0 = \Delta G = \Delta G^{\circ\prime} + RT \ln Q$$

However, at equilibrium the value of Q is simply the equilibrium constant K_{eq}, so we have

$$\Delta G^{\circ\prime} = -RT \ln K_{eq}$$

If we express this equation in terms of base 10 logarithms, it becomes

$$\Delta G^{\circ\prime} = -2.3RT \log K_{eq}$$

or

$$-\frac{\Delta G^{\circ\prime}}{2.3RT} = \log K_{eq}$$

or, by the definition of logarithms,

$$K_{eq} = 10^{-\Delta G^{\circ\prime}/2.3RT}$$

This simple but important relation between the standard free energy of a reaction and the equilibrium constant makes it possible to determine values of $\Delta G^{\circ\prime}$ by simply measuring the concentrations of chemicals at equilibrium, rather than by trying to measure changes in entropy or free energy directly. Note that if $\Delta G^{\circ\prime}$ is negative, then K_{eq} is greater than 1; that is, the formation of products from reactants is favored (Table 2-8).

Although a chemical equilibrium is apparently unchanging and static, it is actually a dynamic state. The two opposing reactions continue to proceed but at the same rate, so that the reactions cancel each other. The point of equilibrium does not depend on the speed of the reactions. When an enzyme or another catalyst speeds up a reaction, it also speeds up the reverse reaction; thus, equilibrium is reached sooner than if the reaction were not catalyzed. However, *the equilibrium constant and thus $\Delta G^{\circ\prime}$ is the same with or without catalysis.*

An equilibrium constant or a $\Delta G^{\circ\prime}$ value is unaffected by any other reactions going on in the same mixture, even

Table 2-8 Values of $\Delta G^{\circ\prime}$ for some values of K_{eq}

K_{eq}	$\Delta G^{\circ\prime}$ (cal/mol)*
0.001	4086
0.01	2724
0.1	1362
1.0	0
10.0	−1362
100.0	−2724
1000.0	−4086

*Calculated from the formula $\Delta G^{\circ\prime} = -2.3RT \log K_{eq}$.

if these change the concentrations of reactants or products. Values of ΔG, however, will be affected by the concentrations of reactants and products, as noted previously (page 38).

If two or more chemical reactions involving the same reactants and products are occurring simultaneously, then the $\Delta G^{\circ\prime}$ values of the individual reactions can be used to predict the equilibrium concentrations of all molecules. For example, if acetic acid (CH_3COOH) is dissolved in water, the hydrogen ions in the solution participate in two dissociation-reassociation reactions:

$$H_2O \rightleftharpoons H^+ + OH^-$$

and

$$CH_3COOH \rightleftharpoons H^+ + CH_3COO^-$$

If enough acetic acid is dissolved to lower the pH to 5, then $[H^+]$ is 10^{-5} M. We can write

$$K_{eq(water)} = \frac{[H^+][OH^-]}{[H_2O]}$$

$$[H^+][OH^-] = 10^{-14}*$$

$$[OH^-] = \frac{10^{-14}}{10^{-5}} = 10^{-9}\ M$$

Thus, $[OH^-]$ is only 1 percent of what it is in pure water.

For the acetic acid reaction, the equation is

$$K_{eq(acetic\ acid)} = \frac{[H^+][CH_3COO^-]}{[CH_3COOH]} = 1.8 \times 10^{-5}\ M$$

which becomes

$$\frac{[CH_3COO^-]}{[CH_3COOH]} = \frac{1.8 \times 10^{-5}}{10^{-5}} = 1.8$$

Evidently, 64 percent of dissolved acetic acid molecules are dissociated at pH 5.

The Generation of a Concentration Gradient Requires an Expenditure of Energy

Cells must often accumulate chemicals, such as glucose and K^+ ions, in concentrations greater than those existing in the cells' environments. Consequently, they must transport these chemicals against a concentration gradient (Figure 2-20). How much energy does this entail? To find the amount of energy required to transfer 1 mol of a substance from outside the cell, where its concentration is C_1, to inside the cell, where its concentration is C_2, we employ the equation

$$\Delta G = RT \ln \frac{C_2}{C_1}$$

*The number of moles of H_2O in a liter of water is 55.6. Because $K_{eq(water)}$ is 1.8×10^{-16} M, we obtain $[H^+][OH^-] = 10^{-14}$ M^2, an equation presented previously (page 31).

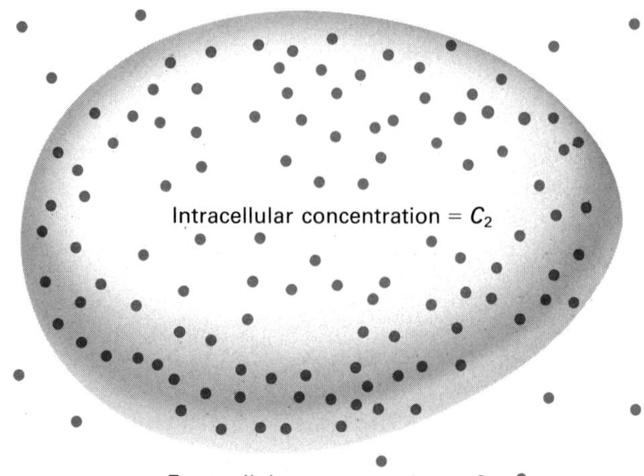

Figure 2-20 In order for a cell to accumulate a molecule in a concentration higher than its environmental concentration, an expenditure of energy is required. If $C_2 > C_1$, then the value of $\Delta G = RT \ln (C_2/C_1)$ will be positive.

where, as in the Boltzmann distribution (page 24), R is the gas constant and T is the absolute temperature. An "uphill" transport of molecules against a concentration gradient ($C_2 > C_1$) clearly requires the input of cellular chemical energy ($\Delta G > 0$). If the concentrations differ by a factor of 10, then $\Delta G = RT \ln 10 = 1.36$ kcal per mole of substance transported. Such calculations assume that a molecule of a given substance inside a cell is identical to a molecule of that substance outside—in other words, that the substance is not sequestered, bound, or chemically changed by the transport. When a glucose molecule is moved across a membrane, the change in heat (ΔH) is near zero. Because $\Delta G = \Delta H - T\,\Delta S$, the main contributor to the change in free energy is the change in entropy.

Many Cellular Processes Involve the Transfer of Electrons in Oxidation-Reduction Reactions

Many chemical reactions result in the transfer of electrons from one atom or molecule to another. This may or may not accompany the formation of new chemical bonds. The loss of electrons from an atom or a molecule is called *oxidation*, and the gain of electrons by an atom or molecule is called *reduction*. Electrons are neither created nor destroyed in a chemical reaction; thus, if an atom or molecule becomes oxidized, another must become reduced. For example, oxygen draws electrons from Fe^{2+} (ferrous) ions to form Fe^{3+} (ferric) ions, a reaction that occurs when energy is generated in cellular particles called mitochondria. Each oxygen atom receives two electrons, one from each of two Fe^{2+} ions:

$$2Fe^{2+} + \tfrac{1}{2}O_2 \longrightarrow 2Fe^{3+} + O^{2-}$$

Fe^{2+} is oxidized and O_2 is reduced. Oxygen accepts electrons in many oxidation reactions in aerobic animal cells. Reduced oxygen (O^{2-}) readily acquires two protons to yield H_2O.

In the transformation of succinate into fumarate, which also takes place in mitochondria, succinate loses two hydrogen atoms (Figure 2-21). This is equivalent to a loss of two protons and two electrons. Thus, succinate is said to be oxidized in its conversion into fumarate, and another molecule (flavin adenine dinucleotide, FAD, which accepts the electrons) is reduced (to $FADH_2$). As in the interconversion of succinate and fumarate, many biologically important oxidation and reduction reactions occur by the removal or the addition of hydrogen atoms (i.e., protons plus electrons) rather than by the transfer of isolated electrons.

When ferrous ion (Fe^{2+}) and oxygen (O_2) combine, the Fe^{2+} is oxidized and the O_2 reduced by the following mechanism:

$$2Fe^{2+} \longrightarrow 2Fe^{3+} + 2e^-$$
$$2e^- + \tfrac{1}{2}O_2 \longrightarrow O^{2-}$$
$$O^{2-} + 2H^+ \longrightarrow H_2O$$

The readiness with which an atom or a molecule takes up an electron is called its *reduction potential.* Represented by E, reduction potentials are measured in volts (V) on a scale whose zero point is arbitrarily set at the reduction potential of the following reaction under standard conditions (25°C, 1 atm, and reactants at 1 M):

$$H^+ + e^- \underset{\text{oxidation}}{\overset{\text{reduction}}{\rightleftharpoons}} \tfrac{1}{2}H_2$$

The value of E for a molecule or an atom under standard conditions is called its *standard reduction potential,* and is written as E_0' (Table 2-9). Standard reduction potentials may differ somewhat from those found under the conditions in a cell, because the concentrations of reactants in a cell are not 1 M. A positive reduction potential means that a molecule or ion (say, Fe^{3+}) has a higher affinity for electrons than does the H^+ ion in the standard reaction just above. A negative reduction potential means that the substance (say, acetate, CH_3COO^-, in its reduction to acetaldehyde, CH_3CHO) has a lower affinity for electrons. In an oxidation-reduction reaction, electrons move spontaneously toward atoms or molecules having more positive reduction potentials. In other words, a compound having a more negative reduction potential can reduce one having a more positive potential.

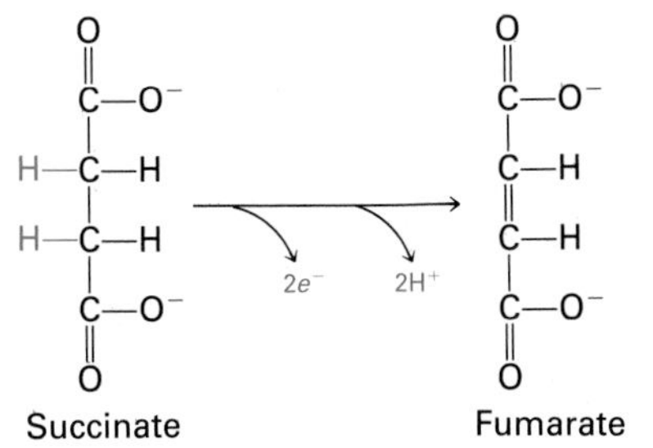

In an oxidation-reduction reaction, the total voltage change (or change in electric potential), termed ΔE, is the sum of the voltage changes (or reduction potentials) of the individual oxidation or reduction steps. Because all forms of energy are interconvertible, we can express ΔE as a change in chemical free energy, ΔG. The charge in 1 mol (6×10^{23} molecules) of electrons is 96,500 coulombs (96,500 joules per volt), a quantity known as the faraday constant ($\mathcal{F}$), named after the British physicist Michael Faraday (1791–1867). We can write

$$\Delta G\ (\text{cal/mol}) = -n\mathcal{F}\,\Delta E = -n\left(\frac{96{,}500}{4.18}\right)\Delta E\ (\text{volts})$$

Table 2-9 Values of E_0', the standard reduction potential, for some important oxidation-reduction reactions (pH 7.0, 25°C)*

Oxidant	Reductant	n	E_0'(V)
α-Ketoglutarate	Succinate + CO_2	2	−0.67
Acetate	Acetaldehyde	2	−0.60
Ferredoxin (oxidized)	Ferredoxin (reduced)	1	−0.43
$2H^+$	H_2	2	−0.42
NAD^+	NADH + H^+	2	−0.32
$NADP^+$	NADPH + H^+	2	−0.32
Glutathione (oxidized)	Glutathione (reduced)	2	−0.23
Acetaldehyde	Ethanol	2	−0.20
Pyruvate	Lactate	2	−0.19
Fumarate	Succinate	2	0.03
Cytochrome *b* (+3)	Cytochrome *b* (+2)	1	0.07
Ubiquinone (oxidized)	Ubiquinone (reduced)	2	0.10
Cytochrome *c* (+3)	Cytochrome *c* (+2)	1	0.22
Fe^{3+}	Fe^{2+}	1	0.77
$\tfrac{1}{2}O_2 + 2H^+$	H_2O	2	0.82

*E_0' refers to the partial reaction written as

$$\text{Oxidant} + e^- \longrightarrow \text{reductant}$$

and n is the number of electrons transferred.

SOURCE: L. Stryer, 1981, *Biochemistry,* 2d ed., W. H. Freeman and Company, p. 310.

Figure 2-21 The conversion of succinate to fumarate is an oxidation reaction: two electrons are released, as well as two protons. This reaction occurs in mitochondria as part of the citric acid cycle, which functions in the final stages of the oxidation of glucose by oxygen to form CO_2.

where n is the number of electrons transferred, and 4.18 is the factor used to convert joules into calories.

The reduction potential is customarily used to describe the electric energy change that occurs when an atom or a molecule gains an electron. In an oxidation-reduction reaction, we also use the oxidation potential—the voltage change that takes place when an atom or molecule *loses* an electron. The oxidation potential is simply the negative of the reduction potential:

$$\textit{Reduction: } Cu^{2+} + e^- \longrightarrow Cu^+ \qquad \Delta E_0' = +0.35 \text{ V}$$
$$\textit{Oxidation: } Cu^+ \longrightarrow Cu^{2+} + e^- \qquad \Delta E_0' = -0.35 \text{ V}$$

The voltage change in a complete oxidation-reduction reaction, in which one molecule is reduced and another is oxidized, is simply the sum of the oxidation and reduction potentials of the atoms or molecules in the partial reactions. Consider, for example, the change in electric potential (and, correspondingly, the change in standard free energy) when succinate is oxidized by oxygen:

$$\text{Succinate} + \tfrac{1}{2}O_2 \rightleftharpoons \text{fumarate} + H_2O$$

In this case, the partial reactions are:

$$\text{Succinate} \rightleftharpoons \text{fumarate} + 2H^+ + 2e^- \qquad \Delta E_0' = -0.03 \text{ V} \quad \Delta G^{\circ\prime} = +1.39 \text{ kcal/mol} \quad (n = 2)$$

$$\tfrac{1}{2}O_2 + 2e^- \rightleftharpoons O^{2-} \qquad \Delta E_0' = +0.82 \text{ V} \quad \Delta G^{\circ\prime} = -37.88 \text{ kcal/mol} \quad (n = 2)$$

$$2H^+ + O^{2-} \rightleftharpoons H_2O$$

$$\text{Sum: Succinate} + \tfrac{1}{2}O_2 \rightleftharpoons \text{fumarate} + H_2O \qquad \Delta E_0' = +0.79 \text{ V} \quad \Delta G^{\circ\prime} = -36.49 \text{ kcal/mol}$$

A positive $\Delta E_0'$ signifies a negative $\Delta G^{\circ\prime}$, and thus, under standard conditions, a reaction that can occur spontaneously.

An Unfavorable Chemical Reaction Can Proceed If It Is Coupled with an Energetically Favorable Reaction

Many chemical reactions in cells have a positive ΔG: they are energetically unfavorable and will not proceed spontaneously. One example is the synthesis of small peptides, such as glycylalanine, from amino acids. How can such a reaction proceed? The cell's solution is to couple a reaction that has a positive ΔG to a reaction that has a negative ΔG of larger magnitude, so that the sum of the two reactions has a negative ΔG. Suppose that the reaction

$$A \rightleftharpoons B + X$$

has a $\Delta G^{\circ\prime}$ of $+5$ kcal/mol and that the reaction

$$X \rightleftharpoons Y$$

has a $\Delta G^{\circ\prime}$ of -10 kcal/mol. The first reaction, the formation of B + X from A, will not proceed spontaneously. However, if any X that is formed in the first reaction is converted into Y, the equilibrium concentration of X will be lowered, and thus the first reaction will tend to proceed in the direction of formation of B + X. In other words, the disappearance of any X that is formed "pulls" the reaction in the direction of B + X.

The overall reaction $A \rightleftharpoons B + Y$ will have a $\Delta G^{\circ\prime}$ that is the sum of the $\Delta G^{\circ\prime}$ values of each of the two partial reactions:*

$$A \rightleftharpoons B + X \qquad \Delta G^{\circ\prime} = +5 \text{ kcal/mol}$$
$$X \rightleftharpoons Y \qquad \Delta G^{\circ\prime} = -10 \text{ kcal/mol}$$
$$\text{Sum: } A \rightleftharpoons B + Y \qquad \Delta G^{\circ\prime} = -5 \text{ kcal/mol}$$

The overall reaction is exergonic. In cells, as we shall see, energetically unfavorable reactions of the type $A \rightleftharpoons B + X$ are often coupled to the hydrolysis of the compound adenosine triphosphate.

Hydrolysis of the Phosphoanhydride Bonds in ATP Releases Substantial Free Energy

Cells extract energy from foods through a series of reactions that have negative free-energy changes. Much of the free energy released is not allowed to dissipate as heat but is captured in chemical bonds formed by other molecules for use throughout the cell. In almost all organisms, the most important molecule for capturing and transferring free energy is *adenosine triphosphate,* or ATP (Figure 2-22).

The useful free energy in an ATP molecule is contained in high-energy *phosphoanhydride bonds.* These bonds are so termed because they are formed from the condensation of two molecules of phosphate by the loss of water:

$$^-O-\overset{O}{\underset{O^-}{\overset{\|}{\underset{|}{P}}}}-OH + HO-\overset{O}{\underset{O^-}{\overset{\|}{\underset{|}{P}}}}-O^- \rightleftharpoons {}^-O-\overset{O}{\underset{O^-}{\overset{\|}{\underset{|}{P}}}}-O-\overset{O}{\underset{O^-}{\overset{\|}{\underset{|}{P}}}}-O^- + H_2O$$

*It is instructive to confirm the direction of $A \rightleftharpoons B + Y$ in a slightly different way, by multiplying together the equilibrium constants for the two partial reactions to obtain the K_{eq} for the overall reaction:

$$K_{eq(1)} = \frac{[B][X]}{[A]} = \exp\left(-\frac{\Delta G_1^{\circ\prime}}{RT}\right) = \exp\left(-\frac{5000}{1.987 \times 298}\right) = 2.15 \times 10^{-4}\,M$$

$$K_{eq(2)} = \frac{[Y]}{[X]} = \exp\left(-\frac{\Delta G_2^{\circ\prime}}{RT}\right) = 2.16 \times 10^{7}\,M$$

$$K_{eq(overall)} = \frac{[B][Y]}{[A]} = \frac{[B][X]}{[A]} \cdot \frac{[Y]}{[X]} = K_{eq(1)} \cdot K_{eq(2)} = 4.65 \times 10^{3}\,M$$
$$= \exp\left(-\frac{-5000}{1.987 \times 298}\right) = \exp\left(-\frac{\Delta G^{\circ\prime}_{overall}}{RT}\right)$$

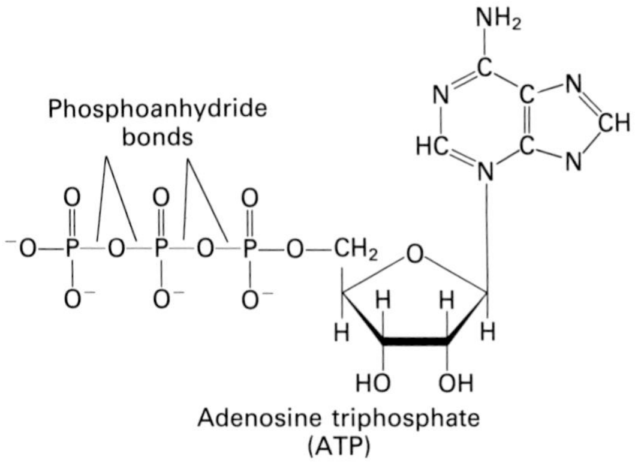

Phosphoanhydride bonds

Adenosine triphosphate (ATP)

Figure 2-22 The structure of adenosine triphosphate. The two high-energy phosphoanhydride bonds that link the three phosphate groups (see text) are shown in color.

An ATP molecule has two phosphoanhydride bonds and is thus often written as

$$\text{Adenosine—p}{\sim}\text{p}{\sim}\text{p}$$

or simply as

$$\text{Ap}{\sim}\text{p}{\sim}\text{p}$$

where p stands for a phosphate group and the symbol ~ denotes a high-energy bond. Although this bond is an ordinary covalent bond, it is referred to as a high-energy bond because it releases about 7.3 kcal/mol of free energy (under standard biochemical conditions) when it is broken, as in hydrolysis:

$$\text{Ap}{\sim}\text{p}{\sim}\text{p} + H_2O \longrightarrow \text{Ap}{\sim}\text{p} + P_i + H^+$$

or

$$\text{Ap}{\sim}\text{p}{\sim}\text{p} + H_2O \longrightarrow \text{Ap} + PP_i + H^+$$

or

$$\text{Ap}{\sim}\text{p} + H_2O \longrightarrow \text{Ap} + P_i + H^+$$

where P_i stands for free inorganic phosphate and PP_i for inorganic pyrophosphate. Removal of a phosphate or a pyrophosphate group from ATP leaves adenosine diphosphate (ADP) or adenosine monophosphate (AMP), respectively (Figure 2-23).

Why is the $\Delta G^{\circ\prime}$ for hydrolysis of a phosphoanhydride bond in ATP so much greater than the $\Delta G^{\circ\prime}$ for hydrolysis of a phosphoester bond, such as that in glucose 6-phosphate, for which $\Delta G^{\circ\prime} = -3.3$ kcal/mol (Figure 2-24)? A principal reason is that ATP and its hydrolysis product ADP are highly charged at pH 7.0. Three of the four ionizable protons in ATP are fully dissociated at pH 7.0, and the fourth, with a pK_a of 6.95, is about 50

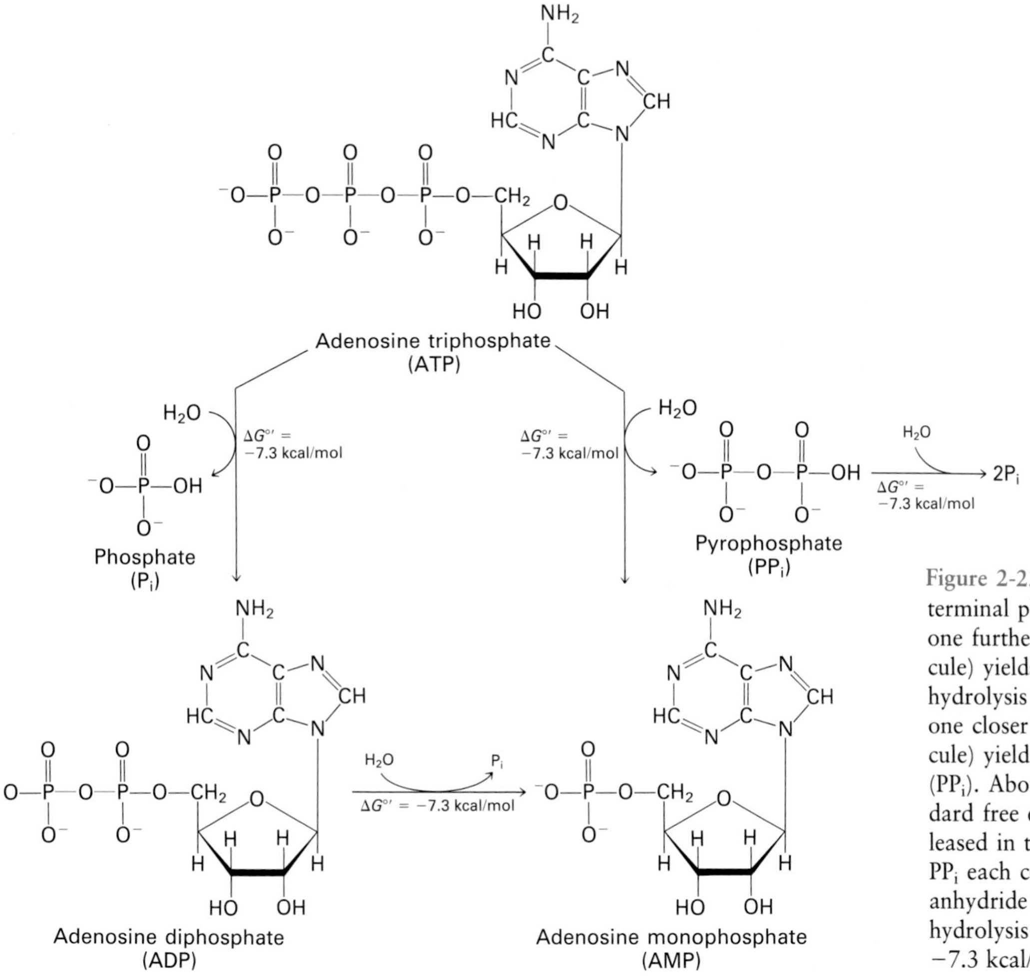

Adenosine triphosphate (ATP)

H_2O $\Delta G^{\circ\prime} = -7.3$ kcal/mol — Phosphate (P_i) — Adenosine diphosphate (ADP)

H_2O $\Delta G^{\circ\prime} = -7.3$ kcal/mol — Pyrophosphate (PP_i) — Adenosine monophosphate (AMP)

Pyrophosphate (PP_i) + H_2O → $2P_i$, $\Delta G^{\circ\prime} = -7.3$ kcal/mol

ADP + H_2O → AMP + P_i, $\Delta G^{\circ\prime} = -7.3$ kcal/mol

Figure 2-23 Hydrolysis of ATP at its terminal phosphoanhydride bond (the one further from the body of the molecule) yields ADP and phosphate (P_i); hydrolysis at the proximal bond (the one closer to the body of the molecule) yields AMP and pyrophosphate (PP_i). About the same amount of standard free energy, −7.3 kcal/mol, is released in these two reactions. ADP and PP_i each contain one phosphoanhydride bond, which has a $\Delta G^{\circ\prime}$ of hydrolysis of approximately −7.3 kcal/mol.

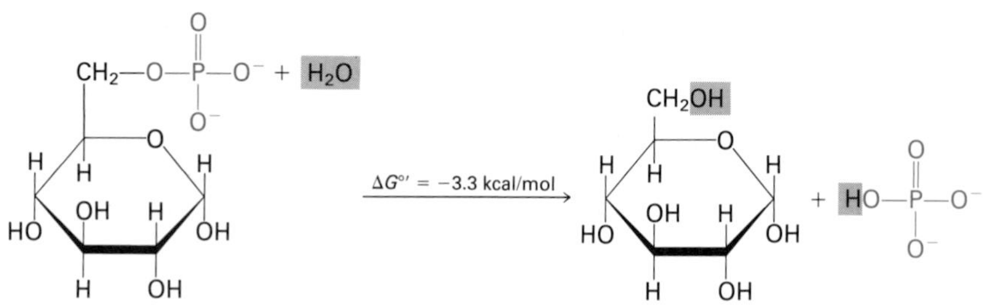

Figure 2-24 The hydrolysis of glucose 6-phosphate to glucose and phosphate has a standard free-energy change of −3.3 kcal/mol, which is less than half the value for hydrolysis of a phosphoanhydride bond in ATP.

percent dissociated.* The closely spaced negative charges in ATP repel one another strongly. When the terminal phosphoanhydride bond is hydrolyzed, some of this stress is removed by the separation of the hydrolysis products ADP^{3-} and HPO_4^{2-}. The separated ADP^{3-} and HPO_4^{2-} anions will repel one another and will tend not to recombine to form ATP. In the phosphate ester glucose 6-phosphate, in contrast, there is no charge repulsion between the phosphate group and the carbon atom to which it is attached. One of the hydrolysis products, glucose, is uncharged and does not repel the negative HPO_4^{2-} ion.

The first step in the metabolism of glucose is the transfer of the terminal phosphate group of ATP to the 6 carbon atom of glucose:

$$ATP^{4-} + \text{glucose} \rightleftharpoons ADP^{3-} + \text{glucose 6-phosphate}$$

The standard free-energy change can be calculated as the sum of the free-energy changes in the hydrolysis of the two compounds.

$$ATP^{4-} + H_2O \rightleftharpoons ADP^{3-} + HPO_4^{2-} \qquad \Delta G^{\circ\prime} = -7.3 \text{ kcal/mol}$$

$$HPO_4^{2-} + \text{glucose} \rightleftharpoons \text{glucose 6-phosphate} + H_2O \qquad \Delta G^{\circ\prime} = +3.3 \text{ kcal/mol}$$

$$\text{Sum: } ATP^{4-} + \text{glucose} \rightleftharpoons \text{glucose 6-phosphate} + ADP^{3-} \qquad \Delta G^{\circ\prime} = -4.0 \text{ kcal/mol}$$

Cells contrive to keep the ratio of ATP to ADP (and AMP) high, often as high as 10:1. Thus, reactions such as the phosphorylation of glucose, in which the terminal phosphate group of ATP is transferred to another molecule, will be driven even further to the right.

In cells, the free energy released by the breaking of high-energy bonds is transferred to other molecules. This supplies them with enough free energy to undergo reactions that would otherwise be impossible. For example, if the reaction

$$B + C \longrightarrow D$$

is energetically unfavorable ($\Delta G > 0$), it can be made favorable by linking it to the hydrolysis of ATP molecules:

$$B + Ap{\sim}p{\sim}p \longrightarrow B{\sim}p + Ap{\sim}p$$
$$B{\sim}p + C \longrightarrow D + P_i$$

Some of the energy in a phosphoanhydride bond is used to transfer a phosphate group to B to form an intermediate molecule, B~p. The intermediate has enough of the remaining free energy to react with C to form D and phosphate. Thus, the overall reaction

$$B + C + Ap{\sim}p{\sim}p \longrightarrow D + Ap{\sim}p + P_i$$

is energetically favorable. Chapter 4 illustrates in detail how the hydrolysis of ATP is coupled to the linking of amino acids to form polypeptides and proteins; in the above equation, B and C would be amino acids, and D would be a dipeptide.

Phosphoanhydride bonds are not the only high-energy bonds. Other bonds, particularly those between a phosphate group and some other substance, have the same high-energy character. As Table 2-10 shows, the interphosphate bond of ATP is neither the most energetic nor the least energetic of such bonds. Why, then, is ATP the most important cellular molecule for capturing and transferring free energy? The free energy of hydrolysis of ATP is sufficiently great that reactions in which the terminal phosphate group is transferred to another molecule have a substantially negative $\Delta G^{\circ\prime}$. However, if the magnitude of the $\Delta G^{\circ\prime}$ for hydrolysis of the phosphoanhydride bond were much higher than it is, cells might require too much energy to form the phosphoanhydride bond in the first place.

ATP Is Used to Fuel Many Cellular Processes

If the terminal phosphoanhydride bond of ATP ruptures by hydrolysis to produce ADP and P_i, the energy released is in the form of heat. In the presence of specific cellular enzymes, however, much of this energy is converted into more useful forms. Some of these uses are summarized in Figure 2-25. Energy is used to synthesize large cellular molecules—for example, to link amino acids to form proteins, or to connect single sugars (monosaccharides) to form oligosaccharides. Energy is also used to synthesize many of the small molecules that are required by the cell. The hydrolysis of ATP supplies the energy needed for cellular movement, both the movement of individual cells

*In cells, ATP and ADP are largely present as the 1:1 Mg^{2+} complexes $MgATP^{2-}$ and $MgADP^{-}$, because of the high affinity of the negatively charged pyrophosphate groups for the Mg^{2+} that is present in high concentrations.

Table 2-10 Values of $\Delta G^{\circ\prime}$ for the hydrolysis of various biologically important phosphate compounds*

Compound	$\Delta G^{\circ\prime}$ (kcal/mol)
Phosphoenolpyruvate 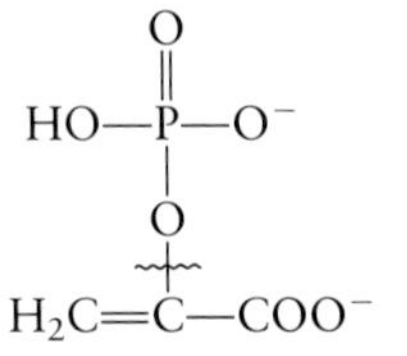	−14.8
Creatine phosphate 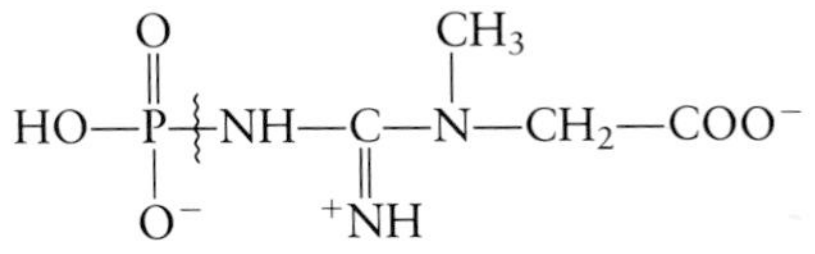	−10.3
Pyrophosphate 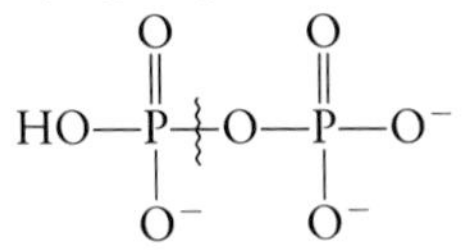	−8.0
ATP (to ADP + P_i)	−7.3
ATP (to AMP + PP_i)	−7.3
Glucose 1-phosphate 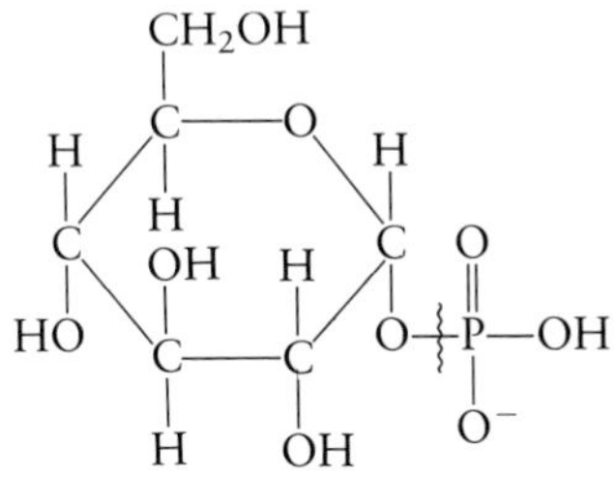	−5.0
Glucose 6-phosphate 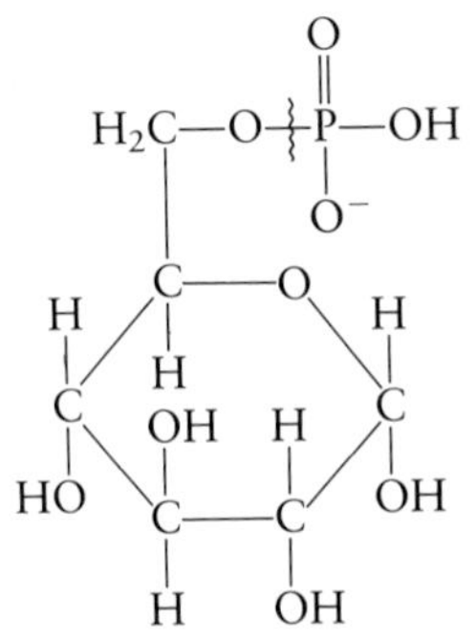	−3.3
Glycerol 3-phosphate 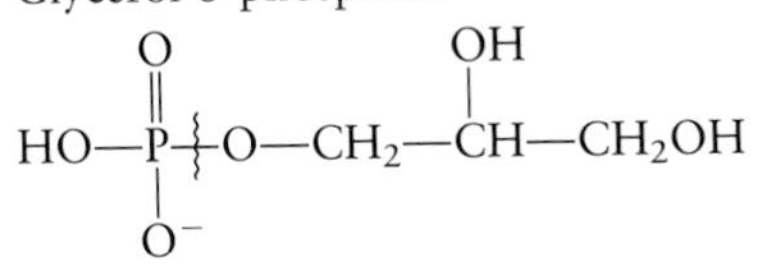	−2.2

*The bond that is cleaved is indicated by the wavy line.

from one location to another and the contraction of muscle cells. An important use of this energy is in the transport of molecules into or out of the cell, usually against a concentration gradient. The generation of gradients of ions, such as Na^+ and K^+, across a cellular membrane is the basis for the electrical activity of cells and, in particular, for the conduction of impulses by nerves.

What are the sources of the energy required to form high-energy bonds? Plants and other organisms obtain this energy by trapping the energy in light through *photosynthesis*. In the chloroplasts of cells and in photosynthetic bacteria, chlorophyll pigments absorb the energy of light and use it to synthesize ATP from ADP and P_i. Our current understanding of the mechanism of these photosynthetic processes is described in Chapter 20. Much of the ATP produced in photosynthesis is used to convert carbon dioxide into glucose:

$$6CO_2 + 6H_2O \xrightarrow{\text{ATP} \rightarrow \text{ADP} + P_i} C_6H_{12}O_6 + 6O_2$$

Animals obtain free energy by oxidizing food molecules through *respiration*. All synthesis of ATP in animal cells and in nonphotosynthetic microorganisms is a result of chemical transformations of energy-rich dietary or storage molecules. The predominant source of energy in cells is the six-carbon sugar glucose. When 1 mol (180 g) of glucose reacts with oxygen under standard conditions according to the following reaction, 686 kcal of energy is released:

$$C_6H_{12}O_6 + 6O_2 \longrightarrow 6CO_2 + 6H_2O \qquad \Delta G^{\circ\prime} = -686 \text{ kcal/mol}$$

If glucose is simply burned in air, all of this energy is released as heat. In the cell, through an elaborate set of enzyme-catalyzed reactions, the metabolism of 1 molecule of glucose is coupled to the synthesis of as many as 36 molecules of ATP from 36 of ADP:

$$C_6H_{12}O_6 + 6O_2 + 36P_i + 36ADP \longrightarrow 6CO_2 + 6H_2O + 36ATP$$

Because one high-energy phosphoanhydride bond in ATP represents 7.3 kcal/mol, about 263 kcal of energy is conserved in ATP per mole of glucose metabolized, which gives an efficiency of about 38 percent (263 ÷ 686). This type of cellular metabolism is termed *aerobic* because it is completely dependent on the oxygen in the air. Aerobic *catabolism* (degradation) of glucose is found in all higher cells—both plant and animal—as well as in many bacterial cells. The overall reaction of glucose respiration,

$$C_6H_{12}O_6 + 6O_2 \longrightarrow 6CO_2 + 6H_2O$$

is an exact reversal of the photosynthetic reaction in which glucose is formed,

$$6CO_2 + 6H_2O \longrightarrow C_6H_{12}O_6 + 6O_2$$

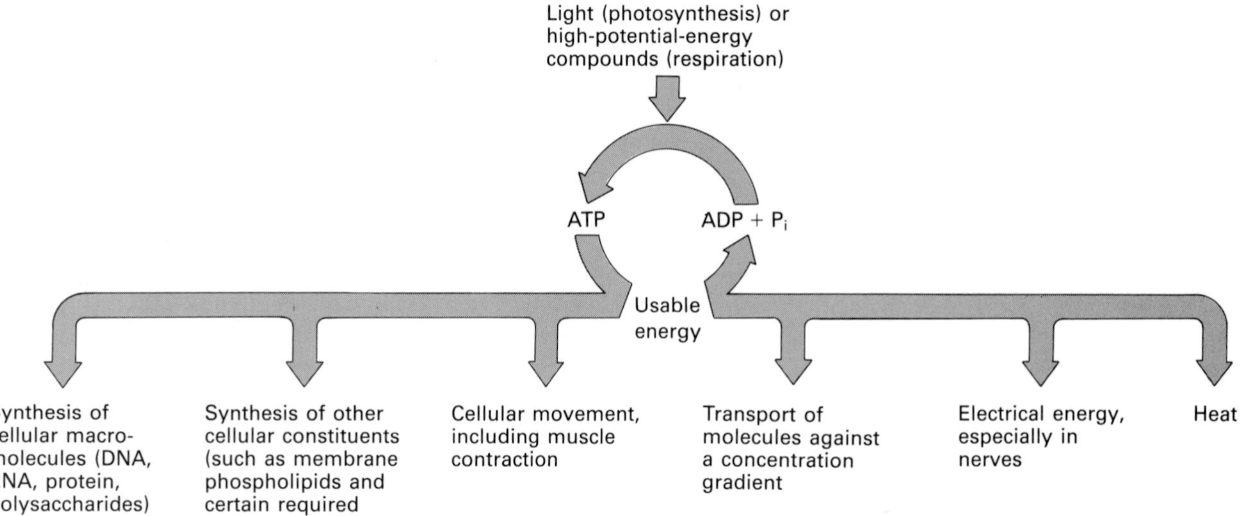

Figure 2-25 The ATP cycle. ATP is formed from ADP and P_i either by photosynthesis or by the metabolism of energy-rich compounds. Hydrolysis of ATP to ADP and P_i is coupled to the performance of many key cellular functions; the free energy released by the breaking of the phosphoanhydride bond is trapped as usable energy to fuel these tasks.

except that the energy of light is essential for the photosynthetic reaction. Respiration and photosynthesis are the two major processes constituting the *carbon cycle* in nature: Glucose and oxygen produced by plants are the raw materials for respiration and the generation of ATP by plant and animal cells alike. The end products of respiration, CO_2 and H_2O, are the raw materials for the photosynthetic production of glucose and oxygen. The only net source of energy in this cycle is sunlight. Directly or indirectly, photosynthesis is the source of chemical energy for all cells.

Activation Energy and Reaction Rate

The Energy Required to Initiate a Reaction Is the Activation Energy

Many chemical reactions that have a negative standard free-energy change do not proceed at a measurable rate. For example, glyceraldehyde 3-phosphate can undergo several different reactions (Figure 2-26), each of which has a negative $\Delta G^{\circ\prime}$. Yet, in normal aqueous solutions, glyceraldehyde 3-phosphate is a fairly stable compound that reacts very slowly or not at all.

Similarly, a mixture of hydrogen and oxygen gases sealed in a flask at room temperature will remain quiescent indefinitely. If, however, the gases are exposed to an electric spark or a flame, the mixture will explode with great force. The hydrogen gas burns vigorously, combining with the oxygen to form water:

$$2H_2 + O_2 \longrightarrow 2H_2O$$

This reaction releases a considerable amount of energy—57.8 kcal per mole of water formed.

To understand this seemingly ambiguous behavior, let us consider the mechanism of the reaction of hydrogen with oxygen. Three molecules, two H_2 and one O_2, are involved. The molecules must come together in such a way that the bonds can rearrange to form the products. At the high temperature in a hydrogen flame, many of the molecules are moving so fast that when two of them collide, one breaks up into single atoms:

$$2H_2 \longrightarrow H_2 + 2H$$

This reaction is *endergonic*—that is, it absorbs energy. Some of the energy of the two colliding molecules is used to break the covalent bond of one of them; thus the products of the reaction do not have as much kinetic energy as the intact molecules did. The kinetic energy used to break the bond has been transformed into chemical potential energy; the single hydrogen atoms react very easily with oxygen molecules*:

$$4H + O_2 \longrightarrow 2H_2O$$

This reaction is *exergonic*—that is, it releases energy. In fact, it releases more energy than was absorbed by the endergonic reaction in which the H_2 was broken apart. If

*The reaction does not actually require a fortuitous meeting between four hydrogen atoms and an oxygen molecule. Instead, it takes place in several steps: $H + O_2 \rightarrow HO + O$; $H + HO \rightarrow H_2O$; $O + H_2 \rightarrow OH + H$; etc. However, this is not pertinent to our discussion.

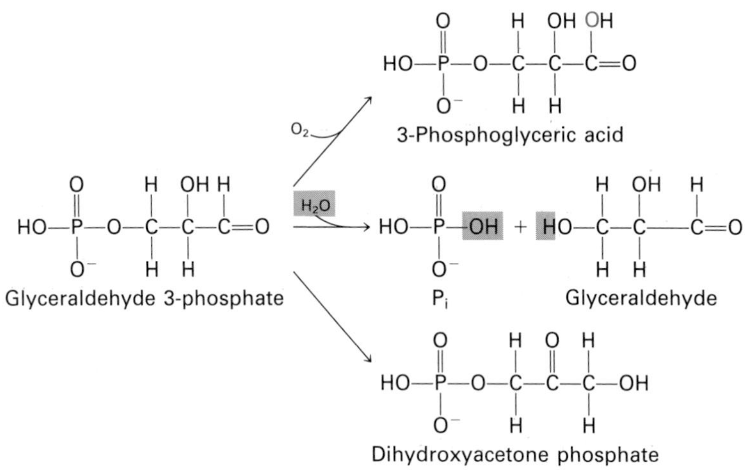

Glyceraldehyde 3-phosphate

O_2 → 3-Phosphoglyceric acid

H_2O → P_i + Glyceraldehyde

→ Dihydroxyacetone phosphate

Figure 2-26 Glyceraldehyde 3-phosphate, like most cellular molecules, can undergo any of several exothermic reactions: oxidation (by O_2) to 3-phosphoglyceric acid, hydrolysis to glyceraldehyde and phosphate, and rearrangement to dihydroxyacetone phosphate. In the absence of enzymes or other catalysts, however, none of these reactions occurs in aqueous solution. Different enzymes catalyze each reaction; the presence of a specific enzyme is required for a particular reaction to proceed.

we consider the two reactions as separate stages of an overall reaction, then because the second stage releases more energy than the first absorbs, the overall reaction is exergonic.

The input of energy required to initiate a reaction is called the *activation energy*. In a burning flame, the kinetic energy of many molecules of H_2 or O_2 is great enough to serve as activation energy. At room temperature, however, the average kinetic energy of a typical gas such as H_2 is about 1.5 kcal/mol. Although many molecules will have more kinetic energy than this average, the chances that an H_2 molecule will have enough kinetic energy to equal the activation energy (~100 kcal/mol) and thereby start the reaction are virtually zero. Thus, a mixture of hydrogen and oxygen will not react until it receives enough energy—say, from a spark or a flame—to achieve the activation energy. The energetic relation between the initial reactants and the products of an exergonic reaction can usually be described by a diagram similar to that in Figure 2-27.

Like the chemical reactions in a flame, exergonic biochemical reactions require activation energy. In some, bonds must be broken before new bonds can be formed; in others, electrons must be excited before they will pair up in a covalent bond; in still others, molecules need only enough energy to overcome the mutual repulsion of their electron clouds so they can get close enough to react. For example, the conversion of dihydroxyacetone phosphate into glyceraldehyde 3-phosphate involves at least one intermediate (Figure 2-28). During formation of the intermediate the following changes take place simultaneously: a proton is removed from one carbon by a basic group in the enzyme catalyst; an acidic group in the enzyme donates a proton to an oxygen, and pairs of electrons move from one bond to another. Each of these partial reactions contributes to the activation energy necessary to form this reaction intermediate.

Thus each stage in a multistep reaction may have its own activation energy (Figure 2-29). For the reaction to proceed, the highest activation energy must be achieved. Because biochemical reactions occur at moderate temperatures, the kinetic energy of colliding molecules is generally insufficient to provide the necessary activation energy; in most cases, reactants must meet this energy requirement in some other way.

A Catalyst Reduces the Activation Energy and Increases the Reaction Rate

The rate at which a chemical reaction occurs depends on several factors. The temperature of the system is impor-

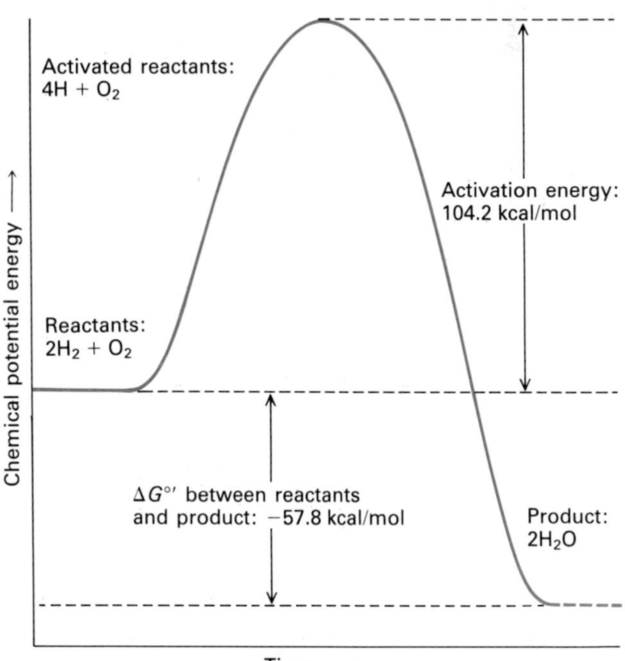

Figure 2-27 The reaction of hydrogen with oxygen requires an initial input of 104.2 kcal/mol—the activation energy—even though the products have a much lower free energy than the reactants do. This diagram depicts the energy content of the reactants at each stage.

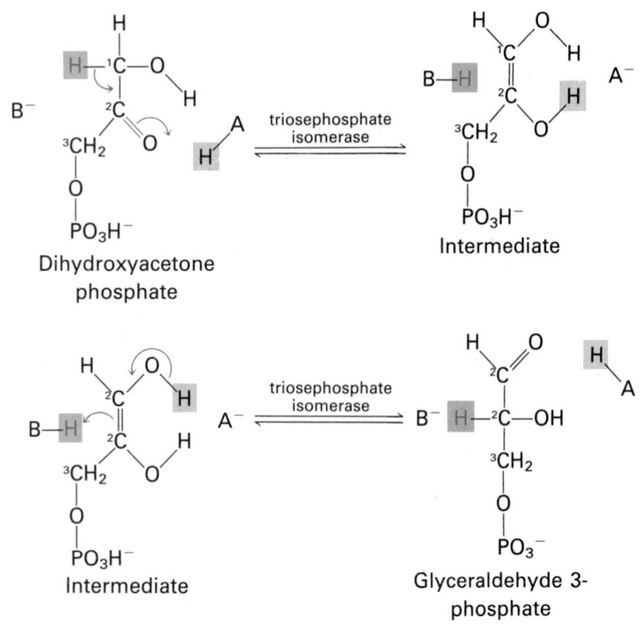

Figure 2-28 The conversion of dihydroxyacetone phosphate into glyceraldehyde 3-phosphate involves an intermediate. Two catalytic groups, a base B^- and an acid HA, are involved; in the enzyme triosephosphate isomerase that catalyzes this reaction, B is a glutamate side chain

$(—CH_2—CH_2—\overset{O}{\overset{\|}{C}}—O^-)$ and HA is probably a lysine side chain $(—CH_2—CH_2—NH_3^+)$. To form the intermediate *(top)* B abstracts a proton *(color)* from the 1 carbon of dihydroxyacetone phosphate, while HA adds a proton to the 2 keto-oxygen. To convert the intermediate into glyceraldehyde 3-phosphate *(bottom)* BH donates its proton to the 2 carbon (regenerating the original B^-), while A^- abstracts a proton from the —OH on carbon 1 (regenerating HA). Shown here also are curved arrows that denote the movements of pairs of electrons that accompany the making and breaking of those bonds. [See D. Straus et al., 1985, *Proc. Nat'l. Acad. Sci.* **82**:2272.]

tant in nonbiochemical reactions, which proceed faster as the temperature increases because an increasing number of molecules in the reaction mixture attain the activation energy. In reactions with two or more reactants, an increase in temperature will also increase the frequency of collisions between molecules or atoms. However, because most cells experience very little variation in temperature, changes in temperature are not important in regulating the rate of biochemical reactions.

For biological systems, two significant rate-regulating factors are the concentrations and the pH of the reactants. A reaction involving two or more different molecules proceeds faster at high concentrations because the molecules are more likely to encounter one another.

However, the most important determinants of biochemical reaction rates are *enzymes,* which are proteins that act as catalysts. A *catalyst* is a substance that brings reactants together but that does not end up among the products of the reaction. This function is aptly reflected in the Chinese term for catalyst: *tsoo mei,* which literally means "marriage broker." A catalyst increases the rate of a reaction but is not itself permanently changed. It does not alter the free-energy change or the equilibrium constant. It accelerates the rates of forward and reverse reactions by the same factor. In fact, whether or not a catalyst is present, the equilibrium constant is equal to the rate of the forward reaction divided by the rate of the reverse reaction:

$$K_{eq} = \frac{\text{rate of forward reaction}}{\text{rate of reverse reaction}}$$

Enzymes and all other catalysts act by reducing the activation energy required to make a reaction proceed (Figure 2-30). Some enzymes bind a *substrate,* the substance on which they act, in a way that strains its bonds and makes it easy for the substrate to react. Other enzymes bind multiple substrates in a way that brings them to-

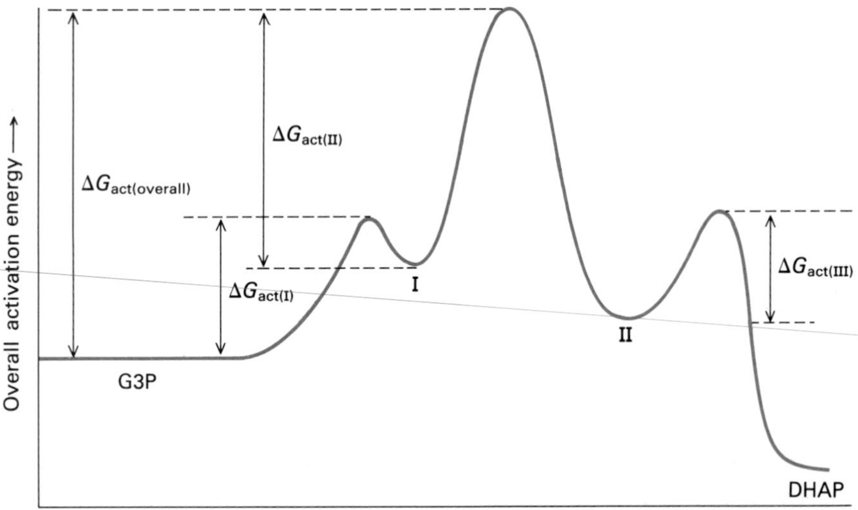

Figure 2-29 Hypothetical energy changes in the conversion of glyceraldehyde 3-phosphate (G3P) into dihydroxyacetone phosphate (DHAP). The troughs in the curve represent stable intermediates in the reaction. The vertical distance from a trough to the succeeding crest represents the activation energy (ΔG_{act}) required for one intermediate to be converted into the next (e.g., $\Delta G_{act(II)}$ represents the activation energy required for the conversion of intermediate I into intermediate II), or for the last intermediate to be converted into product. The total activation energy $\Delta G_{act(overall)}$ is the difference between the free energy of the reactants and that represented by the highest crest along a pathway.

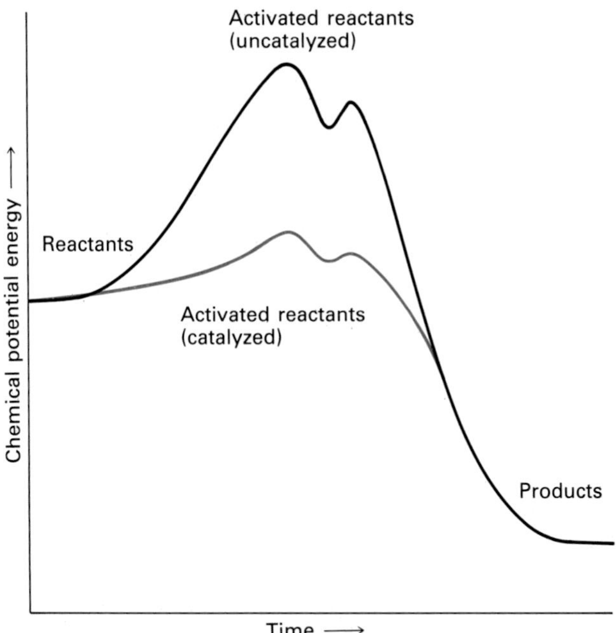

Figure 2-30 A catalyst accelerates the rate of a reaction by reducing the activation energy. It does not alter the free energy of either the reactants or the products and thus does not affect their equilibrium concentrations.

gether so that they react readily with one another. In each case, the overall effect is to reduce the activation energy needed for the reaction to take place. Triosephosphate isomerase, the enzyme that catalyzes the conversion of glyceraldehyde 3-phosphate (G3P) into dihydroxyacetone phosphate, binds G3P as its substrate. Once G3P is bound, all of the requisite movements of its hydrogen atoms, protons, and electrons are facilitated by specific chemical groups on the parts of the enzyme that are adjacent to the bound substrate (see Figure 2-28).

An enzyme binds either a single substrate or a set of similar substrates. Each enzyme catalyzes a single reaction on the bound substrate. Thus, the presence or absence of particular enzymes in a cell or in extracellular fluids determines which of many possible chemical reactions will occur.

Summary

Atoms in a molecule are held together in a fixed orientation chiefly by covalent chemical bonds. Such bonds have relatively high energies of formation (they release from 50 to 200 kcal/mol when they are formed) and consist of the sharing of electrons by two atoms. Many covalent bonds between unlike atoms are dipolar, meaning that the electron spends more time around the more electronegative atom.

A number of weaker chemical bonds and interactions help determine the shape of many large biological molecules and stabilize complexes composed of different molecules. In a hydrogen bond, a hydrogen atom covalently bonded to a donor atom associates with an acceptor atom (whose nonbonding electrons attract the hydrogen). Hydrogen bonds among water molecules are largely responsible for the properties of both the liquid and the crystalline solid (ice) forms. Van der Waals interactions are weak and relatively nonspecific; they are created whenever any two atoms approach each other closely. Ionic bonds result from the electrostatic attraction between the positive and negative charges of ions. In aqueous solutions, all cations and anions are surrounded by a tightly bound shell of water molecules. Hydrophobic interactions occur among nonpolar molecules, such as hydrocarbons, in an aqueous environment. These "bonds" result from the noninteraction of water with the nonpolar molecules and from the resulting van der Waals forces among the nonpolar molecules. Although any single noncovalent bond may be very weak—most release only a few kilocalories per mole when they are formed—several such bonds between molecules, or between the parts of one molecule, can result in very stable structures. The attraction between two large molecules can be quite strong if they have complementary sites on their surfaces.

The energy measure that is most useful in describing chemical reactions in biological systems is the change in free energy, ΔG. This value depends on the change in heat or enthalpy, ΔH (chiefly the overall change in bond energies) and the change in the randomness of molecular motion, or entropy, ΔS: $\Delta G = \Delta H - T\,\Delta S$. Chemical reactions tend to proceed spontaneously only if ΔG is negative.

The direction of a chemical reaction can be predicted if both its standard $\Delta G^{\circ\prime}$ value and the concentrations of the reactants and the products are known. The equilibrium constant K_{eq} bears a simple mathematical relation to $\Delta G^{\circ\prime}$: $K_{eq} = 10^{-\Delta G^{\circ\prime}/2.3RT}$. A chemical reaction having a positive ΔG can proceed spontaneously if it is coupled with a reaction having a negative ΔG of larger magnitude.

In cells, the two phosphoanhydride bonds in ATP are a principal source of chemical potential energy. The hydrolysis of one or both of these bonds is often coupled with another reaction that is energetically unfavorable. Such reactions include those involved in the synthesis of proteins from amino acids, the synthesis of other needed molecules, cellular movement, and the transport of compounds into or out of a cell against a concentration gradient. All of these reactions are fueled by the hydrolysis of ATP. In plant cells, ATP is generated from ADP and P_i through the use of energy absorbed from light; much of the ATP is consumed in the synthesis of glucose from CO_2 and H_2O. In animal cells and nonphotosynthetic microorganisms, most of the ATP is generated during the oxidation (by O_2) of glucose to CO_2 and H_2O. Directly or indirectly, photosynthesis is the source of energy for all cells.

Whether or not a reaction will actually proceed is a function of its activation energy. Sometimes the kinetic energy of the reactant molecules is sufficient to overcome the activation-energy barrier. In other cases, the activation energy is too high to permit the reaction to occur at a measurable rate. Catalysts speed up chemical reactions by decreasing the activation energy. They accelerate both the forward and the reverse reactions to the same extent; they do not change the ΔG or the K_{eq}. Enzymes are biological catalysts that generally facilitate only one of the many possible transformations that a molecule can undergo.

References

BROWN, T. L., and H. E. LEMAY. 1985. *Chemistry: The Central Science,* 3d ed. Prentice-Hall. Chapters 13–16, 18, and 19.

BUTLER, J. M. 1973. *Solubility and pH Calculations.* Addison-Wesley.

CANTOR, C. R., and P. R. SCHIMMEL. 1980. *Biophysical Chemistry,* part 1. W. H. Freeman and Company. Chapter 5.

DAVENPORT, H. W. 1974. *ABC of Acid-Base Chemistry,* 6th ed. University of Chicago Press.

EDSALL, J. T., and J. WYMAN. 1958. *Biophysical Chemistry,* vol. 1. Academic Press.

EISENBERG, D., and W. KAUZMANN. 1969. *The Structure and Properties of Water.* Oxford University Press.

HILL, T. L. 1977. *Free Energy Transduction in Biology.* Academic Press.

KLOTZ, I. M. 1967. *Energy Changes in Biochemical Reactions.* Academic Press.

LEHNINGER, A. L. 1982. *Principles of Biochemistry.* Worth. Chapters 1, 3, 4, 13, and 14.

PAULING, L. 1960. *The Nature of the Chemical Bond,* 3d ed. Cornell University Press.

STRYER, L. 1981. *Biochemistry,* 2d ed. W. H. Freeman and Company. Chapters 1, 6, 11, and 14.

TANFORD, C. 1980. *The Hydrophobic Effect: Formation of Micelles and Biological Membranes,* 2d ed. Wiley.

WATSON, J. D. 1976. *Molecular Biology of the Gene,* 3d ed. Benjamin-Cummings. Chapters 4 and 5.

WOOD, W. B., J. H. WILSON, R. M. BENBOW, and L. E. HOOD. 1981. *Biochemistry: A Problems Approach,* 2d ed. Benjamin-Cummings. Chapters 1, 5, and 9.

ZUBAY, C. 1983. *Biochemistry.* Addison-Wesley. Chapter 7.

3

The Molecules in Cells

All cells are constructed of two different types of chemical substances: polymers and small molecules. The distinction is one of both size and organization. *Small molecules* generally consist of fewer than 50 atoms, and each substance has its own characteristic structure. *Polymers* are composed of many copies of a few small molecules linked in chains by covalent bonds. These subunits of polymers are referred to as *monomers,* or *residues.* The principal cellular polymers—and the primary focus of this chapter—are nucleic acids, the substances that preserve and transmit genetic information, and proteins, the products generated from the transmitted information. Nucleic acids are built of monomers called nucleotides; proteins are constructed from monomeric amino acids. The arrangement of monomers in these biopolymers is linear (Figure 3-1).

Nucleic acids are built of four different nucleotides linked together in chains that can be millions of units long. Because these subunits can be linked in any order, the number of different possible nucleic acids n units long is 4^n (Figure 3-1). A 10-unit nucleic acid has 4^{10} (more than 1 million) possible structures; a 100-unit nucleic acid has 4^{100} (more than 10^{60}) possible structures.

There are 20 different amino acids in proteins. Thus, a 100-unit protein has 20^{100} (more than 10^{130}) possible structures. This enormous variability means that cells and organisms can differ greatly in structure and function

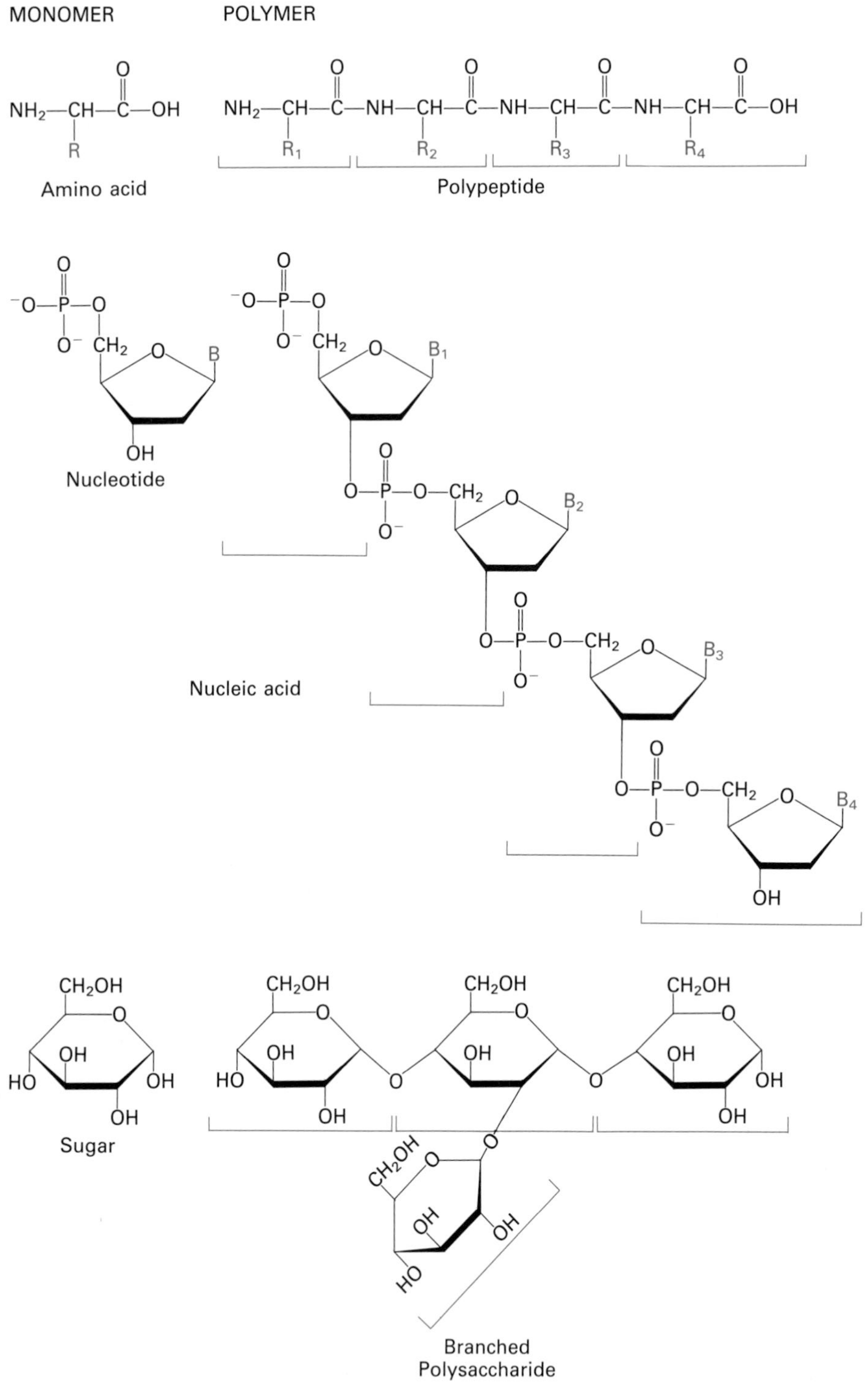

Figure 3-1 Proteins are linear biopolymers formed from monomeric subunits termed amino acids. There are 20 different amino acids, each with a different R group, or side chain. Thus, the polypetide shown here, which is constructed of 4 amino acids, has 20^4 or 160,000 possible structures.

Nucleic acids, which are also linear biopolymers, are formed from four monomers termed nucleotides. Each of the four nucleotides has a different nitrogen-containing base structure (B). Thus, the nucleic acid shown here has 4^4 or 256 possible structures. (The rings are depicted as Haworth projections, so the plane of the rings is perpendicular to the page: the heavy bonds project out of the front of the page, and the oxygen falls behind it. By convention, the ring carbons are not shown. Here, the hydrogens attached to the ring carbons are not shown either.)

Polysaccharides are built of monomeric saccharide (sugar) subunits. Because sugar residues can bind to one another at different positions, nonlinear branching polymers are often formed.

even though they are constructed of the same types of biopolymers produced by similar chemical reactions. The information in living systems is stored in nucleic acids, and the chemical reactions that constitute life processes are directed and controlled by proteins.

Starch (a storage form of glucose in plant cells), cellulose (a constituent of plant cell walls), and glycogen (a storage form of glucose in liver and muscle cells) are examples of another important type of biopolymer: the *polysaccharides*. Polysaccharides are built of monomers called saccharides, or simply sugars (Figure 3-1). At least 15 different sugars are used to form various polysaccharides. The monomeric sugars can be bonded to one another in multiple ways; thus, many polysaccharides are nonlinear, branched molecules.

There are other small molecules besides those that serve as the monomers within polymers. The lipids, for example, are especially important because they form the basic structure of cell membranes. Lipids interact noncovalently in very large nonlinear complexes; the membranes thus formed are as crucial to living systems as are the biopolymers.

This chapter deals with the structures and some of the functions of biopolymers and small molecules; in later chapters, we describe how the polymers are made and consider many of their other functions.

Proteins

The working molecules of the cell are proteins. They catalyze an extraordinary range of chemical reactions, provide structural rigidity, control the permeability of membranes, regulate the concentrations of needed metabolites, recognize and bind noncovalently other biomolecules, cause motion, and control the functioning of genes. As mentioned earlier, these incredibly diverse tasks are performed by molecules synthesized from only 20 different amino acids.

Amino Acids, the Building Blocks of Proteins, Differ Only in their Side Chains

The monomers that make up proteins are called amino acids because, with one exception, each contains an *amino group* ($—NH_2$) and an acidic *carboxyl group* ($—COOH$). The exception, proline, has an *imino group* ($—NH—$) instead of an amino group. At typical pH values in cells, these groups are ionized as $—NH_3^+$, $—COO^-$, and $=NH_2^+$. All amino acids are constructed according to a basic design: a central carbon atom called the α carbon (because it is adjacent to the acidic carboxyl group) is bonded to an amino (or imino) group, to the carboxyl group, to a hydrogen atom, and to one variable group, called a *side chain* or *R* group (Figure 3-2). The side chains are the parts of the amino acids that give them their individuality.

The amino acids can be classified according to their electric charge when they are ionized, and according to the affinity of their side chains for water. Five of the standard amino acids have charged side chains at neutral pH (Figure 3-2). Lysine, arginine, and histidine are positively charged. Essentially all of the side chains of lysine and arginine are protonated at neutral pH. The exact fraction of side chains that are protonated can be calculated from the pK_a value for the ionizing group: this value is 10.5 for lysine and 12.5 for arginine. For the side chain of lysine, we have

$$H_2N—(CH_2)_4— + H^+ \rightleftharpoons {}^+H_3N—(CH_2)_4—$$

$$\log \frac{[H^+][H_2N—(CH_2)_4—]}{[{}^+H_3N—(CH_2)_4—]} = -10.5$$

At pH 7.0, $[H^+]$ is 10^{-7} *M*, so the ratio of protonated to nonprotonated side chains in lysine is 3160:1. For arginine, the ratio is 316,000:1. Histidine is much more weakly ionic; the pK_a of its side chain is 6.0. Thus, at pH 7.0, the ratio of protonated to nonprotonated chains is 1:10.

Aspartic and glutamic acids are negatively charged. The pK_a's for their side chains are 3.9 and 4.2, respectively; thus, at neutral pH, these side chains are virtually all nonprotonated, as shown here for the glutamic acid side chain:

$$—CH_2—CH_2—\overset{\overset{\displaystyle O}{\|}}{C}—OH \longrightarrow H^+ + —CH_2—CH_2—\overset{\overset{\displaystyle O}{\|}}{C}—O^-$$

To emphasize this fact, aspartic and glutamic acids are conventionally referred to as aspartate and glutamate.

Serine, threonine, and tyrosine, which each have an —OH group in their side chains, and cysteine, which has an —SH, can interact strongly with water by the formation of hydrogen bonds. The side chains of asparagine and glutamine have polar amide groups,

$$—\overset{\overset{\displaystyle O}{\|}}{C}—NH_2$$

which render them hydrophilic as well.

The side chains of several amino acids—alanine, valine, leucine, isoleucine, proline, methionine, phenylalanine, and tryptophan—consist only of hydrocarbons, except for the sulfur atom in methionine and the nitrogen atom in tryptophan. The side chains of these amino acids are *hydrophobic*. As a consequence, the amino acids in this group (except for alanine) are only slightly soluble in water, even though the form of the amino acid that exists at neutral pH is a zwitterion (Chapter 2), with its positively charged $—NH_3^+$ and its negatively charged $—COO^-$:

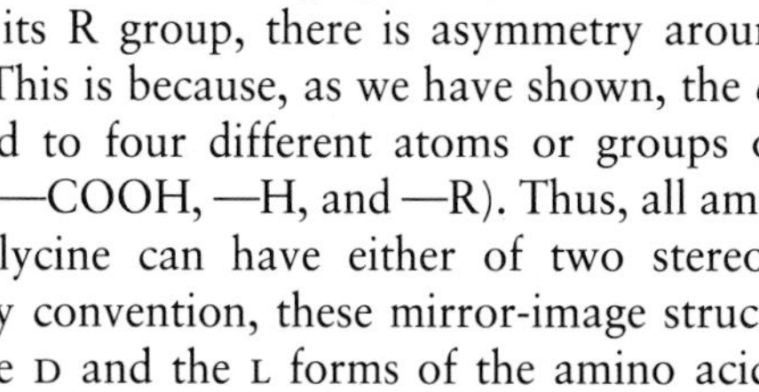

H H
C=C NH₃⁺
HC C—CH₂—CH—COO⁻
C—C
H H

Hydrophobic Hydrophilic

Phenylalanine

In all amino acids except glycine, which has a hydrogen atom as its R group, there is asymmetry around the α carbon. This is because, as we have shown, the α carbon is bonded to four different atoms or groups of atoms ($—NH_2$, $—COOH$, $—H$, and $—R$). Thus, all amino acids except glycine can have either of two stereoisomeric forms. By convention, these mirror-image structures are called the D and the L forms of the amino acid (Figure 3-3). They cannot be interconverted without breaking a chemical bond. With rare exceptions, only the L forms of amino acids are found in proteins.

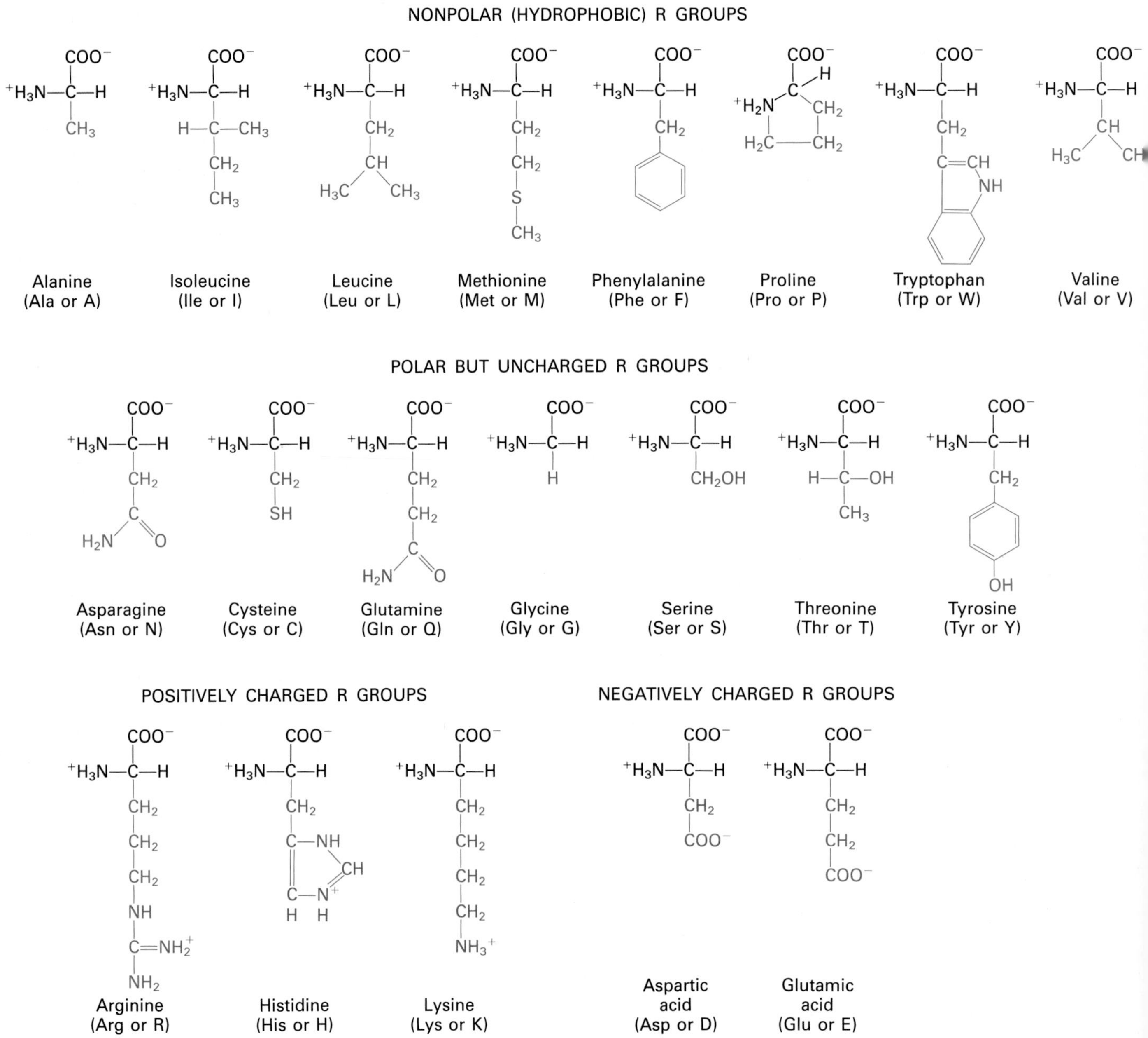

NONPOLAR (HYDROPHOBIC) R GROUPS

Alanine (Ala or A) | Isoleucine (Ile or I) | Leucine (Leu or L) | Methionine (Met or M) | Phenylalanine (Phe or F) | Proline (Pro or P) | Tryptophan (Trp or W) | Valine (Val or V)

POLAR BUT UNCHARGED R GROUPS

Asparagine (Asn or N) | Cysteine (Cys or C) | Glutamine (Gln or Q) | Glycine (Gly or G) | Serine (Ser or S) | Threonine (Thr or T) | Tyrosine (Tyr or Y)

POSITIVELY CHARGED R GROUPS

Arginine (Arg or R) | Histidine (His or H) | Lysine (Lys or K)

NEGATIVELY CHARGED R GROUPS

Aspartic acid (Asp or D) | Glutamic acid (Glu or E)

Figure 3-2 The structures of the 20 common amino acids. All have a central carbon atom (the α carbon) that is bonded to an amino group (or an imino group in proline), a carboxyl group, a hydrogen atom, and an R group. The R groups are shown in color.

Polypeptides Are Polymers Composed of Amino Acids Connected by Peptide Bonds

The *peptide bond,* the chemical bond that connects two amino acids in a polymer, is formed between the amino group of one amino acid and the carboxyl group of another. This reaction, which is called condensation, liberates a water molecule (Figure 3-4a). Because the carboxyl carbon and oxygen atoms are connected by a double bond, the peptide bond between carbon and nitrogen exhibits a partial double-bond character, as shown by the resonance structures

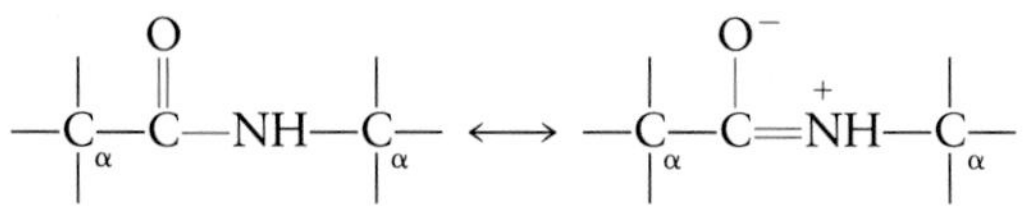

$$-C_{\alpha}-\overset{O}{\overset{\|}{C}}-NH-C_{\alpha}- \longleftrightarrow -C_{\alpha}-\overset{O^-}{\overset{|}{C}}=\overset{+}{N}H-C_{\alpha}-$$

Peptide group

Thus, the carbon-nitrogen peptide bond is shorter than the typical C—N single bond. Because of the resonance bonding, the six atoms of the peptide group—the two α carbons of the adjacent amino acids, and the carbon, oxygen, nitrogen, and hydrogen atoms of the bond—lie in the same plane (Figure 3-4b). However, adjacent peptide

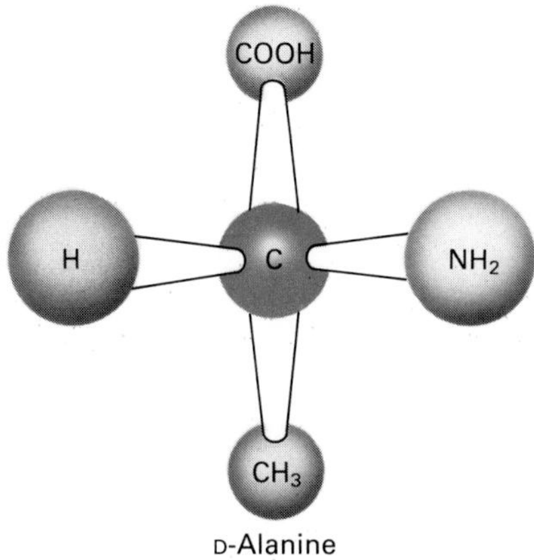

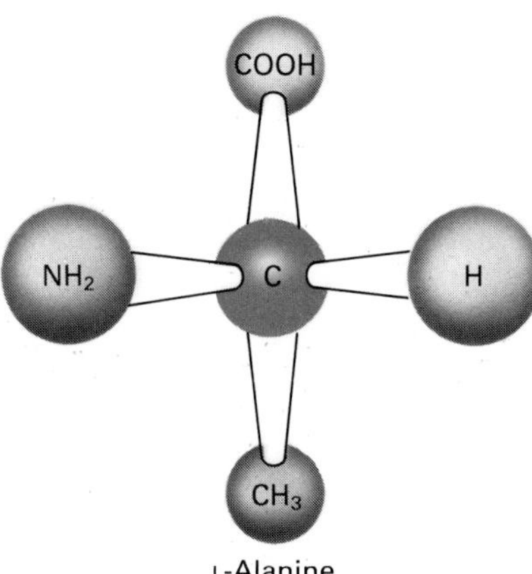

Figure 3-3 Stereoisomers of the amino acid alanine. The α carbon is in color.

groups are not coplanar, due to rotation about the C—C_α and N—C_α bonds (Figure 3-4c).

A single linear array of amino acids is called a *polypeptide.* If the polypeptide is short (fewer than 30 amino acids long), it may be called an oligopeptide or just a peptide. Polypeptides differ greatly in length; they generally contain between 40 and 1000 amino acids, but some, such as glutathione, contain as few as 3. Each polypeptide has a free amino group at one of its ends (the N-terminus) and a free carboxyl group at the other (the C-terminus) (Figure 3-5).

Some proteins are single polypeptides, but many consist of more than one chain. A well-understood multichained protein is *hemoglobin,* the oxygen-carrying protein of red blood cells. Adult human hemoglobin is made up of four polypeptides, or globin chains; two identical *α chains* and two identical *β chains.* The sequence of amino acids in the chains is known, and the amino acids at many positions along the α and β chains are identical or closely related, which suggests that the two chains function similarly in the cell (Figure 3-6). Indeed, that is the case; each chain can bind one oxygen molecule (O_2).

A genetic mutation can lead to the substitution of one amino acid for another at a defined place in the protein. If the substitution inserts an amino acid with a similar side chain—for example, leucine for isoleucine, or aspartate for glutamate—the function of the protein may not be impaired. If, however, an amino acid with a very different side chain is substituted, the protein is often altered to such an extent that it is rendered nonfunctional. A mutation in which valine replaces glutamate at position 6 of the hemoglobin β chain, for example, results in sickle-cell hemoglobin (Figure 3-6), a mutant protein that causes sickle-cell anemia, a serious disease.

The Structure of a Polypeptide Can Be Described at Four Levels

Protein structure has been described at four different levels. *Primary structure* refers to the linear arrangement of amino acid residues along a polypeptide chain, and to the locations of covalent bonds, such as —S—S— bonds, between chains. *Secondary structure* pertains to the folding of parts of these chains into regular structures such as α helixes and β pleated sheets. *Tertiary structure* includes the folding of regions between α helixes and β sheets, and all of the other noncovalent interactions that ensure the proper folding of a single polypeptide chain of a protein, such as hydrogen and hydrophobic bonds and van der Waals interactions. Finally, *quaternary structure* refers to the noncovalent interactions that bind several polypeptide chains into a single protein molecule, such as hemoglobin.

Proteins Fold into Different Shapes; Many Proteins Contain Tightly Bound Prosthetic Groups

Although the primary structure of polypeptides is linear, most fold into compact spherical or ellipsoid forms to yield *globular proteins.* Other proteins adopt more rodlike shapes; these are the *fibrous proteins* such as collagen, the major structural protein in cartilage, tendons, and blood vessels (Figure 3-7). A polypeptide folds into a stable conformation mainly through noncovalent interactions—ionic, hydrogen, van der Waals, and hydrophobic—or through covalent bonding between amino acids in different parts of the chain. Noncovalent forces also hold the separate polypeptide chains together in proteins such as hemoglobin (Figure 3-8).

Another influence on the shape of a folded polypeptide is the presence or absence of a *prosthetic group:* that is, a small molecule, which is not a peptide, that is tightly bound to the protein and that plays a crucial role in its functioning. For example, each of the four chains of hemoglobin binds and enfolds a prosthetic group called a *heme,* which is the oxygen carrier (Figure 3-9). Prosthetic

(a)

$$^{+}H_3N-\overset{H}{\underset{R_1}{C}}-\overset{O}{\overset{\|}{C}}-O^- + {}^{+}H_3N-\overset{H}{\underset{R_2}{C_\alpha}}-\overset{O}{\overset{\|}{C}}-O^- \xrightarrow{H_2O}$$

$$^{+}H_3N-\overset{H}{\underset{R_1}{C_\alpha}}-\overset{O}{\overset{\|}{C}}-\underset{H}{N}-\overset{H}{\underset{R_2}{C_\alpha}}-\overset{O}{\overset{\|}{C}}-O^-$$

Peptide bond

(b)

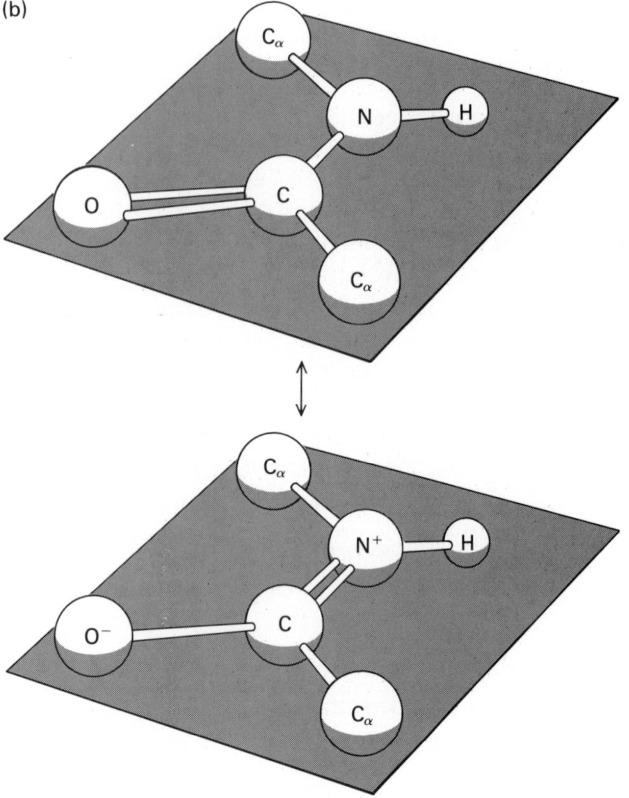

(c)

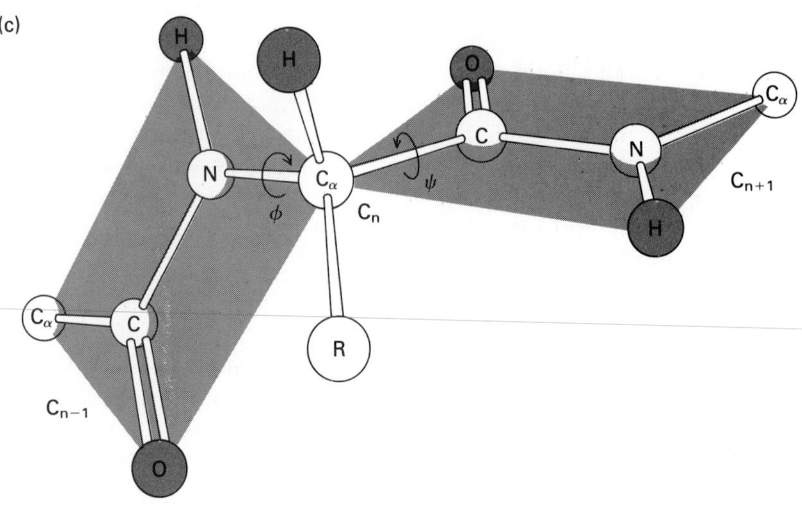

Figure 3-4 (a) The formation of a peptide bond between two amino acids occurs by the elimination of a molecule of water in a condensation reaction. (b) Because the carbon-nitrogen peptide bond has a partial double-bond character, the peptide group is planar; that is, α carbons (C_α) of the connected amino acids and the carbon, oxygen, nitrogen, and hydrogen atoms of the bond are coplanar. (c) However, there is considerable flexibility in the geometry of polypeptides, since rotation is possible about the two covalent single bonds that connect each α carbon to the two adjacent planar peptide units. These rotation angles are labeled ψ (Greek psi), for the C—C_α bond, and ϕ (Greek phi), for the N—C_α bond. There are some restrictions on the values of ψ and ϕ. For example, if the pictured adjacent peptide groups were coplanar, then certain oxygen and hydrogen atoms would be separated by less than their van der Waals radii, so they would repel one another. Thus, a polypeptide can be visualized as having a backbone consisting of a flexible linear arrangement of small planes (the peptide groups); each plane is connected to its two neighboring planes at a single pivot point (the α carbon).

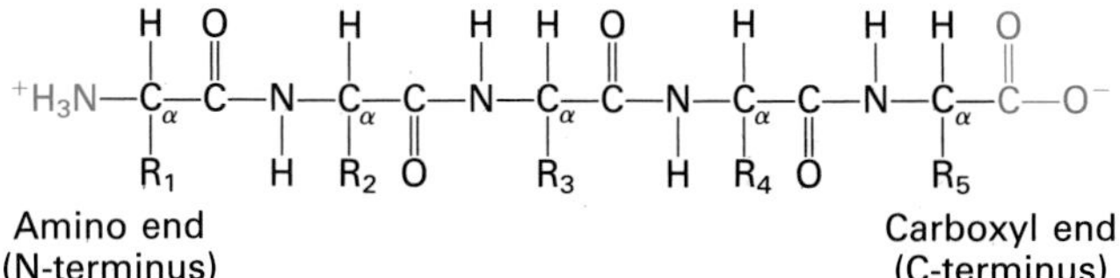

Figure 3-5 A polypeptide has a definite chemical orientation. One end of each chain bears a free amino group; the other end bears a free carboxyl group.

groups can be linked to proteins noncovalently, as in hemoglobin, or covalently. Not all proteins have prosthetic groups.

Covalent Modifications Affect the Structures and Functions of Proteins

An important determinant of the shapes of proteins is the covalent *disulfide bridge* between two cysteine residues in the same polypeptide chain or in different chains. Such —S—S— bonds stabilize folded conformations of proteins. Chemically, the disulfide bridge is formed by an oxidation reaction in which two hydrogen atoms (i.e., two protons and two electrons) are lost (Figure 3-10). Disulfide bridges often play an important role in the structures of extracellular proteins (those that function mainly outside of cells). The hormone insulin, for example, is composed of two polypeptides: A and B chains containing 21 and 30 residues, respectively. One disulfide bridge connects two cysteine residues of the A chain, and two others link the A and B chains together (Figure 3-11).

Intracellular proteins generally contain few disulfide bridges. This is because the concentration of small molecules containing —SH groups is very high inside cells. The tripeptide glutathione (Figure 3-12) is the most important such molecule because it reduces cysteine-cysteine disulfide bridges in intracellular proteins. If we let G stand for the remainder of the glutathione molecule the reaction is

$$2\ \text{G—SH} + \text{—Cys—S—S—Cys—} \longrightarrow \text{G—S—S—G} + 2\ \text{—Cys—SH}$$

[The oxidized glutathione molecules linked in pairs by disulfide bonds (G—S—S—G) are reduced in turn to free molecules (G—SH) in a reaction catalyzed by the enzyme glutathione reductase.]

Proteins undergo other forms of covalent modification. In many cases, a protein is first made as a precursor polypeptide that is then cleaved to form the functional protein. Preproinsulin, for example, is a single polypeptide chain that undergoes several successive cleavages to produce insulin (Figure 3-13).

N-termini										
		α	1 Val		Leu	Ser	Pro	5 Ala	Asp	Lys
		β	Val 1	His	Leu	Thr	Pro	Glu 6	Glu	Lys
α Thr	Asn	10 Val	Lys	Ala	Ala	Trp	Gly	Lys	Val	Gly
β Ser	Ala 10	Val	Thr	Ala	Leu	Trp	Gly	Lys	Val	Asn
α Ala	20 His	Ala	Gly	Glu	Tyr	Gly	Ala	Glu	Ala	Leu
β		Val 20	Asp	Glu	Val	Gly	Gly	Glu	Ala	Leu
30 α Glu	Arg	Met	Phe	Leu	Ser	Phe	Pro	Thr	Thr	40 Lys
β Gly	Arg 30	Leu	Leu	Val	Val	Tyr	Pro	Trp	Thr	Gln
α Thr	Tyr	Phe	Pro	His	Phe		Asp	Leu	Ser	50 His
β Arg 40	Phe	Phe	Glu	Ser	Phe	Gly	Asp	Leu	Ser	Thr 50
α Gly	Ser	Ala						Gln	Val	Lys
β Pro	Asp	Ala	Val	Met	Gly	Asn	Pro	Lys	Val 60	Lys
α Gly	His	Gly	60 Lys	Lys	Val	Ala	Asp	Ala	Leu	Thr
β Ala	His	Gly	Lys	Lys	Val	Leu	Gly	Ala 70	Phe	Ser
α Asn	Ala	70 Val	Ala	His	Val	Asp	Asp	Met	Pro	Asn
β Asp	Gly	Leu	Ala	His	Leu	Asp	Asn 80	Leu	Lys	Gly
α Ala	80 Leu	Ser	Ala	Leu	Ser	Asp	Leu	His	Ala	His
β Thr	Phe	Ala	Thr	Leu	Ser	Glu 90	Leu	His	Cys	Asp
90 α Lys	Leu	Arg	Val	Asp	Pro	Val	Asn	Phe	Lys	100 Leu
β Lys	Leu	His	Val	Asp	Pro 100	Glu	Asn	Phe	Arg	Leu
α Leu	Ser	His	Cys	Leu	Leu	Val	Thr	Leu	110 Ala	Ala
β Leu	Gly	Asn	Val	Leu 110	Val	Cys	Val	Leu	Ala	His
α His	Leu	Pro	Ala	Glu	Phe	Thr	Pro	120 Ala	Val	His
β His	Phe	Gly	Lys 120	Glu	Phe	Thr	Pro	Pro	Val	Gln
α Ala	Ser	Leu	Asp	Lys	Phe	Leu	130 Ala	Ser	Val	Ser
β Ala	Ala	Tyr 130	Gln	Lys	Val	Val	Ala	Gly	Val	Ala
α Thr	Val	Leu	Thr	Ser	Lys	140 Tyr	Arg	C-termini		
β Asp	Ala 140	Leu	Ala	His	Lys	Tyr	His			

Figure 3-6 The sequences of amino acids in the α and β chains of normal adult human hemoglobin. The α chain contains 141 amino acids, and the β chain has 146. The sequences are aligned to show the regions of identity between the chains; this necessitates "gaps" at a few points. Presumably, these gaps represent insertions or deletions of DNA sequences that occurred during the evolution of the α and β globin genes from a common ancestor. Sickle-cell hemoglobin differs from normal hemoglobin only in that the sickle-cell type has a valine instead of a glutamate at position 6 of one or both of the β chains.

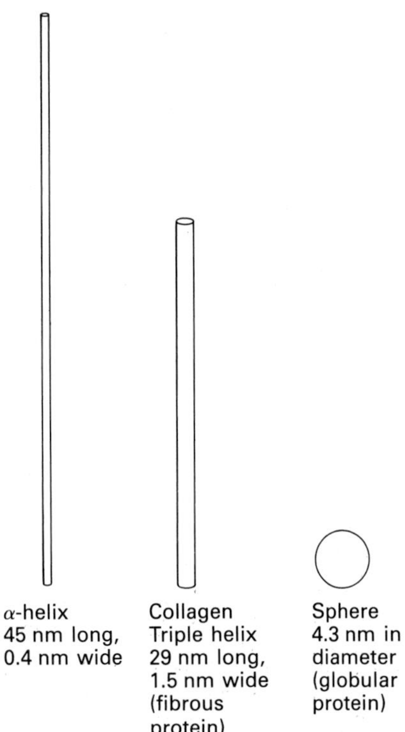

Figure 3-7 Three different shapes that can be assumed by a protein containing 300 amino acids.

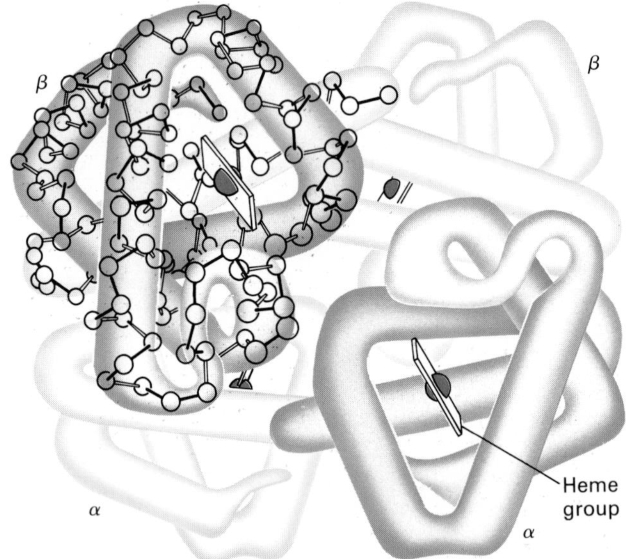

Figure 3-8 The conformations assumed by the two α and two β chains in a molecule of hemoglobin. Each chain forms several α helixes (see Figure 3-16). In this figure, only the backbones formed by the carbon and nitrogen atoms of the chains are shown. A multitude of noncovalent interactions stabilize the conformations of the individual chains and the contacts between them. A heme group is bound to each chain (see Figure 3-9). *After R. E. Dickerson and I. Geis, 1969,* The Structure and Action of Proteins, *Benjamin-Cummings, p. 56. Copyright 1969 by Irving Geis.*

Individual side chains of amino acids in certain proteins are often chemically modified. Because the protein-synthesizing machinery uses only the 20 fundamental amino acids, covalent modification of amino acid residues must take place after they have been polymerized into the polypeptide chain. In collagen, many side chains of proline and lysine are modified by the attachment of hydroxyl groups (Figure 3-14). These modifications are essential for the correct folding of the polypeptide. In addition, the amino group on the lysine side chain can be modified by an acetyl group ($CH_3CO—$) (as can the amino group at the N-terminus of the polypeptide backbone). This modification abolishes the positive charge that the amino group would normally carry at neutral pH:

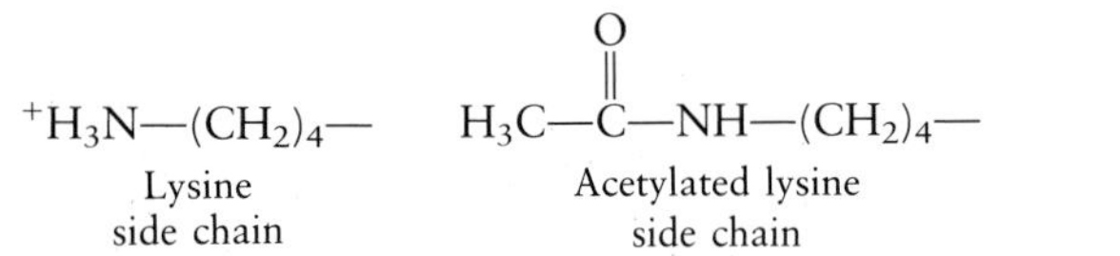

$$^{+}H_3N—(CH_2)_4— \qquad H_3C—\overset{\overset{\large O}{\|}}{C}—NH—(CH_2)_4—$$

Lysine side chain — Acetylated lysine side chain

Phosphate groups can be linked to the hydroxyl groups of tyrosine, serine, or threonine side chains:

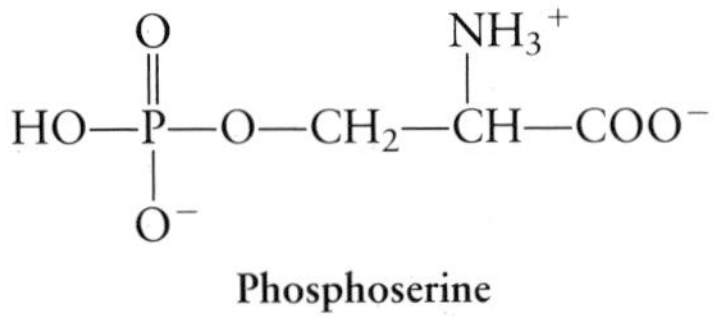

$$HO—\underset{\underset{\large O^-}{|}}{\overset{\overset{\large O}{\|}}{P}}—O—CH_2—\overset{\overset{\large NH_3^+}{|}}{CH}—COO^-$$

Phosphoserine

Phosphorylation (or dephosphorylation) of certain residues can greatly affect the activity of an enzyme. Proteins called glycoproteins have covalently bound carbohydrate groups. Other proteins, the lipoproteins, contain tightly

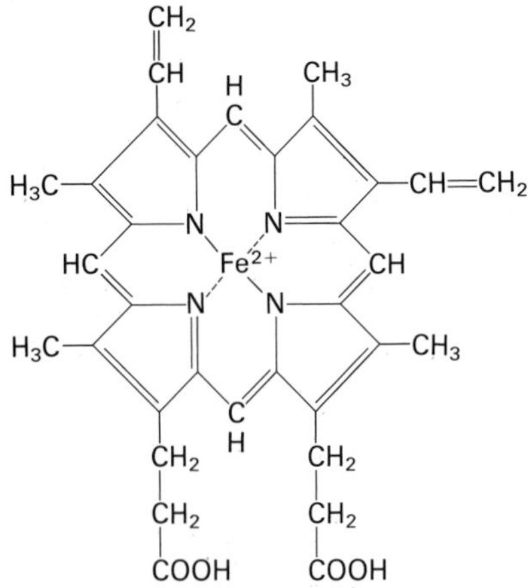

CH_2 CH H C CH_3 H_3C $CH{=}CH_2$ N N HC Fe^{2+} CH N N H_3C CH_3 C H CH_2 CH_2 CH_2 CH_2 COOH COOH

Figure 3-9 The chemical structure of a heme. In the center, an iron ion is bound equally to the four nitrogen atoms of four imidazole groups. The positions of the C=C double bonds vary continuously; the structure depicted is only one of several possible resonance structures. In hemoglobin, a histidine residue lying above the plane of the heme binds to the iron ion, which locks it into place.

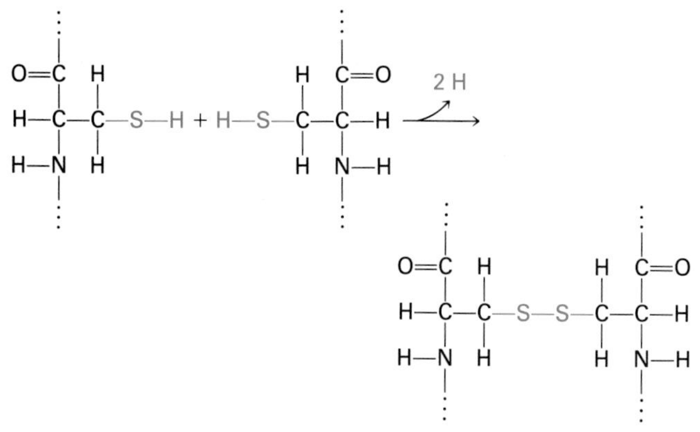

$$\text{—C(=O)—CH(NH—)—CH}_2\text{—S—H} + \text{H—S—CH}_2\text{—CH(NH—)—C(=O)—} \xrightarrow{2\text{H}} \text{—C(=O)—CH(NH—)—CH}_2\text{—S—S—CH}_2\text{—CH(NH—)—C(=O)—}$$

Figure 3-10 The formation of a disulfide bridge between two cysteine residues of a protein.

bound fatty acids, or lipids. In certain membrane lipoproteins, fatty acids are esterified to cysteine groups, which increases the hydrophobicity of the amino acid side chains:

$$H_3C\text{—}(CH_2)_{14}\text{—}\overset{\overset{\displaystyle O}{\|}}{C}\text{—S—}CH_2\text{—}\overset{\overset{\displaystyle NH_3^+}{|}}{C}H\text{—}COO^-$$

Palmitylcysteine

In other lipoproteins, the lipids are tightly bound to amino acid side chains by hydrophobic interactions.

The Native Conformation of a Protein Can Be Denatured by Heat or Chemicals

Even with the constraints imposed by disulfide bridges and other types of cross-links between amino acids, any polypeptide chain can, in principle, be folded into an infinite number of conformations. In general, however, all molecules of any species of protein will adopt the same conformation. This is called the *native state* of the molecule.

A folded protein molecule is most stable in the conformation that has the least free energy in the noncovalent bonds between the amino acids and any prosthetic groups, and between the protein and its environment. The most stable conformation is usually the native state. Most water-soluble proteins, for example, are globular. They are folded so that most of the hydrophobic amino acids are in the interior part of the molecule, away from the water, and so that the amino acids having polar side chains are on the exterior part, in contact with the water. For proteins embedded in membranes, other conformations are more stable (as will be demonstrated in Chapter 14).

The weak bonds that hold a protein together can be disrupted in the laboratory by a number of treatments that cause polypeptides to unfold. When this happens, the protein is said to be denatured. If the native protein is heated, for example, thermal energy can break the weak bonds. Extremes of pH can alter the charges on amino acid side chains, which disrupts ionic and hydrogen bonds. Reagents such as 8 *M* urea,

$$H_2N\text{—}\overset{\overset{\displaystyle O}{\|}}{C}\text{—}NH_2$$

disrupt both hydrogen and hydrophobic bonds. Most denatured proteins precipitate; that is, the molecules clump together because of attractions between hydrophobic groups that would normally be buried inside different molecules.

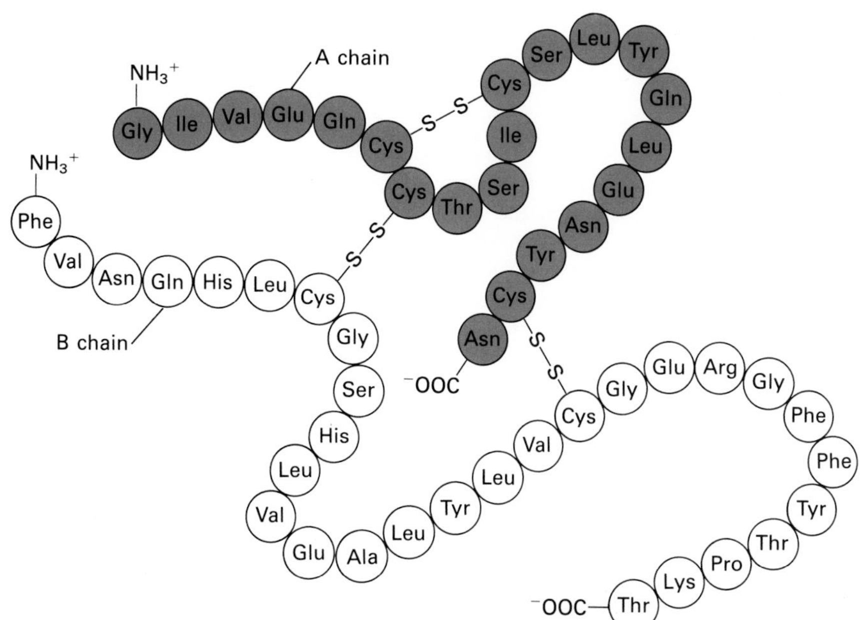

Figure 3-11 Human insulin has two disulfide bridges between cysteine residues of the A and B chains, and one bridge connecting two residues of the A chain.

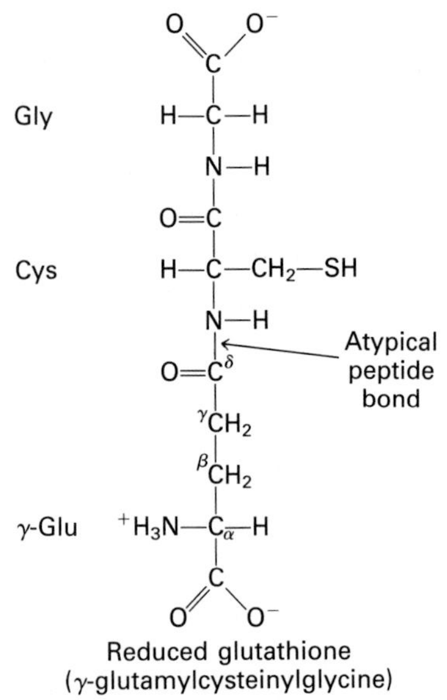

Gly $H-C-H$ (with $C(=O)O^-$ above), $N-H$, $O=C$

Cys $H-C-CH_2-SH$, $N-H$

Atypical peptide bond

$O=C^{\delta}$, ${}^{\gamma}CH_2$, ${}^{\beta}CH_2$

γ-Glu $^{+}H_3N-C_{\alpha}-H$, $C(=O)O^-$

Reduced glutathione
(γ-glutamylcysteinylglycine)

Figure 3-12 The structure of glutathione. The peptide bond is unusual because it is between the carboxyl group at the δ carbon atom of glutamate (rather than the carboxyl group attached to the α carbon) and the nitrogen atom of cysteine.

Some Denatured Proteins Can Renature into Their Native States

Many proteins that have been denatured by other techniques can be completely unfolded and dissolved in 8 *M* urea. If the urea is then slowly removed by dialysis, the denatured proteins will often renature (refold) into their native states. The weak bonds re-form in an ordered series of steps. In this way, the proteins can be carried through a denaturation-renaturation cycle that first destroys and then reestablishes their original structures and functions (Figure 3-15).

In general, renaturation is most complete if a protein is made up of a single polypeptide chain and lacks a noncovalently bound prosthetic group. For example, staphylococcal nuclease, a bacterial enzyme of 149 residues that degrades DNA and RNA, is totally denatured in acid. Yet it regains its native, active conformation within 0.1 s after the solution has been neutralized. The fast renaturation of this protein shows that its three-dimensional architecture is solely a consequence of interactions among its amino acids and with its aqueous environment. In such cases, the genetic program of the cell has only to define the primary structure of the protein—the amino acid sequence—and the tertiary structure will be assured.

For other proteins, the native conformation is not the one having the lowest free energy, and consequently the native form is not completely restored upon renaturation. This is particularly true for multichain proteins. The two chains of insulin, for example, can be separated by a combination of reducing agents (to break the disulfide bridges) and concentrated solutions of chemicals, such as urea, that disrupt hydrogen and hydrophobic bonds. When the insulin is allowed to renature in the presence of oxidizing agents that promote the formation of disulfide bridges, a number of stable multichain aggregates are formed, but the *native* insulin molecule is only a minor component of them.

The reason is, of course, that insulin is formed by the

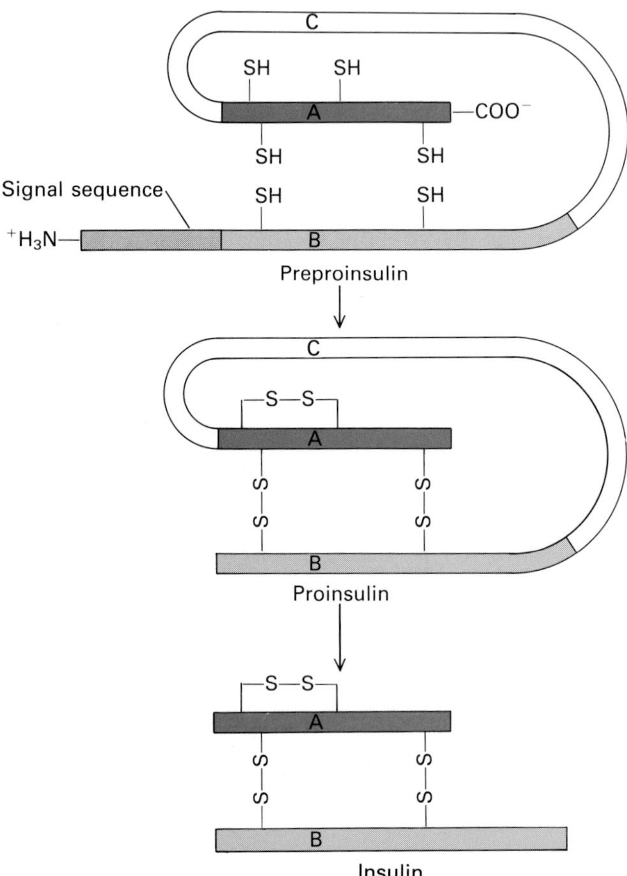

Figure 3-13 The processing of preproinsulin into proinsulin, and of proinsulin into insulin. The cleavage of preproinsulin, which occurs immediately after the synthesis of its chain of 108 amino acids is completed, removes 24 amino acids collectively termed the signal sequence from the amino end of the molecule. This excised polypeptide is degraded. The remaining 84 amino acids constitute proinsulin, a molecule in which all of the correct disulfide bridges have been formed. While the hormone is being packaged for secretion, the 33 residues (the C chain) between the A and the B chains are removed to produce insulin. (Some C chains may be secreted along with the insulin, but the function of these is not known.) *After L. Stryer, 1981,* Biochemistry, *2d ed., W. H. Freeman and Company, p. 848.*

3-Hydroxyproline (mainly in collagen)	$\mathrm{H_2C{-}CH(OH)}$; $\mathrm{H_2C}$ and $\mathrm{CH{-}COO^-}$ joined through $\mathrm{^+NH_2}$ (ring)
4-Hydroxyproline (mainly in collagen)	$\mathrm{HC(OH){-}CH_2}$; $\mathrm{H_2C}$ and $\mathrm{CH{-}COO^-}$ joined through $\mathrm{^+NH_2}$ (ring)
3-Methylhistidine (mainly in actin)	$\mathrm{HC{=}C{-}CH_2{-}CH(NH_3^+){-}COO^-}$; ring $\mathrm{H_3C{-}N{-}C(H){=}N}$
5-Hydroxylysine (mainly in collagen)	$\mathrm{^+H_3N{-}CH_2{-}CH(OH){-}CH_2{-}CH_2{-}CH(NH_3^+){-}COO^-}$
γ-Carboxyglutamate (mainly in prothrombin, an essential blood-clotting factor)	$\mathrm{(^-OOC)_2CH{-}CH_2{-}CH(NH_3^+){-}COO^-}$

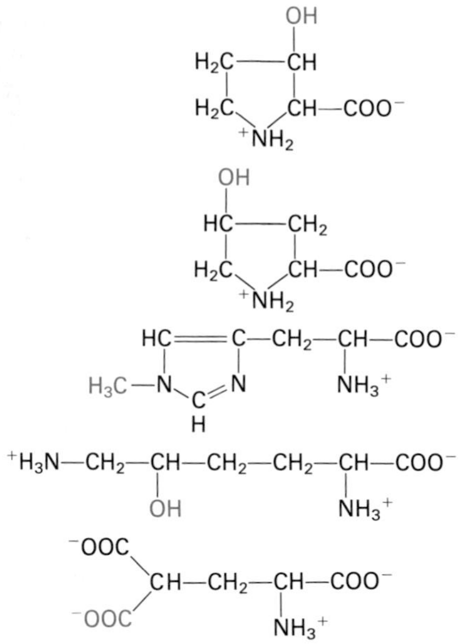

Figure 3-14 The structures of modified amino acids found in specific proteins. In each case, the substituted group (color) replaces a hydrogen atom.

partial proteolysis (breaking down) of proinsulin, its larger precursor. It has been shown, however, that denatured proinsulin can renature to form the native structure of proinsulin with a high efficiency. Presumably, within the cell, either proinsulin or preproinsulin folds in such a way that the state of lowest free energy is the one in which the correct disulfide bridges are formed. By utilizing these intermediate stages, the cell is able to form insulin, whose stable conformation is not the one of lowest free energy.

Certain Structures Are Regular Features of Many Proteins

The α Helix Although many regions of proteins are held together in unique and irregular conformations, there are also a number of regular structures that protein regions may adopt. One such common structure is the α *helix,* which was first described by Linus Pauling and Robert B. Corey in 1951. Through careful model building, these workers came to realize that polypeptide segments composed of certain amino acids tend to arrange themselves in regular helical conformations. In an α helix, the carboxyl oxygen in one peptide bond is hydrogen-bonded to the hydrogen on the amino group of the third amino acid away (Figure 3-16). The resulting helix has 3.6 amino acids per turn. Each amino acid residue represents an advance of about 0.15 nm along the axis of the helix. Every C═O and N—H group in the peptide bonds participates in a hydrogen bond. In this inflexible, stable arrangement of amino acids, the side chains are positioned along the outside of a cylinder. The rigid planarity of the peptide bond contributes to the rigid shape of the α helix. All residues in an α helix have an identical conformation: the ψ angle between adjacent peptide bonds (see Figure 3-4c) is $-50°$, and the ϕ angle is $-60°$. In most proteins, some of the amino acids are organized into α helixes.

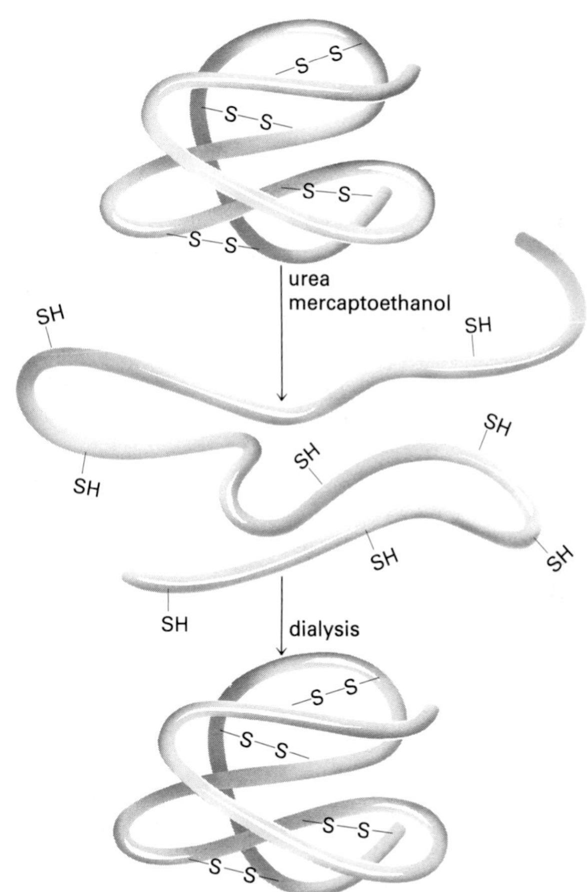

Figure 3-15 Denaturation and renaturation of a protein. Most polypeptides can be completely unfolded by treatment with an 8 *M* urea solution containing mercaptoethanol,

$$\mathrm{HS{-}CH_2{-}CH_2{-}OH}$$

The urea breaks intramolecular hydrogen and hydrophobic bonds, and the mercaptoethanol reduces each cystine —S—S— bridge to two —SH groups. When these chemicals are removed by dialysis, the —SH groups on the unfolded chain oxidize spontaneously to re-form —S—S— bridges, and the polypeptide chain simultaneously refolds into its native configuration.

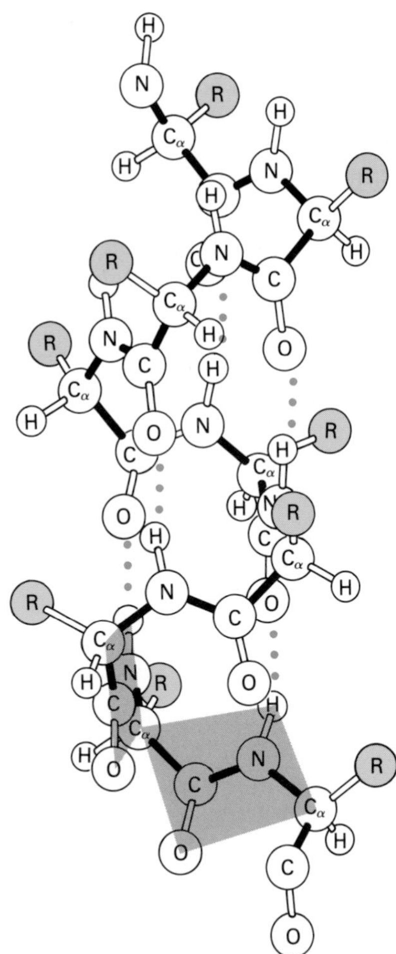

Figure 3-16 A ball-and-stick model of an α helix. The side chains are represented by single balls (color). Some planes of the C_α—CO—NH groups are shaded. Note that there are about 3.6 amino acid residues per turn of the helix, and that the NH···O hydrogen bonds are nearly parallel to the long axis of the helix. The side chains protrude outward from the helix at regular intervals.

The reason why most regions of proteins do not adopt an α-helical structure is that many sequences of amino acids are difficult to incorporate into it. Multiple adjacent amino acids of like charge repel one another too strongly to allow themselves to be bonded into an α helix. For example, in aqueous solutions at pH 7.0, polyalanine (Ala-Ala-Ala- . . .), with its nonpolar side chains, will spontaneously form α-helical coils. On the other hand, polylysine, with its positively charged side chains, does not form an α helix under these conditions; it exists rather as a random coil. At pH 12.0, a condition under which most of the side chains are not protonated, polylysine will form an α helix spontaneously.

Some side chains—that of tryptophan, for example—are too bulky to be accepted easily into the inflexible arrangement of the α helix. Proline is rarely found in the middle of an α helix. When incorporated into a peptide bond, the imino nitrogen atom of proline does not contain a bound hydrogen atom:

```
      O            O
      ‖            ‖
  —C—N—CH—C—NH—
      /    \
   H2C      CH2
      \    /
       CH2
```

Thus, the proline residue cannot participate in the regular array of hydrogen bonds that is characteristic of an α helix. Furthermore, its five-membered ring would cause the α-helical chain to be twisted. However, proline can be found at the beginning or at the end of an α helix, which is where it is found in hemoglobin and myoglobin.

The β Pleated Sheet A second regular structural element of many proteins is the *β pleated sheet.* This structure is created by a series of hydrogen bonds between residues in different polypeptide chains or between residues in different sections of a folded polypeptide. Most adjacent polypeptide chains in pleated sheets are *antiparallel,* which means that they run in opposite directions: the chains are oriented so that the $NH_2 \rightarrow COOH$ direction of one is the $COOH \rightarrow NH_2$ direction of the other (Figure 3-17). However, some adjacent chains are parallel. If many polypeptides participate, the sheet has a rigid wall-like structure. In many structural proteins, multiple pleated sheets provide the requisite toughness and rigidity. Silk fibroin, for example, consists almost entirely of stacks of antiparallel β pleated sheets.

Determination of Three-Dimensional Protein Structure through X-Ray Crystallography We know the detailed three-dimensional structures of numerous proteins because of the painstaking efforts of many workers, notably Max Perutz and John Kendrew, who labored for more than a decade to perfect the methods of x-ray crystallography of proteins. In this procedure, beams of x-rays are passed through a crystalline protein lattice. When any radiation passes through a lattice with openings larger than the wavelengths of the x-rays, the radiation is scattered in a regular pattern, called a diffraction pattern. Because the wavelengths of x-rays are small enough (from about 0.1 to 0.2 nm), they are scattered by the atoms in the protein crystal. From the diffraction pattern produced by the regular array of atoms, the three-dimensional structure of the protein can be deduced. However, the diffraction pattern of a protein crystal is extremely complex; as many as 25,000 diffraction "spots" are obtained from a small protein such as myoglobin. Thus, tedious, elaborate calculations are needed, as well as modified forms of the protein (i.e., molecules

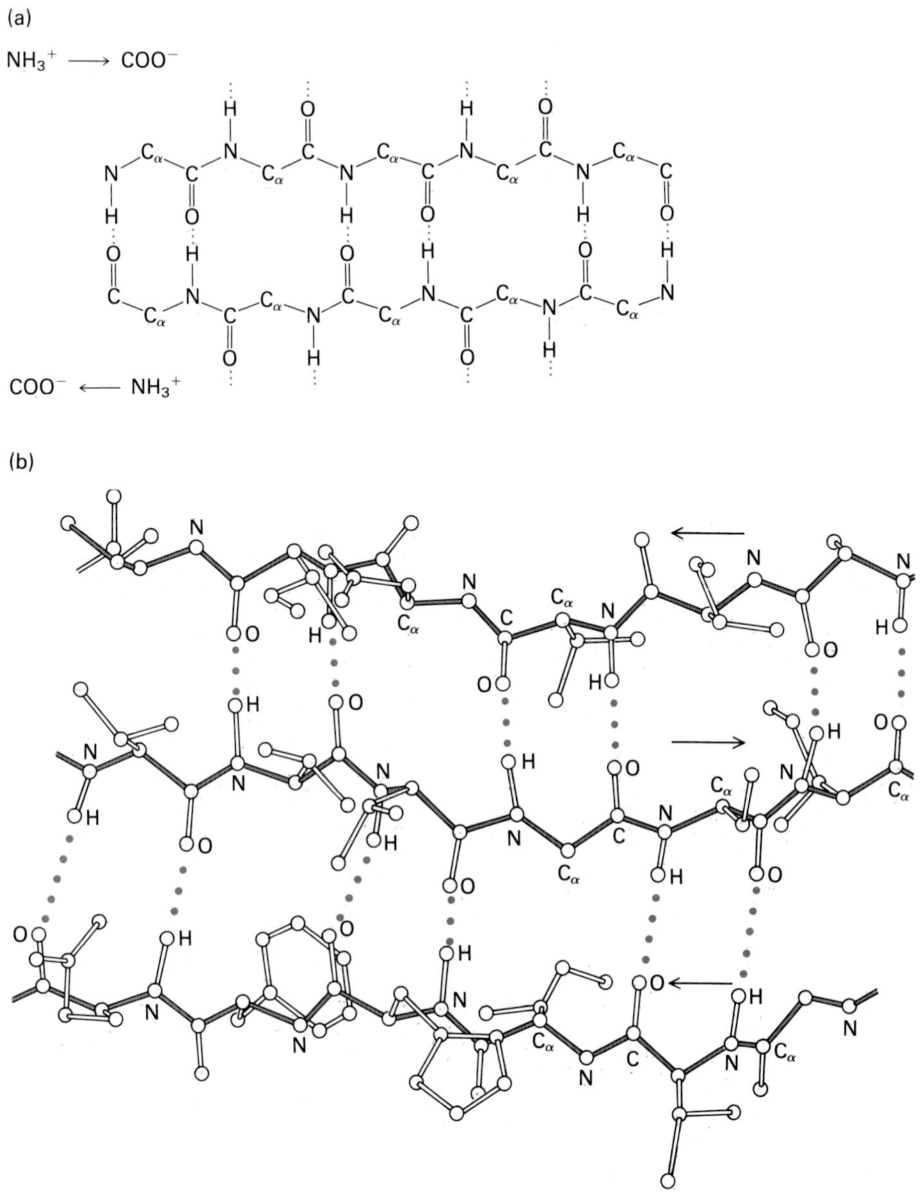

Figure 3-17 Models of an antiparallel β pleated sheet. (a) Only the backbone C_α—CO—NH linkages are shown. (b) The pleats formed by the planar peptide groups are derived from the NH···O hydrogen bonds (dotted lines) between the polypeptide chains and from similar hydrogen bonds to chains not depicted. The side chains project from both sides of the chains. *Part (b) courtesy of Dagmar Ringe.*

containing bound heavy metals), to interpret the diffraction pattern and solve the structure of the protein.

Some proteins—like myoglobin, the oxygen-carrying protein of muscle (Figure 3-18)—are composed of a single polypeptide chain that is folded into one compact unit. Other proteins have obvious separate *domains,* polypeptide regions that fold into globular units that often have independent functions (Figure 3-19). Antibodies typically have several discrete domains.

Recently, it has been found that in many proteins containing multiple polypeptide domains, each domain is encoded by a discrete segment of DNA, and that these segments are often not adjacent to each other. These discoveries, which have important implications for the evolution of proteins, are described in Chapter 25.

Enzymes

Enzymes Catalyze Biochemical Reactions

Protein catalysts called *enzymes* are mediators of the dynamic events of life; almost every chemical reaction in a cell is catalyzed by an enzyme. Like other catalysts, enzymes increase the rates of reactions that are already energetically favorable; more precisely, they increase the rates of forward and reverse reactions by the same factor. Therefore, they do not change the direction of reactions, nor do they change the equilibrium ratio between products and reactants (Chapter 2).

The chemicals that undergo a change in a reaction catalyzed by an enzyme are called the *substrates* of that en-

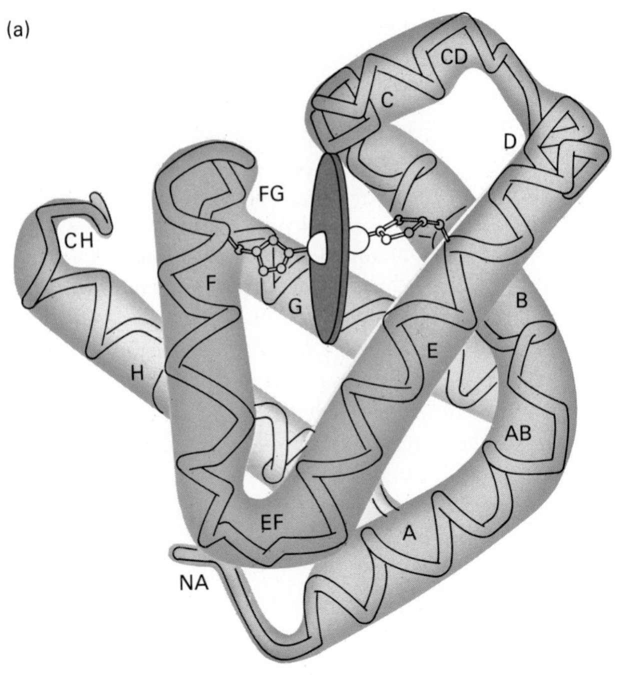

Figure 3-18 (a) An outline of the polypeptide backbone of myoglobin. There are eight α helixes, labeled A through H. Nonhelical segments are labeled by two letters; AB, for example, indicates the segment between the A and the B helixes; NA and HC represent the N-terminal and C-terminal segments, respectively. The heme group is shown in color; the iron atom in the center is bonded to histidine residues in helixes E and F. The oxygen atom at the right of the heme plane is bonded to the iron in the center of the heme. (b) A computer-derived interpretation of the three-dimensional structure of myoglobin. The α helixes are depicted as cylinders (see inset), and the nonhelical segments are represented by bent ribbons tracing the planes of the successive peptide groups. The heme, however, is shown in the form of a space-filling model, and the two histidine residues flanking it are shown as balls-and-sticks. *Part (a) after R. E. Dickerson and I. Geis, 1969,* The Structure and Action of Proteins, *Benjamin-Cummings, p. 47. Copyright 1969 by Irving Geis. Part (b) after A. M. Lesk, 1984,* TIBS *(June 1984), p. v. Copyright 1984 by Elsevier, Amsterdam.*

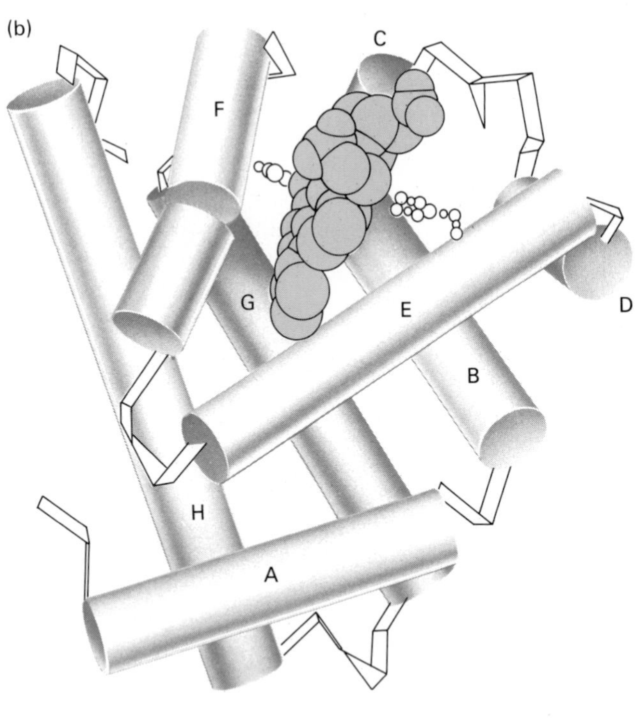

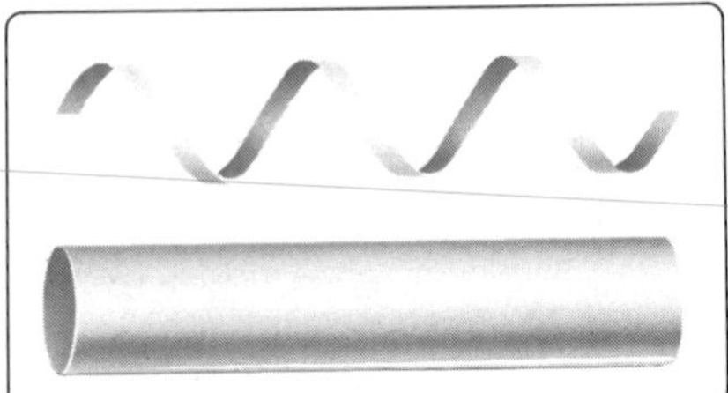

Two ways of depicting the α helix

zyme. Which chemicals are substrates and which are products is often arbitrary; it depends on the direction in which the reaction is written. For instance, triosephosphate isomerase catalyzes the interconversion of glyceraldehyde 3-phosphate and dihydroxyacetone phosphate (see Figure 2-17). Depending on the concentrations of the two compounds in the cell (and thus on the ΔG value), the enzyme will more often use dihydroxyacetone phosphate as the substrate, converting it into glyceraldehyde 3-phosphate, or vice versa.

Most enzymes are found inside cells, but a number are secreted by cells and function in the blood, in the digestive tract, or in other extracellular spaces. In microbial species, some enzymes function outside the organism. The number of different types of chemical reactions in any one cell is very large: a "typical" animal cell, for example, contains from 1000 to 4000 different types of enzymes, each of which catalyzes a single chemical reaction or a single set of closely related chemical reactions. Certain enzyme-catalyzed reactions are common to most cells, and as a result the enzymes involved are found in the vast majority of cells. These enzymes include not only those that catalyze the synthesis of proteins, nucleic acids, and phospholipids, but also those that catalyze the conversion of glucose and oxygen into carbon dioxide and water by a complex set of reactions that produce most of the chemical energy used in animal cells. Any particular type of cell within an organism—such as a liver cell or a nerve cell—carries out chemical reactions that are unique to that cell; consequently, certain enzymes are found only in particular cells. Also, many cells, including erythrocytes (red blood cells) and epidermal (skin) cells, have matured to a stage at which they are no longer capable of

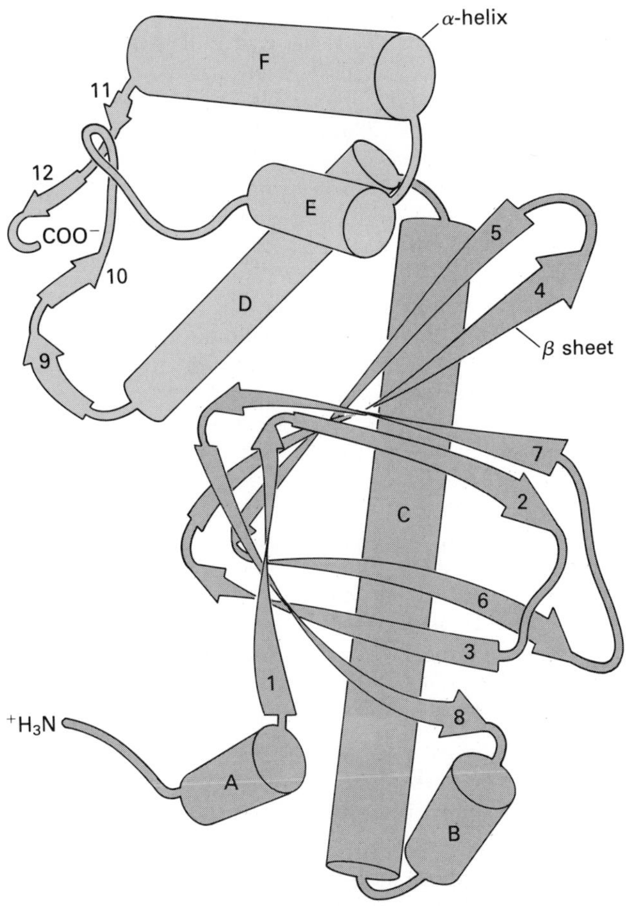

Figure 3-19 A two-domain protein, the DNA-binding CAP regulatory protein from *Escherichia coli.* As in Figure 3-18b, the α helixes are depicted as cylinders. Each β pleated sheet is shown as a flat arrow; the plane of the arrow represents the plane of the peptide backbone. The arrow indicates the N $\rightarrow$ C direction. The carboxyl-terminal domain (color) contains three α helixes (D, E, and F) and four β sheets (9 to 12); this region binds to specific segments of DNA and regulates the expression of adjacent genes. The N-terminal domain (gray; α helixes A to C, β sheets 1 to 8) binds the regulatory molecule 3′,5′-cyclic AMP. The binding of the regulatory molecule to the N-terminal domain induces in the nonordered "hinge" regions a conformational change that allows the C-terminal domain to bind to DNA. *After D. B. McKay and T. A. Steitz, 1981,* Nature *290:746.*

making proteins, nucleic acids, or phospholipids—yet these cells still contain specific sets of enzymes that they synthesized at an earlier stage of differentiation.

Certain Amino Acids in Enzymes Bind Substrates, and Others Catalyze Reactions on the Bound Substrates

Two striking properties characterize all enzymes: their enormous *catalytic power* and their specificity. Quite often, the rate of an enzyme-catalyzed reaction is 10^6 to 10^{12} times that of an uncatalyzed reaction under otherwise similar conditions. The specificity of an enzyme is revealed in the different rates at which it catalyzes closely related chemical reactions or by its ability to distinguish between closely related substrates.

Certain amino acid side chains of an enzyme are important in determining both its ability to accelerate the rate of a reaction and its specificity. The properties of an enzyme are thus functions of its linear arrangement of amino acids and of the proper foldings of the peptide chain. Enzyme molecules have two important regions, or sites: one of these sites recognizes and binds the substrate(s), and the other catalyzes the reaction once the substrate or substrates have been bound. The amino acids in each of these key regions need not be adjacent to one another in the linear polypeptide; they are brought into proximity by the folding of the molecule. The substrate-binding and the catalytic sites are adjacent in the active form of the enzyme; sometimes the catalytic site is part of the substrate-binding site. The two regions are called collectively the *active site.*

The binding of a substrate to an enzyme usually involves the formation of several types of noncovalent bonds: ionic, hydrogen, and hydrophobic bonds, and van der Waals interactions (Figure 3-20). The active site of the enzyme has an array of chemical groups that are precisely arranged so that the specific substrate can be more tightly

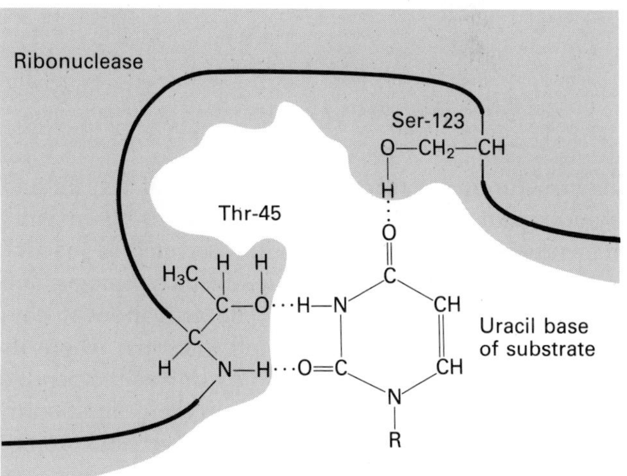

Figure 3-20 The specific binding of a substrate to an enzyme involves the formation of multiple noncovalent bonds. In this example, two amino acid residues of the enzyme ribonuclease bind uracil, part of its substrate, by three hydrogen bonds. Substrates without the two C=O groups and one N—H group in the appropriate positions would be unable to bind, or would bind less tightly. Other regions of the enzyme, not depicted here, bind other parts of the RNA substrate by hydrogen bonds and van der Waals interactions.

bound than any other molecule with the exception of some inhibitors and in such a way that the reaction can readily occur.

In catalysis, covalent bonds between the enzyme and the substrate may be formed (and then broken) as a means of reducing the activation energy for the reaction.

Because the formation of peptide bonds requires an input of free energy, the reverse reaction, hydrolysis, is energetically favorable ($\Delta G^{\circ\prime} = -2$ kcal/mol). Nonetheless, the activation energy for an *uncatalyzed* peptide-bond hydrolysis—say, in a neutral aqueous solution of a protein at room temperature—is so high that there is little or no hydrolysis even after several months. Biochemists can chemically hydrolyze proteins into their constituent amino acids by treating them with a 6 *M* solution of hydrochloric acid in an evacuated tube at 100°C for 24 h. (Complete hydrolysis is the first step in obtaining the amino acid composition of a protein.) Yet enzymes such as chymotrypsin and trypsin catalyze hydrolysis reactions at 37°C and at neutral pH. Depending on the nature of the protein substrate, each enzyme molecule can catalyze the hydrolysis of up to 100 peptide bonds each second.

Enzymes are outstanding examples of large functional proteins that specifically bind small molecules or small regions of large molecules. A molecule that binds specifically to a macromolecule (other than the substrates or products of an enzyme) is often called a *ligand* of that macromolecule. Antibodies are other proteins that bind small molecules, or small regions of large molecules, specifically and tightly.

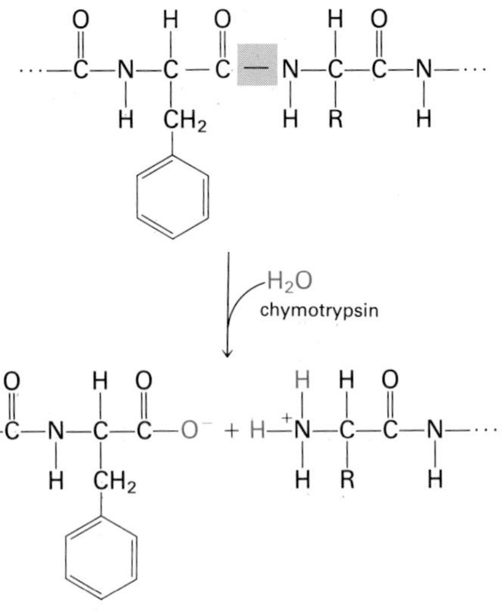

Trypsin and Chymotrypsin Are Well-Characterized Proteolytic Enzymes

The structures and reaction mechanisms of trypsin and chymotrypsin are known in some detail. Both enzymes are synthesized in the pancreas and secreted as inactive precursors, or *zymogens;* these are called trypsinogen and chymotrypsinogen, respectively. The activation of chymotrypsin takes place in the small intestine, where its function, like that of trypsin, is to hydrolyze the peptide bonds of ingested proteins as a step in their digestion to single amino acids (Figure 3-21). Two irreversible proteolytic cleavages activate this enzyme: one cleavage removes serine 14 (i.e., the serine at position 14) and arginine 15 from chymotrypsinogen, and the other removes threonine 147 and asparagine 148 (Figure 3-22). This activation in the intestine serves an important regulatory purpose: it prevents the enzyme from degrading the pancreatic tissue in which it was made.

Chymotrypsin does not hydrolyze all peptide bonds; rather, it is selective for the peptide bond at the carboxyl ends of amino acids such as phenylalanine, tyrosine, and tryptophan, which have large hydrophobic side chains. Trypsin, in contrast, is specific for the peptide bond at the C-termini of lysine and arginine residues.

H_2O
chymotrypsin

Figure 3-21 The hydrolysis of a peptide bond by chymotrypsin.

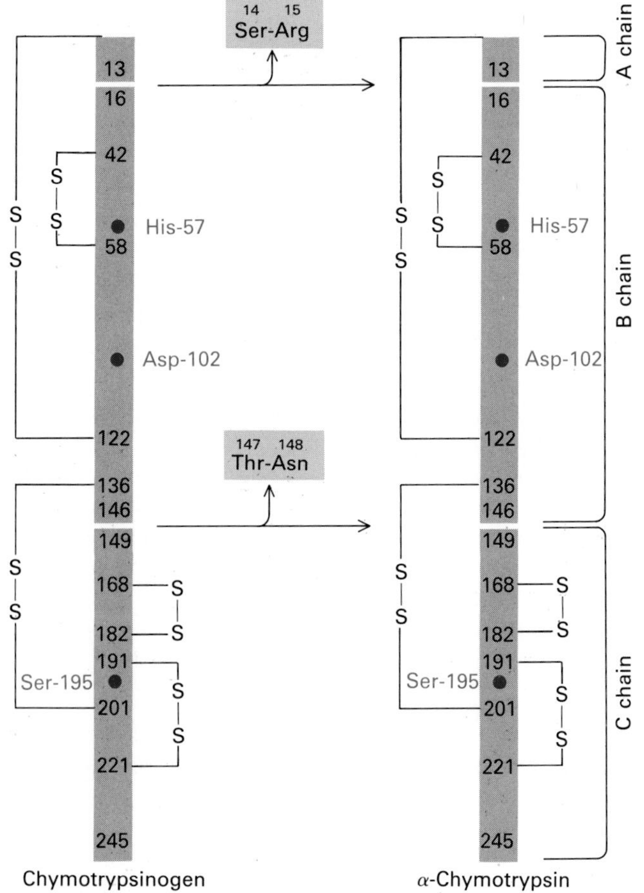

Figure 3-22 A linear representation of the conversion of chymotrypsinogen into chymotrypsin by the excision of two dipeptides. The positions of the disulfide bridges are indicated. In the folded molecule, the histidine 57, aspartate 102, and serine 195 are located in the active site.

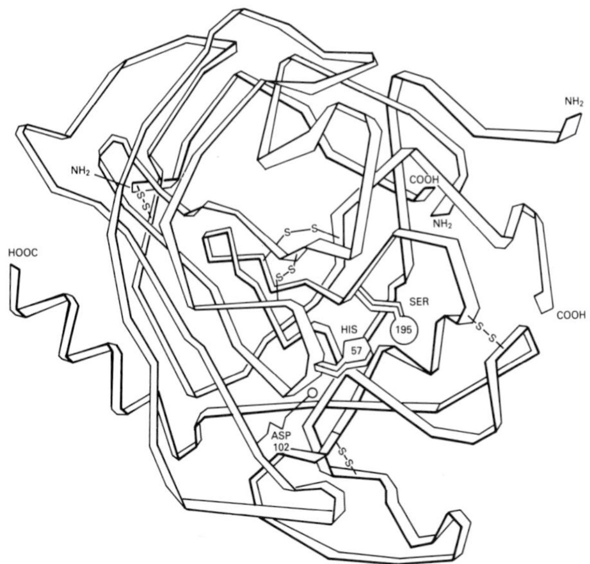

Figure 3-23 A three-dimensional model of α-chymotrypsin, as determined from x-ray analysis. The N- and C-termini of the A, B, and C chains are indicated, as are the —S—S— bridges, the hydrophobic cleft (see text), and the three amino acid residues of the active site (histidine 57, aspartate 102, and serine 195). *After B. W. Matthews et al., 1967,* Nature ***214****:652.*

Substrate Binding by Specific Amino Acid Side Chains of Chymotrypsin The reaction mechanism of chymotrypsin was determined, in part, from the three-dimensional structure obtained by x-ray crystallography (Figure 3-23). The enzyme contains three polypeptide chains—the A, B, and C chains, which have 13, 131, and 97 amino acids, respectively. The chains are interconnected by disulfide bridges (see Figures 3-22 and 3-23). The molecule has two key structural features: the active site and the *hydrophobic cleft*. The latter is a crevice bordered by the side chains of several hydrophobic amino acid residues. The hydrophobic cleft serves as the binding site for specific amino acid residues on the substrate. The conformation of this pocket allows the residues lining it to participate in hydrophobic interactions with the large hydrophobic side chains of phenylalanine, tyrosine, or tryptophan. Neither charged side chains nor small hydrophobic residues on the substrate can participate in these specific interactions.

As mentioned earlier, the hydrophobic residues of most globular proteins are buried in the interior; thus, when such proteins are in their native states, the peptide bonds linking the hydrophobic residues are not accessible to hydrolysis by chymotrypsin. Normally, stomach acids (which, recall, are at pH 1) denature ingested proteins so that proteases in that organ can partially degrade them before they are exposed at neutral pH to further digestion by chymotrypsin in the intestine.

The Role of Other Amino Acid Side Chains of Chymotrypsin in Catalyzing the Hydrolysis of the Bound Substrate The catalytic activity of chymotrypsin depends on three amino acid residues: histidine 57, aspartate 102, and serine 195. These amino acids are distant from one another in the primary structure of the protein (see Figure 3-22), but in the active enzyme molecule the chains are folded in such a way that the three side chains are close together, in the correct position for catalyzing the hydrolysis of a peptide bond in a protein bound to the enzyme (see Figure 3-23). When chymotrypsinogen is proteolytically activated, the polypeptide conformation is altered to bring these three residues into the correct alignment.

The hydrolysis reaction forms a covalent enzyme-substrate complex in several steps. First, the peptide bond is broken and the carboxyl group is transferred to the hydroxyl residue of serine 195:

$$\underset{\text{Enzyme}}{\text{Enz—(Ser-195)—OH}} + \underset{\text{Substrate}}{R_1\text{—NH—}\overset{\overset{\displaystyle O}{\|}}{C}\text{—}R_2} \longrightarrow$$

$$\underset{\text{Acylenzyme}}{\text{Enz—(Ser-195)—O—}\overset{\overset{\displaystyle O}{\|}}{C}\text{—}R_2} + R_1\text{—NH}_2$$

Second, this *acylenzyme* intermediate is hydrolyzed:

$$\text{Enz—(Ser-195)—O—}\overset{\overset{\displaystyle O}{\|}}{C}\text{—}R_2 + H_2O \longrightarrow$$

$$\text{Enz—(Ser-195)—OH} + R_2\text{—}\overset{\overset{\displaystyle O}{\|}}{C}\text{—}O^- + H^+$$

Aspartate 102 and histidine 57 facilitate the acylation reaction by first removing the proton from serine 195 and then adding it to the nitrogen of the departing amino group (Figure 3-24). In a similar manner, aspartate 102 and histidine 57 facilitate the hydrolysis of the acylenzyme. These enzyme-catalyzed steps—transfer of a proton from the enzyme to the substrate, formation of a covalent acylserine intermediate, and hydrolysis of the acylenzyme—all serve to drastically reduce the overall activation energy of the proteolysis reaction.

The hydroxyl group on serine 195 is unusually reactive. The concept of an "active" serine residue at the active site predated the determination of the crystal structure of chymotrypsin. It was already known, for example, that the compound diisopropylfluorophosphate,

$$\begin{array}{l} H_3C\text{—}\underset{\displaystyle CH_3}{\underset{|}{CH}}\text{—O—}\overset{\overset{\displaystyle O}{\|}}{P}\text{—F} \\ \qquad\qquad\qquad\;\; | \\ \qquad\qquad\qquad\;\; O \\ \qquad\qquad\qquad\;\; | \\ \qquad\qquad\quad\;\; HC\text{—}CH_3 \\ \qquad\qquad\qquad\;\; | \\ \qquad\qquad\quad\;\; CH_3 \end{array}$$

is a potent inhibitor of chymotrypsin; it reacts only with

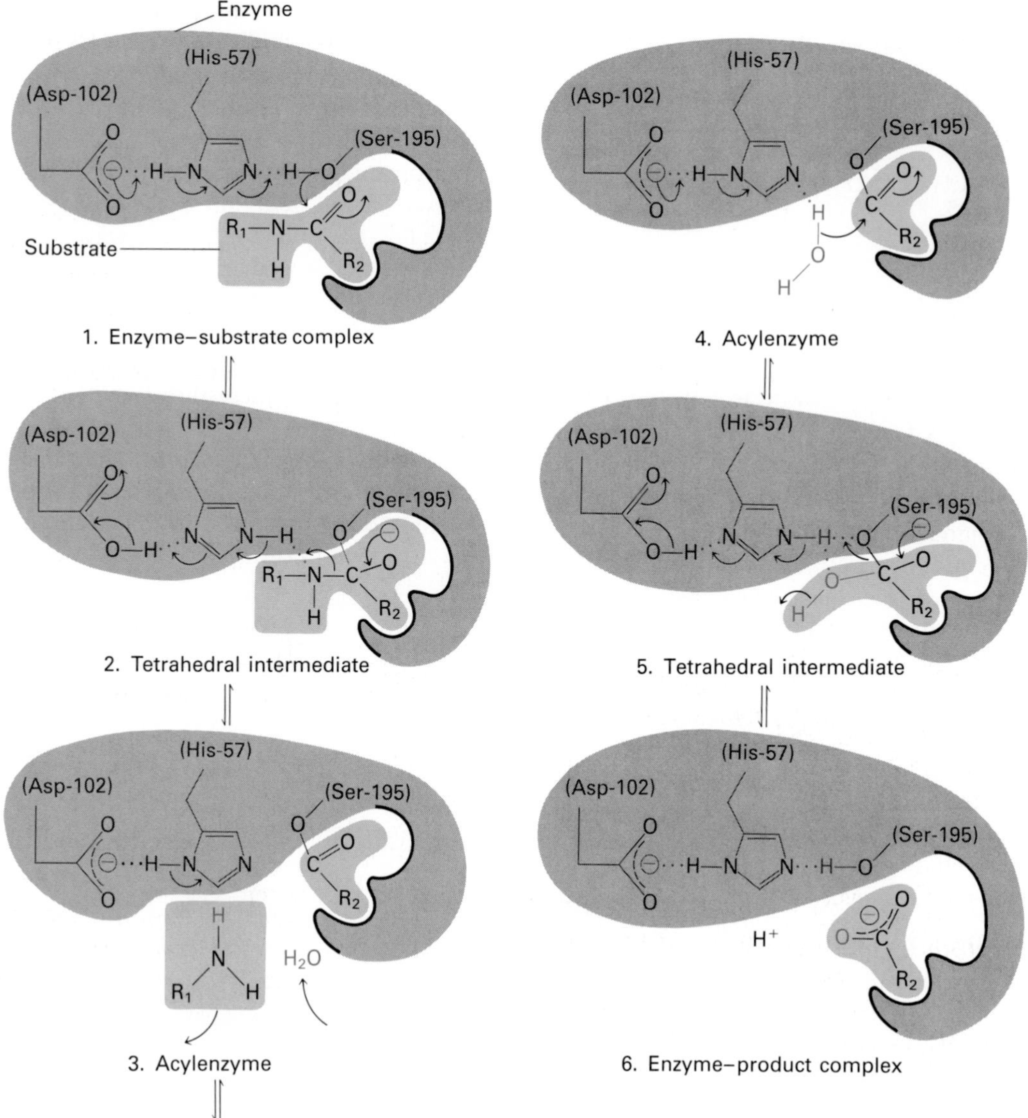

Figure 3-24 The mechanism of hydrolysis of a peptide bond by α-chymotrypsin. (1) The substrate is bound to the enzyme so that the bond to be hydrolyzed is positioned near serine 195. The negative charge surrounding the oxygens in aspartate 102 induces a charge relay system, as shown by the arrows. The relay system is initiated when Asp-102's oxygens attract a proton from one of His-57's nitrogens. When the negative charge reaches the second nitrogen in His-57, the nitrogen removes the proton from the hydroxyl group on Ser-195. The resulting O^- attacks the carbon of the bound substrate to form (2) a tetrahedral intermediate. Then the hydrogen bound to the second nitrogen in His-57 is added to the nitrogen of the substrate. As a result, the C—N bond of the substrate is hydrolyzed to form (3) R_1NH_2 and the acylenzyme intermediate

$$\text{(Ser-195)—O—}\overset{\overset{\displaystyle O}{\|}}{C}\text{—}R_2$$

The R_1NH_2 is discharged from the enzyme and is replaced by water. In the resulting structure (4), a similar charge relay system is induced, and His-57 removes a proton from the hydrogen-bonded H_2O. The OH^- thus generated attacks the carboxyl carbon of the acylenzyme to form (5) another tetrahedral intermediate. The bond between the tetrahedral carbon and the oxygen originally belonging to Ser-195 is hydrolyzed to yield (6) R_2COO^- and the free enzyme bound to the product. *After R. M. Stroud, et al., 1975, in* Proteases and Biological Control, *E. Reich et al., eds. Cold Spring Harbor Laboratory, p. 25.*

the hydroxyl on serine 195 to form the covalent compound

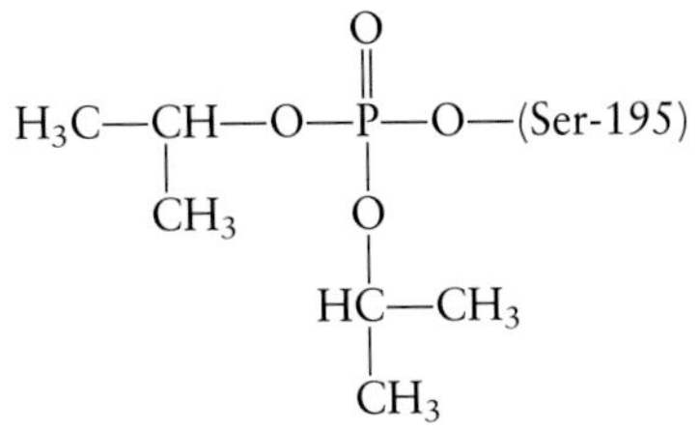

$$\begin{array}{c} O \\ \| \\ H_3C-CH-O-P-O-(\text{Ser-195}) \\ | \qquad\quad | \\ CH_3 \qquad O \\ | \\ HC-CH_3 \\ | \\ CH_3 \end{array}$$

which irreversibly inactivates the enzyme.

The Different Substrate-Binding Sites in Trypsin and Chymotrypsin A comparison of trypsin and chymotrypsin is useful in emphasizing the nature of the specificity of enzyme-catalyzed reactions. About 40 percent of the amino acids in these two molecules are the same; in particular, the amino acid sequences in the vicinity of the key serine residue are identical:

$$\text{-Gly-Asp-}\overset{195}{\text{Ser}}\text{-Gly-Gly-Pro}$$

The three-dimensional structures and the mechanisms of catalysis of the two enzymes are also quite similar, which indicates that they have evolved from a common polypeptide. The major difference is in the side chains of the amino acids that line the substrate-binding site. The negatively charged amino acids in this area of the trypsin molecule facilitate the binding of only positively charged (lysine or arginine) residues, instead of hydrophobic ones.

Active Serine in Other Hydrolytic Enzymes Other, mostly unrelated hydrolytic enzymes also contain an active serine residue that is essential for catalysis. An example is acetylcholinesterase, which catalyzes the hydrolysis of the neurotransmitter acetylcholine to acetate and choline:

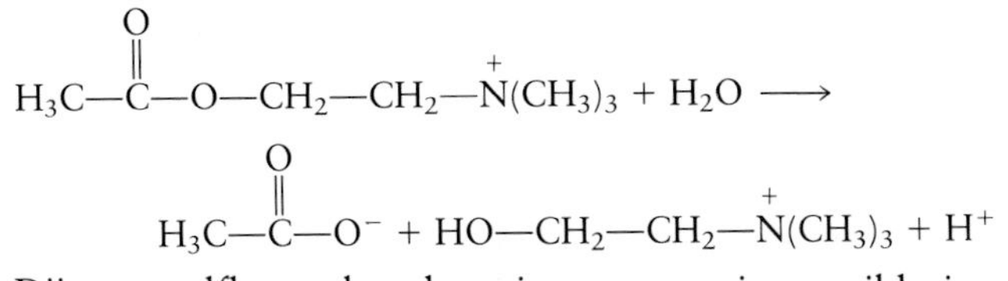

$$H_3C-\overset{\overset{\displaystyle O}{\|}}{C}-O-CH_2-CH_2-\overset{+}{N}(CH_3)_3 + H_2O \longrightarrow$$

$$H_3C-\overset{\overset{\displaystyle O}{\|}}{C}-O^- + HO-CH_2-CH_2-\overset{+}{N}(CH_3)_3 + H^+$$

Diisopropylfluorophosphate is a potent, irreversible inhibitor of acetylcholinesterase as well as of chymotrypsin. The compound is lethal to animals, as it blocks nerve transmission by causing a buildup of the transmitter substance. (The action of this transmitter is discussed further in Chapter 17.)

Coenzymes Are Essential for Certain Enzyme-Catalyzed Reactions

An enzyme often contains a tightly bound small molecule, termed a *coenzyme,* that is essential for the activity of the enzyme. Many vitamins that are required in trace amounts in the diet are converted into coenzymes; coenzyme A, for instance, is derived from the vitamin pantothenic acid, and the coenzyme pyridoxal phosphate is derived from vitamin B_6. To cite just one example of how coenzymes function, consider pyridoxal phosphate. The aldehyde group

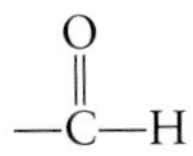

$$-\overset{\overset{\displaystyle O}{\|}}{C}-H$$

can form a covalent complex called a *Schiff base* with an $-NH_2$ group of an amino acid. This facilitates or lowers the activation energy for the breaking of bonds involving the α carbon of the amino acid. Figure 3-25 shows how pyridoxal phosphate catalyzes the decarboxylation of histidine to form histamine, a potent dilator of small blood vessels. Histamine is formed in certain cells as a result of allergenic hypersensitivity.

Substrate Binding May Induce a Conformational Change in the Enzyme

The binding of a substrate to an enzyme may simply involve the fitting together of molecules that have complementary charges, or shapes, or both, into a complex that is stabilized by a variety of noncovalent bonds. Such an interaction resembles the fitting together of a key into a lock and is therefore said to occur by a *lock-and-key* mechanism.

In some enzymes, the binding of the substrate induces in the enzyme a conformational change that causes the catalytic residues to become positioned correctly. Molecules that bind to the enzyme's *recognition site*—substrate-binding site—but that do not induce this conformational change are not substrates of the enzyme. Thus, an enzyme differentiates between substrate and nonsubstrate in two ways: Does the potential substrate bind to the enzyme? If so, does it induce the correct conformational change? When both criteria are met, the enzyme-substrate complex is said to demonstrate *induced fit* (Figure 3-26).

Hexokinase provides an important example of induced fit. This enzyme catalyzes the transfer of a phosphate residue from ATP to a specific carbon atom of glucose (Figure 3-27); this is the first step in the degradation of glucose by cells. X-ray crystallography has shown that hexokinase consists of two lobes. The binding of glucose induces a major conformational change that brings the lobes closer together and creates a functional catalytic site (Figure 3-28). Only glucose and closely related molecules can induce this conformational change, which thus ensures that the enzyme is used to phosphorylate only the correct substrates. Molecules such as glycerol, ribose, or even water may bind to the enzyme at the substrate-binding site, but these cannot induce the requisite conformational change, and so they are not substrates for the enzyme.

The Progress of Enzyme-Catalyzed Reactions Can Be Described by Simple Kinetic Equations

The rate of a simple reaction $S \rightarrow P$ (substrate $\rightarrow$ product) is a function of the concentration of the substrate

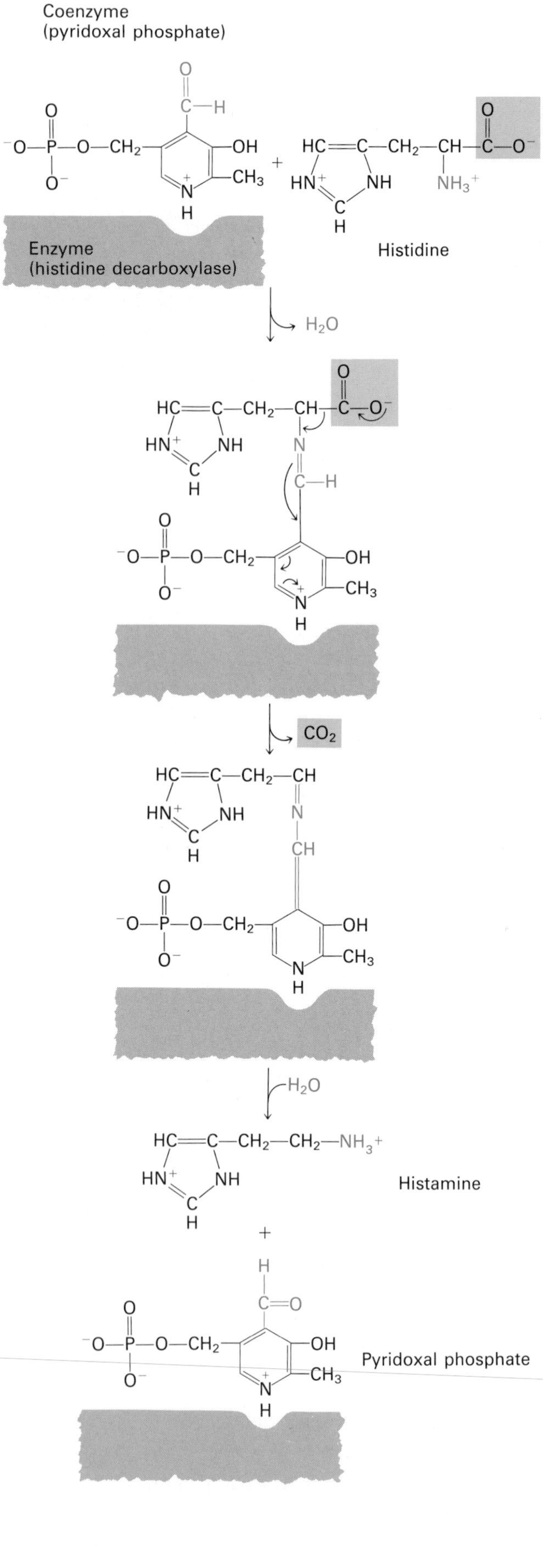

Figure 3-25 Pyridoxal phosphate, a coenzyme, participates in many reactions involving amino acids. In this example, when it is bound to histidine decarboxylase it forms a Schiff base with the α-amino group of histidine. The positive charge on the nitrogen of pyridoxal phosphate then attracts the electrons from the carboxylate group of the histidine, via a charge relay system. This weakens the bond between the α carbon of the histidine and the carboxylate group, causing the release of CO_2. Finally, histamine, the reaction product, is released from the pyridoxal complex.

and of the concentration and properties of the enzyme, E, that catalyzes the reaction. Figure 3-29 shows how, for an enzyme with a single catalytic site, the rate of production of the product depends on the concentration of the substrate when the concentration of the enzyme is kept constant.

At low concentrations of S ($[S] \ll K_m$, the value of [S] at which the reaction rate is one-half of the maximum), the reaction rate is proportional to [S]; as [S] is increased, however, the rate does not increase in proportion to [S] but eventually reaches a plateau where it is independent of [S]. This maximum velocity, V_{max}, is proportional to [E], the concentration of the enzyme and to a rate constant, k_{cat}, that is an intrinsic property of the individual enzyme; halving [E] reduces the rate (at all values of [S]) by half.

To interpret these curves, we recall that all enzyme-catalyzed reactions include at least three steps: (1) the binding of the substrate, S, to the enzyme, E, to form an enzyme-substrate complex, ES; (2) the conversion of ES into the enzyme-product complex, EP; and (3) the release of P from EP, to yield free P:

$$E + S \xrightleftharpoons{\text{Binding}} ES \xrightarrow{\text{Catalysis}} EP \xrightarrow{\text{Release}} P + E$$

In the simplest case, the release of P is so rapid that we can write

$$E + S \underset{k_2}{\overset{k_1}{\rightleftharpoons}} ES \xrightarrow{k_{cat}} E + P$$

The reaction rate, dP/dt, is proportional to the concentration of ES and to the catalytic constant, k_{cat}, for the given enzyme:

$$\frac{dP}{dt} = k_{cat}\,[ES] \qquad (3\text{-}1)$$

In order to calculate [ES], we can assume that the reaction is in a steady-state, that is, the rate of formation of [ES], k_1 [E][S], is equal to the rate of its consumption, either by dissociation of substrate, at a rate k_2[ES], or catalysis, at a rate k_{cat} [ES]:

$$k_1[E][S] = (k_2 + k_{cat})[ES] \qquad (3\text{-}2)$$

If

$$[E]_{tot} = [E] + [ES] \qquad (3\text{-}3)$$

(i.e., if $[E]_{tot}$ is the sum of the free and the complexed enzyme, or the total amount of enzyme), then we can combine Equations 3-2 and 3-3 to obtain

$$[E]_{tot} = [E] + [ES] = \frac{[k_2 + k_{cat}]}{k_1[S]}(ES) + [ES] =$$

$$[ES]\left[1 + \left(\frac{k_2 + k_{cat}}{k_1}\right)\left(\frac{1}{[S]}\right)\right]$$

or, if we define K_m, the Michaelis-Menten constant, as equal to

$$\frac{k_2 + k_{cat}}{k_1}, \quad \text{then} \quad [ES] = \frac{[E]_{tot}}{1 + K_m/[S]} \qquad (3\text{-}4)$$

Thus $$\frac{dP}{dt} = k_{cat}\,[ES] = k_{cat}\,[E]_{tot}\frac{1}{1 + K_m/[S]}$$

$$= k_{cat}\,[E]_{tot}\frac{[S]}{[S] + K_m} \qquad (3\text{-}5)$$

This is the equation that fits the curve shown in Figure 3-29. V_{max}, which is equal to $k_{cat}\,[E]_{tot}$, is the maximum rate of formation of product if all of the substrate-binding sites on the enzyme are filled with substrate. K_m is equivalent to the substrate concentration at which the reaction rate is half-maximal. (If $[S] = K_m$, then from Equation 3-5 we can calculate that the rate of formation of product is $\frac{1}{2}k_{cat}\,[E]_{tot} = \frac{1}{2}V_{max}$.) For most, but not all, enzymes, the slowest step is the catalysis of [ES] to [E] + [P]. In these cases, $k_{cat} \ll k_2$ so that $K_m = \frac{k_2 + k_{cat}}{k_1} \cong \frac{k_2}{k_1}$, is equal to the equilibrium constant for binding of S to E.

Thus the parameter K_m describes the affinity of an enzyme for its substrate. The smaller the value of K_m, the more avidly the enzyme can bind the substrate from a dilute solution (Figure 3-30), and the lower the value of [S] needed to reach half-maximal velocity. The concentrations of the various small molecules in a cell vary widely, as do the K_m values for the different enzymes that act on them. Generally, the intracellular concentration of a substrate is approximately the same as, or is greater than, the K_m value of the enzyme to which it binds.

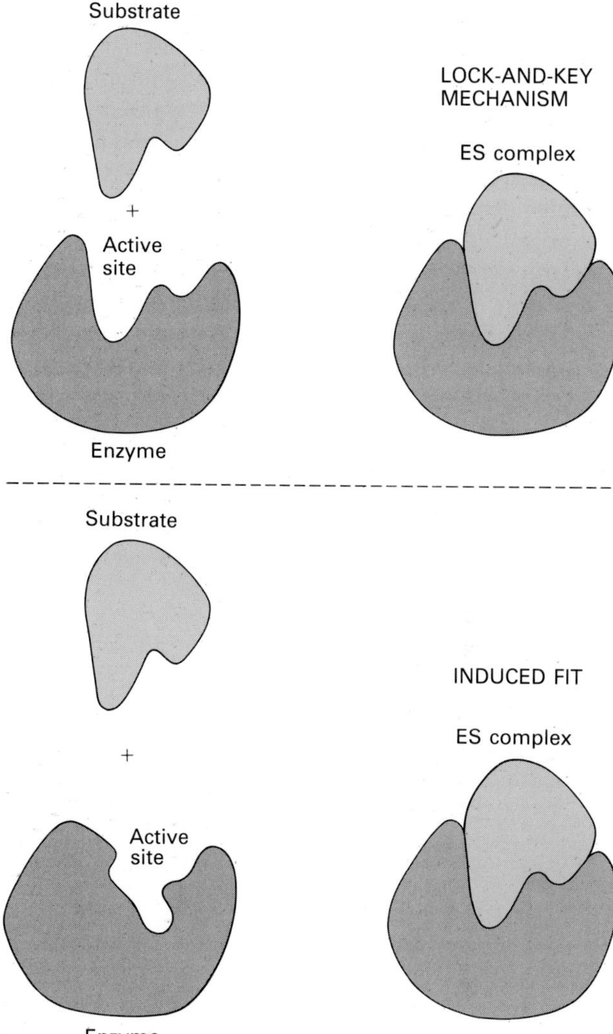

Figure 3-26 Two mechanisms for the interaction of an enzyme and a substrate. In the lock-and-key mechanism (*top*), the substrate fits directly into the binding site of the enzyme. If the binding occurs by induced fit (*bottom*), the substrate induces a conformational change in the enzyme that allows the substrate to be properly positioned for catalysis.

The Actions of Most Enzymes Are Regulated

Many reactions in cells do not occur at a constant rate. This is because the catalytic activity of the enzymes involved is *regulated* so that the amount of reaction product is just sufficient to meet the needs of the cell.

Feedback Inhibition of an Enzyme in a Biochemical Pathway Consider, for example, a set of enzymes that catalyze a series of reactions leading to the synthesis of the amino acid isoleucine. Isoleucine is used primarily as

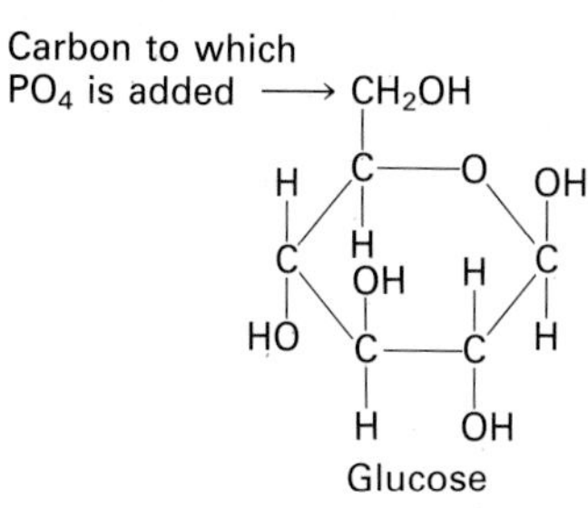

Figure 3-27 The structure of glucose, a substrate for hexokinase.

a monomer in the synthesis of proteins, and the amount of isoleucine needed depends on the rate of synthesis of proteins by the cell. The first step in the synthesis of isoleucine is the conversion of the amino acid threonine into the compound α-ketobutyrate which eliminates an amino group. The enzyme that catalyzes this reaction—threonine deaminase—plays a key role in regulating the level of isoleucine. In addition to its sites for binding the substrate threonine, threonine deaminase contains a binding site for isoleucine. When isoleucine is bound there, the enzyme molecule changes shape so that it cannot function as efficiently. Isoleucine thus acts as an *inhibitor* of the reaction for the conversion of threonine. If the isoleucine concentration in the cell is high, the binding of isoleucine to the enzyme temporarily reduces the rate of synthesis of isoleucine:

$$\begin{array}{c} NH_3^+ \\ | \\ H-C-COO^- \\ | \\ H-C-OH \\ | \\ CH_3 \end{array}$$

Threonine

↓ threonine deaminase (═╪═ ← Feedback inhibition: enzyme inhibited by isoleucine)

$$\begin{array}{c} O \\ \| \\ C-COO^- \\ | \\ CH_2 \\ | \\ CH_3 \end{array} \rightarrow \rightarrow \rightarrow \rightarrow \begin{array}{c} NH_3^+ \\ | \\ H-C-COO^- \\ | \\ H-C-CH_3 \\ | \\ CH_2 \\ | \\ CH_3 \end{array}$$

α-Ketobutyrate **Isoleucine**

This is an example of *feedback inhibition,* a process whereby an enzyme that catalyzes one reaction in a series of reactions is inhibited by their ultimate product (Figure 3-31).

In isoleucine synthesis, as in most cases of feedback inhibition, the final product in the reaction pathway inhibits the enzyme that catalyzes the first step in the pathway. The suppression of enzyme function is not permanent. If the concentration of free isoleucine is lowered, bound isoleucine molecules are released by enzyme molecules, which then resume functioning. The binding of the inhibitor isoleucine to the enzyme and its subsequent release can be described by an equilibrium binding constant, K_i ("*i*" for "*i*nhibition"), which is similar to the K_m used for substrate binding:

$$[E\cdot Ile]_{inactive} \overset{K_i}{\rightleftharpoons} [Ile] + [E]_{active}$$

$$K_i = \frac{[Ile][E]_{active}}{[E \cdot Ile]_{inactive}}$$

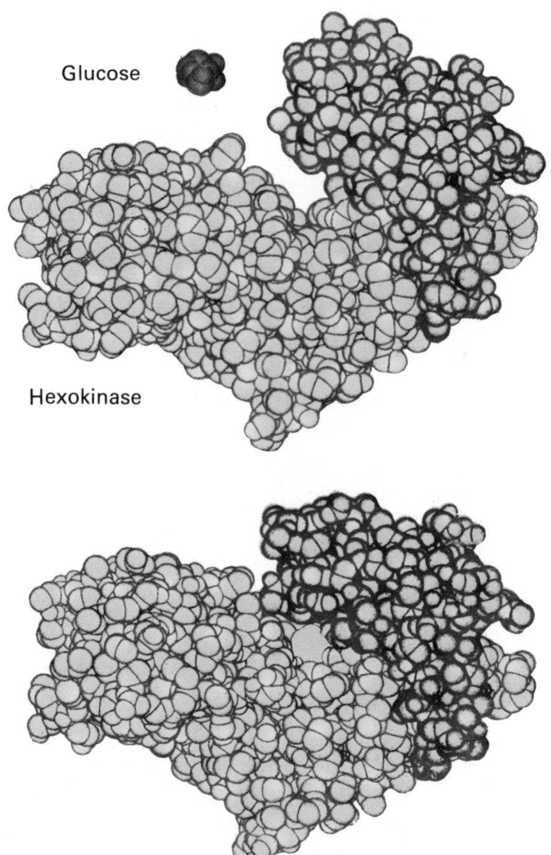

Figure 3-28 The conformation of hexokinase changes markedly when it binds the substrate glucose. The two lobes of the enzyme come closer together to surround the substrate. Molecules such as the five-carbon sugar ribose can also bind to hexokinase by forming specific hydrogen bonds with groups in the substrate-binding pocket of the enzyme, but only glucose can form all of the bonds that cause the enzyme to change its conformation. *Courtesy of Dr. Thomas A. Steitz.*

Multiple Binding Sites for Regulatory Molecules on Many Enzymes Some enzymes have binding sites for small molecules that stimulate their activity; these stimulator molecules are called *activators.* Enzymes may even have multiple sites for recognizing more than one activator or inhibitor. In a sense, enzymes are like microcomputers; they can detect concentrations of a variety of molecules and use that information to vary their own activities. Molecules that bind to enzymes and increase or decrease their activities are said to be *effectors* of enzyme activity. Effectors can modify enzymatic action because enzymes are rather flexible and can assume both active and inactive conformations. Activators are positive effectors, and inhibitors are negative effectors. The sites at which effectors bind are called *regulatory sites,* or *allosteric sites.* The word allosteric is derived from the Greek for "other shape." Thus, the binding of a positive effector

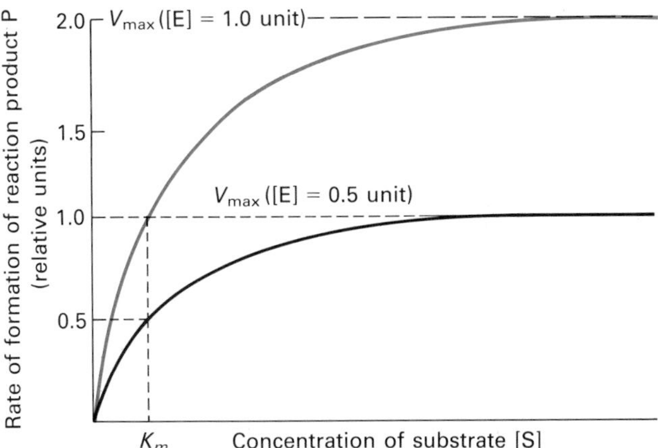

Figure 3-29 The rate of a hypothetical enzyme-catalyzed reaction S → P for two different concentrations of enzyme, [E], as a function of the concentration of substrate, [S]. (The substrate concentration that yields a half-maximal reaction rate is denoted by K_m.) Doubling the amount of enzyme causes a proportional increase in the rate of the reaction, so that the maximum velocity, V_{max}, is doubled. The K_m, however, is unaltered.

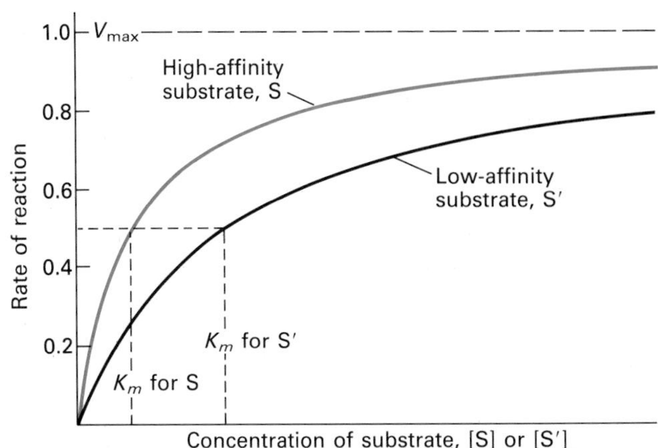

Figure 3-30 The rates of enzyme-catalyzed reactions with a substrate S for which the enzyme has a high affinity, and with a substrate S′ for which the enzyme has a low affinity. The value of K_m is higher for S′ than it is for S. The V_{max} value is the same for S and S′.

to the proper allosteric site on an inactive enzyme changes the enzyme's shape into one that is catalytically active.

Many enzymes, called multimeric enzymes, contain several copies of a single type of polypeptide chain; each copy, or subunit, has its own active site. In many such cases, upon binding an activator, inhibitor, or substrate, a subunit undergoes a conformational change that is transmitted to the other subunits and causes an increase in the affinity of the other chains for the type of molecule bound. Such *cooperative interactions* among the active sites allow an enzyme to respond to small changes in the concentration of an effector or substrate with large changes in the catalytic activity. Figure 3-32 illustrates how a cooperative interaction between the two identical subunits of a hypothetical enzyme might occur. The differences between active and inactive or poorly active enzyme conformations may result in part from small changes in tertiary or secondary structure, but major changes in the quaternary interactions between subunits are often involved. Thus when two subunits interact cooperatively, an increase in the concentration of a substrate causes a larger increase in the rate of the enzymatic reaction than would occur if the two subunits were independent.

The Multiple Types of Regulation of Aspartate Transcarbamoylase A well-studied enzyme that illustrates both cooperative interactions among active sites and the allosteric binding of an inhibitor at different sites is the *aspartate transcarbamoylase* (ATCase) of *Escherichia coli*. ATCase catalyzes the reaction of carbamoyl phosphate with the amino acid aspartate to yield carbam-

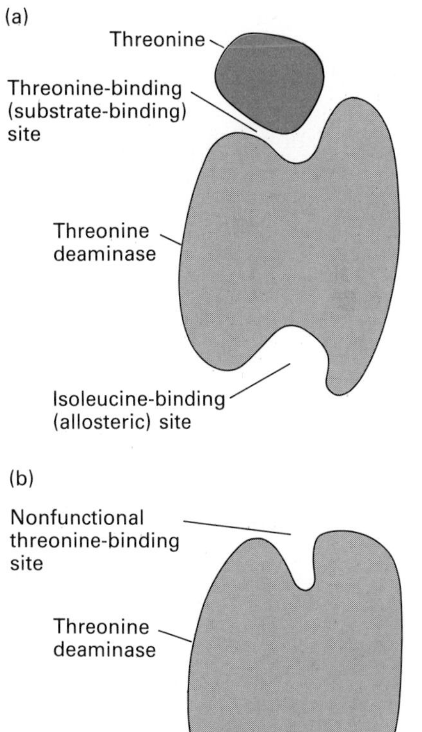

Figure 3-31 Feedback inhibition of threonine deaminase by isoleucine, the end product of the biochemical pathway. (a) In the absence of isoleucine, the enzyme is in the active conformation, and it can bind its substrate, threonine. (b) The binding of isoleucine to the regulatory (allosteric) site causes a conformational change that alters the substrate-binding site; this reduces the affinity of the enzyme for threonine.

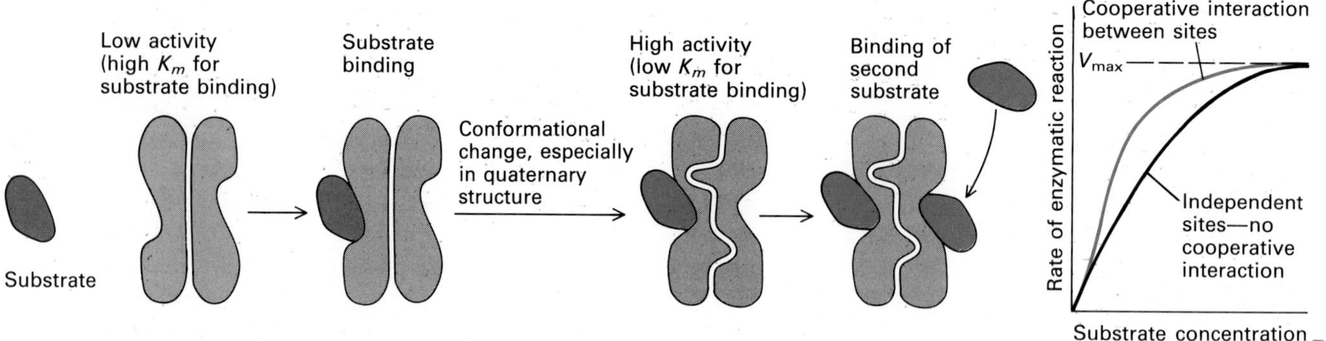

Figure 3-32 A cooperative interaction between active sites. The binding of a substrate to one subunit of a multimeric enzyme induces a conformational change in the adjacent subunit, which lowers the K_m for the binding of the substrate there. Thus, a small change in the substrate concentration can cause a much larger increase in the reaction rate than if there were no cooperative interactions between active sites.

oyl aspartate and phosphate (Figure 3-33). This is the first unique step in the biosynthesis of the pyrimidine nucleotides uridine monophosphate (UMP) and cytidine triphosphate (CTP), which are precursors of DNA and RNA. The enzyme exhibits feedback control; it is inhibited specifically by CTP, the end product of the pyrimidine pathway (Figures 3-33 and 3-34). The binding of CTP to the regulatory subunits of ATCase induces an inhibitory conformational change in the enzyme. ATCase contains six copies each of the catalytic (C) and regulatory (R) subunits; thus, its composition is C_6R_6.

X-ray crystallographic studies have shown that the catalytic subunits form two C_3 *trimers;* these catalytic trimers are positioned on either side of an equatorial plane. The regulatory subunits form three R_2 dimers, each of which is bound to two catalytic subunits, one above and one below the equatorial plane (Figure 3-35).

In the absence of an inhibitor, the native enzyme exhibits cooperative interactions among the substrate-binding sites of the six catalytic subunits. A twofold change in the aspartate concentration, from 0.0025 *M* to 0.005 *M*, causes the reaction rate to increase by a factor of 3.5 (see Figure 3-34). The enzyme is thus extremely sensitive to small changes in the substrate concentration. The binding

Figure 3-33 The biochemical pathway initiated by the reaction catalyzed by aspartate transcarbamoylase (ATCase), and the feedback inhibition of this enzyme by cytidine triphosphate (CTP).

Carbamoyl phosphate + Aspartate —ATCase→ Carbamoyl aspartate + Phosphate (HPO_4^{2-})

Feedback inhibition

Uridine monophosphate (UMP)

Cytidine triphosphate (CTP)

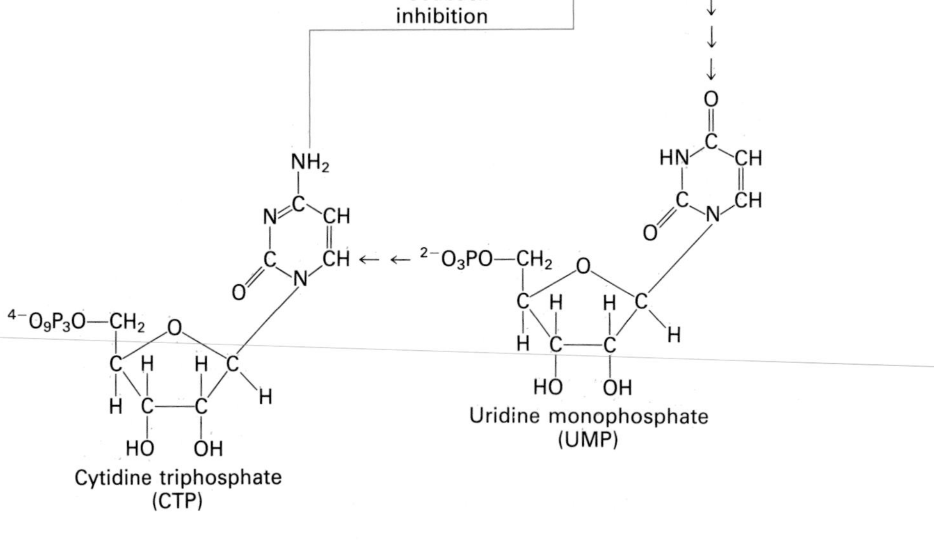

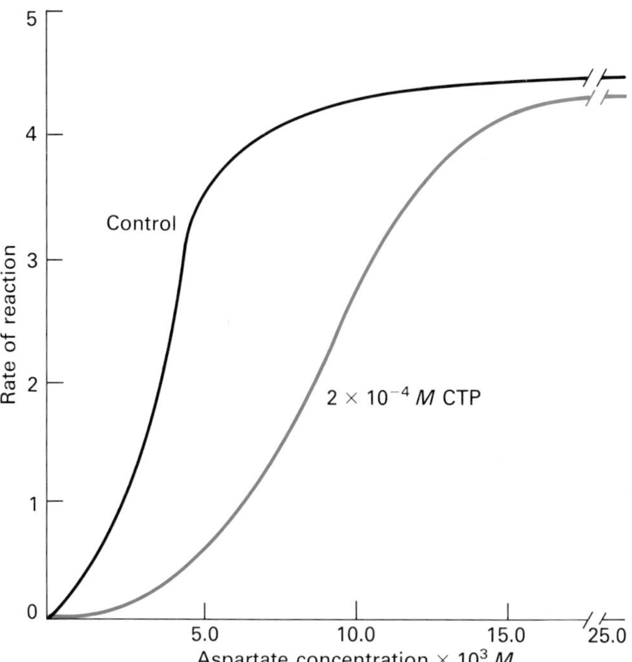

Figure 3-34 The activity of ATCase as a function of the concentration of its substrate, aspartate. The sigmoid saturation curves indicate cooperative interactions between the six catalytic subunits. The binding of the substrate to an active site on one of the catalytic subunits induces a conformational change that lowers the K_m for substrate binding to other catalytic sites. The addition of the allosteric inhibitor CTP causes a decrease in the affinity of ATCase for aspartate, as reflected in the higher concentration of aspartate required to achieve a given reaction rate. In the absence of CTP, an aspartate concentration of 4.0×10^{-3} *M* results in a half-maximal reaction rate; in the presence of CTP, 9.0×10^{-3} *M* aspartate is required. *After J. G. Gerhart, 1962, Ph.D. thesis, University of California, Berkeley.*

of CTP to regulatory subunits decreases the affinity of the catalytic subunits for the substrate, without affecting V_{max}, the maximum rate of catalysis. Thus, a higher concentration of aspartate is required to achieve the reaction rate that would occur in the absence of CTP.

Treatment of ATCase with mercury-containing compounds such as *p*-hydroxymercuribenzoate (*p*HMB), which binds to free SH groups, causes the enzyme to dissociate into two C_3 trimers and three R_2 dimers. The C_3 trimers are catalytically active but, as might be expected, the reaction is not inhibited by CTP. If the *p*HMB is removed by dialysis, the C_3 and R_2 subunits will spontaneously recombine to become a fully active enzyme that is sensitive to allosteric inhibition by CTP.

The Four Interacting Oxygen-Binding Sites of Hemoglobin The four polypeptides in hemoglobin interact with one another in a cooperative fashion, similar to the way the active sites of aspartate transcarbamoylase work together. Each hemoglobin chain binds one O_2 molecule. The binding of a single O_2 to any one of the four chains induces in the molecule a conformational change that increases the affinity of the other oxygen-binding sites for oxygen. The binding of the second O_2 likewise results in a further increase in oxygen affinity. The cooperative interaction between the chains causes the molecule to take up four O_2 molecules, or to lose four O_2 molecules, over a much narrower range of oxygen pressures than it would otherwise. This function ensures that hemoglobin becomes almost completely oxygenated at the oxygen pressure in the lungs and largely deoxygenated at the oxygen pressure in the tissue capillaries (Figure 3-36).

The contrast between hemoglobin and myoglobin is revealing. Myoglobin is a single-chained oxygen-binding protein found in muscle (see Figure 3-18). The oxygen-binding curve of myoglobin (i.e., the plot of the percentage of sites filled with O_2 versus the partial pressure of O_2) has the characteristics of a simple noncooperative protein-ligand equilibrium reaction:

$$E + O_2 \xrightleftharpoons{K_{O_2}} E{-}O_2$$

Myoglobin has a greater binding affinity for O_2 (has a lower K_{O_2}) than hemoglobin has, at all oxygen pressures (Figure 3-36). Thus, at the oxygen pressure in capillaries,

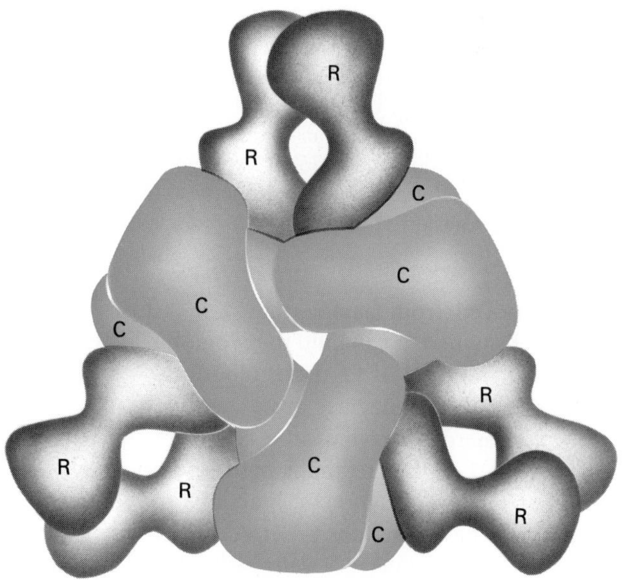

Figure 3-35 The subunit structure of ATCase. Three of the large, catalytic subunits form a catalytic trimer; another catalytic trimer is beneath the one shown. (The equatorial plane is the plane of the page.) The six regulatory subunits (Rs) are in three dimers at the corners of the triangular molecule. *After K. L. Krause et al., 1985,* Proc. Nat'l. Acad. Sci. ***82**:1643.*

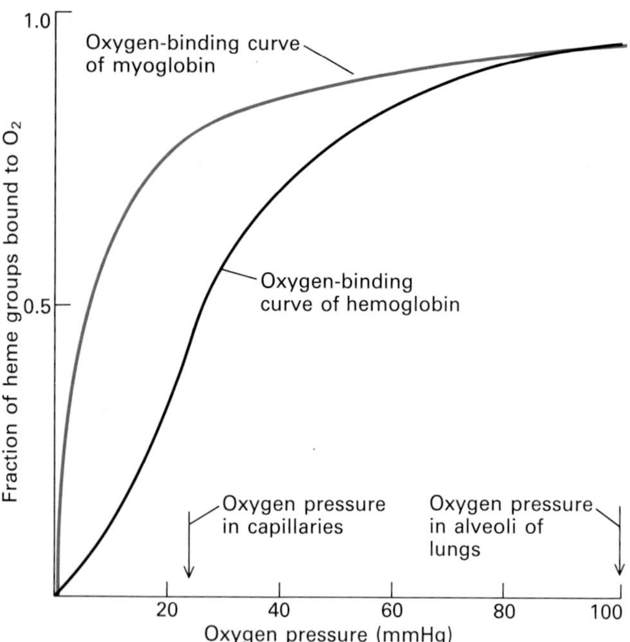

Figure 3-36 The binding of oxygen to hemoglobin depends on cooperative interactions between the four chains. The graph shows the amount of O_2 bound as a function of the oxygen pressure. The colored curve is the oxygen-binding profile of single-chained myoglobin. The black curve illustrates the binding of O_2 to hemoglobin; note that the binding activity increases sharply over a narrow range of oxygen pressures from 20 to 40 mmHg. The oxygen pressures in the lungs and in the tissue capillaries are indicated above the abscissa. Hemoglobin is saturated with O_2 in the lungs, but it releases much of its bound O_2 at the low oxygen pressure in the tissue capillaries. At any oxygen pressure, myoglobin has a higher affinity for O_2 than does hemoglobin. As myoglobin is a principal muscle protein, this property allows oxygen to be transferred from blood to muscle.

O_2 is removed from hemoglobin and diffuses into the muscle cells, where it can bind to myoglobin. This ensures an efficient transfer of O_2 from blood to tissues.

The activities of enzymes are extensively regulated so that the numerous enzymes in a cell work together harmoniously. All of the metabolic pathways are closely controlled at all times. Synthetic reactions occur when the products of these reactions are needed; degradative reactions occur when molecules must be broken down. A host of kinetic controls that affect the activities of key enzymes determine which pathways are going to be used, as well as the rate at which they will function.

Regulation of cellular processes involves more than simply turning enzymes on and off, however. Some regulation is accomplished through compartmentation. Many enzymes are localized in specific compartments of the cell, such as the mitochondria or lysosomes. This restricts the substrates, the effectors, and the other enzymes with which an enzyme can interact. In addition, compartmentation means that reactions that might compete with one another if they were confined to the same solution can occur simultaneously in different parts of a cell. Another means by which cellular processes are regulated is through the control of the rates of synthesis and destruction of many enzymes themselves.

Enzymes Contain Many Regions That Have Pivotal Functions Not Directly Related to Catalysis

We have seen that the enzyme regions involved in substrate binding and catalysis are small in comparison to the overall size of enzymes. For example, glucose, the substrate of hexokinase, occupies a volume that is less than 1 percent of the volume of hexokinase (see Figure 3-28). What role, if any, does the rest of the enzyme play? Years ago, scientists answered this question by assuming that the catalytic site required a large "scaffold" to hold it in the appropriate shape. Indeed, this is the case for α-chymotrypsin (see Figure 3-23), in which the three amino acids that are critical for catalysis must be brought into close proximity near the substrate-binding site. But x-ray crystallographic analysis has since shown that most enzymes have several domains, only a few of which participate directly or indirectly in catalysis; it is often possible to demonstrate that just a fragment of an enzyme molecule is needed for catalytic activity.

Although much remains to be discovered about enzyme domains, some reasons have been found for the large mass of most enzymes. Foremost among the functions of the noncatalytic parts is the binding of effector molecules to allosteric sites. In addition, the extra bulk of an enzyme may be partly attributed to its location within the cell. Some enzymes float freely in the cytosol (the intracellular medium); some remain in specific organelles; others are secreted from the cell; still others are bound to specific cell membranes. There is increasing evidence that the location of an enzyme is a consequence of its structure. In some cases, bulky attachment sites are required. Or, enzymes may contain specific "ticketing" signals (sequences of amino acids or sugar groups) that determine their location within a cell. Other enzymes, such as chymotrypsin, have sequences that keep them in an inactive state until they reach the sites at which they function. Ticketing signals and inactivating fragments are often removed by specific proteases after enzymes have reached their appointed locales.

Proteins that can dissolve in water must fold so that polar, hydrophilic amino acid residues face the outside. Thus, hydrophobic residues tend to be folded into the interior. Yet many of the catalytic and other binding sites of enzymes also require that hydrophobic residues be in "pockets" that are accessible from the exterior. This is the case, for example, with chymotrypsin. The exposed hydrophobic residues on the surface of this enzyme are energetically unfavorable and might tend to cause the protein to aggregate or precipitate. For such a protein to remain soluble, it must have a large number of hydrophilic residues on the surface to counterbalance the hydrophobic ones; this would add bulk to the protein.

There is yet another contributing factor in the large size of enzymes: proteins also include regions that reflect their evolutionary history. Because all proteins have evolved from other proteins, many certainly contain remnants of their evolutionary past in amino acid sequences that are not essential for their present function.

Antibodies

Enzymes are not the only proteins that bind tightly and specifically to smaller compounds. The insulin receptor on the surface of a liver cell, for example, is a protein that can bind to insulin so tightly that the receptors on a cell are half-saturated when the concentration of insulin is only 10^{-9} *M*. This protein, which does not bind to most of the other compounds present in blood, mediates the specific actions of insulin on liver cells. To cite another example, avidin, a protein found in raw egg white, specifically binds the vitamin biotin very tightly. In human beings, biotin is normally made by bacteria in the small intestine and is absorbed through the intestinal wall; avidin binding prevents absorption of biotin, and thus biotin deficiency is found in people who consume large amounts of raw eggs.

The capacity of proteins to distinguish among different molecules is developed more highly in blood proteins called *antibodies,* or immunoglobulins, than in enzymes. Animals produce antibodies in response to an invasion by an infectious agent such as a bacterium or a virus. The antibody has a recognition site that can bind tightly to very specific sites—generally either proteins or carbohydrates—on the surface of the infectious agent. Experimentally, animals produce antibodies in response to the injection of almost any foreign polymer that is larger than a certain size; such antibodies will bind specifically and tightly to the invading substance but, like enzymes, will not bind to unrelated molecules. The binding of an antibody to a bacterium, virus, or virus-infected cell allows certain white blood cells to recognize the invading body as foreign, and they respond by degrading it. In this way, the antibody acts as a signal for the elimination of infectious agents. The specificity of antibodies is exquisite: they can distinguish between proteins that differ by only a single amino acid and between the cells of different individual members of the same species.

All vertebrates can produce a very large variety of antibodies. They can produce antibodies that bind to chemically synthesized molecules that do not exist in nature. Exposure to an antibody-producing agent—called an *antigen,* or an immunogen—causes an organism to make a larger number of different antibody proteins, each of which may bind to a slightly different region of the antigen. These constellations of antibodies may differ from one member of a species to another for a given antigen.

Antibodies Can Distinguish among Closely Related Molecules

We shall use two of many possible examples to illustrate the exquisite specificity of antibodies.

The sequence of bovine insulin is identical to that of human insulin, except for three amino acid replacements. Yet when bovine insulin is injected into people, some individuals respond by synthesizing antibodies that specifically recognize the "altered" amino acids in the bovine molecule, even though human beings generally do not produce antibodies that recognize their own insulin—for obvious reasons.

Animals do not normally synthesize antibodies against any proteins purified from their own species. Thus, injection of mouse albumin—the major serum protein—into mice would not elicit the production of antialbumin antibodies. However, if a small molecule, such as 2,4-dinitrophenol (DNP),

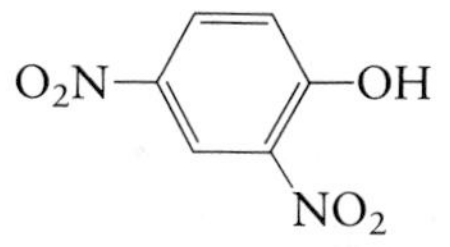

O_2N—(benzene ring)—OH, with NO_2 on the ring

is coupled to the albumin, then the mice will produce antibodies that bind specifically to the modified region of the protein—in this case, to the dinitrophenyl group (Figure 3-37a). A small group capable of eliciting antibody production is called a *hapten.* The anti-DNP albumin antibody will not bind to albumin that is not complexed with DNP, nor will it bind to albumin that has been modified by other haptens, such as phenyl groups with different substituents (Figure 3-37b).

Certain Structural Features Characterize All Antibodies

Despite the enormous variability in the substances to which antibodies will bind, all antibodies have common structural features. Most blood serum antibodies are of the immunoglobulin G (IgG) class. These are made up of four chains: two heavy chains of about 446 amino acids each, and two light chains of about 219 residues each. There are bridges both within and between the chains (Figure 3-38a). The IgG molecule folds into three structural regions, two of which are identical. The identical regions each contain a specific antigen-binding site; thus, they are referred to as the F_{ab} regions. The third region is the common or effector domain; this is often called the F_c

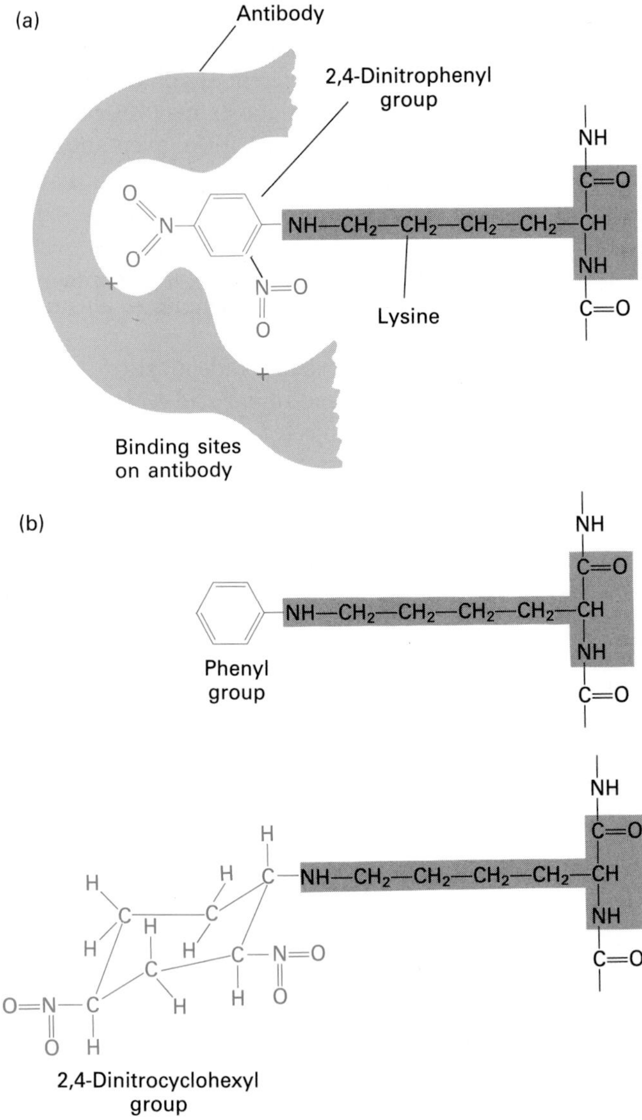

Figure 3-37 (a) A protein that contains a 2,4-dinitrophenyl (DNP) group covalently bound to a lysine residue will bind to an antibody that is specific for the modified region. (b) The antibody for DNP is so specific that it will not bind to either of the haptens shown here. The phenyl hapten lacks the two nitro groups that bind to positive residues on the antibody. Because the cyclohexyl ring is puckered (not planar like the phenyl ring), it cannot position itself into the binding grooves of the antibody.

region because it readily crystallizes (Figure 3-38b). The F_c region is recognized by cells that digest antigen-antibody complexes. The "hinge" region of the heavy chain separates the globular F_{ab} and F_c domains. It is a site for cleavage by several proteases; these generate two F_{ab} fragments and one F_c fragment.

As is described in Chapter 24, each of the F_{ab} and F_c structural regions consists of four polypeptide domains. Each domain is compactly folded independently of the others. The antigen-binding sites are in the variable domains (V_L and V_H) of the F_{ab} regions (see Figure 3-38). The rest of each F_{ab} region is made up of two constant domains (C_L and C_H1). The F_c region comprises four constant heavy-chain domains (two each of C_H2 and C_H3). The polypeptides in the domains of an IgG molecule are folded consistently; each domain contains, for example, a single —S—S— bridge that forms a loop that is the same size in all domains.

In fact, it is believed that all the domains (V_L, V_H, C_L, C_H1, C_H2, and C_H3) have evolved by the duplication of a single ancestral gene that encoded a primordial immunoglobulin region about 100 amino acid residues long. As is true for many other multidomain proteins, the polypeptide segments that form each domain are encoded by discrete, nonadjacent segments of DNA. The rearrangement of these DNA sequences allows the generation of multiple types of antibodies.

The binding of an antigen at the antigen-binding site triggers a conformational change in the F_c effector domain of an IgG. Macrophages and other cells that digest antigen-antibody complexes have surface receptors that bind to the F_c regions of the complexes. This binding is the initial stage in the digestion of immune complexes.

Antibodies Are Valuable Tools for Identifying and Purifying Proteins

Because they bind so selectively to proteins, antibodies can be used experimentally to isolate one protein from a complex mixture. One method of isolating proteins is through the use of *affinity chromatography*. In this technique, the antibody is chemically coupled to tiny plastic beads, which are then placed in a small column. Next, a protein solution is passed through the column. Only the protein to which the antibody is directed will adhere to the column; all of the others will pass through unimpeded. The desired protein can be eluted from the column by adding a solution that disrupts the binding to the antibody (Figure 3-39).

Similarly, antibodies can be used to detect specific proteins in cells or in other biological materials. For example, an antibody to the protein actin can be complexed with a fluorescent dye and then added to a preparation of disrupted tissue culture cells. The antibody binds to the actin filaments, which can be made visible by shining intense visible light on the cells (Figure 3-40).

Structural Proteins

Another important class of nonenzyme proteins are the structural proteins, which play a major role in the lives of cells and of organisms. Certain eukaryotic intracellular proteins (e.g., actin) form filaments or sheets that give the cells their sizes and shapes, and participate as well in cell

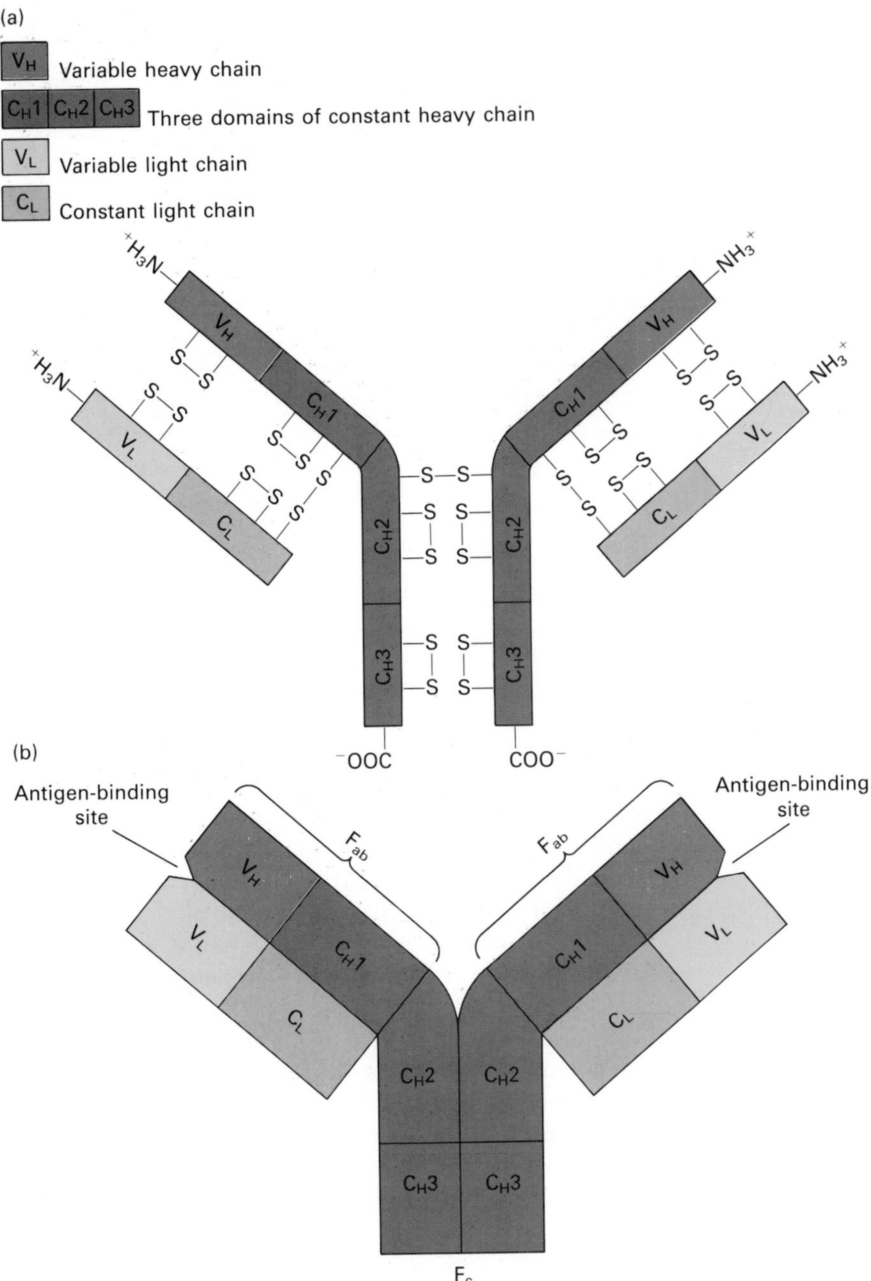

Figure 3-38 The structure of an IgG molecule. (a) The —S—S— bridges run within and between the two heavy chains (denoted by the subscript H) and the two light chains (denoted by the subscript L). The variable and the constant domains are indicated by V and C. (b) The three structural regions of the molecule are the two identical antigen-binding (F_{ab}) regions and the common or effector (F_c) region.

movement and locomotion. Extracellular structural proteins bind cells together in various ways to strengthen tissues or organs and to provide a matrix in which cells can grow. These structural proteins are treated in detail in later chapters; here we wish to describe some of the features that make the filamentous proteins strong and rigid.

Most structural proteins are regular aggregates of single polypeptide chains that form specific covalent or noncovalent bonds with one another. To cite one example: Actin-containing filaments, which are a structural feature of all higher cells, are especially important in the contraction of muscle. Actin, the principal constituent of these fibers, is a two-domain globular protein with a molecular weight of 42,000. In solutions that are weakly ionic, actin is a soluble globular monomer called G-actin. If the salt concentration is increased, G-actin polymerizes into the insoluble fiber F-actin, which is similar in structure to the filaments found in cells. The fibers are made up of a string of G-actin monomers wound in a helix. Each actin monomer interacts with its neighbors in precisely the same way; these regular monomer-monomer interactions are what generate the helical fiber (Figure 3-41).

Collagen, the major extracellular structural protein in all animals, has a very different composition (Chapter 5); yet again, each fiber comprises specific aggregates of smaller protein subunits.

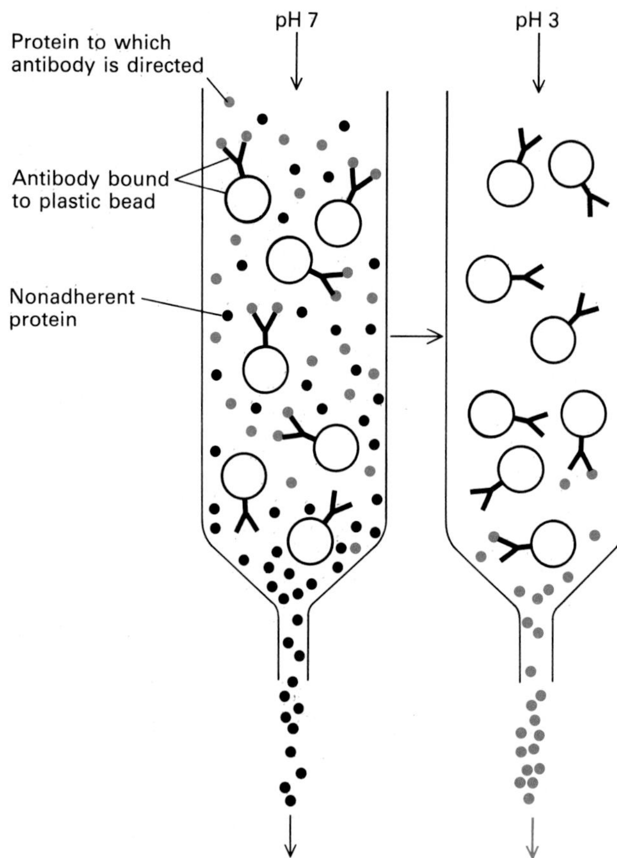

Figure 3-39 The purification of a protein from a mixture by using affinity chromatography. In step 1, (*left*) the mixture is filtered through a column containing antibody molecules that are specific for the desired protein; only that protein binds to the antibody matrix; any others in the mixture are eluted. In step 2, (*right*) a solution such as acetic acid is added to disrupt the antigen-antibody complex. A pure protein is thus eluted from the column.

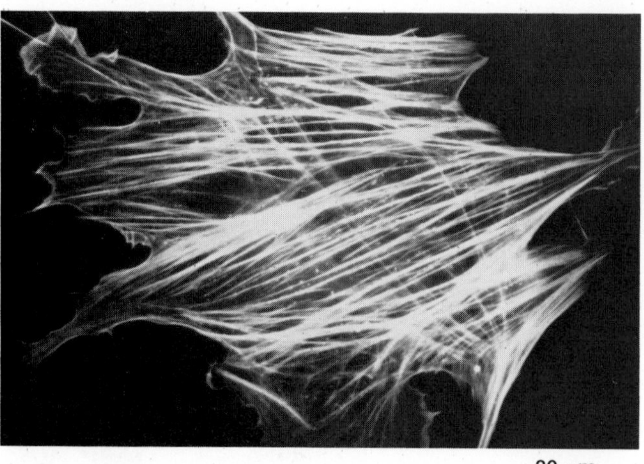

Figure 3-40 A rat mammary cell stained with a fluorescent antibody to actin. Note the strong staining of the long fibers that crisscross the cell, and also the staining at the puckered borders of the cell, where the actin fibers insert into the cell membrane. *Courtesy of M. Osborn and K. Weber.*

Nucleic Acids

Proteins are the catalysts of all chemical reactions in cells, and the determinants of cell movement and shape. Cells receive instructions about which proteins to synthesize, and in what quantities, from nucleic acids, the molecules that store and transmit information in cells. As in any system of communication, the information is stored and transmitted in a code. Just how this code is translated is described in Chapter 4. Here, we deal with the chemical structures of the molecules that store the encoded information.

Nucleic Acids Are Linear Polymers of Nucleotides Connected by Phosphodiester Bonds

There are two closely related information-storing molecules in cells: deoxyribonucleic acid (DNA) and ribonucleic acid (RNA). Like proteins, DNA and RNA are linear biopolymers. However, the number of monomers in a nucleic acid is generally much greater than the number of amino acids in a protein. Cellular RNAs range in length from about 80 to 200,000 units; certain viral RNAs contain 7000 units or more. (The genome of many viruses is DNA, but some viruses have RNA as the genetic material.) The number of units in a DNA molecule can be in the millions.

Both DNA and RNA consist of only four different monomers, called *nucleotides.* A nucleotide has three parts: a phosphate group, a *pentose* (a five-carbon sugar molecule), and an *organic base* (Figure 3-42). In RNA, the pentose is always *ribose;* in DNA, it is *deoxyribose* (Figure 3-43). The only other difference between the monomers of DNA and those of RNA lies in one of their bases.

The base components of nucleic acids are of two kinds, *purines* and *pyrimidines* (Figure 3-44). The structure of purine is a pair of fused rings; a pyrimidine has only one ring. Both are heterocyclic; that is, the rings are built of more than one kind of atom—in this case nitrogen in addition to carbon. The presence of nitrogen atoms gives these molecules their basic character, but none is protonated (ionized) at neutral pH. The acidic character of nucleotides is due to the presence of phosphate, which dissociates, freeing hydrogen ions, under physiological conditions. The names of the bases—adenine, guanine, and cytosine (found in both DNA and RNA); thymine (found only in DNA); and uracil (found only in RNA)—are often abbreviated as A, G, C, T, and U, respectively.*

*Although, strictly speaking, the letters A, G, C, T, and U stand for bases, they are often used in diagrams to represent the nucleotides containing these bases.

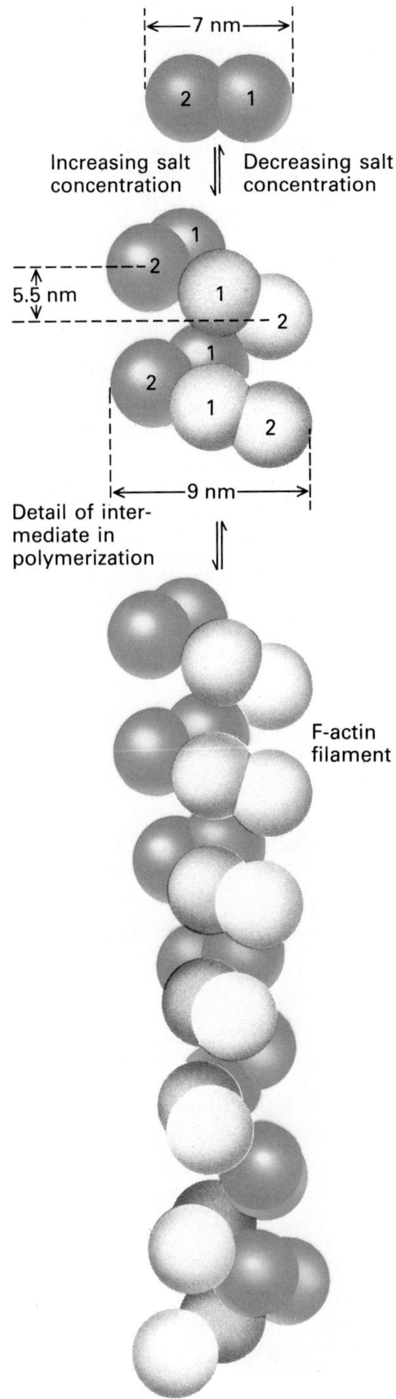

Figure 3-41 The polymerization of globular G-actin to form helical F-actin filaments. G-actin monomeric units, which are about 7 nm in diameter, contain two roughly spherical domains termed 1 and 2. In F-actin, the 2 domain of one unit interacts with the 1 domain of the next. The units form a helix about 9 nm in diameter; each monomer in the helix is displaced about 5.5 nm along the axis from its neighbor. Alternate monomers are colored to make it easier to visualize the structure of the helix, but all subunits are identical. *After E. Egelman and D. DeRosier, 1983,* J. Mol. Biol. ***116****:623–629.*

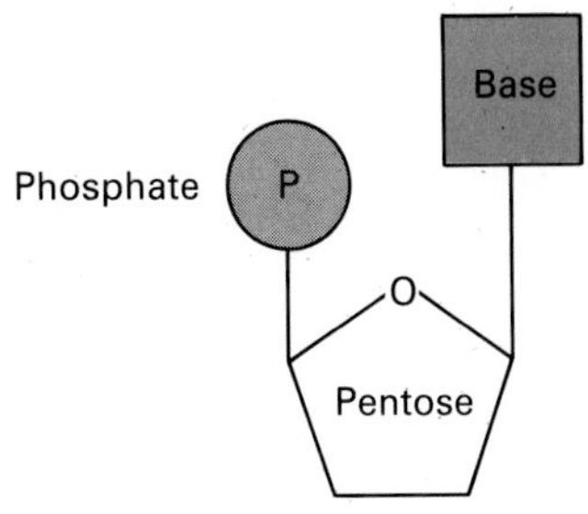

Figure 3-42 A schematic diagram of the structure of a nucleotide.

The sugar component of a nucleotide is the link between the base and the phosphate group. Whether the sugar is ribose or deoxyribose, its 1′ carbon atom is attached to the 9 nitrogen of a purine or the 1 nitrogen of a pyrimidine. The hydroxyl group on the 5′ carbon atom is linked in an ester bond to the phosphate group (Figure 3-45).

Cells contain several types of molecules that differ from nucleotides only in the absence of attached phosphate groups. A combination of a base and a sugar without a phosphate group is called a *nucleoside.* The four *ribo*nucleosides are *adenosine, guanosine, cytidine,* and *uridine.* The four *deoxyribo*nucleosides are *deoxyadenosine, deoxyguanosine, deoxycytidine,* and *deoxythymidine.*

Nucleotides, which have one, two, or three attached phosphate groups, may be referred to as nucleoside phosphates. The phosphates are generally esterified to the 5′ carbon. The nucleoside monophosphates are abbreviated as AMP (adenosine 5′-monophosphate; see Figure 3-45), GMP, CMP, and UMP; and as dAMP (deoxyadenosine 5′-monophosphate), dGMP, dCMP, and dTMP. (Another way of referring to the nucleoside monophosphates is as adenylic acid or adenylate, guanylic acid or guanylate, etc.; deoxyadenylic acid or deoxyadenylate, etc.) Diphosphates contain a pyrophosphate group,

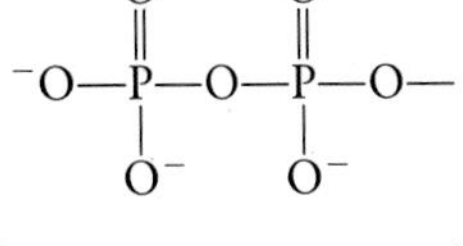

$$^{-}O{-}\overset{\overset{\displaystyle O}{\|}}{\underset{\underset{\displaystyle O^-}{|}}{P}}{-}O{-}\overset{\overset{\displaystyle O}{\|}}{\underset{\underset{\displaystyle O^-}{|}}{P}}{-}O{-}$$

esterified to the 5′ carbon; they are abbreviated as ADP

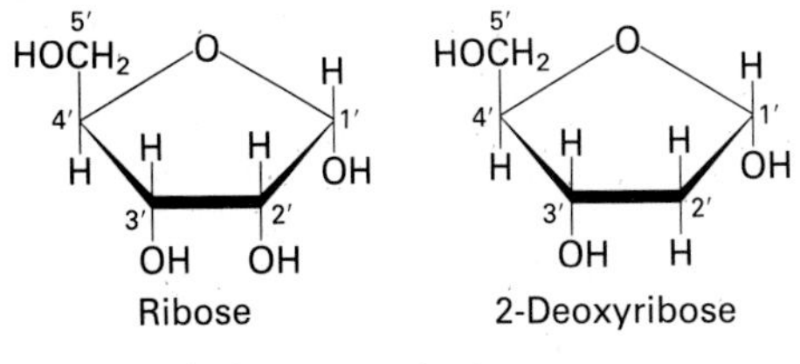

Ribose 2-Deoxyribose

Figure 3-43 Haworth projections (see Figure 3-1) of the structures of ribose and deoxyribose. In nucleotides and nucleic acids, the 5′ carbon is linked in an ester bond to the phosphate, and the 1′ carbon is linked to the base. By convention, the carbon atoms of the pentoses are numbered with primes.

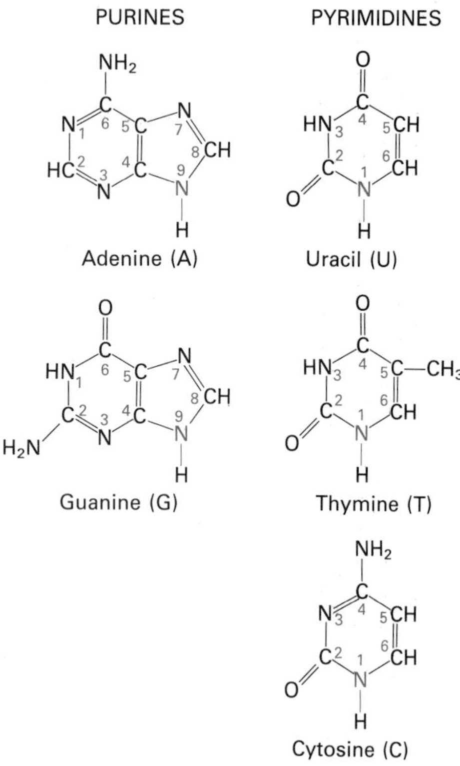

Figure 3-44 The chemical structures of the principal bases in nucleic acids. In nucleic acids and nucleotides, the 9 nitrogen atom of purines and the 1 nitrogen atom of pyrimidines (color) are bonded to the 1 carbon of ribose or deoxyribose.

(adenosine 5′-diphosphate; see Figure 2-23), dADP (deoxyadenosine 5′-diphosphate), etc. Triphosphates are abbreviated as ATP (adenosine 5′-triphosphate; see Figure 2-22), dATP, etc. A supply of nucleoside triphosphates is necessary for the synthesis of nucleic acids; the triphosphate ATP is the most widely used energy carrier in the cell.

When nucleotides polymerize to form nucleic acids, the hydroxyl group attached to the 3′ carbon of a sugar of one nucleotide forms an ester bond with the phosphate of another nucleotide, eliminating a molecule of water:

$$\underset{\displaystyle (\text{Sugar})}{\overset{(\text{Base})_1}{|}}\text{—OH} + \text{HO—}\underset{\displaystyle \text{O}^-}{\overset{\displaystyle \text{O}}{\overset{\|}{\text{P}}}}\text{—O—}\underset{\displaystyle (\text{Sugar})}{\overset{(\text{Base})_2}{|}} \longrightarrow$$

$$\underset{\displaystyle (\text{Sugar})}{\overset{(\text{Base})_1}{|}}\text{—O—}\underset{\displaystyle \text{O}^-}{\overset{\displaystyle \text{O}}{\overset{\|}{\text{P}}}}\text{—O—}\underset{\displaystyle (\text{Sugar})}{\overset{(\text{Base})_2}{|}} + H_2O$$

This condensation is similar to the reaction that forms a peptide bond. Thus, a single nucleic acid chain is a phosphate-pentose polymer (a polyester) with purine and pyrimidine bases as side groups. The links between the nucleotides are called *phosphodiester bonds.* Like a polypeptide, a nucleic acid chain has a chemical orientation: the *3′ end* has a free hydroxyl group attached to the 3′ carbon of a sugar; the other end, the *5′ end,* has a free hydroxyl or phosphate group attached to the 5′ carbon of a sugar (Figure 3-46). The orientation of a nucleic acid chain is an extremely important property of the molecule, as we shall see.

DNA

As was discussed in Chapter 1, the modern era of molecular biology began in 1953, when James Watson and Francis Crick elucidated the double-helical structure of DNA by the detailed analysis of x-ray diffraction patterns

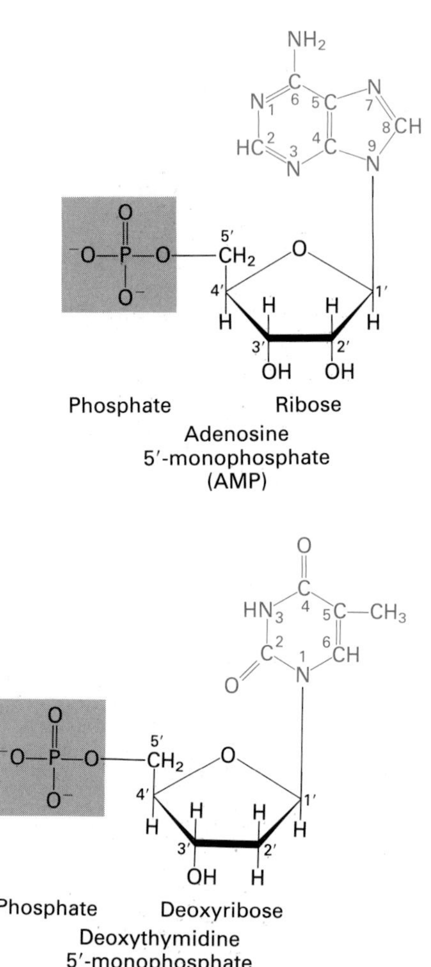

Figure 3-45 The chemical structures of two typical nucleotides, adenosine 5′-monophosphate and deoxythymidine 5′-monophosphate. Note the C—N linkage between the sugar and the base. The C—N bonds shown here are β linkages, meaning that the base lies above the plane of the ribose. (In the α linkage, the base lies below.) Only the β linkage is found in cellular nucleotides.

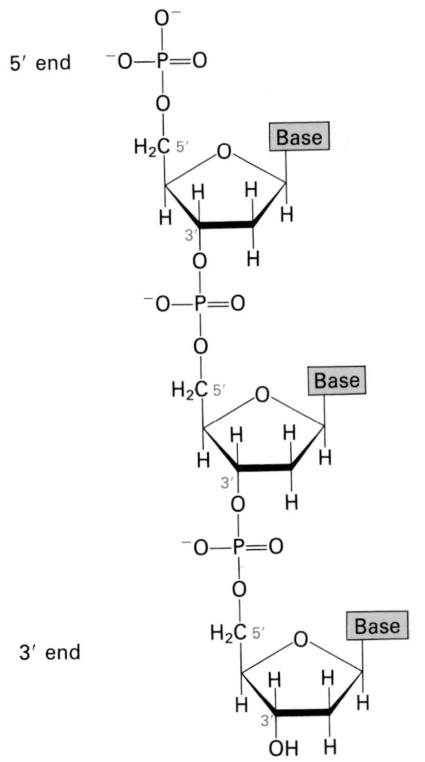

5′ end

3′ end

Figure 3-46 The chemical structure of a trinucleotide, a single strand of DNA containing only three nucleotides. The nucleotide at the 3′ end has a free 3′ deoxyribose hydroxyl group (i.e., a hydroxyl that is not bonded to another nucleotide). Similarly, the 5′ end has a free 5′ hydroxyl or phosphate.

and by careful model building. This event, and the structure of the double helix itself, are quite familiar by now to anyone who reads newspapers and magazines. A closer look at the structure of the "thread of life," as the DNA molecule is sometimes called, will reveal why the discovery was so important.

The Native State of DNA Is a Double Helix of Two Antiparallel Chains That Have Complementary Sequences of Nucleotides

The double helix consists of two intertwined chains of DNA. The orientation of the two chains is antiparallel: one chain runs 5′ → 3′ and the other runs 3′ → 5′. The chains are held together by hydrogen bonds and hydrophobic interactions. The bases on opposite strands are held in precise register by the hydrogen bonds between them: A is paired with T; G is paired with C. The A-T pairs are held together by two hydrogen bonds, and the G-C pairs by three (Figure 3-47). This *base-pair complementarity* is a consequence of the size, shape, and chemical composition of the bases (Figure 3-48).

A double helix that has complementary structure in its component strands has the same informational content in both strands. In DNA replication, the strands of the helix separate, whereupon each strand seeks to replace the hydrogen-bonded complementary nucleotides that it has lost. The newly hydrogen-bonded nucleotides are polymerized, and the result is two double helixes that are identical to the original one (Figure 3-49). This process is described in detail in Chapter 4.

Because of the geometry of the double helix, a purine must always pair with a pyrimidine. The rules are a bit more complicated than that, however. For example, although a guanine residue (a purine) could theoretically form hydrogen bonds with a thymine (a pyrimidine), there would have to be a minor distortion of the helix. A similar distortion would allow the formation of a base pair between two pyrimidines, cytosine and thymine. However, neither G-T nor C-T base pairs are normally found in DNA (Figure 3-50).

Because the bases are planar and form planar pairs, the base pairs stack on top of one another. Van der Waals and hydrophobic interactions between adjacent base pairs in

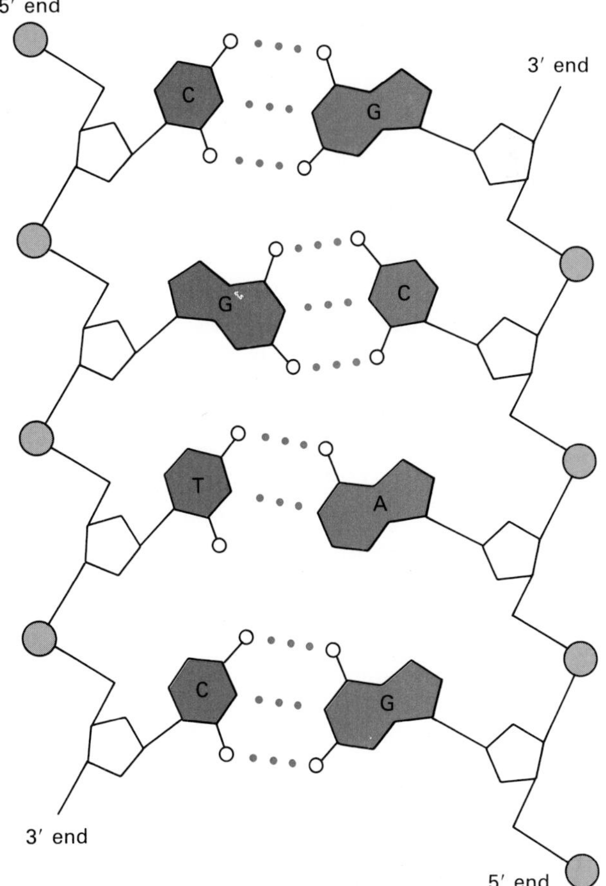

Figure 3-47 A two-dimensional representation of double-stranded DNA.

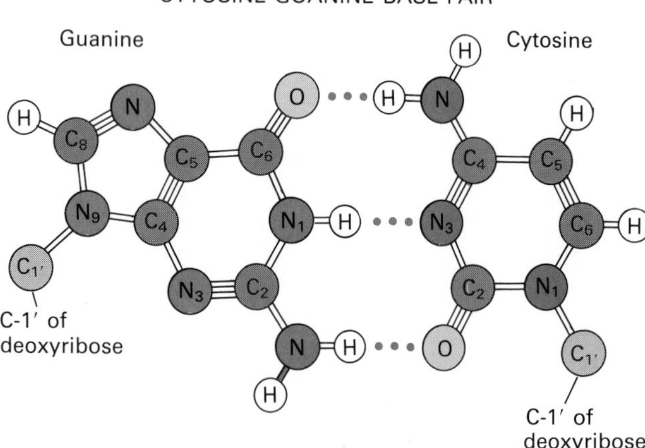

Figure 3-48 Ball-and-stick models of A-T and G-C base pairs.

the stack contribute significantly to the overall stability of the double helix. The two sugar-phosphate chains wind around each other with the base pairs stacked between them (Figure 3-51).

Two polynucleotide strands can, in principle, form either a right-handed or a left-handed helix (Figure 3-52). However, the geometry of the sugar-phosphate backbone is most compatible with a right-handed helix. The x-ray diffraction pattern of DNA has indicated that the stacked bases are regularly spaced, 0.34 nm apart along the axis of the helix. The helix makes a complete turn every 3.4 nm; thus, there are about 10 pairs per turn. The intertwining of the chains results in two helical grooves, of differing widths, along the outside of the molecule (Figure 3-53). Consequently, part of each base is accessible to molecules from outside the helix. The multitude of hydrogen and hydrophobic bonds between the chains gives the structure considerable stability and rigidity. However, the double helix is somewhat flexible because, unlike the α helix in proteins, it has no hydrogen bonds between successive segments to maintain its rigidity.

Recent studies have shown that certain short DNA polymers can adopt, in crystals, an alternative left-handed configuration—the Z-DNA illustrated in Figure 3-53. Whether or not such left-handed helixes occur in nature is a matter of intense current interest; there is growing evidence that certain GC-rich DNA regions within long molecules do adopt a Z-like conformation. Such an alternative structure could be a recognition signal for some important function of DNA.

DNA Is Denatured When the Two Strands Are Made to Separate

When DNA replicates, the two strands unwind and separate from each other. A similar process can be observed experimentally: if a solution of DNA is heated, the thermal energy of molecular motion eventually causes the hydrogen bonds and other forces that stabilize the double helix to break, and the two strands separate, or *denature* (Figure 3-54). This "melting" of DNA can be detected in many ways. One way is to take advantage of the fact that native double-stranded DNA absorbs much less light of 260-nm wavelength (a wavelength of ultraviolet) than does the equivalent amount of single-stranded DNA (Figure 3-55). Thus, as DNA denatures, its absorption of ultraviolet light increases. Only a small increase in temperature is necessary to denature DNA. This is because the double helix is held together by multiple weak cooperative interactions; a small increase in thermal energy destabilizes the entire structure.

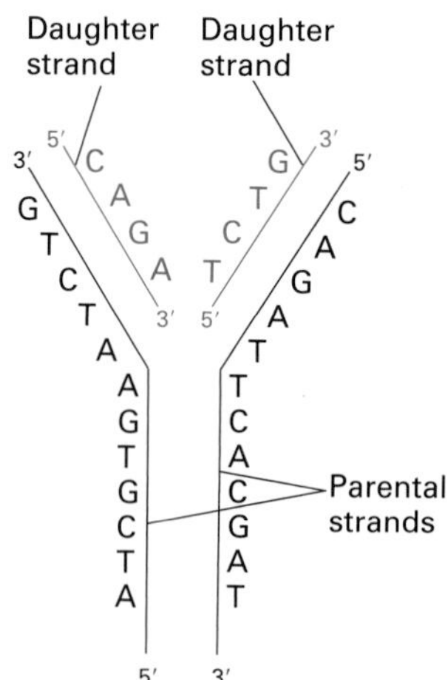

Figure 3-49 A schematic diagram of DNA replication. Because the bases in the two strands of the parental double helix complement one another, the strands carry the same information. When the strands separate in replication, bases complementary to those in the parental strands are added to form the new daughter double helixes; this ensures that the daughter DNA molecules are the same as the parents.

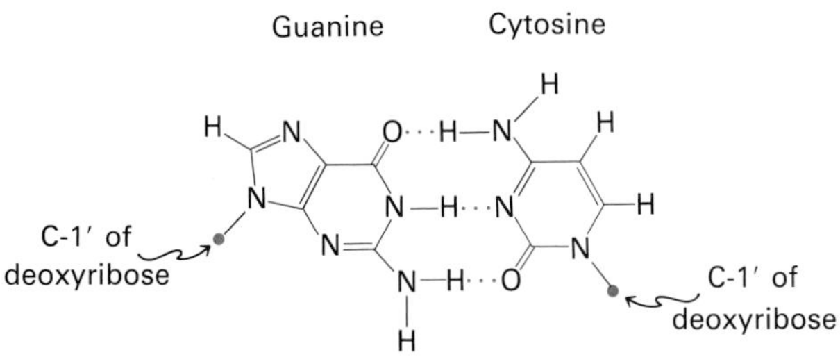

Guanine Cytosine

C-1′ of deoxyribose

C-1′ of deoxyribose

STANDARD G-C BASE PAIR

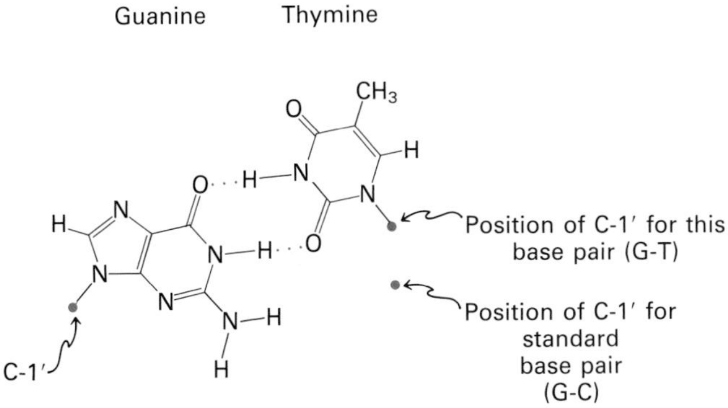

Guanine Thymine

Position of C-1′ for this base pair (G-T)

Position of C-1′ for standard base pair (G-C)

C-1′

NONSTANDARD G-T BASE PAIR

Figure 3-50 Nonstandard base pairs would distort the double helix. Guanine and thymine can form two hydrogen bonds, but the geometry of the G-T base pair is such that if it were incorporated into a standard double helix, the 1′ carbon of the deoxyribose would be 0.2 to 0.3 nm away from its standard position, which would introduce a distortion into the double helix. Chemically plausible hydrogen-bonded structures also exist for the paired pyrimidines T-C and T-T, but these would also introduce a distortion into the helix. No plausible hydrogen-bonded structures can be constructed for A-A, A-G, A-C, G-G, and C-C base pairs.

The temperature at which the strands of DNA will separate depends on several factors. Molecules that contain a greater number of G-C pairs require higher temperatures for denaturation (Figure 3-56), because G-C pairs, which have three bonds, are more stable than A-T pairs, which have two. Solutions with a low salt concentration tend to destabilize the double helix and cause it to melt at lower temperatures. DNA is also denatured by exposure to other agents that destabilize hydrogen bonds: alkaline solutions, or concentrated solutions of formamide or urea.

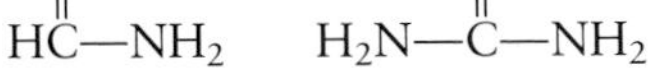

$HC(=O)—NH_2$ (Formamide) $H_2N—C(=O)—NH_2$ (Urea)

Formamide **Urea**

The single-stranded molecules that result from denaturation are stable; even if the temperature is lowered they generally will not undergo renaturation into the native double-stranded molecule. However, under certain well-defined conditions, the two complementary strands can be made to renature. This property is the basis of the powerful technique of nucleic acid hybridization, which is discussed in Chapter 7.

Many DNA Molecules Are Circular

For DNA to denature, the two strands must be free to unwind from each other. Linear DNA molecules, which have free ends, denature quite readily. Some DNA molecules, however, are circular, with the two strands forming

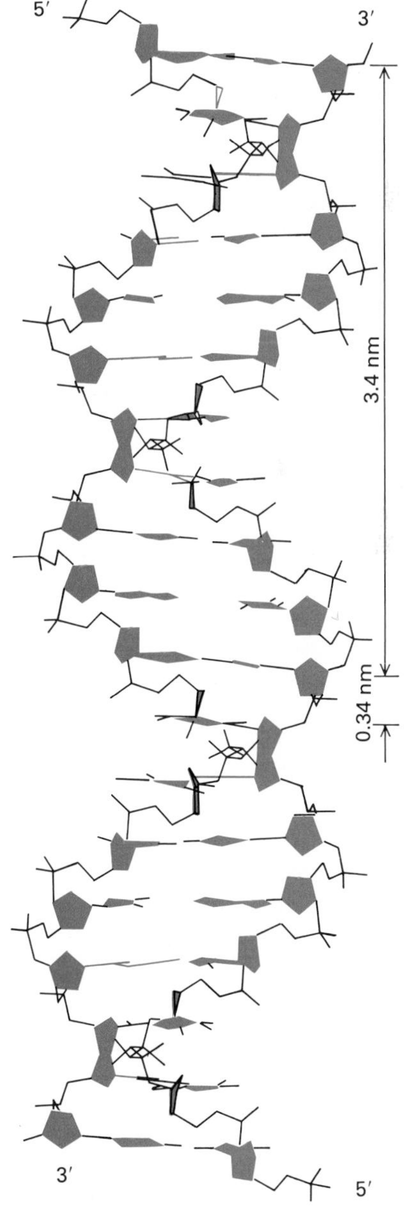

Figure 3-51 A skeletal model of double-helical DNA. The bases and riboses are shown in color. *From L. Stryer, 1981,* Biochemistry, *2d ed., W. H. Freeman and Company, p. 565.*

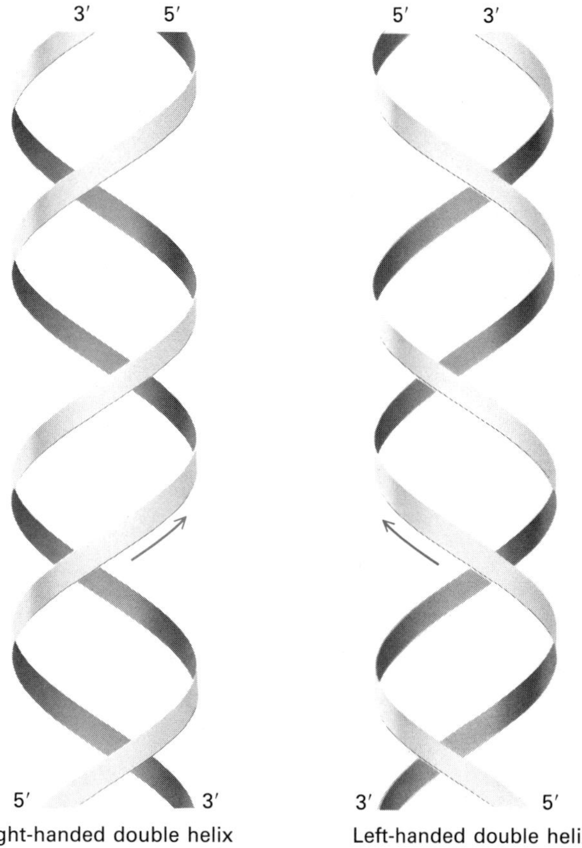

Figure 3-52 Two possible helical forms of DNA. The geometry of the sugar-phosphate backbone of DNA forces the chains into a right-handed double helix. The right-handed and left-handed helixes are defined by convention: As a right-handed helix is viewed from the side, the strand running from the lower left to the upper right (here, 3′ → 5′; see arrow) crosses over the strand running from the lower right to the upper left (here, 5′ → 3′). In a left-handed helix, the strand running from the lower right to the upper left (arrow) crosses over the strand running from the lower left to the upper right. The two helixes are mirror images of each other.

a continuous, closed structure. Bacterial and many viral DNAs are such covalently closed circular molecules. Mitochondria contain a small DNA molecule that encodes several proteins that are localized in mitochondria; this DNA also is a covalently closed circle.

When circular DNA molecules are subjected to elevated temperatures or extremes of pH, the hydrogen bonds and other bonds that stabilize the double helix are destroyed. However, the two strands cannot unwind from each other. The structure that is formed is a knotted, tangled mass of single-stranded DNA in which the two strands are interlocked. When the temperature of the solution is returned to normal, the molecule will spontaneously renature to form the original base-paired double-stranded structure. Why is renaturation of denatured closed circular DNA much faster than renaturation of denatured linear DNA? In the former case, because the two single strands are knotted around each other, they can slide relative to one another until the complementary nucleotide sequences are in perfect alignment. In the latter case, the two strands are separated in solution. The rate-limiting step in renaturation appears to be a collision of the complementary strands in such a way that small parts of the complementary nucleotide sequences are paired together. Once such a nucleating event takes place, however, the rest of the sequences can "snap" or "zip" into alignment.

If the native circular DNA is *nicked*—that is, if one of the strands is cut—the two strands will be able to unwind when the molecule is denatured. Denaturation of a nicked circular DNA results in the separation of the two strands; one will remain circular and the other will become linear (Figure 3-57). The nicking of circular DNA occurs natu-

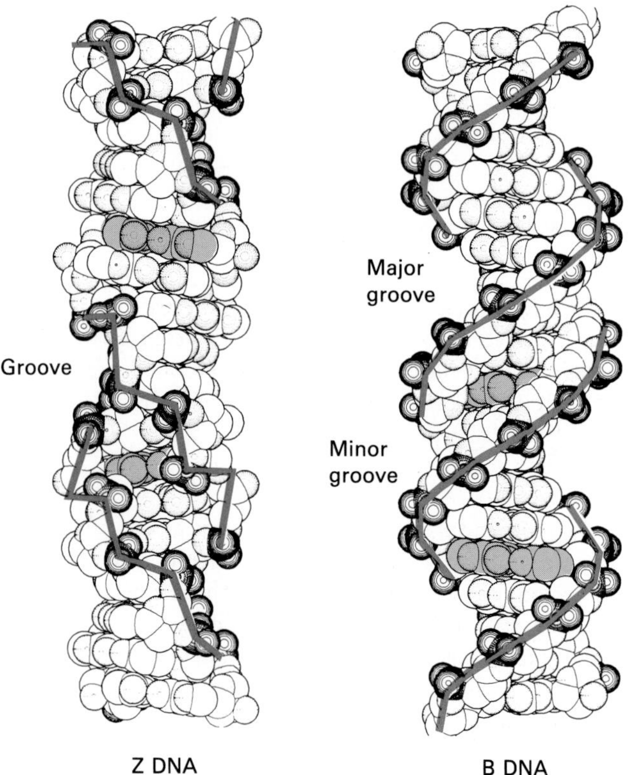

Figure 3-53 Space-filling models of right-handed Watson-Crick B-DNA and of the rarer left-handed Z form. The thick colored lines connect the phosphate residues along the chain. In both forms of DNA, the same base pairs are used; the base pairs are nearly perpendicular to the axis of the helix (two base pairs in each helix are shaded in color). B-DNA has two grooves—one major and one minor. Note the irregularity of the Z-DNA backbone. The single groove of Z-DNA is quite deep: it extends all the way to the axis of the helix. *Courtesy of A. Rich.*

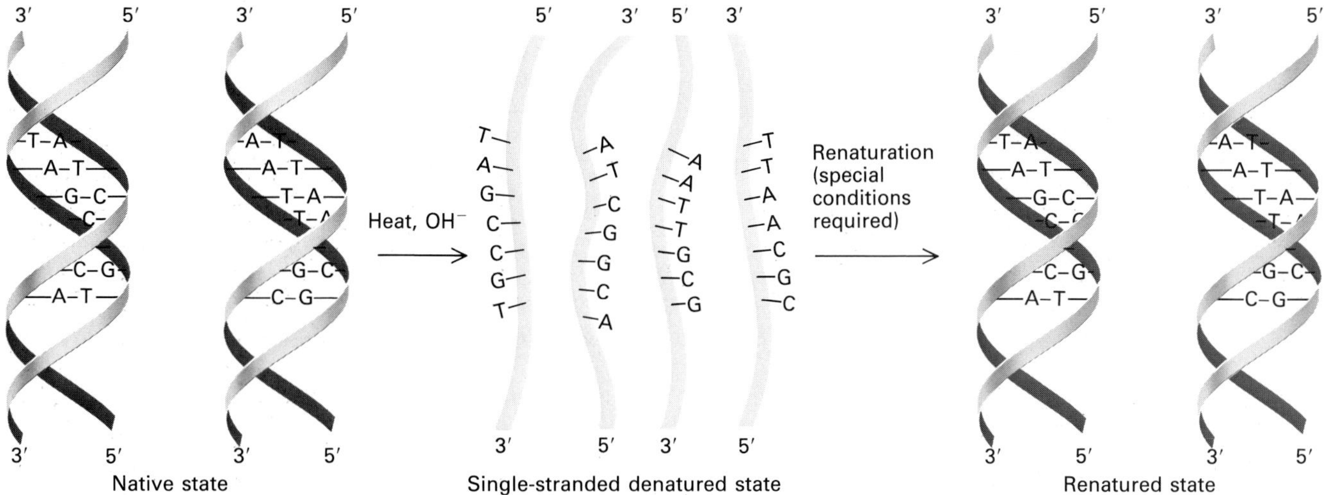

Figure 3-54 The denaturation and renaturation of two double-stranded DNA molecules.

rally during the process of DNA replication. Nicking can also be performed experimentally, by adding to the closed circular DNA a minute amount of deoxyribonuclease (a DNA-degrading enzyme) so that only a single phosphodiester bond is cleaved.

Many Covalently Closed Circular DNA Molecules Are Supercoiled

Covalently closed circular DNA molecules that are carefully isolated in their native forms from mitochondria, viruses, and bacteria are often supercoiled. The configuration of these natural supercoils is right-handed, or, as it is often termed, negative (Figure 3-58a). (Several examples of such supercoiled DNA molecules are depicted in Figure 13-16.)

To understand the origin of these supercoils, we need to recall that the double-helical conformation of DNA in solution, as depicted in Figure 3-51, is one of minimum free energy. If the molecule is bent or twisted, its energy is increased. In particular, if the two strands of the linear molecule are unwound slightly, the molecule spontaneously rewinds to achieve the most stable conformation; that is, one in which all of the bases are in a right-handed double-stranded helical arrangement. Suppose, however, that a molecule of linear DNA were unwound slightly and that the free ends were then linked together to form a covalently closed circle. The unstacked bases would spontaneously reunite in a standard double-helical conformation, but, because the strands would no longer have free ends, they could not easily wind completely around each other to allow base pairing along their entire lengths. The only possible way for the molecule to have the normal double-helical conformation would be for it to twist as a whole. This is how the right-handed (or negative) supercoil is formed. There is one supercoiled twist for each 360° of unwinding (each helical turn) of the original linear DNA (Figure 3-58b).

If a linear DNA helix were instead wound tighter than normal before the free ends were linked together, the resulting closed circular DNA helix would then tend to unwind. This would also generate a supercoil, but one with a left-handed (or positive) twist (Figure 3-58b).

Possibly because of their interaction with proteins such as histones, and possibly because of the ionic composi-

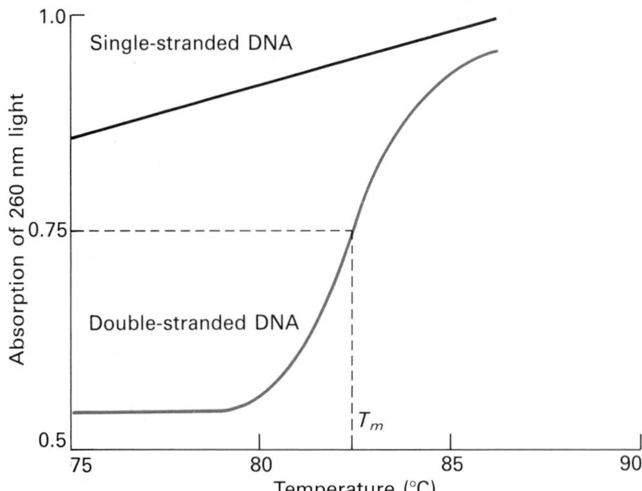

Figure 3-55 A comparison of the amount of absorption of ultraviolet light of 260 nm by single-stranded and double-stranded DNA. As regions of double-stranded DNA unpair, the absorption of light by those regions increases; thus, the increase in absorption is proportional to the extent of denaturation. The temperature at which half of the bases of double-stranded DNA are unpaired is denoted by T_m. Light absorption by single-stranded DNA changes little as the temperature is increased.

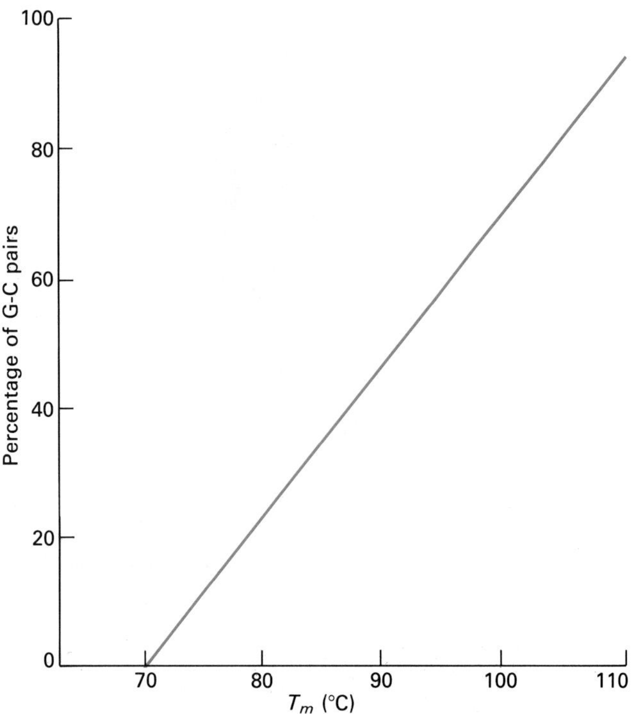

Figure 3-56 A plot of the G-C content of 38 different DNA samples against T_m. An increase in the percentage of G-C base pairs is correlated with increased resistance to thermal denaturation of the double helix. For an unknown DNA sample, the T_m can be used to determine the percentage of G + C.

Figure 3-58 ***(page 89, opposite)*** (a) The structures of three configurations of covalently closed circular double-stranded DNA: a right-handed (or negative) superhelix, a normal circular molecule with no supercoils, and a left-handed (or positive) superhelix. (b) The generation of superhelical DNAs. (*Upper panel*) A right-handed supercoil is formed if the linear double-stranded molecule is slightly unwound before the ends are linked together. When the base pairs are allowed to reunite, the supercoil is formed to compensate for the increase in tension. One supercoil results from the unwinding of one helical turn of 10 base pairs. (*Lower panel*) The converse structure is formed if the linear DNA is wound too tightly before the ends are linked together. When the bases are allowed to unwind so that the normal degree of intrastrand twisting is restored, a left-handed supercoil is formed to compensate for the decrease in tension.

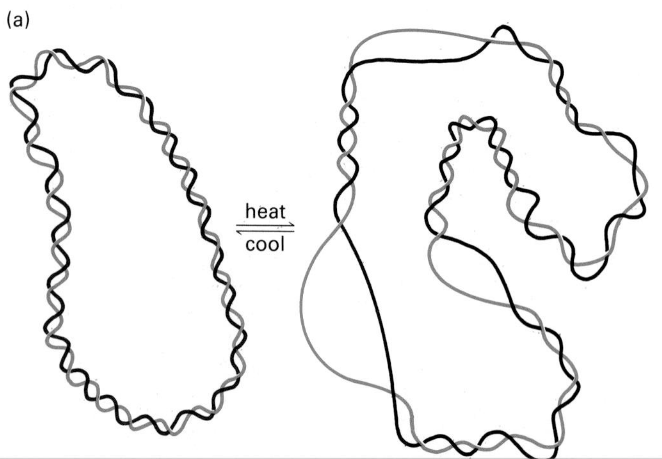

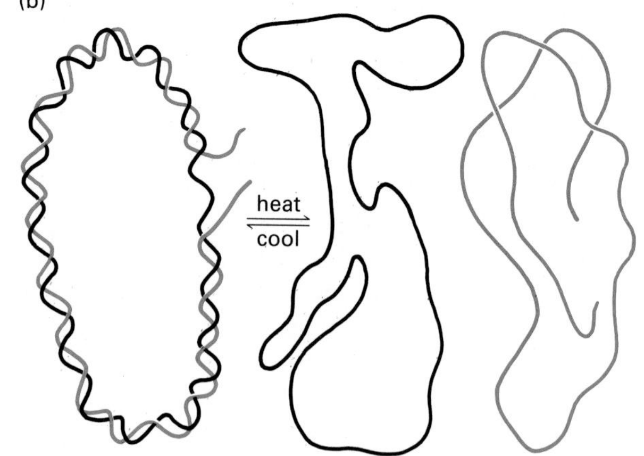

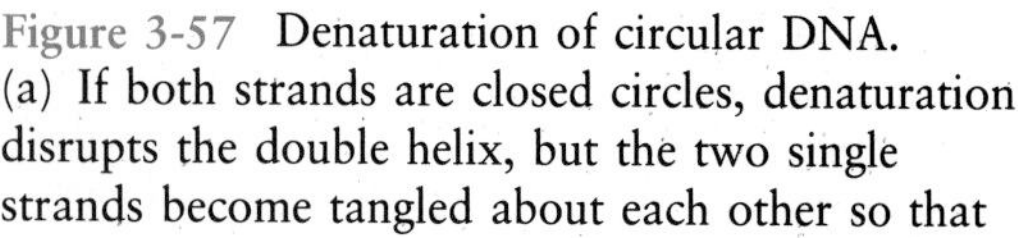

Figure 3-57 Denaturation of circular DNA. (a) If both strands are closed circles, denaturation disrupts the double helix, but the two single strands become tangled about each other so that they cannot separate. (b) However, if one or both strands are nicked, the two strands will separate upon thermal denaturation.

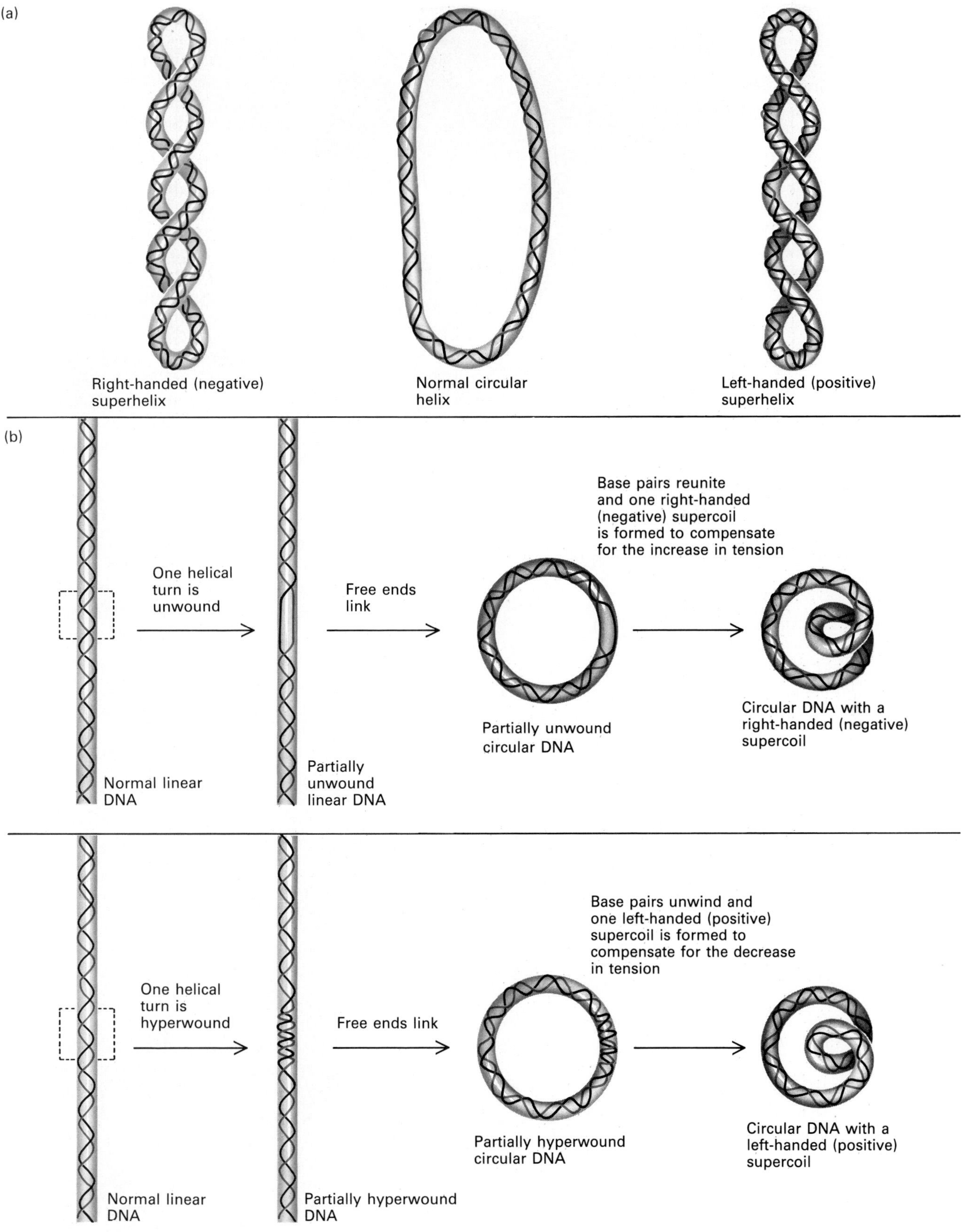
(a)
Right-handed (negative) superhelix
Normal circular helix
Left-handed (positive) superhelix
(b)
One helical turn is unwound
Free ends link
Base pairs reunite and one right-handed (negative) supercoil is formed to compensate for the increase in tension
Normal linear DNA
Partially unwound linear DNA
Partially unwound circular DNA
Circular DNA with a right-handed (negative) supercoil
One helical turn is hyperwound
Free ends link
Base pairs unwind and one left-handed (positive) supercoil is formed to compensate for the decrease in tension
Normal linear DNA
Partially hyperwound DNA
Partially hyperwound circular DNA
Circular DNA with a left-handed (positive) supercoil

tion, DNA helixes within cells appear to be slightly *unwound* with respect to the standard right-handed double-helical conformation of DNA in solution. When such closed circular DNA is removed from the cells and stripped of its protein, it forms right-handed supercoils of the sort depicted in Figure 3-58a. In typical circular DNAs isolated from viruses, there is one supercoiled twist for every 200 bases.

If one strand of a supercoiled DNA is nicked, the two strands will be free to wind (or unwind) around each other. This causes the supercoil to "relax," and thus generates a circular DNA with no supercoils.

RNA Differs from DNA in Its Chemical Structure and in Its Three-Dimensional Folding

Ribonucleic acid (RNA) is as central to the growth and function of cells as is DNA. The molecules in this class vary in structure only slightly from DNA molecules. The function of RNA in a cell is to translate the genetic information stored in DNA into the amino acid sequences of proteins. In some viruses, such as tobacco mosaic virus, the genetic material *is* RNA. In this case, RNA functions in exactly the same way as DNA; that is, it stores genetic information.

The chemical structure of RNA differs from the structure of DNA in that each nucleotide in RNA has a ribose component instead of a deoxyribose; also, RNA has uracil rather than thymine as one of its pyrimidines. Ribose differs from deoxyribose by the presence of a hydroxyl group at the 2′ carbon (see Figure 3-43), and uracil differs from thymine by lacking one methyl group (see Figure 3-44). A major structural difference between DNA and RNA is topological: the RNA in cells exists predominantly as a single strand, whereas DNA exists mainly as a double-stranded helix. However, either nucleic acid can adopt the configuration of the other. (For example, the genomes of some viruses are double-stranded RNA; of others, single-stranded DNA.)

A single chain of RNA can fold back on itself so that the bases form pairs like those in DNA. Thus, an RNA molecule can have a very precise structure with many double-helical regions. (The base pairing in transfer RNA is shown in Figure 4-9, and that in ribosomal RNA in Figure 4-19.)

Although double-stranded RNA forms a right-handed helix similar to that of DNA, the detailed molecular structure is different from the B form of DNA depicted in Figure 3-53. The base pairs, for example, are tilted so that they are not perpendicular to the axis of the helix. Similarly, a DNA-RNA hybrid—a double-stranded structure composed of one DNA and one RNA strand—adopts yet another slightly different right-handed double-helical conformation.

Lipids and Biomembranes

Proteins are the elements that catalyze all of the chemical reactions in a cell, but they would be unable to coordinate their activities were it not for the *biomembranes* that separate a cell from its surroundings, that provide anchoring points for some proteins, and that define the boundaries of the intracellular compartments that are characteristic of eukaryotic cells. The major structural elements of biomembranes are *lipids,* organic molecules that are insoluble or only slightly soluble in water because a large part of each of these molecules is hydrophobic. This hydrophobic part typically contains only hydrogen and carbon atoms; if it were a separate molecule, it would simply be a hydrocarbon. A pure hydrocarbon, such as heptane, is so hydrophobic that it is almost completely insoluble in water (Chapter 2). In an aqueous environment, hydrocarbons segregate into a separate nonaqueous phase, as is apparent to anyone who has tried to mix oil and water.

Fatty Acids Are the Principal Components of Membranes and Lipids

A *fatty acid* molecule contains a long hydrocarbon chain attached to a carboxyl group, which is acidic in character. There are many different fatty acids; they differ in length, in the extent of double bonding, and in the positioning of the double bonds. Most of the fatty acids in cells have 16 or 18 carbon atoms, and from zero to three double bonds are typical (Table 3-1).

Fatty acids with no double bonds are termed *saturated;* those with at least one double bond are *unsaturated.* A double bond in an unsaturated fatty acid can have two possible configurations, cis or trans:

$$\begin{array}{ccc} H_2C & & CH_2 \\ & C{=}C & \\ H & & H \end{array} \qquad \begin{array}{ccc} H_2C & & H \\ & C{=}C & \\ H & & CH_2 \end{array}$$

Cis   Trans

Most unsaturated fatty acids have the cis orientation; the cis double bond introduces a bend in the hydrocarbon side chain (Figure 3-59).

Typical storage forms of fatty acids in cells are the *triacylglycerols,* which are strongly hydrophobic molecules. A triacylglycerol consists of three fatty acid molecules and one molecule of *glycerol,*

$$HOCH_2—CH(OH)—CH_2OH$$

As a free molecule, glycerol has three hydroxyl groups, one bonded to each carbon. In the formation of a triacylglycerol, fatty acids form esters with the three hydroxyl groups of glycerol; three molecules of water are eliminated to form the bonds (Figure 3-60). The fatty acid part

Table 3-1 Structures of some typical fatty acids found in cells

Structure	Systematic name	Common name
SATURATED FATTY ACIDS		
$CH_3(CH_2)_{10}COOH$	*n*-Dodecanoic	Lauric
$CH_3(CH_2)_{12}COOH$	*n*-Tetradecanoic	Myristic
$CH_3(CH_2)_{14}COOH$	*n*-Hexadecanoic	Palmitic
$CH_3(CH_2)_{16}COOH$	*n*-Octadecanoic	Stearic
$CH_3(CH_2)_{18}COOH$	*n*-Eicosanoic	Arachidic
$CH_3(CH_2)_{22}COOH$	*n*-Tetracosanoic	Lignoceric
UNSATURATED FATTY ACIDS		
$CH_3(CH_2)_5CH{=}CH(CH_2)_7COOH$		Palmitoleic
$CH_3(CH_2)_7CH{=}CH(CH_2)_7COOH$		Oleic
$CH_3(CH_2)_4CH{=}CHCH_2CH{=}CH(CH_2)_7COOH$		Linoleic
$CH_3CH_2CH{=}CHCH_2CH{=}CHCH_2CH{=}CH(CH_2)_7COOH$		Linolenic
$CH_3(CH_2)_4(CH{=}CHCH_2)_3CH{=}CH(CH_2)_3COOH$		Arachidonic

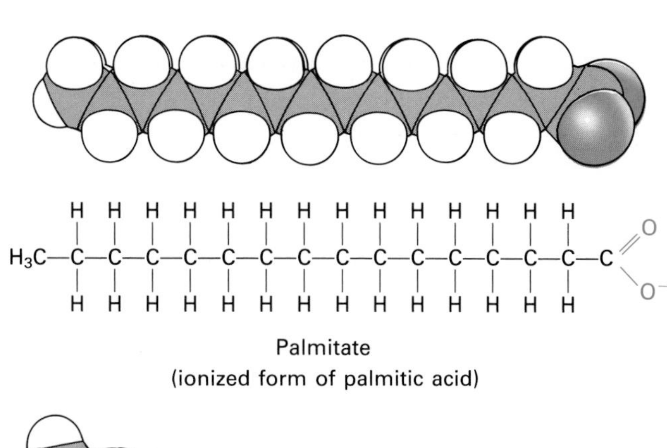

Palmitate
(ionized form of palmitic acid)

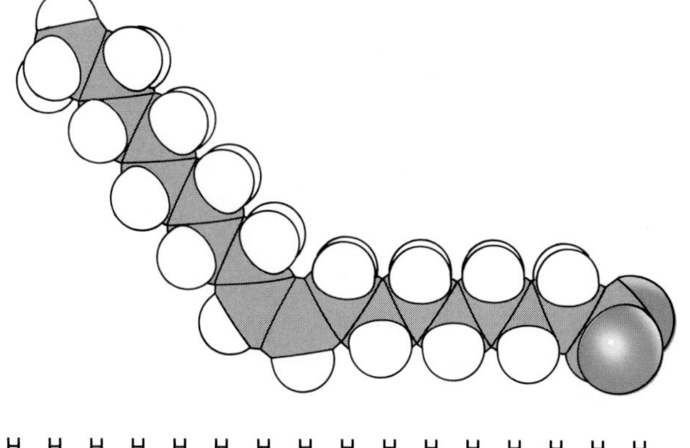

Oleate
(ionized form of oleic acid)

Figure 3-59 Space-filling models and chemical structures of two fatty acids: saturated palmitate and unsaturated oleate. The space-filling models show that the saturated fatty acid forms a linear molecule, whereas the cis double bond in oleate creates a kink in the hydrocarbon chain.

of a triacylglycerol or other ester is termed an *acyl* group.

Triacylglycerols, like pure hydrocarbons, are insoluble in water and in salt solutions; thus, they form lipid droplets in cells. Fat cells, which are also called *adipose cells,* store these triacylglycerols as a source of energy for the body; such cells contain one large lipid droplet (Figure 3-61). When triggered by hormones such as adrenalin, adipose cells hydrolyze the triacylglycerols into free fatty acids. The fatty acids are released into the blood and are used as a source of energy by other cells.

A purely hydrophobic compound such as a triacylglycerol cannot participate in the formation of membranes because it always tends to adopt a configuration that ensures minimal contact with aqueous surroundings, and that thus causes the molecules to cluster in a separate nonaqueous phase. Membrane lipids do contain long-chained fatty acyl groups, but these are linked, usually by an ester bond, to small highly hydrophilic groups. Consequently, membrane lipids do not clump together in droplets but orient themselves in membranes so as to expose their hydrophilic ends to the aqueous environment. Such molecules, in which one end (the head) interacts with

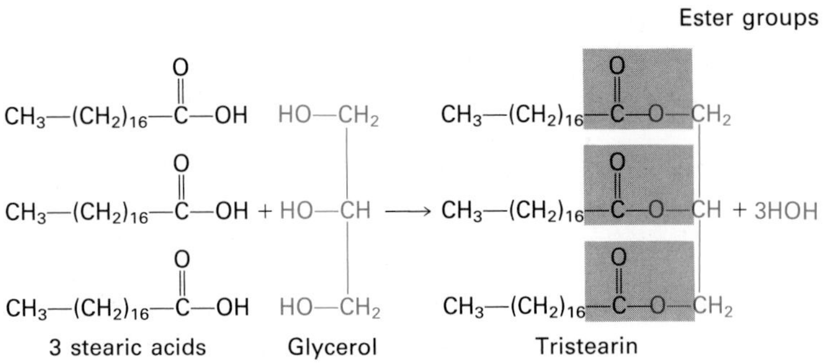

Ester groups

$$3\ CH_3{-}(CH_2)_{16}{-}\overset{O}{\overset{\|}{C}}{-}OH + \begin{matrix} HO{-}CH_2 \\ | \\ HO{-}CH \\ | \\ HO{-}CH_2 \end{matrix} \longrightarrow \begin{matrix} CH_3{-}(CH_2)_{16}{-}\overset{O}{\overset{\|}{C}}{-}O{-}CH_2 \\ | \\ CH_3{-}(CH_2)_{16}{-}\overset{O}{\overset{\|}{C}}{-}O{-}CH \\ | \\ CH_3{-}(CH_2)_{16}{-}\overset{O}{\overset{\|}{C}}{-}O{-}CH_2 \end{matrix} + 3HOH$$

3 stearic acids Glycerol Tristearin

Figure 3-60 The formation of tristearin, a triacylglycerol (a storage form of fatty acids). The molecule is the product of the esterification of each of the three hydroxyl groups of glycerol with stearic acid.

water and the other end (the tail) avoids it, are called amphipathic (from the Greek for "tolerant of both").

Phospholipids Are Amphipathic Molecules That Are Key Components of Biomembranes

The tendency of amphipathic molecules in water to form organized structures spontaneously is the key to the structure of cell membranes, which typically contain a large proportion of amphipathic lipids. The most abundant are the *phospholipids. Phosphoglycerides,* a principal class of phospholipids, contain two fatty acyl side chains that are esterified to two of the three hydroxyl groups of a glycerol molecule. The third hydroxyl group on glycerol is esterified with phosphate. The simplest phospholipid, *phosphatidic acid,* contains only these components (Figure 3-62). In most phospholipids, however, the phosphoric acid is also esterified to a hydroxyl group on a hydrophilic compound such as ethanolamine,

$$HO{-}CH_2{-}CH_2{-}NH_3^+$$

or serine, choline, or glycerol, so that the head group is somewhat larger (Figure 3-63). The structures of several phosphoglycerides commonly found in biological membranes are shown in Figure 3-64. The two fatty acyl side chains may both be saturated or unsaturated, or one may be saturated and the other unsaturated. The negative charge on the phosphate and the charged groups and hydroxyl groups on the alcohol esterified to it will interact strongly with water.

Sphingomyelin is the other principal membrane phospholipid. It, too, is amphipathic. Sphingomyelin contains *sphingosine,* an amino alcohol with a long unsaturated hydrocarbon chain. A fatty acyl side chain is linked to the NH_2 group of sphingosine by an amide bond,

$${-}\overset{O}{\overset{\|}{C}}{-}\overset{H}{\overset{|}{N}}{-}$$

to form ceramide (Figure 3-65). The hydroxyl group of sphingosine is esterified to phosphocholine; thus, the hydrophilic head of sphingomyelin is similar to that of phosphatidylcholine (see Figure 3-64).

Certain Steroids Are Components of Biomembranes

Cholesterol and its derivatives constitute another important class of membrane lipids, the *steroids.* The basic structure of steroids is the four-ring hydrocarbon shown in Figure 3-66a. Cholesterol (Figure 3-66b) is the major steroidal constituent of animal tissues, but other steroids are more important in the membranes of plant cells. Although cholesterol is almost entirely hydrocarbon in composition, it is amphipathic because it contains a hydroxyl group that interacts with water.

Phospholipids Spontaneously Form Micelles or Bilayers in Aqueous Solutions

Phospholipids can exist in three different forms in aqueous solutions: in micelles, in bilayer sheets, and in liposomes (Figure 3-67).

When a solution of phospholipids is violently dispersed in water, the phospholipids typically cluster together in

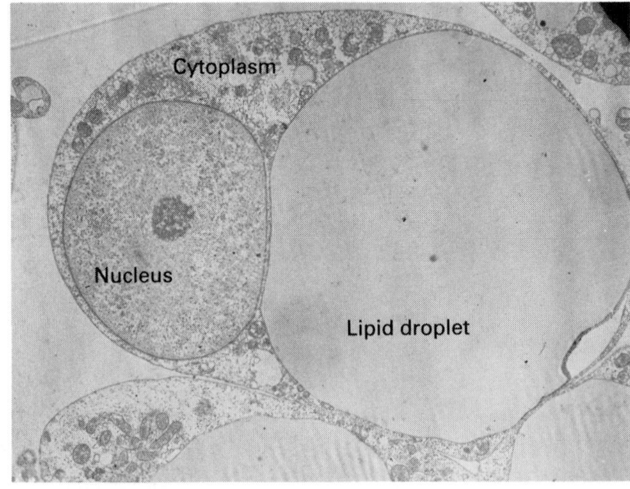

Figure 3-61 An electron micrograph of an adipose (fat) cell showing a single large lipid storage droplet, a small nucleus, and a small cytoplasm. This cell resulted from the differentiation of a line of preadipocyte cells in tissue culture. [See H. Green and M. Meuth, 1974, *Cell* 3:127–133.] *Courtesy of H. Green.*

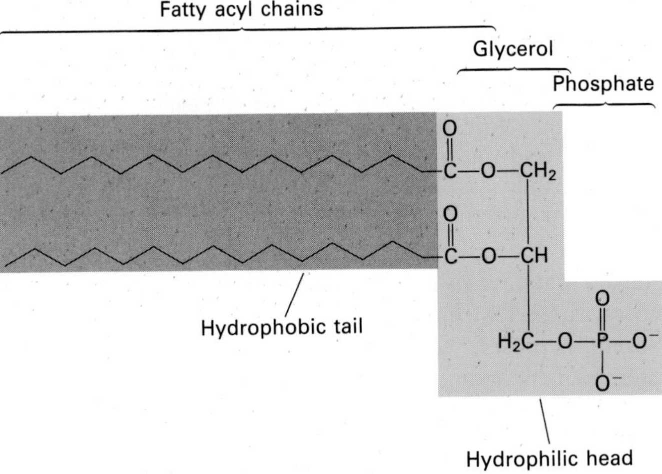

Figure 3-62 The structure of phosphatidic acid, the simplest phospholipid.

micelles. These are spherical structures (about 20 nm or less in diameter) in which the hydrocarbon side chains are sequestered inside and the polar head groups are on the surface, in contact with the water.

Phospholipids also spontaneously form symmetrical sheetlike structures called *phospholipid bilayers.* Bilayers are two molecules thick. The hydrocarbon side chains of each layer minimize contact with water by aligning themselves together in the center of the bilayer. Van der Waals attractions between the hydrocarbon side chains of adjacent phospholipids stabilize the close packing. The interaction of the polar head groups with water is stabilized by both electrostatic and hydrogen bonds. A phospholipid bilayer can be of almost unlimited size—from micrometers to millimeters in length or width.

Liposomes are spherical bilayer structures that contain aqueous interiors. The type of structure formed by a pure phospholipid or a mixture of phospholipids depends on the length and degree of saturation of the fatty acyl chains, on the temperature, on the ionic composition of the aqueous medium, and on the mode of dispersal of the phospholipids in the solution.

For many years it was known that biomembranes are formed from phospholipids, but it was not established whether the membranes are formed from micelles held together by proteins or whether the bilayer sheet is the underlying structure. The bilayer sheet was the more likely alternative, but this notion was widely accepted only after many years of argument and experimentation. Today, all models of membrane structures are based on phospholipid bilayer sheets, but there is still a nagging suspicion that micelles might sometimes play a crucial role.

Proteins, another essential constituent of all biomembranes, allow the various cellular membranes to function in specific ways. In many cases, all or part of a protein molecule interacts directly with the hydrophobic core of a phospholipid membrane bilayer. Such membrane-embedded proteins must adopt conformations quite different from those of water-soluble proteins. This subject is taken up in Chapter 14, which contains a detailed treatment of the structure and function of phospholipids and proteins in biological membranes.

Carbohydrates

As its name implies, a carbohydrate is constructed of carbon (*carbo-*) and hydrogen and oxygen (*-hydrate,* or water). The simplest carbohydrates are the *monosaccharides,* which are often called simple sugars. Monosaccharides have the structure $(CH_2O)_n$, where $n = 3, 4, 5, 6$, or 7. All monosaccharides contain hydroxyl groups and either an aldehyde or a keto group:

Aldehyde   Keto

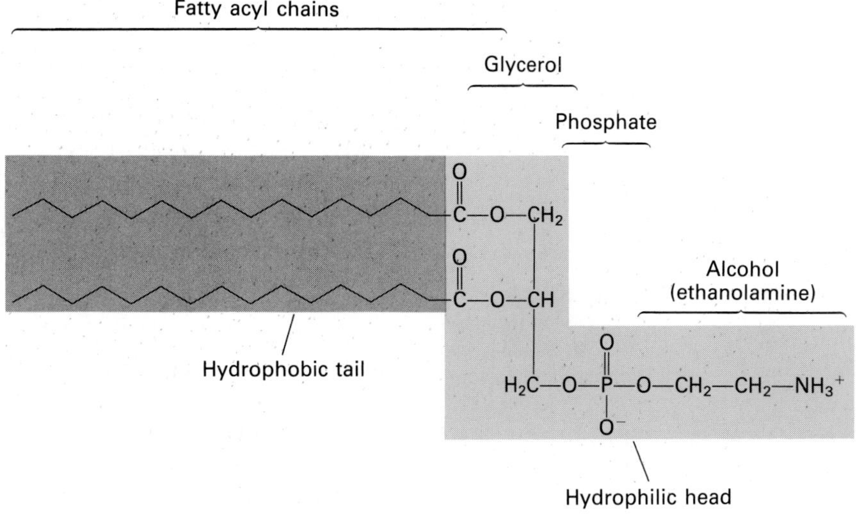

Figure 3-63 The structure of phosphatidylethanolamine, a typical phospholipid.

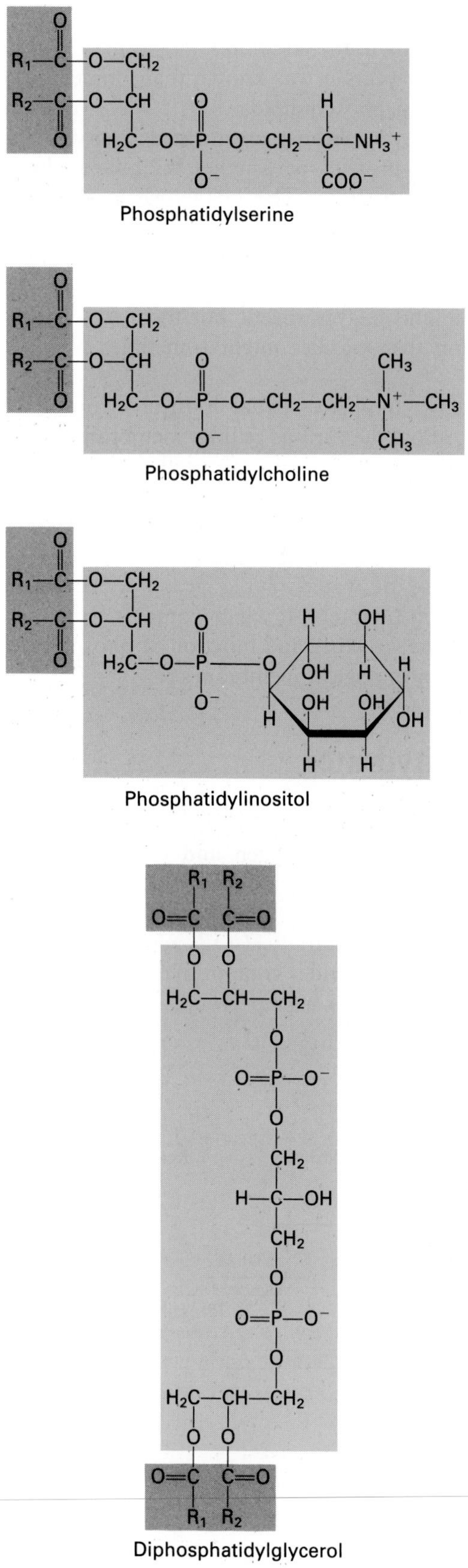

Phosphatidylserine

Phosphatidylcholine

Phosphatidylinositol

Diphosphatidylglycerol

Figure 3-64 The structures of some common phospholipids found in cell membranes. The fatty acyl side chains (represented by R_1 and R_2) can be saturated, or they can contain one or more double bonds. Note that diphosphatidylglycerol contains four fatty acids; it is composed of two molecules of phosphatidic acid linked to the 1 and 3 carbons of a central glycerol.

Sugars that have five carbon atoms ($n = 5$) are called pentoses; those that have six, hexoses; and those that have seven, heptoses. Glucose, $C_6H_{12}O_6$, is an important source of energy in all cells of higher organisms. As mentioned earlier, two pentose sugars, ribose and deoxyribose, are essential constituents of nucleic acids. Certain carbohydrates (often termed complex carbohydrates) consist of polymerized sugar units. Polysaccharides play important roles as energy-storage molecules within cells, and as structural components when they are located on the outer surfaces of cells or between cells. Oligosaccharide chains are often found attached to proteins or lipids, in which cases the modified molecules are called *glycoproteins* or *glycolipids*. These are important constituents of cell surface membranes.

Many Important Sugars Are Hexoses

Because many of the biologically important sugars are hexoses, which are structurally related to D-glucose,* it is important to know some of the structural features of D-glucose (Figure 3-68).

To begin with, glucose can exist in three different forms: it can have a linear structure or one of two different ring structures. The aldehyde group on the 1 carbon can react with the hydroxyl group on the 5 carbon to form a hemiacetal group:

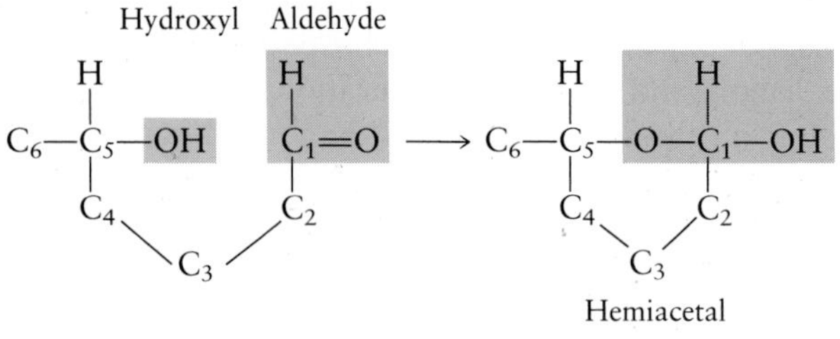

Hydroxyl Aldehyde

$$C_6—C_5(H)—OH \quad H—C_1{=}O \longrightarrow C_6—C_5(H)—O—C_1(H)—OH$$

Hemiacetal

This forms a molecule with a six-membered ring, *glucopyranose*. Similarly, condensation of the hydroxyl group on the 4 carbon with the aldehyde results in the formation of *glucofuranose*, a hemiacetal that contains a five-membered ring. Although all three forms are known to exist in biological systems, the pyranose form is by far the most abundant.

Four of the carbon atoms (specifically, numbers 2, 3, 4,

*D and L sugars are mirror images of one another. The D or L stands for the direction in which a solution of a sugar rotates the plane of polarized light: *dextro* (right) or *levo* (left). Except for L-fucose, sugars found in biological systems are generally of the D type.

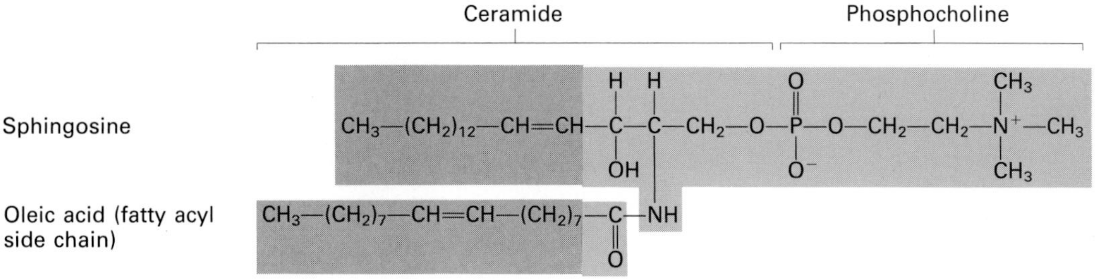

Ceramide   Phosphocholine

Sphingosine: $CH_3—(CH_2)_{12}—CH{=}CH—CH(OH)—CH(NH)—CH_2—O—P(=O)(O^-)—O—CH_2—CH_2—N^+(CH_3)_3$

Oleic acid (fatty acyl side chain): $CH_3—(CH_2)_7—CH{=}CH—(CH_2)_7—C(=O)—NH$

Figure 3-65 The structure of a typical sphingomyelin molecule.

and 5) in the linear form of glucose are *asymmetric*. Each of these carbons is bonded to four different atoms or groups, and the exact spatial orientation of these bonds is of great importance. For example, if the hydrogen atom and the hydroxyl group attached to the 2 carbon were interchanged, a different molecule would result, one that could not be converted to glucose without breaking and making covalent bonds. *Mannose* is the sugar that is identical to glucose except for the orientation of the substituents on the 2 carbon (Figure 3-69). If the pyranose forms of glucose and mannose are shown as planar structures (Haworth projections), the hydroxyl group on the 2 carbon of glucose points downward, whereas that on mannose points upward. This single point of difference results in quite different biological properties for the two molecules. Galactose, another hexose, differs from glucose only in the orientation of the hydroxyl group on the 4 carbon.

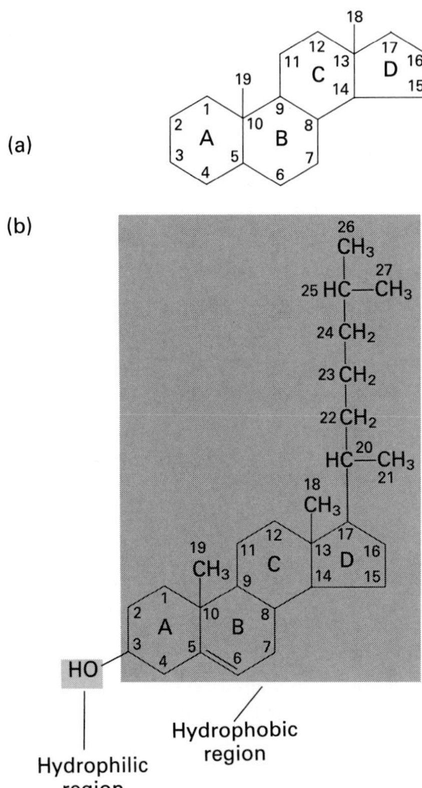

Figure 3-66 (a) The general structure of a steroid. All steroids contain the same four hydrocarbon rings, conventionally labeled A, B, C, and D, with the carbons numbered as shown. (b) The structure of cholesterol.

Another structural feature of hexoses derives from the fact that when the linear form is converted into a ringed pyranose, additional asymmetry is created at the 1 carbon. That is, if the structure of glucopyranose is depicted as a Haworth projection, the hydroxyl group on the 1 carbon can point either downward—in which case the sugar is called α-D-glucopyranose—or upward—which gives a β-D-glucopyranose (Figure 3-70).

However, the Haworth projection is an oversimplification, and the actual conformation of the pyranose ring is not planar. Rather, the molecule adopts a conformation in which each of the ring carbons is at the center of a tetrahedron, just like the carbon in methane (see Figure 2-1). The preferred conformation of pyranose structures is the *chair*. The atoms bonded to the carbon atoms are of two types, *axial* and *equatorial* (Figure 3-71).

Glucose is often present in cells as an independent molecule, whereas mannose, galactose, and the four modified hexoses shown in Figure 3-72 are more commonly found as components of polysaccharides, glycoproteins, or glycolipids. Note that *N*-acetylglucosamine and *N*-acetylgalactosamine contain acetamide groups,

$$—NH—\overset{\overset{\displaystyle O}{\|}}{C}—CH_3$$

in place of the usual hydroxyls. *N*-Acetylneuraminic acid contains not only the acetamide group replacement, but also three extra carbon atoms; because of its carboxyl group, it is negatively charged. The 6 carbon of fucose is a methyl group rather than the CH_2OH group found in most other sugars.

Polymers of Glucose Serve As Storage Reservoirs

Because D-glucose is the principal source of cellular energy, it is important for cells to maintain a reservoir of it.

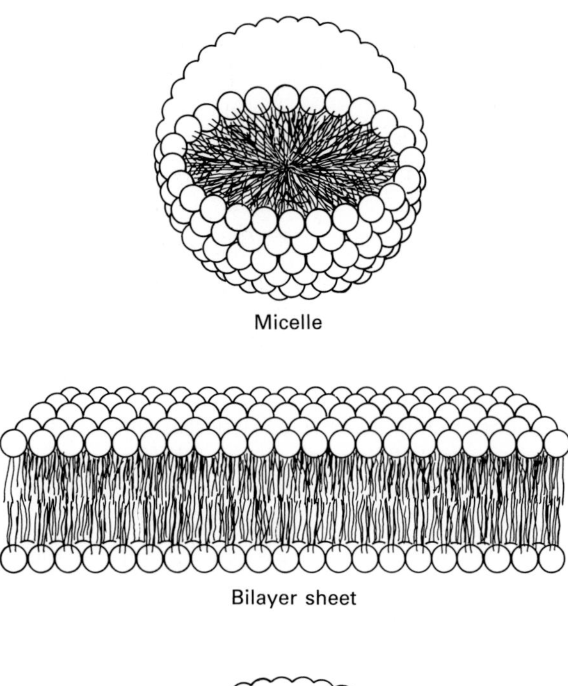

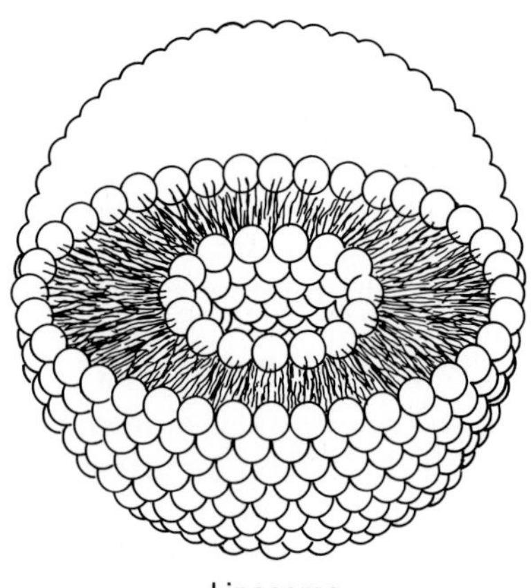

Figure 3-67 Cross-sectional views of the three structures that can be formed by phospholipids in aqueous solutions: spherical micelles with hydrophobic interiors, sheets of phospholipids in a bilayer, and spherical liposomes comprising one phospholipid bilayer. The circles are the hydrophilic heads and the wavy lines are the fatty acyl side chains.

D-Glucose

D-Glucopyranose (common)

D-Glucofuranose (rare)

Figure 3-68 Three alternative configurations of D-glucose. The ring forms are generated from the linear molecule by the reaction of the aldehyde at the 1 carbon with the hydroxyl on the 5 carbon (to form glucopyranose) or with the hydroxyl on the 4 carbon (to form glucofuranose).

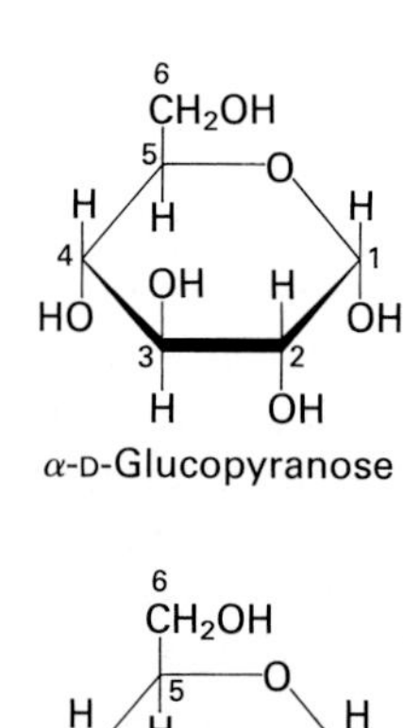

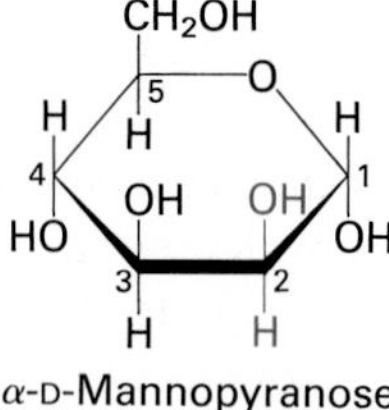

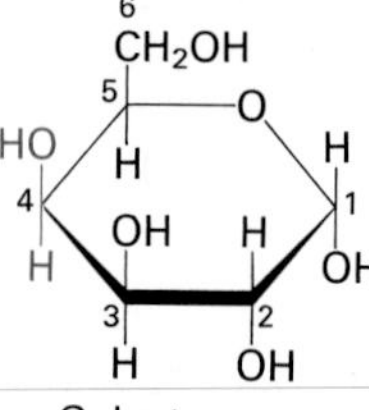

Figure 3-69 Haworth projections of the structures of glucose, mannose, and galactose in their standard pyranose forms. The hydroxyl groups with orientations different from those of glucose are indicated in color.

Some sugars are stored as *disaccharides*, which consist of two monosaccharides linked together by a *glycosidic bond.* In the formation of a glycosidic bond, the 1 carbon atom of one sugar molecule reacts with a hydroxyl group of another; as in the formation of most biopolymers, the linkage is accompanied by the loss of water.

In principle, a large number of different glycosidic bonds can be formed between two sugar residues. Galactose can be bonded to glucose, for example, by any of the following linkages: $\alpha(1 \rightarrow 1)$, $\alpha(1 \rightarrow 2)$, $\alpha(1 \rightarrow 3)$, $\alpha(1 \rightarrow 4)$, $\alpha(1 \rightarrow 6)$, $\beta(1 \rightarrow 1)$, $\beta(1 \rightarrow 2)$, $\beta(1 \rightarrow 3)$, $\beta(1 \rightarrow 4)$, or $\beta(1 \rightarrow 6)$, where α or β specifies the conformation at the 1 carbon in galactose, and the number following the arrow indicates the glucose carbon to which the galactose is bound. The theoretical number of linkages in trisaccharides is much larger, but only a few of them are actually found in nature, owing to the specificities of the enzymes that synthesize oligosaccharides from single sugar residues.

Lactose is the major sugar in milk; sucrose is a sugar found abundantly in most plants. The formation of both disaccharides is depicted in Figure 3-73. The most common storage carbohydrate in animal cells is *glycogen,* a very long, branched polymer of glucose (Figure 3-74). As much as 10 percent of the weight of liver can be glycogen. (The synthesis and utilization of this polymer is described in Chapter 16.) *Cellulose* is a different long-chained polymer of glucose; it is the major constituent of plant cell walls and the major organic chemical on earth. Its important structural and functional features are described in Chapter 5.

Extracellular carbohydrates in mammals do not exist as pure substances, but as complexes with proteins or lipids. They serve to lubricate joints, to fill spaces (e.g., the eye chamber), and to coat the outside of most cells. As water-soluble viscous lubricants, they are important constituents of bone, cartilage, and the umbilical cord. The conjugates usually have many carboxyl and other acid groups on them, and they are thus called *acid mucopolysaccharides.* The most abundant of these is *hyaluronic acid;* other important ones are *chondroitin sulfate* and *keratan sulfate* (Figure 3-75a). These large, negatively charged polymers bind water and cations to form the gel-like ground substance of connective tissue. Like most extracellular polysaccharides, acid mucopolysaccharides consist of repeating disaccharide units.

Many extracellular acid mucopolysaccharides are found as components of complex *proteoglycans*—complexes of proteins and polysaccharides. The proteoglycans from cartilage, for example, contain a very long molecule of hyaluronic acid in their centers. *Link proteins* and *core proteins* noncovalently bind chains of chondroitin sulfate and keratan sulfate to the hyaluronic acid (Figure 3-75b). Other proteins are bound to the extremities of the chondroitin sulfate and keratan sulfate. The entire macromolecular complex can be several micrometers in length—bigger than a bacterial cell.

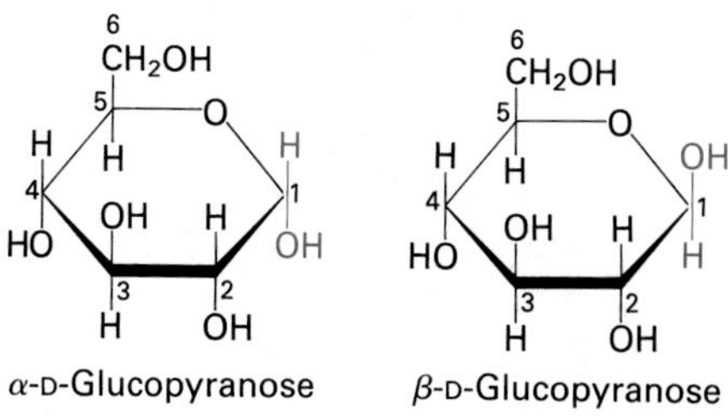

Figure 3-70 Haworth projections of α-D-glucopyranose and β-D-glucopyranose. The substituents that differentiate the two forms are shown in color.

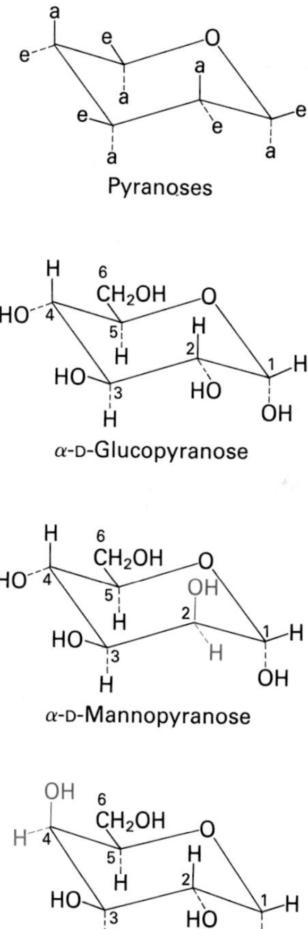

Figure 3-71 Chair conformations of glucose, mannose, and galactose in their pyranose forms. The chair is the most stable conformation of a six-membered ring. In the generalized pyranose ring shown above, a = axial atoms and e = equatorial atoms. The four bonds at each of the ring carbon atoms are tetrahedral. In α-D-glucopyranose, all of the hydroxyl groups except that bonded to the 1 carbon are equatorial. In α-D-mannopyranose, the hydroxyl group bonded to the 2 carbon as well as that bonded to the 1 carbon are axial. In α-D-galactopyranose, the hydroxyl groups bonded to the 1 and 4 carbons are axial.

Glycoproteins Are Composed of Proteins Covalently Bound to Sugars

Many important membrane proteins and serum proteins contain carbohydrate chains; these are termed *glycoproteins.* The carbohydrate chains are often important for the correct folding and hence for the stability of the proteins, as well as for their correct synthesis and positioning within a cell.

Carbohydrates generally play no role in the catalytic

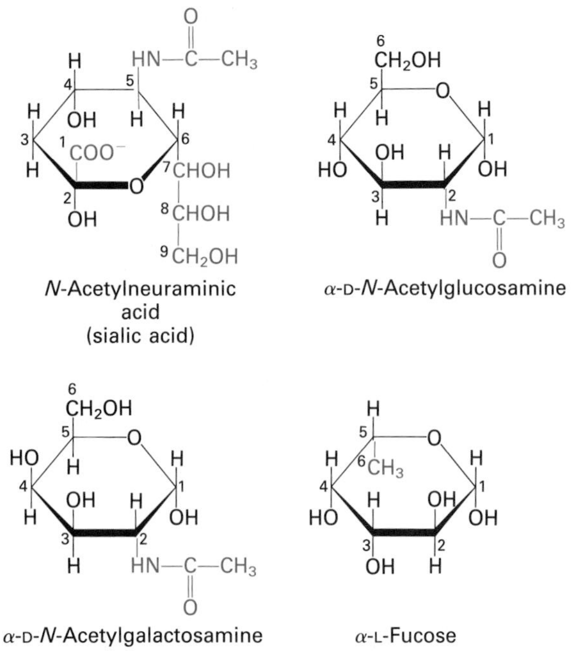

N-Acetylneuraminic acid (sialic acid)

α-D-*N*-Acetylglucosamine

α-D-*N*-Acetylgalactosamine

α-L-Fucose

Figure 3-72 The structures of four sugars commonly found in larger molecules in cells: *N*-acetylneuraminic acid (sialic acid), *N*-acetylglucosamine, *N*-acetylgalactosamine, and fucose. (The color highlights the modifications mentioned in the text.)

functions of proteins; in fact, many synthesized glycoprotein derivatives that lack the carbohydrate chain still function normally. One possible role of carbohydrates in glycoproteins is to increase the solubility of the proteins in water. Often the outermost sugar in a carbohydrate chain attached to a protein is *N*-acetylneuraminic acid, which is negatively charged. This creates a negative charge on the surface of the glycoprotein and, because of charge repulsion, helps to prevent dissolved proteins from clumping together.

Sugar residues in glycoproteins are commonly linked to two different classes of amino acid residues. The sugars are classified as O-linked if they are bonded to the hydroxyl oxygen of serine and of threonine and—in collagen—of hydroxylysine; they are classified as *N*-linked if they are bonded to the amide nitrogen of asparagine. The structures of O-linked and *N*-linked oligosaccharides are very different, and different sugars are usually found in each type. Oligosaccharides that are O-linked are generally shorter and more variable than the *N*-linked structures, and often contain only one to three sugar residues. Figure 3-76 shows the structures of some common O-linked sugars.

Recently, there has been much interest in the *N*-linked oligosaccharides, and particularly in the complex molecules that contain galactose and *N*-acetylneuraminic acid. These chains are quite similar in structure on many membrane and secreted proteins. An example of the oligosaccharide found on serum antibody proteins is shown in Figure 3-77. These structures contain *N*-acetylglucosamine, mannose, fucose, galactose, and *N*-acetylneuraminic acid. The differences among proteins are primarily in the number of branches (two, three, or four), in the number of *N*-acetylneuraminic acid residues (from zero to three), and in the chemical nature of the linkage of *N*-acetylneuraminic acid to galactose. Even oligosaccharides found at the same site in a single protein species are often heterogeneous. (In Chapter 21 we discuss the many steps involved in the biosynthesis of these and related oligosaccharides.)

Glycolipids of Various Structures Are Found in the Cell Surface Membrane

Other abundant constituents of cell membranes are the *glycolipids,* compounds that contain carbohydrate chains

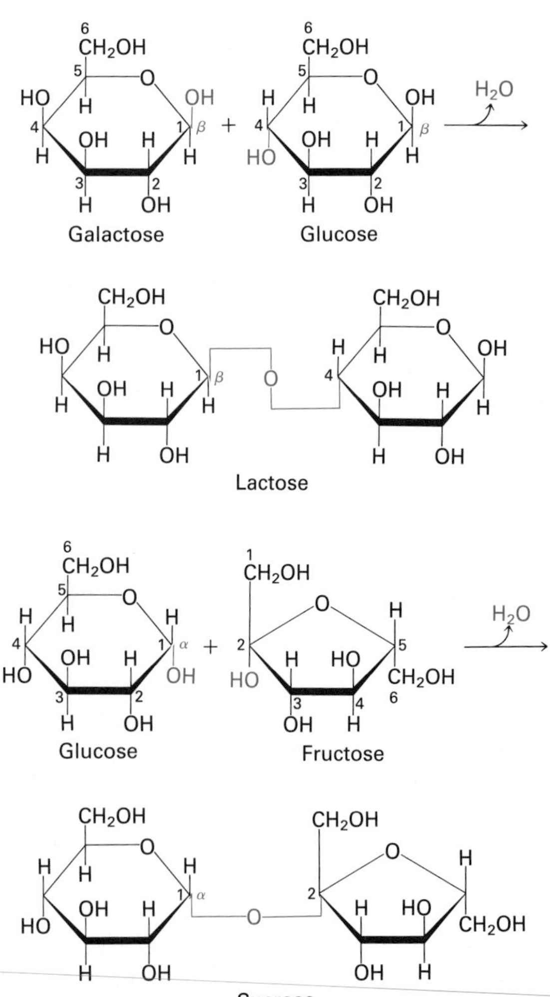

Figure 3-73 The formation of glycosidic linkages to generate the disaccharides lactose and sucrose. The lactose linkage is $\beta(1 \rightarrow 4)$; the sucrose linkage is $\alpha(1 \rightarrow 2)$.

(a)

α(1 → 6) linkage between two glucose chains

α(1 → 4) linkage between two glucose units

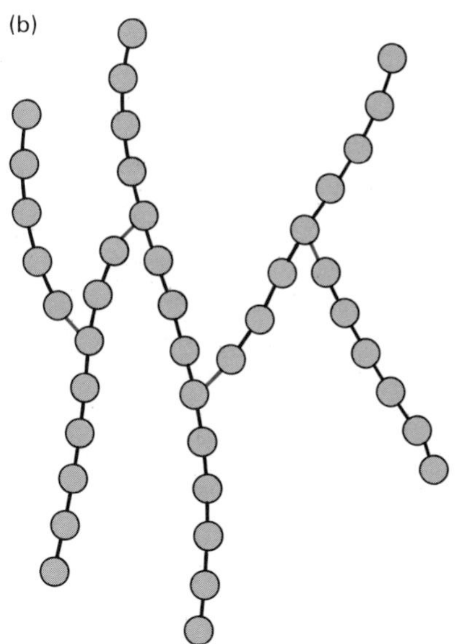

Figure 3-74 The structure of glycogen, a storage polymer of glucose. (a) Haworth projections showing that glycogen consists principally of chains of glucose residues connected by $\alpha(1 \rightarrow 4)$ linkages. The end of one chain is connected to another chain by an $\alpha(1 \rightarrow 6)$ linkage. (b) A schematic drawing of a glycogen molecule with its network of chains. The $\alpha(1 \rightarrow 6)$ linkages are in color; all other linkages are $\alpha(1 \rightarrow 4)$.

covalently linked to lipids. They are similar in structure to sphingomyelin, in that they contain sphingosine and an attached fatty acid (see Figure 3-65). However, glycolipids have a variable number of sugars linked to the hydroxyl group of sphingosine. The structure of the simplest glycolipid—glucosylcerebroside—is depicted in Figure 3-78; it contains a single glucose sugar unit.

Often, one or more of the sugars in glycolipids is *N*-acetylneuraminic acid (see Figure 3-72), which, as we have noted, dissociates to become negatively charged. Glycolipids with *N*-acetylneuraminic acid are called *gangliosides.* Glycolipids are found mainly, but not exclusively, on the surface membranes of cells, and they are situated so that their polar carbohydrate chains face outward, into the environment and away from the cells. The *N*-acetylneuraminic acid residues on the glycolipids (and glycoproteins) give most animal cells a net negative surface charge.

Important human glycolipids and glycoproteins are the *blood group antigens,* which can trigger harmful immune reactions. In human beings and other animals, the exact structures of certain of the carbohydrates linked to membrane lipids or proteins are genetically determined. Thus, when a person receives a blood transfusion containing different antigens, the foreign carbohydrates may stimulate the production of an immune response to the foreign cells.

The carbohydrates of the human blood groups (A, B, and O) have been studied in great detail. The A, B, and O antigens are structurally related oligosaccharides that may be linked either to lipids or to proteins. The O antigen is a chain of fucose, galactose, *N*-acetylglucosamine, and glucose, with a ceramide lipid (or a hydroxyl side chain of a protein) linked to the glucose, as shown in Figure 3-79. The A antigen is identical to the O, except that the A contains a residue of *N*-acetylgalactosamine attached to the outer galactose residue; the B is also similar to the O, except for an extra galactose residue attached to the outer galactose. All people have the enzymes that make the O antigen. People with type A blood also have the enzyme that adds the extra *N*-acetylgalactosamine, whereas those with type B have

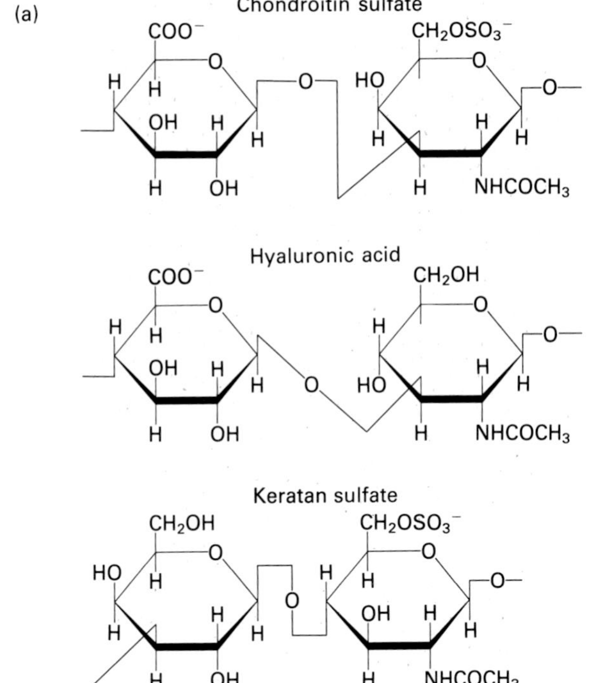

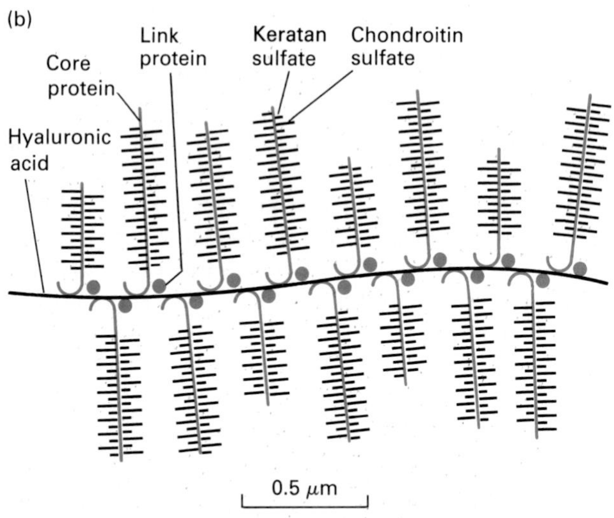

Figure 3-75 (a) The structures of three acid mucopolysaccharides: chondroitin sulfate, hyaluronic acid, and keratan sulfate. All are long polymers of the disaccharide groups shown. (b) A schematic drawing of a typical proteoglycan, such as is found in cartilage. It consists of a central filament of hyaluronic acid several micrometers long. Link proteins are bound to this filament at regular intervals. Long core proteins are bound to the hyaluronic acid by the link proteins. Branching from the core proteins at regular intervals are long chondroitin sulfate and shorter keratan sulfate chains.

the enzyme that adds the extra galactose. Those with type AB synthesize both A and B antigens, whereas those with type O make only the O antigen.

People with blood types A and O normally have anti-B antibodies in their serum; these are specific for the extra galactose in the B antigen. People with types B and O normally have anti-A antibodies, which are specific for the *N*-acetylgalactosamine residue of the A antigen. Thus, when B (or AB) blood cells are injected into a person having type A or O, the person's natural anti-B antibodies bind to the injected cells and trigger their destruction by *phagocytic cells.* Similarly, A or AB cells cannot be safely injected into a person having type B or O. Type O blood can be safely injected into people with type O, A, B, or AB. Because type AB people lack both anti-A and anti-B antibodies, they can receive injections of A, B, AB, or O blood.

The ability of antibodies to distinguish subtle differences in the structures of macromolecules underscores the role of cell surface oligosaccharides in establishing the uniqueness of cells and organisms.

The Primacy of Proteins

Although the major cell constituents have been considered mainly as isolated chemical species in this chapter, in fact they function together. All cellular molecules must be able to interact with other types of molecules, especially with proteins. For example, most nucleic acids in cells have tightly adhering proteins and thus exist as *nucleoproteins;* the DNA of eukaryotes is heavily coated by basic proteins called histones, as well as by a variety of nonhistone proteins that are essential for the proper functioning of DNA in the cell. For another example, cell

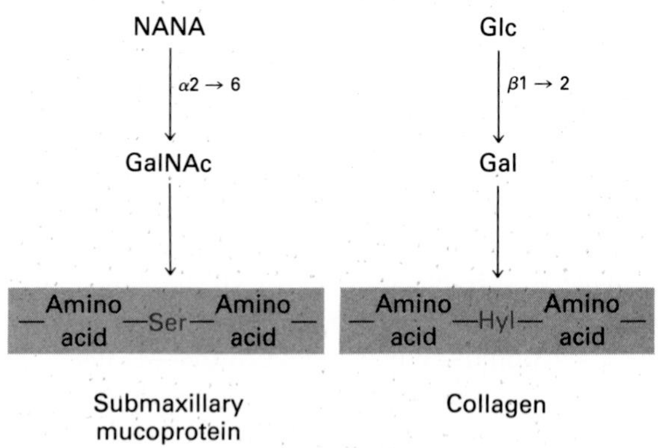

Figure 3-76 The structures of some common *O*-linked oligosaccharides. Abbreviations: NANA = *N*-acetylneuraminic acid (sialic acid), GalNAc = *N*-acetylgalactosamine, Glc = glucose, Gal = galactose, Ser = serine, and Hyl = hydroxyl-lysine. The submaxillary mucoprotein, one of the components of saliva, is secreted by the salivary gland.

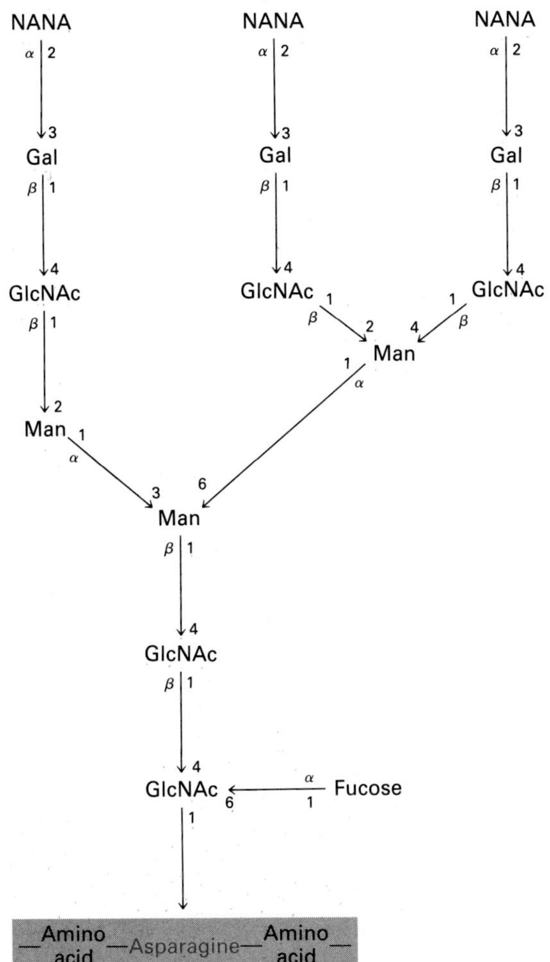

Figure 3-77 The structure of a typical *N*-linked (asparagine-linked) oligosaccharide that is attached to many serum proteins, such as antibodies. Abbreviations: NANA = *N*-acetylneuraminic acid (sialic acid), Gal = galactose, GlcNAc = *N*-acetylglucosamine, and Man = mannose.

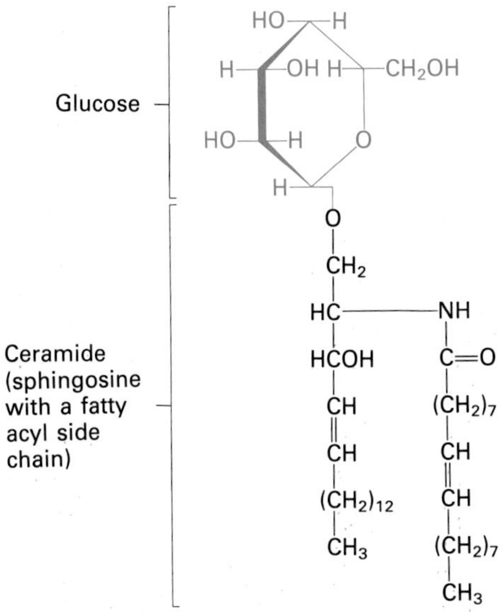

Glucose

Ceramide (sphingosine with a fatty acyl side chain)

Figure 3-78 The structure of glucosylcerebroside, one of the simplest glycolipids.

membranes are never pure phospholipids; attached proteins determine many of the unique features of different cellular and intracellular membranes.

Lipids, nucleic acids, and carbohydrates have fairly uniform chemical structures and physical and chemical behaviors, but proteins can be either hydrophilic or hydrophobic, and either negatively or positively charged. On a single protein molecule there can be both hydrophobic regions and hydrophilic regions, a nucleic acid–binding region, a catalytic region, a region that binds tightly to a sugar, and/or a region that binds tightly to a nucleotide. All parts of a protein fit together to make a functioning unit. It is hard to overestimate the potential variability, specificity, and range of functions of proteins. Although all of the information for synthesizing proteins is found in DNA, DNA is a monotonous molecule com-

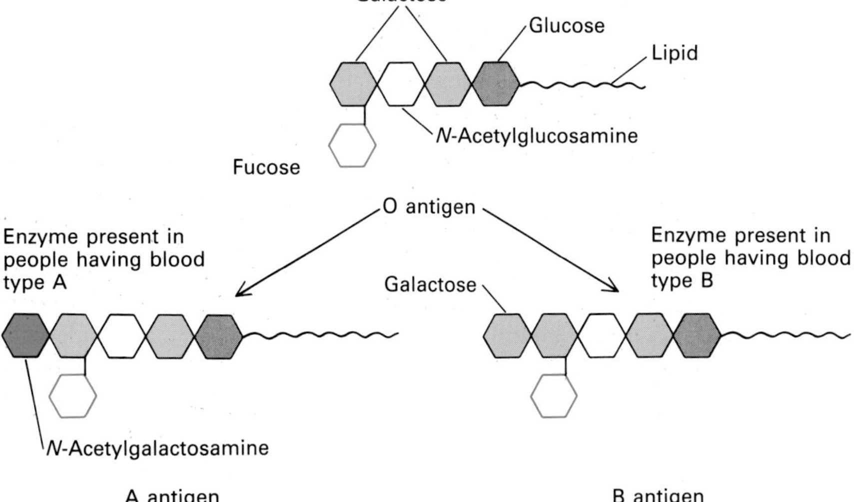

Figure 3-79 The structures of the human blood group antigens. The A antigen differs from the O antigen by the presence of an *N*-acetylgalactosamine attached to the outer galactose residue. The B antigen differs from the O in having an additional galactose residue.

pared with protein. DNA is like a computer tape with building instructions encoded on it, whereas protein has the substantiality and complexity of the final construction.

As we shall see in later chapters, proteins have a hand in every cellular function—the replication and transmission of genes, the building of the cell's physical substance, energy metabolism, cellular and muscular movement, and intracellular communication. It is amazing that a single kind of molecule able to carry out all these functions could have evolved. When we construct a building, we use a variety of substances, each for a specific job. That one type of polymer could be responsible for so many biochemical accomplishments is one of the great wonders of life.

Summary

Proteins are linear polymers fabricated from 20 common amino acids linked by peptide bonds. Amino acids differ from one another in their side chains; some are positively or negatively charged at neutral pH; some are polar; others are hydrophobic. Many proteins contain covalently modified side chains, such as phosphate groups bound to serine, threonine, or tyrosine residues. Glycoproteins contain either short saccharide chains linked to serine or threonine residues or longer oligosaccharides attached to asparagine side chains. Other proteins contain prosthetic groups, such as hemes.

Depending on their size and composition, proteins fold into different shapes: some are compact with one or more globular domains; others are fibrous. Many proteins contain one or more regions of rigid α helix or β pleated sheet. Fibrous proteins such as actin are formed by the aggregation of similar polypeptides in regular, repeating arrays. Water-soluble proteins typically fold so that hydrophobic side chains are internal, away from the aqueous medium. Many proteins maintain their shapes by covalent interactions between the side chains: these interactions take the form of disulfide (—S—S—) bridges between cysteine residues. In addition, most proteins contain multiple polypeptide chains held together by noncovalent bonds. The forces that maintain the native state of a protein can be disrupted by heat or by certain solvents so that the protein becomes denatured. Often the denatured species can renature to restore the native state.

Many proteins are enzymes—catalysts that accelerate the rate of specific chemical reactions. Enzymes have at least two key regions in their active sites: one that tightly binds the substrate(s), and another that catalyzes the chemical reaction on the bound substrate(s). Frequently, a covalent enzyme-substrate complex is an intermediate in catalysis. The binding of a substrate to an enzyme may induce a conformational change in the enzyme that positions the catalytic residues correctly. An example of catalytic specificity is provided by two proteases: trypsin and chymotrypsin. These enzymes have similar catalytic sites, but they require different amino acid side chains in the substrates that they cleave.

Most enzymes in cells do not function at a constant rate. This is because allosteric activators or inhibitors bind at sites different from the substrate-binding site, to modulate the catalytic activities of enzymes. Often, an enzyme that catalyzes one of a series of reactions is inhibited by the ultimate product, in a process called feedback inhibition. A multimeric enzyme can have multiple substrate-binding sites; the binding of a substrate at one site can cause a conformational change that increases the affinity of the other site(s), which makes the enzyme responsive to a narrow range of substrate concentrations. Such cooperativity among binding sites allows a molecule of hemoglobin to bind or release four O_2 molecules over a narrow range of oxygen pressures.

Antibodies, like enzymes, bind specifically and tightly to small regions on molecules. Vertebrates synthesize antibodies in response to the presence of foreign molecules larger than a minimum size. Antibodies can also be synthesized in response to the introduction of chemically modified proteins, in which case the antibodies will be directed to the chemical substituent, a hapten. Typical IgG antibodies contain two identical antigen-binding regions. A third region common to all IgGs is recognized by specific receptors on macrophage cells. Because of their exquisite binding specificity, antibodies can be used to purify a protein in a dilute solution by the technique of affinity chromatography.

A single strand of DNA has a linear backbone of deoxyribose and phosphate. Attached to the $1'$ carbon of the sugar is one of four bases: adenine (A), guanine (G), cytosine (C), or thymine (T). Like a polypeptide, a nucleic acid has a chemical polarity—a $3'$ and a $5'$ end. In their native states, all cellular DNAs are double-stranded: two antiparallel strands are twisted about each other in a right-handed helix with the stacked bases on the inside, perpendicular to the axis of the helix. An adenine residue in one strand is hydrogen-bonded to a thymine residue in the other, and a guanine residue pairs with a cytosine. Some sequences of DNA may form an alternative, left-handed (Z) helix. Heating a linear molecule of double-stranded DNA above a critical temperature causes the two strands to unwind and separate; under appropriate conditions, the molecule can renature to form the native base-paired helix.

Many DNAs are covalently closed circles; these often exist as superhelixes. When a covalently closed circular DNA is denatured, the strands cannot separate, and the molecule will renature spontaneously.

RNA differs from DNA in several ways: in RNA, the sugar is ribose rather than deoxyribose; uracil is found in place of thymine; and the molecule is generally single-stranded, although it may contain a number of double-helical segments.

Long-chained fatty acids and triacylglycerols are, like pure hydrocarbons, insoluble in an aqueous medium, and they will segregate into a separate phase. Adipose cells contain droplets of triacylglycerols, which are storage forms of fatty acids. Phospholipids, sphingomyelin, and cholesterol—the main structural components of cellular membranes—are amphipathic: they contain a large hydrophobic (hydrocarbon) region and a small hydrophilic end. Although micelles and liposomes can be formed from phospholipids, cellular membranes are typically sheets of phospholipid bilayers, which have a hydrophobic central region and two hydrophilic surfaces.

Carbohydrates are important as energy-storage molecules and as structural components in cells. Glucose ($C_6H_{12}O_6$) is the most common sugar in cells. As a linear molecule, it contains four asymmetric carbon atoms. In the formation of the ringed glucopyranose, another asymmetric carbon is created to yield α-D-glucopyranose or β-D-glucopyranose. Other important sugars are mannose and galactose, which differ from glucose only in the positions of substituents on one carbon atom, and modified molecules such as *N*-acetylglucosamine and *N*-acetylneuraminic acid. Storage forms of hexoses include the long-chained polymer glycogen and the disaccharides lactose and sucrose. Sulfated and otherwise modified polysaccharides are principal structural components of extracellular spaces in animals. Short saccharide chains are often bound to lipids or proteins to generate glycolipids or glycoproteins. The compositions and the bonding arrangements of saccharide substituents are genetically determined; as in the case of the A, B, and O blood group antigens, this genetic programming can give rise to harmful immune reactions.

References

General References

SMITH, E. L., R. L. HILL, I. R. LEHMAN, R. J. LEFKOWITZ, P. HANDLER, and A. WHITE. 1983. *Principles of Biochemistry: General Aspects,* 7th ed. McGraw-Hill. Chapters 2–8, 10, and 11.

STRYER, L. 1981. *Biochemistry,* 2d ed. W. H. Freeman and Company. Chapters 2–8, 10, and 24.

ZUBAY, G. 1983. *Biochemistry,* Addison-Wesley. Chapters 1–6, 12, 16, and 18.

Proteins

BLAKE, C. C. F., and L. N. JOHNSON. 1984. Protein structure. *Trends Biochem. Sci.* **9**:147–151.

CANTOR, C. R., and P. R. SCHIMMEL. 1980. *Biophysical Chemistry,* parts 1, 2, and 3. W. H. Freeman and Company. Includes several chapters (2, 5, 13, 17, 20, and 21) on the principles of protein folding and conformation.

CHOTHIA, C., and A. LESK. 1985. Helix movements in proteins. *Trends Biochem. Sci.* **10**:116–118.

CREIGHTON, T. E. 1983. *Proteins: Structure and Molecular Properties.* W. H. Freeman and Company.

DICKERSON, R. E., and I. GEIS. 1969. *The Structure and Action of Proteins.* Benjamin-Cummings. This little book contains many excellent three-dimensional structures of enzymes and other proteins.

DICKERSON, R. E. and I. GEIS. 1983. *Hemoglobin: Structure, Function, Evolution, and Pathology.* Benjamin-Cummings.

DOOLITTLE, R. 1985. Proteins. *Sci. Am.* **253**(4):88–96.

KE, H.-M., R. B. HONZATKO, and W. N. LIPSCOMB. 1984. Structure of unligated aspartate carbamoyltransferase of *Escherichia coli* at 2.6 Å resolution. *Proc. Nat'l Acad. Sci. USA* **81**:4037–4040.

KIM, P.S., and R. L. BALDWIN. 1982. Specific intermediates in the folding reactions of small proteins and the mechanism of protein folding. *Annu. Rev. Biochem.* **51**:459–489. A recent review of experimental results on several proteins.

LESK, A. M. 1984. Themes and contrasts in protein structures. *Trends Biochem. Sci.* **9**:v–vii.

PERUTZ, M. F. 1964. The hemoglobin molecule. *Sci. Am.* **211**(5):64–76 (Offprint 196). This article and the next describe the three-dimensional structure of hemoglobin and the conformational changes that occur during oxygenation and deoxygenation.

PERUTZ, M. F. 1978. Hemoglobin structure and respiratory transport. *Sci. Am.* **239**(6):92–125 (Offprint 1413).

ROSE, G. D., A. R. GESELOWITZ, G. J. LESSER, R. H. LEE, and M. H. ZEHFUS. 1985. Hydrophobicity of amino acid residues in globular proteins. *Science* **229**:834–838.

ROSSMAN, M. G., and P. ARGOS. 1981. Protein folding. *Annu. Rev. Biochem.* **50**:497–532.

THOMAS, K. A., and A. N. SCHECHTER. 1980. Protein folding: evolutionary, structural and chemical aspects. In *Biological Regulation and Development,* vol. 2, R. F. Goldberger, ed. Plenum. A detailed treatment of the variety of three-dimensional structures found in different types of proteins.

Enzymes

BENDER, M. L., R. J. BERGERON, and M. KOMIYAMA. 1984. *The Bioorganic Chemistry of Enzyme Action.* Wiley.

FERSHT, A. 1985. *Enzyme Structure and Mechanism,* 2d ed. W. H. Freeman and Company.

KANTROWITZ, E. R., S. C. PASTRA-LANDIS, and W. N. LIPSCOMB. 1980. *Escherichia coli* aspartate transcarbamoylase. Part 1: Catalytic and regulatory functions. *Trends Biochem. Sci.* **5**:124–128. Part 2: Structure and allosteric interactions. *Trends Biochem. Sci.* **5**:150–153.

PHILLIPS, D. C. 1966. The three-dimensional structure of an enzyme-molecule. *Sci. Am.* **215**(5):78–90 (Offprint 1055). The structure of lysozyme and the nature of the enzyme-substrate complex.

SRERE, P. A. 1984. Why are enzymes so big? *Trends Biochem. Sci.* **9**:387–390.

WALSH, C. 1979. *Enzymatic Reaction Mechanisms.* W. H. Freeman and Company. A detailed discussion of the chemical bases of action of many types of enzymes.

Antibodies

CAPRA, J. D., and A. B. EDMONSON. 1977. The antibody combining site. *Sci. Am.* **236**(1):50–59 (Offprint 1350). A clear description of how an antibody molecule binds to an antigen.

PORTER, R. R. 1973. Structural studies on immunoglobulins. *Science* **180**:713–716.

TONEGAWA, S. 1985. The molecules of the immune system. *Sci. Am.* **253**(4):122–130.

Nucleic Acids

BAVER, W. R., F. H. C. CRICK, and J. H. WHITE. 1980. Supercoiled DNA. *Sci. Am.* **243**(1):118–133 (Offprint 1474). A mathematical model for describing and analyzing DNA supercoiling.

CANTOR, C. R., and P. R. SCHIMMEL. 1980. *Biophysical Chemistry,* part 1. W. H. Freeman and Company. Chapters 22 to 24. The conformation of DNA is described using physical-chemical principles.

DICKERSON, R. E. 1983. The DNA helix and how it is read. *Sci. Am.* **249**(6):94–111.

FELSENFELD, G. 1985. DNA. *Sci. Am.* **253**(4):58–66.

KORNBERG, A. 1980. *DNA Replication.* W. H. Freeman and Company. Chapter 1. A good summary of the principles of DNA structure.

MIN JOU, W., G. HAEGEMAN, M. YSEBAERT, and W. FIERS. 1972. Nucleotide sequence of the gene coding for the bacteriophage MS-2 coat protein. *Nature* **237**:82–88. A vivid demonstration of double helixes within a single-stranded RNA.

RICH, A., and S.-H. KIM. 1978. The three-dimensional structure of a transfer RNA. *Sci. Am.* **238**(1):52–62 (Offprint 1377).

SAENGER, W. 1983. *Principles of Nucleic Acid Structure.* Springer-Verlag. A comprehensive treatise on the structures of RNA, DNA, and their constituents.

WANG, J. C. 1980. Superhelical DNA. *Trends Biochem. Sci.* **5**:219–221.

Lipids and Biomembranes

FINEAN, J. B., R. COLEMAN, and R. H. MICHELL. 1984. *Membranes and Their Cellular Functions,* 3d. ed. Blackwell.

QUINN, P. J. 1976. *The Molecular Biology of Cell Membranes.* University Park Press. Chapter 2. An excellent description of the structures of membrane components.

TANFORD, C. 1980. *The Hydrophobic Effect,* 2d ed. Wiley. Includes a good discussion of the interactions of proteins and membranes.

Carbohydrates

GINSBURG, V., and P. ROBBINS, eds. 1984. *Biology of Carbohydrates.* Wiley.

KORNFELD, R., and S. KORNFELD. 1980. Structure of glycoproteins and their oligosaccharide units. In *The Biochemistry of Glycoproteins and Proteoglycans,* W. J. Lennarz, ed. Plenum.

RODEN, L. 1980. Structure and metabolism of connective tissue proteoglycans. In *The Biochemistry of Glycoproteins and Proteoglycans,* W. J. Lennarz, ed. Plenum. A clear, detailed description of sulfated oligosaccharides.

SHARON, M. 1980. Carbohydrates. *Sci. Am.* **243**(5):90–116 (Offprint 1483). The structures and functions of several oligosaccharides.

4

Synthesis of Proteins and Nucleic Acids

CLASSICAL geneticists from Mendel through Morgan defined the biological units of heredity—the genes—and proved that they are aligned linearly on the chromosomes (Chapters 1 and 10). Then the microbial geneticists showed that genes control the structures of proteins. Since 1953, when James Watson and Francis Crick ushered in the modern era of molecular biology by elucidating the structure of DNA, two questions have dominated research in molecular biology: How is DNA copied? How does DNA control protein synthesis?

The occasional random changes, or mutations, in DNA base sequence, the recombination between chromosomes, the segregation of chromosomes, and the faithful copying of DNA from generation to generation together provide the physical basis for the regular inheritance of characteristics. These mechanisms are also responsible for occasional variations in inheritance—the events that were studied by the classical geneticists.

Many important steps in DNA synthesis have been discovered through analysis of bacterial, or prokaryotic, cells, especially with the aid of bacterial viruses. Nevertheless, how DNA synthesis starts and stops and how recombination is achieved are not entirely known, even for these simple systems. Far less is known about the details of DNA synthesis and recombination in eukaryotic cells, although some important general principles have been discovered.

Since the early 1950s, much successful work has been done on the almost infinitely detailed and interrelated processes of RNA synthesis and protein synthesis in both prokaryotic and eukaryotic cells. Comprehensive discussions of what we have learned about these processes will appear in later chapters. Here we describe the general reactions that occur in the synthesis of proteins and nucleic acids; that is, the reactions by which the monomers—amino acids or nucleotides—are polymerized into long correctly ordered chains.

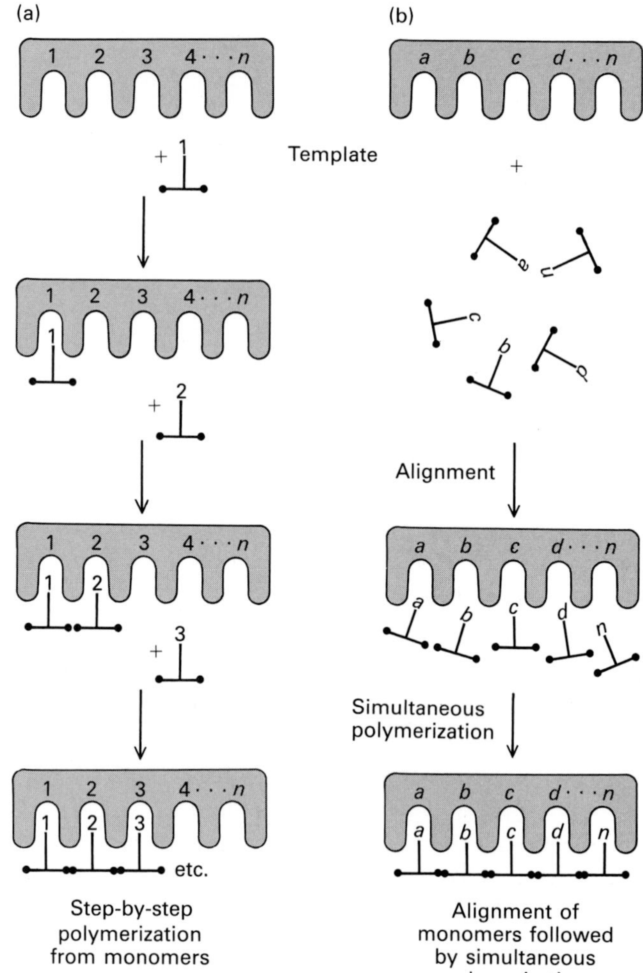

Figure 4-1 (a) Cells produce linear polymers by adding monomers one at a time to a lengthening chain. The order of addition is directed by a template. (b) Cells do not produce polymers by first aligning all the monomers against the template and then fusing them simultaneously.

The Rules for the Synthesis of Proteins and Nucleic Acids

In thinking about the synthesis of all linear polymers—whether they are nucleic acids or proteins—several questions recur. How does synthesis start? How does it progress? How does it end? Four general rules have been drawn from the experimental answers to these questions.

1. *Proteins and nucleic acids are made of a limited number of different subunits.* Although the theoretical number of amino acids is limitless and several dozen have been identified as metabolic products in various organisms, only 20 different ones are used in making proteins. Likewise, only four nucleic acid bases (out of a much larger number of possibilities) are used in cells to construct either RNA or DNA. Cell-free enzyme preparations and cells in culture can be "fooled" into incorporating chemical relatives of these four bases, but in nature this almost never happens.
2. *The subunits are added one at a time in the polymerization of proteins and nucleic acids.* A priori, there is no reason why biological polymers could not have been built by the alignment of all their subunits in the correct order on a template, or mold, followed by the simultaneous fusion of these units. But it does not happen that way. The assembly of proteins and nucleic acids proceeds step by step (Figure 4-1), and in only one chemical direction: protein synthesis begins at the amino (NH_2) terminus and continues through to the carboxyl (COOH) terminus; nucleic acid synthesis begins at the 5′ end and proceeds to the 3′ end (Figure 4-2).
3. *Each chain has a specific starting point, and growth proceeds in one direction to a fixed terminus; this requires start and stop signals.* If the cellular machinery for polymer synthesis did not start and stop the process correctly, a cell would be full of partial, and probably useless, polymers. Elaborate cellular mechanisms ensure correct starts and stops.
4. *The primary synthetic product is usually modified.* The functional form of a nucleic acid or a protein molecule is rarely the same length as the initially synthesized form. The original chain is often inactive or incomplete. By the action of specific enzymes, the original chain is trimmed down, linked to another chain, or even cut apart and reassembled from selected pieces to make a fully active chain (Figure 4-3). Primary chains can also undergo certain chemical additions, either during the formation of the chain or after its synthesis is complete. For example, methyl groups can be added to specific sites in DNA, RNA, and proteins; and phosphate groups and a wide variety of oligosaccharides can be added to proteins, as noted in Chapter 3. Such chemical alterations clearly have important functions, some of which have been elucidated. The details of this macromolecular carpentry will be discussed in Chapters 8 through 13 and in Chapter 21.

H_2O removed as peptide bond forms

Growth of a polypeptide

Growth of a nucleic acid

Figure 4-2 In the formation of proteins in the cell, amino acids are added one after another to the carboxyl end of a growing chain. Thus the growth of a polypeptide chain is said to be from the N-terminus to the C-terminus. In the polymerization of nucleic acids, nucleotides are added one at a time to the 3′ hydroxyl group at the end of a growing chain. The first nucleotide of such a chain retains its phosphates at the 5′ end. Thus, nucleic acids grow from the 5′ end to the 3′ end. (In the polypeptide example, R_1, R_2, etc. denote side chains of amino acids. In the nucleic acid, N_1, N_2, etc. denote nucleic acid bases; Ⓟ represents a phosphodiester bond that connects the 3′ carbon of one sugar to the 5′ carbon of the next.)

Protein Synthesis: The Three Roles of RNA

Describing the process of nucleic acid and protein synthesis in cells is akin to describing a circle. The intricate relation between DNA, RNA, and protein can be diagrammed as follows:

$$DNA \longrightarrow RNA \longrightarrow Protein$$

DNA directs the synthesis of RNA; RNA then directs the synthesis of protein; and special proteins catalyze the synthesis of both RNA and DNA. This cyclic flow of information occurs in all cells, and it has been referred to as the "central dogma" of molecular biology. Thus, a discussion of the synthesis of nucleic acids and proteins can begin with any one of the three kinds of polymers. We shall describe protein synthesis first, because, as we emphasized in Chapter 3, proteins are the active working components of the cellular machinery. In addition, proteins are the targets of evolutionary selection. Whereas DNA stores the information for protein synthesis, and RNA carries out the instructions encoded in DNA, the proteins themselves are responsible for most biological activity, and their synthesis is at the heart of cellular activity.

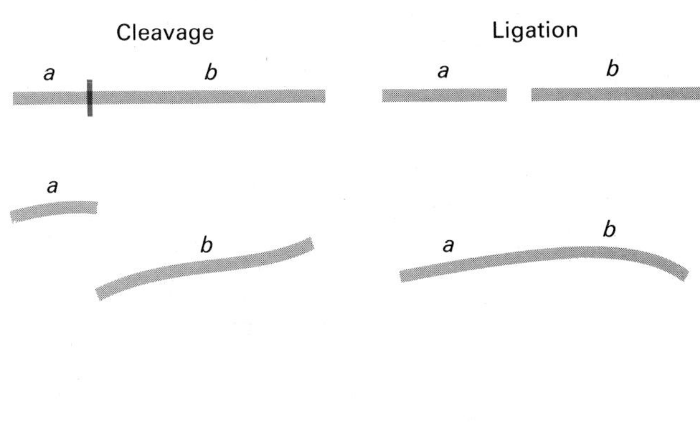

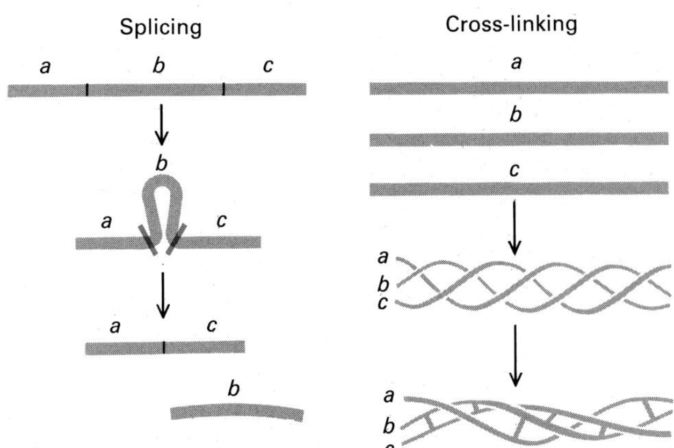

Figure 4-3 Linear biopolymers are typically modified after they are formed. In all kinds of cells, protein and nucleic acid chains can be cleaved into smaller pieces; and short DNA chains can be ligated (linked) to form longer ones. In eukaryotes, many RNA molecules are spliced: a section of the chain is excised and the remaining sections are rejoined. Additionally, many proteins can be chemically modified in more complex ways after synthesis. For example, collagen, the chief protein in connective tissue, has three strands that may be cross-linked by covalent bonds (shown as bars between chains).

The linear order of amino acids in each protein determines its function. Thus, the mechanism for ensuring this order is critical. The design and function of the protein-synthesizing apparatus is similar in all cells. Three kinds of RNA molecules perform different but cooperative roles in linking amino acids together in the correct linear arrangement: *Messenger RNA* (mRNA) encodes genetic information that is copied from DNA; this information is in the form of a sequence of bases that specifies a sequence of amino acids. *Transfer RNA* (tRNA) decodes—translates—the base sequence of the mRNA into the amino acid sequence of the protein. *Ribosomal RNA* (rRNA) combines with many different proteins to form ribosomes, structures that have binding sites for all the interacting molecules necessary for protein synthesis. Ribosomes plus special proteins and attached tRNAs can physically move along an mRNA chain to translate its encoded genetic information. (*Translation* refers to the whole process by which the base sequence of the mRNA determines the amino acid sequence of the protein.) These three types of RNA participate in this essential protein-synthesizing pathway in all cells; and, in fact, the evolution of the three distinct functions of RNA was probably the molecular key to cellular evolution (Chapter 25).

We begin our description of protein synthesis by explaining the genetic code and the role of mRNA in carrying coded information. Next, we describe the structure of tRNA and its elementary biochemistry, to show how the language of nucleic acids is converted into the language of proteins. Finally, we summarize the present understanding of how the ribosome serves as a structure for organizing all the events of protein synthesis.

Messenger RNA Carries Information from DNA in a Three-Letter Genetic Code

As we discussed in Chapter 3, the nucleic acids are linear polymers composed of four mononucleotide units: RNA consists of adenylate (adenosine 5′-monophosphate), cytidylate, guanylate, and uridylate; DNA contains thymidylate instead of uridylate. It is clearly not possible for four nucleotides to specify the linear arrangement of the 20 possible amino acids in a one-to-one fashion. Thus, a *group* of nucleotides is required to symbolize each amino acid. The code employed must be capable of specifying at least 20 "words."

If a group of two nucleotides were used to code for one amino acid, then only 16 (4^2) different code words could be formed, an insufficient number. However, if a group of three nucleotides were used for each code word, then 64 (4^3) code words could be formed. So any code using three or more nucleotides would have more than enough units to encode 20 amino acids. Many different coding systems using three or more bases at a time are mathematically possible, including systems containing "punctuation" (Figure 4-4). For example, only three out of every four bases might be used for coding, with the fourth serving as a "comma" to separate the words. However, the system known to be universal is a "commaless" triplet code.

A recent surprising discovery is that occasionally the DNA sequence contains overlapping information, still in the form of a triplet code. Because it is possible to shift the *reading frame* for any set of triplets by moving the starting point for translation either one or two bases in either direction, two or three different amino acid sequences can be encoded by the same region of the nucleic acid chain (Figure 4-5).

Each triplet is called a *codon*. All but 3 of the 64 possible codons specify individual amino acids (Table 4-1). The meaning of each codon is the same in all known organisms; this is a strong argument that life on earth evolved only once, as we shall discuss in Chapter 25. Recently, however, the genetic code has been found to be different for a few codons in mitochondria and in ciliated protozoans. Because there are 61 codons for 20 amino acids, many of the amino acids have more than one codon; in fact, some amino acids have six (Table 4-2).

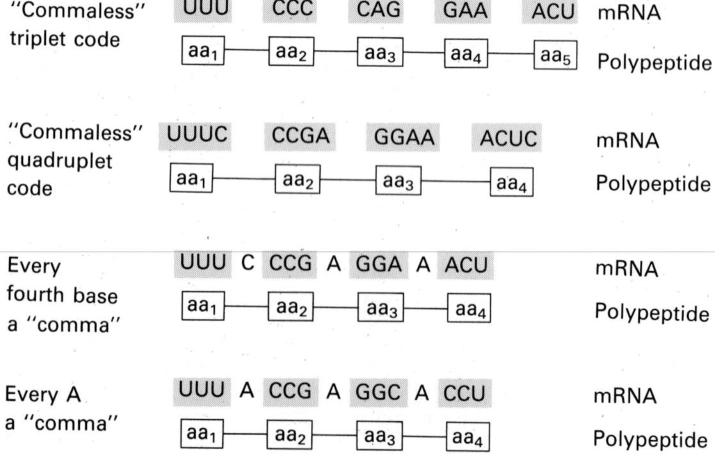

Figure 4-4 Sequences of mRNA bases interpreted by different hypothetical coding systems. The only system known to be used by organisms is the "commaless" triplet code. (The abbreviation aa denotes an amino acid.)

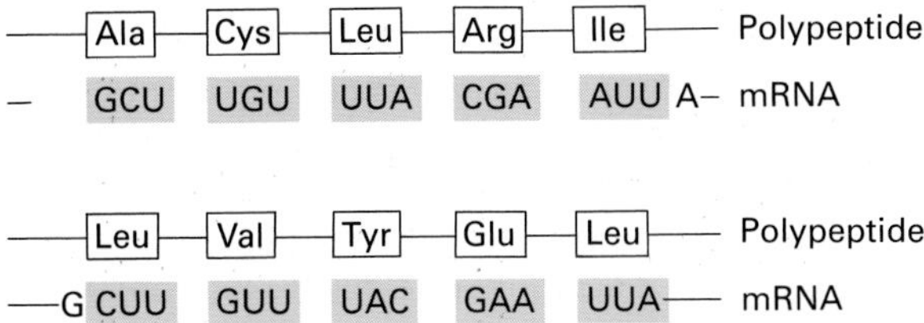

Figure 4-5 An overlapping triplet code that is read in two different frames. The mRNA is the same sequence in both lines but is read in a different "frame." Such codes have been discovered in the DNA of viruses that infect bacterial and mammalian cells.

The different codons for a given amino acid are said to be *synonymous*. The code itself is termed *degenerate*, which means simply that it contains redundancies. However, this degeneracy does not cause ambiguity in translation, because each triplet codes for only one amino acid.

The "start" *(initiator)* codon AUG specifies the amino acid methionine: all protein chains in prokaryotic and eukaryotic cells begin with this amino acid. At the beginning of a few chains, methionine is encoded instead by GUG. The three codons UAA, UAG, and UGA do not specify amino acids but constitute "stop" *(termination)* signals at the ends of protein chains. So a precise linear array of ribonucleotides grouped into threes in the mRNA specifies a precise linear sequence of amino acids in a protein, and also signals to ribosomes where to start and stop synthesis of the protein chain.

Table 4-1 The genetic code*

First position (5′ end)	Second position: U	C	A	G	Third position (3′ end)
U	Phe	Ser	Tyr	Cys	U
	Phe	Ser	Tyr	Cys	C
	Leu	Ser	Stop (och)	Stop	A
	Leu	Ser	Stop (amb)	Trp	G
C	Leu	Pro	His	Arg	U
	Leu	Pro	His	Arg	C
	Leu	Pro	Gln	Arg	A
	Leu	Pro	Gln	Arg	G
A	Ile	Thr	Asn	Ser	U
	Ile	Thr	Asn	Ser	C
	Ile	Thr	Lys	Arg	A
	Met	Thr	Lys	Arg	G
G	Val	Ala	Asp	Gly	U
	Val	Ala	Asp	Gly	C
	Val	Ala	Glu	Gly	A
	Val (Met)	Ala	Glu	Gly	G

*Bases are given as ribonucleotides, so U appears in the table instead of T. "Stop (och)" stands for the ochre termination triplet, and "Stop (amb)" for the amber. AUG is the most common initiator codon; GUG usually codes for valine, but it can also code for methionine to initiate an mRNA chain.

Synthetic mRNA and Trinucleotides Break the Code

The discovery of mRNA and how it functions led to the solution of the genetic code, one of the great triumphs of modern biochemistry. The underlying experimental work on mRNA and the code was largely carried out with the use of cell-free extracts from bacteria. All the necessary components for protein synthesis except mRNA (tRNAs, ribosomes, amino acids, and energy-rich nucleotides—ATP and GTP) were present in these extracts. Upon the addition of chemically defined synthetic mRNAs, the extracts formed specific polypeptides. For example, synthetic mRNA composed only of U residues yielded polypeptides made only of phenylalanine. Thus, it was concluded that UUU codes for phenylalanine. Each of the other three homopolymers likewise coded for a single amino acid (Figure 4-6). Next, synthetic mRNA that has alternating bases was used; for example,

...A C A C A C A C A C A C A...

The polypeptides made in response to this polymer contained alternating threonine and histidine residues. But this result alone was not enough to determine whether threonine was encoded by ACA and histidine by CAC, or

Table 4-2 The degeneracy of the genetic code

Number of synonymous codons	Amino acid	Total number of codons
6	Leu, Ser, Arg	18
4	Gly, Pro, Ala, Val, Thr	20
3	Ile	3
2	Phe, Tyr, Cys, His, Gln, Glu, Asn, Asp, Lys	18
1	Met, Trp	2
Total number of codons for amino acids		61
Number of codons for termination		3
Total number of codons in genetic code		64

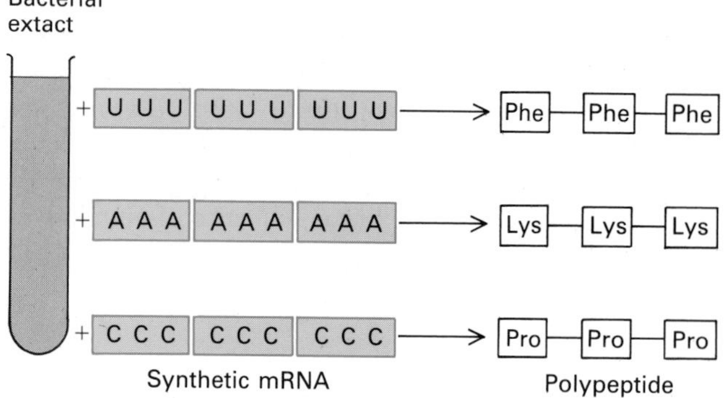

Figure 4-6 The genetic code was worked out largely by using bacterial extracts that contained all the components necessary for protein synthesis except mRNA. When synthetic mRNAs consisting entirely of a single type of nucleotide were added to the extracts, polypeptides composed of a single type of amino acid were formed. For example, polyphenylalanine was formed when polyuridylic acid was added. (Polyguanylic acid, G—G—G— . . . , encodes polyglycine but does so very poorly, because of the tendency of poly G to "stack" or undergo intrachain interactions.) [See M. W. Nirenberg and J. H. Matthei, 1961, *Proc. Nat'l Acad. Sci. USA* **47**:1588.]

vice versa. A further experiment was necessary. An mRNA made of repeated sequences of AAC,

. . . A A C A A C A A C A A C A . . .

stimulated the synthesis of three kinds of polypeptide chains: all asparagine, all threonine, and all glutamine. Apparently, the decoding mechanism could start at any nucleotide, so that it could read the mRNA as three different repeated codons: all AAC, all ACA, or all CAA. The only codon in common between the two-codon mRNA and the three-codon mRNA was ACA, and the only amino acid in common in the polypeptide products was threonine. Therefore, ACA was assigned to threonine (Figure 4-7). Comparisons of the coding capacity of many such mixed polynucleotides revealed a substantial part of the genetic code.

In addition to these experiments, which used synthetic nucleic acids that instructed bacterial extracts to synthesize specific polypeptides, another type of experiment with extracts also was of great importance in the solution of the code. This experiment (Figure 4-8) showed specific trinucleotides would bind individual tRNAs to ribosomes, allowing codon assignment to each trinucleotide.

In all of the experiments in which bacterial extracts were programmed with synthetic mRNAs, the rate of formation of polypeptides was much lower than the rate when natural mRNAs were added to bacterial extracts. This was because the synthetic mRNA lacked start codons and stop codons. It was later appreciated that whereas the coding ability of the synthetic mRNA produced reliable results in experiments designed for the purpose of deciphering the code, it was only when natural mRNAs were added to the bacterial extracts that true proteins were programmed by mRNA. The first successful synthesis of a specific protein occurred when the mRNA of bacteriophage F2 was added to bacterial extracts, and the coat, or capsid, protein (the "packaging" protein that covers the virus particle) was formed.

Transfer RNA Decodes mRNA by Base-Pairing Codon-Anticodon Interactions

There is no evidence for direct chemical recognition between specific nucleic acid bases and specific amino acids. That is, in the synthesis of a polypeptide chain, the trip-

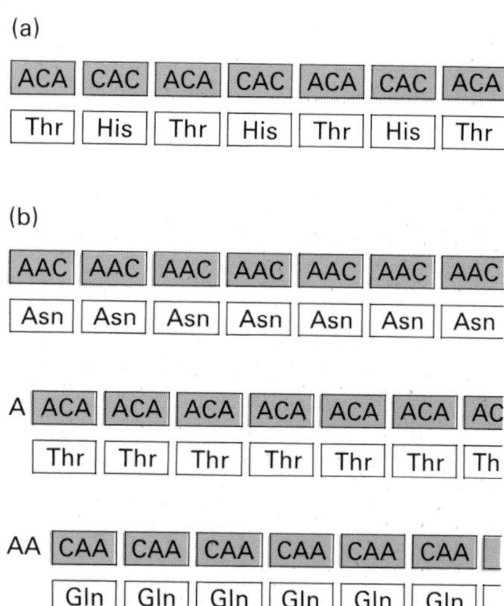

Figure 4-7 When synthetic mRNAs made with alternating A and C residues were added to a bacterial extract, polypeptides made of alternating threonine and histidine residues were formed. The assignment of threonine to ACA and histidine to CAC, as shown in (a), was made possible by another experiment: Synthetic mRNAs made with repeating sequences of

. . . AACAAC . . .

yielded three different kinds of polypeptide chains—all asparagine, all threonine, and all glutamine. Because the only codon in common in the two experiments was ACA, and the only amino acid product in common was threonine, threonine could be assigned to ACA. The other two assignments in (b) (asparagine to AAC and glutamine to CAA) were derived from further experiments. The "cracking" of the genetic code was thus a laborious, step-by-step process. [See H. G. Khorana, 1968, in *Nobel Lectures: Physiology or Medicine (1963–1970)*, Elsevier (1973), p. 341.]

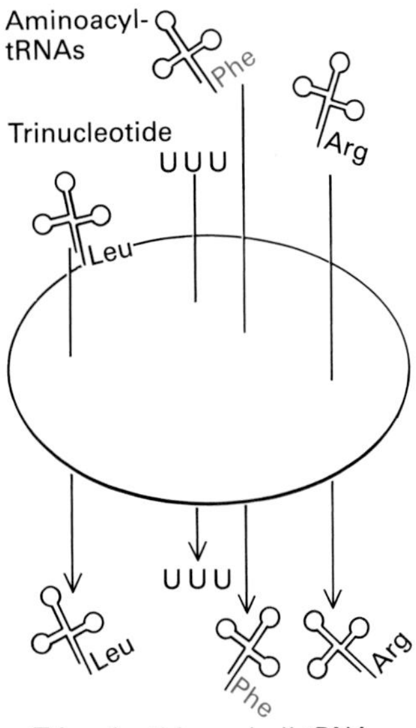

Trinucleotide and all tRNAs pass through filter

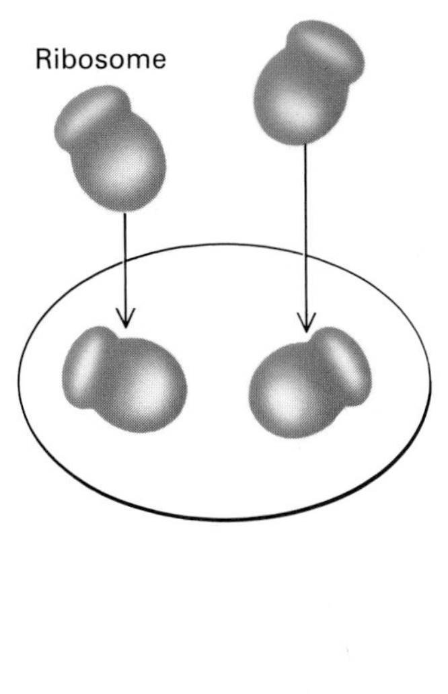

Ribosomes stick to filter

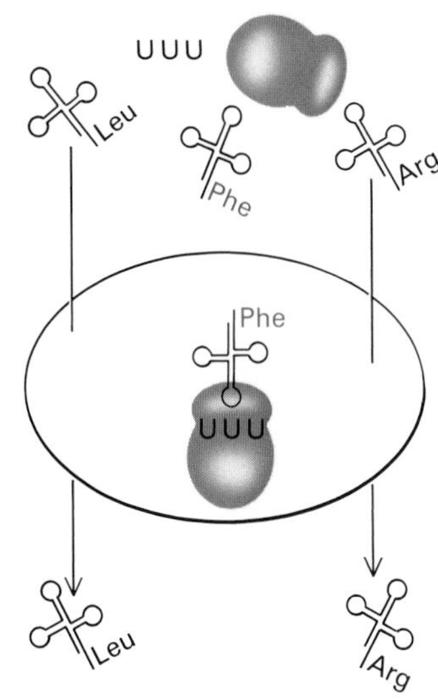

Complex of ribosome, UUU, and Phe-tRNA sticks to filter

Figure 4-8 Marshall Nirenberg and his collaborators used extracts of *E. coli* plus chemically synthesized trinucleotides to decipher the entire genetic code. Twenty mixtures of aminoacyl-tRNAs (tRNAs that have an amino acid attached) were prepared. In each mixture, a different amino acid was radiolabeled (color); the other 19 amino acids were present on tRNAs but they were unlabeled. An aminoacyl-tRNA sample and a trinucleotide would pass through a nitrocellulose filter without binding (left panel). Ribosomes, however, would bind to the filter (center panel). When samples from each of the 20 aminoacyl-tRNA mixtures were mixed with ribosomes and a trinucleotide and then filtered, one of the 20 samples would leave a labeled complex stuck to the filter. In the example shown here, the complex consists of the ribosome, the trinucleotide UUU, and the labeled phenylalanyl-tRNA. This indicates that UUU is recognized by the tRNA for phenylalanine (but not, for example, by the tRNA for leucine or arginine; when these amino acids were labeled, no labeled complex stuck to the filter). Because all possible trinucleotides could be synthesized and tested, this experiment made an enormous contribution to the work on the genetic code. [See M. W. Nirenberg and P. Leder, 1964, *Science* **145**:1399.]

lets in an mRNA molecule do not select amino acids directly. Instead, the protein-synthesizing system uses another molecule to translate, or adapt, the information in each mRNA code word so that the appropriate amino acid is added to the chain. This *adapter molecule* must recognize two things: a codon in mRNA and the amino acid that matches the codon. The adapter function is performed by a transfer RNA (tRNA) molecule to which a single amino acid molecule is attached at one end. The correct aminoacyl-tRNA molecule then binds to the codon in the mRNA strand and transfers its attached amino acid to the polypeptide chain growing there.

Transfer RNA molecules are short—70 to 80 nucleotides long—and varied: there may be as many as 30 or 40 different types in bacteria, and even more in animal cells. Each variety is able to recognize one or more of the several codons that can specify the same amino acid. The structure of the tRNA molecule itself always ends in CCA; the amino acid is attached to the 3′ hydroxyl group of the terminal adenosine.

Transfer RNA molecules in solution are folded into three-dimensional structures. The backbone of the three-dimensional configuration is a stem-loop structure resembling a cloverleaf. The four stems are stabilized by Watson-Crick base pairing; three of the four stems end in loops (Figure 4-9a). The stem-loop structure is then folded into an L-shaped three-dimensional form (Figure 4-9b and c). Hydrogen bonds different from those participating in the standard G-C and A-U base pairs also help to maintain the molecule's shape.

The nucleic acid bases of tRNAs are highly modified after the tRNA is synthesized. The most frequent modification is the addition of a methyl group to specific nucleic acid bases. (Some of the modified structures are shown in Figure 4-10). Most tRNAs are synthesized with a four-base sequence of UUCG near the middle of the molecule. The first uridylate is methylated to become a thymidylate, and the second uridylate is rearranged into a pseudouridylate (abbreviated Ψ), in which the ribose is attached to a carbon instead of to a nitrogen. These modifications produce a characteristic TΨCG segment that is located in an unpaired region at approximately the same

(a)

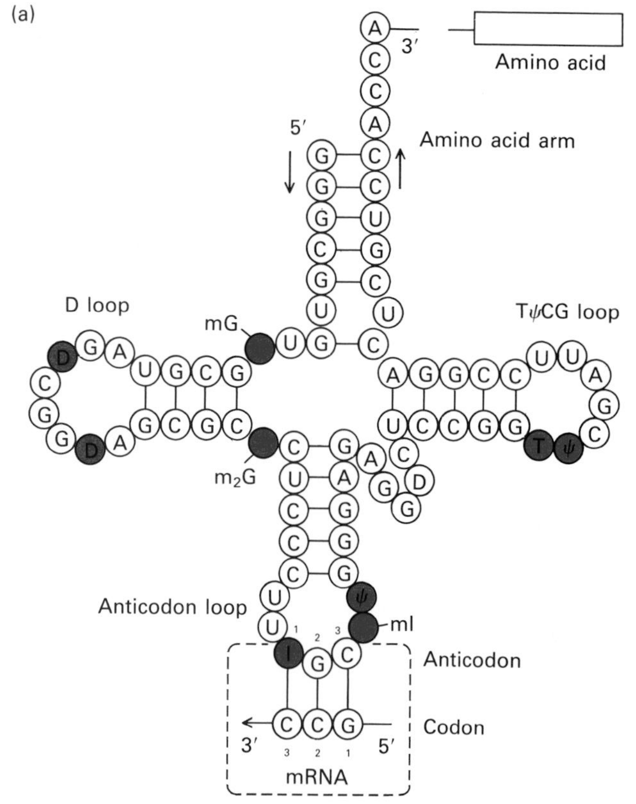

(b)

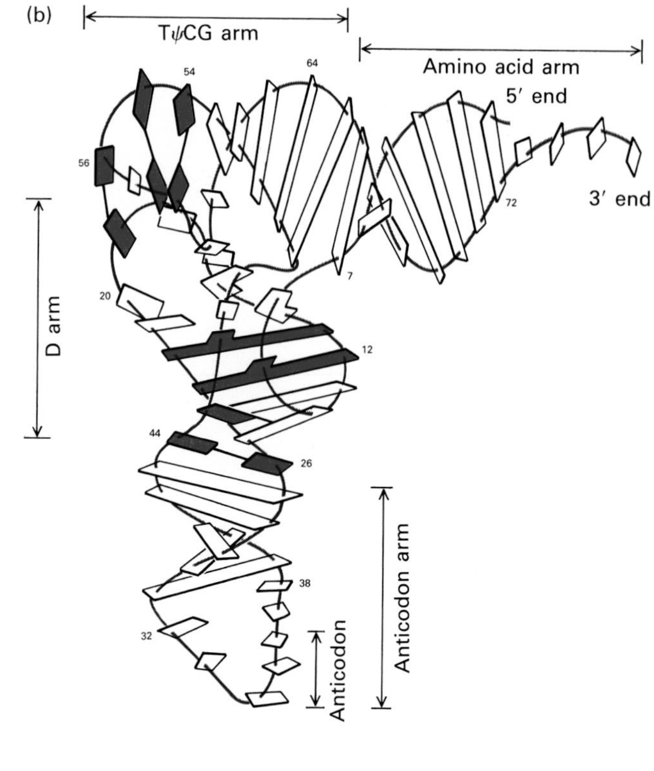

(c)

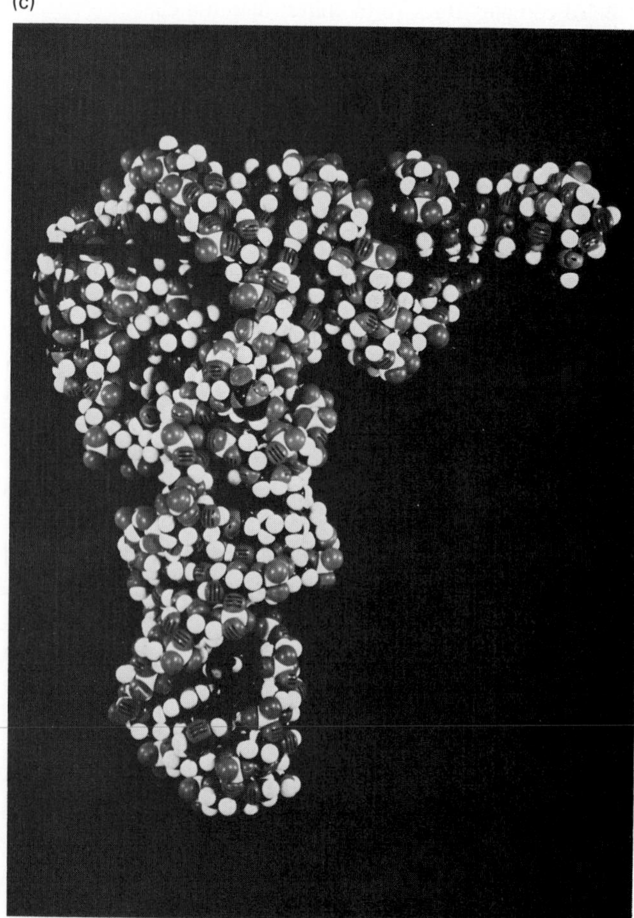

Figure 4-9 (a) The primary structure of yeast alanine tRNA ($tRNA^{Ala}$). This molecule is synthesized from the nucleotides A, C, G, and U, but some of the nucleotides are modified after synthesis: the modified nucleotides are shown in color. (Abbreviations: D = dihydrouridine, I = inosine, T = thymidine, Ψ = pseudouridine, and m = methyl group.) The primary structure is a cloverleaf consisting of four base-paired stems and three loops: the D loop, for dihydrouridine, a virtually constant constituent of this loop; the anticodon loop; and the TΨCG loop, so named because a sequence of thymidylate, pseudouridylate, cytidylate, and guanylate is virtually always present in this loop. (b) This diagram shows the standard Watson-Crick base pairs as rectangular bars connecting segments of the folded molecule of yeast $tRNA^{Phe}$. A number of non-Watson-Crick molecular interactions help stabilize the L-shaped molecule, as indicated in color. (c) A space-filling model of yeast $tRNA^{Phe}$. Parts (b) and (c) are prepared from X-ray crystallographic data. *See J. L. Sussman and S. H. Kim, 1976,* Science ***192****:853.*

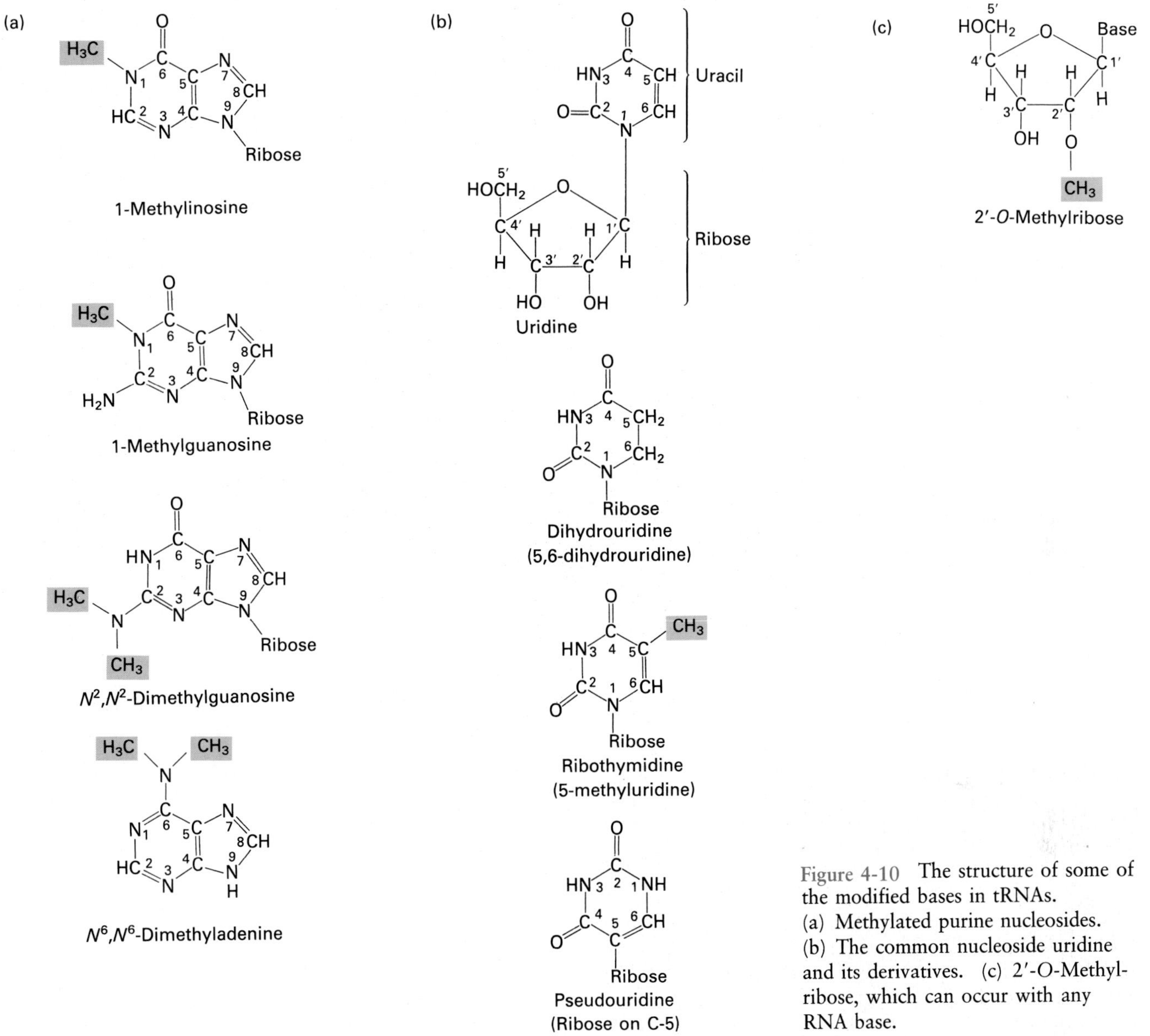

Figure 4-10 The structure of some of the modified bases in tRNAs. (a) Methylated purine nucleosides. (b) The common nucleoside uridine and its derivatives. (c) 2′-O-Methylribose, which can occur with any RNA base.

position in nearly all tRNAs (the TΨCG loop; see Figure 4-9). The exact role of the tRNA modifications has not yet been elucidated. But the fact that certain sites on the tRNA structure are frequently modified in similar ways suggests that these sites have a common role in protein synthesis. This is further supported by the similar two- and three-dimensional structures of tRNAs. The constant features are the D loop, the TΨCG loop, and the anticodon loop (see Figure 4-9).

The anticodon loop contains three nucleotides that can form base pairs with the nucleotides of a specific codon of the mRNA. The three nucleotides in the tRNA molecule are called the *anticodon;* they are *complementary* (not identical) to the three nucleotides in the mRNA codon.

If *perfect* Watson-Crick base pairing were demanded between codon and anticodon, 61 different tRNA species (one for each codon that specifies an amino acid) would be required. However, this is not the case. A pure preparation of tRNA molecules that all have the same anticodon sequence is capable of recognizing more than one codon (but not necessarily every one) corresponding to a given amino acid. One tRNA can recognize multiple codons due to *wobble,* or nonstandard base pairing, between the third position of the codon and its partner in the anticodon. In addition to A-U and G-C, several other combinations of two bases form interactions that are stable enough to allow codon recognition in the wobble position (Figure 4-11). For example, whereas the codon (5′)UUU(3′) in mRNA always calls for phenylalanyl-$tRNA^{Phe}$ (Phe-$tRNA^{Phe}$; $tRNA^{Phe}$ with phenylalanine attached), the anticodon in the Phe-$tRNA^{Phe}$ might be either (3′)AAA(5′), (3′)AAG(5′), or (3′)AAI(5′). (I is the abbreviation for inosine, a modified nucleoside; see below.)

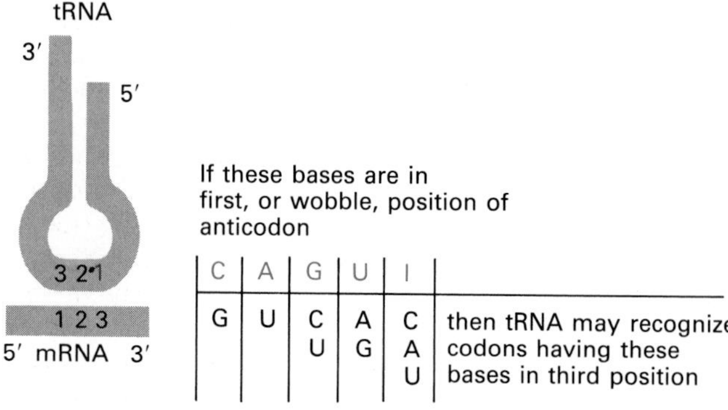

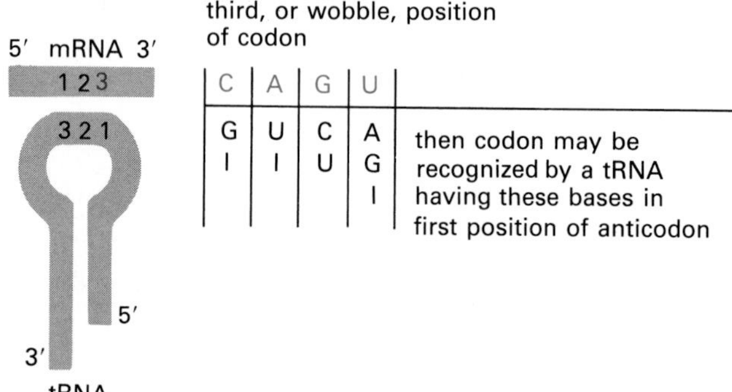

Figure 4-11 The first two bases in an mRNA codon make Watson-Crick base pairs with the second and third bases of a tRNA anticodon. However, the base in the third (or wobble) position of the codon often forms a nonstandard base pair with the base in the first (or wobble) position of the anticodon. Wobble allows a tRNA to recognize more than one mRNA codon; conversely, it allows a codon to be recognized by more than one kind of tRNA, although each tRNA will bear the same amino acid. Note that a tRNA with I (inosine) in the wobble position can "read"—can become paired with—three different codons; a tRNA with G or U in the wobble position can read two codons; but a tRNA with C or A in the wobble position can read only one codon.

This is because bonds, for example, between U and G in the wobble positions of codon and anticodon

(5′)UUU(3)
(3′)AAG(5)

are strong enough to allow recognition of the codon by the anticodon. However, although Phe-tRNA^{Phe} with a (5′)AAG(3′) anticodon can pair with another mRNA codon, (5′)UUC(3′), which also specifies phenylalanine, the anticodon (3′)AAA(5′) does not recognize the codon (5′)UUC(3′). The most versatile tRNAs are those in which the wobble position of the anticodon is occupied by inosine (I), a guanosine analog that lacks an amino group at the 2 carbon. Inosine in the wobble position of the tRNA anticodon can pair satisfactorily with adenine, cytosine, or uracil (Figure 4-12). So, for those amino acids that are encoded by multiple codons differing only in whether they have A, C, or U in the third position (e.g., leucine, proline, serine, valine, and arginine), one tRNA with inosine in the wobble position can decode three different codons. The reason why wobble is allowed in the third anticodon site is still unknown, but its effect may be to speed up protein synthesis by the use of alternative tRNAs.

Aminoacyl-tRNA Synthetases Activate tRNA

How does a tRNA molecule become attached in the first place to the appropriate amino acid? There are at least 20 different amino acid–specific enzymes that recognize amino acids and their compatible, or *cognate,* tRNAs.

Cytosine — Inosine
Adenine — Inosine
Uracil — Inosine
Uracil — Guanine

Figure 4-12 The nonstandard, wobble base pairs C-I, A-I, U-I, and U-G. The heavy bonds signify the positions of attachment of the nitrogenous bases to the 1′ carbons of the riboses.

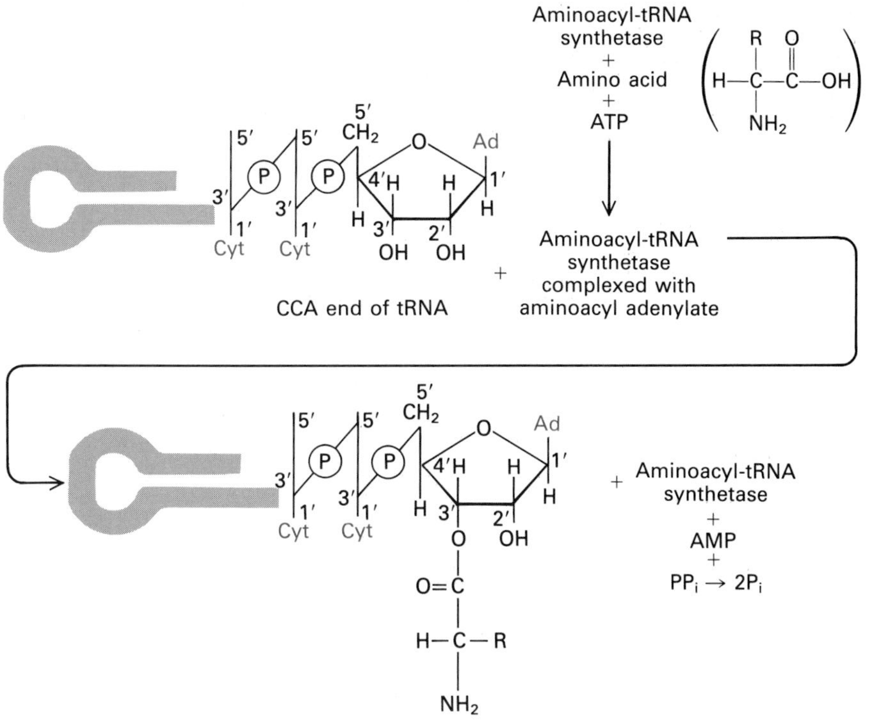

Figure 4-13 To begin the translation of an mRNA strand into a protein, amino acids must be attached to appropriate tRNAs. This is accomplished in two steps by an enzyme that recognizes one kind of amino acid and its cognate tRNAs. The amino acylation of a tRNA occurs in association with the enzyme and requires energy from the hydrolysis of ATP. The amino acid is attached to the 3′ hydroxyl of the terminal adenylate of the tRNA or to the 2′ hydroxyl with prompt rearrangement to the 3′. The equilibrium of the reaction favors products because the pyrophosphate is converted into inorganic phosphate by a pyrophosphatase. (Ad = adenine, Cyt = cytosine.)

Each enzyme can attach one amino acid molecule to the end of a cognate tRNA. A given enzyme is capable of recognizing different tRNAs (i.e., tRNAs with different anticodons) for the same amino acid. These coupling enzymes are called *aminoacyl-tRNA synthetases.* The amino acid is linked to the free 3′ hydroxyl of the ribose of the terminal adenosine of the tRNA. The linkage occurs by a two-step reaction that requires the cleavage of an ATP molecule:

$$\text{Enzyme} + \text{amino acid} + \text{ATP} \xrightarrow{Mg^{2+}} \text{enzyme-(aminoacyl-AMP)} + PP_i \quad (4\text{-}1)$$

$$\text{tRNA} + \text{enzyme-(aminoacyl-AMP)} \longrightarrow \text{aminoacyl-tRNA} + \text{AMP} + \text{enzyme} \quad (4\text{-}2)$$

First the enzyme, the amino acid, and ATP form a complex (Equation 4-1); then the aminoacyl moiety is transferred to the tRNA (Equation 4-2). The resulting aminoacyl-tRNA retains the energy of the ATP, and the amino acid residue is said to have become *activated.* The overall process releases AMP and PP_i (inorganic pyrophosphate) (Figure 4-13). The equilibrium of the reaction is driven further toward activation of the amino acid because *pyrophosphatase* then splits the high-energy phosphoanhydride bond in pyrophosphate. The overall reaction is

$$\text{Amino acid} + \text{ATP} + \text{tRNA} \xrightarrow{\text{enzyme}} \text{aminoacyl-tRNA} + \text{AMP} + 2P_i$$

The basis of the specificity between a tRNA molecule and its cognate tRNA synthetase presumably lies in their three-dimensional structures, but this has not yet been proved. However, the binding site on the tRNA molecule for the synthetase is probably at some distance from the anticodon. Because one enzyme can add the same amino acid to two different tRNAs with different anticodons, the two tRNAs must contain similar binding sites for the synthetase.

The attachment of amino acids to their cognate tRNAs is a critical step in protein synthesis. Once the tRNAs are loaded or "charged" with the correct amino acids, the accuracy of protein synthesis depends only on the base pairing between codons and anticodons. A classic experiment confirmed this hypothesis: A cysteine residue already attached to a $tRNA^{Cys}$ (i.e., to a cysteine-specific tRNA) was chemically changed into alanine, so that the tRNA molecule became an alanyl-$tRNA^{Cys}$ (Figure 4-14). When the alanyl-$tRNA^{Cys}$ was used in the synthesis of a polypeptide, the tRNA added its alanine residue to the growing chain in response to a cysteine codon. Thus, upon completion of synthesis of the polypeptide, it was discovered that all the usual cysteine residues had been replaced with alanine. This proved that only the anticodon of an aminoacyl-tRNA—and not the amino acid—is involved in the recognition step that causes the amino acid to be incorporated during protein synthesis.

Ribosomes Are Protein-Synthesizing Machines

The highly specific chemical interactions of translation do not take place in free solution inside a cell. This critical

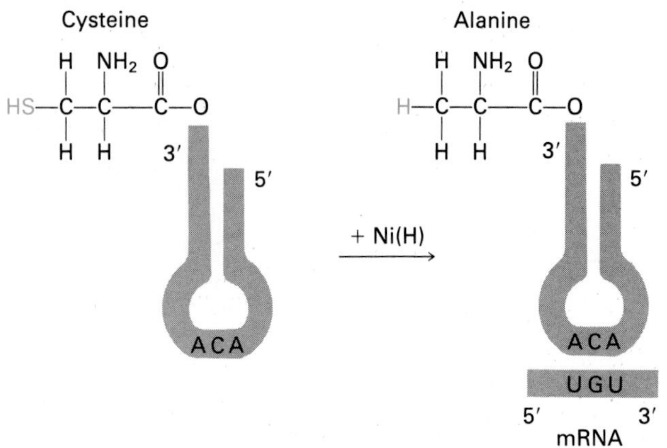

Figure 4-14 After cysteine has been activated by its attachment to a $tRNA^{Cys}$ molecule, it can be chemically altered (by treatment with a nickel compound) into alanine. If the aminoacyl-tRNA is then used in protein synthesis, the resulting polypeptide contains alanine where the mRNA codes for cysteine. [See F. Chappeville, F. Lipmann, G. von Ehrenstein, B. Weisblum, W. J. Ray, and S. Benzer, 1962, *Proc. Nat'l Acad. Sci. USA* **48**:1086.]

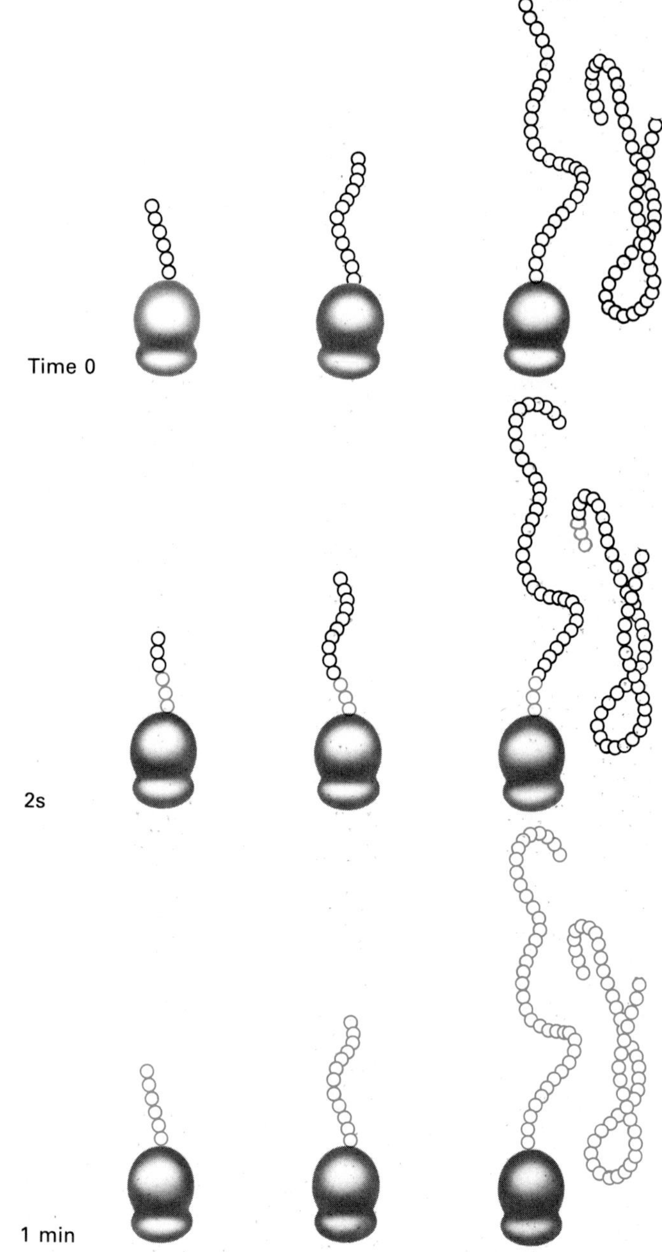

Figure 4-15 An experiment that identified ribosomes as the sites of protein synthesis. In extracts of cells, these discrete structures can be separated from the remainder of the cell. When cells are exposed to radioactive amino acids for very short times—about 2 s—the great majority of newly incorporated labeled amino acids are found in growing protein chains associated with ribosomes. After a longer exposure (about 1 min), completely synthesized proteins that are free of the ribosomes have labeled amino acids along their entire lengths. [See K. McQuillen, R. B. Roberts, and R. J. Britten, 1959, *Proc. Nat'l Acad. Sci. USA* **45**:1437.]

cell function, which results in the formation of more than 1 million peptide bonds each second in an average mammalian cell, requires the participation of a variety of highly specific proteins. Protein synthesis would be very inefficient if each of the many participating proteins had to react in free solution: simultaneous collisions between the necessary components of the reaction would be so rare that the rate of amino acid polymerization would be very slow indeed. Instead, the mRNA with its encoded information and the individual tRNAs loaded with their correct amino acids are brought together by their mutual binding to a complex intracellular structure called a *ribosome*.

Ribosomes were first discovered with the aid of the electron microscope, before it was known that they played a role in protein synthesis. They were seen to be discrete, rounded structures that are prominent in tissues secreting large amounts of protein. Once reasonably pure preparations of ribosomes were achieved, they were found to be very consistent in size. Then newly synthesized polypeptides were shown to be associated with structures the size of ribosomes, and this led to the conclusion that ribosomes are the sites of protein synthesis (Figure 4-15).

Like the structure of tRNAs, the structure of ribosomes is similar but not identical in all cells. This consistency is another reflection of the common evolutionary origin of many of the most basic constituents of living cells. However, as we discuss in Chapter 25, the small differences in ribosomal structure have become important to investiga-

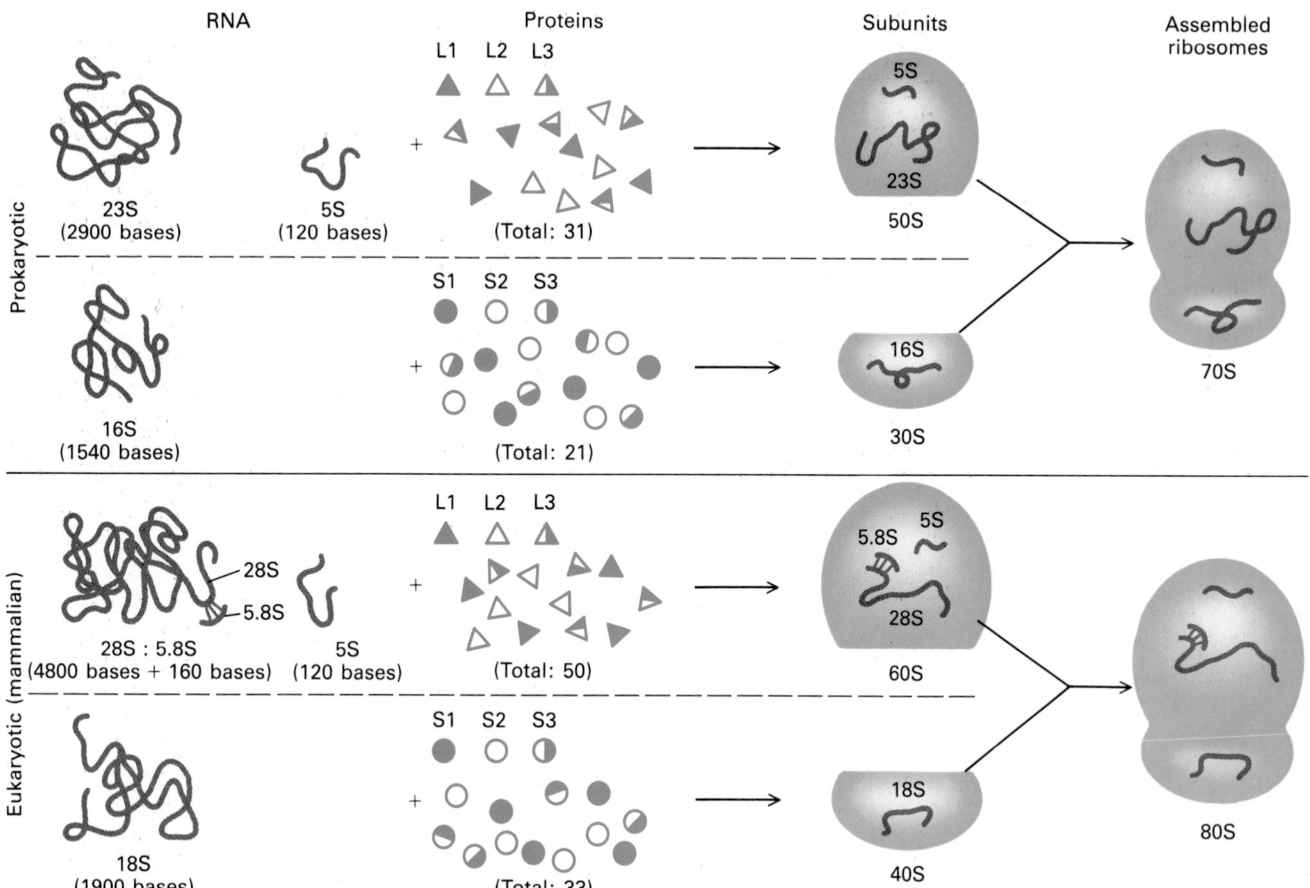

Figure 4-16 The composition of prokaryotic and eukaryotic ribosomes, each consisting of a large and a small subunit. The different subunits contain rRNAs of different lengths and varying numbers and types of proteins (indicated by different shadings). In addition to the two major rRNA molecules, prokaryotic ribosomes have one small 5S rRNA that is about 120 bases long. Eukaryotic ribosomes have two small rRNAs: a 5S molecule similar to the prokaryotic 5S, and a 5.8S molecule that is 160 bases long. The proteins are named L1, L2, etc. and S1, S2, etc., depending on whether they belong to the large or the small subunits. Some cell organelles also have ribosomes; chloroplast ribosomes are similar to prokaryotic ribosomes, whereas ribosomes in mitochondria have smaller RNAs and fewer proteins than prokaryotic ribosomes.

tors attempting to judge the relatedness of various cell types.

The ribosome is a complex composed of individual RNA molecules (the third major type of cellular RNA) and more than 50 proteins, all organized into a large subunit and a small subunit. The proteins in the two subunits differ, as do the molecules of *ribosomal RNA* (rRNA). The large ribosomal subunit contains a large rRNA molecule; the small subunit contains a smaller rRNA. The subunits and the rRNA molecules are commonly designated in terms of svedberg units; e.g., 23S for the large prokaryotic rRNA. Svedberg units are measures of the rates of sedimentation of the various ribosomal components when they are centrifuged under standard conditions. (Centrifugation procedures are discussed in Chapter 7.) The lengths of the major rRNA molecules, the quantity of proteins in each subunit, and consequently the sizes of the subunits differ in prokaryotic and eukaryotic cells (Figures 4-16 and 4-17). At this point, however, we want to emphasize the similarities among all ribosomes, not their differences.

Although the ribosome is a very complicated assembly of proteins and RNA, progress is being made in the study of its structure. Bacterial ribosomal subunits can be disassembled and each constituent protein purified. When the proteins and rRNA are mixed in the correct order, the ribosomal particle can be reassembled. (Such procedures are discussed in Chapter 21.) This has led to experiments that have determined the location of individual ribosomal proteins in the whole particle.

For example, it is possible to take all but one of the proteins of the smaller subunit from the bacterium *E. coli* and replace the similar protein with the same protein from another bacterium (say, a *Bacillus* strain). The ribo-

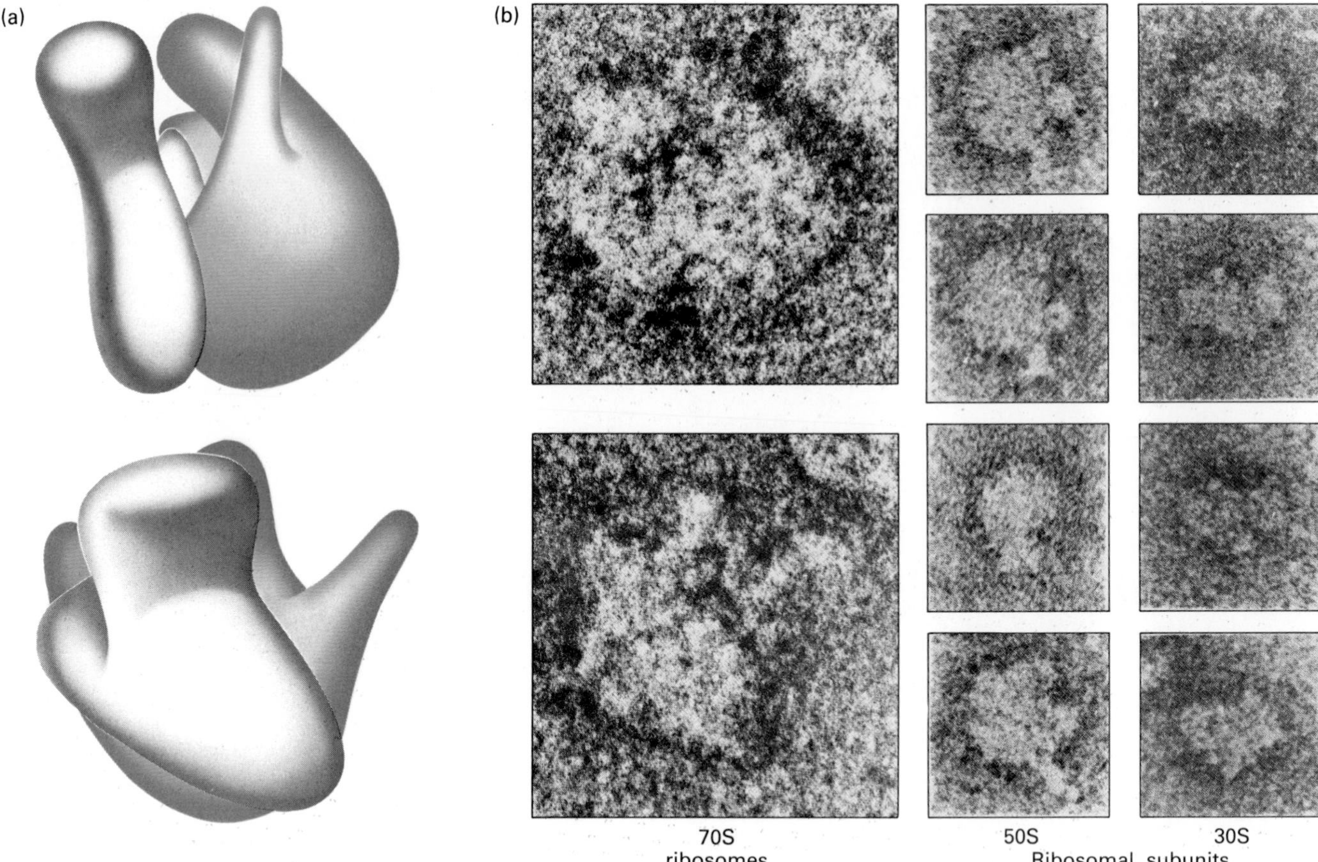

Figure 4-17 (a) Three-dimensional models drawn from electron micrographs of purified *E. coli* ribosomes. (The large subunit is in color.) The exact three-dimensional structure of the ribosome is not yet known, nor is it known what changes in the shape may occur during protein synthesis. (b) Electron micrographs of 70S ribosomes and of 50S and 30S ribosomal subunits. The subunits are viewed from a variety of different angles. *Photographs from J. A. Lake, 1976,* J. Mol. Biol. ***105**:131. Drawings after J. A. Lake.*

somal subunit can be reconstructed in a fully functional form. The position of the *Bacillus* protein on the ribosome can be determined by using antibodies directed against that protein; these will be visible in electron micrographs. Alternatively, a physically marked *Bacillus* or *E. coli* protein (i.e., one in which some hydrogens are replaced by deuteriums) can be used to map its location, because of the difference in the deflection of a neutron beam by the deuterium-containing protein compared with the normal *E. coli* proteins. When two deuterated proteins are used in reassembly, the distance between them can be precisely determined (Figure 4-18). The two techniques—use of antibodies and use of deuterated proteins—together give a high degree of confidence that the positions of the majority of the proteins have been mapped correctly in the bacterial ribosome.

Significant progress on the structure of the small rRNA of *E. coli* has also been made. The entire sequence of nucleotides has been deduced by sequencing the DNA that encodes the rRNA. From the primary sequence, the most probable folding pattern that would produce base-paired stems and the unpaired loops was determined. The folding pattern was confirmed by the use of a chemical reaction that cross-links base pairs, followed by the chemical isolation of the cross-linked regions and comparison of these regions with the total sequence (Figure 4-19).

All the individual proteins of eukaryotic ribosomes have been separated, but reassembly of complete ribosomes has not yet been achieved. It seems most likely that proteins on the surface of the ribosomes, as well as loops of rRNA that are present at the surface, bind the various factors important for the initiation, elongation, and termination of protein synthesis.

The Steps in Protein Synthesis

Protein synthesis is usually considered in three stages—*initiation*, *elongation*, and *termination*—each of which involves distinct and important biochemical events.

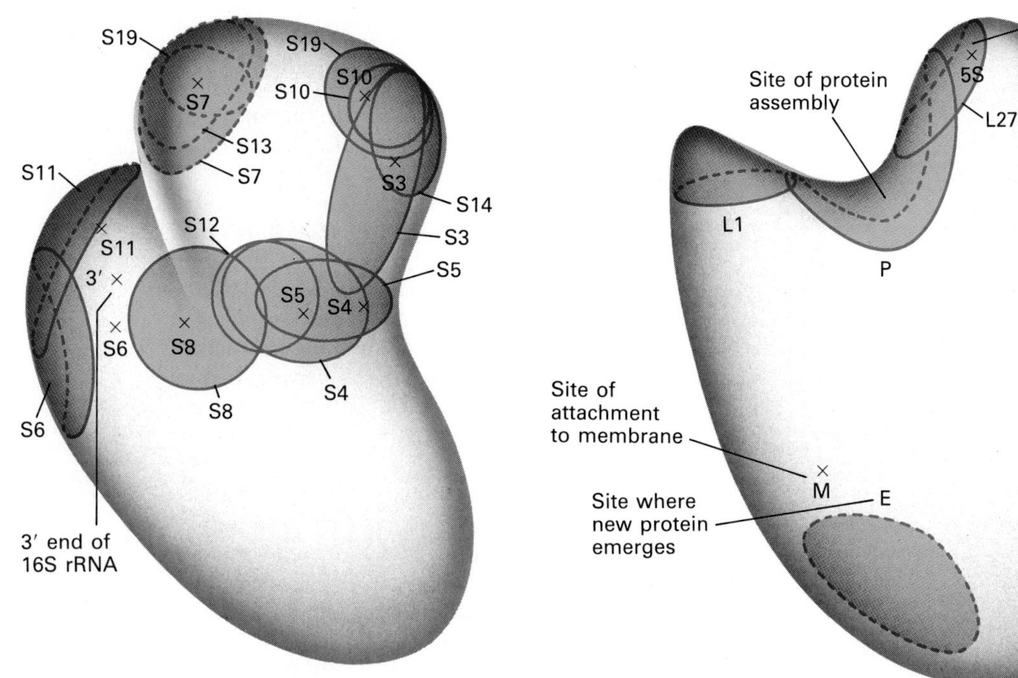

Figure 4-18 Approximate positions of several numbered proteins in the *E. coli* ribosome (S3, S4, etc. in the small subunit; L1, L7, etc. in the large subunit). Several other sites are labeled as well. The colored patches were mapped by viewing, through electron microscopy, the sites on the surface of the ribosome at which specific antibodies bind. The crosses were mapped by neutron diffraction, a technique in which a pair of deuterium-substituted proteins are incorporated into ribosomal subunits and located by neutron bombardment. *After J. A. Lake, 1981,* Sci. Am. ***245****(2):84.*

AUG Is the Initiation Signal in mRNA

The first event of the initiation stage is the attachment of a free molecule of methionine to the end of a tRNA^{Met} by the specific methionyl-tRNA synthetase. There are at least two types of tRNA^{Met}. One is called $\text{tRNA}_i^{\text{Met}}$ because it has the capacity to initiate protein synthesis. Other tRNAs can have methionine attached by the same tRNA synthetase and can be used to incorporate methionine within the protein chain; however, only methionyl-$\text{tRNA}_i^{\text{Met}}$ (methionine attached to $\text{tRNA}_i^{\text{Met}}$) can bind to a small ribosomal subunit to begin the process of protein synthesis. (In bacteria, the amino group of the methionine of the methionyl-$\text{tRNA}_i^{\text{Met}}$ is modified by the addition of a formyl group and is sometimes described as *N*-formylmethionyl-$\text{tRNA}_i^{\text{fMet}}$, but methionyl-$\text{tRNA}_i^{\text{Met}}$ or Met-$\text{tRNA}_i^{\text{Met}}$ is a general designation for the initiator tRNA in all cells.) The Met-$\text{tRNA}_i^{\text{Met}}$ together with a molecule of GTP and a small ribosomal subunit bind to the mRNA, with the aid of proteins called initiation factors, at a specific site that is located near the AUG initiation codon.

Although the methionine codon AUG is present in many places in a long mRNA molecule, protein synthesis always begins at the correct AUG near the ribosomal binding sites. Translation then proceeds in the $5' \rightarrow 3'$ direction along the mRNA. (A variety of experiments with both natural mRNAs and synthetic polynucleotides that have a single distinctive initiation codon have proved the $5' \rightarrow 3'$ direction of translation.) As will be discussed in Chapters 8 and 9, one bacterial mRNA can have several AUG initiation sites, but a eukaryotic mRNA molecule almost always has a single functional AUG near its $5'$ end. The AUG initiation sites are distinguished because the mRNA base sequences just preceding the codons have a high affinity for ribosomes. The $5'$ end of the eukaryotic mRNA is a biochemically modified segment that is recognized by proteins that are important in the binding of ribosomes. This structure is called the mRNA *cap* because a $5' \rightarrow 5'$ pyrophosphate linkage "seals off" the $5'$ end of the mRNA. (The mRNA cap is discussed in more detail in Chapter 9.)

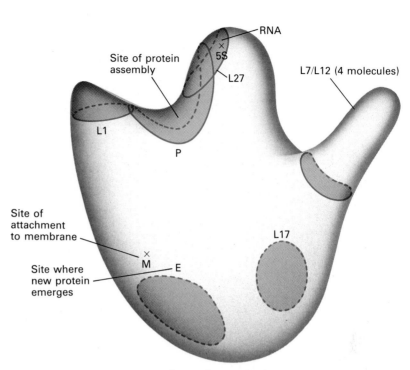

Initiation Factors, tRNA, mRNA, and the Small Ribosomal Subunit Form an Initiation Complex

In addition to the requirement that the mRNA at the initiation site have the proper sequence, a group of proteins called *initiation factors* must be present. Without these proteins, the complex of mRNA, Met-$\text{tRNA}_i^{\text{Met}}$, GTP,

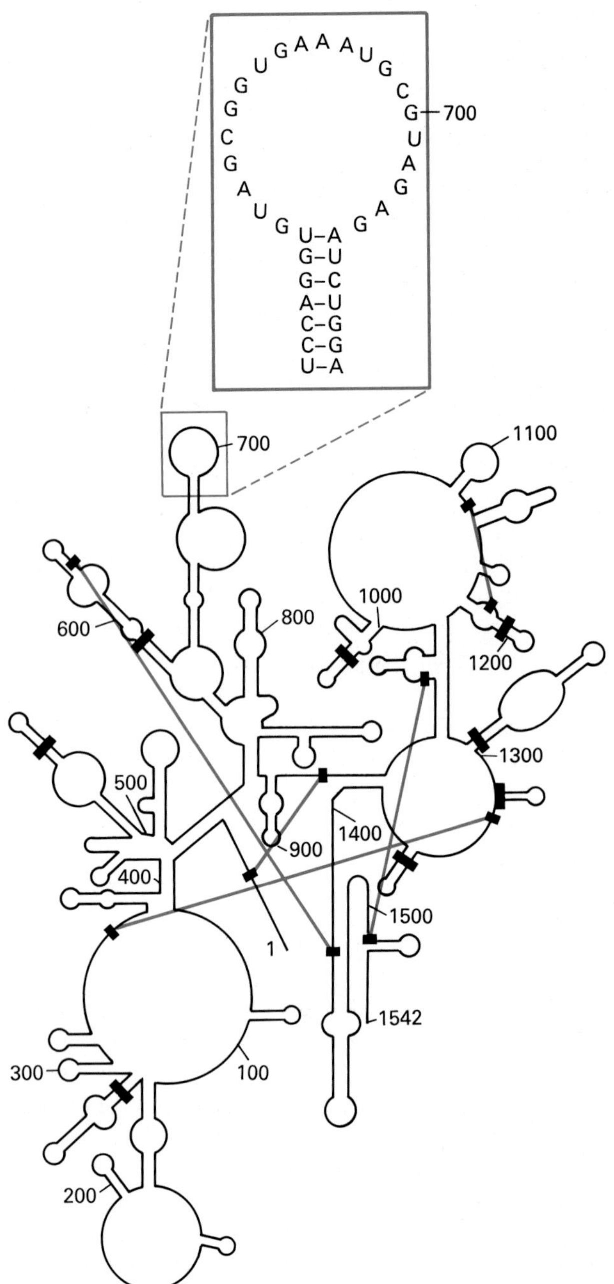

Figure 4-19 Base-paired interactions within *E. coli* 16S rRNA. (The numbers indicate nucleotide positions in the primary sequence; the 5′ end is given the number 1.) From the sequence of 1542 nucleotides in the primary structure of the 16S rRNA, it was possible to draw the most likely secondary structure containing base-paired stems and unpaired loops. The inset shows the nucleotide sequence and the probable stem-loop structure of one region of the molecule. An important chemical proof that the indicated folding is probably correct utilizes a chemical cross-linking reagent that covalently links two base-paired nucleotides together and thus makes the stems of a stem-loop structure chemically more stable. When cross-linked rRNA is digested into pieces by an enzyme that attacks unpaired RNA, nucleotide segments that are not immediately adjacent to one another in the primary structure are found to be linked, thus proving the stemlike interaction between those sites. Most of the interactions are between two nucleotide segments located near each other in the sequence (as indicated by the black bars, which represent the stems identified in this experiment), but some cross-links are between segments that lie at great distances on the molecule (as indicated by the colored lines). *After J. F. Thompson and J. E. Hearst, 1983,* Cell *32:1355. Copyright 1983 by M.I.T; and H. F. Noller and C. R. Woese, 1981,* Science *212:403. Copyright 1981 by the AAAS.*

and the small ribosomal subunit does not form. Three initiation factors have been purified from bacterial cells and at least four are known in eukaryotic cells; each has been shown to play a role in correct protein initiation. The details of the initiation cycle and the discharge of the initiation factors are shown in Figure 4-20a. In general, ribosomes and initiation factors from different bacterial species can be substituted for one another in protein synthesis in the test tube. The same holds true for eukaryotic ribosomes and initiation factors, even when the mixture consists of, for example, extracts of both human cells and yeast cells. But translation of prokaryotic mRNA by eukaryotic ribosomes (and vice versa) is very poor.

Ribosomes Use Two tRNA-Binding Sites, A and P, During Protein Elongation

Once the complex of Met-tRNA$_i^{Met}$, GTP, and the small ribosomal subunit is correctly bound to the mRNA at an initiation site, a large ribosomal subunit joins the complex. For the peptide chain to begin to grow, a second amino acid that is correctly bound to its tRNA must be brought into proper position on the ribosome. (In Figure 4-20b, this second amino acid is phenylalanine.) Two sites on the ribosome are occupied by tRNA molecules. The *A site* accommodates the incoming *aminoacyl*-tRNA that is to contribute a new amino acid to the growing chain. The *P site* contains the *peptidyl-tRNA* complex; that is, the tRNA linked to all the amino acids that have so far been added to the chain. (It is possible that there are alternative types of binding between the ribosome and the tRNAs as the ribosome changes shape during protein synthesis, and that these could be considered intermediate stages between the two major binding conformations.)

Consider the first amino acid addition that is diagrammed in Figure 4-20b. As the contributor of the first amino acid of the chain, the Met-tRNA$_i^{Met}$ enters the P position, and the incoming phenylalanyl-tRNA^{Phe} is then bound to the A position. A peptide bond is created between the carboxyl group of the methionine and the amino group of the Phe-tRNA^{Phe} to form the dipeptide methionyl-phenylalanyl-tRNA^{Phe}; the peptidyl-tRNA^{Phe} vacates the A site and moves to the P site in the process. The hydrolysis of GTP furnishes the energy for this *translocation* of the peptidyl-tRNA. The cycle is repeated for the addition of each amino acid. In our example, the third codon is (5′)CUG(3′), and therefore a leucinyl-tRNA^{Leu} with an anticodon of (3′)GAC(5′) or (3′)GAI(5′) is required. After the Leu-tRNA^{Leu} attaches to the A site, the leucine is incorporated into the new peptidyl chain and the process repeats itself.

For polypeptides, which are longer than the tripeptide depicted in Figure 4-20, this step-by-step process continues until all the amino acids encoded by the mRNA have been added. In each translocation step, the ribosome plus its attached peptidyl-tRNA move three nucleotides closer to the 3′ end of the mRNA. Although the exact mechanism for this movement remains obscure, it seems likely that some protein or proteins in the ribosome as well as the mRNA itself change configuration, using the energy of GTP hydrolysis to propel the mRNA through the ribosome. Because some of the hydrogen bonds that exist in rRNA (see Figure 4-19) are between distant nucleotides, translocation is widely believed to occur through a *contraction-relaxation* cycle in which the folding of the ribosome changes.

The reactivity of a peptidyl-tRNA and an aminoacyl-tRNA with its activated amino acid is such that if the two are brought together, a peptide bond forms spontaneously. Thus, a major role of the ribosome must be to offer binding sites to aminoacyl-tRNA in such a way that the correct codon-anticodon match is made before an activated amino acid is brought close to the peptidyl-tRNA. The selection of the correct aminoacyl-tRNA for elongation of the chain is the most time-consuming part of protein synthesis. The average rate of amino acid addition is about five amino acids per second. In bacterial cells, the proteins that carry out the delivery of aminoacylated tRNAs to the correct site on the ribosome and the proteins that assist in the translocation of the tRNA have been identified (Figure 4-20b).

UAA, UGA, and UAG Are the Termination Codons

In the mRNA for our hypothetical tripeptide being synthesized in Figure 4-20, the three bases following the leucine codon are UAG. This codon, when recognized by protein *termination factors,* signals the release of the peptidyl-tRNA complex (Figure 4-20c). Almost simultaneously the complex divides into an uncharged tRNA molecule (one lacking an attached amino acid) and a newly completed protein chain that can either assume its final shape or combine with additional protein subunits. (Actually, the interaction of two peptide chains can begin while a protein chain is still growing.) After releasing its peptidyl-tRNA, the ribosome disengages from the mRNA and divides into two subunits, whereupon it is ready to start the whole cycle over again.

The peptide synthesis depicted in Figure 4-20 oversimplifies the release process; a tripeptide might, in fact, not even be released. An experiment has shown that a growing peptide chain on a ribosome can be "buried" within the ribosome until it is about 35 residues long (Figure 4-21). In this experiment, brief proteolytic digestion is found to destroy most of each growing peptide chain, but to leave an undigested piece 35 amino acids long associated with each ribosome. The logical place that could accommodate and protect about 35 amino acids is within the area shielded by the two ribosomal subunits. There is some evidence for this: when the protein β-galactosidase is being formed by a ribosome and antibodies to β-galactosidase are introduced, the antibodies react with the newly synthesized protein at a particular location on the periphery of the ribosome—near the junction of the two subunits and on the bottom side—that is, on the side opposite to the side where the projections appear and where protein synthesis occurs (see Figures 4-17 and 4-18). Thus, release of a polypeptide and its first interactions after synthesis may occur once it protrudes beyond an exit site.

Rare tRNAs Suppress Nonsense Mutations

Because each of the codons UGA, UAA, and UAG normally codes for chain termination, it is clear that mutations in a gene could produce an abnormal termination signal (e.g., if UGG mutates to UAG, then a tryptophan codon becomes a stop codon). This type of mutation is called a *nonsense mutation* because the translation apparatus stops too soon. (A *missense mutation* causes the substitution of one amino acid for another.) Nonsense mutations were discovered in bacteria when it was found that shortened forms of bacterial cell or bacterial virus proteins resulted from certain types of mutations.

The chain-terminating mutations were themselves found to be correctable if the mutant gene was experimentally transferred to special bacterial strains. All the mutations that can be corrected by one particular strain are called *amber* mutations, and those that can be corrected by another strain are called *ochre* mutations. The strains that can correct the chain-termination defect are called *suppressor* strains because they suppress the mutation. It is known that the basis for suppression of amber and ochre mutations is the reading of the chain-terminating codon as a codon for an amino acid. This is brought about by a mutation in the anticodon of a tRNA, which changes its reading capacity, or by the presence of a minor, "suppressing" tRNA. Either of these possibilities may produce a low frequency of misinterpretation of stop signals.

For example, the amber codon (5′)UAG(3′) is not normally recognized by a tRNA molecule. However, this codon can be recognized by a minor tyrosine tRNA species that contains (3′)AUC(5′) as its anticodon. Thus, for an mRNA mutation from UGG to UAG, which normally leads to chain termination, a tyrosine residue may be inserted at a low frequency and the nonsense mutation corrected or suppressed; this would allow the chain synthesis to continue. Perhaps because of the existence of suppressor tRNAs, the 3′ ends of coding regions in mRNAs often contain two or more stop codons within a short stretch. In such cases the termination of protein synthesis has a "fail-safe" mechanism.

Suppressor tRNAs have also been discovered in yeast cells. The discovery and characterization of suppressible mutations have provided valuable tools for the molecular geneticist. Any gene that harbors a mutation that is suppressible by a tRNA suppression mechanism presumably codes for a protein. By ascertaining which suppressible site is mutant (e.g., the codon for tyrosine or some other

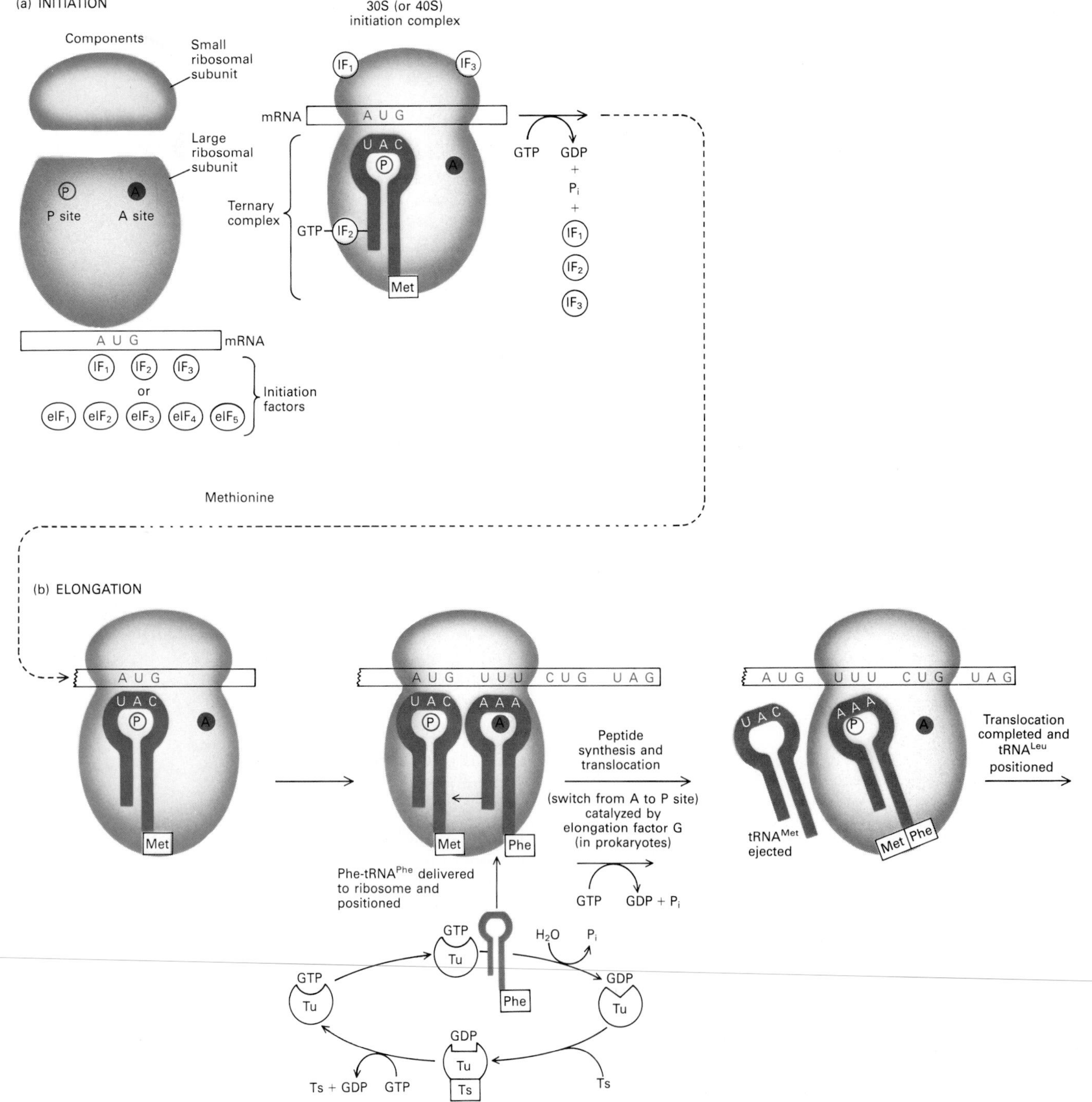

Figure 4-20 ***(opposite and below)*** The three stages of the translation of the genetic message from mRNA into protein.
(a) *Initiation.* An initiation factor (IF_2 in prokaryotes and eIF_2 in eukaryotes) binds a molecule of GTP and a molecule of methionyl-$tRNA_i^{Met}$ to form a ternary complex. This complex binds to mRNA and the small ribosomal subunit (plus other initiation factors) to make an initiation complex (the 30S initiation complex in prokaryotes, and the 40S complex in eukaryotes). The Met-$tRNA_i^{Met}$ is now positioned correctly at the AUG initiation codon. A large ribosomal subunit then joins the complex; the bound GTP is hydrolyzed; and the initiation factors detach. The Met-$tRNA_i^{Met}$ bearing the first amino acid is now bound to the ribosome at the P site. The initiation complex is ready to begin synthesis of the peptide chain.
(b) *Elongation.* The Met-$tRNA_i^{Met}$ is located at the P (for peptidyl-tRNA) site on the ribosome. The growing polypeptide is always attached to the tRNA that brought in the last amino acid. A new aminoacyl-tRNA (Phe-$tRNA^{Phe}$ here) binds to the ribosome at the A site. During elongation in prokaryotes, a protein complex called Tu-Ts catalyzes the binding of each aminoacyl-tRNA to the ribosome. (A protein complex similar in action to Tu-Ts exists in eukaryotic cells.) An activated Tu-GTP complex binds to the TΨCG. The Tu-Ts complex probably binds to the TΨCG loop found in all tRNAs and allows the tRNA to associate with the ribosome. GTP is hydrolized and the Ts protein rejoins GDP-Tu and reactivates it. The complex also binds GTP, whereupon GDP and Ts dissociate from Tu. The remaining GTP-Tu complex then binds to an aminoacyl-tRNA. When the whole elongation factor complex is bound to the ribosome, the GTP is hydrolyzed to GDP, yielding energy to position the aminoacyl-tRNA in the A site. After the incoming aminoacyl-tRNA is correctly placed in the A site–that is, when the codon-anticodon pairing is correct—the peptide chain (or, here, the first methionine) is transferred to the amino group of the newly arrived aminoacyl-tRNA. This generates a peptidyl-tRNA that has acquired an additional amino acid. (In our example, the compound is methionyl-phenylalanyl-$tRNA^{Phe}$.) At this stage, the peptidyl-tRNA is bound to the ribosome at the A site. The ribosome moves one codon down the mRNA chain. (The mRNA codons are illustrated with spaces separating them for convenience; in mRNA there are, of course, no spaces.) The translocation reaction is catalyzed in bacteria by the elongation factor G, using energy from the hydrolysis of GTP. With this movement, the empty tRNA is released from the P site, and the peptidyl-tRNA is shifted to the P site. (In eukaryotes, too, there are special proteins that serve in elongation.) The sequence of events is repeated for every amino acid added to the growing chain. Note that two molecules of GTP are used in the addition of each amino acid.
(c) *Termination.* When the ribosome arrives at the codon UAG, the translation is completed with the aid of a termination factor. Hydrolysis of the peptidyl-tRNA on the ribosome releases the completed polypeptide and the last tRNA, and the two ribosomal subunits separate.

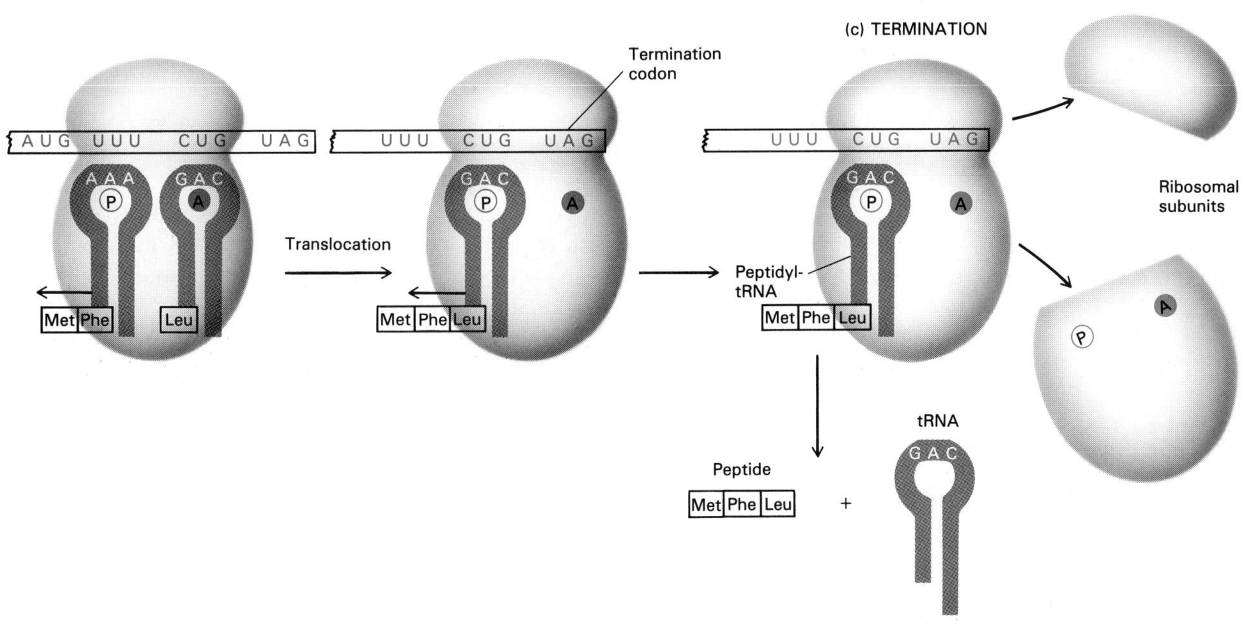

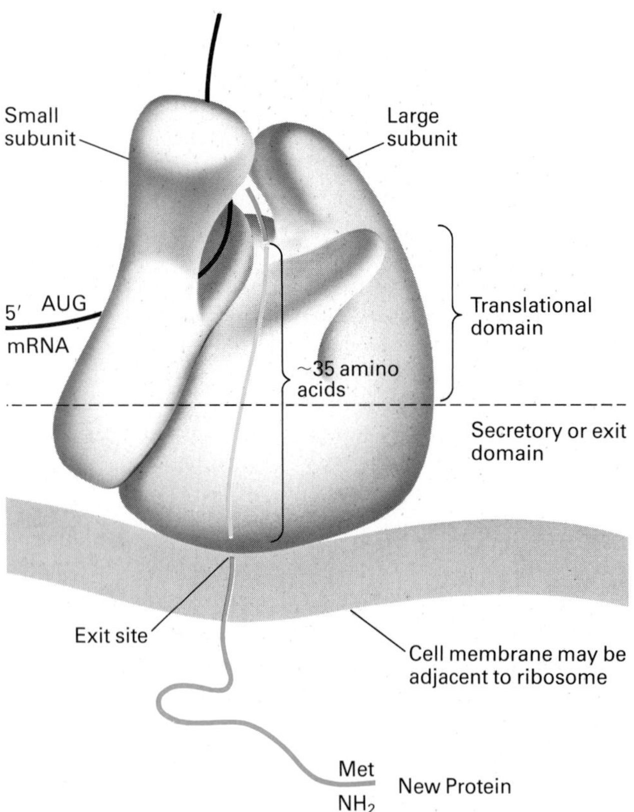

Figure 4-21 Interpretive drawing of a ribosome engaged in protein synthesis. The possible positions of functional domains and the exit site for polypeptides are shown. The top half of the ribosome (as illustrated) includes the binding sites for the initiation, elongation, and termination of peptide chain synthesis. These sites are probably located on both the large and the small subunits. The protein segment labeled "~35 amino acids" is still in the space between the two subunits. The bottom portion of the ribosome contains a region from which the new protein emerges. The proteins that are secreted from the cell may go directly through the cell membrane, which means that the exit site on the ribosome may be bound to the membrane. *Drawing after J. A. Lake.*

amino acid), the experimenter gains some information about the amino acids that are necessary in protein function.

Nucleic Acid Synthesis

The ordered assembly of the basic units in DNA or RNA—deoxyribonucleotides or ribonucleotides—is much simpler than the correct assembly of the amino acids in a protein chain. As Watson and Crick remarked in their renowned paper in which they proposed the double-helical DNA structure: "It has not escaped our notice that the specific pairing we have postulated immediately suggests a possible copying mechanism for the genetic material." Their prediction that the synthesis of nucleic acids would be by the copying of the strands of DNA has proved to be correct. The phenomena associated with the copying of nucleic acids have given rise to two areas of research in molecular biology: (1) DNA replication and its role in cell growth and division, and (2) the synthesis of specific mRNAs and their role in cell adaptation and differentiation.

Nucleic Acid Synthesis Can Be Described by Four Rules

Let us now consider a few general principles governing the synthesis of DNA and RNA chains, and briefly discuss some of the properties of the enzymes that carry out nucleic acid synthesis. The relation of these crucial events to cell growth and differentiation will be elaborated in Chapters 8 through 13 and 21 through 24.

1. *Both DNA and RNA chains are produced in cells by the copying of a pre-existing DNA strand according to the rules of Watson-Crick base pairing.* The DNA from which the new strand is copied is called a *template.* The information (the base sequence) in the template is preserved, although the copy has a complementary sequence and not an identical one. In the replication of a double-helical DNA molecule, or *duplex,* both DNA strands of the original duplex are copied. This complex process is briefly discussed in this chapter and is treated in detail in Chapter 13. In some viruses, RNA molecules are produced by the copying of pre-existing RNA molecules; in one class of viruses, the *retroviruses* (Chapters 6 and 23), DNA is produced by the copying of RNA. However, the vast majority of cellular RNA and DNA synthesis in *cells* is from a pre-existing DNA template.

2. *Nucleic acid strand growth is in one direction: 5′ → 3′.* Because nucleic acids are phosphodiesters (i.e., polymers of nucleotides regularly linked by phosphate groups between adjacent sugars at their 3′ and 5′ positions), each strand has a definite chemical direction: a 5′ (phosphate) end and a 3′ (hydroxyl) end (Chapter 3). All RNA and DNA synthesis, both cellular and viral, proceeds in one chemical direction, from the 5′ to the 3′ end (Figure 4-22). [This directionality has given rise to the convention that polynucleotide sequences are read from left to right in the 5′ → 3′ direction; e.g., the sequence AUG is assumed to be (5′)AUG(3′).] The nucleotides that are used in the construction of nucleic acid chains are 5′-triphosphates of ribo- or deoxyribonucleosides. Strand growth is energetically unfavorable, but it is driven by the energy available in the triphosphates. The α phosphate of the incoming nucleotide is attached to the 3′ hydroxyl of the ribose (or deoxyribose) of the preceding residue to form a phosphodiester bond, with the release of a pyro-

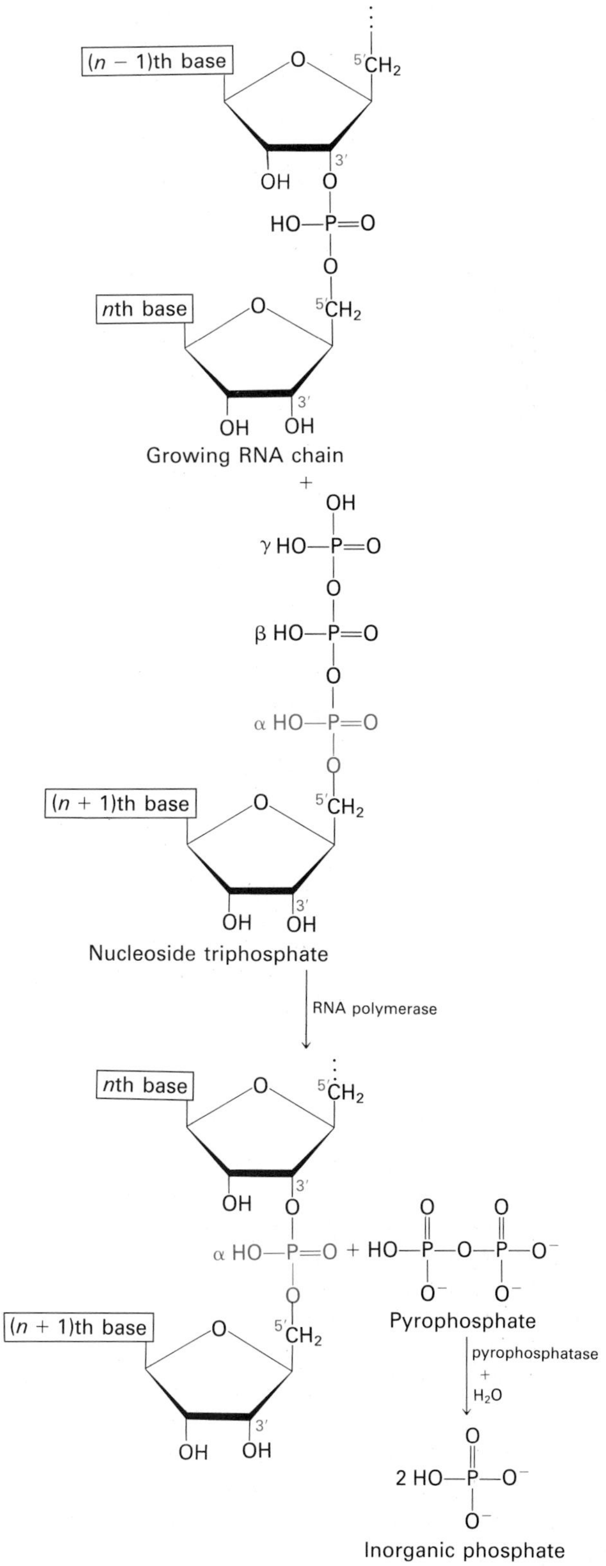

Figure 4-22 Nucleic acids grow one nucleotide at a time and always in the same direction: 5′ → 3′. A nucleoside triphosphate arriving at the 3′ hydroxyl end of the growing strand is bonded to it there, releasing in the process one pyrophosphate ion (PP_i). The remaining phosphate of the triphosphate becomes part of the backbone of the new strand. Both RNA and DNA grow by the same type of reaction, with deoxyribonucleoside triphosphates being substrates for DNA growth and ribonucleoside triphosphates being substrates for RNA growth.

phosphate (PP_i). The equilibrium of the reaction is driven further toward chain elongation by pyrophosphatase, which catalyzes the cleavage of PP_i into two molecules of inorganic phosphate (P_i).

3. *Special enzymes called polymerases elongate RNA or DNA strands.* The enzymes that copy DNA to make more DNA are called *DNA polymerases;* those that copy RNA from DNA are called *RNA polymerases.* An accurate copy of a nucleic acid by a polymerase always requires a template, as noted earlier, and most polymerases do catalyze the synthesis of a copy. However, there are a few polymerases that can catalyze the addition of bases to a chain without a template. One important enzyme of this type is poly A polymerase, discussed in Chapter 9, which adds A's to eukaryotic mRNAs.

The copying of DNA into RNA is called *transcription.* Because the two strands of DNA are complementary rather than identical, they obviously do not encode the same protein chains. For many years it was believed that only one strand of the DNA duplex, when transcribed into RNA, would give rise to usable information; this proved true in a very large number of cases. However, a single case has recently been discovered in which this rule has been violated. In bacteria, DNA molecules called *plasmids* exist free of the major bacterial DNA; a plasmid DNA can be transcribed into mRNA from either strand of DNA in one region. This is the only known DNA that encodes proteins on both strands.

Cells have several different types of DNA polymerases. The physiological role of each of these enzymes is not completely understood, even for viral and bacterial enzymes. This topic is discussed in Chapter 13. Bacterial cells apparently have one type of RNA polymerase that synthesizes mRNA, rRNA, and tRNA. Eukaryotic cells, even the simplest of them, such as yeast cells, have three distinct types of RNA polymerases (discussed in Chapter 9). Each type is responsible for making a different kind of RNA. The product of eukaryotic RNA polymerase I is ribosomal RNA; that of RNA polymerase II is mRNA; and the products of RNA polymerase III are tRNA and other small RNAs. Why the labor of RNA synthesis in eukaryotic cells is divided among three enzymes is not yet known, but the roles of these different RNAs clearly differ, as do the sites of their synthesis.

In both eukaryotic and prokaryotic cells, the RNA strand that is produced initially is often not a biologically active form of RNA. Especially in eukaryotic cells, the ends of the RNA molecules must be chemically modified before the molecules become active. In addition, the primary RNA transcript may undergo shortening and even *splicing:* enzymatic cutting in two or more places, with one or more pieces being discarded and the remaining RNA segments rejoined. Such modifications occur very widely in eukaryotic cells, both in nuclei and in organelles such as chloroplasts and mitochondria.

4. *RNA polymerases can initiate a nucleic acid strand, but DNA polymerases cannot.* A single RNA polymerase can find an appropriate initiation site on duplex DNA, bind the DNA, separate the two strands in that region, and begin generating a new RNA strand (Figure 4-23). The terminal 5′ end of an RNA strand is chemically distinct from the rest of the strand. Unlike the nucleotides within the strand, the nucleotide at the 5′ end retains all three of the phosphate groups of the triphosphate. When each additional nucleotide is added to the 3′ end of the growing strand, only the α phosphate is retained, whereas the β and γ phosphates are lost.

In contrast to RNA polymerases, DNA polymerases cannot initiate a new nucleic acid chain de novo. Rather, they add nucleotides to the hydroxyl group at the 3′ end of a pre-existing RNA or DNA strand. The pre-existing strand is called a *primer*. If RNA is the primer, then the resulting polynucleotide is RNA at the 5′ end and DNA at the 3′ end.

Chemical Differences between RNA and DNA Probably Have Biological Significance

Because duplex DNA consists of two intertwined base-paired strands (see Figure 3-52), the copying of one of the

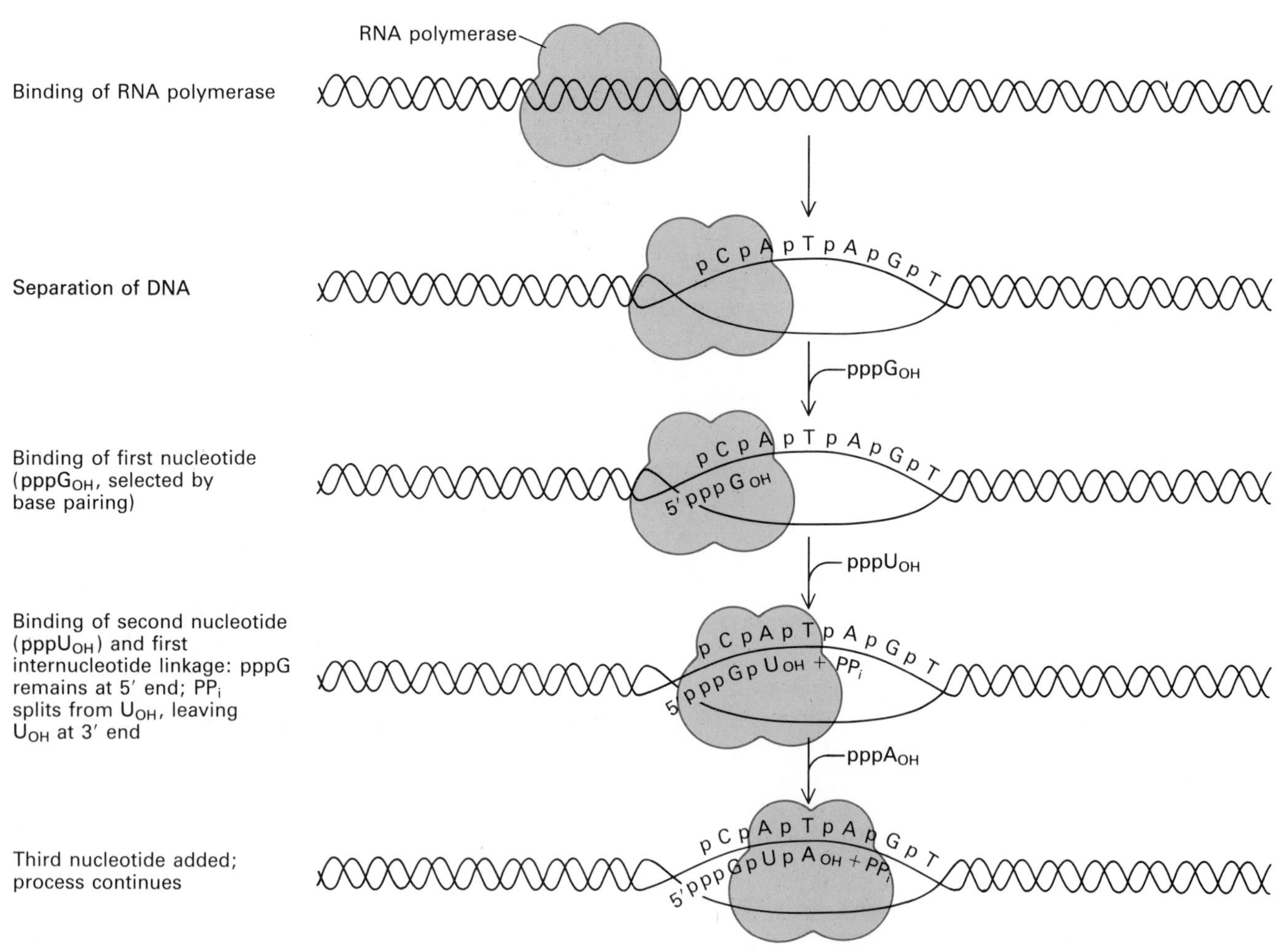

Figure 4-23 The transcription of a DNA strand into an RNA strand is catalyzed by an RNA polymerase, which can initiate the synthesis of strands de novo on DNA templates. The 5′ end of an RNA strand is chemically distinct from the rest of the strand. Unlike the nucleotides within the strand, the nucleotide at the 5′ end retains all three of the phosphate groups of the triphosphate.

two strands by base pairing requires the unwinding of the original duplex *and* a release of the resulting torsional force. This is most likely accomplished by specific "unwinding proteins" (Chapter 13). As mentioned in Chapter 3, the two DNA strands of the duplex are antiparallel; yet all nucleic acid strands grow in the $5' \rightarrow 3'$ direction. How, then, can the two strands of a DNA molecule replicate simultaneously? This is accomplished by the discontinuous synthesis of one of the daughter strands. While synthesis of the daughter strand of one parental strand proceeds continuously in the $5' \rightarrow 3'$ direction, short segments of DNA complementary to the other parental strand are also being synthesized in the $5' \rightarrow 3'$ direction, which is physically opposite to the direction of synthesis of the first daughter strand. In the course of replication, these short, discontinuous segments are linked together to form the second daughter strand (Figure 4-24). The need for a primer and for the ligation of the discontinuous DNA segments makes double-stranded DNA synthesis (either cellular or viral) much more complex than RNA synthesis.

Why should DNA and RNA differ at all, given that the addition of mononucleotides by DNA and RNA polymerases is chemically similar? One possible explanation is that DNA is designed to be stable, but RNA molecules are designed so that they can be changed during cell growth. If this is so, enzymes must be able to degrade RNA selectively. The $2'$ hydroxyl on the ribose renders RNA chemically more labile; it could also be a convenient recognition marker for enzymes that degrade RNA. For example, a low level of hydroxide ion (i.e., a high pH) leads to the cleavage of RNA into $2',3'$-cyclic phosphates; DNA, lacking a $2'$ hydroxyl group, is resistant to such alkaline hydrolysis.

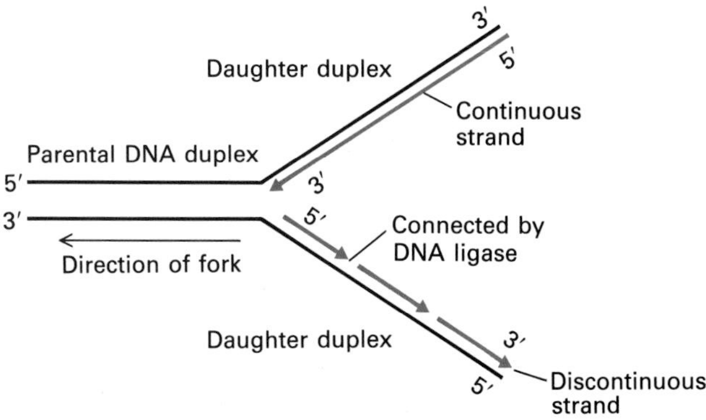

Figure 4-24 Replication of DNA. Nucleic acid chains can grow only in the $5' \rightarrow 3'$ direction. Thus, the new strand in one of the daughter duplexes is replicated continuously, whereas replication in the other daughter duplex is discontinuous. DNA ligase connects these fragments to form a continuous strand. The entire structure is called a *growing fork*.

For another example of how the differences between DNA and RNA relate to their differences in stability, consider that cytosine residues in DNA or RNA can be chemically converted into uracil by the loss of the amino group. This process, called *deamination,* alters the sequence of the molecule. As is detailed in Chapter 13, a uracil residue formed in DNA by the deamination of cytosine can be recognized by cell enzymes as a "mistake," perhaps because thymine, rather than uracil, is normally found in DNA. Certain enzymes can repair this induced base change and convert the uracil back into cytosine. In other words, the enzymes recognize a G-U base pair in DNA and convert it into the normal G-C. An incorrect uracil in RNA, however, cannot be recognized as an error and thus cannot be repaired. In short, then, the presence in DNA of thymine (rather than uracil) and of deoxyribose (rather than ribose) promotes stability.

Summary

The essential biopolymers in cells are the proteins and the nucleic acids. Certain general rules can be applied to the assembly of these polymers. Each is constructed one molecule at a time from a fixed set of subunits (20 amino acids for proteins, and four nucleotides for nucleic acids, with uracil in RNA instead of the thymine in DNA). The synthesis begins and ends at particular sites. The initial product is commonly altered to make the final product.

Most reactions in biology depend on the specificity in protein chains, which resides in the linear order of amino acids in the chains. The sequence of nucleotides in DNA encodes the information for the order of amino acids in proteins. The triplet genetic code is the mechanism that nature has evolved for storing this information in DNA. The translation of genetic information into protein requires the participation of enzymes and three kinds of RNA: messenger RNA (mRNA), which is a copy of the nucleotide sequence from DNA; transfer RNA (tRNA), which carries individual amino acids to the ribosomes, where mRNA is translated into proteins; and ribosomal RNA (rRNA), which functions to bring mRNA and tRNA together so that the protein-synthesizing reactions can occur. Certain protein factors aid in the three steps of protein synthesis: initiation, elongation, and termination. The highlights of these steps are as follows: methionine is attached to $tRNA_i^{Met}$ by methionyl-tRNA synthetase; methionyl-$tRNA_i^{Met}$ locates the AUG initiation codon on the mRNA; the correct amino acids attached to their tRNAs by specific aminoacyl-tRNA synthetases are brought to the ribosome, where one by one the correct tRNA is chosen by codon-anticodon pairing to insert its amino acid into the growing peptide. Finally, a termination codon (UAA, UAG, or UGA) is reached on the mRNA, and the finished peptide chain is released.

Nucleic acid synthesis, which relies on Watson-Crick

base pairing for the selection of the correct nucleotides, is much simpler than protein synthesis. Enzymes termed polymerases elongate nucleic acid chains in the $5' \rightarrow 3'$ direction by using ribo- or deoxyribonucleotide triphosphates as substrates. The synthesis of RNA is called transcription. RNA polymerases can initiate new RNA chains de novo at appropriate sites on DNA. Prokaryotes have one RNA polymerase, whereas eukaryotes have three, each catalyzing the synthesis of either mRNA, rRNA, or small RNAs such as tRNA. RNA primary transcripts are often modified in length before use.

DNA polymerases cannot initiate new chains de novo; they can only elongate from the $3'$ hydroxyl group of pre-existing primer molecules. Because the strands in a molecule of DNA are antiparallel, and because chain elongation proceeds $5' \rightarrow 3'$, DNA replication requires the continuous copying of one strand and the discontinuous copying of the other, followed by ligation.

References

General References

STRYER, L. 1981. *Biochemistry,* 2d ed. W. H. Freeman and Company.

WATSON, J. D. 1977. *Molecular Biology of the Gene,* 3d ed. Benjamin-Cummings.

Protein Synthesis: The Three Roles of RNA

The Genetic Code

Descriptions and explanations of some of the original experiments that broke the code.

CRICK, F. H. C. 1966. The genetic code: III. *Sci. Am.* **215**(4):55–62 (Offprint 1052).

GAREN, A. 1968. Sense and non-sense in the genetic code. *Science* **160**:149–155.

KHORANA, H. G. 1968. Nucleic acid synthesis in the study of the genetic code. In *Nobel Lectures: Physiology or Medicine (1963–1970),* Elsevier (1973).

NIRENBERG, M. W., and P. LEDER. 1964. RNA codewords and protein synthesis. *Science* **145**:1399–1407.

WOESE, C. B. 1967. *The Genetic Code.* Harper & Row.

Messenger RNA

The original demonstrations of mRNA in bacteria and the activity of mRNA in extracts.

BRENNER, S., F. JACOB, and M. MESELSON. 1961. An unstable intermediate carrying information from genes to ribosomes for protein synthesis. *Nature* **190**:576–581.

NATHANS, D., G. NOTANI, J. H. SCHWARTZ, and N. D. ZINDER. 1962. Biosynthesis of the coat protein of coliphage f2 by *E. coli* extracts. *Proc. Nat'l Acad. Sci. USA* **48**:1424–1431.

NIRENBERG, M. W., and J. H. MATTHEI. 1961. The dependence of cell-free protein synthesis in *E. coli* upon naturally occurring or synthetic polyribonucleotides. *Proc. Nat'l Acad. Sci. USA* **47**:1588–1602.

Transfer RNA

Descriptions of the discovery and action of tRNA in protein synthesis, the first tRNA sequences, and their three-dimensional structures.

CHAPPEVILLE, F., F. LIPMANN, G. VON EHRENSTEIN, B. WEISBLUM, W. J. RAY, and S. BENZER. 1962. On the role of soluble ribonucleic acid in coding for amino acids. *Proc. Nat'l Acad. Sci. USA* **48**:1086–1092.

HOAGLAND, M. B., M. L. STEPHENSON, J. F. SCOTT, L. I. HECHT, and P. C. ZAMECNIK. 1958. A soluble ribonucleic acid intermediate in protein synthesis. *J. Biol. Chem.* **231**:241–257.

HOLLEY, R. W., J. APGAR, G. A. EVERETT, J. T. MADISON, M. MARQUISEE, S. H. MERRILL, J. R. PENSWICK, and A. ZAMIR. 1965. Structure of a ribonucleic acid. *Science* **147**:1462–1465.

NISHIMURA, S. 1983. Structure, biosynthesis, and function of queuosine in transfer RNA. In *Progress in Nucleic Acid Research and Molecular Biology,* vol. 28, W. E. Cohn, ed. Academic Press.

PRESS, J., P. BERG, E. J. OFENGAND, F. H. BERGMAN, and M. DRECKMANN. 1959. The chemical nature of the RNA amino acid compound formed by the amino acid activating enzymes. *Proc. Nat'l Acad. Sci. USA* **45**:319–328.

RICH, A., and S.-H. KIM. 1978. The three-dimensional structure of transfer RNA. *Sci. Am.* **240**(1):52–62 (Offprint 1377).

SCHIMMEL, P. R., and D. SÖLL. 1979. Aminoacyl-tRNA synthetases: general features and recognition of transfer RNAs. *Annu. Rev. Biochem.* **48**:601–648.

SCHIMMEL, P., D. SÖLL, and J. ABELSON, eds. 1979. *Transfer RNA: Structure, Properties, and Recognition.* Cold Spring Harbor Laboratory.

SÖLL, D., and P. SCHIMMEL, eds. 1980. *Transfer RNA: Biological Aspects.* Cold Spring Harbor Laboratory.

SUSSMAN, J. L., and S.-H. KIM. 1976. Three-dimensional structure of a transfer RNA in two crystal forms. *Science* **192**:853–858.

Ribosomes

The identification of ribosomal proteins and their function and rudimentary descriptions of rRNA three-dimensional folding.

BERNABEU, C., and J. A. LAKE. 1982. Nascent polypeptide chains emerge from the exit domain of the large ribosomal subunit. *Proc. Nat'l Acad. Sci. USA* **79**:3111–3115.

BRIMACOMBE, R., P. MALY, and C. ZWIEB. 1983. The structure of ribosomal RNA and its organization relative to ribosomal protein. In *Progress in Nucleic Acid Research and Molecular Biology,* vol. 28, W. E. Cohn, ed. Academic Press.

CHAMBLISS, G., G. R. CRAVEN, J. DAVIES, K. KAVIS, L. KAHAN, and M. NOMURA, eds. 1980. *Ribosomes: Structure, Function, and Genetics.* University Park Press.

LAKE, J. A. 1981. The ribosome. *Sci. Am.* **245**(2):84–97 (Offprint 1501).

LAKE, J. A. 1983. Evolving ribosome structure: domains in archaebacteria, eubacteria, and eukaryotes. *Cell* **33**:318–319.

NOLLER, H. F. 1984. Structure of ribosomal RNA. *Annu. Rev. Biochem.* **53**:119–162.

NOMURA, M. 1973. Assembly of bacterial ribosomes. *Science* 179:864–873.

NOMURA, M., R. GOURSE, and G. BAUGHMAN. 1984. Regulation of the synthesis of ribosomes and ribosomal components. *Annu. Rev. Biochem.* 53:75–118.

THOMPSON, J. F., and J. E. HEARST. 1983. Structure of *E. coli* 16S RNA elucidated by psoralen crosslinking. *Cell* 32:1355–1365.

THOMPSON, J. F., and J. E. HEARST. 1983. Structure-function relations in *E. coli* 16S RNA. *Cell* 33:19–24.

WITTMAN, H. G. 1983. Architecture of prokaryotic ribosomes. *Annu. Rev. Biochem.* 52:35–66.

The Steps in Protein Synthesis

Detailed descriptions of the process of translation.

CASKEY, T. H. 1980. Peptide chain termination. *Trends Biochem. Sci.* 5:234–237.

HERSHEY, J. W. B. 1980. The translational machinery: components and mechanisms. In *Cell Biology: A Comprehensive Treatise,* vol. 4, D. M. Prescott and L. Goldstein, eds. Academic Press.

LAKE, J. A. 1981. Protein synthesis in prokaryotes and eukaryotes: the structural bases. In *Electron Microscopy of Proteins,* R. Harris, ed. Academic Press.

MAITRA, U., E. A. STRINGER, and A. CHAUDHURI. 1982. Initiation factors in protein biosynthesis. *Annu. Rev. Biochem.* 51:869–900.

MOLDAVE, K. 1985. Eukaryotic protein synthesis. *Annu. Rev. Biochem.* 54:1109–1150.

Nucleic Acid Synthesis

Complete summaries of the basic mechanics of nucleic acid synthesis.

KORNBERG, A. 1980. *DNA Replication.* W. H. Freeman and Company.

LOSICK, R., and M. J. CHAMBERLIN, eds. 1976. *RNA Polymerase.* Cold Spring Harbor Laboratory.

MCCLURE, W. R. 1985. Mechanism and control of transcription initiation in prokaryotes. *Annu. Rev. Biochem.* 54:171–204.

5

Principles of Cellular Organization and Function

THE discovery that all living matter is built up of smaller units—cells—was one of the notable scientific achievements of the nineteenth century. The pioneering research of Matthias Schleiden and Theodor Schwann (Chapter 1) was extended to an enormous number of plants, animals, and microorganisms, and today this once surprising finding is universally accepted. What perhaps remains surprising is that all cells, in all organisms, show certain common structural features, and that certain complicated metabolic events are carried out in similar ways. For example, the replication of DNA, the synthesis of proteins, and the metabolism of sugars are very similar in all organisms.

This chapter deals with the structural organization of cells, the function of many subcellular structures, and several universal metabolic processes. Some of these topics are covered in greater detail in later chapters. Our purpose here is to present an overview of the key problems and concepts of cell function.

Cells as the Basic Structural Units

All Cells Evolve from Other Cells

It is generally accepted that life first evolved several billion years ago in the form of tiny, one-celled organisms.

Some of the early organisms were photosynthetic and obtained their energy from sunlight. As the environment of the earth changed, so did these primitive cells. However, when multicellular forms of life evolved, their component cells retained the basic organization of free-living single cells. Each cell was distinct and potentially capable of being an autonomous and self-sufficient unit, but cells gradually came to depend more and more on other cells for certain aspects of their existence. By adopting intimate interdependence and cooperation between cells as a way of life, organisms could increase their abilities to grow and survive in different environments without having to evolve any new forms of organization. The new organisms that arose, ranging from the simple sponges to the most complex mammals and plants, consisted of masses of individual cells, with each type of cell capable of carrying out a particular task. The cellular division of labor within large organisms conferred on them advantages of enormous evolutionary importance.

Cells Can Be of Various Sizes

The smallest cells are *bacteria,* single-celled microorganisms that can be less than 1 μm in diameter (Figure 5-1). Some unicellular microorganisms, such as yeast, are slightly larger—about 5 μm in diameter; others, including the photosynthetic algae *Euglena gracilis,* are 50 μm in diameter. Among the largest single-celled organisms are the plant *Acetabularia,* which can be 8 cm in length, and the giant ostrich egg. The average cell in a multicellular animal ranges from 10 to 30 μm in diameter. Erythrocytes (nonnucleated blood cells) are only 7 μm in diameter. At the other extreme, certain nerve cells in mammals have long, slender extensions, called axons, which approach 1 m in length. Growing cells in a multicellular plant—for example, root cells—are from 20 to 30 μm in diameter. The dimensions of some nongrowing cells can be more than 10 times that value; this massive increase in plant cell volume is due mainly to the presence of a large, water-filled vacuole inside the cell.

Diffusion Limits the Sizes of Cells and Unicellular Organisms

Diffusion is the entropy-driven process by which molecules distribute themselves in whatever volume is available to them (see Figure 2-18). Because a cell coordinates its metabolic activities by diffusion alone, the rate at which molecules diffuse throughout the cell limits the typical cell's size to between 30 and 50 μm in diameter. More specifically, in most cells, whatever their shape, no metabolically active interior region is more than 15 to 25 μm from the cell surface.

Both theoretical and experimental analyses of diffusion have established that the rate of diffusion of any molecule over a particular distance is inversely proportional to the square of that distance. For molecules of biological interest, such as sugars, amino acids, and small peptides, equilibration by diffusion over a distance of a few micrometers occurs in several seconds, whereas diffusion over a distance of a few centimeters would require days. Larger molecules, such as proteins, diffuse even more slowly. Thus, if a cell were too large, its metabolism would be compromised by the too slow diffusion of oxygen or other small molecules from the cell surface to the interior, and by the similarly inefficient diffusion of waste or toxic molecules on their way out of the cell. Enzymes and other macromolecules would also require too much time to move across the cell.

Some unicellular organisms, such as the multinucleated slime mold *Physarum polycephalum* and certain algae (e.g., the giant *Acetabularia*), are larger than the optimum size of from 30 to 50 μm; these organisms have had to evolve some means of rapidly moving chemicals from one region of the cell to another. In the mechanism they have developed, which is called *cytoplasmic streaming,* a continuous ribbon of cytoplasm is pushed along one side of the cell, just under the surface membrane, and back along the other side of the cell. The ribbon or stream thus acts as a conveyor belt for the distribution of needed nutrients and for the disposal of wastes.

Extracellular Circulation Enables Cells to Receive Nutrients and to Eliminate Wastes

Because simple diffusion is so slow, all multicellular plants and animals have evolved circulatory systems for rapidly transporting simple chemicals, macromolecules, and chemical signals from one cell to another and for eliminating wastes. Few metabolically active human cells are more than 20 μm away from a *capillary,* the smallest type of blood vessel.

Molecules Enter Cells by Diffusion through the Plasma Membrane or by Transport through Specific Permeases

The rate at which molecules can enter or leave a cell also limits its size. The surface of all cells is a single *plasma membrane* composed largely of a phospholipid bilayer. This membrane defines the boundaries of the cell. A phospholipid bilayer is permeable to certain important gases, such as oxygen and carbon dioxide, and to water, and these substances can diffuse freely across it. However, a simple phospholipid bilayer is virtually impermeable to most of the molecules that a cell needs from its environment, such as sugars, amino acids, proteins, and inorganic ions (e.g., K^+ and Cl^-). To enable these molecules to enter and leave the cell, the plasma membrane utilizes many membrane proteins called *permeases* or *transporters.* These are proteins that are embedded in the

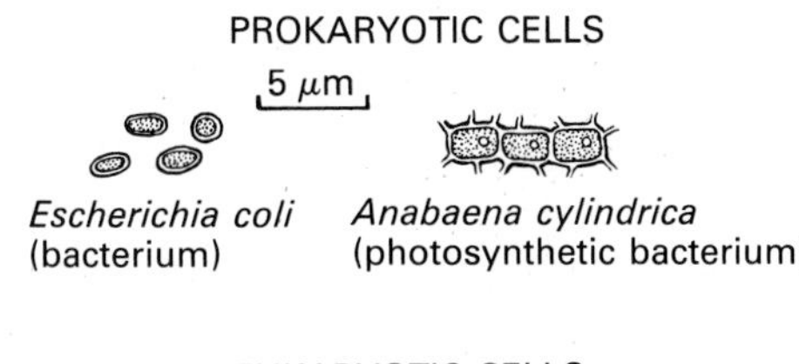

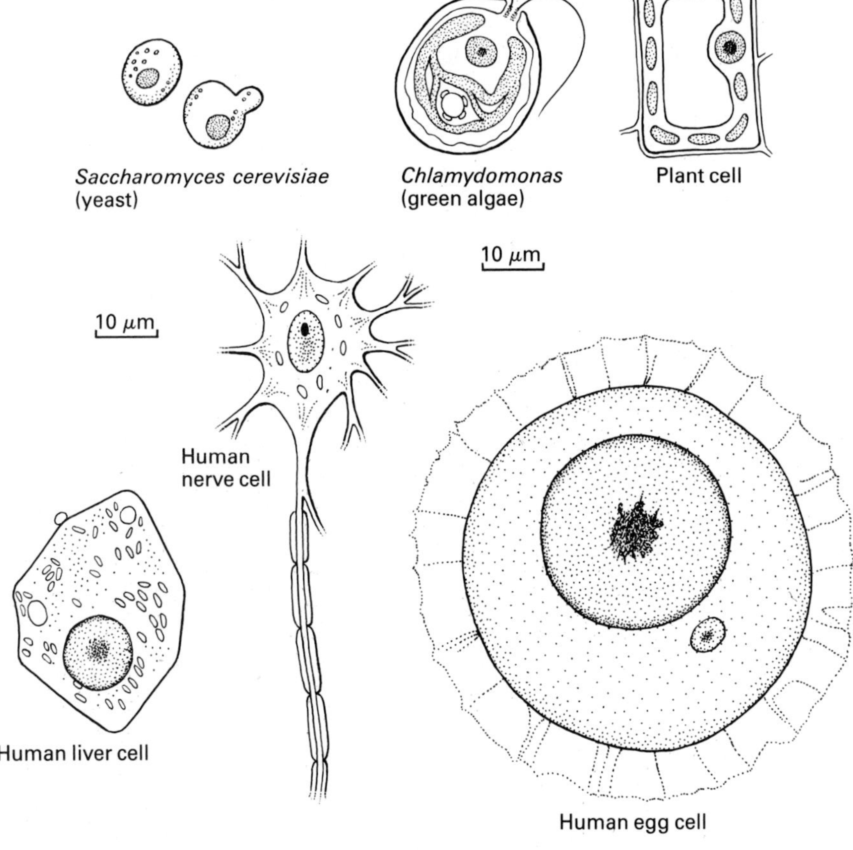

Figure 5-1 Although prokaryotic (bacterial) cells are generally smaller than eukaryotic cells, the sizes and shapes of cells in both classes vary considerably.

phospholipid bilayer, where they form channels that allow certain molecules to enter or leave the cell. (Permeases are discussed in detail in Chapter 15.) For example, some permeases that are present in all animal cells couple the hydrolysis of cellular ATP with the uptake of K^+ ions and extrusion of Na^+. In nerve and muscle cells, other permeases regulate the flow of K^+, Ca^{2+}, and Na^+ ions across the membrane; these ionic fluxes, in turn, are essential for the transmission of nerve impulses. (Ion permeases are discussed in Chapter 15, and nerve conduction is covered in Chapter 17.)

Whether transport is by free diffusion through the lipid bilayer or by permease catalysis, the rate at which molecules enter a cell is proportional to the area of the cell surface, and is thus roughly proportional to the square of the cell's diameter. [For a spherical cell of diameter d, the area is $A = \pi d^2$ (Table 5-1); for a cuboid cell of side s, $A = 6s^2$.] The volume of the cell, in contrast, is proportional to the cube of the diameter. [For a sphere, $V = (\pi/6)d^3$ (Table 5-1); for a cube, $V = s^3$.] If a spherical cell

Table 5-1 Surface-to-volume ratios for spherical cells of various diameters

Diameter (μm)	Surface area (μm^2)	Volume (μm^3)	Surface-to-volume ratio
1	3.14	5.2×10^{-1}	6.0
3	2.83×10	1.41×10	2.0
10	3.14×10^2	5.24×10^2	0.60
30	2.83×10^3	1.41×10^4	0.20
100	3.14×10^4	5.24×10^5	0.06
1000	3.14×10^6	5.24×10^8	0.006

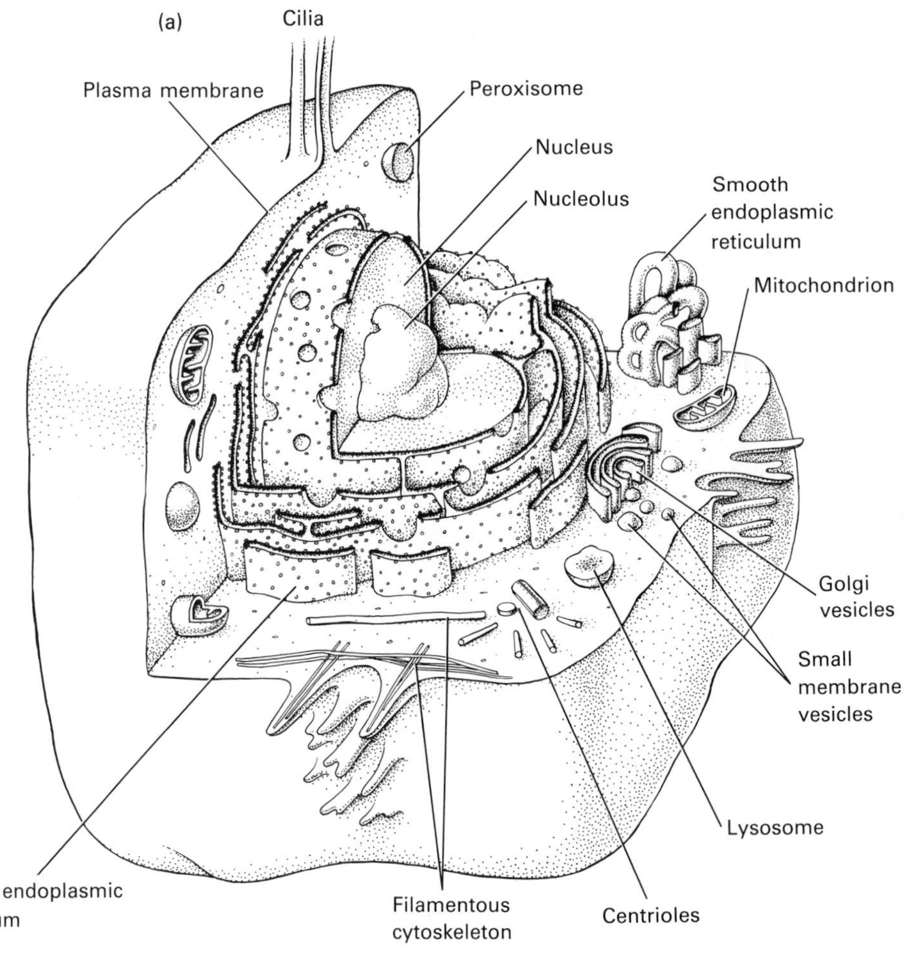

Figure 5-2 ***(right and opposite)*** Typical animal (a), plant (b), and bacterial (c) cells. Not every cell of a given type contains all of the organelles, granules, and fibrous structures shown here, nor are these substructures the only ones that might be present in a cell. Cells also vary considerably in their shapes and in the prominence of their various organelles and substructures.

became too large, the diffusion of metabolites through the surface membrane would be insufficient to supply its internal volume with all of the metabolites it needs. Many cells that are specialized for the absorption of nutrients have highly folded cell surfaces to increase the surface-to-volume ratio; an example is the absorbent cell of the small intestine, which will be discussed later (see Figure 5-49).

The fastest-growing single-celled microorganisms are generally the smallest. These are the ones with the greatest surface-to-volume ratios, which allow a maximum rate of supply of nutrients to the cell interior. Many bacterial populations, for example, can double every 20 min when they are growing aerobically. In general—although there are notable exceptions—the maximum growth rate of a single-celled organism is inversely proportional to its size.

Prokaryotic and Eukaryotic Cells Differ in Many Ways

Before the introduction of electron microscopes, it was generally assumed that all cells shared similar basic principles of organization. True, bacterial cells are much smaller than typical animal, plant, or even fungal cells; but because treatment of some bacteria with a Feulgen stain (a stain for DNA) revealed central masses of nuclear material, it seemed possible that all cells possessed both a defined nucleus and internal compartmentation. In addition, all cells—bacteria, plant, and animal alike—demonstrated certain key metabolic capacities. For example, virtually all cells utilized the same basic pathway for the production of chemical energy via the conversion of glucose into carbon dioxide.

However, the limited resolution of the light microscope (only about 0.5 μm) did not provide an accurate image of the internal structure of bacterial cells. The question of cell structure was not answered until the 1950s, when a variety of biochemical and genetic studies established that there are two radically different types of cells that have persisted independently for perhaps a billion or more years of biological evolution. The two cell types are *eukaryotic cells*—literally, those with a true nucleus—and *prokaryotic cells*—those lacking a defined nucleus (Figure 5-2). However, as is discussed in Chapter 25, prokaryotes and eukaryotes have apparently evolved from the same type of cell, although the nature of this ancestral cell is unknown.

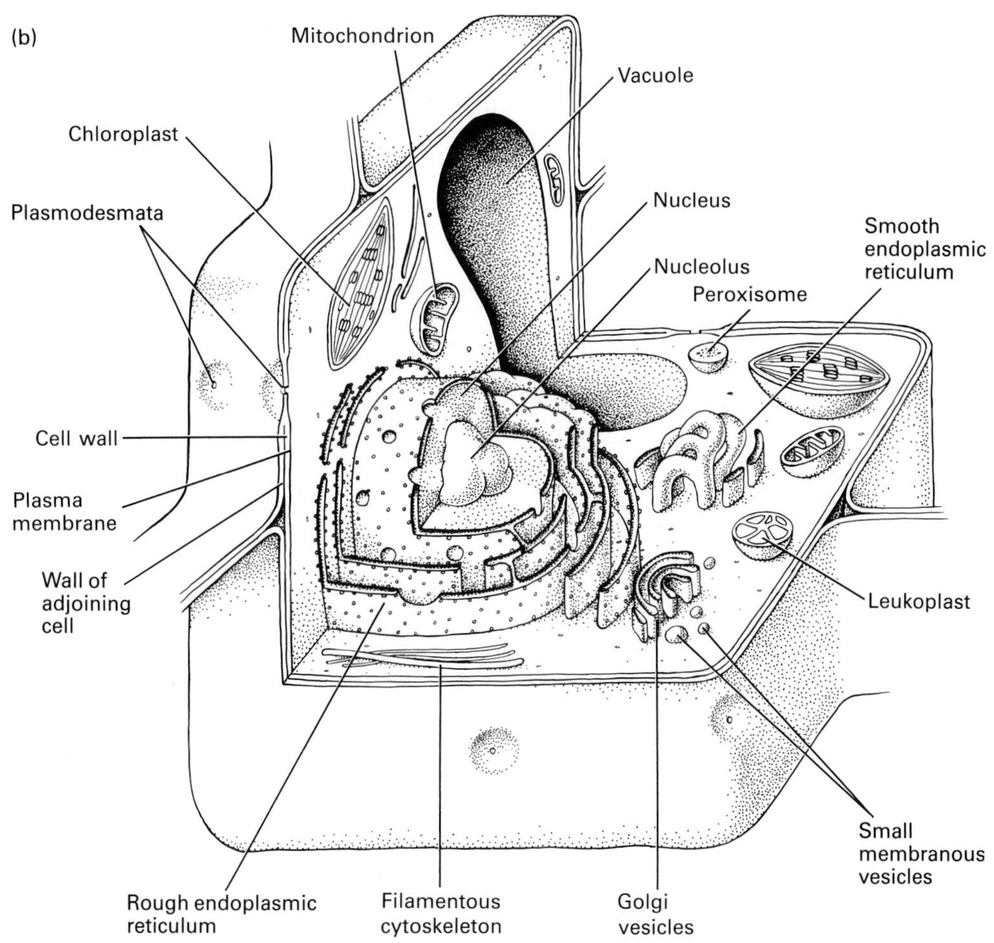

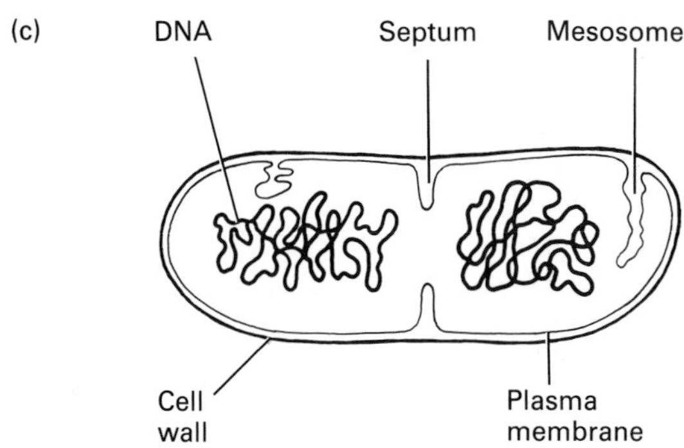

Within the prokaryotes, there are two separate lineages: *eubacteria* and *archaebacteria.* Far more is known of eubacteria than of archaebacteria, and, in general, when bacterial structure or metabolism is discussed, eubacteria are the subject. The eubacteria include most of the bacteria, as well as the organisms formerly known as blue-green algae, which are actually bacterial cells that can obtain energy from photosynthesis. Archaebacteria grow in unusual environments. The methanogens live only in oxygen-free milieus such as swamps; these bacteria generate methane (CH_4), which is also known as swamp gas, by the reduction of carbon dioxide. Other archaebacteria include the halophiles, which require high concentrations of salt to survive, and the thermoacidophiles, which grow in hot (80°C) sulfur springs, where pH values lower than 2 are common.

The eukaryotic nucleus is divided from the remainder of the cell by two phospholipid-containing membranes. Eukaryotes contain other subcellular regions that are bounded by closed phospholipid membranes (Figure 5-2): the *mitochondria,* in which most cellular ATP is generated; the rough and the smooth *endoplasmic reticula,* a network of membranes in which glycoproteins and lipids are synthesized; *Golgi vesicles,* which direct membrane constituents to their appropriate places in the cell; *lysosomes,* which degrade many proteins, nucleic acids, and lipids; *peroxisomes,* which metabolize hydrogen peroxide; and assorted smaller vesicles. Plant cells also contain *chloroplasts,* the sites of photosynthesis, and large fluid-filled organelles termed *vacuoles.*

The cytoplasm of animal and plant cells—the region lying outside of the nucleus—also contains an array of fibrous proteins called, collectively, the *fibrous cytoskeleton.* Among these proteins are the *actin filaments* (see Figure 3-40) and the *centrioles.* Prokaryotic cells generally do not contain a fibrous cytoskeleton nor internal closed membranes, but infoldings of the plasma membrane often form structures termed *mesosomes* (Figure 5-2c).

Macromolecules in Prokaryotic and Eukaryotic Cells

Prokaryotes and Eukaryotes Contain Similar Macromolecules

The volume of a typical mammalian cell is several hundred times that of a typical bacterial cell; yet the composition of all prokaryotic and eukaryotic cells is strikingly similar. By weight, about 70 percent of a typical cell consists of water. Small molecules such as salts, lipids, amino acids, and nucleotides account for another 7 percent (Table 5-2). The other approximately 23 percent is composed of macromolecules, including nucleic acids, proteins, and polysaccharides (Table 5-3).

HeLa cells, which are derived from a human cervical carcinoma, grow very well in cell culture; consequently they are among the most thoroughly studied mammalian cells. (Some properties of HeLa cells are discussed in Chapter 6.) Table 5-3 shows that a HeLa cell contains about 4×10^6 ribosomes and about 7×10^5 mRNA molecules, or roughly 150 times as many of each as is found in *Escherichia coli,* a prokaryotic cell. The number of protein molecules in a HeLa cell is about 5×10^9; these represent from 5000 to 10,000 different polypeptide species. The much smaller *E. coli* cell contains about 3×10^6 protein molecules, representing from 1000 to 2000 different species.

The Amount of DNA per Cell Varies over a Wide Range in Different Organisms

The differences between typical prokaryotic and eukaryotic cells can be seen at many levels. First, there are major differences in the amount of genetic information per cell. The genome of *E. coli* contains 4.4×10^{-15} g [0.0044 picograms (pg)] of DNA, an amount equal to 4×10^6 base pairs, as indicated in Table 5-4. Because three DNA bases encode the position of each amino acid in a protein and because an average protein contains about 400 amino acids, approximately 1200 DNA base pairs are used to encode each protein species. Thus the *E. coli* DNA has a maximum coding capacity of about 3300 different proteins. However, not all of the bacterial DNA encodes proteins, although a large part of it does. A typical bacterial cell actually contains 2000 or so different species of proteins.

All eukaryotic cells contain more DNA than prokaryotic cells do. As can be seen in Table 5-4, yeast cells—

Table 5-2 Some of the chemical components (the water and the small molecules) in a rapidly growing bacterial cell, *Escherichia coli**

Component	Percentage of total cell weight	Average molecular weight	Approximate number per cell	Number of different kinds
H_2O	70	18	4×10^{10}	1
Inorganic ions (Na^+, K^+, Mg^{2+}, Ca^{2+}, Fe^{2+}, Cl^-, PO_4^{3-}, SO_4^{2-}, etc.)	1	40	2.5×10^8	20
Carbohydrates and precursors	3	150	2×10^8	200
Amino acids and precursors	0.4	120	3×10^7	100
Nucleotides and precursors	0.4	300	1.2×10^7	200
Lipids and precursors	2	750	2.5×10^7	50
Other small molecules (hemes, quinones, breakdown products of food molecules, etc.)	0.2	150	1.5×10^7	200

*The small molecules in a HeLa cell—a well-studied human tumor cell—are similar in composition to those in *E. coli.* However, the volume of a HeLa cell is several hundred times larger.

Table 5-3 The main macromolecular components of *E. coli* and HeLa cells

Component	Amount per HeLa cell	Amount per *E. coli* cell
Total DNA	15 picograms (pg)*	0.017 pg†
Total RNA	30 pg	0.10 pg
Total protein	300 pg (5×10^9 molecules, of average m.w. 40,000)	0.2 pg (3×10^6 molecules, of average m.w. 40,000)
Cytoplasmic ribosomes	4×10^6	3×10^4
Cytoplasmic tRNA molecules	6×10^7	4×10^5
Cytoplasmic mRNA molecules‡	7×10^5	4×10^3
Nuclear precursor rRNA molecules	6×10^4	
Heterogeneous nuclear RNA molecules§	1.6×10^5	
Total dry weight	400 pg	0.4 pg

*HeLa cells are hypotetraploid; i.e., they contain about four copies of each chromosome. The normal diploid human DNA complement is about 5 pg per cell.

†A rapidly growing *E. coli* cell contains, on the average, four DNA genomes. Each genomic DNA weighs 0.0044 pg.

‡An average chain length of 1500 nucleotides is assumed.

§An average chain length of 6000 nucleotides is assumed; this group of molecules contains precursor mRNAs.

whose genomes are among the simplest of eukaryotic genomes—have about three times as much DNA as *E. coli* does. The cells of higher plants and animals have from 40 to 1000 times as much DNA as *E. coli,* although the genomes of some amphibians are as large as 200 pg. If all of this DNA were used to encode proteins, then some animal cells would be able to encode as many as 3×10^6 different types of proteins. However, the amount of DNA that actually does encode proteins is usually a small fraction of the total, as is explained in Chapter 10. It is widely believed that invertebrates can make about 5000 proteins and that humans can make about 50,000—far fewer than the theoretical capacities of their respective genomes. However, vertebrates can make *several million* different types of antibody molecules. As is discussed in Chapter 24, these are generated by specific rearrangements of DNA segments encoding immunoglobulins in immune cells and also by specific mutations that occur in these rearranged genes. The functions of noncoding DNA are discussed in Chapter 11.

The Organization of DNA Differs Greatly in Prokaryotic and Eukaryotic Cells

In all prokaryotes studied to date, most or all of the cellular DNA is in the form of a single circular molecule. For example, the *E. coli* genetic information is found in a duplex circular DNA molecule of 4×10^6 base pairs. The cell is said to have a single chromosome, although the arrangement of DNA within this chromosome is very different from that in the chromosomes of eukaryotic cells. Eukaryotic chromosomes are highly structured complexes of DNA and proteins.

Several species of prokaryotes have small circular DNA molecules termed *plasmids* in addition to the chromosomal DNA. Plasmids, which have between 1000 and 30,000 base pairs, generally encode proteins that are not essential for the growth of the cell. Many plasmids encode the proteins that are required for resistance to antibiotics or other toxic materials. The *E. coli* F (male) plasmid determines the difference between male and female strains, and encodes the proteins required for cell-to-cell contact during sexual mating. These plasmids are of great use to experimental molecular biologists—they are essential tools of *recombinant DNA* technology, as we shall see in Chapter 7.

The nuclear DNA of all eukaryotic cells, in contrast, is divided between two or more different *chromosomes.* The number and the size of individual chromosomes vary widely among different eukaryotes (see Table 5-4; Chapter 10 describes eukaryotic chromosomes in more detail). Yeast, for example, has 12 to 18 chromosomes, each of which, on the average, contains only 20 percent of the

Table 5-4 The DNA content of various cells

Organism	Size of DNA genome: Number of base pairs	Size of DNA genome: Total length (mm)	Maximum number of proteins encoded*	Number of chromosomes (haploid)†
PROKARYOTIC				
Escherichia coli (bacterium)	4×10^6	1.36	3.3×10^3	1
EUKARYOTIC				
Saccharomyces cerevisiae (yeast)	1.35×10^7	4.60	1.125×10^4	17
Drosophila melanogaster (insect)	1.65×10^8	56	1.375×10^5	4
Homo sapiens (human)	2.9×10^9	990	2.42×10^6	23

*Assuming 1200 base pairs per protein.

†Haploid refers to a single set of chromosomes. Most insect and human cells are diploid, so they have twice the number of chromosomes shown.

DNA of an *E. coli* chromosome. Human cells, at the other extreme, contain two sets of 23 chromosomes, each of which contains, on the average, about 30 times the amount of DNA present in an *E. coli* cell. Each chromosome is believed to contain a single linear double-stranded DNA molecule. (Some eukaryotic cells also contain plasmid DNAs; these have been well characterized in yeasts and in a few plants.)

As was discussed in Chapter 1, among the most important discoveries in the early history of cell biology were those concerning chromosomes. In most eukaryotic organisms, nearly all the cells are *diploid* ($2n$), which means that they contain two copies of each chromosome. *Haploid* ($1n$) cells contain only one copy of each chromosome. In mammals, virtually all body cells are diploid, whereas sperm and egg cells are haploid. When a sperm fertilizes an egg, the result is a diploid cell called a *zygote*, which grows and divides repeatedly by mitosis to give rise to the somatic, or body, cells of the mammal. In haploid organisms, such as some yeasts, diploid cells are transient; almost as soon as they are formed, they divide into haploid cells by meiosis (see Chapter 6 for a discussion of yeast division).

Almost all chromosomal DNA in eukaryotic cells is associated with a set of five different basic proteins termed *histones*. The interaction between the histones and the DNA is very regular; that is, every sequence of 150 to 180 base pairs of DNA is bound to one molecule of the histone H1 and to two molecules each of the histones H2A, H2B, H3, and H4. The amino acid sequence of some of these histones has been extraordinarily conserved during evolution. The H3 from peas is virtually identical to the H3 from cows (see Figure 10-15). Histones are unknown in prokaryotic cells. Bacterial DNA is apparently associated with a type of protein that does not resemble histones, but its identity is not known.

Microscopy and the Internal Architecture of Cells

A simple light microscope was the instrument used by Schleiden and Schwann when they first identified individual cells, and light microscopy has continued to play a major role in biological research. The development of electron microscopes greatly extended the ability to resolve subcellular particles and yielded much new information on the organization of plant and animal tissue. The nature of the cellular images obtained by these techniques depends on the type of light or electron microscope employed and on the way in which the cell or tissue under observation has been prepared. Each technique is designed to emphasize particular structural features of the cell, and each has its own limitations. Figure 5-3 shows how a typical cell, a human leukocyte, appears when different techniques are used.

The *resolution* of a microscope is its ability to distinguish two objects that are close together. The minimum distance, d, at which two objects can be resolved is a function of λ, the wavelength of the radiation used; n is the refractive index of the air or fluid above the specimen; and α is the aperture angle (Figure 5-4):

$$d \approx \frac{0.6\lambda}{n \sin \alpha}$$

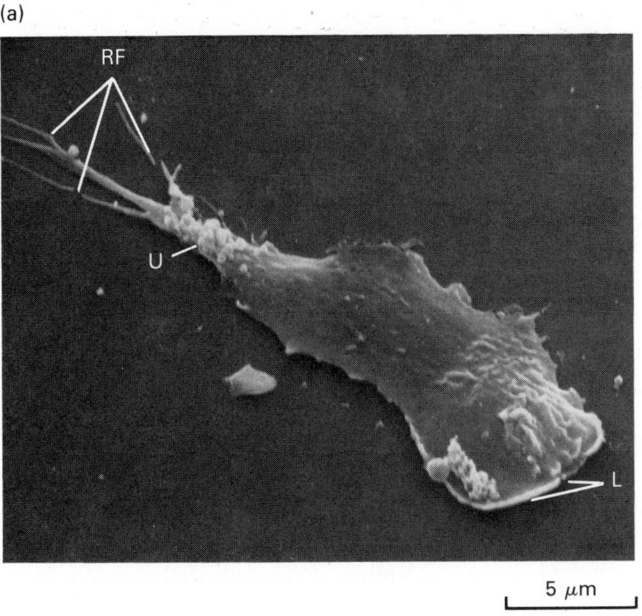

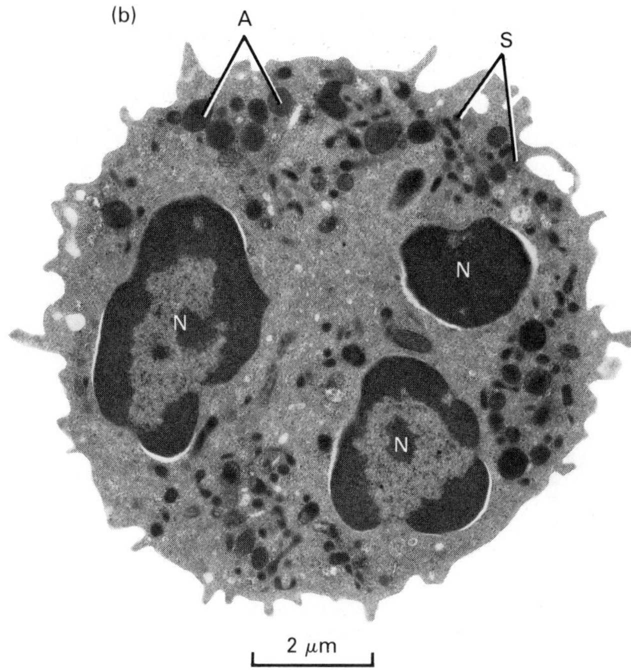

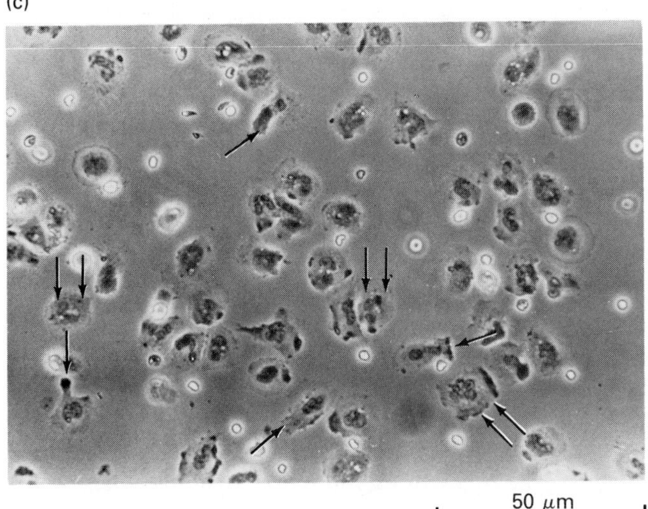

Figure 5-3 Views of human leukocytes derived from three different microscopic techniques. This white blood cell is polymorphonuclear, meaning that its nucleus contains many irregularly shaped lobes. (The microscopic techniques are explained in Figure 5-4 and in the text following.) (a) A scanning electron micrograph. The three-dimensional appearance of the cell surface is characteristic of images obtained by this technique. The shape of this cell indicates that it is a migrating cell. Such a cell has a wide, flattened projection (a lamellipodium, L) at its leading edge, and a narrow tail (the uropod, U) ending in retraction fibers (RF). (b) A thin-section transmission electron micrograph. The leukocyte contains numerous granules that are of two types: the larger, azurophil granules (A) and the smaller, specific granules (S). Also evident in this section are three lobes of the single nucleus (N) and some mitochondria, which are not present in large numbers in these cells. (c) A light micrograph (using phase-contrast optics) of a field of polymorphonuclear leukocytes. These cells have attached to a glass slide. Some of the cells have spread out and have become firmly attached to the glass at many points on their surfaces; these stationary cells are indicated by double arrows. Other cells are migrating along the glass, and their direction of movement is indicated by a single arrow. Note that only the nuclei and a few of the larger cytoplasmic particles are visible with this technique. *Courtesy of M. J. Karnofsky; cells prepared by J. M. Robinson, Department of Pathology, Harvard Medical School.*

The Light Microscope Utilizes the Refraction of Visible Light

The resolution of a light microscope can be increased (that is, *d* in the preceding equation can be decreased) by increasing the refractive index of the medium. This can be done in several ways: for example, oil can be used instead of water between the specimen and the objective lens. With this method, however, the increase in the resolution is small, because the refractive index of water is 1.0 and that of oil is 1.5. The most important variable is the wavelength of the incident light. With visible light of the shortest wavelength possible (0.4 μm, for blue light), the maximum resolution is about 0.2 μm. This is a great improvement over the 100-μm resolution of the unaided human eye. Nonetheless, even the large subcellular struc-

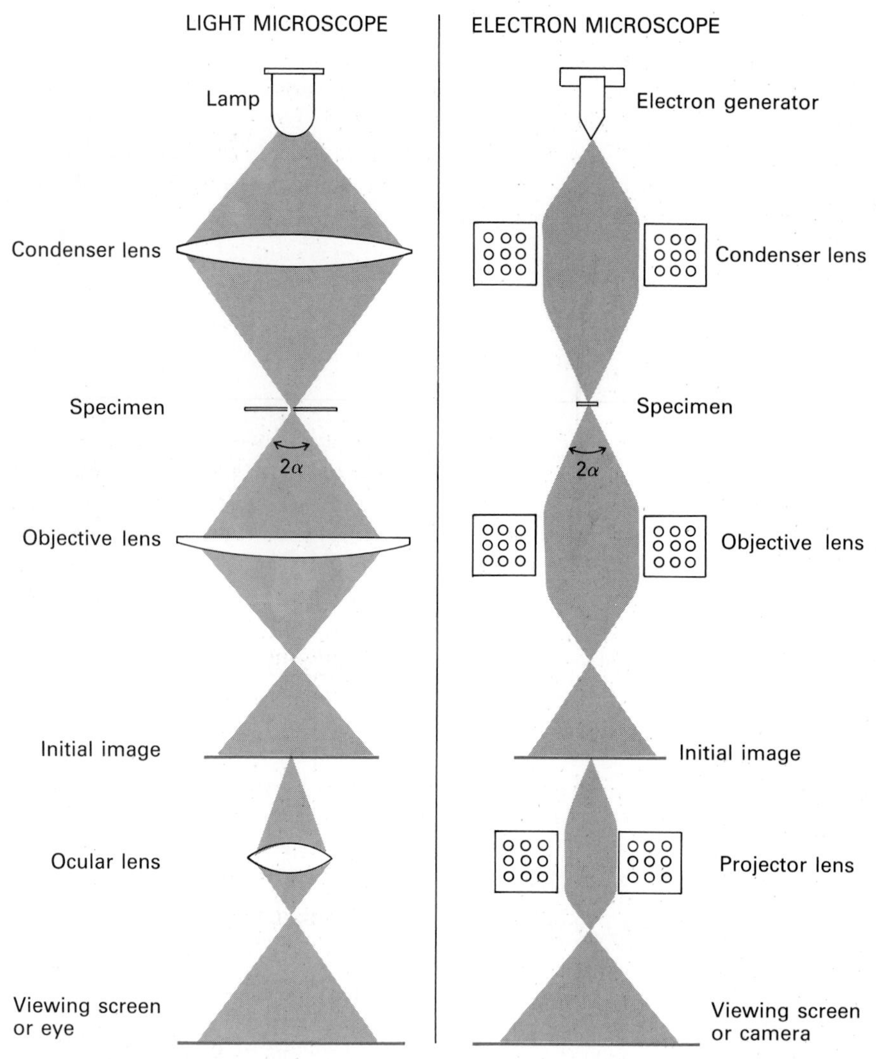

Figure 5-4 A comparative diagram of a light microscope and a conventional (transmission) electron microscope. In the standard light microscope, light is focused by a condenser lens onto the plane of the specimen. Light passing through the specimen is focused by the objective and ocular lenses in such a way that the image is magnified as many as 1000 times. The aperture angle, α (measured from the vertical), is a function of the individual objective lens. When an oil-immersion lens is used, the space between the specimen and the objective lens is filled with oil, which has a relatively high index of refraction, n. This increases the resolution. In the electron microscope, the specimen is exposed to a beam of electrons (rather than a beam of light); the pattern created by the electrons passing through the specimen is focused by two lenses—the objective lens and the lens of a projector—onto either a screen or camera film. The entire column, from the electron generator to the screen, is maintained at a very high vacuum. The aperture angle, α, is actually very small; it is greatly magnified in this drawing.

tures such as mitochondria (which are 0.1 to 0.2 μm in diameter) are barely visible under the light microscope, and smaller structures such as ribosomes and DNA are invisible.

To prepare a specimen for the light microscope, the sample is commonly "fixed" with a solution containing alcohol or formaldehyde; these compounds denature most proteins, and nucleic acids. Formaldehyde also cross-links amino groups on adjacent molecules:

$$R_1{-}NH_2 + H{-}\overset{\overset{\displaystyle O}{\|}}{C}{-}H \longrightarrow R_1{-}NH{-}\underset{\underset{\displaystyle H}{|}}{\overset{\overset{\displaystyle OH}{|}}{C}}{-}H$$

$$R_1{-}NH{-}\underset{\underset{\displaystyle H}{|}}{\overset{\overset{\displaystyle OH}{|}}{C}}{-}H + R_2{-}NH_2 \longrightarrow$$

$$R_1{-}NH{-}\underset{\underset{\displaystyle H}{|}}{\overset{\overset{\displaystyle H}{|}}{C}}{-}NH{-}R_2 + H_2O$$

where R_1 and R_2 are the rest of the molecules. These covalent bonds stabilize protein-protein and protein–nucleic acid interactions and render the molecules insoluble and stable for subsequent procedures. Next, the sample is embedded in paraffin and cut into thin sections; it is then stained to bring out certain features of the cell.

Different cellular constituents (such as the nucleus, the cytoplasm, and the mitochondria) absorb visible light to about the same degree. This absence of contrast makes it difficult to distinguish such entities by using a standard light microscope without staining; often not even the nucleus of a growing cell can be seen. Consequently, certain techniques have been developed to make it possible for an investigator using a light microscope to distinguish particular cellular or subcellular features. Of course, none of these techniques gives a "complete" picture of a cell.

The *phase-contrast microscope* takes advantage of the fact that different regions of a cell have slightly different densities, and thus scatter (or refract) light to different extents (Figure 5-5). This microscope is widely used to study growing tissue culture cells, which are otherwise transparent (Figure 5-6a and b).

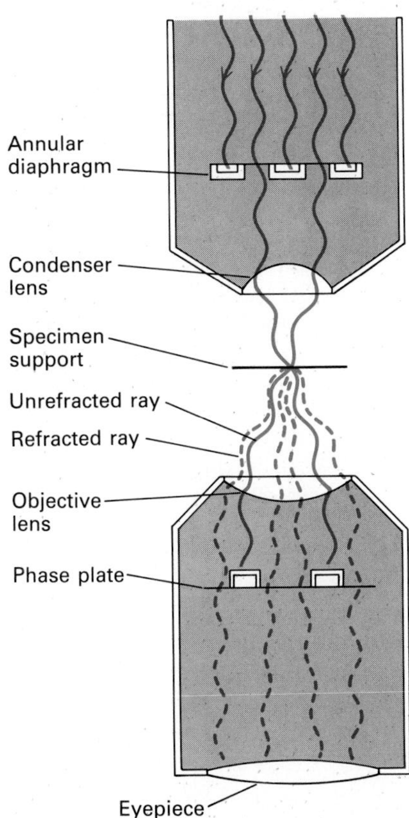

Figure 5-5 A diagram of the phase-contrast microscope. Incident light passes first through an annular diaphragm, which focuses a circular annulus (a ring) of light on the sample. If the incident light is unobstructed in its passage through the specimen, it will be focused onto the annular phase plate in such a way that the sample appears white. However, if, as usually occurs, some light rays are refracted (by a region of the sample with a different refractive index), after passing through the phase plate they will interfere with the unrefracted rays. This produces an image in which the degree of darkness or brightness of a region of the sample depends on the refractive index of that region. The phase-contrast microscope makes it possible to view substructures in unstained cells.

Nomarski microscopy, like phase-contrast microscopy, is a type of interference microscopy, but it uses polarized light. Contrast in Nomarski optics is greatest in regions of the specimen where the light absorption changes dramatically over a small area. Edges of structures, such as the outline of the cell and the nucleus, are clearly seen because of the good contrast (Figure 5-6c).

Many chemical stains bind to molecules that have specific features: for example, *eosin* and *methylene blue* bind to many different kinds of proteins; *benzidine* binds to heme-containing proteins and nucleic acids; and the *fuchsin* used in Feulgen staining binds to DNA. Antibodies that are tagged with fluorescent molecules can be used to detect a wide range of cellular proteins (Figure 5-6b and d). If an enzyme catalyzes a reaction that produces a colored precipitate, the enzyme can be detected in cell sections. This last technique, which is called *cytochemical staining*, can be used with both light and electron microscopes (Figure 5-7).

The Electron Microscope Depends on the Absorption or Scattering of a Beam of Electrons

In the electron microscope, a beam of electrons is substituted for visible light. Electrons accelerated by a decrease in voltage of 50,000 volts have the properties of a wave with a wavelength of only 0.005 nm. The theoretical maximum resolution of the electron microscope is 0.5 nm, about the diameter of a single atom. This would give the electron microscope 400 times the resolution of the light microscope and 200,000 times that of the unaided human eye. The effective resolution of the electron microscope in the study of biological systems is actually considerably less than the ideal; under the best conditions, a resolution of a few angstroms, or a fraction of a nanometer, can be obtained.

Like the light microscope, the conventional (transmission) electron microscope is used for viewing thin sections. It is similarly necessary that samples be fixed before they are sectioned (Figure 5-8). The image in the transmission electron microscope depends on variations in the absorption of the incident electrons by different molecules in the preparation, whereas the image in a light microscope depends on variations in the absorption of light. The composition of cells and tissues is such that without staining, there is little differentiation of components in the electron microscope. Heavy metals, which are very good absorbers of electrons, can easily be seen in the electron microscope; heavy-metal-containing compounds like osmium tetroxide will preferentially stain certain cellular components, such as membranes. (A micrograph showing an osmium-stained membrane appears in Figure 14-5.) A recent advance is the introduction of gold particles and the iron-containing protein ferritin as electron-dense tags; these can be used to detect antibody molecules that are bound to specific target proteins in thin sections (Figure 5-9).

The scanning electron microscope allows the investigator to view the surfaces of unsectioned cells or other specimens. The sample is fixed and dried; then it is coated with a thin layer of a heavy metal, such as platinum, by evaporation in a vacuum. Inside the microscope, a beam of electrons is scanned over the sample; electrons that are scattered by the specimen (rather than those that pass through, as in a transmission microscope) are focused onto a scintillation detector, and the resulting signal is

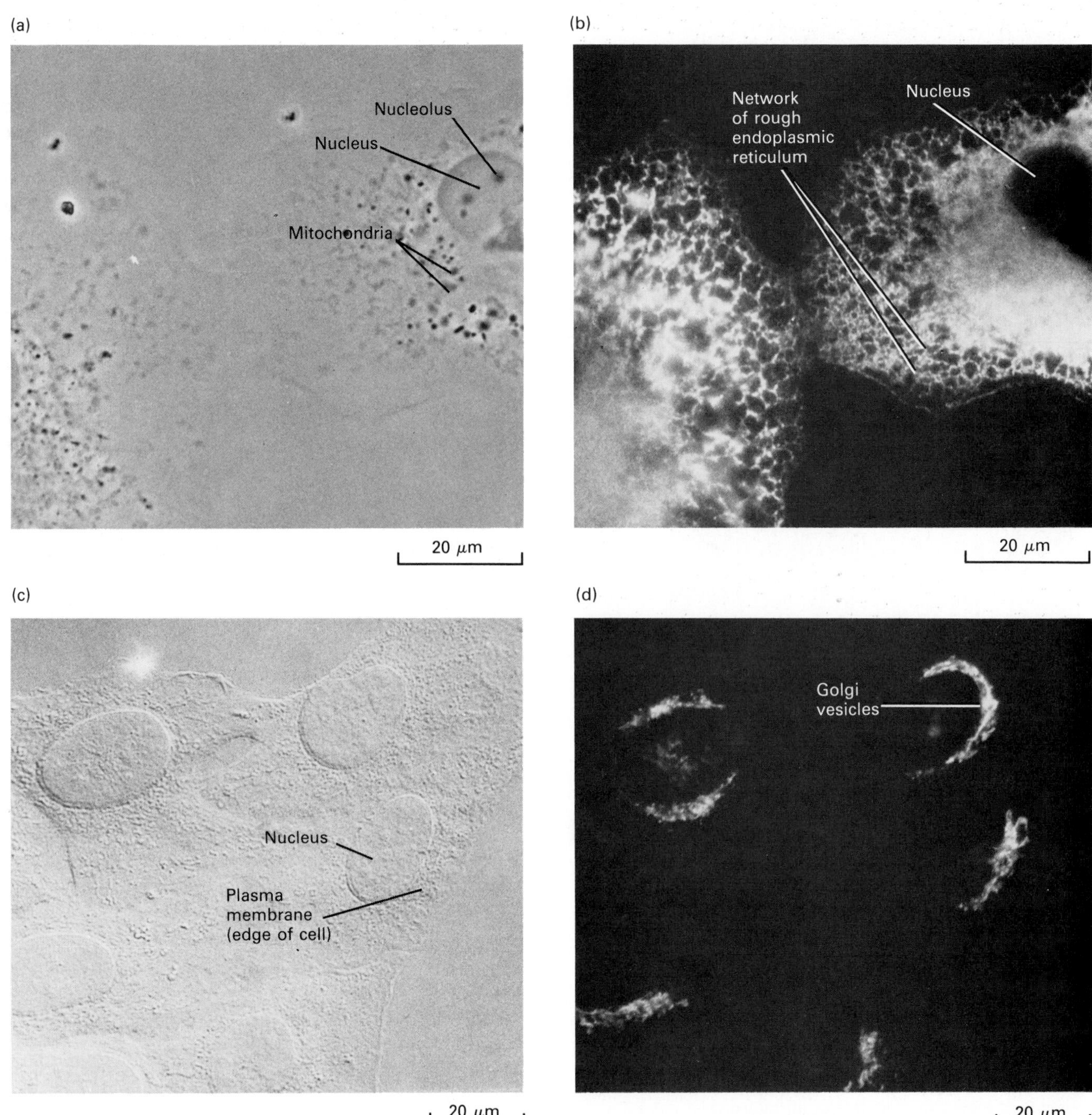

Figure 5-6 Cells viewed by phase-contrast and Nomarski microscopy, and specific organelles detected by staining with antibodies. (a) Baby hamster kidney fibroblasts viewed through a phase-contrast microscope. The nucleus and nucleolus are visible, as are the silhouettes of some of the large cytoplasmic particles, such as mitochondria. (b) The same hamster fibroblasts after they have been fixed, treated with a rabbit antibody that is specific for a rough endoplasmic reticulum protein, and then treated with a fluorescent goat anti-rabbit immunoglobulin, which reveals the bound rabbit antibodies by fluorescence. (c) Rat fibroblasts viewed by Nomarski optics. (d) The same rat fibroblasts after they have been fixed and stained with fluorescent antibodies that are specific for purified Golgi vesicle proteins. Note that the Golgi vesicles are in the region surrounding the nucleus, in contrast with the rough endoplasmic reticulum, which forms a network that fills most of the cytoplasm. [See D. Louvard, H. Reggio, and G. Warren, 1982, *J. Cell Biol.* 92:92.] *Reproduced from the* Journal of Cell Biology *by copyright permission of The Rockefeller University Press.*

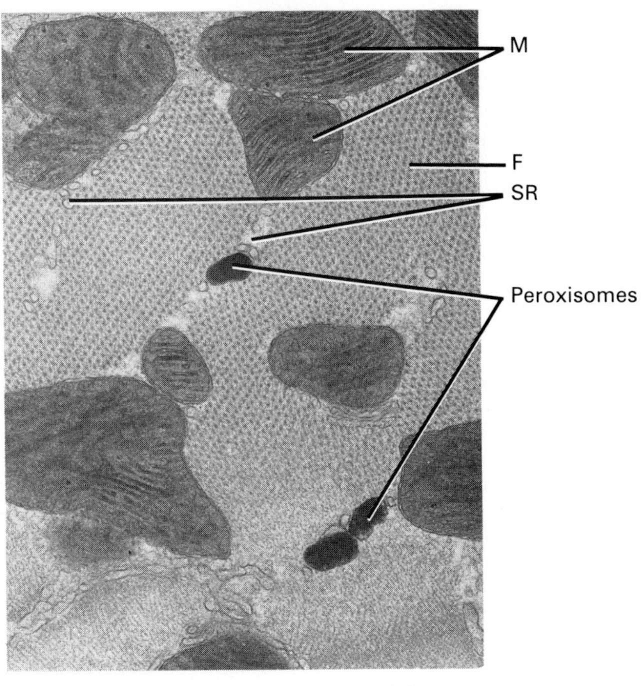

Figure 5-7 Thin sections of rat heart muscle stained for catalase and viewed through an electron microscope. Catalase is found in peroxisomes, small membranous vesicles. It is detected by treating cell sections with hydrogen peroxide and a dye abbreviated DRB; the catalase-catalyzed oxidation of the dye results in an insoluble dense precipitate that is visible as dots scattered throughout the section. Also visible (but unstained) are mitochondria (M); vesicles of the sarcoplasmic reticulum (SR), a set of membranes that store Ca^{2+} ions; and (in cross section) the actin and myosin contractible fibers (F). *From H. D. Fahmi and S. Yokota, 1981,* International Cell Biology 1980–1981, *H. G. Schweiger, ed., Springer-Verlag, p. 640. Courtesy of H. D. Fahmi.*

displayed on a cathode-ray tube. Because the amount of scattering of the beam in any direction is a function of the angle made by the beam with the surface, the specimen has a three-dimensional appearance in the scanning electron micrograph (see Figure 5-3a). The resolving power of scanning electron microscopes is only about 10 nm, much less than that of transmission instruments.

Flow Cytometry Is Used to Sort Cells

By far the most frequent use of optical systems in cell biology is to obtain an image of a cell or a cell fraction. However, one optical technique serves a different function: it can identify particular cells and collect them. An instrument called a *fluorescence-activated cell sorter,* or FACS, can select one cell from many.

Suppose that an antibody to a specific cell surface molecule is chemically conjugated with a fluorescent dye. Cells bearing this molecule will bind the antibody. When these cells are exposed to the correct wavelength of light, delivered by a laser beam, they will fluoresce. In the FACS, a flowing stream of tiny droplets of liquid from a cell suspension is formed, with each droplet containing at most one cell. The droplets flow past a laser beam; the laser light causes the droplets with cells that contain the antibody-dye complex to fluoresce. The fluorescence of each droplet is measured by a detector. Those droplets that contain a fluorescent cell are given a negative electric charge. The stream then passes between two electrically charged plates; droplets that are charged (fluorescent) are separated from those that are not (Figure 5-10). The FACS can select one cell that bears a specific surface marker from thousands that do not; the selected cell can then be grown in culture. A machine with this optical capability can detect any absorbing or emitting characteristic of a cell.

Other uses of *flow cytometry,* as the technique is generally called, include the measurement of a cell's DNA and RNA content and the determination of its general shape and size. Most FACS instruments, for example, contain a detector that measures the amount of laser light scattered by each droplet (Figure 5-10). The amount of scattered light is proportional to the number and size of particles, such as cells, in the droplet. In this way, the instrument can measure simultaneously the size of a cell in the droplet (from the amount of scattered light) and the amount of DNA it contains (from the amount of fluorescence of the DNA-binding dye).

The Structure of the Cell Nucleus

The development of the electron microscope in the early 1950s made possible the study of subcellular organization with a resolving power many times greater than that associated with the light microscope. It immediately became clear that the internal structure of eukaryotes is much more complex than that of the prokaryotes. Especially striking is the intricate structure of the eukaryote nucleus.

The Eukaryotic Nucleus Is Bounded by a Double Membrane; the Nucleus Contains the Nucleolus, a Fibrous Matrix, and DNA-Protein Complexes

The eukaryotic nucleus is surrounded by two phospholipid-containing membranes. The *inner nuclear membrane* defines the nucleus itself. In many cells the *outer nuclear membrane* is continuous with an extensive cyto-

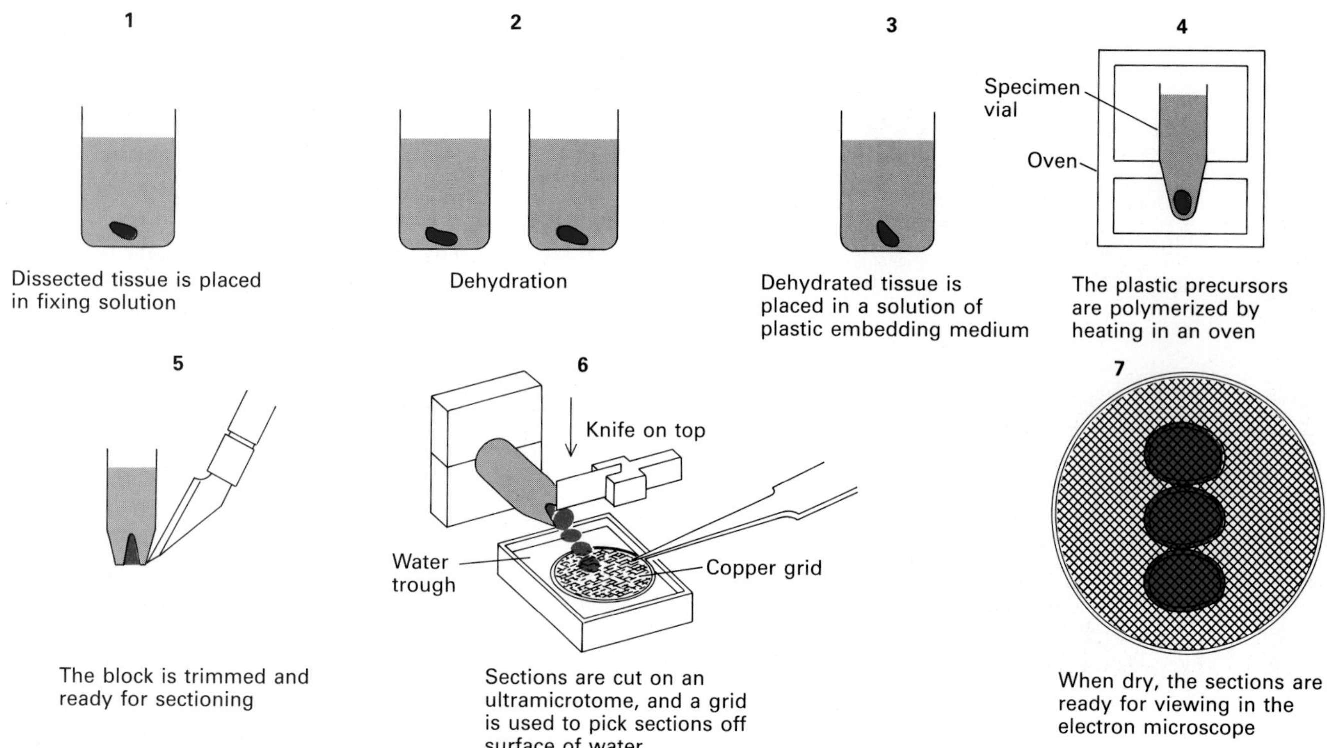
1
2
3
4
Specimen vial
Oven
Dissected tissue is placed in fixing solution
Dehydration
Dehydrated tissue is placed in a solution of plastic embedding medium
The plastic precursors are polymerized by heating in an oven
5
6
7
Knife on top
Water trough
Copper grid
The block is trimmed and ready for sectioning
Sections are cut on an ultramicrotome, and a grid is used to pick sections off surface of water
When dry, the sections are ready for viewing in the electron microscope

Figure 5-8 Steps in preparing a thin section of a tissue for transmission electron microscopy.
(1) Rapid fixation of the tissue is important in order to denature and stabilize the proteins, membranes, and other constituents before degradation of the material begins. The tissues are dissected, cut into small cubes, and plunged into a fixing solution, which cross-links proteins and immobilizes them. Glutaraldehyde is a fixing solution that is frequently used. Another is osmium tetroxide, which stains intracellular membranes and other macromolecules.
(2) The tissue is dehydrated by placing it in successively higher concentrations of alcohol or acetone.
(3) The dehydrated tissue is immersed in a solution of plastic embedding medium.
(4) The specimen is then put in an oven; heat causes the plastic precursors to polymerize into a hard plastic.
(5) The specimen block is trimmed.
(6) Sections less than 0.1 μm thick are cut on an ultramicrotome, a fine-slicing instrument with a diamond blade. The sections are floated off the blade edge onto the surface of water in a trough. A copper grid, coated with carbon or other material, is used to pick up the sections.
(7) After drying, the sections are ready for use.

plasmic membrane system termed the rough (or ribosome-studded) endoplasmic reticulum (ER). The functions of the rough ER will be discussed later, but it should be noted that the space between the inner and outer nuclear membranes is continuous with the *lumen,* or central cavity, of the ER. At the *nuclear pores,* the inner and outer nuclear membranes appear to fuse together (Figure 5-11). The pores are ringlike in shape and appear to be constructed of a specific set of membrane proteins. Pores probably function as channels for the regulated movement of material between the nucleus and the cytoplasm.

In the nondividing, or *interphase,* nucleus, the chromosomes are elongated and they are only about 25 nm thick; they are not observable in the light microscope. However, a suborganelle of the nucleus, the *nucleolus,* is easily recognized in the light microscope (see Figure 5-6). The nucleolus possesses a region of one or more chromosomes termed the *nucleolar organizer.* This region contains many copies of the DNA that directs the synthesis of ribosomal RNA; most of the ribosomal RNA in the cell is synthesized in the nucleolus, as we discuss in Chapter 9. In fact, some of the ribosomal proteins are added to the ribosomal RNAs within the nucleolus as well. The finished (or partly finished) ribosomal subunit passes through a nuclear pore into the cytoplasm.

Although we have some knowledge of how the DNA is packaged in the cell nucleus, we are ignorant about the large-scale organization of the nucleus. The nonnucleolar regions of the nucleus are called the *nucleoplasm.* In the electron microscope, the nucleoplasm can be seen to have regions of high DNA concentration, often closely associated with the nuclear membrane. Fibrous proteins termed

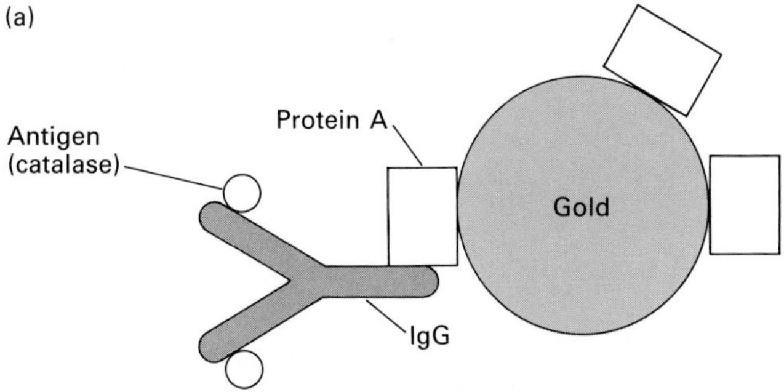

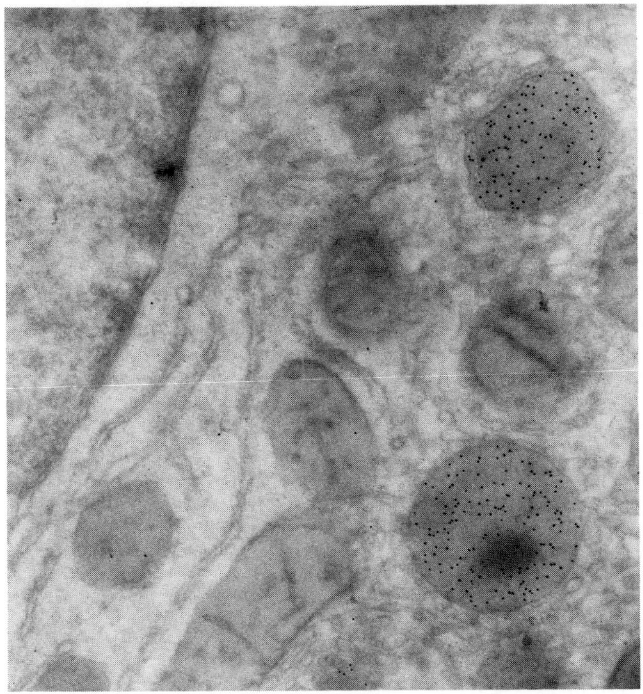

Figure 5-9 Detection of a specific protein, the enzyme catalase, in a cell section. (a) A slice of liver tissue was fixed with glutaraldehyde and then sectioned. An antibody (IgG) to catalase was added to the section. A bacterial protein, the *Staphylococcus aureus* protein A, was complexed with gold particles 5 nm in diameter and then added to the section. The protein A binds tightly to the common Fc domain of the antibody, and the gold particle allows the immune complex to be viewed in the electron microscope. (b) The location of the gold particles in the peroxisomes indicates the presence of the catalase. *From H. J. Geuze, J. W. Slot, P. A. van der Ley, and R. C. T. Scheffer, 1981,* J. Cell Biol. **89**:653. *Reproduced from the* Journal of Cell Biology *by copyright permission of The Rockefeller University Press.*

lamins are associated with the inner surface of the inner membrane, and appear to bind DNA to the nuclear membrane. (Lamins are discussed more extensively in Chapter 21.) Beyond these rudimentary facts, the organization of materials and activities in the nucleoplasm is largely a mystery.

The major physiological function of the nucleus is to direct the synthesis of RNA. In a growing or differentiating cell, the nucleus is the site of vigorous metabolic activity. In other cells, such as resting mast cells (blood-borne cells that release histamine when triggered by allergens) or mature avian erythrocytes, the nucleus is inactive or dormant, and minimal DNA or RNA biosynthesis takes place.

The DNA in Prokaryotic Cells Is Confined to a Nuclear Region

Prokaryotic cells lack a nuclear membrane. Most of the genomic DNA is in the central region of the cell; yet, in many prokaryotes, such as *Bacillus subtilis,* the DNA is apparently attached to the plasma membrane at many points (Figure 5-12). The majority of the ribosomes are found near the periphery of the cell, possibly because they are excluded from the central regions occupied extensively by the DNA. The DNA must be folded back on itself many times: an *E. coli* chromosomal DNA molecule stretched out to its full length would be over 1 mm long, or 1000 times as long as the cell itself.

The Syntheses of Prokaryotic and Eukaryotic Messenger RNAs Differ in Many Important Respects

In eukaryotic cells, the chromosomal DNA is retained within the nucleus. Messenger RNA is fabricated there and then transported to the cytoplasm. Thus, the synthesis of mRNA and its translation into protein by the ribosomes are totally separate events that take place in different regions of the cell.

In prokaryotes, there is no nuclear membrane to separate chromosomal DNA from the ribosomes. As a consequence, ribosomes can and do begin to translate an mRNA molecule into protein while the RNA is still nascent—that is, while it is still being synthesized on the DNA. Thus, *some* ribosomes are translating mRNA that is bound to DNA. Exactly how the bacterial DNA and the growing mRNAs are organized in nuclear regions of the cell is unclear.

The temporal separation of the synthesis of mRNA in the nucleus of eukaryotic cells from its translation allows the mRNA to undergo a number of important postsynthetic modifications (e.g., splicing and addition of poly A; these are covered in detail in Chapter 9) before it is utilized in protein synthesis. Most of these modifications are totally absent in prokaryotic cells, although some, such as RNA splicing, do occur infrequently.

A second difference is that prokaryotic mRNA often begins to be degraded (at its 5′ end) just after its translation has begun, and even before its synthesis is completed.

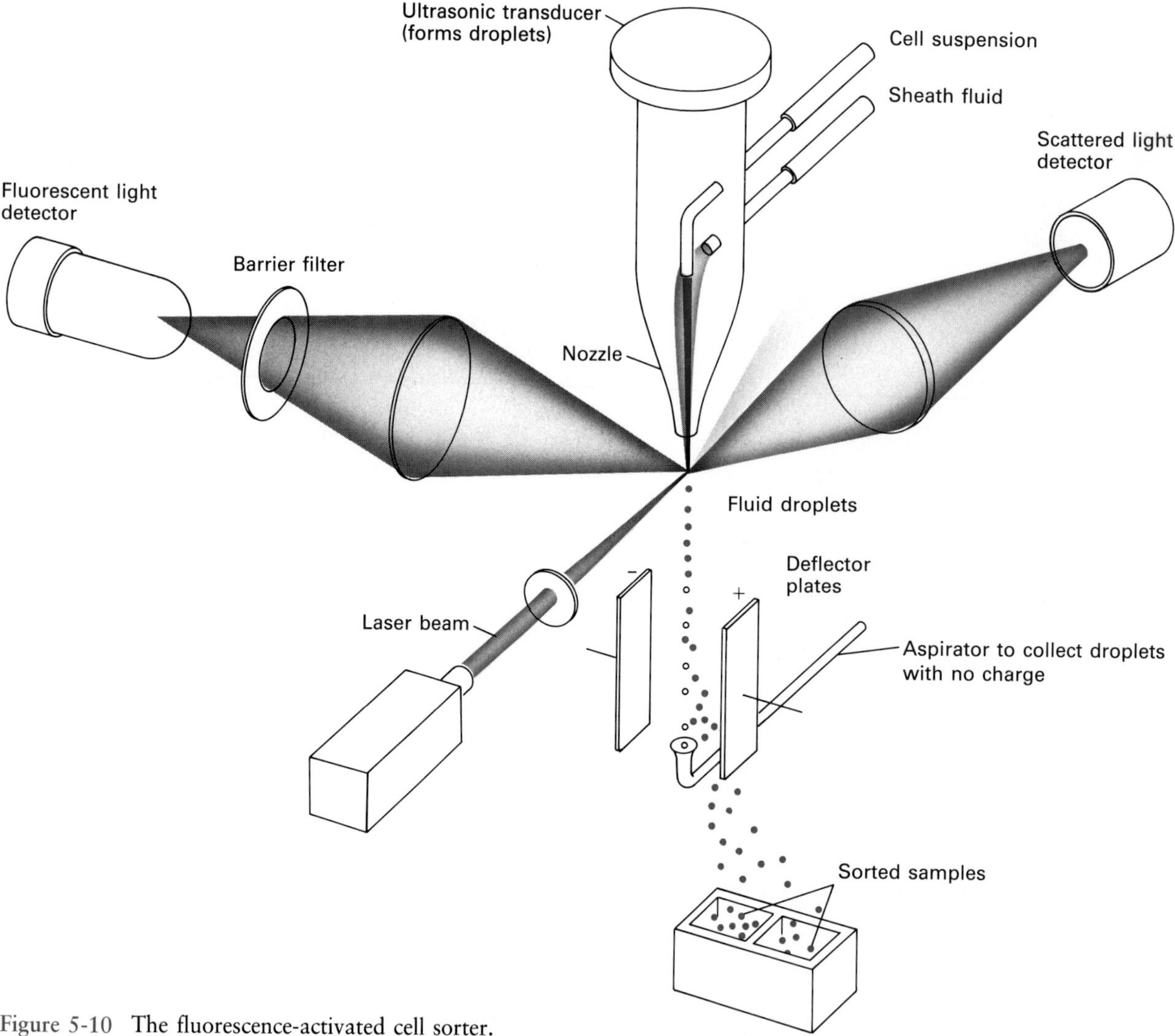

Figure 5-10 The fluorescence-activated cell sorter. To use this technique, a concentrated suspension of cells is allowed to react with a fluorescent antibody, or with a fluorescent dye that binds to a particle or molecule such as DNA. The suspension is mixed with a buffer (the sheath fluid) and forced through a nozzle. Tiny droplets that each contain at most a single cell are formed. The stream of droplets passes through a laser light beam. The fluorescent light emitted by each droplet (due to the fluorescent antibody or dye) is quantified, and the droplet is given an electric charge proportional to the amount of fluorescence. Droplets with no charge and with different electric charges are separated by an electric field and collected. Simultaneously, the light scattered by each droplet is quantified; from this measurement the size and the shape of the cell in the droplet can be determined. It takes only milliseconds to sort each droplet, so many cells can be passed quickly through the machine. In this way, cells that have unusual surface proteins can be separated and then grown. *After D. R. Parks and L. A. Herzenberg, 1982, in* Methods in Cell Biology, *vol. 26, Academic Press, p. 283.*

Prokaryotic mRNA generally has a short half-life—on the order of seconds to minutes. In contrast, cytoplasmic mRNA in eukaryotic cells is relatively stable, with a half-life of hours—although some mRNAs (especially those in single-celled eukaryotes) are much shorter-lived.

Cell Division and DNA Replication

A necessary property of all growing cells—prokaryotic and eukaryotic alike—is the ability to duplicate their genomic DNA and to pass along identical copies of this genetic information to every daughter cell. All growing cells undergo a *cell cycle* that comprises two periods: (1) *cell division* and separation of daughter cells, and (2) *interphase,* the period of cell growth. Prokaryotic and eukaryotic cells differ markedly in their schemes for coordinating DNA synthesis and cell division and for ensuring the continuity of genomic DNA.

(a)

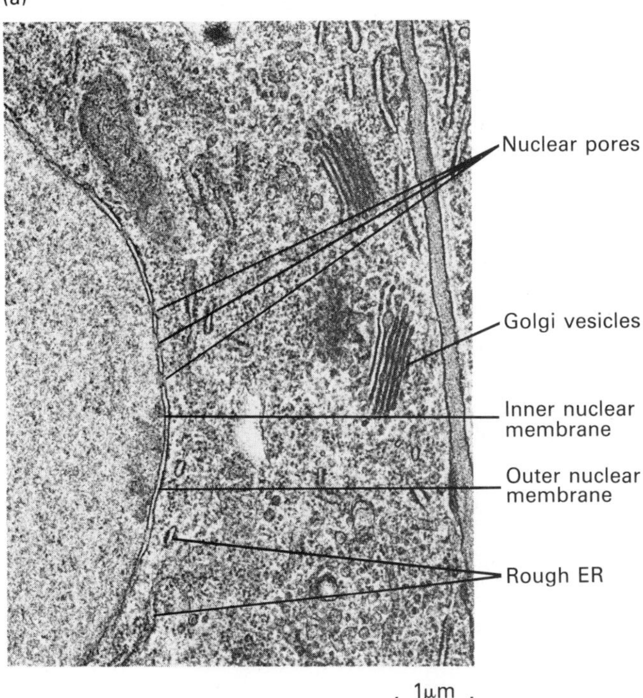

(b)

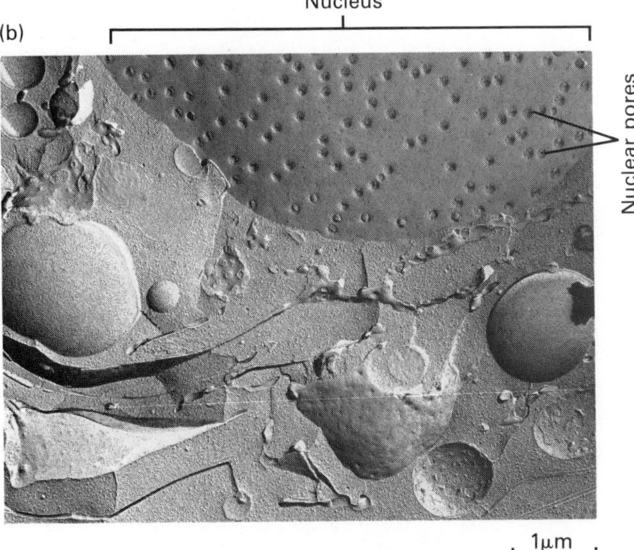

Figure 5-11 Views of the nuclear envelope. (a) A thin section of an *Arabidopsis* (plant) cell, showing the pores between the nuclear membranes. The rough endoplasmic reticulum and stacks of Golgi vesicles are also visible. (b) A freeze-etched preparation of an onion root-tip cell, showing the nucleus with pores in the nuclear membrane. The pores appear to be filled with a granular substance. (The freeze-etching technique is described in Figure 14-26.) *Part (a) from Biophoto Associates/Myron C. Ledbetter/Brookhaven National Laboratory. Part (b) courtesy of D. Branton.*

The Cell Cycle in Prokaryotes Consists of DNA Replication Followed by Cell Division

The genome of a prokaryotic cell is a single circular molecule of DNA. In rapidly growing bacteria, DNA is replicated throughout much of the cell cycle; when replication is complete, *cytokinesis,* or cell separation, follows immediately. No visible apparatus contributes to cell division. Apparently, the two daughter DNA strands are linked at different points on the plasma membrane to ensure that one copy of DNA is delivered to each daughter (see Figure 5-12). In certain bacteria, infolded plasma membrane regions termed *mesosomes* provide anchors for DNA. There is no condensation and decondensation of the DNA, as there is in eukaryotic cells during mitosis.

The Cell Cycle in Eukaryotes Is Resolvable into Several Discrete Stages

Different eukaryotic cells grow and divide at quite different rates. Yeast cells, for example, can divide and double in number in 120 min and so can the cells of sea urchin embryos. Most growing plant and animal cells take from 10 to 20 h to double in number, although some duplicate at a much slower rate. Many cells, such as nerve cells and striated muscle cells, do not divide at all; although such cells continue RNA, protein, and membrane synthesis, their DNA does not replicate. A general scheme describing the cell cycle for all growing and dividing eukaryotic cells is shown in Figure 5-13.

The replication of DNA and the synthesis of the histone proteins occurs only in the *synthetic* (S) phase of the interphase. In this period, each double-helical DNA molecule is replicated into two identical daughter DNA molecules; histones and other chromosomal proteins bind quickly to the newly replicated DNA. The S phase is preceded and followed by two *gap* periods of interphase—G_1 and G_2, respectively—in which there is no net synthesis of DNA, although damaged DNA can be repaired then. During the G_2 period, a cell contains two copies of each of the DNA molecules present in the G_1 cell. Throughout interphase there is continued cellular growth and continued synthesis of other cellular macromolecules, such as RNA, proteins, and membranes. The cell finally divides in the *mitotic* (M) period; identical copies of the cellular DNA are then distributed to each of the daughter cells.

In different growing mammalian cells, the S, G_2, and M periods are roughly constant—about 7 h, 3 h, and 1 h, respectively. The most variable period is the G_1, which can be as short as 2 to 3 h and as long as many days. Tissue culture cells that have exhausted a needed hormone or nutrient cease to grow, and their DNA does not replicate. They are generally stopped in the cell cycle at a stage called G_0. (This stage is not called G_1 because the

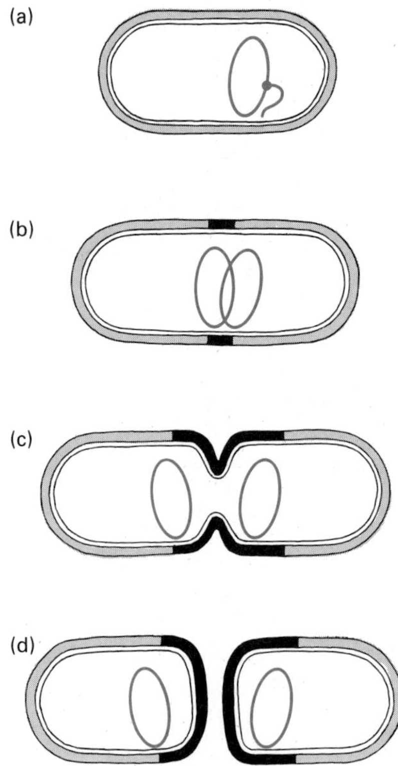

Figure 5-12 The DNA in prokaryotic cells is attached to the cell membrane, and remains attached during cell division. (a) The circular chromosome, which has already begun replication, is attached to the plasma membrane. (b) When chromosomal replication is complete, the new chromosome has an independent point of attachment to the membrane. New membrane and cell wall forms at the center of the cell. (c) More sections of membrane and cell wall form between the points of attachment of the two chromosomes. Part of this growth invaginates to give rise to a septum dividing the cell. (d) Cell division is complete; each of the two daughter cells has a DNA chromosome attached to the membrane.

cells are not prepared or preparing for DNA replication.) The stimulation of such cells to resume growth, such as by the addition of a missing hormone to the medium, induces them to return to the G_1 stage. Similarly, nonreplicating cells such as nerve or striated muscle cells are terminally differentiated, and are generally stopped in the G_0 stage.

Mitosis Is the Complex Process That Apportions Chromosomes Equally to Daughter Cells

During interphase, chromosomes are not visible by light microscopy; DNA-protein complexes called *chromatin* are dispersed throughout the nucleoplasm. The beginning of the cell division, or *mitosis,* is signaled by the appearance of chromosomes as thin threads inside the nucleus. Although the events that follow unfold smoothly, they are conventionally divided into four substages, *prophase, metaphase, anaphase,* and *telophase* (Figure 5-14).

During prophase and metaphase, a chromosome consists of two identical coiled filaments, the *chromatids* (often called *sister chromatids*), each of which contains one of the two identical daughter DNA molecules that were produced in the S phase (Figure 5-15). As prophase progresses, the chromatids become more condensed, owing to the packing of the nucleoprotein fibers. Sister chromatids are held together at their *centromeres,* or constricted regions.

At this time, the small cylindrical particles termed *centrioles* begin to play a key role. As the centrioles move apart, they generate fibrous microtubules that radiate from them in all directions. (Centrioles, which are themselves constructed of microtubules, duplicate during interphase by forming daughter centrioles at right angles to their own positions, as shown in Figure 5-16.) Some of these microtubules connect the centrioles with the *kinetochores,* granular regions that are attached to the centromeres of the chromatids. Other microtubules form a net-

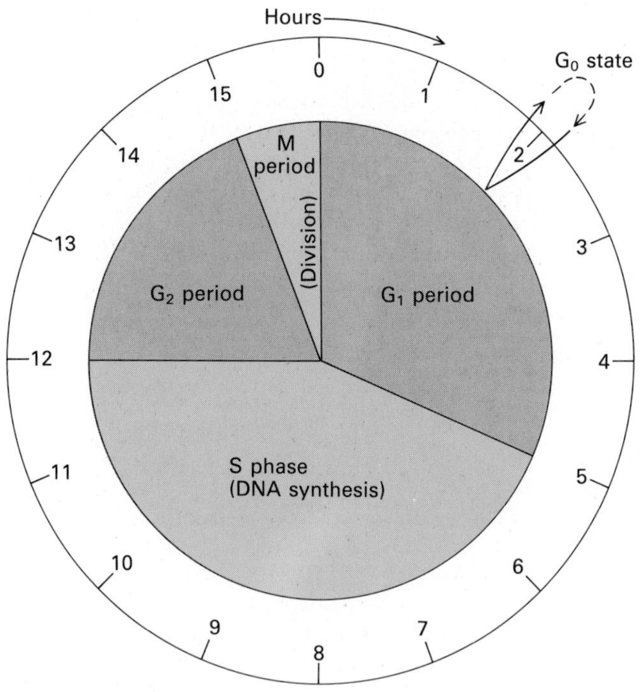

Figure 5-13 The cell cycle in a mammalian cell having a generation time of 16 h. The three periods spanning the first 15 h or so— the G_1 (first gap) period, the S (synthetic) phase, and the G_2 (second gap) period—make up the interphase, during which DNA and other cellular macromolecules are synthesized. The remaining hour is the M (mitosis) period; it is during this period that the cell actually divides.

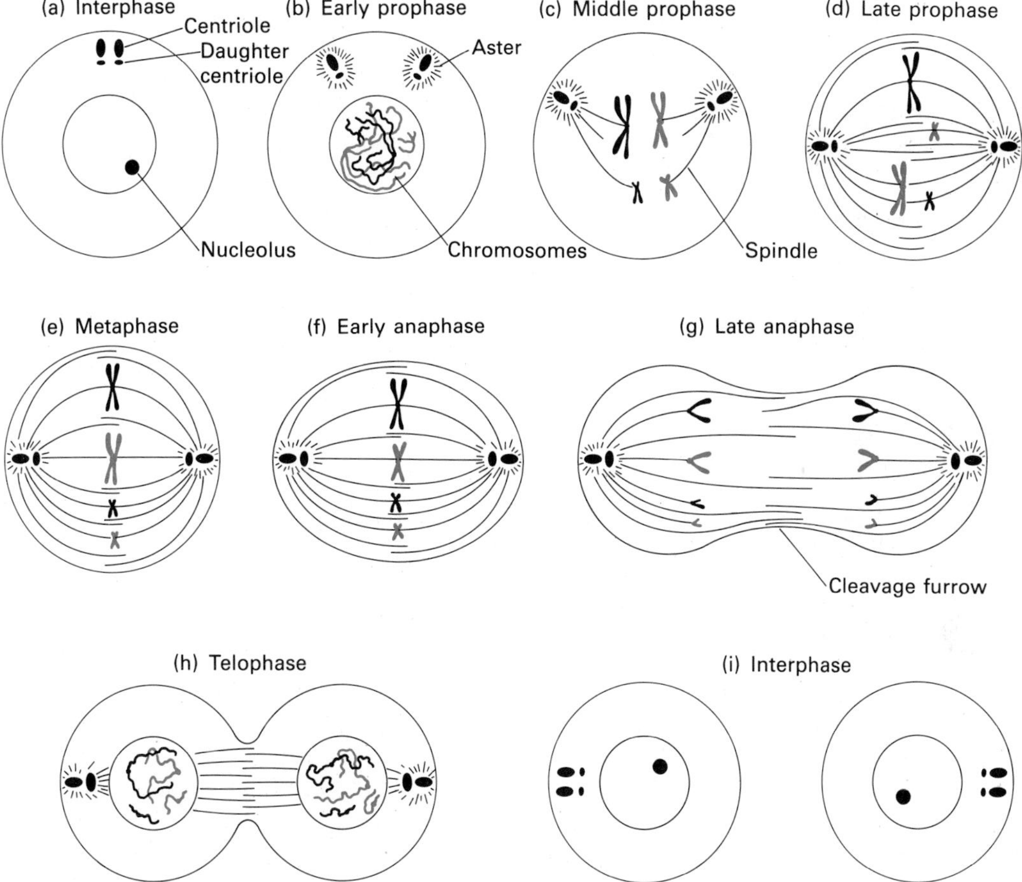

Figure 5-14 The stages of mitosis and cytokinesis in an animal cell.
(a) *Interphase.* During the S phase, chromosomal DNA is replicated and histones and other proteins are bound to the DNA strands, but the chromosomes are not seen as distinct structures. The nucleolus is the only nuclear substructure that is visible under the light microscope. At right angles to each of the two centrioles, a daughter centriole begins to be formed.
(b) *Early prophase.* The centrioles begin moving toward opposite poles of the cell; the chromosomes can be seen as long threads, and the nucleolus is dispersing and becoming less distinct.
(c) *Middle prophase.* Chromosome condensation is completed; each chromosome is composed of two chromatids held together at their centromeres. Each chromatid contains one of the two newly replicated daughter DNA molecules. The microtubular spindle begins to radiate from the regions just adjacent to the centrioles, which are moving closer to their poles.
(d) *Late prophase.* The centrioles reach the poles, and some spindle fibers (black) extend from pole to the center, or equator, of the cell. Other spindle fibers extend from the poles to the chromatids (color) and attach to the kinetochores, which are near the centromeres of the chromosomes. The nuclear membrane begins to disperse and disappear, and the nucleolus is not visible.
(e) *Metaphase.* The chromosomes move toward the equator of the cell, where they become aligned in the equatorial plane. The sister chromatids have not yet separated.
(f) *Early anaphase.* The two daughter chromatids separate. Each contains a centromere that is linked by a spindle fiber to one pole. Led by the centromere, each chromosome begins to move toward the pole to which it is attached. Simultaneously, the cell elongates, as do the pole-to-pole spindles.
(g) *Late anaphase.* Each set of chromosomes (as the daughter chromatids are now called) is nearing its pole, and cytokinesis begins as the cleavage furrow starts to form.
(h) *Telophase.* New membranes form around the daughter nuclei; the chromosomes uncoil and become less distinct; and the nucleolus becomes visible again. Throughout mitosis the "daughter" centriole at each pole has continued to grow until it is full-length. At telophase the duplication of each of the original centrioles is completed, and each of the two centrioles at each pole begins to generate a new daughter centriole at right angles to it. (Elongation of the new daughters will continue throughout interphase and the next mitotic cycle.) Cytokinesis is nearly complete, and the spindle disappears as the microtubules and other fibers depolymerize.
(i) *Interphase.* Upon the completion of cytokinesis, DNA replication begins anew.

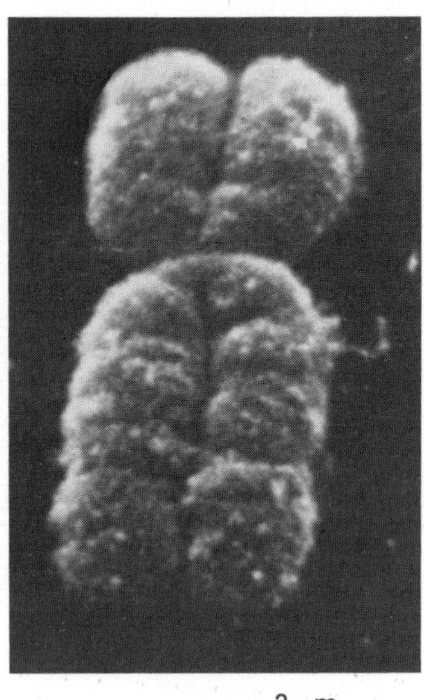

Figure 5-15 A scanning electron micrograph of a human metaphase chromosome number 8, showing the two sister chromatids paired along their axis. *Courtesy of C. J. Harrison, Christie Hospital and Holt Radium Institute, Manchester, U.K.*

work that links the centrioles as they move apart. These microtubules, together with associated fibers and proteins, are called the *spindle*. The regions surrounding the centrioles, from which the microtubules radiate, are termed the *poles* or polar regions of the cell. (Most higher plant cells do not contain centrioles. An analogous region of the cell acts as a microtubule-organizing center, and from it the spindle microtubules radiate.) At the end of prophase, the nuclear membrane disappears.

During metaphase, the condensed sister chromatids, which are connected to each pole of the cell by the microtubules attached to their kinetochores, migrate to the *equatorial plane* of the cell. The microtubules attached to the chromosomes appear to play a role in orienting them.

Anaphase is marked by the separation of the two sister chromatids at their centromeres. This happens simultaneously to the entire set of chromosomes. The members of each chromatid pair—now independent chromosomes—then migrate to opposite poles of the cell, which ensures that each daughter cell receives the same chromosome complement.

The force that propels the members of each pair of chromosomes to opposite poles is a subject of controversy. According to one theory, this movement is due to a shortening of the microtubules attached to the kinetochores. According to another theory, the force is generated by a sliding action between the microtubules attached to the kinetochore and those extending pole to pole in the polar regions of the cell. These ideas and the relevant experiments are discussed more fully in Chapter 18.

In telophase, the chromosomes start to uncoil and become less condensed. Nuclear membrane fragments begin to form at this time; they are then joined together to become two new nuclear membranes surrounding the two sets of daughter chromosomes. Simultaneously, there is separation and segregation of the cell cytoplasm in a process celled *cytokinesis*.

Because a plant cell is surrounded by a rigid cell wall, the shape of the cell does not change greatly in mitosis (Figure 5-17). During telophase, formation of the cell membrane and cell wall in a plane separating the two nuclei is completed.

Meiosis Is the Form of Cell Division in Which Haploid Cells Are Produced from Diploid Cells

In higher animals and plants, the primary cell division that departs from the plan of mitosis is that destined to give rise to gametes—the sperm and the egg cells. Like most cells in higher organisms, those that divide to produce gametes are diploid; that is, they contain two homologous chromosomes of each morphological type. Meiosis involves two separate cell divisions and yields four haploid cells from a single diploid cell (Figure 5-18).

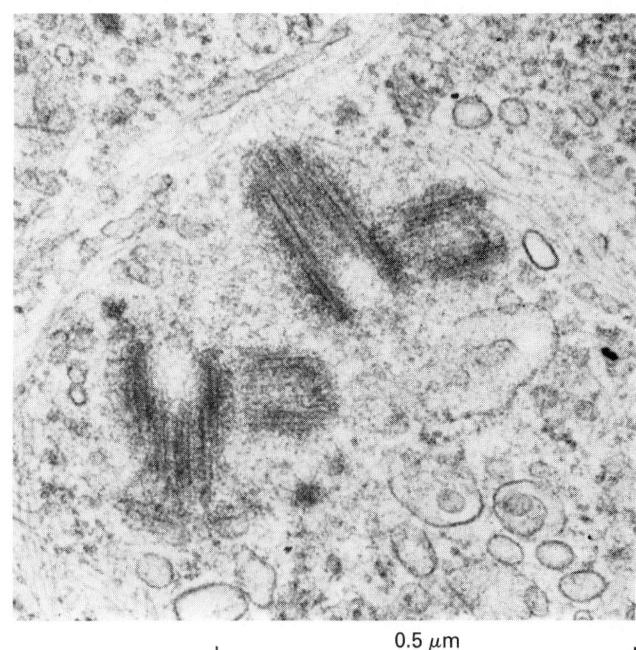

Figure 5-16 Centrioles in a cultured mammalian cell. In this micrograph, taken 4 h after mitosis, a daughter centriole about 0.25 μm long is forming at a right angle to each mature centriole (the daughters are the horizontal structures). During mitosis, the two pairs of centrioles are at opposite poles of the microtubular spindle. *Courtesy of J. B. Rattner.*

(a)

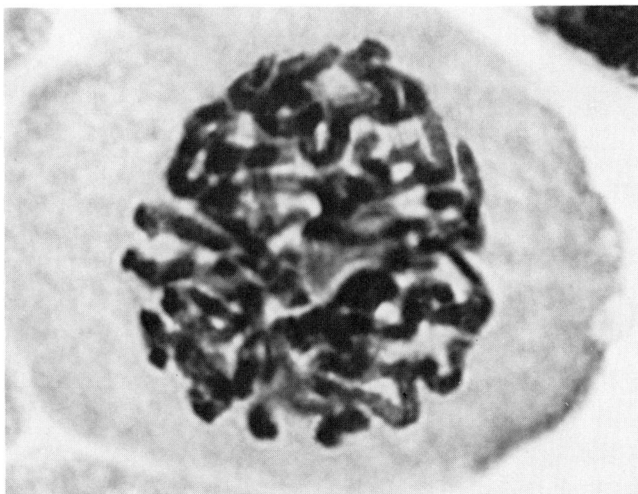

(b)

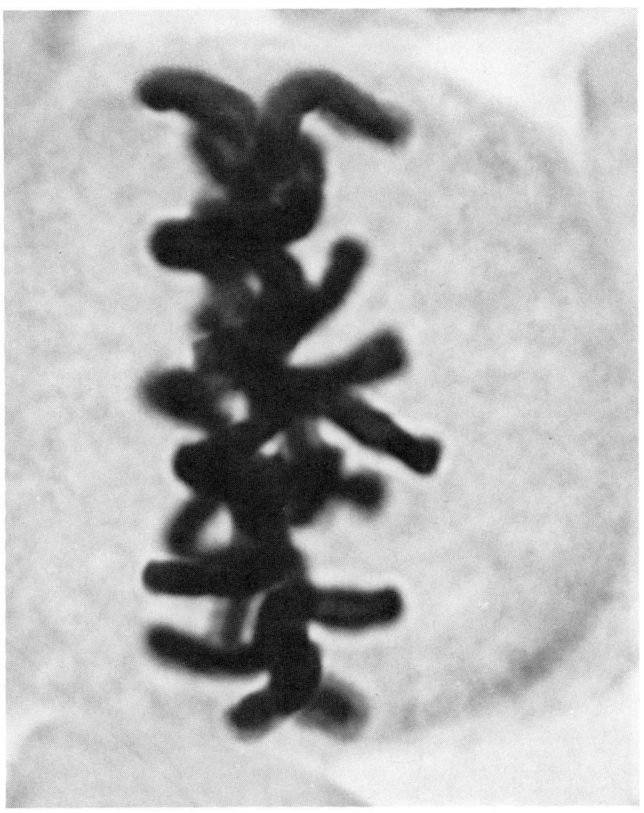

Figure 5-17 Light micrographs of mitosis in a plant.
(a) *Prophase*. The chromosomes are beginning to coil.
(b) *Metaphase*. The chromosomes are aligned on the equatorial plane.
(c) *Anaphase*. The sister chromatids have separated, and the two groups of chromosomes are moving toward opposite poles of the spindle.
(d) *Telophase*. New membranes are forming around the daughter nuclei, and the chromosomes are uncoiling and becoming less distinct.
Courtesy of the Carolina Biological Supply Company.

(c)

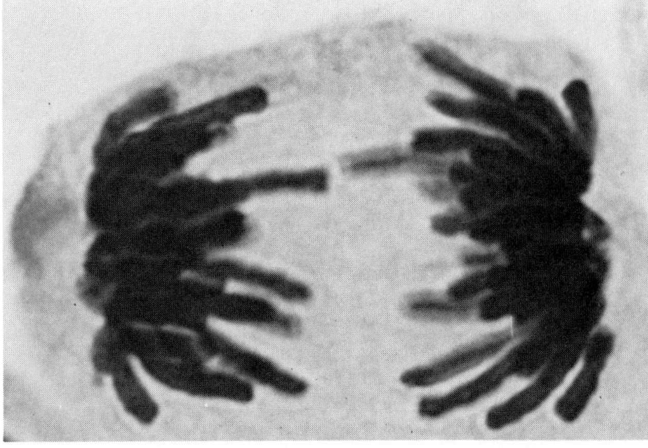

(d)

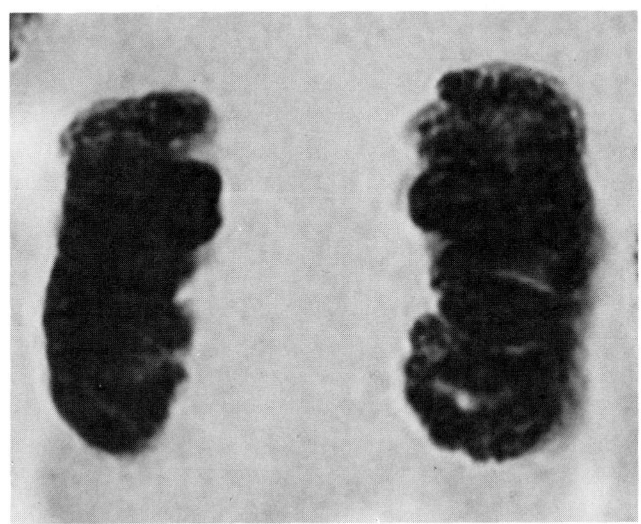

The Assortment of Homologous Chromosomes in the First Meiotic Division Before meiosis begins, the chromosomes in the pregametic cell are replicated to yield two pairs of sister chromatids for each morphological type. Each pair of sister chromatids remains together throughout the first meiotic division. No further DNA replication occurs in meiosis. In prophase I, the homologous chromosomes synapse, or align with each other lengthwise, so that each member of the pair consists of two sister chromatids. As in mitosis, two pairs of centrioles move to opposite sides of the cell; this defines two poles, between which a spindle of microtubules forms. In metaphase I, synaptic pairs of chromosomes line up at the equatorial plane of the cell. In anaphase I, the members of each pair move toward opposite poles so that each half of the cell contains one chromosome (actually one pair of sister chromatids) of each morphological type: a complete haploid set. Unlike the sister chromatids in the anaphase of mitosis, those in anaphase I of meiosis do not uncouple at their centromeres. Thus, the nuclei that form in telophase I are genetically haploid, but contain two copies of each sister chromatid.

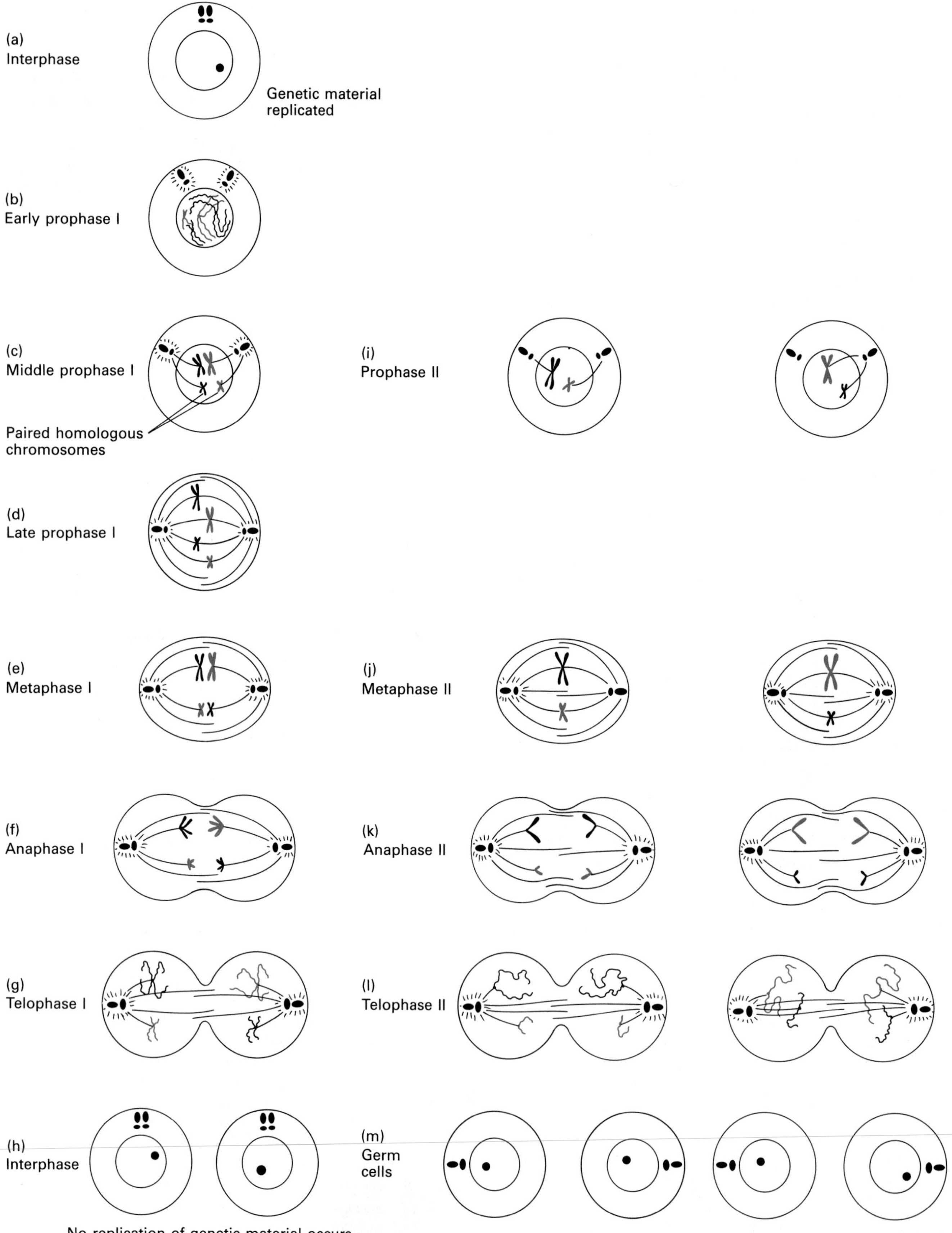
(a)
Interphase
Genetic material replicated
(b)
Early prophase I
(c)
Middle prophase I
Paired homologous chromosomes
(d)
Late prophase I
(e)
Metaphase I
(f)
Anaphase I
(g)
Telophase I
(h)
Interphase
No replication of genetic material occurs
(i)
Prophase II
(j)
Metaphase II
(k)
Anaphase II
(l)
Telophase II
(m)
Germ cells

Figure 5-18 ***(opposite)*** A schematic outline of the two divisions of meiosis in an animal cell.
First meiotic division [parts (a) through (h)]:
(b) *Early prophase I.* The chromosomes have already replicated by this stage, but the chromatids in each pair remain together and are not distinguishable until late prophase I. The chromatid pairs, shown here in condensed form, actually become visible as long, coiled filaments.
(c) *Middle prophase I.* Homologous chromosomes pair, or synapse, along their entire lengths. They also condense and become shorter. Crossing over between chromatids of homologous chromosomes occurs at this stage.
(d) *Late prophase I.* The sister chromatids of each chromosome become visible. The synaptic pairs of chromosomes begin to move to the equator of the cell. The spindle is nearly complete, and the nuclear membrane begins to disappear.
(e) *Metaphase I.* The synaptic pairs align themselves in the equatorial plane of the cell.
(f) *Anaphase I.* The daughter chromatids remain attached to their centromeres. The synaptic pairs separate and the chromosomes from each pair move to opposite poles.
(g) *Telophase I.* A nuclear membrane forms around each group of chromosomes. The chromatids are still attached by their centromeres.
(h) *Interphase.* Division of the haploid cells is complete and there is no further DNA replication. The chromosomes uncoil somewhat and may even become invisible under the light microscope.
Second meiotic division [parts (i) through (m)]: In events similar to those of mitosis, the sister chromatids separate, and one member of each pair is transmitted to each daughter cell.

Interphase is the period between the first and the second meiotic divisions. Unlike the interphase in mitosis, interphase in meiosis does not involve DNA replication.

The second meiotic division is like mitosis. The sister chromatids separate at their centromeres, as in mitosis, and move to opposite poles. The result is four haploid cells.

The Genetic Consequences of Meiosis As mentioned earlier, the pregametic cell contains two copies of each morphological type of chromosome; that is, the cell is diploid. One copy is inherited from the maternal parent and the other from the paternal parent. In the first meiotic division, the homologous chromosomes segregate into the daughter cells randomly with regard to parental origin (Figure 5-19). Thus, the number of different types of cells generated by the first meiotic division of a diploid cell having $2n$ chromosomes is 2^n. The genetic variation thus achieved is further enhanced in prophase I by *crossing over,* the exchanging of genetic material between individual chromatids of homologous chromosomes (Figure 5-20). This makes even sister chromatids different from each other. In the second meiotic division, these chromatids assort randomly into the daughter cells.

In male mammals, each of the four cells produced from a meiotic division differentiates into a sperm cell. In females, only one of the four haploid nuclei survives as the egg nucleus; the other three are discarded. In mammals, as in most animals, the haploid phase is transitory; the fusion of the sperm and the egg regenerates a diploid cell.

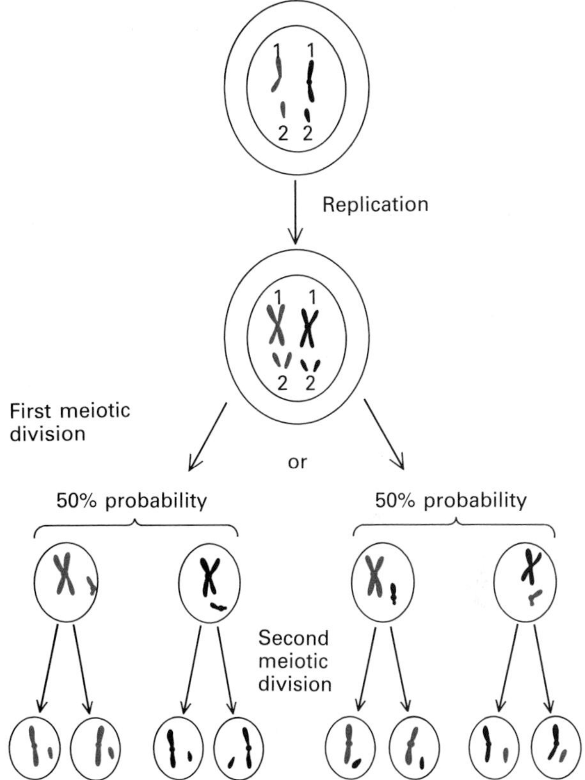

Figure 5-19 The genetic consequences of meiosis are illustrated for a hypothetical organism in which the haploid number of chromosomes is two. Each diploid cell contains two copies of each chromosome, one inherited from each parent. (In this diagram, the maternal chromosomes are distinguished by color.) Before the first meiotic division, the chromosomes replicate, but the daughter pairs of chromatids remain attached at the centromeres. Each daughter cell receives the maternal or paternal pair of homologous replicated chromosomes. Which chromosome is incorporated into which daughter is random, and each pair of chromosomes is segregated independently of the other pairs. Thus, in this simple case, there are two possible outcomes of the first division, each having an equal probability of 50 percent. In the second meiotic division, the chromatids separate to generate a total of four haploid daughter cells, each containing a full set of homologous chromosomes.

This cell, termed a zygote, has a unique genetic constitution; half of its chromosomes are derived from the sperm, half from the egg.

In a number of eukaryotic microorganisms, including many yeast and algae, growing cells are haploid. The diploid stage in yeast is formed by the mating, or fusion, of two haploid cells, as we discuss in Chapter 6. Depending on the environment, the cell in the diploid stage can grow by mitosis, or it can undergo meiosis to yield four haploid cells.

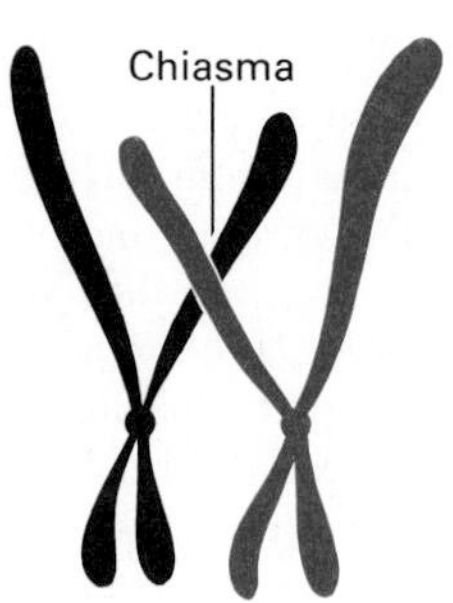

Figure 5-20 Recombination during meiosis. Two chromatids cross over during prophase I and exchange chromosomal segments. The point of attachment of the two chromatids participating in the exchange is a *chiasma*.

Cellular Membrane Systems and Organelles

As we noted in Chapter 3, phospholipid bilayer membranes form an essential structural element of cells. All cellular membranes contain proteins and glycoproteins that give each membrane, or region of membrane, its unique characteristics and enable it to function.

Prokaryotic Cells Have Few Internal Membranes

Prokaryotic cells generally contain no internal membranes that are not connected to the plasma membrane (the surface membrane). Most prokaryotic cells possess only the plasma membrane. However, in some bacteria, the plasma membrane has infoldings called *mesosomes* (see Figure 5-2c). The extensive infoldings of the plasma membrane of photosynthetic bacteria (blue-green algae) contain the proteins that trap light and generate ATP (Figure 5-21). Some of these internal photosynthetic membranes, which are called thylakoid vesicles, may not connect with the plasma membrane and may enclose small regions of the cytoplasm.

Bacterial Cells Contain Complex Cell Walls That Confer Strength and Rigidity

Bacterial species can be divided into two classes, depending on their ability to be stained by the Gram technique. Those in the *gram-negative* class (i.e., those that are not stained by the Gram technique) include the common intestinal bacterium *E. coli.* Gram-negative bacteria contain two surface membranes (Figure 5-22). The inner membrane is the actual plasma membrane. As the major permeability barrier of the cell, it contains proteins that allow certain nutrients and other chemicals to pass into and out of the cell, while preventing others from doing so. The outer membrane is unusual in that it is permeable to many chemicals having a molecular weight of 1000 or more. It contains proteins called *porins,* which line channels large enough to accommodate such molecules (Figure 5-23). A number of lipid-linked oligosaccharides are bound to the outer membrane. The structures of these oligosaccharides can differ extensively even among closely related strains and species. When pathogenic bacteria infect animal cells, the production of antibodies specific for these surface molecules can be important in determining the course of the illness.

Between the two membranes is the *periplasm,* which contains a *peptidoglycan*—a wall-like complex of proteins and oligosaccharides that gives rigidity to the cell—and a space that is generally occupied by proteins secreted by the cell. The peptidoglycan is constructed of linear polymers of the disaccharide *N*-acetylglucosamine–*N*-acetylmuramic acid (Figure 5-24). These chains are cross-linked to one another by small peptides containing both D-amino acids and the L-amino acids normally found in proteins.

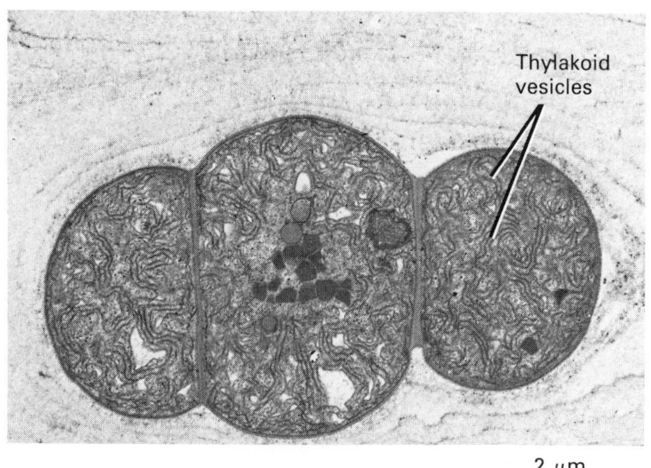

Figure 5-21 An electron micrograph of a thin section through three attached cells of the blue-green alga *Nostoc carneum.* The extensive array of thylakoid vesicles—internal photosynthetic membranes—is characteristic of this prokaryotic group. *Courtesy of T. E. Jensen and C. C. Bowen.*

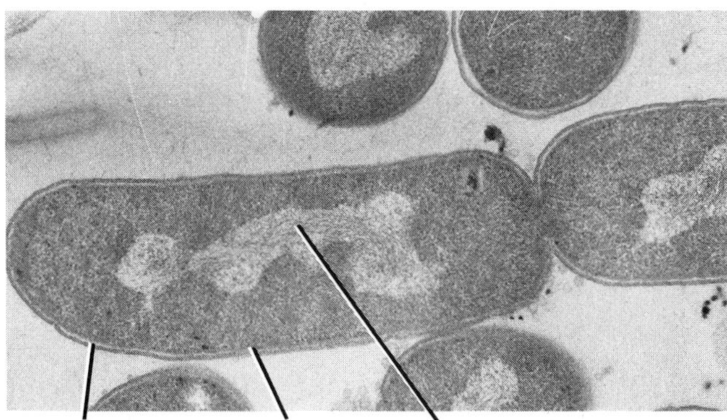

Figure 5-22 A section of *E. coli,* a gram-negative bacterium. The micrograph shows the inner and outer surface membranes and the nucleoid, the DNA-containing fibrous central region of the cell. *Courtesy of I. D. J. Burdett and R. G. E. Murray.*

Gram-positive bacteria, such as *Bacillus polymyxa,* have only a single phospholipid-containing plasma membrane and no outer membrane or periplasmic space. Their peptidoglycan is thicker than that found in gram-negative organisms, and the bacterium secretes proteins into the growth medium rather than into a periplasmic space (Figure 5-25).

Eukaryotic Cells Have Complex Systems of Internal Membranes

Like gram-positive prokaryotes, eukaryotic cells have only the plasma membrane at their surfaces. But unlike most prokaryotes, they also contain extensive internal membranes that are not connected to the plasma membrane. These membranes, which enclose specific regions and separate them from the rest of the cytoplasm, define a collection of subcellular structures called *organelles.*

According to one definition, an organelle is any subcellular entity that can be isolated by centrifugation at a very high speed. This definition would include structures such as ribosomes, particles of glycogen (a polymer of glucose), and large multienzyme complexes. In this book, however, we use the term organelle to refer only to membrane-limited structures. Organelles are found in all eukaryotic cells, whether they be unicellular or of plant or animal origin. In addition to the nucleus, the organelles include the *mitochondria,* the rough and the smooth *endoplasmic reticula, Golgi vesicles, lysosomes, peroxisomes,* and assorted smaller vesicles (Figure 5-26). Plant cells also contain *chloroplasts,* the sites of photosynthesis, and large fluid-filled organelles termed *vacuoles.* Each organelle plays a unique role in the growth and metabolism of the cell, and each contains, accordingly, a specific collection of enzymes that catalyze requisite chemical re-

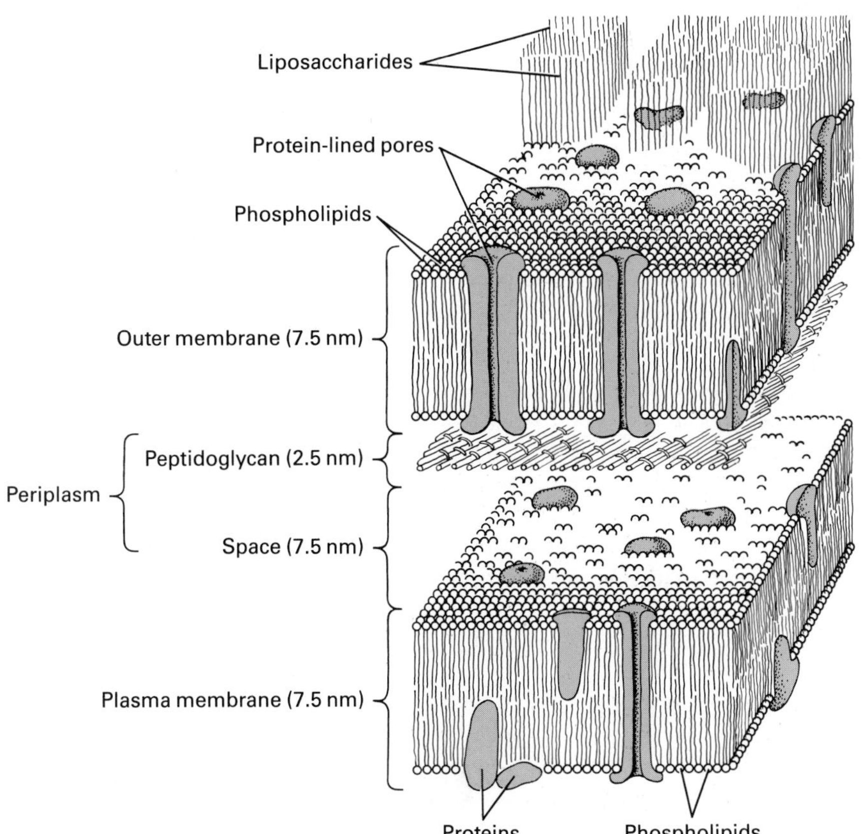

Figure 5-23 A schematic representation of the structure of the gram-negative *E. coli* cell envelope. *After J. DiRienzo, K. Nakamura, and M. Inouye, 1978,* Annu. Rev. Biochem. *47:481.*

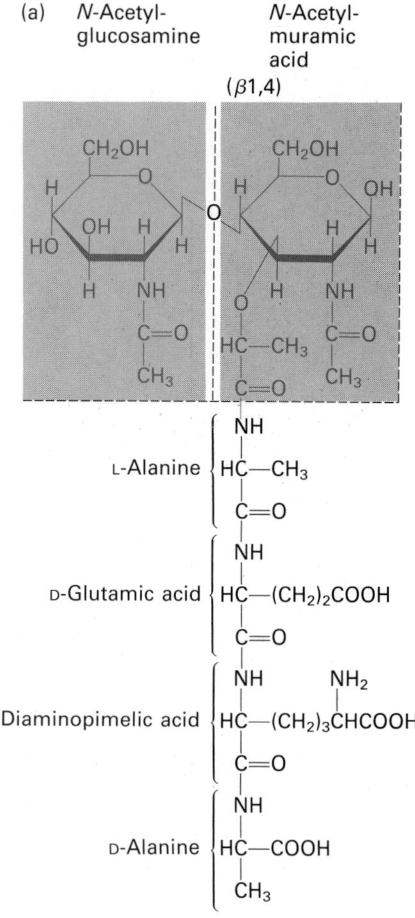

(b) —GlcNAc—MurNAc—GlcNAc—MurNAc—GlcNAc—MurNAc—GlcNAc—
(L-Ala) (D-Glu) DAP (D-Ala)

(c) —GlcNAc—MurNAc—GlcNAc—MurNAc—GlcNAc—MurNAc—GlcNAc—
(L-Ala) GlcNAc—MurNAc—GlcNAc—MurNAc—GlcNAc—
(D-Glu) (L-Ala) GlcNAc—MurNAc—GlcNAc—
DAP (D-Glu) (L-Ala)
(D-Ala)—C(=O)—NH—DAP (D-Glu)
(D-Ala)—C(=O)—NH—DAP
(D-Ala)

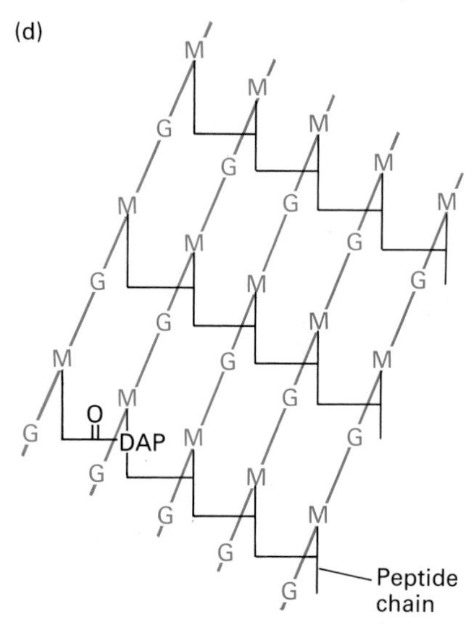

Figure 5-24 The structure of the peptidoglycan. (a) The complete structure of a single subunit, showing the linkage between the two amino sugars and between N-acetylmuramic acid and the four amino acids in the short peptide chain. (b) The subunits are linked into chains formed of alternating N-acetylglucosamine (GlcNAc) and N-acetylmuramic acid (MurNAc). (DAP = diaminopimelic acid.) (c) The chains are linked to one another by bonds between adjacent short peptide chains. (Note that *each* MurNAc has a short peptide chain attached to it, just like the three MurNAcs whose chains are shown.) (d) The resulting two-dimensional sheet has strength and rigidity. (G = N-acetylglucosamine; M = N-acetylmuramic acid.)

actions. Some of this specificity resides in the organelle membranes, to which a number of the enzymes and other proteins are bound.

Organelles Can Be Purified by Various Fractionation Techniques

The purification of each of the major subcellular organelles was a great achievement in cell biology because it made possible the detailed study of their structures and metabolic functions. Rat liver was used in many of the classic studies, but the same principles apply to virtually all cells or tissues.

The initial step for purifying subcellular structures is to rupture the cell wall, the plasma membrane, or both. This is accomplished by first suspending cells in a solution of appropriate pH and salt content, usually isotonic sucrose (0.25 *M*) or a combination of salts similar in composi-

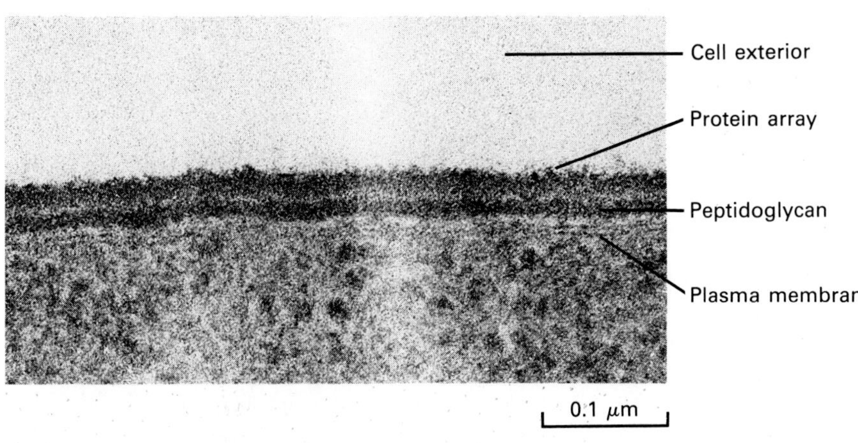

Figure 5-25 A section of the surface of *Bacillus polymyxa,* a gram-positive bacterium. Visible are the thin plasma membrane, a layer composed of a peptidoglycan (similar in structure to that shown in Figure 5-24) and another polymer termed teichoic acid, and the protein array. *Courtesy of R. G. E. Murray.*

tion to those of the cell. Many cells can then be broken by stirring them in a high-speed blender. Plasma membranes can also be sheared by using special tissue homogenizers in which the cells are forced under pressure through a very narrow space between the plunger and the wall of the vessel. Generally, the cell solution is kept at 0°C to best preserve enzymes and other constituents once they are released from the stabilizing forces of the cell.

The various organelles differ both in size and in density. Most fractionation procedures utilize *differential-velocity centrifugation* (sometimes called rate-zonal centrifugation), a technique in which an *ultracentrifuge* is used to generate high centrifugal forces. Large, dense structures form a deposit in a centrifuge tube faster than small, less dense ones. Generally, the cell homogenate is first filtered or centrifuged at relatively low speeds to remove unbroken cells (Figure 5-27). A slightly faster or longer centrifugation selectively deposits the nucleus—the largest organelle (it is usually 5 to 10 μm in diameter). The undeposited material is then transferred to another tube, which is subjected to centrifugation at a higher speed; this results in the deposition of mitochondria (these organelles, which are 1 to 2 μm long, are the sites of ATP production), chloroplasts (these, which also are 1 to 2 μm long, are the sites of photosynthesis), and lysosomes (the sites of degradation of many molecules). A final centrifugation at 100,000 times the force of gravity (*g*) for 60 min results in deposition of the plasma membrane, the endoplasmic reticulum (the site of the synthesis of membrane and secretory components), peroxisomes, and other small materials.

The endoplasmic reticulum, a network of membranous vesicles, spreads through much of the cytoplasm (see Figure 5-6b). The shearing forces required to break the plasma membrane rupture the endoplasmic reticulum into smaller vesicles, often called microsomes. The material that remains undeposited after centrifugation for 1 h at 100,000*g* constitutes the *cytosol,* the soluble aqueous portion of the cytoplasm. Additional centrifugation (not depicted in Figure 5-27) of the cytosol in the ultracentrifuge deposits the ribosomes and such particles as granules of glycogen.

Rate-zonal centrifugation does not yield totally pure organelle fractions, and often some of the fractions must be processed further in a different way. One method that may be employed is *density-gradient centrifugation,* in which organelles are separated by their density rather than by their size. The organelle fraction is layered on top of a solution that contains a gradient of a dense substance such as sucrose or glycerol. The solution is most concentrated and densest at the bottom of the centrifuge tube (where a typical sucrose solution is about 1.23 g/cm^3); the concentration decreases gradually toward the top, so that the solution is least dense at the surface. The tube is centrifuged at high speed (about 40,000 r/min) for several hours, which allows each particle to migrate to a position at which its density is equal to the density of the surrounding liquid. In typical preparations from animal cells, the rough endoplasmic reticulum has a density of 1.20 g/cm^3 and is well separated from the Golgi vesicles (which have a density of 1.14 g/cm^3) and the plasma membrane (which has a density of 1.12 g/cm^3), as indicated in Figure 5-28. The higher density of the rough endoplasmic reticulum is due largely to the ribosomes that are bound to it.

The purity of organelle preparations can be assessed by morphology (through use of the electron microscope) or by their content of *marker enzymes*—enzymes that are known to be concentrated in particular organelles. For example, the protein cytochrome *c* is found only in mitochondria; its presence in other fractions is a measure of their contamination by mitochondria. Similarly, catalase is present only in peroxisomes; acid phosphatase only in lysosomes; and ribosomes only in the rough endoplasmic reticulum or the cytosol.

Frequently, the organelles obtained are not pure even after both rate-zonal and density-gradient centrifugation. The next step will be to take advantage of the fact that the membrane of each organelle contains unique enzymes and other proteins; several groups of investigators are

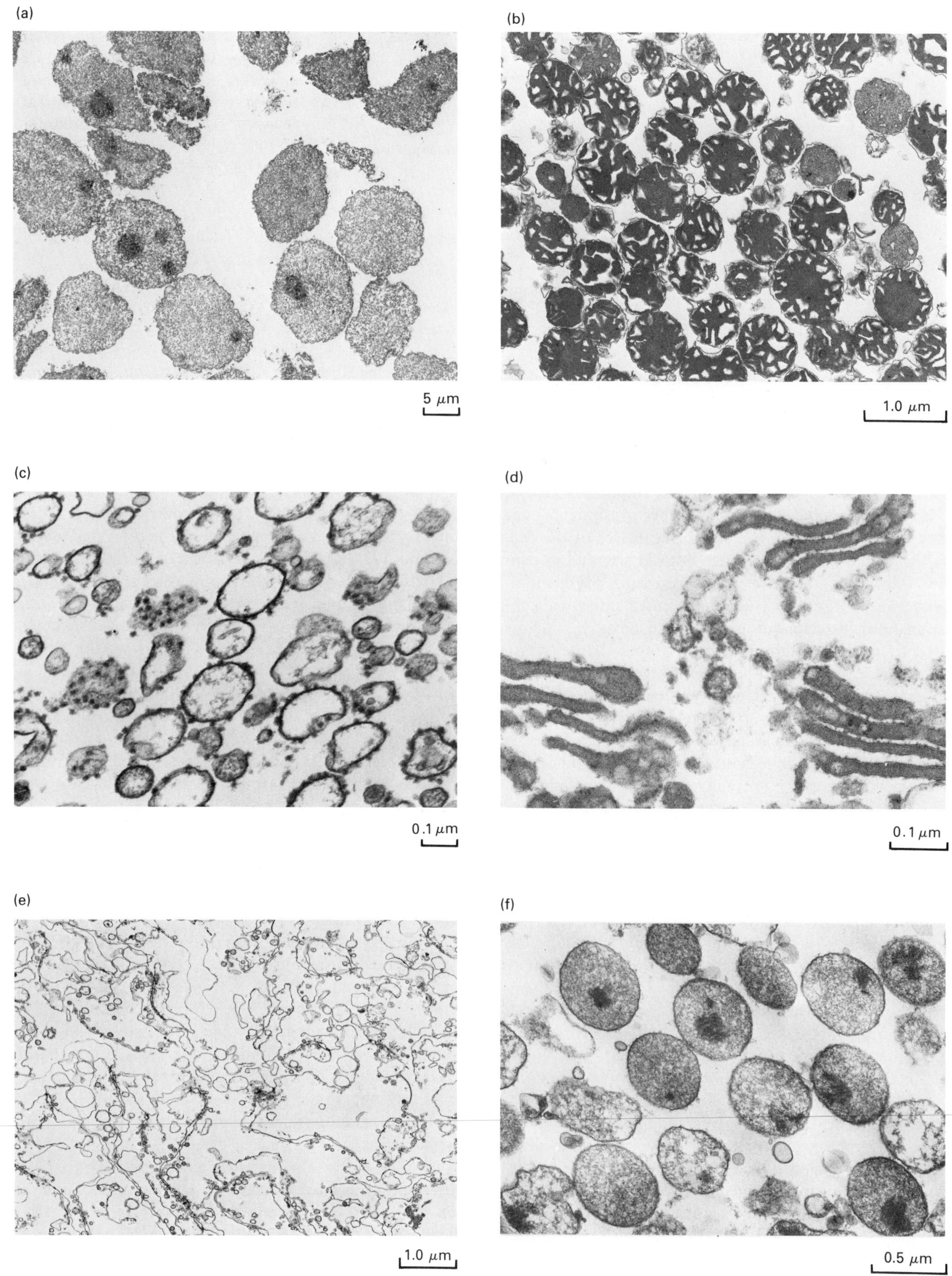
(a)
5 μm
(b)
1.0 μm
(c)
0.1 μm
(d)
0.1 μm
(e)
1.0 μm
(f)
0.5 μm

Figure 5-26 ***(opposite)*** Electron micrographs of purified organelles: (a) nuclei, (b) mitochondria, (c) rough endoplasmic reticulum, (d) Golgi vesicles, (e) plasma membranes, and (f) peroxisomes. *Parts (a) through (e) courtesy of S. Fleischer and B. Fleischer; part (f) courtesy of P. Lazarow.*

now trying to purify organelles by immunological techniques, using antibodies that are specific for these organelle membrane proteins (see Figure 5-6b and d).

The Cytosol Contains Many Cytoskeletal Elements and Particles

The *cytosol* is the nonparticulate region of the cell cytoplasm, that is, the part not contained within any of the subcellular organelles. Initially, the cytosol was thought to be a fairly homogeneous "soup" in which all of the organelles floated. However, it is now known that the cytosol of all cells contains fibers that help to maintain cell shape and mobility and that probably provide anchoring points for the other cellular structures. Collectively, these fibers are termed the *cytoskeleton* (Figure 5-29). At least three general classes of such fibers have been identified. The thickest are the *microtubules* (20 nm in diameter), which consist primarily of the protein tubulin. The other two classes are the *microfilaments* (7 nm in diameter), which are constructed principally of the protein actin, and the intermediate filaments (10 nm in diameter). As is explained in Chapter 19, at least five types of intermediate filaments, each made of a different type of protein, have been identified in various animal cells.

Additionally, the cytosol of many cells contains *inclusion bodies* or inclusion granules—particulate cytoplasmic regions that are not bounded by a membrane. Some cells—specifically, muscle cells and hepatocytes, the principal cell type in the liver—contain cytosolic granules of glycogen, a glucose polymer that functions as a storage form of usable cellular energy (Figure 5-30). In well-fed animals, glycogen can account for as much as 10 percent

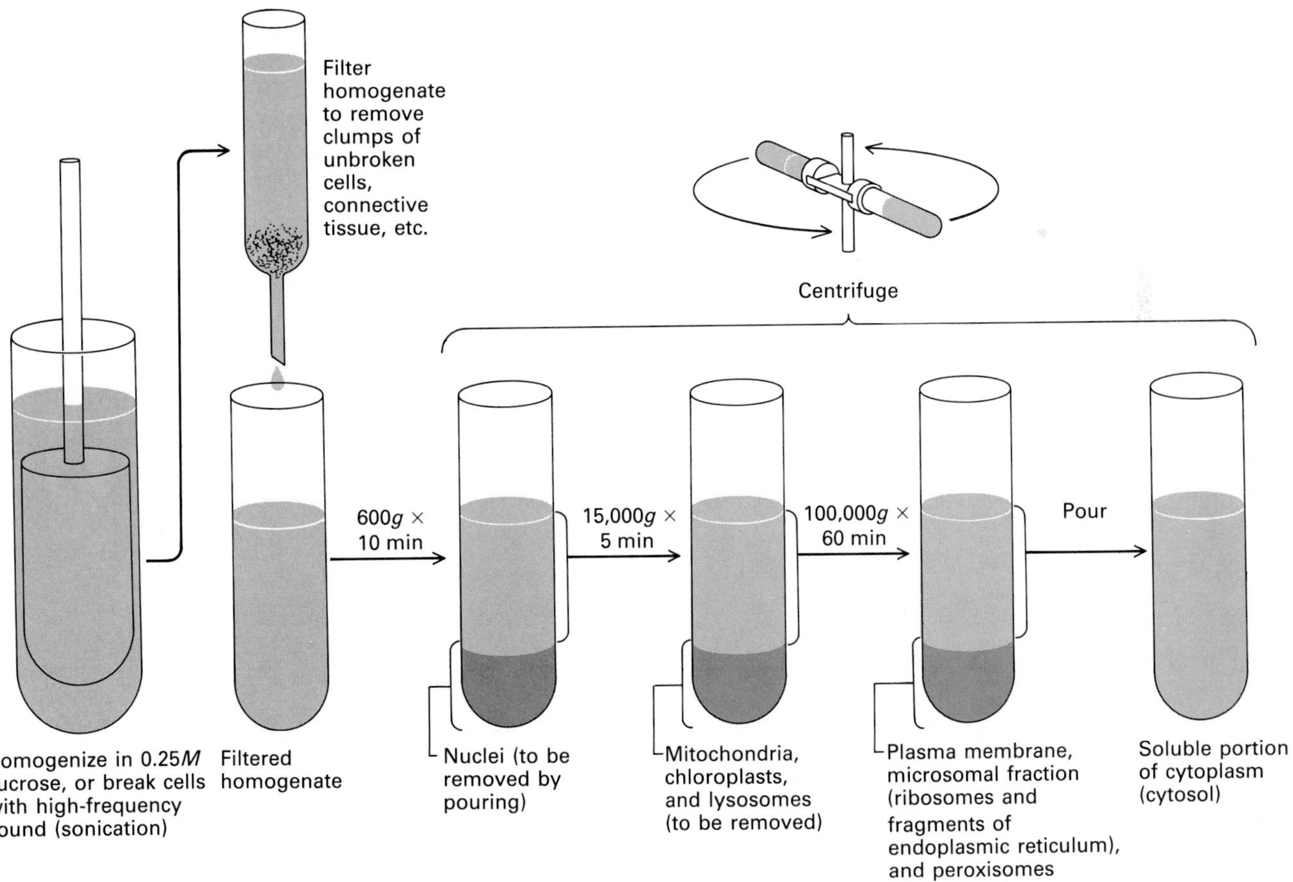

Figure 5-27 Cell fractionation by rate-zonal centrifugation. The different sedimentation rates of various cellular components make it possible to achieve a partial separation of them by centrifugation. Nuclei and viral particles can sometimes be purified completely by such a procedure. Components of the soluble fraction can themselves be further separated by density-gradient centrifugation, as described in Figure 5-28. [The force of gravity is denoted by g. A force of 600g is generated by a typical centrifuge rotor operating at 500 revolutions per minute (r/min). A force of 100,000g requires an ultracentrifuge that can rotate the sample at 50,000 r/min.]

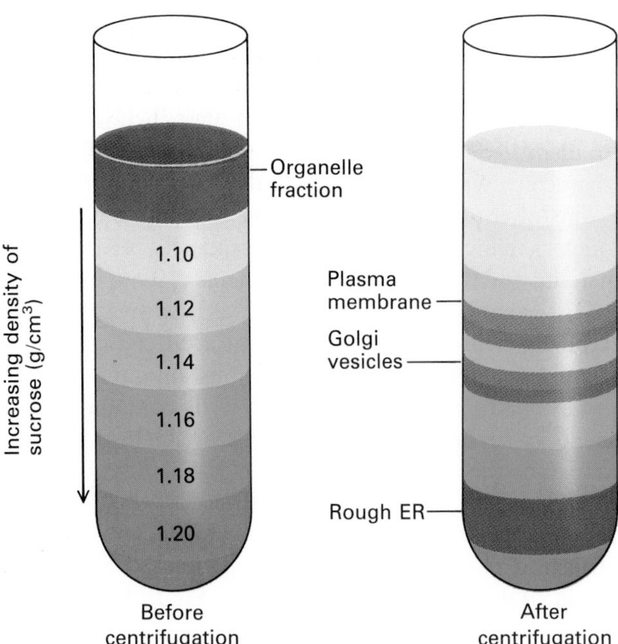

Figure 5-28 The separation of rough endoplasmic reticulum, Golgi vesicles, and the plasma membrane by the density-gradient centrifugation of a crude membrane fraction. The density gradient is set up by layering sucrose solutions that are successively less dense in a centrifuge tube.

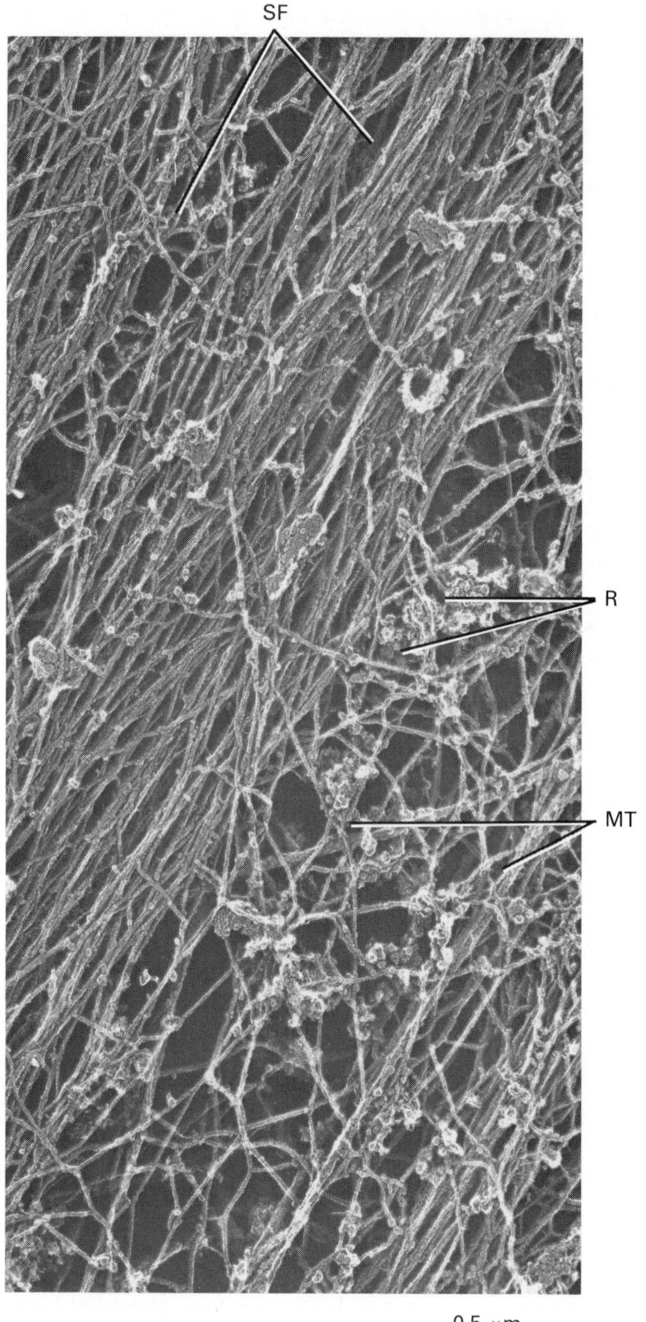

Figure 5-29 An electron micrograph of the cytoskeleton from a fibroblast cell that was extracted with a detergent to destroy the plasma membrane. Prominent are bundles of actin-containing microfilaments that form the cell's stress fibers (SF). Also visible are two thicker microtubules (MT) and a more diffuse meshwork of filaments that are studded with grapelike clusters; these are probably polyribosomes (R). *From J. E. Heuser and M. Kirschner, 1980,* J. Cell Biol. **86**:212. *Reproduced from the* Journal of Cell Biology *by copyright permission of The Rockefeller University Press.*

of the wet weight of the liver. The cytoplasm of the specialized fat cells in adipose tissue contains large droplets of almost pure triacylglycerols, a storage form of fatty acids. Metabolism of these stored compounds generates cellular ATP. In all cells, the cytosol also contains a large number of different enzymes and is a major site of cellular metabolism; often it contains 25 to 50 percent of the total cell protein. Because of the high concentration of protein, the cytosol is a viscous gel through which the cytoskeletal fibers interweave. Many workers believe that the cytosol is highly organized, with most proteins bound to fibers and/or localized in specific regions.

The key roles played by the organelles in eukaryotic cellular metabolism will be discussed in detail in later chapters of this book. In the next section we shall present an overview of their structures and functions.

Internal Membrane Systems of Eukaryotic Cells

The Endoplasmic Reticulum Is an Interconnected Network of Internal Membranes

Generally, the largest membrane in a eukaryotic cell is the *endoplasmic reticulum* (ER), which is actually a network

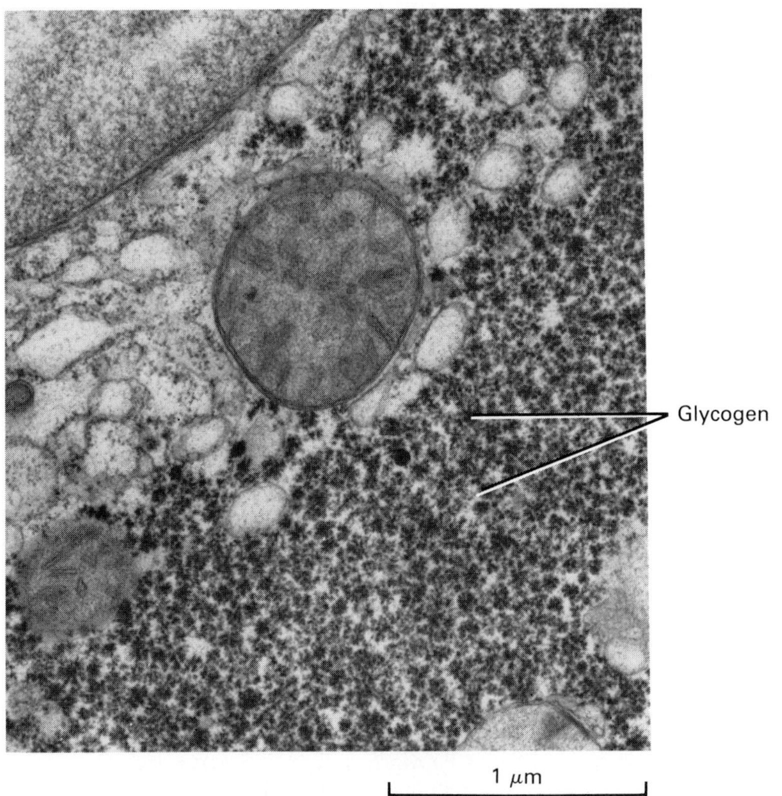

Figure 5-30 An electron micrograph of a portion of the cytoplasm of a rat hepatocyte, showing packed rosettes of glycogen, a mitochondrion, and part of the nuclear envelope. *Biophoto Associates.*

of interconnected closed membrane vesicles. The endoplasmic reticulum has a number of functions in the cell. Notable are those related to the synthesis of many membrane lipids and proteins. Regions of rough endoplasmic reticulum are studded with ribosomes, whereas regions of smooth endoplasmic reticulum lack ribosomes (Figure 5-31).

The Smooth Endoplasmic Reticulum The *smooth endoplasmic reticulum* is the site of the synthesis and metabolism of fatty acids and phospholipids. The amount of smooth ER varies in different cells. Many cells have very little smooth ER. In the hepatocyte, the major cell of the liver, the smooth ER is present in abundance. It contains enzymes that modify or detoxify chemicals such as pesticides or carcinogens by converting them into more water-soluble, conjugated products, which can then be secreted from the body. In experimental animals treated with high doses of such compounds, there is a large proliferation of the smooth ER.

The Rough Endoplasmic Reticulum Ribosomes bound to the *rough endoplasmic reticulum* participate in the synthesis of certain membrane and organelle proteins and also in the synthesis of the proteins that are to be secreted from the cell. Figure 5-32 illustrates the latter: as a growing secretory polypeptide emerges from the ribosome, it passes through the rough ER membrane; specific proteins in the membrane facilitate this transport. The newly made secretory proteins accumulate in the *lumen,* or central cavity, of the ER. The ribosomes that fabricate secretory proteins are bound to the rough ER, in part by

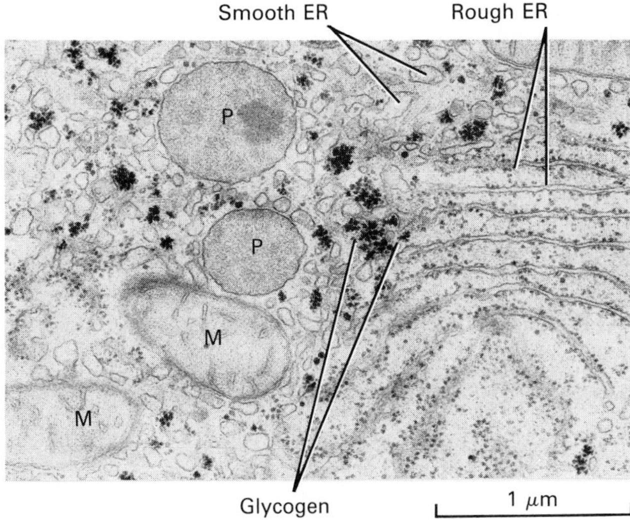

Figure 5-31 A higher-magnification micrograph of a section of a rat hepatocyte, showing two mitochondria (M), two peroxisomes (P), rough and smooth endoplasmic reticula, and glycogen rosettes. *Courtesy of P. Lazarow.*

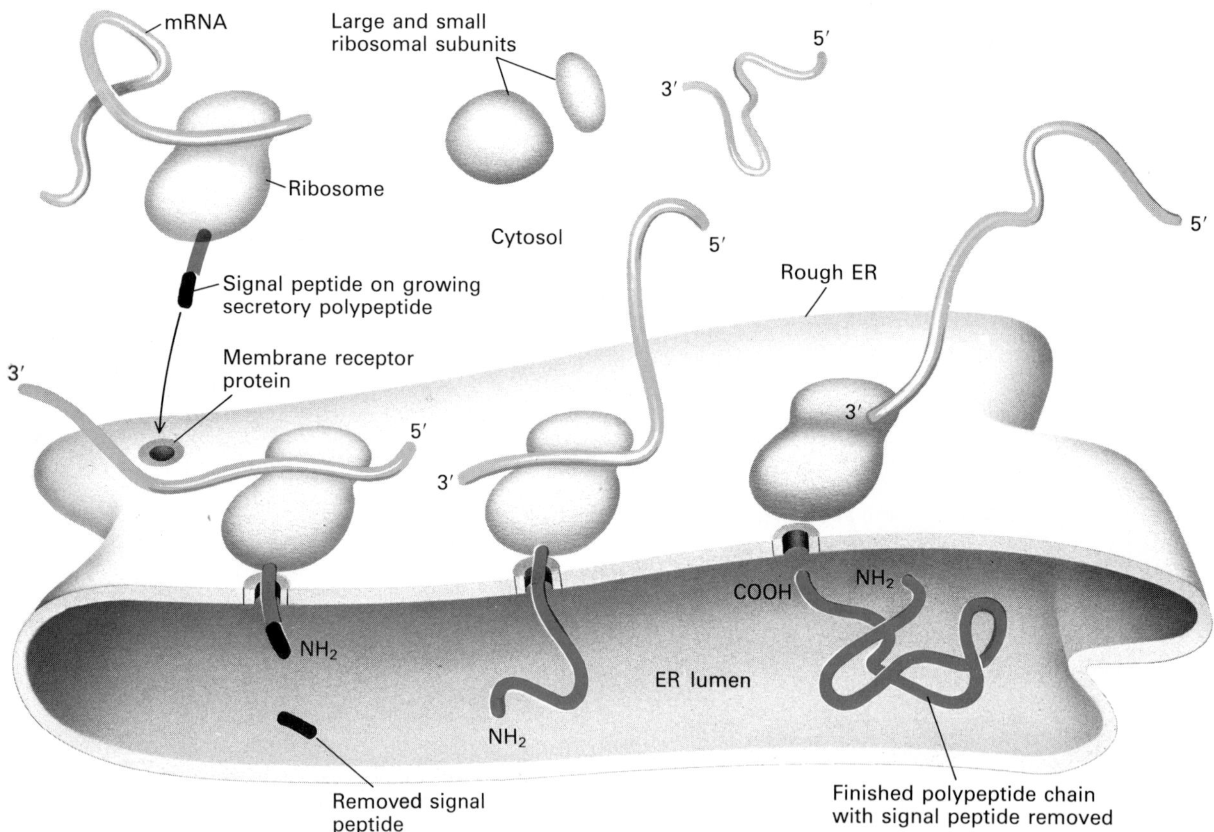

Figure 5-32 The transport of growing secretory proteins through the endoplasmic reticulum membrane. The N-terminus of a protein that is to be secreted from the cell contains a signal peptide, an amino acid sequence that binds to a receptor on the rough ER membrane. This interaction causes the large ribosome subunit to bind tightly to the membrane, and the growing polypeptide is then extruded through it. The signal peptide is removed once it has passed into the lumen of the ER.

the nascent polypeptide chain. Membrane proteins and glycoproteins are also synthesized by ribosomes bound to the ER membrane rather than passed through it.

All eukaryotic cells contain a discernible amount of rough ER because, as mentioned above, it is necessary for the fabrication of some plasma membrane proteins and glycoproteins. However, rough ER is found in great abundance in cells that are specialized for the production of secretory proteins. For example, the pancreatic *acinar cells* (the groups of cells surrounding the ductules) are specialized for the synthesis of digestive enzymes, such as chymotrypsin; and *plasma cells,* for another example, produce serum antibodies. In both types of cells, a large part of the cytosol is filled with rough ER.

Cytoplasmic Ribosomes Other classes of proteins, including cytosolic proteins, are synthesized by ribosomes that appear to be free in the cytosol, but that may be bound to cytoskeletal fibers. Ribosomes that are not bound to the ER membrane synthesize proteins that are incorporated into cytoskeletal fibers (e.g., actin), into mitochondria (e.g., cytochrome *c*), and into peroxisomes (e.g., catalase). Chapter 21 contains a detailed discussion of the problem of *protein sorting*—that is, of how proteins are targeted to their destinations during or after their synthesis.

Golgi Vesicles Process Secretory Proteins and Partition Cellular Proteins and Membranes

Several minutes after their synthesis, secretory proteins in the rough ER move to the lumenal cavity of another group of membrane-limited organelles, the *Golgi vesicles,* which are located near the nucleus in many cells (see Figures 5-6d and 5-26d). Three-dimensional reconstructions of electron micrographs of Golgi vesicles show that they consist of continuous series of flattened membranous sacs. Surrounding them are a number of more or less spherical membrane vesicles that appear to transport proteins to and from the Golgi complex. How a protein moves from the lumen of the ER to the lumen of the Golgi vesicles is controversial. From the available evidence (which is discussed in Chapter 21), it seems that small membranous vesicles containing some of the lumenal ER contents bud off from regions of the rough ER not coated with ribosomes and then fuse with the membranous Golgi sacs. The stack of Golgi sacs has two defined regions—the cis and the trans—and vesicles coming from the ER apparently fuse with the cis region of the Golgi complex. The contents are then transferred to the trans region (Figure 5-33).

Golgi vesicles are often referred to as the "traffic police" of the cell. They play a key role in sorting many of the cell's proteins and membrane constituents, and in directing them to their proper destinations. To perform this function, the Golgi vesicles contain a number of enzymes that react with and modify secretory proteins passing through the Golgi lumen, or membrane proteins and glycoproteins that are transiently in the Golgi membranes as they are en route to their final destinations. For example, a Golgi enzyme may add a "signal" or "tag," such as a carbohydrate or phosphate residue, to certain proteins to direct them to their proper sites in the cell. Or, a proteolytic enzyme in the Golgi secretory vesicles may cut a secretory or membrane protein into two or more specific segments; this is what occurs in the conversion of proinsulin into insulin (see Figure 3-13).

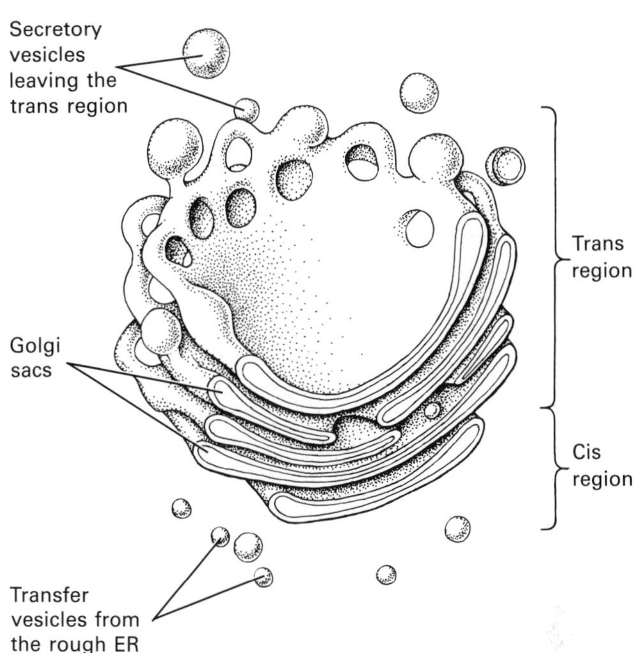

Figure 5-33 A three-dimensional model of the complex of Golgi vesicles in a secretory cell. The transfer vesicles, which have budded off the rough ER, fuse with the cis membranes of the Golgi complex. The secretory vesicles that bud off the sacs on the trans membranes become, in acinar cells, the zymogen granules. *After a model by J. Kephart.*

Proteins Are Secreted by the Fusion of an Intracellular Vesicle with the Plasma Membrane

As shown in Figure 5-34, after secretory proteins are modified in the Golgi vesicles, they are transported out of the complex by membranous *secretory vesicles,* which seem to bud off the trans side of the complex. In many cells, the membranes of these vesicles quickly fuse with the plasma membrane, releasing their contents into the extracellular space. The fusion of the membrane of an intracellular vesicle with the plasma membrane is termed *exocytosis.* In other cells, the Golgi vesicles fuse with secretory vesicles that do not immediately release their contents to the outside. Rather, these secretory vesicles fuse with other similar vesicles and become intracellular membrane-limited storage reservoirs for the secretory proteins (Figure 5-34). When an appropriate signal, such as a hormone, stimulates a cell, the secretory vesicles fuse with the surface membrane, releasing their products into the extracellular fluid. Figure 5-35 shows secretory vesicles containing zymogen granules (particles composed of pancreatic enzyme precursors) just under the apical surface of a pancreatic acinar cell.

Small Vesicles May Shuttle Membrane Constituents from One Organelle to Another

The foregoing discussion of protein secretion illustrates several important principles pertaining to the structure and function of membrane-limited organelles. Each type of organelle has a specific function to perform and contains a specific set of enzymes and other proteins for this purpose. Some organelles (Golgi vesicles) modify proteins; some (secretory vesicles) concentrate and store them. Organelles do not exist in isolation from other organelles, and as they function they are in constant communication with one another. Most likely, this communication is mediated by *small vesicles* that shuttle membrane constituents and lumen contents from one organelle to another. Some of these transport vesicles are termed *coated vesicles* because they contain an outer protein shell composed primarily of the fibrous protein *clathrin.* As is explained in Chapter 15, clathrin may help in the "pinching off" of one vesicle from a larger one. One of the big mysteries in cell biology is how intracellular transport vesicles "know" with which membrane to fuse and thus where to deliver their contents. (Again, Chapter 21 presents current ideas about this sorting process.)

Lysosomes Contain a Battery of Degradative Enzymes That Function at pH 5

Lysosomes are membrane-limited organelles containing *acid hydrolases,* enzymes that degrade polymers into their

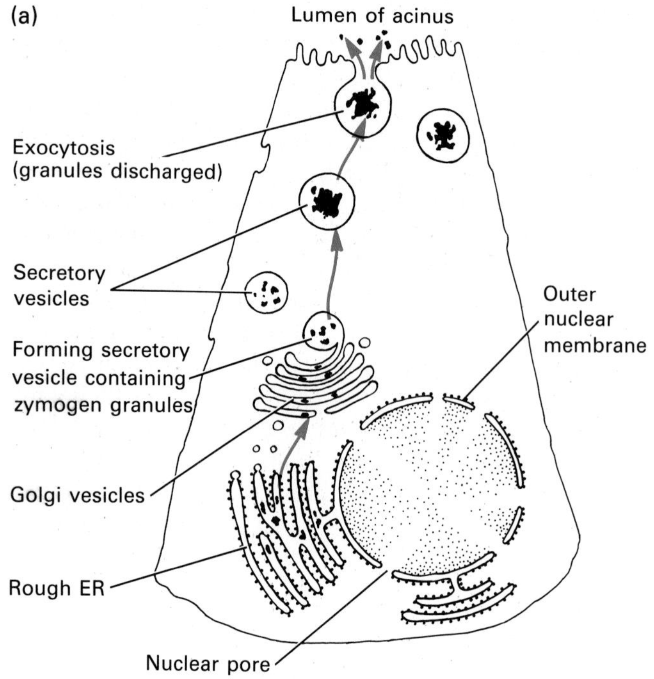

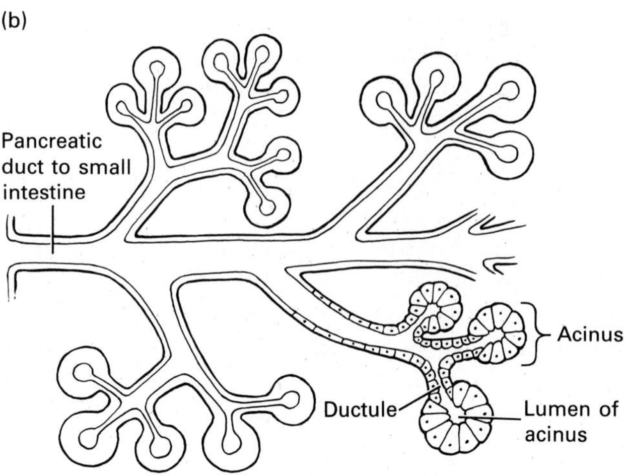

Figure 5-34 (a) The pathway taken by newly synthesized secretory proteins in a rat pancreatic acinar cell. Immediately after synthesis, the secretory proteins are found in the lumen of the rough ER. Transfer vesicles transport them to the Golgi complex (see Figure 5-33). Next they are concentrated in secretory vesicles containing granules of zymogens, or pancreatic enzyme precursors such as chymotrypsinogen. The cell is oriented so that the secretory vesicles are formed under the plasma membrane region that faces a ductule of the acinus. Exocytosis of the vesicles, triggered by hormones or nerve stimulation, releases the contents into the ductule; from there, the inactive precursors move to the intestine, where they are proteolytically activated into digestive enzymes (see Figure 3-22). (b) The organization of the *exocrine* part of the pancreas. This part of the organ synthesizes digestive enzymes and delivers them to the small intestine. Digestive enzymes secreted by the acinar cells pass first into the ductules of the acini (the plural of acinus) and then into the small intestine. Not shown here are the *endocrine* parts of the pancreas. These are small groups of cells, termed islets of Langerhans, that produce hormones such as insulin and glucagon and secrete them directly into the blood.

monomeric subunits (Table 5-5). Lysosomes vary in size and shape (Figure 5-36), and several hundred may be present in a typical cell. They contain ribonucleases and deoxyribonucleases, which degrade RNA and DNA into their mononucleotide building blocks; phosphatases, which remove phosphate groups from mononucleotides and other compounds, such as phospholipids; proteases, which degrade a variety of proteins and peptides; and enzymes that degrade complex polysaccharides and lipids into smaller units.

All of these enzymes work only at acid pH values and are inactive at the neutral pH values of cells and of most extracellular fluids. To enable these enzymes to function, the inside of a lysosome is maintained at about pH 4.8 (or lower) by a hydrogen-ion pump in the lysosomal membrane. The acid pH helps to denature proteins and make them accessible to the action of the lysosomal hydrolases, which themselves have structures that resist acidic denaturation. If an occasional lysosome releases its contents into the cytosol, where the pH is from 7.0 to 7.3, no degradation of cytosolic components takes place. The cytosol of yeast, and probably of other cells as well, contains proteins that can bind and inactivate any leaked lysosomal proteases and other degradative enzymes.

No part of a cell is immortal; even in growing cells, in which macromolecules and organelles are increasing, there is constant degradation of membranes, proteins, and other constituents. Lysosomes degrade many membranes and organelles that have outlived their usefulness to the cell. How aged or defective organelles are marked for degradation and transferred to lysosomes is not known, but occasionally what appears to be a bit of mitochondrion or other membrane can be seen within a lysosome (Figure 5-36b).

Tay-Sachs Disease, the Result of a Defective Lysosomal Enzyme The importance of lysosomes in the degradation of cellular membrane constituents is demonstrated by the existence of mutants that are defective in certain lysosomal hydrolases. Tay-Sachs disease is an in-

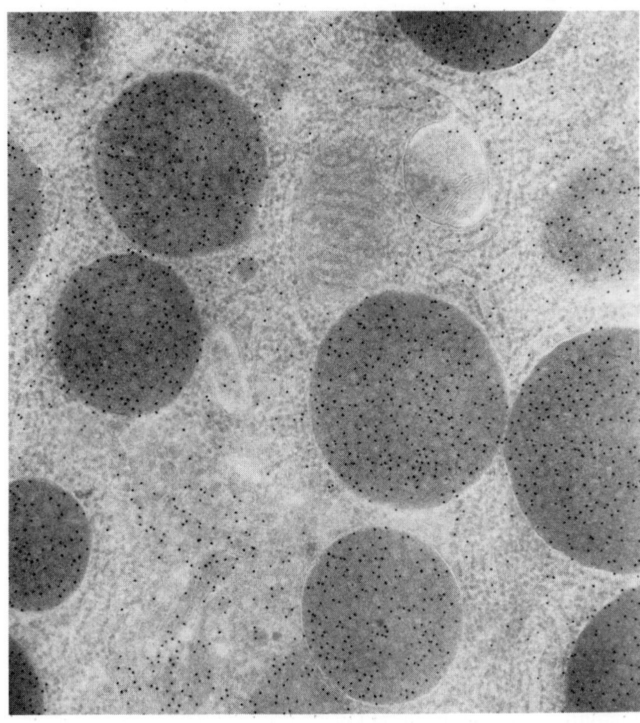

Figure 5-35 An electron micrograph showing secretory vesicles containing zymogen granules in a rat pancreatic acinar cell. To detect amylase, a protein typically found in zymogen granules, sections were treated with an antibody to amylase; then 8-nm gold–protein A complexes were added to reveal the bound antibody (see Figure 5-9). *Courtesy of H. J. Geuze.*

herited recessive disorder that results in mental retardation, derangement of the central nervous system, and death by the age of 5. In normal people, the ganglioside G_{M2}, one of the constituents of the plasma membrane of many mammalian cells (nerve cells in particular), is continually synthesized and degraded; in Tay-Sachs victims, membranes of the brain cells accumulate ganglioside G_{M2}. This accumulation is caused by the absence of a specific lysosomal hydrolase, β-*N*-hexosaminidase A, a key enzyme in the normal turnover of plasma membrane G_{M2} (Figure 5-37). The excess G_{M2} is believed to cause all of the symptoms of Tay-Sachs disease. In humans there are a large number of known *lysosome storage diseases,* similar to Tay-Sachs, in which a missing lysosomal enzyme causes the lysosomes to fill up with partly degraded cellular material.

Endocytosis Enables Proteins and Particles to Enter the Cell

Lysosomes are also important in the degradation of extracellular macromolecules brought into the cell. Many proteins and particles are brought into the cell by *endocytosis,* the progressive invagination of a region of the plasma membrane to form a closed vesicle entirely surrounded by cytoplasm. Extracellular materials are either trapped within the lumen of the vesicle or bound to its lumenal surface. These vesicles are believed to fuse with lysosomes, in which the ingested macromolecules are degraded (Figure 5-38). (Chapter 15 contains a detailed discussion of these processes.)

Certain animal cells called macrophages are specialized for the ingestion of large particulate matter, especially harmful bacteria. Again, a membrane surrounds the ingested particle and creates a vesicle that fuses with the lysosome, in which (one hopes) the bacterium is killed. The lysosomes in macrophages also contain enzymes such as myeloperoxidase, which generates hydrogen peroxide (H_2O_2) and superoxide (O_2^-). These powerful oxidants

Table 5-5 Acid hydrolases that have been located in lysosomes

Enzyme	Natural substrate
PHOSPHATASES	
Acid phosphatase	Most phosphomonoesters
Acid phosphodiesterase	Oligonucleotides and other phosphodiesters
NUCLEASES	
Acid ribonuclease	RNA
Acid deoxyribonuclease	DNA
POLYSACCHARIDE- AND MUCOPOLYSACCHARIDE-HYDROLYZING ENZYMES	
β-Galactosidase	Galactosides
α-Glucosidase	Glycogen
α-Mannosidase	Mannosides, glycoproteins
β-Glucuronidase	Polysaccharides and mucopolysaccharides
Lysozyme	Bacterial cell walls and mucopolysaccharides
Hyaluronidase	Hyaluronic acids; chondroitin sulfates
Arylsulfatase	Organic sulfates
PROTEASES	
Cathepsin(s)	Proteins
Collagenase	Collagen
Peptidases	Peptides
LIPID-DEGRADING ENZYMES	
Esterase(s)	Fatty acyl esters
Phospholipase(s)	Phospholipids

SOURCE: D. Pitt, 1975, *Lysosomes and Cell Function,* Longman.

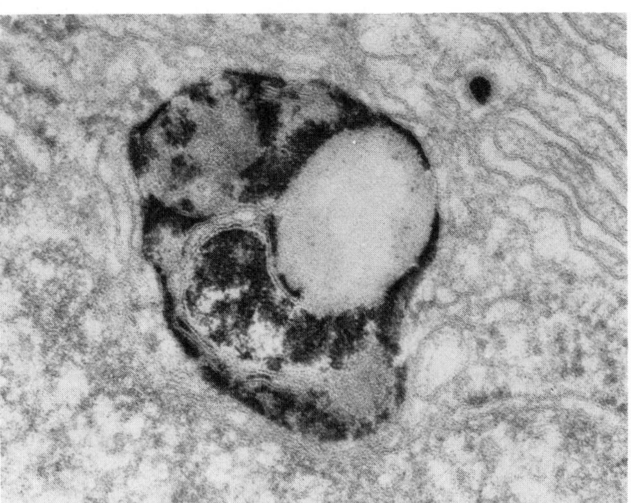

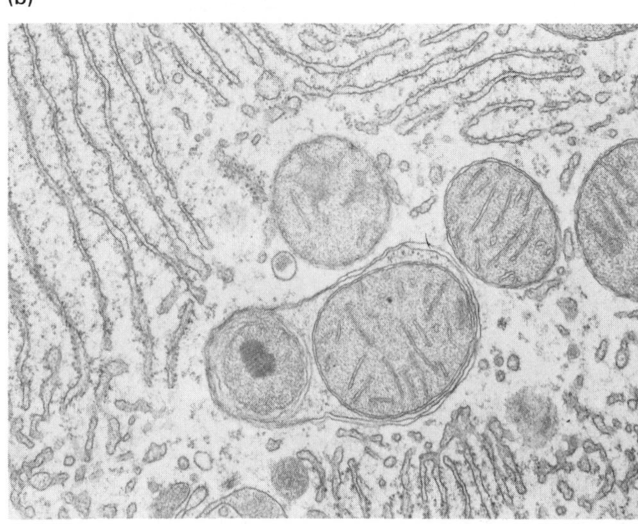

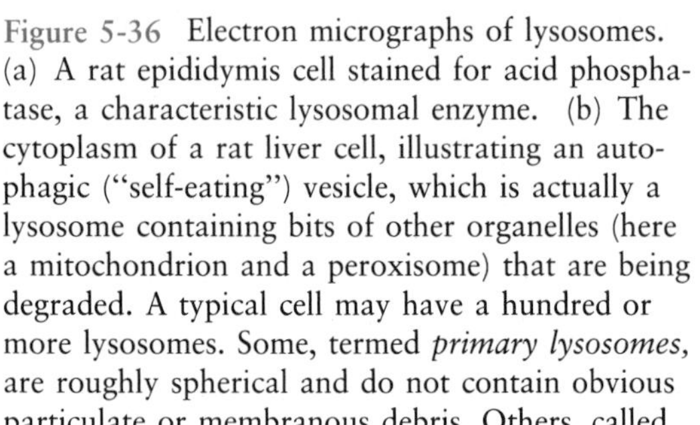
Figure 5-36 Electron micrographs of lysosomes. (a) A rat epididymis cell stained for acid phosphatase, a characteristic lysosomal enzyme. (b) The cytoplasm of a rat liver cell, illustrating an autophagic ("self-eating") vesicle, which is actually a lysosome containing bits of other organelles (here a mitochondrion and a peroxisome) that are being degraded. A typical cell may have a hundred or more lysosomes. Some, termed *primary lysosomes,* are roughly spherical and do not contain obvious particulate or membranous debris. Others, called *secondary lysosomes,* are larger and irregularly shaped and do contain particles or membranes that are being digested. Secondary lysosomes appear to result from the fusion of primary lysosomes with other membranous organelles; the autophagic vesicle depicted here is a secondary lysosome that apparently resulted from the fusion of one or more primary lysosomes with an aged or defective mitochondrion and a similarly disabled peroxisome. *Courtesy of D. Friend.*

are essential for killing bacteria; human genetic defects in these enzymes render a person very sensitive to bacterial infection. Another kind of vesicle present in macrophages contains other antibacterial proteins: for example, lysozyme, an enzyme that degrades the peptidoglycan in bacterial cell walls, and lactoferrin, a protein that binds iron, an essential bacterial nutrient. These vesicles also fuse with vesicles that contain ingested bacteria and participate in their killing.

Peroxisomes and Glyoxysomes Produce and Degrade Hydrogen Peroxide

The *peroxisomes* are a class of small, membrane-limited organelles found in the cytoplasm of all animal cells; *glyoxysomes* are similar organelles found in plants. Because the morphology of these two organelles resembles that of lysosomes, it was thought for a long time that they were in fact lysosomes. However, cell fractionation studies established that the enzymes in peroxisomes and glyoxysomes are very different from those in lysosomes and serve a quite different function. Peroxisomes contain enzymes that degrade fatty acids and amino acids. A byproduct of these reactions is hydrogen peroxide (H_2O_2), a corrosive substance that oxidizes many amino acid side chains, such as methionine (see Chapter 2, page 24). To counter the potential deleterious effects of the hydrogen peroxide, peroxisomes also contain copious amounts of the enzyme catalase, which degrades hydrogen peroxide:

$$2H_2O_2 \xrightarrow{\text{catalase}} 2H_2O + O_2$$

Many peroxisomes have crystalline arrays of catalase molecules, which gives them a very distinctive appearance in the electron microscope (see Figures 5-26f and 5-31). The exact role of peroxisomes in cellular metabolism is a bit mysterious, because many of the enzyme-catalyzed degradative reactions also occur in other organelles without the synthesis and degradation of H_2O_2. Some investigators feel that one function of peroxisomes is the generation of heat, rather than ATP, as a product of the catabolism of energy-rich molecules such as fatty acids.

Plant Cells Contain Vacuoles, Which Store Small Molecules and Enable the Cell to Elongate Rapidly

Most plant cells contain one or more membrane-limited *internal vacuoles* (Figure 5-39). The number and size of

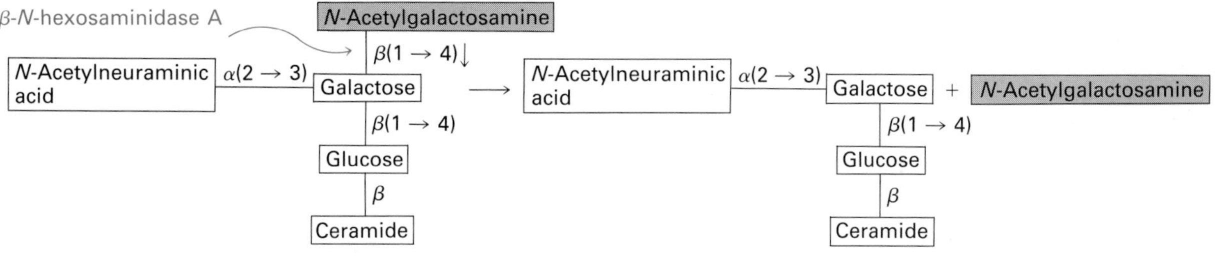

Figure 5-37 The first step in the degradation of the ganglioside G_{M2}. The colored arrow indicates the site of action of β-N-hexosaminidase A, the degradative enzyme that is deficient in Tay-Sachs disease.

vacuoles vary in different cells and in different stages of development; in mature cells, a single vacuole may occupy as much as 80 percent of the cell's volume. Within the vacuoles, plant cells store water, ions, and such food materials as glucose. Vacuoles may also act as receptacles for harmful waste products. The concentration of dissolved materials within the vacuoles is very high—higher

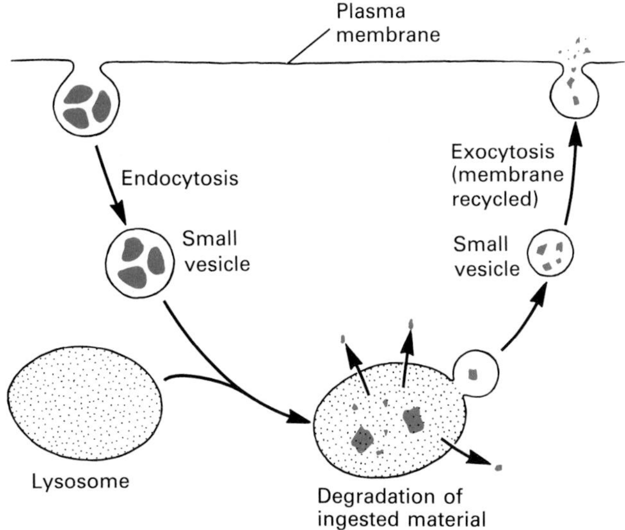

Figure 5-38 The degradation of extracellular material by a lysosome. During endocytosis, a region of the plasma membrane invaginates and forms a vesicle containing material from outside the cell. This vesicle fuses with a primary lysosome, forming a secondary lysosome, where the material is degraded. Many of the products of this degradation, such as amino acids and nucleotides, are transported into the cytoplasm, where they are either incorporated into cellular macromolecules or metabolized. Other products of this degradation are incorporated into vesicles and subjected to exocytosis. The lysosome retains some indigestible material, such as the membranous debris visible in Figure 5-36b.

than that in the cytosol or in extracellular fluids. The vacuolar membrane is semipermeable; that is, it is permeable to water, but not to the small molecules stored within it. Water flows across a semipermeable membrane from a solution having a low concentration of dissolved materials to one having a high concentration. A different way of stating the same point is that water will flow from a solution of high H_2O concentration (one with a low concentration of solutes) to one of low H_2O concentration (one with a high concentration of solutes), in order to attain the same concentration of water on both sides of the membrane. Such movement of water is termed *osmotic flow*.

The entry of water into the vacuole from the cytosol causes the vacuole to expand and creates hydrostatic pressure, or *turgor*, inside the cell. This is balanced by the mechanical resistance of the cellulose-containing cell wall that surrounds all plant cells. Most plant cells have a turgor ranging from 5 to 20 atmospheres, and consequently they must have a wall with sufficient strength to react to this pressure in a controlled way. However, unlike animal cells, plant cells can elongate extremely rapidly—at rates ranging from 20 to 75 μm per hour. Such elongation, which occurs when the somewhat elastic cell wall stretches under the pressure created by water taken into the vacuole, usually accompanies plant growth.

Osmotic Regulation Is an Important Function of Microorganisms and Animal Cells

Like higher plants, many algae and fungi contain cell walls; animal cells and most protozoans do not. Many protozoans, however, contain a *contractile vacuole* which, like the plant vacuole, takes up water from the cytosol. Unlike the plant vacuole, the contractile vacuole discharges its contents periodically through fusion with the plasma membrane (Figure 5-40). Because water continuously enters the cell by osmotic flow, the contractile vacuole prevents too much water from accumulating in the

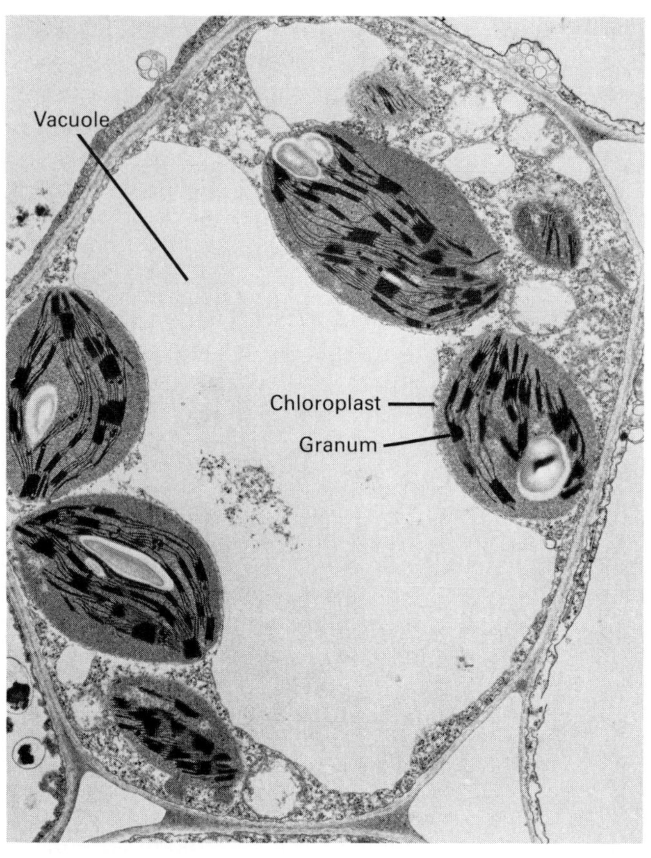

Figure 5-39 An electron micrograph of a leaf cell from *Phleum pratense,* showing a large internal vacuole and parts of five chloroplasts. *Biophoto Associates/Myron C. Ledbetter/Brookhaven National Laboratory.*

cell and swelling it until it bursts. The rupturing of the plasma membrane by a flow of water into the cytosol is termed *osmotic lysis*.

Animal cells do not contain vacuoles for disposing of excess water. If animal cells are placed in a solution in which the concentration of dissolved impermeable molecules, such as salts, is lower than that inside the cells (a *hypotonic* solution), the cells will swell and possibly burst, owing to the osmotic flow of water into them. Immersing animal cells in a *hypertonic* solution (one with a higher than normal concentration of solutes) causes them to shrink, as water leaves by osmotic flow. Consequently, it is essential that animal cells be maintained in an *isotonic* medium—one in which the concentration of solutes is similar to that of the cell cytosol (Figure 5-41).

Certain marine plants and other organisms employ an alternative mechanism for osmotic control. They employ metabolic pathways to synthesize large amounts of molecules termed *osmotic protectants*. These are usually small organic molecules such as glycine, which accumulate in the cell in a high concentration (0.3 *M*) in response to a high-osmotic-strength environment (i.e., an environment high in solutes). This prevents the cells from shrinking.

Because plant cells and bacterial cells are surrounded by rigid cell walls, they do not swell when they are placed in a hypotonic solution. Most bacterial and plant cells do not absorb extra fluid even when they are exposed to pure water.

The Plasma Membrane Has Many Varied and Essential Roles

Some of the essential functions of the *plasma membrane,* the semipermeable barrier that defines the outer perimeter of the cell, are to allow nutrients required by the cell to enter, to filter out unwanted materials in the extracellular milieu, and to prevent metabolites and ions needed by the cell from leaving it. Pure phospholipid bilayer membranes (see Figure 3-67) either are impermeable to most of the small molecules (such as glucose and amino acids) that are needed for cellular metabolism, or they allow these molecules to enter at a rate that is far too slow to be compatible with life. As is discussed in Chapter 15, the plasma membrane and the membranes of organelles con-

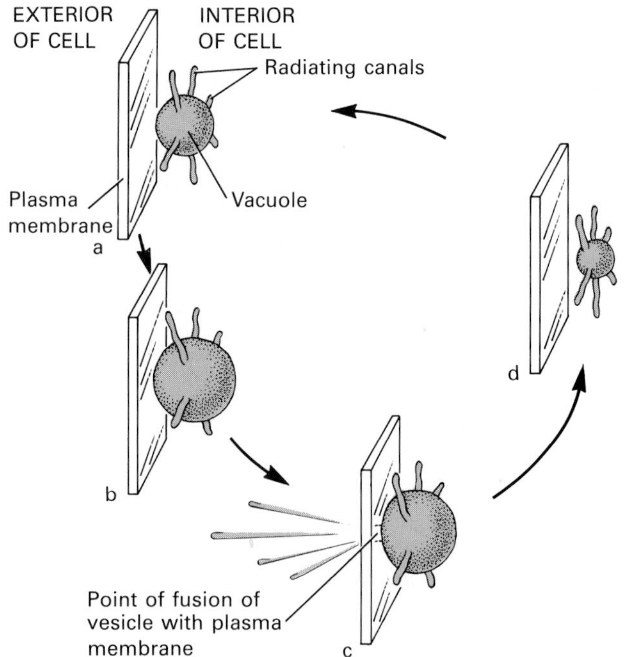

Figure 5-40 The contractile vacuole in *Paramecium caudatuam.* (a),(b) The vesicle in this typical ciliated protozoan is filled by radiating canals that collect fluid from the cytosol. (c) When the vesicle is full, it then fuses for a brief period with the plasma membrane and expels its contents. (d) When the vesicle is nearly empty, the radiating canals begin to refill it with fluid from the cytoplasm.

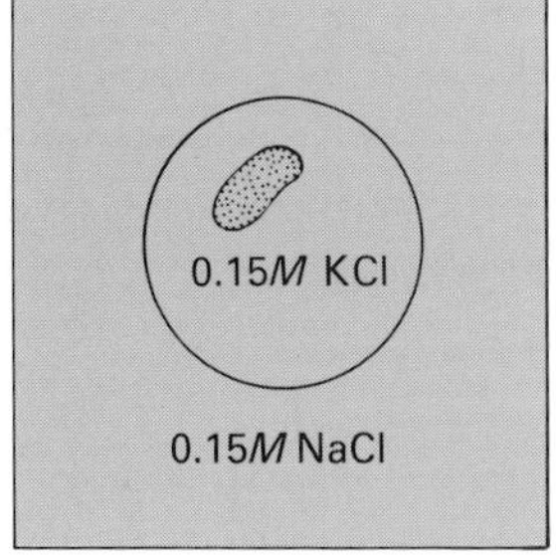

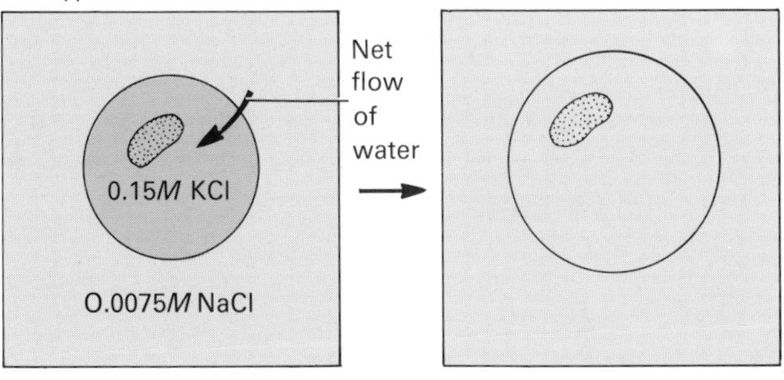

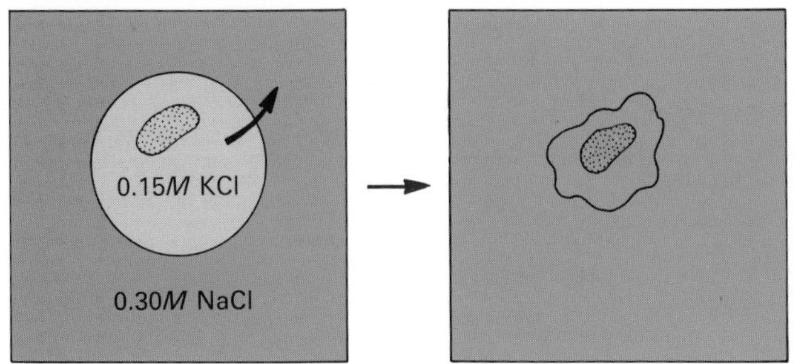

Figure 5-41 Animal cells respond to the osmotic strength of the surrounding medium. Sodium, potassium, and chloride ions do not move freely across the cell membrane, but water does. (a) When the medium is isotonic, there is no net flux of water into or out of the cell. (b) When the medium is hypotonic, water flows into the cell until the total concentration of ions is the same inside and outside. In this example, in which the cytosolic ion concentration is twice the extracellular ion concentration, the cell tends to swell to twice its original volume, at which point the internal and external concentrations of ions are the same. (c) When the medium is hypertonic, water flows out of the cell until the concentration of ions is the same inside and outside. Here the cytosolic ion concentration is half the extracellular ion concentration, so the cell is reduced to half its original volume.

tain specific transport or permease proteins that allow certain molecules to pass through but not others. For example, the plasma membrane of virtually all cells contains a *glucose permease* protein that allows glucose, but not sugars with related structures, to cross it.

We have already seen how particles can enter the cell by endocytosis of the plasma membrane. Many extracellular proteins and other macromolecules enter the cell by a similar process after binding to specific receptors on the cell surface. As mentioned earlier, cellular contents can also be extruded through the plasma membrane as a result of exocytosis, in which intracellular vesicles fuse with the membrane and release their contents into the extracellular fluid. Clearly, regions of the membrane, and the lipids and proteins within them, undergo a great deal of recycling. The plasma membrane also contains anchoring points for many of the fibrous structures that permeate the cytosol.

A major function of the plasma membrane is to communicate and interact with other cells. Most cells in a multicellular animal do not exist as isolated entities; rather, groups of cells with related specializations combine to form *tissues*. Certain designated areas of the plasma membrane make contact with other cells to give strength to tissues and to allow the exchange of metabolites between cells. (The structure of tissues is discussed in Chapter 14.)

Another function of the plasma membrane is to recognize extracellular signals. The prime examples of such signals are *hormones,* chemicals that are secreted by one cell type but that act on another; alternatively, molecules located on cell surfaces may act on neighboring cells. In animals, hormones are usually carried by the blood. They function by means of specific receptors on the surfaces of target cells or within them; these receptors are proteins that recognize the hormones and inform the cell that a signal has been received. Insulin, for example, is a peptide hormone that is secreted into the blood by the β cells of the pancreas. Insulin is required for the growth and metabolism of most mammalian cells, which contain surface receptors for it. In general, a cell may react to a hormonal signal in a variety of ways: by changing its metabolism, by excreting some metabolite or protein, by generating an electric current (in a nerve), or by contracting (in a muscle).

The Generation of Cellular Energy: The Structures and Functions of Mitochondria and Chloroplasts

The principal source of cellular energy in nonphotosynthetic cells is glucose (Chapter 2). The complete aerobic degradation of glucose to CO_2 and H_2O is coupled to the synthesis of as many as 36 molecules of ATP:

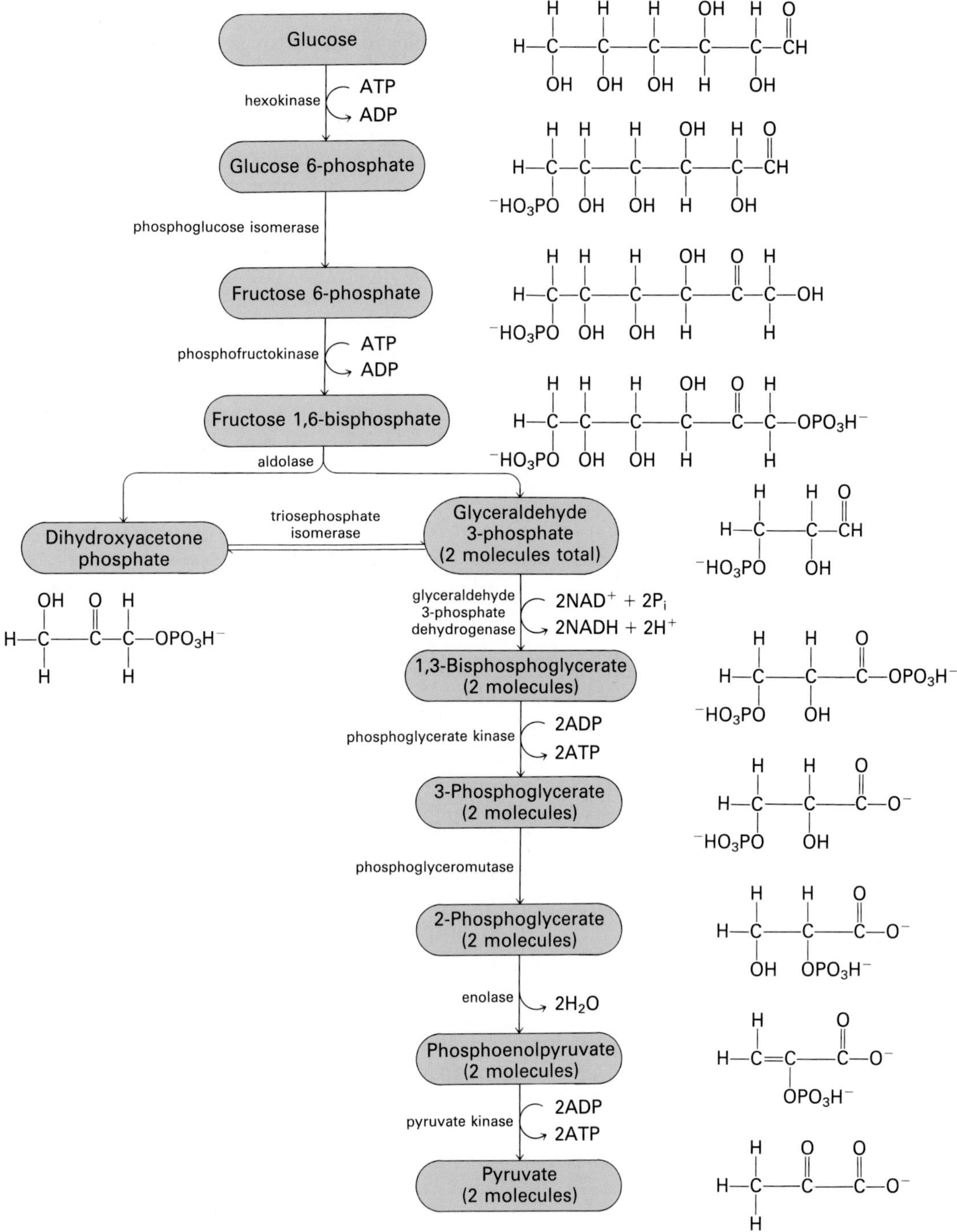

Figure 5-42 The glycolytic pathway, by which glucose is degraded to pyruvic acid.

$$C_6H_{12}O_6 + 6O_2 + 36P_i + 36ADP \longrightarrow 6CO_2 + 6H_2O + 36ATP + 36H_2O$$

In eukaryotic cells, some of the enzyme-catalyzed chemical reactions in the pathway of the aerobic degradation of glucose occur in the cytosol, whereas others occur in the mitochondria. Because of the central importance of this biochemical pathway, we shall describe its predominant features in some detail before turning to the structure and function of the mitochondrion itself.

Glycolysis Is the First Stage in the Metabolism of Glucose and the Generation of ATP

The initial stage of glucose metabolism, glycolysis, is the conversion of each molecule of glucose into two molecules of the three-carbon compound pyruvate, $C_3H_3O_3^-$ (Figure 5-42). Glycolysis takes place in the cytosol, and does not involve molecular oxygen. All of the metabolic intermediates between the initial carbohydrate and the final product, pyruvate, are phosphorylated compounds. As can be seen in Figure 5-42, four molecules of ATP are formed from ADP in glycolysis: two in the step catalyzed by phosphoglycerate kinase, where two molecules of 1,3-bisphosphoglycerate* are converted to 3-phosphoglycerate, and two in the step catalyzed by pyruvate kinase. However, two molecules of ATP were consumed in earlier steps of this pathway: first, by the addition of a phosphate residue to glucose in the reaction catalyzed by hexokinase (see Figure 3-28); and second, by the addition of a second phosphate to fructose 6-phosphate in the reaction catalyzed by phosphofructokinase. Thus, there is a net gain of two molecules of ATP. The balanced chemical equation for this series of reactions shows that four hydrogen atoms (i.e., four protons and four electrons) are also formed:

$$C_6H_{12}O_6 \longrightarrow 2C_3H_4O_3 + 4H$$

These hydrogen atoms are generated in the reaction catalyzed by the enzyme glyceraldehyde 3-phosphate dehydrogenase. The four electrons and two of the protons are transferred to two molecules of nicotinamide adenine dinucleotide (NAD^+), the oxidized form of the compound, to produce NADH, the reduced form (Figure 5-43):

$$2H^+ + 4e^- + 2NAD^+ \longrightarrow 2NADH$$

Thus, the overall reaction for this first part of glucose metabolism is

$$C_6H_{12}O_6 + 2NAD^+ + 2ADP + 2P_i \longrightarrow 2C_3H_4O_3 + 2NADH + 2ATP$$

*The prefix "bis" is used for names of compounds with 2 phosphates attached to 2 different carbon atoms; the prefix "di" is used if the 2 phosphates are attached to the same carbon atom.

Oxidation of Carbohydrates Is Completed in the Mitochondria, Where Most of the ATP Is Produced

The pyruvate formed in glycolysis is transported into the mitochondria, where it reacts with coenzyme A* (CoA) to form CO_2 and the key compound acetyl-CoA:

$$\text{Pyruvate} + \text{HS—CoA} \longrightarrow \underbrace{CH_3\text{—}\overset{\overset{O}{\|}}{C}\text{—S—CoA}}_{\text{Acetyl-CoA}} + CO_2$$

Acetyl-CoA is further metabolized to CO_2 to regenerate free "carrier" CoA:

$$CH_3\text{—}\overset{\overset{O}{\|}}{C}\text{—S—CoA} + O_2 \longrightarrow 2CO_2 + H_2O + \text{HS—CoA}$$

The two molecules of NADH formed in glycolysis reduce NAD^+ molecules within the mitochondria:

$$NADH_{cytosol} + NAD^+{}_{mitochondria} \longrightarrow NAD^+{}_{cytosol} + NADH_{mitochondria}$$

The mitochondrial NADH is reoxidized by O_2:

$$2NADH + 2H^+ + O_2 \longrightarrow 2NAD^+ + 2H_2O$$

These oxidation reactions generate the bulk of the ATP produced from the conversion of glucose into CO_2; 34 of the 36 ATPs generated are produced in the mitochondrion. Actually, 34 ATPs is the maximal number, and even this value is controversial. As is discussed in Chapter 20, much of the energy released in mitochondrial oxidation can be used for other purposes, such as generation of heat and transport of molecules into or out of the mitochondrion. Thus, this organelle deserves to be called the "power plant" of the cell. The details of the process by which ATP is formed in the mitochondrion require a chapter of their own (Chapter 20). Here we will describe only some of the basic structural features of mitochondria.

The Mitochondrion Has an Outer and an Inner Membrane

Most eukaryotic cells contain many mitochondria. They are among the largest organelles in the cell (generally only the nucleus, the vacuoles, and the chloroplasts are larger), and they can occupy as much as 25 percent of the volume of the cytoplasm. They are large enough to be seen under a light microscope, but the details of their structure can be viewed only with the electron microscope. Mitochondria contain two very different membranes, the *outer membrane* and the *inner membrane* (Figure 5-44; see also Figures 5-7 and 5-31).

*The structure of coenzyme A is shown in Figure 14-19.

Figure 5-43 The structures of NAD^+ and NADH. NAD^+ accepts only *pairs* of electrons, so its reduction to NADH requires the transfer of two electrons simultaneously.

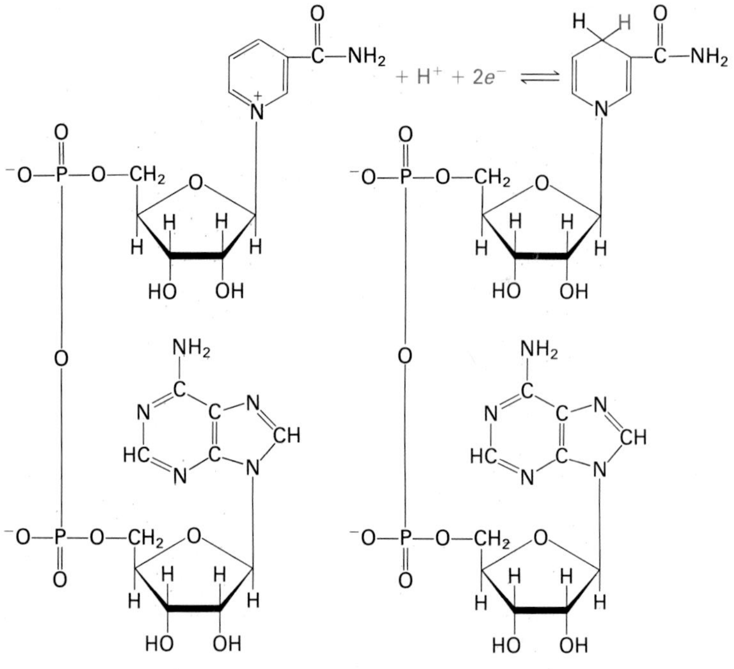

The outer membrane contains porins, proteins that render the membrane permeable to molecules having molecular weights as high as 10,000. In this respect, the outer membrane is uncharacteristic of biological membranes. It is about 50 percent lipids and 50 percent proteins.

The inner membrane is much less permeable. It is distinctive in that it contains about 20 percent lipids and 80 percent proteins, a higher protein-to-lipid ratio than other cellular membranes have. The inner membrane has a large number of infoldings, or *cristae,* that protrude into the *matrix,* or central space. On the matrix side of the inner membrane are a number of short protuberances collectively called the F_0F_1 protein complex; these are made up of about 15 different protein species, and they are important because they constitute the enzyme system that synthesizes ATP from ADP and P_i. The matrix is filled with enzymes that catalyze a number of reactions, including the conversion of acetyl-CoA into CO_2. Several enzymes are bound to the inner membrane itself; they participate in the binding of oxygen and the transport of electrons and protons. A pure phospholipid membrane is impermeable to most of the chemicals that need to move into and out of the mitochondrion, such as citrate, ADP, phosphate, and ATP; *permease* proteins present in the membrane transport these molecules by forming channels through which they can enter and leave. Certain ions, particularly Ca^{2+} and HPO_4^{2-}, are concentrated in the mitochondria; in many cases, the mitochondria represent the major intracellular reservoir of Ca^{2+} ions.

Mitochondria Contain DNA That Encodes Some of the Mitochondrial Proteins

An important discovery was that mitochondria contain several copies of a relatively small DNA molecule, *mitochondrial DNA.* The size of this DNA varies from organism to organism, but it usually does not constitute more than a small fraction of the total DNA in the cell. Mitochondrial DNA contains the genetic information for five or six of the key inner membrane proteins. As is described in Chapter 21, mitochondrial DNA resides in the matrix, where it undergoes transcription to yield mitochondrial mRNAs. These are translated into proteins by mitochondrial ribosomes, which also reside in the matrix, and which are very different from cytosolic ribosomes. However, the genetic information for the vast majority of mitochondrial proteins is located in the nuclear DNA, not in the mitochondrial DNA. These proteins are made on cytoplasmic ribosomes and are transported into the mitochondria after their synthesis.

Chloroplasts Have at Least Three Different Membranes

Except for vacuoles, chloroplasts are the largest and most characteristic organelles in the cells of green plants and green algae. Like mitochondria, chloroplasts do not have fixed positions in cells, and they often migrate from place

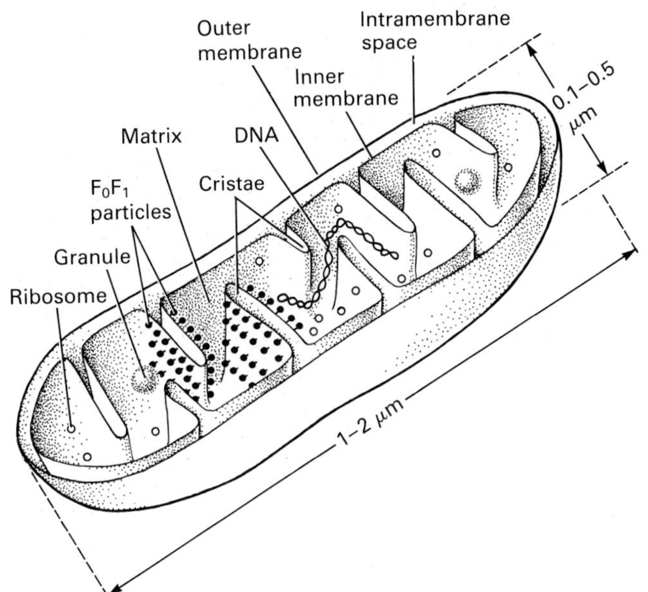

Figure 5-44 A three-dimensional diagram of a mitochondrion cut longitudinally.

to place. They can be as long as 10 μm and they are typically 0.5 to 2 μm thick, but they vary in size and shape in different cells, especially among the algae. On a dry-weight basis, 35 percent of the chloroplast is lipid, 5 percent is protein, and 7 percent is pigment. Predominant among the pigments is chlorophyll *a,* the substance that traps light and gives the chloroplast its green color. In some types of algae, the predominance of other pigments in organelles similar to chloroplasts gives them a different color. For example, rhodoplasts are red, phaeoplasts are brown, and so forth.

The chloroplast is surrounded by two membranes, an outer and an inner membrane (Figure 5-45). Chloroplasts also contain an extensive internal membrane system made up of *thylakoid vesicles;* these are closed membrane vesicles that are flattened to form disks. Thylakoid vesicles are often stacked together to form *grana* (see also Figure 5-39). The *stroma* is the space surrounding the thylakoids and grana.

The outer and inner chloroplast membranes and the space between them constitute a selective barrier to the movement of metabolites and ions. Permease proteins in the chloroplast membrane enable glucose and other molecules formed in the stroma during photosynthesis to exit into the cytoplasm, and also allow certain chemicals from the cytosol to enter the chloroplast. Much of the glucose formed by photosynthesis is stored within the chloroplasts as granules of *starch,* a polymer of glucose similar in structure to glycogen. The pigments and enzymes that participate in trapping the energy of light and generating ATP in a process called *photophosphorylation* are found in the thylakoid vesicles. Like the F_0F_1 proteins in the inner membrane of the mitochondrion, the thylakoid vesicles contain projections that protrude into the inner matrix. As is discussed in Chapter 20, the morphological similarity also extends to function, because these protein particles are the elements that generate ATP from ADP and P_i. The other major product of photophosphorylation is NADPH, the reduced form of nicotinamide adenine dinucleotide phosphate. (NADPH is similar to NADH, but it has an additional phosphate group.) These two molecules, ATP and NADPH, participate in the conversion of CO_2 into intermediates in the synthesis of glucose. This process, which is called carbon dioxide fixation, occurs in the stroma. Chapter 20 treats these topics in detail.

Chloroplasts Also Contain DNA

Chloroplasts, like mitochondria, contain DNA; this DNA is known to encode some of the key chloroplast proteins, but most of these proteins are made in the cytosol and are incorporated into the organelle later. The synthesis of chloroplast proteins is described in Chapter 21.

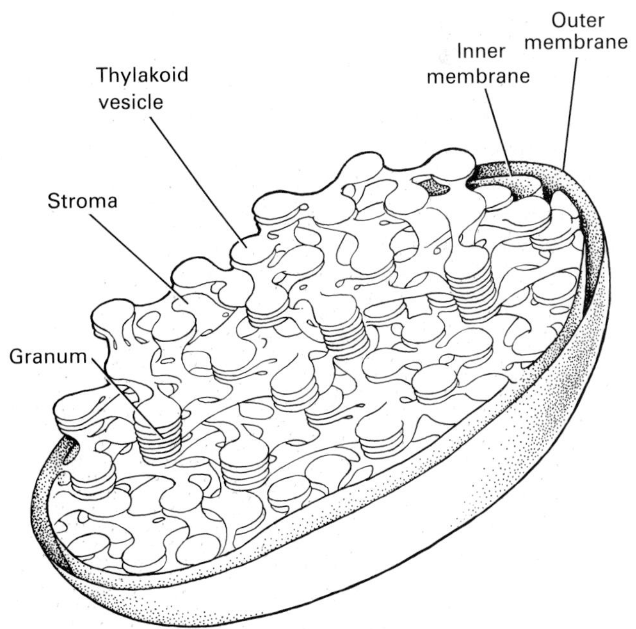

Figure 5-45 The structure of a chloroplast. The internal membrane (thylakoid) vesicles are organized into stacks, or grana, which reside in a matrix termed the stroma. All the chlorophyll in the chloroplast is contained in the membranes of the thylakoid vesicles.

Prokaryotes Oxidize Glucose to Carbon Dioxide

Prokaryotic cells do not contain mitochondria, yet most prokaryotes are capable of the aerobic conversion of glucose into CO_2. Such cells contain only a single plasma membrane, and all of the membrane-bound proteins required for aerobic metabolism and oxidative phosphorylation are contained in one or more regions of the plasma membrane. Likewise, there are certain bacteria—the blue-green "algae"—that are capable of photophosphorylation and photosynthesis, but that do not contain a separate internal chloroplast to hold the requisite membrane proteins and pigments. Rather, the enzymes and pigments are found on membranes that seem to be invaginations of the plasma membrane (see Figure 5-21). For example, bacteriorhodopsin, the purple light–trapping protein, is found in patches of "purple membranes" in photosynthetic bacteria.

Some Eukaryotic and Prokaryotic Cells Metabolize Glucose Anaerobically

Most eukaryotic cells are *obligate aerobes,* which means that they can grow only in the presence of oxygen; they metabolize glucose (or related sugars) completely to CO_2 with the concomitant production of a large amount of ATP. However, most of them can also generate some ATP by anaerobic metabolism. A few eukaryotes, including certain yeasts, are *facultative anaerobes,* meaning that they can grow either in the presence or in the absence of oxygen. Annelids and mollusks can live and grow for many days without oxygen. Many prokaryotes are *obligate anaerobes;* these, of course, cannot grow in the presence of oxygen.

The anaerobic metabolism of glucose does not require mitochondria. In this case, glucose is not converted entirely into CO_2, but rather into one or more two- or three-carbon compounds and, in some cases, CO_2. As a result, much less ATP is produced per mole of glucose. Yeasts, for example, ferment glucose anaerobically to two molecules of ethanol and two of CO_2, with the net production of only two molecules of ATP per molecule of glucose (Figure 5-46; see also Figure 5-42). As pyruvate is converted into ethanol, the NADH produced in the initial stages of glycolysis is reoxidized. This anaerobic fermentation is the basis of the entire beer and wine industry.

During prolonged contraction of mammalian muscle cells, the cells ferment glucose to two molecules of lactic acid—again, with the net production of only two molecules of ATP per molecule of glucose (Figures 5-46 and 5-42). The lactic acid formed is largely secreted into the blood; from there it passes into other tissues, where it is reoxidized into pyruvate and then metabolized further into CO_2. Lactic acid bacteria—the organisms that "spoil" milk—and other prokaryotes also generate ATP by fermentation of glucose to lactate.

In general, nonphotosynthetic eukaryotic cells are totally dependent on an exogenous (added) supply of one of the following: glucose; polymers such as glycogen or sucrose, which can be hydrolyzed to glucose; or other sugars that can be converted into glucose. In most cells, fatty acids also can be a source of ATP. Few other molecules are efficient generators of ATP.

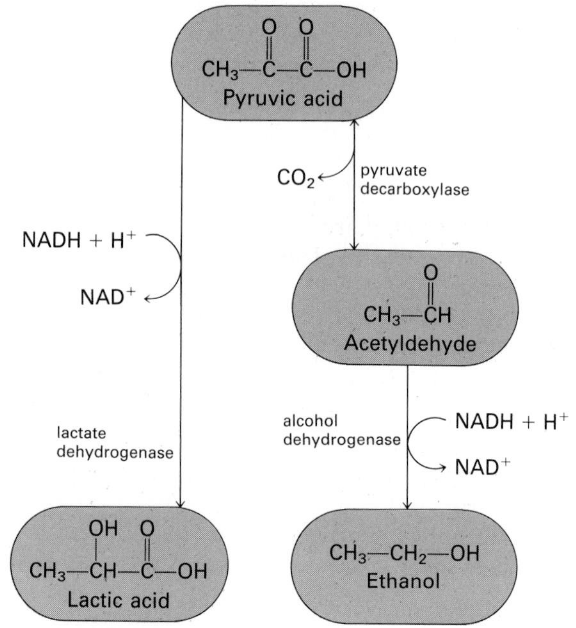

Figure 5-46 The anaerobic metabolism of glucose. In the formation of acid from glucose, one molecule of NAD^+ is reduced to NADH for each molecule of pyruvate formed (see Figure 5-42). If all the NAD^+ in the cell were converted into NADH, glucose metabolism would cease; thus it is necessary to regenerate NAD^+ by the transfer of the two electrons from NADH to an acceptor molecule. If the acceptor is pyruvic acid, as occurs in muscle cells when oxygen supplies are low, lactic acid is formed. In yeasts, acetaldehyde is the acceptor, and ethanol is formed.

Prokaryotic Extracellular Extensions and Eukaryotic Extensions of the Plasma Membrane

Although the plasma membrane forms the principal permeability barrier of the cell, some important parts of the cell lie outside of it. Also, in most eukaryotic cells, long extensions of the plasma membrane surround specialized cytoplasmic regions that serve important functions.

The Flagella and Pili of Prokaryotic Cells Are Constructed of Proteins

The major types of surface appendages on gram-negative bacterial cells are *flagella* and the shorter *pili* (the singular forms are flagellum and pilus; Figure 5-47). These similar long, thin structures originate in the *basal structure,* which is embedded in the outer surface membrane and the peptidoglycan layer. The flagellum is composed of a single species of protein, *flagellin,* which is assembled in helical chains wound around a hollow central core. Flagella play an important role in bacterial movement, as can be demonstrated by a simple experiment: when cells' flagella are sheared off, the cells stop swimming. To propel cells, flagella rotate around their long axes; apparently the terminal parts that are embedded in the inner surface membrane (the plasma membrane) receive energy from the cytosol. The energy for flagellar rotation comes from a proton gradient across the plasma membrane. As is discussed in Chapter 20, this electrochemical gradient is generated during the oxidation of carbohydrates and other molecules.

The functions of pili are less well known. In some pathogenic bacteria, pili participate in the invasion process by aiding the attachment to the host organism's cells (e.g., to cells that line the intestine or the lung airways).

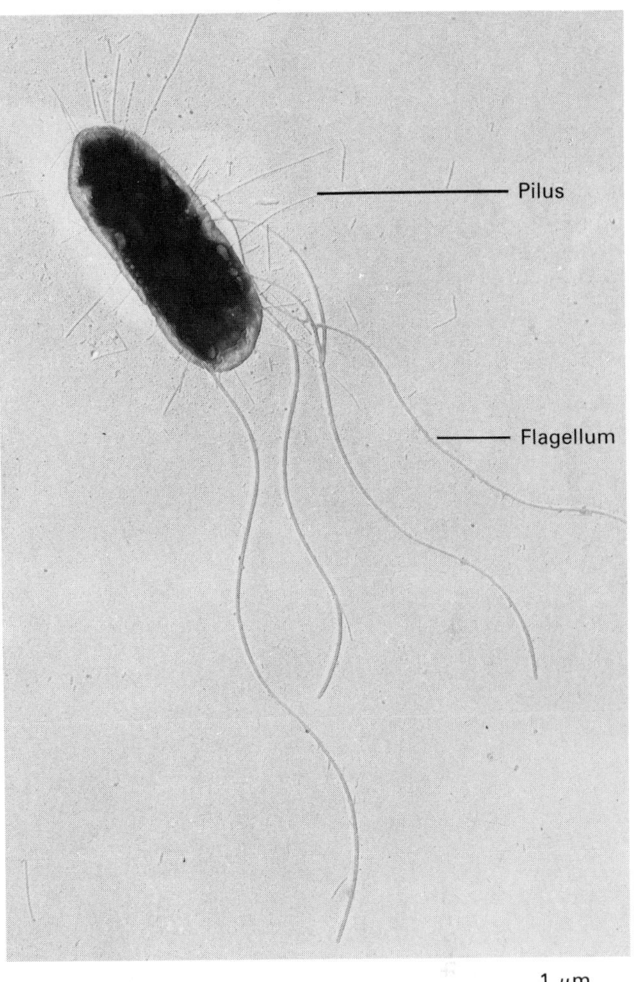

Figure 5-47 An electron micrograph showing flagella and pili protruding from the surface of the gram-negative bacterium *E. coli. Courtesy of H. Berg.*

The Cilia and Flagella of Eukaryotic Cells Are Extensions of the Plasma Membrane

The surfaces of eukaryotic cells contain a number of protuberances and extensions that serve many specific and important functions. *Cilia* and *flagella* are similar long, motile structures that extend from the plasma membranes of many plant and animal cells and of unicellular organisms (Figure 5-48a). The surfaces of these organelles are extensions of the plasma membrane. In this respect, eukaryotic flagella differ from prokaryotic flagella, which consist only of protein and which lie outside the plasma membrane. In addition, eukaryotic cilia and flagella have an entirely different mechanism of movement from prokaryotic flagella. Bunches of filaments called microtubules run the lengths of the central cores of eukaryotic cilia and flagella (Figure 5-48b). These microtubules are constructed of the protein tubulin, and they are similar in structure to the microtubules that participate in the mitotic spindle. Cilia beat backward and forward; flagella, which are longer than cilia, typically rotate in a screwlike manner. Both motions can propel a cell; the beating of the sperm flagellum, for example, is the only means of locomotion of the sperm. The motions of cilia or flagella on an immobile cell can propel extracellular fluids along the surface of the cell. Cilia are present on many

(a)

(b)

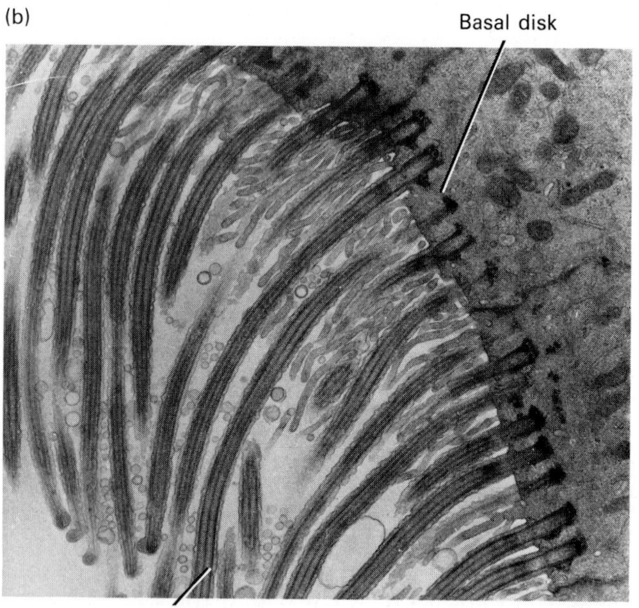

Figure 5-48 (a) A scanning electron micrograph of the surface of hamster trachea cells, showing abundant clusters of cilia protruding into the lumen of the trachea. (b) A micrograph of a thin section through the surface of a rabbit tracheal cell. Note that the surfaces of the cilia are continuous with the plasma membrane and that each cilium contains a core of microtubules that terminate in a basal disk structure just under the plasma membrane. *Courtesy of E. R. Dirksen.*

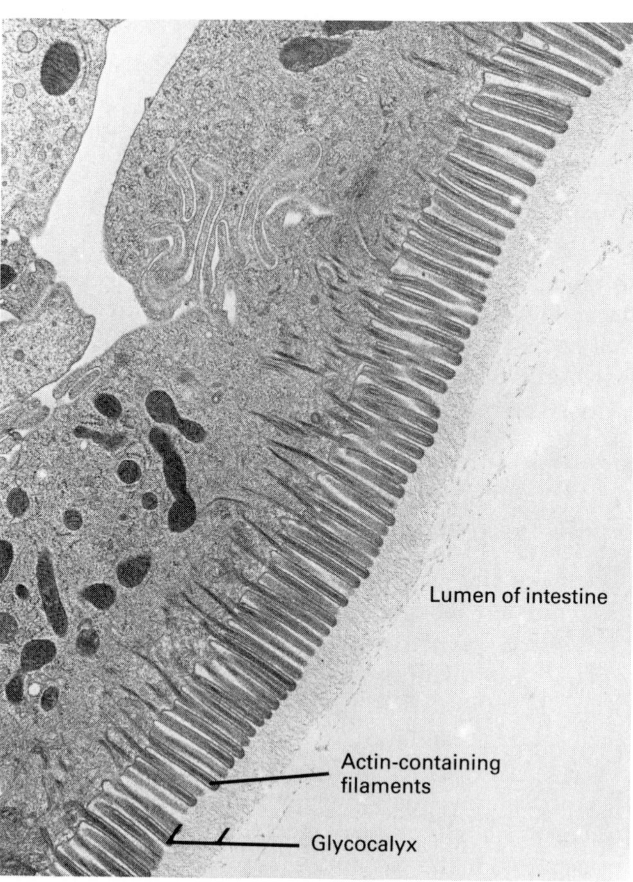

Figure 5-49 A transmission electron micrograph of the brush border of an intestinal epithelial cell, showing the microvilli with their cores of actin-containing microfilaments. Attached to the brush border is a fibrous network of glycoproteins termed the glycocalyx; this network contains glycosidases and peptidases that catalyze the final stages in the digestion of ingested macromolecules. *Courtesy of S. Ito.*

epithelial cells, such as those of the trachea, where the cilia propel small particles along the surface of this sheet of cells that lines the airways.

Microvilli Enhance the Adsorption of Nutrients

On many cells that are specialized for the adsorption of nutrients, the plasma membrane is folded into a large number of fingerlike projections termed *microvilli;* these greatly increase the surface area of the cell and allow a greater rate of adsorption of nutrients. The surfaces of the epithelial cells lining the cavity of the small intestine, which is studded with microvilli, are collectively called the *brush border* (Figure 5-49). The plasma membrane of the brush border contains many enzymes that assist in the degradation of sucrose into glucose and fructose, and of

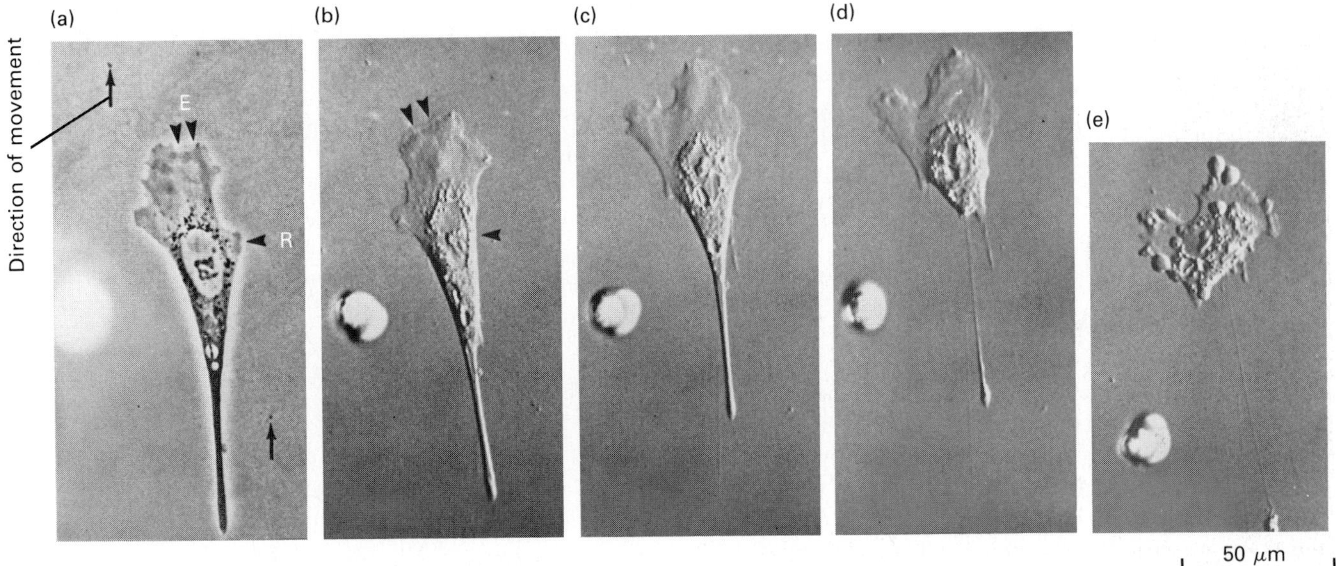

Figure 5-50 The movement of a fibroblast cell along a substratum. (a) The first image, at 0 min, was obtained by phase-contrast microscopy. Successive time-lapse images of the same cell, obtained by Nomarski optics, show that the lamella at the right of the cell is retracting (arrowhead labeled R), and that the lamellipodia at the leading edge of the cell are extending (arrowheads E). Bits of debris adhering to the substratum serve as fixed reference points. (b) In the frame taken at 8 min, the leading edge has moved forward about 9 μm, and the lamellipodia there form a thin flat sheet. (c), (d) At 28 min, the broad leading edge has spread and separated into two lamellae. The thin trailing edge of the cell has begun to retract into the cell body. (e) At 34 min, retraction of the trailing edge is almost complete; only a thin thread of cytoplasm is anchored to the substratum. *From W.-T. Chen, 1981,* J. Cell Sci. ***49**:1.*

proteins into amino acids. It also contains specific permeases that allow the epithelial cell to adsorb these nutrients from the intestine. (Chapter 15 contains a discussion of the transport of amino acids and sugars into and through epithelial cells.)

Lamellipodia Play a Role in the Movement of Microorganisms and Cells in Multicellular Animals

Cellular movement is a dominant feature of many eukaryotic microorganisms and of multicellular animals. Soil amoebae move very rapidly on solid surfaces, searching for food. In the embryonic construction of tissue and organ systems of animals, many cells migrate, individually or in groups, from one area of the embryo to another.

Cell motility has been studied extensively with vertebrate fibroblasts—connective tissue cells—in vitro. Fibroblasts move jerkily along a substratum. As such cells move forward, they develop ruffled *lamellipodia,* flattened extensions that adhere to the substratum (Figure 5-50). Lamellipodia are very thin—from 0.1 to 0.4 μm in diameter—and consist of a meshwork of actin-containing microfilaments (Figure 5-51); their structures are discussed in more detail in Chapter 19. To propel itself, a

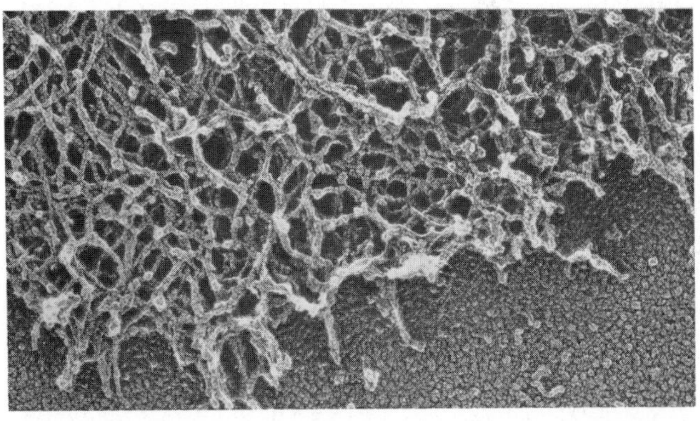

Figure 5-51 A high-magnification electron micrograph of the cytoskeleton beneath a lamellipodium. A fibroblast was treated with the detergent Triton X-100 to dissolve the plasma membrane, and the cell contents were then fixed. A web of actin-containing filaments underlies the plasma membrane in the region of the lamellipodium; these filaments are much more concentrated and interdigitated than filaments in other regions of the cell. *From J. Heuser and M. Kirschner, 1980,* J. Cell Biol. ***86**:212. Reproduced from the* Journal of Cell Biology *by copyright permission of The Rockefeller University Press.*

fibroblast continually establishes new contacts with the substratum at the leading edges of the lamellipodia. The trailing edge of the fibroblast remains attached to the substratum; thus, the tail becomes greatly elongated under the resulting tension. Eventually the tail ruptures, leaving a bit of itself attached to the substratum, whereas the major part of the tail retracts into the cell body (see Figure 5-50).

Extracellular Matrixes

The eukaryotic cellular extensions that we have just described are all formed by the plasma membrane; none lies outside of the permeability barrier defined by this membrane. However, many essential eukaryotic cell entities do lie outside of the plasma membrane. A variety of proteins that are attached to the outer surfaces of animal cells participate in the formation of specific contacts and junctions between cells. Such interactions impart strength and rigidity to multicellular tissues.

Much of the volume of *connective tissue* in animals consists of the spaces between cells. *Loose connective tissue* forms the bedding on which most small glands or epithelial cells lie (Figure 5-52). It contains only a few cells and a number of fibers of various diameters. The blood capillaries that bring oxygen and nutrients to the cells in all parts of the body are confined to loose connective tissue. A principal function of the amorphous spaces between cells in this tissue is to allow such nutrients to diffuse freely into the cells of the epithelia and the glands. *Strong connective tissue* is found in bone, cartilage, and tendons. Like loose connective tissue, it consists largely of extracellular spaces. These contain proteins, proteoglycans, and other substances that are produced and secreted by the relatively few cells present in the tissue.

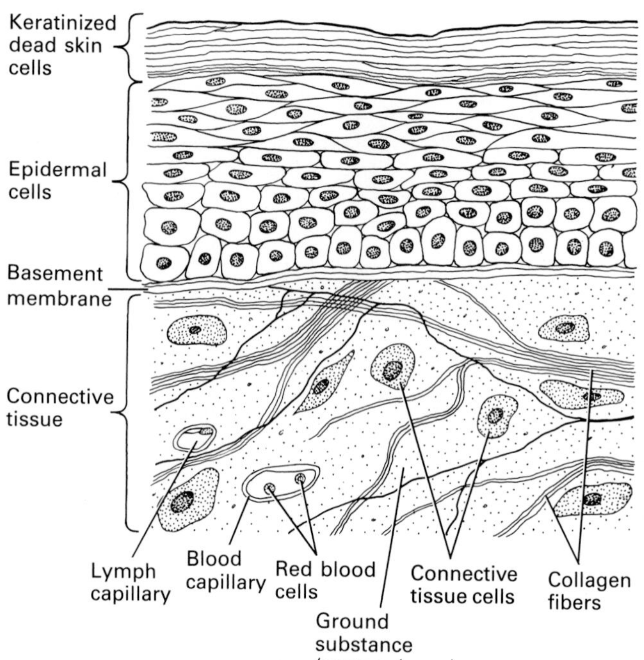

Figure 5-52 A schematic drawing of a section through the outer part of the skin of a pig. The cellular epidermis rests on a layer of loose connective tissue. The latter contains abundant collagen fibers that give it strength and rigidity. Also present are a few cells—mostly fibroblasts—that synthesize the connective-tissue proteins and polysaccharides. The relatively empty areas between the collagen fibers represent sites occupied by the amorphous *ground substance,* which is composed of hybrid molecules called proteoglycans. *Blood capillaries* course through the loose connective tissue and carry nutrients that must cross the capillary walls and diffuse through the connective tissue to nourish the epidermal cells; they also carry away waste products such as CO_2. *Lymph capillaries* are part of a second drainage system, the lymphatic system, which also carries away waste products and debris. The layer of epidermal cells synthesizes *keratin* proteins in abundance. Eventually the cells become filled with this fibrous protein and die; the dead cells form the impermeable keratinized outer layer of skin.

Collagen Forms Much of the Extracellular Framework of Multicellular Animals

Collagen is the major protein in the extracellular spaces of connective tissue. In fact, it is the most abundant protein in the entire animal kingdom. A distinctive characteristic of collagen is that it forms insoluble fibers that have a very high tensile strength. There are at least five types of collagens (Table 5-6); the structural features of each type make it suitable for a particular function in a cell or tissue. Collagens are secreted by cells called *fibroblasts* (Figure 5-53).

The fundamental structural unit of collagen is *tropocollagen,* a long (300-nm), thin (1.5-nm-diameter) protein that consists of three coiled polypeptide subunits called α chains. Each α chain contains 1050 amino acids. The three chains wind around one another in a characteristic right-handed triple helix (Figure 5-54a); hydrogen bonds between the chains hold the helix together. To satisfy the structural constraints of such a helix, every *third* residue in each chain must be glycine, the smallest amino acid. Collagen is also rich in proline and hydroxyproline, amino acids that favor the close packing of the chains (see Figure 3-14). Much of the α chain is, in fact, composed of the following amino acid sequence:

$$(\text{Gly-Pro-X})_n$$

where X can be any amino acid. As shown in Table 5-6,

Table 5-6 Characteristics of different types of collagens

Type	Molecular formula*	Tissue distribution	Distinctive features
I	$[\alpha 1(\text{I})]_2\alpha 2(\text{I})$	Skin, tendon, bone, dentin, fascia (widespread)	Low content of hydroxylysine; few sites of hydroxylysine glycosylation; broad fibers
II	$[\alpha(\text{II})]_3$	Cartilage, notochord, vitreous body	High content of hydroxylysine; heavily glycosylated; fibers usually thinner than in type I
III	$[\alpha 1(\text{III})]_3$	Skin, uterus, blood vessels ("reticulin" fibers generally)	High content of hydroxyproline; low content of hydroxylysine; few sites of hydroxylysine glycosylation; interchain disulfide bridges between cysteines at the carboxyl end of the helix
IV	$[\alpha 1(\text{IV})]_3$ and $[\alpha 2(\text{IV})]_3$	Kidney glomeruli, lens capsule; basement laminae of epithelial and endothelial cells	Very high content of hydroxylysine; almost fully glycosylated; relatively rich in 3-hydroxyproline; low alanine content
V	$[\alpha 1(\text{V})]_2\alpha 2(\text{V})$	Basement laminae of smooth and striated muscle cells, exoskeleton of fibroblasts and other mesenchymal cells (widespread in small amounts)	High content of hydroxylysine; heavily glycosylated; low alanine content

*Each collagen fibril is composed of three intertwined peptide chains. The two kinds of chains that constitute type I collagen are called $\alpha 1(\text{I})$ and $\alpha 2(\text{I})$; each fibril contains two of the former and one of the latter. The different types of collagen are built of different but related collagen polypeptides. Additionally, the chains differ in the extent to which their proline, lysine, and cysteine residues are modified, as indicated in the right-hand column. The structures of these modified amino acids are shown in Figure 3-14.

SOURCE: D. Eyre, 1980, *Science* **207**:1315–1321.

the α chains in the collagen of different tissues differ. The major collagen of skin and bone, type I, contains two chains of the $\alpha 1(\text{I})$ type and one chain of the $\alpha 2(\text{I})$. Because it is a coil of three peptide chains, the tropocollagen triple helix is a much more rigid, inflexible structure than the α helix.

In collagen fibers, tropocollagen molecules pack together side by side, with adjacent molecules being displaced from one another about one-quarter of their length, as shown in Figure 5-54b. This staggered array produces a striated effect that can be seen in electron micrographs of collagen fibers; the characteristic pattern of bands is repeated about every 64 nm (Figure 5-55). The bands are caused by the binding of a metal or a stain to certain sequences of amino acids in the tropocollagen.

As is discussed in Chapter 21, collagen contains a number of modified amino acids. These participate in chemical cross-links both within the tropocollagen triple helix and between adjacent tropocollagen molecules in a fiber. The cross-links make the fibers strong and flexible. Hydroxylysine residues are often glycosylated with a characteristic disaccharide (see Figure 3-76).

Connective tissue is also rich in large, complex molecules termed *proteoglycans,* which consist partly (95 percent) of acidic polysaccharides such as hyaluronic acid, and partly (5 percent) of proteins. Proteoglycans form the *ground substance* in which the collagen and other connective-tissue fibers are embedded.

Cell Walls in Plants Are Constructed of Multiple Layers of Cellulose

Almost all plant cell walls are made of *cellulose,* a polysaccharide formed from $\beta(1 \rightarrow 4)$-linked glucose mono-

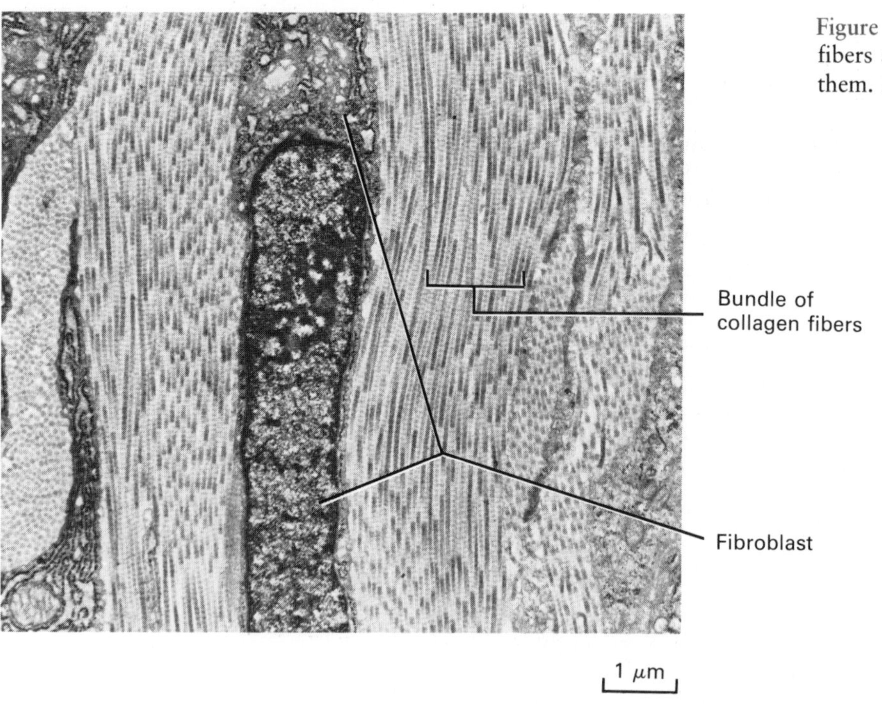

Figure 5-53 An electron micrograph of collagen fibers and the parent fibroblast cells that secreted them. *From D. Eyre, 1980,* Science ***207**:1315.*

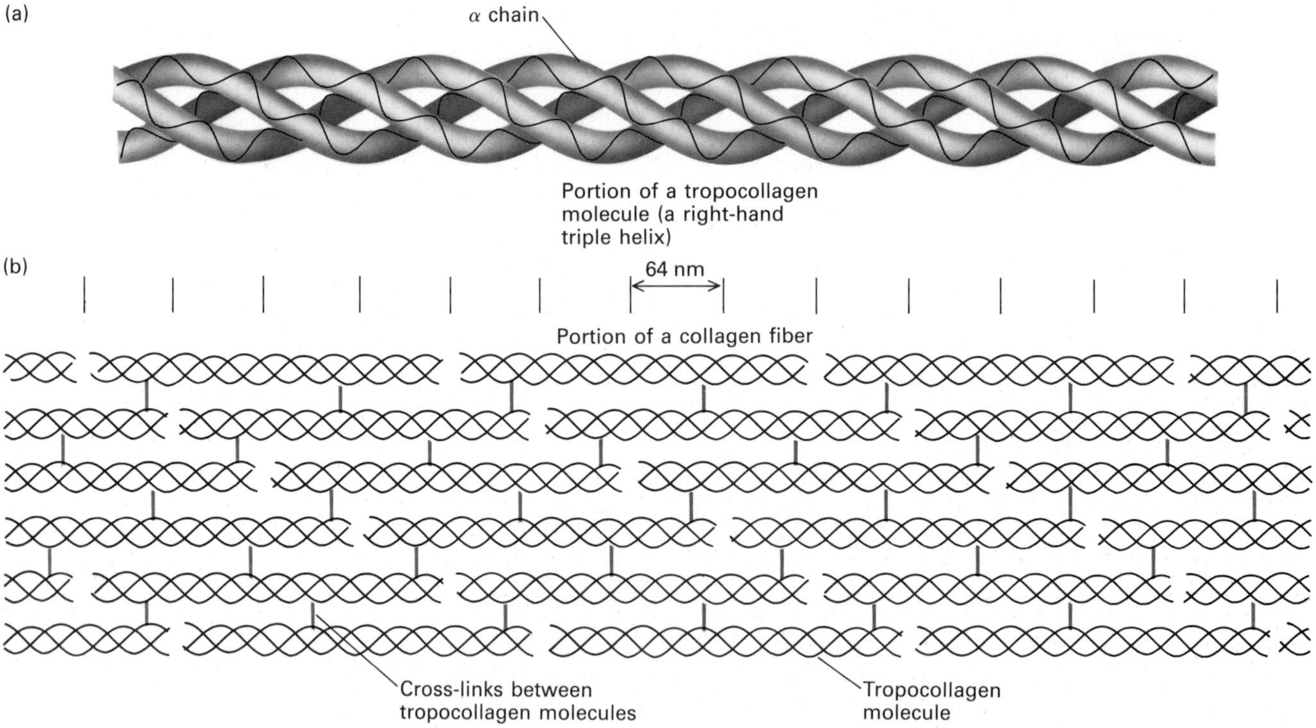

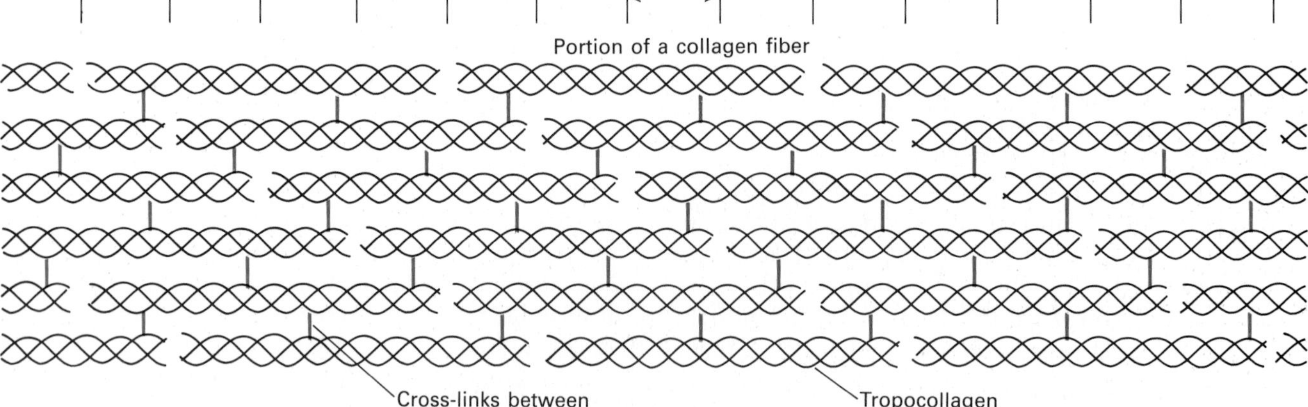

Figure 5-54 The structure of collagen. (a) The basic structural unit is the tropocollagen molecule, a triple-stranded helix. In type I collagen, the helix contains two $\alpha 1(I)$ chains and one $\alpha 2(I)$. Hydrogen bonds (not shown here) connect residues in each strand, or α chain, to the other two strands, which makes the helix rigid. Every third amino acid must be glycine, because a larger amino acid would not fit in the space at the center of the helix. (b) Each tropocollagen molecule is a triple helix 300 nm in length. In collagen, tropocollagen molecules pack together side by side. Adjacent molecules are displaced 64 nm, or slightly less than one-fourth the length of a single molecule. A small gap separates the "head" of one tropocollagen from the "tail" of the next. The side-by-side interactions are stabilized by chemical cross-links (color) between the chains. The mode of packing creates a 64-nm periodic pattern in the structure.

(a)

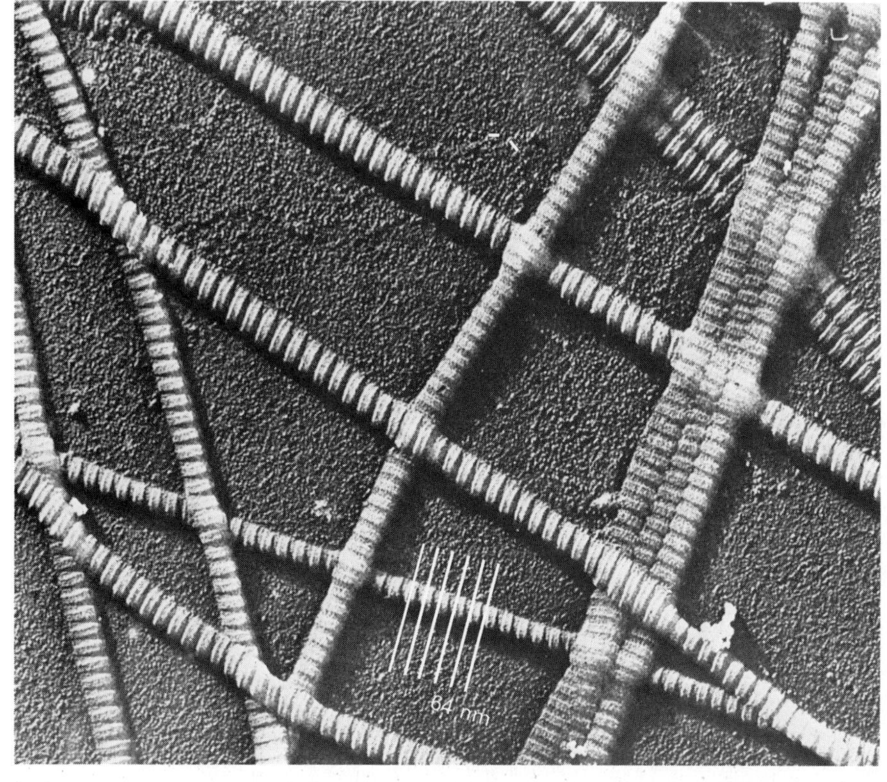

(b)

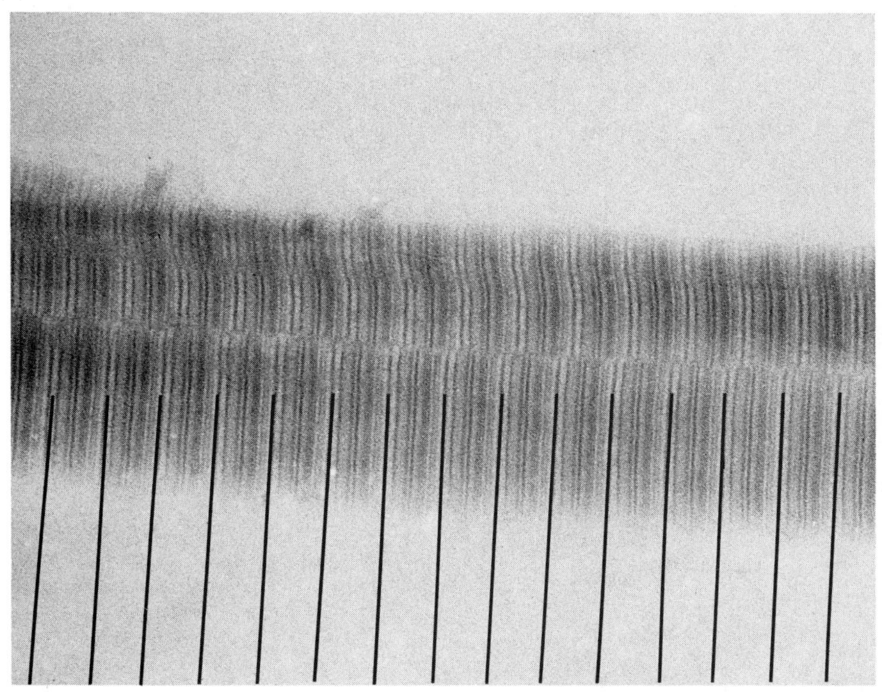

Figure 5-55 (a) A metal-shadowed replica of fibers of calfskin collagen. The 64-nm periodic pattern can be seen along the length of each fiber. In this technique, metallic platinum is evaporated in a vacuum chamber and allowed to settle on the collagen fibers, forming a thin coat. The underlying protein is then dissolved, leaving the metal replica of the surface of the fiber intact. This is picked up on a grid and viewed in the electron microscope. (b) A high-magnification electron micrograph of calfskin collagen fibers stained with phosphotungstic acid. The striations of the 64-nm periodic pattern are clearly visible. *Part (a) courtesy of J. Gross; part (b) courtesy of R. Bruns.*

(a)

$\beta(1 \rightarrow 4)$-Linked D-glucose units (cellulose)

$\alpha(1 \rightarrow 4)$-Linked D-glucose units (glycogen, starch)

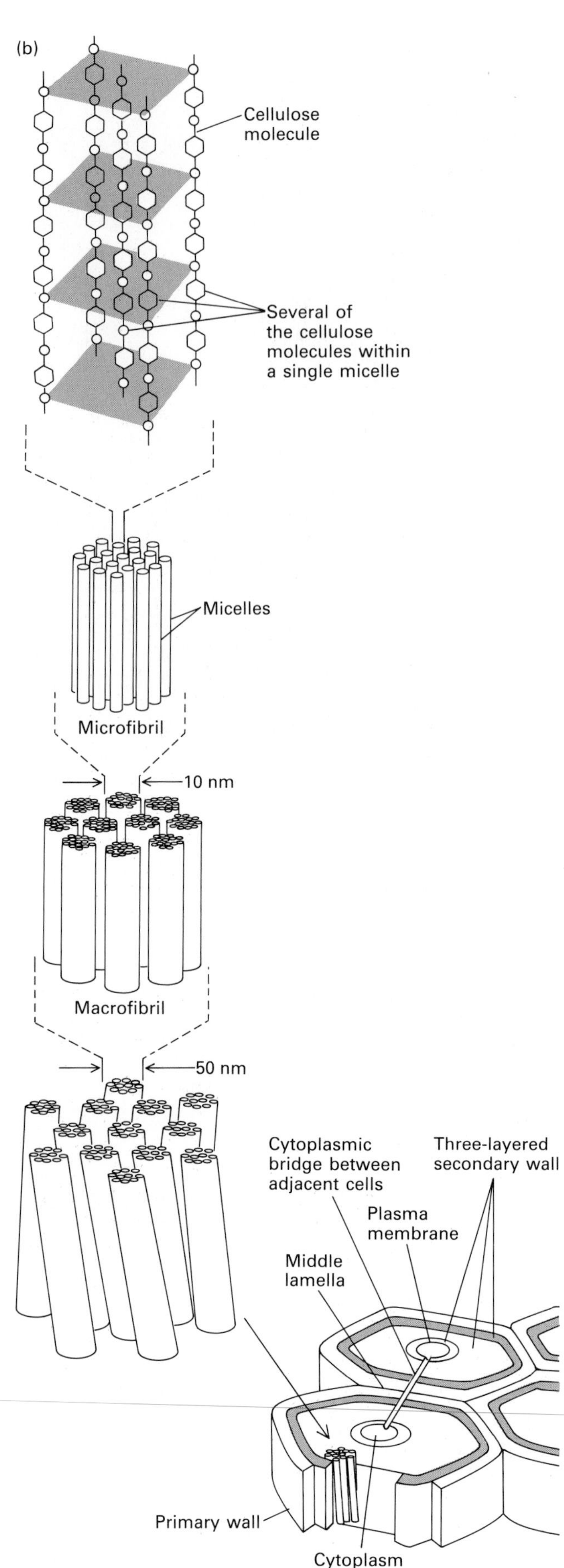

Figure 5-56 The structure of cellulose in the plant cell wall. (a) Cellulose is a linear polymer consisting of glucose residues linked together by $\beta(1 \rightarrow 4)$ glycosidic bonds. The $\beta(1 \rightarrow 4)$ linkages cause the molecule to form straight chains. In contrast, the $\alpha(1 \rightarrow 4)$ linkage in polyglucose molecules such as glycogen and starch causes a turning of the chain; such polyglucose molecules adopt a coiled helical conformation. (b) The linear cellulose molecules pack together to form rodlike structures termed micelles. These are stabilized by hydrogen bonds (not shown here) between the cellulose molecules. Not all of the chains in a micelle are illustrated. Micelles are packed into microfibrils, which are each about 10 nm in diameter and up to several micrometers in length. In turn, about 10 microfibrils are packed together to form macrofibrils, which are about 50 nm in diameter. These can be seen in the light microscope.

The cell wall is built up of a series of layers of cellulose. In each layer the fibers run more or less in the same direction, but the direction varies in the different layers. As plant cells grow, they deposit new layers of cellulose adjacent to the plasma membrane. Thus the oldest layers of cellulose are in the primary wall (the outside wall) and in the middle lamella (the part of the cell wall that is laid down between two daughter cells as they separate during division). Younger regions of the wall—collectively, the secondary cell wall—are laid down as three successive layers of cellulose, each of which is formed adjacent to the plasma membrane. The cytoplasms of adjacent cells are usually connected by tubes that run through the layers of the cell walls.

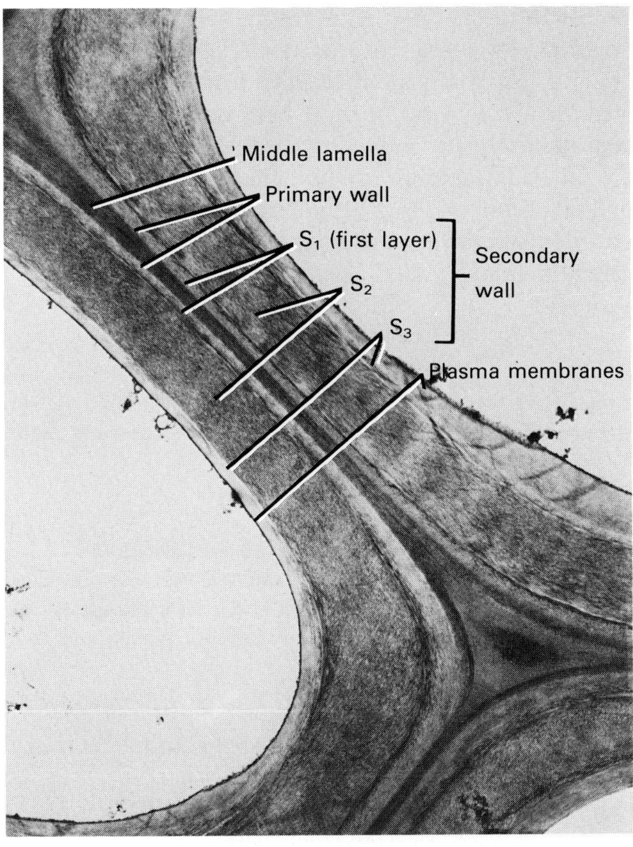

Figure 5-57 An electron micrograph of a thin section showing parts of the cell walls separating three *Taxus canadensis* plant cells. The principal layers of each wall are evident: the middle lamella; the primary wall; and the three layers of secondary wall, S_1, S_2, and S_3. The layers of cellulose fibers are laid down one by one (in the sequence just listed) as the cell matures. The fibers in each layer run in a different direction from those in the preceding layer. The plasma membrane is adjacent to the S_3 layer, the youngest stratum of the cell wall. *Biophoto Associates/Myron C. Ledbetter/ Brookhaven National Laboratory.*

mers. The $\beta(1 \rightarrow 4)$ linkage causes the polyglucose molecule to adopt a fully extended linear conformation; this is accomplished by the flipping of each glucose unit 180° relative to the preceding one (Figure 5-56a). In contrast, glycogen and starch are $\alpha(1 \rightarrow 4)$-linked polyglucose molecules; the $\alpha(1 \rightarrow 4)$ linkage causes the chain to adopt a helical configuration (Figure 5-56a). In plant cell walls, the cellulose molecules aggregate into bundles of fibers termed *microfibrils*, which in turn are clustered into *macrofibrils* (Figures 5-56b and 5-57). Hydrogen bonding of the cellulose molecules within microfibrils makes them very cohesive. Other polysaccharides within the wall cross-link the cellulose fibers. A very different type of complex compound called *lignin* imparts strength and rigidity and is predominant in the walls of woody plants. Other chemicals also are found in the walls of various plant cells; for example, *waxes* prevent the plant tissues and proteins from drying out.

The Walls of Prokaryotic Cells Are Constructed of Peptidoglycan

Like plant cells, many prokaryotic cells have walls. And, like plant cell walls, prokaryotic cell walls are mechanically strong enough to counterbalance the turgor—the osmotic pressure—of the enclosed cells. The chemical structure of the prokaryotic wall, however, is very different. It contains a number of proteins, lipids, and carbohydrates that are not found in plant cell walls; in particular, it invariably contains a *peptidoglycan* layer (see Figure 5-24). Only prokaryotes can synthesize this sheet of cross-linked polymers. The wall is the major determinant of bacterial shape, and its strength gives the cell rigidity.

Summary

Eukaryotic and prokaryotic cells differ in their structural elements, as summarized in Table 5-7, which lists most of the key differences in the structures of plant, animal, and

Table 5-7 Principal differences between prokaryotic and eukaryotic cells*

	Eukaryotes		
Characteristic	Higher plant	Animal	Prokaryotes
Number of chromosomes	>1	>1	1
Nucleoplasm bounded by a membrane	+	+	−
Histones	+	+	−
Nucleolus	+	+	−
Nuclear division by mitosis	+	+	−
ORGANELLES:			
Endoplasmic reticulum	+	+	−
Golgi vesicles	+	+	−
Mitochondria	+	+	−
Chloroplasts	+	−	−
Peroxisomes	+	+	−
Vacuoles	+	−	−
CELL WALL CONTAINING:			
Cellulose	+	−	−
Peptidoglycan	−	−	+

*The presence of a constituent is indicated by +, the absence by −.

prokaryotic cells. In most respects, the internal architecture and central pathways of metabolism are similar in all eukaryotic cells, whether they belong to plants, animals, or unicellular organisms. Eukaryotic cells have intracellular membranes and organelles that are never found in prokaryotes: examples are lysosomes, in which cellular membranes and endocytosed materials are degraded; peroxisomes, in which the synthesis and destruction of hydrogen peroxide takes place; the endoplasmic reticulum, the site of synthesis of secretory proteins, membrane proteins, and lipids; Golgi vesicles, which sort and process proteins—particularly secretory proteins—and membrane constituents; mitochondria, where oxidative phosphorylation takes place; and chloroplasts, where photosynthesis occurs in plants. Endocytosis and intracellular degradation of large particles are unknown in prokaryotes. Both plant and bacterial cells have walls, but these are very different in composition. The peptidoglycan, in particular, is unique to prokaryotes. Depending on the species of bacterium, other proteins, lipids, and carbohydrates are found in the extracellular wall. Plant cell walls are built of cellulose and other polymers. The wall is the major determinant of cell shape, and it gives the cell rigidity.

The metabolic reactions for the conversion of glucose into pyruvate are found in almost all cells, prokaryotic and eukaryotic; reactions for the oxidation of pyruvate to carbon dioxide are found in all aerobic cells. Metabolic pathways that are unique to certain groups of bacteria allow them to grow in unusual environments.

Both bacteria and eukaryotic cells bear flagella, but the structures of these organelles are very different. The flagella of eukaryotes are extensions of the plasma membrane, whereas those of prokaryotes are extracellular. In many eukaryotic cells, other plasma membrane regions, the microvilli and the lamellipodia, facilitate nutrient absorption and cell movement, respectively. The cytoplasm of all animal and plant cells contains a network of fibrous proteins that constitutes the cytoskeleton; this gives structural stability to the animal cell, and it is required for a number of processes, including cell movement. Animal cells are surrounded by an extracellular matrix consisting of collagen, proteoglycans, and other components that give strength and rigidity to tissues and organs.

As will be discussed in detail in subsequent chapters, the structure, replication, transcription, and translation of eukaryotic DNA are similar in plants and animals, but are substantially different in many respects from genetic processes in prokaryotes. The organization of nuclear DNA into several chromosomes and the division of the nucleus by mitosis are common to all eukaryotic cells; these features are absent in prokaryotes. Extranuclear DNA is present in the mitochondria of animals and plants and in the chloroplasts of plants, whereas these organelles are not found in prokaryotes. Because the two types of cells have many biochemical pathways in common, and because most aspects of the translation of mRNA into proteins are similar in all cells, it is believed that prokaryotes and eukaryotes are descended from the same primitive cell. The divergence must have occurred before the separation of plant and animal cells. All extant prokaryotic and eukaryotic cells and organisms are the result of a total of 3.5 billion years of evolution. It is not surprising, then, that cells are so well adapted to their own environmental niches.

References

Many of the topics listed below are treated in more detail in later chapters of this book, where additional articles are cited.

Cells as the Basic Structural Units

MARGULIS, L., and K. V. SCHWARTZ. 1982. *Five Kingdoms*. W. H. Freeman and Company. Electron micrographs and diagrams of representative cells from the principal groups of prokaryotes, eukaryotic microorganisms, plants, and animals.

Plant and Microbial Cells

BROCK, T. D., SMITH, D. W., and M. T. MADIGAN. 1984. *Biology of Microorganisms*, 4th ed. Prentice-Hall.

INGRAHAM, J. L., MAALOE, O., and F. C. NEIDHARDT. 1983. *Growth of the Bacterial Cell*. Sinauer.

RAVEN, P. H., R. F. EVERT, and H. CURTIS. 1981. *Biology of Plants*. 3d ed. Worth. An introduction to the structures and functions of plants and their cells.

Mammalian Cells (Histology Texts)

BLOOM, W., and D. W. FAWCETT. 1975. *A Textbook of Histology*, 10th ed. Saunders. An excellent text containing detailed descriptions of the structures of mammalian organs, tissues, and cells.

CORMACK, D. H. 1984. *Introduction to Histology*. Lippincott. Another excellent histology text.

WEISS, L. and L. LANSING, eds. 1983. *Histology—Cell and Tissue Biology*, 5th ed. Elsevier Biomedical.

Mammalian Cells (Atlases)

The following books are sources of detailed electron micrographs illustrating the structures of most mammalian cells and tissues.

CARR, K. E., and TONER, P. G. 1983. *Cell Structure—An Introduction to Biomedical Electron Microscopy*. Churchill Livingstone.

FAWCETT, D. W. 1981. *The Cell*, 2d ed. Saunders.

KESSEL, R. G., and R. H. KARDON. 1979. *Tissues and Organs: A Text Atlas of Scanning Electron Microscopy*. W. H. Freeman and Company.

PORTER, K. R., and M. A. BONNEVILLE. 1973. *Fine Structure of Cells and Tissues*, 4th ed. Lea & Febiger.

Microscopy and the Internal Architecture of Cells

The following books and reviews cover all of the standard techniques in light and electron microscopy.

EVERHART, T. E., and T. L. HAYES. 1972. The scanning electron microscope. *Sci. Am.* **226**(1):54–69.

HAYAT, M. A., ed. 1980. *Principles and Techniques of Electron Microscopy—Biological Applications,* vol. 1, 2d ed. Wiley.

HEUSER, J. 1981. Quick-freeze, deep-etch preparation of samples for 3-D electron microscopy. *Trends Biochem. Sci.* **6**:64–68.

PEASE, D. C., and K. R. PORTER. 1981. Electron microscopy and ultramicrotomy. *J. Cell Biol.* **91**:287s–292s.

POLAK, J. M. and S. VAN NOORDEN, 1984. *An Introduction to Immunocytochemistry. Roy. Microscopic Soc. Microscopy Handbooks Ser.* Oxford.

SPENCER, M. 1982. *Fundamentals of Light Microscopy.* Cambridge University Press.

WEAKLEY, B. S. 1981. *Beginner's Handbook in Biological Transmission Electron Microscopy,* 2d ed. Churchill Livingstone.

WILLINGHAM, M. C., and PASTAN, I. 1985. *An Atlas of Immunofluorescence in Cultered Cells.* Academic Press.

WISCHNITZER, S. 1981. *Introduction to Electron Microscopy,* 3d ed. Pergamon.

The Structure of the Cell Nucleus

The following books and articles deal with the architecture of the cell nucleus, the chromosomes, and the nuclear membrane.

BOSTOCK, C. J., and A. T. SUMNER. 1978. *The Eukaryotic Chromosome.* Elsevier/North-Holland.

FRANKE, W. W., U. SCHEER, G. KROHNE, and E. D. JARASCH. 1981. The nuclear envelope and the architecture of the nuclear periphery. *J. Cell Biol.* **91**:31s–50s.

GERACE, L., and G. BLOBEL. 1981. Nuclear lamina and the structural organization of the nuclear envelope. *Cold Spring Harbor Symp. Quant. Biol.* **46**:967–978.

GHOSH, S. 1976. The nucleolar structure. *Int. Rev. Cytol.* **44**:1–28.

Cell Division and DNA Replication

The cell cycle and the process of cell division are discussed in the following books and review articles.

INOUE, S. 1981. Cell division and the mitotic spindle. *J. Cell Biol.* **91**:132s–147s.

JOHN, P. C. L., ed. 1981. *The Cell Cycle.* Cambridge University Press.

MITCHISON, J. M. 1972. *Biology of the Cell Cycle.* Cambridge University Press.

PRESCOTT, D. M. 1976. *Reproduction of Eukaryotic Cells.* Academic Press.

The Organization of the Cytoplasm

The following books and articles describe the structure of the cytoplasm, focusing on the cytoskeletal fibers and on cell motility.

FULTON, A. B. 1980. How crowded is the cytoplasm? *Cell* **30**:345–347.

GOLDMAN, R., T. POLLARD, and J. ROSENBAUM, eds. 1976. *Cell Motility.* Cold Spring Harbor Laboratory.

INOUE, S., and R. E. STEPHENS. 1975. *Molecules and Cell Movement.* Raven.

STEBBINGS, H., and J. S. HYAMS. 1979. *Cell Motility.* Harlowe, U. K.: Longman.

Cellular Membrane Systems and Organelles

These articles and books describe the properties of individual fibers and organelles and techniques for purifying them.

BAINTON, D. 1981. The discovery of lysosomes. *J. Cell Biol.* **91**:66s–76s.

BARRANGER, J. A., and R. O. BRADY, eds. 1984. *Molecular Basis of Lysosome Storage Disorders.* Academic Press.

DE DUVE, C. 1975. Exploring cells with a centrifuge. *Science* **189**:186–194. The Nobel prize lecture of a pioneer in the study of cellular organelles. (See also Palade, below.)

FARQUHAR, M., and G. PALADE. 1981. The Golgi apparatus (complex) (1954–1981) from artifact to center stage. *J. Cell Biol.* **91**:77s–103s.

HOLTZMAN, E. 1976. *Lysosomes: A Survey.* Springer-Verlag.

LOUVARD, D., H. REGGIO, and G. WARREN. 1982. Antibodies to the Golgi complex and endoplasmic reticulum. *J. Cell Biol.* **92**:92–107.

MERISKO, E., M. E. FARQUHAR, and G. PALADE. 1982. Coated vesicle isolation by immunoadsorption on *Staphylococcus aureus* cells. *J. Cell Biol.* **92**:846–857.

NOVIKOFF, A. 1976. The endoplasmic reticulum: a cytochemist's view (a review). *Proc. Nat'l Acad. Sci. USA* 73:2781–2787.

PALADE, G. 1975. Intracellular aspects of the process of protein synthesis. *Science* **189**:347–358. The Nobel prize lecture of a pioneer in the study of subcellular organelles. (See also de Duve, above.)

TOLBERT, N. E., and E. ESSNER. 1981. Microbodies: peroxisomes and glyoxysomes. *J. Cell Biol.* **91**:271s–283s.

The Generation of Cellular Energy

The following references contain detailed treatments of the pathways of intermediate metabolism in cells.

BRIDGER, W. A., and J. F. HENDERSON. 1983. *Cell ATP.* Wiley.

EDWARDS, N. A., and K. A. HASSALL. 1980. *Biochemistry and Physiology of the Cell,* 2d ed. McGraw-Hill.

LEHNINGER, A. L. 1982. *Principles of Biochemistry.* Worth.

NEWSHOLME, E. A., and C. START. 1973. *Regulation in Metabolism.* Wiley.

STRYER, L. 1981. *Biochemistry,* 2d ed. W. H. Freeman and Company.

The Extracellular Matrix

These books and articles detail the structures and functions of matrix components, particularly of collagen.

HAY, E. 1981. *The Extracellular Matrix.* Plenum.

MILLER, A. 1982. Molecular packing in collagen fibrils. *Trends Biochem. Sci.* 7:13–18.

PIEZ, K. A., and A. REDDI. eds. 1984. *Extracellular Matrix Biochemistry.* Elsevier/North-Holland.

TRELSTED, R. L., ed. 1984. *The Role of the Extracellular Matrix in Development.* 42d Symposium of the Society for Developmental Biology. Alan R. Liss.

6

Tools of Molecular Cell Biology: Cells and Organisms

IN order to understand current advances in molecular cell biology, the reader requires a familiarity with the most commonly used biological materials, and knowledge of the latest experimental techniques. This chapter deals with the former; the latter are treated in Chapter 7. Particular cells or organisms are chosen for study because they provide good illustrations of an important molecular event or sequence of events—for example, the replication, recombination, and rearrangement of DNA; the control of genes; the construction of membranes; the formation of organelles; the secretion of proteins; the development of cytoskeletal elements; or the differentiation of cells.

Two important techniques in cell biology are endowments from microbiology. First, cultured cells derived from a single cell are valuable in molecular studies because they constitute a homogeneous cell mass in which each cell contains the same constituents in the same proportions. Second, the infection of cells by a virus—whether it be a bacterial, animal, or plant virus—affords an opportunity for studying the action of a limited set of genes designed to carry out a restricted, specific molecular task. Thus a homogeneous cell culture infected by a chosen virus is a popular choice for many molecular studies.

Cultured cells, however, do not generally perform the specialized tasks that differentiated cells in the body perform. Therefore it is often necessary for the investigator to return to the whole organism for material to study. In molecular investigations of specialized cells, the cells of interest within the whole organism are those that perform special tasks such as the synthesis of a large amount of a specific protein or group of related proteins. Tissues such as the liver of vertebrates and the oviduct of birds are rich in a single cell type, and these tissues are often used in molecular analysis. The cells of some tissues (e.g., the spleen, the thymus, the bone marrow, and the blood) are easily separated into single cells and then into individual cell types, even though there may be numerous different cell types in these tissues. The ease with which individual cell types can be purified from these tissues makes them advantageous to biologists—immunologists in particular—and accounts for their widespread use.

Table 6-1 Growth media for microorganisms such as *Escherichia coli* and *Saccharomyces cerevisiae* (baker's yeast)*

MINIMAL MEDIUM†
Carbon source: e.g., glucose or glycerol
Nitrogen source: NH_4^+ (e.g., $NaNH_4HPO_4$) or an organic compound such as histidine
Salts: Na^+, K^+, Mg^{2+}, Ca^{2+}, SO_4^{2-}, Cl^-, and PO_4^{3-}
Trace elements
RICH MEDIUM
Partially hydrolyzed animal or plant protein (rich in amino acids, short peptides, and lipids)
Yeast extract (rich in vitamins and enzyme cofactors, nucleic acid precursors, and amino acids)
Carbon source, nitrogen source, and salts as in minimal medium

*For more detailed information see R. W. Davis, D. Botstein, and J. W. Roth, 1982, *A Manual for Genetic Engineering: Advanced Bacterial Genetics,* Cold Spring Harbor Laboratory.

†Typical for most bacteria and yeast. Some photosynthetic bacteria (e.g., *Rhodospirillum rubrum* and cyanobacteria, or blue-green algae) require CO_2 for the carbon source. Some nitrogen-fixing bacteria (e.g., *Azotobacter*) require atmospheric N_2. Other organisms have special needs: e.g., *Hemophilus* strains require factors found in whole blood.

The Growth of Microorganisms and Cells in Culture

An important characteristic of many cells chosen for study is that a single cell can be readily grown into a colony. The process by which a single cell is grown into a colony of like cells is called *cloning,* and a genetically pure strain of cells is called a *clone.* Microorganisms that grow in nature as single cells are relatively easy to grow in culture dishes—usually on top of *agar,* a semisolid base of plant polysaccharides. The agar is first dissolved in a nutrient medium at an elevated temperature before solidification at a lower temperature. Both prokaryotic and eukaryotic single-celled organisms are routinely grown in this manner. In addition, it is now possible to clone cells from many animals and plants. The techniques of *cell culture* are critically important in modern biological research.

Escherichia coli Is a Favorite Organism of Molecular Biologists

The most commonly used cell in molecular cell biology is the bacterium *Escherichia coli.* A few of the many genetic and molecular experiments that have been performed with *E. coli* are described in Chapter 8. Among the advantages of using prokaryotic cells such as *E. coli* are the simplicity of the growth medium (Table 6-1) and the rapid growth rate of the organism (the division time ranges from 20 min to 1 h, which means that a very large mass of cell material can be obtained quickly). However, the single greatest advantage of using bacteria in experiments is probably the ease with which genetically distinct populations of bacterial cells can be isolated. This advantage is further heightened by concentrating on a single bacterium such as *E. coli.* There are literally tens of thousands of genetically distinct strains of *E. coli* available for study in laboratories throughout the world. Of the maximum estimated number of 3000 genes in *E. coli,* almost half have been located on the circular genetic map of the organism (Figure 6-1).

The basic technique that underlies all of the success in microbiological research is the growth of individual bacterial colonies from single bacterial cells under specific conditions (Figure 6-2). Every cell in such a colony or in a liquid culture started from such a colony can be assumed to be identical in every one of its approximately 3000 genes.

Strictly speaking, however, not all cells in a cultured clone are genetically identical. The probability of a detectable mutation is from 1 in 10^6 to 1 in 10^8 replication of each gene. Given that *E. coli* has about 3000 genes, by the time a single cell has divided to produce a few hundred or a few thousand cells, some individual genes will differ from cell to cell. Because of this genetic impurity, it is a common practice to subject the cultured clone to *selective* conditions to ensure that only cells with a particular desired set of characteristics will continue to survive and grow. For example, a widely available antibiotic such as streptomycin, penicillin, or tetracycline can be added to the medium to identify antibiotic-resistant strains; or a strain requiring a particular nutrient can be identified by the addition of that nutrient to an otherwise growth-limiting medium (Figure 6-3).

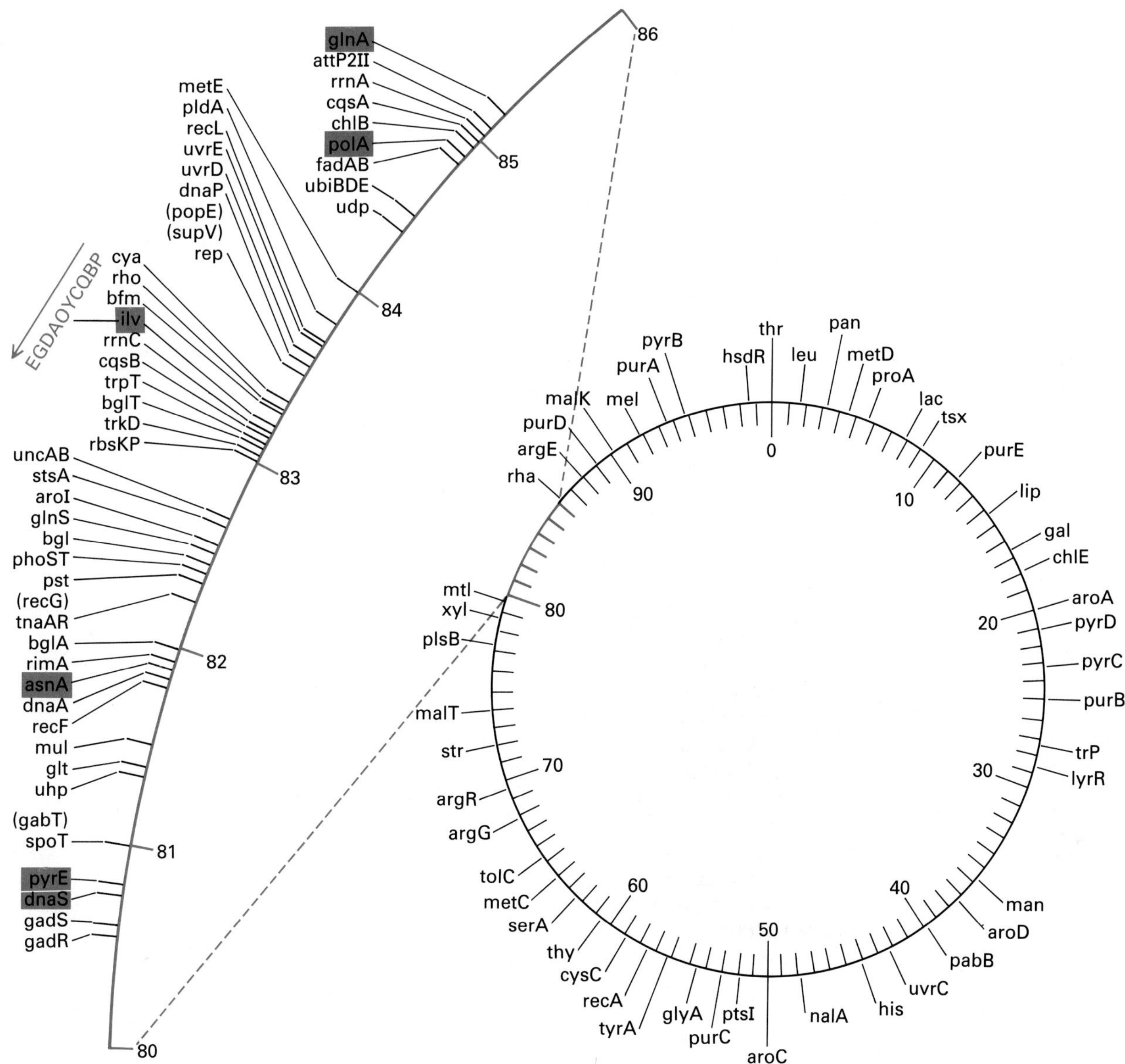

Figure 6-1 A map showing some of the genes on the circular chromosome of *Escherichia coli.* The abbreviations for the genes relate to the metabolic properties of the gene products; for example, *pyr*E, *asn*A, and *gln*A were each identified in a bacterial clone with a gene defect that made the addition of a pyrimidine, asparagine, or glutamine, respectively, necessary for growth. A strain with a genetic deficiency is said to be genetically *marked,* and the resulting genetic trait is termed a *marker* for that strain. A group of such markers can define a strain with great precision.

Some of the regions of *E. coli* have been mapped so completely that every gene is known. An example is *ilv,* the locus for enzymes that catalyze the synthesis of isoleucine, leucine, and valine. Each capital letter is a gene name and represents a known purified enzyme. Some of the markers in this region, such as those resulting from mutations in *dna*S and *pol*A (the gene for a DNA polymerase), represent defects in genes whose products participate in DNA synthesis. The numbers inside the circle indicate the approximate percentage distance around the circular *E. coli* genome from an arbitrary starting point. One percentage point is equal to 3×10^4 base pairs. Although more than 1500 of the estimated 3000 genes of *E. coli* have been mapped, only a few of them are shown here, at two levels of detail. *After B. J. Bachman, K. B. Low, and A. L. Taylor, 1976,* Bacteriol. Rev. ***40****:116.*

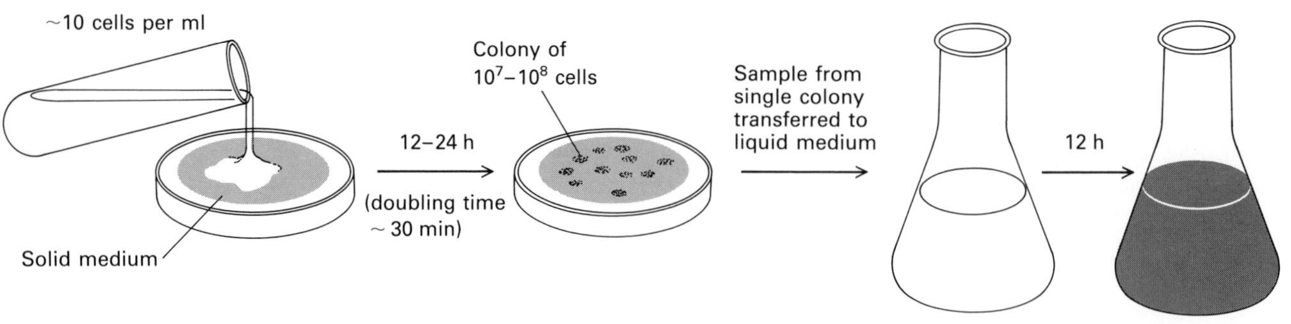

Figure 6-2 The selection and propagation of genetically uniform colonies, or clones, of bacterial cells. A very dilute suspension of bacteria can be spread on a solid medium, which limits the movement of bacteria as individual cells grow and divide. This restriction of movement results in the development of compact colonies, each descended from a single cell. With a doubling time of about 30 min, one cell yields 10^7 to 10^8 cells overnight. A sample from a colony may be cultured further to obtain a large mass of genetically identical cells in a day or two.

Genes Can Be Transferred between Bacteria in Three Ways

In the 1940s and 1950s, the availability of genetically different bacterial strains led to the discovery of a number of means by which genes could be exchanged between bacteria. Each of the strains participating in these transfer experiments was marked because it possessed a different group of genetic traits (e.g., dependence on one or more growth factors, plus a sensitivity or a resistance to a particular chemical or virus). Although the details of this exciting period in the development of molecular genetics are beyond the scope of this book, some of the genetic experiments with *E. coli* are described in Chapter 8. Purified bacterial genes obtained through the use of genetic techniques are also described in Chapter 8; in addition, many of the later chapters explain how the same techniques have been used to purify eukaryotic genes. For these reasons, three mechanisms by which genes can be transferred between two bacterial cells are described here.

Conjugation In *conjugation,* two *E. coli* cells, one designated "male" and one "female," become attached to one another by a conjugation bridge. This allows one strand of the chromosome of the male cell to enter the female cell (Figure 6-4). The newly introduced DNA strand recombines with the DNA of the recipient cell. Subsequent DNA replication and cell division give rise to a new recombinant cell that has characteristics derived from each of the parental cells.

Transduction Viruses that invade bacterial cells are called bacteriophages. Most bacteriophages contain DNA. During growth and reproduction, some bacteriophages incorporate host-cell DNA into their own DNA. If the bacteriophage then infects another cell, the DNA of the first host cell may be transferred to the second. This mechanism of gene transfer by bacteriophages is called *transduction* (Figure 6-5). *Special* or *restricted* transducing bacteriophages pick up specific regions of the host bacterial chromosome; *general* transducing bacteriophages acquire DNA randomly from the bacterial chromosome. Some transducing bacteriophages incorporate so much host DNA that not enough room is left for the viral DNA sequences that are necessary in bacteriophage replication. These *defective* transducing phages are no longer able to kill host bacteria, but they can still invade new hosts and transfer genes to them. (See the section on bacterial viruses later in this chapter for a more detailed description of different types of bacteriophages.)

Transformation *Transformation* is the genetic change of a bacterium after exposure to and recombination with isolated DNA from another, genetically different bacterium (Figure 6-6). This mechanism of gene transfer, which was discovered with *Streptococcus pneumoniae* (Chapter 1), provided a crucial piece of evidence that DNA was the genetic material. However, for many years the transformation of one bacterial cell by the DNA of another was the *only* means of experimental gene transfer, and only a few bacterial strains could be transformed.

The transformation of *E. coli* and other similar bacteria requires that the recipient cells be specially treated to get the DNA inside them. The cells are usually exposed to high concentrations of calcium (Ca^{2+}) ions, which somehow causes the bacterial plasma membranes to admit foreign DNA.

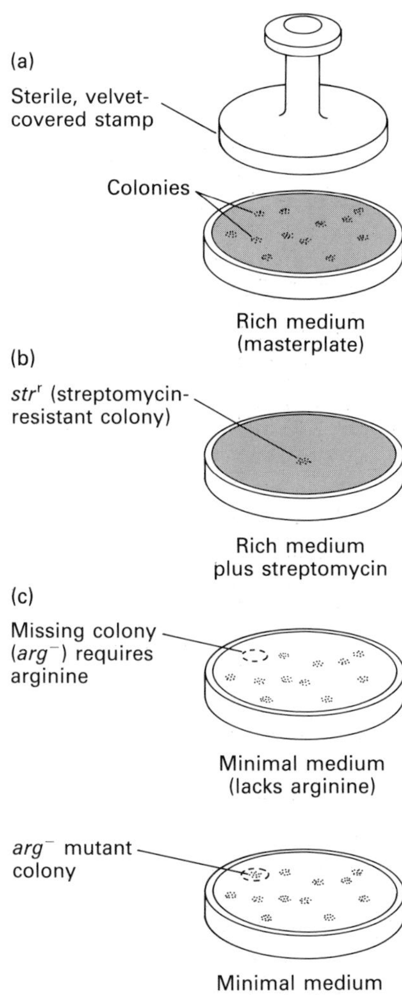

Figure 6-3 Random mutations in bacterial cultures may produce cells that differ from all the other cells of the culture in a single genetic characteristic. It is easy to test the genetic characteristics of colonies growing on a *plate* (a petri dish) by *replica plating,* a technique developed by Joshua and Esther Lederberg in the 1940s. A circular velvet-covered stamp equal in size to the petri dish is pressed into it, so that the stamp picks up a sample of each bacterial colony in this dish (a). The bacteria on the stamp can then be deposited in the same arrangement on new plates. The media in the new plates are chosen to test different genetic characteristics, such as sensitivity or resistance to the antibiotic streptomycin (b), or nutritional auxotrophy (c)—that is, inability to grow without a specific nutrient (e.g., the amino acid arginine). In this manner, colonies arising from single cells that have mutations in particular genes can be identified.

The Yeast Life Cycle Can Include Haploid and Diploid Phases

Genetically homogeneous populations can be obtained with eukaryotic cells as well as with prokaryotic cells. For example, cultures of yeasts and molds can be grown either from a single vegetative cell (a growing cell) or from a single spore (a dormant cell). Subjects that have been particularly well studied with these organisms include the cell's mechanisms for controlling DNA synthesis, the structure of chromosomes, and the rearrangement and replication of chromosomes.

Most species of yeast, including *Saccharomyces cerevisiae* (baker's yeast), grow in nature as single diploid cells. Each cell has from 12 to 18 chromosomes, depending on the species. Yeast cells reproduce asexually by a "budding off" of the cell wall: The bud begins as a small protuberance over a thickened region of the cell wall called a *plaque.* The bud gradually enlarges and finally separates from the parent cell. Nuclear division is not so elaborate in yeast as in most eukaryotic cells; no true mitosis occurs. The chromosomes are attached to a structure called the spindle pole body (SPB). When the chromosomes replicate, the spindle pole body does too, and a spindle apparatus separates the two SPBs so that one of them can move into the bud (Figure 6-7). The duplicated chromosomal material of the parent nucleus is thus partitioned between the two cells.

When a diploid yeast cell encounters adverse conditions (nutrient deprivation, for example) it undergoes a form of differentiation to become a spore. To be precise, the original diploid cell undergoes meiotic divisions to produce four haploid spores. If the required nutrients are restored, the spores may fuse to form two diploid cells; otherwise, each haploid spore may generate a haploid colony (Figure 6-8).

In nature, the fusion of two haploid yeast cells from different strains probably occurs rarely; when it does, it broadens the gene pool available to the species. In the laboratory, the intentional fusion of two different haploid cells is termed a *genetic cross.* This procedure allows the genetic contribution of each parent to be examined.

The fusion of discrete strains of haploid cells can occur because of the capacity of cells to act as *mating types.* Only cells differing in mating type can combine to give rise to diploid cells. (The establishment of different mating types is discussed extensively in Chapters 11 and 12.) When two genetically different haploid strains are mated to produce diploid cells, which are then allowed to undergo meiosis and segregation into new haploid spores, the assortment of chromosomes resulting from the first meiotic division can be studied (Figure 6-9). Such experiments have led to the establishment of *linkage groups*—groups of genes that are located on the same chromosome. A linkage group can be detected by a high

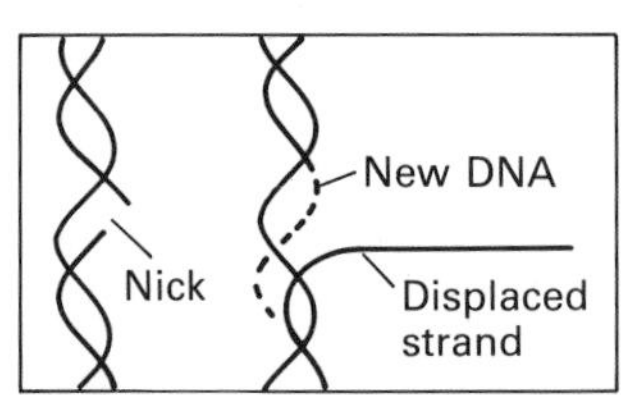
New DNA
Nick
Displaced strand

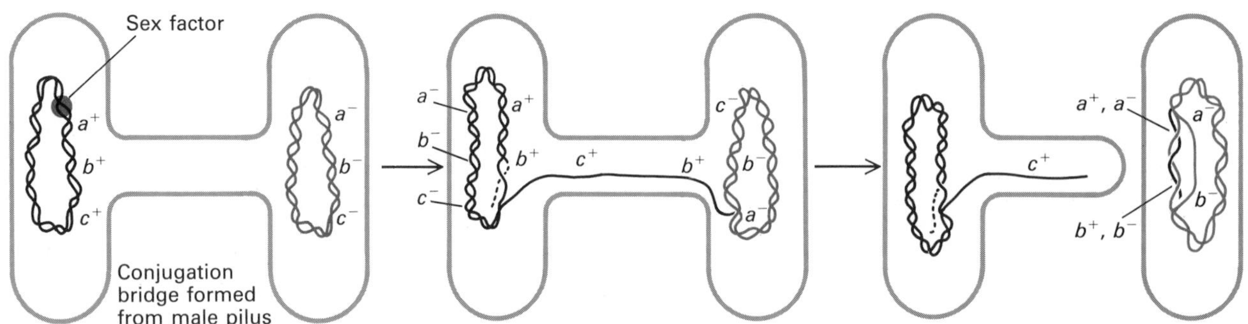
"Male" donor, Hfr
"Female" recipient, F⁻
Sex factor
Conjugation bridge formed from male pilus
Double-stranded bacterial chromosome with functioning a, b, and c genes
Double-stranded bacterial chromosome with nonfunctioning a, b, and c genes
A nick is made in one strand of donor chromosome. DNA synthesis begins at nick, displacing one of the strands. Displaced strand is transferred through conjugation bridge.
The bridge is broken by the experimenter before the entire Hfr DNA is transferred. In this case, conjugation is interrupted before the c⁺ information is transferred. The donor strand joins chromosome of recipient cell at the ab region.

Figure 6-4 Conjugation in *E. coli*. Two bacterial cells can join to exchange genetic information. This is accomplished when the strain Hfr (for "high-frequency recombination"), which is designated a "male" strain because of the presence of a sex factor in the chromosome, is mixed with the strain F^-, a "female" strain. The male cells possess surface projections called pili, which are hollow tubes through which DNA is transferred. Only one strand of the DNA is transferred, and the conjugation bridge can be broken before transfer is complete. (Conjugation bridge is not to scale; actual size of the bridge is only 1 to 2 percent of cell length.)

The figure shows three genes, *a, b,* and *c,* for which the male is positive (i.e., the "plus" genes

frequency of reappearance of parental combinations of alleles in progeny cells. (An allele is one of the several homologous forms that a given gene can take.) Chromosome mapping becomes possible because of crossovers that occur during meiosis, as we shall demonstrate in Chapters 11 and 13.

Many mutant yeast cell strains have been isolated and mapped. Hundreds of nutritional mutants—that is, cells that require nutrients not required by normal cells—are available, as well as strains that are blocked at various steps in the normal progress of the cell cycle.

Animal Cells Share Certain Growth Requirements and Capacities

All animals must obtain from their environments a group of amino acids referred to as the *essential amino acids:* arginine, histidine, isoleucine, leucine, lysine, methionine, phenylalanine, threonine, tryptophan, and valine (Table 6-2). The same is true of all animal cells in culture. In addition, cells in culture also require cysteine, glutamine, and tyrosine. Animals can acquire these latter three amino acids from their own specialized cells; for example, the liver makes tyrosine from phenylalanine, and both the liver and the kidney can make glutamine. The remaining amino acids (glycine, alanine, serine, asparagine, the dicarboxylic amino acids, aspartic acid and glutamic acid) and the imino acid proline can be synthesized by animal cells both within the organism and in culture. The other essential components of a culture medium are vitamins (which the cells cannot make, at least in adequate amounts), salts, glucose, and serum (the noncellular part of blood—essentially a solution of various proteins) (Table 6-3).

It is worth noting that all normal animal cells in culture can synthesize the nucleic acid precursors from the sim-

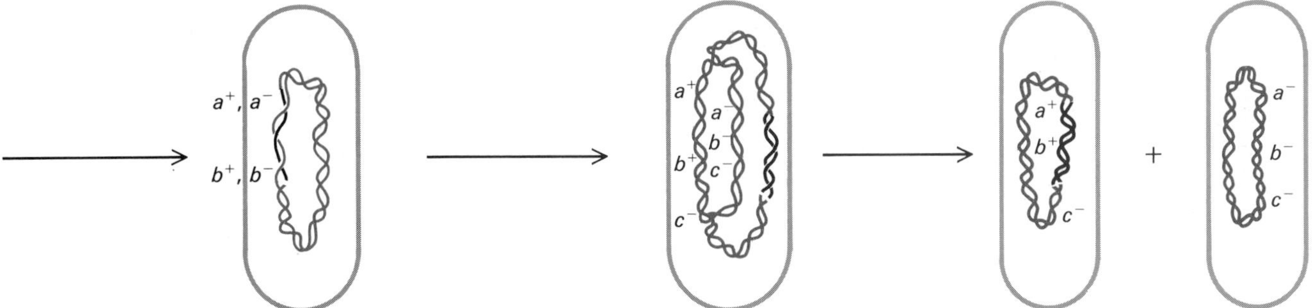

Both strands in recipient cell begin to replicate. One strand produces a double-stranded $a^+b^+c^-$ region, whereas the other produces a double-stranded $a^-b^-c^-$ region. Cell at this stage is merodiploid, because it has two different forms of genes *a* and *b*.

Replication proceeds until complete. After cell division, the two genotypes are segregated.

are wild-type) and the female is negative (i.e., the "minus" genes are mutant). The a^+ and b^+ DNA is transferred, but conjugation is interrupted before the c^+ DNA can be transferred. The synthesis of mRNA from the single transferred DNA strand in half the cases must await chromosome replication within the recipient cell. In other words, the new genes can function within the recipient cell either immediately or after a single cell division. At one point in this process, the recipient cell has two different forms of the same genetic sequence; the cell is then called a *merodiploid* (from the Greek *meros,* meaning "part"). This is a transient state. Segregation of the two chromosomal sequences containing genes *a* and *b* occurs after the chromosome replicates and the cells divide.

pler compounds in the medium. The role of serum, which is required by most cells in a culture medium, is not completely understood, but serum apparently supplies needed trace materials, including protein growth factors. It is now possible to grow a few types of mammalian cells in a completely defined medium supplemented with trace minerals and protein growth factors (Table 6-3). Cells cultured from single cells on a glass or a plastic dish (Figure 6-10a) will form visible attached colonies in 10 to 14 days (Figure 6-10b and c). Some cells can be grown in suspension, which offers a considerable experimental advantage in that equal samples are easier to obtain from suspension cultures.

Primary and Transformed Cultures Two types of cultured animal cells are in general use: primary and transformed. *Primary* cell cultures are derived from animal tissue, probably most often from skin or whole embryos. The cells in such cultures are predominantly *fibroblasts,* which are found in connective tissue in all parts of the body. Fibroblasts produce collagen and other material that lies between the cells of connective tissue. For example, one important role of fibroblasts is to heal wounds in the skin. (In addition to fibroblasts, some epithelial cells have been cultured. Epithelial cells are often highly specialized, which may be why they are more difficult to grow.) Normal human fibroblasts will not grow indefinitely in culture, but will continue doubling for 50 to 100 generations, at which time they reach a "crisis," grow very slowly (if at all) for a few more generations, and then cease to grow altogether. No cause has been found for this limitation in growth capacity, and no nutritional regimen that will cure the problem has yet been discovered.

All mammalian cells that are capable of indefinite growth in culture are derived either from tumor cells taken directly from an animal or from cultured cells that

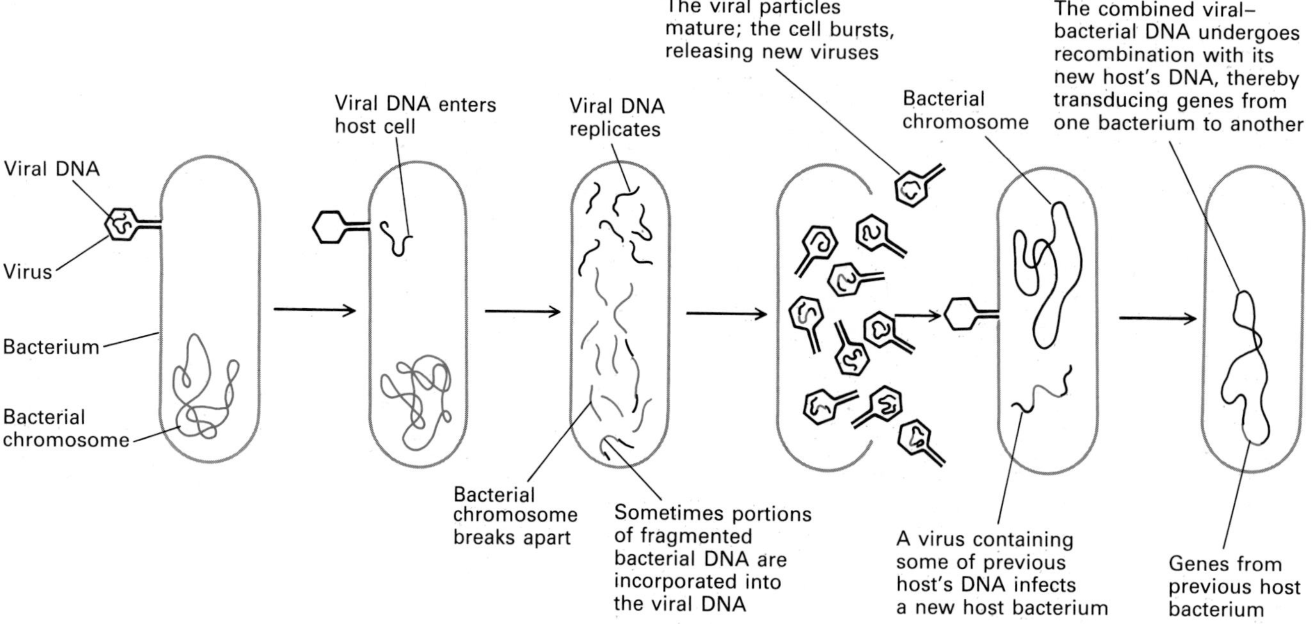

Figure 6-5 Transduction. Sometimes a bacteriophage can incorporate host-cell DNA into its own DNA during a growth cycle. Such a bacteriophage can then carry the host-cell DNA out of the host bacterium and introduce it into a new host, where it can undergo recombination with the chromosomal DNA.

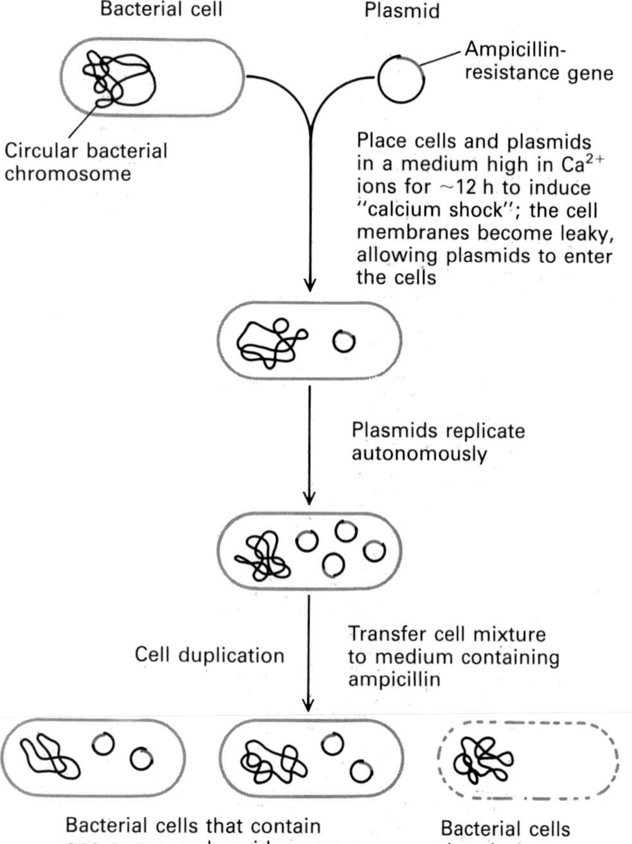

Figure 6-6 (*left*) Transformation. The DNA most commonly used in transformation experiments is in the form of small circular molecules called *plasmids*. These circles of DNA, which replicate autonomously in the bacterial cell, often carry genes that enable a cell to become resistant to an antibiotic. Thus, a cell that has acquired a plasmid can be selected by growing cells in a medium containing the antibiotic. As we shall see in Chapter 7, the development of techniques both for transforming cells with plasmid DNA and for inserting DNA into plasmids has given birth to the field of recombinant DNA research.

Figure 6-7 (**opposite**) Cell division in *Saccharomyces cerevisiae.* (a) Yeast cells do not divide by classic mitosis, but by budding. The distribution of chromosomes to the two new cells is determined by the *spindle pole body* (SPB), located at the nuclear periphery. (b) The beginning of nuclear division is signaled by the division of the SPB. (This section through dividing cells missed the cytoplasmic connection between the bud and the mother cell.) (c) When the two daughter SPBs separate, microtubules connecting them can be seen. (d) A portion of the nucleus containing one SPB then enters the bud. Cell wall formation completes the separation of the two cells. *Courtesy of B. Byers.*

(a)

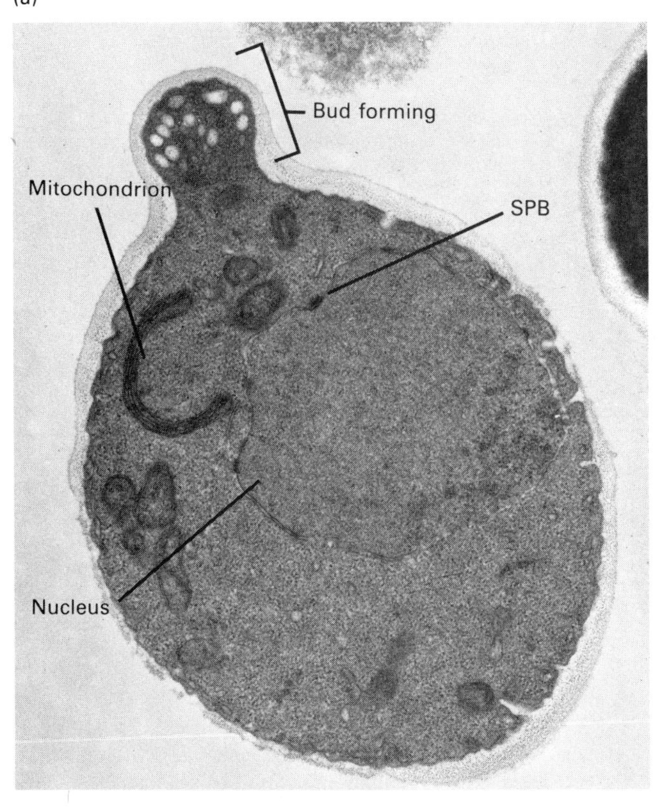
Bud forming
Mitochondrion
SPB
Nucleus

(b)

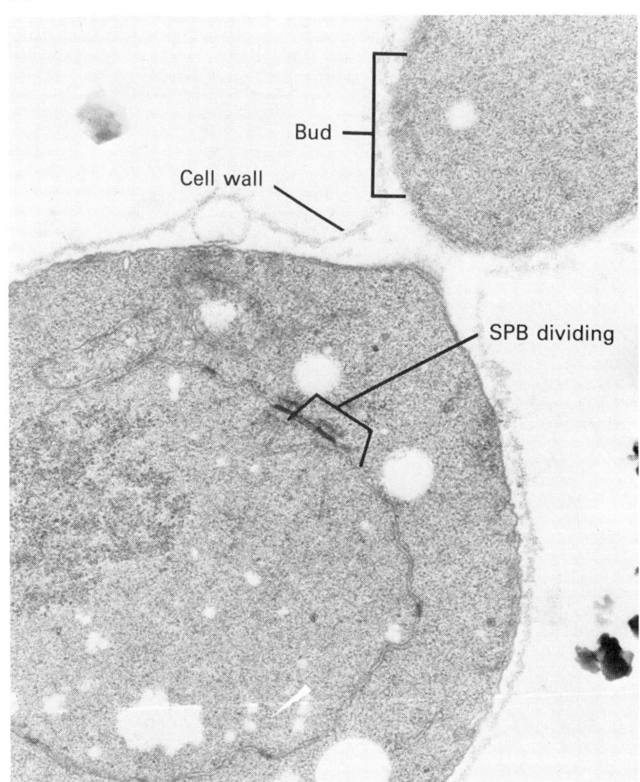
Bud
Cell wall
SPB dividing

(c)

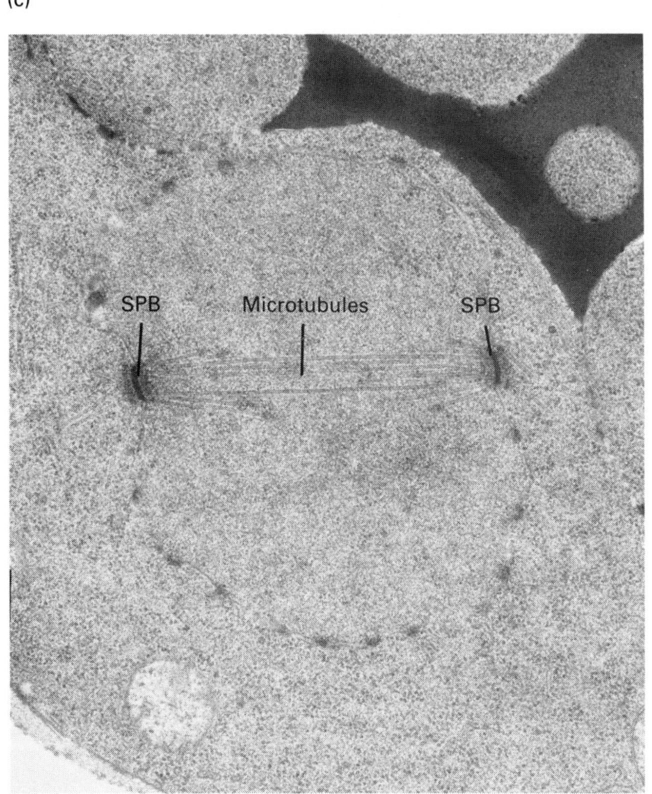
SPB
Microtubules
SPB

(d)

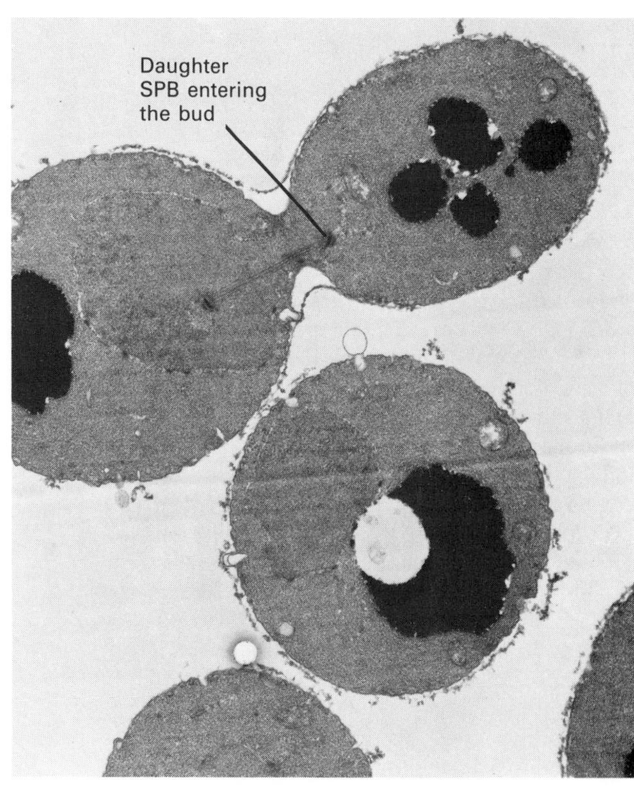
Daughter
SPB entering
the bud

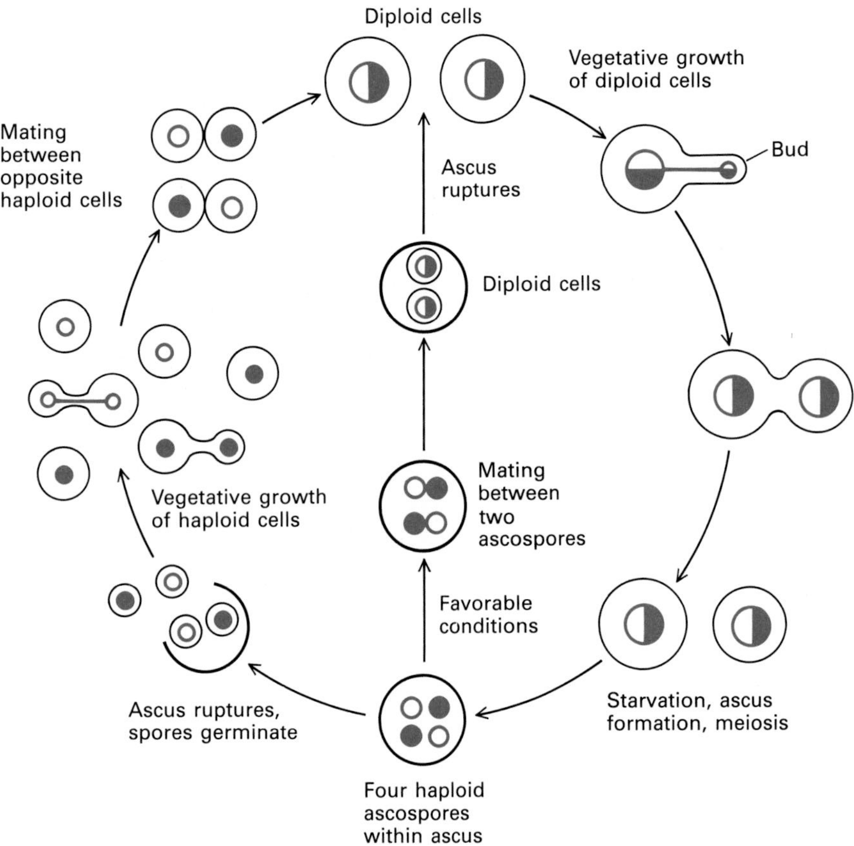

Figure 6-8 The life cycle of *S. cerevisiae*. The two haploid portions of each diploid cell differ in mating type (discussed in text) because mating can occur only between haploid components that are opposite in mating type. Under favorable growth conditions, diploid cells reproduce asexually, by budding. When such a cell is deprived of essential nutrients, however, it forms a tough outer wall, within which a saclike structure called an *ascus* develops. Within the ascus, the cell undergoes meiosis and the yeast chromosomes assort independently in mendelian fashion (Chapter 1; Chapter 5) to yield four haploid *ascospores*. If the ascus is returned to favorable growth conditions, pairs of spores within it may fuse to form diploid cells, which proceed to grow and bud as before; this happens often in nature. On the other hand, each ascospore can remain haploid, so that it can give rise to a haploid colony. Haploid cells from different colonies—that is, haploid cells with different genetic makeups—can fuse (or be fused, in the laboratory) into diploid cells.

Figure 6-9 (***opposite***) (a) A cross between the haploid strains MAT*a* Leu2$^-$ and MATα Leu2$^+$. The individual products of meiosis can be observed in yeast (and in other fungi) because each spore can be independently grown into a colony and tested for different properties. The two properties followed in the diagram are *mating type* (MAT*a* or MATα) and leucine requirement (Leu2$^+$ or Leu2$^-$). The Leu2$^+$ strain has normal enzymatic capacity for leucine synthesis; the Leu2$^-$ strain has a mutant gene that incapacitates the cell for a particular step in leucine biosynthesis. Yeast cells mate heterologously; that is, MAT*a* cells mate with MATα cells but not with other MAT*a* cells. The second meiotic division produces two pairs of haploid ditypes: these are either parental or nonparental.
(b) Tests to prove the genotypes of the meiotic products. To determine the genetic makeup of the meiotic products (here A and B happen to be the parental ditypes), the experimenter first cultures each spore separately in a complete medium. The Leu2$^+$ spores, which can grow without leucine, can be identified from a single replica plate on a medium lacking leucine. Because only the B and C spores grow in this medium, they must be Leu2$^+$.

Determination of mating type requires that an additional marker (such as Met$^-$, indicating the presence of a defective gene that disrupts the cell's ability to grow without methionine) be incorporated into the yeast cells. In addition, two different *tester* strains are introduced into these cultures. The tester strains are Met$^+$ and they, too, have a marker—here, His$^-$. Tester strain 1 (T1) is MATα Met$^+$ His$^-$, and tester strain 2 (T2) is MAT*a* Met$^+$ His$^-$. When the A, B, C, and D cells (which are all Met$^-$ His$^+$) are plated together first with T1 and then with T2 in media lacking methionine and histidine, colonies form only where mating has occurred to produce diploid cells. Because A and C grow on T1, they must be MAT*a*; similarly, B and D grow on T2, so they must be MATα.

have undergone a change that causes them to behave as tumor cells. Such cells are said to be *transformed;* the term is used to describe any cells that are immortal. Unfortunately, the term transformation has two different meanings in cell biology: first, it is the process by which foreign DNA is incorporated and foreign genes are subsequently expressed (see the section "Transformation of Eukaryotic Cells" later in this chapter); and second, as used here, it is the process by which the normal in vivo restraints on cellular growth are abolished, so that the cells no longer die after a limited number of divisions. (Immortal transformed cells are discussed further in Chapter 23.)

Neither primary fibroblasts nor the majority of transformed cells carry out differentiated cell functions such as the production of special proteins. However, undifferentiated cultured cells have been used to advantage in general studies on RNA, DNA, and protein synthesis (as will be seen in Chapters 9 to 13) and in studies of the structural elements that are common to all cells (Chapters 14 to 21).

One of the most popular undifferentiated cultured cells

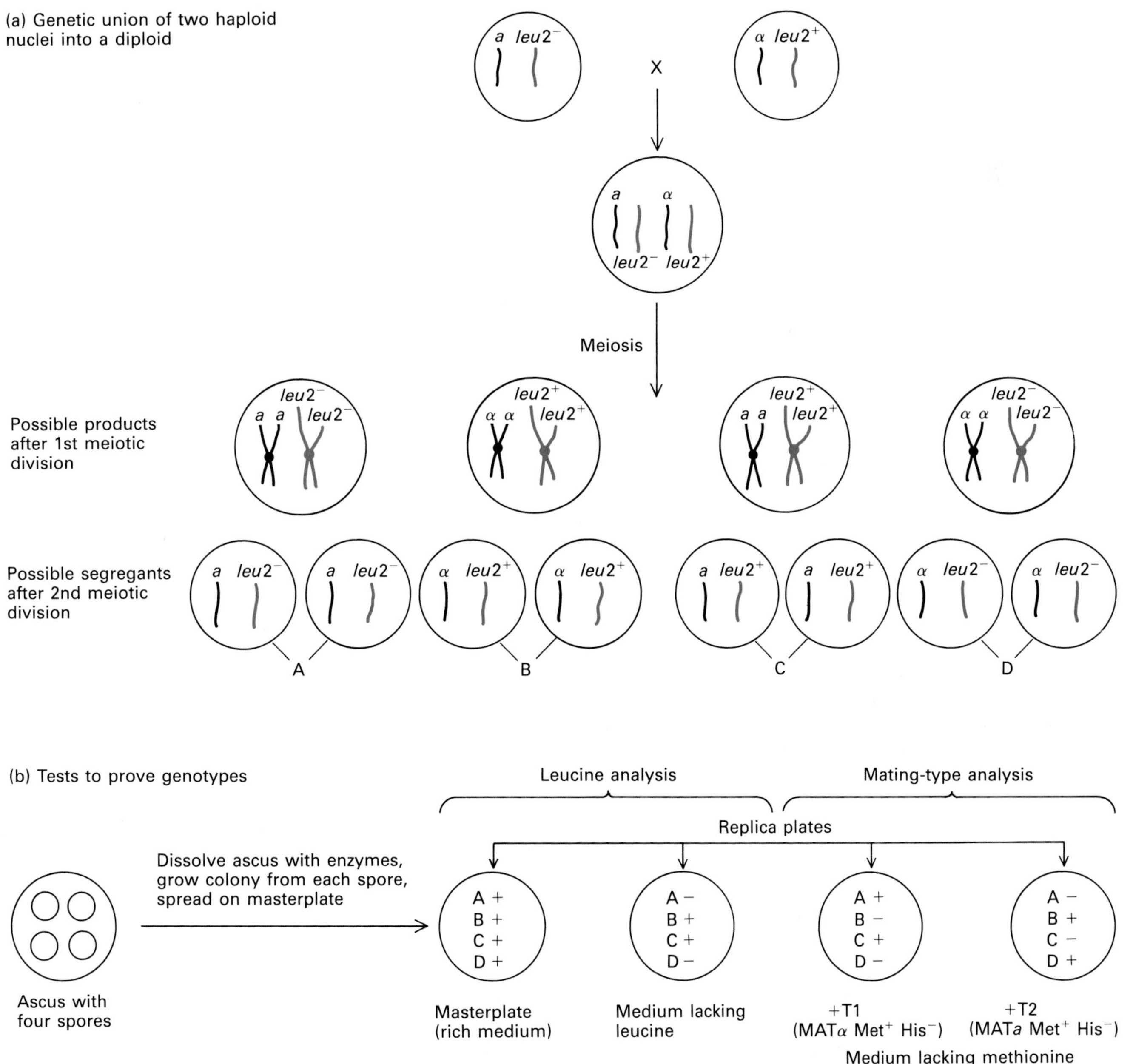

Table 6-2 Daily requirement of essential amino acids (for college-age males)*

Amino acid	Grams
Arginine	0†
Histidine	Unknown‡
Isoleucine	1.30
Leucine	2.02
Lysine	1.50
Methionine	2.02
Phenylalanine	2.02
Threonine	0.91
Tryptophan	0.46
Valine	1.50

*The amino acid requirements of men were established by feeding experiments. All of the listed amino acids are necessary for healthy growth and development. However, arginine and histidine can be omitted from the diet for a short time without damaging body cells.

†Required by infants and growing children.

‡Essential, but the precise requirement is not yet established.

SOURCE: W. C. Rose, R. L. Wixom, H. B. Lockhart, and G. F. Lambert, 1955, *J. Biol. Chem.* **217**:987.

Table 6-3 Mammalian cell media (HeLa cell)

Compound	Concentration (*M*)*
SERUM-CONTAINING MEDIUM: EAGLE'S MEDIUM	
Essential amino acids:	
L-Arginine	6×10^{-4}
L-Cysteine†	1×10^{-4}
L-Glutamine†	2×10^{-4}
L-Histidine	2×10^{-5}
L-Isoleucine	4×10^{-5}
L-Leucine	4×10^{-5}
L-Lysine	4×10^{-5}
L-Methionine	1×10^{-5}
L-Phenylalanine	2×10^{-5}
L-Threonine	4×10^{-5}
L-Tryptophan	5×10^{-6}
L-Tyrosine†	2×10^{-5}
L-Valine	4×10^{-5}
Vitamins:	
Choline	1 mg/l
Folic acid	1 mg/l
Inositol	2 mg/l
Nicotinanide	1 mg/l
Pantothenate	1 mg/l
Pyridoxal	1 mg/l
Riboflavin	0.1 mg/l
Thiamine	1 mg/1
Salts:	
Na^+, K^+, Ca^{2+}, Mg^{2+}, Cl^-, PO_4^{3-}, HCO_3^-	
Glucose	
Serum, dialyzed (5–10% of total volume)‡	
SERUM-FREE MEDIUM	
Amino acids:	
L-Alanine	1.0×10^{-4}
L-Arginine HCl	1.0×10^{-3}
L-Asparagine	1.0×10^{-4}
L-Aspartic acid	1.0×10^{-4}
L-Cysteine HCl	2.0×10^{-4}
L-Glutamic acid	1.0×10^{-4}
L-Glutamine	1.0×10^{-3}
Glycine	1.0×10^{-4}
L-Histidine HCl	1.0×10^{-4}

is the HeLa cell (see Figure 6-10b and c), the first human cell to be grown continuously in the laboratory. The HeLa cell was obtained in 1952 from a malignant tumor, a carcinoma of the uterine cervix. The composition of the HeLa cell as a typical example of an animal cell is compared with that of *E. coli* in Table 5-3. The average animal cell is about 1000 times larger than a bacterial cell, and about 10 times larger than a yeast cell.

Undifferentiated cultured cells readily serve as hosts for viral infections. In a pure cell population, every cell in the culture is equally susceptible to a given virus; thus, in biochemical terms, every cell can be expected to behave in a similar manner.

Differentiated Cells in Culture In addition to undifferentiated normal cells (such as fibroblasts) and undifferentiated cancer cells (such as HeLa cells), investigators study transformed cells that can grow in culture while continuing to perform some of the functions of specialized tissues. For example, cultured *myoblasts* (muscle precursor cells) that have been transformed (to immortality) will still fuse to form multinucleated muscle cells (Figure 6-11). *Erythroleukemia* cells (abnormal precursors of red blood cells) can be induced to produce hemoglobin (Figure 6-12). Cells from a tumor called *teratoma* can

Table 6-3 *Continued*

Compound	Concentration (*M*)*
L-Isoleucine	3.0×10^{-5}
L-Leucine	1.0×10^{-4}
L-Lysine HCl	2.0×10^{-4}
L-Methionine	3.0×10^{-5}
L-Phenylalanine	3.0×10^{-5}
L-Proline	3.0×10^{-4}
L-Serine	1.0×10^{-4}
L-Threonine	1.0×10^{-4}
L-Tryptophan	1.0×10^{-5}
L-Tyrosine	3.0×10^{-5}
L-Valine	1.0×10^{-4}
Vitamins:	
Biotin	3.0×10^{-8}
Calcium pantothenate	1.0×10^{-6}
Choline chloride	1.0×10^{-4}
Folic acid	3.0×10^{-6}
Myoinositol	1.0×10^{-4}
Niacinamide	3.0×10^{-7}
Pyridoxine HCl	3.0×10^{-7}
Thiamine HCl	1.0×10^{-6}
Vitamin B_{12}	1.0×10^{-6}
Salts:	
$CaCl_2 \cdot 2H_2O$	3.0×10^{-4}
$CuSO_4 \cdot 5H_2O$	1.0×10^{-8}
$FeSO_4 \cdot 7H_2O$	3.0×10^{-6}
KCl	3.0×10^{-3}
$MgCl_2 \cdot 6H_2O$	6.0×10^{-4}
NaCl	1.3×10^{-1}
$NaHCO_3$	1.4×10^{-2}
$Na_2HPO_4 \cdot 7H_2O$	1.0×10^{-3}
$ZnSO_4 \cdot 7H_2O$	3.0×10^{-6}
Phenol red (pH indicator)	3.3×10^{-6}
Other additions:	
Glucose	1.0×10^{-2}
Linoleic acid	3.0×10^{-7}
Lipoic acid	1.0×10^{-6}
Hypoxanthine	3.0×10^{-5}
Putrescine dihydrochloride	1.0×10^{-6}

Table 6-3 *Continued*

Compound	Concentration (*M*)*
Sodium pyruvate	1.0×10^{-5}
Thymidine	3.0×10^{-6}
Trace elements:	
$MnCl_2 \cdot 4H_2O$	0.5 n*M*
$(NH_4)_6Mo_7O_{24} \cdot 4H_2O$	0.5 n*M*
$NiSO_4 \cdot 6H_2O$	0.25 n*M*
H_2SeO_3	15 n*M*
$Na_2SiO_3 \cdot 9H_2O$	250 n*M*
$SnCl_2$	0.25 n*M*
$Na_3VO_4 \cdot 4H_2O$	2.5 n*M*
$CdSO_4$	50 n*M*
Hormones and protein growth factors:	
Insulin	0.5×10^{-6} g/ml
Transferrin	$0.5–5 \times 10^{-6}$ g/ml
Hydrocortisone	30–100 n*M*
Fibroblast growth factor	100×10^{-9} g/ml
Epidermal growth factor	$5–30 \times 10^{-9}$ g/ml

*Unless other units are indicated.

†These amino acids are essential for cell cultures, but in an animal they can be formed in specialized tissues and by bacteria in the gut (see Table 6-2).

‡Serum is a mixture of hundreds of proteins with a total protein concentration of 50–70 mg/ml. Albumin is the most plentiful serum protein (30–50 mg/ml). Growth factors and growth hormone are present at 34 ng (nanograms)/ml, and insulin at 0.2 ng/ml.

SOURCE: H. Eagle, 1959, *Science* 130:432; and S. E. Hutchings and G. H. Sato, 1978, *Proc. Nat'l Acad. Sci. USA* 75:901.

generate a variety of different tissue types. Teratomas are thought to be cancerous embryonic cells; during their growth they give rise to many normal cell types, but the differentiated cells are not arranged in an orderly fashion.

In addition to these three cases in which transformed cells have displayed (or have been induced to display) at least partially differentiated cell function, some success has been achieved in culturing normal precursor cells to skin epithelium (Figure 6-13). Under appropriate culture conditions and in the presence of a hormone called epidermal growth factor, the basal cells of the epidermis will divide many times. When the cells have formed a close-packed colony, they will differentiate into the cornified cells that make up the outer surface of the skin. Fat cells have similarly been cultured from precursor cells (probably fibroblasts) found in the subcutaneous tissue.

However, many specialized animal cells cannot be cul-

(a)

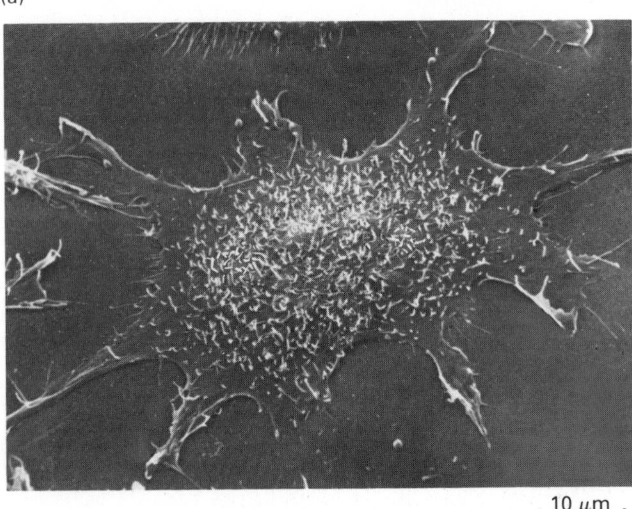

Figure 6-10 Mammalian cells have a life cycle of approximately 16 to 24 h. They can be grown attached to a plate; also, some cells will grow in suspension cultures. When individual cells are plated from a dilute suspension, colonies (clones) arise from them. To separate cells growing as attached cells so that they can be plated individually, a cell culture must be treated with a protease—trypsin, for example. (a) A single mouse cell attached to a plastic petri dish, as viewed through a scanning electron microscope. (b) Macroscopic colonies of human HeLa cells, produced in 10 to 14 days from a single cell. (c) After staining, individual colonies can be seen and counted. [See P. I. Marcus, S. J. Cieciura, and T. T. Puck, 1956, *J. Exp. Med.* **104**:615.] *Photograph (a) courtesy of N. K. Weller; photographs (b) and (c) courtesy of T. T. Puck.*

(b)

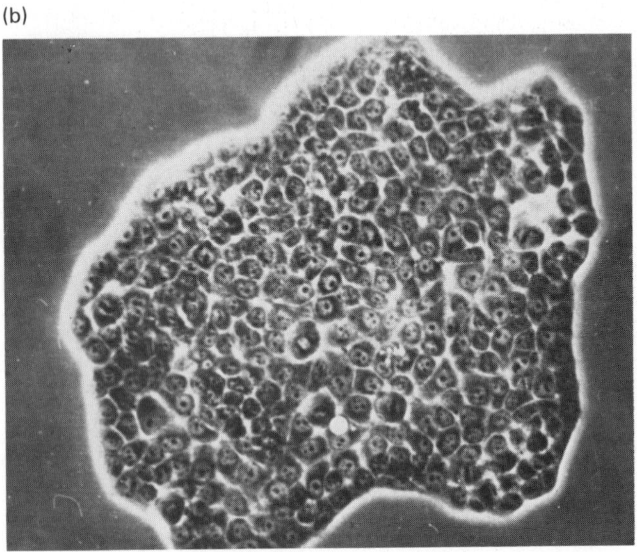

(c)

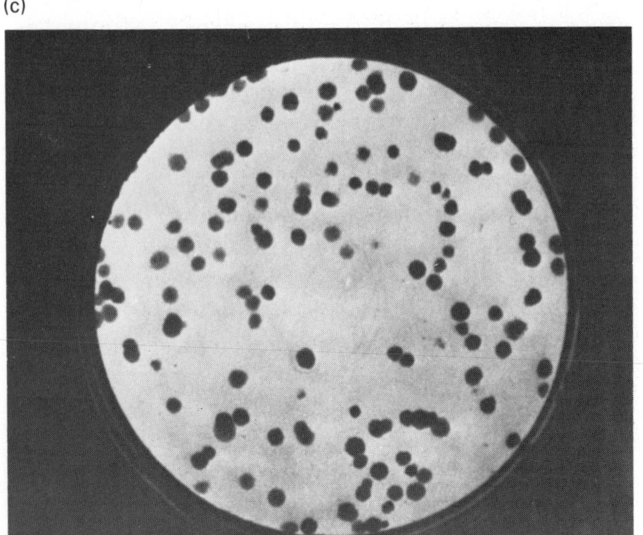

tured at all. This is not altogether surprising, because certain fully differentiated normal cells—for example, muscle, nerve, and kidney cells—do not grow continually in the body. Perhaps structural changes occurring in such cells as they differentiate during the formation of a tissue prevent any further growth. For example, in early development, muscle cells fuse so that many nuclei occupy one cell body; nerve cells develop extensions (axons) that can be several feet long, and that form intricate attachments to other cells. Such structural features would make it difficult for a cell to divide in a way that would yield two equal daughter cells.

Some specialized cells, such as those in the liver, do slowly "turn over"; when they die they may be replaced by an occasional division of remaining liver cells. Moreover, if a large section of liver is removed from a mammal, the remaining cells initiate cell division, which returns the liver to its normal size. Nevertheless, despite much effort, growing cultures of liver cells have not been achieved.

Cell Fusion: An Important Technique in Somatic-Cell Genetics

Because animal cells can be cultured from single cells in a well-defined medium, it is possible to select for genetically distinct animal cells, just as can be done with bacteria and yeast. The chromosomes are large and easily visible after staining, so individual chromosomes can be identified and their distinctive arrangements observed. Moreover, cells can be fused so that two nuclei function in one cell; with new techniques, DNA can be transferred into such growing cultured cells. This branch of cell biology, which is called *somatic-cell genetics,* has as one of its

(a)

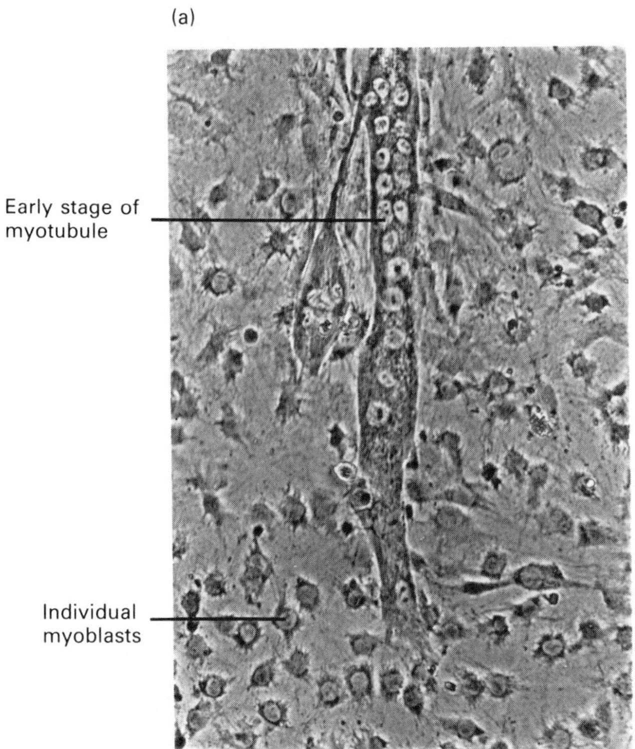

(b)

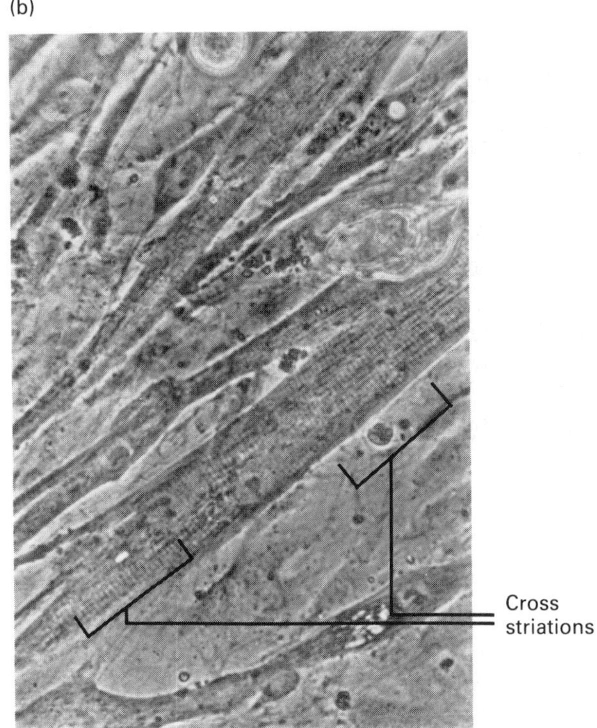

Figure 6-11 (a) A transformed line of rat myoblasts (muscle precursor cells) will grow indefinitely in laboratory cultures as single cells. (b) When the cultured cells are stopped from growing (e.g., by removal of serum from the medium), they fuse and produce *myotubules* that resemble multinucleated muscle cells. The characteristic cross-striations of muscle cells can be seen.

chief goals the study of gene function when genes or whole nuclei are introduced into novel environments.

Several somatic-cell genetic techniques involve the fusion of two cells in such a way that the nuclei from both parents are brought within one joint cytoplasm. If the nuclei remain separate (as occurs in nature for certain molds and fungi), the multinucleated cell is called a *heterokaryon*. Spontaneous fusion of animal cells in culture occurs infrequently, but the rate increases greatly in the presence of certain viruses that have lipoprotein envelopes similar to the plasma membranes of animal cells (see Figure 15-59). Apparently a glycoprotein in the virus membrane promotes cell fusion, but the mechanism is not yet proved. Cell fusion can also be promoted by the addition of polyethylene glycol, which causes cell plasma membranes to adhere to those of any surrounding cells (Figure 6-14). In most fused animal cells, the nuclei eventually also fuse, producing viable cells that contain chromosomes from both "parents." Hybrids between cultured cells from different mammals—for example, between human and rodent cells or between cells of different rodents (rats, mice, and hamsters)—have been widely used in somatic-cell genetics.

As hybrids of human and mouse cells grow and divide, they gradually lose human chromosomes in random order. In a medium that can support both mouse and human cells, the hybrids eventually lose all the human chromosomes. However, in a different medium in which human cells can grow but mouse cells lack one enzyme needed for growth, one human chromosome will remain—the one containing the gene that codes for the needed enzyme (Figure 6-15). The hybrid cells retain the entire chromosome—not just the selected gene.

By using various different media in which mouse cells cannot grow but human cells can, *panels* of hybrid cell lines have been established. Each cell line in a panel contains a different, limited number of human chromosomes (ideally, a single human chromosome) and a full set of mouse chromosomes, each of which can be identified visually through a light microscope. Thus, an individual human chromosome (or a small group of them) can be probed for the presence of a particular gene—for example, by testing a cell line biochemically for a particular enzyme, or immunologically (with an antibody) for a surface antigen; or the methods of DNA hybridization (discussed in Chapter 7) may be used to locate a particular DNA sequence. In this way, many human genes have been mapped to specific human chromosomes. Panels of

(a)

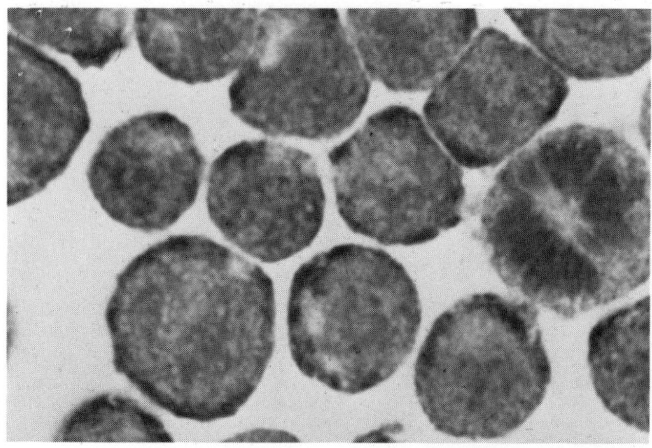

(b)

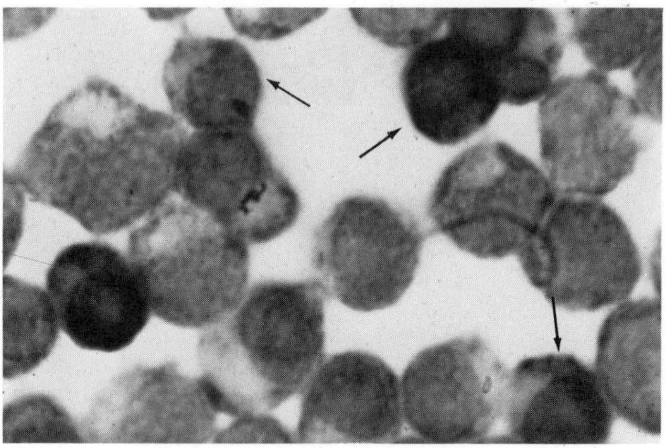

Figure 6-12 (a) Mouse erythroleukemia cells (abnormal erythroblasts, or precursors of red blood cells) that have been transformed by a virus. When these cells are treated with dimethyl sulfoxide or various other agents, they undergo only three or four final divisions. (b) During this time they produce large quantities of hemoglobin, stained dark here (arrows). *Courtesy of R. A. Rifkind.*

mouse-hamster hybrid cells have also been established; in these cells, the majority of mouse chromosomes are lost, which allows the mapping of genes in the mouse genome.

Mutants in Salvage Pathways of Purine and Pyrimidine Synthesis Are Good Selective Markers

One metabolic pathway has been particularly useful in cell fusion experiments. Most animal cells can synthesize the purine and pyrimidine nucleotides de novo—that is, from simpler carbon and nitrogen compounds, rather than from already formed purines and pyrimidines (Figure 6-16). The folic acid antagonists *amethopterin* and *aminopterin* interfere with the ability of the carrier molecule tetrahydrofolate to participate in the early stages of de novo synthesis of glycine, purines, nucleoside triphosphates, and thymidine triphosphate. These drugs are called *antifolates*. However, most cells have enzymes that can use purines and thymidine directly via *salvage pathways* to bypass the antifolates (Figure 6-16).

A number of mutant cell lines that are unable to carry out one of the salvage steps have been discovered. For example, cell cultures lacking thymidine kinase (tk) have been selected, and cultures have been established from humans who lack adenine phosphoribosyl transferase (APRT) and hypoxanthine-guanine phosphoribosyl transferase (HGPRT). These different types of salvage mutants become useful partners in cell fusions with one another or with cells that have salvage pathway

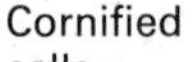

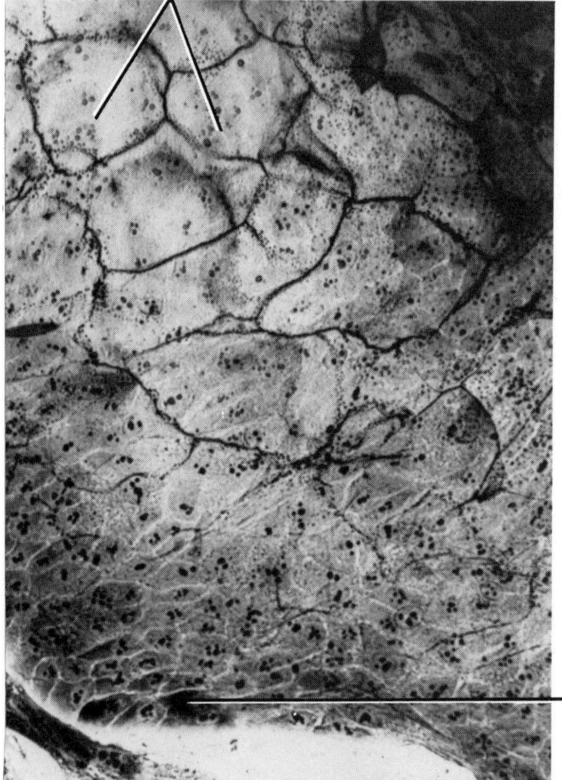

Figure 6-13 The differentiation of skin cells in culture. When the epidermal layer of the skin of newborn animals is placed in culture with irradiated, nongrowing, "feeder" cells, the basal cells of the skin grow and differentiate into cornified epithelial cells. The flattened and tightly packed nature of the epithelial sheet is characteristic of the cells at the body surface. [See T. T. Sun and H. Green, 1976, *Cell* 9:511.]

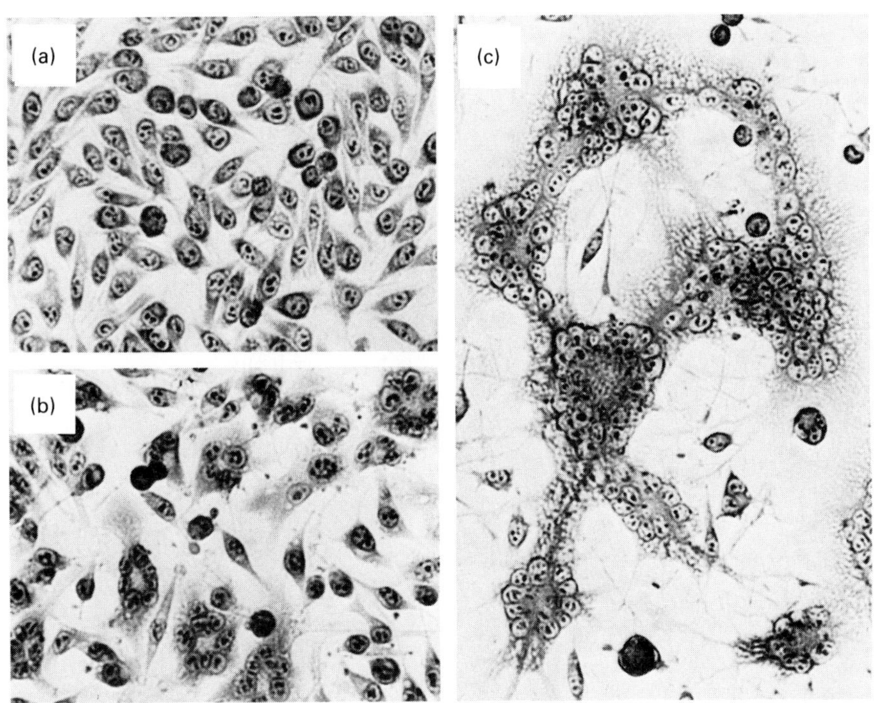

Figure 6-14 Fused cultured cells. Mammalian cells growing in culture can be fused by treating the cells with polyethylene glycol. (a) Unfused growing mouse cells. (b),(c) Mouse cells that have fused as a result of treatment for 1 min with polyethylene glycol. The cells in (b), which were treated with a 45% concentration of polyethylene glycol, show 2 to 5 nuclei per cell; those in (c), which were treated with a 50% concentration, have more than 10 nuclei per cell. By varying the concentrations and the times of exposure, the investigator can maximize the number of heterokaryons containing only two nuclei. *From R. L. Davidson and P. S. Gerald, 1976,* Som. Cell Genet. *2:165.*

enzymes but that cannot grow in culture. The selective medium most often used to culture such fused cells is called *HAT medium,* because it contains hypoxanthine (a purine), amethopterin or aminopterin, and thymidine (Figure 6-17).

Hybridomas Are Fused Cells That Make Monoclonal Antibodies

The technique of cell fusion followed by selection is widely used in the production of monoclonal antibodies. A *monoclonal antibody* is a single pure antibody produced in quantity by a cultured clone of a special type of cell called a B lymphocyte. How is cell fusion used to produce monoclonal antibodies? Each normal B lymphocyte from, say, a rat or mouse spleen is capable of producing a single antibody. If a mouse is injected with an antigen, B lymphocytes that make an antibody that recognizes the antigen are stimulated to grow and produce that antibody. In the animal, this stimulated B cell forms a clone of cells in the spleen or bone marrow. However, normal lymphocytes will not grow continuously in

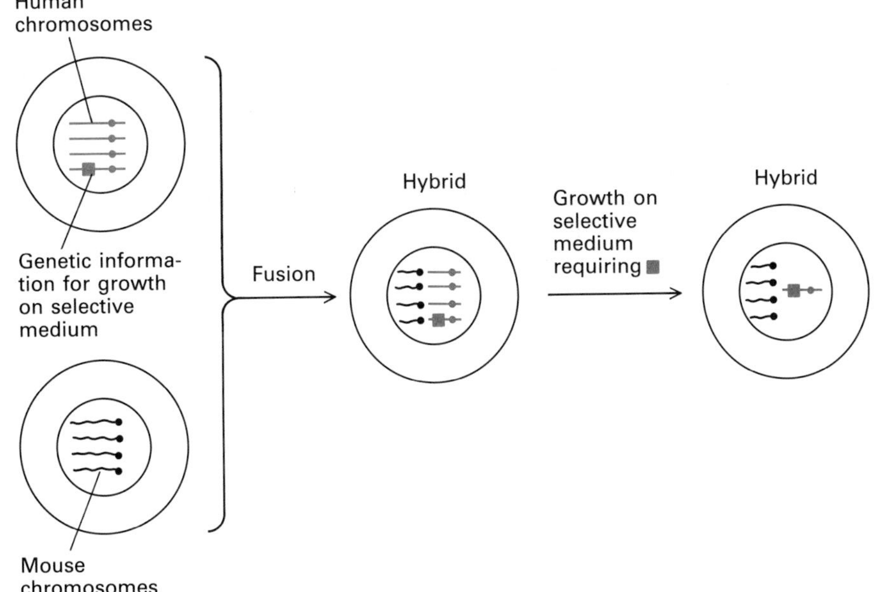

Figure 6-15 If mouse cells are fused with human cells, the hybrids initially contain all of the chromosomes from both cells. With growth and division of the hybrids, the human chromosomes are eventually lost. However, if one of the human chromosomes bears genetic information required for growth in a particular medium, and this information is lacking in the mouse chromosomes, the cells containing that human chromosome will survive and grow in the medium. By using this method, investigators may select cells containing particular human chromosomes.

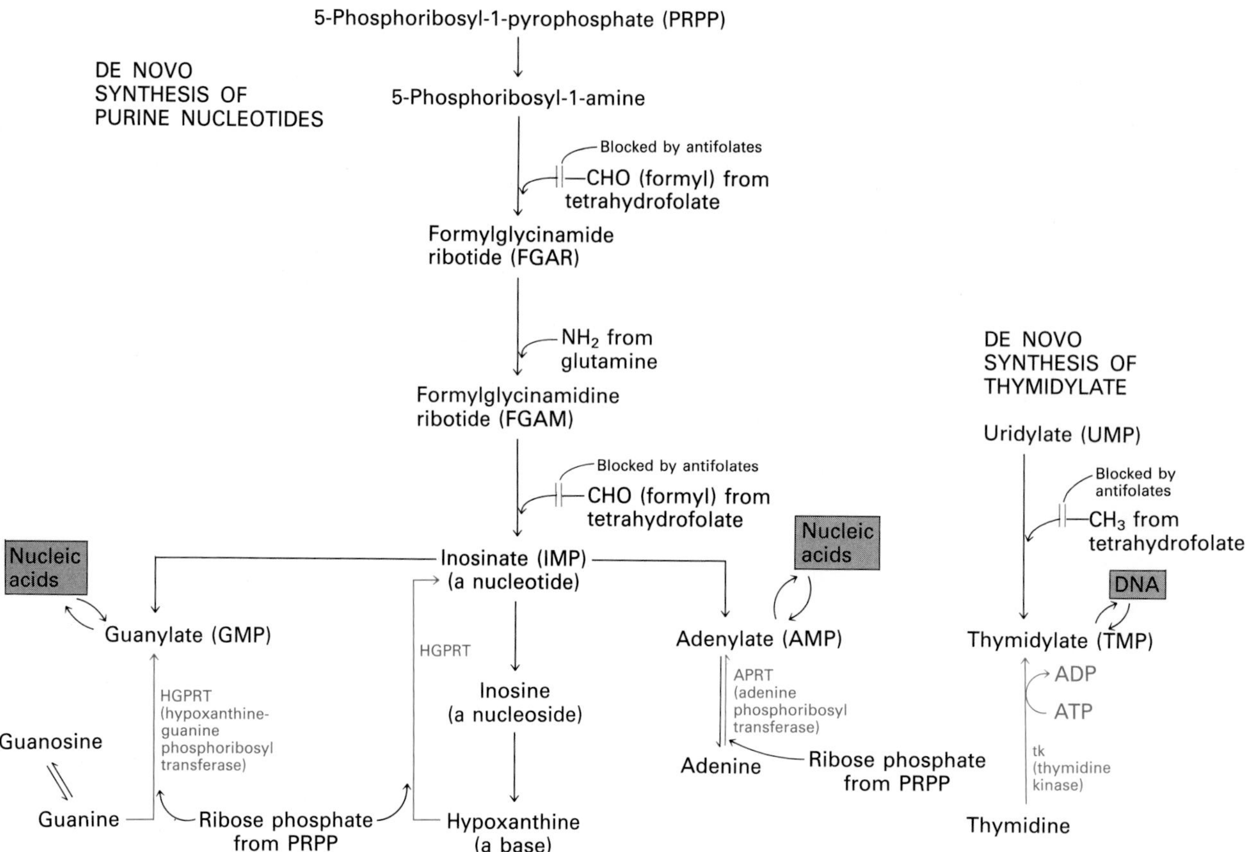

Figure 6-16 Cultured animal cells can obtain purine nucleotides and thymidylic acid in two ways. Under normal circumstances, cells synthesize purine nucleotides and thymidylate by a *de novo pathway* that requires the transfer of a methyl or formyl group from an activated form of tetrahydrofolate (e.g., N^5,N^{10}-methylenetetrahydrofolate) as shown in the top part of the diagram. Antifolates such as aminopterin and amethopterin block the reactivation of tetrahydrofolate, preventing purine and thymidylate synthesis. The second mechanism of obtaining nucleotides is called a *salvage pathway* (bottom part of diagram). These pathways are not blocked by antifolates. If the medium contains purine bases or nucleosides and thymidine, most mammalian cells can use the nucleoside or bases directly to make nucleotides by these pathways. HGPRT, APRT, and tk are the salvage pathway enzymes.

culture and cannot be used to establish an immortal productive clone.

The difficulty is resolved by fusing normal B lymphocytes with cancerous lymphocytes called myeloma cells. The wild, unrestrained growth of lymphocytes is the cause of myeloma and can be induced in mice and rats by various treatments. Unlike normal lymphocytes, myeloma cells are immortal. Many different cultured cell lines of myeloma cells from mice and rats have been established; from these, mutant cell lines that have lost the salvage pathways for purines (as indicated by their inability to grow in HAT medium) have been selected. When these mutant myeloma cells are fused with normal antibody-producing cells from a rat or mouse spleen, *hybridoma* cells result. Like myeloma cells, hybridoma cells grow indefinitely in culture; like normal spleen cells, the fused cells have purine salvage pathway enzymes and grow in HAT medium. If a mixture of fused and unfused cells is placed in HAT medium, the unfused mutant myeloma cells and the unfused spleen cells die, which leaves a culture of immortal hybridoma cells, each producing a single antibody.

Clones of hybridoma cells can be tested separately for the production of a desired antibody, and the clones containing that antibody can be cultured in large amounts to produce it in quantity (Figure 6-18).

Besides being valuable as research reagents, such pure antibodies are already becoming very important in medi-

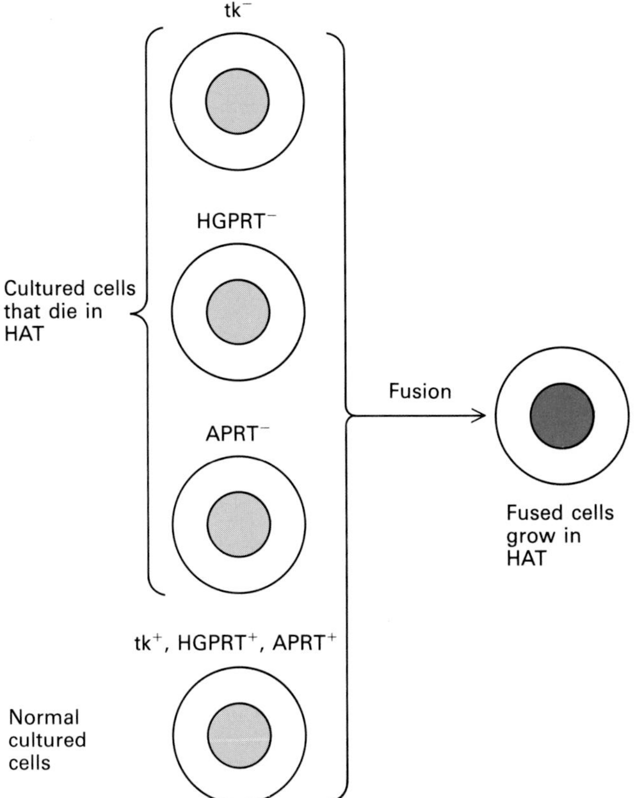

Figure 6-17 Cultured animal cells normally make purines and pyrimidines de novo from simple compounds. They do not use bases previously formed by salvage pathways. Thus, if a cell is mutant in an enzyme of a salvage pathway, it still grows in a normal medium (see Figure 6-16). In HAT medium, de novo synthesis of purines and pyrimidines is blocked, and the cells are called upon to use the salvage pathways. Therefore, cells with mutations in genes coding for salvage pathway enzymes die. If clones of any salvage pathway mutant (e.g., a mutant in the genes for tk or HGPRT) are mixed together with normal cells (primary cells from tissues) that cannot grow in culture, some fusions will take place. Only fused cells are able to grow in HAT medium.

cine for diagnosis; in the future they will probably give rise to new treatments as well. For example, proteins that exist in small quantities in an organ of an organism can be recognized as antigens by antibodies, and thus their locations in specific cells or cell fractions can be identified (Figure 6-19). Once identified, such scarce and, in some cases, valuable proteins can be obtained by the technique of *antibody-affinity chromatography:* The monoclonal antibody is bound to a solid substrate, which is then exposed to a solution containing the desired protein. Only that protein binds to the antibody, and it can be easily purified (see Figure 3-39).

Transformation of Eukaryotic Cells

Pure DNA can be used to transform cultured eukaryotic cells genetically by a procedure similar to that used in the classic experiments of Avery, McLeod, and McCarty with pneumococci. If cultured yeast cells are treated to remove their thick cell walls, they will take up DNA that has been added to the surrounding medium. Cultured mammalian cells also take up DNA directly, particularly if the DNA is converted to a fine precipitate by treatment with calcium (Ca^{2+}) ions. In both kinds of cells, once the foreign DNA

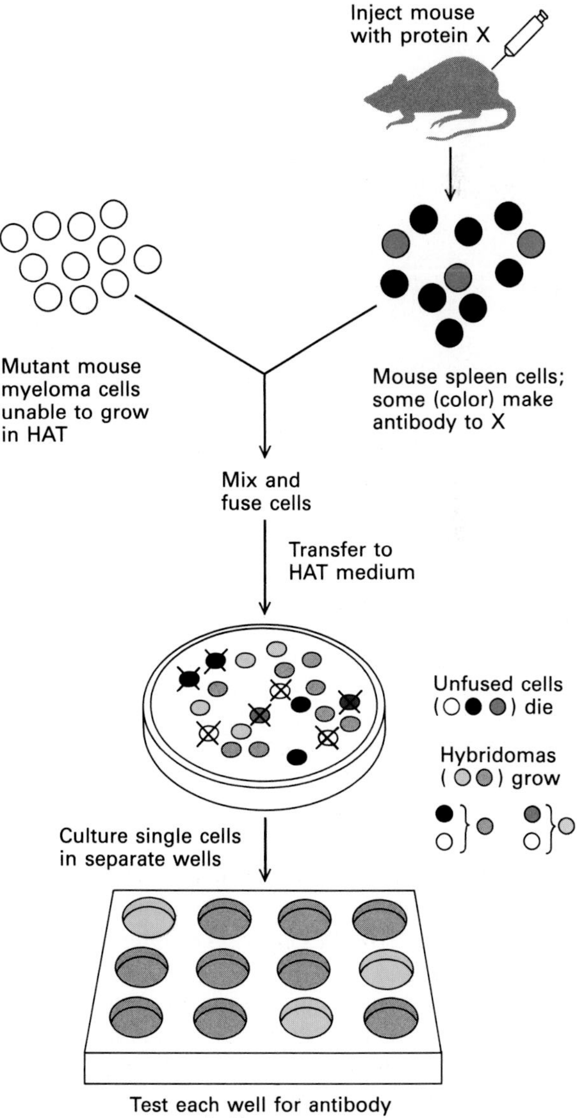

Figure 6-18 The procedure for producing a monoclonal antibody to a protein X. Once a hybridoma clone that produces a desired antibody has been identified, the clone can be cultured in large amounts to yield a large amount of the antibody.

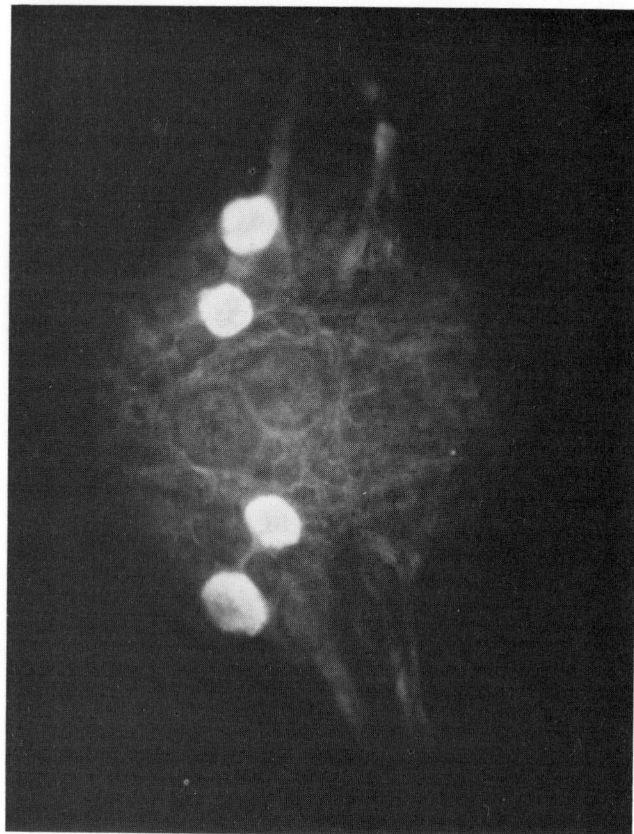

Figure 6-19 Cells in a leech's nervous system have been stained with a fluorescent monoclonal antibody. The distribution of the single protein antigen is shown by the bright spots, which are created by the fluorescence of the individual cells to which the antibody has attached. *From B. Zipser and R. McKay, 1981,* Nature ***289***:*549. Reprinted by permission from* Nature. *Copyright 1981 Macmillan Journals Limited.*

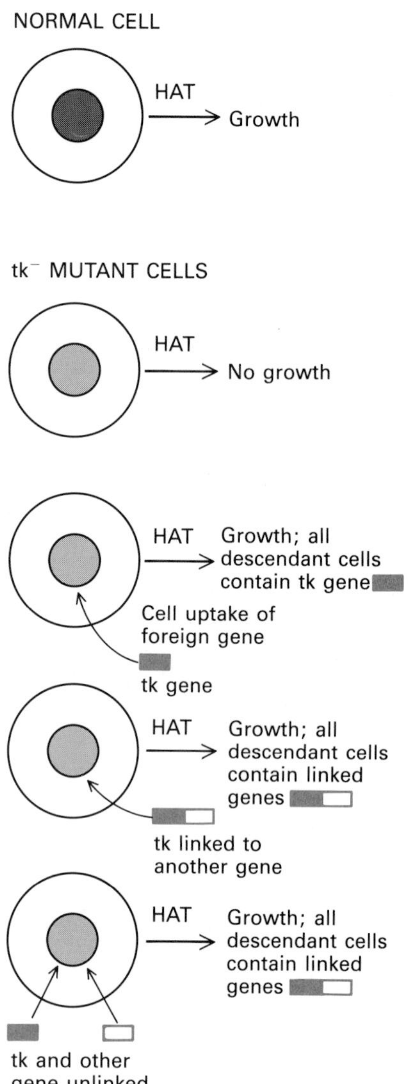

Figure 6-20 The technique of transforming mammalian cells is frequently performed on mutant cells unable to grow on the inhibiting HAT medium because they lack a particular enzyme (see Figures 6-16 and 6-17). The most frequently used selectable markers include tk (thymidine kinase), especially that from the *Herpes simplex* virus; APRT; HGPRT; and GPT (guanosine phosphoribosyl transferase, from bacteria). The gene for each of these markers can be prepared as pure DNA. The mutant cells are transformed by taking up this free DNA, and they are able to grow on the inhibiting medium.

Any other DNA can be linked to DNA encoding the enzyme. Alternatively, the two DNAs can simply be mixed; they will be incorporated together into the chromosomes of the recipient cell. Then, by culturing the mixture of cells in a selective medium, a pure population of transformed cells containing both types of DNA is obtained.

is inside, it combines with the cells' chromosomes. If yeast DNA is used to transform yeast cells, the recombination occurs most often at the chromosomal site that is homologous to the DNA segment that is introduced. In contrast, integration of mammalian DNA in mammalian cells occurs in nonspecific locations. Perhaps this is due to the much larger size of the mammalian genome compared to the yeast genome.

The transformation of cultured cells by pure DNA allows the action of a particular gene to be studied in different cell settings. Individual genes can be prepared as pure DNA (the procedure is explained in Chapter 7). In general, bacterial genes transform only bacterial cells, and yeast genes only yeast cells; but at least some animal genes transform animal cells across wide phylogenetic ranges (e.g., insect genes in human cells). The DNA that codes for a particular selectable trait can be linked with a second stretch of DNA whose action is being studied. After the linked DNA has been mixed with cells, those

(a)

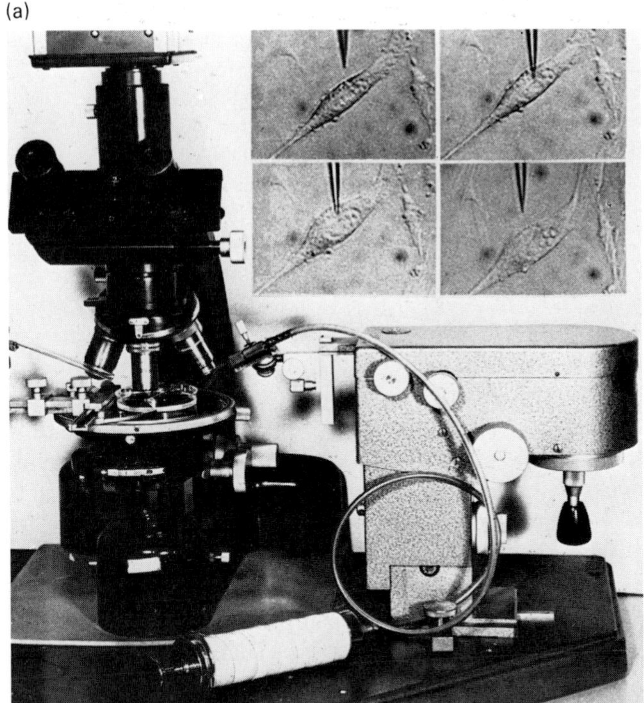

(b)

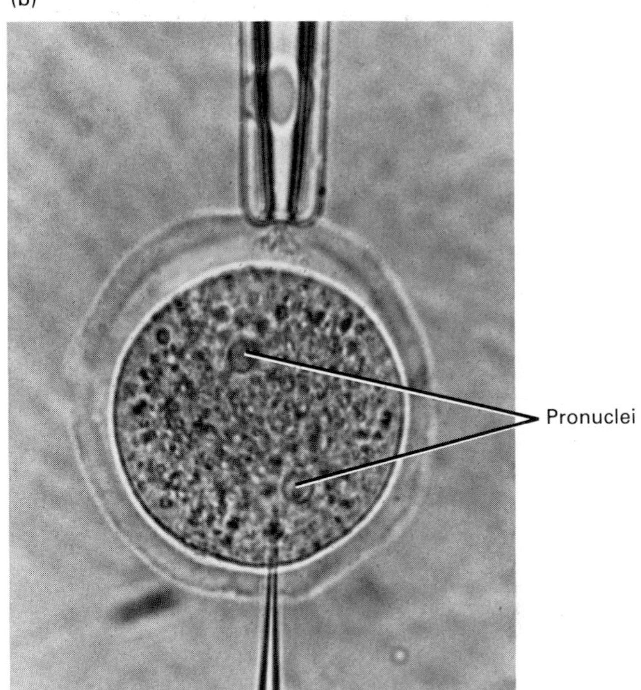

Figure 6-21 (a) DNA can be directly injected into individual cells under the microscope by inserting fine-tipped glass needles for controlled pumping of small volumes (about 2×10^{-9} ml). The four inset photographs show the injection of DNA into a HeLa cell nucleus, with the consequent enlargement of the cell. (b) Injection into a fertilized mouse egg with two pronuclei. The egg is held by the blunt pipette so that a fine-tipped pipette can be inserted into one of the pronuclei. Such injected eggs are viable, and they have grown into mice containing injected DNA sequences in every cell. *Photograph (a) courtesy of A. Graessman, 1968, Ph.D. thesis, Freie University, Berlin; (b) courtesy of R. Brinster.*

that have taken up the DNA can be separated, by means of the selectable trait, from those that have not (Figure 6-20). Even if two unlinked DNA samples are added, the two will often become linked inside the cell, and thus will be integrated simultaneously into the cell's chromosomes. In animal cell cultures, about 1 percent of the cells take up DNA.

DNA can also be injected directly into vertebrate cells and cell nuclei. Frog oocytes are among the most popular targets because of their very large nuclei. The functions of many types of DNA from sea urchins, mammals, and *Drosophila* as well as from frogs have been tested by introducing the DNA into frog nuclei. In addition, the nuclei of several lines of mammalian cells, including HeLa cells, have been injected with DNA with great success (Figure 6-21a).

Foreign DNA Can Be Introduced into the Germ Line of Animals and Plants

Perhaps the most dramatic use of the injection technique is the introduction of DNA into mouse and *Drosophila* eggs in such a way that the DNA becomes part of the chromosomes of the animal. When foreign DNA is injected into one of the two pronuclei (the male and female haploid nuclei contributed by the parents) of a fertilized mouse egg before they are fused (Figure 6-21b), the DNA ultimately becomes incorporated into the chromosomes of the diploid zygote. The injected eggs can be transferred to foster mothers, in which normal cell growth and differentiation occurs. Some of the progeny mice will have the foreign DNA; the mice that do will have equal amounts in *all* tissues (Figure 6-22).

This technique is of great importance in the study of embryogenesis and organ development, because the activity of injected DNA can be observed in various tissues. Similar incorporation of DNA into *Drosophila* can be achieved by injecting the foreign DNA into the region of the embryo giving rise to germ cells. The resulting fly will have some progeny that carry the injected DNA.

Plants Can Be Regenerated from Plant Cell Cultures

Plant cell cultures are not yet as routinely used as animal cell cultures. Established cultures exist for relatively few

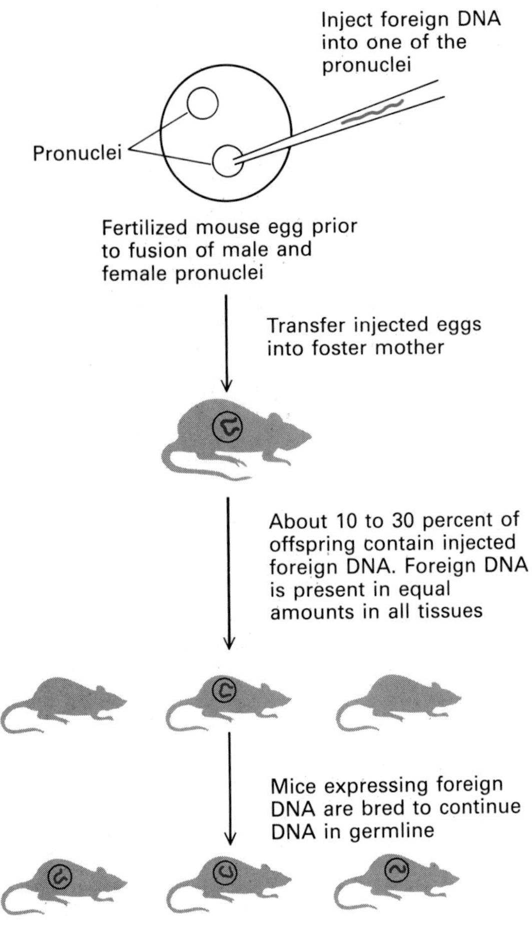
Inject foreign DNA into one of the pronuclei
Pronuclei
Fertilized mouse egg prior to fusion of male and female pronuclei
Transfer injected eggs into foster mother
About 10 to 30 percent of offspring contain injected foreign DNA. Foreign DNA is present in equal amounts in all tissues
Mice expressing foreign DNA are bred to continue DNA in germline

Figure 6-22 The introduction of foreign DNA into the tissues of a mouse. Injected fertilized eggs are inserted in foster mothers (treated appropriately with progesterone and other female hormones), and mice born from the injected eggs are tested for the presence of the foreign DNA. All tissues contain equal amounts of it: from zero to 100 copies. [See R. L. Brinster et al., 1981, *Cell* 27:223.]

plant species, because the growth of a culture from a single plant cell is not yet as reliable as the growth of cultures from single animal cells. However, one striking property of plants and of cultured plant cells makes continued work on plant cell cultures a priority for the future. Whole plants have been reconstructed from single plant cells taken from cultures. The cultured plant cells first grow into *callus tissue,* an undifferentiated lump of cells. Under the influence of plant growth hormones, different plant parts (roots, stems, and leaves) develop to yield whole, fertile plants. The ability to achieve this result predictably and reliably would be a significant advantage to plant biologists. To that end, new techniques of introducing DNA into plant cells have been developed.

Viruses

A virus is a small cellular parasite that cannot reproduce by itself. However, upon infection of a susceptible cell, the virus can direct the cell machinery to produce more viral material. Each virus has either RNA or DNA as its genetic material; no virus has both. The nucleic acid, which may be either single-stranded or double-stranded, is contained within a protein layer. The infectious virus particle is called a *virion.* Some viruses contain only enough RNA or DNA to encode from 3 to 10 proteins, whereas others can encode from 100 to 200. (See Table 6-4 for a list of viruses used in genetic experiments.)

The coat, or *capsid,* that encloses the nucleic acid is composed of one or more proteins that are specific to each kind of virus. The capsid plus the enclosed nucleic acid is called the *nucleocapsid.* There are two ways in which capsid proteins can be arranged: as a deltahedron (a polyhedron having triangular faces) or as a helix (Figure 6-23). At low resolution, the deltahedra (which are frequently 20-sided *icosahedra*) resemble spheres, and the helixes resemble cylinders. Some viruses also have an *envelope* that is external to the nucleocapsid. The envelope consists mainly of lipids, but it also contains some virus-specific proteins (Figure 6-24a). The lipids in the envelope of a virus are similar to those of the plasma membrane of the infected host cell, because the virus envelope is in fact derived from the membrane of its host cell (Figure 6-24b).

Generally, the host range—the group of cell types that the virus will infect—is restricted, and thus serves to initially classify the virus. Viruses that infect only bacteria are called *bacteriophages,* or simply *phages.* Those that infect animal or plant cells are termed, appropriately, *animal viruses* or *plant viruses.* However, a few viruses can grow both in insects and in plants: for example, potato yellow dwarf virus can grow in leafhoppers (a type of insect) as well as in plants. Some other animal viruses have wide host ranges also; for example, *vesicular stomatitis* virus grows in insects and in many different mammalian cells. However, many animal viruses do not cross phyla, and some (e.g., poliomyelitis virus) grow only in closely related species such as primates.

Virology is a highly quantitative science because very accurate methods for counting infectious particles exist. This allows the experimenter to be certain that every cell in a sample is infected, or, alternatively, that only a few cells are. Each virus preparation grown from a single particle represents a clone of virus in which the RNA or DNA constitutes a pure sample of nucleic acid. The number of infectious virus particles in a sample can be determined accurately by counting the local lesions in cells; this is termed the *plaque assay* technique (Figure 6-25).

Table 6-4 Animal viruses commonly used in molecular cell biology

Virus	Known hosts	Type	Nucleic acid class*	Size (kb)†	Lipid membrane?	Research areas in which virus is used
Adenoviruses	Vertebrates	DNA	I	36	No	mRNA production and regulation, DNA replication, cell transformation
SV40	Primates	DNA	I	5.2	No	
Herpes viruses	Vertebrates	DNA	I	150	Yes	
Vaccinia	Vertebrates	DNA	I	200	Yes	Genome structure, mRNA synthesis by virion enzymes
Parvoviruses	Vertebrates	DNA	II	1–2	No	DNA replication
Retroviruses	Vertebrates and (?) invertebrates	RNA/ DNA	VI	5–8	Yes	Cell transformation, "oncogenes"
Reoviruses	Vertebrates	RNA	III	1.2–4.0‡	No	mRNA synthesis by virion enzymes, mRNA translation
Influenza	Mammals	RNA	V	1.0–3.3‡	Yes	Membrane formation, glycoprotein biosynthesis, and intracellular transport
Vesicular stomatitis virus	Vertebrates	RNA	V	12	Yes	
Sindbis virus	Insects and vertebrates	RNA	IV	10	Yes	
Poliomyelitis virus	Primates	RNA	IV	7	No	Viral RNA replication, interruption of host translation, polyprotein cleavage

*Class refers to strategy for mRNA synthesis (see Figure 6-30).

†Size is given in kilobases (1 kb = 1000 nucleotides) for single-stranded nucleic acids, or kilobase pairs for double-stranded molecules.

‡Reoviruses have 10 double-stranded RNA segments, and influenza has 8 single-stranded RNA segments; the length of each segment is in the range indicated.

Viral Growth Cycles Can Be Divided into Stages

A viral infection begins when a virion comes into contact with a host cell and attaches, or adsorbs, to it. In various ways, depending on the specific virus and host, the viral DNA or RNA crosses the plasma membrane into the cytoplasm. The entering viral genetic material is often still associated with inner viral proteins, but sometimes the viral nucleic acid is free of proteins once it has entered the cell cytoplasm. The DNA of most animal viruses (plus any associated proteins) ends up in the cell nucleus, where cell DNA is of course also found. The viral nucleic acid then interacts with the host cell's protein- and nucleic acid–synthesizing machinery (enzymes, ribosomes, tRNA, etc.) to direct its own replication and the synthesis of viral proteins. Some of the protein products of the viral DNA are special enzymes or inhibitory factors that stop cell metabolism, but most virus-encoded proteins are used in the construction of new virions.

Simple viruses such as tobacco mosaic virus (TMV), which contains only one protein plus its RNA, will undergo *self-assembly* from pure components. Other viruses contain dozens of proteins and will not spontaneously assemble. However, within the cell the multiple components may assemble spontaneously in stages, so that subviral particles are formed first and completed particles last.

In most bacterial virus infections and in some plant and animal virus infections, when the new virions formed inside the cell are complete, the cell ruptures (*lyses*) and releases all the virions at once. However, in many plant and animal virus infections no discrete event of lysis occurs; rather, the dead host cell gradually disintegrates, releasing the virions as the disintegration occurs.

These events—adsorption, penetration, replication, and release—describe the *lytic cycle* of virus replication (Figure 6-26). The outcome is the production of a new round of virus particles and the death of the cell. At the other extreme, a virus DNA can enter the cell and become integrated into the host chromosome, where it remains quiescent and replicates as part of the cell. Bacteriophages that become integrated in this manner are called *temperate* phages. A few viruses, both phages and animal vi-

(a) An icosahedral virus

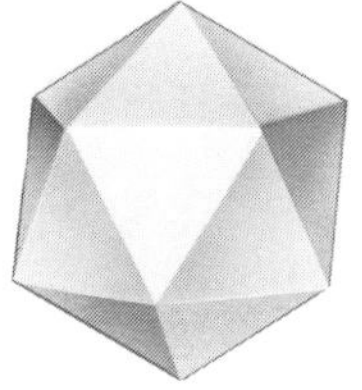

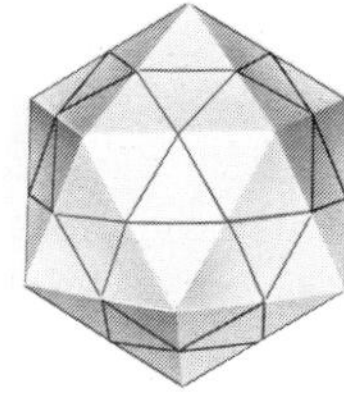

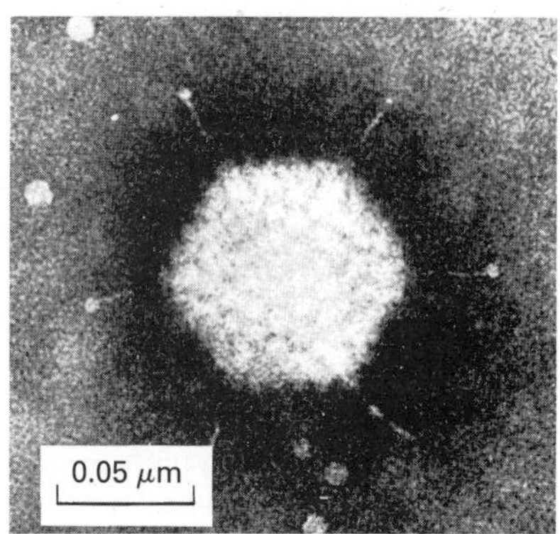

(b) Section of a helical virus

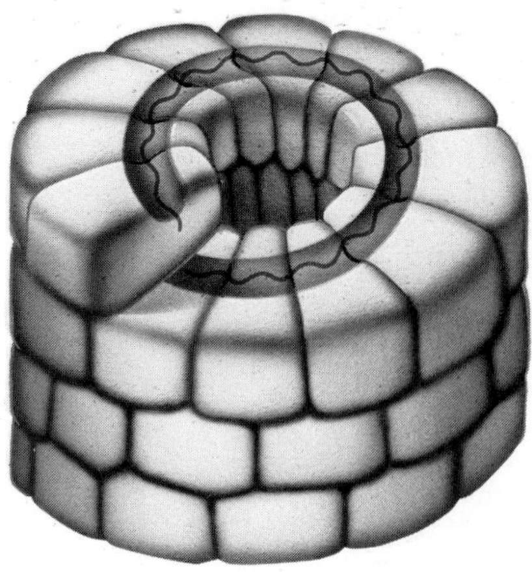

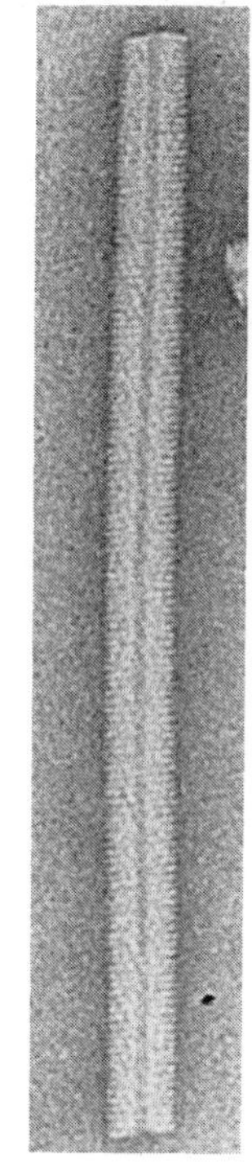

Figure 6-23 The two basic geometrical shapes of virions. (a) Solid, compact viruses are most often assembled into icosahedra, 20-sided deltahedrons. (The actual shape of the protein subunit doesn't conform to a flat triangle, but the overall effect when the subunits are assembled is of a roughly spherical structure with triangular faces.) The electron micrograph is of an adenovirus, a *naked* icosahedral virus (i.e., one having no envelope) with surface projections at its vertexes. (b) Protein subunits can also take the form of helical arrays around an RNA or DNA molecule, with the nucleic acid strand running in a groove within the enclosing proteins. Illustrated here in both the drawing and the electron micrograph is the helical array formed by the subunits of another naked virus, tobacco mosaic virus (TMV), around its RNA (colored in drawing). *After S. E. Luria, J. E. Darnell Jr., D. Baltimore, and A. Campbell, 1978,* General Virology, *3d ed., Wiley, pp. 39–40; photograph (a) courtesy of R. C. Valentine; photograph (b) courtesy of J. Finch.*

ruses, infect cells and cause new virion production without either killing the cell or becoming integrated.

Because pure cultures of susceptible cells can be obtained and large amounts of virus easily prepared, it is possible to deliver virus in such great quantities that every cell in a culture is sure to be infected simultaneously. Therefore the viral genes can be observed acting synchronously on every cell in the culture.

Bacterial Viruses Remain in Wide Use

As we mentioned in Chapter 1, bacterial viruses played an important role in the development of molecular biology. Thousands of different bacteriophages have been isolated, and almost every one is uniquely suited to the investigation of a specific biochemical or genetic event. For our present purposes, however, we will consider only four types in detail:

1. *DNA phages of the T series in E. coli.* These large lytic phages contain a single molecule of DNA. (The DNA molecule is about 2×10^5 base pairs long in T2, T4, and T6, and about 4×10^4 base pairs long in T1, T3, and T7.) A T phage enters an *E. coli* cell through a "tail" (Figure 6-27). The bacteriophage DNA then directs a program of events that produces approximately 100 new bacteriophage particles in about 20 min, at which time the infected cell lyses and releases the new phages. The formation of phage mRNA, its specification of phage proteins, and the multiplication of phage DNA are classic subjects of molecular biological study. Many of the principles described in later chapters of this book were first established from the study of these viruses.
2. *Temperate phages of E. coli.* Typical of this class of phages, which has one of the most well-studied genomes in all biology, is the *E. coli* bacteriophage λ

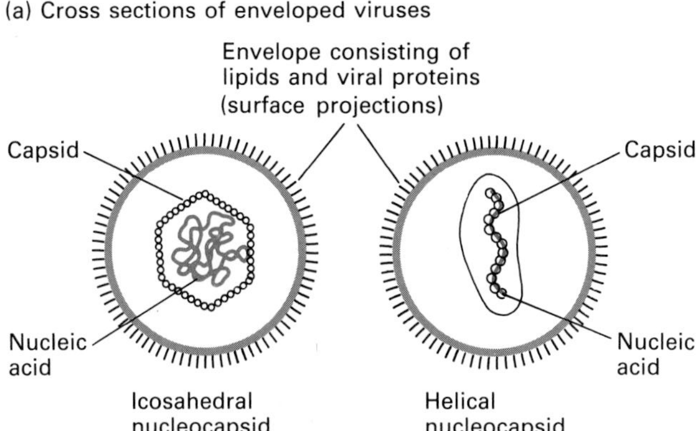

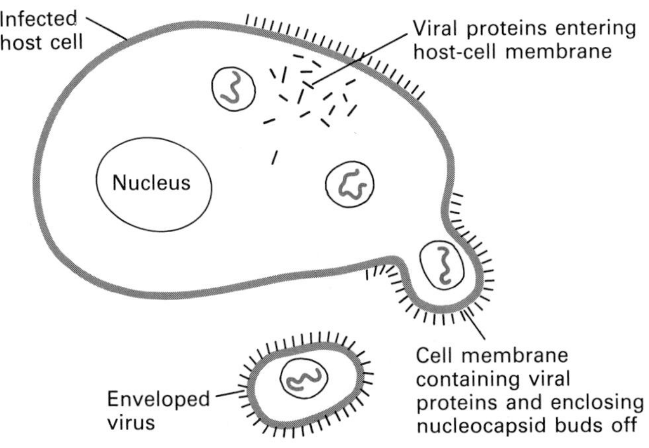

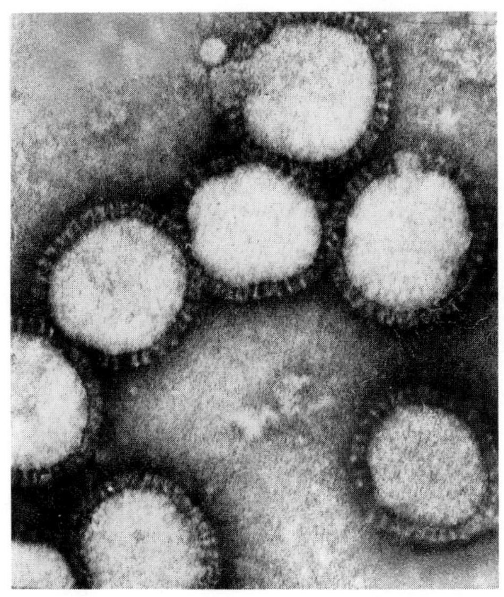

Figure 6-24 (a) In some viruses, an external envelope consisting of lipids and proteins surrounds the nucleocapsid. The influenza virion shown here has a helical nucleocapsid. (b), (c) Once inside a host cell, enveloped viruses encode proteins that enter the host-cell membrane and appear as surface projections on it. When new virions emerge from the cell, they acquire lipid-containing envelopes with embedded viral proteins. *Photograph from S. E. Luria, J. E. Darnell Jr., D. Baltimore, and A. Campbell, 1978,* General Virology, *3d ed., Wiley, p. 280.*

(lambda). Upon entering an *E. coli* cell, the DNA of this virus can take either of two different courses of action: (1) like the T phages, it can direct the production of new phage particles and lyse the cell, or (2) it can integrate into the bacterial chromosome and remain there indefinitely as the host cell grows and divides. The first alternative is the lytic cycle; the second is termed the *lysogenic cycle*. In the latter case, the viral DNA forms a circle and approaches the circular host DNA at a specific site. Enzymes break both circular molecules of DNA and then rejoin the broken ends so that the viral DNA becomes inserted into the host DNA (Figure 6-28). The maintenance of λ DNA as part of the host chromosome is due to the carefully controlled action of viral genes that suppress the lytic functions of the phage. (This and other genetic aspects of λ phage infection are discussed in Chapter 8.)

3. *Small DNA phages of E. coli.* These viruses, which are typified by phage φX174, have been extremely well studied, and the entire DNA sequences of at least five or six such phages have been determined. Because the DNA sequences are known, all of the possible protein products can be read from them; in fact, all of the proteins produced by several of the phages have been identified, and the precise functions of many of these proteins determined. The viruses in this group are so simple that almost every step in the replication of their DNA requires the use of cellular machinery. For this reason they have been useful in the identification of cell proteins involved in DNA replication, and they have also been widely used in the study of DNA synthesis in the test tube, as we shall see in Chapter 13.
4. *RNA phages of E. coli.* Some *E. coli* bacteriophages contain RNA instead of DNA. Because these phages are easy to grow in large amounts and because their RNA genomes serve also as their mRNA, they are a ready source of a pure species of mRNA. Some of the earliest demonstrations that cell-free protein synthesis can be programmed by mRNA (Chapter 4) were made with RNA from these bacteriophages. Also, the first long mRNA molecule to be sequenced was the genome of an RNA bacteriophage.

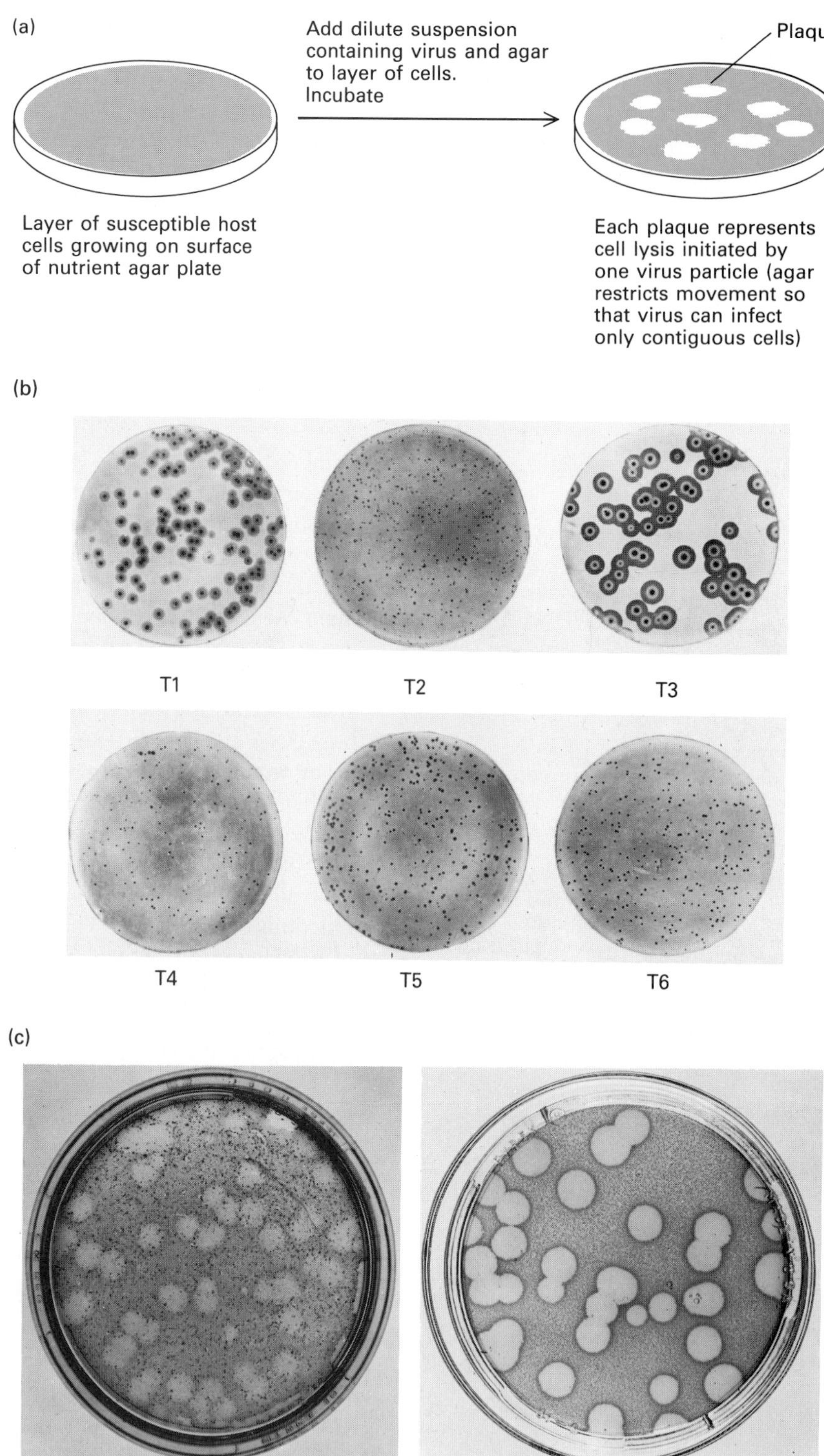

(a)
Add dilute suspension containing virus and agar to layer of cells. Incubate
Plaque
Layer of susceptible host cells growing on surface of nutrient agar plate
Each plaque represents cell lysis initiated by one virus particle (agar restricts movement so that virus can infect only contiguous cells)
(b)
T1
T2
T3
T4
T5
T6
(c)

Figure 6-25 (a) The plaque assay technique, which is used to estimate the number of infectious particles in a virus suspension. Each lesion or *plaque* is caused by one infectious particle, which in turn causes new virus formation with the consequent spread to neighboring cells. The virus in each plaque represents a pure clone of virus. (b) Plates illuminated from behind show plaques formed by different bacteriophages of the T series, plated on *E. coli.* (c) Plaques produced by two animal viruses. *Left:* Western equine encephalomyelitis virus, plated on chicken embryo fibroblasts. *Right:* Poliomyelitis virus plated on HeLa cells. *Photographs in (b) from M. Demerec and R. Fano, 1945,* Genetics *30:119; (c, left) courtesy of R. Dulbecco; (c, right) from S. E. Luria, J. E. Darnell Jr., D. Baltimore, and A. Campbell,* General Virology, *3d ed., Wiley, p. 26.*

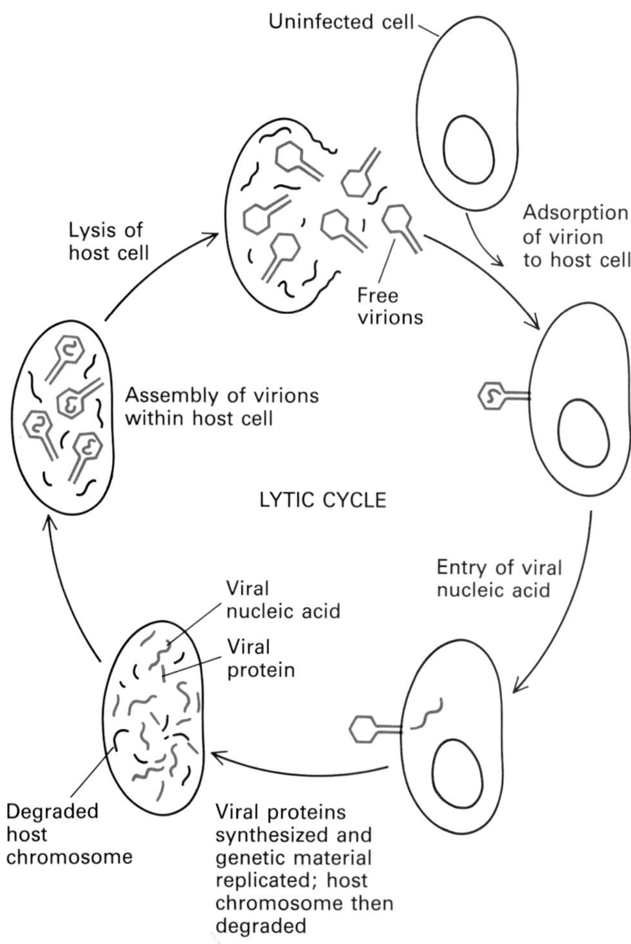

Figure 6-26 A generalized viral lytic cycle.

Plant Viruses Proved That RNA Can Act as a Genetic Material

The study of plant viruses inspired some of the earliest experiments in molecular biology. In 1935, Wendell Stanley purified and partly crystallized tobacco mosaic virus (TMV); other plant viruses were crystallized soon after. Pure proteins had been crystallized only a short time before Stanley's work, and it was at first thought that the TMV crystals were pure protein. Later studies showed that these crystalline preparations also contained RNA. In fact, plant viruses may contain either RNA or DNA, but for many years only RNA-containing plant viruses were known.

Experiments with TMV were very important in establishing nucleic acids as the informational molecules in viruses. For example, the fact that pure RNA from TMV could infect plant cells was demonstrated in 1956 by Alfred Gierer and Gerhard Schramum in Germany, and by Heinz Fraenkel-Conrat and Robley Williams in the United States. Further, it was shown that TMV could be separated into protein and nucleic acid parts and then reassembled into infectious virus. If protein and RNA from different strains of TMV were reassembled, the reassembled virus was infectious, but it was the source of the RNA that determined which of the two strains of virus would be produced in a host cell (Figure 6-29). These classic experiments, plus the study of the self-assembly of the TMV protein and its RNA molecule, made the plant viruses the most popular subjects of biophysical studies throughout the 1940s and 1950s.

Animal Viruses Are Very Diverse

Animal viruses show a bewildering array of shapes, sizes, and genetic strategies. Here we are interested only in viruses that exhibit at least one of two features: (1) they contain or form molecules that mimic cell functions quite exactly, or (2) they can integrate their genomes into those of normal cells. The integration of viral DNA sequences into the host cell frequently results in the induction of cancer. This special case is considered in Chapter 23.

The Classification of Viruses How can viruses be separated into groups? Viruses originally bore the names either of the diseases they caused or of the animals or plants they infected. However, it was soon discovered that viruses of many different kinds can produce the same symptoms or the same apparent disease states; for example, at least a dozen different viruses can cause red eyes, a runny nose, and sneezing. Clearly, this way of classifying viruses obscured many important differences in their structures and in their life cycles.

What *is* central to the viral life cycle is the replication sequence by which mRNA is produced. The relation between the viral mRNA and the nucleic acid of the infectious particle is the basis of the currently used means of classifying viruses. In this system, a viral mRNA is designated as a *plus strand* and its complementary sequence, which cannot function as an mRNA, as a *minus strand.* Likewise, a strand of DNA complementary to a viral mRNA is designated as a *minus strand.* Production of a plus strand of mRNA thus requires that a minus strand of RNA or DNA be used as a template. Six classes of animal viruses have been recognized. (Some bacteriophages and plant viruses also can be classified in this way, but the system has been used more widely in animal virology because representatives of all six classes have been identified.)

The pathway by which the virion nucleic acid becomes the mRNA of the virus is summarized for each of the six classes in Figure 6-30 (see also Table 6-4). Classes I and II are DNA viruses. *Class I* viruses contain a double strand of DNA; examples that will be discussed later in this book include the *adenoviruses* (Figure 6-31), which cause upper respiratory infections in many animals, and *SV40* (simian virus 40), a monkey virus discovered accidently when polio vaccines were prepared in monkey kidney cell cultures made from captured wild monkeys. The DNA of

(a)

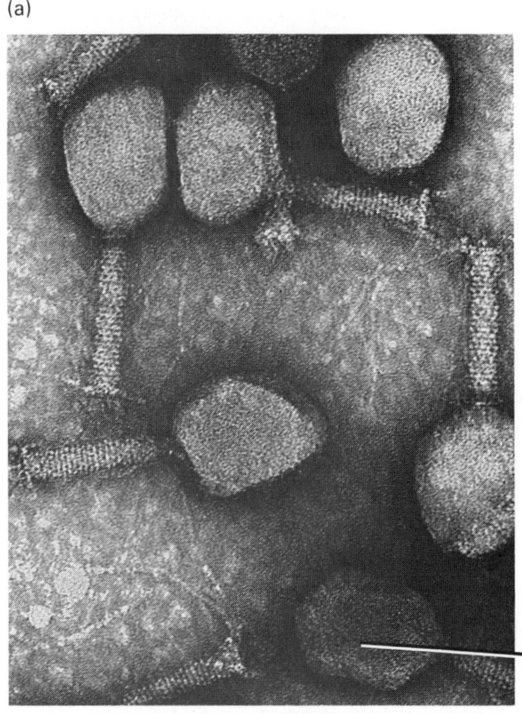

Figure 6-27 The complex structure of bacteriophage T4. The double-stranded DNA is in the polygonal head. (a) A high-magnification electron micrograph. Most heads in this micrograph are saclike because the conditions of preparation (the high salt during drying) triggered the release of DNA; the phage at the lower right has retained the polygonal shape. (b) The phage attaches to a bacterium by the tail fibers; the action of enzymes in the tail plate enables the phage DNA to enter the bacterium through the tail. *Photograph (a) courtesy of J. King and E. Hartwieg, M.I.T.*

(b)

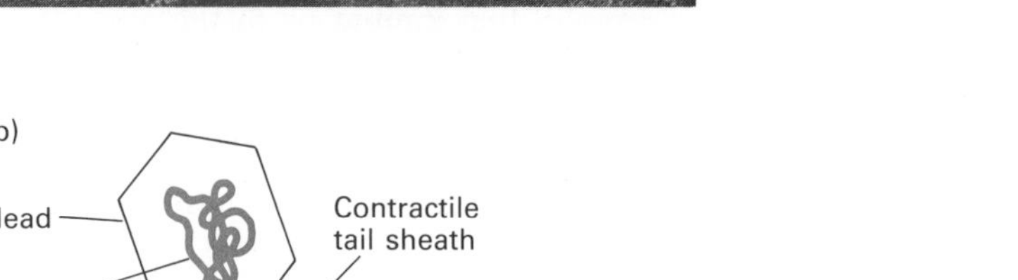

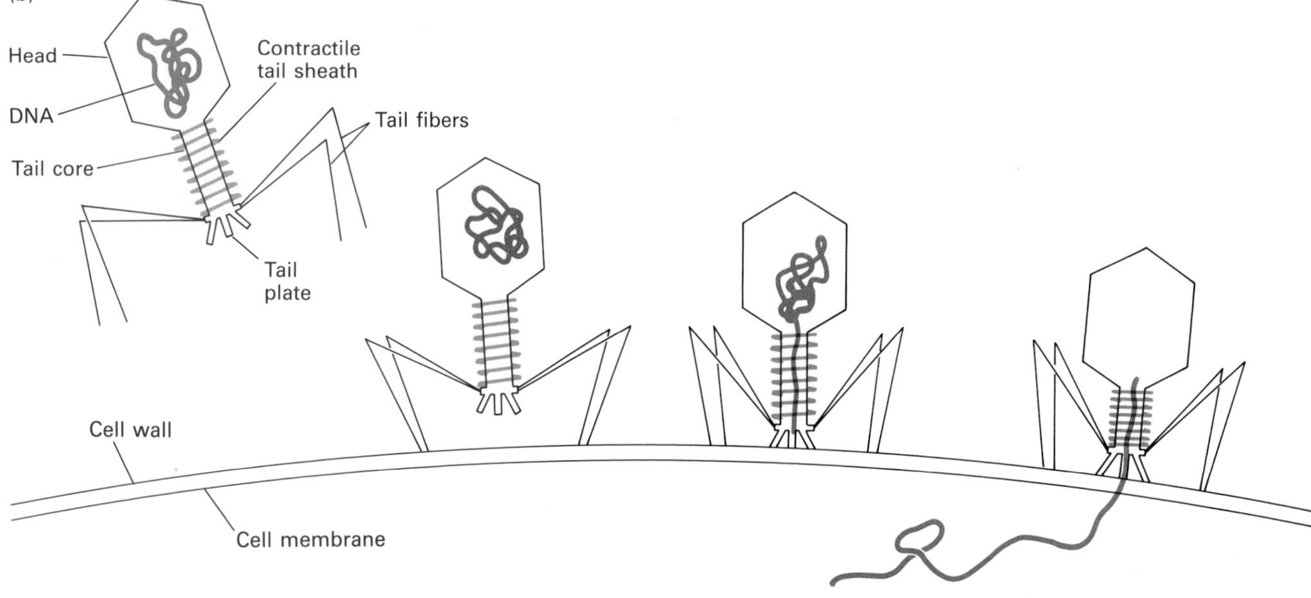

most of these viruses enters the cell nucleus, where enzymes that are normally responsible for producing cellular mRNA are diverted to producing viral mRNA (this mechanism of viral action is discussed further in Chapter 9). Another group of class I viruses, the *poxviruses,* is typified by *variola* (smallpox) and *vaccinia* (Figure 6-31), an attenuated (weakened) poxvirus that is used in vaccinations to induce immunity to smallpox. These very large, brick-shaped viruses ($0.1 \times 0.1 \times 0.2$ μm) carry their own enzymes for making mRNA, and they replicate in the cell cytoplasm.

Class II viruses, called *parvoviruses,* are simple viruses (*parvo-* is from the Latin for "poor") that contain a single strand of DNA. Some parvoviruses encapsidate (enclose) both plus and minus strands of DNA, but in separate virions. Others encapsidate only the minus strand, which is copied inside the cell into double-stranded DNA, which is then itself copied into mRNA.

The other classes of animal viruses, III through VI, contain RNA genomes. A wide range of animals, from insects to human beings, are infected by viruses in each of these classes. *Class III* viruses contain a double strand of RNA.

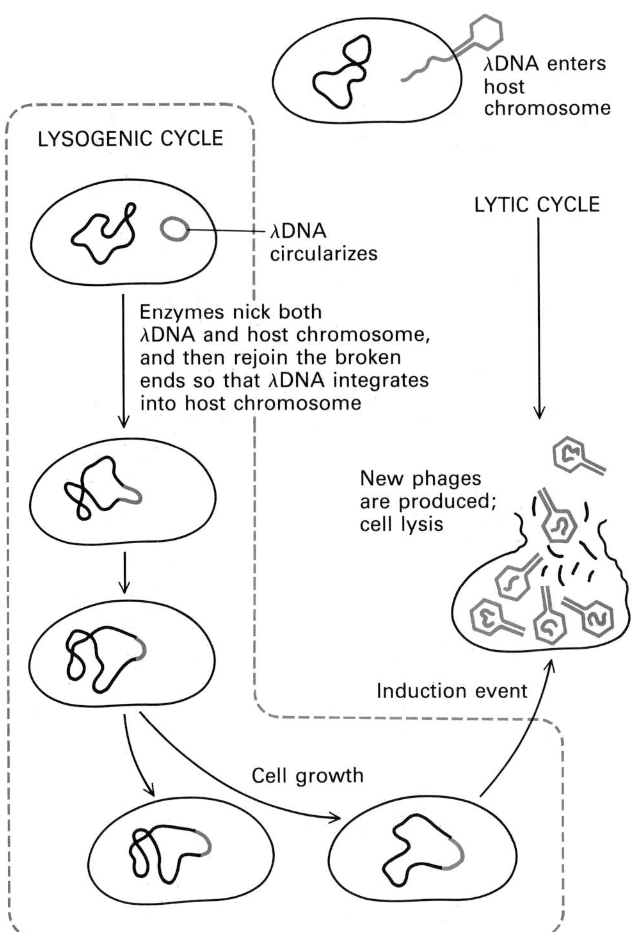
λDNA enters host chromosome
LYSOGENIC CYCLE
λDNA circularizes
LYTIC CYCLE
Enzymes nick both λDNA and host chromosome, and then rejoin the broken ends so that λDNA integrates into host chromosome
New phages are produced; cell lysis
Induction event
Cell growth

Figure 6-28 Temperate bacteriophages, such as λ, do not always produce new virions and lyse their host cells immediately after infection. Sometimes they enter a lysogenic cycle, in which the viral DNA is incorporated into the host-cell DNA and maintained there as the host grows and divides. Under certain conditions (UV light), λ DNA is induced to separate from the host chromosome, and thereby initiate the lytic cycle of viral replication.

Figure 6-29 (***below***) The proof that information for TMV protein is carried in TMV RNA, and not in the protein. The conclusion is based on two experimental results: (1) Only the RNA of viruses reassembled from RNA and protein of different strains determines the specificity of the progeny. (2) The RNA alone is infectious and causes production of the type of virus from which it originated. Both experiments were useful because the infectivity of pure RNA is quite low, whereas the infectivity of the reconstituted whole virus is quite high. [See H. Fraenkel-Conrat, 1958, *The Harvey Lectures, 1957–1958, Ser. 53,* Academic Press, p. 56.]

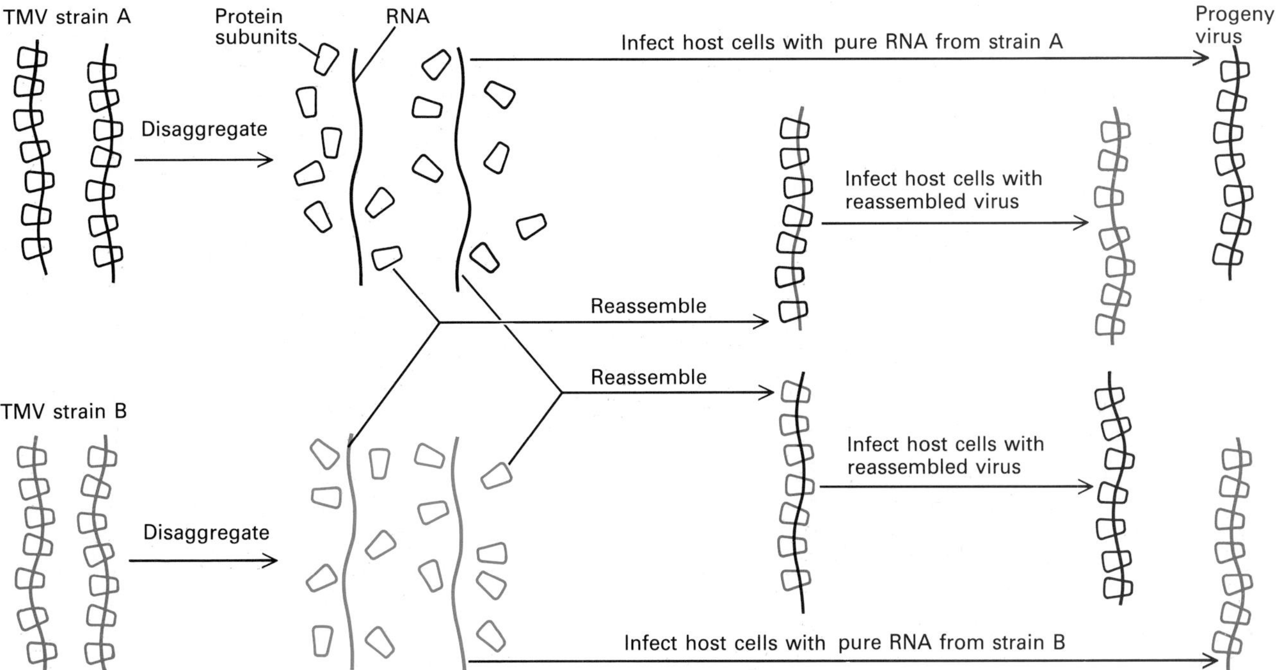
TMV strain A
Protein subunits
RNA
Progeny virus
Infect host cells with pure RNA from strain A
Disaggregate
Infect host cells with reassembled virus
Reassemble
Reassemble
TMV strain B
Infect host cells with reassembled virus
Disaggregate
Infect host cells with pure RNA from strain B

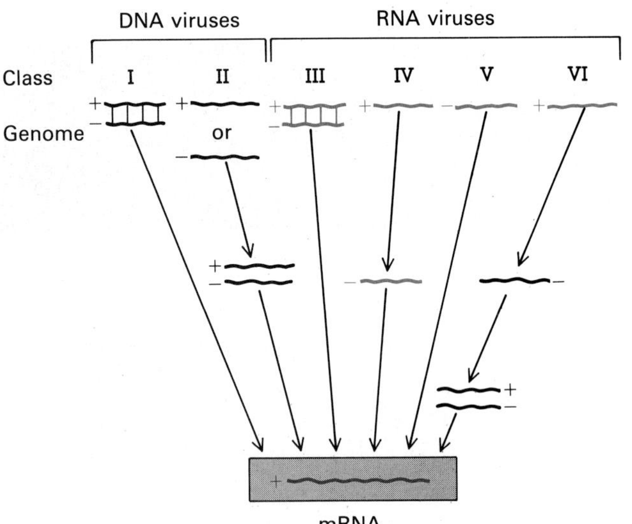

Figure 6-30 Animal viruses can be classified by the composition of their genomes—DNA (black) or RNA (color), single- or double-stranded—and by the pathway of formation of mRNA. If mRNA is designated as the plus strand, six classes of animal viruses can be identified as shown. In classes I and III, the minus strand of the double-stranded genome is copied to make mRNA. In class V, the single-stranded genome is a minus strand that is copied to make mRNA. In class II, the single initial strand of DNA is copied to make a double-stranded DNA template, and the minus strand of this is copied to make mRNA. In class IV, the single plus strand of the virion is copied in the cell into a minus strand, which is then copied to make mRNA. Finally, in class VI (the retroviruses), the single plus strand of RNA is copied into a single DNA strand, which in turn is copied into duplex DNA; the mRNA is then copied from the correct strand of the double-stranded DNA molecule.

The minus RNA strand acts as a template for the synthesis of a plus strand of mRNA. The virions of all class III viruses known to date have segmented genomes containing 8 to 12 double-stranded RNA segments, each of which encodes a single polypeptide. In these viruses, the virion itself contains a complete set of enzymes that can produce mRNA, in the test tube as well as in the cell cytoplasm after infection. A number of important studies have used class III viruses as a source of pure mRNA.

Class IV viruses contain a single plus strand of RNA. Because the genome RNA is identical to the mRNA, the virion RNA by itself is infectious. The mRNA is copied into a minus strand, which then produces more plus strands. Two types of class IV viruses are known. In *class IVa* viruses, typified by *poliomyelitis virus* (Figure 6-31), the RNA molecule in the virion serves as the mRNA to encode all the viral proteins. The individual proteins are first synthesized as a single long polypeptide strand, or *polyprotein,* which is then cleaved to yield the different functional proteins. In these viruses, all the mRNA is the same length as the genome RNA. *Class IVb* viruses, also called *togaviruses* because the virions are surrounded by a lipid envelope (*toga-* is from the Latin for "cover"), synthesize at least two forms of mRNA in the host cell. One of these mRNAs is the same length as the virion RNA, whereas the other corresponds to the third of the virion RNA at the 3′ end. Both mRNAs produce polyproteins. Included in the class IVb group are a large number of rare, insect-borne viruses that cause encephalitis in human beings. These viruses once were called arboviruses (for arthropod-borne viruses).

Class V viruses contain single negative strands of RNA; that is, the virion RNA is complementary in base sequence to the mRNA. Thus, the virion contains a template for making mRNA but does not itself encode proteins. Two subdivisions of class V can be distinguished. The genome of *class Va* viruses is a single molecule of RNA, and a virus-specific polymerase contained in the virion synthesizes several different mRNAs from different parts of this single template strand. Each of the class Va viral mRNAs encodes one protein. *Class Vb* viruses, typified by *influenza virus* (Figure 6-31), have segmented genomes. Each segment is a template for the synthesis of a different single mRNA. As with class Va viruses, the virion contains the virus-specific RNA polymerase necessary to make the mRNA; thus, the minus strands of class V nucleic acids alone—that is, in the absence of the virus-specific polymerases—are not infectious. The influenza virus RNA polymerase initiates the transcription of each mRNA by a unique mechanism. The polymerase begins its mRNA synthesis by stealing 12 to 15 nucleotides from the 5′ end of a cellular mRNA or mRNA precursor in the nucleus. This oligonucleotide acts as a primer for the viral RNA polymerase. The individual mRNAs made by class Vb viruses encode single proteins in most cases, but some of the mRNAs can be read in two different frames to produce two distinct proteins (Chapter 4).

Class VI viruses are also known as *retroviruses* because the RNA of their genome (a single plus strand) directs the formation of a DNA molecule, which ultimately acts as the template for making the mRNA (Figure 6-32). First, the virion RNA is copied into a single strand of DNA, which then forms a complementary strand. This double-stranded DNA is integrated into the chromosomal DNA of the infected cell. Finally, the integrated DNA is transcribed by the cell's own machinery into RNA that either acts as a viral mRNA or becomes enclosed in a virion. This completes the retrovirus cycle. If the retrovirus contains cancer genes, the cell that it infects is transformed into a tumor cell.

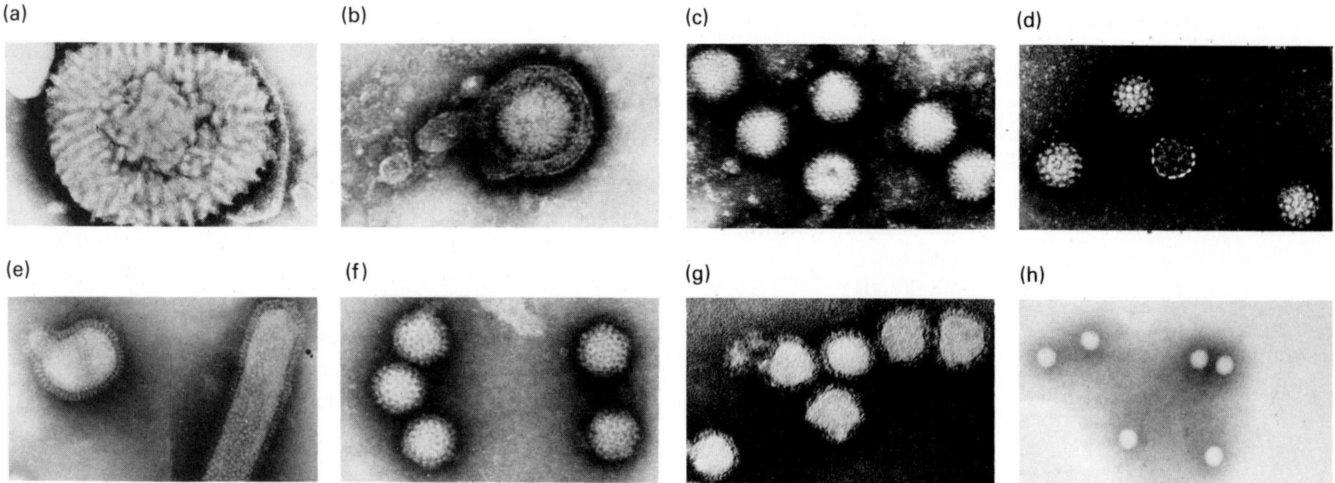
(a)
(b)
(c)
(d)
(e)
(f)
(g)
(h)

Figure 6-31 Electron micrographs of a representative group of animal viruses, all taken at the same magnification and shadowed with phosphotungstic acid. The viral classes are given in parentheses below.

Photos (a) to (d) are DNA viruses: vaccinia (I), herpes (I), adenovirus (I), and human wart virus (I).

Photos (e) to (h) are RNA viruses: influenza (V), reovirus (III), Sindbis virus (IV), and poliomyelitis virus (IV). *Courtesy of P. Choppin; from P. Choppin, 1965,* Viral and Ricketsial Infections of Man, *Lippincott (frontispiece).*

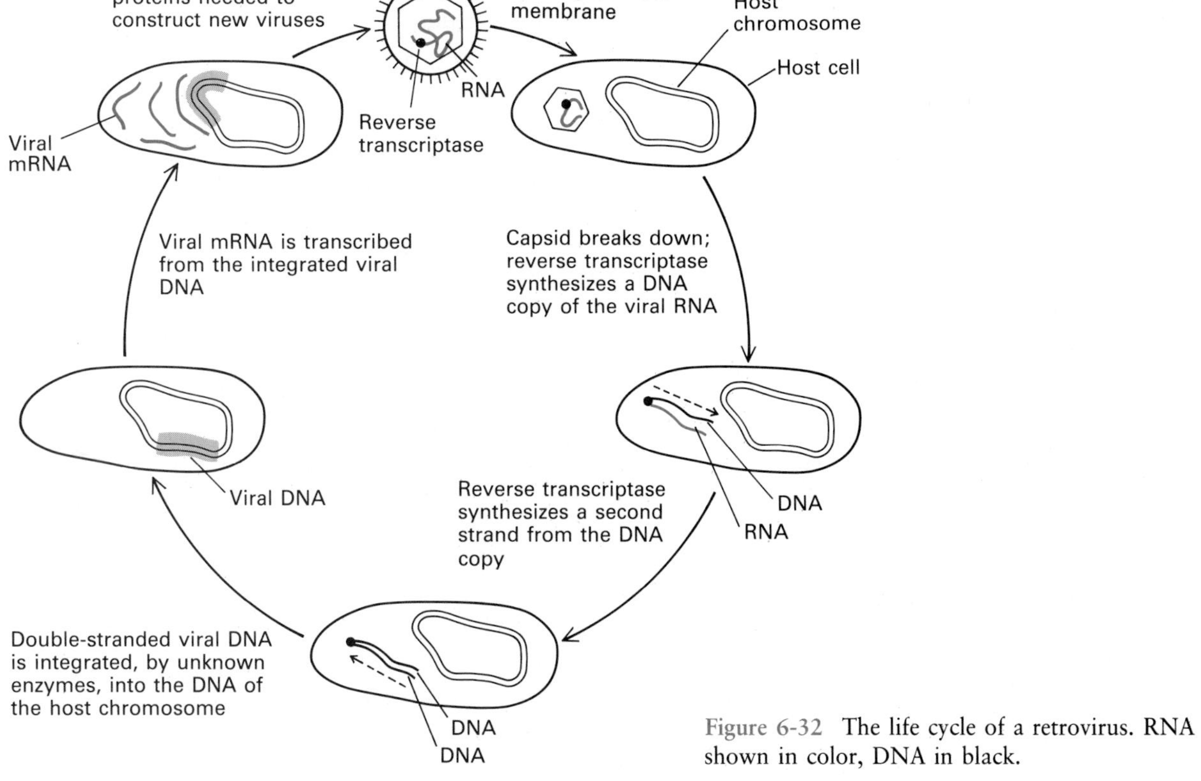
Retrovirus
Capsid enters host cell, leaves envelope on the membrane
Synthesis of viral proteins needed to construct new viruses
Host chromosome
Host cell
RNA
Reverse transcriptase
Viral mRNA
Viral mRNA is transcribed from the integrated viral DNA
Capsid breaks down; reverse transcriptase synthesizes a DNA copy of the viral RNA
Viral DNA
Reverse transcriptase synthesizes a second strand from the DNA copy
DNA
RNA
Double-stranded viral DNA is integrated, by unknown enzymes, into the DNA of the host chromosome
DNA
DNA

Figure 6-32 The life cycle of a retrovirus. RNA is shown in color, DNA in black.

The Use of Viruses in Molecular Cell Biology

Three aspects of virological studies are of particular interest in this book (see Table 6-4):

1. Cellular enzymes synthesize many of the viral molecules, particularly viral mRNAs and virus-specific proteins such as membrane proteins. Thus, viruses can be used to explore the action of the cellular machinery. They have been particularly useful in studying mRNA production in the nuclei of animal cells.
2. Viruses offer a precise means of interrupting host-cell functions, and as a result they can shed light on normal cell processes. For example, the roles of certain factors in protein synthesis have been revealed because viruses interrupt their action.
3. The fact that viral DNA incorporated into host-cell chromosomal DNA is inherited in successive generations of the cell serves as an excellent model for the recombination of foreign DNA with cellular DNA. Such viral genes can also be considered to be representative cellular genes, and studies of their expression can help explain aspects of the expression of normal cell genes (e.g., where transcription begins and ends, and what sequences are required for gene expression). Finally, when certain genes carried by cancer viruses unite with a normal cell, the normal cell becomes a cancer cell. This transformation of the growth potential of cells is one of the most intensively studied processes in all of molecular cell biology, as we shall see in Chapter 23.

Summary

An important factor in the development of molecular cell biology has been the use of a limited number of experimental systems. Different features of organisms—their rapid growth from a single cell, the ease with which their genes can be manipulated, their production of large amounts of a specific substance, and their capacity to carry out developmental changes—govern the choice of which cell or organism is most appropriate in a given study. By far the most popular bacterium is *Escherichia coli;* because genes can be readily transferred in *E. coli* either by conjugation, transduction, or transformation, this organism is the best-understood cell in the world of biology. *Saccharomyces cerevisiae,* a yeast strain, is the most frequently used single-celled eukaryote; yeast genetics also has reached a very advanced state through the use of classical and molecular techniques. Studies with bacteria and yeast established the basic principles of microbiology, including the use of biochemical or immunologic selection of one genetically different cell from among many similar cells. These principles have been adapted to cultured cells from higher organisms.

Many different cultured animal cells are widely used, as are cells from specialized tissues such as bone marrow (blood cells), liver, muscle, and nerves. These tissues are often chosen because of their high content of specific cell types. A primary cell culture is derived from animal cells that will not continue to grow indefinitely, whereas a culture of transformed cells will grow indefinitely. Plant cell cultures are increasing in popularity, especially now that whole plants can be reconstituted from cultured cells.

Among the most important and widely used cell culture techniques are those facilitating gene transfer between cells. This is particularly true now that gene purification has been achieved, as we shall see in Chapter 7. The most common form of gene transfer is the direct uptake of DNA by the recipient cell. In some recipient cells, particularly in large cells like frog eggs but also in certain cultured cells, genes can be transferred by injection. After genes are transferred to growing cells, individual cells that have incorporated the genes can be selected. This procedure is used in the production of hybrid cells that form monoclonal antibodies. In addition to introducing DNA into cultured cells, it is now possible to introduce DNA into germ-line cells. Every cell in an organism produced from such germ cells can have the new DNA.

Viruses play a major role in modern cell biology because they provide an opportunity to study small, reproducible sequences of genes in a single organism, and because viral products are easily purified. An understanding of the basic elements of virology is indispensable to today's biologist. Two kinds of bacterial and animal viruses are in wide use: viruses that grow lytically (i.e., that kill cells) and viruses that are temperate (i.e., that integrate with host-cell genomes).

References

Growth of Microorganisms and Cells in Culture

Bacterial Cells

The following titles cover the logic of selective techniques and genetic transfers, as well as the simple and elegant laboratory procedures used.

BIRGE, E.A. 1983. *Bacterial and Bacteriophage Genetics.* Springer-Verlag.

INGRAHAM, J. L., O. MAALØE, and F. C. NEIDHARDT. 1983. *Growth of the Bacterial Cell.* Sinauer Associates.

MANIATIS, T., E. F. FRITSCH, and J. SAMBROOK. 1982. Propagation and maintenance of bacterial strains and viruses. In *Molecular Cloning* (same authors). Cold Spring Harbor Laboratory.

MILLER, J. H. 1972. *Experiments in Molecular Genetics.* Cold Spring Harbor Laboratory.

STENT, G. S., and R. CALENDAR. 1978. *Molecular Genetics,* 2d ed. W. H. Freeman and Company.

Yeast Cells

The first two entries are papers in two key experimental areas;

the others are up-to-date summaries of techniques and ideas in yeast genetics.

HARTWELL, L. H. 1978. Cell division from a genetic perspective. *J. Cell Biol.* 77:627–637.

STRATHERN, J., J. HICKS, and I. HERSKOWITZ. 1981. Control of cell type in yeast by the mating type locus: the α1-α2 hypothesis. *J. Mol. Biol.* **147**:357–372.

STRATHERN, J., E. JONES, and J. BROACH, eds. 1981. *The Molecular Biology of the Yeast Saccharomyces: Life Cycle and Inheritance.* Cold Spring Harbor Laboratory.

STRUHL, K. 1983. The new yeast genetics. *Nature* **305**:391–397.

Animal Cells: Growth Requirements and Capacities

The following articles explain how to grow animal cells in mass culture and as single cells and describe some specialized functions that can be studied in culture.

BARNES, D. W., D. A. SIRBASKY, and G. H. SATO, eds. 1984. *Cell Culture Methods for Molecular and Cell Biology.* Alan R. Liss.

EAGLE, H. 1955. Nutrition needs of mammalian cells in tissue culture. *Science* **122**:501–504.

EAGLE, H. 1959. Amino acid metabolism in mammalian cell cultures. *Science* **130**:432–437.

FUCHS, F., and H. GREEN. 1981. Regulation of terminal differentiation of cultured human keratinocytes by vitamin A. *Cell* **25**:617–625.

GREEN H., and O. KEHINDE. 1976. Spontaneous heritable changes leading to increased adipose conversion in 3T3 cells. *Cell* **7**:105–113.

HAYFLICK, L., and P. S. MOORHEAD. 1961. The serial cultivation of human diploid cell strains. *Exp. Cell Res.* **25**:585–621.

MARKS, P. A., and R. A. RIFKIND. 1978. Erythroleukemia differentiation. *Annu. Rev. Biochem.* **47**:419–498.

POLLACK, R., ed. 1975. *Readings in Mammalian Cell Culture.* Cold Spring Harbor Laboratory.

PUCK, T. T., and P. I. MARCUS. 1955. A rapid method for viable cell titration and clone production with HeLa cells in tissue culture: the use of x-irradiated cells to supply conditioning factors. *Proc. Nat'l Acad. Sci. USA* **41**:432–437.

SALZMAN, N. P., and E. D. SEBRING. 1959. Utilization of precursors for nucleic acid synthesis by human cell cultures. *Arch. Biochem. Biophys.* **84**:143–150.

SATO, G., ed. 1982. *Hormones and Cell Culture.* Cold Spring Harbor Laboratory.

STICKLAND, S., and V. MAHDAVI. 1978. The induction of differentiation and teratocarcinoma stem cells by retinoic acid. *Cell* **15**:393–403.

SUN, T.-T., and GREEN, H. 1976. Differentiation of the epidermal keratinocyte in cell culture: formation of the cornified envelope. *Cell* 9:511–521.

Cell Fusion: An Important Technique in Somatic-Cell Genetics

These articles describe techniques for fusing cultured cells and for using mutants to select specific cells. Included are explanations of the making of hybridomas for monoclonal antibody production.

DAVIDSON, R. L., and P. S. GERALD. 1976. Improved techniques for the induction of mammalian cell hybridization of polyethylene glycol. *Som. Cell Genet.* **2**:165–176.

D'EUSTACHIO, P., and F. H. RUDDLE. 1983. Somatic cell genetics and gene families. *Science* **220**:919–924.

KOHLER, G., and C. MILSTEIN. 1975. Continuous cultures of fused cells secreting antibody of predefined specificity. *Nature* **256**:495–497.

LITTLEFIELD, J. W. 1964. Selection of hybrids from matings of fibroblasts *in vitro* and their presumed recombinants. *Science* **145**:709–710.

MILSTEIN, C. 1980. Monoclonal antibodies. *Sci. Am.* **243**(4):66–74.

RINGERTZ, N. R., and R. E. SAVAGE. 1976. *Cell Hybrids.* Academic Press.

RUDDLE, F. H. 1982. A new era in mammalian gene mapping: somatic cell genetics and recombinant DNA methodologies. *Nature* **294**:115–119.

SIMINOVITCH, L. 1976. On the nature of heritable variation in cultured somatic cells. *Cell* **7**:1–11.

SORIEUL, S., and B. EPHRUSSI. 1966. Karyological demonstration of hybridization of mammalian cells *in vitro*. *Nature* **190**:653–654.

YELTON, D. E., and M. D. SCHARFF. 1981. Monoclonal antibodies: a powerful new tool in biology and medicine. *Annu. Rev. Biochem.* **50**:657–680.

Mutants in Salvage Pathways

ADAMS, R. L. P., R. H. BURDON, A. M. CAMPBELL, D. P. LEADER, and R. M. S. SMELLIE. 1981. *The Biochemistry of the Nucleic Acids.* London: Chapman & Hall.

CASKEY, C. T., and G. D. KRUH. 1979. The HPRT locus. *Cell* **16**:1–9.

STANBURY, J. B., J. B. WYNGAARDEN, and D. S. FREDERICKSON, eds. 1983. *The Metabolic Basis of Inherited Disease,* 5th ed., part 6: *The Disorders of Purine and Pyrimidine Metabolism.* McGraw-Hill.

Transformation of Eukaryotic Cells

The following titles describe experiments in which DNA has been introduced into cultured cells and into germ-line cells in vivo.

Animal Cells

ANDERSON, W. F., and E. G. DIACUMACOS. 1981. Genetic engineering in mammalian cells. *Sci. Am.* **245**(1):106–121.

BRINSTER, R. L., H. Y. CHEN, M. TRUMBAUER, A. W. SENEAR, R. WARREN, and R. D. PALMITER. 1982. Somatic expression of herpes thymidine kinase in mice following injection of a fusion gene into eggs. *Cell* **27**:223–231.

GRAHAM, F., and A. J. VAN DER EB. 1973. A new technique for the assay of infectivity of human adenovirus 5 DNA. *Virology* 52:456–467.

PALMITER, R. D., G. NORSTEDT, R. E. GELINAS, R. E. HAMMER, and R. L. BRINSTER. 1983. Metallothionein-human GH fusion genes stimulate growth of mice. *Science* **222**:809–814.

PELLICER, A., D. ROBINS, B. WOLD, R. SWEET, J. JACKSON, I. LOWY, J. M. ROBERTS, G. K. SIM, S. SILVERSTEIN, and R. AXEL. 1980. Altering genotype and phenotype by DNA-mediated gene transfer. *Science* **209**:1414–1422.

WIGLER, M., S. SILVERSTEIN, L.-S. LEE, A. PELLICER, Y. CHENG, and R. AXEL. 1977. Transfer of purified herpes virus thymidine kinase gene to cultured mouse cells. *Cell* **11**:223–232.

Plant Cells

CHALEFF, R. S. 1981. *Genetics of Higher Plants: Applications of Cell Culture.* Cambridge University Press.

FRALEY, R. T., S. C. ROGERS, R. B. HORSCH, P. R. SANDERS, J. S. FLICK, S. P. ADAMS, M. L. BITTNER, L. A. BRAND, C. L. FINK, J. S. FRY, G. R. GALLUPPI, S. B. GOLDBERG, N. L. HOFFMAN, and S. C. WOO. 1983. Expression of bacterial genes in plant cells. *Proc. Nat'l Acad. Sci. USA* 80:4803–4807.

Viruses

The following books are useful general references that cover many aspects of virology.

COFFIN, J., N. TEICH, J. TOOZE, H. VARONUS, and R. WEISS, eds. 1982. *Molecular Biology of Tumor Viruses,* 2d ed., part 3: *RNA Tumor Viruses.* Cold Spring Harbor Laboratory.

DENHARDT, D. T., D. DRESSLER, and D. S. RAY, eds. 1978. *The Single-Stranded DNA Phages.* Cold Spring Harbor Laboratory.

FRAENKEL-CONRAT, H., and R. R. WAGNER, eds. 1974–1985. *Comprehensive Virology.* Plenum.

LAUFFER, M. A., and K. MARAMOROSCH., eds. 1953–1985. *Advances in Virus Research,* vols. 1–30. Academic Press.

LURIA, S. E., J. E. DARNELL, D. BALTIMORE, and A. CAMPBELL. 1978. *General Virology,* 3d ed. Wiley.

TOOZE, J., ed. 1980. *Molecular Biology of Tumor Viruses,* 2d ed., part 2: *DNA Tumor Viruses.* Cold Spring Harbor Laboratory.

ZINDER, N. D., ed. 1975. *RNA Phages.* Cold Spring Harbor Laboratory.

7

Tools of Molecular Cell Biology: Molecular Technology

THE greatest advances in molecular cell biology in the recent past have been in the analysis and manipulation of macromolecules, particularly DNA. For years it was clear that many deep biological secrets were locked up in the sequence of DNA. However, obtaining the sequences of long regions of DNA—not to mention altering the sequences at will—seemed a distant dream. An avalanche of technical advances has drastically changed that perspective. First came the discovery of enzymes that cut the DNA from any organism at specific short nucleotide sequences, thus generating a reproducible set of pieces. The availability of these enzymes then greatly facilitated two other important developments, DNA cloning and DNA sequencing.

Through the use of advanced enzymatic and microbiological techniques, pure pieces of any DNA can be inserted into bacteriophage DNA or other carrier DNA to produce *recombinant DNA*. The recombinant molecule can be introduced into bacteria or yeast cells, and cells bearing specific recombinant molecules can be selected. These can then be grown in unlimited quantities. This procedure is referred to as *cloning* a particular DNA sequence; the joining of sequences from two DNA molecules to form a recombinant DNA molecule is accomplished by the use of *recombinant DNA technology*.

In addition to procedures that allow the selection and production of large amounts of pure DNA, a number of other techniques have been developed. For example, *rapid DNA sequencing,* which came into being in the late 1970s, was made possible by advances in certain chemical and enzymatic techniques: fragments of DNA can be labeled with radioactive tracers, and then any fragments containing up to about 500 nucleotides could be separated by gel electrophoresis, a technique that uses an electric field to separate nucleotide chains on the basis of their length. There was no longer any obstacle to obtaining the sequence of a DNA molecule containing 10,000 or more nucleotides. Suddenly, any DNA was accessible to isolation and to sequencing. Techniques were soon developed for modifying and rearranging DNA sequences. These procedures have been coupled with others representing advances in cell biology to test the recombinant DNA molecules for changed biological activity. Finally, selected DNA fragments that encode proteins of particular interest have been transferred to bacteria and to other cells, where the transferred DNA has caused the production of these proteins.

Almost overnight, these techniques of *molecular genetics* have become the dominant approach to the study of many basic biological questions—particularly questions concerning the nature of genes and how they work in eukaryotic cells. The power and the success of the new technology have given birth to many hopes for the practical use of our ever-increasing biological knowledge to benefit human beings.

This chapter outlines the techniques summarized above, as well as some older procedures that are still widely used in molecular experiments.

Radioisotopes: The Universal Modern Means of Following Biological Activity

A major goal of classic biochemistry from the 1930s through the 1950s was to chart the metabolic pathways in cells. This work contributed enormously to our present detailed picture of cellular biochemistry (Chapter 3). Almost all of this seminal biochemistry was performed with chemical tests based on colorimetric or spectrophotometric enzyme assays.

Since World War II, when radioactive materials first became widely available as by-products of work in nuclear physics, chemists and biologists have fashioned radioactive "tracer" molecules of almost limitless variety. Today, labeled precursors of macromolecules have greatly simplified many of the standard biochemical assays and significantly enhanced our ability to follow biochemical events in whole cells. Almost all experimental biology depends on the use of radioactive compounds.

In a labeled molecule, at least one of the atoms is present in a radioactive form, or *radioisotope* (Table 7-1). Labeling does not change the chemical properties of the molecule. For example, an enzyme, whether in a cell extract or in a living cell, does not distinguish between a labeled and an unlabeled molecule when performing a metabolic function (e.g., synthesizing protein, DNA, or RNA). The radioactive atoms can be detected when they emit a particle; in this way, they can be used to trace the activities of the labeled molecules.

Not all labeled materials can be used interchangeably in whole cells and in cell-free systems. One reason for this is that many compounds participating in intermediary metabolism do not enter cells, and therefore cannot be used for labeling cells. For example, labeled ATP may contribute phosphorus 32 (^{32}P) in RNA synthesis in a cell-free system, but ATP does not enter cells. On the other hand, labeled orthophosphate ($^{32}PO_4{}^{3-}$) added to a cell medium will enter cells and will be incorporated into nucleotides and then into RNA in the cells; but $^{32}PO_4{}^{3-}$ is not effective in labeling RNA in a cell extract because it does not get efficiently incorporated into nucleotides in the cell extract.

Amino acids and nucleotides labeled with either carbon 14 (^{14}C) or tritium (^{3}H) are commercially available, as are hundreds of labeled metabolic intermediates. Methionine labeled with sulfur 35 (^{35}S) is widely used as a protein label because of its availability in high specific activities. The *specific activity* is the amount of radioactivity per unit amount of material; for example, commercially available [^{35}S]methionine can have over 10^{11} disintegrations per minute per micromole of methionine. The magnitude of the specific activity depends on the ratio of unstable (potentially radioactive) atoms to stable (nonradioactive) atoms, and on the probability of decay of the unstable atoms as indicated by their half-life.

Because of the availability of phosphate in which every phosphorus atom is ^{32}P, nucleotides labeled with ^{32}P are widely used in various cell-free methods of labeling nucleic acids. Likewise, a radioactive isotope of iodine, ^{125}I, is available in almost pure form, and this tracer atom can be enzymatically or chemically attached to a protein or to a nucleic acid without drastically affecting the macromolecule (see Figure 14-15).

Radioisotopes Are Detected by Autoradiography or by Quantitative Assays

There are two detection schemes in general use for assaying incorporated radioactivity:

1. In autoradiography, a cell or a cell constituent is labeled and then overlaid with a photographic emulsion; development of the emulsion reveals the distribution of labeled material. Autoradiographic studies have been valuable in determining the original sites of synthesis of macromolecules and their subsequent movements within cells. For example, [^{3}H]thymidine

Table 7-1 Commonly used radioisotopes

Radioisotope	Half-life	Energy of emitted particle (MeV)*	Mean path length in water (μm)	Specific activity (mCi/mA)†	Common values for compounds (mCi/mmol)‡
Tritium (hydrogen 3)	12.35 yr	0.0186	0.47	2.92×10^4	10^2–10^5
Carbon 14	5730 yr	0.156	42	62.4	1–10^2
Sulfur 35	87.5 days	0.167	40	1.50×10^6	1–10^5
Phosphorus 33	25.5 days	0.248	—	5.32×10^6	10–10^4
Chlorine 36	3.01×10^5 yr	0.709	—	1.2	10^{-3}–10^{-1}
Phosphorus 32	14.3 days	1.709	2710	9.2×10^6	10–10^5
Iodine 131	8.07 days	0.806	—	1.6×10^7	10^2–10^4
Iodine 125	60 days	0.035	—	2.2×10^6	10^2–10^4

*MeV = 10^6 electronvolts. The maximum energy is given. The particle emitted is a β particle, except in the case of ^{131}I and ^{125}I, which emit γ particles.

†These values are for pure radioisotopes. The unit mCi (millicuries) expresses the number of disintegrations per unit time: 1 mCi = 2.2×10^9 disintegrations per minute. The unit mA (milliatoms) expresses the atomic weight of the element in milligrams.

‡These values are for compounds that are commercially available and that may have many carbon or hydrogen atoms.

SOURCE: New England Nuclear, Boston.

incorporation has identified the nucleus as both the major site of DNA synthesis and the location of most cell DNA (Figure 7-1). Similarly, labeled uridine incorporation into RNA has shown that most RNA is made in the nucleus; however, incorporation of labeled amino acids has revealed that most protein is made in the cytoplasm. The transport pathway of proteins from synthesis to secretion was first documented by electron microscope autoradiography, which makes visible each silver filament resulting from a radioactive disintegration (see Figure 21-28).

2. In the second and more common use of radioisotopes, cells are labeled either in vivo or in vitro and their constituents are isolated and purified in various ways. The amount or type of radioactivity in these constituents is then measured—either by a Geiger counter, which detects ions produced in a gas by the radioactive emissions, or by a scintillation counter. If the latter is used, the sample must first be mixed with a material that will fluoresce after interacting with the particle emitted when a radioactive atom disintegrates. The scintillation counter quantifies the flashes of light.

A combination of labeling and biochemical techniques is often employed. For instance, a cell constituent may be purified before it is chemically or enzymatically labeled for further use. Autoradiography of chemically labeled materials—particularly after separation by paper chromatography or gel electrophoresis (to be discussed later in this chapter)—is very widely used.

The experimental purpose governs the choice of a radioisotope and the method of detection. The specific activity of the labeling material is often a practical restraint. A labeled compound must have a specific activity that is high enough so that when the compound is incorporated into cells there is significant radioactivity in the cell fraction to be studied. For example, ^{3}H-labeled nucleic acid precursors are available in much higher specific activities than ^{14}C-labeled samples; the former will thus allow adequate labeling of RNA or DNA in a shorter time of incorporation or in a smaller cell sample.

For autoradiography, the energy in the particles released by radioactive disintegrations must be considered because this energy affects the experimenter's ability to localize the site of incorporation of the radioactivity. For example, the β particles emitted by ^{32}P are so energetic that the streaks they make on photographic film are about 3 mm long (see Table 7-1)—that is, much longer than the diameters of individual cells. ^{3}H is much preferred for locating radioactive substances or structures in cells: the track created on photographic film by the particle released in ^{3}H atom decay is only about 0.47 μm long. Thus, ^{3}H-labeled structures can be located within cells to an accuracy of about 0.5 to 1.0 μm, or about one-fifth the diameter of the cell nucleus of a mammalian cell.

Pulse-Chase Experiments Must Be Designed with Knowledge of the Cell's Pool of Amino Acids and Nucleotides

In many experiments using radioactive material, a labeled compound is added to cells and then the path of the label is traced as it moves through various cell compartments

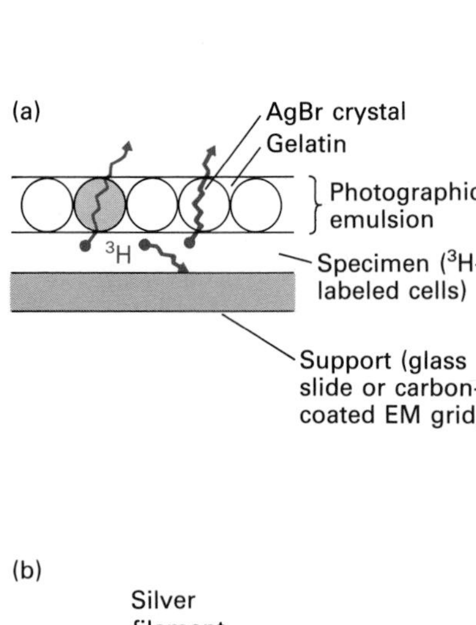

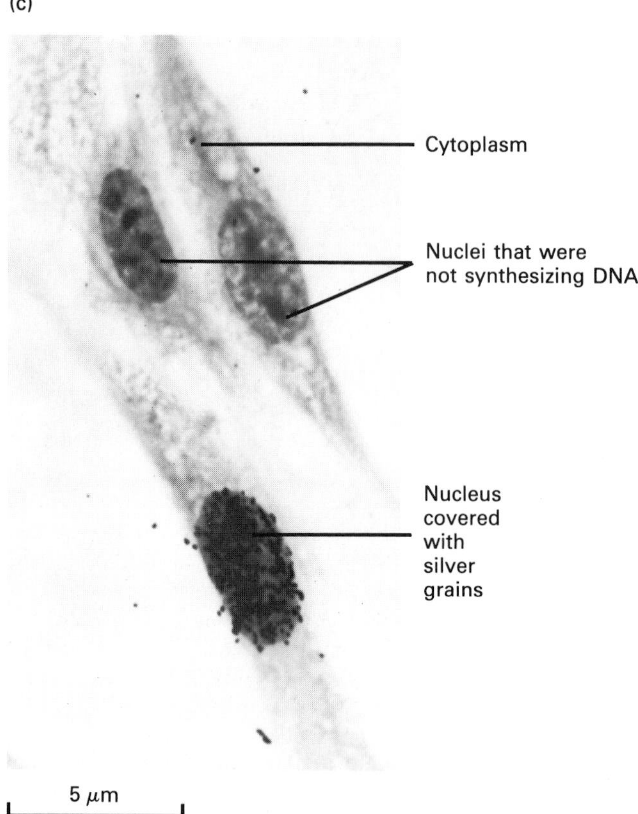

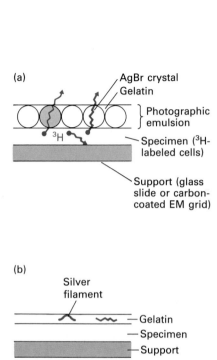

Figure 7-1 The technique of autoradiography (shown in cross section) and an actual autoradiograph. (a) A radiation-sensitive photographic emulsion containing silver salts (AgBr) is placed over labeled cells attached either to a glass slide (for viewing in the light microscope) or to a carbon-coated grid (for viewing in the electron microscope). When the regions of the cell containing the labeled molecules emit radioactive particles, the particles sensitize the salts of the emulsion as they pass through it. (b) The photographic emulsion is developed and the unexposed salts are dissolved and washed away. The exposed portions form silver filaments that appear as dark grains under the light microscope and as curly filaments in the electron microscope. The silver deposits show where the labeled molecules were in the cell. (c) These fibroblasts from Chinese hamsters were labeled with [^{3}H]thymidine for 1 h. Two of the cells were not synthesizing DNA during this time, but one cell was. The grains are almost entirely over the cell nucleus, indicating that the DNA is in the nucleus. *Parts (a) and (b) redrawn from E. D. P. DeRobertis and E. M. F. DeRobertis, 1979,* Cell and Molecular Biology, *Saunders, p. 62; part (c) courtesy of D. M. Prescott.*

or various molecules within the cells. One important type of experiment, called a *pulse-chase* experiment, utilizes the brief addition (a *pulse*) of a labeled compound, followed by its removal and its replacement (the *chase*) by an excess of unlabeled compound; the cells or cell constituents are examined at various times thereafter to see what has happened to the radioactivity that was incorporated during the pulse (Figure 7-2).

Before an amino acid, a nucleoside, or a phosphate ion (for example) is incorporated into a protein or a nucleic acid, it first enters the cell's *pool* of molecular building blocks. This is a collection of small molecules that are probably free to diffuse throughout the cytoplasm and nucleus of the cell, but that are not necessarily free to diffuse freely into or out of membrane-bound organelles such as mitochondria or chloroplasts. The quantities of different components in the pool vary greatly, as do the rates at which different molecules are absorbed, utilized, and secreted by the cell (Figure 7-3).

Because of the rapid exchange of amino acids between the pool and the medium, there is a clear pulse-chase effect in experiments with radioactive amino acids. That is, the acid-soluble pool can be made to contain radioactive amino acids in a few seconds, and then they can be removed just as quickly (Figure 7-4a).

However, as shown in Figure 7-3, ribo- and deoxyribonucleoside precursors to nucleic acids become phosphorylated soon after they enter a cell pool, and phospho-

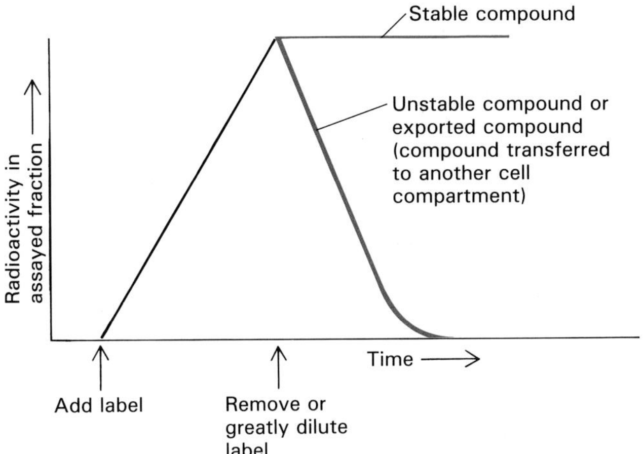

Figure 7-2 The incorporation of radioactivity into a cell sample (whole cells, cell compartments, or a purified compound from cells) during a pulse-chase experiment. The accumulation of label in a sample during the pulse (the brief addition of a radioactive compound) is followed by one of two possibilities during the chase (the replacement of the labeled form of the compound by an excess of unlabeled compound). The sample may retain the radioactivity it gained during the pulse, indicating the labeling of a stable cell constituent; or the sample may lose radioactivity, showing the labeling of a metabolically unstable compound or one that is transferred to another cell fraction.

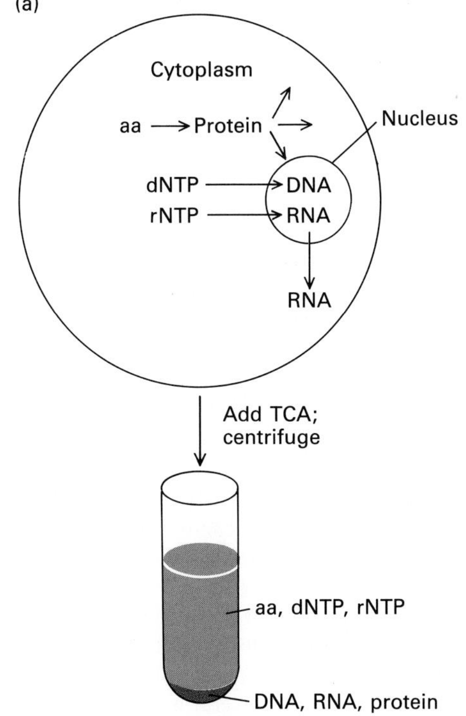

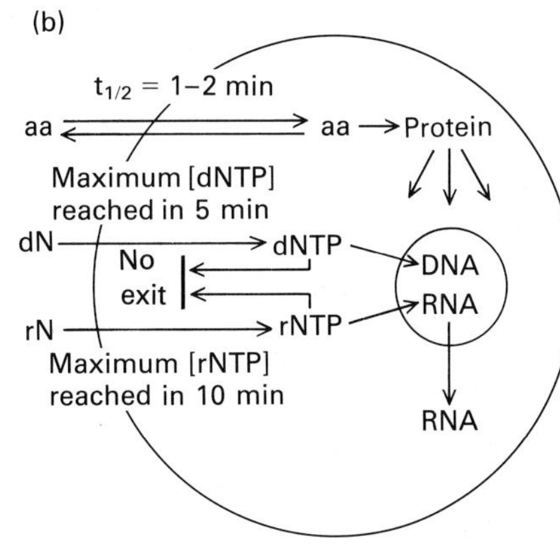

Figure 7-3 (a) A sample of a cell's pool of small soluble molecules [amino acids (aa) and nucleotides (dNTP and rNTP)] and its collection of macromolecules (DNA, RNA, and proteins) may be obtained by adding cold acid, usually trichloroacetic acid (TCA), to the cell population. The acid destroys the cell structure and precipitates all macromolecules. Centrifugation then deposits the macromolecules in a pellet and leaves the amino acids and nucleotides in the supernatant. By taking samples at frequent intervals after labeled amino acids or nucleotide precursors have been added to the medium of a cell culture, the rate at which cells take up the labeled molecules and incorporate them into macromolecules can be determined. (b) Amino acids both enter and leave the cell pool through the plasma membrane very quickly; they approach equilibrium exponentially with a half-time ($t_{1/2}$) of 1 to 2 min. Equilibrium between the labeled amino acids in the cells and those in the medium is therefore reached within about 5 min of the onset of the pulse, even though the concentration of amino acids may be 5 to 20 times higher inside the cells than in the medium. Experiments with labeled deoxyribonucleosides (dN) and ribonucleosides (rN) have revealed that they, too, enter the cells rapidly, but that these compounds are then phosphorylated into deoxyribonucleotides (dNTP) and ribonucleotides (rNTP), which do not leave the cells. Thus labeled precursors to DNA and RNA increase in the cell pool until their maximum concentration is reached; this occurs within about 10 min.

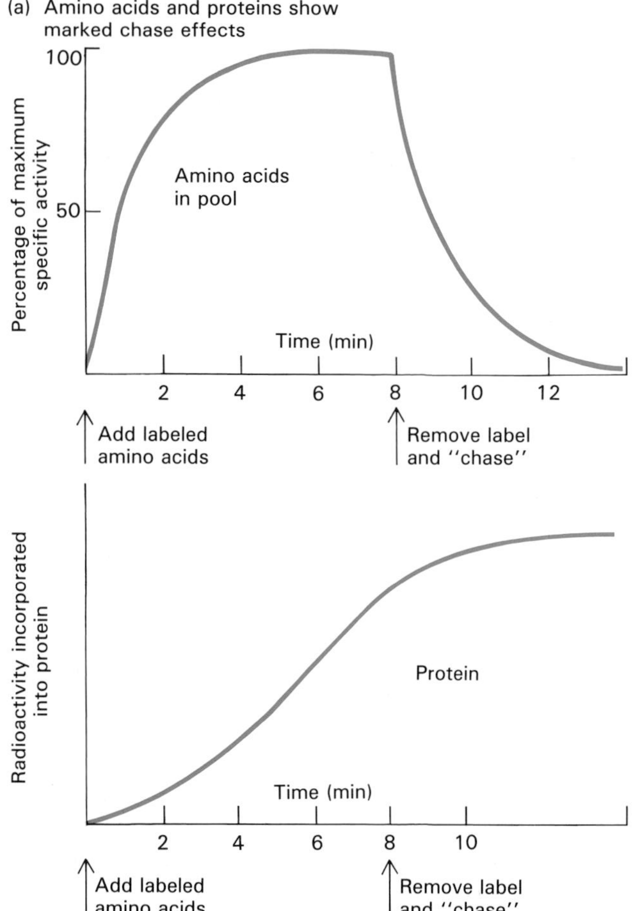

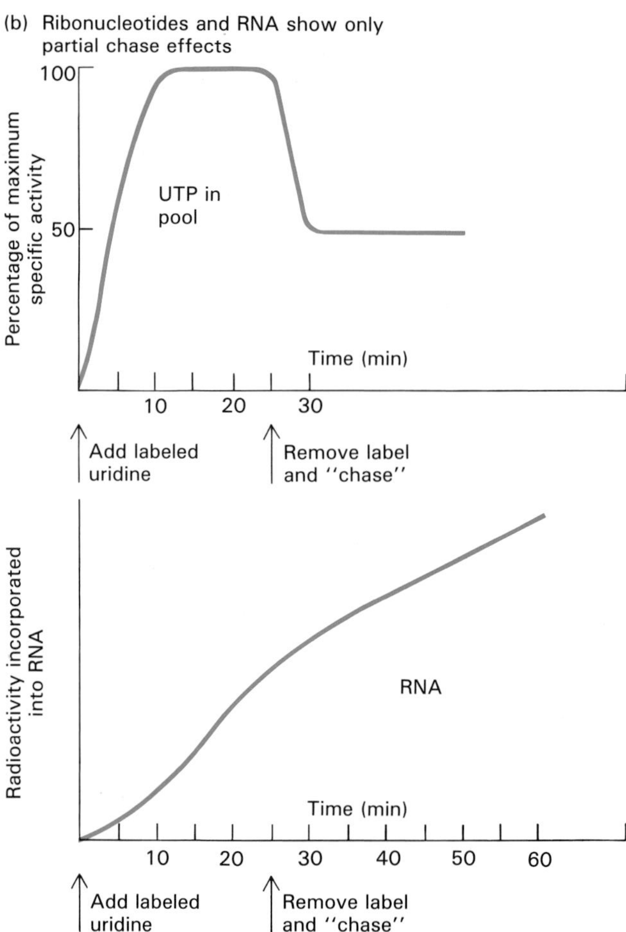

Figure 7-4 (a) If growing cells are exposed to a pulse of labeled amino acids, it takes about 5 min for the amino acids in the cell pool to be completely labeled. The accumulation of radioactivity in proteins starts more slowly, because the label must be incorporated into the amino acid pool first. However, if at 8 min the medium containing labeled amino acids is replaced by a medium containing unlabeled ones, the incorporation of radioactivity into proteins stops within a few minutes, due to the rapid equilibration of amino acids inside the cells with those in the medium. Thus a marked pulse-chase effect is seen with labeled amino acids. (b) Although a pulse of labeled uridine is incorporated into UTP in the cell pool in about 10 min, it takes much longer for a chase with unlabeled uridine to cause a leveling off in the amount of radioactivity incorporated into RNA (i.e., to reduce the ongoing incorporation of radioactivity to zero). This is because the labeled uridine in the pool does not equilibrate with the unlabeled uridine in the medium; all of the labeled uridine must be incorporated into RNA. Because the ribonucleotide content of the cell pool turns over only once in several hours, it takes that long for the labeled uridine to be fully utilized by the cell. Thus no marked chase effect is seen.

rylated compounds do not generally exchange with compounds outside the cell. Thus, no equilibrium is established between the nucleic acid precursors in the medium and the phosphorylated precursors in the cell. Nevertheless, a pulse-chase effect can be obtained in experiments with radioactive *deoxyribonucleosides,* because the deoxyribonucleotide content of the cell pool is sufficient for only a few minutes of DNA synthesis. Labeled thymidine, for example, can be readily chased because it is quickly phosphorylated and taken up by replicating DNA.

On the other hand, there is no immediate and drastic pulse-chase pattern in experiments with labeled *ribonucleosides.* It takes several hours for the ribonucleotide content of animal cell pools to be completely utilized and regenerated. In most cultured cells, the cell pool absorbs a pulse of labeled ribonucleosides within 10 min; however, a marked chase response (i.e., one that occurs within a few minutes) is not possible with RNA precursors. Although the addition of unlabeled ribonucleosides outside the cell may further expand the ribonucleotide content of the pool, dilute the label within it, and decrease the rate

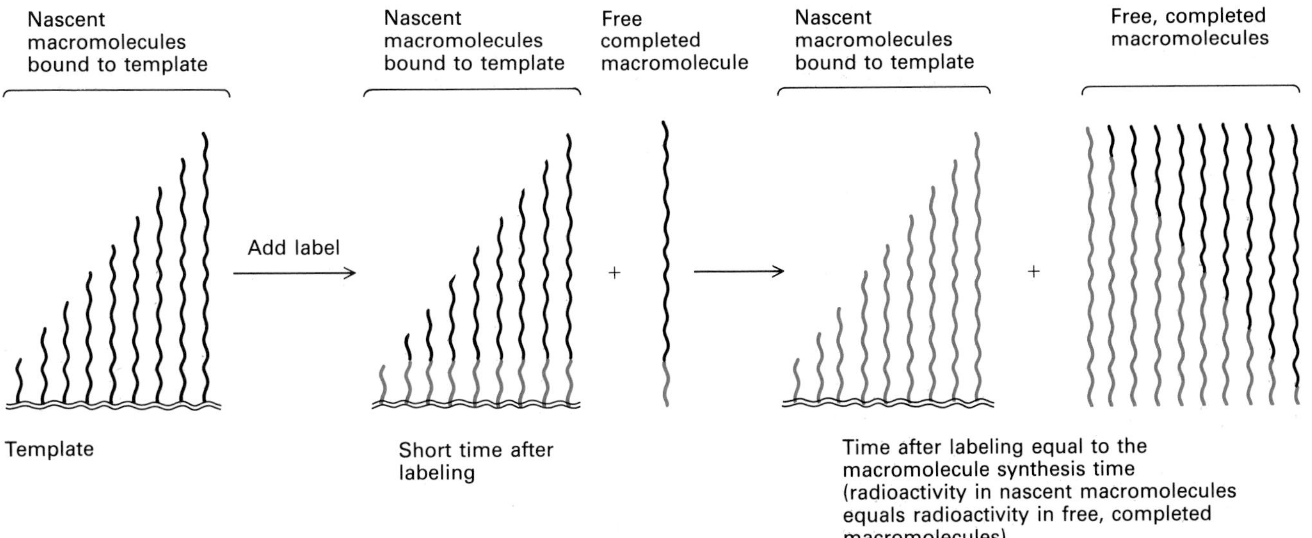

Figure 7-5 The first appearance of labeled radioactive precursors in macromolecules is in nascent macromolecules. As time passes, molecules that contain an increasing amount of radioactivity are completed. At the end of an interval that is equivalent to the synthesis time of the macromolecules, the amounts of radioactivity found in finished and in unfinished molecules are equal.

of labeling of RNA, there is no clear leveling off in the amount of label incorporated for several hours after the chase begins (Figure 7-4b).

In planning and interpreting experiments that use labeled precursors of proteins, DNA, or RNA to study macromolecular synthesis, these characteristics of small molecules in the soluble pool must always be borne in mind.

Labeled Precursors Can Trace the Assembly of Macromolecules and Their Cellular Distribution

The macromolecules that become labeled when a radioactive building block first enters a cell are those that are in the process of being constructed. For example, if a radioactive amino acid is placed in a culture, the *nascent* protein chains (those that are unfinished and still growing) are the first proteins to become labeled. As time passes after the label has been introduced, an increasing number of completed chains contain the radioactive label.

It is possible to estimate the time required to form a specific macromolecule, by sampling a labeled cell culture at very short intervals to assay the relative amounts of radioactivity in the nascent macromolecules (which are still bound to the templates) and in the free completed ones. The first finished chains obtained after the label is added contain only a small amount of the label, because they were almost completed before the label was introduced. Each nascent chain also contains a small amount of label. As time passes, however, finished chains contain more and more label, and, of course, the nascent chains do too. Finally, a point is reached at which some finished chains and nascent chains are labeled along their entire lengths; at this point there is an equal amount of label in the finished and the nascent chains (Figure 7-5). The time elapsed up to this point is equal to the time required for the synthesis of one chain.

The Dintzis Experiment: Proteins Are Synthesized from the Amino End to the Carboxyl End Other very important facts—the origin of synthesis of a macromolecule and the direction of its growth—can be determined by labeling growing chains. Indeed, the analysis of newly finished chains was used by Howard Dintzis in the classic experiment that demonstrated the progressive step-by-step formation of protein chains from the amino terminus to the carboxyl terminus. In his experiment, Dintzis observed the synthesis of globin chains by reticulocytes, which represent the next-to-final stage in the differentiation of red blood cells in the bone marrow of mammals. Over 90 percent of the protein made by reticulocytes consists of the α and β globin chains that form hemoglobin. (Hemoglobin is composed of four globin chains, two α and two β; Chapter 3.) Dintzis exposed reticulocytes to radioactive amino acids, then collected the newly finished labeled chains at short intervals. He separated the α and β chains and digested each with trypsin, an enzyme that attacks at arginine and lysine residues to produce a specific set of fragments for each chain. He knew the sequence of amino acids in both globin chains as well as the sequence in each tryptic fragment, so he knew the position of each fragment within its chain.

Dintzis reasoned that the first completed globin chains to contain the radioactive label would be those that were almost complete when the label was added. Thus, the first portion of the globin chain to become labeled would be the last to be synthesized. By extension, the last portion of the globin chain to be labeled would be the first to be synthesized. His results were consistent with these expectations: the radioactive label always appeared in the fragments in a certain order—in the carboxyl-terminal fragment first and the amino-terminal fragment last, with intermediate fragments becoming labeled consecutively in the order in which they lay between the two termini (Figure 7-6). From this, Dintzis deduced that the amino terminus of each chain is finished first and the carboxyl terminus last, and that synthesis progresses step by step from one end of the chain to the other.

Whereas Dintzis studied the labeling of newly finished molecules, other workers have studied *nascent* molecules. Experiments on the labeling of nascent RNA and DNA will be described in Chapters 9 and 13. The logic for these studies is, however, parallel to that of the Dintzis experiment: the sequences of the shortest labeled molecules in a nascent set will be those near the start site; longer and longer members will contain sequences progressively

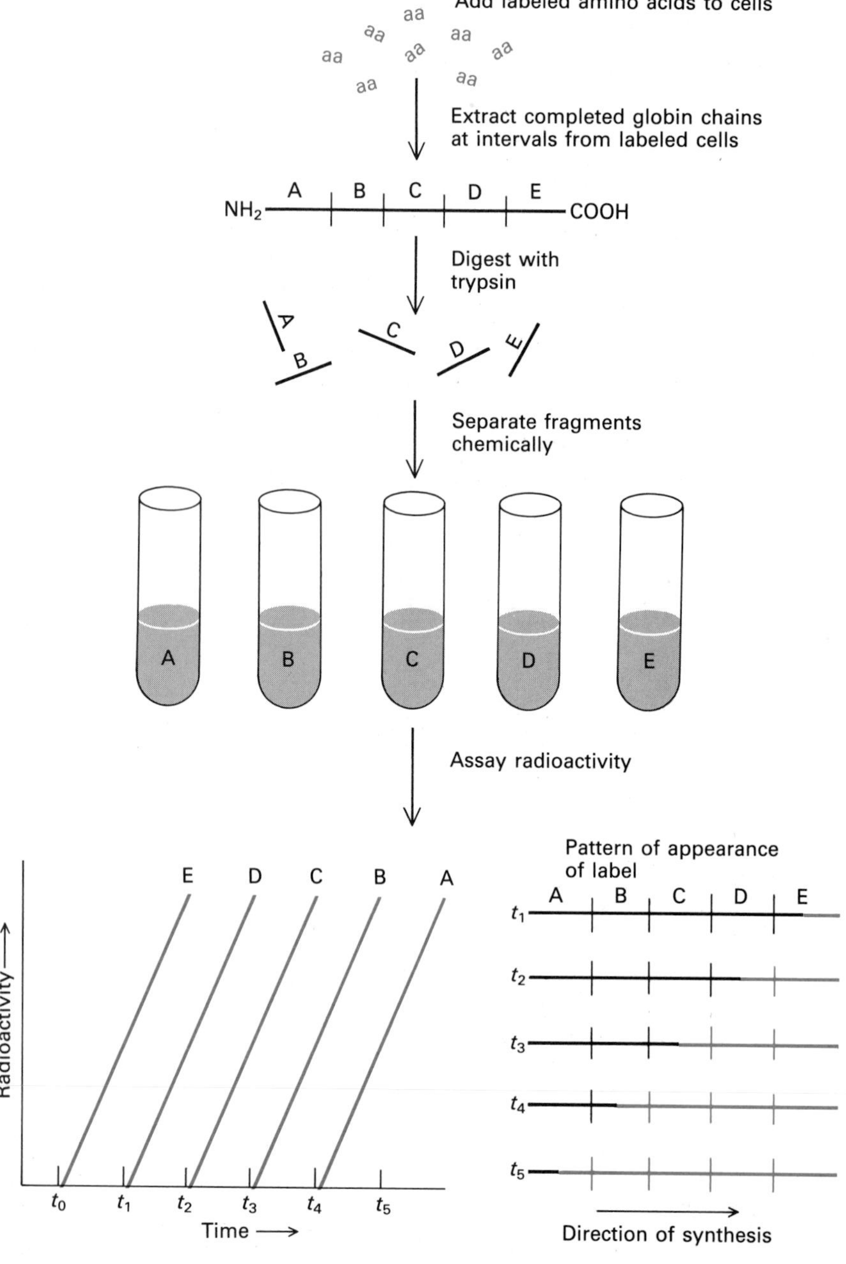

Figure 7-6 A diagram of Howard Dintzis' classic experiment showing the direction of growth of polypeptides. Soon after radioactively labeled amino acids were added to a suspension of reticulocytes, finished labeled globin chains (the protein part of hemoglobin) were released from ribosomes. Samples of released chains were taken at frequent intervals after labeling (t_1 to t_5) and cleaved into fragments by digesting the protein with the protease trypsin; radioactivity in the fragments was assayed. In the first sample of finished chains (t_1), all radioactivity was located in the E fragments—those at the carboxyl end. In the other samples (t_2 to t_5), fragments became labeled in the order D, C, B, and finally A. These results indicated that the E fragment contains the amino acids added last and that the A fragment contains those added first. Thus, Dintzis concluded that synthesis of protein begins at the amino terminus and progresses to the carboxyl terminus. [See H. Dintzis, 1961, *Proc. Nat'l Acad. Sci. USA* **47**:247.]

more remote from the start site; and full-length molecules will complete the set.

Determining Sizes of Nucleic Acids and Proteins

Whereas it is the *sequence* of the subunits in a protein or in a nucleic acid that ultimately determines the functional capacity of the polymer, the most useful physical characteristic of a polymer for many experimental purposes is its unique *length*. This is because it is relatively easy to separate molecules by length, but more difficult (in the case of proteins) or virtually impossible (in the case of nucleic acids) to separate molecules by using chemical differences based on sequence differences. Therefore, the size (or length) of a protein, RNA, or DNA molecule is one of the most frequent measurements in molecular cell biology. In the next few sections we briefly outline the principles of molecular separation according to size and illustrate their use. The newer techniques are so simple and effective that they are frequently not appreciated as "physics in action." The student is encouraged to learn the physical chemistry underlying these crucial techniques. (See, e.g., C. R. Cantor and P. R. Schimmel, 1980, *Biophysical Chemistry,* part II: *Techniques for the Study of Biological Structure and Function,* W. H. Freeman and Company.)

Centrifugation Is Used to Separate Particles and Molecules Differing in Mass and/or Density

Two basic uses of centrifugation recur in the experiments described in this book: (1) separation of particles according to their mass, size, or length, and (2) separation according to their density. We shall discuss both uses in turn.

Rate-Zonal Centrifugation When particles or molecules are layered on top of a column of liquid in a tube and subjected to a centrifugal force, they migrate down the tube at a rate that is controlled by the mass of the particles, the difference between the density of the particles and the density of the suspending medium, and the friction between the particles and the suspending medium (this friction depends on the shape of the particles). For example, a mixture of RNA molecules whose average shape and density are similar would separate in a centrifugal field almost solely according to mass, or chain length. Because different-sized molecules would be found in different zones of a centrifuge tube, this separation technique is commonly called *rate-zonal centrifugation* or more simply *zonal centrifugation* (Figure 7-7a). Samples are centrifuged just long enough to separate the molecules of interest. If they are centrifuged for too short a time, the molecules will not separate sufficiently; if they are centrifuged much longer than necessary, all of the molecules will end up in a pellet at the bottom of the tube.

The column of liquid through which particles are sedimented is often stabilized by a sucrose gradient that is more concentrated (and thus denser) at the bottom than at the top of the centrifuge tube. For this reason the technique is sometimes called "sucrose density-gradient centrifugation," but this terminology is erroneous because the property that is mainly responsible for the separation of particles by zonal centrifugation is *not* density, but mass.

Although the rate of sedimentation is strongly influenced by the mass of the particles, this technique is seldom effective in determining *exact* molecular weights, because even slight variations in shape also affect the rate of sedimentation. These variations in shape are hard to assess exactly, especially for proteins and single-stranded nucleic acid molecules. Nevertheless, zonal centrifugation is very widely used for separating all types of polymers and particles. The ultracentrifuge rotor is held in a high vacuum to reduce friction and control temperature. While an analytical ultracentrifuge (an ultracentrifuge equipped with optical instruments) is in motion, the rate of sedimentation can be measured by photographing moving boundaries of sedimenting layers of molecules. Modern ultracentrifuges reach speeds of 60,000 revolutions per minute (r/min) or greater and generate forces sufficient to sediment particles with masses greater than 10,000 daltons. For a particle located 6 cm from the rotational axis of a centrifuge, 60,000 r/min corresponds to a centrifugal force of 250,000 times gravity (250,000*g*). However, even at such tremendous forces, particles that are quite small (5000 daltons or less) diffuse too freely to be moved uniformly through a centrifugal field.

Equilibrium Density-Gradient Centrifugation In this technique, a density gradient can be established throughout the suspending medium before centrifugation; alternatively, the force of centrifugation can be used to establish a density gradient. The latter case is illustrated in Figure 7-7b. In both cases, the density of the medium should range from less dense than the particles to be separated, to more dense than these particles. During centrifugation, the particles or molecules in the tube float to the level at which the density of the medium is equal to their own density—that is, to the level at which they are *isopycnic* with (equally as dense as) the medium. Even under tremendous centrifugal force, a particle will not sediment through a gradient region that is denser than itself.

The material most commonly used to form density gra-

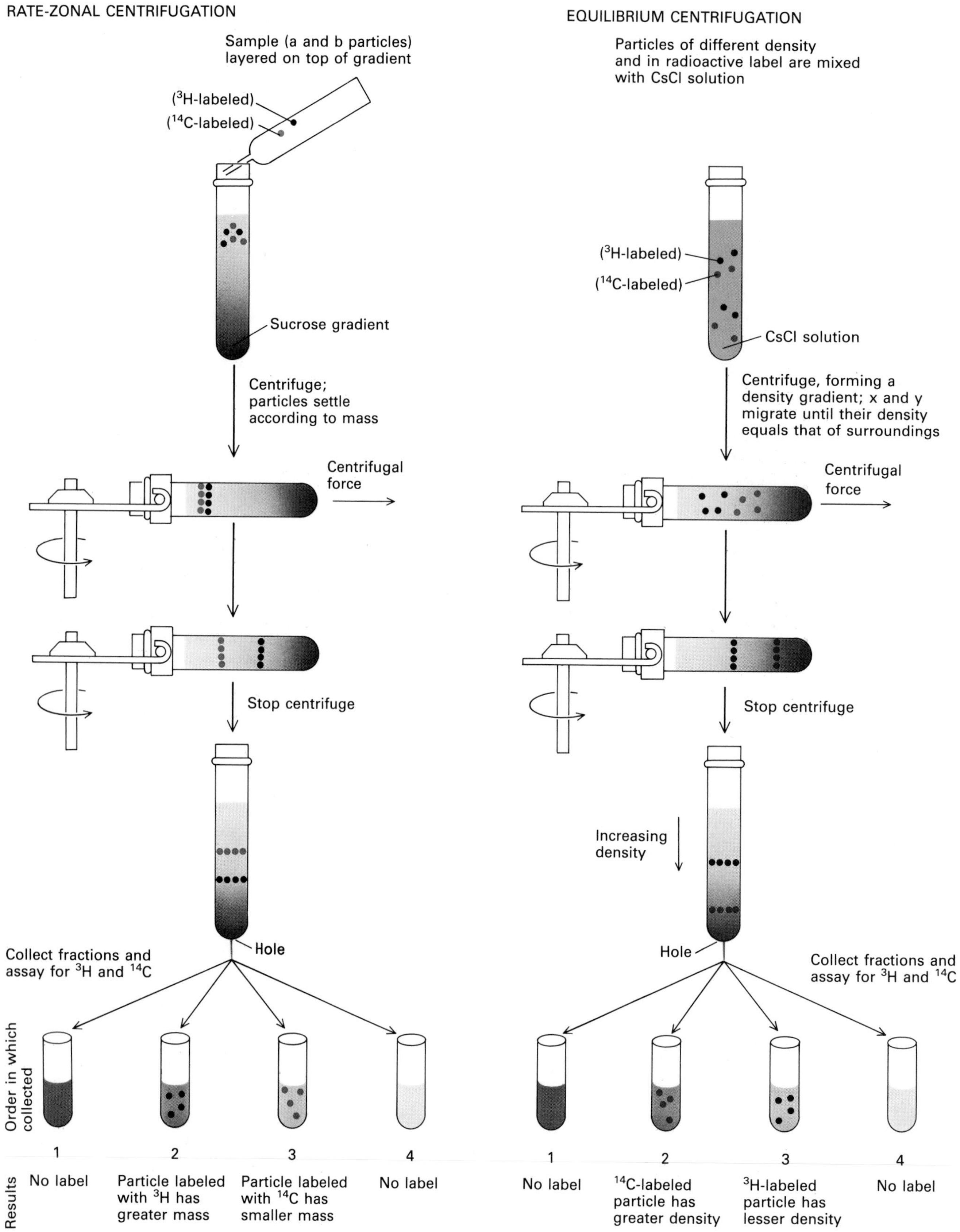

RATE-ZONAL CENTRIFUGATION
Sample (a and b particles) layered on top of gradient
(^{3}H-labeled)
(^{14}C-labeled)
Sucrose gradient
Centrifuge; particles settle according to mass
Centrifugal force
Stop centrifuge
Hole
Collect fractions and assay for ^{3}H and ^{14}C
Order in which collected
1
2
3
4
Results
No label
Particle labeled with ^{3}H has greater mass
Particle labeled with ^{14}C has smaller mass
No label
EQUILIBRIUM CENTRIFUGATION
Particles of different density and in radioactive label are mixed with CsCl solution
(^{3}H-labeled)
(^{14}C-labeled)
CsCl solution
Centrifuge, forming a density gradient; x and y migrate until their density equals that of surroundings
Centrifugal force
Stop centrifuge
Increasing density
Hole
Collect fractions and assay for ^{3}H and ^{14}C
1
2
3
4
No label
^{14}C-labeled particle has greater density
^{3}H-labeled particle has lesser density
No label

Figure 7-7 *(opposite)* Two centrifugation techniques are widely used for separating different subcellular particles or different types of molecules.

Rate-zonal centrifugation is frequently used to separate particles or molecules that differ in mass but that may be similar in shape and density. In the diagram, particles differing in mass have been given distinct radioactive labels and then subjected to centrifugation. When they have separated sufficiently, the centrifuge is stopped and samples are collected from a hole punctured in the bottom of the tube. The different samples are assayed in order to identify them. *Left.* The particles labeled with ^3H have the greater mass and sediment faster than those labeled with ^{14}C. (As an alternative to radioactive labeling, the samples may be assayed on the basis of their differing enzymatic activity.) *Right.* Equilibrium centrifugation allows the separation of particles that differ in density (they may or may not also differ in mass and shape). In this technique, labeled particles move either up or down through a density gradient established by the centrifugal force acting on a dissolved salt such as CsCl. At equilibrium, the particles in the tube collect at levels at which the density of the solution equals their own density. After centrifugation is completed, samples are collected and assayed as in the zonal technique. In this example the particles labeled with ^{14}C have the greater density.

dients for equilibrium centrifugation is probably a water solution of cesium chloride (CsCl). The cesium ion (Cs^+) is so compact that it sediments in the powerful centrifugal fields created in modern ultracentrifuges. A gradient is thereby established, with more Cs^+ (and more Cl^-, which follows the Cs^+ to neutralize the charge) at the bottom of the tube than at the top. In a typical run of an ultracentrifuge, the density gradient formed will be about 0.02 g/ml heavier at the bottom than at the top. Thus, molecules that differ in density by even a fraction of 0.02 g/ml can easily be separated by this technique. The densities of protein, DNA, and RNA in a solution of CsCl are approximately 1.3, 1.6 to 1.7, and 1.75 to 1.8 g/ml, respectively, so these molecules are easily separated from one another. The densities given here are higher than those of the same macromolecules in cells, because ions in a CsCl solution bind to proteins and nucleic acids. The densities of all macromolecules without bound ions are almost equal: 1.25 to 1.3 g/ml. In a CsCl solution, the Cs^+ binds to DNA mainly at phosphate groups; it binds to RNA both at phosphates and at the hydroxyl groups of riboses, thus increasing the density of RNA more than that of DNA.

Proteins or nucleic acids that have ^{13}C or ^{15}N substituted in place of ^{12}C or ^{14}N in amino acids or nucleotides also can be separated from their normal counterparts. For example, since proteins are 14 percent nitrogen and since ^{15}N is 15/14 times more dense than ^{14}N (Table 7-2), a protein substituted completely with ^{15}N is about 1 percent denser than the normal protein; this is a sufficient change in density to allow the normal protein to be completely separated from the substituted one. Thus when cells are grown in a medium containing heavy amino acids or nucleotide precursors, it is possible to physically separate molecules made by the cells before and after the addition of the heavy isotope. (See Figure 13-1, for an illustration of the separation of "heavy" and "light" DNA.)

The Sedimentation Constant When a particle suspended in a medium is subjected to centrifugal force, it will move if its density d is greater than d_0, the density of the surrounding medium. The speed of movement in a stationary medium is proportional to the gravitational acceleration g; in a centrifugal field, g is replaced by the centrifugal acceleration c, which is equal to $(2\pi\omega)^2x$, where ω (Greek omega) is the number of revolutions per unit time and x is the distance of the particle from the axis of rotation.

As the particle moves, it will encounter friction with the medium. As it accelerates, its velocity v increases, which causes an increase in friction. The frictional force ϕ (Greek phi) is given by $\phi = fv$, where f is a frictional coefficient related to the shape of the particle. The value of f for a spherical particle is $6\pi\eta r$, where η (Greek eta) is the viscosity of the medium and r is the radius of the particle. The velocity increases until the frictional force balances the centrifugal force P_c, after which time the particle continues to move at a uniform velocity. Then the following equations hold:

$$\tfrac{4}{3}\pi r^3(d - d_0)c = 6\pi\eta r v$$

$$v = 2cr^2\left(\frac{d - d_0}{9\eta}\right)$$

and

$$r = \sqrt{\frac{9v\eta}{2c(d - d_0)]}}$$

Table 7-2 Commonly used heavy isotopes and their natural (more abundant) counterparts*

Heavy isotope	Atomic mass	Natural isotope	Atomic mass
Deuterium (hydrogen 2)	2.01	Hydrogen 1	1.01
Carbon 13	13.01	Carbon 12	12.00
Nitrogen 15	15.00	Nitrogen 14	14.01
Oxygen 18	18.00	Oxygen 16	16.00

*The greater density of heavy isotopes is brought about by the presence of one or more additional neutrons in their nuclei. The extra neutrons do not affect the chemical bonding properties of the atoms, but do affect their mass.

where, again, d and d_0 are the densities of the particle and the surrounding medium, respectively. These are *Stokes' equations,* which describe the motion of spherical particles in a fluid under ideal conditions (i.e., there is no interaction among the particles; the particles are large compared with the solvent molecules; and there is no disturbance due to convection).

The value of $v/c = s$, the sedimentation constant, is characteristic for a given particle in a given medium at a given temperature. If r and x are expressed in centimeters, g or c in centimeters per second squared, ω in revolutions per second, d in grams per cubic centimeter, and η in grams per centimeter per second, then the sedimentation constant s is expressed in seconds. The value of s is also often expressed in *svedbergs,* S ($1S = 10^{-13}$ seconds), which apply under standard conditions of sedimentation in water at 20°C. (Standard conditions are necessary so that the friction of the sedimenting particles is standardized.) From the first of the preceding equations, s can be calculated as follows:

$$s = \frac{v}{c} = \frac{\frac{4}{3}\pi r^3(d - d_0)}{6\pi\eta r} = \frac{m[1 - (d_0/d)]}{f}$$

where m is the mass (in grams) of the spherical particle. Because the centrifugal force, the density of the medium, and its viscosity can be measured, the preceding equations together serve to estimate the radius and the mass of the spherical particle, provided that we can measure its density and its sedimentation velocity in a centrifuge. Some representative values of s are given in Table 7-3.

Table 7-3 Sedimentation constants and molecular weights for some standard molecules

Molecule or particle	Sedimentation constant (S)*	Molecular weight $\times 10^{-3}$
PROTEINS		
Cytochrome *c*	1.7	13.4
Myoglobin	2.0	16.9
Hemoglobin ($\alpha_2\beta_2$)	4.5	64.5
Fibrinogen	7.6	340
RNA		
Transfer RNA (average)	4.0	25–27
Ribosomal RNA:		
E. coli, small	16	550
E. coli, large	23	1100
Human, small	18	660
Human, large	28	1700
PARTICLES		
Ribosome (human)	80	—
Poliomyelitis virus	150	—
Bacterium	5000	—

*1 svedberg (S) = 10^{-13} seconds.

Electrophoresis Separates Molecules According to Their Charge-to-Mass Ratio

Molecules of RNA, DNA, and protein can be separated according to size by *electrophoresis.* This technique depends upon the fact that dissolved molecules in an electric field move at a speed determined by the ratio of their charge to their mass. For example, if two molecules have the same mass and shape, the one having the greater charge will move faster toward an electrode. Many successful techniques of electrophoresis are in general use. For example, paper electrophoresis is widely used for small molecules such as amino acids and nucleotides. A sample is deposited in a small spot on a filter paper (or other porous support); the paper is then soaked with a conducting solution and an electric field is applied. The small molecules dissolve in the conducting solution and move in the plane of the paper at a rate related to their charge.

Nucleic acids in solution generally have a negative charge because of the ionization of their phosphate groups. Thus, they migrate toward a positive electrode. However, nucleic acid molecules consisting of long chains have almost identical charge-to-mass ratios, no matter what their lengths, because each residue contributes about the same charge and about the same mass. Also, many proteins that differ in shape and mass have almost equal charge-to-mass ratios. Therefore, if electrophoresis of nucleic acids and proteins were simply carried out in solution, or on paper as is done for small molecules, little or no separation of molecules of different lengths would be achieved.

In spite of these difficulties, electrophoretic separation according to chain length has become not only possible but amazingly reliable. First, molecules are now most commonly subjected to electrophoresis not in a liquid solution but in a *gel,* a semisolid material made of agarose (a plant polysaccharide) or of a synthetic polymer such as polyacrylamide. The size of the pores in such gels limits the rate at which molecules can move through them. Nucleic acids that have identical charge-to-mass ratios separate according to length, with the longer ones moving more slowly through the pores of the gel (Figure 7-8). Even very long nucleic acids (chains containing from 10,000 to 20,000 residues) that differ in length by a few percentage points can be separated from one another, and *each individual chain can be isolated* in mixtures containing chains of 500 nucleotides or less.

Protein chains also can be separated according to length. Both before and during electrophoresis, the proteins are continuously exposed to the detergent SDS (so-

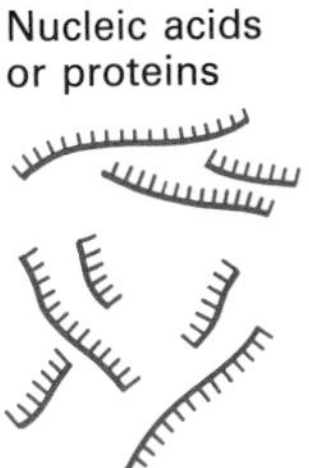
Nucleic acids
or proteins
Negative charges on
phosphate groups
of nucleic acid
or on SDS attached
to protein
Place mixture on an agarose
or polyacrylamide gel
Apply electric field

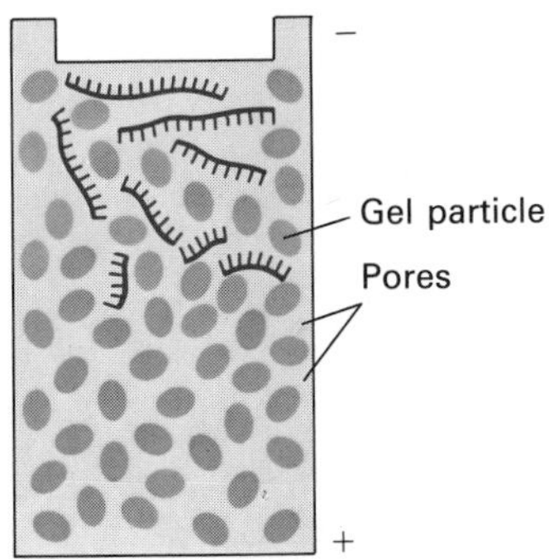
–
Gel particle
Pores
+
The negatively charged
molecules move through the
pores in the gel toward the
positive pole at a rate
inversely proportional to
chain length

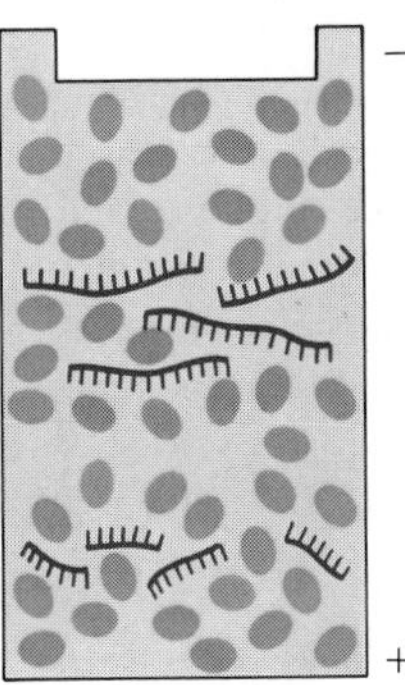
–
+

Figure 7-8 Gel electrophoresis is carried out by pouring a liquid containing either melted agarose or chemically treated polyacrylamide into a cylinder (for a round gel), or between two flat, parallel glass plates 1 or 2 mm apart. As the gel solidifies, it forms interconnected pores, or channels, whose size depends on the concentration of agarose or polyacrylamide. The substances to be separated are then layered on top of the gel (or at one edge of it, if it lies between two plates) and an electric current is passed through the gel. In usual laboratory practice, the migration of RNA or DNA depends on the charges on the phosphates: if the nucleic acid is denoted by

$$N_1pN_2pN_3p \ldots N_np$$

then the charge at neutral pH is $2n$. Proteins can be separated by binding sodium dodecyl sulfate (SDS) residues to their amino acid residues; this will contribute approximately one negative charge per residue. If all of the particles have about the same charge-to-mass ratio, they move through the gel at a rate that is inversely proportional to their chain length.

dium dodecyl sulfate, a common commercial cleaning agent found in toothpaste). Approximately one molecule of detergent binds to each amino acid (Chapter 14). At neutral pH, the detergent is negatively charged so that the adjacent negatively charged SDS molecules repel one another, which forces the proteins with bound detergent into rodlike shapes endowed with similar charge-to-mass ratios. Proteins in this state are said to be *denatured.* As with nucleic acids, chain length (which is equivalent to mass) is the determinant for the separation of proteins during electrophoresis through polyacrylamide gels containing SDS. Again, even chains that differ in molecular weight by less than 1 percent can be separated.

Gel Electrophoresis Can Separate All of the Proteins in a Cell

The traditional biochemical approach to enzyme chemistry has been to detect enzyme activity and then to purify the specific enzyme proteins. Biochemical methods of separating pure proteins from natural mixtures have relied on differences in how their charge changes with varying salt concentrations or pH, or on differences in their sedimentation rates. Much of this protein biochemistry (which was briefly discussed in Chapter 3) remains in day-to-day use.

However, many experiments in molecular biology require an enumeration of the polypeptides formed from a certain genome at a certain time, rather than detection of enzymes or determination of their concentrations. Sometimes all that is desired is the detection of a single given protein within the total cell, so that no purification of the protein is necessary. Frequently, it is important to compare the rate of synthesis of a protein or a set of proteins with the rate of synthesis of all the other proteins in the cell, again without purification or enzymatic detection of the particular protein. Gel electrophoresis can often accomplish these aims.

One-Dimensional and Two-Dimensional Gels Electrophoresis of all of the proteins of cells through a column or a thin rectangular SDS gel reveals only the major proteins that are present. If these are the proteins of interest, or if a cell is producing larger amounts of specific proteins—as occurs, for example, during virus infection—

then one-dimensional analysis may suffice (Figure 7-9).

However, resolution of all of the proteins in the cell can be accomplished in a two-dimensional gel, which separates the proteins in a sample first by charge and then by size (Figure 7-10). The separation by charge is carried out by *isoelectric focusing*. A protein at any given pH has an overall charge on its surface, which will change with pH. When placed in a gradient of pH achieved by small peptides termed *ampholines* and subjected to an electric field a protein will migrate to a point at which its surface charge is balanced and will remain there. This is the isoelectric point of the protein. Proteins separated in a gel of this type can, while still in the gel, be layered on top of another gel soaked with SDS; the proteins can thus be separated by electrophoresis in a second dimension on the basis of size. As many as several thousand different protein chains—virtually the total protein content of a cell—can be detected and separated by this technique. Thus, two-dimensional gels are very useful in studying the expression of various genes in differentiated cells.

There are two widely used methods of detecting proteins in gels:

1. The "steady state" protein content, or the total amount of each type of protein in a mixture, can be estimated with gel electrophoresis by staining the gels with a dye that binds approximately equally to all proteins. This technique therefore indicates the comparative quantities of proteins of different lengths.
2. Gel electrophoresis also provides a way of measuring the *rate* of synthesis of any particular protein without purifying the protein. If cells are radioactively labeled briefly before they are analyzed, each newly synthesized chain can be detected in the gel by autoradiography. However, because proteins may be secreted or may be subject to different rates of metabolic turnover, the total concentration of a protein in a cell may not accurately reflect its rate of synthesis.

In Vitro Protein Synthesis and Gel Electrophoresis Provide an Assay for Messenger RNA

Two general approaches are used to determine what proteins a cell can make; both approaches are actually assays for functional mRNAs. In one method, whole cells are labeled and then their contents are examined for newly synthesized proteins. In the other method, which we shall discuss in this section, mRNA is extracted from the cells and translated in the presence of labeled amino acids by cell-free protein synthesis systems. In either case, the products can be separated and identified by gel electrophoresis. [Figures 7-9 (lanes 1 to 5) and 7-10 show results obtained by the first method; Figures 7-9 (lane 6) and 7-11 show results from the second.]

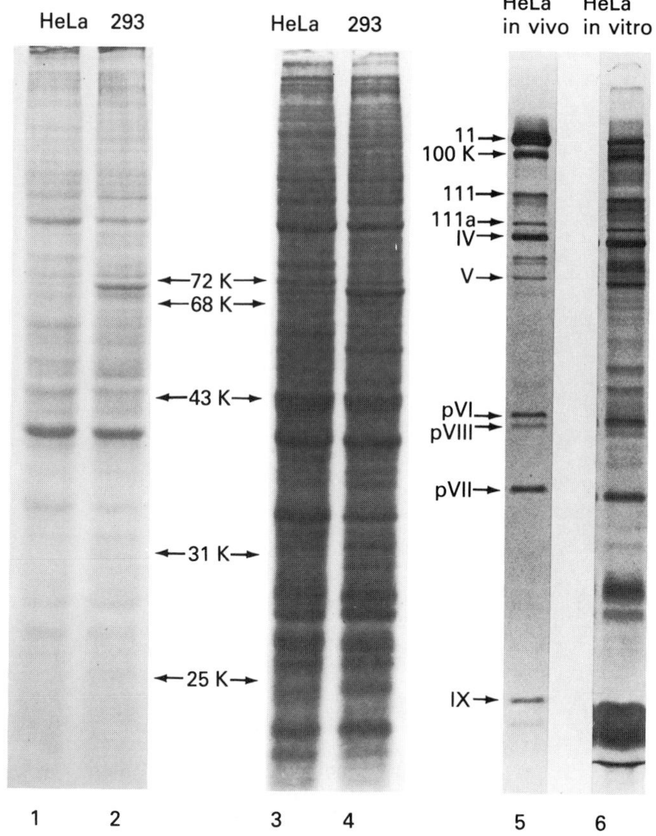

Figure 7-9 Resolution of proteins by one-dimensional gel electrophoresis. The proteins of two human cell lines—HeLa, a human cervical cancer cell, and 293, a virus-transformed embryonic fibroblast—were dissolved in SDS and subjected to electrophoresis. The proteins are made visible in lanes 1 and 2 by autoradiography because the cells were labeled with [^{35}S]methionine, and in lanes 3 and 4 by a dye, Coomassie blue, that stains all proteins. The designations 72 K, 68 K, etc. indicate the positions of marker proteins—proteins of known sizes—that have molecular weights of 72,000, 68,000, etc. The proteins in the two types of human cells are obviously quite similar. Lane 5 is an autoradiograph of the labeled proteins in HeLa cells that have been infected with adenovirus type 2 for 18 h. The total number of labeled proteins is greatly reduced: the host-cell proteins are no longer made, and synthesis is confined to a limited number of virus-specific proteins. The proteins found in virions are labeled with Roman numerals in order of decreasing size. Some of the proteins are synthesized as longer forms and then shortened; these are labeled "p" for "precursor." The autoradiograph in lane 6 results from isolating the mRNA from the infected cells and causing it to be translated by mixing it with extracts of reticulocytes in the presence of [^{35}S]methionine; the labeled proteins now are a mixture of viral and cell proteins. *Photographs courtesy of J. R. Nevins and C. Lawrence.*

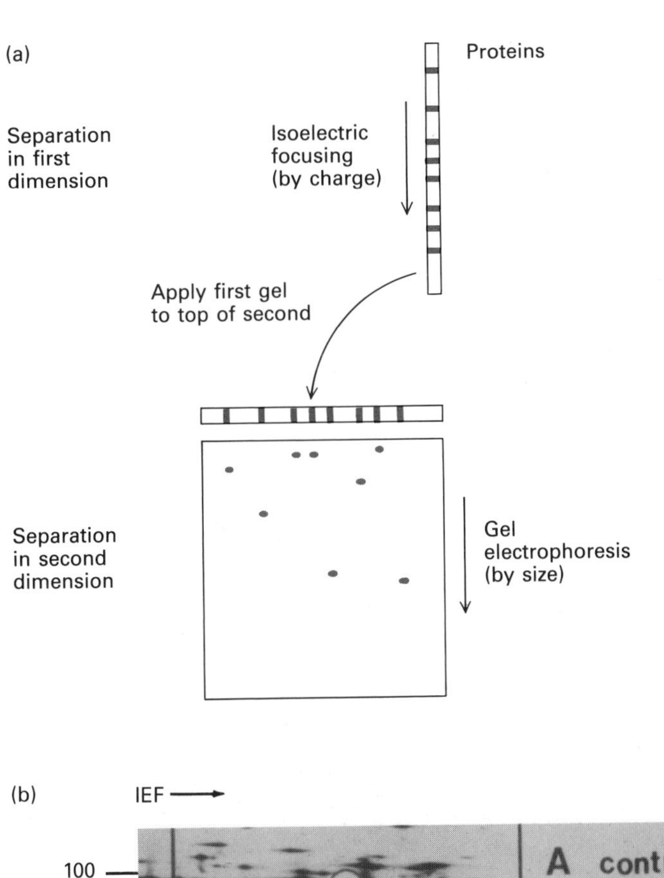

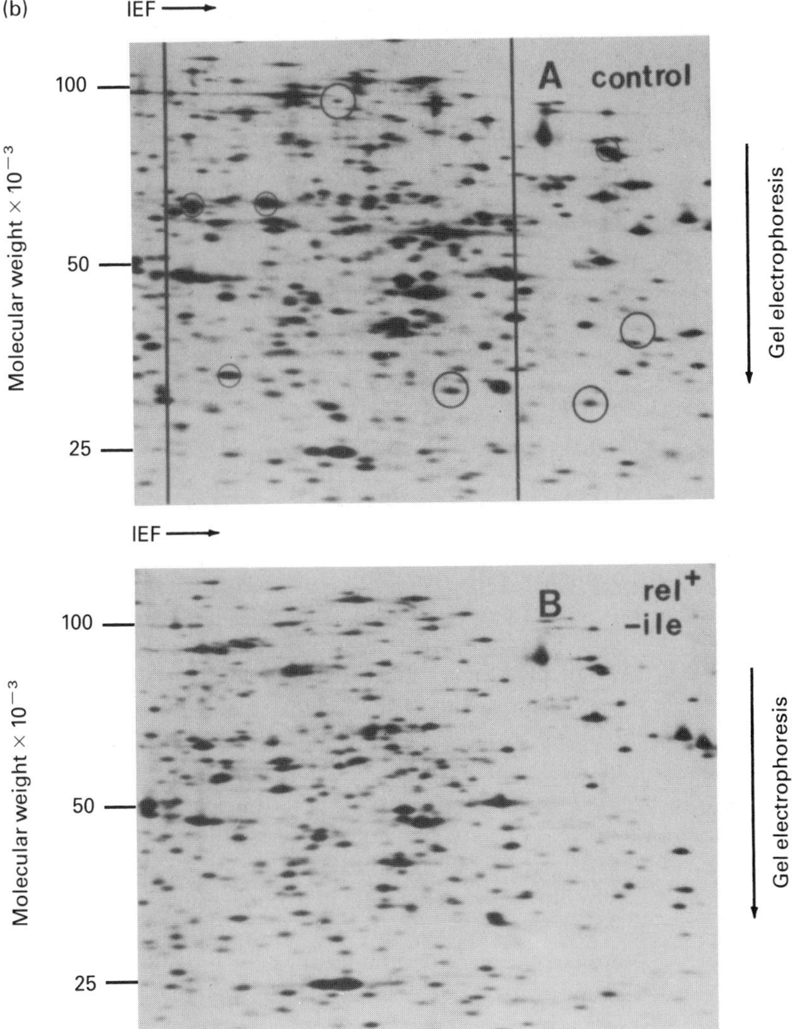

Figure 7-10 Two-dimensional protein gels. (a) All of the proteins in a sample of *E. coli* cells are dissolved and subjected to isoelectric focusing (IEF) in a column filled with a semisolid medium that does not contain SDS and so does not denature the proteins. This procedure separates the native proteins according to their overall charges. The gel containing the charge-separated sample is then applied to the top of a flat gel containing SDS, and the denatured proteins are electrophoretically separated by molecular weight in a second dimension. (b) When the proteins are labeled, their positions in the gel can be detected by autoradiography. Each spot represents a single polypeptide. The spots are elongated horizontally because the average charge on a protein molecule varies somewhat during the isoelectric focusing. These two samples show that the patterns are reproducible and that changes in individual proteins can be detected. The proteins in the top photo are from cells growing on a normal medium supplemented with arginine; the proteins in the bottom photo are from cells growing on a medium that lacks arginine, which is required for this strain. Certain spots *(boxed)* are absent in the bottom photo or are much fainter than in the top photo; one spot *(circled)* is darker. These differences represent changes made in the synthesis pattern of cell proteins in response to amino acid starvation. *From P. H. O'Farrell, 1978,* Cell **14***:545. Copyright M.I.T.*

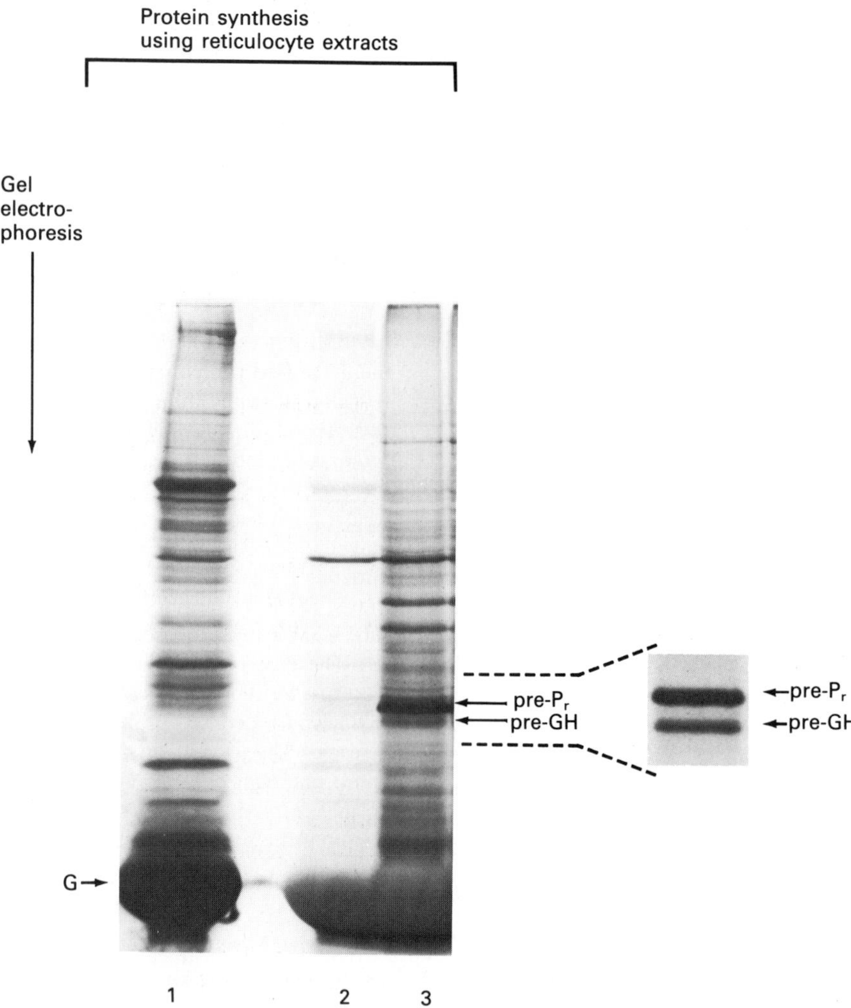

Figure 7-11 Messenger RNA can be translated by mixtures of ribosomes, tRNAs, and protein synthesis factors extracted from reticulocytes. In the photographs shown here, the total protein resulting from such reactions is separated by electrophoresis through a polyacrylamide gel in the presence of SDS. Because [^{35}S]methionine was added to the extract, the newly made proteins are labeled, so they can be made visible by autoradiography. The proteins synthesized from the naturally occurring mRNAs in an untreated reticulocyte extract are shown in lane 1; the large amount of globin synthesis (G) is evident. Lane 2 shows what happens when the reticulocyte extract is treated with a bacterial nuclease (*Micrococcus aureus*) and the nuclease is then inactivated chemically: the extract carries out very little synthesis. The stimulation of protein synthesis by added mRNAs can be observed in lane 3. After all of the mRNAs extracted from rat pituitary cells are added to the treated extract (of lane 2) and translated, several prominent proteins become visible, including two hormone precursors—that of prolactin (pre-P$_r$, 236 amino acids long) and that of growth hormone (pre-GH, 212 amino acids long). (The inset photo was taken under electrophoretic conditions designed to clearly separate pre-P$_r$ and pre-GH.) [See H. R. B. Pelham and R. J. Jackson, 1976, *Eur. J. Biochem.* 67:247.] *Photographs courtesy of D. Anderson.*

A variety of different cell extracts can be used for labeling proteins in vitro (i.e., assaying for active mRNAs). Bacterial cell extracts that can translate homopolymers were first widely used to break the genetic code (Chapter 4) and to examine bacterial and bacteriophage proteins; now extracts of eukaryotic cells are also commonly used. Two of the most popular cell-free systems are extracts of reticulocytes and of wheat "germ," the embryo plant in a fertile wheat seed. Extracts from both of these sources are prepared by treating the cells first with a nuclease that destroys endogenous RNA (RNA from the source cells) and then with a chemical that blocks the nuclease, so that mRNA that is subsequently added will not be destroyed. After this treatment, there is very little protein synthesis by endogenous mRNA (Figure 7-11, lane 2). Thus, whatever mRNA is added is responsible for almost all protein synthesis, and the products of the added mRNA can be easily detected.

Examining the Sequences of Nucleic Acids and Proteins

The first biopolymer to be sequenced was a protein, and this discovery had great historical importance. Before Fred Sanger reported the sequence of human insulin in 1953, some biochemists were not convinced that proteins have specific sequences from end to end. Since that time, the intimate relation of nucleic acid sequences to protein sequences has become clear. Recently, in fact, it has become much easier to obtain long nucleic acid sequences than long protein sequences.

Because the biological effects of nucleic acids depend on the linear sequences of their nucleotides, nucleic acid research relies heavily on techniques that reveal and compare sequences. Experiments vary considerably in the extent and type of sequence information required. In the

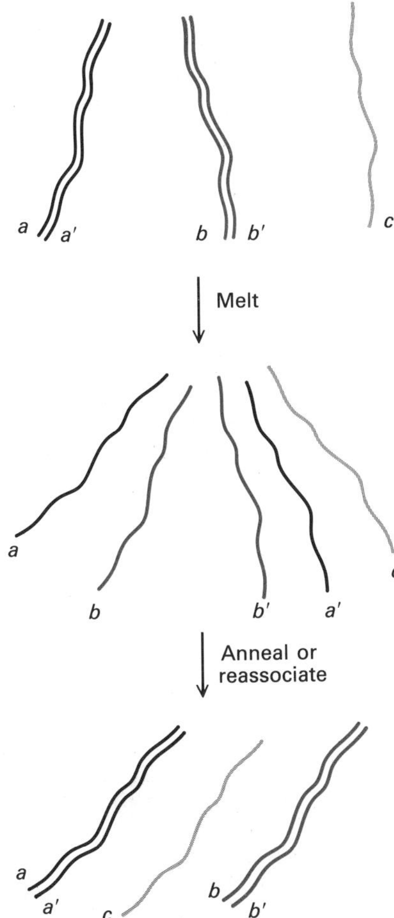

Figure 7-12 Molecular hybridization, or reassociation of the complementary strands of a nucleic acid. Suppose that in a mixture of nucleic acids in solution, *a* and *a′* are complementary strands, *b* and *b′* are complementary strands with different sequences from *a* and *a′*, and *c* has yet a third distinct sequence. Under certain conditions, the duplexes will melt or separate into single strands. After an appropriate change in conditions, they will reassociate: *a* will find *a′*, *b* will find *b′*, and *c* will remain single-stranded. Thus the presence of noncomplementary chains does not affect the rate of recombination of complementary ones.

simplest case, an estimate of the degree of *sequence relatedness,* or similarity, between two nucleic acid or protein samples is all that is required. Often, however, it is necessary to determine whether a particular sequence is present in a given mixture of nucleic acids or proteins. Once the presence of a certain sequence in a mixture of sequences is established, a variety of other questions arise. What is the concentration or amount of the specific sequence? Where within a DNA, RNA, or protein molecule is the sequence located? What are the neighbors of the sequence in the molecule? What is the precise nucleotide or amino acid sequence for the entire molecule under study? A variety of techniques are used to address these questions; naturally, each technique applies better to some questions than to others.

Hybridization of Two Nucleic Acid Strands Can Be Detected in Several Ways

Under the conditions of temperature and ion concentration found in cells, DNA is maintained as a duplex (two-stranded) structure by the many hydrogen bonds of the A-T and G-C base pairs. The two strands of the duplexes can be *melted* (separated) by heating them (usually in a dilute salt solution—e.g., 0.01 *M* NaCl) or by raising the pH to higher than 11. If the temperature is then lowered and the concentration of ions in the solution is raised, or if the pH is lowered, the single strands will *anneal,* or reassociate, to reconstitute duplexes (provided that their concentration in the solution is high enough). Only complementary strands reassociate, and the extent of their reassociation is virtually unaffected by the presence of noncomplementary strands. This specific reassociation is termed *molecular hybridization;* it can take place between complementary strands of either DNA or RNA, or between an RNA strand and a DNA strand (Figure 7-12).

Reassociated molecules can be detected by various techniques. Viewing the hybrid molecules under the electron microscope is the most graphic; single strands of nucleic acid can be distinguished from duplexes. If two single nucleic acid strands that are complementary over only part of their length are allowed to hybridize, the result is a *heteroduplex* (Figure 7-13). Homologous (duplex) regions and noncomplementary (single-stranded) regions can be distinguished from one another. This technique allows an instant determination of the overall relatedness of two samples of nucleic acid. It can be used not only for comparing DNA strands, but also for locating DNA sites that are complementary to RNA molecules. In the latter case, it is possible to distinguish and accurately locate, with a maximum error of 100 bases, which regions of the DNA are transcribed into RNA. This technique is called the *R loop procedure* because regions of RNA-DNA hybridization create loops (R loops) in the nucleic acid molecules. The R loops are formed where the RNA sequence has base-paired with one DNA strand and displaced the other DNA strand; displacement occurs because RNA-DNA hybrid regions are more stable than DNA-DNA hybrids (Figure 7-14). The fact that 1 μm of double-stranded nucleic acid contains about 3000 bases can be used to estimate the number of nucleotides in single- and double-stranded regions; this gives a fairly accurate map of the complementary and noncomplementary portions of the molecules.

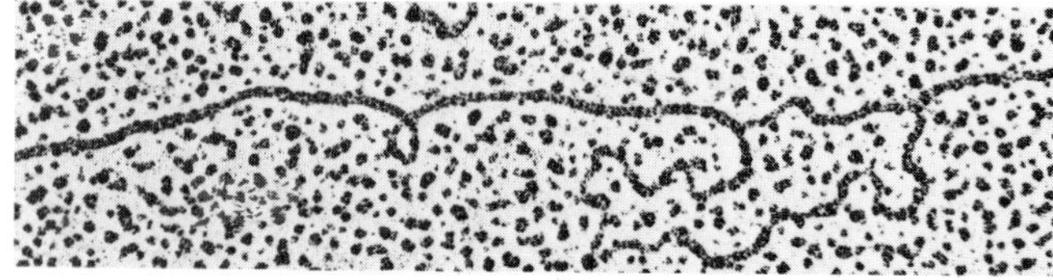

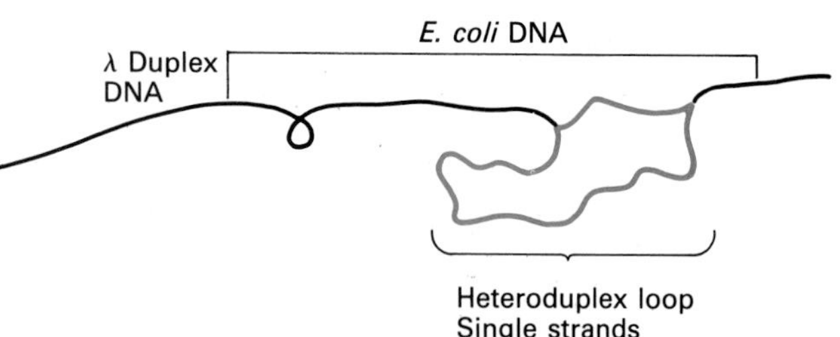

Figure 7-13 (*top*) Electron micrograph of a DNA heteroduplex. DNA molecules on a carbon grid can be distinguished as long threads when they are shadowed with heavy metals (platinum and palladium were used here). The single strands that have partially annealed to form the heteroduplex (*bottom*) are complementary strands from two different λ bacteriophages that have incorporated different but related sequences of *E. coli* DNA (sequences encoding different enzymes needed to make biotin, a vitamin). If the bacteriophage DNA strands are complementary, they form a double-stranded hybrid; where the dissimilar sequences from *E. coli* remain unassociated, the structure contains a heteroduplex loop of single-stranded DNA. *From R. W. Davis and J. S. Parkinson, 1971,* J. Mol. Biol. *56:403.*

A second common method of detecting reassociated nucleic acids is as follows: A particular sequence of a single-stranded nucleic acid (usually DNA) is attached to a solid matrix. Nitrocellulose paper is the most commonly used matrix; it is not known why single-stranded DNA binds to nitrocellulose, but this affinity is enormously useful. Radioactive RNA or DNA that is to be tested for sequences complementary to the bound nucleic acid is allowed to hybridize with it. After an appropriate interval, the unhybridized single strands not associated with the bound DNA are washed away. RNA hybridized to bound DNA is resistant to ribonucleases, whereas unpaired RNA regions are digested by these enzymes; thus any remaining single-stranded RNA regions can be trimmed away. The amount of hybrid formed can then be measured by the amount of bound radioactive label (Figure 7-15). This technique is perhaps the single most common use of molecular hybridization, because one specific RNA (or DNA) sequence in a mixture of many different types of sequences can be detected.

In the procedure called *DNA excess hybridization,* the total RNA from cells (a very complex mixture of sequences) is labeled and exposed to multiple copies of a particular (unlabeled) sequence of DNA. When the DNA is present in excess—that is, when there are more than enough copies to hybridize with all complementary segments of RNA—a characteristic amount of hybrid is formed (Figure 7-16a). This enables the experimenter to measure accurately the amount of the particular mRNA in a mixed sample.

The reciprocal procedure, *RNA excess hybridization,* can be performed with a complex DNA sample to determine what amount of the DNA is complementary to a particular labeled RNA species (Figure 7-16b). This procedure has been used, for example, to determine how many DNA sites encode ribosomal RNA in various organisms.

Finally, it is also possible to test for the presence of a particular RNA and to quantify the amount in different samples by *competition hybridization.* When a labeled RNA species previously tested for by labeled RNA-DNA hybridization is mixed with DNA and an excess of a

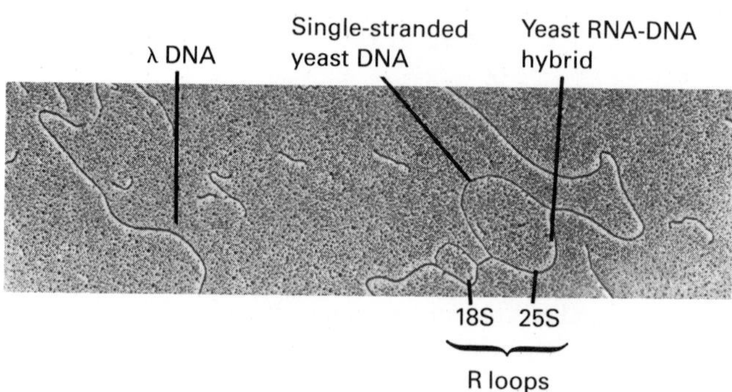

Figure 7-14 If double-stranded DNA is treated with the organic solvent formamide at certain temperatures, some of the hydrogen bonds between the strands of the molecule break, so that although the strands remain in complementary association they are no longer firmly bound. If RNA that is complementary to one strand of the duplex DNA is then introduced, the RNA binds to that region on one DNA strand, displacing the other DNA strand. This is because an RNA-DNA duplex is more stable than a DNA-DNA duplex. The hybrid duplex and the displaced stretch of single-stranded DNA form a loop called an R loop. The two R loops that appear in this electron micrograph result from the hybridization of 18S and 25S ribosomal RNA from yeast with a region of bacteriophage λ DNA that contains an inserted stretch of yeast ribosomal genes. [See M. Thomas, R. L. White, and R. W. Davis, 1976, *Proc. Nat'l Acad. Sci. USA* 73:2294.] *Photograph courtesy of R. W. Davis.*

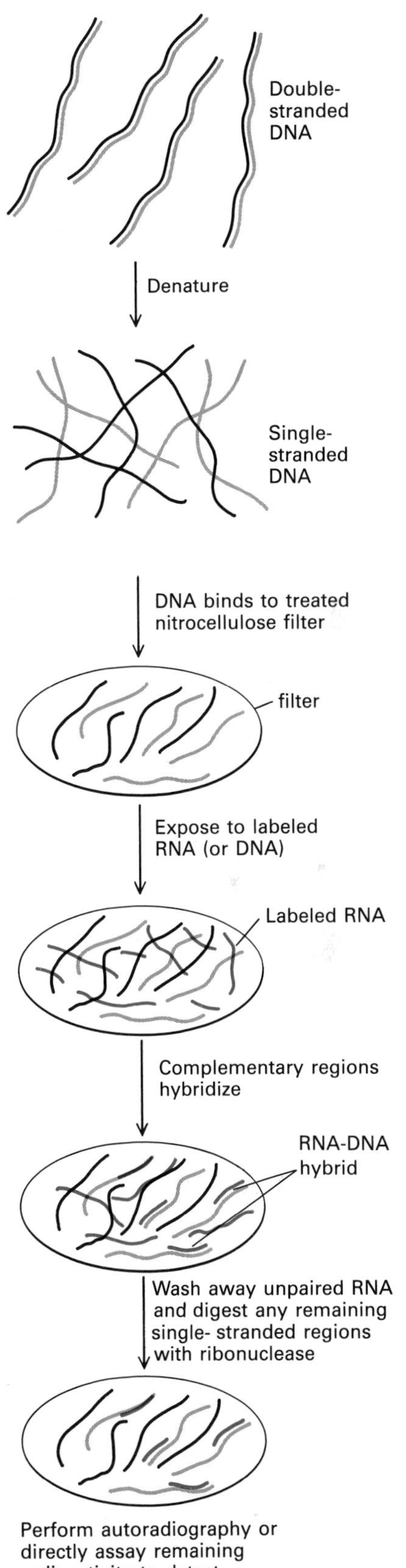

Figure 7-15 The filter-binding assay for RNA-DNA (or DNA-DNA) hybridization is an extremely popular and flexible method of detecting complementary regions. Double-stranded DNA is melted and, while still single-stranded, attached to a solid matrix—usually nitrocellulose. Then, to form molecular hybrids the filter-bound DNA is exposed under the proper conditions of ionic strength and temperature to a labeled RNA (or DNA) sample containing molecules that are complementary to the filter-bound DNA. Unpaired labeled molecules can be washed away; if labeled RNA has been used, in addition to simply washing the filters any residual unpaired RNA can be trimmed away with a ribonuclease that does not attack hybridized RNA. This technique allows as little as 1 part in 10^6 of specific RNA to be detected.

"competing" unlabeled RNA species, the degree of similarity between the two RNAs can be determined. If the competing RNA is identical to the labeled RNA, then increasing the ratio of unlabeled competitor to labeled sample results in a decrease in the amount of labeled RNA hybridized (Figure 7-16c).

The Rate of Nucleic Acid Hybridization as a Measure of Complexity The rate of hybridization between two complementary single-stranded nucleic acids in solution depends on the frequency with which complementary regions collide and *nucleate*—that is, start to form a duplex (Figure 7-17). This collision frequency, in turn, depends on the concentration of the two strands. Suppose that the DNA fragments of two different organisms—say, *E. coli* and yeast—are incubated in two different test tubes in amounts that yield the same *total* DNA concentration. The *complexity* of the DNA (the number of base pairs in the total genome) is about four times greater for yeast than for *E. coli*. A separated strand of *E. coli* DNA would therefore encounter its correct partner four times as often as would a strand of yeast DNA, and the rate of reassociation of the *E. coli* DNA would be faster (Figure 7-18).

The equation for determining the quantitative relation between the reassociation rate and genome complexity is given in Figure 7-18. Thus, the reassociation rate of any DNA sample allows the calculation of the relative complexity of the genome. Because experimentally the initial concentration of DNA (C_0) and the time (t) are varied to measure reassociation rates, the resulting curves are often called C_0t *(cot)* curves.

If all of the DNA sequences of an organism are present once per haploid genome, there is a uniform reassociation curve. If some sequences are repeated, these hybridize

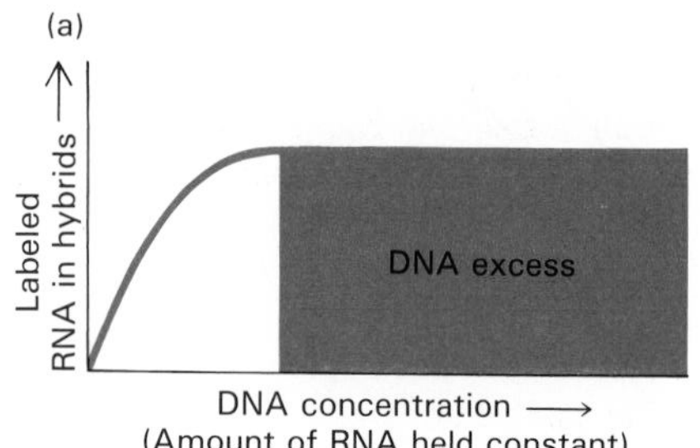

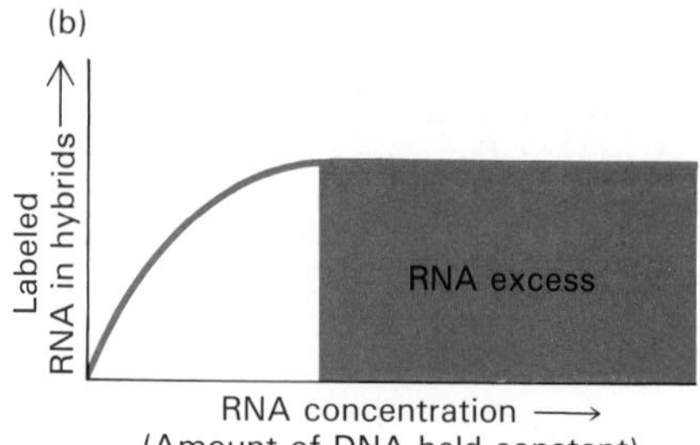

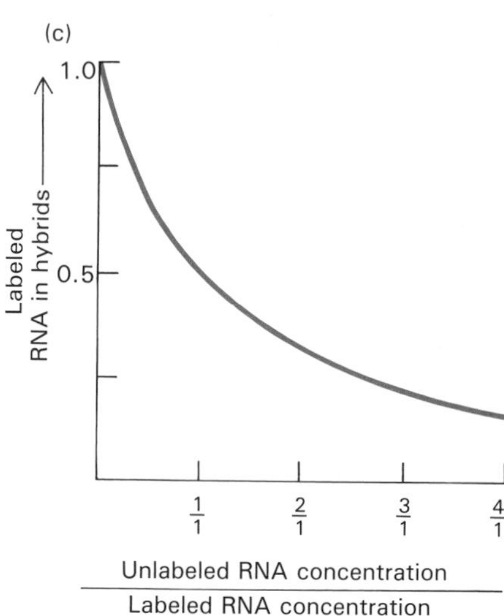

Figure 7-16 Excess and competition hybridization. (a) DNA excess hybridization measures the total amount of RNA in a sample that is complementary to a particular sequence of DNA. A fixed amount of labeled RNA is exposed to greater and greater amounts of DNA bound to a filter paper, as shown in Figure 7-15. When the quantity of radioactive label bound to the filter (i.e., representing hybridized RNA-DNA) reaches a maximum, it indicates that the DNA is in excess. The fraction of the RNA sample that is complementary to the test DNA can therefore be measured. This experiment is often performed with purified viral DNA and RNA from infected cells. (b) RNA excess hybridization measures the total amount of a complex DNA sample that can hybridize with a particular RNA sequence. In this procedure, the filter-bound DNA is held constant, and it is exposed to greater and greater amounts of labeled RNA. When the RNA is in excess, the fraction of DNA that is complementary to the particular RNA can be accurately determined. This type of experiment is performed to measure the number of genomic DNA sites that are complementary to a particular RNA (e.g., rRNA). (c) Competition hybridization measures the diluting effect of an unlabeled RNA species on the amount of labeled hybrid formed. If the unlabeled RNA is identical to the labeled species (as here), the quantity of labeled hybrid drops as the concentration of unlabeled RNA increases. (Note that this effect occurs only when the diluting RNA plus the original RNA exceeds the number of complementary DNA sites.) Competition hybridization is often performed with RNA samples taken from cells under two different conditions—e.g., early and late in virus infection.

more rapidly. Reassociation measurements have been important both in comparisons between different types of organisms (e.g., prokaryotic versus eukaryotic; vertebrate versus invertebrate) and in studies of the degree of repetition of certain sequences within eukaryotic genomes. Characteristics of such repetitious DNA sequences are discussed in detail in Chapter 11.

A variation in the application of reassociation curves to duplex DNA is the use of a pure radioactive DNA sample and unlabeled RNA from a cell of interest. The concentration of the complementary RNA can be estimated in this case because the rate of DNA-RNA hybridization depends on the RNA concentration. Curves from such measurements are referred to as R_0t (RNA concentration) curves; these are discussed in Chapter 12.

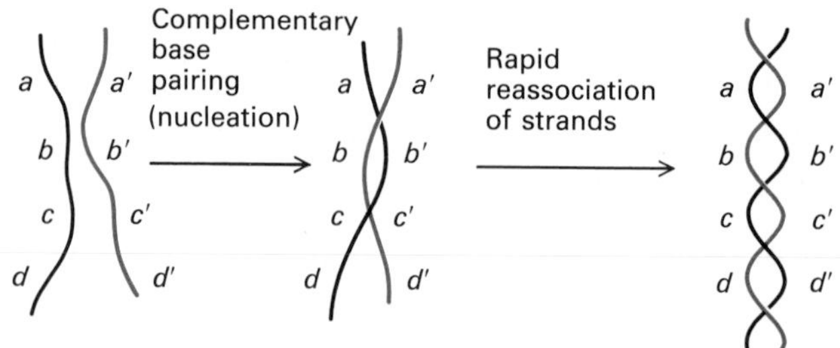

Figure 7-17 The rate of hybridization of two nucleic acid strands depends on the frequency of collision and subsequent nucleation of complementary regions.

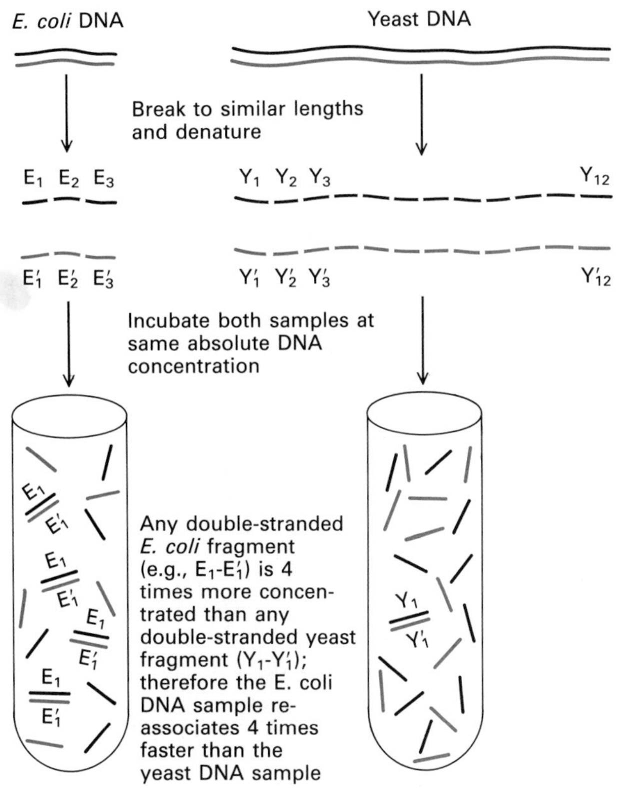
E. coli DNA
Yeast DNA
Break to similar lengths and denature
E1 E2 E3
Y1 Y2 Y3
Y12
E'1 E'2 E'3
Y'1 Y'2 Y'3
Y'12
Incubate both samples at same absolute DNA concentration
Any double-stranded E. coli fragment (e.g., E1-E'1) is 4 times more concentrated than any double-stranded yeast fragment (Y1-Y'1); therefore the E. coli DNA sample reassociates 4 times faster than the yeast DNA sample

Figure 7-18 The complexity of DNA controls the rate of reassociation of denatured DNA. The relative rates of hybridization within two samples of dissolved and melted genomic DNA (here, from *E. coli* and yeast) depend on their relative complexity (the number of DNA base pairs in the genome of each organism), provided that the samples are equal in absolute concentration (i.e., in total nucleotide concentration). For example, *E. coli* DNA has only about one-fourth the number of deoxynucleotides that yeast DNA has. (The reassociation reactions are carried out after breaking the DNA into pieces of 1000 to 2000 bases each, so size plays little role in the comparison. The DNA is melted to single strands and then placed under conditions that encourage reassociation.) The equation for the rate of reassociation is

$$C_t = \frac{1}{1 + KC_0t}$$

where C_0 and C_t are the molar concentrations of single strands at time 0 and time t, respectively, and K is the rate constant for the particular type of DNA (this constant depends on the complexity). [See R. J. Britten and D. E. Kohne, 1968, *Science* **161**:529.]

Partial Sequence Analysis ("Fingerprinting") Allows Quick Comparisons of Macromolecules

The enzymatic fragmentation of proteins and nucleic acids at specific sites provides a means of recognizing particular macromolecules quickly. As we explained in Chapter 3 (and mentioned again in Figure 7-6), the enzyme trypsin digests protein chains, cleaving them after each lysine and each arginine residue to produce, from any given pure protein, a regular and specific set of peptides. Likewise, the enzyme RNase T1 reliably cuts RNA at the 3′ side of each guanylate residue to produce specific fragments ending with a guanylate. The resulting *oligonucleotides* are fairly short: they usually contain from 2 to 20 nucleotides, because consecutive guanylates usually occur within 20 bases of one another. Reliable separation and detection of the different peptides from a pure protein or the different oligonucleotides from a pure RNA can be accomplished by electrophoresis or chromatography, or both.

Because the nucleotides or peptides from a pure RNA or protein are always the same, the pattern of separated fragments is always the same. The characteristic pattern of fragments from a primary sequence is called a *fingerprint* (Figure 7-19). In the case of RNA, labeled molecules are used and detection of oligonucleotides is by autoradiography. On the other hand, peptides in a protein fingerprint are most often detected by reagents that react with the amino groups of the amino acids (Figure 7-20). The fingerprinting technique allows the rapid comparison of two samples of RNA or protein when there is no need to determine the complete nucleotide or amino acid sequence.

A fingerprint of a DNA molecule can also be obtained by treating the DNA with mild acid, which releases purines; the "apurinic" DNA chain can then be broken to yield pyrimidine stretches. Alternatively, the DNA can be copied into RNA so that an RNA fingerprint can be obtained.

The fingerprinting technique was first developed for proteins, which can be cleaved both by enzymes and by chemical reactions. With the historic fingerprint of globin shown in Figure 7-20, Vernon Ingram demonstrated that people suffering from sickle-cell anemia, a genetic disease, have one different amino acid in their β globin. This was the first protein shown to be affected in its function by a change in one amino acid residue.

Restriction Enzymes Allow the Precise Mapping of Specific Sites in DNA

The most flexible, simple, and useful technique for partial sequence analysis of DNA derives from the discovery of bacterial *restriction enzymes* (or *restriction endonucleases*, as they are sometimes called). These enzymes recog-

Spot number	Sequence
1	AG
3	CCCG
4	AAG
5	ACCG
6	AAAG
8	CG
9	AUG
10	UAG
11	UCG
12	CCUG
13	CCUACG
14b	AAUACCG
14c	AUCCAG
15	CCACACCACCUG
16	UUAG
17	UCUG
18	AUCUCG
20	CCG
21	CG
22	pG
G	Gp

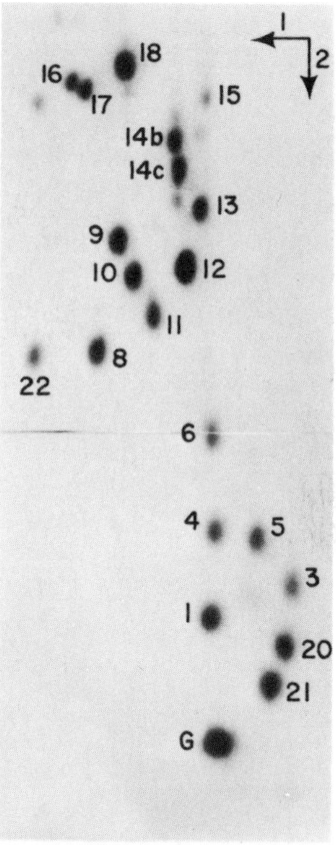

Figure 7-19 A ribonuclease T1 "fingerprint" of 5S ribosomal RNA from oocytes of the frog *Xenopus laevis*. This enzyme cuts RNA on the 3′ side of all guanylate residues:

$$\text{NpGp} \downarrow \text{Np} \longrightarrow \text{NpGp} \quad \text{Np}$$

Therefore the digestion produces fragments that all contain one Gp at their 3′ ends. These are applied to treated paper (cellulose acetate), and two steps of separation are carried out: electrophoresis in one dimension *(arrow 1)*, followed by chromatography in the other *(arrow 2)*. If the starting sample is radioactive (^{32}P-labeled RNA is often used), the oligonucleotides can be identified by autoradiography. "Spots" of RNA can be collected and further analyzed biochemically to determine their sequences. (There are no spots numbered 2, 7, 14a, or 19 because these numbers were given to oligonucleotides that were identified in another type of 5S rRNA, but that are missing from this type.) *From D. D. Brown, D. Carroll, and R. D. Brown, 1977,* Cell ***12****:1045. Copyright M.I.T.*

nize specific short oligonucleotides (from four to six residues long) in DNA and cleave the DNA at all such sites (Figure 7-21a). The name restriction enzyme derives from its function in the bacterium of origin: the enzyme destroys (restricts) incoming foreign DNA (e.g., bacteriophage DNA or DNA accidentally taken up during transformation) by cleaving it at these specific sites, which are called *restriction sites*. Another enzyme protects the bacterium's own DNA from cleavage by modifying it at or near each potential cleavage site: a methylase adds a methyl group to one or two nucleic acid bases—usually to bases within the restriction site. When a methyl group is present there, the restriction endonuclease is prevented from cutting the DNA (Figure 7-21b). Together with the restriction endonuclease, the methylating enzyme forms a *restriction-modification* system that protects the host DNA while destroying foreign DNA.

A restriction enzyme cuts a pure DNA sample into a consistently reproducible set of fragments that can be easily separated by gel electrophoresis (Figure 7-22). Several hundred restriction enzymes are now available (Table 7-4). Digestion of DNA by such enzymes, followed by the application of a simple technique (such as electrophoresis) for separating the fragments, has revolutionized the

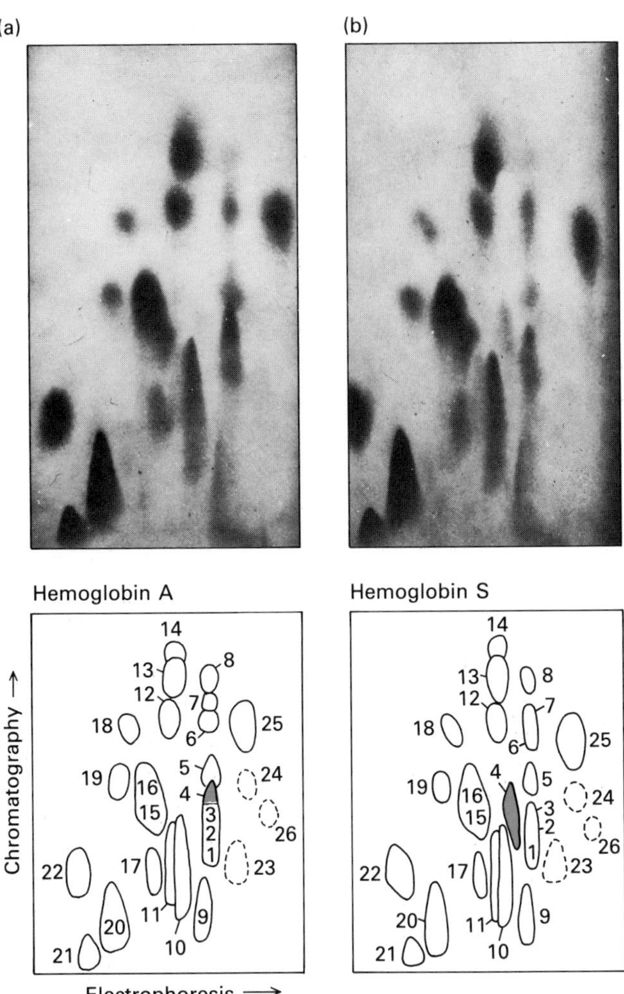

Figure 7-20 Fingerprints of (a) normal and (b) sickle-cell human β globin. Proteins can be subjected to fingerprint analysis by using proteolytic enzymes such as trypsin to break the peptide chain at known amino acid residues (trypsin cuts after each arginine and each lysine). The resulting set of specific fragments can then be separated in two steps, electrophoresis followed by chromatography. Individual peptide spots can be distinguished by spraying the chromatography paper with ninhydrin, a reagent that forms a purple product with free amino groups. (The spots numbered 23, 24, and 26 show up poorly with ninhydrin.) The fingerprints shown here were the first ever made for any macromolecules, and they led to the first proof of the correspondence between a single amino acid change in the β globin and a mutation in the human β globin gene (this mutation was known from genetic studies). The fingerprints are identical, with one exception. Peptide 4 of the β chains of hemoglobin S (the hemoglobin of people with the sickle-cell disease) is found in a location slightly different from that of peptide 4 of normal hemoglobin A. Analysis of the two peptides has shown that hemoglobin S has a valine instead of a glutamic acid at residue 6 in the β globin chain. Thus a single amino acid replacement is the cause of sickle-cell anemia. *From V. Ingram, 1958,* Biochim. Biophys. Acta *28:543.*

mapping of chromosomes. By using two or more restriction enzymes on a pure DNA sample, the order of the restriction sites in the DNA can be determined (Figure 7-23). Also, by partial digestion of terminally labeled DNA, many sites can be located with one enzyme (Figure 7-24). In this way it is possible to produce a map showing the order of the restriction sites in any region of DNA.

Southern Blots Now that we have the ability to divide DNA into reproducible pieces, the following question frequently arises: Where are the restriction sites around a particular sequence of interest? In the laboratory this question becomes: To which restriction fragment(s) will a given purified sequence (a *probe*) hybridize? The latter question can be answered through the use of a technique called the *Southern blot,* after its originator, Edward Southern. DNA restriction fragments are separated by gel electrophoresis; their distribution in the gel is preserved as they are denatured and transferred to a solid substrate (usually a nitrocellulose filter) by blotting; the filter is then exposed to specific radioactive nucleic acid sequences. The blotted DNA fragments that are complementary to the labeled nucleic acids hybridize with them, and their location on the filter can be revealed by autoradiography. This technique is so sensitive that it detects fragments of DNA complementary to a purified nucleic acid sequence at a level of 1 part in 10^6 (Figure 7-25), which makes the mapping of restriction sites in human DNA an easy task (Figure 7-26).

Northern Blots A similar technique is used to detect the presence of specific mRNA molecules. Because this procedure is patterned after the Southern blot, it is often referred to as the *Northern blot,* a bit of laboratory jargon now widely understood because so many experiments of this type have been used to detect specific mRNA. The experimenter first denatures the RNA molecules by mixing them with an agent (e.g., formaldehyde) that breaks the hydrogen bonds between base pairs, and thus ensures that the RNA is in unfolded, linear form. The RNA is then subjected to gel electrophoresis, nitrocellulose blotting, hybridization with a labeled DNA se-

Table 7-4 Some restriction enzymes now in common use

Source microorganism	Enzyme†	Recognized sequence and cleavage site (↓)‡	Number of cleavage sites§			
			λ	Ad2	SV40	φX174
Anabaena cylindrica	*Acy*I	G(Pu) ↓ CG(Py)C	>14	>14	0	7
Anabaena subcylindrica	*Asu*II	TT ↓ CGAA	7	1	0	0
Anabaena variabilis	*Ava*I	C ↓ (Py)CG(Pu)G	8	15	0	1
	*Ava*II	G ↓ G($^{A}_{T}$)CC	>17	>30	6	1
	*Ava*III	ATGCAT	?	?	3	0
Anabaena variabilis^{uw}	*Avr*II	CCTAGG	2	2	2	0
Arthrobacter luteus	*Alu*I	AG ↓ CT	>50	>50	35	24
Bacillus amyloliquefaciens	*Bam*HI	G ↓ GATC*C	5	3	1	0
Bacillus brevis	*Bbv*I	GC*($^{A}_{T}$)GC	>30	>30	23	14
Bacillus caldolyticus	*Bcl*I	T ↓ GATCA	7	5	1	0
Bacillus globigii	*Bgl*I	GCC$(N)_4$ ↓ NGGC	22	12	1	0
	*Bgl*II	A ↓ GATCT	6	12	0	0
Bacillus stearothermophilus ET	*Bst*EII	G ↓ GTNACC	11	8	0	0
Brevibacterium albidum	*Bal*I	TGG ↓ CCA	15	17	0	0
Caryophanon latum L	*Cla*I	AT ↓ CGAT	12	2	0	0
Desulfovibrio desulfuricans, Norway strain	*Dde*I	C ↓ TNAG	>50	>50	19	14
Diplococcus pneumoniae	*Dpn*I	G*A ↓ TC	¶	¶	¶	¶
Escherichia coli RY13	*Eco*RI	G ↓ AA*TTC	5	5	1	0
Escherichia coli R245	*Eco*RII	↓ CC*($^{A}_{T}$)GG	>35	>35	16	2
Fusobacterium nucleatum D	*Fnu*DII	CG ↓ CG	>50	>50	0	14
Haemophilus aegyptius	*Hae*III	GG ↓ C*C	>50	>50	19	11
Haemophilus aphrophilus	*Hpa*II	C ↓ CGG	>50	>50	1	5
Haemophilus gallinarum	*Hga*I	GACGC	>50	>50	0	14
Haemophilus haemolyticus	*Hha*I	GC*G ↓ C	>50	>50	2	18
Haemophilus influenzae R_d	*Hind*II	GT(Py) ↓ (Pu)A*C	34	>20	7	13
	*Hind*III	A* ↓ AGCTT	6	11	6	10
Haemophilus influenzae R_f	*Hinf*I	G ↓ ANTC	>50	>50	10	21
Haemophilus parahaemolyticus	*Hph*I	GGTGA	>50	>50	4	9
Haemophilus parainfluenzae	*Hpa*I	GTT ↓ AA*C	13	6	4	3
	*Hpa*II	C ↓ C*GG	>50	>50	1	5
Klebsiella pneumoniae OK8	*Kpn*I	GGTAC ↓ C	2	8	1	0
Microcoleus species	*Mst*I	TGC ↓ GCA	>10	>15	0	1
Moraxella bovis	*Mbo*I	↓ GATC	>50	>50	8	0
	*Mbo*II	GAAGA	>50	>50	16	11
Moraxella nonliquefaciens	*Mnl*I	CCTC	>50	>50	51	34
Proteus vulgaris	*Pvu*I	CGAT ↓ CG	3	7	0	0
	*Pvu*II	CAG ↓ CTG	15	22	3	0

Continued on next page

Table 7-4 *Continued*

Source microorganism	Enzyme†	Recognized sequence and cleavage site (↓)‡	Number of cleavage sites§			
			λ	Ad2	SV40	φX174
Providencia stuartii 164	*Pst*I	CTGCA ↓ G	18	25	2	1
Rhodospirillum rubrum	*Rru*I	AGT ↓ ACT	?	?	0	0
Serratia marcescens S_b	*Sma*I	CCC ↓ GGG	3	12	0	0
Sphaerotilus natans C	*Sna*I	GTATAC	2	?	0	0
Streptomyces achromogenes	*Sac*I	GAGCT ↓ C	2	7	0	0
	*Sac*II	CCGC ↓ GG	4	>25	0	1
Streptomyces albus G	*Sal*I	G ↓ TCGAC	2	3	0	0
Streptomyces aureofaciens IKA 18/4	*Sau*I	CC ↓ TNAGG	2	7	0	0
Streptomyces phaeochromogenes	*Sph*I	GCATG ↓ C	4	9	2	0
Streptomyces tubercidicus	*Stu*I	AGG ↓ CCT	>10	>6	7	1
Thermus aquaticus YTI	*Taq*I	T ↓ CGA*	>50	>50	1	10
Xanthomonas badrii	*Xba*I	T ↓ CTAGA	1	4	0	0
Xanthomonas holcicola	*Xho*I	C ↓ TCGAG	1	6	0	1
Xanthomonas malvacearum	*Xma*III	C ↓ GGCCG	2	10	0	0

†Enzymes are named with abbreviations of the bacterial strains from which they are isolated; the Roman numeral indicates an enzyme's relative order of discovery in that strain (i.e., *Acy*I was the first restriction enzyme to be isolated from *Anabaena cylindrica*).

‡Recognition sequences are written 5′ → 3′ (only one strand is given) and the point of cleavage is indicated by an arrow (↓). For example, G ↓ GATCC is an abbreviation for

(5′)G ↓ GATCC(3′)
(3′)CCTAG ↑ G(5′)

When no arrow appears, the precise cleavage site has not been determined. The symbol Pu (purine) indicates that either A or G will be recognized; Py (pyrimidine) indicates that either C or T will be recognized. Two bases appearing in parentheses signify that either base may occupy that position in the recognition sequence. Thus, *Ava*II cleaves the sequence GGACC or GGTCC. Where known, the base modified by the corresponding specific methylase is indicated by an asterisk: A* is N^6-methyladenosine; C* is 5-methylcytosine.

§These columns list the number of cleavage sites recognized by the various specific endonucleases on bacteriophage λ DNA (λ), adenovirus type 2 DNA (Ad2), simian virus 40 DNA (SV40), and φX174 R_f DNA (φX174).

¶Only cleaves methylated DNA (one cleavage site per methyl group).

SOURCE: R. J. Roberts, 1981, *Nuc. Acids Res.* 9:75.

quence complementary to the desired mRNA, and autoradiography. Because a Northern blot indicates not only the presence and size of a specific mRNA in a sample but also its amount, the procedure can be used to compare the amounts of an mRNA generated in cells under different conditions (Figure 7-27).

Band Analysis of S1 Digests Another way of detecting complementarity between two specific sequences of nucleic acids employs the nuclease S1, which is an enzyme from the mold *Aspergillus oryzae*. S1 destroys unpaired RNA or DNA but not double-stranded molecules. Like other hybridization procedures, this technique is quite versatile because either the RNA or the DNA in a hybrid may be labeled. For example, the total unlabeled mRNA from a cell can be hybridized to a pure labeled DNA sample (usually consisting of one or more restriction fragments) that includes all or part of the region of DNA that is transcribed to produce one particular mRNA. The labeled DNA-RNA hybrid is then digested with S1, which removes unpaired nucleic acid strands. The remaining hybrid duplexes are all the same size, so when they are subjected to electrophoresis they lie together in a discrete band. The length of the hybrid can be estimated from the positions of bands formed by marker DNA fragments. This technique has been widely used to determine how much of a particular DNA region defined by restriction sites is complementary to an mRNA (Figure 7-28).

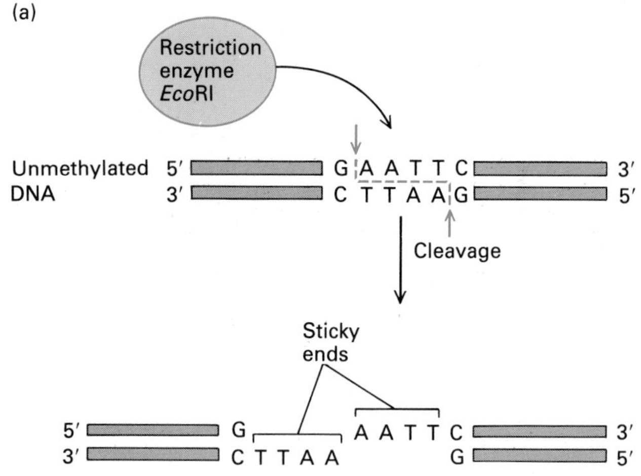

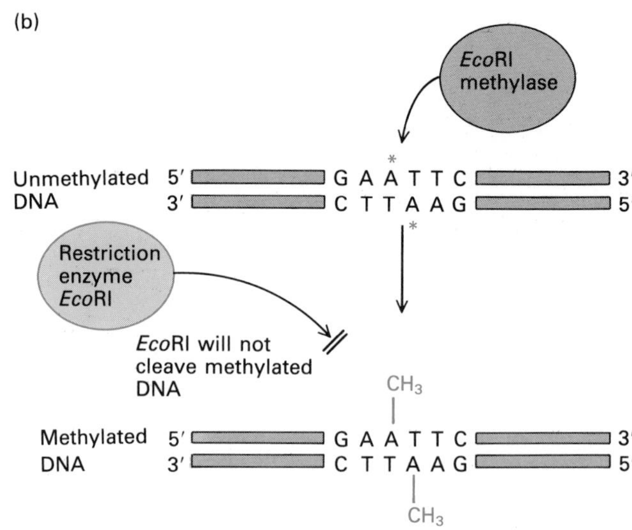

Figure 7-21 (a) *Eco*RI and many other restriction enzymes cleave DNA so that the fragments have short complementary single-stranded segments at the ends. These "sticky ends" are important in recombinant DNA techniques because they readily combine with the ends of other cleavage fragments produced by the same restriction enzyme. *Eco*RI recognizes the six-base-pair sequence

GAATTC
CTTAAG

(b) Most restriction enzymes exist in cells along with modification enzymes. The modification enzyme *Eco*RI methylase catalyzes the methylation of two adenylates *(asterisked)* in the six-base-pair sequence that is normally cleaved by *Eco*RI. If a methylated sample is then exposed to *Eco*RI, the added methyl groups protect the restriction site so that the restriction enzyme does not cut the DNA.

Endonucleases Compared with Exonucleases The enzymes discussed in the previous sections cut DNA or RNA (or both) *within* a chain and are termed *endonucleases*. The restriction endonucleases have a restricted specificity and S1, as well as the widely used pancreatic RNase and DNase, digests almost all nucleic acids. The S1 is a single-stranded endonuclease while the pancreatic enzymes are single- or double-stranded endonucleases. Exonucleases exist that will remove nucleotides one at a time from the ends of RNA or DNA and will do so from single strands only or from one or both strands of duplex DNA. There are too many such enzymes to attempt a comprehensive description of them here and we will simply describe exonucleases as necessary later in this and other chapters.

The Sequence of Nucleotides in Long Stretches of DNA Can Be Rapidly Determined

Molecular cell biology ultimately requires the ability to determine exact nucleotide sequences in DNA or RNA. The discovery of restriction enzymes, which cleave DNA into small, consistently reproducible sets of fragments, was an important step in developing sequencing techniques. A second advance was the development of gel electrophoresis, which has the power to separate fragments that contain from 1 to 500 bases and that differ in length by only one residue.

One highly successful and widely used procedure for DNA sequencing has been named for its inventors, A. Maxam and W. Gilbert. What follows is a simplified description of their method. A quantity of a specific single-stranded DNA fragment is prepared, and the fragment is radiochemically labeled on one of its ends. (A variety of procedures are employed to obtain single-stranded DNA labeled at only one end.) Four samples of the terminally labeled fragment are subjected to four different chemical reactions, each breaking strands specifically at only one base, A, G, C, or T. The reactions are allowed to proceed just long enough for the fragments to be broken only once (on the average). Because the breaks are random, each sample now contains fragments that are labeled at one end and that stretch at their broken ends to *each occurrence* of one of the four bases. Together, the four samples contain fragments that stretch from the labeled end to each base in the original segment.

Parallel gel electrophoresis of the newly created labeled

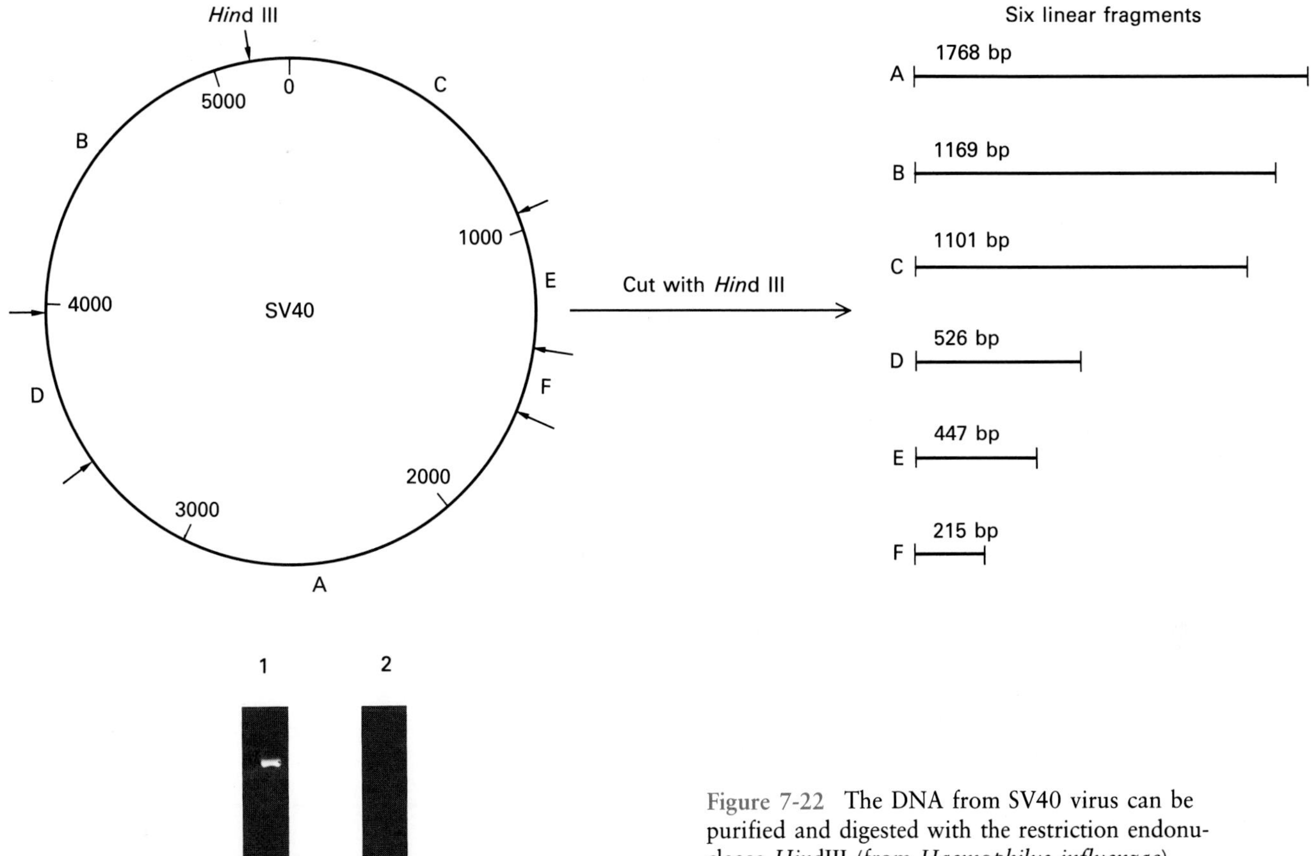

Figure 7-22 The DNA from SV40 virus can be purified and digested with the restriction endonuclease *Hind*III (from *Haemophilus influenzae*). Both the digested and the undigested DNA can then be subjected to electrophoresis. The DNA is made visible in the gel by incorporating ethidium bromide, a molecule that binds to DNA and fluoresces when exposed to ultraviolet irradiation. Lane 1 represents the uncut DNA, and lane 2 the digested DNA. *Hind*III cuts the SV40 molecule six times, producing six fragments. By convention, the pieces of DNA released by a restriction enzyme are labeled A to Z in order of decreasing size; the *Hind*III fragments of SV40 are therefore labeled A to F. The sizes of the various pieces are given in the diagram. *Photograph courtesy of D. Nathans.*

fragments from the four different samples produces four parallel lanes of bands. The bands in each lane correspond to all of the fragments that end in a particular nucleotide, arranged in order of length (Figure 7-29). Because the fragments in different samples end in different nucleotides, the fragments in different columns are never the same length, and the bands formed are never in the same position in any two lanes. Thus, the layout of the bands reveals the order of the nucleotides in the original DNA sample. (In actual practice, the four reactions usually do not include one that breaks only at T, but rather one that breaks at both C and T. A band that appears in the C-and-T lane but not in the C lane is read as T; a band that appears in both lanes is read as C.)

Because sequencing of DNA has proved to be relatively simple and accurate over long stretches of DNA, it is now possible to deduce the sequences of proteins from the sequences of a given DNA and the genetic code. Investigators often obtain protein sequence information by sequencing DNA. Nevertheless, the ability to sequence protein chains directly remains an important tool of molecular biology. To cite one instance, the genome of a higher animal may contain a number of genes that are similar but not identical in sequence. Only by knowing the protein sequence of a product of such related regions can one know which DNA sequence is actually responsible for encoding the protein. Thus, the techniques of direct protein sequencing are still very much in use (Figure 7-30).

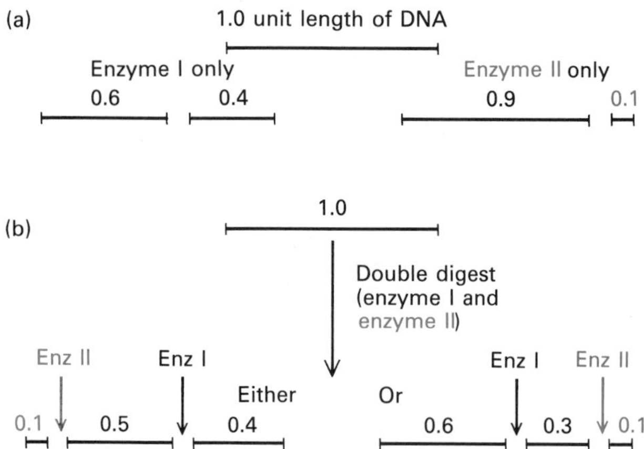

Figure 7-23 Mapping two restriction enzyme sites with respect to one another. (a) When a given piece of DNA is exposed to two restriction enzymes, I and II, each cuts the DNA once. The lengths of the fragments are determined by gel electrophoresis. (b) Digestion with *both* enzymes is used to determine the relative positions of the cuts along the DNA. Analysis of the double digest will give one of the two results shown. The fragment lengths identify the positions of the restriction sites for enzymes I and II with respect to the ends of the DNA and therefore with respect to each other. By continuing this process with different pairs of enzymes, the investigator can construct a detailed and ordered map of restriction sites. The various-sized fragments produced by digestion of any specific DNA are labeled from A to Z,with A the largest fragment.

Recombinant DNA: Selection and Production of Specific DNA

The essence of cell chemistry is purification of sufficient quantities of a substance of interest so that its chemical behavior can be analyzed. Pure samples of specific DNA sequences can be isolated from a simple mixture (e.g., from a virus) by the use of restriction enzymes followed by selection of particular fragments on the basis of size. But if the DNA of interest comes, for example, from a human cell, the total complexity of the mixed DNA sample (4×10^9 base pairs) is so great that the concentration of any specific segment—one, say, from 10^3 to 2×10^3 base pairs long—may be considerably less than 1 part in 10^6. After a restriction digest there may be so many pieces of DNA of about the same size (i.e., more than 10^6 such pieces) that it is impossible to obtain pure genes directly from the cells in which they occur naturally. This obstacle to obtaining pure DNA samples from complex genomes has been overcome by the techniques that are known collectively as *recombinant DNA technology.*

Two widely used types of recombinant DNA preparations are grown or produced in bacterial cells: genomic clones and cDNA clones. A *genomic clone* is a cultured host cell containing a fragment of genomic DNA; a *cDNA clone* is a cultured host cell containing a molecule of *complementary DNA,* DNA that was copied from mRNA in an enzyme-catalyzed reaction (Table 7-5). In both cases, the DNA of interest is linked to a *vector*—either a bacteriophage that can grow in a particular host or a *plasmid,* a small circular DNA molecule capable of

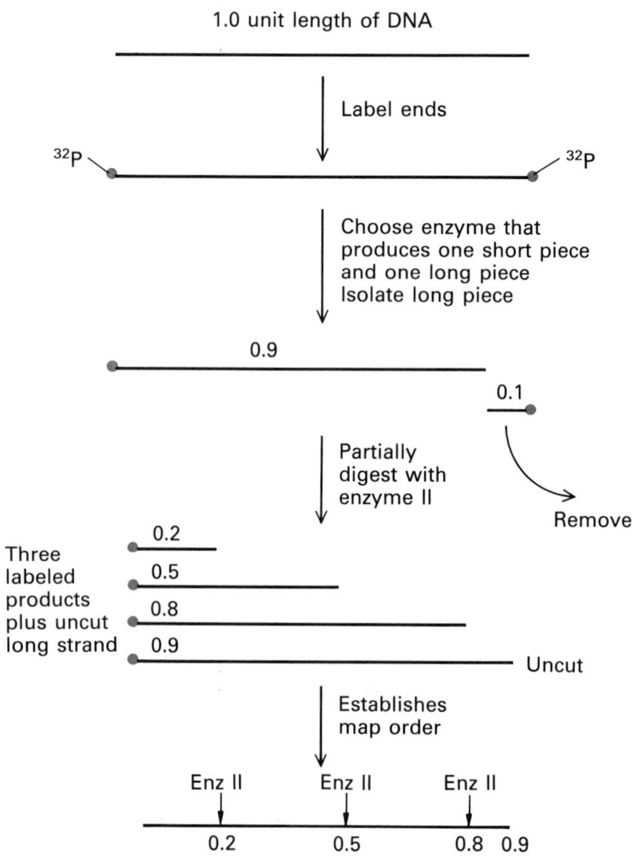

Figure 7-24 Mapping the multiple recognition sites of a restriction enzyme by partial digestion. DNA can be labeled at its termini with ^{32}P, and fragments with *one* labeled terminus can be obtained. In this example, the mapping procedure is applied to the 0.9-unit piece. This piece is cut with an enzyme (enzyme II) that has three sites within the piece. Complete digestion produces one labeled fragment (the 0.2-unit piece), but partial digestion (so that the enzyme cuts each 0.9-unit piece only once) produces three labeled fragments. From the differences in the lengths of the labeled pieces, the positions of enzyme II recognition sites can be inferred. [See H. O. Smith and M. Birnstiel, 1976, *Nuc. Acids Res.* 3:2387.]

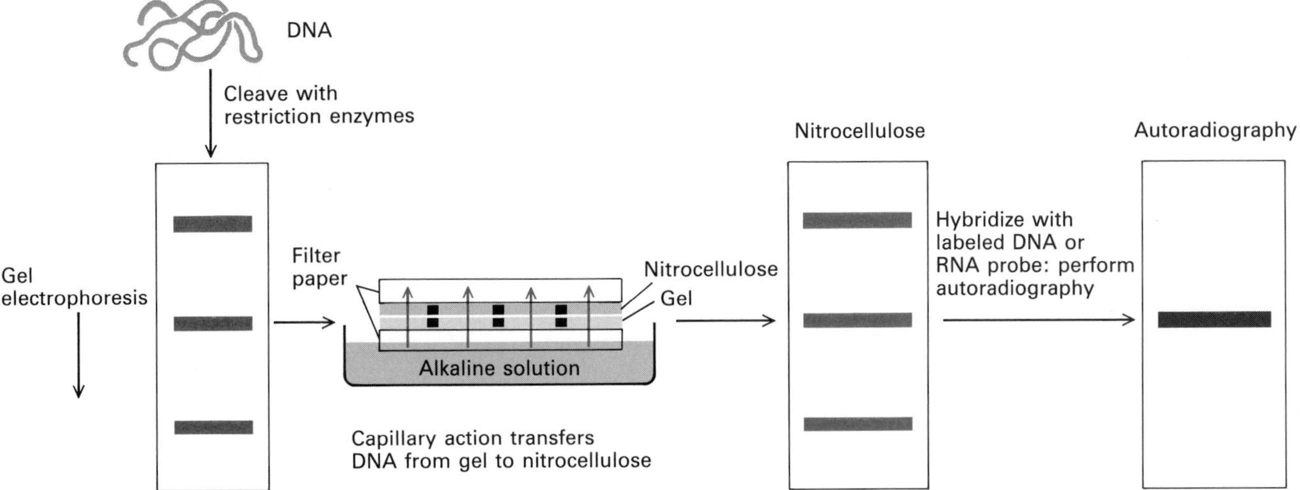

Figure 7-25 The Southern blot technique for detecting the presence of specific DNA sequences. DNA digested with one or more restriction enzymes is separated into fragments by gel electrophoresis. (Marker DNA fragments of known lengths are included in the gel so that their positions can be used to estimate the lengths of the experimental fragments.) The DNA is denatured and transferred from the gel to a nitrocellulose sheet by capillary action; it binds to the sheet. Specific labeled sequences of DNA or RNA can then be hybridized to the bound DNA. Autoradiography locates the hybridized nucleic acids, which indicate the presence of particular known sequences in specific restriction fragments. This technique is sensitive enough that a region of DNA that appears once in the human genome (assuming that the sequence represents at least 1 part in 10^6) can be detected in as little as 5 μg of a DNA sample (the DNA content of about 10^6 cells). [See E. M. Southern, 1975, *J. Mol. Biol.* **98**:508.]

reproducing independently within a bacterial host. (The most widely used *host-vector systems* are *E. coli* with bacteriophage λ or a plasmid.) A *library* consisting of a full set of genomic or cDNA clones can be prepared. The starting material for a library of genomic clones is the total DNA of an organism or cell type; the starting material for a library of cDNA clones is the set of cDNA molecules representing all of the mRNAs in a cell (Figure 7-31).

The preparation and selection of cDNA and genomic clones are illustrated in the following section by a description of how recombinant DNA containing mouse β globin sequences can be obtained.

Complementary DNA (cDNA) Clones Are Copies (Whole or Partial) of Messenger RNA

To prepare cDNA clones with globin-specific sequences (Figure 7-32), the starting material is reticulocytes, red blood cell precursors in which 90 percent of the proteins synthesized are α and β globins. These cells are therefore rich sources of globin mRNA, and extraction and selection of total cell mRNA (a technique made possible by the presence of a 3′ poly A tail on all eukaryotic mRNAs, as is discussed in detail in Chapter 9) yields a product that is almost pure.

Complementary DNA copies of the mRNAs are made by using the enzyme reverse transcriptase (found in retroviruses; Chapters 4 and 6). Like the ordinary DNA polymerases in cells, this enzyme can build a complementary nucleic acid strand on a template, but only by adding nucleotides to a primer. Thus, before the reverse transcriptase can do its work, a short primer strand must be hybridized to the nucleotides near the 3′ ends of the mRNAs. Fortunately, a single oligonucleotide primer—a string of thymidylate residues (poly T)—will serve for most eukaryotic mRNAs, because most end in a string of 50 to 250 adenylate residues (poly A).

When the cDNA strands are complete, the mRNA strands are removed and a complementary DNA strand is built on each cDNA. In one technique for doing this, the 3′ end of each single cDNA strand is elongated by adding several residues of a single nucleotide—say, poly C—through the action of an enzyme that adds bases at free 3′ ends. A poly G primer is then hybridized with the terminal poly C. This primer is elongated by a DNA polymerase into a strand complementary to the cDNA. What results is a complete double-stranded DNA copy of the original mRNA.

The next step is to insert the globin-specific DNA into a

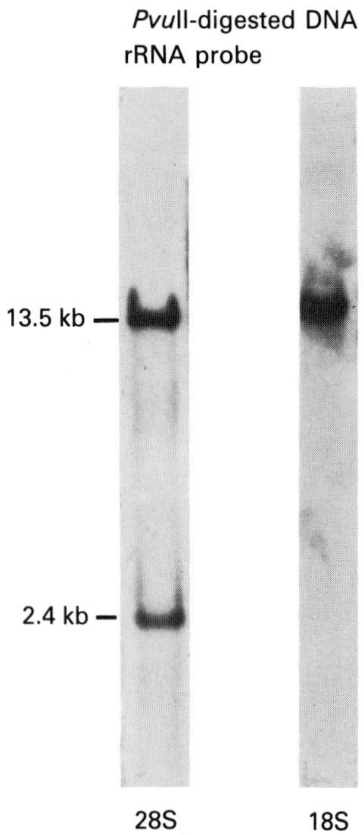

(a)

PvuII PvuII PvuII

45S

18S 28S

Region coding for ribosomal RNAs

(b)

18S 28S

13.5 kb 2.4 kb

Figure 7-26 A Southern blot of human DNA was used to detect the presence of recognition sites for the restriction enzyme *Pvu*II within the gene encoding ribosomal RNA. (a) The map shows the coding regions for the 18S and 28S rRNAs (see Figure 4-15). (As will be explained in Chapter 9, the cell actually transcribes the whole 45S stretch as a single unit and then cleaves the resulting RNA into the smaller 18S and 28S rRNAs.) In this experiment, the *Pvu*II enzyme cut the DNA in and around all of the 45S rRNA transcription unit at the sites indicated by the arrows. (b) After the DNA was transferred to two nitrocellulose sheets (to provide two samples), it was hybridized with two different labeled probes—the 28S and the 18S rRNAs. Autoradiography revealed that the 28S rRNA had hybridized with two different DNA fragments, one 13.5 kb long and one 2.4 kb. (Again, the lengths of the fragments were estimated by using marker DNA samples not shown; kb = kilobases or 1000 bases.) This result indicated that a cut had been made *within* the DNA region encoding the 28S rRNA. The 18S rRNA also hybridized with the 13.5-kb fragment; clearly, then, DNA sequences encoding 18S and 28S rRNA were contained, at least in part, on the same 13.5-kb fragment. *Photograph courtesy of N. Arnheim and A. Franzusoff.*

plasmid, a small circle of double-stranded DNA that can replicate independently in bacteria. Plasmids occur naturally in almost all bacteria, and were originally detected by their ability to transfer genes between bacteria (Chapter 6). It has been shown that a specific region of the circle assures replication in a host bacterium.

The plasmid or vector DNA is cleaved once with a restriction enzyme at a point where cleavage does not affect the vector's ability to grow in its bacterial host. The globin cDNA is then inserted at the cut site and the circle is rejoined. The technique that was first used and that is still the most popular one for carrying out this insertion is called *homopolymeric tailing*. The double-stranded globin cDNA has two 3′ ends, to which a homopolymer—say, poly C—is added. The cut plasmid is also a double-stranded DNA having two 3′ ends, to which a complementary homopolymer (poly G) is added. When the "tailed" plasmid and globin DNA are mixed, their complementary single-stranded tails spontaneously hybridize; the resulting circular recombinant molecule can be resealed with the enzyme DNA ligase (Chapters 4 and 13). Specifically treated *E. coli* cells take up the plasmid, and the recombinant molecule multiplies along with the growing culture of cells.

If the experimenter has chosen a plasmid that encodes the information necessary for resistance to an antibiotic, individual cells that have taken up the plasmid will grow and multiply in the presence of the antibiotic, whereas other cells in the culture will not grow. Thus, each cell in an antibiotic-resistant colony should contain globin cDNA as well. If at the outset the number of plasmids allowed to infect the *E. coli* cells is 1/10 or less of the total number of *E. coli* cells, then it is very unlikely that more than one plasmid will end up in a recipient cell. As a rule,

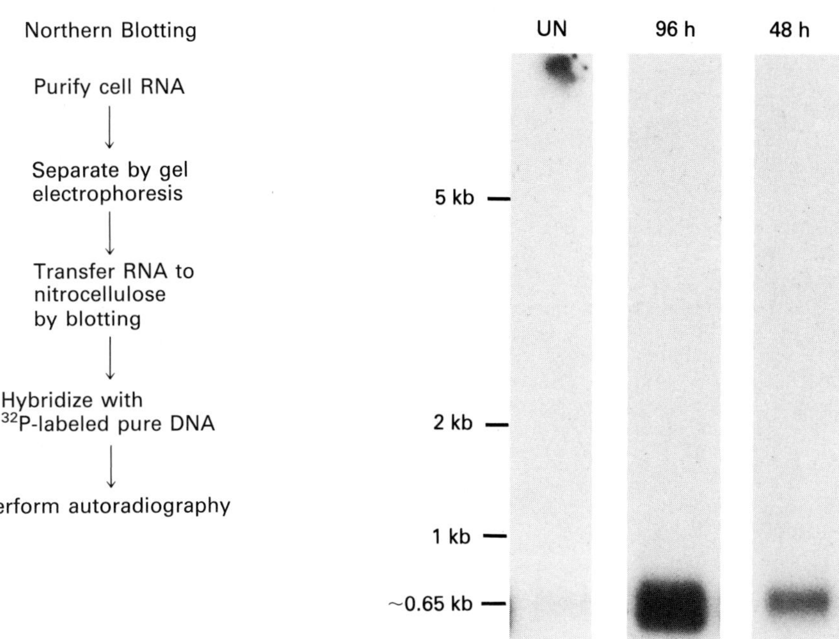

Figure 7-27 The Northern blot is used to locate a specific mRNA molecule in a test sample, and to determine the size of this molecule and the amount of it present. The procedure is similar to the Southern blot, except that the sample to be probed consists of RNA molecules rather than enzyme-digested DNA sequences. The total RNA from cells is extracted, separated into specific sizes by gel electrophoresis, and transferred to a nitrocellulose filter, as in Figure 7-25. The filter-bound RNA is then hybridized with purified labeled DNA that is complementary to the mRNA molecule of interest. An autoradiograph shows the position of that mRNA in the gel, and the intensity of the signal shows the amount of it. The photo indicates the relative quantities of β globin mRNA in erythroleukemia cells at three different times: when cells are growing and have not started to make globin *(lane UN, for "uninduced")*, and 48 and 96 h after they have been treated with dimethyl sulfoxide, a compound that induces them to stop growing and to begin to differentiate. The β globin mRNA is barely detectable in growing cells but increases by a factor of more than 1000 in 96 h of differentiation. *Photograph courtesy of L. Kole.*

then, the recombinant DNA in all the cells of a colony will have descended from a single recombinant DNA molecule. To verify what the plasmid vector contains, the experimenter can sequence the recombinant molecule.

Complementary DNA clones can also be prepared from the unpurified mRNA from any cell type, but this produces a random mixture of individual recombinant clones, which must then be screened to isolate specific clones. Such a screening technique with labeled nucleic acid probes will be described in Figure 7-34. Recently, antibody techniques for identifying the protein products encoded by the cloned cDNA have been worked out. Here, if an experimenter has an antibody that is reactive with a protein (or part of a protein) the antibody can be used directly to detect whether an *E. coli* colony (or a bacteriophage plaque) contains the protein (or part of a protein) reactive with the antibody. If so, then the cDNA sequence corresponding to that protein has been cloned.

Genomic Clones Are Copies of DNA from Chromosomes

The most common procedure for preparing globin-specific clones from mouse genomic DNA—that is, the DNA in mouse chromosomes—makes use of bacteriophage λ. The DNA of the phage is about 50 kb long, but a sequence about 25 kb long is dispensable: this section can be removed and replaced with foreign DNA without impairing the ability of the phage to infect cells and reproduce. This is why bacteriophage λ is so widely used in making libraries of genomic clones (as shown in Figure 7-31), or *genomic libraries.* A genomic library is a collection of recombinant molecules that includes *all* of the DNA sequences of a given species. Once the library has been prepared, it can be screened for the phage that contains the DNA sequence of interest.

The size of a library depends on the haploid amount of DNA in the organism. For example, the human and the

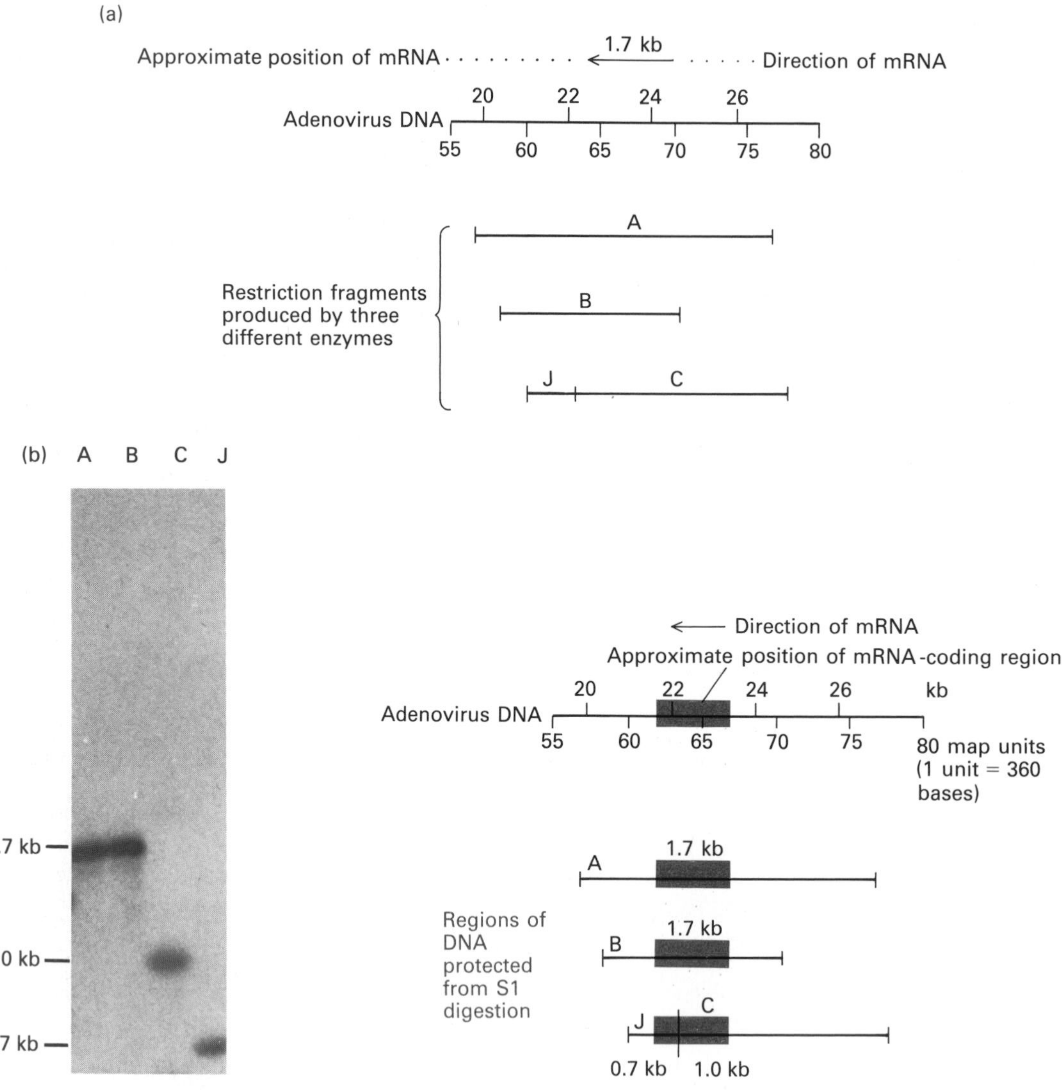

Figure 7-28 The S1 mapping technique is a way of determining the lengths of the complementary sequences in two nucleic acid samples. (S1 is an endonuclease that destroys single-stranded RNA and DNA.) (a) A portion of the map of the DNA of adenovirus is shown. Earlier experiments had established that an mRNA was complementary to the large region spanned here by restriction fragment A. Restriction fragments A, B, C, and J were prepared from ^{32}P-labeled DNA. These four labeled DNA fragments were denatured and hybridized with a large excess of mRNA prepared from virus-infected cells. The mixture was treated with S1 to destroy any unpaired DNA that had not found an mRNA partner and protected DNA fragments were separated by gel electrophoresis. (b) The autoradiograph of the gel shows the lengths of the portions of the DNA fragments that were protected by the mRNA (i.e., that were complementary to the mRNA). Fragments A and B were protected for 1.7 kb, and C and J were protected for 1.0 and 0.7 kb, respectively. Thus the mRNA includes 1.7 kb positioned as indicated. The direction of the mRNA shown by the arrowhead was deduced in additional experiments by separating the DNA strands and determining which of the two was protected by the mRNA. *From A. J. Berk and P. A. Sharp, 1978,* Cell *14:695. Copyright M.I.T.*

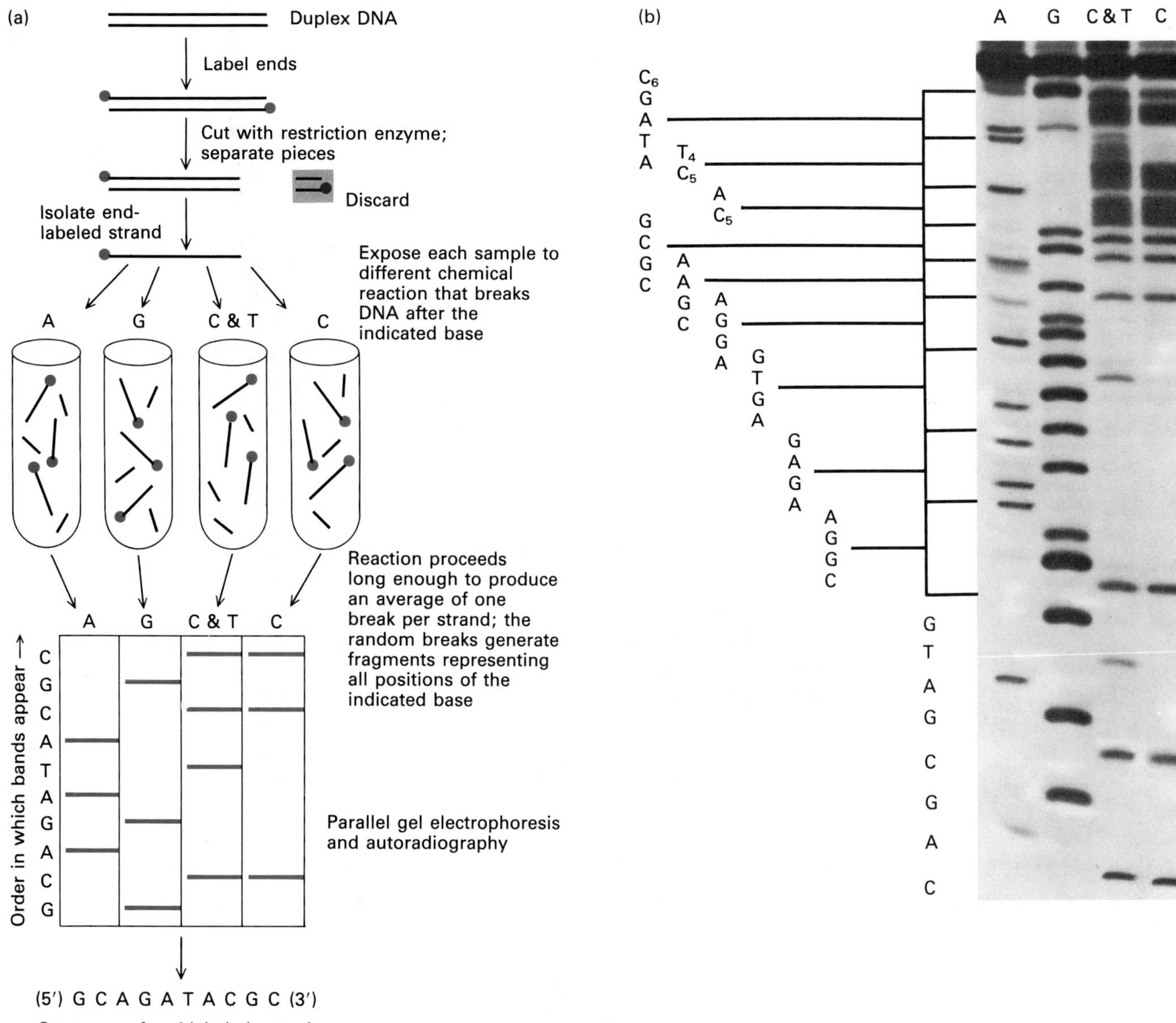

Figure 7-29 DNA sequencing by the Maxam-Gilbert method. (a) A 5′ end-labeled DNA fragment serves as a yardstick for the experimenter to use in determining the position of each base in the sequence. A different chemical reaction in each of four samples breaks the DNA fragment only (or mainly) at A, G, both C and T, and C residues, respectively. The labeled subfragments created by these reactions all have the label at one end and the cleavage point at the other. After gel electrophoresis and autoradiography, the four sets of labeled subfragments together yield one radioactive band for each and every nucleotide in the original fragment. (b) An actual autoradiograph indicating part of the adenovirus genome. This sequence is in the region where most RNA synthesis begins late in adenovirus infection. The variations in the darkness of bands are due to differences in the extent of the breakage reactions. [See A. Maxam and W. Gilbert, 1977, *Proc. Nat'l Acad. Sci. USA* **74**:560.] *Photograph from E. Ziff and R. M. Evans, 1978,* Cell *15:1463 (Fig. 4).*

Table 7-5 Terms used in recombinant DNA research

Genomic DNA	All DNA sequences of an organism
cDNA (complementary DNA)	DNA copied from an mRNA molecule
Plasmid	An extrachromosomal small circular DNA molecule capable of reproducing independently in a host cell
Vector	A plasmid or a viral DNA molecule into which either a cDNA sequence or a genomic DNA sequence is inserted
Host cell	A cell (usually a bacterium) in which a vector can be propagated
Genomic clone	A selected host cell with a vector containing a fragment of genomic DNA from a different organism
cDNA clone	A selected host cell with a vector containing a cDNA molecule from another organism
Library	A complete set of genomic clones from an organism, or of cDNA clones from one cell type

mouse genomes are each about 4×10^9 base pairs long. If one of these genomes is divided into fragments about 2×10^4 base pairs long for insertion into bacteriophage λ, then 2×10^5 different recombinant bacteriophage λ particles are required to constitute a complete library. Because the pieces of DNA are incorporated into recombinant form randomly (as will be obvious below), about 10^6 recombinant phages are necessary to provide a 90 to 95 percent chance that every DNA piece will be included.

The first step in preparing a genomic library is to extract all of the DNA from cells of the organism (Figure 7-33). Often sperm cells or early embryos are used. In preparing mouse globin DNA clones, the DNA of an entire mouse embryo may be extracted. The DNA is then broken into fragments by a restriction enzyme such as *Eco*RI, which cleaves it in a way that produces short single-stranded "sticky" ends (AATT and TTAA) on every fragment (see Figure 7-21). The digestion is not carried to completion but is stopped when the average size of a fragment is approximately 2×10^4 base pairs (20 kb). The bacteriophage λ DNA also can be cut by *Eco*RI to yield a center section of approximately 2.5×10^4 base pairs (25 kb) plus two flanking ends, or *arms*. The center section of the phage DNA is removed by separating it from the two arms.

The arms and the collection of genomic DNA are then mixed in about equal amounts (approximately 2×10^5 DNA fragments and a similar number of pairs of λ arms). Because of the complementarity of the sticky ends, molecules approximately the same length as normal phage DNA will form, but these will include a piece of mouse DNA. DNA ligase, the enzyme that normally joins DNA breaks (Chapter 4), is used to seal the recombinant molecules. The recombinant DNA is then coated with bacteriophage proteins prepared from infected cells. Only DNA molecules of the correct size will be effectively coated (or *packaged*) so that they can give rise to fully infectious λ bacteriophages; these will, of course, contain DNA sequences from the mouse genome. The bacteriophages containing DNA sequences that code for globin can be detected by hybridization of globin-specific cDNA sequences (prepared as described in Figure 7-32) to DNA obtained from each plaque (Figure 7-34).

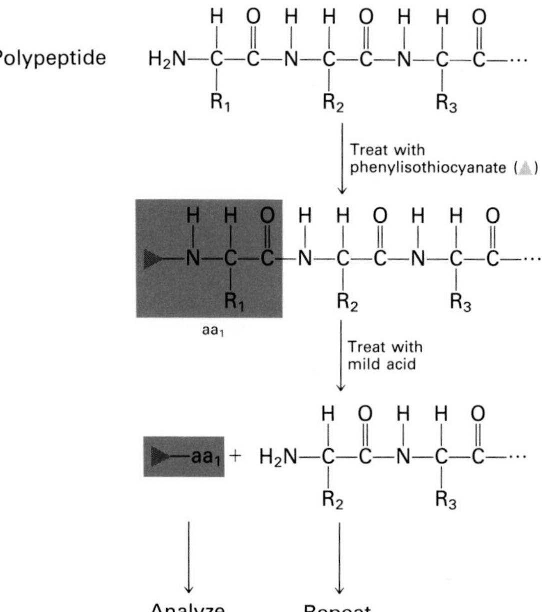

Figure 7-30 Sequencing a protein by the Edman degradation method. The peptide is treated with phenylisothiocyanate, which combines with the amino-terminal residue in the peptide chain, rendering the first peptide bond in the chain labile to treatment with mild acid. The same pair of reactions is carried out repeatedly to remove the amino acids one at a time. After each step, the removed amino acid is chemically identified. In this way, the entire amino acid sequence of the short peptides can be determined. Machines called sequenators can perform this reaction automatically for about 20 to 30 residues of a pure protein, using tiny amounts of it.

Vectors for Recombinant DNA Exist in Many Cell Types

Any gene that can be subjected to a hybridization assay can be purified. Once a bacteriophage or another bacterial vector with the desired gene is prepared, an unlimited amount of the pure gene can be obtained by growing the vector and extracting the DNA. There are also vectors that can carry recombinant DNA molecules in yeast, in higher plant cells, and in human cells. The vectors most frequently used in mammalian cells are the small DNA viruses SV40 and polyoma. In plant cells, the most common vector is the TI plasmid, which is grown first in a bacterium but is then transferred into the plant cell.

For the cell biologist, the availability of unlimited amounts of a pure gene through the use of recombinant DNA technology offers rich opportunities for further chemical and biological study. Access to vectors in yeast and in cultured mammalian cells affords the additional possibility of testing the biological function of particular eukaryotic DNA sequences in a variety of eukaryotic cells.

Industrial microbiologists can use recombinant DNA techniques to engineer bacteria and other easily cultured organisms to make proteins that are useful in medicine, in agriculture, and in research of all kinds. Progress in this area has been astonishingly rapid. A number of virus proteins that are important in immunizations (e.g., against the foot-and-mouth disease virus of cattle, and against hepatitis virus in humans) have already been synthesized in *E. coli,* as have several hormones (e.g., insulin and growth hormone; Figure 7-35). The vectors that direct such programmed protein synthesis are called *expression vectors;* the use of such vectors allows the experimenter to take advantage of bacterial genetic tricks that increase mRNA synthesis to produce large quantities of desired proteins.

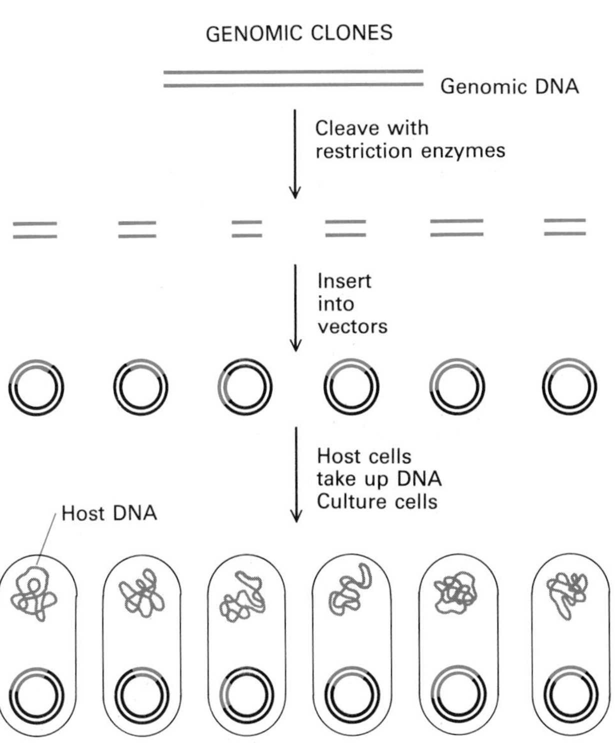

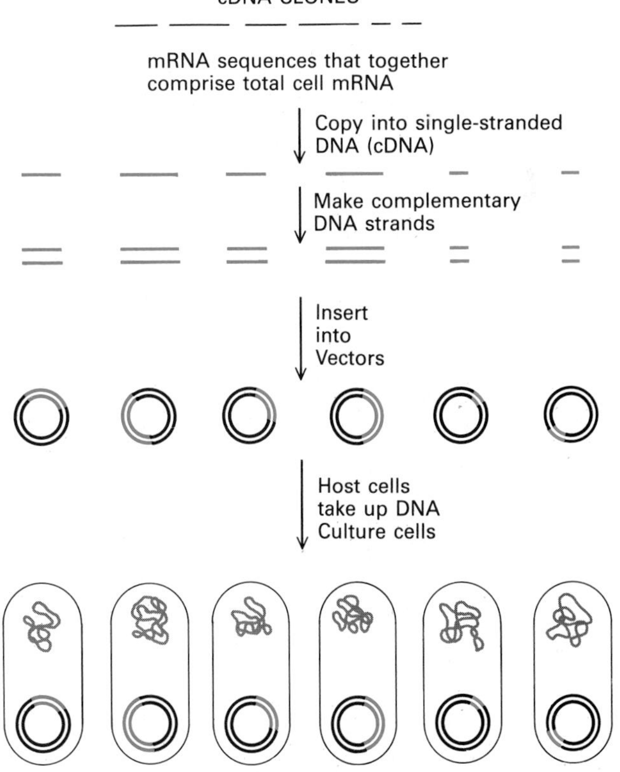

Figure 7-31 A comparison of genomic cloning with cDNA cloning. The major difference is that in genomic cloning, the genomic DNA must be cleaved with restriction enzymes before it can be inserted into vectors; in cDNA cloning, the mRNAs must first be copied into double-stranded DNA molecules.

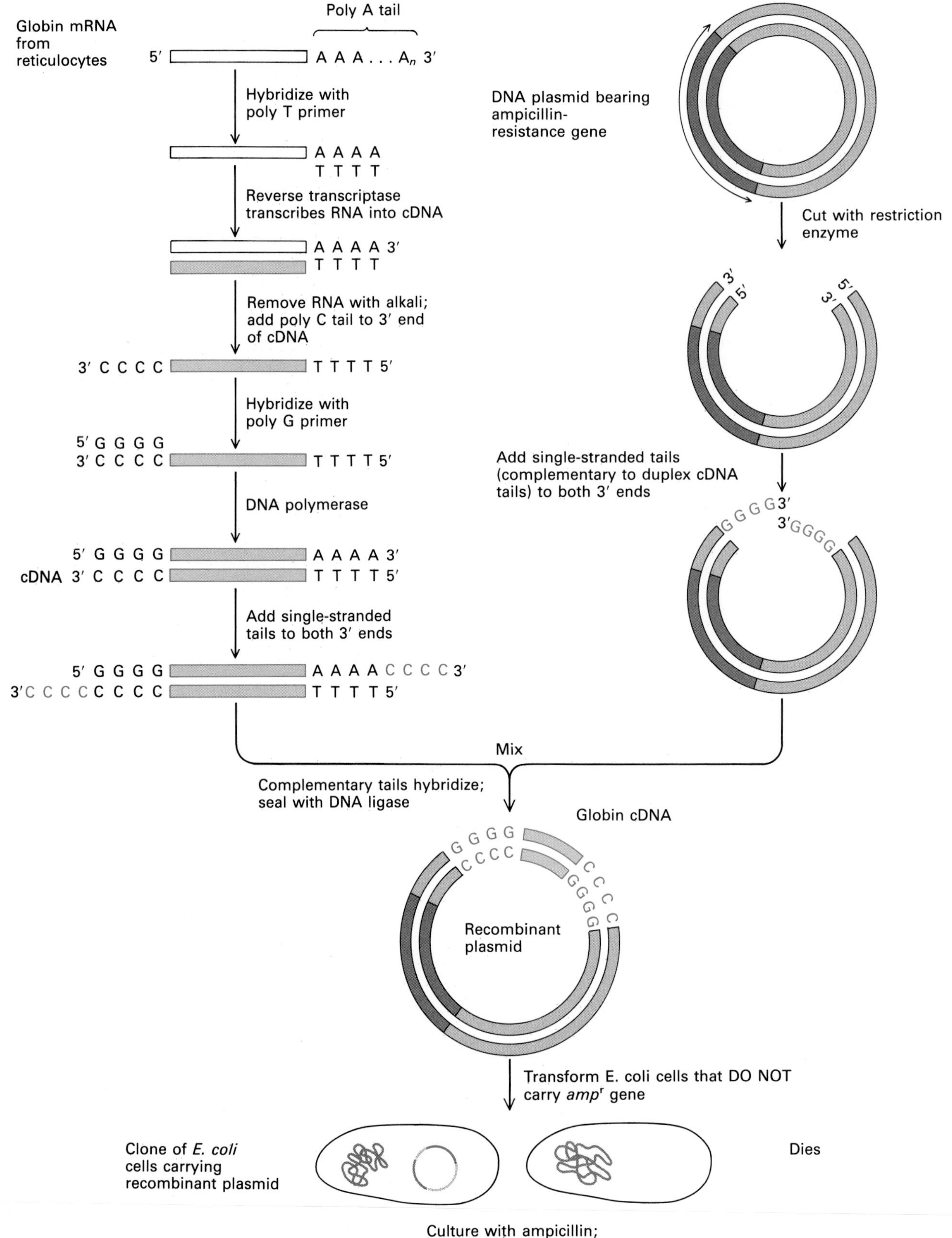

Figure 7-32 The preparation of a cDNA clone containing sequences that code for globin.

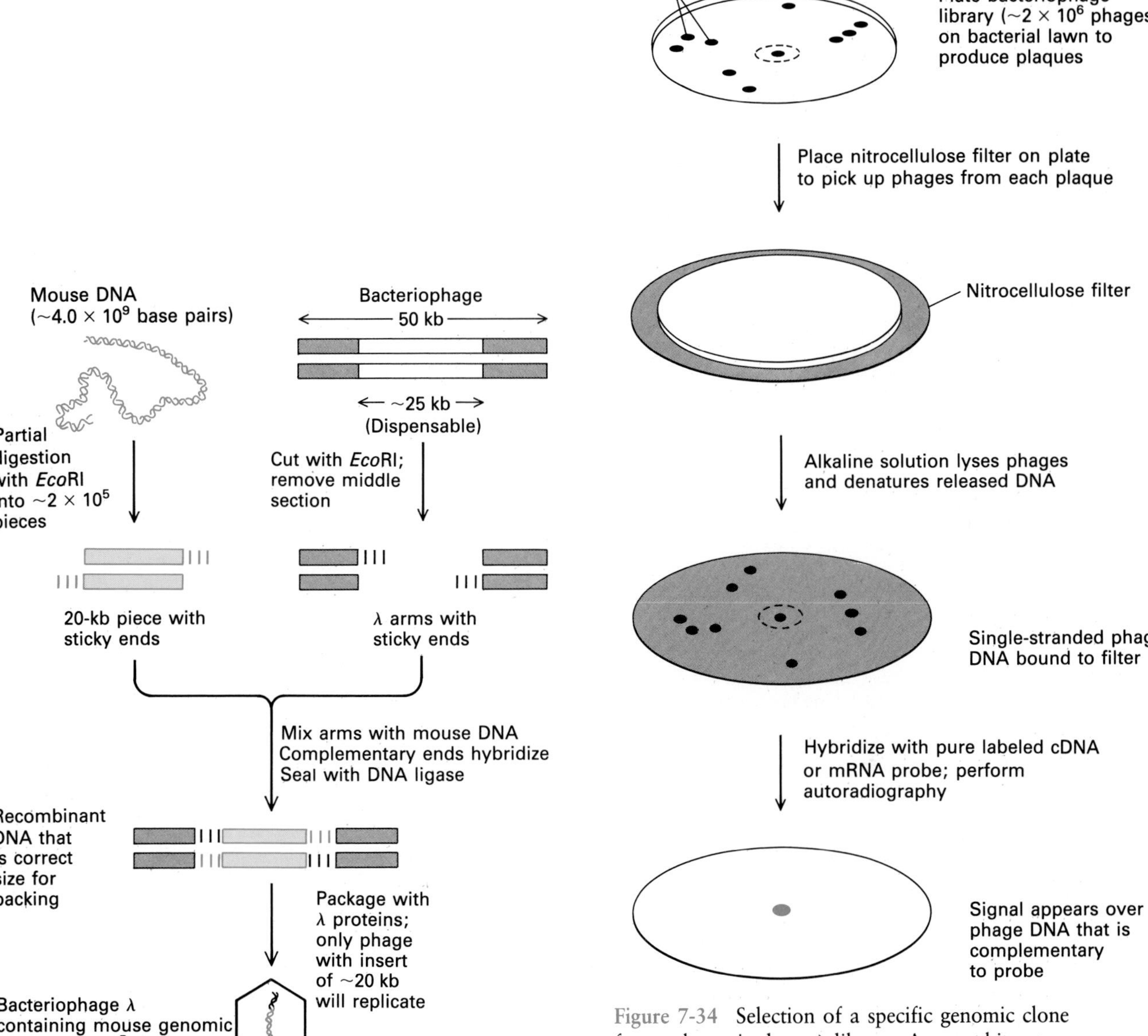

Figure 7-33 The construction of a genomic library of mouse DNA in bacteriophage λ. The total DNA from mouse cells (sperm cells or the cells of embryonic tissue, either of which presumably has a complete set of sequences) is broken into fragments of approximately 20 kb by partially digesting it with endonucleases and purifying the pieces of appropriate size. These are inserted into the DNA of bacteriophage λ, which has two restriction sites that allow removal of the middle section and selection of the two "arms." Only reconstituted phage with approximately 20 kb of foreign DNA will propagate. A single region of the mouse genome, such as that encoding β globin, would occur approximately once in 10^5 particles.

Figure 7-34 Selection of a specific genomic clone from a bacteriophage λ library. A recombinant phage suspension is plated on a large surface of bacterial "lawn," and individual phage plaques appear. (Although about 2×10^5 phages *could* contain all the mouse sequences, 2×10^6 phages are plated to ensure that a phage with the desired sequence is included. This requires an area of 1000 to 2000 cm^2 to accommodate all of the phage plaques.) The phages on the surface will stick to a nitrocellulose filter, and in alkaline solution the DNA will be released and will attach to the nitrocellulose, where it forms an image identical to the plaque distribution on the plate. The filter is then exposed to a pure labeled cDNA or mRNA probe (e.g., the β globin sequence) to identify the complementary phage DNA. The filter is compared with the master λ library plate; the position of the labeled DNA on the filter allows the correct plaque on the plate to be selected. Phage particles from that plaque can then be purified.

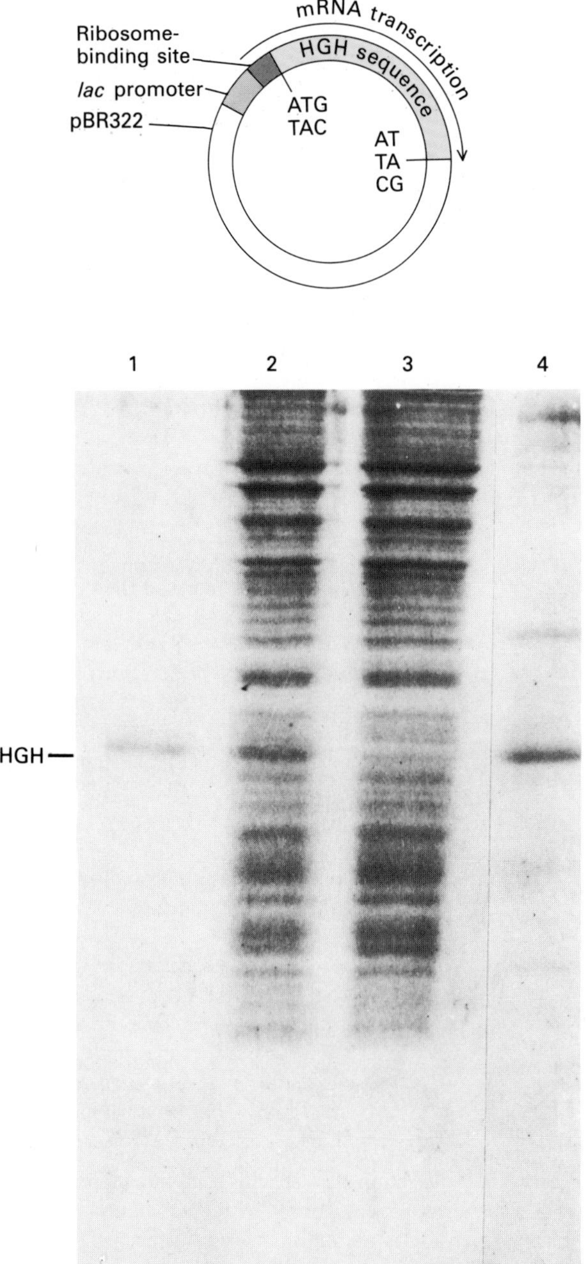

Figure 7-35 The production of human growth hormone (HGH) in *E. coli*. The DNA sequences that encode the 191 amino acids of HGH were inserted by recombinant DNA techniques into the *E. coli* plasmid pBR322. A strong RNA initiation site called the *lac* promoter (promoters are discussed in Chapter 8) causes increased RNA synthesis in the presence of lactose; this sequence also was inserted so that mRNA production could be greatly increased by the presence of lactose in the culture medium.

Messenger RNAs have a site that binds to the ribosome, where, of course, they are translated. In this procedure, the coding region for a site that binds to the *E. coli* ribosome was included, to enable *E. coli* to translate the mRNA made from the inserted HGH DNA. Finally, the sequences TAC and ATC were also inserted. The former is transcribed as AUG, the methionine-coding mRNA triplet that initiates protein synthesis; the latter is transcribed as UAG, one of the termination codons.

The *E. coli* cells were labeled with [^{35}S]methionine, and the labeled proteins were detected by gel electrophoresis. Cells that harbored the recombinant plasmid contained a protein *(visible in lane 2)* that migrated like authentic growth hormone *(lane 1)* during gel electrophoresis. Bacteria containing plasmids that lacked the growth hormone sequences did not make this protein *(lane 3)*. After partial purification of the growth hormone from the bacterial extracts, the band in lane 4 was obtained. *Photograph courtesy of D. V. Goeddel. Reprinted by permission from* Nature, *vol. 281, p. 544. Copyright 1979 by Macmillan Journals Limited.*

Controlled Rearrangement and Site-Specific Mutagenesis in DNA

With the availability of pure DNA in unlimited amounts, a variety of chemical and enzymatic techniques for altering DNA have been developed. The practice of genetics no longer depends on isolating naturally occurring mutant organisms; DNA can be changed in the test tube and reinserted into cells. This technology enables the experimenter to introduce mutations in genes that determine protein structure and to change DNA sequences that function as genetic regulatory or "control" elements (for example, sequences that code for specific protein-binding sites). (Control elements will be discussed in more detail in Chapters 8, 9, and 12.) Two techniques for introducing mutations—the alteration of a single base and the deletion of a short DNA sequence—are illustrated in Figure 7-36. The function of the mutant DNA—whether it be a *deletion mutant* as in Figure 7-36a or a *point mutant* as in Figure 7-36b and c—can be tested by reintroducing it

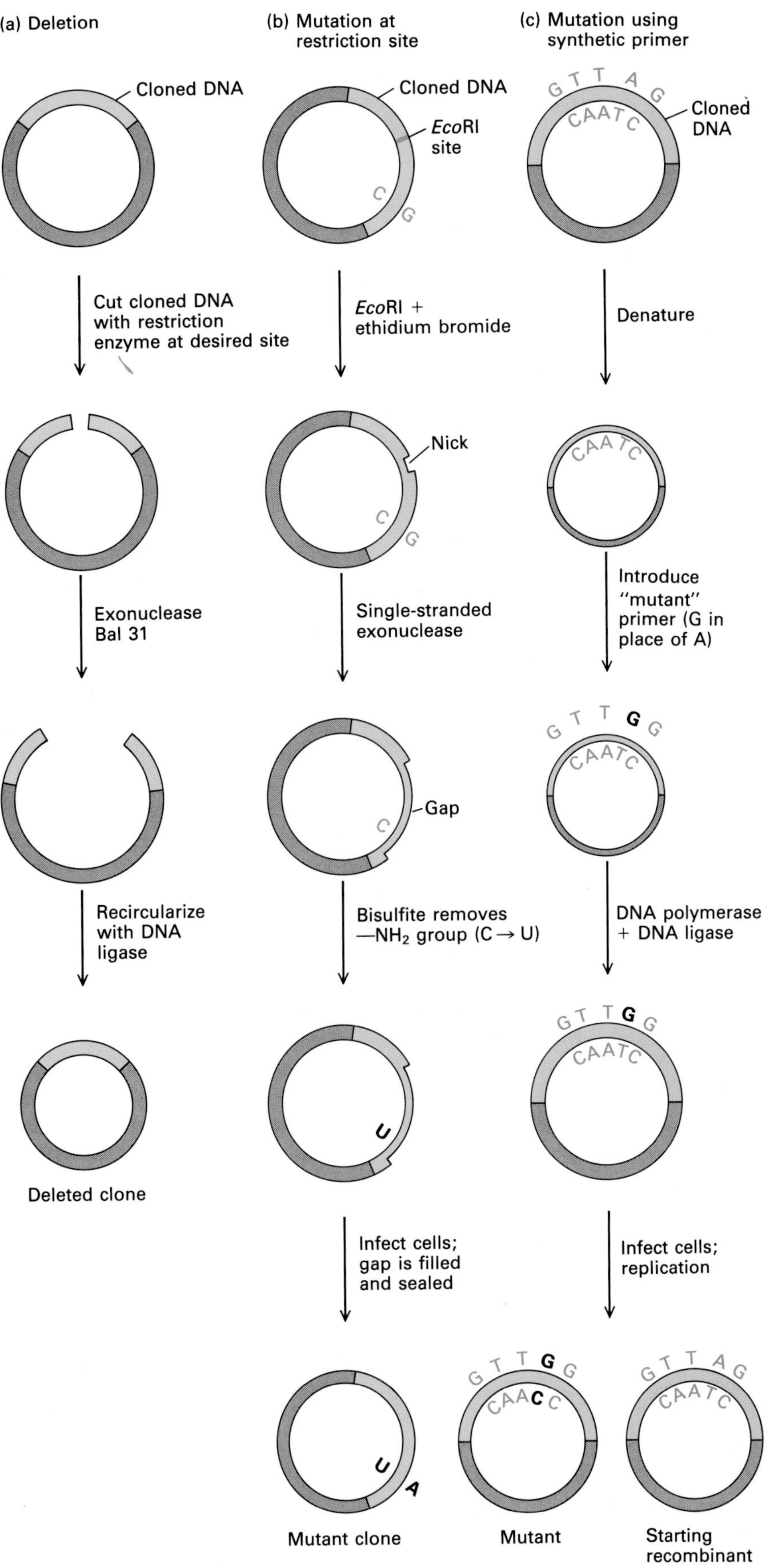

Figure 7-36 In vitro mutagenesis: constructing (a) DNA deletions and (b),(c) DNA point mutations through the use of recombinant DNA techniques. (a) Deletions are made in cloned DNA by removing entire sections of DNA between two restriction sites, or by cutting at a single restriction site and using the exonuclease Bal 31, which removes nucleotides from both ends of a linear double-stranded DNA molecule (as illustrated). From a collection of such deleted molecules, those with deletions of various lengths are chosen. (b) Restriction endonucleases are used under conditions that encourage cleavage of only one strand at a single recognition site (e.g., brief incubation in the presence of ethidium bromide often causes a nick—a single-stranded cut—at one restriction site). Limited single-stranded digestion with an exonuclease produces a gap that exposes only a few bases; treatment with bilsulfite converts a cytosine in the single-stranded region into a uracil. The result is a mutation that will be expressed when the DNA is copied. (c) The two strands of a cloned DNA are separated, and a chemically synthesized oligonucleotide primer (see Figure 7-39) that is *almost* complementary to one of the strands (there is a single mismatched base) is hybridized to that strand. The primer is then extended by a DNA polymerase. The single strands of the double-stranded molecule are each replicated and the resulting molecules are replicated further, to produce two discrete populations of molecules differing in a single base pair.

FROM GENE TO PROTEIN

Take specific cells (e.g., brain or liver)
↓
Isolate mRNA; make cDNA
↓
Select and sequence individual cDNA clones that are tissue-specific
↓
Deduce protein sequence from DNA sequence
↓
Make peptides specified by sequence; inject into animals to produce antibodies
↓
Isolate pure protein by affinity to antibody

FROM PROTEIN TO GENE

Isolate protein (an enzyme or a biologically active protein—e.g., a hormone)
↓
Obtain partial amino acid sequence
↓
Make oligonucleotides that correspond to amino acid sequence
↓
Use labeled oligonucleotides to select genomic DNA clone
↓
Sequence and characterize selected gene

Figure 7-37 With modern techniques it is now possible to identify an mRNA of interest (say, a brain-specific mRNA) and to use it to isolate the protein it encodes without knowing the function of that protein. It is likewise possible to purify a tiny amount of a protein that has a specific function (say, an enzyme or a growth factor) and to use a bit of the protein sequence to isolate the gene.

into a cell by injection or transformation (Chapter 6). The power of this approach is that without knowing beforehand what the role of a particular sequence is, the experimenter can determine its function by altering its structure and reintroducing it into the organism. These practices have been termed "reverse genetics" by Charles Weissman.

Synthesis of Specific Peptide and Nucleotide Sequences: Use in Gene Isolation and Identification

As the primary sequences of proteins and nucleic acids have begun to accumulate, the special importance of certain short sequences—both regulatory "signals" in nucleic acids and functional subsections or "domains" in proteins—have become more apparent. (The former are discussed in detail in Chapters 8 to 13; the latter are treated in Chapters 14 to 21 and 25.) In addition to simply deciphering the sequences of macromolecules, investigators have put great effort into learning to chemically synthesize specific protein and DNA fragments. With such fragments, the function of a part of a protein rather than the whole protein can be tested. Also, "mutant" oligonucleotides can be inserted into normal cloned DNA sequences to study the effects of mutations.

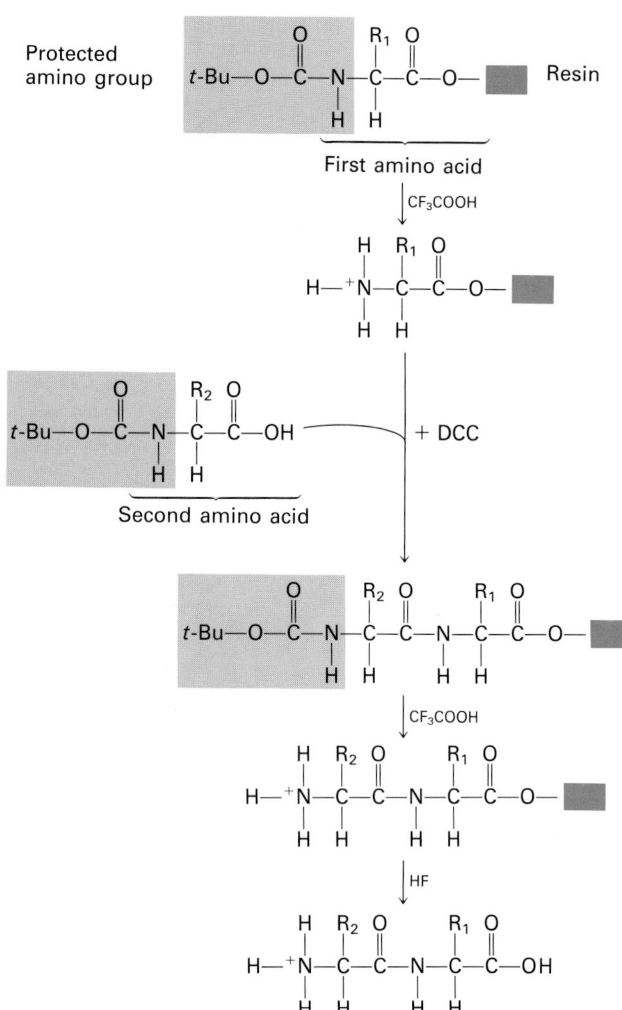

Figure 7-38 Solid-phase peptide synthesis. The first amino acid of the desired peptide is attached by its carboxyl group through a reaction involving esterification to a resin. The amino group of the first amino acid in the peptide under construction is blocked by a reversible organic reaction resulting in the substitution of a *tert*-butyloxycarbonyl, *t*-Bu—O—C(=O)—. This linked structure is shown as the first amino acid. The *tert*-butyloxycarbonyl blocking group is removed by treatment with trifluoroacetic acid (CF_3COOH), and the resulting free amino group forms a peptide bond with a second amino acid, which is presented with a reactive carboxyl group and a blocked amino group together with the coupling agent DCC (dicyclohexylcarbodiimide). The process is repeated until the desired product is obtained, and the peptide is then chemically cleaved (with hydrofluoric acid) from the resin. [See R. B. Merrifield, L. D. Vizioli, and H. G. Boman, 1982, *Biochemistry* **21**:5020.]

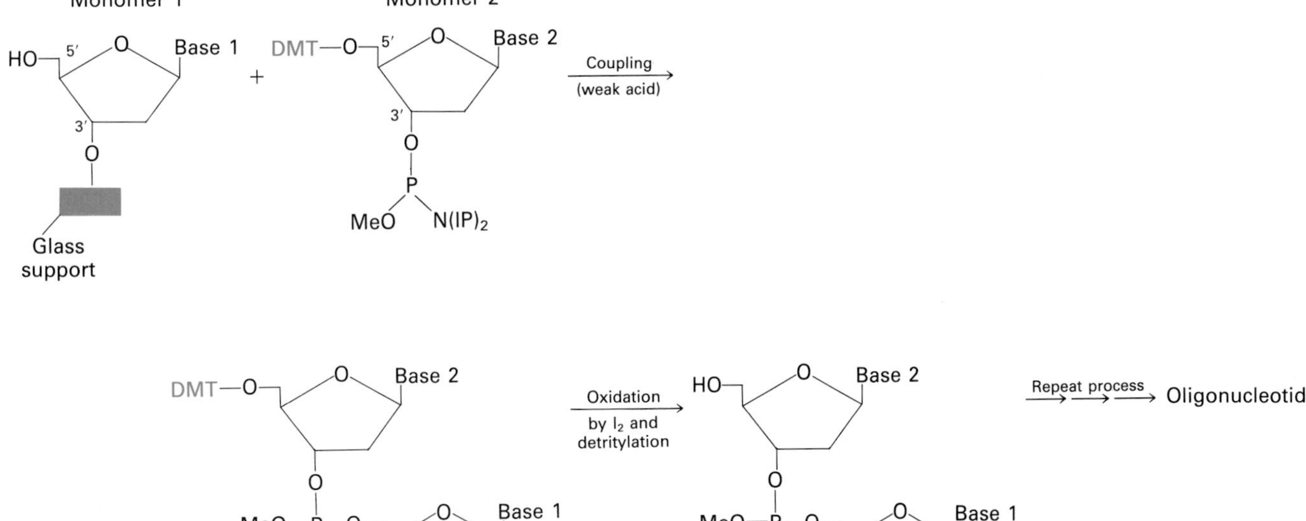

Figure 7-39 Synthesis of oligonucleotides. The first nucleotide *(monomer 1)* is bound to a glass support by its 3′ hydroxyl; its 5′ hydroxyl remains available. The synthesis of the first internucleotide link is carried out by mixing monomer 1 with monomer 2, a reactive 3′-diisopropyl phosphoramidite (with attached methyl group, Me), a nucleotide derivative that has the blocking group DMT (4′,4′-dimethoxytrityl) bound to its 5′ hydroxyl. In the presence of a weak acid, coupling of the two nucelotides occurs as a phosphodiester with phosphorous in a trivalent state. This is oxidized by iodine, giving a phosphotriester where the P is pentavalent; detritylation with $ZnBr_2$ is carried out and the process is repeated. The methyl groups on the phosphates are all removed at alkaline pH at the finish of oligonucleotide synthesis. [See S. L. Beaucage and M. H. Caruthers, 1981, *Tetrahedron Letters* **22**:1859.]

Another extremely valuable aspect of synthetic oligonucleotides and peptides is that they make possible the isolation of whole genes, on the one hand, and pure proteins, on the other. Because the genetic code is universal, a nucleic acid sequence can be used to predict the exact protein sequence it encodes; with less certainty (because of degeneracy in the code), a peptide sequence can be used to predict the approximate nucleic acid sequence that encodes it. Thus it has become feasible to go back and forth between two chemical languages—that of nucleic acids and that of proteins—in order to obtain additional information about a polymer of one type or the other (Figure 7-37).

For example, if a region of an mRNA for a protein that has not yet been studied is cloned and sequenced, a peptide—part of the protein encoded by the mRNA—can be prepared. The synthetic peptide can be used to provoke an antibody that will react with a protein containing that peptide. With such an antibody, the previously unisolated protein corresponding to the already isolated DNA can be identified in cells and purified. A reciprocal selection is also possible: If a protein has been purified and a short region of peptide sequence is available, then oligonucleotides coding for that amino acid sequence can be prepared. The oligonucleotides can be used to screen a genomic or cDNA library for the particular DNA sequence.

Techniques for the chemical synthesis of peptides have been available for some time, and techniques for DNA oligonucleotide synthesis are now also available. The basic logic of both techniques is similar, although the chemistry is different. In the synthesis of peptides, an amino acid in which the amino group has been "blocked" to prevent reaction is attached by its carboxyl group to a solid substrate (a resin), as shown in Figure 7-38. The blocking group at the amino end is removed so that the amino group can react, and a second amino acid residue with a reactive carboxyl group and a blocked amino group is added. Conditions are adjusted to promote peptide-bond formation, which results in a dipeptide. The process is repeated as many times as necessary, and the final product is then cleaved from the substrate. Defined peptide chains of 20 or more amino acids can be synthesized in this manner.

For oligonucleotide synthesis, the first nucleotide is bound by its 3′ hydroxyl to a solid matrix (Figure 7-39). This residue is mixed with a second nucleotide that has a phosphoramidite group on its 3′ hydroxyl and a blocking group on its 5′ hydroxyl; under appropriate conditions, the two residues form a dinucleotide. The blocking group at the 5′ end of the incoming nucleotide is removed and the process is repeated, thus allowing the growth of an oligonucleotide with a specific sequence.

Incidentally, note that in the chemical synthesis of peptides, chain growth is from the carboxyl terminus to the amino terminus; in the chemical synthesis of DNA, chain growth is from the 3′ to the 5′ end. Both of these directions are opposite to the directions in which biosynthetic reactions occur in cells or cell extracts.

Summary

An indispensable adjunct of modern biochemistry is the use of isotopes to label biologically important molecules. The isotopes may be radioactive (the most common examples are ^{3}H, ^{14}C, and ^{32}P) or density-labeled (e.g., ^{15}N and ^{13}C). These tracers are widely used in cell-free biochemical experiments, as well as in the observation of metabolic events within cells. Important considerations in the use of isotopes include the energy of the emitted particle during radioactive decay, the speed of entry of various labeled macromolecular precursors into the cell, and the extent of exchange between compounds in the cell and compounds in the medium. For example, tritiated (^{3}H) compounds give the best autoradiographic images because the β particle emitted has a low energy and the image on the photographic emulsion is therefore better defined. Pulse-chase experiments using labeled amino acids or thymidine to study the synthesis of proteins or DNA can provide valuable information because amino acids exchange between the cell and the medium within a minute or two, and thymidine enters a very small pool that is quickly consumed by cell growth. However, the pulse chase obtained when any labeled RNA precursors are used to study RNA synthesis is much less effective, because RNA nucleotides do not exchange with the medium, and they are part of a large intracellular pool.

Separation techniques for molecules purified from cells have reached the level of a high art. In addition to the many varieties of chromatographic procedures, two basic methods are frequently applied to problems in molecular cell biology: sedimentation and electrophoresis, both of which are most useful in separating molecules according to chain length. Separations of very large molecules that differ by less than 1 percent in size are commonly achieved. For example, the total protein content of cells or the protein translated in vitro from the total cell mRNA can be resolved by two-dimensional gel electrophoresis into more than 5000 individual components. This provides an assay for each mRNA of the cell.

The use of electrophoresis to separate nucleic acids on the basis of size has become one of the most common laboratory procedures. In chains containing between 2 and approximately 500 nucleotides, individual nucleic acids can be separated. These nucleic acid fragments can now be sequenced with such facility that DNA stretches of thousands of nucleotides are typically sequenced within a few weeks. Protein sequencing of shorter peptides has been entirely automated, and chemical synthesis of short oligonucleotides and peptides is also easily accomplished.

Two aspects of nucleic acid biochemistry—molecular hybridization and nucleic acid enzymology used in conjunction with microbial genetics—are responsible for an array of powerful techniques for analyzing and synthesizing nucleic acid sequences. Molecular hybridization (RNA-DNA and DNA-DNA), the fundamental method of testing the identity of a nucleic acid sample, has a wide variety of ingenious applications. To cite one example, the detection of a single gene representing perhaps as little as 1 part in 10^6 in the total human genome is routinely carried out by a hybridization procedure known as the Southern blot.

The success of many hybridization experiments depends on the use of nucleic acid enzymes, a great number of which have been purified and characterized. Among the most important are the restriction enzymes that cut DNA at four- to six-base-pair restriction sites, thus generating reproducible fragments from any genome. A number of enzymes that synthesize DNA and RNA are widely available in highly purified forms, as are enzymes that add to the ends of nucleic acids and enzymes that join DNA segments. The clever use of these enzymes coupled with a deepening understanding of microbial genetics has led to the planned rearrangement and mutation of genes by means of recombinant DNA and other technologies of molecular genetics. Now any gene can be purified and the functional regions of its DNA sequences explored. Succeeding chapters will explain how all of these techniques have been applied to gain a better understanding of gene structure, function, and control.

References

Radioisotopes: The Universal Modern Means of Following Biological Activity

FREEMAN, L. M., and M. D. BLAUFOX. 1975. *Radioimmunoassay.* Grune & Stratton.

HENDEE, W. R. 1973. *Radioactive Isotopes in Biological Research.* Wiley.

QUIMBY, E. H., S. FEITELBERG, and W. GROSS. 1970. *Radioactive Nuclides in Medicine and Biology.* Lea & Febiger.

Autoradiography

PARDUE, M. L., and J. G. GALL. 1969. Molecular hybridization of radioactive DNA to the DNA of cytological preparations. *Proc. Nat'l Acad. Sci. USA* **64**:600–604.

ROGERS, A. W. 1979. *Techniques of Autoradiography,* 3d ed. Elsevier/North-Holland.

Labeling Cells and Pools

DINTZIS, H. 1961. Assembly of the peptide chains of hemoglobin. *Proc. Nat'l Acad. Sci. USA* **47**:247–261.

EAGLE, H., and K. A. PIEZ. 1958. The free amino acid pool of cultured human cells. *J. Biol. Chem.* **231**:533–545.

PUCKETT, L., and J. E. DARNELL JR. 1977. Essential factors in the kinetic analysis of RNA synthesis in HeLa cells. *J. Cell Phys.* **90**:521–534.

Labeling Isolated Molecules

BOLTON, A. E., and W. M. HUNTER. 1973. The labelling of proteins to high specific radioactivities by conjugation to a ^{125}I-containing acylating agent. *Biochem. J.* **133**:529–538.

RIGBY, P. W. J., M. DIECKMANN, C. RHODES, and P. BERG. 1977. Labeling deoxyribonucleic acid to high specific activity *in vitro* by nick translation with DNA polymerase I. *J. Mol. Biol.* **113**:237–251.

Determining Sizes of Nucleic Acids and Proteins

*CANTOR, C. R., and P. R. SCHIMMEL. 1980. *Biophysical Chemistry,* part II: *Techniques for the Study of Biological Structure and Function.* W. H. Freeman and Company.

*KORNBERG, A. 1980. *DNA Replication.* W. H. Freeman and Company. A comprehensive treatment of biochemical and biophysical properties of proteins and nucleic acids.

Centrifugation

BRAKKE, M. K. 1967. Density gradient centrifugation. In *Methods in Virology,* vol. II, K. Maramorosch and H. Koprowski, eds. Academic Press.

DE DUVE, C. 1975. Exploring cells with a centrifuge. *Science* **189**:186–194.

MESELSON, M., and F. W. STAHL. 1958. The replication of DNA in *Escherichia coli. Proc. Nat'l Acad. Sci. USA* **44**:671–682.

SCHEELER, P. 1980. *Centrifugation in Biology and Medical Science.* Wiley.

VINOGRAD, J. 1963. Sedimentation equilibrium in a buoyant density gradient. In *Methods in Enzymology,* vol. VI, S. P. Colowick and N. O. Kaplan, eds. Academic Press.

VINOGRAD, J., R. RASLOFF, and W. BAUER. 1967. A dye-buoyant density method for the detection and isolation of closed circular duplex DNA. *Proc. Nat'l Acad. Sci. USA* **57**:1514–1521.

Electrophoresis

ANDREWS, A. T. 1981. *Electrophoresis.* Oxford University Press.

LAEMMLI, U. K. 1970. Cleavage of structural proteins during the assembly of the head of the bacteriophage T4. *Nature* **227**:680–685.

MAIZEL, J. V. JR. 1971. Polyacrylamide gel electrophoresis of viral proteins. In *Methods in Virology,* vol. V, K. Maramorosch and H. Koprowski, eds. Academic Press.

O'FARRELL, P. H. 1975. High resolution, two-dimensional electrophoresis of proteins. *J. Biol. Chem.* **250**:4007–4021.

RICKWOOD, D., and B. D. HAMES, eds. 1982. *Gel Electrophoresis of Nucleic Acids.* London: IRL Press Ltd.

Examining the Sequences of Nucleic Acids and Proteins

Nucleic Acids: Hybridization

ALWINE, J. C., D. J. KEMP, and G. R. STARK. 1977. A method for detection of specific RNAs in agarose gels by transfer to diazobenzyloxymethyl-paper and hybridization with DNA probes. *Proc. Nat'l Acad. Sci. USA* **74**:5350–5354.

BERK, A. J., and P. A. SHARP. 1977. Sizing and mapping of early adenovirus mRNAs by gel electrophoresis of S1 endonuclease digested hybrids. *Cell* **12**:721–732.

BRITTEN, R. J., and E. D. KOHNE. 1968. Repeated sequences in DNA. *Science* **161**:529–540.

DAVIS, R. W., and N. DAVIDSON. 1968. Electron microscopic visualization of deletion mutations. *Proc. Nat'l Acad. Sci. USA* **60**:243–250.

GILLESPIE, D., and S. SPIEGELMAN. 1965. A quantitative assay for RNA-DNA hybrids with DNA immobilized on a membrane. *J. Mol. Biol.* **12**:829–842.

HALL, B. D., and S. SPIEGELMAN. 1961. Sequence complementarity of T2-DNA and T2-specific RNA. *Proc. Nat'l Acad. Sci. USA* **47**:137–146.

HU, N., and J. MESSING. 1982. The making of single-stranded probes. *Gene* **17**:271–277.

THOMAS, M., R. L. WHITE, and R. W. DAVIS. 1976. Hybridization of RNA to double-stranded DNA: formation of R loops. *Proc. Nat'l Acad. Sci. USA* 73:2294–2298.

THOMAS, P. S. 1980. Hybridization of denatured RNA and small DNA fragments transferred to nitrocellulose. *Proc. Nat'l Acad. Sci. USA* 77:5201–5205.

WETMUR, J. G., and N. DAVIDSON. 1968. Kinetics of renaturation of DNA. *J. Mol. Biol.* **31**:349–370.

Nucleic Acids: Restriction Enzymes

NATHANS, D., and H. O. SMITH. 1975. Restriction endonucleases in the analysis and restructuring of DNA molecules. *Annu. Rev. Biochem.* **44**:273–293.

*Book or review article that provides a survey of the topic.

ROBERTS, R. 1982. Restriction and modification enzymes and their recognition sequences. *Nuc. Acids Res.* **10**:r117–r144.

SMITH, H. O. 1970. Nucleotide sequence specificity of restriction endonucleases. *Science* **205**:455–462.

SMITH, H. O., and M. L. BERNSTIEL. 1976. A simple method for DNA restriction site mapping. *Nuc. Acids Res.* **3**:2387–2398.

SOUTHERN, E. M. 1975. Detection of specific sequences among DNA fragments separated by gel electrophoresis. *J. Mol. Biol.* **98**:503–517.

Nucleic Acids: Sequence Analysis

BROWNLEE, G. 1972. Determination of sequences in RNA. In *Laboratory Techniques in Biochemistry and Molecular Biology,* vol. 3, T. S. Work and E. Work, eds. Elsevier/North Holland.

MAXAM, A. M., and W. GILBERT. 1977. A new method for sequencing DNA. *Proc. Nat'l Acad. Sci. USA* **74**:560–564.

MAXAM, A., and W. GILBERT. 1980. Sequencing end-labeled DNA with base-specific DNA. In *Methods in Enzymology,* vol. 65, L. Grossman and K. Moldave, eds. Academic Press.

SANGER, F. 1981. Determination of nucleotide sequences in DNA. *Science* **214**:1205–1210.

SANGER, F., S. NICKLEN, and A. R. COULSON. 1977. DNA sequencing with chain-terminating inhibitors. *Proc. Nat'l Acad. Sci. USA* **74**:5463–5467.

Proteins: Sequence Analysis

CLEVELAND, D. W., S. G. FISCHER, M. W. KIRSCHNER, and U. K. LAEMMLI. 1977. Peptide mapping by limited proteolysis in sodium dodecyl sulfate and analysis by gel electrophoresis. *J. Biol. Chem.* **252**:1102–1106.

INGRAM, V. M. 1956. A specific chemical difference between the globins of normal human and sickle-cell anaemia haemoglobin. *Nature* **178**:792–794.

SANGER, F. 1952. The arrangement of amino acids in proteins. *Adv. Protein Chem.* **7**:1–67. A description of original sequencing, fingerprinting, and partial-chain-analysis methods.

WALSH, K. A., L. H. ERICSSON, D. C. PARMELEE, and K. TITANI. 1981. Advances in protein sequencing. *Annu. Rev. Biochem.* **50**:261–284. A review of up-to-date methods, including automated sequence analysis.

Recombinant DNA and Mutagenesis

ABELSON, J., and E. BUTZ, eds. 1980. Recombinant DNA. *Science* **209**:1317–1338.

BENTON, W. D., and R. W. DAVIS. 1977. Screening λgt recombinant clones by hybridization to single plaques *in situ. Science* **196**:180–183.

BERG, P. 1981. Dissections and reconstructions of genes and chromosomes. *Science* **213**:296–303.

BOLLUM, F. J. 1974. Terminal deoxynucleotidyl transferase. In *The Enzymes,* 3d ed., vol. 10, P. D. Boyer, ed. Academic Press.

COHEN, S. N. 1975. The manipulation of genes. *Sci. Am.* **233**(1):24–33.

COHEN, S. N., A. C. Y. CHANG, H. W. BOYER, and R. B. HELLING. 1973. Construction of biologically functional bacterial plasmids *in vitro. Proc. Nat'l Acad. Sci. USA* **70**:3240–3244.

EFSTRATIADIS, A., F. C. KAFATOS, A. M. MAXAM, and T. MANIATIS. 1976. Enzymatic *in vitro* synthesis of globin genes. *Cell* **7**:279–288.

GOEDDEL, D. V., D. G. KLEID, F. BOLIVAR, H. L. HEYNEKER, D. G. YANSURA, R. CREA, T. HIROSE, A. KRASZEWSKI, K. ITAKURA, and A. RIGGS. 1979. Expression in *Escherichia coli* of chemically synthesized genes for human insulin. *Proc. Nat'l Acad. Sci. USA* **76**:106–110.

GRUNSTEIN, M., and D. S. HOGNESS. 1975. Colony hybridization: a method for the isolation of cloned DNAs that contain a specific gene. *Proc. Nat'l Acad. Sci. USA* **72**:3961–3965.

HOHN, B. 1979. *In vitro* packaging of λ and cosmid DNA. In *Methods of Enzymology: Recombinant DNA,* vol. 68, R. Wu, ed. Academic Press.

JACKSON, D. A., R. H. SYMONS, and P. BERG. 1972. Biochemical method for inserting new genetic information into DNA of simian virus 40: circular SV40 DNA molecules containing lambda phage genes and the galactose operon of *Escherichia coli. Proc. Nat'l Acad. Sci. USA* **69**:2904–2909.

LOBBAN, P. E., and A. D. KAISER. 1973. Enzymatic end-to-end joining of DNA molecules. *J. Mol. Biol.* **78**:453–471.

MANDEL, M., and A. HIGA. 1970. Calcium dependent bacteriophage DNA infection. *J. Mol. Biol.* **53**:159–162.

MANIATIS, T., R. C. HARDISON, E. LACY, J. LAUER, C. O'CONNELL, D. QUON, D. K. SIM, and A. EFSTRATIADIS. 1978. The isolation of structural genes from libraries of eucaryotic DNA. *Cell* **15**:687–701.

OKAYAMA, H., and P. BERG. 1982. High-efficiency cloning of full-length cDNA. *Mol. Cell. Biol.* **2**:161–170.

SHORTLE, D., and D. NATHANS. 1978. Local mutagenesis: a method for generating viral mutants with base substitutions in preselected regions of the viral genome. *Proc. Nat'l Acad. Sci. USA* **72**:2170–2174.

TEMIN, H., and D. BALTIMORE. 1972. RNA directed DNA synthesis and RNA tumor viruses. *Adv. Virus Res.* **17**:129–186.

VERMA, I. M. 1977. The reverse transcriptase. *Biochim. Biophys. Acta* **473**:1–38.

WILLIAMSON, R., ed. 1979, 1981, 1982. *Genetic Engineering,* vols. 1–3. Academic Press.

Synthesis of Specific Peptide and Nucleotide Sequences

Synthesis of Polypeptides

MARGLIN, A., and R. B. MERRIFIELD. 1970. Chemical synthesis of peptides and proteins. *Annu. Rev. Biochem.* **39**:841–866.

MERRIFIELD, R. B., L. D. VIZIOLI, and H. G. BOMAN. 1982. Synthesis of the antibacterial peptide cecropin A(1–33). *Biochemistry* **21**:5020–5031. The authors were the originators of successful peptide synthesis.

PELHAM, H. R. B., and R. J. JACKSON. 1976. An efficient mRNA dependent translation system from reticulocyte lysates. *Eur. J. Biochem.* **67**:247–256. A description of the most frequently used in vitro technique for protein synthesis.

Synthesis of Oligonucleotides

GOEDDEL, D. V., D. G. YANSURA, and M. H. CARUTHERS. 1977. Studies on gene control regions, part 1: Chemical synthesis of lactose operator deoxyribonucleic acid segments. *Biochemistry* **16**:1765–1772.

ITAKURA, K. 1980. Synthesis of genes. *Trends Biochem. Sci.* **5**:114–116.

ITAKURA, K., and A. D. RIGGS. 1980. Chemical DNA synthesis and recombinant DNA studies. *Science* **209**:1401–1405.

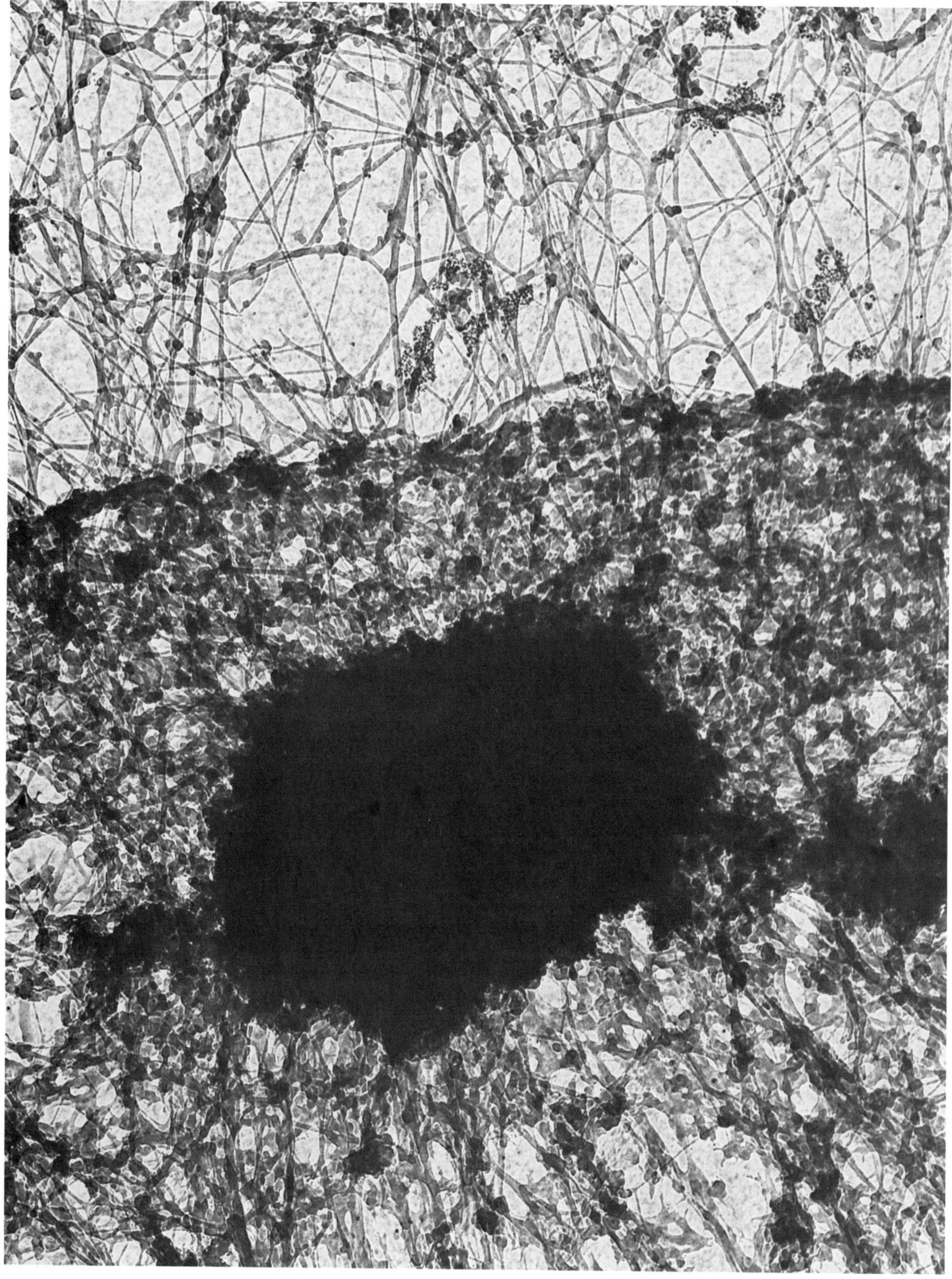

II

Gene Expression, Structure, and Replication

A transmission electron micrograph of a whole mount of a Hela cell, showing a skeletal network within the nucleus. *Courtesy of S. Penman.*

How nucleic acid synthesis and gene control are integrated into a functioning cell is the subject of the next six chapters. The idea governing all such discussions is that *the actions and properties of each cell type are determined by the proteins each contains.* In turn, the amount of each protein in a cell is determined by the concentration of its specific mRNA, by the frequency with which that mRNA is translated, and, finally, by the stability of the protein itself. These three parameters are responsible for which genes are expressed and for the extent to which they are expressed in each cell. Thus the differential activity of different genes determines the properties of cells. The regulation of this differential gene activity is called gene control.

Studies using bacterial cells provide the foundations of our understanding of gene control, and these studies are considered in Chapter 8. In bacteria, most gene control is carried out by regulatory proteins that bind to specific sites on DNA. The binding either prevents or increases the syntheses of specific mRNAs. Knowledge about several of these regulatory proteins has reached an advanced stage and includes crystallographic analyses that reveal how the proteins recognize the DNA double helix.

Molecular biologists were unable to study differential gene control in eukaryotic cells until they had worked out the details of mRNA synthesis. Bacteria transcribe mature mRNA, which requires no modification before it can be used. In contrast, eukaryotic cells transcribe large precursor molecules of RNA. While the precursor RNA is still in the nucleus, it is extensively processed into the mRNA that is used in directing protein synthesis in the cytoplasm. Likewise, precursor molecules are modified to produce rRNA and tRNA. Chapter 9 presents the story of RNA processing, including the proof that large molecules are the precursors. Descriptions of the various processing steps follow and include the surprising fact that RNA must often be spliced before it becomes a functional molecule.

Chapters 10 and 11 examine the anatomy of the eukaryotic chromosomes. Chapter 10 covers their general structure: the distinctive features that can be seen with the microscope, knowledge obtained from classical genetics about the arrangement of the genes on the chromosomes, and the role of proteins in packaging the DNA. Chapter 11 describes the chromosomes at the molecular level, where specific transcription units are found interspersed within long stretches of apparently meaningless DNA. This information about chromosomes follows the chapter on RNA synthesis for two reasons: First, biologists would never have understood chromosome anatomy if they had not known about RNA processing. Second, much of the reason for studying chromosomes is to discover how individual genes are regulated, and this, of course, demands that one first appreciate RNA synthesis.

We then move on, in Chapter 12, to present some current ideas about gene control in eukaryotic cells. It is now known that most eukaryotic gene control lies at the level of transcription but that control of RNA processing and variations in mRNA stability also exist. Thus the foundations have been laid for understanding this subject. Many tantalizing questions remain; for example, what are the details of the interaction of the DNA-binding proteins with the chromatin structure that results in transcriptional control?

The chromosome must, of course, replicate in cell division. And this replication must be controlled to provide multicellular organisms with the right number of cells. In Chapter 13 we deal with the molecular events that allow DNA to replicate. Because of the antiparallel, double-helical structure, DNA replication is a very complicated process requiring many specific proteins. Truly remarkable progress has been made in the understanding of bacterial replication systems, and success in understanding eukaryotic systems appears to be just around the corner. DNA duplication presents similar structural and biochemical problems in all cells; quite possibly, biologists will discover that prokaryotes and eukaryotes have solved these problems in similar ways.

8

RNA Synthesis and Gene Control in Prokaryotes

THIS chapter opens our discussion of gene expression and gene regulation with a look at prokaryotic genes. Several considerations led to this choice. First, prokaryotic gene control is now very well understood. Because eukaryotic gene control is still relatively mysterious, we shall make frequent use of our detailed knowledge of prokaryotic gene control in considering the possible mechanisms of eukaryotic gene control. Furthermore, to grasp the full range of possible eukaryotic control mechanisms, it is most instructive to realize that even in bacteria, where most gene regulation is at the transcriptional level, many different mechanisms of transcriptional control are utilized. Thus, there is probably no cell in which all genes are controlled in the same manner. We shall begin by outlining how the fundamental differences in prokaryotic and eukaryotic cellular organization affect gene expression and gene control.

Transcription of DNA and Translation of mRNA: Fundamental Differences between Prokaryotic and Eukaryotic Cells

During its formation, a bacterial mRNA molecule is completely accessible to ribosomes and to other elements of the protein-synthesis apparatus. Consequently, bacterial protein synthesis begins on an mRNA molecule even while it is still being formed (Figures 8-1 and 8-2). (Recall that the transcription of DNA into RNA proceeds in the $5' \rightarrow 3'$ direction, as does the ribosomal translation of mRNA.) Thus in bacterial cells, mRNA molecules are not chemically modified before they are translated, and *coupled transcription-translation* is the rule. This direct interaction of the transcriptional and the translational machinery is important in certain types of prokaryotic gene control.

In eukaryotic cells, mRNA formation and utilization is not as straightforward. DNA transcription occurs in the nucleus, which contains ribosomal precursors undergoing formation (in the nucleolus), but which does not contain mature ribosomes engaged in protein synthesis. Thus, transcription and translation are not coupled in eukaryotes as they are in prokaryotes. Furthermore, the newly formed eukaryotic mRNA precursor is extensively and specifically modified in the nucleus before it emerges into the cytoplasm to associate with ribosomes. Such modifications include the addition of chemical groups at both ends and, in some cases, the cutting of the molecule into segments; noncontiguous segments from the original molecule may then be spliced together to produce a very different RNA. This biochemical and structural modification of newly formed eukaryotic RNA is one of the greatest distinguishing characteristics between eukaryotic and prokaryotic cells.

The Logic and the Strategy of Prokaryotic Gene Control

Prokaryotes and eukaryotes differ also in the apparent *purpose* of gene control. In bacteria, gene control serves mainly to allow the single cell to adjust to changes in its nutritional environment so that its main function—growth and division—can be optimized. Some genes in metazoan organisms may also respond directly to environmental changes. But in eukaryotes, the most characteristic and biologically far reaching purpose served by gene control is the regulation of a genetic program that underlies embryologic development and tissue differentiation.

Because ribosomes begin translating nascent mRNA in bacteria as soon as the first ribosome-binding site on the mRNA is formed, it follows that *initiation of transcription* is a critical point of gene control in bacterial cells. Somewhat surprisingly, however, transcription can undergo a controlled termination before the coding sequence of a gene has been completely transcribed, and sometimes transcription can terminate between two genes. Thus, *termination of transcription* can be as important as initiation. Presumably these controlled termination reactions spare the bacterium from producing RNA that is not needed.

The concentration of a given mRNA in any cell depends not only on its rate of synthesis, but also on its rate

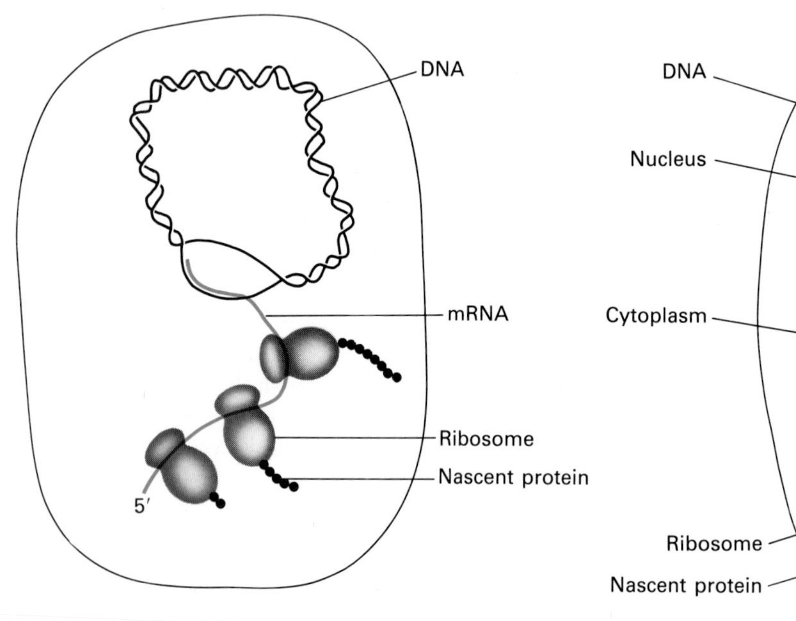

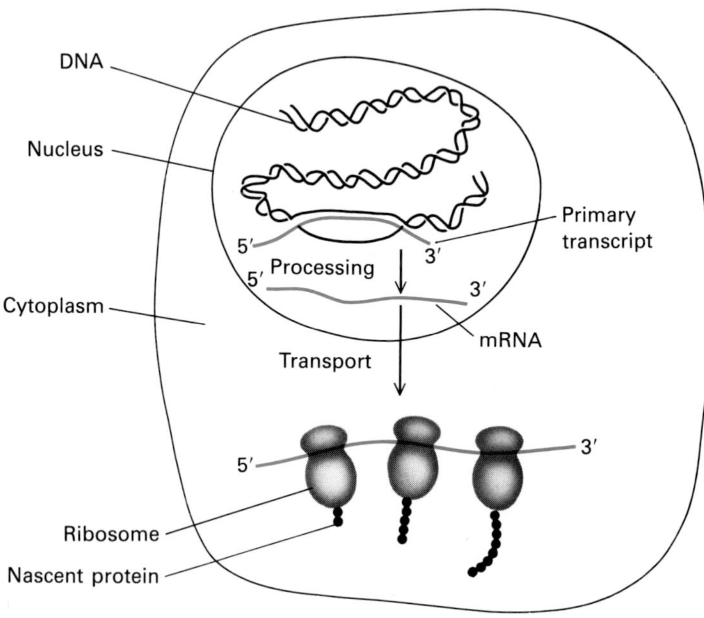

Figure 8-1 The production of functioning mRNA is very different in prokaryotes and eukaryotes. In prokaryotes, the RNA transcript serves directly as the mRNA, and translation begins before transcription is completed; that is, transcription and translation are coupled. In eukaryotes, the primary RNA transcript must be modified in the cell nucleus to form mRNA. Translation takes place only after the completed mRNA is delivered to the cytoplasm.

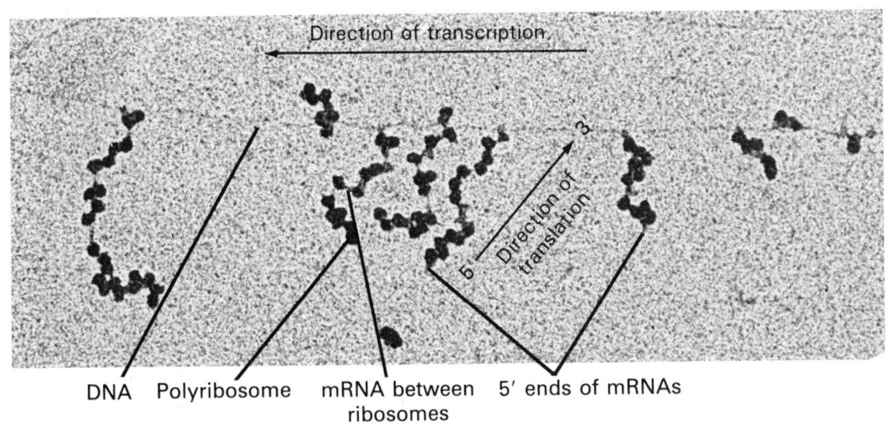

Figure 8-2 An electron micrograph showing coupled mRNA transcription and translation in *E. coli*. An unidentified region of the *E. coli* chromosome is undergoing transcription to produce mRNA. This is translated immediately, forming polyribosomes. The direction of transcription *(arrow)* is inferred because the longer mRNAs (loaded with more ribosomes) nearer completion of transcription are further along in the transcription process. The direction of translation is likewise inferred because the ends of the mRNAs free of the DNA template are the 5′ ends, and mRNA is translated 5′ → 3′. *From O. L. Miller Jr., B. A. Hamkalo, and C. A. Thomas Jr., 1970,* Science **169**:*392.*

of degradation; that is, on its *metabolic stability*. Generally, mRNA in a bacterial cell has a short lifetime compared with the doubling time of the cell so that cessation of the formation of a specific mRNA is rapidly followed by cessation of the synthesis of a specific protein. Only a few bacterial mRNAs are known to have longer lifetimes than the average; even so, these longer-lived mRNAs last only half as long as the doubling time of bacterial cells.

In summary, then, three elements of control—transcriptional initiation, transcriptional termination, and rapid mRNA turnover—constitute the underlying strategy of bacterial gene control.

Initiation of Transcription in Prokaryotes Entails Sequence Recognition by RNA Polymerase

All transcription of DNA in bacteria is catalyzed by a single RNA polymerase (Chapter 4). This enzyme can bind to any region in double-stranded DNA. By binding loosely, releasing momentarily, and then binding again, an RNA polymerase explores the DNA. With the aid of accessory proteins called *σ factors* (sigma factors), the RNA polymerase recognizes specific DNA sites called *promoters*, at which RNA synthesis can begin. The first nucleotide in the RNA chain usually contains a purine base (adenine or guanine), but occasionally a pyrimidine (cytosine or uracil) is found at the 5′ end of an RNA. At the promoter sites, RNA polymerase binds tightly and causes the double-stranded DNA structure to *melt*, or unwind, so that transcription can begin (Figure 8-3). The σ factor is released after a few nucleotides have been joined. About 10 to 12 nucleotides of the growing RNA chains are hydrogen-bonded to the DNA at any one time. As the chain elongates, the RNA "peels off" the DNA. By cross-linking reactions that attach RNA to protein, the RNA can be shown to lie near the surfaces of the β and β′ subunits.

The bacterial promoter region, which is about 40 bases long, includes a common sequence of about 6 bases located upstream (i.e., in the 5′ direction) from the site at which RNA synthesis begins. This sequence is called the *Pribnow box*, for the investigator who recognized it. The position of the nucleotide at the initiation site is designated +1, and the nucleotides that precede this site are designated with negative numbers (there is no zero position). Five to eight bases separate the Pribnow box from the initiation site, so the Pribnow box centers around the −10 position (Figure 8-4). Not all promoters have exactly the same sequence of bases in this region, but a sequence similar to TATAAT is always found. The final thymine residue is present in the Pribnow boxes of all of the many bacterial promoters that have been examined, and the correspondence at each site in the rest of the box is greater than 60 percent. Such a conserved region is called a *consensus sequence*. Another consensus sequence, TGTTGACA, centers around −35; this region is also critical for the accurate and rapid initiation of transcription for most bacterial genes.

Elegant chemical experiments with purified DNA fragments containing promoters have proved that the RNA polymerase touches the DNA at certain bases in the conserved regions. In these chemical experiments, the RNA polymerase is bound to the DNA, which is then subjected to chemical modifications that break the chain at specific unprotected sites, as in a DNA sequence analysis (see Figure 7-29). The bases that are shielded from chemical reaction by being bound to the polymerase cause gaps in the sequence "ladder"—that is, in the array of bands revealed by gel electrophoresis (Figures 8-5 and 8-6). Furthermore, DNAs that have altered promoter activity (either increased or decreased) show changes in the sites

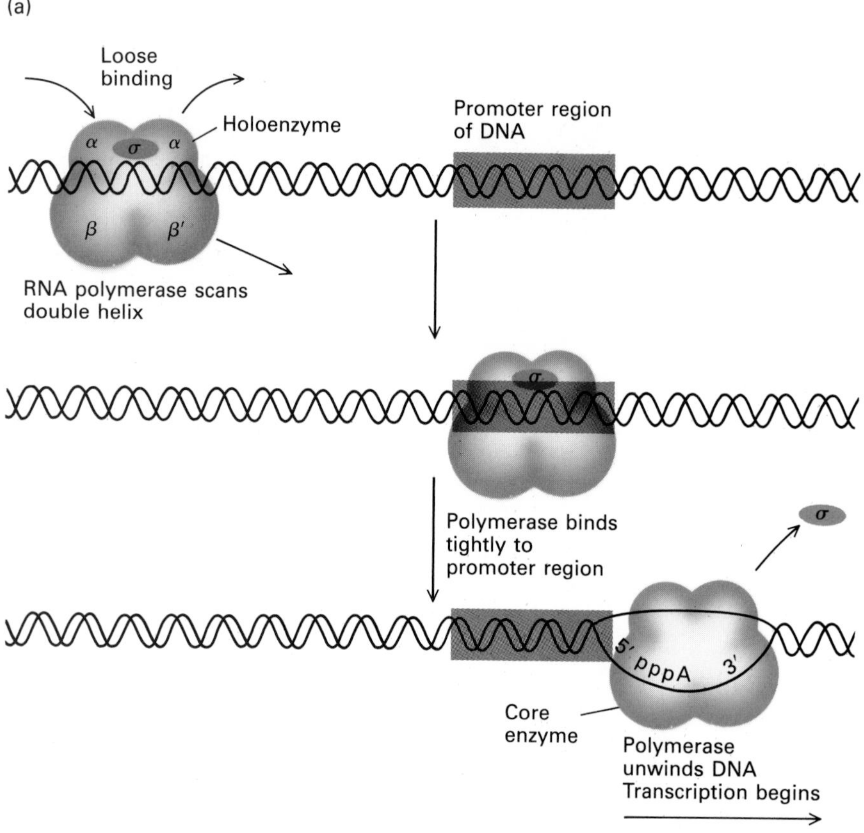

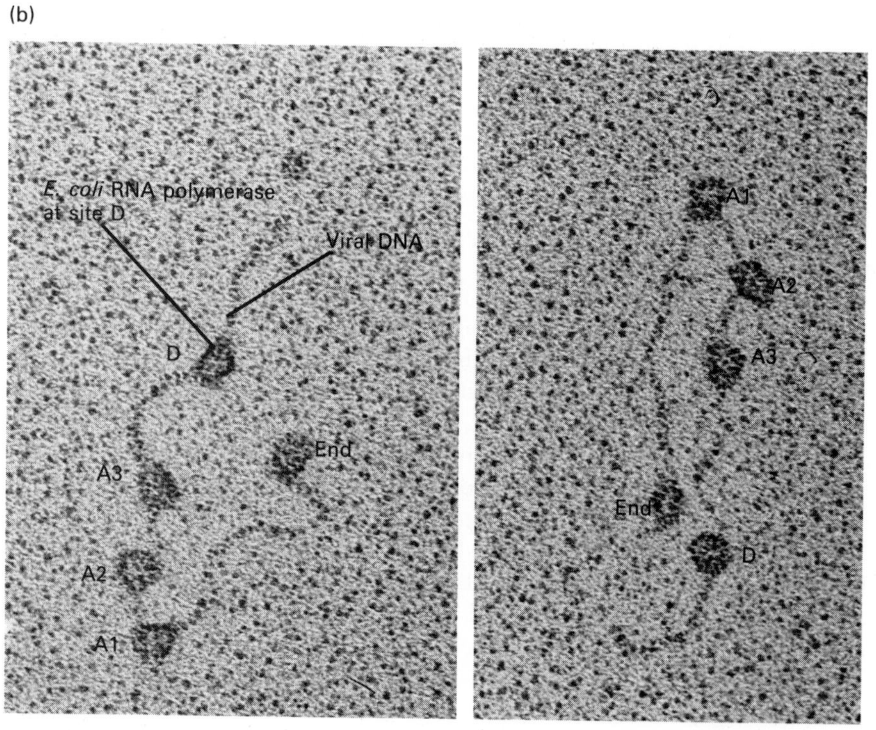

Figure 8-3 The RNA polymerase of *E. coli* is a large enzyme with multiple subunits and a total mass of almost 500 kilodaltons. The *holoenzyme* (the catalytically active form of the enzyme) has five subunits, and the complex is designated as $\alpha_2\beta\beta'\sigma$; the *core enzyme* lacks the σ factor. (a) An RNA polymerase molecule explores a DNA double helix by binding to it weakly, releasing and moving a short distance along the chain, and repeating this process over and over. Eventually the polymerase locates a *promoter,* a sequence of nucleotides to which the enzyme binds tightly, and RNA chain initiation takes place. The σ factor is a protein that assists the polymerase in binding to the correct initiation site. The enzyme then melts (unwinds) a short segment of DNA near the promoter and begins transcribing DNA into RNA. The σ factor is lost in this process. (b) Electron micrographs of *E. coli* RNA polymerase molecules bound to specific regions of two DNA molecules from bacteriophage T7. Viral DNA is used to judge where the polymerase binds because viral molecules are all short and identical in sequence. Thus the binding sites of the polymerase can be identified by measuring lengths along individual molecules. The polymerase reacts with the ends of all DNAs, so the observed interactions at the ends of the phage molecules are nonspecific. However, the other four polymerases are identically spaced on the two molecules; these specific reaction sites on the viral DNA are known promoters, termed A1, A2, A3, and D. [See R. C. Williams, 1977, *Proc. Nat'l Acad. Sci. USA* **74**:2311.] *Photograph courtesy of R. C. Williams.*

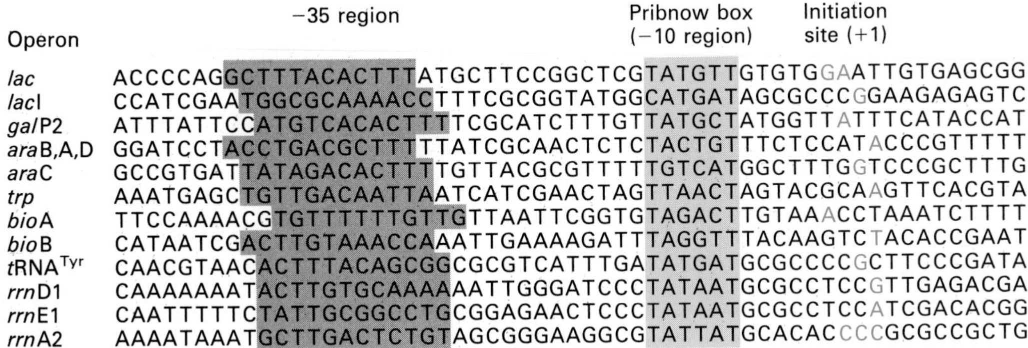

Operon
−35 region
Pribnow box (−10 region)
Initiation site (+1)
lac ACCCCAGGCTTTACACTTTATGCTTCCGGCTCGTATGTTGTGTGGAATTGTGAGCGG
*lac*I CCATCGAATGGCGCAAAACCTTTCGCGGTATGGCATGATAGCGCCCGGAAGAGAGTC
*gal*P2 ATTTATTCCATGTCACACTTTTCGCATCTTTGTTATGCTATGGTTATTTCATACCAT
*ara*B,A,D GGATCCTACCTGACGCTTTTTATCGCAACTCTCTACTGTTTCTCCATACCCGTTTTT
*ara*C GCCGTGATTATAGACACTTTTGTTACGCGTTTTTGTCATGGCTTTGGTCCCGCTTTG
trp AAATGAGCTGTTGACAATTAATCATCGAACTAGTTAACTAGTACGCAAGTTCACGTA
*bio*A TTCCAAAACGTGTTTTTTGTTGTTAATTCGGTGTAGACTTGTAAACCTAAATCTTTT
*bio*B CATAATCGACTTGTAAACCAAATTGAAAAGATTTAGGTTTACAAGTCTACACCGAAT
tRNA^{Tyr} CAACGTAACACTTTACAGCGGCGCGTCATTTGATATGATGCGCCCCGCTTCCCGATA
*rrn*D1 CAAAAAAATACTTGTGCAAAAAATTGGGATCCCTATAATGCGCCTCCGTTGAGACGA
*rrn*E1 CAATTTTTCTATTGCGGCCTGCGGAGAACTCCCTATAATGCGCCTCCATCGACACGG
*rrn*A2 AAAATAAATGCTTGACTCTGTAGCGGGAAGGCGTATTATGCACACCCCGCGCCGCTG

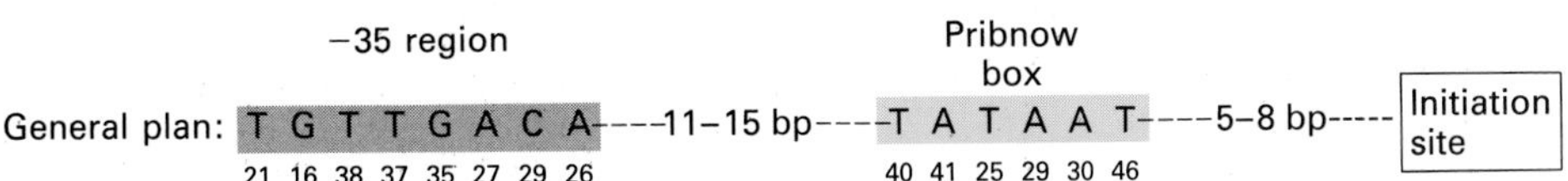

Figure 8-4 A comparison of the nucleotide sequences in different promoter sites that direct transcription by *E. coli* RNA polymerase. Forty-six such sites were sequenced; twelve of the sequences are shown here. Each site belongs to a specific operon—that is, a cluster of genes that are all controlled by the same promoter. Several operons are discussed in detail in this chapter. The Pribnow box *(color)* is a strongly conserved region of 6 bases; an 8- to 12-base sequence around −35 *(gray)* is also conserved. Both of these regions are in contact with RNA polymerase in the initiation complex. The bases in color type are initiation sites. Usually transcription starts at a single nucleotide, a purine; occasionally alternate sites are used (here, wherever more than one base is in color). The general plan of the promoters is shown at the bottom of the diagram; the numbers below each base indicate the frequency with which that base was present in the 46 promoters examined. Note that the initiation site for *bio*B (a gene for the vitamin biotin) is at a thymine (which becomes a uracil in the RNA), and that the site for *rrn*A2 (one of the ribosomal RNA genes) has three cytosines; all of the other initiation sites occur at an adenine or a guanine. [See M. Rosenberg and D. Court, 1979, *Annu. Rev. Genet.* 13:319.]

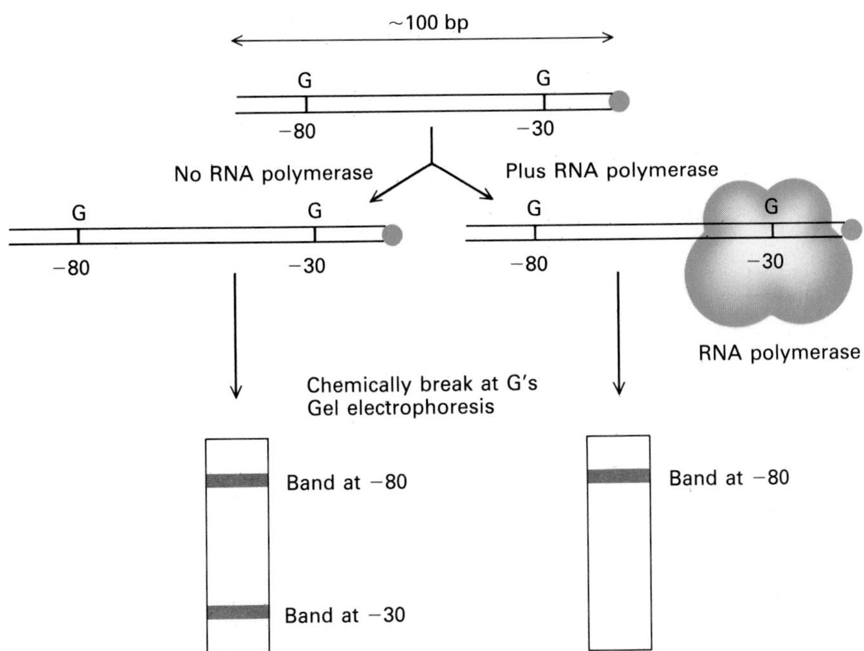

Figure 8-5 A diagram of a footprint experiment: how the sites of interaction of DNA with a protein are determined. A purified DNA fragment that is labeled at the 5′ end of one strand is allowed to interact with a protein—for example, RNA polymerase. The complex is chemically treated in a way that affects certain bases in the DNA (usually by breaking the DNA chain) when they are not in contact with the protein. Gel electrophoresis of the DNA then reveals which of the bases in the DNA have been protected. Only two such sites (two guanines) are shown in this illustration. The results of an actual experiment of this type are shown in Figure 8-6, and the data in Figure 8-7 were derived from several experiments using different kinds of chemical reactions.

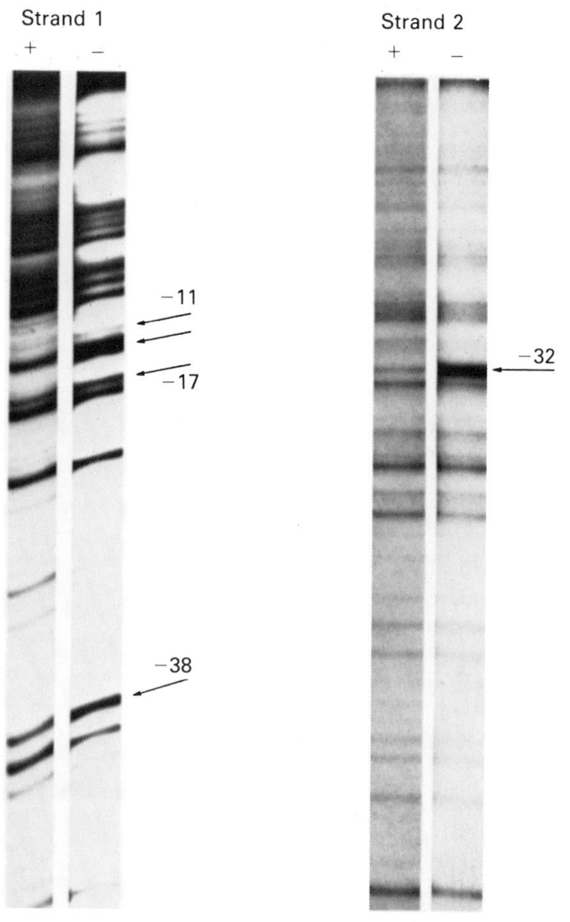

Figure 8-6 End-labeled DNA fragments containing the promoter region of a lactose operon (*lac*UV5) were prepared. Strand 1 (the top sequence shown in Figure 8-7) was labeled at the 5′ end; strand 2 (the complementary sequence, shown just below in Figure 8-7) was labeled at the 3′ end. For each strand, breakage of DNA at purines was carried out on two samples, one with RNA polymerase attached (+) and one without it (−). Bases that were protected from chemical attack by the polymerase resulted in a missing or diminished band in the plus lane, at the positions indicated by the arrows. The strongest protection from reaction occurred at the −32 residue in strand 2. Because the sequence of the promoter region was known (see Figure 8-7), the protected areas could be assigned to specific locations in the region. *From U. Siebenlist, R. B. Simpson, and W. Gilbert, 1980,* Cell *20:269. Copyright M.I.T.*

protected by the polymerase. Such results support the conclusion that the specific contact points between the promoters and the polymerase are important in the formation of an active complex (Figure 8-7).

The polymerase seems to contact the DNA asymmetrically. In other words, from the hypothetical point of view of someone looking down the cylindrical double helix, most of the contact points fall into one 180° sector (Figure 8-8). When the polymerase is bound, the helix is twisted so that it unwinds from about the middle of the Pribnow box to two or three bases past the first base to be transcribed into RNA (see Figure 8-7).

Operons Are Clusters of Genes Controlled at One Promoter Site

It is common in bacteria for one promoter to serve a series of clustered genes; often, gene clusters produce enzymes that are active in a single metabolic pathway. Such a gene cluster, which is called an *operon,* is transcribed into a single messenger RNA. Each gene of the operon is represented in the mRNA, and each section of the mRNA is independently translated. These mRNA molecules are *polycistronic,* meaning that they encode more than one polypeptide chain (Figure 8-9). (The term *cistron* applies to the smallest genetic unit serving a single function—i.e., it encodes one polypeptide. The origin of the term is explained in Chapter 10.)

The genes of a given operon are controlled coordinately: either all are transcribed, or none is transcribed. Often, if several enzymes are required to perform a metabolic function, they are found encoded in one operon. When one such enzyme is formed, so are all the other enzymes in the operon—a remarkably economical control system. However, as we have mentioned, the coding portions of the polycistronic mRNA are independently translated (i.e., there are multiple starts and stops). Consequently, some sections of the mRNA can be translated faster than others; different rates of translation are probably governed largely by folds (stem-loops) in the mRNA around initiation sites.

Not all of the potential promoters in a bacterial cell are available at every instant in the lifetime of the cell. The choice of which promoter sites (and thus which operons) are available to RNA polymerase at any given time is determined mainly by the nutritional content of the medium in which the bacterium is suspended. For example, specific groups of enzymes are needed for the conversion of specific sugars into glucose, a readily metabolized carbon source. As shown in Figure 8-10a, one set of enzymes is needed to convert galactose (a six-carbon monosaccharide) into glucose; an additional enzyme is needed to break down lactose (a disaccharide composed of two six-carbon units—glucose and galactose); and a third set of enzymes is required to convert arabinose (a five-carbon monosaccharide) into a glycolytic intermediate. The genes for these different groups of enzymes are clustered in different operons, called the *gal, lac,* and *ara* operons, respectively (Figure 8-10b). By making only the appropriate promoter available, the bacterium manufactures only

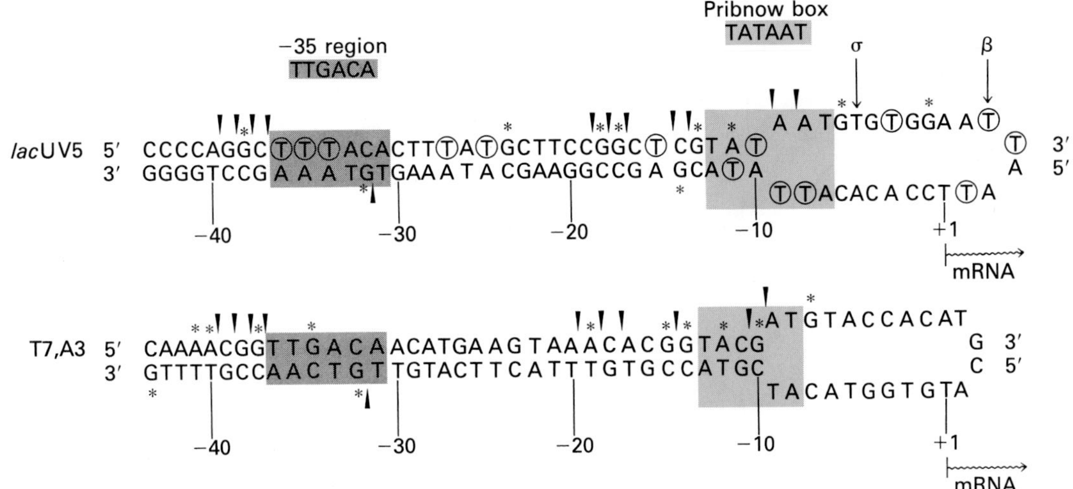

Figure 8-7 DNA sites that interact with the *E. coli* RNA polymerase to begin RNA synthesis: the promoter regions of the *lac*UV5 operon *(top)* and the A3 operon of bacteriophage T7 *(bottom)*. The DNA strands near the RNA initiation sites (+1) are separated, indicating that this DNA region is unwound when RNA polymerase binds. Note that the two promoter sequences are similar but not identical, and that the bases that interact with the polymerase (see the explanation of the symbols below) are similarly spaced in the two promoters. The indicated sites were identified by various chemical reactions with end-labeled promoter DNA sequences in the presence and absence of polymerase, as described in Figures 8-5 and 8-6. The symbols used in the figure have the following meanings:

▾	A phosphate group that may be ethylated, in which event polymerase cannot bind.
* A or * G	A purine at which methylation (with dimethylsulfate) prevents polymerase binding.
Ⓣ	A thymine that polymerase protects.
σ or β	A site at which the σ or β polymerase subunit can be chemically cross-linked to the DNA by a photochemical reaction.

[See U. Siebenlist, R. B. Simpson, and W. Gilbert, 1980, *Cell* 20:269.]

the enzymes needed for the uptake and metabolism of the sugar that it finds in the medium. This specific response is referred to as *enzyme induction,* and synthesis of the enzymes is said to be *induced* by the presence of their substrate, the sugar (Figure 8-11a). Removal of the inducer very quickly stops enzyme synthesis, and the enzymes are diluted as the cell grows and divides.

The lack of a nutrient also can induce the synthesis of enzymes. Many bacteria can synthesize all of the amino acids necessary for growth, but the enzymes required for forming a particular amino acid are present in a bacterial cell only when they are needed to catalyze the production of that amino acid. For example, suppose a bacterial cell can synthesize tryptophan. If, in the laboratory, the bacterium is placed in a medium containing no tryptophan, the cell synthesizes the five enzymes necessary for producing tryptophan at a maximal rate so that it contains a maximal concentration of the enzymes (Figure 8-11b). The genetic information for making these enzymes is clustered in the *trp* operon.

On the other hand, if the medium contains a sufficient quantity of tryptophan, the cell takes up what it needs from the medium and does not make any of the enzymes that catalyze the synthesis of this amino acid. The pre-existing enzyme proteins are diluted by growth. In this case the *trp* operon is said to be *repressed.* This repression of the *synthesis* of new enzyme molecules by a metabolic product is different from feedback inhibition, in which the final product in a metabolic pathway inhibits the *action* of existing enzyme molecules (see Figure 3-31).

Regulatory Proteins Control the Access of RNA Polymerase to Promoters in Bacterial DNA

The induction and repression of bacterial enzymes occur mainly through the control of the transcription of the appropriate genes. How does a bacterium accomplish this task?

The proteins in bacteria and the genes that encode them fall into two groups. The largest group includes all proteins that do not regulate transcription, such as the enzymes, membrane proteins, and ribosomal components. The genes that encode the proteins in this group

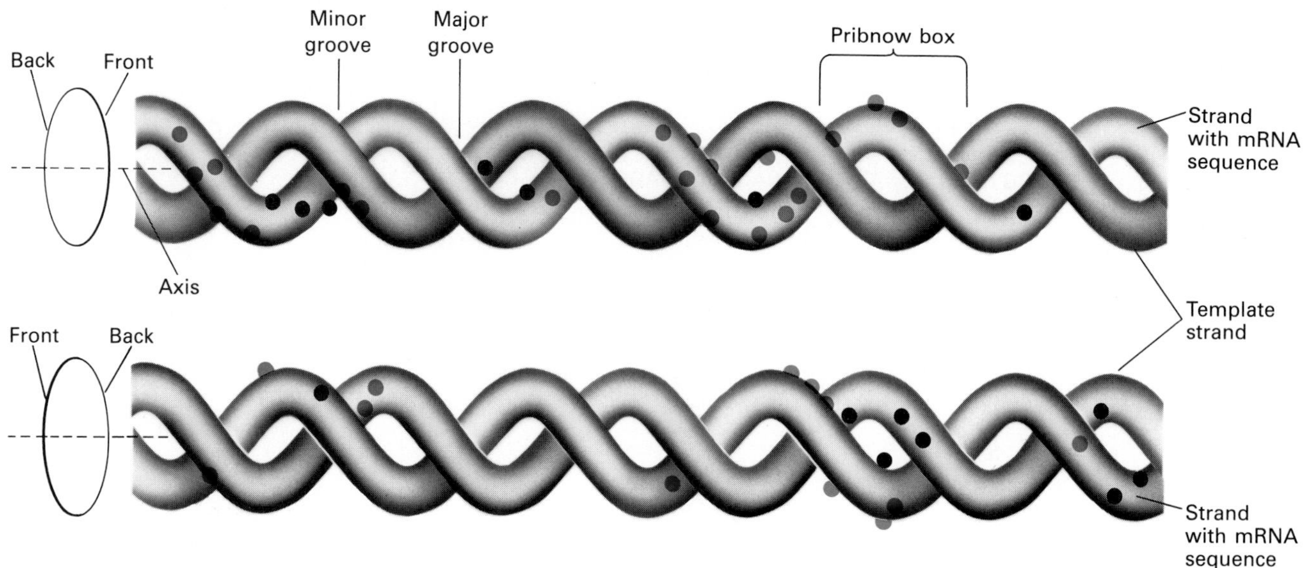

Figure 8-8 A three-dimensional representation of the *lac* promoter, showing the placement of chemical groups that interact with RNA polymerase (see Figure 8-7). The double helix is viewed both from the "front" *(top)* and from the "back" *(bottom)*. The front 180° sector of the cylindrical helix has more binding sites than the back sector. Thus the protein binds mainly to one face of the helix. The red dots are either phosphates or purine bases that are protected by the polymerase, and the black dots are protected thymines. *After U. Siebenlist, R. B. Simpson, and W. Gilbert, 1980,* Cell *20:269.*

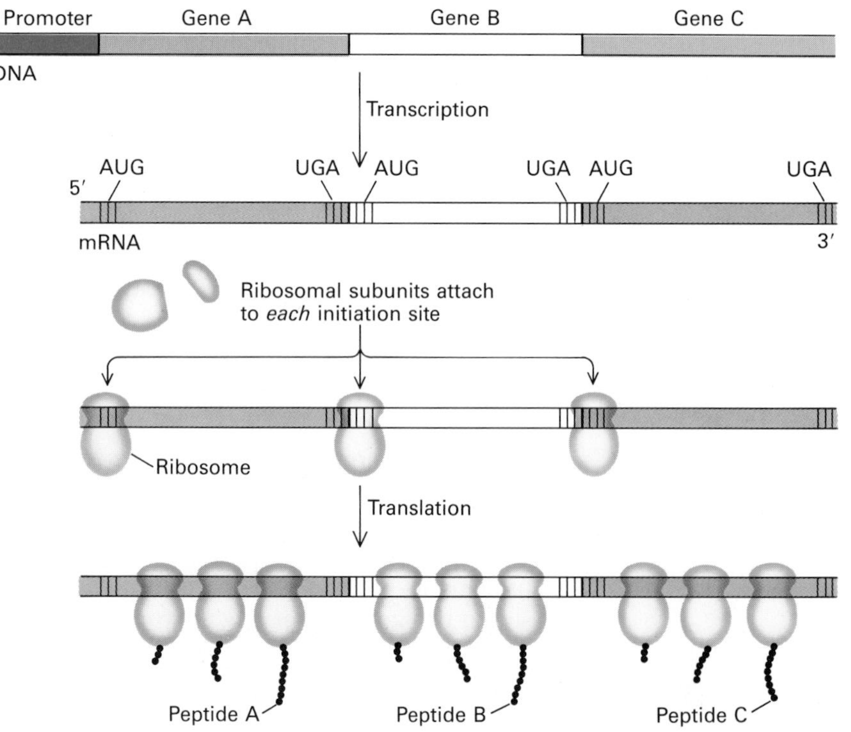

Figure 8-9 A bacterial operon is transcribed into a single polycistronic mRNA molecule. Each segment of the mRNA has its own ribosome-binding site, initiation codon (AUG), and a termination codon (UGA is shown here). The 30S and 50S ribosomal subunits can attach to each initiation site on the mRNA to begin translating the different peptides encoded by the molecule.

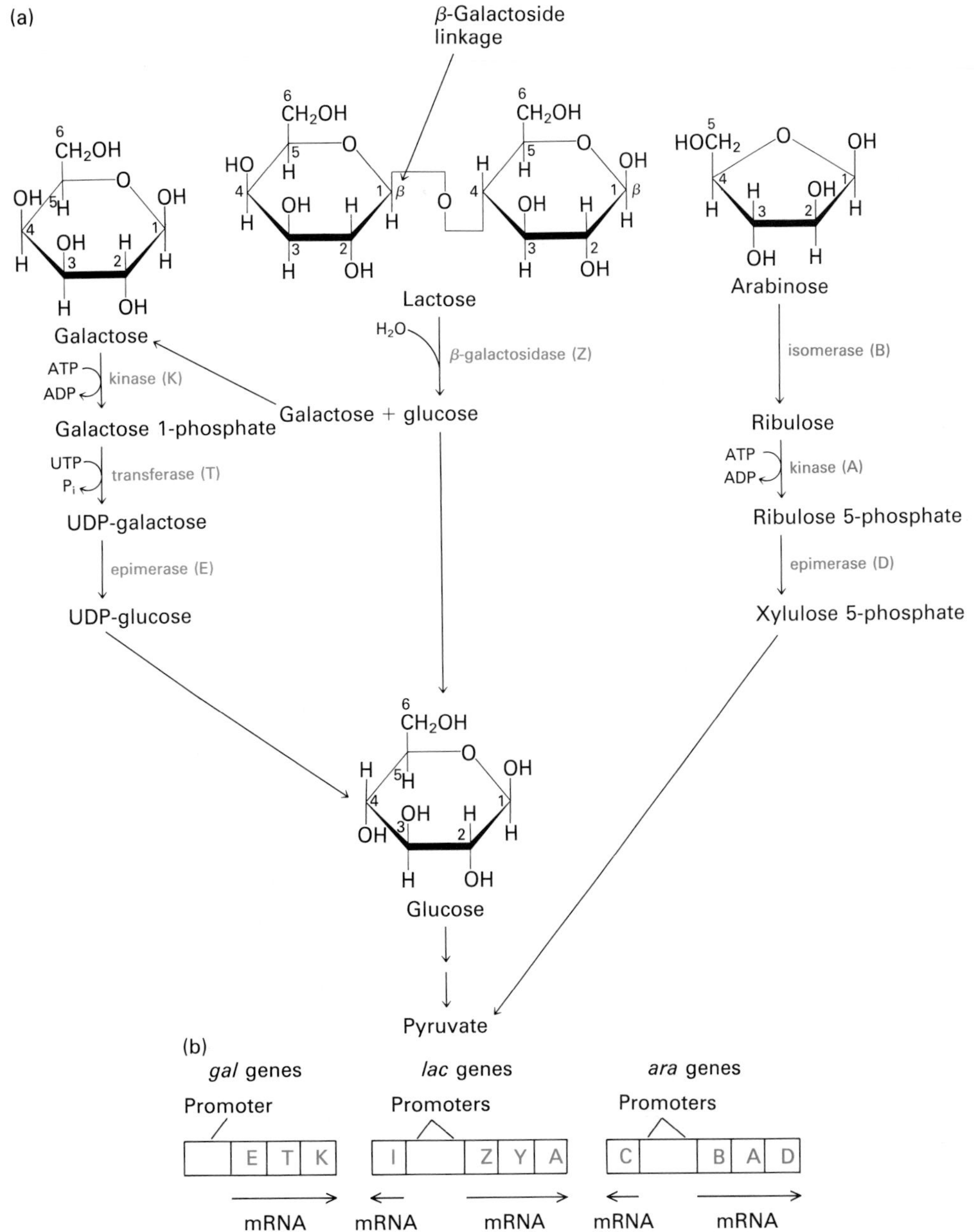

Figure 8-10 (a) Metabolic steps in three well-studied pathways of sugar metabolism in *E. coli*. The conversion of the monosaccharides galactose and arabinose into glucose requires two different sets of enzymes. Lactose is a disaccharide (a β-D-galactoside) that is split by β-galactosidase into galactose and glucose. (The capital letter following each enzyme designates the gene encoding a protein.) (b) Each set of enzymes is encoded by a cluster of genes in a single operon. (The Y and A genes clustered with the Z gene for β-galactosidase are discussed later in this chapter, as are the regulatory *lac*I gene and the regulatory *ara*C gene.) The arrows show the direction of transcription, which varies at different locations on the circular *E. coli* chromosome.

are called *structural genes*. The rate of synthesis of RNA for some structural genes is carefully controlled or regulated. However, many structural genes are transcribed continually at a rate that is more or less constant. Such unregulated genes are said to exhibit *constitutive* function.

Regulatory proteins are the other important category of proteins. Their job is to help the cell sense the environment and regulate the rate of transcription of structural genes by binding to DNA. The genes encoding the regulatory proteins are referred to as *regulatory genes*. It was originally thought that the active product of regulatory genes might be RNA, but almost all regulatory genes studied so far encode a regulatory protein.

There are two types of regulatory proteins, negative-acting and positive-acting proteins. *Negative-acting proteins* can repress a gene or an operon by binding to DNA

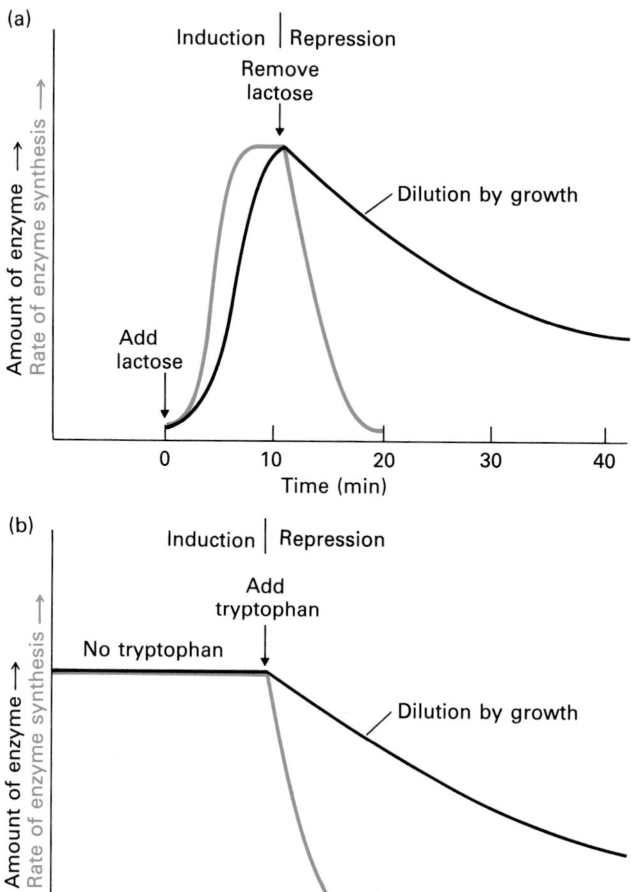

Figure 8-11 (a) The enzymes needed for metabolizing lactose are induced by addition of that sugar. Their rates of synthesis and their concentrations within the cell rise to a maximum within approximately 10 min after the lactose is added. When the sugar is removed, enzyme synthesis is quickly repressed (i.e., synthesis stops). The concentration of the enzymes is then diluted by the growth of the organism. (b) The enzymes that produce a required nutrient, such as tryptophan, are induced when the nutrient is lacking. Maximal enzyme synthesis and enzyme concentrations are reached in such cultures. When the nutrient is added to the medium, enzyme synthesis is immediately repressed, and the enzyme concentration decreases as the cells grow and divide.

at or near a promoter site and physically preventing mRNA synthesis by denying RNA polymerase access to that gene or operon. These regulatory proteins are called *repressors,* and the DNA site at which the repressor binds is called the *operator.* The operator sequence is always close to the promoter sequence and often overlaps it.

Repressors can combine with small molecules called *effectors,* which greatly affect the binding affinity of repressors for their operator sites. Two types of effectors exist. One type, the *inducers,* combine with repressors to decrease their binding affinity to the DNA. For example, a repressor protein for the genes that encode enzymes for metabolizing a particular sugar—say, lactose—will bind to its operator in the DNA until lactose or a metabolic product of lactose combines with the repressor. In this case, the lactose or its metabolic product acts as the inducer. The binding of the inducer produces a change in the shape of the represssor protein, and this is what decreases its binding affinity for its operator. When the repressor disengages from the operator, the genes for the sugar-metabolizing enzymes such as β-galactosidase are transcribed (Figure 8-12a).

Some effector molecules have the opposite role. *Corepressors* combine with repressors that are not functional when the corepressor is absent. For example, when the amino acid tryptophan (the product of the enzymes encoded by the *trp* operon) is readily available, it combines with the *trp* repressor protein. Only then can the *trp* repressor bind tightly to the operator and inhibit transcription of the genes for the enzymes that make tryptophan. Thus, the amino acid is acting as a corepressor, and the addition of tryptophan to an *E. coli* culture stops the synthesis of tryptophan-producing enzymes (Figure 8-12b).

The second type of regulatory proteins, the *positive-acting proteins,* bind to DNA at or near a promoter site and increase the efficiency with which RNA polymerase binds to the promoter. Positive-acting proteins are also called *activators,* and the DNA sites to which they bind are called *activator sites* (Figure 8-13). Among the best-studied activators is a protein termed the AraC protein; AraC is necessary in the transcription of genes responsible for arabinose utilization.

The complexity of the control of bacterial gene transcription has only recently come to be appreciated. Not only are there negative-acting proteins for some operons and positive-acting proteins for others, but many operons have *both* positive- and negative-acting proteins associated with them. For such doubly regulated genes, the maximum possible rate of mRNA synthesis is achieved only when the negative-acting protein is rendered ineffective by an inducer *and* the positive-acting protein is available to bind to the DNA.

Considerable progress has been achieved lately in studies of the physical mechanisms by which regulatory proteins from bacteria and bacteriophages bind to DNA. Three proteins that have regulatory functions have been crystallized (more discussion about these functions will follow): the *cro* protein, a negative regulator (i.e., a negative-acting protein) in bacteriophage λ; the CAP protein, a positive regulator for a number of *E. coli* genes; and the *c*I repressor, a bacteriophage λ protein that has both a positive regulatory role on the transcription of its own gene and a negative regulatory role on the transcription of other λ genes. The three-dimensional shapes of all

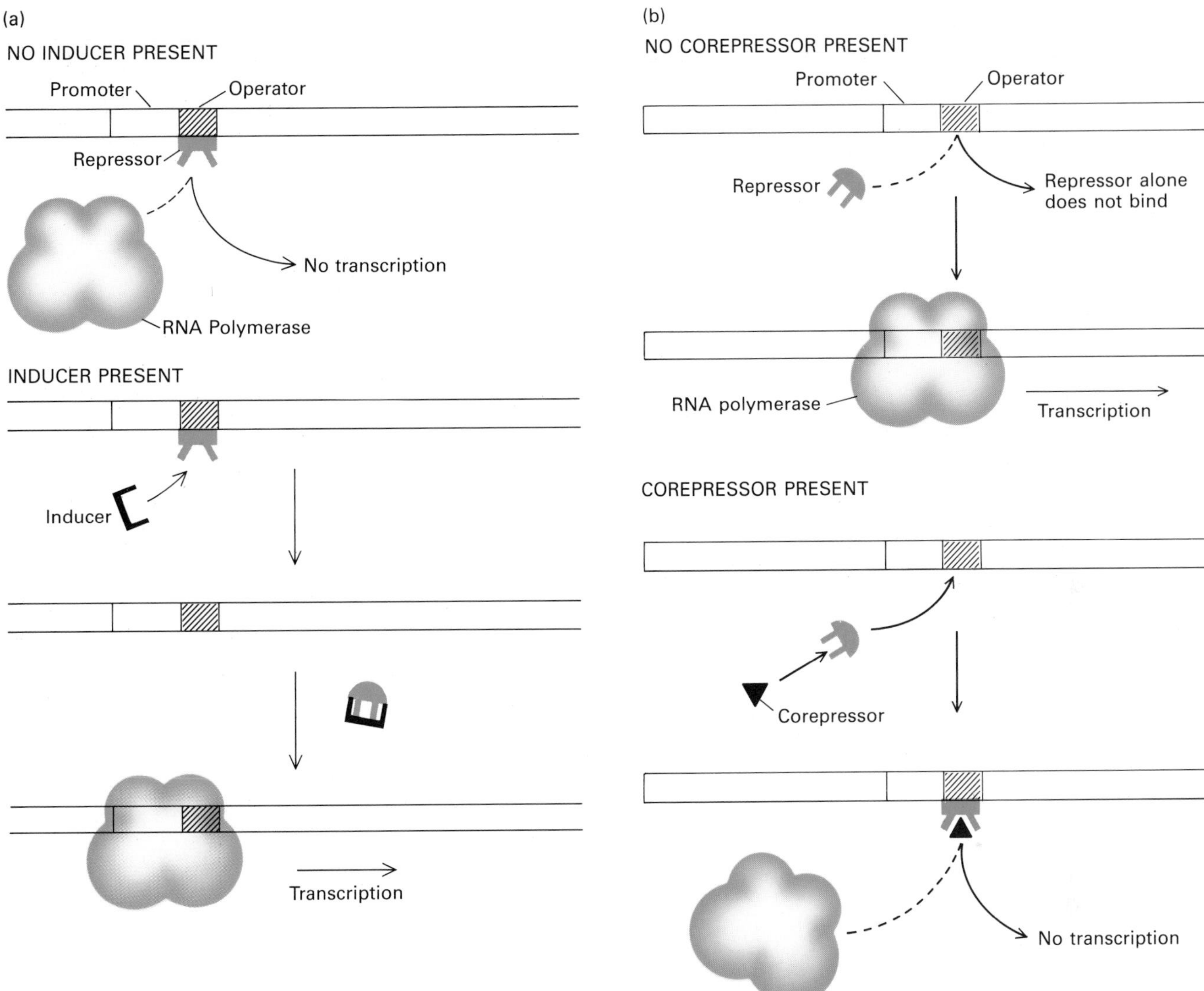

Figure 8-12 Two kinds of effectors: inducers and corepressors. (a) Some repressor proteins bind to DNA in their native states. Therefore, with no inducer present, such a repressor binds to its operator. This blocks the access of RNA polymerase to the promoter region, and mRNA synthesis cannot take place. When an inducer is present, it combines with the repressor, which causes the repressor to change shape. As a result the repressor disengages from the DNA molecule, and the promoter region becomes accessible to RNA polymerase. (b) Another type of repressor protein is ineffective unless a corepressor is present. The repressor alone does not bind tightly to its operator, so it does not prevent mRNA synthesis. When a corepressor combines with the repressor, the complex has the right shape for binding tightly to the operator; thus mRNA synthesis does not occur.

three of these proteins have striking overall similarities (Figure 8-14). All of the proteins bind as dimers to DNA, and in each monomer of the dimers there are three α-helical protein regions that have conformations similar to those in the other monomers.

Model building with DNA and the protein structures has shown that one of the helixes in each monomer of these proteins fits very comfortably into the major groove of the DNA structure. This helix is labeled α_3 in *cro* and α_F in CAP. A space-filling model of the structure of *cro* bound to DNA shows how closely the *cro* dimer can bind to the double helix of DNA by fitting into two adjacent major grooves (Figure 8-15). Studies of bacteriophage λ mutants that have "incorrect" amino acids in the binding regions of *cro* have revealed that these mutations block the function of the regulatory protein. Thus the models

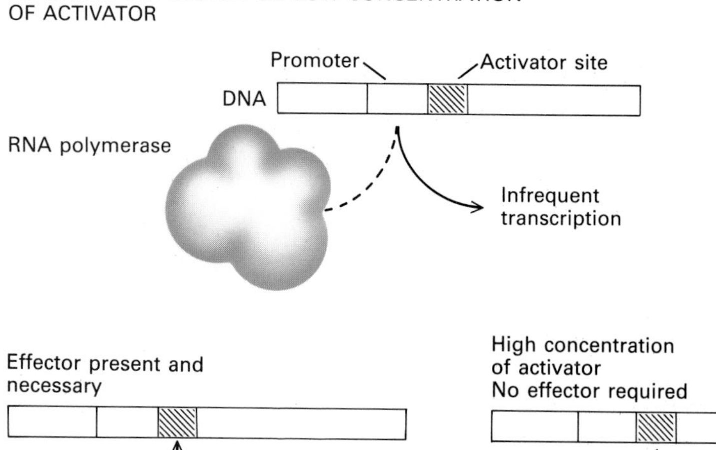

Figure 8-13 Regulation can depend on positive-acting proteins, or activators, some of which require effectors. When an effector binds to the activator (or, in cases where no effector is required, when the activator is present in a high enough concentration), the activator binds to the DNA, attracting RNA polymerase to the gene and increasing the transcription.

probably do describe how this group of regulatory proteins do their job.

In the following sections we shall trace the historical development of our knowledge of gene regulation, and describe several operons in detail to illustrate how gene control participates in bacterial physiology.

Negative Control of Transcription: The Lactose Operon

Mutations in structural genes of bacteria alter enzyme activity; mutations in regulatory genes, on the other hand, alter the response of the bacteria to environmental changes. The identification of mutant regulatory genes in bacteria was indispensable in achieving an understanding of normal bacterial gene regulation. The techniques of conjugation, transduction, and transformation allowed gene transfers to be engineered in *E. coli* so that the effects of mutations in structural and regulatory genes could be examined (Chapter 6).

Historically, the most important experiments on bacterial gene regulation concerned *E. coli* and its metabolism of lactose, a disaccharide found in mammalian milk (see Figure 8-10). If *E. coli* is grown on lactose (or on a chemically similar β-galactoside), three proteins increase in concentration in the cells: the enzyme β-galactosidase, which splits lactose into glucose and galactose; permease, an inner-bacterial-membrane protein that increases the amount of lactose taken into the cells (the function of permease is discussed in detail in Chapter 15); and transacetylase, an enzyme whose exact role in lactose metabolism is still unclear. Together, the three genes coding for these proteins constitute the structural part of the *lactose operon*. They are arranged in the following order on the *E. coli* chromosome: the Z gene (for β-galactosidase), the Y gene (for permease), and the A gene (for transacetylase). Lactose or a metabolic product of lactose induces the transcription of the operon. The classic genetic and biochemical studies on the lactose operon yielded the first example of *negative control* (i.e., control by a negative-acting protein) of the initiation of mRNA synthesis.

The genetic makeup of the lactose operon of a normal bacterium can be written $Z^+ Y^+ A^+$, to indicate the normal production of active forms of all three proteins. A mutation in one of the genes can result in a bacterium that produces an inactive form of the corresponding protein, or none at all. A mutant $Z^- Y^+ A^+$, for example,

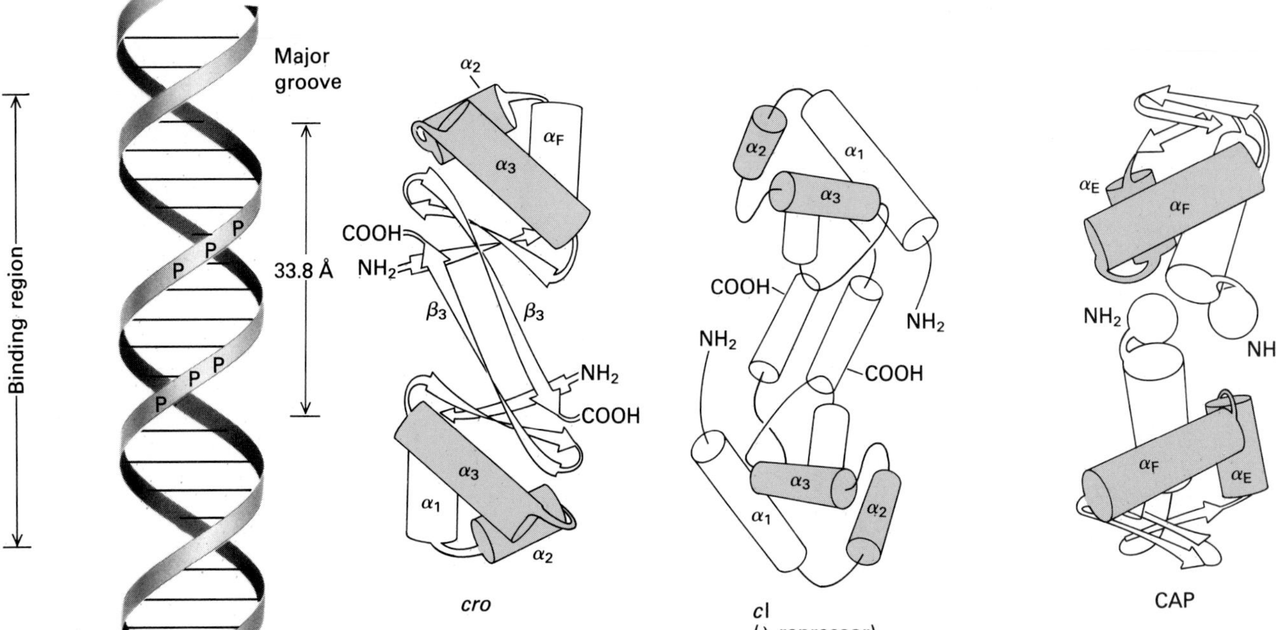

Figure 8-14 Models of three regulatory proteins. The *cro* protein and λ repressor are encoded in the genome of bacteriophage λ. Their physiological roles in the bacteriophage λ life cycle are discussed on pages 296 to 299. *Cro* is a negative-acting protein and the λ repressor can act both negatively and positively. The CAP protein is an *E. coli* protein that activates a wide array of genes. All act by binding to DNA. The diagram of the three-dimensional structure of each protein was determined from x-ray crystallographic analysis. The cylindrical forms are α-helical and the lines or arrows contain regions of β pleated sheet structures (see Chapter 3). On the left of the diagram is the double helix of DNA drawn to the same scale. Note for example the distances between the two helixes in the *cro* protein labeled α3 are very similar to the distances in the DNA binding region between the two major (wider) grooves in DNA.

would respond to the presence of lactose by producing permease and transacetylase but no effective β-galactosidase. Some mutant bacteria have normal structural information for the three proteins but do not have the normal ability to regulate any of the three genes. Many of these regulatory mutants make the three proteins even in the absence of an inducer. Such mutants are *constitutive* mutants, in contrast to normal, *inducible* bacteria, which make the proteins only in the presence of lactose. Some of the mutations that cause constitutive production of β-galactosides were found in genetic tests to lie in the gene for inducibility. This is termed the I gene (Figure 8-16). Thus a normal, inducible bacterium can be represented by I^+ and a constitutive mutant by I^-.

The nature of the I gene function was first suggested by conjugation experiments performed in the late 1950s in the laboratories of François Jacob and Jacques Monod at the Pasteur Institute in Paris (Figure 8-17). Normal donor bacteria (I^+ Z^+) were crossed with mutant recipient bacteria (I^- Z^-). After conjugation, the recipient bacteria contained both sets of genes; they had become *merozygotes* (from the Greek for "joined parts"). Their genetic constitution can be represented by I^+ Z^+/I^- Z^- (donor/recipient).

Before conjugation began, the donor cells were not producing β-galactosidase because no inducer (lactose) was present in the medium. The recipient cells also produced no β-galactosidase: although they required no inducer, their Z gene was defective. Shortly after DNA transfer began, β-galactosidase began to be formed whether or not inducer was present. However, if no inducer was present, enzyme synthesis stopped again after about 2 h (as indicated in Figure 8-18). Because the donor cells had been killed as soon as enough time had passed for the transfer of the I and Z genes, the recipient bacteria (the merozygotes) were the ones that had produced β-galactosidase and then stopped.

How were these facts to be explained? To the experimenters, it seemed likely that in the donor cells, in the absence of inducer, a product of the I^+ gene prevented expression of the Z^+ gene. This product was absent in the recipient cells because their I genes were inactive. Thus, when the Z^+ gene entered the recipient cells from the donors, it suddenly found itself in surroundings that

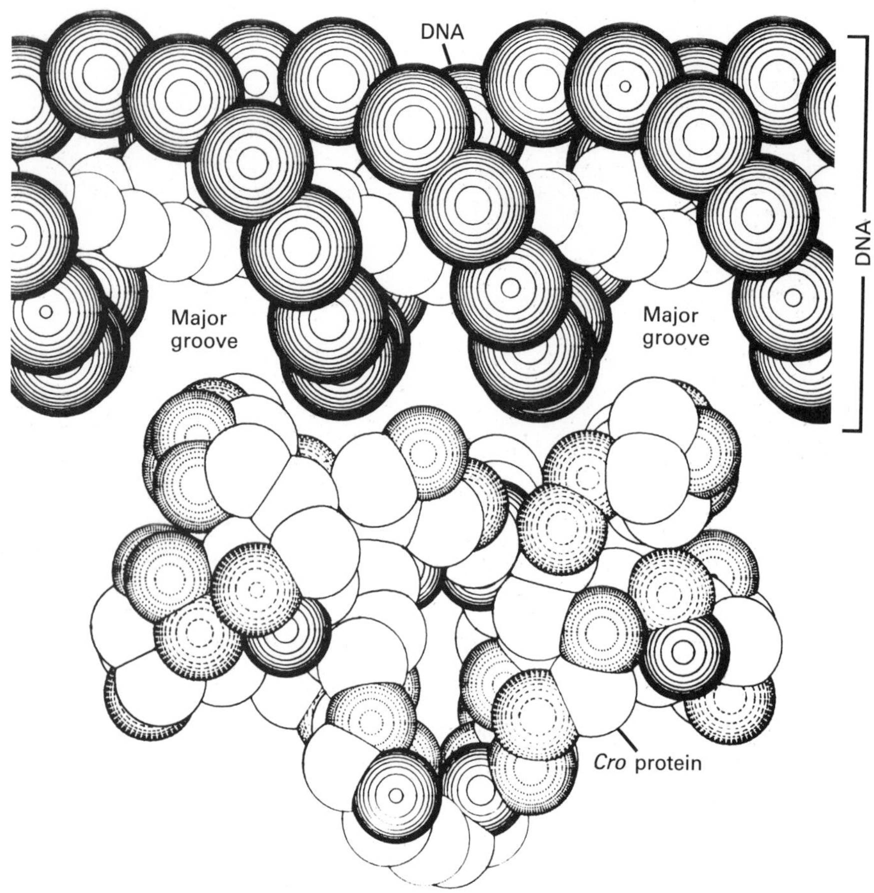

Figure 8-15 Three-dimensional computer model of *cro* protein bound to DNA. The crystallographic data of the protein and DNA structure are used to make a computer model of best fit for the binding of protein to DNA. [See Y. Takeda et al., 1983, *Science* **221**:1020.]

lacked the I^+ product. Released, Z^+ began to direct the production of β-galactosidase. However, I^+ also had been donated to the recipient cells; in time, in the absence of inducer, the product of I^+ accumulated in the recipients and the activity of Z^+ was again suppressed. Thus arose the concept of a gene product that acted as a repressor (the I^+ product).

Additional genetic experiments with lactose genes greatly illuminated the function of the repressor. Different combinations of mutants in the lactose operon and in the I gene could be inserted into plasmids, which were then introduced into *E. coli* cells that already had one lactose operon and an I gene in their chromosomes. (Recall that a plasmid is a small circular DNA molecule capable of independent replication within a bacterium.) The experiments with these partial diploid cells suggested that the repressor protein had to bind to a particular site on the DNA in order to function, and that this site was physically contiguous to the Z gene being regulated.

The results of these experiments are shown in Table 8-1, which lists the activities of various bacterial strains containing combinations of mutant *lac* genes. First, in cells that were $I^- Z^+/I^+ Z^-$, inducer was still required for β-galactosidase synthesis. This suggested that the repres-

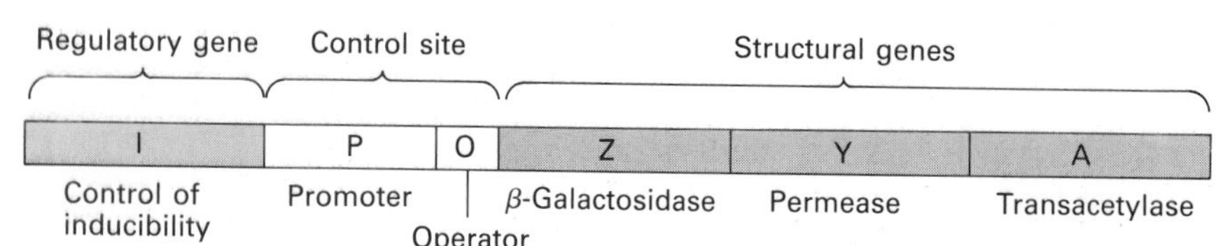

Figure 8-16 The *E. coli* genes that govern lactose metabolism: the regulatory I gene for inducibility, the promoter-operator control site, and the structural genes of the lactose operon. All of these regulatory elements for lactose utilization were detected by genetic analysis of bacterial mutants.

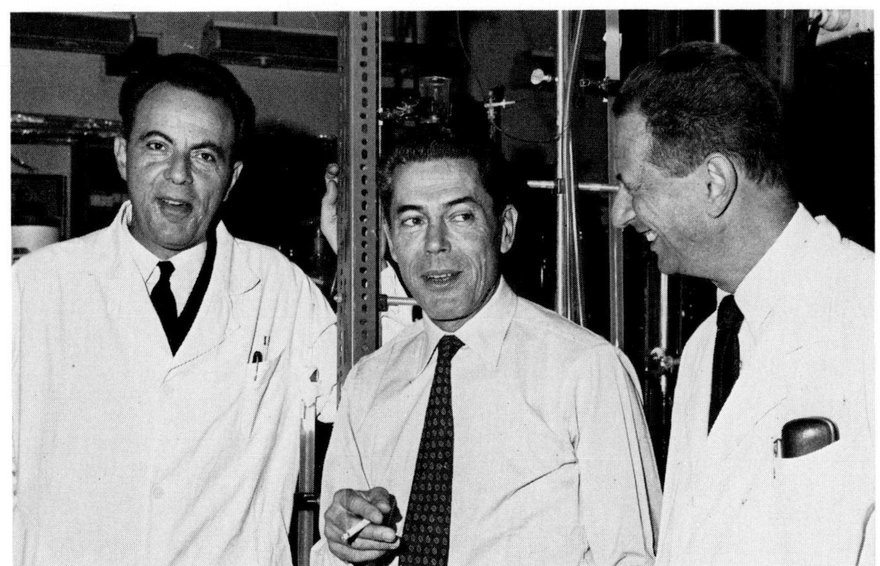

Figure 8-17 François Jacob (1920– , *left*) and Jacques Monod (1919–1976, *center*) at the Pasteur Institute in Paris, together with André Lwoff (1902– , *right*), with whom they shared the Nobel Prize in 1965. *Photograph by René Saint-Paul, Paris; courtesy of F. Jacob.*

resided at another chromosomal location. Then the discovery of mutations in sequences called O (for "operator") suggested that the repressor acted on a particular DNA site; this site was shown by further genetic analysis to be located very close to the Z gene. The mutations in O sequences caused the neighboring Z^+ genes to be active even without inducer, so these mutations were called O^c, for "operator constitutive." When an O^c mutant with an active Z^+ gene was used to create a partial diploid ($O^c\ I^-\ Z^+/O^+\ I^+\ Z^+$), the cells were still partially constitutive. Thus O^c continued to cause the neighboring Z^+ to be active, even in the presence of the I gene product.

At this point, two types of mutations that result in constitutive β-galactosidase production were recognized—mutations in the I gene and mutations in the operator. Mutations in the operator were *cis-active;* that is, they affected only the Z gene on the same chromosome. However, mutations in the I gene were *trans-active,* meaning that they could affect either a Z gene on the same chromosome or one on another chromosome. Thus it was concluded that a diffusible product (the I gene product) acted at a site next to the Z gene (the operator). Then temperature-sensitive mutations in the I gene product were detected, which suggested that the repressor was a protein because proteins are far more sensitive to heat than nucleic acids. In addition, the I gene was also subject to suppressible mutations (i.e., mutations that were correctable by suppressor tRNAs; see the discussion of cor-

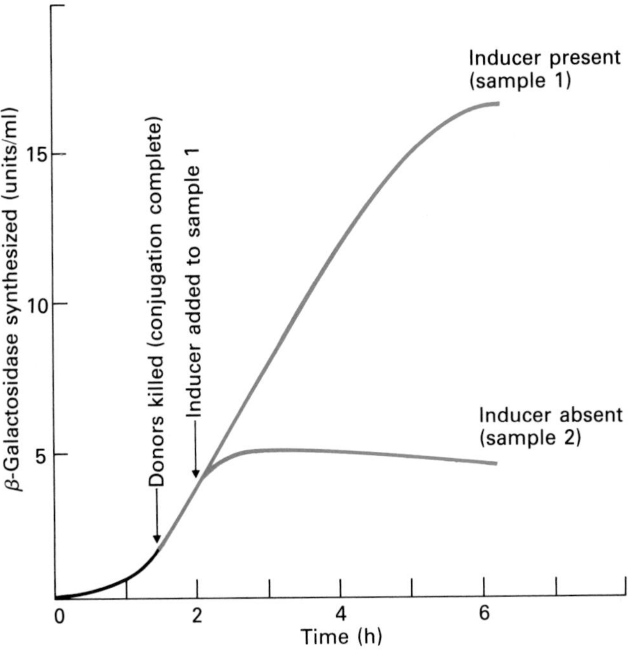

Figure 8-18 The enzyme β-galactosidase is produced by $I^+\ Z^+$ *E. coli* cells only in the presence of inducer; it is not produced at all by $I^-\ Z^-$ cells. After the conjugation of $I^+\ Z^+$ (donor) cells with $I^-\ Z^-$ (recipient) cells, the donors were killed. Initially the merozygotes ($I^+\ Z^+/I^-\ Z^-$) produced β-galactosidase whether or not inducer was added. However, when no inducer was present (sample 2), synthesis of the enzyme ceased about 2 h after conjugation began. When inducer was added (sample 1) enzyme synthesis continued. [See A. B. Pardee, F. Jacob, and J. Monod, 1959, *J. Mol. Biol.* **1**:165.]

Table 8-1 β-Galactosidase levels in various genetically different *E. coli* strains

Strain	Enzyme level: With inducer	Enzyme level: Without inducer	Conclusion
$O^+ I^- Z^+$	Maximal	Maximal	Constitutive enzyme synthesis
$O^+ I^- Z^+/O^+ I^+ Z^-$	Maximal	Low	Diffusible, trans-active* repressor; I^+ dominant over I^-
$O^c I^+$ (or I^-) Z^+	Maximal	Near maximal	Operator constitutive and dominant over I^+; repressor unable to function
$O^c I^+$ (or I^-) $Z^+/O^+ I^+ Z^+$	Maximal	Half maximal	O^c dominant and cis-active*
$O^c I^+$ (or I^-) $Z^+/O^+ I^+ Z^-$	Half maximal	Half maximal	O^c dominant and cis-active

*"Trans-active" means capable of activity on another chromosome. "Cis-active" means active only on the same chromosome; for a mutation to be cis-active, it must be located close to the affected gene. See Figure 10-24.

rection of "nonsense" mutations in Chapter 4). This was strong evidence that the repressor was a protein. How did it work?

Jacob and Monod proposed in 1961 that the protein produced by the I gene specifically regulated the DNA that encoded all three enzymes of the lactose operon. Jacob and Monod further suggested that an unstable intermediate molecule—an mRNA—conveyed the information in DNA to the protein-synthesizing machinery, and that the function of the repressor was to regulate the amount of mRNA (Figure 8-19).

The critical test for this theory came with the measurement of mRNA synthesized from the *lac* genes. When normal cells are induced by lactose to form β-galactosidase, they contain a large amount of mRNA that can be measured by hybridization with the lactose operon DNA (Figure 8-20). In the absence of lactose, these same bacteria contain less than 1 percent as much of the mRNA that is complementary to the lactose operon DNA.

The lactose repressor protein has since been purified, and its entire sequence of 360 amino acids is known. The repressor binds to a 25- to 30-base DNA region that includes several bases before and 20 bases after the initiation site for the synthesis of β-galactosidase mRNA. The RNA polymerase also binds in this region: the promoter overlaps the operator. The overlap explains why the binding of the repressor to the operator prevents the

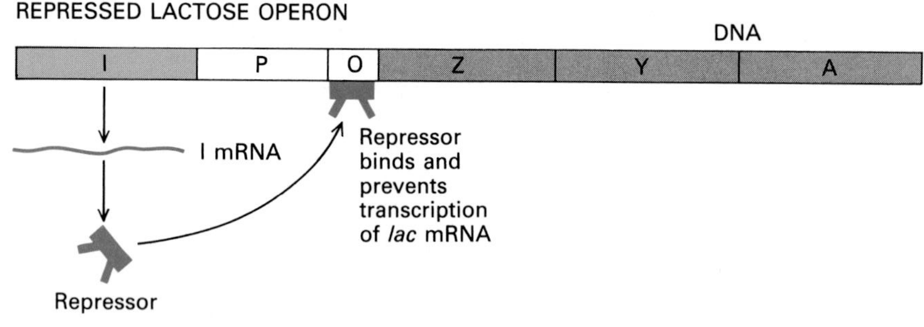

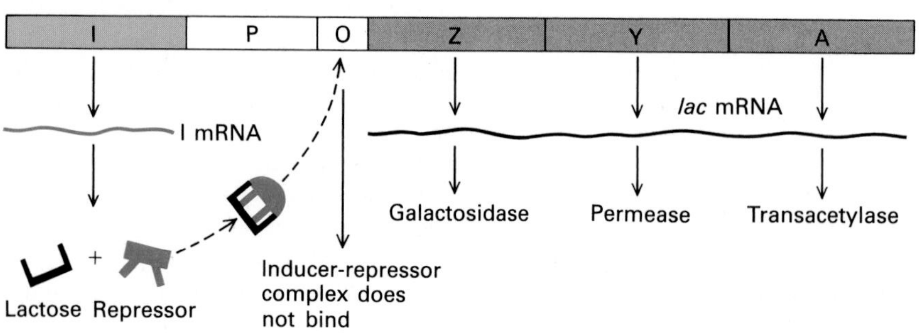

Figure 8-19 The original Jacob-Monod model of negative regulation of the *lac* operon. The *lac*I gene produces a repressor protein that binds to the operator sequence (O) of the *lac* operon and thus blocks the synthesis of a *lac* mRNA. This prevents the synthesis of the lactose-metabolizing enzymes encoded by the Z, Y, and A genes. When lactose (the inducer) is present, the repressor protein combines with it. The complex is unable to bind to the operator sequence, so the synthesis of the lactose enzymes proceeds. [See F. Jacob and J. Monod, 1961, *J. Mol. Biol.* 3:318.]

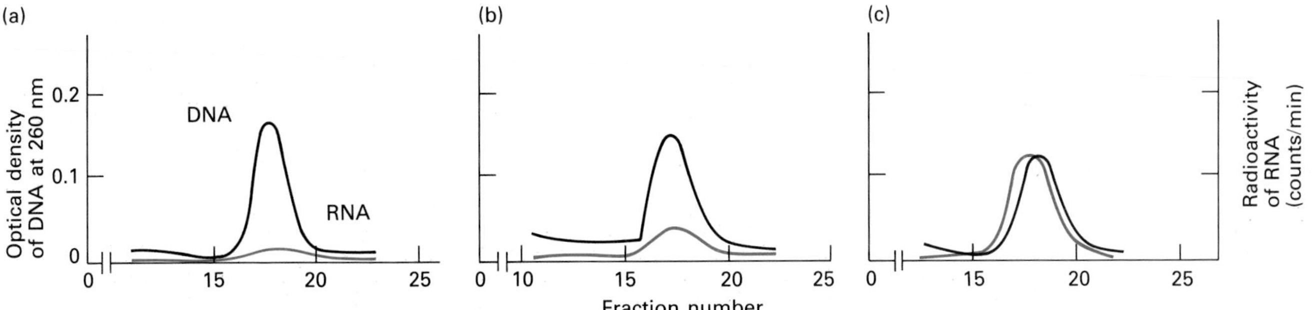

Figure 8-20 DNA was prepared from a transducing bacteriophage that carried the *lac* genes from *E. coli*. This DNA was hybridized with labeled RNA (color) from three different *E. coli* cultures. The amount of RNA that formed a molecular hybrid with the DNA was measured by separating the DNA by equilibrium density-gradient centrifugation in cesium chloride (Chapter 7), and then collecting samples (fractions) of the gradient from bottom to top (fraction 1 is from the bottom of the gradient). The optical density revealed the amount of DNA, and the radioactivity represented the amount of labeled RNA. Three RNA samples were tested: (a) one from cells that had no *lac* genes; (b) one from cells that had *lac* genes, but that were cultured in a medium containing no inducer; and (c) one from cells that had *lac* genes, and that were cultured with inducer. Only in the cells in the (c) group was there substantial labeled *lac*-specific mRNA. [See M. Hayashi et al., 1962, *Proc. Nat'l Acad. Sci. USA* **49**:729.]

binding of RNA polymerase to the promoter and the initiation of β-galactosidase mRNA synthesis (Figure 8-21).

Because both mutant and normal DNA sequences for the promoter-operator region of the lactose operon have been known for some time, it has been possible to document the importance of contact points between the operator and the repressor molecule. The techniques used were similar to those in which the binding of RNA polymerase with promoter sites was explored (see Figures 8-5 and 8-7). The normal or *wild-type* repressor protein has contact points in the region in which mutations are known to cause constitutive β-galactosidase synthesis. Thus mutations in DNA sequences can cause the repressor to be unable to bind to the DNA.

Positive Control of Transcription: Arabinose Metabolism and the *ara*C Gene

When *E. coli* cells are grown on arabinose as an energy source, they produce the three enzymes needed to convert arabinose into xylulose-5-phosphate (see Figure 8-10). (Xylulose 5-phosphate can be converted into an intermediate in the glycolytic pathway, where oxidation of its carbon skeleton supplies energy for growth; see Figures 8-10 and 5-42.) These three enzymes—an isomerase, a kinase, and an epimerase—are the products of three genes—B, A, and D—that belong to a single operon, the

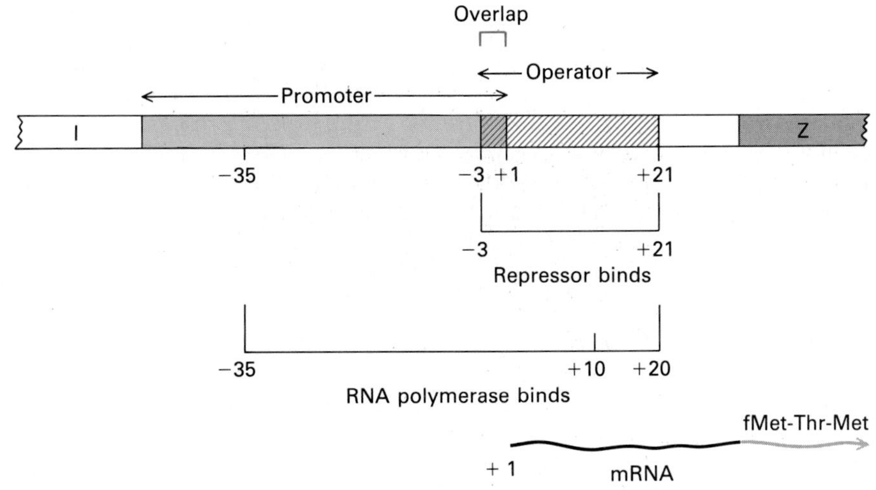

Figure 8-21 The arrangement of the promoter and operator sequences that control the *lac* operon in *E. coli*. The end of the I gene and the beginning of the Z gene are also shown. The promoter and operator sequences overlap, which accounts for the exclusion of RNA polymerase from the promoter when the repressor occupies the operator. [See W. S. Reznikoff and J. N. Abelson, 1978, in *The Operon*, J. H. Miller and W. S. Reznikoff, eds., Cold Spring Harbor Laboratory.]

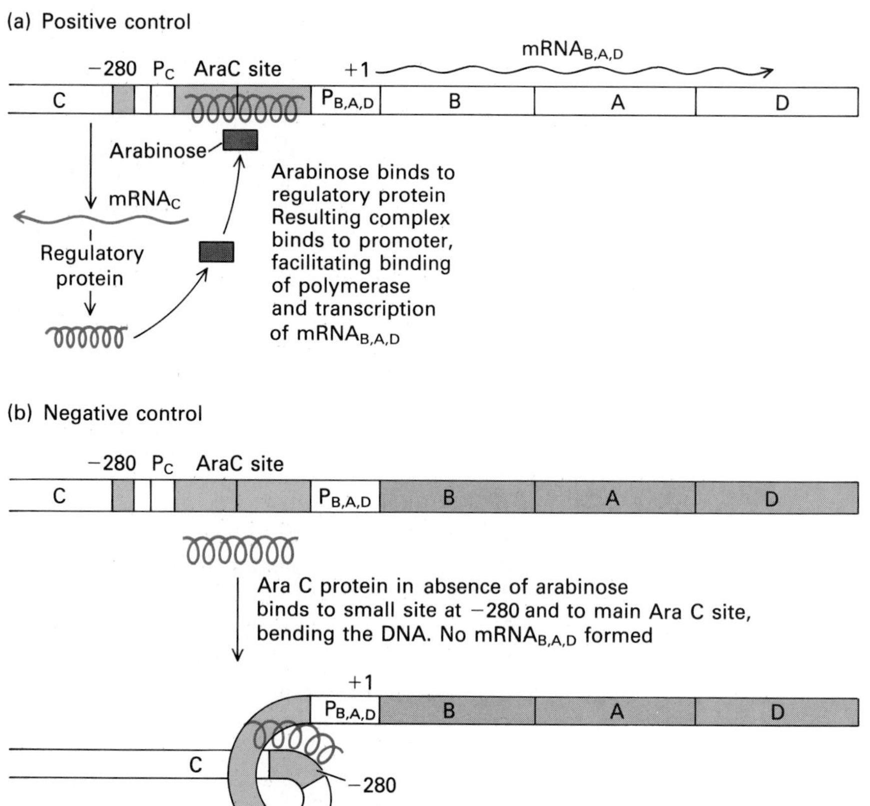

Figure 8-22 The positive and negative regulation of the *ara* operon by the AraC protein. (a) The *ara*C gene of *E. coli* produces a regulatory protein that binds to the indicated sequence in the *ara*B,A,D operon. When arabinose is present, it binds to the regulatory protein AraC, which causes the protein to assume a shape that assists in the binding of RNA polymerase and so stimulates the transcription of the structural genes (B, A, D) of the operon. (b) When arabinose is absent, the binding of this regulatory protein blocks the access of RNA polymerase to the promoter and thus prevents the transcription of the structural gene. This negative action apparently involves not only binding at the major *ara*C site, but at a site that lies 280 nucleotides upstream (−280 site) of the transcription start site for the mRNA of the B,A,D operon. These two binding sites must be brought together by bending the DNA. [See T. M. Dunn et al., 1984, *Proc. Nat'l Acad Sci. USA*, 8:5017.]

arabinose *(ara)* operon. Thus new enzymes are made to metabolize arabinose just as enzymes are made to metabolize lactose, but it is now known that the regulatory "circuits" of the two operons are very different.

The original examinations of mutant *E. coli* cells that were incapable of utilizing arabinose revealed three different classes of mutants, each of which contained one of the three enzyme proteins mentioned above in an inactive form. Workers subsequently discovered a fourth cell type that is mutant in a regulatory gene termed *ara*C. The *ara*C gene product (the protein AraC*) is different from the *lac* repressor. Mutations in the *lac* repressor cause cells to be constitutive, but mutations in *ara*C make cells incapable of responding to arabinose, even when they contain the normal structural genes (B, A, D) of the *ara* operon. The B, A, and D genes were shown to be normal by transduction experiments.

The AraC protein, which has been purified, is a regulatory protein that binds to DNA. In the absence of arabinose, this protein exercises negative control over the three structural genes for arabinose enzymes by binding to an operator site and repressing the synthesis of their mRNA. However, in the presence of arabinose, the AraC protein exercises *positive control* over the *ara* genes (i.e., becomes positive-acting) and binds to the B,A,D promoter as a necessary element in efficient transcription (Figure 8-22). When AraC is bound to arabinose, the protein probably changes its shape—that is, undergoes an *allosteric* change—so that it interacts differently with the DNA of the *ara* promoter. The function of the AraC protein thus contrasts with that of the Lac repressor protein, which is simply removed in the presence of lactose.

An experiment that was important in confirming the positive-acting nature of the *ara*C gene product employed cell-free protein synthesis. Cell extracts containing all of the elements of the protein-synthesizing system, plus RNA polymerase and the DNA purified from a transducing bacteriophage carrying the *ara* operon, manufactured the B, A, and D enzymes, but *only* if the AraC protein plus arabinose itself were present. This proved that a sugar-specific, positive-acting protein can participate in a cell-free system to regulate the metabolism of arabinose. Exactly how the AraC protein increases transcription is not yet known. Other bacterial genes are similarly regulated by positive-acting elements.

Most of the interest in the AraC protein has been generated by its positive-acting function when arabinose is present. Study of its negative action when arabinose is absent has also proved extremely informative. The negative action of the AraC protein is apparently not carried out by binding immediately upstream from the transcription start site. Rather, approximately 280 bases upstream is a binding site (an operator, O_{p2}), which is thought to

*Note that the *products* of certain genes (e.g., those in the *ara, lac, gal,* and *trp* operons) are differentiated from the genes themselves by setting the name of the product or protein in roman type with an initial capital letter.

bend the DNA so that the AraC protein without arabinose binds to two regions, stopping transcription of B,A,D mRNA. This is a very important result because it shows that all DNA sites involved in regulation do not have to lie immediately adjacent to transcription start sites.

Compound Control of Transcription in Several Operons

In the years when negative control of the lactose operon and positive control of the arabinose operon were being established, a conscious oversimplification was made by investigators who worked on these problems. The sugar normally used to grow bacteria is glucose, the simplest, most directly utilized sugar because it enters cell metabolism without requiring the induction of any new enzymes. However, if glucose is present in a medium to which lactose, arabinose, or any of a number of other sugars is added, enzymes needed for metabolizing the other sugar cannot be induced. Apparently, glucose or a breakdown product of glucose, a *catabolite,* prevents the synthesis of mRNA for a wide variety of sugar-metabolizing enzymes. This phenomenon is now referred to as *catabolite repression.* In the original studies of the induction of the *lac* and *ara* operons, a medium lacking glucose was used to prevent catabolite repression.

Interest in catabolite repression revived, however, when biochemists discovered that bacteria starved for glucose show a marked increase in the synthesis of an unusual nucleotide, cyclic adenosine 3′,5′-monophosphate (cyclic AMP, or cAMP; Figure 8-23). It appears that the increased level of cyclic AMP is an "alert" signal for dire metabolic stress. In addition, cyclic AMP has been linked to gene regulation. When cyclic AMP was added to the medium of cells grown in both glucose and lactose, the induction of β-galactosidase was *not* suppressed as it was when no cyclic AMP was added. The same result was obtained for arabinose enzyme induction. Apparently the cyclic AMP played a direct role in overcoming the catabolite repression of several different enzymes. It is now known that cyclic AMP does, in fact, participate in initiating the transcription of a wide variety of genes, not only in bacterial cells but in eukaryotes as well.

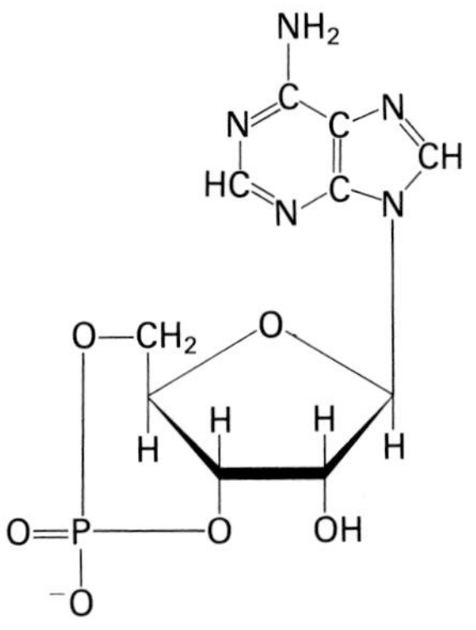

NH_2 N C C N CH HC N C N O—CH_2 O H H H H O=P O OH $^-$O

3′,5′-Cyclic AMP

Figure 8-23 The chemical structure of 3′,5′-cyclic AMP. The phosphate is attached to the oxygens on both the 3′ and the 5′ carbons by a cyclic linkage, rather than being attached to a single oxygen, as is the case in most nucleotides.

A Single Protein, CAP, Has a Positive Regulatory Function in the *lac* and *ara* Operons

Once the transcriptional effects of cyclic AMP were discovered, it was hypothesized that proteins must mediate or "interpret" connection between high cyclic AMP levels (glucose starvation) and the need to induce other sugar-metabolizing enzymes. To test this theory, laboratory workers sought mutant cells that could not grow in the presence of any of a wide variety of sugars even if glucose were absent from the medium. The workers isolated individual mutant *E. coli* colonies that could use neither lactose, galactose, maltose, nor arabinose. By a second mutation, these cells regained the ability to use all of these sugars. In other words, a mutation in a single gene appeared to be the reason for the mutant cells' inability to utilize any of the sugars.

About half of the mutant cells lacked the ability to make cyclic AMP, and thus they apparently could not detect that they were undergoing glucose starvation. The other half could make cyclic AMP, but did not make more of the enzymes for metabolizing any of the sugars mentioned above. It was hypothesized that the second group of mutants contained a defective protein that could not bind to cyclic AMP. A protein that could bind to cyclic AMP was then isolated from normal cells, and this protein was indeed found to be missing in the second group of mutant cells. The final proof that the lactose operon is controlled by cyclic AMP and the cyclic AMP–binding protein—now called the *catabolite activator protein,* or CAP—came from in vitro transcription of the *lac* operon. Purified DNA containing that operon was transcribed by RNA polymerase only about 5 percent as well without either cyclic AMP and CAP as it was with both. Thus, the *lac* operon is under both the positive control of the CAP–cyclic AMP complex and the negative control of the Lac repressor protein (Figure 8-24).

In the in vitro transcription of the arabinose operon, *two* positive-acting proteins are required for maximal activity, the AraC protein and the CAP protein.

Although the compound control of the transcription of the *lac* and the *ara* operons is complicated, it is advantageous to the *E. coli* cell. As long as glucose is plentiful, little cyclic AMP is produced; thus, CAP is not activated

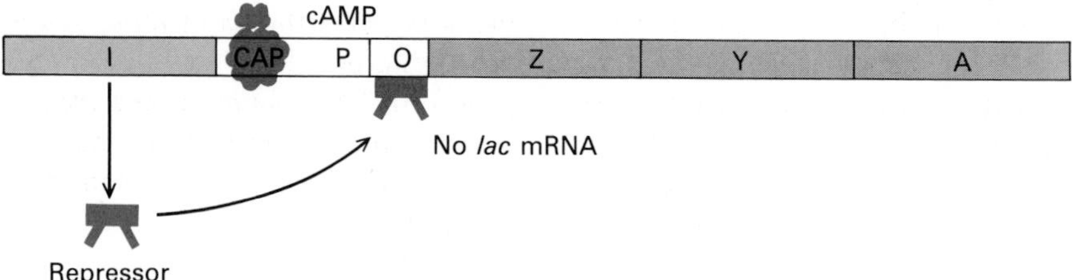

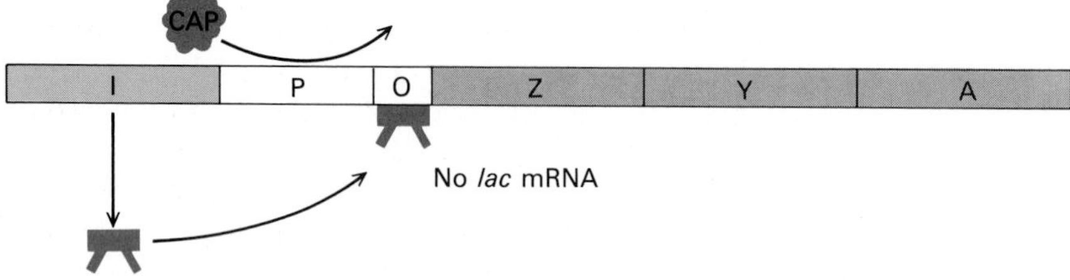

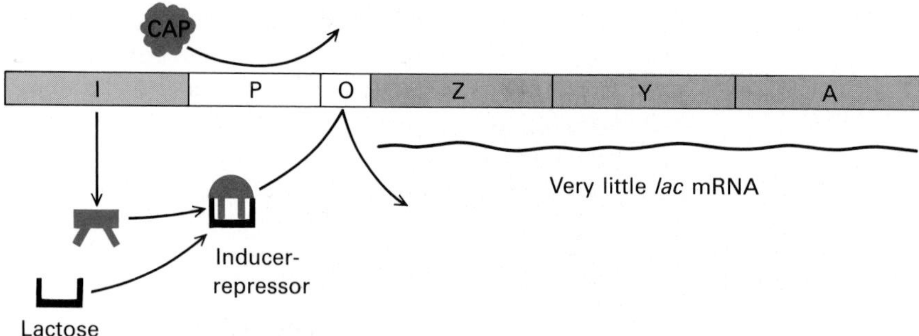

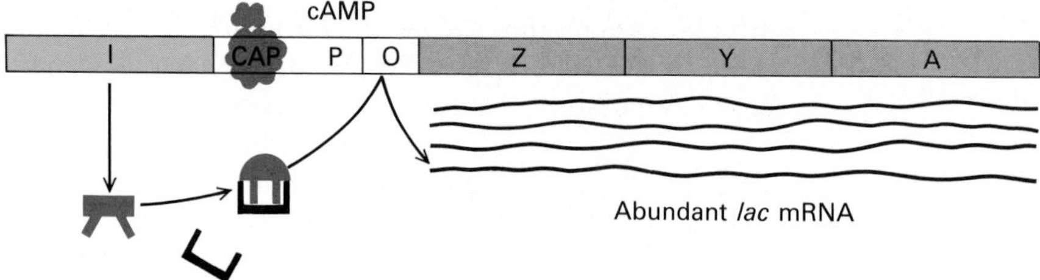

Figure 8-24 The control of the *lac* operon by two proteins: the LacI repressor and CAP, the cyclic AMP–binding protein. (a),(b) Without lactose, no *lac* mRNA is formed whether or not glucose is present, because the repressor protein is bound to the operator. (c) The presence of lactose sugars removes the repressor, but when the level of cyclic AMP is low (i.e., when glucose is present) the CAP protein does not bind to the promoter, and a minimal amount of *lac* mRNA is synthesized. (d) Maximal transcription of the *lac* operon is achieved when lactose is present and glucose is absent. This is because the presence of lactose removes the repressor protein from the DNA, and the absence of glucose causes an increase in cyclic AMP, which binds to the CAP protein. The resulting CAP–cyclic AMP complex binds to the promoter region and activates transcription of the operon.

and the induction of enzymes that digest other sugars is not necessary. When glucose is absent *and* another sugar is present, transcription is initiated and the sugar can be metabolized.

The physical basis for the action of CAP has been elucidated by DNA sequencing and by DNA-protection studies in the presence of CAP. In the best-studied genes, those of the lactose operon, the binding sites for CAP and RNA polymerase (see Figure 8-8) have been found to be contiguous and on the same side of the DNA helix. It seems likely, therefore, that the CAP protein somehow attracts the RNA polymerase or otherwise facilitates transcriptional initiation.

The Galactose Operon Has Both a Regulated Promoter and a Constitutive One

Growth of *E. coli* on the six-carbon monosaccharide galactose leads to an increase of galactose-metabolizing enzymes, which are encoded by the *gal* operon (see Figure 8-10). (The enzymes are a kinase, a transferase, and an epimerase.) However, the level of the three galactose enzymes produced by fully induced transcription of the *gal* operon in the presence of galactose and in the absence of glucose is only 10 to 15 times as high as that produced without induced transcription. In contrast, the maximally induced level of β-galactosidase may be 1000 times the uninduced level. The presence of glucose prevents the induced increase of the galactose-metabolizing enzymes, as it does for β-galactosidase. However, the basal, noninduced level of the enzymes is not decreased by glucose. These findings may be related to the fact that one of the enzymes, the epimerase, produces a direct precursor for cell wall biosynthesis. This enzyme is needed by the cell whether or not glucose is in the medium; perhaps this is why the epimerase is continuously synthesized at a modest rate.

The basis for this complex control has been illuminated by recent studies. In the *gal* operon, two initiation sites for RNA synthesis are recognized by the *E. coli* RNA polymerase. One site, S_2, binds the polymerase and allows the transcription of genes for the galactose-metabolizing enzymes at low levels of cyclic AMP or CAP; that is, this site is responsible for the constitutive level of these enzymes. The other binding site, S_1, requires high levels of both cyclic AMP and CAP in order for RNA polymerase to bind and transcribe the genes for the three galactose enzymes (Figure 8-25). It appears that the S_1 polymerase-binding site has a much higher affinity for polymerase than does S_2, so when cyclic AMP is high and S_1 is used, the rate of RNA chain initiation is greater and more *gal* mRNA is produced.

The Arginine Repressor: One Repressor Acting at Multiple Sites

The regulation of the three operons *lac, ara,* and *gal* in *E. coli* illustrates the flexibility and subtlety of the control of transcriptional initiation in bacteria. In all three of the operons mentioned above, the genes for the necessary sugar-metabolizing enzymes are controlled as one transcription unit. In eukaryotes, genes that contribute to one pathway are not clustered into operons. The *E. coli* genes that encode the enzymes for arginine biosynthesis are not clustered either. For this reason, the genes for the arginine biosynthetic enzymes are a valuable model for transcriptional control in eukaryotes.

These genes are scattered over a number of nonadjacent sites on the *E. coli* chromosome (Figure 8-26a). When arginine is in plentiful supply, transcription of all of these genes is repressed by a single repressor molecule. This scattered group of coordinately regulated genes is referred to as the arginine *regulon*. The DNA sequences near the initiation site for mRNA synthesis from two of the arginine genes are shown in Figure 8-26b. It is clear that the presumed operator sequences around the RNA initiation site are very similar, which may explain why one repressor recognizes these distant sites in DNA. The action of a single regulatory protein at multiple sites in *E. coli* is a possible model for eukaryotic cells, in which different chromosomes may contain genes for coordinated functions (e.g., galactose metabolism in yeast, and liver function in mammals; the genes for both of these complex functions are discussed in Chapter 12).

The Control of Regulatory Proteins

The discovery of bacterial regulatory proteins immediately raised the question of whether the regulatory proteins themselves were regulated, and if so, how. In *E. coli*, two situations were found: (1) the regulatory proteins are formed constitutively, or (2) the regulatory proteins control their own synthesis by a mechanism called autogenous regulation.

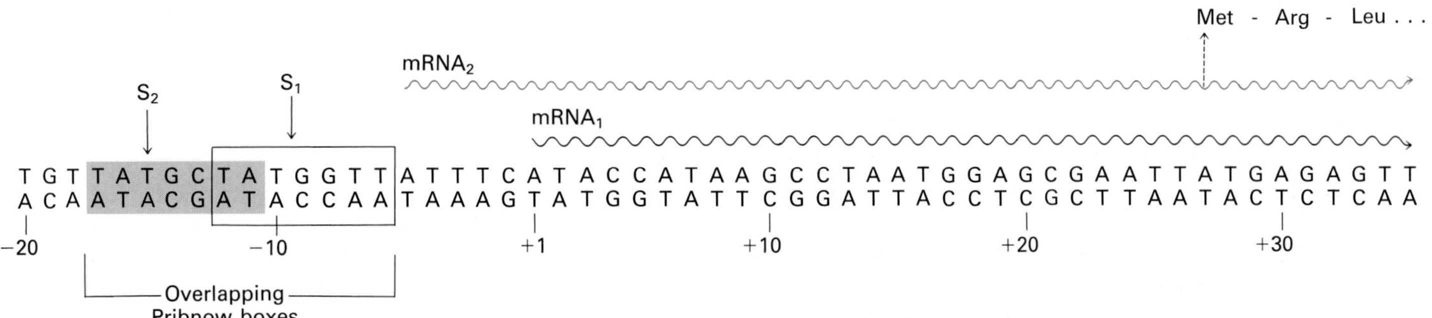

Figure 8-25 The galactose operon has two RNA initiation sites, each of which has its own binding site for RNA polymerase. One binding site (S_2) is active constitutively (i.e., even when glucose is present), so it does not require CAP or cyclic AMP. The second binding site (S_1) requires galactose, CAP, and high levels of cyclic AMP (brought about by the absence of glucose). [See B. de Crombrugghe and I. Pastan, 1978, in *The Operon*, J. H. Miller and W. S. Reznikoff, eds., Cold Spring Harbor Laboratory.]

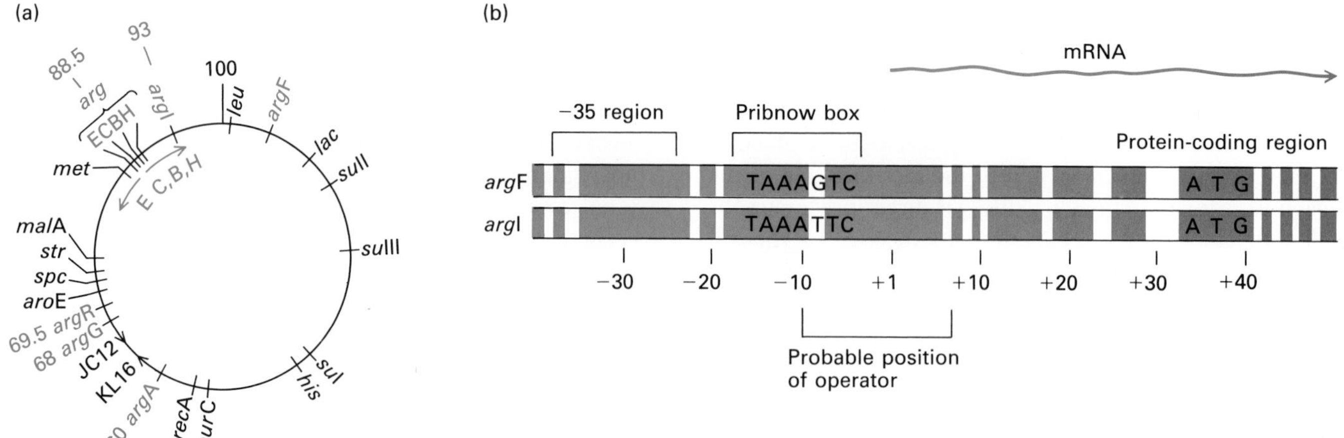

Figure 8-26 (a) Some of the genes that encode the enzymes for arginine synthesis in *E. coli*—the *arg* F, A, G, R, E, C, B, H, and I genes—are widely separated from one another on the *E. coli* circular chromosome. Each gene in the arginine synthetic pathway was located by means of a separate mutation that rendered cells incapable of growing without arginine. The numbers represent units of the *E. coli* chromosome (actually, minutes required to transfer a gene during conjugation; see Chapter 6). The arrows indicate directions of transcription. (b) The sequences of two genes *arg*F and *arg*I were determined and are compared here. Regions of identity are shown in color. There is a much greater conservation of upstream sequences in these two genes than when a diverse group of genes is compared (see Figure 8-4). The similarity continues in the first part of the coding region. Presumably the single repressor for arginine genes can specifically recognize the highly conserved operator regions. *Data for part (b) courtesy of E. James and S. Moore.*

Some Regulatory Proteins Are Synthesized Continually

The lactose repressor is not formed from genes located within the *lac* operon (see, for example, Figure 8-19 or 8-24). Rather, the *lac*I gene is situated in a neighboring region of DNA; this region has its own promoter, which is always available for binding RNA polymerase. The synthesis of mRNA for the lactose repressor from this separate transcription unit is not regulated, but occurs at a constant rate at all times. It seems likely that many negative-acting repressors are made constitutively at a low level. Their negative action is overcome when excess effector molecules become available from the medium.

Some Regulatory Proteins Control Their Own Synthesis: Autogenous Regulation

One example of autogenous control of the synthesis of regulatory proteins is provided by the repressor for the histidine-utilization *(hut)* operon. The gene that encodes this repressor protein is one of the three genes in the operon, so the protein is considered to be self-regulating. A second design for autogenous control of regulatory genes is exemplified by the major repressor for the bacteriophage λ genes and by the AraC protein. Each of these proteins is encoded by a single gene with its own promoter-operator (i.e., by a one-protein operon), and each exerts control over the synthesis of its own mRNA; that is, the proteins are *autogenously regulated.* We shall describe the *hut* operon first. The bacteriophage λ regulatory system will be discussed in a later section. Because virtually all of the known control mechanisms are used in λ, it is a useful example of an integrated system of transcriptional control.

The *hut* Operon *Klebsiella aerogenes* and *Salmonella typhimurium,* two bacterial species closely related to *E. coli,* can degrade the amino acid histidine into ammonia, glutamic acid, and formamide. These breakdown products are a usable source of both carbon and nitrogen, so histidine can serve as the only nutrient in the medium for these cells. Two groups of enzymes called *hut* enzymes are necessary for histidine degradation (Figure 8-27).

The *hut* enzymes are encoded in two separate operons, each of which has its own operator and promoter. (These two operons are completely distinct from the operon that controls the enzymes for histidine biosynthesis.) In the absence of histidine, a single repressor protein prevents the transcription of both *hut* operons. Thus, identical repressor molecules bind at two different operator sites. Histidine acts as an inducer by combining with the repressor molecules, which can then no longer bind to the operons, and transcription is initiated. The *hut* repressor protein is itself encoded by a gene in one of the two operons, so the synthesis of the *hut* repressor is under the control of histidine, the inducer. When the inducer becomes scarce, free repressor accumulates, which shuts

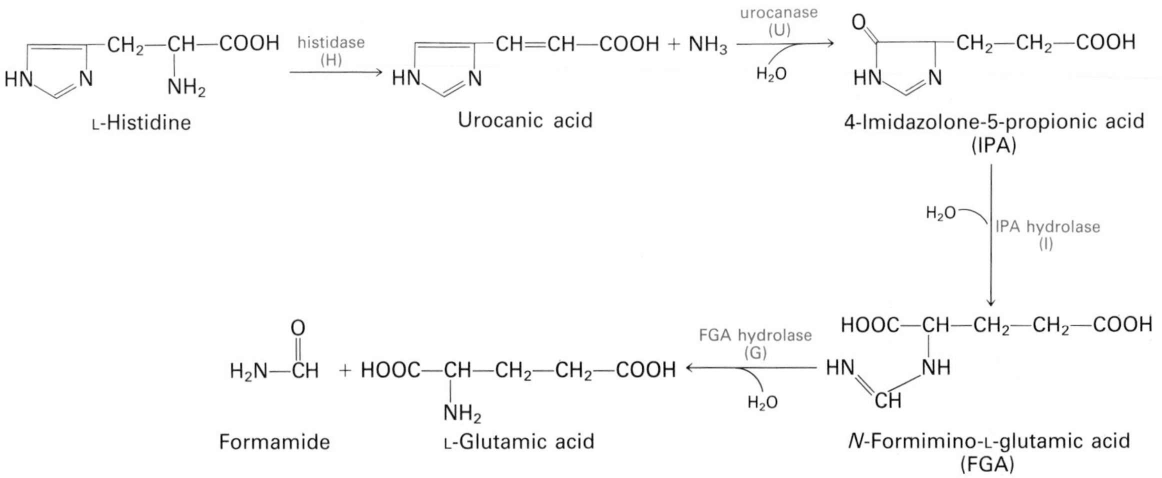

L-Histidine —histidase (H)→ Urocanic acid + NH_3 —urocanase (U), H_2O→ 4-Imidazolone-5-propionic acid (IPA)

4-Imidazolone-5-propionic acid (IPA) —H_2O, IPA hydrolase (I)→ N-Formimino-L-glutamic acid (FGA)

N-Formimino-L-glutamic acid (FGA) —FGA hydrolase (G), H_2O→ Formamide + L-Glutamic acid

Figure 8-27 The metabolic pathway of histidine degradation in bacteria. Histidase and urocanase are encoded in one operon, and IPA hydrolase and FGA hydrolase are encoded in another. Glutamic acid can be used directly in protein synthesis or metabolized as an energy source.

down the synthesis of more repressor (Figure 8-28). The synthesis of the repressor is therefore controlled by the repressor itself. This is an example of *negative autogenous control,* which provides the bacterium with a regulatory circuit that ensures a controlled production of proteins from the separate *hut* operons.

The *hut* genes not only respond to the presence of histidine, but they are also sensitive to the presence of glucose, just as the *lac, ara,* and *gal* genes are. Evidence for this sensitivity is found in the fact that cyclic AMP and CAP are needed for in vitro synthesis of *hut* mRNA.

Control of the Termination of Transcription in Bacteria

The initiation of DNA transcription is, a priori, the most logical point for the exercise of gene control in bacteria. Clearly, control of transcriptional initiation does play an important role in the flexible adaptation of bacteria to the environment. A possible second level of transcriptional control occurs in the form of regulated termination (or nontermination) of RNA synthesis. Three possible termination/nontermination events are as follows:

1. If transcription were terminated prematurely, before RNA polymerase had reached the structural genes in an operon, this could constitute secondary regulation of mRNA production in the form of a "veto."
2. Controlled termination between genes in an operon could cause more transcription of the DNA nearer the promoter than of the DNA farther from it; this would result in a "polar effect" (because less RNA would be synthesized as the distance from the promoter increased).
3. If an RNA polymerase were allowed to continue transcription—that is, to "read through"—from one operon to the next, then regulated initiation of the second operon would be overridden (Figure 8-29).

Each of these possible termination mechanisms is thought to play at least some role in the control of various bacterial and bacteriophage genes.

The molecular basis for mRNA chain termination by RNA polymerase is considerably less well understood than the molecular events in initiation. There are apparently several contributing factors in chain termination, and two general types of chain termination are recognized: *rho*-dependent and *rho*-independent.

Rho-dependent termination requires the presence of a protein called the *rho factor.* Sites for *rho*-dependent termination in *E. coli* were first recognized during in vitro studies of transcription of the DNA of bacteriophage λ. If *E. coli* polymerase is mixed with λ DNA, RNA synthesis begins in a central region of the DNA at the promoter sites P_L and P_R (for "left" and "right," respectively), which are known to function in infected cells. (The P_L and P_R promoters and the regulation of λ transcription are discussed in detail in the next section.) Messenger RNA is transcribed by the RNA polymerase in two directions (to the left of P_L and to the right of P_R). The resulting transcripts are several thousand nucleotides long; mRNAs this long are not present inside cells during the early stages of infection.

If extracts of uninfected cells are added to the cell-free transcription system, it forms two discrete products—one containing 500 nucleotides and one containing 1000 nucleotides—that are ordinarily found inside infected cells. These products extend from the P_L and P_R promoters to specific *termination sites.* The *rho* factor, a cell protein that can be purified from the cell extract, is responsible

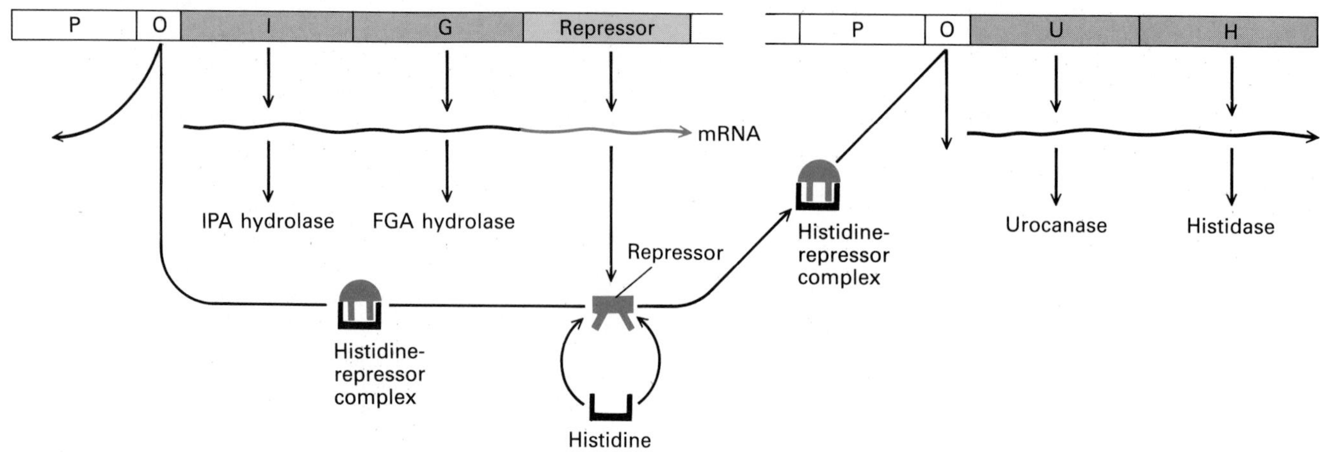

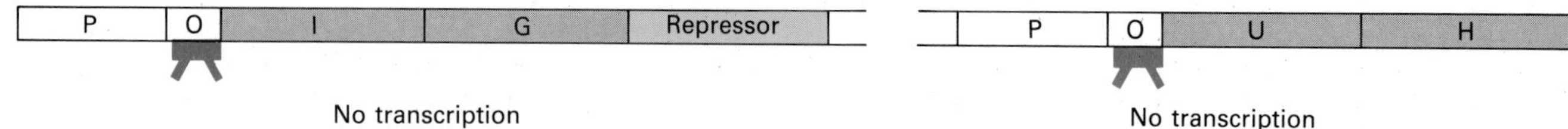

Figure 8-28 Certain bacterial cells can use histidine as a source of either nitrogen or carbon when their medium lacks any other nitrogen source or glucose. Two separate operons encode histidine-utilizing *(hut)* enzymes, and one repressor controls both operons. Without histidine, this repressor is bound to DNA; with histidine present, the repressor does not bind. The repressor protein is itself encoded by a gene in one of the operons. Thus the *hut* repressor negatively controls its own synthesis: when the medium lacks histidine, repressor synthesis quickly declines. [See B. Magasanik, 1978, in *The Operon,* J. H. Miller and W. S. Reznikoff, eds., Cold Spring Harbor Laboratory.]

for this correct chain termination (Figure 8-30). Several *rho*-dependent chain-terminating regions have been identified in λ and *E. coli* DNA. Comparisons of the sequences of these regions yield no obvious similarities. Mutations that have been discovered in the *rho* gene itself cause RNA polymerase to read through from one operon to the next, so it is clear that the *rho* protein plays an important role in the termination of RNA chains inside the cell. However, the exact mechanism of *rho* termination is not yet clear.

Rho-independent termination also takes place, both inside the cell and in vitro. For example, synthesis of the *trp* mRNA ceases at a site 36 bases beyond the last coding region in the mRNA. At this site there are four consecutive U residues preceded by 22 nucleotides in a sequence that is largely self-complementary; that is, it can fold back on itself and base-pair to form a perfect "hairpin." Within this hairpin are eight base pairs (Figure 8-31). At least 30 such *rho*-independent sites have been compared, and these three features—high GC content, dyad symmetry (self-complementarity), and several U residues in a row—are common to almost all of these sites. The effi-

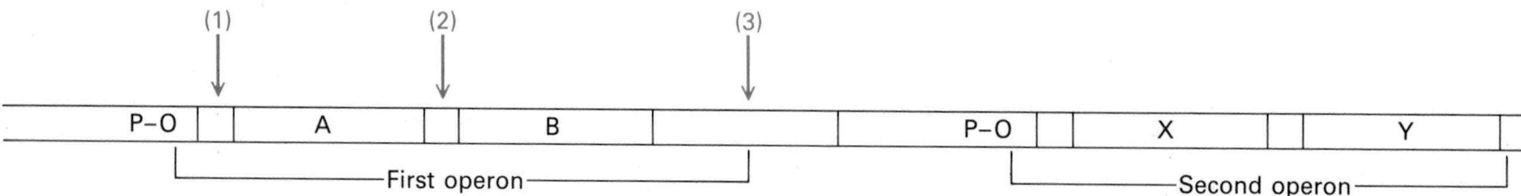

Figure 8-29 Bacteria can terminate transcription in three different places on DNA: (1) before the structural genes in an operon, (2) between genes in an operon, which creates a polar effect, and (3) at the end of an operon. If an mRNA molecule is not terminated at the third possible locus, then the polymerase can "read through" to the next operon.

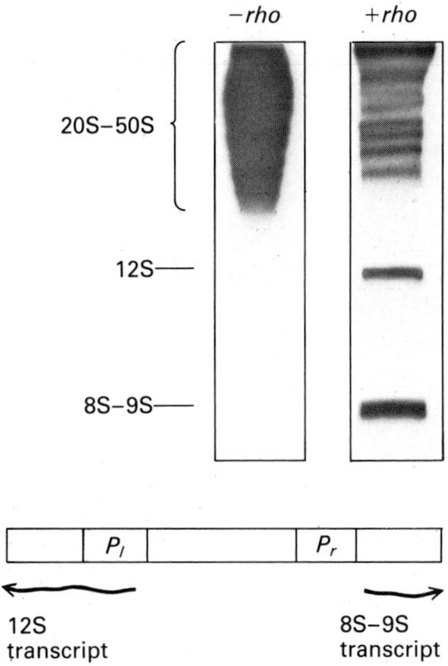

Figure 8-30 The products of the transcription of bacteriophage λ DNA by the *E. coli* RNA polymerase in vitro in the presence and absence of purified *rho* factor. RNA was labeled during transcription with ^{32}P-labeled nucleotides, and the resulting products were analyzed by gel electrophoresis and autoradiography. Note that without *rho*, only very large RNA molecules (20S to 50S chains containing more than 2000 nucleotides) are formed; but when *rho* is present, two discrete short bands, one at 12S (i.e., about 1000 nucleotides long) and one at 8S to 9S (i.e., about 500 nucleotides long), appear. Experiments of this type yielded the first indication of a specific termination factor in RNA synthesis. *From M. Rosenberg et al., 1978, in* The Operon, *J. H. Miller and W. S. Reznikoff, eds., Cold Spring Harbor Laboratory, p. 348.*

ciency of termination at *rho*-independent sites varies from less than 25 percent to more than 75 percent.

Although it is uncertain how these regions cause chain termination, a reasonable hypothesis has been advanced. When the self-complementary regions of a growing mRNA chain are synthesized, they base-pair with one another, which causes the polymerase to pause. Perhaps because the DNA is now free to reanneal behind the polymerase, the polymerase is forced off the template and the RNA chain is released (Figure 8-32). Experimental evidence supports these ideas. For example, mutations that weaken the dyad symmetry or that result in fewer consecutive U residues decrease termination at these sites.

Attenuation Is a Secondary Control System for the *trp* Operon

We now consider the role of chain termination in the regulation of gene expression in bacteria. As we mentioned earlier, when *E. coli* cells are grown in a medium lacking the amino acid tryptophan, they produce five different enzymes that catalyze the five steps in the synthesis of tryptophan. All of these enzymes are encoded in the *trp* operon. In normal cells, tryptophan acts as a corepressor; it combines with a repressor molecule and together they bind to the operator that precedes the *trp* operon, which thus represses the synthesis of mRNA^{Trp}. The more tryptophan in the medium, the less mRNA^{Trp} is produced, so the primary level of control is at the level of transcriptional initiation.

However, the discovery of *E. coli* mutants that manufactured greater-than-normal amounts of tryptophan-synthesizing enzymes provided bacterial geneticists with experimental material that illuminated yet another basic mechanism of prokaryotic gene control. One group of these mutants appeared to make a normal *trp* repressor and to have a normal *trp* operator, although their basal level of *trp* enzymes was higher than normal, and they

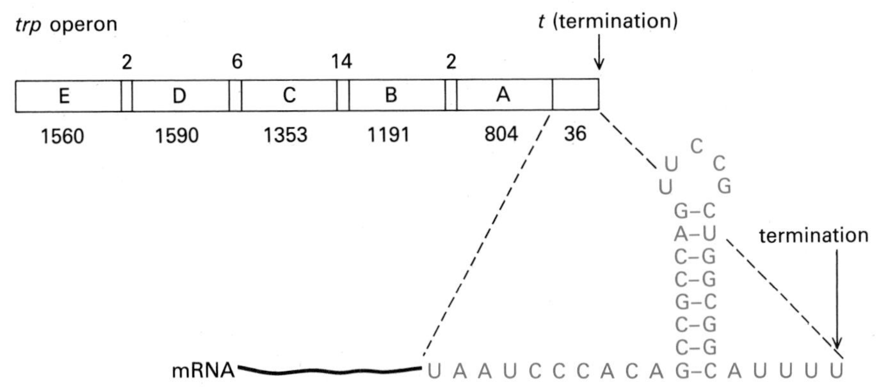

Figure 8-31 Termination at the *trp t* site, a *rho*-independent site. The *trp* operon is composed of five genes separated from one another by very short "spacer" sequences. (The number of nucleotides in each segment of the operon is indicated above or below.) The genes are followed by the sequence shown, at the end of which termination occurs. The stem-loop ("hairpin") structure in the mRNA preceding the final four U residues is characteristic of *rho*-independent termination sites. [See T. Platt, 1981, *Cell* 24:10.]

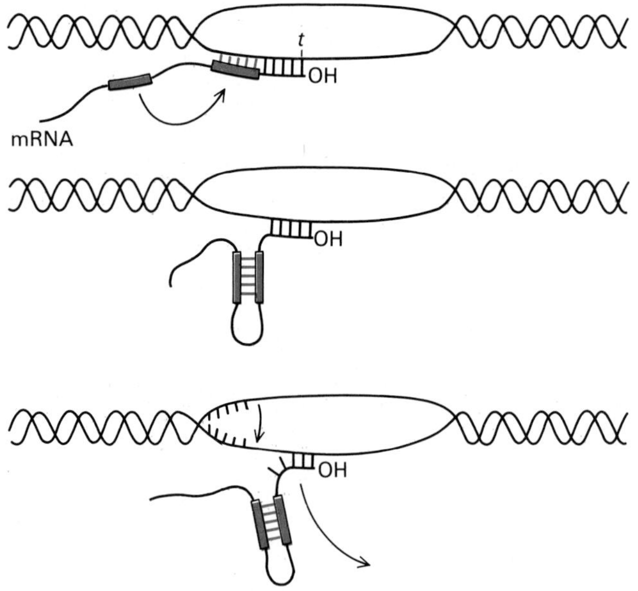

Figure 8-32 A model for RNA chain termination. RNA polymerase has reached a termination site (*t*); the transcript preceding this site is characterized by a string of U residues (see Figure 8-31) preceded by a region of high GC composition with dyad symmetry *(color)*. The bases in the symmetrical region of the mRNA form base pairs, because RNA-RNA hybrids are more stable than RNA-DNA hybrids. This removes the RNA from the template and allows the unpaired DNA to reanneal. The RNA chain and the polymerase are displaced in the process of DNA reannealing. [See T. Platt, 1981, *Cell* **24**:10.]

responded to the absence of tryptophan with an even higher rate of transcription. The tryptophan genes from this group of mutants were isolated in transducing bacteriophages to compare the sequences of these genes with the sequences of the wild-type genes. It was found that the mutants had deletions of short stretches—fewer than 50 base pairs—of DNA; the deletions appeared just after the promoter-operator region in the normal *trp* operon and before the coding sequences for the *trp* enzymes.

How could the deletion of a short region of DNA after the RNA initiation site cause unregulated synthesis of mRNA for the tryptophan-synthesizing enzymes? In the hope of answering this question, experimenters sequenced the beginning of the mRNA molecule transcribed from the *trp* operon and the DNA region near the *trp* promoter in both normal and mutant *E. coli* strains. This led to the discovery of a new mechanism in the regulation of bacterial transcription. (Several other operons concerned with amino acid synthesis have a similar regulatory mechanism.)

The mRNA transcribed from the *trp* operon contains 162 nucleotides upstream (i.e., toward the 5′ end of the molecule) from the AUG codon that constitutes the protein initiation site for the first of the five tryptophan enzymes. This stretch of mRNA is called the *leader sequence*. In normal *E. coli* cells, when tryptophan is abundant only small amounts of any $mRNA^{Trp}$ sequences are made, and these consist entirely of the leader RNA. When tryptophan is scarce, however, the full-length $mRNA^{Trp}$ (about 7000 nucleotides) is transcribed, including the leader sequence and the entire coding sequence for the *trp* enzymes. Thus control is exercised at

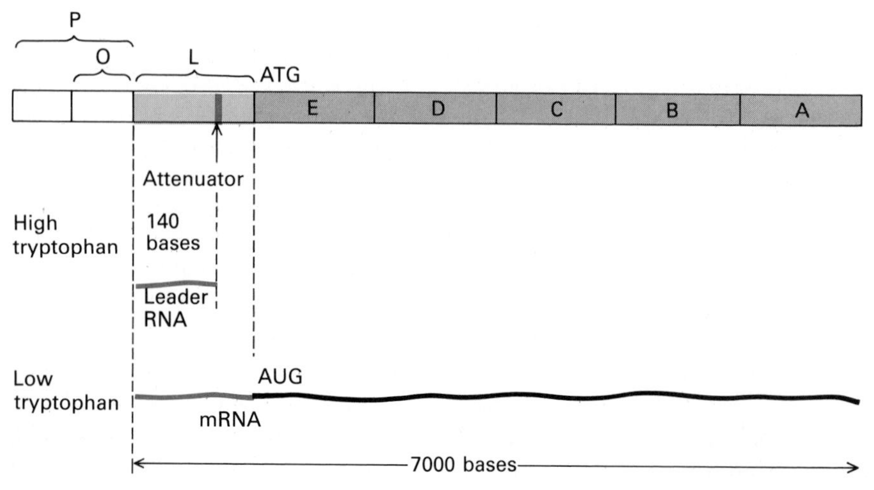

Figure 8-33 Attenuation in the *trp* operon. Genes E, D, C, B, and A encode tryptophan-synthesizing enzymes. The leader sequence (L) is a stretch of nucleotides between the operator (O) and the structural genes. When the supply of tryptophan is low, the entire operon, including the leader sequence, is transcribed into mRNA. When the supply of tryptophan is high, only 140 bases (the part of the leader sequence that precedes the attenuator) are transcribed into mRNA; the structural genes are not transcribed. [See T. Platt, 1978, in *The Operon*, J. H. Miller and W. S. Reznikoff, eds., Cold Spring Harbor Laboratory.]

Tryptophan leader	Met - Lys - Ala - Ile - Phe - Val - Leu - Lys - Gly - Trp - Trp - Arg - Thr - Ser - Stop 5′~~AUG AAA GCA AUU UUC GUA CUG AAA GGU UGG UGG CGC ACU UCC UGA~~3′
Phenylalanine leader	Met - Lys - His - Ile - Pro - Phe - Phe - Phe - Ala - Phe - Phe - Phe - Thr - Phe - Pro - Stop 5′~~AUG AAA CAC AUA CCG UUU UUC UUC GCA UUC UUU UUU ACC UUC CCC UGA~~3′
Histidine leader	Met - Thr - Arg - Val - Gln - Phe - Lys - His - His - His - His - His - His - His - Pro - Asp - 5′~~AUG ACA CGC GUU CAA UUU AAA CAC CAC CAU CAU CAC CAU CAU CCU GAC~~3′

Figure 8-34 The short polypeptides encoded by the leader sequences of the tryptophan, phenylalanine, and histidine operons. In each case, the polypeptide is fairly rich in the amino acid *(color)* synthesized by the enzymes translated from the specific mRNA.

the level of termination, which occurs at the 140th nucleotide of the $mRNA^{Trp}$ when tryptophan is abundant.

Even when the total $mRNA^{Trp}$ is being transcribed in large amounts, more leader sequences than whole mRNA molecules are transcribed. This situation has been interpreted as indicating that the leader sequence contains an *attenuator,* a site that exercises a "veto" over transcription after a certain distance (Figure 8-33). When tryptophan is scarce, about 25 to 50 percent of the RNA polymerase molecules continue transcribing past the attenuator. When tryptophan is abundant, some initiation of transcription takes place but virtually all of the transcripts are cut short.

Thus, when tryptophan is in plentiful supply, the *trp* operon is controlled not only by a repressor-operator function, but also by attenuation of RNA synthesis at a site 140 nucleotides from the start site. The number of RNA polymerase molecules that pass the attenuator depends precisely on the concentration of tryptophan in the medium; this finely tuned control balances the number of tryptophan-synthesizing enzymes formed with the need for tryptophan. Mutant cells that lack the attenuator produce more $mRNA^{Trp}$ under all conditions.

The phenomenon of attenuation apparently depends on the intramolecular interactions of the mRNA leader sequence—that is, on its base-paired, stem-loop structure. This structure depends in turn on the rate of ribosomal translation of the nucleotide sequence. The leader sequence of each of the three mRNAs that are translated into enzymes for the synthesis of tryptophan, phenylalanine, and histidine encodes a remarkable set of signals. The first half of the leader contains from 14 to 16 codons that specify a short polypeptide. This short polypeptide is relatively rich in the very amino acid synthesized by the enzymes translated from that mRNA (Figure 8-34). A ribosome engages the short translatable mRNA sequence soon after it is synthesized. Whether the ribosome efficiently translates the leader sequence into protein depends on the supply of amino acids encoded by the sequence—in particular, on the supply of the amino acid that is the product of the enzymes encoded by that mRNA.

The case of the tryptophan leader has been analyzed most thoroughly. If tryptophan is present in sufficient quantity, the ribosome moves quickly along the leader transcript. The portion of the transcript that has not yet been translated folds itself into a particular stem-loop structure (the 3–4 pairing shown in Figure 8-35) that allows the termination site in the mRNA to be recognized. Attenuation is therefore maximized when tryptophan is present in abundance.

If, on the other hand, the supply of tryptophan is low, the ribosome will pause at each codon for the scarce polypeptide component. Because the leader sequence has many such codons, translation of the sequence is repeatedly halted. While the ribosome is stalled, the "waiting" mRNA transcript folds itself into a different stem-loop structure (the 2–3 pairing shown in Figure 8-35) which blocks recognition of the termination signal (the 3–4 pair), so that RNA polymerase continues transcription. Thus attenuation is minimized when tryptophan is scarce.

Antitermination Proteins Prevent Termination and Allow "Read-Through" Control

Termination at an attenuator site is a *rho*-independent event that is spontaneous unless a ribosome prevents formation of the secondary 3–4 RNA-RNA structure. In contrast, *rho*-dependent termination can be regulated by specific antitermination proteins. Perhaps the best-studied case concerns early events in the infection of *E. coli* by bacteriophage λ.

As we mentioned previously, in the early stages of infection, transcription begins in a central region of the λ DNA and proceeds to the left and to the right of two promoters, P_L and P_R. Because they are terminated at specific sites, the RNA transcripts from each promoter region are at first quite short, representing about 1000 and 500 bases from P_L and P_R, respectively (see Figure 8-30). The P_L transcript encodes a bacteriophage protein called N. The N protein is an *antiterminator* which allows transcription to continue into additional coding regions so that additional phage genes will be expressed (Figure 8-36). Because the *rho* protein of the bacterial cell is an active agent of early termination, the bacteriophage N gene product must somehow combat the effect of the *rho*

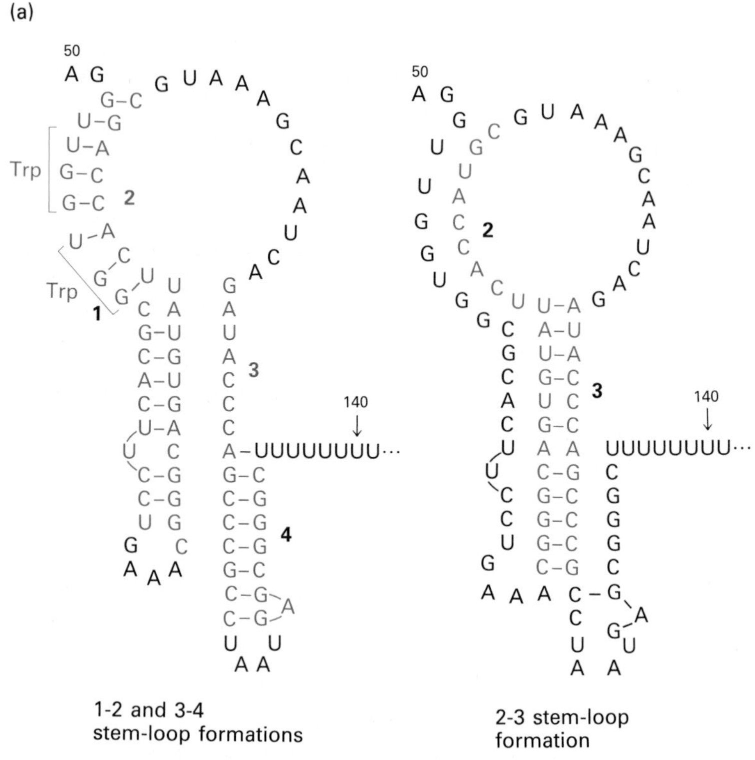
(a)
50
Trp
Trp
1
2
3
4
140
1-2 and 3-4 stem-loop formations
2-3 stem-loop formation

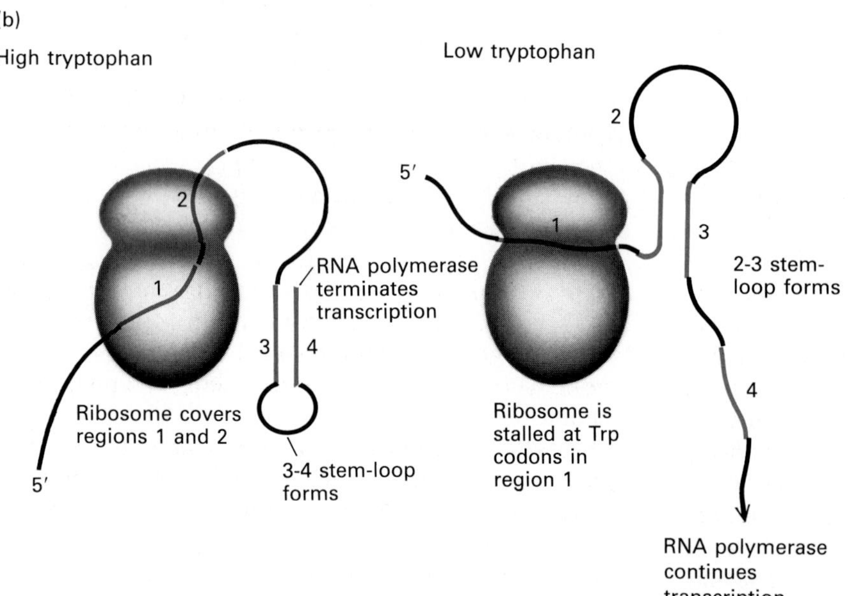
(b)
High tryptophan
Low tryptophan
5′
Ribosome covers regions 1 and 2
RNA polymerase terminates transcription
3-4 stem-loop forms
Ribosome is stalled at Trp codons in region 1
2-3 stem-loop forms
RNA polymerase continues transcription

Figure 8-35 (a) The leader region in mRNA^{Trp}, showing the possible stem-loop structures *(color)* formed by RNA sequences in regions 1 and 2 and in regions 3 and 4 *(left)* and in regions 2 and 3 *(right)*. (b) Ribosomes apparently play a role in preventing the formation of the stem-loop structure between regions 3 and 4, which leads to RNA chain termination (for the reasons suggested in Figure 8-32). When translation of the leader sequence stalls, the 3-4 pair does not form and transcription continues. [See C. Yanofsky, 1981, *Nature* **289**:751.] *Part (b) from T. Platt, 1981,* Cell *24:10.*

protein. Although the molecular mechanism of antitermination has not been completely worked out, it is clear that the N protein binds to λ DNA sites (called *nut* sites for "N utilization"), and it is likely that several bacterial proteins also participate in this antitermination. Some of these bacterial proteins may bind to the polymerase, for example, to prevent its termination at *rho* sites.

In summary, then, a balance between terminators and antiterminators can control genes at the level of transcription just as effectively as operators and repressors can.

The Early Events of Bacteriophage λ Infection: A Transcriptional Program Leads to Alternate Physiological States

In our discussion of the known transcriptional control mechanisms of bacteria—positive and negative regulation of initiation, autogenous control, and regulation of termination—bacteriophage λ has often been mentioned. The entire DNA sequence of over 50,000 base pairs is known; more than 50 different genes have been identified and located on the genetic map. Among these are the

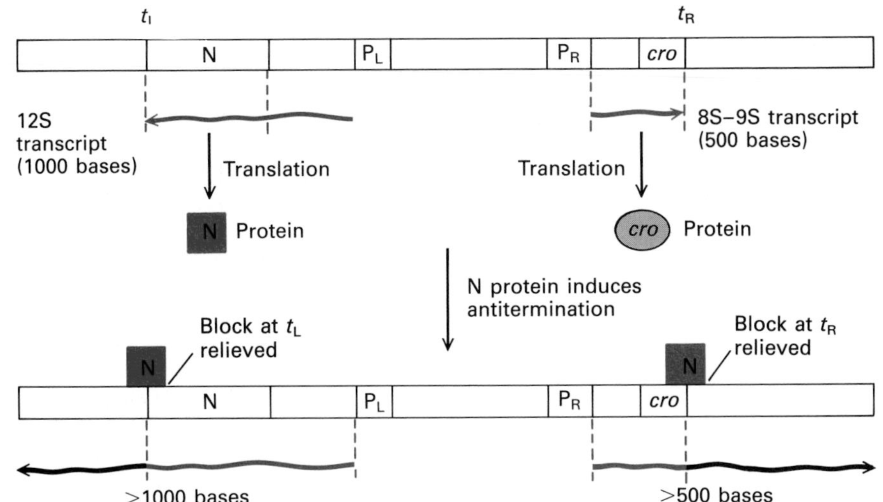

Figure 8-36 N protein is an antiterminator. Early in bacteriophage λ infection, mRNA is made from the promoters P_L and P_R. The left-hand transcript terminates at t_L as a 1000-nucleotide RNA and the right-hand transcript terminates at t_R as a 500-nucleotide RNA that encodes the *cro* protein. The *rho*-factor is required in both cases. However, translation of the left-hand mRNA produces the N protein, which binds to sites on the DNA that block early termination. This allows transcription of longer molecules.

genes that encode the phage-specific proteins that regulate transcription of the phage's own DNA. From this detailed knowledge, much has been learned about how an integrated set of transcriptional controls is responsible for physiological choices in a cell containing λ DNA. There are two different pathways in the λ life cycle: lytic growth and maintenance of the lysogenic state. The factors that determine which pathway is chosen are fairly well understood. Such choices may not reflect the exact molecular mechanisms of alternate choices in eukaryotic cells, but they are most instructive as models for our thinking about how transcriptional control networks might function when cells choose different pathways—as occurs, for example, in developing organisms.

Bacteriophage λ consists of 50,000 base pairs of double-stranded DNA enclosed in a protein coat. When the DNA has entered a host cell, it may be transcribed and translated into the proteins required to make more DNA and more coat protein. Such a cell soon lyses, releasing about 100 new phage particles. This course of events is called the *lytic cycle*. In contrast, if the *lysogenic pathway* is followed, no new bacteriophage is formed. Rather, the λ DNA becomes incorporated into the bacterial chromosome (Chapter 6), where it behaves as a quiescent set of bacterial genes and replicates, along with the bacterial cell, through hundreds of generations. The lysogenic state is stable as long as no unusual event disrupts the cell-virus combination. However, a number of events that damage DNA—for example, ultraviolet irradiation—cause the virus to come out of its lysogenic state and enter the lytic state, whereupon it produces the 100 new phage particles and kills its host cell. How is the delicate balance of lysogeny established and maintained?

The concentration of a master transcriptional control protein decides the issue. The *c*I gene product is a repressor protein that regulates transcription at two key promoters (P_L and P_R), and in so doing plays the central role in suppressing lytic growth of bacteriophage λ. If enough *c*I λ repressor is synthesized just after λ enters a host cell, the cell is pushed into lysogeny; if this synthesis is interrupted and the supply of repressor diminishes, lytic growth can begin. The level of *c*I λ repressor is determined in turn by a set of genes whose complex interactions are reasonably well understood. It is our understanding of how this network operates that makes bacteriophage λ such an instructive model.

When λ infects a host cell, no *c*I λ repressor is present, and therefore unregulated transcription begins at the two promoter sites P_L and P_R. The products of this early transcription are mRNAs for the N and *cro* proteins (see Figure 8-36). The most important physiological effect is that of the antiterminator N protein, which allows the transcription of the *c*II and *c*III genes, as well as genes to the left of *c*III and genes to the right of *c*II (Figure 8-37). (Transcripts of the genes downstream from *c*III set the stage for integration of λ DNA; those downstream from *c*II cause λ DNA replication and phage production.)

The *c*II gene product is of crucial importance. By a mechanism not yet thoroughly studied, this protein increases the production of *c*I λ repressor, probably by instituting transcription of the *c*I gene from the promoter for repressor establishment (P_E). The *c*I λ repressor mRNA that is initiated at P_E may also be translated more efficiently than the *c*I mRNA that is initiated at P_{RM}. In addition, the *c*III gene product interferes with the functioning of H_fl, a host-cell protein (probably a protease) that decreases *c*II gene activity.

The overall result of these early interactions is to increase *c*II activity and thereby cause a burst of *c*I λ repressor synthesis after a few minutes of infection. This supply of λ repressor then curtails all synthesis from P_L and P_R; if such synthesis were to continue after the initial antitermination by N, it would lead to the lytic cycle. The *c*I λ repressor is bound near P_{RM} (the promoter for repressor maintenance), where it stimulates the synthesis of its own

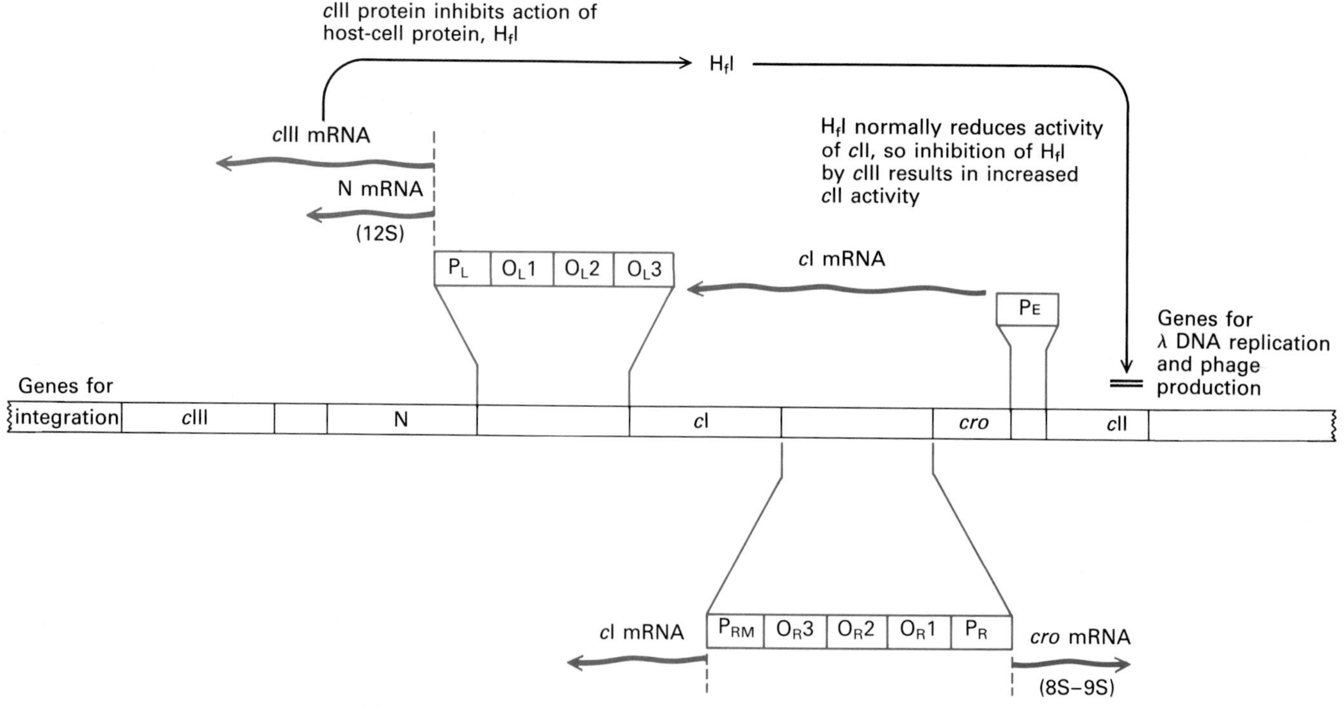

Figure 8-37 Bacteriophage λ: the "early" gene region (i.e., the genes that are transcribed first). This DNA controls repressor establishment and the other critical events that lead to lysogeny. The genes *c*I, *c*II, *c*III, N, and *cro* encode proteins important in the regulation of λ DNA transcription. The symbols P_L, P_R, and P_{RM} indicate promoters that operate either early in infection (P_L and P_R) or later, when λ DNA has integrated into the host-cell chromosome and repressor maintenance has been established (P_{RM}). The symbols O_R1, O_R2, and O_R3 indicate operator sites that bind *c*I λ repressor. When transcription is initiated at P_L, it leads to production of the N protein, the antiterminator. The promoter P_E is not fully understood, but it probably functions in the establishment of lysogeny.

The events controlling the function of λ DNA are summarized in the chart. The arrows pointing up indicate increases, and those pointing down, decreases.

Left	Right	Major result
FIRST STAGE		
↑ N (antiterminator)	↑ *cro* (represses P_L and P_R)	Antitermination of 12S and 8S–9S transcripts
SECOND STAGE		
↑ *c*III (inhibits H_fL)	↑ *c*II (stimulates P_E)	Increases *c*II and *c*I λ repressor
↑ Integration enzymes		
THIRD STAGE		
*c*I dominates	*c*I dominates	Lysogeny established
↓ N	↓ *cro*	
↓ Integration enzymes	↓ Replication	

(*c*I) mRNA from that site. Thus the *c*I protein is its own positive regulator, and the cell has a supply of repressor that keeps the remainder of the λ genes in a repressed state.

During the initial establishment of lysogeny, while this supply of repressor is accumulating, the genes to the left of *c*III and those to the right of *c*II are being transcribed. A group of genes downstream from the *c*III gene encode proteins that enable the λ DNA to become integrated into the bacterial DNA. Some mRNA products that lead to the lytic cycle are also formed, but apparently in insufficient quantities to trigger that cycle.

Once a sufficient supply of *c*I λ repressor protein is established, lysogeny can be maintained for a long time. Elegant biochemical studies on the interactions between λ repressor and three operator regions of the DNA—O_R1, O_R2, and O_R3—have elucidated the molecular events that underlie the control of λ DNA in the lysogenic state. Each operator consists of a DNA sequence 17 base pairs long, and three of these 17-base-pair sites lie in front of each promoter, P_L and P_R (Figures 8-37 and 8-38). The *c*I gene is located between P_L and P_R; and P_{RM}—the autogenously regulated promoter for the *c*I gene—is close to the O_R3 site.

The repressor protein dimerizes; in this form it has a high affinity for DNA. The repressor dimers bind sequen-

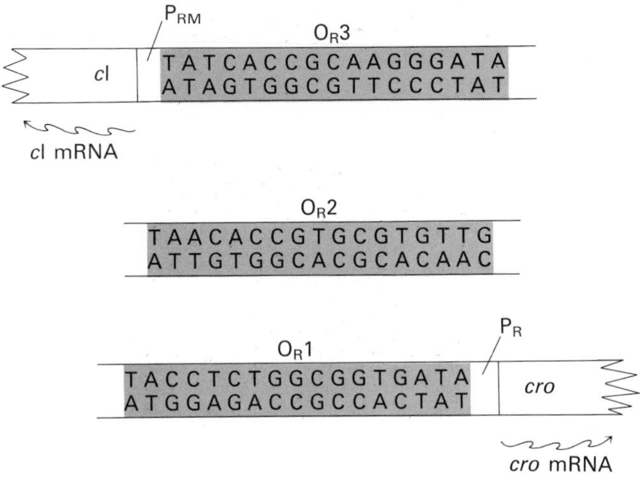

Order of binding of regulatory proteins
*c*I: $O_R1/O_R2/O_R3$
cro: $O_R3/O_R2/O_R1$

Figure 8-38 A diagram of part of the genome of bacteriophage λ: O_RP_R, the right-hand region to which the *c*I λ repressor and the *cro* protein bind. The λ repressor binds about 10 times as strongly to the three operator sites as *cro* does. The base sequences indicate that the operator sites are similar but not identical. Mutations in these sites upset the balance of the lysogenic state and can result in phage capable only of lytic growth. [See M. Ptashne et al., 1980, *Cell* **19**:1.]

tially in the order O_R1, O_R2, and O_R3. The repressor bound to O_R1 cooperatively increases the binding to O_R2 but not to O_R3. Thus, O_R1 and O_R2 are largely saturated by bound repressor in lysogenic cells. When repressors occupy O_R1 and O_R2, transcription of the *c*I gene is significantly stimulated at the nearby P_{RM} promoter (Figure 8-39a). (At very high repressor concentrations all three sites are filled, which eventually decreases repressor mRNA synthesis.) When the first repressor dimer occupies O_R1, it represses RNA polymerase interactions with P_R and shuts down the transcription of *cro*. Thus, in the lysogenic state, *c*I λ repressor continues to be synthesized because of autogenous positive control, and a permanent *epigenetic* effect of repression is maintained. For example, the *cro* gene is not changed, but its repression is nevertheless long-lasting.

However, as we mentioned earlier, DNA-damaging events can release the quiescent phage DNA. When a cell is subjected to ultraviolet irradiation, it undergoes a number of biochemical reactions that allow it to repair DNA. (These reactions are outlined in Chapter 13.) One of these biochemical changes is the activation of a protease, which cleaves a number of proteins, including the λ repressor. The cleaved repressor no longer dimerizes, and thus loses its ability to bind to DNA. This allows the initiation of transcription of the *cro* gene from P_R; the *cro* protein, also a DNA-binding repressor protein, competes favorably with the cleaved *c*I λ repressor protein for binding to the O_L and O_R regions (Figure 8-39b). The *cro* protein binds in the order $O_R3 > O_R2 > O_R1$ (see Figure 8-38), and in so doing shuts down transcription of the *c*I gene at P_{RM} (Figure 8-39c). This allows transcription of the N gene. All of the other transcripts necessary for the synthesis of proteins in the lytic pathway can then be made, due to the absence of an effective amount of *c*I λ repressor, and the lytic cycle ensues. Thus, whether a λ-infected *E. coli* cell enters the lytic cycle or remains stable in a lysogenic state depends on the differential rates of transcription of a remarkable set of interacting genes.

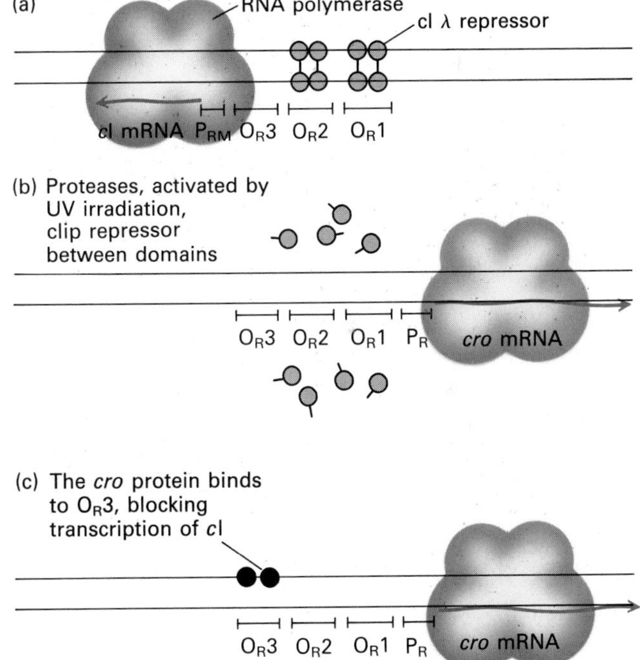

Figure 8-39 (a) A model of the configuration of the *c*I λ repressor and RNA polymerase after the establishment of lysogeny. The λ repressor first binds to O_R1, and then cooperative effects enhance the binding of the repressor to O_R2. The presence of λ repressor at O_R1 and O_R2 stimulates the binding of RNA polymerase at P_{RM}, which results in the transcription of the *c*I gene and produces a large supply of *c*I protein. (b) After ultraviolet irradiation, activated proteases clip the *c*I λ repressor, decreasing its ability to form dimers and its efficiency of binding to DNA. The P_R promoter is activated, and the *cro* protein is produced to shut off P_{RM}. [See A. D. Johnson et al., 1981, *Nature* **294**:217.] (c) The *cro* protein binds strongly to O_R3, and the synthesis of *c*I mRNA is shut down.

Table 8-2 RNA polymerase σ factors and conserved promoter sequences in *Bacillus subtilis**

Cell state	σ factor	Consensus sequences	
		−35 region	Pribnow box (−10 region)
Growing	σ^{55}	T T G A C A	T A T A A T
Infected	σ^{gp28}	T N A G G A G A N N A	T T T N T T T
Growing	σ^{37}	G G N T N A A A	T A T T G T T T
Growing	σ^{28}	C T A A A	C C G A T A T

*Growing *Bacillus subtilis* cells have three different σ factors (with masses of 55, 28, and 37 kilodaltons), and bacteriophage-infected cells have an additional σ factor (the product of gene 28). Each of the σ factors causes the RNA polymerase to recognize a different set of genes. The consensus sequences in the −35 and −10 regions for genes recognized by the σ^{55} are similar to the corresponding *E. coli* sequences, but both the −35 and −10 sequences vary for the other σ factors. (The N's indicate positions in which any base can be present.)

SOURCE: R. Losick and J. Pero, 1981, *Cell* **25**:582.

Gene Control by Proteins That Act Directly on RNA Polymerase

All of the cases of transcriptional control that have been discussed so far are mediated by a change in the rate of initiation or the frequency of termination by RNA polymerase. DNA transcription can also be affected by a change in the ability of the RNA polymerase to recognize the common features exhibited by most bacterial promoters. Such an event might be expected to have a broader range of action than individual repressors and activators do.

In *Bacillus subtilis,* several proteins that bind directly to the RNA polymerase itself have been discovered. As mentioned earlier in the description of the tight binding of the *E. coli* RNA polymerase to DNA, a protein called the σ (sigma) factor interacts with the four-subunit ($\alpha_2\beta\beta'$) RNA polymerase core to aid in the selection of the correct initiation sites on DNA. *B. subtilis* has not one but three different σ factors in growing cells. Moreover, sporulating *B. subtilis* cells have at least one other σ factor. In sporulating cells, the new σ factor redirects the normal cell RNA polymerase to bind to certain DNA sites and begin the transcription of genes whose products allow sporulation (this mechanism is discussed in Chapter 22). Bacteriophages that infect *B. subtilis* also encode their own σ factors. In bacteriophage-infected cells, the new phage-specific σ factor results in the transcription of specific phage genes. Certain DNA sequences in the promoters of the different types of genes—that is, in the promoters of the genes in growing vegetative cells, of the genes for sporulation, and of the bacteriophage genes—are more alike within gene types than between gene types. These conserved sequences presumably enable RNA polymerases with different σ factors to recognize their specific sites (Table 8-2).

In *E. coli* cells infected by bacteriophage T4, at least six different modifications of the RNA polymerase occur. For example, the enzyme may be phosphorylated, or it may have chains of poly ADP-ribose attached to it. The chemically altered polymerase then selects a different set of promoter sites. Thus, it is not uncommon for gene control in bacteria to depend, at least in part, on chemical modifications of the polymerase. It is not yet known whether chemical modifications of the polymerase occur in eukaryotes, and whether such changes are more or less frequent than regulation by DNA-binding proteins such as repressors and activators.

The Stability of Biopolymers in Bacterial Cells

As we explained earlier in this chapter, if a bacterium is placed in a solution lacking an essential nutrient, the enzymes necessary for synthesizing that nutrient are induced. Furthermore, if the nutrient is added to the medium, the mRNA for the enzymes will no longer be made. What happens to this mRNA and to the enzymes that synthesize the nutrient when they are no longer needed?

Turnover Rates of mRNAs Can Differ

In general, bacterial mRNAs have short life spans. Unknown nucleases destroy them so quickly that their half-lives range from 3 to 5 min, whether or not inducers are still present. Even in bacterial cells that grow and divide every 20 to 30 min, mRNA must be renewed from 5 to 10 times in each generation. Thus, the finely tuned mechanisms for the control of mRNA initiation plus the rapid degradation of mRNA molecules allow very efficient and flexible control of metabolism. When the bacterial cell needs certain enzymes, it makes them immediately; when the need is gone, mRNA synthesis stops and the existing mRNA is destroyed quickly. Although the enzymes responsible for mRNA turnover are not known, experiments have suggested that destruction begins near the 5′

end of the mRNA. There is also the possibility that polycistronic mRNA is cleaved between the individual coding segments before destruction begins.

A few mRNAs of bacterial cells have half-lives that are longer than the average. For example, the mRNA that encodes a cell membrane protein has a half-life of about 10 min. The basis for the increased longevity of this mRNA is unknown, but it could be because the mRNA product is a membrane protein that enters the membrane and anchors the mRNA there.

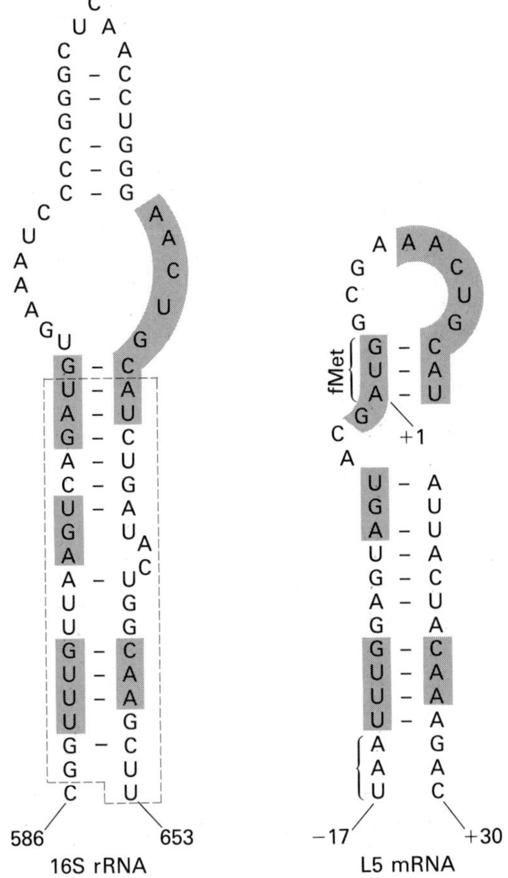

Figure 8-40 A comparison of primary and secondary structure in an rRNA and in an mRNA for a ribosomal protein. The sequence from the mRNA for the ribosomal protein L5 begins 17 nucleotides before the AUG initiation codon. In the 16S rRNA, the binding site for the S8 protein is indicated by the colored dashed lines. Gray boxes highlight base sequences that are identical in the two RNAs. The striking similarities in the sequences and in the stem-loop structures indicate why these regions presumably compete for free S8 protein. [See P. O. Olins and M. Nomura, 1981, *Nuc. Acids Res.* 9:1757.]

The Synthesis of Some Ribosomal Proteins Is Regulated by Control of Translation

The production of ribosomal proteins and the cellular content of ribosomal RNAs closely parallel the rate of cell growth. In cell-free systems, specific mRNAs for ribosomal proteins are synthesized from added DNA and translated into proteins. In the presence of excess ribosomal proteins, the genes that encode these proteins can still be transcribed into mRNAs, but some of the mRNAs are not translated. This indicates that the translation of specific mRNAs into ribosomal proteins is regulated by the amounts of the ribosomal proteins present.

An excess of certain ribosomal proteins can inhibit the translation of other types of ribosomal proteins. For example, excess ribosomal protein S8 (i.e., protein number 8 associated with the small ribosomal subunit) inhibits the translation of the mRNA for protein L5 (protein number 5 associated with the large subunit), but not that for L14 or L24. A comparison of the sequences of the 16S ribosomal RNA and of the mRNA encoding the beginning of the L5 gene suggests a physical basis for the translational arrest caused by excess ribosomal proteins. In the 16S rRNA, there is a stem-loop structure whose base sequence has many similarities with that at the beginning of the mRNA region encoding L5 proteins (Figure 8-40). This region of 16S rRNA binds tightly to the S8 protein. But if the amount of S8 protein exceeds that of rRNA (as occurs, for example, in bacteria in growth-limiting conditions), the S8 protein will be free to bind to the L5 mRNA and directly inhibit its translation. Untranslated mRNA is destroyed; thus the level of L5 mRNA is held at a low level in nongrowing cells, even though this mRNA is being synthesized.

Synthesized Proteins Become Diluted When They Are No Longer Needed

Although the mRNAs in bacterial cells disappear quickly when their products are no longer needed, this is not true of the bacterial proteins themselves. Once the inducing condition for an enzyme is removed, most enzyme molecules are simply diluted by the growth and division of cells; they are not destroyed. At one time it was thought that there was absolutely no intracellular turnover of bacterial proteins. However, if bacteria are severely deprived of nutrients, they can break down particular proteins to synthesize other proteins. For example, cells that are acutely starved for tryptophan can make the enzymes needed for the synthesis of tryptophan by breaking down other proteins. It has recently been demonstrated that small quantities of proteins are probably being broken down constantly into amino acids for the purpose of reusing these components. However, this process is not normally thought to be an important element of metabolic regulation in bacterial cells.

Summary

RNA synthesis and its control in bacteria are among the most thoroughly understood subjects of molecular cell biology. A single RNA polymerase is responsible for locating promoter sites in DNA. Promoters contain two consensus sequences, one centered about 10 base pairs before the start site for RNA, and one centered about 35 base pairs before this site. When the RNA polymerase makes contact with the promoter and binds to it, RNA synthesis is initiated. Some bacterial genes are transcribed at a constant rate (usually slowly) without any regulation.

For many other genes, the rate at which RNA synthesis is initiated is governed by site-specific proteins that interact with DNA regions termed operators. Operator sequences lie near promoter sites and sometimes overlap the promoters (i.e., include some of the same base pairs). The regulatory proteins that bind to operators are called repressors if they inhibit initiation, and activators if they increase it. Some genes are under the complex control of both repressor and activator proteins; the repressor must be removed and the activator must be present in order for these genes to be transcribed. Such genes often have two neighboring control sites, one recognized by the positive-acting regulatory protein and one recognized by the negative-acting protein. In general, the promoter-operator sites in prokaryotic DNA control the transcription of regions encoding more than one protein. The entire DNA transcription unit is called an operon. A prokaryotic mRNA therefore often contains information for more than one protein.

Even while the mRNA is being formed—that is, before the polymerase has reached any signals for RNA termination—ribosomes begin to translate the mRNA. The synthesis of each protein is initiated independently at a particular site in the mRNA. Although most regulation takes place through control of the rate of *transcription* initiation, translation can be regulated by the binding of specific proteins at *translation* initiation sites in the mRNA.

Prokaryotic gene control may also occur at the level of transcription termination. During the process of mRNA synthesis (and concomitant translation), the RNA polymerase does not always complete the transcription of all coding information in an operon. Thus the amounts of some mRNAs in bacteria are controlled by the regulated termination of transcription. Such regulation can occur between two genes in an operon, either by termination at that site or by antitermination—prevention of termination by the binding of a specific protein (an antiterminator). Regulated termination can also occur before the coding part of a gene is reached. This premature termination, which is called attenuation, is widely used to regulate the production of mRNAs that encode enzymes for the synthesis of amino acids.

References

General Mechanisms of Prokaryotic Gene Control

BIRGE, E. A. 1981. *Bacterial and Bacteriophage Genetics.* Springer-Verlag.

DUBNAU, D., ed. 1982. *The Molecular Biology of the Bacilli,* part I: *Bacillus subtilis.* Academic Press.

GLASS, R. E. 1982. *Gene Function: E. coli and Its Heritable Elements.* University of California Press.

PABO, C. T., and R. T. SAUER. 1984. Protein–DNA interactions. *Annu. Rev. Biochem.* 53:293–321.

STENT, G. S., and R. CALENDAR. 1978. *Molecular Genetics,* 2d ed. W. H. Freeman and Company.

WATSON, J. D. 1977. *Molecular Biology of the Gene,* 3d ed. Benjamin.

RNA Polymerase

CHAMBERLIN, M. J. 1976. RNA polymerase, an overview. In *RNA Polymerase,* R. Losick and M. J. Chamberlin, eds. Cold Spring Harbor Laboratory.

HANNA, M. M., and C. F. MEARES. 1983. Topography of transcription: path of the leading end of nascent RNA through the *E. coli* transcription complex. *Proc. Nat'l Acad. Sci. USA* 80:4238–4242.

LOSICK, R., and M. J. CHAMBERLIN, eds. 1976. *RNA Polymerase.* Cold Spring Harbor Laboratory.

LOSICK, R., and J. PERO. 1981. Cascades of sigma factors. *Cell* 25:582–584.

RABUSSY, D. 1982. Changes in *Escherichia coli* RNA polymerase after bacteriophage T4 infection. *ASM News* 48:398–403.

VON HIPPEL, P. H., D. G. BEAR, W. D. MORGAN, and J. A. MCSWIGGEN. 1984. Protein–nucleic acid interaction in transcription. *Annu. Rev. Biochem.* 53:389–446.

Operons

MILLER, J. H., and W. S. REZNIKOFF, eds. 1978. *The Operon.* Cold Spring Harbor Laboratory.

Promoters

PRIBNOW, D. 1975. Bacteriophage T7 early promoters: nucleotide sequences of two RNA polymerase binding sites. *J. Mol. Biol.* 99:419–443.

PRIBNOW, D. 1975. Nucleotide sequence of an RNA polymerase binding site at an early T7 promoter. *Proc. Nat'l Acad. Sci. USA* 72:784–789.

ROSENBERG, M., and D. COURT. 1979. Regulatory sequences involved in the promotion and termination of RNA transcription. *Annu. Rev. Genet.* 13:319–353.

SIEBENLIST, U., R. B. SIMPSON, and W. GILBERT. 1980. *E. coli* RNA polymerase interacts homologously with two different promoters. *Cell* 20:269–281.

Negative Control of Transcription

CUNIN, R., T. ECKHARDT, J. PIETTE, A. BOYEN, A. PIERARD, and N. GLANSDORFF. 1983. Molecular basis for modulated regulation of gene expression in the arginine regulon of *E. coli* K-12. *Nuc. Acids Res.* **11**:5007–5019.

DUNN, T. M., S. HAHN, S. OGDEN, and R. F. SCHLEIF. 1984. An operator at −280 base pairs that is required for repression of the BAD operon promoter. *Proc. Nat'l Acad. Sci. USA* **81**:5017–5020.

GILBERT, W., and B. MÜLLER-HILL. 1967. Isolation of the *lac* repressor. *Proc. Nat'l Acad. Sci. USA* **58**:2415–2421. The first report of the isolation of a transcriptional regulatory protein.

JACOB, F., and J. MONOD. 1961. Genetic regulatory mechanisms in the synthesis of proteins. *J. Mol. Biol.* **3**:318–356. A historic presentation of the hypothesis that a short-lived messenger, probably an RNA, exists.

KAISER, A. D., and F. JACOB. 1957. Recombination between related temperate bacteriophages and the genetic control of immunity and prophage localization. *Virology* **4**:509–521. This paper was the first to suggest that genes could encode negative-acting regulatory elements.

PARDEE, A. B., F. JACOB, and J. MONOD. 1959. The genetic control and cytoplasmic expression of inducibility in the synthesis of β-galactosidase by *E. coli. J. Mol. Biol.* **1**:165–178. Another landmark paper showing that genes could encode negative regulatory elements.

TAKEDA, Y., D. H. OHLENDORF, W. F. ANDERSON, and B. F. MATTHEWS. 1983. DNA-binding proteins. *Science* **221**:1020–1026.

ZIPSER, D., and J. BECKWITH, eds. 1970, 1977. *The lac Operon.* Cold Spring Harbor Laboratory.

Positive Control of Transcription

ENGLESBERG, E., and G. WILCOX. 1974. Regulation: positive control. *Annu. Rev. Genet.* **8**:219–242. A review of the work on positive-acting regulation, by its major discoverer.

GREENBLATT, J., and R. SCHLEIF. 1971. Arabinose C protein regulation of the arabinose operon *in vitro. Nature New Biol.* **223**:166–169.

LEE, N. 1978. Molecular aspects of *ara* regulation. In *The Operon,* J. H. Miller and W. S. Reznikoff, eds. Cold Spring Harbor Laboratory.

RAIBAUD, O., and M. SCHWARTZ. 1984. Positive control of transcription initiation in bacteria. *Annu. Rev. Genet.* **18**:173–206.

ZUBAY, G., D. SCHWARTZ, and J. BECKWITH. 1970. Mechanism of activation of catabolic-sensitive genes: a positive control system. *Proc. Nat'l Acad. Sci. USA* **66**:104–110. The first presentation of definitive evidence for positive-acting proteins in transcriptional regulation.

Compound Control of Transcription

DE CROMBRUGGHE, B., and I. PASTAN. 1978. Cyclic AMP, the cyclic AMP receptor protein, and their dual control of the galactose operon. In *The Operon,* J. H. Miller and W. S. Reznikoff, eds. Cold Spring Harbor Laboratory.

GOTTESMAN, S. 1984. Bacterial regulation: global regulatory networks. *Annu. Rev. Genet.* **18**:415–442.

Control of Regulatory Proteins

AIBA, H. 1983. Autoregulation of the *Escherichia coli crp* [or CAP] gene. CRP is a transcriptional repressor for its own gene. *Cell* **32**:141–149.

IRANI, M. H., L. OROSZ, and S. ADHYA. 1983. A control element within a structural gene: the *gal* operon of *E. coli. Cell* **32**:783–788.

MAGASANIK, B. 1978. Regulation in the *hut* system. In *The Operon,* J. H. Miller and W. S. Reznikoff, eds. Cold Spring Harbor Laboratory.

Termination Control of Transcription in Bacteria

HOLMES, W. M., T. R. PLATT, and M. ROSENBERG. 1983. Termination of transcription in *E. coli. Cell* **32**:1029–1032.

PLATT, T. 1978. Regulation of gene expression in the tryptophan operon of *Escherichia coli.* In *The Operon,* J. H. Miller and W. S. Reznikoff, eds. Cold Spring Harbor Laboratory.

PLATT, T. 1981. Termination of transcription and its regulation in the tryptophan operon of *E. coli. Cell* **24**:10–32.

ROBERTS, J. W. 1975. Transcription termination and late control in phage *lambda. Proc. Nat'l Acad. Sci. USA* **72**:3300–3304.

YANOFSKY, C. 1981. Attenuation in the control of expression of bacterial operons. *Nature* **289**:751–758. A comprehensive review of the proof of attenuation by its discoverer.

The Early Events of Bacteriophage λ Infection

HERSHEY, A. D., ed. 1971. *The Bacteriophage Lambda.* Cold Spring Harbor Laboratory.

HERSKOWITZ, I., and D. HAGEN. 1980. The lysis-lysogeny decision of phage λ: explicit programming on responsiveness. *Annu. Rev. Genet.* **14**:399–446. An exceedingly readable treatment of a complex subject.

HOCHSCHILD, A., N. IRWIN, and M. PTASHNE. 1983. Repressor structure and the mechanism of postive control. *Cell* **32**:319–325.

JOHNSON, A. D., A. R. POTEETE, G. LAUER, R. T. SAUER, G. K. ACKERS, and M. PTASHNE. 1981. λ repressor and cro—components of an efficient molecular switch. *Nature* **294**:217–223.

LWOFF, A. 1953. Lysogeny. *Bact. Rev.* **17**:269–337. A summary of the phenomenon of lysogeny, written by its major discoverer before the structure of DNA was known.

PABO, C. O., and M. LEWIS. 1982. The operator-binding domain of λ repressor: structure and DNA recognition. *Nature* **298**:443–447.

The Stability of Biopolymers in Bacterial Cells: Translational Control

The following three papers beautifully illustrate interlocking events in ribosomal protein synthesis.

DEAN, D., J. L. YATES, and M. NOMURA. 1981. *Escherichia coli* ribosomal protein S8 feedback regulates part of *spc* operon. *Nature* **289**:89–91.

NOMURA, M., D. DEAN, and J. L. YATES. 1982. Feedback regulation of ribosomal protein synthesis in *Escherichia coli. Trends Biochem. Sci.* 7:92–95.

OLINS, P. O., and M. NOMURA. 1981. Translational regulation by ribosomal protein S8 in *Escherichia coli:* structural homology between rRNA binding site and feedback target on mRNA. *Nuc. Acids Res.* 9:1757–1764.

9

RNA Synthesis and Processing in Eukaryotes

In prokaryotic cells, all RNA synthesis—that of ribosomal RNA, transfer RNA, and messenger RNA—is catalyzed by a single RNA polymerase. Control of the action of this enzyme determines which mRNAs are made and therefore which proteins are produced, because the translation of mRNA begins shortly after the initiation of transcription. The expression of genes in eukaryotes differs significantly from that in prokaryotes (Figure 9-1):

1. Most eukaryotic DNA is packaged with proteins termed *histones* in such a way as to produce a highly ordered *chromatin* structure. The association of the DNA with histones makes it less accessible or even inaccessible to transcription. It is widely believed that changes in the architecture of chromatin are necessary for the activation of transcription of eukaryotic genes. This contrasts with the accessibility of prokaryotic DNA; in bacteria, the DNA is not bound to protein in a compact form, and the transcription of specific genes can be started or stopped in seconds. The same quick response of starting or stopping transcription may also apply to many genes in single-cell eukaryotes and even a few genes in multicellular organisms, but it is generally true that gene control in multicellular organisms is not characterized by quick response to outside stimuli.
2. When a gene is first transcribed in eukaryotes, the initial transcript is, in most cases, *not* a functional rRNA,

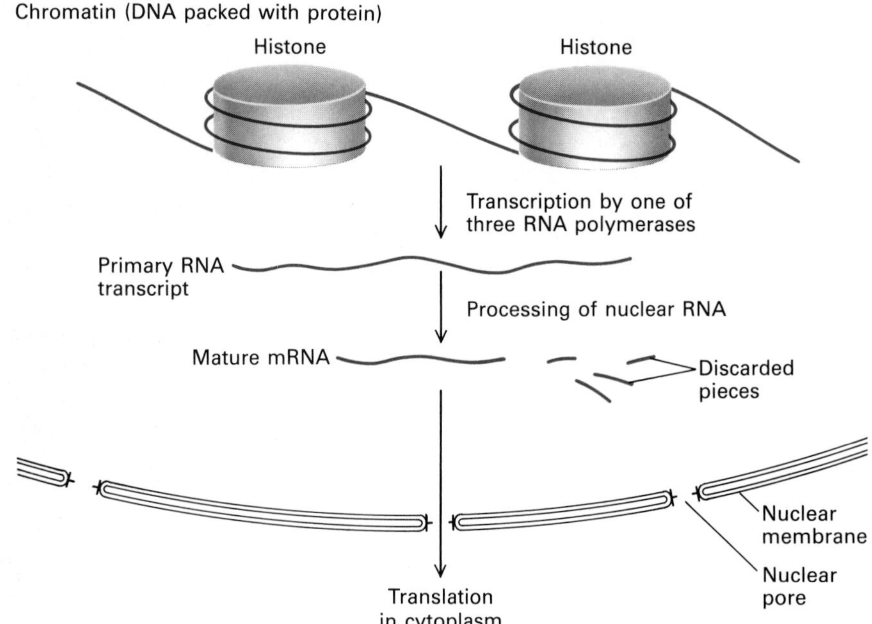

Figure 9-1 Four distinguishing features of mRNA production in eukaryotes are not found in prokaryotes: (1) The DNA to be transcribed is wound around a histone core. (2) It is transcribed by one of three RNA polymerases. (3) The primary RNA transcript is not a finished rRNA, tRNA, or mRNA; it must be processed within the nucleus into the finished RNA. (4) Only after transport of the mRNA—presumably through a nuclear pore—does translation occur in the cytoplasm.

tRNA, or mRNA. Rather, a precursor RNA molecule is produced in the eukaryotic nucleus, where it undergoes several biochemical modifications that usually reduce it in size before it enters the cytoplasm as an rRNA, tRNA, or mRNA molecule. (Actually, it is now known that rRNA and tRNA are formed from precursor RNA molecules even in bacteria.) The modifications in eukaryotic mRNA formation frequently include splicing, in which distant regions of the primary RNA molecule are joined together and the intervening sequence is eliminated. Splicing in the formation of mRNA is necessary because the gene sequences encoding a single protein are frequently not contiguous in eukaryotic DNA.

3. Eukaryotic RNA synthesis is carried out by three separate RNA polymerases. The precursor to three of the four rRNAs (the 28S, 5.8S, and 18S rRNAs) is produced by RNA polymerase I in the nucleolus of the cell. The precursors to mRNAs are made by RNA polymerase II; and the small RNAs (tRNAs and the 5S rRNAs) or their precursors are made by RNA polymerase III.
4. Transcription and translation are not coupled as they are in prokaryotic cells. In eukaryotic cells, the site of synthesis of mRNA is in the nucleus, whereas the translation of the mRNA into protein occurs in the cytoplasm.

These fundamental differences between prokaryotes and eukaryotes in chromosome structure and in RNA production suggest that the two types of cells also differ in their mechanisms of gene control. It is now known that many eukaryotic genes are regulated by their transcription. Some genes are also regulated by differential processing of their primary RNA transcripts, by differential rates of translation of their mRNAs in the cytoplasm, and by differential rates of degradation of these mRNAs.

Because our first information about the expression of eukaryotic genes came from biochemical studies of RNA synthesis, and because a knowledge of the biochemistry of nuclear RNA aids greatly in understanding the structure of eukaryotic genes, RNA synthesis and processing are described in this chapter. Chapters 10 and 11 deal with the structure and organization of genes and chromosomes in eukaryotes. These subjects pave the way for the discussion of gene control in eukaryotes in Chapter 12.

Eukaryotic RNA: Nuclear Synthesis but Cytoplasmic Residence

A central problem in the study of gene expression in eukaryotic cells arises from the fact that whereas proteins are synthesized in the cytoplasm, and most of the RNA molecules in the cell are found in the cytoplasm, the genes (DNA molecules) that control protein synthesis are located in the nucleus, where the RNA is synthesized.

That most of the eukaryotic RNA is in the cytoplasm was revealed by studies done with RNA-specific staining and spectrophotometric scanning of individual cells before and after treatment with RNase and DNase. This fact was established long before biologists recognized the existence of different types of RNA that each have specific biological functions. The idea that RNA had something to do with protein synthesis was suggested origi-

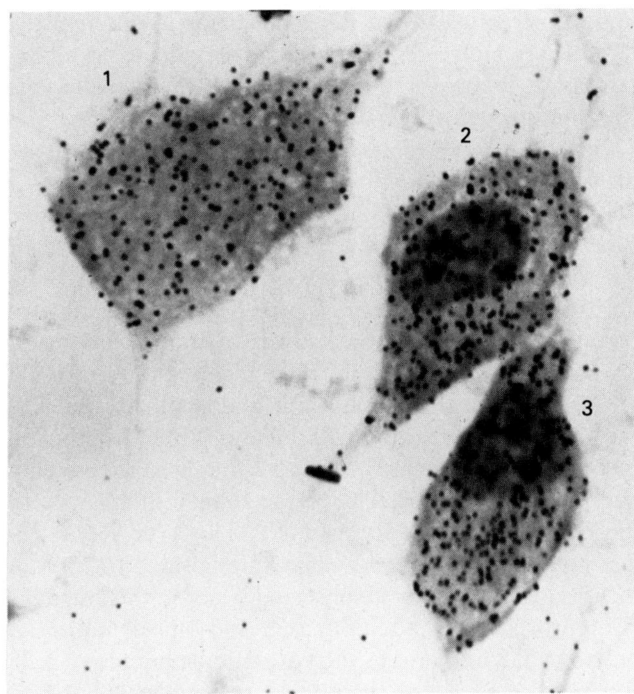

Figure 9-2 Cytoplasmic protein synthesis in cultured Chinese hamster cells. The cells were exposed to amino acids labeled with tritium (^{3}H) for 10 min, and then autoradiographs were made. Most of the silver grains *(black dots)* indicating the incorporation of ^{3}H are over the cytoplasm. The culture has been treated with cytochalasin, a drug that does not affect protein synthesis but that causes some cells to lose their nuclei. Cells 2 and 3 each have a dark-stained central nucleus; cell 1 has lost its nucleus but still incorporates amino acids into proteins, thus illustrating clearly that proteins are formed in the cytoplasm. *Courtesy of D. Prescott.*

nally because of these histological studies. Cells that make a large amount of protein for export (e.g., pancreatic cells, which secrete digestive enzymes, and liver cells, which secrete blood proteins) contain the most cytoplasmic RNA. One of the earliest discoveries of electron microscopy was the presence of RNA-containing granules, which were later termed ribosomes, in the cytoplasm of pancreatic and liver cells.

When radioactively labeled amino acids became available, the cytoplasmic location of protein synthesis was confirmed directly (Figure 9-2). But geneticists from Morgan through McClintock, Beadle, and Tatum had concluded that the control of protein synthesis somehow resided in the genes that were located on the chromosomes in the nucleus. How was molecular communication achieved between the nucleus and the cytoplasm?

By the 1950s, the demonstration of the infectiousness of pure viral RNA without protein had established an important biological role for RNA (see Figure 6-29). Investigators of cell function began to think of RNA as a possible information courier between the nucleus and the cytoplasm, and the first attempts to study RNA synthesis were made. When cells were exposed to radioactive building blocks of RNA such as [^{3}H]uridine, the nucleus became labeled long before the cytoplasm did (Figure 9-3). Thus it seemed that RNA synthesis took place in the nucleus, and that RNA might have an informational role to play.

Before studies of nuclear RNA had progressed very far, the separate roles of rRNA, tRNA, and mRNA in bacterial protein synthesis had been determined (Chapters 4 and 8), and ribosomes and tRNA had been shown to exist in animal cells. Clearly, an understanding of the control of eukaryotic genes in molecular terms depended on establishing the relation between nuclear RNA and cytoplasmic RNA. In pursuit of this goal, researchers separated the newly formed nuclear RNA into classes that could be identified and followed as the RNA migrated from the nucleus to the cytoplasm. They characterized first the nuclear RNA precursors of rRNA and then the precursors of tRNA. Finally, the details of the formation of mRNA from its nuclear precursors were worked out.

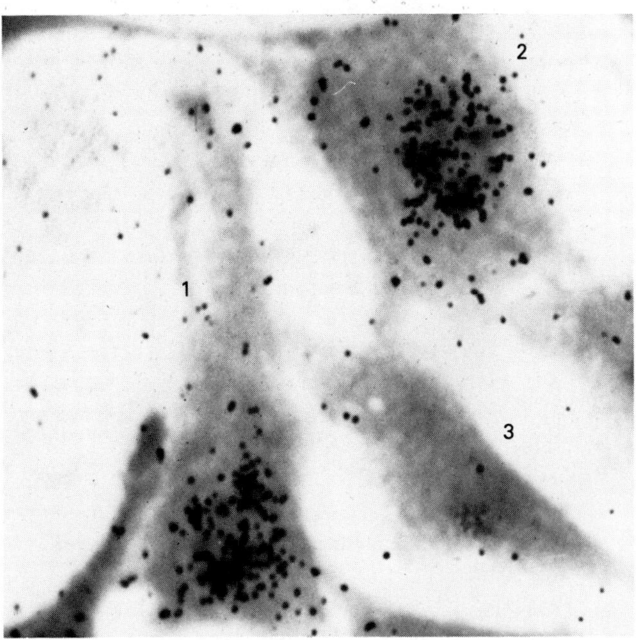

Figure 9-3 Nuclear RNA synthesis in cultured Chinese hamster cells. The cells were exposed to [^{3}H]uridine for 10 min, and then autoradiographs were made. Like the cells shown in Figure 9-2, these were treated with cytochalasin. Cells 1 and 2 are whole cells with normal nuclei, and most of the silver grains indicating incorporated [^{3}H]uridine appear over the nuclei. Cell 3 has no nucleus, and it has incorporated almost no label. *Courtesy of D. Prescott.*

The fractionation schemes and experimental approaches that developed from this early work on RNA synthesis are still in current use, and we shall describe them in the next few sections.

Analyzing Pulse-Labeled RNA in Cell Fractions: The Discovery of RNA Processing

Understanding the relationship of nuclear RNA to cytoplasmic RNA requires two kinds of fractionation. In one, the nucleus is physically separated from the cytoplasm so that RNA in each fraction can be examined (Figure 9-4). The second kind of fractionation involves time; this procedure allows the kinetic relationship between nuclear and cytoplasmic RNA to be followed. The sequential appearance of a labeled RNA precursor in different nuclear RNA molecules before its appearance in cytoplasmic products points to the initial, the intermediate, and the final products in a precursor-product pathway.

Cultured mammalian cells, particularly HeLa cells, are ideal for cell fractionation and the separation of nuclear and cytoplasmic RNAs. Centrifugation of the purified total cell RNA separates molecules according to their size. Each size class in the total cell RNA is detected by its absorption of ultraviolet light; the RNA species that are present in the largest amounts are, of course, the easiest to observe. By this analysis, three major species of RNA were identified: tRNA plus large and small rRNAs. The 28S rRNA, which is about 5 kilobases (kb) in length, and the 18S rRNA, which is about 2 kb, appear to make up approximately 80 percent of the total cell RNA, with most of the remainder being tRNA. However, RNA species such as mRNA, which constitutes a small fraction of the total RNA, cannot be observed by using this fractionation procedure. In addition, each ribosome contains not only one 28S and one 18S RNA molecule but also a 5.8S RNA and a 5S RNA; these latter two molecules constitute only about 2 percent of the total RNA in a ribosome, and thus escape detection with this method.

Of the rRNA and tRNA molecules extracted from fractionated cells, almost all (more than 90 percent) are recovered from the cell cytoplasm. (Again, they are detected by their absorption of UV light.) However, after cells are exposed for 5 min or less to a labeled RNA precursor

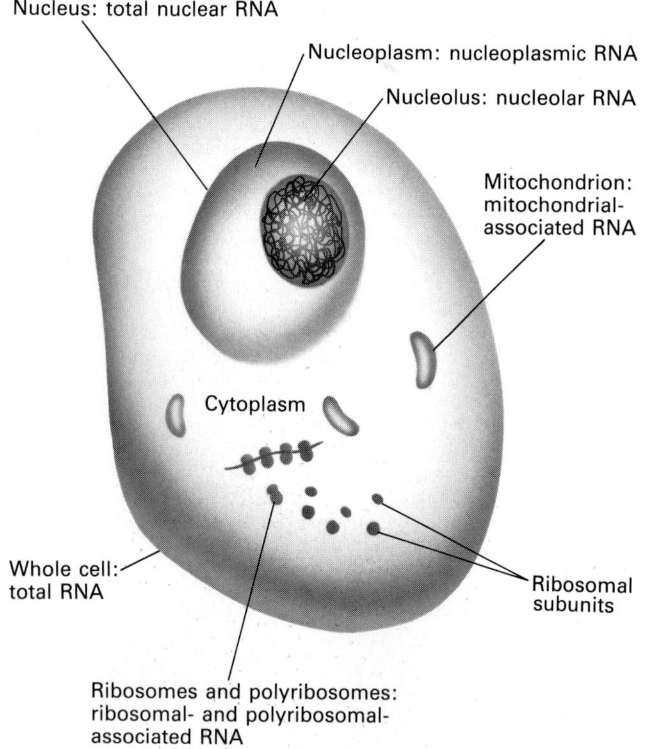

Figure 9-4 The mammalian cell can be fractionated and RNA can be purified from each fraction. Commonly used experimental preparations are total cell RNA, total nuclear RNA, cytoplasmic RNA, nucleolar RNA, and extranucleolar or nucleoplasmic RNA. The mitochondria and the polyribosomes can be separated from the remainder of the cytoplasmic fraction, and their RNA can then be extracted. In plant cells, chloroplasts also can be separated from the cytoplasm, and their RNA, too, can be extracted.

Figure 9-5 ***(opposite)*** The amounts of labeled and unlabeled RNA components from separated fractions of HeLa cells after three different labeling intervals. The graphs show the distribution of the RNA components within the whole cell and within three cell fractions. The extracted RNA is separated by rate-zonal sedimentation (Chapter 7) on the basis of size. Larger molecules are to the left, which represents the bottom of the centrifuge tube. The pre-existing or "old" RNA is monitored by its absorption of ultraviolet light (260 nm); the resulting curves are black. The color curves indicate the relative amounts of radioactive or "new" RNA from cells labeled for 5, 15, or 60 min before fractionation and RNA extraction. Equal-sized cell samples were taken at each interval. Clearly, more and more radioactivity is incorporated with time. Also, the predominating species of labeled RNA changes. For example, the first discrete labeled RNA to appear is the 45S precursor to ribosomal RNA (pre-rRNA); 45S pre-rRNA is followed by the 32S pre-rRNA, and then by the 28S and 18S rRNAs. Note, too, that the nucleolar and nucleoplasmic RNAs are labeled before the cytoplasmic rRNA is. [See J. E. Darnell, 1968, *Bacteriol. Rev.* **32**:262.]

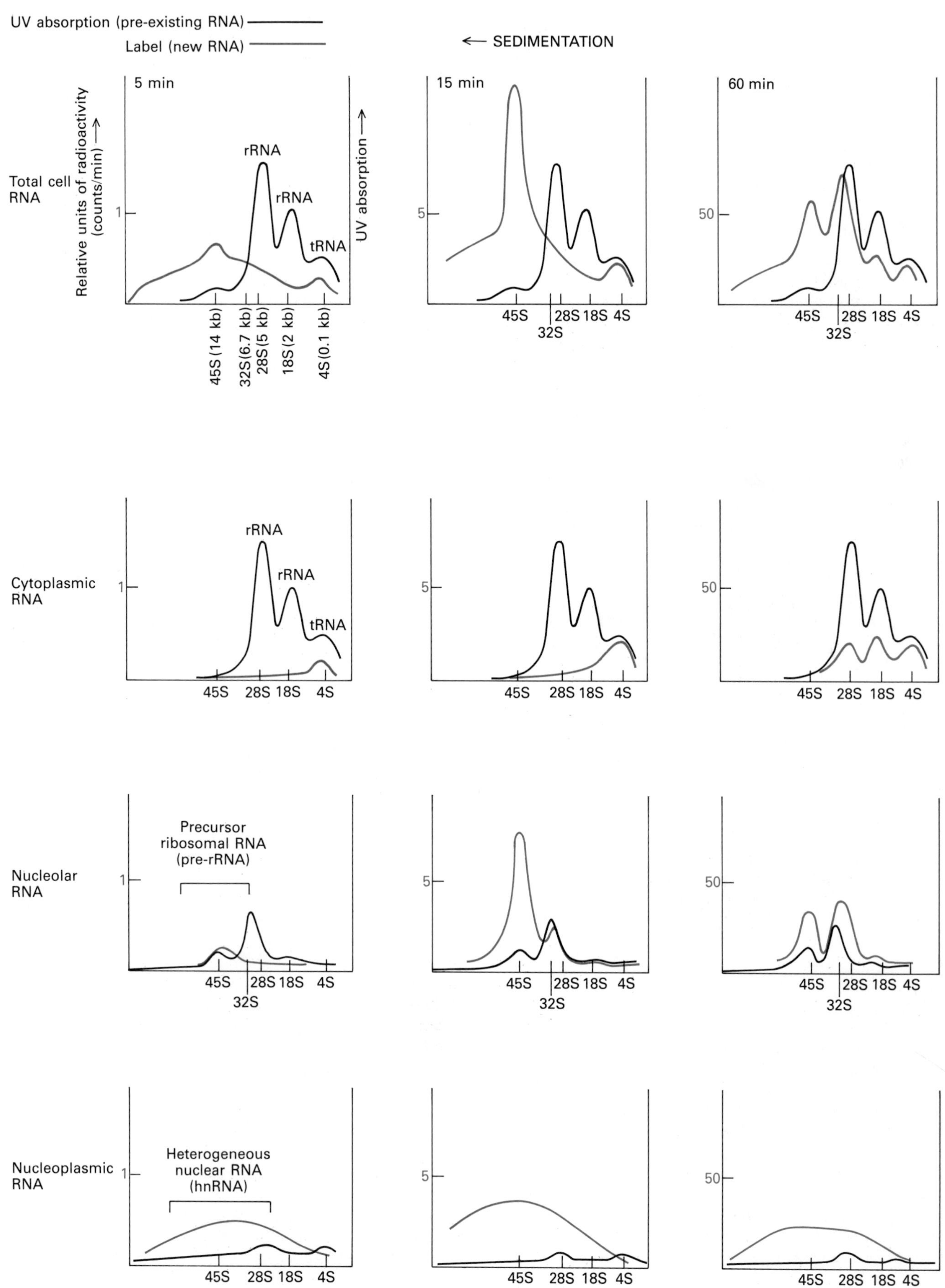
UV absorption (pre-existing RNA)
Label (new RNA)
← SEDIMENTATION
5 min
15 min
60 min
Relative units of radioactivity (counts/min) →
UV absorption →
Total cell RNA
rRNA
rRNA
tRNA
45S (14 kb)
32S (6.7 kb)
28S (5 kb)
18S (2 kb)
4S (0.1 kb)
45S 28S 18S 4S
32S
Cytoplasmic RNA
rRNA
rRNA
tRNA
45S 28S 18S 4S
Nucleolar RNA
Precursor ribosomal RNA (pre-rRNA)
45S 28S 18S 4S
32S
Nucleoplasmic RNA
Heterogeneous nuclear RNA (hnRNA)
45S 28S 18S 4S
1
5
50

such as [^{3}H]uridine almost all of the *labeled* RNA is in the nucleus. Furthermore, the size of the newly labeled nuclear RNA does not match that of the majority species in the cytoplasm. The briefly labeled nuclear molecules are quite heterogeneous in size: they vary in length from a few hundred nucleotides to more than 30 kb, and most of the label is found in molecules that sediment as fast or faster than the 28S rRNA that is found in the cytoplasm (Figure 9-5).

This result suggested that the cytoplasmic RNAs might be formed initially as precursor molecules. If the cells are labeled for longer periods before fractionation (e.g., for 15 and 60 min), labeled nuclear molecules of specific sizes begin to emerge against the background of molecules of many sizes. The first labeled species to be detected in any quantity is a 45S RNA that is about 14 kb long; soon another labeled species, a 32S RNA, about 6.7 kb long appears. Molecules in these two size classes constitute a large percentage of the total newly labeled nuclear RNA. A short time after the labeled 6.7-kb species appears, label begins to appear in the cytoplasmic 28S and 18S rRNAs that are about 5 kb and 2 kb respectively. As we shall see, the 45S and 32S RNAs are nucleolar precursors to the 28S, 18S, and 5.8S rRNAs. The specific-sized molecules are also referred to by their sedimentation values: 45S (14 kb), 32S (6.7 kb), 28S (5 kb), and 18S (2 kb).

The 45S RNA of HeLa Cells Is a Precursor to Ribosomal RNA

In fractionation studies, the nucleolus can be separated from the remainder of the nucleus. When RNA is then extracted from each of these fractions, the labeled 45S RNA is found in the nucleolus. Chemical experiments have shown that this nucleolar molecule is related to rRNA. Its nucleotide composition is similar to that of rRNA, and it has the same methylated oligonucleotide sequences as does rRNA. In addition, labeled 45S RNA is converted into smaller RNA molecules. Cells that have been labeled for a short time (for about 10 to 20 min) can be prevented from synthesizing further RNA by the addition of a drug (actinomycin D) that immediately stops transcription by binding to DNA. The amount of labeled 45S RNA decreases, and the amounts of 28S and 18S rRNAs increase (Figure 9-6). The 45S precursor rRNA molecule, or *pre-rRNA,* is cleaved in several steps to make the 28S and 18S rRNA molecules. (As we shall discuss later, it is cut to yield the 32S pre-rRNA, which is then cut again to form the 28S and the 5.8S rRNAs.) The experiments showing a flow of label from 45S to ribosomal RNAs were the first indications that nuclear RNA molecules are refashioned, or *processed,* into cytoplasmic RNAs.

The characterization of the pre-rRNA gave rise to two

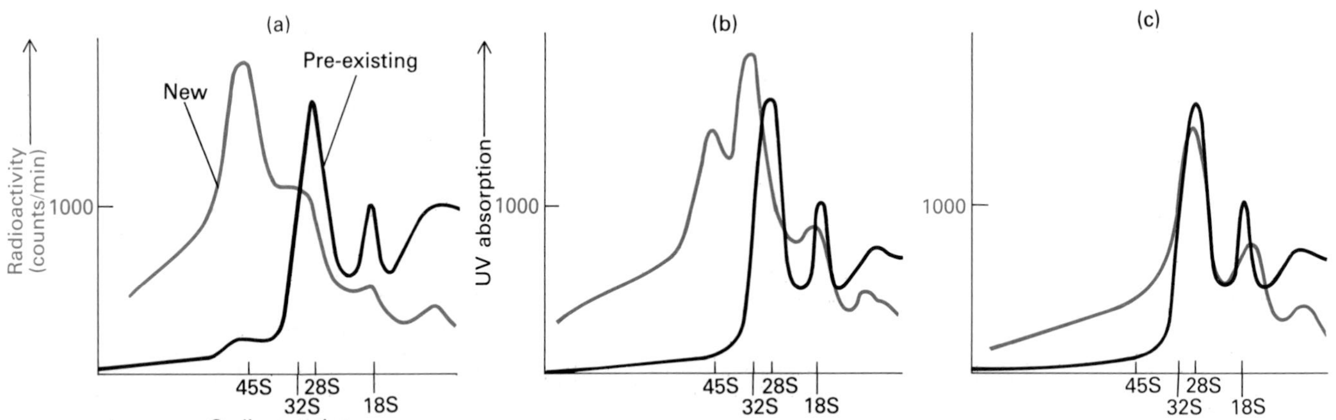

Figure 9-6 An actinomycin pulse-chase experiment, showing by the sedimentation analysis of extracted RNA that 45S nucleolar RNA is a precursor to rRNA. HeLa cells were labeled for 25 min, and the RNA was extracted and examined. The distribution of the labeled RNA molecules at that time is shown by the color curve in (a). RNA synthesis was then stopped by the addition of actinomycin D, a compound that can enter cells and tightly bind DNA to prevent further transcription. A sample taken 10 min after the actinomycin treatment generated the color curve shown in (b): there is a decrease in labeled 45S RNA and an increase in labeled 32S RNA and 18S rRNA. A further sample taken 35 min after the actinomycin treatment yielded the results in (c): there is a further decrease in labeled 45S RNA, a marked decrease in 32S RNA, and a marked increase in 28S rRNA. These results provided evidence that 45S RNA is a precursor to 32S RNA and 18S rRNA, and that 32S RNA is in turn a precursor to 28S rRNA. [See J. R. Warner et al., 1966, *J. Mol. Biol.* **19**:349.]

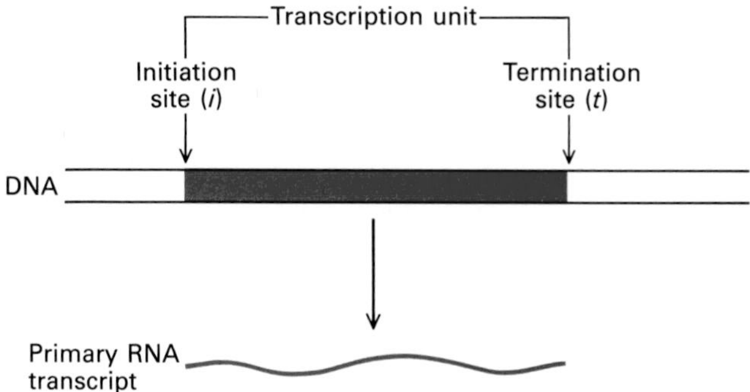

Figure 9-7 Transcription units in DNA have initiation sites and termination sites. Primary RNA transcripts are the initial unmodified RNA products.

terms that will be used repeatedly in this book: a newly synthesized RNA molecule that has not yet been modified in length is a *primary RNA transcript*. The *transcription unit* is the DNA region that is transcribed to produce the primary transcript (Figure 9-7).

Heterogeneous Nuclear RNA Is a Separate Class of Primary Nuclear RNA Transcripts

The early studies with HeLa cells also identified a second important group of nuclear RNA molecules. As we mentioned earlier, when HeLa cells are allowed to incorporate [^{3}H]uridine for 5 min or less (see Figure 9-5), most of the labeled nuclear RNA is heterogeneous in size. Thus, it is referred to as *heterogeneous nuclear RNA* (hnRNA). These molecules range in length from 200 bases to 30 kb, with an average length of about 5 kb. Most of the hnRNAs are found outside of the nucleolus, so they are not precursors to rRNAs. These nuclear RNA molecules include precursors to mRNAs, but not all hnRNAs function as mRNA precursors. The hnRNAs are discussed extensively later in this chapter and again in Chapter 12.

The RNA Polymerases and General Transcription Factors of Eukaryotic Cells

We mentioned in the introduction to this chapter that unlike prokaryotic cells, in which a single RNA polymerase makes all types of RNA, all eukaryotic cells examined so far (e.g., human, rodent, frog, fruit fly, yeast, and plant cells) have three different polymerases that make different classes of RNA in the cell nucleus. Each of the three RNA polymerases has many more protein subunits than the bacterial RNA polymerase does. We still lack detailed information on how RNA synthesis is begun by the eukaryotic polymerases, but before describing in the remainder of this chapter what we do know about eukaryotic RNA synthesis, we pause here to describe briefly how the three enzymes have been characterized.

The synthesis of pre-rRNA is catalyzed in the nucleolus by RNA polymerase I; hnRNA, containing mRNA precursors, is made by RNA polymerase II; and 5S rRNA and precursors to tRNA and other small nuclear RNAs are synthesized by RNA polymerase III. Each of these enzymes is complicated in structure (at least 10 major polypeptide chains are typically found in a highly purified preparation of an individual polymerase). The functions of the different polypeptide chains in the enzymes have not yet been determined. In addition to the polymerases, several auxiliary proteins called *transcriptional factors* are required to assist eukaryotic polymerases in starting RNA synthesis at correct sites on the DNA. This very active area of research will be dealt with mainly in Chapter 12.

The eukaryotic RNA polymerases are designated I, II, and III because, in one of the steps of chromatographic purification, most of the proteins responsible for RNA synthesis elute at three different salt concentrations (Figure 9-8a). The polymerases can be distinguished further by their sensitivity to α-amanitin, a product of the poisonous mushroom *Amanita phalloides:* polymerase I is very insensitive, polymerase II has the greatest sensitivity, and polymerase III is intermediate in its sensitivity to α-amanitin (Figure 9-8b). In addition to the multiple subunit proteins of RNA polymerases, there are several generalized transcription factors necessary for RNA chain initiation. At least three such factors exist for RNA polymerases III and II. The factors are different for the two polymerases. It is thought likely that all cells possess these nonspecific but necessary factors. There are, it is widely assumed, specific transcription factors as well, which we discuss in detail in Chapter 12.

Three Methods for Mapping Transcription Units

To learn which sequences in DNA may play a role in starting and stopping transcription, it is necessary to first define the transcription unit—that is, where on DNA the RNA polymerase starts and stops. In eukaryotic cells, the general boundaries of transcription units were originally defined by examining the primary RNA transcripts themselves; recombinant DNA technology subsequently provided the genetic manipulations for identifying exact transcription initiation sites on DNA. The first transcription unit to be studied was that for the 45S pre-rRNA in mammalian cells. Three experimental designs were used: kinetic labeling and nascent chain analysis, electron mi-

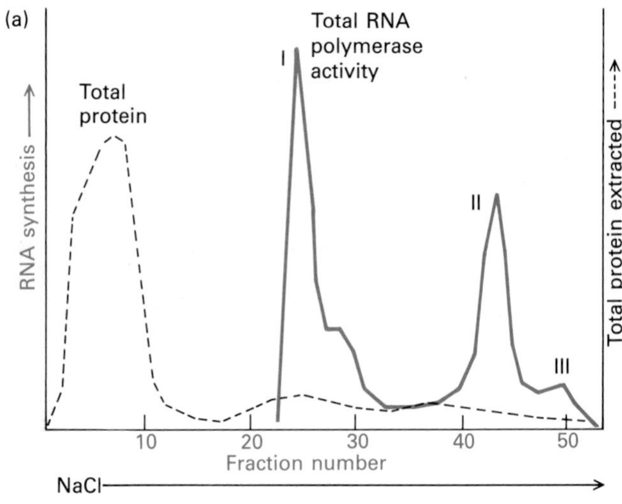

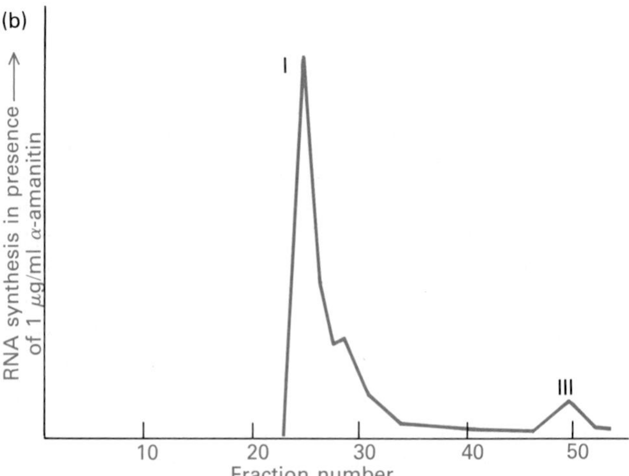

Figure 9-8 The identification and separation of the three RNA polymerases of frog cells by chromatographic analysis of proteins. (a) A protein extract from the nuclei of cultured frog cells was passed through an adsorbent column (DEAE Sephadex, a column to which charged proteins adsorb differentially), and adsorbed proteins were eluted *(dashed curve)* with an increasing concentration of NaCl. Successive fractions of the eluted proteins were assayed for the ability to transcribe DNA *(color curve)* in the presence of the four nucleoside triphosphates (including ratioactive UTP). Most of the proteins did not bind to the column, but the enzymes did. (b) The sensitivity of the RNA synthesis to 1 μg/ml of α-amanitin was measured in a second procedure, which demonstrated that polymerases I and III are insensitive to the compound at that concentration, whereas polymerase II is sensitive (i.e., it ceases RNA synthesis). (Polymerase III is sensitive to 10 μg/ml of α-amanitin, however, whereas polymerase I remains unaffected.) [See R. G. Roeder, 1974, *J. Biol. Chem.* **249**:241.]

croscopy, and ultraviolet light irradiation. These techniques have since proved useful in the mapping of transcription units for various mRNAs. Because most of our ideas about RNA synthesis, processing, and control require knowledge of the boundaries of transcription units, we shall explain each of the three mapping techniques.

Kinetic Labeling and Nascent Chain Analysis Provide "Snapshots" of RNA Molecules during and Just after Synthesis

The underlying assumption in a kinetic labeling analysis is that macromolecules are formed by a sequential pathway that can be outlined by observing the flow of radioactive label (derived from monomers). Label accumulation can be divided into three stages:

1. Over very short intervals, only nascent molecules are labeled. The nascent chains emerging from a single transcription unit range in length from one nucleotide to one less than the number included in the finished chain. Experimentally, this was observed first for the 14-kb, 45S pre-rRNA from HeLa cells. Nucleolar RNA from cells labeled for only a few minutes is heterogeneous in size, but between 5 and 10 min the 45S primary transcript becomes the dominant labeled molecule (Figure 9-9). Mixtures of nascent chains from different transcription units also can be examined to determine the size range of those units. Although all transcription units will contribute smaller labeled nascent chains, only the longest transcription units will contribute the longest nascent chains. Recognition of this distribution enabled the size range of transcription units encoding HeLa cell hnRNA to be estimated (Figure 9-10). About half of the hnRNA comes from transcription units smaller than 4 to 5 kb, and half from larger transcription units that we now know range up to 50 kb and more.
2. When labeling time exceeds that necessary to form a complete molecule, most of the radioactivity will no longer be in nascent molecules but will accumulate in the primary product, as shown in Figure 9-9 and in the Dintzis experiment on globin protein synthesis (see Figure 7-6).
3. As labeling time increases, the radioactivity will appear successively in the family of RNA molecules derived from the initial transcripts (see Figure 9-6).

In addition to the use of kinetic labeling analysis to establish the size of an initial RNA precursor and to follow its fate, another experimental design involving analysis of labeled nascent RNA chains has been very important in mapping transcription units. The first step is to obtain a sufficient quantity of pure DNA from which a particular RNA is transcribed in cells (e.g., viral DNA can be used; or other DNA can be cloned in a bacterial vec-

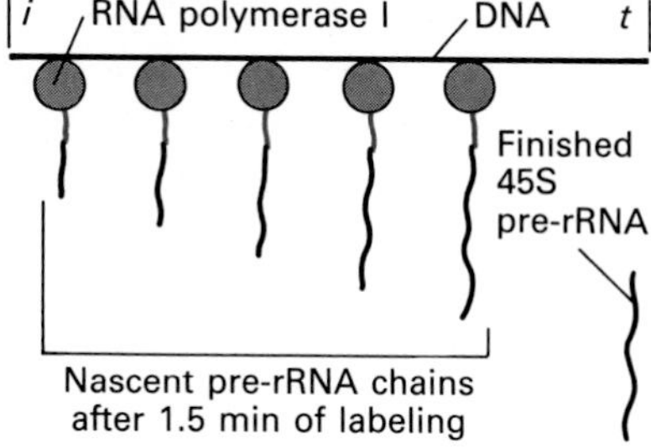

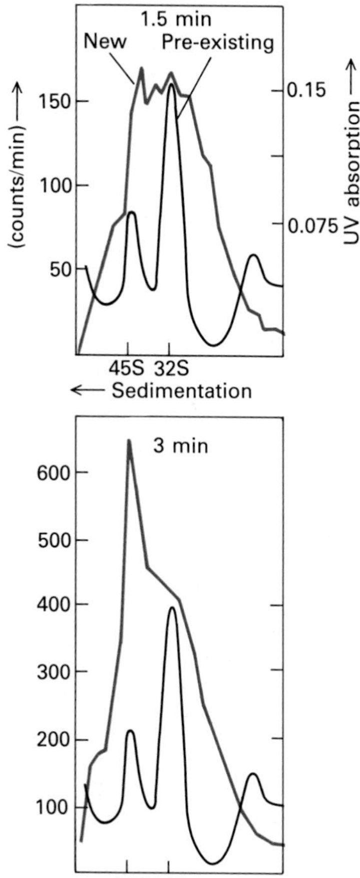

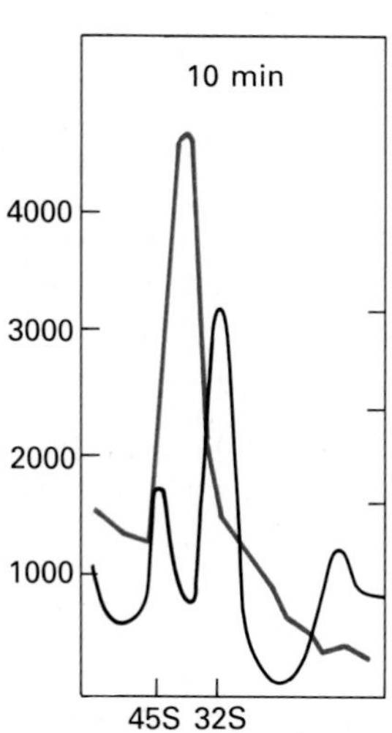

Figure 9-9 Nascent chains of precursor ribosomal RNA, detected by the brief labeling of HeLa cells. A radioactive label was incorporated into pre-rRNA for 1.5, 3, and 10 min; nucleolar RNA obtained by fractionation was sedimented through a sucrose gradient. Note increasing scales for radioactivity. The pre-existing nucleolar RNA was detected by absorption of ultraviolet light at 260 nm *(black curve)*. After 1.5 min of labeling, the pattern of newly synthesized RNA, as detected by radioactivity *(color curve)*, was indistinct (i.e., the RNA that had become labeled was of varying lengths, with a maximum at 45S). This indicated that most of the radioactivity was in nascent chains, as shown in the diagram at the top. After 3 min of labeling, 45S pre-rRNA began to appear in larger amounts; after 10 min, it was unambiguously the dominant species. From experiments of this type, it has been estimated that 45S pre-rRNA is synthesized in about 2 to 4 min. [See H. Greenberg and S. Penman, 1966, *J. Mol. Biol.* **21**:527.]

tor). The availability of specific restriction endonucleases (Chapter 7) allows the DNA to be cut up into pieces whose successive order in the genome is known, and these DNA pieces can be used in RNA-DNA hybridization experiments. DNA fragments that lie within the transcription unit will hybridize to labeled nascent RNAs synthesized from that transcription unit; the DNA that lies on either side of the transcription unit will not hybridize to the labeled RNAs. The sizes of the labeled nascent RNAs that hybridize to various members of the set of ordered DNA fragments will indicate the direction of RNA chain growth: the shortest labeled nascent RNA will hybridize to the DNA nearest the promoter; longer and longer nascent RNA chains will hybridize to successively more distant DNA fragments (Figure 9-11). This technique has been found to be very useful in mapping transcription units for mRNAs in animal cells, as will be discussed later in this chapter.

The Electron Microscope Is Used to View Transcription Units in Action

The first transcription units to be viewed in the electron microscope were ribosomal DNA from the nucleolus of frog oocytes (cells on the pathway to becoming egg cells). Because these cells have greatly increased amounts of rDNA and increased numbers of nucleoli, all making pre-rRNA, they are excellent material with which to observe pre-rRNA synthesis in the electron microscope. If oocyte nuclei are suspended in a detergent solution, the nuclear contents are released. When the suspension is spread as a thin film on the surface of an underlying solution having a greater density, such as a concentrated protein solution, the long threads of nuclear DNA with their associated

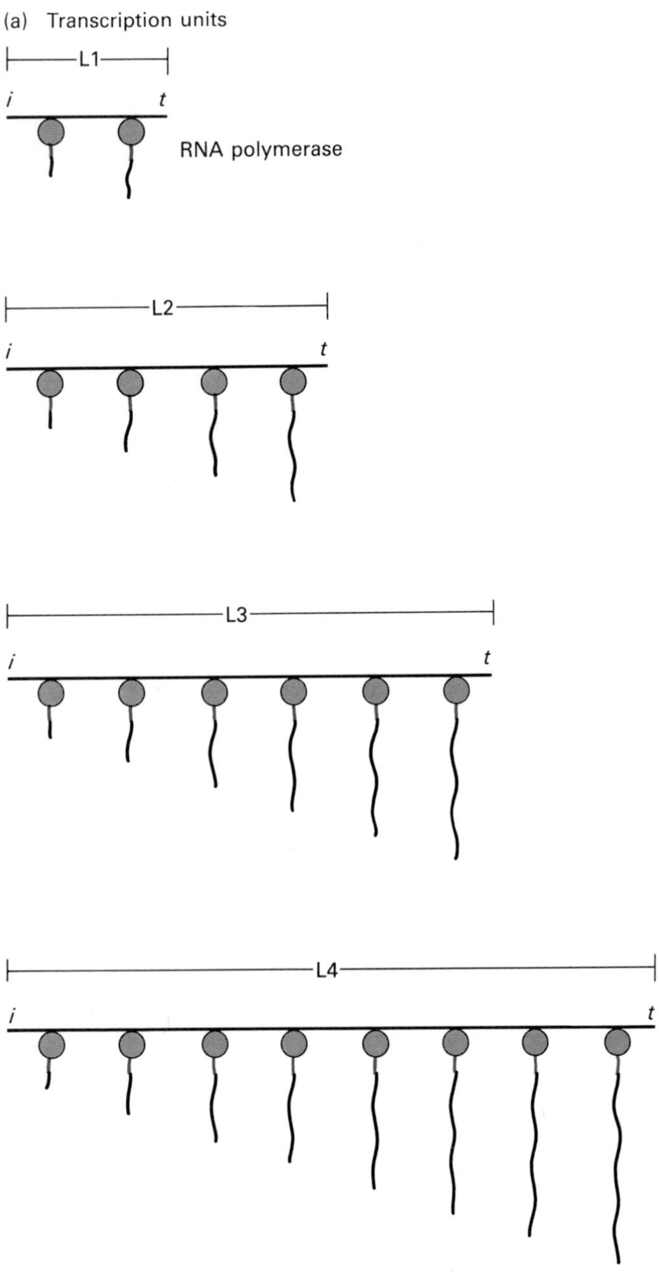

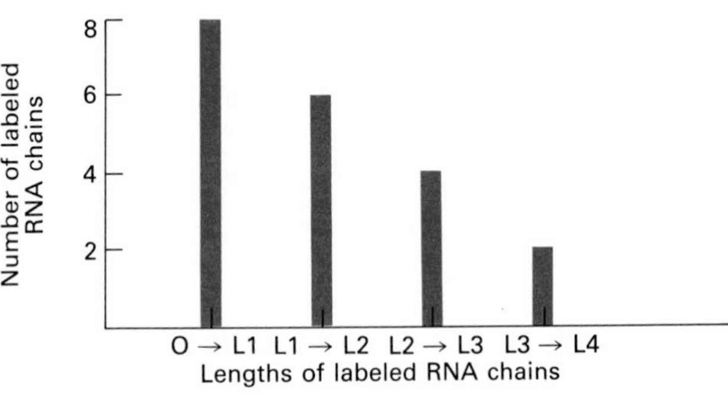

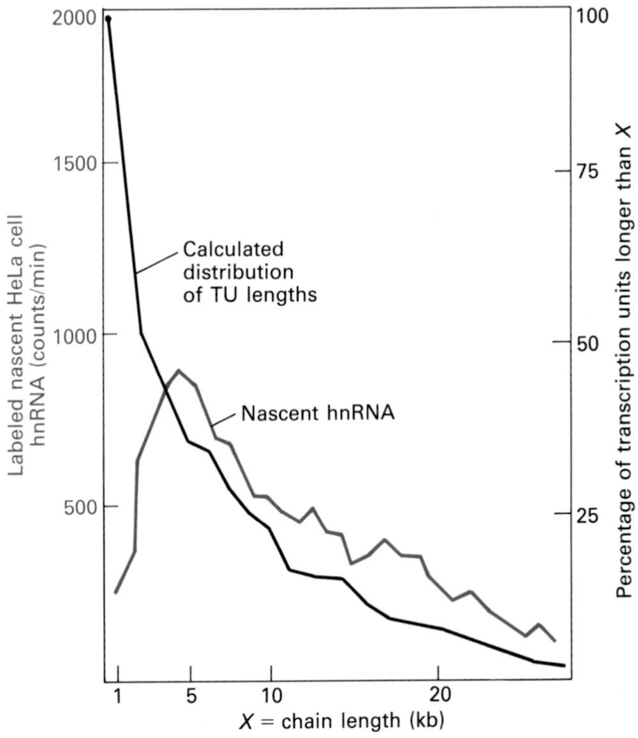

Figure 9-10 Estimation of the size range of transcription units of HeLa cells by analysis of nascent hnRNA chains. (a) A mixture of very briefly labeled hnRNA molecules derived from transcription units of various lengths would be composed of shorter molecules from all transcription units (L1 to L4) and longer molecules only from the longer transcription units (say, L3 and L4). (b) The ratio of the number of labeled chains in the longest category (L3 to L4) to the number in the shortest (0 up to L1) predicts the percentage of molecules that will be in the longest category when chain synthesis is finished (2:8, or 25 percent). (c) An actual experiment in which nascent Hela cell hnRNA that had been labeled for 30 min was separated according to size by rate-zonal sedimentation; the numbers of pulse-labeled chains in size classes ranging from 1 to 30 kb were estimated *(color curve)*, as in part (b). Then the percentage of primary transcripts—or, likewise, of transcription units—longer than each size class was calculated *(black curve)*. About half of the transcription units are longer than 5 kb and at least 10 percent are 20 kb or larger. [See E. Derman, S. Goldberg, and J. E. Darnell, 1976, *Cell* 9:465.]

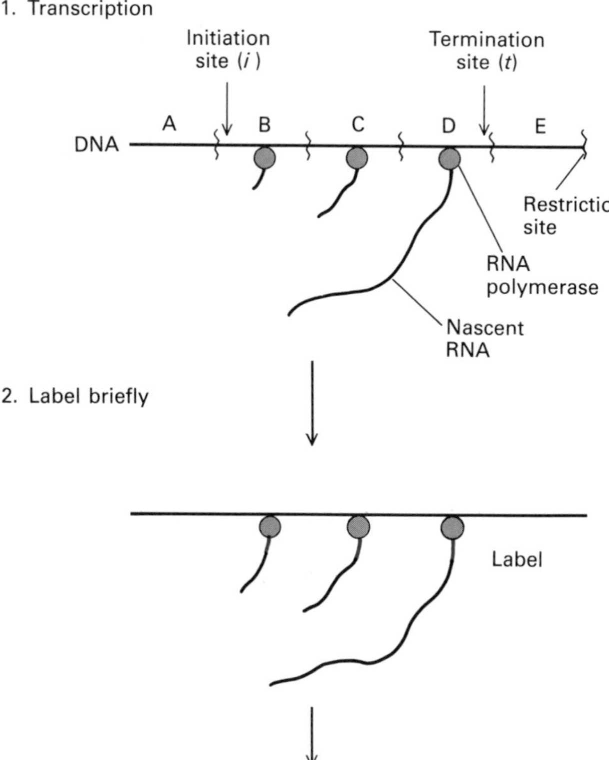

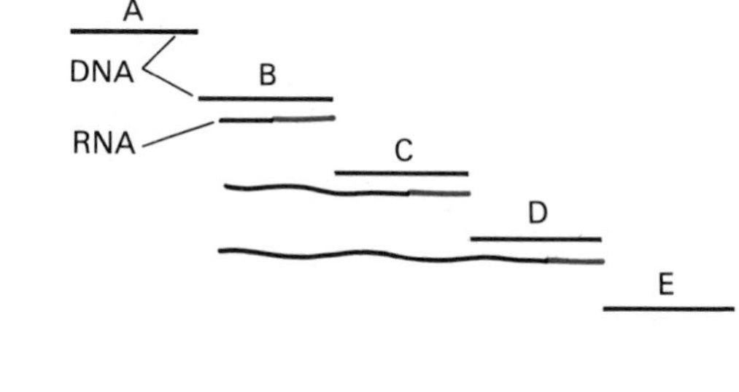

Figure 9-11 The mapping of a transcription unit by hybridization of labeled nascent RNA. A section of DNA with transcription start and stop sites is illustrated in the act of being transcribed. The introduction of radioactive label for a brief time produces labeled nascent RNA molecules. A sufficient quantity of the purified DNA (e.g., the DNA of a virus or cloned, recombinant DNA can be used; see Chapter 7) is then divided into an ordered set of fragments (A to E) by cutting at defined sites with a restriction endonuclease. Hybridization of the labeled nascent RNA to the DNA pieces defines which fragments lie within the transcription unit and in which order: the labeled regions of the shortest RNAs will hybridize to fragment B in the diagram, and the labeled regions of the longer RNAs will hybridize to succeeding fragments, showing that RNA synthesis begins in the B fragment and continues through fragments C and D.

proteins spread out; they can then be captured on a grid for viewing in the electron microscope. The transcribing RNA polymerases, trailing fibrils of pre-rRNA, can be easily seen (Figure 9-12). Such stunning electron micrographs have shown as many as 50 polymerases transcribing pre-rRNA from a single transcription unit. The transcription units for pre-rRNA were seen to exist in repeated tandem copies with nontranscribed regions separating them.

Other eukaryotic transcription units have been viewed in the electron microscope. The resulting micrographs have so far been used mainly to assess the average size of these other transcription units in comparison with ribosomal transcription units.

The electron microscope can also aid in assessing the relationship between primary transcription products and ultimate products. Views of 45S pre-rRNA have revealed that the molecule contains reproducible and characteristic stem-loop structures (caused by intramolecular complementarity). These structures are also visible in the final products, the 28S and 18S cytoplasmic rRNAs; thus the final products can be identified within their precursors (Figure 9-13).

Damage Caused by Ultraviolet Light Can Be Used to Map Transcription Units

A third means of measuring the transcription units that encode primary RNA transcripts is a physical-biological technique in which cells or viruses are irradiated by ultraviolet light. The technique, originally devised for the study of bacterial and bacteriophage genes, was first adapted for the study of eukaryotic genes in an examination of the formation of ribosomal RNA in cultured mouse cells. The irradiation of DNA by UV light causes a chemical reaction in the absorbing nucleic acid bases so that they form intrachain pyrimidine dimers, mainly between adjacent thymidylate residues. RNA polymerases terminate transcription at these dimers.

Any transcription unit that has been damaged by UV light behaves as if its termination site has been moved closer to its initiation site. Thus, the damaged transcription units produce short, incomplete RNA transcripts. RNA synthesis from regions near the promoter decreases, but not as much as synthesis from regions farther away (Figure 9-14). Measurement of the synthesis of rRNA in cultured, irradiated cells by the incorporation of labeled uridine shows that more 18S rRNA is synthesized than 28S rRNA (Figure 9-15). The difference indicates that the 18S rDNA segment is closer to the 5′ (promoter) end of the transcription unit than is the 28S segment. On the other hand, the amount of 28S rRNA synthesized is only slightly greater than the amount of full-length 45S pre-rRNA. Because any UV damage to the transcription unit will stop 45S synthesis, and because 28S synthesis is al-

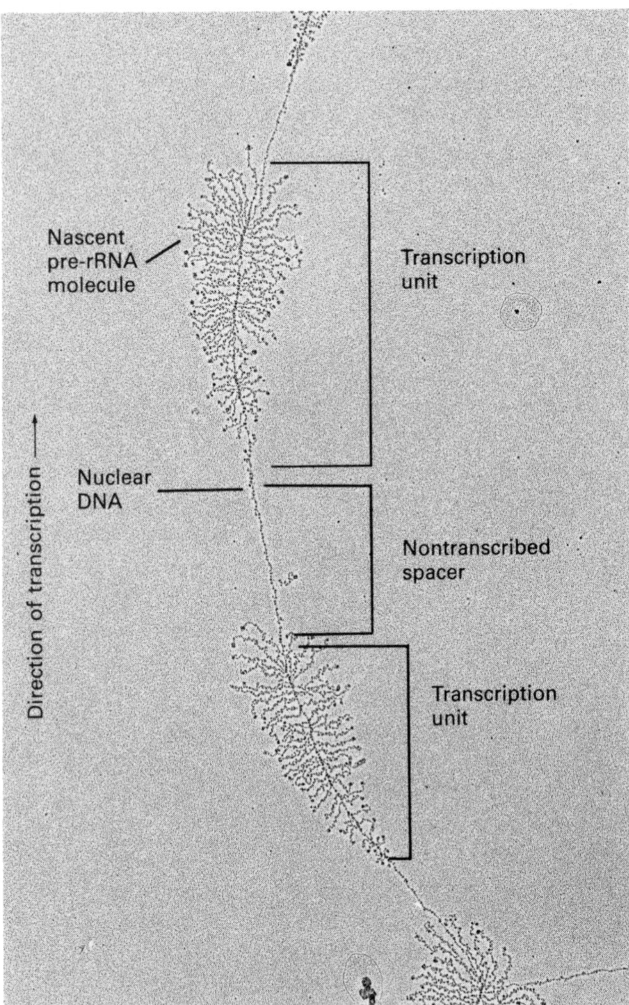

Figure 9-12 Ribosomal DNA transcription as viewed by electron microscopy. Frog oocytes were lysed and the nuclear contents were captured on the surface of an electron-microscope grid; this material was then dried and shadowed. Each "feather" represents nascent pre-rRNA molecules of varying lengths emerging from a single rDNA transcription unit. The tandemly arranged transcription units are separated by nontranscribed spacer regions. *Courtesy of Y. Oshein and O. L. Miller Jr.*

most as sensitive as 45S synthesis, the 28S rDNA must be near the end of the 45S transcription unit.

Another more subtle but equally important point was derived from the differential sensitivity of 18S and 28S rRNA synthesis to UV light. The transcription units for ribosomal RNA in vertebrate cells are arranged in long tandem arrays (see Figure 9-12). Suppose that the RNA polymerase could begin transcription only at the beginning of one of these long arrays of multiple ribosomal genes and not at the beginning of each individual transcription unit. After UV damage, all genes following the damage site would be knocked out, and both 28S and 18S synthesis would decline in parallel. But this does not happen: after any UV dose, more 18S than 28S rRNA is synthesized. Thus, many (maybe all) individual rDNA transcription units in a tandem cluster undergo transcription independently.

The Synthesis and Processing of Pre-rRNA

The ribosomal DNA from many different eukaryotic cells has been purified and cloned as recombinant DNA. We have, then, detailed information about these genes in many species. This information, coupled with that derived from studies of nucleolar RNA synthesis in a wide variety of plants, animals, and single-celled eukaryotes, has produced a comprehensive picture of ribosomal RNA formation.

The rDNA sequences in all eukaryotic cells show great similarities. Yet, important differences also distinguish the ribosomal genes from species to species. Of special interest are the nucleotide sequences around the initiation site for the pre-rRNA. Human rDNA has been compared with rat and mouse rDNA. Although the first 19 bases of

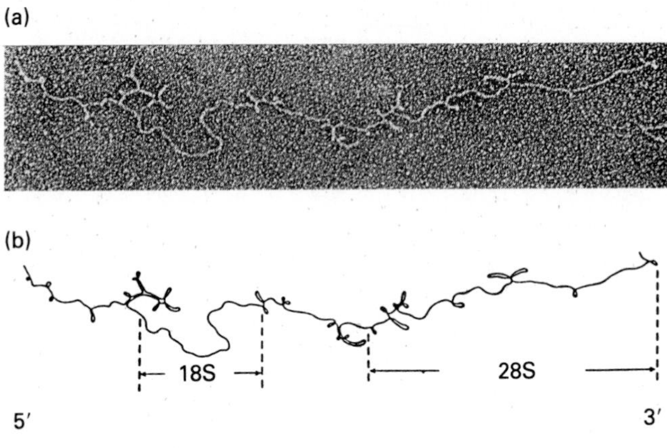

Figure 9-13 A 45S pre-rRNA molecule from a HeLa cell nucleolus. (a) The electron-microscopic image is obtained by spreading out the isolated RNA, as described in the text, and shadowing the molecules with a heavy metal. (b) The patterns of stems and loops in the 45S RNA are reproducible, and some of the same patterns can be seen in isolated 28S and 18S cytoplasmic rRNAs. By matching up characteristic regions of secondary structure in the 28S and 18S rRNAs to regions in the 45S RNA, a map of the positions of the product rRNAs within the precursor molecule was made. [See P. K. Wellauer and I. B. Dawid, 1979, *J. Mol. Biol.* **128**:289.] *Courtesy of P. Wellauer.*

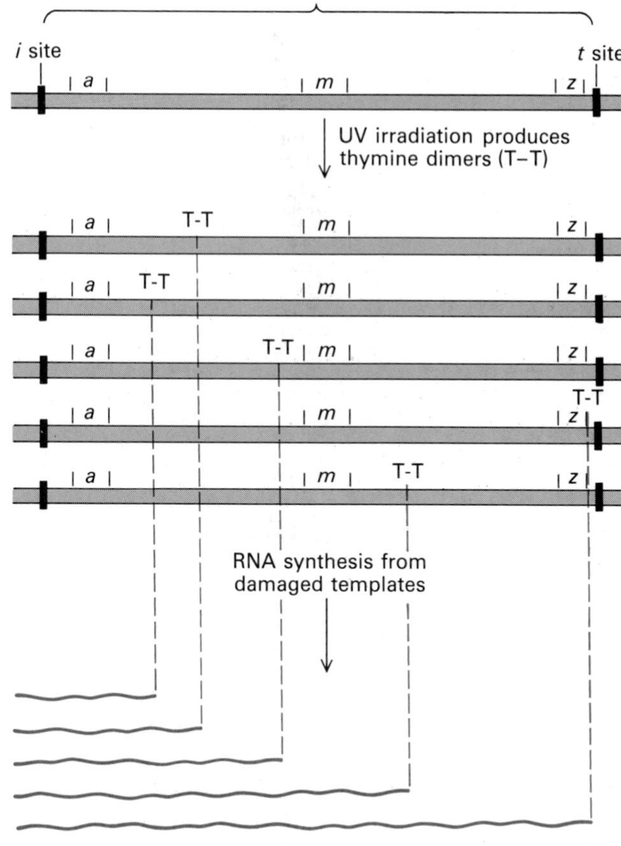

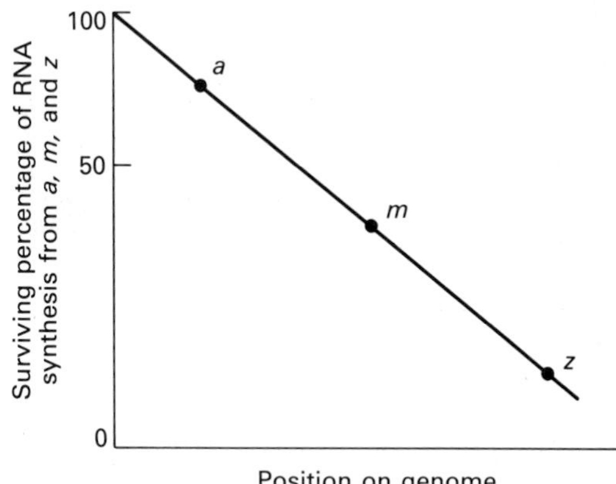

Figure 9-14 Transcription unit mapping by UV irradiation. Chemical changes induced in DNA by the UV light result in the formation of thymine dimers (T-T). In a long DNA molecule, this damage is distributed approximately randomly. Five damaged templates are illustrated. When a polymerase reaches a thymine dimer, the synthesis of RNA stops. RNA synthesis in region *a*, which is near an initiation site, is less likely to be stopped than RNA synthesis in regions *m* or *z*, which are farther from the initiation site; the chances that the polymerase will encounter a T-T dimer increase exponentially as it transcribes regions more and more distant from the initiation site. The equation relating the surviving RNA synthesis to the number of damage sites *(D)* in the transcription unit is

$$\frac{\text{Rate of RNA synthesis after irradiation}}{\text{Original rate of RNA synthesis}} = \frac{1 - e^{-D}}{D}$$

the pre-rRNA transcripts are very similar in the cells of all three animals, the sequences on the 5′ side of the initiation site are less well conserved between the human and the rodent genomes (Figure 9-16). This change in sequence is correlated with results of in vitro pre-rRNA synthesis. The polymerase I of mouse or human cells will transcribe either mouse or human DNA, but a species-specific transcriptional factor (a protein) is required. The binding site of this protein on the DNA is thought to reside upstream (to the 5′ side) of the initiation site, in the region where the sequences are not the same in rodent compared with human DNA.

Despite the differences in certain regions of the ribosomal genes, the similarities in ribosome biosynthesis in all eukaryotic cells remain striking. The synthesis of pre-rRNA by polymerase I always occurs within the nucleolus and always yields three cleavage products—28S, 5.8S, and 18S—which appear in ribosomes. Both the pre-rRNA and the final 28S and 18S molecules vary in size in different eukaryotic cells (Table 9-1). However, in all eukaryotic cells and even in bacteria, the transcription unit is longer than the sum of the two major finished rRNA molecules (Figure 9-17; Table 9-1).

In human cells, pre-rRNA is about 13.7 kb long, whereas 28S rRNA is 5.1 kb, 5.8S is 160 bases, and 18S is 1.9 kb. Thus only about half of the primary transcript appears in the final ribosomal products. Because the other half of the primary transcript does not accumulate, it must be subject to a rapid turnover or destruction. There is no known function for this *transcribed spacer RNA*.

Within the pre-rRNA of many different eukaryotes, the arrangement of the regions that are to be preserved and those that are to be discarded (the transcribed spacers) is very similar. A variable-sized region at the 5′ end is discarded; it is followed by the preserved 18S region. Spacer regions separate the 18S, 5.8S, and 28S regions.

In the ribosome, the small 5.8S rRNA fragment is hydrogen-bonded to the 28S rRNA. Nucleolytic enzymes digest the spacer sequences that separate the 5.8S and 28S regions and leave the two preserved regions hydrogen-bonded to each other. Recent results show that transcription does not terminate at the 3′ end of the 28S region but continues for at least a few hundred bases. Thus a cleavage is also presumably required to create the 3′ end of 28S rRNA.

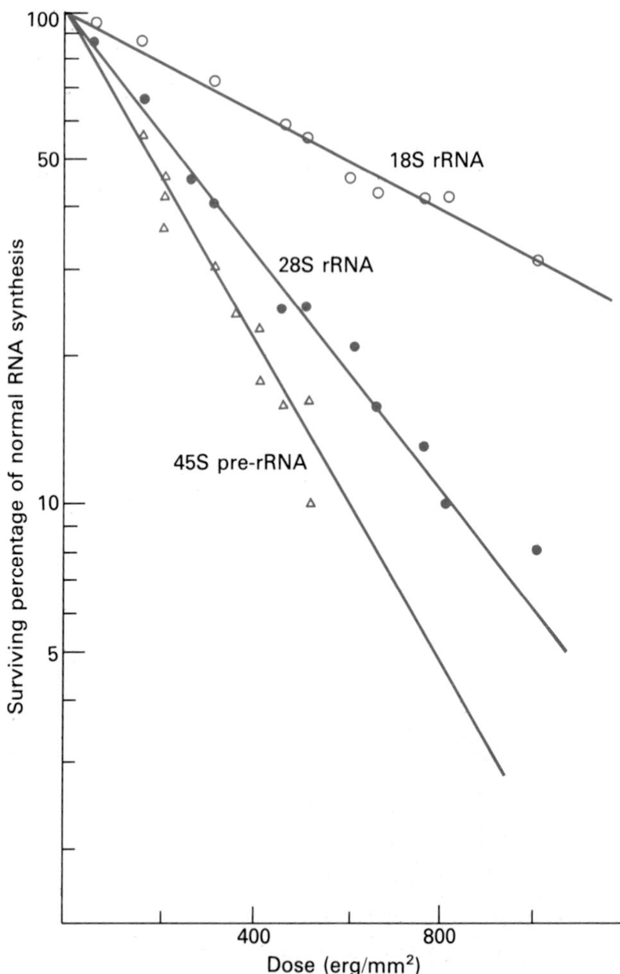

Figure 9-15 The mapping of ribosomal DNA (rDNA) transcription units in mouse cells by UV irradiation. Cultured mouse cells labeled with [^{3}H]uridine were exposed to no irradiation or to increasing doses of irradiation. The amounts of labeled 45S pre-rRNA, 28S rRNA, and 18S rRNA were measured in the irradiated and nonirradiated cultures, and the changes in the rates of synthesis after irradiation were calculated. The synthesis of 45S RNA was the most sensitive to irradiation, followed by the synthesis of 28S and then of 18S. Thus, the 18S rDNA is closest to the start site for RNA synthesis. [See P. B. Hackett and W. Sauerbier, 1975, *J. Mol. Biol.* **91**:235.]

As we shall discuss later in this chapter, in a few organisms the 28S rRNA molecules are spliced together from primary transcript regions that are separated by intervening sequences. So far as is known, this process occurs only in single-cell organisms (e.g., in *Tetrahymena pyriformis*) and in the mitochondrial rRNA of some fungi (e.g., *Neurospora crassa*). In any case, a very small proportion of rRNA biosynthesis involves RNA splicing. It is important to note that the spacer sequences that separate the 18S, 5.8S, and 28S regions are not in any way related to sequences removed by splicing.

Synthesis of Pre-rRNA Is Complete Before Processing

For any molecular precursor to accumulate in cells, it must be synthesized faster than it is processed. In mammalian cells, the synthesis of pre-rRNA by polymerase I requires from 3 to 4 min. The molecule is not cleaved, however, until the entire chain has been finished. The first cleavage removes only a few hundred nucleotides at the 5′ end; the first cleavage that markedly reduces the size occurs on the 3′ side of the 18S RNA and requires between 5 and 10 min. About 30 min are required to complete the cleavages (Figure 9-18). The small (40S) ribosomal subunit containing the 18S rRNA is finished and delivered to the cytoplasm within minutes of cleavage of the primary transcript—much faster than the large (60S) subunit is processed. The pool of nuclear precursors to large riboso-

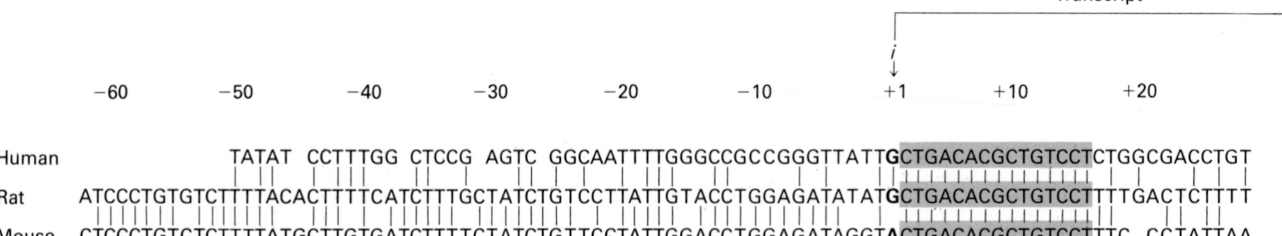

Figure 9-16 Base sequences around the starting sites for pre-rRNA synthesis in human, rat, and mouse DNA. (The index numbers refer to the rat sequences; the human and mouse sequences have been aligned with those of the rat in order to show the regions of consensus. The vertical bars indicate bases conserved between adjacent chains.) The sequences from position 2 through position 16 are identical in all three transcription units. However, upstream of the initiation site, the conservation of the DNA sequence is less striking, particularly between the human and the rodent sequences. [See L. I. Rothblum, R. Reddy, and B. Cassidy, 1982, *Nuc. Acids Res.* **10**:7345.]

Table 9-1 The ribosomal RNA molecules

	Primary transcript*		Ribosomal RNA length*		
	S value	Length (kb)	26S–28S (kb)	16S–18S (kb)	Percentage of precursor preserved
Escherichia coli (prokaryote)	30	6.0	3.0	1.5	75
Saccharomyces cerevisiae (yeast)	37	6.6	3.8	1.7	77
Dictyostelium discoideum (slime mold)	37	7.4	4.1	1.8	80
Drosophila melanogaster (fruit fly)	34	7.7	4.1	1.8	76
Xenopus laevis (frog)	40	7.9	4.5	1.9	81
Gallus domesticus (chicken)	45	11.2	4.6	1.8	57
Mus musculus (mouse)	45	13.7	5.1	1.9	51
Homo sapiens (human)	45	13.7	5.1	1.9	51

*The lengths of the various RNA molecules are estimates based on gel electrophoresis and direct measurements of electron micrographs.
SOURCE: B. Lewin, 1980, *Gene Expression*, vol. 3, Wiley, p. 867.

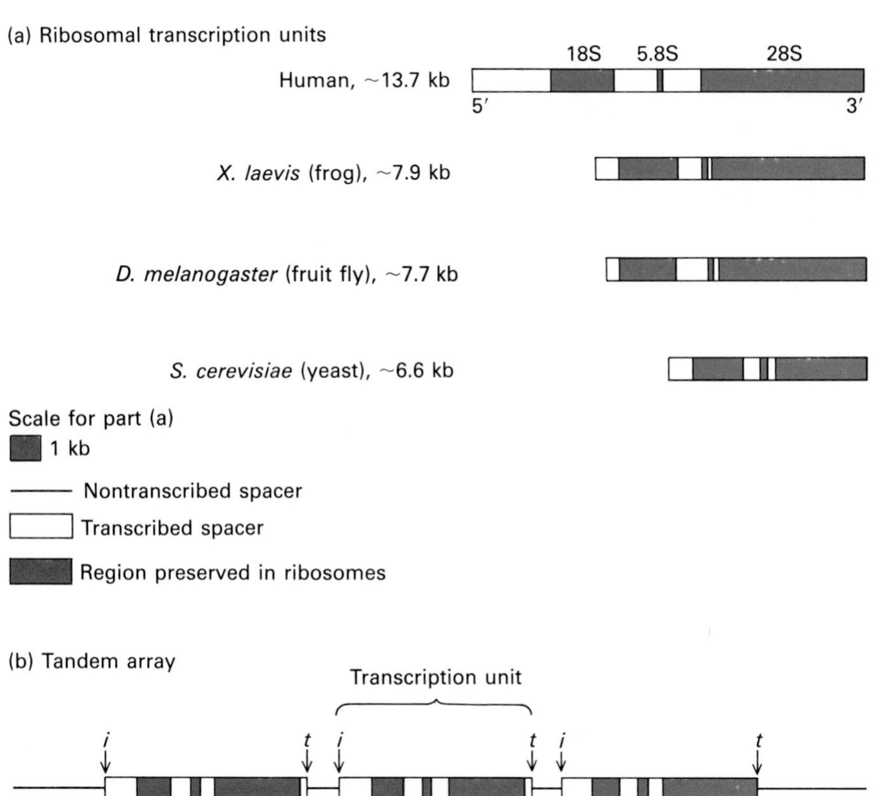

Figure 9-17 (a) The ribosomal transcription units of four eukaryotes. Variations in the lengths of the transcribed spacer regions *(white bars)* account for the major differences in the lengths of the transcription units. The regions that appear in ribosomes are always 18S, 5.8S, and 26-28S. (b) There are also wide variations in length *(not indicated)* in the nontranscribed spacer regions between transcription units. The nontranscribed regions can vary from 2 kb in frogs to 30 kb in humans. The genomes of all animals contain multiple tandem copies of the transcription units (as viewed in Figure 9-12).

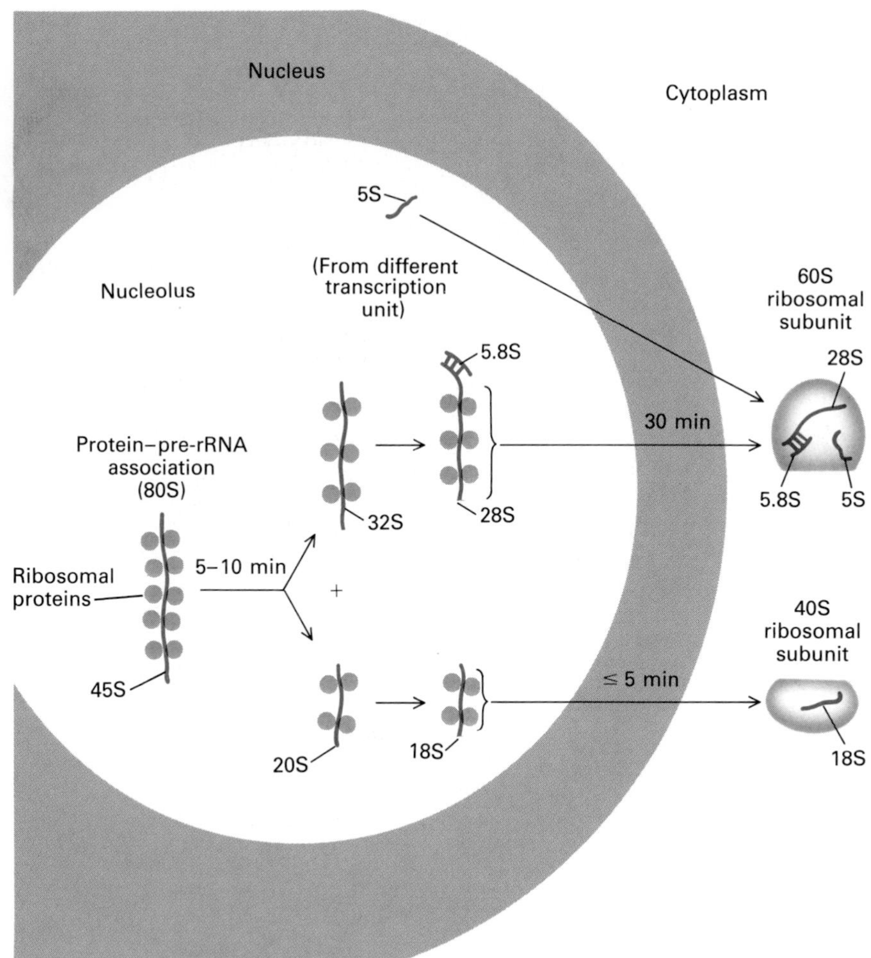

Figure 9-18 The mammalian ribosome contains four rRNAs: a 28S, a 5.8S, an 18S (these three are cleaved from the 45S precursor), and a 5S (this one is synthesized from an entirely separate transcription unit). The diagram shows the paths by which the 45S RNA is processed into three of the four rRNAs and the sedimentation values of the preribosomal and ribosomal particles.

Nucleoprotein particles that can be extracted from nucleoli contain uncut 45S pre-rRNA associated with ribosomal proteins. These nucleolar particles are larger (80S) than the finished ribosomal subunits. The overall time of assembly of the small (40S) subunit is 10 min or less, a much shorter interval than the 35- to 40-min synthesis time of the large (60S) subunit. Thus, the pool of precursors to large subunits (i.e., of 32S pre-rRNAs) is larger than that of precursors to small subunits (i.e., of 20S pre-rRNAs). Note, for example, that in Figure 9-5 there is a substantial nucleolar peak for 32S pre-rRNA, whereas there is no peak for 20S pre-rRNA. [See J. Warner and R. Soeiro, 1967, *Proc. Nat'l Acad. Sci. USA* 58:1984.]

mal subunits is thus larger than that for small ribosomal subunits. In fact, the most prominent RNA component in the nucleolus is the 32S pre-rRNA, the intermediate precursor to the 28S rRNA molecule. In growing HeLa cells, there are about five times as many 32S molecules as there are intact 45S primary transcripts.

Although pre-rRNA is not cut before it is finished, some steps in processing take place before the end of transcription. In human cells, more than 100 methyl groups are added to specific bases and to the riboses of specific ribonucleotides; most of these methyl groups can be detected even in nascent pre-rRNA molecules. In addition, a few methyl groups are added to 45S pre-rRNA molecules after synthesis is finished, and four methyl groups are added to the 18S rRNA in the cytoplasm—that is, after cleavage. All of the methyl groups added to the primary transcript are preserved during processing (i.e., they are found in 28S and 18S rRNA in the cytoplasm).

The reason for rRNA methylation is not known with certainty. Perhaps the methylases that add the methyl groups have access only to sequences in the preserved region. In any case, it seems likely that the added methyl groups play a role in the preservation of those sequences. If cells are deprived of methionine, an intermediate in the transfer of methyl groups, the processing of pre-rRNA is interrupted. The sequences in the methylated regions are highly conserved among different species, as is the methylation pattern. This, too, suggests that methyl groups play a role in pre-rRNA processing.

The pre-rRNA apparently binds proteins, including most of the ribosomal proteins found in completed ribosomes, while it is being synthesized in the nucleolus. The RNA is probably never completely free of associated proteins. In electron micrographs of the nucleolus, both granular and fibrillar structures can be seen (Figure 9-19). Which structures represent pre-rRNA is not clear. Despite extensive attempts, ribosomal proteins and rRNA from eukaryotic ribosomes have not been successfully separated and recombined in the test tube.

Very possibly the proper rRNA-protein interaction will occur only if the proteins are added in a certain order during pre-rRNA synthesis. The assembly of bacterial ribosomal subunits from purified, separated components does require such a step-by-step addition (see Figures 4-18 and 21-10).

When the synthesis of ribosomal particles in the nucleolus is complete, they move—by mechanisms as yet unknown—to the cytoplasm, where they appear first as free subunits (see Figure 9-18). Since the nuclear pores (to be discussed later in this chapter) are only about 10 to 20 nm in diameter and the large ribosomal subunit is about the same size, there may be a change in the shape of one of the two structures to admit the subunit to the cytoplasm.

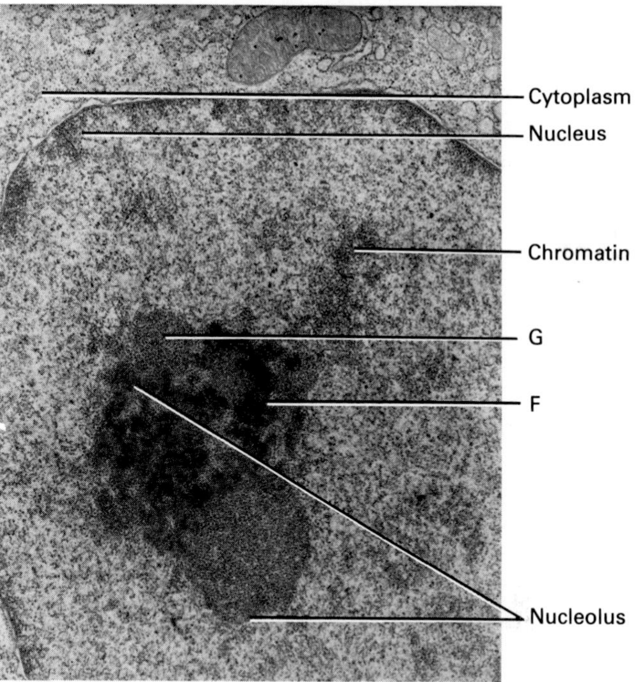

Figure 9-19 An electron micrograph of a nucleolus, showing granular *(G)* and fibrillar *(F)* structures. Which structures correspond to biochemically defined nucleolar structures is not yet clear. *Courtesy of K. Porter.*

Transcription by RNA Polymerase III

Eukaryotic cells contain many dozens of RNA molecules 300 nucleotides or less in length. Most of these molecules are synthesized by RNA polymerase III outside of the nucleolus of the cell. Although the functions of many of these short molecules are not yet known, we do know that the most common RNAs in the cell, the tRNAs, are included in this group, as is the 5S rRNA.

The Synthesis of 5S rRNA Is Well Understood

In addition to the 18S, 5.8S, and 28S rRNAs, which come from pre-rRNA, eukaryotic ribosomes also contain a short 120-nucleotide molecule known as 5S rRNA. The 5S rRNA is transcribed from a gene that is separate from the pre-rRNA gene. The sequence and probably the secondary structure of the 5S rRNA are highly conserved in many different eukaryotic cells. In fact, the 5S rRNA molecules are similar in *all* cells (Figure 9-20) and in mito-

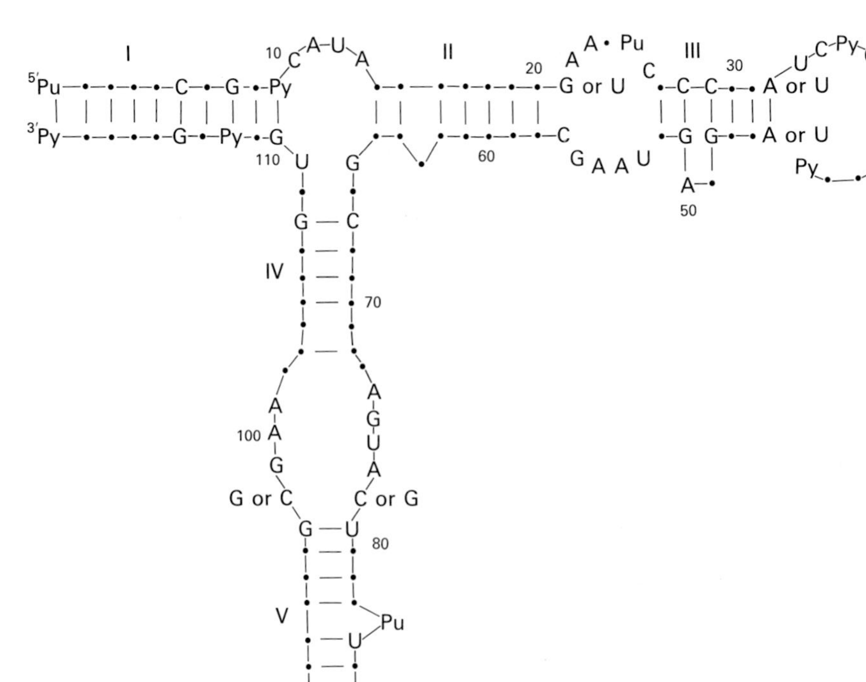

Figure 9-20 A generalized structure of 5S rRNA obtained from an analysis of over 30 5S molecules of animals, plants, and single-celled organisms. There are five stem-loop regions (I to V); lines connect bases that are hydrogen-bonded. (Pu = a purine, and Py = a pyrimidine.) Where dots appear for bases, the sequence may vary but the base pairing is maintained. [See N. Delihas and J. Andersen, 1982, *Nuc. Acids Res.* **10**:7323.]

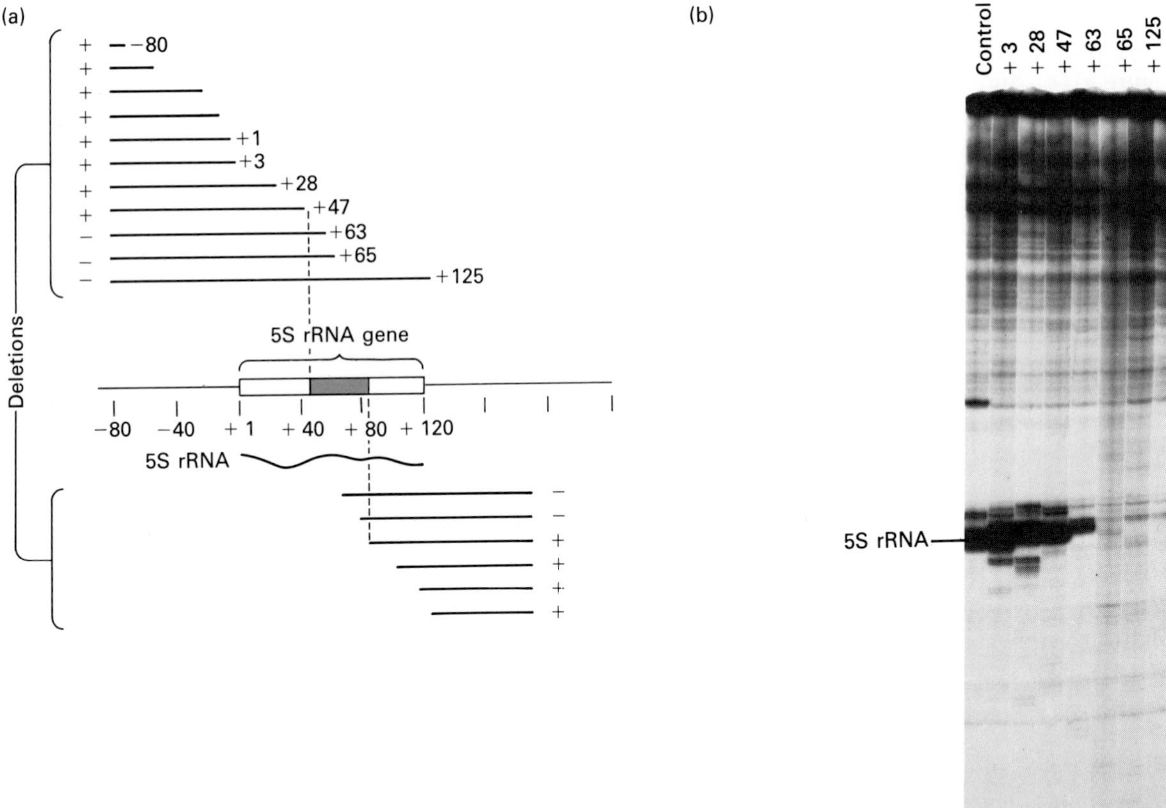

Figure 9-21 (a) A schematic diagram showing the frog 5S gene region *(color)* that is necessary for accurate initiation of transcription in extracts of frog cells. The 120 bases encoding the 5S rRNA are represented by the box. The plus signs on the extreme right or left of the diagram indicate deletions that do not affect transcription; minus signs indicate deletions that stop transcription. The horizontal lines above and below the gene indicate the region deleted in each of a series of deletion mutations (Chapter 7). For example, the top line indicates a deletion extending from upstream to −80 [80 bases preceding the first nucleotide (+1) of the 5S gene]. The numbers +1, +3, +28, etc. indicate the endpoints of the deletions that start upstream. The deletions in the bottom part begin within the gene and the sequences upstream are intact. (b) An autoradiograph of electrophoretically separated 5S rRNA transcripts from normal 5S genes and from deletion mutants. The dark band labeled 5S is the major labeled RNA product from the normal gene and from some of the mutants. The mutant gene products represent some of the deletions diagrammed in part (a); transcription is not affected by deletion of the first 47 bases of the gene, but it is greatly decreased by deletions after that. (In the deletions to +3, +28, and +47, RNA polymerase III must start at DNA sequences that have been inserted to replace the normal start site.) *Photograph from S. Sakonju, D. F. Bogenhagen, and D. D. Brown, 1980,* Cell ***19**:27.*

chondria and chloroplasts, as we shall discuss further in Chapter 25. In eukaryotic chromosomes, the genes for 5S rRNA are not contiguous with the other ribosomal genes, but are clustered in long tandem arrays elsewhere in the chromosomes. (The arrangement of the various ribosomal genes is described in more detail in Chapter 11.)

The transcription of 5S rDNA is catalyzed by RNA polymerase III. According to the findings from studies of frog cells, the RNA is not processed. The 5′ end of the RNA is a pppG nucleotide, which indicates that the RNA polymerase III starts with this guanylate residue; the structure of the 3′ end of the molecule likewise indicates that it is formed by termination of an RNA polymerase. Thus, the primary transcript of 5S rRNA in frog cells is a functional molecule, which may make it unique among eukaryotic RNA molecules.

The 5S rRNA can be accurately synthesized in vitro by purified RNA polymerase III and chromatin from frog cells. Also, if purified 5S rDNA is injected into frog oocytes, or if recombinant DNA that includes 5S rDNA is added to extracts of frog oocyte nuclei, endogenous RNA polymerase III catalyzes the transcription of the 5S genes.

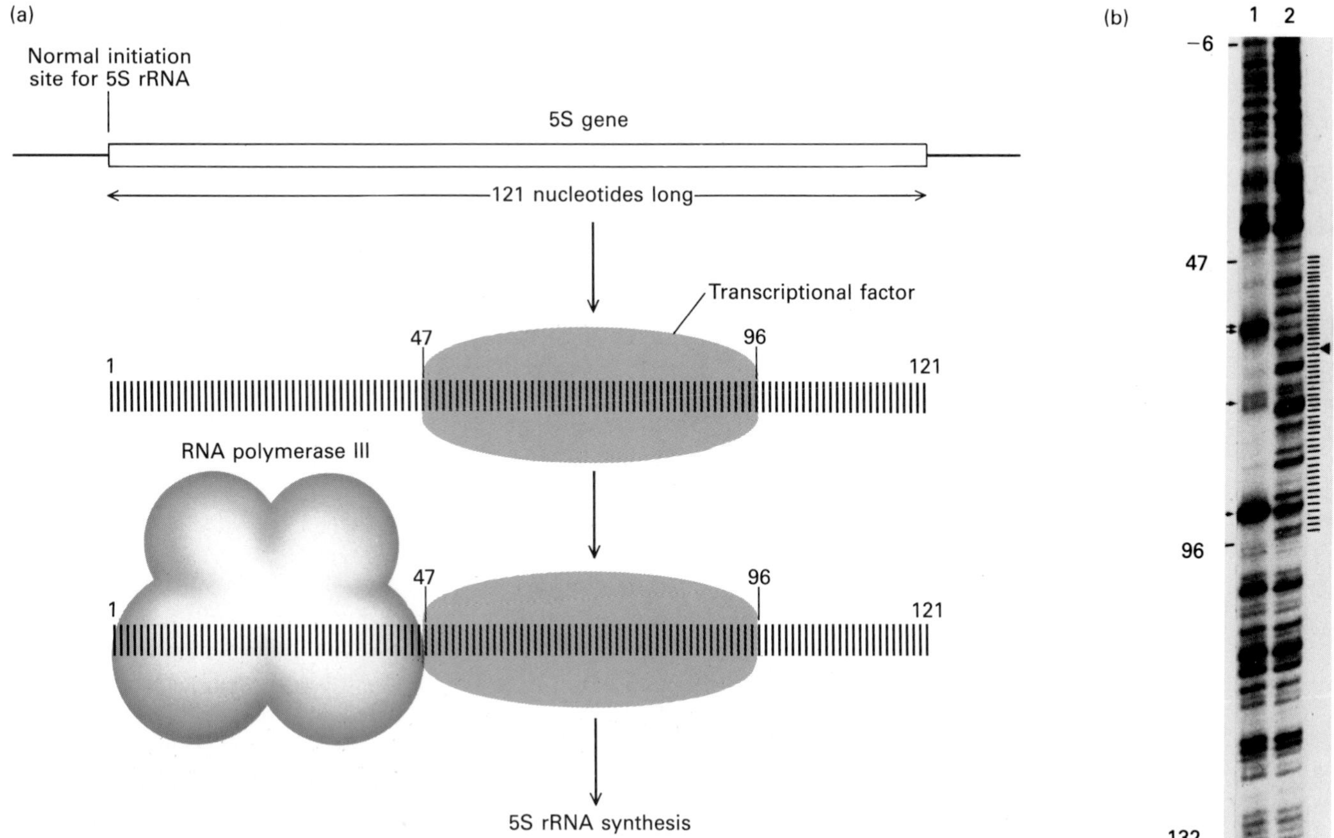

Figure 9-22 (a) A schematic diagram showing the relation between the binding site for the transcriptional factor *(gray)* and that for RNA polymerase III *(color)* on the 5S rRNA gene. (b) An autoradiograph demonstrating protection by the transcriptional factor. The protein was purified and mixed with 5′ end-labeled 5S rDNA. This mixture *(lane 1),* as well as end-labeled 5S rDNA by itself *(lane 2),* was digested briefly with DNase, which resulted in about one cut per 150 bases. Each cleavage produced an end-labeled band stretching from the end to the cleavage site. The sample with the bound protein displayed a lack of bands in the region from base 47 to base 96 on the 5S gene sequence; this "footprint" indicated that these bases were protected from digestion. The four bands that do appear in the protected region in lane 1 *(small arrowheads)* are not present in the absence of the transcriptional factor; these "hypersensitive sites" represent DNA regions that become *more* sensitive to DNase as a result of the binding of the transcriptional factor. [See R. G. Roeder, 1980, *Cell* **19**:717.] *Photograph courtesy of R. G. Roeder.*

Such assays have proved useful in defining the sequences necessary for the correct initiation and termination of transcription of 5S rDNA.

A series of recombinant DNA plasmids that contain either the entire 5S gene or varying deletions in the region of the 5S gene can be assayed for transcription. Surprisingly, the removal of any part of a sequence of 40 to 50 bases *within* the gene greatly inhibits the initiation of transcription (Figure 9-21). A transcriptional factor—a DNA-binding protein necessary for the action of RNA polymerase III—binds to this region, which begins about 47 bases past the initiation site for the transcription of the 5S gene. Almost any recombinant DNA containing an intact binding region from within the 5S gene will be transcribed by RNA polymerase III, although transcription will not be at the maximum rate if the normal 5S start sequences are missing from the 5′ end. With reconstructed DNA molecules lacking the normal RNA initiation site but containing the binding site for the transcriptional factor, RNA initiation starts 47 bases upstream from the first base of the protein-binding site, regardless of what sequences are present there.

The accessory protein that is required for 5S rRNA synthesis was the first transcriptional activation protein to be purified from eukaryotic cells. When this purified protein is allowed to interact with purified DNA containing the 5S gene, the DNA is protected from chemical damage within the 40- to 50-base-pair region that has been shown to be important for the initiation of transcription (Figure 9-22).

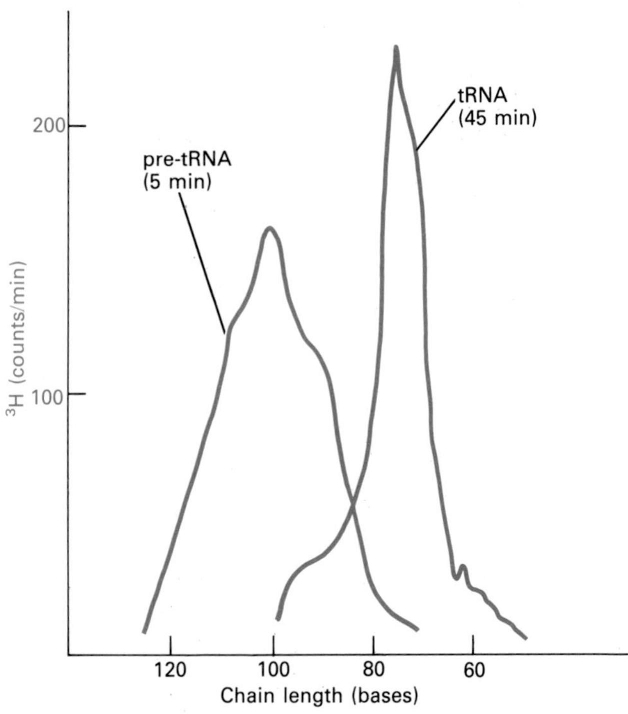

Figure 9-23 The conversion of precursor tRNA into tRNA. HeLa cells labeled for 5 min with [^{3}H]uridine contain short RNA molecules about 80 to 120 nucleotides long; these molecules can be separated by gel electrophoresis and then assayed for radioactivity. Other experiments have shown these molecules to have some of the base modifications (methylations and pseudouridines) characteristic of tRNA. Addition of actinomycin D to stop RNA synthesis followed by a 40-min incubation results in the disappearance of the 80- to 120-nucleotide class and the appearance of labeled RNA in the tRNA size class (70 to 80 nucleotides). [See J. E. Darnell, 1968, *Bacteriol. Rev.* **32**:262; and D. Bernhardt and J. E. Darnell Jr., 1969, *J. Mol. Biol.* **42**:43.]

Several interesting but as yet unexplained facts about 5S rRNA are known. First, the 5S rRNA is not formed coordinately with the other RNA molecules found within the ribosome. Not only are the 5S genes separate from the precursor rRNA genes on chromosomes, but in human cells, for example, the rate of synthesis of 5S rRNA is about five times that of pre-rRNA. Because only one 5S molecule is associated with each ribosome, an excess nuclear pool of 5S rRNA exists. The unused 5S molecules in the nucleus are degraded back into nucleotides.

Some animals—perhaps most—have several sets of 5S genes whose sequences differ in only a few nucleotides out of the 120. For example, frogs have two major sets and one minor set of tandemly repeated 5S genes. One major set is transcribed during the formation of the oocyte, which produces a huge number of ribosomes from its multiple tandem arrays of ribosomal genes (such amplified genes are discussed in Chapter 11). The other major set of 5S genes is transcribed in all other cells of the animal. One of the experimental goals of developing an accurate cell-free synthesis system for 5S rRNA is to define the factors that control each set of genes in transcription. The protein factor required for 5S transcription binds to both the somatic and the oocyte 5S genes, and therefore cannot by itself be the controlling element that determines which 5S genes are to be active.

The Synthesis of tRNA Reveals a Dual Role for Conserved tRNA Regions

Next to ribosomal RNA, transfer RNA is the most plentiful RNA by weight; it makes up about 15 percent of the total RNA of the cell, compared to 80 percent for rRNA. The tRNAs average only 75 to 80 nucleotides in length,

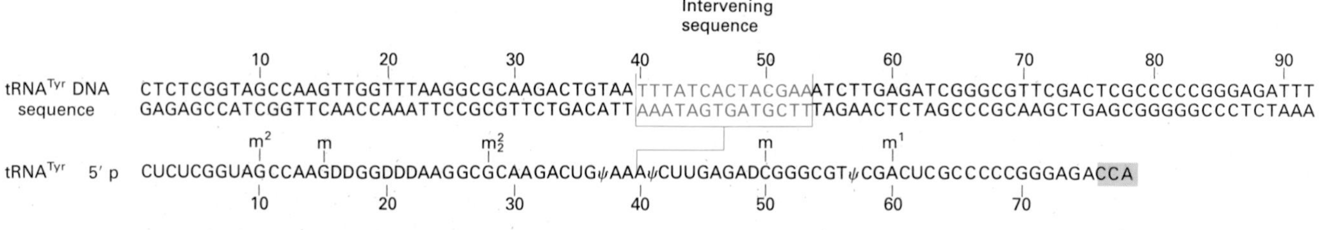

Figure 9-24 The sequence of a tyrosine tRNA gene from yeast and the tRNA^{Tyr} sequence found in the cytoplasm of yeast cells. In the DNA there is a 14-base intervening sequence that does not appear in the tRNA. (Note that the CCA 3′ terminus of the mature tRNA is not encoded by the DNA.) The modified bases in the tRNA are as follows: D = dihydrouridine, ψ = pseudouridine, m = methyl group on ribose, m^2 = methyl on position 2 of base, m_2^2 = dimethyl on position 2 of base, and m^1 = methyl on position 1 of base. [See H. M. Goodman, M. V. Olson, and B. D. Hall, 1977, *Proc. Nat'l Acad. Sci. USA* **74**:5453.]

however, so there are actually many more tRNAs than rRNAs. Many tRNA genes exist in multiple copies in chromosomal DNA, although they are not so highly repeated as 5S rRNA genes are. All of the primary transcription products of tRNA genes studied so far are, like pre-rRNA, larger than the final products. Thus, precursor tRNA molecules require shortening as they mature. They also require many other chemical modifications, such as the addition of methyl and isopentenyl groups and the conversion of uridines into pseudouridines (Chapter 4). In fact, about 1 base in every 10 of a tRNA molecule is modified in some way.

In the first experiments that defined precursor tRNA and its relation to product tRNA, the drug actinomycin D was used to stop all RNA synthesis without affecting the processing of existing precursors into products. The precursors were identified because they were longer than tRNA and contained some of the same modified bases, and because in the presence of actinomycin D these longer molecules gradually decreased in number while the amount of tRNA increased (Figure 9-23).

Some tRNA genes in eukaryotic cells do not exist as a consecutive stretch of DNA. For example, to form the yeast phenylalanine and tyrosine tRNAs, a segment must be removed from the middle of the primary transcripts, and the two ends must be spliced together (Figure 9-24). The enzymes for tRNA splicing have been partially purified from yeast cells. The action of these enzymes is described later in this chapter. (RNA splicing was first discovered in studies of mRNA; the enzymatic events in the splicing of different types of molecules are discussed together to facilitate comparison; see page 353.)

The transcription of tDNA (the genes that encode tRNA) is catalyzed by RNA polymerase III. As has been done with 5S rDNA, the tDNA sequences from a number of different eukaryotic organisms have been cloned as recombinant DNA molecules in *E. coli*. After injection into the nuclei of frog oocytes, recombinant DNA molecules containing the yeast tyrosine tDNA sequence can be transcribed by the frog RNA polymerase III in the oocyte nucleus. Moreover, an incompletely processed RNA transcript can be isolated and injected, whereupon it can be processed into a finished, spliced molecule. The processing steps include the addition of CCA to the 3′ ends of the yeast tRNAs, and the addition of various other chemical structures such as methyl and isopentenyl groups. Thus the recognition of the tRNA by the appropriate enzymes extends over a wide evolutionary range.

Because the nucleus can be manually removed from an intact oocyte, it can be shown that only mature tRNA and not precursor tRNA enters the cytoplasm (Figure 9-25). Even if the precursor molecules are injected directly into the cytoplasm of an oocyte with a nucleus, they are not processed. Thus, the location of tRNA processing is in the nucleus. A similar conclusion was

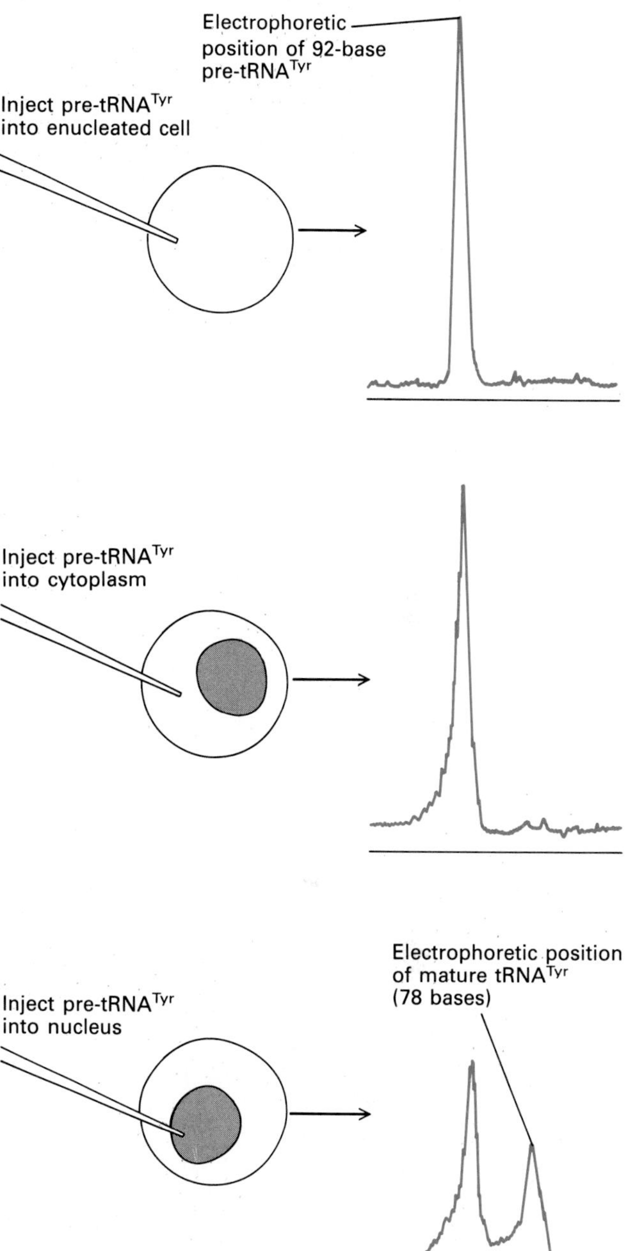

Figure 9-25 An experiment demonstrating that pre-tRNA is spliced in the nuclei of frog oocytes. The purified 92-base-long unspliced precursor of yeast tRNA^{Tyr} was injected into frog oocytes. In one sample, the oocyte nucleus had been removed; in two other samples, the precursor was injected into the cytoplasm or the nucleus, respectively, of a normal oocyte. The oocyte RNA was extracted within 1 h and any possible processing of the labeled RNA was then detected by gel electrophoresis. Only when the precursor was injected into the nucleus was the shortened spliced form—that is, a molecule of 78 nucleotides, the same size as mature tRNA—detected. [See D. A. Melton, E. M. DeRobertis, and R. Cortese, 1980, *Nature* 284:143.]

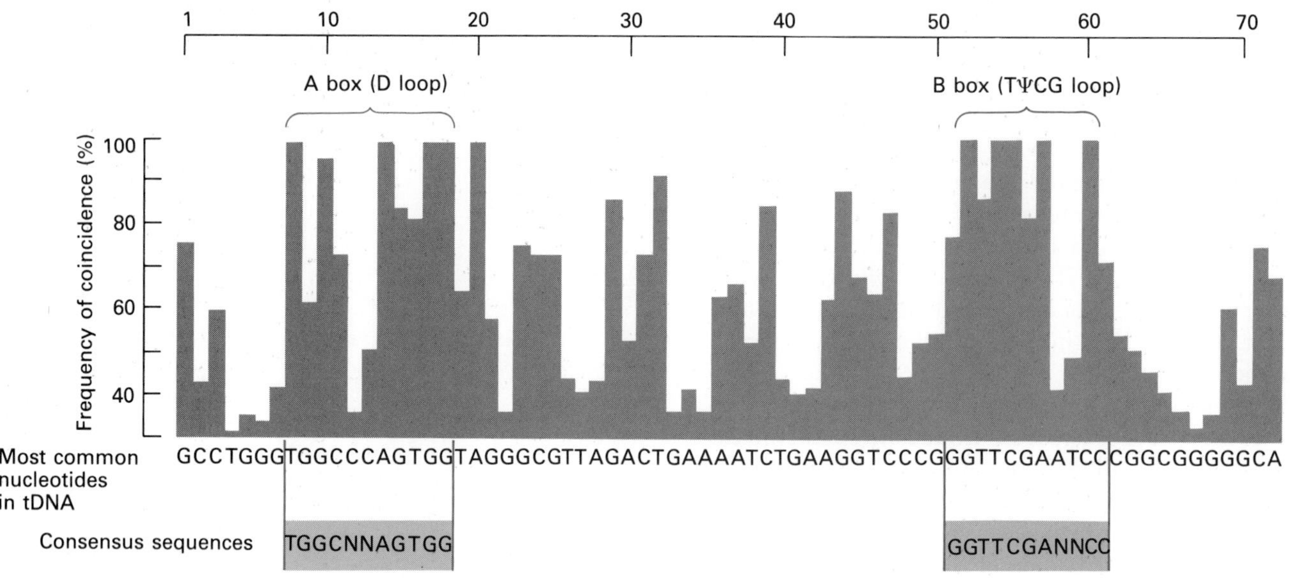

Figure 9-26 The consensus sequences in the +10 to +20 and +50 to +60 regions of 80 different eukaryotic tRNA genes show the strong conservation of nucleotides in these regions, which are often referred to as the A box and the B box, respectively. The A box encodes the D loop of the tRNA, and the B box encodes the TΨCG loop; both loops are constant features of tRNA. The A and B boxes also serve as parts of the promoter in tRNA genes. [See D. H. Gauss and M. Sprinzl, 1981, *Nuc. Acids Res.* 9:r1; and G. Galli, H. Hofstetter, and M. L. Birnstiel, 1981, *Nature* **294**:626.]

reached when cultured mammalian cells were fractionated in nonaqueous solvents in which nuclear contents did not "leak out" during the fractionation. Precursor tRNA was found only in the nucleus, and completed tRNA only in the cytoplasm.

With the availability of an assay for the production of tRNA in frog oocytes, the sequences required for the transcription of several tRNA genes have been determined. The removal of either of two regions of a cloned tRNA gene greatly diminishes transcription. These two regions—located between positions +10 and +20 and between +50 and +60 of the mature tRNA molecule—correspond to the D and the TΨCG stem-loop structures, respectively, in the tRNA (see Figure 4-9). The corresponding DNA sequences are highly conserved (Figure 9-26). Apparently these conserved sequences not only encode the structural components needed for the transfer of amino acids by tRNA molecules, but also constitute protein-binding sites in the DNA. Thus a dual role is now known for these highly conserved regions: one role in structuring functional tRNA and one role in promoting the transcription of the tRNA genes.

The Nucleus and the Cytoplasm of Eukaryotic Cells Contain Many Small RNAs with Uncertain Functions

Many types of small RNA molecules other than tRNA and 5S rRNA have been detected in a variety of eukaryotic cells, from *Drosophila melanogaster* cells to mammalian cells, but no clear physiological role has yet been defined for many of them (Figure 9-27). In the cell, most (if not all) of the small RNAs interact with specific proteins to form ribonucleoprotein particles (RNPs). These are called snRNPs or scRNPs, for small nuclear (or cytoplasmic) ribonucleoprotein particles. People with the disease *lupus erythematosus* develop antibodies against their own cell constituents. These antibodies are often directed against the proteins contained in the sn- and scRNPs. When blood serum from different affected individuals is mixed with cell extracts, different classes of snRNPs precipitate (Table 9-2; Figure 9-28). The availability of these antibodies has been of great assistance in characterizing the small ribonucleoproteins. Many of these small RNA molecules are probably formed by RNA polymerase III, because they contain at their 5′ ends the unmodified pppN structure that is characteristic of polymerase III products, including 5S rRNA, as we have mentioned. However, some of the small RNAs may be formed by polymerase II, as indicated by their different type of 5′ end structure. Several of the latter, such as U1, U2, and U3, have been sequenced; their 5′ termini contain a 2,2,7-methylguanylate residue that is part of a structure called a *cap*: $m^{2,2,7}$GpppN (Table 9-2). The guanylate residue, G, is in a 5′ → 5′ linkage with the first nucleotide, N, of the RNA chain. The cap structure is also found in nonribosomal nuclear RNA (hnRNA) and in mRNA. Synthesis of these latter groups of molecules is catalyzed by RNA polymerase II, which is why it is suggested that

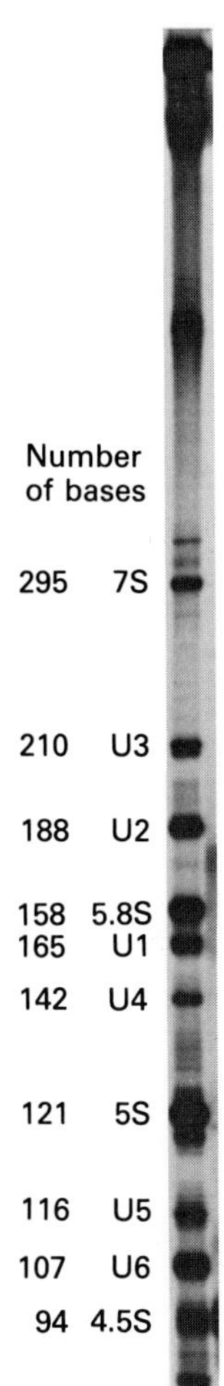

Figure 9-27 The small RNA molecules from cultured Chinese hamster cells. RNA labeled with $^{32}PO_4$ was purified from the nuclei of cultured cells and subjected to gel electrophoresis. Autoradiography has revealed the various small RNAs. Almost all of the RNAs identified here have been sequenced. The species labeled U1 to U6 are rich in uridylates. U3 is found mainly in the nucleolus. The 5.8S and 5S are the ribosomal molecules, and the 7S belongs to a particle that assists in transporting proteins through membranes. These last three molecules are present mainly in the cell cytoplasm. The 4.5S RNA, which is found in both the nucleus and the cytoplasm, may function in the initiation of protein synthesis. *Courtesy of W. Jelinek and S. Haynes.*

the capped small nuclear molecules also are polymerase II products (see pages 333–334).

It is very likely that at least several of the small nuclear RNA molecules play a role (probably as parts of ribonucleoprotein particles) in processing larger molecules. The U3 snRNA, which is found in the nucleolus, is thought to be associated with pre-rRNA. The U1 and U2 snRNPs are associated with high-molecular-weight hnRNA that is processed to yield mRNA. (The role of these snRNPs is discussed later in this chapter.) Yet another small nuclear RNA—labeled 4.5S in Figure 9-27—is found associated with mRNA that has just entered the cytoplasm; this snRNA may be involved in initiating protein synthesis. Perhaps the most clearly identified role for any of the RNPs is in protein secretion. The 7S cytoplasmic RNA, a chain of 294 nucleotides, combines with a set of nonnuclear proteins to function in transporting nascent proteins across membranes during their secretion (see Figure 21-35). Further study will certainly uncover direct evidence about the roles of these small RNA molecules.

The First Isolation of Eukaryotic mRNA

Despite the great interest in the studies on the synthesis of rRNA and tRNA, it was considered of even greater importance to learn how mRNA is made in eukaryotic cells, because the array of mRNAs in a cell determines which proteins the cell makes. The discoveries in the 1960s that tRNA and most rRNAs were formed by the cleavage of precursor RNAs suggested that nuclear precursors to mRNA might also exist. Newly labeled nuclear RNA had been divided into pre-rRNA molecules (located in the nucleolus) and the hnRNA molecules (located in the nucleoplasm) whose lengths ranged from 2 to 30 kb. Also the hnRNA fraction had an average base composition like that of genomic DNA, suggesting that hnRNA was made up of many different molecules, as also might be expected of the total mRNA of a cell. Were these hnRNA molecules precursors to mRNAs? Early in the 1960s, cell-free translation of eukaryotic mRNA had not been achieved, so direct tests of mRNA function could not be attempted. In their place, however, experiments designed to detect mRNA as a labeled RNA species were begun, so that the sizes and average base compositions of mRNA and of hnRNA could be compared.

Polyribosomes Contain the Cell's mRNA

The first step in studying the origin of eukaryotic mRNA was the application of a cell fractionation procedure that would yield purified mRNA. Because active mRNA molecules should be associated with ribosomes in the process of making proteins, cells pulse-labeled with radioactive amino acids were fractionated to isolate ribosomes containing new protein chains. Almost all of the newly labeled protein was associated with groups of ribosomes, called *polyribosomes* or *polysomes*. If the cell extracts were treated briefly with ribonuclease, the polyribosomes were broken down into single intact ribosomes with nascent proteins still attached. This result suggested that the

Table 9-2 RNA content of small ribonucleoprotein particles

	Ribonucleoprotein			
	RNA portion			
Antibody*	**Components†**	**Chain length (bases)**	**5′ end‡**	**Molecular weights of antigenic proteins × 10^{-3}**
Anti-Sm	U2	196	m_3Gppp	32, 28, 16, 13
	U4	125	m_3Gppp	(12, 11 also present
	U5	118	m_3Gppp	on Sm snRNPs)
	U6	106	?	
Anti–U1 RNP	Human U1	165	m_3Gppp	68, 33, 22
Anti-*rho*	Mouse Y1	~110	pppG	Not characterized
	Y2	~95	pppG	
	Human Y1, Y2	~110	pppG	Not characterized
	Y3	~110	pppG	
	Y4	~95	pppG	
	Y5	~90	pppA	
Anti-La	Many cellular RNAs, including:			Not characterized
	Mouse $4.5S_I$, $4.5S_H$	96, 94	pppG	
	tRNA precursors	80–100	ppp$\binom{A}{G}$	
	5S rRNA	~122	pppG	

*Each antiserum is from a patient with lupus erythematosus (see text). Such antibodies are in wide use to specifically precipitate different classes of small ribonucleoprotein particles.

†All of the RNAs except those in the Y series are nuclear; the Y series is cytoplasmic.

‡m_3G = 2,2,7-methylguanylate.

SOURCE: M. R. Lerner and J. A. Steitz, 1981, *Cell* 25:298.

groups of ribosomes were attached to a separate RNA molecule.

The sizes of the polyribosomes varied among cells making different proteins (Figure 9-29). In reticulocytes, which synthesize mainly globin, which is about 140 amino acids long, the polyribosomes contained three to five ribosomes. In cultured HeLa cells, which make a broad array of proteins with an average size of 400 amino acids, the size of the polyribosomes varied from 3 to about 20 ribosomes. HeLa cells infected with poliovirus, the RNA of which is 7000 nucleotides long, had very large polyribosomes with 40 or more ribosomes. The sizes of these various polyribosomes as determined by the speed of sedimentation were confirmed by electron microscopy (Figure 9-30). It seemed very likely that each polyribosome contained a single mRNA molecule being translated by many ribosomes and that the size of the polyribosome was dictated by the length of the mRNA.

Messenger RNA Has the Same Average Base Composition as hnRNA

Reticulocytes from mammals are "dead" cells without a nucleus, so the globin mRNA they produce cannot be easily studied by labeling techniques. However, in growing HeLa cell cultures, the mRNA can be labeled.

A polyribosome-associated class of RNA ranging in length from 0.5 to 3 kb becomes labeled before any newly labeled rRNA reaches the cytoplasm. This RNA species could thus be studied as a pure labeled RNA in the presence of unlabeled pre-existing RNA. Although the average size of the labeled polyribosomal RNA was smaller than the 18S rRNA (2 kb), and considerably smaller than hnRNA, it had a base composition very much like that of hnRNA and unlike that of rRNA. Moreover, in poliovirus-infected cells making only virus-specific proteins, the whole 7-kb poliovirus RNA replaced the rapidly labeled cellular RNA in the polyribosomes. Thus, the labeled, polyribosome-associated nonribosomal RNA appeared to be cellular mRNA.

The message-bearing nature of the rapidly labeled polyribosome-associated RNA has since been proved in many different cases by isolating it and translating it in cell-free extracts: the products have been specific proteins. (See Figure 7-11, for an earlier example of cell-free protein synthesis.) In recent years, techniques have been developed for partially purifying specific polyribosomes so that their mRNA content can be isolated. Purification

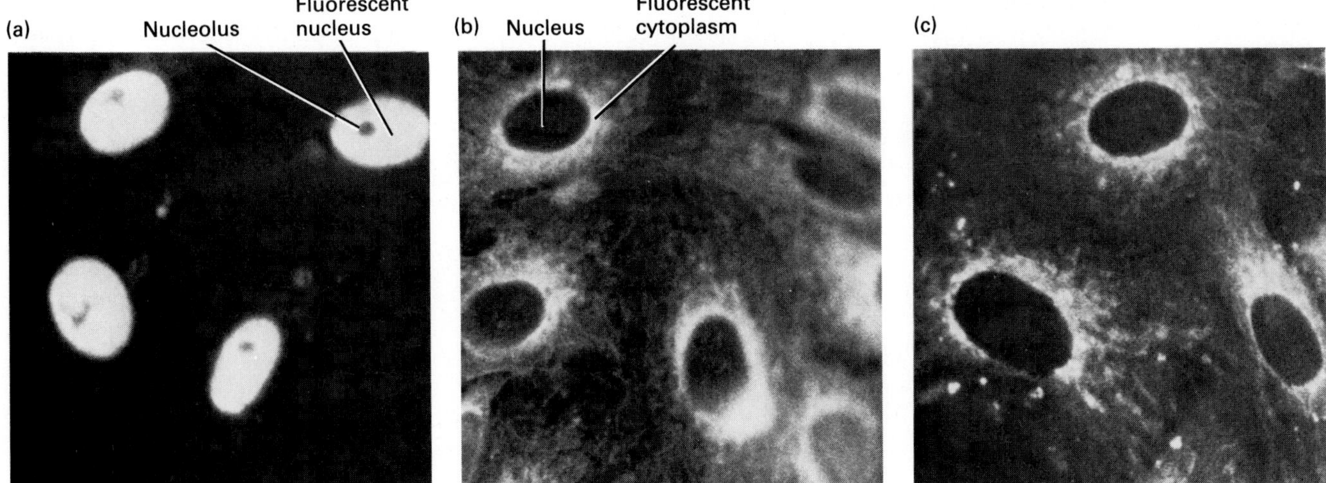

Figure 9-28 The location of ribonucleoprotein particles in cultured human cells. Diseases such as lupus erythematosus, in which people form antibodies to proteins in their own cells, are called *autoimmune* diseases. Blood serum samples from different autoimmune individuals will react with different cell constituents. When the antibodies are tagged with a fluorescent dye, treatment of cultured cells with the antibody-containing serum allows the cellular localization of particular antigens. (a) Cells exposed to a fluorescent antibody that precipitates small ribonucleoprotein particles. The distribution of the fluorescence *(light areas)* indicates that this class of ribonucleoproteins is localized in the nuclei and does not appear in the nucleoli *(small black dots)* or in the cytoplasm. The nuclear particles include the U1, U2, U4, and U6 RNAs. (b) The cytoplasmic distribution of an antigen that reacts with a serum containing a ribosomal antibody. Cells attached to plates are thicker in the center around the nucleus, so most of the reactivity is concentrated there, but ribosomes are quite evenly distributed in the cytoplasm. (c) The distribution of *rho*-type cytoplasmic ribonucleoproteins (see Table 9-2). These antigens, which are not associated with ribosomes, are generally widely distributed in the cytoplasm. *Courtesy of J. Steitz and M. Rose.*

is accomplished by using antibodies specific to the growing protein chains on the polyribosome, and then binding the antibody-polyribosome complex to a matrix that binds antibodies.

Because the average size of eukaryotic cellular mRNA was 1 to 2 kb, it was thought that each eukaryotic mRNA encodes only one polypeptide chain. Recent mRNA sequence studies have established that this is indeed the case. Thus, there is a fundamental difference between eukaryotes and prokaryotes in the information content of their mRNA. Almost all mRNAs in eukaryotes are *monocistronic* (i.e., they encode a single polypeptide chain), as evidenced by their single start site for protein synthesis, whereas many (but not all) mRNAs in prokaryotes are *polycistronic* (i.e., they encode multiple polypeptide chains), as indicated by their multiple protein start and stop signals (Figure 9-31). (Note that even in monocistronic mRNAs the start and stop sites for protein synthesis do not correspond to the ends of the mRNA. Also, recent studies have shown that eukaryotic mRNAs, particularly those from animal viruses, may have two initiation sites and produce two different proteins from overlapping stretches of mRNA.)

The Structure and the Biosynthesis of Eukaryotic mRNAs

Once radioactive cellular mRNA free of other labeled RNA was obtained from polyribosomes, it became possible to identify distinctive features of eukaryotic mRNA that were not found in tRNA and rRNA. In each case, the modifications found in mRNA were also found in hnRNA, which pointed to the derivation of mRNA from some portion of the hnRNA population. In this section we describe these modifications and explain how their study revealed the pathway of mRNA biosynthesis.

Eukaryotic mRNAs and hnRNAs Have Modifications at Both Ends: 5′ Caps and 3′ Poly A's

The first chemically distinct feature of mRNA to be discovered was a 3′ terminal sequence of adenylate residues

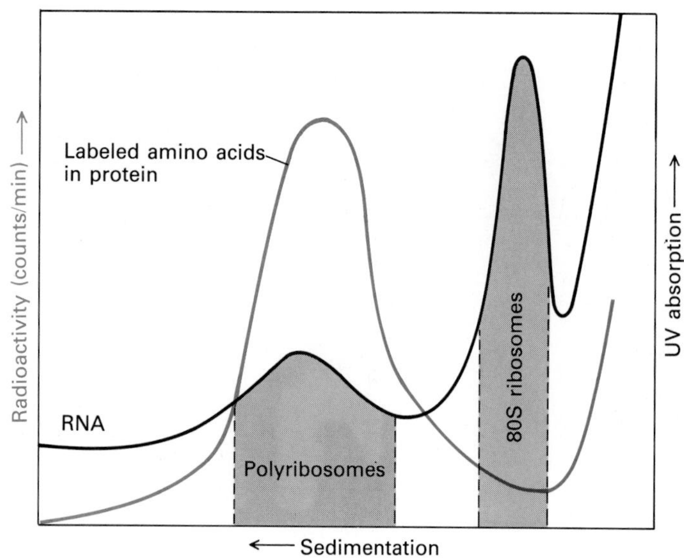

Cultured
HeLa cells

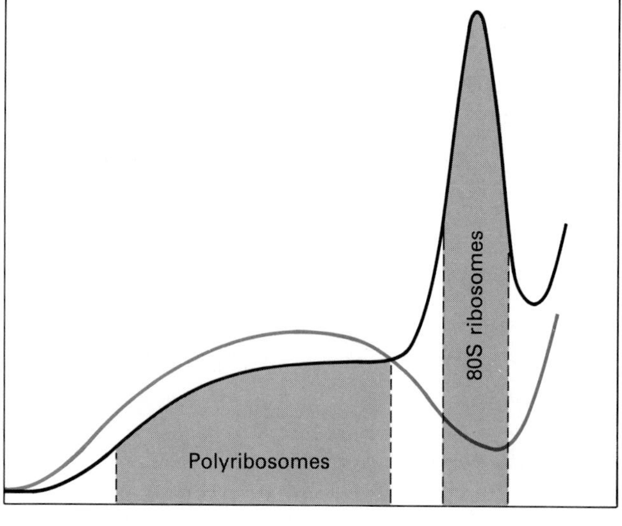

Poliovirus-
infected
HeLa cells

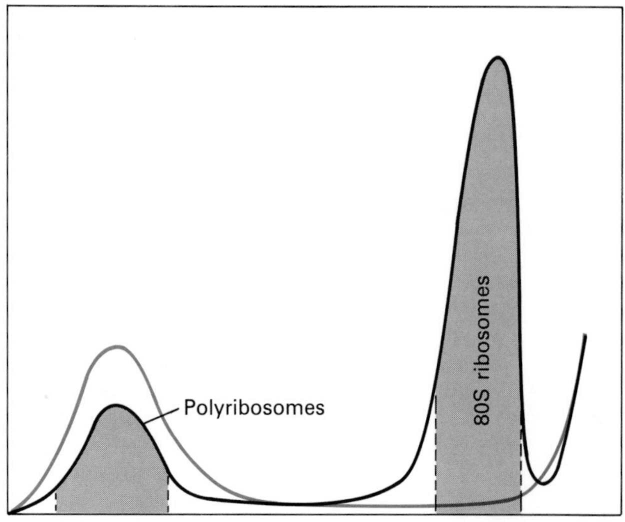

Figure 9-29 Polyribosomes are the sites of protein synthesis in mammalian cells. Three different cell samples were pulse-labeled for about 1 min with radioactive amino acids. The cells were then broken and the cell extracts were sedimented through a sucrose gradient to separate structures of various sizes. The distribution of RNA-containing material in the gradient was detected by the absorption of UV (260-nm) light *(black curve);* most of this material was rRNA. The shaded areas show the relative amounts in single 80S ribosomes and in polyribosomes. The newly formed nascent proteins still associated with ribosomes were detected by radioactivity *(color curve).* In the three samples, almost all of the protein synthesis had occurred in polyribosomes, but the sizes of the polyribosomes varied, the largest (here, those in poliovirus-infected HeLa cells) sedimenting last. [See J. Warner, P. Knopf, and A. Rich, 1963, *Proc. Nat'l Acad. Sci. USA* **49**:122; and S. Penman, K. Scherrer, and J. E. Darnell, 1963, *Proc. Nat'l Acad. Sci. USA* **49**:654.]

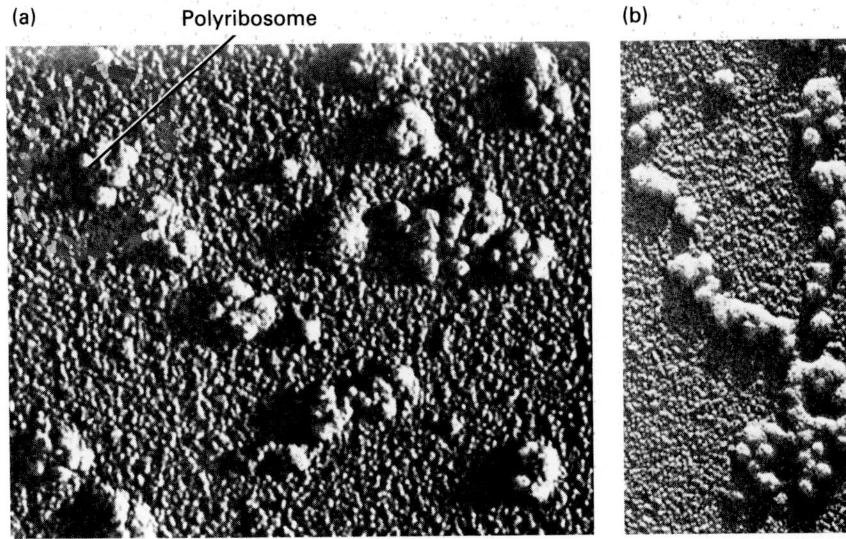

Figure 9-30 Polyribosomes were prepared from rabbit reticulocytes and from poliovirus-infected HeLa cells by sedimentation analysis, as shown in Figure 9-29. The polyribosomes were then adsorbed to electron-microscope grids and photographed. (a) The reticulocyte polyribosomes, which synthesize α and β globin, consist of three to five ribosomes each. The mRNA is now known to be approximately 650 nucleotides long. (b) The polyribosomes that synthesize poliovirus are made up of more than 40 ribosomes. The 7000-nucleotide RNA is translated into one long polypeptide of approximately 2200 amino acids. *Courtesy of A. Rich.*

that is now referred to simply as *poly A*. In the early 1960s, an enzyme termed poly A polymerase had been isolated; this enzyme did not copy a template, but simply added adenylate residues to the ends of RNA chains. Also, adenylate-rich RNA was found in cytoplasmic extracts of mammalian cells. Then, as has so often been the case in the study of mammalian gene expression, a key observation was made in work with viruses.

Vaccinia, a large DNA virus that grows in the cytoplasm of mammalian cells (Chapter 6), carries with it all of the enzymes necessary for producing mRNA from its own DNA. All of the vaccinia mRNA molecules produced by these viruses have a string of adenylate residues at their 3′ ends. This fact was discovered because the ApA linkage in poly A is relatively indigestible by many ribonucleases, and thus segments of poly A were left over after digestion of the mRNA. Labeled polyribosome-associated cellular mRNA was then shown to contain this segment of poly A, as was the mRNA made from DNA viruses whose DNA enters the cell nucleus.

The Addition of Poly A to hnRNA in the Cell Nucleus Although *all* eukaryotic cells contain mRNA with a 3′ terminal poly A, the feature has been studied most often in mammalian cells. It was discovered that about one-fourth of the hnRNA molecules in such cells have this repetitious terminal segment, which subsequently appears in all mRNA molecules. This chronology strongly suggested a link between the mRNA and at least the fraction of the hnRNA that contains the poly A. It was later shown that the poly A is added to hnRNA in the nucleus immediately after transcription (Figure 9-32). (As we shall discuss in more detail later in the chapter, the poly A is probably added to the hnRNA after an endonuclease cuts the RNA chain.)

The nuclear synthesis of poly A was proved by exposing cells for a very brief time (2 min or less) to [^{3}H]adenosine. At the beginning of exposure, the only labeled poly A was in the nucleus and was associated with hnRNA. Later, most of it was in cytoplasmic polyribosomal RNA. However, cellular DNA does not contain long

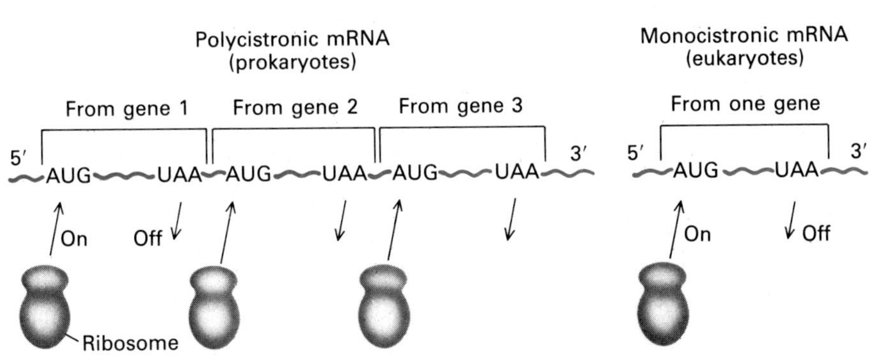

Figure 9-31 The average size of mammalian mRNAs (approximately 1.5 kb) first suggested that eukaryotic mRNA is monocistronic, in contrast to prokaryotic mRNA, which includes RNA from entire operons. This fundamental difference has generally held true for the many specific mRNAs that have been studied. A few cases of eukaryotic mRNAs with multiple protein start sites and even multiple reading frames (and therefore multiple protein products) have been discovered.

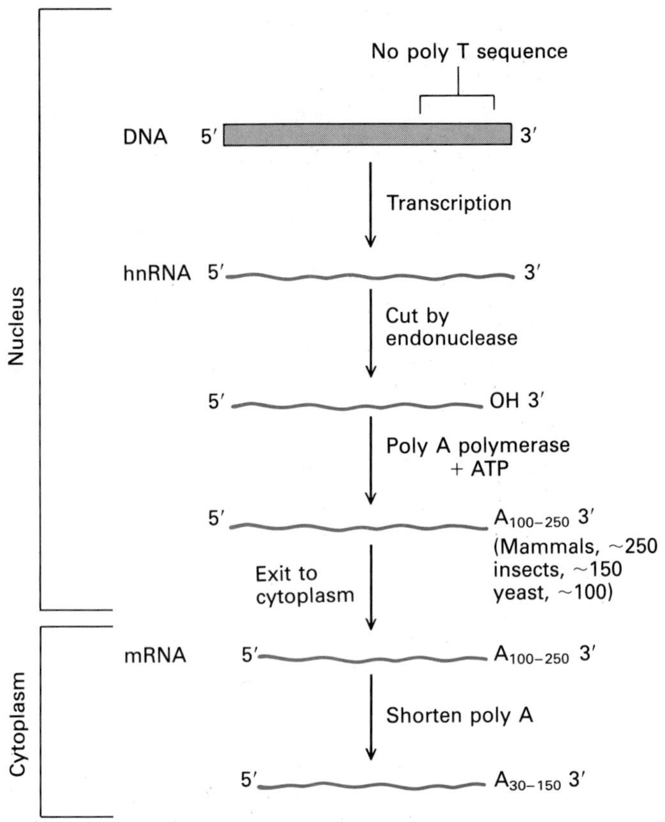

Figure 9-32 Poly A in eukaryotic RNA. In all eukaryotic cells, from 100 to 250 adenylate residues are added to the nuclear RNA (hnRNA) that will become mRNA. This segment—the poly A—is not copied from DNA, but is added one A at a time by a transferase reaction at the 3′ hydroxyl of the hnRNA. The enzyme responsible for this addition is called poly A polymerase, and the 3′ hydroxyl end is probably created by cutting the hnRNA chain. In the cytoplasm, the poly A becomes shorter as the mRNA ages.

poly T sequences, so poly A could not be copied from DNA. Also, poly A is not added to tRNA or rRNA. Thus, poly A was identified as a posttranscriptional addition product that is added specifically to hnRNA (RNA polymerase II transcripts) and that subsequently accumulates in polyribosomal mRNA.

The length of a poly A segment in the nucleus is quite consistent: in cultured mammalian cells and probably in all vertebrate cells, it is between about 200 and 250 nucleotides. In the cells of lower animals and plants, the initial length of the poly A may be shorter than in mammalian cells, but the size of the segment appears to be fairly uniform in each species. This specificity in poly A length—a feature that is not determined by copying a template—is only one example of the nuclear RNA processing events that have been clearly demonstrated, but that are not yet understood. After the poly A–containing mRNA reaches the cytoplasm, the poly A segment is shortened so that its size is no longer uniform, but varies between 30 and 250 nucleotides.

The Absence of Poly A in Histone mRNAs Very soon after the discovery of poly A, one class of mRNAs was shown to lack it. These were the histone mRNAs, short molecules that encode DNA-packaging proteins. Thus, in order to enter the cytoplasm and direct protein synthesis, an mRNA does not need to have a poly A.

In most cell samples, a few other individual cytoplasmic mRNA molecules associated with polyribosomes are found either to lack poly A or to have a very short poly A.

(a) Poly A^+

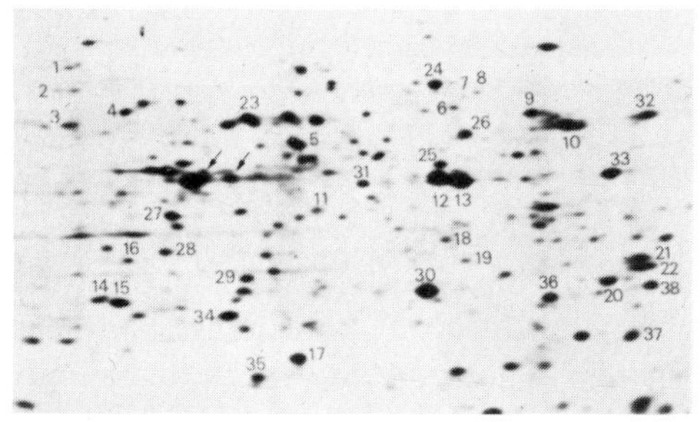

(b) Poly A^-

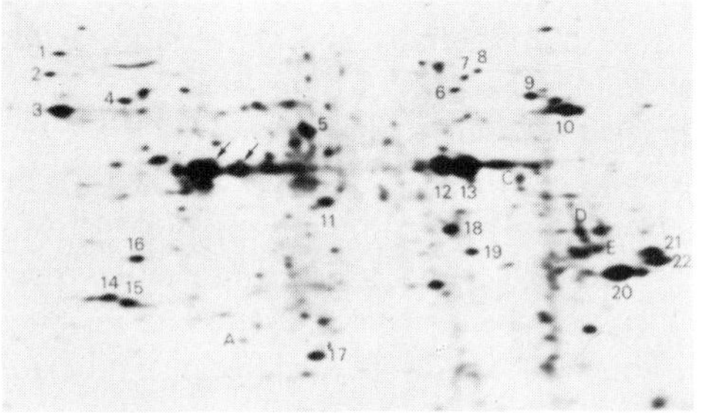

Figure 9-33 Translation products of poly A^+ and poly A^- mRNAs. RNA containing poly A can be purified from RNA either lacking it or having very short poly A (see Figure 9-34). This figure shows autoradiographs of translation products synthesized in vitro from labeled amino acids and separated into individual protein components by two-dimensional gel electrophoresis (the procedure in which both size and charge are used to separate the polypeptides). The individual proteins are numbered for comparison. Virtually every polypeptide that results from translating the poly A^- mRNA fraction (b) is also present in the poly A^+ fraction (a). Thus no unique translation products of the poly A^- fraction were found. (Histone proteins are detected in another gel system as products of the poly A^- fraction.) The two spots marked by arrows are not due to translation of added mRNA, but all of the other spots are. [See A. J. Minty and F. Gros, 1980, *J. Mol. Biol.* **139**:61.] *Courtesy of A. Minty.*

However, the in vitro translation products of such *poly A minus* (poly A^-) mRNAs are extremely similar to the translation products of mRNAs that contain poly A (i.e., poly A^+ mRNAs), as shown in Figure 9-33. Quite possibly, aside from the histones, no specific *class* of cytoplasmic mRNAs lacks poly A. Most poly A^- mRNA probably derives from poly A^+ molecules by cytoplasmic shortening of the poly A segments.

The Purification of mRNA and the Discovery of the Methylated Cap Structure A practical benefit was derived from the discovery of poly A in mRNA molecules. The presence of poly A enables eukaryotic mRNA to be purified away from rRNA and tRNA by affinity chromatography. When homopolymers of either poly rU (polyuridylic acid) or poly dT (polythymidilic acid) are bound to a solid matrix, such as paper fibers, they can selectively hybridize to the 3′ poly A segment of mRNA. All non–poly A–containing RNA, such as rRNA and tRNA, can be removed, and then the pure mRNA can be released from the bound homopolymer (Figure 9-34).

The purification of eukaryotic mRNA led to the discovery of a second modification of eukaryotic mRNA molecules. Examination of purified mRNA revealed that methyl groups were present in both mRNA and hnRNA. About 1 percent of the adenylate residues in the mRNA of mammalian cells are methylated at position 6 of the adenine, to give 6-methyladenylate. But most attention has centered on a complex methylated structure at the 5′ ends of the mRNA and hnRNA molecules. This structure, called a *cap* (like that we mentioned earlier for U snRNAs), was first characterized in the products of enzymes that synthesize viral mRNAs in virus particles. The 5′ cap was then also isolated from the poly A–containing polyribosomal mRNA of cells. The cap consists of a terminal nucleotide, 7-methylguanylate (m^7G) in a $5' \rightarrow 5'$ linkage with the initial nucleotide of the mRNA chain. Thus there is no free 5′ end of an mRNA molecule; in fact, it has 3′ hydroxyl groups on the ribose rings of the nucleotides at both ends (Figure 9-35a). As we shall discuss later, the cap probably aids in translation after a mRNA molecule is completely synthesized, but the structure is made in the nucleus by the series of enzymatic steps outlined in Figure 9-35b.

With the finding that both mRNA and hnRNA molecules have poly A at their 3′ ends and caps at their 5′ ends, it seemed very likely that mRNA must be processed from hnRNA.

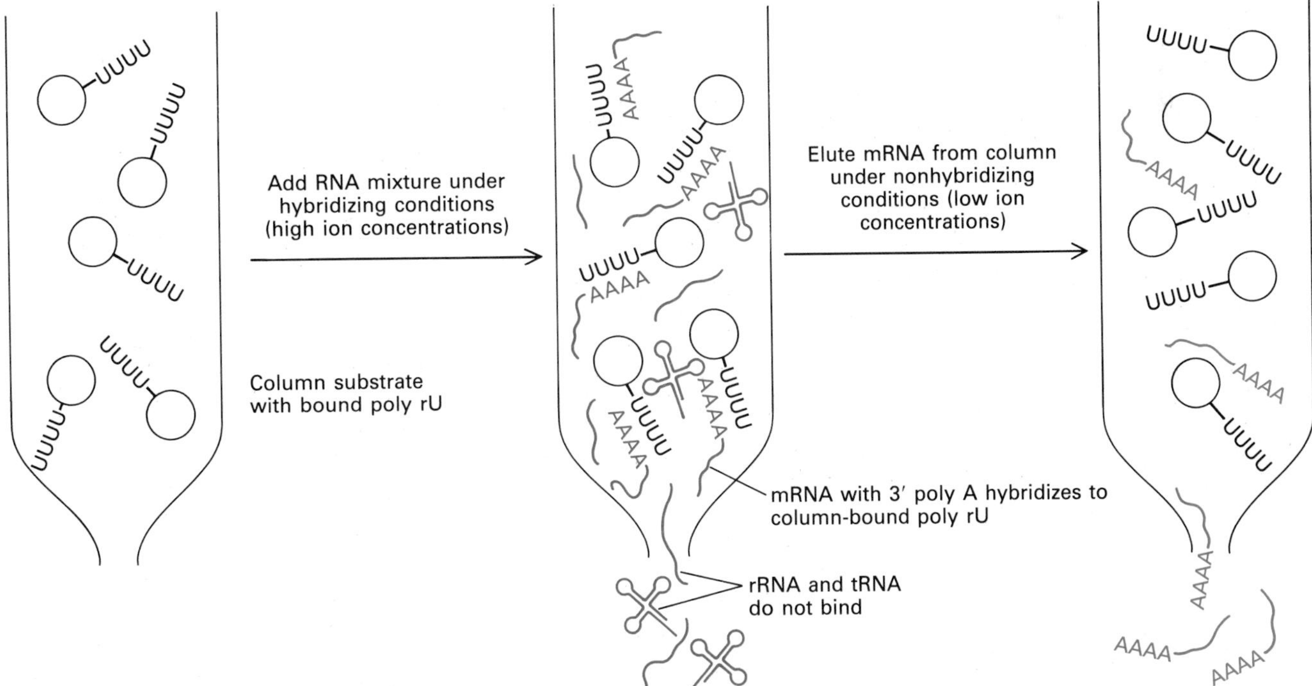

Figure 9-34 The purification of eukaryotic mRNA. Because eukaryotic mRNAs (except histones) contain a 3′ terminal poly A segment, it is possible to purify the molecules by affinity chromatography. The mRNA is allowed to interact with a substrate (e.g., cellulose) to which poly rU or poly dT is attached. A hybrid forms between the poly A and the column-bound poly rU or poly dT, and all RNA that does not contain poly A, such as rRNA and tRNA, is not bound. Under nonhybridizing conditions (e.g., very low ion concentrations) the pure mRNA can be eluted.

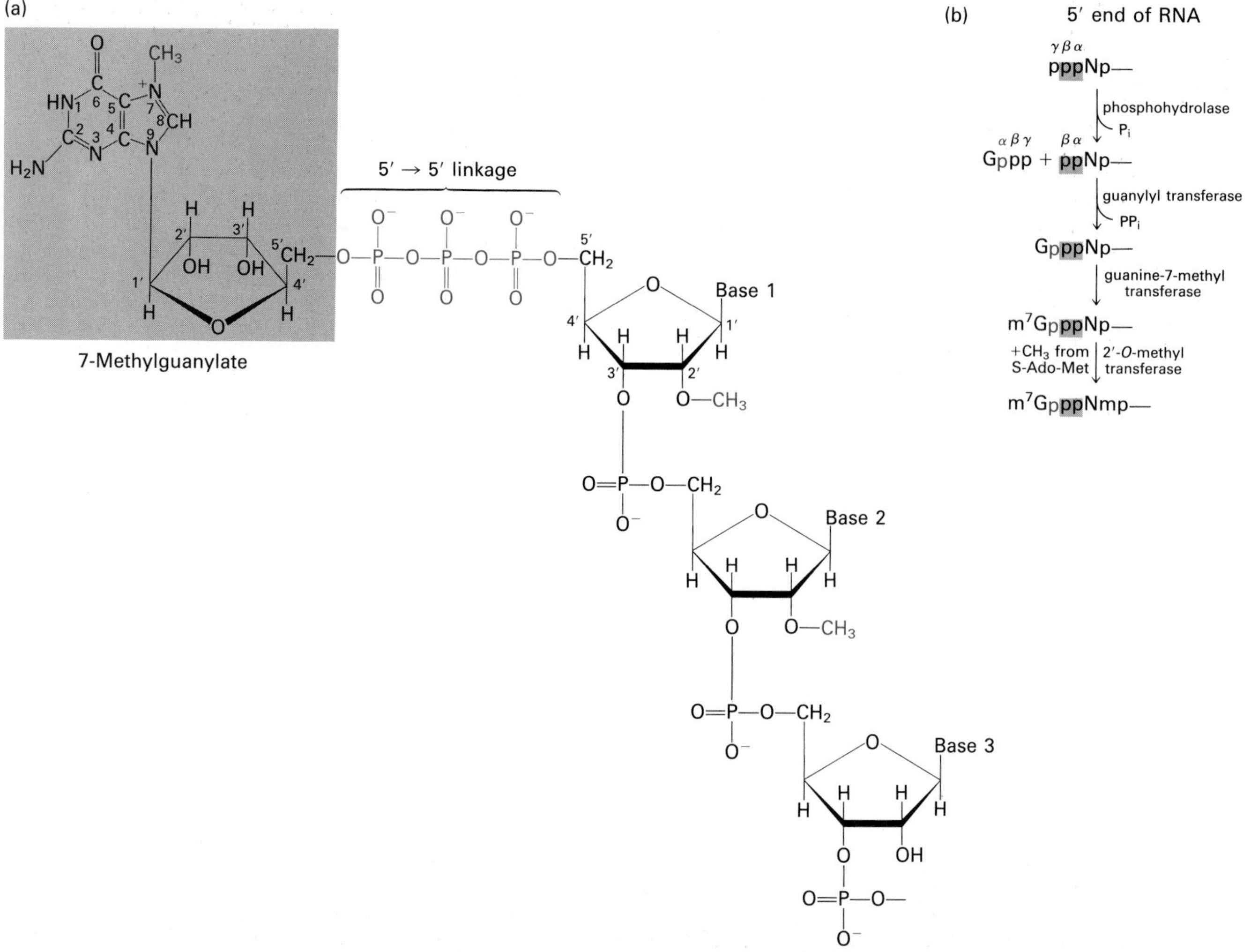

Figure 9-35 (a) The structure of the 5′ methylated cap of eukaryotic mRNA. The distinguishing chemical features are the 5′ → 5′ triphosphate linkage of 7-methylguanylate to the initial nucleotide of the mRNA molecule, and the methyl group at the 2′ hydroxyl of the ribose of the first nucleotide (base 1) in all animal cells and in cells of higher plants. Yeasts lack this methyl group. The ribose of the second nucleotide (base 2) also is methylated in vertebrates. [See A. J. Shatkin, 1976, *Cell* 9:645.] (b) Steps in the biosynthesis of the cap structure in the nuclei of mammalian cells. The last three enzymes can catalyze the addition of guanylate only to diphosphate (ppN) ends. The methyl (CH_3) donor, S-Ado-Met is S-adenosylmethionine. [See S. Venkatesan and B. Moss, 1982, *Proc. Nat'l Acad. Sci. USA* 79:304.]

The Study of Adenovirus mRNA Formation Proved the Steps in mRNA Processing

The evidence that was needed to clinch the precursor relationship of hnRNA to mRNA was proof of an obligatory larger primary transcript for a specific mRNA. It had been easy to demonstrate that the 45S pre-rRNA was a precursor to 28S and 18S rRNA, because such a large portion of the total cellular RNA was rRNA. Both the precursor and the product could be easily purified so that their chemical similarities could be proved. However, normal cells do not make such large amounts of a single mRNA, so purification of a single pure nuclear transcript was impractical. The solution to this problem lay once again in using viruses: cells infected with a virus provide a source of nuclei in which a large fraction of the RNA synthesis comes from only one or a few transcription units.

The Adenovirus Infectious Cycle The investigation of nuclear precursors to mRNA was therefore undertaken by using HeLa cells infected by adenovirus, a DNA virus. A subviral particle of the virus gains entry into the cell, and the double-stranded viral DNA enters the cell nucleus to direct viral replication (Figure 9-36). Late in adenovirus infection, about half of the total nonribosomal nuclear RNA synthesis is adenovirus-specific. Also, the only new mRNA molecules that arrive in polyribosomes are virus-specific. Thus the cell becomes a factory for making

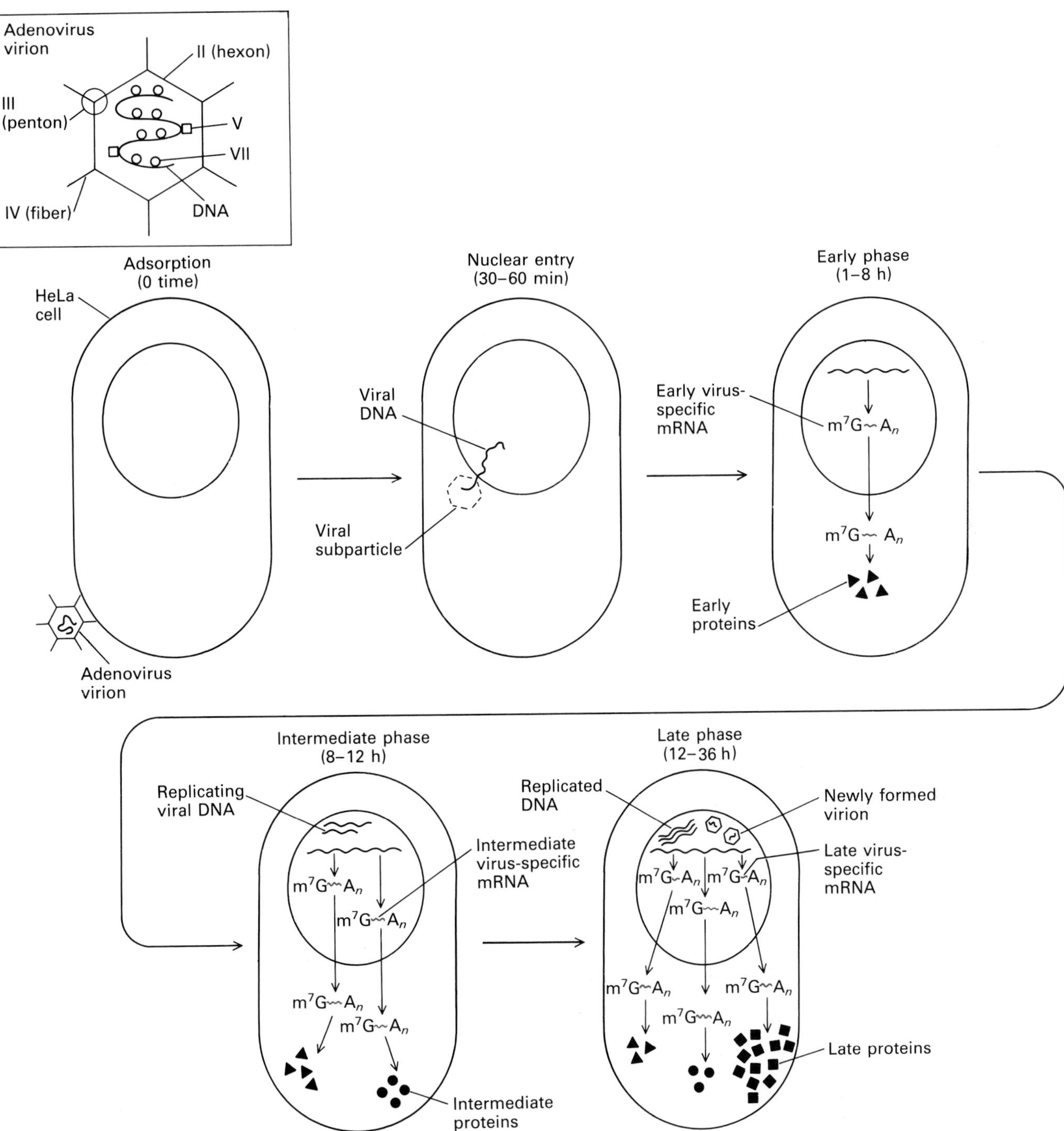

Figure 9-36 The adenovirus infectious cycle in HeLa cells. The viral particle (the virion) is composed of about 10 different proteins, some of which are indicated. The virion proteins are numbered with Roman numerals in order of decreasing size; for example, II, the hexon protein, has the highest molecular weight and is also present in the largest quantity. To enter the cell, the virion first attaches by the fiber protein (IV) to specific cell receptor proteins. A portion of the virion containing both DNA and proteins (at least II, III, V, and VII) gains entry into the cell. The viral DNA molecule, about 36 kb in length, is released at or near the nuclear border, and is transcribed within the nucleus by pre-existing cellular enzymes. During each of three phases of infection—early, intermediate, and late—different viral mRNAs accumulate and undergo translation; the three types of mRNAs are distinguishable from one another by hybridization tests and by cell-free translation to yield different products. The virus-specific mRNA contains methylated 5′ caps (m^7G) and 3′ poly A (A_n), just like cellular mRNA. The early phase of infection precedes any DNA synthesis; the late phase is one of rapid viral DNA synthesis; and the intermediate phase overlaps the two. Adenovirus late mRNA is synthesized in larger amounts than any other known eukaryotic mRNA.

adenovirus mRNA. Studies of the production of adenovirus late mRNAs yielded the first clear evidence of the processing of specific large nuclear molecules to make specific mRNAs. When the adenovirus DNA enters the cell nucleus, the cell forms a group of mRNAs that encode virus-specific proteins not found in virions. These are the early mRNAs; another group, the intermediate viral mRNAs, are formed around the time viral DNA synthesis is initiated. Finally, the late mRNAs appear and encode the 10 *capsid* proteins that are found in the coats of the mature virus particles, the virions.

Like normal cellular mRNAs, virus-specific mRNAs contain caps at their 5′ ends and poly A at their 3′ ends. Adenovirus early mRNAs have the same modifications. Furthermore, these mRNAs can be made in infected cells before any new viral proteins have been synthesized. Consequently, the enzymatic machinery for forming adenovirus mRNA is very likely to be the same as that for forming cellular mRNA.

Mapping the Adenovirus Late mRNAs on the Genome The first step in the study of adenovirus mRNA synthesis in the nucleus was to locate the viral DNA regions that encode each viral mRNA molecule. The discovery of DNA restriction enzymes afforded a method for cutting the viral DNA at specific sites. Once the order of fragments in the linear DNA molecule was determined, the fragments could be hybridized with adenovirus late mRNA molecules obtained from the polyribosomes of infected cells to see which of the DNA fragments would hybridize with the mRNAs (Figure 9-37). Specific DNA fragments were also used to select by RNA-DNA hybridization specific mRNAs, which could then be translated into proteins. With these two techniques, the genomic regions encoding particular proteins could be mapped. Finally, the two strands of the viral DNA could be separated, as could the two strands of specific DNA fragments. The DNA strand that was copied into mRNA at any site on the genome could therefore be determined. A number of adenovirus late mRNAs were all found to hybridize to the same DNA strand, which meant that they are synthesized from the DNA in the same orientation (left to right in the conventional diagram; see Figure 9-37). Almost the entire right-hand two-thirds of the viral genome—more than 20 kb of it—encodes a series of rightward-reading late mRNAs (Figure 9-37).

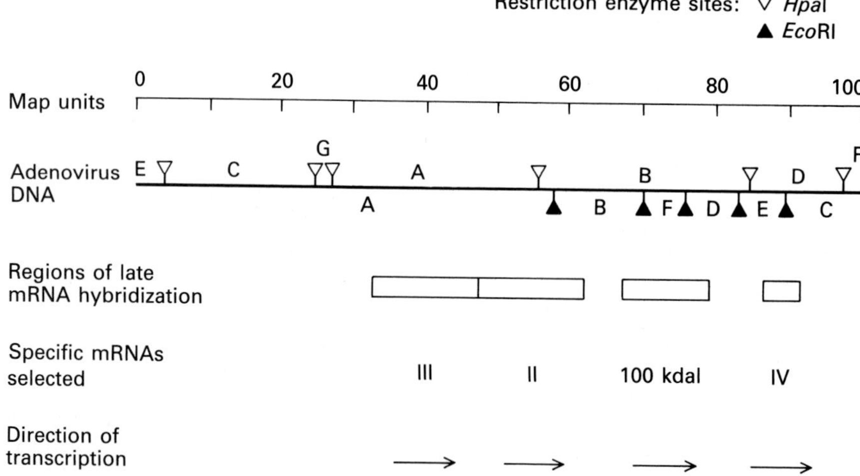

Figure 9-37 The map positions of some adenovirus late mRNAs. This partial map of the 36-kb adenovirus DNA molecule (1 map unit = 360 base pairs) shows the regions that encode three of the major virion protein molecules (II, III, and IV) and the 100-kdal (-kilodalton) protein, which helps assemble the virion proteins. The sites of action for two restriction endonucleases (*Hpa*I and *Eco*RI) are shown on a conventional physical map that reads from left to right. The genomic positions of the DNA fragments (A–G and A–F) created by these restriction enzymes can be ordered from left to right, and each fragment can be isolated for use in hybridization experiments (see Figures 7-23 and 7-24). The diagram indicates regions of the adenovirus DNA *(white bars)* to which particular late mRNAs hybridize. For example, *Eco*RI DNA fragment E(84 to 89) and the *Hpa*I fragment D (83 to 98) both hybridize to the mRNA that translates into the fiber protein (protein IV). Because this distance (about 5 kb) is longer than the mRNA itself, the bar is placed at the junction of the two fragments. The direction of transcription *(arrows)* was found by determining the 5′ → 3′ direction of each of the two complete DNA strands and then determining which strand hybridized to any particular mRNA. All of the four mRNAs hybridized to the same strand; the 3′ end of this strand is the left end on the conventional map. Because mRNA is formed in the 5′ → 3′ direction and each of the four mRNAs are complementary to this strand, all four are represented as synthesized from left to right. [See J. Flint, 1977, *Cell* **10**:153.]

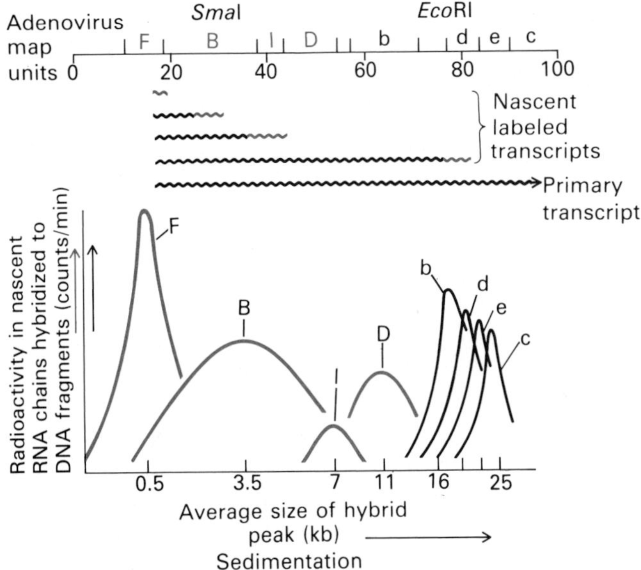

Figure 9-38 Mapping the single adenovirus transcription unit that encodes many different late mRNAs. Nascent nuclear RNAs labeled only briefly so that their 3′ growing ends carried the label (see Figure 9-11) were sedimented through a sucrose gradient to separate the molecules according to size, and the size-fractionated RNA samples were hybridized to a series of DNA pieces from the adenovirus genome. (Two restriction endonucleases, *Sma*I and *Eco*RI, were used to fragment the genome.) The shortest labeled RNA (averaging 0.5 kb in length) was complementary to the region between 11 and 18 on the genome—that is, to the *Sma*I F fragment, a DNA fragment of over 2 kb. Since the nascent RNA complementary to this 2-kb fragment was only 0.5 kb long, the initiation site for RNA synthesis was about 0.5 kb to the left of the right end of the F fragment. Longer RNAs had labeled segments complementary to fragments representing successive regions to the right of the F fragment. The adenovirus DNA that is transcribed to make this series of longer and longer nascent RNAs encodes a number of different protein products (see Figure 9-37). Short nascent RNAs corresponding to each of the mRNAs encoding these products were not observed. Thus, it was concluded that the long nuclear RNA molecules are processed to yield shorter mRNAs. [See J. Weber, W. Jelinek, and J. E. Darnell Jr., 1977, *Cell* **10**:611; and R. M. Evans et al., 1977, *Cell* **12**:733.]

The Definition of a Single Adenovirus Late Transcription Unit Were these adenovirus late mRNAs, which encode different capsid proteins, the products of separate transcription units, or did all of the mRNAs come from one transcription unit through the processing of a long nuclear RNA transcript? The nuclear RNA made from this region of the adenovirus genome was studied by two different techniques to define the boundaries of the transcription unit or units responsible for late mRNA production. Both pulse-labeled nascent chain analysis (Figure 9-38) and transcription mapping by ultraviolet irradiation (Figure 9-39) pointed to only *one* start site (near 16 on the map) for nuclear RNA synthesis from the 25-kb stretch on the right side of the adenovirus genome. Initiation of transcription for each of the different mRNAs that map between 35 and 95 was near 16. Thus, the late mRNA molecules had to be derived by some type of cleavage from the large nuclear adenovirus-specific RNA. An unexpected discovery revealed the final step in the synthesis of the mRNA.

The Discovery of Splicing in the Formation of Adenovirus Late mRNA Throughout the earlier work on mRNA formation and control in bacterial cells, the sequence of any given gene was shown to be represented in a continuous fashion in the mRNA and in proteins encoded by that gene. Also, in mammalian and amphibian

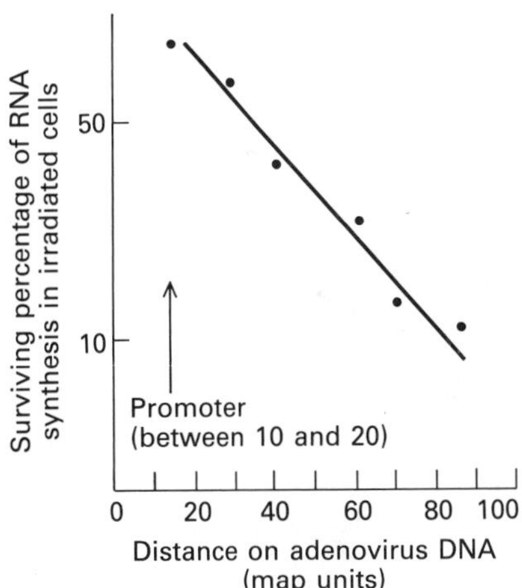

Figure 9-39 Mapping the adenovirus late transcription unit by using UV irradiation (see Figure 9-14). HeLa cells infected with adenovirus were exposed to UV light; infected cells not exposed served as controls. The nuclear RNA in both cultures was labeled with [^{3}H]uridine. The effect of the irradiation on nuclear RNA synthesis at various sites in the adenovirus genome was determined by hybridizing labeled nuclear RNA to these sites and comparing the results for irradiated and nonirradiated cells. The most sensitive region was at the far right end of the genome, and the most resistant was at the left end. When the amount of RNA synthesis that survived irradiation in each section of the genome was plotted on a semilogarithmic scale, an exponential increase in damage (i.e., an exponential decrease in RNA synthesis) was observed between about 20 and 100 map units on the genome. This indicated that synthesis began at or before 20 and progressed to 100. [See S. Goldberg, J. Weber, and J. E. Darnell Jr., 1977, *Cell* **10**:617.]

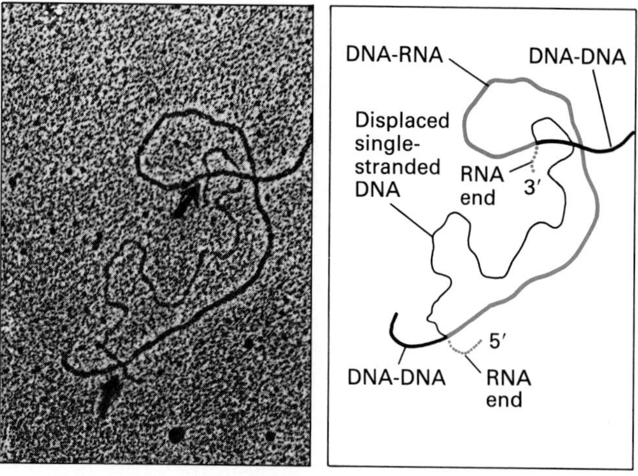

Figure 9-40 An electron micrograph and tracing of an adenovirus mRNA-DNA hybrid. The mosaic nature of viral mRNA (i.e., its derivation from noncontiguous DNA sequences) was proved by micrographs such as this one and that in Figure 9-41. In this figure, double-stranded DNA spanning the region from 50 to 73 map units was hybridized to purified mRNA encoding capsid protein II (the hexon) by the R loop procedure, in which complementary RNA displaces DNA (Chapter 7). Besides the R loop, *two* protruding RNA ends (5′ and 3′) are visible, which indicates that these ends are not homologous with adenovirus DNA. One end (the 3′ end) is poly A; the homology of the other end to distant sites in the adenovirus genome was tested by experiments of the type shown in Figure 9-41. [See S. Berget et al. 1977, *Proc. Nat'l Acad. Sci. USA* **74**:3171.] *From S. Berget et al., 1977.*

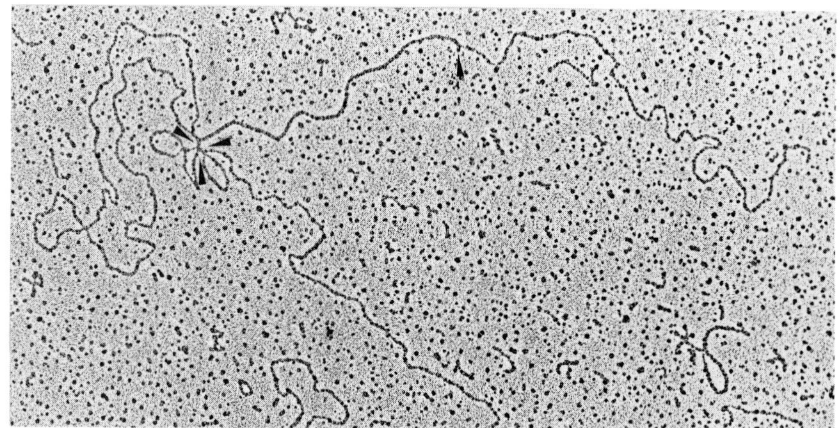

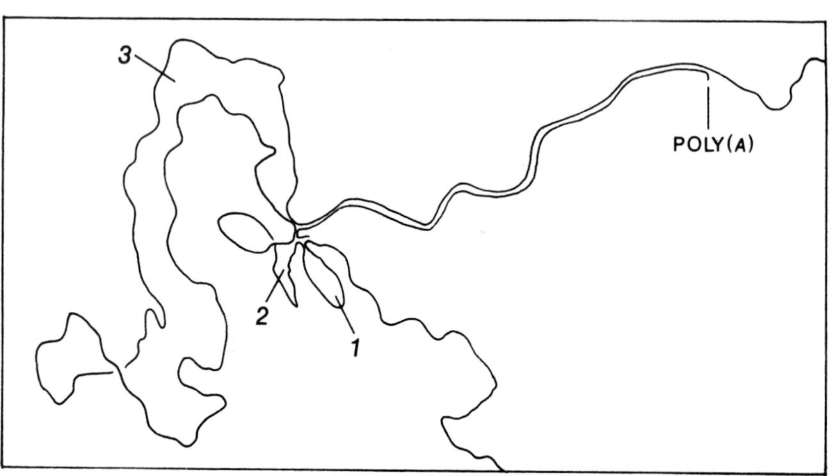

Figure 9-41 Microscopic identification of the regions of adenovirus DNA that are complementary to a late mRNA. The *single* strand of the adenovirus genome that is known to be transcribed in the indicated direction and to be complementary to late mRNA molecules (see Figures 9-38 and 9-39) was purified and hybridized to adenovirus mRNA for the 100-kdal protein. The major part of the coding sequence for this protein lies between 67 and 79 map units. The 5′ end of the mRNA hybridizes to three regions whose distances from the left (3′) end of the single strand are 16.4, 20, and 27. Thus the mRNA has components from *four* regions of the DNA: When the four separated regions hybrized to the DNA, loops were formed of a distinct length. Loop 1 stretched from 16 to 19 map units; loop 2 from 19 to 27 map units, and loop 3 stretched from 27 to 67 map units. The RNA-DNA hybrid at the 16.4 region corresponds to the region predicted by nascent chain analysis (Figure 9-38) and UV mapping (Figure 9-39) to contain the initiation site for the primary transcript. [See L. T. Chow et al., 1977, *Cell* **12**:1.] *Photograph courtesy of L. T. Chow and T. R. Broker.*

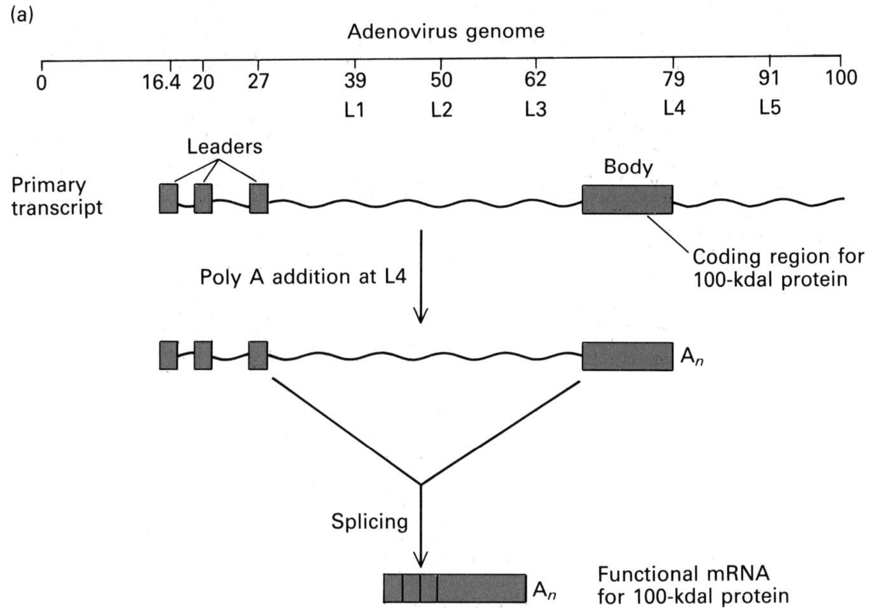

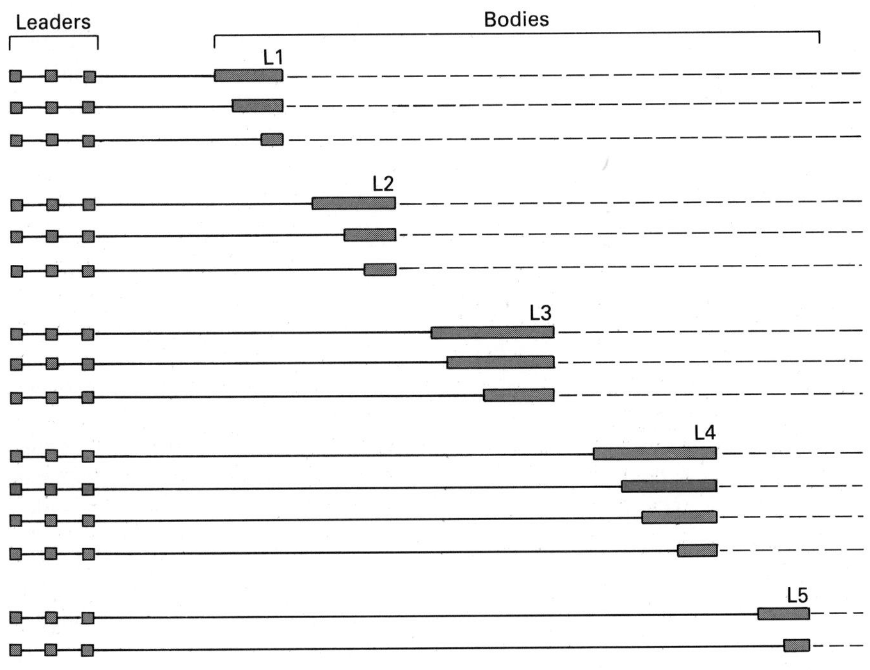

Figure 9-42 Different sites for poly A addition and for splicing allow the assembly of different adenovirus late mRNAs. (a) The primary RNA transcript from adenovirus late DNA contains five potential poly A addition sites labeled L1 to L5 (late poly A sites 1 to 5). The formation of a particular mRNA molecule begins by cleavage at one of these sites. For example, the primary transcript yields the mRNA for the 100-kdal protein only when the RNA is cleaved and polyadenylated at the L4 site. Splicing must also occur between the leaders, so that the first is joined to the second and the second to the third. Finally, splicing permits the third leader to be joined to a site near 67 on the map, and the formation of the 100-kdal mRNA is complete. (b) Each poly A addition site, including L4, is preceded by several different cleavage sites to which the third leader can be joined; in effect, then, there are several different bodies associated with each poly A site. Fifteen different mRNAs can be produced from the primary transcript; which mRNA is produced depends on which poly A site is chosen and on where the third leader is spliced to the body. The splicing alternatives for these mRNAs are schematically represented. The components of the 100-kdal protein are shown in color. [See L. T. Chow and T. R. Broker, 1978, *Cell* 15:487; reviewed in J. R. Nevins and S. Chen-Kiang, 1981, in *Advances in Virus Research,* vol. 26, Academic Press, p. 1.]

rRNA formation from precursor rRNA, the 28S and 18S sequences represented separate but continuous stretches within the larger precursor molecule. Therefore, it was most surprising when the individual adenovirus late mRNA molecules were discovered to be complementary to *noncontiguous sites* in the DNA.

The mosaic nature of the adenovirus late mRNAs was first revealed by electron microscopy of mRNA-DNA hybrids. All of the mRNAs that derive from the long nuclear RNA (i.e., the mRNAs for the II, III, IV, and 100-kdal proteins) form hybrids with adenovirus DNA in a long region, a *body,* and in short regions, called *leaders,* at three separate genomic sites (16.4, 20, and 27 map units; Figures 9-40 and 9-41). The leaders in the mRNA molecule are at or near the 5′ end. It is now known from nucleotide sequencing that the leaders are not translated, and that translation begins within the body of the molecule. Because the adenovirus nuclear primary RNA transcript contains all of the sequences that make up the individual mRNA (i.e., the body and each of the short leader sequences), the only logical explanation is that the primary transcript is cut and spliced to give functional mRNA.

A point of great interest is that the splicing event can assemble different mRNAs encoding different proteins by acting on different regions. In fact, when the mapping of the adenovirus late mRNAs was completed, at least 15 different mRNAs were found to be formed from the late transcription unit (Figure 9-42). There are five possible poly A sites that define five different 3′ termini, and preceding each of the five termini there is more than one body sequence that can be joined to the three leaders. The production of multiple mRNAs from the same primary transcript allows gene control at the level of RNA processing. We shall consider such events in detail in Chapter 12.

Splicing in Other Virus Transcription Units The concept of RNA splicing was promptly applied to the mRNAs produced from the genome of SV40, a small DNA virus that is of particular interest because it can cause cells to become cancerous. The entire 5250-base sequence of this virus was already known. When the sequences of individual SV40 mRNA molecules were located on the SV40 genome, they, too, were found to derive from separate, noncontiguous regions of the SV40 DNA. Again, different splicing patterns between the cap and the poly A of the early and late nuclear RNA transcripts produced different SV40 mRNAs (Figure 9-43). Also, a number of different adenovirus early transcription units give rise to primary transcripts that are spliced to form mRNAs. The splicing of one of the adenovirus early mRNAs was then demonstrated both in isolated nuclei and in cell extracts (by adding the RNA). Progress in clarifying the in vitro splicing reaction has been most rapid in studies using viral sequences and is discussed on page 353.

Cellular mRNAs Also Are Spliced

The splicing of primary nuclear transcripts to make finished mRNAs is not an aberration of viral infection. This fact was soon demonstrated with a great variety of cellular genes. One particularly well-studied case illustrates the point.

The β chain of mammalian hemoglobin is encoded in an mRNA that is about 0.7 kb long. A 1.5-kb nuclear RNA molecule that is complementary to cloned β globin DNA can be detected in the nuclei of cultured mouse erythroleukemia cells that produce hemoglobin. The relation of this nuclear RNA molecule to hemoglobin mRNA can be illustrated through the use of a recombinant DNA

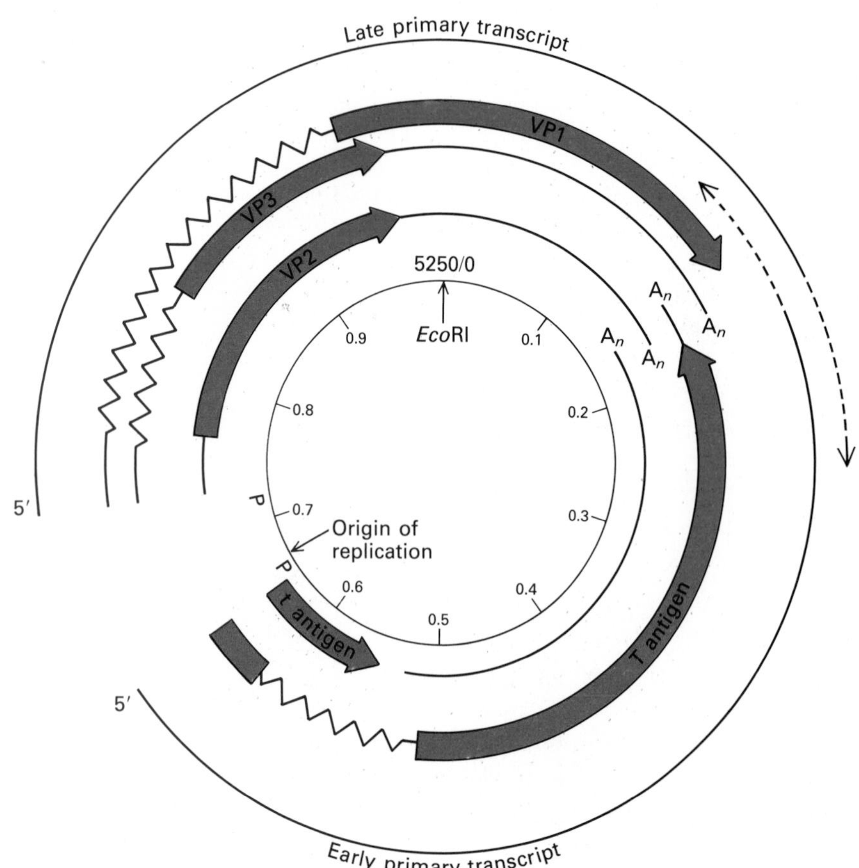

Figure 9-43 A map of SV40 mRNAs. The genome is a circular DNA whose 5250-base sequence has been entirely determined. The two regions that encode the early and late viral proteins are shown in color. [VP1, VP2, and VP3 are virion proteins produced late in infection; the T (large T) and t (small t) antigens are produced early in infection.] The zigzag lines show primary transcript regions that are removed by RNA splicing, and the solid black lines show untranslated regions. The two P's indicate the promoter regions for the early and late transcripts, and the A_n's designate poly A sites. The dashed parts of the arrows show the primary transcripts extending beyond the poly A sites, because the termination sites for the primary transcripts are not known. Note that the T antigen is encoded by two noncontiguous segments. [See W. Fiers et al., 1978, *Nature* **273**:113.] *After Fiers et al., 1978.*

molecule containing the genomic DNA segment that includes the entire mouse β globin gene.

The cloned β globin genomic DNA can be hybridized to either the 1.5-kb globin-specific nuclear molecule or to the shorter β globin mRNA molecule. The 1.5-kb nuclear molecule hybridizes to a continuous sequence of the cloned genomic DNA, whereas the globin mRNA molecule hybridizes to the genomic DNA in three separate regions, producing two unhybridized loops (only one of which is large enough to see easily; Figure 9-44). A comparison of the nucleotide sequence of genomic DNA and the sequence of the mRNA confirmed that the DNA contains two intervening sequences that are not present in the mRNA (Figure 9-45). Again, the logical explanation for the hybridization results and the absence of the intervening sequences in the mRNA is that the primary RNA transcript is spliced to form the mRNA.

The number of specific mRNAs known to be derived from transcription units with intervening sequences is now very large. Splicing is required not only to form mRNAs in vertebrates, but also those in invertebrates, in plants, and in single-celled organisms such as yeast.

Although the adenovirus late mRNAs are not spliced within the translated sequences of the mRNA, in some of the viral mRNAs and in many of the cellular mRNAs the splicing junctions *are* located within the mRNA region that is translated into protein. For example, the β globin gene is divided between the codons for amino acids 31 and 32 and again between the codons for amino acids 105 and 106 (see Figure 9-45). Some mRNAs are spliced within a codon. Clearly, precise rejoining is necessary to ensure that the completed mRNA is translatable. In many cases, when protein-coding portions of an mRNA are in separate locations in the DNA (and in the primary RNA transcript), the separated sequences represent distinct *domains* (functional regions) of the final protein product. The important implications of these arrangements for genome structure and evolution are discussed in Chapters 10, 11, and 25.

Introns and Exons Walter Gilbert suggested the term *intron* to refer to a part of a primary transcript (or the DNA encoding it) that is not included in a finished mRNA, rRNA, or tRNA. An *exon* is a primary transcript region (or, again, the DNA encoding it) that exits the nucleus and reaches the cytoplasm as part of an RNA molecule. The terms intron and exon are meaningful when they are applied to a primary transcript that gives

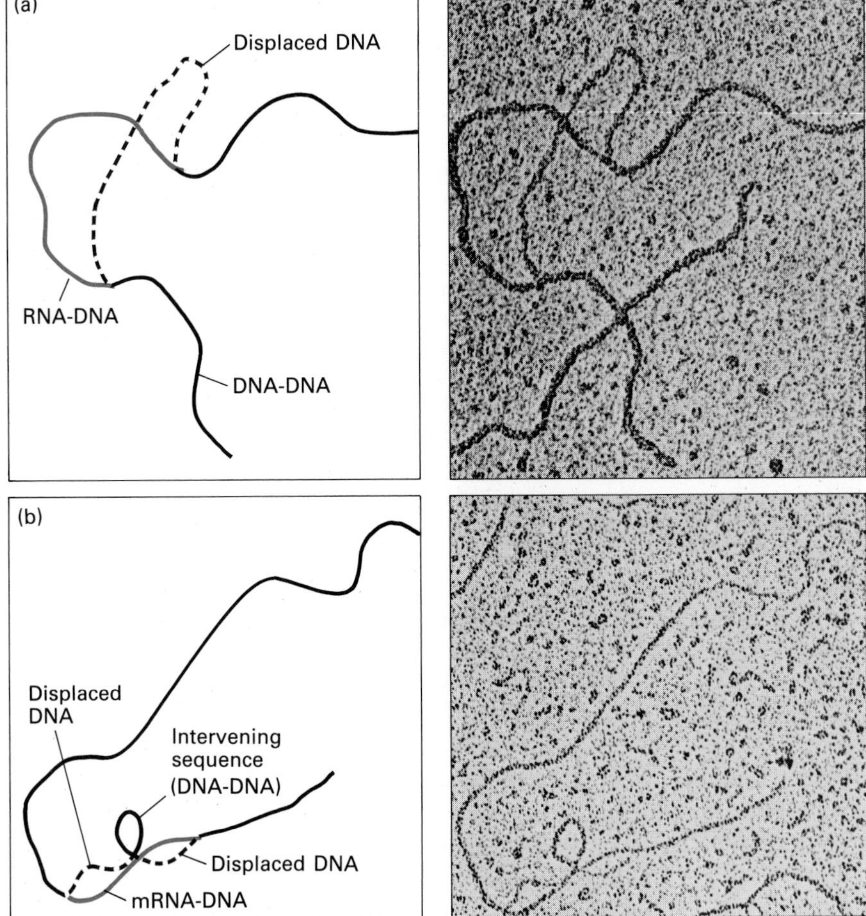

Figure 9-44 Electron micrographs and tracings of R loop structures between cloned genomic β globin DNA and (a) 1.5-kb globin-specific nuclear RNA and (b) cytoplasmic globin mRNA. The entire 1.5-kb sequence of nuclear RNA hybridizes in a continuous fashion to the β globin DNA. The presence of the loop in the hybridized mRNA-DNA in (b) indicates that part of the DNA sequence is absent from the mRNA molecule. [See S. M. Tilghman et al., 1978, *Proc. Nat'l Acad. Sci. USA* 75:1309.] *Photograph courtesy of P. Leder.*

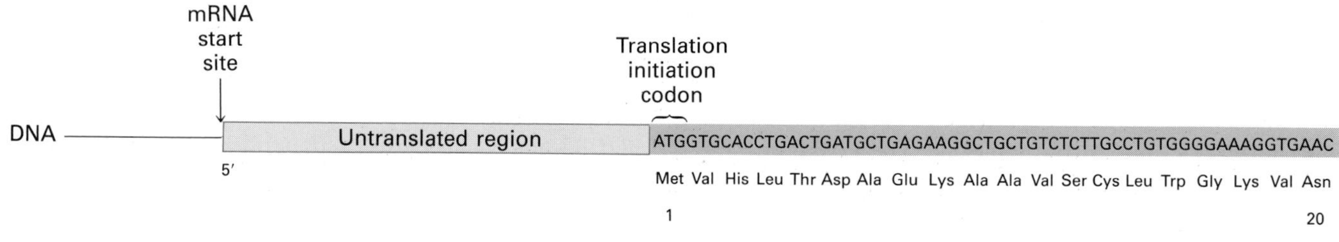

Figure 9-45 The DNA sequence that encodes mouse β globin mRNA. From the genetic code and the known sequence of the β globin protein, the protein-coding regions can be located. Two intervening sequences separate the three protein-coding portions of the mRNA. These are not present in the cDNA copy of the mRNA, which has also been sequenced. The 3′ poly A addition site and the nucleotide to which the 5′ methylated cap is added are also indicated. The sequences between the mRNA start site and the initiation codon, and between the termination codon and the poly A site, are transcribed but not translated. [See D. A. Konkel, S. Tilghman, and P. Leder, 1978, *Cell* 15:1125.]

rise to only one mRNA molecule. However, as we have mentioned (and as will be discussed in more detail later in this chapter), complex transcription units like that for the adenovirus late genes have multiple splicing arrangements. Here the definitions of intron and exon become blurred. A sequence that is an intron (i.e., that is destroyed in the nucleus) in the processing of one RNA transcript may become an exon (i.e., may become part of the mRNA) in the processing of another primary transcript from the same transcription unit. However, the terms intron and exon are widely used, and so is "intervening sequence." The terms intron and intervening sequence are used interchangeably in this book.

A Summary of the Structure of Eukaryotic mRNA and Its Pathway of Formation With the discovery of RNA splicing, the understanding of events leading to the formation of eukaryotic mRNAs was greatly enhanced. The mystery of the existence of long nuclear RNA and short mRNA, both of which have 5′ caps and 3′ poly A, was solved. Gene cloning and sequencing have shown that there is only one pathway for mRNA formation in

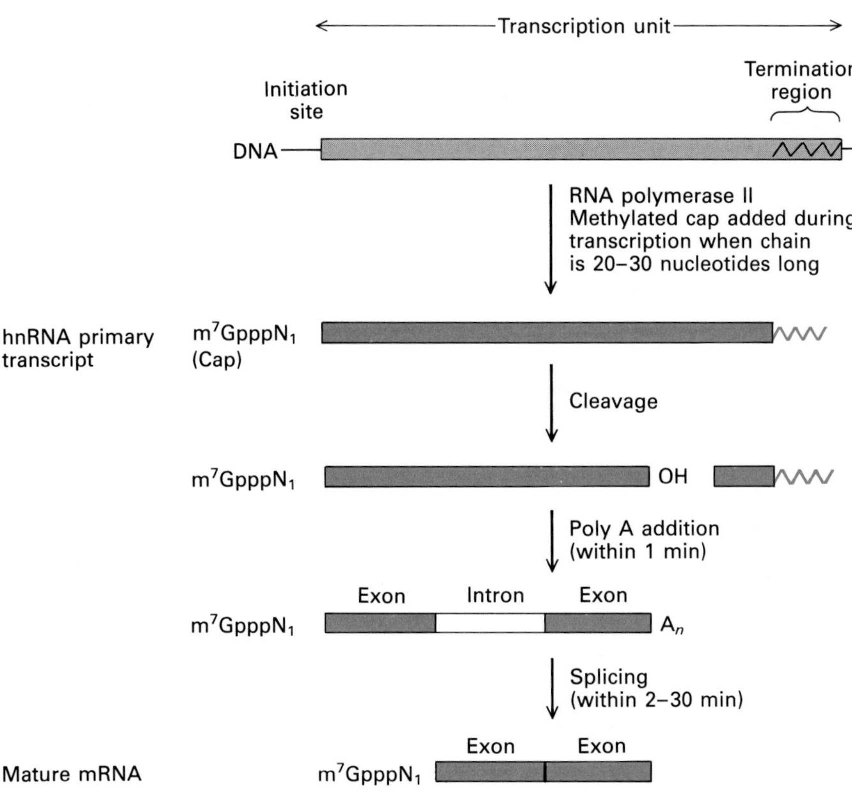

Figure 9-46 The general pathway of eukaryotic mRNA formation. Not all steps are necessary for all mRNAs. Termination regions in the DNA are known, but the exact termination sites are not. The methylation of adenylate residues, which is common in vertebrates but uncommon in single-celled eukaryotes and invertebrates, is not shown. Most of the time measurements have been made on viral mRNA molecules.

the nucleus of all eukaryotes; it differs for different mRNAs only in the extent to which primary transcripts are modified (Figure 9-46).

The central features of the pathway of mRNA biosynthesis include:

1. The initiation of transcription by RNA polymerase II. The first nucleotide of the primary RNA transcript becomes the first nucleotide of the mRNA.
2. The addition of a methylated cap structure at the 5′ end of each primary transcript when it is about 20 to 30 nucleotides long.
3. The continuation of transcription past the poly A addition site, where cleavage by an endonuclease exposes a 3′ hydroxyl end to which poly A is promptly added.
4. The splicing of primary transcripts, if necessary, within the nucleus to complete the mRNA. Some splicing reactions are complete in a few minutes, and others require from 20 to 30 min. (Because the unused RNA pieces do not accumulate, it can be assumed that they are destroyed.)

The finished mRNA molecules have distinct features (Figure 9-47). Between the capped 5′ end and the first AUG translation initiation codon, there are usually from 10 to as many as several hundred nucleotides that are not translated (see also Figure 9-45). Likewise, between the translation termination codon and the poly A at the 3′ end of the mRNA molecule, there is another untranslated region that is usually at least 50 nucleotides long and that can be as long as 2000 nucleotides. Among the functions suggested for these regions is the binding of proteins that play a role in efficient translation or in stabilization of the mRNA, but thus far these roles have not been proved by experiment.

The Biochemical Signals for the Steps in mRNA Biogenesis

Although we know the central events in the processing of primary RNA transcripts, important questions remain.

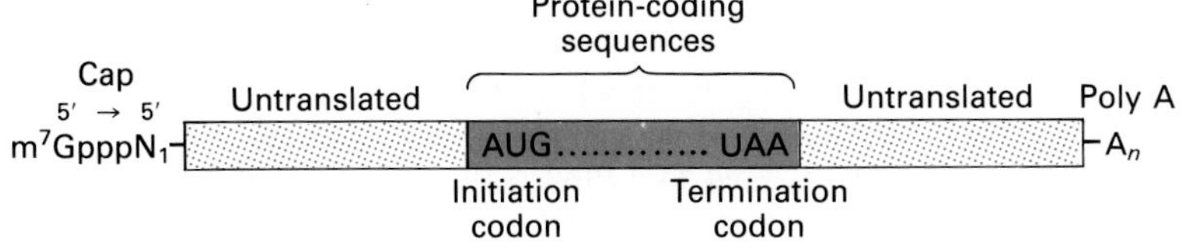

Figure 9-47 Regions of eukaryotic mRNAs, showing the terminal cap and poly A structures, the translation initiation and termination sites, and the untranslated regions at both the 5′ and the 3′ ends.

What guides an RNA molecule through the steps of processing? Are there specific molecular signals for each event, and what proteins and nuclear structures participate? What roles do the biochemical modifications play in the cell? The answers to these and other questions about the formation and functioning of mRNAs are critical to an understanding of gene structure (the subject of Chapters 10 and 11) and gene control (the subject of Chapter 12).

Questions that are perhaps even more basic concern how initiation of the transcription of eukaryotic genes occurs and how this step in mRNA biosynthesis is regulated. Because the initiation of RNA synthesis is so critical, we shall deal with it now in some detail.

RNA Polymerase II Binds to Promoter Sites and Begins mRNA Synthesis at Cap Sites

Early experiments using nascent chain hybridization (see Figure 9-11) and ultraviolet irradiation (see Figure 9-14) indicated that the RNA initiation site in the adenovirus late transcription unit was within several hundred nucleotides of the site to which the 5′ ends of all of the adenovirus late mRNAs hybridized. However, the exact start site for RNA synthesis was not known with certainty. Subsequent in vivo and in vitro experiments with transcription units of both viruses and cells supported the conclusion that RNA polymerase II starts transcription at the nucleotide to which the cap is added. This site in DNA is frequently referred to as the *cap site*. Several experimental results contributed to this conclusion.

First, no sequences complementary to the DNA sequences upstream (on the 5′ side) of the cap site have been found in nuclear precursor RNA for adenovirus late mRNAs (Figure 9-48), or for any cellular mRNAs so far examined.

Second, studies of the nucleotides involved in capping have suggested that the first nucleotide laid down by RNA polymerase receives the cap. The cap structure,

$$m^7GpppNmp$$

could theoretically be derived in two different ways:

1. Cleavage of an RNA chain preceding addition of the cap:

$$—NpNpNpNpNp—$$
$$\downarrow$$
$$—NpNp\quad NpNpNp— \text{ or } —NpN\quad pNpNpNp—$$
$$\downarrow \text{Gppp addition and methylation}$$
$$m^7GpppNmpNpNp \text{ or } m^7GpppNmpNpNp + P_i$$

2. Addition of the cap to the first nucleotide of a chain, so that two 5′-terminal phosphates of the chain are retained. For example:

$$pppNpNpNpNp$$
$$\downarrow \text{Add Gppp; allow cap reaction and methylation}$$
$$m^7GpppNmpNpNpNp + P_i + PP_i$$

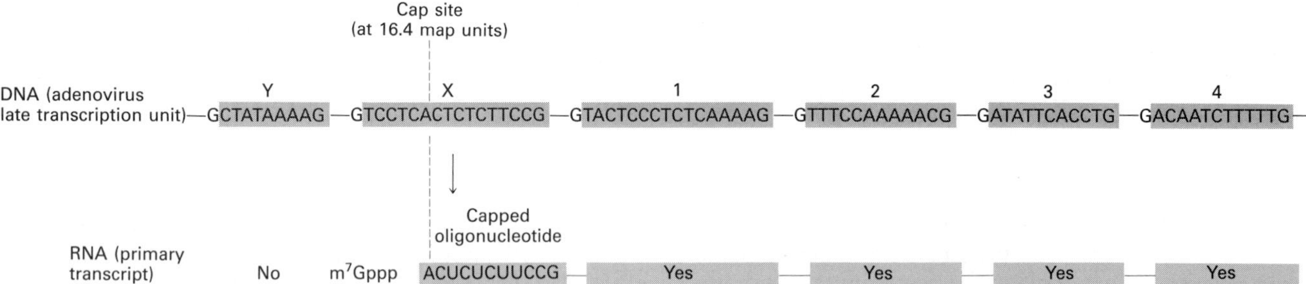

Figure 9-48 Proof of the initiation site of late adenovirus RNA. The DNA sequence known to contain the proposed RNA start site (see Figures 9-38 and 9-39) and the site to which the first leader hybridized and the binding site for the first leader (see Figure 9-41) in the mRNA was obtained (top line shows the DNA strand equivalent to mRNA). In this region several long stretches that lie between guanylate residues were identified. If such a region is transcribed into RNA, oligonucleotides ending in G's can be released by digestion with the ribonuclease T1, which cuts on the 3′ side of all G's. If *E. coli* RNA polymerase is used to nonspecifically copy this region of the DNA, all of the indicated oligonucleotides (X, Y, and 1 through 4) can be identified in the copied RNA. However, in the nuclear RNA formed in adenovirus-infected cells, oligonucleotides 1 through 4 are found, but no X or Y is found. In place of X an oligonucleotide with a cap on it is detected. Thus there is no RNA synthesis upstream (on the 5′ side) of the cap site. [See E. Ziff and R. Evans, 1978, *Cell* **15**:1463.]

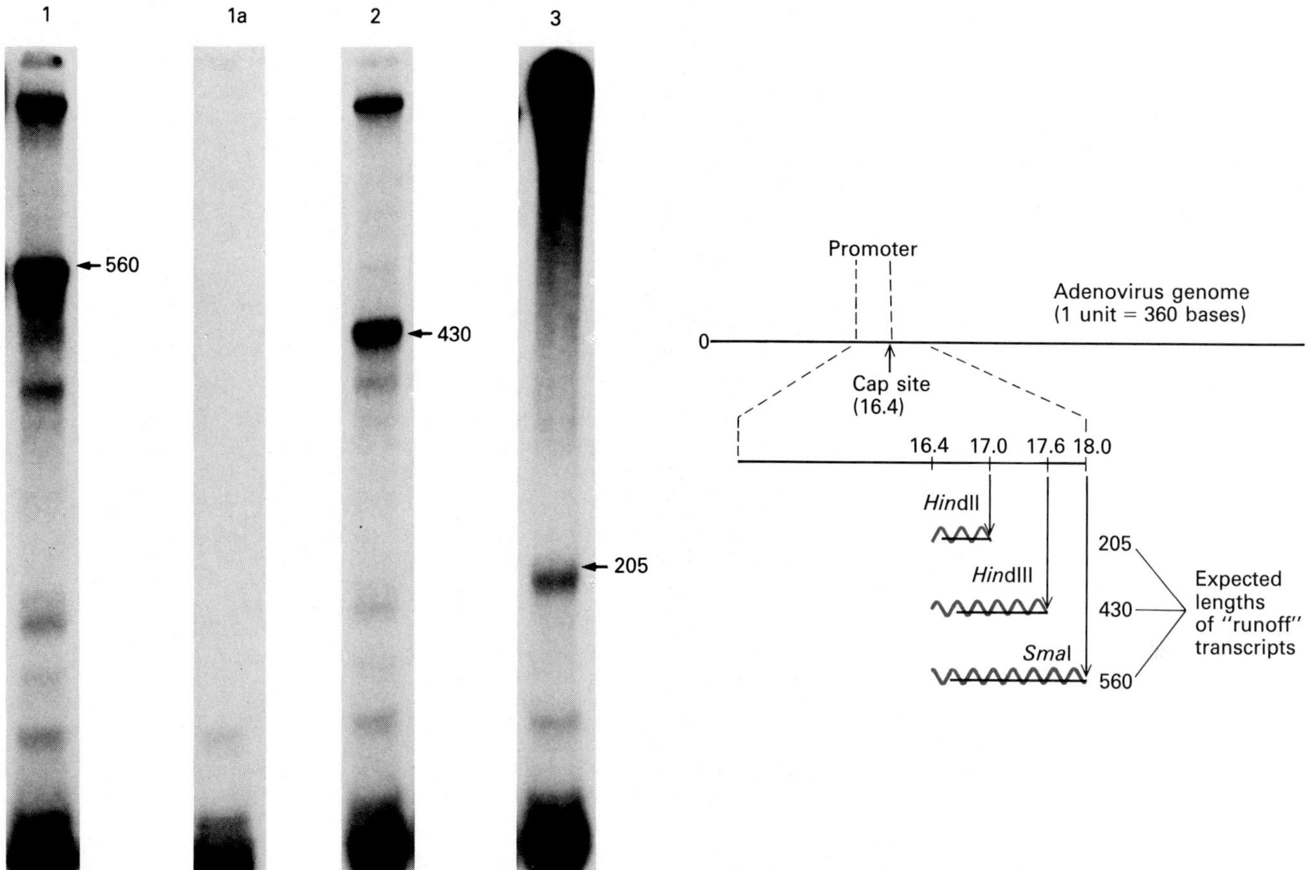

Figure 9-49 Correct in vitro initiation of RNA synthesis by RNA polymerase II. Three samples of purified adenovirus DNA that encodes the cap site (located at approximately 16.4 map units) were digested with one of three restriction enzymes that cut at 17.0, 17.6, and 18.0. The cut DNA templates were mixed and incubated with RNA polymerase II plus extracts of HeLa cells and labeled ribonucleoside triphosphates to synthesize labeled RNA. When the RNA polymerase reaches a cut end, it "runs off" the template. If the start site is the cap site, the RNA products should stretch from the cap site to the cut end; this prediction was verified by subjecting the labeled RNA to gel electrophoresis and autoradiography. Lanes 1, 2, and 3 show RNA made from DNA cut at 18.0, 17.6, and 17.0, respectively. The sample analyzed in lane 1a is the same as that in lane 1, but α-amanitin, the inhibitor of polymerase II, is included at 1 μg/ml. Thus the in vitro starting point of RNA polymerase II is the cap site. [See P. A. Weil et al., 1979, *Cell* 18:469.] *Photograph courtesy of R. G. Roeder.*

Experiments with labeled triphosphates (e.g., where N = A) have shown that the second mechanism occurs in cell nuclei.

Third, and most important, in vitro RNA synthesis with purified polymerase II starts at cap sites. In prokaryotes, the initiation site for mRNA synthesis was worked out in the 1960s and early 1970s because the single bacterial RNA polymerase would start RNA synthesis in vitro at the correct sites on bacteriophage DNA or on purified bacterial DNA. For many years after the discovery of the three polymerases in eukaryotic cells, and even after it had become clear that polymerase II makes hnRNA, it was not possible to obtain evidence of the correct initiation of mRNA in vitro because the initiation sites were not known. Finally, after the initiation region of the adenovirus late transcription unit was identified, attempts to detect site-specific initiation of viral RNA were successful, and the in vitro start site was the same as the cap site (Figure 9-49).

After the initial success with adenovirus DNA, crude extracts of cells mixed with RNA polymerase II and many different specific cloned DNA segments were shown to initiate RNA synthesis at the cap site. In addition to starting RNA chains at the cap site, the cell extracts also add the complete methylated cap structure. In fact, no chains started by RNA polymerase II either in cell extracts or inside the cell lack a cap. Even prematurely terminated chains are capped. As is the case in the cell nucleus, the

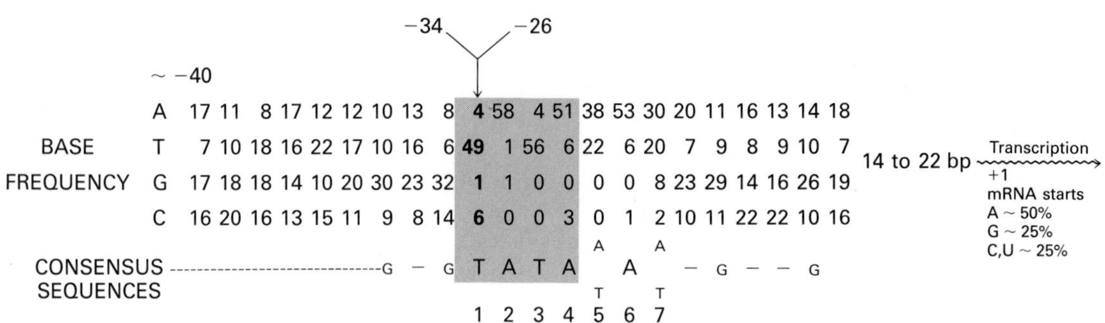

Figure 9-50 The TATA box in eukaryotic DNA (color) is located upstream of the 5′ end of protein-coding genes. DNA strands containing sequences of the same 5′ → 3′ polarity as mRNAs from 60 different protein-coding genes were compared, beginning at about −40 (40 bases upstream of the +1 mRNA start sites). Each sequence was aligned to yield maximum homology in the region from −35 to −20. The tabulated numbers give the frequency of each base at each position out of the total of 60. The maximum homology occurs over a seven-base consensus sequence in which the first four bases are TATA. The TATA box begins about 30 bases from the start site. The most frequent mRNA start is an A but a pyrmidine makes up about 25% of the starts. [See R. Breathnach and P. Chambon, 1981, *Annu. Rev. Biochem.* 50:349.]

middle phosphate of the cap added in vitro derives from the first nucleotide of the RNA chain (see possibility 2 above). The conclusion that mRNA synthesis is initiated at the cap site is now widely accepted.

Promoter Sequences for RNA Polymerase II What are the signals that direct RNA polymerase II to begin synthesis at the cap site? Examination of the DNA sequences of a variety of different transcripton units has revealed a short consensus sequence that starts about 30 bases upstream of the cap site. This highly conserved sequence has been named the *TATA box* (Figure 9-50).

The importance of the TATA box in correctly positioning the RNA polymerase II in vitro was demonstrated by experiments with recombinant molecules. If the sequences downstream from the TATA box of the adenovirus late transcription unit are replaced by other sequences, the polymerase still begins transcribing at a nucleotide approximately 30 nucleotides downstream (Figure 9-51). However, a single base change within the TATA box (e.g., G or A substituted for the second T) drastically decreases transcription (Figure 9-52). After the TATA box directs the binding of the RNA polymerase II, the nucleotide selected as the start site for RNA synthesis

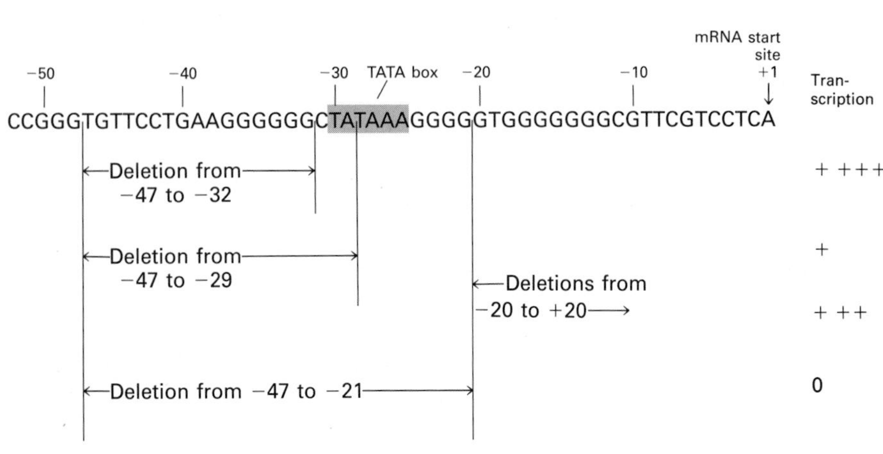

Figure 9-51 The TATA box as a functional element of the adenovirus major late promoter. The promoter region can be deleted at a number of sites and replaced with other DNA (bacterial plasmid DNA, for example). In vitro synthesis by RNA polymerase II normally begins 30 nucleotides downstream of the TATA box. Removal of all or part of the TATA box (as in two of the deletions shown here) stops synthesis. When the TATA box is left intact and sequences are removed either upstream (the deletion from −47 to −32) or downstream (various deletions from −20 to +20), RNA synthesis continues to begin 20 to 30 bases from the TATA box. [See J. Corden et al., 1980, *Science* 209:1406.]

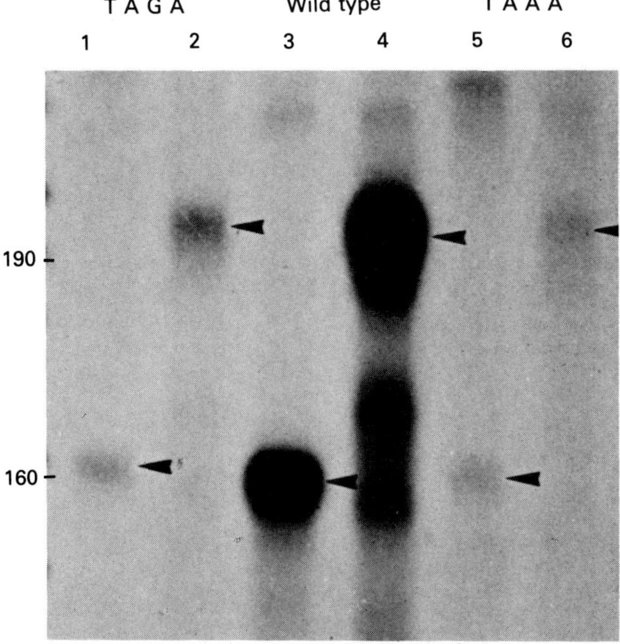

Figure 9-52 Transcription of recombinant DNA with sequence variations in the TATA box. The second T in the wild-type TATA box of the chicken conalbumin gene was replaced, by in vitro mutagenesis, with a G (to yield TAGA) or with an A (to yield TAAA). When wild-type DNA that had been cut with two different restriction enzymes was transcribed, the products were two RNA bands at about 160 and 190 nucleotides, respectively *(lanes 3 and 4)*. When the mutated DNA was transcribed, these products were greatly decreased *(lanes 1, 2, 5, and 6)*. [See B. Wasylyk and P. Chambon, 1981, *Nuc. Acids Res.* 9:1813.] *Courtesy of P. Chambon.*

varies somewhat. The polymerase II usually starts with a purine, but a pyrimidine can be the first nucleotide (25 percent of the time in mammalian cells; less frequently in lower cells).

The eukaryotic TATA box is reminiscent of the Pribnow box, the conserved oligonucleotide centered at about −10 in bacterial genes (see Figure 8-4). Recall that in bacteria, a second site centered about 35 bases upstream of the start site also facilitates polymerase binding. The in vitro transcription of eukaryotic DNA has shown the importance of the TATA box; sequences farther upstream have much less effect on the in vitro reaction. However, when DNA is introduced into cells via a plasmid or a virus—and when viral DNA is integrated into chromosomes of cultured cells—additional sequences are required to achieve mRNA production.

For example, transcription of the yeast *his*4 gene (which encodes an enzyme active in histidine biosynthesis), of the thymidine kinase gene of the herpes virus, and of the human β globin gene depends in each case on sequences located from about −110 to −40 (Figure 9-53). Presumably, this means that transcription is initiated in vivo only when these upstream sequences are present, although direct measurements of RNA synthesis have been made in just a few cases. The upstream sequences of the large family of globin genes (those for both α and β globin) contain the sequence CAAT around −80. When many different genes are examined, the sequence conservation around −80 is not as impressive as that in the TATA box, but there is a recognizable consensus region. In general, the segment from −110 to −40 contains two elements (designated "first" and "second" in Figure 9-53) that are both necessary for in vivo transcription. It is very likely that the TATA box and the sequences from −110 to −40 form at least part of the promoter site for RNA polymerase II.

Enhancers The promoter elements just described are thought to act in binding the RNA polymerase II and in locating the RNA initiation site. Another type of DNA sequence element that has a strong but poorly understood effect on transcription by RNA polymerase II has been discovered. An *enhancer* sequence is a DNA sequence that somehow, without regard to its position or its orientation in the DNA, increases the amount of RNA synthesized from DNA introduced into cells. Enhancers can be several thousand base pairs away from—and can appear in either orientation with respect to—the transcription unit in which they increase transcription.

The enhancer effect was first found in studies of deletion mutants of the SV40 virus. In the DNA sequence that precedes the cap site for the SV40 early transcription unit there is a 144-nucleotide region consisting of two 72-base-pair repeats in tandem. The entire viral DNA, including the upstream region, was cloned in an *E. coli* plasmid so that it could be grown in large amounts. In some experiments DNA sequences such as those encoding for the β globin mRNA, were inserted into the viral DNA just after the cap site and in place of the SV40 early region for use as a probe to study the effect on transcription of various changes in the upstream tandem repeats. (Figure 9-54). Removal of one of the 72-base-pair elements did not interrupt the production of β globin mRNA or SV40 early mRNA for T antigen, but removal of both did. Moreover, the insertion of a single 72-base-pair segment at any site, regardless of its distance from the RNA initiation site, restored activity. In addition, the enhancer element could be inserted in either a $5' \rightarrow 3'$ or a $3' \rightarrow 5'$ orientation and still stimulate mRNA production. To summarize, then, two properties define enhancer elements: action over long distances and action in either orientation in the genome.

Subsequent experiments have shown that other viruses (including other small circular DNA viruses such as polyoma virus and bovine papilloma virus, as well as retrovi-

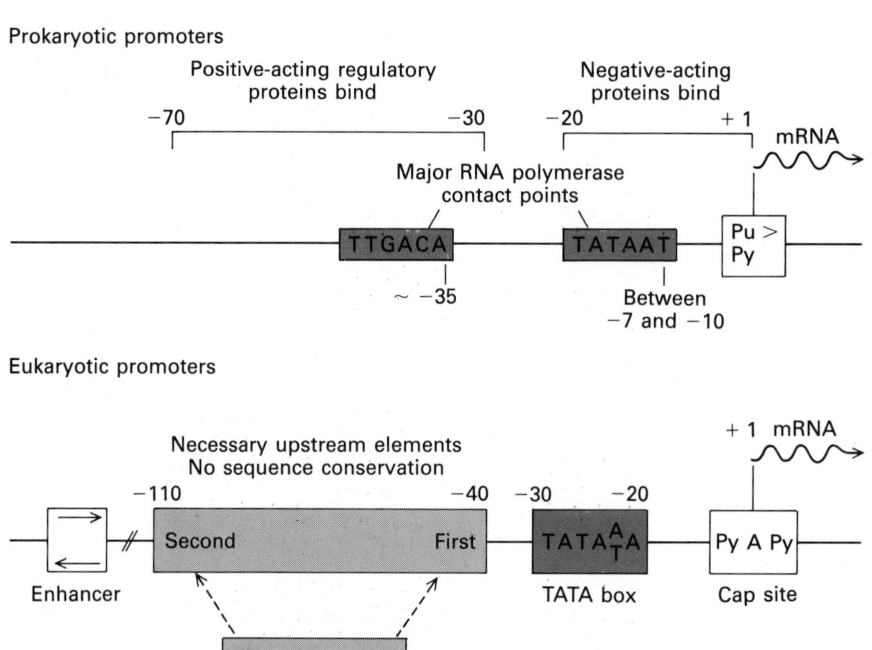

Figure 9-53 A comparison of the DNA elements that affect transcription in prokaryotes and eukaryotes.

Prokaryotic promoters have two sequence elements, one centered about 35 bases and one centered about 10 bases upstream of the start site for mRNA. The RNA polymerase makes contact with the DNA in both of these regions. Messenger RNA synthesis begins more often with purines than with pyrimidines (Pu > Py). Regulatory proteins bind as indicated in the regions from −70 to −30 to increase transcription (positive regulation) and in the region from −20 to +1 to decrease transcription (negative regulation).

In eukaryotes, the DNA sequences necessary for transcription are found in at least three regions close to the RNA initiation site: the TATA box and two upstream elements located between −110 and −40. In many genes, a GC-rich region associated with a CAAT sequence may serve as one of the upstream elements. Finally, an enhancer region (see text, next section, and Figure 9-54) may be present at greatly varying distances from the initiation site. How the various necessary eukaryotic promoter sites participate in regulation is not well known at present. [See S. L. McKnight and R. Kingsbury, 1982, *Science* **217**:316; and R. Breathnach and P. Chambon, 1981, *Annu. Rev. Biochem.* 50:349.]

ruses and adenoviruses) contain enhancer elements. Some cell transcription units also have regions that behave like enhancer elements. In at least one case—the immunoglobulin heavy chain—a sequence with the properties of an enhancer lies *within* the transcription unit. Recombinant plasmids containing this enhancer and the 5′ sequences of the heavy chain gene produce mRNA in cultured myeloma cells that make immunoglobulin but not in other cell types. The extent of the presence of enhancers in cell genes and their degree of tissue specificity is just beginning to be studied. Comparisons of sequences from different viruses that contain enhancer elements have not revealed striking sequence homology, so in most cases the exact functional sequences are not known.

The role of enhancer elements in stimulating transcription has not yet been determined. It is known that the requirement for enhancer sequences is sometimes eliminated by the presence of certain viral proteins [e.g., an early protein product (E1A) of adenovirus]. Two general explanations of enhancers may be offered: (1) they bring DNA into a form that enables it to be actively transcribed (or they direct it to a cellular location at which the chance of transcription is increased); or (2) enhancers may simply help to sequester transcriptional factors that may be

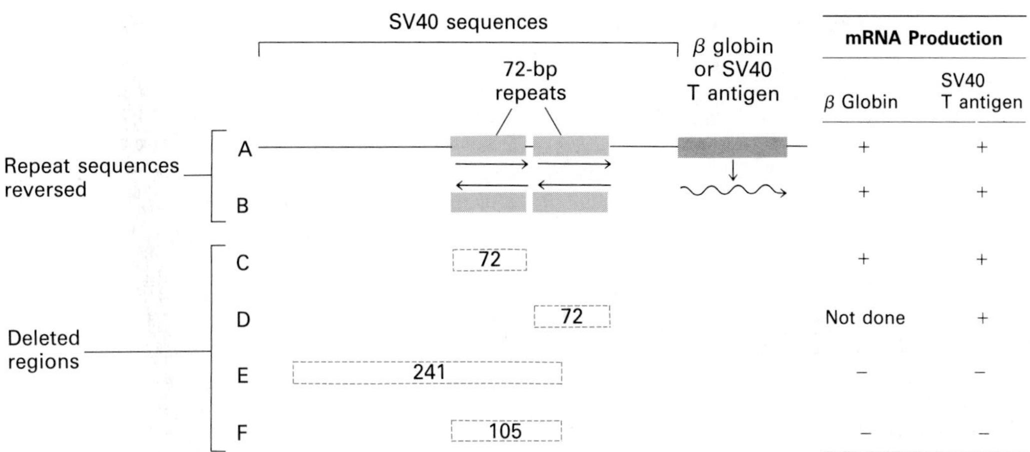

	β Globin	SV40 T antigen
A	+	+
B	+	+
C	+	+
D	Not done	+
E	–	–
F	–	–

Figure 9-54 The demonstration of enhancer sequences in SV40 DNA. The diagram shows a portion of a recombinant DNA molecule containing SV40 DNA sequences. This segment lies between the two RNA initiation sites for early and late RNA synthesis (at about 8 o'clock on the circular genome depicted in Figure 9-43). The SV40 sequences were joined to the protein-coding portion of the β globin or to the SV40 T antigen. The synthesis of the mRNA for β globin and SV40 T antigen were measured after introduction of different recombinants. The table at the right indicates whether or not each mRNA was synthesized when recombinant plasmids of various designs were introduced into cells. The structures of the plasmids are shown at the left. The 72-base-pair repeat stimulated mRNA production when it was inserted in either orientation *(samples A and B)*. Deletion of one or the other of the 72-base-pair repeats did not affect mRNA production *(samples C and D)*, but an additional deletion of a portion of the remaining 72-base-pair repeat *(samples E and F)* did abolish mRNA production. [See J. Banerji, S. Rusconi, and W. Schaffner, 1981, *Cell* **27**:299; and M. Fromm and P. Berg, 1982, *J. Mol. App. Genet.* **1**:457.]

in limiting supply. The possible roles of enhancers in transcriptional regulation are discussed in more detail in Chapter 12.

Transcriptional Termination by RNA Polymerase II Occurs Downstream of Poly A Addition Sites

Once transcription has been initiated by RNA polymerase II, where does it end? We have already mentioned that the primary transcript continues past the poly A addition site before terminating. This discovery was made by hybridizing labeled nascent nuclear RNA to DNA regions both upstream and downstream of sites to which poly A is attached (Figure 9-55). The uniform result, obtained first with adenovirus and SV40 transcription units, and subsequently with a number of vertebrate transcription units (e.g., mouse β globin, mouse α amylase, and rat calcitonin), was that transcription regularly passes the poly A site. For all of these cellular genes, transcripts appear to extend about 0.5 to 2 kb past the poly A site. However, no single preferred site for termination has been detected. The sequences responsible for termination after the β globin gene in the mouse chromosome also cause termination when they are transferred to another gene (an adenovirus gene). Thus there are probably conserved sequences whose job it is to ensure termination in the eukaryotic cell.

If termination is not responsible for creating the 3′ end of the mRNA, how is it formed? Experiments using recombinant DNA to study the sequences at the ends of mRNA have provided a convincing answer. About 15 to 30 nucleotides upstream from the poly A site appears a strongly conserved consensus sequence, AAUAAA (Figure 9-56). Removal of this sequence prevents synthesis of mRNA from the transcription unit. Mutation within the AAUAAA (for example, to AAGAAA) allows transcription but prevents correct mRNA formation. Finally, removal of sequences starting about 12 nucleotides downstream from the poly A addition site also prevents correct mRNA formation. In the latter two cases (mutation of AAUAAA to AAGAAA and removal of downstream sequences), long RNA molecules that stretch across the poly A site can be detected (Figure 9-57). Only very occasionally do these transcripts yield mRNAs containing poly A added at the correct site. Such experiments have suggested that the AAUAAA sequence and the sequences downstream from the poly A site cooperate to signal the location of the poly A site, which results in cleavage of the primary transcript and addition of the poly A.

These two events, cleavage at the poly A site and addition of about 250 adenylate residues, occur very close in

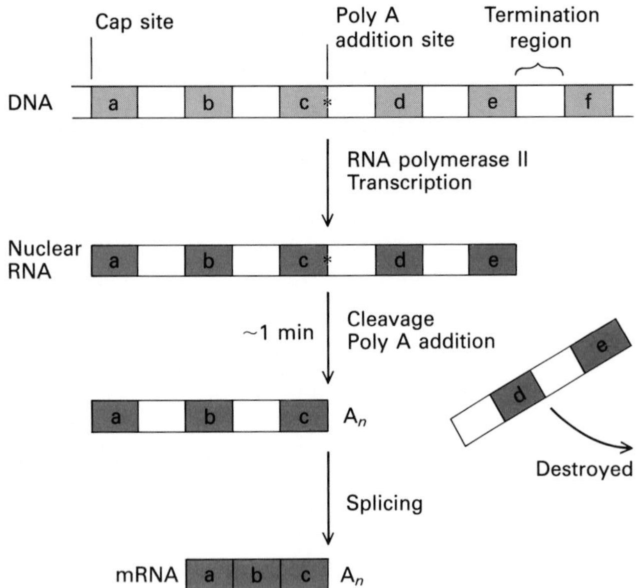

Figure 9-55 Poly A addition is distinct from transcriptional termination in mammalian cells. The diagram shows a section of DNA from which restriction fragments a to f have been purified. Segments a, b, and c represent exons, portions of the primary transcript found in mRNA. The hybridization of pulse-labeled RNA to the DNA fragments was used to derive the following findings and conclusions: (1) Labeled nascent nuclear RNA hybridizes equally well to a, b, c, d, and e: transcription passes the poly A site, and cleavage is necessary for poly A addition. (2) Nuclear RNA does not hybridize to f: termination occurs in the vicinity of e to f. (3) Sequences d and e do not accumulate in the cell: they are destroyed. (4) ^{3}[H]Uridine appears in the c segment of poly A^+ RNA no more than 1 min after it appears in a: poly A addition occurs very soon after transcription. (5) RNA with newly labeled poly A is not spliced: splicing generally occurs after poly A addition. [These experiments are reviewed in J. E. Darnell Jr., 1982, *Nature* **297**:365.]

time in the cell, but they can be separated. The addition of 3′-deoxyadenosine to a cell culture completely stops the addition of poly A in cells, but the primary transcript is still cut correctly. Even in histone mRNAs that do not receive a poly A, the 3′ end is created by cleavage of a primary transcript. Again, sequences downstream from the cleavage site are required and the site is located with the aid of a specific small nuclear ribonucleoprotein (different from U1, U2, and so forth).

Methylation of Adenylate Residues Is a Prominent Feature of Vertebrate mRNAs

Many mRNAs of higher organisms (mostly mammalian cells have been studied) are methylated at the 6 nitrogen of adenylate residues within the mRNA chain. Some lower organisms, such as yeasts and slime molds, seem to have few, if any, methylated adenylates. The role of these methyl groups has not been determined, but it is known that in the course of adenovirus infection, the methyl groups are added to nuclear RNA precursors that have not yet been spliced. Also, methyl groups are added within the adenovirus primary transcript region that remains part of the mRNA. As in pre-rRNA molecules, then, the methyl groups may play some role in protecting the portion of the primary transcript that is to be preserved.

The Splicing Reaction Probably Requires Interactions between Ribonucleoproteins

All splicing of protein-coding mRNAs occurs in the nucleus. This is known because unspliced mRNA precursors are not found in the cytoplasm, and, as previously mentioned, isolated nuclei can carry out the splicing reaction. However, the enzymes and other factors responsible for the splicing of mRNA have not yet been purified. Cells exhibit only a limited species specificity in the splicing of primary transcripts. For example, a virus that requires splicing to produce its mRNA may grow in a variety of different mammalian cells. In addition, recombinant molecules containing rabbit or chicken genomic DNA can be transcribed, and their transcripts can be correctly spliced, in mouse cells or frog oocytes. Therefore, the biochemical

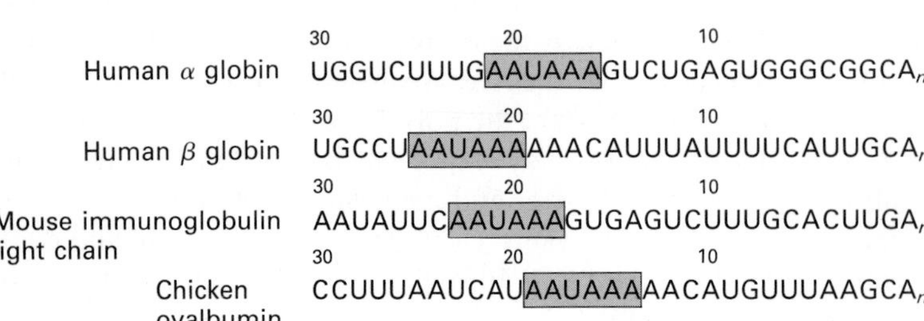

Figure 9-56 A comparison of sequences adjacent to poly A in four eukaryotic mRNAs (A_n is poly A.) The sequence AAUAAA is present in each mRNA, and it is located about 20 bases upstream of the poly A addition site. Well over 100 sequences are now known to conform to this rule, and only a few mRNAs fail to show the conserved sequence. [See N. J. Proudfoot and G. G. Brownlee, 1976, *Nature* **263**:211.]

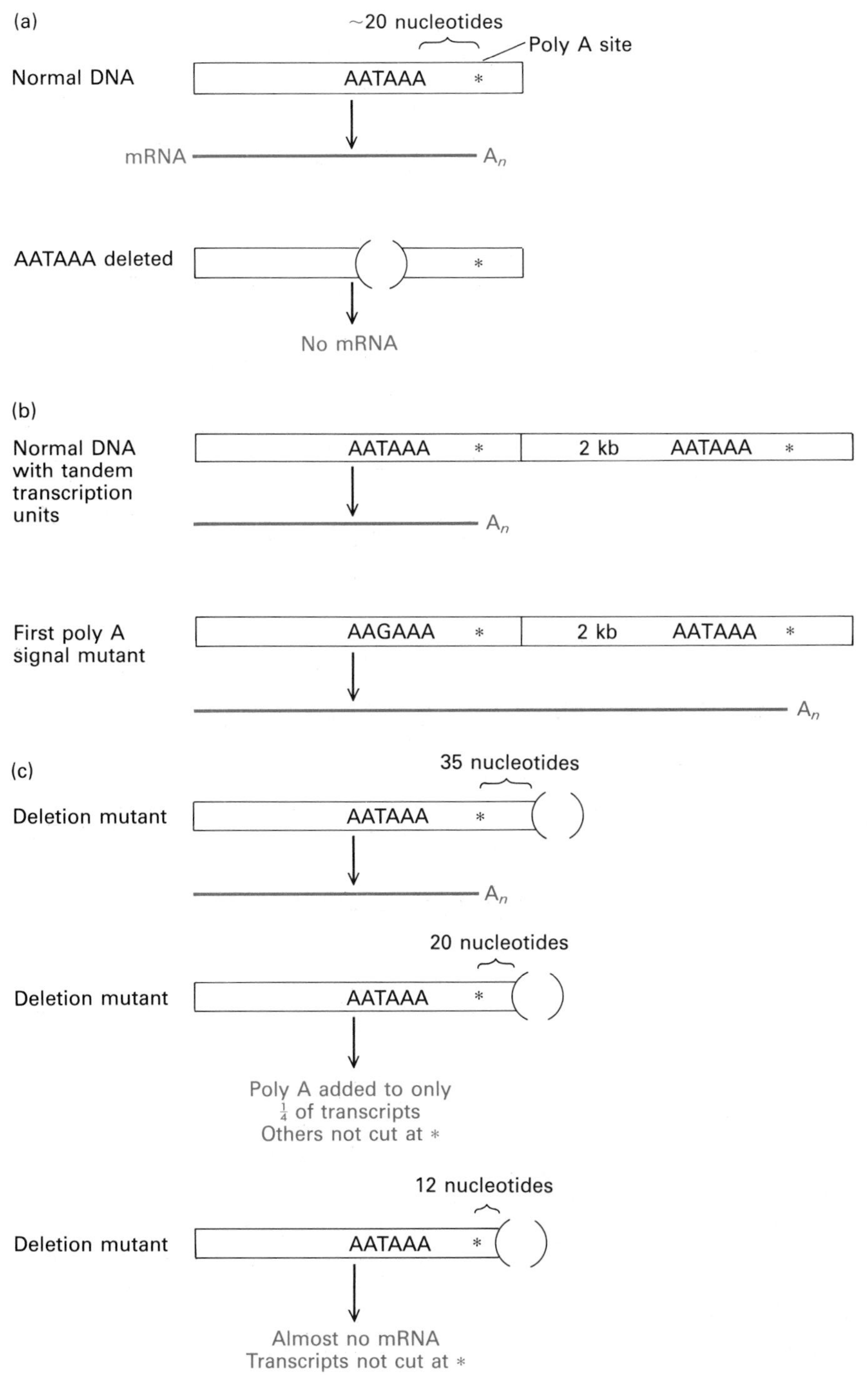

Figure 9-57 Sequences involved in poly A addition. Recombinant viruses and recombinant plasmids containing transcription units with a variety of changes upstream and downstream from the poly A sites have been constructed. (AAUAAA, the poly A signal in RNA is shown as AATAAA in the DNA strand equivalent to mRNA.) (a) Transcription of deletion mutants lacking the AATAAA sequence shows that this sequence is necessary for poly A addition. (b) A single base change (from T to G) in the consensus sequence prevents the signal from working, presumably because the primary RNA transcript is not cut at the proper site. In the example here, a recombinant molecule containing two neighboring transcription units normally yields approximately equal quantities of two mRNAs—one from the first transcription unit and one (not shown) from the second. When the first poly A signal is mutant, cutting of the primary transcript only occurs at the second poly A site. (c) Some DNA deletions following the poly A site reduce or prevent poly A addition. The downstream sequences involved probably cooperate with the AAUAAA in the primary RNA transcript to signal the correct cut site. [See M. Fitzgerald and T Shenk, 1981, *Cell* **24**:251; C. Montell et al., 1983, *Nature* **305**:600; and M. A. McDevitt et al., 1984, *Cell* **37**:993.]

mechanisms of splicing are probably fairly universal, at least in vertebrate cells.

Analyses of the sequences adjacent to the splice sites for a number of different mRNAs from the same organism and from different organisms have revealed only a moderately conserved short consensus sequence at the splice junctions (Table 9-3). The only universally conserved nucleotides in the junction sequence are the first two and the last two in the intron:

(5′) Exon ⦚ GT . . . AG ⦚ Exon (3′)

It is possible to prepare recombinant DNA containing the 5′ exon-intron junction of one transcription unit (e.g., the SV40 late region) and the 3′ intron-exon junction of another transcription unit (e.g., the mouse β globin gene). When such artificial DNA is introduced into cells, the result is a spliced mRNA molecule that joins the two exon sequences with the loss of the chimeric intron. This outcome shows that the 5′ and 3′ splicing recognition sequences are equally compatible when they come from different transcription units. Other experiments with recombinant DNA have shown that as few as 20 to 30

Table 9-3 A summary of DNA sequence analyses at boundaries between exons and introns in protein-coding genes

EXON-INTRON JUNCTIONS (5′ splice junctions)	(5′) Exon				: Intron							
Position of nucleotide*	−4	−3	−2	−1	+1	+2	+3	+4	+5	+6	+7	+8
Number of cases with each nucleotide:												
A	42	56	89	12	0	0	86	94	12	23	53	33
T	28	10	18	17	0	139	9	16	7	87	30	36
C	42	60	16	8	0	0	3	13	3	17	28	40
G	27	13	16	102	139	0	41	16	117	12	25	27
Number of cases analyzed†	139	139	139	139	139	139	139	139	139	139	136	136
Consensus‡	—	A/C	A	G	G	T	A/G	A	G	T	—	—

INTRON-EXON JUNCTIONS (3′ splice junctions)	Intron														: Exon (3′)		
Position of nucleotide*	−15	−14	−13	−12	−11	−10	−9	−8	−7	−6	−5	−4	−3	−2	−1	+1	+2	+3
Number of cases with each nucleotide:																		
A	17	11	11	19	8	19	14	24	15	4	13	33	5	130	0	29	22	25
T	58	50	57	67	75	62	62	57	57	73	75	38	40	0	0	11	48	37
C	21	28	35	27	30	38	42	35	46	46	36	28	84	0	0	23	28	42
G	17	24	11	13	13	7	9	11	9	6	6	31	1	0	130	67	32	26
Number of cases analyzed†	113	113	114	126	126	126	127	127	127	129	130	130	130	130	130	130	130	130
Consensus‡	T/C	T/C	T/C	T/C	T/C	T/C	T/C	T/C	T/C	T/C	T/C	—	C/T	A	G	—	—	—

*The positions (−4, −3, etc., and +1, +2, etc.) signify the distance before (−) or after (+) the junction.

†The examples studied were from diverse organisms, including viruses, soybeans, sea urchins, fruit flies, mice, and humans.

‡Summary of results: The strongest consensus sequences are at the ends of the introns. At the 5′ end, the first two bases are GT in every case, and the 3d through the 6th positions are strongly conserved. At the 3′ end, a distinctive stretch of pyrimidines is followed by AG, the last two bases in each case. Elsewhere, the consensus sequence is much weaker; the accordance rate in exons drops to less than 50 percent within two bases of the junctions.

SOURCE: S. M. Mount, 1982, *Nuc. Acids Res.* **10**:459.

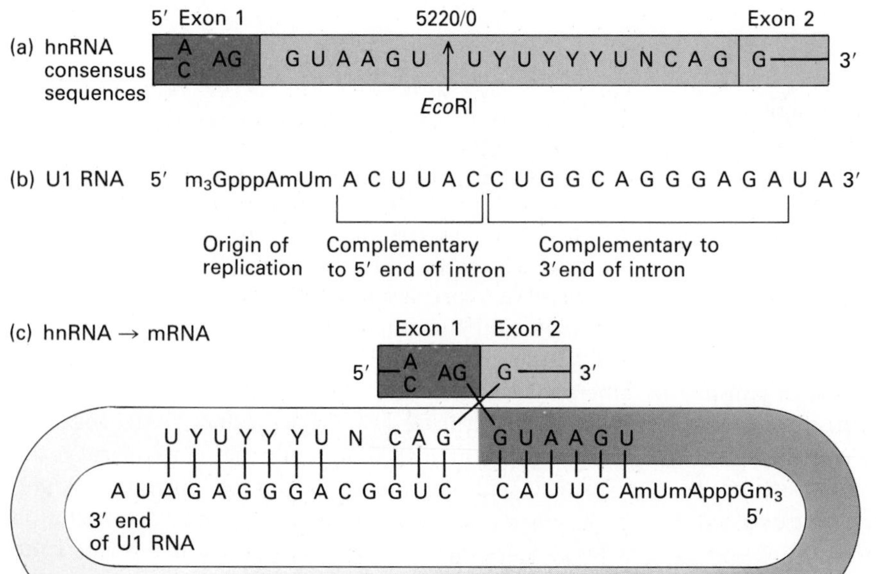

Figure 9-58 Possible base-pairing interactions between the U1 RNA and splice-junction sequences. (a) More than 100 cloned DNA sequences that encode mRNA were analyzed to determine the consensus sequences at exon-intron and intron-exon junctions (as in Table 9-3). Y is a pyrimidine; N is any base. (b) Brackets indicate the regions of U1 that are postulated to base-pair with the 5′ and 3′ ends of the introns. (c) A model of the base-paired structure. The two guanylates in the exons at the splice junctions would be joined after excision of the intron to make an mRNA. Experiments strongly support the model for the 5′ interaction with the 5′ end of U1 RNA. [See M. R. Lerner et al., 1980, *Nature* **283**:220.]

nucleotides of the intron sequence at each splice junction suffice for the splicing reaction to occur correctly.

Because the sequences near the ends of introns are highly conserved over only a very short region, a popular hypothesis is that some additional cellular elements must play a role in holding the two splice junctions in the correct position for splicing. Promising candidates are the small ribonucleoprotein particles containing U1 or U2 RNAs. At least some of these are associated with hnRNA in the nuclei of metazoan cells. The sequence of the U1 RNA is very similar—particularly at its 5′ end—in a variety of vertebrates and in *Drosophila;* this observation indicates that the U1 RNA particle has undergone little evolutionary change. A region of complementarity lies between the 5′ end of the U1 RNA and the hnRNA consensus sequence at the 5′ end of the intron (Figure 9-58).

The role of the U1 and U2 ribonucleoproteins in splicing has been tested by using both isolated nuclei and extracts of whole nuclei, either of which can carry out the correct splicing reaction. The addition of antibodies to ribonucleoprotein particles containing U1 or U2 stops the in vitro splicing reaction. Clearly a good deal of experimental support exists for the idea that U1 and U2 ribonucleoprotein particles help to align the hnRNA sequences so that splicing can occur.

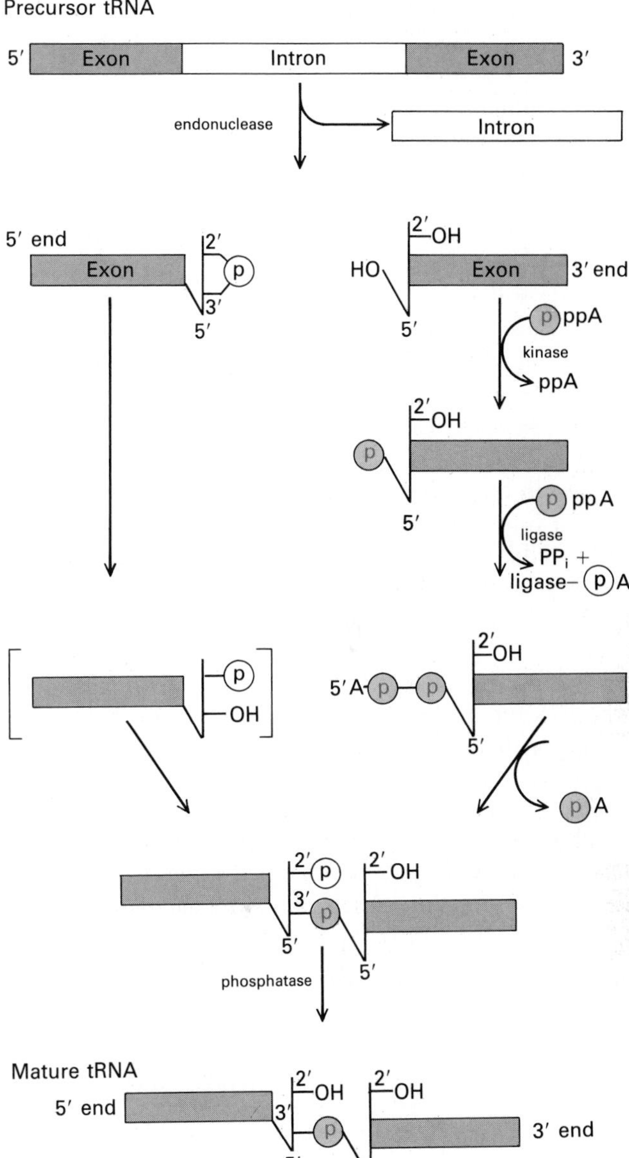

Figure 9-59 A model for the removal of an intron in tRNA. Partially purified extracts of yeast carry out tRNA splicing in two steps: (1) the precise excision of introns by cleavage of the precursor at two sites, and (2) the ligation of the two exons of the tRNA with the introduction of a phosphate not previously in the chain. The cleavage leaves a 5′ exon with a 2′,3′-cyclic phosphomonoester on its exposed end and a 3′ exon with an exposed 5′ hydroxyl. A kinase and a ligase use two ATP molecules to form a phosphoanhydride (Ap—p—) linkage on the exposed end of the 3′ exon. The nature of the interaction of the two exons is not certain, but the resulting 2′-phosphomonoester and the 3′ → 5′ linkage at the ligation site have been verified. The fact that the 2′-phosphomonoester does not appear in the finished molecule implies that a 2′-phosphatase exists. [See C. L. Greer et al., 1983, *Cell* **32**:537.]

The Biochemistry of RNA Splicing

We have mentioned that rRNAs are spliced in a few organisms, and that some tRNAs are spliced in most, if not in all, eukaryotes. We have also just dwelt in detail on the widespread occurrence of splicing in eukaryotic mRNAs. In this section we shall discuss the chemical details of splicing. Much progress has been made recently in understanding the biochemical events that occur during splicing, even though all of the necessary enzymes and cofactors have not yet been identified. Three different types of splicing reactions have been recognized and exemplified in studies of yeast tRNA, of mammalian cell or virus mRNAs, and of rRNA from the protozoan *Tetrahymena pyriformis.* A description of how each of the three different splicing reactions occurs in nature and a consideration of the evolutionary role of splicing may be found in Chapter 25.

Possibly, the best understood of these three reactions results in the removal of the short intron from the anticodon loop of tRNA (see Figure 9-24). There are two steps in the reaction: the cutting of the primary transcript twice to remove the intron, and the sealing together of the ends. These steps can be carried out separately by partly purified extracts. The first step takes place in the absence of ATP, but the second step requires it (Figure 9-59). The phosphate group that links the two halves of the finished tRNA comes from one of the two ATP molecules used during the reaction. Particularly important is the unusual 2′-phosphomonoester that exists together with the 3′ → 5′ phosphodiester linkage at the splice junction just after ligation is completed.

The enzymes involved in mRNA splicing have not yet been isolated. However, the reaction pathways have recently been outlined through the use of cell extracts that correctly splice short synthetic RNA substrates containing introns. The most popular substrates are stretches of adenovirus and globin RNA. The first step in the reaction is cleavage at the 5′ exon-intron junction, which leaves a 3′ hydroxyl on the 3′ nucleotide of the first exon and a 5′ phosphate on the guanylate that is always the 5′ nucleotide in the intron (Figure 9-60). Finding and cutting the 5′ exon-intron junction is the step that requires the U1 snRNP. The next step that requires the U2 snRNP involves the formation of a branched nucleotide structure similar to that described above for tRNA. The nucleotide at the branch point not only has a conventional 3′ → 5′ linkage to the next nucleotide in the chain, but the branch-point nucleotide also participates in a 2′ → 5′ linkage with the guanylate on the 5′ end of the intron (Figure 9-60). This structure was first characterized in hnRNA in the nucleus of cultured cells.

The nucleotide at the branch point is almost always an adenylate that lies 10 to 30 nucleotides from the 3′ intron-exon junction. Adenylate is the branch point in both the adenovirus and the globin RNAs. A consensus sequence exists in the branch-point region:

$$\mathrm{CU}\binom{\mathrm{G}}{\mathrm{A}}\mathbf{A}\binom{\mathrm{C}}{\mathrm{U}}$$

where **A** is the branch point. This sequence occurs in all vertebrate mRNAs so far examined. Note that the vertebrate branch-point consensus sequence is similar to the last five nucleotides, *CUAAC*, of the yeast sequence:

UA*CUAAC*

which is found in all mRNA introns about 50 nucleotides upstream of the 3′ intron-exon junction in yeast. Thus it may be that all introns are removed from mRNA precursors by a reaction involving a similar recognition sequence.

The mechanism which locates the 3′ intron-exon junction and completes the final step of cleavage and ligation has not yet been worked out. In the mRNAs studied so far, the first AG dinucleotide downstream from the branch point is the site of the second cleavage. Because almost all introns are pyrimidine-rich at their 3′ ends, they lack AG sequences in this stretch. Therefore, it is plausible that, having formed the branch point within the CU(Pu)**A**(Py) sequence cited above, the splicing mechanism scans downstream to find the next AG and com-

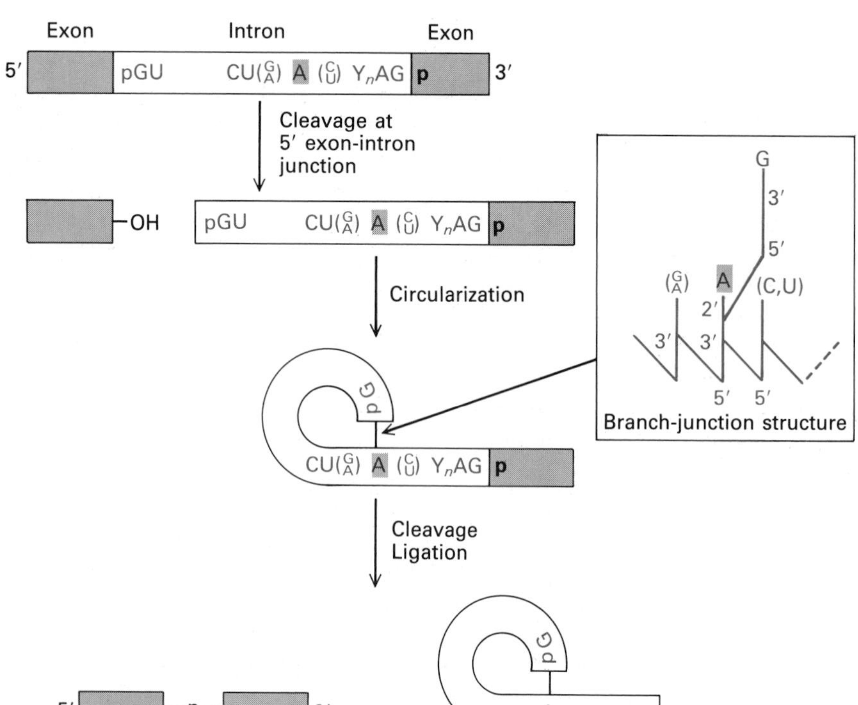

Figure 9-60 The pathway of mRNA splicing. The diagram illustrates the steps that have been shown to occur during in vitro splicing of adenovirus and globin RNA sequences by cell extracts. The consensus nucleotides GU and AG that begin and end the intron are indicated, as is the pyrimidine-rich stretch (Y_n) near the 3′ end. After the cleavage of the first (5′) exon-intron junction, a circularization occurs to create a branched structure in which the guanylate at the cut 5′ end of the intron forms an unusual linkage (5′ → 2′) with an adenylate located near the 3′ end of the intron *(inset)*. (The adenylate also retains its normal 3′ → 5′ and 5′ → 3′ linkages to adjacent nucleotides.) In a number of different hnRNAs, the adenylate at the branch point is surrounded by similar sequences:

$$\mathrm{CU}^{\mathrm{G}}_{\mathrm{A}}\mathrm{A}^{\mathrm{C}}_{\mathrm{U}}\mathrm{Y}_n\mathrm{AG}$$

How the cleavage and ligation occur at the 3′ junction is not yet known, but the phosphate linking the two exons is derived from the first nucleotide of the 3′ exon. [See B. Ruskin et al., 1984, *Cell* 38:317; and R. A. Padgett et al., 1984, *Science* 225:898.]

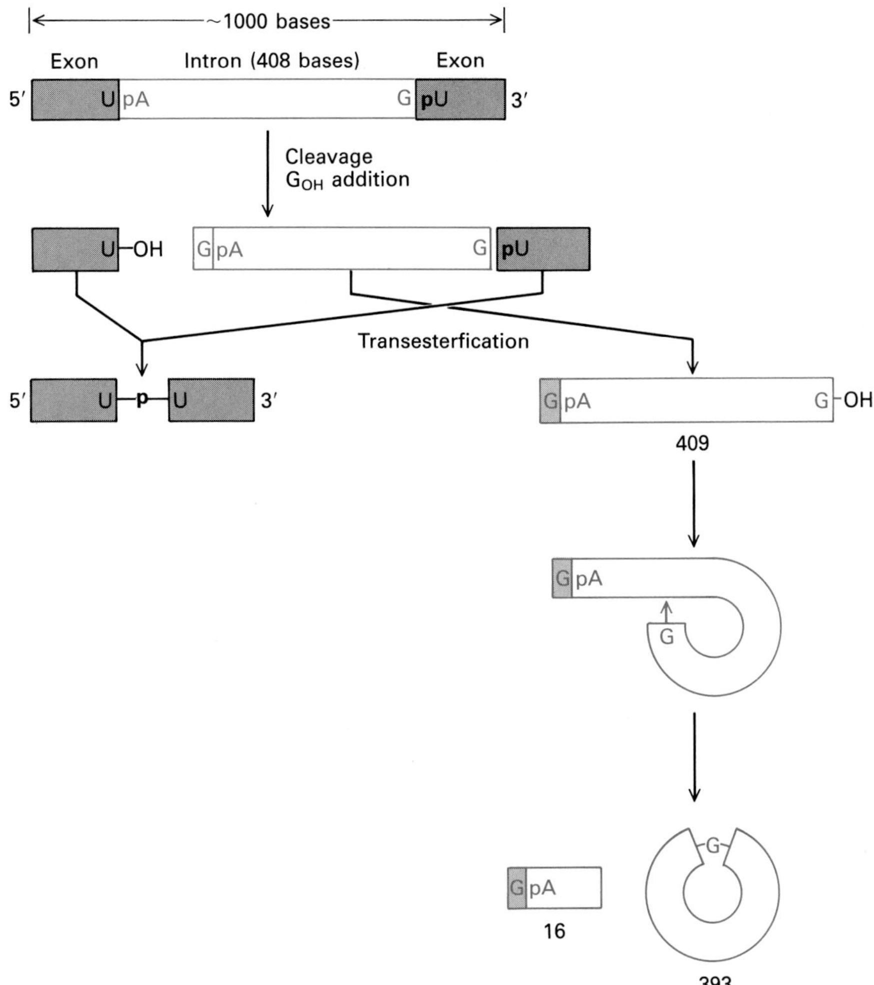

Figure 9-61 The self-splicing of *Tetrahymena pyriformis* ribosomal RNA sequences without protein. A recombinant DNA molecule containing the coding sequences for a portion of the rRNA precursor is prepared and transcribed. The resulting RNA molecule contains about 1000 nucleotides of the *Tetrahymena* sequence (two exon regions plus a 408-base intron). This transcript can undergo splicing without assistance from a protein. Guanosine (G_{OH}) is a necessary cofactor in the cutting-and-splicing reaction. The hydroxyl group of the guanosine attacks the indicated site (the U—pA bond) at the 5′ exon-intron junction, and during the cleavage at this site the guanosine is added to the 5′ cut end of the intron. Cleavage at the 3′ end of the intron is accomplished during a transesterification that links the two U residues by a phosphodiester linkage; the phosphate used in this linkage is the one that was already present between the G and U residues at the 3′ intron boundary. The excised intron undergoes a second cutting-and-splicing reaction that produces a 393-nucleotide circular RNA molecule and a 16-nucleotide fragment. [See T. R. Cech, 1983, *Cell* **34**:713.]

pletes the reaction. The 3′ hydroxyl of the first exon is then linked to the 5′ phosphate of the first nucleotide after the AG (Figure 9-60). Thus the phosphate linkage that is involved in the final ligation already existed in the RNA chain. The source of any energy needed to drive the cleavage and ligation reaction has not been determined.

The third RNA splicing reaction involves the removal of introns from certain ribosomal RNAs. This reaction was actually discovered before the details of mRNA splicing were worked out, and it can occur in purified RNA without the participation of proteins. This very surprising result was first achieved with the ribosomal RNA precursor from *Tetrahymena pyriformis*. The most extensive experiments have utilized the protein-free incubation of a *Tetrahymena* rRNA segment of about 1000 nucleotides, including a 408-base-pair intron (Figure 9-61). The first step in the removal of the intron—a cleavage at the 5′ exon-intron junction—is mediated by a guanosine cofactor (probably specifically by the chemical activity of the 2′ and 3′ hydroxyl groups on its ribose). As the cut in the RNA is induced, the free 3′ hydroxyl of the guanosine is linked to the 5′ phosphate of the nucleotide at the 5′ end of the intron. The cut leaves a 3′ hydroxyl on the 3′ nucleotide of the first exon. This hydroxyl acts in a transesterification with the phosphate that joins the 3′ nucleotide of the intron and the 5′ nucleotide of the second exon. The transesterification step, which is not yet fully understood, thus releases the intron and joins the two exons by a phosphate that was already present in the chain. The intron itself undergoes a second cleavage-and-splicing reaction that releases a small oligonucleotide and circularizes the remaining segment.

The three types of RNA splicing mechanisms are clearly different in detail. There are, however, superficial similarities that might indicate some relatedness at an early stage of evolution. For example, in both mRNA and tRNA splicing (and possibly in the *Tetrahymena* splicing mechanism as well), a single nucleotide that has both a 2′ → 5′ and a 3′ → 5′ linkage is involved. And in both mRNA and in cell-free rRNA splicing, circular intermediates are formed and the phosphate that links the exon is already in the RNA.

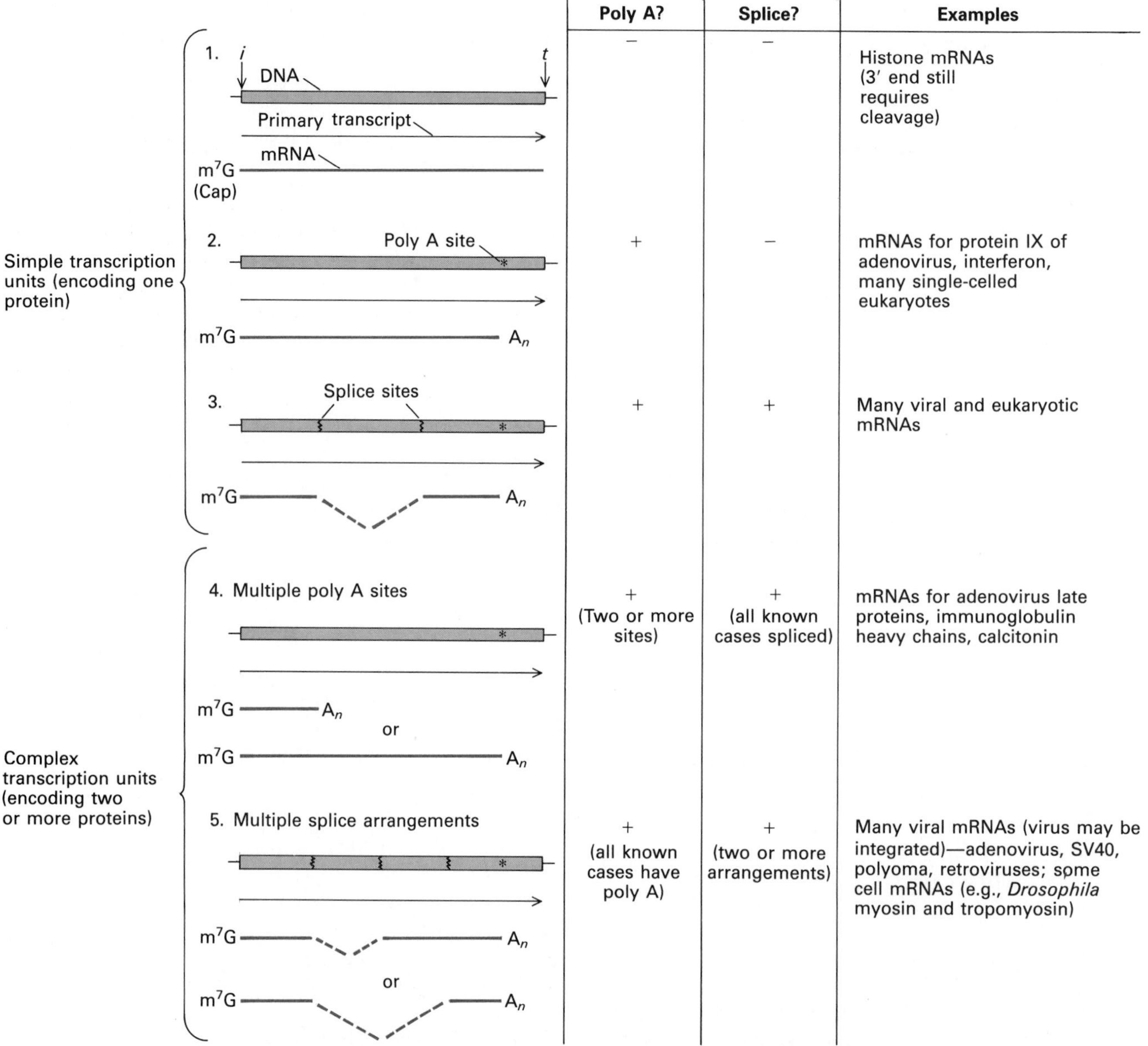

Poly A?
Splice?
Examples
Simple transcription units (encoding one protein)
1. i t
DNA
Primary transcript
mRNA
m7G
(Cap)
−
−
Histone mRNAs (3′ end still requires cleavage)
2. Poly A site
m7G
An
+
−
mRNAs for protein IX of adenovirus, interferon, many single-celled eukaryotes
3. Splice sites
m7G
An
+
+
Many viral and eukaryotic mRNAs
Complex transcription units (encoding two or more proteins)
4. Multiple poly A sites
m7G
An
or
m7G
An
+ (Two or more sites)
+ (all known cases spliced)
mRNAs for adenovirus late proteins, immunoglobulin heavy chains, calcitonin
5. Multiple splice arrangements
m7G
An
or
m7G
An
+ (all known cases have poly A)
+ (two or more arrangements)
Many viral mRNAs (virus may be integrated)—adenovirus, SV40, polyoma, retroviruses; spme cell mRNAs (e.g., Drosophila myosin and tropomyosin)

Figure 9-62 Diagrams and examples of simple and complex transcription units encoding mRNAs. The diagrams show an initiation and a termination site in DNA; the direction and extent of the primary transcript are indicated by an arrow. The mRNA derived from the primary transcript is shown in color. Poly A sites appear as asterisks, and splice sites as zigzags. For each type of transcription unit, an example is given. The most common categories are probably the simple transcription units 2 and 3.

The Design of Simple vs. Complex Transcription Units for Eukaryotic mRNAs

We have now considered all of the steps known to occur during mRNA processing. Not all transcription units produce primary transcripts that require every step in processing in order to become a functional molecule. The most useful division of the types of transcription units focuses on the protein-coding capacity of the primary transcripts.

From comparisons of the gene structures of cloned DNA sequences, two types of transcription units have been distinguished. *Simple transcription units are those whose primary RNA transcripts give rise to only one mRNA, which encodes only one protein* (Figure 9-62). The nuclear RNA products of simple transcription units may or may not require all of the steps of RNA processing. All such products acquire caps, but many mRNAs are not spliced, and most histone mRNAs undergo neither splicing nor addition of poly A. There are a number of mRNAs (e.g., those of interferon, some viruses, and most yeast species) that have caps and poly A, but that do not require splicing. Globin and ovalbumin genes are examples of simple transcription units whose primary transcripts undergo both poly A addition and splicing.

Complex transcription units are those whose primary transcripts can produce two or more different mRNAs, each encoding a different protein (Figure 9-62). Several types of complex transcription units can be distinguished. The primary transcripts may have two or more poly A sites, each of which can lead to a nuclear RNA molecule that requires splicing for completion as an mRNA. Or, the primary transcripts may have one poly A site but two or more possible splicing arrangements, each resulting in a different mRNA. It is also possible to have a mixture of the two—that is, multiple poly A sites and multiple splicing arrangements. Most adenovirus transcription units and all of those in SV40 and the retroviruses are complex. In addition, a growing number of complex transcription units encoding cell proteins are being reported in both vertebrates and invertebrates; these studies are described in Chapters 11 and 12. Selection mechanisms that cause a particular mRNA to be produced from a complex primary transcript are a form of genetic regulation, which we shall discuss in Chapter 12.

Proteins Associated with hnRNA

As the foregoing discussion has indicated, a great deal has been learned about how a nuclear RNA molecule is processed before it becomes an mRNA molecule and enters the cytoplasm to function in protein synthesis. However, much less is known about how nuclear structures and nuclear enzymes carry out the processing reactions.

Because RNAs are highly charged, the assumption has been that newly formed hnRNA becomes associated with proteins. Extracts of isolated nuclei have been prepared to search for ribonucleoprotein complexes. Most of the hnRNA recovered is found associated with a complex mixture of proteins: three groups of proteins (termed A, B, and C in Figure 9-63a) are most evident when the ex-

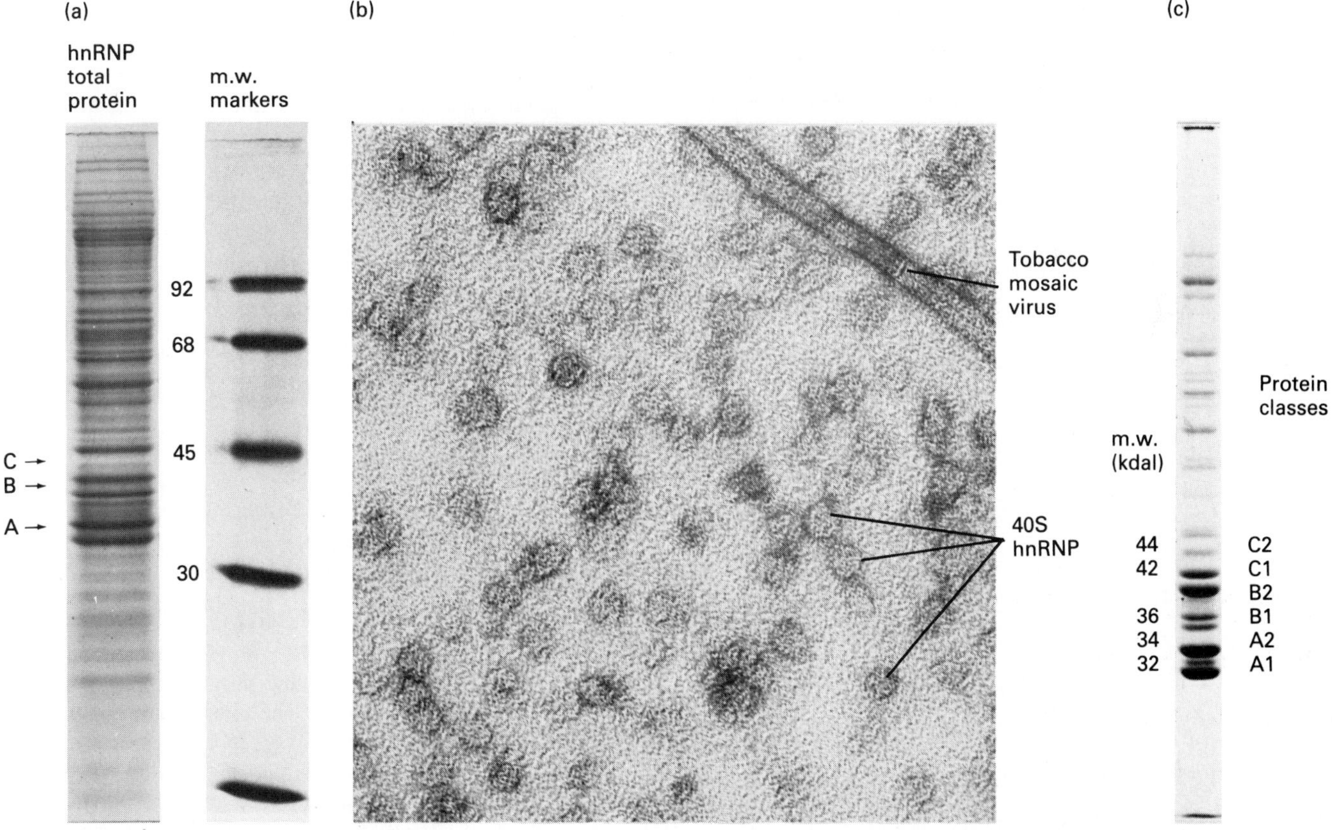

Figure 9-63 The hnRNA-associated proteins of HeLa cell nuclei. (a) Purified nuclei were extracted and a fraction containing hnRNPs—high-molecular-weight hnRNAs plus proteins—was obtained. The total protein content of the hnRNP fraction is seen in the left lane. The major protein classes A, B, and C are marked. The m.w. (molecular weight) markers are four well-characterized proteins of 30 to 92 kdal. (b) An electron micrograph of 40S hnRNPs shows regular units with a diameter of about 20 nm in uranyl acetate–stained preparations. (The rodlike structure is tobacco mosaic virus, included as a size marker.) These particles are thought to be regular monomers in which the proteins A, B, and C are contained in the ratio 6:2:6. (c) The protein composition of pure 40S hnRNPs is seen by acrylamide gel electrophoresis to be quite simple. *Part (a) courtesy of T. Pederson; parts (b) and (c) courtesy of W. LeSturgeon.*

tract is examined by gel electrophoresis. When such extracts are exposed briefly to a nuclease, a 40S particle containing hnRNA fragments is released; this particle can be obtained in an almost pure form (Figure 9-63b). The three protein groups A, B, and C make up the majority of the proteins in the 40S particle (Figure 9-63c). These three groups contain the unusual modified amino acid residues di- and trimethylarginine, and are therefore highly distinctive. Antibodies directed against these proteins show that they are contained entirely within the nucleus.

When the contents of any nuclei are spread for examination in the electron microscope, particles that presumably contain protein can be seen in association with growing nascent hnRNA chains (Figure 9-64). In addition, the RNA chain between obvious particles is probably coated evenly with proteins. The protein particles visible in spread nuclei are not randomly distributed along the hnRNA chains; rather, on each transcript from a single transcription unit the particles appear to be distributed in approximately the same way (Figure 9-65). However, the distribution of particles varies in the products of different transcription units.

A striking example of the reproducibility of the distributions of RNA and protein in nascent RNA strands comes from examining the transcription of two new daughter DNA strands that have just been replicated from a single parental strand. In this case, the transcriptional activity on the two new molecules should be identical. In fact, microscopic observations have shown that the distribution of nascent transcripts on two such DNA strands is indeed similar, and that the associated protein likewise forms a very similar pattern. These observations suggest that regular processing events may be responsible for the consistent particle distribution on nascent hnRNA.

In addition to the particles that are associated with hnRNA, some of the small ribonucleoproteins mentioned earlier in the discussion of processing may be recovered in association with the large hnRNPs. Functions have not yet been determined for most of these protein structures, but they might be involved in any step of mRNA processing: methylation, poly A addition, splicing, or transport of the mRNA from the nucleus to the cytoplasm.

Nuclear Pores: The Presumed Exit Sites for mRNA

The internal architecture of the nucleus is a subject of great interest. We know a great deal about the biochemistry of the nucleus but very little about the physical organization of nuclear components and processes. Are the active genes arranged in a particular structure, and does any such structure help to get hnRNA processed and mRNA transported to the cytoplasm? There is not yet any generally accepted procedure for examining the internal structure of the nucleus. Thus it is not even certain whether the fibrous protein network that can be viewed

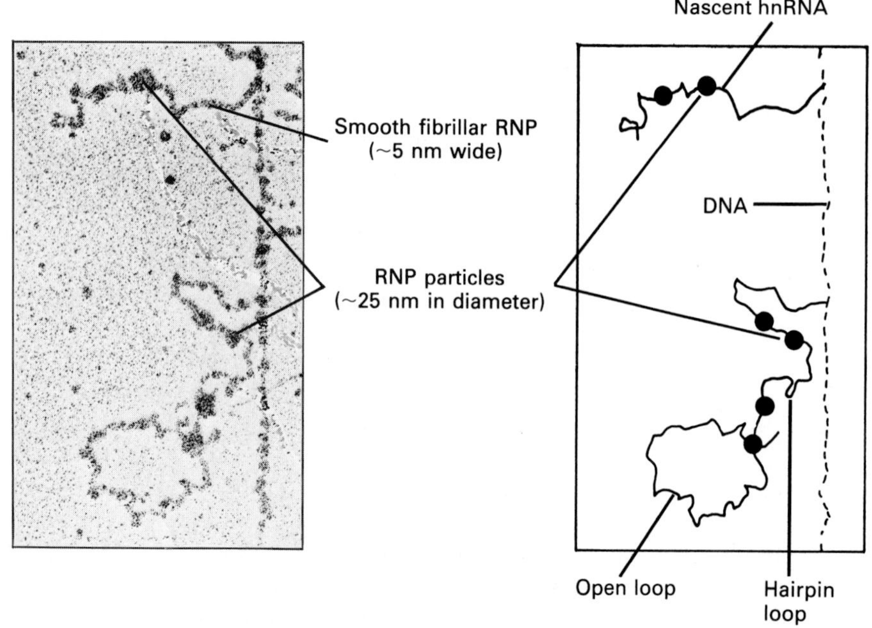

Figure 9-64 An electron micrograph and a tracing showing transcription products. The photograph is a highly magnified (×125,000) view of a nonribosomal transcription complex (presumed to be an RNA polymerase II product) from the nuclei of *Drosophila melanogaster* cells. Associated with the nascent RNAs are particles similar in diameter to the 40S nucleoprotein particles shown in Figure 9-63. The 5-nm-diameter RNA chain without particles is apparently also associated with protein ("smooth fibrillar RNP"), or it would not be visible. *Photograph courtesy of A. Beyer.*

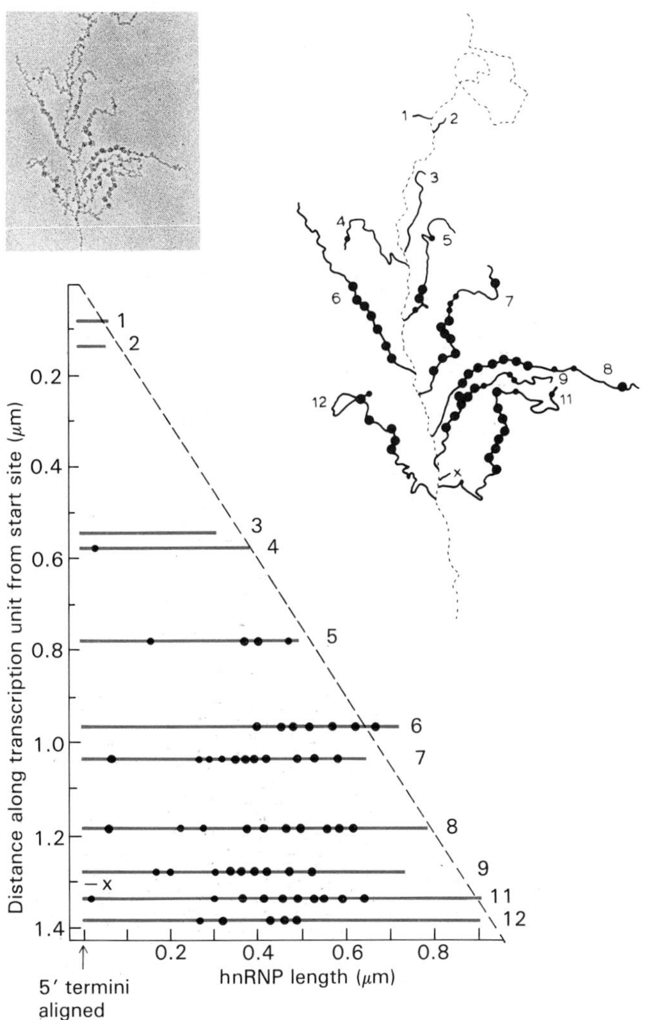

Figure 9-65 The hnRNPs from a single transcription unit look similar. *Drosophila* cells were lysed and examined in the electron microscope, which revealed proteins still attached to the RNA. The central thread is DNA, and a series of individual nascent strands of hnRNA are emerging from it. (The tracing at the right shows DNA as a black dashed line, hnRNA as solid lines, and protein particles as black dots.) The distribution of the 25-nm-diameter protein particles along the hnRNAs *(color)* is shown in the plot at the bottom. Note that the particles are not distributed randomly, but that their locations are similar on each of the presumably similar RNA chains. [See A. L. Beyer, O. L. Miller Jr., and S. L. McKnight, 1980, *Cell* **20**:75.] *From Beyer et al., 1980.*

(Figure 9-66) after the destruction of nuclear RNA and DNA is real, or whether it is an artifact. However, at the surface of the nucleus there are distinctive structures, called *nuclear pores,* that are embedded in the double-layered nuclear envelope (Figures 9-66 and 9-67). It has been assumed for years—although it remains to be proved—that communication between the nucleus and the cytoplasm is achieved either by diffusion or by specific transport through the nuclear pores. Specific RNA transport could possibly be important in gene regulation.

Nuclear pores exist in all eukaryotic cells. The number of pores ranges from several thousand in somatic cells to several million in large cells such as amphibian oocytes. Just inside the nucleus, underlying the double-layered nuclear envelope, is a layer of interacting fibrous proteins called *lamins.* The pores appear to be embedded in the lamins (see Figure 9-67). In electron micrographs the pores appear octagonal, a shape conferred by a crystal-like array of similar structural proteins surrounding a central cavity. The entire pore complex is quite large; estimates made by physical techniques suggest that 10^5 kdal of protein is present in each pore. In some pores, the central cavity is filled with a protein "plug" (Figure 9-68), the physiological significance of which is unknown at present.

The diameter of the cavity in the pore is approximately 10 to 20 nm, which is smaller than the diameter of a large ribosomal subunit. If a material such as mRNA or a ribosome enters the cytoplasm through the cavity in a pore, either the pore or the material passing through it may have to change shape. However, almost nothing is known at present about this critical phase in the life cycle of an mRNA molecule. It remains unproved that an active process of mRNA transport exists; it is not even certain whether the mRNA enters the cytoplasm through the nuclear pores.

The Appearance of mRNA in the Cytoplasm and the Initiation of Protein Synthesis

To understand the life cycle of an mRNA molecule and the events that may determine its functional concentration in the cytoplasm, not only should its formation and entry into the cytoplasm be considered, but also its use there. Nuclear events produce a finished mRNA molecule in 5 to 20 min; within at most another few minutes, the mRNA is part of a cytoplasmic polyribosome.

A key question about eukaryotic cells is still unanswered after two decades of study: Does a new mRNA molecule enter the eukaryotic cytoplasm by associating with a ribosome or a ribosomal subunit at the nuclear border? Many studies have been done with drugs that stop translation (cycloheximide, emetine, puromycin); such drugs therefore eventually prevent the formation of new ribosomes. However, even when the drugs are present, new mRNA can still be synthesized and can still enter the cytoplasm. Thus *new* ribosomes are not required to accompany mRNA to the cytoplasm. In addition, cell fractionation studies have shown that almost no old ribosomes are present in isolated nuclei. Therefore, the first

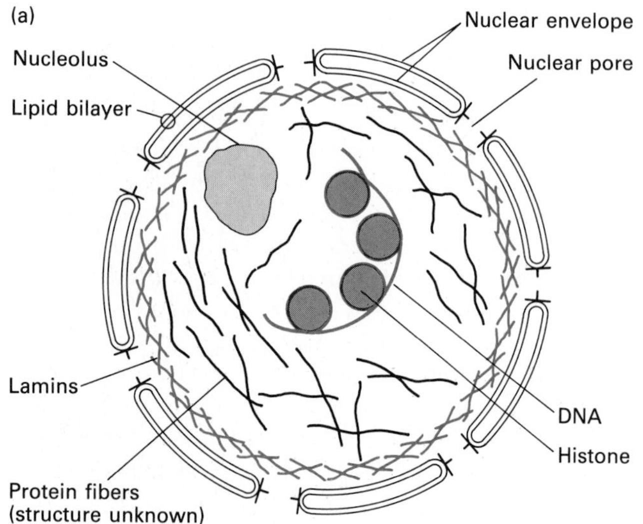

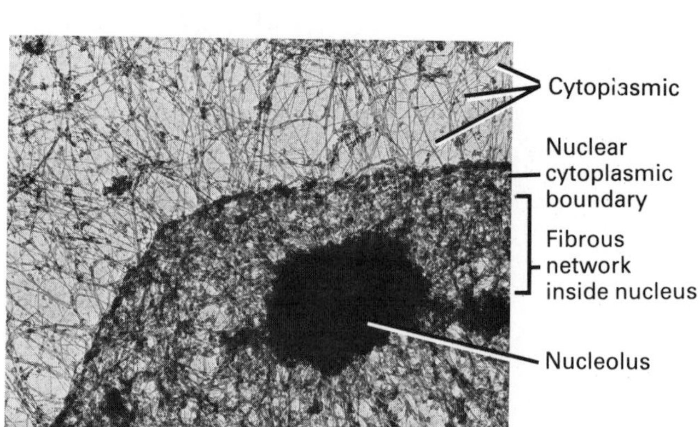

Figure 9-66 The eukaryotic nucleus. (a) A generalized diagram. The double-layered envelope, nucleolus, nuclear pores, and fibrous lamins are prominent nuclear structures. The DNA forms complexes with nonfibrous proteins called histones; these complexes and their structural organization will be discussed in Chapter 10. Biochemical experiments have identified other proteins within the nucleus, and electron micrographs like the one in (b) show a fibrous network there. However, it is not known whether the proteins exist as such a network in living tissues. (b) A transmission electron micrograph of a whole mount of a HeLa cell, showing a skeletal network within the nucleus. The cell was prepared by removing lipids and soluble factors with a mild detergent. The remaining skeletal structure was then treated to remove most of the RNA and DNA. The sample was fixed with glutaraldehyde, but no heavy-metal shadowing was done. [See S. Penman et al., 1982, *Cold Spring Harbor Symp. Quant. Biol.* **46**:1013.] *Photograph courtesy of S. Penman.*

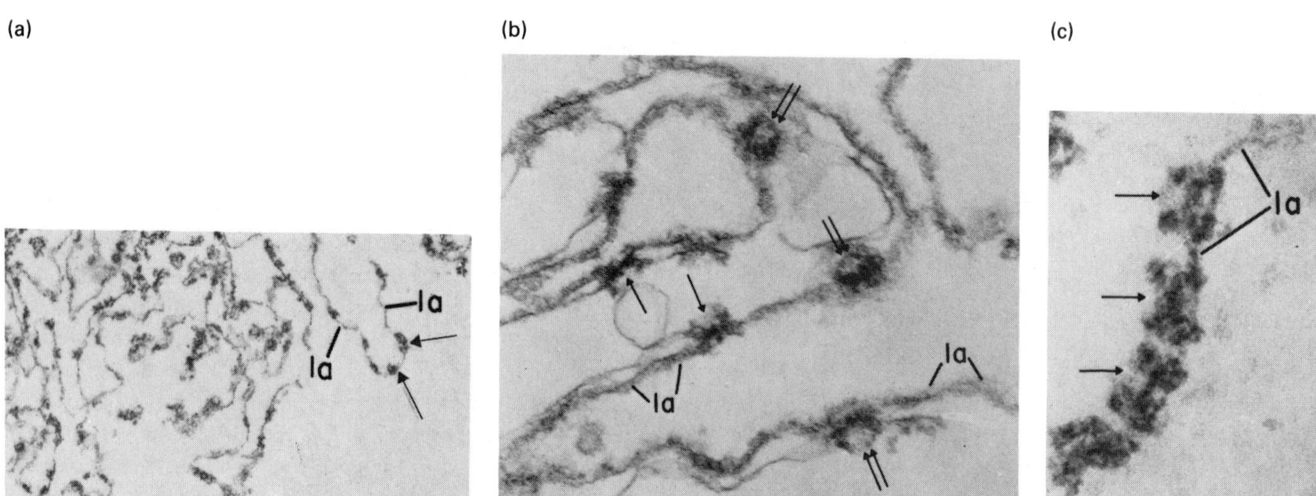

Figure 9-67 Electron micrographs of nuclear pore–lamin complexes. Rat liver nuclear envelopes were prepared from lysed nuclei. The fibrous lamin proteins (la) make up the majority of the sinuous fibers in the photographs, and nuclear pores *(arrows)* are embedded in the fibers. The magnifications are as follows: (a) ×5000, (b) ×20,000, and (c) ×50,000. [See N. Dwyer and G. Blobel, 1976, *J. Cell Biol.* **70**:581.] *Courtesy of N. Dwyer. Reproduced from the* Journal of Cell Biology, *1976, by copyright permission of the University of Rockefeller Press.*

(a)

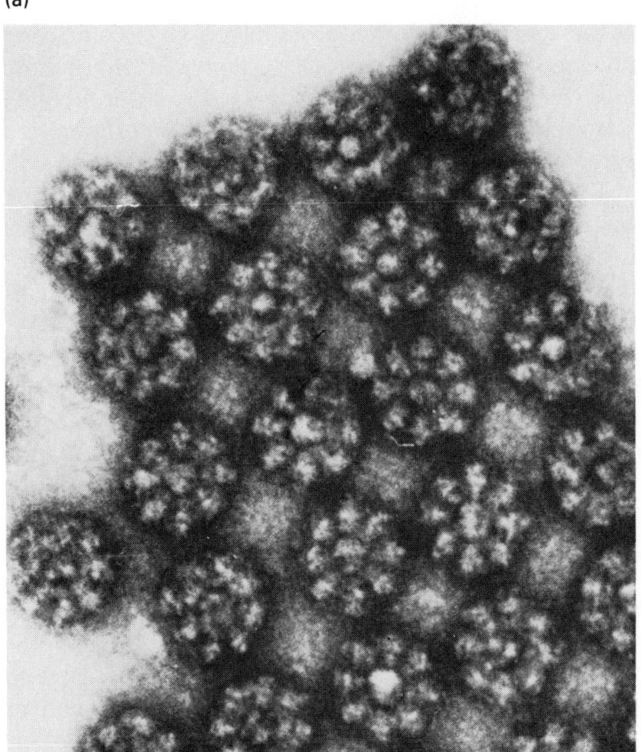

(b)

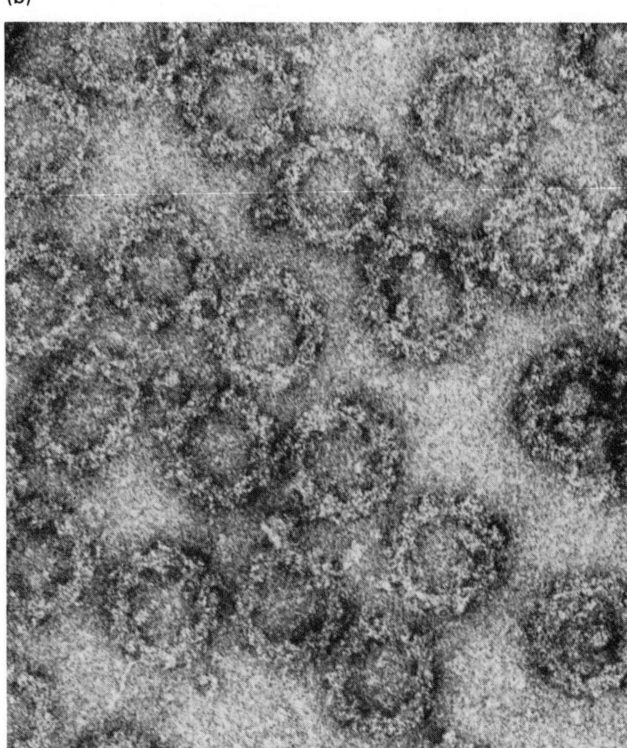

(c)

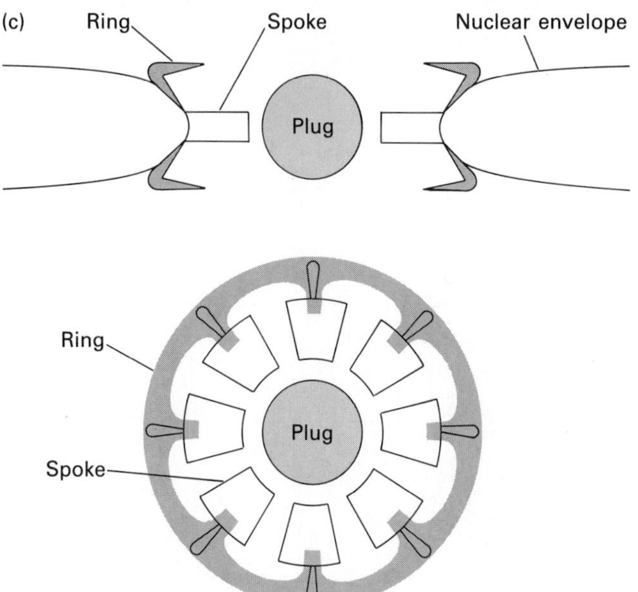

Figure 9-68 Nuclear pores from *Xenopus laevis* oocytes, and a model of pore structure. The micrographs show purified pores (a) with central cores or "plugs," and (b) without plugs. Because of the regularity of their structure, these nuclear pores could be subjected to image processing (a computer-assisted analysis of the average structure of many similar objects). In (c), a physical model of a pore within the nuclear envelope is viewed in cross section *(top)* and face on *(bottom)*. [See P. N. T. Unwin and R. A. Mulligan, 1982, *J. Cell Biol.* 93:63.] *Photographs (a) and (b) from Unwin, 1982; part (c) redrawn from same source.*

possible association of a ribosomal subunit with a newly emerging mRNA molecule would be at the nuclear border, perhaps at a nuclear pore.

If the new mRNA does not enter the cytoplasm by association with a ribosomal subunit, the possibility remains that the mRNA enters by association with one or more of the proteins that initiate translation. A *cap-binding protein* that could have such a role in first the transport and then the utilization of mRNA has been characterized. This protein, which has a molecular weight of 24,000, can be purified by virtue of its affinity for the cap structure at the 5′ end of mRNA (m^7GpppN; the methyl group is important for recognition). Probably in association with other initiation factors, this protein aids in binding the capped end of mRNA to the 40S ribosomal subunit. The rate of translation of mRNA in vitro is higher if it has a cap; the rate of translation is inhibited by methylated guanylic acid (m^7GMP), which may act as a competitive inhibitor of the cap-binding protein.

The biochemical mechanisms of all of the steps in the initiation of protein synthesis in eukaryotic cells have not yet been worked out. It is known that there are a number of initiation factors (called eIF, for "eukaryotic initiation factors"), some of which are complexes of several different polypeptide chains. For example, as noted above, the cap-binding protein itself is probably part of a complex. Because the entire cycle of ribosome binding and the roles of individual initiation factors have been studied only in cell-free extracts (most intensively in extracts of mammalian reticulocytes), the relation of these factors to cell structures is unknown at present. However, it is possible that the initiation factors and the cap-binding protein are the first proteins to come into contact with new mRNAs in the cytoplasm, and they might even aid in transport to the cytoplasm.

Is There a Specific mRNA Transport System?

An alternative scenario is that mRNA enters the cytoplasm by association with one or more specialized, not-yet-identified transport proteins, rather than by association with any part of the translation apparatus. However, no labeling experiment—neither with cells from animals nor with cultured cells—has successfully demonstrated the transfer of new mRNA into the cytoplasm in a ribosome-free form, followed by association with ribosomes.

In cultured cells, within a few minutes most new mRNA is found in polyribosomes. In cytoplasmic extracts, a minority mRNA fraction (10 to 20 percent) is unassociated with ribosomes. For example, in mammalian and avian reticulocytes, a small proportion of globin mRNA is found unattached to ribosomes and in association with a set of proteins different from those in polyribosomes. However, because these enucleated cells are no longer synthesizing mRNA, the unattached mRNA must represent older molecules no longer associated with ribosomes—not newly emerging mRNA. Cultured mammalian cells also have some cytoplasmic mRNA unassociated with ribosomes. But, as noted earlier, attempts to show that new mRNA passes through a cytoplasmic stage on the way to polyribosomes have not succeeded. Thus, the cytoplasmic mRNAs that are unassociated with ribosomes could simply be the normal fraction that is temporarily unoccupied; alternatively, these could result from factors that restrain translation more permanently.

The Possible Association of mRNA with the Cytoskeleton

Most of the above-mentioned experiments on mRNA movement have employed cell fractionation techniques to examine the presence of mRNA in cytoplasmic extracts. It is very possible, however, that the translation of mRNA in the cell cytoplasm does not take place in free solution. For example, polyribosomes can be intimately associated with portions of the endoplasmic membrane (the rough endoplasmic reticulum). The proteins that are synthesized at these sites are exported from the cell or become components of new membranes (see Chapter 21). Even those mRNAs not making proteins destined to associate with membranes may be associated with the *cytoskeleton,* a cytoplasmic network of diverse fibers (see Figure 5-29).

The association of some mRNAs with the cytoskeleton can be demonstrated by cell fractionation. Cultured cells treated briefly with detergent can be stripped of lipids; the cell border becomes porous, and many proteins (about 75 percent of the total) and other materials (e.g., more than 90 percent of the tRNAs) wash out of such preparation. The polyribosomes, however, are completely retained (Figure 9-69).

Before they are subjected to the procedures outlined above, polyribosomes can be disaggregated by treating cells with a drug. If the cells are now subjected to detergent extraction, some of the released ribosomes wash out of the cells, but the mRNAs remain within the cytoskeletons. Thus the mRNAs or proteins associated with them may act as anchors for the polyribosomes. One protein that may have such a function is the poly A–binding protein, a 75,000-molecular-weight molecule found in polyribosomes that are bound to the poly A segment. Interestingly, the mRNA that is unassociated with polyribosomes in cell fractionation experiments is depleted of this poly A–binding protein. However, the actual basis for the retention of the polyribosomes in detergent-extracted

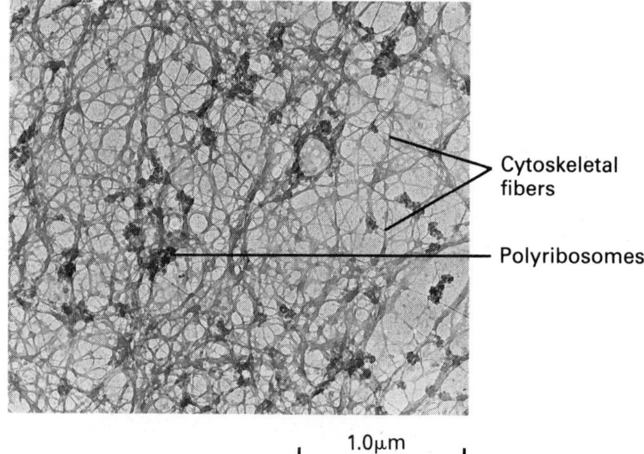

Figure 9-69 A transmission electron micrograph of a whole mount of a human diploid fibroblast. The cell was prepared by removing lipids and most proteins with mild detergents. The remaining cytoskeletal structure was then fixed with glutaraldehyde. This area of the cytoplasm reveals many fibers and dark groups of objects whose approximate diameters are equal to those of ribosomes. [See S. Penman et al., 1982, *Cold Spring Harbor Symp. Quant. Biol.* **46**:1013.] *From Penman, 1982.*

cells has not been proved to be a true functional attachment due to particular proteins.

Likewise, whether these possible associations of mRNA-protein complexes with cytoskeletal elements play a role in the movement of mRNA into the cytoplasm is as yet unknown. However, a reasonable hypothesis is that most of the complicated reactions of mRNA synthesis in the cell nucleus occur in association with a nuclear matrix, and that when the mRNA is finished it is ushered into the cytoplasm and immediately becomes attached to structural elements of the cells.

The Metabolic Turnover of mRNA in the Cytoplasm

Unlike rRNA and tRNA, which are synthesized in growing cells at a rate that allows these molecules to double in number with each cell generation, mRNA is apparently synthesized faster than is required for maintenance of its steady-state amount. The higher rate of synthesis is necessary because mRNA is constantly being degraded. This process was detected, first in bacteria and later in animal cells, when it was observed that labeled uridine or PO_4^{3-} appeared in the mRNA fraction sooner than it did in the ribosomal RNA. For example, when a steady state is reached, mRNA may amount to 5 percent of the total labeled RNA in polyribosomes. However, after a shorter time—an interval equal to, say, one-fourth of the generation time or less—the mRNA may represent 15 to 20 percent of the total label; the shorter the label time, the higher the percentage of mRNA. From such observations, it has become clear that mRNA is unstable; it is "turned over" (renewed) as often as 10 times per generation.

Measuring the half-life of mRNA in animal cells is not an easy experimental task. The simplest technique would seem to be to use a pulse-chase: to add label briefly, remove the label, and measure the retention of the labeled mRNA (Chapter 7). However, because the ribonucleotide pool that supplies precursors for RNA synthesis is quite large, and because once a labeled precursor enters that pool as a phosphorylated intermediate it cannot be chased out of the cell (Chapter 7), a simple pulse-chase cannot be achieved within a short period of time. Nevertheless, an effective dilution of label in the pool can sometimes mimic a perfect chase well enough that the stability of mRNA can be studied and conclusions drawn about mRNA turnover, at least for relatively long-lived molecules (Figure 9-70, *top*).

Another approach to studying mRNA turnover is to label cells continuously and study the curve of accumulation of a specific mRNA as a function of time. Because achievement of the maximum rate of label accumulation in cytoplasmic mRNA requires time for the ribonucleotide pool to equilibrate and for the nuclear RNA precursor to become fully labeled, the accumulation technique is not accurate for measuring the turnover time of very short-lived species. (As we have seen, the pulse-chase technique also fails to accurately measure short-lived species.) However, with longer labeling times (say, times longer than 1 h in mammalian cells), the accumulation curves of specific mRNAs can be used to obtain fairly accurate estimates of mRNA half-lives (Figure 9-70, *bottom*). With cultured mammalian cells, individual mRNAs with half-lives ranging from about 3 to 24 h can be observed through the use of one of these two techniques (Table 9-4).

Most of the individual mammalian mRNAs studied have been found to have long half-lives. This may be sim-

Table 9-4 Half-lives of messenger RNAs

		mRNA half-lives*	
Cell	**Cell generation time**	**Average**	**Range known for individual cases**
Escherichia coli	20–60 min	3–5 min	2–10 min
Saccharomyces cerevisiae (yeast)	3 h	22 min	4–40 min
Cultured human or rodent cells	16–24 h	10 h	30 min or less (histone mRNAs) 30 min (virus mRNAs) 3–24 h (specific mRNAs of cultured cells)

*For information on specific mRNA half-lives for *E. coli,* see A. Hirashima, G. Childs, and M. Inouye, 1973, *J. Mol. Biol.* **119**:373; for yeast, see L.-L. Chia and C. McLaughlin, 1979, *Mol. Gen. Genet.* **170**:137; and for mammalian cells, see M. M. Harpold, M. Wilson, and J. E. Darnell, 1981, *Mol. Cell. Biol.* **1**:188.

PULSE-CHASE

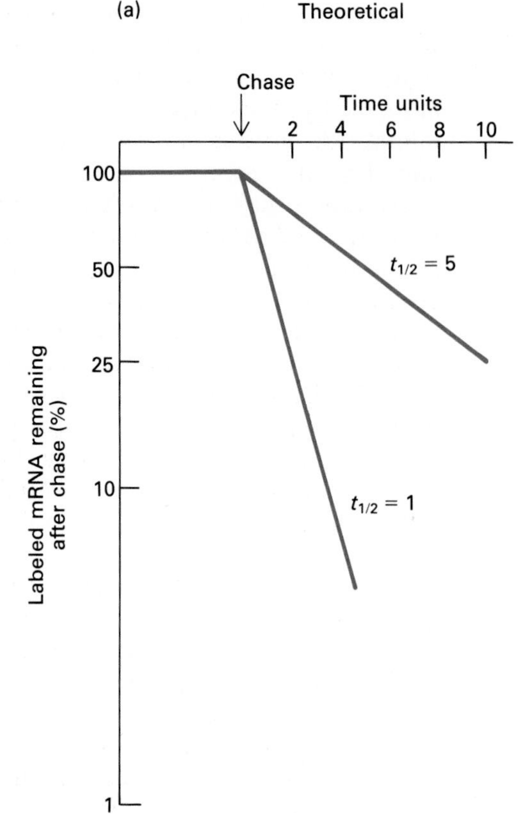

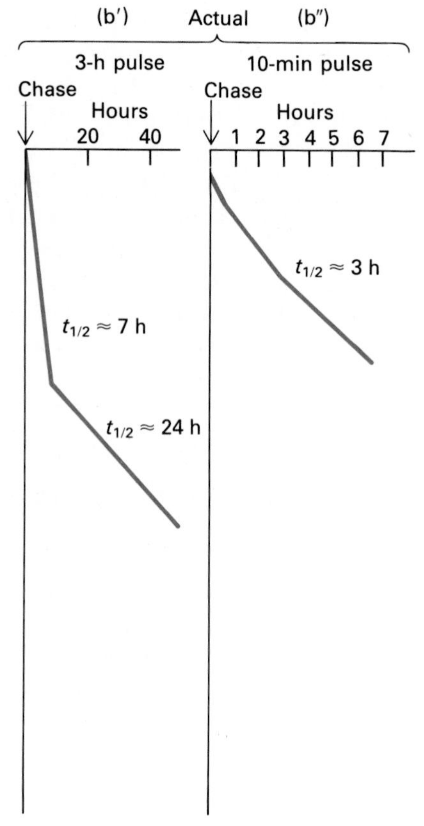

LABELING TO STEADY STATE

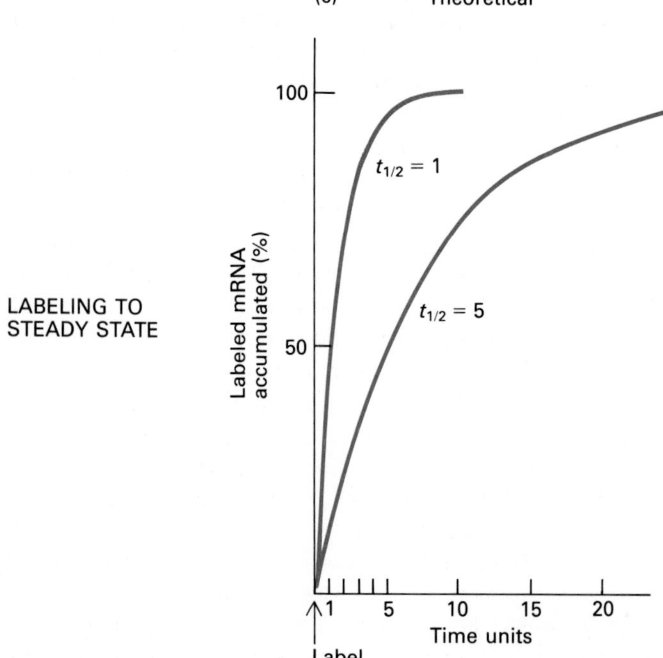
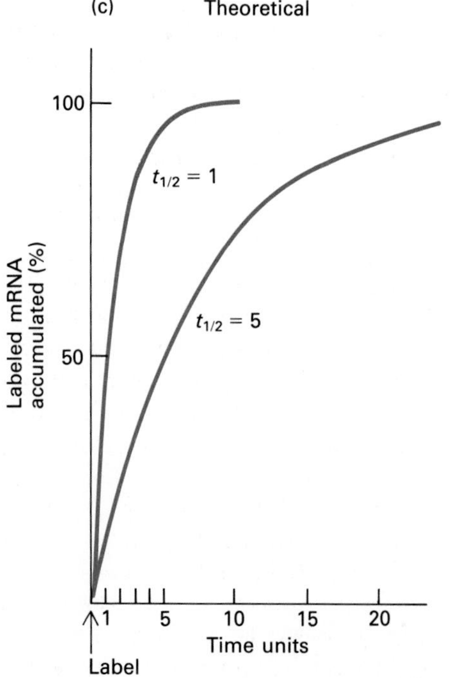

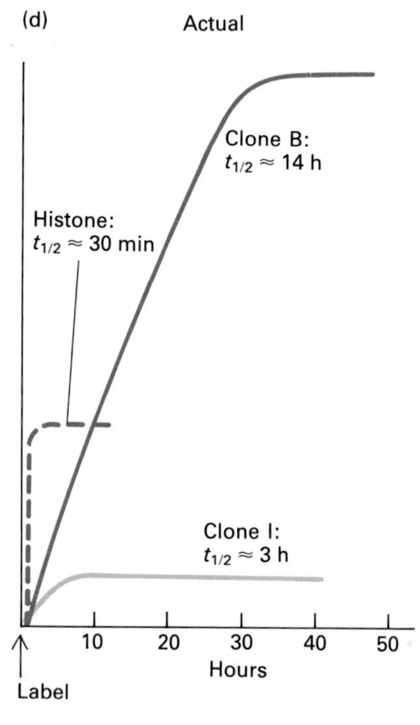

Figure 9-70 ***(opposite)*** Two methods of assessing mRNA half-lives: pulse-chase and labeling to steady state.
(a),(b′),(b″) *The pulse-chase method.* In this type of experiment, cells are labeled briefly and then the source of label is abruptly removed. Theoretically, this would lead to a 50 percent reduction in the remaining labeled mRNA with the passage of each half-life. Theoretical curves for the behavior of two mRNAs with half-lives ($t_{1/2}$) of 1 and 5 time units are shown in (a). Actual data are shown in (b′) and (b″), where the decline of the total HeLa cell mRNA is measured after a 3-h label (b′) and after a 10-min label (b″). These two decay curves are very different. It is clear that the initial decay rate of the mRNA depends on the length of the prelabeling period. In the short pulse (10 min), at least 75 percent of the labeled RNA has a half-life of 3 h or less, whereas the RNA labeled in the longer pulse (3 h) exhibits two decay rates with 7-h and 24-h half-lives. (The curve for the 24-h half-life shows decay from approximately 17 perent to 7 percent in 40 h.) Thus, the pulse-chase method fails to easily detect short-lived mRNAs and is most useful for measuring mRNAs with a long half-life.
(c),(d) *Labeling to steady state.* The approach to steady state theoretically begins at the moment of label addition, as in (c). The data in such an experiment can be used to calculate the half-life of the mRNA from the following equation:

$$\frac{A}{A_\infty} = 1 - e^{-\ln 2(1/K_D + 1/K_{1/2})t}$$

in which A and A_∞ represent the radioactivity in mRNA per cell sample at times t and t_∞, respectively; K_D is the growth constant of the cell culture; and $K_{1/2}$ is the turnover rate constant of the mRNA. The time necessary to reach equlibrium is controlled by the turnaround rate. Actual data appear in (d). Here the radioactivity of three specific mRNAs of cultured cells is plotted per cell through 50 h. One mRNA (I) is fully labeled within 8 h or less, whereas the other (B) requires at least 24 h. Histone mRNA reaches a plateau value within 1 h. From the equation above, the half-life calculated for clone B mRNA is 14 h and for clone I is 3 h, and the histone mRNA has a half-life of 30 min or less. [See J. R. Greenberg, 1972, *Nature* **240**:102; R. H. Singer and S. Penman, 1972, *J. Mol. Biol.* **78**:321; L. Puckett and J. E. Darnell Jr., 1977, *J. Cell Phys.* **90**:521; M. M. Harpold, C. Wilson, and J. E. Darnell Jr., 1981, *Mol. Cell. Biol.* **1**:188; and N. Heintz, H. L. Sive, and R. G. Roeder, 1983, *Mol. Cell. Biol.* **3**:538.]

ply because mRNAs that have long half-lives are more likely to be abundant and therefore more likely to be studied. There is also evidence for a considerable number of mRNAs with half-lives of 1 h or less, but only a few specific mRNAs with half-lives less than 3 h have been studied to date.

The variation in half-life between different mRNAs is not understood at present, nor has the molecular damage that causes mRNA to undergo turnover been identified. Experiments with adenovirus-infected HeLa cells, cultured Chinese hamster cells, and cells injected with hemoglobin mRNA have emphasized one point about mRNA turnover: the presence of the 3′ poly A tail is necessary for a normal half-life. For example, injection of frog eggs with rabbit hemoglobin mRNA that has a minimum of 20 to 30 adenylate residues at its 3′ end results in hemoglobin production for at least 24 h. Eggs or HeLa cells injected with hemoglobin mRNA that has a very short poly A or none at all either do not synthesize hemoglobin or do so for only a short time. Likewise, adenovirus mRNA made in the presence of an inhibitor of poly A synthesis is correctly spliced and enters polyribosomes, but has a very short (30-min) cytoplasmic half-life. Finally, histone mRNA, which normally lacks poly A when it enters the cytoplasm of mammalian cells, has a very short half-life (see Figure 9-70). Thus there is strong circumstantial evidence that mRNA stability is governed by the presence of poly A, but the mechanism by which poly A confers stability is unknown.

Summary

The RNA of eukaryotic cells is made in the nucleus by one of three distinct RNA polymerases. In almost all cases, the primary RNA transcript is not a functional RNA molecule; biochemical modification of the primary transcript is the rule.

The precursor to eukaryotic ribosomal RNA, the product of RNA polymerase I, is made in the nucleolus. Pre-rRNA varies in length from about 11 to 14 kb. In the majority of species, it is cut into three pieces: an 18S small rRNA of 1.8 to 2.0 kb, a 26S-to-28S large rRNA of 4 to 5 kb, and a 160-nucleotide 5.8S rRNA that is base-paired to the 28S rRNA. Thus in all species, only about half of the pre-rRNA appears in cytoplasmic ribosomes. The 5′ end and at least two middle sections of the precursor are discarded; the 18S is positioned nearer to the 5′ end than is the 26S-to-28S rRNA. In a few species, portions of the large rRNA are divided by an intervening sequence and must be spliced together after transcription and endonucleolytic cleavage. The large and the small rRNA sequences found in the cytoplasm are moderately well conserved within various eukaryotic species, and the two-dimensional structure of the rRNA is highly conserved.

However, species specificity is found just upstream of the initiation sites of ribosomal genes from different mammals, and species-specific factors that participate in the initiation of pre-rRNA synthesis by polymerase I have been identified. Ribosomal RNA processing begins with the addition of methyl groups (more than 100 in human pre-rRNA), and the nascent RNA combines with specific proteins as it is being formed and before cleavage occurs. The entire process of assembly of the ribosomal subunits in the nucleolus requires about 10 min for the small ribosomal subunit and about 30 to 40 min for the large subunit. The large and small subunits with their ribosomal proteins (numbering approximately 50 and 30, respectively) are delivered independently to the cytoplasm, where they enter into protein synthesis.

The products of RNA polymerase III are small RNA molecules, including the two major types of small RNA molecules of the cell, the tRNAs and the 5S rRNA. The primary transcript of the 5S rDNA may be unmodified in most cases, but the tRNA primary transcripts are highly modified. Most of the well-known tRNA modifications occur in the nucleus, and all tRNA transcripts are shortened from longer precursor molecules. About one-fourth of the primary transcripts of the eukaryotic tRNAs must be spliced. The sequences that are responsible for directing RNA polymerase III to begin proper transcription lie *within* both the 5S and the tRNA genes, and not upstream of them. A specific factor required for the transcription of 5S genes in frogs binds to nucleotides in positions 47 to 96 of the 120-nucleotide 5S gene. The sequences necessary for the initiation of tRNA gene transcription encode those that fold into the D loop and the TΨCG loop in the finished tRNA. Thus these highly conserved sequences play a role in transcription and in cytoplasmic tRNA functioning as well.

The synthesis and processing of messenger RNA is carried out by RNA polymerase II and a multitude of nuclear factors. At least three types of recognition elements are found in DNA that play a role in initiation of transcription by RNA polymerase II: (1) the TATA box lies 30 bases from the start site; (2) the GC-containing sequences that, often featuring a CAAT sequence, are found between −50 and −100; and (3) the enhancer sequences, transcriptional activators that can exist on either side of the initiation site and that can carry out their function when oriented in either direction in the DNA.

The assembly of mRNA is more complex than that of any of the other RNAs. Heterogeneous nuclear RNA (hnRNA), the primary product of polymerase II, varies greatly in size, from shorter than 2 kb to longer than 30 kb. Some hnRNA molecules, but probably not all, are converted into functioning mRNAs by a series of modifications: the 5′ end of the new RNA chain acquires a *cap*, a 5′ → 5′ linkage between a methylated guanylate residue and the initial nucleotide of the transcript. Within the RNA chains of all pre-mRNA molecules except for most histone mRNAs, cleavage by an endonuclease precedes the addition of poly A, a chain of 100 to 250 adenylate residues, to the 3′ end of the molecule. This process does not use a DNA template. The final step in the processing of many mRNAs is splicing: the removal of one or more intervening RNA sequences, or *introns*, and the joining of the remaining pieces, the *exons*, which will appear in the finished mRNA product. In mammalian cells, over three-fourths of the mRNAs are spliced; probably fewer are spliced in invertebrate cells; and perhaps only 10 to 20 percent in single-celled eukaryotes. In protein-coding primary transcripts, the sequences at the ends of the introns are always (5′)GU . . . AG(3′), but the intron sequences farther from the splice junctions are not highly conserved. It is very likely that splice junctions are brought together for correct cutting and splicing by a small ribonucleoprotein structure containing the U1 RNA, a 165-base nuclear species of RNA. The U1 RNA has a continuous sequence that is approximately homologous to the intron sequences at most splice junctions, and antibodies to the U1 ribonucleoprotein inhibit cell-free splicing. In fact, all events of RNA processing and even cytoplasmic transport could involve ribonucleoproteins, because hnRNA can be recovered in association with specialized nuclear proteins whose functions are not yet known. The role of the various mRNA modifications is not completely understood but most evidence suggests that the cap plays a role in translation initiation and poly A has an effect on stabilizing mRNAs that are known to be less stable than rRNA and tRNA.

From the work on purified genes from animal viruses and cells and on the primary RNA transcripts that come from them, we recognize two types of protein-coding transcription units. Simple transcription units encode one protein, and their primary transcripts may or may not have poly A or require splicing. Complex transcription units encode two or more proteins; these genes have one initiation site, but they contain multiple poly A addition sites and/or multiple variations in the sequences spliced out to make mRNA. The selection of a particular mRNA from the several possibilities encoded by a complex transcription unit constitutes gene control at the posttranscriptional level.

Once completed, mRNA, like new ribosomes and new tRNA, quickly enters the cytoplasm to participate in protein synthesis. The mechanism(s) of the transport of mRNA to the cytoplasm are unknown at present. The nuclear pore is a specialized, octagonal protein array that is embedded in the nuclear envelope; its central hole is the likely means of passage of molecules between the nucleus and the cytoplasm. There is little evidence as yet about the internal structure of the nucleus, but protein fibers that can be seen in whole mounts of the nucleus may represent a scaffolding around which nuclear events—including RNA synthesis, processing, and transport—may be organized.

References

The RNA Polymerases and Associated Factors of Eukaryotic Cells

CHAMBON, P. 1975. Eucaryotic nuclear RNA polymerases. *Annu. Rev. Biochem.* **44**:613–638.

DYNAN, W. S., and R. TJIAN. 1985. Control of eukaryotic messenger RNA synthesis by sequence-specific DNA-binding proteins. *Nature* **316**:774–778.

MATSUI, T., J. SEGALL, P. A. WEIL, and R. G. ROEDER. 1980. Multiple factors required for accurate initiation of transcription by purified RNA polymerase II. *J. Biol. Chem.* **255**:11992–11996.

ROEDER, R. G. 1975. Multiple forms of deoxyribonucleic acid–dependent ribonucleic acid polymerase in *Xenopus laevis*. *J. Biol. Chem.* **249**:241–248.

SEGALL, J., T. MATSUI, and R. G. ROEDER. 1980. Multiple factors are required for the accurate transcription of purified genes by RNA polymerase II. *J. Biol. Chem.* **255**:11986–11991.

Transcription Unit Mapping: The Synthesis and Processing of rRNA

DARNELL, J. E. JR. 1968. Ribonucleic acids from animal cells. *Bacteriol. Rev.* **32**:262–290.

GRUMMT, I., H. SORBAZ, A. HOFMANN, and E. ROTH. 1985. Spacer sequences downstream of the 28S RNA coding region are part of the mouse rDNA transcription unit. *Nuc. Acids Res.* **13**:2293–2304.

HACKETT, P. B., and W. SAUERBIER. 1975. The transcriptional organization of the ribosomal RNA genes in mouse L cells. *J. Mol. Biol.* **91**:235–250.

MILLER, O. L. 1981. The nucleolus, chromosomes, and visualization of genetic activity. *J. Cell Biol.* **91**:15s–27s.

REEDER, R. H. 1984. Enhancers and ribosomal gene spacers. *Cell* **38**:349–351.

ROTHBLUM, L. I., R. REDDY, and B. CASSIDY. 1982. Transcription initiation site of rat ribosomal RNA. *Nuc. Acids Res.* **10**:7345–7362.

SALIM, M., and B. E. H. MADEN. 1981. Nucleotide sequences of *Xenopus laevis* 18S ribosomal RNA inferred from gene sequence. *Nature* **291**:205–208.

SAUERBIER, W., and K. HERCULES. 1978. Gene and transcription unit mapping by radiation effects. *Annu. Rev. Genet.* **12**:329–363.

Transcription Unit Mapping: The Synthesis of 5S rRNA and tRNA

BOGENHAGEN, D. F., S. SAKONJU, and D. D. BROWN. 1980. A control region in the center of the 5S RNA gene directs specific initiation of transcription, II: The 3′ border of the region. *Cell* **19**:27–35.

BROWN, D. D., and J. B. GURDON. 1977. High-fidelity transcription of 5S DNA injected into *Xenopus* oocytes. *Proc. Nat'l Acad. Sci. USA* **74**:2064–2068.

BROWN, D. D., and J. B. GURDON. 1978. Cloned single repeating units of 5S DNA direct accurate transcription of 5S RNA when injected into *Xenopus* oocytes. *Proc. Nat'l Acad. Sci. USA* **75**:2849–2853.

DELIHAS, N., and J. ANDERSON. 1982. Generalized structures of the 5S ribosomal RNAs. *Nuc. Acids Res.* **10**:7323–7344.

ENGELKE, D. R., S.-Y. NG, B. S. SHASTRY, and R. G. ROEDER. 1980. Specific interaction of a purified transcription factor with an internal control region of 5S RNA genes. *Cell* **19**:717–729.

GALLI, G., H. HOFSTETTER, and M. L. BIRNSTIEL. 1981. Two conserved sequence blocks within eukaryotic tRNA genes are major promoter elements. *Nature* **294**:626–631.

LASSAR, A. B., P. L. MARTIN, and R. G. ROEDER. 1983. Transcription of class III genes: formation of preinitiation complexes. *Science* **222**:740–748.

SAKONJU, S., D. F. BOGENHAGEN, and D. D. BROWN. 1980. A control region in the center of the 5S RNA gene directs specific initiation of transcription, I: The 5′ border of the region. *Cell* **19**:13–25.

Transcription Unit Mapping for hnRNA and mRNA Precursors

DERMAN, E., S. GOLDBERG, and J. E. DARNELL. 1976. hnRNA in HeLa cells: distribution of transcript sizes estimated from nascent molecule profile. *Cell* **9**:465–472.

FIERS, W., R. CONTRERAS, G. HAEGEMAN, R. ROGIERS, A. VAN DE VOORDE, H. VAN HEUVERSWYN, J. VAN HEEREWEGHE, G. VOLCKAERTAND, and M. YSEBAER. 1978. Complete nucleotide sequence of SV40 DNA. *Nature* **237**:113–118.

FLINT, J. 1977. The topography and transcription of the adenovirus genome. *Cell* **10**:153–166.

KONKEL, D. A., S. M. TILGHMAN, and P. LEDER. 1978. Sequence of the chromosomal mouse β-globin major gene: homologies in capping, splicing and poly(A) sites. *Cell* **15**:1125–1132.

TILGHMAN, S. M., P. J. CURTIS, D. S. TIEMEIER, P. LEDER, and C. WEISSMAN. 1978. The intervening sequence of a mouse β-globin mRNA precursor. *Proc. Nat'l Acad. Sci. USA* **75**:1309–1313.

The Processing of mRNA

DARNELL, J. E., JR. 1982. Variety in the level of gene control in eukaryotic cells. *Nature* **297**:365–371.

DARNELL, J. E., JR. 1983. The processing of RNA. *Sci. Am.* **249**:90–100.

NEVINS, J. R. 1983. The pathway of eukaryotic mRNA formation. *Annu. Rev. Biochem.* **52**:441–466.

NEVINS, J. R., and J. E. DARNELL JR. 1978. Steps in the processing of Ad2 mRNA: poly$(A)^+$ nuclear sequences are conserved and poly(A) addition precedes splicing. *Cell* **15**:1477–1493.

Capping

COPPOLA, J. A., A. S. FIELD, and D. S. LUSE. 1983. Promoter-proximal pausing by RNA polymerase II *in vitro:* transcripts shorter than 20 nucleotides are not capped. *Proc. Nat'l Acad. Sci. USA* **80**:1251–1255.

SHATKIN, A. J. 1976. Capping of eukaryotic mRNAs. *Cell* 9:645–654.

VENKATESAN, S., A. GERSHOWITZ, and B. MOSS. 1980. Purification and characterization of mRNA guanyl transferase from HeLa cell nuclei. *J. Biol. Chem.* 255:2829–2834.

VENKATESAN, S., and B. MOSS. 1980. Donor and acceptor specificities of HeLa cell mRNA guanyl transferase. *J. Biol. Chem.* 255:2834–2842.

Poly A Addition and Termination

BIRNSTIEL, M. L., M. BUSSLINGER, and K. STRUB. 1985. Transcription termination and 3′ processing: the end is in site! *Cell* 41:349–359.

BRAWERMAN, G. 1981. The role of the poly(A) sequence in mRNA. *Crit. Rev. Biochem.* 10:1–38.

FALCK-PEDERSON, E., J. LOGAN, T. SHENK, and J. E. DARNELL, JR. 1985. Transcription termination within the E1A gene of adenovirus induced by insertion of the mouse β-major globin terminator element. *Cell* 40:897–905.

FITZGERALD, M., and T. SHENK. 1981. The sequence 5′—AAUAAA—3′ forms part of the recognition site for polyadenylation of late SV40 mRNAs. *Cell* 24:251–260.

HENIKOFF, S., and E. H. COHEN. 1984. Sequences responsible for transcription termination on a gene segment in *S. cerevisiae*. *Mol. Cell Biol.* 4:1515–1520.

HIGGS, D. R., S. E. Y. GOODBOURN, J. LAMB, J. B. CLEGG, D. J. WEATHERALL, and N. J. PROUDFOOT. 1983. α-Thalassaemia caused by a polyadenylation signal mutation. *Nature* 306:398–400.

MCDEVITT, M. A., M. J. IMPERIALE, H. ALI, and J. R. NEVINS. 1984. Requirement of a downstream sequence for generation of a poly(A) addition site. *Cell* 37:993–999.

MONTELL, C., E. F. FISHER, M. H. CARUTHERS, and A. J. BERK. 1983. Inhibition of RNA cleavage but not polyadenylation by a point mutation in the mRNA 3′ consensus sequence AAUAAA. *Nature* 305:600–605.

ZEEVI, M., J. R. NEVINS and J. E. DARNELL JR. 1982. Newly formed mRNA lacking polyadenylic acid enters the cytoplasm and polyribosomes but has a shorter half-life in the absence of polyadenylic acid. *Mol. Cell. Biol.* 2:517–525.

Splicing

BERGET, S. M., C. MOORE and P. A. SHARP. 1977. Spliced segments at the 5′ terminus of adenovirus 2 late mRNA. *Proc. Nat'l Acad. Sci. USA* 74:3171–3175.

BLACK, D. L., B. CHABOT, and J. A. STEITZ. 1985. U2 as well as U1 small nuclear ribonuclearproteins are involved in premessenger RNA splicing. *Cell* 42:737–750.

CECH, T. R. 1983. RNA splicing: three themes with variations. *Cell* 34:713–716.

CHOW, L. T., R. E. GELINAS, T. R. BROKER, and R. J. ROBERTS. 1977. An amazing sequence arrangement at the 5′ ends of adenovirus 2 messenger RNA. *Cell* 12:1–8.

HERNANDEZ, N., and W. KELLER. 1983. Splicing of *in vitro* synthesized mRNA precursors in HeLa cell extracts. *Cell* 35: 89–99.

PADGETT, R. A., M. M. KONARSKA, P. J. GRABOWSKI, S. F. HARDY, and P. A. SHARP. 1984. Lariat RNAs as intermediates and products in the splicing of messenger RNA precursors. *Science* 225:898–903.

PADGETT, R. A., S. M. MOUNT, J. A. STEITZ, and P. A. SHARP. 1983. Splicing of messenger RNA precursors is inhibited by antisera to small ribonucleoproteins. *Cell* 35:101–107.

RUSKIN, B., A. R. KRAINER, T. MANIATIS, and M. R. GREEN. 1984. Excision of an intact intron as a novel lariat structure during pre-mRNA splicing *in vitro*. *Cell* 30:317–331.

SHARP, P. A. 1981. Speculations on RNA splicing. *Cell* 23:643–646.

Transcription by RNA Polymerase II

Definition of Promoter Sites

BAKER, C. C., and E. B. ZIFF. 1980. Biogenesis, structures, and sites of encoding of the 5′ termini of adenovirus-2 mRNAs. *Cold Spring Harbor Symp. Quant. Biol.* 44:415–428.

BENOIST, C., and P. CHAMBON. 1981. *In vivo* sequence requirements of the SV40 early promoter region. *Nature* 290:304–310.

CONTRERAS, R., and W. FIERS. 1981. Initiation of transcription by RNA polymerase II in permeable SV40-infected or noninfected CV1 cells: evidence for multiple promoters of SV40 late transcription. *Nuc. Acids Res.* 9:215–236.

CORDEN, J., B. WASYLYK, A. BUCHWALDER, D. SASSONE-CORSI, D. KEDINGER and P. CHAMBON. 1980. Promoter sequences of eukaryotic protein coding genes. *Science* 209:1406–1413.

DAVISON, B. L., J-M. EGLY, E. R. MULVIHILL, and P. CHAMBON. 1983. Formation of stable preinitiation complexes between eukaryotic class B transcription factors and promoter sequences. *Nature* 301:680–686.

GROSVELD, G. C., E. DE BOER, C. K. SHEWMAKER, and R. A. FLAVELL. 1982. DNA sequences necessary for transcription of the rabbit β-globin gene *in vivo*. *Nature* 295:120–126.

MCKNIGHT, S. L., and R. KINGSBURY. 1982. Transcriptional control signals of a eukaryotic protein-coding gene. *Science* 217:316–324.

WASYLYK, B., and P. CHAMBON. 1981. A T to G base substitution and small deletions in the conalbumin TATA box drastically decrease specific in vitro transcription. *Nuc. Acids Res.* 9:1813–1824.

WEIL, P. A., D. S. LUSE, J. SEGALL, and R. G. ROEDER. 1979. Selective and accurate initiation of transcription of the Ad2 major late promoter in a soluble system dependent on purified RNA polymerase II and DNA. *Cell* 18:469–484.

Definition of Enhancer Sites

BANERJI, J., S. RUSCONI, and SCHAFFNER, W. 1981. Expression of a β-globin gene is enhanced by remote SV40 DNA sequences. *Cell* 27:299–308.

FROMM, M ., and P. BERG. 1982. Deletion mapping of DNA regions required for SV40 and early region promoter function *in vivo*. *J. Mol. App. Genet.* 1:457–481.

GLUZMAN, Y., and T. SHENK. 1983. *Enhancers and Eukaryotic Gene Expression.* Cold Spring Harbor Laboratory.

GROSSCHEDL, R., and M. L. BIRNSTIEL. 1980. Spacer DNA sequences upstream of the TATA sequence are essential for promotion of H2A gene transcription *in vivo*. *Proc. Nat'l Acad. Sci. USA* 77:7102–7106.

KHOURY, G., and P. GRUSS. 1983. Enhancer elements. *Cell* 33:313–314.

Proteins Associated with hnRNA

BEYER, A. L., M. E. CHRISTENSEN, B. W. WALKER, and W. M. LESTOURGEON. 1977. Identification and characterization of the packaging proteins of core 40S hnRNP particles. *Cell* **11**:127–138.

BEYER, A. L., O. L. MILLER JR., and S. L. MCKNIGHT. 1980. Ribonucleoprotein structure in nascent hnRNA is nonrandom and sequence-dependent. *Cell* **20**:75–84.

LERNER, M. R., and J. A. STEITZ. 1981. Snurps and scyrps. *Cell* **25**:298–300.

PEDERSON, T. 1983. Nuclear RNA-protein interactions and messenger RNA processing. *J. Cell Biol.* **97**:1321–1326.

REDDY, R., and H. BUSCH. 1981. U snRNA's of nuclear snRNP's. In *The Cell Nucleus,* vol. 8. Academic Press.

Nuclear Pores: The Presumed Exit Sites for mRNA

GERACE, L., and G. BLOBEL. 1980. The nuclear envelope lamina is reversibly depolymerized during mitosis. *Cell* **19**:277–287.

MAUL, G. G. 1977. The nuclear and the cytoplasmic pore complex: structure, dynamics, distribution, and evolution. In *International Review of Cytology, Suppl. 8,* Academic Press.

UNWIN, P. N. T., and R. A. MILLIGAN. 1982. A large particle associated with the perimeter of the nuclear pore complex. *J. Cell Biol.* 93:63–75.

The Appearance of mRNA in the Cytoplasm and the Initiation of Protein Synthesis

MAITRA, U., E. A. STRINGER, and A. CHAUDHURI. 1982. Initiation factors in protein biosynthesis. *Annu. Rev. Biochem.* 51:869–900.

THIMMAPPAYA, B., C. WEINBERGER, R. J. SCHNEIDER, and T. SHENK. 1982. Adenovirus VAI RNA is required for efficient translation of viral mRNAs at late times after infection. *Cell* **31**:543–551.

The Possible Association of mRNA with the Cytoskeleton

CERVERA, M., G. DREYFUSS, and S. PENMAN. 1981. Messenger RNA is translated when associated with the cytoskeletal framework in normal and VSV-infected HeLa cells. *Cell* 23:113–120.

10

Eukaryotic Chromosomes and Genes: General Structure and Definition

THE preceding chapter described how mRNAs, rRNAs, and tRNAs are produced from the transcription units in eukaryotic DNA. In the next two chapters we shall examine the structures that contain the DNA—the eukaryotic chromosomes. Viewed in the light microscope, the chromosomes are morphologically distinctive, species-specific structures. At the molecular level, chromosomes can be regarded as an assembly of transcription units that are precisely duplicated in each cell generation. Ever since Sutton hypothesized (in 1902) that the chromosomes were the carriers of Mendel's independently segregating genes, studies of how the structure of the chromosome is related to its function have been at the heart of biological research. Modern molecular genetic research is answering the following long-standing questions about the relation of chromosome structure to gene function:

What is the relation of eukaryotic transcription units to the traditional functional chromosome unit, the gene?

In chromosomal DNA, how are the transcription units arranged in relation to other sequences that are not transcribed?

Which genes are present in multiple copies?

What is the function (if any) of DNA sequences that are not transcribed?

What is the nature of the so-called repetitive DNA?

How stable is the arrangement of DNA in chromosomes?

What proteins are associated with the DNA in chromosomes, and what is this DNA-protein structure like?

The other traditional problems that have attracted students of chromosomes concern replication and repair of DNA:

At what site or sites in the chromosomes does replication begin?

What enzymes and other proteins participate in replication?

How is a chromosome repaired after it suffers damage from chemicals or from irradiation?

What are the molecular events in the DNA at the site of recombination?

This chapter provides an introduction to the structure of eukaryotic chromosomes and a definition of eukaryotic genes. Molecular studies that are largely based on the cloning and sequencing of segments of eukaryotic chromosomes are described in Chapter 11. In that chapter, we also discuss the surprising finding that eukaryotic chromosomes have a marked instability due to transposable elements. Chapter 12 integrates the present facts about eukaryotic transcription with what we know about gene and chromosome structure to develop a detailed picture of eukaryotic gene control. Chapter 13 deals with the dynamic events of DNA biochemistry—replication, repair, and recombination.

The Morphology of Chromosomes

Today, the fine details of chromosome structure, including the nucleotide sequences of individual genes, are being determined through the use of modern molecular technology. However, many important general principles of chromosome structure were derived from earlier cytologic studies (Chapter 1). In this chapter we shall discuss chromosome morphology before proceeding to questions about specific gene structure.

Chromosomes Are Species-Specific in Number and Shape

During mitosis and meiosis, chromosomes are visible in the light microscope (Chapter 5). In cells not undergoing mitosis or meiosis, the chromosomes are not visible, even with the aid of histological stains for DNA (e.g., Feulgen or Giemsa stains). Therefore, almost all cytogenetic work has been done with condensed metaphase chromosomes obtained from dividing cells—either somatic cells in mitosis or dividing gametes during meiosis. During cell division, chromosomes are duplicated structures: each metaphase chromosome consists of two *sister chromatids* that are separated along their lengths except at one point of attachment, the *centromere.* The number, sizes, and shapes of the metaphase chromosomes constitute the *karyotype,* a set of chromosomal features that is distinctive for each organism (Figure 10-1). In most organisms, all cells have the same karyotype. However, species that appear quite similar can have very different karyotypes (Figure 10-2; Table 10-1).

Cellular DNA Content Does Not Correlate with Phylogeny

The total amount of DNA in the set of chromosomes from different animals or plants does not vary in a consistent manner with the apparent complexity of the organisms, as we have already seen in Figure 10-2c and d. Yeasts, fruit flies, chickens, and humans have successively larger amounts of DNA in their haploid chromosome sets (0.015, 0.15, 1.3, and 3.2 picograms, respectively), in keeping with what we perceive to be the increasing complexity of these organisms. Yet the vertebrate species with the greatest amount of DNA per cell are amphibians (Figure 10-3), which are surely less complex than humans in their structure and behavior. And tulips have 10 times the DNA per cell that humans have (Figure 10-3 and Table 10-2).

In addition, there is considerable intragroup variation in DNA content. For example, all insects or all amphibians would appear to be similarly complex, but the amount of haploid DNA within each of these phyla varies by a factor of 100 (Figure 10-3). The same wide variation in DNA content per cell is common within groups of plants that have similar structures and life cycles, such as bean plants or cereal grain plants (Table 10-2). From these facts it would seem that some of the DNA in certain organisms is "extra" or expendable; apparently all DNA in all organisms does not serve to encode proteins. The total amount of DNA per haploid cell in an organism is referred to as the *C value,* and the failure of C values to correspond to phylogenetic complexity has been termed the *C value paradox.* As we shall see in Chapter 11, this perplexing variation in genome size occurs, at least in

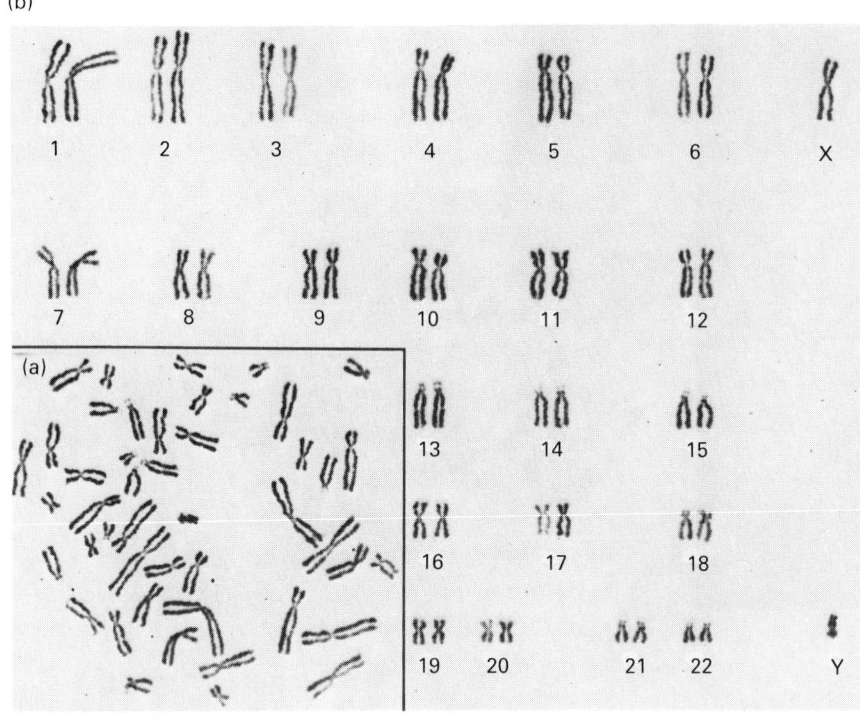

Figure 10-1 Human male metaphase chromosomes. (a) The chromosomes from one mitotic cell, a lymphocyte, were stained with orcein (a dye that reacts with DNA), and the cell was "squashed" on a microscope slide. (b) After identification of each pair of homologues, photographs of the chromosomes were mounted in sequence by size. Each number indicates a homologous pair, and the entire set is the *karyotype*. The unnumbered, unpaired X and Y chromosomes are the sex chromosomes; note that they are morphologically distinct.

A metaphase chromosome consists of two identical chromatids; their point of association is the centromere. According to the position of the centromere, the chromosome is termed metacentric (centromere in the middle; e.g., chromosome 1), acrocentric (asymmetrical centromere; e.g., chromosome 4), or telocentric (centromere at an end; chromosome 14 is the best example here). *Courtesy of J. German.*

Table 10-1 Chromosome numbers of various species*

Common name	Species	Diploid number	Common name	Species	Diploid number
ANIMALS			DIPLOID PLANTS AND FUNGI		
Human	*Homo sapiens*	46	Yeast	*Saccharomyces cerevisiae*	36±
Rhesus monkey	*Macaca mulatta*	42	Green algae	*Acetabularia mediterranea*	20±
Cattle	*Bos taurus*	60	Garden onion	*Allium cepa*	16
Dog	*Canis familiaris*	78	Barley	*Hordeum vulgare*	14
Cat	*Felis domesticus*	38	Rice	*Oryza sativa*	24
Horse	*Equus calibus*	64	Corn	*Zea mays*	20
House mouse	*Mus musculus*	40	Snapdragon	*Antirrhinum majus*	16
Rat	*Rattus norvegicus*	42	Tomato	*Lycopersicon esculentum*	24
Golden hamster	*Mesocricetus auratus*	44	Tobacco	*Nicotinana tabacum*	48
Guinea pig	*Cavia cobaya*	64	Kidney bean	*Phaseolus vulgaris*	22
Rabbit	*Oryctolagus cuniculus*	44	Pine	*Pinus* species	24
Chicken	*Gallus domesticus*	78±	Garden pea	*Pisum sativum*	14
Alligator	*Alligator mississipiensis*	32	Potato	*Solanum tuberosum*	48
Frog	*Rana pipiens*	26	Broad bean	*Vicia faba*	12
Carp	*Cyprinus carpio*	104			**Haploid number**
Silkworm	*Bombyx mori*	56	HAPLOID PLANTS AND FUNGI		
House fly	*Musca domestica*	12			
Fruit fly	*Drosophila melanogaster*	8	Slime mold	*Dictyostelium discoideum*	7
Flatworm	*Planaria torva*	16	Mold (fungus)	*Aspergillus nidulans*	8
Freshwater hydra	*Hydra vulgaris attenuata*	32	Pink bread mold	*Neurospora crassa*	7
Nematode	*Caenorhabditis elegans*	11/12 (male/female)	Penicillin mold	*Penicillium* species	4
			Green algae	*Chlamydomonas reinhardi*	16

*In most organisms the chromosome number does not vary, but in some species (those with numbers listed as ±) either there is a natural variation in strains or the correct number has not been determined. In a few organisms the sexes differ; in nematodes, the male is haploid for the sex chromosome (X,0), and the female is diploid (X,X).

SOURCE: M. Strickberger, 1976, *Genetics*, 2d ed., Macmillan, p. 16.

(a)

(b)

(c)

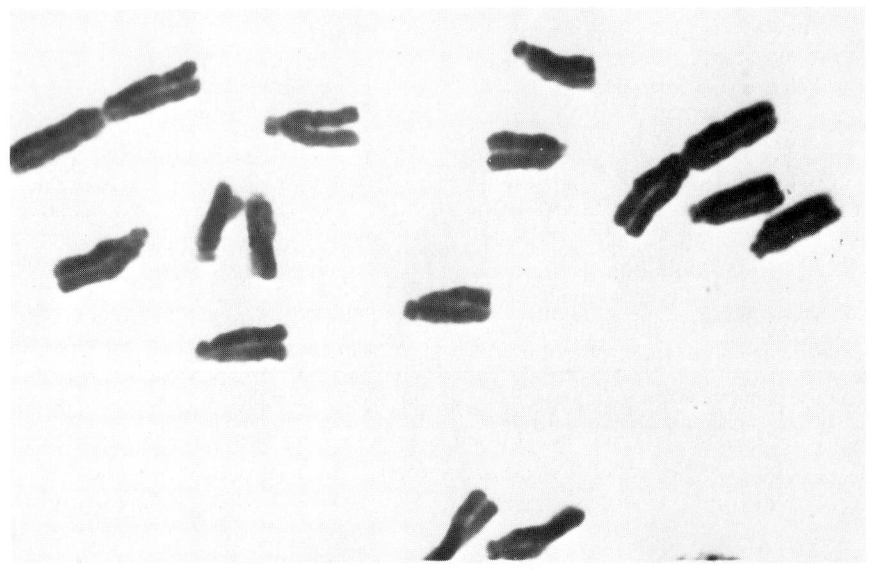

(d)

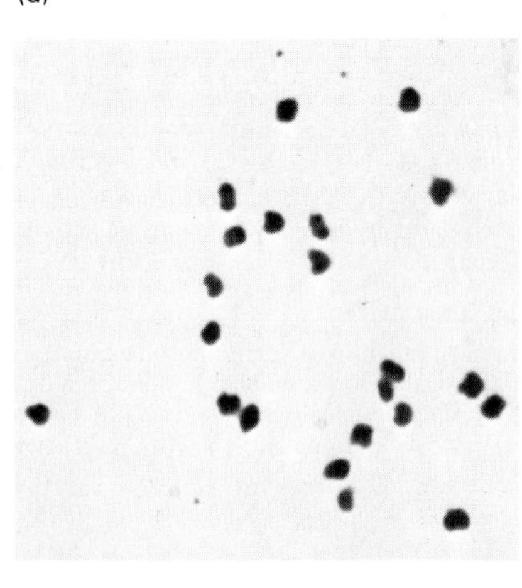

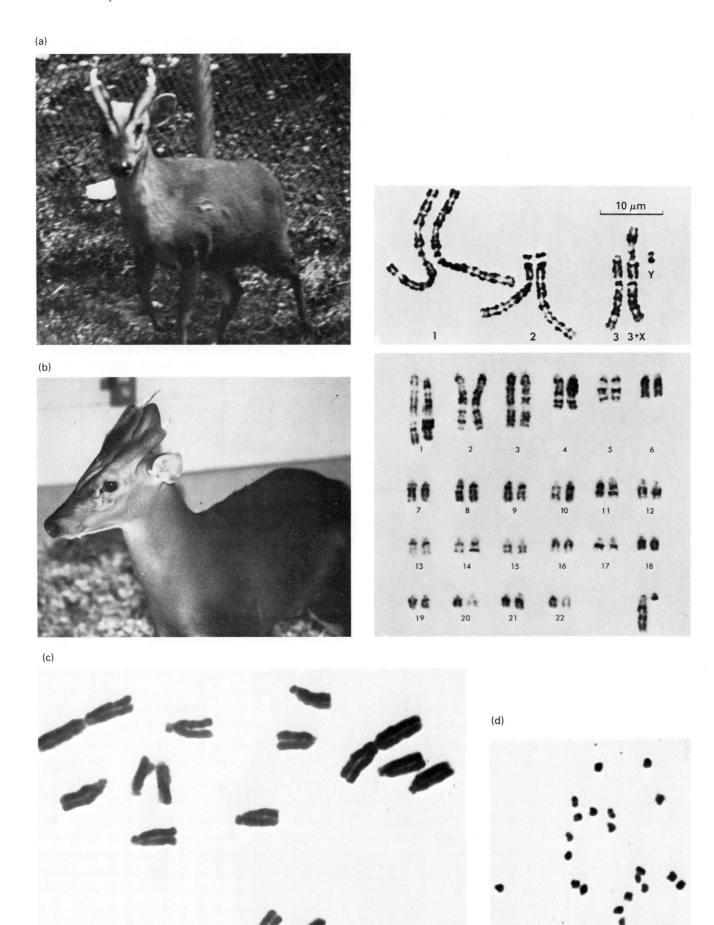

Figure 10-2 *(opposite)* Chromosomes of related organisms. The Reeves muntjac (a) and the Indian muntjac (b) are two species of small deer that are quite similar in size, appearance, and behavior, but they do not interbreed. As the karyotypes show, the genomes of the two species are arranged quite differently: one has 4 chromosomes; the other, 23. However, the two genomes contain about the same amount of total DNA. [The magnification is the same in (a) and (b). The chromosomes were stained after G banding, a procedure explained in Figure 10-4.]

Similar plants can also have strikingly different karyotypes. The broad bean *Vicia faba* (c) has about half the number of chromosomes of the kidney bean plant *Phaseolus vulgaris* (d). However, the broad bean has about three to four times as much DNA per cell as the kidney bean. (Again, the chromosomes in these two photographs are shown at the same magnification.) *Parts (a) and (b) courtesy of R. Church. Parts (c) and (d) courtesy of M. D. Bennett; from M. D. Bennett, 1976,* Environmental and Experimental Botany *16:93.*

part, because eukaryotic chromosomes contain variable amounts of repeated DNA stretches, some of which are never transcribed and many of which are probably dispensable.

Stained Chromosomes Show Characteristic Banding Patterns

A variety of microscopic techniques are used to stain certain regions or *bands* of chromosomes more intensely than other regions. These banding patterns are specific for individual chromosomes. They serve as landmarks along the length of each chromosome and also help to distinguish among chromosomes with similar sizes and shapes.

Metaphase Chromosome Bands and Chromosome Rearrangements Quinacrine, a fluorescent dye that inserts, or *intercalates,* itself into the DNA helix, produces *Q bands* when stained chromosomes are viewed microscopically. However, because the Q bands fade with time, other techniques are generally preferred in the laboratory. For example, chromosomes may be treated briefly in various ways (e.g., they may be subjected to mild heat or proteolysis) and then stained with permanent DNA dyes (such as Giemsa) to produce a pattern of *G bands* (Figure 10-4). Treatment of the chromosomes with a hot alkaline solution before staining with Giemsa produces *R bands* in a pattern that is the reverse of the G band pattern. Because of the distinctiveness of the banding patterns, cytologists are able to identify the specific parts of a chromosome. A standardized map of bands in human chromosomes has been developed (Figure 10-5). With this map, the sites of breaks and rearrangements in chromosomes have been accurately located, and partial chromosome duplications have been discovered (Figure 10-6).

The molecular basis for the regularity of chromosomal bands remains unknown. Because the chromosomal bands are not visible in isolated DNA, they are presumed to be due to folding produced by the interaction of DNA and proteins. Human chromosomes contain from 10^8 to 3×10^8 base pairs; a single band represents about 5 to 10 percent of a chromosome, or about 10^7 base pairs. The constancy of banding patterns over these large DNA regions (the patterns are nearly identical in almost every copy of a chromosome, in almost every tissue) implies a constancy of DNA-protein interactions at specific sites within the regions.

Interphase Polytene Chromosomes For over 50 years, cytologists have observed bands on interphase chromosomes in the salivary glands of *Drosophila melanogaster* and other dipteran insects. Although the bands in human chromosomes apparently represent very long linear stretches of DNA, the bands in *Drosophila* chromosomes represent much shorter stretches of only 50,000 to 100,000 base pairs. A fortunate structural feature of these insect chromosomes accounts for the fact that such short segments can be resolved in the light microscope. The presumably single duplex DNA molecule in insect salivary chromosomes is repeatedly copied (as many as 1000 times) into parallel arrays of identical DNA molecules. This amplification without separation, termed *polytenization,* results in thick bundles of parallel DNA molecules that all have the same banding pattern across the width of the bundle (Figure 10-7). Recent molecular cloning experiments plus mRNA mapping studies have suggested that each band contains a limited number of transcription units, perhaps only one unit in some cases.

Heterochromatin Consists of Chromosome Regions That Do Not Uncoil

As cells exit from mitosis and the chromosomes uncoil, certain sections of the chromosomes remain dark-staining. The dark-staining areas, termed *heterochromatin,* are regions of condensed chromatin, or DNA plus proteins. Heterochromatin appears most frequently—but not exclusively—at the centromere and telomeres (ends) of the chromosome. The light-staining, less condensed portions are called *euchromatin.* Because the heterochromatin regions apparently remain condensed throughout the life of the cell, they have long been regarded as sites of inactive genes. In recent years, molecular studies have shown that most of the DNA in heterochromatin is

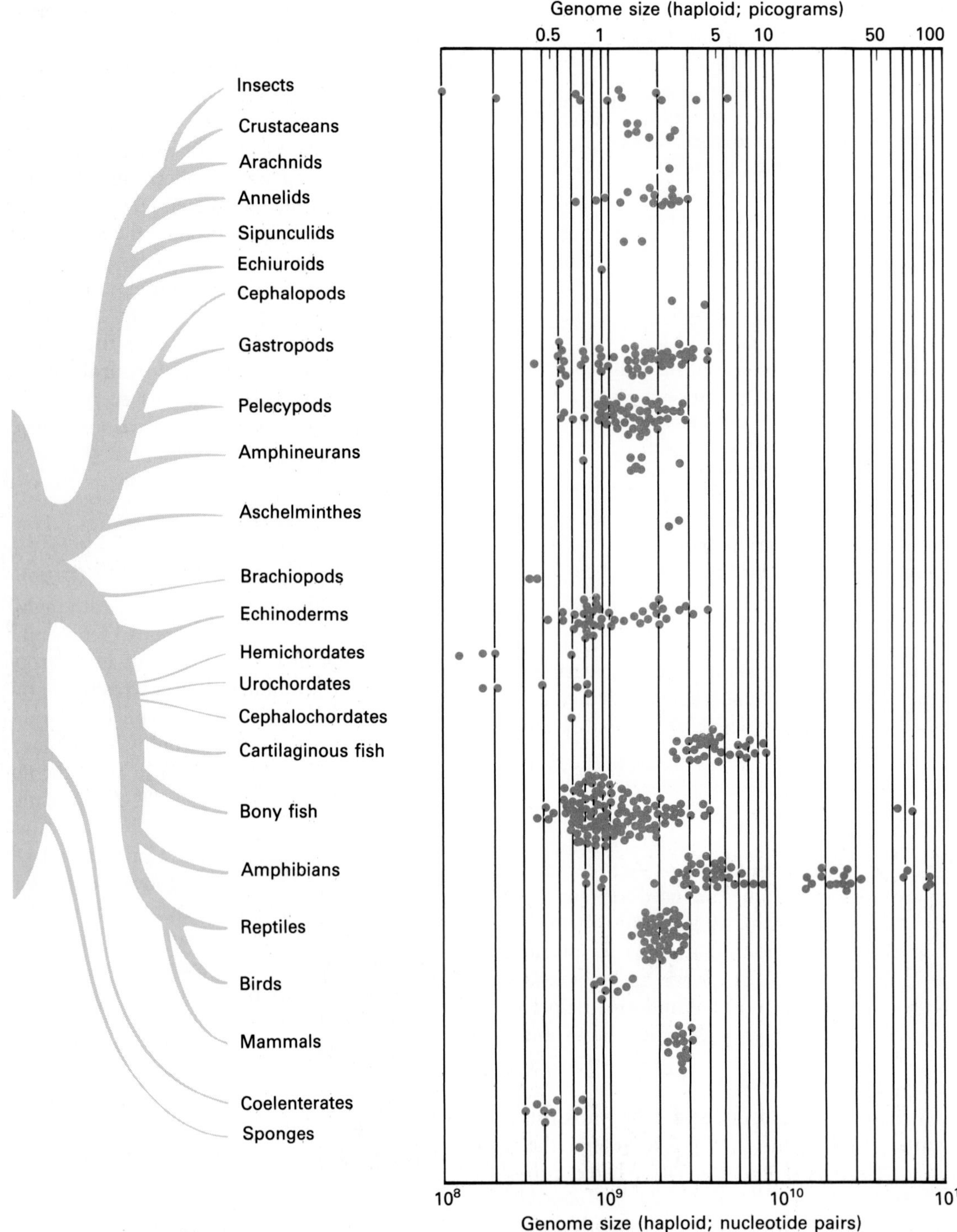

Figure 10-3 The amount of DNA per haploid chromosome set in a variety of animals. Variation within certain phyla is quite wide (e.g., see insects and amphibians). [See R. J. Britten and E. Davidson, 1971, *Quart. Rev. Biol.* **46**:111.]

Table 10-2 Nuclear DNA content of plants

Type of plant	Species	Haploid nuclear DNA content (picograms)
Grass	*Poa trivialis*	3.5
Garden pea	*Pisum sativum*	5.0
Sunflower	*Helianthus annuus*	5.3
Corn	*Zea mays*	5.5
Barley	*Hordeum vulgare*	6.7
Wheat	*Triticum monococcum*	7.0
Rye	*Secale cereale*	9.4
Broad bean	*Vicia faba*	14.6
Garden onion	*Allium cepa*	16.8
Tulips	*Tulipa kaufmanniana*	31.2

SOURCE: R. B. Flavell et al., 1974, *Biochem. Genet.* **12**:257.

indeed highly repeated DNA that is never (or very seldom) transcribed. However, all inactive genes and nontranscribed regions of DNA are surely not visible as heterochromatin, nor is it necessarily true that no transcription occurs within heterochromatin.

The Inactive X Chromosome in Mammalian Females One important case of heterochromatinization does correlate with gene inactivation. The random inactivation and condensation of one of the two female sex chromosomes (the X chromosomes) occurs in virtually all of the somatic cells of female mammals. The inactive X, which appears as heterochromatin throughout the cell cycle, is visible during interphase as a dark-staining, peripheral nuclear structure called the *Barr body,* after its discoverer, the Canadian cytologist M. L. Barr. The process of X inactivation is often termed *lyonization,* after the British cytogeneticist Mary Lyons, who first recognized that the random inactivation of one X chromosome in females can lead to important genetic consequences when the individual has two X chromosomes, one of which carries a defective gene.

Each cell in a female has two X chromosomes, one of maternal origin (X_m) and one of paternal origin (X_p). Early during embryologic development (when a few million cells have been formed), inactivation of either the X_m or the X_p chromosome occurs in each cell. The embryo thus becomes a mosaic of cells in which about half have an inactive X_m and the other half have an inactive X_p. All subsequent daughter cells maintain the same inactive X chromosomes as do their parent cells. As a result, the adult female is a patchwork of clones, some expressing the genes on the X_m chromosome and the rest expressing the genes on the X_p.

Such an arrangement is exhibited in the fur color of cats. Male cats, which have only one X chromosome, can be black or yellow, but they are almost never both. In contrast, certain females, called *calicos,* have coats that are patches of black and yellow. The two colors have been traced to a gene on the X chromosome. The males have either the black (Bl^+) or the yellow (Bl^-) allele of

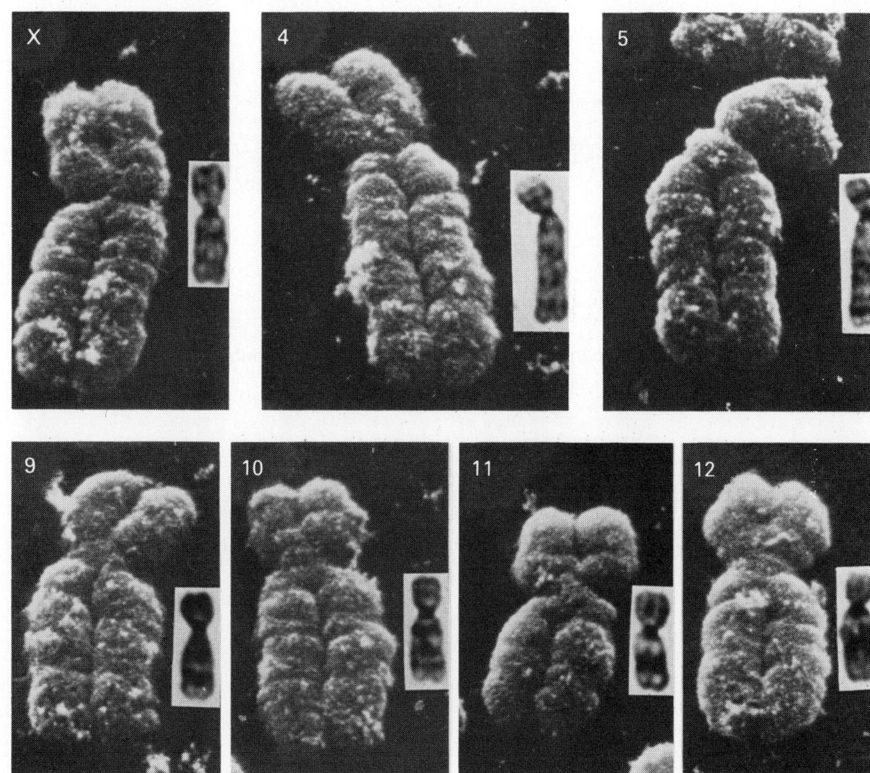

Figure 10-4 Human male chromosomes after G banding. The chromosomes have undergone brief proteolytic treatment followed by staining with the Giemsa reagent. After this procedure, which does not stain all regions of the chromosome equally, distinctive bands appear at characteristic places. The bands enable investigators to distinguish chromosomes of similar lengths from one another. The light micrographs *(insets)* show the bands; the scanning electron micrographs show constrictions at the sites where the bands are observed. *Courtesy of C. J. Harrison; from C. J. Harrison et al., 1981,* Exp. Cell Res. ***134**:141.*

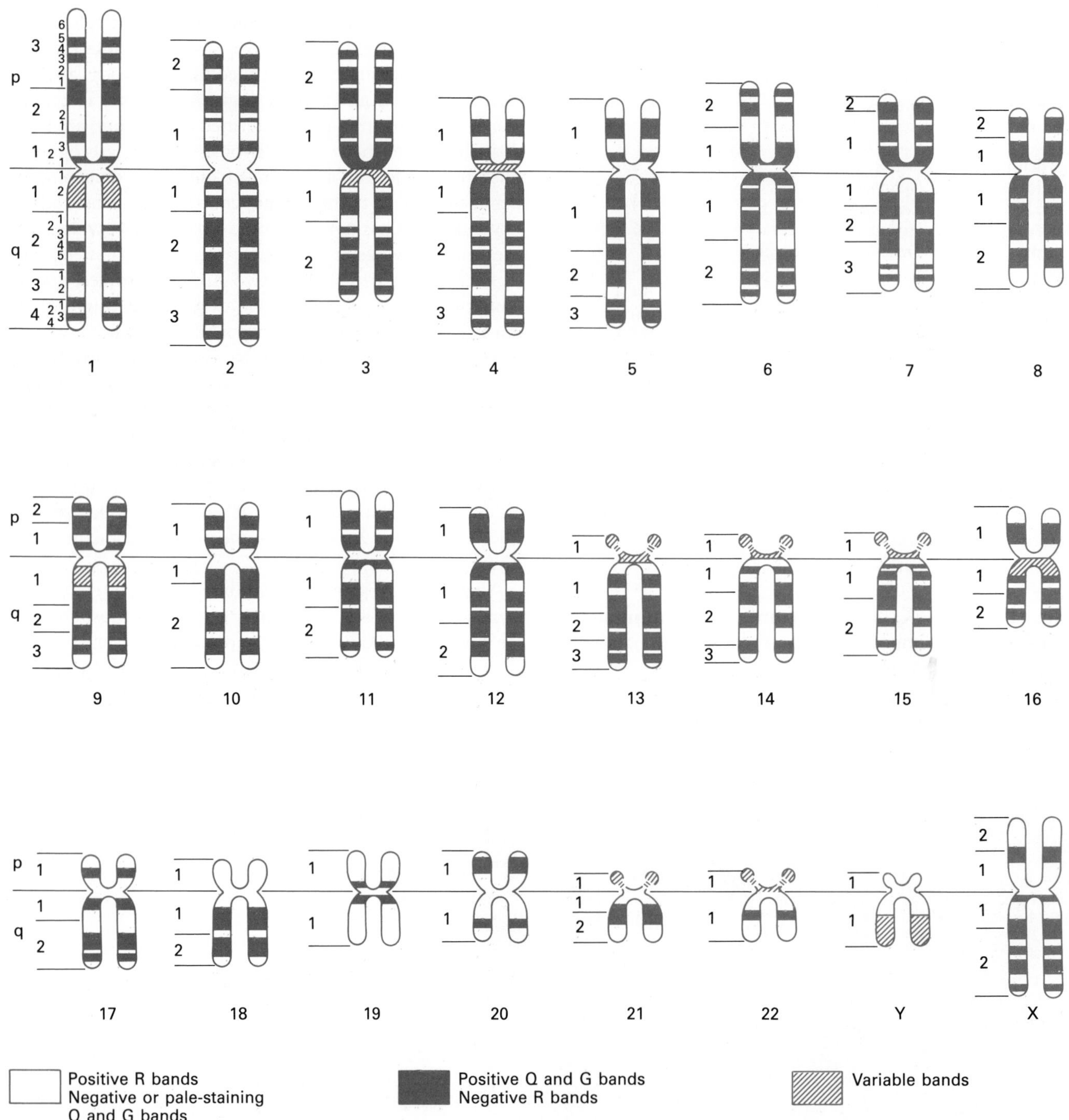

Figure 10-5 Diagrams of the standard Q, G, and R bands in human chromosomes, as recognized by an international group of cytologists. The parts of the chromosomes are labeled according to convention: the two arms are labeled p and q, with q being the longer arm. Each arm is divided into sections (such as p1, p2, and p3 on chromosome 1), and then into subsections defined by the light and dark bands (for example, p35 on chromosome 1; note that the numbers are written serially, without separation). Variable bands, which are seen particularly in the centromere regions (e.g., on chromosomes 1, 3, 4, etc.) and on the Y chromosome, may be due to variable amounts of repeated DNA. *After Paris Conference, 1971.*

this gene, whereas the calicos are females that are heterozygous (Bl^+/Bl^-). The black patches on the heterozygotes are produced by clones in which the active X chromosome carries the Bl^+ allele, and the yellow patches mark clones in which the active X has the Bl^- allele (Figure 10-8). The rare calico male has a sex chromosome constitution of XXY, in which one X is also inactivated.

Recent studies have shown that the genes on an inactive X can be reactivated by preventing the methylation of the DNA. From this result, it has been inferred that a high degree of methylation is one of the changes related to the inactivation of X chromosomes. (This topic is discussed in detail in Chapter 12.)

Chromosomes Probably Have One Linear DNA Molecule

How many individual DNA molecules does a chromosome represent? All DNA viruses seem to possess one molecule of DNA, and bacterial cells contain a single chromosomal DNA molecule. The general belief—which has not yet been completely proved—is that most eukaryotic chromosomes also represent a single long DNA molecule. As we have mentioned, the DNA content of a single human chromosome varies between about 10^8 and 3×10^8 base pairs. Such DNA molecules are almost 10 cm long and are difficult to handle experimentally without breaking them, so the upper limit of DNA size for human chromosomes remains unknown. However, in lower eukaryotic cells, the sizes of the largest DNA molecules that can be extracted are consistent with the hypothesis that a chromosome is a single DNA molecule. For example, sedimentation analysis of the largest DNA molecules from several genetically different *Drosophila* species and strains has shown that they are from 6×10^7 to 1×10^8 base pairs long. These sizes match the DNA content of single stained metaphase chromosomes of *Drosophila melanogaster,* as measured by the amount of DNA-specific stain absorbed (Table 10-3). Therefore, each chromosome is probably a single DNA molecule.

The largest DNA molecules from yeast cells also correlate with the entire DNA content of the longest chromosomes. The length of yeast chromosomes ranges from approximately the length of *Escherichia coli* chromosomes (3×10^6 base pairs) to only 5 to 10 percent of that figure.

The Linearity of Genetic Maps and Visible Chromosome Exchanges Mitochondrial DNA, chloroplast DNA, bacterial chromosomal DNAs, and many viral DNAs are circular molecules. But it is very unlikely that the DNA in eukaryotic chromosomes is circular. First, genetic maps of all normal eukaryotic chromosomes are

Table 10-3 **Comparison of chromosome size with DNA extracted from *Drosophila* species and strains**

Species	Karyotype*	Molecular weight of largest DNA†	DNA content of largest chromosome‡
Drosophila melanogaster			
Wild type		$41 \pm 3 \times 10^9$ 6.2×10^7 base pairs	43×10^9
Translocation		$58 \pm 6 \times 10^9$	59×10^9
Drosophila hydei			
Wild type		$40 \pm 4 \times 10^9$	39×10^9
Deletion-breakage		$24 \pm 4 \times 10^9$	26×10^9
Drosophila americana			
Wild type		$79 \pm 10 \times 10^9$	79×10^9

* The lengths of the lines indicate approximate lengths of various chromosomes and the largest chromosomes in each set are identified by color. Notice that translocation, deletion, and breakage change the visible sizes of chromosomes, and thus the sizes of the largest sedimenting DNAs.

† The molecular weights of the chromosomal DNAs were estimated by comparing their sedimentation coefficients with the coefficients of bacteriophage DNAs of known molecular size.

‡ The DNA content was measured by spectrophotometry of metaphase chromosomes and is expressed as molecular weight of one chromosome.

SOURCE: R. Kavenoff, L. C. Klotz, and B. H. Zimm, 1973, *Cold Spring Harbor Symp. Quant. Biol.* **32**:2.

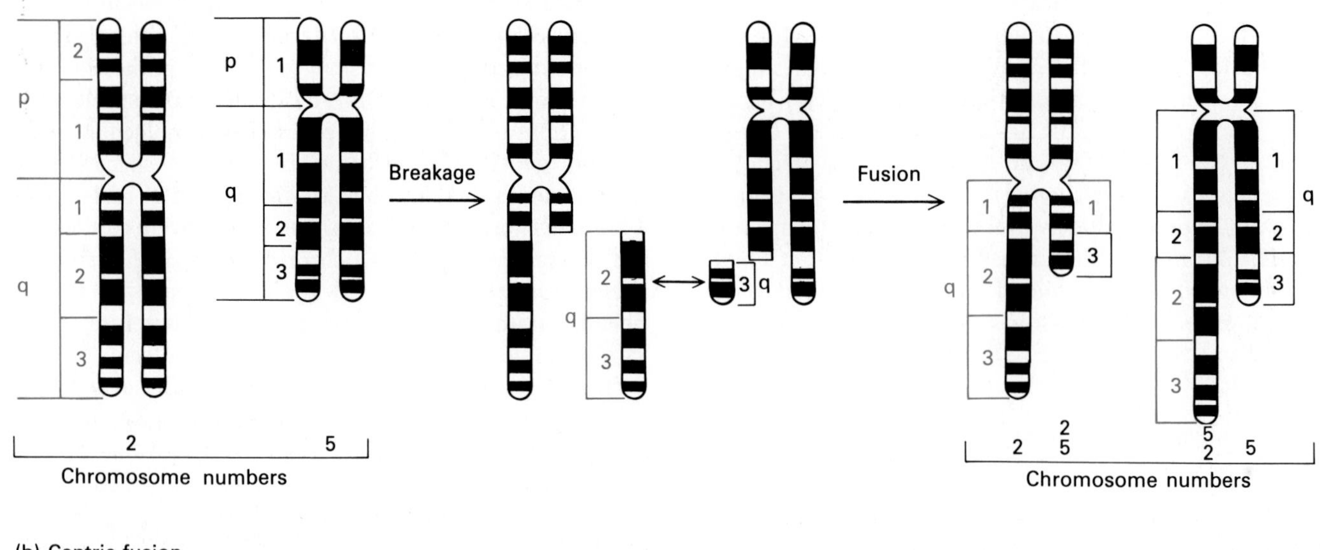
(a) Reciprocal translocation
p
q
2
1
1
2
3
p
1
q
1
2
3
Breakage
q
2
3
3 q
Fusion
q
1
2
3
1
3
1
2
2
3
1
2
3
q
2
5
2 5
2 5
Chromosome numbers
Chromosome numbers

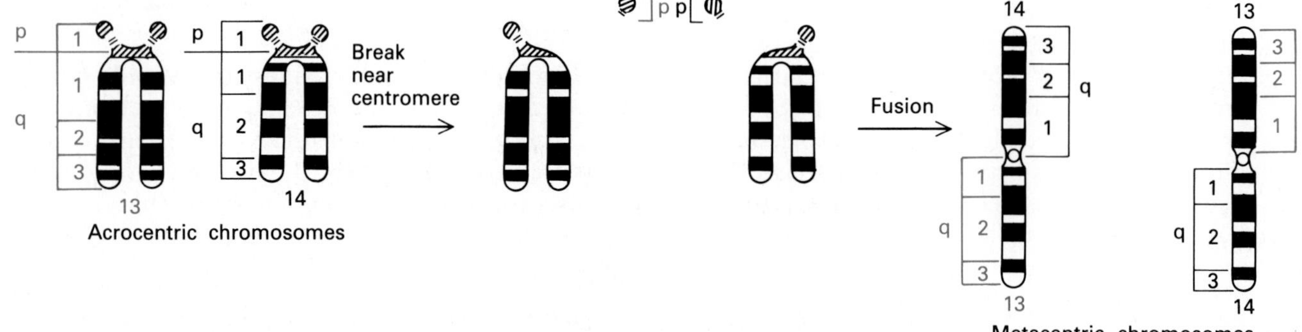
(b) Centric fusion
p
q
1
1
2
3
p
q
1
1
2
3
13
14
Acrocentric chromosomes
Break
near
centromere
p p
Fusion
14
3
2
1
q
1
2
3
q
13
13
3
2
1
1
2
3
q
14
Metacentric chromosomes

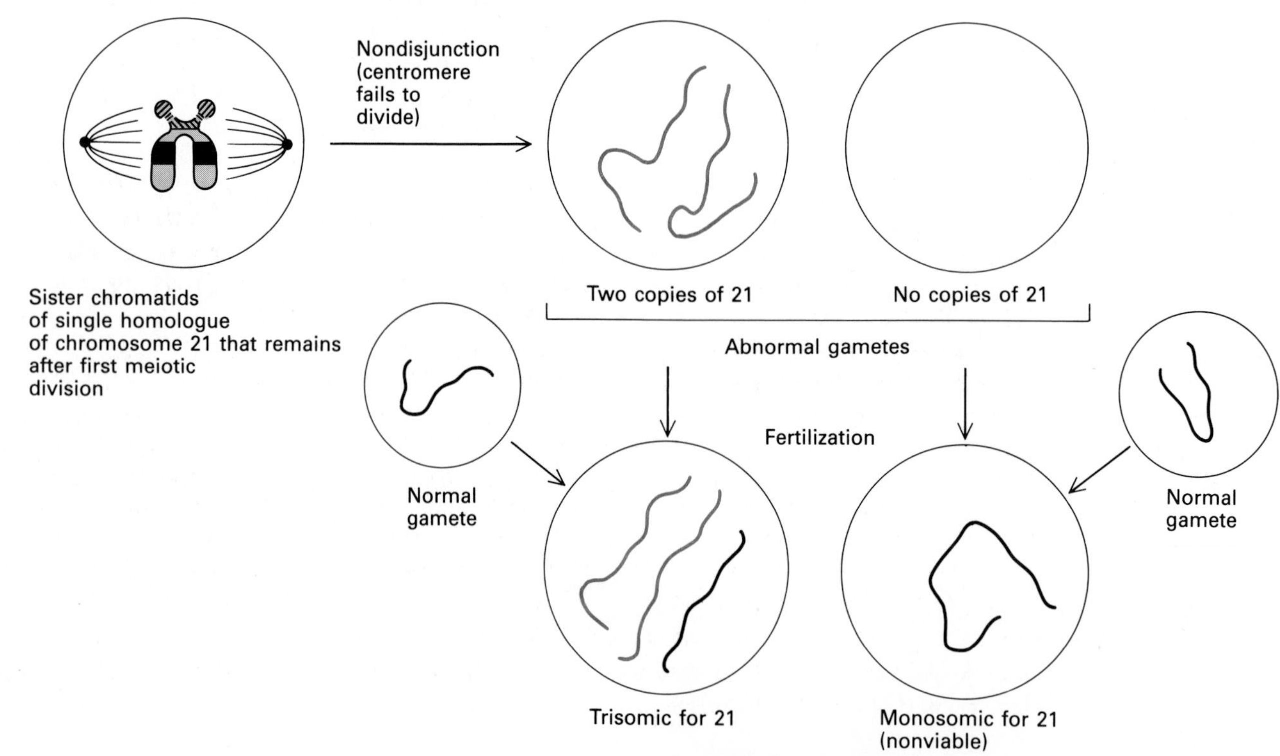
(c) Nondisjunction during second meiotic division
Nondisjunction
(centromere
fails to
divide)
Sister chromatids
of single homologue
of chromosome 21 that remains
after first meiotic
division
Two copies of 21
No copies of 21
Abnormal gametes
Fertilization
Normal
gamete
Normal
gamete
Trisomic for 21
Monosomic for 21
(nonviable)
Abnormal zygotes

Figure 10-6 ***(opposite)*** Chromosomal abnormalities have been found in about 0.1 to 1 percent of the human population by studying karyotypes and analyzing bands. (a) A *translocation* is a rearrangement in which a piece breaks off of one chromosome and attaches to another. Most translocations are *reciprocal,* which means that each of the chromosomes involved is both a donor and a recipient. Reciprocal translocations in which no part of either chromosome is lost in the exchange are seldom associated with any known abnormalities of phenotype. (b) A fairly common type of translocation is *centric fusion* (also called Robertsonian fusion). Centric fusions always involve acrocentric chromosomes, such as the human 13 and 14, which are illustrated here. The process occurs as follows: Breaks appear at or near the centromere of one chromatid in each of the metaphase chromosomes. The long arm and centromere of one broken chromatid fuse to the long arm of the other to form a metacentric chromosome. [It is likely that the regions near the centromere contain repetitive sequences (discussed in Chapter 11), and that the fusions occur within these repetitive sequences.] The unbroken chromatids remain normal at division. The short arms of the broken chromatids may be lost, or they may fuse with the other centromere to become a *dot chromosome.* (c) Another kind of chromosomal abnormality is *nondisjunction,* which is either the failure of sister chromatids to separate at mitosis during the second meiotic division, or the failure of homologous chromosomes to separate during the first meiotic division. The nondisjunction of the sister chromatids of chromosome 21 during meiosis results in trisomy (three copies) of that chromosome in one of the zygotes. Trisomy of chromosome 21 is the most common serious human chromosomal defect, and 95 percent of the cases result from nondisjunction. The abnormality is expressed as Down's syndrome, a condition involving mental retardation and characteristic physical deformities. [See J. J. Yunis, 1976, *New Chromosomal Syndromes,* Academic Press.]

linear. These maps are based on recombination frequencies (in meiosis) between genes on homologous chromosomes. (Mapping will be described later in the chapter.) Second, mitotic chromosomes, which appear as sister chromatids joined at the centromere, *look* linear. Furthermore, as indicated earlier, nonhomologous chromosomes occasionally undergo *translocations*—that is, they break and exchange large pieces (see Figure 10-6a and b). The effects of translocations are visible in the light microscope. The specific Q, G, and R banding patterns are linearly distributed along the lengths of stained chromosomes; translocation between two nonhomologous chromosomes exchanges a set of linearly distributed bands on one chromosome with a set of linearly distributed bands on the other. Exchange can also occur between sister chromatids in single chromosomes of cells undergoing mitosis; this phenomenon is readily seen in cultured cells. Such exchanges occur in a reciprocal linear fashion (Figure 10-9). Although it is possible that the two ends of the chromosome come together to form a circle at some stage during the life cycle of the cell, the eukaryotic chromosome behaves genetically, and appears visually, as a linear structure containing a single DNA duplex.

Telomeres, Centromeres, and Autonomously Replicating Sequences Are Required for Replication and Stable Inheritance of Chromosomes

So far we have emphasized that eukaryotic chromosomes are linear structures probably composed of single DNA molecules. Although chromosomes are different in length and number in different organisms, they all behave similarly at the time of cell division. For example, at mitosis the attached sister chromatids become aligned on the metaphase plate; they then separate at the centromeres, and one chromatid of each metaphase chromosome is distributed to each daughter cell (Chapter 5; see also Chapter 18 for a discussion of microtubules, the proteins that participate in mitosis). Are there regular chromosomal structures that are responsible for these regular events? Recent recombinant DNA research with yeast cells has identified all of the chromosomal elements that are necessary to perform this feat of equal segregation to daughter cells. These experiments were based on the assumption that several regions of the chromosome might be special: (1) the centromere, (2) the two ends, or telomeres, and (3) any special sequences involved in the initiation of DNA replication.

The Cloning of Autonomously Replicating Sequences and Centromere Regions from Yeast Cells In yeasts, a circular plasmid can enter a host cell and grow along with the cell, as long as the plasmid contains *autonomously replicating sequences (ARS)* (Figure 10-10a). These *ARS* sequences and their role as initiation sites for DNA replication are discussed in Chapter 13. If the host cell lacks a particular gene—for example, a gene responsible for synthesizing an amino acid such as leucine—the necessary DNA (the normal Leu gene) from a normal cell can be cloned into plasmids, and the growth of the mutant (leu, leucine requiring) cell can be made dependent on the presence of a replicating Leu^+ plasmid.

In any culture of yeast with such a circular Leu^+ plasmid, only about 5 to 20 percent of the cells contain the plasmid, because mitotic segregation of the plasmids is faulty. However, if random bits of genomic DNA are

(a)

Chromocenter

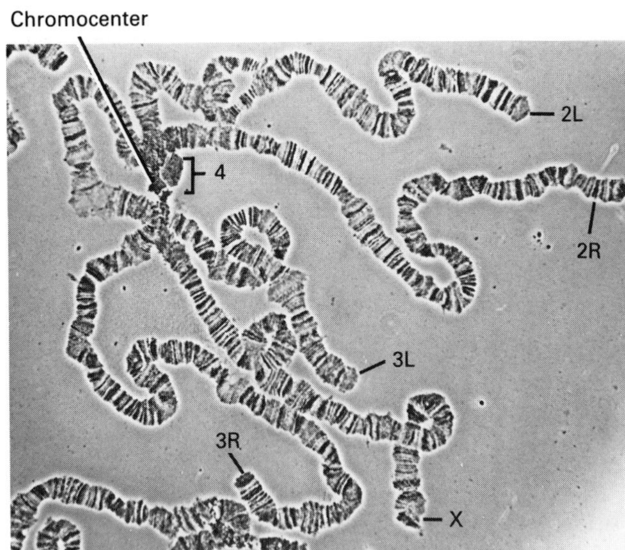

(b)

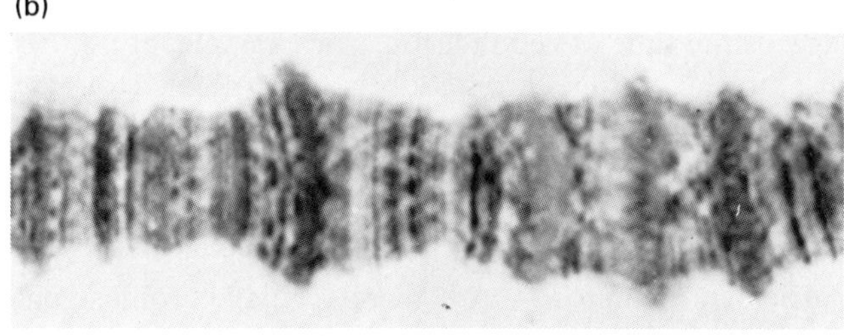

(c)

Parallel strands of DNA

Segment of polytene chromosome

Figure 10-7 Light micrographs of polytene insect chromosomes stained to reveal the very reproducible banding pattern. (a) In the four salivary gland chromosomes (X, 2, 3, and 4) of *Drosophila melanogaster,* a total of approximately 5000 bands can be distinguished. The centromeres of all four chromosomes often appear fused at the *chromocenter;* the smallest chromosome (4), a dot chromosome, is also associated with the chromocenter. The tips of the metacentric 2 and 3 chromosomes are labeled (L = left arm; R = right arm). The tip of the acrocentric X chromosome is also labeled. (b) A higher-power magnification of a section of chromosome C, one of the four chromosomes of the fly *Rhynchosciara americana.* (c) The DNA in salivary polytene chromosomes is repeated about 1000 times; the duplicated DNA fibers are thought to remain parallel. Therefore, any staining property in a chromosome with one DNA duplex would be amplified 1000 times to produce a transverse band. *Part (a) courtesy of J. Gall; (b) courtesy of F. Lara.*

added to such circular plasmids, specific centromere regions called *CEN* sequences are found to confer equal segregation at mitosis (Figure 10-10b). Such cloning experiments have led to the recovery of sequences that improve mitotic segregation to the extent that over 90 percent of the cells in a culture contain the Leu^+ plasmid. These cells and their descendants therefore grow well in a medium lacking leucine.

The Function of Telomeres in Linear Chromosomes Yeast plasmids are circular molecules, whereas the genetic maps of yeast chromosomes are linear. If the circular plasmids are cut once with a restriction enzyme to make them linear, they are unable to replicate, and cut recombinant plasmids no longer supply a necessary function for the cell (e.g., leucine synthesis). Experiments have been done to identify the portion of the chromosome that gives its linear DNA the ability to replicate. The entire yeast genome can be cut into pieces and randomly recombined at the end of a Leu^+ plasmid that has been made linear. This procedure allows the selection of pieces of the yeast chromosome that, when attached to the ends of the plasmid, enable it to replicate as a linear molecule. Each yeast cell yields 30 to 40 pieces of yeast DNA that are able to carry out this terminal function—or about 2 pieces per chromosome. The cloned terminal fragments all have similar sequences. These cloned sequences are believed to be those at the termini of the yeast chromosomes (Figure 10-10c).

Building an Artificial Chromosome The research on circular and linear plasmids in yeast has provided all of the basic components of an artificial chromosome (Figure

10-10d). The telomere sequences from yeast cells or from the protozoan *Tetrahymena* can be combined with yeast centromere sequences; to these are added DNA with selectable yeast genes and enough DNA from any source to make a total of more than 50 kb. (Smaller DNA segments do not work as well.) This artificial chromosome replicates in yeast cells and segregates almost perfectly (approximately 1 daughter cell in 1000 to 10,000 fails to receive an artificial chromosome). During meiosis, the two sister chromatids of the artificial chromosome appear to separate correctly to produce haploid spores. Such studies strongly support the conclusion that yeast chromosomes, and probably all eukaryotic chromosomes, are linear, double-stranded DNA molecules with special sequences—including centromere (*CEN*), telomere (*TEL*), and replication origins (*ARS*)—that ensure replication and proper segregation.

Molecular and Cytologic Characteristics of Yeast Centromeres Although, as we have seen, telomeres allow linear yeast plasmids to be propagated, the yeast centromeres are the regions that confer mitotic stability (as in Figure 10-10b). These sequences have been cloned and analyzed, and a common sequence pattern has emerged (Figure 10-11). In addition, proteins that bind to centromere regions have been purified but not yet characterized.

Finally, it is known that during yeast mitosis, microtubules appear to bind to a dark-staining structure called the *spindle pole body*. This structure, which is replicated during interphase, is embedded in the nuclear envelope. When budding occurs, the two spindle pole bodies separate and one enters the bud. The microtubules extend from the spindle pole body to the chromatin (the DNA plus its associated proteins) in the center of the mother cell (Figure 10-12; see also Figure 6-7). Approximately 15 to 20 individual microtubules (each 20 nm in diameter) bind to the spindle pole body, which makes it likely that one microtubule exists for each of the approximately 17 chromosomes of yeast. Therefore, the yeast centromere DNA may serve as the nucleation site for microtubular assembly (Figure 10-13; see also Chapter 18).

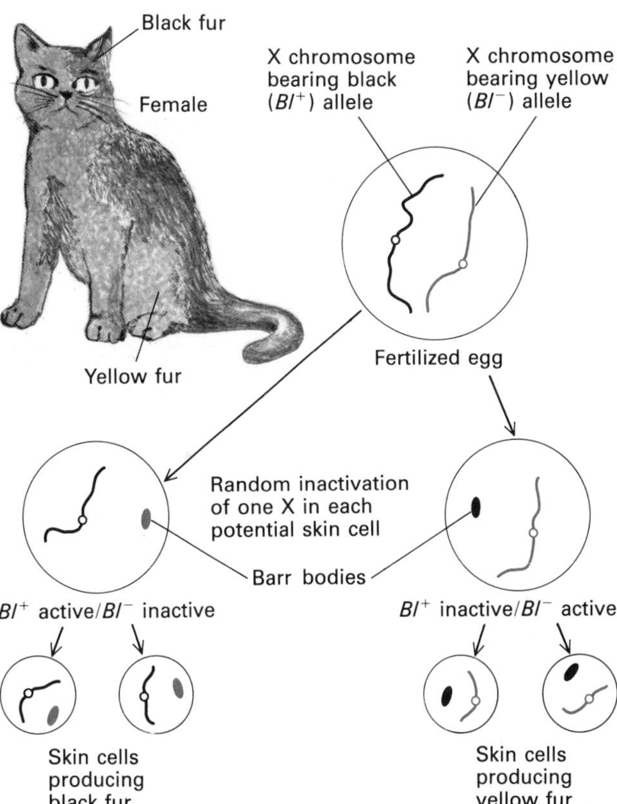

Figure 10-8 The derivation of coat-color variation in a calico cat. The alleles (alternate forms of a gene) for coat color exist on the X chromosome: one (Bl^+) produces black hair, and the other (Bl^-) produces yellow hair. In a heterozygous female, the random inactivation of one or the other of the two X chromosomes in the precursors to skin cells results in patches of yellow and black fur.

Chromosomal Proteins: Histones and Nucleosomes

What material other than DNA does the eukaryotic chromosome contain? When metaphase chromosomes are isolated from cells undergoing mitosis, or when the DNA from interphase nuclei is isolated in isotonic buffers, twice as much protein as DNA is recovered. This mixture of DNA and protein constitutes the *chromatin*. The general structure of chromatin has been found to be remarkably similar in all eukaryotic cells.

Histones Are the Most Abundant Proteins in Chromatin

The most prominent proteins associated with metaphase chromosomes and with interphase DNA are the *histones*, a family of basic proteins found in all eukaryotic nuclei. There are five major types of histone chains; these are termed H1, H2A, H2B, H3, and H4 (Figure 10-14a). The amino acid side chains of histones are frequently modified by posttranslational additions of phosphate, methyl, or acetate groups, which cause the major species to separate into subspecies that are detectable by high-resolution gel electrophoresis (Figure 10-14b).

The amino acid sequences of four histones (H2A, H2B, H3, and H4) are now known for a wide variety of organisms. There is a remarkable conservation of the sequences among distantly related species. For example, the sequences of histone H3 from sea urchin tissue and of H3 from calf thymus are identical exept in a single amino acid. And there are only four amino acid differences be-

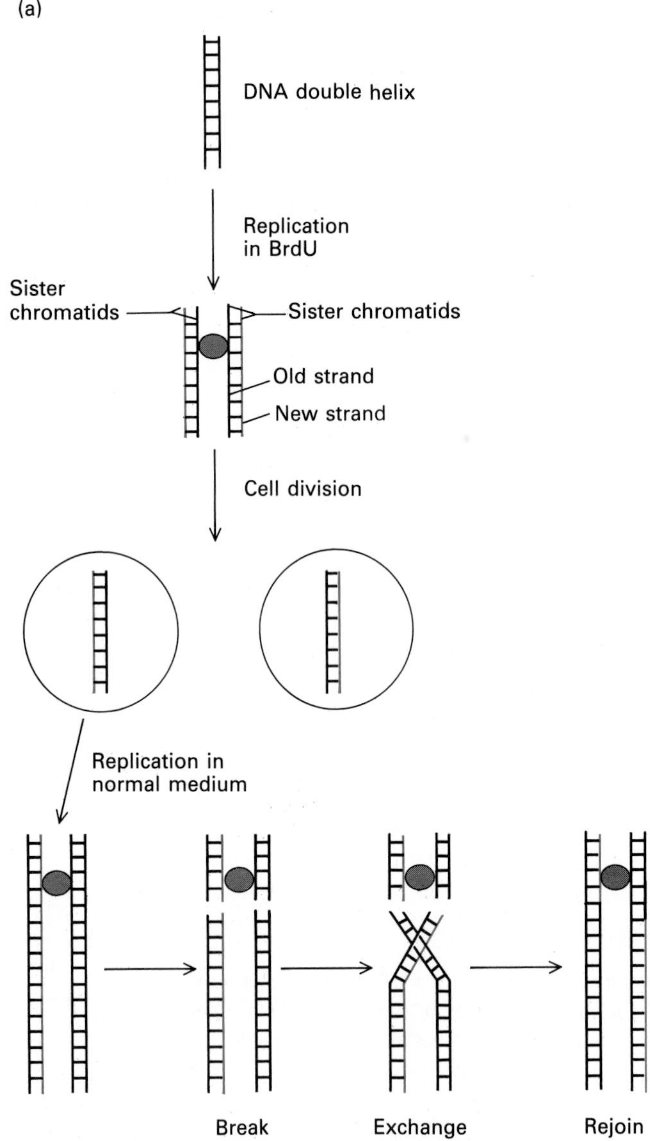

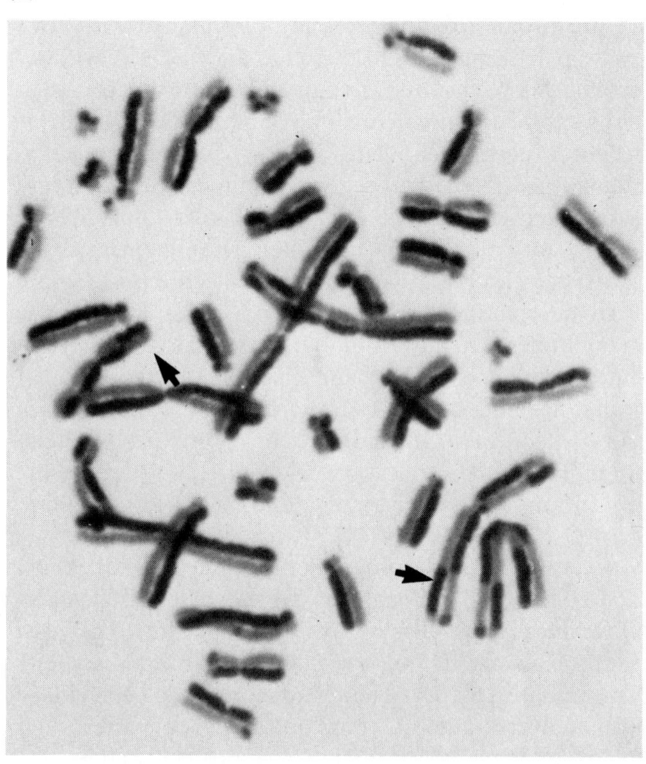

Figure 10-9 Sister chromatid exchange. (a) Cells are placed in a medium with the thymidine analog bromodeoxyuridine (BrdU) just long enough for the DNA to replicate once (the newly synthesized strands are indicated in color); then the cells are returned to a normal medium for a second replication. This procedure yields cells in which one chromatid of each pair contains BrdU and the other lacks it. The distribution of the BrdU allows breakage and exchange between sister chromatids to be detected, as shown in (b). (b) When stained with Giemsa (a DNA stain) and a fluorescent dye, the BrdU-containing chromatids fluoresce more brightly (i.e., look lighter) than the unsubstituted chromatids. Regions of interchange between the labeled and the unlabeled linear sister chromatids can be seen *(arrows). Part (b) courtesy of S. Wolff; from S. Wolff and P. Perry, 1975,* Exp. Cell Res. *93:23.*

tween the H3 of the garden pea and that of the calf thymus (Figure 10-15).

The amino acid sequence of H1 varies more from organism to organism than do the other major histone species; the H1 is replaced in certain tissues by special histones. For example, in the nucleated red blood cells of birds, a histone termed H5 is present instead of H1 (see Figure 10-14a). In addition, very careful analyses of the histones of individual organisms have revealed the existence of minor histone families. But the similarity of the histones among all eukaryotes still remains impressive.

Nucleosomes, the Primary Structural Units of Chromatin When chromatin is extracted from the nuclei in a solution of low ionic strength, it stretches out in a formation that resembles beads on a string under the electron microscope. In this form the DNA is a thin fiber connecting the 10-nm-diameter "beads." If the ionic concentration used in the extraction is raised (or if small amounts of divalent cations are added), the fibers become thicker, about 30 nm in diameter (Figure 10-16). The "beads" are structures termed *nucleosomes,* and in the thicker fiber the nucleosomes are packed into a spiral or *solenoid.*

A nucleosome consists of a disk-shaped histone core plus a segment of DNA that winds around the core like thread around a spool. The core is an octomer constructed of two copies each of histones H2A, H2B, H3, and H4 (Figure 10-17a). The DNA wrapped around the histone core is about 140 base pairs in length; this length of DNA makes slightly less than two turns around the histone octomer (Figure 10-17b). Histone H1 binds to the DNA just next to the nucleosomes (Figure 10-18). When chromatin is extracted in the extended beads-on-a-string configuration, the H1 is often released.

The isolation of individual nucleosomes for physical and chemical analysis can be carried out by partial nuclease digestion either of DNA within the intact nucleus or of DNA in extracted chromatin. In addition to single nucleosomes, gentle digestion also releases nucleosome dimers, trimers, and higher oligomers, which can be isolated by various physical methods (Figure 10-19). Extensive nuclease treatment finally digests all the DNA between the individual nucleosomes to release single nucleosomes. The DNA content of a single nucleosome before all connecting DNA is digested varies between 160 and 200 base pairs, but after trimming the figure is very close to 140 base pairs in all species.

Higher Levels of Packing: Solenoids and Loops Most (maybe all) DNA in the eukaryotic nucleus interacts with histones to form nucleosomes. The long strings of nucleosomes themselves form coils that produce higher-order chromosomal structures. Because all nucleosomes have the same basic structure—an octomeric histone core with loops of DNA around it—the same chemical groups likely reside at the surfaces of all (or at least most) nucleosomes, and allow for regular bonds to form between nucleosome surfaces that come into contact. Adjacent nucleosomes are packed into a helical secondary DNA-protein structure termed a *solenoid.* When chromatin is released from the nucleus still in the solenoidal form, it appears in electron micrographs as the thick, 30-nm-diameter fiber (see Figure 10-16), in comparison to the 10-nm-diameter thin fiber or beaded string. The internal arrangement of a solenoid is probably a chromatin fiber coiled into a helix containing six nucleosomes per turn (Figure 10-20). The solenoids may be further organized into giant supercoiled loops. Our knowledge of nucleosome and solenoid structure has been drawn from an array of physical studies, including x-ray crystallography and neutron diffraction.

The Structure of Metaphase Chromosomes Is Maintained by a Protein Scaffold

Electron micrographs of metaphase chromosomes have provided evidence that DNA associated with histones is attached at intervals to a structural protein. The complete removal of histones from metaphase chromosomes of HeLa cells releases long DNA loops (10 to 90 kb) that are, at their bases, still anchored to a nonhistone protein scaffolding. This scaffolding actually reproduces the shape of the metaphase chromosome (Figure 10-21) and persists even if the DNA is digested by nucleases. (As we shall see in Chapter 13, 10 to 90 kb is the size range of units of replication in human DNA.) A very limited set of proteins is in this scaffold: in the purest preparations of metaphase chromosomes, only two scaffold proteins have been found.

Whether these loops are, in fact, units that are replicated and/or transcribed, and whether the attachment sites have fixed positions along the chromosomal DNA, are questions of the utmost importance. It has been found that during interphase, long loops of DNA are attached to some nuclear protein structure at regular intervals. This conclusion was reached by treatment to remove all histones and digesting the DNA with restriction enzymes. Specific DNA sequences in known regions of the genome were found to be released free of the residual nuclear protein structure; other specific fragments remained bound (Figure 10-22). Thus chromosomal DNA may be bound at specific sites in the nucleus, and the binding sites may be retained when the metaphase chromosome assembles. Further studies of the function of such binding sites in replication and transcription will be of great interest.

Nonhistone Proteins May Serve Regulatory Functions

The total mass of the set of histones associated with a DNA molecule is about equal to that of the DNA. Interphase chromatin and metaphase chromosomes also contain smaller amounts of other proteins. Because the regulation of transcription is thought to be controlled by specialized proteins associated with interphase chromatin, there has been great interest in attempts to isolate, characterize, and eventually determine the function of nonhistone chromosomal proteins. Relatively few of these proteins have been purified. One nonhistone class called HMGs (for "high mobility group"—a reference to the behavior of these proteins during electrophoretic separation) is thought to be associated mainly with actively transcribed genes. HMGs are not specific for individual genes, but appear to be associated with many—perhaps with all—actively transcribed genes.

The Biological Definition of a Gene

The reason for the intense interest in chromosome structure throughout the last century is, of course, that the genes reside on the chromosomes. Before going any further in our disussion of genes and chromosomes, we shall struggle with a semantic problem that is now 85 years old: what is a gene?

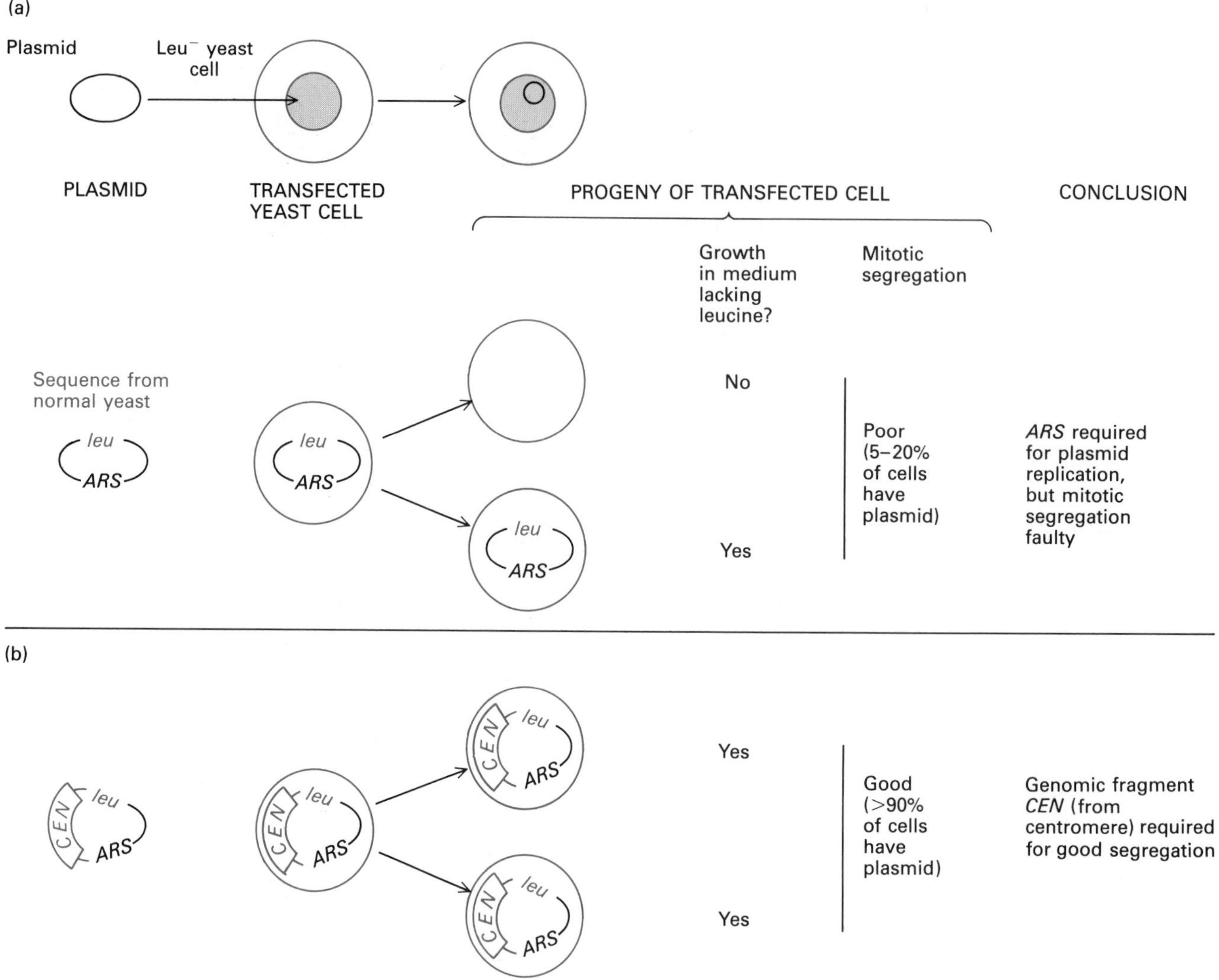

Figure 10-10 ***(above and opposite)*** Functional chromosomal elements from yeast. Leu$^-$ yeast cells lack an enzyme that is required for leucine synthesis. In the experiments shown, the *leu* gene for that enzyme is inserted into yeast plasmids, which are introduced into the Leu$^-$ cells by transfection. (Recall that transfection of yeast cells involves the removal of their outer cell walls, which induces the uptake of DNA and its incorporation into the nucleus.) The plasmids may then replicate along with the nuclear DNA of the cells. Because only the plasmids carry the information for leucine synthesis, only cells containing a plasmid will grow in a medium that is not supplemented with leucine. (a) The plasmid must possess sequences that allow autonomous replication *(ARS)*, or it will not replicate in the cell. However, even plasmids with *ARS* exhibit poor segregation during mitosis, and therefore do not appear in each of the daughter cells. (b) Randomly broken pieces of genomic yeast DNA are inserted into the plasmids containing *ARS* and *leu*. Some of the transfected cells pro-

Recombination Tests Separated Linked Genes

The original definition of a gene came from breeding experiments in which individual traits, or "characters," were followed. In crosses of genetically distinct strains of garden peas, Mendel chose simple traits that could be easily followed: stem length, seed color, seed texture, etc. (Chapter 1). Each trait was determined by one of two alternative expressions of a "gene"—the dominant or the recessive form. These gene forms were seen to be stable through many generations. Alternative forms of a gene encoding alternative forms of a single trait are called *alleles.*

Later, T. H. Morgan and his coworkers carried out an extensive genetic study of *Drosophila melanogaster.* Be-

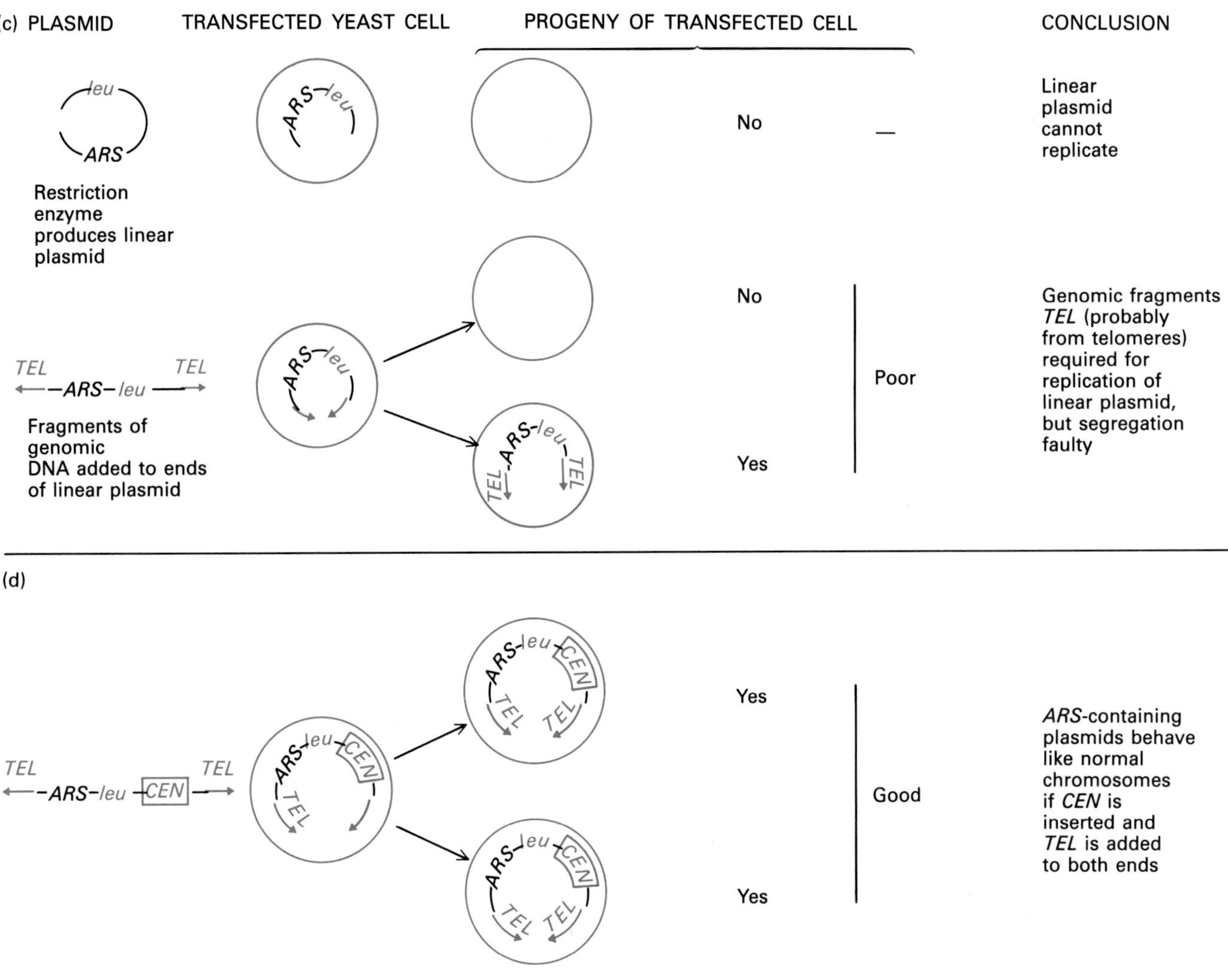

duce large colonies, indicating that a high rate of mitotic segregation among their plasmids is facilitating the continuous growth of daughter cells. The DNA recovered from plasmids in these large colonies contains yeast sequences called *CEN* sequences, which are believed to have originated in the centromere region. The *CEN* sequences improve the segregation efficiency of the plasmid and the plasmid segregates as a normal chromosome. (c) When a circular plasmid is made linear before transfection, no colonies grow. However, if fragments of the total yeast cell DNA are inserted at the ends of the linear plasmids, some are able to replicate and provide for leucine synthesis. The fragments that give linear plasmids the ability to replicate are called *TEL;* these are thought to be telomere sequences. Linear plasmids containing *TEL, ARS,* and *leu* do not segregate well, though. (d) A combination of all of the above elements produces an artificial chromosome that behaves very much like a normal chromosome in both mitosis and meiosis. [See A. W. Murray and J. W. Szostak, 1983, *Nature* 305:89.]

tween 1910 and 1925, careful visual observations of *Drosophila* uncovered well over 100 physical abnormalities due to gene mutations. A sizable number of these were variations in eye color; other mutations affected body parts such as the wings, legs, antennae, and bristles. Morgan et al. determined whether all of the mutants that produced variations in such phenotypic traits as, say, eye color were alleles of one eye-color gene. Recall that the first important result of the studies with *Drosophila* (mentioned in Chapter 1) was that each of the genetically determined abnormalities of whatever type segregated during Mendelian crosses into one of four groups, and that this number of groups coincided with the presence of four chromosomes. The mutant gene loci that segregated together were said to be "linked," meaning that they were all on the same chromosome. The various eye-color mu-

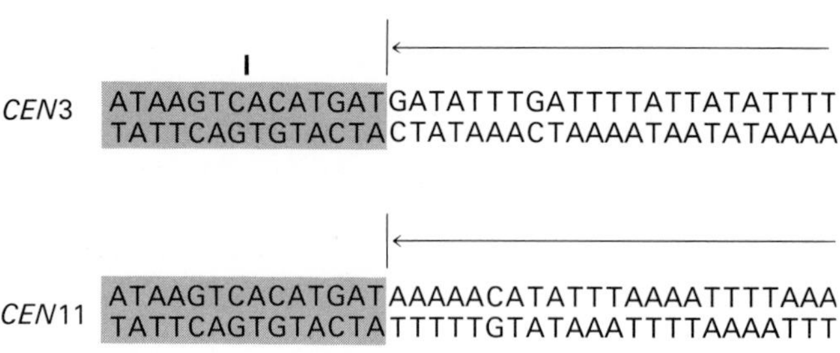

Figure 10-11 *(right)* Sequences of yeast centromere DNA. Two DNA fragments were independently cloned: one from chromosome 3 and the other from chromosome 11 of *Saccharomyces cerevisiae*. The fragments were selected as in Figure 10-10b because they were known to provide centromere DNA function (i.e., they improved plasmid segregation). A comparison of these centromere sequences reveals two elements, I and III, that are identical at both sites. Between I and III lies element II, an 85- to 90-base-pair stretch of very AT-rich DNA. [See M. Fitzgerald-Hayes, M. L. Clarke, and J. Carbon, 1982, *Cell* **29**:235.]

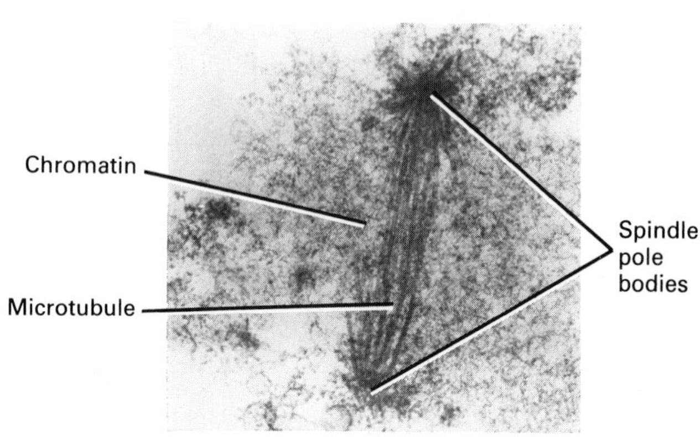

Figure 10-12 An electron micrograph of yeast spindle pole bodies and attached microtubules. The yeast cell has been fixed and sectioned, so some of the microtubules in the cell cannot be seen. Through observation of serial sections, however, the number of microtubules can be shown to be approximately equal to the number of chromosomes. *Courtesy of H. Ris; from J. B. Peterson and H. Ris, 1976,* J. Cell Sci. *22:219.*

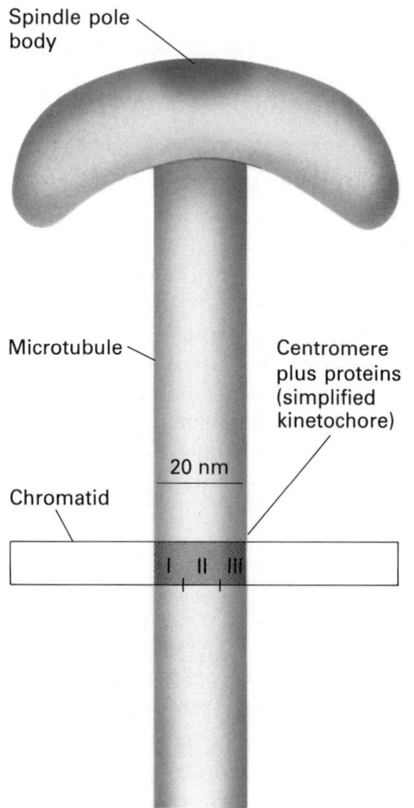

Figure 10-13 A model of the binding of the yeast centromere to a microtubule. The DNA sequence shown in Figure 10-11, with its three elements (I, II, and III), is approximately the correct length for the binding of a single microtubule 20 nm in diameter; this sequence is known to bind to certain cell proteins and to be resistant to DNase digestion. In cells from multicellular plants and animals, the microtubules are much more numerous than the 15 to 20 that are present in yeast cells; in higher organisms the microtubules appear to attach to the chromosome at the *kinetochore,* a granular region at or near the centromere (see Chapter 18). The yeast centromere plus its proteins *(color)* could be a simplified kinetochore. [See K. S. Bloom, M. Fitzgerald-Hayes, and J. Carbon, 1982, *Cold Spring Harbor Symp. Quant. Biol.* **47**:1175.]

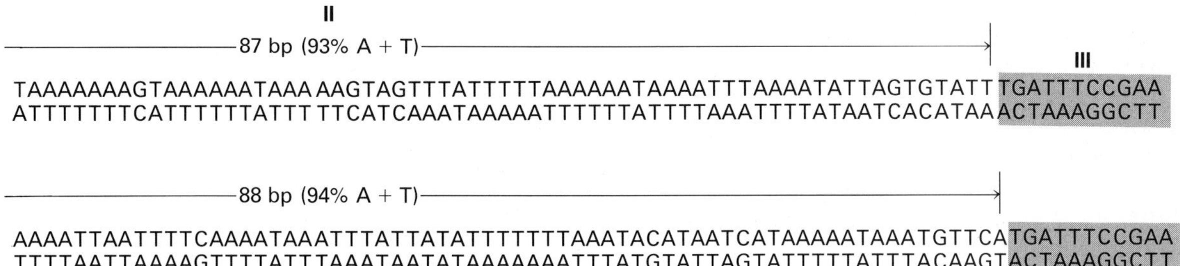

tants were mapped to three of the four chromosomes, which illustrated that different genes can contribute to one phenotypic trait (Figure 10-23). Clearly, all of the eye-color mutations were not alleles of the same gene, but represented a series of genes.

Although linkage to one of four groups was the rule for the various *Drosophila* genes, it was soon realized that linkage does not occur 100 percent of the time. This is due to recombination in gametes during meiosis; the phenomenon can be observed by following two linked genes during a genetic cross. For example, consider one eye-color gene and one gene for wing morphology. The wild-type or normal genes for dark red eyes (sc^+) and straight wings (cr^+) are dominant alleles. Scarlet eyes (sc^-) and curled wings (cr^-) result from recessive mutations. By continuous inbreeding of mutant flies, geneticists can produce homozygous strains of flies that are either $sc^-\ cr^-/sc^-\ cr^-$ or $sc^+\ cr^+/sc^+\ cr^+$. (The slash separates the genetic types contributed by each of the two parental gametes.) Both the sc^- and the cr^- mutations segregate with all of the other mutations assigned to chromosome III. Mating of the two pure strains produces $sc^-\ cr^-/sc^+\ cr^+$ heterozygotes in the F_1 generation (Figure 10-24a).

Suppose now that a heterozygous female is crossed with a male that is homogyzous for both recessive traits. If linkage between the eye-color and wing-shape genes was 100 percent, all offspring with dark red eyes would also have straight wings, and all scarlet-eyed flies would have curled wings. In fact, about 90 percent of the offspring follow this pattern, establishing the linkage of these traits. However, such a mating also produces a minority of flies with straight wings and scarlet eyes and flies with curled wings and dark red eyes (Figure 10-24b). These results indicate that some of the gametes of the heterozygotes have a chromosome III in which the linkage between the eye-color and wing-shape alleles has been disrupted by recombination. Recombination separates the alleles of different genes on the same chromosome, frequently in some cases and rarely in others.

The explanation for recombination has come from cytology. In the formation of gametes, the homologous chromosomes pair up during the first meiotic division. This process is called *synapsis*. The paired (maternal and paternal) homologous chromosomes are actually broken and reunited during synapsis. The crossed chromosomes participating in recombination are visible as *chiasmata* (Figure 10-25a; see also Figure 13-48).

A quantitative tabulation of crossovers between genes on genetically marked homologous chromosomes allows the mapping of genes along the chromosome. The maps are based on the laws of chance: the farther apart two linked genes lie, the greater the chance that a recombination will occur to disrupt the linkage between alleles of these genes. In the chromosome in Figure 10-25b, a crossover would occur most frequently between *a* and *e*, less frequently between *a* and *d*, etc. A crossover between genes *a* and *b* is least likely.

Thus, the work of the early *Drosophila* geneticists led to the recognition of three properties of a gene: (1) it is a chromosomal site that controls an observable characteristic; (2) it can be changed or "mutated"; and (3) it can recombine with a homologous site on another chromosome. According to these properties, if two genetic traits could be separated by recombination, they were considered to be encoded by separate genes.

Complementation Tests Improved Gene Definition

The definition of a gene by recombination tests has several drawbacks. Measuring the frequency of recombination can be laborious; genes that lie very close together require the screening of large numbers of offspring. Furthermore, recombination analyses place genes on a genetic map but do not define the function of a gene. Analyses of inherited traits that depend on gene function therefore were developed. Complementation tests rely on gene function and are designed to bring into the same cell two sets of genes required to carry out one function. If either gene set contains a mutation in a gene not mutant in the other set then the cell will function and complementation will have occurred. Even if such functionally related genes are side by side in a chromosome they can be detected easily as separate genes in such a test. These have been widely used, especially in microbial genetics.

For example, the bread mold *Neurospora crassa* can grow from a single haploid spore into a *mycelium,* a

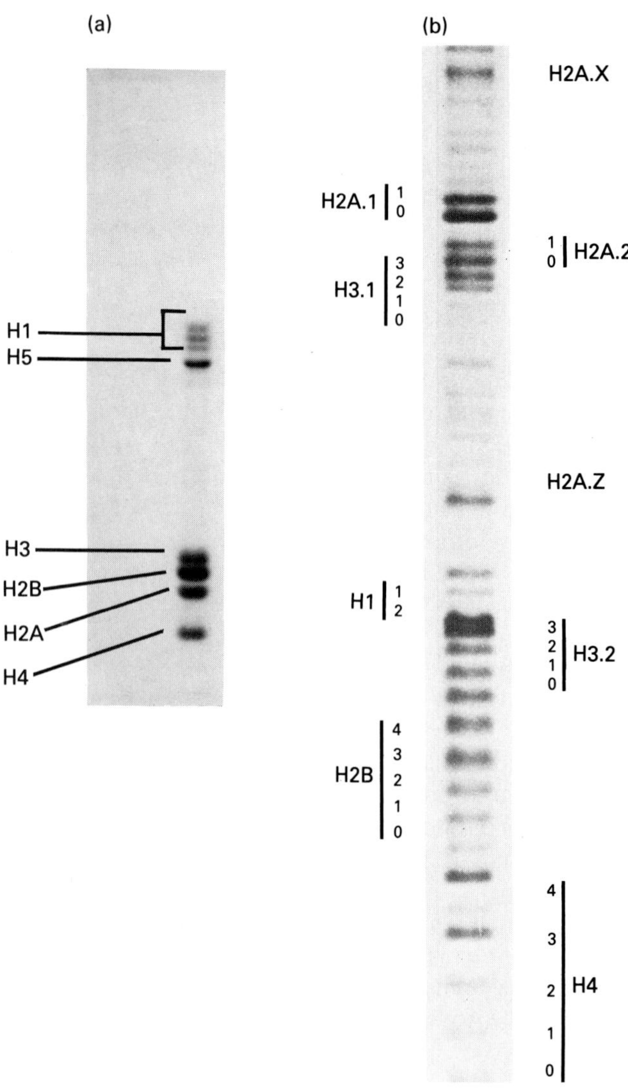

Figure 10-14 Histones separated by gel electrophoresis. (a) This sample was extracted from the nuclear chromatin of chicken blood cells and examined by low-resolution gel electrophoresis. The major species H2A, H2B, H3, and H4 are present in about equal amounts. The other major histone, H1 (from white blood cells), can also be seen, as can H5 (a species that is present in the red blood cells of birds). (b) A high-resolution separation of histones extracted from calf thymus. The proteins were subjected to electrophoresis through an acrylamide gel at an acid pH in the presence of urea and in the absence of SDS. Amino acid sequence variants are indicated by the large-type numbers and letters following the decimals (e.g., H3.1 is slightly different in sequence from H3.2). The small-type numbers indicate charge variations within each major class. Variation within the H1 class is due mainly to phosphorylation, whereas charge variants within H2A, H2B, H3, and H4 are caused primarily by different degrees of acetylation. *Part (a) courtesy of V. Allfrey; (b) courtesy of M. Plesko and R. Chalkley.*

CALF THYMUS H3 SEQUENCE

$Me^{O\text{-}2}$ (Lys 9) Ac (Lys 14)

H_2N-Ala-Arg-Thr-Lys-Gln-Thr-Ala-Arg-Lys-Ser-Thr-Gly-Gly-Lys-Ala-Pro-Arg-Lys-Gln-Leu- (10, 20)

Ac (Lys 23) $Me^{O\text{-}2}$ (Lys 27)

-Ala-Thr-Lys-Ala-Ala-Arg-Lys-Ser-Ala-Pro-Ala-Thr-Gly-Gly-Val-Lys-Lys-Pro-His-Arg- (30, 40)

-Tyr-Arg-Pro-Gly-Thr-Val-Ala-Leu-Arg-Glu-Ile-Arg-Arg-Tyr-Gln-Lys-Ser-Thr-Glu-Leu- (50, 60)

-Leu-Ile-Arg-Lys-Leu-Pro-Phe-Gln-Arg-Leu-Val-Arg-Glu-Ile-Ala-Gln-Asn-Phe-Lys-Thr- (70, 80)

-Asp-Leu-Arg-Phe-Gln-Ser-Ser-Ala-Val-Met-Ala-Leu-Gln-Glu-Ala-Cys/Ser-Glu-Ala-Tyr-Leu- (90, 100)

-Val-Gly-Leu-Phe-Glu-Asp-Thr-Asn-Leu-Cys-Ala-Ile-His-Ala-Lys-Arg-Val-Thr-Ile-Met- (110, 120)

-Pro-Lys-Asp-Ile-Gln-Leu-Ala-Arg-Arg-Ile-Arg-Gly-Glu-Arg-Ala-COOH (130)

SEQUENCE CHANGES BETWEEN CALF THYMUS AND OTHER SPECIES

Chicken, carp, and shark	Cys/Ser ⟶ Ser at 96
Pea	Tyr ⟶ Phe at 41
	Arg ⟶ Lys at 53
	Met ⟶ Ser at 90
	Cys/Ser ⟶ Ala/Ser at 96
Sea urchin	Asp ⟶ Glu at 81

Figure 10-15 The sequence conservation of histone H3. The entire sequence of H3 from calf thymus is compared with H3 from four other species. Changes occur at the colored amino acids. Histones are modified after synthesis by addition of acetyl groups (Ac) and methyl groups ($Me^{O\text{-}2}$, dimethyl). The Cys/Ser in position 96 indicates that two different H3 histones, one with cysteine and one with serine, exist in the calf. [See B. Lewin, 1980, *Gene Expression*, Wiley, p. 305.]

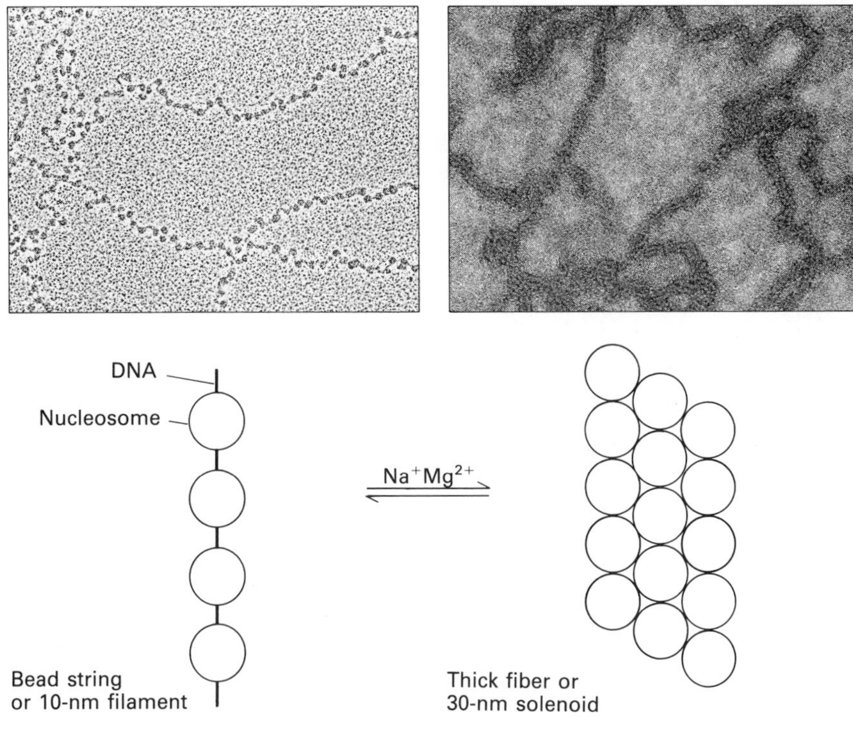

Figure 10-16 Chromatin fibers. *Left:* An electron micrograph of chromatin fibers from the nuclei of *Drosophila melanogaster* cells shows their characteristic beads-on-a-string appearance at low ionic strength. The "beads" are nucleosomes that are about 10 nm in diameter. *Right:* Chromatin is converted into a more compact arrangement by an increase in the ionic strength or by the addition of divalent cations. In this form the chromatin consists of thick, 30-nm-diameter fibers; the nucleosomes are in a solenoidal arrangement. *Left photograph courtesy of D. S. McKnight and O. L. Miller Jr.; right photograph courtesy of B. Hamkalo and J. B. Rattner.*

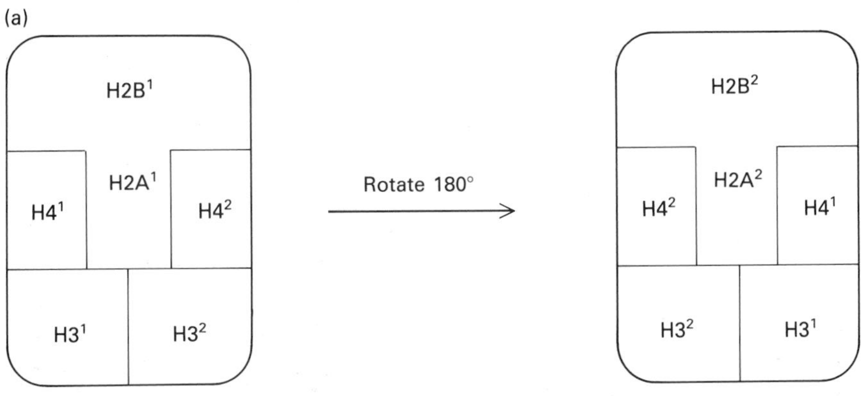

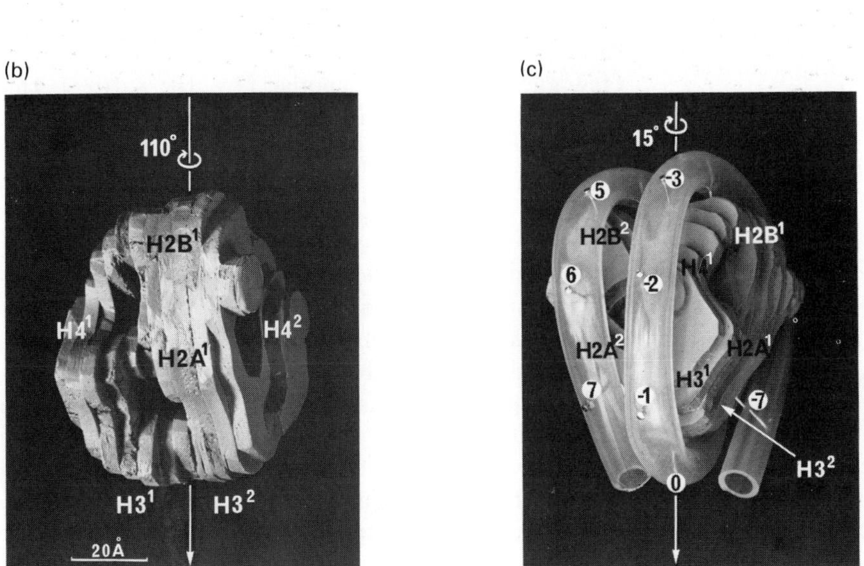

Figure 10-17 A model of a nucleosome. (a) Two copies each of the histones H2A, H2B, H3, and H4 compose an octomeric histone core. (b) From electron microscopy and x-ray diffraction studies, a fairly precise representation of the general locations of the various histones in the core can be constructed. The representation of the mass of the core as "slices" is taken from electron density maps. The indicated rotation angle is for purposes of orientation; it is based on a 0° position not shown. [Note that one copy of H2A (H2A²) and one of H2B (H2B²) are not visible from this vantage point.] (c) The DNA *(clear tube)* wraps around the histone core. Each number on the DNA indicates 10 base pairs; 70 base pairs appear in each turn of the DNA, or about 140 base pairs in the nucleosome. Again, the indicated rotation angles are for orientation purposes only. [The position in (c) differs slightly from that in (b).] [See A. Klug, 1981, in *Nucleic Acid Research: Future Development,* K. Mizobuchi, I. Watanabe, and J. D. Watson (eds.), Academic Press, p. 91; and A. Klug et al., 1980, *Nature* **287**:509.] *Parts (b) and (c) courtesy of A. Klug; reprinted by permission from* Nature. *Copyright 1980 Macmillan Journals Limited.*

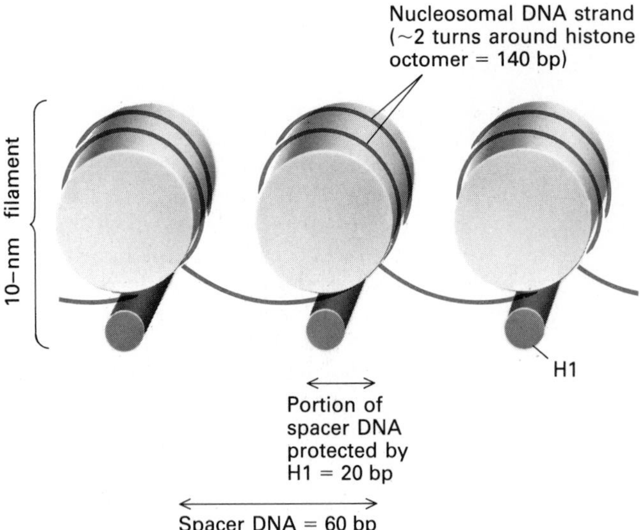

Figure 10-18 The chromatin fiber: a drawing of nucleosomes with histone H1 in place. H1 is the histone that is the least tightly bound; it is released first from the DNA when nuclei are treated with increasing concentrations of salt. The H1 may also be released in regions undergoing transcription.

spongy clump of branching chains called *hyphae*. Each hypha is segmented, with each segment containing one haploid nucleus. Neighboring hyphae originating from two different spores occasionally fuse, bringing into the same cytoplasm haploid nuclei from different starting spores. This structure containing two nuclei is called a *heterokaryon*. It can grow and divide while maintaining two separate nuclei. The formation of heterokaryons allows for tests that determine whether mutants resulting in the same phenotype are affected in the same gene.

For example, multiple mutants of *Neurospora* cannot grow without histidine. When heterokaryons consisting of nuclei from His^- mutants are tested, some of the paired nuclei do not grow without histidine, but other pairs do. This difference results because there are several steps in the enzymatic synthesis of histidine, and a mutation can damage any one of the steps. When a pair of mutants that fail at two different steps is tested, growth occurs; if mutants affecting the same step are crossed, no growth occurs (Figure 10-26). When two mutants correct the phenotypic defect, they are said to *complement* each other. The complementation test defines separate genes. Each of the complementing groups of mutants is now known to be deficient in one of the enzymes necessary in histidine synthesis.

The Cistron Is a Complementing Genetic Unit That Encodes One Polypeptide

The complementation test and extensive recombination analyses were both used by Seymour Benzer to show that a small region of the T4 bacteriophage chromosome contains two distinct genes that control one function—namely, the ability of phage to grow on the host bacterium *E. coli*, strain K. Normal T4 phage can grow on both the B and the K strains, but mutants unable to grow on K were found. A large group of these bacteriophage mutants, called *r*II mutants, was collected.

When bacteria are infected with two different mutant phages either recombination or complementation are possible. Complementation can be distinguished from recombination in two ways: (1) complementation occurs with greater frequency and (2) the individual phage particles produced by complementation still have one or the other of the original mutations, whereas recombinants differ from their parents.

To test for genetic identity between the different mutant *r*II bacteriophages, strain B cells were infected with two independently isolated *r*II mutants. Recombination between two *r*II mutant phages to yield wild-type bacteriophage could be detected by plating any new phage on strain K. Recombination between *r*II mutants was relatively infrequent compared to recombination of other genetic markers in the bacteriophage genome; this finding showed that all of the *r*II mutations lay close together in the bacteriophage chromosome. The recombination frequency between the most closely spaced *r*II mutants was only about 10^{-5} of the frequency between other bacteriophage mutants that were spaced farther apart. Considering the size of the T4 genome—2×10^5 base pairs—recombination frequencies of 10^{-5} indicated that the least frequently recombining *r*II mutations must lie within a few nucleotides of one another ($2 \times 10^5 \times 10^{-5}$). This impressive result indicated that recombination tests, if they are sensitive enough and if they are carried out on enough organisms, can detect single-base differences in mutants.

However, Benzer's work with the *r*II mutants is probably even more well known because of his complementation tests. All of the *r*II mutants shared a single characteristic—the inability to grow on *E. coli* strain K—and all were capable of being complemented by a normal function; for example, they would all replicate in strain K together with normal, wild-type phage. When *E. coli* strain K cells were infected with pairs of the *r*II mutants, the mutants appeared to fall into two classes. The two groups *r*IIA and *r*IIB were distinguished because

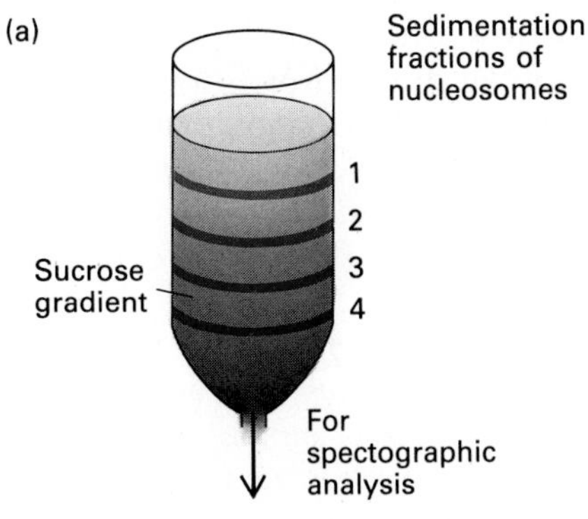

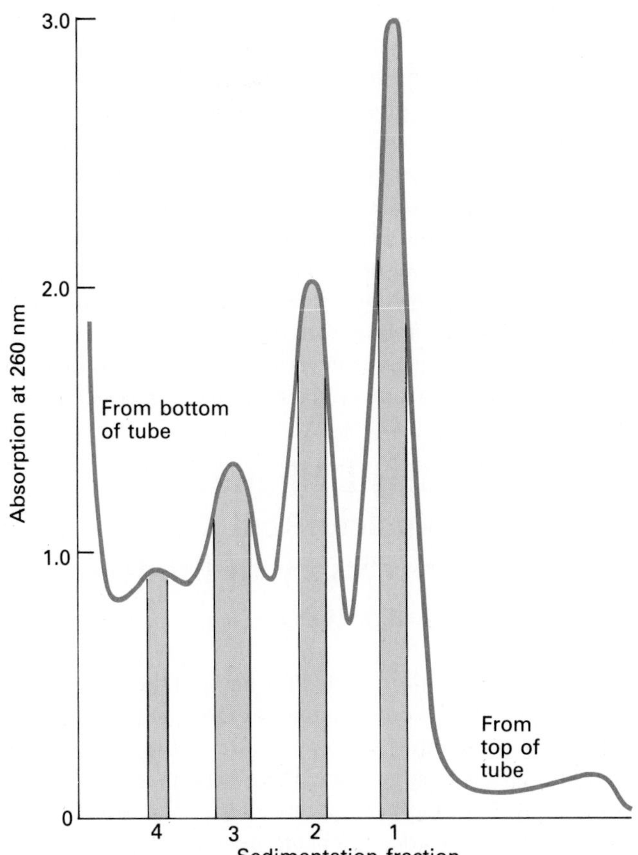

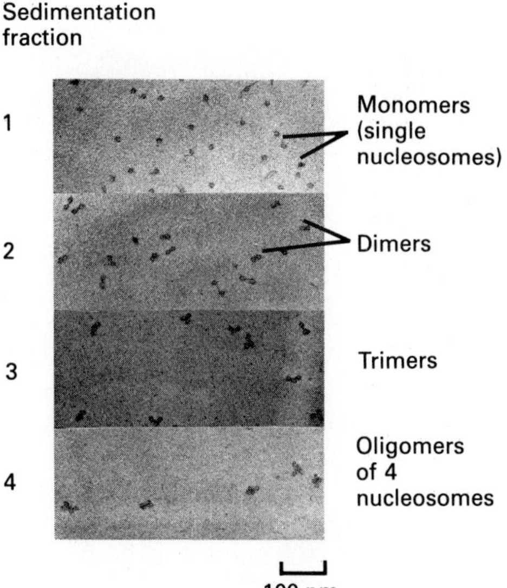

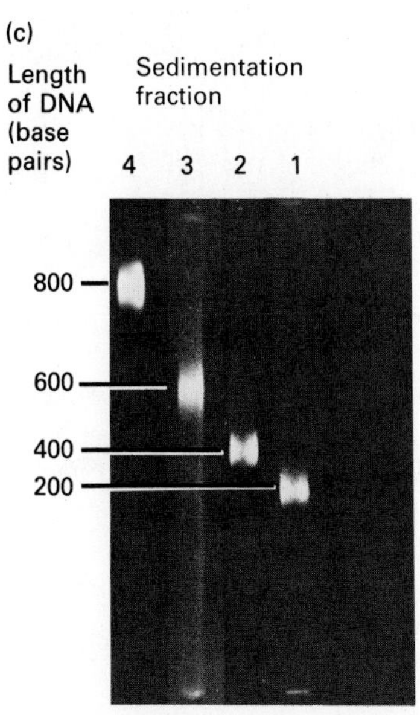

Figure 10-19 The spacing of nucleosomes in chromatin. (a) Rat liver nuclei briefly digested with DNase release DNA-containing structures that sediment through sucrose gradients as discrete peaks. (b) Electron microscopy reveals that each of the various sedimenting fractions contains one, two, three, or four units (nucleosomes). (c) When DNA is extracted from each fraction and analyzed, the approximate lengths are 200, 400, 600, and 800 base pairs, corresponding to nucleosome monomers, dimers, trimers, etc. From data of this type, the model of nucleosomes in Figure 10-18 was formulated. [See J. T. Finch, M. Noll, and R. D. Kornberg, 1975, *Proc. Nat'l Acad. Sci. USA* 72:3320.] *Parts (b) and (c) from Finch, 1975.*

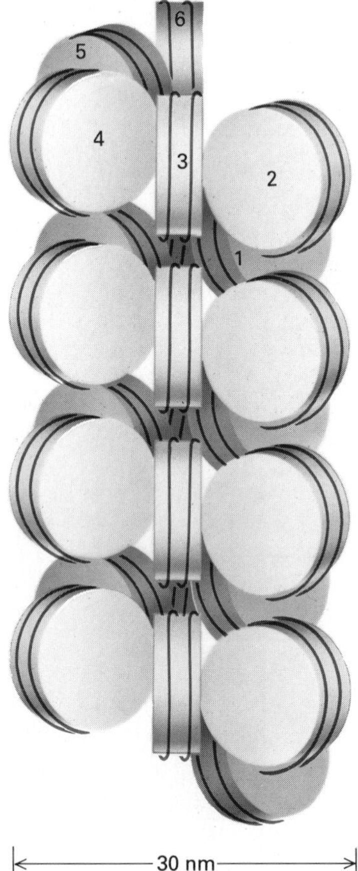

Figure 10-20 The packing of nucleosomes in a solenoid. When chromatin is extracted under the proper ionic conditions, a compact form is isolated (see Figure 10-16). Physical studies have shown that the structure is a helix with six nucleosomes per turn, and that the individual nucleosomes are oriented as shown here. [See J. D. McGhee et al., 1980, *Cell* **22**:87.]

growth of phage occurred when one *r*IIA and one *r*IIB infected a cell, but not when two *r*IIAs or two *r*IIBs infected a cell. Note, however, that complementation occurred only when *r*IIA and *r*IIB were on different copies of the bacteriophage chromosome (i.e., when the mutations were in the trans position). (If a single copy of the chromosome carried both the *r*IIA and the *r*IIB mutations—i.e., if the mutations were in the cis position—they would not complement.) Thus *r*IIA and *r*IIB, although they lay very close together on the chromosome, acted as if they were independent genetic functions. Benzer called these two independent trans-active complementation units *cistrons*. He concluded that the genetically inherited trait of growth of T4 bacteriophage on *E. coli* strain K depended on the two cistrons *r*IIA and *r*IIB (Figure 10-27).

Benzer's definition implied that a cistron is a unit that encodes one polypeptide chain. Initially, the characterization of a cistron as a DNA region encoding one polypeptide chain was thought to be sufficiently precise that a cistron could be considered the equivalent of a gene. However, as we shall see, the complicated arrangements of the information in many DNA molecules do not allow every separate polypeptide to be detected by a complementation or recombination test; further, many mutations affect more than one polypeptide. The two terms gene and cistron are no longer deemed equivalent.

The Molecular Definition of a Gene: All Nucleic Acid Sequences Necessary to Produce a Peptide or an RNA

Modern genetics has traveled far beyond the classic studies just described. Some changes in DNA sequence within a polypeptide-coding region can alter more than one genetic function. Furthermore, DNA sequence changes not located in a coding region can alter genetic function(s). For example, consider the bacterial operon. It consists of a single multigene transcription unit that produces polycistronic mRNA; such mRNA has multiple start and stop signals for the translation of a series of independent polypeptides. This operon can contain independent mutations in different regions; each of the different polypeptide-coding sequences defined by independent mutations can be scored as an independent cistron in a complementation test. However, every mutation that affects the operon does not necessarily affect only one cistron. Mutations in the promoter or the operator affect the entire group of cistrons in the operon; these are called *pleiotropic mutations*. For example, a mutation in the operator of the lactose operon renders all genes in the operon constitutive—that is, continuously functioning, not subject to regulation (see Table 8-1). Therefore, a genetically defined

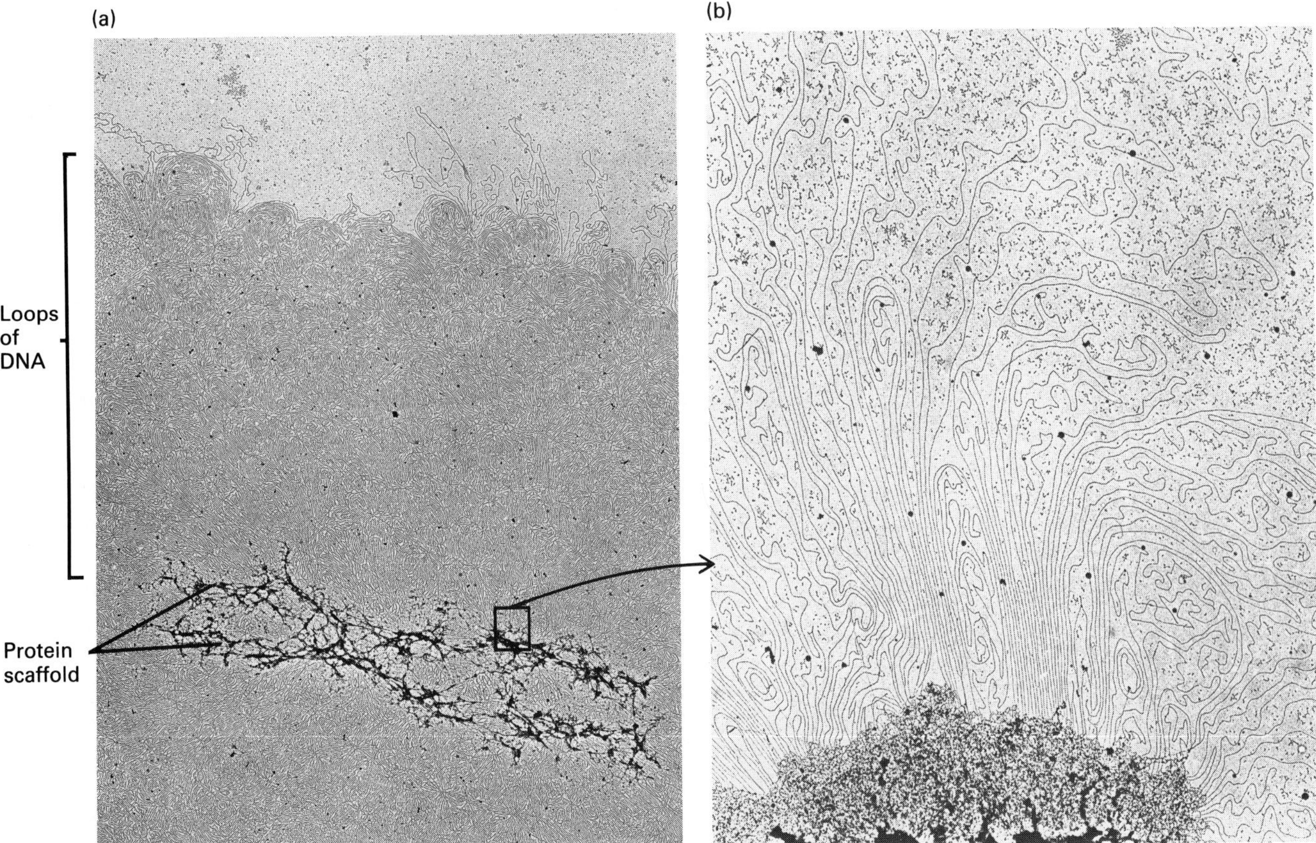

Figure 10-21 Electron micrographs of a histone-depleted metaphase chromosome. (Metaphase chromosomes from HeLa cells were prepared and treated with a mild detergent to release the histones.) (a) The mitotic chromosome maintains its general shape by a nonhistone protein scaffolding *(dark structure)* from which long loops of DNA protrude (50,000 ×). (b) A higher magnification of a section of the micrograph in (a) shows the apparent attachment of the loops to the scaffold (150,000 ×). *From J. R. Paulson and U. K. Laemmli, 1977,* Cell **12**:*817. Copyright 1977 M.I.T.*

cistron is not the "whole" gene; the controlling sequences, which are not transcribed, are also part of the gene.

If we attempt to encompass all of the recent knowledge about eukaryotic nuclear RNA transcription and mRNA processing (Chapter 9), the definition of a gene becomes even more problematic.

Transcription Units Are Not Necessarily Single Genes

In Chapter 9 we defined simple eukaryotic transcription units as those producing only one mRNA, and therefore encoding only one protein. Many eukaryotic transcription units are simple, and therefore many eukaryotic "genes" and transcription units are coextensive and might be scored as individual cistrons in a complementation test.

Complex transcription units such as those of adenovirus, SV40 virus, and certain eukaryotic cellular transcription units can produce more than one mRNA. Here we must differentiate between the transcription unit and the gene or potential complementation unit. A complex transcription unit may contain two or more poly A sites, or two or more splicing variations. This in turn can lead to two or more mRNAs, and each mRNA can encode a separate protein. Thus, if complex transcription units are considered to contain as many genes as the different protein-coding mRNAs that they produce, they contain more than one gene. However, genetic tests to prove the separate existence of each gene could be difficult. Mutations within an exon of the transcription unit that is shared by all mRNAs from that transcript can affect all encoded proteins. In such a case, one mutation affects the formation of two or more proteins, but the protein-coding functions are not separable by complementaton. Two such "genes" cannot be identified as cistrons (Figure 10-28a).

In a complex transcription unit, mutations can also

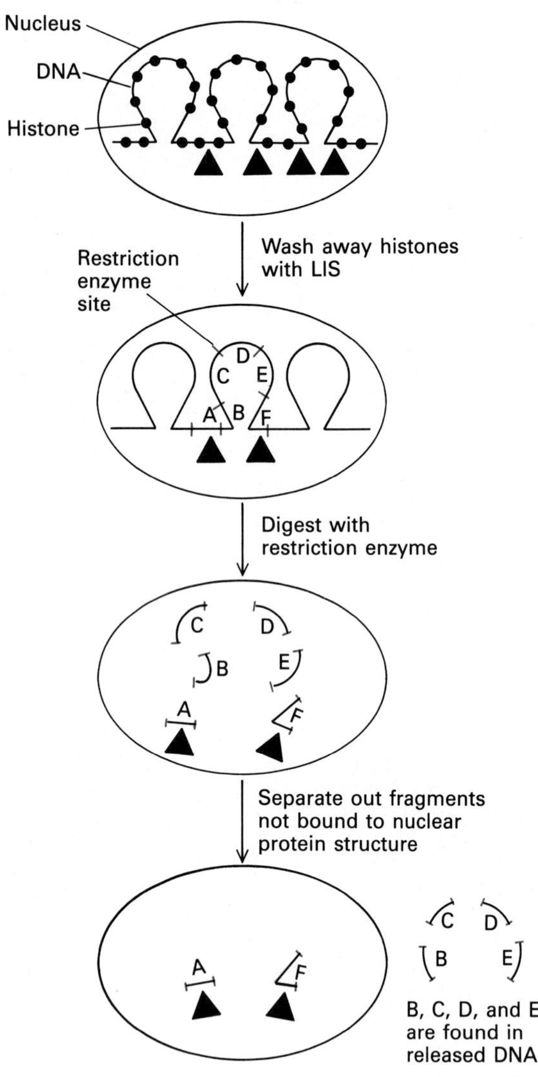

Figure 10-22 Organized loops of chromatin are bound to nuclear proteins. Nuclei from *Drosophila melanogaster* cells can be washed free of all histones by treating them with the detergent lithium diiodosalicylate (LIS), which exposes the DNA to digestion with added enzymes. Regions of the *Drosophila* genome have been cloned and sequenced, so specific sites for restriction endonucleases are known. One such region is diagrammed. After it is digested with a restriction enzyme to yield fragments A to F, the fragments that remain associated with nuclear proteins and the fragments that are released can be detected by the technique called Southern blotting (Chapter 7), in which hybridization of labeled cloned DNA samples is used to select fragments of specific sizes. Of the six DNA fragments produced by the enzyme, four—B, C, D, and E—are released and recovered outside of the nucleus. Extraction of the DNA that remains bound to the residual nuclear proteins yields the two fragments A and F. [See J. Mirkovitch, M.-E. Mirault, and U. K. Laemmli, 1984, *Cell* 39:223.]

occur in two or more different exons that each appear in only one of the possible mRNAs. In such cases, the mutation sites can be identified as separate complementing genes (Figure 10-28b and c). Thus the following rule applies, at least theoretically, to genetic tests of complex transcription units: if mRNAs contain different, nonoverlapping portions of a primary transcript, then complementation should be possible between at least some mutants in these distinct regions.

Finally, keep in mind that only one set of promoter sequences serves a complex eukaryotic transcription unit, and that a mutation in one of these nontranscribed but essential DNA sequences affects the expression of all functions encoded in the transcription unit.

Genes in Two Reading Frames In addition to the possible differences in the processing of a primary transcript, other molecular complications prevent or make very difficult the definition of a gene by genetic tests. For example, within the DNA of animal and bacterial viruses, there are overlapping genes that produce an mRNA whose triplet code is "read" (translated by ribosomes) in different frames (Figure 10-29). Because multiple codons exist for each amino acid, it is theoretically possible that two mutations in such overlapping genes might each affect only one of the possible proteins encoded by that region. In such cases, the two mutants in the same region of the DNA could be shown by complementation tests to contain two genes. However, mutations that damage both genes could not be complemented by any mutant in that region. Therefore, identifying overlapping genes by complementation does not always work. Identification of two such genes may require instead a molecular demon-

Figure 10-23 ***(opposite)*** (a) The four chromosomes of *Drosophila melanogaster,* with the map positions of the approximately 110 mutants identified by T. H. Morgan's group by 1926. The names beginning with capital letters represent dominant mutant alleles; the names beginning with lower-case letters represent recessive mutant alleles. Eighteen separate gene loci, each affecting eye color, are indicated in color. The arabic numbers are map distances derived from the crossover frequencies. Group I is the X chromosome, and group IV is the small dot chromosome shown in (b). [See T. H. Morgan, 1926, *The Theory of the Gene,* Yale University Press.] (b) A diagram of the four chromosomes of *Drosophila ampelophila,* a close relative of *Drosophila melanogaster.* The original drawing of Calvin Bridges clearly indicates that the male has two unequal chromosomes (the X and the Y, *bottom*), whereas the female has an equal pair (two X's). The other three pairs of chromosomes, the *autosomes,* are the same size and shape in males and females. *Part (b) from C. Bridges, 1916,* Genetics *1:3.*

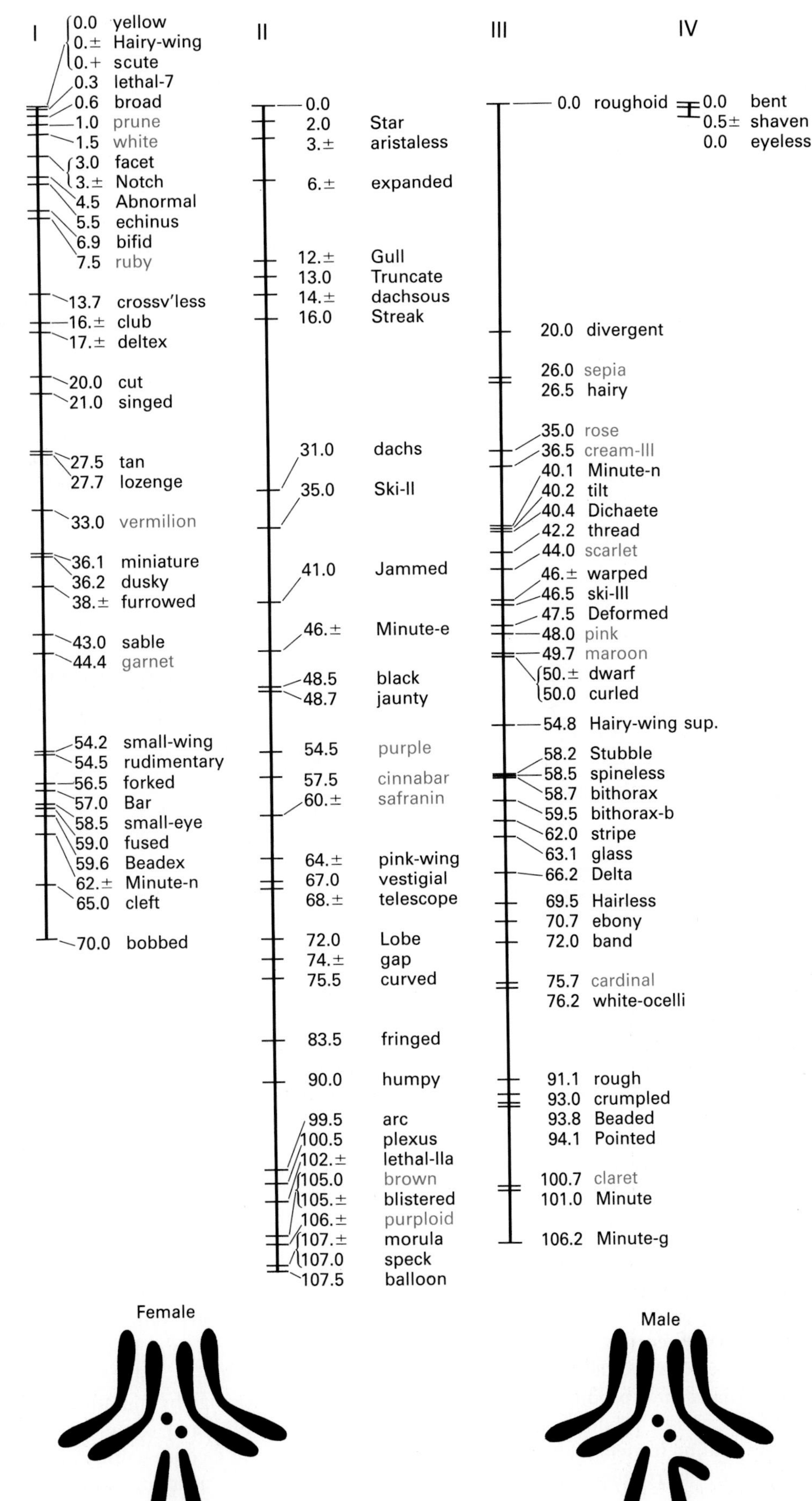
(a)
I
0.0 yellow
0.± Hairy-wing
0.+ scute
0.3 lethal-7
0.6 broad
1.0 prune
1.5 white
3.0 facet
3.± Notch
4.5 Abnormal
5.5 echinus
6.9 bifid
7.5 ruby
13.7 crossv'less
16.± club
17.± deltex
20.0 cut
21.0 singed
27.5 tan
27.7 lozenge
33.0 vermilion
36.1 miniature
36.2 dusky
38.± furrowed
43.0 sable
44.4 garnet
54.2 small-wing
54.5 rudimentary
56.5 forked
57.0 Bar
58.5 small-eye
59.0 fused
59.6 Beadex
62.± Minute-n
65.0 cleft
70.0 bobbed
II
0.0
2.0 Star
3.± aristaless
6.± expanded
12.± Gull
13.0 Truncate
14.± dachsous
16.0 Streak
31.0 dachs
35.0 Ski-II
41.0 Jammed
46.± Minute-e
48.5 black
48.7 jaunty
54.5 purple
57.5 cinnabar
60.± safranin
64.± pink-wing
67.0 vestigial
68.± telescope
72.0 Lobe
74.± gap
75.5 curved
83.5 fringed
90.0 humpy
99.5 arc
100.5 plexus
102.± lethal-IIa
105.0 brown
105.± blistered
106.± purploid
107.± morula
107.0 speck
107.5 balloon
III
0.0 roughoid
20.0 divergent
26.0 sepia
26.5 hairy
35.0 rose
36.5 cream-III
40.1 Minute-n
40.2 tilt
40.4 Dichaete
42.2 thread
44.0 scarlet
46.± warped
46.5 ski-III
47.5 Deformed
48.0 pink
49.7 maroon
50.± dwarf
50.0 curled
54.8 Hairy-wing sup.
58.2 Stubble
58.5 spineless
58.7 bithorax
59.5 bithorax-b
62.0 stripe
63.1 glass
66.2 Delta
69.5 Hairless
70.7 ebony
72.0 band
75.7 cardinal
76.2 white-ocelli
91.1 rough
93.0 crumpled
93.8 Beaded
94.1 Pointed
100.7 claret
101.0 Minute
106.2 Minute-g
IV
0.0 bent
0.5± shaven
0.0 eyeless
(b)
Female
Male

(a) Preparation of mating types

Strain	**Phenotype**	**Genotype**	**Gametes (assuming that *sc* does not segregate from *cr*)**
Strain 1 (homozygous recessive)	Scarlet eyes (sc^-) Curled wings (cr^-)	$\frac{sc^-\ cr^-}{sc^-\ cr^-}$	All $sc^-\ cr^-$
Strain 2 (homozygous dominant)	Dark red eyes (sc^+) Straight wings (cr^+)	$\frac{sc^+\ cr^+}{sc^+\ cr^+}$	All $sc^+\ cr^+$
F_1 (heterozygous offspring of strain 1 × strain 2)	Dark red eyes Straight wings	$\frac{sc^+\ cr^+}{sc^-\ cr^-}$	50% $sc^+\ cr^+$ 50% $sc^-\ cr^-$

(b) Experimental cross

F_1 female × Strain 1 male ⟶ Expected Offspring (F_2)

$\frac{sc^+\ cr^+}{sc^-\ cr^-} \times \frac{sc^-\ cr^-}{sc^-\ cr^-}$ ⟶ 50% $\frac{sc^+\ cr^+}{sc^-\ cr^-}$ Dark red eyes, Straight wings; 50% $\frac{sc^-\ cr^-}{sc^-\ cr^-}$ Scarlet eyes, Curled wings

Actual

⟶ 45% $\frac{sc^+\ cr^+}{sc^-\ cr^-}$ Dark red eyes, Straight wings; 45% $\frac{sc^-\ cr^-}{sc^-\ cr^-}$ Scarlet eyes, Curled wings; 5% $\frac{sc^+\ cr^-}{sc^-\ cr^-}$ Dark red eyes, Curled wings; 5% $\frac{sc^-\ cr^+}{sc^-\ cr^-}$ Scarlet eyes, Straight wings

Conclusion: Linkage between *sc* and *cr* is usually maintained, but recombination occurs in about 10% of gametes

Actual F_1 gametes:

45% $sc^+\ cr^+$
45% $sc^-\ cr^-$
5% $sc^+\ cr^-$
5% $sc^-\ cr^+$

Figure 10-24 An illustration of linkage and recombination: two closely linked markers followed through a genetic cross. (a) Two pure strains ($sc^-\ cr^-/sc^-\ cr^-$ and $sc^+\ cr^+/sc^+\ cr^+$) of *Drosophila* are achieved by inbreeding the two phenotypes; the strains are then crossed. The F_1 progeny are all heterozygotes, and the dominant alleles—dark red eyes and straight wings—are expressed. (b) Linkage and recombination are both evident when F_1 females are crossed with males of strain 1 (i.e., males that are homozygous for both recessive traits). If hundreds of progeny are counted, about 90 percent of the offspring demonstrate the expected linkage: dark red eyes remain associated with straight wings, and scarlet eyes with curled wings. However, in the other 10 percent of the offspring the linkage is disrupted: flies with dark red eyes have curled wings, and scarlet-eyed flies have straight wings. This is due to a crossover that occurs on chromosome III during the formation of gametes (see text). (A female F_1 animal is selected because recombination in male gametes of *Drosophila* is very rare.) Quantitative experiments of this type established the genetic maps of the four chromosomes shown in Figure 10-23a.

stration that there are two coding functions in the same DNA sequence.

Multiple Genes in One Polypeptide A final complication in defining genes of eukaryotic cells arises because not all mRNAs are translated into products that act as single polypeptides. Some viruses produce *polyproteins* that are cleaved after translation to yield as many as 10 to 12 different polypeptides of diverse functions. The same is true for cell polyproteins such as proopiomelanocortin, which upon cleavage can give rise to four or more different hormonally active polypeptides, some of whose sequences overlap (Figure 10-30). Theoretically each of these final functioning polypeptide units is individually susceptible to change by mutation of the appropriate region of the DNA, and complementation tests between two mutants could conceivably indicate that the peptides are separate. Such a finding would define a nucleic acid

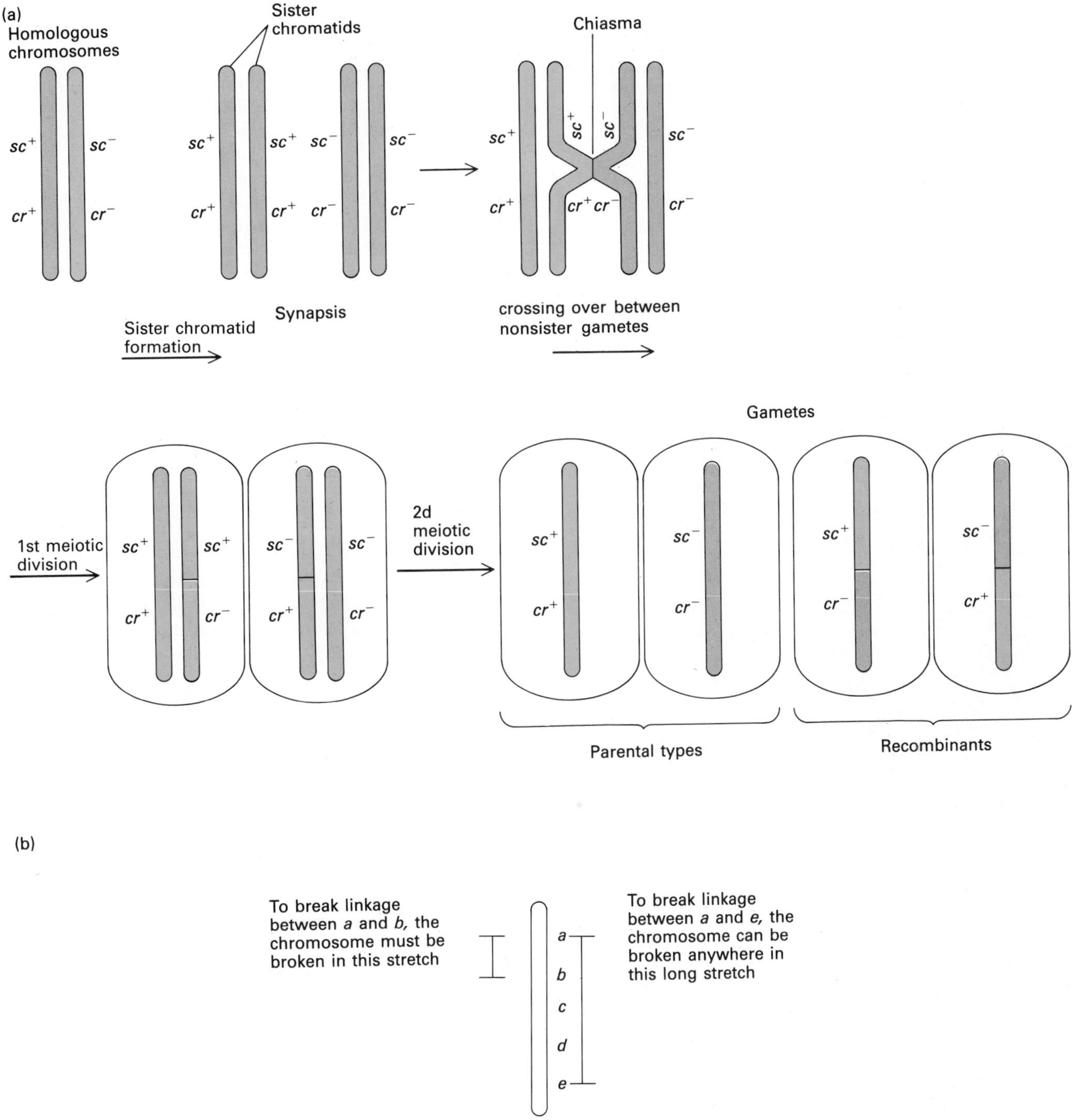

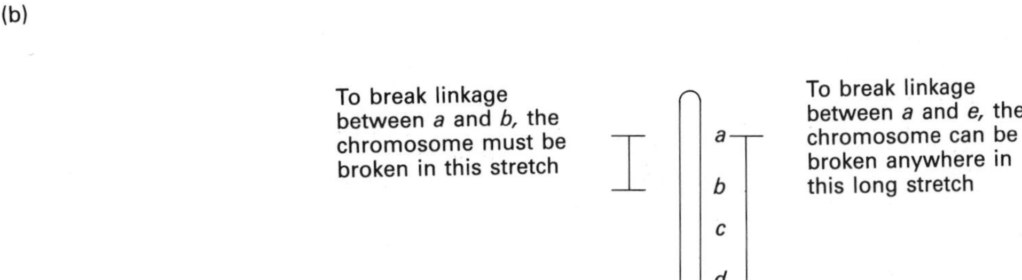

Figure 10-25 The cytologic basis of recombination. (a) A diagram of crossover at meiosis. After sister chromatid formation, the two bivalent chromosomes align (synapsis), and crossover events can occur. When the crossover is between sister chromatids, no genetic effect is detectable; but when a crossover occurs between nonsister chromatids, genetic recombination results. The meiotic divisions yield two parental and two recombinant gametes. (b) The "strength" of the linkage between an allele of one gene and an allele of another depends on the relative positions of the genes of the chromosome. By choosing one gene (say, *a*) and determining the frequency of its recombination with each of the other genes on the chromosome (*b*, *c*, *d*, and *e*), the order of the genes on the chromosome can be obtained. Also, in genetic crosses in which three *factors* (genetic traits) are followed—for example, the factors encoded by *a*, *b*, and *c*—recombination between *a* and *b* or between *b* and *c* would be less frequent than between *a* and *c*; this would establish the gene order *a-b-c*.

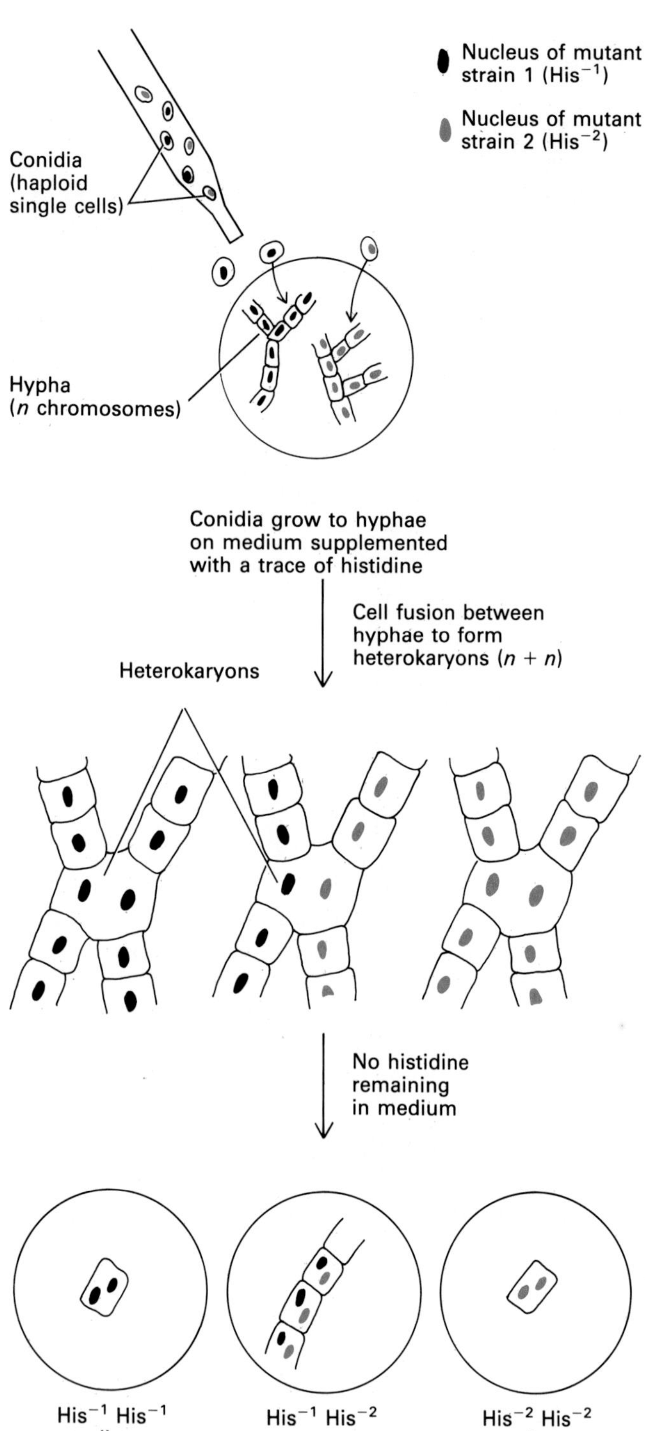

Figure 10-26 The complementation test in *Neurospora crassa.* Mutant *conidia* (haploid cells) will germinate and grow into hyphae if the medium supplies any needed compound that the cell cannot synthesize. Conidia from two different mutant strains that require histidine (His^{-1} and His^{-2}) are grown together on a medium supplemented with a trace of histidine, in order for minimal growth to occur. When the hyphae of the mutant cells touch, some cells fuse to produce heterokaryons. If one nucleus in a heterokaryon can compensate for the genetic defect in the other nucleus (here, the defect in the pathway for histidine biosynthesis), the heterokaryon will continue to grow in the same medium (which has by now been depleted of histidine) until the cell attains a visible size. The His^{-1} and His^{-2} mutant conidia complement each other because the two mutations are in different enzymes of histidine biosynthesis, and each mutant contributes a normal allele that is lacking in its partner.

sequence that encodes one mRNA and one translated peptide product as two genes.

Genes That Do Not Encode Proteins Although most genes are transcribed into mRNAs that encode proteins, clearly some transcribed RNA, including tRNA and rRNA, does not encode proteins. However, the DNA that encodes pre-tRNA and pre-rRNA can be mutated and detected in genetic experiments. We thus speak of tRNA genes and rRNA genes, even though the final products of these genes are RNA molecules and not proteins. Many other RNA molecules—for example, the small nuclear RNAs found in eukaryotic nuclei—also do not encode proteins and therefore are transcribed from "genes" that do not encode proteins.

How the Term "Gene" Is Used in the Era of Known DNA Sequences

Because "gene" may be the single most frequently used word in biology, we want to finally offer a definition that takes into account all of the newly accumulated molecular evidence. Respecting the term's heritage from genetic experiments, we say that any heritable trait that can be identified in a mutant form and separated from any other mutant form by a complementation test represents a separate gene. Thus we speak of a gene or genes for blue or brown eyes in humans, for coat color in mice, or for sucrose utilization in yeast. And we speak of the multiple genes for histidine formation in, for example, bacterial or yeast cells. A gene defined in this manner—whether it be prokaryotic or eukaryotic—usually encodes a single polypeptide chain. However, some distinct polypeptides encoded within the same DNA (or RNA) stretch are difficult or impossible to detect as separate genes by either a recombination or a complementation test. *Therefore, we define a gene in molecular terms simply as a nucleic acid sequence (usually a DNA sequence) that encodes a functional polypeptide or RNA sequence.* Nucleic acid stretches without demonstrable functions should not be referred to as genes.

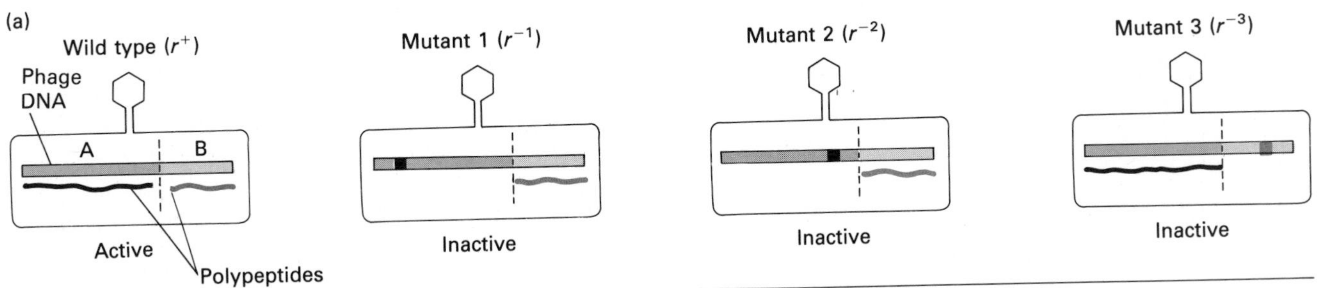

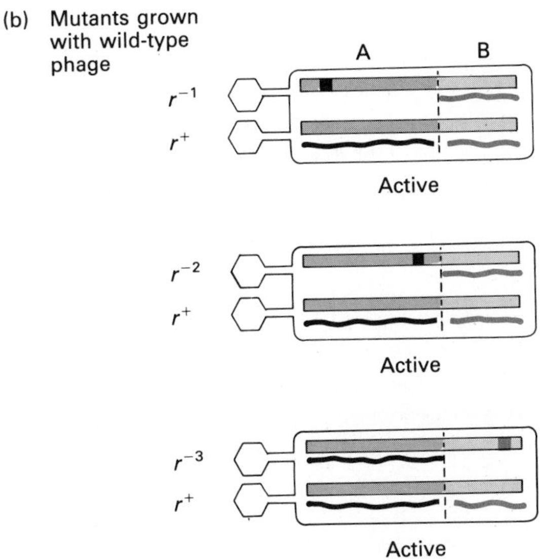

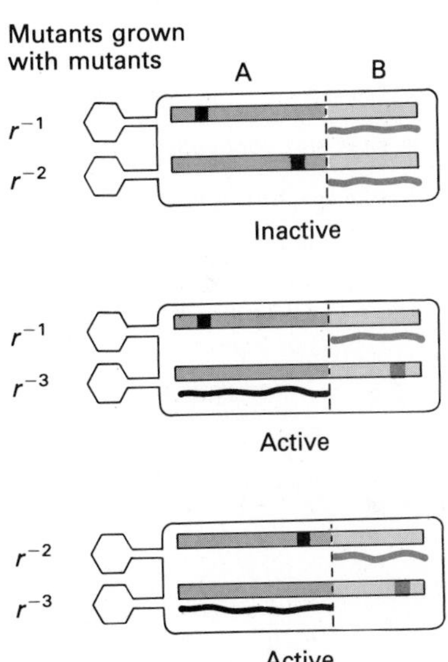

Figure 10-27 Benzer's cis-trans test of *r*II mutants of bacteriophage T4 in *E. coli* strain K. (a) The A and B genes were intact in the wild-type (r^+) phage genome, so both A gene and B gene polypeptides were formed. All mutants were inactive alone, and therefore any combination of r^{-1}, r^{-2}, and/or r^{-3} on one chromosome (i.e., any combination of mutants in the cis position) was also inactive. (b) When grown on strain K together with normal phage, all mutants could grow new phage, which showed that the growth of r^{-1}, r^{-2}, and r^{-3} was possible when the normal genes were present on another chromosome (i.e., when the mutant and the wild-type genes were trans). Likewise, any mutant in the A group could rescue any mutant in the B group when the two mutants were trans. No A's were transactive with one another, and no B's were transactive with one another; but any A was transactive with any B. The two transactive groups of mutants for the *r* function defined two cistrons, cistron A and cistron B, both of which are necessary for the *r* function. [See S. Benzer, 1962, *Sci. Am.* **206**(1):70.]

From studies in bacterial genetics (Chapter 8) and from recombinant DNA work with eukaryotic DNA, we know that gene function can be altered not only by mutations in the DNA sequences that are expressed in the primary RNA transcript, but also by mutations in the immediately surrounding DNA. In addition, in eukaryotic genes, sequences at a great distance from the start site for transcription can play a role in regulation—for example, by affecting chromatin structure. Until the potential functions of such very distant sequences are completely characterized, the final dimensions of a "gene" remain uncertain.

Summary

Eukaryotic chromosomes are visible in the light microscope during mitosis and meiosis. The karyotype—the collective term for chromosome number, size, and shape—is species-specific. Organisms with similar structures and degrees of complexity can differ greatly in their karyotypes and in the total amount of DNA in their genomes; these variations suggest that many eukaryotic genomes contain "extra" or unused DNA. The staining of chromosomes with dyes after various treatments produces light and dark bands at characteristic places along the lengths

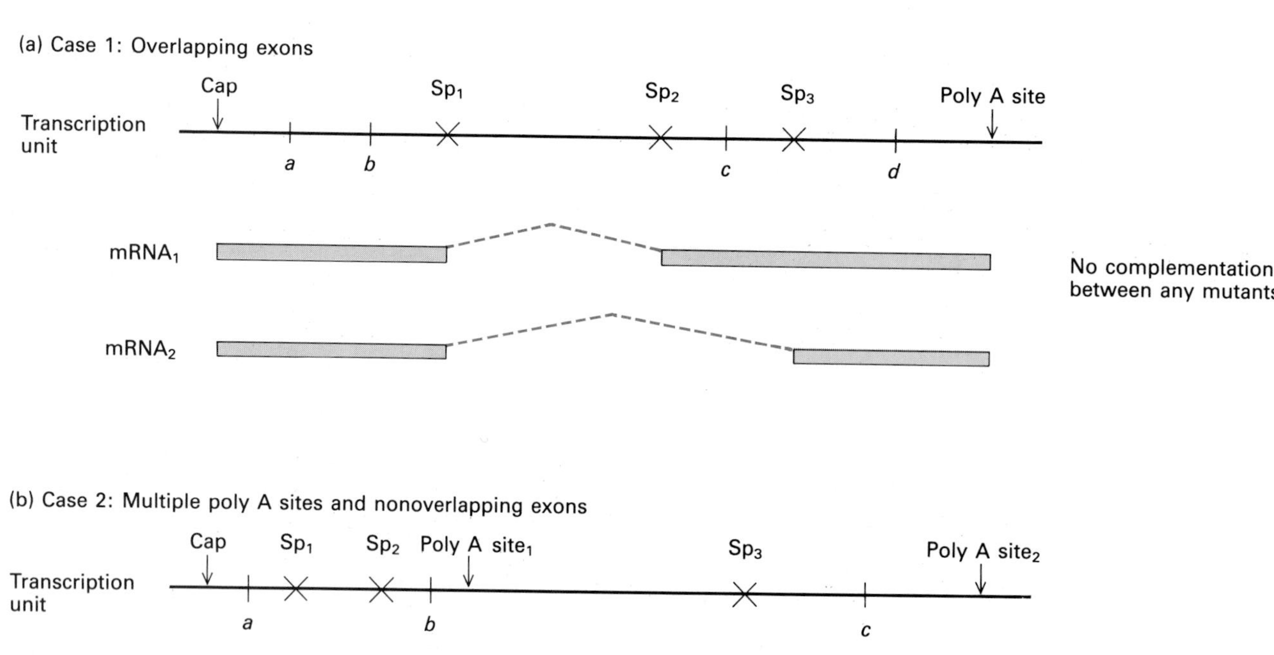

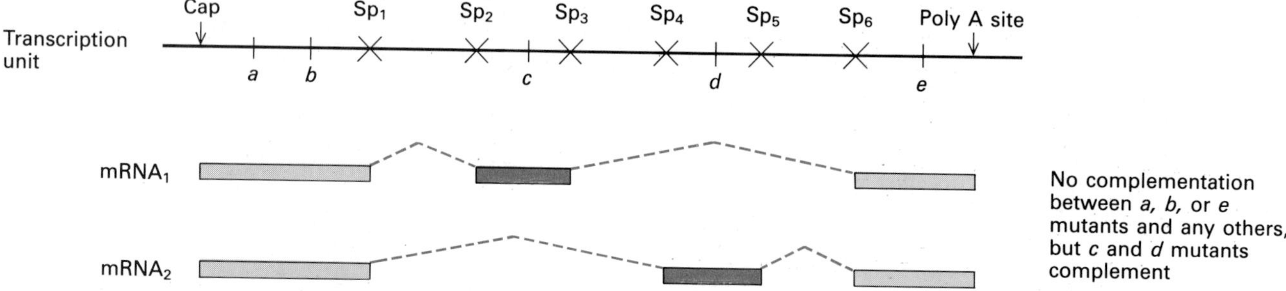

Figure 10-28 Complex transcription units and the genetic detection of coding capacity for different proteins. The three cases illustrated demonstrate that mutations in nonoverlapping exons could complement each other, but that mutations in shared exons cannot. The poly A sites, splice sites (Sp), and mutation sites (*a* through *e*) are marked. The mRNAs derived from transcription are shown below the transcription unit with nonoverlapping exons in dark color. (a) *Case 1:* Each of two mRNAs is formed from two exons that overlap with the exons in the other mRNA. A mutation at any one of the three sites *a*, *b*, or *d* prevents the formation of a normal copy of *either* mRNA. Furthermore, no complementation is possible between two chromosomes bearing individual mutations at any of these sites. For example, *a* and *b* do not complement each other, nor do *b* and *d*. (b) *Case 2:* A transcription unit contains two poly A sites and therefore two nonoverlapping 3′ exons. Mutants at *b* and *c* can complement one another. As in case 1, a mutation at site *a* would be in the shared 5′ exon, and no complementation with mutants at *b* or *c* could occur. (c) *Case 3:* A transcription unit has six splice sites and produces two mRNAs, each containing an exon that is spliced out of the other. Mutants at *c* and *d* can complement one another. Again, mutants at *a*, *b*, or *e* do not complement any other mutants.

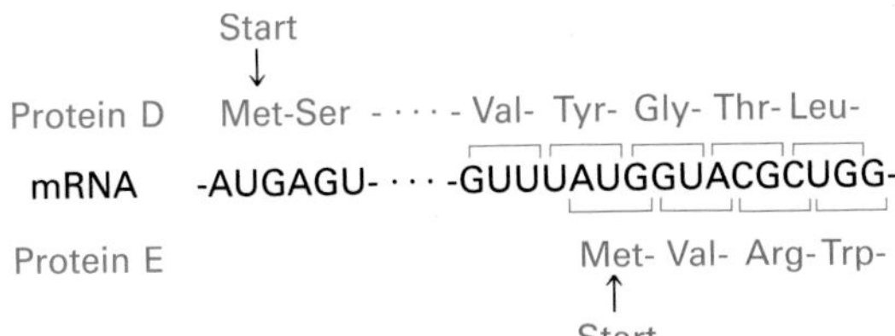

Figure 10-29 Overlapping reading frames in the same mRNA: part of the translated portion of an mRNA sequence from the bacteriophage ϕX174. Translation of the D protein of the virus begins at the first AUG codon; translation of the E protein begins in a different reading frame in the same mRNA. [See B. G. Barrell, J. M. Air, and C. A. Hutchinson, 1976, *Nature* **264**:34.]

of the chromosomes. These bands, which can be resolved in the light microscope, are the landmarks by which cytogeneticists have prepared physical maps of the chromosomes of many species.

Each eukaryotic chromosome is apparently a long linear DNA molecule. And each has two specialized areas: the ends, or telomeres, which somehow facilitate replication of a linear molecule; and the centromere, to which the spindle attaches at nuclear division (mitosis or meiosis).

By far the most prominent proteins in the nucleus are basic proteins called histones, which have very similar amino acid sequences in all eukaryotes. Octomers containing two each of histones H2A, H2B, H3, and H4 form the core around which helical DNA is wrapped approximately twice to form nucleosomes. Almost all of the DNA in the nucleus is packaged in this form. It is possible that the regular packing of DNA around the histone occurs in all DNA that is not being transcribed, and that at least some structural changes in the DNA-histone packing are necessary when transcription occurs. A higher-order packing of chromatin is achieved by a helical assembly of nucleosomes into a solenoid, a compact arrangement of six nucleosomes per turn of the helix. Finally, there are probably sequences that divide the chromatin into long loops, the bases of which are attached to a protein scaffold. During mitosis, the long loops of chromosomal DNA still associated at their base with an underlying protein-containing "scaffold" are condensed into a visible form.

From studies of linkage groups and recombination during meiosis, the early geneticists developed recombination mapping and a definition of the gene. The maps showed that the eukaryotic chromosome is a linear array of genes. A gene was defined as any trait that could be separated from a neighboring trait by recombination.

Complementation analysis, raised to its zenith in studies of the *r*II locus of bacteriophage T4, increased the efficiency and precision of genetic analysis and defined a gene as a unit of DNA encoding one polypeptide. As more and more was learned about regulatory elements in the control of gene function in prokaryotes, it became clear that the expression of individual "genes" often depends on sequences that lie at some distance from them. And when modern molecular genetics arrived and demonstrated multiple polyadenylation sites, multiple splicing sites, translation reading in two frames, and other complexities, it became clear that no single simple molecular definition of a gene suffices for all cases.

The concept of the gene as a biological entity remains intact: a gene is still considered a heritable function detected by observing the effect of a mutation. However, according to the current molecular definition, a gene consists of all the DNA sequences necessary to produce a single peptide or RNA product. Thus, the gene is no longer thought of as a single contiguous stretch of DNA.

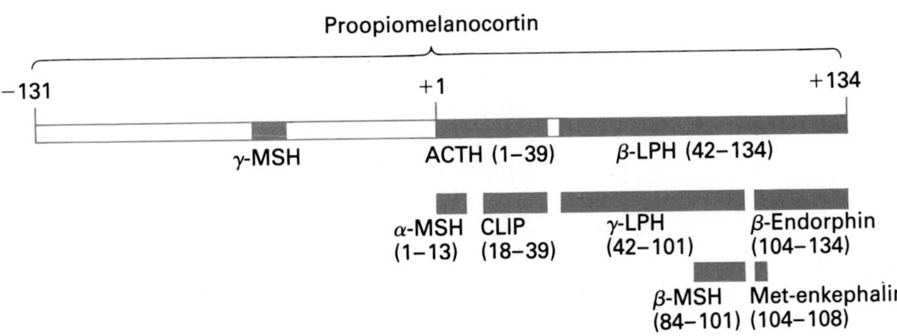

Figure 10-30 A number of polypeptides that have hormonal activity are contained within the primary translation product proopiomelanocortin, a polyprotein produced by the anterior pituitary. Proopiomelanocortin is cleaved to yield these hormonal peptides, which include adrenocorticotropic hormone, ACTH (amino acids 1 to 39; because ACTH was sequenced first, its first amino acid is numbered +1); lipotropic hormones, LPHs (amino acids 42 to 134); melanocyte-stimulating hormones (MSHs); and enkephalins, which are neuroactive hormones. CLIP is a short polypeptide whose function is not yet known. [See S. Nakanishi et al., 1979, *Nature* 278:423.]

References

General References

Collections of Articles

Chromosome Structure and Function. 1974. *Cold Spring Harbor Symp. Quant. Biol.*, vol. 38.

Chromatin. 1978. *Cold Spring Harbor Symp. Quant. Biol.*, vol. 42.

DNA Structures. 1983. *Cold Spring Harbor Symp. Quant. Biol.*, vol. 47.

Books

BOSTOCK, C. J., and A. T. SUMNER. 1978. *The Eukaryotic Chromosome.* Elsevier/North-Holland.

LEWIN, B. 1980. *Gene Expression,* vol. 2: *Eucaryotic Chromosomes,* 2d ed. Wiley.

STRICKBERG, M. 1976. *Genetics,* 2d ed. Macmillan.

SUZUKI, D. T., A. J. F. GRIFFITHS, J. H. MILLER, and R. C. LEWONTIN, eds. 1985. *An Introduction to Genetic Analysis.* W. H. Freeman and Company.

YUNIS, J. J., ed. 1977. *New Chromosomal Syndromes.* Academic Press.

Classical Genetics

BEADLE, G. W., and E. L. TATUM. 1941. Genetic control of biochemical reactions in *Neurospora. Proc. Nat'l Acad. Sci. USA* **27**:499.

BENZER, S. 1961. Genetic fine structure. In *The Harvey Lectures, 1961, Ser. 56.* Academic Press.

BENZER, S. 1962. The fine structure of the gene. *Sci. Am.* **206**(1):70–84.

MORGAN, T. H. 1926. *The Theory of the Gene.* Yale University Press.

PETERS, J. A., ed. 1959. *Classic Papers in Genetics.* Prentice-Hall.

VOELLER, B. R., ed. 1968. *The Chromosome Theory of Inheritance: Classic Papers in Development and Heredity.* Appleton-Century-Crofts.

The Morphology of Chromosomes

General Structure: DNA Content, Banding Patterns, Heterochromatin, Loops

BENDER, W., P. SPIERER, and D. S. HOGNESS. 1983. Chromosomal walking and jumping to isolate DNA from the *Ace* and *rosy* loci and the bithorax complex in *Drosophila melanogaster. J. Mol. Biol.* **168**:17–33.

BOSSY, B., L. M. C. HALL, and P. SPIERER. 1984. Genetic activity along 315 kb of the *Drosophila* chromosome. *EMBO J.* **3**:2537–2541.

GALL, J. G. 1981. Chromosome structure and the C-value paradox. *J. Cell Biol.* **91**:3s–14s.

GILLIES, C. B. 1975. Synaptonemal complex and chromosome structure. *Annu. Rev. Genet.* **9**:91–109.

KAVENOFF, R., L. C. KLOTZ, and B. H. ZIMM. 1974. On the nature of chromosome-sized DNA molecules. *Cold Spring Harbor Symp. Quant. Biol.* **38**:1–8.

LYON, M. F. 1972. X-chromosome inactivation and developmental patterns in mammals. *Biol. Rev.* **47**:1–35.

MACGREGOR, H. C., and J. M. VARLEY. 1983. *Working with Animal Chromosomes.* Wiley.

MARTIN, G. R. 1982. X-chromosome inactivation in mammals. *Cell* **29**:721–724.

MILLER, O. L. 1981. The nucleolus, chromosomes, and visualization of genetic activity. *J. Cell Biol.* **91**:15s–27s. (A review.)

MIRKOVICH, J., M.-E. MIRAULT, and U. K. LAEMMLI. 1984. Organization of the higher-order chromatin loop: specific DNA attachment sites on nuclear scaffold. *Cell* **39**:223–232.

Paris Conference 1971, 1972. *Birth Defects,* Orig. Art. Ser. 3, no. 7.

SPIERER, P., A. SPIERER, W. BENDER, and D. S. HOGNESS. 1983. Molecular mapping of genetic and chromometric units in *Drosophila melanogaster. J. Mol. Biol.* **168**:35–50.

YUNIS, J. J. 1976. High resolution of human chromosomes. *Science* **191**:1268–1270.

Telomeres, Centromeres, and Autonomously Replicating Sequences

BLACKBURN, E. H., and J. W. SZOSTAK. 1984. The molecular structure of centromeres and telomeres. *Annu. Rev. Biochem.* **53**:163–194.

BLOOM, K. S., M. FITZGERALD-HAYES, and J. CARBON. 1982. Structural analysis and sequence organization of yeast centromeres. *Cold Spring Harbor Symp. Quant. Biol.* **47**:1175–1185.

CARBON, J. 1984. Yeast centromeres: structure and function. *Cell* **37**:351–353.

MURRAY, A. W., and J. W. SZOSTAK. 1983. Construction of artificial chromosomes of yeast. *Nature* **305**:189–193.

SHAMPAY, J., J. W. SZOSTAK, and E. H. BLACKBURN. 1984. DNA sequences of telomeres maintained in yeast. *Nature* **310**:154–157.

STRUHL, K. 1983. The new yeast genetics. *Nature* **305**:391–397.

SZOSTAK, J. W., and E. H. BLACKBURN. 1982. Cloning yeast telomeres on linear plasmid vectors. *Cell* **29**:245–255.

WALMSLEY, R. W., C. S. M. CHAN, B.-K. TYE, and T. D. PETES. 1984. Unusual DNA sequences associated with the ends of yeast chromosomes. *Nature* **310**:157–160.

Chromosomal Proteins: Histones and Nucleosomes

CHAMBON, P. 1977. The molecular biology of the eukaryotic chromosome is coming of age. *Cold Spring Harbor Symp. Quant. Biol.* **42**:1209–1234.

GOODWIN, G. H., J. M. WALKER, and E. W. JOHNS. 1978. The high mobility group (HMG) non-histone chromosomal proteins. In *The Cell Nucleus,* vol. 6, H. Busch., ed. Academic Press.

HEWISH, D. R., and L. A. BURGOYNE. 1973. Chromatin substructure. The digestion of chromatin at regularly spaced sites by a nuclear deoxyribonuclease. *Biochem. Biophys. Res. Commun.* **52**:504–510.

HOWARD, G. C., S. M. ABMAYR, L. A. SHINEFELD, V. L. SATO, and S. C. R. ELGIN. 1981. Monoclonal antibodies against a specific nonhistone chromosomal protein of *Drosophila* associated with active genes. *J. Cell Biol.* **88**:219–225.

KORNBERG, R. D. 1974. Chromatin structure: a repeating unit of histones and DNA. Chromatin structure is based on a repeating unit of eight histone molecules and about 200 base pairs of DNA. *Science* **184**:868–871.

KORNBERG, R. D., and A. KLUG. 1981. The nucleosome. *Sci. Am.* **244**(2):52–64.

KORNBERG, R. D., and J. O. THOMAS. 1974. Chromatin structure: oligomers of the histones. *Science* **184**:865–868.

MCGHEE, J. D., and G. FELSENFELD. 1980. Nucleosome structure. *Annu. Rev. Biochem.* **59**:1115–1156.

MCGHEE, J. D., D. C. RAU, E. CHARNEY, and G. FELSENFELD. 1980. Orientation of the nucleosome within the higher order structure of chromatin. *Cell* **22**:87–96.

OLINS, J. D., and D. E. OLINS. 1974. Spheroid chromatin units (nu bodies). *Science* **183**:330–332.

RICHARD, T. J., J. T. FINCH, B. RUSHTON, D. RHODES, and A. KLUG. 1984. Structure of the nucleosome core particle at 7Å resolution. *Nature* **311**:532–537.

11

Eukaryotic Chromosomes and Genes: Molecular Anatomy

THE techniques of DNA cloning and sequencing are only about a decade old, but they have already provided a wealth of detail about the types of sequences found in various eukaryotic genomes and about the distribution of these sequences. As we discussed in Chapter 9, the surprising interruption of protein-coding genes by intervening sequences (introns) was proved by sequencing cloned genes. Cloning and sequencing have also confirmed the widespread existence of "families" of protein-coding genes; such gene families had already been suggested by the discovery of groups of proteins whose sequences were similar.

Many other structural features of eukaryotic chromosomes have been illuminated as well. For example, sequence analysis has shown that a large fraction of the eukaryotic genome does not encode precursors to mRNAs or any other RNAs. No function for the majority of this "extra" DNA has yet been found. In addition, many regions of DNA in multicellular organisms contain stretches of sequences that are very similar but not identical. For most of this repetitive DNA no role in the life cycle of organisms is known. Finally, the widespread distribution of apparently nonfunctional repetitious sequences has been linked to another surprising discovery: some repetitious DNA sequences are not found in constant positions in the DNA of organisms of the same (or similar) species. Such "mobile" DNA segments, which are present in both prokaryotic and eukaryotic organisms, may play an important role in evolution, even though they seem to have no role in the life cycle of most organisms.

Classifying Eukaryotic DNA

In Chapter 10 we considered the large-scale structure of eukaryotic chromosomes and defined eukaryotic genes and transcription units. Here we shall describe the present picture of eukaryotic chromosomes at the molecular level.

Eukaryotic DNA Includes Solitary Genes and Duplicated Genes

According to mendelian genetics, a single copy of each gene is passed from each parent to the offspring when the germ cells unite to form a zygote. According to the rules of classical genetics, therefore, each protein-coding DNA segment would occur just once in a germ cell.

The cloning of many different genes from the germ cells of animals and plants has indicated that some genes are solitary, but that others are duplicated. The extent of duplication varies for different genes in the same organism and for the same gene in different organisms (Table 11-1). Why some genes remain solitary and others become duplicated is unknown. With few exceptions, the pattern of duplication in the germ-cell DNA of an organism is identical to that in its somatic-cell DNA.

In multicellular organisms, probably a minority of the protein-coding genes are represented only once in the haploid nucleus. Most of the protein-coding genes within each organism are members of families of two or more genes that have very similar but not identical sequences and that encode similar but not identical proteins. These gene families are apparently duplicates of an original gene that diverged during evolution. Thus, the sequences of certain duplicated, functioning genes change over time—a phenomenon known as drift.

Table 11-1 Classes of eukaryotic DNA

Protein-coding genes
- Solitary
- Duplicated (most duplicates are no longer exact)

RNA-coding genes (most are tandemly duplicated)

Repetitious DNA
- Simple sequences
- Dispersed intermediate repeats (short, 150–300 bp; long, 5–7 kb)
  - Mobile genetic elements
    - Transposons
    - Reverse transcription copies

Unclassified spacer or connecting DNA

The sequences of other duplicated genes are somehow kept "correct." In addition to the families of related protein-coding genes, eukaryotic DNA contains duplicated genes that have remained exact (or almost exact) copies of one another. These genes include all sequences encoding ribosomal RNAs (both the pre-rRNA and the 5S), some of the sequences encoding tRNAs, and some of the histone genes. A distinguishing feature of these families of identical genes is that the multiple copies exist in tandem (i.e., they remain localized in the genome, with one copy next to another).

Eukaryotes Have Repetitious DNA Sequences and "Extra" DNA That May Be Superfluous

As we have mentioned, the genomes of multicellular organisms contain many long DNA stretches whose function is unclear: they do not seem to encode protein or RNA. Short sequences within these noncoding regions probably serve as sites to initiate DNA replication. However, much of the chromosomal DNA apparently serves only to connect genes and complete the integrity of chromosomes. Interspersed throughout this connecting or *spacer* DNA are many repeated DNA sequences. Some of these repeated sequences are short (from 100 to 300 nucleotides in length), and some are long (up to 10,000 nucleotides). Certain repeated sequences are transcribed into RNA, but much of this RNA has no known function. Possibly some of the RNA copies of the longer repeated sequences do encode proteins.

For several reasons considerable interest has focused on determining the possible functions of the repeated sequences in the chromosomes of higher cells. First is the popular darwinian notion that all features preserved during evolution must have a function for the organism. Such an idea is not necessarily true. If DNA is lost in evolution by multicellular organisms very slowly, as is probably the case, then the existence of a stretch of noncoding DNA is not evidence that it has been *selectively* preserved. Second was the thought that some of these repeated sequences might function as binding sites for specific proteins to regulate gene expression or replication. However, no evidence has been obtained to support this possibility for these repetitious DNA sequences.

A third reason for the interest in repeated elements in eukaryotic DNA is that most classes of repeated DNA have been found at different sites and/or in different amounts in different individuals of the same or closely related species. Because of the variable locations of such repeated sequences, they have been termed *mobile genetic elements*. In a few instances this variability in chromosome structure determines changes in cell physiology. In most cases, though, mobile genetic elements apparently play little or no role in the life of an individual organism. By affecting genes nearby, the repetitive mobile elements probably do play a role in evolution. Because these sequences may have mechanisms for their own propagation and dispersal within a genome, even though they do not seem useful from the organism's point of view, they are sometimes called "selfish DNA."

Solitary Protein-Coding Genes

Most mutant genes that are studied by geneticists are simple transcription units; that is, they encode single polypeptides. We shall begin our detailed consideration of types of eukaryotic chromosomal DNA by examining such protein-coding genes.

As we explained in Chapters 9 and 10, the structure of the eukaryotic transcription unit is fairly consistent (Figure 11-1). The transcription unit includes promoter sequences that control the initiation of RNA synthesis, followed by the DNA region that is copied into the primary transcript. The protein-coding portion of the gene may be segmented by one or more introns. The first exon contains a 5′ untranslated region; the final exon contains a 3′ untranslated region extending to a poly A addition site. What is known about the sequences that flank such transcription units? Not enough, as yet; but DNA sequence data show that many of these flanking DNA regions cannot encode protein, because they have translation stop codons in all three reading frames.

An example of a protein-coding gene that is solitary at least in chickens is the gene for lysozyme. Lysozyme is an enzyme that cleaves certain polysaccharides, including those in the cell walls of some bacteria. It is found in human tears and is an abundant component of egg-white protein. Neither the flanking sequences upstream nor those downstream of this gene appear to code for protein. The flanking sequences do contain DNA elements that are repeated elsewhere in the genome (Figure 11-2). Thus, the transcription unit for lysosyme is a 15-kb sequence containing a single gene, and the solitary nature of this gene has been demonstrated experimentally: no protein-coding information occurs for about 20 kb on either side of it.

The most extensively studied eukaryotic DNA region in which the gene density has been mapped is in *Drosophila melanogaster*. A 315-kb region of chromosome 3 was isolated in a series of overlapping recombinant DNA clones in bacteriophage λ. This region corresponds to the regions designated 87D and 87E in the map of salivary gland polytene chromosomes; 87D/E has about 12 *chromomeres,* or dark-staining bands (Figure 11-3a; see Figure 10-7 for a photograph of stained chromosomes). By genetic analysis 12 different lethal complementation groups have been shown to lie within this chromosomal region, so it includes at least 12 genes. The entire region was tested for the presence of mRNAs in embryonic, larval, and adult fly tissue. Forty-three mRNAs of distinct sizes and chromosomal locations were detected in these three types of tissues (Figure 11-3b and c). The mRNA-coding sequences accounted for a total of about 33 percent of the 315 kb examined. In some regions the density of sequences encoding mRNA was much greater than in other regions. For example, in the center of the 315-kb segment, a 156-kb stretch encodes only seven mRNAs, whose primary transcripts total about 14 kb in length.

This 315-kb region of chromosome 3 is thought to be typical in its distributions of both lethal mutations and chromomeric bands. The region may therefore also have a typical distribution of genes. If so, we can use our estimate for the number of genes in this stretch—between 12 and 43 (the number of nonoverlapping mRNAs would be the maximum possible number of genes)—to estimate the number of genes in the entire *Drosophila* genome. Because 315 kb is 0.25 percent of the total DNA, we calculate that the genome should contain between 4800 and

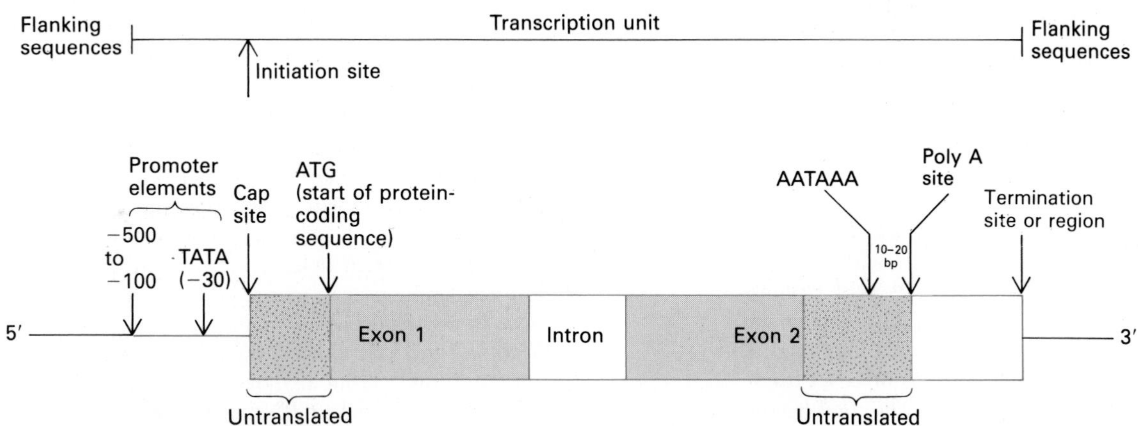

Figure 11-1 The recognized elements of a typical protein-coding transcription unit in eukaryotes. All of the designated features are not present in all transcription units. For example, some genes have no introns, whereas others have as many as 50. The cap site or initiation site is +1.

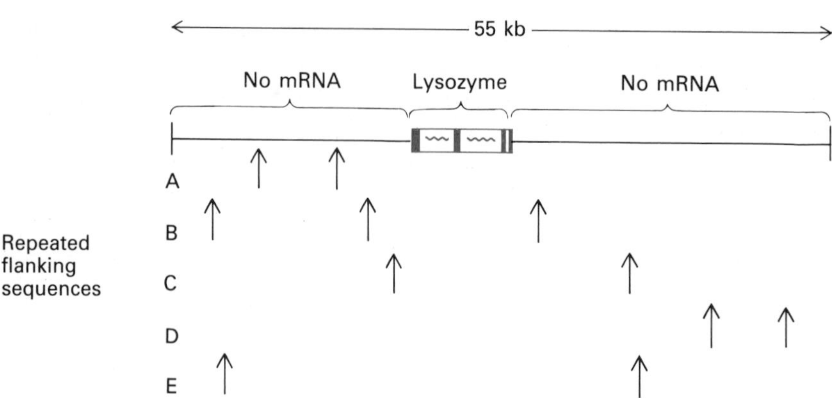

Figure 11-2 The chicken lysozyme gene and its surrounding region. This gene is one of several that is active in producing egg-white proteins in the oviducts of laying hens. By recombinant DNA techniques, several DNA clones from a chicken genomic DNA library were selected. These clones included the four exons (*color-shaded segments*) present in the lysozyme mRNA and the three introns (*white segments*) separating them. Together these sequences occupy about 15 kb. In addition, flanking sequences constituting about 40 kb have been mapped.

The lysozyme mRNA accounts for about 3 percent of the mRNA of the hen oviduct. To determine whether any of the genomic sequences flanking the lysozyme-coding sequence encodes another mRNA, a sensitive hybridization assay for mRNAs that constitute about 0.1 percent or less of the total mRNA was made. No such sequence was found. Sequences at the sites marked by arrows (A to E) were discovered to be repeated many times in the chicken genome, but were not found in mRNA. [See P. Balducci et al., 1981, *Nuc. Acids Res.* 9:3575.]

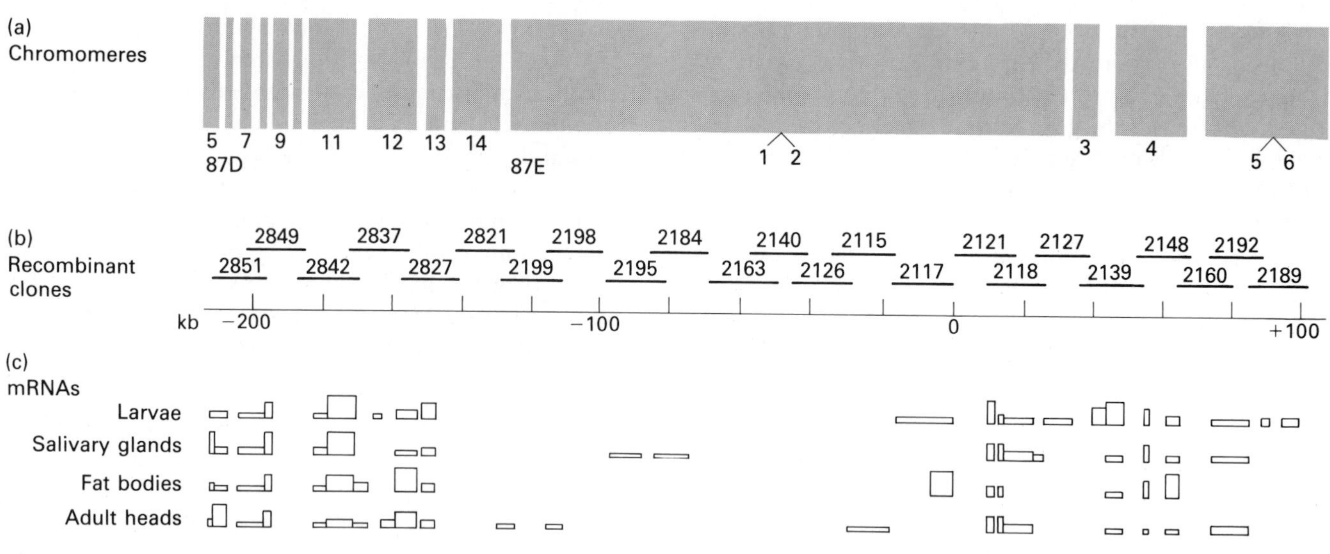

Figure 11-3 The distribution of mRNA-coding regions in the 87D/E region of the salivary gland chromosome 3 of *Drosophila melanogaster*. (a) The pattern of bands seen in the stained polytene chromosome are aligned here with the genomic map in (b) (see Figure 10-7 for a photograph of bands). (b) A series of 23 recombinant λ bacteriophages (shown as numbered lines) containing partially overlapping portions of DNA from the 87D/E region were selected. The total length of the region is 315 kb. (c) RNA preparations from *Drosophila* larvae, salivary glands, fat bodies (an organ with primitive liver-like function), and adult heads were tested for mRNAs that hybridized to each of the λ clones. The bars plotted show about 30 different positions to which mRNAs hybridized. Each bar represents a separate mRNA detected either by its unique size, or by its presence in different tissues, or by its presence in different amounts (i.e., in amounts different from those of other mRNAs of similar size). (The heights of the bars indicate the relative mRNA concentrations.) From analyses of this type, 43 different nonoverlapping mRNAs have been detected. [See B. Bossy, L. M. C. Hall, and P. Spierer, 1984, *Eur. Mol. Biol. Org. J.* 3:2537.]

17,000 genes. This estimate agrees fairly well with that of the classical geneticists who counted chromosome bands and lethal mutations to propose 5000 genes as the total complement of the fly. The hybridization studies have indicated that in the fruit fly, about two-thirds of the DNA may not contain genes encoding mRNA.

Families of Protein-Coding Genes

All protein-coding genes are not solitary like the gene for lysozyme. Frequently, the DNA that lies within 5 to 10 kb of a known gene contains sequences that are close copies of the gene. These duplicated protein-coding genes have been found so often in the genomes of vertebrates (e.g., chickens, mice, and humans) that they seem to be the rule rather than the exception. The discovery of such repeated chromosomal regions was not unexpected, because families of proteins had already been recognized by similarities in amino acid sequences. For example, it was known that all vertebrate hemoglobin contains two α-type and two β-type globin polypeptides (see Figures 3-6 and 3-8). Humans (and all other mammals examined) produce several slightly different "β-like" globin chains during embryogenesis and in adult life. In humans, the β-like globins include β, δ (delta), ${}^{A}\gamma$ and ${}^{G}\gamma$ (A- and G-gamma), and ϵ (epsilon). The proteins in this group have sequences that are similar but not identical (Figure 11-4). The individual globins are formed successively during embryogenesis, but only the β and δ chains continue to be formed after 6 months of age.

Many other families of proteins with similar but nonidentical amino acid sequences are known. Moreover, clusters of gene families with similar nucleic acid sequences have been identified. Fairly commonly these families contain 10 or more members; examples are the genes in the actin, tubulin, and keratin families in vertebrates (Table 11-2). Hundreds—perhaps even a thousand—nearly duplicate genes exist for the variable portions of

Table 11-2 Protein families common to many vertebrates and invertebrates

Family	Number of duplicate genes
COMMON PROTEINS	
Actins	5–30
Tubulins (α and β)	5–15
Myosin, heavy chain	5–10
Histones	100–1000
Keratins	>20
Heat-shock proteins	3
INSECTS	
Eggshell proteins (silk moth and fruit fly)	>50
VERTEBRATES	
Globins (many species)	
α	1–3
β-like	⩾5
Ovalbumin (chicken)	3
Vitellogenin (frog, chicken)	5
Immunoglobulins, variable regions (many species)	>500
Transplantation antigens (mouse and human)	50–100

```
Human β globin  ------VHLTPEEKSAVTALWGKV----NVD
       δ globin  ------VHLTPEEKTAVNALWGKV----NVD
      Gγ globin  ------GHFTEEDKATITSLWGKV----NVE

EVGGEALGRLLVVYPWTQRFFESFGDLSTPDAVMGNPKVK
AVGGEALGRLLVVYPWTQRFFESFGDLSSPDAVMGNPKVK
DAGGETLGRLLVVYPWTQRFFDSFGNLSSASAIMGNPKVK

AHGKKVL-GAFSDGLAHLDN---LKGTFATLSELHCDK--LHV
AHGKKVL-GAFSDGLAHLDN---LKGTFSQLSELHCDK--LHV
AHGKKVL-TSLGDAIKHLDD---LKGTFAQLSELHCDK--LHV

DPENFRLLGNVLVCVLAHHFGKEFTPPVQAAYQKVVAGV
DPENFRLLGNVLVCVLARNFGKEFTPQMQAAYQKVVAGV
DPENFKLLGNVLVTVLAIHFGKEFTPEVQASWQKMVTGV

ANALAHKYH------
ANALAHKYH------
ASALSSRYH------
```

Figure 11-4 The amino acid sequences of three β-like globin chains from humans. (Figure 3-2 provides a key to the single-letter amino acid symbols.) The β chain is produced by adults, and is part of the hemoglobin tetramer ($\alpha_2\beta_2$) in adult red blood cells. The β-like ${}^{G}\gamma$ and ${}^{A}\gamma$ chains are produced during fetal life; the β-like δ chain is produced late in fetal life and at a low percentage throughout adult life (see Figure 11-7). The amino acid sequence homology between β and δ approaches 90 percent; between β and ${}^{G}\gamma$ the homology is over 80 percent. (Amino acids differing from those in the β chain are shown in color. Spaces are apparent at sites where a β globin from some other animal differs for a few amino acids.) [See M. Dayhoff, ed., 1976, *Atlas of Protein Sequence and Structure*, Suppl. 2, National Biomedical Research Foundation, Washington, D.C., p. D-218.]

mouse and human immunoglobulin. (How all these gene copies participate in immunoglobulin diversity is explained in Chapter 24.)

It is necessary to keep in mind that the genomes of present-day organisms are the result of evolution. That the DNA of even such simple organisms as insects and echinoderms contains large gene families implies that gene duplication was an early evolutionary mechanism important in genome expansion. Some present-day single-celled eukaryotes have been examined for duplicated genes (i.e., that belong to families in higher organisms). Most yeasts, including *Saccharomyces cerevisiae,* contain one or two copies of the genes for actin or tubulin. Other simple organisms, however, contain multiple copies; for example, the slime mold *Dictyostelium discoideum* has 17 actin genes that are near-duplicates. These findings make it unsafe to conclude that *S. cerevisiae* represents a single-celled organism that is an evolutionary bridge to a past in which there was no gene duplication. Extra duplicates in yeast may simply have not been retained.

The evidence from protein sequences and DNA sequences of cloned genes from gene families shows clearly that gene duplication is a common feature of eukaryotes. Furthermore, each original "duplicate" apparently has evolved independently, explaining why the copies are not identical. No completely satisfactory molecular model for gene duplication during DNA synthesis has been proposed, much less validated by experiments. Probably, however, unequal crossovers during meiosis have produced gene duplication by chance. Because the DNA appears to be duplicated over regions longer than the transcription unit, the unequal crossover perhaps occurs between regions that lie outside of the transcribed region of the duplicated gene (Figure 11-5).

The β-Like Globin Cluster Is Representative of Gene Families

The β-like globin family is worth examining in more detail because its features typify many duplicated genes and

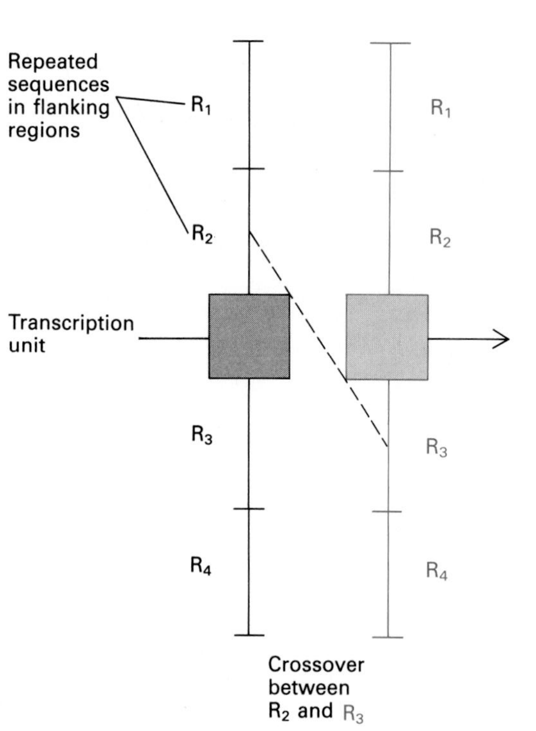

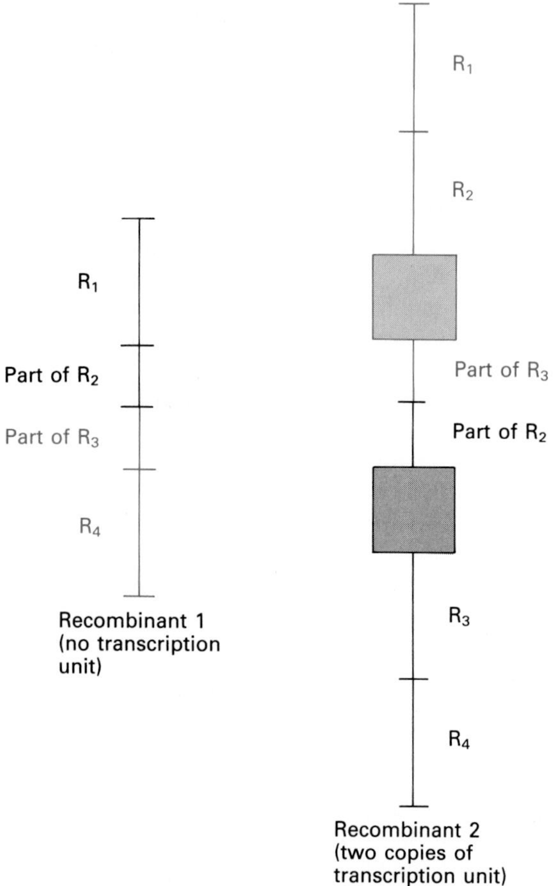

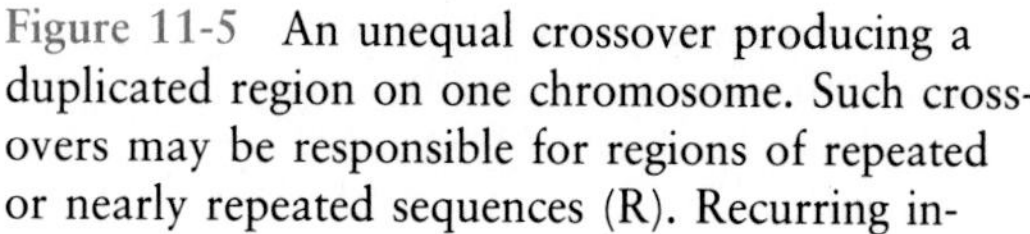

Figure 11-5 An unequal crossover producing a duplicated region on one chromosome. Such crossovers may be responsible for regions of repeated or nearly repeated sequences (R). Recurring instances of unequal crossovers in the descendants carrying recombinant 2 could theoretically produce multiple copies of the transcription unit and its immediate flanking sequences.

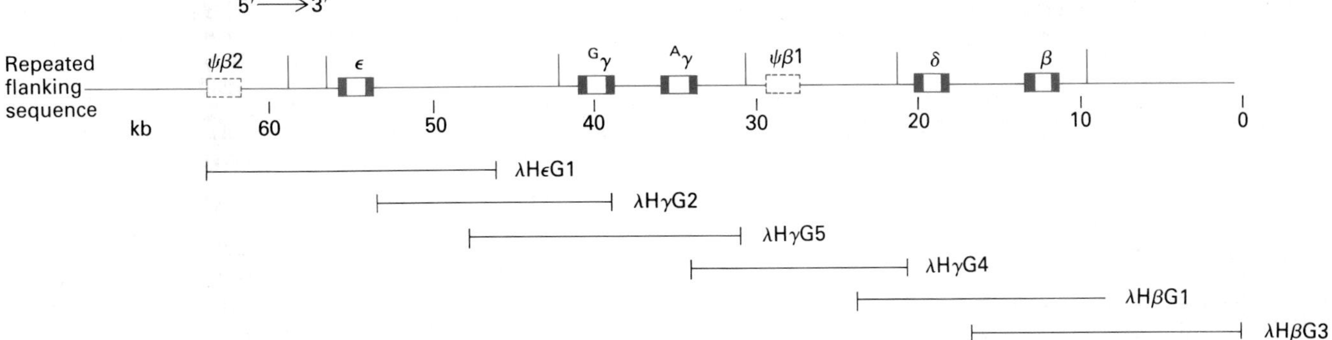

Figure 11-6 The β-like globin gene cluster on human chromosome 11. The segments labeled λHεG1 through λHβG3 represent overlapping fragments of human genomic DNA cloned in bacteriophage λ; these pieces were selected by their complementarity to cloned cDNA sequences made from β-like globin mRNAs. The total human DNA represented by this group of overlapping genomic clones is more than 60 kb. The positions of ε, $^G\gamma$, $^A\gamma$. δ, and β—the five genes known to encode β-like globin protein products—are indicated on the map. The direction of transcription is from left to right in all cases. Two nonfunctional pseudogenes, ψβ1 and ψβ2, are also shown. The vertical bars mark the sites of repetitious DNA sequences present at many other sites in the human genome. [See E. F. Fritsch, R. M. Lawn, and T. Maniatis, 1980, *Cell* **19**:959.]

may shed some light on their evolutionary history. Several mammalian β-like globin gene clusters, including human ones, have been isolated by recombinant DNA techniques. All the β-like genes in humans lie within a region of about 60 kb on chromosome 11 (Figure 11-6). In the map of this DNA region the genes are arranged from left to right in the same order in which they are expressed in development (Figure 11-7). (The δ and β genes are actually expressed around the same time, but activity from δ is initially low.) The β-like globin genes appear in similar long DNA regions in rabbits, mice, sheep, chickens, and frogs.

Not only are the sequences within the coding portions of the duplicated genes similar, but their transcription units have very similar designs. Each transcription unit has two introns that occur in the same position in each gene. The design of this transcription unit is preserved for all globin genes—not only for the β-like genes but for the α genes as well—in all species (Figure 11-8). Although the α group and the β-like group are now located on

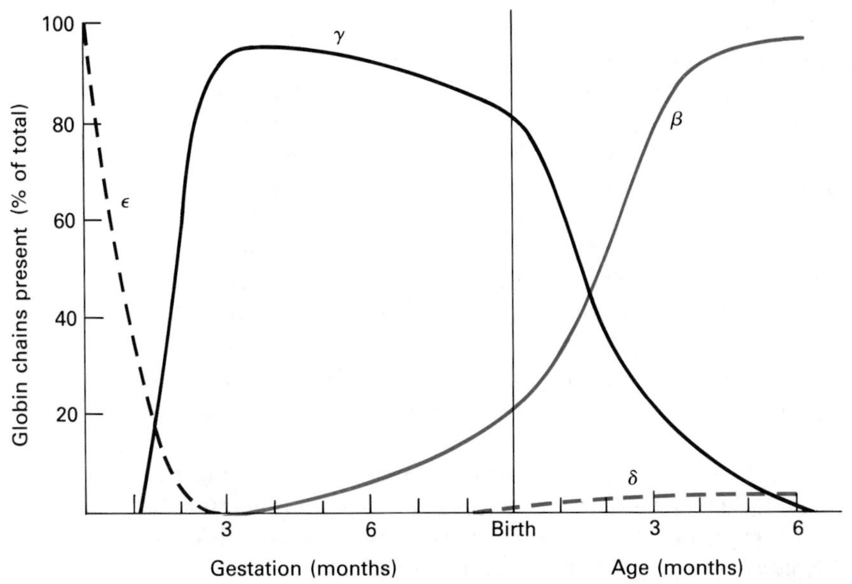

Figure 11-7 The relative expression of individual β-like globin genes in human red blood cells of a fetus (*left of vertical line*) and of an infant (*right of line*). The γ curve is a mixture of $^G\gamma$ and $^A\gamma$. [See R. M. Winslow and W. F. Anderson, 1983, in *The Metabolic Basis of Inherited Disease,* 5th ed., J. G. Stanbury et al., eds., McGraw-Hill, p. 1666.]

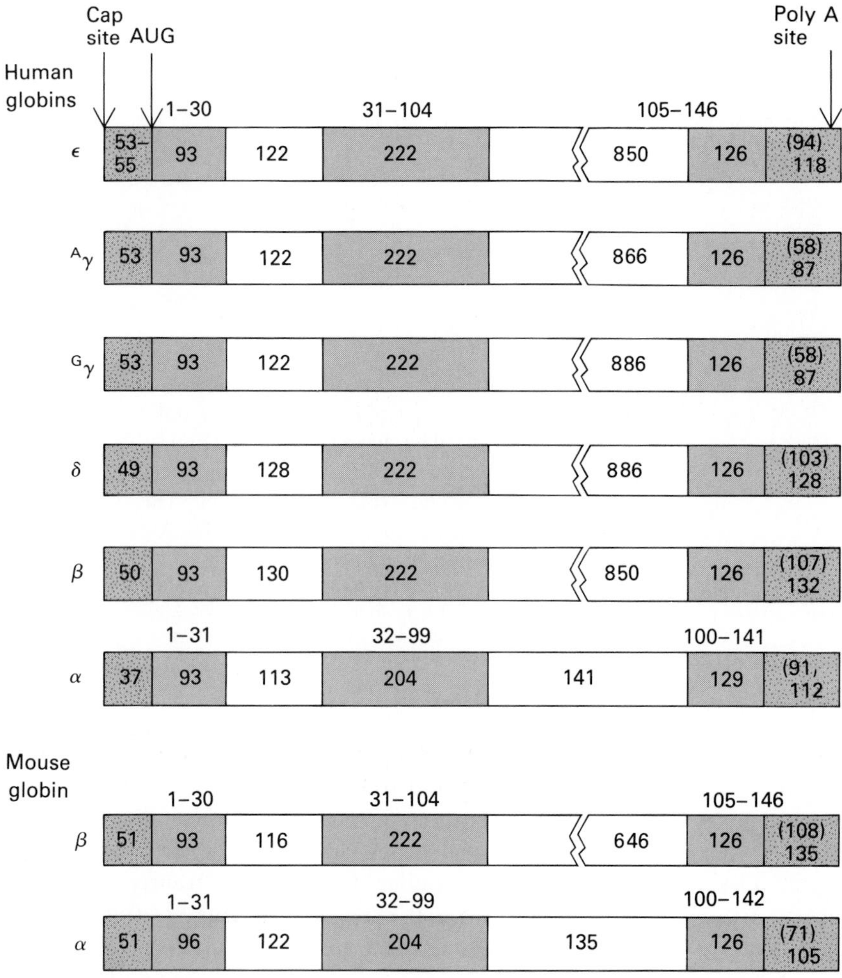

Figure 11-8 The remarkable similarity in the design of the transcription unit of the β-like and α globin genes from humans and mice. The positions of the transcribed portions of each globin gene are compared, beginning with the cap site and extending to the poly A site. Each gene has two introns (*white boxes*) that divide the protein-coding sequence in very similar places. (The second introns of the β-like genes are broken to indicate that these introns, in order to align the exons, are not shown in their proportionate lengths.) The stippled areas represent untranslated regions of the exons (*color boxes*). The numbers within the boxes indicate the number of nucleotides present in each region of the primary transcript, whereas the numbers above the boxes designate the corresponding amino acid positions in the resulting polypeptides. (Because the methionine encoded by the AUG is removed from the finished proteins, the numbering of the amino acids begins after the methionine.) In the final box, the number in parentheses indicates the length of the 3′ untranslated region up to the first A of the AAUAAA sequence; the other number gives the length of the entire 3′ untranslated region, up to the site of poly A addition. [See A. Efstratiadis et al., 1980, *Cell* **21**:653.]

different chromosomes, they are presumably duplicates of a primitive globin gene. The implications of structural conservation of the globin genes for the early evolutionary history of cells are discussed in Chapter 25.

As Figure 11-9 shows, the DNA sequences before the 5′ end of the globin transcription units are also similar for more than 100 bases preceding the cap site, with the strongest regions of conservation centering on −80 and −30. These regions are known to be necessary for accurate and frequent initiation of RNA synthesis by RNA polymerase II. Thus, the duplications of DNA that occurred and became fixed in the population of mammals included not only the sequences encoding the transcription unit but also regions flanking it.

Other Gene Families Exhibit a Variety of Duplication Patterns

To show the general applicability of the principles derived from studies of the globin family, we shall briefly describe several additional examples of protein-coding gene families. These examples will illustrate that differences as well as similarities exist in the structure of duplicated genes.

Ovalbumin, X, and Y in Chickens One very well-studied example of duplicated genes occurs in the chicken genome. Three contiguous structurally similar genes, the ovalbumin gene and two genes termed X and Y, are located within a 40-kb segment in chicken DNA (Figure 11-10). Ovalbumin is the major protein of egg white. The mRNAs for X and Y have been detected, but their protein products have not yet been characterized. The X and Y proteins are made at the same time as ovalbumin, but in much smaller quantities.

The primary sequences of significant parts of the three genes have been determined; they reveal very great similarities in design among the three transcription units. Each gene contains exactly the same number of introns (seven) and exons (eight) (Table 11-3). Although the introns in corresponding positions in the three genes vary

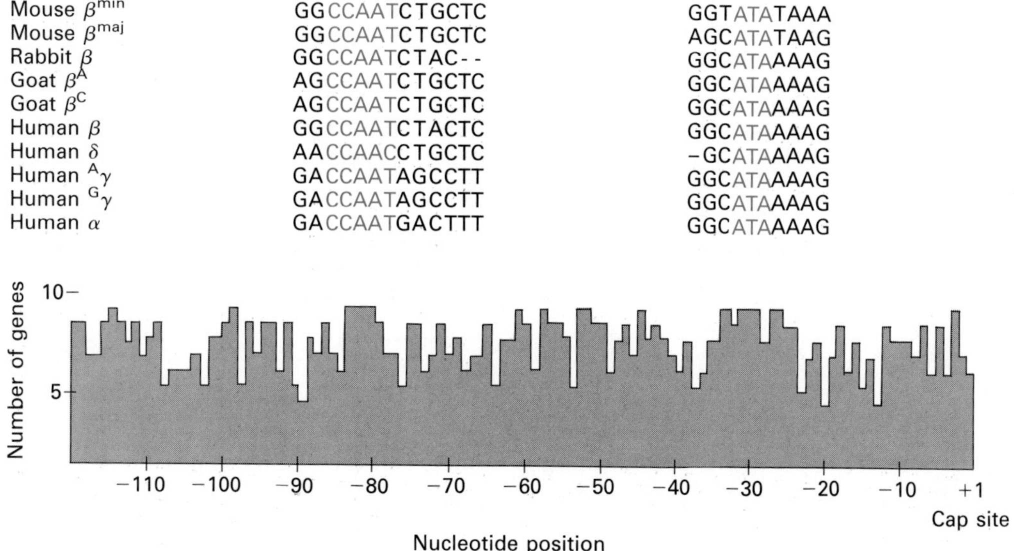

Figure 11-9 Similarities in sequences upstream of the coding portion of globin genes in mammals. The sequences preceding the mRNA cap site of 10 different mammalian β-like globin genes and a human α gene are aligned. Dashes indicate gaps introduced to achieve the most consistent sequence alignment; in a few cases some deletions were made to give the best alignment of the sequences on either side. Two highly conserved regions common to all globin genes sequenced to date—the CCAAT and the ATA boxes—are indicated in color. A histogram shown below illustrates the extent of homology at each nucleotide position. The ordinates indicate the frequency of occurrence of the predominant nucleotide at each position; for example, where the height of the graph is 5, 5 of the 10 genes have the same base at that position. [See A. Efstratiadis et al., 1980, *Cell* **21**:653.]

in length, the corresponding exons—which, of course, code for protein—are approximately equal in length. The last exon, which contains the 3′ untranslated region, is the only one that varies substantially in length. In addition to the constancy in length, the exons exhibit many fewer base changes than the introns do. Thus the X and Y genes appear to be duplicates that function like the ovalbumin gene, but at a greatly reduced rate.

Albumin and α Fetoprotein in Mammals For an even more striking example of the duplication of a lengthy transcription unit without alteration of the basic structure, we turn to the albumin and α fetoprotein genes. Examples of these proteins have been studied in many animals; the genes encoding the proteins have been examined most extensively in mice. Albumin, a protein secreted into the blood by liver cells, binds to many small

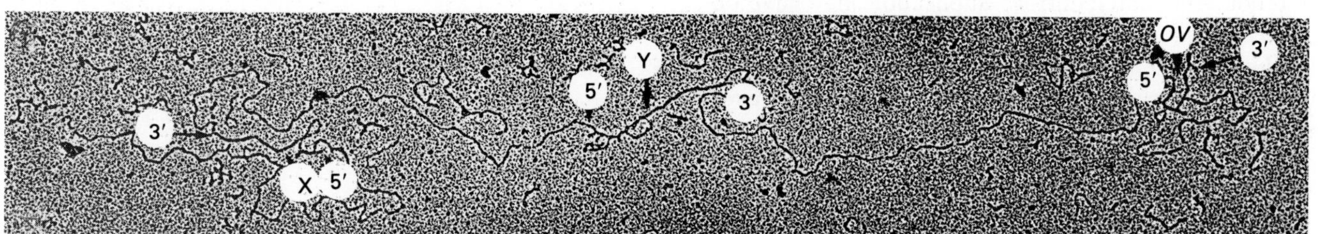

Figure 11-10 RNA-DNA hybrids showing the locations of coding portions of the ovalbumin family genes. A series of genomic clones of chicken DNA were shown to have three genes—the gene for ovalbumin (*ov*) and two others designated X and Y—clustered within a 40-kb region. The entire 40-kb stretch was cloned in a *cosmid* vector, which is a large recombinant plasmid designed to carry inserts from 35 to 45 kb in length. The electron micrograph shows the RNA-DNA hybrids formed in the X, Y, and *ov* regions when the cosmid DNA is hydribized with RNA from oviduct cells. Detailed analyses of such micrographs plus the sequencing of genomic fragments and cDNA molecules have established the structure of all three transcription units. [See A. Royal et al., 1979, *Nature* **279**:125.] *Courtesy of P. Chambon.*

Table 11-3 Lengths (in base pairs) of regions of chicken egg-white genes

Gene	Exons								Introns						
	1	2	3	4	5	6	7	8	a	b	c	d	e	f	g
ovalbumin	47	185	51	129	118	143	156	1043	1560	251	582	401	1029	323	1614
X	?	195	59	137	127	145	167	1503	1587	604	894	523	778	261	898
Y	?	199	54	142	128	148	165	1146	1619	153	357	808	235	50	886

SOURCE: R. Heilig et al., 1980, *Cell* 20:625.

molecules that are only slightly soluble in the blood serum. Because the bound form is soluble, albumin provides a way of transporting the small molecules through the bloodstream. The protein α fetoprotein, which during development may play a role similar to albumin, is synthesized in fetal liver and other cells. Folding the chains of these two proteins allows disulfide linkages to form between cysteine residues. Three very similar protein domains result in the two proteins isolated from several different animals (Figure 11-11).

Not surprisingly, the albumin gene in mice is very similar in structure to the α fetoprotein gene. Both genes have 14 introns and 15 exons. Although the sizes of the corresponding introns in the two genes vary by as much as a factor of 3, the lengths of the corresponding exons are quite similar. Twelve of the exons fall into three subgroups, which generate the three domains seen in the protein chains (Table 11-4). The total lengths of the four exons in each of these subgroups are similar, both within the same gene and between the two genes. This structure suggests that an earlier, primordial form of the gene had four exons that encoded a single domain, and that this region was triplicated to produce a protein containing three domains. Later, further duplication may have occurred to produce the present-day albumin and α fetoprotein (Figure 11-12). However the pattern of duplications occurred, the primordial gene contained an intron-exon pattern that has been retained during duplication.

Chorion Proteins in Silk Moths Silk moths provide a final example of duplicated protein-coding genes. As insect eggs develop they form an outer shell, called a *chorion,* that is composed of organized layers of fibrous proteins. Many of these proteins have similar amino acid sequences. The similar chorion proteins are encoded by a very large family of genes, whose relatedness has been proved both by amino acid sequencing of the proteins and by DNA sequencing of isolated genes.

The genes fall into three classes termed A, B, and C. The similarities within each class are greater than those between classes. All the classes, however, have a single intron. In the chromosome, the A class genes and the B class genes exist in A/B pairs; several such pairs can reside in a chromosomal segment of 10 to 15 kb (Figure 11-13). Each pair of genes represents a duplicated DNA segment, with perhaps as many as 50 such segments in the genome. Interestingly, the A and B genes of a pair are transcribed in opposite directions.

In each duplicated set, the two characteristic features—the divergent transcription and the single intron—are maintained. Seemingly, to arrive at the present structure, a primordial gene must have been duplicated, the copy (or the original) must have inverted, and then the pair of genes must have been duplicated as a unit. Duplication events are apparently not limited to single genes (i.e., to DNA encoding single proteins); rather, they may involve whole areas of a chromosome.

Table 11-4 The lengths of the exons in the mouse genes α fetoprotein and serum albumin

Domain encoded	Exon	Base pairs	
		α Fetoprotein	Albumin
	1	114 ± 32	110 ± 29
	2	53 ± 14	75 ± 27
I	3	148 ± 37	121 ± 48
I	4	218 ± 52	240 ± 49
I	5	144 ± 26	170 ± 70
I	6	104 ± 22	118 ± 33
II	7	133 ± 29	137 ± 35
II	8	230 ± 23	218 ± 65
II	9	154 ± 67	136 ± 30
II	10	125 ± 35	125 ± 28
III	11	135 ± 45	153 ± 31
III	12	280 ± 70	222 ± 42
III	13	175 ± 50	122 ± 36
III	14	69 ± 27	62 ± 26
	15	149 ± 33	112 ± 31

SOURCE: D. Kioussis et al., 1981, *J. Biol. Chem.* 256:1960.

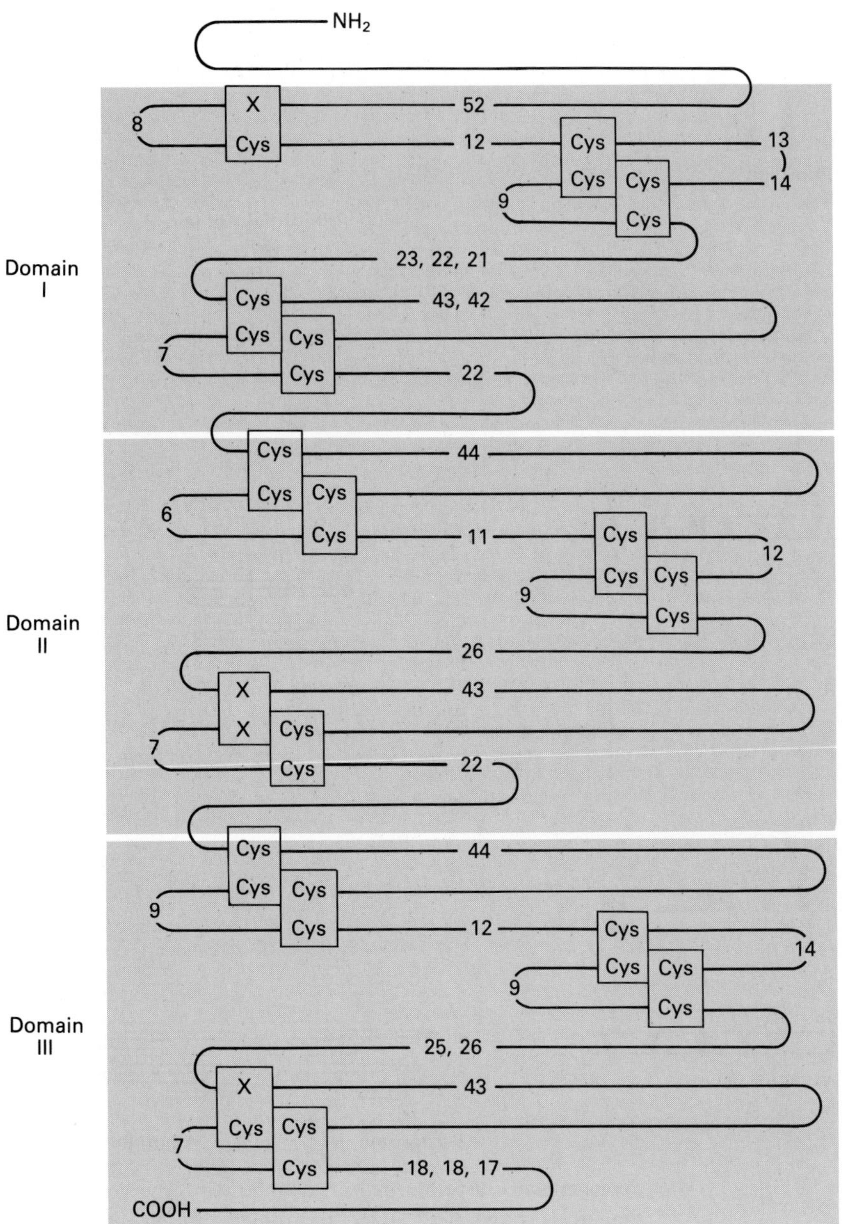

Figure 11-11 The similarity in the protein domains of albumin and α fetoprotein. The switchback line represents the amino acid sequence of mouse α fetoprotein, human albumin, and bovine albumin. The sequences are similar in the last 80 percent of the chains (the sequences in the first 20 percent precede the first domain). The chain is folded to allow the formation of cysteine-cysteine disulfide bridges; the numbers indicate the distances, in amino acids, between the Cys-Cys bridges. (A single number indicates that all three proteins have segments of the same length between these bridges; where two numbers are given, the first refers to the α fetoprotein and the second to the two albumins; three numbers represent the α fetoprotein, the human albumin, and the bovine albumin, in that order. Wherever an X is shown, α fetoprotein is lacking a cysteine residue that is present in the albumins.) The distances between the corresponding disulfide bridges in the three proteins are remarkably similar. The repeating pattern divides the proteins into three very similar domains. [See M. B. Gorin et al., 1981, *J. Biol. Chem.* 256:1954.]

Pseudogenes Are Duplications That Have Become Nonfunctional

At least two regions in the human β-like globin gene cluster and three regions in the mouse β-like globin cluster have nonfunctional sequences similar to those of the functional β-like globin genes. Because no known protein corresponds to these regions, they are called *pseudogenes* (see Figure 11-6). These copies originally may have been functional gene duplicates; if so, *sequence drift* has apparently rendered them nonfunctional. In some pseudogenes stop codons would block translation of any RNA copied from that region. In other pseudogenes mutations arise that would prevent RNA processing if a transcript were produced. Thus pseudogenes presumably represent duplicated genes that are no longer active because they have drifted in sequence.

Some of the presently active β-like globin genes (e.g., the human δ globin gene) produce very little mRNA; a mutation that completely halted the activity of such an infrequently used gene duplicate might well be tolerated by the organism. Such a "silencing" genetic event apparently occurred in the δ gene of gibbons some 5 to 10 million years ago. The initial mutation was probably in the promoter region and this gene has now accumulated a number of mutations that would interrupt its coding function as an RNA. Gibbons have survived perfectly well on one adult β-like globin gene.

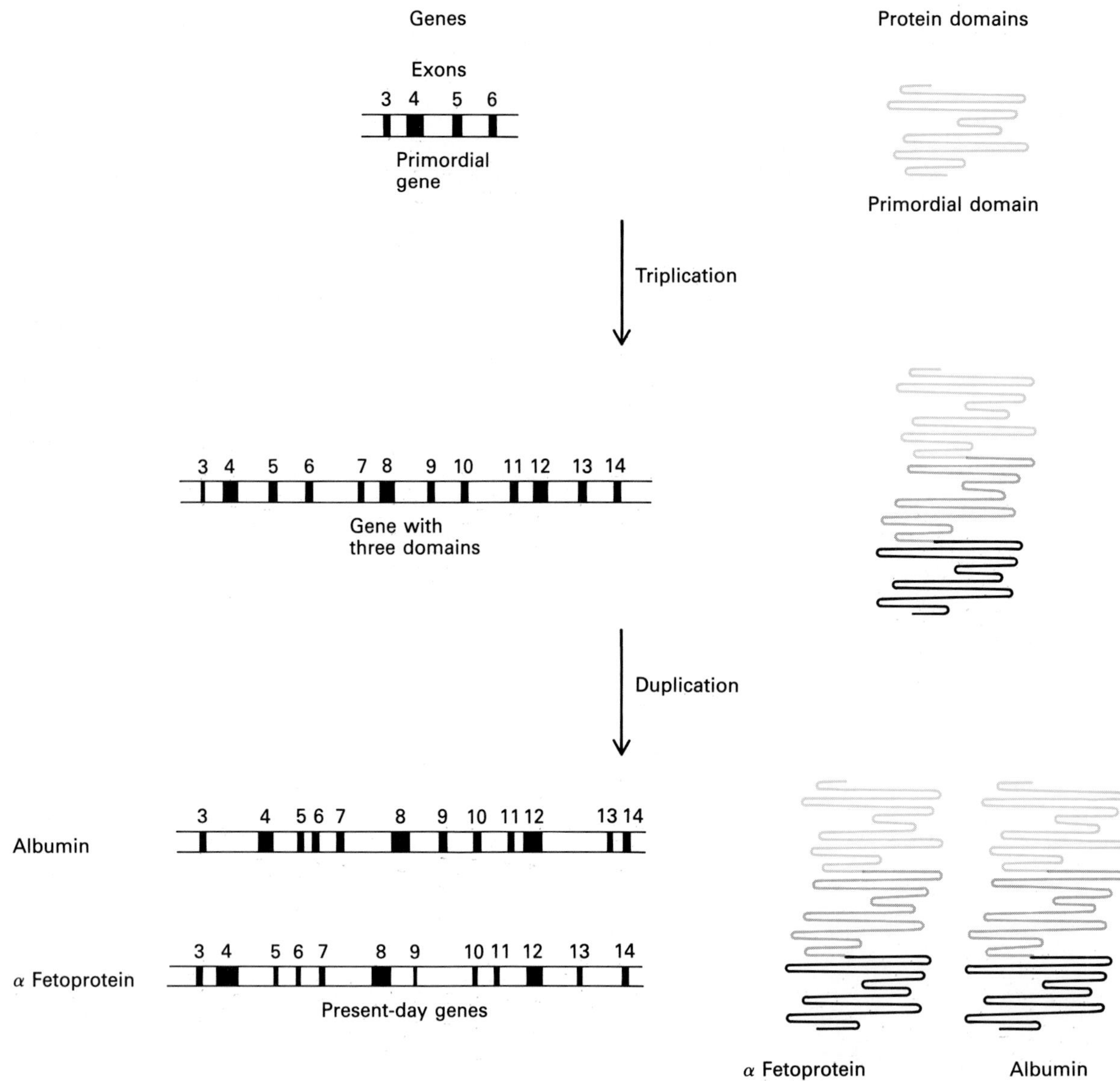

Figure 11-12 Duplication of a primordial gene to build a present-day gene. The 12 exons in the protein-coding portions of mouse albumin and α fetoprotein have very similar structures. (These exons are labeled 3 to 14 because the first two exons do not encode protein.) Within each gene, a repeating exon pattern is clear: three groups of four exons are present, and the pattern in each group is short/long/short/short. Each group of four exons encodes one of the domains in albumin or α fetoprotein (see Figure 11-11). This repeating pattern suggests that a primordial gene composed of four exons encoded a primordial protein domain. The domain is shown here folded according to the pattern in Figure 11-11. The primordial exons for this domain apparently triplicated during evolution to give a larger protein similar to albumin and α fetoprotein. Probably, the subsequent duplication of the larger gene produced the present-day α fetoprotein and albumin genes. The actual spacing of the exons in the two present-day genes is shown at the bottom. [See M. B. Gorin et al., 1981, *J. Biol. Chem.* **256**:1954.]

Besides the pseudogenes in the globin family, there are numerous other instances of extra near-copies of protein-coding genes. These occur, for example, in the tubulin and actin families of many different organisms (see Chapters 18 and 19). In some species, analysis of mRNAs has shown that only one or a few of the duplicated genes are active. Sequencing the genomic copies of these genes to determine whether they are able to encode proteins is perhaps the easiest means of determining whether a gene duplicate is functional.

In addition to the full but nonfunctional gene copies that constitute pseudogenes, partial copies of some genes have been identified. For example, fragments from the 5′ and 3′ ends of the tubulin genes in humans are quite common.

Protein-coding transcription units are not the only DNA sequences that can give rise to useless copies; extra unused copies of RNA-coding genes have also been found. For example, pseudogenes that are similar in sequence to 5S rDNA are common, but no RNA corresponding to these regions is found in ribosomes. Many extra near-copies of the DNA encoding the U series of small nuclear RNAs (snRNAs) (U1, U2, U4, etc.; Chapter 9) have also been identified; these extra sequences do not correspond to the known sequences of the RNAs either. As will be discussed later in this chapter, these spurious, silent copies of small RNA genes may have been derived by the copying of the snRNA back into DNA (reverse transcription) followed by the reintegration of the new DNA copy. Because the reintegrated DNA is a nonfunctional gene, the sequence has drifted.

Summary: Distribution of Genes and Gene Families

So many duplicated genes have been found that the following question has been raised: Are most genes in higher organisms members of gene families that arose through duplication? The examination of cloned duplicate genes has provided much important new information.

First, in vertebrates there seem to be very great distances between functional transcription units—not only between solitary genes such as the chicken lysozyme, but even between duplicated genes. In the human β-like globin gene region, only 20 percent of the region is occupied by active transcription units. In invertebrates such as *Drosophila melanogaster,* as much as two-thirds of the genome may not encode any mRNAs. Although the absence of a known gene in a region of DNA does not prove that no gene is present, where sequence data exist stop codons are often found in all possible reading frames. The

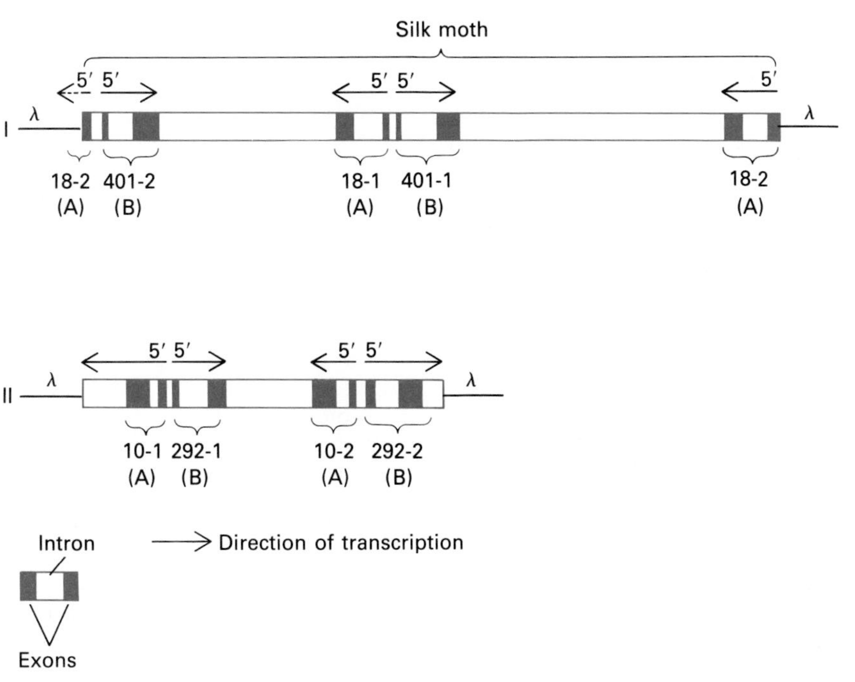

Figure 11-13 Maps of chorion genes in the silk moth. Regions of silk moth genomic DNA (*color*) that encode chorion proteins were selected from a genomic library prepared in bacteriophage λ. The chorion genes in two λ clones (I and II) are shown and named 18, 401, 10, and 292. All these genes have similar structures (e.g., one intron and two exons). Every 18 gene is paired with a 401, and every 10 gene is paired with a 292. The chorion proteins exist in three related but distinct families termed A, B, and C; the proteins encoded by the 18 and the 10 genes are in the A class, and those encoded by the 401 and the 292 genes are in the B class. Two or three examples of such gene pairs can be found within a 10- to 15-kb region of genomic DNA. The paired genes are arranged head to head; that is, their 5′ ends abut, so that the genes are transcribed in opposite directions. During egg formation, the pairs in any genomic region are expressed together. For example, the 10/292 pairs shown here are expressed midway in egg formation, and the 18/401 pairs are expressed later. [See C. W. Jones and F. Kafatos, 1980, *Cell* **22**:855.]

prevalence of apparently unused stretches of DNA or infrequently used copies of known genes in the genomes of higher animals indicates the relative sluggishness with which "extra" sequences are lost from these genomes. Unused duplicate genes, together with large regions of intragene spacer DNA, obviously account for much of the noncoding DNA that is such a prominent feature of eukaryotes.

Second, although the duplication of sequences may lead to an accumulation of unused DNA, clearly great evolutionary usefulness derives from gene duplication. Duplicated genes may continue to function although they are no longer exactly like the originals. Their continuing expression would allow the evolution of multiple similar proteins that may be used at different times during the life cycle of an organism or that may be used for multiple functions in the different tissues of an adult. For example, some of the genes in the β-like globin cluster are functional duplicates used at special times in the life cycle; other duplicate sequences are now pseudogenes. The extra copies of the chorion genes in silk moths and fruit flies are active at different times in the construction of eggs. And different actin and tubulin genes are expressed in different tissues (see Chapters 18 and 19). Gene duplication would appear to be a critical evolutionary mechanism that allows expanded and specialized gene function.

Tandemly Repeated Genes Encoding rRNA, tRNA, and Histones

The genes for 45S pre-rRNA, the 5S rRNA, various tRNAs, and one family of protein-coding genes, the histone genes, occur in *tandemly repeated arrays.* In a tandem array, copies of a sequence appear one after the other, in a head-to-tail fashion, down a long stretch of DNA. Only nonfunctional spacer DNA separates the individual copies. Within a tandem array of rRNA or tRNA genes, each copy is exactly—or almost exactly—like all the others. The arrays of tandemly repeated histone DNA are somewhat more complex; however, each histone gene, too, has multiple identical copies. The tandem arrangement in ribosomal genes was the first to be recognized (Figure 11-14). Although the transcribed portions of ribosomal genes are the same in a given individual, the nontranscribed spacers can vary.

Retention of Exact Copies May Be Related to a Need for Plentiful Transcripts

The High Demand for rRNA and tRNA As we have noted before, most of the cell RNA consists of tRNA and rRNA. A single gene must have a limit on the number of RNA copies it can provide during one generation, even if it is fully loaded with RNA polymerases. Therefore, multiple copies of the genes for structural RNAs may be required for adequate rates of RNA formation during cell growth. Figure 11-15 demonstrates the need for multiple copies of such genes in ribosome formation in HeLa cells. Given the cell's requirement for 5 to 10 million ribosomes in each generation, it is not surprising that human cells possess over 100 copies of 45S pre-rRNA genes, most of which must be close to maximally active for the cell to grow. In fact, the genes for all the structural RNAs of the cell—45S pre-rRNA, 5S rRNA, and at least some tRNAs—exist in multiple copies in eukaryotic cells (Table 11-5). The pre-rRNA genes are present in 100 or more copies in all species, and the 5S genes can be present as many as 20,000 times. The range of repetitions is from 10 to 100 for individual tRNA genes. The multiple copies of all these genes appear in tandem arrays.

Levels of mRNA Production from Nonrepeated Genes For the great majority of mRNAs, single genes are the rule. These single genes suffice for the synthesis of enough RNA to produce all the individual protein re-

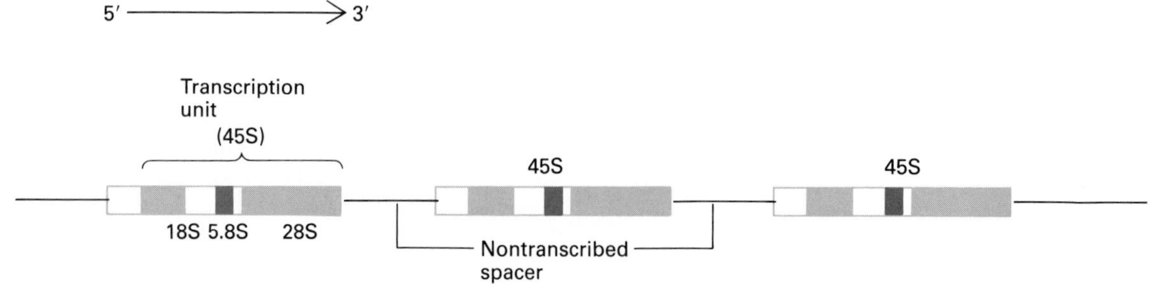

Figure 11-14 The tandem arrangement of human ribosomal genes. The transcription units (which are about 14 kb in length) encode the 45S pre-rRNAs, which after cleavage yield the 18S, 5.8S, and 28S rRNAs. Between each of the approximately 200 to 300 copies of the transcription unit, there is a nontranscribed spacer region that varies from 20 to 30 kb in length. [See P. K. Wellauer and I. B. Dawid, 1979, *J. Mol. Biol.* **128**:289.]

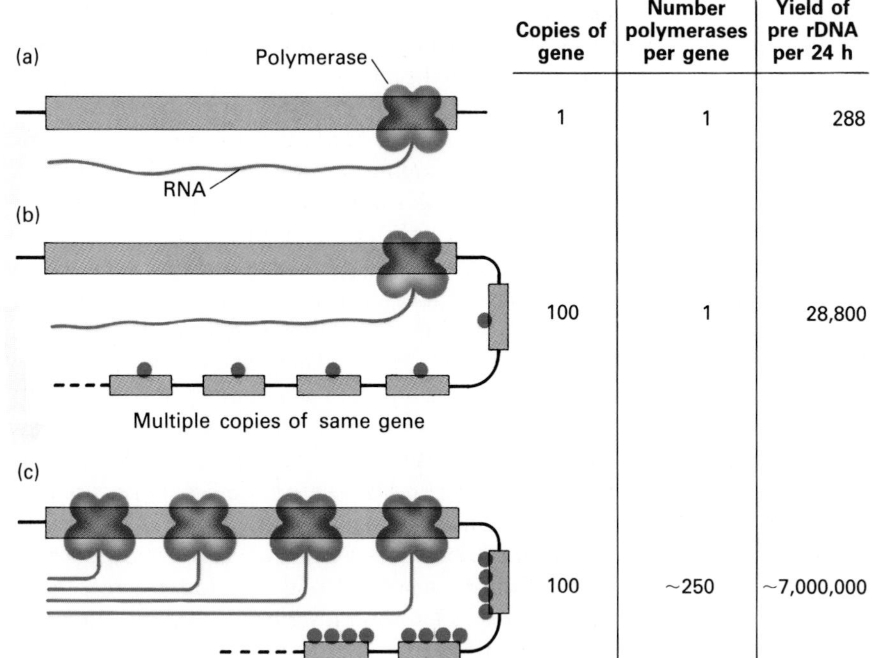

Figure 11-15 HeLa cells, which have a doubling time of 24 h, have between 5×10^6 and 1×10^7 ribosomes per cell. (a) Because RNA polymerase I takes about 5 min to make the 45S pre-rRNA primary transcript, one pre-rRNA gene with one polymerase per gene would make about 288 copies in 24 h. (b) Here 100 copies of the gene are available but each is still being transcribed by one polymerase at a time (colored dot), so the number of transcripts produced is still insufficient. (c) Again, 100 gene copies are available, but because each is being transcribed by many polymerases at any given time, ribosomes are produced in the necessary quantity. To generate the needed copies of rRNA cells must have multiple gene copies that are near maximally loaded with RNA polymerase I.

quired. For example, a single protein-coding gene about 5000 nucleotides long occupied by 20 polymerases that each make an mRNA molecule every 3 min could turn out about 10,000 mRNAs in 24 h, the doubling time of most animal cells in culture. Many mRNA molecules are not stable for long periods, but even if the instability of the mRNA were such that an average of, say, 1000 mRNA molecules (i.e., 1/10 of what is produced per cell generation) existed per cell at any given time, this number would still suffice to synthesize a large amount of a single protein. The average mRNA is associated with 5 to 10 ribosomes, and a peptide chain takes about 1 min

Table 11-5 Repeated genes for eukaryotic rRNAs and tRNA*

Species	Number of copies of pre-rRNA genes	Number of copies of 5S rRNA genes	Number of tRNA genes†
Saccharomyces cerevisiae	140	140	250
Dictyostelium discoideum	180	180	?
Tetrahymena pyriformis			
Micronucleus	1	325	800
Macronucleus	200	323	800
Drosophila melanogaster			
X chromosome	250	165	860
Y chromosome	150	165	860
Xenopus laevis	450	24,000	1150
Human	~250	2000	1300

*The gene numbers in this table were estimated by hybridizing saturating amounts of labeled RNA to DNA.

†The tRNA numbers include all tRNA sites and therefore represent more than 50 different tRNA genes in some organisms.

SOURCE: B. Lewin, 1980, *Gene Expression,* vol. 2, Wiley, p. 876.

to synthesize. Thus 1000 mRNAs could direct the synthesis of up to 1.4×10^7 protein chains per generation (1000 mRNAs $\times 10^1$ ribosomes $\times 1440$ min $= 1.4 \times 10^7$).

This amount of protein would represent about 0.5 to 1 percent of the total protein of a cell (Figure 11-16), and very few single proteins make up such a high proportion of the total. Moreover, if the mRNA had a longer half-life or if the density of polymerases transcribing the gene were higher than that in the example above, a single protein-coding gene could be responsible for even more protein. These factors almost surely account for the high productivity of the silk fibroin gene in silk moths. This single gene is responsible for one mRNA that encodes over 30 percent of the total protein of the silk gland.

Histone Genes One group of mRNA molecules is needed in especially large amounts in all eukaryotic cells. These are the mRNAs that encode the histones (H1, H2A, H2B, H3, and H4; Chapter 10), whose combined weight in the cell nucleus is equal to that of the DNA. Each of the major species of histones makes up at least 0.5 to 1 percent of the total cell protein. The histone genes are present in multiple copies (from 50 to 500) in all cells of multicellular organisms. (Yeast cells, however, have only two copies of the various histone genes, and one copy suffices for growth.)

In a number of invertebrates (e.g., sea urchins) and in some vertebrates (e.g., frogs and newts), all the members of the histone gene family are clustered together, and the entire group is repeated in tandem arrays that cover approximately 5 to 10 kb (Figure 11-17). The lengths of the mRNAs from genes of a given type vary somewhat from species to species, but the variation in length exists mainly in the untranslated regions.

Perhaps the most striking aspect of histone gene structure within a given species is the constancy of the base sequences in the coding portions of the tandem arrays. Just as with the ribosomal genes, the histone DNA regions copied into mRNAs of a particular type are very nearly identical in sequence. Variations in the spacer sequences between genes or gene clusters are found in individual animals. Even though the sequences of histone proteins are highly conserved among different species, the mRNA-coding regions do vary among species; differences occur most often in the third bases of the codons.

The histone genes of different animals are oriented differently on their respective chromosomes. In sea urchins, fruit flies, and newts, the histone genes are all clustered in tandem arrays; however, in the sea urchin, transcription of all five histone genes is in one direction, whereas in the fruit fly, three genes are transcribed in one direction and two are transcribed in the opposite direction. In the newt, all but one of the genes are transcribed in the same direction. Genes transcribed in opposite directions have opposing coding (template) strands; the changes in transcriptional direction within some chromosomal histone-coding segments imply that each histone gene is an independent transcription unit.

The gene arrangement of histones in mammals has not yet been completely determined because the distances between the copies are greater in mammals than in invertebrates, and the repetition frequency is somewhat lower. For example, histone genes are repeated from 20 to 50

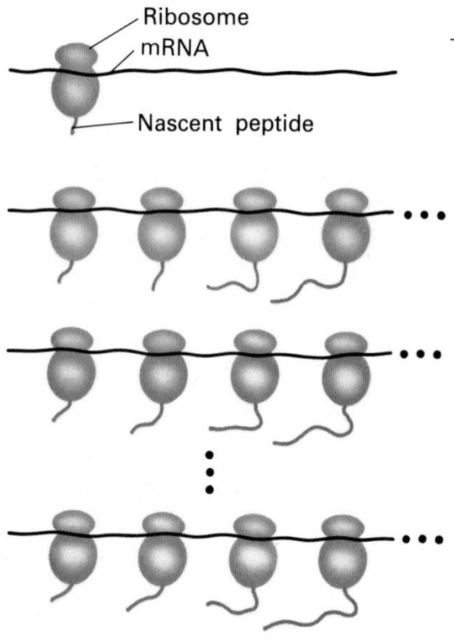

Number of mRNAs and ribosomes	Yield of single type of protein chain per 24 h	Percentage of total cell protein (given 2.8×10^9 protein chains per cell)
1 mRNA 1 ribosome	1.4×10^3	0.00005
1 mRNA 10 ribosomes	1.4×10^4	0.0005
1000 mRNAs 10 ribosomes each	1.4×10^7	0.5

Figure 11-16 The relation between number of mRNA copies, extent of ribosomal loading of mRNAs, and protein output. HeLa cells contain from 2×10^9 to 4×10^9 polypeptide chains (2.8 is used to simplify the calculation). For a specific protein to be synthesized in a concentration that is 0.5 percent of the total protein (the amount of actin in many cultured cells), there must be about 1000 copies of the specific mRNA, each filled with 10 ribosomes.

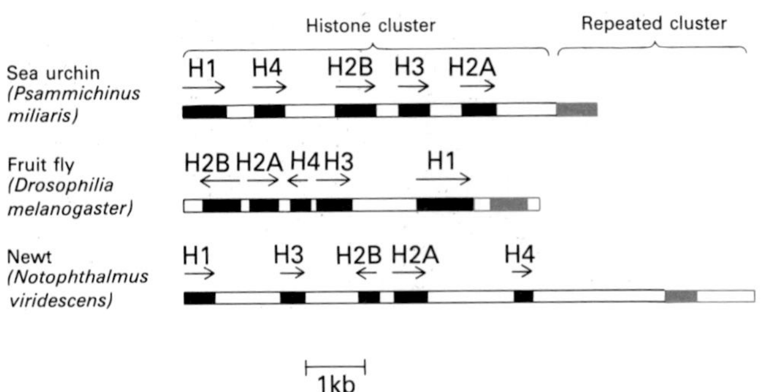

Figure 11-17 Histone gene maps for three animals. The five histone genes encoding H1, H2A, H2, H3, and H4 occur grouped together within 5 to 6 kb in each of the three genomes represented here. In other vertebrates, such as birds and mammals, the genes are not clustered within such short repeat regions. The direction of transcription is shown by the arrows. [See C. C. Hentschel and M. L. Birnstiel, 1981, *Cell* **25**:301.]

times in human cells, and no two histone genes are any closer than about 10 kb. Still, the distant copies of the same histone (e.g., H3) are identical, or nearly so.

Spacer Length Between Tandem Genes Varies

The base sequences of individual copies of the genes encoding the 45S pre-rRNA, the 5S rRNA, or single histone mRNAs are exactly (or very nearly exactly) alike, both within any individual organism and, to a considerable extent, from one individual to another within a species. In this respect, the rRNA and histone gene clusters are strikingly different from the families of near-duplicate protein-coding genes such as mammalian globin genes and insect chorion genes. However, the nontranscribed spacer sequences between individual members of tandem gene families can vary. Between some genes—for example, between the 5S tandem repeats—the spacer sequence is composed of a series of repeated (or nearly repeated) oligonucleotides about 15 bases long. The number of oligonucleotide repeats is not constant, leading to variation in the length of the spacer DNA between 5S genes (Figure 11-18). Between species, the spacer sequences have completely diverged (e.g., between *Xenopus laevis* and *Xenopus borealis,* two frog species), whereas the DNA portions encoding the 5S rRNAs and the pre-rRNAs remain quite similar.

Two Mechanisms Can Maintain Identical Gene Copies

The apparent maintenance of tandem arrays of repeated sequences without the accumulation of frequent mutations has presented an interesting molecular puzzle. Population geneticists formerly insisted that some special mechanism must exist to keep a whole set of genes free of mutations in a population of organisms. This notion spawned speculation about a "master-slave" relationship among the gene copies within each organism. According to one hypothesis, the repeated genes would be regenerated in each gamete through amplification of one copy of the repeated gene family. This mechanism would indeed ensure uniformity in the copies of the genes subjected to it. The simplest model of such a master-slave mechanism was ruled out, however, when the lengths of the spacer sequences between individual 5S genes were shown to vary within a single individual. This individual variation implied that the 5S gene copies could not arise by the copying of a single 5S gene and its spacers in each gamete.

The maintenance of sequence identity in tandem genes within a species is now believed to occur through one of two mechanisms: (1) frequent unequal meiotic crossovers or (2) gene conversion, which "corrects" one gene sequence against another.

Sequence Maintenance by Unequal Crossovers Let us consider first the effects of frequent *unequal* crossovers on the rRNA genes. The assumption is that rRNA molecules are highly conserved because their structures have gradually been perfected, and that no change is easily tolerated by the protein-synthesizing apparatus. An individual with a large proportion of mutant rRNA would thus be unlikely to survive. One model proposes that during meiosis, regions of the long tandem array of rRNA genes frequently undergo misaligned pairing between sister chromatids in such a way that unequal crossovers can occur (Figure 11-19). If unequal crossovers were frequent, then any one of the original gene copies could be raised to a higher frequency. A mutant copy of a tandem gene could, like any other copy, increase or decrease in number by this mechanism. If mutant copies were to increase, the organism containing them would be placed at a selective disadvantage, and the mutant copies would be lost from the population. However, individuals with extra copies of a nonmutant gene would survive. The chromosome would thereby be constantly "purified" of deleterious mutations as successive generations passed.

Because spacer DNA (at least in the 5S ribosomal genes) is made up of short, repeated oligonucleotides, and because the lengths of spacers vary, they could very well be the actual sites at which the proposed frequent unequal crossovers occur.

Recombinant DNA studies with yeast rDNA have shown that frequent unequal crossovers do, in fact, occur within the rRNA genes. To demonstrate this, two haploid

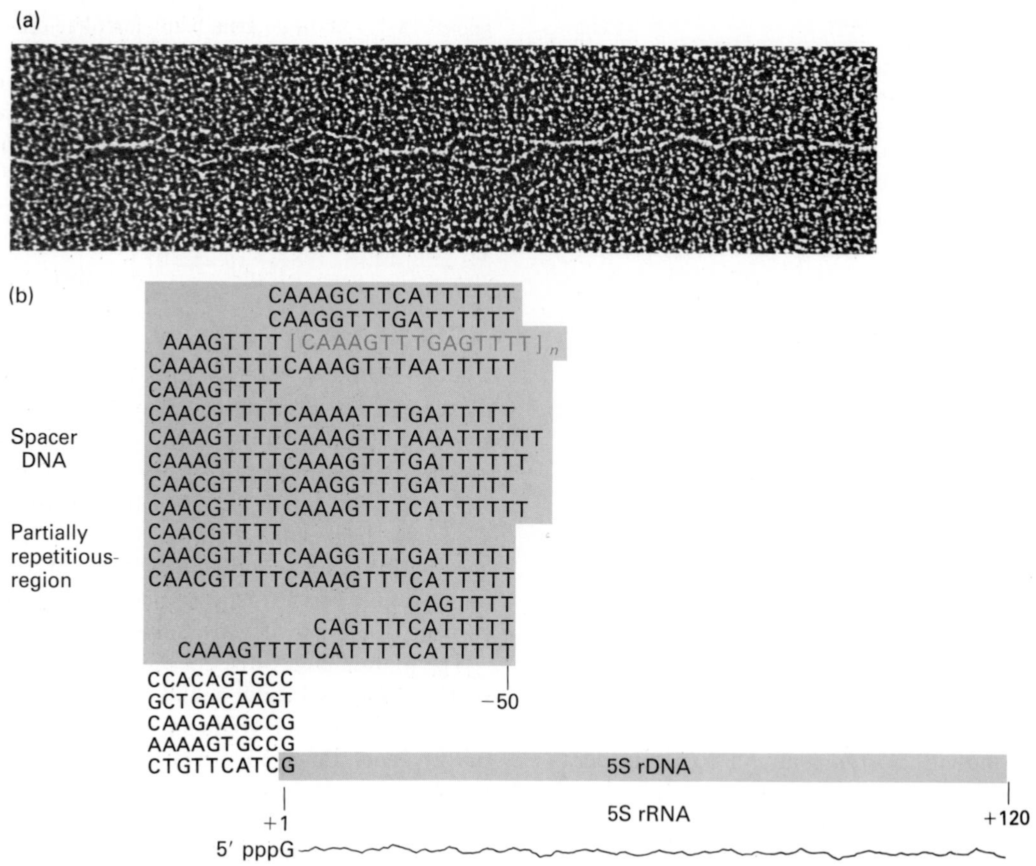

Figure 11-18 The tandem array of 5S rRNA genes in the frog *Xenopus laevis.* (a) Partially denatured purified 5S rDNA presents loops in the spacer regions; these serve as structural markers when the molecules are examined in the electron microscope. The loops are formed because the spacer is rich in A and T residues, which cause it to melt more easily. The spacer loops vary in length, whereas the nondenatured 5S coding regions are all the same size. (b) The spacer sequence before the 5S gene of *X. laevis.* The sequence to the 5′ side of the G that begins the transcribed portion of the rDNA is shown. About 50 bases upstream, the end of the partially repetitious AT-rich region can be found. This sequence is shown in a stacked arrangement (the last nucleotide in each row is actually adjacent to the first nucleotide in the row below) to emphasize the repetitious portions. About 200 bases upstream of the coding portion of the gene is a sequence (*color*) that appears a variable number of times (n = 2 to 12) in each spacer. This short variably repeated sequence is responsible for all of the length variation seen in the spacers of this gene cluster. [See N. V. Federoff, 1979, *Cell* 16:697.] *Part (a) from D. Brown, P. C. Wensink, and E. Jordan, 1971,* Proc. Nat'l Acad. Sci. USA *68:3175; courtesy of D. D. Brown.*

strains mutant in *leu2* (a gene for the production of one of the enzymes needed in leucine synthesis) were transformed by insertion of cloned rDNA containing the cloned *leu2* gene. This recombinant DNA within the yeast cell recombined into the cluster of rRNA genes and served to mark the rDNA. The frequency of unequal crossovers within the rDNA region could be followed by the retention of the *leu2* gene. The two haploid strains were mated to produce a diploid strain, which was then sporulated to form four haploid spores from each cell. If no crossover had occurred, each spore should have retained the *leu2* marker; however, a small percentage of the spores had lost the gene (Figure 11-20).

The physical basis for this loss was determined in several cases. Each analysis showed that meiotic recombination between the rDNA of sister chromatids had caused the loss of the leucine marker (Figure 11-21). Such studies have supported the possibility that frequent unequal crossovers play a role in maintaining the identity of tandemly repeated sequences.

Sequence Maintenance by Gene Conversion The original description of gene conversion stems from genetic experiments with yeast cells in which meiotic recombination between homologous chromosomes was under study. Reciprocal exchange during chromosomal

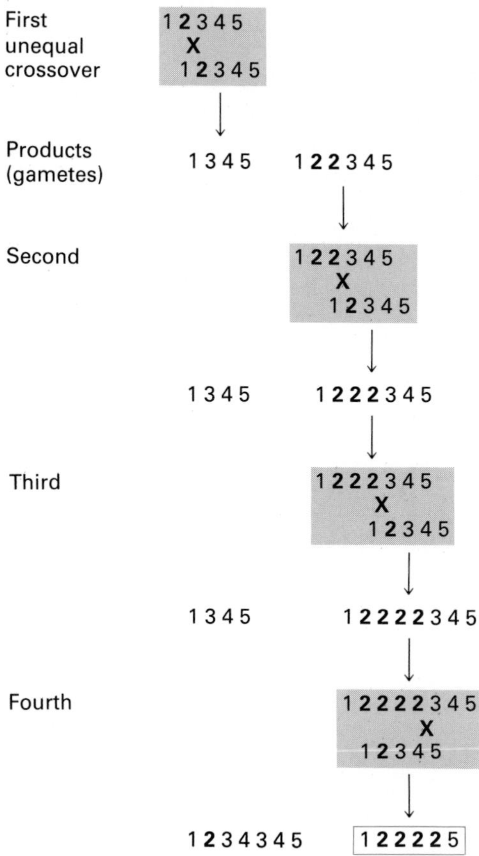

Figure 11-19 The results of unequal crossovers in repetitious DNA regions. 1, 2, 3, 4, and 5 are repeats of the same unit of, say, rDNA. The diagram follows the result of a series of unequal crossovers between sister chromatids during meiosis, leading to the cell (gamete) with the 1-2-2-2-2-3-4-5 array at the end of the third crossover. At the fourth crossover, the descendant loses repeats 3 and 4, but retains the four copies of repeat 2 from the original array.

recombination leads to the equal transmission of all alleles to the progeny; however, often during recombination genetic markers that lay *between* reciprocally exchanged alleles were not recovered in progeny spores in equal amounts, but rather in 3:1 ratios. This unequal inheritance of alleles has been attributed to *gene conversion* within the recombinant region. A model for the molecular events of a gene conversion associated with recombination is presented in Figure 13-42. For our purposes in this section, we wish to consider gene conversion in a less strict sense, that is, as a set of events that do not necessarily depend on meiotic recombination but that result in erasing (or decreasing) mutations in tandem gene arrays.

A diagram of a simple form of gene conversion appears in Figure 11-22. A nick (a single-strand cut) in a DNA duplex allows the single-stranded region near one of the cut ends to hybridize to a homologous region in another duplex. This hybridization displaces part of the complementary strand in the second duplex. After the first duplex is filled in and the displaced strand in the recipient duplex is excised, three DNA strands have the same genetic information. Replication of these duplexes produces three copies of one allele of each gene at the site of strand invasion, and one copy of the other allele.

Whether this simple molecular mechanism actually occurs is not known, but such localized invasions would explain why sequence identity is maintained in some regions of a tandem array of genes better than in other regions of the same array. This mechanism is true, for example, of ribosomal gene and histone gene arrays, where the sequences of the transcription units are maintained more consistently than the flanking sequences between the transcription units.

Gene conversions that are more frequent than mutations would tend to erase (or decrease) the mutated sites in tandem gene arrays. When a mutation occurs in a tandem array of identical genes, there are many "correct" repeated sequences and only one mutant one. Therefore, the hypothetical nick starting a gene conversion would occur most often in a wild-type region. Most gene conversions would thus tend to remove mutations, and DNA replication would result in progeny with fewer mutant genes. (In the example given in Figure 11-22, suppose that site I is "correct" and site II is mutant; the result of conversion is to produce three wild-type progeny DNAs and one mutant progeny molecule.) Continuous repetition of this process would lead to the suppression of mutations in favor of the numerically greater number of wild-type genes. Many events that are best explained by gene conversion of the type outlined here are known to occur in a number of different organisms.

Repetitious DNA

Besides the tandemly repeated genes (like rRNA, tRNA, and histone genes) and the families of near-duplicate genes (such as the globin genes), eukaryotic cells contain other regions of DNA that are generally referred to as repeated sequences or as *repetitious DNA*. These regions represent several different types of sequences. Some consist of a repetitious oligonucleotide (i.e., a sequence of 5 to 10 nucleotides repeated in tandem many times); some have a repetitious unit of 150 to 300 nucleotides; and in some, the repeated unit is as long as 5000 to 6000 nucleotides. For the latter two types of repetitious sequences, a region may contain no two units that are *exact* repeats. The existence of repetitious DNA regions was first recognized when some of the denatured DNA of mammals was observed to renature more rapidly than the bulk of the cell DNA. We shall briefly review these reassociation experiments.

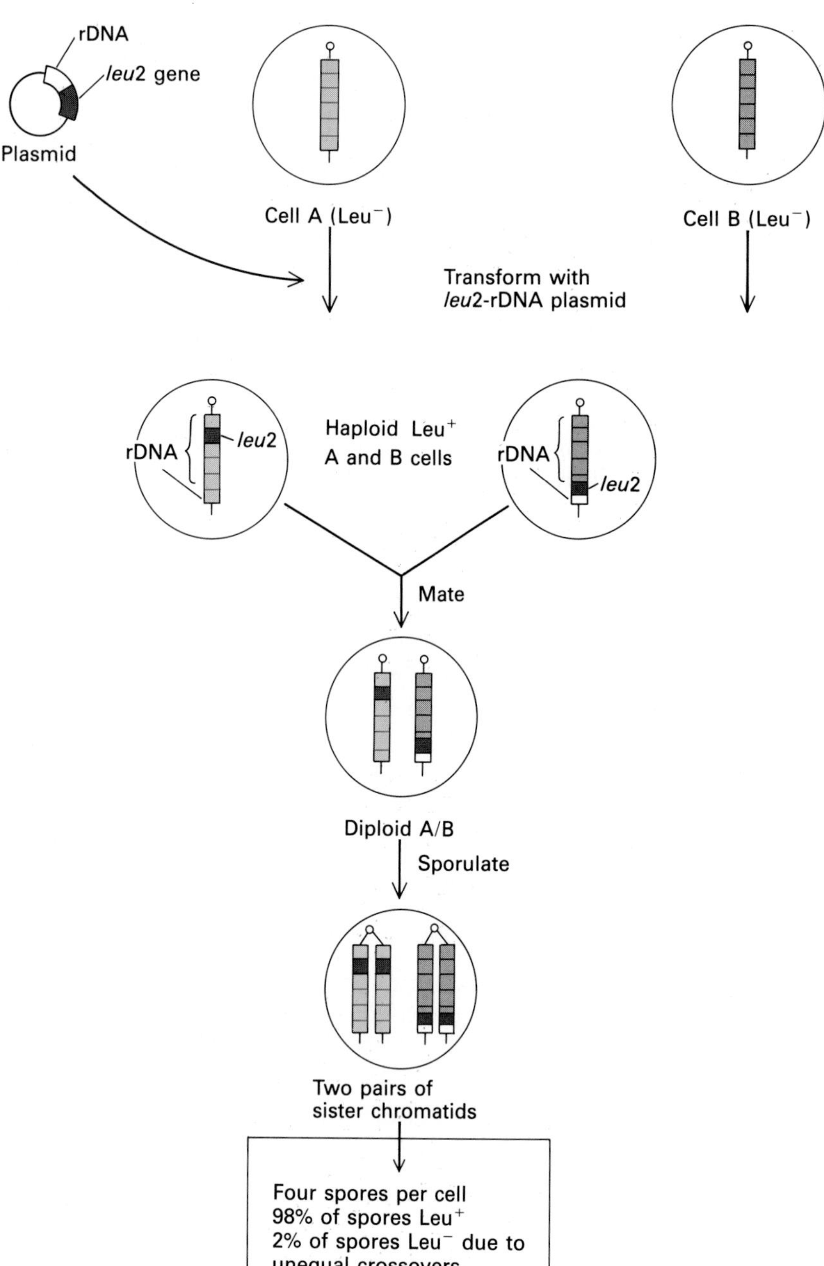

Figure 11-20 Marking yeast DNA to detect unequal crossovers. Two haploid Leu$^-$ yeast strains (A and B) were chosen; the strains could be distinguished from each other because the restriction sites in the spacers in their rDNA differed. Otherwise, the strains were identical, and neither could grow in a medium without leucine. (They lacked the *leu2* gene, which encodes an enzyme needed in the biosynthesis of leucine.) The strains were transformed with recombinant DNA that contained the *leu2* gene linked to a short stretch of rDNA. Each strain acquired the *leu2* gene by recombination at the rDNA locus. The two haploid strains were mated to produce diploids with identical chromosomes except at the rRNA genes. These diploids were then caused to sporulate (i.e., each chromosome was copied to become a pair of sister chromatids). During meiosis, four haploid spores were formed from each diploid cell. If no crossover had occurred during meiosis, each spore would have contained the *leu2* gene still embedded in the rRNA gene cluster. However, about 2 percent of the spores had no *leu2* gene and could not produce colonies on a leucine-free medium. Thus unequal crossovers occur quite frequently in the rDNA region.

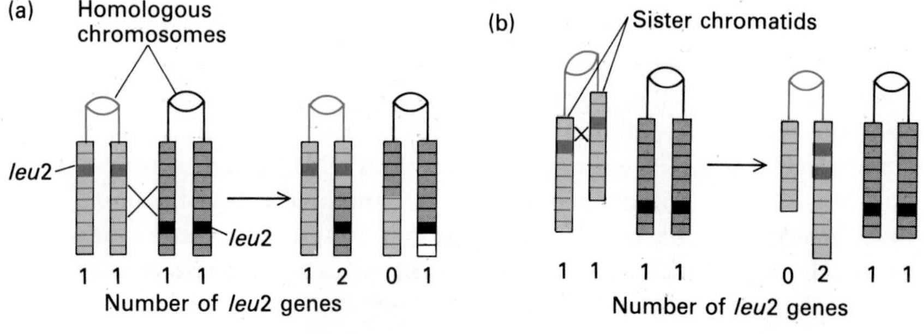

Figure 11-21 Two mechanisms that might cause the loss of *leu2* in a tandem array of rDNA are diagrammed: (a) unequal crossover between homologous chromosomes, and (b) unequal crossover between sister chromatids. Analysis of six clones that had lost *leu2* showed that the second mechanism (b) had occurred in all six cases. [See T. D. Petes, 1980, *Cell* **19**:765.]

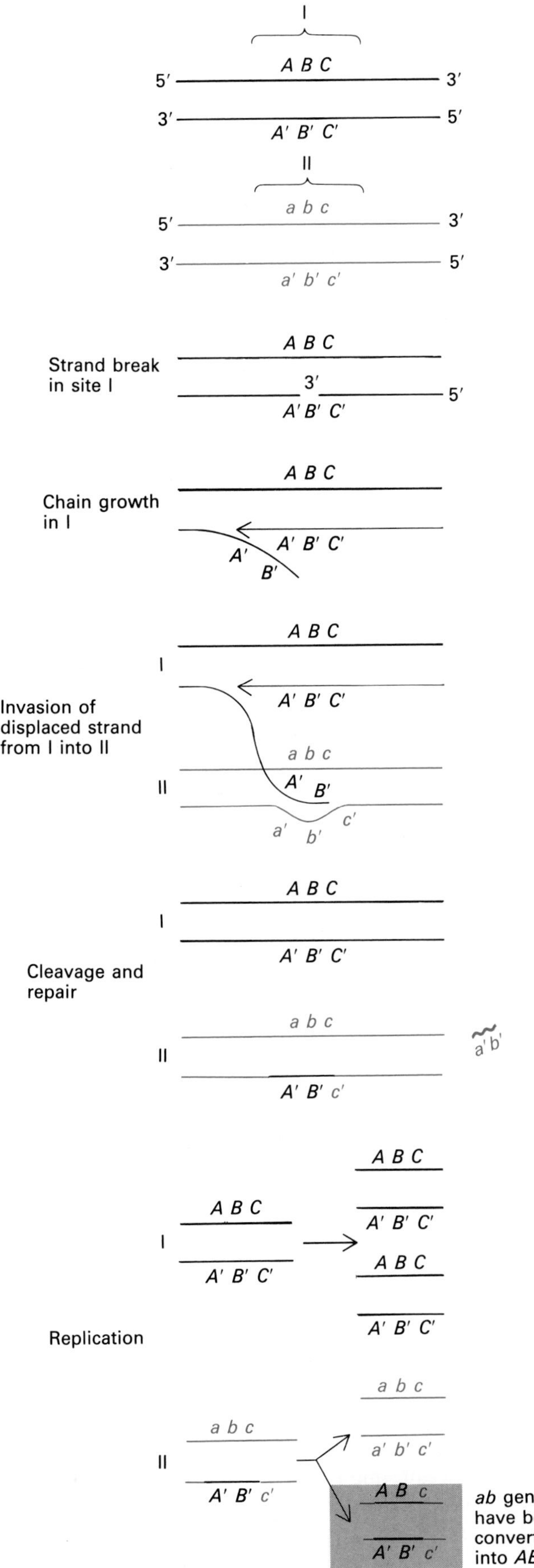

Figure 11-22 A diagram illustrating gene conversion between two homologous DNA sites, I and II. (*A* and *a* are different alleles of the same gene, as are *B* and *b*, and *C* and *c*. Primes designate the corresponding sequences on the complementary strands.) The two genetically distinct sites can be on the same chromosome (e.g., within a tandem repeat) or on different chromosomes.

Reassociation Experiments Reveal Repetitious DNA Fractions

Suppose that the total DNA of an organism is broken into fragments with an average length of about 1000 bases. The DNA is then dissociated into single chains and placed under conditions that allow strand reassociation to occur (e.g., a favorable ionic concentration and a favorable temperature; Chapter 7). All the DNA segments would re-form duplexes at about the same speed if none contained sequences that were repeated in the genome. A strand containing a repeated sequence would find a complementary partner more quickly and thus reassociate faster than a strand without repeated sequences. For example, the DNA encoding 45S pre-rRNA and that encoding 5S rRNA reassociate faster than the average DNA does.

The parameters that affect the degree to which a sequence reassociates are its initial concentration (C_0) and the time (t) allowed for the reaction. The C_0t of a reaction is the product of the molar concentration of the DNA and the reaction time in seconds. A term convenient for comparing the reassociation rates of different DNA fractions is the $C_0t_{1/2}$ value—the C_0t that allows one-half of a given fraction to renature. The lower the $C_0t_{1/2}$, the higher the reassociation rate. The $C_0t_{1/2}$ value of any particular DNA fraction can be compared with the $C_0t_{1/2}$ value of a "standard" nucleic acid, such as a homopolymer or a viral or bacterial DNA of known length (both of which are assumed to have no repeated DNA). Such a comparison allows an estimation of the approximate frequency of repeats within the fraction of interest.

When a reassociation experiment is performed with the total DNA of a mammalian genome, about 10 to 15 percent of the DNA reassociates almost immediately (Figure 11-23), and has a $C_0t_{1/2}$ value of 0.01 or less. The rapidly reassociating fraction ($C_0t_{1/2} \leq 0.01$) has proved to be the simplest DNA to analyze. (This portion was easily purified away from the majority of the cell DNA even beore reassociation experiments or DNA cloning were in use.) The rapidly reassociating fragments are largely composed of several different sets of repeated short oligonucleotides arranged in long tandem arrays; this fraction is referred to as *simple sequence DNA*.

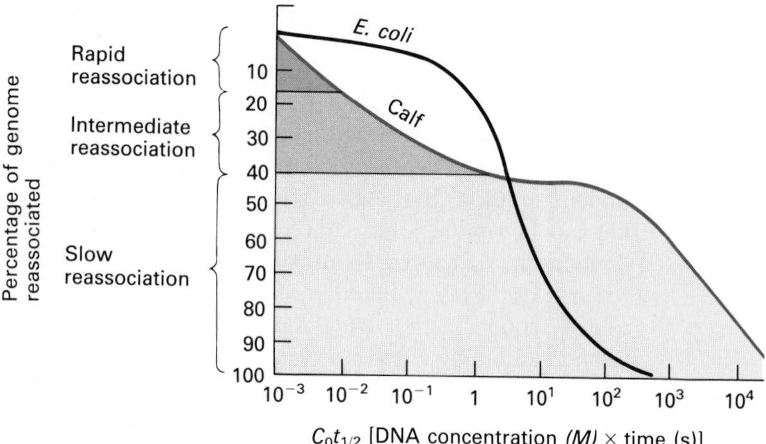

Figure 11-23 The reassociation curve of calf DNA compared with that of *E. coli* DNA. Samples of DNA from the calf thymus and from *E. coli* cells were purified, broken into pieces of about 1000 base pairs, denatured into single strands, and reassociated for various periods of time and at various DNA concentrations. The reassociated fraction was measured by a chromatographic technique and plotted against C_0t (molar concentration × time in seconds). The most rapidly reassociating calf DNA renatures almost immediately. Two other broad reassociation fractions are commonly noted: a fraction that reassociates at an intermediate rate and a slowly reassociating fraction. The *E. coli* DNA has a much narrower range of $C_0t_{1/2}$ values, which indicates that the *E. coli* fragments all tend to reassociate at about the same speed. The calf DNA that reassociates at intermediate and rapid rates is repetitious DNA. The slowly reassociating fraction of calf DNA renatures about 500 to 1000 times slower than *E. coli* DNA and is thought to represent DNA present only once per diploid calf genome. [See R. J. Britten and D. E. Kohne, 1968, *Science* **161**:529.]

Another 25 to 40 percent of the DNA reassociates at an intermediate rate, and over a broad range of $C_0t_{1/2}$ values from 0.01 to 10 (Figure 11-23). This *intermediate repeat DNA* has been more difficult to analyze. On the basis of reassociation measurements, it was once thought that in most invertebrates and in all vertebrates there were hundreds of distinct families of different kinds of repeated sequences of 100 to 1000 nucleotides in length. However, sequence analyses of cloned examples of intermediate repeat DNA from many different animals (e.g., fruit flies, frogs, rodents, and humans) has failed to distinguish such large numbers of distinct families of sequences. In fact, only one major class of frequent intermediate repeats and several classes of less frequent intermediate repeats are found in mammalian genomes.

A great deal is known about intermediate repeats in fruit fly cells and yeast cells. But because during reassociation these sequences do not appear to be repeated as often as the repeated elements in mammals, their possible similarity to the mammalian elements was only recently recognized. In fruit flies and yeasts these repeated elements have been termed *transposition elements* or *mobile DNA,* because they are found in different places in chromosomes of different strains of the same species. Transposition elements are described more fully under the next major heading. It now seems likely that most, if not all, of the intermediate repeat DNA in higher animals is a form of mobile DNA.

The remaining 50 to 60 percent of the mammalian DNA sample reassociates at a much slower rate and has $C_0t_{1/2}$ values of 100 to 10,000 (Figure 11-23). The most slowly reassociating sequences in mammalian cells reanneal about 500 times more slowly than *E. coli* sequences (Figure 11-23). Because the haploid amount of DNA in mammalian cells is about 700 times the amount in *E. coli,* and because almost all *E. coli* DNA is thought to be present in a single copy only, the most slowly reassociating fraction of mammalian DNA is assumed to be *single-copy DNA*. According to mendelian genetics, only one copy of each gene is contained in the haploid DNA set; thus the single-copy DNA fraction is expected to contain most of the genes encoding mRNA. (The reverse, of course, is not necessarily true: every single-copy DNA sequence does not necessarily perform a genetic function.)

Simple Sequence (Satellite) DNA Is the Most Rapidly Reassociating DNA

As we have mentioned, the most rapidly reassociating DNA is *simple sequence DNA,* most of which is composed of short (5- to 10-base) oligonucleotides that are tandemly repeated. A few instances of tandem repeats of DNA sequences of 100 to 200 nucleotides long are also known. Obviously, the repeated units of simple sequence DNA are not nearly long enough to encode rRNA or tRNA, and any mRNA synthesized from them would make a highly repetitious protein that has never been found.

Most examples of simple sequence DNA occur in very long stretches (of at least 10^5 base pairs). This type of DNA was originally termed "satellite" DNA because some varieties of simple sequence DNA separate from the majority of the DNA during equilibrium centrifugation in CsCl. This separation as a "satellite band" occurs because DNAs of different average composition bind different amounts of Cs^+ ions and thus have a different buoyant density (Figure 11-24). All simple sequence DNA does not have a different average base composition and does not separate from the main-band DNA to form satellites. Therefore the term simple sequence DNA is preferred to satellite DNA.

Each animal or plant cell has a number of different types of simple sequence DNAs based on different repeating oligonucleotides (Table 11-6). In some cases the repeating oligomeric units of simple sequence DNA have very similar sequences, which suggests an ancestral relationship between two or more different simple sequence types. In *Drosophila virilis* cells, for example, the seven-base repeats constituting three of the simple sequence types are related to one another by a single base change (see Figure 11-24).

In humans, at least 10 varieties of simple sequence DNA exist, many of which are short oligonucleotide tandem repeats of the type already discussed. A single type of simple sequence can account for as much as 0.5 to 1 percent of the total human DNA, an amount equivalent to approximately 10^7 base pairs, or three times the total genome size of *E. coli.* Exactly how much of the simple sequence DNA exists at any one chromosomal location is not known. However, a large part of the simple sequence DNA appears to be located near centromeres. This location was discovered in mice by hybridizing fixed, stained chromosomes to labeled simple sequence DNA (Figure 11-25). The use of this technique marked the first time chemically identified DNA was localized by in situ hybridization.

The locations of several specific simple sequence DNAs within distinct chromosome regions have been established in *D. melanogaster.* As in mammalian chromosomes, *Drosophila* simple sequence DNAs are concentrated in the centromere region. However, some of the simple sequence DNA is located within chromosome arms and appears to constitute some telomeres (chromosome tips) (Figure 11-26; see also Table 11-6). None of the individual simple sequence DNAs is confined to a single chromosome. In *D. melanogaster,* the total amount of heterochromatin, as measured by photometric scanning of the DNA in stained chromosomes, approximately equals the amount of isolated, chemically characterized simple sequence DNA. Thus all—or certainly most—of the heterochromatin regions must consist of simple sequence DNA.

As we have mentioned, the simple sequence DNAs are generally not transcribed into RNA. Their function in chromosomes is unknown, but some workers have tended to assign these sequences a structural or organizational role, because they are located mainly in the centromeres and telomeres and do not encode proteins or RNAs. If the simple sequences serve as binding sites for chromosomal proteins, then proteins that attach to specific simple sequence DNAs might be expected to exist. However, no such proteins have yet been isolated. Furthermore, the centromere sequences that probably bind the single microtubule of spindle proteins in yeast DNA are only 100 to 200 nucleotides long. The several dozen microtubules in a spindle of a metazoan cell would certainly not require the huge stretches of simple sequence DNA that are present in centromeres of higher cells. It seems unlikely, therefore, that all the simple sequence DNA at the centromere is related to kinetochore function.

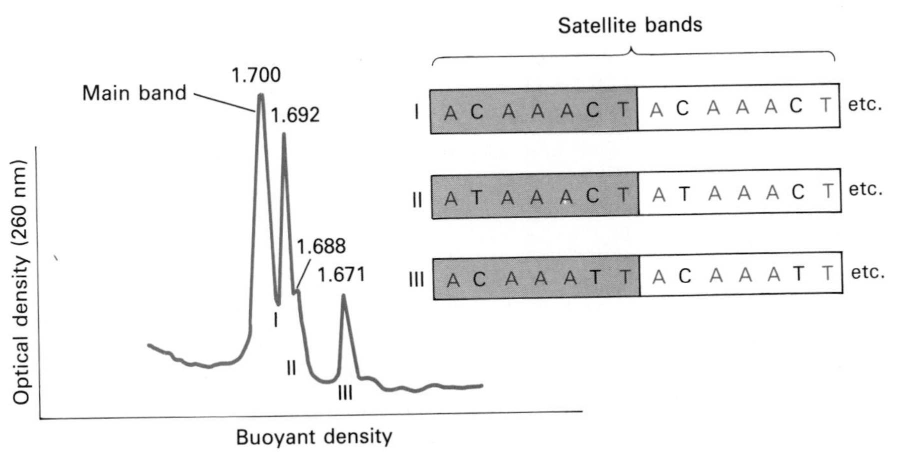

Figure 11-24 Satellite DNA in *Drosophila virilis.* The DNA from embryonic tissue was extracted and subjected to equilibrium density-gradient sedimentation in cesium chloride to separate DNAs differing in buoyant density (Chapter 7). The DNA content of different zones in the gradient was monitored by measuring the absorption of ultraviolet light at 260 nm (i.e., the optical density). The main band of the DNA—that is, the greatest part of the DNA—has a density of 1.700. The three satellite bands, I, II, and III, are less dense: DNA sequence analyses show that each of the satellite bands is composed of a DNA that is a long tandem repeat of a seven-base sequence. [See J. G. Gall, E. H. Cohen, and D. D. Atherton, 1973, *Cold Spring Harbor Symp. Quant. Biol.* 38:417.]

Table 11-6 Some simple sequence DNAs

Organism	Base pairs per repeat	Sequence of one repeat unit*	Location
Drosophila melanogaster (fruit fly)	5 7	AGAAG (polypurine) ATAAT ATATAAT	Arms of Y chromosome; centromeric heterochromatin of chromosome 2; long arm of 2, near end
	10	AATAACATAG AGAGAAGAAG	Centromeric heterochromatin of all chromosomes; tip of long arm of 2
Drosophila virilis	7	ACAAACT (band I) ATAAACT (band II) ACAAATT (band III)	Centromeric heterochromatin
Cancer borealis (marine crab)	2	AT	?
Pagurus pollicaris (hermit crab)	4 3	ATCC CTG	? ?
Cavia poriella (guinea pig)	6	CCCTAA	Centromeric heterochromatin
Dipodomys ordii (kangaroo rat)	10	ACACAGCGGG	Centromeric heterochromatin
Cercopithecus aethiops (African green monkey; α sequences)*	172		Throughout chromosomes
Homo sapiens (human; alphoid sequences)*	171		Throughout chromosomes

*The eukaryotes listed have simple sequences other than the ones shown, but most repeats are between 2 and 10 bases in length. Exceptions are the primate α and alphoid sequences, which are much longer. All species have more than one kind of simple sequence DNA; for example, humans have at least 10, many of which are near centromeres.

SOURCE: K. Tartoff, *Annu. Rev. Genet.* 9:355.

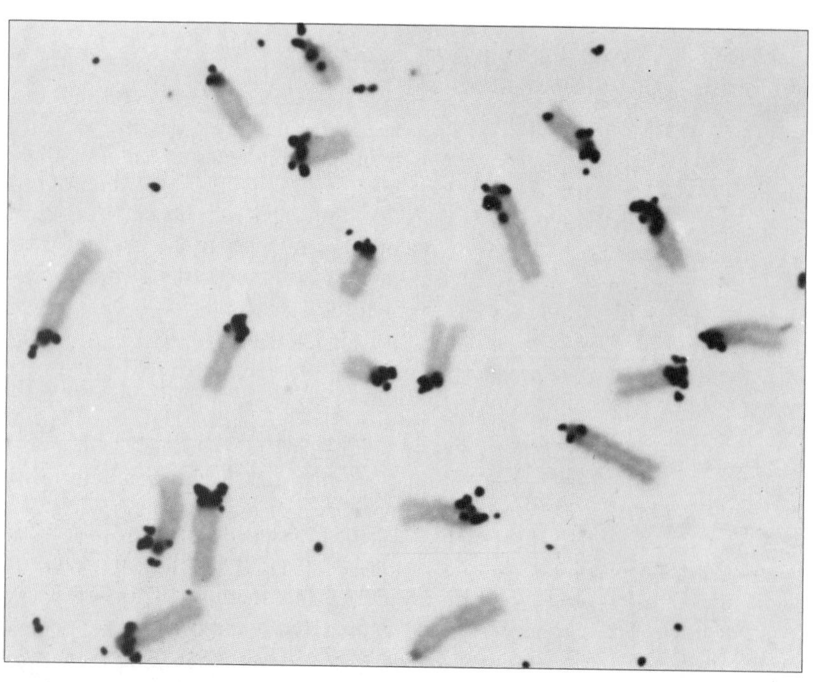

Figure 11-25 The localization of simple sequence DNA. Purified simple sequence DNA from mouse cells was randomly transcribed by *E. coli* RNA polymerase to make labeled RNA. The mouse chromosomes from cultured cells were fixed and denatured on a microscope slide, and then the chromosomal DNA was hybridized in situ to the labeled RNA. Autoradiography of the telocentric mouse chromosomes shows that most of the complementary simple sequence DNA lies close to the centromere. *From M. L. Pardue and J. G. Gall, 1970,* Science ***168:1356;*** *courtesy of J. G. Gall.*

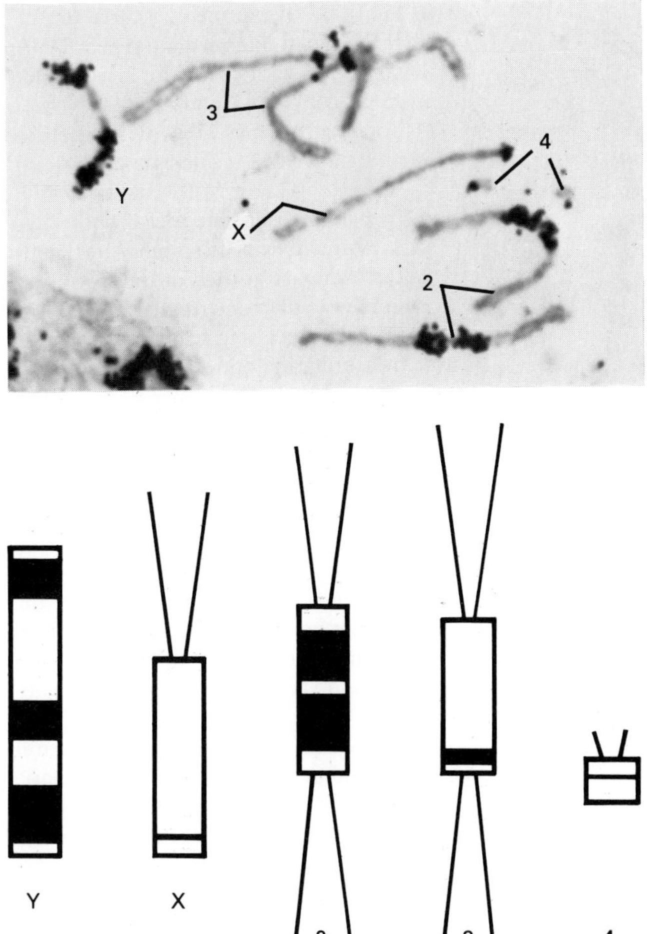

Figure 11-26 The localization of simple sequences within the heterochromatin of metaphase chromosomes of *D. melanogaster*. Labeled simple sequence DNA was hybridized in situ to fixed denatured chromosomes. The dark spots in the resulting autoradiograph show where this DNA is abundant. In the diagram below, the shaded chromosomal areas represent regions of heterochromatin (i.e., regions that are always condensed and dark-staining). Comparison of the autoradiograph to the diagram reveals that simple sequence DNA occupies a large part of the heterochromatin. *Photograph from W. J. Peacock et al., 1977,* Cold Spring Harbor Symp. Quant. Biol. *42:1121; courtesy of W. J. Peacock.*

The conservation of sequence identity in the simple sequences of any one organism is impressive. Very likely the mechanism of unequal crossovers, previously described for DNA encoding pre-rRNA and 5S rRNA, plays a key role in maintaining the long stretches of simple sequence repeats with infrequent base changes. Occasional mutations in a simple sequence would probably be "spread" by unequal crossovers, particularly if there were no strong selective pressure against such base changes. For example, a fairly recent set of unequal crossovers could account for the existence of the three different but closely related simple sequence types of *D. virilis* (see Figure 11-24).

Intermediate Repeat DNA Consists of Dispersed Blocks of Related Nonidentical Sequences

We consider now the *intermediate repeat DNA,* which reassociates faster than single-copy DNA but slower than the highly repetitious simple sequence DNA (see Figure 11-23). Intermediate repeat DNA has several distinguishing structural features: the repeated units are similar to one another but not identical. Unlike the clusters of duplicated genes and the tandem arrays, the intermediate repeats are scattered individually throughout the genome. Finally, the most common intermediate repeat sequences in most vertebrates are between 150 and 300 bases long—longer than the units of most simple sequence DNA but shorter than the sequences that encode most mRNAs.

All eukaryotes contain some intermediate repeat DNA, although the amount varies from a few percent of total yeast DNA, to about 20 percent in most invertebrate species, to about 40 percent in mammals. Although a number of examples of this type of DNA have been sequenced and much information about its structure has resulted, very little is known as yet about its functions. However, because these sequences were originally thought to be candidates for regulatory sites in DNA, they have attracted great attention.

Sequence Similarity, Not Identity, in Intermediate Repeats One prominent property of intermediate repeat sequences was recognized very early: they are not exact repeats. DNA that lacks repeated sequences (e.g., viral, bacterial, and single-copy mammalian DNA) has approximately the same melting temperature, T_m, after dissociation and reassociation as does freshly isolated, native DNA (DNA that has never been denatured). This indicates near perfect reconstitution of the duplex during reassociation. In native DNA, any intermediate repeat DNA is presumably perfectly paired like all the rest of the DNA. However, if DNA containing intermediate repeats is broken, denatured, and reassociated, the intermediate repeat fraction has no sharply defined T_m like that of native DNA or reassociated single-copy DNA (Figure 11-27). This difference is attributable to imperfect base pairing in the reassociated intermediate repeat fraction: during reassociation, these DNA strands do not find their exact partners, but instead form inexactly base-paired hybrids with related but nonidentical sequences. Addi-

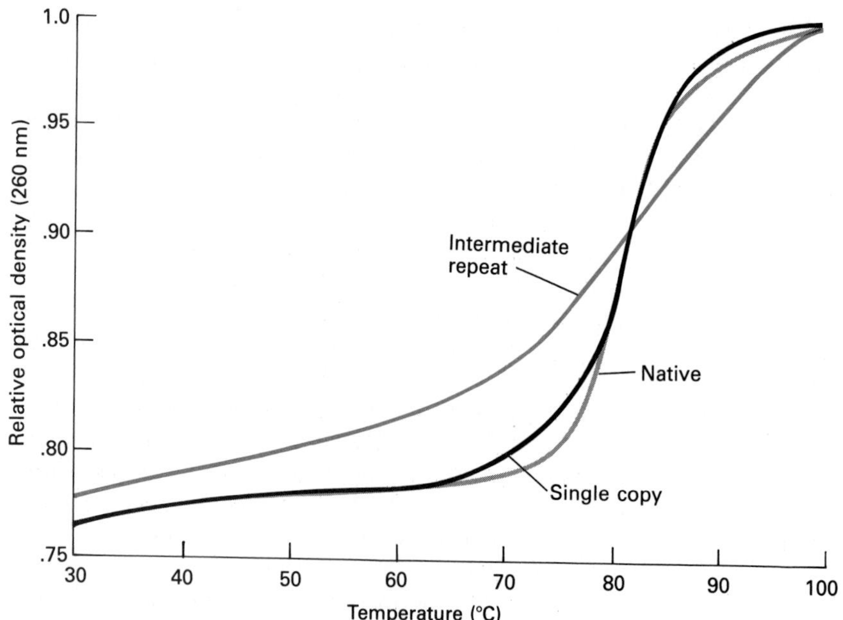

Figure 11-27 A melting curve for reassociated intermediate repeat DNA compared with curves for reassociated single-copy DNA and native DNA. Total calf thymus DNA was denatured and allowed to reassociate as described in Figure 11-23, in order to prepare intermediate repeat and single-copy fractions. The stability of the DNA duplex in each of the three fractions was then determined by exposing them to increasing temperatures and measuring the optical density at each temperature. (The higher the optical density, the more denatured the DNA. See Figure 3-55.) The reassociated intermediate repeat DNA is more dissociated at each of the lower temperatures, and at the temperature at which the other two fractions melt abruptly, the intermediate repeat fraction is still melting gradually. The sharply defined T_m (melting temperature) for the native and single-copy duplexes indicates good base pairing over the length of the duplex; the gradual melting seen in the intermediate repeat fraction signifies poor base pairing, which indicates that the repetitious fraction does not consist of exact copies. [See R. J. Britten and D. E. Kohne, 1968, *Science* 161:529.]

tional evidence that intermediate repeats are nonidentical lies in the sensitivity of reassociated intermediate repeats to enzymes that digest mismatched DNA.

The Distribution of Intermediate Repeats in the Genome A second characteristic property of the intermediate repeat DNA is its interspersion throughout long stretches of nonrepetitious DNA: in virtually all vertebrates examined, intermediate repeats are not clustered like the simple sequences but are widely distributed at different genomic sites. For example, if human DNA is broken into pieces with an average length of 5 kb and if the pieces are denatured and allowed to reassociate to an intermediate C_0t value, well over one-half of the individual DNA strands will exist, in at least a portion of their lengths, as DNA-DNA hybrids. (Hybrids are detected by their binding to material that selects double strands.) Such experiments have indicated that intermediate repeats are present at many different locations throughout the genome. Similar results are obtained with the DNA of other mammals and with that of frogs and sea urchins.

Another indication of the widespread distribution of these repetitive sequences comes from the frequency with which recombinant DNA clones contain repetitive DNA sequences. If human DNA is used to make a set of genomic clones with an average length of 20 kb, more than 90 percent of all randomly chosen clones contain an intermediate repeat sequence.

From these two simple experiments (reassociation of 5-kb pieces and cloning of 20-kb pieces), the number of human genomic sites harboring repetitious DNA has been estimated to be in the hundreds of thousands. (For example, if 3×10^9 base pairs comprise the human genome, then breaking it into fragments each of 5×10^3 base pairs would yield 6×10^5 fragments; half of this number, 3×10^5 fragments, are known to contain an intermediate repeat sequence.)

Intermediate Repeat Lengths To gain more information on the distribution and nature of intermediate repeat sequences, denatured DNA fragments have been allowed to reassociate for limited times (so that *only* repeated elements have a chance to reassociate), and the interacting molecules have then been viewed in the electron microscope. DNA that remains single-stranded can be distinguished from reassociated double-stranded DNA. In the electron micrographs, short base-paired regions (duplex DNA) are interspersed with long single-stranded regions (Figure 11-28). Most animals, including many different vertebrates and invertebrates, show this pattern of widely scattered and rather short (300-base-pair) regions of intermediate repeat DNA. When a human DNA sample before denaturation consists of fragments about 5 to 10 kb long, reassociated regions between two single strands (Figure 11-28a) occur once or twice, on the average, along each strand. Estimates of the distance between such points of reassociation range from 1 to 20 kb, with an average of about 5 kb.

In addition to reassociation between two different single strands of DNA, reassociation *within* single strands can also be observed in the electron microscope (Figure 11-28b). Intrastrand reassociation implies a repetitious inverted sequence: *abc* . . . *c'b'a'*, where *a* and *a'*, *b* and *b'*, and *c* and *c'* are complementary sequences. The unpaired region within the inverted repeat is called the *turnaround* (Figure 11-28b). Most inverted repeats (i.e., most duplex regions created by intrastrand base pairing) are about 300 bases long, like the other intermediate repeats described above. Some of the repetitious inverted sequences lie very close together. These form *hairpins* during renaturation (Figure 11-28b). The two halves of an inverted repeat can also be located at distances of up to several thousand base pairs; however, when the DNA is melted, the sequences are kept close together because they are on the same molecule and can therefore base-pair very quickly. These sequences are described as *snap-back* sequences.

Thus, from the frequency with which cloned sequences contain intermediate repeat DNA and from its distribution in electron micrographs, intermediate repeats have been established as widely dispersed repetitious elements about 300 nucleotides long. The sequencing of some intermediate repeat DNA has greatly clarified its nature.

The *Alu* Sequence Family in Mammals Segments of human, mouse, and hamster genomic DNA containing the 150- to 300-base-pair intermediate repeat sequences have been isolated, cloned, and sequenced. Within many of these repetitious sequences in human DNA lies a recognition sequence for the restriction enzyme *Alu*I. The entire family has been referred to as the *Alu* family. This term is a bit misleading because many of the regions belonging to the family lack the *Alu* site; however, the name is convenient and widely used by workers in the field.

Over 50 such regions have been sequenced in human DNA. No two are exactly alike, but almost all members are sufficiently similar that they will cross-hybridize with one another. A feature of the 300-base-pair human repeat is that the left-hand (5') 130 or so bases are quite similar to the right-hand (3') 170 or so bases. The right-hand portion has an added insert of about 30 bases and a few extra bases at the end (Figure 11-29). The total number of such repeats is estimated at 5×10^5 per genome. At 300 base pairs each, this constitutes between 5 and 10 percent of the human genome.

The human genome contains not only hundreds of thousands of copies of full-length *Alu* family sequences, but also many partial *Alu*-like sequences. Numerous blocks of 10 to 20 nucleotides clearly related to the *Alu* family have been found scattered between genes and within introns. Thus the fraction of the genome related to this particular sequence family is greater than 10 percent.

In mouse, Chinese hamster, monkey, and chicken cells, the intermediate repeat sequence is only about 150 base pairs long, and it is fairly homologous with both halves of the human *Alu* family sequence (Figure 11-29). However, even though the human and rodent sequences are similar by sequence analysis and clearly belong to the same ancestral family, they do not cross-hybridize between species, or they do so very poorly. Therefore, only by sequence comparisons can the interrelations of this widely repeated family be established.

Possible Functions of *Alu* Sequences The discovery of a high concentration of repetitive sequences in heterogeneous nuclear RNA (hnRNA) provided one of the earliest indications of the existence of repeated sequences in animal cell DNA. Their presence was originally detected in reassociation experiments when some sequences within long hnRNA molecules were observed to hybridize more rapidly than other sequences. In addition, two-dimensional oligonucleotide fingerprints of the rapidly hybridizing hnRNA segments showed a simple pattern indicative of low nucleotide complexity in the repetitious fraction. The sequences of these oligonucleotides from hnRNA were compared to *Alu* sequences when the *Alu* family was described, demonstrating clearly that *Alu*-type sequences constituted a prominent part of hnRNA. However, the repetitious *Alu*-type oligonucleotides of hnRNA are not found, or are less abundant by a factor of at least 20, in mRNA. With the discovery of intervening sequences (introns) came the logical conclusion that the repetitious part of the hnRNA must lie mainly in intervening sequences, although no function for them in hnRNA is known.

The cloning and sequencing of genomic DNA segments containing identified genes has yielded a long list of *Alu* sequences that occur within and around transcription units. For example, *Alu* sequences are found within the transcription units for mouse α fetoprotein, mouse dihydrofolate reductase, human insulin, and rat growth hormone. In almost every case, the *Alu* sequences within a transcription unit lie within introns (Figure 11-30). This finding explains the much greater content of *Alu*-like sequences in hnRNA than in mRNA. About 5 percent of mRNAs are known to contain *Alu* sequences. In these cases, the sequence lies either in the 5' or the 3' untranslated portion of the mRNA. Thus, the *Alu* sequences within transcription units all lie in areas with no assigned function. That is, no function has been recognized for *Alu* or for any other repetitious sequence in starting transcription, in locating poly A sites, or in splicing. (The repeated sequences that do serve a signaling function for protein-coding regions are short oligonucleotides. For example, the six- to eight-base TATA box helps position RNA polymerase II molecules at the start site for transcription, and the RNA hexanucleotide AAUAAA signals the location of a poly A site downstream.)

Genomic mapping has shown that many of the *Alu* sequences lie far outside the presumed boundaries of RNA

(a)

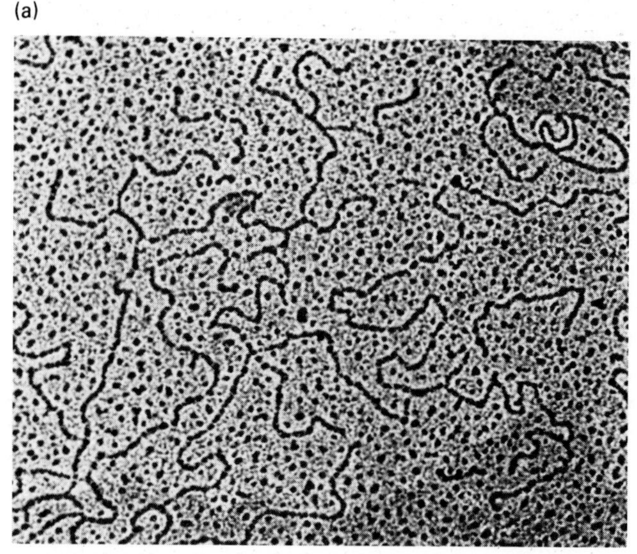

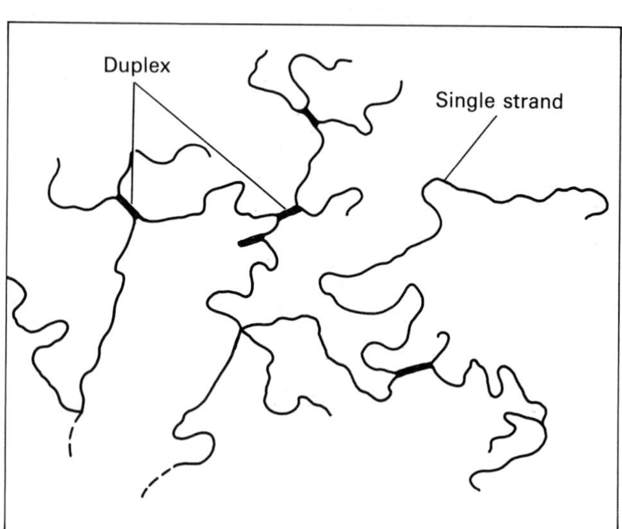

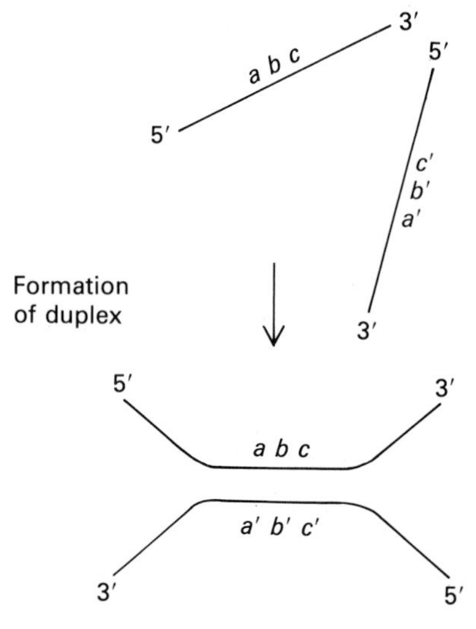

Interstrand pairing

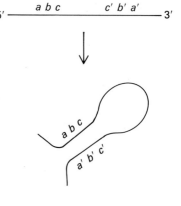

(b)

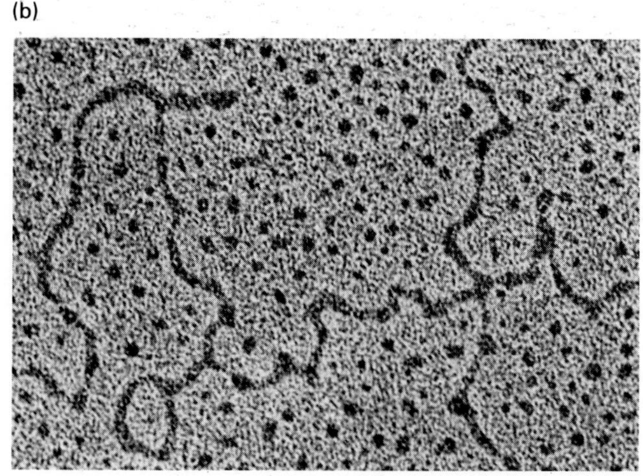

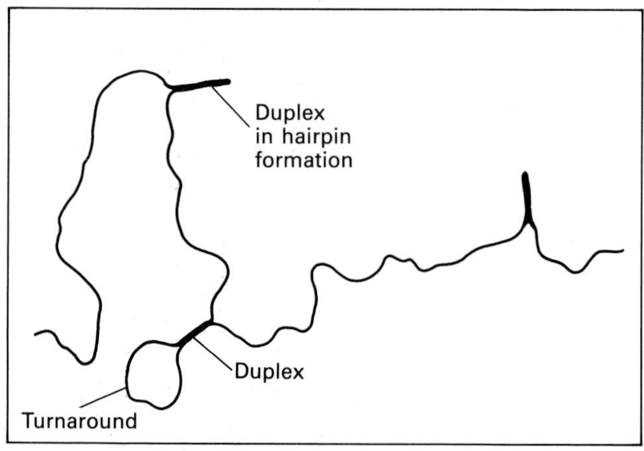

5′ a b c c′ b′ a′ 3′

a b c
a′ b′ c′

Intrastrand pairing

Figure 11-28 (*opposite*) The reassociation of repetitious sequences in human DNA. Human DNA is fragmented, dissociated, and allowed to reassociate just long enough for only the repetitious regions to form hybrids. When observed in the electron microscope, the reassociated DNA is thicker than the still denatured unpaired regions. Two hybrid structures are commonly seen: (a) pairing between two different strands and (b) intrastrand pairing. By measuring the lengths of the fragments and of the hybrid portions, the repeated sequences can be estimated to be about 300 bases long and to occur an average of once every 5000 bases. The tracings show the repetitious element as *abc* and the corresponding complementary element as *a'b'c'*. *Photographs from P. L. Deininger and C. W. Schmid, 1976,* J. Mol. Biol. ***106****:773; courtesy of C. W. Schmid.*

polymerase II transcription units. For example, at least seven or eight *Alu* sequences are located in the 65-kb section that spans the human β-like globin genes. However, comparisons among the globin genomic regions of different animals have revealed no regular positioning of *Alu* or any other repeated sequence with respect to globin transcription units.

One possible role for *Alu* sequences is in the initiation of DNA synthesis. *Alu* contains a 14-base-pair stretch that is also present in the replication origins of a number of DNA viruses. In addition, some of the scattered *Alu* sequences are transcribed as small RNA molecules that might act as primers from which DNA chain synthesis could begin. The possibility that *Alu* plays a role in the initiation of DNA synthesis will be discussed further in Chapter 13 (see also Figure 11-29).

The Mobility of *Alu* Sequences in the Genome The best suggestion about the nature of the *Alu* family—and by extension, perhaps, of all intermediate repeat DNA—has come from an analysis of the general structure of *Alu* sequences compared to other repetitious sequences that have been characterized in yeast and fruit flies. As we

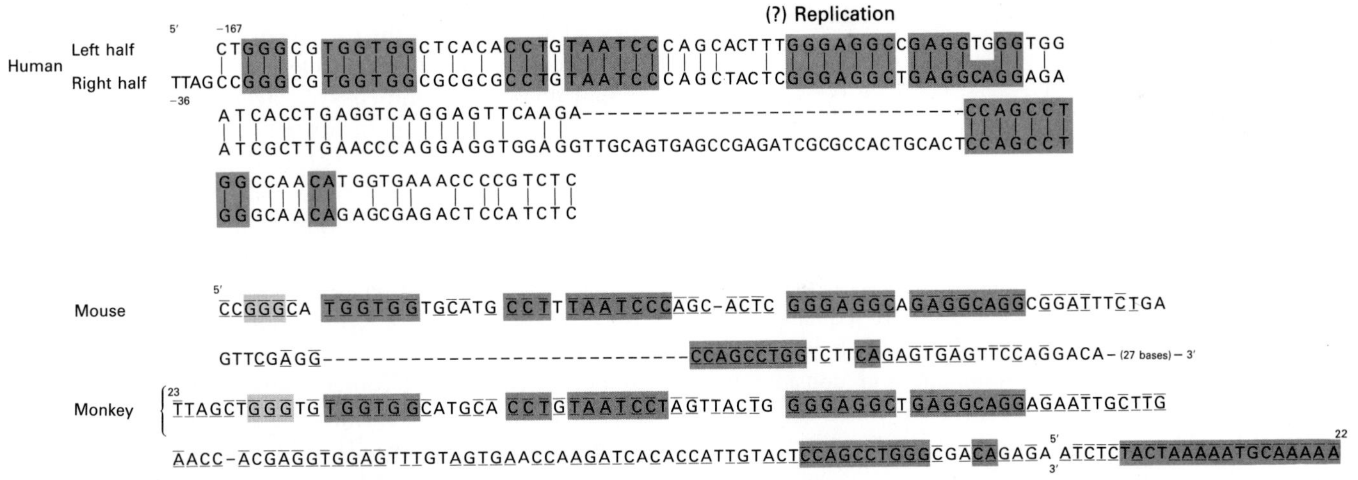

Figure 11-29 Interspecies comparisons of the sequence of the most prominent repetitious DNA family in mammals, the *Alu* family. Regions of strong sequence homology (colored regions) were found in cloned samples of human, mouse, and monkey DNA that had been shown by hybridization tests to contain sequences repeated many times in the genome. In humans there is a 300-base region whose left- and right-hand portions are similar. (The human *Alu* sequence is stacked; the first nucleotide in the second line follows the last nucleotide in the first line, and identical bases are connected by lines. The right-hand portion has an insert that is not present in the left-hand portion, where dashes appear.) When different members of the human *Alu* family of repeats are compared, there is about 85 percent homology. In mice and monkeys (and in other vertebrates such as chickens), the *Alu* family members are only about 150 bases long. The region marked (?) *Replication* is similar to a 14-nucleotide conserved region in several DNA viruses and is discussed in Chapter 13 as a possible replication origin for mammalian cells. [See P. L. Deininger et al., 1981, *J. Mol. Biol.* **151**:17.]

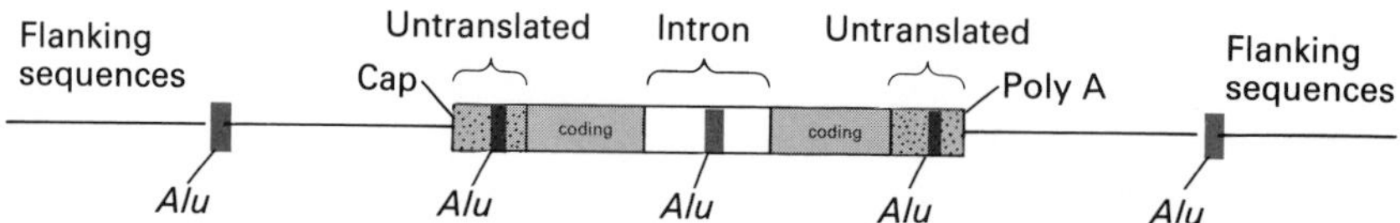

Figure 11-30 Positions of *Alu* sequences in and around a typical protein-coding transcription unit. There is no fixed or regular location of *Alu* sequences within transcription units. The repetitious sequences may be contained within introns or within the noncoding regions at the 5′ and 3′ ends of mRNAs. In many cases, *Alu*-type sequences have been found flanking transcription units, but they do not occur in regular positions.

shall discuss in the next section, regions of DNA that can apparently "move" from place to place over the course of many generations have been found in various animal, yeast, and bacterial genomes. These elements make up most, and possibly all, of the intermediate repeat DNA of yeast and fruit fly cells. An outstanding characteristic of all these mobile repeated elements is the presence of short (5- to 10-base) oligonucleotides on each side of the sequence. Each *Alu* site likewise is flanked by a repeated oligonucleotide; these sequences, consisting of 6 to 20 bases, vary at different sites (Figure 11-31). However, many of the mobile elements discovered in all cell types have structural features that are lacking in *Alu,* and *Alu* has some consistent sequence characteristics not found in the common mobile elements. (A comparison of *Alu*-like sequences and mobile elements is made later, in the section "All Mobile Elements in Eukaryotes May Use RNA Intermediates.")

Mammals Also Have Families of Long Interspersed Repeats

The *Alu* family of mammals is by far the most common intermediate repeat DNA. However, other families of repeats have been cloned, isolated, and sequenced, and additional examples will be found as more sequence studies are completed. Both mice and humans have one related family that is very prominent and considerably longer than the *Alu* family. It has been called the R (repeated) family in mice; in humans it has been termed the *Kpn* family because the restriction enzyme *Kpn*I releases the repeats in a fairly uniform size class. This repeated DNA segment can be as long as 5 to 6 kb, although in many genomic DNA segments only a portion of the entire repeat has been found.

Sequence analyses of the DNA of many of these *long interspersed repeats* have shown frequent translation stop codons in all three reading frames, so it is likely that few of these sequences encode protein. Transcripts from them are seldom found in mRNA. The frequency with which these elements (and probably relatives of them) appear in DNA is on the order of 10^4 times per genome. Thus, long interspersed repeats may make up as much as 1 to 2 percent of the total DNA in mammalian genomes: (10^4 copies) × (5×10^3 bases per copy) = 5×10^7 bases; about 3×10^9 base pairs constitute the human genome.

Rearrangements in Chromosomal DNA

The stability of genetic traits from generation to generation has long implied the existence of some stable chemical structure. Once the general size and shape of each chromosome in a species was found to be characteristic of that species, and DNA was shown to be the hereditary principle present in all chromosomes, then the stability of the DNA structure in the germ cells (gametes) was assumed.

However, for many years it has been known that not all somatic cells obey the rule of chromosomal constancy. In the cells of particular tissues of some lower animals (e.g., roundworms and cockroaches), part or all of some chromosomes are lost. In other lower animals (particularly dipteran insects such as fruit flies), part or all of the chromosomes are repeatedly copied or amplified to different degrees in different tissues. The cells in which these radical chromosomal changes occur no longer divide.

The vertebrates differ from the invertebrates cited above in that in each vertebrate somatic cell, the properties of the two sets of chromosomes (i.e., their number, shape, average DNA content, and fluorescent banding patterns) are similar to those of the two gametes that formed the fertilized egg. The biological capacities of sets of chromosomes from at least some differentiated cells of higher animals also appear to be identical. This conclusion is based in part on nuclear transplantation experiments in frogs. The haploid nucleus of a frog egg (from *X. laevis* or *Rana pipiens*) can be physically removed and a diploid nucleus from another frog cell—for example, an intestinal cell of a tadpole—can be injected in its place. An egg treated in this manner can undergo development directed by the egg cytoplasm plus the injected nucleus (see Figure 22-16). An analogous experiment has been performed in plants. A single cell from a root tip can be physically isolated and regrown in culture into a whole plant. These experiments support the long-held belief that in most multicellular organisms, differentiated cells con-

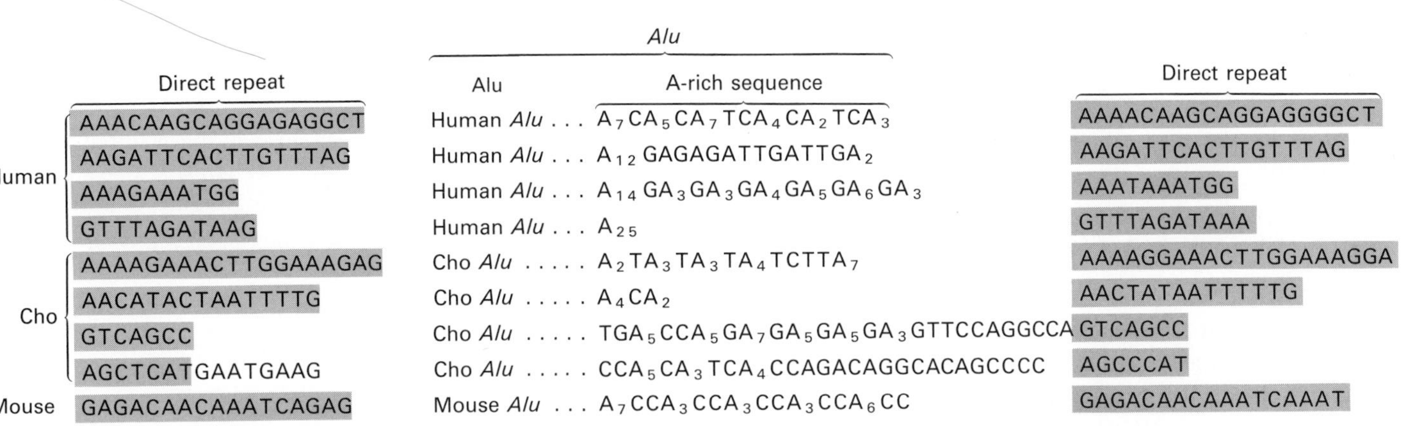

Figure 11-31 The general structure of the *Alu* region, including the A-rich sequence which joins the flanking direct repeats, is illustrated in nine independent examples of *Alu* family DNA from humans, Chinese hamsters (Cho), and mice. [See S. R. Haynes et al., 1981, *Mol. Cell. Biol.* **1**:573.]

tain the same DNA as undifferentiated cells—or, if changes do exist, they cannot be so extensive as to be irreversible.

But a very important question has remained: Is the primary structure of DNA in the chromosomes of cells in different animal tissues *exactly* the same? In particular, are some protein-coding genes amplified in certain tissues, and/or do sequence rearrangements sometimes occur during development? The transplantation experiments in frogs and the plant regeneration experiments do not rule out a few key changes in DNA structure that might, under the conditions of the experiments, be reversible. Armed with the techniques for isolating and sequencing specific genes from different cells, biologists have been able to answer some questions about gene number and gene arrangement in chromosomes from different cells of the same organism. (The rearrangements considered here are to be distinguished from recombination, which occurs mainly during gamete formation and tends to result in reciprocal exchanges between the chromosomes of two gametes. Recombination is discussed in Chapter 13.)

How Might Chromosomal Rearrangements Affect Organisms?

Specific rearrangements of chromosome structure could theoretically affect cells and organisms at two distinct levels: (1) at the level of gene expression and cellular differentiation, and (2) at an evolutionary level. The chromosomal rearrangements that might play a role in differentiation must occur in specific cells in each generation and must be tightly integrated into a developmental pathway. Only a few examples of such programmed rearrangements are known at present. The DNA rearrangements that may have an evolutionary impact occur infrequently (say, once in 10 generations, at the most) and, it appears thus far, randomly. However, because such rearrangements accumulate in the chromosomes, many examples have been recognized in natural populations. In fact, many naturally occurring mutations are due not to simple base changes but to DNA rearrangements.

The question of whether DNA rearrangements are the basis for developmental gene control was reopened when transcription units for cellular mRNAs were discovered to contain introns. The finding renewed speculation that perhaps differentiation consisted of gene rearrangements that eliminated the introns. Subsequent analyses of many transcription units have shown that this is not the case. The active transcription units for a great number of tissue-specific proteins contain the introns. Only one type of rearrangement—that occurring in cells of the immune system—has been related to differentiation in higher organisms. (Chapter 24 describes these events in detail.) In several different experiments, genes that were expressed in only one tissue were cloned both from germ cells and from somatic cells of the tissue in which the genes were expressed. No differences in sequence were found within 1000 bases of the RNA initiation site.

Thus, at least for most genes in multicellular organisms, rearrangements of DNA probably do not occur prior to transcriptional activation. If any such changes do happen, either they must be very subtle or they must be confined to sites located a considerable distance from the transcription unit. One frequent biochemical change has been linked to differentiated gene expression: genes are less methylated when they are undergoing transcription than when they are inactive. (This change will be discussed in Chapter 12.) However, methylation does not involve any change in the primary DNA sequence.

Although most DNA that codes for protein in multicellular eukaryotic organisms is probably stable, new techniques in DNA biochemistry have uncovered an increasing number of differences in the DNA of different members of the same species. Most of these changes involve mobile chromosome elements whose functions, as we shall see, are unknown; the mobile elements usually do not alter the biology of the organism during each life cycle, but they do appear to have greatly affected evolution.

Four Major Classes of Chromosomal Rearrangements Are Known

Chromosomal rearrangements of four general types have been discovered in recent years (Figure 11-32). *Insertion* and *transposition* involve the movement of mobile genetic elements—*insertion sequences* and *transposons,* respectively—into, or less often, out of, a chromosomal site. Both insertion sequences and transposons may exist outside the cell chromosome in plasmids, and enter the chromosome by recombination; transposons may also move from one chromosomal site to another. In addition, specific chromosomal segments can undergo *amplification,* or repeated localized copying. Finally, *deletions* of specific sequences can bring together parts of a transcription unit and thus activate genes.

A fifth mechanism for the rearrangement of chromosomes—gene conversion—has already been discussed (see Figure 11-22). However, this mechanism appears to be limited primarily to single-celled organisms and to germ cells of multicellular organisms.

Bacterial Insertion Sequences and Transposons Promote Evolutionary Change

Although our main consideration in this chapter is DNA sequences found in eukaryotic chromosomes, the earliest molecular understanding of insertion sequences and transposable elements came from studies with bacteria. Because these bacterial sequences are so similar in design to many elements found in eukaryotes, we shall briefly describe the bacterial elements before turning to their eukaryotic counterparts.

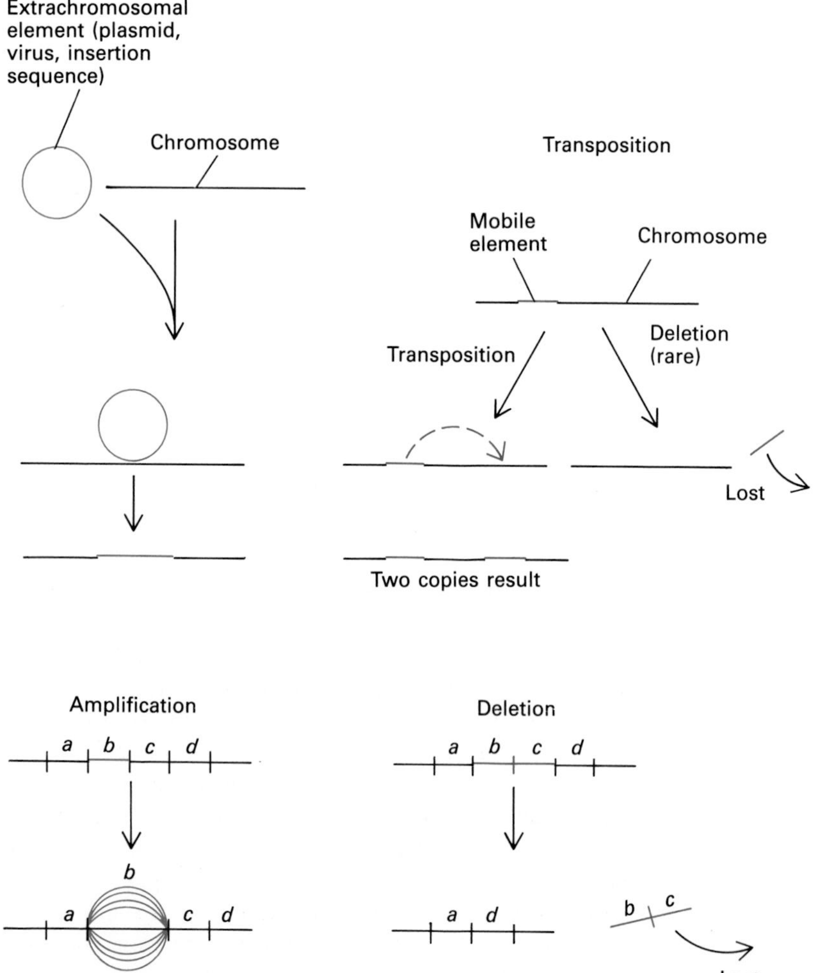

Figure 11-32 The four known types of somatic-cell DNA rearrangements. *Insertion* is the introduction of foreign DNA. *Transposition* is the movement of sequence information from one site to another, usually without the information disappearing from the original site. In prokaryotes, the movement is within the same chromosome; in eukaryotes, it may occur either within one chromosome or between two different chromosomes. *Amplification* is the localized copying of a particular DNA sequence. *Deletion* (as it will be considered in this chapter) involves the loss of a sequence at a specific site. Deletion may occur as the reverse of insertion or transposition.

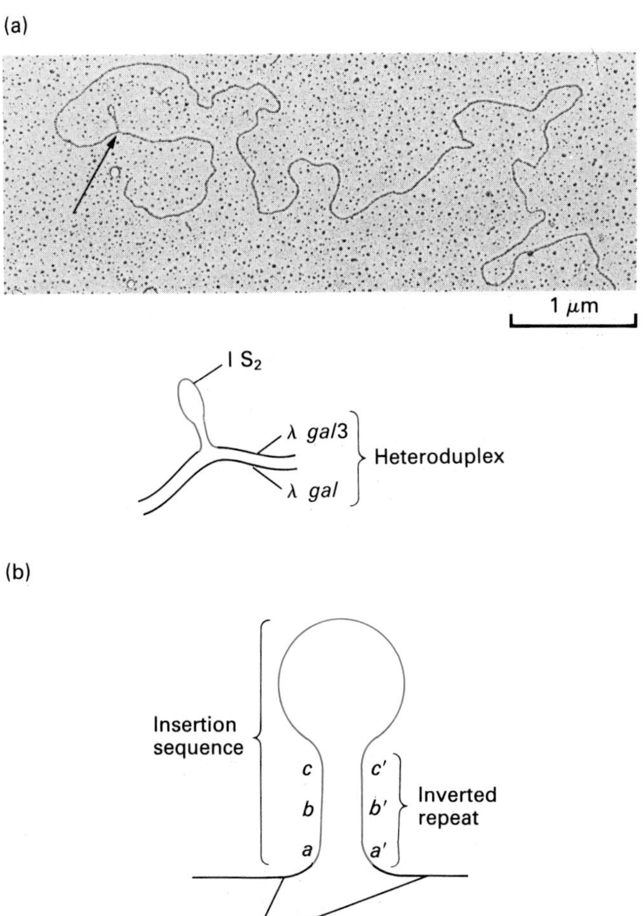

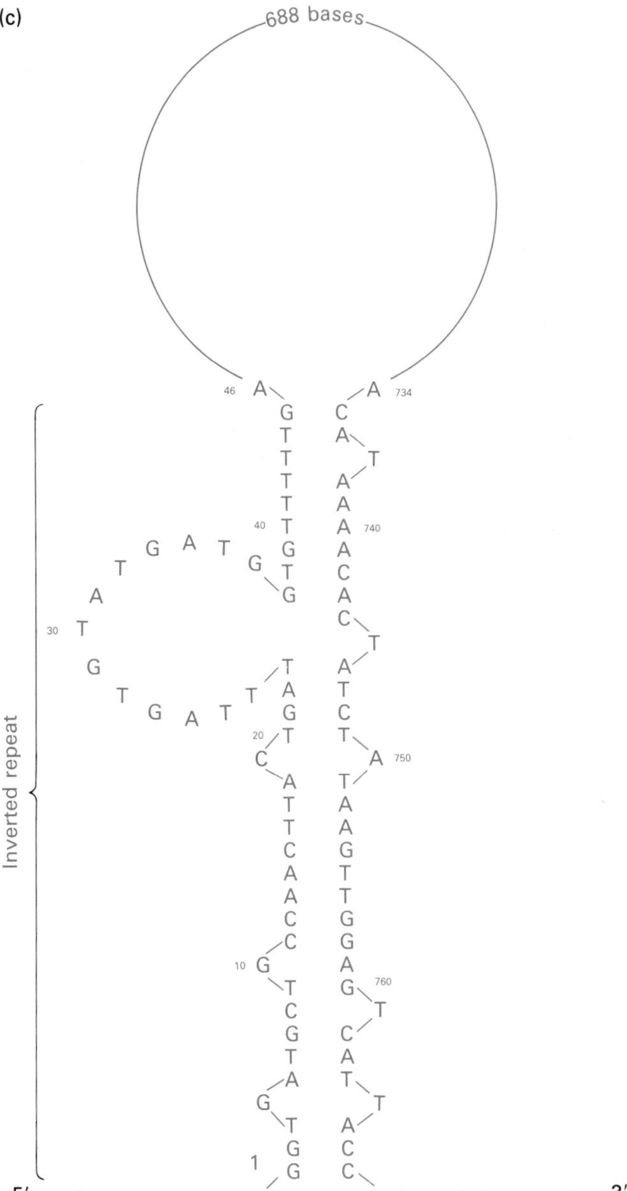

Figure 11-33 Bacterial insertion sequences. (a) IS_2, an insertion element in the galactose gene of *E. coli*. The photograph shows a heteroduplex between two λ bacteriophages, one carrying the normal *E. coli* gene (*gal*) that encodes an enzyme for the metabolism of galactose, and one carrying a "mutant" (i.e., defective) form of this gene (*gal*3). The *gal*3 gene has an extra piece of DNA, not present in *gal*, that forms an *insertion loop (arrow)*. Bacteria with the *gal*3 gene "mutation" occasionally revert to the normal Gal^+ phenotype. When λ phage DNA carrying the revertant *gal* gene is examined in the microscope, the insertion loop is gone. Thus, the presence of the insertion loop appears to inactivate the *gal* gene, whereas the removal of the loop restores the gene's function. (b) The general structure of IS_1, another insertion element in *E. coli*. The terminal regions of the element are inverted repeats. The DNA sequence of one strand is shown. (c) The DNA sequence at the ends of IS_1, showing that the inverted repeated structure is not perfectly matched. [See A. Ahmed and D. Scraba, 1975, *Mol. Gen. Genet.* **136**:233; and H. Ohtsubo and E. Ohtsubo, 1978, *Proc. Nat'l Acad. Sci. USA* **75**:615.] *Part (a) courtesy of A. Ahmed.*

Bacterial Insertion Sequences Bacterial *insertion sequences* (or IS elements, as they are commonly called) were first detected through their effect on gene function: when they are inserted into the middle of a gene, they inactivate the gene. When they are removed from the gene, it regains its activity. For example, *E. coli* loses the ability to grow on galactose after the insertion of an element called IS_2, but the cell regains the ability to grow on galactose upon the removal of IS_2. The loss and the subsequent recovery of the ability to metabolize galactose was shown to be due not to a mutation in a single base, but rather to the insertion and removal of an extra piece of DNA in the galactose gene region (Figure 11-33a).

At least six different IS elements have been found for *E. coli* alone, and other bacteria also contain them. Although they do not provide any known function in the bacterial cell, they are widely distributed in bacteria in nature. Characteristically, one end of a bacterial IS element is an inverted repeat of the other end (Figure 11-33b and c). The length of the section between the inverted repeats can vary from a few hundred bases to a few thousand, but it is generally shorter than 1500 nucleotides. The IS elements do not show long open reading frames (i.e., sequences that can encode protein), so they can be detected only by their damaging effects.

Bacterial Transposons The *transposons* are a group of larger mobile bacterial elements that contain genes. Most transposons encode proteins that confer antibiotic resistance and often exist in nature in plasmids (circular DNA molecules that replicate independently in bacteria). Transposons can enter the bacterial chromosome, often damaging a resident bacterial gene in the process, just as IS elements do. Because transposons contain genes that encode antibiotic resistance proteins, bacteria with chromosomes that have undergone transposition can be selected by growing them in the presence of an antibiotic in which bacteria normally do not grow. Bacteria that have not incorporated the antibiotic-resistance DNA into their chromosomes will die.

Comparisons of the DNA sequences of IS elements and transposons reveal a close relationship between the two. Transposons contain direct repeats several hundred nucleotides long at each end. In fact, in a number of transposons the direct repeats are IS elements. Figure 11-34 shows the general structure of Tn-9, an *E. coli* transposon that contains a gene encoding resistance to the antibiotic chloramphenicol as well as a gene for a protein needed to promote transposition of Tn-9 DNA.

A Model for the Mechanism of Bacterial Transposition The molecular mechanism for the movement of bacterial transposons is under active study and can now be induced to operate in cell extracts. Genetic work has indicated that at least several host enzymes are necessary for the recognition and integration of transposons, and, as mentioned above, some transposons encode a protein necessary for transposition. This protein is called a *transposase;* its precise biochemical role is unknown. For example, the transposon Tn-3, which is similar in design to Tn-9, encodes both a transposase and a repressor protein that controls production of the transposase. The transposase likely recognizes the inverted repeats at the ends of the transposon within the donor chromosome or plasmid and cuts the DNA at that site to begin the transposition. A proposed mechanism for transposition is shown in Figure 11-35.

This model of transposition accounts for a number of constant properties of transposition and transposons. The direct repeats (often IS elements) at the ends of the transposons would serve as recognition sites to trigger the cuts in the donor DNA. How the staggered cut would be made in the DNA of the recipient chromosome remains speculative. The generation of a direct repeat of the target site and an extra copy of the transposon would result from DNA synthesis from the two free 3′ hydroxyl ends of the DNA adjacent to the target sites. A successful transposition requiring DNA synthesis has recently been achieved in a cell-free bacterial system. With this advance, the details of how transposition occurs may be worked out soon.

As we shall see in later sections, the design of transposons in eukaryotic cells seems to be identical to the design of prokaryotic transposons; however, the eukaryotic mechanism of transposition may be quite different.

Loss of Transposons or IS Elements Homologous recombination between the IS elements at the end of a transposon can lead to the deletion of the entire transposon. Insertion sequences can also be deleted, perhaps by recombination between target-site repeats. However, transposition and insertion appear to occur much more often than deletion. Probably, the sequences of mobile elements must eventually drift randomly and become unrecognizable, or they must somehow undergo partial deletion; otherwise, the entire genome would be overrun by these sequences of the type we referred to earlier as "selfish DNA."

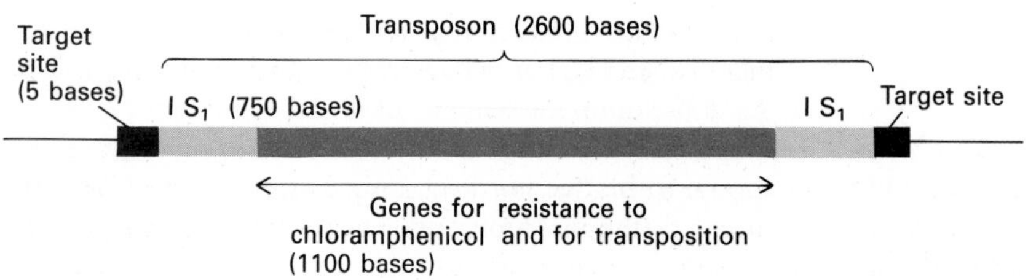

Figure 11-34 Elements of a bacterial transposon: Tn-9 of *E. coli.* The termini of the transposon are IS_1 sequences, which can also be found independently in the *E. coli* chromosome. Sequencing of the integration site before and after the entry of transposons has shown that a short chromosomal sequence termed the *target site* becomes duplicated during transposition. As a result, identical target sites flank the integrated transposon. The target-site duplication for Tn-9 is five bases long.

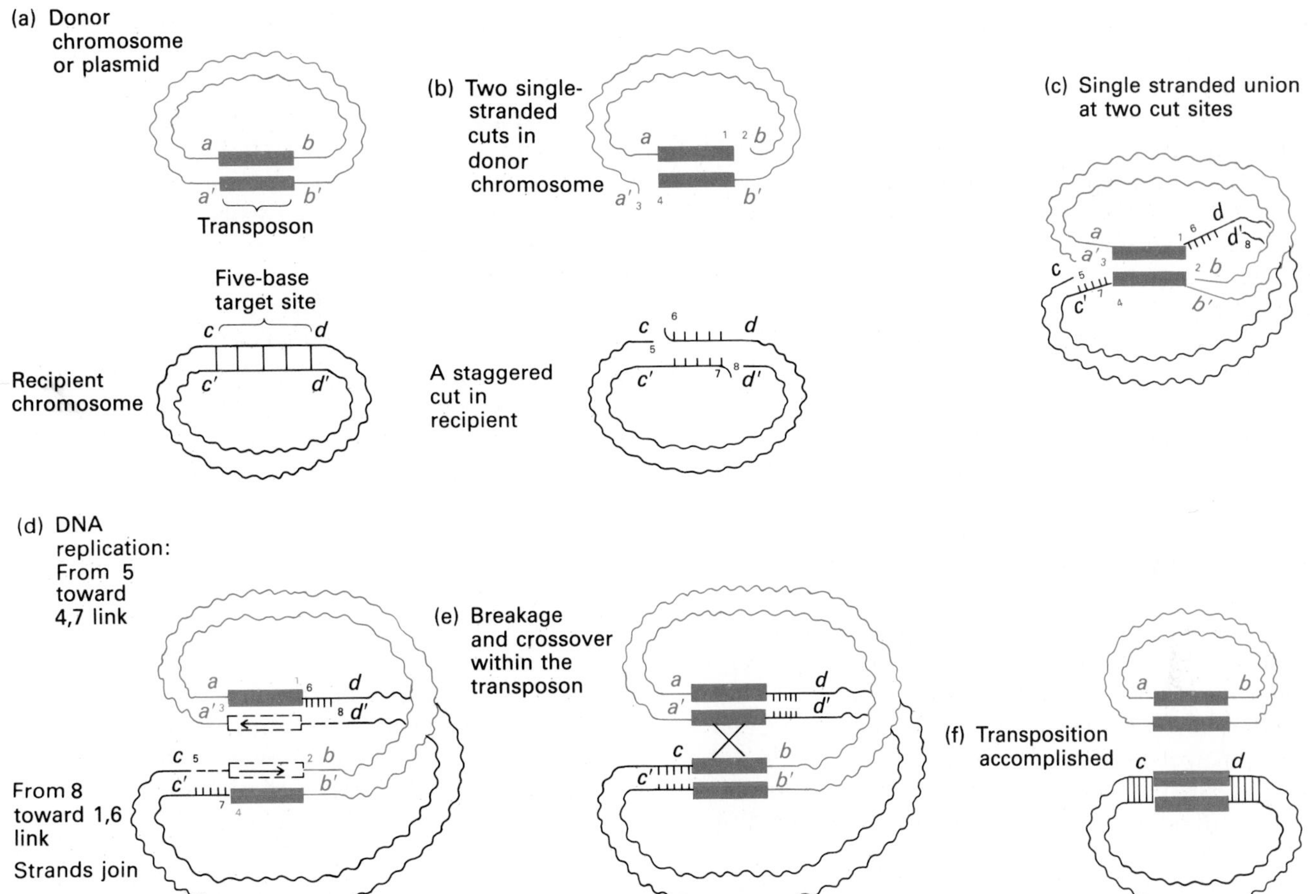

Figure 11-35 A model for bacterial transposition. (a) The initial state. (The *a* and *b* sequences flank the transposon; they are indicated to make it easier to follow subsequent events.) (b) A single-strand cut is made at each end of the transposon, on opposing strands to produce the free ends 1, 2, 3, and 4. In the recipient DNA, a staggered cut is made between the *c* and *d* sequences, which are five nucleotides apart. This cut produces the free ends 5, 6, 7, and 8. The staggered cut, like one made by a restriction enzyme, leaves single-stranded protruding ends at the integration site. (c) Each end of the transposon is attached to one of the single-stranded protruding ends resulting from the staggered cut (1 is attached to 6, and 4 is attached to 7). Note that the target site is now present as two single-stranded regions. Let us suppose that the remaining cut ends in the recipient chromosome (the 5 and 8 ends) have exposed 3′ hydroxyl groups; DNA synthesis can begin at such sites. The resulting two replication forks copy first the target site and then the transposable element. (d) Elongation of the DNA chains from sites 8 and 5 has generated a target-site repeat and a duplicate of the original transposon. Because the donor and the recipient chromosomes were each originally parts of independent circles, the synthesis across the transposons and the resealing of the DNA form a large fused circle called a *cointegrate.* (e),(f) A crossover within the transposons (indicated by an X) resolves the cointegrate into two parts: the donor chromosome, which retains a copy of the original transposon, and the recipient chromosome, which now contains a copy of the transposon flanked by a short direct repeat (the duplicated target site). [See J. Shapiro, 1979, *Proc. Nat'l Acad. Sci. USA* **76**:1933.]

Some Bacterial Gene Rearrangements Are Not Due to Transposons

Not all gene rearrangements in bacterial cells are due to transpositions of sequences from plasmids or sequences from another chromosomal site. One DNA rearrangement that does not involve a duplication, as occurs during transposition, deserves particular discussion. This was, in fact, the first bacterial chromosomal rearrangement to be discovered.

Phase variation in *Salmonella,* an intestinal bacterium, is a well-studied mechanism by which an invertible (and therefore mobile) chromosomal segment controls the alternate expression of two different genes. *Salmonella* are motile bacteria whose locomotion depends on the beating of a flagellum. The flagellar proteins in *Salmonella* consist of subunits called *flagellin,* which are designated either H1 or H2 according to their antigenic properties. The flagellin in the same bacterial strain can change from H1 to H2 and back again. The DNA segments containing the genes responsible for the H1 and H2 proteins have been cloned as recombinant DNA molecules. No changes in the cloned H1 DNA are ever detectable, whether the strain from which the DNA is cloned is expressing H1 or H2. However, in the DNA of the H2 region, a section of about 900 base pairs can exist in either of two inverted orientations. If this DNA is in one orientation, the cell produces H2 flagellin; if the DNA is in the other orientation, the cell expresses H1.

This phenomenon results because the invertible DNA contains promoter sequences. In one orientation both an mRNA for the H2 flagellin and an mRNA encoding a repressor protein for the H1 locus are made. Inversion of the DNA containing these promoters prevents the formation of both mRNAs encoded by this segment, thereby allowing the synthesis of H1 mRNA (Figure 11-36). Thus a chromosomal rearrangement has a profound effect on the character of the bacterial cell.

The inversion of the H2 promoter segment occurs randomly, an average of once in every 10 generations. It is not a programmed part of the life cycle of individual cells. However, *Salmonella* and any other strains of bacteria harboring this ability to change their antigenic properties would be at a selective advantage in the presence of antibodies.

Another case of gene rearrangement in prokaryotes was described recently. The cyanobacteria *Anabaena* (blue-green algae) cluster together in long filaments. When these organisms are deprived of a nitrogen source, about 1 in every 10 cells develops cysts and becomes dormant. In such encysted cells an 11-kb piece of DNA is excised. The removal of this sequence brings nitrogen-fixing genes close together, probably so that they will function better to provide nitrogen-containing nutrients (e.g., amino acids) to the cells.

Transposons in Eukaryotic Cells Probably Move by Using an RNA Intermediate

Although perhaps the most intensively studied cases of transposition are in bacteria, the first indication of the existence of mobile genetic elements was in the angiosperm plant *Zea mays* (maize). Over 30 years ago Barbara McClintock described genetically unstable loci in corn, which led her to conclude that genes were "jumping

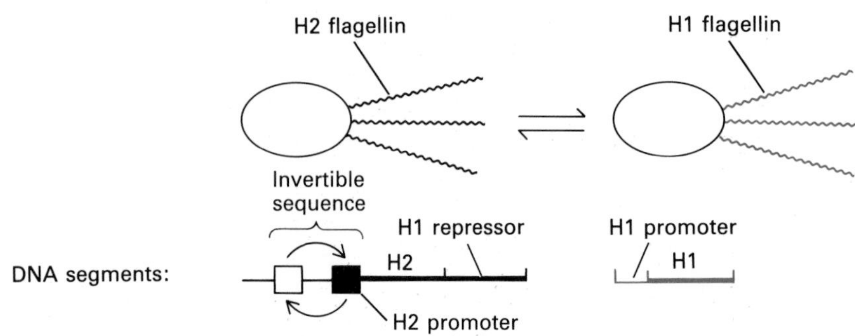

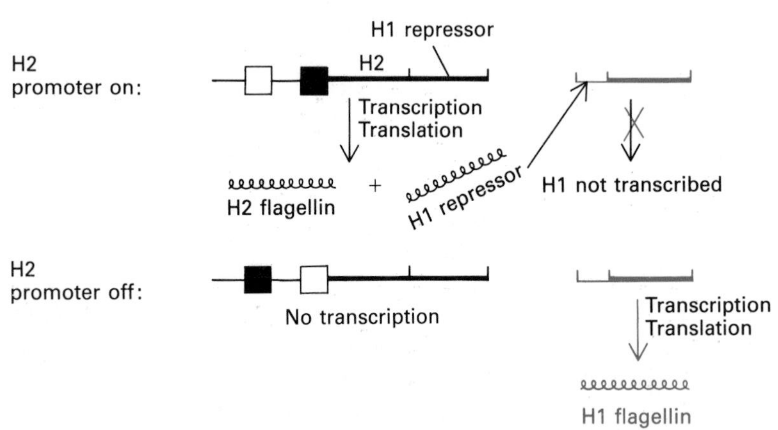

Figure 11-36 A model for the alternate expression of the H1 and H2 flagellin genes in *Salmonella* through phase variation. Expression of the type of flagellin is regulated by the orientation of an invertible DNA sequence adjacent to the H2 operon. When this promoter is in the "on" orientation, the bacterium is in the H2 phase and the H2 operon is transcribed: the H2 protein and the H1 repressor protein are made. (The H1 repressor prevents transcription of the H1 gene.) When the promoter is inverted, it is "off" and the bacterium is in the H1 phase: the H2 operon is not transcribed, and the H1 gene is expressed. [See M. Silverman and M. Simon, 1980, *Cell* **19**:845.]

around." As was later the case with bacteria, the first sign of mobile elements in corn were certain unexplained gene inactivations. These inactivations were distinguished from single point mutations, which were much rarer and usually left permanently altered products. A class of these inactivated genes could spontaneously reactivate, often in association with the spontaneous inactivation of other genes. Recent results obtained with recombinant DNA techniques have revealed that the genetic events McClintock discovered in corn were in fact due to transposons.

Guided by the pioneering studies in corn and the subsequent clear description of transposons in bacteria, workers have searched for transposons in eukaryotic cells. Such elements have been found in yeast, insects, worms, frogs, birds, and mammals. There is a striking similarity in the general design of many of the transposons in these organisms (Figure 11-37). A short direct repeat appears at the site of insertion, and a second direct repeat, a *long terminal repeat* (LTR), is present at the ends of each transposon. Also, sequences similar to TGT and ACA, respectively, are found at the ends of each LTR (one trinucleotide per end) (Table 11-7).

Although the general structure of transposons is similar in bacteria and eukaryotes, the mechanism by which transposons move from site to site may be quite different, as we have mentioned. Recall that the bacterial mechanism involves the making of an extra DNA copy from the already resident copy as the insertion process proceeds. In eukaryotes, the DNA to be moved is copied first into RNA and then back into DNA. This mechanism was not a complete surprise because retroviruses, which are widely distributed in vertebrate cells, have a design exactly as described in Figure 11-37. These viruses have RNA as their virion genetic material, but the particle possesses the enzyme reverse transcriptase, which makes first a single-stranded and then a double-stranded DNA copy of the virion RNA (Chapters 4 and 6; details of retrovirus structure and action are described in Chapter 23). When this double-stranded copy is inserted into the genome, it is flanked by an LTR on both ends and a 5- to 10-base-pair repeat at the target site (Table 11-7). It now seems likely that most movement in eukaryotic genomes occurs in a similar fashion.

Ty Elements of Yeast Certain DNA sequences isolated from yeast cells have been found to be repeated 25 to 50 times in the genome. Aside from the genes for the major RNA molecules (the rRNAs and the tRNAs), these elements are the only long repeated sequences in yeast DNA. When restriction enzymes were used to fragment the DNA from three different yeast strains, these repeated DNA segments were not present in exactly the same set of fragments in each strain (Figure 11-38a). Thus the positions of repeated sequences in the yeast chromosomes were shown to vary among different strains. By growing a culture from a single yeast cell, dividing this culture, and growing the separated cultures for extended periods, changes in the chromosomal positions of repeated sequences were detected in the separated cultures (Figure 11-38b). The repeated sequences are apparently mobile.

The yeast mobile elements, which are called *Ty* elements (for transposon, yeast), are about 5 kb long. Analysis of Ty elements has revealed that they have an LTR—a δ (delta) sequence—of about 300 base pairs (see Figure 11-37 and Table 11-7). Just as with bacterial transposons, Ty elements generate a 5-base-pair repeat at their chromosomal target sites. Ty elements apparently do not move in a coordinated manner; rather, they seem to move in a random fashion at a fairly low rate, perhaps once in 20 or more generations. These mobile elements do not have a regular programmed function in yeast physiology, and some strains have many more of them than others. However, an RNA transcript is produced from at least some Ty elements, and this RNA is used to make DNA copies during the transposition.

As happens in bacteria, the insertion of a mobile element in the yeast genome can damage pre-existing genes.

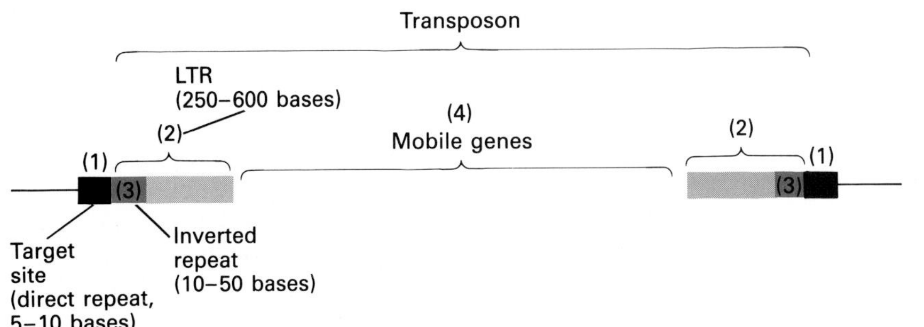

Figure 11-37 The common elements of eukaryotic transposons. (1) A direct repeat (5 to 10 bases) of the genomic DNA occurs at the target site; the sequence may vary for different transposons, but it tends to be the same length at each target site for a particular transposon. (2) A long terminal repeat (LTR) of 250 to 600 bases occurs within the transposon. There may be slight variations in different transposons. In yeast the LTR is called the δ (delta) sequence. (3) Inverted repeats (10 to 50 bases) occur at the outside end of each LTR and may be analogous to IS sequences in bacterial transposons. (4) Any gene or genes encoded within the transposons are located between the repeats. Some transposons encode proteins.

Table 11-7 Prokaryotic and eukaryotic transposons with the same basic design

Transposon	Transposon size (bp)	LTR size (bp)	Inverted repeat at LTR ends?	Sequence at LTR ends*	TATA box and AATAAA in LTR?†	Size of direct repeat at target site (bp)
Escherichia coli						
Tn-9	2638	768 (IS_1)	Yes	—	No	5
Saccharomyces cerevisiae						
Ty-1	5600	334 (δ sequence)	Yes	TGT . . . TCA	Yes	5
Drosophila melanogaster						
Dm mdg-1	7200	422	Yes	TGT . . . ACA	Yes	4
Dm mdg-3	5600	268	Yes	TGT . . . CAG	Yes	5
Dm *copia*	5000	276	Yes	TGT . . . ACA	Yes	5
VERTEBRATE RETROVIRUSES						
Molony murine (mouse) sarcoma provirus	5900	588	Yes	TGT . . . ACA	Yes	4
Molony murine leukemia provirus	8800	515	Yes	TGT . . . TCA	Yes	4
Avian (chicken) spleen necrosis provirus	8300	569	Yes	TGT . . . ACA	Yes	5

*The highly preserved trinucleotide sequences at the ends of all eukaryotic LTRs are occasionally found in bacteria [e.g., in phage μ (mu)].
†The TATA box is found in eukaryotic promoters, and AATAAA is part of the signal for poly A addition.
SOURCE: W. Jelinek and C. Schmid, 1982, *Annu. Rev. Biochem.* **51**:813.

In other cases, insertion can cause genes to be more active or to become *constitutive*—that is, continuously active instead of active only in response to induction. For example, ADH_{II}, one of two genes that encode alcohol dehydrogenase in yeast, is normally induced greatly by ethanol in the absence of glucose; however, no mRNA is made from this gene when normal cells are grown on glucose, with or without ethanol. When Ty inserts into the 5′ region of the ADH_{II} gene, the gene is activated in such a way that the mRNA and the enzyme are produced in great quantities in the presence of glucose and without ethanol. In these cases of inappropriate activation, the Ty element is often inserted so that its transcription is in the opposite direction from that of the gene it affects (Figure 11-39). Perhaps the effect of Ty insertion and the introduction of a directional change in transcription is to "loosen" the chromatin structure in the region and thereby increase transcription of neighboring genes.

Mobile Elements of *Drosophila* Species Most of the repetitious DNA in *Drosophila* species is now known to consist of mobile elements (see Table 11-7), some of which have structures similar to those of the transposons we have already described. These repetitious sequences clearly change positions in the chromosomes, but they require many generations to do so. The function of these sequences is not known; they were discovered with the use of cloned genomic DNA segments that hybridize to *Drosophila* chromosomes at multiple sites (Figure 11-40). At least several different types of repeated sequences occur in each species studied. Each type of repetitious sequence is present in multiple copies ranging in number from 10 to 100 per cell, and each is distributed throughout the four *Drosophila* chromosomes. As much as 15 percent of the total *Drosophila* DNA, and all the intermediate repeat DNA, appears to consist of various members of these repeated sequence classes.

From DNA cloning and sequencing experiments, three types of mobile repetitious elements have been recognized (Figure 11-41). These repetitious sequences in *Drosophila* remain constant in number and stationary in position through many generations in a given strain of flies. However, if the chromosomes of related *Drosophila* species that have been separated for lengthy periods (hundreds or thousands of years) are compared, the incidence and positioning of each of the repetitive elements vary, whereas the general band pattern in the chromosomes is fairly consistent (Figure 11-42). It appears, therefore, that these

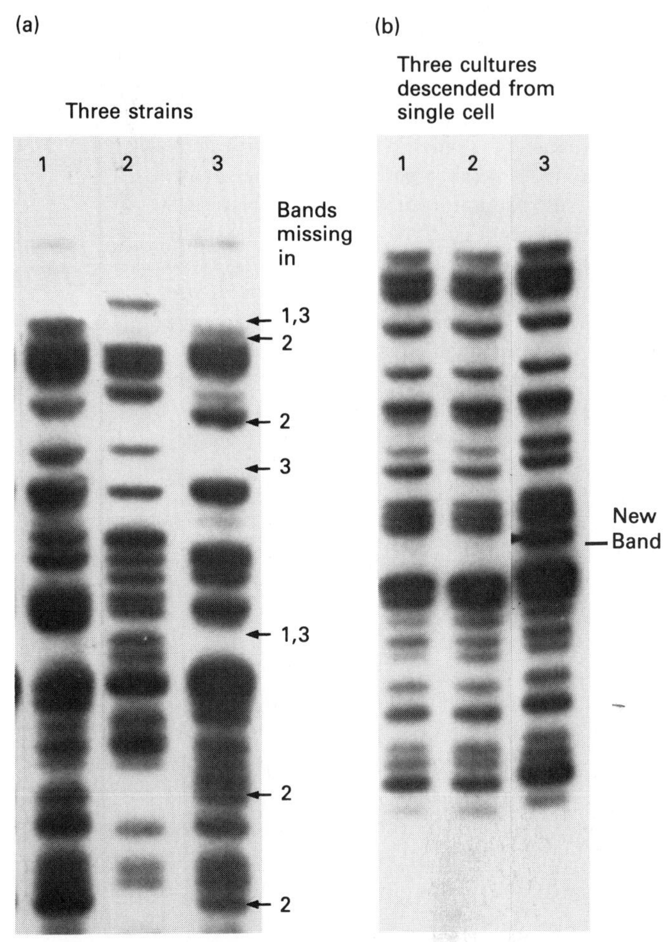

Figure 11-38 A demonstration of the existence of a transposon in yeast cells by a Southern blot analysis of yeast cell DNA. (a) A recombinant DNA clone that hybridizes to many DNA segments in yeast cells was used as a labeled probe. When the DNA of three strains is compared by Southern blot analysis (Chapter 7), they are found to share some—but not all—DNA fragments that react with the probe. Thus, the repeated sequence is not distributed in the same way in each strain. (b) The ability of the repeated sequence to move within the genome of a single strain was tested by taking a single yeast cell, growing a culture overnight, dividing the culture into several parts, and growing these for 30 days. The DNA from each culture was then tested by Southern blot analysis with the labeled probe. Lanes 1 and 2 are alike, showing no movement, but lane 3 shows a new band that is caused by movement of the repetitive element. The element was named Ty-1. [See J. R. Cameron, E. Y. Loh, and R. W. Davis, 1979, *Cell* **16**:739.] *Courtesy of R. W. Davis.*

elements can move about the genome. Two examples will illustrate the effects of insertion of mobile sequences on known *Drosophila* genes.

Variation in eye color is a prominent visible trait studied by the earliest *Drosophila* geneticists. One group of changes—originally thought to be point mutations—occurs at a site termed the *white* locus. A wide variety of changes at this locus modify the normal brick-red eye color to various paler shades, including pure white. Several known enzymes that contribute to making eye pigment are encoded outside the *white* locus. Exactly what protein is encoded at the *white* locus is not known, but its gene product somehow affects pigment production. The various mutations resulting in different intensities of eye color therefore correspond to the production of different amounts of pigment.

The DNA from a number of the mutants has been examined, and repetitive mobile elements have been found to be inserted at the *white* locus (Figure 11-43). These insertions are the cause of the variations in eye color, even though the inserted sequences do not encode protein and may, in fact, lie outside of the protein-encoding region. How the insertions cause the variations in eye pigment is not at all clear. However, it is now estimated that perhaps

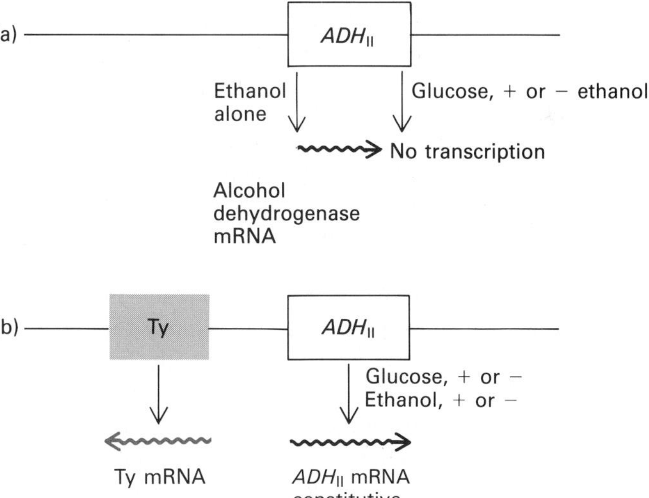

Figure 11-39 Yeast Ty insertion activates the ADH_{II} gene. (a) Normally, alcohol dehydrogenase of the ADH_{II} type is produced in the presence of ethanol, but is inhibited when glucose is present in the medium or when no ethanol is present. (b) When the Ty element is inserted next to the ADH_{II} gene in such a way that Ty is transcribed in the opposite direction from the gene, ADH_{II} mRNA is formed constitutively. [See V. M. Williamson, E. T. Young, and M. Ciriacy, 1981, *Cell* **23**:605.]

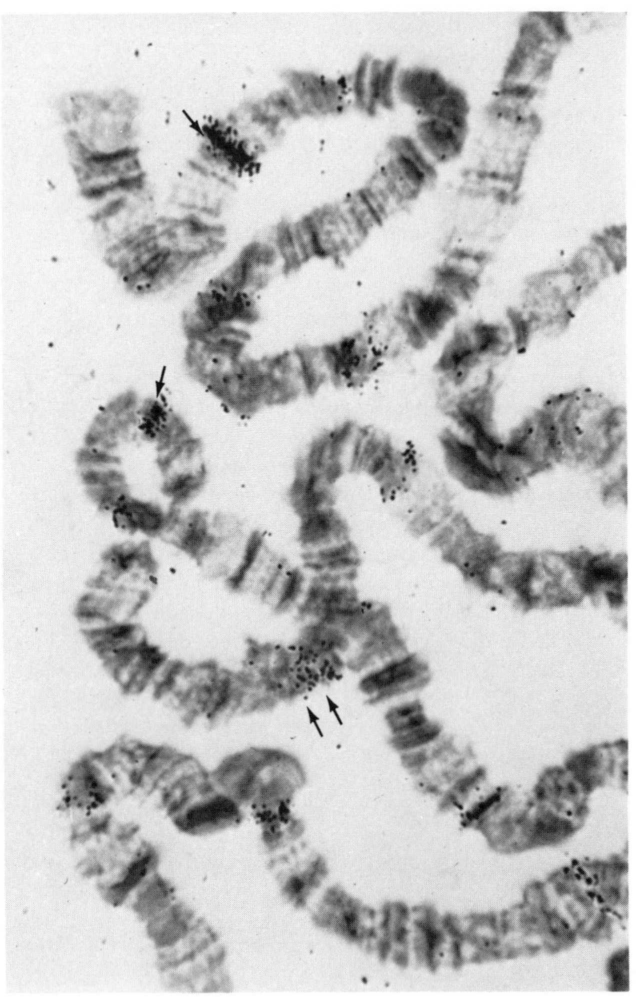

Figure 11-40 The distribution of an intermediate repeat sequence on a chromosome from *D. melanogaster.* Reassociation kinetics were used to show that several different cloned DNA sequences occur approximately 50 times in the *Drosophila* genome. One example of this repetitive DNA was named *copia* (from the Latin for "plentiful") because so much RNA from cultured *Drosophila* cells was complementary to this clone. When *copia* DNA was labeled, hybridized to fixed denatured chromosomes in situ, and then made visible by autoradiography, prominent silver grains appeared in more than 20 chromosomal regions (*arrows show several*), indicating the presence of *copia* DNA at many sites in the genome. [See M. W. Young, 1979, *Proc. Nat'l Acad. Sci. USA* **76**:6274.] *Courtesy of M. W. Young.*

as many as half of all the spontaneous mutations collected from *Drosophila* species are not point mutations at all, but rather changes in chromosome structure brought about by movements of mobile elements.

In spite of the prevalence of mobile elements in the *Drosophila* chromosomes, it seems unlikely that mobile sequences play an essential role in either the early development of the fly or its later life cycle. For example, mobile elements are about 20 times more prevalent in *D. melanogaster* than in the almost identical *D. simulans,* a sibling species. And *D. simulans* lacks altogether some of the types of repetitive elements seen in *D. melanogaster.* Apparently the repetitive sequences are unnecessary to "build a fly." However, because of the dramatic effects the mobile elements can have on gene functions, they probably do participate in evolution.

One type of mobile element in *Drosophila,* the P element (see Figure 11-41), shows how important these sequences can be in promoting the evolution of distinct species by forcing a separation between strains and therefore potentially contributing to mating isolation. P elements exist in some strains (P strains) of *D. melanogaster,* but they are absent in other strains (M strains). When males of the P strain are crossed with females of the M strain, the progeny are defective heterozygous flies; it is as if the two genomes P and M cannot merge. The term *hybrid dysgenesis* is used to describe the unsuccessful cross. (The reverse mating—an M strain male with a P strain female—is not nearly so damaging.) The defects that the P/M heterozygotes display include greatly decreased fertility and increased mutations in any offspring; these induced mutations are frequently associated with chromosomal rearrangements.

It is considered highly likely that the P × M mating triggers the transcriptional activation of the P element, and that this activation is followed by the transposition of the P element, which in turn causes hybrid dysgenesis. The favored model supposes that the eggs of the M strain do not contain a repressor for the P element, which is carried in the incoming sperm (Figure 11-44). The model draws an analogy to zygotic induction, which occurs when two bacteria mate and the male introduces bacteriophage λ without a repressor into the female cytoplasm.

The P element is known to be responsible for many mutations that have been recovered through the years in *Drosophila,* both in laboratory strains and in nature. For instance, at least five or six *white* locus mutations are due to P element insertion. The mobility of the P element has been neatly confirmed by a *white* mutant that reverted to normal eye color. When the DNA of the *white* locus of both the mutant and the revertant was sequenced, the P element was found to have been excised exactly, leaving only a single copy of the eight-base-pair duplication that previously had flanked the P element. Clearly, mobile elements like the P element would work in nature to prevent the productive interbreeding of certain strains, and this type of separation might lead eventually to speciation.

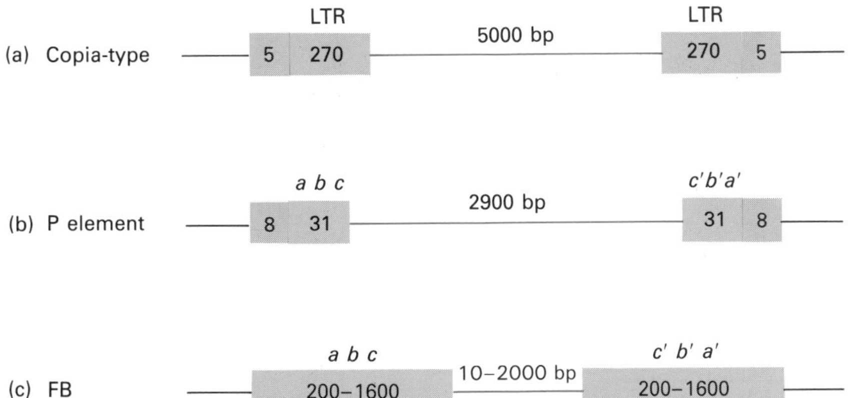

Figure 11-41 The three distinct types of dispersed repetitious transposons in *Drosophila* species. In flies in nature, these elements vary in number from 10 to 100 copies per cell. (a) The first type to be characterized was *copia* (see Figure 11-40). This sequence has the classic transposon structure, including LTRs similar to those of the Ty elements in yeast and the bacterial transposons. The LTRs bracket a 5-kb central region highly consistent in sequence at each of its approximately 50 genomic sites. The chromosomal target site is duplicated for 5 base pairs. (The Dm mdg-1 and Dm mdg-3 transposons listed in Table 11-7 are *copia*-type transposons.) (b) The P element exemplifies a different type of transposon in the *Drosophila* genome. It lacks LTRs but contains a 31-base-pair inverted repeat at its termini and produces an 8-base-pair duplication at the chromosomal target site. The sequence between the terminal structures is 2.9 kb long; this region has long open reading frames that could encode protein. (c) A third type of mobile element, the FB (fold-back) element, also has terminal inverted repeats and no LTRs. The inverted sequences of FB elements range from 200 to 1600 nucleotides in length. Because of this unusual structure, these elements can be detected in the electron microscope by hybridization of the inverted repeats within a single strand. The inverted repeats in the FB elements tend to contain tandemly repeated sequences. The region enclosed between the repeats is quite variable in length, from a few nucleotides to 2000, and thus the elements do not regularly encode protein.

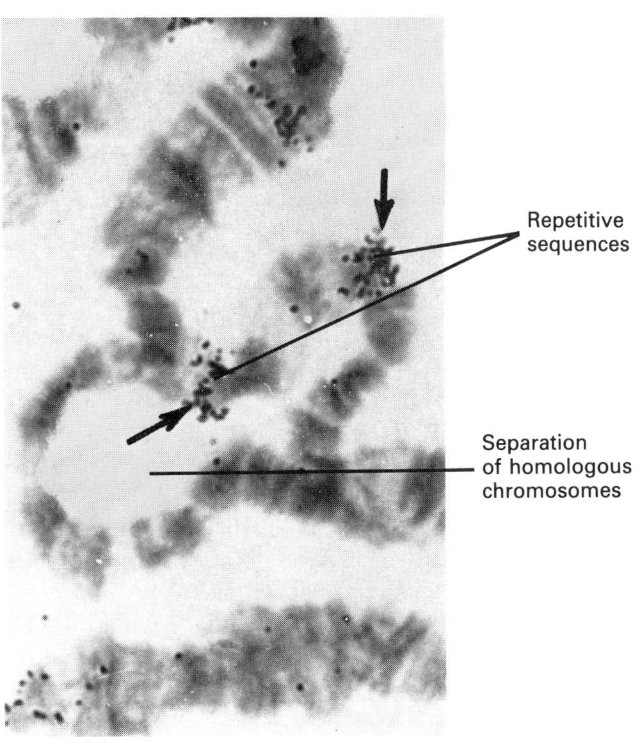

Figure 11-42 A demonstration that repetitive sequences in *Drosophila* species are mobile. A cloned intermediate repeat DNA (like the *copia* sequence in Figure 11-41) was hybridized to salivary gland chromosomes of a hybrid fly. Both polytene chromosomes of a homologous pair are aligned in register, but occasionally they separate along their longitudinal axis. The micrograph clearly shows the repetitive sequence is present in two sites in one of the parents, but absent in the other parent. When the experiment is carried out in many strains and sibling species (very closely related species), it is found that there is not a single fixed position in which a particular repetitive DNA occurs in all strains. [See M. W. Young, 1979, *Proc. Nat'l Acad. Sci. USA* **76**:6274.] *Courtesy of M. W. Young.*

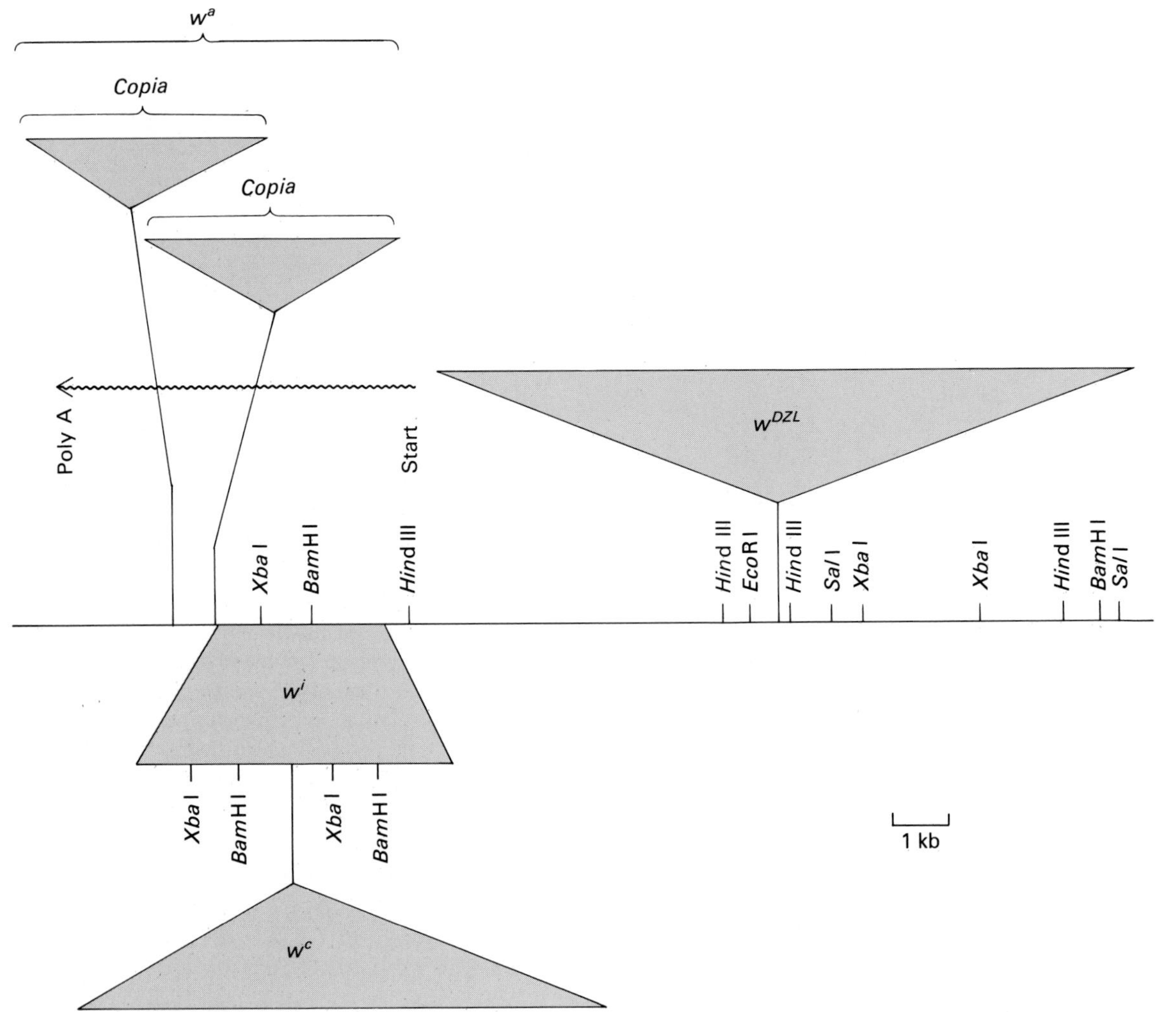

Figure 11-43 A restriction-site map of the *white* locus on the *Drosophila* chromosome and DNA arrangements responsible for mutations. The start site for transcription and poly A site are indicated on the white locus by the wavy line. The protein encoded by this locus has an unkown role in producing eye pigment; thus mutations result in loss of the normal brick-red color. Colored wedges in the figure represent insertions detected in some of the large number of *white* locus mutants. The colored area labeled w^i (ivory eyes) represents a mutant with almost no eye pigment. The mutation is a tandem 3-kb duplication identified by the repetition of the restriction enzyme sites (*Xba*I, *Bam*HI). Insertion of a 10-kb fold-back mobile element (see Figure 11-41) within the duplicated region of w^i results in a pale-pink-eyed fly called w^c; the mutant gene produces at least some eye pigment. Thus the w^i rearrangement, a DNA duplication, is partially corrected by the w^c rearrangement, a DNA insertion. The w^a mutation, which produces a slightly lighter eye color than w^c does, is associated with an insertion of two *copia* elements near the same site. The w^{DZL} mutant, another pale-eyed fly, has a long fold-back element inserted as shown. This sequence is often excised precisely, which accounts for the instability of the mutation from one generation to another. [See G. M. Rubin, 1983, in *Mobile Genetic Elements*, J. Shapiro, ed., Academic Press, p. 329; R. Levis, K. O'Hare, and G. M. Rubin, 1984, *Cell* 38:471].

All Mobile Elements in Eukaryotes May Use RNA Intermediates

As we have mentioned, it now seems likely that most mobile DNA elements in eukaryotes move in the genome through an RNA intermediate. During the years when the structure and the distribution of mobile DNA elements were being worked out—first in bacteria and then in yeast and fruit flies—the function of reverse transcriptase as a copying enzyme in retroviruses had already been discovered. Then came recognition of the similarities between the LTRs and the target-site duplications of retroviruses, Ty elements, and some of the *Drosophila* mobile elements (the *copia* class; see Table 11-7). Now it is

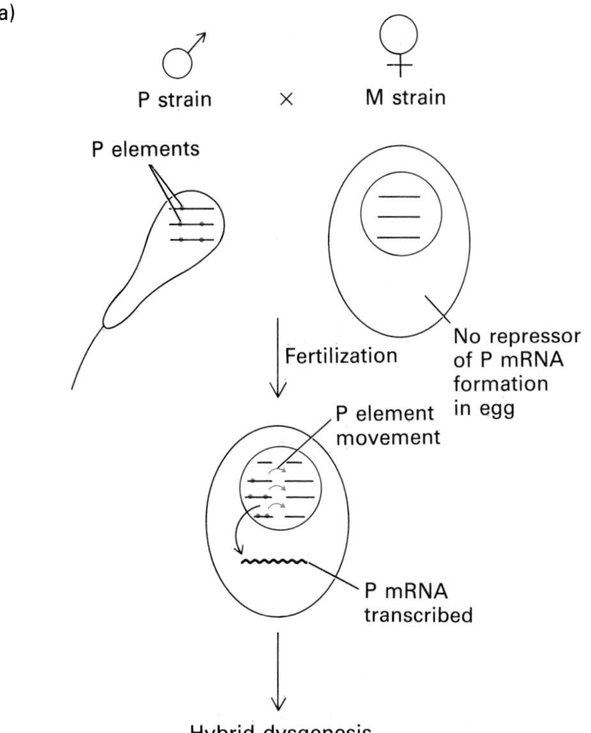

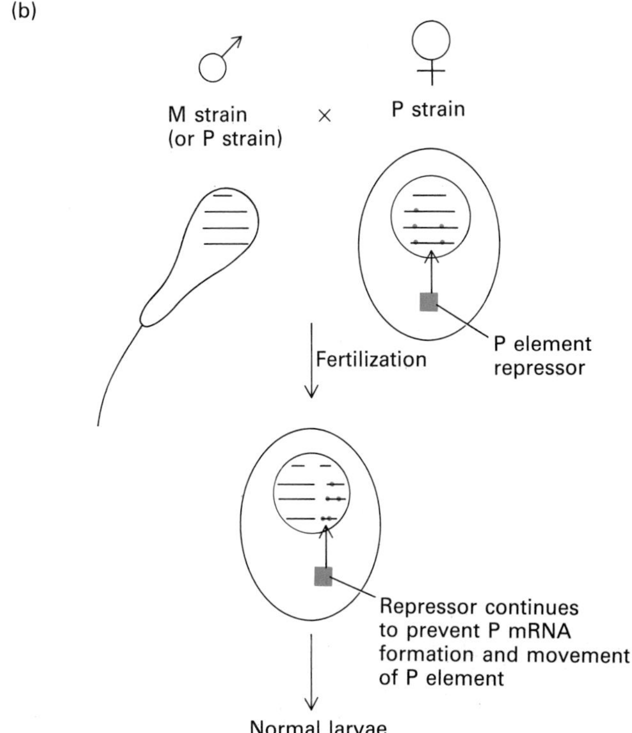

Figure 11-44 A model for hybrid dysgenesis in *D. melanogaster*. (a) The offspring produced by the cross of a P strain male (♂) with an M strain female (♀) are defective, probably due to an induction of P element mobility. (b) The reverse cross is not so damaging, perhaps because there is a repressor of the P element in the cytoplasm of the P strain egg. Likewise, a P male and a P female do not produce dysgenic offspring. [See K. O'Hare and G. M. Rubin, 1983, *Cell* **34**:25.]

known that *Drosophila* cells in culture produce retrovirus-like particles with reverse transcriptase activity, and that some of these particles harbor single-stranded *copia*-like RNA sequences. Moreover, double-stranded plasmid-like DNA molecules with *copia* sequences also exist in cultured *Drosophila* cells. These results clearly support the notion of sequences made mobile through an RNA intermediate.

A recent set of experiments with yeast cells have all but completely confirmed transposon mobility via the RNA route in eukaryotes. A galactose-sensitive transcriptional promoter (i.e., a promoter made more active by the presence of galactose) was inserted into a recombinant form of a yeast Ty element. Yeast cells were transfected with plasmids containing the galactose-sensitive Ty element. If RNA were an intermediate in the mobility of this transposon, increased transposition should have occurred when these cells were grown on a galactose-containing medium. Cells transformed with this DNA did, in fact, show increased transposon movement in the presence of galactose (Figure 11-45).

Next, a second Ty recombinant was constructed. An intron taken from another cloned yeast gene was inserted into the putative protein-coding region of the Ty transposon. The galactose-sensitive promoter was left in this second recombinant. When yeast cells were transformed with the intron-containing Ty element and then exposed to galactose, transposition occurred at a high frequency, as before, but *none* of the newly transposed Ty sequences contained the intron (Figure 11-45b). This result strongly implied that RNA was transcribed, that the intron was removed by RNA-RNA splicing, and that a DNA copy of the intronless RNA was produced and integrated into the yeast chromosome at many different sites. Thus transposition of Ty elements in yeast apparently does involve an RNA intermediate.

As we have mentioned, this interpretation had already been suggested by the known activity of retroviruses. In addition, the structures of other repetitious sequences in the mammalian genome strongly supported the notion of transposon mobility through RNA intermediates. Researchers had cloned and sequenced the partial inactive copies of many different mammalian genes (e.g., globin, tubulin, immunoglobulin) in which the introns and the 5′ and 3′ flanking sequences found at the active gene site were missing (Figure 11-46). Each of these copies even contained a poly A stretch, as if the whole mRNA had been copied. The inevitable conclusion was that the mRNAs, already completely processed, had been copied into DNA and then the DNA copy had been inserted into

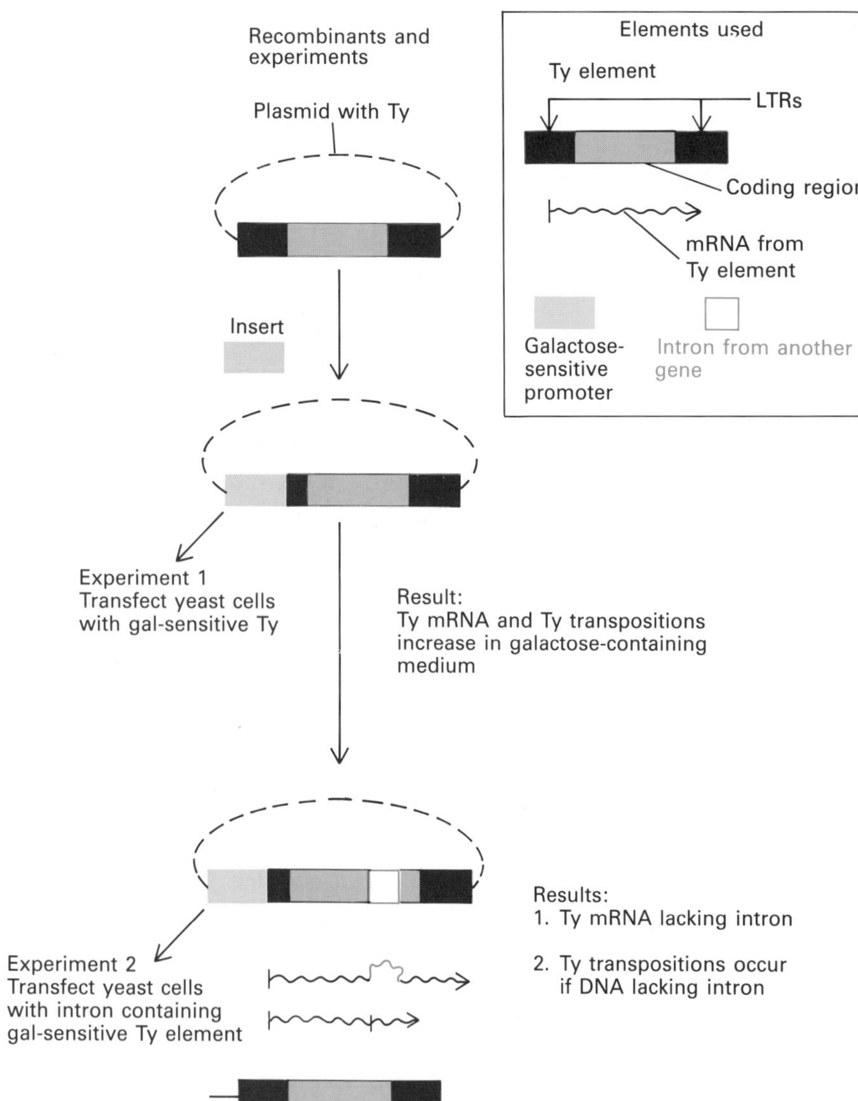

Figure 11-45 A scheme for demonstrating Ty transposition through RNA intermediates. A Ty element encoding an mRNA molecule can be cloned in a yeast plasmid. The yeast plasmid can enter the cell and the Ty element can integrate into the yeast chromosome and then transpose to other sites. By genetic tests the transpositions within the genome can be detected. In the first experiment a portion of the left LTR was removed and replaced by a promoter region of a gene encoding a galactose-metabolizing enzyme. When cells are grown on galactose, transcription of the galactose gene is known to increase. The recombinant DNA molecule was used to transfect yeast cells. After integration of the new DNA into the host chromosome, the cells in the presence of galactose produced much more Ty mRNA than before and transpositions were greatly increased. In a second experiment, an intron was inserted in the galactose-sensitive Ty element, and transfection of yeast cells again resulted in the integration and transposition of the intron-containing Ty element into the genome. Not only did all Ty mRNAs analyzed lack the intron, but all the transposed copies of the Ty element that were detected lacked the intron as well. [See J. Boeke et al., 1985, *Cell*, **40**:491.]

the genome. The resulting "pseudogenes" or "processed genes" were nonfunctional because they did not contain promoters; furthermore, in many cases their sequences had drifted to include translation termination signals, so no protein could be encoded.

A similar mechanism involving reverse transcription has also been proposed for the movement of *Alu* sequences and the numerous pseudogenes similar but not identical to the genes encoding the small nuclear RNAs (U1, U2, U4, etc.). Both *Alu* sequences and the U-like pseudogenes are transcribed by RNA polymerase III. The genomic insertion sites for *Alu,* for the pseudogenes for globin, tubulin, and immunoglobulin, and for the U-like pseudogenes all have a similar structure. There is a 5- to 35-base target-site repeat that varies at different genomic sites (Figure 11-31).

Toward the 3′ end of each of the *Alu* sequences and U-like pseudogenes lies a 20- to 50-nucleotide-long A-A-rich region. A short distance beyond this A-rich segment in the genome, a region rich in T residues is often found. It was hypothesized that if an RNA were copied from such a site, the resulting U-rich region could pair with the A-rich region of the copied RNA sequence, and that this pairing could produce a hairpin primer that a reverse transcriptase–type enzyme could use to make a DNA copy of the entire sequence. This hypothesis was based entirely on DNA sequence analyses plus the knowledge that RNA molecules were in fact transcribed from the U-like pseudogenes and from *Alu,* beginning just at the 5′ end of the first target-site duplication.

These suggestive findings and the results from the experiments with recombinant Ty elements make it highly

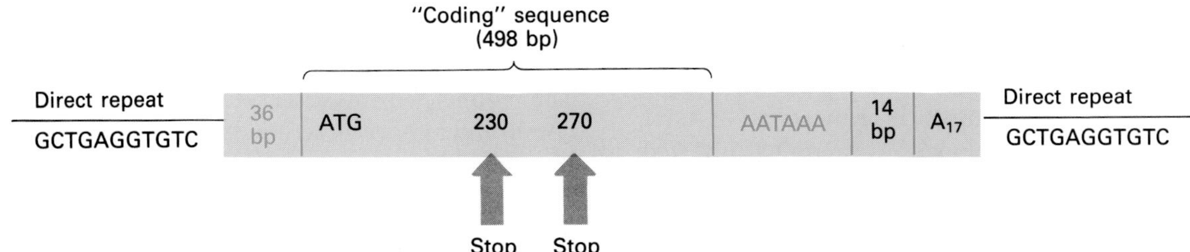

Figure 11-46 A human β tubulin pseudogene that probably arose by transposition through an RNA intermediate. The genomic DNA region diagrammed here was detected because of its complementarity to a tubulin cDNA. The region is flanked by an 11-base-pair direct repeat. Reading from left to right, there is a 36-base-pair stretch, followed by a 498-base-pair region that could encode the known amino acid sequence of β tubulin. However, base changes at nucleotides 230 and 270 have introduced translation stop codons. After the "coding" region, there is an AATAAA (a poly A signal) and after 14 bases a poly A tract of 17 bases before the direct repeat is encountered again. Because the true β tubulin gene contains several introns, this pseudogene probably arose by reverse transcription of an mRNA and insertion of the resulting DNA into the chromosome. The translation stop codons, which render any product mRNA nonfunctional, represent accumulated mutations. [See C. D. Wilde et al., 1982, *Nature* **297**:83.]

likely that most, if not all, of the mobility of DNA within eukaryotic chromosomes is a result of RNA transcription followed by reverse transcriptase–type activity and integration of the new DNA copy. For such events to have an impact in natural populations, they must occur in germ cells or in the precursors to germ cells. The enzyme with the reverse transcriptase activity would act much like the viral enzyme. Regions in both the protein-coding sequences of *copia* DNA from *Drosophila* and Ty elements from yeast are similar to the reverse transcriptase–coding sequences found in several retroviruses. The presence of these specific sequences in two such different eukaryotes strongly implies that this enzyme is encoded by an ancient gene with a history that precedes the evolutionary separation of unicellular and multicellular organisms.

A Summary of Transposable Elements

The DNA of all cells, from bacteria to mammals, contains elements that can move from site to site in the genome by

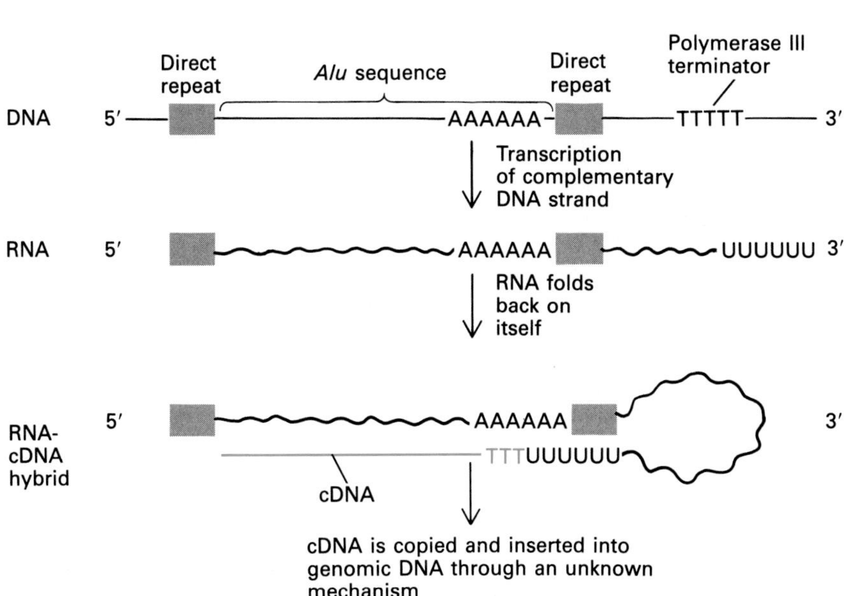

Figure 11-47 A proposed mechanism for the mobility of *Alu* genomic DNA segments through an RNA intermediate. The *Alu* sequence is known to produce a short RNA through polymerase III transcription, beginning at the first nucleotide of the *Alu* sequence. The polymerase reads through to a nearby termination signal known to be T-rich. Thus, the 3′ ends of *Alu* transcripts in vivo contain a stretch of U's. It has been suggested that the *Alu* transcript folds so that the A-rich region and the terminal U-rich region form a hybrid. Reverse transcription might then be initiated at the 3′ hydroxyl end of the U-rich region to make a cDNA copy of the entire *Alu* region. By unknown mechanisms, the cDNA copy might become integrated into the chromosomes. [See P. Jagadeeswaran, B. G. Forget, and S. M. Weissman, 1981, *Cell* **26**:141; and S. W. Van Arsdell et al., 1981, *Cell* **26**:11.]

mechanisms very different from reciprocal recombination. In bacteria these elements move by a DNA-copying mechanism that occurs simultaneously with the transposition event. Insertion sequences (IS elements) in bacteria are about 500 nucleotides long, have terminal inverted repeats, and may not encode any information. Two insertion elements can bracket a segment of DNA to form a transposon. The entire transposon—that is, the bracketing sequences and the DNA between—can move in the genome. At least some of the enzymatic machinery for such movement in bacteria is encoded within certain transposons. A characteristic result at the site of insertion of IS elements or transposons is a duplication of 5 to 10 genomic nucleotides.

Eukaryotic transposons such as the yeast Ty elements, the *copia* elements in *Drosophila,* and the retroviruses in birds and mammals all have the same general structure. However, the mechanism for movement in eukaryotic cells is probably through an RNA intermediate. This is known to be true for the retroviruses, and it is highly likely for much of the intermediate repeat DNA in yeast (Ty elements) and *Drosophila* (*copia*-like elements). Other types of repetitious sequences—*Alu* elements in mammals and pseudogenes that appear to be copies of mRNAs or snRNAs—also seem to move in the genome through an RNA intermediate.

Some Genomic Rearrangements Have Definite Functions

The mobile elements discussed so far seem to move randomly and to have either no effect or damaging effects on the life cycles of organisms. However, as we have mentioned, there are a number of chromosomal rearrangements that apparently occur according to programs and provide genetic flexibility to organisms. The random and the programmed rearrangements may have similar enzymatic mechanisms, but their effects on cells and organisms are very different.

Yeast Mating-Type Switching: A Well-Controlled Gene Conversion *Saccharomyces cerevisiae* undergoes a regular chromosomal rearrangement that controls an important cell feature—the mating type. *Saccharomyces cerevisiae* grows in culture as a haploid cell. Two haploids can "mate" or fuse to form a diploid cell. The cells of opposite mating types are called *a* and *α*. Fused cells are designated *a*/*α*. The fusion is promoted by oligopeptides termed *a* and *α* factors; each of these factors is produced by its homologous cell type, and each factor arrests the growth of the other cell type to enable fusion to occur. When faced with starvation, the fused diploid cell undergoes two meiotic divisions to produce four spores (Figure 11-48). A cell sample can be tested for mating type by

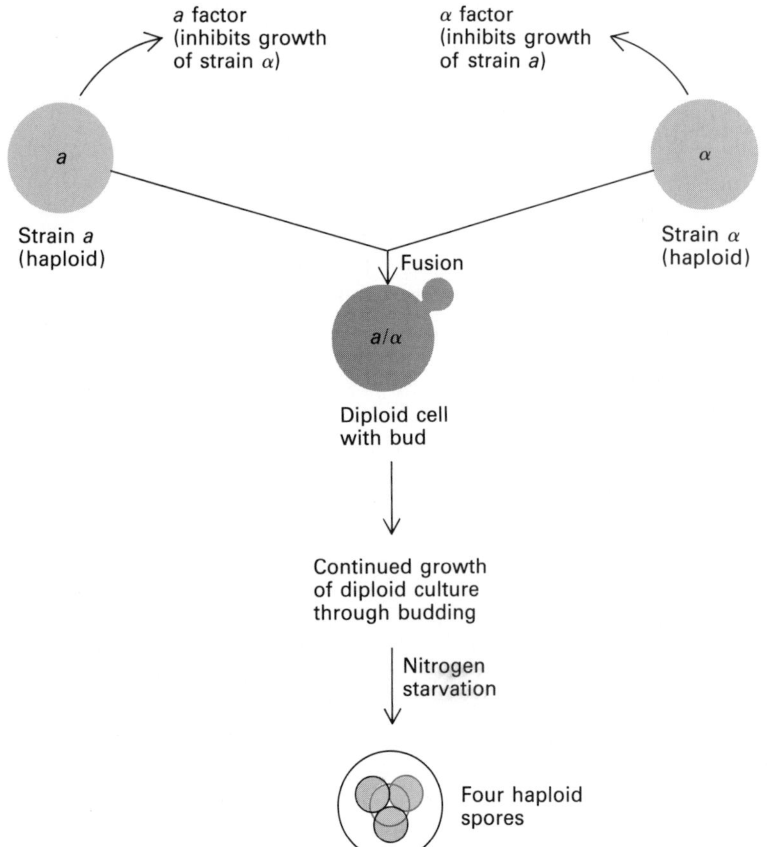

Figure 11-48 The yeast mating cycle: the two mating types *a* and *α* produce proteins called *a* factor and *α* factor, respectively; these reciprocally inhibit the growth of cells of the opposite mating type and thereby encourage the fusion of heterologous haploid cells. The resulting diploid cells can grow as such, until starved of nitrogen (or other nutrients) when they undergo meiosis to produce four spores each.

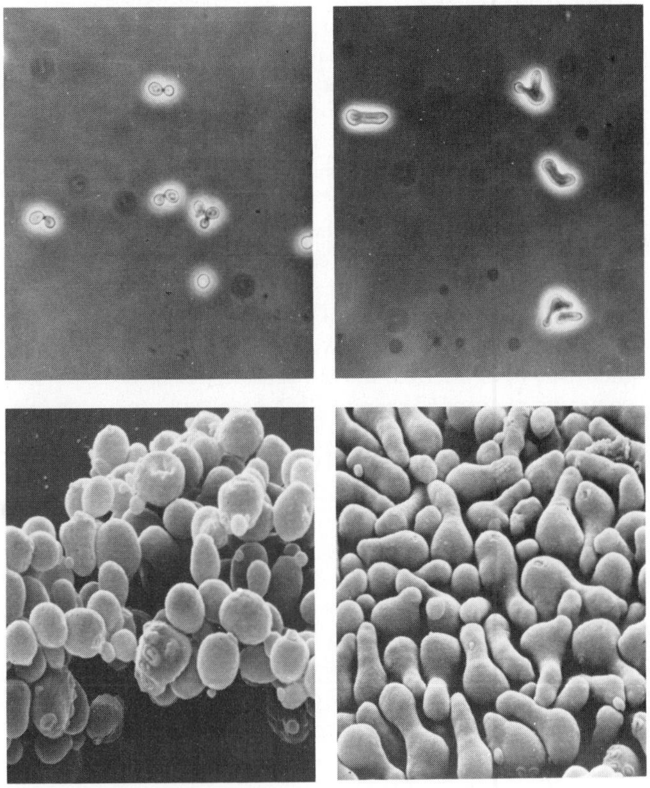

Figure 11-49 Detection of *a* and α phenotypes in yeast. Cells of an unknown mating type were separated into samples 1 and 2. Sample 1 was not exposed to α factor, but sample 2 was. *Top:* The light (phase contrast) micrograph of sample 1 shows buds and completed daughter cells; sample 2 shows inhibited cell division. Therefore the cells are *a,* because they are inhibited by α factor. *Bottom:* The scanning electron micrograph of sample 1 shows normal budding cells; a similar micrograph of sample 2 shows inhibition of cell division. (Yeast geneticists refer to the inhibited cells as "shmoos" after the "Lil Abner" cartoon characters that are plastic in shape.) [See J. Thorner, 1980, in *The Molecular Genetics of Development,* T. A. Leighton and W. A. Loomis, eds., Academic Press.] *Courtesy of J. Thorner, R. Kunisawa, and M. Davis.*

adding one of the two factors to yeast cells in culture (Figure 11-49).

A normal haploid cell can grow into a colony that contains both *a* and α mating types. This means that either mating type can be expressed from a single genome. In fact, a normal haploid cell has the ability to switch its mating type with nearly each generation. This ability depends on the *homothallic (HO)* locus in the genome: mutants in the *HO* locus no longer switch mating type. If the *HO* locus is normal, mating-type switching will be frequent, and cells expressing different mating types will be brought into proximity so that fusion can occur.

This remarkable switching in mating types has been traced to DNA rearrangements on chromosome 3 of *S. cerevisiae.* Three genetic loci directly involved in mating-type switching have been located. The central locus is termed MAT, or the mating-type locus. The DNA found at the MAT locus is actively transcribed into mRNA, and the protein encoded by this mRNA acts to regulate genes not directly linked to the MAT locus. These other genes encode proteins that give the cell its *a* or α phenotype. Two additional "silent" (nontranscribed) copies of the genetic information found at MAT are stored at loci termed HML and HMR (homothallic copies to the left and right of MAT, respectively, as the genetic map is usually drawn). One of the silent loci has the information for type *a* and the other for type α. Some of the sequences from HML and HMR is transferred alternately (between cell generations) into MAT, from which expression can then occur (Figure 11-50).

The DNA from the three mating-type loci in yeast has been cloned and sequenced. The diagram in Figure 11-51 indicates the structures of these sequences; the figure shows that blocks of sequence are similar, but that one block appears in α DNA only. The *a* DNA and the α DNA are quite similar in the X and Z regions, but they differ in the Y region. The mRNAs from the MAT locus include the differing Y sequences and variable parts of the X and Z regions. The transcriptional control of these mRNAs will be discussed in Chapter 12.

The variation in sequence content at MAT is probably accomplished by an act of gene conversion. A site-specific endonuclease isolated from HO^+ strains is absent in HO^- strains; the *HO* product is probably the endonuclease. Analysis of the DNA at MAT in HO^+ strains has shown that the endonuclease cleaves the α or *a* DNA specifically at the boundary between the Z and the Y sequences, as indicated in Figure 11-52a. About 3 percent of the cells in an HO^+ culture have a cut in MAT, but never one in the HML or HMR locus. This enzyme was the first sequence-specific endonuclease to be isolated from eukaryotic cells.

The gene conversion at the MAT locus presumably involves the free cut end of the DNA. This end can pair with the DNA of either HMR or HML. Because the switching of the information at MAT is the rule in HO^+ strains, the pairing to HMR or HML is somehow directed to the site containing information different from that presently residing at MAT (i.e., to HMLα in Figure 11-52b). DNA synthesis proceeding from the free 3′ hydroxyl on the Z_1 sequence of MAT starts a growing fork (see Figure 13-11 for an explanation of how a growing fork is initiated), and this results in the copying of the information at HMLα (Figure 11-52b). A recombination between the new copy of HMLα and MAT puts the α information at MAT and displaces the *a* information (Fig-

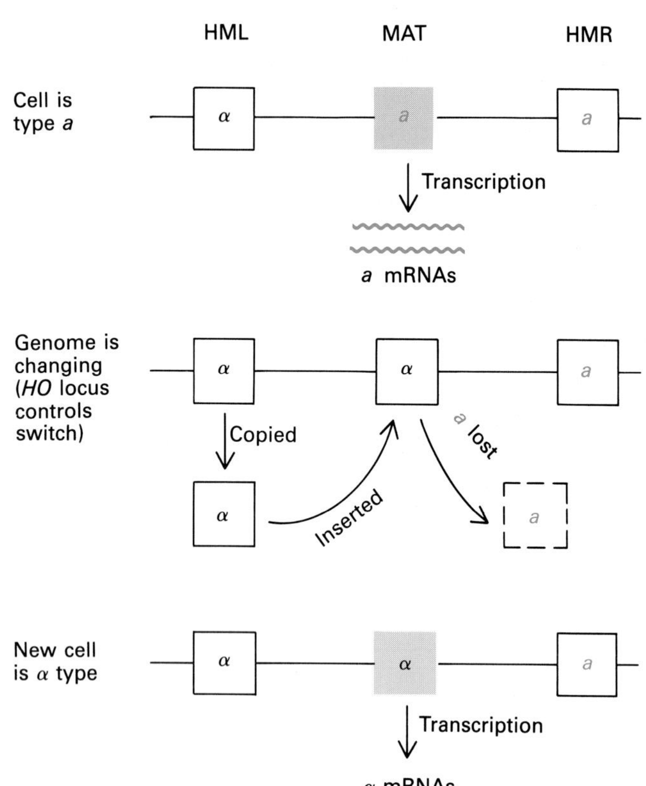

Figure 11-50 The cassette model of mating-type interconversion. Three copies of mating-type information exist on chromosome 3 in each yeast cell. The active copy at the MAT locus can be either *a* or *α*. In the diagram, MAT has the *a* information originally. Two "silent" (unexpressed) copies of mating-type information also exist, one at HML and one at HMR. In normal strains the information at HMR is opposite to that at HML. Here, *α* is at HML and *a* is at HMR. The mRNAs produced from MAT regulate other genes that determine mating type. When the cell switches to the opposite mating type, a replica of the new information is inserted at MAT (like a cassette tape is inserted into a tape player). As the new cassette (here, the *α*) is inserted, the old cassette (the *a*) is somehow removed. The model has been very useful in clarifying events during the process of mating-type interconversion. However, recent molecular studies have shown that the change does not actually involve free-floating pieces of DNA acting as cassettes. [See P. J. Kushner, L. C. Blair, and I. Herskowitz, 1979, *Proc. Nat'l Acad. Sci. USA* **76:**5264.]

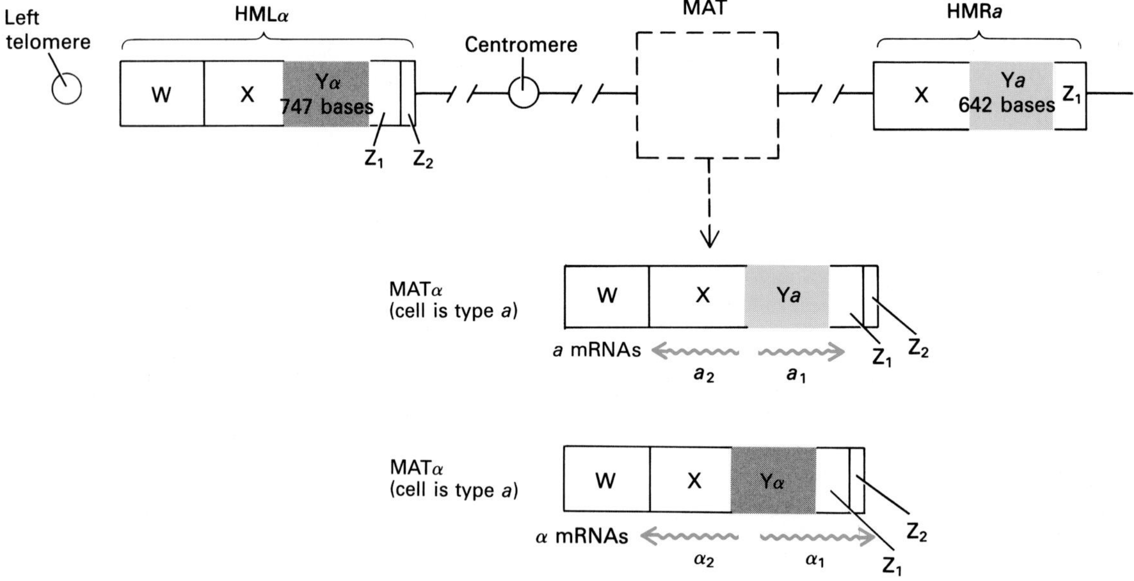

Figure 11-51 A comparison of cloned mating-type regions, using recombinant DNA from HML*α*, MAT*a*, MAT*α*, and HMR*a*. Regions X and Z_1 are present at all four sites; W and Z_2 are present at HML*α*, MAT*a*, and MAT*α*, but not at HMR*a*. The mRNAs from MAT*a* and MAT*α* differ most in their Y sequences, which are the main protein-coding regions. These conclusions were reached both by DNA sequencing and by electron microscopic analyses. Two different mRNAs are formed at MAT for each mating type. When a switch occurs at MAT, it is mainly the change in the Y sequence that differentiates the *a* and *α* protein products. The positions of the various loci with respect to the centromere and the telomere on the short arm of chromosome 3 are shown. [See K. A. Nasmyth, 1983, *Nature* **302:**670.]

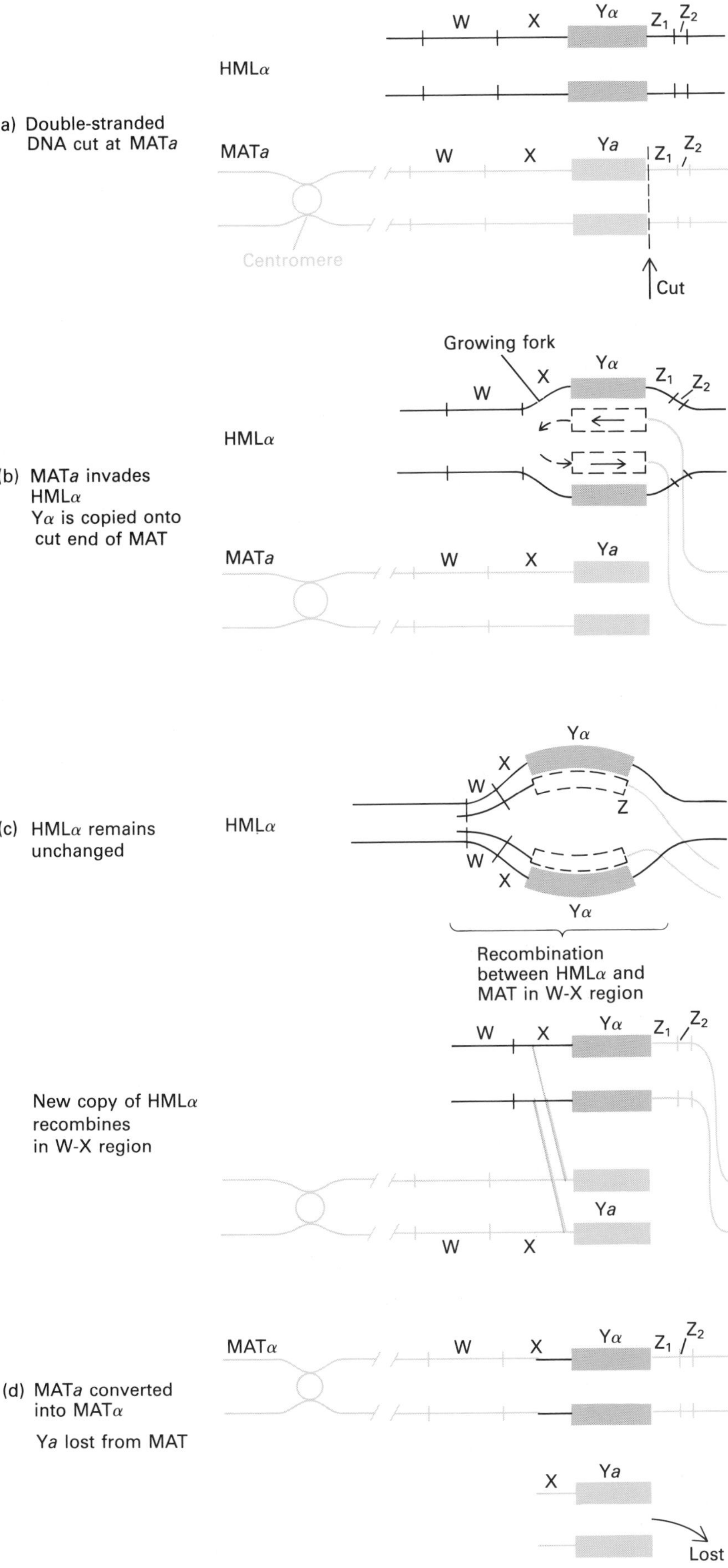

Figure 11-52 Gene conversion after an endonuclease cut at MAT. (a) An endonuclease formed by the *HO* locus can cut the DNA duplex specifically at the Y-Z junction. In vivo, only MAT is cut. (b) The free cut ends on the Z side of the duplex MAT*a* DNA invade (hybridize with) HMLα. DNA synthesis (*dashed lines*) initiated at one of the invading ends of MAT results in the copying of the Yα information onto the Z region of MAT. (c),(d) A recombination within the W-X region completes the transfer of the Yα information into MAT. [See J. N. Strathern et al., 1982, *Cell* **31**:183.]

ure 11-53d). The information in HML and HMR remains unchanged throughout this procedure.

Antigenic Variation in Trypanosomes Trypanosomes are protozoans that can cause severe illness (e.g., African sleeping sickness) in animals and humans. The best-studied of these organisms is *Trypanosoma brucei,* which can be grown in rats. This organism contains a highly antigenic surface glycoprotein. The glycoprotein is also variable: the organism has a mechanism that frequently changes the protein to a similar but distinct antigen. This change enables the protozoan to avoid host immune responses. The proteins involved in this switching are called *variable surface glycoproteins* (VSGs). The mRNAs responsible for several different VSGs were partially purified and used to make cDNA clones that were then sequenced. In addition, the genomic sequences complementary to the cDNA were examined. A number of interesting things were discovered.

Clusters of DNA sequences called *basic copies* (BCs) were found to encode the VSGs. Presumably, each VSG is encoded by a different BC. However, the sequence at the 3′ ends of all cDNAs tested showed that the BCs all had 3′ end regions different from the 3′ end of the mRNAs. In addition, the BCs did not have the 5′ end sequence found in the mRNAs. Apparently, the mRNAs were not synthesized directly from the BCs. By Southern blot analysis, the active site of transcription was determined to differ from the BCs. This site was found to contain a transposed and duplicated copy—the *expression-linked copy* (ELC)—of the active VSG gene, inserted next to a 3′ end sequence that was the same as the 3′ region found in the mRNAs. Upstream of the ELC are clusters of sequences that probably represent multiple start sites for transcription of the 5′ ends (approximately 35 bases in length) for the VSG mRNAs. These may be spliced from a primary transcript, or they may be ligated in a bimolecular reaction (Figure 11-53).

The chromosomal rearrangements giving rise to differing mRNAs for VSGs in trypanosomes are *not* programmed changes that occur before each cell division, rather they are probably random events. These variations in DNA structure are similar to yeast mating-type changes to the extent that one copy is maintained and one copy is activated, and a gene conversion may be responsible. Because the trypanosome circumvents the effects of antibodies produced by the infected animal, the changes have an evolutionary value for the trypanosome. It is possible that a similar rearrangement of DNA occurs in *Paramecium,* another protozoan that changes its surface proteins often.

Ribosomal DNA Amplification in Amphibian Oocytes All of the DNA rearrangements we have discussed so far involve sequences that change positions within the genome. However, the first proven case of a programmed change in genomic DNA did not involve a change in the position of a sequence, but rather the amplification of a specific DNA—namely, the DNA encoding ribosomal RNA. Amplification is the result of site- or locus-specific DNA synthesis. The amplified copies may remain in tandem at the site where the increase takes place, or they may be released as free-floating extra copies. Specific gene amplification was first discovered in the rRNA genes of frogs and beetles; it probably occurs also in many other animals with highly developed egg cells.

The eggs of frogs are 2 to 3 mm in diameter and contain many stored ribosomes. During the development of the egg, the number of nucleoli increases dramatically, and so does the amount of ribosomal DNA. This increase is brought about by *rolling circle* DNA synthesis (Figure 11-54), which amplifies the normal level of ribosomal DNA by a factor of 2000. Active transcription of this amplified DNA produces the large number of stored ribosomes. (See Figure 9-12 for a view of rDNA transcription in frog oocytes.)

Polytenization and Site-Specific Amplification in Insects The amplification of DNA in the salivary gland chromosomes of *Drosophila* species has been observed and exploited for many years. The bands on these enlarged chromosomes are the basis for all detailed cytogenetic studies of the organisms. The enlargement of chromosomes in the salivary glands, and in other tissues as well, occurs when the DNA repeatedly replicates but the daughter strands fail to separate. The result is a *polytene* chromosome composed of as many as 1000 parallel copies of itself. Although nearly the whole chromosome participates in polytenization, certain sequences, such as the simple sequence DNA near the centromere, do not participate in this increase. Furthermore, the ribosomal genes tend to be amplified less than the other amplified sequences during polytenization (Figure 11-55).

The molecular basis for the varying extent of replication along the presumably linear chromosomal DNA molecules remains unknown, as does the reason for the differences in the degree of polytenization in different cell types. Polytenization does vary according to tissue, with the salivary glands possibly achieving a 1000-fold increase whereas abdominal tissue contains cells that amplify DNA about 100 times only.

In spite of the rather widespread occurrence of polytenization (some of it unequal, as noted), until recently it was thought that localized amplification of specific genes happened very rarely, if in fact at all. However, in flies of the genus *Rhynchosciara,* isolated cases of chromosomal enlargements or "puffs" have been attributed to site-specific DNA amplifications (Figure 11-56). Recent evidence has also identified increased synthesis of a pro-

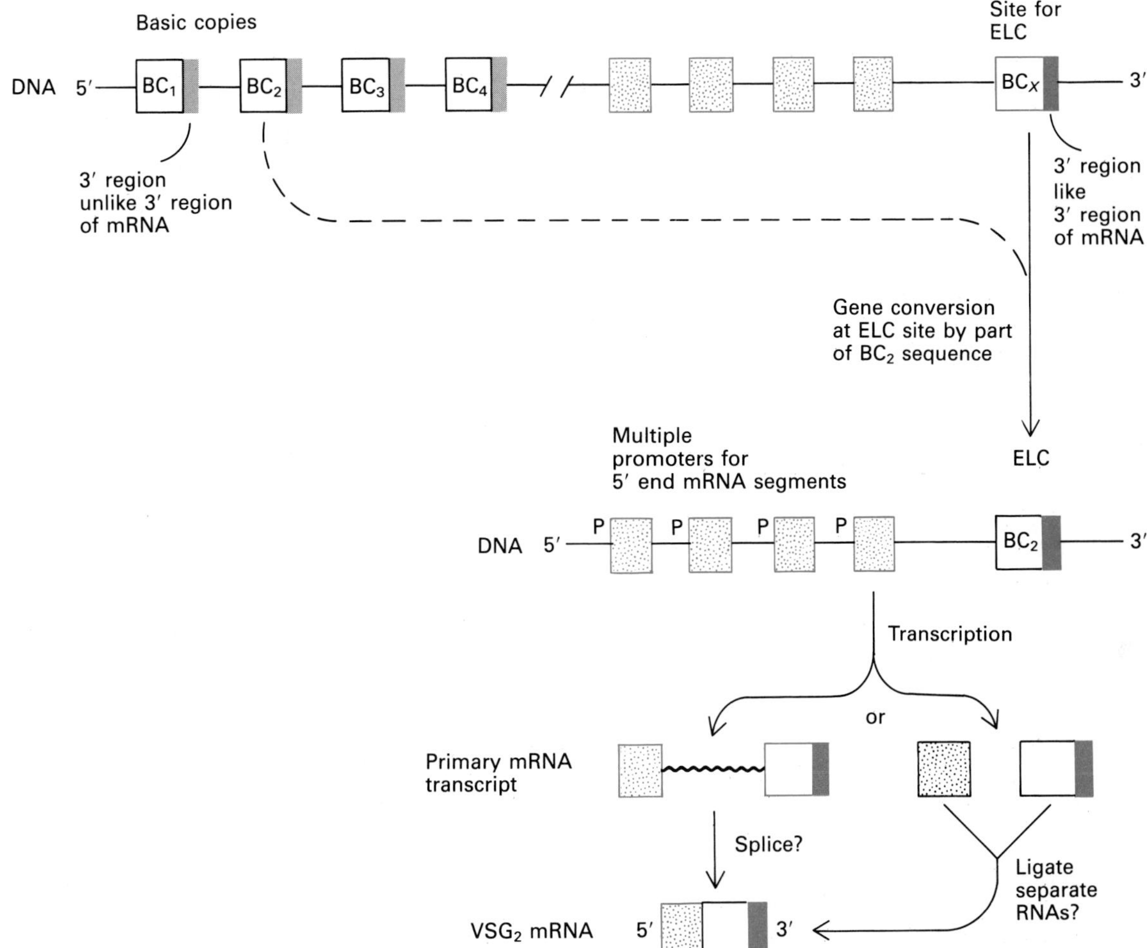

Figure 11-53 The structure and arrangement of DNA that encodes variable surface glycoproteins (VSGs) in trypanosomes. After the discovery that cDNA clones prepared from VSG mRNAs had different end sequences than the genomic BC (basic copy) clones to which the cDNAs hybridized, Southern blot analysis was used to locate an active site of transcription elsewhere in the genome. Gene conversion was inferred as the mechanism by which BCs become ELCs (expression-linked copies). [See P. Borst and G. M. Cross, 1982, *Cell* 29:291.]

tein that might be encoded in the amplified DNA of *Rhynchosciara*.

In addition to the cytologic results with *Rhynchosciara*, it is known that genes encoding eggshell (chorion) proteins in the ovarian follicle cells of *D. melanogaster* are amplified. The amplification occurs only in the ovarian follicle tissue (Figure 11-57). A DNA region of about 100 kb participates in this amplification. In insects, gene-specific amplification may employ the same enzymes as does the generalized polytenization of the chromosomes.

Insect genes called upon to program the production of large amounts of protein do not necessarily become amplified. For example, the gene that encodes silk fibroin, the major protein in silk, is not amplified in the silk moth. Many vertebrate genes that produce large amounts of mRNA are also known to be unamplified. Examples are the hemoglobin, ovalbumin, and serum albumin genes. However, now that protein-coding gene amplification has been discovered in insects, renewed attempts will undoubtedly be made to find cases of this phenomenon in vertebrates.

DNA Amplification in Cultured Mammalian Cells When mammalian cells are grown in culture, specific sites on their DNA can become amplified. However, this amplification probably does not reflect a capacity of normal mammalian cells to engage in the kind of amplification we have described for frog and fruit fly cells. Sequence amplification in cultured mammalian cancer cells occurs when cells are kept in the presence of toxic substances such as the drug *methotrexate*, which inhibits the action of dihydrofolate reductase (DHFR), an enzyme necessary for the transfer of methyl groups. This enzyme

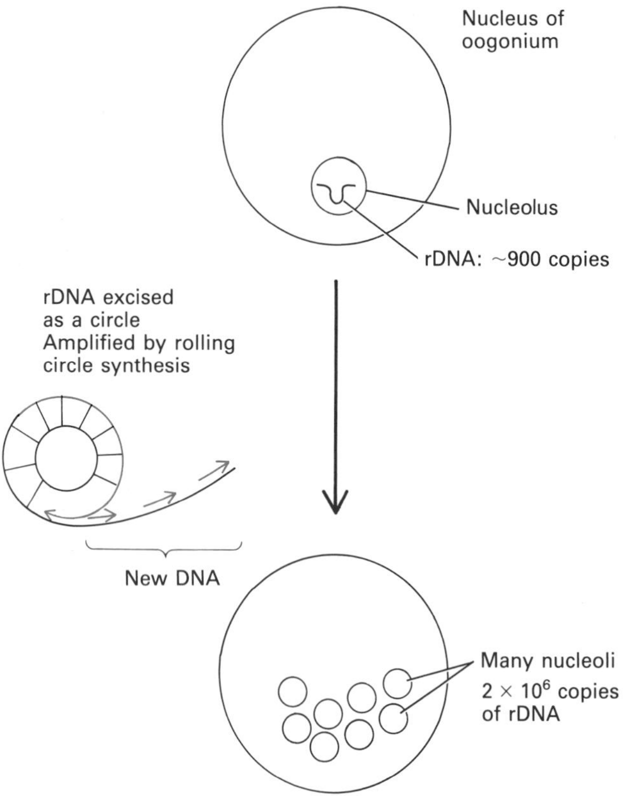

Figure 11-54 Ribosomal DNA amplification in frog oocytes. At the beginning of oocyte development, the oogonia (small oocytes) have a single nucleolus and approximately 900 copies of rDNA, the genes encoding rRNA. By the end of prophase I, the oocyte has as many as 3000 nucleoli and 2×10^6 copies of rDNA. During oocyte development, many copies of extrachromosomal rDNA are found as circular molecules that somehow must have been cut out of the chromosome. These circles are the source of the amplified rDNA. The replication of the new DNA takes place by a mechanism called rolling circle replication (described in Chapter 13). The replicative molecules are illustrated in the electron microscope autoradiograph shown here. The cells were exposed to [^{3}H]thymidine, a DNA precursor, and DNA was isolated. Circles of DNA were observed; the long DNA strand extending from the circle is newly made DNA because it contains [^{3}H]thymidine (detected by silver grains in the photographic emulsion). All of this extra rDNA is lost after the egg matures and is fertilized. [See J. G. Gall and M. L. Pardue, 1969, *Proc. Nat'l Acad. Sci. USA* **63**:378; and J.-D. Rochaix, A. Bird, and A. Bakken, 1974, *J. Mol. Biol.* **87**:473.] *Photograph courtesy of A. Bird.*

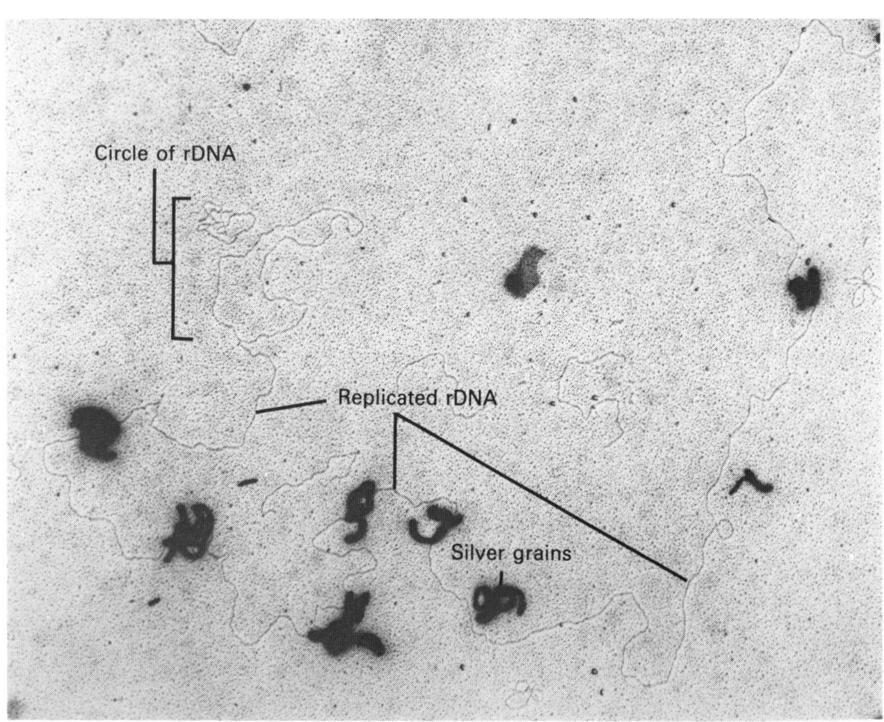

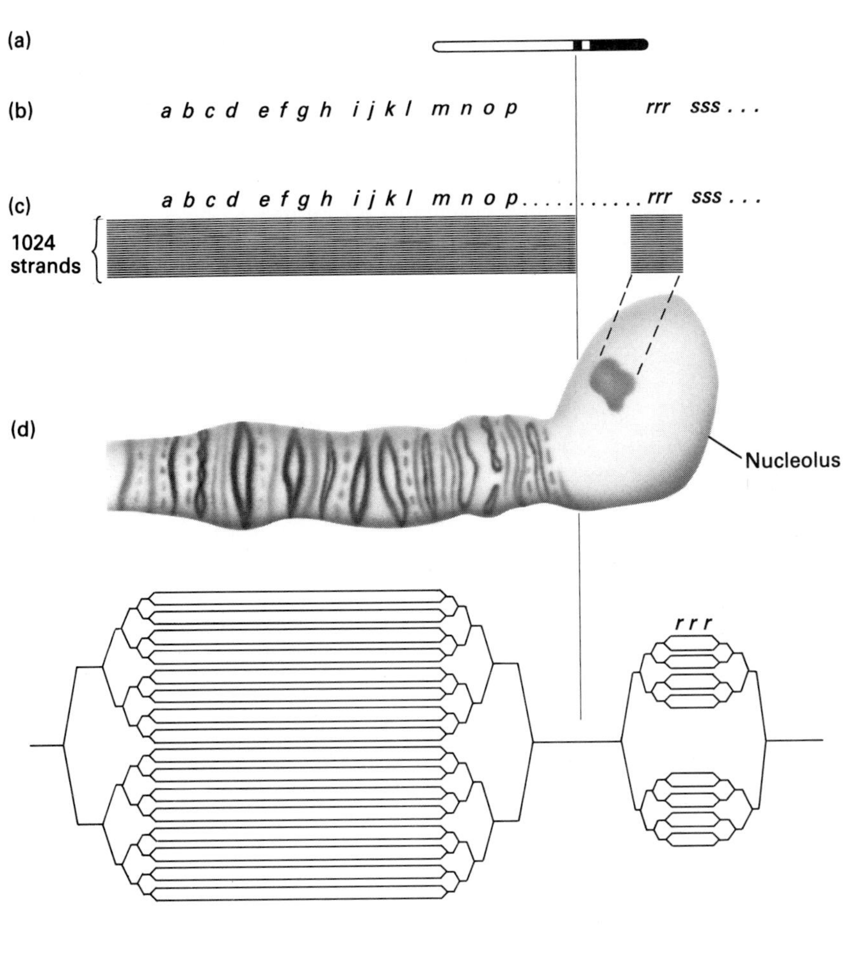

Figure 11-55 A model of a *Drosophila* polytene chromosome. (a) The chromosome pictured is the *Drosophila hydei* X chromosome, in which the entire left arm is euchromatic (lightly staining). (b) The left arm contains many sequences, some of which are represented by the string of letters from *a* to *p*. The heterochromatic region near the centromere is known to contain simple sequence DNA (*sss*) as well as ribosomal genes (*rrr*). (c) All of the DNA in the left arm is replicated in parallel many times. The number 1024 is suggested since 2^{10} duplications would give 1024. The *sss* DNA does not appear to increase at all; the ribosomal genes do increase, but less than the DNA in the left arm. (d) The polytenization produces a banding pattern that amplifies the visual variation along each strand. The *rrr* genes are localized in the nucleolus. (e) The diagram shows the layout of the replicated strands in a polytene chromosome. [See C. D. Laird et al., 1973, *Cold Spring Harbor Symp. Quant. Biol.* **38:**311.]

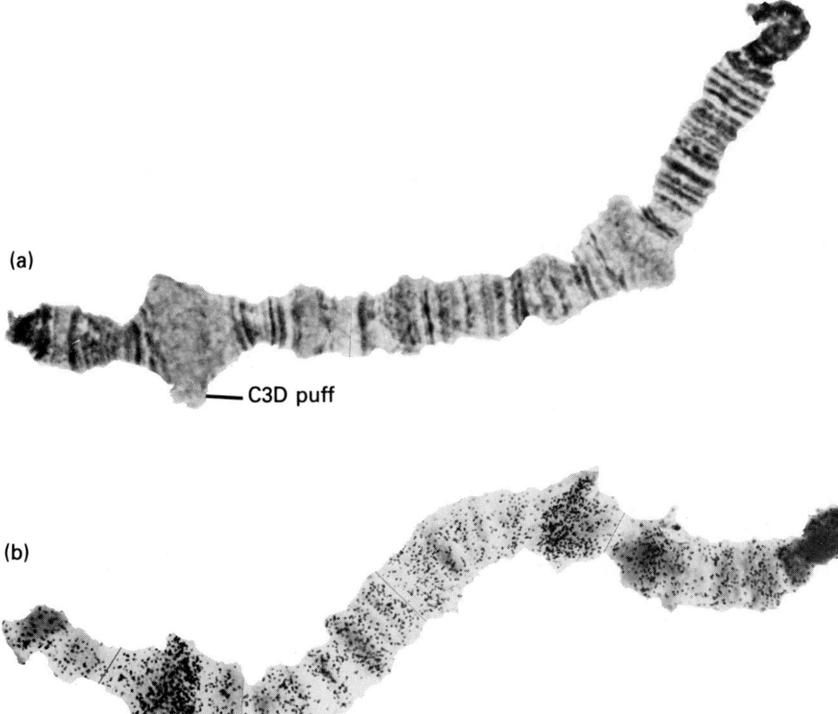

Figure 11-56 DNA "puffs" in *Rhynchosciara* salivary gland chromosomes. (a) An enlarged chromosomal region termed C3D is easily located microscopically in a stained chromosome. (b) The same region is shown by [^{3}H]thymidine incorporation and autoradiography to be a site of intense DNA synthesis. *Courtesy of F. Lara.*

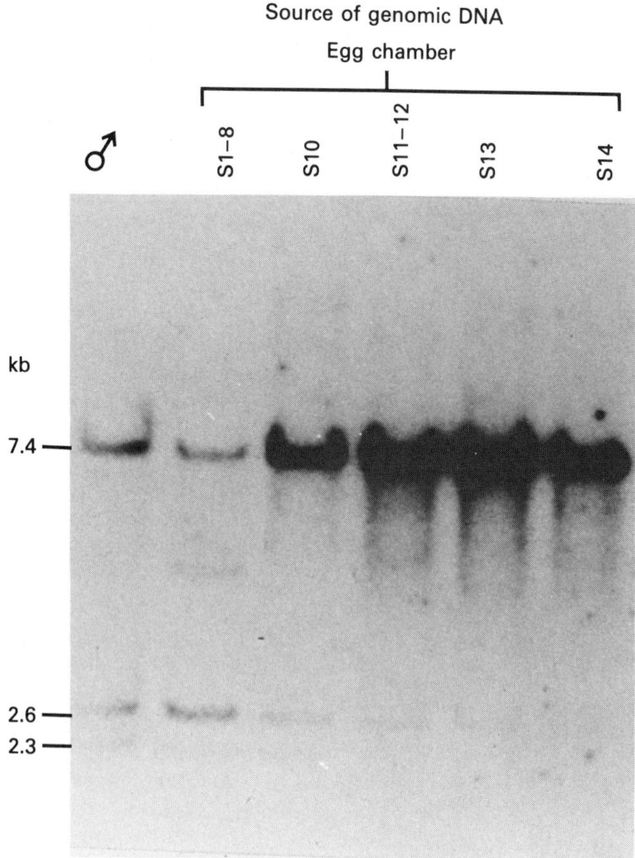

Figure 11-57 Amplification of specific DNA during *Drosophila* development. As *Drosophila* eggs develop in their egg chambers, they become coated with a thick fibrous chorion produced by ovarian follicle cells. The mRNAs for chorion proteins have been cloned as cDNA copies. The genomic DNAs complementary to several of the cDNA chorion clones have been found to be amplified in follicle cells during development.

The results shown here are from Southern blot analysis of equal total amounts of DNA from normal adult males and from the egg chamber at several different recognizable stages of larval development (these are labeled S1–8 and S10). A 7.4-kb DNA band complementary to the cloned cDNA is observed to increase by a factor of 50 during the stages from S10 to S14. [See A. C. Spradling and A. P. Mahowald, 1980, *Proc. Nat'l Acad. Sci. USA* 77:1096.] *Courtesy of A. C. Spradling.*

is required in the synthesis of nucleic acid bases and in a number of other synthetic reactions.

When mammalian cell cultures are exposed to methotrexate, they develop drug-resistant colonies. As the concentration of the drug is gradually increased, the average level of dihydrofolate reductase in the culture becomes 1000 times higher than normal. Along with this increase in enzyme activity is an increase in the DNA encoding the enzyme. On the surface, it would appear that specific gene amplification occurs in response to need.

Although the precise mechanism for mammalian DNA amplification is not known, methotrexate treatment likely leads to chromosomal instability. Unequal sister chromatid exchanges resulting in donation of the DHFR site to one chromosome would be favored during methotrexate treatment. In addition, chromosomal fragments termed *minute chromosomes* are created by the treatment; these fragments could be, upon re-entry into normal chromosomes, a source of "amplification" (Figure 11-58). Repetition of either of these events would, in the continuing presence of methotrexate, lead to increased DHFR genes.

It is of considerable interest that all cultured cells known to undergo drug-selected amplification are abnormal cancer cells, and that normal fibroblasts in culture do not undergo amplification. Therefore, normal cells in the mammalian body may not possess special amplification mechanisms in response to a need for the gene product. Of course, if certain cells within a mammal can directly amplify genes in response to such a need, the amplification mechanisms of DHFR genes in cultured cells could be extremely important models. Although no direct gene amplification due to environmental stress has yet been found in normal cells, in cancer patients treated with methotrexate, cancer cells with amplified DHFR genes have been observed. Thus in abnormally growing cells DNA can undergo amplification in vivo as well as in culture.

Deletions Produce Immunoglobulin Transcription Units The greatest interest in DNA rearrangement during differentiation in mammals has centered on one class of genes: the immunoglobulins. In all vertebrates, a series of interacting cells collectively called the *immune system* cooperate to protect the animal from noxious outside influences—for example, viruses and bacteria. Perhaps the most well-studied event of immunity is the production and secretion of specific antibodies (immunoglobulins), which are produced by lymphoid cells. While these cells are undergoing differentiation, specific deletions occur in

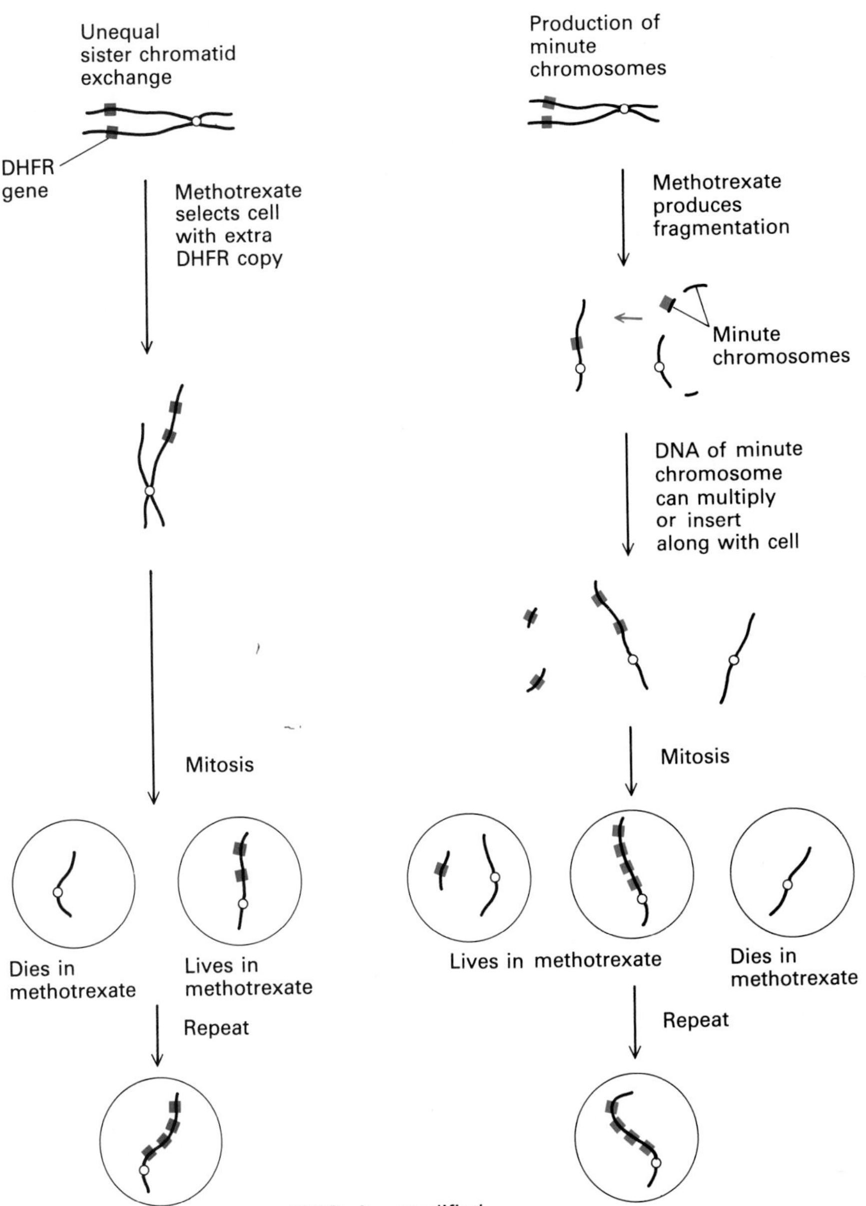

Figure 11-58 Mechanisms for gene amplification in cultured mammalian cells. The distribution of minute chromosomes is irregular, but they can insert themselves into full-sized chromosomes to form stable associations.

their DNA. The deletions bring together regions of DNA encoding the L ("light") protein chain of immunoglobulin as well as regions encoding the H ("heavy") chain. (Figure 3-38 shows the structure of an immunoglobulin.) Apparently these DNA rearrangements occur randomly in the lymphoid cell population of the animal. Only a few cells in any individual survive to become successful antibody producers. Chapter 24 discusses the lymphoid system in detail.

Summary

In recent years, a wealth of new information about the DNA in the chromosomes of many different organisms has become available. Although generalizations are risky at this point, some summary statements may be helpful.

1. The genome of bacteria, exemplified by *E. coli,* has little wasted space. Many of the protein-coding genes are

arranged in operons, which when transcribed usually result in polycistronic mRNAs (Chapter 8). Apparently little nonfunctional DNA is contained both within and between operons. The only repetitive DNAs are for ribosomal genes, insertion sequences (IS elements), and transposons. The only *E. coli* chromosomal DNA that may serve no useful function is that of the IS elements and some of the transposons.

2. Eukaryotic single-celled organisms such as yeast have only three to five times as much DNA per cell as prokaryotes; they, too, appear to have little "extra" DNA aside from transposons and the relatively few introns.

3. The protein-coding portion of eukaryotic cells of all metazoans and higher plants makes up a small fraction—possibly as low as 1 percent—of the total DNA of the cell. Even if transcription units (exons plus introns) average ten times the length of mature mRNA, only 10 percent of the genome would be involved in encoding transcription units. The protein-coding genes seem to be islands floating in a sea of meaningless DNA. Distinct functional transcription units are most often separated by great distances.

4. The "extra" DNA in eukaryotes consists of introns, pseudogenes, simple sequence DNA, transposable elements, and unclassified spacer DNA between transcription units. The amount of simple sequence DNA varies greatly among genomes that are otherwise comparable in size, which suggests that even if some simple sequence DNA serves an important function, there may be vastly more than is required. Pseudogenes of many types have been identified; sequence analysis has shown that these genes were formerly normal but have lost critical nucleotide sequences. Introns are present in some transcription units from all eukaryotic cell types, and they abound in vertebrate genes, where they may make up 80 to 90 percent of all DNA contained within transcription units. Introns perform no known function, and, like other extra DNA, they appear to be dispensable. Finally, clearly identified transposons exist in many cells. In yeast and in fruit flies, all (or nearly all) of the dispersed intermediate repeat DNA is contributed by such elements. Most transposons appear to be no more useful to the eukaryotic cell than to the prokaryotic cell.

5. The discovery of gene conversion in yeast and trypanosomes, of specific gene amplification in insects and frogs, and of DNA rearrangement in the differentiation of immunoglobulin-forming cells suggests that DNA rearrangements might play a wider role in gene expression than has been suspected. Many pathways of differentiation are thought to proceed systematically toward greater and greater cell specialization. In these cases, it is tempting to think that permanent (or near-permanent) gene rearrangements at the DNA level might underlie some "irreversible" developmental changes. However, the gene rearrangements in immunoglobulin formation, in trypanosome surface protein variation, and in most other cases are not strictly programmed and involve much cell death before a successful rearrangement is selected. Perhaps only in cases where cell death is a prominent part of embryogenesis is gene rearrangement likely. For example, considerable cell death occurs during the development of the nervous system. Before the importance of DNA rearrangements in differentiation can be ascertained, much work remains to be done: many developmentally important genes must be isolated and studied to determine whether their structures are altered during development.

References

Protein-Coding Genes and Gene Families: Nature and Distribution

BOSSY, B., L. M. C. HALL, and P. SPIERER. 1984. Genetic activity along 315 kb of the *Drosophila* chromosome. *EMBO J.* 3:2537–2541.

DAYHOFF, M., ed. 1976. *Atlas of Protein Sequence and Structure,* Suppl. 2. National Biomedical Research Foundation, Washington, D. C.

EFSTRATIADIS, A., J. W. POSAKONY, T. MANIATIS, R. M. LAWN, C. O'CONNELL, R. A. SPRITZ, J. K. DERIEL, B. G. FORGET, S. M. WEISSMAN, J. L. SLIGHTOM, A. E. BLECHL, O. SMITHIES, F. E. BARALLE, C. C. SHOULDERS, and N. J. PROUDFOOT. 1980. The structure and evolution of the human β-globin gene family. *Cell* **21**:653–668.

EIGERMAN, F. A., P. R. YOUNG, R. W. SCOTT, and S. M. TILGHMAN. 1981. Intragenic amplification and divergence in the mouse α fetoprotein gene. *Nature* **294**:713–718.

FIRTEL, R. A. 1981. Multigene families encoding actin and tubulin. *Cell* **25**:6–7.

GORIN, M. B., D. L. COOPER, F. EIFERMAN, P. VAN DE RIJN, and S. M. TILGHMAN. 1981. The evolution of α-fetoprotein and albumin, I: A comparison of the primary amino acid sequences of mammalian α-fetoprotein and albumin. *J. Biol. Chem.* **256**:1954–1959.

HENTSCHEL, C. C., and M. L. BIRNSTIEL. 1981. The organization and expression of histone gene families. *Cell* **25**:301–313.

JONES, C. W., and F. KAFATOS. 1980. Structure, organization and evolution of developmentally regulated chorion genes in a silk moth. *Cell* **22**:855–867.

KAFATOS, F. C. 1983. Structure, evolution and developmental expression of the chorion multigene families in silk moths and *Drosophila.* In *Gene Structure and Regulation in Development,* S. Subtelny and F. C. Kafatos, eds. Alan R. Liss.

PIATIGORSKY, J. 1984. Lens crystallins and their gene families. *Cell* **38**:620–621.

ROYAL, A., A. GARAPIN, B. CAMI, F. PERRIN, J. L. MANDEL, M. LEMEUR, F. BRÉGÉGÈRE, F. GANNON, J. P. LEPENNEC, P. CHAMBON, and P. KOURILSKY. 1979. The ovalbumin gene region: common features in the organisation of three genes expressed in chicken oviduct under hormonal control. *Nature* **279**:125–132.

STEINEET, P. M., A. C. STEVEN, D. R. ROOP. 1985. The molecular biology of intermediate filaments *Cell* **42**:411–419.

SUBTELNY, S., and F. C. KAFATOS, eds. 1983. *Gene Structure and Regulation in Development.* Alan R. Liss.

Tandemly Repeated Genes Encoding rRNA

ARNHEIM, N., and E. M. SOUTHERN. 1977. Heterogeneity of the ribosomal genes in mice and men. *Cell* **11**:363–370.

FEDEROFF, N. 1979. On spacers. *Cell* **16**:697–710.

OHTA, T. 1983. On the evolution of multigene families. *Theoretical Population Biol.* **23**:216–240.

PETES, T. D. 1980. Unequal meiotic recombination within tandem arrays of yeast ribosomal DNA genes. *Cell* **19**:765–774.

WELLAUER, P. K., and I. B. DAWID. 1979. Isolation and sequence organization of human ribosomal DNA. *J. Mol. Biol.* **128**:289–303.

Mechanisms of Sequence Maintenance

ARNHEIM, N. 1983. Concerted evolution of multigene families. In *Evolution of Genes and Proteins,* M. Nei and R. K. Koehn, eds. Sunderland, Mass.: Sinauer Associates.

PETES, T., and G. R. FINK. 1982. Gene conversion between repeated genes. *Nature* **300**:216–217.

SMITH, G. P. 1976. Evolution of repeated DNA sequences by unequal crossovers. *Science* **191**:528–537.

SPRADLING, A. C., and G. M. RUBIN. 1981. *Drosophila* genome organization: conserved and dynamic aspects. *Annu. Rev. Genet.* **15**:219–264.

Repetitious DNA

BRITTEN, R. J., and D. E. KOHNE. 1968. Repeated sequences in DNA. *Science* **161**:529–540.

DOOLITTLE, W. F., and C. SAPIENZA. 1980. Selfish genes, the phenotype paradigm and genome evolution. *Nature* **284**:601–603.

GALL, J. G. 1981. Chromosome structure and the C-value paradox. *J. Cell Biol.* **91**:3s–14s.

GEBHARD, W., and H. G. ZACHAU. 1983. Organization of the R family and other interspersed repetitive DNA sequences in the mouse genome. *J. Mol. Biol.* **170**:255–270.

LONG, E. O., and I. B. DAWID. 1980. Repeated genes in eucaryotes. *Annu. Rev. Biochem.* **49**:727–764.

ORGEL, L. E., and F. H. C. CRICK. 1980. Selfish DNA: the ultimate parasite. *Nature* **284**:604–607.

SINGER, M. F. 1982. SINES and LINES: highly repeated short and long interspersed sequences in mammalian genomes. *Cell* **28**:433–434.

The *Alu* Sequence Family and Genomic Re-Entry via RNA Copying

HAYNES, S. R., and W. R. JELINEK. 1981. Low molecular weight RNAs transcribed *in vitro* by RNA polymerase III from *Alu*-type dispersed repeats in Chinese hamster DNA are also found *in vivo*. *Proc. Nat'l Acad. Sci. USA* **78**:6130–6134.

JAGADEESWARAN, P., B. G. FORGET and S. M. WEISSMAN. 1981. Short interspersed repetitive DNA elements in eucaryotes: transposable DNA elements generated by reverse transcription of RNA Pol III transcripts? *Cell* **26**:141–142.

JELINEK, W. R., and C. W. SCHMID. 1982. Repetitive sequences in eukaryotic DNA and their expression. *Annu. Rev. Biochem.* **51**:813–844.

SCHMID, C. W., and W. R. JELINEK. 1982. The *Alu* family of dispersed repetitive sequences. *Science* **216**:1065–1070.

SHARP, P. A. 1983. Conversion of RNA to DNA in mammals: *Alu*-like elements and pseudogenes. *Nature* **301**:471–472.

VAN ARSDELL, S. W., R. A. DENISON, L. B. BERNSTEIN, A. M. WEINER, T. MANSER, and R. F. GESTELAND. 1981. Direct repeats flank three small nuclear RNA pseudogenes in the human genome. *Cell* **26**:11–17.

Rearrangements in Chromosomal DNA: Transposition and Transposable Elements

BOEKE, J. D., D. J. GARFINKEL, C. A. STYLES, and G. R. FINK. 1985. Ty elements transpose through an RNA intermediate. *Cell,* **40**:491–500.

BURR, B., and F. A. BURR. 1982. *Ds* controlling elements of maize at the *shrunken* locus are large and dissimilar insertions. *Cell* **29**:977–986.

CALOS, M. P., and J. H. MILLER. 1980. Transposable elements. *Cell* **20**:579–595.

FINCHAM, J. R. S., and G. R. K. SHASTRY. 1974. Controlling elements in maize. *Annu. Rev. Genet.* **8**:15–50.

GARFINKEL, D., J. BOEKE, and G. R. FINK. 1985. Ty element transposition: reverse transcriptase and virus-like particles. *Cell* **42**:507–517.

GOLDEN, J. W., S. J. ROBINSON, and R. HASELKORN. 1985. Rearrangement of nitrogen fixation genes during heterocyst differentiation in the cyanobacterium *Anabaena. Nature* **314**:419–423.

KLECKNER, N. 1981. Transposable elements in prokaryotes. *Annu. Rev. Genet.* **15**:341–404.

MCCLINTOCK, B. 1956. Controlling elements and the gene. *Cold Spring Harbor Symp. Quant. Biol.* **21**:197–216.

O'HARE, K., and G. RUBIN. 1983. Structures of P transposable elements and their sites of insertion and excision in the *Drosophila melanogaster* genome. *Cell* **34**:25–35.

ROEDER, G. S., and G. R. FINK. 1983. Transposable elements in yeast. In *Mobile Genetic Elements,* J. Shapiro, ed. Academic Press.

RUBIN, G. M. 1983. Dispersed repetitive DNAs in *Drosophila.* In *Mobile Genetic Elements,* J. Shapiro, ed. Academic Press.

RUBIN, G. M., and A. C. SPRADLING. 1982. Genetic transformation of *Drosophila* with transposable element vectors. *Science* **218**:348–353.

SHAPIRO, J. A. 1979. Molecular model for the transposition and replication of bacteriophage Mu and other transposable elements. *Proc. Nat'l Acad. Sci. USA* **76**:1933–1937.

SHAPIRO, J. A., ed. 1983. *Mobile Genetic Elements.* Academic Press.

SILVERMAN, M., and M. SIMON. 1980. Phase variation: genetic analysis of switching mutants. *Cell* **19**:845–854.

SIMON, M., and I. HERSKOWITZ, eds. 1985. Genome rearrangement. In *UCLA Symposia on Molecular and Cellular Biology,* vol. 20. Alan R. Liss.

SPRADLING, A. C., and G. M. RUBIN. 1982. Transposition of cloned P elements into *Drosophila* germ line chromosomes. *Science* **218**:341–347.

TRUETT, M. A., R. S. JONES, and S. S. POTTER. 1981. Unusual structure of the FB family of transposable elements in *Drosophila*. *Cell* 24:753–763.

WILLIAMSON, V. M., E. T. YOUNG, and M. CIRIACY. 1981. Transposable elements associated with constitutive expression of yeast alcohol dehydrogenase II. *Cell* 23:605–614.

YOUNG, M. W. 1982. Differing levels of dispersed repetitive DNA among closely-related species of *Drosophila*. *Proc. Nat'l Acad. Sci. USA* 79:4570–4574.

Functional Genomic Rearrangements

Yeast Mating-Type Switching

HERSKOWITZ, I., and O. OSHIMA. 1982. Control of cell type in *S. cerevisiae:* mating type and mating type interconversions. In *Molecular Biology of the Yeast* Saccharomyces, J. N. Strathern, E. W. Jones, and J. R. Broach, eds. Cold Spring Harbor Laboratory.

KOSTRIKEN, R., J. N. STRATHERN, A. J. S. KLAR, J. B. HICKS, and F. HEFRON. 1983. A site-specific endonuclease essential for mating-type switching in *Saccharomyces cerevisiae*. *Cell* 35:167–174.

KUSHNER, P. J., L. C. BLAIR, and I. HERSKOWITZ. 1979. Control of yeast cell types by mobile genes: a test. *Proc. Nat'l Acad. Sci. USA* 76:5264–5268.

NASMYTH, K. A. 1983. Molecular analysis of a cell lineage. *Nature* 302:670–676.

STRATHERN, J. N., E. W. JONES, and J. R. BROACH, eds. 1982. *Molecular Biology of the Yeast* Saccharomyces. Cold Spring Harbor Laboratory.

STRUHL, K. 1983. The new yeast genetics. *Nature* 305:391–397.

Antigenic Variation in Trypanosomes

BORST, P., and G. A. M. CROSS. 1982. Molecular basis for trypanosome antigenic variation. *Cell* 29:291–303.

DNA Amplification

ALT, F. W., R. E. KELLEMS, J. R. BERTINO, and R. T. SCHIMKE. 1978. Selective multiplication of dihydrofolate reductase genes in methotrexate resistant variants of cultured murine cells. *J. Biol. Chem.* 253:1357–1370.

GOLDSMITH, M. R., and F. C. KAFATOS. 1984. Developmentally regulated genes in silkmoths. *Annu. Rev. Genet.* 18:443–488.

SCHIMKE, R. T. 1980. Gene amplification and drug resistance. *Sci. Am.* 243(5):60–69.

SPRADLING, A. C., and A. P. MAHOWALD. 1980. Amplification of genes for chorion proteins during oogenesis in *Drosophila melanogaster*. *Proc. Nat'l Acad. Sci. USA* 77:1096–1100.

STARK, G. R., and G. M. WAHL. 1984. Gene amplification. *Annu. Rev. Biochem.* 53:447–491.

12

Gene Control in Eukaryotes

THE molecular basis of gene control in eukaryotic cells, particularly in animal cells, is currently one of the most actively studied areas in all of biology. Throughout the development of classical genetics, a major aim was to discover how genes participate in controlled programs that result in the development of an animal or a plant. Although that broad challenge still exists, considerable progress has been made in our understanding of the control of gene expression, not only in single-celled eukaryotes but also in animal cells, both normal and virus-infected. In this chapter we shall review that progress and discuss some of the problems that remain. In Chapter 22 we shall return briefly to these issues, particularly as they apply to the development of animals.

The Scope of Studies of Eukaryotic Gene Control

The complex set of interactions that is broadly referred to as gene control is responsible for differences in the rate of synthesis of various proteins by different cells. Such differential gene expression results from the regulation of the formation and use of mRNA at a number of different molecular steps, or *levels of control. Differential tran-*

scription is probably the most frequent basis of differential protein synthesis. We shall see, however, that *differential processing* of RNA transcripts in the cell nucleus, *differential stabilization* of mRNA in the cytoplasm, and *differential translation* of mRNA into protein can also be factors in gene control. Furthermore, as we discussed in Chapter 11, chromosomal segments can be amplified or rearranged so as to affect gene expression. Inquiries into the control of specific genes normally begin with experiments to determine the level at which gene control operates.

In bacteria, the mechanisms and levels of gene control were elucidated by the discovery and analysis of mutations in regulatory genes (Chapter 8). The most important technical contributions to the understanding of bacterial regulatory genes were experimental procedures that enabled workers to rearrange the mutated regulatory genes by transduction and conjugation and more recently by recombinant DNA methods. These rearrangements allowed tests to determine how different mutations in the regulatory apparatus affected gene activity.

Some regulatory mutations acted only in *cis;* that is, the mutation had to be very close to the regulated gene. Such mutations were interpreted as affecting key sequences in the DNA involved in the control of RNA synthesis. Other regulatory mutations acted in *trans;* that is, the mutation did not have to be close to the regulated gene. The *trans-active* regulatory elements are separate genes encoding *regulatory proteins* that can diffuse through the cell to act upon a susceptible gene or genes. Biochemical experiments on isolated trans-active proteins (e.g., the *lac* and the λ repressors) to determine their effects on isolated genes have been done only relatively recently. In eukaryotes it is not possible to conduct such simple and informative genetic transfers as conjugation and transduction, which have been so useful with haploid cells; hence steady progress in research on the molecular basis for eukaryotic gene control had to await the advent of recombinant DNA technology.

Molecular details of the control—at several different levels—of a sizable number of individual eukaryotic viral and cellular genes are now becoming known. As we mentioned, the most frequent type of control is transcriptional, and a few of the regulatory proteins that affect transcription have been discovered. Some of these proteins in vertebrate and invertebrate cells have been isolated by biochemical techniques. In addition, genetic analysis in single-celled eukaryotes such as yeast has identified a number of regulatory genes that have been cloned. Thus, the proteins encoded by these genes can now be prepared and studied.

Before we turn to specific mechanisms of eukaryotic gene control, we want to consider two general questions: What is the "purpose" of gene control? How many eukaryotic genes are, in fact, under specific control?

The "Purpose" of Gene Control in Unicellular vs. Multicellular Organisms

Most of our understanding of the molecular details of gene control and of the logic of the regulatory circuits comes from studies of prokaryotes (Chapter 8). However, gene control obviously serves a limited function in prokaryotes, in keeping with their limited life cycles in comparison to eukaryotes. Gene control in prokaryotes functions to adjust the enzymatic machinery of the cell to its immediate nutritional and physical environment. Within a *single* bacterial cell, genes are reversibly turned on and off by transcriptional control in order to allow growth and division, the raison d'être of bacterial existence.

Some eukaryotic cells, such as yeast cells, also seem to be designed only, or mainly, for the purpose of growth; such cells possess many genes that are controlled to respond to the environment (Table 12-1). Even in the organs of higher animals—for example, in mammalian liver—some genes can respond reversibly to external stimuli. However, the hallmark of existence of metazoan cells is protection from immediate outside influences. Most cells in metazoans experience a tightly controlled environment, which is why relatively few genes in multicellular organisms may be devoted to responses to environmental changes.

The most characteristic and presumably the most demanding aspect of gene control in multicellular organisms is the execution of a controlled developmental pathway so that the right gene is activated in the right cell at the right time. In most cases, these developmental decisions cannot be reversed or repeated within a single cell. Many differentiated cells (e.g., skin cells, red blood cells, the lens cells of the eye, and antibody-producing cells) march down a pathway to final cell death in carrying out their genetic programs, and they leave no progeny behind. Thus, fixed patterns of gene control that lead to differentiation serve the need of the whole organism. This plan is clearly different from the environmentally stimulated gene control that allows the propagation of the individual responding cell.

How Different Are Proteins and mRNAs from Cell to Cell?

If what makes one cell different from another is the kind and the amount of the proteins contained in each, then the *complexity* of the mRNA (i.e., the number of different types of mRNA molecules), the abundance of each mRNA, and the number of times each mRNA is used before it is destroyed determine what a cell is or can be-

Table 12-1 Possible differences in types of regulatory genes in rapidly growing unicellular and multicellular organisms

Number of regulatory genes	Process regulated	
	General	**Specific**
BACTERIA AND YEASTS		
~200–500	Growth and division	Amino acid synthesis Purine and pyrimidine synthesis Sugar metabolism and energy production Cell wall and membrane protein formation rRNA and ribosomal protein synthesis Initiation of DNA synthesis and repair
~50	"Differentiation"	Spore formation and activation Changes in cell shape and mating behavior
MULTICELLULAR ORGANISMS		
<100?	Growth and division	Some changes occur in response to starvation or heat, but many nutritional responses present in single-celled organisms are missing in multicellular ones
		Factors that control the cell cycle may be encoded by regulatory genes and possibly "oncogenes"
Hundreds to thousands?	Developmental choices (determination)*	Proteins in this class are largely unknown; they are responsible for the existence of various cell types, and for differences in the shapes and sizes of organisms
	Specialized cell functions (differentiation)*	Several hundred cell types are probably characterized by many thousands of cell-specific proteins

*Determination sets a cell and its descendants on a particular developmental course. Differentiation, which is often recognized by the production of cell-specific proteins, occurs when the determined cell receives the correct molecular signal.

come. How much do these factors actually vary, and how different are the proteins in various differentiated cells? Does each differentiated cell contain many different specialized proteins, or do variations in the amounts of shared proteins contribute most to the differentiated state?

The Number and Kinds of Proteins in a Cell Can Be Estimated by Two-Dimensional Gel Analysis

One experimental method has been widely used to estimate the number of different proteins and their concentrations in a particular cell. In two-dimensional gel analysis, the entire protein content of a sample of cells is dissolved and separated by two different physical methods: isoelectric focusing, which separates the proteins on the basis of their charge, and gel electrophoresis in the presence of sodium dodecyl sulfate (SDS), which further separates the proteins on the basis of their chain length (see Figure 7-10). By staining the proteins or subjecting the gel to autoradiography (if the proteins have been labeled radioactively), several thousand, perhaps 5000, different polypeptide chains can be detected within almost any vertebrate cell.

Whether a great many more proteins exist in quantities too small to be detected is a matter of debate. There could be as many as 5000 additional proteins present at very low concentrations in most cells. If a gel analysis is made using proteins from different tissues, such as liver, kidney, and muscle, a large number of the individual protein "spots" are the same. However, up to 100 proteins that are not found in one tissue or cell type may be found in another (Figure 12-1). These cell-specific proteins can be presumed to be among those that give the different cells their specific character. In addition, it is also common for the same (or apparently the same) protein to be found in many different cell types in different amounts (Figure 12-1). These results confirm and extend the classical biochemical findings that some enzymes are common to many tissues (either in similar or varying concentrations), whereas other enzymes and some specialized secretory products are formed only in single cell types. Whether

(a)

2
3
4 5
6 7
8
9
10
K
1

(b)

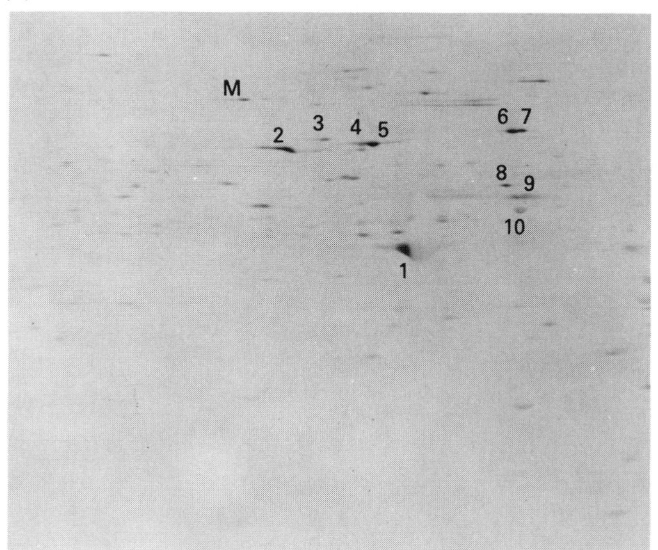
M
2
3
4 5
6 7
8
9
10
1

(c)

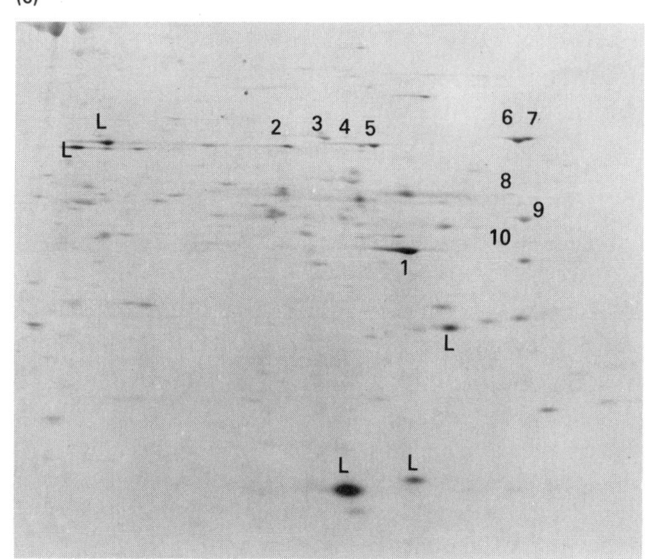
L
L
2
3 4 5
6 7
8
9
10
1
L
L
L

(d)

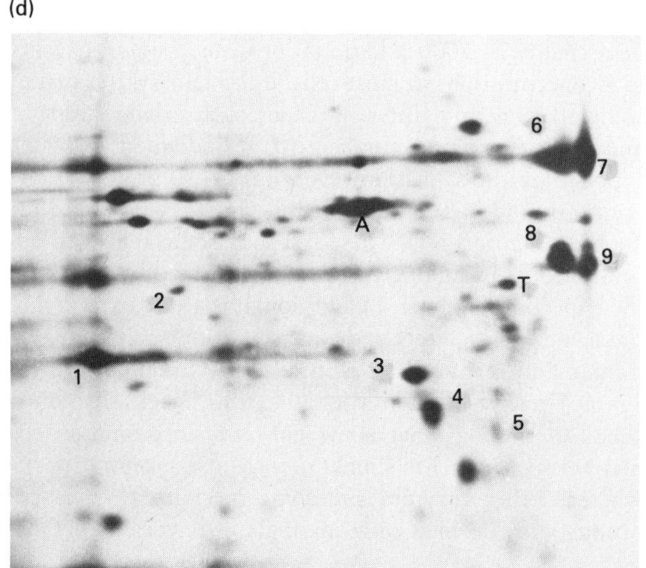
6
7
A
8
9
2
T
1
3
4
5

(e)

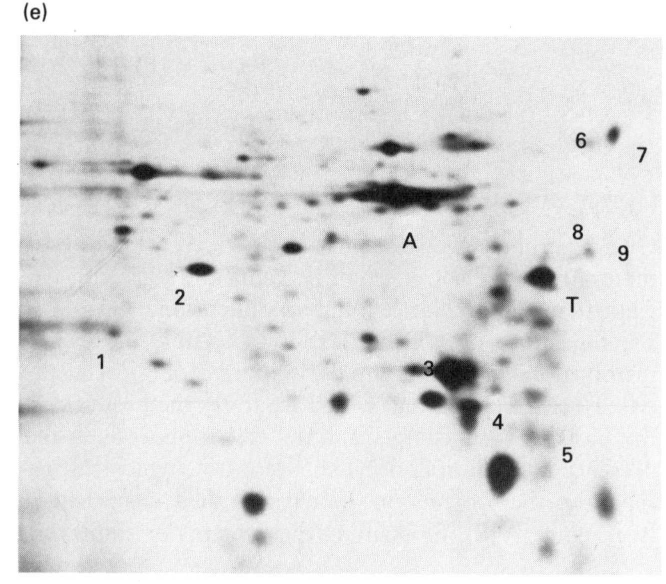
6
7
A
8
9
2
T
1
3
4
5
9

Figure 12-1 ***(opposite)*** Two-dimensional gel analysis of proteins from different cells. (a),(b),(c) Labeled proteins from mouse kidney (a), muscle (b), and liver (c). Animals were sacrificed and tissues were minced, washed free of blood, and homogenized in neutral buffer; individual proteins from each tissue were separated by isoelectric focusing (in the horizontal dimension) followed by SDS gel electrophoresis (in the vertical dimension). The proteins were detected by staining them with the dye Coomassie blue, which reacts with all proteins. The spots labeled with numbers indicate proteins that are present in all tissues; there are quantitative variations from tissue to tissue. Proteins labeled with letters appear to be present only in the specific tissue. (d),(e) A comparison of the proteins in an embryonic mouse heart at 14 days of gestation (d) and in an adult mouse heart (e). The proteins were labeled with $[^{35}S]$methionine during in vitro translation of the extracted mRNA, and the spots were detected by autoradiography. (A = actin; T = tropomyosin.) Many similarities can be seen in the patterns from the two sources, as well as some differences—both quantitative (e.g., compare spots 3 to 7) and qualitative (note that spots 8 and 9 are missing in the adult sample). *Parts (a) to (c) courtesy of P. O'Farrell; (d) and (e) courtesy of A. Ouelette.*

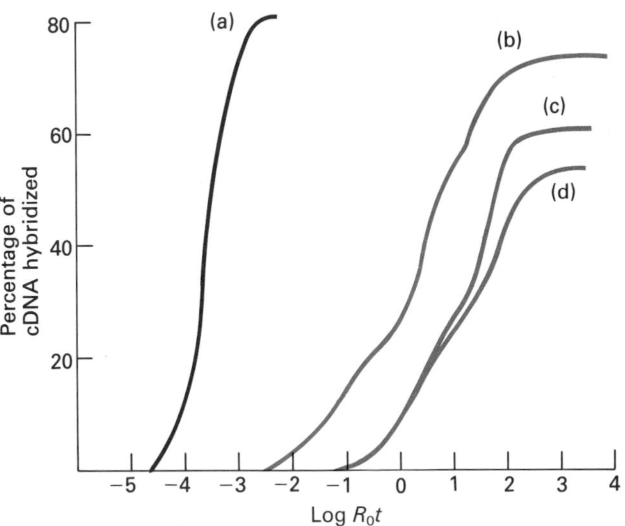

Figure 12-2 Hybridization reactions of labeled cDNA with excess mRNA. Labeled cDNA prepared by reverse transcriptase copying of mRNA was hybridized with increasing concentrations of mRNA or for increasing intervals of time: the log R_0t is the logarithm of the mRNA concentration R (in M, or moles per liter) × time t (in seconds). The higher the R_0t necessary for a cDNA to hybridize, the lower the proportion of the corresponding mRNA in the total sample. The extent of hybridization of the labeled cDNA with the mRNA sample is monitored by its resistance to a nuclease that digests only single-stranded DNA. Curve (a) shows the reaction of labeled globin cDNA with its own mRNA; this provides a standard profile of the rate of reaction of a single cDNA with its mRNA. The reaction is half-complete at a log R_0t of −3.5. Curve (b) traces the reaction of labeled total mouse liver cDNA with total mouse liver mRNA. The most rapid part of the reaction with the liver sample (the first 20 percent) has a midpoint at about log R_0t −1.5, which makes it slower than the pure globin reaction by a factor of approximately 100. Such a result is consistent with the presence of about 20 different mRNAs, each constituting about 1 percent of the total mRNA. The remainder of the liver reaction curve shows that most mRNAs are present at much lower concentrations (causing a reaction that is slower than the globin reaction by a factor varying from 10^3 to 10^6). Curves (c) and (d) show that the majority of the labeled liver cDNA that reacts with liver mRNA also reacts with mRNA from kidney (c) and brain (d) at rates similar to those of the reaction with liver mRNA. Thus there is extensive sharing of mRNA sequences in different tissues. [See N. D. Hastie and J. O. Bishop, 1976, *Cell* 9:761.]

these common quantitative variations in proteins or the less frequent qualitative variations are more important in determining cell specificity is not clear.

The Concentrations of mRNAs Can Be Measured by Kinetics of Hybridization

A method that is used to compare the *genes* expressed in different cell types is based on the hybridization of mRNA sequences. When denatured complementary nucleic acids are reassociated, the higher the concentration of two complementary strands, the faster their reassociation rate (see Figures 7-12 and 11-23). Both a purified sample of a specific mRNA and the total mRNA from cells of one type may serve as templates for the manufacture of radioactive single-stranded cDNA (complementary DNA); the reverse transcriptase of retroviruses is used to make the copies (Chapters 6 and 7). The rates at which the labeled cDNA copied from the purified mRNA associates with known concentrations of its homologous mRNA provide known reaction rates for a single mRNA. The cDNA copied from the total cell mRNA can then be mixed with known concentrations of the total cell mRNA, and the two sets of reaction rates can be compared to estimate the range of the concentrations of the various mRNAs within the total cell sample (Figure 12-2).

The association reaction with total cell mRNA (using either cultured cells or cells from specialized tissues) occurs over a wide range of concentrations of total mRNA, which indicates that eukaryotic cells contain a very broad range of mRNA concentrations. The most abundant mRNAs may be present in 10^4 to 10^5 times the quantity of the scarce mRNAs. About 10 to 20 percent of the cDNA reacts very rapidly; perhaps 30 to 50 percent associates at an intermediate rate; and the remainder reacts at a very slow rate. These results indicating the widely differing quantities of different mRNAs confirm what is also apparent in the different intensities of protein spots in the two-dimensional gel patterns. A liver cell contains about 10^6 total mRNA molecules. From the rates of hybridization of the total mouse liver cDNA with its mRNA, it has been estimated that there exist about 100 species of abundant mRNA with 5000 to 50,000 copies per cell, several hundred species with 100 to 1000 copies per cell, and perhaps as many as 10,000 different mRNAs with as few as 0.1 to 10 copies per cell.

When the cDNA prepared from the mRNA of one tissue—say, liver—is allowed to react with the mRNA from another tissue—say, kidney or brain—there is substantial hybridization for all but the most abundant mRNA species. This result is expected because the most abundant products of the liver, such as serum albumin, are not produced in the brain or kidney, and the mRNAs for these products are not present in either the brain or the kidney. The partial cross-reaction between the cDNAs from one tissue and the large number of moderately abundant and scarce mRNAs in another tissue indicate that many proteins are shared, but that some of the mRNAs are specific for the second tissue.

From two-dimensional gel analyses of proteins and from hybridization experiments with mRNAs and mRNA copies from different tissues, it has been postulated that the production of most proteins may be subject to quantitative gene controls that regulate abundance over a range of 10-fold to perhaps 100-fold. In this view, very few genes are subject to absolute on-off control. Before accepting this hypothesis, which may or may not be correct, we should reflect on a problem that was mentioned earlier. Some proteins (e.g., the actins and the tubulins) are common to many tissues. Many of the corresponding proteins in different tissues may actually be encoded by different members of multigene families whose mRNAs might not cross-hybridize and thus would be counted as a single species in cDNA hybridization experiments. If this proves true, more gene-specific as opposed to quantitative regulation will be indicated.

It is, however, possible although not yet proved that the same gene may be active to different extents in different cell types. How are such variations brought about? For example, can one and the same gene be transcribed at a variety of different, controlled rates, or are changes in the quantity of the same gene product brought about by differential mRNA stabilization or translation?

How Important Are Scarce mRNAs and Proteins?

We should mention one final difficulty that complicates the issue of which mechanisms are really most important in establishing or maintaining differentiation of cell types. Although major proteins like globin in red blood cells and albumin in liver cells are the most conspicuous indicators of cell differentiation, the proteins that are most important in controlling cell specification may be scarce proteins that have not yet been identified. For example, the amounts of some enzymes constitute only 10^{-4} to 10^{-5} of the total cell protein; these proteins would hardly be noticed in a two-dimensional gel. Among the most important cell-specific proteins may be cell surface proteins, some of which (e.g., the insulin receptor) represent only 10^4 out of the total of 10^9 to 10^{10} protein molecules in a cell.

Whether all or even most scarce mRNAs serve specific functions in the cells in which they are found is still an open question. It may be that in a given cell almost any mRNA may be synthesized occasionally and that many of the very scarce mRNAs (say, those with less than one copy per cell) do not contribute to cell functions. If much of cell specificity does, in fact, lie in the regulation of the myriad very scarce mRNAs and proteins, then discovering how differentiation and specific cell function are brought about will be extraordinarily difficult.

The Three Components of Gene Control: Signals, Levels, and Mechanisms

Three separate aspects of cell function must be analyzed for an understanding of the molecular control of the concentration and use of mRNA in eukaryotes (Figure 12-3):

1. What are the *signals* to which a specific gene responds?
2. At which *level* (i.e., at which step or steps) in the chain of events leading from the transcription of DNA to the use of mRNA in protein synthesis is control exerted on the specific gene?
3. What are the *molecular mechanisms* of each level of gene control? That is, which individual molecules or cell structures are involved in the control of the gene, and how do they exert their effects?

The complete answers to these questions are not yet known, but some generalizations have been formulated from current research. In the following sections we shall

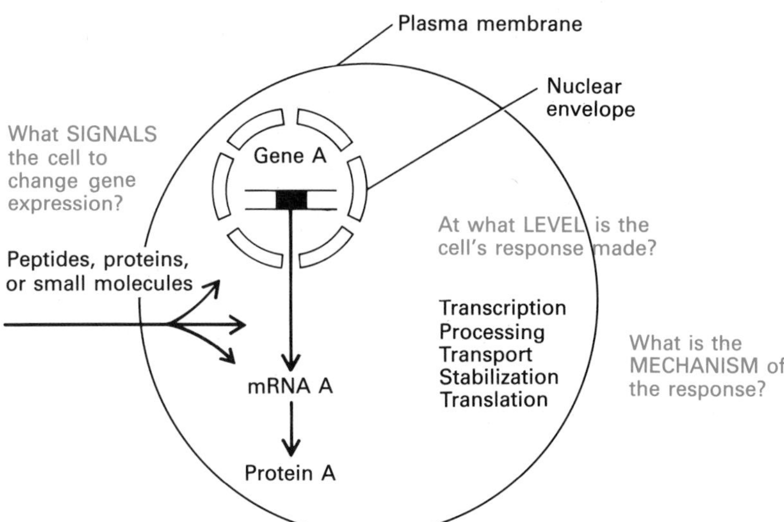

Figure 12-3 The three questions in studying gene control.

describe the types of experiments that will continue to be necessary in finding answers to the questions posed above.

Signals for Gene Control in Eukaryotic Cells

Hormones, Both Small Molecules and Polypeptides, Can Cause Differential Gene Expression

Dozens of substances produced within the body bring about differential gene expression (Table 12-2). The largest group of known gene-signaling substances in animals is the *hormones*. Hormones have far-reaching effects on cells, both in adult organisms and during growth and development. A few hormones are known to operate by controlling individual genes or gene sets, but hormones may also affect the action of already existing enzymes or cell structures, as we shall see in Chapter 16. Each hormone is produced in one cell type and transported, usually through the circulatory system, to target cells.

Some target cells are scattered in many tissues; others are concentrated in a single tissue. Hormones such as the thyroid hormones and the glucocorticoids (adrenal steroids) affect groups of perhaps 10 or more genes in their target tissues, rather than single genes (see Figure 16-31). Many different cell types can receive signals from these hormones. Protein hormones like insulin and growth hormone may also have similar effects on many different cell types. In contrast, prolactin stimulates only acinar cells in breast tissue to produce milk proteins. Still other hormones have one effect on one cell type and a different effect on another, as is detailed in Chapter 16.

In addition to the long-recognized group of circulating proteins and steroids that are generally included in any list of vertebrate hormones, there exist many proteins that are released from one cell type but that exert dramatic effects on a variety of other cells, either neighboring cells or distant cells. One class of these proteins is known as growth factors. For example, *nerve growth factor* causes cells of neural origin to extend long, axon-like processes. Another such protein is the *epidermal growth factor* or EGF, so named because it was originally discovered by its capacity to increase the speed of epidermal development when injected into rodents late in fetal development. It is now known that many cells besides skin cells have surface receptors for EGF and are stimulated by it. Apparently a variety of other protein factors have similar generalized effects of stimulating or retarding cell growth. The mechanisms by which these effects may be mediated are discussed in Chapters 16 and 23.

A few protein factors have been discovered to have more specific functions. For example, the *interferons*, a family of 20 to 25 low-molecular-weight proteins, cause cells to become resistant to the growth of a wide variety of viruses. This change in viral susceptibility is apparently due to the ability of interferon-treated cells to make a set of proteins that they were unable to make before. Because many cultured cells respond to the interferons, this cell-protein interaction is a particularly good model for studying gene activation in response to a single protein added to cells. In addition to new protein synthesis and the establishment of the antiviral state, interferon causes some cells to stop or slow down division.

As the chemical nature of various hormones and growth factors was deciphered, it became clear that two general types of signaling molecules exist: (1) small molecules such as the steroids and thyroid hormones, which actually enter cells to cause their effect, and (2) polypep-

Table 12-2 Gene-activating signals in eukaryotes

Classes	Examples	Targets
Hormones:		
Proteins	Growth hormone	Many cells
	Prolactin	Secretory cells in breast tissue
Steroids	Estrogens	Liver, brain, reproductive organs
	Testosterone	Muscle, bone, skin, reproductive organs
Circulating or secreted protein factors	Nerve growth factor	Axons (differentiating nerve cells)
	Epidermal growth factor	Many surface tissues (skin, eye, etc.) as well as cultured cells of all types
	Interleukins (lymphokins)	White blood cells
	Erythropoietin	Red blood cell precursors
	Interferon	Most epithelial cells, white blood cells
	PDGFs (platelet-derived growth factors)	Many fibroblast cell types
Environmental	Nutritional signals:	
	Lower eukaryotes	Most genes for synthetic activity (amino acids, nucleic acid components) and for hydrolytic functions (phosphatases, saccharidases)
	Animal cells	Gluconeogenic enzymes* in starvation Some genes for synthetic activity (there is some repression by excess products)
	Heat shock	Genes for specific proteins (induced); most other mRNAs (general translation suspended); widespread in vertebrate and invertebrate cells
	Toxic substances:	
	Drugs, carcinogens	Cytochrome P-450 proteins in liver
	Heavy metals	Metallothioneins in liver, kidney, and other tissues, and in single-celled eukaryotes
	Products of hemorrhage or inflammation	White blood cells

*Liver enzymes that convert amino acids into glucose.

tides or proteins that bind to specific receptors on cell surfaces; in most cases these exert their action before entering cells. Because it is frequently found in biology that some entities use two or more common mechanisms to achieve a single end, it has come as no surprise that some proteins that appear to bind to cell surfaces are subsequently transported inside, where they or their cleavage products may be active factors in signaling genes. The details of cell surface binding and intracellular signaling are covered in Chapter 16.

Cell-to-Cell Contacts Can Act as Signals to Control Genes

Most hormones and some growth factors are released into the circulatory system to travel to their target cells. Other gene-activating signals may require direct contact between cells. For example, during the embryonic determination of most specialized cells, two or more different cell types must make contact. Usually the interacting cells come from different primordial cell layers. The most extensively studied cases involve contact between mesenchymal cells (cells derived from the mesoderm) and either endodermal or ectodermal cells. The nature of the information passed in such cell-to-cell contacts is unknown, but it may be that a surface signal mediated by the contact between cell surface proteins leads to gene control. The interaction may also take place more indirectly: a cell of one type may form an extracellular matrix, and contact with this matrix may be the signal for the cell of the other type. Assuming that at least one of these mechanisms does occur, the means by which such extracellular information is communicated to the cell interior (especially to the nucleus) is one of the most profound mysteries of present-day cell biology.

Needless to say, the identification of signals for eukaryotic gene control is far from complete. Probably only the signals with the most dramatic effects have been uncovered so far. Regarding the signals that have been identified, there is still much to learn about how gene control occurs after the interaction of the signal molecule with the target cell.

Environmental and Nutritional Signals for Gene Control Are Used More by Bacteria Than by Multicellular Organisms

The eukaryotic gene control signals mentioned so far—hormones, growth factors, and contacts between cells—are very different from the environmental signals that free-living cells like bacteria encounter and respond to. However, many genes in lower eukaryotes such as yeasts and molds respond to such environmental signals as nutritional stress. A number of inducible enzyme systems for sugar metabolism, use of organic sulfate, phosphate, and nitrate, and biosynthesis of amino acids have been documented in lower eukaryotes. However, starvation of yeast cells does not induce biosynthetic enzymes in the cells to nearly the same extent as starvation of bacterial cells does. For example, yeast cells increase the levels of enzymes for the synthesis of tryptophan, histidine, isoleucine, and valine by a factor of only 2 to 10 when the cells are starved for these amino acids. In contrast, when *Escherichia coli* is starved for the same amino acids it changes the levels of biosynthetic enzymes for them by as much as a factor of 100.

Gene Responses of Mammalian Cells to Changes in Culture Conditions Although some genes in higher eukaryotic cells can respond to nutritional stress, most do so in a limited fashion, and many genes cannot be called into action by nutritional demand. Perhaps the relative lack of response to nutritional stress is related to the fact that cells in multicellular organisms are seldom called upon to respond to nutritional stress.

Cultured mammalian cells also have an uneven response to varying nutritional conditions. For example, most cultured mammalian cells can make purines and pyrimidines, so these nucleic acid bases do not have to be added to the culture medium. However, if adenine is added to a culture of mammalian cells, after several generations the cells derive almost 100 percent of the adenine and guanine in their RNA and DNA from the adenine in the medium. On the other hand, if uridine is added as a pyrimidine source, only about 50 percent of the uracil or cytidine residues come from the medium. It appears that adenine, or a derivative of it, is a signal that leads to the inhibition of the synthesis of at least one enzyme in the purine synthesis pathway. In contrast, the capacity to make pyrimidines continues in a relatively unrepressed fashion (Figure 12-4). The same is true for endogenous synthesis of the amino acids serine and glycine: even

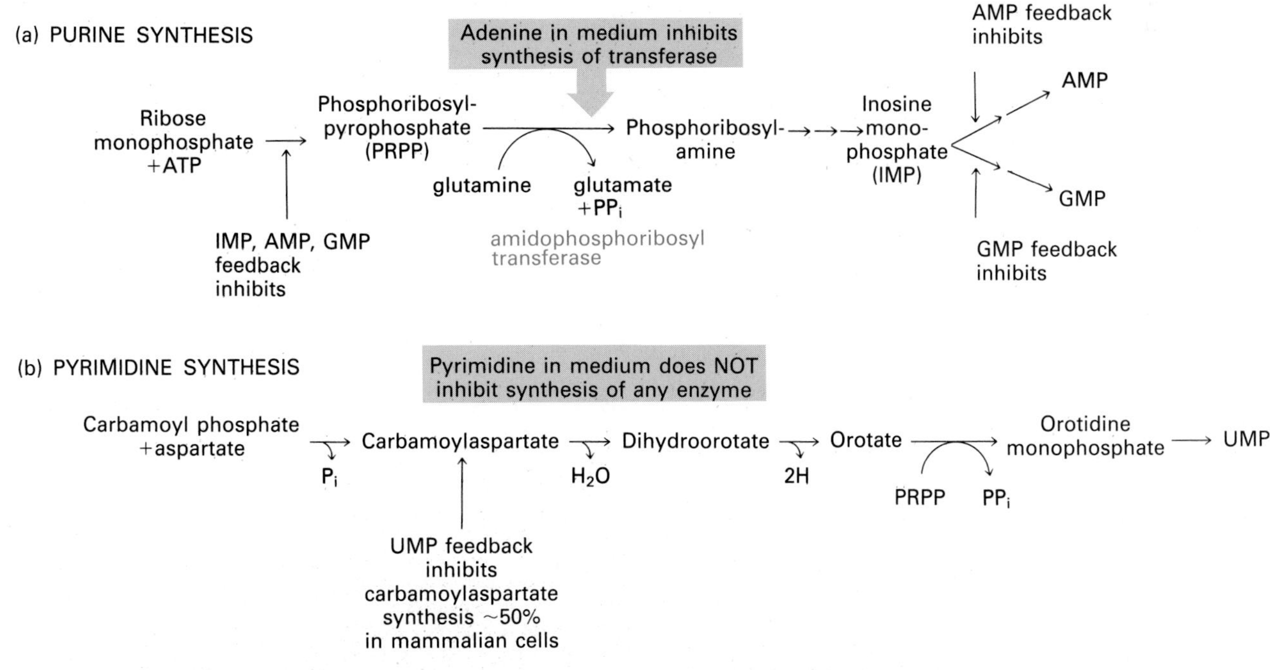

Figure 12-4 The control of purine and pyrimidine pathways in mammalian cells. (a) If cultured cells are grown in the presence of adenine, their overall rate of purine synthesis decreases by more than 95 percent, partly through inhibition of the synthesis of at least one enzyme. (b) Cells grown in the presence of an excess of uridine (or cytidine) continue to synthesize pyrimidines at about 50 percent of the normal rate.

when they are present in the medium, the cell continues to synthesize them. Thus, cultured mammalian cells cannot always suppress the synthesis of unneeded enzymes.

Nor can such cells always respond to metabolic needs even when they possess the requisite genes. For example, liver cells contain the enzyme phenylalanine hydroxylase, which converts phenylalanine into another amino acid, tyrosine. If all body cells have the same DNA as liver cells, the genetic capacity to perform this biochemical step exists in all of the cells of an animal. Yet other body cells cannot perform this function; nor can any cultured mammalian cells, including liver cells. Even when they are starved for tyrosine, cultured cells cannot convert phenylalanine into tyrosine.

Finally, nearly all mammalian cells can make certain necessary nutrients—cysteine, for example—in very limited quantities only, and a culture medium must be supplemented with these nutrients if the cells are to grow. When the cells are placed in a medium that lacks cysteine, they do not increase their production of the amino acid and they stop growing. Thus, nutritional conditions that cause bacteria and yeast to adjust their biosynthetic machinery to produce nutrients that are needed and to repress production of nutrients that are not needed are not necessarily signals for mammalian cells.

Although the nutritional responses of eukaryotic cells seem limited, all types of cells, including those of eukaryotes, can respond to an increase in temperature, or *heat shock,* by synthesizing a specific group of proteins. This generalized response is probably related to the occasional encounter with high temperatures that all cells experience, even those of multicellular organisms.

Gene Responses in Animals to Environmental Changes To what extent do the cells in an animal respond to nutritional variables by differential protein synthesis? If a rat is placed on a low-carbohydrate diet, the level of gluconeogenic enzymes (enzymes that convert amino acids into glucose) in the liver increases dramatically, mainly because of the synthesis of new enzyme molecules (Figure 12-5). If the animal is returned to a normal diet, a more ready source of glucose, then the concentration of gluconeogenic enzymes will decrease. Many of the responses of the liver in meeting these metabolic demands are regulated by hormones that affect the synthesis and the stability of specific mRNAs in liver cells (Figure 12-6), as we shall discuss later in this chapter.

The liver receives part of its blood supply directly from the gut via the large portal vein, which picks up not only nutrients but also toxic substances that are ingested. Hepatocytes, the main cell in the liver, are responsible for detoxifying many noxious chemicals. To metabolize some of these compounds, special proteins are required. For example, the presence of cadmium or other metals causes the liver to increase production of the metallothioneins, two small metal-binding proteins (each with 61 amino acid residues, including 20 cysteines) that bind heavy metal ions and thus protect against their toxic effects. In the membranes of liver cells a group of enzymes termed cytochrome P450, or simply P450, can be increased by a factor of 100 or more to cope with the organism's need to oxidize and then excrete toxic substances such as phenobarbital, codeine, morphine, and carcinogens such as methylcholanthrene. Different genes in the large family that encodes the P450 enzymes respond to different toxic substances. The molecule or molecules that recognize the toxic substance and that thus constitute the active signal for gene control during these metabolic responses in the liver cell are not yet known.

Unlike liver cells, most cells in the animal body are shielded from changing metabolic fortunes. Some cells not only do not adapt metabolically if presented with unusual compounds but show damage if their nutritional environment changes greatly. For example, cells in the brain use glucose for energy; normally, glucose is the major hexose delivered to the brain. In a human genetic disease called *galactosemia,* brain cells are damaged by being exposed to galactose, which they cannot use. Lactose, the major sugar in milk, is a disaccharide of glucose and galactose. The cells of the intestines and the liver split the disaccharide and convert the galactose into glucose when an infant ingests milk. In homozygous galactosemic individuals, the inability to metabolize galactose leads to an elevated concentration of galactose in the blood, which in turn leads to severe brain damage (Figure 12-7).

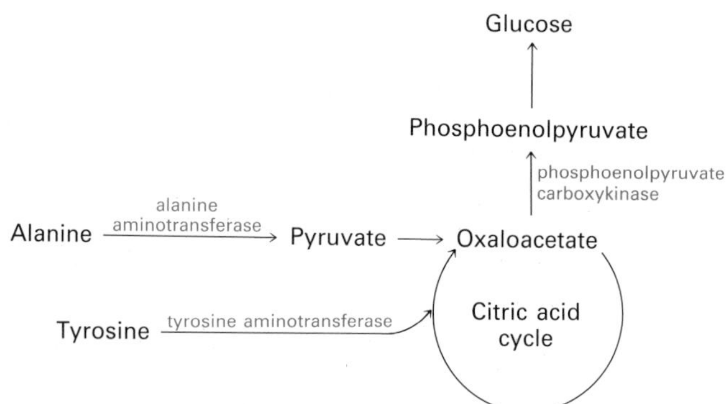

Figure 12-5 The control of glucose production in the liver. When food intake is limited, the concentration (and hence the activity) of a number of liver enzymes increases to allow gluconeogenesis, the production of glucose from other metabolites; this enables the cell to maintain normal blood levels of glucose. The increases in the enzyme activity are due to increased synthesis of enzyme proteins brought about by higher concentrations of mRNAs. Antibody precipitation of specific labeled proteins is used to illustrate their increased synthesis. (More details about these interconversions are given in Figure 20-33.)

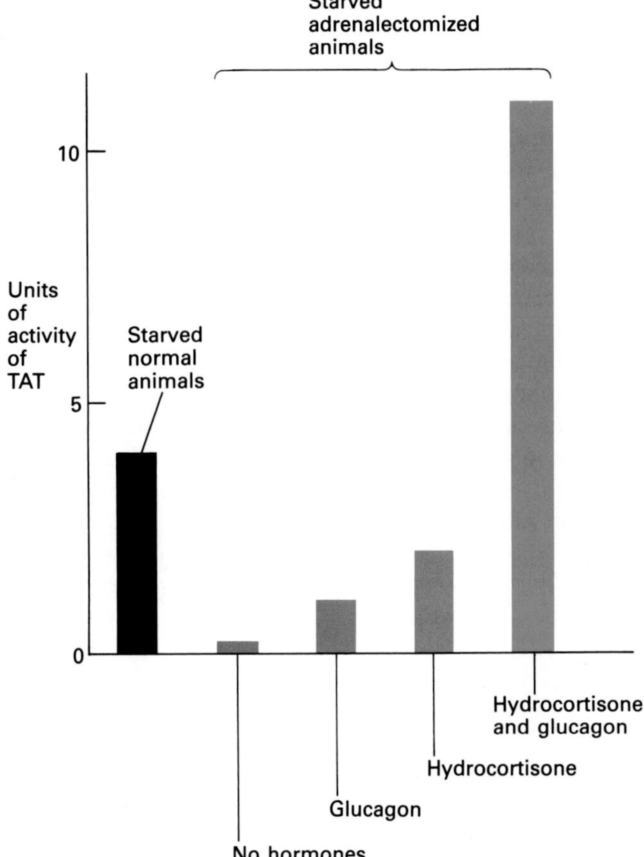

Figure 12-6 Liver enzyme levels in response to starvation or hormone treatment. Many of the responses that result in changing liver enzyme levels during starvation are a result of stimulation of the liver by hormones. This was demonstrated by monitoring tyrosine aminotransferase (TAT) activity in animals that had been adrenalectomized (their adrenal glands had been removed). These animals did not respond to starvation alone, but they did respond when they were given the adrenal hormone hydrocortisone. They responded even more when given hydrocortisone plus glucagon, a pancreatic hormone that helps the liver turn out more glucose. [See O. Greengard and G. T. Baker, 1966, *Science* 209:146; see also Figure 16-27.]

These few examples illustrate that some genes in cultured cells and some genes in cells of whole animals may be controlled in response to environmental signals, but that not all genes in all tissues are equally responsive to such signals.

Summary: Signals That Cause Changes in Eukaryotic Gene Expression

Many simple environmental signals (e.g., a change in nutrient supply) that result in changes in gene expression in bacteria also cause changes in single-celled eukaryotic organisms. However, the cells of multicellular organisms do not, in general, respond to such signals. Perhaps this is because in whole organisms a constant environment is normally maintained for individual cells. Although even in mammals there are occasional instances of gene control in response to environmental changes, the rule is that conditions such as amino acid deprivation or a change in the sugar supply do not result in a change in gene control to accommodate to the new situation.

Most of the signals that do bring about differential control of genes in multicellular organisms have a different purpose: to cause cells to perform specialized tasks that, when integrated with the tasks performed by many other cell types, will lead to the formation of organs and whole organisms. The signals producing such differentiation include molecules such as the hormones and the

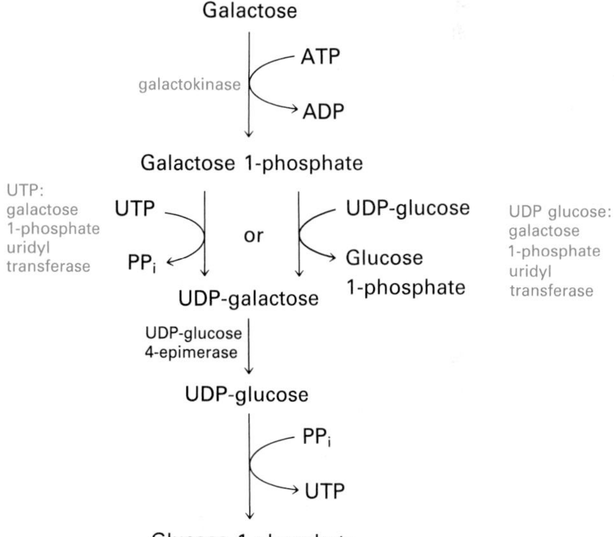

Figure 12-7 The metabolic pathways that convert galactose to glucose in humans. (UDP = uridine diphosphate, a carrier of hexose groups; UTP = uridine triphosphate.) Two different reactions can convert galactose 1-phosphate to UDP-galactose; the reaction on the left is present in adults but lacking in newborns. Therefore mutations that inactivate either the galactokinase or the UDP glucose:galactose 1-phosphate uridyl transferase (reaction on the right in the second step) cause the disease galactosemia in infants. Babies who are homozygous for mutations in either of these two enzymes have high blood galactose concentrations and suffer brain damage because their brain cells cannot metabolize the blood galactose that comes from milk. Close to 1 percent of people in the United States have one mutant UDP glucose:galactose 1-phosphate uridyl transferase gene (i.e., they are heterozygous for the mutation); galactokinase mutations are less frequent.

growth factors, which have long been familiar. Many additional molecules that serve as signals must remain to be discovered, including specific molecules on the surfaces of two interacting cell types.

Levels of Regulation in mRNA Metabolism

Eukaryotic cells respond to molecular signals by synthesizing different proteins in different amounts, as we know from observing the differential protein composition of cells from different tissues and of cells grown under different conditions. The next task in determining the molecular basis for gene control is to establish the *levels* at which gene control occurs.

Using pure, cloned DNA that encodes specific mRNA, measurements of RNA-DNA hybridization can distinguish control at three levels: (1) nuclear RNA synthesis, (2) use of primary transcripts, and (3) stabilization (or destabilization) of mRNAs in the cytoplasm (see Figure 12-3).

Enough individual examples have been analyzed to show that each of these levels of mRNA regulation operates in eukaryotes. Of course, not every gene is—or can be—controlled at all three of these levels. Furthermore, some genes are controlled at the level of mRNA *translation,* as we shall see later. In the following sections we shall describe examples of each of the three levels of control listed above and explain the types of experiments that have brought them to light.

Transcriptional Control and How It Is Demonstrated in Eukaryotes

Transcriptional control of a eukaryotic gene means that controlled increases or decreases in the synthesis of a primary RNA transcript in the cell nucleus are the main cause of changes in the rate of synthesis of a particular protein. Transcriptional initiation and transcriptional termination are both potentially subject to control, but control of initiation appears to be most important. After the initiation and termination sites for a primary RNA transcript have been determined, the rate of synthesis of the primary transcript can be measured.

The simplest and most direct method of measuring transcription rates would seem to be to expose the cells in question for a brief time (e.g., 5 min or less) to a labeled RNA precursor and measure the incorporation. This technique is not practical, however, in whole animals, and even with cultured cells it is often easier to use another method of labeling RNA. Nuclei isolated from cells incorporate ^{32}P from labeled nucleoside triphosphates directly into nascent (growing) RNA chains to produce highly labeled RNA preparations. Fortunately, it appears that all polymerases that are active at the time nuclei are taken from cells remain active and add a few hundred nucleotides to the already initiated RNA chains during the incubation of nuclei. RNA labeled in such a fashion can be hybridized to specific DNA, and the differential rates of transcription of different genes can thus be assayed accurately.

See, for example, Table 12-3, which shows the results of hybridization experiments with labeled RNA prepared by both methods (pulse labeling of whole cells and labeling of isolated nuclei). Both transcripts that are abundant and transcripts that are rare are formed in equivalent proportions in nuclei and in whole cells. Thus isolated nuclei provide a remarkably direct means of measuring transcription rate.

Transcriptional Control of Many Genes Has Been Proved

DNA Viruses in Mammalian Cells The production of virus-specific proteins (and mRNAs) during the course of infection with several different DNA-containing animal viruses follows an orderly pattern: first, a group of *early* proteins (and mRNAs) is detected; then viral DNA replication occurs; and finally a group of *late* proteins (and mRNAs), including the structural or *capsid* proteins that form virus particles, are made in large amounts.

Using both pulse-labeled RNA from whole cells and RNA synthesized in vitro by isolated nuclei, a differential pattern of usage of virus transcription units has been proved to be the basis for this program of early and late gene expression. These specific transcriptional patterns in DNA viruses, and more particularly in adenovirus, can be used to illustrate several points.

First, one transcription unit (that for protein 1A) is activated immediately upon the entry of the virus (Figure 12-8a). No new protein synthesis is necessary for this activation, so it must be carried out by pre-existing cell proteins. A second group of early transcription units (primarily those labeled 1B, 2, 3, and 4) is activated only if the protein product of the 1A transcription unit is formed. (How the 1A protein may act to cause increased transcription is discussed later in this chapter.) In addition, synthesis of the 1A protein causes a low rate of transcription initiating at P16, the major late promoter.

Second, there are also instances of specific decreases in ongoing transcription during adenovirus infection. Transcription unit 4 is first strongly activated and then deactivated (Figure 12-8a). However, viruses that are mutant in the protein encoded by transcription unit 2 do not deactivate transcription unit 4. Therefore, at least part of the overall transcriptional control program results from an interaction between the products of the various transcription units.

Table 12-3 A comparison of two methods of labeling nuclear RNA for use in hybridization experiments to assay transcription rates*

DNA sequences used in hybridization	Percent of total labeled RNA hybridized	
	Whole cells	Isolated nuclei
Adenovirus genome:		
Cells early in infection	0.75	0.58
Cells late in infection	16.6	14.4
β globin	0.01	0.01
cDNA made from 8 different Chinese hamster mRNAs	0.0001–0.001	0.0001–0.001
Ovalbumin cDNA†	0.00018	0.00024
Conalbumin cDNA†	0.00015	0.00022

*Labeled nuclear RNA was prepared both by labeling whole cells with [^{3}H]uridine for 1 to 3 min and by adding four ribonucleoside triphosphates (including [^{32}P]UTP) to isolated nuclei. Because approximately the same fraction of labeled RNA hybridizes in both cases, chain elongation of RNA in isolated nuclei must be carried out about as faithfully as it is in whole cells.

†For the two cDNAs representing chicken egg-white proteins, the source of labeled RNA from whole cells was cultured minced chicken oviducts labeled for 20 min.

SOURCE: J. E. Darnell Jr., 1982, *Nature* **297**:365; and G. S. McKnight and R. D. Palmiter, 1979, *J. Biol. Chem.* **254**:9050.

Third, DNA replication is necessary for the activation of transcription of specific promoters. For example, consider the "left" end of the adenovirus genome, where within a 4-kb region there are three transcription units, those for proteins 1A, 1B, and IX (Figure 12-8b). As we have already mentioned, transcription of 1A precedes transcription of 1B and is necessary for the activation of 1B. The 1B transcription unit is fully active by about 5 h after infection; however, the protein IX transcription unit, which lies within the same DNA stretch, is not active at the early time.

The protein IX transcription unit does not become active until after DNA replication begins, at which point it becomes very active (Figure 12-8c). DNA replication appears to be required to prepare a DNA template for this restricted and very productive "late" mode of transcription. The need for template replication to allow transcription was shown clearly by introducing a second strain or type of adenovirus *after* the original virus DNA had begun to replicate. Because the late mRNAs of the second virus were distinguishable from those of the original virus, it could be determined that only the already replicated DNA molecules yielded high levels of synthesis of protein IX RNA and the RNA that initiates at P16 (the major late promoter). The DNA of the second virus had to replicate before its IX and 16 promoters became highly active.

Thus adenovirus offers an array of transcriptional regulatory events for study, including specific increases and decreases in transcription rates and differential transcriptional activation depending on the state of the template (pre- or postreplication). The investigative utility of the adenovirus system is that the proteins responsible for the transcriptional changes can be identified and studied.

Tissue- or Cell-Specific Gene Control When recombinant DNA techniques became available, investigators used them to clone mRNA sequences encoding proteins that were known to be prominent products of particular differentiated cells. Through the use of cDNA clones complementary to a variety of such mRNAs and labeled nuclear RNA produced by isolated nuclei, it has now been proved beyond a doubt that transcription is the primary level of gene control in the production of a variety of mRNAs that are found only in specific differentiated tissues (Table 12-4).

Globin Genes Specific individual genes can become very active during the last stages of differentiation. The globin genes of vertebrate erythroblasts (the precursors of red blood cells) are a classic example of the phenomenon. In adults, the differentiation of erythroblasts occurs mainly in the bone marrow. In embryos, it occurs in extraembryonic tissues called blood islands and in the embryonic liver, where circulating erythroblasts can lodge. The cells that differentiate into red blood cells do not form globin until the last four or five divisions before they become mature nondividing red blood cells.

Globin gene control during this differentiation process occurs primarily at the level of transcriptional initiation. Different globin genes are known to be active during embryogenesis and in adults (see Figure 11-4). For exam-

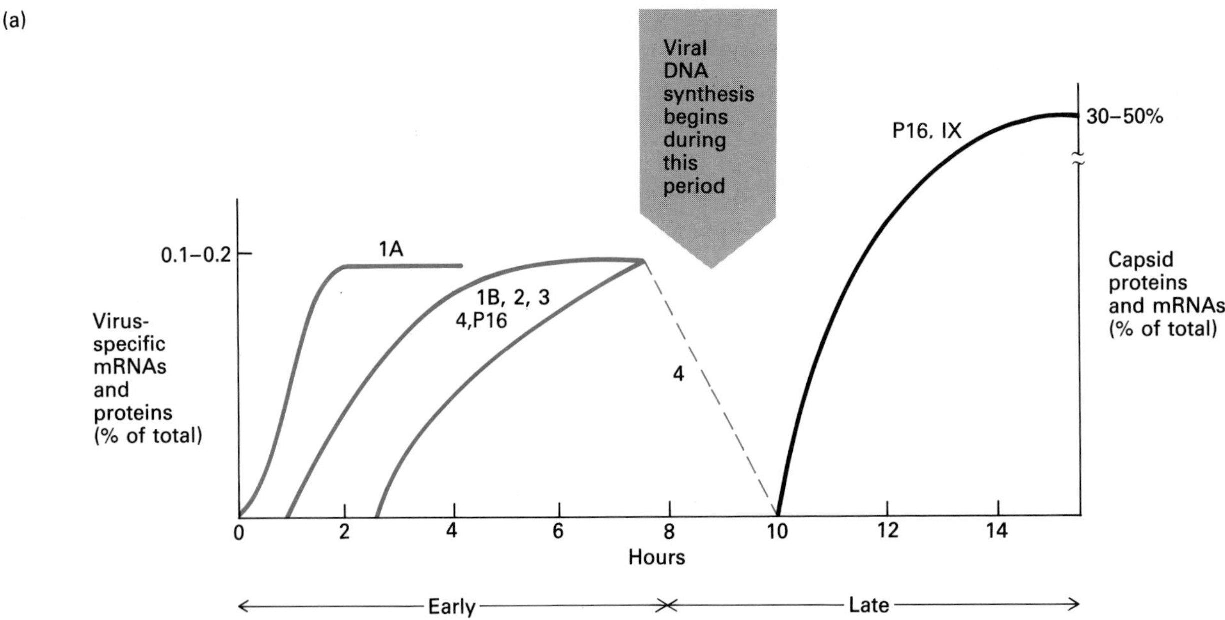

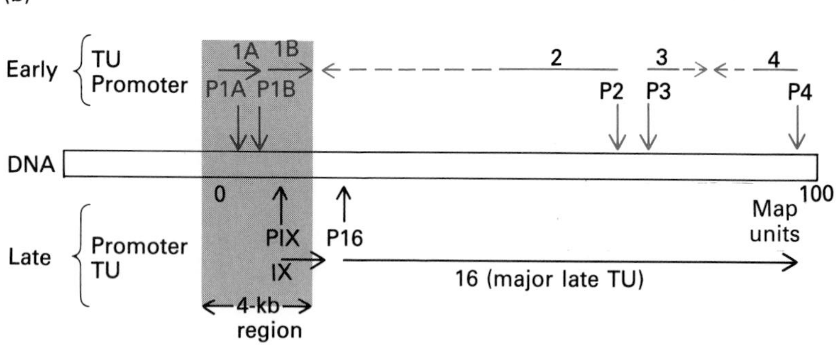

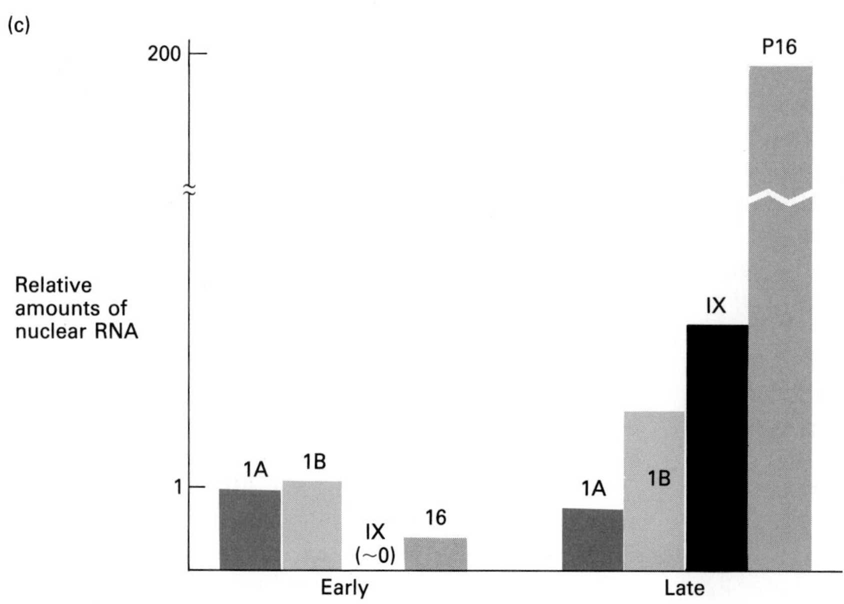

Figure 12-8 A simplified version of the adenovirus life cycle and the program of viral gene expression. (a) The curves show the accumulation of early and late virus proteins and their mRNAs. DNA synthesis, which begins at about 8 h, roughly divides early and late events. Virus-specific RNA synthesis reaches 0.1 to 0.2 percent of total transcription early in infection, and 30 to 50 percent late in infection. (b) A map of the 36-kb adenovirus genome, showing the locations of seven of the promoters (P) for viral genes. The arrows indicate the respective transcription units (TU) and their directions of transcription. The exact transcription termination sites are not known (as indicated by the dashed portions of the arrows). (c) The activity of the three promoters on the left end of the virus genome is graphed along with the activity from P16, the major late promoter, which is weakly active early in infection. These comparisons are based on an analysis of pulse-labeled nuclear RNA. [For genome structure and mRNA identification and production, see E. Ziff, 1980, *Nature* 287:491. For adenovirus gene control, see J. E. Darnell Jr., 1982, *Nature* 297:365; and J. R. Nevins, 1982, *Cell* 28:1.]

Table 12-4 Specific instances of control of transcription by RNA polymerase II in chicken and mouse cells*

Protein	Cell	Signaling agent
CHICKEN		
Egg-white proteins: ovalbumin, conalbumin, ovomucoid, lysosyme	Oviduct	Estrogen
Vitellogenin*	Liver	Estrogen
Muscle proteins	Myoblasts	Cell fusion
Globins (embryonic and adult, α and β chains)	Erythroblasts	?Erythropoietin ?Terminal cell division
MOUSE		
Liver: many serum proteins, gluconeogenic and other enzymes	Hepatocytes†	?Cell contact, plus hormones in some cases
Globin, α and β	Mouse erythroleukemia cells	?Terminal cell division
Mammary tumor virus	Tumor cells	Glucocorticoid (steroid)
Muscle proteins	Rat myoblasts	Cell contact or terminal cell division

*The control mechanisms of many of the products listed in this table probably have a component of mRNA stabilization in addition to their major level, transcriptional control. Vitellogenin mRNA is particularly well studied: in chicken and frog hepatocytes, it is clearly controlled by estrogens at both the transcriptional level and the level of mRNA stabilization.

†Some mouse liver–specific mRNAs are known to be present from mid-gestation through adulthood; others become present in late fetal or immediate postnatal life; still others appear only in adulthood. Most changes appear to be transcriptionally regulated.

ple, in 5-day-old chick embryos, nuclei from red blood cell precursors make only embryonic globin mRNA sequences; however, nuclei taken from red blood cell precursors after 12 days make a different set of globin sequences that are identical to those made by nuclei from adult chickens (Figure 12-9). Nuclei from other tissues do not make globin RNA in detectable amounts. Notice that the same precursor cell does not differentiate to form first embryonic and then adult globin; rather, all of the precursors before 5 days differentiate to form embryonic globin, and all (or most) after 12 days produce the adult globin. A final important point is that although globin synthesis can make up 50 to 90 percent of total protein output in the late stages of red blood cell development, the rate of globin gene transcription never rises higher than 0.01 to 0.04 percent of total hnRNA and the synthesis of all (or almost all) other mRNA sequences continues in the cell nucleus. Thus, the accumulation of most nonglobin RNAs specifically declines during red blood cell maturation.

Mouse Liver mRNAs The vertebrate liver produces dozens of proteins—both enzymes and proteins secreted into the blood—that are not made in other organs. Some of these represent products that are very prominent (e.g., serum albumin constitutes 1 percent of total liver cell protein synthesis), but many liver-specific proteins are produced in far lower amounts. The mRNAs complementary to cDNA clones representing both prominent and scarce liver mRNAs have been prepared. Most of these mRNAs are not found in detectable amounts in, for example, brain, spleen, or kidney cells. At what level is this tissue-specific expression of a large number of different products varying in concentration controlled?

Nuclear synthesis of RNA sequences complementary to 12 different mouse cDNA clones made from liver mRNAs was studied to answer this question (Figure 12-10). The amounts of nascent RNA produced by liver nuclei range from 20 to 500 times those produced by nuclei from other tissues. Therefore transcription is the major level of control for a wide variety of genes that are expressed in a tissue-specific fashion. Also included in the hybridization analysis were cDNAs complementary to actin and tubulin; the synthesis of nascent RNA comple-

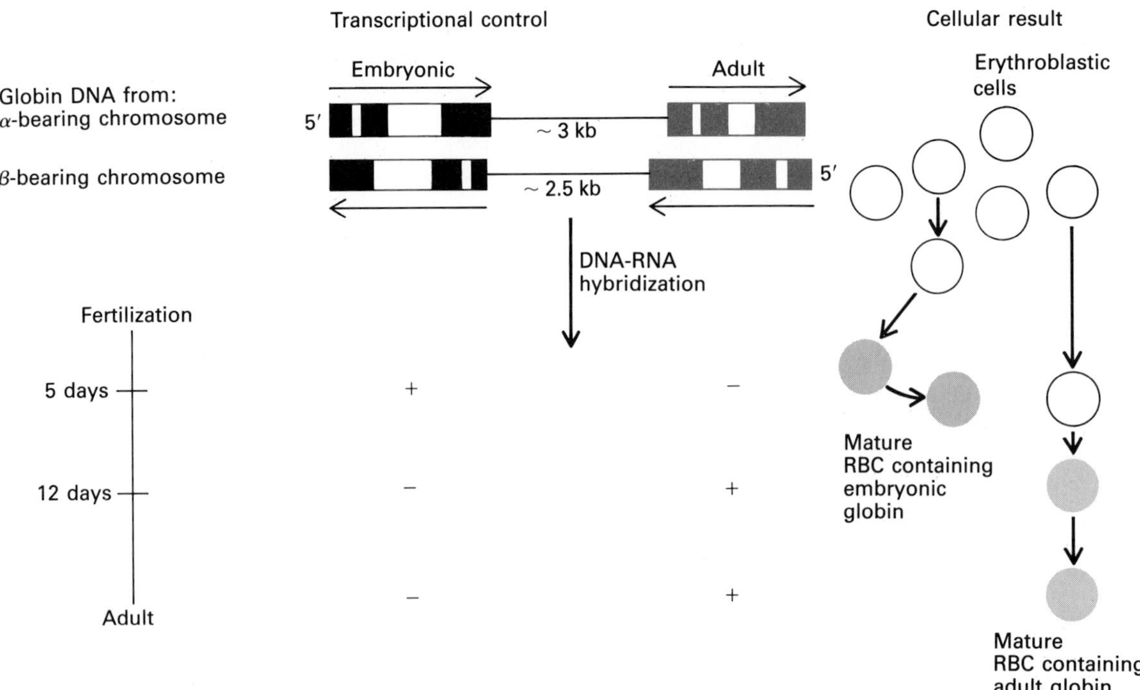

Figure 12-9 Red blood cell (RBC) differentiation in the chick. The α and β globin genes of chicks are on different chromosomes. In the α globin cluster the embryonic gene is on the 5′ side of the adult gene, whereas the reverse is true in the β globin cluster. Hybridization of RNA labeled in nuclei in vitro to cloned DNA of embryonic and adult DNA regions shows transcription of embryonic genes at 5 days; at 12 days and thereafter, the adult genes are the site of all detectable globin transcription. However, note that no one cell first makes embryonic and then adult globin. The switch from transcription of embryonic to adult genes occurs within the erythroblast population as a whole, but not within individual cells. [See J. D. Engel and J. B. Dodgson, 1980, *Proc. Nat'l Acad. Sci. USA* 77:2596; J. B. Dodgson, J. Strommer, and J. D. Engel, 1979, *Cell* 17:879; H. Weintraub, A. Larsen, and M. Groudine, 1981, *Cell* 24:333; and M. Groudine, M. Peretz, and H. Weintraub, 1981, *Mol. Cell. Biol.* 1:281.]

mentary to these cDNAs was similar in a variety of tissues. Because the concentrations of these proteins (and their mRNAs) vary in different tissues, some posttranscriptional mechanism must be responsible for controlling these "common" genes.

Egg-White Proteins of Chickens In some cases the signal that induces a transcriptional increase is known. For example, the transcriptionally controlled genes that encode the egg-white proteins of chicken eggs have been well studied. Estrogens cause the epithelial cells that line the chick oviduct (the organ that produces eggs) to increase in number and size and to secrete proteins. During these cell divisions, the cells begin to produce six or seven egg-white proteins, the most prominent of which is ovalbumin. When the process is completed, as much as 50 percent of the total protein being synthesized by the oviduct cells is egg-white protein. Estrogens also stimulate the synthesis of vitellogenin, a yolk protein normally formed in the female liver and transported through the serum to the oviduct and from there into the egg.

The rate of transcription of all of these genes (both oviduct and liver) increases dramatically in cells of estrogen-treated young chicks compared to untreated animals. As a result of this transcriptional increase, nuclear RNA precursors for a number of the egg-white proteins can be detected. Figure 12-11 shows the major nuclear mRNA precursors of one egg-white protein, ovomucoid.

As much as 0.05 percent of the total nuclear RNA in estrogen-stimulated oviduct cells is synthesized from the ovalbumin gene. However, as much as 30 percent of the total protein made by these cells is ovalbumin (20% is other egg-white proteins). It seems likely that the transcriptional increase caused by the steroid estrogen is not the only determinant of the very large increase in ovalbumin mRNA. In some way, the accumulation of other mRNAs must be decreased; the half-life of the ovalbumin mRNA is also increased during estrogen stimulation. This stabilization of mRNAs by steroids has been recognized in several cases, as we shall discuss later in this chapter.

"Puffs" on Insect Chromosomes During the normal development of insects, microscopically visible enlarge-

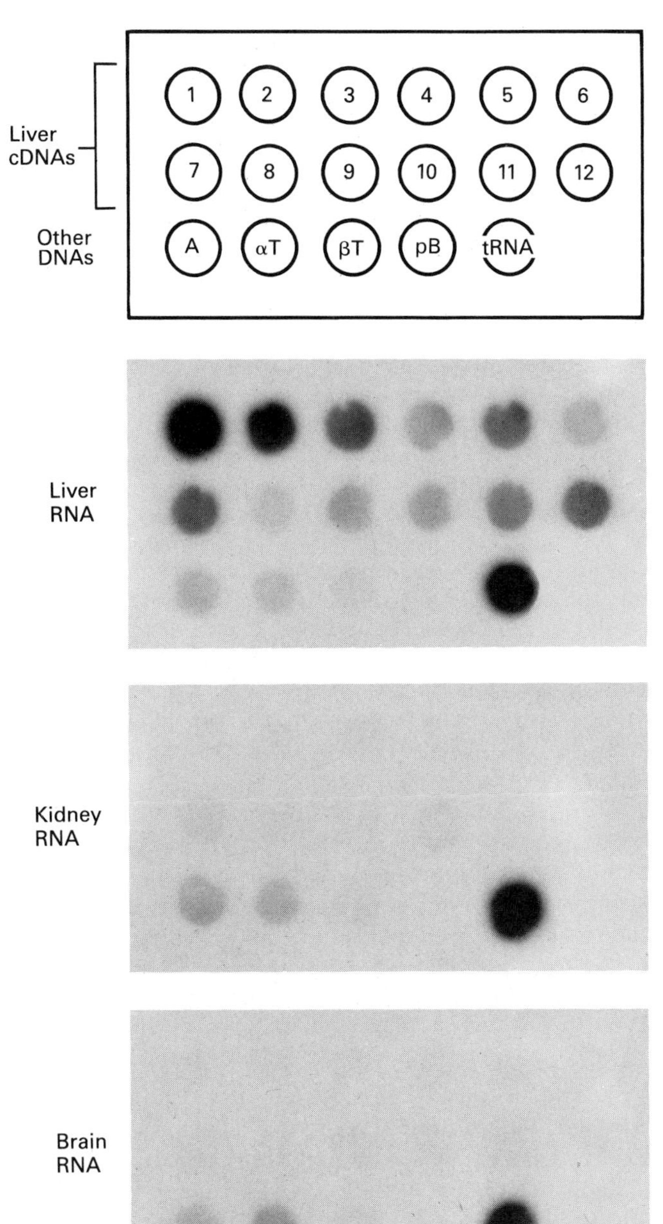

Figure 12-10 Tissue-specific transcriptional control for a variety of genes in the same tissue. The dark dots in positions 1 through 12 demonstrate RNA synthesis in liver nuclei that did not occur in kidney and brain samples. Nuclei from mouse liver, kidney, and brain cells were incubated in vitro with [^{32}P] nucleoside triphosphates to prepare nascent labeled RNA. The RNA was hybridized to samples of filter-bound cloned cDNAs made from mRNAs found in mouse liver cells (dots 1 to 12). The cDNA dots 1, 3, 5, 6, and 12 contain coding sequences for the major urinary protein, α1 antitrypsin, prealbumin, transferrin, and albumin, respectively—all proteins secreted into blood serum by liver cells. The other clones (2, 4, and 7 through 11) contain sequences complementary to specific liver mRNAs but the proteins encoded are not known. Cloned cDNAs complementary to actin (A) and to α and β tubulin (αT and βT) mRNA were also included, as was a sample of DNA encoding a methione-tRNA. All of the mouse DNA sequences were cloned in the *Escherichia coli* plasmid pBR322, and therefore pBR322 DNA by itself (pB) was also included as a control. (This plasmid DNA was not expected to hybridize to any mouse RNA.)

The dots that hybridized to labeled RNA were detected by autoradiography. The labeled liver nuclear RNA hybridized to all 12 liver-specific cDNA probes with varying intensities, signifying varying rates of RNA synthesis. Also as expected, there was no hybridization with the *E. coli* plasmid. *All* of the samples of RNA hybridized to the actin, tubulin, and tRNA probes. The kidney nuclear RNA reacted weakly with clones 7 and 10, which are known to be complementary to mRNAs in kidney as well as in liver but not with any other cDNAs. The brain RNA did not react in any quantity with the 12 liver-specific cDNAs. Thus the major level of control in the production of the 12 liver-specifc mRNAs tested is transcriptional. [See E. Derman et al., 1981, *Cell* **23**:731; and D. J. Powell et al., 1984, *J. Mol. Biol.* **179**:21.]

ments called "puffs" appear on their chromosomes. The puffs occur at sites of active [^{3}H]uridine incorporation, and they have long been interpreted as evidence of differential transcriptional activity at these locations. Some of the puffs appear only upon treatment of the insect larvae with the hormone ecdysone, which causes the larvae to molt; other puffs are formed after particular environmental shocks, such as excessive heat (Figure 12-12a and b).

The heat-shock genes of *Drosophila melanogaster* have been isolated as recombinant DNA clones. Increased transcriptional activity from the heat-shock genes of cells subjected to heat has been detected by hybridization of nuclear RNA to the purified DNA (Figure 12-12c). Thus it seems highly likely that most puffs do in fact represent transcriptionally controlled regions of *D. melanogaster* chromosomes.

The Transcriptional and Posttranscriptional Control of Histone Production The histone genes represent a case in which trancriptional control plays an important role but not an exclusive one. First of all, there are at least two different major sets of histone genes in

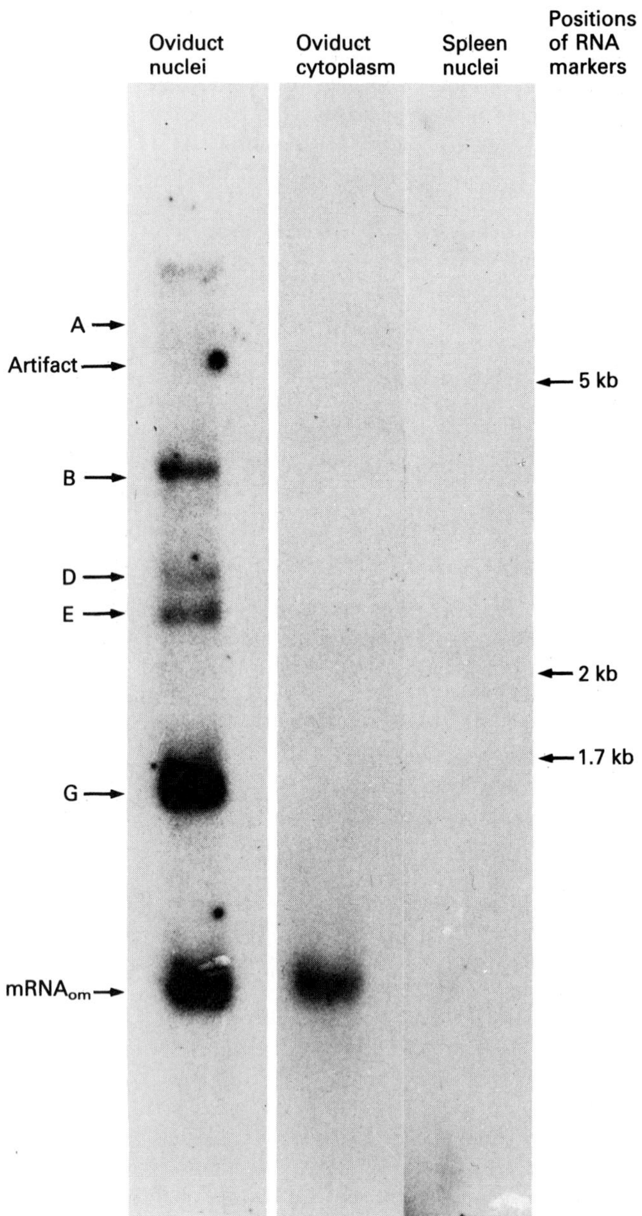

Figure 12-11 Ovomucoid mRNA precursors in nuclei of oviduct cells from estrogen-treated animals were detected by the Northern blot procedure (see Figure 7-27). Nuclear RNA was separated by electrophoresis in a gel and adsorbed to nitrocellulose filters. The RNA on the filters was then hybridized with cloned labeled DNA containing all of the ovomucoid sequences (5450 nucleotides) between the cap and the poly A sites, so that any RNA complementary to any of the labeled DNA would react. The oviduct nuclei yield ovomucoid RNA molecules that are equal in length to the entire distance from the cap to the poly A site *(first lane, band A)*. The nuclei also contain RNAs of intermediate length *(bands B, D, E, and G)*; these bands represent partially processed mRNA precursors that are missing intervening sequences. The experiment was also performed with RNA from the oviduct cytoplasm *(second lane:* the ovomucoid mRNA precursor molecules are absent) and from other nuclei (e.g., spleen nuclei, *third lane:* no ovomucoid RNA is present). [See J. L. Norstrom, D. R. Roop, M.-J. Tsai, and B. W. O'Malley, 1979, *Nature* **278**:328; and E. C. Lai et al., 1979, *Cell* **18**:829.] *Courtesy of B. W. O'Malley.*

animals such as sea urchins. Both of these gene sets occur in tandemly repeated arrays (see Chapter 11). The histone mRNAs from one set, the early embryonic set, are synthesized in eggs, but soon after fertilization the mRNAs from the other set appear. These latter mRNAs continue to be synthesized into adult life, when they represent the only histone mRNAs that are present in either the nucleus or the cytoplasm (Figure 12-13a). Thus these different sets of genes are subject to transcriptional control during early and late embryogenesis.

The model in Figure 12-13a is a simplified version of sea urchin histone gene expression; the adult "set" of histone genes actually contains at least three slightly different clusters of genes encoding late histones. In addition, there is preferential translation of early histone mRNAs during early embryogenesis. Nevertheless, it is fair to conclude that a major level of control resides at transcription for the early and late groups of histones.

Another level of control has also been demonstrated for histone genes. In cultured mammalian cells, histone mRNAs are found mainly during the S phase of the cell cycle, and histone synthesis itself occurs at this time. However, the transcription of histone genes is found to occur at a low rate throughout the cell cycle. During the S phase there is about a 4-fold increase in histone transcription, but in the same period there is about a 16-fold increase in the amount of histone mRNAs (Figure 12-13b). Consequently the stability of histone mRNAs must also be increased during the S phase. In this one family of genes, therefore, a variety of modes of gene control operate in different situations.

Transcriptional Control in Single-Celled Eukaryotes It has long been assumed that transcriptional control underlies most changes in the rate of protein synthesis in yeast. Because the nuclei of yeast cells are difficult to isolate without RNA being destroyed and because the isolated nuclei do not function well during in vitro transcription, direct measurements of nuclear RNA synthesis in yeast have not been made. However, briefly labeled total yeast cell RNA has been hybridized to cloned genes (e.g., the gene for orotic acid decarboxylase, an enzyme in the pathway of pyrimidine synthesis, and genes

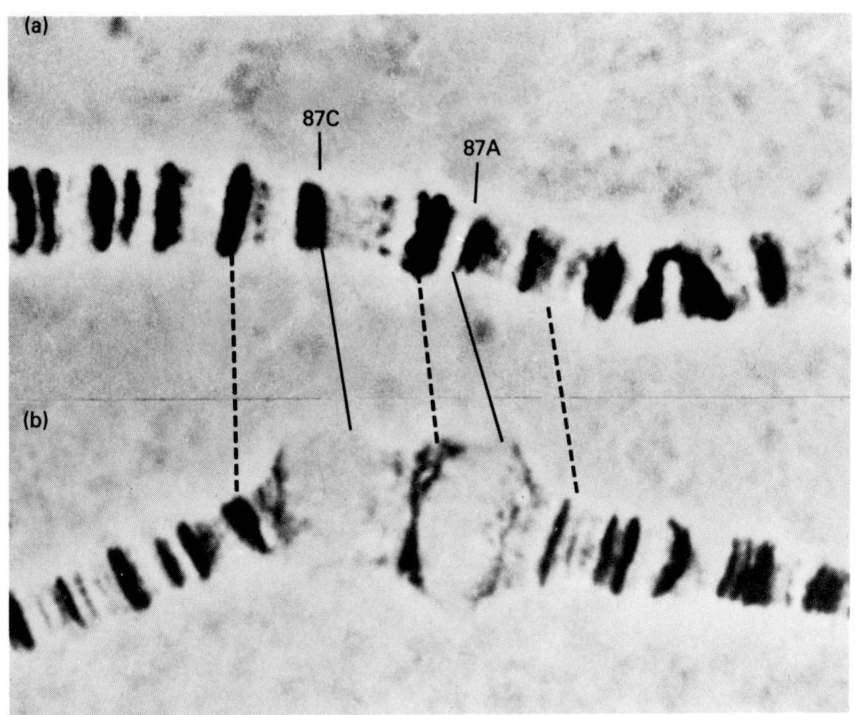

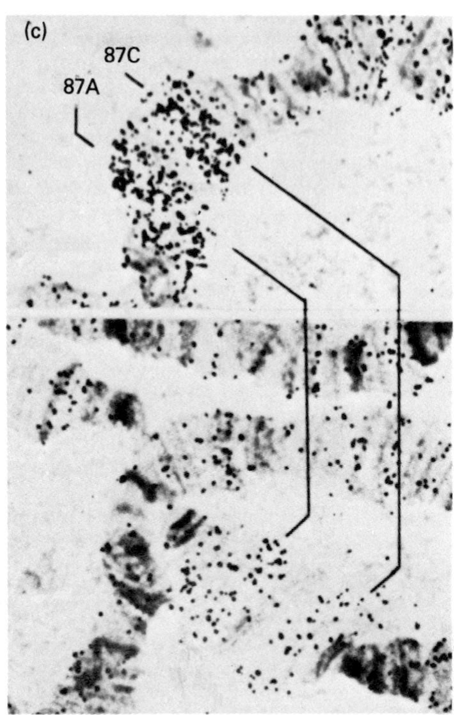

Figure 12-12 Response to heat shock in *Drosophila melanogaster* chromosomes. In nature, these flies normally grow at temperatures below 37°C. When the animals are exposed to 37°C for 40 min, "puffs" appear on the salivary gland chromosomes at characteristic sites. The two here are designated 87C and 87A. Photos (a) and (b) compare this region in chromosomes taken from normal and heat-shocked animals, respectively. Active RNA synthesis can be observed by incorporation of [^{3}H]uridine and autoradiography, as demonstrated in (c). The top photo shows incorporation of label after heat shock, and the bottom photo shows label incorporation when normal temperatures were maintained. [See M. Ashburner and J. J. Bonner, 1979, *Cell* 17:241; and J. J. Bonner and M. L. Pardue, 1976, *Cell* 8:43.] *Courtesy of J. J. Bonner.*

for ribosomal proteins). These hybridization experiments show that when the mRNA concentration is increased, there is increased [^{3}H]uridine incorporation into specific RNA. Thus it is very likely that most gene control in yeast cells is vested at the level of transcription. As we shall see later in this chapter, many yeast genes are regulated by proteins whose probable mode of action is binding to DNA.

For the slime mold *Dictyostelium discoideum,* isolated nuclei have been used to show transcriptional control of a variety of genes that function at different times in the life cycle of the organism (see Chapter 22). Once again, then, transcription appears to be the major level of gene control in simpler eukaryotes.

Control at the Level of RNA Processing

Ever since the initial discovery of the hnRNA and the suggestion of its role as precursor to mRNA, molecular biologists have speculated that selective RNA processing might be a factor in gene control. Now that the design of eukaryotic transcription units and the nature of posttranscriptional modifications (capping, methylation, poly A addition, and splicing) are well known, the mechanisms of selective processing of primary transcripts can indeed be specified exactly. A number of cases of gene control at the level of RNA processing have been documented.

Recall from Chapters 9 and 10 that there are at least two types of primary RNA transcripts from complex transcription units (those encoding two or more proteins):

1. Primary transcripts that have two or more alternative poly A sites, only one of which can be used in any one molecule in the production of an mRNA
2. Primary transcripts that have one poly A site, but that can be spliced into two or more different mRNAs

The cell's ability to make variable choices between poly A sites or between splice sites within the same primary transcript constitutes gene control by *differential RNA proc-*

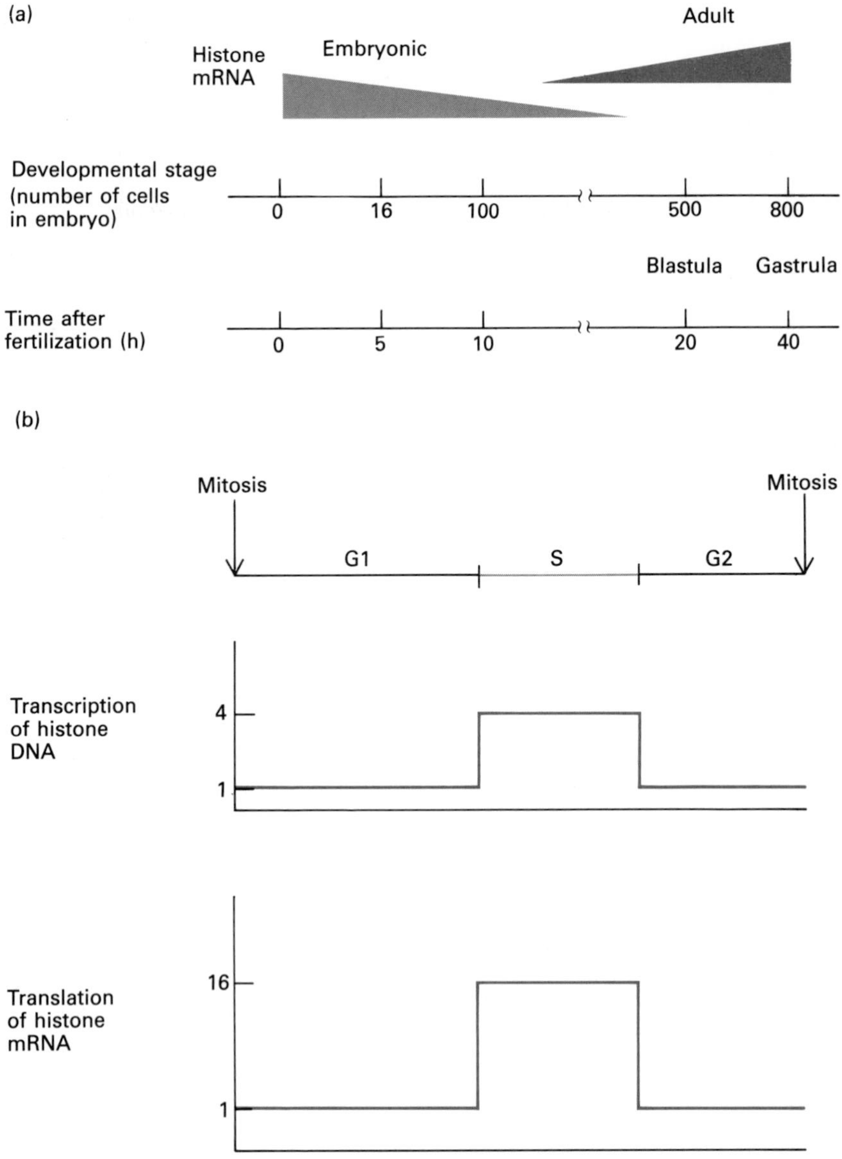

Figure 12-13 Control of histone genes. (a) Different histone genes—embryonic and adult—are used by sea urchin cells at different times in the animal's life cycle; this is true for other animals as well. The production of embryonic or adult histone mRNAs is controlled at the level of transcription. [See G. Childs, R. Maxson, and L. H. Kedes, 1979, *Devel. Biol.* 73:153.] (b) In cultured HeLa cells, analysis of histone RNA synthesis and histone protein synthesis at different times in the cell cycle shows about a 4-fold increase in transcription during the S phase, but about a 16-fold increase in histone mRNA concentration and histone protein synthesis during the same period. Thus there is also control of histone mRNA at the level of mRNA accumulation. [See N. Heintz, H. L. Sive, and R. G. Roeder, 1983, *Mol. Cell. Biol.* 3:539.]

essing (Figure 12-14a). The primary transcripts of several cellular and viral transcription units have been shown to undergo differential processing.

A second type of regulatory decision may also operate at the posttranscriptional level. Suppose that primary RNA transcripts are correctly processed under one physiological condition or in one cell type, but that they are discarded under other conditions or in a different cell type. Such decisions could be described as *process-versus-discard decisions* (Figure 12-14b).

One reason for the speculation about process-versus-discard decisions is that many nuclear RNA molecules apparently are not used to make mRNA; that is, their sequences are not represented in cytoplasmic mRNA. For example, in one experiment with cultured hamster cells, the sequences found in a number of different mRNAs (eight were tested) were contained mostly in the 25 percent of the total nuclear RNA that contained poly A (Table 12-5). The 75 percent of the total hnRNA without poly A therefore had a very low concentration of the sequences found in mRNA. As was discussed in Chapter 9, except for histone mRNAs, no specific mRNAs that lack poly A are known. The notion of process-versus-discard decisions could help to account for the existence of nuclear hnRNAs lacking poly A: these hnRNAs could be transcripts that are in a discard mode. Process-versus-discard decisions would thus serve as an on-off form of processing control. The discarded transcripts could be

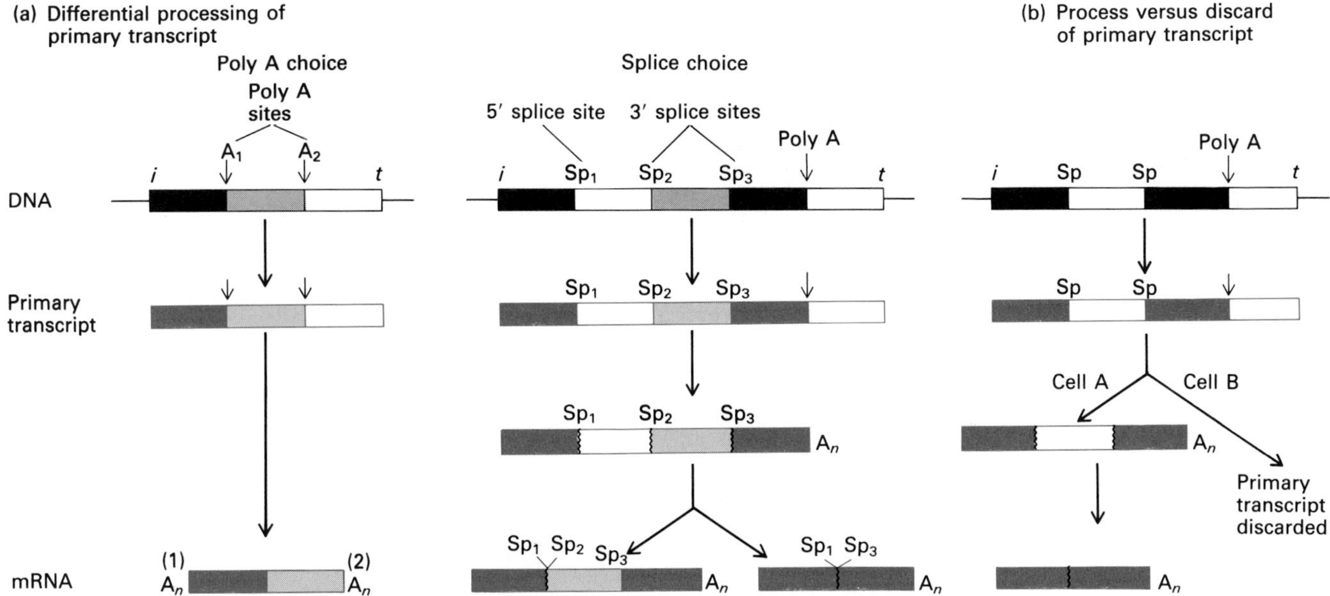

Figure 12-14 Two possible mechanisms of RNA processing control in eukaryotic gene expression. (a) Differential processing by the choice of poly A sites and/or splice sites is known to occur. (b) Process-versus-discard decisions are a hypothetical regulatory mechanism.

turned off only temporarily, or they could represent genes that have become pseudogenes through evolution.

Two apparent examples of hormones controlling process-versus-discard decisions have recently been described. Adrenal steroid hormones cause liver cells to accumulate the mRNA for a particular glycoprotein (α1), and thyroid hormone causes liver cells to accumulate the mRNA for a protein of unknown function (the protein is called spot 12, for its electrophoretic behavior). Because no difference in transcription rate was observed in either of these cases, but because there was an increase in the steady-state amount of nuclear RNA, the increase in mRNAs is thought to have been due to the conservation of nuclear transcripts that would otherwise have been discarded. Whether the addition of poly A was the "decisive" mechanism is not yet known.

Differential Nuclear RNA Processing Is Widely Used

Studies of adenovirus mRNA formation led the way in showing posttranscriptional regulation of mRNA levels: the first cases of differential poly A choice and differential splicing were observed during the adenovirus infectious cycle. A sizable number of cases of differential processing in cellular genes are now known (Table 12-6). One such example is diagrammed in Figure 12-15: differential poly A choice is exercised on the same primary transcript of the rat calcitonin gene in cells from the thyroid and brain. The two prehormones that result are similar in their amino-terminal portions but different at their carboxyl termini. Proteolytic cleavage releases either calcitonin, a peptide active in calcium retention, or the neuropeptide CGRP (calcitonin gene–related peptide). It was important to show that transcription continues past both poly A sites in both cell types. This finding proved the differential poly A choice in the processing of calcitonin transcripts. While the differential poly A choice leads to a different splicing arrangement it is the differential choice of poly A sites that dictates the different splicing arrangement.

Table 12-5 The composition of hnRNA in cultured hamster cells

Class of hnRNA	Percentage in poly A$^+$ fraction	Percentage in poly A$^-$ fraction
Total labeled hnRNA	25	75
Cap-containing hnRNA*	25	75
hnRNA with mRNA sequences (determined by hybridization to 8 cDNAs)	70–90	10–30

*According to these percentages, caps are present on all hnRNA molecules. This implies that the major posttranscriptional modification that distinguishes sequences destined to become stable cytoplasmic mRNAs from sequences that will not is the presence of poly A.

SOURCE: M. Salditt-Georgieff and J. E. Darnell Jr., 1981, *Mol. Cell. Biol.* 2:701.

Table 12-6 Genes controlled by differential processing

Gene products	Notes about the different products	Basis of differential control: Poly A choice	Basis of differential control: Splicing variation
MAMMALS			
Calcitonin (rat)	Two proteins, calcitonin (Ca^{2+}-regulating hormone) and a calcitonin-related neuropeptide that probably functions in taste, differ in their carboxyl termini	+	
Kininogens (cow)	Two prehormones that both yield bradykinin, which controls blood pressure, differ in their protease susceptibility	+	
Immunoglobulin heavy chains (mouse)	These differ at their carboxyl ends in different antibody molecules	+	
Troponin (rat)	Two different muscle proteins appear in different rat skeletal muscles		+
Myosin light chains (rat)*	Two proteins 140 and 180 amino acids long, respectively, appear in different "fast twitch" muscles		+
Preprotachykinin (cow)	Two prehormones that release substance P, a neuropeptide that affects smooth muscle, are formed; the larger prehormone also releases substance K, a neuropeptide of unknown function		+
Fibronectin (rat)	The stuctural protein has three forms		+
DROSOPHILA			
Tropomyosin	Two muscle proteins differ between embryos and adults		+
Myosin heavy chains	Two muscle proteins differ between embryos and adults; a third form is found in pupae and adults	+	
Ubx	Different gene products that control thoracic segment development are found in early embryos and larvae	+	
Glycinamide ribotide transformylase	This enzyme is encoded by a long mRNA; a shorter mRNA that shares several exons encodes a second polypeptide	+	

*The rat myosin light-chain proteins are processed from two different overlapping primary transcripts with different 5′ ends. The resulting mRNAs also contain different exons within the region of overlap, plus four exons that are similar. An almost identical situation exists in chicken myosin genes.

A clear-cut case of differential splicing occurs in the formation of the mRNA for the muscle protein tropomyosin I in *Drosophila* species (Figure 12-16). Here the 5′ and the 3′ ends of the mRNA are identical in embryos and adults, but the adult mRNA includes an exon that is not present in the embryonic mRNA. Because translation is terminated in the third exon in each case, there is a 27-amino-acid difference at the carboxyl termini of the two proteins.

Where different cells have been shown to contain mRNAs with different exons or different poly A sites from the same transcription unit, it has generally been impossible to study nuclear RNA directly to show that the differences are based on differential nuclear processing and not, for example, on differential mRNA stabilization. Many of the cases listed in Table 12-6 seem almost certain to be due to differential processing, however, because tissues contain only one of the two or more possible products. For differential mRNA stabilization to produce these results it would operate in an all-or-none fashion, which seems unlikely.

Possible Mechanisms of Differential Processing The factors that govern differential poly A choice and differential splicing are still unknown. Recall that the primary transcript terminates downstream from the poly

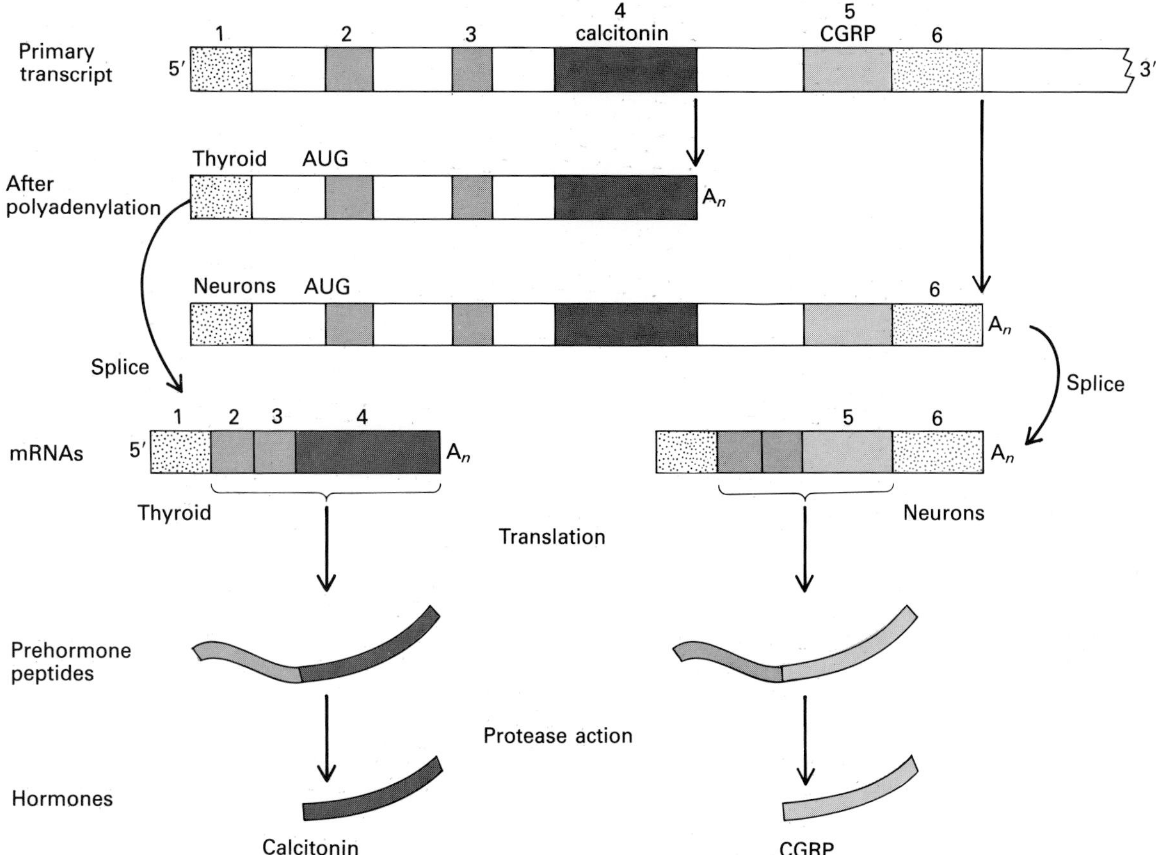

Figure 12-15 Poly A choice in the differential expression of the calcitonin gene in rats. The primary transcript has two poly A sites. During RNA processing, the first poly A site is used in the thyroid, and the second is used in the brain and other nerve tissue. (Coding exons are shaded; untranslated exons are stippled.) Translation of the different mRNAs produces two prehormones; proteolytic cleavages release the two different polypeptides, calcitonin and calcitonin gene–related peptide (CGRP), respectively. [See M. G. Rosenfeld et al., 1983, *Nature* **304**:129.]

A site, but the RNA is cut soon after the RNA polymerase passes the poly A site. We still do not know whether the hypothetical recognition factor(s) locate the poly A by associating with the newly formed nuclear RNA or whether they associate with the DNA to provide the signals for use of a particular poly A site.

In the case of differential splicing, nuclear RNA molecules (snRNA) associated with protein (snRNPs) may play a role. As we discussed in Chapter 9, the U1 snRNA has sequences that are complementary to at least the 5′ junction at many splice sites, so the snRNPs may act to bring the two parts of the transcript close together. If different snRNPs have different specificities, then variations in their concentrations could lead to differential splicing.

Another possible mechanism of differential processing is differential methylation of nuclear RNA. For example, the m^6A residues (adenylate residues methylated at the 6 carbon of the base) that are added to nuclear RNA when it is still unspliced might play a role in the choice of poly A sites; or such methylation could protect some splice sites and thereby cause others to be favored.

These puzzles are extremely interesting to researchers at present. Most of the components of processing events seem to have been identified, because processing in vitro is now possible. Details about mechanisms of differential processing will probably come to light much more quickly than the long-sought proof that differential processing of nuclear RNA actually exists.

Overlapping Transcription Units: Transcriptional Control Not Processing Control

One transcription unit arrangement leads to a subtle type of transcriptional control that is easily confused with processing control. The control of the enzyme α amylase,

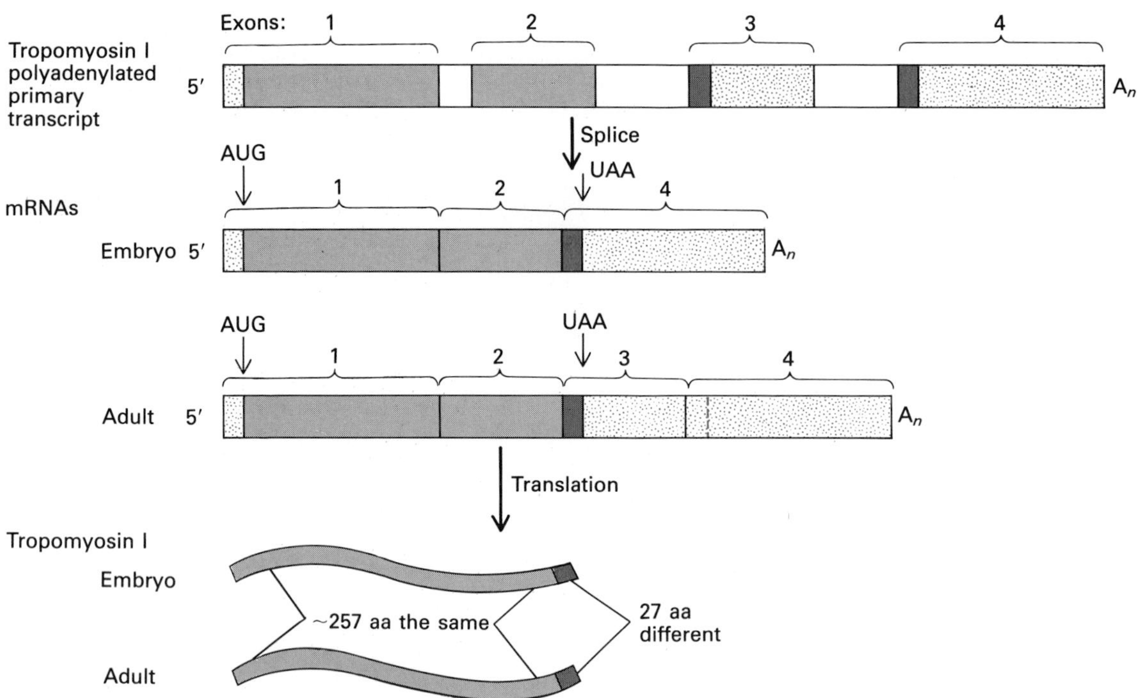

Figure 12-16 Splicing in the differential expression of the tropomyosin I gene in *Drosophila* species. The tropomyosin I transcription unit has four exons. Embryo muscle contains an mRNA with exons 1, 2, and 4, whereas thoracic muscle in adults has an mRNA with all four exons. The stippled areas represent untranslated regions. The proteins translated from the two mRNAs share the first 257 amino acids, whereas the final 27 amino acids are different. Note the position of the UAA termination codon in each 3′ exon. [See G. S. Basi, M. Boardman, and R. V. Storti, 1984, *Mol. Cell Biol.* 4:2828.]

which is produced in the liver and in the salivary glands for the digestion of starches, is a case in point. The enzyme has an identical amino acid sequence in the two tissues, but it is found in a concentration that is 100 times higher in the salivary gland. The sequences in the coding regions of the amylase mRNAs are the same. However, the 5′ ends of the two mRNAs are different. Examination of the genomic DNA encoding these mRNAs reveals how they are derived and suggests the basis for the different rate of enzyme synthesis in the two tissues. The two mRNAs have different cap sites, which are separated in the genome by about 2.8 kb (Figure 12-17). The different cap sites produce different 5′ exons (which are not translated in either case.) Thus both the starting site and the rate of transcription are tissue-specific. The first 3′ splice junction is the same in both mRNAs, which are also identical throughout the rest of their lengths. The two mRNAs are actually from two different but overlapping transcription units; the transcriptional initiation of each is a tissue-specific choice.

Yeast invertase, the disaccharidase that splits sucrose, provides a similar example. The transcription unit for this enzyme has two different transcription start sites, one that operates constitutively and one that operates when the enzyme is induced by sucrose. It is not known whether the inducible site is negatively or positively controlled, but yeast invertase is an attractive case for further analysis of this type of problem.

Gene Control in the Cytoplasm

Thus far in this chapter we have discussed nuclear events that regulate the production of a given mRNA. Nuclear events probably account for most eukaryotic gene control, but cytoplasmic events are also important. The rate of protein synthesis is affected by the rate at which mRNA molecules are transported into the cytoplasm, as well as by the cytoplasmic lifetime of the mRNA and by the frequency of its translation. Finally, even posttranslational events can function to regulate gene expression. Proteins are not completely stable, and differential protein turnover may occur. Also, differential processing of the same primary *translation* product in different cells can lead to different protein products. (See the discussion of proopiomelanocortin processing in the pituitary gland; Figure 10-30.) In this section, however, we shall confine our discussion to immediate posttranscriptional control:

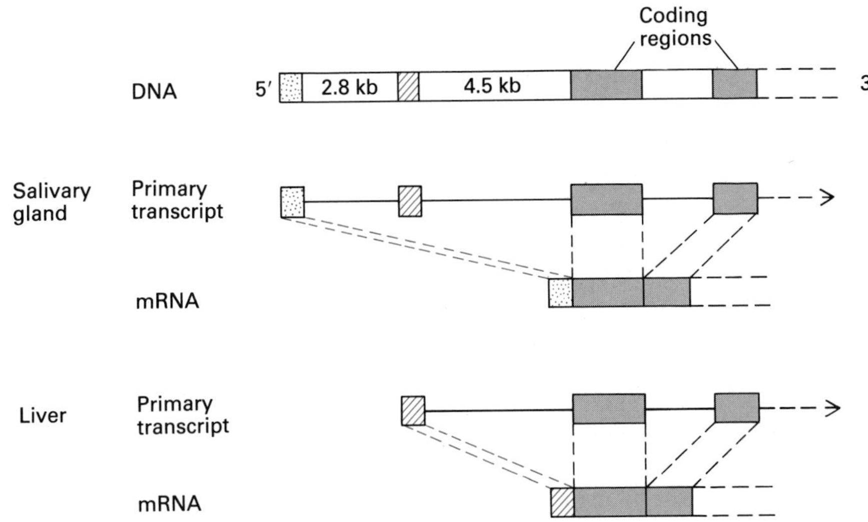

Figure 12-17 Transcription of the α amylase gene in the mouse salivary gland and liver. The 5′ ends of the mRNAs for the α amylase enzyme differ in the two tissues because the primary transcripts begin in different places. The coding portions of the mRNAs are identical. Synthesis of the salivary gland primary transcript is about 100 times more frequent than synthesis of the liver primary transcript. [See R. A. Young, O. Hagenbüchle, and U. Schibler, 1981, *Cell* 23:451.]

differential mRNA stabilization and differential translation.

Gene Control Can Occur by Changes in mRNA Stability

Hormones acting on tissue-specific mRNAs are responsible for most of the known cases of differential stabilization of mRNAs in vertebrates. This type of control (like that already described for the histone mRNAs) often accompanies increases in transcription, with the result that the overall increase in the production of a specific protein is magnified. We do not yet understand why mRNAs are metabolically unstable, so we know almost nothing about why mRNA stability changes.

The Role of Prolactin in Prolonging the Half-Life of Casein mRNA Casein is the most abundant protein in milk. It is produced in the epithelial cells of breast tissue in response to hormones, including the polypeptide hormone prolactin. Small pieces of breast tissue that are cultured in the absence of prolactin contain only about 300 molecules of casein mRNA per cell, whereas cells cultured in the presence of prolactin contain about 30,000 casein mRNA molecules per cell. However, the nuclei of the cultured breast cells synthesize only about three times as many casein mRNA sequences with prolactin as without.

The majority of the 100-fold increase in casein mRNA concentration with prolactin occurs because the half-life of the casein mRNA is increased by a factor of from 30 to 50 in the presence of the hormone. This increase in mRNA half-life was measured by introducing an RNA label into the culture medium for each sample, then removing the label from the medium and observing the loss of radioactivity over a 48-h period (Figure 12-18). The results strongly suggested that most of the prolactin-

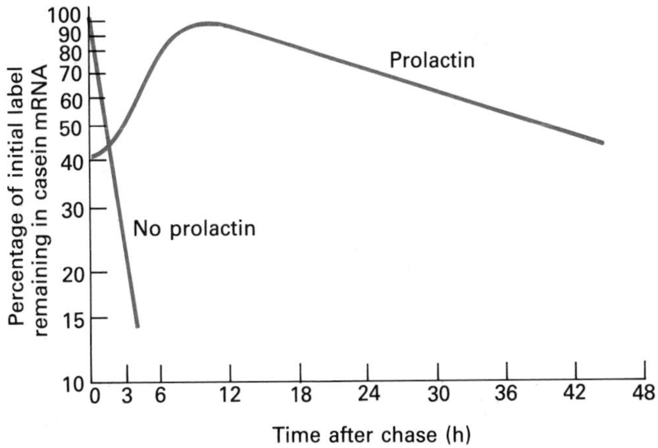

Figure 12-18 The hormone prolactin causes stabilization of casein mRNA in breast cells. Breast tissue from lactating rats was placed for 4 h into a culture medium lacking prolactin. One sample was then labeled with [^{3}H]uridine in the presence of prolactin, and another sample was labeled without the hormone. After 3 h the label was removed, a "chase" of unlabeled nucleosides was added, and RNA samples were taken at intervals to measure the levels of casein mRNA by a hybridization procedure. The highest value in the culture without prolactin was found immediately after the label was removed, and the decay of labeled mRNA was very rapid (a 50 percent loss occurred in less than 1 h). The culture with prolactin accumulated labeled mRNA for several hours because of the ineffective chase of RNA precursors (see Figure 7-2). The decay of labeled mRNA in this culture was much less rapid (a 50 percent loss required 40 h). Thus prolactin greatly reduces the turnover of casein mRNA, and this stabilization is mainly responsible for the high level of casein mRNA in breast cells. [See W. A. Guyette, R. J. Matusik, and J. M. Rosen, 1979, *Cell* **17**:1013.]

induced increase in the casein mRNA of the initial sample was due to the increased stability of the mRNA rather than to its increased synthesis.

The Estrogen-Dependent Stability of a Liver mRNA As we have mentioned, vitellogenin is a protein that is made in female chicken and frog liver in response to estrogens. After its synthesis it is transported to the oviduct. In male chickens also, the liver is responsive to experimental addition of estrogens (such as estradiol), and the synthesis of vitellogenin in males not treated experimentally with estrogen is very low. In nuclei from liver cells, a very large and prompt transcriptional increase (by a factor of at least several hundred) follows the treatment of animals with estrogen. The accumulation of vitellogenin mRNA has been plotted, and from such curves a half-life of about 24 h has been estimated. (See Chapter 9 for an explanation of how the accumulation of mRNA to a plateau value can be used to estimate mRNA half-life.) After acute withdrawal of estrogens, the vitellogenin mRNA concentration falls abruptly, indicating a half-life of less than 3 h. Thus the hormonal control of this protein is quite significant at two levels—transcriptional initiation and mRNA stabilization.

Changes in the Half-Lives of Other mRNAs We have seen that mRNA stability can be regulated by proteins or steroid hormones. Other factors may increase or decrease mRNA stability (Table 12-7). The precise signals for the changes in half-lives are often not known. Viral proteins that accumulate late in infection may be responsible for stabilizing some early adenovirus mRNAs.

The mRNAs for the cytoskeletal elements are known to change in concentration even when no difference in transcription rate is detectable. For example, actin and tubulin mRNAs increase in liver cells wihout an increase in transcription during the cell regeneration that occurs after surgical removal of part of the liver. Likewise, the concentrations of these mRNAs vary between cells of different tissues without significant changes in transcription rates. Such changes are likely caused by differential mRNA stabilization of RNA processing.

In particular, tubulin protein subunits may have a role in regulating the stability of tubulin mRNA. When tubulin fibers (microtubules) are disaggregated by the treatment of cells with colchicine, a drug that binds to monomers and prevents microtubules from forming, the tubulin mRNA rapidly declines (see Figure 18-6). The finding that the disaggregation of the microtubules apparently destabilizes the tubulin mRNA implies that the tubulin normally has a role in stabilizing its own mRNA. Such a mechanism constitutes autoregulation of a gene in the cytoplasm.

Purified DNA segments have been used to perform accurate kinetic analyses of mRNA formation and stabilization, and such experiments have already proved a number of instances of changes in mRNA half-lives. It seems likely that mRNA stabilization is an important basis for at least a secondary form of gene control.

The Translation Efficiency of Specific mRNAs Can Be Controlled

As we have mentioned, differential turnover of mRNA is only one type of cytoplasmic gene control; differential

Table 12-7 Biological systems that exhibit regulation of mRNA stability*

mRNA	Tissue or cell	Regulatory signal	Half-life of mRNA	
			With effector	Without effector
Vitellogenin	Frog liver	Estrogen	500 h	16 h
Vitellogenin	Rooster liver	Estrogen	~24 h	<3 h
Ovalbumin, conalbumin	Hen oviduct	Estrogen, progesterone	>24 h	2–5 h
Casein	Rat mammary gland	Prolactin	92 h	5 h
Prostatic steroid-binding protein	Rat prostate	Androgen	Increases 30×	
Histones	HeLa cells	DNA replication	30–60 min	<10 min
Histones	Yeast	DNA replication	i.e., during replication: >15 min	i.e., after replication: 5 min
Adenovirus 1A and 1B (early mRNAs)	HeLa cells	Late viral proteins	i.e., late in infection: 60–100 min	i.e., early in infection: 6–10 min
Cytoplasmic actin	Liver	Regenerating liver	Increases 10×	
Cytoplasmic β tubulin	Liver	Regenerating liver	Increases 10×	
Cytoplasmic β tubulin	Mouse cells in culture	Disaggregated tubulin	Decreases >10×	

*For all mRNAs listed except those of adenovirus, actin, and tubulin, the effector (the agent responsible for stabilization) also causes an increase in transcription.

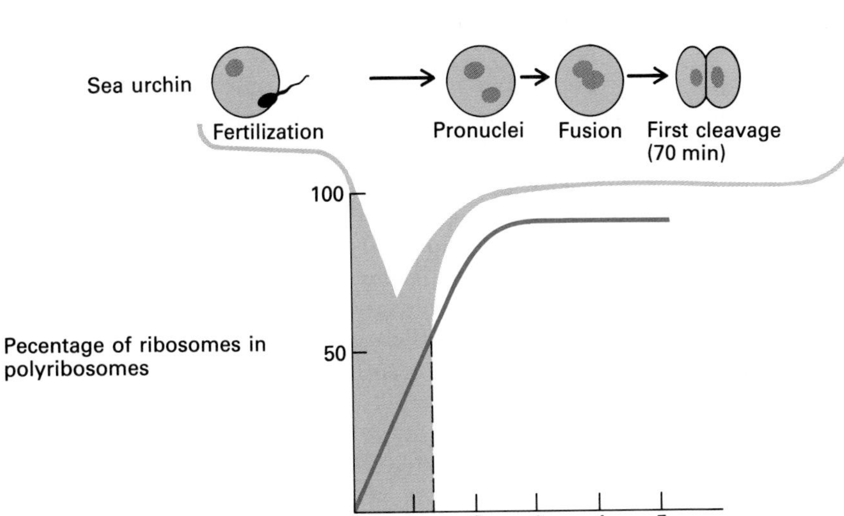

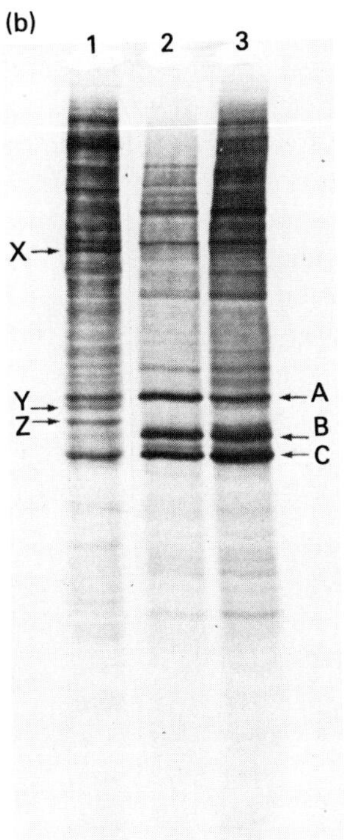

Figure 12-19 Translational control of protein synthesis in the early cleavage stages of embryogenesis. Eggs from marine invertebrates such as sea urchins and clams are easy to collect, and they can all be fertilized at the same time by adding excess sperm. (a) Protein synthesis increases in most marine eggs just after fertilizaton by the association of ribosomes with preexisting mRNA. (The percentage of ribosomes in the polyribosomes is a good index of the extent of translation taking place.) The delayed use of the mRNA illustrates translational control. [See B. Brandhorst, 1976, *Devel. Biol.* **52**:310.] (b) In eggs of the clam *Spisula solidissima,* protein synthesis is sufficiently active before fertilization to label proteins for examination by gel electrophoresis and autoradiography. A comparison of electrophoretic patterns obtained from eggs before and after fertilization reveals changes in protein synthesis. Lane 1 shows the proteins made by unfertilized eggs. Lane 2 shows egg proteins made after fertilization but before the first cleavage, and lane 3 shows a similar sample extracted after the first cleavage. X, Y, and Z are proteins that are prominent synthetic products before fertilization but not after; the reverse is true for A, B, and C. However, when mRNA was extracted before and after fertilization and translated in vitro, the protein patterns were nearly identical in the two samples: all six bands were present in each case. Because the proteins synthesized differed so much more than the mRNA composition in fertilized versus unfertilized eggs, the experiment demonstrated translational control. [See E. T. Rosenthal, T. Hunt, and J. V. Ruderman, 1980, *Cell* **20**:487.] *Part (b) courtesy of J. V. Ruderman.*

translation can also greatly affect gene expression. Translational control was first discovered in connection with the delayed use of the so-called "maternal" mRNA that is stored in egg cells. RNA molecules that have all of the molecular hallmarks of mRNAs—the correct length, 5′ caps, and 3′ poly A—are stored in the cytoplasm of unfertilized eggs in many vertebrate and invertebrate animals. While the egg remains unfertilized, protein synthesis is slow; after fertilization, protein synthesis increases dramatically. Because this increase can occur without any new mRNA being formed, translational control must be responsible (Figure 12-19a).

The changes in protein synthesis may be qualitative as well as quantitative. For example, in the surf clam *Spisula solidissima,* certain proteins that are formed before fertilization are not formed nearly so rapidly after fertilization. Other proteins appear prominently only after fertilization. However, when the total mRNA is extracted from both fertilized and unfertilized eggs and translated by cell extracts, the same mRNAs are found to be present before and after fertilization (Figure 12-19b). The translational step that is suspended in vivo for certain proteins at certain stages of embryogenesis appears to be initiation, because the untranslated mRNAs of both the fertilized and the unfertilized eggs are not associated with ribosomes.

In Chapter 8 we noted that translational control operates for a number of bacterial ribosomal proteins (those proteins associated with ribosomal RNA in ribosomes). Translational control also occurs for ribosomal proteins in mammalian cells. When growth is suspended by placing cultured cells in a protein-free medium, the mRNAs for ribosomal proteins are not associated with polyribosomes and are no longer translated. When serum is returned to the medium, these mRNAs again become engaged by the translation initiation apparatus and are translated.

Another very interesting case of translational control occurs late after the adenovirus infection of HeLa cells. The pre-existing cell mRNA is still present but it is not translated; translation of the late viral mRNAs accounts for over 90 percent of the proteins made. In addition to encoding virus mRNAs, the virus genome also encodes a small (160-nucleotide) RNA that is synthesized by RNA polymerase III. This RNA is called VA RNA, for virus-associated RNA. Inside the cell, the VA RNA is part of a ribonucleoprotein particle (RNP) found both in the nucleus and in the cytoplasm. If the viral sequences that encode the VA RNA are deleted, the virus mRNAs synthesized are identical to those in a normal virus infection. However, there is about a 10-fold decrease in virus mRNA translation. The low rate of translation of virus mRNA in cells infected by the VA^- mutants suggests that the small RNPs that contain VA have a direct role in translation. By extension, it seems possible that all mRNAs require some small RNPs for assistance in efficient translation.

The Overall Rate of Translation Is Controlled under Various Conditions

Translational control of specific proteins has been proved in only a few cases. In contrast, known mechanisms of translational control for all mRNAs operate in all cells to change the *average* rate of protein synthesis. For example, when cultured mammalian or insect cells are heated above 40°C, most translation initiation is suppressed, but the formation of heat-shock proteins is stimulated. As we mentioned earlier, an increase in the transcription of the heat-shock genes (see Figure 12-12) accompanies the favored translation of heat-shock mRNAs. Mammalian cells adapt to the new elevated temperature (as long as the temperature is not higher than about 42°C) and resume normal protein synthesis and growth within 2 h after a heat shock.

Another occasion for the general inhibition of protein synthesis occurs during each cell cycle. When cells enter mitosis the translation initiation rate is inhibited, which causes a fall in the rate of protein synthesis to 30 percent of normal. The decline in initiation is indicated by a decrease in the size of polyribosomes; that is, there are fewer ribosomes per mRNA, a sign that initiation is occurring less frequently even though polypeptide elongation and termination are continuing. If the rate of chain elongation is inhibited by a drug to restore the balance between elongation and initiation, the polyribosomes regain normal size. There is also a great decrease in RNA synthesis during mitosis, but the decline in the number of mRNAs is not sufficient to account for the decrease in protein synthesis; translational control is clearly also a factor.

Thus mechanisms for the inhibition of total protein synthesis are available to cells. One possible mechanism is the phosphorylation of initiation factors by protein kinases. These enzymes add phosphates to serine, threonine, and tyrosine residues (see Figure 16-12). The biochemistry of translation initiation reactions in mammalian cells is most well known in reticulocytes because a number of the components of their translational machinery have been purified. When the protein initiation factor eIF_2 (eukaryotic initiation factor 2) is phosphorylated, it is inactive. In the experimental system described in Figure 12-20, the trigger for phosphorylation by a protein kinase can be either the absence of heme (a component of hemoglobin) or the presence of a segment of double-stranded RNA. Since hemoglobin contains both heme and the globin chains, a balance of the two is assured by this translational control; a decrease in heme leads to a decrease in globin mRNA translation. It is likely that all cells possess these protein kinases, and that many factors other than heme and double-stranded RNA may affect their activity.

Control of rRNAs and tRNAs in Different Cells

All of the cases of gene control discussed so far have involved protein-coding genes. There are, of course, many more protein-coding genes than other types, and most cell specificity resides in the control of the protein-coding genes. Nevertheless, an important element of cellular growth and development is the control of the ribosomal genes transcribed by RNA polymerase I, and the genes for small RNAs transcribed by RNA polymerase III.

Ribosomal RNA synthesis is not controlled at the same level in every type of cell and in every organism. Bacteria placed in a poor medium (e.g., one lacking amino acids) grow more slowly than bacteria in a rich medium. This slower growth is accompanied by an almost complete cessation in pre-rRNA synthesis. Control of pre-rRNA synthesis is mediated in bacteria by the polyphosphate 5′ppGpp3′, which increases in concentration in non-

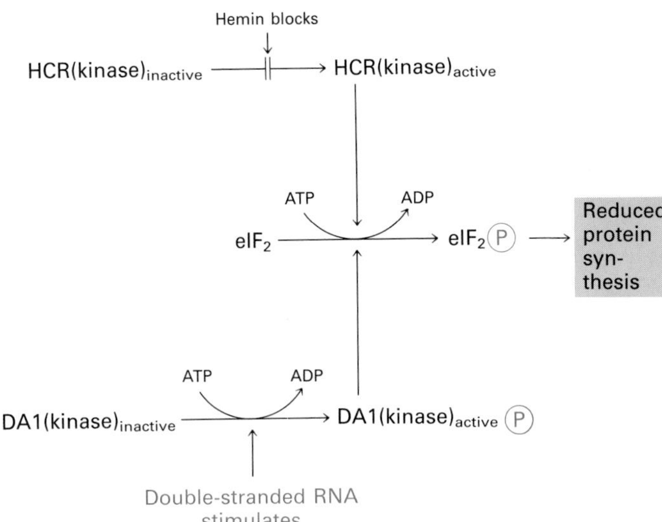

Figure 12-20 The mechanism of generalized translational control in the reticulocyte protein synthesis system. The diagram shows two possible mechanisms by which eukaryotic initiation factor 2 (eIF_2) may become phosphorylated, and thus be rendered much less active in promoting initiation of protein synthesis. HCR (hemin control reactant) is an inhibitory complex containing a kinase protein that is activated by the absence of hemin, which is a derivative of heme. DA1 is a double-strand-activated protein kinase. Thus either the presence of double-stranded RNA (during, say, RNA virus infection) or the absence of hemin (conceivably due to iron starvation) leads to phosphorylation of eIF_2 and decreases the initiation of protein synthesis. [See P. J. Farrell et al., 1977, *Cell* **11**:187.]

growing cells. The ppGpp plays a direct role in repressing the initiation of ribosomal gene transcription. This so-called "stringent" response immediately causes a great decrease in ribosome production. Yeast cells also have a stringent response that stops ribosome formation; in the face of starvation or during sporulation, ribosomal RNA synthesis decreases by at least 80 percent (Figure 12-21).

Cultured mammalian cells, however, continue to make pre-rRNA at 30 to 50 percent the normal rate when they are starved of an essential amino acid. No new ribosomal proteins are formed under these conditions, so new ribosomes cannot be assembled and the pre-rRNA is simply degraded. Some control of ribosomal gene transcription is known to occur in vertebrate cells. For example, frog embryos do not begin pre-rRNA synthesis until about 16 h after fertilization (Figure 12-21). Special preparations for the suspension of pre-rRNA synthesis early in frog development are made at the time of oogenesis. During egg formation the ribosomal genes are amplified over 1000 times, and pre-rRNA synthesis and ribosome formation increase accordingly (Chapter 11). These ribosomes are stored in the egg and parceled out during the early embryonic divisions until pre-rRNA synthesis is activated after about 12 to 14 divisions. Mammalian eggs, however, do not have large ribosomal stores, and transcription of pre-rRNA begins at the two-cell stage.

It is possible that most mammalian cells exercise little transcriptional control over pre-rRNA synthesis. The cells in a variety of tissues all tend to synthesize pre-rRNA at similar rates, but they differ in their rate of cell turnover. In several cases the efficiency of processing pre-rRNA clearly regulates ribosome formation. For example, in mammals in which one kidney is removed, there is compensatory hypertrophy (growth) of the remaining kidney; its mass increases by at least 50 percent. New ribosome formation accompanies the increase in mass, but there is no increased labeling of nucleolar RNA. Apparently pre-rRNA is processed more efficiently.

A similar situation exists in regenerating liver. When mammals lose a large amount of liver tissue—for example, during surgery, infection, or chemical poisoning—the remaining liver tissue regenerates a normal-sized liver within a few days. There is an increase in the appearance of new ribosomes in the cytoplasm, but no increase in nuclear preRNA synthesis. Thus differential processing of pre-rRNA seems to be a major means of controlling rRNA formation in the cells of adult mammals. In addition the new ribosomal proteins required for regeneration arise from an increased translation of the same amount of mRNA for ribosomal proteins.

One case in which transcriptional control does seem to occur in a ribosomal component is in the selection of 5S rRNA genes in frogs. As we discussed in Chapter 9, RNA polymerase III transcription of both 5S rRNA and tRNA genes is known to require specific transcriptional factors that bind to sequences within the transcribed regions of these short genes. Once the 5S rDNA is bound to its cognate transcriptional factor, the resulting active transcriptional complex is maintained for long periods. In the case of 5S genes in frogs, a particular set of 5S genes (e.g., somatic or oocyte genes) is selected for expression, and the selection is almost surely regulated at the level of transcription. The factors governing selection are not yet clear, because the same transcriptional factor shows binding with similar affinity (within a factor of 2 or 3) to both somatic and oocyte genes.

Mechanisms of Transcriptional Control: Site-Specific DNA Binding and Changes in Chromosomal Topology

Most experimentation and speculation about mechanisms of eukaryotic gene control have concerned transcriptional control. This focus is appropriate because the initiation of transcription is the first step in gene control, and more protein-coding genes are probably controlled at this level than at any other level.

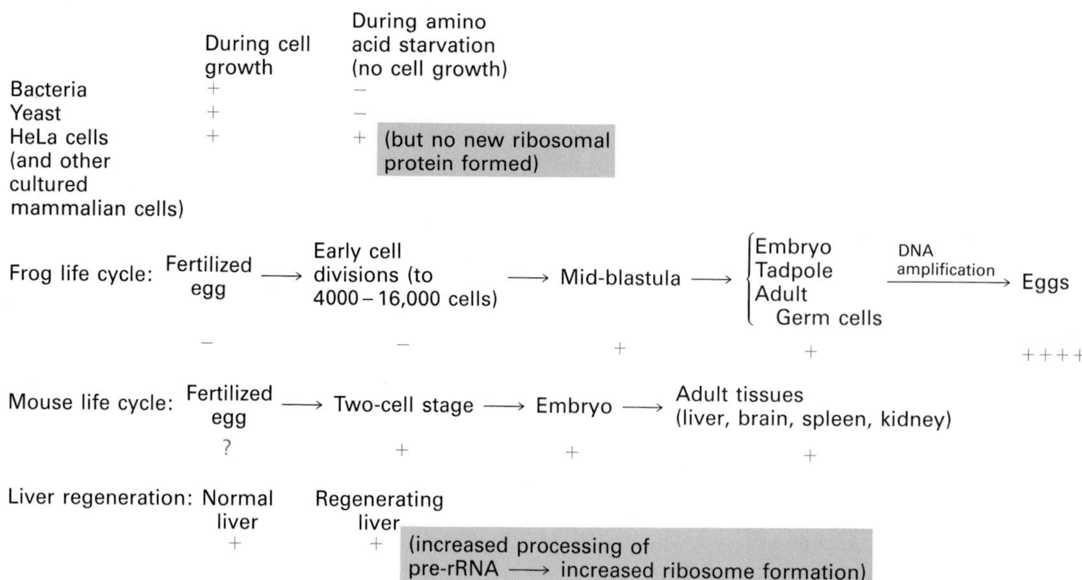

pre-rRNA Synthesis
During cell growth
During amino acid starvation (no cell growth)
Bacteria + −
Yeast + −
HeLa cells (and other cultured mammalian cells) + + (but no new ribosomal protein formed)
Frog life cycle: Fertilized egg → Early cell divisions (to 4000 – 16,000 cells) → Mid-blastula → Embryo Tadpole Adult Germ cells → DNA amplification → Eggs
− − + + ++++
Mouse life cycle: Fertilized egg → Two-cell stage → Embryo → Adult tissues (liver, brain, spleen, kidney)
? + + +
Liver regeneration: Normal liver + Regenerating liver +
(increased processing of pre-rRNA → increased ribosome formation)

Figure 12-21 Patterns of control of pre-rRNA synthesis in various organisms. Starved bacterial and yeast cells stop pre-rRNA synthesis (the "stringent" response). HeLa cells and other cultured cells do not exhibit this response. Frog eggs have an enormous store of ribosomes that are produced during oogenesis, when ribosomal genes are amplified and actively transcribed (Chapter 11). Transcription of rDNA ceases in ripe frog eggs, and ribosome synthesis does not resume after fertilization until mid-blastula. Mouse eggs do not have abundant stored ribosomes, and pre-rRNA and ribosome synthesis begin as early as the two-cell stage. Many mouse cells that differ in their life spans have equal rates of pre-rRNA synthesis. For example, mouse liver and brain cells grow very slowly, whereas spleen cells grow quite rapidly; but pre-rRNA synthesis is the same for all. During liver regeneration in animals ribosome formation increases but pre-rRNA synthesis does not; this observation implies increased pre-rRNA processing.

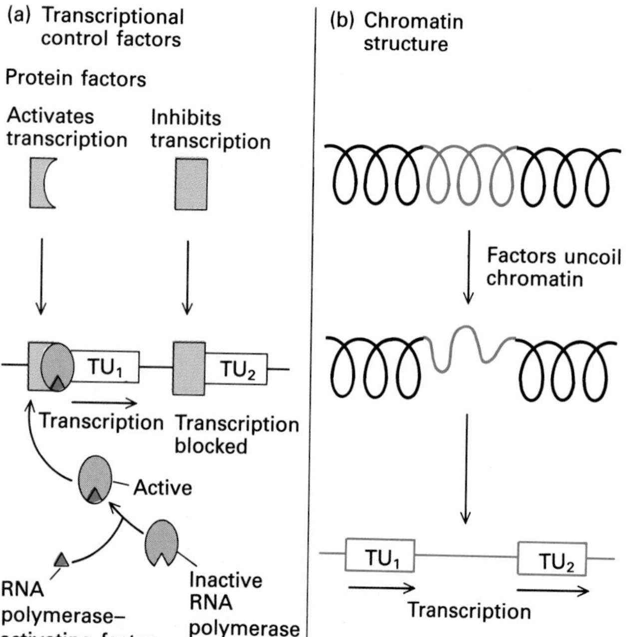

Figure 12-22 Two possible contributing factors to gene control: transcriptional factors and chromatin structure. (a) Positive- or negative-acting transcriptional factors might bind to DNA to effect a transcription unit (TU). Other factors might interact with the RNA polymerase (or, more broadly, with any part of a transcription complex).
(b) Tightly coiled chromatin is probably inactive for transcription, so some means of localized uncoiling may accompany transcriptional activation. Transcription may be automatic after uncoiling, or it may be subject to the control factors in (a).

Two main assumptions are often made in considering the mechanisms of eukaryotic transcriptional control. The first is that eukaryotic cells, like prokaryotic cells, contain direct-acting transcriptional factors—presumably proteins—that control transcription by interacting with DNA sites very near the target gene. Proteins that bind directly to the polymerases and change either their rates of activity or their specificity are also possible (Figure 12-22a). A few proteins that are thought to directly affect the initiation of transcription of specific genes have been isolated, although direct assays for the activity of such factors are still difficult. For example, mixing purified DNA plus purified RNA polymerase II plus a putative positive-acting factor does not allow for transcription. In vitro transcription from pure DNA still requires cell extracts containing a number of proteins other than the polymerases themselves for correct initiation to occur. Thus a straightforward demonstration of the activity of a positive-acting factor is difficult.

A second assumption often made about eukaryotic transcriptional control is that site-specific transcription in eukaryotic chromatin depends on a "loosening" or uncoiling of the chromatin structure to free the DNA in a particular region (Figure 12-22b). Such changes might include the removal of nucleosomes, alterations in their structure, or local structural changes in the DNA itself (e.g., an unwinding of the helix, the conversion of a normal right-handed helix to the left-handed type known as Z DNA, or the local supercoiling of the DNA; Chapter 3) (Figure 12-23). Many of these changes in the topology of DNA may act through the agency of topoisomerases, enzymes that change the extent of the supercoiling of DNA. (Topoisomerases and DNA superhelicity are discussed in Chapter 13.) Structural changes in the chromatin might lead to automatic transcription of uncovered genes; alternatively, the uncovered genes might be further subject to positive or negative control by specific factors (see Figure 12-22). As we shall see, there is direct evidence that chromatin structure differs for active genes and inactive ones, although many of the details about the types of changes that occur and the ways they are brought about are still unknown.

The Control of Eukaryotic Regulatory Proteins Raises Several Questions

Site-specific regulatory proteins like those that act in bacteria almost surely exist in eukaryotes. However, a problem arises in how the eukaryotic genome can contain information for many different regulatory genes. Tran-

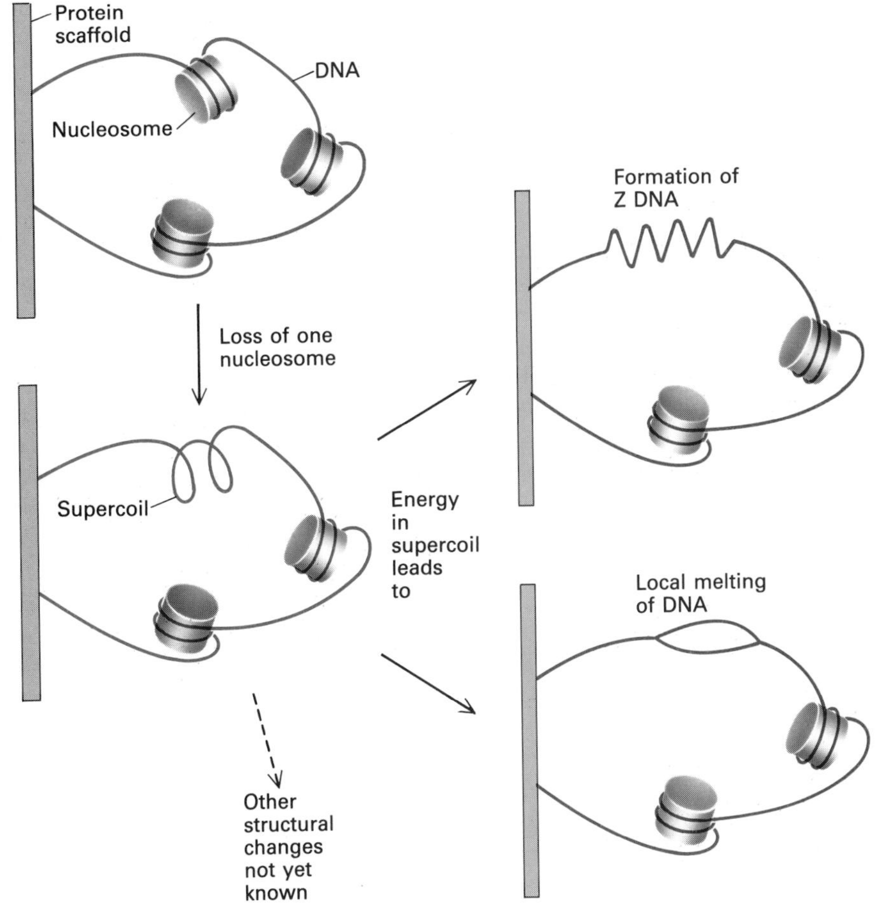

Figure 12-23 Hypothetical events in chromatin uncoiling. DNA is depicted as a loop or *domain* that is bound to a protein scaffold; this structure has been suggested by studies of *D. melanogaster* chromosomes. Loss of a nucleosome within a domain would leave extra supercoils in the DNA (see Figure 3-53). The energy of the supercoiling could cause regions of the DNA to assume the Z configuration, to undergo localized untwisting, or to change its configuration in some way that has not yet been discovered.

scription units in eukaryotes are mostly simple; that is, they produce only a single mRNA. It therefore seems unlikely that each and every transcription unit has its own unique site-specific regulatory protein whose only function is to serve in regulation. Such a scheme would require as many regulatory proteins as there are genes to be regulated; it would also leave open the question of how synthesis of the regulatory proteins themselves would be regulated. If the transcription of many or most eukaryotic genes is in fact controlled, then we can anticipate the following possibilities: (1) autogenous regulation of transcription, and/or (2) regulation of multiple genes by a single protein, and/or (3) dual-functioning proteins—that is, enzymes or structural proteins that would also act in gene regulation.

Autogenous regulators do exist in bacteria (e.g., for bacteriophage λ and for the *hut* operon; Chapter 8) and they could exist to control eukaryotic transcription units also.

Regarding the second possibility, an individual regulatory protein might operate on a large number of different genes scattered throughout the chromosomes (e.g., all of the genes necessary to make a red blood cell, a skin cell, or a liver cell). The idea of a single regulatory protein acting on multiple sites has been popular because genes for proteins that are tissue- or cell-specific are usually scattered on different chromosomes. Again, there is a possible model in *E. coli.* The genes for the enzymes of arginine biosynthesis are located at several distant sites in the *E. coli* chromosome, and all are controlled by the presence of arginine plus a single repressor protein (see Figure 8-26).

Finally, dual-functioning proteins, if they exist, could not only circumvent the need for each regulated transcription unit to have its own separate regulatory protein, but they could also connect metabolic events directly to gene control.

Positive- and Negative-Acting Regulatory Proteins Exist in Yeast Cells

Geneticists working with yeasts have discovered that these cells contain genes encoding regulatory proteins for a number of metabolic pathways, for determination of mating type, and for sporulation. This group of regulatory proteins includes both positive- and negative-acting proteins, some of which can affect more than one gene in a pathway. We shall describe several yeast regulatory proteins to illuminate issues raised in the previous section.

Galactose Regulation In yeasts, three enzymes that are required to convert galactose into glucose phosphate are encoded by genes that lie on chromosome II. A fourth gene, which is for galactose transport, is on chromosome XII. A fifth gene located on still another chromosome encodes α galactosidase, which converts melibiose to galactose. All five genes are induced by a metabolic product of galactose—but only if the protein product of a sixth gene, *gal*4, is present. Thus the Gal4 protein is a positive regulator (Figure 12-24). The *gal*4 gene has been cloned and sequenced, so the sequence of its protein is now known. The Gal4 protein apparently causes the increase in mRNAs for the galactose-metabolizing enzymes by binding to specific DNA sequences from 150 to 250 bases upstream from the site of initiation of the various mRNAs. This conclusion is inferred from the protection of certain DNA bases from chemical reaction by the presence of the Gal4 protein. The Gal4 protein itself is negatively regulated by the product of another gene (*gal*80) on another chromosome. It is not yet known if the Gal80 protein binds to DNA.

Transcription of the Mating-Type Locus The expression of the yeast mating type involves another sophisticated network of interacting genes that is governed by the action of separate regulatory genes. As we discussed in Chapters 6 and 11, haploid yeast cells exhibit either *a* or α mating types. Each mating type produces an oligopeptide (an *a* or an α factor, respectively) that binds to cells of the opposite type and stops their growth. Each cell thus tends to aggregate with a cell of the opposite type, and the result is diploid cells that are designated *a*/α. The *a*/α diploids can undergo meiosis and sporulation to produce haploid *a* and α cells once again. The mating type of a haploid cell is determined by which of the sequences, *a* or α, is present at the MAT locus; two different pairs of mRNAs—an *a*-specific pair and an α-specific pair—are produced from MAT. Storage copies of *a* and α DNA reside at two silent (nonexpressed) loci, HML and HMR (sites to the left and right, respectively, of MAT on chromosome III) (see Figure 11-50). No RNA is normally transcribed from the sequences at HML and HMR.

The product of a gene called *SIR* (for silent information repressor) has been identified as the protein that is responsible for suppressing the transcription of the silent DNA copies at HML and HMR. Mutations in the *SIR* genes cause the yeast cells to transcribe abnormally the DNA sequences at these sites. It is believed that the *SIR* gene product does not simply bind to the DNA at the start site for RNA synthesis, but rather acts at a distance from the start site and prevents transcription by changing the chromosomal structure.

The mode of action of the *SIR* gene product was determined in an experiment in which plasmids of various types were prepared. Some contained the DNA found at a silent mating-type locus (HMR); others contained the sequences from the active MAT locus. The plasmids were then individually introduced into normal yeast cells and into cells bearing mutations in the *SIR* gene. When the MATα plasmid was introduced into either the wild-type

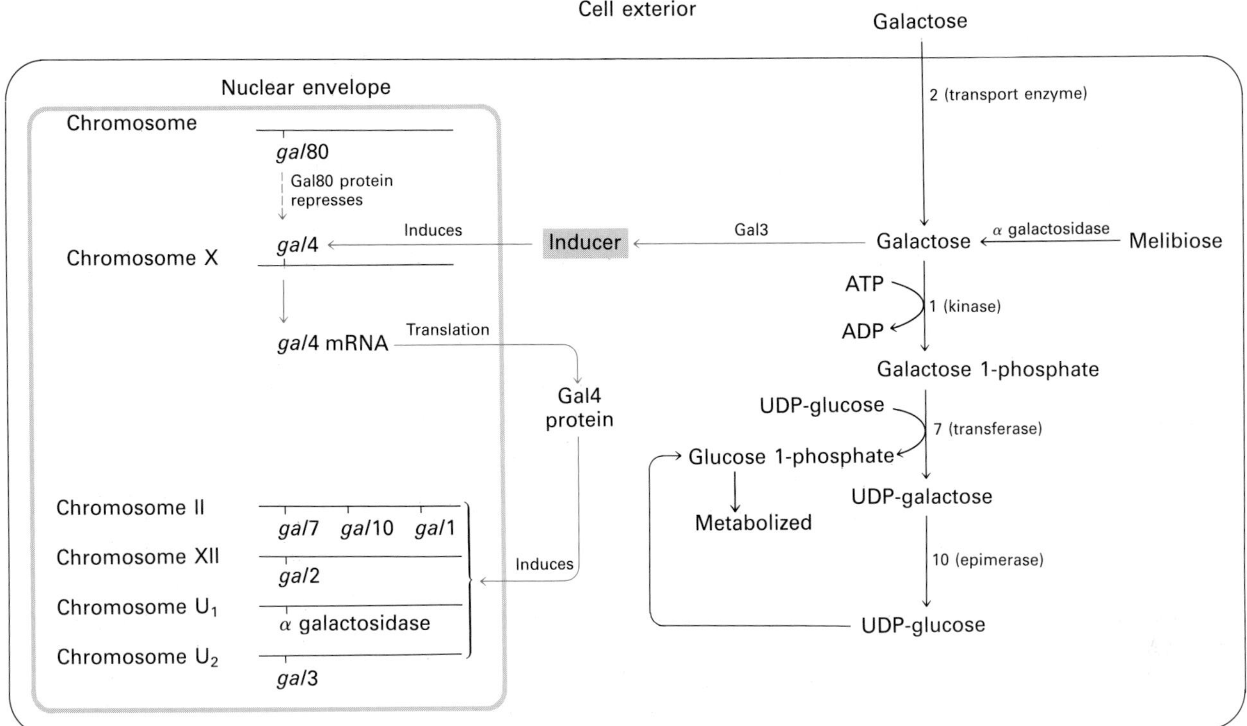

Figure 12-24 Induction of galactose enzymes in yeast. Enzymes responsible for galactose metabolism are shown. The actual inducer is a metabolic product of galactose formed by the enzyme encoded by the *gal*3 gene. This product has not yet been identified. The inducer causes the regulatory locus, *gal*4, to produce the mRNA for a protein that has been identified. Defective mutants in *gal*4 do not make any of the other products of the galactose operon—that is, neither enzymes from *gal* genes 1, 2, 3, 7, and 10, nor α galactosidase. The enzymes 7, 10, and 1 are encoded by separate transcription units on a single chromosome (II). The other genes lie on chromosome XII and on two additional unassigned (U_1 and U_2) chromosomes. The *gal*4 gene itself is under the negative control of the product of the *gal*80 gene. [See J. Hopper et al., 1978, *Proc. Nat'l Acad. Sci. USA* **75**:2878; and J. R. Broach, 1979, *J. Mol. Biol.* **131**:41.]

strain (SIR^+) or a mutant strain (SIR^-), active transcription of MATα sequences occurred. The plasmids containing HMRα were not transcribed in the SIR^+ strain, as would have been the case if the HMR sequences were located in the yeast chromosome. But the HMRα sequences *were* transcribed in the SIR^- mutants (Figure 12-25).

Because plasmids are discrete small DNA molecules, they can be isolated from cells into which they have been introduced, and their superhelicity (supercoiling) can be tested. This was done in the above experiment. The MATα-bearing plasmid had the same degree of superhelicity in both the SIR^+ and the SIR^- strains. In the SIR^+ strain (the one in which transcription of HMRα DNA was repressed), the HMRα plasmids were more highly supercoiled than they were in the SIR^- cells (Figure 12-26a). Thus transcriptional repression is associated with a physical change—in this case an increase in the superhelicity of the plasmid DNA (see Figure 3-58).

The HMRα sites on which the *SIR* gene product acts have been located by deleting portions of the plasmid. The major DNA region that is necessary for transcriptional repression by the *SIR* product (and the concomitant increase in superhelical density) is over 1000 bases from the site where RNA synthesis begins. A second DNA region whose deletion causes a partial loss of *SIR* activity is likewise distant (in the other direction) from the transcription initiation site. Thus the two regions on which *SIR* products act bracket a domain in which the two α transcription units lie (Figure 12-26b). Such experiments have suggested that regulatory proteins may act outside of transcription units in such a way as to change the DNA structure within a domain, and thereby change the accessibility of promoter sites to RNA polymerase.

Interacting Transcriptional Controls in Determination of Yeast Mating Types Control of the mRNAs from the silent copies of the MAT locus is only one aspect of a

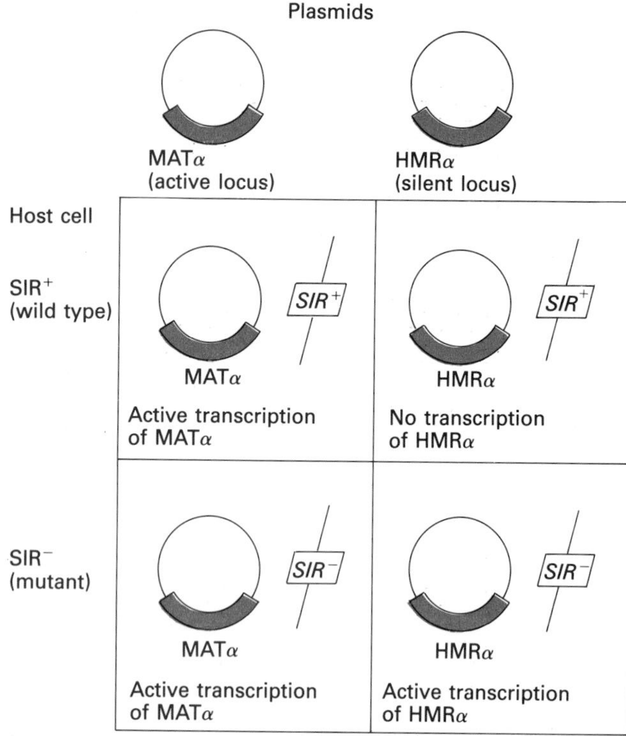

Figure 12-25 A diagram of experimental results establishing that the *SIR* gene product in yeast exerts negative transcriptional control over HMRα. Mating-type sequences from HMR (which contains a silent copy) and MAT (which contains the active copy) were inserted into yeast plasmids. The plasmids were propagated in two different strains—one that was the wild type (+) for the *SIR* genes, and one that was mutant (−) in these genes. The MATα DNA in the plasmid was actively transcribed in both strains. The SIR^+ strain effectively suppressed transcription of the HMRα DNA, whereas SIR^- allowed transcription. Thus the *SIR* gene products are responsible for the suppression of transcription of DNA in HMR (and, by extension, in HML).

whole cascade of events leading to the expression of yeast mating type. Genetic studies and molecular cloning studies have begun to describe the activities of an interlocking set of transcriptionally controlled genes that account for functions that are specific to either α cells, *a* cells, or diploid *a*/α cells. These interrelations are outlined in Figure 12-27.

In this model the α1 mRNA encodes the α1 protein, which has positive transcriptional control over α-specific genes. For example, the α factor (the secreted mating hormone) is positively controlled by the α1 protein. (At least three other known genes are controlled in the same manner.) The α2 protein is a negative regulator for *a*-specific genes. (At least three such genes are known, one of which is for the *a* factor.) These two regulatory actions account for many of the specific properties of the α cell: α functions are expressed and *a* functions are repressed.

In an *a* cell, both *a*1 and *a*2 mRNAs are formed, but their effects have not yet been studied in detail. It is known that *a* cells lack the α2 mRNA (which encodes the repressor protein mentioned above) and that they consequently express *a*-specific genes. Certain haploid-specific genes are equally expressed in *a* and α cells.

In the *a*/α diploids the character of the α2 protein is changed by an interaction with the *a*1 protein. These combined proteins repress the transcription of the haploid-specific genes, and they also repress α1 mRNA synthesis. Without the α1 protein (the positive regulator) the cells cannot perform α-specific functions, but because α2 mRNA is still being made, the negative regulator, the α2 protein, can repress *a*-specific genes; therefore the *a*/α diploid does not express either the *a* or the α phenotype.

This sophisticated system provides a look at a set of intricately interrelated eukaryotic genes. Just as the study of bacterial genes has revealed positive and negative regulatory gene actions, the yeast mating-type system should be of great use as a model of gene networks in eukaryotic gene function.

Site-Specific Regulation Occurs in Mammalian and *Drosophila* Cells

Animal cell and virus DNA sequences that are necessary for a maximal rate of transcription by RNA polymerase II have been identified through the use of recombinant DNA techniques (Chapter 7). In several cases, specific proteins that bind to the apparent regulatory sites in DNA also have been identified. We shall discuss three examples of such regulatory factors: the negative-acting SV40 early protein, a mouse steroid-binding protein, and a *Drosophila* protein. The latter two have positive transcriptional activity.

The SV40 T Antigen SV40 DNA contains only two transcription units, an early one and a late one. Just after SV40 virus infects susceptible cells, the predominant transcription products come from the early transcription unit; the reverse is true late in infection. The protein involved in this shift in transcription is the T (or large T) antigen of SV40. The T antigen is known to participate also in viral DNA synthesis because viruses that are mutant in this protein do not make viral DNA. The T antigen is encoded by the early transcription unit and has three specific binding sites on the SV40 DNA. (This region of the DNA also contains the origin of DNA replication, as we shall discuss in Chapter 13.)

Because of the close proximity of these binding sites to the major initiation site for early RNA synthesis, when the T antigen is present in abundance it stops (or slows)

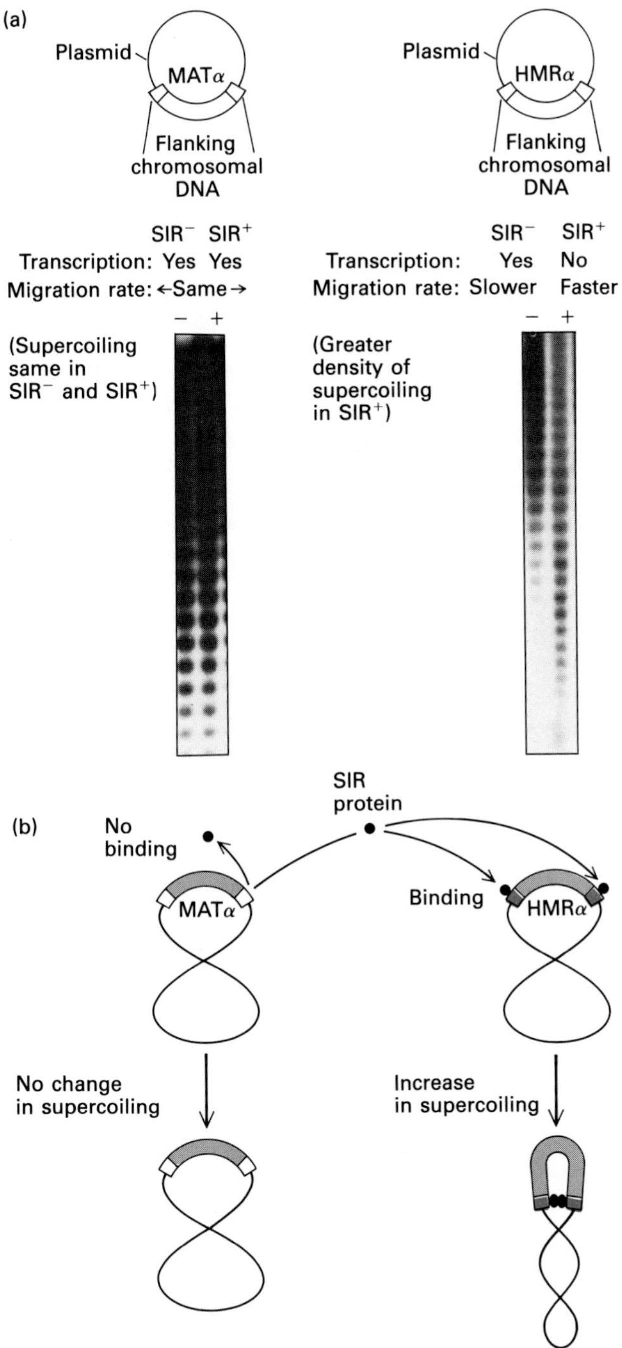

Figure 12-26 How the SIR protein might work: by changing the supercoiling of DNA. (a) Yeast plasmids with MATα sequences or HMRα sequences and some of the flanking chromosomal DNA were introduced into yeast cell strains that were SIR^+ or SIR^- (i.e., wild-type or mutant, respectively, for the SIR protein). The plasmid DNA was later extracted from cells and examined by gel electrophoresis. The migration of the plasmid DNA during electrophoresis is sensitive to the degree of supercoiling of the DNA: the more supercoils, the faster the migration (see Figure 13-19). The MATα plasmid, which is transcribed in both SIR^+ and SIR^- cells, has the same degree of supercoiling in both cell types. The HMRα plasmid has more supercoils in SIR^+ cells, where it is not transcribed, than in SIR^- cells, where it is transcribed. (b) The DNA sites that are necessary for the effects of the SIR protein on HMRα (increased supercoiling and decreased transcription) were identified by deletion mutations. Apparently the binding of SIR to sequences flanking the protein-coding sequences increases the supercoiling. [See K. A. Nasmyth et al., 1981, *Nature* **289**:244.] *Photographs courtesy of K. A. Nasmyth.*

transcription from the early transcription unit, the same region that encodes the T antigen itself (Figure 12-28). This effect on RNA initiation can be demonstrated in cell-free transcription systems that contain RNA polymerase plus other necessary proteins. When bound to the SV40 DNA, the T antigen prevents RNA polymerase II from initiating SV40 RNA synthesis at the early promoter site. Thus the T antigen appears to be an autoregulatory protein; it blocks the synthesis of its own mRNA. It was the first known example of a eukaryotic protein that inhibits transcription by binding to a specific DNA site.

This repressor action of the T antigen is limited to blocking *initiation* at the SV40 promoter. It will not block the passage of an RNA polymerase that initiates upstream from the T antigen–binding site. For example, recombinant DNA procedures were used to insert the adenovirus major late promoter segment (the P16 promoter; see Figure 12-8) only 50 bases upstream from the early SV40 promoter region. The binding of the SV40 T antigen to its sequences downstream of the adenovirus promoter did not affect transcription beginning at the adenovirus promoter (Figure 12-29). Apparently the transcriptional apparatus can either pass or displace a bound protein if it starts transcription before encountering the protein.

Steroid-Binding Proteins as Positive Activators of Transcription As we noted earlier, the transcription of many genes depends on the presence of steroid hormones (see Table 12-2). In cells that are responsive to steroids, proteins exist that specifically bind to the effective hormones. An estrogen-binding *receptor* is found in estrogen-sensitive cells, and a glucocorticoid *receptor protein* is found in cells responsive to glucocorticoid hormones. Chemically inactive analogs of naturally occurring steroids do not bind to the receptor proteins. This and other tests make it very likely that the identified receptors do function in the cell (see Chapter 16). Because

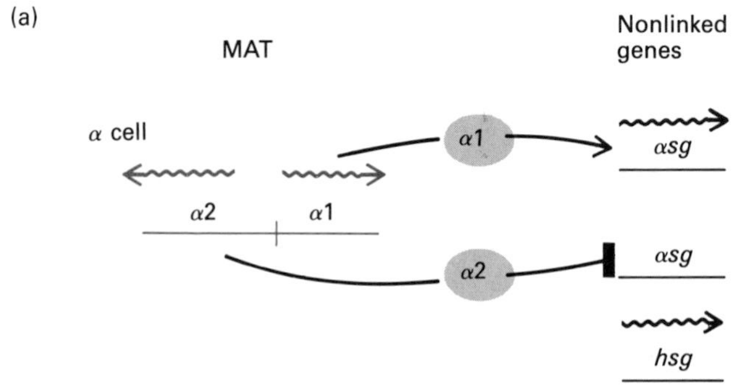

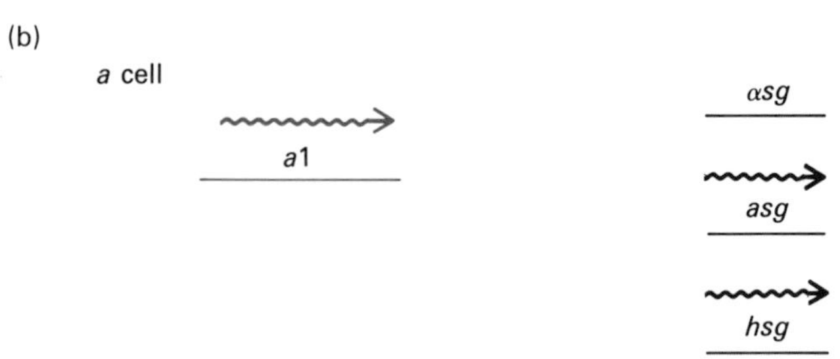

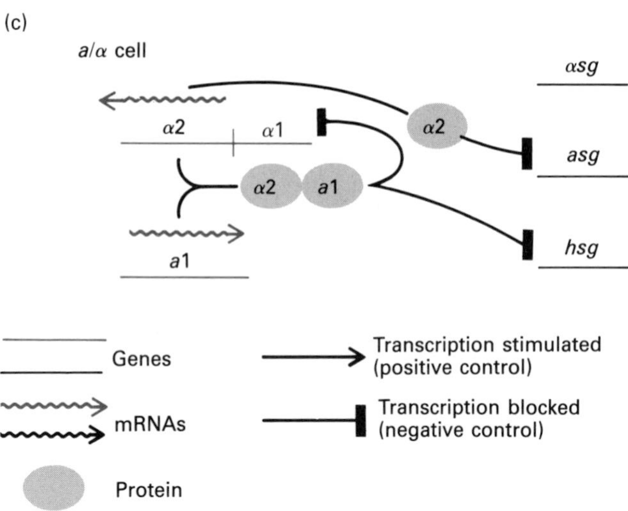

Figure 12-27 The regulation of *a*-, *α*-, and haploid-specific genes (*asg, αsg,* and *hsg*) in yeast cells. Depending on whether the cell is *a* or *α* at the MAT locus, either *a* or *α* mRNAs are transcribed from the active MAT genes (only *a*1 mRNA is shown because *a*2 has not been well studied). The *a*1, *α*1, and *α*2 mRNAs code for regulatory proteins whose actions on a set of nonlinked genes are diagrammed. (a) In *α* cells the *α*-specific genes (*αsg*) are stimulated by the *α*1 proteins and the *a*-specific genes (*asg*) are repressed by *α*2 proteins. (b) The *a*-specific genes are not repressed in *a* cells because no *α*2 mRNA is formed. No regulatory effect of *α*1 has yet been found in *a* cells. Both *α* and *a* cells transcribe haploid-specific genes (*hsg*). (c) In diploid *a*/*α* cells, an interaction between *α*2 and *a*1 proteins forms a new repressor that prevents the synthesis of haploid-specific mRNA as well as *α*1 mRNA; the absence of the *α*1 protein in turn prevents *α*-specific gene transcription. Because *α*2 mRNA is still formed, transcription of *a*-specific mRNA is blocked. Thus the diploid cell loses all *a*-, *α*-, and haploid-specific characteristics. *After K. L. Wilson and I. Hershkowitz, 1984,* Mol. Cell. Biol. *4:2420.*

steroids are soluble in lipids, they easily diffuse through the lipid of the plasma membrane; there the steroid molecule encounters and binds to its specific receptors. ^{3}H-labeled steroids can be observed to enter the cell nucleus apparently still bound to the cytoplasmic receptor. It has long been assumed that the receptor-steroid complex somehow affects transcription directly.

One case of transcriptional control of a gene by a receptor-steroid complex has been particularly well analyzed at the molecular level. Mammary tumor virus of mice (MMTV) is a retrovirus that is produced at elevated levels in cells when cortisone-like (glucocorticoid) steroids are present. The hormone acts by increasing transcription. The MMTV DNA (the complete DNA copy of the viral RNA that is integrated into cells) has been recovered in a genomic clone. The promoter for the virus DNA was identified by in vitro transcription and by sequence analysis. When plasmids containing this virus promoter are introduced into cells, added steroids cause increased transcription of the plasmid DNA. Maximal transcriptional activity from the introduced plasmids demands the presence of a glucocorticoid hormone in the culture medium.

The sequences necessary for the glucocorticoid-induced maximal transcription have been pinpointed by deletion mutations. These sequences are positioned within the 305 bases upstream of the RNA initiation site (Figure 12-30a). The purified glucorticoid receptor can actually be seen by electron microscopy to bind to this region (Figure 12-30b). Finally, the exact sites of contact of the receptor-steroid complexes have been determined by footprint analysis (Figure 12-30c). The binding sites on the DNA occur in two main regions. Thus it appears that the steroid activation of transcription—at least for MMTV—does require entry of the steroid, binding to its receptor, and subsequent binding of the receptor-steroid complex to the DNA.

A Positive Activator for a *Drosophila* Heat-Shock Gene When *Drosophila* cells (or in fact any metazoan cells) are subjected to elevated temperatures, they are stimulated to produce a class of proteins that have been termed heat-shock proteins. As we have mentioned, the increase in these proteins is caused by the increased tran-

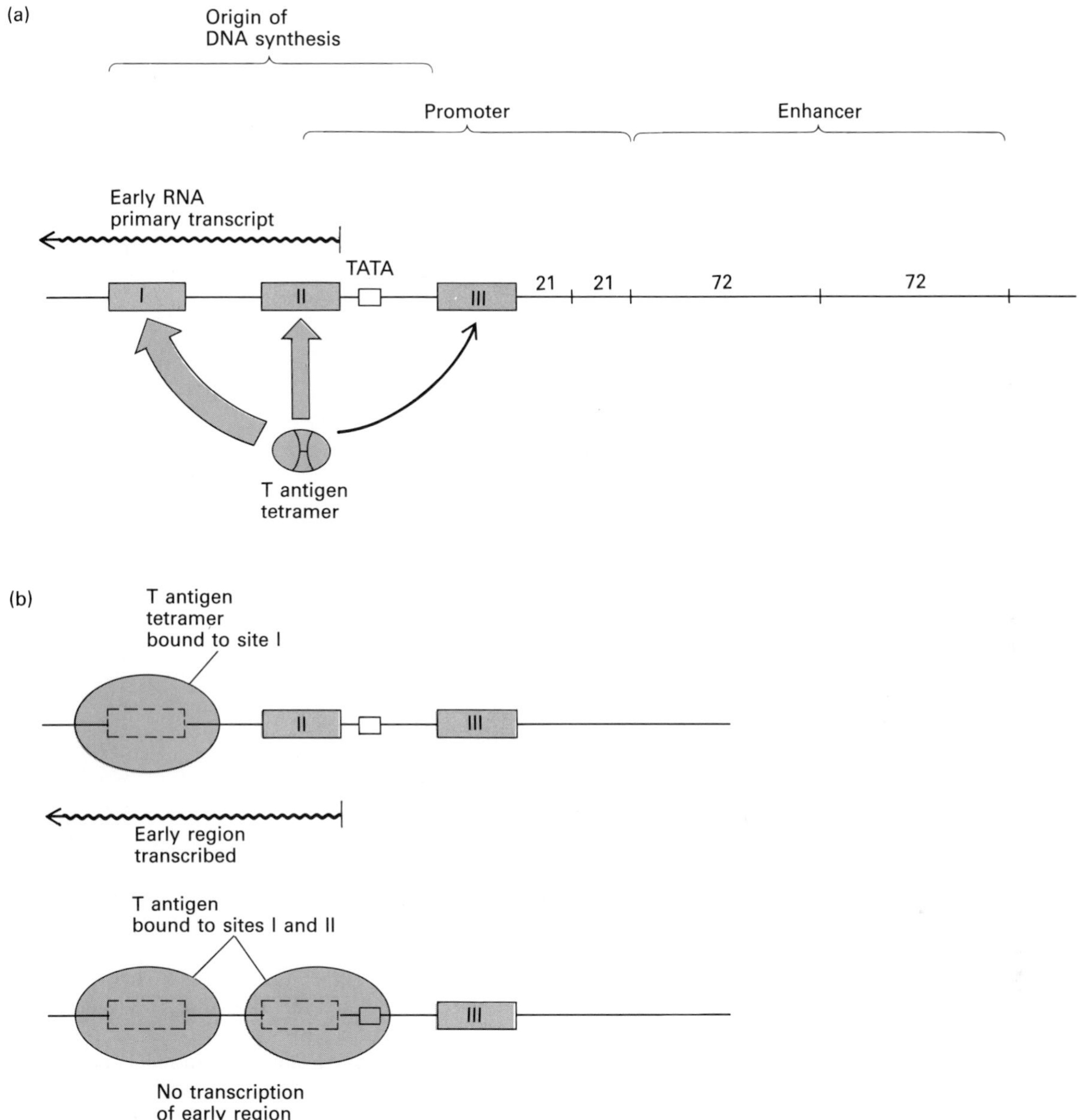

Figure 12-28 The control region of SV40 early RNA synthesis. (a) The control region contains three 26-bp DNA sites—I, II, and III—to which the T antigen binds. Other sites that play a role in the initiation of early RNA synthesis are nearby. The early RNA encodes the T antigen itself. The promoter region includes two 21-bp repeats that are GC-rich, as well as 72-bp repeats that have enhancer function. A cellular protein factor (called sp1) binds to the 21-bp repeat and is required for transcription. The TATA box and RNA start site lie between sites II and III, and are presumably blocked by the T antigen. The thickness of the arrows indicates the order of binding affinity: the T antigen binds preferentially to site I, but if the antigen is at higher concentrations it also binds to sites II and III. The binding at site I may increase the binding at site II. (b) In vitro, binding at both sites I and II is necessary to stop transcription that initiates at the early promoter. Thus a variation in the amount of T antigen in the cell will exert a variable control on its own synthesis. [See R. M. Myers et al., 1980, *Cell* **25**:373; W. S. Dynan and R. Tjian, 1985, *Nature* 316:774.]

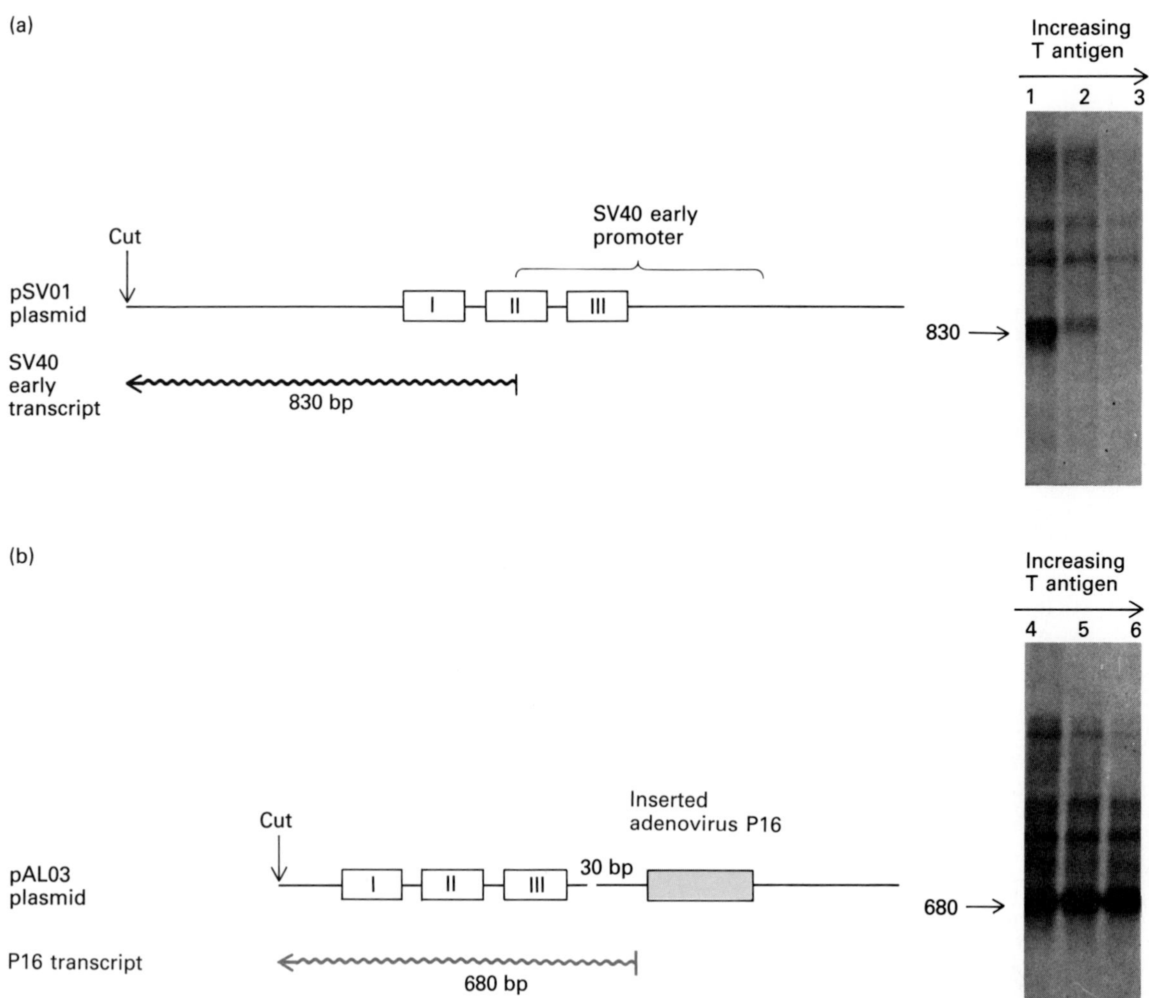

Figure 12-29 Control of in vitro SV40 transcription by the T antigen. In vitro RNA synthesis is carried out on viral DNA cloned in either of two plasmids, pSV01 or pAL03. The first has the normal SV40 early promoter region, and the second has the adenovirus major late promoter (P16) inserted as indicated. The template DNA within the plasmids is cut with restriction enzymes downstream of the promoters so that RNA molecules reaching from the initiation site to the cut ends will be formed; the lengths of these molecules can be predicted. The transcription products are labeled and detected by autoradiography after gel electrophoresis. The transcript of pSV01 is 830 nucleotides long, and that of pAL03 is 680. The two products formed in the absence of the T antigen appear in lanes 1 and 4. When the T antigen is added in increasing amounts, it gradually stops transcription from pSV01 *(lanes 2 and 3)* but not from pAL03 *(lanes 5 and 6)*. The action of the T antigen in stopping initiation is therefore specific to its binding region, and a polymerase that starts upstream of the binding sites is not prevented from reading through (transcribing) the binding-site region. [See R. M. Myers et al., 1981, *Cell* 25:373.] *Photographs courtesy of R. Tjian.*

scription of genes that have been identified (see Figure 12-12). Another protein (i.e., one not encoded by the heat-shock genes themselves) has been purified from the nuclei of cultured *Drosophila* cells by its selective binding affinity to the upstream sequences of a cloned *Drosophila* heat-shock gene. This protein greatly stimulates RNA polymerase II transcription of the cloned heat-shock gene in an in vitro transcription system. The effect of this protein seems to repesent a clear-cut case of positive transcriptional factor acting at a particular site on DNA. A number of genes are transcriptionally activated by heat shock, but it is not yet known whether this same protein is involved in the activation of any of these other genes.

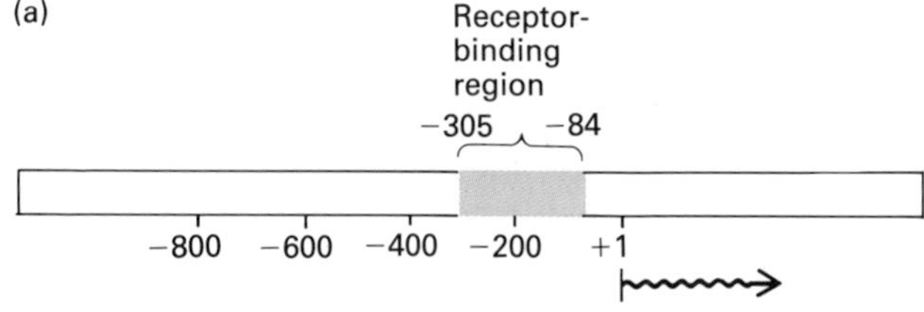

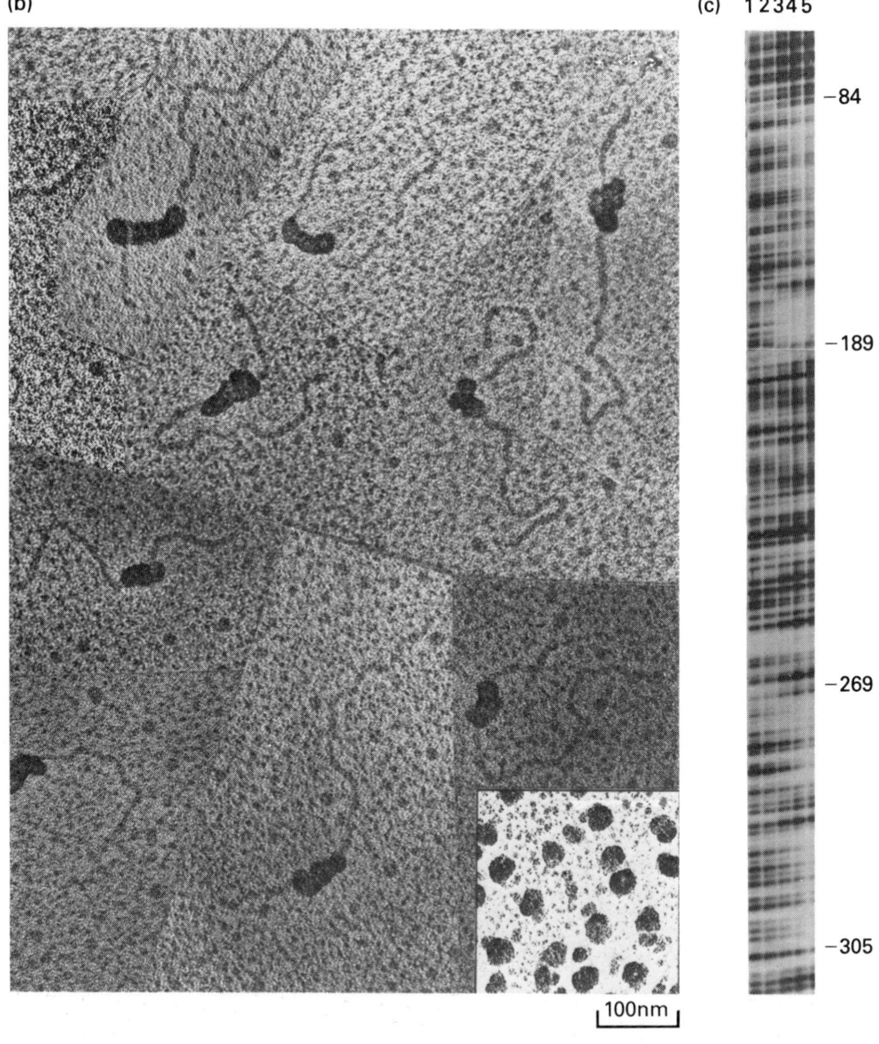

Figure 12-30 The binding of the glucocorticoid receptor to mouse mammary tumor virus (MMTV) promoter sequences. (a) The region of the MMTV genome containing the start site (+1) for RNA synthesis is shown. When MMTV virions are introduced into cells, RNA synthesis beginning at +1 is greatly stimulated by glucocorticoids (cortisone-like steroids). The color shading indicates a 220-bp region that is responsible for this steroid-stimulated transcription. (b) Electron micrographs of glucocorticoid receptor-hormone complexes bound to copies of the 1.2 kb fragment of MMTV diagrammed in (a). The thin threads are DNA and the dark objects are protein. The receptor protein probably forms tetramers when it is binding. The position of the protein-binding region indicated in (a) is the statistical average calculated from EMs such as this one. The inset at the lower right shows receptor-hormone complexes without DNA. (c) An autoradiograph footprint showing that the glucocorticoid receptor protects the promoter segment of MMTV from DNase digestion. DNA was end-labeled and exposed to brief DNase I treatment without receptor protein *(lane 1)* or with increasing amounts of glucocorticoid receptor-hormone complex *(lanes 2 to 5)*. The DNA fragments resulting from the DNase treatment show up as dark bands in the gel electrophoretic pattern. The more cuts, the more dark bands. Because in samples exposed to hormone-receptor complex the bands grow progressively fainter in the regions marked −84 to −189 and −269 to −305, it was concluded that the receptor-hormone complex binds to the DNA and prevents the DNase from cutting in those regions. These numbers indicate the positions of nucleotides upstream from the RNA start site (+1). [See F. Payvar et al., 1983, *Cell* 35:381.] *Parts (b) and (c) courtesy of K. Yamamoto.*

Generalized Transcriptional Activation Is Achieved by Viral Proteins

The transcriptional regulatory proteins we discussed in the preceding sections appear to have a high affinity for specific sequences near the initiation sites of eukaryotic transcription units; they apparently act to stimulate or inhibit the frequency of transcriptional initiation. In this section we shall describe the activity of two viral gene products, the 1A protein of adenovirus and the immediate early protein of herpes virus. These proteins, which have been identified by mutations, have the ability to increase transciption from a *variety* of genes. They differ from the T antigen and the heat-shock regulatory protein in that they do not have a high affinity for sequences just upstream of the genes whose transcription they stimulate. The effects of the viral proteins on transcription are similar to the effects of enhancer sequences in DNA (Chapter 9); in fact, these proteins may be able to eliminate the need of some genes for enhancer elements.

As we have described (see Figure 12-8), the adenovirus early transcription units 1B, 2, 3, and 4 are not transcribed rapidly in infected cells if the 1A protein is not made. However, in the sequences upstream of these early transcription units there is no obvious sequence homology that might indicate binding sites for the 1A protein; nor does purified 1A protein bind to specific sites on the purified viral DNA in footprint tests. During herpes virus infection, the *immediate early protein* (the first virus-encoded protein) is another protein that is necessary for transcription of later virus transcription units.

One interesting experimental observation is that the herpes virus product will support transcription of early adenovirus transcription units 1B, 2, 3, and 4. In fact, it stimulates early adenovirus transcription even better than the adenovirus 1A protein does. This interchangeability between different viruses was surprising, because there is no detectable DNA homology in promoter regions between the two viruses. The 1A protein and the herpes immediate early protein were also found to stimulate transcription of certain cellular genes, which further implied that the proteins do not act directly by binding to the specific DNA promoter regions. How might we explain the fact that the 1A and the herpes proteins assist entering virus genomes to become actively transcribed?

A clue to the possible mechanism of this activation has come from studies of the infectious cycle of 1A deletion mutants. Although the 1A region is necessary for maximal transcription of the adenovirus early transcription units in a normal infectious cycle, the deletion mutants in the 1A region can eventually enter a state in which they are equally actively transcribed in the host cell without the assistance of 1A. Thus the cell can transcribe the early adenovirus genes (1B, 2, 3, and 4) without 1A, but the process of formation of an active transcription complex may be aided by 1A (or by the herpes immediate early protein) without those proteins actually binding to DNA. It appears that the adenovirus 1A protein (and the herpes immediate early protein) may augment a normal function that creates active transcription complexes.

Enhancers and Proteins May Provide Interchangeable Functions

An interesting connection has been established between the adenovirus 1A protein and enhancer sequences in DNA. Recall from Chapter 9 that enhancer sequences can increase the transcriptional activity of genes in newly introduced recombinant DNA, and that many viral enhancers can function effectively in a number of different transcription units. In some cases either the 1A protein or the enhancer sequence can stimulate gene function.

Cell strains that have the adenovirus 1A DNA integrated into their chromosomes will produce the 1A protein. When genes normally requiring an enhancer function are transferred to such 1A-producing cells, the enhancer is no longer necessary. For example, plasmids containing the mouse or rabbit β globin gene do not stimulate the formation of globin mRNA when they are introduced into human cells unless an enhancer is present in the plasmid DNA. The enhancer can come from SV40 DNA or from any of several other sources. If plasmid DNA lacking an enhancer is introduced into human cells that are producing 1A protein, the β globin mRNA is synthesized just as efficiently as if an enhancer element were present. Thus the 1A protein in this case replaces the requirement for the enhancer.

Because both enhancers and the 1A protein affect a variety of genes that do not appear to be related in sequence, the mechanism by which transcription is increased may be to increase the chance the a gene will successfully encounter the specific transcription factors that allow it to become an active transcriptional complex. One idea that couples enhancers and proteins that have 1A-type activity hinges on the fact that 1A-type transcriptional factors are normally present at rate-limiting amounts in mammalian cells; an enhancer might trap and bring the 1A-type protein into the vicinity of an initiation site for RNA polymerase II and thereby activate that site. Apparently once such a site is activated it remains so.

However, the mechanisms by which the enhancers and the 1A protein increase transcription are not yet understood. Specifically, it is unclear whether a higher transcription rate of newly introduced test genes derives from an increase in the rate of initiation of a fixed number of templates that would have a basal level of activity or from the recruitment of an increased number of templates into active transcription complexes whose rate of initiation is then an intrinsic property of the complex. In the first case, an enhancer function could be continuously necessary; in the second, enhancer function would be required only to establish an active transcription complex (Figure 12-31).

The Tissue Specificity of Some Enhancers One final note about RNA polymerase II enhancers and transcriptional initiation: there is some cell specificity to enhancer action. For example, enhancers obtained from some mouse viruses do not activate transcription of enhancer-dependent genes in primate cells, and the reverse is also true. In addition, enhancers for mouse immunoglobulin genes work in mouse lymphoid cell lines (tumor cells derived from lymphoblastic cells) and possibly in similar cells in humans, but not in some nonlymphoid cell lines. Thus enhancers have some tissue and species specificity. If enhancers do serve to sequester transcriptional

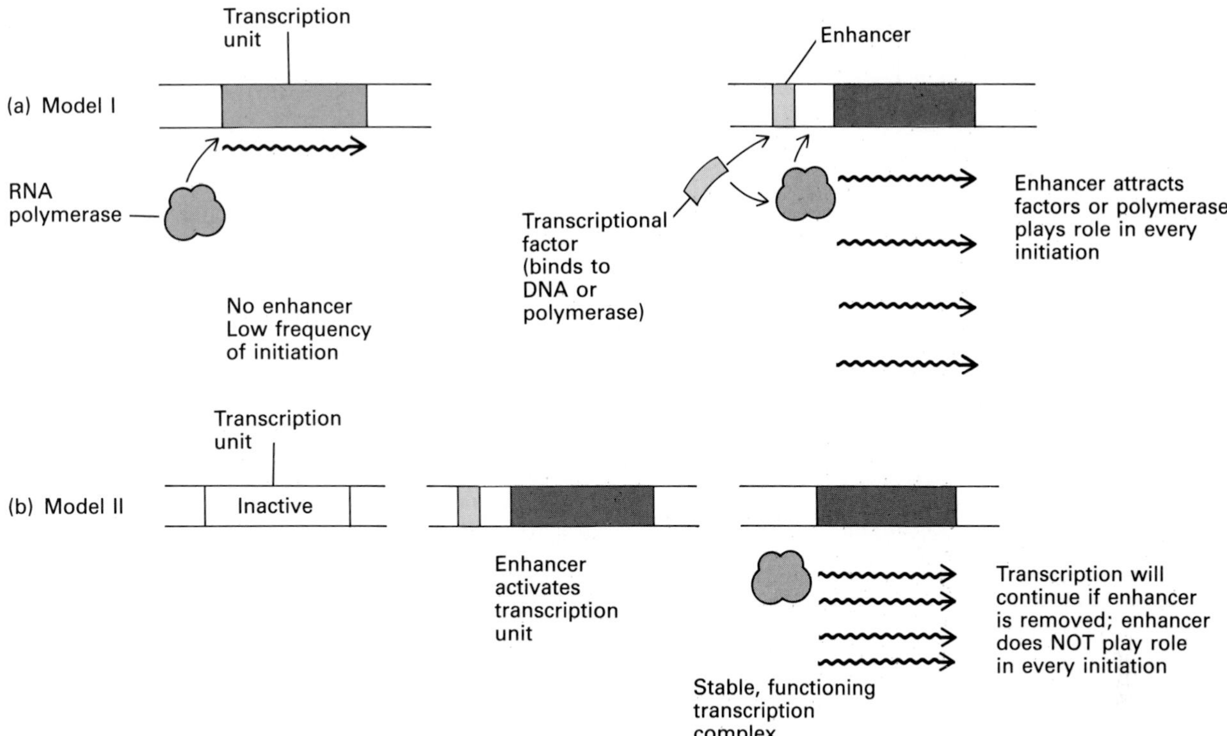

Figure 12-31 Two possible models for enhancer activity in the stimulation of RNA synthesis. (a) In Model I, the enhancer element would attract RNA polymerase (or other components of the transcription initiation system) continuously. (b) In Model II, the enhancer sequence would help establish an active transcription complex that would then be stable.

factors, as we suggested earlier, tissue-specific transcriptional factors may operate by recognition of enhancer sequences.

Ribosomal Enhancers Recruit More Templates to Active Transcription

Recent work on RNA polymerase I transcription of the ribosomal genes in frogs has suggested that enhancers operate in ribosomal DNA transcription. The initiation site for RNA synthesis of the frog pre-rRNA transcription units is preceded by blocks of repeated sequences of 60 to 80 nucleotides. *Xenopus laevis* rDNA has about 15 such blocks, and *Xenopus borealis* has about 5. Recombinant DNA plasmids containing the start site for RNA polymerase I either with or without the upstream blocks have been constructed and reintroduced into frog egg nuclei. Transcription is markedly increased when the upstream blocks are present. In common with the enhancer sequences for RNA polymerase II, these rDNA upstream sequences stimulate transcription when they are inserted into plasmids in *either* orientation. They are thus termed ribosomal enhancers (Figure 12-32).

Perhaps the most important conclusion to come from studying the ribosomal enhancers is that they do not work by increasing the initiation rate of transcription of an already active template, but rather by causing a DNA molecule to become an active transcription unit. Recall that when a ribosomal gene in a cell is active, polymerase initiation is so rapid that the transcribed gene appears in the electron microscope as a "Christmas tree" or a "feather," with many nascent chains of pre-rRNA simultaneously emerging from it (see Figure 9-12). With recombinant circular plasmids containing genes for ribosomal RNA, the same thing is true: electron micrographs show fully loaded genes with the polymerases spaced very close together (Figure 12-33). The number of such active transcription units varies depending on the presence of the enhancing element, but the rate of initiation from a single active transcription unit (as indicated by the "Christmas tree" structure) is the same with or without the enhancers. Thus, whenever a pre-rRNA transcription unit becomes active it is maximally active, but its chances of becoming active are greatly increased by having an enhancer. This could also be true for protein-coding genes transcribed by RNA polymerase II.

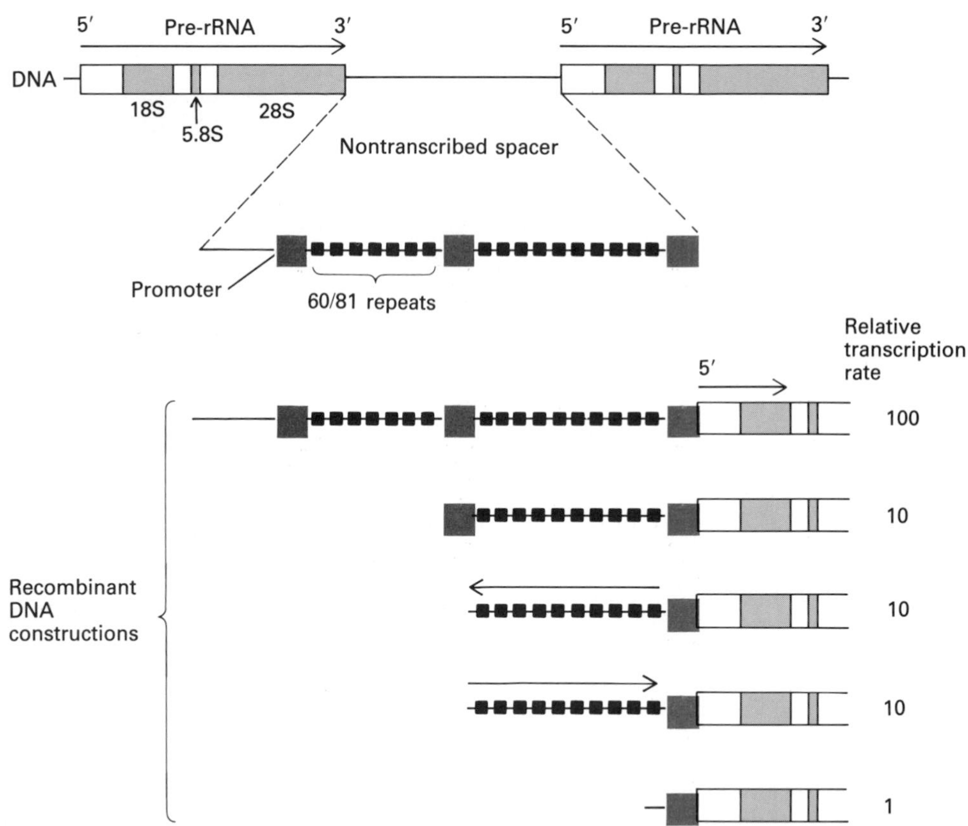

Figure 12-32 A diagram of ribosomal genes from *Xenopus laevis,* showing the transcriptional effects of sequences upstream from the RNA initiation site. The rRNA genes of *X. laevis,* like those of all other animals, are repeated in tandem *(top).* Transcription of the genes produces long pre-rRNA transcripts that are processed to yield 18S, 5.8S, and 28S rRNAs. The ribosomal transcription units are separated by nontranscribed spacers. Sequence analysis and recombinant DNA experiments have shown that within each nontranscribed spacer there is a promoter element about 100 nucleotides long *(right-hand color box)* just before the adjacent initiation site. Upstream there are additional repeated elements with the same sequence as the promoter, and a large number of other repeated stretches 60 or 81 nucleotides long *(black boxes).* Transcription of the gene requires the proximal *(right-hand)* promoter sequence and a string of 60/81 repeated sequences. This is indicated in the diagram, which shows five different recombinant DNA constructions that were injected into oocyte nuclei. Extra promoters do not increase transcription, but extra 60/81 repeats do. Because the 60/81 repeats function equally well when they are oriented in either direction *(arrows),* they are analogous to RNA polymerase II enhancer sequences. [See R. H. Reeder, J. G. Roan, and M. Dunaway, 1983, *Cell* 35:449.]

Chromatin Structure and Transcriptional Control

In view of the condensed nature of the chromatin in many regions of chromosomes, it has long been considered likely that transcription requires a "loosening" in the chromatin packing. For example, it has been known for years that histones will inhibit any in vitro transcription of DNA in a nonspecific manner. More recent studies show nucleosomes to be stacked in ordered helical arrays (see the solenoidal structure in Figure 10-20) which might prevent access of the transcriptional apparatus to the initiation sites. These ideas have provoked widespread investigations of the accessibility of specific regions of chromosomal DNA to various nucleases, on the assumption that active genes might be more vulnerable to attack than inactive genes. Such studies have been remarkably successful in demonstrating the existence of physical differences between active and inactive genes, apparently either in the packing of protein along the DNA or in the folding and supercoiling of the DNA-protein structure.

Active Sites of Transcription Have Increased DNase Sensitivity

In one experiment that has demonstrated a change in DNA-protein structure when genes become active, whole

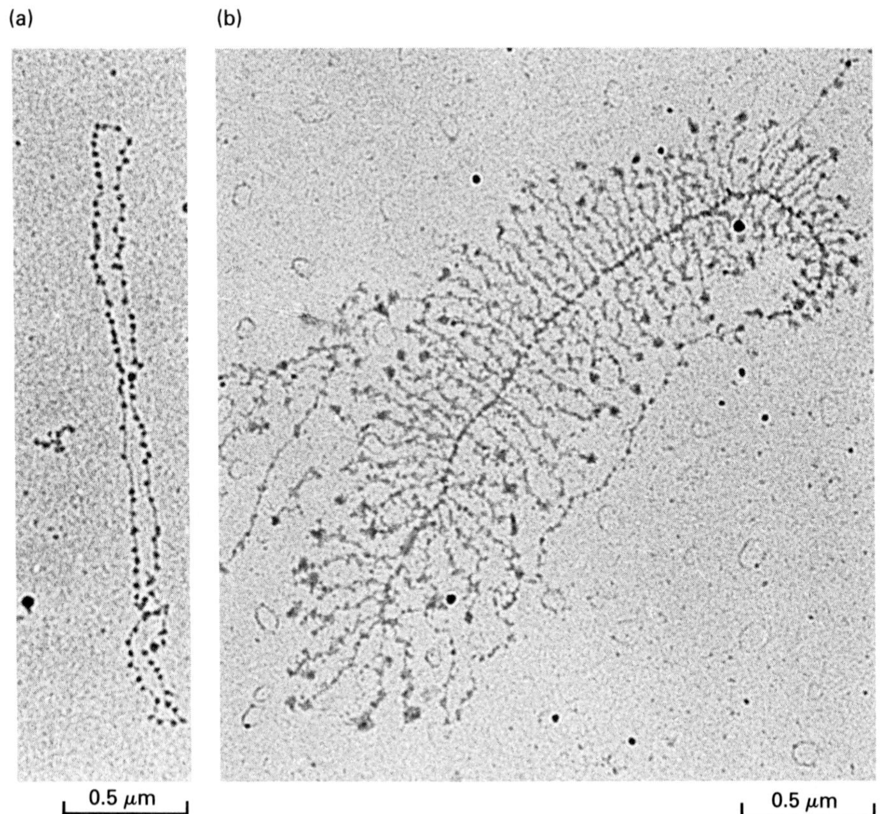

Figure 12-33 Electron micrographs of active and inactive transcription complexes of recombinant DNA containing ribosomal promoters. (a) The chromatin of this inactive recombinant ribosomal transcription unit is visible as a "beaded string." (b) The initiation rate of ribosomal transcription units is so rapid that many nascent pre-rRNAs can be observed. The active transcription complex shown here came from a recombinant ribosomal transcription unit inserted into a circular plasmid and injected into a frog egg. The active complexes look the same with and without the upstream enhancer sequences (see Figure 12-32), but they are much more numerous when the recombinant DNA contains the enhancer region. [See G. T. Morgan, R. H. Reeder, and A. H. Bakken, 1983, *Proc. Nat'l Acad. Sci. USA* 80:6490.] *Courtesy of A. H. Bakken and R. H. Reeder.*

nuclei from different tissues are treated with an enzyme such as pancreatic DNase. This enzyme cuts most regions of pure DNA equally, but a different result is obtained in whole nuclei. For example, in the nuclei of red blood cell precursors (erythroblasts), the DNA sequences encoding globin mRNA are more susceptible than other sequences to attack by DNase. On the other hand, in hen oviduct nuclei, in which ovalbumin mRNA is produced, the ovalbumin DNA sequences are more susceptible than the globin sequences. The greater sensitivity of the active genes to the nuclease indicates that their associated protein is in a less protective configuration than the protein on the inactive genes.

A specific technique that has revealed the increased DNase sensitivity of active genes uses Southern blotting (see Figure 7-25) before or after DNase treatment of nuclei. For example, after nuclei from chicken erythroblasts are treated with a range of concentrations of DNase I, the DNA may be purified and subjected to restriction enzyme digestion. In chicken DNA not exposed to previous DNase action, the restriction enzyme releases specific-sized globin DNA fragments that are detected by hybridization to cloned labeled globin sequences. If the DNase treatment has cut the DNA within the borders of the fragments released by restriction enzyme digestion, the unit-sized globin-specific DNA pieces will not be observed. In practice the latter result occurs in nuclei from cells that are transcribing the gene in question: given sufficient concentrations of DNase, only smaller globin DNA fragments are observed (Figure 12-34a). Nuclei from cells not making globin do not show this sensitivity. In addition, genes that are activated in other tissues, such as ovalbumin in oviduct cells, are not sensitive in the erythroblasts (Figure 12-34b), but they *are* sensitive in the oviduct cells.

This test has shown that almost all transcriptionally active genes have an increased DNase sensitivity. The specific region that is DNase-sensitive during transcription varies with different genes: it may include only a few thousand base pairs around the gene, or it may extend to include as much as 20 kb of the flanking sequences.

Other experiments have supported the notion that changes in DNase sensitivity in active genes are due to changes in the structure of the chromatin rather than to another aspect of the transcriptional activity. For example, in laying hens that are withdrawn from estrogens, the oviduct ceases to make ovalbumin mRNA. However, the chromatin from such cells retains the increased DNase sensitivity even though transcription is drastically reduced. Thus the altered chromatin structure may be necessary but not sufficient for active transcription.

Changes in Chromatin Structure Can Be Site-Specific: DNase Hypersensitivity

If transcriptionally active genes are generally more sensitive to DNase, perhaps some DNA sites—for example, those to which the polymerase attaches—are exquisitely

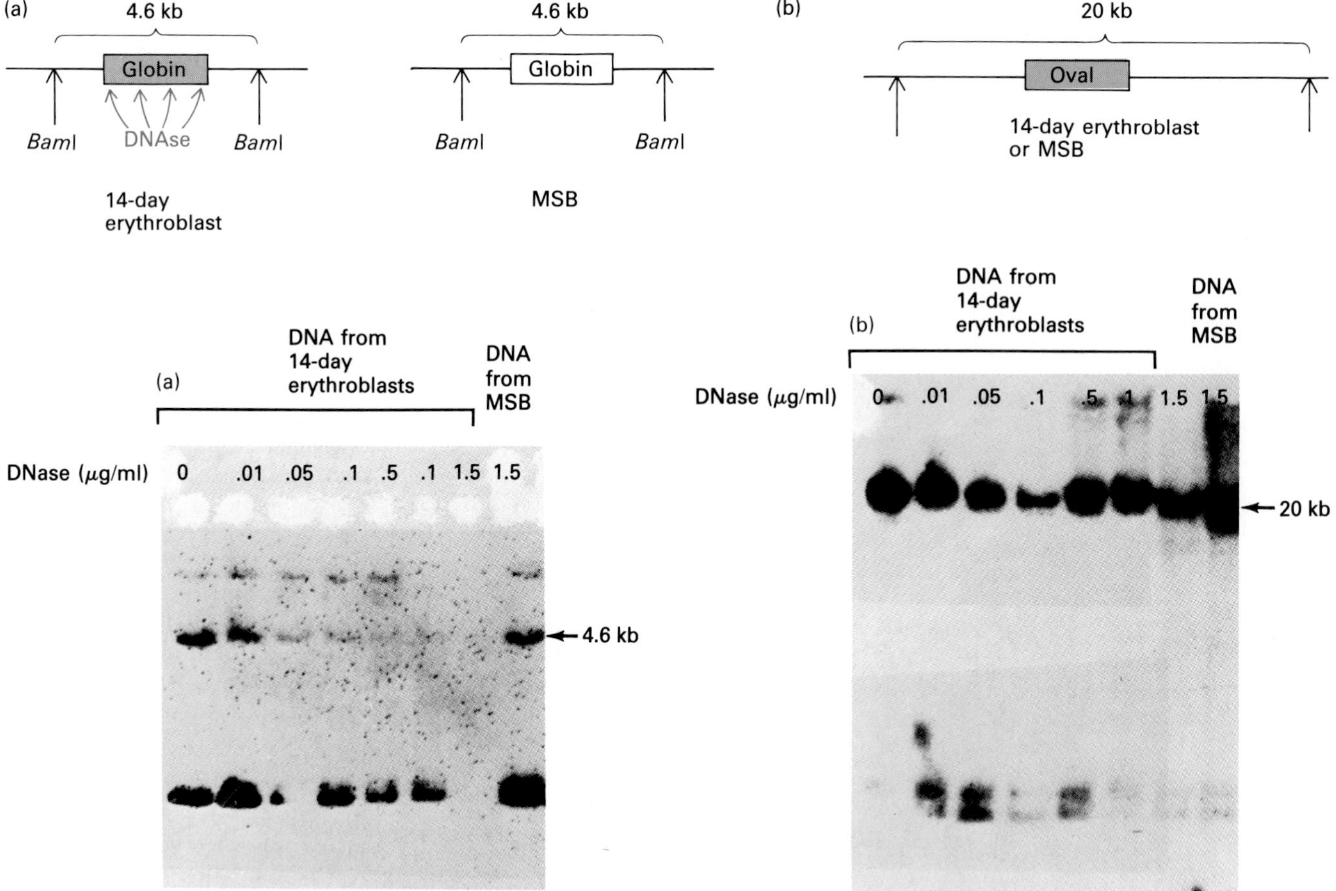

Figure 12-34 The sensitivity of globin DNA sequences to DNase increases when chick embryos begin to make adult globin at 14 days of incubation. (a) Nuclei from red blood cell precursors (erythroblasts) were extracted at 14 days and exposed to increasing concentrations of DNase (0 to 1.5 μg/ml). The nuclear DNA was then extracted free of protein. Each DNA sample was then digested by the restriction enzyme *Bam*I, which cleaves the DNA *around* the globin sequences; *Bam*I digestion normally releases a 4.6-kb globin fragment. The DNase- and *Bam*I-digested DNA was subjected to Southern blot analysis, and the DNA corresponding to the adult globin sequence was detected by hybridization to labeled clones of adult globin DNA. The 4.6-kb globin region in the nuclear DNA from the 14-day cells *was* sensitive to DNase digestion, as was indicated by the gradual disappearance of the 4.6-kb band with increasing concentrations of DNase. As a control to the experiment, the globin DNA sequences from the nuclei of cultured cells not making globin (MSB cells) were shown to be insensitive to DNase digestion. (b) As a further control, DNA from 14-day erythroblasts was exposed to DNase and then to a restriction enzyme that cleaves around the ovalbumin sequences. The ovalbumin-specific bands were stable even for DNA subjected to the highest DNase concentration. Thus in the whole nucleus an actively transcribed gene is more sensitive to DNase treatment than a gene that is not actively transcribed. [See J. Stalder et al., 1980, *Cell* **19**:973.] *Photographs courtesy of H. Weintraub.*

sensitive. With this logic in mind, investigators have sought and found evidence of *hypersensitive* sites in chromatin. A hypersensitive site or region would be the first to be digested by DNase. Therefore a hypersensitive site that lies between two restriction sites can be detected in a Southern blot of DNA from DNase-treated nuclei as a specific band (or two bands) that is (are) smaller than the restriction fragment from nuclei not digested with DNase.

Hypersensitive regions have been located at the 5′ ends of globin genes and a variety of other genes (Figure 12-35) in cells where the genes are being actively transcribed. Many—but not all—DNase-hypersensitive sites are in the region near the start of transcription, where transcriptional factors are most likely to be bound.

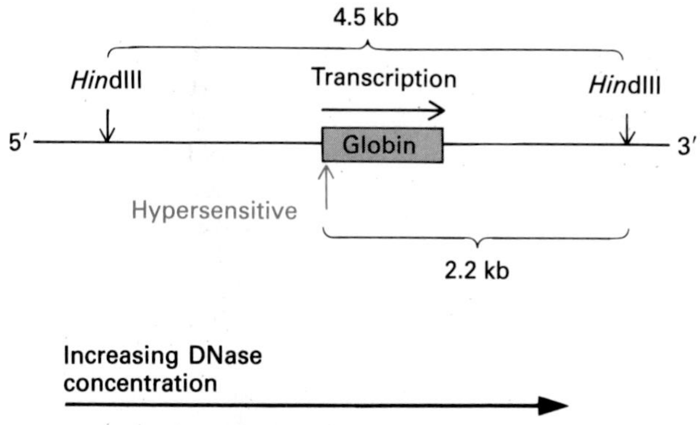

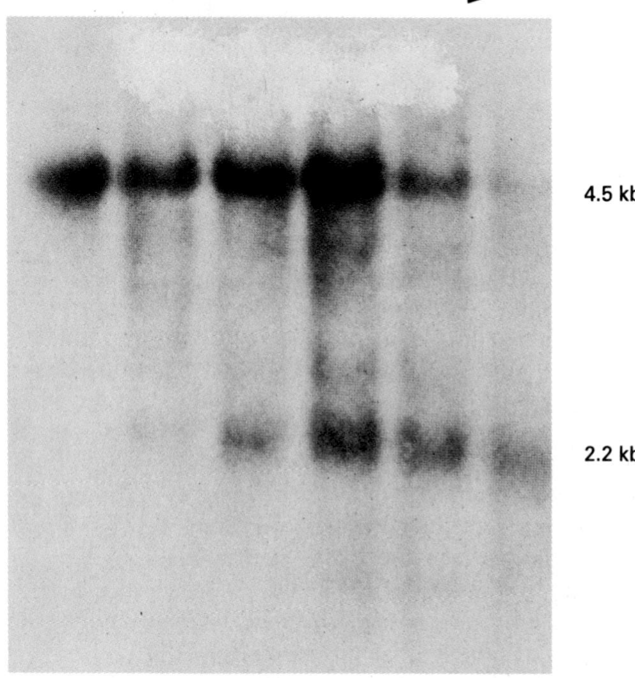

Figure 12-35 A DNase-hypersensitive site in globin DNA. The nuclei of erythroblasts from 5-day chick embryos were exposed to increasing concentrations of DNase (as in Figure 12-34). DNA was then isolated and digested with the restriction endonuclease *Hin*dIII. The resulting DNA fragments were subjected to Southern blot analysis for embryonic globin sequences. The expected 4.5-kb band generated by *Hin*dIII gradually disappeared; as the 4.5-kb band disappeared, a new band of 2.2 kb appeared. This was due to a DNase-hypersensitive site that was cut about 2.2 kb from the right-hand *Hin*dIII site. Digestion with other restriction enzymes has confirmed the position of the hypersensitive site. The location of this hypersensitive site is very close to the 5′ end of the globin transcription unit. [See J. Stalder et al., 1980, *Cell* **20**:451.] *Photograph courtesy of H. Weintraub.*

What Causes DNase Sensitivity in Chromatin?

It seems likely that DNase-sensitive sites in chromatin are caused by the nature of the protein-DNA interactions at those sites. A specific DNA region that is active in transcription might be stripped of proteins that would otherwise protect it from digestion by DNase; or the region might be bound to proteins in such a way that it is presented more easily to a nuclease.

A protein that has been purified from the nuclei of chicken erythrocytes specifically binds to DNA about 200 nucleotides upstream from the start of initiation of the chicken β globin gene. This DNA region is known to be particularly sensitive to nucleases in erythroblasts. When the purified protein attaches to its binding site in cloned DNA before histones are bound to the DNA, the binding site becomes hypersensitive. It thus seems quite likely that hypersensitive sites generally do point to regions of specific protein-DNA interaction.

Unfolded Histone Cores An important issue that has not yet been resolved is the state of nucleosomes in the region of transcribed genes. Are they totally removed? Are only nucleosomes at the 5′ end removed? Electron micrographs of gently spread chromatin with attached nascent RNA chains show nucleosomes both just before and just after the polymerase. If removal occurs, it must be immediately reversible. However, in chromatin containing ribosomal genes—which, recall, are distinct, identifiable transcription units on which polymerases are densely packed—the nucleosomes are less obvious than in rarely transcribed regions.

In at least one case, the histone proteins associated with ribosomal genes are now known to change in conformation when the genes become active. In one stage of the life cycle of the slime mold *Physarum polycephalum,* the ribosomal gene replicates independently of chromosomes and can be purified as free active chromatin. Both the transcribed ribosomal gene region and the nontranscribed spacer region are associated with histones. The protein of the chromatin can be treated with iodoacetaminofluorescein, a fluorescent reagent that binds to —SH groups of cysteine if the —SH groups are not involved in the internal structure of the protein. The —SH groups of the histone H3 in the nucleosomes associated with the transcribed region of the ribosomal genes do react, but those of the H3 in the nucleosomes of the nontranscribed spacer region of the gene do not react. This result and further physical measurements have indicated that the nucleosomes of the transcribed region are more relaxed in their structure (Figure 12-36). Thus the nucleosomes are still present in the transcribed portion of the ribosomal DNA, but they may have fewer histone-histone interactions and a less compact configuration.

Other studies have also suggested that histones are not dissociated from transcribed regions. For example, after

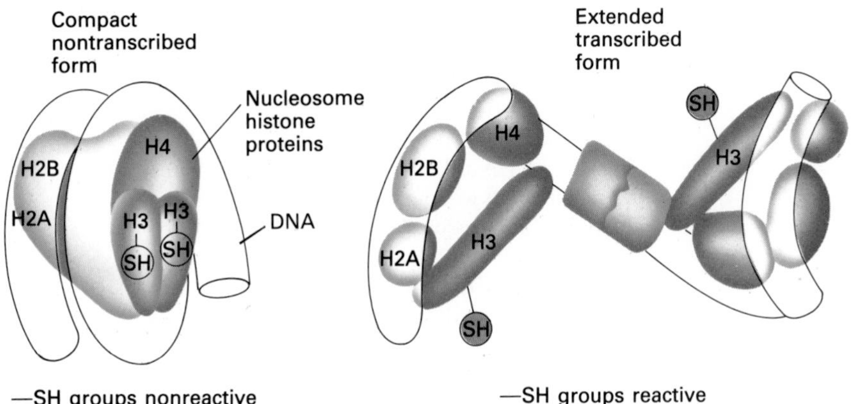

Figure 12-36 Models for active and inactive nucleosome structure, based on experiments with ribosomal transcription units of *Physarum polycephalum*. These transcription units can be isolated as active chromatin. The nucleosomes in the transcribed region can be separated from those in the nontranscribed spacer. The sulfhydryl (—SH) groups on the H3 histones in the nucleosomes of the nontranscribed spacer are not chemically reactive, whereas those on the transcribed region are reactive. This experiment and others have suggested that the histone core unfolds where transcription occurs. [See C. P. Prior et al., 1983, *Cell* **34**:1033.] *After Prior, 1983.*

brief nuclease digestion of whole erythroblast nuclei, large chunks of chromatin still exist in the solenoidal form. These large solenoidal chromatin pieces contain the same proportion of actively transcribed globin genes as the total DNA of the starting nuclei. Thus solenoidal chromatin is not necessarily inactive.

Active Genes Are Undermethylated

The structural changes we have been describing for nucleosomes in active genes are well established, but the bases for the changes remain largely unknown. Another difference—a chemical one—between active and inactive chromosomal DNA has been discovered, but its cause and its functional significance likewise remain mysterious. In the DNA of vertebrates and many plants, there are less than half as many (5′)CG(3′) sequences as would be predicted on the basis of chance. Moreover, it has long been known that the cytidylate residue of this uncommon CG sequence is frequently methylated to yield 5-methyldeoxycytidylic acid, m^5C. A CG in one DNA strand must be matched by a GC in the complementary strand: when a C of a CG is methylated, its dinucleotide partner in the other strand is nearly always methylated as well:

pmCpG
| |
GpCmp

Because methylation is known to occur soon after DNA replication, it seems likely that the signal for the methylation of one strand is a resident methyl group in the other strand:

Old strand (1) pmCpG
New strand (2) GpCp

$\downarrow +CH_3$

(1) pmCpG
(2) GpCmp

This establishes a mechanism whereby the presence of a methyl group could be passed along to succeeding generations. Prevention of methylation at a specific site would, after DNA replication, then permanently uncover the site.

It has been shown that actively transcribed genes generally contain many fewer methylated CG's than the average number in DNA. This undermethylation of specific genes can be demonstrated by the Southern blot technique; restriction enzymes can be used to assay for the presence of specific methyl groups. Some restriction enzymes that have CG as part of their recognition site are inhibited if the C is methylated. However, other restriction enzymes with the same sequence in their recognition site are not inhibited by the presence of m^5C.

For example, the enzyme *Msp*I (from a *Moraxella* species) cuts both CCGG and CmCGG; the enzyme *Hpa*II (from *Hemophilus parainfluenzae*) recognizes and cuts nonmethylated CCGG, but not the methylated sequence. Thus DNA samples from two different types of differentiated tissue (e.g., chicken red blood cells and cells from brain or liver) or from a differentiated tissue and cultured undifferentiated cells can be digested with *Hpa*II and *Msp*I and analyzed by the Southern blot procedure to determine where methyl groups have prevented cleavage of the *Hpa*II-treated samples (Figure 12-37).

By this test various genes undergoing transcription (globin, ovalbumin, genes for enzymes such as thymidine kinase, and integrated virus genes) have all shown a decrease or an absence of methylation at sites that are methylated when the same genes are inactive. The undermethylated sites that are critical for transcription appear to be in the region immediately before the transcription initiation site. In plasmids bearing virus transcription units, DNA methylation by bacterial enzymes has been carried out within 200 bases of initiation sites. When the methylated DNA is transferred into cells, transcription is blocked. When the methylation is more than 500 nucleotides upstream, no effect on transcription is noted.

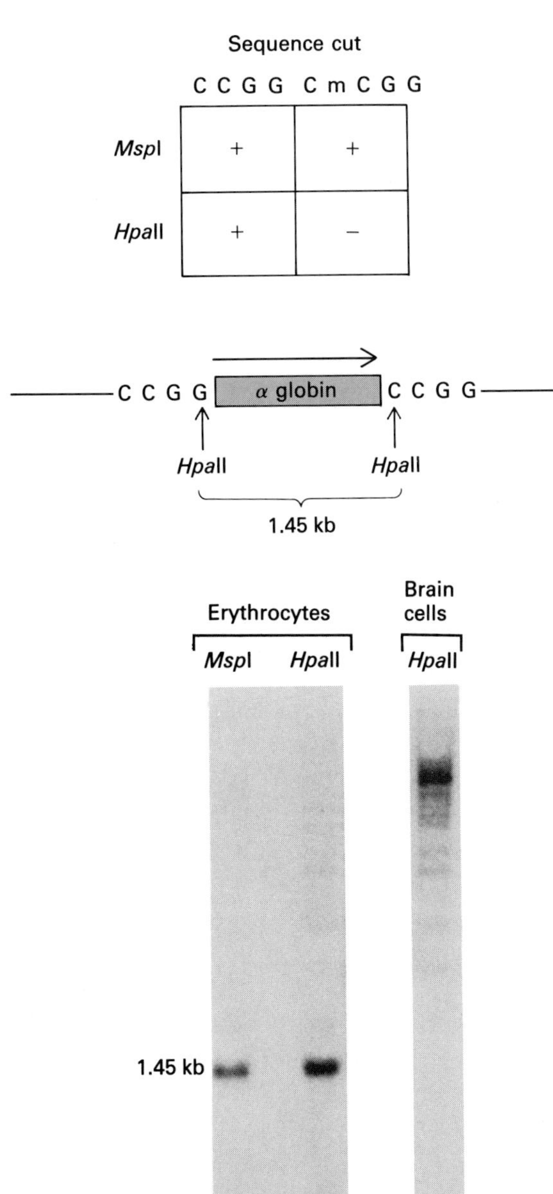

Figure 12-37 Methylation of DNA around an α globin gene in chick tissues. DNA samples from erythrocytes and brain were prepared. One sample of erythrocyte DNA was cleaved with *Msp*I, a restriction endonuclease that cuts at both CCGG and CmCGG (mC is 5-methyldeoxycytidylic acid). The other sample of erythrocyte DNA was exposed to the restriction endonuclease *Hpa*II, which cuts at CCGG only. The DNA from the brain was exposed to *Hpa*II also. The DNA samples were then subjected to Southern blot analysis and hybridized to an α globin DNA probe. The erythrocyte DNA was digested by *Hpa*II at these sites, but the brain DNA was not. (DNAs from six other tissues were also not digested.) Thus methyl groups that are present in the vicinity of the α globin gene in many tissues are absent in erythrocytes, where the α globin gene is actively transcribed. [See H. Weintraub, A. Larsen, and M. Groudine, 1981, *Cell* **24**:333.] *Photograph courtesy of H. Weintraub.*

Similar results are obtained when the upstream sequences of a globin gene are either methylated or left unmethylated at various sites and then introduced into cells. The newly introduced methylated globin DNA is incorporated into the chromosome, where it maintains its methylation pattern. When the 500 bases upstream from the start site for RNA synthesis are unmethylated, transcription is active; however, methylation in this region results in very little transcription. Transcription of the globin gene is insensitive to methylation in any other parts of the gene, either within it or farther than 500 bases upstream.

Inactive Genes Can Be Activated by 5-Azacytidine, a Nonmethylatable Analog of Cytidine

How are methyl groups added or removed when tissue-specific gene regulation is imposed? A drug has been used to investigate this question. The nucleoside analog 5-azacytidine contains a nitrogen instead of a carbon at the 5 position.

5-Methylcytidine	5-Azacytidine
NH_2 at C4; N3; C5—CH_3; O=C2; 6CH; N1—Ribose	NH_2 at C4; N3; 5N; O=C2; 6CH; N1—Ribose

The analog therefore cannot be methylated at the 5 position, but it can exist as part of a nucleotide and be incorporated into DNA. The product, 5-azacytidine-containing DNA, is undermethylated. When a DNA duplex containing the analog in one strand is copied, one of the resulting new duplexes will be totally unmethylated. Thus a cell that has incorporated 5-azacytidine briefly will have random unmethylated sites.

Undifferentiated mouse cells that originated from skin tissue and that were probably derived from fibroblasts have been widely used in cell culture experiments. These cells produce a small amount of collagen, a product of fibroblasts, but normally grow continuously without differentiating. When a group of such cells was exposed to 5-azacytidine for 24 h and then placed in a normal medium, differentiated colonies of contractile muscle cells and chondroblasts (cartilage-forming cells) developed within 2 to 3 weeks. It was already known that cell growth is necessary for the appearance of the differentiated phenotype; this experiment suggested that loss of methylation at specific sites occurs during the DNA replicative cycle.

In another study, loci on an inactivated human X chromosome were reactivated by exposure to 5-azacytidine. This experiment and the one described above have been used to support the hypothesis that a maintenance methylase is responsible for continually propagating the methylated state when a gene is to remain suppressed in a particular cell lineage. Thus an initial step in the activation of genes may be to block the methylation that normally occurs shortly after DNA replication.

Preparation for Transcriptional Activity May Occur at the Time of DNA Replication

The idea that an active site of transcription may be built into DNA at the time of replication is an appealing possibility. As we shall discuss in Chapter 13, the formation of nucleosomes is an event that occurs soon after DNA replication; methylation of all appropriate C residues also occurs just after replication. Site-specific binding proteins might be added around the same time, before nucleosome formation. Many steps in genetic determination (programming) and differentiation (the execution of a program) occur only if cells are allowed to continue multiplying, as we shall show in Chapter 22. Thus it is possible that gene control events fall into two broad categories: (1) the construction of chromatin in the proper form at the site of DNA replication, and (2) the activity of site-specific regulatory proteins on properly prepared chromatin.

Summary

The genes in the cells of multicellular organisms are regulated by a set of signals, many of which we do not yet understand. The environmental and nutritional signals that dominate the control of prokaryotes probably make up a very minor fraction of the controlling signals of eukaryotic cells. Circulating substances such as hormones are important in gene signaling, but many of the signals that are responsible for tissue specificity have not yet been identified.

We now know that gene control in eukaryotic cells can occur at several levels: transcription, RNA processing, mRNA stabilization, and translation. Observations of differential transcription rates, differential choices in nuclear RNA processing, and differential rates of mRNA turnover have been confirmed and reconfirmed, for both viral and cellular genes. Control usually exists at the level of transcription, and probably at initiation; however, establishing this fact has been much more difficult for eukaryotic cells than it was for bacteria.

From genetic studies with yeast, it is clear that regulatory genes govern both metabolic pathways and proteins, such as the mating type, that change cell behavior. A number of yeast loci that encode regulatory proteins have been cloned. Although the mechanisms by which these proteins change transcription rates are not yet established with certainty, some (e.g., the Gal4 protein) may operate directly as positive activators. Others (e.g., the *SIR* gene product) may act by changing the superhelicity of the DNA.

From studies with mammalian viruses and a few cellular genes, it is known that transcriptional regulatory proteins also exist in higher cells, although only a few such proteins have been isolated. Most of these are positive regulatory proteins; their mode of action is currently being explored. Information about which DNA sequences are important for the action of regulatory proteins of individual transcription units is beginning to accumulate. Most of these sequences lie upstream of the initiation site for transcription, but some tissue-specific enhancer sequences that lie within the transcription unit have been discovered. We know that active genes are characterized by chromosomal changes that include undermethylation and a looser packing of the nucleosomal DNA; the latter change was revealed by the increased nuclease sensitivity of the active genes. Regulatory proteins may be responsible for creating nuclease-hypersensitive sites.

We do not yet know precisely how the chromatin structure is changed during transcription. Can fixed loops of chromatin containing transcription units be subject to changes in superhelicity? If so, where do the sequences that allow such changes in large DNA loops reside? Finally, are transcriptional events that must be coordinated—for example, those in the tissue-specific expression of many combined genes—regulated at the level of loops of DNA, at the level of sequences near each gene, or both?

The mechanisms of gene regulation that occurs *after* transcription remain entirely unknown, but such regulation certainly exists. Primary transcripts from complex transcription units can be processed in different cells in different ways by the differential choice of poly A sites or splice sites. In addition, cells may make "decisions" (process-versus-discard decisions) about whether to use or discard primary transcripts. The regulatory molecules in differential choices appear to be either protein, RNA, or ribonucleoprotein particles. Finally, there are many clearly established cases of cytoplasmic regulation both of mRNA half-life and differential translation of mRNAs. Here also the mechanisms of regulation are not yet known.

References

Signals for Gene Control

ANDERSON, J. E. 1983. The effect of steroid hormones on gene transcription. In *Biological Regulation and Development,* R. F. Goldberger and K. R. Yamamoto, eds. Plenum.

ASHBURNER, M. 1980. Chromosomal action of ecdysone. *Nature* **285**:435–436.

LENGYEL, P. 1982. Biochemistry of interferons and their actions. *Annu. Rev. Biochem.* **51**:251–282.

LITWACK, G., ed. 1984. *Biochemical Actions of Hormones,* vol. 14. Academic Press.

STILES, C. D. 1983. The molecular biology of platelet-derived growth factor. *Cell* **33**:653–655.

Levels of Regulation in mRNA Metabolism

DARNELL, J. E. JR. 1982. Variety in the level of gene control in eucaryotic cells. *Nature* **297**:365–371.

DARNELL, J. E. JR. 1983. The processing of RNA. *Sci. Am.* **249**(10):90–100.

NEVINS, J. R. 1982. Adenovirus gene expression: control at multiple steps of mRNA biogenesis. *Cell* **28**:1–2.

Transcriptional Control

ASHBURNER, M. 1973. Temporal control of puffing activity in polytene chromosomes. *Cold Spring Harbor Symp. Quant. Biol.* **38**:655–662.

DERMAN, E., K. KRAUTER, L. WALLING, C. WEINBERGER, M. RAY, and J. E. DARNELL JR. 1981. Transcriptional control in the production of liver-specific mRNAs. *Cell* **23**:731–739.

DYNAN, W. S., and R. TJIAN. 1985. Control of eukaryotic messenger RNA synthesis by sequence-specific DNA-binding proteins. *Nature* **316**:774–778.

EDLUND, T., M. D. WALKER, P. J. BARR, and W. J. RUTTER. 1985. Cell-specific expression of the rat insulin gene: evidence for role of two distinct 5′ flanking elements. *Science* **230**:912–916.

LANDFEAR, S. M., P. LEFEBVRE, S. CHUNG, and H. F. LODISH. 1982. Transcriptional control of gene expression during development of *Dictyostelium discoideum. Mol. Cell. Biol.* **2**:1417–1426.

MCKNIGHT, G. S., and R. D. PALMITER. 1979. Transcriptional regulation of the ovalbumin and conalbumin genes by steroid hormones in chick oviduct. *J. Biol. Chem.* **254**:9050–9058.

NEVINS, J. R. 1981. Mechanism of activation of early viral transcription by the adenovirus E1A gene product. *Cell* **26**:213–220.

NEVINS, J. R., M. J. IMPERIALE, H-T. KAO, and L. T. FELDMAN. 1984. Role of the adenoviral E1A gene product in transcriptional activation. In *Oncogenes and Viral Genes,* G. F. Van de Woude et al., eds. Cold Spring Harbor Laboratory.

PARKER, C. S., and J. TOPOL. 1984. A *Drosophila* RNA polymerase II transcription factor binds to the regulatory site of the heat-shock protein 70 gene. *Cell* **37**:273–283.

PAYVAR, F., D. DEFRANCO, G. L. FIRESTONE, B. EDGAR, O. WRANGE, S. OKRET, J.-A. GUSTAFSSON, and K. R. YAMAMOTO. 1983. Sequence-specific binding of glucocorticoid receptor to MTV DNA at sites within and upstream of the transcribed region. *Cell* **35**:381–392.

SHENK, T. 1981. Transcription control regions: nucleotide sequence requirements for initiation by RNA polymerase II and III. *Curr. Topics Microbiol. Immun.* **93**:25–40.

TIJAN, R. 1981. T antigen binding and the control of SV40 gene expression. *Cell* **26**:1–2.

WALKER, U., and M. ASHBURNER. 1981. The control of ecdysterone-regulated puffs in *Drosophila* salivary glands. *Cell* **26**:269–277.

Transcriptional and Posttranscriptional Control of Histone Production

BORUN, T. W., F. GABRIELLI, K. AJIRO, A. ZWEIDLER, and C. BAGLIONI. 1975. Further evidence of transcriptional and translational control of histone mRNA during the Hela S3 cycle. *Cell* **4**:59–68.

CHILDS, G., R. MAXSON, and L. KEDES. 1979. Histone gene expression during sea urchin embryogenesis: isolation and characterization of early and late messenger RNAs of *S. purpuratus* by gene-specific hybridization and template activity. *Devel. Biol.* **73**:153–173.

HEINTZ, N., H. L. SIVE, and R. G. ROEDER. 1983. Regulation of histone gene expression. *Mol. Cell. Biol.* **3**:539–550.

HEINTZ, N., and ROEDER, R. G. 1984. Transcription of human histones genes in extracts from synchronized HeLa cells. *Proc. Nat'l Acad. Sci. USA* 81:2713–2717.

HENTSCHEL, C., and M. L. BIRNSTIEL. 1981. The organization and expression of histone gene families. *Cell* **25**:301–313.

Control at the Level of RNA Processing

BREITBART, R. E., H. T. NGUYEN, R. M. MEDFORD, A. T. DESTREE, V. MAHDAVI, and B. NADAL-GINARD. 1985. Intricate combinational patterns of exon splicing generate multiple regulated troponin T isoforms from a single gene. *Cell* **41**:67–82.

EARLY, P., J. ROGERS, M. DAVIS, K. CALAME, M. BOND, R. WALL, and L. HOOD. 1980. Two mRNAs can be produced from a single immunoglobulin μ gene by alternative RNA processing pathways. *Cell* **20**:313–319.

NEVINS, J. R., and M. WILSON. 1981. Regulation of adenovirus-2 gene expression at the level of transcriptional termination and RNA processing. *Nature* **290**:113–118.

ROSENFELD, M. G., J. J. MERMOD, S. G. AMASA, L. W. SWANSON, P. E. SAWCHENKO, J. RIVIER, W. W. VALE, and R. M. EVANS. 1983. Production of a novel neuropeptide encoded by the calcitonin gene via tissue-specific RNA processing. *Nature* **304**:129–135.

ROZEK, C. E., and N. DAVIDSON. 1983. *Drosophila* has one myosin heavy-chain gene with three developmentally regulated transcripts. *Cell* **32**:23–34.

SCHWARZBAUER, J. E., J. W. TAMKUN, I. R. LEMISCHKA, and R. O. HYNES, 1983. Three different fibronectin mRNAs arise by alternative splicing within the coding region. *Cell* **35**:421–431.

Overlapping Transcription Units

CARLSON, M., R. TAUSSIG, S. KUSTU, and D. BOTSTEIN. 1983. The secreted form of invertase in *Saccharomyces cerevisiae* is synthesized from mRNA encoding a signal sequence. *Mol. Cell. Biol.* 3:439–447.

HAGENBÜCHLE, O., M. TOSI, U. SCHIBLER, R. BOVEY, P. K. WELLAUER, and R. A. YOUNG. 1981. Mouse liver and salivary gland α-amylase mRNAs differ only in 5′ non-translated sequences. *Nature* **289**:643–646.

WILSON, M. C., N. W. FRASER, and J. E. DARNELL. 1979. Mapping of RNA initiation sites by high doses of UV irradiation: evidence for three independent promoters in the left 11% of the adeno-2 genome. *Virology* **94**:175–184.

YOUNG, R. A., O. HAGENBÜCHLE, and U. SCHIBLER. 1981. A single mouse α-amylase gene specifies two different tissue-specific mRNAs. *Cell* **23**:451–458.

Gene Control in the Cytoplasm

BROCK, M. L., and D. J. SHAPIRO. 1983. Estrogen stabilizes vitellogenin mRNA against cytoplasmic degradation. *Cell* **34**:207–214.

DANI, CH., J. M. BLANCHARD, M. PRECHACZYK, S. EL SABOUTY, L. MARTZ, and PH. JEANTEUR. 1984. Extreme instability of myc mRNA in normal and transformed cells. *Proc. Nat'l Acad. Sci. USA* **81**:7046–7050.

DOUGLASS, J., O. CIVELLI, and E. HERBERT. 1984. Polyprotein gene expression: generation of diversity of neuroendocrine peptides. *Annu. Rev. Biochem.* **53**:665–716.

FRIEDMAN, J. M., E. V. CHEUNG, and J. E. DARNELL. 1984. Gene expression during liver regeneration. *J. Mol. Biol.* **179**:37–53.

GEYER, P., O. MEYUHAS, R. P. PERRY, and L. F. JOHNSON. 1982. Regulation of ribosomal protein mRNA content and translation in growth-stimulated mouse fibroblasts. *Mol. Cell. Biol.* 2:685–693.

GUYETTE, W. A., R. J. MATUSIK, and J. M. ROSEN. 1979. Prolactin mediated transcriptional and post-transcriptional control of casein gene expression. *Cell* **17**:1013–1023.

ROSENTHAL, E. T., T. HUNT, and J. V. RUDERMAN. 1980. Selective translation of mRNA controls the pattern of protein synthesis during early development of the surf clam, *Spisula solidissima. Cell* **20**:487–494.

SHAPIRO, D. J., and M. L. BROCK. 1984. Messenger RNA stabilization and gene transcription in the estrogen induction of vitellogenin mRNA. In *Biochemical Action of Hormones,* G. Litwack, ed. Academic Press.

Mechanisms of Transcriptional Control

Positive- and Negative-Acting Regulatory Proteins in Yeast Cells

ABRAHAM, J., J. FELDMAN, K. A. NASMYTH, J. N. STRATHERN, A. J. S. KLAN, J. R. BROACH, and J. B. HICKS. 1983. Sites required for position-effect regulation of mating-type information in yeast. *Cold Spring Harbor Symp. Quant. Biol.* 47:989–998.

GINIGER, E., S. M. VARNUM, and M. PTASHNE. 1985. Specific DNA of Gal4, a positive regulatory protein of yeast. *Cell* 40:767–774.

JOHNSON, A. D., and I. HERSKOWITZ. 1985. A repressor (MAT α2 product) and its operator control of expression of a set of cell-type specific genes in yeast. *Cell* **42**:237–247.

JOHNSTON, S. A., and J. E. HOPPER. 1982. Isolation of the yeast regulatory gene GAL4 and analysis of its dosage effects on the galactose/melibiose regulon. *Proc. Nat'l Acad. Sci. USA* 79:6971–6975.

JONES, E. W., and G. R. FINK. 1982. Regulation of amino acid and nucleotide biosynthesis in yeast. In *The Molecular Biology of the Yeast* Saccharomyces: *Metabolism and Gene Expression,* J. N. Strathern, E. W. Jones, and J. R. Broach, eds. Cold Spring Harbor Laboratory.

LAUGHON, A., R. DRISCOLL, N. WILLS, and R. F. GESTELAND. 1984. Identification of two proteins encoded by the *Saccharomyces cerevisiae* GAL4 gene. *Mol. Cell. Biol.* **4**:268–275.

MILLER, A. M., V. L. MACKAY, and K. A. NASMYTH. 1985. Identification and comparison of two sequence elements that confer cell-type specific transcription in yeast. *Nature* **314**:598–603.

NASMYTH, K. A. 1982. The regulation of yeast mating-type chromatin structure by SIR: an action at a distance affecting both transcription and transposition. *Cell* **30**:567–578.

NASMYTH, K. A. 1983. Molecular analysis of a cell lineage. *Nature* **302**:670–676.

OSHIMA, Y. 1982. Regulatory circuits for gene expression: the metabolism of galactose and phosphate. In *The Molecular Biology of the Yeast* Saccharomyces: *Metabolism and Gene Expression,* J. N. Strathern, E. W. Jones, and J. R. Broach, eds. Cold Spring Harbor Laboratory.

WARNER, J. R. 1982. The yeast ribosome: structure, function and synthesis. In *The Molecular Biology of the Yeast* Saccharomyces: *Metabolism and Gene Expression,* J. N. Strathern, E. W. Jones, and J. R. Broach, eds. Cold Spring Harbor Laboratory.

WILSON, K. L., and I. HERSKOWITZ. 1984. Negative regulation of STC 6 gene expression by the α2 product of *Saccharomyces cerevisiae. Mol. Cell Biol.* 4:2420–2427.

Enhancers

BANERJI, J., L. OLSON, and W. SCHAFFNER. 1983. A lymphocyte-specific cellular enhancer is located downstream of the joining region in immunoglobulin heavy chain genes. *Cell* 33:729–740.

GILLIES, S. D., S. L. MORRISON, V. T. OI, and S. TONEGAWA. 1983. A tissue-specific trancription enhancer element is located in the major intron of a rearranged immunoglobulin heavy chain gene. *Cell* **33**:717–728.

GLUZMAN, Y., and T. SHENK, eds. 1983. *Enhancers and Eukaryotic Gene Expression.* Cold Spring Harbor Laboratory.

KHOURY, G., and P. GRUSS. 1983. Enhancer elements. *Cell* 33:313–314.

QUEEN, C., and D. BALTIMORE. 1983. Immunoglobulin gene transcription is activated by downstream sequence elements. *Cell* 33:741–748.

Chromatin Structure and Transcriptional Control

BROWN, D. D. 1984. The role of stable complexes that repress and activate eucaryotic genes. *Cell* **37**:359–365.

CARTWRIGHT, I. L., S. M. ABMAYR, G. FLEISCHMANN, K. LOWENHAUPT, S. C. R. ELGIN, M. A. KEENE, and G. C. HOWARD. 1982. Chromatin structure and gene activity: the role of nonhistone chromosomal proteins. *CRC Crit. Rev. Biochem.* **13**:1–86.

ELGIN, S. C. R. 1981. DNase I-hypersensitive sites of chromatin. *Cell* **27**:413–415.

EMERSON, B. M., and G. FELSENFELD. 1984. Specific factor conferring nuclease hypersensitivity at the 5′ end of the chicken adult β globin gene. *Proc. Nat'l Acad. Sci. USA* **81**:95–99.

KMIEC, E. B., and A. WORCEL. 1985. The positive transcription factor of the 5S RNA gene induces a 5S-DNA–specific gyration in *Xenopus* oocyte extracts. *Cell* **41**:955–963.

KOHWI-SHIGEMATSU, T., R. GELINAS, and H. WEINTRAUB. 1983. Detection of an altered DNA conformation at specific sites in chromatin and supercoiled DNA. *Proc. Nat'l Acad. Sci. USA* **80**:4389–4393.

KORNBERG, R. 1981. The location of nucleosomes in chromatin: specific or statistical? *Nature* **292**:579–580.

MCGHEE, J. D., W. I. WOOD, M. DOLAN, J. D. ENGLE, and G. FELSENFELD. 1981. A 200 base pair region at the 5′ end of the chicken adult β-globin gene is accessible to nuclease digestion. *Cell* **27**:45–55.

PRIOR, C. P., C. R. CANTOR, E. M. JOHNSON, V. C. LITTAU, and V. G. ALLFREY. 1983. Reversible changes in nucleosome structure and histone accessibility in transcriptionally active and inactive states of rDNA chromatin. *Cell* **34**:1033–1042.

SMITH, R. D., R. L. SEALE, and J. YU. 1983. Transcribed chromatin exhibits an altered nucleosomal spacing. *Proc. Nat'l Acad. Sci. USA* **80**:5505–5509.

TUAN, D., W. SOLOMON, Q. LI, and I. M. LONDON. 1985. The "β-like" globin gene domain in human erythroid cells. *Proc. Nat'l Acad. Sci. USA.* **82**:6328–6388.

WEINTRAUB, H. 1985. Assembly and propagation of repressed and derepressed chromosomal states. *Cell* **42**:705–711.

WEISBROD, S. 1982. Active chromatin. *Nature* **297**:289–295. (A review.)

WEISBROD, S., M. GROUDINE, and H. WEINTRAUB. 1980. Interaction of HMG 14 and 17 with actively transcribed genes. *Cell* **19**:289–301.

Undermethylation of Active Genes

BUSSLINGER, M., T. HURST, and R. A. FLAVELL. 1983. DNA methylation and regulation of globin gene expression. *Cell* **34**:197–206.

DOERFLER, W. 1983. DNA methylation in gene expression. *Annu. Rev. Biochem.* **52**:93–124.

JONES, P. A., S. M. TAYLOR, T. MOHANDAS, and L. J. SHAPIRO. 1982. Cell cycle specific reactivation of an inactive X-chromosome locus by 5-azadeoxycytidine. *Proc. Nat'l Acad. Sci. USA* **79**:1215–1219.

KRUCZEK, I., and W. DOERFLER. 1983. Expression of the chloramphenicol acetyltransferase gene in mammalian cells under the control of adenovirus type 12 promoters: effect of promoter methylation on gene expression. *Proc. Nat'l Acad. Sci. USA* **80**:7586–7590.

RAZIN, A., and A. D. RIGGS. 1980. DNA methylation and gene function. *Science* **210**:604–610.

STEIN, R., N. SCIAKY-GALLILI, A. RAZIN, and H. CEDAR. 1983. Pattern of methylation of two genes coding for housekeeping functions. *Proc. Nat'l Acad. Sci. USA* **80**:2422–2426.

WIGLER, M. H. 1981. The inheritance of methylation patterns in vertebrates. *Cell* **24**:258–286.

13

DNA Synthesis, Repair, and Recombination

BEFORE Watson and Crick discovered the duplex structure of DNA, one of the most mysterious aspects of biology was how genetic material was exactly—or almost exactly—duplicated from one generation to the next. Recognition of the base-paired nature of the DNA duplex immediately gave rise to the notion that a template was involved in the transfer of information between generations, but many structural and biochemical questions soon followed. When replication has been completed, are old strands paired with new, or old with old and new with new? How does replication begin, and how does it progress along the chromosome? What enzymes take part in DNA synthesis, and what are their roles? How does duplication of the long helical duplex occur without the strands becoming tangled? What mechanisms ensure exact replication?

DNA Synthesis: The General Problems

The functions that assemble new DNA chains are often collectively referred to as DNA synthesis. DNA replication (or chromosomal replication) is a more comprehensive term that encompasses not only DNA chain synthesis but also its initiation and termination. Studies of replication, therefore, tend to be concerned with how DNA synthesis starts and stops in such a way that each chromosome is duplicated exactly. Another issue in replication studies is how the two new chromosomes become separated.

With these definitions in mind, we can begin to consider the details of DNA synthesis and replication. Two general principles apply to all cells. The first of these is that each new DNA strand pairs with one old strand; that is, DNA synthesis is semiconservative. The second principle is that regions of DNA chain growth called growing forks move for thousands of bases along the chromosome; the single duplex ahead of the fork becomes a pair of duplexes behind the fork. The enzymatic events that are responsible for DNA chain growth at the growing fork are now fairly well understood in bacteria and are probably very similar in all cells. The action of a variety of different enzymes and assisting proteins can account for the cell's ability to accomplish the biochemically complex task of copying the two antiparallel DNA strands.

After discussing semiconservative replication and the mechanics of the growing fork, we turn to mechanisms for the initiation of DNA synthesis; here again we use bacterial models. We also describe some of the experimental work that has contributed to the present understanding of replication origins. Next, because chromosomes in higher cells are not pure DNA molecules but rather complexes of DNA with protein, we must deal with the packaging of newly formed DNA together with the chromosomal proteins. The processes of initiation of DNA synthesis and chromosome packaging are not yet fully understood in any eukaryotic cell, but they are among the more important aspects of cell division.

For DNA to serve as the genetic link between generations, not only must the base sequence be copied correctly during replication, but the integrity of the sequence must be maintained continuously for accurate protein synthesis in all cells. Consequently, it was not a surprising discovery that cells possess enzymatic "repair" functions to keep DNA sequences accurate. Some of the enzymatic events necessary for repair of damaged DNA also play a role in genetic recombination, a major factor in evolution. Therefore, we shall discuss both repair and recombination briefly toward the end of this chapter.

DNA Replication Is Semiconservative

The Watson-Crick model suggested that new DNA chains were copied from old. This raised a new question: Was replication *conservative;* that is, did each new duplex contain two new strands, which would mean that the old duplexes remained intact? Or was replication a *semiconservative* process in which each old chain paired with a new chain copied from it? This question was settled for *Escherichia coli* with proof that each newly formed duplex contained one old and one new DNA chain.

When *E. coli* cells are grown in ammonium salts containing ^{15}N instead of ^{14}N, the heavy atoms are incorporated into the deoxyribonucleotide precursors and then into DNA. The resulting DNA is about 1 percent denser than the DNA containing only the normal isotope. Heavy DNA can be separated from normal, light DNA by its buoyant density after equilibrium density-gradient (isopycnic) centrifugation in solutions of CsCl (Chapter 7): the ^{15}N- and ^{14}N-containing DNAs form discrete bands. Bacteria grown for several generations in a "heavy" medium will contain almost completely substituted (i.e., heavy) DNA. Cells containing all heavy DNA can then be transferred to a normal (light) medium and allowed to double their DNA. At this point, the density of the DNA in the CsCl solution is intermediate between the densities of heavy and light DNAs. After the cells undergo another doubling, this "heavy-light" DNA is replaced by a mixture whose density is one-half heavy-light and one-half completely light. This experiment has supported the model of semiconservative replication (Figure 13-1).

The experiment can be refined further by denaturing and then separating the long strands of heavy-light DNA into single strands. These single strands can then be subjected to equilibrium centrifugation to show that each of the long strands is either all heavy or all light (because two clear bands are formed). Thus, semiconservative replication extends over long stretches of DNA.

Experiments of a different design first demonstrated semiconservative DNA replication in eukaryotic chromosomes. Cells were labeled during DNA synthesis with [^{3}H]thymidine. The *mitotic* chromosomes were then examined by autoradiography beginning at the first division after labeling, which occurred within a few hours. The autoradiographs revealed that all mitotic chromosomes were labeled in *both* chromatids (Figure 13-2a). Presumably each chromatid represented one double-stranded DNA molecule, and one DNA strand in each chromatid was new (labeled) and one was old.

This interpretation was greatly strengthened by autoradiographs made by the same experimenters after one further round of cell division in the absence of labeled thymidine. At this point one chromatid in each mitotic pair was labeled and one was totally unlabeled (Figure 13-2b)—a result consistent with the semiconservative model; the labeled chromatid contained the old labeled strand, and the unlabeled chromatid, the old unlabeled strand. Thus each labeled strand retained its label and would be paired after each replication with a new unlabeled strand. Evidence of semiconservative replication has been obtained with both plant and animal chromosomes. Apparently all cell DNA, both prokaryotic and eukaryotic, is replicated by a semiconservative mechanism.

DNA Synthesis Occurs before—Not during—Mitosis

Each cell must have exactly the correct amount of DNA, which means that the synthesis of DNA must be closely

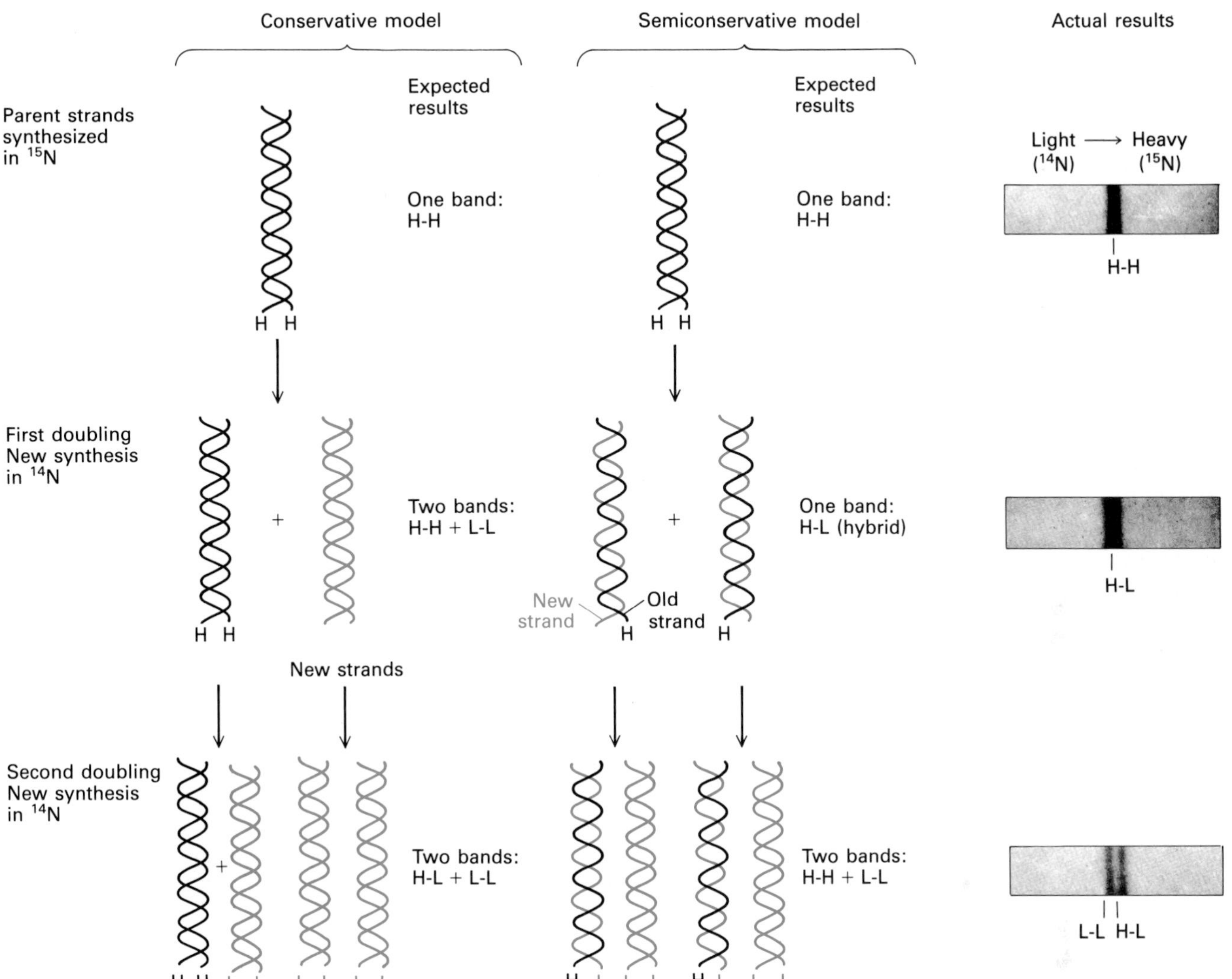

Figure 13-1 Semiconservative DNA replication: the Meselson-Stahl experiment. The diagram shows the origin of each strand of duplicated DNA duplexes when replication is conservative and when it is semiconservative. The experimental proof of semiconservative replication in *Escherichia coli* is at the right. *Escherichia coli* cells grown in a "heavy" medium (one with ammonium salts containing ^{15}N instead of ^{14}N) yield duplex DNA that is "heavy-heavy" (H-H); this determination was made upon banding the DNA in a CsCl solution by equilibrium centrifugation (Chapter 7). After one DNA doubling, the DNA is heavy-light; after two doublings, it is half heavy-light and half light-light. [See M. Meselson and F. W. Stahl, 1958, *Proc. Nat'l Acad. Sci. USA* **44**:671.] *Photographs from A. Kornberg, 1980,* DNA Replication, *W. H. Freeman and Company, p. 348.*

integrated into the division cycle of the cell. In rapidly growing bacteria, the length of the DNA replication cycle essentially determines the time of cell doubling; very soon after the chromosome is duplicated, the cell divides. During the process of division, the two daughter chromosomes segregate to two sections formed within the cell, and a wall is built between them (see Figure 5-12). Physical division of the cell into two daughter cells requires a relatively short time compared to the duplication of the chromosome. When cell division is completed, chromosome duplication begins anew and the cycle goes on. When a growing culture of bacterial cells is exposed for a few minutes to [^{3}H]thymidine and subjected to autoradiography, nearly all the cells become labeled. This means that DNA synthesis is active in all bacterial cells almost all the time.

Continual DNA synthesis is not the pattern for eukaryotic cells. In 1953 it was discovered that when root tips, in which cells divide rapidly, were exposed to [^{3}H]thymidine for a few minutes, a fraction of the cells incorporated high levels of label into DNA but the remainder were not labeled at all. Most of the cells in root tips, however, divide at least once every 15 h or so. It thus appeared that DNA synthesis did not occur throughout the cell division cycle but was perhaps restricted to a part of it.

Thorough analyses of the relationship between eukaryotic DNA synthesis and cell division have been made in

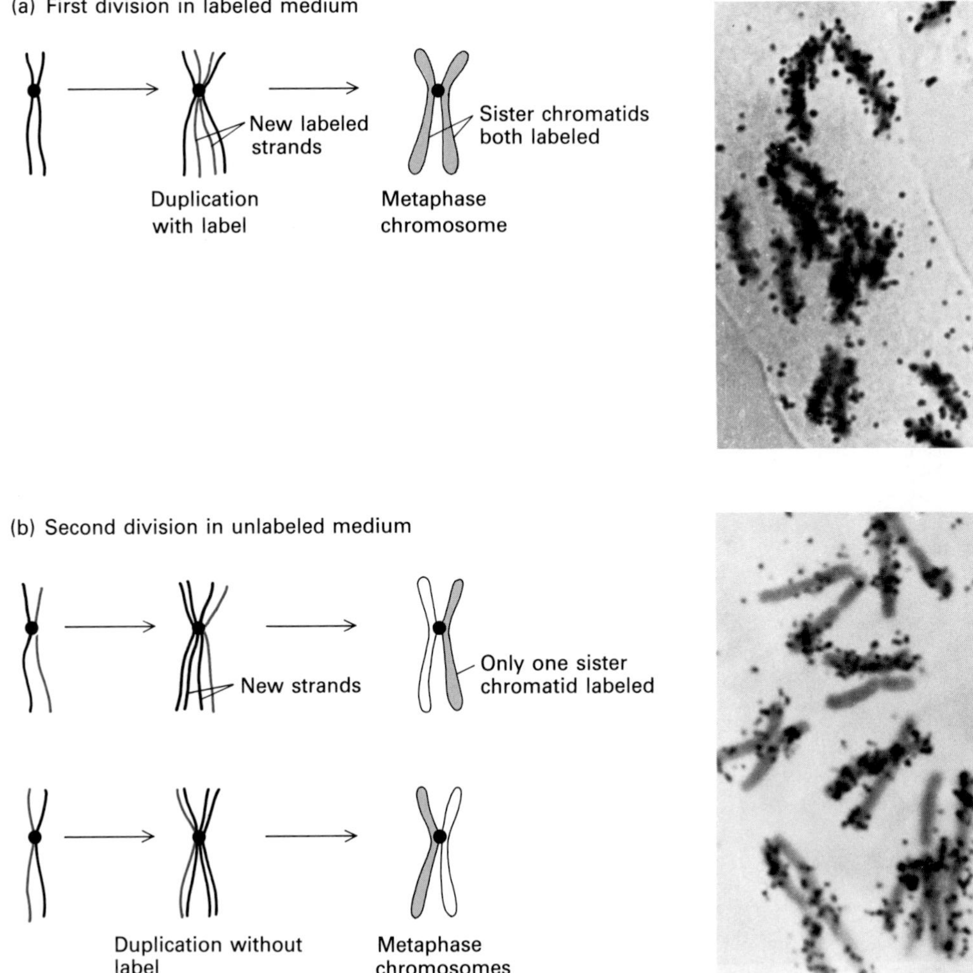

Figure 13-2 Autoradiographs indicating semiconservative replication in plant chromosomes. (a) DNA from the growing root cells of a lily plant *(Bellavalia romana)* was labeled by a pulse of [^{3}H]thymidine, and cells were mounted and stained at intervals. Mitotic chromosomes were then located microscopically and autoradiographed. The first labeled mitotic cells to appear after labeling (within approximately 8 h) showed both chromatids to be labeled. (b) At the second mitosis (within approximately 24 h), individual chromatids tended to be either all labeled or all unlabeled.

The experimenters who originally obtained these results were not certain that each chromatid represented one DNA double helix, as is now thought to be the case. However, their work clearly demonstrates semiconservative DNA synthesis: each initially labeled chromatid had one labeled and one unlabeled strand, and after a second mitosis some chromatids were unlabeled. (Some of the chromosomes have engaged in sister chromatid exchange of whole segments of DNA; this accounts for contiguous patches of grains in a few of the otherwise unlabeled chromosomes that underwent a second mitosis.) [See J. H. Taylor, P. S. Woods, and W. L. Hughes, 1957, *Proc. Nat'l Acad. Sci. USA* **43**:122.] *Photographs courtesy of J. H. Taylor, 1957.*

mammalian cell cultures in which each cell was capable of growth and division. Cells were pulse-labeled briefly with [^{3}H]thymidine to label those that were synthesizing DNA. At hourly intervals after the period of labeling, cell samples were stained so that each mitotic chromosome would be visible. The samples were then exposed to autoradiography so that the labeled mitotic chromosomes could be counted. (In a culture of growing human cells, about 3 percent of the total cells are in mitosis at any given time. The time required to complete mitosis is about 30 to 45 min.) The first cell samples examined after the labeling showed no mitotic chromosomes that contained labeled DNA. After about 6 or 7 h, all of the mitotic chromosomes contained labeled DNA. In about 18 h, unlabeled mitotic chromosomes began to recur and by 26 h after exposure to [^{3}H]thymidine, very few labeled mitotic cells were observed (Figure 13-3).

This experiment led to the conclusion that DNA synthesis in eukaryotic cells occurs only during one phase of the cell cycle: the S (synthetic) phase. A gap of time occurs

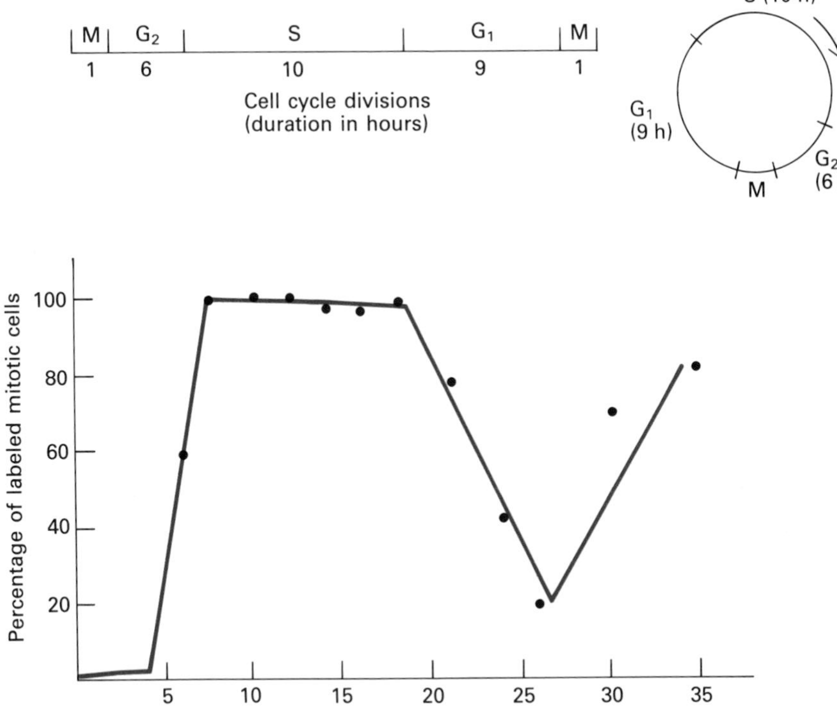

Figure 13-3 An experimental derivation of the existence and duration of the divisions of the cell cycle in cultured mammalian cells. HeLa cells were exposed to a pulse of [^{3}H]thymidine, and samples taken at intervals to analyze to determine how many of the cells in the mitotic (M) phase were labeled. The mitotic cells in the first sample were unlabeled, as were those in succeeding samples up to 6 or 7 h after labeling; at this point, the number of labeled mitotic cells increased dramatically. It was reasoned that the first group of mitotic cells to show the label were those that were in the last part of the synthesis (S) phase during the pulse. The gap (G_2) between the S phase and the M phase was thus estimated to be about 6 h long (the duration of mitosis was visually observed to be less than 1 h). The same reasoning was applied to infer a timetable for the rest of the cycle: About 17 h after the pulse, the frequency of labeled mitotic cells began to decrease; hence the duration of the S phase = 17 − (6 + 1) = 10 h. The decline in labeled mitosis after 17 h signified another DNA synthesis gap (G_1). The cell doubling time was known to be 26 h, so G_1 = 26 − 17 = 9 h. The completed cycle is shown at the right. [See R. Baserga and F. Weibel, 1969, *Int. Rev. Exp. Pathol.* 7:1.]

after DNA synthesis and before mitosis; another gap of time occurs after mitosis and before the next S period. The cell growth cycle thus consists of the M (mitosis) phase, a G_1 phase (the first gap), the S phase, a G_2 phase (the second gap), and back to M. Many nondividing cells in tissues (for example, all resting fibroblasts in the body) suspend the cycle after mitosis and just prior to DNA synthesis. Such "resting" cells, which have exited from the cell cycle before the S phase, are described as being in G_0. The regulatory events guiding the cell from phase to phase are unknown, and they remain a major experimental challenge to modern cell biology.

Most DNA Replication Is Bidirectional

Let us consider several possible mechanisms of chain copying when semiconservative DNA synthesis begins at a given site on a chromosome. Model 1 in Figure 13-4 shows one of the simplest possibilities: one new strand derives from one origin and the other new strand derives from another origin. Only one strand of the duplex grows at each growing point. This model does, in fact, describe replication of the linear DNA occurring in viruses such as adenovirus. In these cases the ends of the DNA molecules serve as fixed sites for the initiation and termination of replication.

In contrast, while the point of chain growth—the *growing fork*—moves along the DNA in one direction, both DNA strands might somehow be copied at the same time (Model 2 in Figure 13-4). A third possibility is that synthesis might start at a single site and proceed in both directions, so that both strands are copied at each of *two* growing forks (Model 3 in Figure 13-4). The available evidence suggests that the third alternative is the most common mode of DNA replication. DNA replication proceeds in both directions—that is, *bidirectionally*—from a given starting point, with both strands being copied. Thus, two growing forks emerge from the start site.

In a circular chromosome (the form found in bacteria and some viruses), one origin often suffices, and the two resulting growing forks merge on the opposite side of the circle to complete the replication. In the long linear chromosomes of eukaryotes there are multiple origins, and the two growing forks continue until they meet the advancing

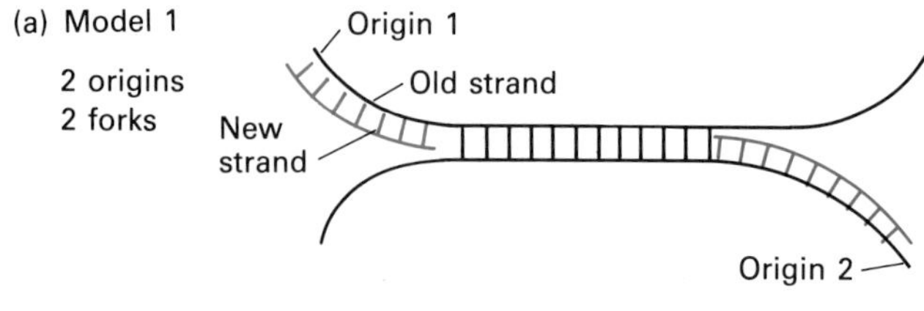

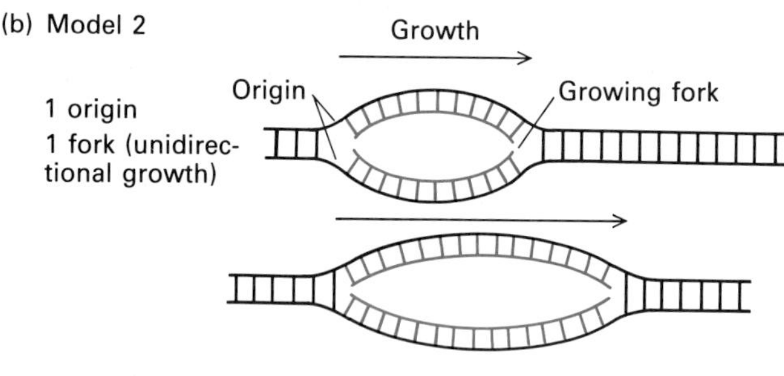

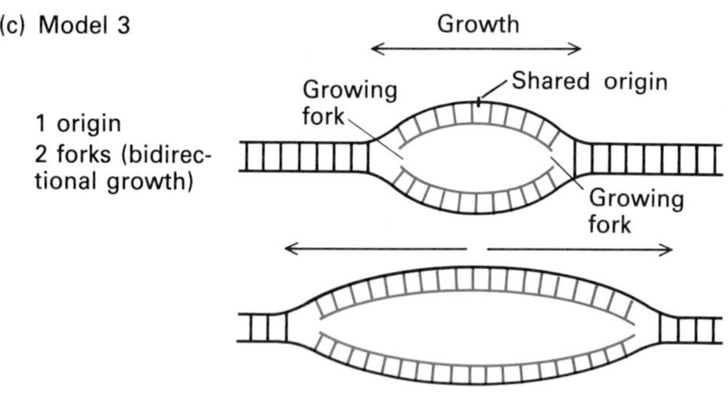

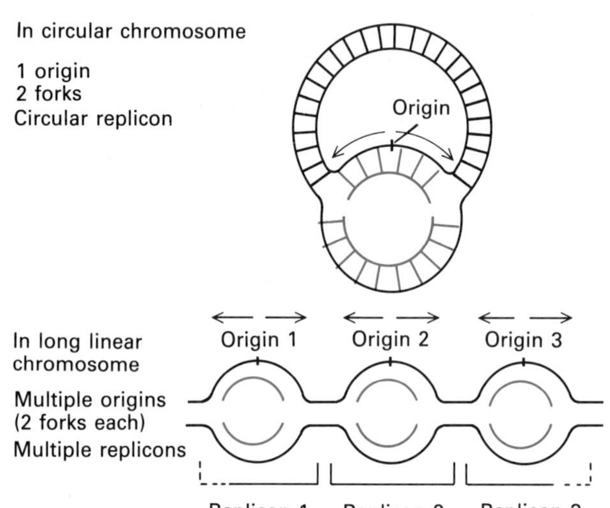

Figure 13-4 Possible mechanisms of DNA strand synthesis from origins of replication. (a) Model 1 occurs in a few viruses. (c) Model 3 (bidirectional growth) is the most common mechanism in cells, although (b) model 2 (unidirectional growth) probably occurs occasionally. Model 3 is found in circular chromosomes with one origin; in such cases the whole circle is one replicon. In long linear chromosomes, multiple origins are used, and each DNA stretch served by one origin is a replicon.

growing forks from a neighboring region. Each region served by a DNA origin is called a *replicon.*

Bidirectional replication was first detected by autoradiography of labeled DNA molecules from cultured mammalian cells. DNA extracted from cells and dried on a microscope slide becomes aligned in parallel fibers. Stretched DNA fibers from newly made [^{3}H]thymidine-labeled DNA will show autoradiographic tracks; these will be dark if the DNA is heavily labeled and lighter if it is not (Figure 13-5). Analysis of such alternating tracks has revealed bidirectional growth in a wide variety of both eukaryotic and prokaryotic cells. It is also true that many virus DNA molecules replicate bidirectionally and therefore serve as excellent models for the study of cellular DNA replication.

Replication "Bubbles" or "Eyes" Electron microscopy has revealed multiple origins of DNA replication with many adjacent "bubbles" or "eyes." In Figure 13-6, each of the bubbles represents either one or two growing forks that have double-stranded DNA both in front of them and behind them. Differentiating between uni- and bidirectional fork movement in cell DNA is not easy, and the observed replication bubbles could theoretically be the result of either mode of replication (see Figure 13-4).

However, viral genomes identical in size and sequence have identical sites for restriction endonucleases. By cutting replicating circular viral molecules at identical sites with a restriction endonuclease, the positions of the centers of the replication bubbles can be determined with respect to the restriction site (Figure 13-7). The most common result from such analyses is a series of ever larger bubbles whose centers are always mapped to the same site, indicating bidirectional replication of both DNA strands from that site. Thus, both fiber autoradiography and electron microscopy have indicated that bidirectional DNA replication is the general rule.

Once DNA synthesis has begun, the rate of chain growth appears to be constant in each cell type of a given organism. Bacterial chromosomes of 4×10^6 base pairs duplicate in 20 min with one origin of replication and two growing forks. The rate in each fork is therefore about 1600 base pairs per second. In human cells the rate of chain growth estimated from fiber autoradiography

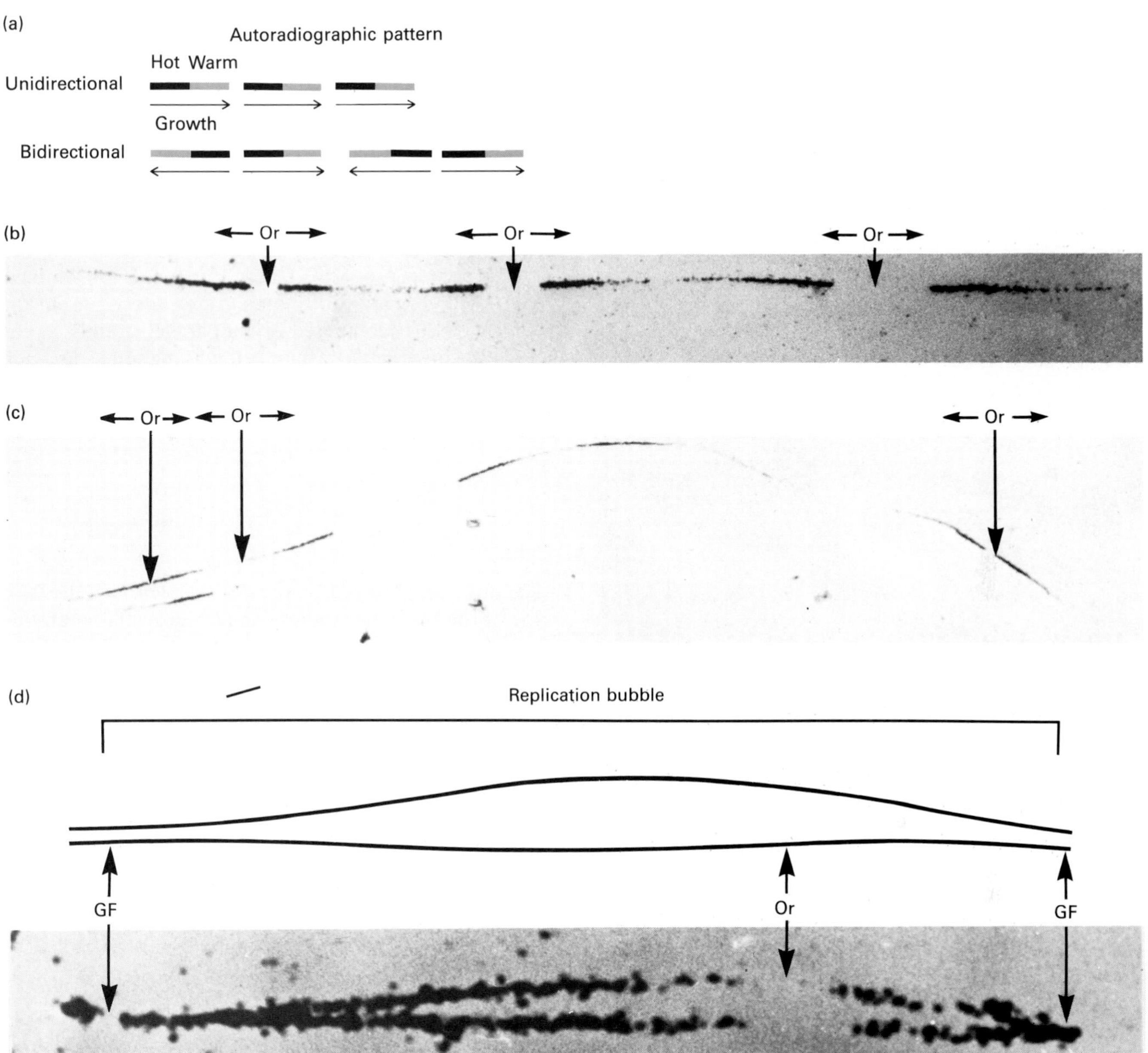

Figure 13-5 Fiber autoradiography of replicating DNA. Exposure of DNA to first a high level and then a low level of [^{3}H]thymidine or to unlabeled medium should produce heavily labeled ("hot") and then lightly labeled ("warm") DNA regions. Because of its physical properties, DNA tends to dry as long linear molecules on a microscope slide. Autoradiography of "hot-warm" DNA should reveal dark regions closer to the origin and light regions farther away. (a) The diagram indicates the expected label distributions for uni- and bidirectional DNA synthesis. (b) and (c) The autoradiographs clearly show origins (Or) with growing forks moving in *two* directions. (d) In this photo, the replication "bubble" created by the two growing forks (GF) has not collapsed as it has in (b). Bidirectional DNA synthesis was first diagnosed from experiments of this type. [See J. A. Huberman and A. D. Riggs, 1968, *J. Mol. Biol.* **32**:327; and J. A. Huberman and A. Tsai, 1973, *J. Mol. Biol.* **75**:5.] *Photographs courtesy of J. A. Huberman.*

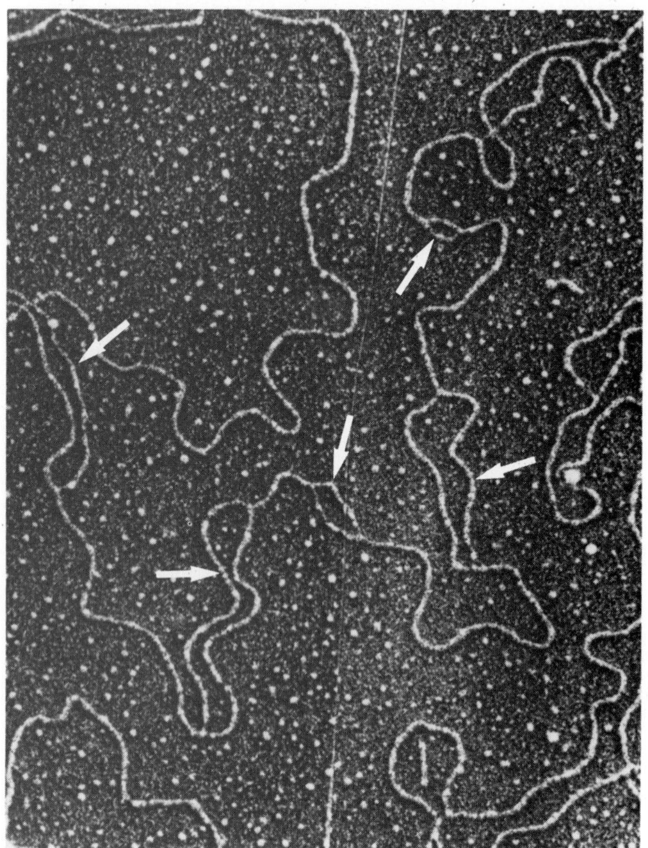

Figure 13-6 Multiple origins of replication are visible in this electron micrograph of replicating DNA from *Drosophila melanogaster*. DNA from rapidly dividing nuclei in early *D. melanogaster* embryos was extracted and spread for the EM examination. All regions of the DNA strands have the appropriate diameter for double strands. The centers of replication bubbles *(arrows)* are spaced about 5 kb apart. [See A. B. Blumenthal, H. J. Kreigstein, and D. S. Hogness, 1973, *Cold Spring Harbor Symp. Quant. Biol.* 38:205.] *Courtesy of D. Hogness.*

after different label times is considerably lower, about 100 to 150 base pairs per second at each fork. The growth rate is approximately the same in most other growing eukaryotic cells. In cases in which the doubling of the total DNA is faster—for example, in the earliest mitotic cycles of embryonic development—the rate of fork movement is similar, and the faster doubling is due to the activation of additional origins of replication.

Properties of DNA Polymerases and DNA Synthesis in Vitro

What types of enzymes and what other factors participate in the synthesis of DNA chains at the growing fork? Enzymatic studies in *E. coli* have been chiefly responsible for answering this question. The DNA polymerases in *E. coli* share two properties with all other such enzymes isolated to date. First, DNA polymerases cannot initiate synthesis of a DNA chain; they can only elongate a pre-existing *primer* strand of DNA or RNA (Chapter 4). Second, all of the DNA polymerases catalyze nucleotide addition to the 3′ hydroxyl end of the primer and thus direct growth in the 5′ → 3′ direction.

This second property creates a dilemma. As we have seen, each growing fork moves in *one* direction; but the duplex DNA is composed of antiparallel strands copied in a semiconservative fashion. How is the single direction of the growing fork maintained while the two strands are replicated in opposite directions?

A Growing Fork Has a Continuous Leading Strand and a Discontinuous Lagging Strand

The problem is solved by the *discontinuous* synthesis of one of the two new strands. At each growing point, one new strand may grow continuously in the 5′ → 3′ direction that coincides with the overall movement of the growing point. This strand is referred to as the *leading strand.* To form the other new strand, the *lagging strand,* short segments are synthesized in the 5′ → 3′ direction that is *opposite* to the overall direction of movement of the growing fork (Figure 13-8). The discontinuous segments are then linked together. The primer for the leading strand can be a free end created within the parental DNA, but what is the primer for the lagging strand?

RNA Primers Are Used in Lagging-Strand Synthesis

In the replication of most viral DNAs and of all cell DNAs studied, the primer for lagging-strand synthesis is RNA. As the synthesis of the leading strand progresses, sites uncovered on the unreplicated parental strand are copied into short RNA oligonucleotides. The RNA primers are then elongated in the 5′ → 3′ direction by DNA polymerase to make the lagging strand. These short (1000 nucleotides or less) segments of RNA plus DNA are called *Okazaki fragments,* named after their discoverer Reiji Okazaki.

As each segment of the lagging strand approaches the 5′ end of the adjacent Okazaki fragment (the one synthesized just previously), a second enzymatic capacity of the DNA polymerase comes into play. The polymerase has a 5′ → 3′ exonuclease activity that chews away the RNA primer at the 5′ end of the neighboring fragment; the gap is then filled by the polymerase. Finally another critical enzyme, *DNA ligase,* joins the two fragments together (Figure 13-9).

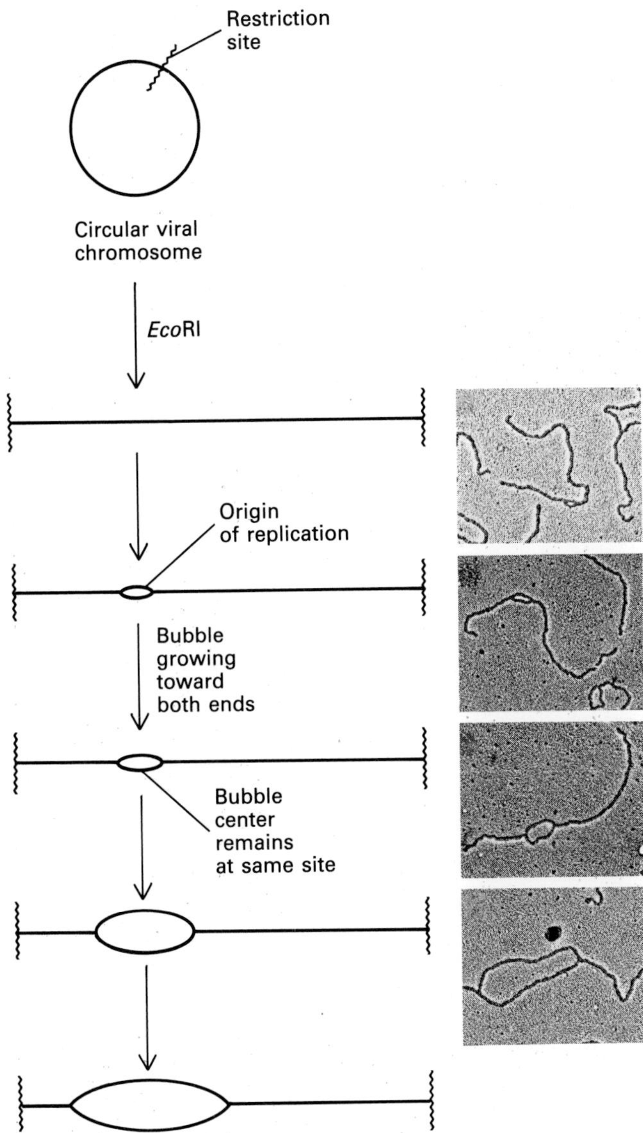

Figure 13-7 Bidirectional growth from a single origin in SV40 DNA replication. Electron micrographs of replicating SV40 DNA were made after the replicating circular DNA was cut at one site by the restriction enzyme *Eco*RI. Bubbles of various sizes represent molecules in various stages of completion of replication. All of the bubbles have their centers at the same position with respect to the ends of the cut molecules, indicating bidirectional DNA synthesis from one origin. [See G. C. Fareed, C. F. Garon, and N. P. Salzman, 1972, *J. Virol.* 10:484.] *Photographs courtesy of N. P. Salzman.*

The formation of RNA primers is an important key to the dilemma of replicating antiparallel strands. Bacterial cells possess a specific RNA-synthesizing enzyme, called a *primase*, that is responsible for RNA primer synthesis. This enzyme works by binding to DNA together with two additional proteins to form a structure called a *primosome*. All of these proteins—the primase and its two associated proteins—were identified in *E. coli* by mutations in their genes. The primase gene is called *dna*G. The genes encoding the other two proteins are *dna*B and *dna*C, and their products are simply called the dnaB and dnaC proteins (Figure 13-10).

Inside cells, the locations (and the sequences) of the RNA primers formed by the primase are not random. Special recognition proteins called N proteins somehow choose sites at which the primase can act (Table 13-1). For example, the products of the bacteriophage T7 gene 4 protein, which serves as a primase (and as an unwinding protein), are the RNA oligonucleotides

(5′)ACCA(3′) and (5′)ACCC(3′)

(Table 13-1). Probably any complementary DNA site (TGGT or GGGT) could serve as a binding site for the T7 enzyme and allow priming of DNA synthesis from that point. As the leading strand progresses from 5′ to 3′, additional sites for the initiation of lagging-strand synthesis are exposed, and short discontinuous DNA segments are initiated opposite them by new primer RNAs.

As we have mentioned, Okazaki fragments have been found in replicating DNA in animal cells as well as in bacteria. The study of the replication of small DNA viruses such as the polyoma virus of rodents and SV40 virus of monkeys has yielded very clear evidence of primer function in DNA formation in animal cells. All the primers made on polyoma virus DNA are about 10 bases long, and all begin with an adenylate or a guanylate; however, the primers are not identical in sequence (Table 13-1).

Although the mechanism of RNA priming in the initiation of DNA synthesis is somewhat different for the class of RNA viruses termed retroviruses (Chapters 6 and 11), an RNA primer is still required for DNA synthesis. These virus particles contain a DNA polymerase that copies

Table 13-1 RNA primers (primase products) used in DNA chain initiation

Replicating DNA	RNA oligonucleotide*
Bacteriophage T4	pppAC $(N)_3$
Bacteriophage T7	pppACCA pppACCC
Mouse polyoma virus	pppA $(N)_9$ pppG $(N)_9$
Lymphoblastoid cells	pppA $(N)_8$ pppG $(N)_8$

*N = any ribonucleotide. The indicated primer lengths for the mouse polyoma virus and the animal cells are averages.

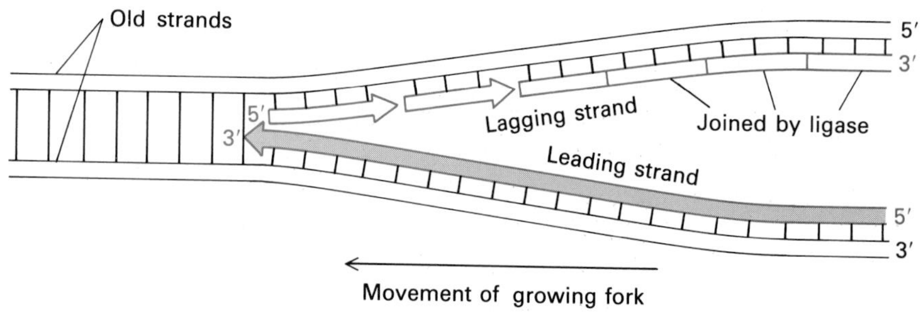

Figure 13-8 Discontinuous DNA synthesis is necessary in one strand of DNA at each growing fork. All known DNA polymerases synthesize DNA by adding nucleotides to a 3′ hydroxyl (here, arrowheads are at 3′ ends), thereby causing growth in a 5′ → 3′ direction. The fact that the overall movement of a growing fork is in one direction means that only one new DNA chain, the leading strand, can be made continuously in a 5′ → 3′ direction. The other new DNA strand, the lagging strand, is made by discontinuous, piecemeal synthesis (also 5′ → 3′ but opposite to the direction of movement of the growing fork) and eventual ligation of the pieces.

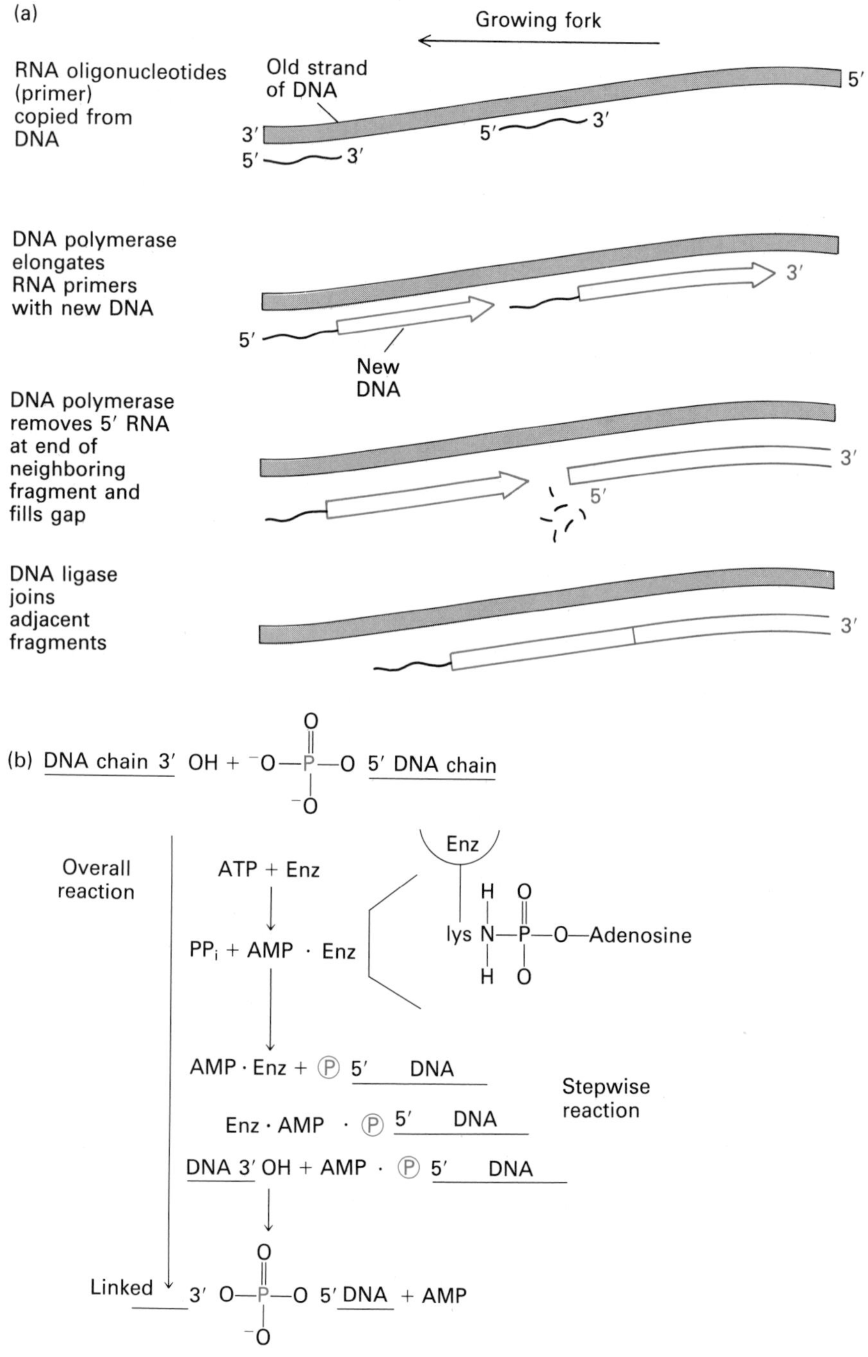

(a)

Growing fork

RNA oligonucleotides (primer) copied from DNA — Old strand of DNA; 3′, 5′; 5′—3′; 5′—3′

DNA polymerase elongates RNA primers with new DNA — New DNA; 5′; 3′

DNA polymerase removes 5′ RNA at end of neighboring fragment and fills gap — 3′; 5′

DNA ligase joins adjacent fragments — 3′

(b) DNA chain 3′ OH + $^{-}$O—P(=O)(O^{-})—O 5′ DNA chain

Overall reaction

ATP + Enz
↓
PP_i + AMP · Enz [Enz—lys N(H)(H)—P(=O)(O)—O—Adenosine]
↓
AMP · Enz + Ⓟ 5′ DNA
Enz · AMP · Ⓟ 5′ DNA
DNA 3′ OH + AMP · Ⓟ 5′ DNA
↓

Stepwise reaction

Linked: 3′ O—P(=O)(O^{-})—O 5′ DNA + AMP

Figure 13-9 Lagging-strand synthesis. (a) DNA polymerases require a free 3′ hydroxyl end on a nucleic acid chain in order to add nucleotides; in the synthesis of the lagging strand of DNA at a growing fork, this free 3′ end is supplied by RNA primers. As each growing Okazaki fragment (each segment of RNA plus DNA) meets the RNA primer of the adjacent Okazaki fragment, DNA polymerase digests the primer and fills the gap with DNA. (The substrate for DNA synthesis is the deoxyribonucleoside triphosphate, and PP_i is split off; see Chapter 4.) Another enzyme, DNA ligase, then joins the lagging-strand fragments. (b) DNA ligase carries out the overall reaction linking the 3′ OH end of one DNA chain to the 5′ phosphate at the end of another chain. The reaction involves attaching an AMP residue covalently to the enzyme and consumes an ATP in the process. The activated AMP-enzyme complex is momentarily covalently attached to the 5′ phosphate of the DNA, activating this Ⓟ; the AMP is then liberated when the 3′ OH end of the DNA is joined to the activated phosphate.

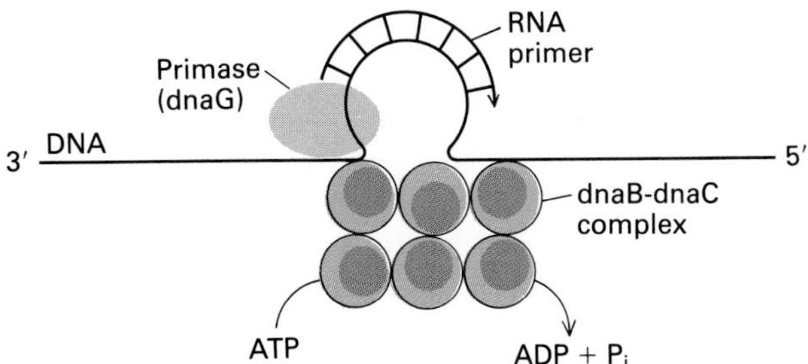

Figure 13-10 The *E. coli* primosome. The primase (the dnaG protein) plus a complex of the dnaB and dnaC proteins can bind to DNA and synthesize RNA primers on any single-stranded DNA. This structure is called a primosome. Additional factors called N proteins (not shown in the diagram) somehow direct the primosome to specific sequences. Hydrolysis of ATP generates the movement of the primosome along the DNA as it searches for specific sequences. *After A. Kornberg, 1982,* 1982 Supplement to DNA Replication, *W. H. Freeman and Company, p. S105.*

their RNA genome into DNA. This enzyme, a reverse transcriptase, begins deoxyribonucleotide polymerization on the 3′ end of a tRNA molecule that is bound to the virus RNA. Different retroviruses use different tRNA molecules as primers.

Bidirectional Replication Bubbles May Form after Initiation of Synthesis on One DNA Strand

The mode of DNA chain growth at a growing fork—continuous elongation of the leading strand and discontinuous, primer-assisted synthesis of the lagging strand—holds the key to the formation of replication bubbles. Any mechanism that starts the synthesis of one new leading strand will automatically generate sites on the opposite DNA strand at which RNA priming and lagging-strand synthesis can begin. Examples of such mechanisms include a nick (a single-strand cut) that exposes a 3′ hydroxyl end of DNA or the initiation of an RNA primer. Once both leading- and lagging-strand synthesis are underway, one growing fork has been created.

A second growing fork moving in the opposite direction will then automatically arise from the same site. For example, synthesis of the leading strand at one growing fork (the rightward-growing fork in the diagram in Figure 13-11) might begin by extension of an RNA chain (either one produced at that site or one hybridized to that site). The growth of the leading strand makes available a new single-stranded region of DNA. Eventually, a site appears in this region that can be primed for synthesis of the lagging strand. When the lagging-strand segment closest to the origin extends past the origin, this lagging strand for the rightward-growing fork can continue as a leading strand for a second, leftward-growing fork. As this leading strand for the leftward-growing fork grows, lagging-strand sites are exposed. By appropriate "patching" (cutting away of primers and gap filling) followed by ligation, a new copy of each nucleotide in the original duplex is created, including the nucleotides at the point of origin.

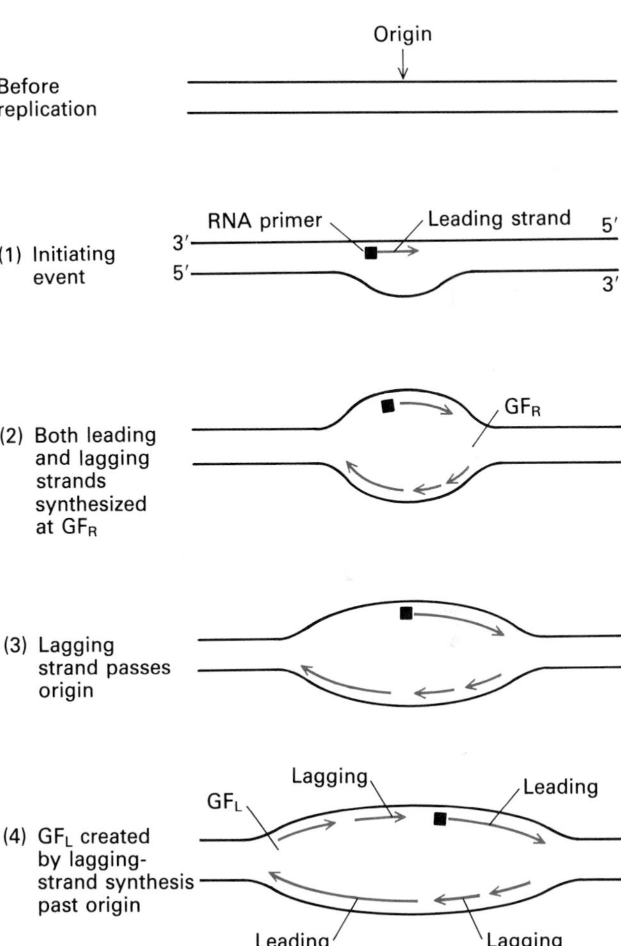

Figure 13-11 The opening of DNA for bidirectional replication: DNA chain initiation in one direction can lead to two growing forks. (GF$_R$ = rightward-growing fork; GF$_L$ = leftward-growing fork.)

(1) An RNA primer initiates synthesis of the leading strand of GF$_R$ at the origin.
(2) Segments of the lagging strand of GF$_R$ are initiated by short RNA primers.
(3) The lagging strand of GF$_R$ then passes the origin; it becomes the leading strand of GF$_L$.
(4) Segments of the lagging strand of GF$_L$ are initiated by short RNA primers. Ligation (not shown) of all pieces completes duplication of DNA.

Cells Have Multiple DNA Polymerases

Escherichia coli has three different DNA polymerases: I, II, and III. All of them may play important roles in correctly synthesizing DNA chains during replication or during the repair of damaged DNA chains (Table 13-2). One of them, however, *polymerase III,* is most conspicuous in DNA synthesis during genome replication. This enzyme has the fastest rate of chain elongation in vitro, approaching 100 percent of the in vivo rate. Mutations in the gene encoding polymerase III can result in the production of temperature-sensitive enzymes. When bacteria carrying a mutation of this type are raised to a high temperature, DNA replication stops immediately, showing clearly that this enzyme has a central role in *E. coli* DNA replication.

Both DNA polymerases I and III have the $5' \rightarrow 3'$ exonuclease activity that can be used to remove RNA primers. Polymerase I mutations that disable only the $5' \rightarrow 3'$ exonuclease function are also lethal to the bacterial cell. Thus it is thought that polymerase III is responsible mainly for fork movement and that polymerase I is most active in gap-filling on the lagging strand. In any case, the two enzymes are not simply interchangeable (Table 13-2).

Animal cells, too, contain three different DNA polymerases: α, which is found in the nucleus only; β, which appears in both the nucleus and the cytoplasm; and γ, which is located primarily in the mitochondria. The activity or number of α polymerase molecules fluctuates throughout the cell cycle, in proportion to the rate of DNA synthesis; therefore the α polymerase may be the most important enzyme in DNA replication. The drug aphidicolin, which inhibits partially purified α polymerase but not β or γ polymerase, stops DNA synthesis in growing cells. Mutant cells that are able to grow in aphidicolin have an α polymerase that is resistant to the drug. These experiments do not prove that the β and γ polymerases have no role in DNA replication, but they have established that the α polymerase is a necessary enzyme for DNA replication. The γ DNA polymerase may be responsible for mitochondrial DNA synthesis.

Like the bacterial polymerases I and III, each of the three animal cell enzymes requires a primer—a free 3′ hydroxyl on one strand of a duplex nucleic acid—in order to synthesize DNA.

Table 13-2 Properties of DNA polymerases

E. coli	I	II	III	
Polymerization: $5' \rightarrow 3'$	+	+	+	
Exonuclease activity: $3' \rightarrow 5'$	+	+	+	
$5' \rightarrow 3'$	+	−	+	
Synthesis from:				
Intact DNA	−	−	−	
Primed single strands	+	−	−	
Primed single strands plus single-strand-binding protein	+	−	+	
In vitro chain elongation rate (nucleotides per minute)	600	?	10,000	
Molecules present per cell	400	?	10–20	
Mutation lethal?	+	−	+	
Animal cells	α	β	γ	**Retrovirus**
Polymerization: $5' \rightarrow 3'$	+	+	+	+
Exonuclease activity: $3' \rightarrow 5'$ (editing function)	−	−	+	−
Requires primer?	+	+*	+	+
DNA primer, RNA template	−	−	−	+
RNA primer, DNA template	+	−	−	+
DNA primer, DNA template	−	−	+	+
RNA primer, RNA template				+
Increases during S phase?	+	−	−	−
Sensitive to aphidicolin (inhibitor of cell DNA synthesis)?	+	−	−	−
Cell location	Nuclei	Nuclei, cytoplasm	Mitochondria	Virion

*Polymerase β is most active on DNA molecules with gaps of about 20 nucleotides and is thought to play a role in DNA repair.

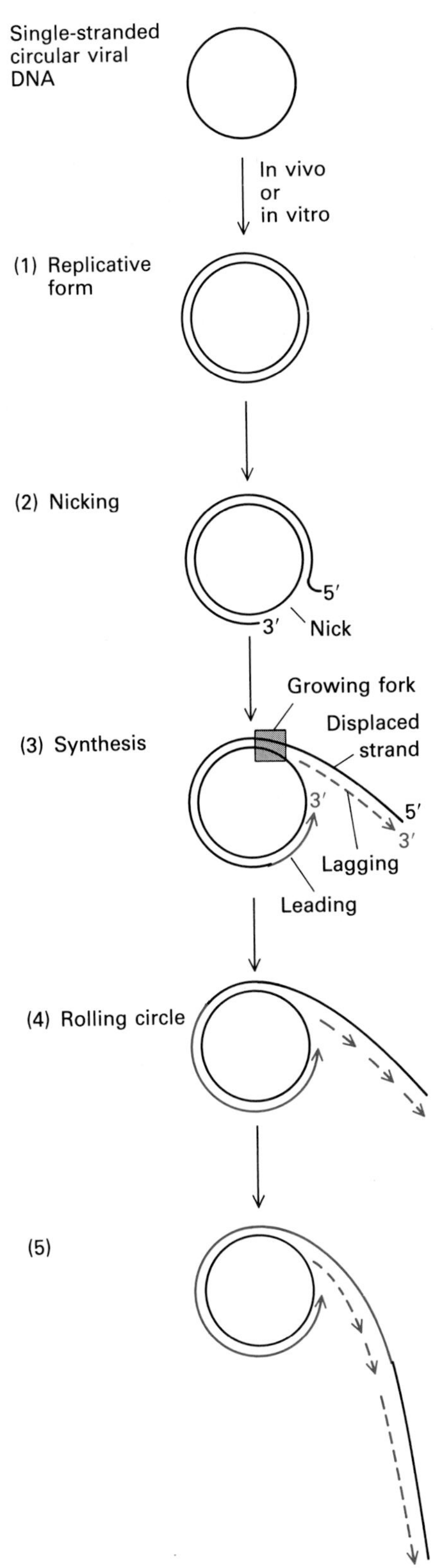

Figure 13-12 The single-stranded circular DNA of small *E. coli* viruses such as ϕX174 and M13 is converted into replicative forms, which are covalently closed circular duplexes. One nick (a single-strand cut) is made with a DNase (ϕX174 encodes a protein that nicks its replicative form at a particular site and becomes covalently attached to the DNA at that site). A new strand (the leading strand) grows from the 3′ hydroxyl at the nick and displaces the other end of the nicked strand, whereupon lagging-strand synthesis can begin on the displaced end. Thus a growing fork with leading- and lagging-strand synthesis is created by rolling circle replication.

A Growing Fork Can Be Constructed from Many Proteins in the Test Tube

The entire apparatus for the synthesis of DNA at a growing fork has been reconstructed in the test tube with purified proteins from both bacteriophage-infected and uninfected bacterial cells. The intricate details of the functions of most of the 10 to 20 proteins in each of these systems have been resolved, and the cooperative interactions of the proteins can be observed. Many of the events at the growing fork are probably carried out by a multienzyme complex sometimes referred to as a *replisome*.

The whole *E. coli* chromosome and even the large bacteriophage chromosomes are not easy to work with in such enzymatic experiments; the DNA used most often in the study of growing forks is the double-stranded replicative form of a small bacterial virus such as M13 and ϕX174. The DNA of these viruses exists in the virion as a single-stranded circular molecule that is converted by host-cell enzymes into a DNA duplex, the *replicative form* (step 1 in Figure 13-12). This first step of in vivo replication can now be carried out completely in vitro. The double-stranded replicative form is a convenient, fairly small (approximately 6-kb) template for the in vitro analysis of DNA synthesis at the growing fork.

The replication of such circular phage DNA in *E. coli* is carried out by *E. coli* enzymes. Therefore, purified proteins that copy the circular duplex DNA can be obtained from uninfected *E. coli* cells. In addition, *E. coli* cells infected with the large bacteriophage T4 or the medium-sized bacteriophage T7 can serve as sources of DNA-replicating enzymes. These enzymes are entirely different from the *E. coli* enzymes, because these larger bacteriophages encode proteins that can replicate their own DNA (Chapter 6).

Each of the proteins involved in DNA synthesis at a growing fork was identified and purified by first isolating mutant *E. coli* cells or mutant T4 bacteriophages that could not make DNA. A series of such mutants was collected (the mutant bacterial genes were named A, B, C, D, etc.; the mutant T4 phage genes were numbered 41, 43, etc.). Extracts of mutant cells or cells infected with mutant virus fail to carry out DNA synthesis, but extracts of normal cells or cells infected with normal T4 phage can. Furthermore, the normal extract or protein fractions from it can "complement" (in the manner of genetic complementation; Chapter 10) the inactive extracts. Thus protein fractionation techniques plus the in vitro complementation assay have allowed the purification of the proteins necessary in the synthesis of DNA.

The replication system assembled from proteins encoded by bacteriophage T4 was the first to be completely reconstructed; this system is somewhat simpler than the one derived from *E. coli* proteins. However, each of the events that occurs in the T4 system has a counterpart in *E. coli* DNA replication, though the proteins differ.

The purified T4 proteins are added to the mixture containing the replicative form of the M13 or ϕX174 DNA. As shown in the second step of Figure 13-12, the DNA is then nicked (cut in one strand). No synthesis is observed without the nick, although the nick site can be random for in vitro synthesis. (In the true intracellular replication of the virus ϕX174, in contrast, a nick is placed at a specific site of one strand of the replicative form DNA by a specific viral protein.) Next (step 3), the free 3′ hydroxyl end at the nick site is elongated by the T4 DNA polymerase, using the unnicked inner circle as the template; this forms the leading strand of the growing fork. Simultaneously, a complex group of proteins synthesizes the lagging strand from the old strand that has been displaced. This completes the growing fork; the double-stranded *linear* product (step 5) has one new and one old strand. The growing fork can continue to move around and around the circle, generating longer and longer DNA. This process is called *rolling circle replication.*

Keep in mind that the sequence of events we have just described is an artificial system (M13 DNA and T4 enzymes) in which the mechanics of a growing fork are under study; the product is not a finished, viable DNA molecule. Rolling circle replication does occur in nature—for example, in infection by a number of different bacteriophages and in the amplification of frog rDNA (Chapter 11). To make use of the long ribbons of DNA produced in this manner, the infected bacterial cells or the frog cells must have specific enzymes that recognize and cut specific sites on the DNA so that regular units of new DNA result.

The Seven Proteins Required for Replication of Bacteriophage T4 DNA At the growing fork in the reconstructed phage T4 DNA replication system, seven polypeptides are required for DNA synthesis (Figure 13-13). The T4 genes that encode these polypeptides are known, and they have been given numbers on the bacteriophage genetic map. The product of gene 41 is a DNA *helicase,*

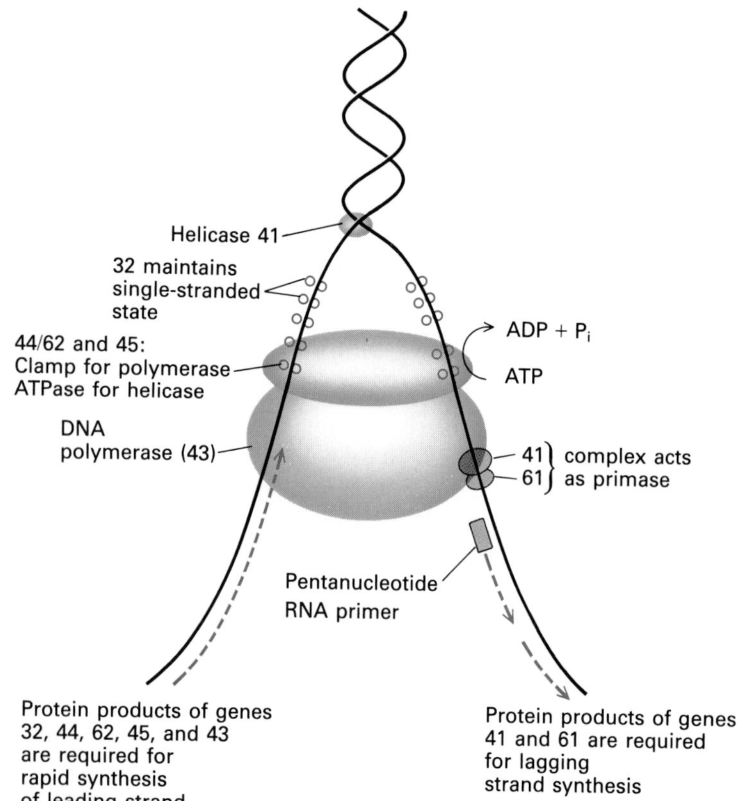

Figure 13-13 An in vitro replication fork created by proteins encoded by seven bacteriophage T4 genes. [See B. M. Alberts et al., 1980, *ICN-UCLA Symp.* **19**:449; and C. C. Huang, J. E. Hearst, and B. M. Alberts, 1981, *J. Biol. Chem.* **256**:4087.]

which uses energy derived from the hydrolysis of ATP to unwind the DNA helix so that the resulting single strands can be copied. The product of gene 32 is a single-stranded DNA–binding protein that keeps unwound chains single-stranded. The gene 43 protein is the T4 DNA polymerase molecule. The helicase from gene 41 also complexes with the gene 61 product; both are necessary for lagging-strand synthesis. The gene 61 protein is thought to have the primase activity that produces pentanucleotides beginning with AC residues by transcription of DNA. These serve as primers for lagging-strand synthesis (see Table 13-1). (A DNA ligase is encoded by gene 30, but its action is not necessary for the in vitro fork movement.)

Accessory proteins enable the polymerase (the gene 43 product) to do an efficient job. In addition to the gene 32 protein (the single-stranded DNA–binding protein), two other accessory proteins are the products of genes 44 and 62. These two proteins form a complex that binds to DNA along with the product of gene 45 and the DNA polymerase. The combination of the 44/62 complex and the 45 protein has ATPase activity. That is, these three proteins catalyze the hydrolysis of ATP to ADP, which provides the energy for the helicase function. The ATPase activity depends on the presence of free ends in DNA chains. The three ATPase proteins plus the gene 32 protein markedly stimulate the rate of DNA synthesis by the DNA polymerase. Apparently the 44/62 and 45 combination acts as a "clamp" to maintain the continuous activity of the polymerase; when this group of proteins is present, the polymerase does not disengage from the template.

The whole assembly of proteins probably remains intact throughout rapid DNA synthesis. When all the proteins necessary for leading-strand synthesis (43, 44/62, 45, and 32) are present, the rate of synthesis is about 800 nucleotides per second. When the complex of 41 and 61 is added, both strands are formed at a similar rate; this is as fast as DNA synthesis occurs inside cells.

The Replisome in *E. coli* As we mentioned above, many small bacteriophages that do not have sufficient information to encode their own DNA-duplicating machinery use the host-cell enzymes. In a replication system that has been constructed, the replicative forms of these small viruses are used as templates by a complex of about 20 different proteins extracted from uninfected bacterial cells. These proteins include two DNA polymerases, I and III. In addition, the replicative proteins include the primosome (primase, dnaB, and dnaC), which initiates primer synthesis, and primer recognition factors (the N proteins), which cause priming at specific sites. All the proteins necessary in vitro for replication of the M13 virus also appear to have a role in vivo in the replication of the *E. coli* chromosome, because mutations in the genes encoding these proteins cause various defects in *E. coli* DNA replication.

From information of this sort, a generalized scheme linking all necessary activities for *E. coli* DNA replication has been devised (Figure 13-14). At the growing fork there is a giant multiprotein complex containing at its advancing edge an unwinding protein plus a primosome—the collection of proteins that starts an RNA primer for the lagging strand. The *replisome,* the suggested collective name for all of the associated proteins needed for replication, also includes DNA polymerase III at the site of deoxynucleotide polymerization. It has been proposed that a loop formed in the denatured lagging-strand template when it passes around the replisome brings the two sites of deoxynucleotide polymerization—one for the leading strand and one for the lagging strand—close together. (The two polymerization sites therefore are not at equivalent points on the leading and lagging strands.) By this mechanism, the same polymerase III enzyme dimer could carry out elongation of both leading and lagging strands without dissociating from the growing fork. Such a *processive* (continuously engaged) enzyme complex might account for the great speed of movement at the growing fork.

Conclusion

In spite of the brilliant demonstrations that T4, T7, and *E. coli* enzymes can replicate small bacteriophage templates, several central issues in DNA replication have not been completely resolved. As yet, most of the systems reconstructed in vitro do not start synthesis of DNA at the correct site or in the correct manner. In fact, how the start site is located is not clear in most cases. Furthermore, the mechanisms for producing two *untangled* daughter molecules are just beginning to be studied. Finally, the events regulating the initiation of DNA replication are only vaguely understood.

Superhelicity in DNA; Topoisomerases

Thus far in our consideration of DNA replication we have dealt with "local" issues concerning how enzymes might manage to establish and then to propagate a growing fork. Newer information has allowed two additional aspects of chromosome replication to be considered: (1) the importance of chromosome folding and twisting—that is, of the "higher-order" DNA structure; and (2) the means by which replication might be initiated inside cells. As we have discussed in Chapter 10, chromosomes within eukaryotic cells are not simple, free-floating bits of DNA; rather, they have proteins bound to them to form nucleosomes and solenoidal stacks of nucleosomes. The impor-

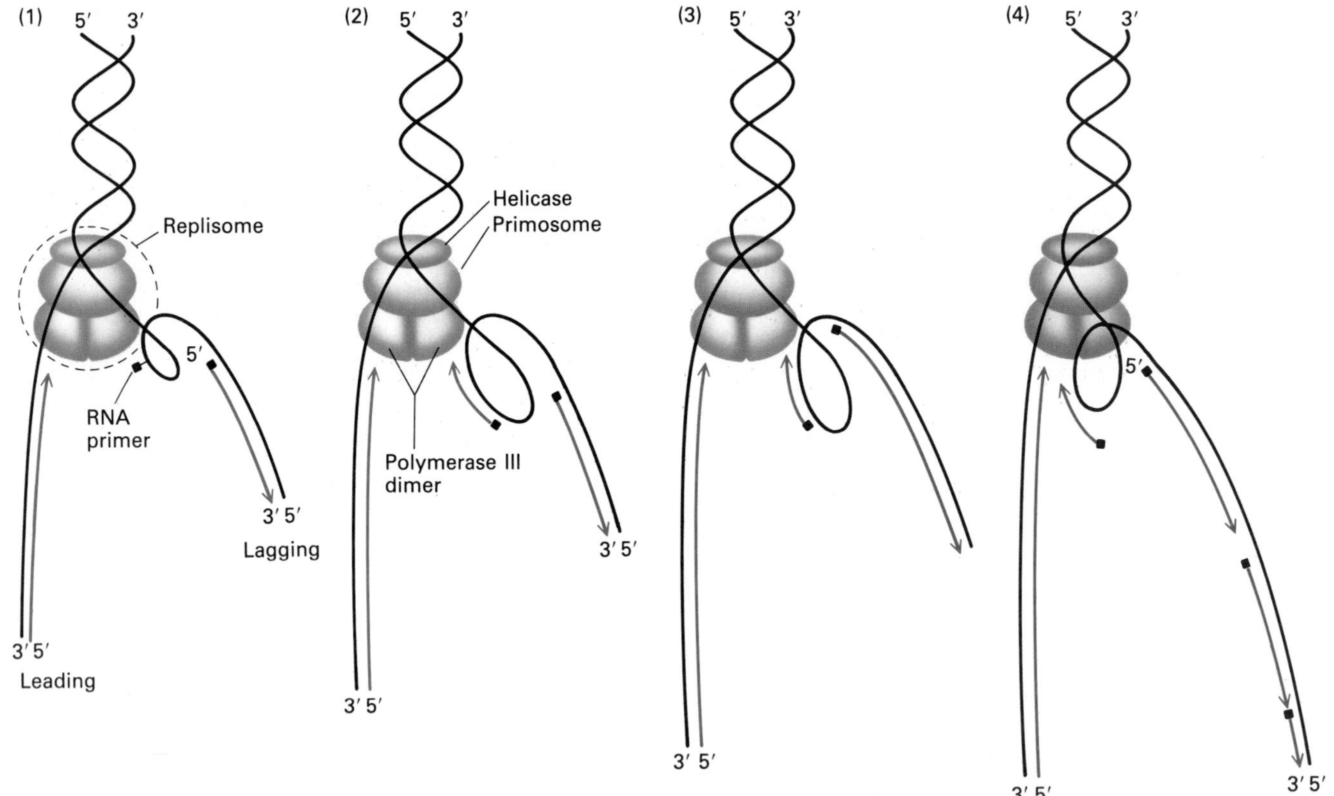

Figure 13-14 A scheme for DNA synthesis by a single advancing multienzyme complex, the replisome. Note that the loop in the lagging strand makes it possible for one enzyme complex to add nucleotides to both growing strands even though the template strands are antiparallel in the duplex. The steps in synthesis are described in the text. *After A. Kornberg, 1982,* 1982 Supplement to DNA Replication, *W. H. Freeman and Company, p. S125.*

tant effects of these proteins on the twisting and packing of DNA must be taken into account in any consideration of the functions of chromatin inside cells.

Soon after the discovery of the double helix, it was recognized that the structure complicated replication. Replication of the two interwound DNA strands would be impossible without a "swivel" point of some sort. A similar topological problem would possibly exist during transcription if strand separation were required. The necessity for a swivel can be visualized by thinking of pulling apart the twisted strands of a piece of rope or twine (Figure 13-15). If the two ends are attached to fixed structures or to each other, as in a circle, it would be impossible to separate the strands. Partial separation in one region would be accompanied by tighter twisting in another region.

If circular strands are being separated, or if the two anchor points of a linear structure cannot rotate, the structure will have an additional response: it will rotate on itself to redistribute the strain. Such a rotation produces *supercoils* or *supertwists* (see Figure 3-58). Supercoiling of DNA, which occurs widely in nature, first came to light in the study of the small DNA animal viruses polyoma and SV40. Studies of small viral chromosomes continue to contribute important informaton about the superstructure of chromosomes, and we shall discuss recent findings in some detail.

If the DNA is both helical and supercoiled, how is tangling prevented during replication? A newly discovered class of enzymes termed *topoisomerases* have the capacity to deal with this problem. Some topoisomerases (type I) can produce changes in the supercoiling by catalyzing the nicking (single-strand cutting) of the DNA, which allows the rotation of a free end, followed by the resealing of the nick. Other topoisomerases (type II) alter the supercoiling in order to facilitate chromosome folding and twisting. An enzyme of this type cuts both strands of one DNA double helix so that a neighboring region of helix can pass through the cut ends. The enzyme then reseals the ends. By this maneuver, knots that may have accumulated can be untied. Very likely the topoisomerases are involved in DNA fork movement, in the initiation of DNA synthesis, and in the final release of the two completed chromosomes after replication is finished.

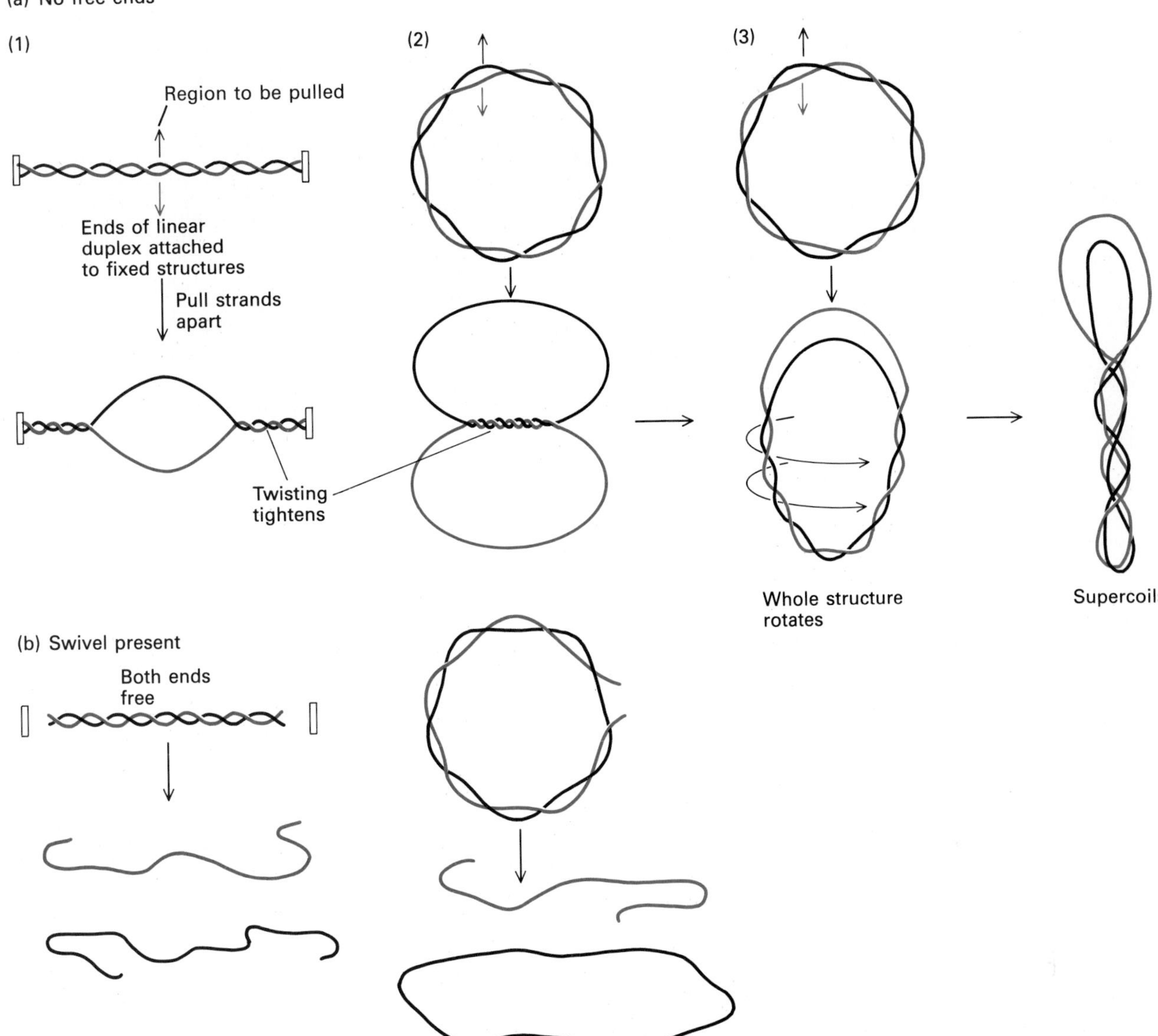

Figure 13-15 (a) In duplex DNA with no free ends, twisting increases in one region when strands are pulled apart in another region *(frames 1 and 2)*. An alternative response to the separation of strands is for the whole structure to rotate on itself to relieve the strain *(frame 3)*. (b) In the linear molecule, both ends are free, so that separation can occur. A break in one strand of the circular structure allows the free end *(colored strand)* to swivel around its partner *(black strand)*, eventually freeing both (see also Figures 3-57 and 3-58).

Superhelicity Was Discovered in SV40 and Polyoma DNA

The *superhelicity* of DNA (another term for supercoiling) was discovered by electron microscopy and by sedimentation analysis of the small DNA molecules of SV40 and polyoma, viruses that occur naturally in monkeys and rodents. When DNA from these viruses is freed from protein very carefully and examined in the electron microscope, most of the DNA molecules are supercoiled (Figure 13-16). Occasional molecules are not supercoiled but appear rather as open circles. A very brief exposure of supercoiled molecules to a quantity of DNase sufficient to place one nick in one strand of the DNA results in molecules that all appear as open, double-stranded circles. These are called *relaxed* circles. The unnicked, native molecule is said to be in form I, and the nicked, relaxed molecule in form II. The molecule is called form III when the circle is cut across the double helix to form a linear duplex.

Sedimentation analysis of the three types of DNA, first at neutral pH and then at increasingly alkaline values, helped researchers to understand the structures of the three forms (Figure 13-17). The linear form III in neutral

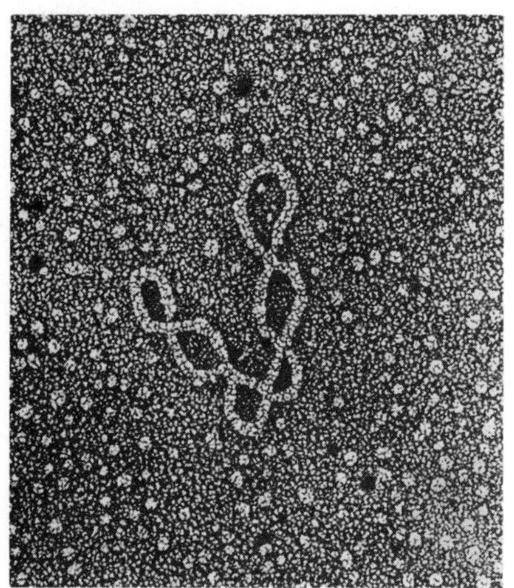
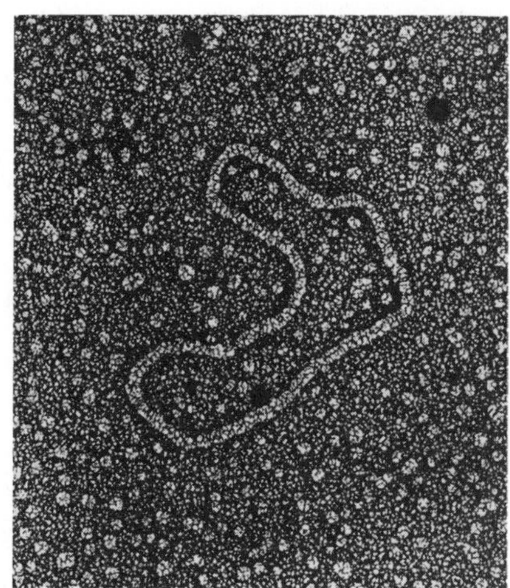
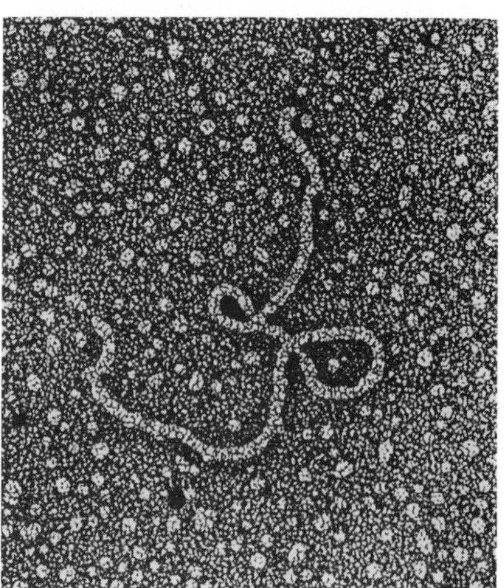

Figure 13-16 Electron micrographs of different SV40 molecules. Form I (native state): supercoiled circle. Form II (relaxed): open duplex circle produced by a nick in one strand. Form III: linear duplex produced by cleavage of both strands. *Courtesy of M.-T. Hsu.*

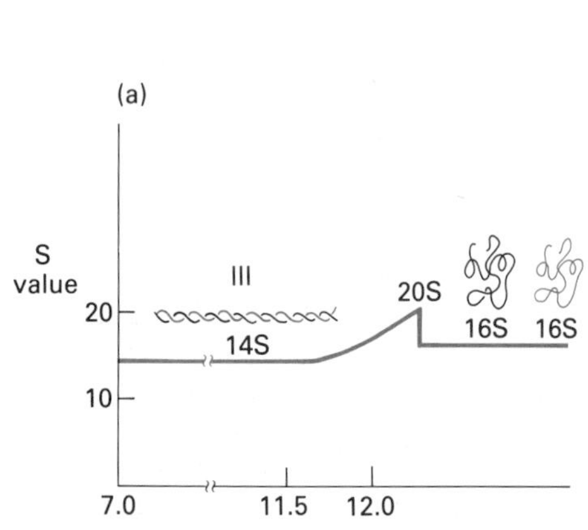

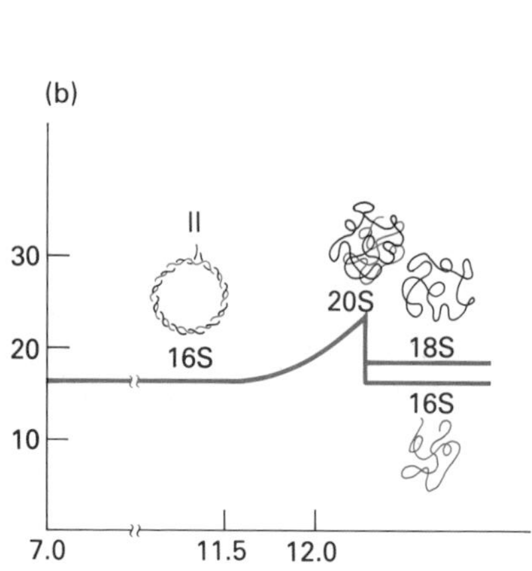

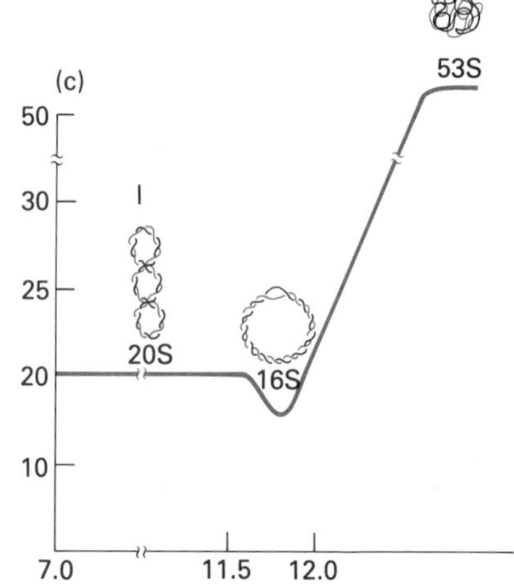

Figure 13-17 The various forms of SV40 DNA as deduced from centrifugation and microscopic studies. (a) Form III is broken once in both strands; its linear structure sediments at 14S and denatures in alkaline solution into two strands. As the strands denature the single-stranded portions become more and more compact; just before the duplex completely separates the highest sedimentation rate (20S) is reached. The denatured strands form such a compact structure that they sediment faster (16S) than the rigid linear double strand, which actually has twice the molecular weight of one single strand. (b) Form II has a nick in one strand; therefore it, too, denatures in alkaline solution into two strands. Again the highest S value occurs just before the two strands separate. The circular single strand sediments slightly faster than the nicked, linear single strand. (c) Form I is freshly purified SV40 DNA. It has no nicks, and its sedimentation rate is 20S in neutral solution (pH 7.0). As the pH is raised, the S value of the molecule first decreases and then increases rapidly. The initial decrease is due to the unwinding of a few bases in the double-stranded helix, which is compensated for by the loss of the supercoils. The subsequent increase occurs because the covalently closed double strand becomes more tangled and compact the more it denatures. [See J. Vinograd et al., 1965, *Proc. Nat'l Acad. Sci. USA* **53**:1104.]

solutions has a sedimentation rate (an S value) of about 14S; the nicked circular form II has an S value of 16S. When either of these two types of DNA is exposed to an alkaline solution (pH 12.3 or greater), it denatures, releasing two single-stranded molecules. When form III denatures in this manner, two linear strands with sedimentaton rates of 16S are released. When form II is denatured, the products are a linear 16S and a circular 18S. Recall that the S value of a molecule depends both on molecular weight and on shape (Chapter 7), which explains the similar S values of the rather rigid initial double-stranded molecules and the single-stranded products, which can assume compact shapes and thereby be subject to less viscous drag during sedimentation. Likewise, the slight difference in shape between the circular single-stranded product and the linear single-stranded (nicked) product explains the slight difference between their sedimentation rates (18S versus 16S, respectively).

Form I, the predominant molecule in DNA freshly isolated from virions, has a higher S value, 20S, than the nicked form II. Because the two molecules have the same molecular weight, form I must be more compact than form II. The greater compactness is evident in the electron micrographs showing that form I is supercoiled (see Figure 13-16).

Careful analysis of the sedimentation value of form I in increasingly alkaline solutions has shown an initial drop in sedimentation rate at about pH 11.8, indicating a conversion into a less compact structure. The change has been interpreted as a relaxation of the supercoils; the energy for this relaxation apparently came from a "loosening" of the double helix as the denaturation pH approached. Thus a connection between the degree of winding of the duplex and its supercoiled nature was suggested (see Figure 3-58).

With further increases in pH, the sedimentation rate of form I rose rapidly. This occurred because the covalently closed double strand could not completely denature into two strands, but denatured instead into a tangled mass which became quite compact.

Linking Number, Twist, and Writhe Describe DNA Superstructure

The topology of a DNA molecule can be described by three parameters. First is the *linking number* of the DNA. This number, always an integer, is a property of the whole molecule. The linking number is equal to the number of times one strand of the helix crosses the other within the boundaries being considered (i.e., within a circular molecule, within a linear molecule with anchored ends, or within a linear region of a molecule between fixed sites).

The second parameter is *twist*, which is related to the frequency or periodicity of the winding of one strand about the other. Twist can vary from segment to segment within a molecule; it is a complicated function that cannot be described by a simple number. In purified, isolated DNA at normal physiological conditions of salt and temperature, an *average* of one right-handed helical turn (and one crossover) occur every 10.4 bases. Possibly the twist of DNA inside cells is not always uniform over short distances, because the DNA is associated with protein.

A circular DNA molecule like SV40, which is about 5200 bases long, would be expected to have a linking number of about 500 under the salt and temperature conditions found in cells; if this were so, the circle would be "flat" (relaxed) (Figure 13-18a). SV40 is supercoiled however when extracted from virions.

The accepted explanation for the supercoiling of SV40 is that the DNA is *underwound* by about 25 turns; that is, its linking number is about 475. In effect, an untwisted region should be left when the DNA is purified from association with protein (Figure 13-18b). There are several possible adjustments the molecule could then make: if one chain could be broken, the helix would be wound another 25 turns. Without a chain break, however, either the average twist in the molecule would have to change, or the molecule would coil on itself, producing supercoils (Figure 13-18c and d). Supercoils occur when the temperature and salt concentrations are normal, because DNA stabilizes at an average twist of 10.4 bases per turn under these conditions. The linking number would not change during the formation of the supercoils, nor would the average twist change. The supercoils that form to compensate for the effects of underwinding in a helix are said to be *negative* supercoils. An overwound SV40 helix—that is, one with more than 10.4 base pairs per turn and a linking number greater than 500—would tend to form supercoils in the opposite direction. These would be called *positive* supercoils (see Figure 3-58).

The supercoiling phenomenon, which is reciprocally related to twist when there is no change in linking number, is an expression of the third parameter that governs the overall structure of the helix. This parameter is called *writhe*. The exact definitions of twist and writhe would require a complex topological or geometric statement, and we shall not attempt any mathematical description here. In simplified terms, L (the linking number) = T (twist) + W (writhe). Briefly, then, twist is related to the frequency of turns around the central axis of the helix within any given region, and writhe to the pathway in space of the axis of the helix. Unlike the linking number, which is a property of the whole molecule, twist and writhe can vary in different parts of the molecule, but they are always in a reciprocal relationship (Table 13-3).

The relationship between linking number and the degree of superhelicity of molecules is experimentally demonstrated in Figure 13-19. Fully supercoiled SV40 DNA can be slowly digested with a topoisomerase to remove supercoils one at a time. When the mixture is sampled at various intervals in the course of the reaction and the

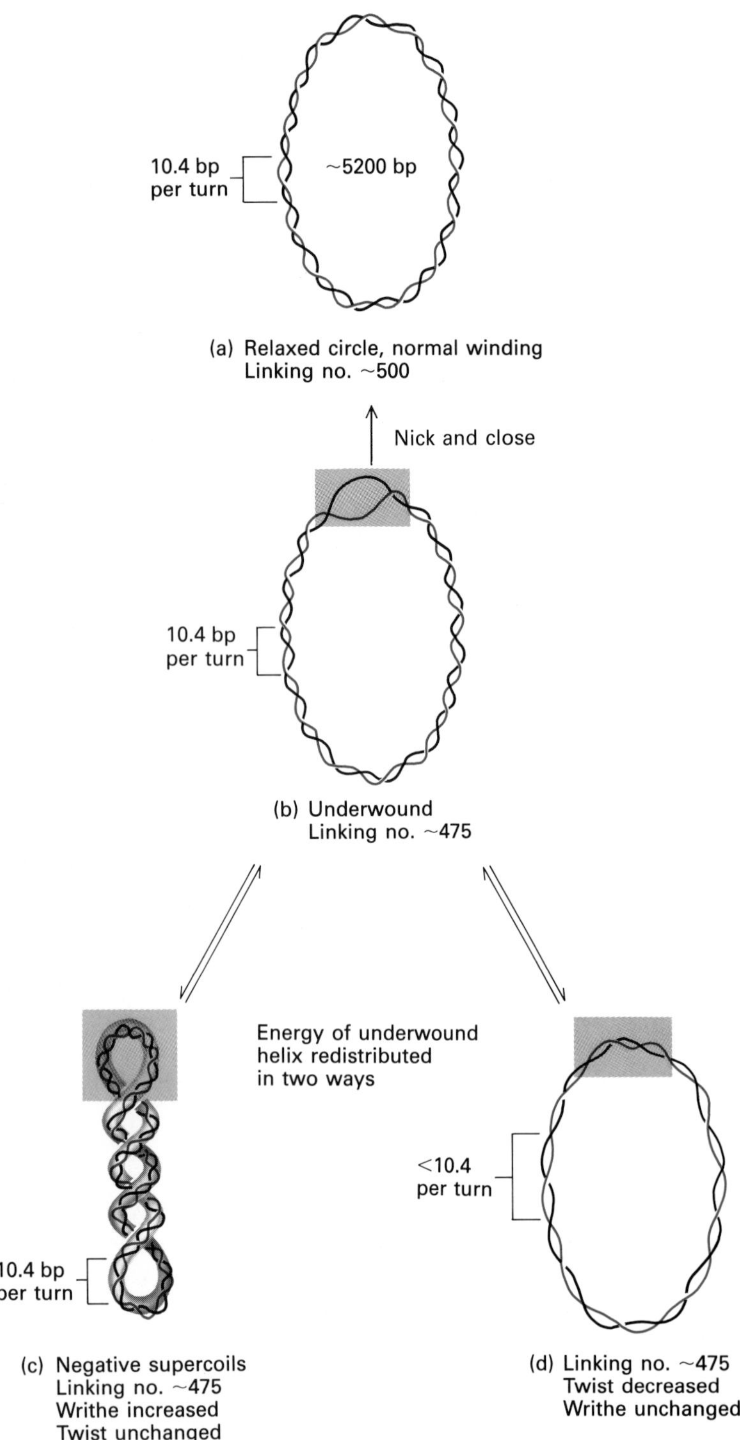

Figure 13-18 Different states of SV40 DNA show the interrelationship of twist and writhe (see text). Because the DNA, which is about 5200 bases long, forms a helix with about 10.4 base pairs per turn, it would have a linking number of about 500 (5200 ÷ 10.4) in the relaxed state shown in a. However, as we saw in Figure 13-16, the DNA from the virions is supercoiled.

The explanation can be found in (b) and (c). The virion DNA is in fact underwound so that its linking number is about 475. [To convert (b) into (a) would require the breaking of a DNA chain, additional winding of the broken chain about the intact chain, and resealing of the broken chain.] The underwinding is represented here as a nonhelical space in the DNA *(shaded)*. Molecular forces cause this underwound double helix to assume the structure shown in (c). Because the nonhelical region in (b) does attempt to form a helix with 10.4 bases per turn, the strain causes the axis of the DNA backbone to curve in space, and negative supercoils result (see also Figure 3-58).

A different model that would relieve the situation in (b) appears in (d): the underwinding would be distributed evenly throughout the molecule, so that the number of bases per turn would be slightly less than 10.4. No change in writhe would occur, and the average twist would decrease. Both examples (c) and (d) are consistent with the known reciprocal relationship between twist and writhe; however, under normal salt and temperature conditions, the supercoiling in (c) results.

DNA molecules are separated by gel electrophoresis, molecules with between zero and about 25 supercoils are found. All viruses with circular DNA, whether they be bacterial, plant, or animal, contain supercoiled DNA, and all DNA from natural sources is negatively supercoiled. An important point to remember is that increased negative supercoiling can be converted into unwinding of the helix, and vice versa. Thus the melting of short stretches of DNA may occur to enable proteins that recognize single strands to bind.

What is the basis for the supercoiling observed when SV40 DNA is freed from association with proteins? When the SV40 molecule is released gently from infected cells it is associated with histone octomers acquired during its replication. Like the chromosome of its eukaryotic host, the SV40 DNA is wound around the histones to

Table 13-3 Physical and biochemical effects on supercoiling*

Change twist (T), leave linking number (L) constant; supercoils (W) change: If $\Delta L = 0$, then $\Delta T = -\Delta W$	Change linking number (L), leave twist (T) constant; supercoils (W) change: If $\Delta T = 0$, then $\Delta L = \Delta W$
CAUSES	CAUSES
Change salt concentration or temperature, or add intercalating agents (chloroquine, ethidium bromide, etc.)	Topo I (normally decreases linking number) Topo II (can increase or decrease linking number) DNA damage followed by resealing

*The degree of supercoiling is indicated as W in the expression $L = T + W$.

form nucleosomes. (In virions the histones are replaced by viral proteins with supercoils remaining in the DNA.) The winding of the helix in the nucleosome produces a *toroidal* supercoil (loops around the circle) which interconverts with an *interwound* supercoil when the protein is removed (Figure 13-20). The nucleosome structure contains 1.75 coils of DNA, and SV40 has about 20 nucleosomes that produce the 25 supercoils in the protein-free molecule. The number of supercoils per nucleosome has not yet been settled; much depends on the exact twist of the DNA in nucleosomes, and on the pathway of the DNA axis between nucleosomes.

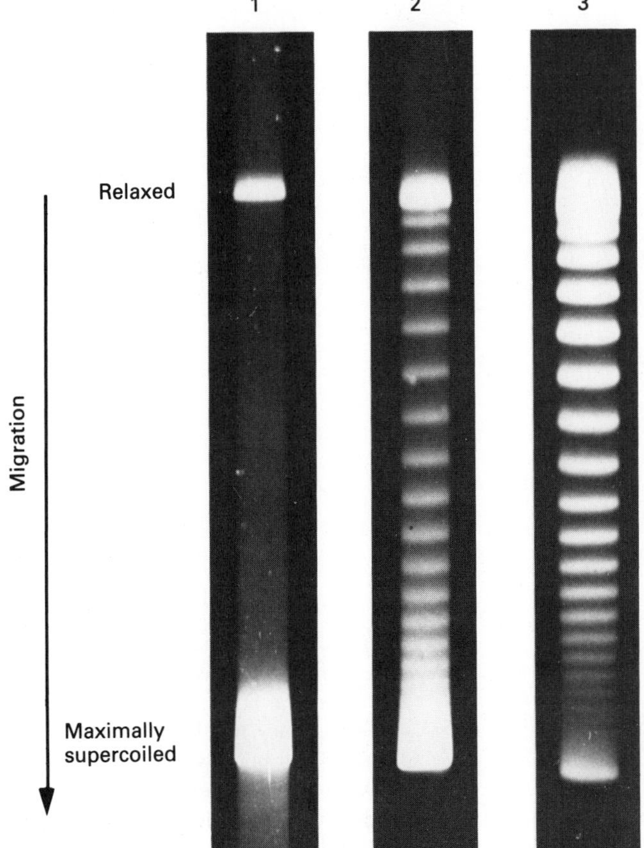

Topoisomerases Can Change the Linking Number

Once it had become clear that DNA replication and packaging can be very demanding structural exercises, biochemists began searching for enzymes that would have the capacity to affect the overall structure of the helix without digesting the molecule to pieces. A number of topoisomerases with such abilities have been purified from both bacterial and animal cells. Some of these enzymes are undoubtedly involved in DNA replication.

Type I Topoisomerases The first topoisomerase discovered was the ω (omega) protein in *E. coli*. This enzyme has the capacity to completely remove negative supercoils without leaving nicks in the DNA molecule (Figure 13-21). The enzyme binds to the DNA, cuts one strand, and the free phosphate on the DNA becomes covalently attached to a tyrosine residue in the enzyme. The complex rotates, relieving a supercoil, and the DNA is resealed.

Figure 13-19 Separation of SV40 DNA with different numbers of supercoils. Maximally supercoiled DNA was obtained from SV40 virus; DNA in this state migrates fastest during gel electrophoresis. Completely relaxed DNA was obtained by nicking duplex circles (see Figure 13-16) and closing the nick with DNA ligase. These two samples are shown in lane 1. In lanes 2 and 3, supercoiled DNA was acted upon by a topoisomerase (type I) that alternately nicks and closes the DNA, gradually relaxing the supercoiled circles. The DNA in lane 2 was digested for 0.5 min, and that in lane 3 for 30 min. The different bands represent different states, or topoisomers, of SV40 DNA with different degrees of supercoiling. About 25 bands are observed in lanes 2 and 3, so the maximally supercoiled molecule has 25 supercoils. [See W. Keller and I. Wendel, 1974, *Cold Spring Harbor Symp. Quant. Biol.* 39:199.] *Courtesy of W. Keller.*

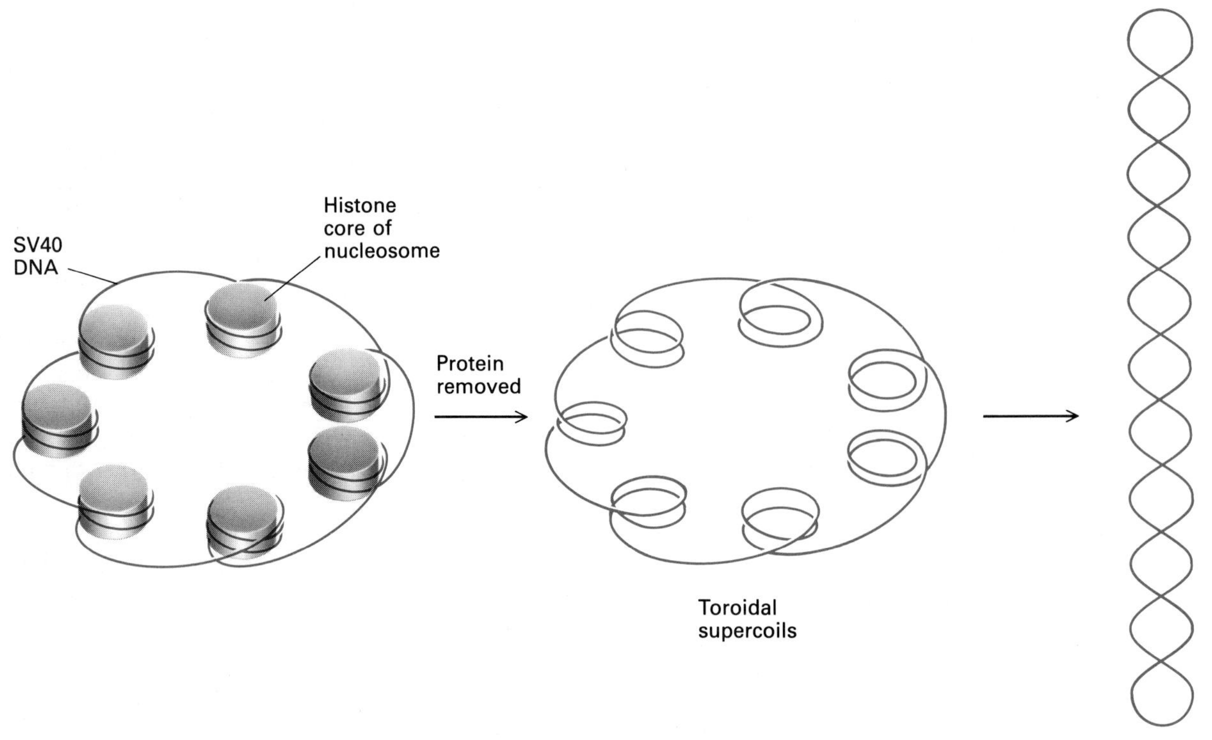

Figure 13-20 SV40 DNA that has been released gently from cells is found in nucleosomes, in which the DNA forms toroidal supercoils around histones. When the DNA is separated from the histones, the toroidal supercoils interconvert with interwound supercoils. (In virions the same structure exists but the proteins are viral.)

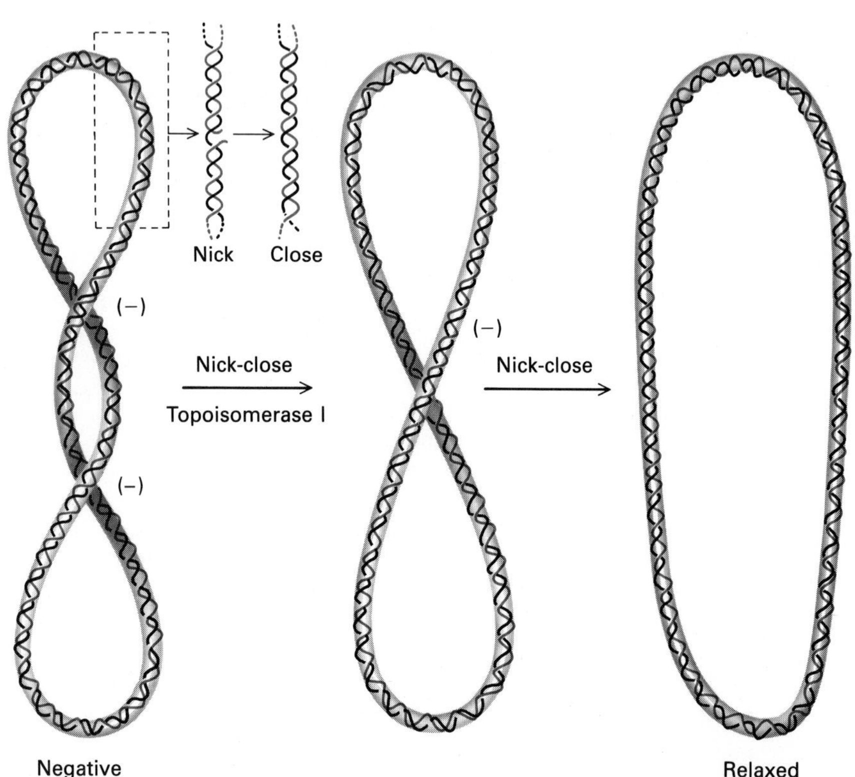

Figure 13-21 The action of topoisomerase I enzymes. This enzyme nicks and closes the DNA. If a helix has supercoils, they will be relieved in the presence of topo I, which allows the DNA to achieve a relaxed state of winding (about 10.4 base pairs per turn of the helix with no supercoils).

The enzyme is much less active on positively supercoiled molecules. Animal cells have similar *nicking-closing* enzymes that also become attached to DNA by a phosphotyrosine linkage during relaxation of supercoils. Because the relaxation of DNA supercoils by the nicking-closing enzymes is energetically favorable, the reaction proceeds without an energy requirement. Such enzymes are termed *topoisomerase I* (or topo I) enzymes. Genetic evidence from bacteria suggests that topoisomerase I may participate in the movement of replication forks and in DNA transcription, but the precise role of these enzymes in replication has not yet been demonstrated in vitro.

Type II Topoisomerases A second type of topoisomerase, *topoisomerase II* (topo II), is illustrated by the *E. coli* enzyme termed *gyrase*. This enzyme has the ability to cut a double-stranded DNA molecule, pass another duplex DNA through the cut, and reseal the cut (Figure 13-22). Because this maneuver has the effect of changing a positive supercoil into a negative supercoil, it changes the linking number of the DNA by 2, as can be shown through gel electrophoresis. ATP is consumed in the process to yield the energy that fuels the change in superhelicity.

There are two protein subunits in the *E. coli* gyrase. One subunit apparently covalently attaches briefly to the enzymatically broken ends of the DNA (possibly at sequence-specific regions). The other subunit (an ATPase) is responsible for the energy transduction necessary to change the supercoiling. These two activities can be inhibited separately by two drugs: novobiocin, an antibiotic that binds to the protein subunit attached to DNA, and oxolinic (or nalidixic) acid, which binds to the ATPase subunit (Figure 13-23).

Gyrase is necessary for DNA replication in *E. coli* cells; a temperature-sensitive gyrase protein renders a bacterium unable to grow at elevated temperatures. It seems likely that the normal enzyme functions in DNA replication and in the movement of the growing fork. The transcription of many genes is also slowed considerably in cells with mutant gyrase. The topo I (ω) and topo II (gyrase) enzymes in *E. coli* compete to balance the level of supercoiling in the *E. coli* chromosome. Measurements of the degree of supercoiling in *E. coli* cells have suggested that there is one negative supercoil for each 15 turns of the DNA helix.

Both type I and type II topoisomerases have been found in eukaryotic cells—for example, in embryonic tissue of *Drosophila melanogaster* and in cultured mammalian cells. The topo II activity from these sources can result in a variety of changes in the form of the DNA (Figure 13-24). The enzymes can produce catenated dimers (two interlocked circular double-stranded DNA molecules, as diagrammed in reaction 5 of Figure 13-23) and can then re-separate the two molecules. Another DNA configuration is the knotted circle formed by one type of viral DNA; the topo II activity can knot or unknot such molecules.

The Role of Topoisomerase II in Releasing Final Products during SV40 Replication Topoisomerase II activity is apparently needed in the completion of SV40 replication. As noted above, SV40 DNA is a circular supercoiled molecule. In replication, the parental strands remain intact and retain their superhelicity (Figure 13-24). As the two replication forks approach each other, how does the as yet unreplicated parental duplex region become unpaired to allow replication to finish? This situation also arises in the meeting of two replication forks in a cell chromosome. If the replication is completed without the parental helix becoming broken and then unpaired, the finished molecules will be catenated dimers, that is, covalently closed linked circles. Such forms can be recovered, in various states of interlocking, from SV40-infected cells that have been treated with abnormally high NaCl concentrations. This treatment probably inhibits topo II activity. For such molecules to become released as single supercoiled circles, there must be topoisomerase II activity.

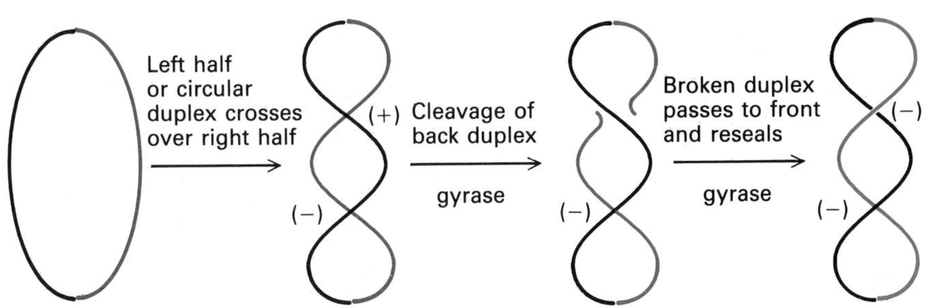

Figure 13-22 The action of topoisomerase II: folding the circle as in the second frame gives a positive (+) and a negative (−) supercoil. (The assignment of positive and negative is by convention: if the helix is stood on its end, in a positive supercoil the "front" strand is falling from left to right as it crosses over the "back" strand; in a negative supercoil the front strand is rising from left to right.) The diagrammed action of gyrase has the effect of removing a supercoil of one sign (+) and replacing it with a supercoil of the opposite sign (−). Thus each action changes the linking number of the DNA by 2. [See N. R. Cozzarelli, 1980, *Science* 207:953.]

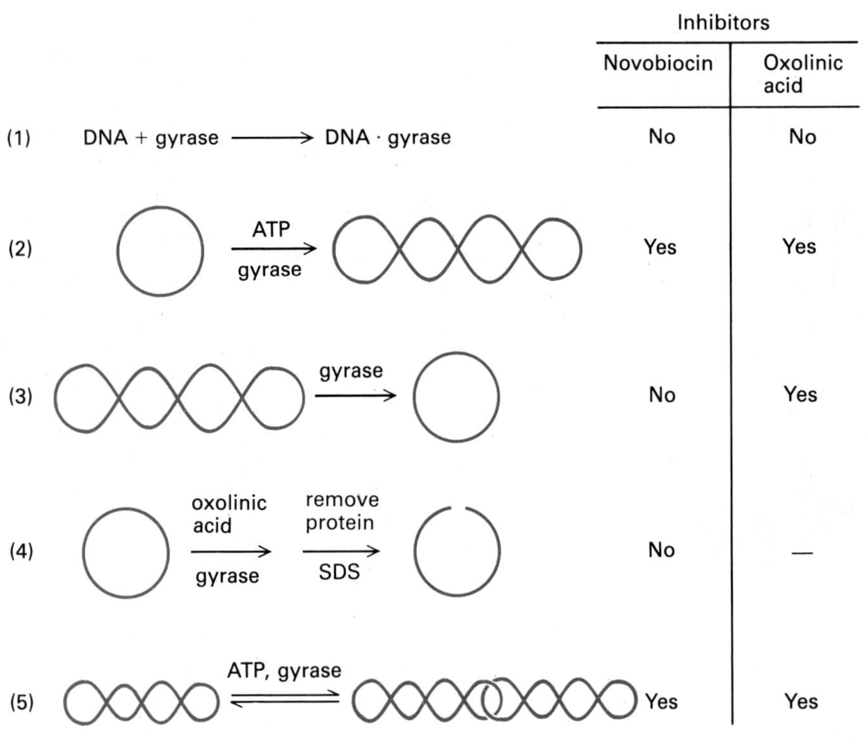

Figure 13-23 The activities and properties of *E. coli* DNA gyrase. The steps in gyrase activity and the participation of two different protein subunits were determined in part because of its susceptibility to the chemical inhibitors novobiocin and oxolinic acid. The reactions are:
(1) The simple binding of gyrase to DNA.
(2) The introduction of negative supercoils (ATP required).
(3) The relaxation of negative or positive supercoils (no ATP required).
(4) The double-strand cleavage of DNA without resealing in the presence of oxolinic acid. This shows that re-ligation but not cutting is due to the oxolinic acid–sensitive part of the gyrase.
(5) The catenation of two (or more) circular supercoiled DNA molecules.
[See N. R. Cozzarelli, 1980, *Science* 207:953.]

Mechanisms and Sites of Initiation in DNA Synthesis

If DNA is to be replicated exactly, then not a single extra nucleotide may be inserted or omitted during each replicative cycle. These exacting requirements pose several major questions: At each replicative cycle, does DNA synthesis begin at the same point? If so, how is that site chosen? How can it be determined that replication has proceeded from the initiation site to the termination site? Although the details of initiation are not yet clear, a variety of experiments with viruses, bacteria, and cultured mammalian cells have suggested that synthesis does begin at or near the same nucleotide in each replicative cycle.

In Bacteria, Strand Nicking, Topoisomerase Melting, or Primer Synthesis Can Initiate Replication

It is generally assumed that the origin for DNA synthesis contains a molecular signal, probably a particular base sequence. For the copying of one strand to begin, one of two events must occur at the origin:

1. The cleavage of a previously intact strand to provide a 3′ hydroxyl end from which chain growth can begin.
2. The synthesis of an RNA primer at or near the origin. A site-specific melting of the duplex may occur so that the primer can be synthesized at that site; alternatively, the primer could be transcribed at another place and hybridize at the origin.

In any case, a specific protein must play a role in recognizing the proposed signal sequence. Only a few known proteins can perform such a role.

For example, the gene A protein in ϕX174, the small DNA virus of *E. coli*, recognizes one site and nicks the double-stranded replicative form of the virus DNA to allow the initiation of leading-strand replication. This begins the duplication of the double-stranded replicative form of the single-stranded virus by a mechanism similar to that shown in Figure 13-12.

A type II topoisomerase encoded by the bacteriophage T4 is a protein that may play a role in the initiation of DNA replication by selectively binding to the origin region and unwinding it. Bacteriophages with temperature-sensitive mutations in the gene encoding this protein do not initiate DNA replication at the nonpermissive temperature. The enzyme could bind to two sites on the DNA and bring them together, at which point new supercoils might be introduced. Increased negative supercoiling would result in this unwinding of the duplex, which might expose the origin as a single-stranded region and thus favor RNA primer formation (Figure 13-25). Whether such a mechanism actually operates in T4 DNA initiation is unknown, but the scheme is plausible. Mutations in the DNA gyrase of *E. coli* do render the cell incapable of initiating DNA synthesis, so a major role for topoisomerases in bacterial DNA initiation seems possible.

A third possible mechanism for starting DNA chain

(a)

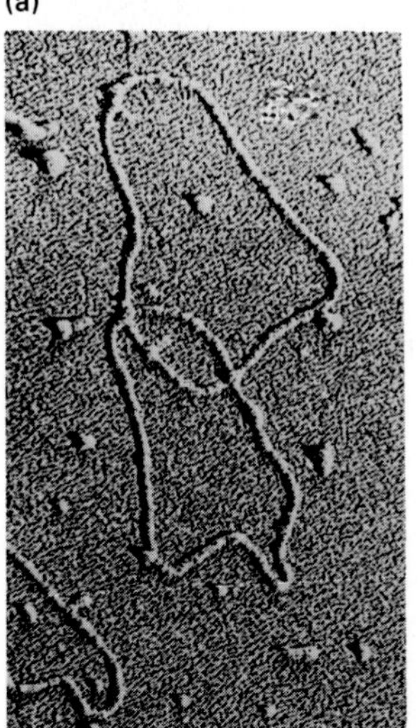

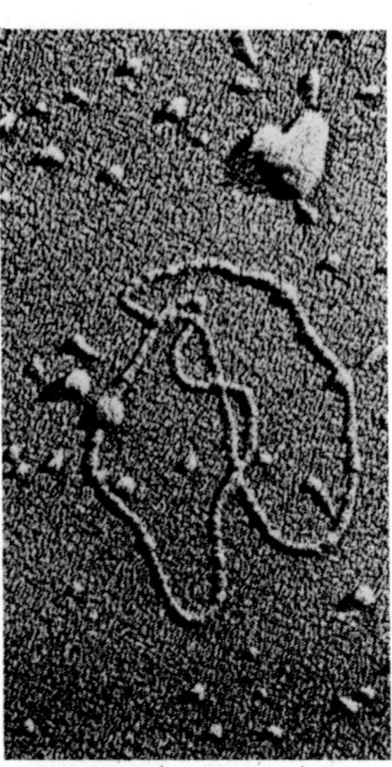

(b)

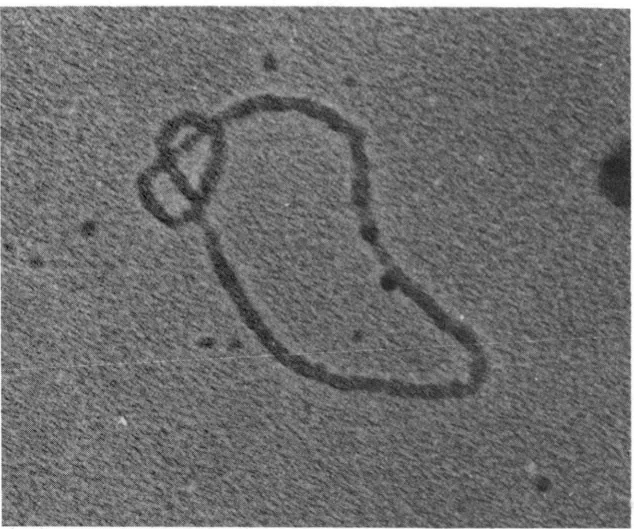

Figure 13-24 Electron micrographs of changes resulting from topo II activity. (a) Catenated dimers of SV40 DNA *(interlocked circles)*. The two circles at the left are interlocked once; the two at the right are interlocked twice. Each circle is a single SV40 molecule which must be broken to release it from the other molecule. (b) "Knotted" DNA molecules of the *E. coli* plasmid pBR322. The knotted circles can be caused or released by topo II activity, depending on incubation conditions. [See O. Sundin and A. Varshavsky, 1981, *Cell* **25**:659; and L. F. Liu, C.-C. Liu, and B. M. Alberts, 1980, *Cell* **19**:697.] *Part (a) courtesy of A. Varshavsky; (b) courtesy of B. M. Alberts. Copyright 1980 M.I.T.*

growth at particular sites involves neither nicking nor site-specific melting followed by primase action. Rather, RNA polymerase—the enzyme that is active in normal transcription—could synthesize an RNA segment near the origin of replication, and a part of the resulting transcript might act as an RNA primer. Such RNA primer molecules are formed by RNA polymerase in the replication of colicin E1 bacterial plasmid, and probably also in the replication of the medium-sized bacteriophages λ and T7. This mechanism may also operate at the origin of replication of the circular chromosome of *E. coli*.

In the case of colicin E1, the RNA that eventually serves as a primer is initiated by the RNA polymerase (and not by a primase) at a site 555 nucleotides before the origin of DNA synthesis (Figure 13-26). When this RNA chain, called RNA II, is elongated toward the origin, it forms an RNA-DNA hybrid (an *R loop*) upstream of the origin. The RNA of the RNA-DNA duplex is then cleaved by an enzyme called *RNase H* (for its ability to digest only hybridized RNA). The cleaved 5′ fragment of this RNA is elongated at its 3′ hydroxyl end by DNA polymerase I and a replication bubble is established.

A remarkable recent discovery is that the frequency of initiation of colicin E1 replication is controlled by a second RNA strand called RNA I (Figure 13-26). This strand is transcribed in the direction opposite to that of RNA II, and is therefore complementary to RNA II. When RNA I is present in abundance, it hybridizes to the RNA II and somehow prevents the RNA-DNA hybrids (the R loops) from forming. Therefore no RNase H action occurs, and no RNA primer or replication bubble is formed.

Specific start sites for DNA replication are thus recognized in a variety of ways in bacterial viruses and in plasmids containing bacterial cell DNA.

Eukaryotic DNA Probably Has Specific Initiation Sites

In eukaryotic cells, too, a number of different experiments have indicated that DNA synthesis begins at specific sites, or at least in specific regions. Because these experiments have exhibited such a wide range of molecular biological ideas, we shall review them here. Just as bacterial DNA replication systems were first understood by studying viruses, small animal viruses that use host-cell enzymes have provided important models. We shall begin our discussion of eukaryotic DNA initiation by describing three different modes of DNA replication exhibited by viruses.

The Single Replication Origin of SV40 DNA The small DNA viruses like SV40 and polyoma may be the best analogs of host DNA replication, because the virus chromosome is packaged in nucleosomes and because the resulting structure is a single replicon that undergoes bidirectional replication. The replication bubbles that pro-

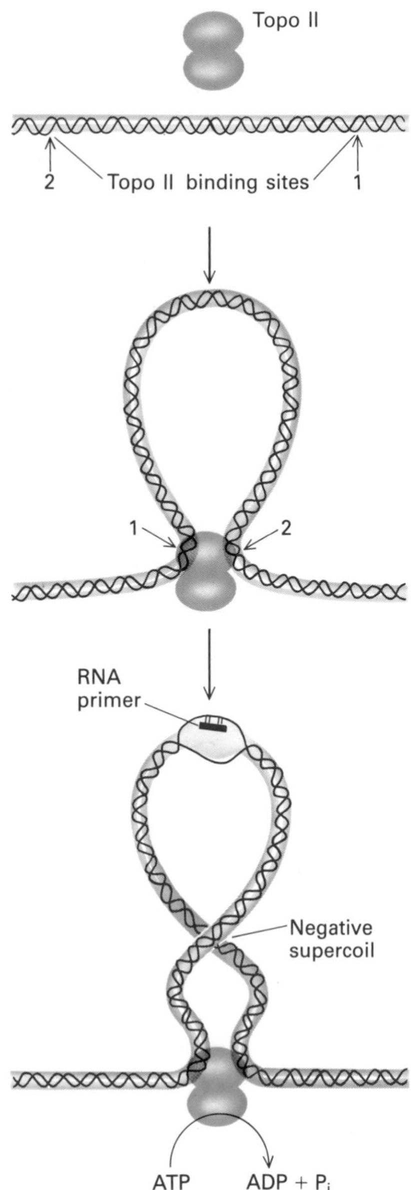

Figure 13-25 A mechanism for the participation of topoisomerase in the initiation of DNA synthesis. Binding sites near the origin of DNA replication would attract the enzyme. The ATP-dependent introduction of negative supercoils could lead to the melting of the DNA in that region, allowing an RNA primer to initiate replication. [See L. F. Liu, C.-C. Liu, and B. M. Alberts, 1979, *Nature* **281**:456.]

ceed from a single site in SV40 DNA are shown in Figure 13-7.

A biochemical experiment actually provided the first evidence of this mode of replication. Labeled completed viral DNA from SV40-infected cells was collected at intervals after the beginning of labeling. The most briefly labeled DNA molecules, that is, those *finished* in the shortest pulse time, all contained label in one limited segment of the genome—the region in which *termination* of replication occurred (Figure 13-27). With longer label times, longer stretches on either side of the region of termination become labeled, indicating that termination is approached bidirectionally from *one* origin.

The single origin of SV40 DNA replication has been confirmed by introducing mutations that prevent replication. A number of single-base changes within the origin prevent replication, many by affecting the binding of the T antigen (Figure 13-28). (The T antigen, which is encoded in the early SV40 transcription unit, binds at the origin to help in DNA initiation.) Clearly, then, there is little question that the small DNA viruses that replicate in the nuclei of eukaryotic cells have restricted sites within which DNA replication originates. The proteins that participate in this replication are, except for the T antigen, encoded by the host cell, and it seems very likely that these proteins also begin DNA replication at discrete sites in the host-cell chromosomes.

Initiation of Adenovirus DNA by a Protein-Nucleotide Complex The linear adenovirus DNA molecule is replicated one strand at a time, beginning at its ends. The mode of DNA chain initiation is unusual. Instead of using a nucleic acid primer, the virus uses a protein that is covalently linked to the 5′ end of each strand of the adenovirus DNA molecule (Model 1 in Figure 13-4).

Replication can be carried out in vitro with a series of purified proteins. The terminal protein attached to the parental viral DNA has a mass of 55 kdal (kilodaltons); a larger (80-kdal) "preterminal" protein also exists in infected cells. The preterminal protein is associated with three other proteins: a host factor that specifically recognizes the sequence at the end of the adenovirus DNA molecule, a single-stranded DNA–binding protein encoded by the virus, and the viral DNA polymerase. The preterminal complex adds a dCMP residue covalently to the preterminal protein; this nucleotide is the first nucleotide in the new DNA chain. Polymerization then occurs at the 3′ hydroxyl of the dCMP, and chain growth proceeds by base-pair copying of the template strand until about half of the molecule is copied; at this point the reaction halts (Figure 13-29a). Continuation of the DNA synthesis reaction to produce a full-length adenovirus single strand requires the addition of a topoisomerase from the cell nucleus. This result suggests that the ends of the virus molecule are not free to rotate in solution. Perhaps the terminal proteins associate with one another to prevent such rotation (Figure 13-29b), and the topoisomerase acts to relieve the torsional strain that develops during replication.

It is not known whether eukaryotic cells make use of a mechanism of DNA synthesis similar to that of adenovirus, but it is very interesting that a host-cell protein has the ability to recognize adenovirus terminal sequences. In addition, phage ϕ29 of *Bacillus subtilis*, plasmids from

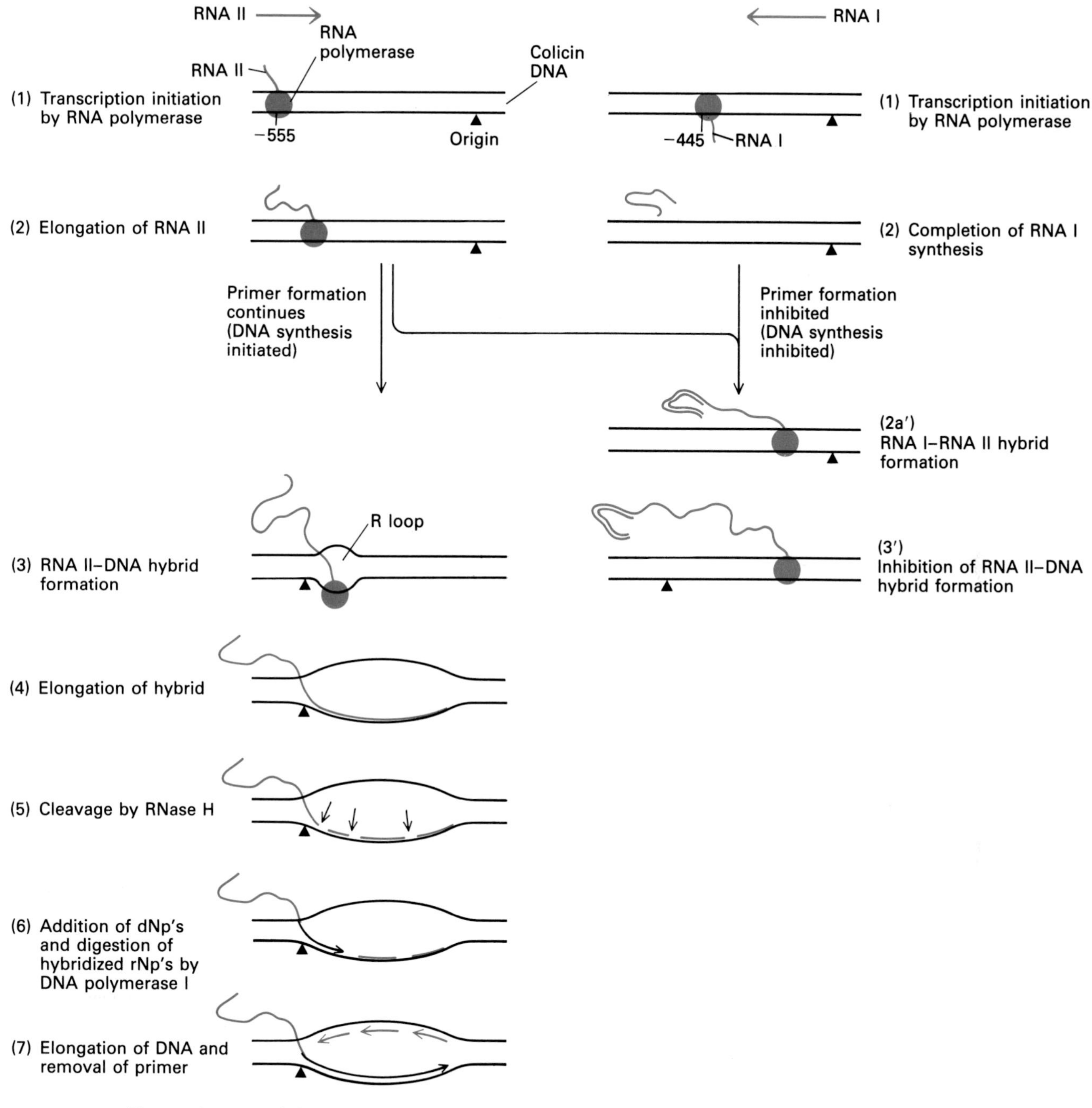

Figure 13-26 The mechanism of the initiation of DNA synthesis and its control in the *E. coli* plasmid colicin E1. When RNA I is formed in high amounts, it prevents the formation of an RNA primer for DNA synthesis. [See T. Itoh and J. Tomizawa, 1982, *Nuc. Acids Res.* **10**:5949; and H. Masukata and J. Tomizawa, 1983, *Cell* **36**:513.]

yeast and *Streptomyces rochei,* and a mitochondrial DNA from maize cells use proteins as initiators in DNA replication.

Parvoviruses as Models for Telomere Replication

Studies of small single-stranded DNA viruses called parvoviruses have illustrated yet another means by which DNA synthesis is initiated: *self-priming.* This viral mechanism of initiation and replication provides a model for how the termini of chromosomes might replicate (Figure 13-30). The sequences at each end of these small single-stranded DNAs are self-complementary. (This does not mean that the two ends are complementary to each other.) Self-complementary sequences can fold into *hairpins*—complementary regions with as few as five or six bases forming a "turnaround" in the middle. A DNA poly-

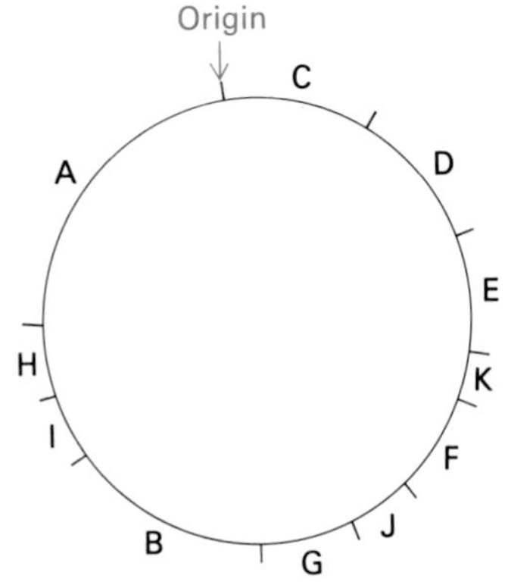

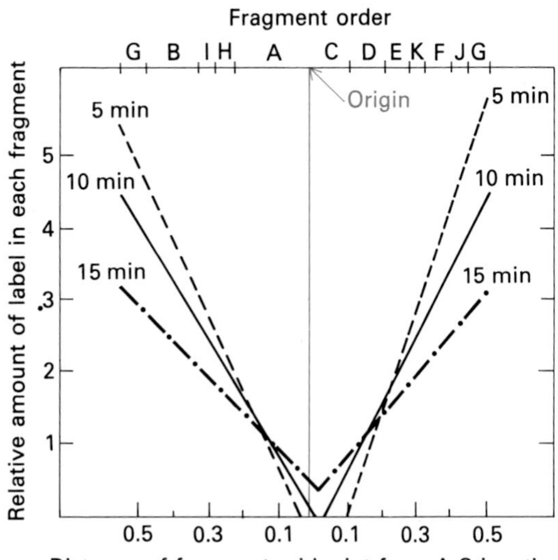

Figure 13-27 The growth of SV40 DNA from one origin. Completed DNA from SV40-infected cells labeled for 5, 10, and 15 min was digested with restriction enzymes, and the relative amount of radioactivity in each SV40 DNA fragment (i.e., the amount per unit length) was measured. The map positions of fragments A through K are shown on the circle. In a newly finished circle with one origin, the label should appear first in a fragment most distant from the origin. The most highly labeled fragment at 5 min was G, and the most lightly labeled fragments were A and C; thus, a single origin could be located near the A-C junction. With longer times, the fragments were labeled more equally. [See K. J. Danna and D. Nathans, 1972, *Proc. Nat'l Acad. Sci. USA* 69:3097.]

merase can begin base-pair copying by chain elongation at the 3′ hydroxyl end of the single strand; the molecule may then be copied all the way to the 5′ end. The two ends at this stage may be separated, whereupon they can both form hairpins; continuation of the copying by extending the 3′ hydroxyl end results in a complete duplication of the original sequence. Thus, without any RNA primer, the DNA molecule has become duplicated.

This idea is important because the ends of linear DNA molecules cannot be copied by a mechanism such as a replication bubble, which employs RNA primers. As the bubble approached the end of the molecule the leading strand would be complete, but the lagging strand, which requires a primer, could not be finished. Apparently, the ends of chromosomes (telomeres) contain nicked hairpin sequences (Chapter 10), so their mode of replication is probably similar to that of parvoviruses.

The Cloning of Replication Origins A recently developed—and a very elegant—genetic test for the capacity of a DNA region to serve as an origin uses recombinant DNA. Independently replicating plasmids are carried by many bacteria, by many yeast strains, and possibly by other eukaryotic cells as well. The replication origin of the plasmid can be removed by restriction enzymes and replaced with any DNA suspected of possess-

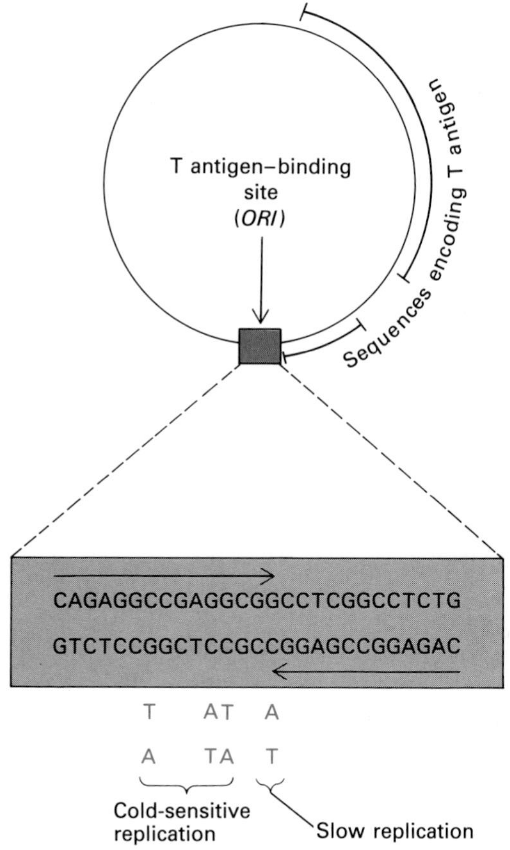

Figure 13-28 The sequence to which the SV40 T antigen binds is located at or near the origin *(ORI)* of SV40 DNA replication. The binding sequence is a region of perfect dyad symmetry *(arrows)*. The indicated base changes at four sites cause DNA replication to be either cold-sensitive or slow (as shown by the small size of plaques produced in vitro). [See D. R. Shortle, R. F. Margolskee, and D. Nathans, 1979, *Proc. Nat'l Acad. Sci. USA* 76:6128.]

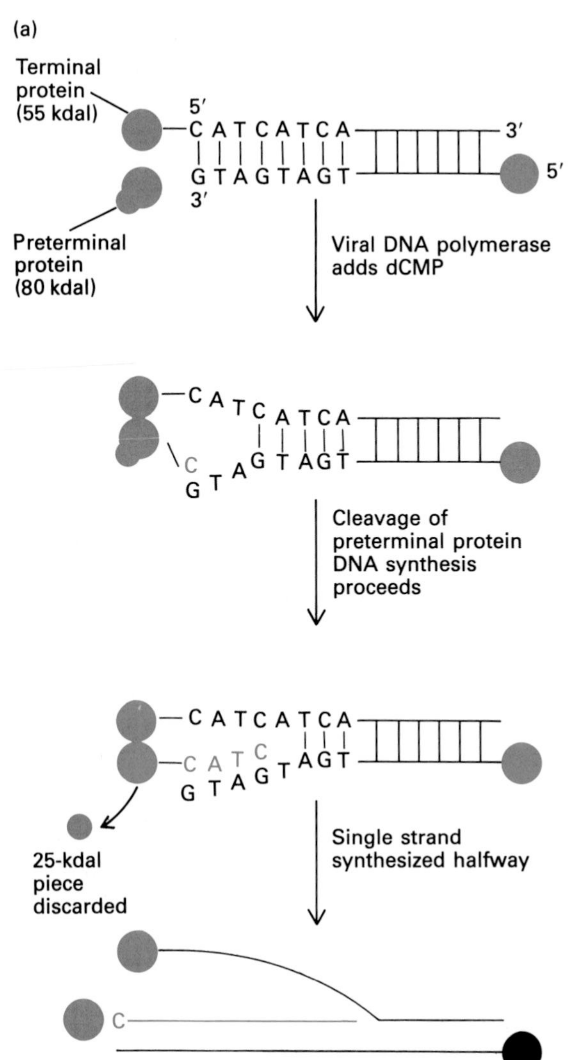

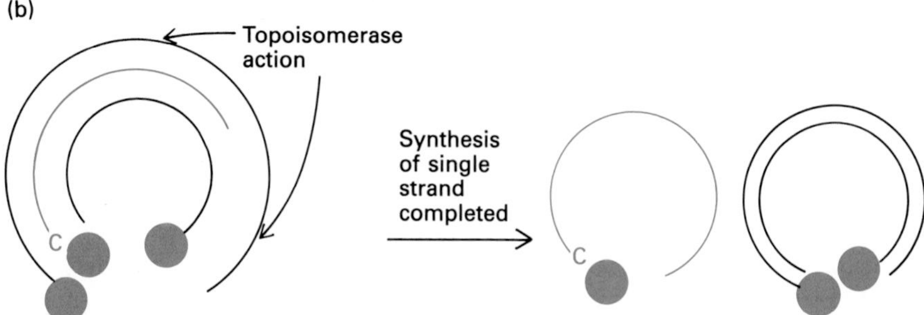

Figure 13-29 Adenovirus DNA replication can be carried out in vitro by purified proteins. (a) The reaction starts with adenovirus DNA that contains a 55-kdal terminal protein attached to each 5′ end of the duplex DNA. An 80-kdal preterminal protein plus a host factor (I, 47 kdal) recognize the end of the DNA; in the presence of the single-stranded DNA–binding protein, the adenovirus DNA polymerase adds a dCMP residue (from dCTP) to the preterminal protein, which is then cleaved to become a 55-kdal terminal protein. Synthesis of a single strand of DNA proceeds about halfway to the end of the molecule. (b) Completion of the single strand requires a topoisomerase. The need for a topoisomerase implies that the two ends of the parent molecule are associated with one another so that the relief of torsional stress can occur (see Figure 13-15). [See K. Nagata, R. A. Guggenheimer, and J. Hurwitz, 1983, *Proc. Nat'l Acad. Sci. USA* 80:4266 and 80:6177.]

ing the function of a replication origin (see Figure 10-10). If the recombinant plasmid is replicated when it is reintroduced into the bacterial or yeast cell, then a replication origin has been inserted. In this manner a single origin of replication has been isolated for *E. coli,* and numerous origins termed autonomous replicating sequences have been found in yeast cells. This general approach, making the replication of an autonomous plasmid DNA depend on a selected portion of chromosomal DNA, should help locate specific sequences that function as replication origins in any cell type for which a selectable plasmid is available.

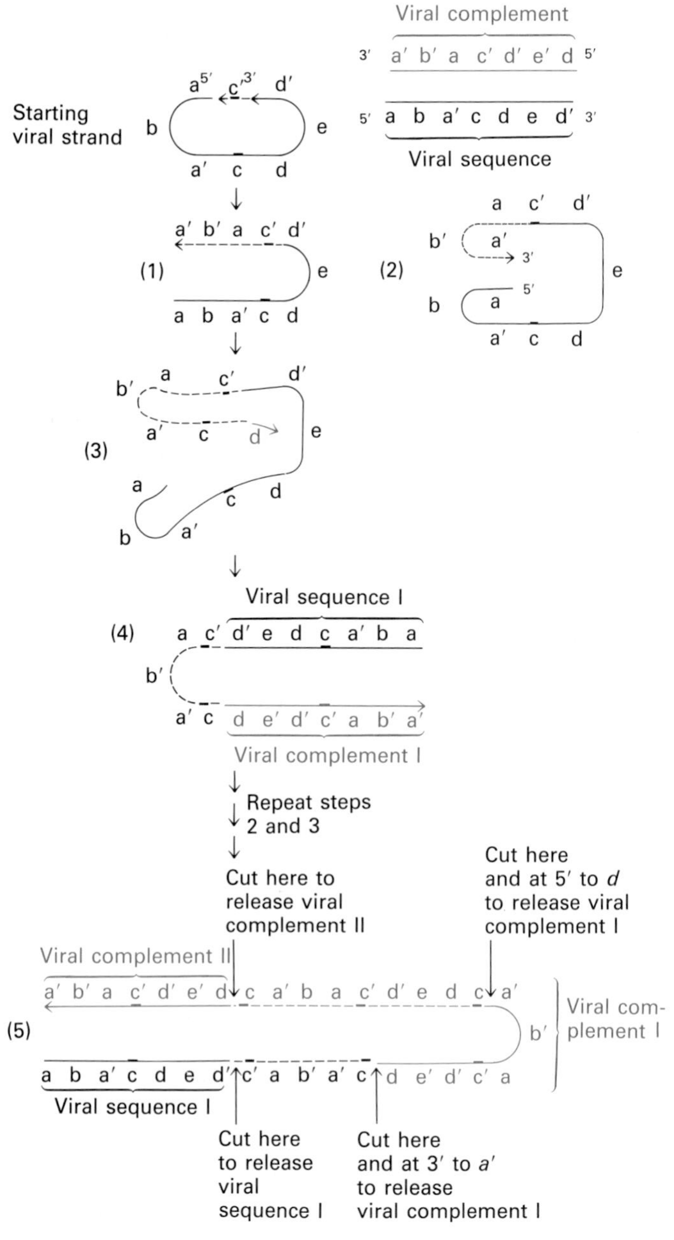

Figure 13-30 The rolling hairpin model of DNA replication. Small mammalian viruses called *parvoviruses* have single-stranded DNA molecules with ends that form self-complementary hairpins. (The end sequences *a* and *a′*, *d* and *d′* are complementary.) By extending the 3′ hydroxyl end *(arrows)* of the parent molecule *(black)*, a DNA polymerase can eventually copy the entire molecule (step 1). After step (1) the ends form self-complementary structures (2) and the elongation continues (3). An appropriate nick is required to release the completed molecule. After step 4 the viral sequence at the 5′ end could be liberated by a cleavage between *d′* and *c′*; replication could continue because a full complement with a hairpin end is present at the 3′ end. Alternatively, *each* full sequence—be it the viral sequence or the viral complement—could be chopped off the 5′ end of the hairpin once a full (template) sequence has been replicated at the 3′ end. The latter situation must occur in some parvovirus infections, because both viral strands and their complements are found in virions. [See E. A. Faust and D. C. Ward, 1979, *J. Virol.* 32:293.]

Early- and Late-Replicating DNA in Cell Nuclei Suggest Order in Replication

In considering the replication of eukaryotic DNA—for example, the replication of the 3×10^9 base pairs of human DNA distributed in 23 different sets of chromosomes—we must answer a number of general questions before we can deal with specific problems at any given site. One such question is the following: As a cell goes through the growth cycle, are the same regions of DNA replicated in the same order in each cycle? It appears that this is generally true, but the exact start sites have not yet been determined for even one cell replicon. Therefore, the detailed organization of chromosome replication remains unknown.

In cultured mammalian cells, the same regions of each chromosome replicate early in each S (synthesis) phase of the cell cycle, and other fixed regions replicate late. This is consistent with the idea of fixed origins of replication. By including either [^{3}H]thymidine or bromodeoxyuridine (BrdU), which substitutes for thymine, in the culture medium and examining the mitotic cells that occur at intervals thereafter, researchers have found different times of replication for different chromosomal regions.

DNA that has incorporated BrdU is revealed by staining chromosomes with a fluorescent dye that binds poorly to the BrdU-substituted DNA; incorporated thymidine is detected by autoradiography (Figure 13-31). The late-replicating chromosome sections are the first to appear labeled with [^{3}H]thymidine, and the first to bind more dye. Although these techniques are not precise, they do show that there is a general organization of chromosomal replication.

Not only can large segments of mammalian chromosomes be demonstrated to replicate at the same time, but so can individual DNA sequences. DNA with bromouracil substituted for thymine is more dense (heavier) than unsubstituted DNA. Cells in a culture can be synchronized so that all are in the same stage of the cell cycle. As the synchronized cells go through the S phase, bromouracil can be introduced at intervals. By using cloned DNA sequences as probes to detect specific DNA molecules, it can be readily shown that certain DNAs become heavy and thus are replicated at specific times during S phase.

Experiments of this type have indicated that in cultured (undifferentiated) cells, most of the genes that are tran-

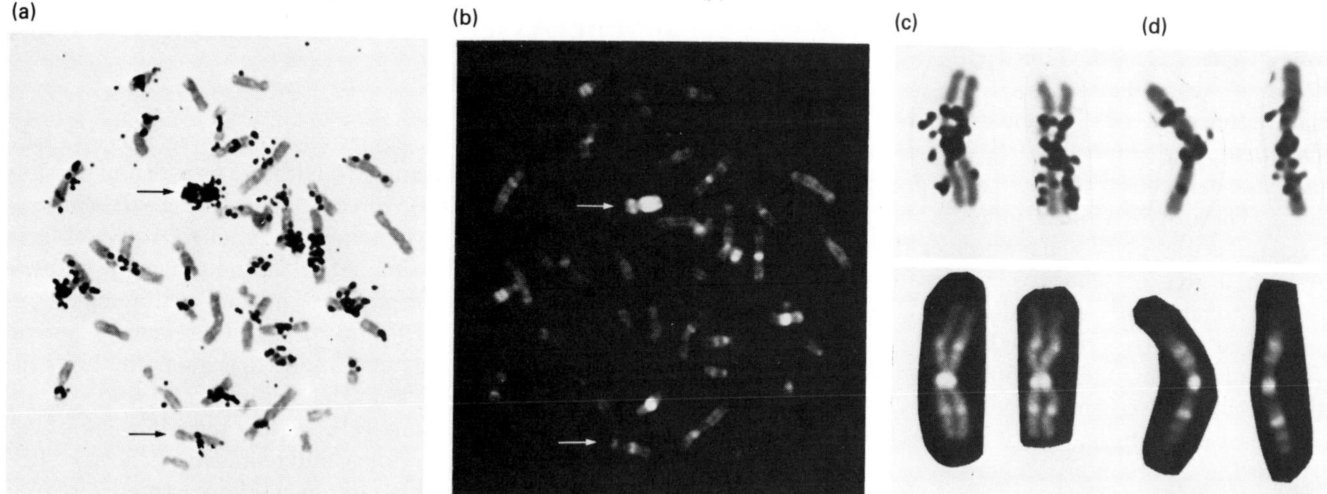

Figure 13-31 Late-replicating DNA in human cells. White blood cells, which can be made to divide a few times outside the body, were cultured in bromodeoxyuridine (BrdU) for 48 h and then placed in a medium containing [^{3}H]thymidine. It was assumed that the last parts of individual chromosomes to replicate would be the first parts to appear with the [^{3}H]thymidine label. Within a few hours, labeled mitotic cells were observed. (a) The regions of [^{3}H]thymidine incorporation were detected by autoradiography *(black dots)*. (b) These same regions were also bright-staining. [The arrows in (a) and (b) are simply for purposes of orientation.] The stain is a fluorescent dye that binds well to DNA containing thymidine, and poorly to DNA substituted with BrdU.

The experiment shows that certain regions of certain chromosomes replicate late in the S phase. Each dividing cell has two homologous chromosomes of each morphological type; the late-replicating regions behave similarly in each homologue. This result is shown below in higher magnifications of chromosome 1. (c) Two pairs of number 1 chromosomes, each pair taken from a different cell, are compared. (d) An unpaired chromosome 1 taken from one cell is compared to a single chromosome 1 from another cell. The pairs of homologues show the same autoradiographic and staining pattern. [See S. A. Latt, 1973, *Proc. Nat'l Acad. Sci. USA* 70:3395.] *Courtesy of S. A. Latt.*

scriptionally active are those that replicate early. The late-replicating DNA contains many individual genes that are not transcribed in cultured cells; many repetitive satellite DNAs that are not transcribed also replicate late in the S phase. In a few cases, cultured cells can be induced to express tissue-specific genes—for example, globin genes have been expressed in erythroleukemia cells. The globin genes replicate early in erythroleukemia cells, but late in undifferentiated HeLa cells (which do not express these genes). These findings certainly support the conclusion that the transcription units that are active in cultured cells are contained in the replicons activated the earliest.

The body of work on replication origins and on early- and late-replicating DNA constitutes impressive evidence that chromosomal replication proceeds according to a regular plan during each DNA synthesis cycle.

Eukaryotic Chromosomes Have Multiple Origins of Replication

Although bacteria, plasmids, and most DNA viruses contain only a single origin of DNA replication, all eukaryotic cells have multiple origins of replication in each chromosome. The human genome (23 chromosomes in its haploid state) is estimated to have between 2×10^4 and 1×10^5 independent origins of replication, and other eukaryotes have comparable numbers. Even yeast chromosomes, each of which is smaller than the *E. coli* genome, have one replication origin for every 20,000 to 40,000 base pairs, or about five origins in the smallest chromosomes. Multiple sites of origin in replicating eukaryotic DNA are easily observed either by electron microscopy or by autoradiography of stretched DNA fibers from cells labeled with [^{3}H]thymidine (see Figures 13-5 and 13-6).

The number of replication origins used by a eukaryotic cell may vary in different cell types. During the cleavage stages of embryogenesis in insects and amphibians, the most rapid DNA synthesis and nuclear divisions occur. The entire chromosomal DNA of *D. melanogaster* (about 2×10^8 base pairs) is replicated in less than 5 min. Probably all available origins are used in these early developmental cleavages. During slower cell growth—for example, in the proliferation of somatic cells or in the growth of cells in culture—perhaps only 10 percent of the available origins are used to initiate replication.

Characteristics of Sequences at DNA Replication Origins The sequences near origins of replication do not have particularly distinctive nor diagnostic characteristics. The origins from a number of different types of bacterial cells, however, have been cloned and sequenced. A high degree of sequence conservation is maintained over a stretch of 50 nucleotides. One characteristic of the conserved regions is the repetition of the oligonucleotide GATC nine times within a 150-nucleotide stretch; four of the repeats are shown in Figure 13-32a.

Several yeast chromosomal regions that are capable of supporting autonomous replication of plasmids in yeast cells have been examined; again, common oligonucleotides are present in these regions, but the conserved sequences are different from those found in bacteria (Figure 13-32b).

In multicellular eukaryotes, the functional origins of replication have not been clearly identified, as we have mentioned. But the origin regions have been identified in a number of DNA viruses, and these regions also contain a highly conserved sequence. Furthermore, this same sequence is present in the *Alu* family of repetitious sequences: these appear in 200,000 to 500,000 copies in the genomes of all mammals, frogs, and chickens (Figure 13-32c).

Because there are 10,000 to 100,000 potential origins of DNA initiation, the *Alu* sequences seem to be reasonable candidates for initiation sites. One other fact about the *Alu* sequences suggests their possible role in initiating DNA replication: some (approximately half) of the *Alu* sequences are transcribed by RNA polymerase III (Chapters 9 and 11). These short RNA transcripts, which do not accumulate in the cell, could conceivably act as RNA primers in initiation. Recent experiments show that plasmids containing transcribed *Alu* sequences do replicate in cultured monkey cells, whereas the same plasmids lacking the *Alu* region do not replicate.

Finally, it should be noted that only for viruses that start DNA synthesis at defined nick sites (e.g., ϕX174) or at the ends of linear DNAs (e.g., adenovirus) is there any evidence of DNA initiation at a precisely defined nucleotide. For example, nascent strands of SV40 DNA were collected and the 5′ nucleotide of origin of these strands were determined (Figure 13-33). Although the great majority of the origins were mapped within the region known to contain the sequences necessary for initiation of DNA synthesis, no one site accounted for more than about 20 percent of the total initiating nucleotides.

Origin regions therefore may not function to get DNA synthesis started at the same nucleotide each time. But within the origin region, gap filling and patching by DNA polymerase are then called upon to make the duplication precise.

The Assembly of DNA into Nucleosomes

Two aspects of the regulation of eukaryotic gene expression appear to be connected with cell growth and possibly with DNA replication directly. First, during early embryogenesis, cell fate is somehow determined many cell generations before the final differentiation of cell phenotype occurs. (A full description of changing cell potential appears in Chapter 22.) Thus there is inheritance of an *epigenetic* state, or a potential for a specific pattern of gene expression. Second, after the application of a specific inducing stimulus, many cells require several cycles of cell division to fully express a differentiated phenotype. Red blood cell precursors, for example, can undergo four or five divisions after stimulation by the hormone erythropoeitin before they all make globin.

Both determination of cell fate and differentiation are widely believed to be controlled by the deposition of proteins and possibly by the failure of some DNA to become methylated during cell replication. A logical time for such effects to occur is during or just after replication. Because the most prominent proteins associated with cell DNA are histones, great interest has focused on the assembly of chromatin in newly replicated DNA. The majority of histone protein is synthesized during the S phase of the cell cycle, whereupon it quickly enters the nucleus to become associated with DNA. (Small amounts of histones are probably synthesized throughout the cell cycle, perhaps for repair purposes.)

New DNA Quickly Associates into Nucleosomes

Within a few minutes of its synthesis, new DNA becomes associated with histones in nucleosomal structures. This has been shown by DNase digestion within nuclei of new, [^{3}H]thymidine-labeled DNA. Both the new, labeled DNA and the old, unlabeled DNA are released into fragments containing multiples of 200 base pairs; these are characteristic of chromatin arranged in nucleosomal arrays (Figure 13-34a).

Old Histones and New Histones Do Not Mix; New Histone Octomers Go to New DNA

Recall that each nucleosome consists of about 140 base pairs of DNA wrapped around an octomer of eight histone chains. When the new chromatin is being formed just after DNA duplication, do old and new histone monomers mix, or do old histones remain together and new histones likewise become associated only with other new histones to form new octomers? Experiments with isotopically labeled dense amino acids showed that the old his-

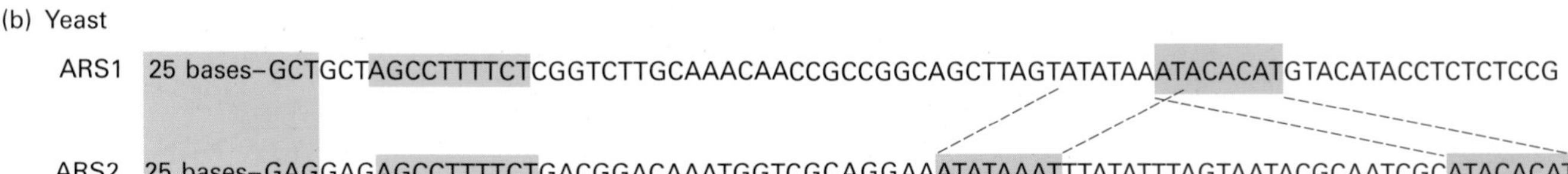

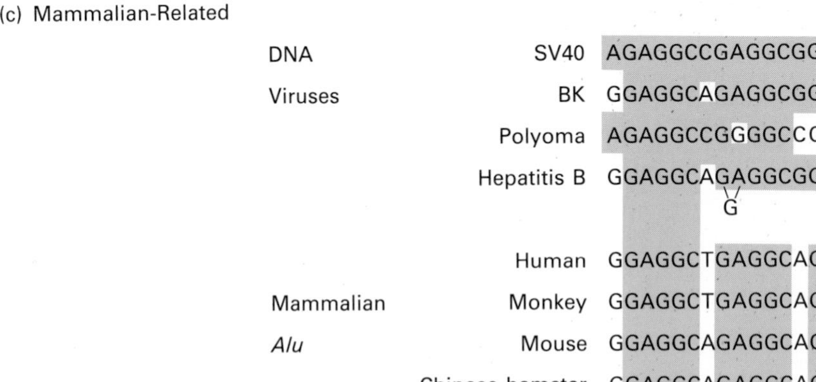

Figure 13-32 Conserved sequences at origins for DNA synthesis. (a) The single origins of replication of five bacterial strains *(E. coli, Salmonella typhimurium, Enterobacter aerogenes, Klebsiella pneumoniae,* and *Erwinia carotavora)* were cloned and sequenced. The most highly conserved nucleotides (in a stretch of about 50) are shown. Dashes indicate positions in which different bases occur in different strains. The frequent occurrence of GATC in the sequence is highlighted. (b) The two yeast chromosomal regions ARS1 and ARS2, which are capable of supporting autonomous replication of plasmids, show blocks of homology. There is no homology in the first 25 bases of the ARS region. (c) The replication origins in four animal DNA viruses contain the conserved sequence shown. SV40 is a monkey virus, BK and hepatitis B are human viruses, and polyoma is a rodent virus. The repetitious *Alu* sequences in the chromosomes of four species of mammals include a stretch that closely resembles the conserved sequence in the viral origins of replication. [See A. Kornberg, 1982, *1982 Supplement to DNA Replication,* W. H. Freeman and Company, p. S148; G. Tschumper and J. Carbon, 1982, *J. Mol. Biol.* **156**:293; and C. W. Schmid and W. R. Jelinek, 1982, *Science* **216**:1065.]

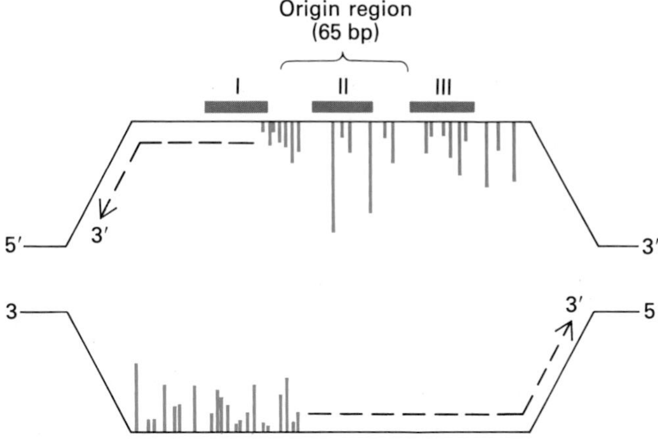

Figure 13-33 The sites of initiation of new DNA chains in the origin region of SV40. (The three horizontal bars I, II, and III indicate binding sites for the T antigen, which is required in DNA replication.) Nascent short DNA was recovered from replicating SV40 molecules within infected cells, its ends were labeled, and the endpoints were mapped by a hybridization procedure. Each site at which a chain started is marked by a colored bar, and the height of the bar indicates the frequency of starts at that site. [See R. T. Hay and M. L. DePamphilis, 1982, *Cell* **28**:767.]

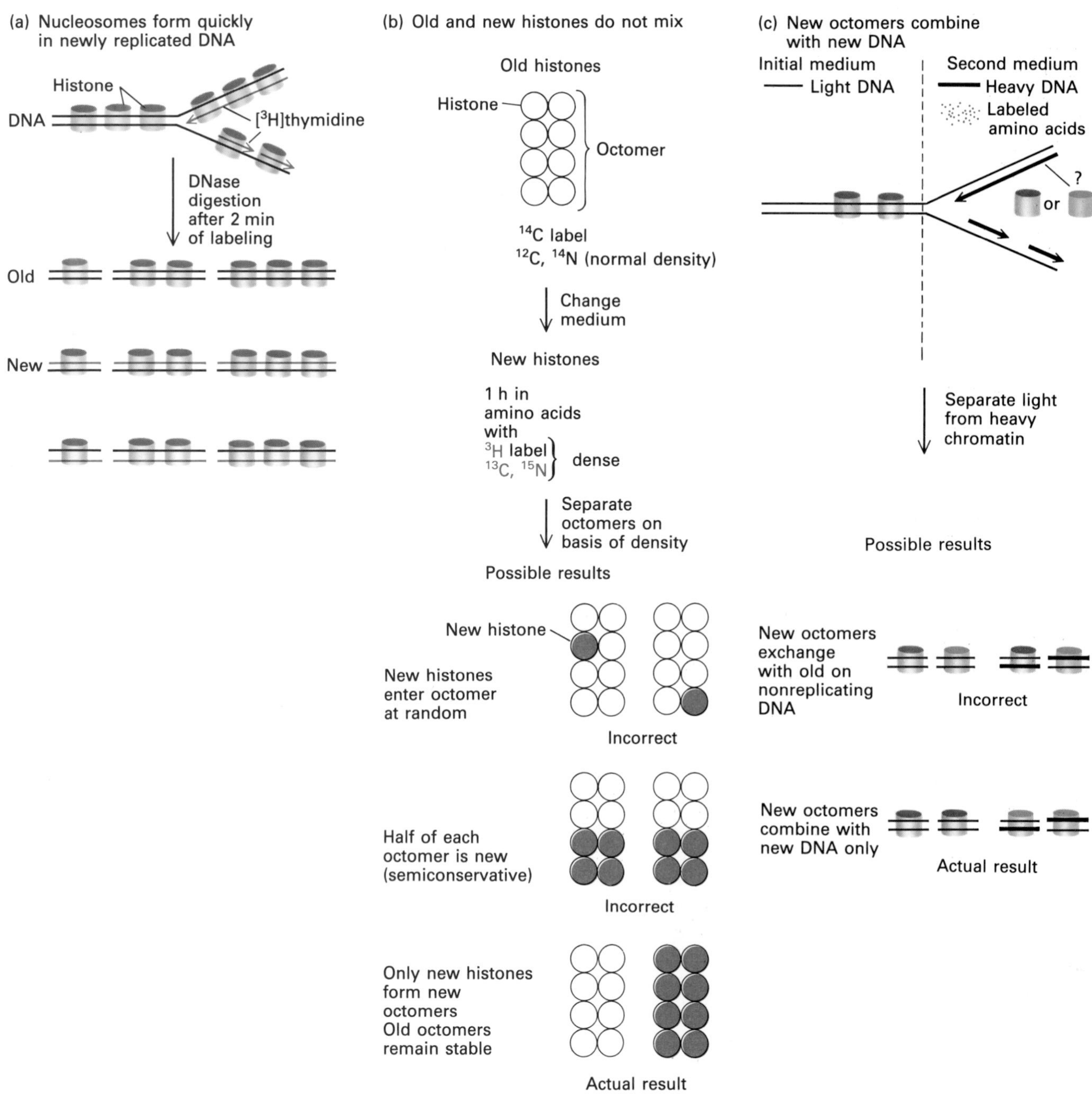

Figure 13-34 Chromatin formation after DNA synthesis. (a) Newly labeled DNA associates with histone octomers within 2 min after synthesis. This was determined by labeling cells with [^{3}H]thymidine and showing that all the new DNA liberated by DNase digestion of nuclei was in nucleosomes. (b) Old histones and new ones are not found in the same octomers. Cells grown for 1 h in amino acids containing radioactive label (^{3}H) and density label (^{13}C and ^{15}N) produced histone chains that were "hot and heavy" *(color dots)*. The octomers were separated on the basis of density. Each octomer had the same density as either the normal or the "hot and heavy" histone chains, and no octomers that were intermediate in density were detected. (c) New histone octomers combine with new DNA. Histones were labeled by growing cells in labeled amino acids, and at the same time a density label (iododeoxyuridine) was added to make the DNA heavy. Chromatin was digested into nucleosomes and separated according to density. The dense nucleosomes contained the new histone octomers. [See I. M. Leffak, R. Grainger, and H. Weintraub, 1977, *Cell* **12**:837; and G. Russev and R. Hancock, 1982, *Proc. Nat'l Acad. Sci. USA* **79**:3143.]

tone octomers in pre-existing nucleosomes remained associated and that new octomers were formed entirely of new histones. Thus the nucleosome core (the histone octomer) apparently emerges intact from replication (Figure 13-34b).

Another issue is whether the new octomers are added mainly to newly replicated DNA, or whether they can exchange with old octomers on DNA that is not replicating. This question has also been answered with density experiments. A heavy analog of thymidine (iododeoxyuridine), when incorporated into DNA, makes the new DNA heavy. When the chromatin containing heavy DNA is separated from that containing light DNA, new histones are found in octomers associated with heavy DNA. Therefore, the new octomers associate mainly with new DNA, and do not frequently exchange with nonreplicating DNA (Figure 13-34c).

Old Histone Octomers May Preferentially Associate with One Daughter Duplex

Finally, we consider the distribution of new and pre-existing histone octomers to two DNA duplexes newly created by a growing fork. This distribution could be random; or all the old octomers might go to one daughter duplex and all the new to the other daughter duplex. The solution of this problem might be especially critical if inheritance of a determined state by each daughter cell were controlled by the spacing or by the specificity of proteins bound to a region of chromatin.

It is a matter of controversy at present whether the deposition of histones on daughter duplexes is selective. The basis for differentiation of the daughter duplexes might lie in the fact that one daughter duplex contains the newly made leading strand and the other daughter the newly made lagging strand. Thus the two daughter duplexes might be distinguished biochemically. If so, the distinction is probably an important one, because DNA modifications that affect transcription (e.g., methylation) could be decided by elements that recognize whether a leading or a lagging strand has just been synthesized (Figure 13-35). Further research should clarify the connection between the direction of leading-strand growth and subsequent transcriptional activity.

Repair and Recombination of DNA

Errors in DNA sequence can be induced by environmental factors, such as radiation and mutagenic chemicals; errors are also occasionally committed by DNA polymerases during replication. If these errors were left totally uncorrected, both growing and nongrowing somatic cells might accumulate such great genetic damage that they could no longer function. In addition, the DNA in germ cells might be subject to far too many mutations for viable offspring to be formed. Thus, the correction of DNA sequence errors in all types of cells is important for survival.

Proofreading Nullifies Errors of DNA Copying

The enzymatic basis for the maintenance of correct base sequences during DNA replication is most completely understood in *E. coli*. First of all, the DNA polymerases demand a very accurate base pairing in the copying of a DNA strand. The error frequency of the *E. coli* DNA polymerase I is about 1 incorrect base in 10^4 internucleotide linkages. Because an average *E. coli* gene is about 10^3 bases long, an error frequency of 1 in 10^4 base pairs would still cause a potentially harmful mutation in every 10th gene during each replication, or 10^{-1} mutations per gene per generation. However, the measured mutation rate in bacterial cells is much less, about 10^{-5} to 10^{-6} mutations per gene per generation.

The mystery of this increased accuracy was cleared up when the proofreading function of DNA polymerases was discovered. A DNA template-primer complex containing a mismatched base at the 3′ hydroxyl end of the primer was synthesized. When the mismatched complex was supplied to either of the *E. coli* DNA polymerases I or III, the incorrectly hydrogen-bonded base (plus some additional bases as well) was removed by a 3′ → 5′ exonuclease activity that is a built-in function of the DNA polymerase molecule (Figure 13-36; also see Table 13-2). Therefore, it is likely that if an incorrect base were accidentally incorporated during DNA synthesis, the enzyme would excise it before going on.

This corrective activity, called *proofreading,* is a property of almost all bacterial DNA polymerases. The purified forms of animal cell DNA polymerases (α, β, and γ) do not have a 3′ → 5′ exonuclease capacity, but a newly discovered enzyme, polymerase δ, may have such an activity. Because the rate of mutation of coding regions is thought to be about the same in animal cells as in bacteria, proofreading is presumably used in higher cells also, but the enzyme or enzymes responsible for proofreading are not known.

Environmental DNA Damage Can Be Repaired

Many cells that divide very slowly or not at all (e.g., liver and brain cells) must use the information in their DNA for weeks, months, or even years. However, as we shall see, the DNA in all cells, even the nongrowing ones, is inherently unstable, both physically and chemically. Uncorrected base changes in the DNA of nongrowing cells would result in the production of faulty proteins at an unacceptable rate. The evolutionary reponse to this problem has been the development of DNA repair systems.

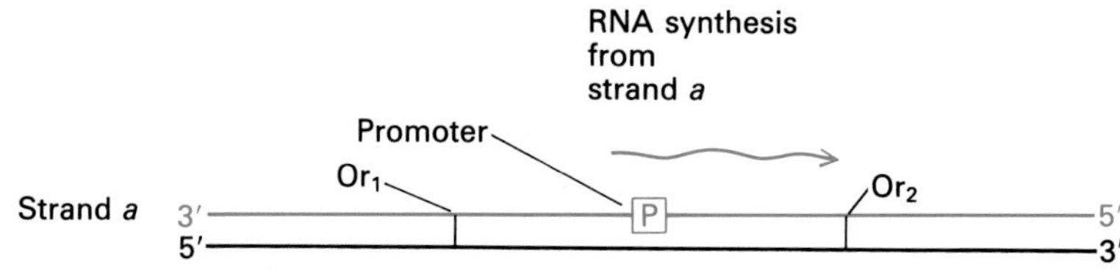

Fork arriving from Or₁ at P, *a* is template for leading strand

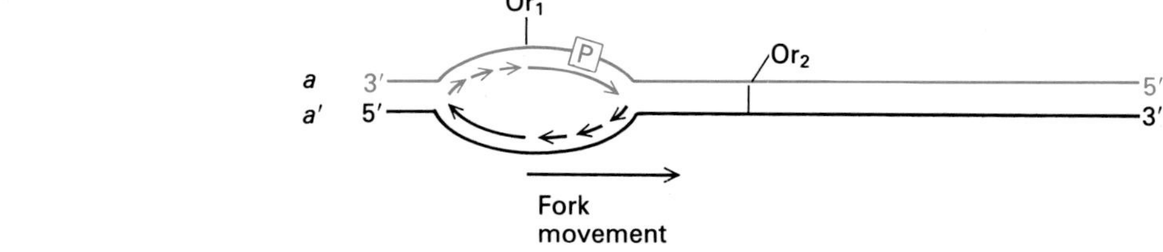

Fork arriving from Or₂ at P, *a* is template for lagging strand

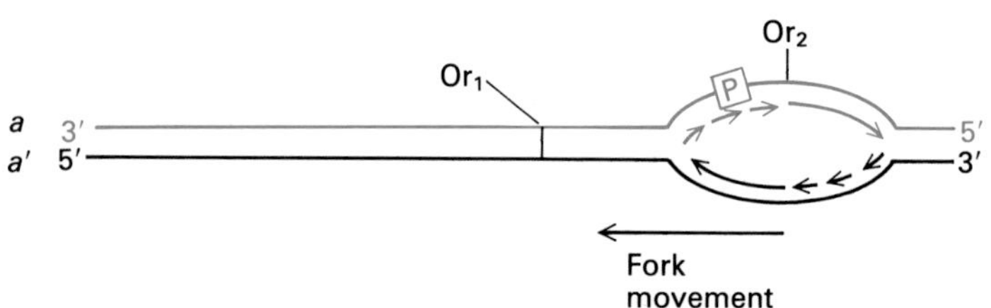

Figure 13-35 RNA synthesis from a template strand may be affected by the manner in which that strand is replicated in DNA synthesis. Strand *a* is the template strand for RNA synthesis from the promoter P. In the promoter region, the *a* strand is also the template for the DNA leading strand when the growing fork is moving from left to right, but the *a* strand is the template for the lagging strand when the growing fork is moving in the other direction. Biochemical events such as DNA methylation (or prevention of methylation) or deposition of modified proteins (histones or others) might depend on this directionality. [See M. M. Seidman, A. Levine, and H. Weintraub, 1979, *Cell* **19**:439.]

Common Types of DNA Damage Table 13-4 lists the general types of DNA damage and their causes. A prominent cause of damage in exposed cells is ultraviolet light, which stimulates the formation of pyrimidine dimers, most frequently between adjacent thymine bases in the same DNA strand (Figure 13-37). RNA synthesis ceases at pyrimidine dimers, and DNA synthesis may also stop.

Purines are susceptible to various types of damage. One is *spontaneous depurination*, which is due to the intrinsic instability of the nucleic acid linkage between the base and the deoxyribose at normal cell pH (Figure 13-38). It has been calculated that as many as 10,000 purine-sugar (*N*-glycosidic) bonds are cleaved per 24-h period in each mammalian cell, leaving as many *apurinic sites* in the DNA. Chemical exposure to many agents can also modify the purines. Some of these compounds, such as the nitrosoureas and complex organic molecules like aflatoxins (mold products), bind irreversibly to purines or break the purine ring. In addition, ionizing radiation can break the imidazole ring of the purine.

Table 13-4 DNA lesions that require repair

DNA lesion	Cause
Missing base	Acid and heat remove purines ($\sim 10^4$ purines per day per cell in mammals)
Altered base	Ionizing radiation; alkylating agents
Incorrect base	Spontaneous deaminations: C $\rightarrow$ U, A $\rightarrow$ hypoxanthine
Deletion/insertion	Intercalating agents (e.g., acridine dyes)
Cyclobutyl dimer	UV irradiation (see Figure 13-37)
Strand breaks	Ionizing radiation; chemicals (bleomycin)
Cross-linking of strands	Psoralen derivatives (light-activated); mitomycin C (antibiotic)

SOURCE: A Kornberg, 1980, *DNA Replication,* W. H. Freeman and Company, p. 608.

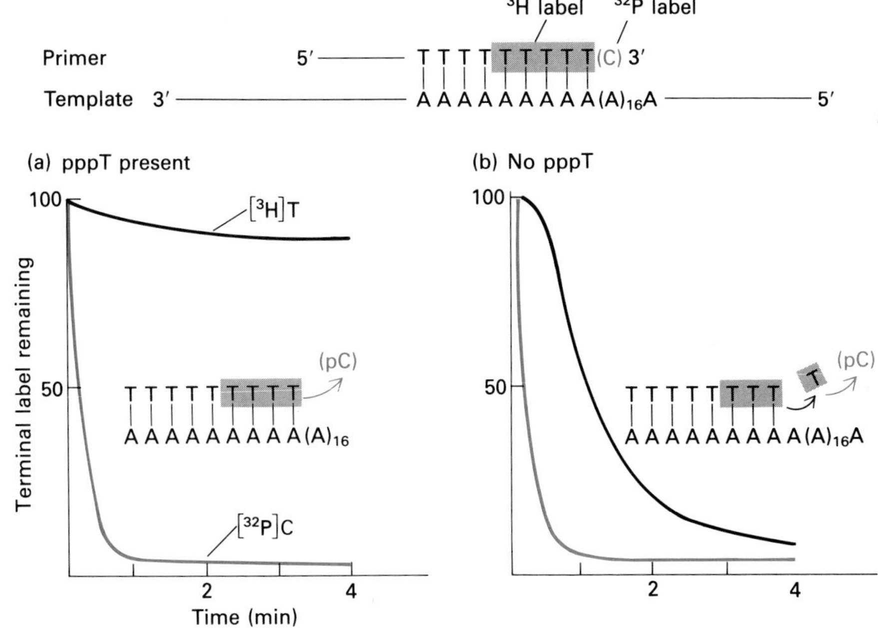

Figure 13-36 The proofreading function of the *E. coli* DNA polymerase I. A template (poly dA) and a primer were constructed. The 3′ end of the primer was labeled with [^{3}H]deoxythymidylate *(shaded)* and [^{32}P]deoxycytidylate *(color)*. The template-primer complex was then incubated with purified DNA polymerase I under each of two sets of conditions: (a) DNA chain elongation could occur (pppT, thymidine triphosphate, was present in the medium), or (b) chain elongation could not occur (no pppT available). In both cases, the 3′ → 5′ exonuclease proofreading activity promptly removed the C residues (releasing pC, deoxycytidylic acid). In (a), where polymerization of the deoxythymidylate was possible, the terminal T's were not digested. But in (b), where the enzyme had no thymidine triphosphates to polymerize, it digested the C's and then proceeded to remove the T's as well. [See A. Kornberg, 1980, *DNA Replication,* W. H. Freeman and Company, p. 128.]

Pyrimidine-glycoside linkages are much more stable than purine-glycoside linkages. The amino group of cytosine, however, is susceptible to loss at 37°C, resulting in the conversion of cytosine into uracil (Figure 13-38).

The Repair Cycle Similar events repair the DNA after various kinds of DNA damage. The DNA at or near the lesion is first *incised* by the correct enzyme. The signal for incision is often a mismatch or a disruption in the base pairing. For example, after a depurination, an unpaired pyrimidine is left; or after cytosine is deaminated, it no longer forms three hydrogen bonds with guanine. The lesion in the DNA must be removed by a nuclease; frequently, some surrounding bases also are removed. The gap in the DNA must then be filled and ligated (Figure 13-39). The enzymes responsible for each of these steps have been identified in bacterial cells that harbor mutations in one of the repair enzymes. In *E. coli,* pyrimidine-dimer–specific and apurinic–site-specific endonucleases and *N*-glycosidases have been purified. Many of the enzymes that function in repair also function during recombination of DNA.

The DNA in eukaryotic cells must also be repaired after chemical or physical damage, although most of the enzymes involved in this type of repair have not yet been identified. Naturally occurring mutations may help to

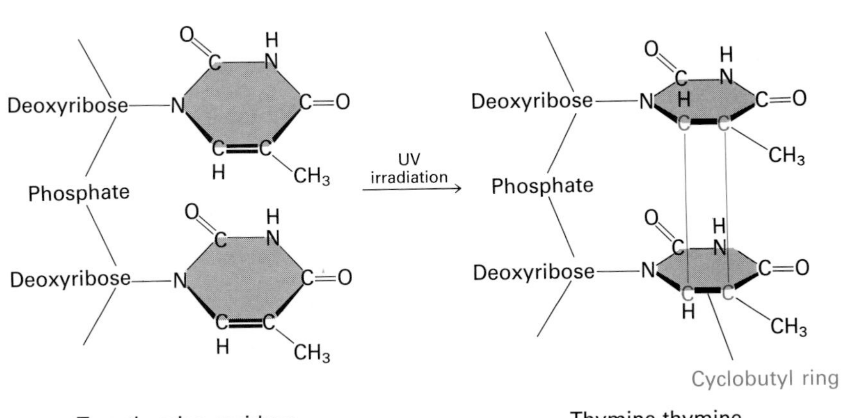

Figure 13-37 The effects of UV irradiation on DNA: production of thymine-thymine dimers. Two adjacent thymine residues in a single DNA chain are joined after irradiation by a cyclobutyl ring. Because the damage occurs in *one* chain of the duplex, it can be repaired by excising the dimer and recopying the missing bases from the other chain.

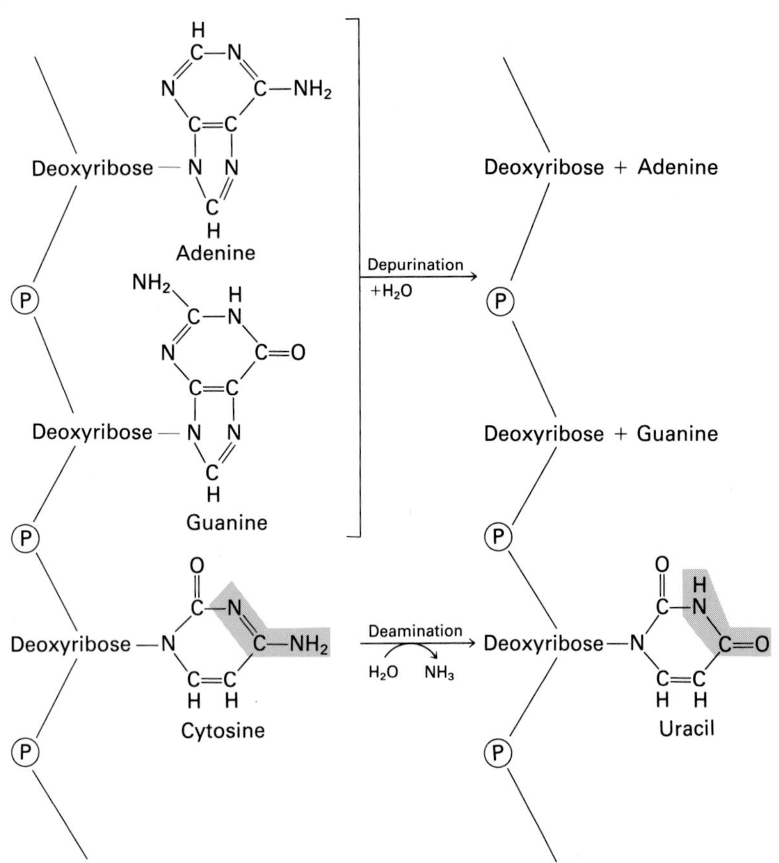

Figure 13-38 Susceptible bonds in DNA. The *N*-glycosidic bond that attaches purine bases to the sugar-phosphate backbone of DNA can break spontaneously, and it is also subject to enzymatic attack. In addition, cytosine can be deaminated to uracil.

elucidate the DNA repair pathways. For example, some humans who are mutant in the ability to perform UV repair have a disease known as xeroderma pigmentosum. This disease is associated with a very high risk of skin cancer, which is probably related to exposure of the skin to the UV rays in sunlight. Cells from various patients with xeroderma pigmentosum have shown evidence of several different defects in UV repair, suggesting that several different enzymes are required for DNA repair after UV irradiation. At least several other disease syndromes in humans seem to be associated with DNA repair defects (Table 13-5). Further analysis of the enzymatic deficiencies in cells from mutant individuals should illuminate both the particular liabilities these patients suffer and the normal mechanisms of DNA repair in eukaryotic cells.

Table 13-5 Human diseases associated with DNA repair defects

Disease	Sensitivity	Cancer susceptibility	Symptoms
Xeroderma pigmentosum	UV irradiation, alkylation	Skin carcinomas and melanomas	Skin and eye photosensitivity
Ataxia telangiectasia	γ irradiation	Lymphomas	Unsteady gait (ataxia); dilation of blood vessels in skin and eyes (telangiectasia); chromosomal aberrations
Fanconi's anemia	Cross-linking agents	Leukemias	General decrease in numbers of all blood cells; congenital anomalies
Bloom's syndrome	UV irradiation	Leukemias	Photosensitivity; facial changes

SOURCE: A. Kornberg, 1980, *DNA Replication,* W. H. Freeman and Company, p. 622.

(a)

Undamaged: C A T T C C N

UV irradiation

T—T dimer

Damaged: C A T T C C N

UV endonuclease

C A OH — T T P OH — P C C N

exonuclease

C A OH — C P OH — P OH N

DNA polymerase copies complementary strand

C A T T C C OH N

DNA ligase

Repaired: C A T T C C N

(b)

Undamaged: C A T T C G N

Deamination

Damaged: C A T T U G N

Uracil–DNA glycosylase

C A T T U G N

AP endonuclease (AP = apurinic- or apyrimidinic-site specific)

C A T T OH — G N

exonuclease

C A T T OH — OH — G OH — N

DNA polymerase copies complementary strand

C A T T C G OH N

DNA ligase

C A T T C G N

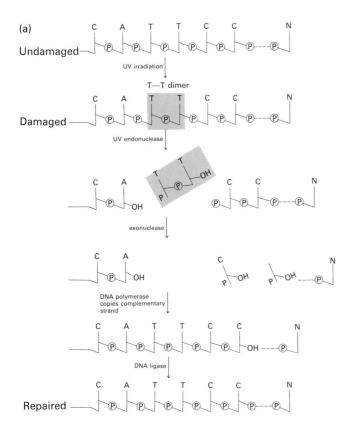

Figure 13-39 Steps in the repair of DNA lesions. (a) The removal of a thymine-thymine dimer by a specific endonuclease is followed by repair at the damaged site. (b) A similar process can repair a deamination of cytosine to uracil. The removal of uracil is also carried out by a specific enzyme. In both cases, a DNA polymerase (e.g., DNA polymerase I in *E. coli*) is necessary to fill the gap, which is then sealed by DNA ligase. [See T. Lindahl, 1979, *Prog. Nuc. Acid Res.* **22**:135.]

Recombination

Soon after Mendel's rules of independent gene segregation were rediscovered, and the segregation of linked groups of genes on individual chromosomes was widely recognized, another great genetic discovery was made in *D. melanogaster*. Blocks of genes from homologous chromosomes could be exchanged by the process of crossing over or recombination (Figure 13-40). (See Chapter 10 for descriptions of classical recombination studies and their terminology.)

Recombination, which takes place during meiosis in sexually reproducing organisms, is the basis for the classical genetic maps. Recombination between homologous

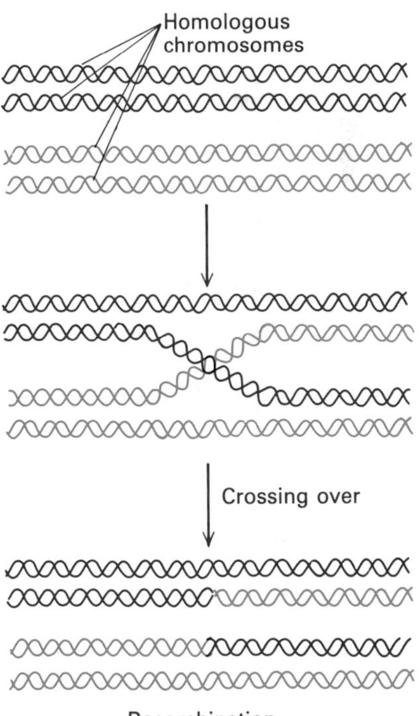

Figure 13-40 The molecular event of recombination at meiosis may be schematically represented by two double-stranded molecules breaking and rejoining. However, no enzyme systems that carry out such a neat recombination are known in meiotic cells.

chromosomes occurs not only in animals and plants but also in prokaryotes, viruses, plasmids, and even in the DNA of cell organelles such as mitochondria. The events in a reciprocal recombination are equivalent to the breakage of two duplex DNA molecules representing homologous but genetically distinguishable chromosomes, an exchange of *both* strands at the break, and a resolution of the two duplexes so that no tangles remain. How such a double-stranded cleavage at two precisely analogous sites can take place, followed by the swapping of duplex regions and the rejoining of ends, is not at all clear. Closer examinations of DNA molecules at the sites of recombination have shown that recombination is most often not such a simple sequence of events.

Models of Recombination Must Explain Gene Conversion

The most popular current models for the molecular events of recombination are outcomes of attempts to explain the behavior of genetic markers near the site of reciprocal recombination between two chromosomes. Although markers at some distance from the crossover point are exchanged in a reciprocal fashion during recombination, there is frequently evidence of a nonreciprocal exchange at or near the crossover point (Figure 13-41). This nonreciprocal exchange has been termed *gene conversion* because, as is shown in Figure 13-41b, one allele is apparently "converted" into another. In the figure, two chromatids bearing the *D* sequence and two bearing *d* enter meiosis, but in the spores there is a 3:1 relationship between *D* and *d* sequences at this locus.

The Holliday Model To explain gene conversion during the recombination event in which distant markers are exchanged reciprocally, we consider the following model. First, suppose that a nick occurs in one strand of each of the two homologous chromosomes that are going to recombine. Strand exchange might then occur at the site of the nick (Figure 13-42, steps 1 through 4). The point of

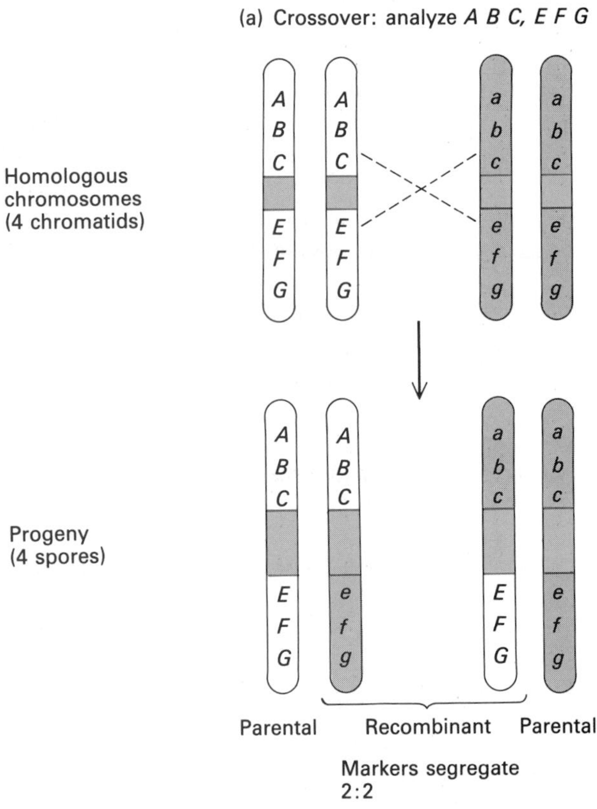

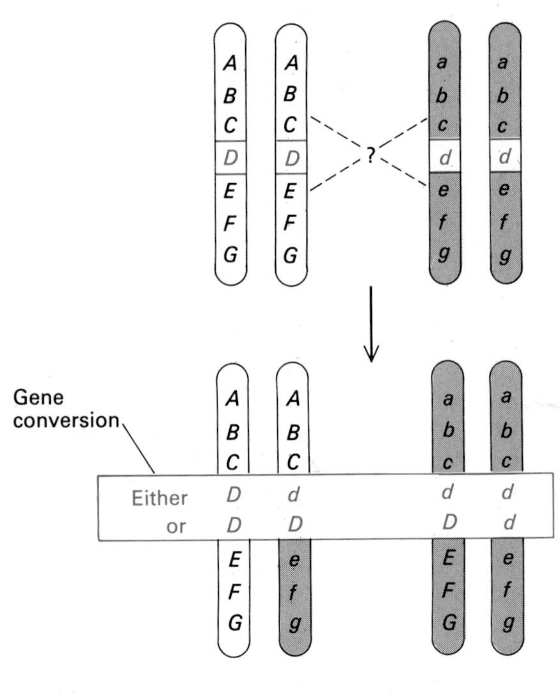

Figure 13-41 Examination of recombination or crossover events in yeast reveals frequent gene conversion near crossover points. One chromatid in each of the two chromosomes in a homologous pair can recombine during meiosis. The individual crossover events are recorded in the four spores that result, and each spore can be recovered so that its genotype can be examined. (a) When examination is restricted to fairly distant markers, for example, *ABC*/*abc* and *EFG*/*efg*, recombination will yield two recombinants and two parental types—that is, a 2:2 segregation. (b) However, if closer examinations are made—say, at the locus *D*/*d*—there might be one *D* and three *d*'s, or three *D*'s and one *d*.

connection of the two chromosomes could then be moved by *branch migration:* the breaking of hydrogen bonds within the parental duplexes in chromosomes I and II, followed by the exchange of strands and the remaking of hydrogen bonds (Figure 13-42, step 5). A region of heteroduplex is created by these maneuvers, but no recombination has yet occurred. A cut is made across all four strands at the site of the crossed strands; the left side of chromosome I is joined to the right side of chromosome II, and vice versa. This exchange produces the reciprocal recombination of all markers to the left and right of the crossed-strand region and leaves a heteroduplex in the region through which the branch migration occurred (Figure 13-42, steps 6 and 7). Because of the mismatches in this region, the heteroduplexes might be subject to DNA repair mechanisms, during which the *D* or the *d* information in one of the strands of the heteroduplex would be lost or "corrected" by the copying of the other strand generating the 3:1 or 1:3 ratios.

This scheme of events, which was originally proposed by Robin Holliday, would account for reciprocal recombination on either side of the recombination site and gene conversion at or near the recombination site. The crosssed-strand structure (i.e., that in step 5 of Figure 13-42) is referred to as the Holliday structure.

A refinement in the original model simplifies the enzymatic cutting that would be necessary to disentangle the recombining DNA molecules. The two chromosomes in the Holliday structure can be rotated at the site of the connecting strands (Figure 13-42, steps 8 and 9); the rotated version is an isomer of the original crossed-strand version. It is now possible to release the two connected duplexes from each other by cutting only two strands. The cutting can occur in two ways (across the page or up and down the page, as shown in step 9 of Figure 13-42). This type of reaction, which releases entangled DNA helixes, is said to *resolve* two connecting duplex molecules. Sealing the cut sites produces the two alternative results shown in Figure 13-42, frames 10a and 10b. The cut leading to frame 10a produces heteroduplex regions at the crossover site but no recombination; the cut yielding frame 10b produces recombination and heteroduplexes. Correction of the heteroduplexes resulting from either alternative could result in gene conversion.

The Meselson-Radding Model: Single-Strand Uptake A further modification of the recombination scheme was suggested by Matthew Meselson and Charles Radding. A single invading DNA strand can be responsible for creating a crossed-strand Holliday structure (Figure 13-43). The single-strand crossover becomes a two-strand crossover by isomerization (steps 8 and 9, Figure 13-42), thus eliminating the need for a nick in both chromosomes. After branch migration of the crossed-strand region, rotation and resolution produce results similar to those described in Figure 13-42. This finding is an important addition to the thinking about the mechanisms of recombination, because it is experimentally demonstrable that a single strand can be used in cell-free enzymatic experiments to initiate a recombination event as described in the next section (see Figure 13-45).

Virus and plasmid DNA molecules in the act of recombining can be extracted from both bacterial and animal cells. Many of these molecules have the linked strands predicted by Holliday structures in the rotated or isomeric form (Figure 13-44). Thus, however the initial event of recombination occurs, the final connection between the unresolved chromosomes seems to involve branch migration and chromosomal rotation.

Many of the *E. coli* Proteins Involved in Recombination and Repair Are Known

Most enzymatic studies of recombination in *E. coli* have tended to focus on strand nicking and initial uptake of one marked DNA strand. Recombination in *E. coli* may proceed largely by this mechanism. Of particular interest is the RecA protein, the product of the *rec*A gene. Mutations in this gene render recombination infrequent in the mutant cells. The protein is capable of carrying out some of the proposed steps in recombination. It causes single-stranded DNA to be "taken up" in a hybrid structure at either the end of a duplex or at a gap in a duplex (Figure 13-45). This uptake of a single strand can lead, as shown in Figure 13-43, to recombination. ATP is a cofactor in the uptake reaction, but it is not hydrolyzed. The RecA protein will then promote branch migration so that the incoming single strand displaces one of the strands of the original helix. This latter reaction is driven by the consumption of ATP. (Note in the theoretical diagram of Figure 13-43 that when two double-stranded chromosomes are linked originally by one strand, helix rotation can result in two-strand crossover. RecA protein action after such a rotation would cause the branch migration in the crossed-strand structure.)

SOS Repair A second fascinating aspect of the RecA protein is the series of gene control functions in which the protein participates. When *E. coli* cells are irradiated with UV light, a whole series of genes are activated, including the *rec*A gene itself. Apparently UV light also puts the RecA protein into an active state (RecA′), probably by its interaction with oligonucleotides formed as a result of DNA degradation after irradiation. The RecA′ form has proteolytic activity that attacks various DNA-binding proteins (Figure 13-46). The protein repressor of lysogenic phages such as λ can be cleaved by the RecA′ protease in vitro; this incapacitation of the repressor is presumably why the dormant phage is induced in irradiated cells (Chapter 8).

Another target of the RecA′ protease activity has helped elucidate its role in UV repair. This protein, which

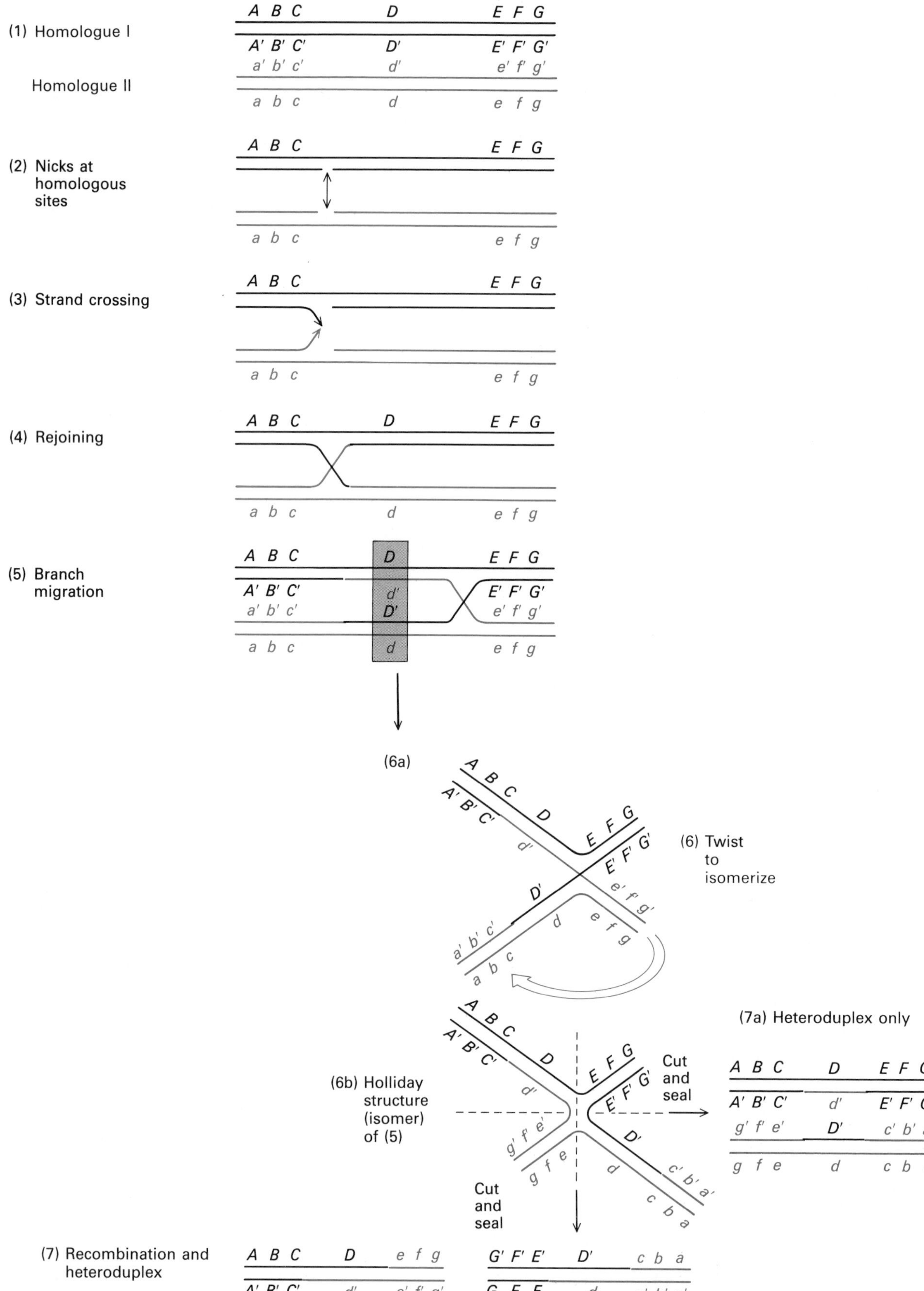
(1) Homologue I
Homologue II
A B C D E F G
A' B' C' D' E' F' G'
a' b' c' d' e' f' g'
a b c d e f g
(2) Nicks at homologous sites
(3) Strand crossing
(4) Rejoining
(5) Branch migration
(6a)
(6) Twist to isomerize
(6b) Holliday structure (isomer) of (5)
Cut and seal
(7a) Heteroduplex only
(7) Recombination and heteroduplex
A B C D e f g
A' B' C' d' e' f' g'
G' F' E' D' c b a
G F E d c' b' a'
A B C D E F G
A' B' C' d' E' F' G'
g' f' e' D' c' b' a'
g f e d c b a

Figure 13-42 (***opposite***) The Holliday model of genetic recombination.
(1) Two genetically distinct homologous chromosomes, I and II, are aligned. Consider strand with letter without ′ to be one polarity (e.g., $5' \rightarrow 3'$) and the other strand the opposite polarity.
(2) A nuclease cuts one strand in each duplex at homologous sites in strands of same polarity.
(3),(4) The two cut ends are exchanged and rejoined.
(5) Branch migration results in heteroduplex formation.
(6a) In order to uncross the strands, the bottom half of the connected structure is rotated, leaving a center with no crossed strands.
(6b) This is the Holliday structure.
(7a), (7b) Two planes of strand cutting can occur. As diagrammed, a cut in the horizontal plane and resealing yields no recombination, but because of the branch migration two heteroduplexes now exist at *D*. A cut in the vertical direction followed. Any resealing yields reciprocal recombination for *A*, *B*, *C*, *E*, *F*, and *G* and a heteroduplex at *D*. [Re steps 1 to 5, see R. Holliday, 1964, *Genet. Res.* 5:282. Re steps 6 to 7, see D. Dressler and H. Potter, 1982, *Annu. Rev. Biochem.* **51**:727. Also see M. Meselson and C. M. Radding, 1975, *Proc. Nat'l Acad. Sci. USA* **72**:358; and N. Sigal and B. Alberts, 1972, *J. Mol. Biol.* **71**:769.]

is called LexA (it is the product of the regulatory gene *lex*A), has negative control over a series of other genes. Some of these have no apparent connection with DNA repair but others do; for instance, both the *rec*A gene itself and the genes for the uvrA nuclease, which excises thymine dimers, are expressed when the LexA protein is digested. This elaborate response, termed *SOS repair*, sheds light on why UV irradiation not only induces repair synthesis but also, because of the increased RecA production, increases recombination.

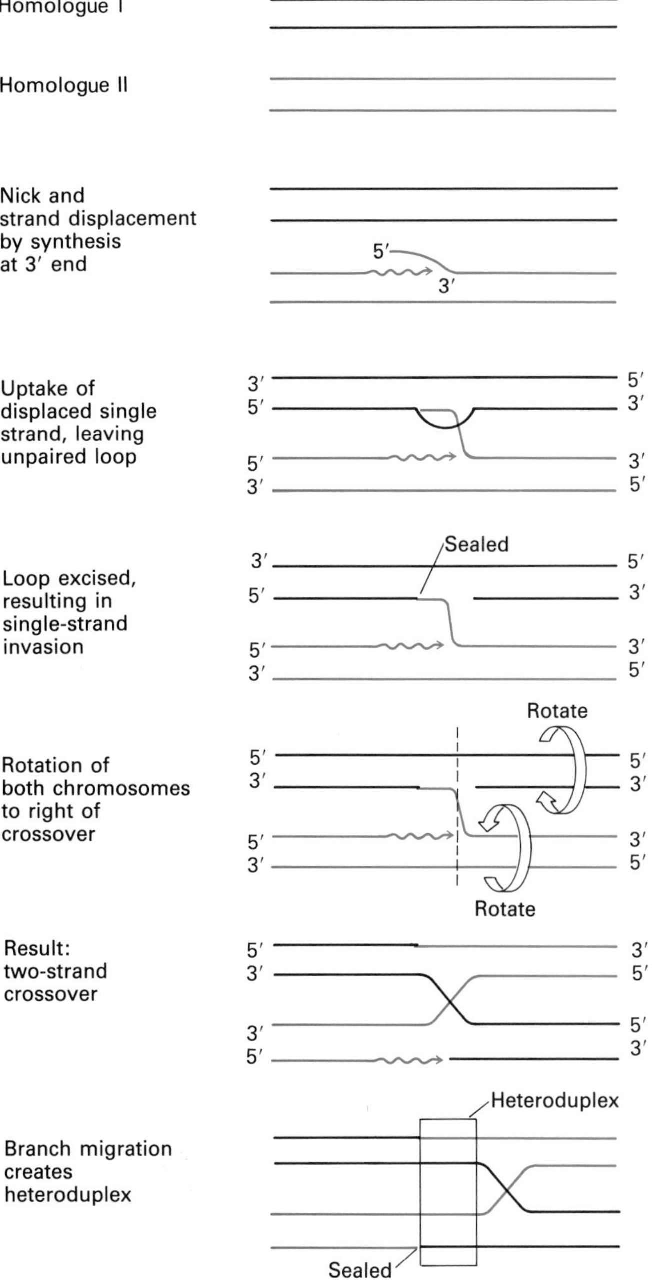

Figure 13-43 The Meselson-Radding model for recombination begins with a single nick. A free 5′ end in chromosome II is shown pairing with its homologous site in chromosome I. The free 3′ hydroxyl end that is created at the nick allows displacement synthesis of the DNA chain from that end. Hybridization of the invading region of chromosome II to chromosome I creates a loop (an unpaired region) in chromosome I. The loop is cleaved and removed, leaving a heteroduplex region with a single strand joining the two chromosomes. Rotation of the two chromosomes at the point of connection could result in a two-strand linkage between them. Branch migration could then occur, followed by resolution (further rotation and cutting of strands), as shown in Figure 13-42, steps 8 to 10. [See M. Meselson and C. Radding, 1975, *Proc. Nat'l Acad. Sci. USA* 72:358.]

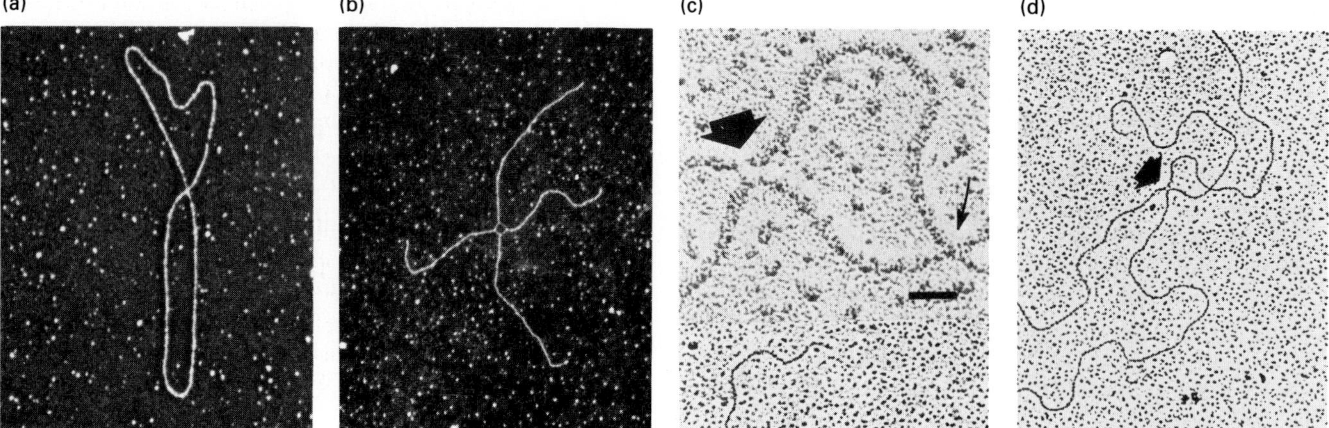

Figure 13-44 Electron micrographs of recombining DNA molecules. (a) A plasmid dimer in the process of recombination. (b) In this more highly magnified plasmid dimer, the single-stranded ring in the Holliday structure is visible. (c),(d) These recombining adenovirus type 2 molecules released from human cells show a similar type of structure [*wide arrows;* the thin arrow in (c) simply designates where one strand crosses over the other]. Part (c) is a higher magnification of the same molecules shown in (d). [See H. Potter and D. Dressler, 1978, *Cold Spring Harbor Symp. Quant. Biol.* **43**:969; and D. J. Wolgemuth and M.-T. Hsu, 1980, *Nature* **287**:168.] *Parts (a) and (b) courtesy of D. Dressler. Parts (c) and (d) courtesy of M.-T. Hsu. Reprinted by permission from* Nature. *Copyright 1980 Macmillan Journals Limited.*

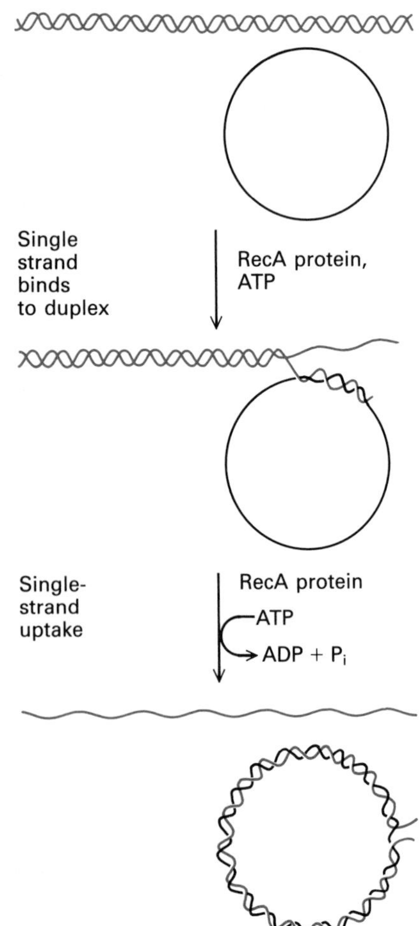

Figure 13-45 Strand uptake catalyzed by the *E. coli* RecA protein. Linear duplex DNA plus purified RecA protein will bind to single-stranded DNA at a region of homology between the single-stranded molecule and the end of the duplex. ATP is required for this step, but it is not hydrolyzed. ATP hydrolysis accompanies single-strand uptake in a second stage of the reaction. [See M. M. Cox and I. R. Lehman, 1981, *Proc. Nat'l Acad. Sci. USA* 78:3433.]

Recombination in Yeast Probably Involves Double-Strand DNA Breaks

The mechanisms for recombination originating from single-strand breaks have been well studied in bacteria, but some recombination may not be the result of single-strand breaks. For example, in yeast cells, double-strand cuts in the DNA may participate actively in mitotic, and possibly in meiotic, recombination. Yeast cells can be transformed with plasmids containing a selectable yeast gene with a known sequence and therefore with known restriction enzyme sites. Such transforming plasmid DNA is found recombined into the yeast genome at its homologous site on yeast chromosomes. Recombination is much more frequent if before transformation *double-strand breaks* are introduced within the yeast gene contained in the plasmid (Figure 13-47).

Even if a section of the gene in the plasmid is removed, the recombination still occurs; all the recombinants contain full-sized genes, which implies that a gene conversion

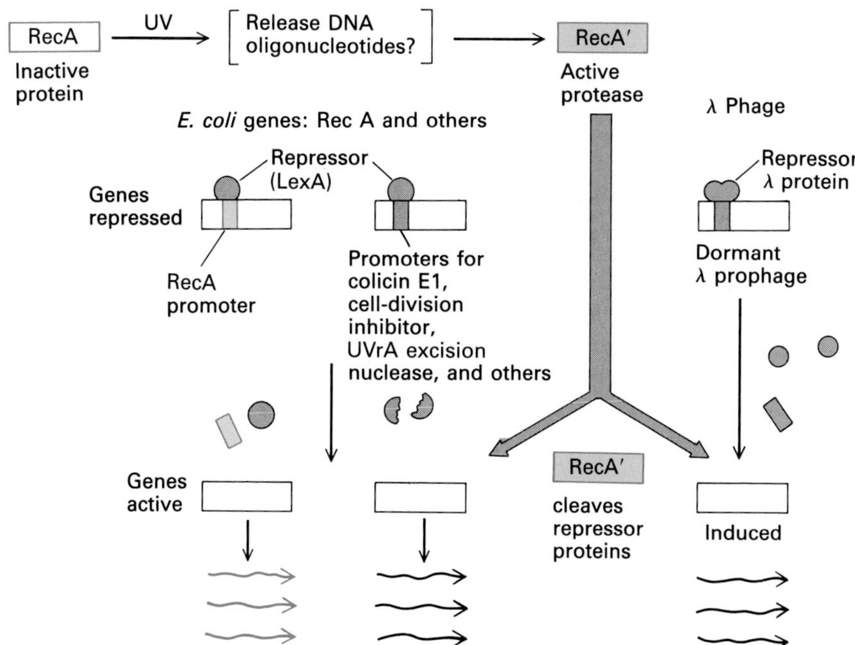

Figure 13-46 Genes that are activated during the SOS response to UV irradiation of bacteria. The bars represent genes for the functions indicated; shaded areas designate operator (repressor-binding) sites. Activation of the RecA′ protease results in the increased synthesis of a number of proteins, including proteins that function in the repair of damaged DNA. The collective name for this sequence of events following UV damage is the SOS repair response. [See J. W. Little and D. W. Mount, 1982, *Cell* **29**:11.]

has corrected the gap. The recombination of the plasmid into a yeast chromosome is dependent on a gene called *RAD52*. Mutations in *RAD52* render cells very sensitive to irradiation and incapable of carrying out such recombination. All these experiments suggest that at least some recombination in yeasts is dependent on double-strand breaks rather than on single-strand nicks with strand invasion.

To determine exactly how recombination occurs at the molecular level, we shall need additional enzymatic studies involving recombination proteins and probably also topoisomerases, which would resolve tangles as they arose. Quite likely different enzymatic schemes for recombination occur in different cells.

Recombination Requires Special Proteins: The *red* Genes in Bacteriophage λ and Synaptonemal Complexes in Eukaryotes

Recombination does not occur at the same rate in all genomes in all cells at all times. For example, bacteriophage λ has genes (the *red* genes) that are dispensable for phage growth, but that when active produce proteins that greatly increase the rate of recombination between λ genomes. Also, DNA sequences called *chi* (X) increase recombination frequency in their vicinities about fivefold. Thus recombination should be thought of as a metabolic process requiring the right enzymes at the right time and not as an ongoing process for all somatic cells at any given time. Recently purified proteins such as the *int* gene product (*int* for integration) of bacteriophage λ have been shown capable of cutting and rejoining DNA within a 12-base-pair recognition sequence leading to the insertion of one DNA circle into another. This type of recombination or integration occurs in lysogenization of *E. coli* by λ bacteriophage.

Some recombination can occur in somatic cells. Sister chromatid exchange is well documented in cultured mammalian cells, but breaks and reunions between sister chromatids in somatic cells do not result in genetic recombination because the exchanging chromosomes are identical. Mitotic recombination might arise during metaphase, if the homologues could pair and exchange DNA before centromere division. The DNA biochemistry that would allow any such recombination events in eukaryotic cells remains mysterious at present.

It is important to keep in mind that at the molecular level, recombination is not equally frequent at all sites on chromosomes. In yeast, for example, the frequency of meiotic recombination can vary among different specific sites by a factor of 10. The overall average is about a 0.35 percent chance of recombination when two genes are 1 kb apart.

The pairing of homologous chromosomes during meiosis may begin at similar points that become aligned when the chromosomes are close together at the nuclear periphery. This is the phenomenon called *synapsis*. A specialized structure called the *synaptonemal complex*, which is visi-

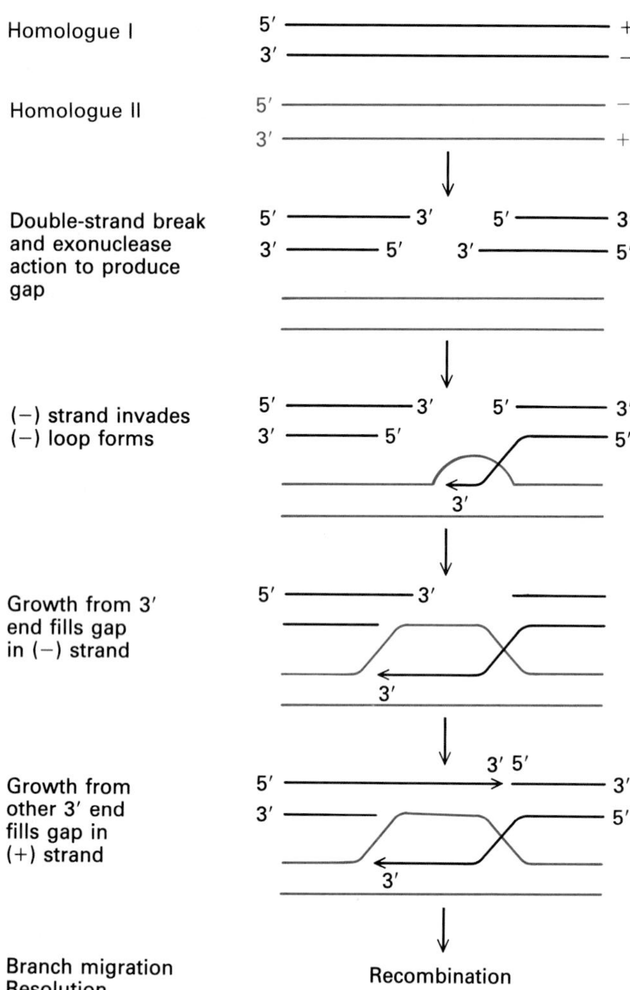

Figure 13-47 A double-strand break model of genetic recombination. Molecules with double-strand breaks can initiate recombination in yeast, even if the strands have gaps in them. This diagram is a model illustrating how such gap filling might initiate recombination. The gap is created in chromosome I, leaving 3′ protruding ends that can invade chromosome II. By extension of the 3′ ends, the gap in first one strand and then the other is repaired. Branch migration creates one of the Holliday structures shown in Figures 13-42 and 13-43, and resolution of the crossed strands leads to recombination. [See J. W. Szostak, T. L. Orr-Weaver, and R. J. Rothstein, 1983, *Cell* 33:25.]

ble in electron micrographs (Figure 13-48a), is assembled in many eukaryotic organisms to function as a scaffold on which meiotic recombination might occur (Figure 13-48b). Mutant animal and plant strains lacking fully formed synaptonemal complexes have greatly decreased frequencies of recombination.

Little biochemical detail about meiotic recombination is available as yet. Increases in the concentrations of endonucleases and DNA-binding proteins during meiosis in plants have been described. In addition, the incorporation of [^{3}H]thymidine into meiotic DNA has been used to detect the patching and gap filling that are presumed to occur during recombination.

Our best hope for learning the details of meiotic recombination comes from yeast, where a number of mutants that cannot undergo meiosis are known; in some of these cases the genes have been cloned. Through combined genetic and biochemical analyses, discovery of the proteins involved in meiosis and their actions in recombination should be possible.

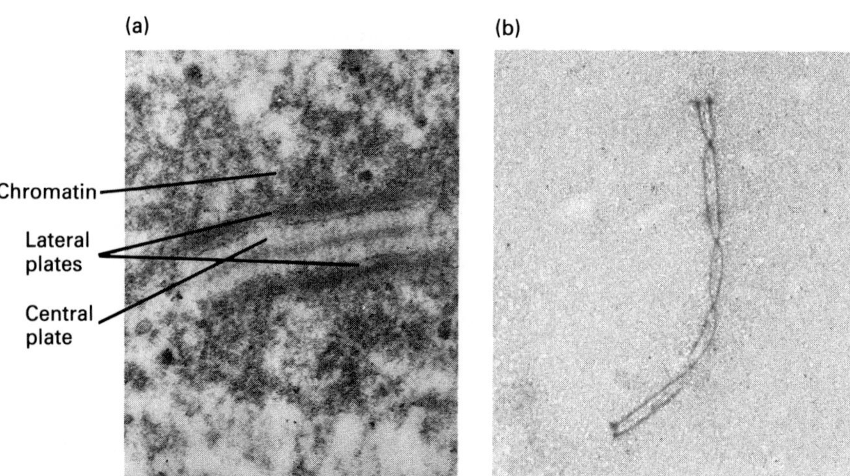

Figure 13-48 (a) Electron micrograph of a thin section from a rooster spermatocyte in the pachytene stage (the meiotic stage in which chromatids divide). Dense amorphous chromatin surrounds the synaptonemal complex (SC), a rigid structure defined by two lateral plates connected to a central plate by filamentous layers. (b) Two meiotic chromosomes (human, probably numbers 19 or 20) stained with phosphotungstic acid, which darkens only the SC so that the chromatin does not show. The SC ends are attached to the nuclear membrane. [See M. J. Moses, 1968, *Annu. Rev. Genet.* 2:363; and M. J. Moses, S. J. Counce, and D. F. Paulson, 1975, *Science* 187:363.] *Courtesy of M. J. Moses. Reproduced with permission, from the* Annual Review of Genetics. *Copyright 1968 by Annual Reviews Inc.*

Summary

The general principles of DNA replication seem to apply, with little modification, to all cells. Viewed at the level of whole chromosomes, DNA synthesis initiates at regions called origins; bacterial chromosomes have single origins, and eukaryotic chromosomes have multiple origins. Synthesis usually moves bidirectionally away from an origin via two growing forks proceeding in opposite directions; this produces a replication bubble. The copying of the DNA duplex at the growing fork is semiconservative; that is, each daughter duplex contains one old strand and its newly made complementary partner.

In bacterial cells, the generation of the new chromosome occupies almost the entire doubling time, so that each cell is in almost continuous DNA synthesis. In eukaryotic cells, DNA synthesis is confined to a fraction of the cell cycle, the S (synthesis) phase. The control of the entry of cells into the S phase is not well understood.

In contrast, much progress has been made in understanding the enzymes and the other factors that perform necessary functions at a growing fork. These proteins have been most extensively studied in *E. coli,* but many such proteins are known in eukaryotic cells; the problems of DNA replication are likely similar—and solved in a similar manner—in all cells. Most of the enzymatic events at the growing fork are a consequence of two properties of the DNA double helix and one property of DNA polymerases, the enzymes that copy DNA. First, the DNA in the helix is an antiparallel duplex; that is, the $5' \rightarrow 3'$ direction of one strand is opposite to the $5' \rightarrow 3'$ direction of the other. Second, the strands of the duplex are interwound and cannot simply be pulled apart along their entire lengths all at once. Third, DNA polymerases cannot initiate a new DNA chain de novo; they must use a nucleic acid primer—that is, they need a 3′ hydroxyl end of a DNA or an RNA molecule—to begin synthesis. All DNA chain growth consequently proceeds in the $5' \rightarrow 3'$ direction. The growing fork proceeds in one overall direction, however, causing one of the two strands at each growing fork to be synthesized in a discontinuous fashion.

The strand synthesized continuously in the direction of movement of the growing fork is called the leading strand. It can start from the 3′ hydroxyl end created by a nick in the DNA. For synthesis of the other strand, the lagging strand, a series of short RNA pieces must be synthesized on the remaining parental template strand, in the direction opposite to the overall direction of the growing fork. Enzymes called RNA primases make the RNA chains. DNA chains then grow from the RNA primers; the resulting segments of RNA plus DNA are called Okazaki fragments. After a nuclease that is part of DNA polymerase removes the RNA primer, the enzyme DNA ligase seals the neighboring Okazaki fragments together. Thus the growing fork moves along in one direction and the principle of $5' \rightarrow 3'$ DNA chain synthesis is preserved.

Many protein factors assist DNA polymerase, primase, and DNA ligase in the construction of a new chromosome. Two of these factors are helicases, which can unwind the DNA, and topoisomerases. The latter can change the form of DNA in a number of ways by cutting it; the DNA may be allowed to swivel so that it can be unwound; or, with an energy expenditure, the topoisomerase can pass one DNA duplex through another to introduce supercoils or to remove knots. Topoisomerases are probably important in starting DNA replication and in resolving any tangles remaining after DNA duplication has been finished. The entire set of proteins responsible for replication appear to form a giant complex, the replisome, which moves the growing fork along at great speed.

How DNA synthesis is initiated is less well understood than the events at an already functioning growing fork. Specific sites of initiation have been identified in bacterial cells, in many viruses, and in yeast cells; such sites probably exist in all cells. The means by which the DNA polymerase locates an initiation site and begins to replicate the first leading strand of a replication bubble may involve a nick in the DNA, or, in some cases, RNA synthesis at or near the origin. Lagging-strand synthesis follows, and a replication bubble is begun.

In eukaryotic cells, chromosomal replication involves more than just making the DNA. The new DNA is packaged promptly into nucleosomes. Most of the old histone octomers are retained in nucleosomes, and new histone octomers are added to new DNA shortly after its synthesis. The addition of new histone octomers and the retention of old histone octomers may follow patterns of nucleosome packing established in previous cell generations; these patterns may affect the transcription of the DNA. The means by which such effects are transmitted by chromatin packing, however, is not yet clear.

Environmental hazards damage bases in DNA and necessitate repair of the DNA so that mutations do not occur at an intolerable rate. This repair is achieved by a battery of enzymes capable of recognizing damaged or mismatched base pairs and excising them. The enzymes capable of single-stranded DNA synthesis then often widen the single-strand gap before filling it in and religating the free ends of the DNA.

The enzymes that take part in the repair function may also participate in recombination, the event in which two DNA helixes are broken and exchanged. Probably at least two initiating events start recombination: (1) invasion of a duplex from the cut end of one strand of a neighboring similar duplex and (2) invasion from the end of the duplex with both strands cut. These initial events are followed by rotation of the interconnected DNA molecules and finally by enzymatic resolution of the recombining molecules so that a double-strand exchange occurs. Special enzymes and, in eukaryotes, special nuclear meiotic structures called *synaptonemal complexes* are used in recombination; it is not an event that is equally probable in all cells at all times.

References

General References

DNA: Replication and Recombination. 1979. *Cold Spring Harbor Symp. Quant. Biol.*, vol. 43.

KORNBERG, A. 1980. *DNA Replication.* W. H. Freeman and Company.

KORNBERG, A. 1982. *1982 Supplement to DNA Replication.* W. H. Freeman and Company.

DNA Replication

HAND, R. 1978. Eukaryotic DNA: organization of the genome for replication. *Cell* **15**:317–325.

HUBERMAN, J. A., and A. D. RIGGS. 1968. On the mechanism of DNA replication in mammalian chromosomes. *J. Mol. Biol.* **32**:327–341.

HUBERMAN, J. A., and A. TSAI. 1973. Direction of DNA replication in mammalian cells. *J. Mol. Biol.* **75**:5–12.

KREIGSTEIN, H. J., and D. S. HOGNESS. 1974. Mechanisms of DNA replication in *Drosophila* chromosomes: structure of replication forks and evidence for bidirectionality. *Proc. Nat'l Acad. Sci. USA* **71**:135–139.

MESELSON, M., and F. STAHL. 1958. The replication of DNA in *E. coli. Proc. Nat'l Acad. Sci. USA* **44**:671–782.

SUNDIN, O., and A. VARSHAVSKY. 1980. Terminal stages of SV40 DNA replication proceed via multiple intertwined catenated dimers. *Cell* **21**:103–114.

SUNDIN, O., and A. VARSHAVSKY. 1981. Arrest of segregation leads to accumulation of highly intertwined catenated dimers: dissection of the final stages of SV40 DNA replication. *Cell* **25**:659–669.

TATTERSALL, P., and D. WARD. 1976. Rolling hairpin model for replication for parvovirus and linear chromosomal DNA. *Nature* **263**:106–109.

TAYLOR, J. H. 1958. The duplication of chromosomes. *Sci. Am.* **198**(6):36–42.

Properties of DNA Polymerases and DNA Synthesis in Vitro

ALBERTS, B. M., and R. STERNGLANZ. 1977. Recent excitement in the DNA replication problem. *Nature* **269**:655–661.

ARIGA, H. 1984. Replication of cloned DNA containing the *Alu* family sequence during cell extract–promoting simian virus 40 DNA synthesis. *Mol. Cell. Biol.* **4**:1476–1482.

HAY, R. T., and M. L. DEPHAMPHILIS. 1982. Initiation of SV40 DNA replication *in vivo:* location and structure of 5′ ends of DNA synthesized in the *ori* region. *Cell* **28**:767–779.

HIBNER, U., and B. M. ALBERTS. 1980. Fidelity of DNA replication catalyzed *in vitro* on a natural DNA template by the T4 bacteriophage multi-enzyme complex. *Nature* **285**:300–305.

HUBERMAN, J. A. 1981. New views of the biochemistry of eucaryotic DNA replication revealed by aphidicolin, an unusual inhibitor of DNA polymerase α. *Cell* **23**:647–648.

ITOH, T., and J.-I. TOMIZAWA. 1980. Formation of an RNA primer for initiation of replication of ColE1 DNA by ribonuclease H. *Proc. Nat'l Acad. Sci. USA* **77**:2450–2454.

LI, J. J., and T. J. KELLY. 1984. Simian virus 40 DNA replication *in vitro. Proc. Nat'l Acad. Sci. USA* **81**:6973–6977.

MASUKATA, H., and J. TOMIZAWA. 1983. Effects of point mutations on formation and structure of the RNA primer for ColE1 DNA replication. *Cell* **36**:513–522.

NAGATA, K., R. A. GUGGENHEIMER, and J. HURWITZ. 1983. Adenovirus DNA replication *in vitro:* synthesis of full-length DNA with purified proteins. *Proc. Nat'l Acad. Sci. USA* **80**:4266–4270.

NOSSAL, N. 1983. Prokaryotic DNA replication systems. *Annu. Rev. Biochem.* **52**:581–616.

OGAWA, T., T. A. BAKER, A. VAN DER ENDE, and A. KORNBERG. 1985. Initiation of enzymatic replication at the origin of the *E. coli* chromosome:contributions of RNA polymerase and primase. *Proc. Nat'l. Acad. Sci. USA* **82**:3562–3566.

RICHARDSON, C. C. 1983. Bacteriophage T7: minimal requirements for the replication of a duplex DNA molecule. *Cell* **33**:315–317.

SUGIMOTO, K., A. OKA, H. SUGISAKI, M. TAKANAMI, A. NISHIMURA, Y. YASUDA, and Y. HIROTA. 1979. Nucleotide sequence of *Escherichia coli* K-12 replication origin. *Proc. Nat'l Acad. Sci. USA* **76**:575–579.

TABOR, S., and C. C. RICHARDSON. 1981. Template recognition sequence for RNA primer synthesis by gene 4 protein of bacteriophage T7. *Proc. Nat'l Acad. Sci. USA* **78**:205–209.

TSCHUMPER, G., and J. CARBON. 1982. Delta sequences and double symmetry in a yeast chromosomal replicator region. *J. Mol. Biol.* **156**:293–307.

TSENG, B. Y., J. M. ERICKSON, and M. GOULIAN. 1979. Initiator RNA of nascent DNA from animal cells. *J. Mol. Biol.* **129**:531–545.

Superhelicity in DNA

BAUER, W. R., F. H. C. CRICK, and J. H. WHITE. 1980. Supercoiled DNA. *Sci. Am.* **243**(1):118–133.

CRICK, F. H. C. 1976. Linking numbers and nucleosomes. *Proc. Nat'l Acad. Sci. USA* **73**:2639–2643.

KELLER, W. 1975. Determination of the number of superhelical turns in simian virus 40 DNA by gel electrophoresis. *Proc. Nat'l Acad. Sci. USA* **72**:4876–4880.

VINOGRAD, J., J. LEBOWITZ, R. RADLOFF, R. WATSON, and P. LAPIS. 1965. The twisted circular form of polyoma viral DNA. *Proc. Nat'l Acad. Sci. USA* **53**:1104–1111.

Topoisomerases

COZZARELLI, N. R. 1980. DNA gyrase and the supercoiling of DNA. *Science* **207**:953–960.

GELLERT, M. 1981. DNA topoisomerases. *Annu. Rev. Biochem.* **50**:879–910.

KIKUCHI, A., and K. ASAI. 1984. Reverse gyrase—a topoisomerase which introduces positive superhelical turns into DNA. *Nature* **309**:677–681.

LIU, L. F., C.-C. LIU, and B. M. ALBERTS. 1980. Type II DNA topoisomerases: enzymes that can unknot a topologically knotted DNA molecule via a reversible double-strand break. *Cell* **19**:697–707.

WANG, J. C. 1982. DNA topoisomerases. *Sci. Am.* **247**(1):94–109.

WANG, J. C. 1985. DNA topoisomerases. *Annu. Rev. Biochem.* **54**:665–698.

The Assembly of DNA into Nucleosomes

LASKEY, R. A., and W. C. EARNSHAW. 1980. Nucleosome assembly. *Nature* **286**:763–767.

LEVY, A., and K. M. JAKOB. 1978. Nascent DNA in nucleosome-like structures from chromatin. *Cell* **14**:259–267.

RUSSEV, G., and R. HANCOCK. 1982. Assembly of new histones into nucleosomes and their distribution in replicating chromatin. *Proc. Nat'l Acad. Sci. USA* **79**:3143–3147.

SEIDMAN, M. M., A. LEVINE, and H. WEINTRAUB. 1979. The asymmetric segregation of parental nucleosomes during chromosome replication. *Cell* **18**:439–449.

WEINTRAUB, H., S. J. FLINT, M. LEFFAK, M. GROUDINE, and R. H. GRAINGER. 1978. The generation and propagation of variegated chromosome structures. *Cold Spring Harbor Symp. Quant. Biol.* **42**:401–407.

Repair of DNA

FREIDBERG, E. A., and P. E. HANAWALT. 1983. *DNA Repair: A Laboratory Manual of Research Procedures.* M. Dekker.

GOTTESMAN, S. 1981. Genetic control of the SOS system in *E. coli. Cell* **23**:1–2.

LINDAHL, T. 1979. DNA glycosylases, endonucleases for apurinic/apyrimidinic sites, and base excision-repair. *Prog. Nuc. Acid Res. Mol. Biol.* **22**:135–192.

LITTLE, J. W., and D. W. MOUNT. 1982. The SOS regulatory system of *Escherichia coli. Cell* **29**:11–22.

Recombination

DRESSLER, D., and H. POTTER. 1982. Molecular mechanisms in genetic recombination. *Annu. Rev. Biochem.* **51**:727–761.

GONDA, D. K., and C. M. RADDING. 1983. By searching processively RecA protein pairs DNA molecules that share a limited stretch of homology. *Cell* **34**:647–654.

HOLLIDAY, R. 1964. A mechanism for gene conversion in fungi. *Genet. Res.* **5**:282–304.

HOWARD-FLANDERS, H. 1981. Inducible repair of DNA. *Sci. Am.* **245**(3):72–80.

MESELSON, M., and C. M. RADDING. 1975. A general model for genetic recombination. *Proc. Nat'l Acad. Sci. USA* **72**:358–361.

ORR-WEAVER, T. L., and J. W. SZOSTAK. 1983. Yeast recombination: the association between double-strand gap repair and crossing-over. *Proc. Nat'l Acad. Sci. USA* **80**:4417–4421.

ORR-WEAVER, T. L., J. W. SZOSTAK, and R. J. ROTHSTEIN. 1981. Yeast transformation: a model system for the study of recombination. *Proc. Nat'l Acad. Sci. USA* **78**:6354–6358.

SMITH, G. R. 1983. Chi hot spots of generalized recombinations. *Cell* **34**:709–710.

SZOSTAK, J. W., T. L. ORR-WEAVER, R. J. ROTHSTEIN, and F. W. STAHL. 1983. The double-strand-break repair model for recombination. *Cell* **33**:25–35.

TAYLOR, J. H. 1958. Sister chromatid exchanges in tritium-labeled chromosomes. *Genetics* **43**:515–529.

VON WETTSTEIN, D. S., W. RASMUSSEN, and P. B. HOLM. 1984. The synaptonemal complex in genetic segregation. *Annu. Rev. Genet.* **18**:331–414.

III

Cell Structure and Function

Freeze fracture of a vesicle of the mitochondrial inner membrane, showing random distribution of intramembrane protein particles. *Courtesy of A. E. Sowers.*

GENES are the master controllers of cell behavior and specialization. But most of the events of cell life take place outside the nucleus and in the cytosol or within the many subcellular compartments, such as the many organelles and small vesicles.

Each of these compartments as well as the cell itself is surrounded by a phospholipid bilayer membrane. Embedded in the membrane and attached to its surfaces are proteins, which catalyze reactions, control the passage of certain substances into and out of the cell and its compartments, and anchor the cytoskeleton. Because this membrane is so important to the structure and function of a cell, it is the subject of the first two chapters of this section of the book. Chapter 14 describes the structure of cell membranes, with special attention to the plasma membrane, which is the portion that encloses the entire cell. Chapter 15 examines the role of membrane proteins in transporting molecules into and out of the cell.

In multicellular organisms, a complex network of cell-to-cell communication is required to coordinate differentiation, growth, metabolism, and behavior. Cellular secretions, which include hormones, act as the communication signals. Chapter 16 discusses the receptor proteins that govern the cell's response to chemical signals. And Chapter 17 takes a look at the important topic of nerve-cell communication. The plasma membrane of a nerve cell is specialized to conduct electrical signals along its length; in addition, a chemical called a neurotransmitter is usually required to relay the message to the next nerve cell or muscle cell.

Chapters 18 and 19 turn to the complex system of fibers that lies within the cell. Collectively, these fibers are called the cytoskeleton, and they are responsible for the cell's shape and motility. Fibers protrude into the microvilli and thereby support these thin projections of the plasma membrane. Some fibers exist in cilia and flagella. Fibers are also responsible for the contraction of muscle, the movement of chromosomes during cell division, and the migration of cells along a substratum.

Cellular events require energy, and this energy is usually supplied by hydrolysis of the high-energy phosphoanhydride bonds in ATP. Chapter 20 examines the formation and conversion of high-energy bonds during photosynthesis and during the oxidation of carbohydrates, fatty acids, and amino acids. The generation of phosphoanhydride bonds in ATP occurs on membranes in the chloroplast and mitochondrion and depends on ionic gradients across those membranes.

Finally, Chapter 21 describes the assembly of the components that make a cell: multiprotein aggregates, ribosomes, membranes, and organelles. The assembly of viruses and the secretion of cellular proteins are discussed as well.

14

The Plasma Membrane

UNTIL the advent of the electron microscope, scientists were unable to appreciate fully the complexity of cytoplasmic organization. By light microscopy, they could not unambiguously identify the plasma membrane, the limiting membrane around the cell, and they debated at length about its existence. Although even electron microscopy cannot make visible the exact manner in which the individual molecules of the cytoplasm are distributed, it does clearly show that a limiting membrane surrounds every eukaryotic cell and, moreover, that the cell contains a multitude of internal membranous structures as well as a large number of separate fibrous systems coursing through it. Membranes close off specific regions of the eukaryotic cell that perform specialized tasks: oxidative phosphorylation in the mitochondrion, degradation of macromolecules in the lysosome, and so forth. Membrane-bound enzymes catalyze reactions that would occur with difficulty in an aqueous environment. Other proteins in the membrane provide anchors for the cytoskeletal fibers that give the cell its shape. Still other proteins provide a passageway across the membrane for certain molecules or regulate the fusion of the membrane with others in the cell. Thus, far from being the mere bag of soluble components, the cell as we now comprehend it is a highly organized entity with many functioning subcompartments.

The plasma membrane of the cell is a highly differentiated structure. Every cell type has in its outer membrane specific proteins that help control the intracellular milieu and that interact with specific signals to influence the cell's behavior. In this chapter we first discuss the basic principles governing the organization of proteins and phospholipids in all biological membranes. Then we consider three examples of specialized plasma membranes: those of the erythrocyte (red blood cell), the pancreatic acinus, and the intestinal epithelial cell.

The relative simplicity of the erythrocyte membrane allows us to examine the ways in which it interacts with a submembranous cytoskeleton to give the cell its shape and flexibility. The membranes of pancreatic acinar and intestinal epithelial cells demonstrate how different regions of the same plasma membrane can specialize to perform different tasks. The membrane of the intestinal epithelial cell, for example, has two or more distinct parts, each responsible for different processes necessary in transporting products of digestion, such as sugars and amino acids, across the epithelium and into the blood. In the last part of the chapter, we focus on junctions, the various specialized regions of the plasma membrane that allow adjacent cells in a tissue to bind to each other or to exchange small molecules.

The Architecture of Lipid Membranes

All Membranes Contain Proteins and Lipids; Many Contain Glycoproteins and Glycolipids

By a combination of subcellular fractionation techniques (detailed in Figures 5-27 and 5-28) one can separate from many kinds of cells several important biological membranes, such as the plasma membrane and the mitochondrial membranes. Often these preparations are contaminated with membranes from other organelles. But the plasma membrane of human erythrocytes can be isolated in near purity because the cells contain no other membranes. The myelin membrane that surrounds certain nerves can also be obtained without contamination.

All membranes, regardless of source, contain proteins as well as lipids (Table 14-1). The ratio of protein to lipid varies enormously: the inner mitochondrial membrane is 76 percent protein, the myelin membrane only 18 percent. As Table 14-2 shows, the composition of lipids also varies greatly among different membranes. All membranes contain a substantial proportion of phospholipids, such as phosphatidylcholine, phosphatidylserine, phosphatidylethanolamine, sphingomyelin, and phosphatidylinositol (see Figure 3-64). Many contain cholesterol (Figure 3-66). Cholesterol is especially abundant in the plasma membrane of mammalian cells and is absent from most prokaryotic cells. Cardiolipin (diphosphatidylglycerol) is restricted to the inner mitochondrial membrane.

Table 14-1 Chemical composition of some purified membranes (in percentages)

Membrane	Protein	Lipid	Carbohydrate
Myelin	18	79	3
Plasma membrane			
Human erythrocyte	49	43	8
Mouse liver	44	52	4
Ameba	54	42	4
Halobacterium purple membrane	75	25	0
Mitochondrial inner membrane	76	24	0
Chloroplast			
Spinach lamellae	70	30	0

SOURCE: Adapted from G. Guidotti, 1972, *Annu. Rev. Biochem.* **41**:731.

Carbohydrates are an important constituent of many membranes. They are bound either to proteins as constituents of glycoproteins (see Figures 3-76 and 3-77) or to lipids as constituents of glycolipids (Figure 3-78). Carbohydrates are especially abundant in the plasma membranes of eukaryotic cells. They are absent from the inner mitochondrial membrane, the chloroplast lamellae, and several other intracellular membranes; glycoproteins are absent from all prokaryotic membranes (Table 14-1).

The Phospholipid Bilayer Is the Basic Structural Unit of Biological Membranes

Despite their variable composition, all biological membranes are thought to be constructed on a common pattern (Figure 14-1). They all contain a phospholipid bilayer as the basic structural unit. Let us recall that all phospholipids are amphipathic molecules: they have a hydrophobic portion and a hydrophilic portion. The primary physical forces for organizing biological membranes are the hydrophobic interactions between the fatty acyl chains of lipid molecules. These interactions result in the formation of a phospholipid bilayer, a sheet containing two layers of phospholipid molecules whose polar head groups face the surrounding watery surface while the fatty acyl chains form a continuous hydrophobic interior (Figures 14-1 and 14-2). Each layer of phospholipid is called a *leaflet*.

All phospholipids except cardiolipin contain two fatty acyl chains. Nearly all fatty acyl found in the membranes of eukaryotic cells have an even number of carbon atoms, usually 16, 18, or 20. Unsaturated fatty acyl usually have one double bond, but some have two, three, or four. In general, all such double bonds are of the cis configura-

Table 14-2 Lipid composition of membrane preparations (in percentages)*

Source	Cholesterol	PC	SM	PE	PI	PS	PG	DPG	PA	Glycolipids
RAT LIVER										
Cytoplasmic membrane	30.0	18	14.0	11	4.0	9.0	—	—	1	—
Endoplasmic reticulum (rough)	6.0	55	3.0	16	8.0	3.0	—	—	—	—
Endoplasmic reticulum (smooth)	10.0	55	12.0	21	6.7	—	—	1.9	—	—
Mitochondria (inner)	3.0	45	2.5	24	6.0	1.0	2.0	18.0	0.7	—
Mitochondria (outer)	5.0	50	5.0	23	13.0	2.0	2.5	3.5	1.3	—
Nuclear membrane	10.0	55	3.0	20	7.0	3.0	—	—	1.0	—
Golgi	7.5	40	10.0	15	6.0	3.5	—	—	—	—
Lysosomes	14.0	25	24.0	13	7.0	—	—	5.0	—	—
Myelin	22.0	11	6.0	14	—	7.0	—	—	—	12
RAT ERYTHROCYTE	24.0	31	8.5	15	2.2	7.0	—	—	0.1	3
E. coli CYTOPLASMIC MEMBRANE	0	0	—	80	—	—	15.0	5.0	—	—

*PC, phosphatidylcholine; SM, sphingomyelin; PE, phosphatidylethanolamine; PI, phosphatidylinositol; PS, phosphatidylserine; PG, phosphatidylglycerol; DPG, diphosphatidylglycerol (cardiolipin); PA, phosphatidic acid.

SOURCE: Adapted from M. K. Jain and R. C. Wagner, 1980, *Introduction to Biological Membranes,* Wiley.

tion; this introduces a bend in the otherwise straight chain. A major difference among phospholipids concerns the polar heads. At neutral pH, the polar head group may have no net negative charge (phosphatidylcholine, phosphatidylethanolamine) or it may have net negative charges (phosphatidylglycerol, cardiolipin, phosphatidylserine). Rarer phospholipids have a net positive charge. Nonetheless, all the polar head groups can pack together into a phospholipid bilayer. Sphingomyelin, though having a different chemical structure from phospholipids, has a similar appearance in space-filling models, and can form mixed bilayers with phospholipids. Glycolipids are also amphipathic and can pack together with phospholipids to form bilayers.

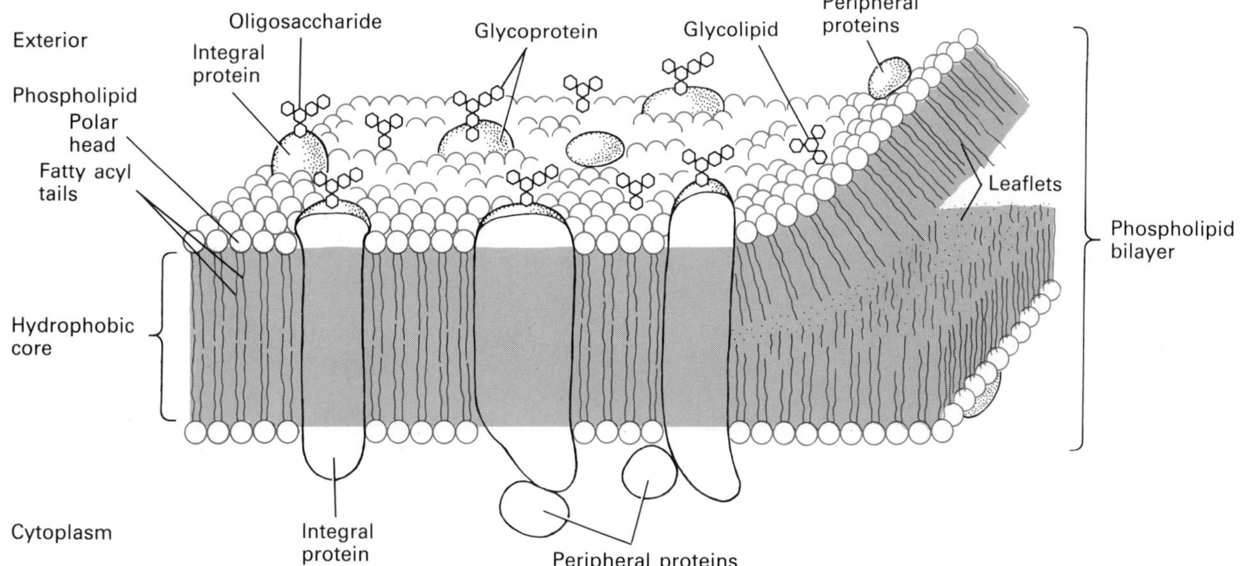

Figure 14-1 A general model of the structure of biological membranes. A phospholipid bilayer constitutes the basic structure. The hydrophobic tails of the fatty acyls form the middle of the bilayer, while the polar, hydrophilic heads of the phospholipids line both surfaces. Each half of the bilayer constitutes a leaflet. The integral proteins have one or more of their regions embedded in the lipid bilayer, interacting with the fatty acyl chains in the hydrophobic cores. Peripheral membrane proteins are merely associated with the membrane by specific protein-protein interactions. Oligosaccharides are bound mainly to membrane proteins, forming glycoproteins; some are bound to lipids, forming glycolipids. Because proteins are visualized as free to move laterally in the two-dimensional lipid fluid, this model is often termed the fluid mosaic model of membranes.

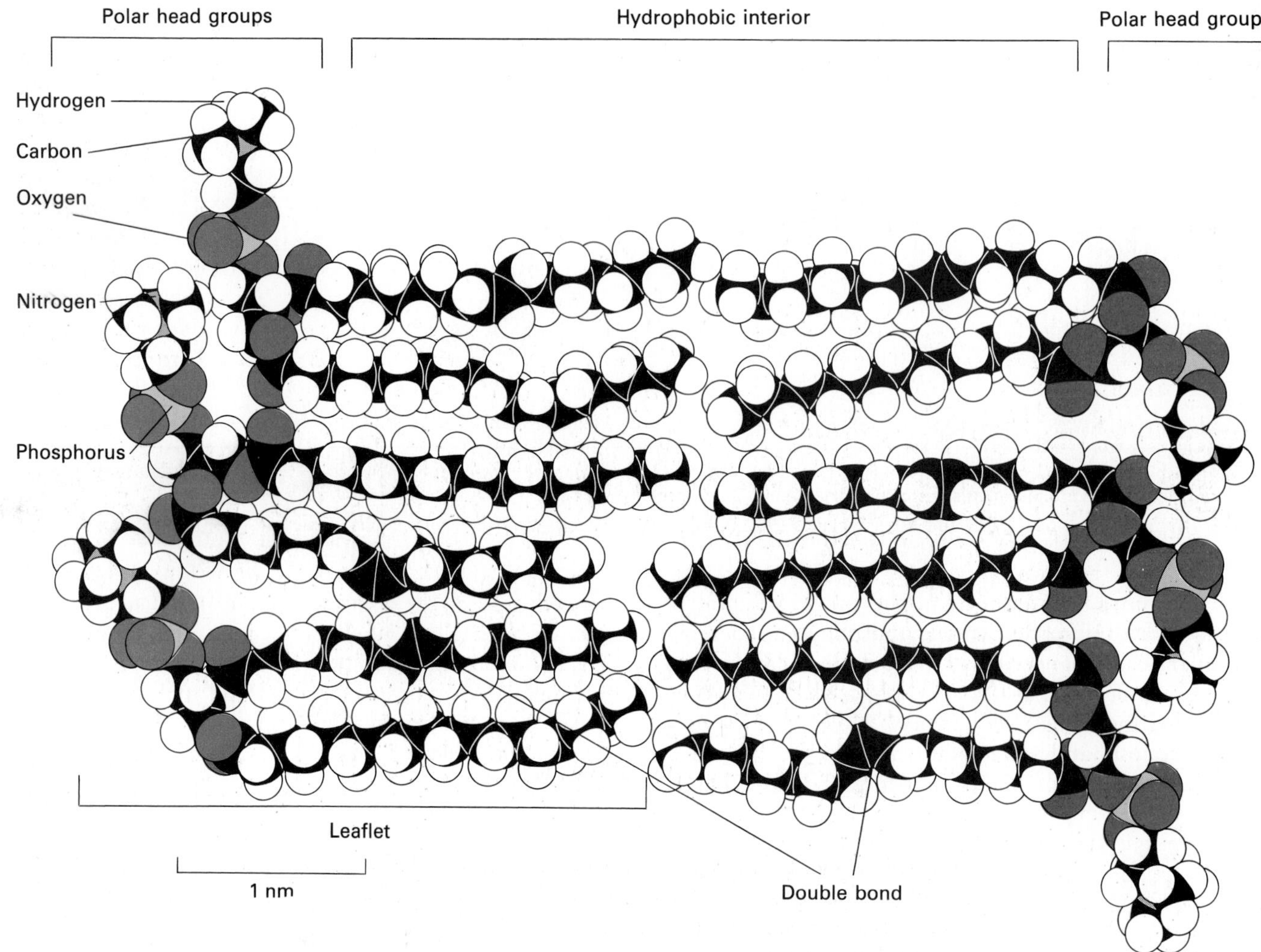

Figure 14-2 A space-filling model of a typical phospholipid bilayer membrane. The hydrophobic interior is generated by the fatty acyl side chains. Some of these chains have bends, caused by the double bonds. Although the polar head groups of different phospholipids are of different sizes and bear different charges, they all lie on the outer, aqueous surface of the membrane. [See L. Stryer, 1981, *Biochemistry,* page 213, W. H. Freeman and Company.] *Courtesy of L. Stryer.*

Might some specialized membranes or regions of membranes be built of a structure other than a phospholipid bilayer? Although there is persuasive evidence that the bilayer structure is prevalent, certain alternative structures may exist. A major conceptual question raised years ago was whether lipids alone are responsible for membrane structure. We know that they are not the only component, because all biological membranes, no matter how carefully purified, are found upon examination to contain proteins (Table 14-1). The percentage and exact nature of the adhering proteins vary with the type of membrane, but there is no question that proteins are present. As Figure 14-1 shows, some membrane proteins contain regions that are bound to the hydrophobic fatty acyl "core" of the bilayer, while other proteins are bound to the membrane primarily by protein-protein interactions. The issue, however, is whether proteins are an essential part of the structure of membranes or whether they are present only as participants in membrane function. If proteins do contribute to the structural organization of membranes, then we might expect to find some common structural protein in all membranes. No such protein has been found. Instead, the data suggest that proteins are specialized constituents of the membrane, each fulfilling one or more specific membrane function, and that the membrane owes its integrity to the properties of its constituent lipids.

Several Types of Evidence Point to the Universality of the Phospholipid Bilayer

If the general architecture of membranes is in fact determined by the interactions of lipid molecules, then the bilayer structure is the most likely one. This likelihood does not rest on theory alone; direct experimental evidence exists. In 1925, E. Gorter and F. Grendel extracted the

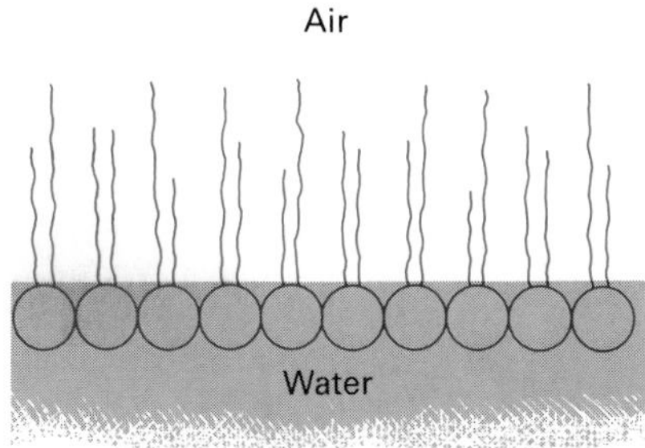

Figure 14-3 A monolayer is formed when phospholipids are floated on the surface of an aqueous solution.

lipids from erythrocytes and floated them on the surface of a water solution. Under such conditions, phospholipids are known to form a unimolecular film: the hydrophilic head groups face the water, and the hydrophobic tails point upwards into the air (Figure 14-3). The area of this monolayer was approximately twice that of the surface of the original erythrocytes. Since erythrocytes contain no internal membranes, these investigators concluded that the lipids are arranged in the membrane as a continuous bilayer.

Later experiments established that phospholipids, when dispersed in aqueous solutions, spontaneously form either sheets of bilayers or closed vesicles that contain a wall of a phospholipid bilayer. These results are obtained both with pure species of phospholipids and with mixtures such as are found in natural membranes.

But are *natural* membranes based on the phospholipid bilayer? Perhaps the best evidence comes from low-angle x-ray structural analysis, a technique by which one can determine the density of electrons, and thus of matter, in stacked layers or spherically symmetrical arrays. A conveniently stacked membrane, used in many such studies, is the multimembrane myelin sheath that covers and insulates many mammalian nerve cells (Figure 14-4). As illustrated in Figure 14-4a, myelin is formed from the plasma membrane of a Schwann cell that wraps around the axon, or elongated part, of a neuron. The Schwann cell becomes little more than a series of stacked membranes, and myelin becomes the major membranous component of such nerves. Myelin can be separated from other cellular membranes in a pure state.

Analyses of these stacked plasma membrane units have shown a very low density in the middle of each membrane, suggesting that in this region the amount of lipid significantly exceeds that of protein (Figure 14-4b). Such a distribution of matter implies a bilayer organization in which protein is located on either side of a membrane whose central region is almost pure lipid. Parts of proteins do pass through the lipid bilayer, but these polypeptide segments make up only a small part of the inner mass of the membrane, generally less than 10 percent.

Possibly the most convincing argument for the universality of the bilayer structure is furnished by the electron microscope. If an electron-opaque stain could bind to the polar head groups of phospholipids, then a bilayer membrane exposed to such a stain and viewed with an electron microscope would look like a railroad track: two thin lines (the stain-phosphate complex) with a uniform space of about 2 nm (the lipid) between them. In fact, membranes stained with osmium tetroxide look just this way (Figure 14-5), but to which part of the phospholipid the electron-opaque agent is bound is uncertain. In all probability it is bound to the polar head groups, and possibly to double bonds in the fatty acyl chains. In any case, most but not all osmium-stained membranes resemble railroad tracks. It may be that in certain regions of membranes that do not exhibit such a uniform appearance, phospholipids are arranged differently, especially where specialized proteins play a large role in the structure. There may even be regions of membrane that lack lipid.

Phospholipid Membranes Form Closed Compartments

Biophysical studies of pure phospholipids have revealed a number of properties that are important for an understanding of biological membranes. Perhaps the most important finding is that phospholipid membranes spontaneously seal to form closed structures. A lipid bilayer sheet would be an unstable structure if it had a free edge at which the hydrophobic region of the bilayer were in contact with water. All cellular membranes surround closed compartments of the cell, and all bilayers have an *internal face,* the side oriented towards the interior of the compartment, and an *external face,* the side presented to the surrounding environment. Because most organelles are surrounded by a single bilayer membrane, it is also useful to speak of the *cytoplasmic* and *extracytoplasmic* faces of the membrane, the extracytoplasmic face being the side directed away from the cytoplasm. For the plasma membrane, the external or extracytoplasmic face defines the outer limit of the cell (Figure 14-6). Some organelles, such as the nucleus and mitochondrion, are surrounded by two membranes, in which case the extracytoplasmic surfaces are those that face the lumen between the two membranes.

Phospholipid Bilayers Form a Two-Dimensional Fluid

The motions of lipid molecules within membranes have been studied most intensely in pure phospholipid bilayers. Two systems of pure bilayers have been especially useful in such studies: liposomes and planar bilayers. Liposomes are spherical vesicles of up to 1 μm in diameter that one can form by mechanically dispersing phospho-

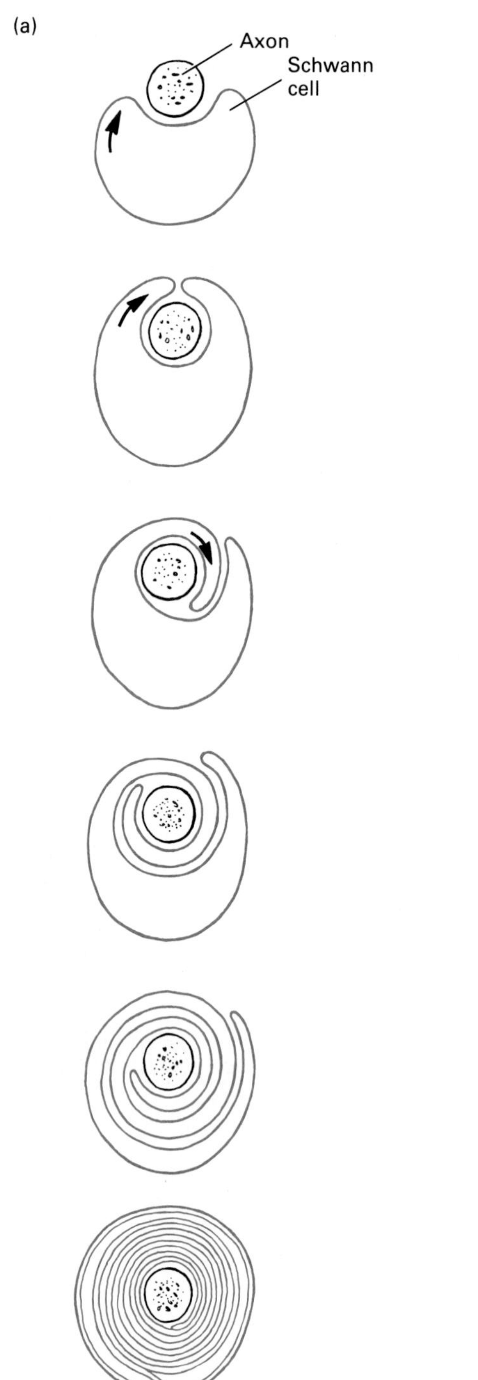

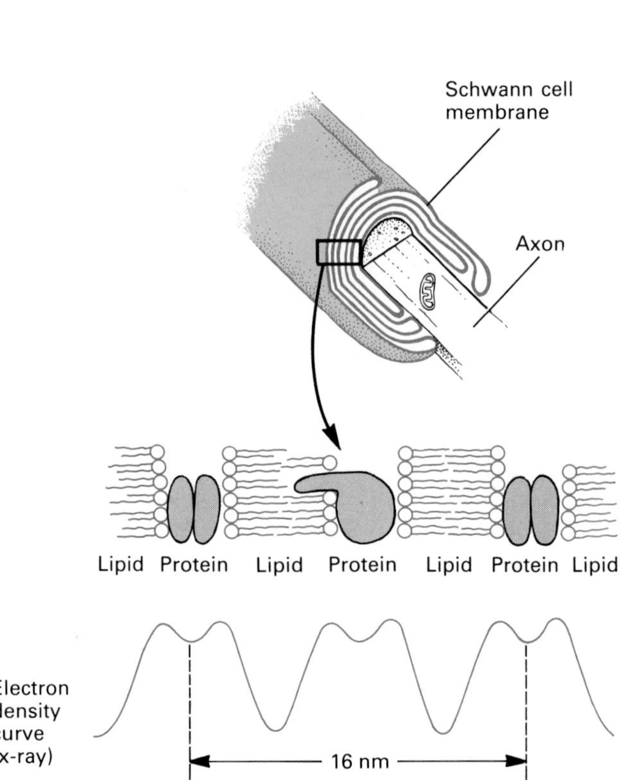

Figure 14-4 Low-angle x-ray analysis of lipid-protein distribution in membranes. (a) The structure and embryonic development of the myelinated nerve fiber. During development of the nervous system, a large cell, called the *Schwann* cell envelops the *axon,* or elongated part, of a neuron. The continuous growth of the Schwann cell membrane into its own cytoplasm, together with rotation of the nerve axon, results in a laminated spiral of double plasma membranes around the axon. Mature myelin is thus a stack of plasma membranes of the Schwann cell. (b) The profile of electron density, and thus of matter, obtained by x-ray diffraction studies on fresh nerve, and the relation of this profile to the protein and lipid components of the myelin membranes. [See W. T. Norton, 1981, in *Basic Neurochemistry,* 3d ed., G. J. Siegel et al., eds., p. 68, Little, Brown.]

lipids in water; planar bilayers are formed across a hole in a partition that separates two aqueous solutions (Figure 14-7). In such membranes, phospholipids and glycolipids are free to rotate around their long axis and to diffuse laterally within the membrane leaflet. Such movements are caused by the natural thermal motion of molecules. Because the movement is lateral or rotational, the fatty acyl chains remain in the hydrophobic interior of the membrane. A typical lipid molecule exchanges places with its neighbors in a leaflet about 10^7 times per second and will diffuse several micrometers per second at 37°C. Thus a lipid can diffuse the 1 μm length of a bacterial cell in only 1 second and the distance equal to the length of an animal cell in about 20 seconds.

In pure phospholipid bilayers, phospholipids do not migrate, or flip-flop, from one face of the membrane to the other. In some natural membranes they occasionally do. Such movements are somehow catalyzed by one or

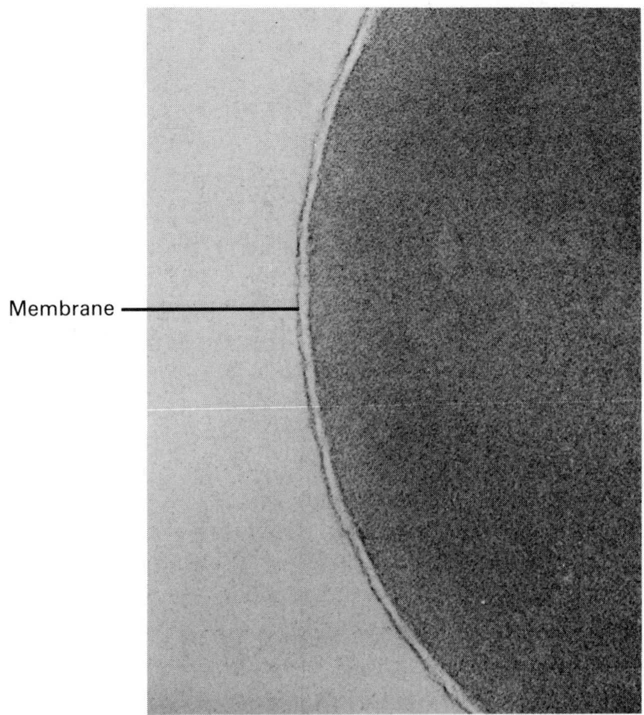

Figure 14-5 Electron micrograph of osmium-stained erythrocyte membrane. Note the "railroad track" appearance of the phospholipid bilayer at the cell surface. *Courtesy of J. D. Robertson.*

more membrane proteins. They are energetically extremely unfavorable, as they require the transport of the polar head of a phospholipid through the hydrophobic interior of the membrane (Figure 14-8).

One technique used for measuring the mobility of membrane lipids is electron spin resonance (ESR) spectroscopy. Synthetic phospholipids that contain a nitroxide group (Figure 14-9) are introduced into otherwise normal phospholipid membranes. ESR measures the energy absorbed by the unpaired electron of the nitroxide group as a function of the magnitude of an external magnetic field. The energy absorption is a complex function of, among other factors, the viscosity of the membrane. Such studies, on both synthetic and natural membranes, show that membranes have a low-viscosity, fluid-like consistency.

The Fluidity of a Bilayer Depends on Its Lipid Composition, Cholesterol Content, and Temperature

When a suspension of liposomes made up of a single type of phospholipid is heated, it undergoes an abrupt change in physical properties over a very narrow temperature range. This phase transition is due to a change in the organization of the fatty acyl side chains: the chains pass from a highly ordered or gel-like state to a more mobile state (Figure 14-10). This change is accompanied by increased motion about the C—C bonds of the fatty acyl chains, allowing them to assume a more random conformation. *Differential scanning calorimetry* is a common technique for measuring these transitions. It measures the heat absorption as a function of temperature; during the gel→fluid transition a relatively large amount of heat is absorbed.

In general, lipids with short or unsaturated fatty acyl chains undergo the phase transition at lower temperatures than lipids with long or saturated chains. Short chains have less surface area with which to form van der Waals interactions with each other. Unsaturated fatty acyl chains have kinks, thus tend to adopt a more random, fluid state and form less stable van der Waals contacts with other lipids. They tend to prefer a more fluid phase.

Liposomes prepared from mixtures of phospholipids

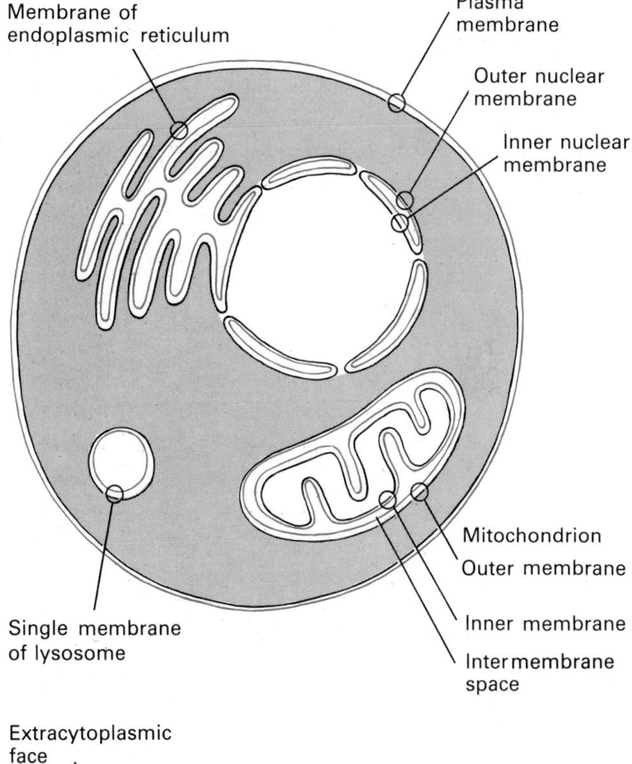

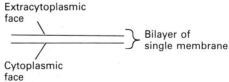

Figure 14-6 Faces of cellular membranes. The extracytoplasmic surfaces are in color. For organelles such as the nucleus, chloroplast, and mitochondrion, which are enclosed in two phospholipid bilayers, the extracytoplasmic face is the one that faces the space between the inner and outer membrane.

undergo phase transitions at intermediate temperatures that average out the properties of the individual side chains. In all cells the membranes contain a mixture of different fatty acyl chains and are fluid at the temperature at which the cell is grown. All animal and bacterial cells adapt to a decrease in growth temperature by increasing the proportion of unsaturated to saturated fatty acids in the membrane; this tends to maintain a fluid bilayer at the reduced temperature.

Cholesterol is a major determinant of membrane fluidity. Too hydrophobic to form a sheet structure on its own, cholesterol intercalates among the phospholipids. Its polar hydroxyl group is in contact with the aqueous solution, near the polar head groups of the phospholipids, while the steroid ring interacts with and tends to immobilize the fatty acyl chains of the phospholipids (Figure 14-11). The net effect of cholesterol on membrane fluidity

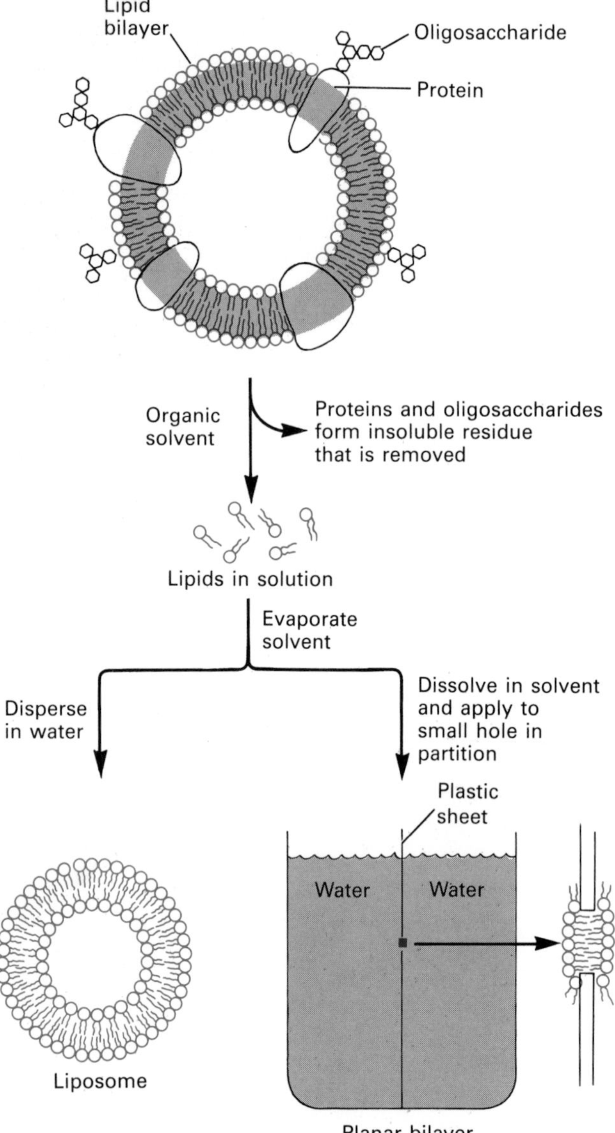

Figure 14-7 Construction of pure phospholipid bilayers. A preparation of biological membranes is treated with an organic solvent, such as a mixture of chloroform and methanol (3:1), which selectively solubilizes the phospholipids and cholesterol. Proteins and carbohydrates remain in an insoluble residue. The solvent is removed by evaporation. *Left:* If the lipids are mechanically dispersed in water, they spontaneously form a liposome. *Right:* A planar bilayer can be formed over a small hole in a partition that separates two aqueous phases. Such bilayers are often termed "black lipid membranes" because of their appearance.

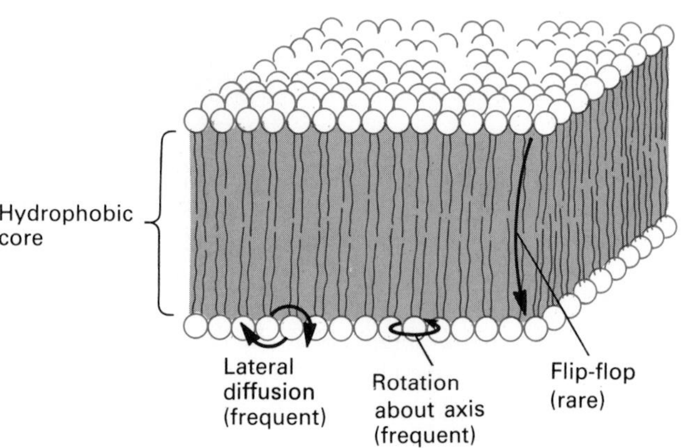

Figure 14-8 In a pure phospholipid bilayer, lipid molecules rotate rapidly about their long axis and rapidly (10^7 times per second) exchange places with a neighbor in the same leaflet. But they rarely migrate from one leaflet to another. An individual lipid molecule might flip-flop across the hydrophobic region less than once a week.

Figure 14-9 A spin-label analog of phosphatidylcholine, a compound used to probe the fluidity of natural and synthetic phospholipid membranes. The nitroxide "spin-label" is colored.

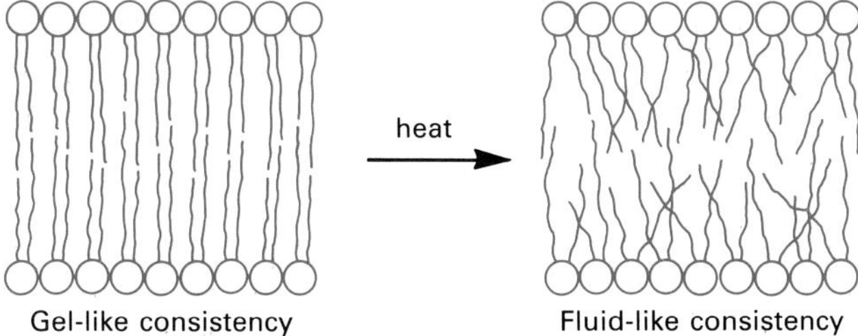

Figure 14-10 Alternative forms of the phospholipid bilayer. Absorption of heat induces the gel → fluid transition over a temperature range of only a few degrees. The fluid phase is also favored by the presence of short fatty acyl chains and by a double bond in the acyl chains.

varies, depending on the lipid composition. Cholesterol restricts the random movement of the part of the fatty acyl chains lying closest to the outer surface of the bilayer. But it separates and disperses the tails of the fatty acyls and causes the inner regions of the bilayer to become slightly more fluid. At the high concentrations found in eukaryotic plasma membranes, cholesterol tends to make the membrane, overall, less fluid at growth temperatures near 37°C. By preventing the hydrocarbon fatty acyl chains of the membrane lipids from binding to each other, cholesterol prevents the fluid→gel transition and the consequent drastic reduction in fluidity that would otherwise occur if the membrane were placed at a lower temperature.

Membrane Proteins

Membranes vary in behavior from one cell type to another and from one organelle to another by the variation of their complement of attached proteins. The mitochondrial membrane proteins differ markedly from the plasma membrane proteins, and the plasma membrane components of a liver cell are strikingly different from those of an intestinal cell. How specific proteins are deposited in specific membranes is the subject of Chapter 21. Here we are concerned with their general roles in basic membrane structure.

Membrane proteins serve a wide range of functions. They transport molecules into and out of cells. They receive signals from hormones and other chemicals in the surrounding fluid and transmit those signals to the cell interior. They act as anchors for cytoskeletal components and for components of the extracellular matrix. Because some membrane-bound proteins are located on the outer faces of the cells, they are responsible for endowing cells with the individuality that allows them to assort appropriately during differentiation and to form specific connections with other cells. Finally, the various enzymes bound to the plasma membrane as well as to the membranes surrounding specific organelles allow different chemical reactions to be catalyzed in different parts of the cell.

Proteins Attach to Membranes in Many Different Ways

The ways proteins can attach to membranes reflect their different functions in the cell. Some proteins are bound only to the surface. Others have one region buried within the membrane and another protruding outside of it. Still others span the membrane from face to face. Because it is difficult to purify most membrane proteins and because many proteins are present in cells in very tiny amounts, we have only a rudimentary understanding of the molecular interactions that hold membrane proteins in place. Nevertheless, it is useful to classify membrane proteins according to the types of interaction that maintain the protein-membrane relationship, using as a guide what is now known about a few proteins (Figure 14-12).

Integral membrane proteins, also called *intrinsic proteins,* have one or more segments interacting directly with the hydrophobic "core" of the phospholipid bilayer.

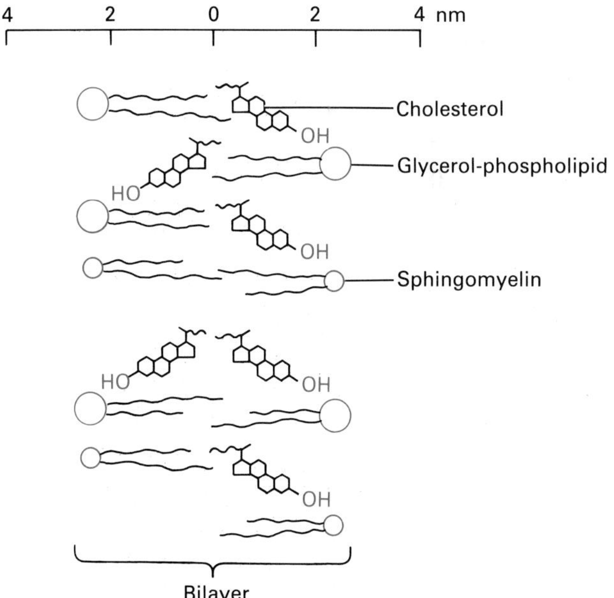

Figure 14-11 Arrangement of phospholipids and cholesterol in a typical myelin membrane. The polar OH group of cholesterol faces the aqueous surface of the membrane, and the hydrophobic hydrocarbon portion nestles among the fatty acyl side chains. In color are the polar OH groups of cholesterol and the polar head groups of sphingomyelin (*small circles*) and glycerol-phospholipids (*large circles*). *After D. Caspar and D. A. Kirschner, 1971,* Nature **231**:46.

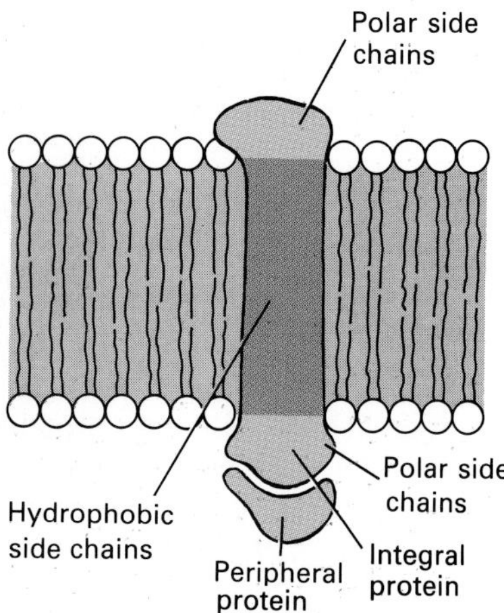

Figure 14-12 Schematic outline of the interaction of a typical integral membrane protein and peripheral protein with a lipid bilayer. Most integral proteins span the bilayer, as depicted here, and contain polar regions on both sides of the membrane. Most peripheral proteins interact with lipid polar head groups or, as depicted here, with integral proteins; they are not in contact with the hydrophobic core. Peripheral proteins are also called extrinsic proteins.

Most integral proteins, but not all, contain polar regions on both sides of the membrane, as is depicted in Figure 14-13. Such proteins span the bilayer and thus are called *transmembrane* proteins. Integral membrane proteins can be removed from the membrane only by the action of detergents, which displace the lipids bound to the hydrophobic side chains of the proteins (see "Detergents Are Used to Solubilize and Study Integral Membrane Proteins").

Peripheral proteins, or *extrinsic proteins,* do not interact directly with the hydrophobic core of the phospholipid membrane. They are usually bound to the membrane either indirectly by interactions with integral membrane proteins, as shown in Figure 14-12, or directly by interactions with lipid polar head groups. The proteins spectrin and ankyrin, cytoskeletal proteins that are bound to the inner face of the red blood cell membrane, are examples of peripheral proteins. They are discussed later on in this chapter. Other peripheral proteins are localized to the outer surface of the plasma membrane, such as certain proteins of the glycocalyx (page 593).

Most peripheral proteins are soluble in aqueous solution and are bound to specific integral membrane proteins by ionic or other weak interactions. Characteristically, many peripheral proteins can be removed from the membrane either by solutions of high ionic strength (high concentrations of salt), which disrupt ionic bonds, or by chemicals that bind divalent cations such as Mg^{2+}. Most peripheral proteins are not solubilized by detergents, since they are not bound directly to the hydrophobic core of the membrane.

Integral Membrane Proteins Contain Regions That Are Embedded in the Phospholipid Bilayer

Most integral membrane proteins are not soluble in water; segments of most such proteins (from 10 to 20 residues long) contain hydrophobic amino acids that enter the interior of the membrane lipid bilayer and bind the protein to the lipid core. A few integral proteins are anchored to the membrane mainly by a glycophospholipid that is attached covalently to the carboxyl-terminus. Three basic types of interaction keep integral proteins embedded in membranes: ionic interactions with the polar head groups of the lipids, hydrophobic interactions with the lipid interior of the membrane, and specific interactions with defined structures of the lipid (such as regions of cholesterol or of complex glycolipids). Both ionic and hydrophobic interactions are believed to function in the anchoring of glycophorin, a major erythrocyte membrane protein, to the phospholipid bilayer (Figure 14-13).

Glycophorin Spans the Membrane Once The membrane-binding region of glycophorin consists of 34 residues (numbers 62–95), including one sequence of 23 residues (73–95) that contains only uncharged, hydrophobic amino acids as phenylalanine, leucine, isoleucine, valine, tryptophan, and tyrosine. Serine and threonine residues are, as in other such proteins, found near the end of a membrane-binding region. It is thought that the membrane-embedded regions form an α helix. The polar amide groups would be in the center of the helix, anchored by hydrogen bonds, and therefore would not be directly exposed to the fatty acyl interior of the membrane. The hydrophobic side chains would protrude from the core of the helix and form hydrophobic bonds with the lipid (Figure 3-16). Charged amino acids (such as arginine, lysine, glutamate, histidine, and aspartate) are often found bordering the hydrophobic segments; perhaps they interact with the polar head groups of the phospholipids and hold the hydrophobic region firmly in the lipid. Ionic interactions with polar head groups could be direct, such as a lysine bound to a negatively charged phosphate or a glutamate residue bound to a positively charged choline. Or they could be indirect, such as a glutamate or aspartate residue bound to a negatively charged phosphate by a bridging Mg^{2+} ion. As each amino acid residue adds 0.15 nm to the length of an α helix, a helix with 25 residues would be 3.75 nm long, just sufficient to span the hydrocarbon core of a phospholipid bilayer.

Bacterial Rhodopsin Spans the Bilayer Seven Times Bacterial rhodopsin is found in the membrane of the photosynthetic bacterium *Halobacterium,* where it

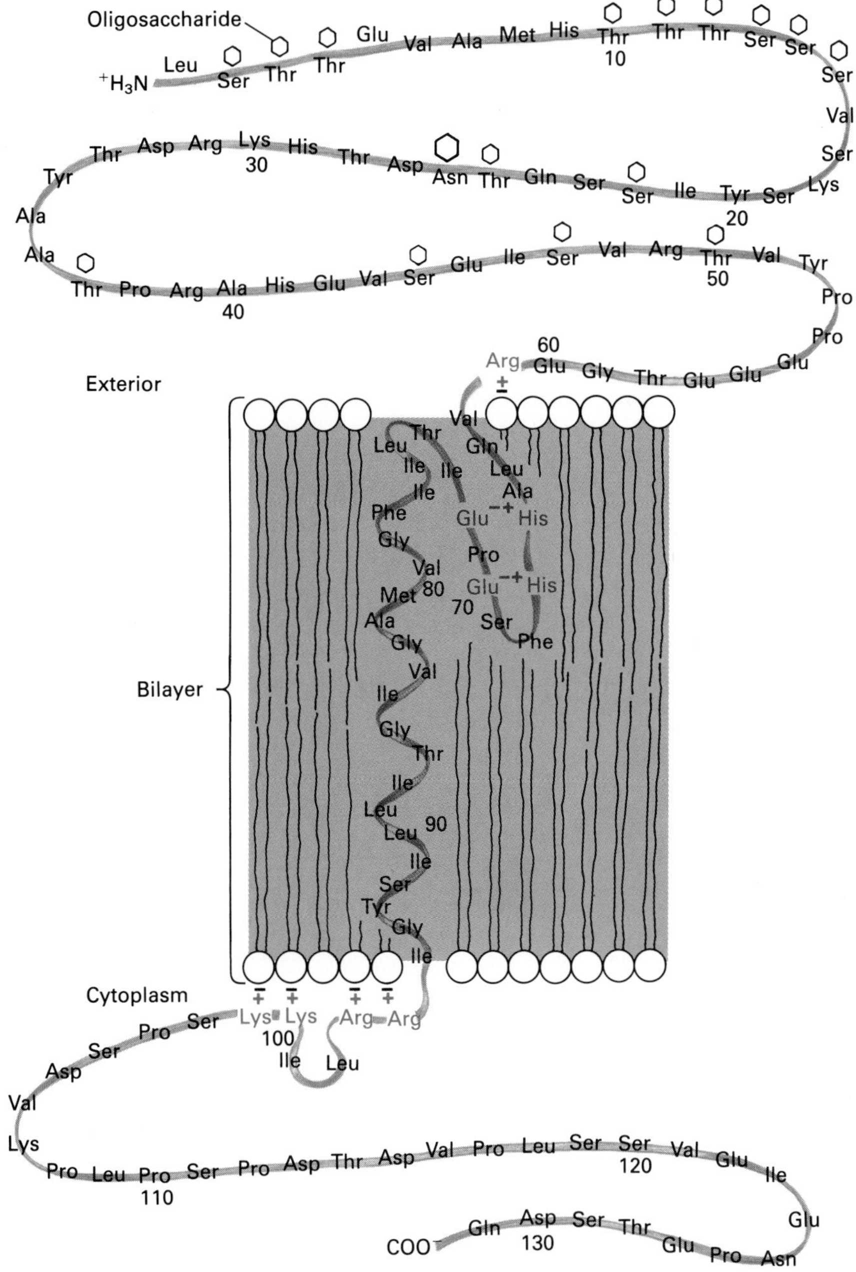

Figure 14-13 Amino acid sequence and transmembrane disposition of glycophorin A from the red blood cell membrane. The amino-terminal portion of the molecule lies outside the cell and has 16 carbohydrate units attached to it: 15 are attached to serine or threonine residues, and one is attached to an asparagine side chain. These oligosaccharides are rich in negatively charged sialic acid groups. The carboxyl-terminal part of the molecule, located inside the cell, is rich in negatively and positively charged amino residues. A hypothetical ionic interaction is indicated between the positively charged arginine and lysine residues at positions 96, 97, 100, and 101 with negative phospholipid head groups at the cytosolic surface of the membrane as well as one between the original arginine at 61 and a polar head group. Residues 62–95 are buried in the bilayer and are hydrophobic except for the charged groups at 66, 67, 70, and 72. Recent work suggests that the negatively charged glutamic acid residues at positions 70 and 72 are bound in ionic linkage with the positively charged histidine residues at positions 66 and 67, allowing residues 62–72 to insert into the hydrophobic core of the bilayer. [See V. T. Marchesi, H. Furthmayr, and M. Tomita, 1976, *Annu. Rev. Biochem.* **45**:667; A. H. Ross et al., 1982, *J. Biol. Chem.* 257:4152.]

traps the energy of light and uses it to pump protons across the cell membrane. Because bacterial rhodopsin forms a major part of the plasma membrane (it generates the color that gives it the name "purple membrane"), it has been studied in detail. It contains 247 amino acids and a covalently linked prosthetic group termed *retinal.* It is composed of seven α-helical segments, each of which contains between 26 and 32 amino acids and each of which spans the phospholipid membrane (Figure 14-14a). Short nonhelical segments of as many as 11 amino acids link the α-helical segments together (Figure 14-14b).

Many of the amino acids in the nonhelical segments are hydrophilic, some with charged side chains; they interact with the aqueous solution on each side of the membrane. Some of the membrane-spanning helixes contain positively charged lysine or arginine residues at positions near the outer surface of the membrane. As in glycophorin, the erythrocyte membrane protein, these residues may interact with the negatively charged phosphates on the head groups of the phospholipids. The vast majority of the amino acid side chains in the membrane-spanning regions are hydrophobic; they are in direct contact with the fatty acyl tails of the phospholipid bilayer. There are, however, several lysine, arginine, glutamine, and aspartic acid side chains in what appears to be the middle of the membrane-spanning regions. Most likely, these side chains are charged but are not embedded in the lipid matrix; they

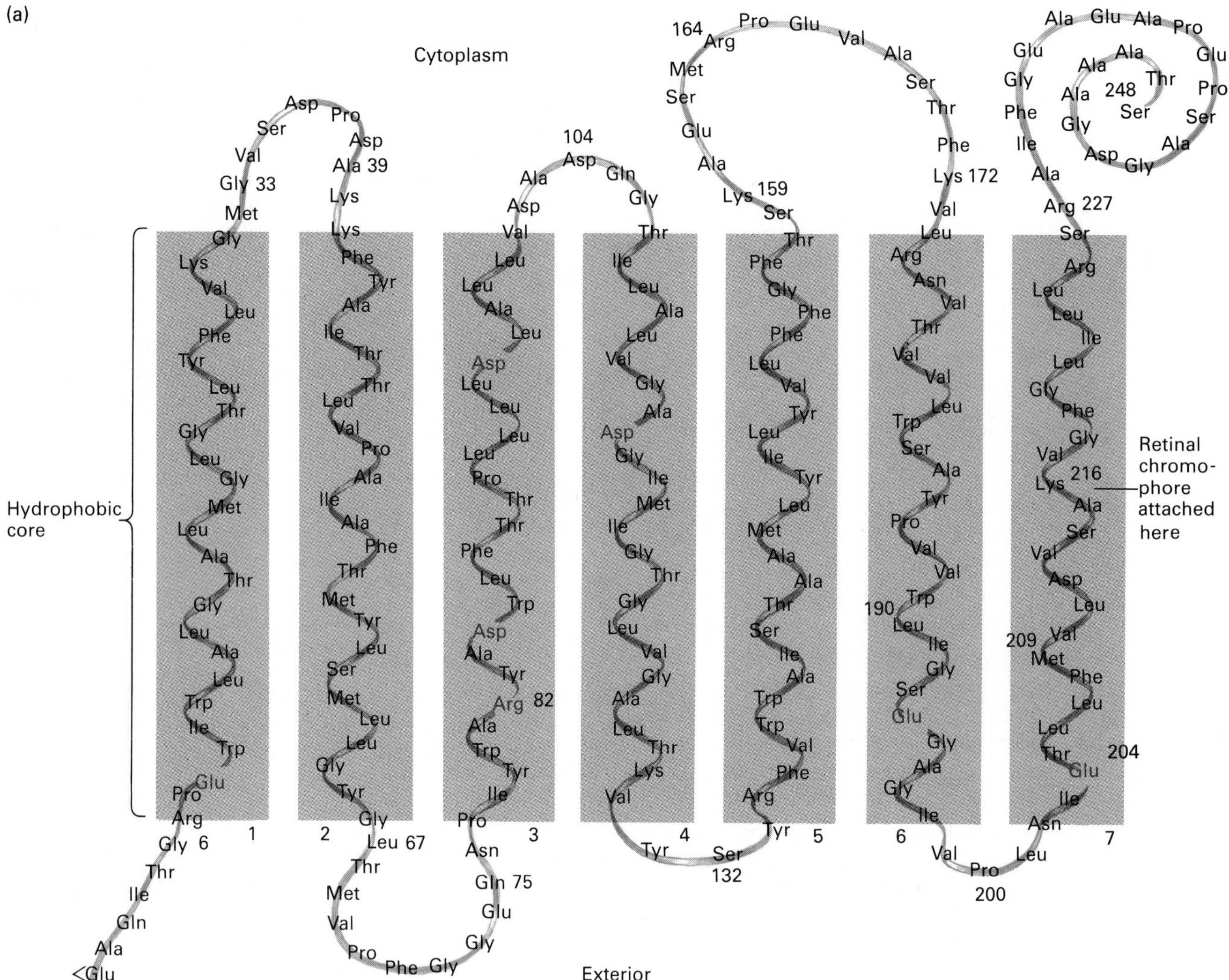

(b)

Cytoplasm

Bacteriorhodopsin

Exterior

Figure 14-14 The bacteriorhodopsin molecule. (a) A linear representation. The molecule contains seven α helixes that span the plasma membrane, labeled 1–7, from the NH_2-terminus. The boxed sequences are those believed to be in the hydrophobic core of the phospholipid bilayer. The retinal chromophore is attached to lysine 216. (b) As this three-dimensional model shows, the seven helixes fold in the membrane to generate an overall cylindrical shape. The polar residues in the membrane-spanning regions, shown in (a) in color, are believed to participate in bonding the helixes to each other. [See D. M. Engelman et al., 1980, *Proc. Nat'l Acad. Sci. USA* 77:2023.] *Part (a) after diagram modified by H. G. Khorana from a structure in Engelman et al.; part (b) after N. Unwin and R. Henderson, 1984,* Sci. Am. *250(2):78.*

form hydrogen bonds or ionic bonds with the polar residues of different helixes. For example, arginine 82 in helix 3 is probably hydrogen-bonded to glutamate 204 in helix 7. As might be expected, very few charged groups are actually embedded in the hydrocarbon core.

We recall that hydrophilic side chains are preponderant on the surface of aqueous proteins and hydrophobic groups tend to be confined to the interior. In proteins such as rhodopsin, the situation is reversed: hydrophobic side chains are on the outside of the protein molecule, exposed to the hydrocarbon interior of the membrane. Hydrophilic side chains are in the interior of the molecule, away from the hydrophobic lipids, or in regions of the molecule that are either exposed to the polar head groups of lipids or that project into the aqueous exterior. Thus, such membrane-embedded proteins are "inside-out" with respect to an aqueous protein.

Because of their exposed hydrophobic regions, membrane-binding proteins present a special problem to the protein chemist. In general, the exterior regions of water-soluble proteins contain a preponderance of hydrophilic groups; in aqueous solutions, therefore, they remain individual. But the exposed hydrophobic regions of membrane-binding proteins cause the protein molecules to aggregate and precipitate from water solutions. Such proteins can be solubilized in the presence of detergents that have affinity for both hydrophobic groups and water. Because many important membrane proteins are present in only tiny amounts, it has been difficult to obtain enough pure protein to determine its sequence, let alone its membrane interactions. The sequence of many membrane proteins is now being determined from the sequence of the cloned cDNA. Very recently, the three-dimensional structure of the first membrane-imbedded protein—a bacterial-photosynthetic reaction center—has been determined by x-ray crystallography.

The Orientation of Proteins in Membranes Can Be Determined by Chemical or Enzymatic Reactions

The direction in which a protein is oriented within the plasma membrane can be determined in several ways. One is by using a covalent labeling reagent that cannot penetrate membranes and will thus bind covalently only to those regions of the polypeptide that are exposed on the outside of the membrane. A commonly used reagent is the enzyme lactoperoxidase. In the presence of hydrogen peroxide and radioactive iodine [^{125}I], this enzyme oxidizes the iodine to an enzyme-bound active intermediate and iodinates tyrosine and histidine side chains of proteins. Because lactoperoxidase, with its bound activated [^{125}I], cannot penetrate membranes, it will react only with proteins that are exposed on the cell surface (Figure 14-15).

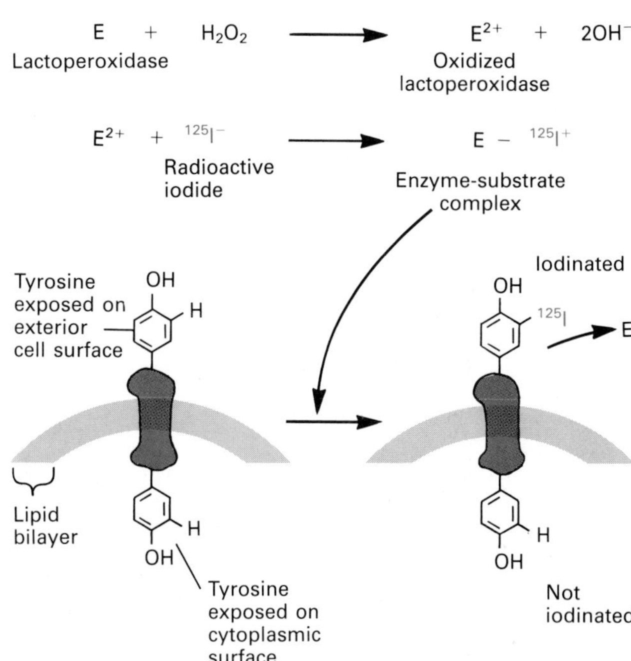

Figure 14-15 Selective labeling of proteins on the outside of the plasma membrane by lactoperoxidase-catalyzed iodination. All reactions in which iodine is a participant require enzyme-bound intermediates.

Similarly, proteases will digest only those regions of cell surface proteins that are exposed to the extracellular medium.

Suppose, however, that the plasma membrane is disrupted by ultrasonic vibration or by detergents. Now added proteases can penetrate to the cytoplasm; they will digest regions of the plasma membrane protein that are exposed to the cytoplasm as well as the regions that face the outside.

Experiments have shown that the NH_2-terminus of glycophorin and all of the oligosaccharides are on the extracytoplasmic surface, as depicted in Figure 14-13. Only this extracytoplasmic segment is digested by proteases added to the outside of the cell, and only this segment is reactive with extracellular lactoperoxidase. The COOH-terminal residues 106–131 face the cytoplasm. They are reactive with such reagents only if the permeability barrier of the plasma membrane is disrupted. The hydrophobic segment (residues about 64–94) is not reactive with any aqueous enzymes or agents. It is believed to span the membrane. Thus, the experimenters concluded that glycophorin is a transmembrane protein.

Detergents Are Used to Solubilize and Study Integral Membrane Proteins

Detergents, as amphipathic molecules, have made important contributions to the study of biological membranes.

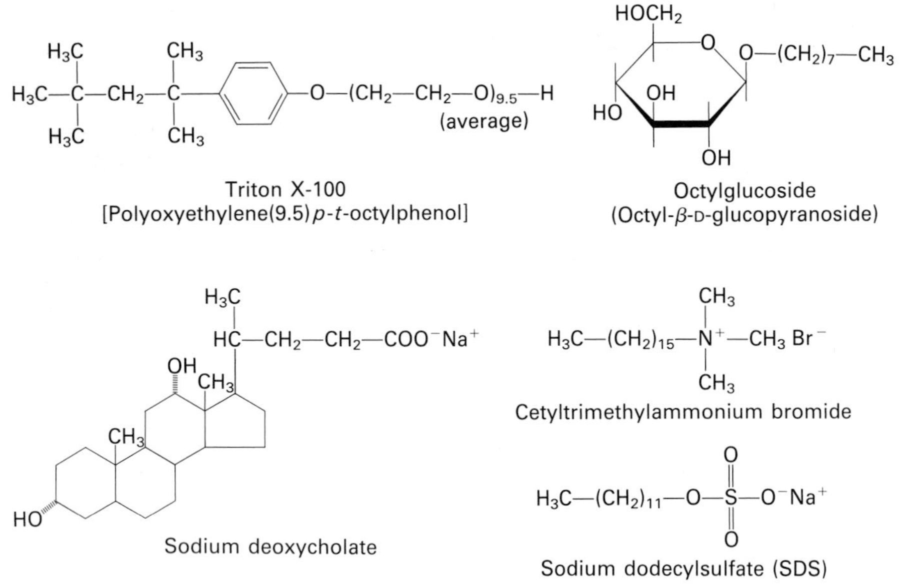

Triton X-100
[Polyoxyethylene(9.5)*p*-*t*-octylphenol]

$H_3C-C(CH_3)_2-CH_2-C(CH_3)_2-C_6H_4-O-(CH_2-CH_2-O)_{9.5}-H$ (average)

Octylglucoside
(Octyl-*β*-D-glucopyranoside)

Sodium deoxycholate

$H_3C-(CH_2)_{15}-N^+(CH_3)_2-CH_3\ Br^-$

Cetyltrimethylammonium bromide

$H_3C-(CH_2)_{11}-O-SO_2-O^-\ Na^+$

Sodium dodecylsulfate (SDS)

Figure 14-16 Structures of some commonly used detergents.

Detergent molecules disrupt membranes by intercalating into phospholipid bilayers and solubilizing lipids and proteins. Some detergents are natural products, such as the bile salt sodium deoxycholate, but most are synthetic molecules developed by the chemical industry for cleaning and for dispersing mixtures of oil and water (Figure 14-16). The hydrophobic part of a detergent molecule is attracted to hydrocarbons, such as oil, and mingles with them readily, whereas the hydrophilic part is strongly attracted to water. Thus, in a mixture of oil and water, the detergent forms a monomolecular film at the boundary between the oil and water.

Agitation of oil-water mixtures breaks large volumes of oil into smaller droplets (Figure 14-17). In the absence of detergent, these droplets coalesce once again into their original aggregate. In the presence of detergent, however, each droplet is surrounded by a single layer of detergent molecules with outward-pointing hydrophilic ends. This arrangement makes the droplet as a whole a hydrophilic object. It is easily suspended in water and can be rinsed away.

At very low concentrations in pure water, detergents are dissolved as isolated molecules. As the concentration increases, the molecules begin to form micelles: small, spherical aggregates having their hydrophilic parts facing outward and their hydrophobic parts clustering in the center (Figure 14-17). The size of a micelle is limited by the need for all its molecules to have access to the hydrophilic surface. The concentration at which micelles form, called the *critical micelle concentration* (CMC), is characteristic of the detergent and is a function of the structures of the hydrophobic and hydrophilic parts of the detergent.

Ionic detergents, such as those whose structures are given in Figure 14-16, contain a charged polar head. They bind to the hydrophobic regions of proteins. They also alter the conformation of the hydrophilic regions and disrupt such noncovalent bonds as ionic and hydrogen bonds. Sodium dodecylsulfate (SDS), for example, completely denatures proteins at high concentrations—every side chain will have one or more SDS molecules bound to it. The mobility of SDS-solubilized proteins in gel electro-

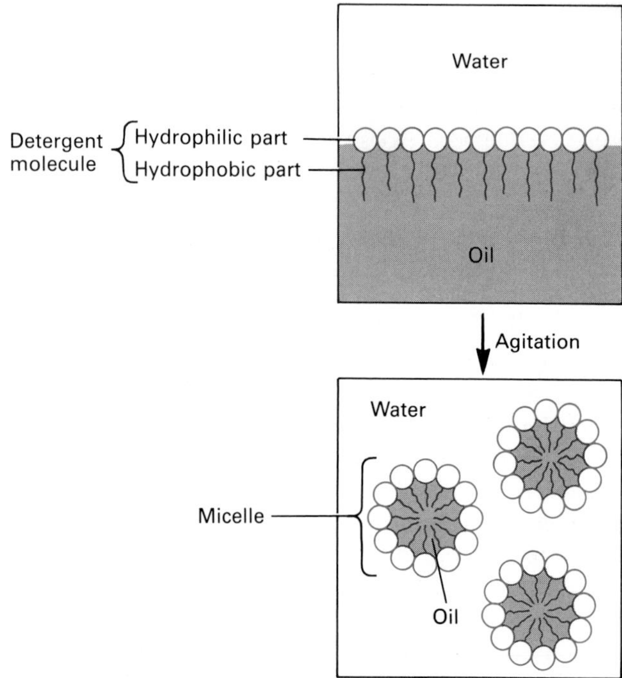

Figure 14-17 Detergents can solubilize hydrocarbons by forming micelles.

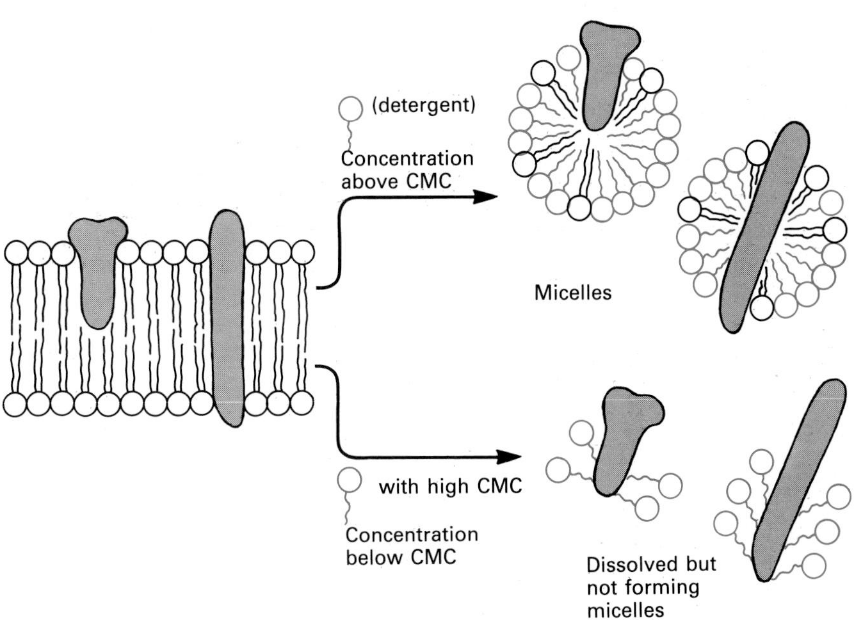

Figure 14-18 Behavior of detergents. The addition of detergents to biological membranes at concentrations higher than the critical micelle concentrations (CMC) results in the solubilizing of lipids and integral membrane proteins and the forming of mixed micelles of detergents, protein, and lipid. If a detergent has a high CMC, it may dissolve membrane proteins at concentrations well below its CMC—that is, without forming micelles.

phoresis is a good measure of their molecular weight (Chapter 7).

Nonionic detergents, such as Triton X-100 and octylglucoside (see Figure 14-16), are without a charged group. Usually they solubilize biological membranes by forming mixed micelles of detergent, phospholipid, and integral membrane proteins (Figure 14-18). These detergents generally will not denature proteins. Detergents such as octylglucoside have a high CMC but will solubilize most membrane proteins at concentrations well below the CMC. When present in low concentrations, octylglucoside binds to the hydrophobic regions of the protein that are normally embedded in the membrane but does not enclose the hydrophobic regions in a micelle. Such solubilized preparations of membrane proteins constitute the first step in their purification.

Membrane-Bound Enzymes Catalyze Reactions with Amphipathic Substances

The important interrelation between the structure and function of membrane-bound proteins is best explained by example. An appropriate one is furnished by an enzyme of the endoplasmic reticulum that catalyzes an important step in the synthesis of phospholipids. Stearoyl coenzyme A (stearoyl CoA), one of the substrates of the reaction, is an amphipathic molecule. Its stearic acid constituent is hydrophobic and is dissolved in the lipid core of a membrane. Its CoA constituent is hydrophilic and water soluble and protrudes into the cytosol (Figure 14-19). The other substrate is glycerol phosphate, a small water-soluble compound. The product of the reaction, stearoyl-glycerol phosphate, is another amphipathic molecule. The enzyme itself, then, should also be amphipathic. Its detailed structure is not known, but it appears to have one segment embedded in the hydrophobic core of the membrane so as to interact with the stearoyl part of the substrate, and another segment that is hydrophilic and interacts with glycerol phosphate and CoA. Possibly the hydrophobic region plays a role in holding the protein in the endoplasmic reticulum membrane by binding to endoplasmic reticulum-specific proteins or lipids.

Viral Envelope Proteins Are Important Examples of Membrane-Bound Proteins

The glycoproteins that cover the surfaces of many types of viral particles are well-studied proteins known to have hydrophobic segments that span a phospholipid bilayer. *Enveloped viruses* acquire an outer phospholipid bilayer as they bud from the host cell surface (see Chapter 6). This outer envelope surrounds an internal complex of viral nucleic acids and viral proteins. Embedded in the viral phospholipid membrane are one or more species of virus-specific glycoproteins (Figure 14-20) that are essential for the infectivity of the virus because they determine the cells to which the virus can bind. The structure of the vesicular stomatitis virus (VSV) glycoprotein, called the G protein, is shown in Figure 14-21. Most viral glycoproteins are believed to be anchored in the membrane in a manner similar to that of glycophorin. A single segment of about 20 very hydrophobic residues spans the phospholipid membrane. Most likely, the amino acids in this segment fold into an α helix. These proteins have as important substituents several covalently attached fatty acids (Figure 14-22) that appear to be linked to cysteine

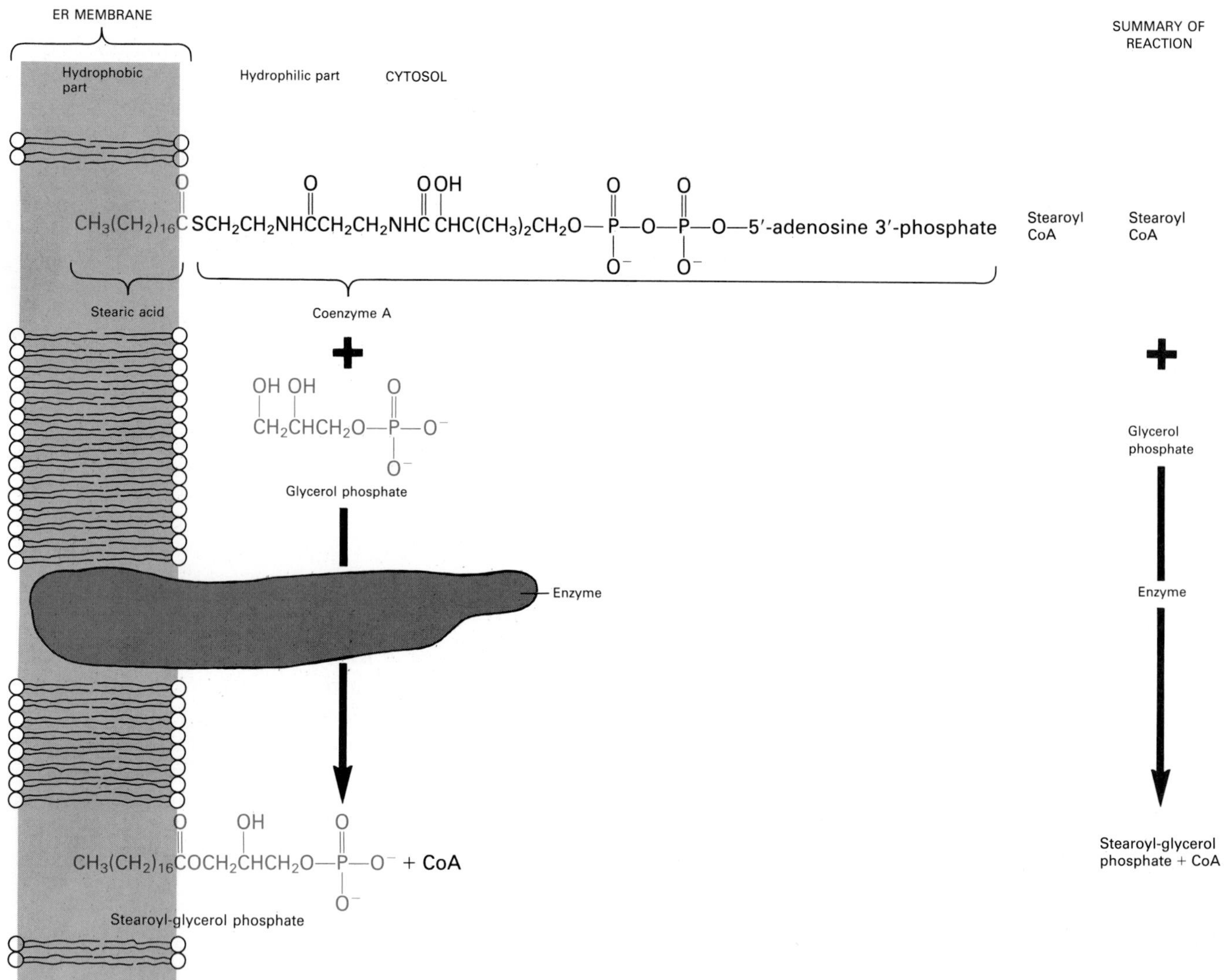

Figure 14-19 Hypothetical structure of the endoplasmic reticulum (ER) enzyme that catalyzes the reaction of stearoyl CoA and glycerol 3-phosphate, forming stearoyl-glycerol phosphate and CoA. Hydrophobic and hydrophilic parts of the stearoyl CoA and stearoyl-glycerol phosphate are indicated in relation to the membrane.

residues in or near the membrane-spanning part and help anchor the protein in the lipid bilayer. A small region of the polypeptide protrudes into the interior of the viral particle. This region contains basic amino acids (lysine and arginine) that may bind to the negative polar head groups of the phospholipids. Like that of glycophorin (Figure 14-13), the bulk of the polypeptide chain protrudes from the surface into the surrounding medium and is the region that binds the virus to the plasma membrane of target cells. Almost all of the carbohydrates in VSV are attached to the G protein; the two "complex" oligosaccharide chains, whose structure is depicted in Figure 3-77, are attached to asparagine residues on the extracytoplasmic segment of the protein. Interestingly, the carbohydrate residues are not required for viral infectivity; the specificity of virus infection is due only to the protein part of the G protein.

Cell Surface Antigens Are Important in Cell-to-Cell Recognition

Like all large molecules, the proteins bound in the plasma membrane can serve as antigens; that is, they can stimulate the production of antibodies. If, for example, live mouse cells are injected into a rat, the immune system of the rat recognizes the surface of the injected cells as foreign and makes antibodies that react with surface proteins on the mouse cells.

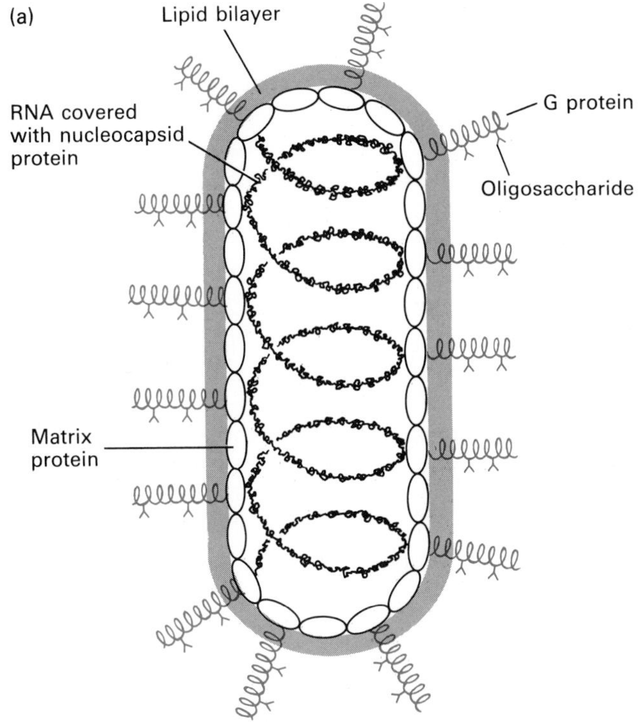

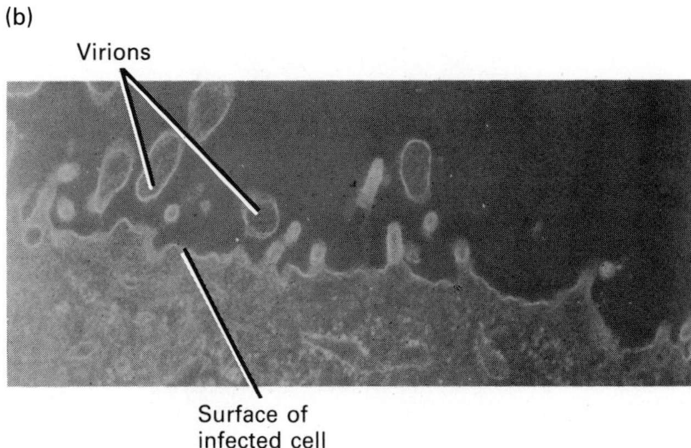

Figure 14-20 (a) Diagram of vesicular stomatitis virus, one of the viruses surrounded by a phospholipid bilayer membrane. (b) Vesicular stomatitis virus particles budding from the surface of an infected cell. The "fuzz" on the outer surface of the particles is due to the virus's glycoprotein (G protein) spikes. *Photograph courtesy of D. Knipe.*

Some surface protein molecules that react with such antibodies have known functions. Most have unknown functions and are simply designated *cell surface antigens.* A major antigen called *H-2* is found on most mammalian cells. In humans, the analogous antigen is called HLA. It is constructed of two different polypeptides. The smaller of the two chains (β_2-microglobulin) is the same in all animals of a given species. The larger chain is capable of so many slight structural differences that the HLA type antigen of each individual organism in a natural population is different from that of almost every other (Figure 14-23). The variability of HLA effectively prevents tissue transplants, because the receiving organism's immune system recognizes the foreign cells and rejects them. The A, B, and O blood-group carbohydrates are also cell surface antigens.

Principles of Membrane Organization

Thus far we have discussed primarily the structures of pure phospholipid vesicles and the conformations of a few relatively well understood membrane proteins. We now turn to more complex questions of the architecture of lipids and proteins in cellular membranes, particularly in the plasma membrane. How are the lipids and proteins actually organized in such membranes? How mobile are they? Let us first consider their asymmetry, which is essential to the function of all biological membranes.

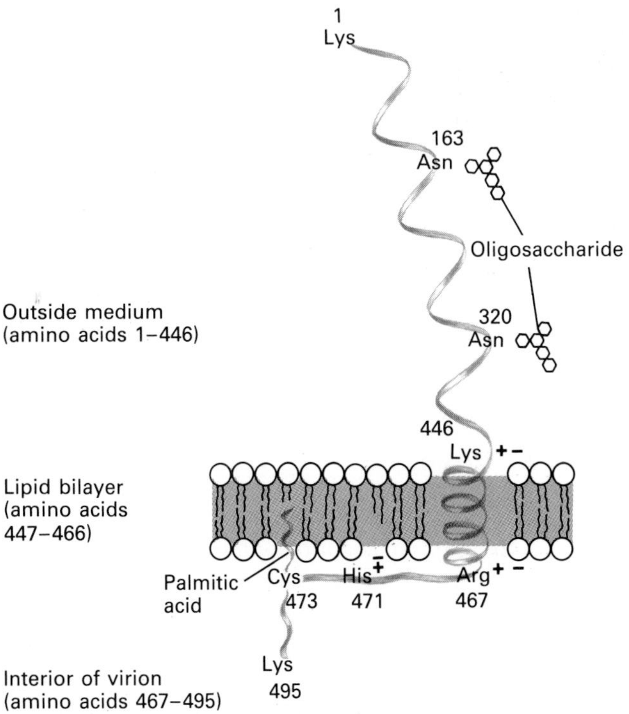

Figure 14-21 Structure of VSV glycoprotein with specific regions indicated. [See J. K. Rose et al., 1980, *Proc. Nat'l Acad. Sci. USA* 77:3884.]

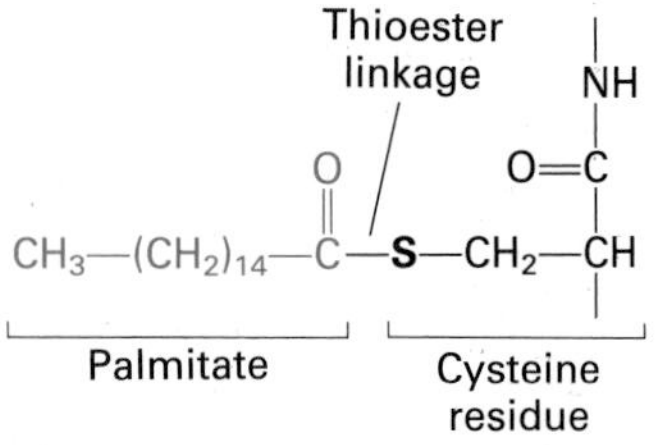

Thioester linkage

$$CH_3{-}(CH_2)_{14}{-}\overset{O}{\overset{\|}{C}}{-}S{-}CH_2{-}CH(-C{=}O)(-NH)$$

Palmitate Cysteine residue

Figure 14-22 Structure of a fatty acyl (palmitate) bound in a thioester linkage to a cysteine side group in a membrane glycoprotein. The palmitate, shown in color, is believed to help anchor the protein in the membrane.

All Membrane Proteins Bind Asymmetrically to the Lipid Bilayer

Each type of membrane protein has a specific orientation within its phospholipid membrane, causing an asymmetry of the two membrane faces that gives them different characteristics. When glycophorin (Figure 14-13) or bacteriorhodopsin (Figure 14-14b) or a viral glycoprotein (Figure 14-21) spans the phospholipid membrane, all molecules of that protein species lie in the same direction. When a species of enzyme is bound to a membrane, every molecule of that enzyme species has its active site on the same face of the membrane. When a cytoskeletal protein binds to one face of a membrane, all molecules of that species are found on the same face of the membrane. The absolute nature of this asymmetry can easily be seen for glycoproteins (Figures 14-13 and 14-21). In all cases, the oligosaccharide residues are on the extracytoplasmic face of the membrane (see Figure 14-6); they are never to be found on the cytoplasmic surface. Proteins have never been observed to flip-flop across a membrane; such a movement would be energetically unfavorable because it would require a transient movement of hydrophilic amino acid and sugar residues through the hydrophobic core of the membrane.

The Two Membrane Leaflets Have Different Lipid Compositions

The phospholipid compositions of the two leaflets of a membrane are quite different, at least in the plasma membranes that have been analyzed. In the human erythrocyte, for example, all the glycolipids, such as sphingomyelin, and most of the phosphatidylcholine are found in the extracytoplasmic leaflet whereas phosphatidylserine and phosphatidylethanolamine are preferentially located on the cytoplasmic face (Figure 14-24). In general, the oligosaccharide side chains of glycolipids are found exclusively on the extracytoplasmic surface of both plasma membranes and internal membranes. In the plasma membrane, all the sugars, both on glycoproteins and glycolipids, are exposed to the outside of the cell. In the endoplasmic reticulum, all the protein- and lipid-bound oligosaccharides are on the lumenal surface.

In contrast with the absolute asymmetry of glycolipids, most kinds of phospholipids, as well as cholesterol, are generally found in both leaflets of membranes, though often enriched in one or the other. In the erythrocyte plasma membrane, phosphatidylcholine is preferentially in the outer leaflet; in other cells it is preferentially in the inner.

One way to determine the presence of specific phospholipids on one or the other face of the plasma membrane is to treat intact cells with various phospholipases, enzymes that hydrolyze the ester or phosphodiester bonds (Figure 14-25). Phospholipids on the outer face of the

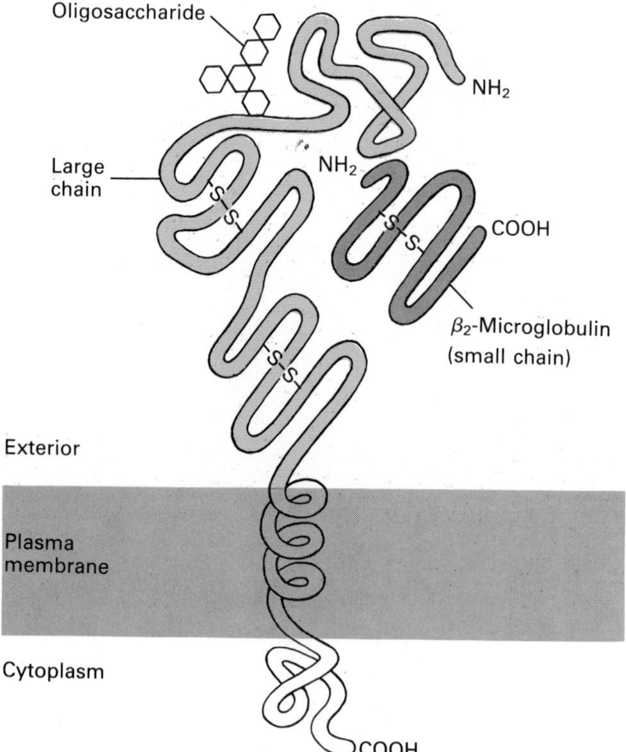

Figure 14-23 Conformation of the HLA cell surface antigen, which is found on the surface of virtually all nucleated human somatic cells. The larger of the two polypeptides that form this antigen is folded into three domains (shown in color), two of which are stabilized by disulfide bonds. It is bound noncovalently to the smaller polypeptide, β_2-microglobulin (shown in gray). It is the large chain that is highly variable among different members of the species. *After H. L. Ploegh, H. T. Orr, and J. L. Strominger, 1981,* Cell **24**:287.

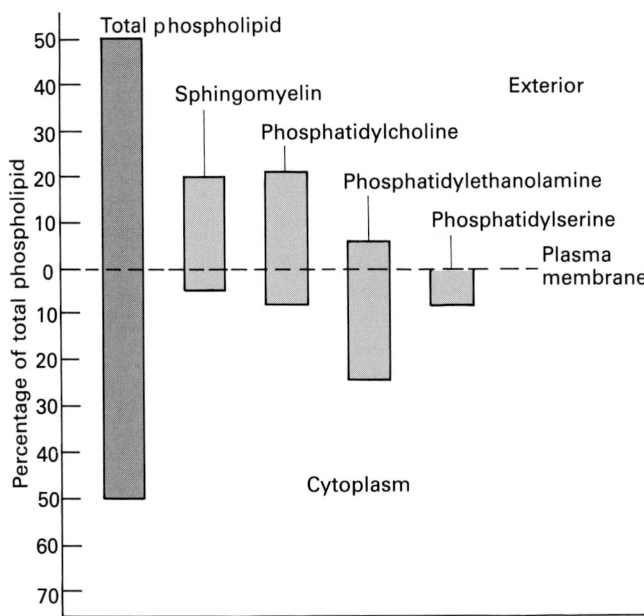

Figure 14-24 Distribution of phospholipids on the two faces of the erythrocyte surface membrane. The values are expressed as a percentage of total membrane lipids. Note that 50 percent of the total phospholipid is found in each face. [See J. E. Rothman and J. Lenard, 1977, *Science* **195**:743.]

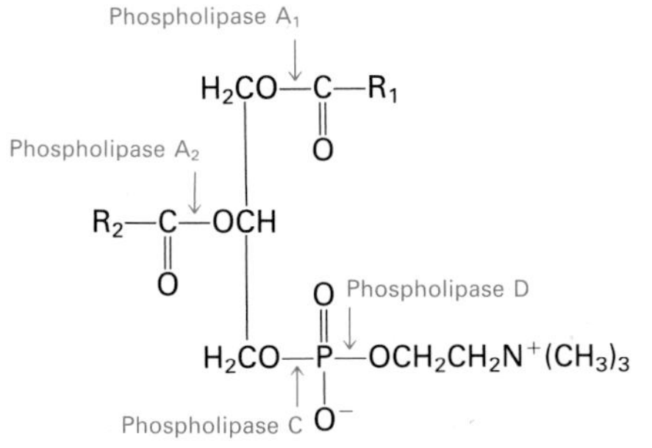

Phospholipase A_1

$H_2CO—C—R_1$ (C=O)

Phospholipase A_2

$R_2—C—OCH$ (C=O)

Phospholipase D

$H_2CO—P—OCH_2CH_2N^+(CH_3)_3$ (P=O, P–O^-)

Phospholipase C

Figure 14-25 Sites of action of phospholipases on phosphatidylcholine. R_1 and R_2 are fatty acyl side chains.

erythrocyte membrane are, in general, substrates for such enzymes. Phospholipids on the cytoplasmic face are not exposed to the enzymes and therefore do not react with them unless the integrity of the membrane is disrupted with detergents or other reagents.

It is not clear how the differences in the lipid composition in the two leaflets arise. One possibility is that certain lipids bind to certain membrane-embedded proteins that occur preferentially on one leaflet of a membrane, so that these lipids will be present in greater quantity in that leaflet.

The Freeze-Fracture Technique Reveals the Two Membrane Faces

The two faces of any cellular membrane can be separated and studied by the freeze-fracture and freeze-etching technique (Figure 14-26). When a frozen specimen is fractured by a sharp blow, the fracture line frequently runs through the hydrophobic interior of the membrane, separating the two phospholipid leaflets. Part of the surface ice is removed by sublimation in a vacuum, which further exposes the sample. The specimen is then coated with a thin layer of carbon and shadowed with platinum. When the organic matter is removed with acid, a metal replica remains for examination in the electron microscope. The fractured surfaces of many membranes reveal numerous particles, most of which are membrane proteins (Figure 14-27). In freeze-etching studies, the cytoplasmic face of a membrane is customarily called the P (for protoplasmic) face, and the extracytoplasmic, or exoplasmic, one, the E face. It is not unusual for most or all of the intramembranous particles to be on one of the two surfaces and their mirror images to be on the other in the form of pits or holes. Most likely the intramembranous proteins are bound more tightly to the lipids in one leaflet than in the other.

Most But Not All Membrane Proteins and Lipids Are Laterally Mobile in the Membrane

From experimental evidence it seems that many membrane proteins do float quite freely in the lipid and diffuse on the membrane surface. Two different cells, say a mouse fibroblast and a human fibroblast, can be fused for the purpose of observing the movement of their distinct surface antigens. At different times after incubation at 37°C, the cells are chilled on ice and immediately fixed with compounds, such as glutaraldehyde ($CHOCH_2CH_2CH_2CHO$), that cross-link lysine side chains of proteins and thus prevent further protein movement. The surface antigens of one of the cells can then be detected by using a specific antibody, such as that for the mouse H-2 antigens, tagged with a fluorescent dye. Immediately after fusion the mouse antigens are grouped in one area of the cell, but they quickly diffuse into the human area. Soon both halves of the fused cells are equally fluorescent, demonstrating that most of the surface H-2 proteins on the original mouse cells were not rigidly held in place (Figure 14-28). Such experiments have encouraged the notion that many membrane proteins are free to diffuse in the two-dimensional space of the membrane. The concept of proteins floating freely in a sea of lipid has been popularized as the *fluid mosaic model* of membrane structure (see Figure 14-1); the membrane is a two-dimensional mosaic of phospholipid and protein.

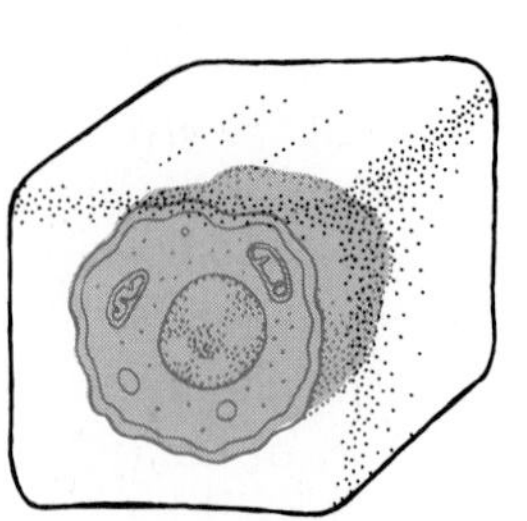

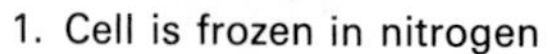
1. Cell is frozen in nitrogen

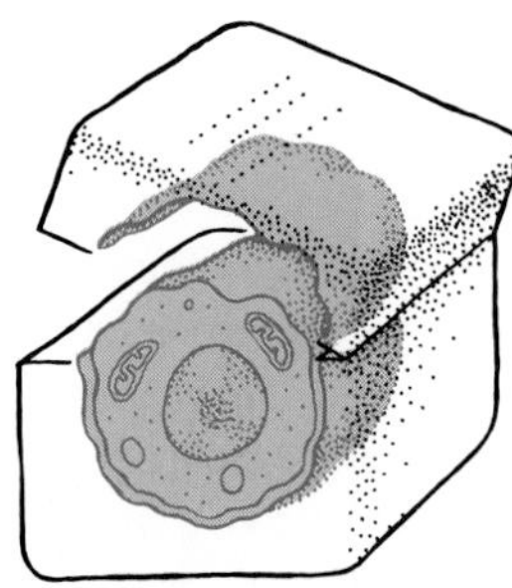
2. Fracture splits the plasma membrane

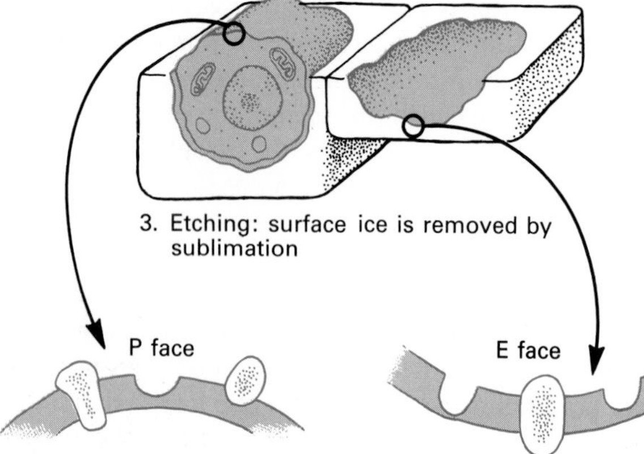
3. Etching: surface ice is removed by sublimation
P face
E face
Detail of the two exposed membrane leaflets

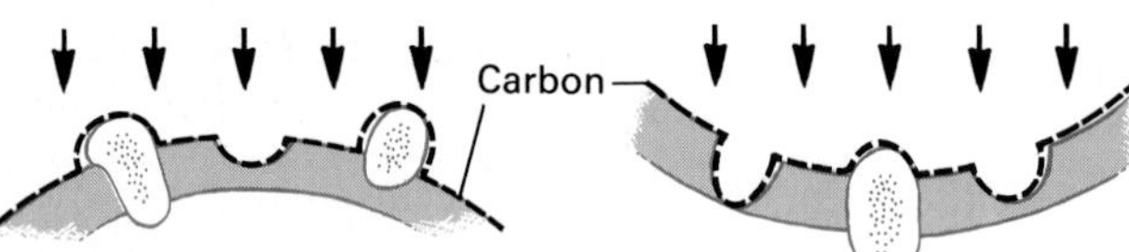
Carbon
4. Carbon is added to form a continuous surface

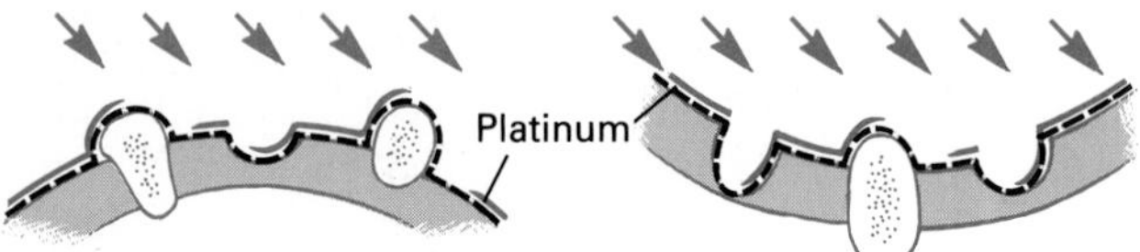
Platinum
5. Surface is shadowed with a thin layer of platinum

6. Tissue is dissolved with acid, carbon-metal replica can be viewed under the electron microscope

Figure 14-26 The production of a freeze-etch image of a cell membrane. A preparation of cells or tissues is quickly frozen in liquid nitrogen at −196°C, which instantly immobilizes cell components without forming ice crystals. The block of frozen cells is fractured with a sharp blow, occasionally separating the two leaflets along the fracture surface. Membrane proteins, the main constituents of intramembrane particles, are bound to one or the other leaflet. Next the specimen is placed in a vacuum and the surface ice removed by sublimation. A thin layer of carbon is evaporated vertically onto the surface to produce a carbon replica. Then a thin layer of platinum is deposited at an angle to the exposed surface to bring out the surface features by producing a shadow effect. The organic material is removed by acid, leaving the carbon-metal replica of the tissue ready for examination with the electron microscope. When prints of the photograph are made the image usually is reversed so that coated areas are light and shadowed areas dark.

Many integral membrane proteins form stable, noncovalent interactions with other such proteins. Glycophorin, for instance, is a dimer of two identical chains, depicted in Figure 14-13. In such cases the entire "island," or aggregate, of protein is thought to float in the lipid.

A more quantitative estimate of the rate and extent of lateral movements of surface proteins and lipids can be obtained from studies called *fluorescence recovery after photobleaching* (FRAP). Surface proteins or lipids are labeled with a fluorescent dye, which initially is uniformly distributed on the surface. A laser light source is used to irreversibly bleach the dye, and thus inhibit the fluorescence, in a small patch of the cell. The fluorescence of the bleached area increases with time, owing to the diffusion into it of unbleached fluorescent surface molecules (Figure 14-29). The extent of recovery is proportional to the fraction of labeled molecules that are mobile in the membrane. The rate of recovery can be used to calculate the diffusion coefficient, or the rate at which the molecules diffuse.

Studies using phospholipids labeled with fluorescent dyes show that all lipids in plasma membranes are freely mobile and that their average rate of lateral diffusion is characteristic of lipid diffusion in a protein-free liposome vesicle. Although membrane phospholipids might bind to the hydrophobic regions of an integral membrane protein, they do not form a permanent attachment. Were they to do so, their movement would be considerably restricted, yet the diffusion coefficient, D, for lateral motion of membrane phospholipids is about 10^{-8} cm^2/s, the same as in pure phospholipid bilayers.

Depending on the cell under study and the particular protein or class of proteins labeled, from 30 to 90 percent of surface proteins are freely mobile; their diffusion rate is characteristic of the same protein embedded in synthetic liposomes. Because of their larger size, proteins diffuse more slowly than lipids. Typically a surface protein takes from 10 to 60 min to diffuse from one pole of a 20-μm-diameter cell to the other.

Yet another technique for demonstrating the mobility of plasma membrane proteins takes advantage of the fact that antibody molecules are bivalent, that is, they have

(a)

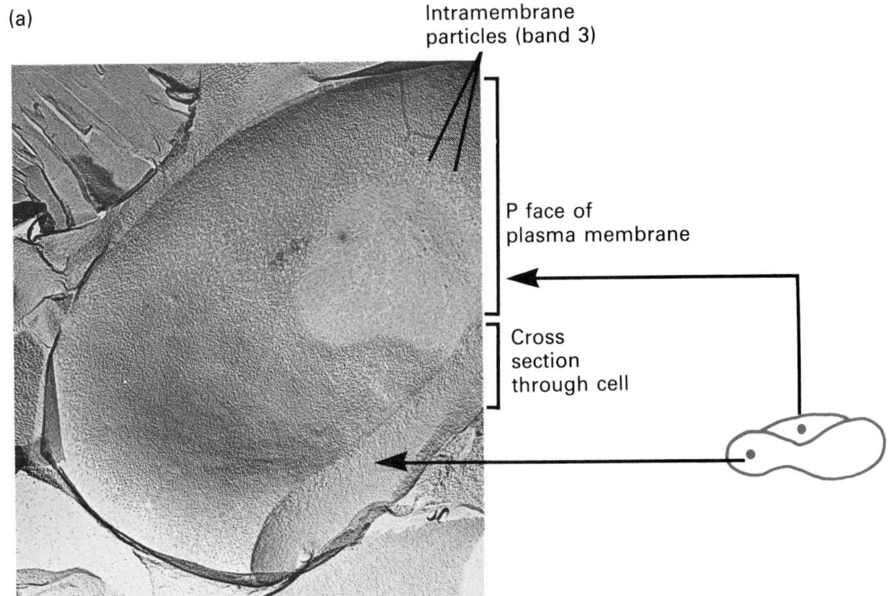

(b)

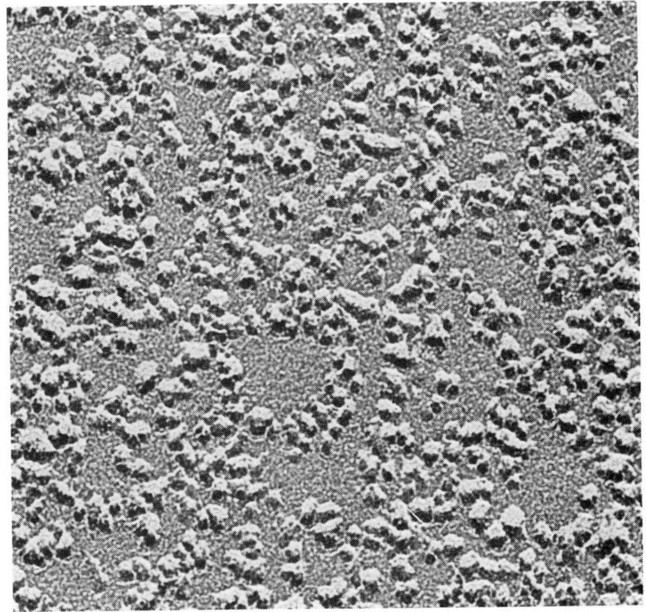

Figure 14-27 Freeze-etch image of an erythrocyte plasma membrane. (a) The P face of the plasma membrane and a cross section through the cell are illustrated. The particles in the cross section are hemoglobin. (b) The intramembrane particles are shown at higher magnification. These particles are composed mainly of band 3, the major intramembrane protein. *Photographs courtesy of D. Branton.*

two identical binding sites for an antigen. Different antibodies directed toward a single protein species will often bind to different sites on the protein. When cells such as lymphocytes are exposed to antibodies specific for a particular cell surface antigen (say HLA), the antibodies cross-link the surface protein molecules. Since the HLA proteins can diffuse laterally in the membrane, they eventually accumulate in antibody–cross-linked networks called patches. This phenomenon, called patching, can be observed experimentally on most types of nucleated cells (Figure 14-30).

Analogous clustering of surface glycoproteins is induced by *lectins,* plant proteins that have two or more specific binding sites for a particular sugar. Wheat germ agglutinin, for example, binds to glucose and mannose. Most plasma membrane glycoproteins contain multiple exposed sugar groups; thus lectins cross-link glycoproteins that have multiple appropriate sugar substituents.

Lateral diffusion of protein occurs in intracellular membranes as well. Inner mitochondrial membranes, for example, contain stable aggregates of integral proteins that are revealed as intramembranous particles by freeze fracturing. If isolated vesicles prepared from the inner mitochondrial membrane are placed in a strong electric field, all the intramembranous particles (which bear a negative electric charge) move to one end of each vesicle (Figure 14-31). When the field is turned off, the particles rapidly return to their original random distribution. This demonstrates that such particles, which are complexes of as many as 15 membrane proteins, are laterally mobile.

The diffusion coefficient for proteins in cell membranes varies from 5×10^{-9} cm^2/s for rhodopsin in the disk membranes of vertebrate retinal cells to about 10^{-11} cm^2/s for many other cell surface proteins. In comparison, the diffusion coefficient for hemoglobin in water is 7×10^{-7} cm^2/s. Thus, the diffusion of membrane proteins is 100 to 10,000 times slower in membranes than the diffusion of water-soluble proteins in water. Apparently the viscosity (resistance to movement) of the phospholipid core of biological membranes is 100 to 10,000 times that of water.

Cytoskeletal Interactions Affect the Organization and Mobility of Surface Membrane Proteins

Fluorescence photobleach studies suggest that not all cell surface proteins conform to the fluid mosaic model.

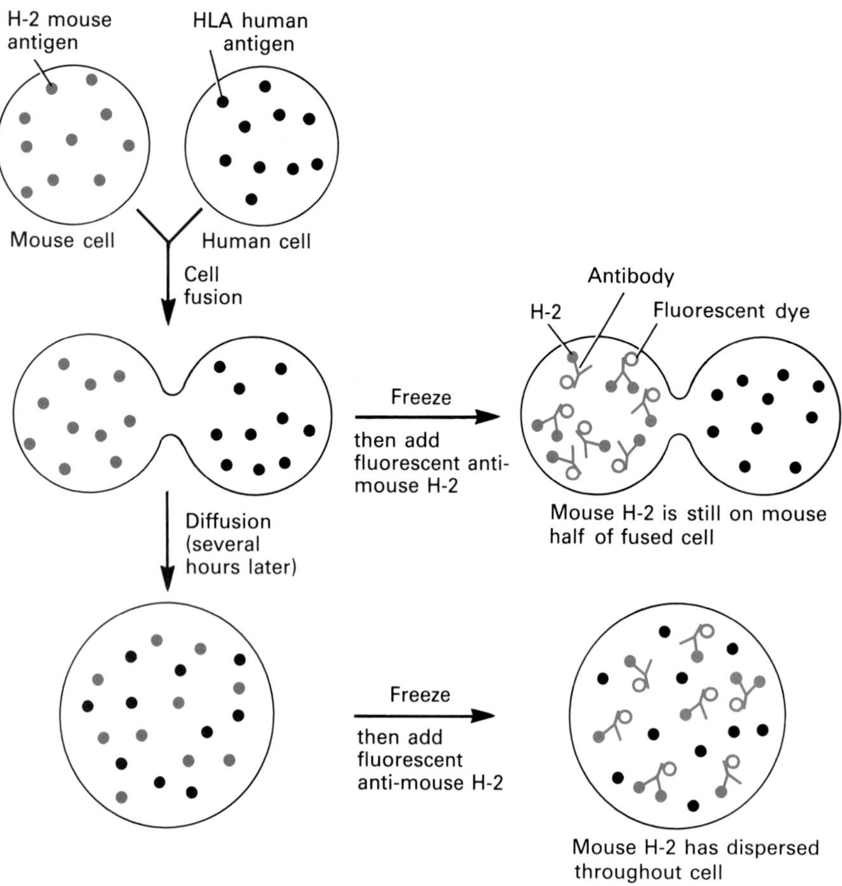

Figure 14-28 The mobility of cell surface proteins is shown by an experiment in which human and mouse cells are fused in culture. Initially, the human and mouse cell antigens remain on their respective halves of the fused cell. After several hours of incubation, the two sets of antigens are thoroughly intermingled. The mouse (H-2) antigens are detected by fluorescent antibodies to them. [See L. D. Frye and M. Edidin, 1970, *J. Cell Sci.* 7:319.]

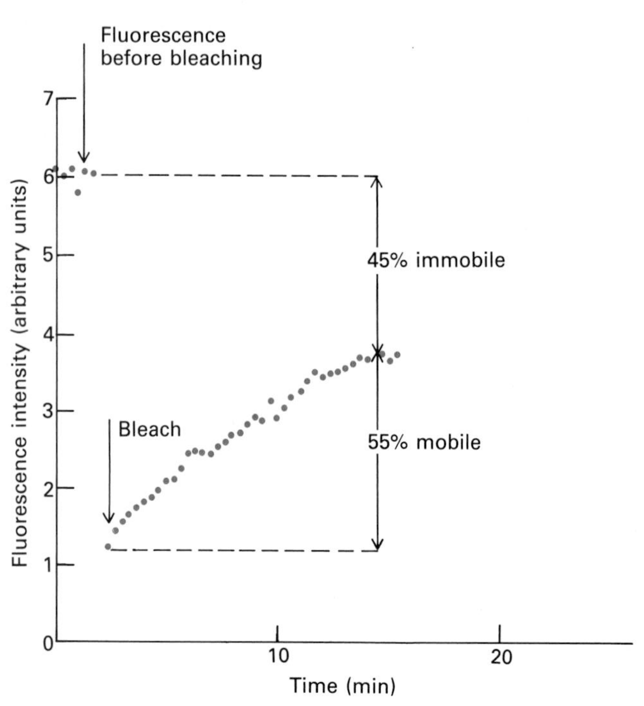

Figure 14-29 A fluorescence photobleach recovery experiment. Myoblasts (precursors of skeletal muscle cells) are treated with fluorescent concanavalin A, a protein that binds tightly to the sugar residues of surface glycoproteins. When a small patch of surface is bleached with a laser light and the myoblasts incubated at 37°C, 55 percent of the fluorescence eventually returns to the bleached area. This indicates that 55 percent of the glycoproteins in this patch of membrane are mobile and 45 percent are immobile. The rate of recovery of fluorescence is proportional to the rate of diffusion of proteins into the region of membrane, and thus to the diffusion constant. [See J. Schlessinger et al., 1976, *Proc. Nat'l Acad. Sci. USA* 73:2409.]

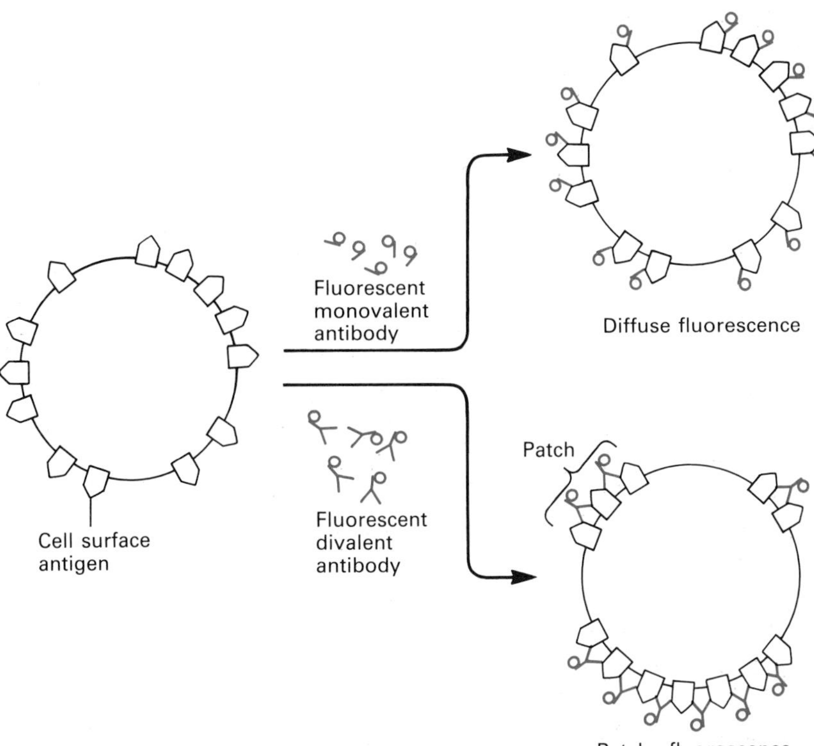

Figure 14-30 Diagram of antibody-induced patching of antigens on the surface of a cell. The antibody is tagged with a fluorescent dye so that it can be detected in the fluorescence microscope. Normal divalent antibodies cross-link the antigens and cause them to cluster into patches. The digestion of antibody molecules by the proteolytic enzyme papain produces monovalent fragments, which have only a single antigen-binding site (see Figure 3-38). Although these fragments still bind to the target protein on the cell surface, they fail to form the cross-linked clusters and thus bind diffusely over the entire surface.

(a)

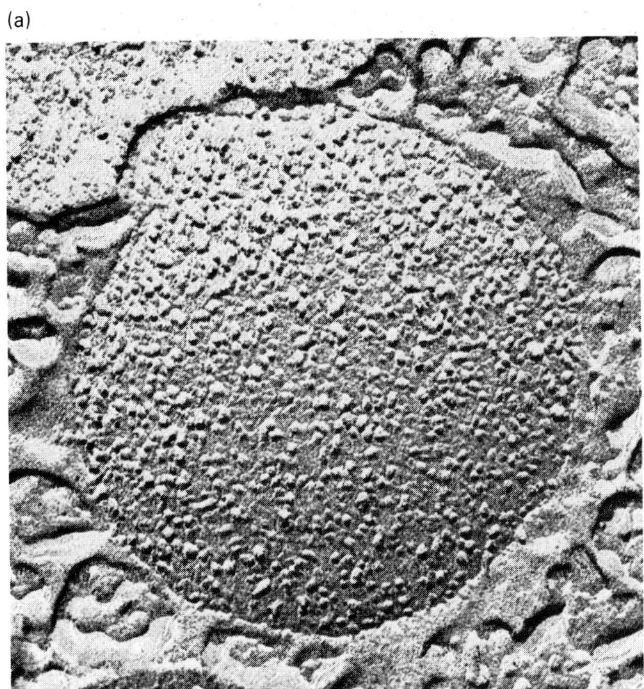

(b)

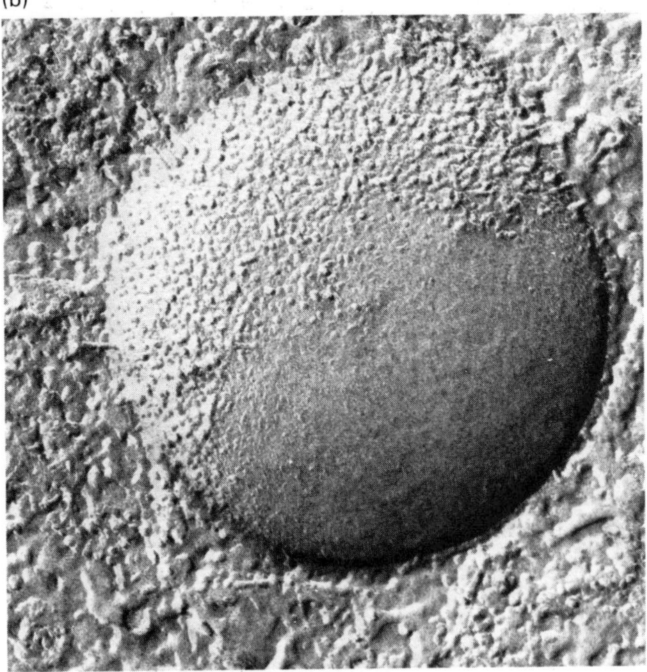

Figure 14-31 (a) Freeze fracture of a vesicle of the mitochondrial inner membrane, showing random distribution of intramembrane protein particles. (b) After the mitochondria are subjected to a strong electric field and then rapidly frozen, all the particles cluster at one end, which shows that the particles have moved within the plane of the membrane under this voltage gradient. Measurement of the rate of this movement indicates that it is similar to the diffusion of many other proteins in a fluid mosaic membrane. [Part (b), see A. E. Sowers and C. R. Hackenbrock, 1981, *Proc. Nat'l Acad. Sci. USA* 78:6246.] *Part (a) courtesy of A. E. Sowers; part (b) courtesy of A. E. Sowers and C. R. Hackenbrock.*

Many of them are immobilized by contacts with other membrane proteins and with the intracellular network that makes up the fibrous *cytoskeleton*. A major component of the cytoskeleton is the long fibers of actin that lie just under the plasma membrane and appear to connect with it at several points. Microtubules and tonofilaments (also called intermediate filaments) also contribute to the cytoskeleton network. The mobility of many plasma membrane proteins is increased when parts of the cytoskeleton are disrupted by drugs, which strongly suggests that those particular membrane proteins are linked to cytoskeletal fibers. By bleaching a two-dimensional grid on the surface of fibroblasts, experimenters have even shown that proteins are more mobile in the direction parallel to the actin fibers that run just under the plasma membrane than in the direction perpendicular to them (see Figure 3-40). This implies that the fibers may serve as tracks along which the proteins tend to move.

Moreover, antibody-induced patching of certain cell surface proteins often causes an analogous redistribution of the submembranous actin cytoskeleton, which results in "copatching" of the surface molecule and the actin—additional suggestive evidence of an interaction between the cell surface proteins and the actin-containing cytoskeleton.

Another striking piece of evidence is that treatment of intact cells with a nonionic detergent at physiological salt concentration releases all of the lipids but only a fraction of the integral membrane proteins, even though these proteins, when purified, are freely soluble in the detergent. Therefore, many of these integral proteins must be tightly bound to the cytoskeleton, perhaps to certain fibrillar elements. Chapters 18 and 19 contain a detailed discussion of these cytoskeletal fibers and their interactions with membranes.

The Glycocalyx Is Made Up of Proteins and Oligosaccharides Bound to the Outer Surface of the Cell

An important element of membrane structure, especially of the plasma membrane, is the glycocalyx, a coat formed by the oligosaccharide side chains of membrane lipids and proteins, among which are found additional proteins bound to the external face of the membrane (Figure 14-32). Most of the proteins of the glycocalyx are not bound into the lipid but rather to proteins embedded in the lipid bilayer (Figure 14-33). Because many of the glycoproteins and glycolipids in mammalian cells contain several negatively charged sialic acid residues, the glycocalyx imparts a negative charge to the surface of most cells.

A major constituent of the cell coat is a fibrous protein called *fibronectin* (see Chapter 19) which has binding sites for many other extracellular proteins and is thus a nexus between the cell and the extracellular proteins that make a tissue rigid. Foremost among the extracellular proteins that maintain tissue architecture is collagen, to which fibronectin binds tightly.

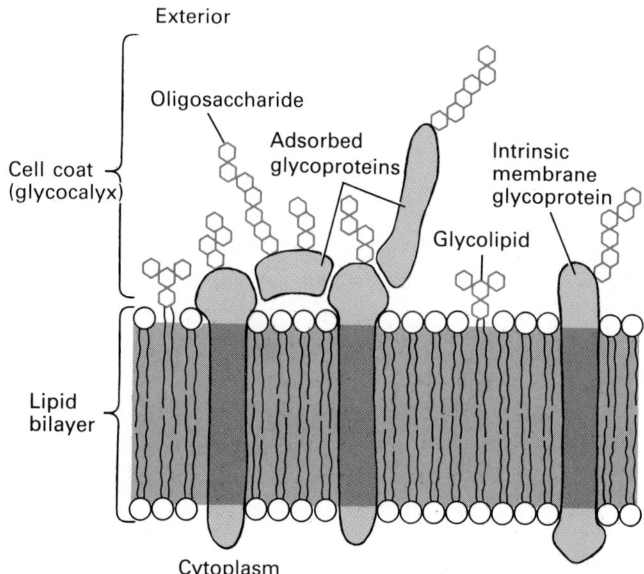

Figure 14-32 Schematic diagram of the glycocalyx, the cell coat made up of the oligosaccharide side chains of intrinsic membrane glycolipids and glycoproteins and of adsorbed glycoproteins and proteoglycans. Note that all the oligosaccharides are on the extracytoplasmic surface of the membrane.

The Erythrocyte Membrane: Cytoskeletal Attachment

The best-characterized cell membrane is the one surrounding the mammalian red blood cell, or erythrocyte (Figure 14-34). The erythrocyte has no nucleus and contains no intracellular membranes. Essentially a bag of hemoglobin with relatively few other proteins, its function is to transport oxygen from the lungs to the tissues and carbon dioxide back to the lungs. We will return to this cell often because its simplicity makes possible detailed analysis.

The erythrocyte normally adopts the shape of a biconcave disk 7 μm in diameter. It is very flexible, however, and often squeezes through capillaries much thinner than 7 μm. Aged or deformed cells that do not possess this flexibility are trapped in the capillaries of the spleen and ingested by macrophages. The erythrocyte membrane must also be durable. In its lifetime of 120 days, a typical human erythrocyte makes half a million circuits of the arteries and veins, a journey of about 300 miles.

The erythrocyte membrane is not representative of

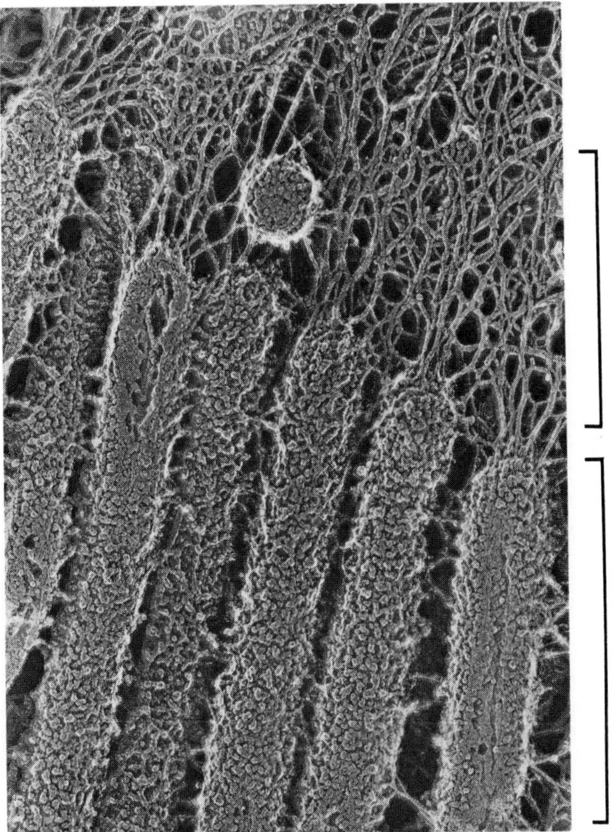

Figure 14-33 The glycocalyx on the surface of microvilli of intestinal epithelial cells, as viewed by the deep-etch technique. The surface of each microvillus is covered with a series of bumps, presumably integral membrane proteins. The glycocalyx, covering the apexes, or tips, of the microvilli, is composed of a network of anastomosing strands. Many of these strands are glycoproteins; some are digestive enzymes such as sucrase-isomaltase (Figure 14-47b) and other glycosidases and peptidases that complete the degradation of proteins and carbohydrates. The halved sphere in the upper center of the photo is the tip of another microvillus. *Courtesy of N. Hirokawa and J. E. Heuser, 1981,* J. Cell Biol. *91:399.*

most cell membranes because it is homogeneous. Its proteins appear to be more or less uniformly distributed in the plane of the membrane, without large specialized patches. The red cell is also unusual among mammalian cells because its cytoskeleton forms a shell that lies under the entire plasma membrane and is attached to it at many points (Figure 14-35). This structure gives the membrane great strength and flexibility. In contrast, most mammalian cells have a cytoskeleton that courses through the cytoplasm and that is anchored to the membrane at relatively few points. Nevertheless, the erythrocyte membrane provides important examples of membrane organization and cytoskeleton attachment.

The Erythrocyte Membrane Retains the Size and Shape of the Erythrocyte

Because the erythrocyte basically consists of a membranous bag containing hemoglobin, the analysis of red cell membrane proteins is relatively straightforward. The cells are placed in distilled water, and the ensuing influx of water by osmosis causes them to swell. Eventually the plasma membranes rupture and release the hemoglobin and other internal proteins. Because the membrane in the depleted cell retains the overall size and shape of the intact cell (proving that the plasma membrane plus associated proteins largely determines the red cell's shape), it is called a *ghost* (Figure 14-36).

In solutions of appropriate ionic strength, ghosts will re-form into sealed vesicles of the size and shape of the original erythrocytes. In solutions of normal ionic strength, the ghosts reseal with their original "outside out" orientation: the extracytoplasmic face is outwards. By changing the composition of the medium, one can obtain sealed inside-out ghosts, with the cytoplasmic face outwards (Figure 14-37). With inside-out preparations, one can study the cytoplasmic surface of the membrane.

Overall, the ghost is 52 percent protein, 40 percent lipid, and 8 percent carbohydrate; 93 percent of the carbohydrate (as oligosaccharides) is attached to protein, forming glycoproteins, and 7 percent to lipids, forming glycolipids.

The Erythrocyte Has Two Main Integral Membrane Proteins

To study the various protein constituents of the erythrocyte plasma membrane, one dissolves the ghost in the negatively charged detergent sodium dodecylsulfate (SDS) and analyzes it by gel electrophoresis. By convention, the major proteins are numbered in order of de-

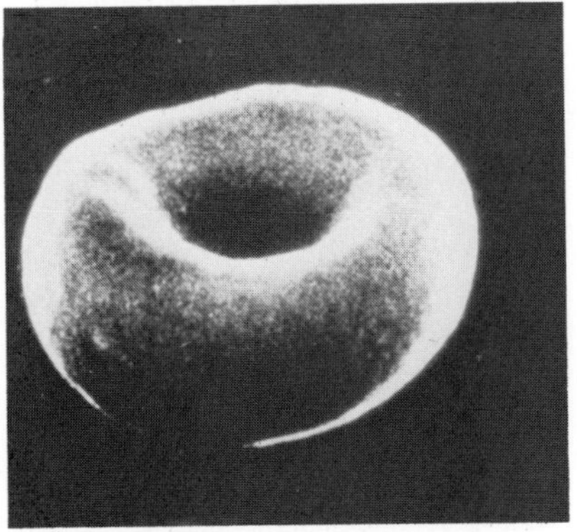

Figure 14-34 Scanning electron micrograph of a normal disk-shaped human erythrocyte; the surface not visible is also concave. *Courtesy of S. E. Lux.*

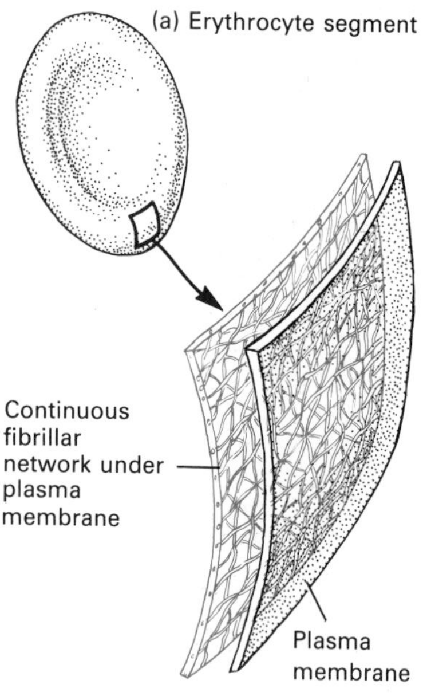

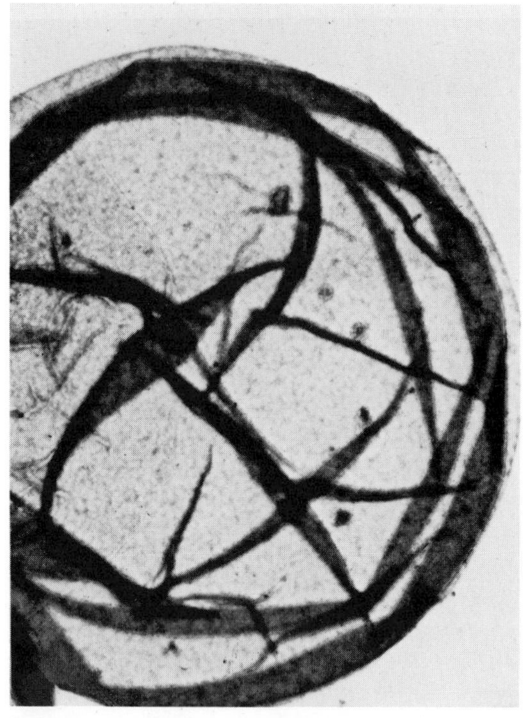

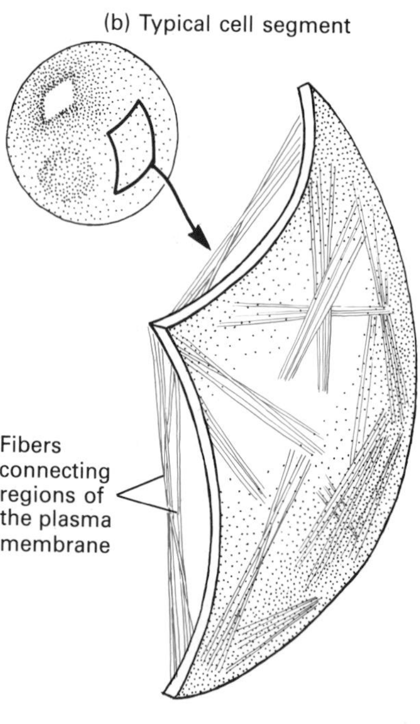

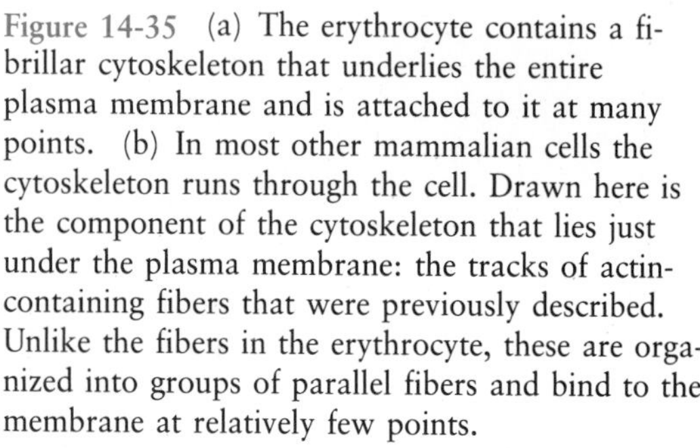

Figure 14-35 (a) The erythrocyte contains a fibrillar cytoskeleton that underlies the entire plasma membrane and is attached to it at many points. (b) In most other mammalian cells the cytoskeleton runs through the cell. Drawn here is the component of the cytoskeleton that lies just under the plasma membrane: the tracks of actin-containing fibers that were previously described. Unlike the fibers in the erythrocyte, these are organized into groups of parallel fibers and bind to the membrane at relatively few points.

Figure 14-36 Electron micrographs of negatively stained erythrocyte ghosts. (a) An intact ghost showing folds (*dark bands*) in the membrane surface. (b) A region at higher magnification. [See J. Yu, A. Fischman, and T. L. Steck, 1973, *J. Supramol. Struct.* **1**:233.] *Courtesy of J. Yu, A. Fischman, and T. L. Steck.*

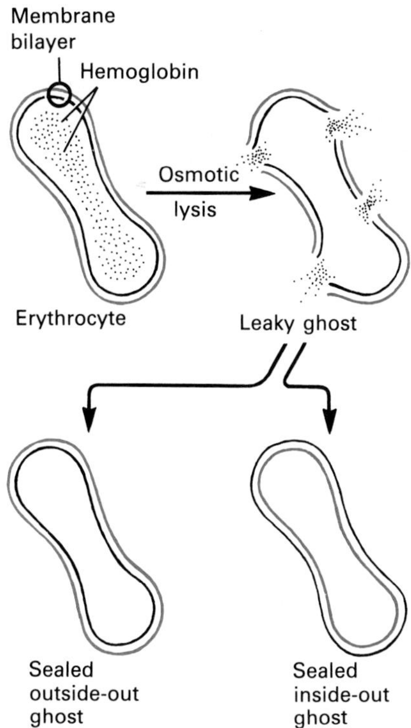

Figure 14-37 Depending on the composition of the medium, leaky erythrocyte membranes can form sealed outside-out or inside-out ghosts.

creasing molecular weight (Figure 14-38). Virtually the same principal proteins are found in erythrocyte membranes from all vertebrates, a result attesting to the generality and importance of this membrane structure.

Two of the erythrocyte membrane proteins, glycophorin and band 3, are glycoproteins. Of the major proteins they are the only two that have exposed regions on the outer surface of the cell. They are also the two main integral membrane proteins. The other proteins depicted in Figure 14-38 are confined to the inner, or cytoplasmic, face of the membrane. They are peripheral proteins that form the cytoskeleton of the erythrocyte.

Glycophorin, one of the two predominant glycoproteins, is not easily detectable in the electrophoretic profile because it stains poorly. Its external segment is heavily glycosylated with both *O*- and *N*-linked types of carbohydrate chains (see Figure 14-13). The *N*-linked oligosaccharides are found on many membranes and soluble glycoproteins. The arrangement of the 15 sugar residues in this type of chain is shown in Figure 3-77. The *O*-linked chains are much shorter, and glycophorin is unusual in having these chains as its major oligosaccharide substituents. Carbohydrate constitutes 64 percent of the weight of glycophorin, an unusually high sugar content.

The other major integral membrane protein in the erythrocyte, band 3, is present in about 1.2 million copies per cell. The orientation of band 3 in the membrane is quite different from that of glycophorin. It is a dimer of two identical chains, each of about 929 amino acids (Figure 14-39). The carboxyl-terminal segment is tightly bound to the lipid membrane and makes multiple passages through it (probably 12 or 13). This part of the molecule contains an oligosaccharide chain facing, as usual, the outside of the cell. The amino-terminal component of the molecule is largely water soluble and protrudes into the cytoplasm. It is this domain of band 3 that anchors the cytoskeleton to the membrane.

The Erythrocyte Extrinsic Proteins Affect Cell Shape and the Mobility of the Intrinsic Proteins

The integral glycoproteins of the erythrocyte membrane are unusual in that they are immobile and cannot diffuse laterally in the plane of the membrane. But they are free to rotate along the axis perpendicular to the plane of the membrane. When erythrocyte ghosts are placed in a solu-

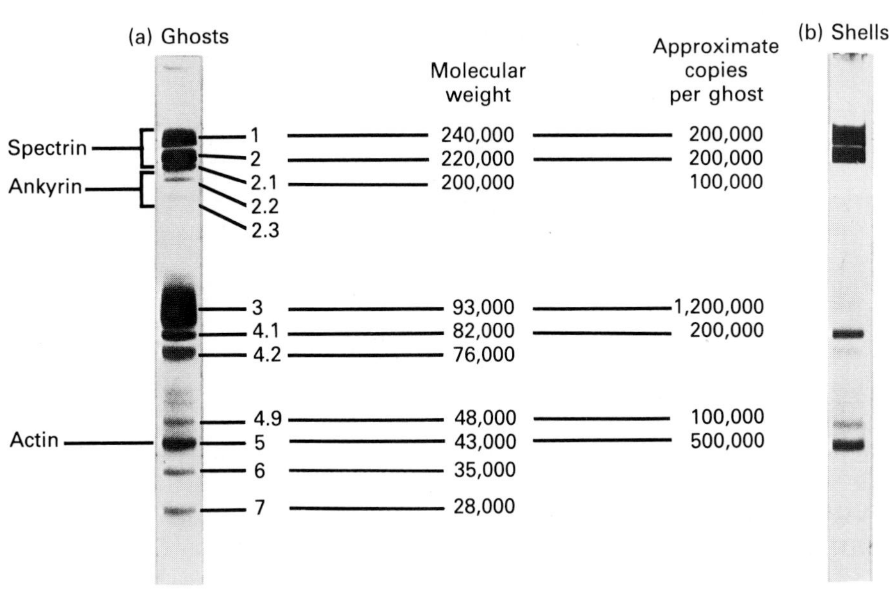

Figure 14-38 SDS-polyacrylamide gels of the polypeptide (a) erythrocyte ghosts and (b) of the shells remaining after the ghosts were treated with 1% Triton X-100 detergent in hypotonic buffer. This treatment removes the band 3 integral membrane protein and certain peripheral proteins but leaves the cytoskeletal proteins in the shell: bands 1 and 2 (spectrins), band 2.1 (ankyrin), band 4.1, and band 5 (actin). Glycophorin, though present in the ghosts, does not stain with the dye that was used and thus is not visible. [See D. Branton, C. M. Cohen, and J. Tyler, 1981, *Cell* **24**:24.]

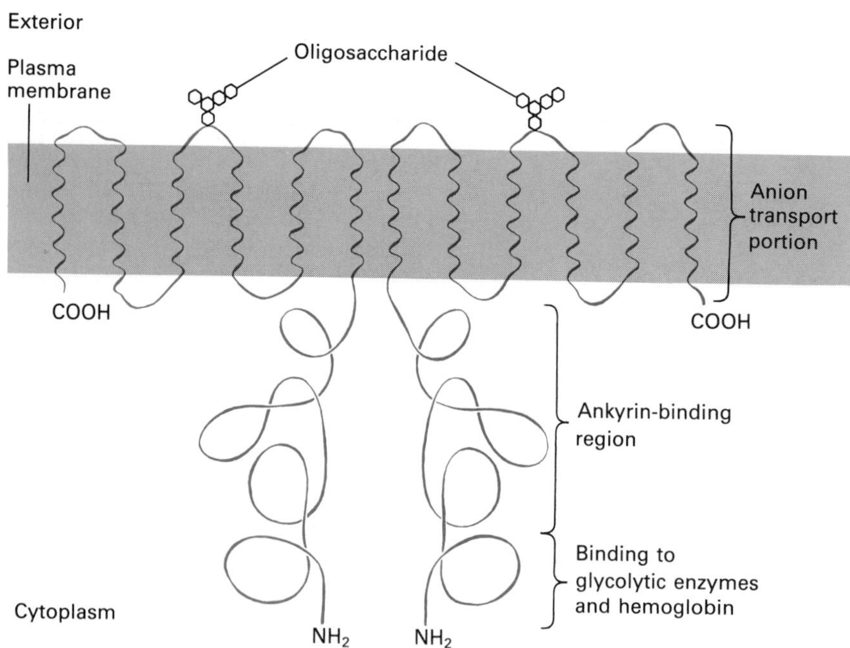

Figure 14-39 The arrangement of band 3 protein (color) in the erythrocyte membrane. Only 6 of the 12 or 13 membrane-spanning helixes per peptide are shown. [See W. A. Braell and H. F. Lodish, 1981, *J. Biol. Chem.* **256**:11337; R. Kopito and H. F. Lodish, 1985, *Nature* **316**:234.]

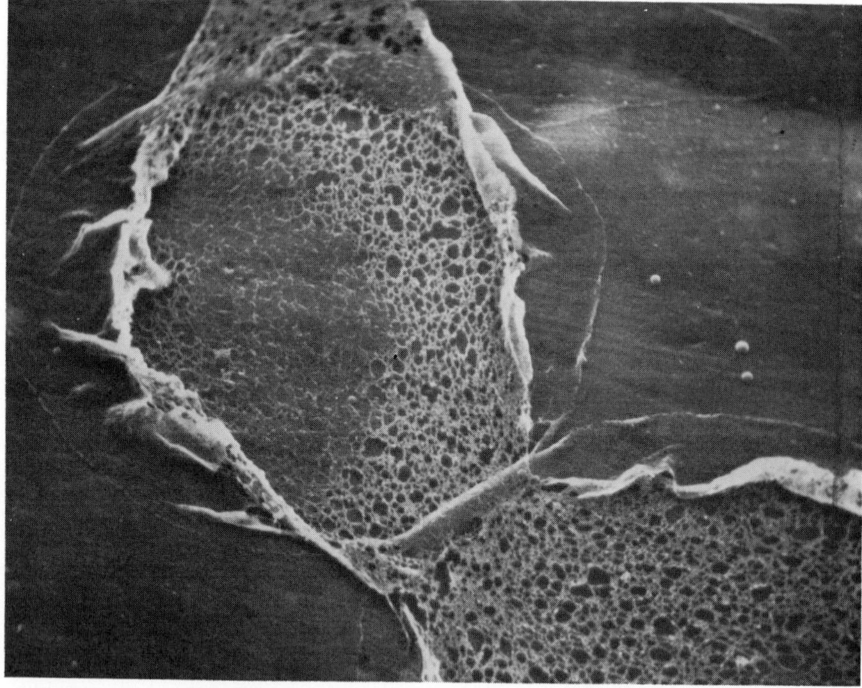

Figure 14-40 Electron micrograph of an erythrocyte shell. The ghost was treated with Triton X-100 detergent. [See J. F. Hainfeld and T. L. Steck, 1977, *J. Supramol. Struct.* **6**:301–311.] *Courtesy of J. F. Hainfeld and T. L. Steck.*

tion of low ionic strength, the principal extrinsic proteins, bands 1 and 2 (spectrins), and band 5 (actin) are solubilized. These three proteins are the major components of the cytoskeleton and removing them has two consequences: the ghost loses its rigid biconcave shape, and the membrane glycoproteins acquire lateral mobility. It appears that the cytoskeleton is the major determinant of the shape of the red blood cell and that it acts to restrict the lateral motion of the membrane glycoproteins.

The Erythrocyte Cytoskeleton Is Constructed of a Network of Fibrous Proteins Just Beneath the Surface Membrane

Ghosts are stripped to their cytoskeletons by treatment with a nonionic detergent at a physiological ionic strength similar to that found in the cell. All the lipid and glycophorin, and about 80 percent of band 3, are solubi-

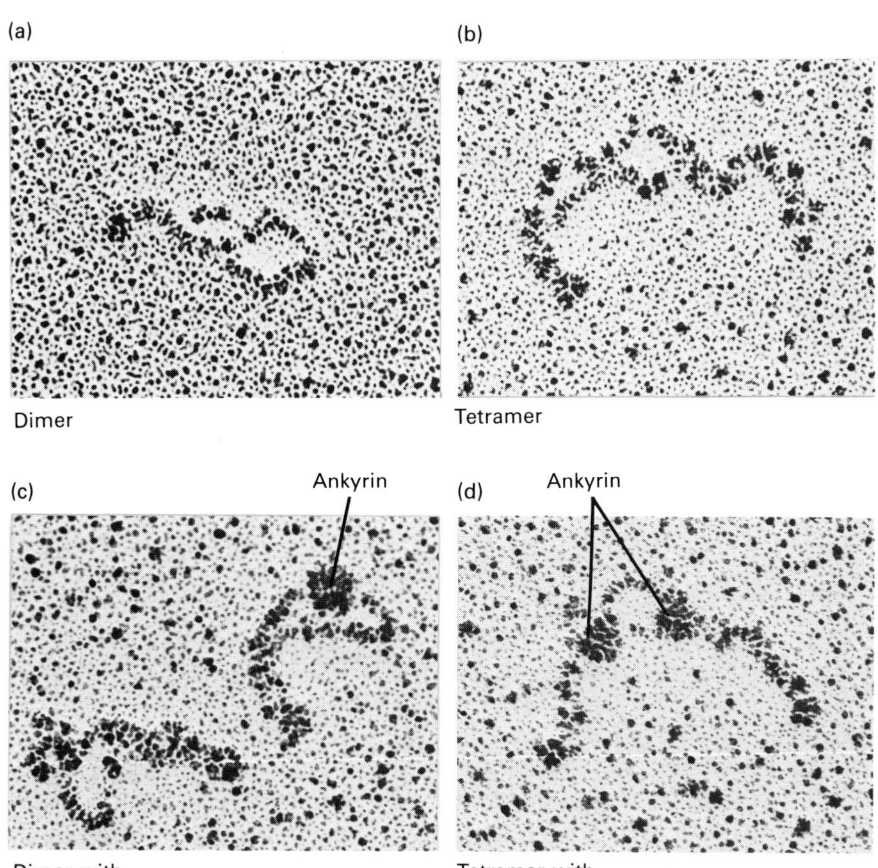

Figure 14-41 Electron micrographs of $\alpha\beta$ dimers and $(\alpha\beta)_2$ tetramers of spectrin. The proteins shown in (c) and (d) are bound to ankyrin; those in (a) and (b) are not. Ankyrin is bound near the middle of the tetramers. *Courtesy of D. Branton.*

lized. What remains in the insoluble fraction is the cytoskeleton, the fibrous shell retaining the size and shape of the ghost (Figure 14-40) and containing the same predominant extrinsic proteins: bands 1 and 2 (spectrin), band 2.1 (ankyrin), band 4.1, and band 5 (actin) (see Figure 14-38). Electron micrographs show that the shell is "an anastomosing lacelike framework like a net woven by a myopic fisherman."* These five proteins are held together by noncovalent bonds and they form the strong but flexible framework responsible for the shape and pliability of the red blood cell.

The major constituents of the cytoskeleton are α and β spectrin (bands 1 and 2), the polypeptides having the highest molecular weight. The two spectrin chains combine to form $\alpha\beta$ dimers that are long (100 nm), slender (5-nm diameter), wormlike molecules in which the subunits coil about each other. Two spectrin $\alpha\beta$ dimers combine head to head to form an $(\alpha\beta)_2$ tetramer (Figure 14-41); this interaction can take place in an aqueous solution. Each erythrocyte ghost contains about 200,000 spectrin dimers. They form the lacelike fibers of the cytoskeleton and are bound to the cytoplasmic face of the erythrocyte membrane by two different types of protein-protein interactions.

*L. Tilney and R. Detmers, 1975, *J. Cell Biol.* **66**:508–520.

In the cytoskeletal network, the free ends of several spectrin tetramers are held together by the short fibrillar chains of the protein actin, band 5 (Figure 14-42). Actin, a major constituent of all cells, normally forms quite long chains—several micrometers long (see Figures 3-40 and 3-41)—and plays crucial roles in cell movement and muscle contraction. Actin is actually a family of proteins encoded by different genes. The short chains of actin found in red blood cells are different from the actin in other cells. The fourth cytoskeletal protein, called band 4.1, lies at the actin-spectrin junction and helps join the actin to the spectrin. Because spectrin binds to the sides of actin filaments, a single fiber of actin has many potential binding sites for spectrin. It affords junction points for three or more spectrin molecules and thus enables a network to form.

The interactions described thus far produce the lacelike cytoskeleton but do not attach it to the erythrocyte membrane. This attachment is mediated through the protein called ankyrin, band 2.1. There are about 100,000 molecules of ankyrin in a cell—on the average, one per spectrin tetramer. Ankyrin has two domains, each of which forms a tight interaction with a different key protein. One binds tightly and specifically to a site on one of the spectrin chains near the center of the tetramer (as shown in Figure 14-41), and the other domain binds tightly to a

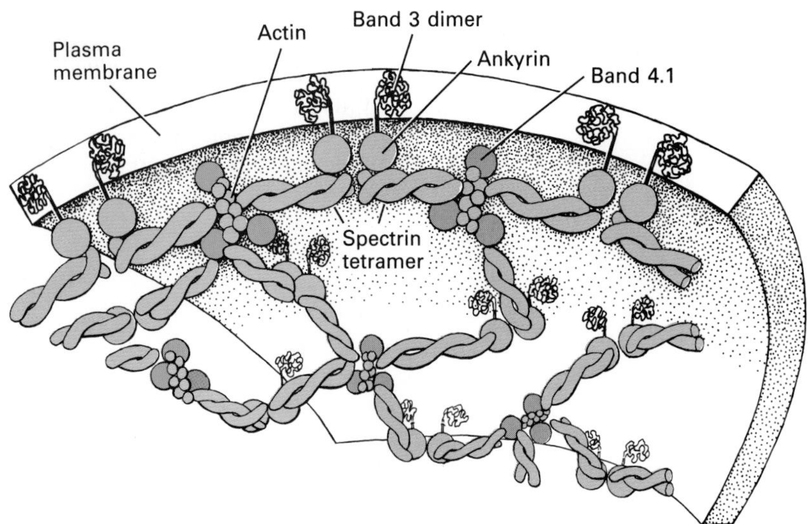

Figure 14-42 Postulated organization of the major erythrocyte cytoskeletal proteins. *After S. E. Lux, 1979,* Nature ***281****:426; and B. W. Shen, R. Josephs, and T. L. Steck, 1984,* J. Cell. Biol. *99:810.*

region on the band 3 protein that protrudes into the cytoplasm (as shown in Figure 14-42). This bridgelike property of ankyrin binds the cross-linked spectrin network to the membrane. Spectrin will not bind to the membrane without it. Thus, the cytoskeleton of the erythrocyte is a large network of band 3, ankyrin, spectrin, band 4.1, and actin. Additionally, band 4.1 binds to the carboxyl-terminal piece of glycophorin and also to band 3; these interactions might also anchor the spectrin skeleton to the membrane. Ankyrin-like, spectrin-like, and band 4.1-like proteins are found in many other cell types, where they may serve as anchors for other cytoskeletal protein.

Only about 20 percent of the band 3 protein is directly bound to an ankyrin molecule and, thus, to the spectrin cytoskeleton. But all the band 3 molecules in the cell seem to be caught up in the cytoskeletal network directly underneath the phospholipid bilayer, because their lateral mobility is restricted by the cytoskeleton.

Although the rigidity of the plasma membrane is due to the interactions of a complex of cytoskeletal proteins, formation of the biconcave disk shape requires additional cytoskeletal elements. Presumably, minor cytoskeletal proteins cause the skeleton to curve inward in a biconcave fashion, and the identification of these morphogenetic proteins would be a major advance.

Several Hereditary Diseases Affect the Cytoskeleton

Hereditary spherocytosis and elliptocytosis are human diseases resulting from genetic mutations. In both diseases the erythrocytes are abnormally shaped (Figure 14-43). In some cases the defect is attributable to an abnormal spectrin polypeptide that is unable to interact correctly with other cytoskeletal proteins to produce a rigid cytoskeleton. The abnormal spectrin either fails to form head-to-head tetramers or binds defectively to either the band 4.1 protein or ankyrin. In other cases, the band 4.1 protein is absent. The consequence of any of these defects is an unstable cytoskeleton and thus an abnormal cell shape, with fragments of membrane occasionally budding

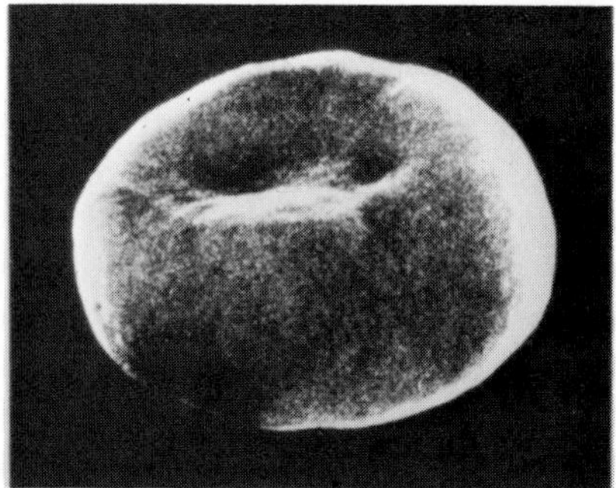

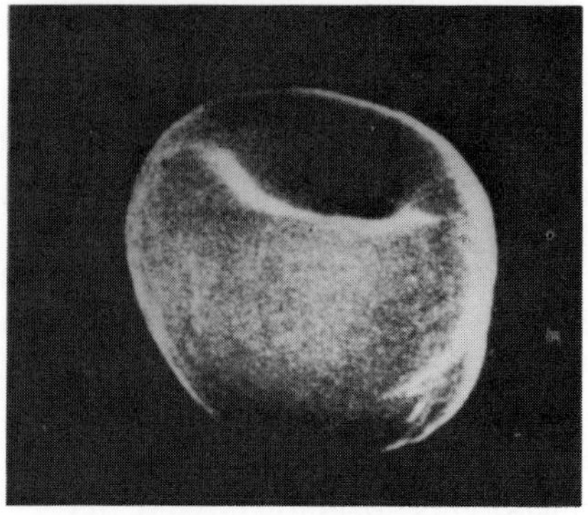

Figure 14-43 Scanning electron micrographs of spherocytes (rounded erythrocytes) from patients with hereditary spherocytosis. *Courtesy of S. E. Lux.*

from the surface. Because these abnormal erythrocytes are degraded by the spleen more rapidly than are normal ones, the persons affected have fewer circulating erythrocytes and are said to be anemic.

The Erythrocyte Is Generated from a Precursor Cell by Loss of a Nucleus and Specific Membrane Proteins

It is interesting to examine the structure of the erythrocyte membrane in light of the process of cell differentiation that produces it. Adult red blood cells are manufactured in the bone marrow from nucleated stem cells. The stem cells grow, divide, and begin synthesizing hemoglobin, at which stage they are called *erythroblasts* (hematologists divide this stage into several substages of development). Erythroblasts contain hemoglobin as well as spectrin and other characteristic erythrocyte membrane proteins. Next, the part of the cell that contains the nucleus is pinched off and eventually degraded. The remaining non-nucleated cell, now called the *reticulocyte,* continues to synthesize hemoglobin and other erythrocyte proteins but eventually loses its ribosomes and acquires the biconcave disk structure of the erythrocyte. At the stage when the cell is losing its nucleus, all the spectrin and glycophorin stay behind in the nascent reticulocyte (Figure 14-44). Most of the other surface glycoproteins, though not band 3 or glycophorin, are lost in the fragment that contains the nucleus. Although the details of the process by which the nucleus is lost are not known, it does appear that the spectrin-rich cytoskeleton "selects" certain integral membrane proteins for inclusion in the reticulocyte and specifically discriminates against others, which are discarded with the segment containing the nucleus.

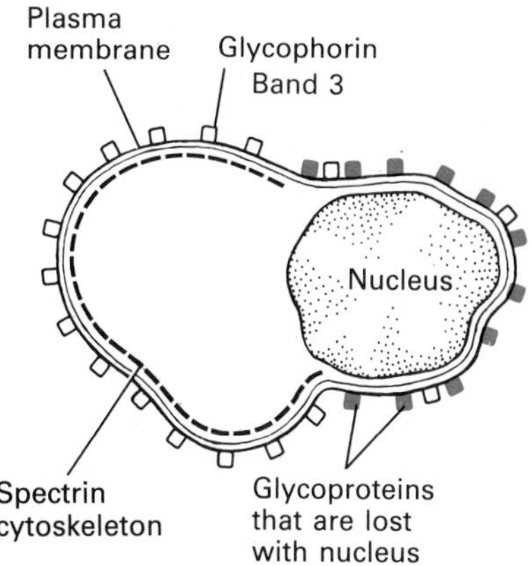

Figure 14-44 Distribution of membrane proteins in the enucleating erythroblast. *After J. B. Geiduschek and S. J. Singer,* Cell *16:149; A. H. Sarris and G. E. Palade, 1982,* J. Cell. Biol. *93:591.*

Specialized Regions of the Plasma Membrane

Unlike blood cells, which are independent, free-floating units, the majority of cells in the mammalian body are organized into multicellular arrays forming solid tissues. Most cells in tissues can carry out their designated function only if their plasma membrane is nonhomogeneous; it must be organized into at least two discrete regions, each with very different tasks. Such cells are termed *polarized.* Moreover, for a tissue to have strength, its cells must be able to form tight, strong contacts with their neighbors. To illustrate the importance and variety of the plasma membrane's specialized regions, and explore the nature of its cell-to-cell junctions, we examine two cell types in detail: the acinar cell of the pancreas, which secretes a number of digestive enzymes, and the epithelial cell that lines the lumen of the small intestine.

The Pancreatic Acinus Is an Aggregate of Cells Having Two Very Different Regions of Plasma Membrane

A pancreatic acinus is a more or less spherical aggregate of a dozen or so cells (Figure 14-45). The lumen (central cavity) of the acinus is connected to a little duct, or ductule. This ductule merges with ductules from those of other acini to form a larger duct, eventually leading into one of several main ducts. The main pancreatic ducts open into the lumen of the small intestine (Figure 5-34b). Acinar cells synthesize enzymes—amylases, proteases, ribonucleases, and so on—that function in the intestine to degrade most food macromolecules. In the cell, these enzymes are stored as inactive precursors in membrane-limited secretory vesicles. The vesicles containing the secretory enzymes are clustered under the *apical* membrane of the cell—that is, under the region of the plasma membrane that is adjacent to the ductule. It is with this part of the plasma membrane, and only this part, that the vesicles will fuse (see Figure 5-34a), which ensures the release of the digestive enzymes into the ductule. On the sides of the cell, the *basolateral* membrane has two different functions. Nutrients from the surrounding blood are transported through this plasma membrane region into the cell, and it is also the site of mechanisms for signaling the secretion of enzymes. The stomach and intestine respond to the presence of food by releasing into the blood several peptide hormones that act on the acinar cells to trigger the secretion of their digestive enzymes. The receptors for these peptide hormones reside on the basolateral membrane of the acinar cell.

(a)

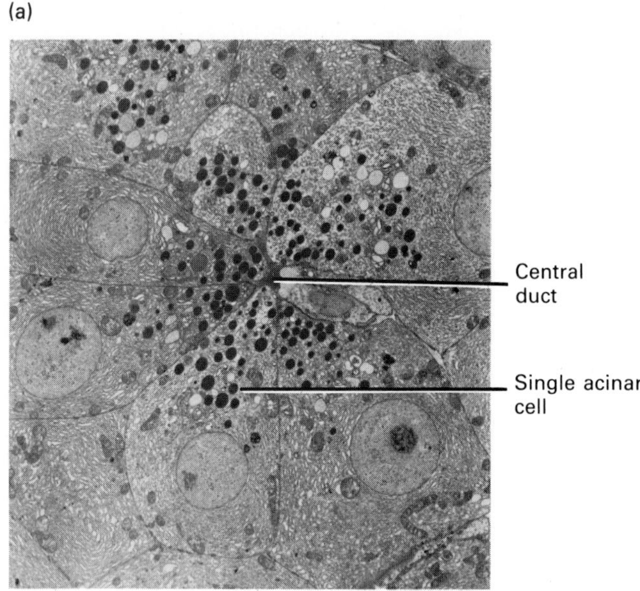

(b)

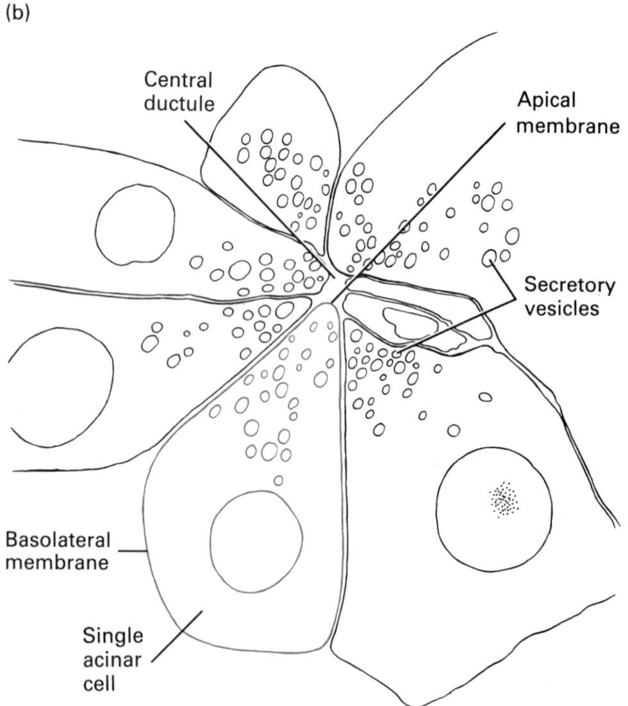

Figure 14-45 (a) A low-magnification (×5000) electron micrograph of a rat pancreatic acinus, showing the overall arrangement of the cells surrounding the central duct. The nuclei are close to the base of the cells. The secretory vesicles fill those parts of the cell that are near the lumen. Fusion of the membrane of the secretory vesicle with the part of the plasma membrane facing the lumen causes exocytosis of the contents of the vesicles into the lumen. (b) Tracing of the electron micrograph seen in (a). *From Biophoto Associates.*

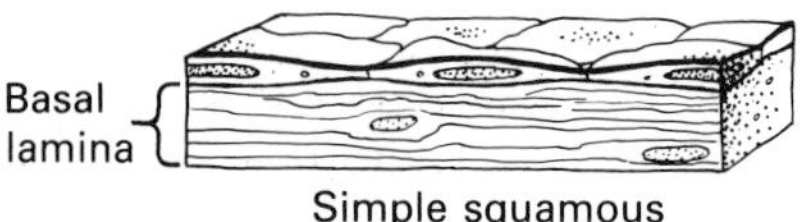

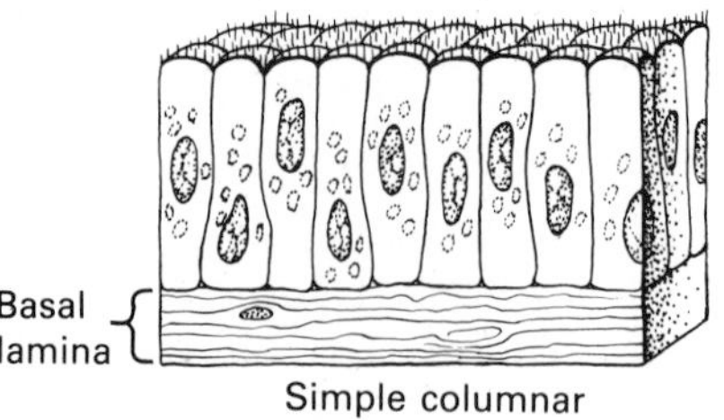

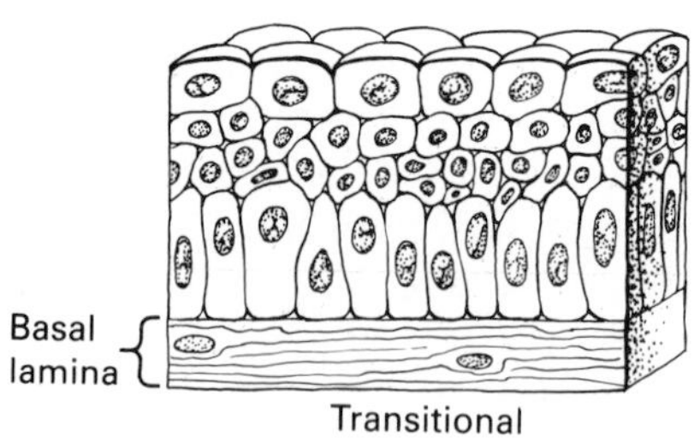

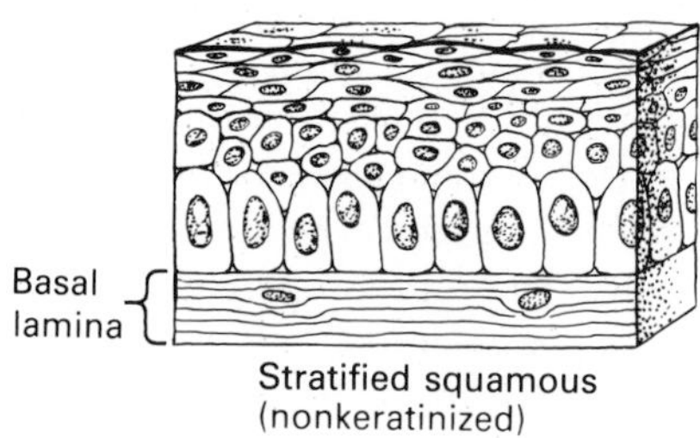

Figure 14-46 Principal types of epithelial membranes that line the surfaces of body cavities. *Simple squamous* membranes are thin cells such as the endothelial cells lining the blood vessels and the mesothelial cells that line body cavities such as the pleural cavity surrounding the lungs. *Simple columnar* membranes are elongated cells that include mucus-secreting cells, such as those lining the stomach and cervical tract, and absorptive cells, such as those lining the small intestine—the latter containing microvilli at their apical surface. *Transitional* membranes are composed of several layers of cells of different shapes lining certain cavities that are subject to expansion and contraction, such as the urinary bladder. *Stratified squamous nonkeratinized* membranes line surfaces such as the mouth and the vagina; such a lining is resistant to abrasion and generally does not participate in either secretion or absorption of materials into or out of the cavity. The structure of the stratified keratinized cells that form skin is detailed in Chapter 19.

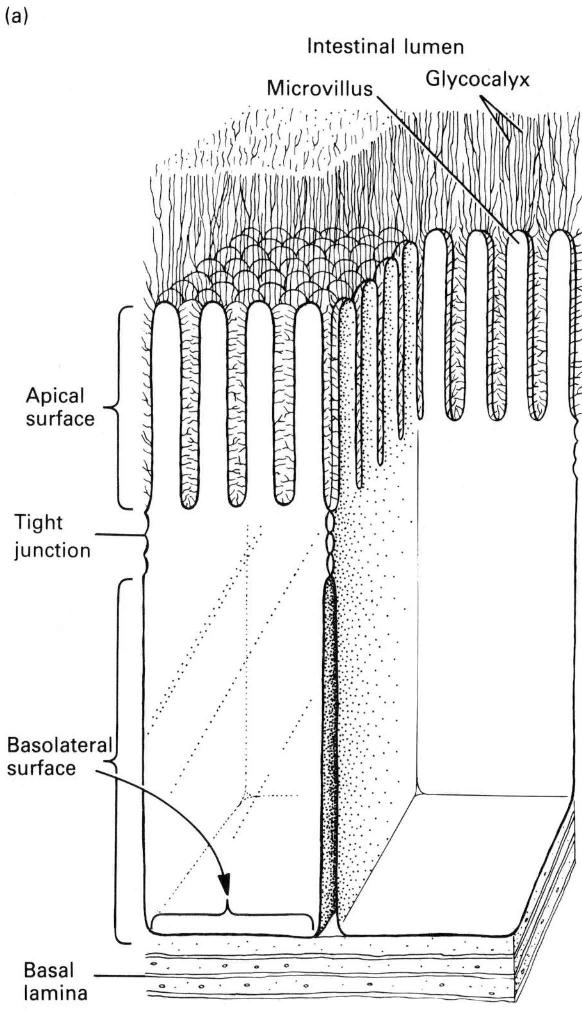

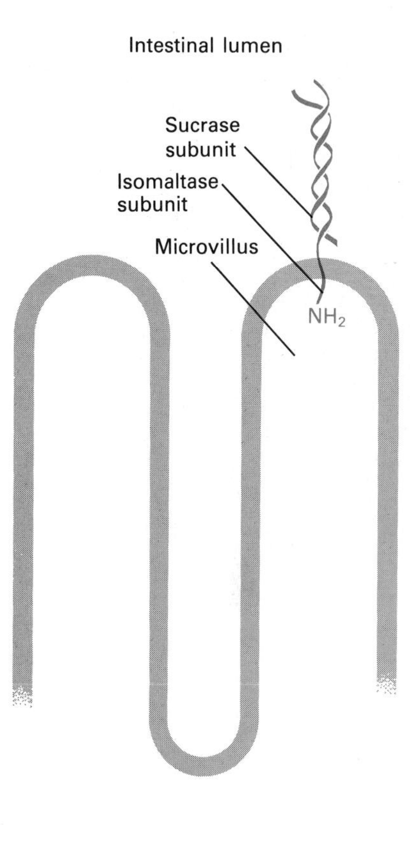

Figure 14-47 Intestinal epithelial cells. (a) The brush border, made up of microvilli, faces the intestinal lumen. This region of the cell is called the *apical surface*. The cells rest on a basal lamina, a fibrous network of collagen and proteoglycans that supports the epithelial cell layer. The *basolateral surface* refers to the membrane on the sides of the cell below the microvilli and on the base of the cell, where it is attached to the basal lamina. (b) Structure of sucrase-isomaltase in the plasma membrane of the intestinal epithelial cell. The major sugar in human diets is sucrose, a disaccharide; to be absorbed from the intestine, it has to be hydrolyzed to its individual sugar units, fructose and glucose, by sucrase-isomaltase. This enzyme is a two-chained, rather elongated protein found on the lumenal surface of the intestinal epithelial cells. It is anchored to the membrane by a hydrophobic region consisting of 30 amino acids, but the bulk of the protein protrudes into the intestinal lumen. Thus, sucrase-isomaltase produces glucose and fructose very close to the cell membrane. [See J. Brunner et al., 1979, *J. Biol. Chem.* **254**:1821; W. Hunziker et al., unpublished sequence data.]

All Body Surfaces Are Lined with Epithelial Cells

With few exceptions, all internal and external body surfaces are covered with a polarized layer of cells called an *epithelium*. An epithelial membrane consisting of a single layer of thin cells, such as the layer that lines most blood vessels (Figure 14-46), is called a *simple squamous epithelium*. Single layers of columnar cells, such as the epithelia that line the cervical canal, the stomach, and the small intestine, are called *simple columnar epithelium*. Membranes that are several cells thick, the *stratified epithelial membranes*, can withstand more wear and tear than those only one cell thick. The inside of the mouth and esophagus are lined with a stratified squamous epithelium. Skin is composed of a stratified squamous epithelium that has undergone the process of keratinization: the fibrous protein keratin fills the most superficial cells, which then metamorphose into a rough, nonliving layer. The structural origin of a keratinized epithelium is detailed in Chapter 19. All epithelial layers rest upon a fibrous basal lamina that is usually composed of collagen and proteoglycans.

The Plasma Membrane of Intestinal Epithelial Cells Is Divided into Two Regions of Different Structure and Function

A well-studied example of a very polarized epithelial cell is the cell that lines the lumen of the small intestine (Figure 14-47). The intestinal epithelial cell has two major functions: to absorb from the lumen of the small intestine the nutrients produced from digested food (sugars, amino acids, lipids, and the like) and to transfer these nutrients across the single cell layer and into the blood. The *lumenal,* or apical, surface of the cells is highly specialized for absorption. This region, often called the *brush border* because of its appearance, consists of large numbers of fingerlike projections (100 nm in diameter) called *microvilli.* These extensions of the cell surface greatly increase the membrane area and enhance the rate of absorption into the cells. The microvillar membrane contains transport proteins that allow glucose, amino acids, and other compounds to pass into the cell. Digestive enzymes are bound to the microvillar surface as well. After proteins and carbohydrates are degraded into small peptides and oligosaccharides in the intestine, mostly by pancreatic enzymes, they must be broken down further into monosaccharides, such as glucose, and amino acids, before they can be absorbed by the epithelial cells. This is the function of the peptidases and glycosidases that are bound to the microvillar surface. The orientation in the membrane of one of these enzymes, sucrase-isomaltase, is depicted in Figure 14-47b. This rodlike enzyme catalyzes the hydrolysis of the disaccharide sucrose (see Figure 3-73) into glucose and fructose. The other surface enzymes have similar structures. In electron micrographs of microvilli these sets of hydrolytic enzymes appear as a "fuzz," or glycocalyx, on the outer surface of the membrane (Figure 14-33). After the nutrients that have been broken down by enzymes are absorbed through the lumenal surface, they must be transported out of the cell and into the blood. This task is accomplished by a discrete set of proteins on the basolateral surface, as is detailed in Chapter 15.

Microvilli Have a Rigid Structure

Various components of the plasma membrane and cytoskeleton contribute to the structure of the microvilli. Structural proteins maintain their shape and uniform diameter. A bundle of actin microfilaments runs down the center of each microvillus, anchoring itself to proteins on the inner surface of the microvillar membrane and intersecting with a network of actin-myosin filaments that transverse the cell just beneath the microvilli (see Figure 5-49). These microfilaments give rigidity to the microvilli and possibly enable them to move backward and forward in the intestinal lumen. The transverse fibers of actin-myosin insert into the belt desmosome, a type of cell junction that is described later in this chapter (Figure 14-48).

Microvilli also occur on other types of cells. In the tubules of the kidney, for example, they serve to resorb into the blood material that would otherwise pass into the urine.

Types of Cell Junction

Tissues like the pancreatic acinus and the intestinal epithelium are aggregates of individual cells. Several types of cell-to-cell interconnections, or *junctions,* are necessary to give any such tissue the architecture and integrated cellular activity it needs to carry out its particular functions.

In some parts of an organism, adjacent cells must be

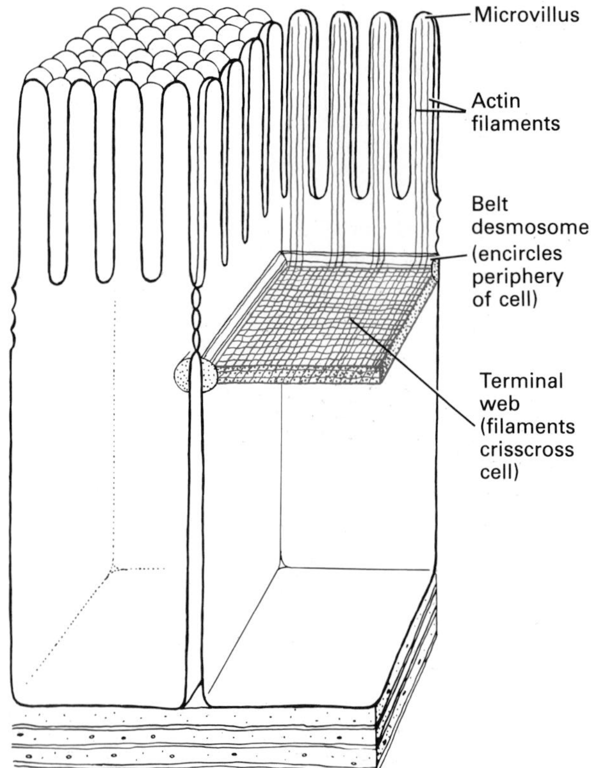

Figure 14-48 Bundles of actin fibers run down the cores of the microvilli and intersect with a layer of fibers called the terminal web. The terminal web transverses the intestinal epithelial cell just below the microvilli and inserts into the belt desmosome.

sealed together to prevent the passage of fluids through the cell layer. Seals are normally formed between the epithelial cells that line all the cavities of the mammalian body. These impermeable junctions, termed *tight junctions* or *zonulae occludens,* seal off, for example, the lumen of the intestine from the fluid that flows past the basolateral surface of the epithelial cells, and they seal off the lumen of the pancreatic acinus from the surrounding blood. Various adhering junctions, termed *desmosomes* or *zonulae adherens,* bind cells tightly to each other not to prevent the passage of fluid but to give the tissue strength. Possibly related junctions called *hemidesmosomes* cause the cell to adhere to the lamina of collagen and other proteins that surround it. Similar junctions called *adhesion plaques* enable motile cells such as fibroblasts to adhere to the substratum. Finally, *gap junctions* form channels of communication between cells; they allow cells to exchange small molecules and help to integrate the metabolic activities of all of the cells in the tissue. The three principal types of junctions—the tight junction, the gap junction, and the desmosomes—are illustrated in Figure 14-49.

Electron micrographs of tissue sections have shown that in nonjunctional regions there is ordinarily a space of about 20 nm between the plasma membranes of two neighboring cells. This space contains extracellular glycoproteins that cover the cells, and interactions between these glycoprotein coats are probably also important to intercellular adhesion.

Tight Junctions Seal Off Body Cavities

Tight junctions are composed of thin bands that completely encircle a cell and are in contact with thin bands of adjacent cells. In epithelial cells, tight junctions are usually located just below the apical microvillar surface (see Figure 14-49). When thin sections through a tight junction are viewed in the electron microscope, the plasma membranes of adjacent cells appear to touch each other at intervals and even to fuse (Figure 14-50). That these junctions indeed are impermeable to most substances is illustrated in Figure 14-51. Here lanthanum hydroxide, a colloid of high molecular weight, is injected into the pancreas of an experimental animal, and the pancreatic acinar cells are analyzed by electron microscopy. The lanthanum, an electron-dense material, covers the basolateral surface of the cell but cannot penetrate past the outermost of the tight junctions. Other studies have shown that these junctions are impermeable to salts and possibly even to water.

We know that all nutrients are absorbed from the intestine into one side of the epithelial cell and then released from the other side into the blood because tight intercellular junctions do not allow small molecules to diffuse directly from the intestinal lumen into the blood. In pancreatic acinar tissue the tight junction has a similar function. It prevents the leakage of pancreatic secretory proteins, including digestive enzymes, into the blood.

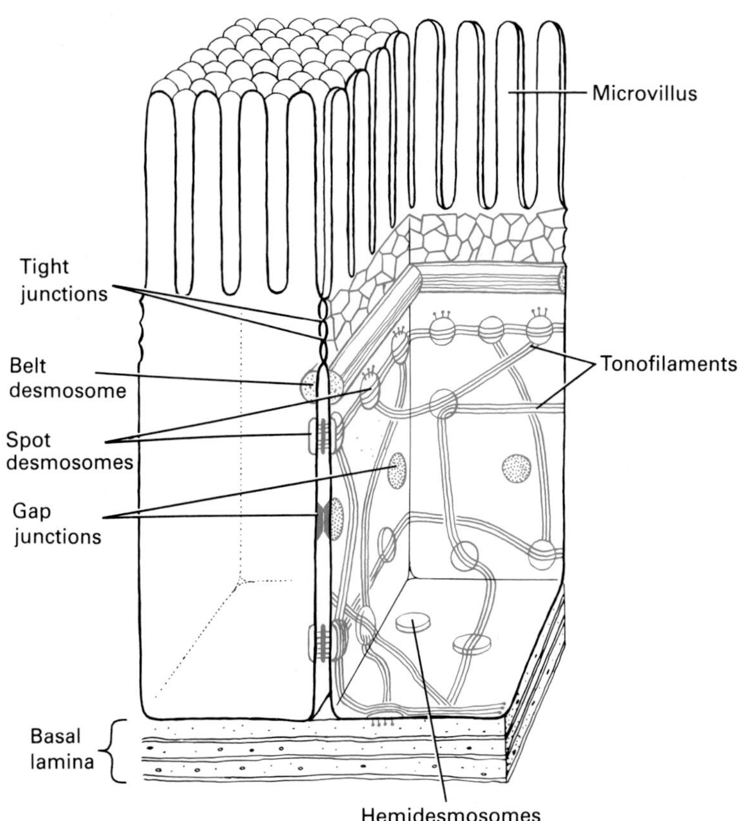

Figure 14-49 Schematic diagram of the principal types of cell junctions, as found in the intestinal epithelial cell. The tight junctions are located just under the microvilli; they act to seal off the lumen of the epithelium from the basolateral surfaces. Below these junctions is the belt desmosome, a layer of fibers encircling each cell that aids in the adhesion of adjacent cells. Spot desmosomes are localized regions of adhesion of adjacent cells. Hemidesmosomes are points of adhesion of the cell to the basal lamina. Bundles of tonofilaments course through the cell and interconnect the spot desmosomes and hemidesmosomes. Gap junctions, distributed along the lateral surface of the cells, allow small molecules to pass from cell to cell.

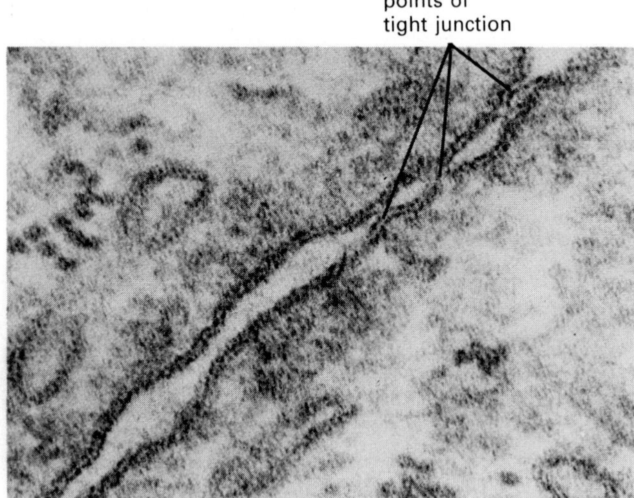

Figure 14-50 A tight junction between two hepatocytes (liver cells). In this electron micrograph of a thin section of fixed cells, the junction is seen as three points of contact (*arrows*) between the plasma membranes of the adjacent cells. *Courtesy of D. Goodenough.*

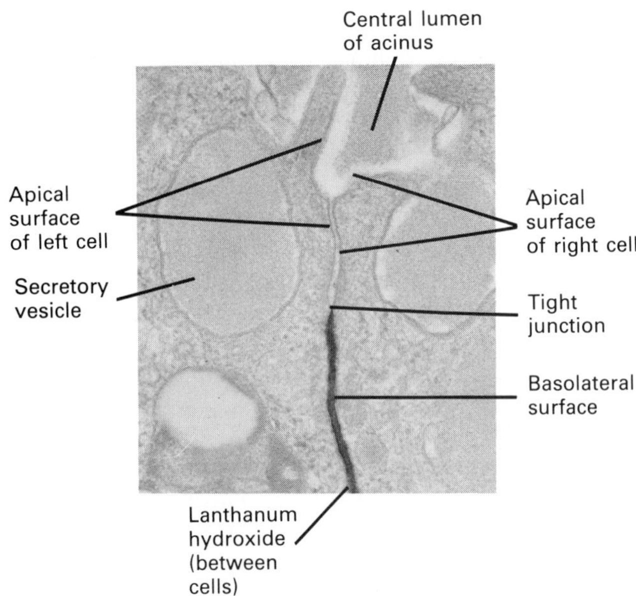

Figure 14-51 Tight junctions block passage of fluids between cells, as shown in this electron micrograph of two pancreatic acinar cells. Lanthanum hydroxide, a colloid, was added to the outside of the acinus a few minutes before the material was fixed and processed for microscopy. The electron-opaque lanthanum hydroxide penetrates between the two acinar cells but is arrested at the level of the tight junction. The tight junction appears as a point of fusion between the contiguous membranes. *Courtesy of D. Friend.*

Freeze-fracture electron microscopy affords a different view of the tight junction. In Figure 14-52 the junction appears to comprise an interlocking network of ridges in the plasma membrane. More specifically, there appear to be ridges on the cytoplasmic face of the plasma membrane of each of the two contacting cells. Although not shown in the figure, there are corresponding grooves on the external face. High magnification reveals these ridges to be made up of particles, possibly of protein, from 3 to 4 nm in diameter. The junction is formed by a double row of these particles, one row donated by each cell. Although the molecular structure of these junctions is not known, it is possible that the protein particles on the two cells form extremely tight links with each other, essentially fusing the two plasma membranes and creating an impenetrable seal! Treatment of an epithelium with the protease trypsin destroys the tight junctions, implicating proteins as an essential structural component. Figure 14-53 depicts one version of protein links in a typical tight junction.

Another view is that the intramembrane particles in tight junctions are lipid micelles, not proteins, and that the sealing junction between two cells is built of a long cylindrical lipid micelle (Figure 14-54). Proteins might be essential structural components of the junctions, but the lipid micelles would form the actual seal.

Interestingly, a single junction ridge does not make a

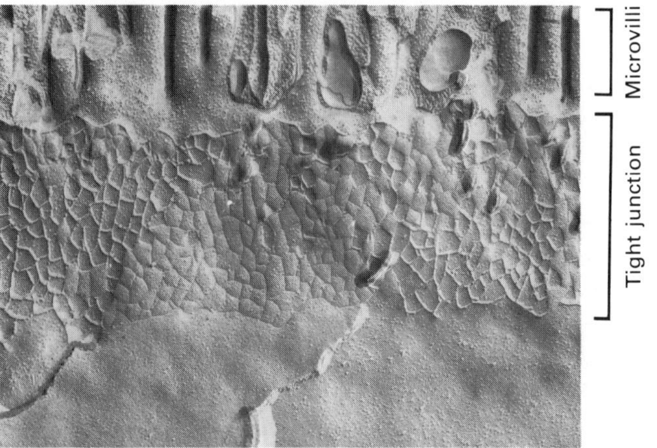

Figure 14-52 Freeze fracture of a tight junction between two intestinal epithelial cells. The fracture plane passes through the plasma membrane of one of the two adjacent cells. The microvilli are visible at the top of the micrograph. The honeycomb-like network of ridges immediately below forms the tight junction. These ridges appear to be built of rows of globular particles; in the intact junction they are intramembraneous particles—possibly intramembrane proteins—that also bridge the space between the adjacent cells. *Courtesy of L. A. Staehelin.*

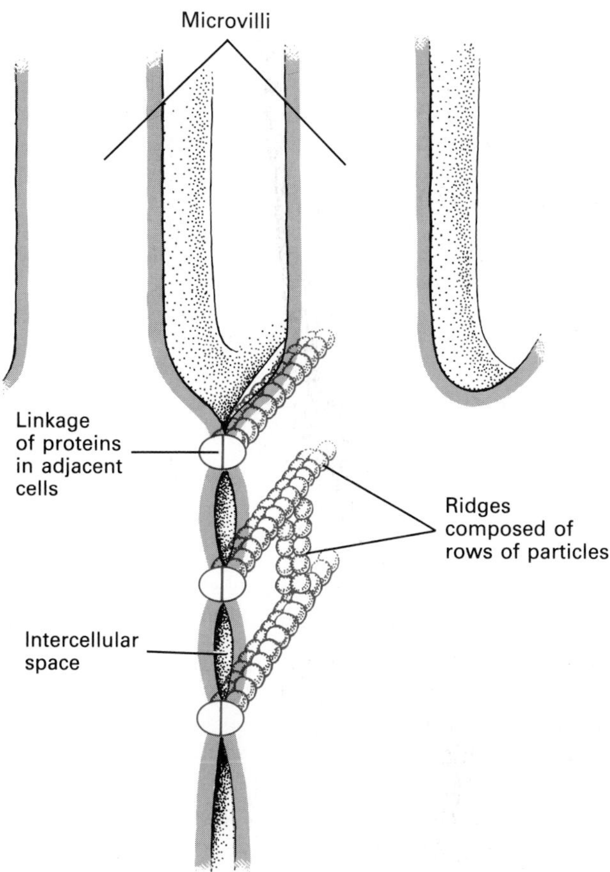

Figure 14-53 One model of the arrangement of membrane proteins in the sealing strands of a tight junction. In this version the junction was formed by linkage of proteins in the plasma membrane of adjacent cells. *After L. A. Staehelin and B. E. Hull, 1978,* Sci. Am. *238(5):140.*

perfect seal. The number of these ridges in different tissues is more or less proportional to the differences in composition of the fluids on the two sides of the epithelium. For example, the function of the epithelial cells in the proximal tubule of the kidney is to absorb glucose, amino acids, and certain salts from the filtrate of the blood that will become the urine and return them to the bloodstream. The concentrations of salt and nutrients on either side of this cell layer are not very different. As a consequence the tight junction is made up of only one or two strands of ridges. In contrast, three or four strands of ridges connect the cells of the pancreatic acinus, and six or more connect those of the epithelium of the small intestine.

The geometric organization of the junctional ridge is also coordinated with the properties of the individual epithelium. The absorptive cell layer of the large intestine is subjected to a great deal of stretching and compressing as large boluses of semisolid, partly digested food are transported through it. The tight junctions between these cells form a loosely interconnected pattern, permitting the network of junctions to stretch under tension and still maintain the tight seals (Figure 14-55). The lumen of the small intestine, in contrast, is filled with the mostly liquid partial products of digestion, thus the epithelium is not subjected to tension-producing forces. The tight junctions there form a set of fairly regular polygons that cannot readily expand or contract.

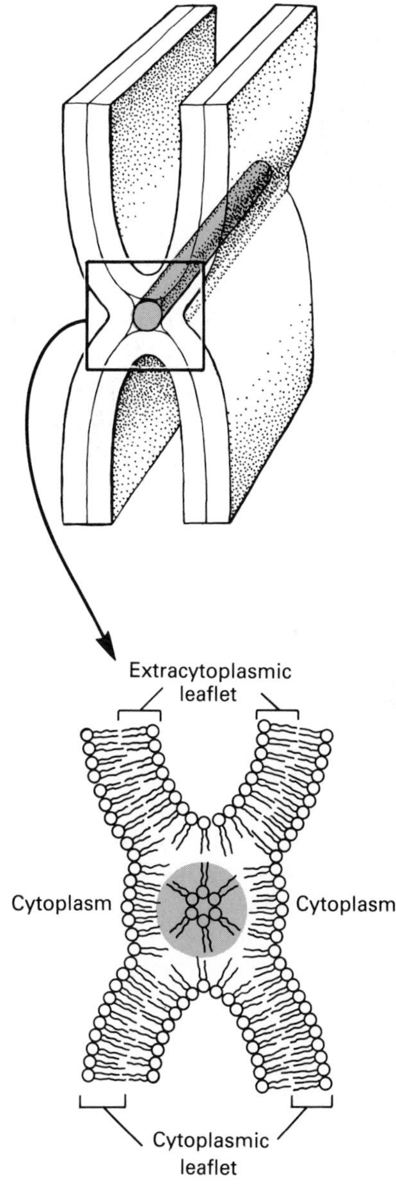

Figure 14-54 An alternative hypothesis of tight junction structure, in which the extracytoplasmic leaflets of two plasma membranes fuse into a single continuous leaflet. In the core of the junction is an inverted cylindrical micelle of phospholipid, with its hydrophilic head groups facing the center of the membrane. These polar heads would form the intramembranous particles visualized upon freeze fracture. *After P. P. da Silva and B. Kachar, 1982,* Cell *29:441.*

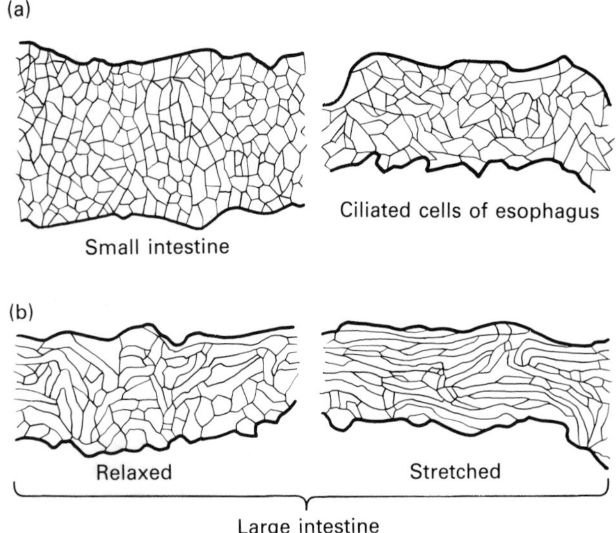

Figure 14-55 Tracings of representative micrographs of tight junctions summarizing the network patterns found in cells lining the alimentary tract of the frog *Xenopus laevis*. These tracings illustrate the two basic types of geometric organization found in tight-junction networks. (a) The first type, a stable, evenly interconnected pattern composed of fairly regularly shaped polygons, seals the absorptive cells of the small intestine and the ciliated cells of the esophagus. (b) The second type, a loosely interconnected pattern found between the absorptive cells of the large intestine, allows the network to stretch under tension. [See B. E. Hull and L. A. Staehelin, 1978, *J. Cell Biol.* **68**:688.] *Courtesy of B. E. Hull and L. A. Staehelin.*

The tight junction also appears to be the demarcation between the apical and basolateral surfaces of the intestinal epithelial cells. The composition of the plasma membrane on one side of a tight junction differs greatly from that on the other side. Recent work suggests that the tight junction blocks the lateral diffusion of membrane proteins and lipids and maintains the integrity of the two principal regions of membrane on the epithelial cell.

Desmosomes Are Junctions That Bind Adjacent Cells

Desmosomes are thickened regions of the plasma membrane where cells are tightly attached to their neighbors. They increase the rigidity of tissues by holding the cells securely together and they act as anchorages for intracellular fibers. There are several very different types (see Figure 14-49). These junctions have only recently been purified, and little is yet known of their composition or specific protein interactions. Thus the description of desmosomes must be based mainly on what has been revealed by electron microscopy.

Spot desmosomes are buttonlike points of contact between cells; they are often thought of as a "spot weld" (Figure 14-56a). At a spot desmosome, the distance between plasma membranes of adjacent cells is about 30 nm. The central stratum between the two membranes consists of specific desmosomal proteins, and running from it to both membranes are many thin filaments constructed of protein and carbohydrate. Presumably the function of these filaments is to bind the cells together. Lying just beneath each plasma membrane and connected to it is a disk-shaped plaque from 15 to 20 nm thick. Tonofilaments (intermediate filaments), a principal type of cytoskeletal fiber, course into and out of these plaques. Sets of these 10-nm-thick fibers run parallel to the cell surface and also penetrate and traverse the cytoplasm. They are thought to be part of the internal structural framework of the cell, giving it shape and rigidity. If so, the spot desmosomes also transmit shearing forces from one region of a cell layer to the epithelium as a whole. Examination of the faces of fractured cell membranes by electron microscopy reveals that spot desmosomes are groups of intramembrane particles. The nature of the particles is not known, but treatment of an epithelium with trypsin, which destroys tight junctions, also destroys the spot desmosomes and breaks the epithelium up into single cells. Therefore, the desmosome must contain an essential protein (Figure 14-56b).

Hemidesmosomes, structures morphologically similar to the desmosomes, are found in regions of epithelial cells in contact with the basal lamina, and it is thought that they anchor extracellular protein networks to the cell.

The *belt desmosome,* or intermediate junction, is a fundamentally different type of desmosomal structure. Characteristically, it is located just below the tight junction in an epithelial cell, such as the columnar cell of the intestinal epithelium (see Figure 14-49), and forms a belt of cell-to-cell adhesion that surrounds each cell. At a belt junction, the plasma membranes of the adjacent cells are parallel and 15 to 20 nm apart, and the space between them is filled with an amorphous material (Figure 14-57a). The plasma membrane in the region of these junctions is thicker than usual; it is the site of attachment for a large number of 7-nm actin filaments. Bands of these filaments lie just under the plasma membrane and encircle the cell like a belt. Other filaments, containing actin and actin-binding proteins, traverse the cell at the level of the belt desmosome to form the *terminal web* (Figure 14-57b). These terminal web fibers also bind to the actin fibers that extend down from the cores of the microvilli. It is believed that, like spot desmosomes, belt

(a)

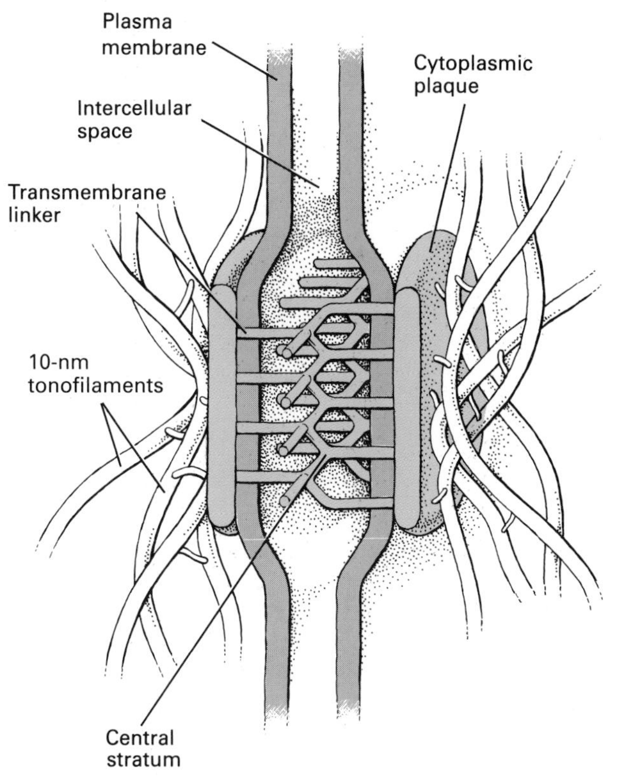

(b)

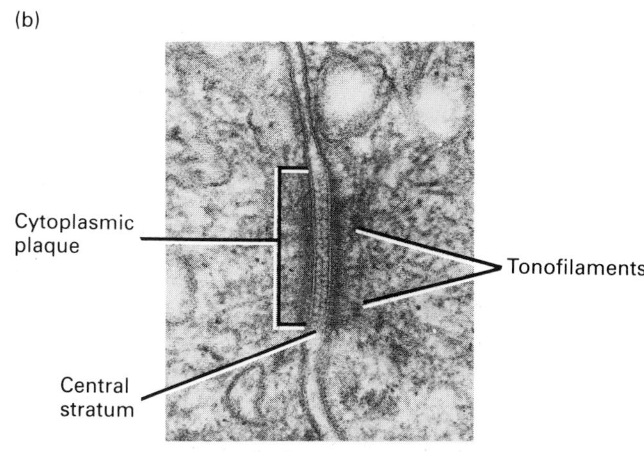

Figure 14-56 A spot desmosome. (a) The 10-nm-diameter tonofilaments, or intermediate filaments, form a tensile network that crisscrosses the interior of the cell. They are attached to the plaques of the spot desmosome through poorly defined filamentous structures. Other filaments, called transmembrane linkers, connect the plaques of the spot desmosomes across the intercellular space. The junction apparently couples the intermediate filament networks of adjacent cells, allowing shearing stresses to dissipate throughout the tissue. (b) Electron micrograph of a thin section of a spot desmosome. Bundles of tonofilaments radiate from the two darkly staining plaques that line the inner surfaces of adjacent cell membranes. Where the tonofilaments have been cut in cross section they appear as black dots. Thinner filaments originating within the plaques appear to extend through the plasma membrane and connect in a staggered configuration to the central stratum, a stained structure between the two cell membranes. *Courtesy of B. E. Hull and L. A. Staehelin.*

desmosome structures transmit forces between cells in an epithelium. But much remains to be learned about them.

Adhesion Plaques Anchor Cells to the Substratum

The *adhesion plaque,* another form of junction complex, is the one most apparent in cultured cells. Adhesion plaques attach cultured cells to the glass or plastic dish on which they are grown. In tissues, they make cell-to-cell contacts. Certainly in culture, and probably in tissues, they serve as reversible anchoring points that allow cells to move over surfaces.

Adhesion plaques are found where the bundles of actin-containing microfilaments that pass through the cell's cytoplasm meet the plasma membrane. In this sense they are similar to belt desmosomes. They contain a major protein, *vinculin,* that probably helps to bind the microfilaments to the membrane but is not an integral membrane protein. The integral protein that serves as the final anchoring point for the filaments is not known.

Gap Junctions Allow the Passage of Small Molecules between Adjacent Cells

Almost all cells that come in contact with each other have regions of junctional specialization characterized by a gap (2 to 3 nm) between the plasma membranes that is filled by a well-defined set of particles (Figure 14-58). Morphologists named these regions *gap junctions,* a poorly chosen name in retrospect, because the gap is not their most important feature. The "particles" in the gap make this junction unique; they are actually tiny channels that directly link the cytoplasms of the two cells, allowing the free passage of very small molecules between a cell and its neighbor. This is not a sealing junction. It does not form a barrier to the passage of extracellular fluid between the two cell membranes.

The size of the intracellular channels can be measured by injecting a cell with a fluorescent dye covalently linked to molecules of various sizes and observing with a fluo-

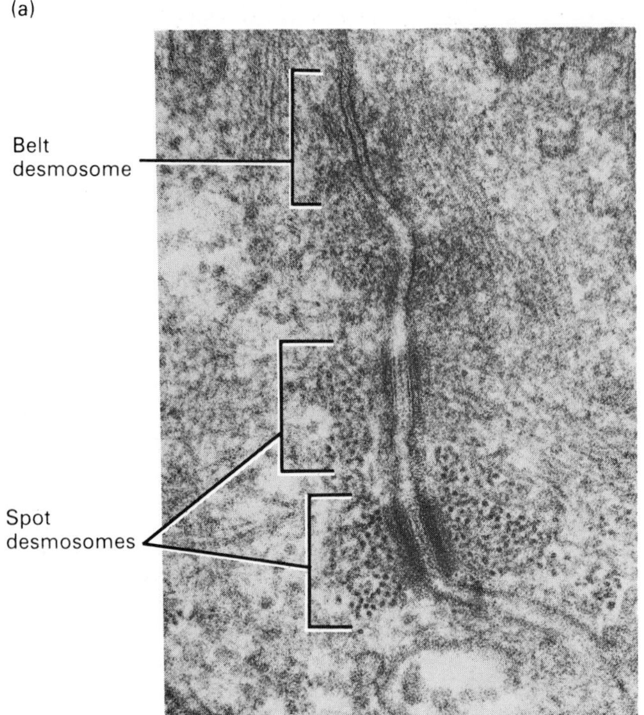

Figure 14-57 Desmosomes. (a) Electron micrograph of a belt desmosome and two spot desmosomes. The fibers associated with the belt desmosome, 7-nm filaments made of actin, are of smaller diameter than the 10-nm tonofilaments associated with the spot desmosomes. The belt desmosomes lack the thick cytoplasmic plaque characteristic of spot desmosomes, and the two plasma membranes are closer together in the belt desmosome. (b) Model of the belt desmosome of an intestinal epithelial cell. In the region of the belt desmosome the plasma membranes of the adjacent cell are separated by only 15 nm, but how they are connected is not known. Bands of actin filaments encircle the cell just under the plasma membrane in the region of the belt desmosome. These actin filaments, in turn, connect with the terminal web, a layer of filaments, crisscrossing the cell at the level of the belt desmosome. Actin filaments also form the core of the microvilli that lie above the terminal web; these filaments extend into the web and appear to be anchored there. *Photograph courtesy of B. E. Hull and L. A. Staehelin.*

rescence microscope whether they are able to pass into neighboring cells (Figure 14-59). Gap junctions between mammalian cells permit the passage of molecules as large as 2 nm in diameter. In insects the junctions are permeable to slightly larger molecules, as much as 3 nm in diameter. Generally speaking, molecules having a molecular weight lower than 1200 pass freely, those of 2000 or more do not pass, and the passage of intermediate-sized molecules is variable and limited. Thus ions and many of the low-molecular-weight building blocks of cellular macromolecules, such as amino acids and nucleoside phosphates, pass freely from cell to cell.

A vivid example of this cell-to-cell transfer is the phenomenon of *metabolic coupling,* or *metabolic cooperation.* Cells can transfer to neighboring cells molecules that the recipients are unable to synthesize. For example, adenosine mono-, di-, or triphosphate can pass through gap junctions. In normal fibroblasts, hypoxanthine (adenine, with the C—NH_2 group replaced by a C=O) can serve as a precursor of DNA. By the pathway depicted in Figure 14-60, hypoxanthine is converted first to inosine 5′-phosphate, in a reaction catalyzed by the enzyme hypoxanthine-guanine phosphoribosyltransferase (HPRT), and thence to dATP, the immediate precursor of DNA. Fibroblasts of a mutant line lacking HPRT are unable to incorporate radioactive hypoxanthine into their DNA. But if cells lacking HPRT are cocultured with cells that have this enzyme, and labeled hypoxanthine is added to the medium, then radioactivity is frequently found in the nuclear DNA of the mutant cells (Figure 14-61). (The two cells can be differentiated by their distinct morphologies or by feeding one of the cell lines carbon particles before mixing it with the other line.) dATP derived from hypoxanthine is incorporated into DNA of the mutant cells only if they are in direct or indirect contact (i.e., through an intermediate cell) with wild-type cells. It is thought that

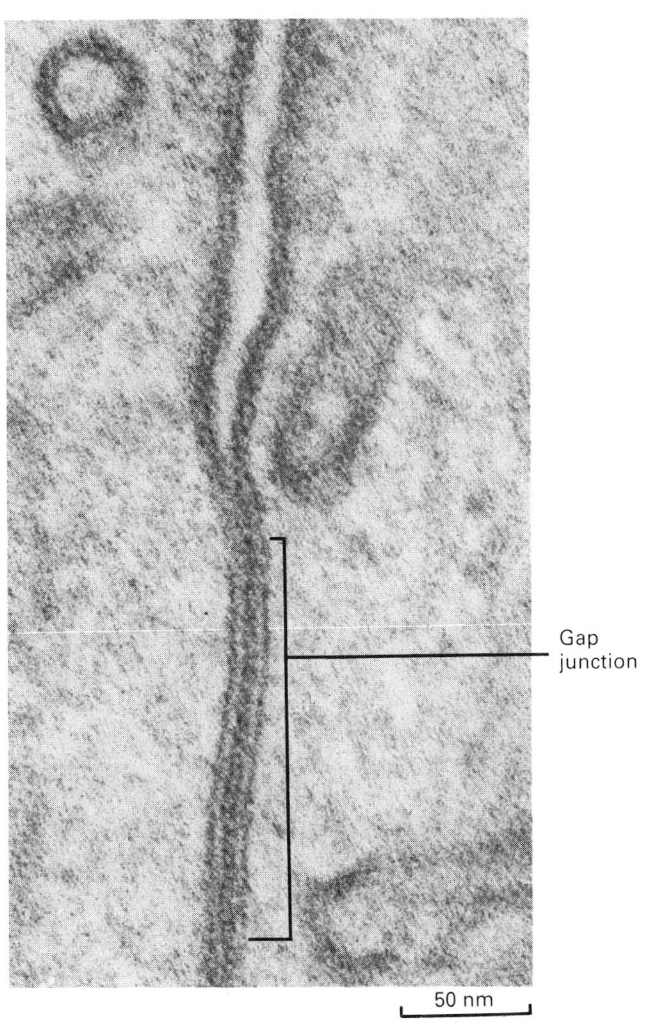

Figure 14-58 Electron micrograph of a gap junction connecting two mouse liver cells. The two plasma membranes are closely associated for a distance of several hundred nanometers, separated by a "gap" of 2 to 3 nm. This space is occupied by a series of cylindrical particles that form channels which allow small molecules to flow from the cytoplasm of one cell to another. *Courtesy of D. Goodenough.*

labeled adenosine mono-, di-, or triphosphate is synthesized from the labeled hypoxanthine by wild-type cells and then passed through gap junctions to the mutant cells.

Another important compound transferred from cell to cell through gap junctions is cyclic AMP (cAMP). The amount of cellular cAMP increases in response to the treatment of cells with many different hormones (cAMP acts as a type of intracellular messenger and regulates a number of cellular metabolic activities). The fact that cAMP can pass through gap junctions means that hormonal stimulation of just one or a few cells in an epithelium will initiate a metabolic reaction in many of them.

Because all the cells of a pancreatic acinus are linked by gap junctions, the hormonal stimulation of secretion in one cell in the acinus is transmitted immediately to all the other cells, ensuring a synchronized secretory response. Specifically, secretory hormones bind to receptors on the basal plasma membranes of these cells and increase the

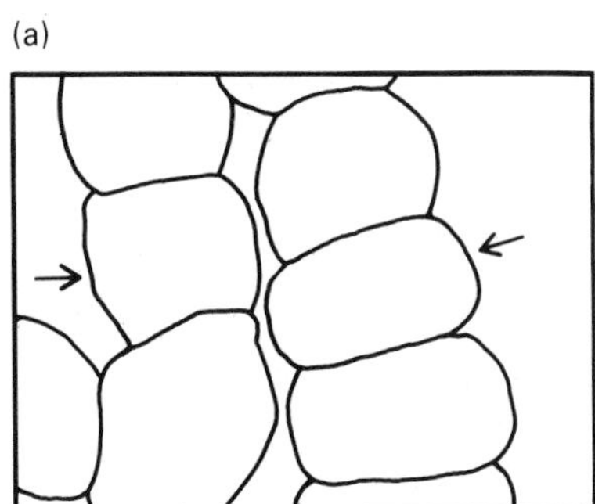

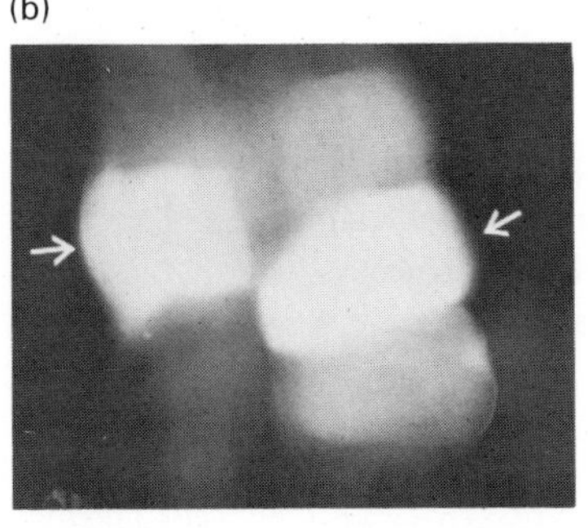

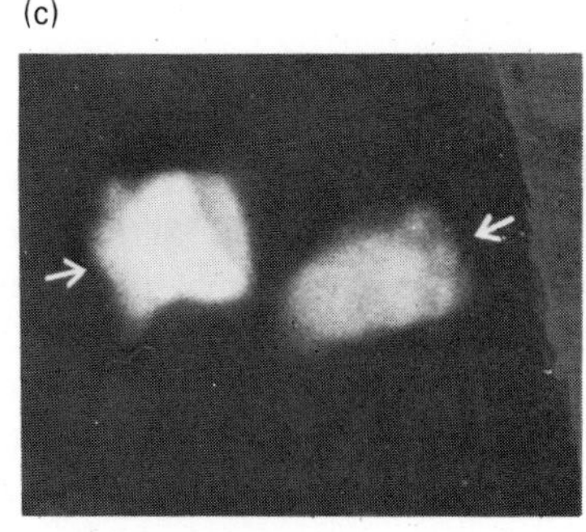

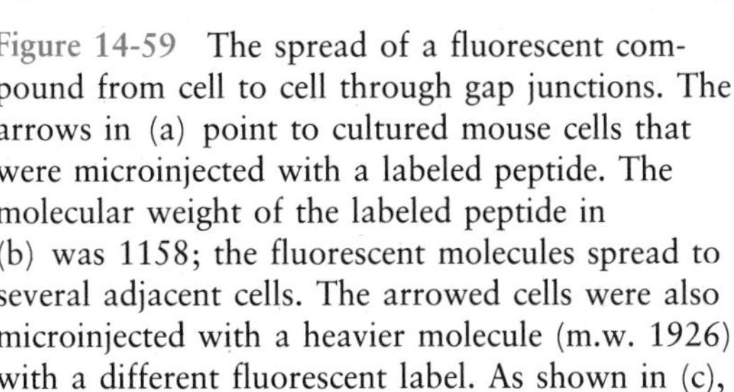

Figure 14-59 The spread of a fluorescent compound from cell to cell through gap junctions. The arrows in (a) point to cultured mouse cells that were microinjected with a labeled peptide. The molecular weight of the labeled peptide in (b) was 1158; the fluorescent molecules spread to several adjacent cells. The arrowed cells were also microinjected with a heavier molecule (m.w. 1926) with a different fluorescent label. As shown in (c), this molecule remained in the injected cells, an indication that it is probably too large to pass through gap junctions. The fluorescent compounds can be distinguished from one another because each gives off light of a different wavelength. [See I. Simpson, B. Rose, and W. R. Loewenstein, 1977, *Science* **195**:294.] *Courtesy of I. Simpson, B. Rose, and W. R. Loewenstein.*

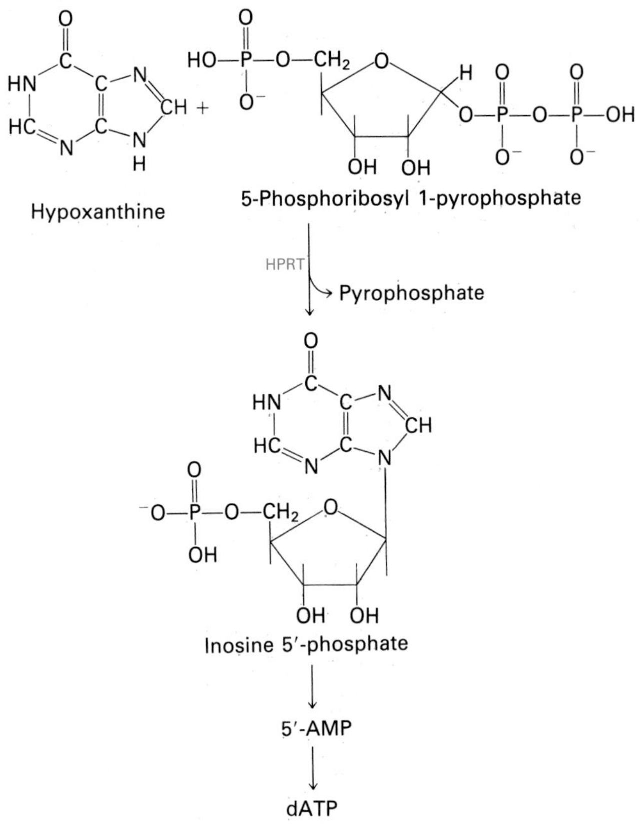

Hypoxanthine + 5-Phosphoribosyl 1-pyrophosphate

HPRT → Pyrophosphate

Inosine 5′-phosphate

↓

5′-AMP

↓

dATP

Figure 14-60 Hypoxanthine, deaminated adenine, can be converted to dATP, a normal precursor of DNA. The enzyme that catalyzes the first step in this pathway, hypoxanthine-guanine phosphoribosyltransferase (HPRT), is missing in mutant HPRT$^-$ cells (see page 202). These mutants are unable to use hypoxanthine as a precursor of dATP for DNA synthesis.

intracellular concentration of either cAMP or Ca^{2+} ions, substances that trigger the secretion of the contents of the secretory vesicles. Both "intracellular messengers" can pass through the intercellular gap junctions, so hormonal stimulation of one cell triggers secretion by many.

An important aspect of gap junction physiology is that the channels close in the presence of high concentrations of Ca^{2+} ion. The Ca^{2+} concentration in extracellular fluids is quite high, from 1×10^{-3} *M* to 2×10^{-3} *M*, whereas inside the cytoplasm the concentration of free Ca^{2+} is lower than 10^{-6} *M*. If the membrane of one cell in an epithelium is ruptured, Ca^{2+} enters the cell, traveling down its concentration gradient, and the concentration of Ca^{2+} in the cytoplasm rises markedly. This closes the channels that connect the cell with its neighbors and prevents leakage of the low-molecular-weight contents of the cytoplasm in all the cells in the epithelium.

Even slight increases in the level of cytoplasmic Ca^{2+} ions, or changes in cytoplasmic pH, can cause the permeability of gap junctions to decrease. Thus cells may modulate the degree of coupling with their neighbors, but precisely why and how they accomplish this is a matter of debate.

In many tissues, the liver for example, large numbers of individual junctions are clustered together in an area about 0.3 μm in diameter (Figure 14-62). This clustering enables researchers to separate gap junctions from other parts of the plasma membrane. To accomplish this, purified plasma membrane is sheared into small fragments. Those containing gap junctions have a higher density than the bulk of the plasma membrane, owing to their relatively high protein content, and can be purified on an

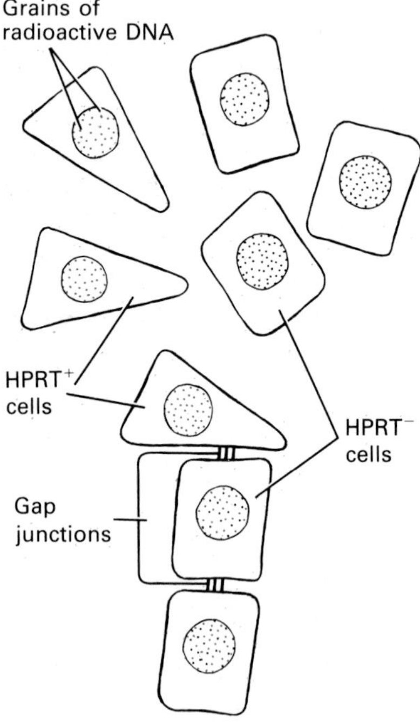

Figure 14-61 Diagram of an autororadiograph of a mixed culture of cells having hypoxanthine-guanosine phosphoribosyltransferase (HPRT$^+$) and cells lacking it (HPRT$^-$), demonstrating the transfer of metabolites from cell to cell through gap junctions. After the cells were mixed, [^{3}H]hypoxanthine was added. After 4 h, the HPRT$^-$ cells in contact with HPRT$^+$ cells are labeled, but those that are isolated from HPRT$^+$ cells remain unlabeled. Labeled inosine monophosphate (IMP) or other metabolic products of hypoxanthine are able to pass from HPRT$^+$ cells when the cells are adjacent.

(a)

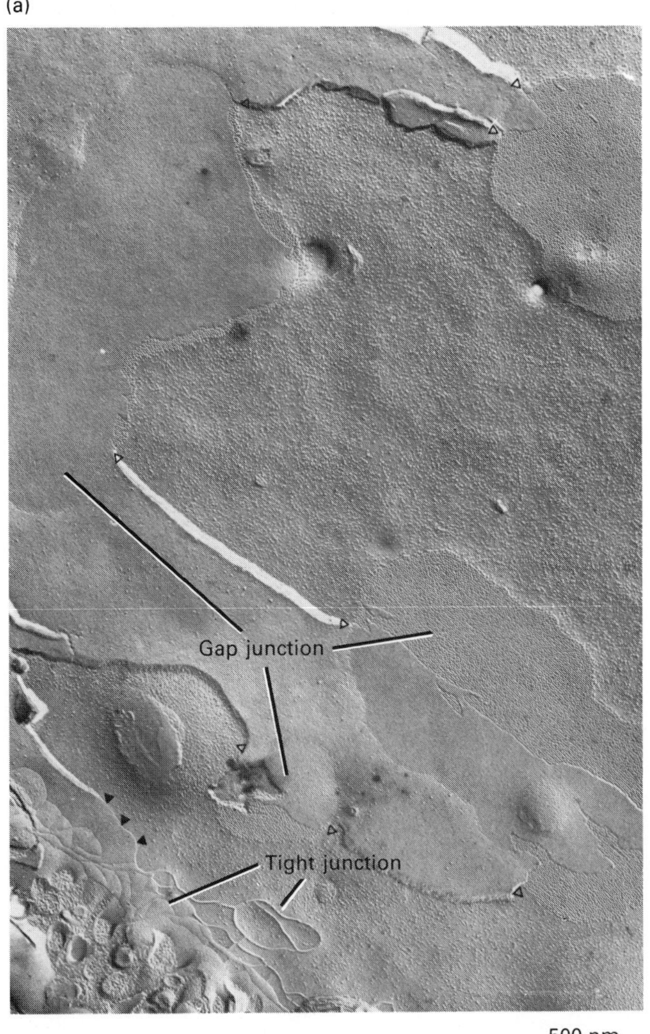

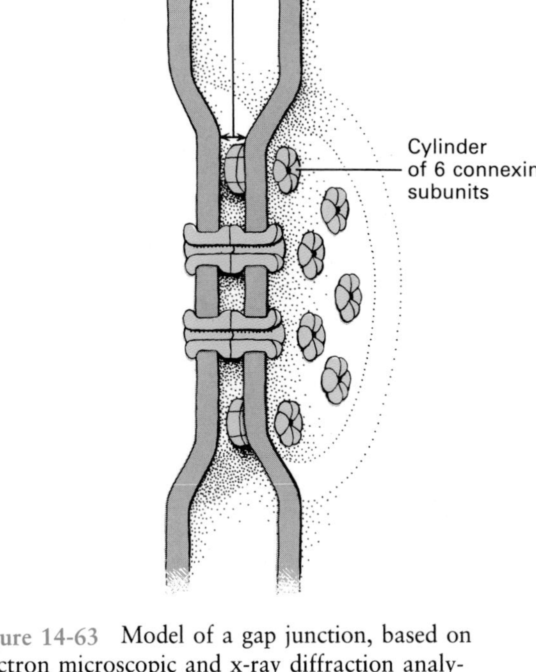

Figure 14-63 Model of a gap junction, based on electron microscopic and x-ray diffraction analyses. Two plasma membranes are separated by a gap of about 15 nm. Both membranes contain cylinders of six dumbbell-shaped protein subunits called connexin. Two such cylinders join in the gap between the cells to form a channel, about 1.5 to 2.0 nm in diameter, that connects the cytoplasm of the two cells.

(b)

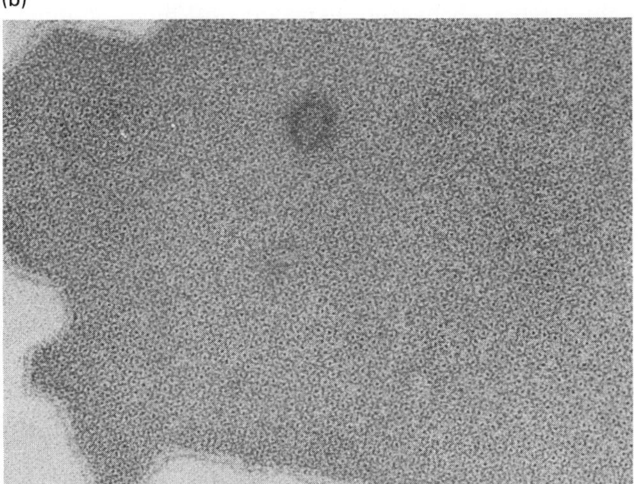

Figure 14-62 (a) Freeze-fracture replicas of gap junctions in liver cells. Part (b) is a higher magnification of negatively stained purified junctions, showing the tiny holes in the center of the hexagonal channels. *Courtesy of D. Goodenough.*

equilibrium gradient (see Figure 5-28). Electron micrographs of stained isolated gap junction reveal a lattice of hexagonal particles with hollow cores as the intercellular channels (Figures 14-62 and 14-63). The purified gap junction material consists of a single major protein (of m.w. 25,000) called connexin. Each hexagonal particle consists of 12 connexin molecules, 6 formed in a hexagonal cylinder in one plasma membrane joined to 6 arranged in the same array in the adjacent cell membrane. As deduced from image analysis of the electron micrographs, the six connexin subunits that make up a single cylinder can interact in two different but related ways, one resulting in an open channel and the other in a closed channel (Figure 14-64). It has been suggested that the transition from one form of interaction to the other might be mediated by the level of Ca^{2+} ion in the cell (Figure 14-64).

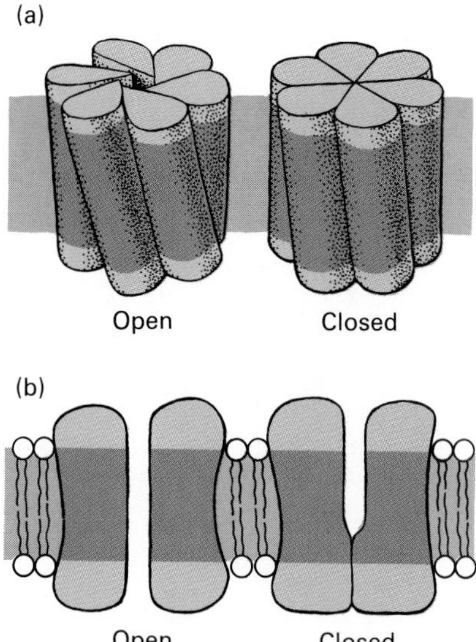

Figure 14-64 Two models of a gap junction derived from analyses of electron micrographs. (a) In one model a rotation of the six connexin subunits about a central axis mediates the transition from an open to a closed state. (b) In the other, a smaller conformational change implements the transition. *Part (a) after P. T. Unwin and G. Zampighi, 1980,* Nature *283:545; part (b) after N. Hirokawa and J. Heuser, 1982,* Cell *30:395.*

Summary

The basic structure of all biological membranes is the closed phospholipid bilayer. The lipid composition of the bilayer varies among the diverse cellular membranes: glycolipids and cholesterol are abundant in the plasma membrane, and large quantities of cardiolipin are found in the inner mitochondrial membrane. The phospholipid composition of the two leaflets in a single membrane may also differ, as in the erythrocyte plasma membrane, but such variations as these do not explain the unique properties of different biological membranes. Each type of membrane—or even patch of membrane—owes much of its individuality to the distinctive properties of its protein species.

The membrane of the mammalian erythrocyte, for example, is different from that of most tissue cells in that it contains a homogeneous array of surface proteins unable to diffuse laterally in the membrane. Just under the plasma membrane, a lacelike cytoskeletal network constructed mostly of spectrin, a fibrous protein, is bound to one of the major integral membrane glycoproteins by many highly specific protein-protein interactions. As no cytoskeleton courses through the erythrocyte cell, apparently the submembrane spectrin network is uniquely responsible for its shape.

Membrane proteins can be classified into two broad groups. An integral protein interacts directly with the phospholipid bilayer and usually contains one or more long α-helical sequences of hydrophobic amino acids that form bonds within the hydrophobic middle of the bilayer. Examples of integral proteins are the viral glycoproteins and the two major erythrocyte glycoproteins, glycophorin and band 3. The second group, peripheral proteins, are bound to the membrane primarily, if not exclusively, by protein-protein, rather than protein-lipid, interactions. Examples are spectrin and other erythrocyte skeletal proteins, and also many proteins of the glycocalyx.

Biological membranes are highly asymmetric. All integral membrane proteins bind asymmetrically to the lipid bilayer. All molecules of any one kind, as glycophorin or a viral glycoprotein, lie in the same direction. All the molecules of a particular membrane-bound enzyme face the same surface of the membrane, and all membrane oligosaccharides, both in glycolipids and glycoproteins, face the extracytoplasmic surface of the bilayer. This asymmetry is necessary for all aspects of membrane function.

Proteins are also essential in the formation of junctions between cells. Gap junctions allow the passage of small molecules between adjacent cells. They are constructed of 12 copies of a single protein that form into a transmembrane channel interconnecting the cytoplasm of the two cells. Belt desmosomes and spot desmosomes are junctions that bind the membranes of adjacent cells in such a way as to give strength and rigidity to the entire tissue. On the cytoplasmic surface of a cell, desmosomal complexes connect with specific cytoskeletal fibers. Such interactions contribute to the cell's rigid shape and allow shearing forces applied to one group of cells to be transmitted to the entire tissue. Other surface proteins form tight junctions between cells, sealing off fluids on different sides of a cell layer.

In epithelial cells, such as those that form the pancreatic acinus or intestinal epithelium, tight junctions demarcate two regions of the plasma membrane, the apical and basolateral surfaces. Each region contains unique proteins that allow it to perform specialized functions, such as binding hormones or fusing with intracellular vesicles that contain secretory vesicles. Most proteins and all lipids in the plasma membrane, as well as in internal membranes, are laterally mobile. Some proteins are immobile, most likely because they are anchored to parts of the cytoskeleton.

References

General References

BRETSCHER, M. S., 1985. The molecules of the cell membrane. *Sci Am.* **253**:(4)100–109.

FINEAN, J. B., R. COLEMAN, and R. H. MICHELL. 1984. Membranes and their cellular functions. Blackwell Scientific Publications.

QUINN, A. J., and D. CHAPMAN. 1980. The dynamics of membrane structure. *CRC Crit. Rev. Biochem.* 8:1–117.

SINGER, S. J., and G. L. NICOLSON. 1972. The fluid mosaic model of the structure of cell membranes. *Science* **175**:720–731.

*TANFORD, C. 1980. *The Hydrophobic Effect,* 2d ed. Wiley. Includes a good discussion of the interactions of proteins and membranes.

The Architecture of Lipid Membranes

CHAPMAN, D. 1975. Lipid dynamics in cell membranes. In *Cell Membranes: Biochemistry, Cell Biology, and Pathology,* G. Weissmann and R. Claiborne, eds. Hospital Practice, pp. 13–22.

*NORTON, W. T. 1981. Formation, structure, and biochemistry of myelin. In *Basic Neurochemistry,* 3d ed., G. J. Siegel, R. W. Albers, B. W. Agranoff, and R. Katzman, eds. Little Brown.

Membrane Proteins

Attachment of Integral Membrane Proteins

CHAPMAN, D., J. C. GOMEZ-FERNANDEZ, and F. M. GONI. 1982. The interaction of intrinsic proteins and lipids in biomembranes. *Trends Biochem. Sci.* 7:67–70.

DEISENHOFER, J., O. EPP, K. MIKI, R. HUBER, and H. MICHEL. 1985. Structure of the protein subunits in the photosynthetic reaction centre of *Rhodopseudomonas viridis* at 3 Å resolution. *Nature* **318**:618–624.

ENGLEMAN, D. M., and G. ZACCAI. 1980. Bacteriorhodopsin is an inside-out protein. *Proc. Nat'l Acad. Sci. USA* 77:5894–5898.

SCHMIDT, M. F. G. 1982. Acylation of proteins: a new type of modification of membrane glycoproteins. *Trends Biochem. Sci.* 7:322–324.

TSE, A. G. D., A. N. BARCLAY, A. WATTS, and A. F. WILLIAMS. 1985. A glycophospholipid tail at the carboxyl terminus of the Thy-1 glycoprotein of neurons and thymocytes. *Science* 230:1003–1008.

UNWIN, N., and R. HENDERSON. 1984. The structure of proteins in biological membranes. *Sci. Am.* **250**(2):78–94.

Detergents

HELENIUS, A., and K. SIMONS. 1975. Solubilization of membranes by detergents. *Biochim. Biophys. Acta* **415**:29–79.

Principles of Membrane Organization

Membrane Asymmetry

BRANTON, D. 1966. Fracture faces of frozen membranes. *Proc. Nat'l Acad. Sci. USA* 55:1048–1056.

HIRANO, H., B. PARKHOUSE, G. L. NICOLSON, E. S. LENNOX, and S. J. SINGER. 1972. Distribution of saccharide residues on membrane fragments from a myeloma-cell homogenate: Its implication for membrane biogenesis. *Proc. Nat'l Acad. Sci. USA* **69**:2945–2949.

LUFT, J. H. 1976. The structure and properties of the cell surface coat. *Int. Rev. Cytol.* **45**:291–382.

ROTHMAN, J., and J. LENARD. 1977. Membrane asymmetry. *Science* **195**:743–753.

Mobility of Membrane Proteins and Lipids

DE PETRIS, S., and M. C. RAFF. 1973. Normal distribution, patching, and capping of lymphocyte surface immunoglobulin studied by electron microscopy. *Nature New. Biol.* **241**:257–259.

DRAGSTEIN, P., P. HENKART, R. BLUMENTHAL, J. WEINSTEIN, and J. SCHLESSINGER, 1979. Lateral diffusion of surface immunoglobulin, Thy-1 antigen, and a lipid probe in lymphocyte plasma membranes. *Proc. Nat'l Acad. Sci. USA* **76**:5163–5167.

FRYE, L. D., and M. EDIDIN. 1970. The rapid intermixing of cell surface antigens after formation of mouse-human heterokaryons. *J. Cell Sci.* 7:319–335.

*HACKENBROCK, C. R. 1981. Lateral diffusion and electron transfer in the mitochondrial inner membrane. *Trends Biochem. Sci.* **6**:151–154.

*JACOBSON, K., E. ELSON, D. KOPPEL, and W. WEBB. 1982. Fluorescence photobleaching in cell biology. *Nature* **295**:283–284.

KORNBERG, R. D., and H. M. MCCONNELL. 1971. Lateral diffusion of phospholipids in a vesicle membrane. *Proc. Nat'l Acad. Sci. USA* **68**:2564–2568.

POO, M., and R. A. CONE. 1974. Lateral diffusion of rhodopsin in the photoreceptor membrane. *Nature* **247**:438–441.

SOWERS, A. E., and C. R. HACKENBROCK. 1981. Rate of lateral diffusion of intramembrane particles: Measurement by electrophoretic displacement and rerandomization. *Proc. Nat'l Acad. Sci. USA* **78**:6246–6250.

*VAZ, W., F. GOODSAID-ZALDVONDO, and K. JACOBSON. 1984. Lateral diffusion of lipids and proteins in bilayer membranes. *FEBS Lett.* **174**:199–207.

The Erythrocyte Membrane

Integral Membrane Proteins

*CABANTCHIK, Z. I., P. A. KNAUF, and A. ROTHSTEIN. 1978. The anion transport system of the red blood cell: The role of membrane protein evaluated by the use of "probes." *Biochim. Biophys. Acta* **515**:239–302.

GOLAN, D. E., and W. VEATCH. 1980. Lateral mobility of band 3 in the human erythrocyte membrane studied by fluorescence photobleaching recovery: Evidence for control by cytoskeletal interactions. *Proc. Nat'l Acad. Sci. USA* 77:2537–2541.

JENKINS, J. D., F. J. KEZDY, and T. L. STECK. 1985. Mode of interaction of phosphofructokinase with the erythrocyte membrane. *J. Biol. Chem.* **260**:10426–10433.

JENNINGS, M. L. 1984. Oligomeric structure and the anion transport function of human erythrocyte Band 3 protein. *J. Mem. Biol.* **80**:105–117.

KOPITO, R. R., and H. F. LODISH. 1985. Primary structure and transmembrane orientation of the murine anion exchange protein. *Nature* **316**:234–238.

*A book or review article that provides a survey of the topic.

TSAI, I.-H., S. N. P. MURPHY, and T. L. STECK. 1982. Effect of red cell membrane binding on the catalytic activity of glyceraldehyde 3-phosphate dehydrogenase. *J. Biol. Chem.* 257:1438–1442.

VIITALA, J., and J. JARNEFELT. 1985. The red cell surface revisited. *Trends Biochem. Sci.* **10**:392–395.

WALDER, J. A., R. CHATTERJEE, T. L. STECK, P. S. LOW, G. F. MUSSO, E. T. KAISER, P. H. ROGERS, and A. ARNONE. 1984. The interaction of hemoglobin with the cytoplasmic domain of band 3 of the human erythrocyte membrane. *J. Biol. Chem.* **259**:10238–10246.

YU, J., D. A. FISCHMAN, and T. L. STECK. 1973. Selective solubilization of proteins and phospholipids from red blood cell membranes by nonionic detergents. *J. Supramol. Struct.* 1:233–248.

The Cytoskeleton

* BRANTON, D., C. M. COHEN, and J. TYLER. 1981. Interaction of cytoskeletal proteins on the human erythrocyte membrane. *Cell* **24**:24–32.

BENNETT, V. 1985. The membrane skeleton of the human erythrocyte and its implications for more complex cells. *Annu. Rev. Biochem.* **54**:273–304.

BYERS, T. M., and D. BRANTON. 1985. Visualization of the protein associations in the erythrocyte membrane skeleton. *Proc. Nat'l Acad. Sci. USA* **82**:6153–6157.

COHEN, C. M., J. M. TYLER, and D. BRANTON. 1980. Spectrin-actin associations studied by electron microscopy of shadowed preparations. *Cell* **21**:875–883.

HAEST, C. W. M. 1982. Interactions between membrane skeleton proteins and the intrinsic domain of the erythrocyte membrane. *Biochim. Biophys. Acta* **694**:331–352.

PASTERNACK, G., R. ANDERSON, T. LETO, and V. MARCHESI. 1985. Interactions between protein 4.1 and band 3. *J. Biol. Chem.* **260**:3676–3683.

SHEN, B. W., R. JOSEPHS, and T. L. STECK. 1984. Ultrastructure of unit fragments of the skeleton of the human erythrocyte membrane. *J. Cell Biol.* **99**:810–821.

SIEGEL, D. and D. BRANTON. 1985. Partial purification and characterization of an actin-bundling protein, band 4.9, from human erythrocytes. *J. Cell Biol.* **100**:775–785.

SPEICHER, D. W., and V. T. MARCHESI. 1984. Erythrocyte spectrin is comprised of many homologous triple helical segments. *Nature* **311**:177–180.

Hereditary Diseases

AGRE, P., J. CASELLA, W. ZINKHAM, C. MCMILLAN, and V. BENNETT. 1985. Partial deficiency of erythrocyte spectrin in hereditary spherocytosis. *Nature* 314:380–383.

* RICE-EVANS, C. A., and M. J. DUNN. 1982. Erythrocyte deformability and disease. *Trends Biochem. Sci.* **6**:282–286.

TOMASELLI, M. B., K. M. JOHN, and S. E. LUX. 1981. Elliptical erythrocyte membrane skeletons and heat-sensitive spectrin in hereditary elliptocytosis. *Proc. Nat'l Acad. Sci. USA* 78:1911–1915.

Generation of the Erythrocyte Membrane

GEIDUSCHEK, J. B., and S. J. SINGER. 1979. Molecular changes in the membranes of mouse erythroid cells accompanying differentiation. *Cell* **18**:149–163.

SARRIS, A. H., and G. E. PALADE. 1982. Immunofluorescent detection of erythrocyte sialoglycoprotein antigens on murine erythroid cells. *J. Cell Biol.* **93**:591–603.

* ZWEIG, S. E., K. T. TOKUYASU, and S. J. SINGER. 1981. Membrane-associated changes during erythropoiesis. On the mechanism of maturation of reticulocytes to erythrocytes. *J. Supramol. Struct.* **17**:163–181.

SIMONS, K., and S. D. FULLER. 1985. Cell-surface polarity in epithelial cells. *Annu. Rev. Cell Biol.* **1**:243–288.

Specialized Regions of the Plasma Membrane

BARTLES, J. R., L. BRAITERMAN, and A. HUBBARD. 1985. Endogenous and exogenous domain markers of the rat hepatocyte plasma membrane. *J. Cell Biol.* **100**:1126–1138.

* KENNY, A. J., and A. G. BOOTH. 1978. Microvilli: Their ultrastructure, enzymoloy, and molecular organization. *Essays Biochem.* **14**:1–44.

NISHI, Y., and Y. TAKESUE, 1978. Localization of intestinal sucrase-isomaltase complex on the microvillous membrane by electron microscopy using nonlabeled antibodies. *J. Cell Biol.* **79**:516–525.

SEMENZA, G. 1976. Small intestinal disaccharidases: Their properties and role as sugar translocators across natural and artificial membranes. In *The Enzymes of Biological Membranes,* vol. 3, A. Martinosi, ed. Plenum, pp. 349–382.

ZIOMEK, C. A., S. SCHULMAN, and M. EDIDIN. 1980. Redistribution of membrane proteins in isolated mouse intestinal epithelial cells. *J. Cell Biol.* **86**:849–857.

Cell Junctions

GILULA, N. B. 1974. Junctions between cells. In *Cell Communication,* R. P. Cox, ed. Wiley, pp. 1–29.

STAEHELIN, L. A. 1974. Structure and function of intercellular junctions. *Int. Rev. Cytol.* **39**:191–283.

* STAEHELIN, L. A., and B. E. HULL. 1978. Junctions between cells. *Sci. Am.* **238**(5):140–152.

Tight Junctions

DA SILVA, P. P., and B. KACHAR. 1982. On tight-junction structure. *Cell* **28**:441–450.

HULL, B. E., and L. A. STAEHELIN. 1976. Functional significance of the variations in the geometrical organization of tight junction networks. *J. Cell Biol.* **69**:688–704.

KACHAR, B., and T. S. REESE. 1982. Evidence for the lipidic nature of tight junction strands. *Nature* **296**:464–466.

Desmosomes

FRANKE, W. W., S. WINTER, C. GRUND, E. SCHMIC, D. L. SCHILLER, and E. D. JARASCH. 1981. Isolation and characterization of desmosome-associated tonofilaments from rat intestinal brush border. *J. Cell Biol.* **90**:116–127.

GORBSKY, G., S. COHEN, H. SHIDA, G. GUIDICE, and M. STEINBERG. 1985. Isolation of the non-glycosylated proteins of desmosomes and immunolocalization of a third plaque protein: desmoplakin III. *Proc. Nat'l Acad. Sci. USA* **82**:810–814.

Gap Junctions

BENNETT, M., and D. SPRAY, eds. 1985. Gap Junctions. Cold Spring Harbor Laboratory.

FRASER, S. E. 1985. Gap junctions and cell interactions during development. *Trends Neurosci.* **8**:3–4.

GILULA, N. B., O. R. REEVES, and A. STEINBACH. 1972. Metabolic coupling, ionic coupling, and cell contacts. *Nature* **235**:262–265.

GOODENOUGH, D. A., and J. P. REVEL. 1970. A fine structural analysis of intercellular junctions in the mouse liver. *J. Cell Biol.* **45**:272–290.

* HERTZBERG, E. L., T. S. LAWRENCE, and N. B. GILULA. 1981. Gap junctional communication. *Annu. Rev. Physiol.* **43**:479–491.

HIROKAWA, N., and J. HEUSER. 1982. The inside and outside of gap-junction membranes visualized by deep etching. *Cell* **30**:395–406.

LAWRENCE, T. S., W. H. BEERS, and N. B. GILULA. 1978. Transmission of hormonal stimulation by cell-to-cell communication. *Nature* **272**:501–506.

* LOWENSTEIN, W. R. 1979. Junctional intercellular communication and the control of growth. *Biochim. Biophys. Acta* **560**:1–75.

NEYTON, J., and A. TRAUTMANN. 1985. Single-channel currents of an intercellular junction. *Nature* **317**:331–335.

REVEL, J.-P., B. J. NICHOLSON, and S. B. YANCEY. 1984. Molecular organization of gap junctions. *Fed. Proc.* **43**:2672–2677.

ROSE, B., I. SIMPSON, and W. R. LOEWENSTEIN. 1977. Calcium ion produces graded changes in permeability of membrane channels in cell junctions. *Nature* **267**:625–627.

SCHWARZMANN, G., H. WIEGANDT, B. ROSE, Z. ZIMMERMAN, D. BEN-HAIM, and W. R. LOEWENSTEIN. 1981. Diameter of the cell-to-cell junctional membrane channels as probed with neutral molecules. *Science* **213**:551–553.

UNWIN, P. N. T., and G. ZAMPIGHI. 1980. Structure of the junction between communicating cells. *Nature* **283**:545–549.

15

Transport across Cell Membranes

THE plasma membrane acts as a semipermeable barrier between the cell and the extracellular environment. This permeability must be highly selective if it is to ensure that essential molecules such as glucose, amino acids, and lipids can readily enter the cell, that these molecules and metabolic intermediates remain in the cell, and that waste compounds leave the cell. In short, the selective permeability of the plasma membrane allows the cell to maintain a constant internal environment. Similarly, the organelles within the cell often have a different internal environment from that of the surrounding cytosol, the soluble portion of the cytoplasm, and organelle membranes maintain this difference. For example, within the lysosome, the organelle involved in digestive and degradative processes, the concentration of protons (H^+) is 100 to 1000 times that of the cytosol. This gradient is maintained solely by the lysosomal membrane.

An artificial membrane composed of pure phospholipid, or of phospholipid and cholesterol, is permeable to water, to gases such as oxygen and carbon dioxide, and to small, relatively hydrophobic but water-soluble molecules such as ethanol (CH_3CH_2OH). It is essentially impermeable to most water-soluble molecules such as glucose, glucose phosphates, ATP, nucleosides, amino acids, and proteins, and to hydrogen, sodium, calcium, and potassium ions (Figure 15-1). The entry of any of these molecules and ions into a cell must be facilitated by specific

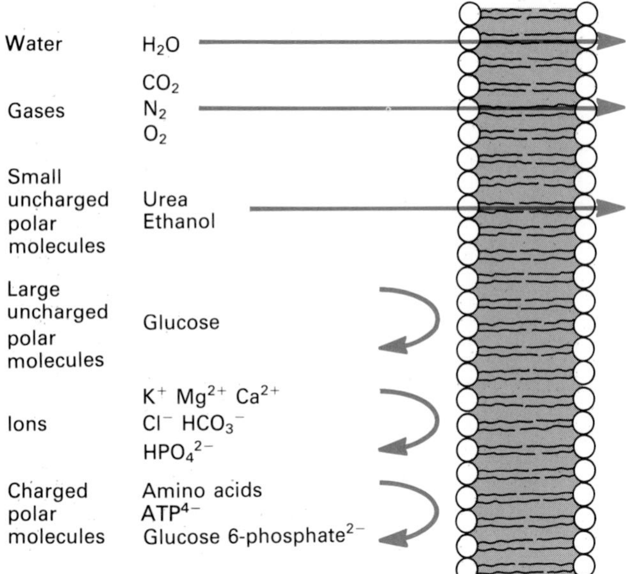

Figure 15-1 A pure artificial phospholipid bilayer is permeable to water, to small hydrophobic molecules, and to small uncharged polar molecules. It is not permeable to ions or to large uncharged polar molecules. Figure 14-7 demonstrates how such artificial membranes are prepared. Using an apparatus in which a small phospholipid bilayer separates two aqueous compartments (Figure 14-7), one can easily measure the permeability of the membrane to various substances: simply add a small amount of the material to one chamber and measure its rate of appearance in the other. The use of radioactive substances such as [^{14}C]glucose or radiolabeled Na^+ greatly facilitates such experiments.

transport molecules within the plasma membrane. Two different types of mechanisms have evolved to allow such molecules to enter or leave the cell.

First, ions and small molecules, including sugars and amino acids, are transported across the lipid of the plasma membrane. Each of various specific integral membrane proteins, termed *permeases,* facilitates the transport of only a limited range of molecules. Because different cell types require different mixtures of low-molecular-weight compounds, the plasma membrane of each cell type contains its own individually tailored battery of permease proteins. Similarly, the membrane surrounding each type of subcellular organelle contains a specific set of permeases that allow only certain molecules to cross it. Indeed, it has become evident in recent years that virtually all the permeability of membranes to small molecules is both facilitated and regulated in various ways by proteins within the membrane. These various mechanisms are discussed extensively in the first part of this chapter.

Second, protein molecules and larger particles enter the cell by endocytosis and phagocytosis (see Figure 5-38). Small regions of the plasma membrane surround the macromolecule or particle required by the cell, then the membrane and its contents are internalized by the cell, forming an intracellular vesicle. A broad array of nutrients, viruses, and particles enter the cell this way. The processes are discussed in detail in the last section of this chapter.

Transport and the Intracellular Ionic Environment

An important function of the plasma membrane is to maintain an ionic composition in the cytosol very different from that of the surrounding fluid. In both vertebrates and invertebrates, for example, the concentration of sodium ion is about 10 to 20 to 40 times higher in the blood than within the cell. The concentration of potassium ion is the reverse, generally 20 to 40 times higher inside the cell (Table 15-1). The generation and maintenance of such gradients on either side of a semipermeable membrane requires the expenditure of a great deal of energy.

Transport across a membrane may be passive or active. *Passive transport* is a type of diffusion in which an ion or molecule crossing a membrane moves down its electrochemical or concentration gradient. No metabolic energy is expended in passive transport. In *simple diffusion,* pas-

Table 15-1 Typical ionic concentrations in invertebrates and vertebrates

	Cell	Blood (mM)
SQUID AXON*		
K^+	400 mM	20
Na^+	50 mM	440
Cl^-	40–150 mM	560
Ca^{2+}	0.3 μM	10
X^-†	300–400 mM	
MAMMALIAN CELL		
K^+	139 mM	4
Na^+	12 mM	145
Cl^-	4 mM	116
HCO_3^-	12 mM	29
X^-†	138 mM	9
Mg^{2+}	0.8 mM	1.5
Ca^{2+}	<1 μM	1.8

*The large nerve axon of the squid is chosen as an example of an invertebrate cell, as it has been used widely in studies of the mechanism of conduction of electrical impulses.

†X^- represents proteins, which have a net negative charge at the neutral pH of blood and cells.

sive transport takes place unaided: a molecule crosses the membrane without the help of permeases. Gases such as oxygen and carbon dioxide, and small molecules such as ethanol enter cells by simple diffusion. In *facilitated diffusion,* a special type of passive transport, ions or molecules cross the membrane rapidly because specific permeases in the membrane facilitate their crossing.

Active transport uses metabolic energy to move ions or molecules against an electrochemical gradient. For example, low sodium concentration inside the cell is maintained by the *sodium potassium ATPase* (Na^+K^+-ATPase), a specific active transport mechanism located in the membrane that transports sodium from the interior of the cell to the outside.

Ion gradients are utilized in driving many biological processes. The transmembrane concentration gradients of sodium and potassium ions are essential for the conduction of an electrical impulse down the axon of a nerve cell. The differences between the concentration of free calcium ion within a cell and the blood are also important. Blood calcium concentration is generally in the millimolar range, while inside the cell, cytosol concentrations are lower than 1 μM. In many cells an increase in the concentration of calcium ion is an important regulatory signal. In muscle cells, for example, it initiates contraction; in the exocrine cells of the pancreas it triggers secretion of digestive enzymes.

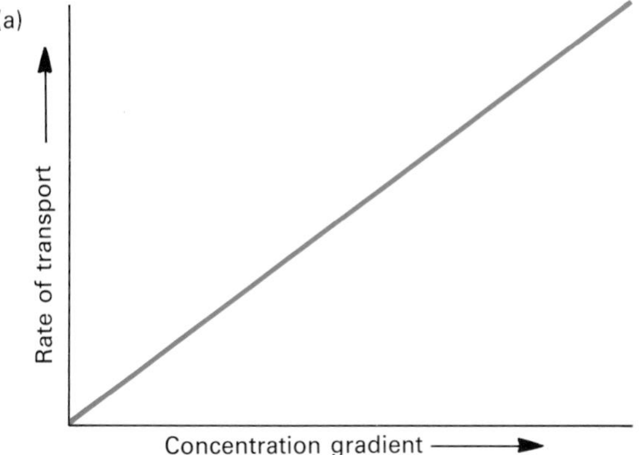

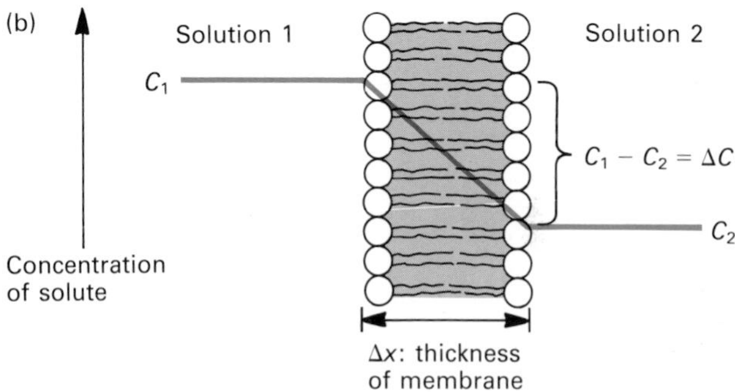

Figure 15-2 (a) For simple diffusion, the rate of transport is directly proportional to the concentration gradient across the membrane. Small hydrophobic molecules or hydrophilic molecules, such as ethanol, cross the plasma membrane by traversing the phospholipid bilayer directly, without aid of a permease protein. (b) The concentration of a solute, C, is different in the two chambers that are separated by a membrane.

Passive Transport across the Cell Membrane

Some Small Molecules Cross the Membrane by Simple Diffusion

In simple diffusion, a small molecule in aqueous solution dissolves into the phospholipid bilayer, crosses it, and then dissolves into the aqueous solution on the opposite side. There is little specificity to the process. To a first approximation, the relative rate of diffusion of the molecule across the phospholipid bilayer will be proportional to the concentration gradient across the membrane (see Figure 15-2a). The rate-limiting step in simple diffusion across the membrane is the movement of the substance from the aqueous solution into the hydrophobic interior of the phospholipid bilayer (Figure 15-3). The rate of diffusion of the molecule will be proportional to its hydrophobicity. A measure of hydrophobicity is the *partition coefficient,* the equilibrium constant for partition of the molecule between oil and water. The partition coefficient is a measure of the relative affinity of the molecule for lipid versus water.

Let us now consider more quantitatively this simple passive diffusion of small molecules through a membrane. Suppose a membrane of area A and thickness Δx separates two solutions of concentrations C_1 and C_2, where $C_1 > C_2$ (Figure 15-2b). In this case, the rate of diffusion is given by a modification of Fick's law, which states that the rate of diffusion, dn/dt (in mol/s) is directly proportional to the concentration gradient dC/dx across the membrane (in mol/cm^4) where C is concentration in mol/cm^3 and x is distance across the membrane, in centimeters:

$$\frac{dn}{dt} = J = -DA\frac{dC}{dx} = -DA\frac{\Delta C}{\Delta x} = \frac{-DA(C_1 - C_2)}{\Delta x}$$

where D is the diffusion constant and J is the rate of transport across the membrane. The concentration gradient across the membrane, dC/dx, is equal to

$$\frac{\Delta C}{\Delta x} = \frac{C_1 - C_2}{\Delta x}$$

In the case of diffusion of a species across a pure phospholipid bilayer, the diffusion constant, D, will be proportional to the ability of the species to dissolve in the

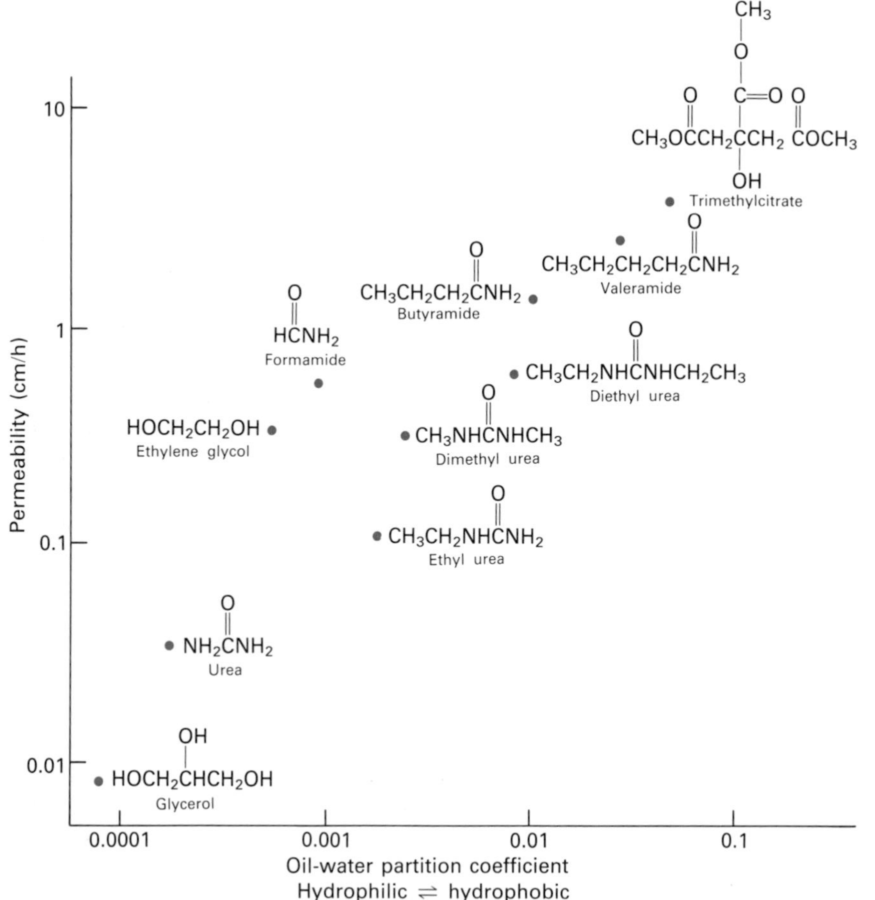

Figure 15-3 Permeability of plasma membranes of the algae *Chara* to small molecules, plotted against the partition coefficients of these molecules in an oil-water mixture. Abscissa: the oil-water partition coefficient, the ratio of equilibrium concentrations of the molecule in oil versus water. Ordinate: the rate of entry of the molecule into the cell. Small molecules enter the cell by diffusing through the phospholipid bilayer; they do not utilize specific permease proteins. The graph shows that molecules of increasing hydrophobicity enter the cell at an increasingly faster rate. Hydrophobic molecules are more capable of dissolving in the fatty center of the phospholipid bilayer, and thus are more capable of crossing the plasma membrane. The rate-limiting step for such molecules to enter the cell is their passage through the center of the bilayer. Note, for instance, the series of molecules urea, ethyl urea, dimethyl urea, and diethyl urea. Increasing the number of CH_3 or CH_2 groups increases the hydrophobicity and also increases the ability of the molecule to cross the plasma membrane. *After R. Collander, 1954,* Physiol. Plant. *7:420.*

strongly hydrophobic interior of the membrane, and thus, to its hydrophobicity (see Figure 15-3).

The values of D for diffusion across a phospholipid membrane are 100 to 1000 times lower than for diffusion of the same molecule in water because the typical cell membrane is 100 to 1000 times more viscous than water. Experiments on simple cells—erythrocytes, for example—show that the transport of small neutral molecules such as urea and ethanol follows Fick's law and is proportional to the concentration gradient across the membrane (Figures 15-2*a*, and 15-3).

Diffusion across a membrane down a concentration gradient has a large negative free energy (see Chapter 2). For example, the movement of 1 mol of substance from 1 *M* solution to 0.1 *M* solution, a 10-fold concentration gradient, releases 1359 cal of free energy (25°C):

$$
\begin{aligned}
\Delta G &= -RT\ln\frac{C_2}{C_1}\\
&= -[1.98\ \text{cal/(degree}\cdot\text{mol)}](298\ \text{K})\left(\ln\frac{1.0}{0.1}\right)\\
&= -1359\ \text{cal/mol}
\end{aligned}
$$

Conversely, to transport 1 mol of a substance "uphill" against a 10-fold concentration gradient would require an input of 1359 cal.

Fick's law does not apply to charged molecules. The diffusion of charged molecules across a membrane (even one that is permeable to the ion) is determined not only by the concentration gradient but also by any electric potential gradient that might exist across the membrane. In addition, each charged ion transported by diffusion (e.g., Na^+) must be moved with an ion of opposite charge (e.g., Cl^-).

Membrane Proteins Speed the Diffusion of Specific Ions or Molecules across the Membranes

The characteristics of facilitated diffusion by membrane proteins are very different from simple diffusion (Figures 15-2 and 15-4). First, the rate of transport of the molecule across the membrane is far greater than would be expected from a simple diffusion model based on solubility of the molecules in a phospholipid bilayer and Fick's equation. Second, the process is specific; each facilitated diffusion protein transports only a single species of ion or molecule, or a single group of closely related molecules. Third, there is a maximum rate of transport (Figure 15-4). That is, when the concentration gradient of molecules across the membrane is low, an increase in the concentra-

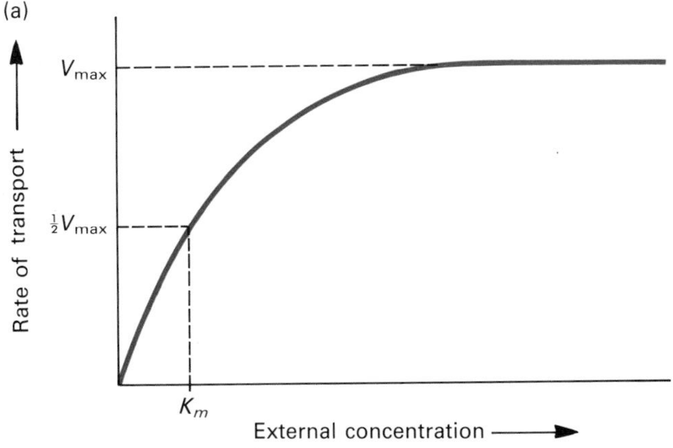

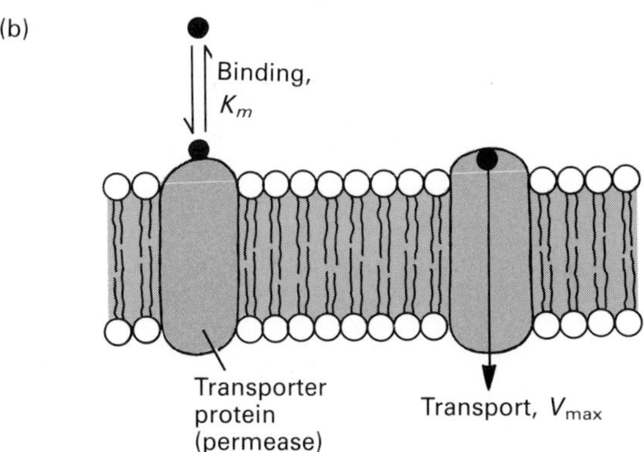

Figure 15-4 The characteristics of facilitated diffusion across a membrane are very different from those of simple diffusion. (a) There is a maximum rate of transport for a facilitated diffusion process V_{max}; the reaction is said to be saturable. The transport is highly specific for different types of molecules. (b) The molecule binds to the transporter protein and is then transported across the membrane. Binding of the molecule to the transporter protein is described by the constant K_m, which is the concentration that gives one-half the maximal velocity of transport.

tion gradient results in a corresponding increase in the rate of transport. The rate of uptake reaches a maximum when the concentration gradient across the membrane becomes large.

The kinetics of a facilitated diffusion process can be described by the same type of equation used for a simple enzyme-catalyzed chemical reaction (see page 70). Let us consider, for simplicity, the situation in which a substance, S, is present initially only on one side of the membrane, the "outside." We can write

$$S_{out} + \text{permease} \xrightleftharpoons{K_m} \underset{\text{complex}}{(S \cdot \text{permease})} \xrightarrow{V_{max}} S_{in}$$

in which K_m is the binding constant for the substance to the permease and V_{max} is the maximum rate of transport of S into the cell. If we set C as the concentration of S outside the cell (initially the concentration inside is zero), then, by exactly the same derivation used for the Michaelis-Menten equation (see page 70) we can write

$$V = \frac{V_{max}}{1 + C/K_m}$$

in which V is the rate of transport of the species into the cell. The meaning of K_m and V_{max} is exactly the same as in the Michaelis-Menten equation. Here K_m is the concentration of substance at which there is half-maximal transport across the membrane. V_{max} is the rate of transport if all molecules of permease were to contain a bound S. This would occur, of course, at high S concentrations. The above equation describes the curve in Figure 15-4.

For many years membrane transport proteins were pictured as shuttles, moving from one side of the membrane to the other, or rotating an active site from one face of the membrane to another. At one face the protein bonded to the molecule while at the other face it released the molecule. Such models are called *carrier* models.

Carrier molecules do explain the properties of certain small peptide antibiotics that greatly facilitate the diffusion of certain ions. For example, the antibiotic valinomycin increases potassium ion transport across biological membranes. Valinomycin forms a sphere around the potassium, its hydrophobic amino acid side chains making up the outer surface while six or eight oxygen atoms on the inside coordinately bind to the ion (Figure 15-5). The hydrophobic exterior makes the ion-carrier complex soluble in the lipid interior of the membranes and facilitates its diffusion across the interior. But the bilayer structure of all cellular membranes, with distinct hydrophobic and hydrophilic planes, makes such movement of proteins energetically expensive, and carrier models are not likely to apply to protein-mediated transport in biological membranes. In fact, no evidence of physical rotation or face-to-face translocation of a permease has been found. Investigators now believe that membrane transport proteins form channels through the membrane that permit certain ions or molecules to pass across.

Transport proteins can be extracted, purified, and then reincorporated into artificial membranes or into vesicles made from natural membranes. In the most elegant studies vesicles called liposomes are formed from pure lipids (see Figure 14-7) and purified transport proteins are incorporated into them. In liposomes the functional properties of individual components can be studied without ambiguities (Figure 15-6). But pure systems are technically difficult to handle. The main problem is that not all of the incorporated transport proteins have the same orientation. Some may be facing "right side" out in the vesicle while others are "wrong side" out. But such problems are being overcome, and transport studies are rapidly developing the precision that has characterized the studies of soluble enzymes.

The human erythrocyte contains two of the best char-

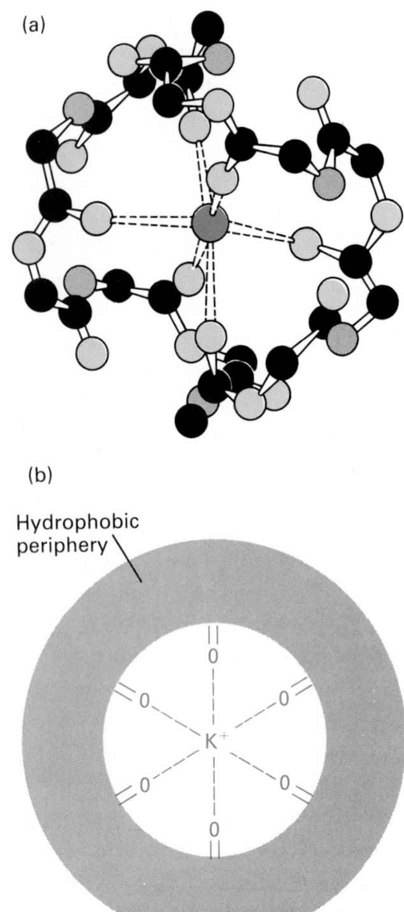

Figure 15-5 (a) Model of valinomycin complexed with one K^+ ion. (b) Schematic diagram of a K^+-valinomycin chelate, showing the hydrophobic periphery and, on the inside of the complex, the peptide C=O bonds binding to K^+.

acterized examples of these transport proteins, one for the facilitated diffusion of glucose and the other for anions such as chloride and bicarbonate. These systems are examined in the sections that follow.

Facilitated Diffusion Transports Glucose into Erythrocytes

Researchers in the past used tissue slices or whole cells for transport studies, but the many competing reactions were hard to separate. The use of red blood cell ghosts was therefore a major advance. Because these ghosts are pure membranes, any transport properties they demonstrate can be attributed solely to membrane-bound proteins. And because ghosts spontaneously reseal into closed vesicles, researchers can seal substrates into the ghosts and then measure both outward and inward transport.

The rate of entry of glucose into erythrocytes is described by the curve in Figure 15-7, a plot of the initial velocity of entry for different external concentrations of glucose. This experiment makes several important points. First, the rate is far greater than one would expect from a free diffusion model based on the simple solubility of glucose into phospholipids and Fick's equation. Second, the

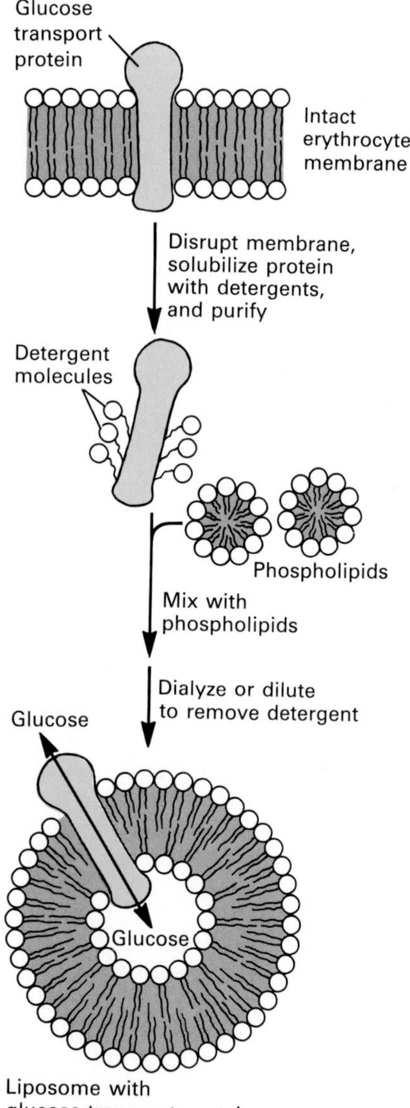

Figure 15-6 The glucose permease from erythrocytes is solubilized by a nonionic detergent such as octylglucoside (see Figure 14-16), and then the protein-detergent complex is purified. To incorporate the protein into liposomes, the detergent-protein complex is mixed with a large excess of phospholipid. If necessary, excess detergent can be removed by dialysis. The resultant liposome-protein complexes can transport glucose by facilitated diffusion.

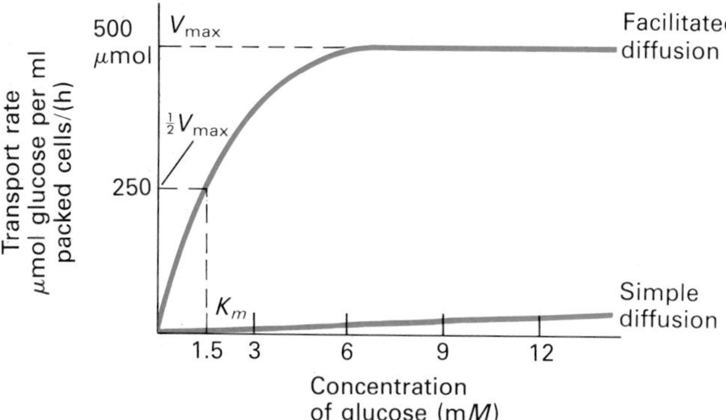

Figure 15-7 Rate of transport of glucose into erythrocytes as a function of glucose in the extracellular medium. Graph shows saturation kinetics that implicate a transport protein (permease) and define facilitated diffusion. The initial concentration of glucose in the erythrocyte is very low, less than 0.5 mM, so the gradient of concentration of glucose across the membrane is effectively the same as the external concentration. The rate of uptake of glucose in the first few seconds is plotted against the glucose concentration. K_m is the concentration at which the rate of glucose uptake is half maximal; V_{max} is the maximum rate of uptake at saturating levels of extracellular glucose. The dashed line is the calculated curve for rate of glucose uptake if the compound were to enter solely by simple diffusion through a phospholipid bilayer rather than by facilitated diffusion. The rate of uptake would be proportional to the extracellular concentration, and thus to the concentration gradient across the membrane. It would *not* show a diminished rate at high concentrations (saturation). The K_m for glucose uptake, 1.5 mM, is less than the glucose concentration in human blood, which is about 4 to 6 mM. This allows glucose uptake into the erythrocyte to occur at near maximal rate.

rate of entry of glucose into the cells does not increase linearly as the concentration gradient across the erythrocyte membrane is increased. The rate of uptake reaches a maximum at 500 μmol (90 mg) per milliliter of packed cells per hour. This defines the V_{max} for the glucose transport. The K_m for glucose transport is 1.5 mM (Figure 15-7); thus at 1.5 mM glucose half of the permease molecules will be expected to contain a bound glucose. The kinetics of glucose uptake are described by the equation for facilitated diffusion on page 621.

The specificity of the transport process is indicated by the fact that the nonbiological L-isomer of glucose does not enter the red blood cell at a measurable rate (Table 15-2): the K_m is greater than 3000 mM. Other sugars structurally related to D-glucose, such as D-mannose or D-galactose, also are transported but somewhat higher concentrations are needed to half-saturate the transport reaction, implying a lower affinity (higher K_m) of the permease for these substrates. The presence of one of the three D-sugars in the medium competitively inhibits the uptake of the others. To accomplish this selectivity, a specific protein must be involved that, like an enzyme, recognizes structural features of its substrate. Such a protein is termed a *glucose transporter* or a *glucose permease,* or, more properly, a D-*hexose permease.* This or a similar glucose transporter is present in most mammalian cells.

After glucose is transported into the erythrocyte, it is rapidly phosphorylated, forming glucose 6-phosphate. This reaction, the first step in the glycolytic conversion of glucose to pyruvate, is catalyzed by hexokinase and requires ATP (see Figure 5-42). Once phosphorylated, the glucose can no longer leave the cell; moreover, the concentration of the simple glucose in the cell is lowered. Its concentration gradient across the membrane is therefore increased, allowing the facilitated diffusion system to continue importing glucose.

The impermeability of red blood cells to glucose phosphate is an example of the general phenomenon that eukaryotic plasma membranes do not contain permeases for phosphorylated compounds. A cell can ensure that molecules will be retained internally by phosphorylating them. ATP and other phosphorylated nucleosides are never released from cells having a normal intact plasma membrane. However, ATP and ADP transport proteins do exist in the mitochondrial membrane, and they allow these molecules to move across it.

The erythrocyte glucose transport protein has been purified and sequenced. It is an integral and probably transmembrane protein of m.w. 45,000. It accounts for 2 percent of the erythrocyte membrane protein. Because different molecules of the glucose permease bear different numbers of oligosaccharide chains, the permease yields a diffuse band during sodium dodecylsulfate gel electrophoresis; it is not visible on the gel depicted in Figure 14-38. The pure protein has been inserted into artificial liposomes, where it dramatically increases their permea-

Table 15-2 Transport of sugars into the erythrocyte by facilitated diffusion

Sugar	K_m (concentration required for a half-maximal rate of transport) (mM)
D-Glucose	1.5
L-Glucose	>3000
D-Mannose	20
D-Galactose	30

SOURCE: P. G. LeFevre, 1961, *Pharmacol. Rev.* 13:39.

bility to D-glucose (see Figure 15-6). All the properties of glucose permeation into erythrocytes are retained in the artificial system. In particular, D-glucose, D-mannose, and D-galactose are transported and L-glucose is not.

The mechanism by which this permease facilitates transmembrane movement of glucose is not known, but it probably does not contain a permanent narrow tunnel in the membrane through which only glucose and closely related sugars can move. If it did, any molecule similar to but smaller than glucose (say glycerol) would also be able to use this "hole." A more likely scenario is depicted in Figure (15-8): the binding of glucose to a site on the exterior surface of the permease induces a conformational change in the polypeptide, generating a passageway that accommodates only the protein-bound sugar.

Transport of Chloride and Bicarbonate Anions across the Erythrocyte Membrane Is Catalyzed by Band 3, the Anion Exchange Protein

The glucose permease allows glucose molecules to enter the cell one at a time. The second erythrocyte permease we examine, the anion exchange protein, catalyzes a different type of transport—a one-for-one exchange of anions such as Cl^- and HCO_3^- across the membrane. Chloride and bicarbonate are the predominant anions in most eukaryotic cells (see Table 15-1). The plasma membrane of most cells is relatively impermeable to Cl^- anions. Typically, their rate of movement through membranes is one-tenth or less that of K^+ or Na^+. The erythrocyte is an exception. Its plasma membrane is highly specialized for the transport of anions both into and out of the cell, and its permeability to Cl^- is about 100,000 times that of most other plasma membranes. The *anion transporter* that enables anions to penetrate the erythrocyte membrane is the band 3 polypeptide, the predominant integral glycoprotein of the erythrocyte (see Figures 14-38 and 14-39).

The rapid flux of anions in the erythrocyte facilitates the transport in the blood of CO_2 from the tissues to the lungs. Waste CO_2 released from cells into the capillary blood diffuses across the erythrocyte membrane. In its gaseous form, CO_2 dissolves poorly in aqueous solutions such as blood plasma, but inside the erythrocyte the potent enzyme carbonic anhydrase converts it to a bicarbonate anion:

$$CO_2 + H_2O \xrightleftharpoons{\text{carbonic anhydrase}} H^+ + HCO_3^-$$

This process occurs while the hemoglobin in the erythrocyte is releasing its oxygen into the blood plasma. The removal of oxygen from hemoglobin induces a change in its conformation that enables a globin histidine side chain to bind the proton produced by carbonic anhydrase. The bicarbonate anion formed by carbonic anhydrase is transported out of the erythrocyte in exchange for an entering Cl^- anion

$$HCO_3^-{}_{in} + Cl^-{}_{out} \rightleftharpoons HCO_3^-{}_{out} + Cl^-{}_{in}$$

(see Figure 15-9). As the total volume of the blood plasma is about twice that of the total erythrocyte cytoplasm, this exchange triples the amount of bicarbonate that can be carried by blood as a whole. Without an anion exchange protein, bicarbonate anions generated by carbonic anhydrase would remain within the erythrocyte and blood would be unable to transport all of the CO_2 produced by tissues. The entire exchange process is completed within 50 ms, during which time 5×10^9 HCO_3^- ions are exported from the cell. The process is reversed in the lungs: HCO_3^- diffuses into the erythrocyte in exchange for a Cl^-. The HCO_3^- combines with H^+ to yield H_2O and CO_2. Oxygen binding to hemoglobin causes release of the proton from hemoglobin. The CO_2 diffuses out of the erythrocyte and is eventually expelled in breathing.

The behavior of band 3 has been studied extensively. It catalyzes a sequential one-for-one exchange of anions on opposite sides of the membrane precisely as required to preserve electroneutrality in the cell, but it does not allow anions to flow unidirectionally from one side to another. It also serves other functions in the erythrocyte membrane, such as the binding of ankyrin (see Figure 14-42). Band 3 is constructed of two domains, one embedded in the membrane, the other on the cytoplasmic face, each of about 450 amino acids (see Figure 14-39). Experiments with protease-generated fragments of band 3 have shown that the membrane-bound domain in the protein catalyzes anion transport. This region of the polypeptide is folded to span the membrane at least 12 times. It is not

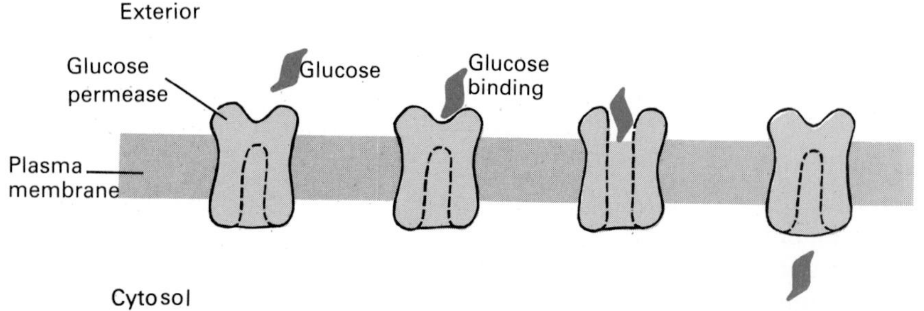

Figure 15-8 Hypothetical scheme for the operation of the glucose permease. The binding of glucose triggers a conformational change in the permease. This change generates a pore in the protein that allows the bound glucose—and only this molecule—to pass through the membrane.

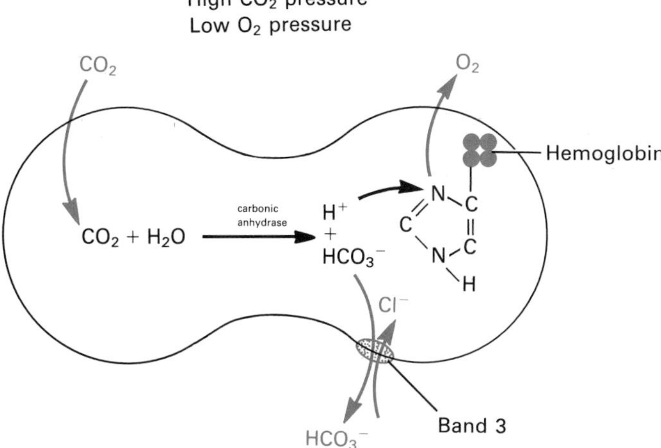

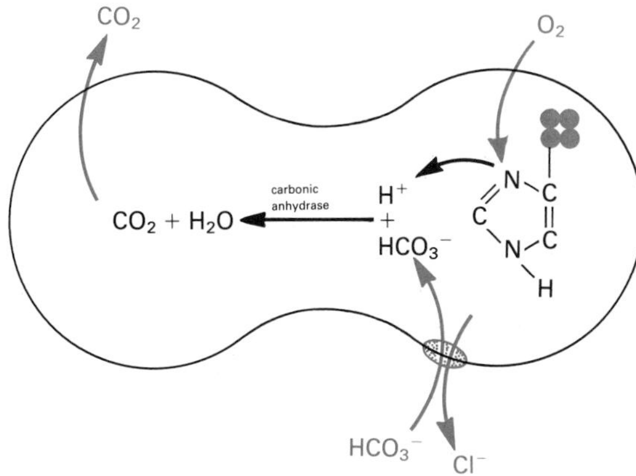

Figure 15-9 Schematic drawing showing anion transport through the erythrocyte membrane in the capillaries and in the lungs. Band 3 catalyzes the exchange of the anions Cl^- and HCO_3^- across the erythrocyte membrane. This reaction is essential for transport of CO_2 from the tissues to the lungs. It allows an HCO_3^-, generated from CO_2 by carbonic anhydrase, to leave the cell in exchange for a Cl^- ion, greatly increasing the ability of the blood to transport HCO_3^- ions.

known exactly how the anion channel is formed nor how it functions; Figure 15-10 suggests a possible model.

Active Transport of Ions and ATP Hydrolysis

When the production of ATP in a cell is inhibited experimentally (treatment with 2,4-dinitrophenol will do this), the concentration of ions inside the cell gradually approaches that of the exterior environment. This is caused by a slow leak of ions across the membrane down their electrical and concentration gradients. Eventually the cell dies, in part because many intracellular enzymes are specialized to function in a solution of low Na^+ and high K^+ ions. In every cell a significant fraction of available energy goes into maintaining the concentration gradients of ions, such as Na^+, K^+, and Ca^{2+}, across the plasma membrane and across the membranes of intracellular compartments. In the human erythrocyte, up to 50 percent of the energy produced (i.e., energy stored in ATP molecules) is used for this purpose. Thus a central issue of cellular metabolism is how permeation systems use energy.

In at least three types of enzyme systems, the hydrolysis of ATP is directly coupled to the transport of ions. One of these systems, the Na^+K^+-ATPase, transports Na^+ out of a cell and K^+ in. A second transports Ca^{2+} ions out of the cell or, in muscle cells, from the cytosol into the compartment of the sarcoplasmic reticulum. A third transports protons. The coupling of interphosphate bond energy to transport is direct—an enzyme system that can split ATP into ADP and phosphate is part of the transport systems for these ions. It is called an *ATPase* but is, in fact, a system for collecting the free energy released in the hydrolysis of ATP and using it to move ions up an electrical or concentration gradient.

We first discuss these three systems of active transport in some detail, then turn to another type of transport against a concentration gradient where the movement of one molecule, say Na^+, *down* its electrical or concentration gradient is coupled to the movement of another, say glucose, *up* its concentration gradient, thus using the energy stored in an Na^+ (or H^+) gradient to power the transport of other molecules.

Sodium Potassium ATPase Maintains the Intracellular Concentration of Sodium and Potassium Ions

Because of its importance to cellular metabolism, the Na^+K^+ transport system has been studied in considerable detail. It has been solubilized and purified from the membranes of several types of cells, including those in the mammalian kidney and the electric organ of eels (a tissue very rich in this enzyme; see Table 15-3). In the membrane, the enzyme is probably a dimer, each subunit containing two different polypeptide chains. One polypeptide is a 50,000-m.w. glycoprotein and the other is a 120,000-m.w. nonglycosylated polypeptide that has a catalytic site for ATP hydrolysis (Figure 15-11a). The overall process of transport involves three Na^+ ions moved out of the cell and two K^+ ions moved into the cell per ATP molecule split (Figure 15-11b).

Several lines of evidence indicate that the Na^+K^+-ATPase is responsible for the coupled movement of K^+

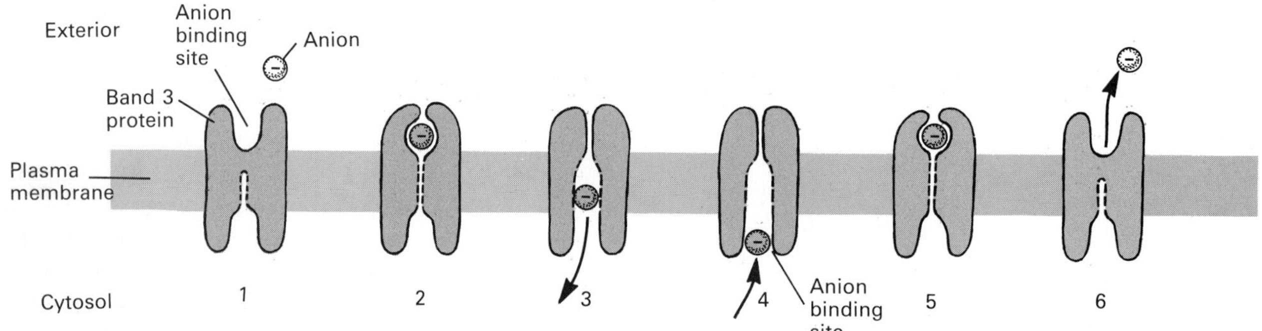

Figure 15-10 Schematic diagram of anion transport by band 3 protein. The protein catalyzes a one-for-one exchange of anions, probably shuttling between two conformations. In one form it binds an anion at the outside face (form 1), causing a conformational change that not only generates a tunnel through which the anion crosses the membrane (form 2) but also creates an anion-binding site on the cytoplasmic surface (form 3). This allows an anion to bind on the cytoplasmic side (form 4), leading to reestablishment of the original conformation, passage of the second anion across the membrane in the opposite direction (form 5) and re-creation of the binding site on the outer surface (form 6). Experiments supporting this model have shown that band 3 can have an anion binding site either on the cytoplasmic or extracytoplasmic face, but not on both faces simultaneously.

and Na^+ into and out of the cell. When a number of tissues are compared (Table 15-3), there is a good correspondence between the flux of cations across the plasma membrane and the activity of the Na^+K^+-ATPase. In addition, the cardiac glycoside ouabain specifically inhibits Na^+K^+-ATPase and also prevents cells from maintaining their Na^+/K^+ balance. Ouabain (Figure 15-12) binds to a specific region of the protein on the outer (extracytoplasmic) surface of the cell membrane. It is ineffective when injected inside cells. Mutant cells resistant to the action of ouabain have an altered Na^+K^+-ATPase. Any doubt that the Na^+K^+-ATPase is responsible for ion movement has been dispelled by the demonstration that when the enzyme is purified from the membrane and then reinserted into liposomes it will drive K^+ and Na^+ transport when supplied with ATP (Figure 15-13).

The Na^+K^+-ATPase is oriented in the membrane with part of the protein exposed at both faces. The region of the protein that faces the cytoplasm has a site for binding ATP and three high-affinity sites for binding Na^+ ions (see Figure 15-11a). The K_m, or binding constant, for the binding of Na^+ to the ATPase is 0.2 m*M*. This value is well below the intracellular Na^+ concentration, so Na^+ ions are pumped out of the cell at the maximum rate. On the outer face the ATPase has two high-affinity binding sites for K^+ ions as well as a binding site for ouabain. The binding constant for K^+ ions is about 0.05 m*M*, much lower than the extracellular K^+ concentration, indicating

Table 15-3 Comparison of cation fluxes and Na^+K^+-ATPase activities

Tissue	Temperature (°C) at which investigation conducted	Cation flux [10^{-14} mol/(cm^2 · s)]	Na^+K^+-ATPase activity [10^{-14} mol/(cm^2 · s)]	Ratio
Human erythrocytes	37	3.87	1.38	2.80
Frog toe muscle	17	985	530	1.86
Squid giant axon	19	1,200	400	3.00
Frog skin	20	19,700	6,640	2.97
Toad bladder	27	43,700	17,600	2.48
Electric eel, noninnervated membrane	23	86,100	38,800	2.22

Data are tabulated as ion flux or enzyme activity (in 10^{-14} mol/s) per square centimeter of surface membrane. Cells that pump a great deal of K^+ and Na^+ have a proportionately higher activity of the Na^+K^+-ATPase.

SOURCE: S. L. Bonting and L. L. Caravaggio, 1963, *Arch. Biochem. Biophys.* **101**:37.

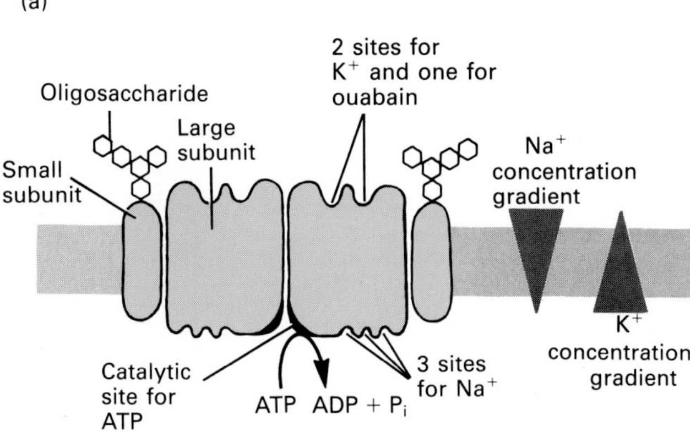

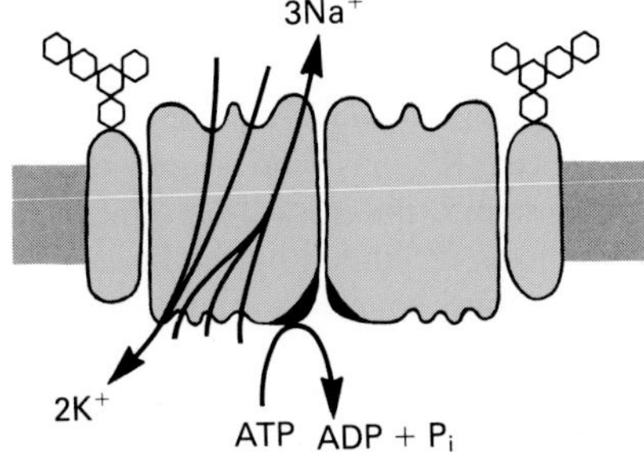

Figure 15-11 Structure and function of the Na^+K^+-ATPase. (a) The enzyme contains two copies each of a glycosylated small subunit (m.w. 55,000) and a large subunit (m.w. 120,000). The large subunit performs ion transport and has sites for two K^+ ions and also the inhibitor ouabain (see Figure 15-12) on its extracytoplasmic surface and sites for three Na^+ ions and one ATP molecule on its cytosolic side. (b) Hydrolysis of one molecule of ATP to ADP and P_i is coupled to the outward transport of three Na^+ ions and inward transport of two K^+ ions.

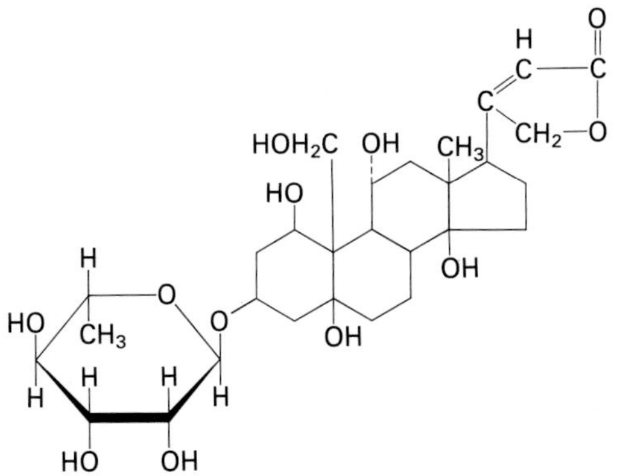

Figure 15-12 The structure of ouabain, a potent inhibitor of the Na^+K^+-ATPase. Ouabain binds to a specific site on the exterior of the large subunit.

that K^+ ions will be pumped inward at the maximum rate. But the activity of this and other cellular ion pumps is closely regulated (by mechanisms that are presently not known), so that the ionic balance of the cell is kept at a constant level. Since binding of Na^+ and K^+ to sites on the opposite faces occurs independently of each other, it appears that the protein neither rotates nor shuttles back and forth within the membrane. Using the energy derived from hydrolysis of a phosphoanhydride bond in ATP, the ATPase moves these ions through itself. The protein must contain one or more ion channels, and conformational changes in the protein, powered by ATP hydrolysis, might allow it to pump Na^+ and K^+ ions against their electrochemical gradients.

The mechanism of coupling between the Na^+ and K^+ transport reactions and ATP cleavage is not known in detail but can be divided into several steps (see Figure 15-14). First, with the protein in one conformational state, termed E_1, three Na^+ ions and ATP bind to their respective sites on the inner surface of the cell membrane.

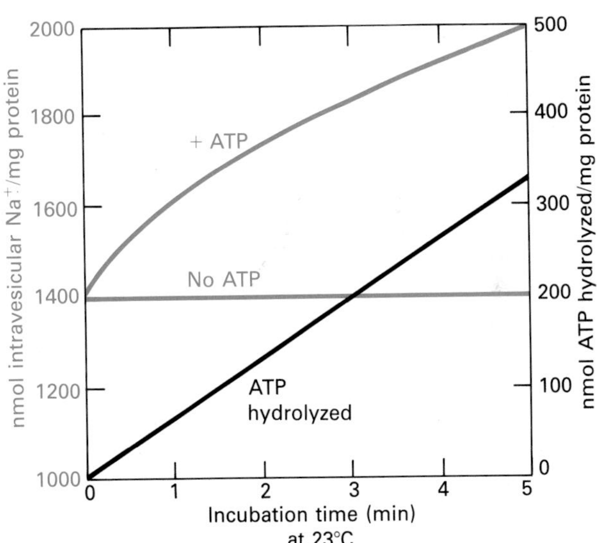

Figure 15-13 Function of the Na^+K^+-ATPase when reconstituted in pure phospholipid vesicles. Purified kidney Na^+K^+-ATPase was incorporated into liposomes with the Na^+ binding site on the outside (inside out with respect to the normal arrangement in a cell). The vesicles were loaded with K^+Cl^- solution, and radioactive Na^+Cl^- was placed on the outside. The increase in intravesicular Na^+ was measured in parallel with the hydrolysis of external ATP to ADP. Note that at early times 3 nmol of Na^+ are incorporated into vesicles per hydrolysis of 1 nmol of ATP: in 2 min, 125 nmol of ATP are hydrolyzed, and 350 nmol of Na^+ (1750 − 1400) are pumped into the vesicles. In the absence of ATP, no Na^+ accumulated in the vesicles. [See S. M. Goldin, 1977, *J. Biol. Chem.* **252**:5630.]

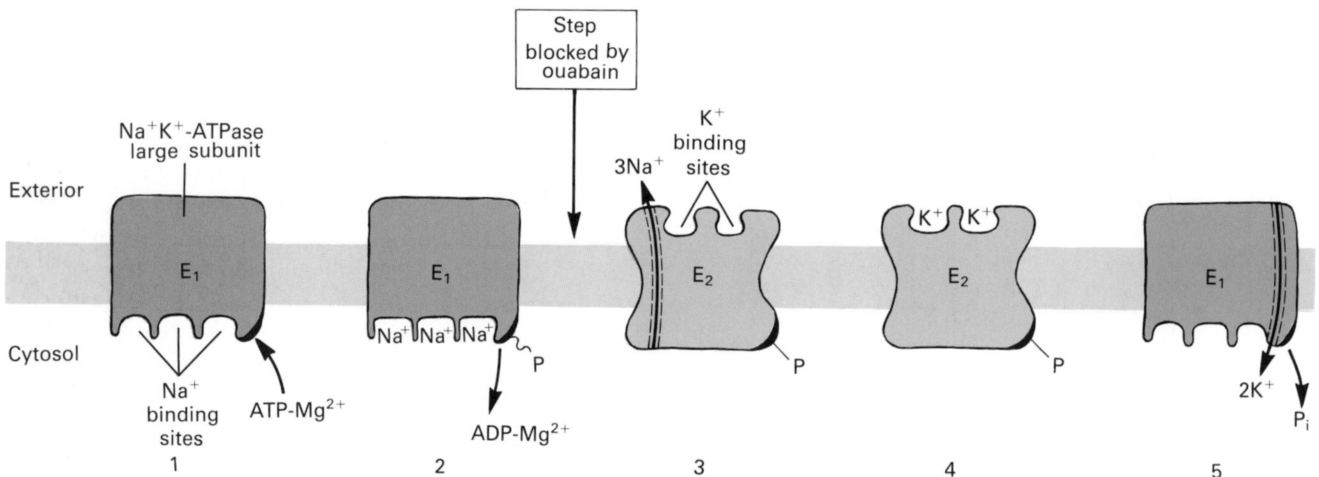

Figure 15-14 Mechanism of action of Na^+K^+-ATPase. E_1 and E_2 are conformational isomers of the ATPase. In cells each ATP and ADP molecule is in a tight complex with an Mg^{2+} ion. In step 1, three Na^+ ions and ATP (complexed to Mg^{2+}) bind to sites on the cytoplasmic surface. In step 2, the γ-phosphate of ATP is transferred to an aspartyl residue, forming a high-energy acyl phosphate bond, shown as E~P. The protein then undergoes a conformational change from E_1 to E_2 (step 3), during which the energy of the phosphate bond is reduced; it is now depicted as E-P. Simultaneously the three Na^+ ions are pumped outward. Possibly a channel is created in the ATPase through which the three bound Na^+ ions can move. In step 4, two K^+ ions bind to sites on the outer surface. In step 5, the inward movement of the bound K^+ ions is facilitated by the $E_2 \rightarrow E_1$ conformational change, coupled to the hydrolysis of the acyl-phosphate bond. The E_2-to-E_1 conversion re-creates the Na^+ binding sites on the inner surface.

In a reaction that requires the Mg^{2+} ion tightly complexed to the ATP, the bound ATP is hydrolyzed to ADP while the liberated phosphate is transferred to a specific aspartate residue in the protein, forming a high-energy acyl phosphate bond. The protein then changes its conformation to E_2, propelling the three Na^+ ions to the outside of the cell. Two K^+ ions subsequently bind to the outside face of the protein and the acyl phosphate is hydrolyzed to aspartate and free phosphate. (In the absence of K^+, the aspartyl-phosphate bond is stable, and the phosphorylated protein can be isolated and characterized.) Finally, the enzyme changes back to its original E_1 conformation, releasing the two K^+ ions into the cell. This scheme is still hypothetical, but most investigators believe that the ATP hydrolysis somehow drives conformational changes in the Na^+K^+-ATPase that allow it to transport Na^+ out and K^+ in.

Calcium ATPase Pumps Calcium Ions Out of the Cytoplasm, Maintaining a Low Concentration

The Na^+K^+-ATPase is a prototype *ion pump*. Another pump important to the proper function of cells is the calcium pump, or Ca^{2+}-ATPase. This transporter holds the Ca^{2+} concentration in cells at a low level. In some cells, such as the erythrocyte, it is located in the plasma membrane and functions to transport Ca^{2+} out of the cell. In muscle cells, a similar activity is found in the membrane of the internal organelle termed the sarcoplasmic reticulum. The Ca^{2+}-ATPase transports Ca^{2+} from the cytosol to the interior of the sarcoplasmic reticulum (see Chapter 19), which concentrates and stores Ca^{2+} ions. Release of Ca^{2+} from the sarcoplasmic reticulum into the muscle cytosol causes contraction, and rapid removal of Ca^{2+} by the Ca^{2+}-ATPase is needed for relaxation.

The Ca^{2+}-ATPase constitutes more than 80 percent of the integral membrane protein of the sarcoplasmic reticulum, a circumstance that has facilitated its purification and characterization. The calcium pump consists of a 100,000-m.w. polypeptide that goes through a phosphoaspartyl intermediate state in a mechanism closely analogous to that of the Na^+K^+-ATPase. Two Ca^{2+} ions are transported per ATP hydrolyzed. There may be a countertransport of one Mg^{2+} ion. In any case, the Ca^{2+}-ATPase requires Mg^{2+} for its function, probably to complex the ATP. The very high affinity of the cytoplasmic face of this enzyme for Ca^{2+} ion (K_m of 10^{-7} M) allows it to transport Ca^{2+} very efficiently from the cytosol ($Ca^{2+} = 10^{-7}$ to 10^{-6} M) into the sarcoplasmic reticulum, where the concentration can be as high as 10^{-2} M. The activity of this and other calcium pumps is regulated so that if the cytosolic Ca^{2+} concentration rises too high, the rate of calcium pumping increases until the concentration is reduced to less than about one micromolar. The erythrocyte plasma membrane Ca^{2+}-ATPase contains, as an essential subunit, the Ca^{2+}-binding regulatory protein termed calmodulin (see Chapter 16). A rise in cytosolic Ca^{2+} induces binding of Ca^{2+} to calmodulin; this triggers an allosteric activation of the Ca^{2+}-ATPase, accelerating the export of Ca^{2+} from the cell.

In the interior of sarcoplasmic reticulum vesicles are two soluble proteins that bind Ca^{2+} ions. One of these, calsequestrin (m.w. 44,000), is extremely acidic; 37 percent of its residues are aspartic and glutamic acid. Each molecule of calsequestrin binds 43 Ca^{2+} ions, amply justifying its name. The second protein, known as the "high-affinity Ca^{2+}-binding protein," has a somewhat lower capacity for Ca^{2+} but binds it at a higher affinity (K_m = 3 to 4 μM; K_m is the concentration of solute for half-maximal binding). These proteins serve as a reservoir for intracellular Ca^{2+}. They also reduce the concentration of free Ca^{2+} ions in the sarcoplasmic reticulum vesicles and thus decrease the energy needed to pump Ca^{2+} into them from the cytosol.

The Lysosomal Membrane Contains an ATP-Dependent Proton Pump

The third type of active transport of an ion, ATP-dependent transport of protons across a membrane, occurs in several cellular systems. It is of major importance in mitochondria and chloroplasts, where it participates in the generation of ATP from ADP. A discussion of this transport is deferred until Chapter 20. Physiological experiments suggest that an ATP-dependent proton transport system is present in lysosomal membranes and is responsible for keeping the interior of lysosomes very acidic. The pH of the interior of lysosomes in growing tissue culture cells is usually about 4.5 to 5.0. This can be measured precisely. In one technique, phagocytic cells such as macrophages are fed particles of dextran, a cross-linked glucose polymer, to which is coupled the dye fluorescein. The particles ingested by phagocytosis are transferred to the lysosomes. Fluorescein is highly fluorescent, and the ability of different wavelengths of ultraviolet light to excite this fluorescence is highly dependent on pH. By shining beams of ultraviolet light of different wavelengths on the intact cell and determining the amount of emitted fluorescence by the ingested fluorescein-dextran particles, one can measure the intralysosomal pH. It is usually between pH 4.5 and pH 5.0. The pH of the cytosol is about 7.0. The maintenance of this proton gradient of more than 100-fold between the lysosome interior and the cytosol depends on ATP production by the cell.

For investigators to show that isolated lysosomes contain an ATP-dependent proton pump, they first had to determine the pH inside a population of isolated lysosomes. An indirect method had to be found because lysosomes are less than 0.1 μm in diameter, far too small for insertion of a pH electrode. The pH of a population can be determined by adding to the suspension a weak base such as methylamine, which can cross a biological membrane in the unprotonated state (CH_3—NH_2) but not in the protonated state (CH_3—NH_3^+). As is detailed in the legend of Figure 15-15, vesicles with an interior of low pH will concentrate weak bases from the surrounding medium. The higher the concentration of methylamine in the organelle, the higher the H^+ concentration, or the lower the pH. When radioactive methylamine is added to a preparation of freshly isolated lysosomes, the lysosomes concentrate the methylamine. By measuring the concentration of methylamine in isolated organelles, it was observed that the pH of freshly isolated lysosomes is about 4.5 to 5.0 but gradually increases to pH 7.4, the pH of the medium. The addition of Mg^{2+} and ATP to the medium causes an immediate drop in the lysosomal pH to less than 5.0, as measured by a concentration of methylamine 100 times that of the medium. Thus, it was shown that isolated lysosomes contain an ATP-dependent proton pump (Figure 15-15).

A Proton Pump Causes Acidification of the Stomach Contents

The mammalian stomach contains a 0.1 *M* solution of hydrochloric acid (H^+Cl^-). This strongly acidic medium denatures many ingested proteins, facilitating their degradation by proteolytic enzymes, such as pepsin, that function at acid pH. Hydrochloric acid is secreted into the stomach by the *parietal cells* (also known as *oxyntic cells*) in the gastric lining. The details of the process are controversial, but they appear to involve an ATP-dependent proton pump in the apical membrane of the cell, the portion of the cell surface that faces the stomach lumen (Figure 15-16). Hydrolysis of ATP is coupled to the transport of H^+ ions out of the cell, but since the enzyme has not been purified it is not clear how many protons are transported per ATP hydrolyzed.

Active transport of H^+ from the cell into the lumen is apparently accompanied by a passive transport of Cl^- ion, preserving electroneutrality. The net result is a concentration of H^+Cl^- in the stomach lumen 1 million times greater than in the cell cytoplasm (pH 1.0 versus pH 7.0). The proton secreted by the parietal cell into the lumen of the stomach is derived from hydration of CO_2 within the cell by carbonic anhydrase:

$$H_2O + CO_2 \rightleftharpoons H^+ + HCO_3^-$$

When HCl is being secreted into the lumen of the stomach, the pH of the cytosol must be maintained near neutrality. Bicarbonate ion (HCO_3^-) diffuses out of the basolateral membrane of the cell in exchange for an incoming Cl^- ion by means of an anion exchanger similar in function to the erythrocyte band 3. Tight junctions

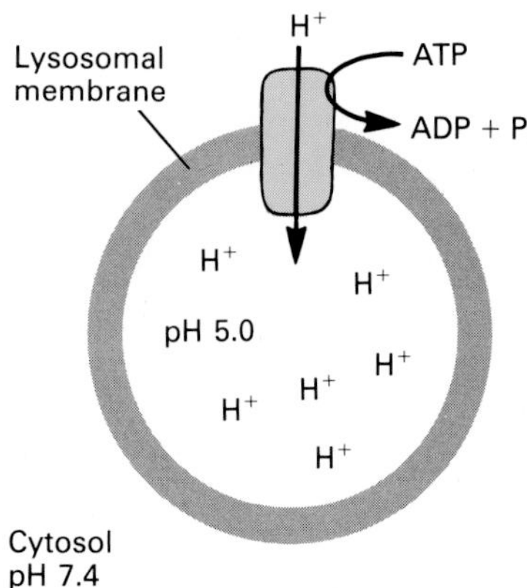

Figure 15-15 The membrane of a lysosome contains an ATP-dependent pump that transports protons from the cytosol into the lumen of the organelle.

The lysosome pH can be determined from the distribution of methylamine ($CH_3—NH_2$) across the lysosomal membrane. A weak base such as methylamine (Me) can exist in either a protonated or unprotonated form:

$$\underset{\text{Me}}{CH_3—NH_2} + H^+ \rightleftharpoons \underset{\text{MeH}^+}{CH_3NH_3{}^+}$$

$$K_a = \frac{[H^+][Me]}{[MeH^+]}$$

where K_a is the dissociation constant of methylamine. Only the *unprotonated* form can diffuse across the lysosomal membrane, and its concentration will be the same inside and outside the lysosome.

Lysosomal Membrane ↓

Medium or Cytosol | Lysosome

$$MeH^+_{out} \overset{K_a}{\rightleftharpoons} H^+_{out} + Me_{out} \rightleftharpoons \Big|\Big| \rightleftharpoons Me_{in} + H^+_{in} \overset{K_a}{\rightleftharpoons} MeH^+_{in}$$

If $[Me]_{out}$ and $[Me]_{in}$ are the concentrations of the unprotonated form in the outside and inside of the lysosome, then

$$[Me]_{out} = [Me]_{in}$$

However, the amount of the *protonated* form on each side of the membrane, $[MeH^+]_{out}$ and $[MeH^+]_{in}$, will be different if the pH and thus the proton concentration is different:

$$K_a = \frac{[H^+]_{out}[Me]_{out}}{[MeH^+]_{out}} = \frac{[H^+]_{in}[Me]_{in}}{[MeH^+]_{in}}$$

or we can write the equations as:

$$[MeH^+]_{out} = \frac{[Me]_{out}}{K_a}[H^+]_{out}$$

$$[MeH^+]_{in} = \frac{[Me]_{in}}{K_a}[H^+]_{in}$$

The total concentration of methylamine $[Me_t]$ in the cytosolic or lysosomal fluid will then be

$$[Me_t]_{out} = [Me]_{out} + [MeH^+]_{out}$$

$$= [Me]_{out} + \frac{[Me]_{out}}{K_a}[H^+]_{out}$$

$$= [Me]_{out}\left(1 + \frac{[H^+]_{out}}{K_a}\right)$$

and similarly,

$$[Me_t]_{in} = [Me]_{in}\left(1 + \frac{[H^+]_{in}}{K_a}\right)$$

But, since $[Me]_{out} = [Me]_{in}$, we have

$$\frac{[Me_t]_{in}}{[Me_t]_{out}} = \frac{1 + [H^+]_{in}/K_a}{1 + [H^+]_{out}/K_a} = \frac{K_a + [H^+]_{in}}{K_a + [H^+]_{out}}$$

An increased concentration of methylamine inside the organelle ($[Me_t]_{in} > [Me_t]_{out}$) is indicative of a high internal proton concentration ($[H^+]_{in} > [H^+]_{out}$) or a low internal pH. Quantitatively, the K_a of methylamine is about 10^{-9}. If the lysosomal pH is 5.0 and the cytosolic pH is 7.0, then $[H^+]_{in} = 10^{-5}$ and $[H^+]_{out} = 10^{-7}$.

$$\frac{[Me_t]_{in}}{[Me_t]_{out}} = \frac{10^{-9} + 10^{-5}}{10^{-9} + 10^{-7}} = 99$$

or a 99-fold concentration of methylamine inside the lysosome.

connect all of the epithelial cells, preventing H^+ ions in the lumen from diffusing into the fluids at the basolateral surface and preventing $HCO_3{}^-$ from diffusing into the lumen.

In all cells—of animals, plants, and microorganisms—the normal operation of metabolism results in a net production of H^+ ions. This arises primarily from the metabolic generation of CO_2, which is accompanied by the production of H^+ and $HCO_3{}^-$ or lactic acid or other organic acids. In many plants and microorganisms, an ATP-dependent transport of protons out of the cell appears to be of major importance in the regulation of ionic balance, and involves a proton pump whose sequence is similar to that of vertebrate Na^+ K^+ and Ca^{2+} pumps.

All ATP-Driven Ion Pumps Are Reversible

An important characteristic of all ion pumps is their reversibility. The following example involving the Ca^{2+}-ATPase of the muscle sarcoplasmic reticulum demonstrates this reversibility. During its isolation from the cell, the sarcoplasmic reticulum fragments into a series of small vesicles, each of which, like the sarcoplasmic reticulum itself, contains a high internal Ca^{2+} concentration.

(a)

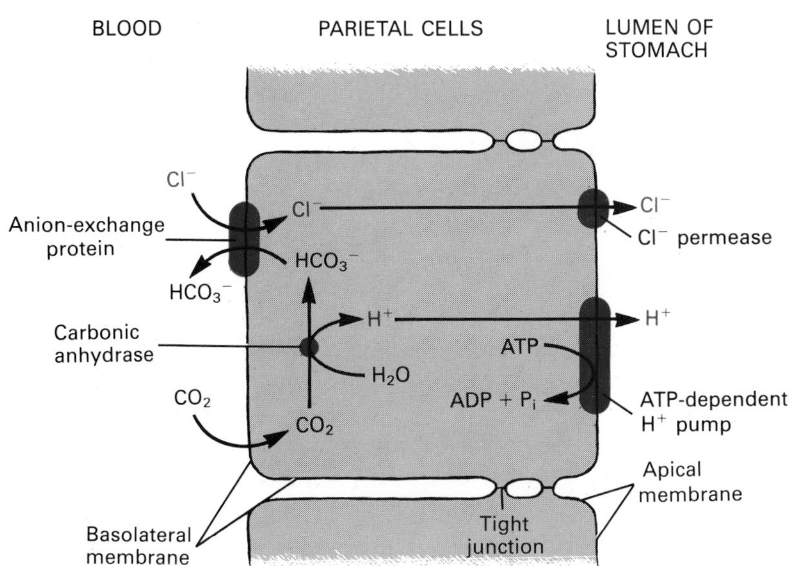

(b)

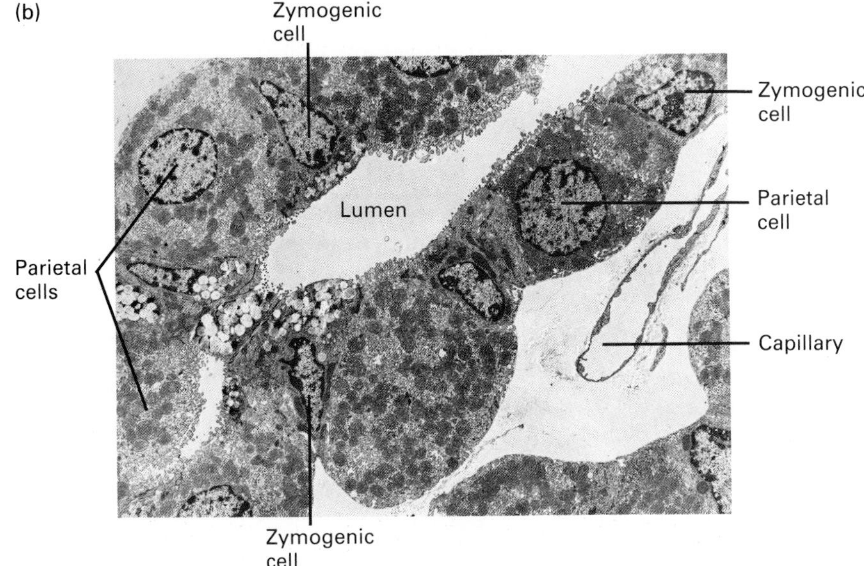

Figure 15-16 (a) Schematic representation of hydrochloric acid (H^+Cl^-) secretion by gastric epithelial cells (parietal cells). CO_2 diffuses into the cell from the blood across the basolateral membrane. Inside the cell, CO_2 is hydrated by carbonic anhydrase to H^+ and bicarbonate (HCO_3^-). A band 3–like anion exchange protein in the basolateral membrane catalyzes the exchange of HCO_3^- for Cl^- ions. The proton generated by carbonic anhydrase is pumped into the lumen of the stomach by an ATP-driven proton pump localized to the apical surface of the cell. (Possibly this is an ATP-driven H^+/K^+ exchange protein.) How Cl^- ions are transported into the lumen of the stomach is not clear, but the process presumably involves a Cl^- permease. (b) Electron micrograph of a region of the stomach wall, showing acid-secreting parietal cells as well as zymogenic cells that secrete the protease pepsinogen. Parietal cells contain microvilli on the lumenal face; the basal surface rests on a basement membrane that, in turn, is fed by a small blood capillary. Parietal cells contain abundant mitochondria that provide energy for the transcellular transport of H^+ and Cl^- ions for production of hydrochloric acid. *Part b from R. Kessel and R. Kardon, 1979,* Tissues and Organs: A Text-Atlas of Scanning Electron Microscopy, *W. H. Freeman and Company, p. 170.*

Suppose these vesicles are placed in a solution without Ca^{2+} or ATP but containing ADP and P_i. The Ca^{2+}-ATPase will transport Ca^{2+} outwards, down its concentration gradient, while ADP and P_i are combined to form ATP. This reaction

$$Ca^{2+}_{in} \xrightarrow{ADP + P_i \;\to\; ATP} Ca^{2+}_{out}$$

is a reversal of the ordinary transport function of the Ca^{2+}-ATPase:

$$Ca^{2+}_{out} \xrightarrow{ATP \;\to\; ADP + P_i} Ca^{2+}_{in}$$

When the ions run through the pump *down* the concentration gradient, the pump transfers energy from the gradient to ATP. The Na^+K^+-ATPase pump also can be driven, by concentration gradients of Na^+ and K^+ ions, to synthesize ATP. Reversal of the normal mode of action of these proteins is an important model of how ATP is synthesized in mitochondria from oxidation of carbohydrates and in chloroplasts from the energy of light. In both organelles a transmembrane gradient of protons supplies the energy for the formation of ATP from ADP and phosphate.

Cotransport: Symport and Antiport

The only known substances whose transport into or out of a cell is directly coupled to ATP hydrolysis are Na^+, K^+, Ca^{2+}, and H^+. Yet cells often must import other molecules such as glucose and amino acids against a concentration gradient. To do this, the cell utilizes the energy stored in the transmembrane gradient of Na^+ or H^+ ions. The movement of an Na^+ (or H^+) ion (the "cotransported ion") across the membrane, *down* its concentration gradient, is obligatorily coupled to movement of the "transported" molecule *up* its concentration gradient. As depicted in Figure 15-17, the transported molecule and cotransported ion can move in the same direction, in which case the process is called *symport.* When they move in opposite directions it is called *antiport.* The entry of glucose into erythrocytes, liver, and certain other cells by facilitated diffusion, uncoupled to the movement of any other ion or molecule, is called a *uniport* transport process.

Transport of Amino Acids and Glucose into Many Cells Is Directly Linked to Sodium Entry (Symport)

Importing glucose and amino acids from the lumen of the small intestine against their concentration gradients is a function of the specialized epithelial cells studded with microvilli on the side that faces the intestinal lumen (Figure 14-47a). Similar epithelial cells that line the kidney tubules absorb glucose and amino acids from the forming urine. Glucose and amino acids are transported from the lumen of the intestine into these epithelial cells and then

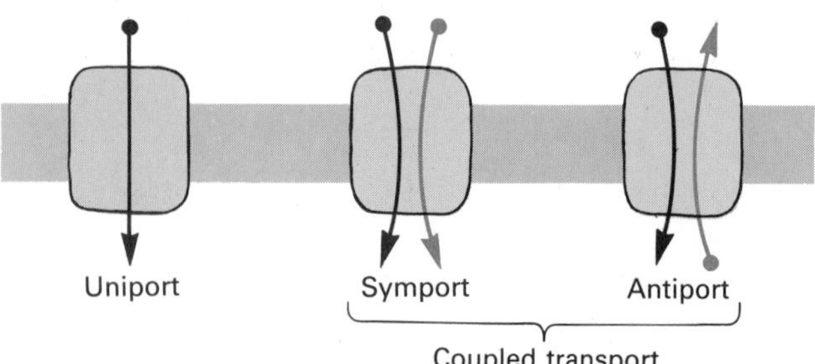

Figure 15-17 A schematic diagram of uniport, symport, and antiport membrane transport proteins. The transported molecule is shown in black; the cotransported ion is shown in color.

into the blood (Figure 15-18). Movement from the intestinal lumen to the blood is a transcellular transport process that occurs in two steps involving two sets of permeases, one in the microvilli on the apical surface membrane and the other on their basolateral surfaces. In fact, a necessary condition for transcellular transport of glucose and amino acids is that the epithelial cell be polarized, with different sets of transport proteins localized to the basolateral or apical surfaces. In particular, the Na^+K^+-ATPase is localized exclusively to the basolateral surface. To demonstrate this, radiolabeled ouabain was added to a preparation of epithelial cells. As shown by radioautography it bound only to the basolateral membranes, indicating that the Na^+K^+-ATPase is confined to that region.

The transport of glucose from the intestinal lumen across the apical surface of the brush border cells is accomplished by a specific glucose-Na^+ symport protein located in the microvillar membranes. In this system, transmembrane movement of glucose is *obligatorily* coupled to the transport of one Na^+ ion (Figure 15-18):

$$Na^+_{out} + glucose_{out} \rightleftharpoons Na^+_{in} + glucose_{in}$$

Because the concentration of Na^+ is much higher in the intestinal lumen than in the cell, the overall direction of this reaction is toward the right, even though the concentration of glucose is higher inside the cell than in the lumen. The cell utilizes the Na^+ gradient generated by the Na^+K^+-ATPase to transport glucose against its concentration gradient. A similar amino acid–Na^+ symport system uses the Na^+ gradient to take up amino acids from the lumen into the cell.

We noted that the Na^+K^+-ATPase is found exclusively in the basolateral region of the plasma membrane. In the steady state, all of the Na^+ that leaks from the lumen into the cell during symport of glucose or amino acids is pumped out across the basolateral membrane. This membrane is often called the serosal (blood-facing) membrane. Glucose and amino acids concentrated within the cell by symport also diffuse outward through the serosal membrane, presumably by facilitated diffusion. The net result is a movement of Na^+ ions, amino acids, and glucose from the intestinal lumen across the cellular epithelium and into the blood. The tight junctions between the epithelial cells prevent these molecules from diffusing back into the intestinal lumen.

In summary, the transepithelial movement of glucose and amino acids is driven by cellular hydrolysis of ATP. The coupling of ATP hydrolysis to the entry of glucose and Na^+ is indirect. ATP hydrolysis is used directly to generate an Na^+ gradient, and that gradient is used as a source of energy to drive the movement of glucose and amino acids.

How a glucose-Na^+ symport protein functions is not known. Figure 15-19 illustrates one possible mechanism.

In many body cells other than erythrocytes, and in tissue culture cells, Na^+-dependent symport is used for up-

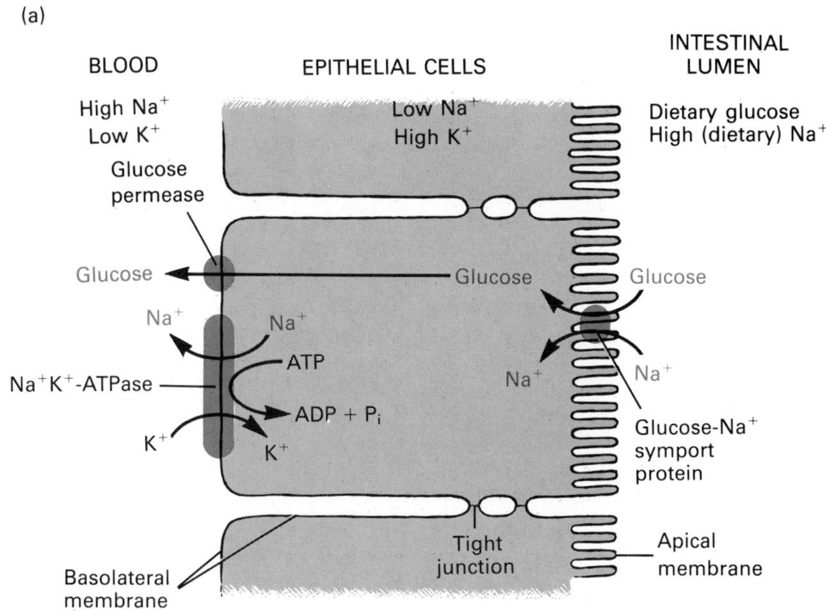

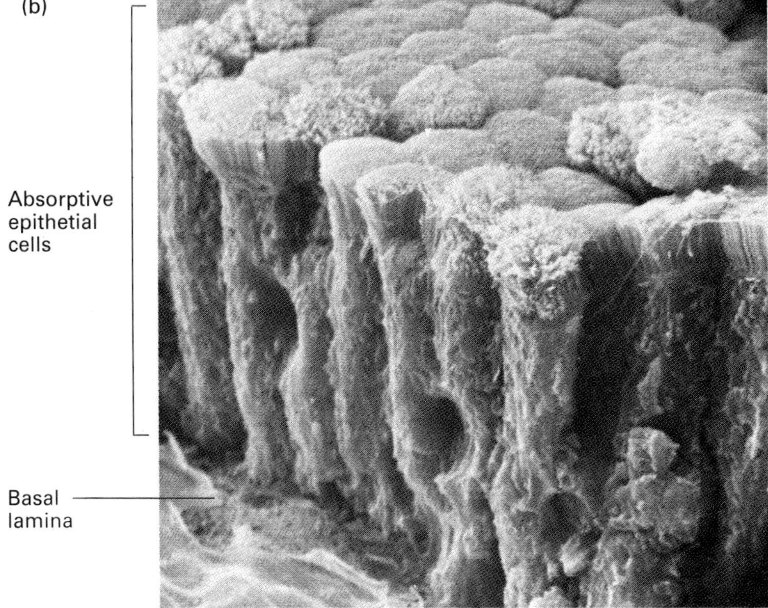

Figure 15-18 (a) Entry of glucose from the intestinal lumen to the epithelial cells is catalyzed by a glucose-Na^+ symport protein located in the apical surface membrane. The Na^+ gradient that provides the energy for glucose uptake is generated by the Na^+K^+-ATPase in the basolateral surface membrane. It pumps out the Na^+ ions that enter by glucose-Na^+ and amino acid–Na^+ symports. Glucose leaves the cell by a facilitated diffusion process across the basolateral membrane. (b) Scanning electron micrograph of the absorptive epithelial cells of the small intestine. The cells reside on a layer of basal lamina and contain abundant microvilli on their apical surface. *Part b from R. Kessel and R. Kardon, 1979,* Tissues and Organs: A Text-Atlas of Scanning Electron Microscopy, *W. H. Freeman and Company, p. 176.*

take of many amino acids, particularly those with small side chains, such as alanine.

Transport of Calcium and Protons Out of Some Cells Is Linked to the Entry of Sodium (Antiport)

In symport, the movement of one substance (usually Na^+) into the cell down a concentration gradient is coupled to the inward movement of a different substance, such as glucose, against a lesser gradient. Antiport is a similar process (see Figure 15-17) except that the inward movement of Na^+ is obligatorily coupled to the *outward* movement of a different molecule.

$$Na^+_{out} + X_{in} \rightleftharpoons Na^+_{in} + X_{out}$$

where X is exported from the cell against a concentration gradient.

An important antiport system, studied in some detail, is used by cardiac muscle cells to export Ca^{2+} ions (Figure 15-20):

$$Ca^{2+}_{in} + Na^+_{out} \rightleftharpoons Ca^{2+}_{out} + Na^+_{in}$$

This antiport reaction reduces the Ca^{2+} concentration in the heart cells. As in most muscle cells, a rise in the intra-

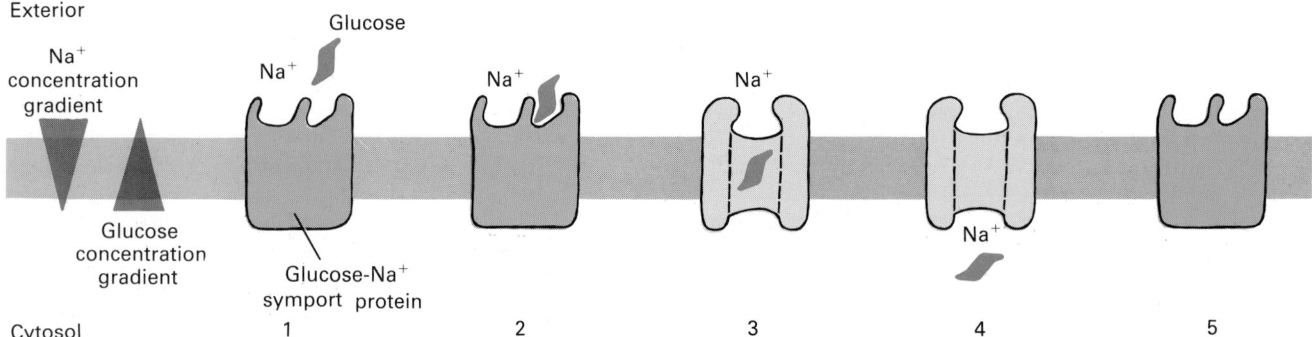

Figure 15-19 Hypothetical scheme for the operation of a glucose-Na^+ symport. In conformational tional form 1, the protein has a binding site for Na^+ and for glucose on its external (extracytoplasmic) face. Na^+ and glucose bind to these sites (form 2). Simultaneous binding induces a conformational change in the permease, generating a transmembrane pore or "tunnel" that allows both the Na^+ and glucose to pass into the cytosol (forms 3, 4). After this passage, the protein reverts to its original conformation (form 5).

cellular Ca^{2+} concentration in cardiac muscle triggers contraction. Thus, the operation of the Na^+Ca^{2+} antiport reduces the frequency of contraction of the heart muscle. The Na^+K^+-ATPase in the plasma membrane of cardiac cells, as in other body cells, creates the Na^+ gradient that is used to power Ca^{2+} export. Both the Na^+K^+-ATPase and the Na^+Ca^{2+} antiport appear to be distributed throughout the plasma membrane of cardiac muscle cells, unlike the localized distribution of the Na^+K^+-ATPase in polarized epithelial cells.

Drugs such as ouabain and digoxin are of great clinical significance. They increase the force of contraction of heart muscle and are widely used in the treatment of congestive heart failure. The primary effect of these drugs is to inhibit the Na^+K^+-ATPase. As a result, the intracellular Na^+ concentration rises. With a lower Na^+ concentration gradient the Na^+Ca^{2+} antiport system functions less efficiently, therefore fewer Ca^{2+} ions are exported, the intracellular level of Ca^{2+} ion is raised, and the muscle contracts more often and more strongly.

The plasma membrane of many animal cells contains an Na^+H^+ antiport that may be important in regulating cytosolic pH. Entry of an Na^+ ion into the cell (down its concentration gradient) is obligatorily coupled to the export of a proton, or H^+ ion (Figure 15-20). Since the intracellular pH and that of the surrounding medium are, for animal cells, generally about the same (pH ~7.0), expenditure of energy is usually required to pump the protons out of the cell. We recall that the anaerobic metabolism of glucose, a nonacidic compound, yields lactic acid, and aerobic metabolism yields carbon dioxide that is hydrated to carbonic acid (H_2CO_3). These weak acids

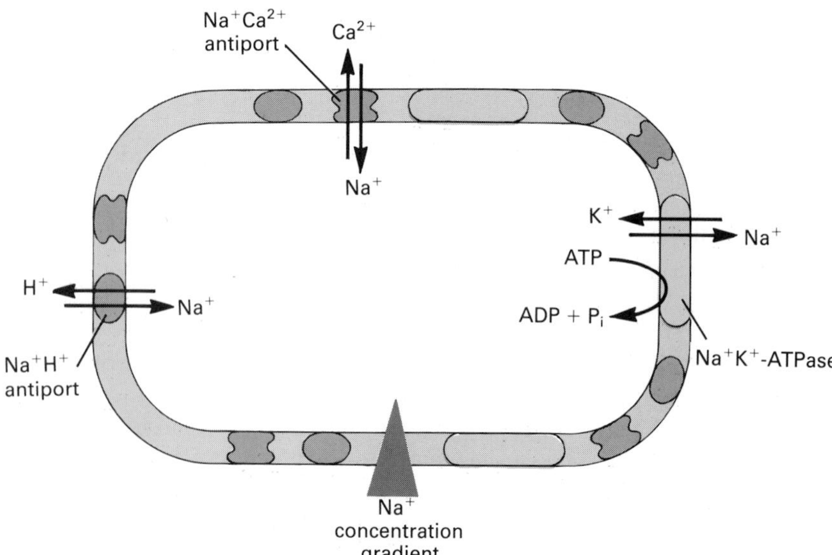

Figure 15-20 Na^+Ca^{2+} and Na^+H^+ antiports in the membranes of many animal cells are used to pump Ca^{2+} and H^+ out of the cytosol. They use energy stored in the Na^+ concentration gradient generated by the Na^+K^+-ATPase. These transporters are believed to be distributed uniformly in the plasma membrane of fibroblasts and muscle cells.

dissociate, yielding protons. Were these protons not exported from the cell, the pH of the cytoplasm would drop precipitously. The Na^+H^+ antiport is believed to be important in removing some of the "excess" protons generated during metabolism.

Small changes in the cytoplasmic pH may have profound effects on the overall rate of cellular metabolism. For instance, primary fibroblast cells grown to confluence in tissue culture generally become quiescent. DNA synthesis stops; the rates of glucose catabolism, RNA synthesis, and protein synthesis are reduced, and the cytoplasmic pH drops from the 7.4 characteristic of growing cells to about 7.2. Contact with other cells or a reduction in the level of serum growth factors may cause these metabolic reductions in cells that are in a stationary phase. Treatment of quiescent cells with a mixture of growth factors from serum restimulates cell growth and DNA synthesis. One of the earliest effects of these growth factors is a marked increase in the cytoplasmic pH to 7.4. This represents a 40 percent decrease in the cytoplasmic concentration of protons, and it occurs within minutes. This dramatic change is apparently caused by a stimulation of the Na^+H^+ antiport, which expels protons into the medium. The rise in cytoplasmic pH is believed to be important in triggering the activation of cellular metabolic pathways required for cell growth and division.

Transport into Prokaryotic Cells

Transport of nutrients into bacterial cells is a more formidable problem than transport into mammalian cells for two main reasons. First, the blood that bathes mammalian cells has a finely regulated concentration of glucose, amino acids, and other nutrients, whereas bacteria are often subjected to media of widely differing compositions. The common intestinal bacterium *Escherichia coli,* for instance, also grows in soil and in freshwater lakes, environments with very different nutrient compositions. Second, bacteria often must concentrate nutrients such as sugars, amino acids, and vitamins against very high concentration gradients. Bacteria have evolved systems for nutrient uptake that can function even at a 100-fold (or more) concentration gradient. Generally these permeases are inducible; the quantities of a transport protein in the cell membrane are regulated according to the concentrations of the nutrient in the medium and according to the metabolic needs of the cell.

Bacteria use two very different systems for nutrient uptake. One is symport, but an H^+ gradient across the bacterial membrane is used to power nutrient uptake instead of the Na^+ gradient used in mammalian cells. The other is a system unique to bacteria (so far as is known), in which phosphorylation of a sugar is part of the uptake mechanism.

Proton Symport Systems Import Many Nutrients into Bacteria

Many aerobic bacteria derive their energy from the oxidation of glucose to CO_2. During this process electrons are transferred from NADH (nicotinamide adenine dinucleotide, reduced) and pyruvate, key metabolic intermediates, to oxygen, the ultimate electron acceptor. The electron transporters are integral proteins in the bacterial plasma membrane. As electrons move along the transport chain, protons are pumped out of the cell. The result can be a 100-fold increase in the H^+ concentration in the surrounding medium relative to the cytoplasm—the pH of the medium will become 5.0 to 5.5, while the cytoplasm pH remains about 7.3. The energy stored in this proton gradient is used for many purposes. One principal use is the generation of ATP from ADP and P_i (detailed in Chapter 20). Another is the import of nutrients from the medium. In *Escherichia coli,* the sugar lactose is concentrated from the medium by a pump called the *lactose permease,* a 30,000-m.w. integral protein encoded by the *y* gene of the *lac* operon (Figure 8-16). A flow of protons from the outside through the membrane (down the proton concentration gradient) is coupled to lactose uptake (Figure 15-21). Thus, the lactose permease is an H^+-lactose symport.

This coupled transport can be demonstrated experi-

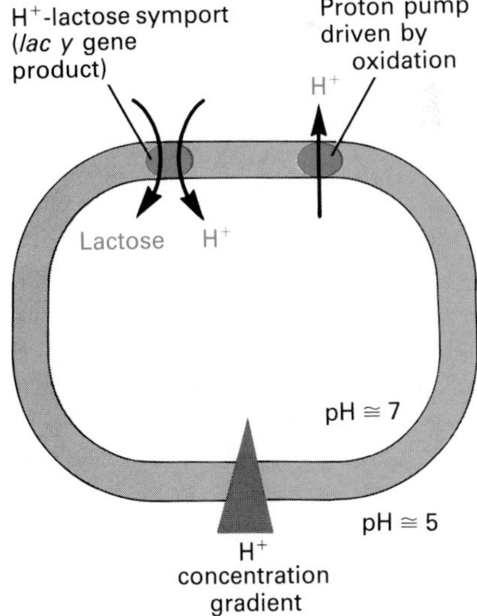

Figure 15-21 The import of lactose into *Escherichia coli* is fueled by the energy stored in a proton gradient across the plasma membrane. Oxidation of pyruvate to CO_2 is coupled to the pumping of protons out of the cytosol, generating as much as a 100-fold H^+ gradient across the membrane. Lactose is imported by an H^+-lactose symport process (catalyzed by the *lac y* gene product). Inward movement of protons is obligatorily coupled to inward movement of lactose.

mentally. *E. coli* cells poisoned with inhibitors of oxidative phosphorylation, such as cyanide, are unable to acidify the external medium and do not concentrate lactose from the medium. The pH of the medium and of the cytosol come to be the same. When the medium is acidified by the addition of HCl, the cells can again concentrate lactose, using the energy of this artificially imposed proton gradient.

Certain Molecules Are Phosphorylated during Passage across the Cell Membrane

When eukaryotic cells take up glucose or amino acids, either by facilitated diffusion or by antiport, the molecules are not usually modified chemically during the transport process. After entering the cell, however, sugars are phosphorylated and amino acids are bound to tRNAs. These reactions lower the intracellular concentration of the unmodified substance and facilitate inward movement of more of the compound. Bacterial cells, by contrast, use transport systems in which the nutrient is chemically modified during transport. This modified substance accumulates in the cytoplasm and cannot pass across the plasma membrane into the medium. This process is called *group translocation.* Since the concentration of the *unmodified* nutrient within the cell is always very low, such systems are not, strictly speaking, active transport, although energy in the form of the phosphoanhydride bond is used and the concentration of the modified nutrient in the cell greatly exceeds that of the unmodified form in the medium.

One well-studied example of a group translocation reaction is the phosphotransferase system used by *E. coli* and other bacteria for the concentration of sugars such as glucose, mannose, *N*-acetylglucosamine, fructose, and maltose (though not lactose), and for the concentration of the three-carbon compound glycerol. At least four proteins participate in this process (Figure 15-22). Two are soluble proteins used for the uptake of all such compounds: *enzyme I* and a small protein called *HPr.* In the first step of glucose transport, enzyme I catalyzes the transfer of a phosphate group from phosphoenolpyruvate (PEP) to a histidine residue in the HPr protein. This system is unusual in that it uses PEP rather than ATP as a phosphate donor, but the free energy of hydrolysis of the phosphoanhydride bond in PEP is greater than that of ATP, and PEP is an intermediate in the catabolism of glucose to pyruvate (see Figure 5-42).

The other two proteins, *enzyme II* and *enzyme III,* mediate the second step of glucose transport, the actual transmembrane uptake of sugars. First, HPr-PO_3H^- transfers its phosphate to enzyme III. Enzyme III-PO_3H^- in turn phosphorylates the appropriate sugar. Enzyme II apparently forms the transmembrane channel, and phos-

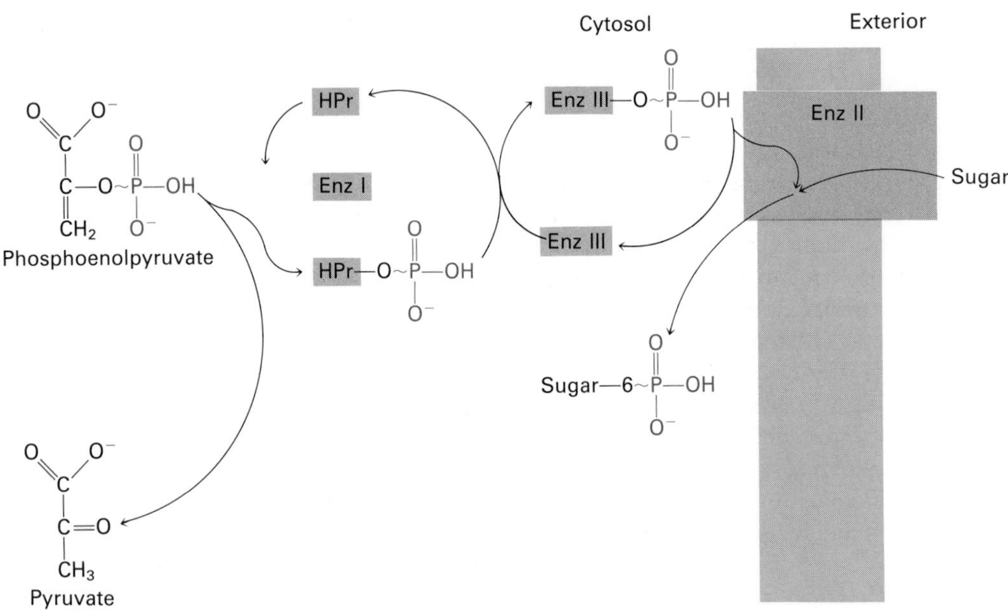

Figure 15-22 Import of sugars by *E. coli:* the phosphotransferase system. Each type of sugar, say glucose and fructose, is imported by a different type of enzyme II, an integral protein in the plasma membrane. As the sugars are transported into the cell they are phosphorylated on the 6 carbon atom. The donor of the phosphate is the phosphorylated form of enzyme III. A different enzyme III species is used to phosphorylate each sugar. As is described in the text, enzyme III's are phosphorylated using phosphoenolpyruvate (PEP) as the phosphate donor and the proteins enzyme I and HPr as intermediates.

phorylation of the sugar by enzyme III somehow occurs concomitantly with transport of the sugar into the cell.

Enzymes II and III are specific for particular sugars, whereas HPr and enzyme I are involved in the transport of all sugars. For instance, different enzymes II and III transport glucose and fructose. Genetic studies on sugar transport were essential to establish these conclusions. Bacterial mutants defective in the transport of just one sugar, such as glucose, lack functional enzyme II_{glu} or enzyme III_{glu}. Mutants that lack functional HPr or enzyme I protein are defective in the transport of several sugars.

Movement of Water across Cell Layers

The bulk flow of water across a layer of cells is a feature of many biological systems. For example, in humans there is a net movement of water from the blood, across the layers of epithelial cells that line the stomach, into the lumen of the stomach. In these and similar systems, movement of water in a specific direction depends on the metabolic activity of the epithelial cells. These epithelial cells, in effect, "secrete" water into the stomach lumen.

The net movement of water across a layer of cells is caused by either of two circumstances. One is a difference in the concentration of dissolved substances in the two solutions, since water moves by osmosis from a solution of lower solute concentration to one of higher concentration (see Figure 5-41). A difference in hydrostatic pressure between the two solutions will also cause water flow. Both factors contribute to the flow of water into the stomach lumen through the surrounding epithelial cells.

Osmotic Pressure Causes Movement of Water across One or More Membranes

Osmotic pressure, a main cause of the movement of water across cell membranes and cell layers, is defined as the hydrostatic pressure required to stop the net flow of water across a membrane when the membrane separates solutions of different composition. In this context, the membrane may be a layer of cells or a plasma membrane. The osmotic pressure of a dilute solution is given approximately by the van't Hoff equation

$$\Pi = RT(C_1 + C_2 + C_3 + \cdots + C_n)$$

where Π is the osmotic pressure (in units of atmospheres, or millimeters of mercury), R is the gas constant, T is the absolute temperature, and $C_1, \ldots, C_n$ are the molar concentrations of all the solutes—ions or molecules—in the solution. It is the total number of solute molecules that is important. For example, a solution of 0.5 M NaCl is actually 0.5 M Na^+ ions and 0.5 M Cl^- ions, and it has approximately the same osmotic pressure as a 1 M glucose or lactose solution.

If the membrane is permeable to water but not to solutes, the osmotic pressure across the membrane is simply

$$\Pi = RT\,\Delta C$$

where ΔC is the difference in total solute concentration on the two sides (Figure 15-23). A concentration gradient of 10 mM sucrose (or 5 mM NaCl) produces a water flow across a semipermeable membrane that is balanced by a hydrostatic pressure of 0.22 atm, or 167 mmHg.

Movement of Water Accompanies Transport of Ions or Other Solutes

To illustrate net flow of water across a layer of cells, let us consider again the epithelial cells that line the stomach. The parietal cells actively secrete H^+ into the lumen. This causes a flow of Cl^- ions from the cell to the lumen, which is necessary for preserving electroneutrality. Thus, there is a net flow of H^+Cl^- from the cell to the lumen, a circumstance that generates a gradient of osmotic pressure across the cell membrane and causes water to flow from the cell into the lumen. These osmotic effects are quite large. About 300 to 400 water molecules must be

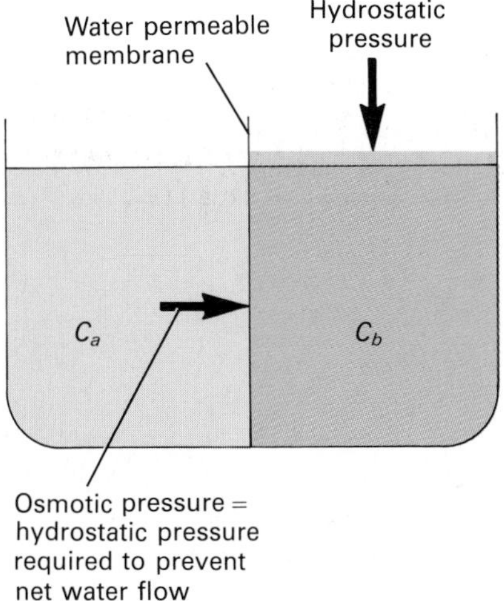

Figure 15-23 Osmotic pressure. In this example, solutions a and b are separated by a membrane that is permeable to water but not to any of the solutes. Suppose C_b, the total concentration of solutes in solution b, is greater than C_a. Water will tend to flow across the membrane from solution a to solution b. The osmotic pressure Π between the solutions is the hydrostatic pressure that one would have to apply to solution b to prevent this water flow. From van't Hoff's equation, this pressure is $\Pi = RT(C_b - C_a)$.

moved per H^+ or Cl^- ion pumped in order to maintain equal osmotic pressure on both sides of the apical (lumen-facing) cell membrane.

A major force pushing water from the blood *into* the parietal cell is a pressure gradient. The hydrostatic pressure in capillaries varies from 15 to 35 mmHg, while in the cell the pressure is nearly zero. Thus, water moves into the parietal cell by hydrostatic pressure and out of it into the lumen by osmotic pressure following the secretion of H^+ and Cl^- ions.

The Internalization of Macromolecules and Particles

Cells exchange with their environment not only small molecules, such as inorganic ions or sugars, but also macromolecules, particularly proteins, and even particles several micrometers in size. Cells secrete various proteins (e.g., peptide hormones, serum proteins) and extracellular structural proteins (e.g., collagen) by a process called *exocytosis,* the fusion of an intracellular vesicle with the plasma membrane and the release of the contents of the vesicle into the environment (see Figure 5-34). We examine exocytosis in Chapter 21. The processes by which cells bind and internalize macromolecules and particles from the environment involve endocytosis and phagocytosis (Figure 15-24).

In older usage, endocytosis referred to any process by which a region of the plasma membrane enveloped an external particle or a sample of the external medium, forming an intracellular vesicle. Recognizing that phagocytosis is a fundamentally different process, we now use the term endocytosis to refer to the situation where only a small region of the plasma membrane folds inward, or *invaginates,* until it has formed a new intracellular membrane-limited vesicle about 0.1 μm in diameter. There are two types of endocytic processes. *Pinocytosis* is the nonspecific uptake of small droplets of extracellular fluid by such vesicles. Any material dissolved in the extracellular fluid is internalized in proportion to its concentration in the fluid. In *receptor-mediated endocytosis,* a specific receptor on the surface of the membrane "recognizes" an extracellular macromolecule and binds with it. The substance bound with the receptor is referred to as the *ligand.* The region of plasma membrane containing the receptor-ligand complex undergoes endocytosis. The same endocytic vesicle can be used for both pinocytosis and receptor-mediated endocytosis; whether a macromolecule is said to enter the vesicle by one process or the other depends on whether it first binds to a specific receptor on the cell surface.

Phagocytosis is the intake of large particles such as bacteria or parts of broken cells—or in experimental situations, even tiny plastic beads. Phagocytosis is used by many protozoans to ingest food particles, and by certain white blood cells called *macrophages* to take in and destroy bacteria. First the target particle is bound to the cell's surface, then the plasma membrane expands along the surface of the particle and eventually engulfs it. Vesicles formed by phagocytosis are typically 1 to 2 μm or

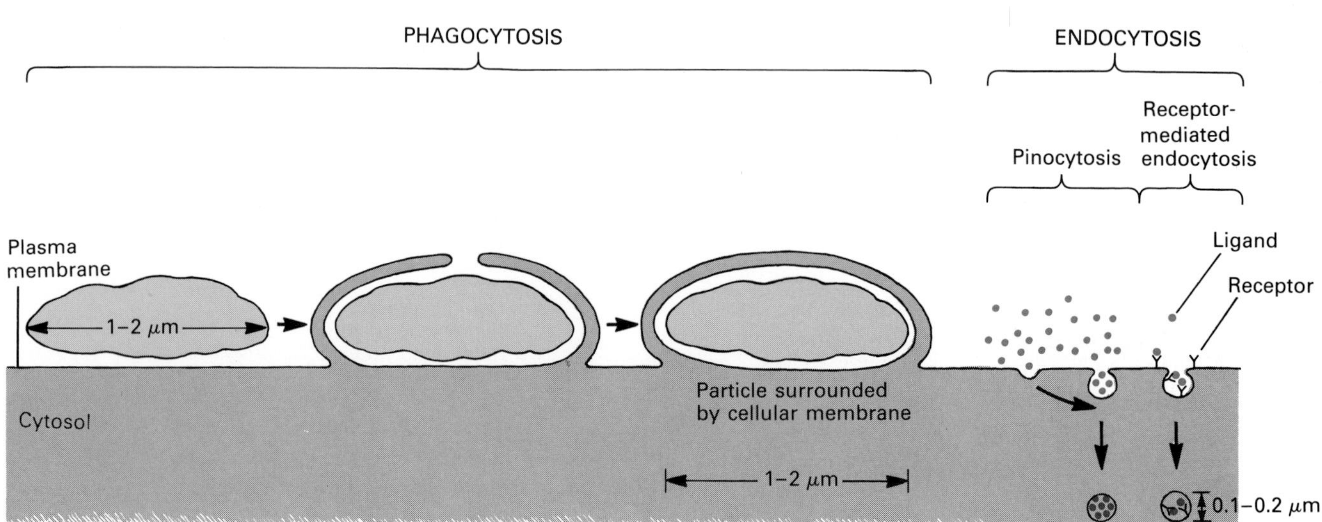

Figure 15-24 The processes of phagocytosis and endocytosis. Endocytosis may involve the specific internalization of macromolecules that are bound to cell surface receptors, a process termed *receptor-mediated endocytosis. Pinocytosis* is the nonspecific uptake of extracellular molecules in an endocytic vesicle; the rate of uptake of molecules is simply proportional to their concentration in the extracellular solution. *Phagocytosis* involves engulfing the particles, which are bound to the plasma membrane surface, by an expansion of the membrane around them.

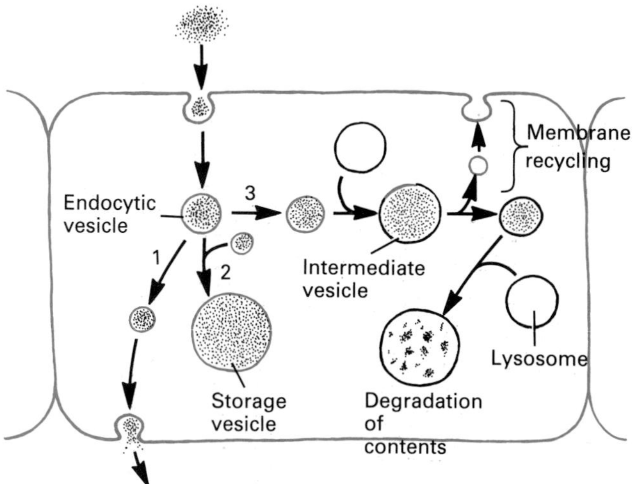

Figure 15-25 Possible fates of endocytic vesicles. The vesicle can be transported to the other side of the cell and exocytose its contents intact (pathway 1). Or it can fuse with other vesicles (pathway 2) and store its contents in the cell. Most often the endocytic vesicle fuses with other cellular vesicles (pathway 3) to form an intermediate transport vesicle. The region containing the plasma membrane receptor (color) buds off to form a separate vesicle (color) and is recycled to the cell surface, while the remainder of the vesicle, with the ingested material, fuses with a lysosome. In the lysosome, hydrolytic enzymes degrade the ingested material to small molecules such as amino acids.

larger in diameter, much larger than those formed by endocytosis. Another difference is that in phagocytosis the expansion of the plasma membrane around the phagocytosed particle requires the active participation of actin-containing microfilaments that lie just under the cell surface; endocytosis does not involve these filaments.

Depending on its structure and content, the new intracellular vesicle formed by endocytosis or phagocytosis may either fuse with an organelle, such as the Golgi complex or a lysosome, or return to the plasma membrane (Figure 15-25; see also Figure 5-38). Usually the vesicles ultimately fuse with lysosomes, where the ingested materials are degraded. There are many important exceptions to this outcome, but before plunging into the specifics of endocytosis and phagocytosis we need to discuss the mechanisms of a process that is basic to an understanding not only of vesicle formation but the formation and fate of all cellular membranes: the fusion of two membranes.

Membrane Fusions Occur in Endocytosis, Exocytosis, and Many Other Cellular Phenomena

Exocytosis, endocytosis, and phagocytosis all involve the fusion of membranes. In exocytosis a small vesicle must fuse with a region of the plasma membrane. In endocytosis and phagocytosis one part of a membrane region fuses with another part of the region to "bud off" an intracellular vesicle (Figure 15-26). Membrane fusion is involved in other crucial cellular events: cell separation during mitosis, for example, or the fusion of individual myoblasts to form a multinucleated muscle cell (Figure 15-26).

The fusion of membranes is not a simple process, as membranes do not spontaneously coalesce. Liposomes do not fuse, nor do cells in the absence of special treatment. If organelles are to maintain their integrity (and they generally do), they must be protected from fusing randomly with each other. One obstacle to membrane fusion is the

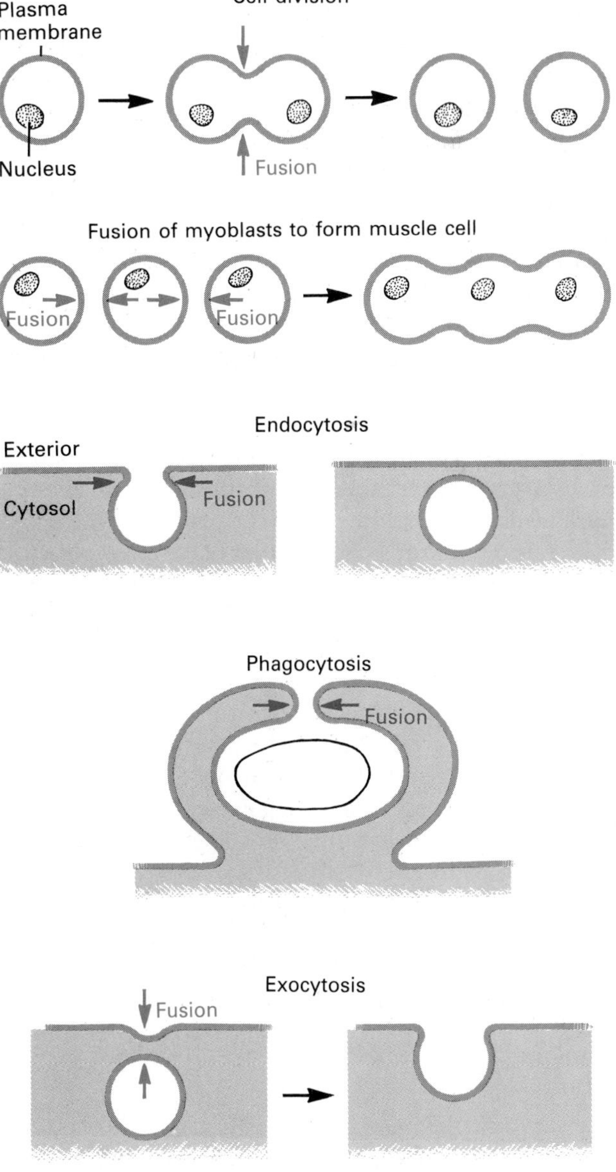

Figure 15-26 Examples of membrane fusion events. The colored arrows indicate the points of fusion of two membrane regions.

high negative surface charge of the phosphate groups on the phospholipids and on the sialic acid residues found in cell surface glycolipids, which causes membranes to repel one another. Another obstacle is that the charges on surface proteins of membranes also inhibit intimate contact. A third obstacle is the absence of free edges in membranes—for fusion to occur, continuous sheets or spheres of membrane must be made to open (Figure 15-27).

Membranes can be made to fuse by artificial means. Experiments with fusing live cells have shown that two different processes are involved: the coming into close proximity, and the actual fusion. Plant lectins, multivalent proteins that bind to carbohydrates, can induce proximity by binding to surface glycoprotein and glycolipid molecules on cells, thus holding the membranes together. Several proteins purified from cells have, along with Ca^{2+}, caused membranes to adhere tightly together. Whether these proteins actually participate in membrane fusion is not known. One reason for supposing that they do is that Ca^{2+} ion triggers many membrane fusion events, such as exocytosis.

When a second substance is added, the adhering membranes somehow fuse. Polyethylene glycol can act as such a *fusion factor,* but how it does so is not known. Viral glycoproteins also act as fusion factors.

The mechanism of fusion is still obscure. Electron micrographs taken during natural fusion events such as exocytosis have detected intermediates with a trilaminar structure, interpreted as partly fused membranes in which the opposing layers of the two membranes are in close contact (step 2, Figure 15-27). Freeze-fracture studies have suggested that these regions of contact are bereft of intramembranous protein particles, and that the subsequent membrane fusion would actually take place between regions of protein-depleted membranes. Such a trilaminar structure could be an intermediate for either fusion or membrane separation. This view of fusion is not universally accepted. In studying stimulated exocytosis of secretory granules in mast cells, some researchers find that the initial region of contact between vesicle and membrane is through a single narrow-necked pore. This pore gradually widens, allowing exocytosis of the vesicle content. There appears to be no depletion of proteins in this region of fusion.

Let us now return to our discussion of the internalization of macromolecules and particles, starting with phagocytosis.

Phagocytosis Depends on Specific Interactions at the Cell Surface

Phagocytosis, the ingestion of particles up to several micrometers in diameter, is a property of many types of cells in the living world. Many unicellular eukaryotic organisms, such as amebas and slime molds, grow by ingesting bacteria (and occasionally eukaryotic microorganisms). Degradation of such foodstuffs in the lysosomes provides the organism with sugar, amino acids, lipids, and other nutrients that it utilizes for biosynthetic reactions. Animals contain several types of cells called phago-

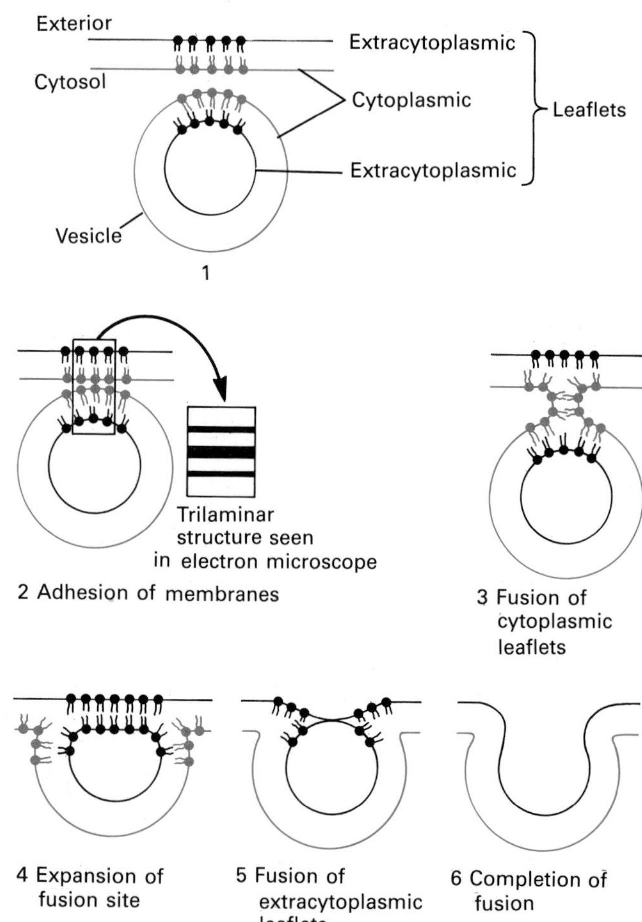

Figure 15-27 Hypothetical scheme for membrane fusion, as deduced from electron microscopic images of exocytosis of secretory granules. Step 1 is the situation before fusion, illustrating the faces of the two membranes. In step 2, the vesicle moves close to the plasma membrane and adheres to it. In the transmission electron microscope this image may appear trilaminar, as if the two opposing cytoplasmic leaflets merge into one. Next (step 3) the cytoplasmic leaflets of the plasma membrane and vesicle membrane fuse, forming a continuous leaflet. The point of contact between the two membranes widens slightly (step 4), so that the hydrophobic regions of the extracytoplasmic leaflets of the two membranes contact each other. Finally the two extracytoplasmic leaflets fuse (step 5) and the membrane fusion event is completed (step 6). A similar series of fusions would occur during endocytosis or cell-to-cell fusion, except that the *extracytoplasmic* leaflets of the two membranes would fuse first (see Figure 15-26).

cytes; their main function is to rid their own organism of such particles as antigen-antibody complexes and such harmful intruders as bacteria and viruses (Figure 15-28). Some macrophages line the sinusoids, the narrow blood channels of the liver, spleen, and other organs, where they "filter out" particles from the passing blood. Such macrophages also devour aging erythrocytes. Each human erythrocyte survives about 120 days; about 10^{11} are destroyed each day by macrophages. Other phagocytic white blood cells called monocytes migrate into tissue areas of infection or inflammation, where they differentiate into macrophages that ingest and degrade broken cells, killed microorganisms, and other debris.

Macrophages are "gourmets" in that they bind and ingest only certain of the particles available to them. Ingestion is very much dependent on the charge, hydrophobicity, and chemical composition of the particle's surface, but the exact requirements are not totally understood. Many pathogenic bacteria, such as *Streptococcus,* have carbohydrate surface structures, called capsules, that inhibit their being bound to and ingested by macrophages. Nonpathogenic strains of the same bacteria lack these capsules and are readily ingested.

In response to infection by pathogenic bacteria, the host organism usually produces serum antibodies (see Chapter 3) that bind to specific proteins or carbohydrates on the bacterial surface, coating the bacteria. An antibody coat on a bacterium or other particle will stimulate its ingestion by macrophages, because macrophages have specific surface receptors that bind to a region common to all antibody molecules, the stem of the Y-shaped structure called the Fc region (see Figure 3-38). Other serum proteins, collectively called *complement,* coat foreign particles nonspecifically. They stimulate ingestion by binding to specific receptors on the macrophage called *complement receptors.*

After the particle has been bound to the cell surface, ingestion commences with the continuous apposition of a segment of the plasma membrane to the particle's surface, excluding most if not all of the surrounding fluid. Binding of a particle to the macrophage does not trigger its immediate engulfment. Rather, the envelopment of the particle proceeds by the sequential and circumferential interaction of receptors on the surface of the macrophage with ligands distributed uniformly over the particle's surface. The effect of the receptors and ligands linking the two

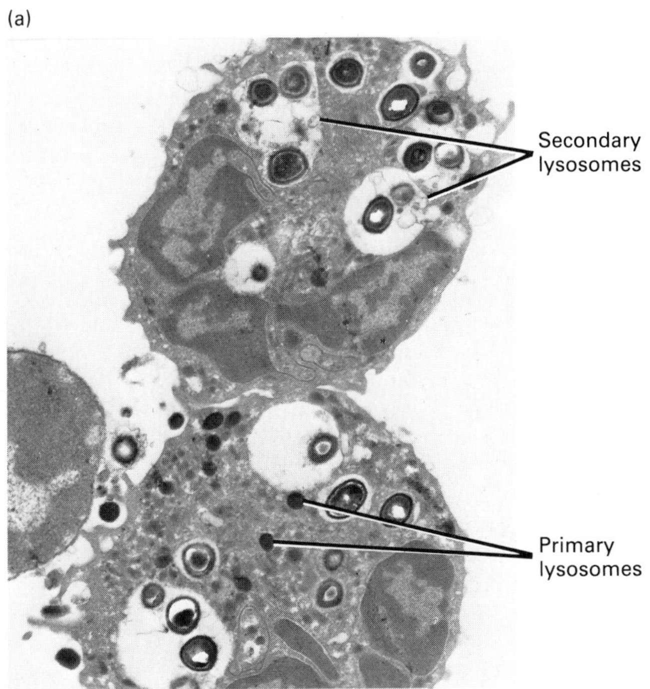

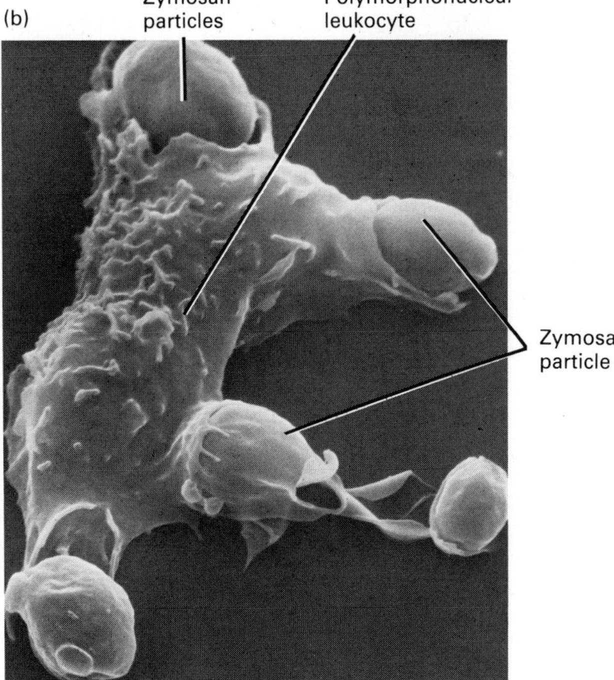

Figure 15-28 Human polymorphonuclear leukocytes (neutrophils), a class of phagocytic white blood cells. (a) A thin-section micrograph. The dark bodies are primary lysosomes. The white bodies are secondary lysosomes containing degradation products of ingested staphylococci. The secondary lysosomes arise by fusion of primary lysosomes with the phagocytic vesicles containing the bacteria. The surface of the neutrophil contains many thin folds, or ruffles, that were immobilized by the fixative in various phases of their phagocytic activity. (b) A scanning electron micrograph of a human polymorphonuclear leukocyte ingesting small particles of zymosan, pieces of yeast cell wall. *Part (a) Biophoto Associates; part (b) courtesy of M. J. Karnofsky.*

surfaces is like that of a zipper closing, and it is called the *zipper interaction* (Figure 15-29a).

Phagocytosis is a localized response of one region of the macrophage plasma membrane and its underlying cytoskeleton. The surface proteins of lymphocytes can be modified in such a way that they bind tightly to macrophages, but the lymphocyte cells will not be ingested. Antibody-coated bacteria subsequently fed to the macrophages will be actively engulfed, but the macrophages still will not ingest the lymphocytes, even if a lymphocyte happens to be bound next to an antibody-coated bacterium. Furthermore, if the target particle is only partly covered with antibodies (or other molecules that stimulate phagocytosis), the macrophage plasma membrane will envelop only the coated part of the particle (Figure 15-29b).

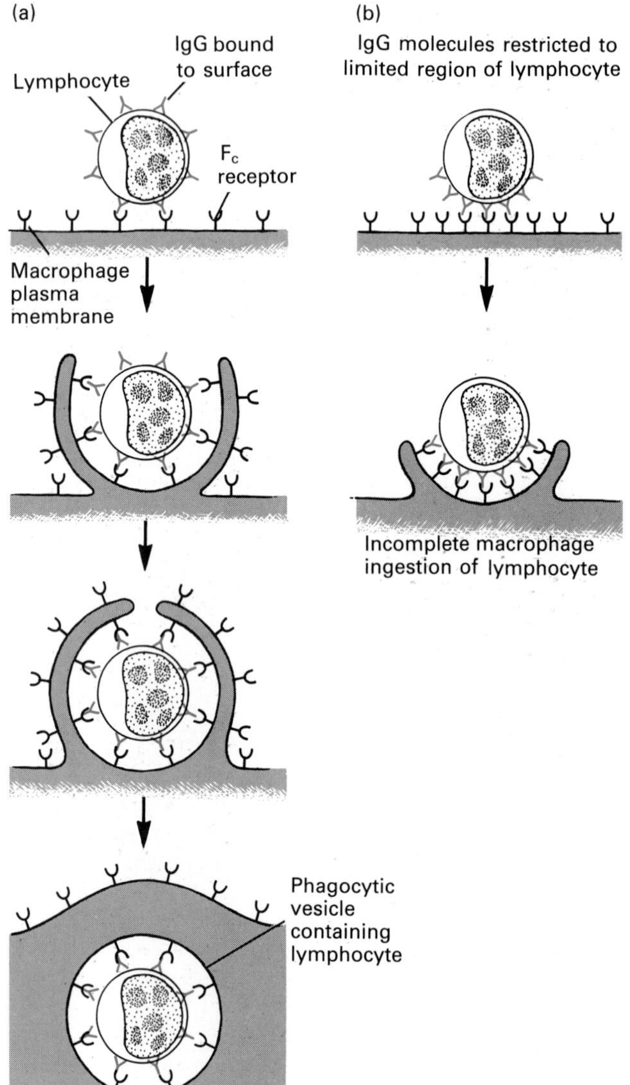

Figure 15-29 Diagram of the zipper interaction. (a) If the target lymphocyte has IgG antibody all over its surface, the macrophage plasma membrane will completely engulf it. (b) If the antibody molecules are localized by "capping" to one region of the surface, the plasma membrane will spread only over the antibody-coated region and the lymphocyte will not be ingested.

Once ingested, the membrane-enveloped particles are delivered to lysosomes by fusion of the phagocytic vesicle with the lysosome membrane, and they are usually destroyed. After fusion with one or more phagocytic vesicles, lysosomes are customarily called "secondary lysosomes" to distinguish them from the smaller, unfused "primary lysosomes." Some parts of the ingested material, such as the cell walls of microorganisms, are resistant to lysosomal hydrolases. They accumulate within the secondary lysosomes as residual bodies. Accumulation of these residual bodies may be one reason why macrophages have a very short lifetime—usually less than a few days. In addition to the normal set of lysosomal hydrolases (Chapter 5), macrophage lysosomes contain enzymes that generate hydrogen peroxide (H_2O_2) and other toxic chemicals that aid in killing bacteria. When antibody ligand binds to the macrophage Fc receptor, it triggers the activation of these enzymes and also increases the rate of nonspecific pinocytosis as well as phagocytosis.

The Fc receptor has recently been purified and studied in detail. It is a permease for cations, activated when an antibody Fc region is bound to it. The influx of Na^+ or K^+ into the cell, triggered by ligand binding, appears to activate many specific macrophage functions: phagocytosis, cell movement, and the generation of hydrogen peroxide and other antibacterial compounds in lysosomes.

Some bacteria are ingested but not killed by macrophages. *Mycobacterium leprae,* the causative agent of leprosy, and parasites of the genus *Leishmania,* the protozoa that cause leishmaniasis, actually grow only in the endocytic vesicles of macrophages, and for this reason they are extremely difficult to kill (Figure 15-30). *Legionella pneumophila,* the bacterial agent of Legionnaire's disease, also multiplies intracellularly in human phagocytes. It inhibits the acidification of the phagocytic vesicles and thus the activation of lysosomal hydrolases that function only at acidic pH and that would otherwise kill the bacterium.

Pinocytosis Is the Nonspecific Uptake of Extracellular Fluids

Pinocytosis is the nonspecific form of endocytosis in which any extracellular material is taken up at a rate that is simply proportional to its concentration in the extracellular medium. To trace the fate of pinocytic vesicles, a solution of horseradish peroxidase, which does not bind to cell surfaces, is added to the extracellular fluid. This

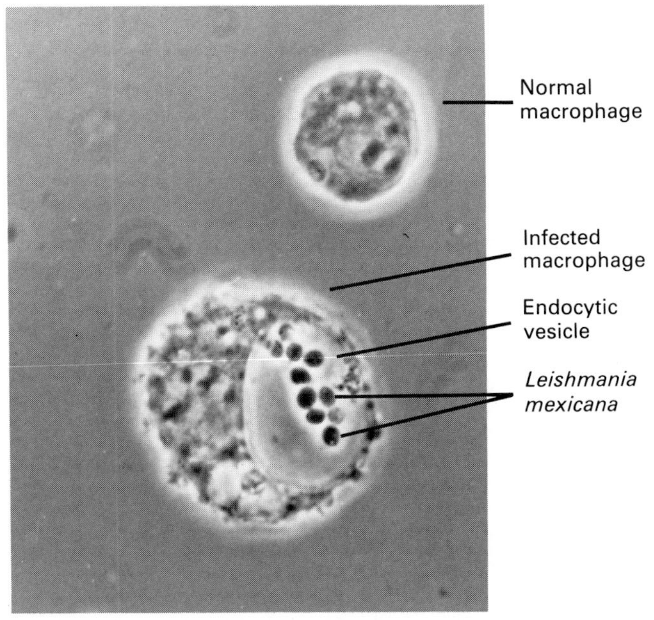

Figure 15-30 Replication of the unicellular parasite *Leishmania mexicana* inside an endocytic vesicle of a macrophage. Shown for comparison is an uninfected macrophage, lacking the expanded vesicle. *Courtesy of D. McMahon Pratt and J. Caulfield.*

protein can be readily distinguished in the electron microscope by a histochemical procedure (Figure 15-31). Endocytic vesicles containing the peroxidase are initially found just under the cell surface, and are small—about 0.1 μm in diameter. These endocytic vesicles fuse with each other, forming larger vesicles. The larger vesicles usually fuse with lysosomes, where the internalized protein is presumably destroyed by lysosomal proteases.

Receptor-Mediated Endocytosis Enables Cells to Internalize Specific Extracellular Proteins and Other Compounds

Receptor-mediated endocytosis allows the selective uptake of extracellular proteins and small particles. Receptor proteins on the cell surface bind certain molecules with a high degree of specificity, after which the region of the plasma membrane containing the receptor-ligand complexes is internalized. Endocytosis of these specific complexes usually occurs at *coated pits,* specialized depressions on the cell surface. A visible proteinaceous layer on the cytoplasmic side of the membrane surface gives a coated appearance to these parts of the plasma membrane and, initially, to the vesicles that form from them (Figure 15-32). Coated vesicles lose their coat after endocytosis, forming a smooth-surfaced vesicle. However, both coated and uncoated pits exist on the plasma membrane, and it is not always apparent whether a given endocytic process involves one or the other.

Different Cell Types Bind and Ingest Different Macromolecules by Receptor-Mediated Endocytosis

Different types of cells have different cell surface receptors and therefore they internalize different molecules by receptor-mediated processes (Table 15-4). Cells lining the yolk sac of the developing mammalian fetus specifically bind and internalize maternal antibodies, as do the intestinal epithelial cells of the young animal. These cells contain a surface receptor protein that binds to the Fc region of IgG-type antibodies (Figure 3-38). Internalized antibodies are transported inside vesicles across the cell and are ultimately exocytosed and delivered into the circula-

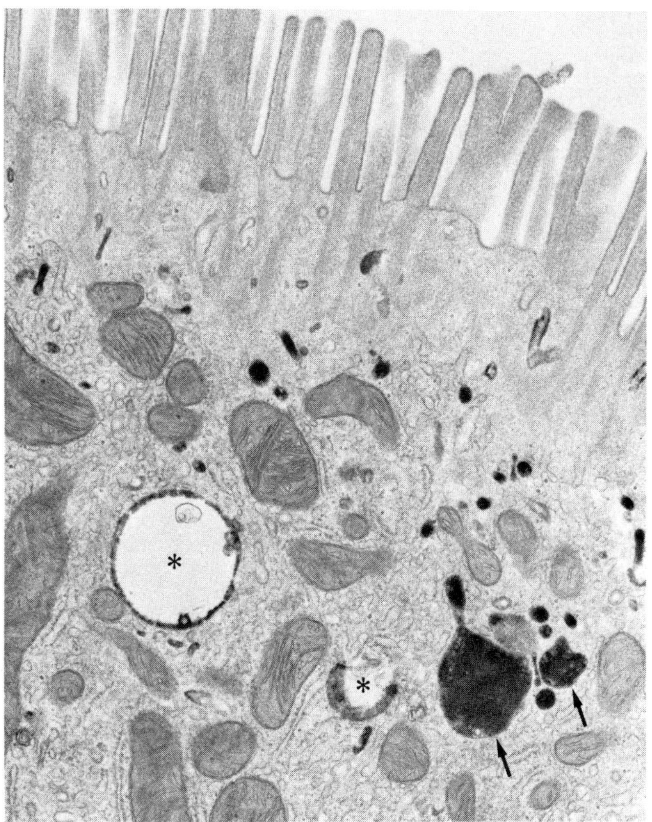

Figure 15-31 Uptake of horseradish peroxidase by intestinal epithelial cells. A solution of horseradish peroxidase is added to the lumen of a rat small intestine. Two hours later the tissue is fixed. The histochemical stain for this enzyme in tissue sections makes a dense precipitate. The peroxidase is visible in a variety of endocytic vesicles (the small dark objects) and inside the periphery of larger vesicles (labeled with an asterisk). High concentrations of peroxidase are also seen with lysosome-like bodies (arrows). [See D. R. Abrahamson and R. Rodewald, 1981, *J. Cell Biol.* **91**:270.] *Courtesy of R. Rodewald. Reproduced from the* Journal of Cell Biology *by copyright permission of the Rockefeller University Press.*

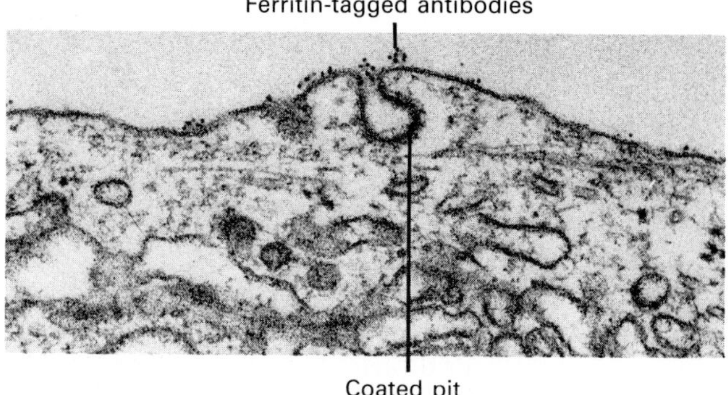

Figure 15-32 Electron micrograph of a coated pit on a mouse fibroblast cell. The cell was stained with ferritin-tagged antibodies to a cell surface antigen, designated theta, θ. Ferritin is an electron-opaque iron-containing protein that is visualized as a dense "dot" in electron micrographs. The tagged antibody labels the outer surface of the cell; it is not found in coated pits. The coat on the inner surface of the pit is readily visible. [See M. S. Bretscher, N. H. Thomson, and B. M. F. Pearse, 1980, *Proc. Nat'l Acad. Sci. USA* 77:4156.] *Courtesy of M. S. Bretscher.*

tion of the fetus or nursing animal. The offspring acquires the maternal antibodies and is able to respond to infectious agents.

Many cells bind and internalize potentially injurious or toxic extracellular proteins, such as proteases. In these cells the vesicles eventually fuse with lysosomes and the proteins are destroyed. Hepatocytes, the predominant cells in the liver, contain a potent surface receptor, the asialoglycoprotein receptor, that binds galactose-terminal glycoproteins. As we saw in Figure 3-77, galactose residues in serum proteins are usually covered by sialic acid residues. Abnormal glycoproteins lacking terminal sialic acids—that is, having exposed galactose residues—are rapidly removed from the circulation by binding to these hepatocyte receptors (Figure 15-33). After receptor-mediated endocytosis, asialoglycoproteins are destroyed by hepatocyte lysosomes. Other cells, especially macrophages, contain specific receptors for different abnormal or foreign glycoproteins.

Cell Surface Receptors Bind Ligands Specifically and Tightly

The binding of a molecule to its surface receptor can be separated experimentally from the internalization of the receptor-ligand complex. Internalization requires an expenditure of energy—it will not occur at 4°C or less, nor at 37°C in cells that have been poisoned with inhibitors of ATP production. But these conditions do not interfere with receptor-ligand binding. At 4°C, then, one can measure the amount of ligand bound to cell receptors and measure the affinity of the receptor for its ligand. Figure 15-34 shows that a hepatocyte will bind about 500,000 molecules of asialo-orosomucoid, a galactose-terminal glycoprotein, to surface asialoglycoprotein receptors, and that the concentration for half-maximal binding (K_m) of the ligand is 8×10^{-9} *M* of asialo-orosomucoid.

Surface receptors such as the one for galactose-terminal glycoproteins are exquisitely specific and have a high affinity for their ligands. Nonglycosylated proteins, for instance, will not bind to the asialoglycoprotein receptor with any measurable affinity, nor will glycoproteins that contain terminal sugars with structures similar to galac-

Table 15-4 Materials taken up by receptor-mediated endocytosis in animal cells

Ligand	Function of receptor-ligand complex	Cell type
Low-density lipoproteins	Supplies cholesterol	Most
Transferrin	Supplies iron	Most
Glucose- or mannose-terminal glycoproteins	Removes injurious agents from circulation	Macrophage
Galactose-terminal glycoproteins	Removes injurious agents from circulation	Hepatocyte
Immunoglobins	Transfers immunity to fetus	Fetal yolk sac; intestinal epithelial cells of neonatal animals
Phosphovitellogenins	Supplies protein to embryo	Developing oocyte
Fibrin	Removes injurious agents	Epithelial

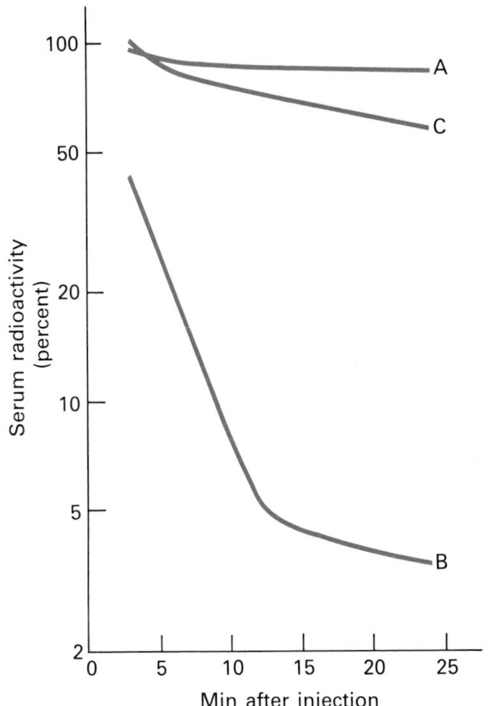

Figure 15-33 Disappearance from serum of ^{64}Cu-labeled native and modified rabbit ceruloplasmin injected into the bloodstream of rabbits.

(A) *Ceruloplasmin:* the normal serum protein with the oligosaccharide

```
                           Man-GlcNAc-Gal-SA
                          /
Protein-GlcNAc-GlcNAc-Man
                          \
                           Man-GlcNAc-Gal-SA
```

where GlcNAc = *N*-acetylglucosamine
Man = Mannose
Gal = Galactose
SA = Sialic acid

(B) *Asialoceruloplasmin:* protein with the sialic acid removed and galactose exposed

```
                           Man-GlcNAc-Gal
                          /
Protein-GlcNAc-GlcNAc-Man
                          \
                           Man-GlcNAc-Gal
```

(C) *Asialoagalactoceruloplasmin:* protein with both galactose and sialic acid removed

```
                           Man-GlcNAc
                          /
Protein-GlcNAc-GlcNAc-Man
                          \
                           Man-GlcNAc
```

Only (B), the protein with terminal galactose residues, is lost from the circulation. Further studies showed that it is removed by hepatocytes in the liver. The galactose receptor on the surface of these cells binds the asialoglycoproteins. After receptor-mediated endocytosis, the asialoglycoproteins are degraded in the secondary lysosomes.

[See G. Ashwell and A. E. Morrell, 1974, *Adv. Enzymol. Related Areas Mol. Biol.* **41**:99–128.]

tose (such as mannose). Binding constants of about 10^{-8} *M* are typical for many receptors that participate in endocytosis. These receptors will be half-saturated with ligand at a ligand concentration of 10^{-8} *M*, or 0.4 μg/ml for a typical protein of molecular weight 40,000. Since the total protein concentration in blood and other extracellular fluids is on the order of 100 mg/ml, these surface receptors are able to work efficiently in the presence of a 100,000-fold excess of unrelated and undesired ligands!

For experimental purposes, proteins bound to cell surface receptors can be removed by proteases or by agents that disrupt the receptor-ligand interaction but cannot penetrate the cells. Internalized ligands would not be dislodged by these treatments. Using this procedure to differentiate bound and internalized ligands, one can show that asialo-orosomucoid, bound to the asialoglycoprotein receptor on the surface of hepatocytes at 4°C, is rapidly internalized once the cells are warmed to 37°C (Figure 15-35).

In summary, the expenditure of energy is not required for cell surface receptors to bind their ligands. Like binding of a substrate to an enzyme, the formation of a receptor-ligand complex utilizes many specific noncovalent interactions between the two molecules. Internalization

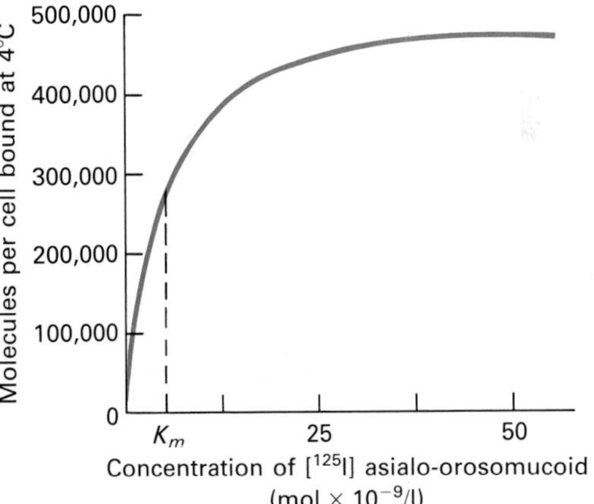

Figure 15-34 Binding of ^{125}I-labeled asialo-orosomucoid to rat liver hepatocytes at 4°C as a function of the concentration of ligand. At saturation, 500,000 molecules of ligand are bound to the cell, thus there are 500,000 surface receptors for asialoglycoprotein. Half of these sites are filled at a ligand concentration of 8×10^{-9} *M* (8 n*M*), thus the K_m of the receptor for this ligand is 8×10^{-9} *M*. This binding curve fits the binding equation with an equilibrium constant $K_m = 8 \times 10^{-9}$ *M*, where *L* is the concentration of free ligand and *R* and *RL* the amount of receptor and receptor-ligand concentrations, respectively: $L + R \rightleftharpoons (RL)$. [See A. L. Schwartz, D. Rup, and H. F. Lodish, 1980, *J. Biol. Chem.* **255**:9033.]

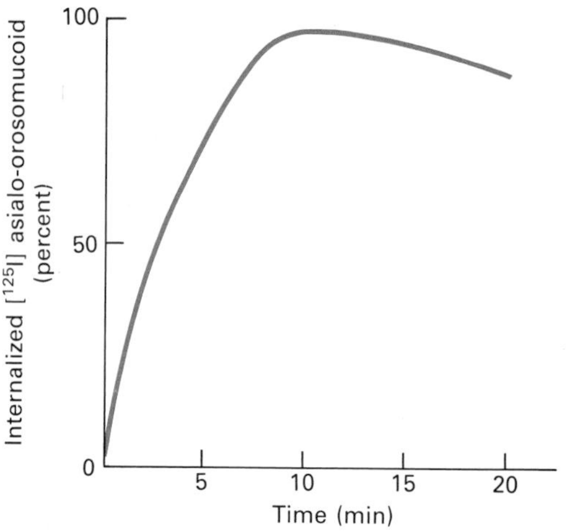

Figure 15-35 Internalization of protein bound to cell surface asialoglycoprotein receptor. ^{125}I-tagged asialo-orosomucoid was bound for several hours at 4°C to the surface of hepatoma (liver tumor) cells, as in Figure 15-34. At 0 min on this graph, the cells were warmed to 37°C. Internalized [^{125}I]protein was measured by treating the cultures with ethylenediaminetetraacetic acid (EDTA), a chemical that cannot penetrate inside the cells but that binds Ca^{2+} and thus removes ligand bound to the cell surface. At 0 min, therefore, all of the radioactivity is accessible to the reagent, and thus is on the cell surface. By 10 min after warming, all of the [^{125}I]ligand is resistant to removal by EDTA because it has been internalized. [See A. L. Schwartz, S. Fridovich, and H. F. Lodish, 1982, *J. Biol. Chem.* 257:4230.]

of the ligand, bound to its surface receptor, does require energy.

Clathrin, a Fibrous Protein, Forms a Lattice Shell around Coated Pits and Vesicles

Coated pits make up about 2 percent of the surface of cells such as hepatocytes and fibroblasts (Figure 15-36). The internalization of ligands begins when the receptor-ligand complex is localized over coated pits, which then pinch off from the membrane to form separate coated vesicles. Indeed, many endocytosed ligands have been visualized in such intracellular vesicles (Figure 15-41 shows one example) and most researchers believe that coated vesicles are an intermediate in the endocytosis of all ligands bound to cell surface receptors. Coated vesicles are readily purified from all eukaryotic cells studied (the brain is a particularly rich source). They probably func-

(a)

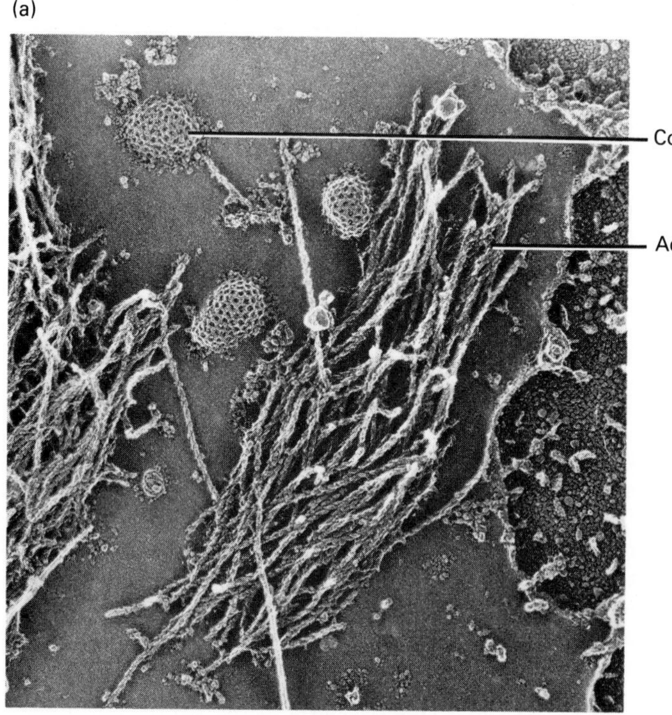

(b)

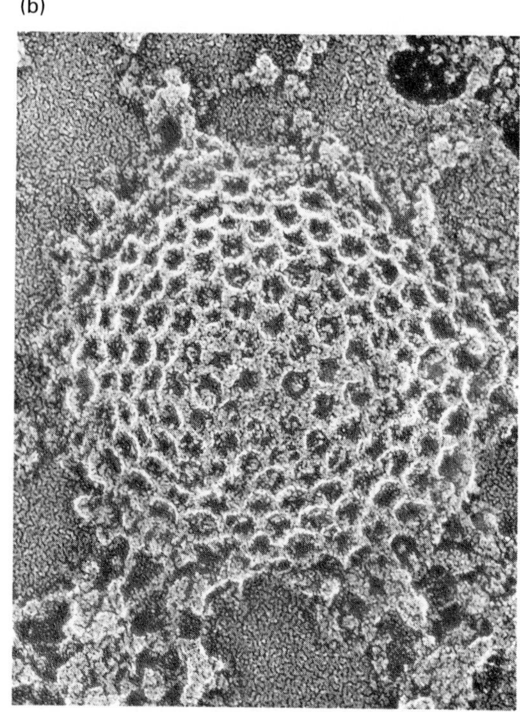

Figure 15-36 Views of coated pits on the cytoplasmic surface of the plasma membrane of a fibroblast. (a) The cells were rapidly frozen in liquid helium, freeze-fractured, and then treated by the deep etch procedure. Note the filaments of actin that also line the inside of parts of the plasma membrane. (b) Higher magnification of the polygonal network in a forming coated vesicle. *Courtesy of J. Heuser.*

tion in processes other than endocytosis—they might, for instance, be intermediates in transporting proteins from the Golgi to the cell surface.

The coat material of the coated vesicles contains a protein called *clathrin.* It is composed of a heavy chain (a large polypeptide of m.w. 180,000) and several light chains (peptides of m.w. 20,000 to 40,000). These molecules can form a cagelike lattice around the entire vesicle (Figure 15-37). Clathrin can be purified as a soluble protein free of membranes. The purified protein has the form of a triskelion. It is a three-armed trimer of the 180,000-m.w. subunit (Figure 15-38), also containing three clathrin light chains. The triskelions can combine, or polymerize, into a cagelike structure even in the absence of membranous vesicles. Indeed, it is thought that the polymerization of clathrin into a lattice along the cytoplasmic surface of a membrane pit causes the pit to expand and ultimately pinch off from the membrane, thus completing the clathrin cage structure (see Figure 15-37) and forming a coated vesicle.

Typical coated vesicles are 50 to 100 nm in diameter, with a membrane vesicle inside. The ~20-nm space between the membrane and the clathrin coat is filled with several proteins that apparently serve to bind the phospholipid bilayer and the proteins within it to the clathrin shell.

Coated vesicles are stable at the pH and ionic composition of the cell cytosol, yet coated vesicles normally lose their clathrin coat just after endocytosis. How does this happen? Recently an enzyme from liver was purified that depolymerizes coated vesicles to clathrin triskelions—the depolymerization is dependent on the hydrolysis of ATP to ADP + P_i. An input of energy is required to disrupt, in some manner, the multiple clathrin-clathrin interactions that stabilize the coated vesicle. Both the formation and depolymerization of coated vesicles must be highly regulated in the cell, as both processes occur simultaneously. But much remains to be learned about the molecular details of these processes. It is possible that different clathrin light chains are associated with various populations of

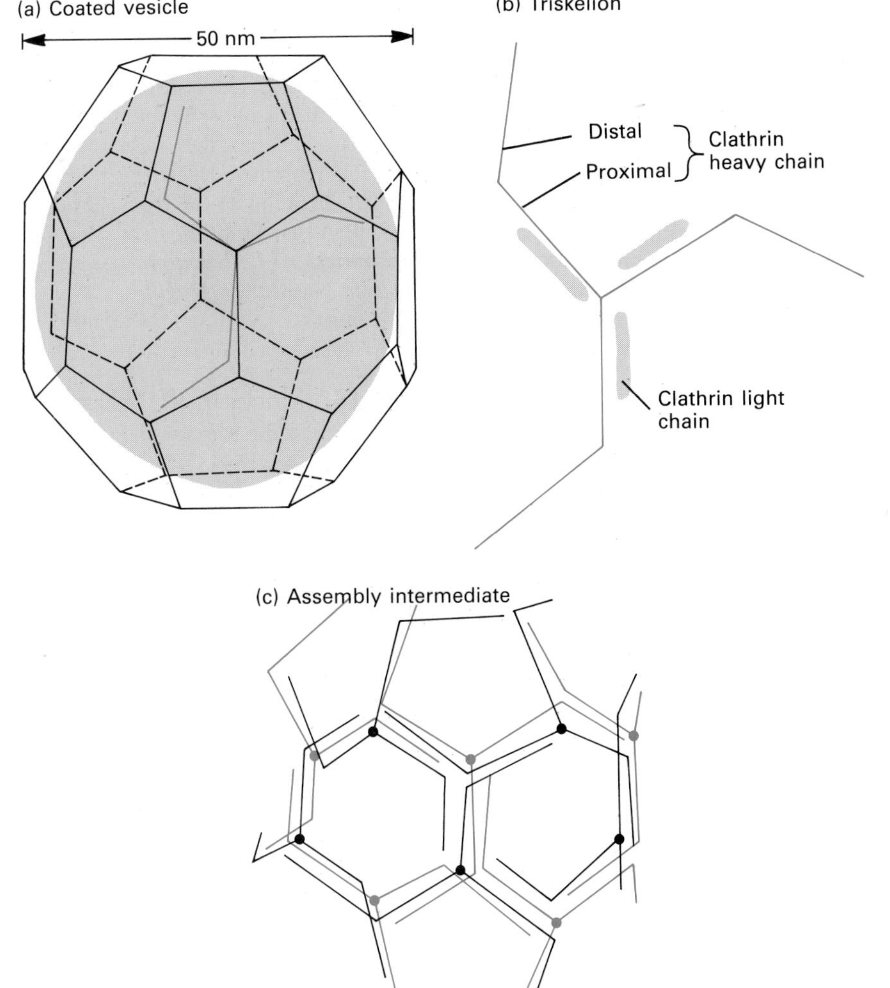

Figure 15-37 Structure and assembly of a coated vesicle. (a) A typical coated vesicle contains a membrane vesicle about 40 nm in diameter surrounded by a fibrous network of 12 pentagons and 8 hexagons. The fibrous coat is constructed of 36 clathrin triskelions, one of which is shown in color. One clathrin triskelion is centered on each of the 36 vertices of the coat. Coated vesicles having other sizes and shapes are believed to be constructed similarly: each vesicle contains 12 pentagons but a variable number of hexagons. [See R. A. Crowther and B. M. F. Pearse, 1981, *J. Cell Biol.* **91**:790.] (b) Detail of a clathrin triskelion. Each of three clathrin heavy chains is bent into a proximal arm and a distal arm. A clathrin light chain is attached to each heavy chain, most likely near the center. (c) An intermediate in the assembly of a coated vesicle, containing 10 of the final 36 triskelions, illustrates the packing of the clathrin triskelions. Each of the 54 edges of a coated vesicle is constructed of two proximal and two distal arms intertwined. The 36 triskelions contain 36 × 3 = 108 proximal and 108 distal arms, and the coated vesicle has precisely 54 edges.

(a)

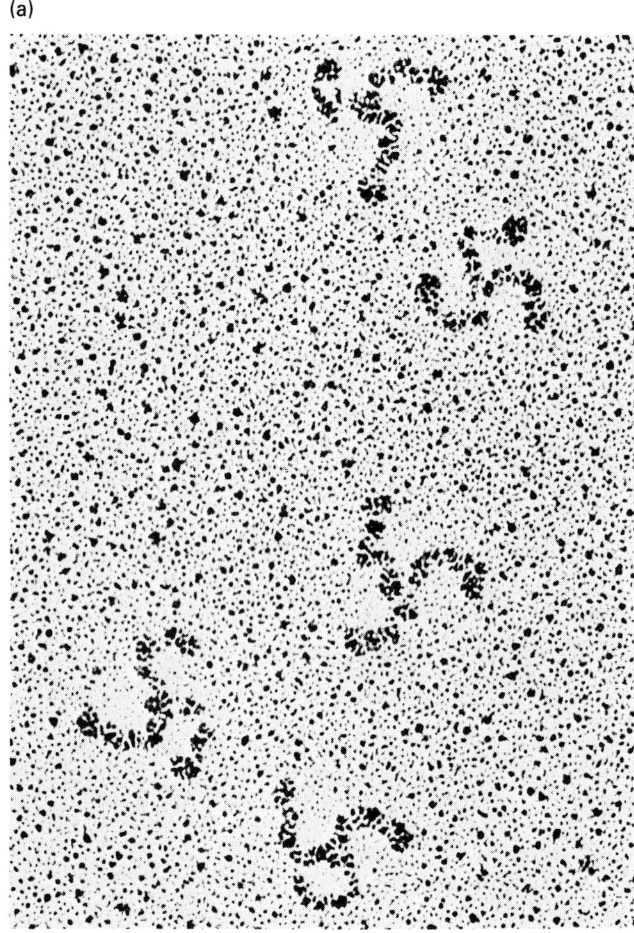

(b)

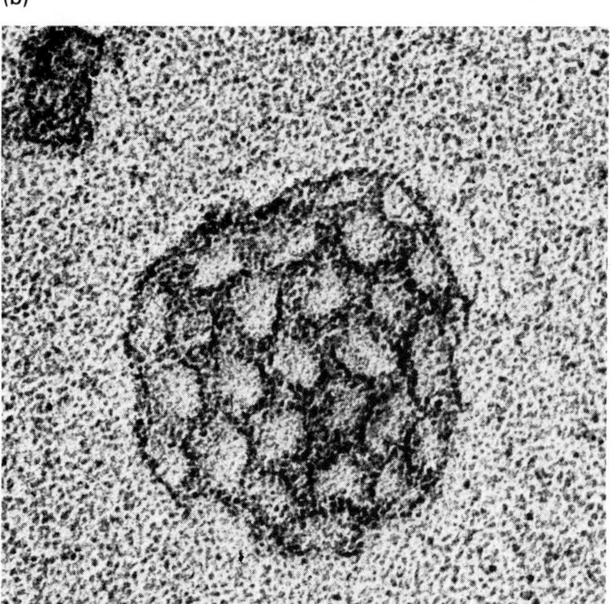

Figure 15-38 Electron micrographs of purified clathrin triskelions. (a) At pH 8.3, triskelions will not polymerize further. (b) The pH is lowered to 6.5 in the presence of Mg^{2+} ions, and the triskelions spontaneously polymerize into spherical structures of the same geometry as coated vesicles. This experiment shows that neither a phospholipid vesicle, ATP, nor any proteins other than the clathrin heavy and light chains present in the triskelion are required for polymerization of the triskelions. [See E. Ungewickell and D. Branton, 1981, *Nature* **289**:420.] *Courtesy of D. Branton. Reprinted by permission from* Nature. *Copyright 1981 Macmillan Journals Limited.*

coated vesicles within a cell and regulate the assembly or disintegration of their clathrin coats.

Low-Density Lipoprotein Receptor Mediates the Uptake of Cholesterol-Containing Particles

The most intensively studied system of receptor-mediated endocytosis is the system by which cholesterol is accumulated in cells. We shall use it to illustrate many important aspects of receptor function. Cholesterol is insoluble in body fluids. Whether taken in with the diet or synthesized in the liver, it must be transported by a carrier. Low-density lipoprotein (LDL) is one of a variety of complexes that carry cholesterol through the bloodstream, and most cells manufacture receptors that specifically bind LDL. The LDL particle is a sphere 20 to 25 nm in diameter (Figure 15-39). Its outer surface is a phospholipid monolayer membrane in which is embedded a single species of hydrophobic protein, called apo-B. Inside is an extremely apolar core of cholesterol esterified through the single hydroxyl group to a long-chain fatty acid. For most cells, serum LDL particles are the major source of cholesterol.

The internalization of LDL particles by fibroblasts, like most systems of receptor-mediated endocytosis, occurs in several stages (Figures 15-40 and 15-41). Receptors for LDL particles are located in coated pits. Bound LDL particles are also seen clustered over these pits (Figure 15-41). As in other ligand-receptor systems, the binding of LDL to its surface receptor does not require energy but the subsequent step, internalization of the receptor-containing vesicle, does—it will not occur at 4°C or in the presence of energy poisons. Normally, internalization occurs 2 to 5 min after binding, at which time LDL particles are incorporated into clathrin-coated vesicles.

LDL receptors are located in coated pits even when the LDL particles are not bound to them (Figure 15-42). Other receptors, such as the one for galactose-terminal glycoproteins, enter coated pits only if the receptors are occupied with ligand. Clearly the interaction between receptors and coated pits is complicated, and whether the receptors have transmembrane affinity directly with clathrin or whether some other interaction holds the re-

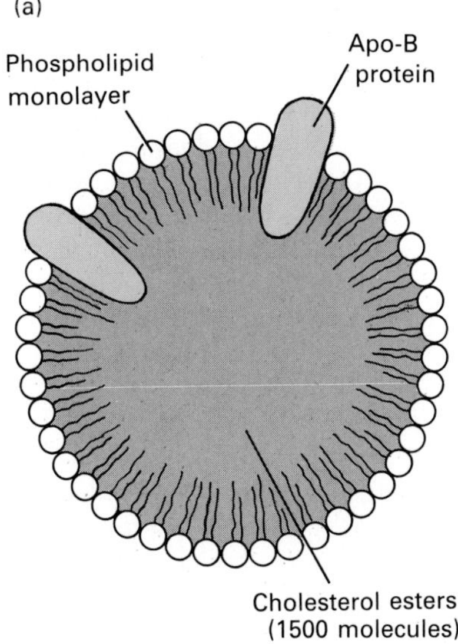

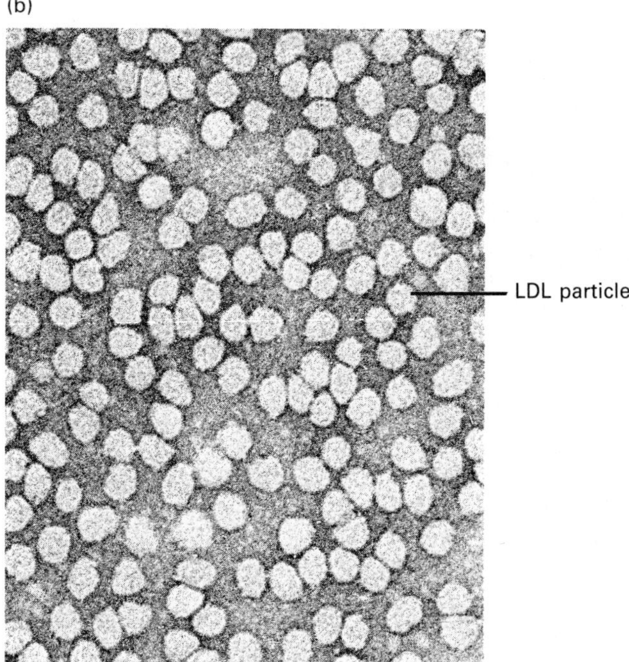

(c)

$CH_3(CH_2)_nCH_2\overset{O}{\overset{\|}{C}}O$–cholesterol

Fatty acid ester of cholesterol

H_2O → cholesterol esterase (lysosomal)

$CH_3(CH_2)_nCH_2\overset{O}{\overset{\|}{C}}OH$ Fatty acid

HO–cholesterol: Cholesterol

Figure 15-39 Structure of a low-density lipoprotein particle (LDL). (a) The surface of the particle is a monolayer of phospholipid with the hydrophilic ends facing outward and the hydrophobic ends facing the hydrophobic core rich in fatty acid esters of cholesterol. Several copies of the hydrophobic apo-B protein are embedded in the phospholipid membrane. (b) Electron micrograph of a negatively stained preparation of LDL particles. (c) Fatty acid esters of cholesterol are found in the interior of LDL particles. The lysosomal enzyme cholesterol esterase hydrolyzes the ester bond, forming a free fatty acid and cholesterol. Cholesterol is then incorporated into cell membranes, and the fatty acids are used to form phospholipids. [See R. Anderson, 1979, *Nature* **279**:679.] *Photograph courtesy of R. Anderson. Reprinted by permission from* Nature. *Copyright 1979 Macmillan Journals Limited.*

ceptors in the pits is not known. Certain surface proteins, such as the theta, H-2, or HLA surface antigens, are excluded from the coated pits (Figure 15-32); the system selectively internalizes some plasma membrane proteins and specifically excludes others.

Degradation of Low-Density Lipoprotein Particles Takes Place in the Lysosomes Once inside cells, the coated vesicles appear to lose the clathrin coat, becoming smooth-surfaced vesicles variously termed *endosomes* or *receptosomes* (Figure 15-40). Vesicles just under the plasma membrane surface are small ($\sim$0.2 μm in diameter). Larger LDL-containing vesicles farther removed from the surface apparently derive from fusion either with other LDL-containing vesicles or with another kind of smooth vesicle. Vesicles containing LDL particles eventually fuse with lysosomes to form secondary lysosomes (Figures 15-40 and 15-41). But before this point in the process, the LDL particles dissociate from the LDL receptors. Vesicles, enriched in LDL receptors, bud off and mediate the recycling of the LDL receptors to the plasma membrane. The LDL receptors, like most cell surface receptors that participate in endocytosis, apparently never reach the lysosome and so are spared degradation by potent lysosomal proteases.

The apo-B protein of the LDL is degraded inside the lysosome (Figure 15-43). Fatty acid is split off from the

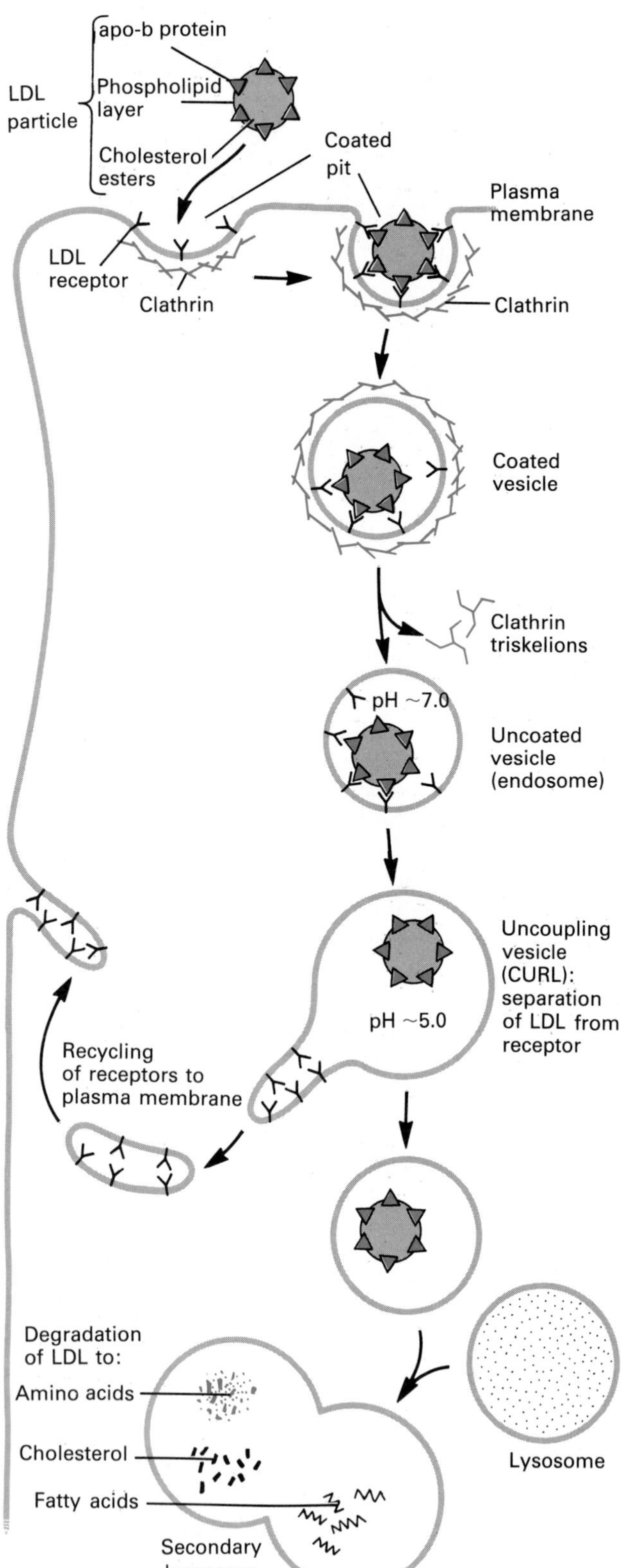

Figure 15-40 Fate of LDL particles and LDL receptors after endocytosis. The same pathway is followed by other ligands that are internalized by receptor-mediated endocytosis and degraded in the lysosome, such as asialoglycoproteins.

After an LDL particle binds to an LDL receptor on the plasma membrane, the receptor-ligand complex is internalized in a clathrin-coated pit that pinches off to become a coated vesicle. Then the clathrin coat depolymerizes to triskelions, resulting in an uncoated (smooth-surfaced) vesicle, often called an endosome. This then fuses with an uncoupling vesicle called CURL (compartment of uncoupling of receptor and ligand) that is characterized by an internal pH of about 5.0. The low pH causes the LDL particles to dissociate from the LDL receptors. A receptor-rich region buds off to form a separate vesicle that recycles the LDL receptors back to the plasma membrane. The vesicles containing the LDL particles fuse with lysosomes, forming a large secondary lysosome. In this lysosome the apo-B protein is degraded to amino acids, and the cholesterol esters are hydrolyzed to fatty acids and cholesterol. Cholesterol is incorporated into cell membranes. Abundant imported cholesterol inhibits synthesis by the cell of both cholesterol and LDL receptor protein.

core cholesterol, releasing cholesterol for use in the cell membrane.

The cells of the adrenal cortex synthesize many important steroid hormones, all of which are made from cholesterol. Because they must import much more cholesterol than most other mammalian cells, the adrenal cells are much richer in LDL receptors, and researchers often use them as a source of purified LDL receptors. The LDL receptor is a single-chain glycopeptide of 839 amino acids. A sequence of 22 hydrophobic amino acids spans the plasma membrane once (Figure 15-44), presumably as an α helix. About 50 amino acids at the carboxyl-terminus face the cytoplasm. Perhaps this latter region is involved in binding the receptor to clathrin in the coated pits. The most striking feature of the receptors is the 322 amino-terminal amino acids. This segment is extremely rich in disulfide-bonded cysteine residues; it includes an eightfold repeat of a sequence of 40 amino acids that may contain the LDL binding site.

The Syntheses of Low-Density Lipoprotein Receptor and Cholesterol Are Tightly Regulated The LDL uptake system is part of a complicated mechanism by which cells regulate cholesterol synthesis. All cells can make cholesterol, but if it is provided to them from the outside they shut down their own cholesterol manufacture. Exogenous cholesterol released from its LDL matrix in the lysosome causes at least three regulatory changes in the cell: (1) inhibition of a key enzyme of cholesterol bio-

(a)

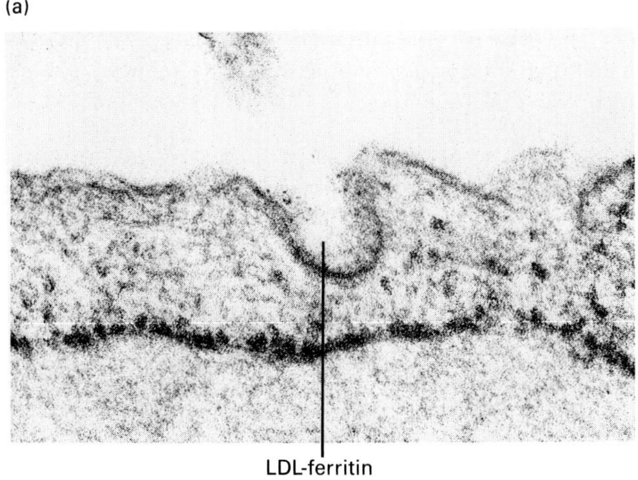

(b)

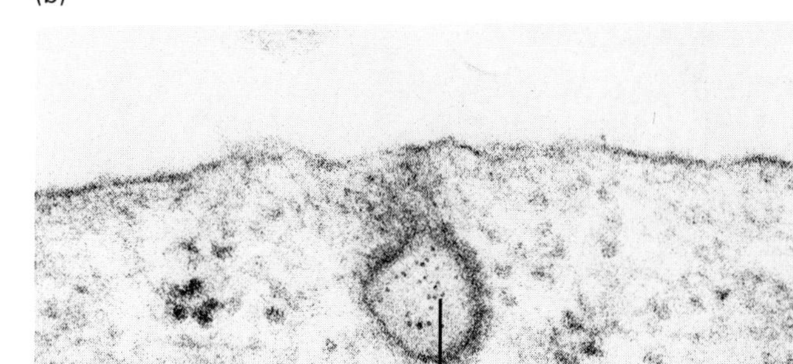

(c)

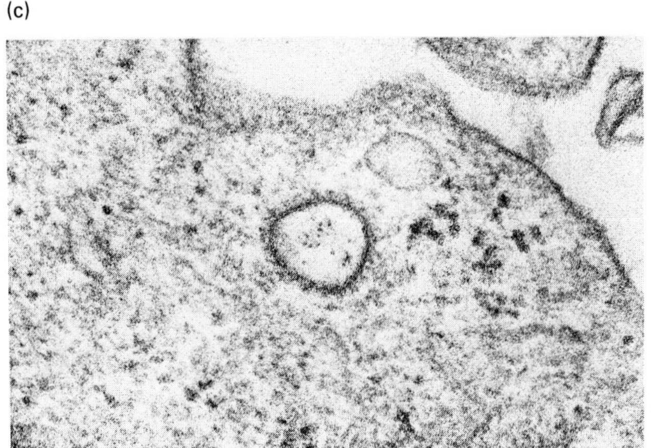

(d)

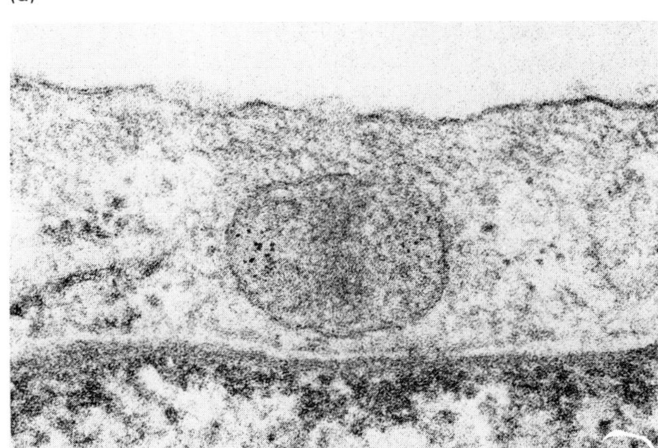

Figure 15-41 The initial stages of receptor-mediated endocytosis of LDL particles, revealed by electron microscopy. (a) A coated pit. The small dots over the pit are LDL particles labeled with ferritin. (b) A pit containing LDL apparently closing on itself to form a coated vesicle. (c) A coated vesicle containing LDL particles. (d) Ferritin-tagged particles in a smooth-surfaced endosome 6 min after they were added to cells. [See J. Goldstein, R. Anderson, and M. S. Brown, 1979, *Nature* **279**:679.] *Courtesy of R. Anderson. Reprinted by permission from* Nature. *Copyright 1979 Macmillan Journals Limited.*

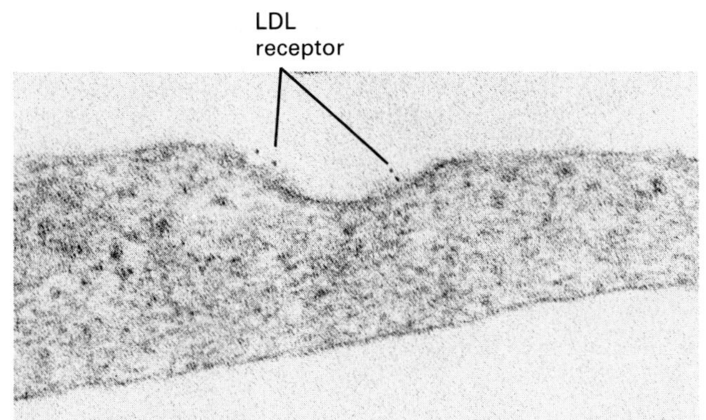

Figure 15-42 LDL receptors in coated pits. A monoclonal antibody to the LDL receptor protein was coupled to ferritin, an electron-dense protein. The antibody binds to LDL receptors on the plasma membrane surface. Electron microscopy reveals that they are located in coated pits on the membrane surface, indicating that LDL receptors are concentrated in coated pits in the absence of LDL particles. When ferritin-tagged LDL particles are bound to the cell, they are also localized in coated vesicles (see Figure 15-41), further confirming that LDL receptors are in the coated pit regions. [See R. Anderson et al., 1982, *J. Cell Biol.* **93**:523.] *Courtesy of R. Anderson. Reprinted by permission from* Nature. *Copyright 1979 Macmillan Journals Limited.*

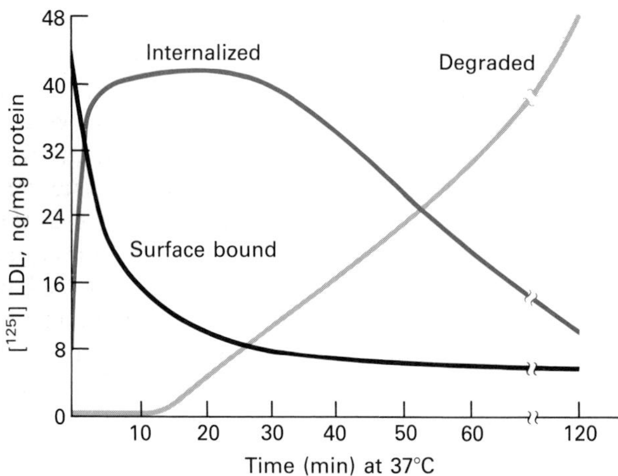

Figure 15-43 Internalization and degradation of LDL bound to normal human fibroblasts. The cells were exposed to particles containing radioactive apo-B protein at 4°C. The unbound LDL was washed away, and at time 0 the cells were placed in a 37°C incubator. At various times plates of cells were removed and the radioactive LDL on the cell surface and inside the cells was determined, as was the LDL that had been degraded to an acid-soluble form. Internalization of the surface LDL begins immediately upon warming the cells, but degradation begins only after 10 min, when the LDL particles have reached the lysosomes. [See M. Brown and J. Goldstein, 1979, *Proc. Nat'l Acad. Sci. USA* 76:3330.]

synthesis, (2) activation of an enzyme used in cholesterol storage, and (3) inhibition of synthesis of the LDL receptor itself. As the existing LDL receptors are slowly degraded, the plasma membrane will contain fewer receptors and will endocytose much less cholesterol in LDL particles. This regulation of the numbers of surface receptors is important, as it ensures that cells import only the cholesterol they need. Fibroblasts grown in medium containing high levels of cholesterol possess one-tenth the receptors of fibroblasts grown in a low-cholesterol medium.

A Hereditary Disease Is Due to a Genetic Defect in the Low-Density Lipoprotein Receptor The human LDL system has been invaluable for analyzing receptor-mediated endocytosis because of the discovery that some persons carry mutant forms of the LDL receptor protein. They have an inherited disorder called *familial hypercholesterolemia* ("too much cholesterol in the blood"). Persons homozygous for the mutant alleles often die at an early age of heart attacks caused by *atherosclerosis,* a buildup of cholesterol deposits that ultimately blocks the arteries. The genetic defect specifically affects the LDL receptor. In some persons with this disorder the receptor is simply not produced; in others it binds LDL poorly if at all. The gene for one mutant receptor has a deletion for the exons that encode the membrane-spanning segment. Although the mutant receptor binds LDL normally it can-

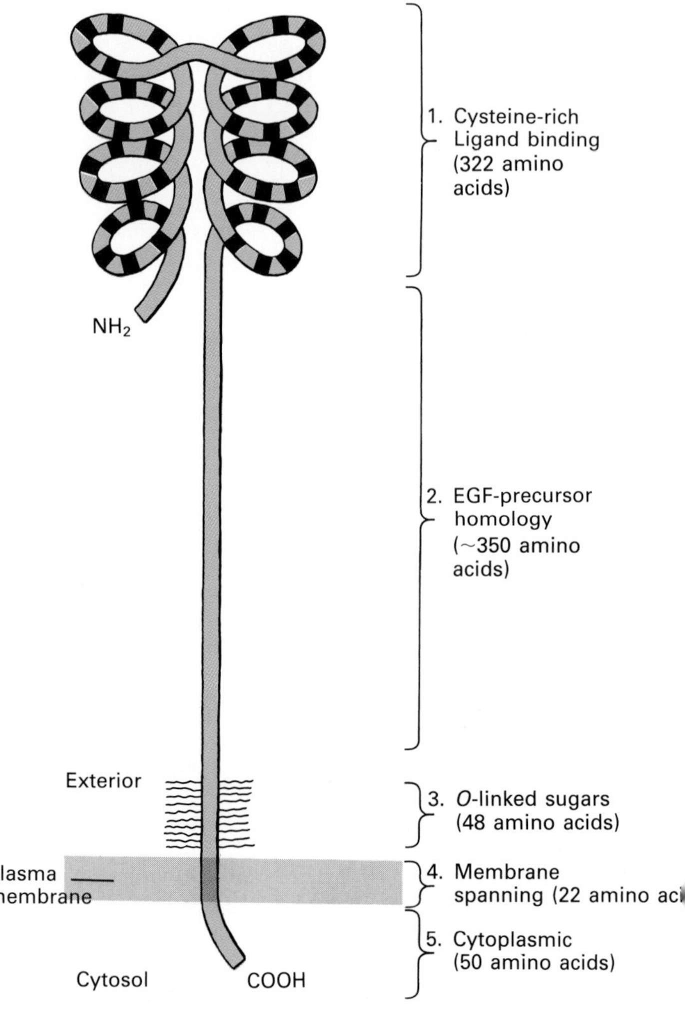

Figure 15-44 The five domains in the structure of the human LDL receptor. The sequence of the protein was deduced from the sequence of the cloned cDNA. The receptor is a dimer of two identical 839-residue polypeptides. From the amino-terminus, these domains are: (1) a 322 amino acid segment that is extremely rich (15 percent) in disulfide-bonded cysteine and that contains the ligand binding site; (2) 350 residues which are (unexpectedly) homologous to the precursor of the growth hormone epidermal growth factor; (3) a region just outside the plasma membrane rich in serine and threonine residues, and the site of O-linked glycosylation; (4) a membrane-spanning segment; (5) and the 50 COOH-terminal residues that project into the cytoplasm. [See T. Yamamoto et al., 1984, *Cell* 39:27.]

not be anchored to the plasma membrane, and it is secreted from the cell. In one especially instructive case, the mutant receptor binds LDL normally but the LDL-receptor complex cannot be internalized by the cell because, rather than being confined to coated pits, it is distributed evenly over the cell surface. Other receptors on this person's plasma membranes are internalized normally by the coated pit system. The problem seems to be that the mutant receptor is not recognized by the coated pit. The mutant receptor has a single amino acid change in the cytoplasmic domain; this residue must be part of the "pit-binding" site of the receptor.

Individuals who are heterozygous for familial hypercholesterolemia possess one mutant allele for LDL receptor and one wild-type allele. Their cells contain half the normal amount of functional LDL receptors. Their serum cholesterol level is higher than normal but lower than that of persons homozygous for the defect.

Most Surface Receptors Are Recycled

The LDL system illustrates an important characteristic of receptor-mediated endocytosis of ligands: Most surface receptors recycle, and they repeatedly mediate the internalization of ligand molecules. For example, a cultured hepatocyte contains about 500,000 asialoglycoprotein receptor molecules, each of which can bind one ligand. In 2 h, a hepatocyte can internalize more than 3 million asialoglycoprotein molecules, even if the synthesis of all cellular protein, including that of new receptors, is totally inhibited. This rate of internalization suggests that each receptor can recycle about every 20 min.

The recycling of receptors raises several questions: Precisely where in the cell do receptors and their internalized ligands dissociate? What causes a receptor to dissociate from the ligand it had bound so very tightly on the cell exterior?

During recent studies on the asialoglycoprotein receptor system, a novel organelle was discovered within which, it turns out, receptors and their ligands dissociate. It has been named CURL, for the compartment of uncoupling of receptor and ligand. Asialoglycoprotein receptor molecules are distributed uniformly on the plasma membrane surface of the hepatocyte. When ligand is added to a culture medium, some of the surface receptors are internalized together with the ligand and are found in small coated and uncoated vesicles immediately under the surface. Within 15 min after internalization, the bound ligands (asialoglycoproteins) are transferred to the lysosomes—but the receptors themselves are never found in these organelles. When both asialoglycoprotein and its receptor are found associated with the membrane of small endocytic vesicles just under the cell surface, it is presumed that the ligand is bound to the membrane receptor. But in the larger vesicles that contain both receptor and ligand, free ligand occurs within the lumen, presumably unbound to the receptor (Figure 15-45). Indeed, the spherical part of these vesicles contains little receptor at all. But attached to the large vesicle are tubular membranous structures that are rich in receptor and rarely contain ligand. One may assume, therefore, that these tubules contain receptor that dissociated from its ligand. Thus, it was concluded that this organelle, the CURL (see Figure 15-40), is the site of uncoupling of the receptor from its ligand. It is believed that receptor-rich elongated membrane vesicles bud off from CURL and then mediate the recycling of receptors back to the cell surface.

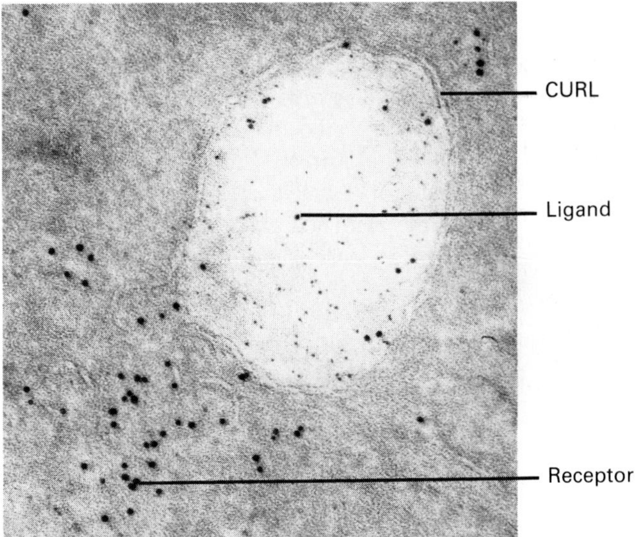

Figure 15-45 CURL, the compartment of uncoupling of receptor and ligand. Both asialoglycoprotein receptor and a ligand (asialofetuin) are localized simultaneously in the same thin section of a hepatocyte after receptor-mediated endocytosis. Liver cells were perfused with asialofetuin, then fixed and sectioned for electron microscopy. The sections were stained with antibodies specific for the receptor, and these antibodies were tagged with protein A linked to 8-nm-diameter particles of gold (the large grains) in the micrograph (see Figure 5-9 for a discussion of this technique). Localization of ligand was then detected by reaction of the sections with antibody to asialofetuin, followed by protein A linked to 5-nm-diameter particles of gold. The ligand particles (small grains) are in the lumen of the vesicle, while the receptor is clustered on tubular membranous structures that communicate with the vesicle. These membranous tubules are believed to be an intermediate in the recycling of receptor to the plasma membrane (Figure 15-40). [See H. J. Geuze et al., 1983, *Cell* 32:277.] *Courtesy of H. J. Geuze. Copyright, 1983, M.I.T.*

The Membrane Phospholipids That Form Endocytic Vesicles Are Recycled to the Plasma Membrane

The overall rate of endocytic internalization of the plasma membrane is quite high. Cultured fibroblasts regularly internalize 50 percent of their cell surface each hour even in the absence of any specifically internalized ligands. This extent of internalization of the plasma membrane requires that some process mediate the return of membrane material to the cell surface. It would be wasteful of the cell to degrade all this lipid and protein material only to synthesize new membrane constituents. We have already mentioned evidence that many cell surface receptor proteins recycle. The same is true for most other internalized plasma membrane proteins and for most of the endocytosed phospholipid as well. Presumably some membrane vesicle buds off from the endocytic vesicle or a lysosome and eventually returns to the plasma membrane to fuse with it. As described in Chapter 21, the Golgi complex is involved in directing certain newly made membrane and secretory proteins to the cell surface, and there is considerable evidence that the Golgi is also involved in recycling at least some endocytosed vesicles.

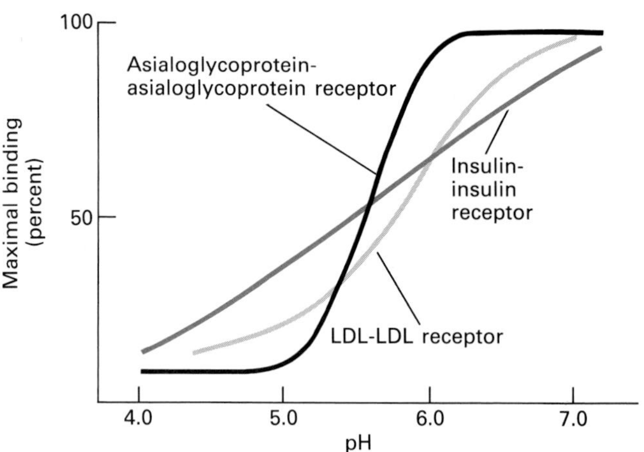

Figure 15-46 Effect of pH on the binding of proteins to their respective receptors. In all three situations shown—LDL binding to the LDL receptor, asialoglycoprotein to the asialoglycoprotein receptor, and insulin to the insulin receptor—the internalized ligands are transported to lysosomes where they are degraded, while the receptors recycle. Dissociation of receptor and ligand is believed to be triggered by the low pH of an endosome or CURL vesicle. [See A. Dautry-Varsat, A. Ciechanover, and H. F. Lodish, 1983, *Proc. Nat'l Acad. Sci. USA* 80:2258.]

Ligands Are Uncoupled from Receptors by Acidification of Endocytic Vesicles

It had been known for many years that the pH of lysosomes is acidic (about 4.5 to 5.0), but it came as a surprise that the pH of small endocytic vesicles—prelysosomal vesicles such as CURL—is also acidic. To establish this, fluorescein-tagged α_2-macroglobulin was internalized into endocytic vesicles by fibroblasts. The excitation spectrum of fluorescence emitted by the dye attached to these endocytosed molecules indicated that the environment of the vesicles was about pH 5.0, compared with a lysosomal pH of about 4.5.

These vesicles contain an ATP-dependent proton (H^+) pump. Pumping only 25 protons into the lumen of a vesicle 0.2 μm in diameter would be sufficient to lower its pH from 7.0 to 5.0 providing there is no buffer in the vesicle. Endosomes have been partially purified, and they do appear to contain an ATP-dependent proton pump different from the one in lysosomes. Surprisingly, coated vesicles may also contain a proton pump. Thus, acidification may begin even at the coated vesicle stage of endocytosis.

The acidity of endocytic vesicles answers the longstanding question we have asked concerning receptor-mediated endocytosis: How can a receptor be persuaded to let go of a tightly bound ligand? Most receptors, such as the asialoglycoprotein receptor and the LDL receptor, bind their ligand tightly at neutral pH but release the ligand if the pH is lowered to less than 5.0 (Figure 15-46). Acidification of endocytic vesicles may serve another useful purpose in the cell: any potentially hazardous endocytosed material stands a good chance of being inactivated.

Proteins Internalized by Receptor-Mediated Endocytosis Undergo Various Fates

We have observed that internalized LDL particles, and also potentially hazardous galactose-terminal serum proteins, are transferred to lysosomes, where they are destroyed. But sometimes endocytosed material simply remains in the cells and is minimally processed. Developing insect and avian egg cells (oocytes), for example, internalize yolk proteins and other proteins from the blood or surrounding cells. (Coated pits were first discovered in insect eggs, where they occupy a large portion of the plasma membrane.)

A hen's egg is a single cell containing several grams of protein, virtually all of which is imported from the bloodstream by endocytosis. *Vitellogenin*, a precursor of yolk proteins (principally of lipovitellin and phosvitin), is synthesized by the liver and secreted into the bloodstream, from which it is endocytosed into the developing egg. Yolk proteins remain in storage granules within the egg and are used as a source of amino acids and energy by the developing embryo after fertilization. Egg white proteins, such as ovalbumin, lysozyme, and conalbumin, are secreted by cells lining hen oviduct (see Figure 16-30).

(a)

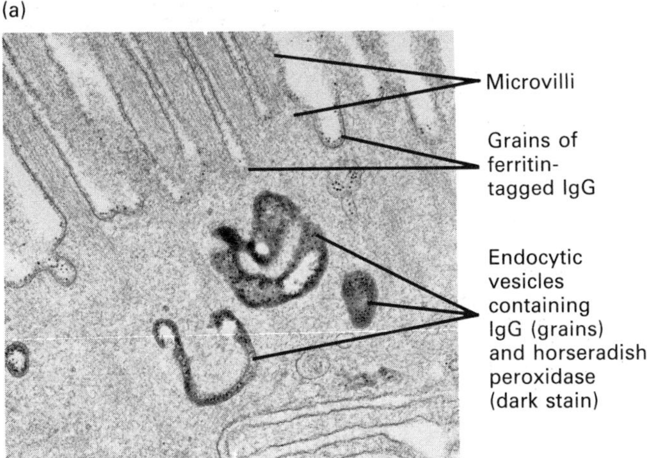

(b)

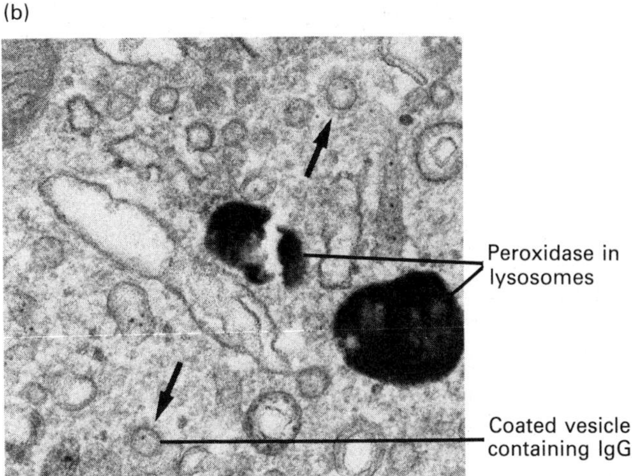

Figure 15-47 Sorting of the contents of the endocytic vesicle into different compartments. A mixture of ferritin-tagged IgG antibody and horseradish peroxidase was applied to the lumen of the small intestine of a newborn rat. (a) Both the IgG (dots of ferritin) and peroxidase (diffuse strain) are found together in endocytic vesicles in the apical region of the cell. The IgG, but not peroxidase, is specifically bound to the surface of the microvilli of the cells; presumably it is bound to an IgG receptor. (b) A view of the basal region of the cell. The two tracers appear in distinctly different vesicles. Horseradish peroxidase is only in large, dense bodies, presumably lysosomes. IgG is found in small tubules and coated vesicles (*arrows*) that will fuse with the basolateral surface membrane of the cell. The horseradish peroxidase has apparently been sorted and sequestered for destruction in lysosomes, while the IgG has been sorted for transcellular transport. [See D. R. Abrahamson and R. Rodewald, 1981, *J. Cell Biol.* **91**:270.] *Courtesy of R. Rodewald. Reproduced from the* Journal of Cell Biology *by copyright permission of the Rockefeller University Press.*

In other cases, endocytosed material passes all the way through the cells and is exocytosed, or secreted, from the plasma membrane at the opposite side (see Figure 15-25). An example of this transcellular transport is the movement of maternal antibodies across mammalian yolk sac cells into the fetus, or across the intestinal epithelial cells of the newborn offspring.

Some sorting of internalized macromolecules occurs after endocytosis to determine their different fates. When a mixture of immunoglobulins (conjugated to ferritin) and horseradish peroxidase is added to the intestinal cells of the newborn rat, both molecules are found initially at the apical surface in the same endocytic vesicles. The horseradish peroxidase enters by pinocytosis, the immunoglobulins by receptor-mediated endocytosis. The contents of these endocytic vesicles are sorted: the antibody but not the peroxidase is found in vesicles near the lateral and basal surfaces of the cell; the antibody escapes degradation in lysosomes and is subsequently released on the opposite side of the cell (see Figure 15-25). The horseradish peroxidase is transported to lysosomes, where it is degraded (Figure 15-47). Thus, two molecules destined for different fates—antibody for transcellular transport and horseradish peroxidase for lysosomal degradation—are sorted into different vesicles after endocytosis in a common vesicle. How and where such sorting occurs is not yet known. However, in other epithelial cells different endocytic vesicles are employed for pinocytosis and receptor-mediated endocytosis; sorting would thus occur at the internalization step.

Transferrin Delivers Iron to Cells by Receptor-Mediated Endocytosis

Transferrin is a major serum glycoprotein that transports iron to all tissue cells from the liver, the main site of its storage in the body, and from the intestine, the site of iron absorption. The iron-free form, *apotransferrin,* binds two Fe^{3+} ions very tightly forming ferrotransferrin. All growing cells contain surface transferrin receptors that avidly bind ferrotransferrin ($K_m = 6 \times 10^{-9}$ M), at neutral pH (Figure 15-48), after which receptor-bound ferrotransferrin is subjected to endocytosis. But there the similarity with other endocytosed ligands, such as LDL, ends: the two bound iron atoms remain in the cell, but the apotransferrin is secreted from the cell within minutes (Figure 15-49). The secreted apotransferrin is carried in the bloodstream to the liver or intestine, where it is reloaded with iron.

What properties of this receptor-ligand system cause the retention of iron within the cell and the secretion of apotransferrin? The answer lies in the unique ability of apotransferrin to remain bound to the transferrin receptor at the low pH (5.0) of the endocytic vesicles. At a pH

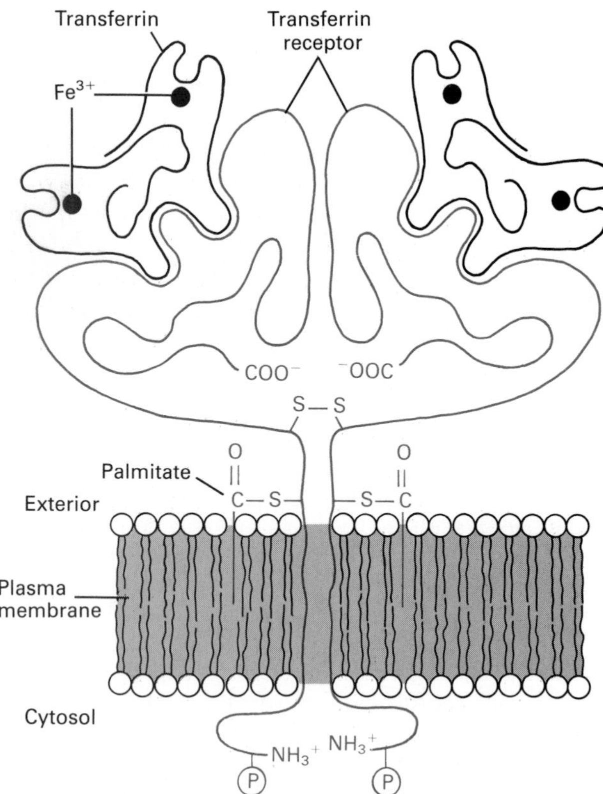

Figure 15-48 Structure of the transferrin receptor as deduced from the sequence of the cloned cDNA. The receptor contains two identical polypeptide chains. Most of the polypeptide is on the outer surface of the plasma membrane. A segment of about 26 hydrophobic residues spans the lipid core of the membrane, and a small segment at the amino-terminus faces the cytoplasm. The two chains are linked by a disulfide bond. Each chain is linked to the fatty acid palmitate, and the polypeptides are phosphorylated. The functions of the phosphate and palmitate are not known. [See C. Schneider et al., 1984, *Nature* **311**:675.] *From A. Dautry-Varsat and H. F. Lodish, 1984, Sci. Am.* ***250****(5):52. Copyright by Scientific American, Inc.*

of 6.0 or less the two bound Fe^{3+} atoms dissociate from ferrotransferrin. The iron remains in the endocytic vesicle or the CURL, and from there it is transported into the cytoplasm (in an unknown manner). The apotransferrin formed by the dissociation of the iron atoms remains bound to the transferrin receptor at the pH of these vesicles. As with the LDL receptor, the transferrin receptor recycles back to the surface, but the transferrin receptor takes the apotransferrin along with it! Remarkably, although apotransferrin binds tightly to its receptor ($K_m = 6 \times 10^{-9}\ M$) at pH 5.0, it has no measurable binding at neutral pH (Figure 15-50). Apotransferrin will dissociate from its receptor when the recycling vesicles fuse with the plasma membrane and the receptor-ligand complex encounters the neutral pH of the extracellular interstitial fluid or growth medium. The surface receptor will then be free to bind another molecule of ferrotransferrin. Figure 15-51 summarizes this pathway.

Endocytic Vesicles Increase the Permeability of Blood Vessel Walls

One of the more puzzling problems of biology has been the permeability of small blood vessels. Endothelial cells

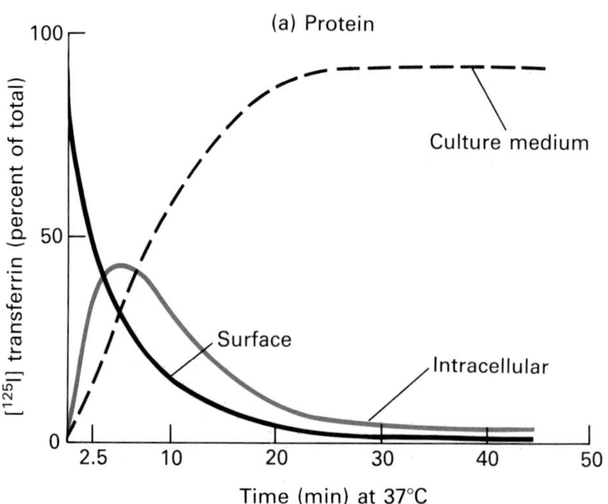

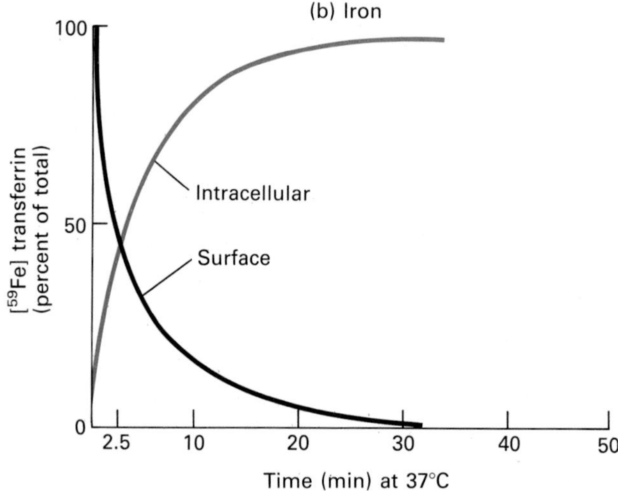

Figure 15-49 After endocytosis, the fate of (a) ferrotransferrin and (b) its two bound iron atoms. Ferrotransferrin labeled both in the protein [^{125}I] and iron [^{59}Fe] is bound to the surface of hepatoma cells at 4°C. After being warmed to 37°C, ferrotransferrin is rapidly internalized. The iron dissociates and remains in the cell, while the iron-free transferrin (apotransferrin) is rapidly secreted. [See A. Ciechanover et al., 1983, *J. Biol. Chem.* **258**:9681.]

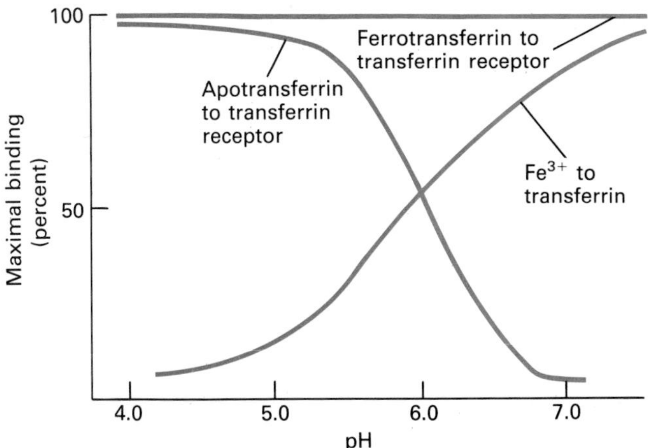

Figure 15-50 Effect of pH on the binding of Fe^{3+} to transferrin and on the binding of ferrotransferrin and apotransferrin to the transferrin receptor. Ferrotransferrin binds equally well to the transferrin receptor at pH values from 4.0 to higher than 7.0. Apotransferrin binds only at pH 5.0 and not at neutral pH. Lowering the pH from 7.0 to 5.0 causes release of the two bound Fe^{3+} atoms from transferrin:

$$\text{Ferrotransferrin} \underset{\text{pH } 7.0}{\overset{\text{pH } 6.0}{\rightleftharpoons}} 2Fe^{3+} + \text{apotransferrin}$$

[See A. Dautry-Varsat, A. Ciechanover, and H. F. Lodish, 1983, *Proc. Nat'l Acad. Sci. USA* 80:2258.]

lining the capillaries make tight junctions with one another, and—at least in certain organs—there is no space between the cells. Nevertheless, the vessels are highly permeable, allowing many of the macromolecular constituents of the blood to pass into the tissue space.

The solution to the paradox of high tissue permeability without intercellular holes was the discovery of *plasmalemmal vesicles*. These vesicles shuttle constantly from the blood side of a blood vessel endothelial cell to the tissue side and back, capturing blood fluid (plasma) on one side by endocytosis and releasing it at the other side by exocytosis (Figure 15-52). In plasmalemmal vesicles, molecules can traverse the blood vessel cells in about 1 min, and the total volume of vesicles is so large that the permeability of the vessels is 100 to 1000 times what would be expected in their absence. In certain cells, plasmalemmal vesicles even fuse to form channels that directly connect the blood vessel space with the exterior.

Infection by Many Membrane-Enveloped Viruses Is Initiated by Endocytosis

Many animal viruses contain an outer phospholipid bilayer membrane surrounding the viral genetic material and protein coat. The virus is usually formed by budding from the host cell's plasma membrane (see Figure 14-20), and its phospholipids derive from that membrane. The viral membrane also contains one or more virus-specific glycoproteins (see Figure 14-21), without which the virus would not be able to infect a target cell. Infection by this type of virus is almost an exact reversal of exocytosis, in that the vesicle—the virus particle—approaches the membrane from the outside and fuses itself into the cell. In this way the genetic material from the inside of the virus is released into the cytoplasm of the host cell. How

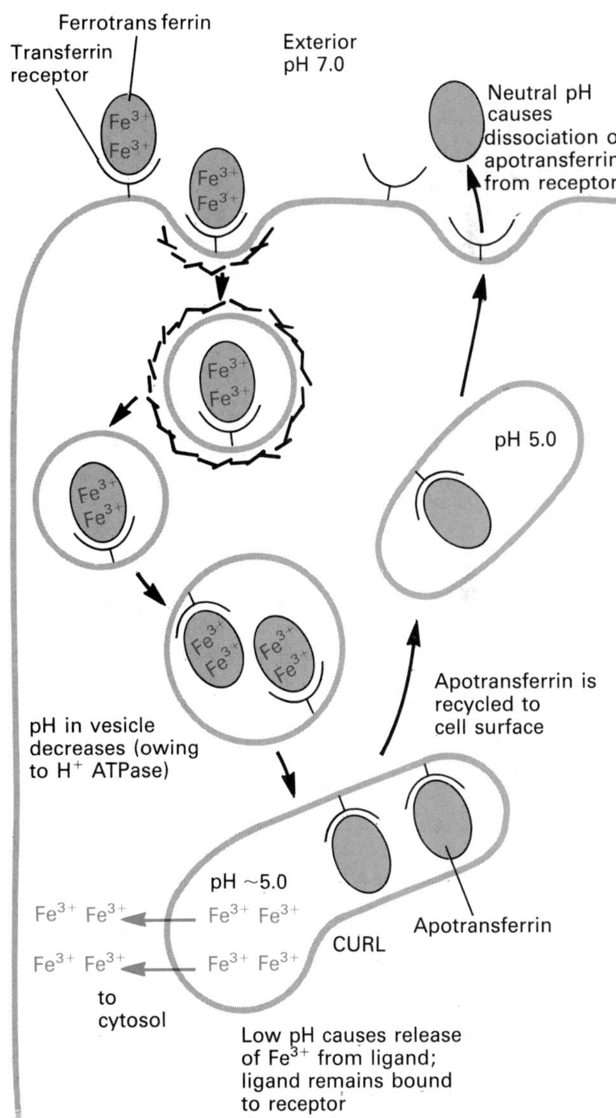

Figure 15-51 The transferrin cycle. After endocytosis, iron is released from the transferrin-receptor complex in the acidic CURL compartment. The transferrin protein remains bound to its receptor, and they cycle to the cell surface together. When the receptor-transferrin complex encounters the neutral pH of the exterior medium, the iron-free transferrin is released.

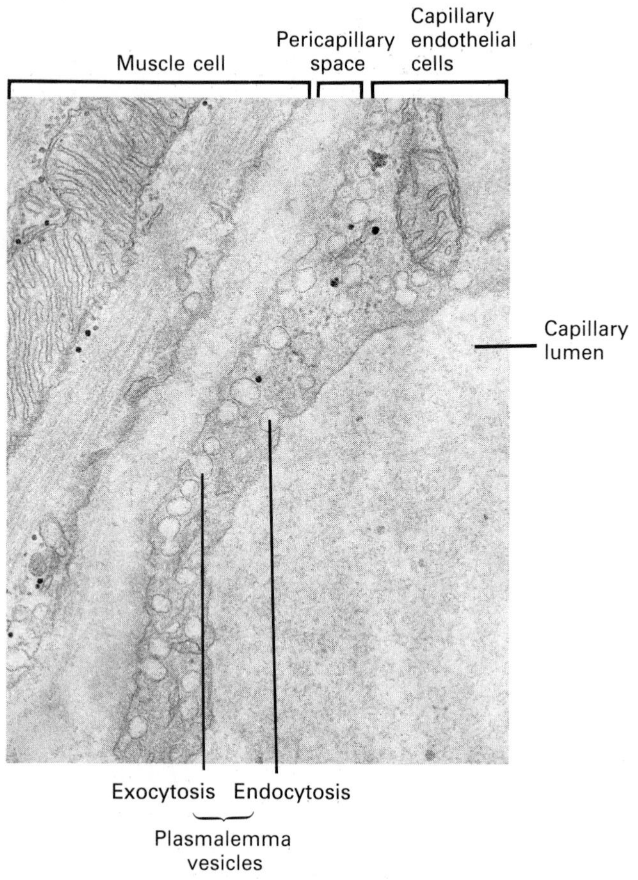

Figure 15-52 Electron micrograph of a blood capillary in the rat diaphragm, showing many plasmalemmal vesicles in the endothelial cells that line the capillary. These vesicles shuttle fluids and macromolecules from the blood to the pericapillary space that surrounds the adjacent muscle cells. *Courtesy of G. Palade and R. Bruns.*

the membrane of an animal virus fuses with the membrane of the host cell has been studied in some detail. It is an important system to know about because, although it has yet to be fully elucidated, it is likely to be the first membrane fusion system that is understood.

The myxovirus group of viruses has taught us most about fusion. The influenza virus is a good example. Its interior is made up of an RNA genome core bound to a structural protein coat, or capsid. This arrangement, called the nucleocapsid, is surrounded by a phospholipid bilayer membrane in which are embedded two different viral glycoproteins (Figure 15-53). The predominant glycoprotein of influenza virus, the hemagglutinin, or HA, protein is grouped in trimers that form the larger spikes on the virus surface. The HA spikes bind to receptors on the cell surface. *N*-acetylneuraminic acid (sialic acid) is a major component of these receptors. Figure 15-54 shows the three-dimensional structure of the HA spike.

Adsorption of influenza virus to the plasma membrane is not usually followed by membrane fusion. Endocytosis of the bound virus is essential but it does not always occur. In fact, the second viral glycoprotein, the neuraminidase, cleaves off cell surface *N*-acetylneuraminic acid residues, destroying the receptor. This causes the virus to desorb from the cell if endocytosis does not occur, freeing it to infect another cell.

(a)

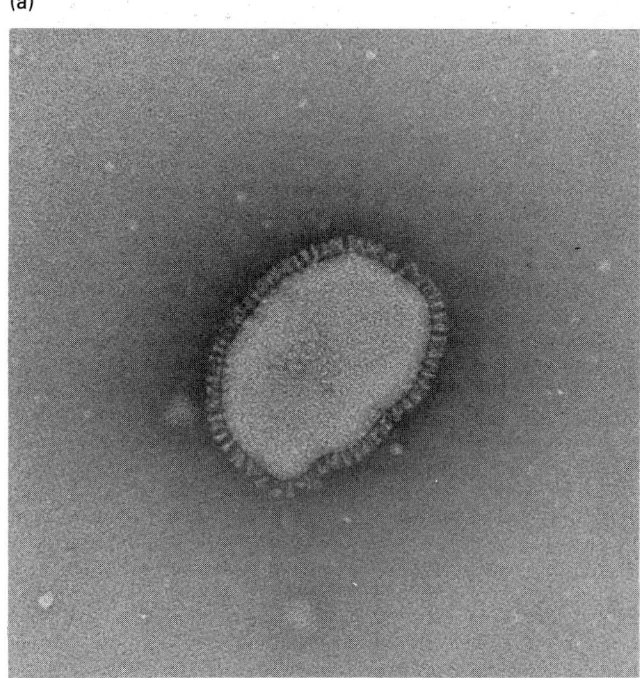

(b)

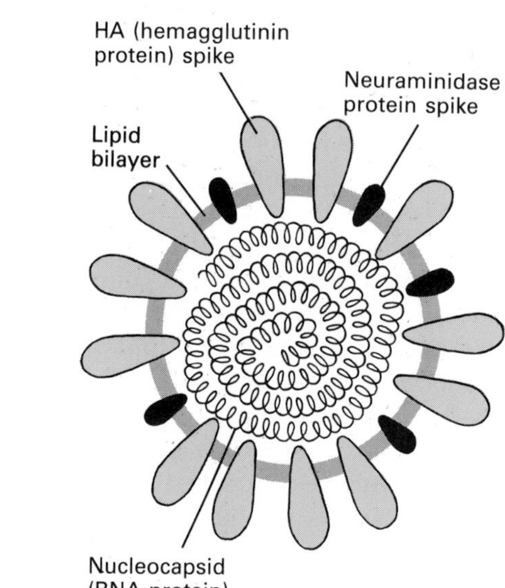

Figure 15-53 (a) Electron micrograph of a negatively stained influenza virus particle. The large spikes on the surface are constructed of trimers of the HA (hemaglutinin) protein; smaller spikes are constructed of multimers of the neuraminidase proteins. (b) Diagram of a cross section of an influenza virus particle. *Photograph courtesy of A. Helenius and J. White.*

(a)

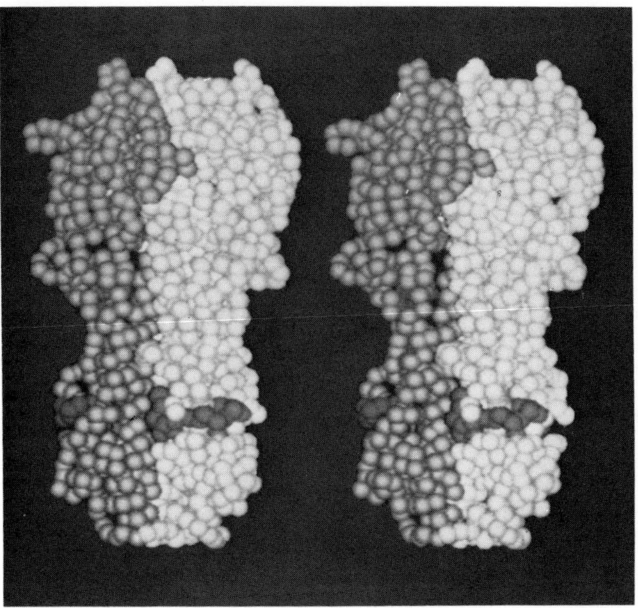

(b)

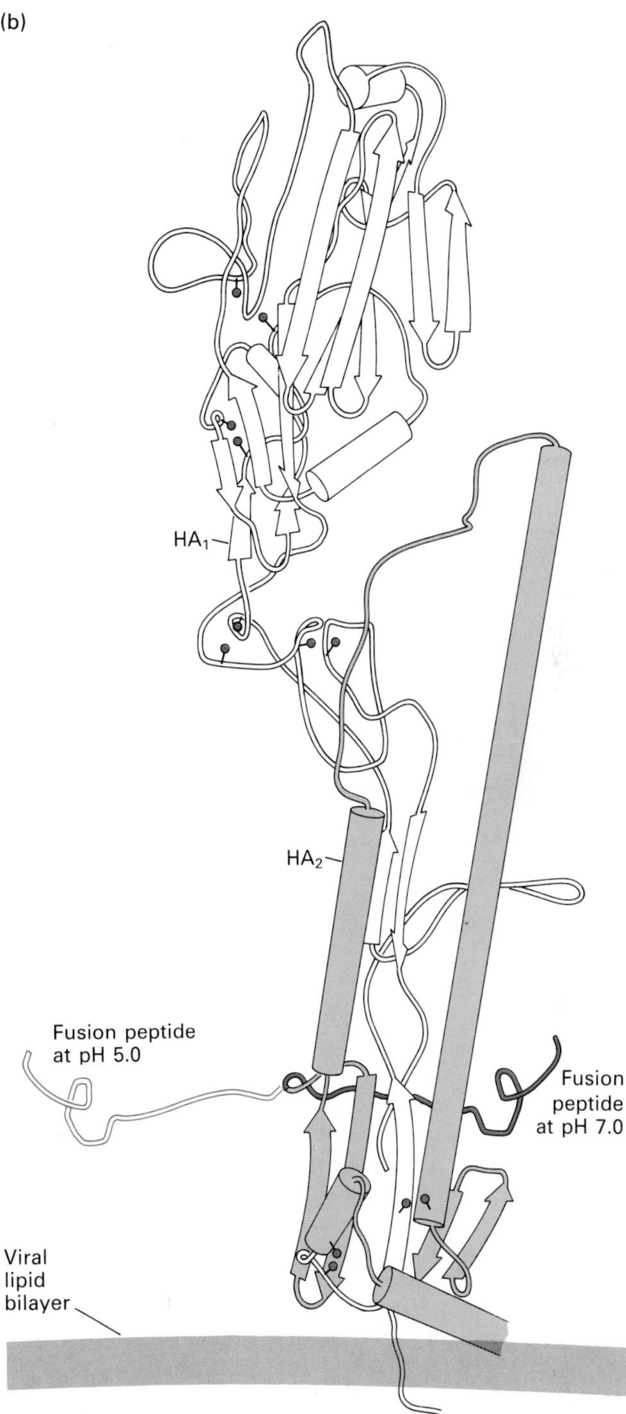

Figure 15-54 Influenza virus HA spike. (a) Three-dimensional structure. The virus membrane would be at the bottom of the picture and the spike protrudes outward (see Figure 15-53). The spike is composed of three $HA_1 + HA_2$ heterodimers. One dimer is white, the other dark grey; the third, mostly hidden in back, is black. The "fusion peptides" are in black, in the conformation at pH 7.0 where they are tucked into the spike. (b) A more detailed structure of one of the three identical $HA_1 + HA_2$ dimers that form the large spike (the HA_2 peptide is colored). The "fusion peptide" at the amino-terminus of the HA_2 protein is shown in its alternative configurations: its location at pH 7.0, tucked into the interface with the other HA chains in the spike, and the hypothetical exposed conformation of the fusion peptide at pH 5.0, when it is capable of mediating fusion with the membrane of the endocytic vesicle. *Part a courtesy of Richard Feldman and Thomas Porter; part b courtesy of D. C. Wiley.*

Low pH Triggers Fusion of the Viral and Cellular Membranes

Bound virus particles are endocytosed into acidic intracellular vesicles. It is at this stage that the virus and vesicle membranes fuse, releasing the virus nucleic acids into the cytosol and initiating virus replication (Figure 15-55). There is considerable evidence that the low pH of the enclosing endocytic vesicle triggers its fusion with the viral membrane. For one thing, virus infection is inhibited by agents that raise the pH of lysosomes or acidic endosomes such as CURL—provided these agents are added at the time of infection. They include lipid-soluble bases such as ammonia (NH_3) and chloroquine (Figure 15-56).

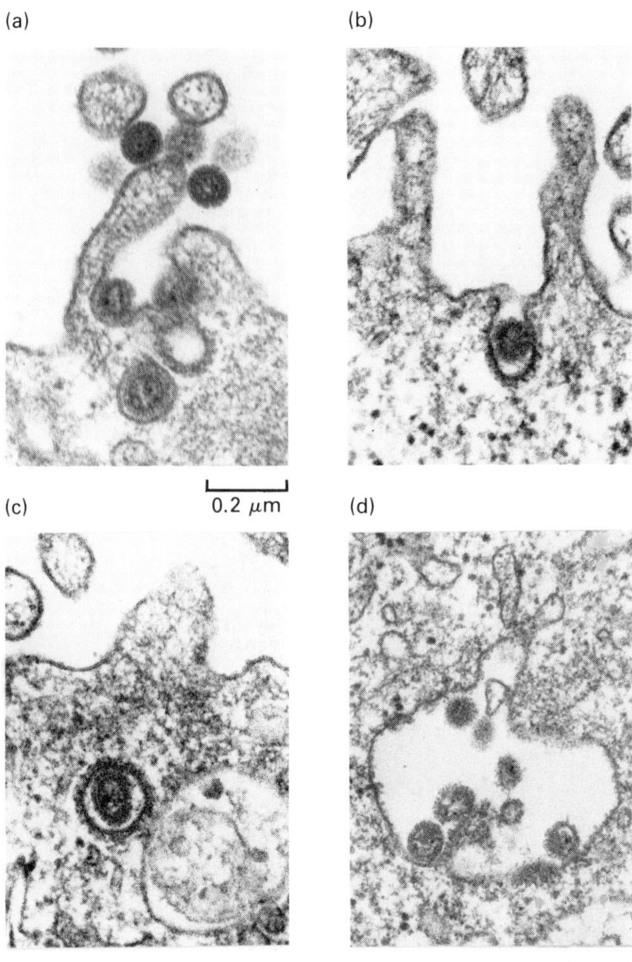

Figure 15-55 Entry of fowl plague virus, an avian influenza virus, into cultured dog kidney cells. Virus was bound to the surface of cells at 4°C. The cells were warmed to 37°C for different lengths of time to initiate virus penetration, then fixed with glutaraldehyde. Initially the virus is only on the surface of the cells. Within 5 min at 37°C, virus particles are seen in (a) smooth-surfaced pits and vesicles, (b) coated pits, and (c) coated vesicles. After 10 min, viruses are found in (d) endosomes and multivesicular lysosome-like bodies. It is believed that these CURL-like vesicles are acidic and that fusion of the viral and cellular membranes occurs in them. [See K. S. Matlin et al., 1981, *J. Cell Biol.* **91**:601.] *Courtesy of A. Helenius. Reproduced from the* Journal of Cell Biology *by copyright permission of the Rockefeller University Press.*

Surface-bound virus particles can fuse with the plasma membrane if the virus-cell complex is treated briefly with acid, thus simulating the lysosomal environment. If cells are treated with large amounts of influenza virus or Simliki forest virus, kept at 4°C to block endocytosis, and transiently exposed to acid pH, some viruses will fuse with two cells simultaneously. Large numbers of adjacent cells fuse together, forming giant multinucleated cells (Figure 15-57).

The molecular events of the membrane fusion process are only partly understood. They depend on a conformational change in the HA spike protein that is induced at pH 5.0. Each HA spike is composed of three copies of each of two chains, HA_1 and HA_2. The amino-terminus of HA_2 contains a strongly hydrophobic amino acid sequence that is highly conserved among many types of viral glycoproteins: Glu-Leu-Phe-Gly-Ala-Ile-Ala-Gly-Phe-Ile-Glu. This sequence is often termed the "fusion peptide." Crystallographic studies on the HA spike show that at pH 7.0 the fusion peptide is tucked into a crevice in the spike. pH 5.0 induces a conformational change in which the fusion peptide swings outward (Figure 15-54)! It is believed that the three exposed fusion peptides insert into the cellular membrane (Figure 15-58) concomitant with yet another change in conformation of the HA spike, bringing the viral and cellular membranes yet closer together. But the last stages in this process—the actual fusion of the viral and cellular membranes—are still not well understood.

Other types of virus, such as the paramyxoviruses Sendai virus and Newcastle disease virus, normally fuse with the cell surface membrane without undergoing endocytosis. These viruses also have two glycoproteins on their surface. One is a hemagglutinin, or binding protein; the other is a fusion factor. If both the hemagglutinin and the fusion protein are active on a paramyxovirus particle, the virus will bind to the cell and its membrane will fuse with the cell's plasma membrane. These viruses usually bind to *N*-acetylneuraminic acid residues on surface glycoproteins, and thus the hemagglutinin behaves very much like a lectin (a carbohydrate-binding protein).

Viruses of this sort have been used to fuse cells together. When a paramyxovirus, such as Sendai virus, is mixed with a cell suspension, many virus particles bind to

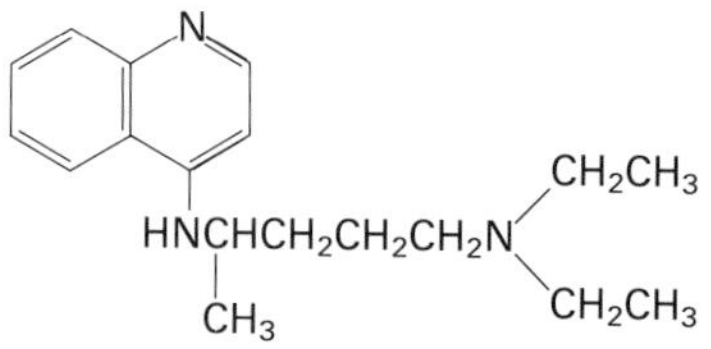

N
CH_2CH_3
$HNCHCH_2CH_2CH_2N$
CH_3
CH_2CH_3

Figure 15-56 Structure of chloroquine, a weak base that can diffuse readily across phospholipid membranes. When added to animal cells, it can diffuse into lysosomes or into acidic endosomes such as CURL, where it binds some of the protons:

$$(\text{Chl}) + H^+ \rightleftharpoons (\text{Chl})H^+$$

thus raising the pH, or reducing the concentration of protons in the lysosome or endosome.

more than one cell, agglutinating the cell suspension. The virus may then try to fuse into both cells simultaneously, resulting in a fused cell pair (Figure 15-59). When the nucleic acid of the virus has previously been inactivated, making the virus incapable of replication or cell damage, such a mixture results in fused but undamaged cells. In a fused cell both nuclei often go into division simultaneously, and after re-forming a single nuclear membrane, a viable *cell hybrid* arises (Chapter 6). It contains chromosomes from both cells in a single nucleus. Thus Sendai virus is a widely used fusagenic agent.

Endocytosis Internalizes Bacterial Toxins

Other unwanted guests also enter animal cells by receptor-mediated endocytosis. Certain pathogenic bacteria synthesize and secrete *toxins,* proteins that poison or kill susceptible mammalian cells and are the causative agents of the bacterial diseases. An important and well charac-

(a)

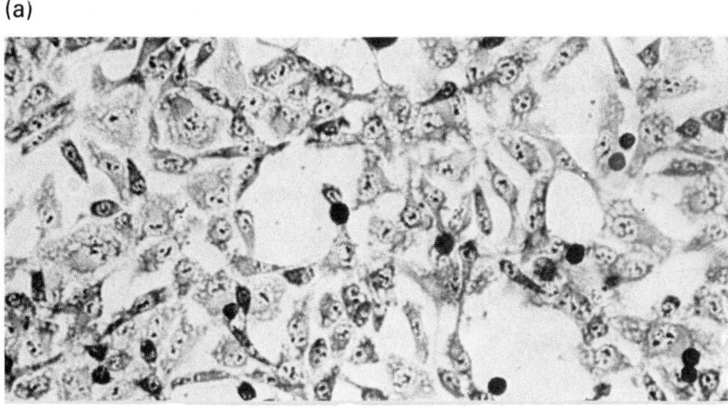

(b)

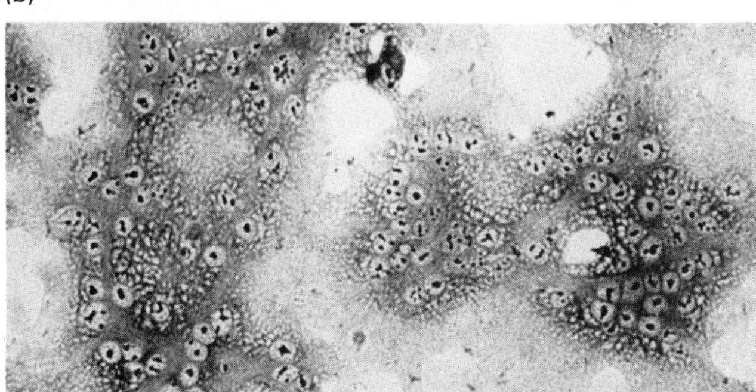

Figure 15-57 BHK-21 cells fused by added influenza virus. Virus was added to the cells at 4°C, and the cells were treated for 1 min at (a) pH 7.0 or (b) pH 5.0. Subsequently they were incubated at 37°C for 1 h, fixed, and stained. Transient exposure to acid pH allows surface-adsorbed virus to fuse with the surface of two cells simultaneously, generating large multinucleated syncytia. [See J. White, K. Matlin, and A. Helenius, 1981, *J. Cell Biol.* 89:674.] *Courtesy of J. White and A. Helenius. Reproduced from the* Journal of Cell Biology *by copyright permission of the Rockefeller University Press.*

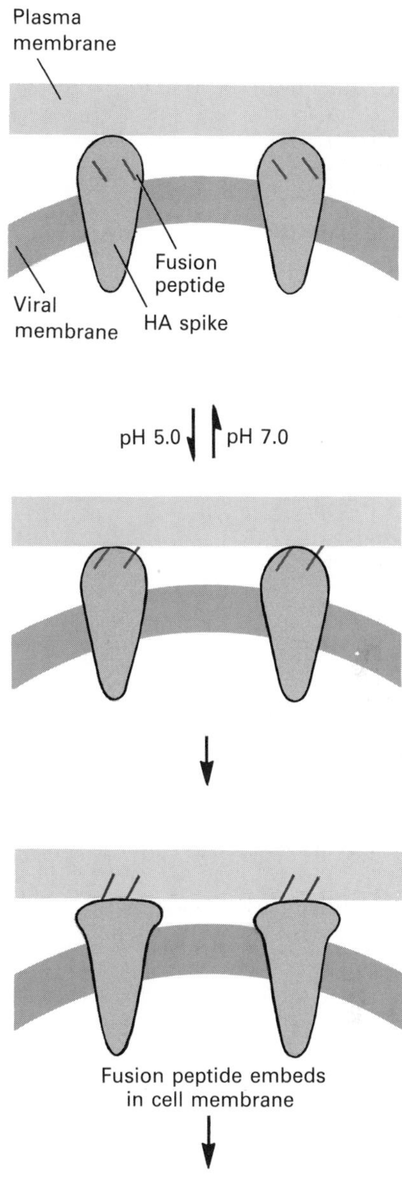

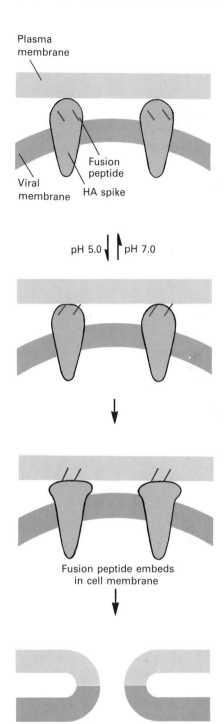

Figure 15-58 Hypothetical scheme by which low pH induces fusion of viral and cellular membranes. At pH 7.0, the HA spike binds to receptors on the cellular membrane. A change to pH 5.0 induces a conformational change in HA, in which the fusion peptide swings outward (see Figure 15-54). The hydrophobic fusion peptide may then embed into the cellular membranes simultaneously with another, as yet hypothetical, change in the HA conformation. Thus, the two membranes would be brought even closer together—but how the actual fusion occurs is still mysterious.

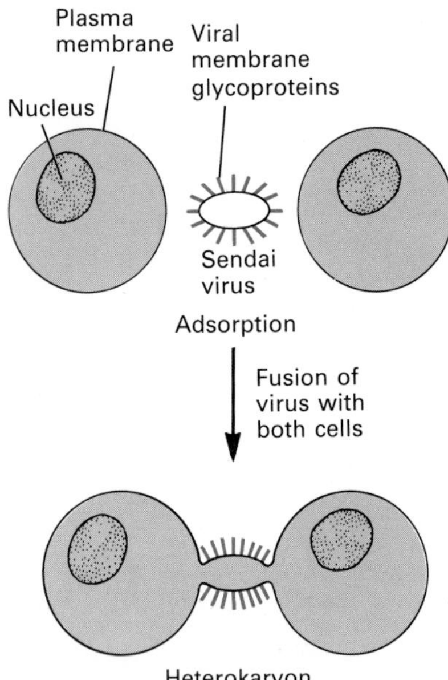

Figure 15-59 Fusion of cells by ultraviolet-light-inactivated Sendai virus. The virus and cellular membranes fuse immediately after adsorption of Sendai virus to the surface of susceptible cells. If two cells are close together, a Sendai virus particle can adsorb to and fuse with both cells simultaneously. A heterokaryon, a cell containing two nuclei, is formed. Provided the viral genome is activated by UV irradiation before the virus is added to the cells, the heterokaryon can grow.

terized example is diphtheria toxin, secreted by *Corynebacterium diphtheriae.* It kills cells by irreversibly inhibiting protein synthesis. The toxin contains two polypeptide chains serving very different functions. The B chain binds the toxin to a receptor on the cell surface; the A chain actually enters the cytoplasm and inhibits protein synthesis by inactivating elongation factor 2, the factor that translocates ribosomes along mRNA concomitant with hydrolysis of GTP (see Chapter 4). Current evidence suggests that entry of the A peptide into the cytoplasm is initiated by receptor-mediated endocytosis of the toxin-receptor complex. In the acid environment of the endosome or lysosome, the B fragment undergoes a conformational change and makes pores in the vesicular membrane. These channels are large—at least 2 nm in diameter—and they allow the A peptide to cross the vesicle membrane and enter the cytoplasm.

Summary

The plasma membrane regulates the traffic of molecules into and out of the cell. Ions, sugars, and amino acids cannot diffuse across the membrane's phospholipid bilayer at rates sufficient for the cell's needs. They are transported by a class of integral membrane proteins called transporters, or permeases. Protein-catalyzed entry into the cell of some molecules is driven only by a concentration gradient. For example, for glucose to enter an erythrocyte a glucose permease forms a transmembrane passage allowing only glucose and closely related molecules to cross.

Some molecules require active transport to enter a cell, a process in which energy is expended. The Na^+K^+-ATPase, for example, pumps three Na^+ ions out of the cell and two K^+ in (against their concentration gradients), utilizing the energy of hydrolysis of a phosphoanhydride bond of ATP. The Ca^{2+}-ATPase pumps Ca^{2+} ions out of the cell or, in muscle, into the sarcoplasmic reticulum. These pumps create an intracellular ionic milieu of high K^+, low Ca^{2+}, and low Na^+, very different from the extracellular fluid milieu of high Na^+, high Ca^{2+}, and low K^+. An ATP-dependent H^+ pump in the lysosomal and endosomal membranes is responsible for maintaining a lower pH inside the organelles than in the surrounding cytosol.

The import or export of a small molecule against a concentration gradient may also be achieved by coupling its movement to that of another molecule or ion, usually H^+ or Na^+, down its concentration gradient. Glucose and amino acids enter many cells by coupling to an influx of Na^+ down its concentration gradient, the process termed symport. In cardiac muscle cells, the export of Ca^{2+} ion is coupled to the import of Na^+; this is termed antiport. In yet other cases, the uptake of nutrients is driven by a proton gradient across the cell membrane; one example is the uptake of lactose by the bacterium *E. coli* catalyzed by a sugar H^+ symport.

The plasma membranes of many cells are differentiated so that specific membrane segments contain different permeases and other proteins and that therefore carry out quite different transport processes. In the intestinal epithelial cell, for example, the glucose-Na^+ and amino-Na^+ acid–symporters are located in the apical portion facing the intestinal lumen while the Na^+K^+-ATPase and the glucose and amino acid permeases are in the basolateral portion facing the blood capillaries. This combination allows the cells to transport amino acids and glucose from the lumen into the blood. Bacterial cells import some nutrients by a different group-translocation reaction; the phosphorylation of sugars is coupled to their uptake.

There are many examples of the bulk flow of water across cell layers. Flow is driven either by hydrostatic pressure (as from a blood vessel into a cell) or by osmotic pressure. Where a cell epithelium secretes ions, such as H^+Cl^- secreted by parietal cells lining the stomach, the movement of water out of the cells is driven by the resulting osmotic gradient.

Proteins, bacteria, and viruses are imported into cells by processes of a fundamentally different nature, requiring regions of the plasma membrane to detach and form intracellular vesicles. One such process is phagocytosis, the means by which very large particles, such as bacteria, are internalized. Phagocytosis is usually initiated by binding of the particle to specific receptors on the cell surface. Then the plasma membrane gradually envelops the bound particle, eventually forming an intracellular membrane-limited vesicle. The vesicle fuses with one or more lysosomes, triggering the degradation of the particle within. Macrophages are cells specialized for ingesting potentially hazardous foreign objects by phagocytosis. Antibodies bound to foreign particles facilitate their ingestion by macrophages, which contain specific receptors for the Fc region of antibody that all IgG antibodies have in common. The killing of bacteria by macrophages is facilitated by the generation of hydrogen peroxide within the lysosomes. But some parasitic organisms, such as mycobacteria and leishmanias, not only escape destruction by macrophage lysosomes, they are actually specialized to grow only within them.

In nonspecific pinocytosis, small regions of the plasma membrane are internalized, forming small intracellular vesicles that contain a bit of the extracellular fluid. In a related process termed receptor-mediated endocytosis, particles such as viruses, small proteins, and oligosaccharides first bind to specific receptor proteins on the plasma membrane, and the regions of the plasma membrane containing these receptor-ligand complexes are selectively internalized. Such endocytic vesicles are often coated with a basketlike network composed of clathrin, a fibrous protein. After internalization, the vesicles are uncoated by depolymerization of the clathrin coat to clathrin triskelions. The smooth vesicles fuse with others; concomitantly, their interior is acidified.

The fate of the endocytosed ligand can vary considerably. Endocytosed membrane-coated viruses, such as the influenza virus, fuse with the membrane of the endocytic vesicle, enabling the viral nucleic acid to enter the cytosol, where it can replicate. The low pH induces a conformational change in the viral HA spike, triggering fusion of the viral membrane with a cellular membrane.

The diphtheria toxin and some other bacterial toxins have a different means of escaping degradation in the acidic endocytic vesicles. One subunit of the toxin forms a transmembrane channel across the vessel. The other subunit crosses the vesicle membrane and enters the cytoplasm, where it catalyzes the reactions that kill the cell. Other endocytosed ligands, such as LDL particles and galactose-terminal glycoproteins, are transported via a series of membrane-limited vesicles to the lysosome. Most endocytosed proteins are degraded in this organelle, but cholesterol is released from the LDL particle and is used for the synthesis of cell membranes.

In most cases of receptor-mediated endocytosis, the receptors are not degraded upon internalization. Rather, they release their internalized ligand and return to the cell surface, where they are recycled. The low pH of the CURL organelle triggers the dissociation of most receptors and ligands. Each transferrin protein transports two Fe^{3+} ions into the cell. Following binding to the transferrin receptor and endocytosis, the Fe^{3+} ions are released in the CURL and are transported to the cytosol, while the transferrin protein is rapidly secreted.

References

Transport and the Intracellular Ionic Environment

*CHRISTENSEN, H. N. 1975. *Biological Transport*, 2d ed. Benjamin.

GRAVES, J. S., ed. 1985. Regulation and development of membrane transport processes. *Soc. Gen. Physiol.* and John Wiley and Sons. An excellent collection of recent papers.

WILSON, D. B. 1978. Cellular transport mechanisms. *Annu. Rev. Biochem.* 47:933–965.

Passive Transport

Facilitated Diffusion of Glucose

ALLARD, W. J., and G. E. LIENHARD. 1985. Monoclonal antibodies to the glucose transporter from human erythrocytes. *J. Biol. Chem.* 260:8668–8675.

*JONES, M. N., and J. K. NICKSON. 1981. Monosaccharide transport proteins of the human erythrocyte membrane. *Biochim. Biophys. Acta* 650:1–20.

MUECKLER, M., C. CARUSO, S. A. BALDWIN, M. PANICO, I. BLENCH, H. R. MORRIS, W. J. ALLARD, G. E. LIENHARD, and H. F. LODISH. 1985. Sequence and structure of a human glucose transporter. *Science* 229:941–945.

Anion Transport Protein

*CABANTCHIK, Z. I., P. A. KNAUF, and A. ROTHSTEIN. 1978. The anion transport system of the red blood cell: The role of membrane protein evaluated by the use of "probes." *Biochim. Biophys. Acta* 515:239–302.

FALKE, J. J., and S. I. CHAN. 1985. Evidence that anion transport by Band 3 proceeds via a ping-pong mechanism involving a single transport site. *J. Biol. Chem.* 260:9537–9544.

*JENNINGS, M. L. 1984. Oligomeric structure and the anion transport function of human erythrocyte band 3 protein. *J. Membrane Biol.* 80:105–117.

Active Transport

*KYTE, J. 1981. Molecular considerations relevant to the mechanism of active transport. *Nature* 292:201–204.

*TANFORD, C. 1983. Mechanism of free energy coupling in active transport. *Annu. Rev. Biochem.* 52:379–409.

WALDERHAUG, M., R. POST, G. SACCOMANI, R. LEONARD, and D. BRISKIN. 1985. Structural relatedness of three ion-transport adenosine triphosphatase around their active sites of phosphorylation. *J. Biol. Chem.* 260:3852–3859.

*Book or review article that provides a survey of the topic.

Na^+K^+-ATPase

* CANTLEY, L. C. 1981. Structure and mechanism of the (Na, K)-ATPase. *Curr. Topics Bioenerget.* **11**:201–237.

SHULL, G. E., A. SCHWARTZ, and J. B. LINGREL. 1985. Amino-acid sequence of the catalytic subunit of the (Na^+,K^+)-ATPase deduced from a complementary DNA. *Nature* **316**:691–695.

* SKOU, J. C., and J. G. NORBY, eds. 1979. *$Na^+K^+ATPase$: Structure and Kinetics.* Academic Press.

* SWEADNER, K. J., and S. M. GOLDIN. 1980. Active transport of sodium and potassium ions: Mechanism, function and regulation. *N. Engl. J. Med.* **302**:777–783.

TANIGUCHI, K., K. SUZUKI, D. KAI, I. MATSUOKA, K. TOMITA, and S. IIDA. 1984. Conformational change of sodium- and potassium-dependent adenosine triphosphatase. *J. Biol. Chem.* **259**:15228–15233.

Ca^{2+}-ATPase

* HASSELBACH, W., and H. OETLIKER. 1983. Energetics and electrogenicity of the sarcoplasmic reticulum calcium pump. *Ann. Rev. Physiol.* **45**:325–339.

* MACLENNAN, D. H., C. J. BRANDL, B. KORCZAK, and N. M. GREEN. 1985. Amino-acid sequence of a Ca^{2+},Mg^{2+}-dependent ATPase from rabbit muscle sarcoplasmic reticulum, deduced from its complementary DNA sequence. *Nature* **316**:696–700.

* SCHATZMAN, H. J. 1983. The red cell calcium pump. *Ann. Rev. Physiol.* **45**:303–312.

Proton Pumps

* FORTE, J. G., T. E. MACHEN, and K. J. OBRINK. 1980. Mechanism of gastric H^+ and Cl^- transport. *Annu. Rev. Physiol.* **42**:111–126.

GLUCK, S., C. CANNON, and Q. AL-AWQATI. 1982. Exocytosis regulates urinary acidification in turtle bladder by rapid insertion of H^+ pumps into the luminal membrane. *Proc. Nat'l Acad. Sci. USA* **79**:4327–4331.

MUALLEM, S., C. BURNHAM, D. BLISSARD, T. BERGLINDH, and G. SACHS. 1985. Electrolyte transport across the basolateral membrane of the parietal cells. *J. Biol. Chem.* **260**:6641–6653.

OHKUMA, S., Y. MORIYAMA, and T. TAKANO. 1982. Identification and characterization of a proton pump on lysosomes by fluorescein isothiocyanate-dextran fluorescence. *Proc. Nat'l Acad. Sci. USA* **79**:2758–2762.

RABON, E., R. D. GUNTHER, A. SOUMARMON, S. BASSILIAN, M. LEWIN, and G. SACHS. 1985. Solubilization and reconstitution of the gastric H,K-ATPase. *J. Biol. Chem.* **260**:10200–10207.

* SACHS, G., B. WALLMARK, G. SACCOMANI, E. RABON, H. B. STEWART, D. R. DIBONA, and R. BERGLINDH. 1982. The ATP-dependent component of gastric acid secretion. *Curr. Topics Membrane Transp.* **16**:135–160.

Cotransport: Symport and Antiport

Amino Acids and Glucose

* GUNN, R. B. 1980. Co- and counter-transport mechanisms in cell membranes. *Annu. Rev. Physiol.* **42**:249–259.

* SEMENZA, G., M. KESSLER, M. HOSANG, J. WEBER, and U. SCHMIDT. 1984. Biochemistry of the Na^+, D-glucose cotransporter of the small-intestinal brush border membrane. *Biochim. Biophys. Acta* **779**:343–379.

* ULLRICH, K. J. 1979. Sugar, amino acid, and Na^+ cotransport in the proximal tubule. *Annu. Rev. Physiol.* **41**:181–195.

* WEST, I. C. 1980. Energy coupling in secondary active transport. *Biochim. Biophys. Acta* **604**:91–126.

Calcium and Protons

* BUSA, W. B., and J. H. CROWE. 1983. Intracellular pH regulates transitions between dormancy and development of brine shrimp (*Artemia salina*) embryos. *Science* **221**:366–368.

MOOLENAAR, W. H., R. Y. TSIEN, P. T. VAN DER SAAG, and S. W. DE LAAT. 1983. Na^+/H^+ exchange and cytoplasmic pH in the action of growth factors in human fibroblasts. *Nature* **304**:645–648.

POUYSSEGUR, J., C. SARDET, A. FRANCHI, G. L'ALLEMAIN, and S. PARIS. 1984. A specific mutation abolishing Na^+/H^+ antiport activity in hamster fibroblasts precludes growth at neutral and acidic pH. *Proc. Nat'l Acad. Sci. USA* **81**:4833–4837.

* SCHULDINER, S., and E. ROZENGURT. 1982. Na^+/H^+ antiport in swiss 3T3 cells: Mitogenic stimulation leads to cytoplasmic alkalinization. *Proc. Nat'l Acad. Sci. USA* **79**:7778–7782.

Transport into Prokaryotic Cells

BROKER, R., T. H. WILSON. 1985. Isolation and nucleotide sequences of lactose carrier mutants that transport maltose. *Proc. Nat'l Acad. Sci. USA* **82**:3959–3963.

POSTMA, P. W., and J. W. LENGELER. 1985. Phosphoenolpyruvate:carbohydrate phosphotransferase system of bacteria. *Microbiol. Rev.* **49**:232–269.

SAIER, M. H. 1985. *Mechanisms and regulation of carbohydrate transport in bacteria.* Academic Press.

Movement of Water

KREGENOW, F. M. 1981. Osmoregulatory salt transporting mechanisms: Control of cell volume in anisotonic media. *Annu. Rev. Physiol.* **43**:493–505.

Internalization of Macromolecules and Particles

DAVEY, J., S. HURTLEY, and G. WARREN. 1985. Reconstitution of an endocytic fusion event in a cell-free system. *Cell* **43**:643–652.

* DAUTRY-VARSAT, A., and H. F. LODISH. 1984. How receptors bring proteins and particles into cells. *Sci. Am.* **250**(5):52–58.

* HELENIUS, A., I. MELLMAN, D. WALL, and A. HUBBARD. 1983. Endosomes. *Trends Biochem. Sci.* **8**:245–250.

* PASTAN, I., and M. C. WILLINGHAM. 1983. Receptor-mediated endocytosis: Coated pits, receptosomes and the Golgi. *Trends Biochem. Sci.* **8**:250–254.

* STEINMEN, R. M., I. S. MELLMAN, W. A. MULLER, and Z. A. COHN. 1983. Endocytosis and the recycling of plasma membrane. *J. Cell Biol.* **96**:1–27.

* WILSCHUT, J., and D. HOEKSTRA. 1983. Membrane fusion: from liposomes to biological membranes. *Trends Biochem. Sci.* 9:474–483.

Phagocytosis

AGGELER, J., and Z. WERB. 1982. Initial events during phagocytosis by macrophages viewed from outside and inside the cell: Membrane-particle interactions and clathrin. *J. Cell Biol.* **94**:613–623.

GRIFFIN, F. M., JR., J. A. GRIFFIN, J. E. LEIDER, and S. C. SILVERSTEIN. 1975. Studies on the mechanism of phagocytosis: I. Requirements for circumferential attachment of particle-bound ligands to specific receptors on the macrophage plasma membrane. *J. Exp. Med.* **142**:1263–1282.

GRIFFIN, F. M., JR., J. A. GRIFFIN, and S. C. SILVERSTEIN. 1976. Studies on the mechanism of phagocytosis: II. The interaction of macrophages with anti-immunoglobulin IgG-coated bone marrow–derived lymphocytes. *J. Exp. Med.* **144**:788–809.

HORWITZ, M. A. 1984. Phagocytosis of the Legionnaires' disease bacterium (*Legionella pneumophila*) occurs by a novel mechanism: engulfment within a pseudopod coil. *Cell* **36**:27–33.

HORWITZ, M. A. and F. R. MAXFIELD. 1984. *Legionella pneumophila* inhibits acidification of its phagosome in human monocytes. *J. Cell Biol.* **99**:1936–1943.

PFEFFERKORN, L. C. 1984. Transmembrane signaling: an ion-flux-independent model for signal transduction by complexed Fc receptors. *J. Cell Biol.* **99**:2231–2240.

YOUNG, J. D.-E., J. C. UNKELESS, T. M. YOUNG, A. MAURO, and Z. A. COHN. 1983. Role for mouse macrophage IgG Fc receptor as ligand-dependent ion channel. *Nature* **306**:186–189.

Pinocytosis

ADAMS, C. J., K. M. MAUREY, and B. STORRIE. 1982. Exocytosis of pinocytic contents by Chinese hamster ovary cells. *J. Cell Biol.* **93**:632–637.

GONELLA, P. A., and M. R. NEUTRA. 1984. Membrane-bound and fluid-phase macromolecules enter separate prelysosomal compartments in absorptive cells of suckling rat ileum. *J. Cell Biol.* **99**:909–917.

MELLMAN, I. S., R. M. STEINMAN, J. C. UNKELESS, and Z. A. COHN. 1980. Selective iodination and polypeptide composition of pinocytic vesicles. *J. Cell Biol.* **86**:712–722.

Clathrin

AGGELER, J., R. TAKEMURA, and Z. WERB. 1983. High-resolution three-dimensional views of membrane-associated clathrin and cytoskeleton in critical-point-dried macrophages. *J. Cell Biol.* **97**:1452–1458.

GOUD, B., C. HUET, and D. LOUVARD. 1985. Assembled and unassembled pools of clathrin: A quantitative study using an enzyme immunoassay. *J. Cell. Biol.* **100**:521–527.

* HARRISON, S. C., and T. KIRCHHAUSEN. 1983. Clathrin, cages and coated vesicles. *Cell* **33**:650–652.

* PEARSE, B. M. F., and M. S. BRETSCHER. 1981. Membrane recycling by coated vesicles. *Annu. Rev. Biochem.* **50**:85–101.

SCHMID, S. L., A. K. MATSUMOTO, and J. E. ROTHMAN. 1982. A domain of clathrin that forms coats. *Proc. Nat'l Acad. Sci. USA* **79**:91–95.

SCHMID, S. L., and J. E. ROTHMAN. 1985. Enzymatic dissociation of clathrin in a two-stage process. *J. Biol. Chem.* **260**:10044–10049.

ZAREMBA, S., and J. H. KEEN. 1983. Assembly polypeptides from coated vesicles mediate reassembly of unique clathrin coats. *J. Cell Biol.* **97**:1339–1347.

Low-Density Lipoprotein Receptor and LDL Uptake

ANDERSON, R. G. W., M. S. BROWN, U. BEISIEGEL, and J. L. GOLDSTEIN. 1982. Surface distribution and recycling of the low density lipoprotein receptor as visualized with antireceptor antibodies. *J. Cell Biol.* **93**:523–531.

* BROWN, M. S., and J. L. GOLDSTEIN. 1979. Receptor-mediated endocytosis: Insights from the lipoprotein receptor system. *Proc. Nat'l Acad. Sci. USA* **76**:3330–3337.

* BROWN, M. S. and J. L. GOLDSTEIN. 1984. How LDL receptors influence cholesterol and atherosclerosis. *Sci. Am.* **251**(5):58.

GOLDSTEIN, J. L., M. S. BROWN, G. W. ANDERSON, D. W. RUSSELL, and W. J. SCHNEIDER. 1985. Receptor-mediated endocytosis. *Annu. Rev. Cell Biol.* **1**:1–40.

LEHRMAN, M. A., J. L. GOLDSTEIN, M. S. BROWN, D. W. RUSSELL, and W. J. SCHNEIDER. 1985. Internalization-defective LDL receptors produced by genes with nonsense and frameshift mutations that truncate the cytoplasmic domain. *Cell* **41**:735–743.

LEHRMAN, M. A., W. J. SCHNEIDER, T. C. SUDHOF, M. S. BROWN, J. L. GOLDSTEIN, and D. W. RUSSELL. 1985. Mutation in LDL receptor: Alu-Alu recombination deletes exons encoding transmembrane and cytoplasmic domains. *Science* **227**:140–146.

SCHNEIDER, W. J., U. BEISIEGEL, J. L. GOLDSTEIN, and M. S. BROWN. 1982. Purification of the low density lipoprotein receptor, an acidic glycoprotein of 164,000 molecular weight. *J. Biol. Chem.* **257**:2664–2673.

SÜDHOF, T., J. GOLDSTEIN, M. S. BROWN, and D. W. RUSSELL. 1985. The LDL receptor gene. A mosaic of exons shared with different proteins. Science **228**:815–822.

YAMAMOTO, T., C. G. DAVIS, M. S. BROWN, W. J. SCHNEIDER, M. L. CASEY, J. L. GOLDSTEIN, and D. W. RUSSELL. 1984. The human LDL receptor: a cysteine-rich protein with multiple Alu sequences in its mRNA. *Cell* **39**:27–38.

Recycling of Receptors

DICKSON, R. B., L. BEGUINOT, J. A. HANOVER, N. D. RICHERT, M. C. WILLINGHAM, and I. PASTAN. 1983. Isolation and characterization of a highly enriched preparation of receptosomes (endosomes) from a human cell line. *Proc. Nat'l Acad. Sci. USA* **80**:5335–5339.

GEUZE, H. J., J. W. SLOT, G. J. A. M. STROUS, H. F. LODISH, and A. L. SCHWARTZ. 1983. Intracellular site of asialoglycoprotein receptor-ligand uncoupling: Double-label immunoelectron microscopy during receptor-mediated endocytosis. *Cell* **32**:277–287.

MERION, M., and W. S. SLY. 1983. The role of intermediate vesicles in the adsorptive endocytosis and transport of ligand to lysosomes in human fibroblasts. *J. Cell Biol.* **96**:644–650.

* SCHWARTZ, A. L. 1984. The hepatic asialoglycoprotein receptor. *CRC Crit. Rev. in Biochem.* **16**:207–233.

SCHWARTZ, A. L., S. E. FRIDOVICH, and H. F. LODISH. 1982. Kinetics of internalization and recycling of the asialoglycoprotein receptor in a hepatoma cell line. *J. Biol. Chem.* **257**:4230–4237.

STAHL, P., P. H. SCHLESINGER, E. SIGARDSON, J. S. RODMAN, and Y. C. LEE. 1980. Receptor-mediated pinocytosis of mannose glycoconjugates by macrophages: Characterization and evidence for receptor recycling. *Cell* **19**:207–213.

Acidification of Endocytic Vesicles

FORGAC, M., L. CANTLEY, B. WIEDENMANN, L. ALTSTIEL, and D. BRANTON. 1983. Clathrin-coated vesicles contain an ATP-dependent proton pump. *Proc. Nat'l Acad. Sci. USA* **90**:1300–1303.

MERION, M., P. SCHLESINGER, R. M. BROOKS, J. M. MOEHRING, T. J. MOEHRING, and W. S. SLY. 1983. Defective acidification of endosomes in Chinese hamster ovary cell mutants "cross-resistant" to toxins and viruses. *Proc. Nat'l Acad. Sci. USA* **80**:5315–5319.

TYCKO, B., and F. R. MAXFIELD. 1982. Rapid acidification of endocytic vesicles containing α_2-macroglobulin. *Cell* **28**:643–651.

YAMASHIRO, D. and F. MAXFIELD. 1984. Acidification of endocytic compartments and the intracellular pathway of ligands and receptors. *J. Cell. Biochem.* **26**:231–246.

Fates of Internalized Proteins

MOSTOV, K. E., and N. E. SIMISTER. 1985. Transcytosis. *Cell* **43**:389–390.

HOPPE, C. A., T. P. CONNOLLY, and A. L. HUBBARD. 1985. Transcellular transport of polymeric IgA in the rat hepatocyte:biochemical and morphological characterization of the transport pathway. *J. Cell Biol.* **101**:2113–2123.

ROTH, T. F., and K. R. PORTER. 1964. Yolk protein uptake in the oocyte of the mosquito *Aedas aegypti*. *J. Cell Biol.* **20**:313–332.

Transferrin and Iron Uptake

DAUTRY-VARSAT, A., A. CIECHANOVER, and H. F. LODISH. 1983. pH and the recycling of transferrin during receptor-mediated endocytosis. *Proc. Nat'l Acad. Sci. USA* **80**:2258–2262.

* NEWMAN, R., C. SCHNEIDER, R. SUTHERLAND, L. VODINELICH, and M. GREAVES. 1982. The transferrin receptor. *Trends Biomed. Sci.* 7:397–400.

ROUAULT, T., K. RAO, J. HARFORD, E. MATTIA, and R. D. KLAUSNER. 1985. Hemin, chelatable iron, and the regulation of transferrin receptor biosynthesis. *J. Biol. Chem.* **260**:14862–14866.

SCHNEIDER, C., M. J. OWEN, D. BANVILLE, and J. G. WILLIAMS. 1984. Primary structure of human transferrin receptor deduced from the mRNA sequence. *Nature* **311**:675–678.

WATTS, C. 1985. Rapid endocytosis of the transferrin receptor in the absence of bound transferrin. *J. Cell. Biol.* **100**:633–637.

YAMASHIRO, D. J., B. TYCKO, S. R. FLUSS and F. R. MAXFIELD. 1984. Segregation of transferrin to a mildly acidic (pH 6.5) para-Golgi compartment in the recycling pathway. *Cell* **37**:789–800.

Blood Vessel Walls

* BUNDGAARD, M. 1983. Vesicular transport in capillary endothelium: Does it occur? *Fed. Proc.* **42**:2425–2430.

SIMIONESCU, N., M. SIMIONESCU, and G. E. PALADE. 1975. Permeability of muscle capillaries to small heme-peptides: Evidence for the existence of patent transendothelial channels. *J. Cell Biol.* **64**:586–607.

Penetration of Viruses into Cells

DANIELS, R., J. DOWNIE, A. HAY, M. KNOSSOW, J. SKEHEL, M. WANG, and D. WILEY. 1985. Fusion mutants of the influenza virus hemagglutin glycoprotein. *Cell* **40**:431–439.

DOXSEY, S. J., J. SAMBROOK, A. HELENIUS, and J. WHITE. 1985. An efficient method for introducing macromolecules into living cells. *J. Cel. Biol.* **101**:19–27.

EDWARDS, J., E. MANN, and D. T. BROWN. 1983. Conformational changes in Sindbis virus envelope proteins accompanying exposure to low pH. *J. Virol.* **45**:1090–1097.

* GAROFF, H., K. SIMONS, and A. HELENIUS. 1982. How an animal virus gets into and out of its host cell. *Sci. Am.* **246**(2):58–66.

HELENIUS, A., J. KARTENBECK, K. SIMONS, and E. FRIES. 1980. On the entry of Semliki Forest virus into BHK-21 cells. *J. Cell Biol.* **84**:404–420.

HSU, M.-C., A. SCHEID, and P. W. CHOPPIN. 1982. Enhancement of membrane-fusing activity of Sendai virus by exposure of the virus to basic pH is correlated with a conformational change in the fusion protein. *Proc. Nat'l Acad. Sci. USA* 79:5862–5866.

Penetration of Bacterial Toxins into Cells

* CLEMENS, M. 1984. Enzymes and toxins that regulate protein synthesis. *Nature* **310**:727.

KAGAN, B. L., A. FINKELSTEIN, and M. COLOMBINI. 1981. Diphtheria toxin fragment forms large pores in phospholipid bilayer membranes. *Proc. Nat'l Acad. Sci. USA* **78**:4950–4954.

MOYA, M., A. DAUTRY-VARSAT, B. GOUD, D. LOUVARD, and P. BOQUET. 1985. Inhibition of coated pit formation in Hep_2 cell blocks the cytotoxicity of diphtheria toxin but not that of ricin toxin. *J. Cell Biol.* **101**:548–559.

SANDVIG, K., and S. OLSNES. 1981. Rapid entry of nicked diphtheria toxin into cells at low pH: Characterization of the entry process and effects of low pH on the toxin molecule. *J. Biol. Chem.* **256**:9068–9076.

16

Cell-to-Cell Signaling: Hormones and Receptors

No cell lives in isolation. An elaborate cell-to-cell communications network coordinates the growth, differentiation, and metabolism of the multitude of cells in the diverse tissues and organs that make up a fruit fly, a frog, or a human being.

Within small groups of cells, communication is often by direct cell-cell contacts. As we saw in Chapter 14, gap junctions permit adjacent cells to exchange small molecules and coordinate metabolic responses. Adhesive junctions between the plasma membranes of adjacent cells determine the shape and rigidity of many tissues. Moreover, the establishment of specific contacts between different types of cells is a necessary step in the differentiation of many tissues.

But cells also have to communicate over distances longer than those facilitated by chains of cell-cell contacts. For this purpose extracellular products act as signals: a substance released by the *signaling cell* is recognized by the *target cell,* in which it induces an appropriate response. Cells use an enormous variety of chemicals and signaling mechanisms to communicate with each other.

In many microorganisms, such as yeast, slime molds, and protozoans, secreted molecules coordinate the aggregation of free-living cells for sexual mating or for differentiation in altered environmental conditions. Chemicals released by one organism that can alter the behavior of other organisms of the same species are called *phero-*

mones. Some animals release pheromones, usually dispersing them into the air or water, to attract members of the opposite sex.

Within plants and animals, extracellular signals control the growth of most tissues, govern the synthesis and secretion of proteins, and regulate the composition of body fluids. In animals, internal extracellular signals affect overall growth and behavior, provide a stimulus for digestive enzymes after a meal, and regulate wound healing, to cite a few examples. A convenient classification of these extracellular signals is based on the distance over which the signal must act.

In *endocrine signaling,* the cells of endocrine organs release *hormones,* signaling substances that act on a distant set of target cells. In animals, an endocrine hormone is usually carried by the blood from its site of release to its target (Figure 16-1).

In *paracrine signaling,* the target cell is close to the signaling cell, and the signaling compound affects only the group of target cells adjacent to it. The conduction of an electrical impulse from one nerve cell to another or from a nerve cell to a muscle cell (inducing or inhibiting muscle contraction) involves signaling by extracellular chemicals called *neurotransmitters.* Neurotransmitters and *neurohormones* are examples of paracrine signals. Because of the special importance of the nervous system, nerve conduction is considered separately in Chapter 17.

In the third type of signaling, *autocrine signaling,* cells respond to substances that they themselves release. Autocrine signaling is usually confined to pathologic conditions. Certain tumor cells, for instance, synthesize and release *growth factors* that are required for normal cellular growth and division. But these growth factors stimulate the inappropriate, unregulated growth of the tumor cell itself, as well as adjacent nontumor cells, and can cause a tumor to form.

The same compound sometimes acts in two or even three types of signaling. Epinephrine and certain small peptides, for instance, function both as neurotransmitters (paracrine) and as systemic hormones (endocrine).

Communication by extracellular signals usually involves six steps: (1) *synthesis* and (2) *release* of the chemical by the signaling cell; (3) *transport* to the target cells; (4) *detection* of the signal (often by a specific receptor protein); (5) *transduction* of the signal from the receptor to other proteins; (6) *response.*

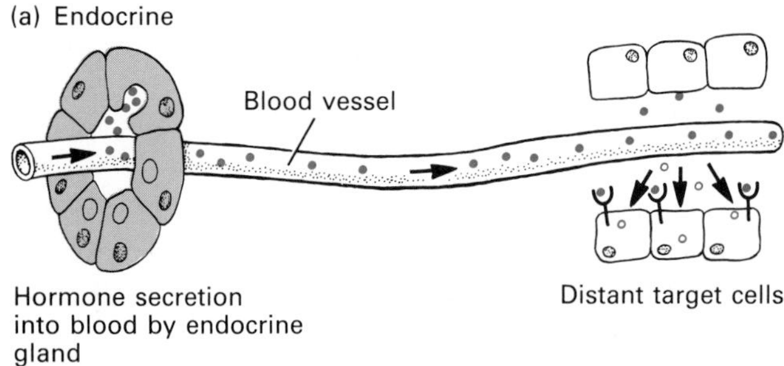

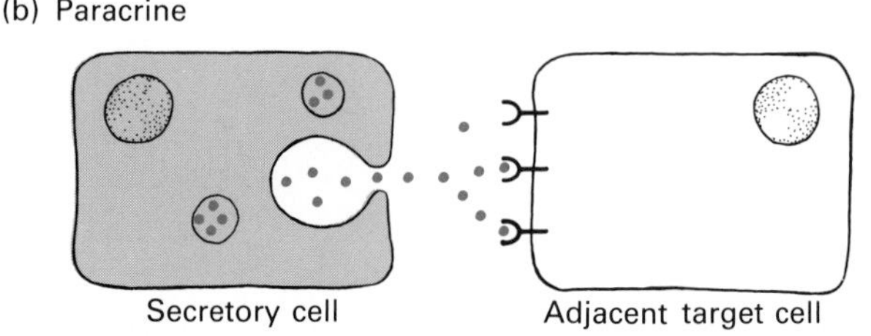

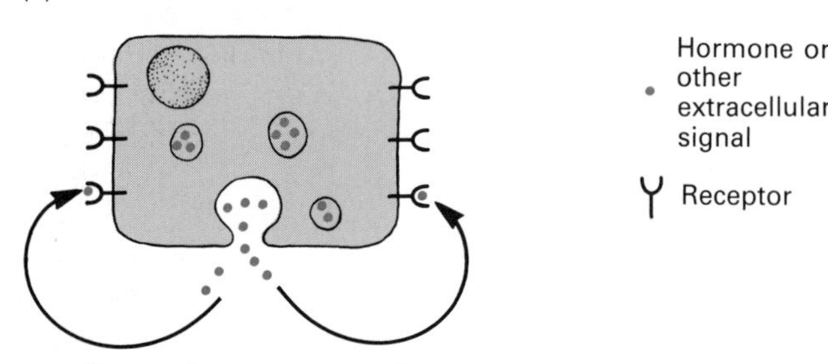

Figure 16-1 Three general schemes in animals for cell-to-cell signaling by extracellular chemicals. Such signaling can occur over very small to very large distances, from a few micrometers in paracrine or autocrine secretions to several meters in endocrine secretions.

The Role of Extracellular Signals in Cellular Metabolism

Some signals induce a modification in the activity of one or more enzymes already present in the target cell. This type of reaction allows the cell to respond quickly—within minutes or seconds. Most signaling molecules that induce such rapid changes are water soluble and bind to receptors located in the plasma membrane.

Other signaling molecules primarily alter the pattern of gene expression. Generally, molecules of this class are poorly soluble in aqueous solutions but soluble in lipids. In comparison to water-soluble molecules, lipid-soluble molecules induce in their target cells responses that are slower and longer lasting. Such prolonged interactions are crucial during growth and differentiation. Steroid hormones, the best-known examples of this class, are believed to interact with intracellular receptors to cause induction of specific genes by binding to controlling regions in DNA.

Hormones may regulate the synthesis, release, and degradation of other hormones. Complex feedback loops involving several hormones coordinate the metabolic response of many, if not most, cells in a multicellular organism.

There are many important hormones, pheromones, and neurotransmitters—more than we can possibly cover in this chapter. We will examine examples of those whose cellular and molecular mechanisms are reasonably well understood.

Specific Receptors Mediate the Response of Cells to Extracellular Signals

On the surface of the target cell, or in its cytoplasm, is a *receptor protein* that has a binding site with high affinity for a particular signaling substance (a hormone, pheromone, or neurotransmitter). When the signaling substance binds to the receptor, the receptor-ligand complex (a ligand is a substance that binds to, or "fits," a site) initiates a sequence of reactions that changes the function of the cell.

The response of a cell or tissue to a group of hormones is dictated both by the particular set of hormone receptors it possesses and by the intracellular reactions initiated by the binding of any one hormone to its receptor. One cell may have two or more receptor types for the same ligand, each type initiating a different response. Or different cells may have different sets of receptors for the same hormone, each receptor inducing a different response. Or the same receptor may occur on various cells and binding of the hormone trigger different responses. Clearly, cells respond in a variety of ways to the same hormone. For instance, acetylcholine receptors are found on the surface of both striated muscle cells and pancreatic acinar cells. Release of acetylcholine from a neuron adjacent to a muscle cell triggers contraction whereas release adjacent to an acinar cell triggers exocytosis of secretory granules. In some systems, different receptor-hormone complexes induce the same cellular response. In liver cells, either the binding of glucagon to glucagon receptors or the binding of epinephrine to epinephrine receptors can induce the degradation of glycogen and the release of glucose into the blood. Thus two aspects to receptor function are: *binding specificity* of the receptor for the ligand, and *effector specificity* for the resulting change in cellular behavior.

In most receptor-ligand systems, the ligand appears to have no function other than binding to the receptor. The ligand is not metabolized to useful products, is not itself an intermediate in any cellular activity, and has no enzymatic properties. The ligand's only function appears to be to change the properties of the receptor, which signals to the cell the presence of a specific product in the environment. Often target cells modify or degrade the ligand. In doing so, they can modify or terminate their response (or the response of neighboring cells) to the signal.

Many Types of Chemicals Can Be Used as Hormones

Hormones fall into three broad categories: (1) small lipophilic molecules that diffuse through the plasma membranes and interact with cytosolic or nuclear receptors; (2) hydrophilic molecules that act by binding to cell surface receptors; (3) large lipophilic molecules that bind to plasma membrane receptors (Figure 16-2).

(a)

Surface receptors

Ligands on surface receptors

Ligands

Low concentration of cytosolic cAMP or Ca^{2+}

High concentration of cytosolic cAMP or Ca^{2+}

(b)

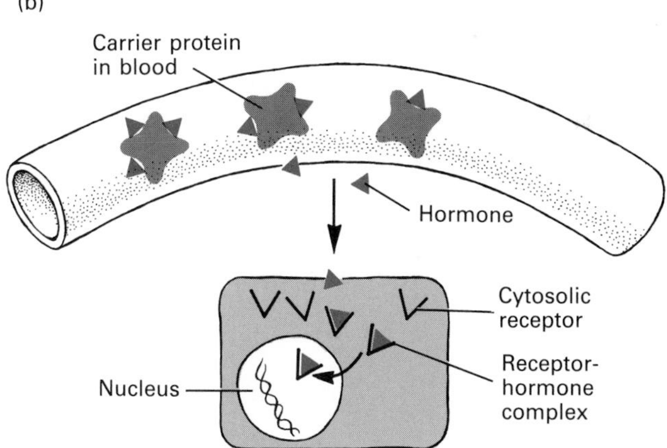

Figure 16-2 Some hormones bind to cell surface receptors, others to receptors within the cell. (a) Cell surface receptors. Peptide and protein hormones, prostaglandins, and epinephrine and its related compounds bind to cell surface receptors. Binding triggers an increase (or decrease) in the cytosolic concentration of cAMP, Ca^{2+}, or some other substance. (b) Cytosolic or nuclear receptors. Steroids and thyroxine, being very hydrophobic, are transported by carrier proteins in the blood. Dissociated from the carrier, the hormones enter the cell, bind to specific receptors in the cytosol or nucleus, and act on nuclear DNA to increase transcription of specific genes.

Table 16-1 Some hormones that interact with cytosolic or nuclear receptors

Hormone	Structure	Origin	Major effects
STEROIDS			
Progesterone	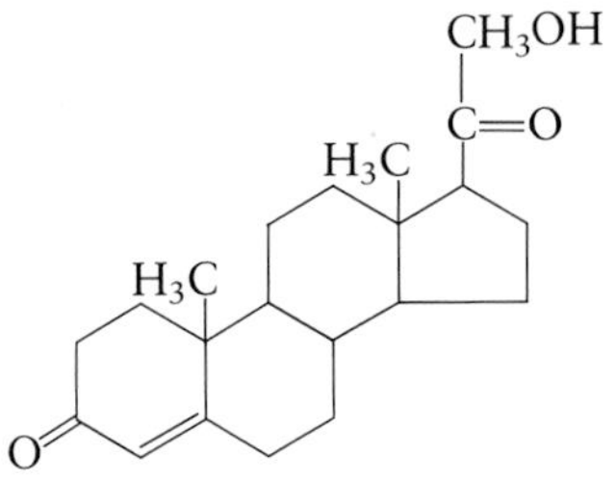 CH_3OH, C=O, H_3C, H_3C, O	Ovary (corpus luteum); placenta	Differentiation of the uterus in preparation for implantation of the early embryo; maintenance of pregnancy; development of the alveolar system in mammary glands
Estradiol (one of three estrogens)	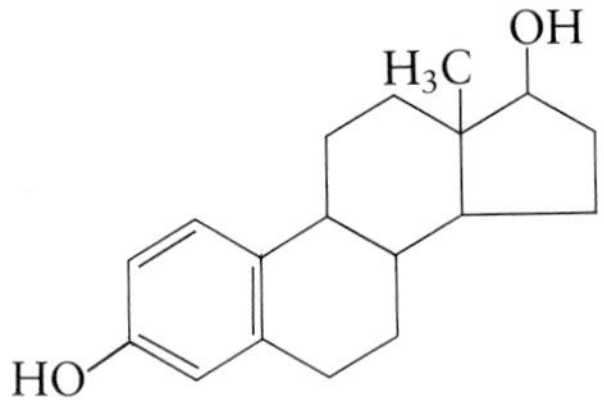 OH, H_3C, HO	Ovary (corpus luteum); placenta	Differentiation of the uterus and other female sex organs; maintenance of secondary female sex characteristics and of the normal cyclic function of accessory sex organs; development of the duct system in mammary glands
Testosterone	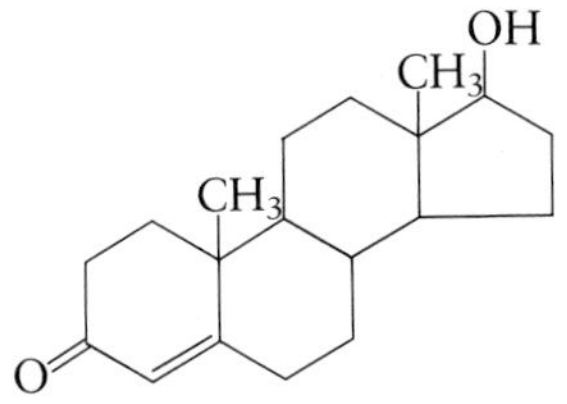 OH, CH_3, CH_3, O	Testis	Maturation and normal function of accessory male sex organs; development of male sex characteristics
Cortisol	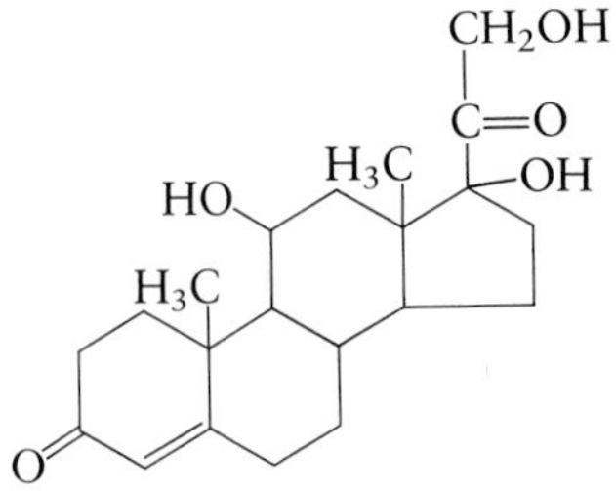 CH_2OH, C=O, OH, HO, H_3C, H_3C, O	Adrenal cortex	Effect on metabolism of carbohydrates, lipids, and proteins; reduction of inflammation, enhanced immunity, and thus resistance to infection
Corticosterone	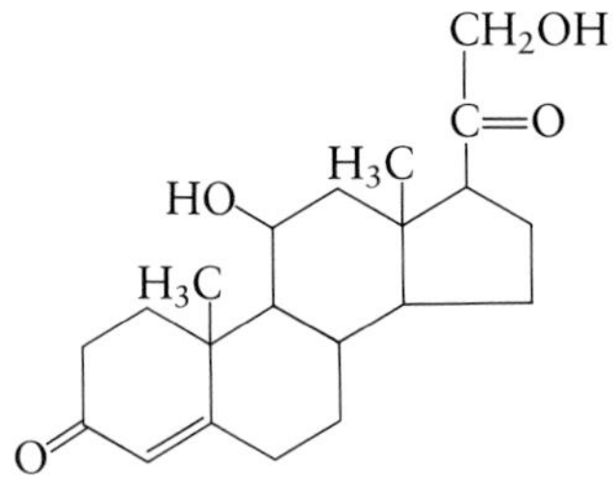 CH_2OH, C=O, HO, H_3C, H_3C, O	Adrenal cortex	Effect on metabolism of carbohydrates, lipids, and proteins; enhanced production of glucose by the liver

Continued on next page

Most Lipophilic Hormones Interact with Cytosolic or Nuclear Receptors The principal representatives of the group of hormones interacting with intracellular receptors are the steroids and thyroxine (Table 16-1). All steroids are synthesized from cholesterol and thus have a similar chemical skeleton. After crossing the plasma membrane they interact with receptor proteins in the nucleus or cytosol. These receptor-steroid complexes are believed to accumulate in the nucleus where, directly or indirectly, they modify DNA transcription (see Chapter 12); they may also affect the stability of specific mRNAs. Steroid effects are of relatively long duration—hours or days—and they often affect growth and differentiation of specific tissues. For example, the steroid sex hormones estrogen and progesterone stimulate the production of egg-white proteins in chickens and are neces-

Table 16-1 *Continued*

Hormone	Structure	Origin	Major effects
Aldosterone	HO, H_3C, O, CH, CH_2OH, C=O	Adrenal cortex	Maintenance of water and ion balance
Cholesterol (not a hormone)	HO, CH_3, CH_3, CH_3, CH, CH_2, CH_2, CH_2, CH, CH_3, CH_3	Liver; cholesterol-containing food	Manufacture of steroid hormones; maintenance of cell membranes
STEROID-LIKE HORMONES			
α-Ecdysone	HO, HO, O, OH, CH_3, CH_3, CH_3, CH, CH, OH, CH_2, CH_2, C—OH, CH_3, CH_3	Endocrine glands (insects, crustaceans)	Differentiation and maturation of larvae
OTHER HORMONES			
Thyroxine [tetraiodothyronine (T4)]	HO—(I, I)—O—(I, I)—CH_2—C(H)(NH_3^+)—COO^-	Thyroid	Increase of oxidation of sugars by most body cells; induction of some enzymes
Triiodothyronine (T3)	HO—(I)—O—(I, I)—CH_2—C(H)(NH_3^+)—COO^-		

sary for cell proliferation in the growth of the hen oviduct. In mammals, estrogens stimulate growth of the wall of the uterus in preparation for implantation of the embryo. In insects and crustaceans, ecdysone, although not classified chemically as a steroid, triggers the differentiation and maturation of larvae; like estrogens, it induces the transcription of specific gene products. Thyroxine, the principal iodinated compound in the body, causes an increase in metabolism and oxidation in many body tissues. Although not a steroid, it is believed to act in a similar fashion. By binding to nuclear receptors it affects the transcription of specific mRNAs and may also act to stabilize certain mRNAs.

Because steroids and thyroxine are poorly soluble in aqueous solutions, they are transported in the blood tightly bound by carrier proteins. They typically have a

Table 16-2 Some hormones that interact with cell surface receptors*

Hormone	Structure	Origin	Major effects
Epinephrine	3,4-(HO)$_2$-C$_6$H$_3$—CH(OH)—CH_2—NH—CH_3	Adrenal medulla	Increase in pulse rate and blood pressure; contraction of most smooth muscles; glycogenolysis in liver and muscle; lipid hydrolysis in adipose tissue
Norepinephrine	3,4-(HO)$_2$-C$_6$H$_3$—CH(OH)—CH_2—NH_2	Adrenal medulla	Contraction of arterioles; decrease of peripheral circulation
Glucagon	Peptide, 29 aa†	Pancreas α cells	Glycogenolysis in liver; lipid hydrolysis in adipose tissue
Insulin	Polypeptide A chain, 21 aa B chain, 30 aa	β Cells of pancreas	Stimulation of glucose uptake into cells and carbohydrate catabolism; stimulation of lipid synthesis by adipose tissue; general stimulation of protein synthesis and cell proliferation
Gastrin	Polypeptide, 17 aa	Intestine	Secretion of HCl and pepsin in stomach
Secretin	Polypeptide, 27 aa	Small intestine	Secretion of HCO_3^--rich pancreatic juice
Cholecystokinin	Polypeptide, 23 aa	Small intestine	Secretion of pancreatic digestive enzymes; emptying of gallbladder
Adrenocorticotropic hormone (ACTH)	Polypeptide, 39 aa	Anterior pituitary	Lipid hydrolysis from adipose tissue; stimulation of adrenal cortex to produce cortisol, corticosterone, and aldosterone
Somatotropin (growth hormone)	Polypeptide, 191 aa	Anterior pituitary	Stimulation of liver to produce somatomedins, which, in turn, cause growth of bone and muscle

* Molecules that are primarily used as neurotransmitters or neurohormones are depicted in Table 17-1.
†Amino acids.

Continued on next page

lifetime in the blood plasma of several hours (steroids) to several days (thyroxine), and their effects on target cells last from hours to days.

Water-Soluble Hormones Interact with Cell Surface Receptors Many hormones are not soluble in lipids. They cannot diffuse across the plasma membrane and interact with intracellular receptors, but instead must interact with cell surface receptors. This class of hormone includes both large polypeptides such as insulin, growth hormones, and glucagon and small charged compounds such as the catecholamines epinephrine and norepinephrine (Table 16-2). The effects of these surface-bound hormones are usually immediate and very short term; the cell reacts in milliseconds or at most a few seconds. The presence of a receptor-ligand complex on the surface of a cell triggers a short-lived increase in the concentration of an intracellular compound termed a *second messenger*, which is often either 3′,5′-cyclic AMP (cAMP) or a calcium ion (Ca^{2+}). The elevated intracellular concentration of either of these two substances triggers a rapid alteration in the activity of one or more enzymes or nonenzymatic proteins. The metabolic functions controlled by these hormones include uptake and utilization of glucose, storage and mobilization of fat, and secretion of cellular products. Some receptors, such as the insulin receptor,

Table 16-2 *Continued*

Hormone	Structure	Origin	Major effects
Epidermal growth factor (EGF)	Polypeptide, 53 aa	Salivary and other glands?	Growth of epidermal and other body cells
Follicle-stimulating hormone (FSH)	Protein: α chain, 92 aa β chain, 118 aa	Anterior pituitary	Stimulation of growth of oocyte and ovarian follicles; stimulation of estrogen synthesis by follicles
Luteinizing hormone (LH)	Protein: α chain, 92 aa β chain, 115 aa	Anterior pituitary	Maturation of oocyte; stimulation of estrogen and progesterone synthesis by ovarian follicles
Thyroid-stimulating hormone (TSH)	Protein: α chain, 92 aa β chain, 112 aa	Anterior pituitary	Release of thyroxine by thyroid cells
Parathyroid hormone	Protein, 84 aa	Parathyroid	Increase in blood Ca^{2+} and phosphate owing to resorption of bone and resorption in kidney
Somatomedins 1, 2 (insulin-like growth factors 1, 2)	Peptides	Liver	Growth of bone, muscle, and other cells
Vasopressin	Peptide, 9 aa	Posterior pituitary	Increase in water absorption from urine by kidney tubules; constriction of small blood vessels and rise in blood pressure
Histamine	HC═C—CH_2—CH_2—NH_3^+ (imidazole ring: N, NH, C, H)	Mast cells	Dilation of blood vessels
Prostaglandins PGE_2	O, COOH, OH, OH	Most body cells	Contraction of smooth muscle

may not utilize a second messenger but instead act directly to modify the activity of cytoplasmic proteins by phosphorylating them.

Prostaglandins Prostaglandins, lipid-soluble hormone-like chemicals produced in most body tissues, are believed to bind to cell surface receptors (Table 16-2). They contain a cyclopentane ring and are synthesized from arachidonic acid, a 20-carbon fatty acid with four double bonds. There are at least 16 different prostaglandins in nine different chemical classes designated PGA, PGB, . . . , PGI. Continuously synthesized and secreted by many types of cells in both vertebrates and invertebrates, they are continuously and rapidly destroyed by enzymes in body fluids. Many act as local mediators (paracrine signaling) and are rapidly destroyed near their site of synthesis. They modulate the responses of other hormones and can have profound effects on many cellular processes. Certain prostaglandins cause blood platelets to aggregate and adhere to the walls of blood vessels. Because platelets play a key role in clotting blood and plugging leaks in blood vessels, these prostaglandins can affect the course of vascular disease and wound healing. Aspirin and other anti-inflammatory agents inhibit the synthesis of these prostaglandins. Other prostaglandins initiate the contraction of smooth muscle cells. They accumulate in the uterus at the time of childbirth and are believed to be important in inducing uterine contraction—they are now often used to induce abortion. Prostaglandins may act by changing the cytosolic concentration of cAMP, Ca^{2+}, or other second messengers.

The Synthesis, Release, and Degradation of Hormones Are Regulated

Organisms must be able to respond instantly to many changes in the internal or external environment. Typically, responses of seconds or minutes in duration are mediated by peptide hormones or the catecholamines. The signaling cells typically have one to several days' supply of hormone stored in secretory vesicles just under the plasma membrane. All peptide hormones, such as insulin and adrenocorticotropic hormone (ACTH), are synthesized as part of a longer polypeptide that is proteolytically processed to the mature and active molecule before or just after it is transported into the secretory vesicle.

Stimulation of the signaling cell by a neurotransmitter or hormone causes immediate exocytosis of the peptide hormone into the blood. The signaling cells are also stimulated to synthesize the hormone and replenish the cell's supply. The released peptide hormones have a lifetime in the blood of only seconds or minutes and are rapidly degraded by blood and tissue proteases. Released catecholamines are inactivated by different enzymes. The initial actions of these signaling substances on target cells—activation or inhibition of specific enzymes—similarly lasts only seconds or minutes (Table 16-3). Thus the catecholamines and some peptide hormones mediate short-lasting responses in which termination of the response is caused by their own degradation.

The steroid hormones and thyroxine, by contrast, persist in the blood for hours or days, bound to carrier proteins (Table 16-3). Target cells thus are exposed to the signal for long periods of time, and their responses last for hours or longer. Release of steroids and thyroxine by the signaling cells is closely regulated. For thyroxine release, thyroid-stimulating hormone (TSH) triggers endocytosis of the iodinated precursor protein thyroglobulin and its proteolysis to thyroxine. Steroids are synthesized from cholesterol by pathways involving 10 or more enzymes. Steroid-producing cells, such as those in the adrenal cortex, typically store a few hours' supply of precursors. When stimulated, the cells convert the precursor to the finished hormone, which then diffuses across the plasma membrane into the blood.

The K_m Values for Hormone Receptors Approximate the Concentration of Circulating Hormone

Hormone receptors, whether on the plasma membrane or within the cell, bind their ligands with great specificity and high affinity. Binding of a hormone to a receptor involves the same types of weak interactions—ionic and van der Waals bonds and hydrophobic interactions—as those characterizing the specific binding of a substrate to an enzyme (Chapter 3). Binding can usually be described by a simple equation:

$$R + H \rightleftharpoons [RH] \qquad K_m = \frac{[R][H]}{[RH]}$$

where [R] and [H] represent the concentrations of free receptor and hormone, respectively, and [RH] is the con-

Table 16-3 Characteristic features of principal types of mammalian hormones

Feature	Steroids	Thyroxine	Peptides and proteins	Catecholamines
Feedback regulation of synthesis	Yes	Yes	Yes	Yes
Storage of preformed hormone	Few hours	Several weeks	One day	Several days, in adrenal medulla
Mechanism of secretion	Diffusion through plasma membrane	Proteolysis of thyroglobulin	Exocytosis of storage vesicles	Exocytosis of storage vesicles
Plasma binding proteins	Yes	Yes	Rarely	No
Lifetime in plasma	Hours	Days	Minutes	Seconds
Time course of action	Hours to days	Days	Minutes to hours	Seconds or less
Receptors	Cytosolic or nuclear	Nuclear	Plasma membrane	Plasma membrane
Mechanism of action	Transcriptional control and mRNA stability	Transcriptional control and mRNA stability	Second messenger in cytosol	Second messenger in cytosol

SOURCE: E. L. Smith et al., 1983, *Principles of Biochemistry: Mammalian Biochemistry,* 6th ed., McGraw-Hill, p. 358. Reproduced by permission of McGraw-Hill.

centration of the receptor-hormone complex. K_m is the dissociation constant of the receptor-ligand complex and measures the affinity of the receptor for the ligand.

In general, the K_m values for hormone receptors approximate the concentration of the hormone in the circulation. Changes in hormone concentration will be reflected in proportional changes in the fraction of receptors occupied. Suppose, for instance, that the normal (unstimulated) concentration of a hormone in the circulation is 10^{-9} *M*, and the K_m of the receptor for the hormone is 10^{-7} *M*. We calculate from the above equation that 1.01 percent of the receptors would be bound with hormone, that is, $[RH]/\{[RH] + [R]\} = 0.0101$. Suppose the concentration of hormone rises 10-fold, to 10^{-8} *M*. The concentration of RH complex rises about 10-fold, to 11.1 percent. If the induced cellular response is proportional to the amount of receptor-hormone complex, as is often the case, then the cellular responses will also increase about 10-fold.

But consider a hypothetical case where the K_m of a cell for hormone *X* is 10^{-9} *M* and the normal concentration of the hormone is 10^{-7} *M*, a hundred times greater. Then 99 percent of the receptor sites would normally be bound with hormone, and twofold changes in hormone concentration would result in changes of less than 2 percent in the fraction of receptors that contain bound hormone.

Frequently cells must respond to small changes in the concentrations of signaling molecules with large reactions. One way to do this is to bind two or more molecules of ligand, where an effect requires that two ligands be bound simultaneously to a receptor,

$$R + H \xrightleftharpoons{K_{m1}} [RH]$$

$$[RH] + H \xrightleftharpoons{K_{m2}} [RH_2] \longrightarrow \text{effect}$$

Binding of the two ligands can be noncooperative, which means that binding of the first H molecule does not affect the binding of the second (i.e., $K_{m1} = K_{m2}$). Noncooperative binding sharpens the response curve (see curve B of Figure 16-3.)

Receptors often bind more than one molecule of hormone in a *positive cooperative interaction:* the binding of one molecule increases the affinity of the receptor for a second molecule ($K_{m1} > K_{m2}$), much in the way binding of one O_2 molecule to hemoglobin increases the affinity of binding of additional O_2 molecules (Chapter 3). In this way, cells are more sensitive to small changes in hormone levels over a restricted range of hormone concentrations than they could be if binding were noncooperative (curve C, Figure 16-3).

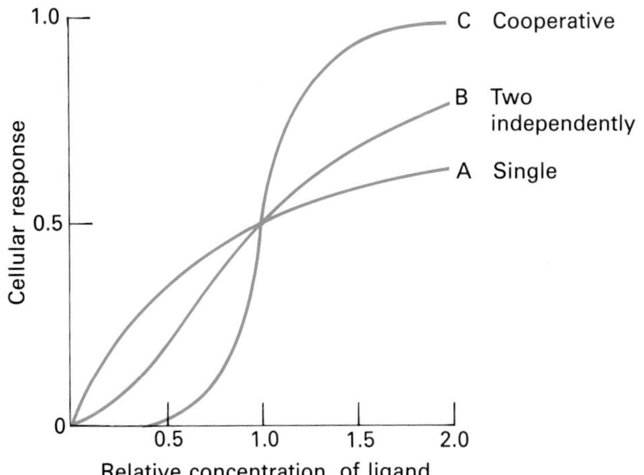

Figure 16-3 Graph shows how a sharp activation curve can result if two effector molecules (hormones) must bind simultaneously to a receptor in order to induce a response. Curve A shows the response if binding of only a single ligand is required to induce the response, and curve B the response if two ligands binding to a receptor independently of each other are required. In the latter case, a relatively small increase in ligand concentration causes a proportionately greater cellular response. Curve C shows the situation if there is a positive cooperative interaction between the two sites: binding of one molecule of hormone to a receptor causes a conformational alteration in that receptor or in an adjacent receptor that increases affinity for a second hormone molecule. Small changes in hormone concentration (e.g., from 0.8 to 1.0 unit) can cause very large changes in the fraction of receptors that are occupied and, as a result, very large changes in the cellular response. Sigmoidal curves, such as B or C, are always indicative of an effect in which multiple ligands bind simultaneously to a single receptor.

Hormone Receptors Are Detected by Their Binding to Radioactive Hormones

The identification and purification of hormone receptors is a difficult task, mainly because they are present in such minute amounts. Typically, the surface of a cell will bear 10,000 to 20,000 receptors for a particular hormone, but this quantity is only about 10^{-6} of the total protein in the cell or about 10^{-4} of the plasma membrane protein. As a consequence, it is difficult to purify and characterize most receptor proteins. Usually, receptors are detected and their amounts measured by a functional assay—their ability to bind radioactive hormone to a cell or to fragments of a cell (Figure 16-4). The development of chemical syntheses of radioactive hormones that retain normal hormone activity was crucial for the identification of many receptors. Once bound to a receptor, the radioactive hormone provides a means to identify and follow the receptor through isolation and purification procedures.

To illustrate the general principles of receptor-hormone interaction, let us examine several well-studied systems. We will focus on several key issues: (1) the detection and

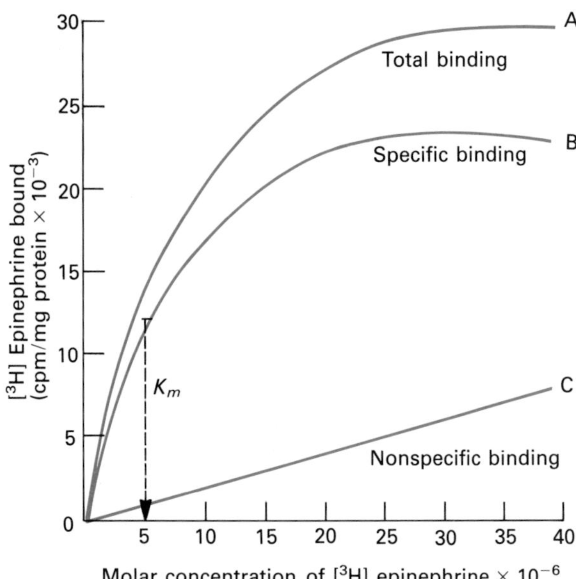

Figure 16-4 The identification of a specific epinephrine receptor on the surface of frog erythrocytes. A suspension of cells is incubated for 1 h at 0°C with increasing concentrations of $[^3H]$hormone. To measure the amount bound to the cells, the suspension is layered on a tube of mineral oil and centrifuged; the cells with their surface-bound hormone are recovered in the pellet. Total binding (curve A) consists of both the specific binding of hormone to high-affinity receptors and a nonspecific sticking to other molecules on the cell surface. The contribution of nonspecific binding to the total (curve C) can be determined by measuring binding of $[^3H]$hormone in the presence of a 100-fold excess of unlabeled hormone. This large excess of unlabeled hormone ensures that all the specific sites are occupied by unlabeled ligand. Specific binding (curve B) is calculated from the difference of curves A and C. Curve B fits the simple binding equation:

$$K_m = \frac{[R][H]}{[RH]} \quad \text{or} \quad [RH] = \frac{R_T}{1 + K_m/[H]}$$

where R_T is the total amount of receptor. $R_T = [R] + [RH]$. The K_m for this receptor is 5×10^{-6} *M*. [See R. J. Lefkowitz, L. E. Limbird, C. Mukherjee, and M. G. Caron, 1976, *Biochim. Biophys. Acta* **457**:1.]

quantitation of a receptor; (2) the means by which ligand binding is transmitted into a metabolic response; (3) the regulation of receptor levels.

Epinephrine Receptors and the Activation of Adenylate Cyclase

The best understood mammalian cell surface receptor system is that found on many cells for the hormones epinephrine and norepinephrine (Table 16-2). These hormones were originally recognized as products of the medulla, or core, of the adrenal gland, and they are also known as adrenaline and noradrenaline. Nerve cells derive embryologically from the same tissue as the adrenal medulla cells, and epinephrine and norepinephrine are also secretory products of differentiated nerve cells. Norepinephrine is employed most often as a neurotransmitter. Epinephrine is usually used as an endocrine hormone but is also a neurotransmitter for some neurons. Both hormones are charged compounds that belong to the catecholamines, which are active amines containing the compound catechol:

[catechol structure: benzene ring with two adjacent OH groups]
OH
OH

In times of stress, as in fright or during heavy exercise, all tissues have an increased need for glucose and fatty acids. These principal metabolic fuels can be supplied to the blood within seconds by rapid breakdown of glycogen in the liver (glycogenolysis) and of triacylglycerol in the adipose storage cells (lipolysis). In mammals, the liberation of glucose and fatty acids can be triggered by the binding of epinephrine or norepinephrine to *β-adrenergic receptors* on the hepatic (liver) and adipose cells. Binding of epinephrine to similar β-adrenergic receptors on heart muscle cells increases the rate of contraction and thus increases blood supply to the tissues. Binding to β-adrenergic receptors on smooth muscle cells of the intestine causes a relaxation of the bowel. A quite different type of epinephrine receptor, the *α-adrenergic receptor,* is found on smooth muscle cells lining the blood vessels in the intestinal tract, skin, and kidneys. Binding of epinephrine to these receptors causes the arteries to constrict, cutting off circulation to these peripheral organs. All of these diverse effects of one hormone are directed to a common end—supplying energy for the rapid movement of the body's major locomotor muscles in response to stress.

Despite the very different tissue-specific effects induced by the binding of epinephrine to β-adrenergic receptors, all are mediated by a rise in the intracellular level of cAMP. Cyclic AMP acts as a signal transducer, or second messenger, modifying the rates of different enzyme-catalyzed reactions in different tissues (Table 16-4). Before discussing the role of cAMP in cellular metabolism, let us investigate some key properties of the β-adrenergic receptors and the mechanism by which hormone binding triggers an increase in the level of cAMP.

Table 16-4 Some metabolic responses to a rise in intracellular cAMP (all are mediated by cAMP-dependent protein kinase)

Tissue	Hormone inducing a rise in cAMP	Metabolic response
Adipose	Epinephrine; ACTH; glucagon	Increase in hydrolysis of triglyceride; decrease in amino acid uptake
Liver; muscle	Epinephrine; norepinephrine	Increase in conversion of glycogen to glucose; inhibition of synthesis of glycogen; increase in amino acid uptake; increase in gluconeogenesis (synthesis of glucose from amino acids) by liver
Ovarian follicle	FSH; LH	Increase in synthesis of estrogen, progesterone
Adrenal cortex	ACTH	Increase in synthesis of aldosterone, cortisol
Cardiac muscle cells	Epinephrine	Increase in rate of beating and increase in strength of contraction
Smooth muscle (intestine, peripheral capillaries)	Epinephrine; norepinephrine	Decrease in tension
Thyroid	TSH	Secretion of thyroxine
Bone cells	Parathyroid hormone	Increase in resorption of calcium from bone

SOURCE: E. W. Sutherland, 1972, *Science* 177:401.

β-Adrenergic Receptors Can Be Detected by Ligand Binding

The binding of epinephrine to sensitive cells serves as a good example of the method for identifying and characterizing a surface hormone receptor. When increasing amounts of radioactive epinephrine are added to a cell, the amount binding to the surface increases at first and then approaches a plateau value (see Figure 16-4). How can one determine what fraction of this binding is specific to the surface receptor and what fraction is nonspecifically bound to the multitude of other proteins and phospholipids on the cell surface? Nonspecific binding of hormone may be misinterpreted to mean there are more receptor sites than in fact there are. One can measure nonspecific binding of labeled hormone by conducting the binding assay in the presence of a large excess of unlabeled hormone (Figure 16-4). The large excess ensures that all of the specific, high-affinity binding sites are filled by unlabeled hormone. Specific binding is calculated as the difference between total binding and nonspecific binding.

The specific binding is saturable. From the saturation level one can calculate the number of epinephrine-binding sites per cell. As is shown by curve B in Figure 16-4, there are about 1500 receptor molecules on each frog erythrocyte. The binding curve can be described by the simple binding equilibrium, also seen on Figure 16-4. In the erythrocyte epinephrine receptors depicted in Figure 16-4, $K_m = 5 \times 10^{-6}$ *M*; in other words, the receptor is half-saturated with hormone when the hormone is present at only 5×10^{-6} *M*. The binding affinities of other receptors can be even greater; the dissociation constant of the receptor-insulin complex is in the range of $1\text{–}5 \times 10^{-9}$ *M*.

But how do we know that specific binding of epinephrine to the cell surface actually represents binding to the physiologically important receptor, the receptor that triggers the cell's response? One indication is that the concentration of epinephrine half-saturating the receptor, about 5 μM, is the same as the concentration causing the level of cAMP inside the cell to increase to half its maximum (Figure 16-5).

Further evidence has come from studies of purified β-adrenergic receptors. Many receptors can be solubilized from plasma preparations by detergents and still fully retain their hormone-binding activity. As Figure 16-6 shows, the receptor can be purified by affinity chromatography. The receptor is a single polypeptide chain of m.w. about 64,000. The pure receptor can be incorporated into liposomes, and the liposomes fused with a cell that contains adenylate cyclase (the enzyme that synthe-

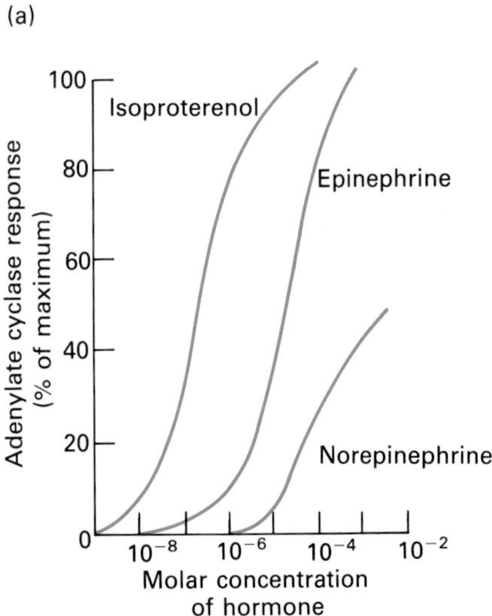

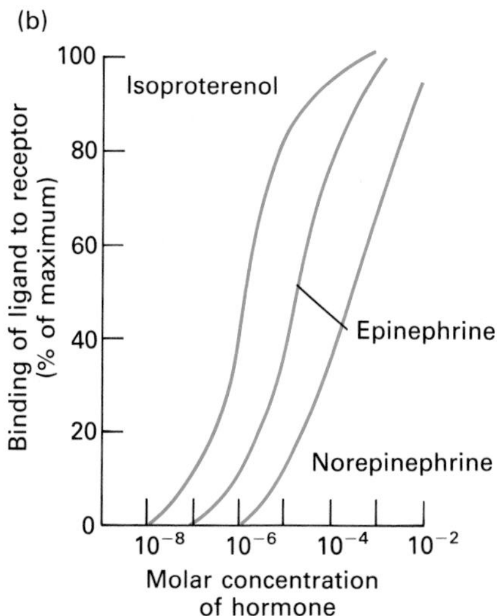

Figure 16-5 Comparison of the ability of three catecholamines to activate synthesis of cAMP in frog erythrocytes (a) and to bind to the cell surface β-adrenergic receptor (b). In (a), different concentrations of the agonists isoproterenol, epinephrine, and norepinephrine were incubated with a cell suspension at 37°C; the cells were then broken and the activity of adenylate cyclase (the enzyme that synthesizes cAMP) measured. The binding of the agonists to the receptor (b) was measured by an indirect competition assay. Cells were incubated with a constant low concentration of the [^{3}H] antagonist alprenolol, and variable concentrations of each of the unlabeled agonists isoproterenol, epinephrine, or norepinephrine. Binding of the unlabeled agonist to the receptor inhibits the binding of the [^{3}H]alprenolol, allowing the estimation of the binding affinities of unlabeled agonists for the receptor. The curves show that each agonist induces adenylate cyclase proportionate to its ability to bind to the receptor: the concentration of agonist required for half-maximal binding to the receptor is the same as that required for activation of adenylate cyclase. These and similar studies using other hormone agonists indicate that elevation of cAMP levels is a consequence of the binding of agonists to the same surface β-adrenergic receptor.

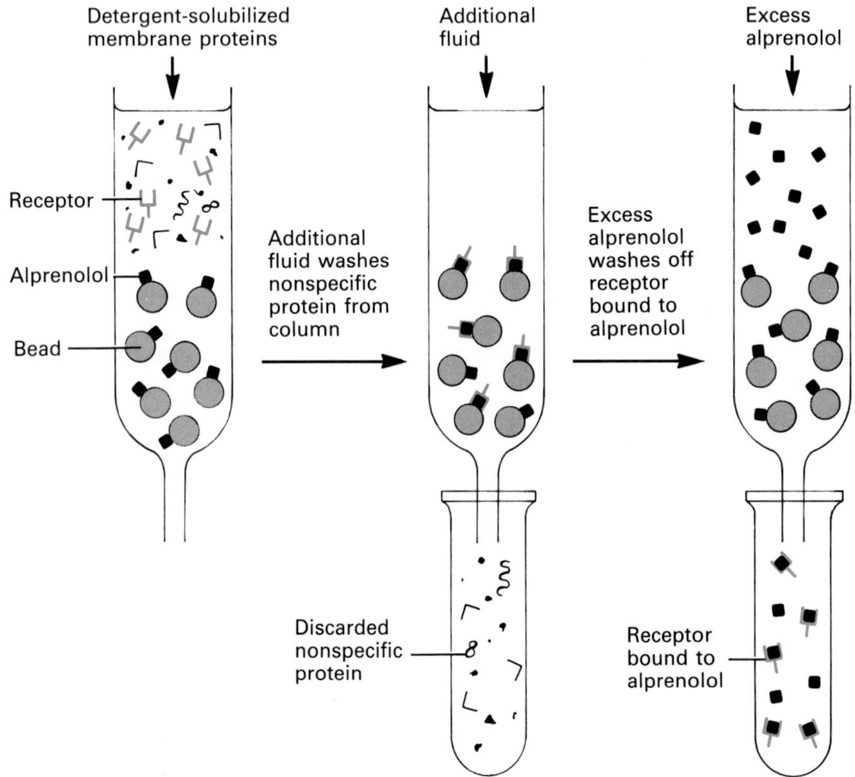

Figure 16-6 Purification of the β-adrenergic receptor by affinity chromatography. The potent receptor-binding antagonist alprenolol (see Table 16-5) is chemically linked to polystyrene beads. A crude, detergent-solubilized preparation of erythrocyte membranes is passed through a column containing these beads. Only the receptor binds to the beads; the other proteins are washed through by excess fluid. With the addition of an excess of alprenolol to the column, the bound receptor is displaced from the beads and eluted. [See J. L. Benovic et al., 1984, *Biochemistry* **23**:4510.]

sizes cAMP), but not β-adrenergic receptor (Figure 16-7). (The cell does contain receptors for other hormones that activate adenylate cyclase.) These receptor-laden vesicles confer cAMP responsiveness of the target cells to epinephrine, a proof that the receptor is indeed involved in inducing synthesis of cAMP.

Hormone Analogs Are Important in the Study of Receptor Action

Studies of a wide array of chemically synthesized analogs of natural hormones such as epinephrine provide additional evidence that the binding of a hormone to a saturable cell surface receptor is physiologically relevant. These analogs fall into two classes: *agonists* that mimic hormone function, binding to the receptor and causing the normal response, and *antagonists* that bind to the receptor but do not activate the hormone-induced functions. A bound antagonist competes with the binding of the natural hormone (or of an agonist) and blocks its physiological activity.

By studying the molecular structure and parameters of these agonists and antagonists, we can define those parts of the epinephrine molecule that are necessary for binding and those parts that are necessary for the subsequent induction of the cellular response. In general the catechol ring is essential for a compound to elevate the level of cAMP, that is, for a compound to be an agonist. The NH-containing side chain determines the affinity of the agonist or antagonist to the β-adrenergic receptor. As Table 16-5 shows, some agonists—isoproterenol is an example—are more potent than their natural counterparts in that they bind to the receptor and activate synthesis of cAMP at a 10-fold to 100-fold lower concentration than epinephrine. Similarly, antagonists differ in their affinities for the β-adrenergic receptor.

In general, the K_m of the binding of an agonist to the receptor is the same as the concentration required for half-maximal elevation of cAMP. This relationship indicates that the activation of cAMP accumulation is proportional to the number of surface receptors that are filled with the agonist (Figure 16-5).

There are at least two kinds of β-adrenergic receptors. In humans, β_1 receptors on cardiac muscle cells promote increased heart rate and contractility. They bind catecholamines with the rank order of affinities isoproterenol > norepinephrine > epinephrine. The β_2 receptors, found on the smooth muscle cells that line the bronchial passages, mediate the relaxation of these passages. They bind β agonists with the order of affinities isoproterenol >> epinephrine > norepinephrine. The discovery of these two types of receptors led to the development and clinical use of β_1-selective antagonists, such as practolol, to slow down heart contraction in the treatment of cardiac arythmias and angina. Such "beta blockers," as they are called, generally have little effect on the β-adrenergic receptors of other cell types. Similarly, β_2-selective agonists, such as terbutaline, are used in the treatment of asthma because they specifically cause the opening of the bronchioles, the small airways of the lung.

Binding of Hormone to β-Adrenergic Receptors Activates Adenylate Cyclase

β-Adrenergic receptors are present on many kinds of mammalian cells. The binding of epinephrine to such receptors on different cells will, in general, trigger distinctly different cellular events. But the initial response is always

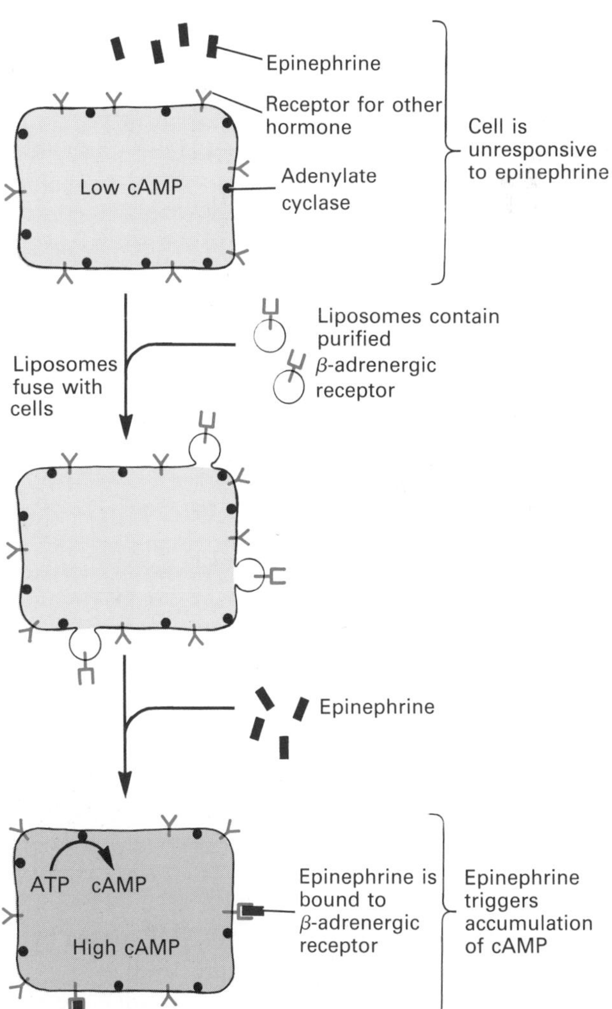

Figure 16-7 The β-adrenergic receptor mediates the induction of epinephrine-initiated cAMP synthesis. The target cell lacks any receptors for epinephrine but contains adenylate cyclase and other proteins required for the synthesis of cAMP. The purified β-adrenergic receptor (Figure 16-6) can be incorporated into liposomes and fused with the target cell. The resultant cell responds to the addition of epinephrine by synthesizing high levels of cAMP. [See R. A. Cerione et al., 1983, *Nature* 306:562.]

Table 16-5 Structure of typical agonists and antagonists of the β-adrenergic receptor

Structure	Compound	K_m for binding to the receptor on frog erythrocytes
HO, HO–(ring)–CH(OH)–CH_2–NH–CH_3	Epinephrine	5×10^{-6} *M*
AGONISTS		
HO, HO–(ring)–CH(OH)–CH_2–NH–CH(CH_3)–CH_3	Isoproterenol	0.4×10^{-6} *M*
HO, HO–(ring)–CH(OH)–CH_2–NH–C(CH_3)(CH_3)–CH_2–(ring)–OH		0.06×10^{-6} *M*
ANTAGONISTS		
CH_2=CH–CH_2–(ring)–O–CH_2–CH(OH)–CH_2–NH–CH(CH_3)–CH_3	Alprenolol	0.0034×10^{-6} *M*
(naphthalene)–O–CH_2–CH(OH)–CH_2–NH–CH(CH_3)–CH_3	Propranolol	0.0046×10^{-6} *M*
CH_3–C(=O)–NH–(ring)–OCH_2–CH(OH)–CH_2–NH–CH(CH_3)–CH_3	Practolol	21×10^{-6} *M*

the same: an elevation in the level of cAMP caused by an activation of adenylate cyclase (Figure 16-8).

To trace the mode of action of epinephrine we must first consider how surface binding results in an activation of adenylate cyclase and then consider how the diverse sets of responses are all mediated by cAMP.

Adenylate cyclase is a membrane-bound enzyme, with its ATP-binding site on the cytoplasmic face of the plasma membrane. The receptor protein has its binding site on the extracellular face. A third protein, tightly bound to the cytoplasmic face, functions as a communicator between the receptor and adenylate cyclase; this protein is called G because it binds guanosine phosphates (GTP and GDP).

G Protein Cycles between Active and Resting Forms

The G protein binds GTP noncovalently but very tightly. In fact, G can be purified from detergent-solubilized membranes by affinity chromatography on columns that contain chemically bound GTP. G is composed of three peptide chains, α, β, and γ of m.w. about 42,000, 35,000, and 10,000, respectively. The α subunit binds

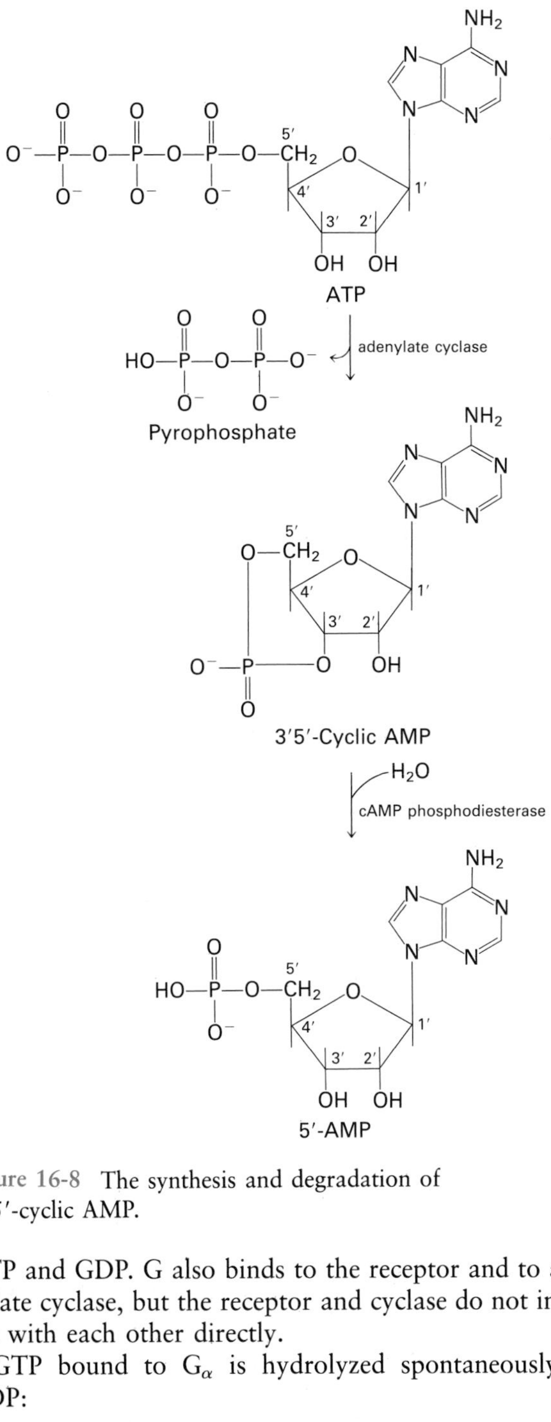

ATP

adenylate cyclase

Pyrophosphate

3′5′-Cyclic AMP

H_2O

cAMP phosphodiesterase

5′-AMP

Figure 16-8 The synthesis and degradation of 3′,5′-cyclic AMP.

GTP and GDP. G also binds to the receptor and to adenylate cyclase, but the receptor and cyclase do not interact with each other directly.

GTP bound to G_α is hydrolyzed spontaneously to GDP:

$$G_\alpha \cdot GTP \longrightarrow G_\alpha \cdot GDP + P_i$$

When GTP is bound to G_α, the G_α and a complex of the G_β and G_γ subunits dissociate. The G_α subunit (with its bound GTP) apparently undergoes a conformational change that enables it to bind to and activate adenylate cyclase (Figure 16-9). When G_α has bound GDP rather than GTP, it binds to the $G_\beta G_\gamma$ subunit—it cannot bind to or activate adenylate cyclase.

In an unstimulated cell—a cell to which no hormone is bound—most G molecules contain GDP. G_α is bound to the $G_\beta G_\gamma$ complex and thus adenylate cyclase is not active. Binding of a hormone or agonist to the β-adrenergic receptor changes the conformation of the receptor, causing it to bind to G in such a manner that GDP is displaced and GTP is bound (Figure 16-9). This reaction results in dissociation of the G_α and $G_\beta G_\gamma$ subunits, followed by binding of G_α to the adenylate cyclase and the activation of cAMP synthesis. The activation is short-lived, however, because GTP bound to G hydrolyzes quickly to GDP, leading to the association of G_α with $G_\beta G_\gamma$ and the inactivation of adenylate cyclase. Therefore, a GTP-GDP cycle (Figure 16-10) is crucial to the hormone-dependent activation and inactivation of adenylate cyclase.

The G_α protein functions as a shuttle between two membrane proteins, the hormone receptor and the adenylate cyclase. G_α is also a signal transducer since it relays to the cyclase enzyme the conformational change in the receptor that has been triggered by hormone binding.

Important evidence supporting this model came from studies using a nonhydrolyzable analog of GTP, called GMPPCP,

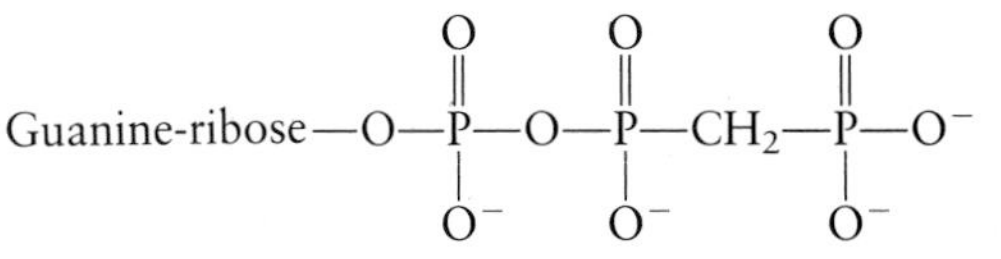

$$\text{Guanine-ribose}-O-\underset{O^-}{\overset{O}{\overset{\|}{P}}}-O-\underset{O^-}{\overset{O}{\overset{\|}{P}}}-CH_2-\underset{O^-}{\overset{O}{\overset{\|}{P}}}-O^-$$

in which a —P—CH_2—P bond replaces the terminal phosphodiester bond in GTP. Although this analog cannot be hydrolyzed it binds to G protein as well as does GTP. Addition of GMPPCP together with agonist to an erythrocyte membrane preparation results in an activation of adenylate cyclase that is much larger and longer lived than that with agonist and GTP. Once the GDP bound to G_α is displaced by GMPPCP, the GMPPCP will remain permanently bound to G_α. Because the G_α·GMPPCP complex is as functional as the normal G_α·GTP complex in activating cyclase, adenylate cyclase will be "locked" into the active state.

The G·GTP complex apparently interacts with the hormone receptor as well as with adenylate cyclase. The interaction is important in terminating the hormone response. Replacement of GDP bound to G by GTP results in a *decrease* in the affinity of the β-adrenergic receptor for hormone (i.e., K_m is increased). This shifts the receptor-hormone equilibrium toward dissociation and leads to inactivation of the cyclase.

What is the purpose of all the complex interactions among these proteins? First, the hormone signal is amplified substantially. A single receptor-hormone complex causes the conversion of several G_α·GDP·$G_\beta G_\gamma$ complexes to the active G_α·GTP form, which in turn activates several adenylate cyclase molecules. Each of these enzymes catalyzes the synthesis of many cAMP molecules during the time that G_α·GTP is bound to it. Although the exact extent of this amplification is difficult to measure, it

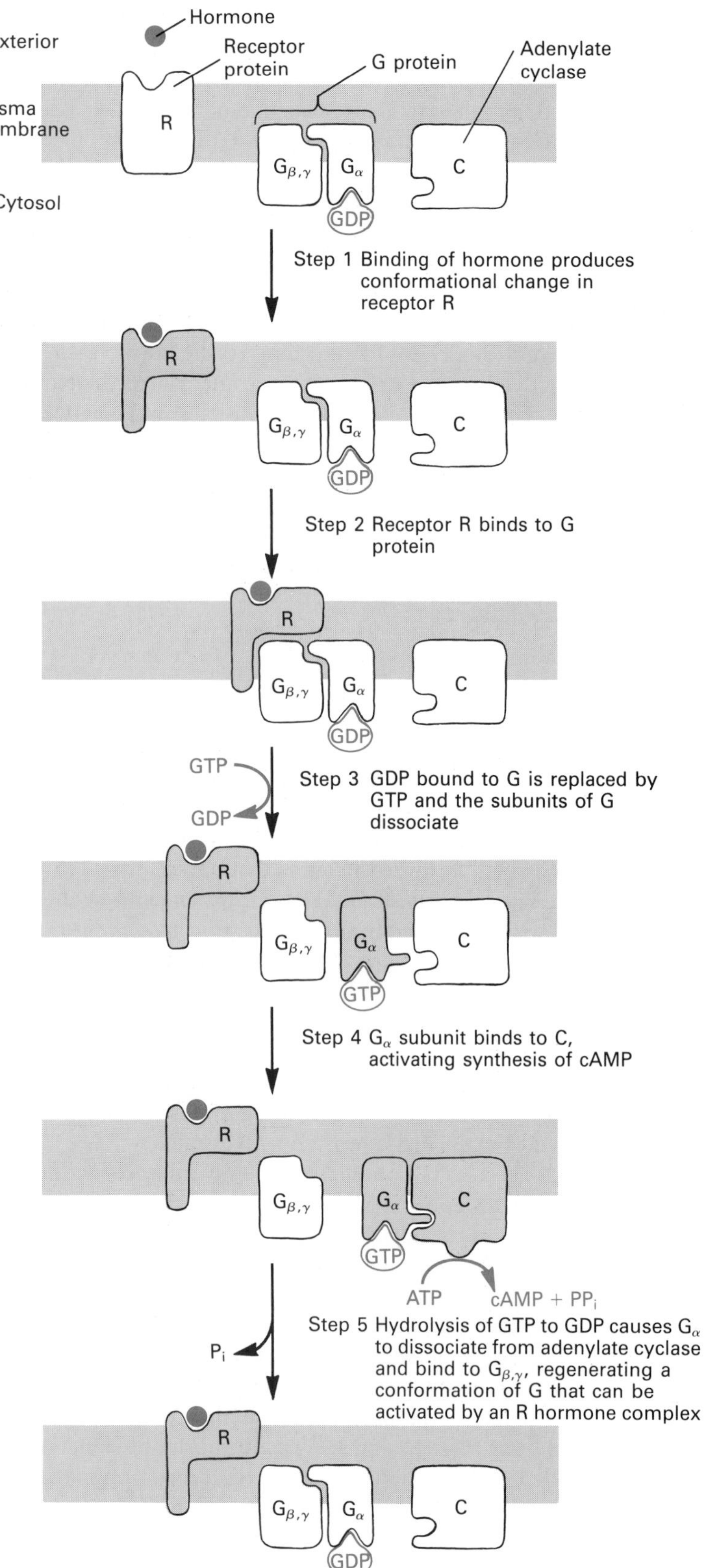

Hormone
Exterior
Receptor protein
G protein
Adenylate cyclase
Plasma membrane
Cytosol
R
$G_{\beta,\gamma}$
G_α
C
GDP
Step 1 Binding of hormone produces conformational change in receptor R
Step 2 Receptor R binds to G protein
GTP
GDP
Step 3 GDP bound to G is replaced by GTP and the subunits of G dissociate
GTP
Step 4 G_α subunit binds to C, activating synthesis of cAMP
ATP
cAMP + PP_i
P_i
Step 5 Hydrolysis of GTP to GDP causes G_α to dissociate from adenylate cyclase and bind to $G_{\beta,\gamma}$, regenerating a conformation of G that can be activated by an R hormone complex

Figure 16-9 The activation of adenylate cyclase by binding of a hormone to its receptor. A patch of cell membrane is depicted, which contains on its outer surface a receptor protein (R) for a hormone (colored circle). On the inside surface of the membrane is adenylate cyclase protein (C) and the transmitter protein G. In the resting state GDP is bound to the α subunit of G. When a hormone binds to R, a conformational change occurs in R (step 1). Activated R binds to G (step 2). This activates G so that it releases GDP and binds GTP, causing the α and the complex of β and γ subunits of G to dissociate (step 3). Free G_α subunit binds to C and activates it so that it catalyzes the synthesis of cAMP from ATP (step 4); this step may involve a conformational change in G_α. When GTP is hydrolyzed to GDP, a reaction most likely catalyzed by G_α itself, G_α is no longer able to activate C (step 5), and G_α and $G_{\beta,\gamma}$ reassociate. The hormone dissociates from the receptor and the system returns to its resting state.

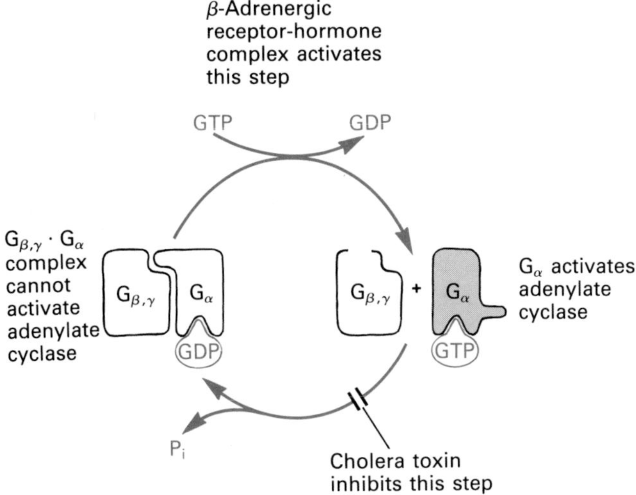

Figure 16-10 The cycling of the G protein between GTP and GDP is coupled to the activation and inactivation of adenylate cyclase. Hydrolysis of GTP to GDP is believed to be catalyzed by the G_α subunit itself.

is clear that the binding of a hormone molecule to one receptor can result in the synthesis of at least several hundred cAMP molecules before the hormone dissociates from the receptor. Second, the series of proteins enables the cell to rapidly terminate its response to a hormone when the concentration of hormone is reduced. Hydrolysis of the GTP bound to G reverses the activation of adenylate cyclase; the continuous presence of epinephrine is required for continuous activation of adenylate cyclase. Hydrolysis of GTP also results in an increase in the affinity of the receptor for epinephrine, making it better able to respond to any further increases in epinephrine concentration.

Several Receptors Interact with a Single Type of Adenylate Cyclase

A third reason for the involvement of three proteins in this system is that it permits several hormone receptor-ligand complexes to interact with the same adenylate cyclase enzyme. In the liver, for instance, glucagon (a peptide hormone) and epinephrine bind to different receptors but both activate the same adenylate cyclase. Both types of receptor probably bind to and activate the same G protein and convert the inactive G_α·GDP·$G_\beta G_\gamma$ complex to the active G_α·GTP. The concentration of G_α·GTP, and thus the activity of adenylate cyclase, will be the sum of that caused by the binding of the individual hormones to their own receptors.

Certain hormones cause a decrease in the level of cAMP and others cause an increase. In the adipose cell, for instance, epinephrine, glucagon, ACTH, and vasopressin all stimulate adenylate cyclase whereas prostaglandin PGE_1 and adenosine inhibit it. Interaction of these "inhibitory" receptors with adenylate cyclase is mediated by an "inhibitory" G, or coupling protein, called G_i, which contains the same β and γ subunits as the stimulatory G_s and a different G_α, a $G_{i\alpha}$ subunit that also binds GTP or GDP. In response to hormone binding to its receptor, G_i binds GTP and dissociates into $G_{i\alpha}$·GTP and a $G_\beta G_\gamma$ complex. This dissociation causes an inhibition, rather than an activation, of adenylate cyclase activity (Figure 16-11). How the inhibition is accomplished is not known. One possibility is that $G_{i\alpha}$·GTP binds to adenyl-

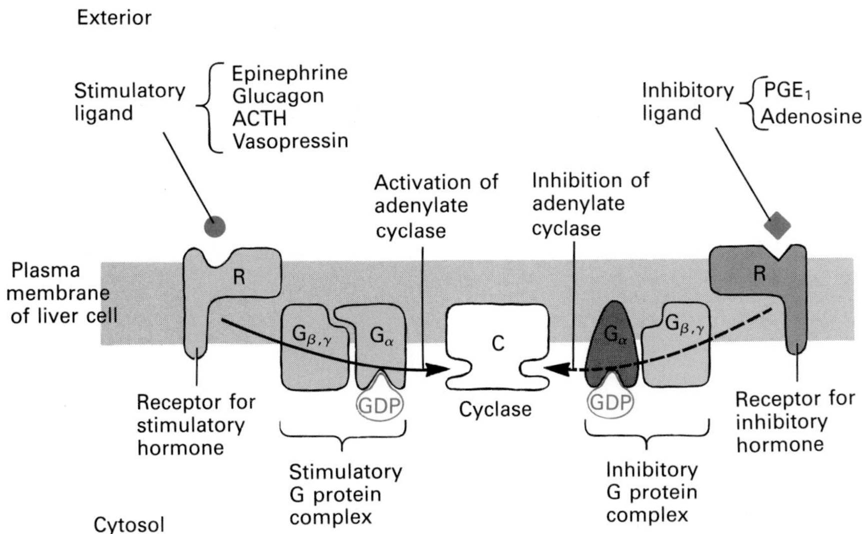

Figure 16-11 Different G proteins mediate activation and inhibition of adenylate cyclase. As depicted in Figure 16-9, many receptor-hormone complexes bind to a stimulatory G protein, causing the replacement of bound GDP with GTP and the generation of the free G_α·GTP subunit, which binds to and activates adenylate cyclase. But other receptor-hormone complexes bind to a different, inhibitory G protein, also constructed of an α subunit that binds GDP or GTP and β and γ subunits. In a way that is not yet known, the inhibitory G protein is modified so that it binds to and inhibits adenylate cyclase. The G_β and G_γ subunits are the same in both the stimulatory and inhibiting G proteins. The G_α subunit and the receptors differ. *After A. G. Gilman, 1984,* Cell *36:577.*

ate cyclase and inactivates it. A second is that the $G_\beta G_\gamma$ complex, formed by dissociation of G_i, binds the stimulatory α subunit, $G_{s\alpha}$, and prevents it from activating adenylate cyclase.

Whatever the detailed mechanism, such a system allows a cell to fine-tune its level of cAMP by integrating the response of a number of hormones, each of which binds to its own surface receptor. For instance, the simultaneous addition of epinephrine and PGE_1 to an adipocyte results in an activation of adenylate cyclase, but to a level less than that stimulated by epinephrine alone.

Cholera Toxin Alters the G Protein

Confirmation of the GTP cycle came from a seemingly unlikely source, a study of cholera toxin. This toxin, a peptide, is produced by the bacterium *Vibrio cholerae.* The classic symptom of cholera is massive diarrhea, caused by water flow from the blood through the epithelial cells into the small intestine. Death is often due to dehydration. The study showed that the toxin *irreversibly* activates adenylate cyclase in the intestinal epithelial cells, which causes a high level of cAMP and therefore an increased water flow out of the intestinal cells into the lumen. Later studies showed that the toxin irreversibly activates adenylate cyclase in a large number of cell types. Like diphtheria toxin (Chapter 15), cholera toxin consists of two types of peptide chains. One chain is an enzyme that penetrates the cell surface membrane and enters the cytosol, where it catalyzes the covalent addition of an ADP ribosyl group from intracellular NAD^+ to the α subunit of G_s protein (see below).

This irreversibly modified $G_{s\alpha}$ subunit can activate adenylate cyclase normally but cannot hydrolyze bound GTP to GDP (Figure 16-10). Thus GTP remains bound to $G_{s\alpha}$, and G_α is continuously in the "activation" mode. Adenylate cyclase is continuously turned on. As a result, the level of cAMP in the cytosol rises 100-fold or more. In the intestinal epithelial cell, this rise apparently causes certain membrane proteins to permit a massive flow of water from the blood into the intestinal lumen.

cAMP Activates a Protein Kinase

The next step in tracing the effect of hormones that elevate the level of cAMP is to see how cAMP affects enzyme activity. Recall that cAMP is the second messenger for many hormones and that the effects of elevated cAMP differ markedly in cells of different types (Table 16-4). All of cAMP's effects, however, are believed to be mediated in a similar manner in all cells: cAMP modifies the activity of a specific group of enzymes through the action of cAMP-dependent enzymes called *protein kinases.*

Protein kinases transfer the terminal phosphate group

Nicotinamide

ADP-ribose

+ $G_{s\alpha}$ protein

cholera toxin

Nicotinamide + $G_{s\alpha}$

ADP-ribosylated $G_{s\alpha}$ subunit

of ATP to serine, threonine, or tyrosine residues of substrate proteins (Figure 16-12). The phosphorylated form of many enzymes is much more active than the unphosphorylated form; in other enzymes the phosphorylated form is less active. Cyclic AMP binds to one type of inactive protein kinase, the *cAMP-dependent protein kinase,* and increases its ability to phosphorylate specific acceptor proteins. This protein kinase contains four subunits, two regulatory (R) and two catalytic (C). When cAMP binds to the R subunits (Figure 16-13), they dissociate from the C subunits and the kinase activity of C increases:

$$R_2C_2 \text{ (inactive)} + \text{cAMP} \longrightarrow R_2 \text{ (dimer)} \cdot \text{cAMP} + 2\text{C (active monomer)}$$

$$\text{ATP} + \text{acceptor protein} \xrightarrow{\text{C}} \text{ADP} + \text{acceptor protein—P}$$

The same C subunit of the cAMP-dependent kinase enzymes is present in most tissues, but different R subunits are found in different cell types. The substrate proteins for the cAMP-dependent kinases vary widely among the cell types. Depending on the tissue, cAMP may activate or inhibit different enzyme systems.

Of the many effects of the cAMP-dependent protein kinase, the one that was first discovered and that has received the most detailed study is the release of glucose from glycogen. This reaction occurs both in muscle and liver cells treated with epinephrine or with agonists of β-adrenergic receptors. Glucagon causes a similar effect only in liver. We shall focus on these systems in some detail, as they illuminate several key mechanisms of action of cAMP.

Glycogen Synthesis and Degradation Is Controlled by cAMP

Glycogen, a polymer of glucose (Figure 3-74), is the major storage form of glucose in the body. Glycogen is found principally in the liver and in muscle cells. Glucose generated from liver glycogen is secreted into the blood,

$$\text{ATP} + \text{protein—OH} \xrightarrow{\text{protein kinase}} \text{ADP} + \text{protein—O—}\overset{\overset{\displaystyle O}{\|}}{\underset{\underset{\displaystyle O^-}{|}}{P}}\text{—O}^-$$

$$H_2O + \text{protein—O—}\overset{\overset{\displaystyle O}{\|}}{\underset{\underset{\displaystyle O^-}{|}}{P}}\text{—O}^- \xrightarrow[H_2O]{\text{protein phosphatase}} \text{protein—OH} + \text{HO—}\overset{\overset{\displaystyle O}{\|}}{\underset{\underset{\displaystyle O^-}{|}}{P}}\text{—O}^-$$

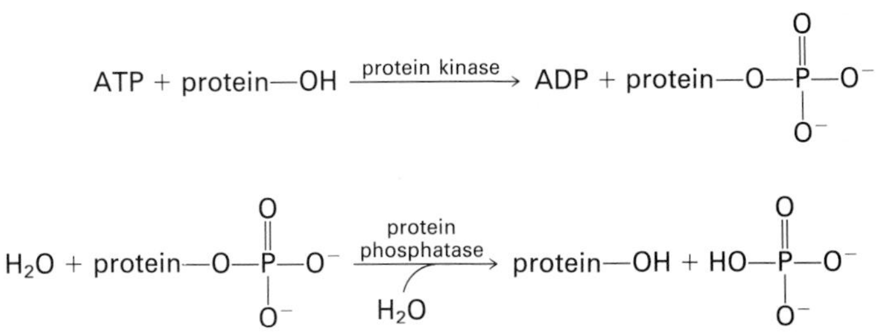

Figure 16-12 Reactions catalyzed by protein kinase *(top)* and protein phosphatase *(bottom).* Often the phosphorylated and dephosphorylated forms of the protein differ markedly in enzyme reactivity.

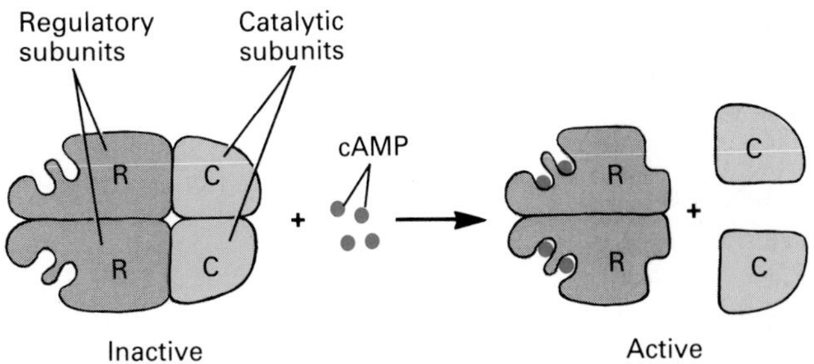

Figure 16-13 Cyclic AMP activation of the AMP-dependent protein kinase. Binding of cAMP to the regulatory (R) subunits causes dissociation of the catalytic (C) subunits. Only when dissociated from the R subunits is a C subunit an enzymatically active protein kinase. Each R subunit binds two cAMP molecules in a positive cooperative interaction. As curve C in Figure 16-3 shows, this enhances cellular response. Small changes in the level of cytosolic cAMP can cause proportionately large changes in the amount of dissociated C subunits and thus in the activity of the protein kinase.

where it serves as an energy source for many body tissues that do not store energy reserves, such as the brain. Glucose produced from glycogen in muscle cells is oxidized quickly to CO_2, yielding ATP as a source of energy for contraction.

Like most polymers, glycogen is synthesized by one set of enzymes and degraded by another (Figures 16-14, 16-15).

The intermediate in glycogen synthesis is uridine diphosphate glucose (UDP-glucose), which is synthesized from glucose by three enzyme-catalyzed reactions:

$$\text{Glucose} + \text{ATP} \xrightarrow{\text{hexokinase}} \text{glucose 6-P} + \text{ADP}$$

$$\text{Glucose 6-P} \xrightarrow{\text{phosphoglucomutase}} \text{glucose 1-P}$$

$$\text{Glucose 1-P} + \text{UTP} \xrightarrow{\text{UDP-glucose pyrophosphorylase}} \text{UDP-glucose} + PP_i$$

Then glycogen synthase transfers the glucose residue from UDP-glucose to a free 4OH of a glucose residue at the end of a glycogen chain (Figure 16-14):

$$\text{Glycogen}_{n \text{ residues}} + \text{UDP-glucose} \xrightarrow{\text{glycogen synthase}} \text{glycogen}_{n+1 \text{ residues}} + \text{UDP}$$

The degradation of glycogen involves the stepwise removal of glucose residues from the same end, a reaction catalyzed by glycogen phosphorylase (Figure 16-15):

$$\text{Glycogen}_{n \text{ residues}} + P_i \xrightarrow{\text{glycogen phosphorylase}} \text{glycogen}_{n-1 \text{ residues}} + \text{glucose 1-P}$$

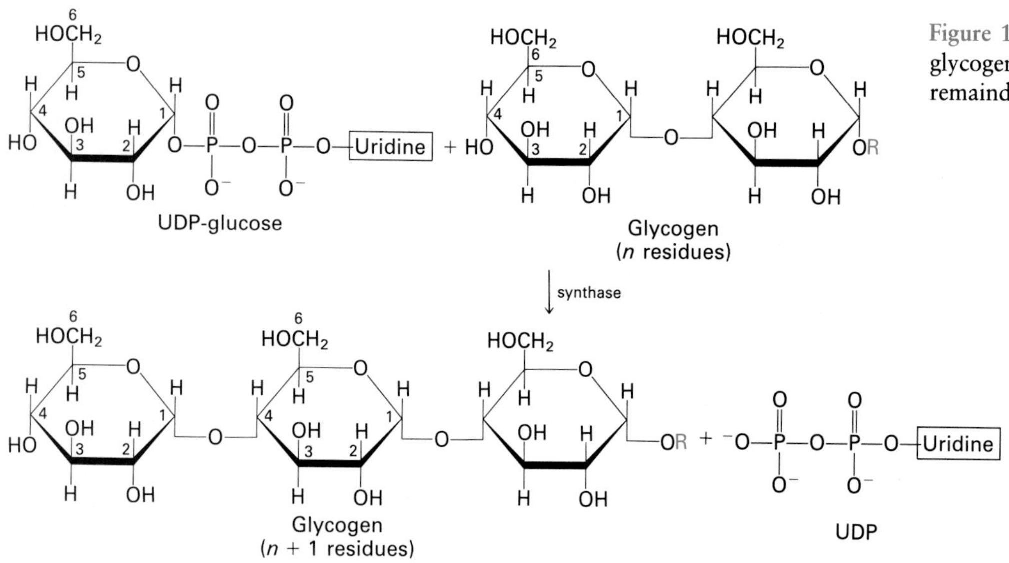

Figure 16-14 Synthesis of glycogen by glycogen synthase. R stands for the remainder of the glycogen chain.

cAMP-Dependent Kinases Regulate the Enzymes of Glycogen Metabolism

Both synthesis and degradation of glycogen are regulated by cAMP. Elevation of the cAMP level enhances the conversion of glycogen to glucose by inhibiting glycogen synthesis and increasing glycogen breakdown. The active form of the cAMP-dependent protein kinase—the C subunit free of the R subunit—phosphorylates glycogen synthase and in so doing converts it to a much less active molecule. This reaction directly inhibits synthesis of glycogen. The cAMP-dependent kinase also activates glycogen phosphorylase but does so by an indirect route involving a multienzyme cascade (see Figure 16-16). The C enzyme phosphorylates and activates another protein kinase, called *glycogen phosphorylase kinase,* which, in turn, phosphorylates and activates a third enzyme, *glycogen phosphorylase.* Active glycogen phosphorylase degrades glycogen to glucose 1-phosphate.

The entire process is reversed when hormone is removed and the level of cAMP drops. The reversal is mediated by a set of *protein phosphatases,* which remove the phosphate residues from glycogen synthase (thus activating it) and from glycogen phosphorylase kinase and glycogen phosphorylase (thus inactivating them). Actually, a single specific protein phosphatase enzyme may catalyze the removal of phosphate from all three enzymes. An inhibitor of the phosphatase is a substrate of the cAMP-dependent protein kinase. As shown in Figure 16-17, phosphorylation of the inhibitor allows it to bind to the phosphatase and inhibit its activity. Thus, the protein phosphatase is activated when the level of cAMP drops (see Figure 16-16). As a result of a lowered level of cAMP, the synthesis of glycogen by glycogen synthase is enhanced and phosphorolysis of glycogen by glycogen phosphorylase is inhibited.

The degradation of glycogen involves *phosphorolysis*—breakdown by adding phosphate rather than water. In muscle, glucose 1-P is converted by phosphoglucomutase to glucose 6-P, which is then metabolized by the Embden-Meyerhof pathway (Figure 16-18).

In liver, much glucose 6-P formed from glycogen is hydrolyzed to glucose and secreted into the blood:

$$\text{Glucose 6-P} + H_2O \longrightarrow \text{glucose} + P_i$$

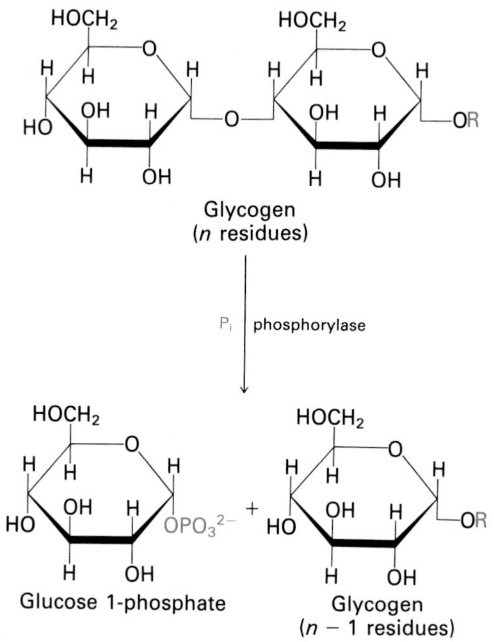

Figure 16-15 Degradation of glycogen by glycogen phosphorylase. R stands for the remainder of the glycogen chain.

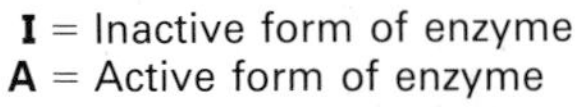

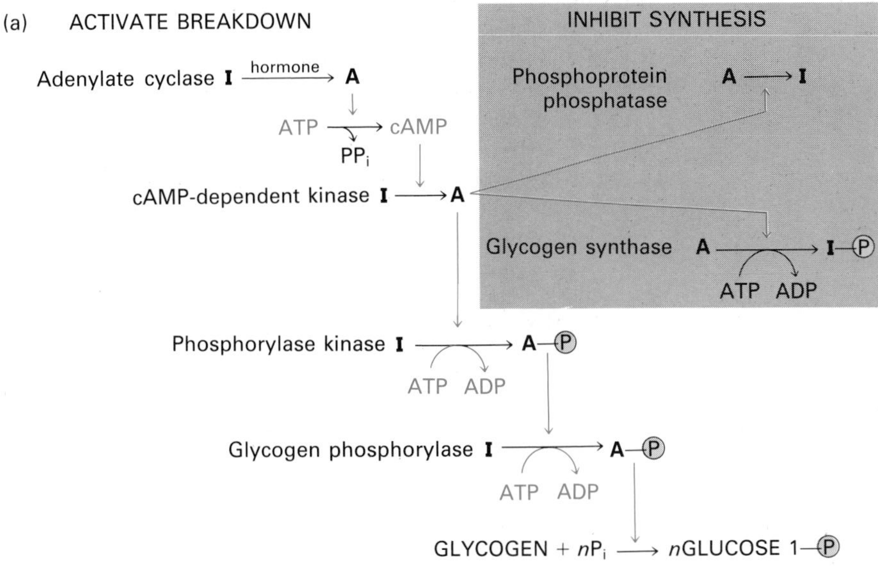

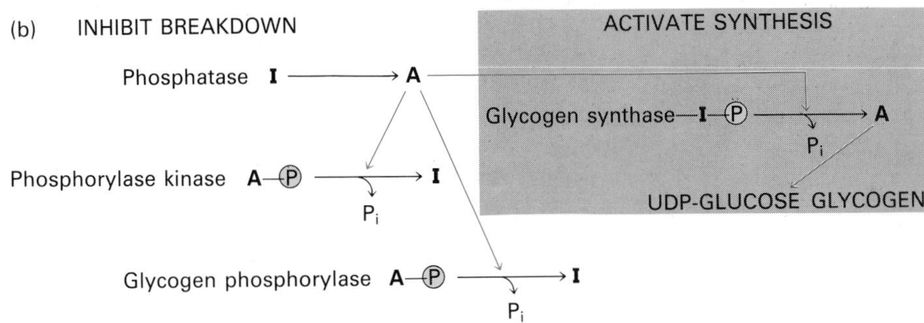

Figure 16-16 Cyclic AMP control of glycogen breakdown and synthesis in liver and muscle. (a) Increased level of cAMP causes an increase in glucose level by (1) activating glycogen breakdown and (2) inhibiting glycogen synthesis. Phosphorylation of enzymes by the cAMP-dependent protein kinase causes an increase in the level of glucose: phosphorylation activates glycogen phosphorylase and inhibits glycogen synthase. Also, the cAMP-dependent protein kinase phosphorylates an inhibitor of the phosphoprotein phosphatase, causing (3) inactivation of phosphoprotein phosphatase. Thus, the phosphate groups added to the other enzymes are not removed. (b) Decreased level of cAMP decreases the level of glucose by (1) inhibiting glycogen breakdown and (2) activating glycogen synthesis. This is accomplished by (3) activation of phosphoprotein phosphatase by removing the phosphate from the inhibitor, inactivating it. The active phosphatase removes the phosphate residue from glycogen phosphorylase kinase and glycogen phosphorylase, thus inhibiting degradation of glycogen. Removal of the phosphate from glycogen synthase, by contrast, activates the enzyme, causing synthesis of glycogen.

Degradation of cAMP Is Also Controlled by Ca^{2+}

The cascade of reactions triggered by cAMP is usually regulated by adenylate cyclase at the site of cAMP synthesis. Other points of regulation are the *cAMP phosphodiesterases;* these enzymes degrade cAMP, terminating the effect of hormonal stimulation (see Figure 16-8):

$$3'5'\text{-cAMP} + H_2O \longrightarrow 5'\text{-AMP}$$

Many cAMP phosphodiesterases are activated by an increase in cytosolic Ca^{2+}, another intracellular second messenger. Increases in cytosolic Ca^{2+} are often induced by neuronal or hormonal stimulation. Calcium ions bind to *calmodulin,* a ubiquitous binding protein. When a Ca^{2+}-calmodulin complex binds to cAMP phosphodiesterase, the enzyme begins to hydrolyze cAMP (Figure 16-19).

Both the synthesis and degradation of cAMP can be subject to complex regulation by multiple hormones. Some cells also modulate the level of cAMP by secreting it

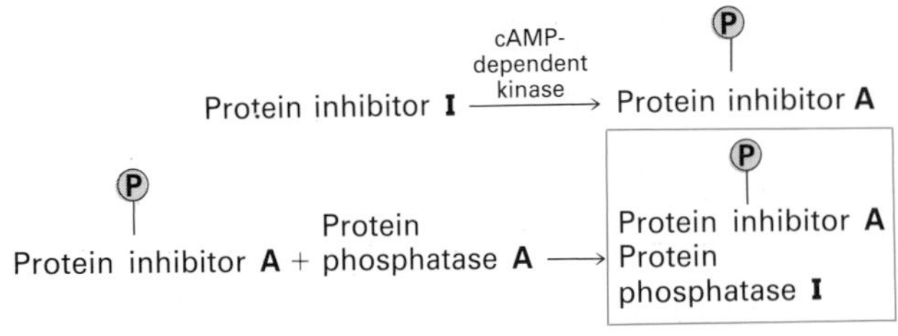

Figure 16-17 Inhibition of phosphoprotein phosphatase by cAMP. Phosphoprotein phosphatase is enzymatically active except when an inhibitor is bound to it. Only when the inhibitor is phosphorylated by the cAMP-dependent protein kinase can the inhibitor bind to and inhibit the phosphoprotein phosphatase. Thus, in the presence of a high level of cAMP the phosphoprotein phosphatase is inactive. It is active only when the level of cAMP is low. [See P. Cohen, 1982, *Nature* **296**:613.]

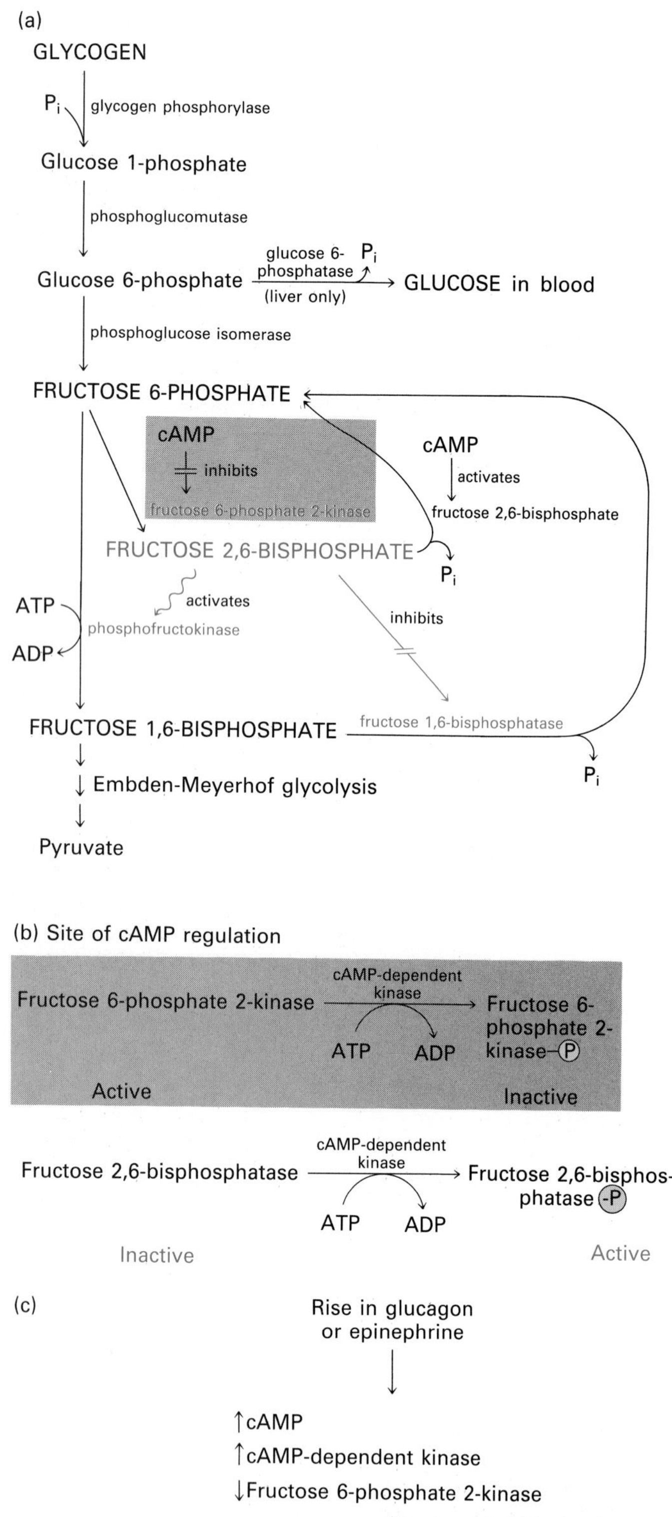

Figure 16-18 Control of the glycolytic pathway in the liver by cAMP. The fate of glucose 1-phosphate is different in liver and muscle. In muscle cells, glucose 1-phosphate generated from glycogen is used immediately for the generation of ATP. Phosphoglucomutase converts the compound to glucose 6-phosphate, an intermediate in the Embden-Meyerhof glycolytic pathway.

The liver stores and releases glucose primarily for use by other tissues, muscle and brain in particular. Glucose is not a major source of ATP in the liver. Liver cells, unlike muscle cells, contain a glucose 6-phosphatase that converts glucose 6-phosphate to glucose, which is immediately released into the blood. How is the flow of glucose 6-phosphate through the glycolytic pathway blocked in the liver? (a) The key regulatory enzymes in the liver are *phosphofructokinase,* which catalyzes the phosphorylation of fructose 6-phosphate to fructose 1,6-bisphosphate, and *fructose 1,6-bisphosphatase,* which catalyzes the hydrolysis of the 1-phosphate of fructose 1,6-bisphosphate to form fructose 6-phosphate. Together these enzymes serve to regulate the flow of metabolites from glucose to pyruvate, or in the reverse direction. The recently discovered molecule fructose 2,6-bisphosphate is an allosteric *activator* of phosphofructokinase and an allosteric *inhibitor* of fructose 1,6-bisphosphatase. Fructose 2,6-bisphosphate is synthesized by an enzyme, fructose 6-phosphate 2-kinase, whose activity is *inhibited* when phosphorylated by the cAMP-dependent protein kinase. Fructose 2,6-bisphosphate is hydrolyzed back to fructose 6-phosphate by the enzyme fructose 2,6-bisphosphatase whose activity is *increased* when phosphorylated by the cAMP-dependent protein kinase. (b) The control of fructose 6-phosphate 2-kinase is shown in more detail. (c) A rise in blood glucagon or epinephrine level triggers an elevation in the activity of the cAMP-dependent protein kinase and thus a *reduction* in the level of fructose 2,6-bisphosphate. This causes an *inhibition* of phosphofructokinase and an *activation* of fructose 1,6-bisphosphatase, resulting in a block to the metabolism of glucose 6-phosphate through the Embden-Meyerhof pathway. Glucose 6-phosphate is diverted to free glucose by the action of glucose 6-phosphatase and is released into the blood. [See M. R. El-Maghrabi et al., 1982, *Proc. Nat'l Acad. Sci. USA* **79**:315; and M. R. El-Maghrabi and S. S. Pilkus, 1984, *J. Cell Biochem.* **26**:1.]

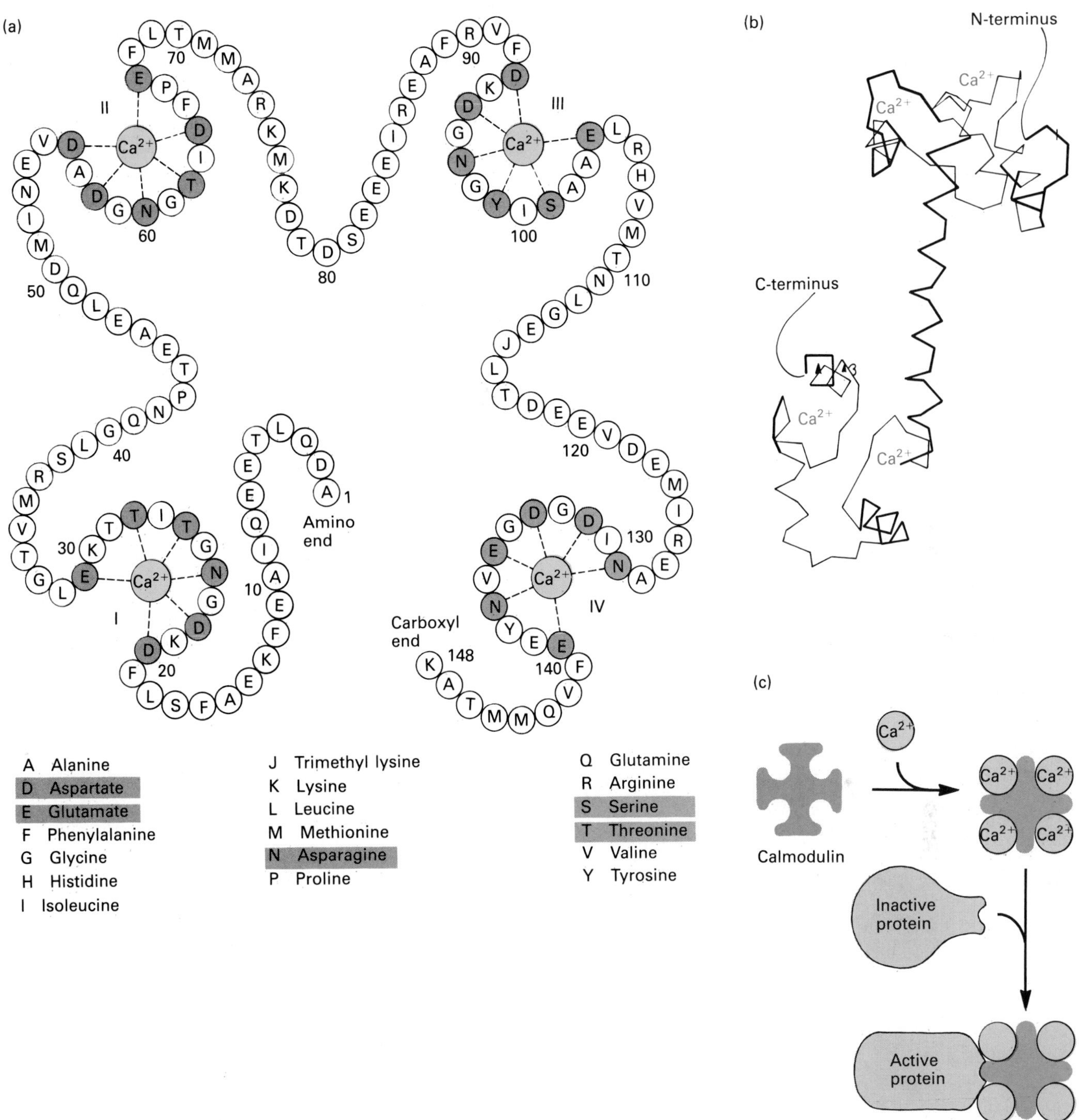

Figure 16-19 (a) Calmodulin is a cytoplasmic protein of 148 amino acids that modulates most of the regulatory functions of Ca^{2+} ions. The chain has four domains (numbered I through IV), each binding one Ca^{2+} ion. Each binding site is a loop containing aspartate and glutamate side chains that form ionic bonds with the Ca^{2+}. The oxygen atoms on the side chains of threonine, serine, tyrosine, and asparagine residues also participate in Ca^{2+} binding. (b) Drawing of the backbone of α-carbon atoms of calmodulin, as deduced from the 3-D structure. The four Ca^{2+} ions are represented by circles. (c) On binding Ca^{2+}, calmodulin undergoes a major conformational change that allows it to bind to other proteins, modifying their enzymic activity. The Ca^{2+}-calmodulin complex, for instance, binds to and activates cAMP phosphodiesterase. Calmodulin contains only 148 amino acids and is smaller than most proteins with which it interacts. Because binding of Ca^{2+} to calmodulin is cooperative, a small change in the level of cytosolic Ca^{2+} causes a large change in the level of "active" calmodulin. *Part (a) from W. Y. Cheung, 1982,* Sci. Am. **246***(6):62. Copyright 1982 by Scientific American, Inc.; part (b) after Y. S. Babu et al., 1985,* Nature *315:37.*

into the extracellular medium. This variability allows cells to integrate responses to many types of changes in the internal and external environment.

One Function of the Kinase Cascade Is Amplification

The set of protein phosphorylations and dephosphorylations we have described is a *cascade*—a series of stages in which the protein catalyzing each step is activated (or inhibited) by the product of the preceding step. The cascade may seem overcomplicated but it has several rationales. First, it allows an entire set of enzyme-catalyzed reactions to be regulated by the level of a single species of molecule, cAMP, and it allows a rapid reversal of the effects when the level of cAMP drops. Second, the cascade of events provides a huge amplification of an initially tiny signal (Figure 16-20). For example, the concentration of epinephrine needed in the blood to stimulate glycogenolysis in liver and muscle can be as low as 10^{-10} *M*. This stimulus generates a concentration of more than 10^{-6} *M* cAMP in the cell. Because three more catalytic steps precede the release of glucose, another 10^4 amplification can occur, so that blood glucose levels ultimately increase by as much as 50 percent. In striated muscle, the concentrations of the three successive enzymes in the glycogenolytic cascade—cAMP-dependent protein kinase, glycogen phosphorylase kinase, and glycogen phosphorylase—are in the ratio 1:10:240. This ratio supports the view that the system amplifies the effects of epinephrine and cAMP.

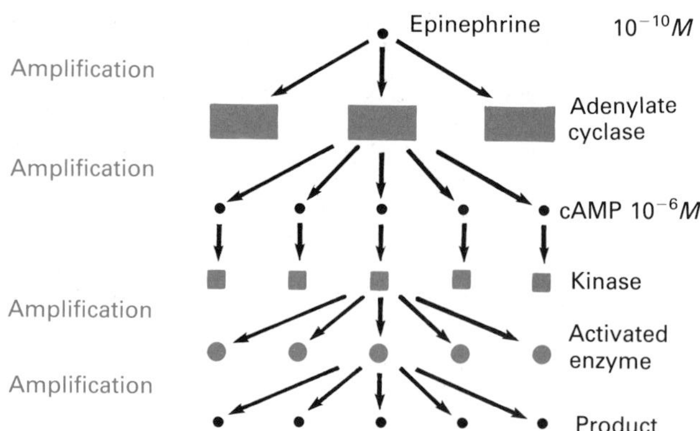

Figure 16-20 Cellular transduction and amplification of an extracellular signal. In this example, binding of a single epinephrine molecule results in the synthesis of a large number of cAMP molecules which, in turn, activate multiple enzyme molecules.

cAMP Operates in All Eukaryotic Cells

The effects of cAMP on synthesis and degradation of glycogen are confined mainly to liver and muscle, cells that store glycogen. But cAMP also mediates the intracellular response of many other cells to a wide variety of hormones (see Table 16-4). In virtually all eukaryotic cells that have been studied, the action of cAMP appears to be mediated through one or two cAMP-dependent protein kinases. The response of different cells to an elevation of cAMP varies according to which enzymes are activated or inhibited by the cAMP-dependent kinase.

In adipocytes, the fat-storage cells, an epinephrine-induced cascade regulates the synthesis and degradation of triacylglycerols, the storage form of fatty acids (see Figure 3-60). Oxidation of a gram of triacylglycerols to CO_2 generates 2.5 times as many molecules of ATP as does oxidation of a gram of glycogen. Thus triacylglycerols are a more potent source of ATP than glycogen (or glucose). Activation of the β-adrenergic receptor on adipose cells triggers an increase in cytosolic cAMP and activation of the cAMP-dependent protein kinase. The *lipase* that hydrolyzes triacylglycerols to fatty acids is activated by phosphorylation by this kinase. The released fatty acids are transferred to the blood, where they are bound to albumin, a major serum protein. In this form fatty acids are transported to other tissues, particularly the heart, muscles, and kidneys, where they are used as a source of ATP.

Still other cells respond to an increased level of cAMP in a different fashion. The cells of the adrenal cortex are the principal ones that have surface receptors for adrenocorticotropic hormone (ACTH), an anterior pituitary peptide. Binding of ACTH activates adenylate cyclase, which elevates the level of cAMP, which in turn activates the cAMP-dependent protein kinase. The protein kinase activates enzymes that synthesize several steroid hormones, such as cortisone, corticosterone, and aldosterone (see Table 16-1). Ovarian cells respond to follicle-stimulating hormone (FSH), another anterior pituitary peptide, by activating adenylate cyclase. Via the cAMP-dependent protein kinase, FSH triggers the synthesis of estradiol and progesterone, two steroids crucial in the development of female sex characteristics. Thyroid-stimulating hormone (TSH), yet another product of the anterior pituitary, stimulates—via a cAMP-mediated cascade—the production and release of thyroxine by the follicle cells of the thyroid gland.

FSH, TSH, and luteinizing hormone (LH), all synthesized by the anterior pituitary, are all dimers consisting of an α and β chain (see Table 16-2). The same α chain is found in all three proteins but the β chain varies. Most likely the β chain binds to the hormone receptor on the different target cells and the common α chain mediates activation of the G protein that, in turn, activates adenylate cyclase.

Ca^{2+} Ions and Other Intracellular Second Messengers

Ca^{2+} Ions Also Control Hydrolysis of Muscle Glycogen

The multiple steps in the β-adrenergic system permit other signals to activate the cascade at many different points. In muscle cells, for instance, stimulation by nerve impulses causes the release of Ca^{2+} from the sarcoplasmic reticulum and an increase in Ca^{2+} concentration in the cytosol. The rise in Ca^{2+} triggers contraction and also degradation of glycogen to glucose 1-phosphate, which fuels prolonged contraction. How does Ca^{2+} interact with the enzyme systems that catalyze glycogen synthesis and hydrolysis? The key regulatory enzyme is glycogen phosphorylase kinase. Its activity depends not only on phosphorylation but on Ca^{2+} as well (Figure 16-21). The enzyme has the subunit structure $(\alpha\beta\gamma\delta)_4$, in which the γ subunit is the catalytic protein. The regulatory α and β subunits are phosphorylated by cAMP-dependent protein kinase. The δ subunit is the Ca^{2+}-binding *calmodulin* (see Figure 16-19). The binding of Ca^{2+} to the δ subunit activates glycogen phosphorylase kinase, which is maximally active if at least the α subunit is phosphorylated (by the cAMP-dependent protein kinase) and Ca^{2+} is bound. In fact, the binding of Ca^{2+} may be essential for enzyme activity. Phosphorylation of the β and α subunits (by the cAMP-dependent kinase) increases the affinity of the calmodulin δ subunit for Ca^{2+}, making it possible for Ca^{2+} to be bound to the enzyme at the Ca^{2+} concentrations found in cells that are not stimulated by nerves. Thus, activity of glycogen phosphorylase kinase is increased in an incremental fashion by increases in the cytosolic concentration of either Ca^{2+} or cAMP, or both together. As a consequence of the elevated Ca^{2+} level after neuronal stimulation, glycogen phosphorylase kinase will be active even if it is unphosphorylated. When nerve stimulation ceases, Ca^{2+} is pumped back into the sarcoplasmic reticulum, the cytosolic Ca^{2+} level falls, activity of glycogen phosphorylase kinase is reduced, and less glycogen is converted to glucose phosphate.

Triphosphoinositol and Diacylglycerol Are Second Messengers for Certain Cellular Signals

Many hormones bind to receptors and cause an increase in cytosolic Ca^{2+} (Figure 16-22), largely by triggering release of Ca^{2+} from the endoplasmic reticulum. Examples include binding of acetylcholine to receptors on smooth muscle, nerve, and pancreas cells. Recent work suggests that the compound triphosphoinositol might be the intracellular transducer, or second messenger, that couples receptor-hormone binding to Ca^{2+} release. Triphosphoinositol is generated by hydrolysis of the glycolipid phosphatidylinositol 4,5-bisphosphate, one of the inositol glycolipids located in the inner leaflet of the plasma membrane (Figure 16-23). Binding of most hormones that lead to an increase in cytosolic Ca^{2+}, such as acetylcholine to pancreatic acinar cells or smooth muscle cells, enhances the rate of this hydrolysis. The resultant triphosphoinositol somehow triggers Ca^{2+} release from the endoplasmic reticulum. Indeed, direct addition of triphosphoinositol to endoplasmic reticulum leads to Ca^{2+} release, presumably by increasing the activity of a Ca^{2+} permease in the endoplasmic reticulum membrane.

1,2-Diacylglycerol, the other product of phosphatidylinositol 4,5-bisphosphate hydrolysis, may also have a biological function. It can diffuse in the membrane and activate a protein kinase, termed protein kinase C, that is bound to the cytoplasmic face of the plasma membrane. Although the substrates of protein kinase C are not

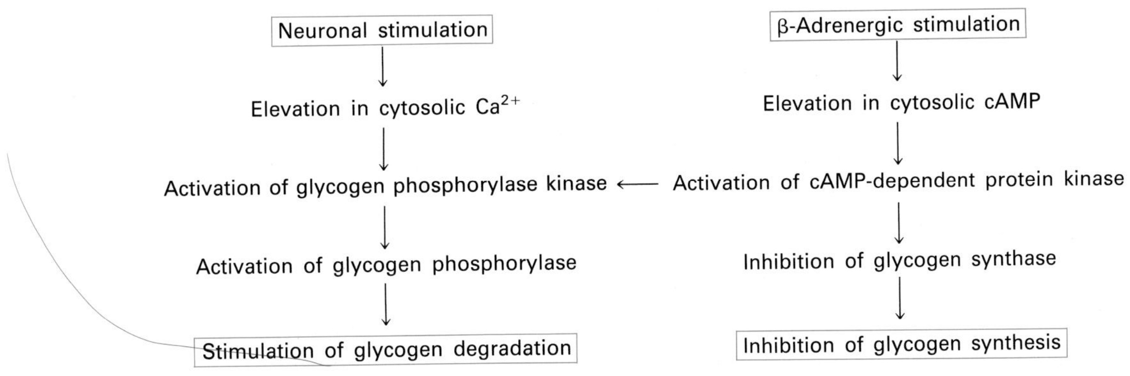

Figure 16-21 Activation of glycogen phosphorylase kinase in striated muscle. Neuronal stimulation and epinephrine use different second messengers, Ca^{2+} and cAMP, but both can activate glycogen phosphorylase kinase and thus accelerate the breakdown of glycogen.

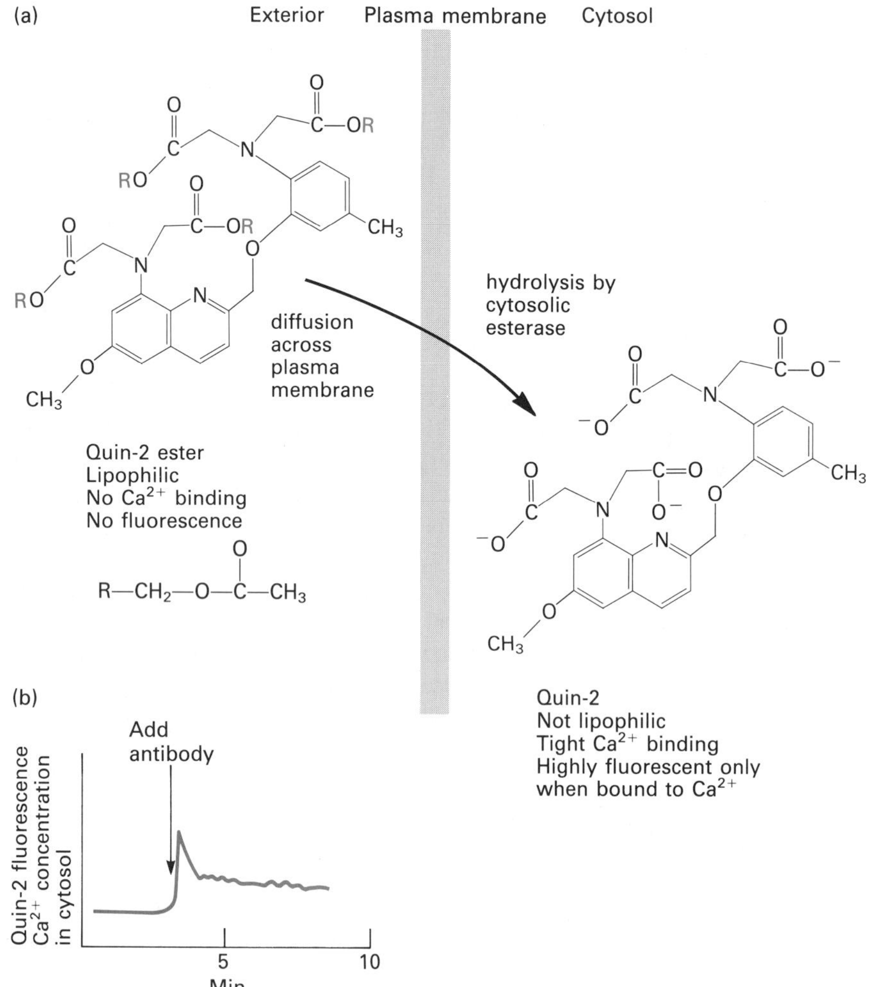

Figure 16-22 (a) Detection of cytosolic Ca^{2+} by fluorescence of quin-2. The vast majority of cellular Ca^{2+} ions are sequestered in the endoplasmic reticulum and mitochondria, but it is the cytosolic concentration of free Ca^{2+} that is critical for the function of Ca^{2+} as a second messenger. The cytosolic concentration of Ca^{2+} can be monitored continuously by the fluorescence of Ca^{2+}·quin-2 complexes. When added to cells the lipophilic quin-2 ester (*left*) diffuses across the plasma membrane and is hydrolyzed to quin-2 by cytosolic esterases. Quin-2 *(right)* cannot cross cellular membranes and remains in the cytosol. In the absence of Ca^{2+}·quin-2 is not fluorescent, but Ca^{2+}:quin-2 complexes are highly fluorescent, and the fluorescence is proportional to the concentration of Ca^{2+} ion in the cytoplasm. (b) The graph shows, by way of example, that addition of an antibody to a cell surface immunoglobin on the surface of lymphocytes preloaded by quin-2 causes an increase in quin-2 fluorescence and thus an increase in the concentration of cytosolic Ca^{2+} ion. *Part (a) from R. Y. Tsien et al., 1982,* J. Cell Biol. ***94****:325; part (b) after T. Pozzan et al., 1982,* J. Cell Biol. ***94****:335.*

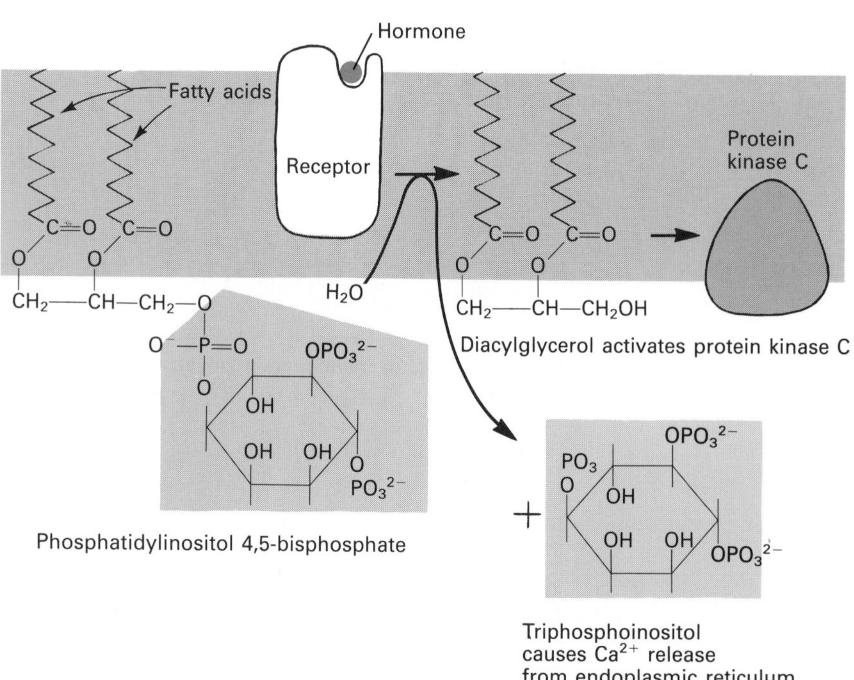

Figure 16-23 Receptor-hormone complexes can trigger production of triphosphoinositol by hydrolysis of phosphatidylinositol 4,5-bisphosphate. Hydrolysis is induced, in an unknown manner, by formation of many plasma membrane receptor-ligand complexes. One of the products, triphosphoinositol, causes release of Ca^{2+} from the endoplasmic reticulum and thus an elevation of cytosolic Ca^{2+}. Diacylglycerol activates protein kinase C and thus causes phosphorylation of several cytosolic proteins. [See M. J. Berridge and R. F. Irvine, 1984, *Nature* **312**:315.]

known with certainty, some are thought to be proteins essential to controlled growth of the cell (see Chapter 23).

Insulin and Glucagon: Hormonal Regulation of Blood Glucose Levels

Epinephrine, as we have seen, induces a wide array of responses to stress or exercise; among these responses is an increase in the concentration of glucose in the blood. During normal day-to-day living, regulation of blood glucose levels is the function of two other hormones, insulin and glucagon, which are produced by the islets of Langerhans, clusters of cells scattered throughout the pancreas. Insulin contains two polypeptide chains, called A and B, linked by disulfide bonds (see Figure 3-11); synthesized by the B cells in the islets, insulin acts to reduce the level of blood glucose (Figure 16-24). Glucagon, a single-chained 29-amino-acid peptide produced by the A cells, has the opposite effect, causing an increase in blood glucose by stimulating glycogenolysis in the liver. Each islet functions as an integrated unit delivering into the blood an amount of both hormones that is appropriate for the metabolic needs of the animal. The secretion of the hormones is regulated by a combination of neuronal and hormonal signals. Of the hormones that affect blood glucose, insulin is by far the most important, and its receptor is among the most extensively studied.

Insulin Controls the Level of Blood Glucose

Insulin acts on many cells in the body. Its effects are of two kinds, immediate and long term. Its immediate action is to increase the rate of glucose uptake from the blood into the cell by increasing the activity of glucose permease in the plasma membrane (see Chapter 15). Insulin's effects on glucose uptake and catabolism occur within minutes and do not require new protein synthesis. Continued exposure to insulin also has longer lasting effects, causing an increase in the activity of liver enzymes that synthesize glycogen and an increase in the enzymes in adipose cells that synthesize triacylglycerols. In these respects, the actions of insulin are opposite to those of epinephrine.

Within 15 min of the addition of insulin to adipocytes there is a dramatic increase in permeability to glucose. Is this due to an increase in the affinity of existing permeases for glucose or to an increase in the number of permeases in the surface membrane? Experiments have shown the latter to be correct. Insulin-stimulated cells contain 10 times more functional glucose permeases on their plasma membrane than do nonstimulated cells, and these permeases have the same binding affinity (K_m) for glucose as do permeases in unstimulated cells.

Unstimulated cells have been found to contain, just under the plasma membrane, vesicles with abundant glucose permeases. The binding of insulin to its receptor is thought to induce rapid fusion of these vesicles with the plasma membrane, greatly increasing the number of permeases in the plasma membrane (Figure 16-25). Other

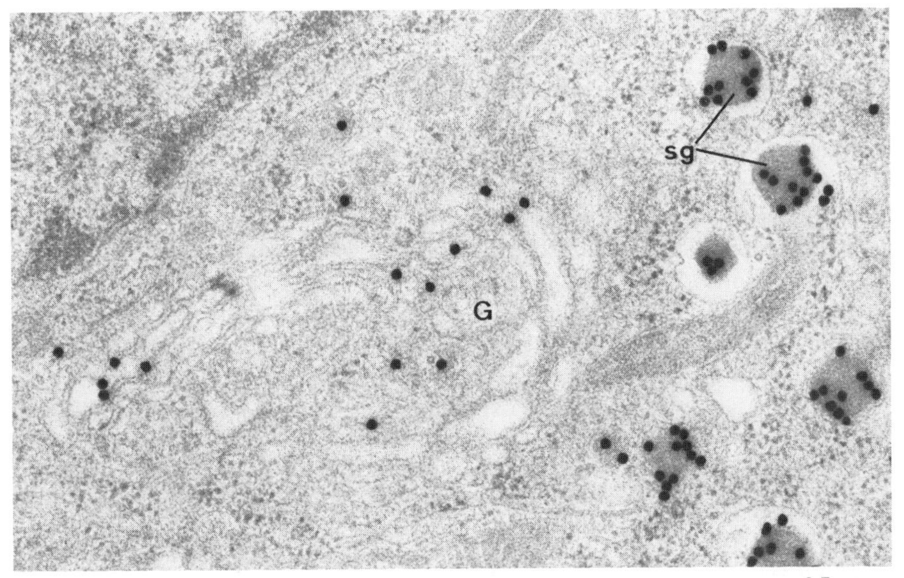

Figure 16-24 Electron micrograph of an insulin-producing B cell in the islets of Langerhans of a rat pancreas. Insulin is stored in secretory granules. Exocytosis of these granules, with release of insulin into the blood, is triggered by a rise in the level of blood glucose. The cell is stained with anti-insulin antiserum and the bound antibodies are detected by the protein A–gold technique (see Figure 5-9). Gold particles revealing the sites of binding of the antiserum with insulin antigens appear selectively concentrated over the dense core of the secretory granules (sg). A moderate degree of labeling is present over the cisternae of the Golgi apparatus (G); these molecules represent newly synthesized proinsulin molecules en route from the rough endoplasmic reticulum to the secretory vesicles (see Figure 5-34). *Courtesy of L. Orci.*

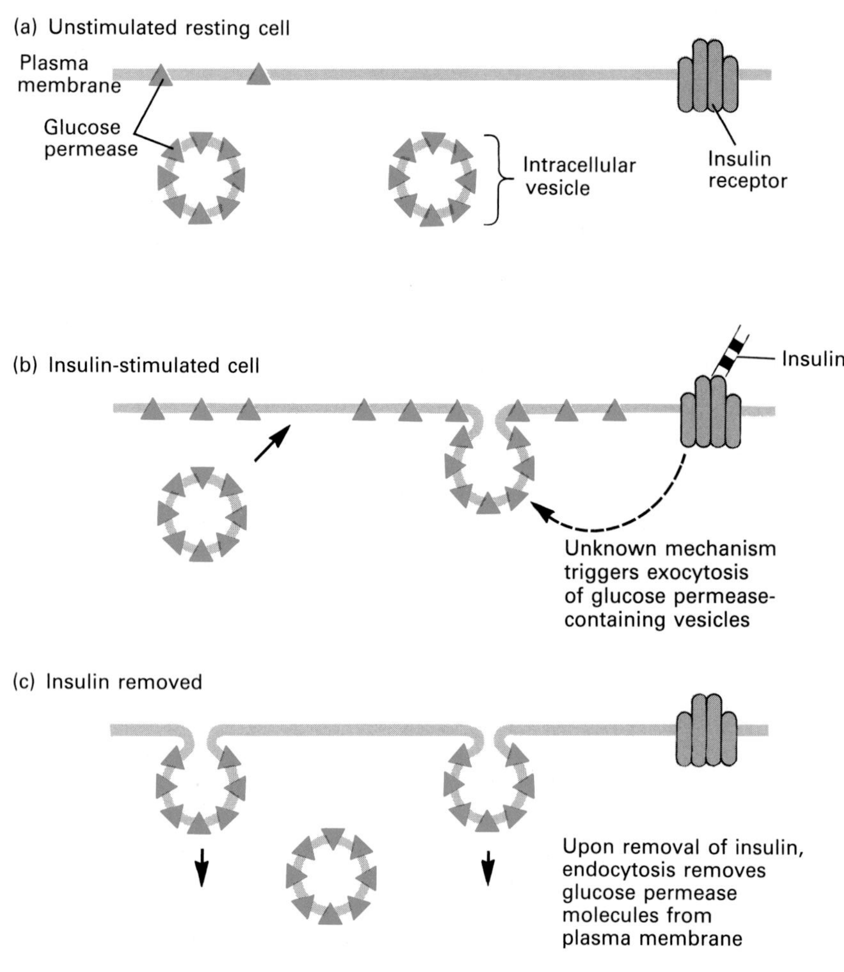

Figure 16-25 A schematic depiction of the regulation of the glucose transport system in adipocytes by exocytosis and endocytosis. (a) Binding of insulin to its receptor on resting cells. (b) This triggers—by an unknown mechanism possibly involving a protein kinase—exocytosis of vesicles that contain abundant glucose permeases, causing a rapid increase in the rate of glucose uptake by the cells. (c) Following the loss or removal of insulin, regions of the plasma membrane rich in glucose permeases are subjected to endocytosis, depleting the plasma membrane of the permeases. [See G. E. Lienhard, 1983 *TIBS* 8:125.]

hormones that trigger rapid increases in membrane permeability may do so by causing the exocytosis of vesicles that contain other cryptic permease proteins.

The Insulin Receptor Has Been Purified, Cloned, and Sequenced

Although insulin receptors are easily demonstrated on the surface of most cells, relatively little is known about their mode of action. They can be detected on most cells by the criteria of saturability and specificity of binding of labeled insulin. The insulin receptors have been identified by *affinity labeling*. This technique involves the binding of radioactive insulin to cells followed by treatment with a chemical agent that covalently cross-links the labeled insulin to the hormone-binding subunit of the receptor (Figure 16-26). Such cross-linking agents contain two reactive sites for free amino groups in proteins. The labeled insulin was found to be specifically bound to a protein of m.w. 84,000, a subunit of the receptor itself.

The insulin receptor, present in only 1 part per 10,000 to 1 million, can be purified by affinity chromatography, similar to the way in which β-adrenergic receptor is purified (see Figure 16-6). The receptor is solubilized from cell membranes by the addition of nonionic detergents. The solubilized receptor adheres tightly to an affinity column containing an inert matrix, such as agarose, to which insulin is covalently bound. Other proteins will not bind to the insulin or the matrix and will pass through the column. The receptor can be recovered from the column by elution with an acidic urea solution, which partially denatures the receptor. The recovered receptor can be purified by dialysis to remove the urea. By raising the pH, the receptor is renatured. Purification of the receptor can be about 200,000-fold by a single step! The insulin receptor contains two copies each of an α (m.w. 84,000) and β (m.w. 70,000) chain, all linked together by S—S bonds (Figure 16-26).

The insulin receptor can itself be shown to have protein kinase activity. When researchers incubated highly purified receptors with labeled ATP in the presence of insulin, they found that a phosphate group was transferred from ATP to the β subunit of the receptor. Little phosphorylation was observed if insulin was omitted. They concluded that the receptor can phosphorylate itself. Presumably, in the cell, it phosphorylates other proteins as well, but the identity of these proteins is not known. Exposure of muscle cells to insulin results in the activation of glycogen

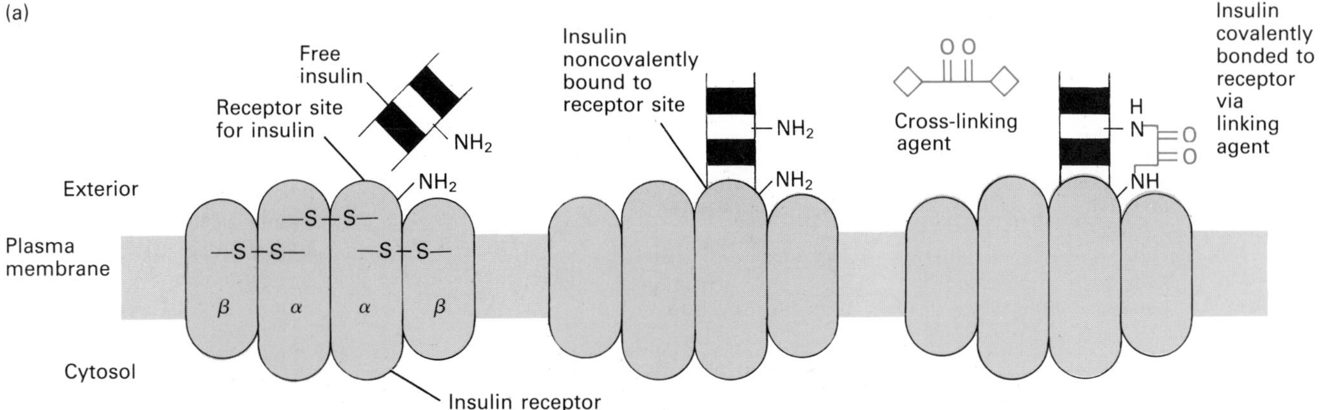

(b)

Sites of reaction with free amino groups

Receptor protein —N(H)—C(=O)—$(CH_2)_6$—C(=O)—N(H)—Insulin + 2 N-hydroxysuccinimide (N—OH)

Figure 16-26 For investigative purposes insulin can be covalently cross-linked to its receptor. (a) Binding of insulin to its receptor brings free NH_2 groups on the two proteins close together. (b) The chemical cross-linking agent reacts with free amino groups on insulin and the receptor protein, forming a covalent bond which ensures that radioactively labeled insulin will remain attached to its receptor throughout isolation and purification procedures. The insulin receptor contains two α subunits (m.w. 84,000) that bind to insulin, and two β subunits (m.w. 70,000), all linked by disulfide bonds. [See J. Massague and M. P. Czech, 1982, *J. Biol. Chem.* **257**:729; A. Ullrich et al., 1985, *Nature* **313**:756.]

synthase and, therefore, glycogen synthesis. Activation is due to removal of phosphates from 3 of the 10 phosphorylated sites on the glycogen synthase. It is now thought that insulin binding to the insulin receptor triggers phosphorylation of a protein phosphatase, thus activating it. The active phosphatase would then remove the three phosphate residues from glycogen synthase, activating the enzyme and triggering glycogen synthesis.

Insulin and Glucagon Balance Glucose Blood Levels

Glucagon is synthesized by the A cells of the islets of Langerhans and released in response to low blood glucose levels. Glucagon primarily affects liver cells. Like epinephrine, it induces adenylate cyclase and the cAMP cascade, causing a degradation of glycogen and an increase in blood glucose. Glucose availability for metabolism is regulated during periods of abundance (following a meal) or scarcity (following fasting) by the adjustment of insulin and glucagon concentrations in the circulation (Figure 16-27). Epinephrine use is regulated only under stressful conditions.

When, after a meal, blood glucose rises above its normal level of 80 to 90 mg per 100 ml, insulin is released into the blood from secretory vesicles in the B cells in the islets of Langerhans of the pancreas. The islet cells themselves respond to the rise in level of glucose or amino acid by releasing insulin into the blood, which transports it throughout the body. By binding to cell surface receptors, insulin causes removal of glucose from the blood and its storage as glycogen. If glucose falls below about 80 mg per 100 ml, then the A cells of the islets start secreting glucagon. The glucagon binds to a glucagon receptor on liver cells, activating adenylate cyclase and the cAMP cascade (a reaction similar to that of epinephrine). The result is the degradation of glycogen and the release of glucose into the circulation.

Receptor Regulation

The amount of functional hormone receptor on a cell's surface is not constant. The receptor level is modulated up (*up regulation*) or down (*down regulation*), permitting the cell to respond optimally to small changes in the hormone level. Prolonged exposure of a cell to a hormone usually results in a reduction of functional surface receptor molecules. This desensitizes the cell to the high level of hormone present. Up to a point, an increase in hormone concentration can still cause typical rather than excessive hormone-induced responses.

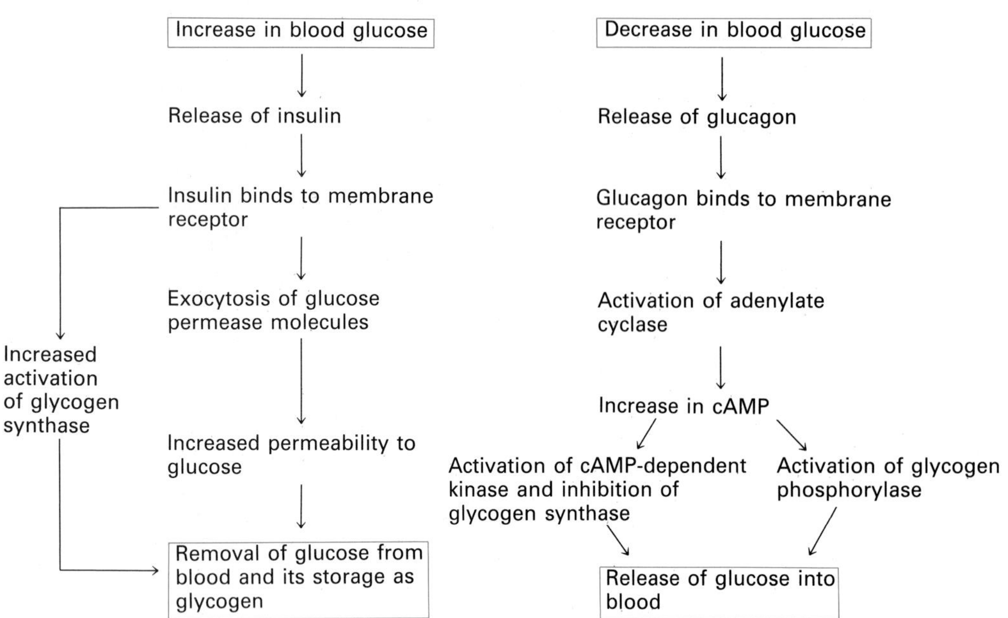

Figure 16-27 The level of blood glucose is regulated by the opposing effects of insulin and glucagon. Insulin causes an increase in glucose uptake mainly in muscle and adipocyte cells; glucagon acts mainly on liver cells to cause glycogen degradation.

Down Regulation of Receptors Is a General Response to a High Level of Circulating Hormone

We know a least four ways in which the level of cell surface hormone receptors can be reduced. (1) They can be endocytosed and destroyed. (2) They can be internalized by endocytosis and remain stored in an intracellular vesicle; since only cell surface receptors can bind extracellular hormone, this would also result in down regulation. (3) The receptors can remain on the cell surface but change in a way that makes them unable to bind ligand. (4) They can bind ligand but the receptor-ligand complex is unable to induce the normal hormone response.

On turkey erythrocytes, several hours' exposure of epinephrine receptors to epinephrine causes a nonfunctional binding of hormone to receptor. The receptor can bind epinephrine but cannot activate adenylate cyclase. Its inactivation is reversible. After incubation of the cells for several hours in the absence of epinephrine, the surface receptors are reactivated. Prolonged exposure to epinephrine causes phosphorylation of several serine and threonine residues of the β-adrenergic receptor. At least some of the phosphorylation is catalyzed by the cAMP-dependent protein kinase, whose activity is enhanced by the high level of cAMP induced by the epinephrine. Phosphorylated receptor binds ligand normally but cannot activate adenylate cyclase. For instance, when phosphorylated receptor is purified and implanted into a cell (see Figure 16-7), it is not able to stimulate cAMP synthesis after the addition of epinephrine. Enzymatic removal of the phosphate groups reactivates the ability of the receptor to stimulate cyclase. Thus, these cells utilize a feedback loop to modulate the activity of the surface β-adrenergic receptor. High intracellular levels of cAMP result in phosphorylation and thus of desensitization of the receptor, so that less cAMP will be induced by extracellular epinephrine.

A different type of down regulation of surface receptors appears to follow the binding of peptide hormones such as insulin, glucagon, and epidermal growth factor (Table 16-2). The receptor-hormone complex is brought into the cell by receptor-mediated endocytosis. The internalized hormone is subsequently degraded in lysosomes—a fate similar to that of other endocytosed proteins such as low-density lipoprotein (LDL) and asialoglycoproteins (see Chapter 15). Internalization and degradation most likely terminate the hormone signal. Ligand-induced endocytosis of receptor causes a transient reduction in the level of surface hormone receptors (Figure 16-28). Even if the hormone receptors recycle to the cell surface by exocytosis, as they do in most cases, at any one time a substantial fraction will be in the internal membrane compartments involved in receptor cycling and will be unavailable to bind extracellular hormone.

Most internalized cell surface receptors for peptide hormones do recycle back to the cell surface. The average half-life of an insulin receptor on a chicken liver cell is about 9 h; during its lifetime it will mediate the binding, internalization, and degradation of about 200 insulin molecules, recycling back to the cell surface after each discharge of ligand (see Figure 16-28). But each time an

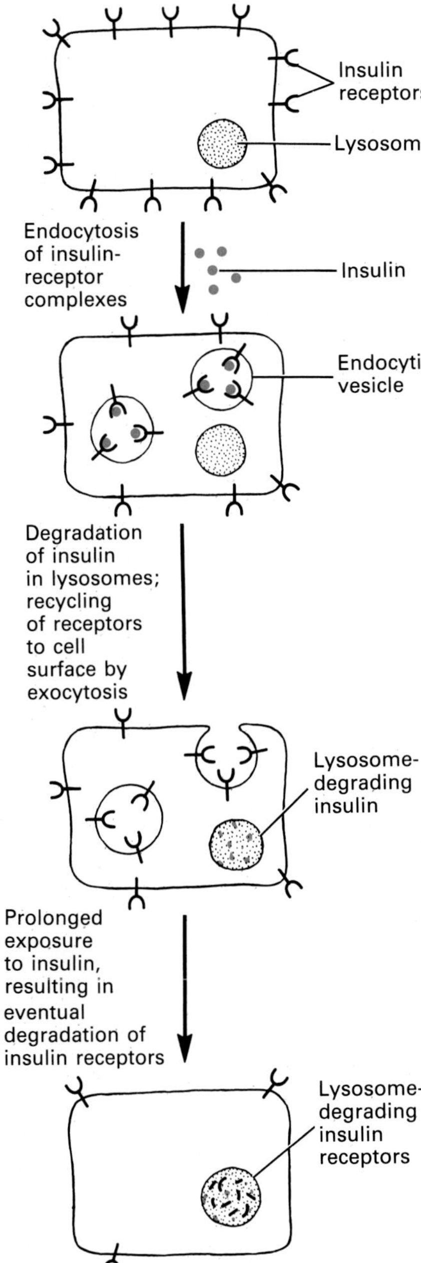

Figure 16-28 Two mechanisms for reduction in the level of cell surface insulin receptors by exposure to insulin. The first is receptor internalization. Surface insulin-receptor complexes are internalized by endocytosis. Insulin is degraded in lysosomes while 90 percent of the receptors escape degradation and recycle to the cell surface by exocytosis. Since many of the receptors are transferred, transiently, to the various intracellular organelles involved in receptor recycling, fewer remain on the cell surface. The second mechanism is receptor degradation. Upon prolonged exposure to insulin, endocytosis and recycling of insulin receptors occurs continuously. During each cycle of endocytosis a small portion (less than 10 percent) of the internalized receptors are degraded. After several hours of endocytosis, the total number of insulin receptors is reduced substantially. [See G. V. Ronnett et al., 1983, *J. Biol. Chem.* **258**:283.]

insulin receptor is internalized, it has a slight probability (less than about 10 percent) of being degraded in the lysosome. Exposure of a liver cell to high levels of insulin for several hours will induce many rounds of endocytosis and exocytosis of insulin receptor. After many hours most of the receptors will be degraded. If the concentration of extracellular insulin is then reduced, the level of cell surface insulin receptors will recover, but only after 12 to 24 h. Synthesis of new receptors is needed to replace those degraded by endocytosis, but this is a slow process and may take more than a day.

Abnormal Regulation of Hormone Receptors Is One Cause of Diabetes

Abnormal function or regulation of insulin receptors has been demonstrated in some persons with diabetes, a disease caused by insufficient insulin action in the body. Childhood, or early onset, diabetes is caused by deficient insulin synthesis by the B cells of the pancreatic islets. Most such persons produce antibodies against insulin-producing B cells in their own pancreas, with subsequent autoimmune reactions destroying the cells. In most of these circumstances injections of insulin can overcome the problem. The deficiency can also be caused by the absence of normal insulin synthesis due to the production of a structurally abnormal insulin, or a defect in the conversion of proinsulin to insulin.

In most nonobese persons with adult onset diabetes, the insulin that is produced is normal but there is a defect in its secretion from the B cells. Many of these diabetics respond well to regular insulin injection, but some secrete normal levels of insulin—which focuses attention on defective receptors as a possible cause of disease. Rather rare receptor defects have been uncovered, including decreased numbers of receptors and defective receptor function.

Some diabetics make antibodies to the insulin receptor, which affects receptor function. Antireceptor antibodies that bind to receptors may prevent insulin binding or can trigger an insulin-like response. Such diabetics do not respond to insulin therapy. In other insulin-resistant diabetics, the target cells have low levels of insulin receptors (down regulation) together with a high chronic excess of insulin secretion. The cells are unable to respond to the high level of insulin in the blood, with severe metabolic consequences. The condition is often associated with chronic obesity, but whether obesity is a cause or an effect is not known.

Growth Factors

The *growth factors,* another class of polypeptide signaling substances, affect the growth and differentiation of specific types of animal cells, both in the developing organism and in tissue culture. Many growth factors fulfill the general definition of a hormone. Insulin is a growth factor for many cells in culture. Its growth-promoting actions require about 10^{-7} *M* insulin, whereas much lower concentrations, 10^{-9} to 10^{-10} *M*, are sufficient to induce very rapid actions, such as glucose uptake.

The first growth factor to be discovered, *nerve growth factor* (NGF), is needed if explanted sympathetic nerve cells are to differentiate in culture. Antibodies to NGF injected into a fetal mouse specifically inhibit differentiation of sympathetic neurons, establishing the role of NGF in the animal. The discovery of NGF was closely followed by the discovery of *epidermal growth factor* (EGF), a protein required for the growth of epithelial cells in culture. Epidermal growth factor is misnamed because many other cell types react to its growth-stimulating effects.

More recently the peptides *fibroblast growth factor* (FGF) and *platelet-derived growth factor* (PDGF) have been described, along with a class of proteins called *somatomedins* that are often required for the growth of specific cells (Table 16-2). Add to these the other hormones needed for cell growth—insulin, in particular—and we begin to understand why growth of most mammalian cell cultures requires a medium containing 10 to 20 percent animal serum: the serum provides all the hormones and growth factors (many of which have not yet been identified or purified). Increasingly, the substances affecting the growth of many cell types is being expanded (see Chapter 6), and many cells can now be grown in a medium of a defined composition that includes only a few peptide hormones.

The Receptor for Epidermal Growth Factor Is a Protein Kinase

Epidermal growth factor is a small protein, found at high concentration in the mouse submaxillary gland. In culture, it has dramatic stimulatory effects on cell growth. Its action depends on the presence of a specific surface receptor but its effector function is not known. An important clue to its effector activity is the presence in purified receptor preparations of a protein kinase that is activated by the binding of EGF to its receptor. Substrate proteins for this kinase are phosphorylated on tyrosine residues rather than on the serine or threonine residues phosphorylated by most protein kinases (such as the cAMP-dependent kinase). The receptors for insulin and other growth hormones also appear to be tyrosine-specific protein kinases, as are certain viral and cellular proteins that, in abundance, may cause a cell to become cancerous. Most likely, hormone binding to receptor triggers phosphorylation of certain cytosolic proteins by the EGF (or insulin) receptor and in this way is the basis of growth stimulation. The identity and function of the proteins phosphorylated are not yet known.

Among the types of cells that respond to EGF in culture are keratinocytes, cells that grow and differentiate into skin. EGF markedly stimulates the proliferation of these cells and leads to enhanced synthesis of keratins, the major structural proteins of skin. EGF also appears to be a fetal growth hormone responsible for the proliferation and differentiation of epithelial cells in the embryo. Infusion of EGF into fetal lambs, for instance, markedly stimulates the growth and differentiation of the epithelial tissue lining the air passages in the lung.

Steroid Hormones and Their Cytosolic Receptors

So far we have discussed hormones and growth factors that bind to cell surface receptors and cause a signal to be transmitted across the plasma membrane. Now let us consider the steroids, the lipid-soluble hormones that pass freely through the plasma membrane and interact with receptors in the nucleus or cytosol (see Figure 16-2).

Steroids Affect Gene Expression

Typical cells contain 10,000 to 100,000 steroid receptor molecules; each can bind one molecule of a steroid hormone with high affinity. For instance, the estrogen receptor in target organs such as the uterus and mammary glands has a dissociation constant for estrogen of about 3×10^{-10} *M*. It is controversial whether, in cells prior to steroid treatment, the receptors are localized in the cytosol or nucleus. Upon subcellular fractionation, most receptors fractionate with the cytosol. Thus, it was thought that unoccupied receptors are in the cytosol, and that hormone binding triggers a conformational change in the receptor that causes the receptor-hormone complex to migrate to the nucleus. In the nucleus it can bind to specific regions of DNA and modify the synthesis of specific mRNAs (Figure 12-30). However, recent studies on immunocytochemical localization of the receptor in sections of cells indicate that unoccupied, as well as occupied, receptors are in the nucleus. Except for glucocorticoid receptors, which are widespread, receptors for steroids are found in only a limited number of target tissues. Nonetheless, the same hormone—estrogen, for example—will induce different changes of gene expression in different tissues, as Table 16-6 shows. Not all the effects of steroids

Table 16-6 Specific proteins induced by steroid hormones

Hormone	Tissue	Protein	
Estrogen	Chick oviduct	Ovalbumin Conalbumin Ovomucoid Lysozyme	Egg-white proteins
	Chick liver	Vitellogenin Transferrin	
	Frog liver	Vitellogenin	
	Rat pituitary	Prolactin	
Progesterone	Chick oviduct	Avidin Ovomucoid Conalbumin Lysozyme	Egg-white proteins
	Rat uterus	Uteroglobin	
Testosterone	Rat liver	α-2u Globulin	
	Rat prostate	Aldolase	
Glucocorticoids	Rat liver	Tyrosine aminotransferase Tryptophan oxygenase	
	Rat kidney	Phosphoenolypyruvate carboxykinase	
	Mouse mammary cells	Mammary tumor virus RNA	
	Rat pituitary	Growth hormone	

SOURCE: L. Chan and B. W. O'Malley, 1978, *Ann. Intern. Med.* 89:649.

are on gene transcription. We noted in Chapter 12 that steroids also affect the stability of certain mRNAs.

Mutants Demonstrate the Role of the Steroid Receptor

How can one be certain that the steroid receptor, whether cytosolic or nuclear, indeed mediates the initial effects of steroid hormones? One way is to examine a series of agonists to a steroid hormone such as cortisol. In general, the ability of different agonists to affect gene expression correlates with their binding affinity for the receptor.

A second approach would be to isolate cell mutants that lack the receptor or have a defective receptor. But how can one select such *nonresponsive* mutants when, in the absence of hormone, a gene product would generally not be made? One way is to take advantage of the finding that treating an animal with cortisol kills almost all the cells in the thymus gland as well as the lymphocyte-related cells of the bone marrow. Many cultured lines of lymphatic tumor cells can also be killed by cortisol. Thus, cell lines *resistant* to killing by cortisol might contain defective cortisol receptors. Indeed, most such cortisol-resistant mutants lack the glucocorticoid receptors, a result establishing that these receptors mediate the cell-killing action of cortisol, be it at the induction of new mRNAs or at another level. Other cortisol-resistant mutants contain receptors that bind cortisol normally, but then the receptor-cortisol complexes cannot accumulate in the nucleus. These mutants lend support to the hypothesis that cortisol exercises a principal function in the nucleus.

Mutants in the testosterone receptor occur naturally in many animals as well as in the human male. Males with the *testicular feminization syndrome* have testes that secrete testosterone normally but the testosterone receptor is absent in all of the normal target organs, such as the vas deferens and seminal vesicles, that are essential for sexual development of the male. Since all mammals develop along female lines unless testosterone is active during embryonic differentiation, affected males appear to be female. This disorder reveals that the steroid receptor is indeed essential for homone function. It also shows that the same hormone receptor is present in many cells, even though the effects of the hormone are different in the various cell types.

The Levels of Steroid Hormones Are Regulated by Complex Feedback Circuits

Estrogen and progesterone stimulate the growth and differentiation of cells in the endometrium, the tissue that lines the interior of the uterus. The changes in the endo-

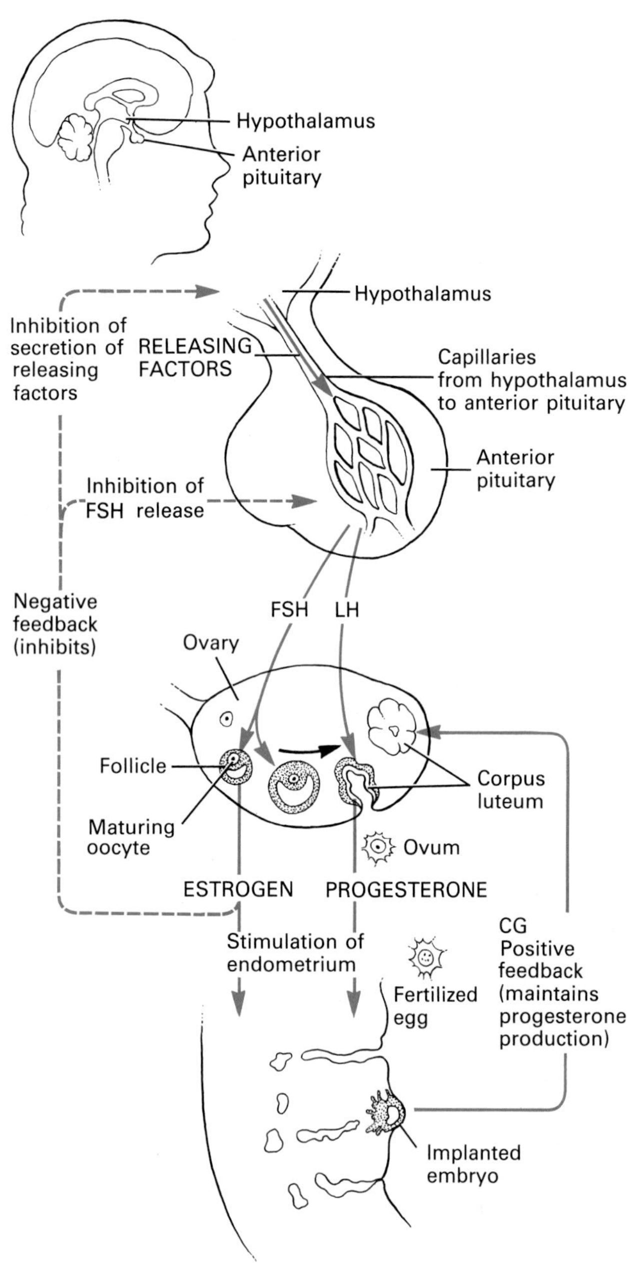

Figure 16-29 The levels of estrogen and progesterone in the circulation of female mammals are regulated by complex feedback systems. A key role is played by the anterior pituitary gland, an organ at the base of the brain that is separate from but controlled by the brain (see inset). The anterior pituitary is connected directly to the hypothalamus, a part of the brain, by a special set of blood vessels. Nerve cells in the hypothalamus discharge hypothalamic peptide-releasing factors that enter these vessels and bind to receptors on the anterior pituitary cells.

Each developing mammalian egg, called an oocyte, matures into an ovum inside an ovarian follicle made up of many cells. Under the influence of follicle-stimulating hormone (FSH) released by the anterior pituitary, the follicle grows in size and number of cells. The follicular cells secrete estrogens that, in turn, stimulate the growth of the uterine wall and its glands in preparation for implantation of an embryo should fertilization occur. Estrogens also act on both the anterior pituitary and the hypothalamus to reduce the secretion of FSH, which then lowers the level of estrogens (negative feedback).

As a follicle approaches maturity, the output of FSH decreases. The hypothalamus releases luteinizing hormone–releasing hormone (LHRH), which triggers the anterior pituitary to secrete LH. The LH completes the maturation of the follicle and ovum; the mature follicle releases the ovum into the oviduct and is transformed into a temporary endocrine gland, the corpus luteum. Under the continuing stimulation of LH, the corpus luteum secretes progesterone. Progesterone, in turn, acts to induce the further growth of the uterine wall, preparing it to receive an embryo. In the absence of fertilization of the ovum and pregnancy, the corpus luteum degenerates. The resulting decrease in the level of circulating estrogens and progesterones causes degeneration of the uterine wall and the start of menstruation. However, if fertilization and implantation occur, the placental tissues produce a peptide hormone called chorionic gonadotropin (CG) that has a structure and function similar to LH. The CG prevents degeneration of the corpus luteum, in particular maintaining its continued synthesis of progesterone (positive feedback). Progesterone, in turn, maintains the cells in the endometrial lining of the uterus, and a good blood supply forms to nourish the implanted embryo.

metrium prepare the organ to receive and nourish an embryo. The level of both hormones is regulated by a complex feedback circuit involving several other hormones (Figure 16-29). Control of estrogen and progesterone levels is typical of the way all hormones are closely regulated. Upon stimulation by two pituitary hormones, FSH and LH, the follicle cells that surround the developing oocyte secrete estrogen and progesterone. FSH and LH are themselves secreted by cells of the anterior pituitary when triggered by specific releasing hormones produced by the hypothalamus, a part of the brain. Estrogen, in turn, acts on the hypothalamus to *reduce* secretion of releasing hormones and on the anterior pituitary directly to *inhibit* the release of FSH. This *negative feedback* by estrogen serves to modulate the level of circulating estrogen in the nonpregnant female (Figure 16-29).

During pregnancy, a high level of progesterone is required to maintain the endometrium and provide an adequate blood supply for the embryo embedded in the uterine wall. A *positive feedback* circuit maintains this high level. Either the placenta or the embryo secretes a peptide, chorionic gonadotropin (CG), that causes the *corpus luteum* (a matured and enlarged ovarian follicle) to continue synthesizing progesterone. By maintaining the endometrial cell layer, progesterone enables the embryo to grow and synthesize additional CG, which then induces the synthesis of additional progesterone.

Estrogen and Progesterone Induce Protein Synthesis in the Chicken Oviduct

The stimulatory effects of estrogen and progesterone have been studied extensively in the chicken oviduct. The molecular mechanisms by which these hormones increase the levels of certain mRNAs have been discussed in Chapter 12. Here we emphasize the cellular events that follow gene activation and contribute to cellular differentiation.

The cells lining the major segment of the mature hen oviduct are of three types. Predominant are the *tubular gland cells* that, in response to estrogen, synthesize and secrete the major egg-white proteins ovalbumin, conalbumin, ovomucoid, and lysozyme. The *goblet cells* synthesize and secrete avidin, another egg-white protein, in response to progesterone. These proteins are taken up by receptor-mediated endocytosis into the developing egg cell as it makes its way down the oviduct. During growth of the embryo, these proteins are degraded to amino acids and used for the synthesis of chick proteins. The third type of cells are *ciliated cells;* the cilia, beating in synchrony, propel the egg down the oviduct.

In the sexually immature chick, the oviduct is lined with a single layer of undifferentiated epithelial cells (Figure 16-30a). These do not synthesize any of the specific egg-white proteins. Sexual differentiation can be induced in an immature female chick by implanting a small, estrogen-containing pellet under the skin. The estrogen slowly dissolves, generating an inducing level in the blood that lasts for many days.

Within 2 days of implantation there is an increase in the activity of all three nuclear RNA polymerases and a marked increase in ribosome and tRNA accumulation in the epithelial cells. Concomitantly the rate of cell growth and division accelerates and the proliferating epithelial cells lining the oviduct differentiate into recognizable tubular gland, goblet, and ciliated cells (Figure 16-30b). The tubular gland cells begin to accumulate large amounts of rough endoplasmic reticulum membranes, essential for the secretion of proteins. Beginning at 2 to 3 days after estrogen implantation, the tubular gland cells accumulate mRNA for ovalbumin and other egg-white proteins, and they begin to synthesize and secrete large amounts of these proteins into the oviduct. Ovalbumin synthesis peaks about 10 days after estrogen implantation. At all times the amount of ovalbumin mRNA per cell is proportional to the rate of ovalbumin synthesis.

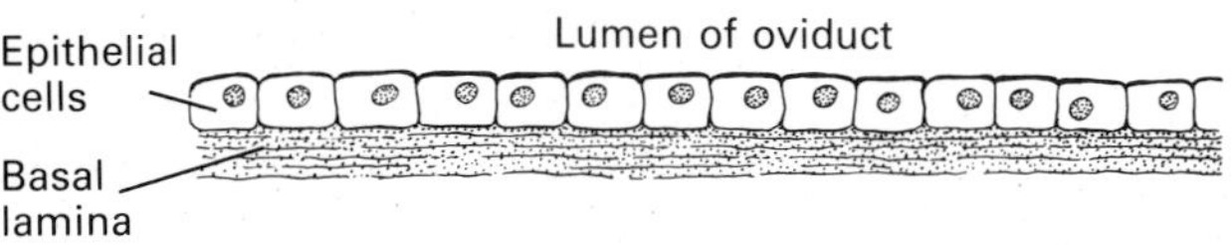

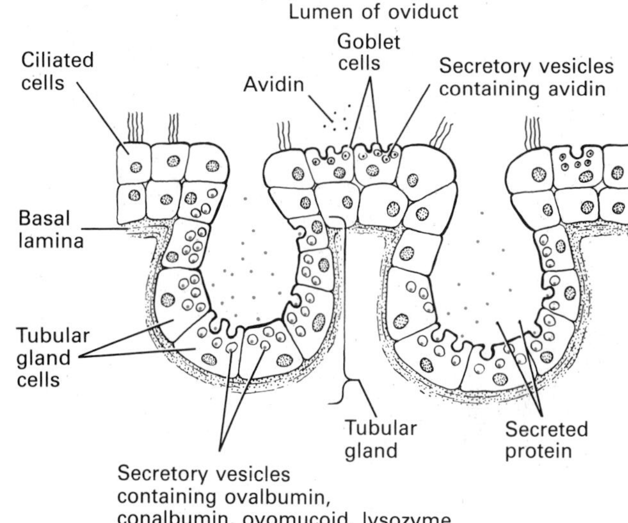

Figure 16-30 Estrogens and progesterone induce proliferation and differentiation of the epithelial cells lining the hen oviduct. (a) In the sexually immature chick the oviduct is lined with a single layer of columnar epithelial cells resting on a collagen-rich basal lamina (basement membrane). (b) Treatment with estrogen and progesterone for six days causes marked cell proliferation of the epithelium and differentiation of three types of cells: *tubular gland cells* that synthesize and secrete ovalbumin and other egg-white proteins, *goblet cells* that secrete avidin, an egg-white protein that binds the vitamin biotin, and *ciliated cells* that help move the ovum through the duct. Cell proliferation has increased the thickness of the epithelial layer and caused pouches called tubular glands to develop. As a result of cell proliferation, the epithelium is now several cell layers thick in places.

The Differentiated State Is Maintained and Reactivated by Estrogen

In experimental systems such as the chick oviduct, it is often difficult to uncouple the effects of steroid hormones on tissue growth from their effects on the synthesis of specific proteins. What is the initial action of the hormone and what are the secondary consequences of hormone action? Estrogen directly increases the synthesis of many egg-white protein mRNAs, including ovalbumin mRNA. If the estrogen pellet is removed from immature chicks after 2 to 3 weeks of induction, synthesis of ovalbumin mRNA ceases almost immediately and the levels per cell of ovalbumin mRNA and the other egg-white protein mRNAs drop precipitously. Most of the other metabolic activities of the oviduct cells are relatively unaffected. One week after estrogen withdrawal, the level of ovalbumin mRNA in oviduct cells has fallen to less than 1 percent that of stimulated animals.

The reintroduction of estrogen to estrogen-withdrawn animals leads to a rapid synthesis of ovalbumin mRNA. Within 3 h the rate of synthesis of ovalbumin mRNA is already about one-third that of a fully differentiated tissue. This rate is much faster than the 2 to 3 days that elapse between initial estrogen implantation and ovalbumin mRNA synthesis. The cells have retained a kind of "memory" that they once were stimulated with estrogen; they can respond much more quickly during restimulation with the hormone. Part of the memory may result from an elevation of the number of estrogen receptors in the stimulated cells, facilitating the response to even low levels of estrogen.

Fast response to secondary estrogen stimulation can also be seen in cultured explants of oviduct tissue from estrogen-withdrawn chicks. In such cells, the addition of estrogen leads to a rapid elevation in the rate of synthesis of ovalbumin mRNA and conalbumin mRNA, followed, as might be expected, by an accumulation of both mRNAs in the cultured cells. In Chapter 12 we discussed the evidence that estrogen acts at least in part to initiate transcription of the ovalbumin gene. Estrogen also appears to increase the half-life of conalbumin mRNA and ovalbumin mRNA, and the two effects cause the accumulation of these mRNAs in stimulated cells. The mechanism of this stabilization is not known.

Glucocorticoids Stimulate the Induction of Specific mRNAs

Corticosterone, the predominant glucocorticoid hormone, has many actions on different body cells (Table 16-1). One principal overall effect of the hormone is to increase the supply of glucose to the blood, ensuring an adequate supply to critical tissues such as the brain. A few hours after administration of corticosterone, or the related steroids cortisol or cortisone, glucose release from the liver as well as deposition of glycogen increases. Cortisol inhibits glucose uptake by peripheral tissues, such as muscle and adipose tissues, most likely by blocking glucose transport. It also causes an increase in the release of fatty acids from adipose cells. Especially important, cortisol, in a process called *gluconeogenesis,* acts on hepatocytes to enhance the synthesis of glucose from precursors such as amino acids. Cortisol increases the synthesis of hepatic enzymes, such as tyrosine aminotransferase and tryptophan pyrrolase, that speed the conversion of tyrosine and tryptophan to acetyl CoA and then to glucose. In this fashion dietary protein can be converted to glucose for transport, metabolism, or storage as glycogen.

The induction of tyrosine aminotransferase by glucocorticoids has received extensive attention, since it can be demonstrated in intact liver, in cultured liver cells, and in continuous cell lines of liver tumor (hepatoma). All these cells respond to treatment with glucocorticoids by a rapid induction of enzyme synthesis, yielding a 5-fold to 15-fold increase in enzyme concentration in 10 h. The induction in hepatoma cells involves a selective synthesis of only a few mRNAs, including that for tyrosine aminotransferase, without a general increased RNA synthesis. Glucocorticoids have no effect, however, on the synthesis of the vast majority of hepatoma (or liver) proteins (Figure 16-31). This result shows that the receptor-steroid complex can have selective and dramatic effects on gene transcription.

In summary, the initial action of steroid hormones on target cells increases (or occasionally decreases) synthesis of a very small number of mRNAs and proteins. The elevated levels of specific enzymes can induce a variety of secondary effects, for instance gluconeogenesis in the liver caused by enzymes induced by cortisol. Presumably some of the proteins induced by hormones such as estrogen trigger the growth and eventual differentiation of target tissues.

The Actions of Thyroxine Resemble Those of Steroid Hormones

The thyroid gland secretes a mixture of two iodinated hormones, thyroxine, or tetraiodothyronine (T4), and triiodothyronine (T3) (Table 16-1). They are formed by intracellular proteolysis of the iodinated protein thyroglobulin, produced by the thyroid cells themselves, and released immediately into the circulation. They accelerate the rates of oxygen consumption and heat production in most tissues of the body, principal exceptions being the spleen, adult brain, and gonads. Thyroid hormones stimulate catabolism of glucose, fats, and proteins by increasing the levels of many enzymes that catalyze these metabolic reactions, such as liver glucose 6-phosphatase, hexokinase, and mitochondrial enzymes for oxidative phosphorylation.

Thyroxine, like steroid hormones, diffuses freely across the plasma membrane and binds specifically and with high affinity to receptors in the nucleus. Hormone-responsive cells such as liver have several thousand thyroxine receptors, while few, if any, are found in nonresponsive tissues. Much evidence suggests (but does not prove) that binding of thyroxine to receptors induces

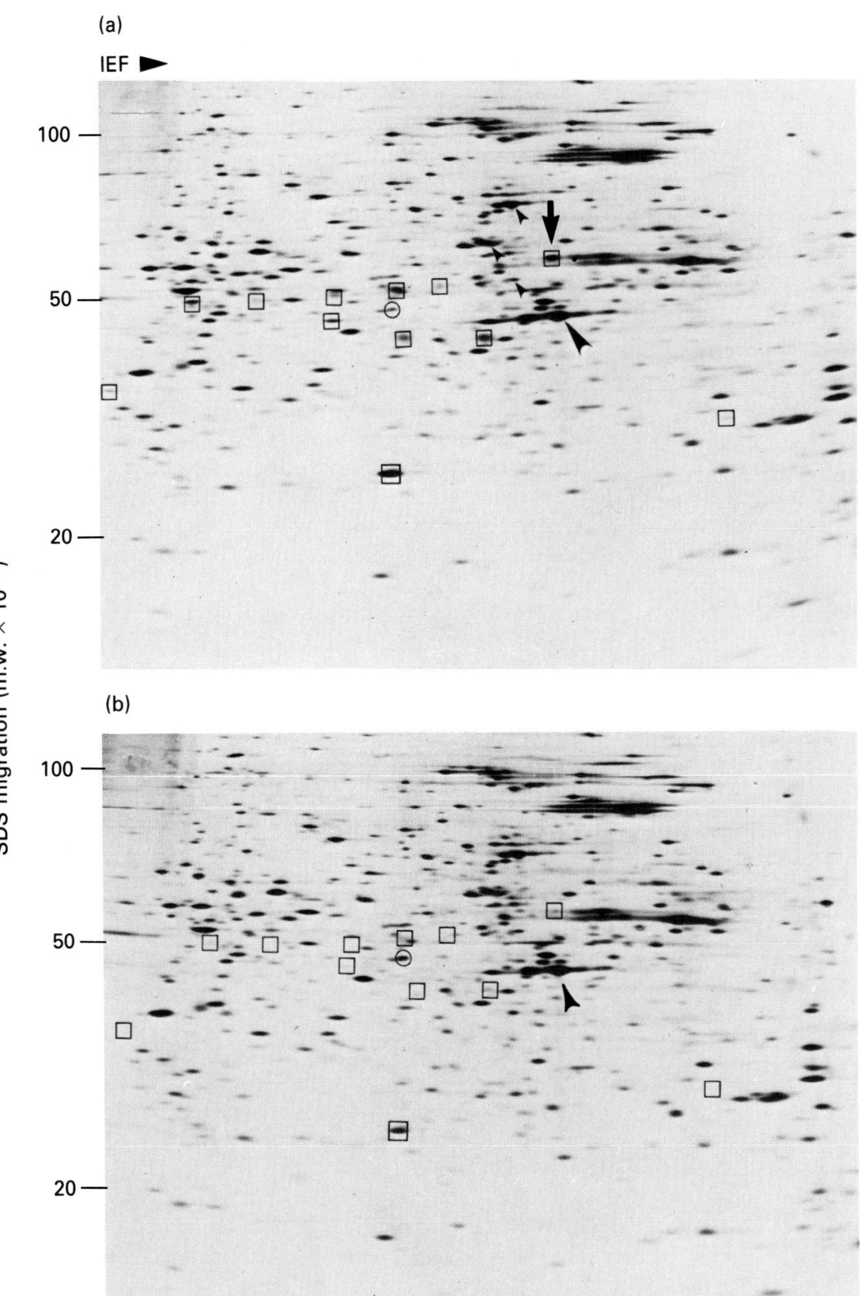

Figure 16-31 Demonstration of highly specific alterations in the pattern of protein synthesis following treatment of hepatoma cells with dexamethasone, a cortisol analog. Growing cells were (a) treated with 10^{-6} *M* dexamethasone for 18 h, or (b) not treated, then labeled for 30 min with [^{35}S]methionine. The labeled proteins made by the cells were resolved by a two-dimensional gel electrophoresis technique. The gel is then subjected to autoradiography. The squares around some of the spots indicate polypeptides whose synthesis is induced by dexamethasone. The large arrow points to tyrosine aminotransferase. The circles indicate a spot whose synthesis is repressed. Note that synthesis of most of the proteins is unaffected by hormone, indicating that its effect is very specific. The small arrows point to actin, a typical protein whose synthesis is unaffected by glucocorticoids. The increase in protein synthesis is due to increases in the levels of specific mRNAs, as shown by translating mRNAs in vitro and examining the pattern of newly synthesized proteins on similar gels. [See R. D. Ivarie and P. H. O'Farrell, 1978, *Cell* **13**:41.] *Courtesy of P. H. O'Farrell.*

transcription of specific mRNAs. In the kidney, for instance, ornithine aminotransferase (OAT) is induced by thyroxine. Using a cloned DNA probe, researchers have shown that thyroxine addition causes a rapid elevation in the rate of transcription of the OAT gene. Also, with various thyroid hormone agonists a good correlation exists between their ability to induce enzyme synthesis and their affinity for nuclear receptors. Triiodothyronine, for instance, is threefold to fivefold more potent than thyroxine and binds with a correspondingly higher affinity to the thyroxine receptors. Under normal conditions the nuclear receptors, with a K_m of 1 to 5 × 10^{-10} *M*, are half-occupied with triiodothyronine.

Thyroid hormones are important for growth and differentiation. This is seen most dramatically in thyroidectomized tadpoles; they are unable to undergo metamorphosis into adult frogs. Metamorphosis to normal tadpoles is induced when they are placed in water containing thyroxine. Growth and development of young mammals stops if the thyroid is removed or ceases to function. But it is not yet known which effects of thyroxine are immediate and which are a consequence of the increased levels of the enzymes that are induced by thyroxine.

Hormones and Cell-to-Cell Signaling in Microorganisms

Cell-to-cell communication by hormonal signaling is not confined to multicellular animals. It is widely used by eukaryotic microorganisms and also by higher plants. Eukaryotic microorganisms such as yeasts and slime molds can propagate asexually as solitary cells but need to aggregate with other cells of the species for sexual mating and recombination or for initiating the developmental stage of the life cycle to form spores. Indeed, many species of yeast and protozoa have evolved systems involving a secreted, diffusible peptide or other molecule as an intercellular signal. These pheromones often attract one gamete cell to another or induce sexual differentiation.

A Pheromone Attracts Yeast Cells for Mating

The mating that occurs between two yeast cells of opposite mating types (see Chapters 6 and 11) is controlled by two secreted peptide pheromones termed *a* factor and α factor. An *a* haploid cell type secretes the *a* oligopeptide "mating factor," or pheromone; an α cell type secretes the "α factor." These extracellular hormones bind to haploid cells of the opposite mating type and trigger at least three major biochemical events: (1) alterations in the cell surface that enhance the ability of cells to bind strongly and selectively to cells of opposite mating type—possibly involving synthesis of new surface glycoproteins; (2) arrest of the growth of the target cells—specifically blocking the initiation of new DNA synthesis and thus synchronizing the cell cycles of the mating partners in the G1 stage; and (3) alterations in the cell wall and in the cell membrane macromolecules that facilitate the fusion of the two mating cells and the eventual fusion of the two nuclei. It appears that *a* cells have surface receptors only for the α factor and not for the *a* factor, and vice versa—but the definitive identification of these specific surface receptors has not yet been achieved.

It is not known how the binding of the hormone to the receptor triggers all three of the above biochemical changes. Initial experiments suggest that the effects of α factor on *a* cells is mediated by an inhibition of adenylate cyclase, a reduction in the level of cellular cAMP, and presumably a reduction of phosphorylation of certain proteins.

Aggregation in Cellular Slime Molds Is Dependent on Cell-to-Cell Signaling

Cell-to-cell signaling is an essential prerequisite for developmental changes in the life cycle of many organisms. The best characterized are the slime molds such as *Dictyostelium discoideum,* unicellular free-living amebas that aggregate to form a motile multicellular organism. *Dictyostelium* amebas grow in soil and feed on bacteria and other organic matter. The differentiation process, depicted in Figure 16-32, is triggered by starvation. Beginning at about 6 h, cells begin streaming toward aggregation centers. By about 10 h mounds have formed, each

Figure 16-32 **(*opposite*)** (a) The differentiation of *Dictyostelium discoideum.* Amebas divide by mitosis, feeding on bacteria and other organic matter. The differentiation cycle is triggered by starvation. During differentiation DNA synthesis and cell division cease, and cells generate ATP by catabolism of cellular proteins and carbohydrates. Chemotaxis toward aggregation centers begins at about 6 h; as Figure 16-33 shows, cells are attracted to cAMP released by neighboring cells. Eventually the aggregating cells form into streams that flow into centers. By 10 h, mounds have formed, each containing about 100,000 cells. The cells do not fuse with each other; they retain their individual identities. About 2 h later a discrete "tip" is formed. The tip contains the cells (prestalk cells) that will form the stalk cells (color) in the mature fruiting body; those at the rear will become spore cells. Under certain environmental conditions, the mound will form into a motile, wormlike creature called a slug that migrates toward light or warmth; the prestalk cells form the "leading edge" of the structure *(lower part of figure).* The migrating slug stage can last for several days. Culmination, the final stage of differentiation, is triggered by overhead light. The stalk cells elongate and vacuolate, pushing down through the mass of differentiating spore cells *(inset).* This motion hoists the mass of spore cells up along the stalk. Simultaneously the spore cells secrete glycoproteins and polysaccharides that form a rigid spore coat around each spore. The spores lose water and become metabolically inactive. The mature "fruiting body" contains about 70,000 spores supported by a stalk built up of about 30,000 cells. Mature spores are much more resistant to desiccation, ultraviolet light, and other toxic treatments than are growing cells. Upon exposure to a suitable supply of nutrients, the spores germinate. Rupture of the spore coat releases viable, free-living amebas. (b) Scanning electron micrographs of fields of aggregating slime molds: 1, a field of *Polysphondylium pallidum* cells streaming toward aggregation centers; 2, a forming aggregate of *D. discoideum;* 3, a higher micrograph of aggregating *Dictyostelium,* showing the polarized cells forming end-to-end and side-by-side contacts as they stream toward the aggregation center. *Part (b1) courtesy of M. Claviez and G. Gerisch; parts (b2) and (b3) courtesy of R. Guggenheim and G. Gerisch.*

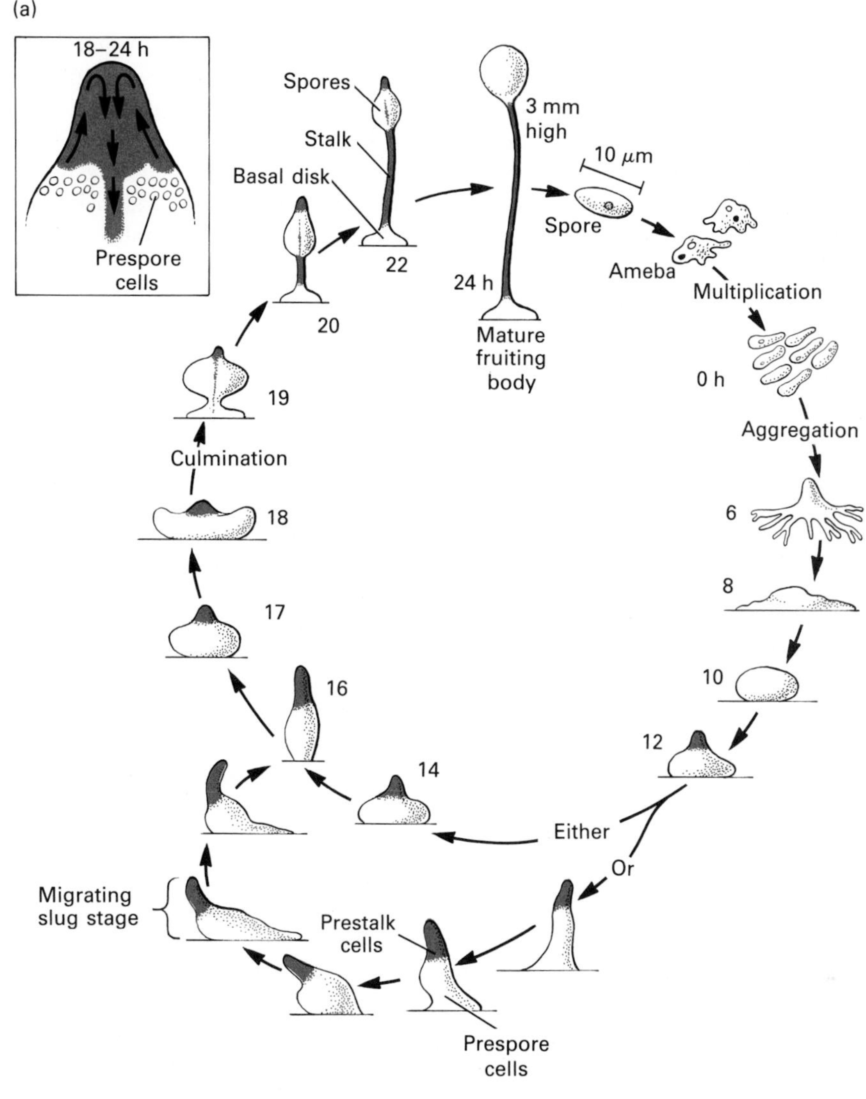
(a)
18–24 h
Prespore cells
Spores
Stalk
Basal disk
22
20
19
Culmination
18
17
16
14
3 mm high
10 μm
Spore
24 h
Mature fruiting body
Ameba
Multiplication
0 h
Aggregation
6
8
10
12
Either
Or
Migrating slug stage
Prestalk cells
Prespore cells

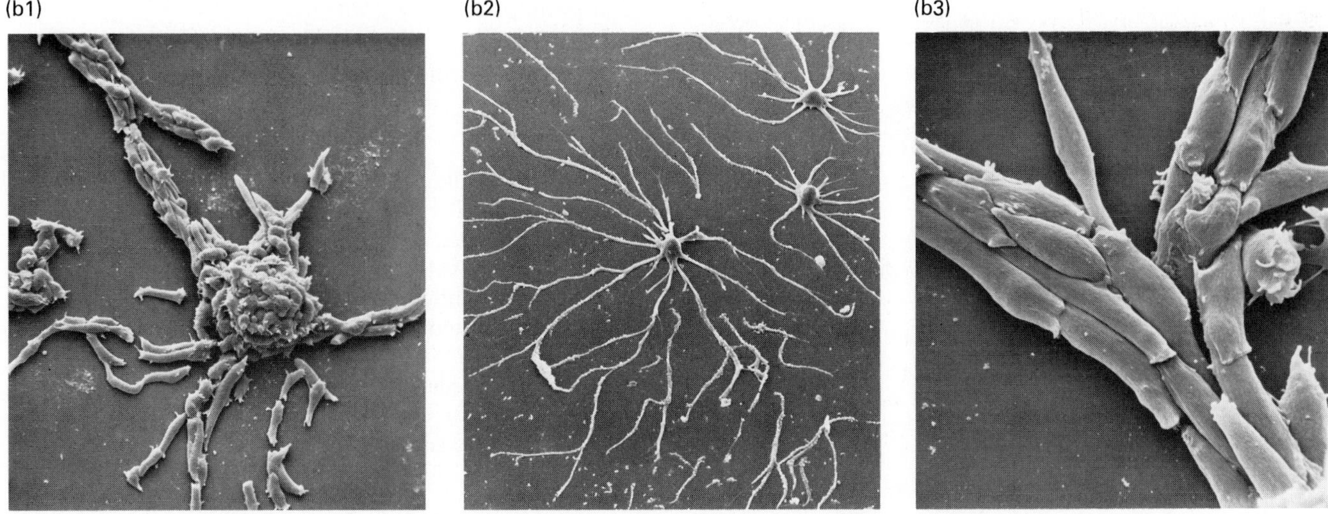
(b1)
(b2)
(b3)

containing about 100,000 cells. Differentiation of cells within the newly formed aggregates generates the two major distinct cell types that contribute to the mature fruiting body: spore cells and stalk cells. A very few cells, less than 10 percent, form basal disks. The fruiting body contains a bag of about 70,000 spores hoisted on a stalk made up of about 30,000 dead vacuolated stalk cells surrounded by a cellulose sheath. Under suitable environmental conditions the spores germinate, each releasing a free-living ameba that proceeds to feed.

The developmental program of the slime mold involves many alterations in the pattern of gene expression; these changes are examined in detail in Chapter 22. Here we discuss two aspects of cell signaling during the formation of the multicellular aggregates: *chemotaxis,* the streaming of cells toward local centers, and *cellular adhesion,* the formation of specific, tight cell-cell connections. Amebas, the growing cells, are incapable of either response. Both processes require the elaboration of specific cell surface macromolecules during the 6-h preaggregation phase.

Each slime mold species uses a different chemical as its chemotactic agent. *Dictyostelium discoideum,* which has received the most study, uses the nucleotide 3′, 5′-AMP as a chemotactic signal. This is the only known organism in which cAMP itself is a pheromone.

Aggregation in *D. discoideum* involves cell-cell signaling by periodic synthesis and secretion of cAMP. An individual ameba that synthesizes and secretes a pulse of cAMP creates a gradient of cAMP around itself. A neighboring ameba responds to such a gradient with several actions (Figure 16-33). It moves some micrometers up the gradient toward the source of cAMP, and it synthesizes and releases its own pulse of cAMP, attracting neighboring cells. It then becomes refractory for a period of several minutes and cannot move or respond to cAMP signals during this time. Under most natural conditions the cAMP signal is amplified by each cell, which means that the cell synthesizes and releases more cAMP than was present in the original stimulus. This relaying results in cell-to-cell propagation of the cAMP signal.

Proper functioning of this response requires a number of macromolecules: adenylate cyclase to synthesize 3′,5′-cAMP from ATP; a surface cAMP receptor; and both a secreted and surface cAMP phosphodiesterase to degrade cAMP to 5′-AMP. The phosphodiesterase is essential to prevent the extracellular hormone building up to a level that swamps out any gradients.

Because cell-cell signaling by extracellular cAMP can extend over distances of only 10 to 100 μm, cells are at-

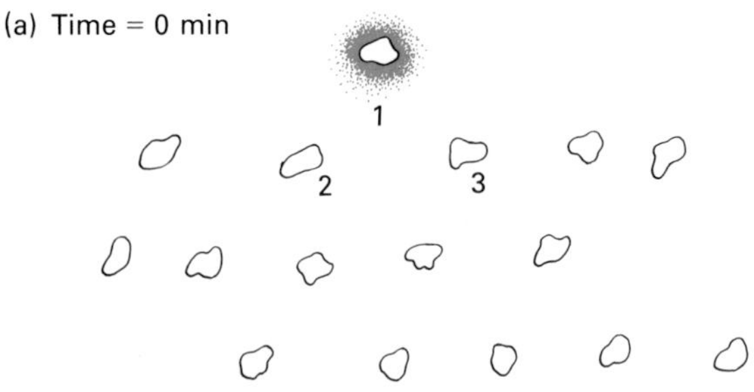

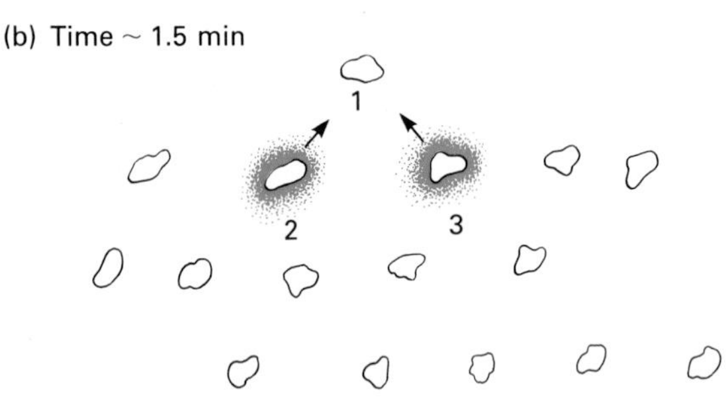

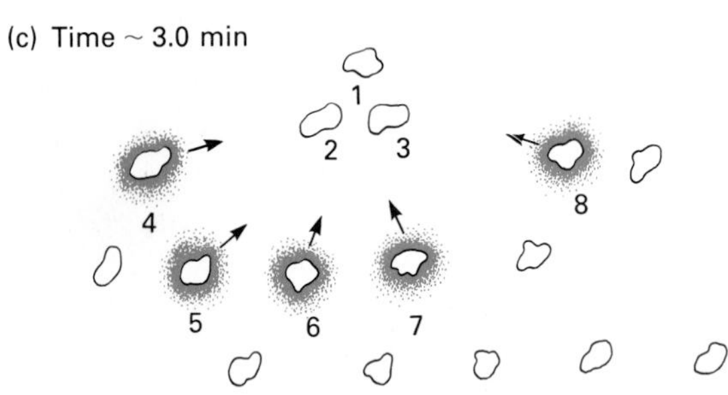

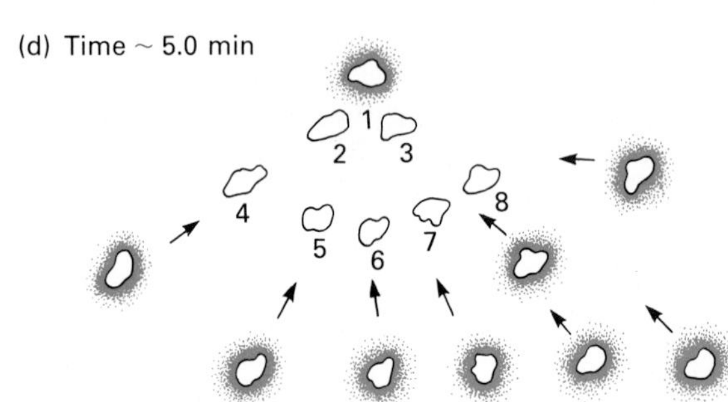

Figure 16-33 Cell-cell signaling with cAMP in *D. discoideum.* (a) $t = 0$ min: One ameba, cell 1, spontaneously releases a pulse of cAMP. This finite amount of cAMP diffuses away from the cell. (b) $t \cong 1.5$ min: Influenced by the cAMP, two adjacent cells, 2 and 3, move a few micrometers toward the source of cAMP, release their own pulse of cAMP, and become refractory to further stimuli. Meanwhile, the cAMP released by cell 1 has been destroyed by cAMP phosphodiesterase. (c) $t \cong 3.0$ min: Cells 4 and 5 move toward cell 2 and release cAMP; cells 6, 7, and 8 move toward cell 3, secrete cAMP, and become refractory. (d) $t \cong 5.0$ min: Other cells move inward toward cells 4 to 8 and relay the cAMP pulses. Meanwhile, the refractory period of cell 1 has ended. It synthesizes and releases a pulse of cAMP, repeating the cycle.

tracted primarily to the cAMP released by adjacent cells. No specialized cell serves as the sole source of chemoattractant to which all other cells are attracted. Aggregation is initiated by random cells releasing pulses of cAMP. Pulsatile waves of cAMP signals radiate outward from these "initiator" cells every 3 to 5 min; concomitantly there occurs a pulsatile movement of the cells inwards toward the centers. An initially homogeneous array of amebas rapidly breaks up into aggregation centers, each containing about 100,000 cells.

A key unanswered question is how the pulses of cAMP are amplified. How does the signal of cAMP bound to the surface of a cell cause the intracellular synthesis and then the release of an additional quantity of cAMP? Intracellular 3′,5′-cyclic GMP has been proposed to be one such intracellular signal. Binding of cAMP to the surface cAMP receptor is followed, within a few seconds, by a transient rise in the level of intracellular cyclic GMP. Possibly the cAMP receptor is coupled to guanylate cyclase, the enzyme that synthesizes 3′,5′-cyclic GMP from GTP. The cyclic GMP could then directly or indirectly activate the enzymes that synthesize and secrete cAMP. And binding of cAMP to the cell surface receptor triggers a transient rise in the level of intracellular Ca^{2+}. Calcium ions might also act as activators of cAMP synthesis and release.

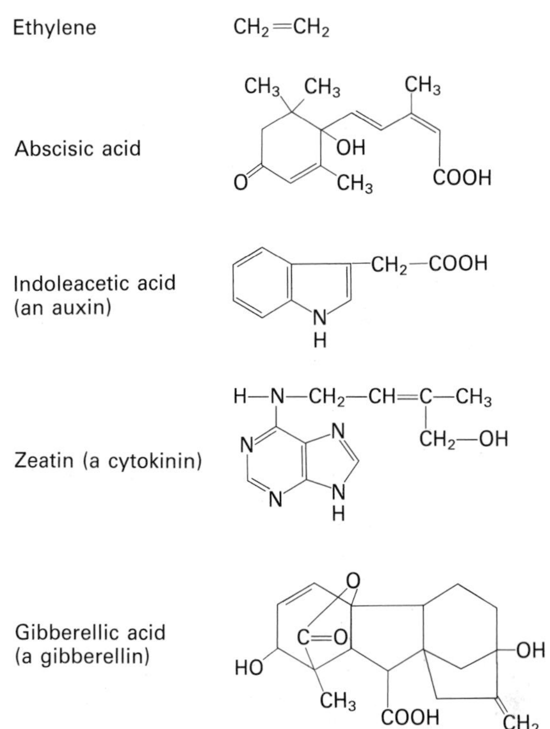

Figure 16-34 Examples of the five classes of plant hormones.

Plant Hormones

Extracellular hormones regulate many biological processes in higher plants, such as cellular growth and enlargement, differentiation of specialized tissues, and the induction of protein synthesis. Plant hormones can be divided into five principal classes: auxins, gibberellins, cytokinins, abscisic acid, and the gas ethylene (Figure 16-34). Frequently, two or more hormones must act together to induce a specific process. The molecular basis of action of plant hormones is being elucidated only slowly. In some cases the mode of action is strikingly different from those of animals and microorganisms. Let us examine the effects on target cells of two plant hormones, auxin and gibberellin.

Auxin Affects the Growth of Higher Plants

Many higher plants grow more by cell enlargement than cell proliferation. The size and shape of a plant are very much determined by the amount and direction of such enlargement. We saw in Chapter 5 that plant cells are surrounded by rigid cell walls constructed of cellulose fibers embedded in a matrix of protein, other polysaccharides such as hemicellulose, and pectin. The tensile strength of the cell walls allows the plant cell to develop a considerable internal pressure, or *turgor*. This outward pressure is caused by the higher osmotic pressure in the cytosol than in the surrounding extracellular fluids. Individual plant cells can increase in size very rapidly by loosening the wall on one or two sides and pushing the cytosol and plasma membrane outward against it. During this elongation the amount of cytosol remains constant; the increase in cell volume is due only to expansion of the intracellular vacuole (Figure 16-35). The magnitude of this phenomenon can be appreciated by the following example: if all of the cells in a redwood tree were enlarged only to the size characteristic of a typical liver cell, about 20 μm in diameter, the tree would have a maximum height of only 1 meter! In higher plants the region of cell division, the meristem, is separate from the region of cell elongation (Figure 16-36) and elongation takes place mainly after cell division is completed.

The mode of action of auxin has been investigated intensely for many years, but no satisfactory explanation exists yet for the molecular basis of any auxin effect. The reported effects are diverse, ranging from enhancement of cell elongation to stimulation of cell division to modification of cellular differentiation. Auxin has both primary and secondary effects, which are difficult to separate from one another. Moreover, primary and secondary effects may differ according to plant tissue and species.

The effects of auxin on cell elongation are fairly well understood. In typical cases, such as a pea stem, cell enlargement is restricted to a zone just below the tip. Elongation and growth of the stem are caused by elongation of the cells, primarily in one direction. Auxin stimulates cell elongation by inducing a localized H^+ secretion from

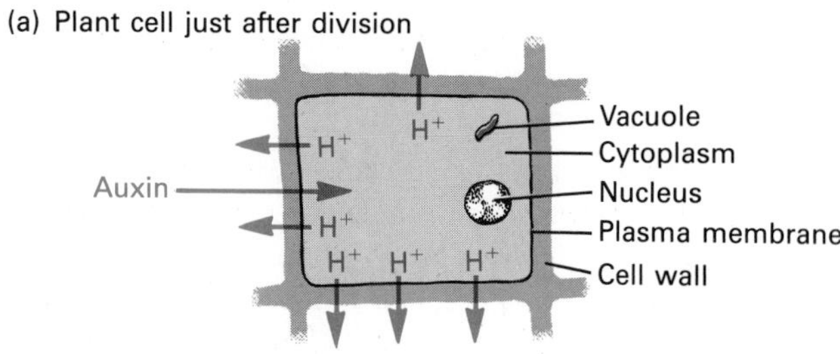

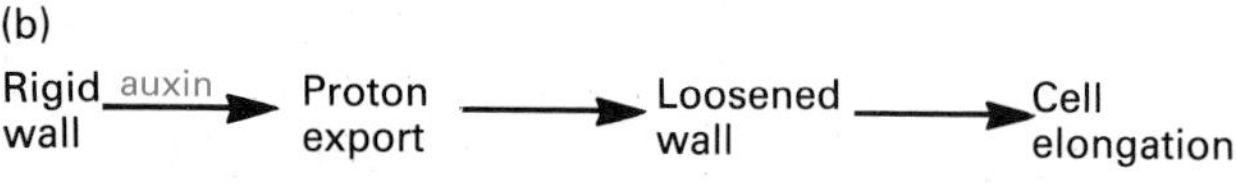

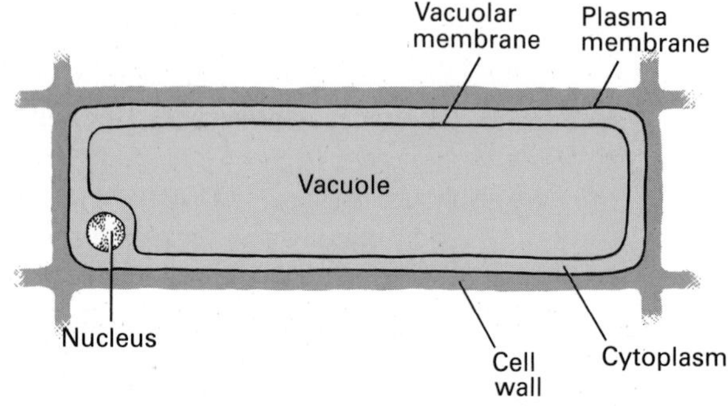

Figure 16-35 Stimulation of cell elongation by auxin. The hormone causes local loosening of the cell wall. Uptake of water by the vacuole creates a state of pressure, or turgor, that causes the cell membrane to expand against the loosened wall, elongating the cell. The increase in cell size is mainly due to an increase in the vacuole; the amount of cytoplasm is constant.

cells, which causes the walls to undergo localized loosening (see Figure 16-35). Auxin apparently activates (directly or indirectly) a membrane-bound proton pump (Chapter 15), with the result that the pH of the region of the cell wall near the plasma membrane is lowered, possibly to as low as pH 4.5 from the normal pH 7.0. As a consequence of the lowered pH, hydrogen bonding between cellulose fibrils of the wall is reduced. The lowered pH may also activate enzymes in the cell wall that degrade various protein or polysaccharide constituents, contributing to the localized softening and loosening of the wall and thus the elongation and enlargement of the cell. Unfortunately, we do not know the identity of these enzymes, nor the means by which auxin activates proton extrusion from the cells.

Is this softening of the cell wall the only effect of auxin? Auxin appears to affect RNA and protein synthesis in several systems but, again, it is not known what steps intervene between the addition of auxin and the observed changes in gene expression (see Figure 16-36). In other words, are these primary or secondary effects? Indeed, it is not known whether the "receptor" for auxin is located on the membrane or is cytosolic, nor whether the hormone acts by more than one mechanism. Auxin treatment, for instance, will activate massive synthesis of both protein and RNA in the differentiated hypocotyl region of a soybean seedling (see Figure 16-36). Some evidence indicates that auxin suppresses the synthesis of a few proteins of unknown function (and their mRNAs) while specifically inducing the synthesis of others. Many of these altered mRNA levels can be detected within 4 h of auxin addition—relatively early in the context of other auxin-induced changes, including cell division, which are not detected until 12 h. But 4 h still would allow ample time for secondary effects to become manifest, such as those that might be triggered by changes in the intracellular composition of ions or metabolites. An important question remains: Does auxin ever have a primary action on DNA, inducing mRNA and protein synthesis?

Gibberellic Acid Induces Specific mRNAs

Gibberellic acid (see Figure 16-34) is an example of a class of plant hormones, the gibberellins, with actions

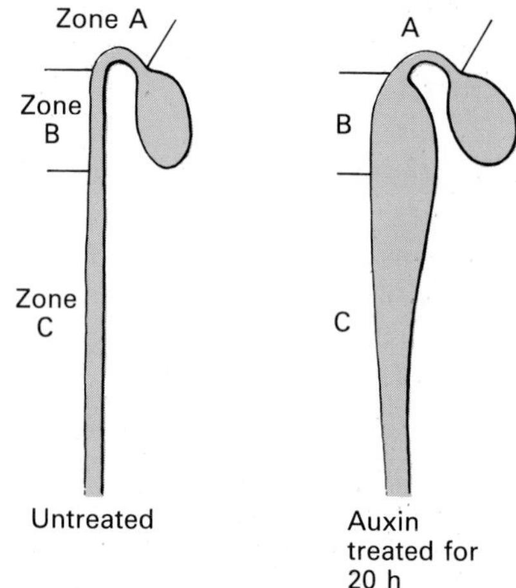

Figure 16-36 Regions of growth of a soybean seedling and the effects of auxin. Zone A: In untreated tissue, zone A is the meristematic zone, the zone of cell division, where levels of RNA and protein synthesis are high. Following auxin treatment, cell growth and division and macromolecular synthesis are inhibited in this region of the seedling. Zone B: Normally, zone B is the zone of maximal cell elongation but not of cell division. Following auxin treatment, the level of protein synthesis is maintained and RNA synthesis increases severalfold. Cell elongation causes widening of the seedlings. Zone C: The fully differentiated hypocotyl. Protein synthesis is low. Auxin activates massive synthesis of protein and RNA, and cell division is initiated.

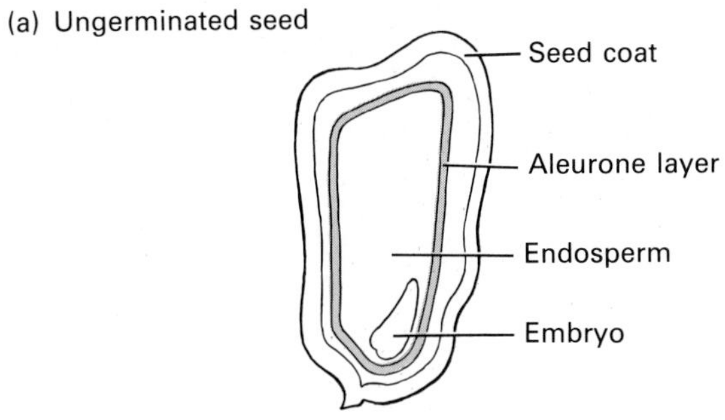

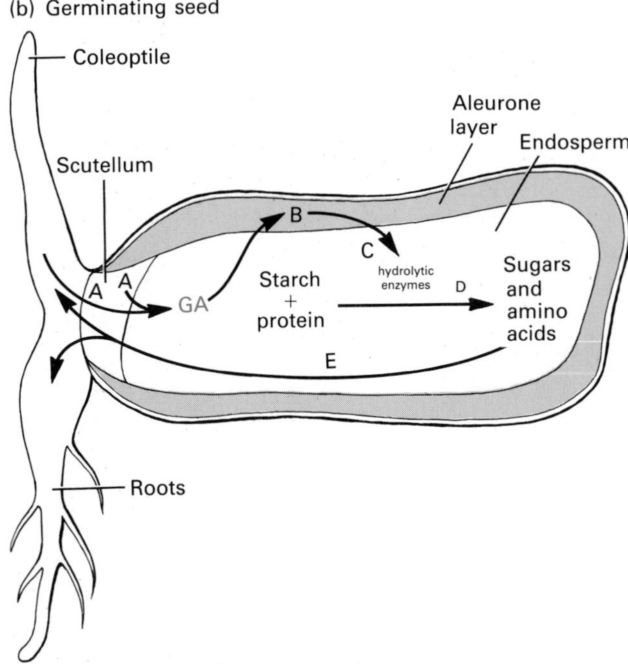

Figure 16-37 Diagrammatic representation of the relationship between gibberellic acid (GA) production, α-amylase production, and sugar accumulation in germinating barley seeds. (a) The ungerminated seed with its cell layers. The embryo—the cells that will give rise to the mature plant—forms a small part of the seed. (b) After germination, the embryo grows and differentiates into several cell types. Gibberellic acid produced by the coleoptile and scutellum regions of the embryo (A) diffuses into the aleurone layer (B), where its presence induces hydrolytic enzyme synthesis and release (C). These enzymes, such as proteases and amylases, serve to hydrolyze the protein and carbohydrate reserves in the endosperm, (D), producing the glucose and amino acids that nourish the growing embryo (E). [See J. E. Varner and D. T.-H. Ho, 1976, in *The Molecular Biology of Hormone Action,* Academic Press, p. 173.]

similar to those of steroid hormones in mammals, although their chemical structures and properties are very different. Gibberellic acid induces enzyme synthesis and is responsible for dramatic changes in plant cell form and function (cell differentiation). The germinating barley seed is a well-worked-out cereal grain system in which we can follow the role of gibberellic acid. The embryo is a small part of the barley seed (Figure 16-37a). After germination, the embryo cells grow and divide rapidly, forming the sheath (coleoptile) and roots of the young plant. The *endosperm* within the seed is a store of proteins, carbohydrates, and minerals whose digestion products—sugars and amino acids—are used for early plant cell growth. Surrounding the endosperm is the *aleurone layer,* composed mostly of nondividing cells.

The degradation and mobilization of these endosperm reserves are controlled by the hormone gibberellin. Gibberellic acid is secreted by the growing embryo cells (Figure 16-37b) and diffuses through the endosperm into the aleurone layer. In the aleurone layer it induces the synthesis and secretion into the endosperm of several hydrolytic enzymes. For example, proteases and α-amylase, which degrade stored proteins to amino acids and starch to glucose, provide support for the early growth of the barley seedling. In parallel, the level of mRNA for α-amylase is increased; these findings and others suggest that gibberellic acid initiates the synthesis of the α-amylase mRNA (Figure 16-38). Gibberellic acid also causes an enhancement of phospholipid synthesis and a proliferation of the rough endoplasmic reticulum and Golgi membrane sys-

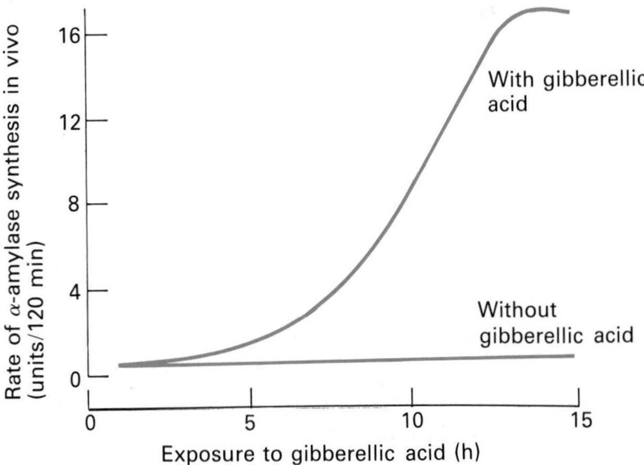

Figure 16-38 The time course of α-amylase synthesis in isolated aleurone layers from germinating barley seedlings in the presence or absence of added gibberellic acid. Although not shown here, the level of α-amylase mRNA increases in parallel to the synthesis of the α-amylase protein. This experiment indicates that gibberellic acid—directly or indirectly—induces synthesis of α-amylase mRNA. [See T. J. V. Higgins, J. A. Zwar, and J. V. Jacobsen, 1976, *Nature* **260**:166.]

tems in the aleurone layer. These changes appear to be essential for the secretion of α-amylase into the endosperm.

In many respects, therefore, the action of gibberellic acid on barley aleurone cells is similar to that of estrogen on the chick oviduct. Despite much effort, however, a cellular receptor for gibberellic acid has not been identified, and the molecular details of its mode of action remain unclear.

Summary

Hormones and other molecules used for cell-to-cell signaling can be categorized into two broad groups. Lipid-soluble hormones, principally thyroxine and the steroids and their derivatives, diffuse across the plasma membrane and interact with protein receptors in the cytosol or nucleus. Peptide hormones and small lipid-insoluble hormones such as epinephrine bind to protein cell surface receptors.

The binding of several hormones, such as glucagon and epinephrine, to cell surface receptors triggers an activation of adenylate cyclase and a consequent elevation in the intracellular concentration of 3′,5′-cyclic AMP. Interaction between the receptors and the cyclase is mediated by a transducing protein termed G, which binds GTP in response to hormone binding to ligand. The G·GTP complex, in turn, dissociates into its subunits and activates adenylate cyclase. The mode of action of cAMP is mediated by a cAMP-dependent protein kinase. In the well-understood case of epinephrine-induced glycogenolysis in liver or muscle, the cAMP-dependent kinase phosphorylates glycogen synthase, inactivating it. Cyclic AMP also phosphorylates another protein kinase in the first step in a cascade of protein phosphorylation that results in activation of glycogen phosphorylase. Ca^{2+} ion also activates glycogen phosphorylase in muscle cells by binding to calmodulin—a ubiquitous Ca^{2+} binding protein that is also a subunit of the phosphorylase.

The receptors for several hormones, such as insulin receptor and β-adrenergic receptor, can be purified and studied in isolation. Receptors can be identified and quantified on the cell by cross-linking to ligands. The level of receptors is often subject to regulation, and the continued exposure of a cell to a hormone results in a reduction of surface receptors. Some receptors, such as those for insulin, are endocytosed with the bound hormone. Internalized hormone is degraded by lysosomes, while the receptor usually (but not always), escapes degradation and recycles to the surface. Receptors for insulin, EGF, and other growth hormones are ligand-activated tyrosine-specific protein kinases, although which intracellular phosphorylations are important is unknown.

Steroid hormones effect long-term changes in cell growth and proliferation. In several well-studied cases they induce the synthesis of new mRNAs. Examples include the estrogen induction of egg-white proteins in the hen oviduct and the glucocorticoid induction of synthesis of specific liver proteins and mammary tumor virus proteins. These effects are mediated by a receptor-hormone complex that accumulates in the cell nucleus.

Microorganisms also utilize cell-cell signaling to coordinate aggregation, differentiation, and sexual reproduction. Yeast strains use diffusible peptide pheromones, or mating factors, termed *a* and *α*, to coordinate the fusion of two haploid cells into a diploid cell during sexual reproduction. Slime molds use extracellular signaling to coordinate chemotaxis of cells into aggregates. *Dictyostelium discoideum* cells attract each other by pulsatile synthesis and release of cAMP. The binding of cAMP to cell surface receptors triggers movement of cells toward the source of cAMP, which is followed by the synthesis and release of cAMP.

Certain plant hormones, gibberellic acid in particular, also affect the pattern of gene expression. Auxin induces the elongation of plant cells by causing a localized loosening of the cell walls. Internal turgor due to water uptake then expands the cell.

References

General Properties of Hormone Systems

BERRIDGE, M. 1985. The molecular basis of communication within the cell. *Sci. Am.* **253**(4):142–150.

*COHEN, P., ed. 1984. *Molecular Aspects of Cellular Regulation,* vol. 3: *Enzyme Regulation by Reversible Phosphorylation—Further Advances.* Elsevier.

SMITH, E. L., R. L. HILL, I. R. LEHMAN, R. J. LEFKOWITZ, P. HANDLER, and A. WHITE. 1983. *Principles of Biochemistry: Mammalian Biochemistry,* 6th ed. McGraw-Hill. Chapters 11 through 20 describe in detail the biochemistry of the endocrine systems.

*WILSON, J. D., and D. W. FOSTER. 1985. *Williams Textbook of Endocrinology,* 7th ed. Saunders.

Prostaglandins and Leukotrienes

*HAMMARSTROM, S. 1983. Leukotrienes. *Annu. Rev. Biochem.* **52**:355–378.

*HARRIS, R. H., P. W. RAMWELL, and P. J. GILMER. 1979. Cellular mechanisms of prostaglandin action. *Annu. Rev. Physiol.* **41**:653–668.

*SAMUELSON, B. 1983. Leukotrienes: Mediators of immediate hypersensitivity reactions and inflammation. *Science* **220**:568–575.

*Book or review article that provides a survey of the topic.

Epinephrine Receptors and the Activation of Adenylate Cyclase

*GILMAN, A. G. 1984. G proteins and dual control of adenylate cyclase. *Cell* **36**:577–579.

HOUSLAY, M. 1983. Dual control of adenylate cyclase. *Nature* **303**:133.

*KREBS, E. G., and J. A. BEAVO. 1979. Phosphorylation-dephosphorylation of enzymes. *Annu. Rev. Biochem.* **48**:923–959.

LEFKOWITZ, R. J., L. E. LIMBIRD, C. MUKHERJEE, and M. G. CARON. 1976. The beta-adrenergic receptor and adenylate cyclase. *Biochim. Biophys. Acta* **457**:1–39

LEFKOWITZ, R. J., J. M. STADEL, and M. CARON. 1983. Adenylate cyclase-coupled beta-adrenergic receptors: Structure and mechanisms of activation and desensitization. *Annu. Rev. Biochem.* **52**:159–186.

RODBELL, M. 1980. The role of hormone receptors and GTP-regulatory proteins in membrane transduction. *Nature* **284**:17–22.

Coupling of Receptors to Adenylate Cyclase

ARAD, H., J. P. ROSENBUSCH, and A. LEVITZKI. 1984. Stimulatory GTP regulatory unit N_s and the catalytic unit of adenylate cyclase are tightly associated: Mechanistic consequences. *Proc. Nat'l Acad. Sci. USA* **81**:6579–6583.

BENOVIC, J. L., R. G. L. SHORR, M. G. CARON, and R. J. LEFKOWITZ. 1984. The mammalilan $beta_2$-adrenergic receptor: Purification and characterization. *Biochemistry* **23**:4510–4518.

CERIONE, R. A., D. R. SIBLEY, J. CODINA, J. L. BENOVIC, J. WINSLOW, E. J. NEER, L. BIRNBAUMER, M. G. CARON, and R. J. LEFKOWITZ. 1984. Reconstitution of a hormone-sensitive adenylate cyclase system. *J. Biol. Chem.* **259**:9979–9982.

COOPER, D., and K. SEAMAN, eds. 1985. Dual regulation of adenylate cyclase. *Adv. Cyclase Nucleotide Protein Phosphorylation Res,* vol. 19.

EIMERL, S., G. NEUFELD, M. KORNER, and M. SCHRAMM. 1980. Functional implantation of a solubilized beta-adrenergic receptor in the membrane of a cell. *Proc. Nat'l Acad. Sci. USA* 77:760–764.

HILDEBRANDT, J. D., J. CODINA, and L. BIRNBAUMER. 1984. Interaction of the stimulatory and inhibitory regulatory proteins of the adenylyl cyclase system with the catalytic component of *cyc*$^-$ S49 cell membranes. *J. Biol. Chem.* **259**:13178–13185.

HILDEBRANDT. J. D., R. D. SEKURA, J. CODINA, R. IYENGAR, C. R. MANCLARK, and L. BIRNBAUMER. 1983. Stimulation and inhibition of adenylyl cyclases mediated by distinct regulatory proteins. *Nature* **302**:706–709.

LIMBIRD, L. E., D. M. GILL, and R. J. LEFKOWITZ. 1980. Agonist-promoted coupling of beta-adrenergic receptor with the guanine nucleotide regulatory protein of the adenylate cyclase system. *Proc. Nat'l Acad. Sci. USA* 77:775–779.

NORTHUP, J. K., M. D. SMIGEL, P. C. STERNWEIS, and A. G. GILMAN. 1983. The subunits of the stimulatory regulatory component of adenylate cyclase. *J. Biol. Chem.* **258**:11369–11376.

PFEUFFER, E., R. DREHER, H. METZGER, and T. PFEUFFER. 1985. Catalytic unit of adenylate cyclase: Purification and identification by affinity cross-linking. *Proc. Nat'l Acad. Sci. USA* 82:3086–3090.

Cholera Toxin

CASSEL, D., and T. PFEUFFER. 1978. Mechanisms of cholera toxin action: Covalent modification of the guanyl nucleotide-binding protein of the adenylate cyclase system. *Proc. Nat'l Acad. Sci. USA* **75**:2669–2673.

HOLMGREN, J. 1981. Actions of cholera toxin and the prevention and treatment of cholera. *Nature* **292**:413–417.

KAHN, R. A., and A. G. GILMAN. 1984. ADP-ribosylation of G_s promotes the dissociation of its alpha and beta subunits. *J. Biol. Chem.* **259**:6235–6240.

*LAI, C-Y. 1980. The chemistry and biology of cholera toxin. *CRC Crit. Rev. Biochem.* **9**:171–206.

cAMP-Dependent Protein Kinase and Cascades of Protein Phosphorylation-Dephosphorylation

*COHEN, P. 1981. The role of protein phosphorylation in the neural and hormonal control of intermediary metabolism. In *Cellular Controls in Differentiation,* C. W. Lloyd and D. A. Rees, eds. Academic Press, pp. 81–105.

*COHEN, P. 1982. The role of protein phosphorylation in neural and hormonal control of cellular activity. *Nature* **296**:613–620.

*COHEN, P. 1983. *Control of Enzyme Activity,* 2d ed. Chapman & Hall.

*EL-MAGHRABI, M. R., and S. J. PILKIS. 1984. Rat liver 6-phosphofructo 2-kinase/fructose 2,6-bisphosphatase: A review of relationships between the two activities of the enzyme. *J. Cell Biochem.* **26**:1–17.

HOPPE, J., and K. G. WAGNER. 1979. Cyclic AMP-dependent protein kinase I, a unique allosteric enzyme. *Trends Biochem. Sci.* **4**:282–285.

PILKUS, S. J., M. R. EL-MAGHRABI, M. MCCRANE, J. PILKIS, and T. H. CLAUSE. 1982. Regulation by glucagon of hepatic pyruvate kinase, 6-phosphofructose-1-kinase, and fructose-1,6-bisphosphatase. *Fed. Proc.* **41**:2623–2628.

SHORAIN, V. S., B. S. KHATRA, and T. R. SODERLING. 1982. Hormonal regulation of skeletal muscle glycogen synthase through covalent phosphorylation. *Fed. Proc.* **41**:2618–2622.

Ca^{2+} Ions and Other Intracellular Second Messengers

Calcium and Calmodulin

BABU, Y. S., J. SACK, T. GREENHOUGH, C. BUGG, A. MEANS, and W. COOK. 1985. Three-dimensional structure of calmodulin. *Nature* **315**:37–40.

*CHEUNG, W. Y. 1982. Calmodulin. *Sci. Am.* **246**(6):48–56.

COX, J. A. 1984. Sequential events in calmodulin on binding with calcium and interaction with target enzymes. *Fed. Proc.* **43**:3000–3004.

*KLEE, C. B., T. H. CROUCH, and P. G. RICHMAN. 1980. Calmodulin. *Annu. Rev. Biochem.* **49**:489–515.

*KRETSINGER, R. H. 1981. Mechanisms of selective signalling by calcium. *Neurosci. Res. Program Bull.* **19**:213–328.

KRUSKAL, B. A., C. H. KEITH, and F. R. MAXFIELD. 1984. Thyrotropin-releasing hormone–induced changes in intracellular

[Ca^{2+}] measured by microspectrofluorometry on individual quin-2-loaded cells. *J. Cell Biol.* 99:1167–1172.

*MEANS, A. R., and J. G. CHAFOULEAS. 1982. Calmodulin in endocrine cells. *Annu. Rev. Physiol.* 44:667–682.

*MEANS, A. R., and J. R. DEDMAN. 1980. Calmodulin: An intracellular calcium receptor. *Nature* 285:73–77.

POZZAN, T., G. GATTI, N. DOZIO, L. VICENTINI, and J. MELDOLESI. 1984. Ca^{2+}-dependent and -independent release of neurotransmitters from PC12 cells: A role for protein kinase C activation? *J. Cell Biol.* 99:628–638.

*TSIEN, R. Y., T. POZZAN, and T. J. RINK. 1984. Measuring and manipulating cytosolic Ca^{2+} with trapped indicators. *Trends Biochem. Sci.* 9:263–266.

Inositol Triphosphate, Diacylglycerol, and Protein Kinase C

*BERRIDGE, M. J., and R. F. IRVINE. 1984. Inositol trisphosphate, a novel second messenger in cellular signal transduction. *Nature* 312:315–321.

*HOKIN, L. E. 1985. Receptors and phosphoinositide-generated second messengers. *Annu. Rev. Biochem.* 54:205–236.

*JOSEPH, S. K. 1984. Inositol trisphosphate: An intracellular messenger produced by Ca^{2+} mobilizing hormones. *Trends Biochem. Sci.* 9:420–421.

MAJERUS, P., D. WILSON, T. CONNOLLY, T. BROSS, and E. NEUFELD. 1985. Phosphinositide provides a link in stimulus–response coupling. *Trends Biochem. Sci.* 10:168–171.

MOOLENAAR, W. H., L. G. J. TERTOOLEN, and S. W. DE LAAT. 1984. Phorbol ester and diacylglycerol mimic growth factors in raising cytoplasmic pH. *Nature* 312:371–374.

ROZENGURT, E., A. RODRIGUEZ-PENA, M. COOMBS, and J. SINNETT-SMITH. 1984. Diacylglycerol stimulates DNA synthesis and cell division in mouse 3T3 cells: Role of Ca^{2+}-sensitive phospholipid-dependent protein kinase. *Proc. Nat'l Acad. Sci. USA* 81:5748–5752.

VOLPE, P., G. SALVIATI, F. DI VIRGILIO, and T. POZZAN. 1985. Inositol 1,4,5-trisphosphate induces calcium release from sarcoplasmic reticulum of skeletal muscle. *Nature* 316:347–349.

Receptors for Insulin and Glucagon

CZECH, M. P., ed. 1985. *Molecular Basis of Insulin.* Plenum.

CZECH, M. P. 1980. Insulin action and the regulation of hexose transport. *Diabetes* 29:399–409.

FEHLMANN, M., J.-L. CARPENTER, A. L. CAM, P. THAMM, D. SAUNDERS, D. BRADENBURG, L. ORCI, and P. FREYCHET. 1981. Biochemical and morphological evidence that the insulin receptor is internalized with insulin in hepatocytes. *J. Cell Biol.* 93:82–87.

KAHN, B. B., and S. W. CUSHMAN. 1985. Subcellular translocation of glucose transporters: Role in insulin action and its perturbation in altered metabolic states. In *Diabetes/Metabolism Reviews*, vol. 1.

KAHN, C. R. 1982. Autoimmunity and the aetiology of insulin-dependent diabetes mellitus. *Nature* 299:15–16.

KASUGA, M., Y. FUJITA-YAMAGUCHI, D. L. BLITHE, and C. R. KAHN. 1983. Tyrosine-specific protein kinase activity is associated with the purified insulin receptor. *Proc. Nat'l Acad. Sci. USA* 80:2137–2141.

*LIENHARD, G. E. 1983. Regulation of cellular membrane transport by the exocytotic insertion and endocytic retrieval of transporters. *Trends Biochem. Sci.* 8:125–127.

PETRUZZELLI, L. M., S. GANGULY, C. J. SMITH, M. H. COBB, C. S. RUBIN, and O. M. ROSEN. 1982. Insulin activates a tyrosine-specific protein kinase in extracts of 3T3-L1 adipocytes and human placenta. *Proc. Nat'l Acad. Sci. USA* 79:6792–6799.

ULLRICH, A., et al. 1985. Human insulin receptor and its relationship to the tyrosine kinase family of oncogenes. *Nature* 313:756–761.

Receptor Regulation

Down Regulation by Covalent Modification

NAMBI, P., J. R. PETERS, D. R. SIBLEY, and R. J. LEFKOWITZ. 1985. Desensitization of the turkey erythrocyte β-adrenergic receptor in a cell-free system.

SIBLEY, D. R., R. H. STRASSER, M. G. CARON, and R. J. LEFKOWITZ. 1985. Homologous desensitization of adenylate cyclase is associated with phosphorylation of the β-adrenergic receptor. *J. Biol. Chem.* 260:3883–3886.

SIBLEY, D. R., and R. J. LEFKOWITZ. 1985. Molecular mechanisms of receptor desensitization using the β-adrenergic receptor-coupled adenylate cyclase system as a model. *Nature* 317:124–129.

STRULOVICI, B., R. A. CERIONE, B. F. KILPATRICK, M. G. CARON, and R. J. LEFKOWITZ. 1984. Direct demonstration of impaired functionality of a purified desensitized beta-adrenergic receptor in a reconstituted system. *Science* 225:837–840.

Endocytosis and Receptor Degradation

CRETTAZ, M., I. JIALAL, M. KASUGA, and C. R. KAHN. 1984. Insulin receptor regulation and desensitization in rat hepatoma cells. *J. Biol. Chem.* 259:11543–11549.

FEHLMAN, M., J-L. CARPENTIER, E. VAN OBBERGHEN, P. FREYCHET, P. THAMM, D. SAUNDERS, D. BRANDENBURG, and L. ORCI. 1982. Internalized insulin receptors are recycled to the cell surface in rat hepatocytes. *Proc. Nat'l Acad. Sci. USA* 79:5921–5925.

KNUTSON, V. P., G. V. RONNETT, and M. D. LANE. 1983. Rapid, reversible internalization of cell surface insulin receptors. *J. Biol. Chem.* 258:12139–12142.

RONNETT, G. V., V. P. KNUTSON, and M. D. LANE. 1982. Insulin–induced down-regulation of insulin receptors in 3T3-L1 adipocytes: Altered rate of receptor inactivation. *J. Biol. Chem.* 257:4285–4291.

RONNETT, G. V., G. TENNEKOON, V. P. KNUTSON, and M. D. LANE. 1983. Kinetics of insulin receptor transit to and removal from the plasma membrane: Effect of insulin-induced down-regulation in 3T3-L1 adipocytes. *J. Biol. Chem.* 258:283–290.

Receptors for Growth Factors

BUHROW, S. A., S. COHEN, and J. V. STAROS. 1982. Affinity labeling of the protein kinase associated with the epidermal growth factor receptor in membrane vesicles from A431 cells. *J. Biol. Chem.* 257:4019–4022.

*CARPENTER, G. 1984. Properties of the receptor for epidermal growth factor. *Cell* **37**:357–358.

*CARPENTER, G., and S. COHEN. 1979. Epidermal growth factor. *Annu. Rev. Biochem.* **48**:193–216.

*CARPENTER, G. and S. COHEN. 1984. Peptide growth factors. *Trends Biochem. Sci.* **9**:169–171.

COOPER, J. A., D. F. BOWEN-POPE, E. RAINES, R. ROSS, and T. HUNTER. 1982. Similar effects of platelet-derived growth factor and epidermal growth factor on the phosphorylation of tyrosine in cellular proteins. *Cell* **31**:263–273.

*CZECH, M. P. 1982. Structural and functional homologies in the receptors for insulin and the insulin-like growth factors. *Cell* **31**:8–10.

DOWNWARD, J., P. PARKER, and M. D. WATERFIELD. 1984. Autophosphorylation sites on the epidermal growth factor receptor. *Nature* **311**:483–485.

*HUNTER, T. 1984. The epidermal growth factor receptor gene and its product. *Nature* **311**:414–416.

*ROTH, J., and S. I. TAYLOR. 1982. Receptors for peptide hormones: Alternatives in diseases of humans. *Annu. Rev. Physiol.* **44**:639–652.

SCHREIBER, A. B., T. A. LIBERMAN, I. LAX, Y. YARDEN, and J. SCHLESSINGER. 1983. Biological role of epidermal growth factor–receptor clustering: Investigation with monoclonal anti-receptor antibodies. *J. Biol. Chem.* **258**:846–853.

Steroid and Thyroid Hormones and Their Receptors

*ASHBURNER, M. 1980. Chromosomal action of ecdysone. *Nature* **285**:435–436.

ASHBURNER, M., C. CHICHARA, P. MELTZER, and G. RICHARD. 1974. Temporal control of puffing activity in polytene chromosomes. *Cold Spring Harbor Symp. Quant. Biol.* **38**:655–662.

GEHRING, U., and G. M. TOMKINS. 1974. A new mechanism for steroid unresponsiveness: Loss of nuclear binding of acivity of a steroid hormone receptor. *Cell* **3**:301–306.

HOLLENBERG, S. M., C. WEINBERGER, E. S. ONG, G. GERELLI, A. ORO, R. LEBO, E. B. THOMPSON, M. G. ROSENFELD, and R. M. EVANS. 1985. Primary structure and expression of a functional human glucocorticoid receptor cDNA. *Nature* **318**:635–641.

IVARIE, R. D., and P. H. O'FARRELL. 1978. The glucocorticoid domain: Steroid-mediated changes in the rate of synthesis of rat hepatoma proteins. *Cell* **13**:41–55.

KATZENELLENBOGEN, B. S., and J. GORSKI. 1975. Estrogen actions on synthesis of macromolecules in target cells. In *Biochemical Actions of Hormones,* vol. 3, G. Litwach, ed. Academic Press.

*MEANS, A. R., and B. W. O'MALLEY. 1975. Oestrogen-induced differentiation of target tissues. In *Biochemistry of Cell Differentiation,* Biochemistry Series 1, vol. 9. J. Paul, ed. ITP International Review of Science, pp. 161–180.

NARAYAN, P., C. W. LIAW, and H. C. TOWLE. 1984. Rapid induction of a specific nuclear mRNA precursor by thyroid hormone. *Proc. Nat'l Acad. Sci. USA* **81**:4687–4691.

SHEPERD, J. H., E. R. MULVIHILL, P. S. THOMAS, and R. D. PALMITER. 1980. Commitment of chick oviduct tubular gland cells to produce ovalbumin mRNA during hormonal withdrawal and restimulation. *J. Cell Biol.* **87**:142–151.

SIBLEY, C. H., and G. M. TOMKINS. 1974. Mechanisms of steroid resistance. *Cell* **2**:221–227.

WELSHONS, W. V., M. E. LIEBERMAN, and J. GORSKI. 1984. Nuclear localization of unoccupied oestrogen receptors. *Nature* **307**:747–749.

*YAMAMOTO, K. R., and B. M. ALBERTS. 1976. Steroid receptors: Elements for modulation of eukaryotic transcription. *Annu. Rev. Biochem.* **45**:721–746.

Hormone Signaling in Microorganisms

Yeast

JENNESS, D. D., A. C. BURKHOLDER, and L. H. HARTWELL. 1983. Binding of alpha-factor pheromone to yeast *a* cells: Chemical and genetic evidence for an alpha-factor receptor. *Cell* **35**:521–529.

*THORNER, J. 1981. Pheromonal regulation of development in *Saccharomyces cerevisiae.* In *The Molecular Biology of the Yeast* Saccharomyces: *Life Cycle and Inheritance,* J. N. Strathern, E. W. Jones, and J. R. Broach, eds. Cold Spring Harbor Laboratory.

Cellular Slime Molds

*BONNER, J. T. 1983. Chemical signals of social amoebae. *Sci. Am.* **248**(4):114–120.

*DEVREOTES, P. N. 1982. Chemotaxis. In *The Development of Dictyostelium discoideum,* W. F. Loomis, ed. Academic Press, pp. 117–168.

GERISCH, G. 1982. Chemotaxis in *Dictyostelium. Annu. Rev. Physiol.* **44**:532–552.

KLEIN C., J. LUBS-HAUKNESS, and S. SIMONS. 1985. cAMP induces a rapid and reversible modification of the chemotactic receptor in *Dictyostelium discoideum. J. Cell Biol.* **100**:715–720.

THEIBERT, A., and P. N. DEVREOTES. 1983. Cyclic 3′,5′-AMP relay in *Dictyostelium discoideum:* Adaptation is independent of activation of adenylate cyclase. *J. Cell Biol.* **97**:173–177.

Plant Hormones

HAGEN, G., and T. J. GUILFOYLE. 1985. Rapid inducton of selective transcription by auxins. *Mol. Cell Biol.* **5**:1197–1203.

JACOBSEN, J. V., and L. R. BEACH. 1985. Control of transcription of α-amylase and rRNA genes in barley aleurone protoplasts by gibberellin and abscisic acid. *Nature* **316**:275–277.

MOZER, T. J. 1980. Control of protein synthesis in barley aleurone layers by the plant hormones gibberellic acid and abscisic acid. *Cell* **20**:479–485.

*RAYLE, K. L., and R. CLELAND. 1977. Control of plant cell enlargement by hydrogen ions. *Curr. Top. Devel. Biol.* **11**:187–214.

*VARNER, J. E., and D. T-H. HO. 1976. The role of hormones in the integration of seedling growth. In *The Molecular Biology of Hormone Action,* 34th Symposium of the Society for Development Biology. Academic Press, pp. 173–194.

*WAREING, P. F., and I. D. J. PHILLIPS. 1981. *Growth and Differentiation in Plants.* Pergamon. Chapters 3–5 cover most plant hormones.

17

Nerve Cells and the Electrical Properties of Cell Membranes

THE nervous system regulates all aspects of bodily function. It is staggering in its complexity. Millions of specialized nerve cells sense changes in both the external and internal environments—light, touch, sound, pain, the stretching of muscles. They transmit this information to other nerve cells for processing and storage. Millions more nerve cells regulate the contraction of muscles and the secretion of endocrine or exocrine glands. The brain—the control center that stores, computes, integrates, and transmits information—contains about 10^{12} nerve cells, each apparently with a defined separate function and each forming as many as a thousand connections with other nerve cells. Moreover, nerve cells can be classified into at least 1000 different types. Whether we shall ever be able to understand such human qualities as thought, emotion, or learning in terms of the function of individual nerve cells is a matter of intense discussion.

Despite the complexity of the nervous system as a whole, the structure and function of individual nerve cells is understood in great detail, perhaps in more detail than any other type of cell. We know, for instance, that electrical impulses are conducted along the length of every nerve cell. We can now explain this conduction in terms of specific plasma membrane proteins that regulate the flow of ions into and out of the cell and cause changes in the electric potential across the plasma membrane. We know a great deal about the membrane proteins of certain nerve cells, such as the photoreceptor cells in the eye, that sense environmental signals and convert them into electrical impulses. We have been able to identify many of the receptor proteins that enable a nerve cell to receive an electrical or a chemical signal from another cell and convert it into an electrical impulse for transmission along its length to the next cell. Even some simple aspects of memory and learning can now be explained in terms of modifications of specific neurons.

Our focus in this chapter is on how individual nerve cells function and how small groups of cells function together. Many of the nerve systems of greatest interest and importance, such as the mammalian brain, are too complex to be studied at this level with current techniques. A great deal of information has been gleaned, however, from simpler nervous systems. Squids, sea slugs, and nematodes contain large neurons that are relatively easy to identify and manipulate experimentally. Moreover, only a few neurons may be involved in a specific task; thus, their function can be studied in some detail. Detailed genetic studies of *Drosophila* have also yielded important insights into the structure and function of nerve cells. Although many of our examples will be taken from such invertebrates, the principles involved are basic and are applicable to complex nervous systems as well, including that of humans.

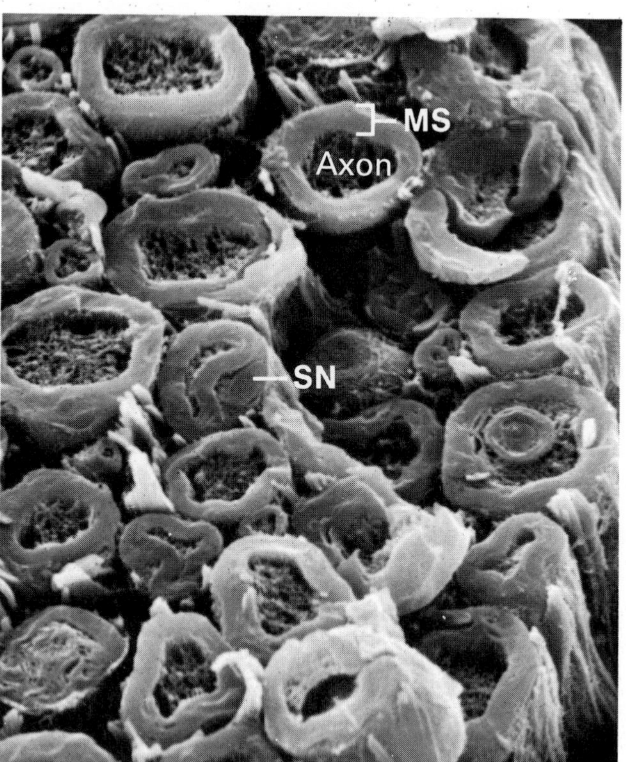

Figure 17-1 Freeze-fracture preparation of a rat sciatic nerve viewed in a scanning electron microscope. Each nerve axon is surrounded by a myelin sheath (MS) formed from the plasma membrane of a Schwann cell (SN). The axonal cytoplasm contains abundant filaments—mostly microtubules and intermediate filaments—that run longitudinally and serve to make the axon rigid. *From R. G. Kessel and R. H. Kardon, 1979,* Tissues and Organs: A Text-Atlas of Scanning Electron Microscopy, *W. H. Freeman and Company, p. 80.*

Neurons, Synapses, and Nerve Circuits

The nervous system contains two principal classes of cells: the *neurons,* or nerve cells, and the *neuroglia* (or *glia*), cells that fill the spaces between the neurons, nourishing them and modulating their function. Neurons exist in a bewildering array of sizes and shapes, but they all possess the ability to conduct an electrical impulse along their length. They make specific contacts with other cells at specialized sites called *synapses,* across which signals are passed. Groups of interconnecting neurons frequently form nerve circuits for the passage of electrical signals. In the peripheral nervous system the long axons of neurons are bundled together to form a *nerve* (Figure 17-1).

The Neuron Is the Fundamental Unit of All Nervous Systems

Most neurons contain four distinct regions, which carry out specialized functions of the cell: the cell body, the dendrites, the axon, and the specialized axon terminals (Figure 17-2).

The *cell body* is the region containing the nucleus and most of the ribosomes and endoplasmic reticulum. It is the site of synthesis of virtually all neuronal proteins and membranes. Here newly synthesized macromolecules are assembled into membranous vesicles or multiprotein particles and transported to other regions of the neuron.

Most neurons contain an *axon,* a single fiber that conducts the nerve impulse away from the cell body. In humans, axons may be a meter or more in length. In the giraffe the axons that reach from the spinal cord to the hoof may be as long as 10 meters. The diameter of an axon may vary from a micrometer in certain nerves of the human brain to a millimeter in the giant fiber of the squid. The speed with which an axon conducts an impulse is roughly proportional to its diameter—it can be as fast as 100 meters per second. Some axons are surrounded by a sheath of myelin (see Figure 14-4) made up of a stacked array of plasma membranes of a Schwann cell, a type of glial cell. Myelinated axons conduct electrical impulses faster than nonmyelinated ones of the same diameter.

Fibers of 10-nm-diameter intermediate filaments and microtubules (called neurofilaments) run the length of the

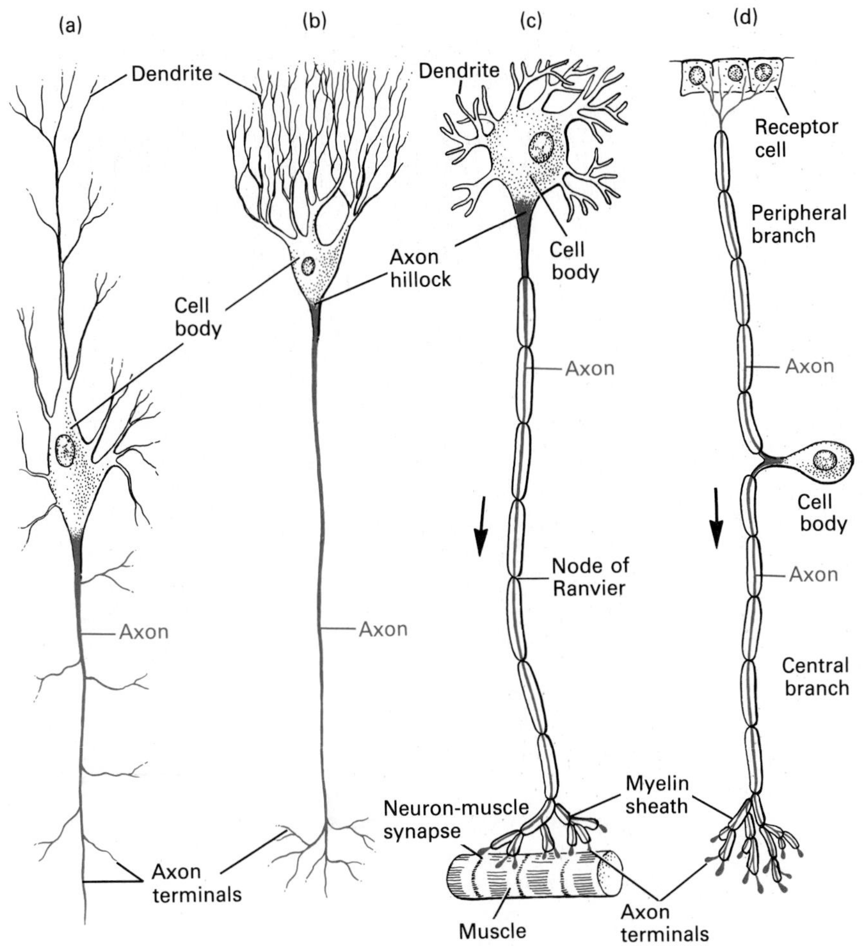

Figure 17-2 Structure of typical mammalian neurons. Axons and axon terminals are in color. (a) A neuron with multiple spiny-looking dendrites and a single axon from which branch laterally a number of axon terminals. (b) A brain neuron with profusely branched dendrites (which form synapses with several hundred other neurons) and a single long axon that branches at its terminus. (c) A motor neuron that innervates a muscle cell. Typically, motor neurons have a single long axon that runs from the cell body, located in the central nervous system, to the effector cell. Motor neurons in mammals usually are myelinated. Myelin covers all parts of the axon except at the nodes of Ranvier and the synaptic junctions. (d) A vertebrate sensory neuron. This type, the most common, has only a single process, or fiber, that branches just after it leaves the cell body. One branch carries the nerve impulse from the receptor site to the cell body located in the dorsal root ganglion near the spinal cord; the other branch carries the impulse from the cell body to the spinal cord or brain. Both branches are structurally and functionally axons, except at their terminal portions, even though the peripheral branch conducts impulses toward, rather than away from, the cell body.

axon; they help to transport proteins, membrane vesicles, and other macromolecules from the cell body down the length of the axon to the terminal. This movement, called *axoplasmic* (or *orthograde*) *transport,* is essential for the renewal of membranes and enzymes in the nerve terminal. During differentiation of the nervous system, axoplasmic transport enables axons to elongate by the addition of new membrane material at their distal tips. The axonal fibers also appear to be involved in *retrograde transport,* the movement of membranes and organelles up the axon toward the cell body. Retrograde transport enables damaged membranes and proteins to be returned to the cell body and degraded, most likely in lysosomes.

Axons are specialized for the conduction of an electrical impulse, called an *action potential,* down their length without diminution. Action potentials originate at the *axon hillock,* the junction of the axon and cell body, and travel to the small branches of the axon, the axon terminals, from which point electrical signals are passed on to other cells: to the next neuron in a nerve circuit, to a muscle cell at a neuromuscular junction, or to any of various other types of cells.

Dendrites are thinner fibrous projections extending outward from the cell body (Figure 17-2). Dendrites contain regions that receive signals from sense organs or from the axons of other neurons, convert these signals into electrical impulses, and transmit them to the cell body. The cell body receives signals independently as well. Electrical disturbances that are generated in the dendrites or cell body spread passively to the axon hillock. There, if the disturbance is great enough, an action potential—a larger electrical signal—is generated and is actively conducted down the axon.

Synapses Are Specialized Sites Where Neurons Communicate with Other Cells

Synapses generally conduct signals in only one direction: An axon terminal from the *presynaptic* cell sends signals that are picked up by the *postsynaptic* cell (Figure 17-3). The two types of synapses, electrical and chemical, differ in both structure and function.

Cells communicating by *electrical synapses* are connected by gap junctions (Figure 17-3b). This allows an

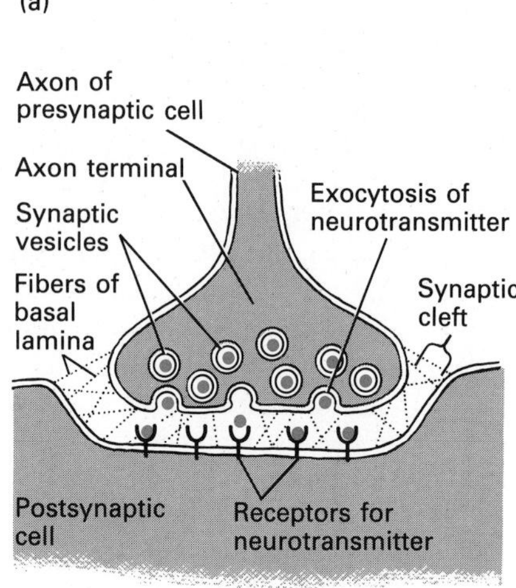

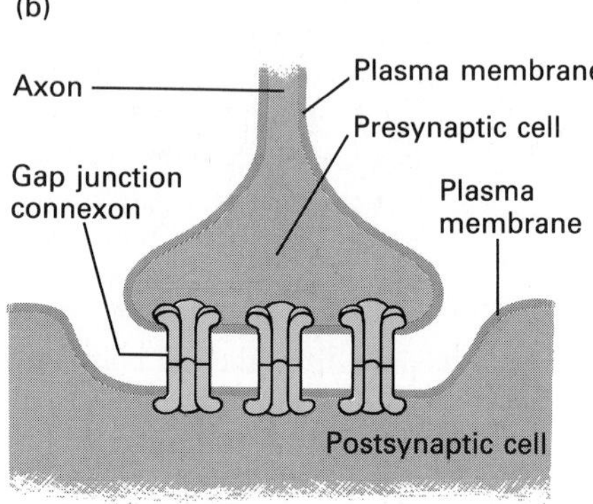

Figure 17-3 Two types of synapses: (a) chemical, and (b) electrical. In the chemical synapse a narrow region, the *synaptic cleft,* separates the plasma membranes of the presynaptic axon and the postsynaptic cell. The latter might be a dendrite, a cell body, the axon of a neuron, or a muscle or gland cell. When the postsynaptic cell is a muscle, the synapse is called a *neuromuscular junction* or *motor end plate.* The axon terminal contains synaptic vesicles filled with one or more neurotransmitters. An electrical impulse triggers exocytosis of these vesicles. The released neurotransmitter binds to receptors on the postsynaptic cell and induces a response. The synaptic cleft may contain a network of fibrous proteins, such as collagen, that binds the two cells together. In some cases, enzymes attached to this network destroy the chemical signal after it has functioned. In other cases the chemical signal diffuses away or is reincorporated into the presynaptic cell. In the electrical synapse, the plasma membranes of the presynaptic and postsynaptic cells are linked by gap junctions. The flow of ions through these channels allows the electrical impulse to be transmitted directly from one cell to another.

electrical impulse to pass directly from a presynaptic cell to a postsynaptic one. Electrical synapses allow an action potential to be generated in the postsynaptic cell with greater certainty than chemical synapses and without a lag period.

In *chemical synapses* (Figure 17-3a), by far the more common, the axon terminal of the presynaptic cell contains vesicles filled with a particular neurotransmitter substance (Figure 17-4), such as epinephrine or acetylcholine. When the nerve impulse reaches the axon terminal, these vesicles are exocytosed, releasing their contents into the *synaptic cleft,* the narrow space between the cells. The transmitter diffuses across the synaptic cleft and, after a lag period of about 0.5 ms, binds to receptors on the postsynaptic cells. Upon binding, it induces a change in the ionic permeability of the postsynaptic membrane that results in a disturbance of the electric potential at this point. This electrical disturbance may be sufficient to induce an action potential or, depending on the type of cell, a muscle contraction or the release of hormones.

Many nerve-nerve and most nerve-muscle chemical synapses are *excitatory.* The chemical signal released by the presynaptic cell causes a change in the plasma membrane of the postsynaptic cell that tends to induce an action potential. Often, however, a nerve impulse in a presynaptic neuron will affect the electrical properties of the postsynaptic membrane in such a way as to prevent the generation of an action potential. This type of synapse is termed *inhibitory.* A single neuron can be affected simultaneously by excitatory and inhibitory stimuli from synapses with many axons. Generally these excitatory and inhibitory stimuli are transmitted passively along the cell membrane to the cell body. Whether a nerve cell generates an action potential in the axon hillock depends on the balance of the timing, amplitude, and localization of all the various excitatory and inhibitory inputs. Action potentials are generated when the membrane potential at the axon hillock reaches a certain level called the *threshold potential.* In a sense, each neuron is a tiny computer that averages all the electrical disturbances on its membrane and makes the yes-or-no decision whether to trigger an action potential and conduct it down the axon. Particularly in the central nervous system, neurons have extremely long dendrites with complex branches. This allows them to form synapses with and receive signals from a large number of other neurons, perhaps up to a thousand. A single axon in the central nervous system can synapse with many neurons and induce responses in all of them simultaneously.

Similarly, the axon of a neuron may branch and form synapses with several muscle cells. An electrical impulse from the neuron will reach each of these muscle cells at the same time and cause them to contract simultaneously.

Often one axon terminal will synapse with another. As Figure 17-5 shows, such a synapse may modulate the ability of an axon terminal to exocytose its synaptic vesi-

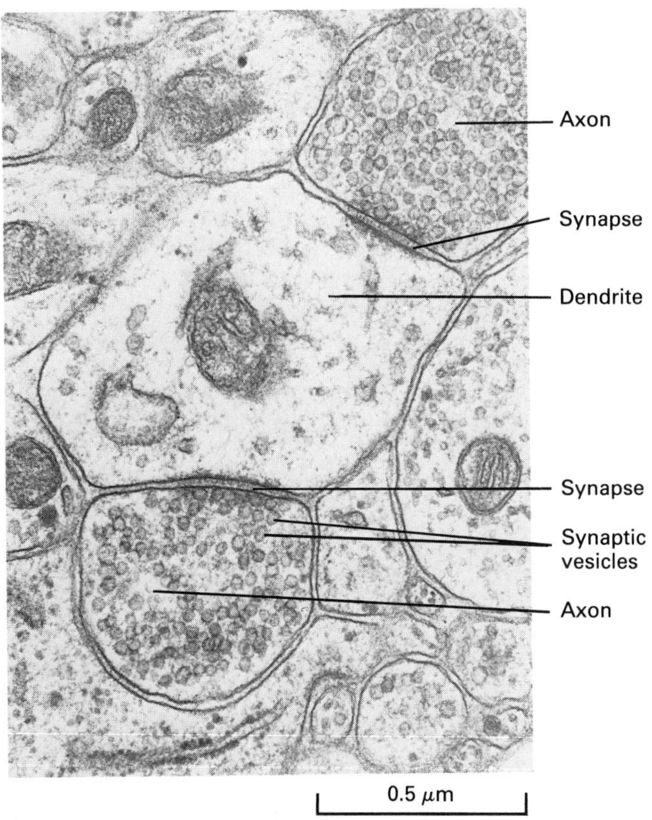

Figure 17-4 Electron micrograph showing a cross section of a dendrite surrounded by two axon terminals filled with synaptic vesicles. In the region of the synapse note the apparent thickening of the plasma membrane both in the presynaptic axon terminal and the postsynaptic dendrite. This region of plasma membrane in the presynaptic cell is specialized for fusion and exocytosis of synaptic vesicles; the opposing membrane of the postsynaptic cell contains receptors for the neurotransmitter. Certain protein fibers just under the plasma membrane may interact with the receptor proteins to keep them from diffusing out of the region of the synapse. *From C. Raine, 1981, in* Basic Neurochemistry, *3d ed., G. J. Siegel et al., eds., Little, Brown, p. 32. Copyright 1981, Little, Brown and Company.*

cles and signal to a postsynaptic cell. We shall see later how such synapses can enable an animal to learn.

Neurons Are Organized into Circuits

Sponges, the most primitive multicellular animals, contain contractile muscle cells that surround their pores and are in direct contact with the sea water. Whenever the muscles are excited by a noxious stimulus in the water, they contract and close the pores. Sponges have no specialized cells for sensory reception or transmission of an excitation.

In more complex multicellular animals, muscle cells lie deep in the interior. This requires that sensory cells be located at or near the surface of the animal and be able to conduct impulses to the interior muscles. In the tentacles of the sea anemone (phylum Cnidaria), this connection is direct. One of the outer epidermal cells, specialized to receive sensory information from the environment, extends to a deeper muscle cell (Figure 17-6). Appropriate stimulation of the sensory receptor cell causes contraction of the underlying muscle. In more advanced animals such

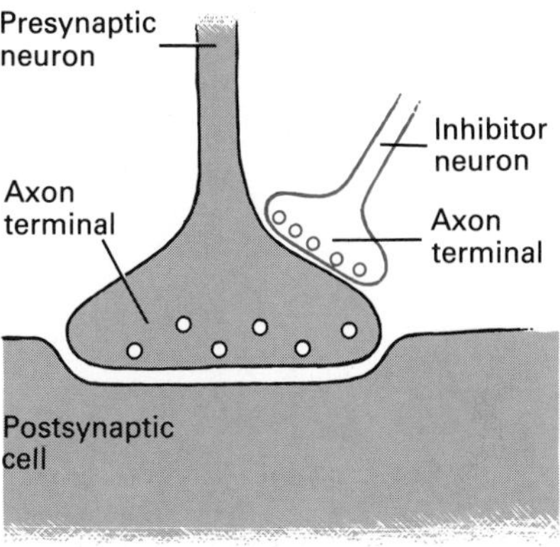

Figure 17-5 Diagrammatic outline of a modulatory synapse. An inhibitor neuron forms a synapse on the terminal of the presynaptic neuron. When the inhibitor neuron is stimulated and releases its transmitter substance, it reduces the presynaptic axon's ability to transmit a signal to the postsynaptic cell. This type of presynaptic inhibition leaves unchanged the ability of the postsynaptic cell to respond to signals from other cells.

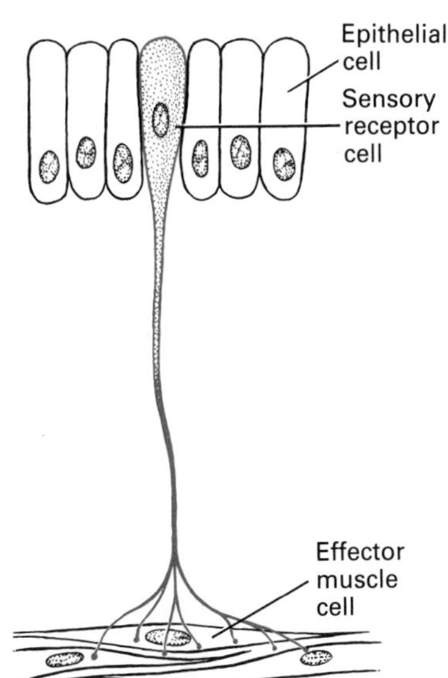

Figure 17-6 Diagram of a simple receptor-effector system, as seen in the tentacles of the sea anemone. The receptor cell transmits an impulse directly to muscle cells.

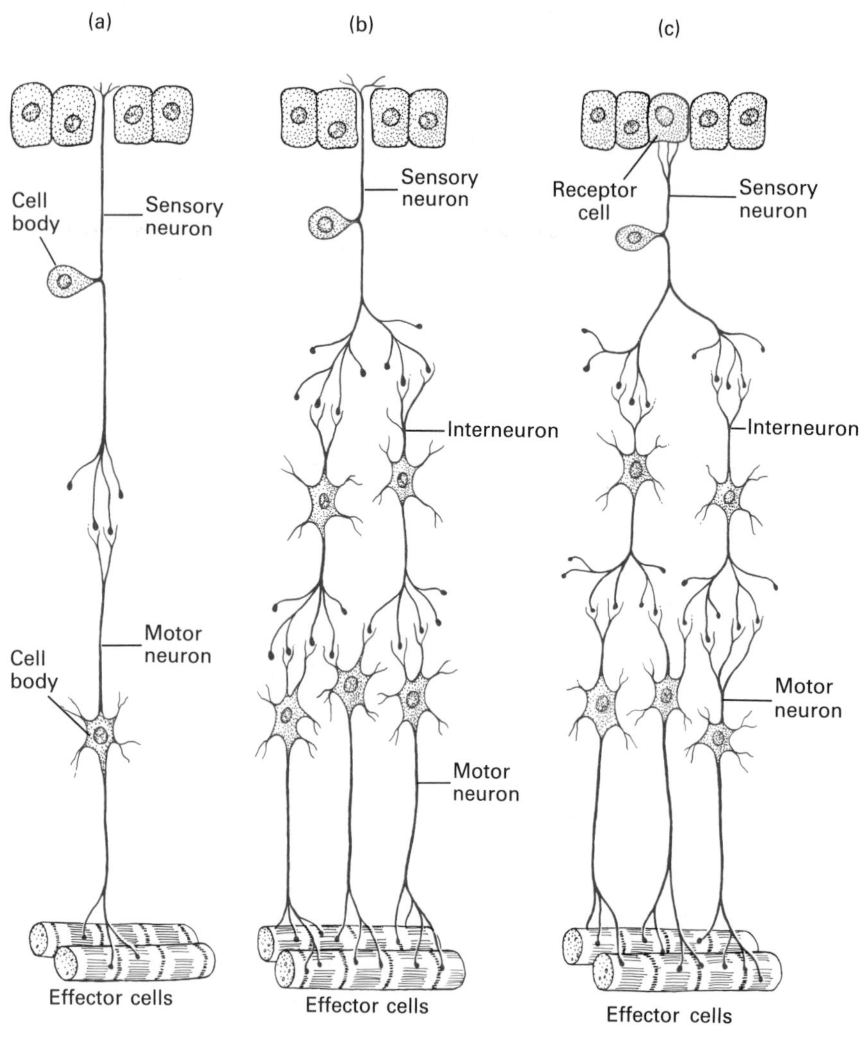

Figure 17-7 Neural pathways consisting of several neurons are common in animals. (a) In many invertebrates, such as the earthworm, part of a sensory neuron such as a stretch receptor has its cell body on the exterior of the animal, where it receives sensory stimuli. The axon forms a synapse directly with the motor neuron, which, in turn, stimulates muscle contraction. (b) In higher animals interneurons are usually interposed between sensory and motor neurons, allowing the nerve impulse to follow multiple routes and to affect multiple effector cells. In this diagram the sensory neuron acts both as a receptor and as a conductor. (c) In other cases a separate nonneural receptor cell relays signals to a sensory neuron.

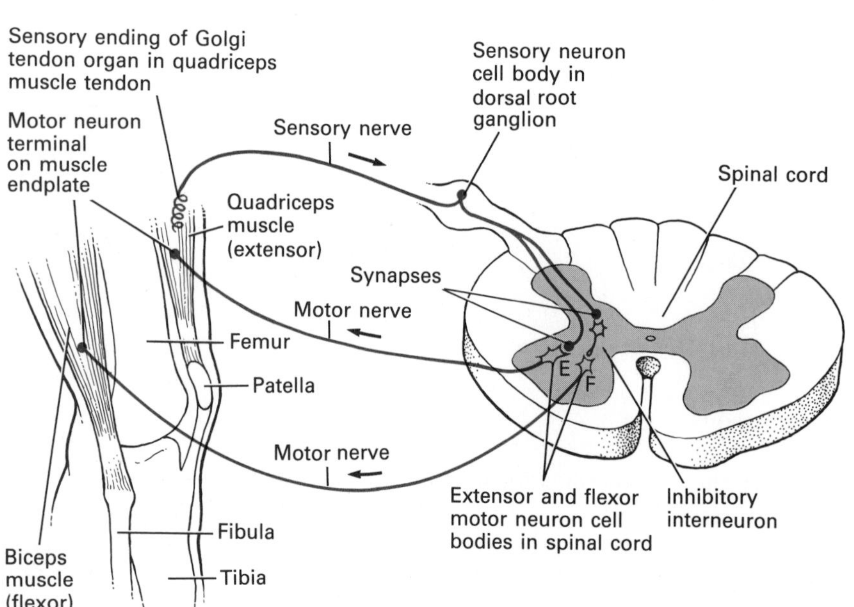

Figure 17-8 The knee-jerk reflex arc in the human. Position and movement of the knee joint is accomplished by two muscles that have opposite actions. The quadriceps muscle inserts in the patella (kneecap); it is termed an *extensor* because its contraction straightens the leg. The biceps muscle is termed a *flexor* because its contraction flexes, or bends, the leg. It inserts into the tibia, the long bone in the lower leg. The knee-jerk response, a sudden extension of the leg, is stimulated by a blow just below the patella. The blow stimulates sensory neurons of the Golgi tendon organ, located in the tendon of the quadriceps muscle. (Each cell illustrated here actually represents a nerve, that is, a population of neurons.) An axon from the sensory neuron extends from the Golgi tendon organ to the spinal cord, where it synapses with two neurons. One is a motor neuron (E) that causes contraction of the extensor quadriceps muscle. The other is an inhibitory interneuron that synapses with the motor neuron innervating the flexor biceps muscle (F). Stimulation of the sensory neuron causes a contraction of the quadriceps and, via the inhibitory neuron, a simultaneous inhibition of contraction of the biceps muscle. The net result is an extension of the leg at the knee joint.

signaling pathways consist of two or more neurons and highly specialized receptor cells.

Sensory receptor cells respond to specific stimuli in the environment: to light, heat, stretching, pressure, osmolarity, and so forth. *Sensory neurons* lead from receptor cells to other neurons, *motor neurons* lead to muscle cells, and *interneurons* connect neurons to each other (Figure 17-7).

In the earthworm, a sensory neuron synapses with one or more motor neurons, which in turn stimulate the appropriate muscle. This constitutes a simple reflex arc (Figure 17-7) and is an example of the organization of neurons into circuits. In vertebrate reflex arcs, interneurons between the sensory and motor neurons integrate and enhance the reflex. The knee-jerk reflex in humans is a complex reflex arc system in which one muscle is stimulated to contract while another is inhibited from contracting (Figure 17-8). Such circuits allow an organism to respond to a sensory input by the coordinated action of sets of muscles that together achieve a single purpose.

The nervous system of higher animals is divided into the *central nervous system,* comprising the brain and spinal cord, and the outlying *peripheral nervous system* (Figure 17-9). Each peripheral nerve is a bundle of axons. Some axons are extensions of motor neurons that stimulate specific muscles or glands, while others are sensory neuron endings that convey information back to the central nervous system. These sensory neurons have their cell bodies clustered in *ganglia,* masses of nerve tissues that lie just outside the spinal cord. Also constituting ganglia are the cell bodies of the motor neurons that make up the *autonomic nervous system,* a division of the peripheral system that controls the involuntary responses of glands and smooth muscles. Most internal glands, such as the liver and pancreas, and most smooth muscles, such as those that surround the digestive tract and the heart muscle, are innervated by two classes of these autonomic nerves: one class stimulates the muscle or gland and the other inhibits it. The cell bodies of the motor neurons that stimulate voluntary muscles are located inside the central nervous system, either in the brain or the spinal cord. However, most of the 10^{12} neurons in the central nervous system are interneurons.

Having surveyed the general features of neuron structure, interactions, and circuits, let us turn to the mechanism by which a neuron generates and conducts an electrical impulse. We begin with the electric potential that exists across the plasma membrane of nerve cells.

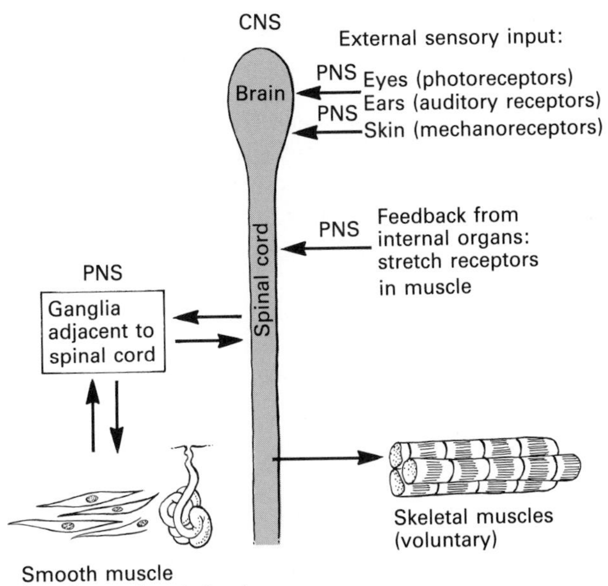

Figure 17-9 A highly schematic diagram of the vertebrate nervous system. The central nervous system (CNS) comprises the brain and spinal cord and is composed mainly of interneurons. It receives sensory input from receptors that react to the outside environment and also from receptors on muscles and other internal organs. The cell bodies of most of the motor neurons that innervate voluntary muscles are in the central nervous system; their axons exit from the central nervous system to stimulate the muscles. The peripheral nervous system (PNS) comprises three sets of neurons with cell bodies lying in ganglia adjacent to the spinal cord: sensory neurons that relay information to the CNS on the state of internal organs, motor neurons that innervate smooth involuntary muscles such as those that line the intestine, and autonomic neurons that innervate glands such as the liver and pancreas.

The Origin of the Electric Potential

An ionic gradient exists across the plasma membrane of virtually all metazoan cells: the concentration of potassium ions inside the cell is about 10 times that in the extracellular fluids, while sodium ions are present in much higher concentrations outside the cell than inside (see Table 15-1). We saw in Chapter 15 how these concentration gradients are maintained by the Na^+K^+-ATPase with the expenditure of cellular energy. Another important property of the plasma membrane is that it is selectively permeable to different cations and anions, including the principal cellular ions Na^+, K^+, and Cl^-, so that the different ions tend to move down their concentration gradients through the plasma membrane at different rates. These two properties, *selective permeability* and *ionic concentration gradients,* lead to a difference in electric potential between the inside and the outside of a cell.

Simple Models Explain the Potential across the Cell Membrane

The situation outlined in Figure 17-10 is similar to that between a cell and its aqueous environment. A 150-m*M* solution of NaCl on the right is separated by a membrane from a solution of 15-m*M* NaCl on the left. A potentiometer (voltmeter) is connected to the solution on each side of the membrane to measure any electric potential difference across it. If the membrane is *impermeable* to both Na^+ and Cl^- ions, no ions will flow across it and there will be no potential difference across the membrane (Figure 17-10a).

Suppose, however, that the membrane is permeable only to Na^+ ions—it contains permeases or pores that accommodate Na^+ but exclude Cl^- ions. Na^+ ions then tend to move down the concentration gradient of Na^+ ions from the right chamber to the left. This leaves the right chamber with an excess of negatively charged Cl^- ions compared with Na^+ ions, and it generates an excess of positive Na^+ ions compared with Cl^- in the left chamber. There is now a separation of charge across the membrane. With a potentiometer it can be measured as an electric potential, or voltage, across the membrane. The right side is *negative* with respect to the left (Figure 17-10b). As more and more Na^+ ions move across the

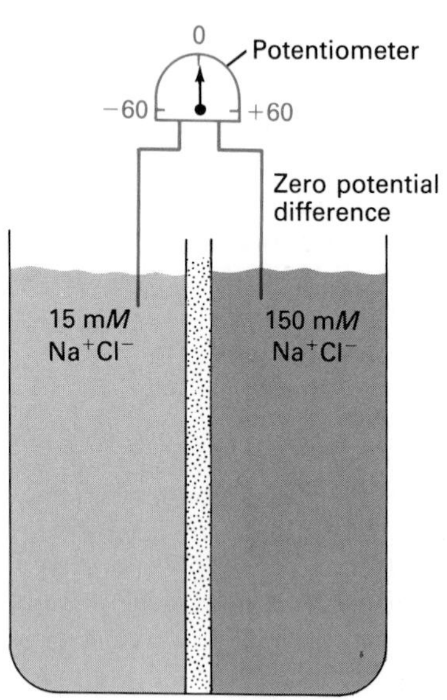

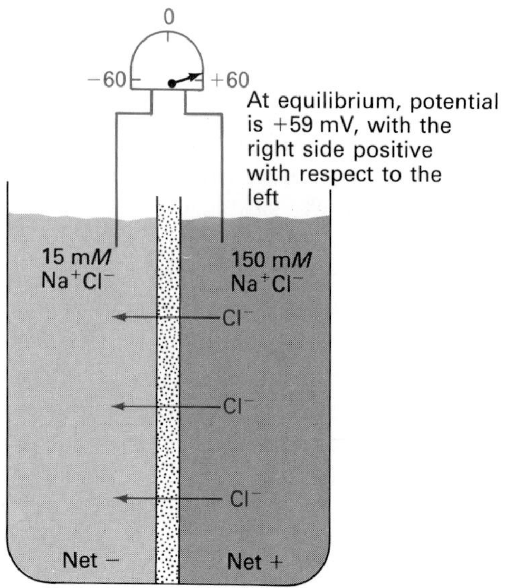

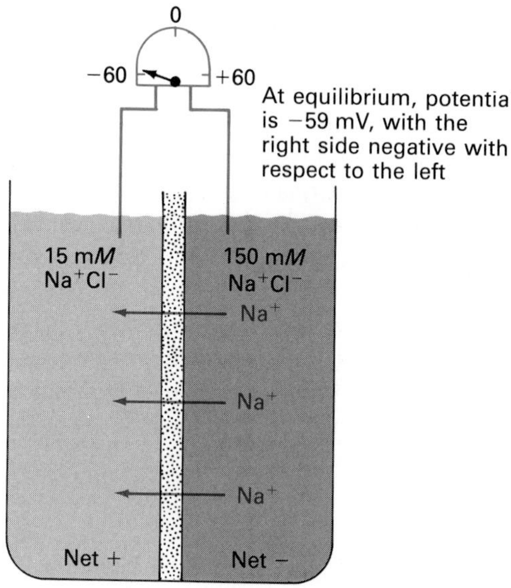

Figure 17-10 A voltage potential is created by the selective permeability of a membrane to different ions. (a) An impermeable membrane separates a 150-m*M* NaCl solution from a 15-m*M* solution. No ions move across the membrane and no electric potential is registered on the potentiometer that connects the two solutions. (b) The membrane is permeable only to Na^+ ions. Na^+ diffuses from right to left down the concentration gradient. The anion Cl^- cannot cross the membrane, so a net positive charge builds up on the left side and a negative charge on the right. At equilibrium, the membrane potential caused by the charge separation becomes equal to the Nernst potential E_{Na} as registered on the potentiometer, and the movement of Na^+ in the two directions becomes equal. (c) In this case, the membrane is selectively permeable only to Cl^- ions. Cl^- diffuses from right to left down the concentration gradient. A net negative charge builds up on the left side and a net positive charge on the right. At equilibrium, the membrane potential becomes equal to E_{Cl}.

membrane, the magnitude of this charge difference increases, but continued movement of the Na^+ ions eventually is inhibited by the excess of positive charges accumulated on the left side of the membrane and by the attraction of Na^+ ions to the excess negative charge built up on the right side. Soon the system reaches an equilibrium in which the two opposing factors that determine the movement of Na^+ ions—the membrane potential and the concentration gradient—balance each other out. At equilibrium there is no net movement of Na^+ ions across the membrane.

The magnitude of the resultant membrane potential (equilibrium potential) is given by the Nernst equation, which is derived from basic principles of physical chemistry:

$$E_{Na} = \frac{RT}{Z\mathscr{F}} \ln \frac{Na_L}{Na_R} \tag{17-1}$$

where R is the gas constant [1.98 cal/(degree · mol) or 8.28 joules/(degree · mol)], T is the absolute temperature (293 K at 20°C), $\mathscr{F}$ is the Faraday constant [23,062 cal/(mol · V) or 96,000 coulombs/(mol · V)], and Z is the valency (here +1). Na_L and Na_R are the Na^+ equilibrium concentrations in the left and right chambers. E_{Na} is the *sodium equilibrium potential* measured in volts. It is the potential across a membrane permeable only to Na^+ ions. The Nernst equation is similar to those used for calculating the voltage change associated with oxidation or reduction reactions—reactions that also involve movement of electric charges (see Chapter 2).

At 20°C this equation reduces to

$$E_{Na} = 0.059 \log_{10} \frac{Na_L}{Na_R} \tag{17-2}$$

In the above example, $Na_L/Na_R = 0.1$ and $E_{Na} = -0.059$ V or -59 mV, with the right side *negative* with respect to the left.

If the membrane is permeable only to Cl^- ions, and not to Na^+, the calculation is the same:

$$E_{Cl} = \frac{RT}{Z\mathscr{F}} \ln \frac{Cl_L}{Cl_R} \tag{17-3}$$

except that $Z = -1$. The *magnitude* of the potential is the same, 59 mV, except that the right chamber is now *positive* with respect to the left (Figure 17-10c). This is precisely the opposite polarity to that obtained with selective Na^+ permeability.

If the membrane is permeable to Na^+ and Cl^- to the same degree, then Na^+ and Cl^- can move together from the right chamber to the left, down their concentration gradients. In this case, no membrane potential is expected and none is observed.

The situation, of course, can also be intermediate between these two extremes. The membrane could be permeable to both Na^+ and Cl^- ions but more permeable to Na^+. Then initially the right side would have a negative potential relative to the left, but the magnitude of the potential would be somewhat less than E_{Na} of -59 mV. Eventually, of course, owing to diffusion of Na^+Cl^-, there will be an equal concentration of ions on both sides of the membrane, and no membrane potential.

The movement of ions such as K^+, Cl^-, or Na^+ across selectively permeable membranes is thus governed by two forces, the membrane electric potential, and the concentration gradient of the ions across the membrane. These forces may act in the same or opposite direction. Consider K^+, the predominant cellular cation. At the potassium equilibrium potential these two forces acting on K^+ balance each other, and $E = E_K$. For any particular potential across the plasma membrane E, the electrical force acting on K^+ ions driving them one way or another across the membrane is proportional to the difference $E - E_K$. Similarly, the electrical force on Na^+ ions will be $E - E_{Na}$ and on Cl^- ions $E - E_{Cl}$. The actual number of ions that flow through the membrane depends not only on this value but also on the permeability of the membrane for that ion. The permeability (P) for K^+ or Na^+ or Cl^- ions, P_K or P_{Na} or P_{Cl}, is a measure of the ease with which they flow through a unit area of membrane. Thus, the flow of K^+ ions across a unit area of cell membrane driven by the electrical forces will be $P_K(E - E_K)$, and similarly for the other ions. The units of the permeability constant P are in centimeters per second. P is the measure of permeability of an ion across a membrane of unit area (1 cm^2) driven by a 1-M difference in concentration.

The Electric Potential of the Membrane Is a Function of Its Relative Permeabilities to Potassium, Sodium, and Chloride Ions

The situation in typical cells is more complex than in the above example because there are three principal ions to consider, Na^+, K^+, and Cl^- (see Table 15-1), and three permeability constants, P_{Na}, P_K, and P_{Cl}. Cells, of course, contain other ions, such as HPO_4^{2-}, Ca^{2+}, SO_4^{2-}, and Mg^{2+}. But their membrane permeabilities are very small relative to K^+, Na^+, and Cl^-. Furthermore, in electrically active cells such as nerves and muscles it is only K^+, Na^+, and Cl^- (and occasionally Ca^{2+}) that affect the membrane potential. These three main ions are the only ones we need consider here.

In particular plasma membranes, P_K, P_{Cl}, and P_{Na} will be dependent on the amounts and activities of K^+, Cl^-, and Na^+ channel proteins; in general, permeability values will vary. Channel proteins have many of the characteristics of permeases, but in neurophysiology the term channel has become established, and that convention will be followed here.

The resultant membrane potential across a cell surface

membrane is given by a more complex version of the Nernst equation in which the concentrations of the ions are weighted in proportion to their permeability constants:

$$E = \frac{RT}{F} \ln \frac{P_K K_o + P_{Na} Na_o + P_{Cl} Cl_i}{P_K K_i + P_{Na} Na_i + P_{Cl} Cl_o}$$
$$= 59 \log_{10} \frac{P_K K_o + P_{Na} Na_o + P_{Cl} Cl_i}{P_K K_i + P_{Na} Na_i + P_{Cl} Cl_o} \quad (17\text{-}4)$$

The "o" and "i" denote the concentrations outside and inside the cell, and by convention, the membrane potential is expressed as inside relative to outside. Because of their opposite charges [Z value in Equations (17-1) and (17-2)] K_o and Na_o but Cl_i are placed in the numerator; K_i and Na_i and Cl_o are placed in the denominator. This equation will adequately describe the potential at any point where the concentrations and permeabilities are known. If the concentrations are equilibrium concentrations, then the potential will be the equilibrium potential and will remain until outside energy is applied or the concentration of an ion is altered. If the concentrations are not equilibrium ones, then the equation will describe the present potential but not the potential at any future time.

Note that if $0 = P_{Na} = P_{Cl}$, then the membrane is permeable only to K^+ ions, and Equation (17-4) reduces to

$$E = \frac{RT}{F} \ln \frac{K_o}{K_i} = 59 \log_{10} \frac{K_o}{K_i} = E_K \quad (17\text{-}5)$$

which is the same as Equation (17-1) with $Z = 1$. Similarly, if $0 = P_K = P_{Cl}$, then the membrane is permeable only to Na^+ and

$$E = \frac{RT}{F} \ln \frac{Na_o}{Na_i} = E_{Na} \quad (17\text{-}6)$$

The potential across the plasma membrane can be measured by the use of two electrodes. One is a microelectrode made by filling a glass tube of extremely small diameter with a conducting fluid. It is inserted inside the cell in such a way that the surface membrane seals itself around the microelectrode. The other electrode is in the extracellular fluid. The two are connected to a voltmeter capable of measuring small potential differences (Figure 17-11). In excitable tissues (nerve and muscle) the potential difference maintained across the cell membrane in the absence of stimulation is called the *resting potential* and the cell is said to be in the *resting state.*

Changes in Ion Permeability Can Depolarize or Hyperpolarize the Membrane

Suppose a membrane separates two solutions. Solution A contains 140 mM of K^+, 12 mM of Na^+, and 4 mM of Cl^-. This solution is similar to that of the cytosol of mammalian cells. Solution B contains 4 mM of K^+, 150 mM of Na^+, and 120 mM of Cl^- and is similar to the extracellular fluid (Figure 17-12). Further suppose that the membrane is permeable only to K^+, Na^+, and Cl^- ions and that $P_K = 10^{-7}$ cm/s, $P_{Na} = 10^{-8}$ cm/s, and $P_{Cl} = 10^{-8}$ cm/s; thus $P_K > P_{Na}$ or P_{Cl}. These values are similar to the values in most cells, as determined directly by the permeability of the plasma membrane to radioactive ions.

(a)

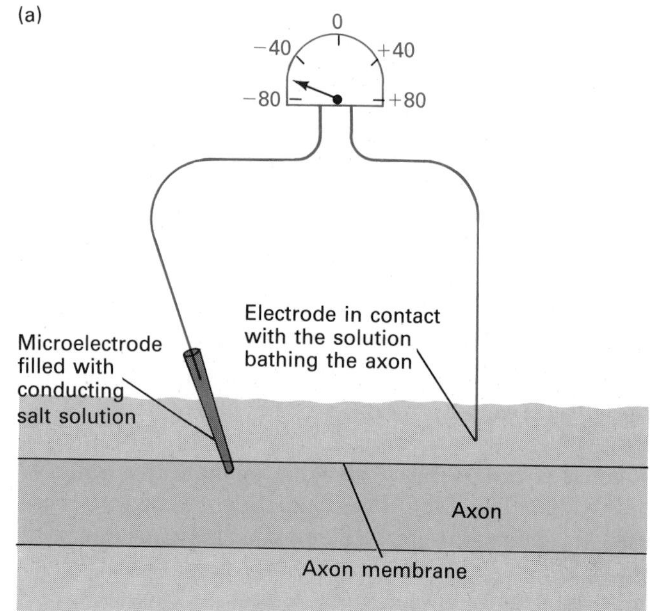

(b)

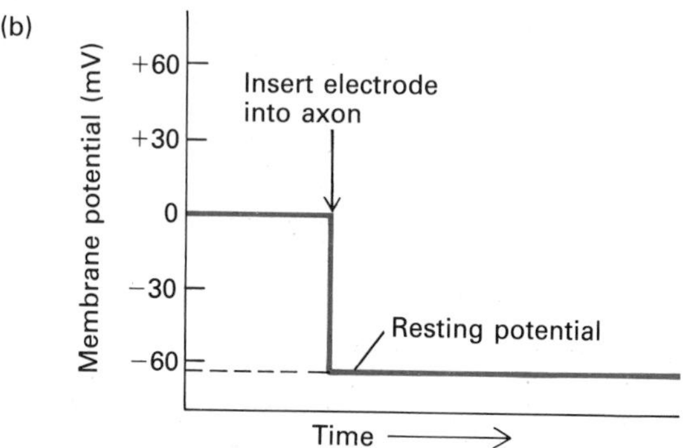

Figure 17-11 Diagram of an axon impaled with a microelectrode. (a) The potentiometer measures the voltage difference between the inside of the cell and the solution surrounding the cell. The fine glass tube microelectrode holds conducting fluid and serves as a microelectrode when inserted into the axon; it insulates the conducting fluid from the extracellular fluid bathing the axon. The potentiometer shows a resting potential for this cell of −60 mV. (b) There is no potential if the tip of the microelectrode is simply in the bathing medium; a potential is registered only when the tip is inserted into the axon.

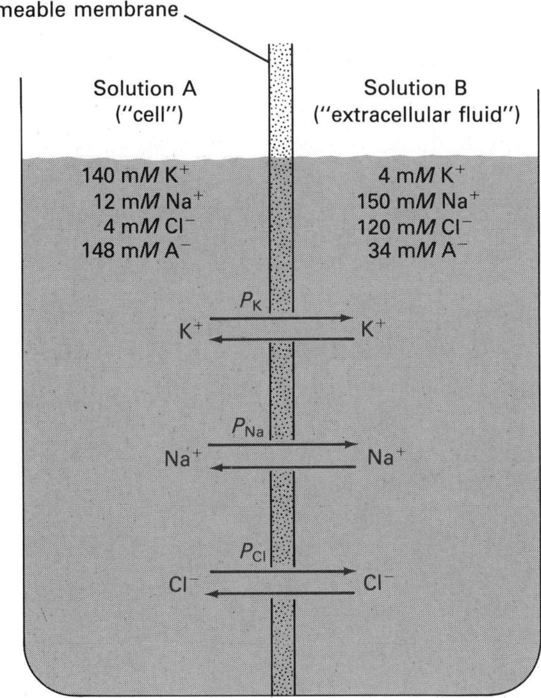

Figure 17-12 A semipermeable membrane separates two solutions. Solution A has a composition similar to that of a typical nerve cell cytoplasm; B is similar to blood. A^- signifies negative charges on proteins that neutralize the excess of positive charges contributed by Na^+ and K^+.

From Equation (17-4) we have

$$E = 59 \log_{10} \frac{(10^{-7})(0.004) + (10^{-8})(0.15) + (10^{-8})(0.004)}{(10^{-7})(0.14) + (10^{-8})(0.012) + (10^{-8})(0.12)} = -52.9 \text{ mV}$$

$$E_K = -91.1 \text{ mV}$$
$$E_{Na} = +64.7 \text{ mV}$$
$$E_{Cl} = -87.2 \text{ mV}$$

The potential is close to but less than the potassium potential E_K. This situation, of course, is not at equilibrium. Eventually enough K^+ and Na^+ ions will diffuse down their respective concentration gradients for both sides of the membrane to have the same concentration of all ions. The equilibrium potential, of course, will then be 0 mV. In cells, the Na^+ and K^+ gradients are maintained by Na^+K^+-ATPase.

Let us return to the original situation, where $E = -52.9$ mV. Suppose the permeability for sodium is suddenly increased 200-fold, to 2×10^{-6} cm/s. Keeping all of the other original values constant, then:

$$E = 59 \log_{10} \frac{(10^{-7})(0.004) + (2 \times 10^{-6})(0.15) + (10^{-8})(0.004)}{(10^{-7})(0.14) + (2 \times 10^{-6})(0.012) + (10^{-8})(0.12)} = +52.2 \text{ mV}$$

Thus, merely by changing P_{Na} so that it is much greater than P_K or P_{Cl}, the potential is shifted to inside positive, close to but less than E_{Na}.

Let us again return to the original set of concentrations and permeabilities, but now increasing P_K 200-fold to $P_K = 2 \times 10^{-5}$ cm/s. Then

$$E = 59 \log_{10} \frac{(2 \times 10^{-5})(0.004) + (10^{-8})(0.15) + (10^{-8})(0.004)}{(2 \times 10^{-5})(0.14) + (10^{-8})(0.012) + (10^{-8})(0.12)} = -90.6 \text{ mV}$$

The membrane is now hyperpolarized, with E almost identical to the E_K of -91.1 mV.

If instead, we increase P_{Cl} 200-fold to 2×10^{-6} cm/s, we find that

$$E = 59 \log_{10} \frac{(10^{-7})(0.004) + (10^{-8})(0.15) + (2 \times 10^{-6})(0.004)}{(10^{-7})(0.14) + (10^{-8})(0.012) + (2 \times 10^{-6})(0.12)} = -83 \text{ mV}$$

Again, the membrane becomes hyperpolarized.

In summary, increasing the permeability of Na^+ ions depolarizes the membrane. Increasing either the K^+ or Cl^- permeability hyperpolarizes it. In neuron function, excitation, inhibition, and synaptic transmission, changes in the permeability constants for certain ions play an essential role.

Most Cells Are More Permeable to Potassium Ions than to Sodium or Chloride Ions

The surface membrane of most animal cells is much more permeable to K^+ ions than to Na^+ or Cl^- ions. Thus:

$$P_{Na} \text{ and } P_{Cl} < P_K$$

This is true for nerve and muscle cells as well as for most cells growing in tissue culture. For example, in fibroblasts growing in tissue culture, direct measurements of the flow of radioactive ions across the plasma membrane were used to calculate permeability constants:

$$P_K = 5 \times 10^{-7} \text{ cm/s}$$
$$P_{Na} = 5 \times 10^{-9} \text{ cm/s}$$
$$P_{Cl} = 1 \times 10^{-8} \text{ cm/s}$$

The permeability of the membrane to K^+ is 50-fold that of Cl^- and about 100-fold that of Na^+. Using these values for the permeability constants and the ion concentrations listed in Table 15-1, we calculate using Equation (17-5) that for this mammalian cell $E = -81$ mV, the inside is *negative* relative to the outside of the membrane. This is close to but less than the potassium potential E_K, the membrane potential of −89 mV that would occur if the membrane were permeable only to K^+ ions, because of the contribution of Cl^- and Na^+ permeabilities to the membrane potential.

The plasma membrane potential should be responsive primarily to internal and external K^+ ions. From the Nernst equation we calculate that any change in either the external or internal K^+ ion concentration by a factor of 10 should change the membrane potential by 59 mV. Such experiments were performed on the giant axon of the squid. In one, the external K^+ concentration was varied and the membrane potential measured. The experimental points lie close to the theoretical line derived from the Nernst equation except in the region of low external K^+ concentration where, according to Equation (17-4), permeabilities to Cl^- and Na^+ affect the potential to a greater extent (Figure 17-13).

The squid axon is so long and wide (1 mm in diameter) that it is possible to squeeze out the contents of the cytoplasm and replace it with various solutions of defined K^+ ionic concentrations, permitting the membrane potential to be studied at different internal K^+ concentrations. In this case also, the resultant membrane potential was that predicted by the modified Nernst equation (17-4).

Similar principles apply in other cells, but the permeabilities of the ions are different. In nonstimulated skeletal muscle cells Cl^- ion permeability is substantial. Erythrocytes, as we noted in Chapter 15, have a very much greater permeability to Cl^- than to K^+ or Na^+. The glial cells in the brain that surround certain neurons have a permeability to K^+ vastly larger than to Na^+ or Cl^-. Glial cells have a membrane potential near that of E_K.

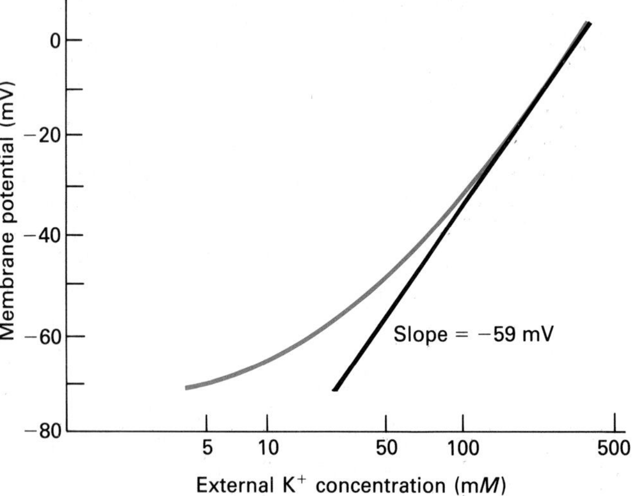

Figure 17-13 Membrane potential of the squid giant axon as a function of the external potassium concentration *(colored curve)*. A tenfold change in external K^+ results in a change in potential of 59 mV. At low external K^+ concentration the potential varies from the theoretical *(black curve)* because of significant contributions from permeabilities to Cl^- and, especially, Na^+ ions. [See A. L. Hodgkin and R. D. Keynes, 1956, *J. Physiol.* 131:592.]

Electrical Activity Is a Result of Permeability Changes to Specific Ions

The potential across the surface membrane of most animal cells generally does not vary with time. The cells are said to be electrically inactive; the membrane potential exists as a consequence of the concentration gradient and the differential permeability of the plasma membrane to K^+, Na^+, and Cl^- ions. As we noted in Chapter 15, a gradient of ions across the cell membrane is important for many cellular activities, and much of the cell's energy production goes into its maintenance.

Neurons and muscle cells are the principal types of electrically active cells. Controlled changes in their membrane potential are central to their function. In muscle cells, transient reduction in the membrane potential, or *depolarization* of the surface membrane, is often induced by an impulse from a nerve. This depolarization triggers contraction of the muscle. Nerve cells and many muscle cells conduct electrical impulses along their membrane by controlling sequential changes in the permeability of the surface membrane to Na^+ and K^+ ions.

The Action Potential

When an impulse passes down a neuron it is observed as a movement of negative charges along the axon. One way to measure this is to place a series of electrodes on the surface of an axon and record the electric potential at various points on the surface relative to the bulk extracellular fluid. Each electrode will register a transient negative potential on the surface of the neuron as the impulse moves along the axon. The magnitude of the electric potential detected is generally several millivolts. Depending on the neuron, the rate of conduction of the nerve impulse varies from 1 to 100 meters per second.

The Action Potential Reflects the Sequential Depolarization and Repolarization of a Region of the Nerve Membrane

It would appear from the outside that negative charges—that is, electrons—simply are propelled along the outer surface of the neuron, but the situation is much more complex. What actually occurs can be followed by measuring changes in the membrane potential of a very small region of the nerve cell plasma membrane as the electrical impulse passes along it (Figure 17-14). Most of the classic studies have been done on the giant axon of the squid, in large part because multiple microelectrodes can be inserted into it without damage and thus the potential across any region of the membrane can be measured. The microelectrodes are connected to an amplifier, and often to a computer. The resultant electrical signals are stored and displayed on an oscilloscope. Electrical changes occurring within fractions of a millisecond can be recorded.

In the resting state, the potential across a small region of the squid axon membrane is −60 mV, typical for most neuronal cells. As the action potential traverses the region, the membrane suddenly becomes depolarized and then the membrane potential becomes about +35 mV. The inside is now *positive* relative to the outside, and there is a net potential change of almost 90 mV. Quite rapidly thereafter, the potential returns to a negative value but to a potential (−75 mV) slightly more negative than the resting value. Gradually the potential returns to the resting value. Thus the action potential is a cycle of depolarization, hyperpolarization, and return to the resting value. The cycle lasts 1 to 2 ms.

The apparent movement of negative charges along the surface of the neuron is, in fact, the successive depolarization and repolarization of adjacent regions of the plasma membrane (Figure 17-15). The action potential is propagated unidirectionally along the axon, from the axon hillock to the axon terminals.

All of these changes in the membrane potential—depolarization, hyperpolarization, and unidirectional

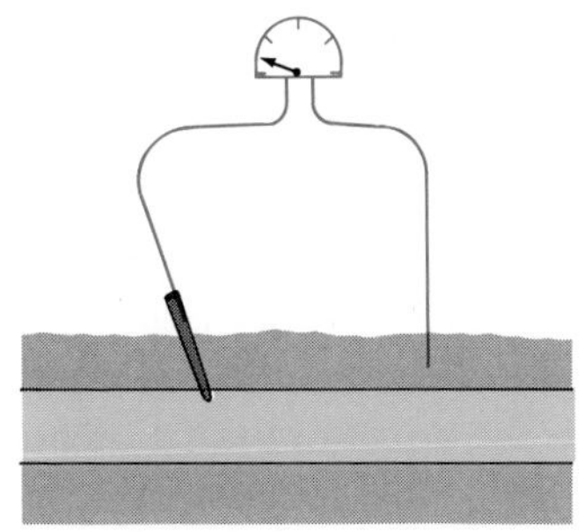

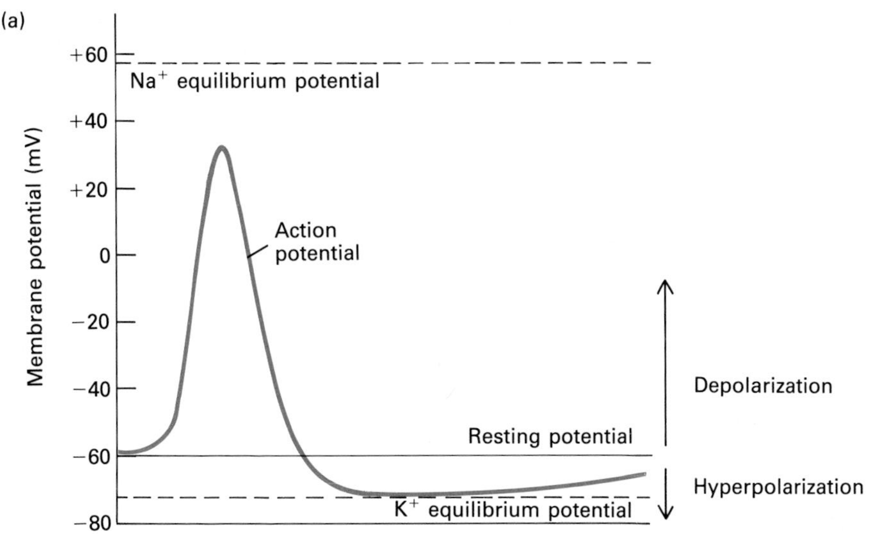

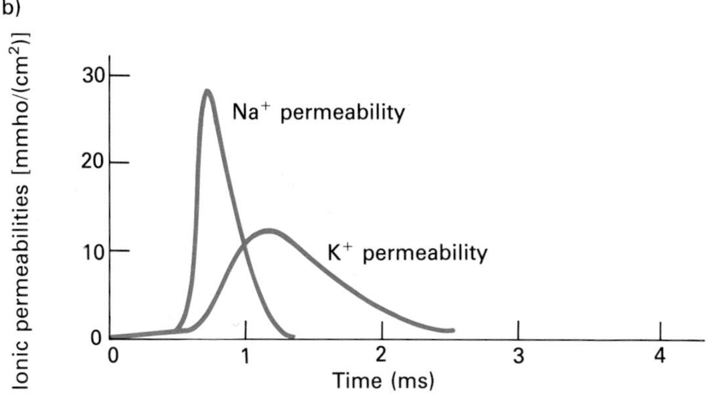

Figure 17-14 The action potential and accompanying changes in sodium and potassium permeability (conductance). The time scale is measured in milliseconds and the unit of electrical conductance is the mho. The resting potential is −60 mV. The action potential depicted in (a) is obtained by implanting a microelectrode within the giant axon of a squid while another electrode is in contact with the bathing solution. Rapid depolarization is followed by hyperpolarization, then a slow return to the resting potential. (b) The changes in permeability (conductance) for Na^+ and K^+. Increased Na^+ permeability occurs at the same time the membrane is depolarized and precedes increased K^+ permeability, which occurs when the membrane is hyperpolarized. [See A. L. Hodgkin and A. F. Huxley, 1952, *J. Physiol.* **117**:500.]

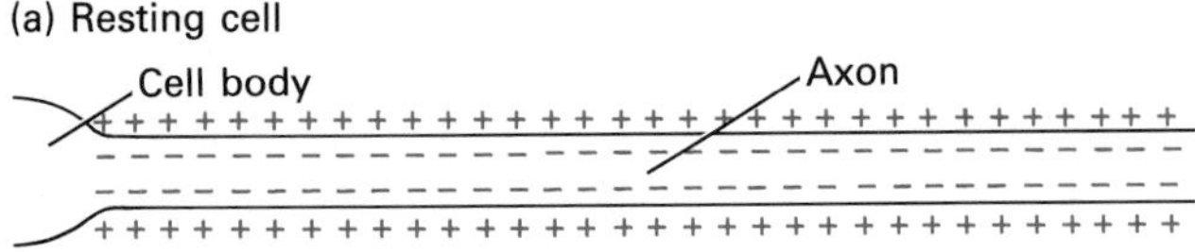

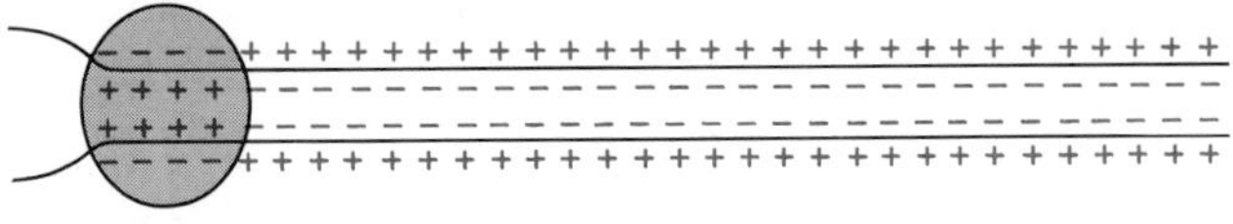

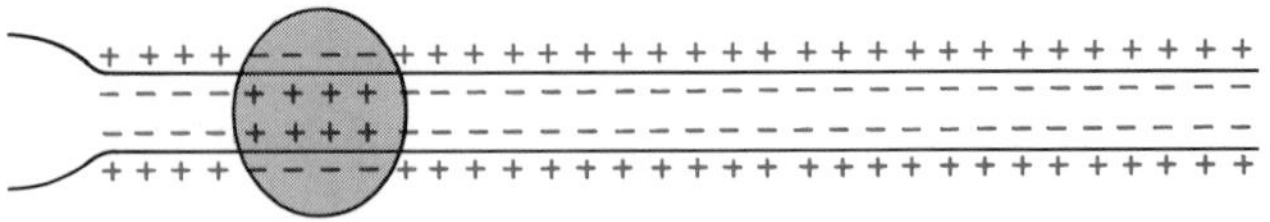

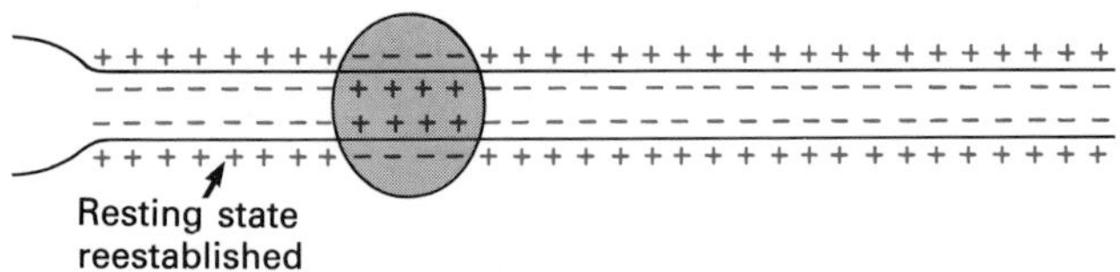

Figure 17-15 (a–d) Movement of an electric current down the axon of a neuron is accomplished by the successive depolarization of regions of the membrane. (c and d) Following the depolarization, the resting state is reestablished.

propagation—can be ascribed to sodium and potassium channels across the plasma membrane that open and shut in response to stimulation of the membrane. These channels are transmembrane proteins whose ion permeability is dependent on the electric potential across the membrane. Voltage-dependent Ca^{2+} channels are important for contraction of muscle as well as for triggering the rise of intracellular Ca^{2+} ions that causes exocytosis of synaptic vesicles in the axon terminals. Voltage-dependent ion channels are, in fact, critical for all aspects of the function of electrically active cells. Let us see how such channels and the coordinate changes in ion permeabilities can explain the action potential.

An Increase in Sodium Permeability Depolarizes the Cell

The sudden depolarization of a region of the surface membrane during the action potential is caused by a sudden massive increase in the permeability of that region to Na^+ ions (see Figure 17-14). This increase in Na^+ permeability is caused by the opening of voltage-dependent Na^+ channels. The channels open briefly when the membrane is depolarized—more channels open as the depolarization of the membrane is increased.

During conduction of an action potential the depolarization of a region of membrane spreads passively to the adjacent downstream region of membrane and depolarizes it slightly, which opens a few of the downstream voltage-dependent Na^+ channels. Na^+ ions are driven into the cell by a combination of two forces acting in the same direction. One is the concentration gradient of Na^+ ions: the concentration is greater outside than inside. The other is the initial potential gradient of the cell—inside negative—which also tends to attract Na^+ ions into the cell. As more and more Na^+ ions enter the cell, the inside of the cell membrane becomes more positive and the membrane is depolarized further. This depolarization causes the opening of more Na^+ channels (Figure 17-16). If the initial depolarization of the region of membrane exceeds a certain minimum value called the *threshold potential,* it sets into motion an explosive entry of Na^+ ions that is completed within a fraction of a millisecond. This is the origin of the action potential. For a fraction of a millisecond, at the peak of the depolarization, P_{Na} becomes vastly greater than P_K (or P_{Cl}), and thus the membrane potential approaches that of E_{Na}:

$$E_{Na} = \frac{RT}{Z\mathcal{F}} \ln \frac{Na_o}{Na_i} = 59 \log_{10} \frac{Na_o}{Na_i} \tag{17-7}$$

At the point where the membrane potential reaches E_{Na}, further net inward movement of Na^+ ions ceases, since the concentration gradient of Na^+ ions (outside >

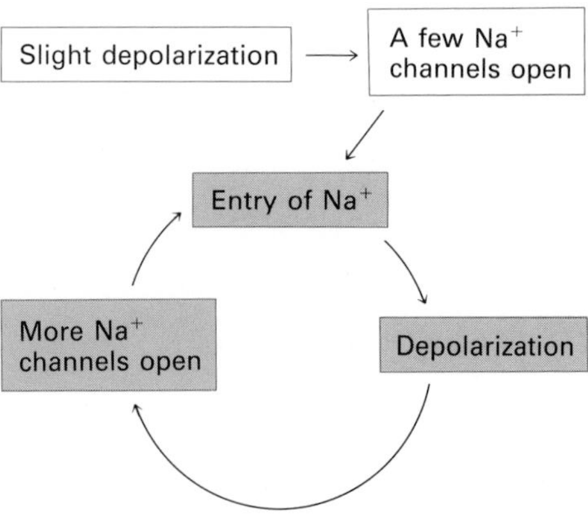

Figure 17-16 The consequences of voltage-dependent Na^+ channels for the generation of an action potential. An initial depolarization triggers a massive opening of Na^+ channels, causing the membrane potential to become, for a short time, close to that of E_{Na}.

inside) is balanced by the membrane potential E_{Na} (inside positive).

This analysis predicts that the membrane potential at the peak of the action potential will equal E_{Na}. Indeed, for the squid's giant axon, $Na_o = 440\ mM$ and $Na_i = 50\ mM$ (see Table 15-1) and thus $E_{Na} = +55$ mV, close to the peak value of the axon's action potential of +35 mV (see Figure 17-14). The relationship between the magnitude of the action potential and the concentration of Na^+ ions inside and outside the cell has been confirmed experimentally. For instance, if the concentration of Na^+ ions in the solution bathing the squid axon is reduced to one-third of normal, the magnitude of the depolarization is reduced by 40 mV, exactly as predicted (Figure 17-17).

The Movements of Only a Few Sodium Ions Generate the Action Potential

Voltage changes are not caused by bulk movements of ions across the membrane. Calculations indicate that the movement of very few Na^+ ions per square centimeter of membrane are sufficient to cause the membrane potential to change from inside negative to inside positive, from near E_K to near E_{Na}. Equation (17-5) and the discussion on page 725 support this conclusion—large changes in the voltage across the membrane will result from changes in the *permeability constants* for Na^+ or K^+ ions even when the *concentration* of ions on the two sides does not change at all. If $P_K \gg P_{Na}$, as in the resting state, then $E \sim E_K$. If $P_{Na} \gg P_K$, then $E \sim E_{Na}$. Measurements of the amount of radioactive sodium entering and leaving single squid axons and other axons during a single impulse show that, depending on the size of the neuron, only about one K^+ ion in 3,000 to 300,000 in the cytosol (0.0003 to 0.03 percent) needs to be replaced with Na^+ to generate the reversal of membrane polarity.

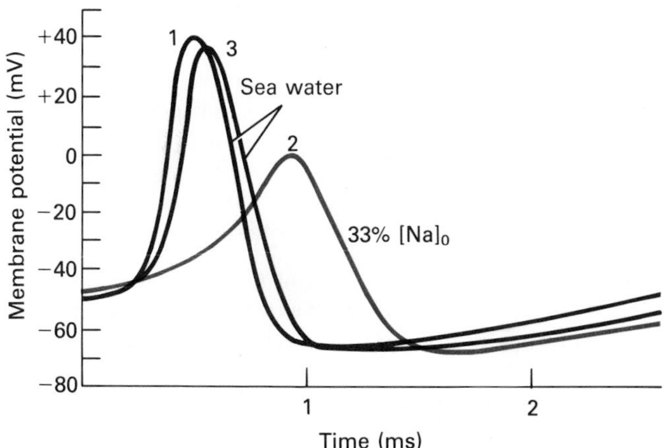

Figure 17-17 Effect on the magnitude of the action potential in the squid axon of changing the external Na^+ concentration. In curve 2 the external NaCl concentration is reduced to 150 m*M*, one-third of normal. Curves 1 and 3 are the control values of normal sea water before and after testing the low NaCl solution. [See A. L. Hodgkin and B. Katz, 1949, *J. Physiol.* **108**:37.]

Na^+K^+-ATPase Plays No Direct Role in Nerve Conduction

In any cell, be it nerve, muscle, or fibroblast, the existence of a membrane potential is dependent primarily on a gradient of K^+ ions. The gradient is generated and maintained by Na^+K^+-ATPase. If dinitrophenol or another inhibitor of ATP production is added to cells, the membrane potential gradually falls to zero as all the ions equilibrate across the membrane. The process is extremely slow in most cells, requiring hours. From this and similar experiments we conclude that the membrane potential is essentially independent of the supply of ATP over the short time-spans required for nerve or muscle cells to function. Nerve cells, in particular, normally fire thousands of times in the absence of an energy supply because the ionic movements during each discharge involve only a minute fraction of the cell's K^+ and Na^+ ions.

The Sodium Channel Is Voltage-Dependent

Electrophysiological studies on the squid axon and other axons have established some of the remarkable properties of the voltage-dependent Na^+ channels that help explain the generation and propagation of an action potential.

The neuronal plasma membrane contains a number of individual Na^+ channels that can be either open or closed. There are two distinct closed conformations. One, in the resting state, can be opened upon depolarization of the membrane. The other is inactivated, incapable of opening (Figure 17-18). In the unstimulated cell most of the channels are closed. Depolarization increases the probability that any channel will open. The greater the depolarization the greater the chance that channels will open. Once opened the channels stay open for about 1 ms, during which time they allow the passage of about 6000 Na^+ ions. The channels then close spontaneously and enter the inactive state. They will not reopen as long as the membrane is depolarized. When the membrane potential returns to the resting—inside-negative—state the channels enter the closed, resting state and will reopen in response to a depolarization. The inactive state of the sodium channel is the explanation for the *refractory period,* the period of time after firing when it is impossible for a neuron to generate or conduct an action potential. The inability of an Na^+ channel to reopen for several

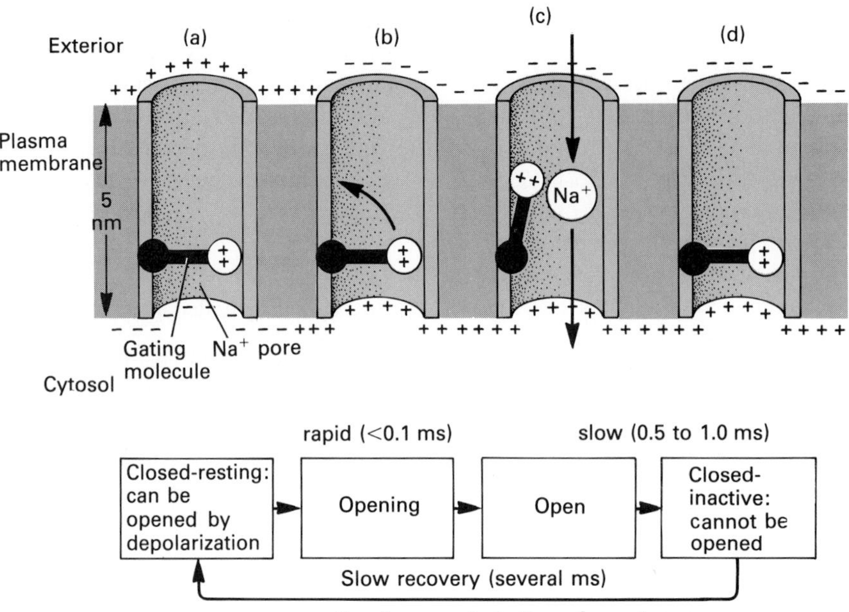

Figure 17-18 Schematic mechanism of the voltage-gated Na^+ channel in the invertebrate neuronal plasma membrane, as deduced from electrophysiological studies. The average open time of a channel is 0.5 to 1.0 ms, and then it enters the closed, inactivated state. As long as the membrane is depolarized the channel cannot reopen. When the membrane potential returns to its original inside-negative state the channel enters the closed, resting conformation; it can be opened again if the membrane is depolarized. An interpretation of the depolarization-induced opening of the channel is depicted here. (a) In the resting stage there is a potential difference of 70 mV across the 5-nm-thick plasma membrane; the resultant voltage gradient is 140,000 V/cm. Thus all intramembrane proteins, such as the Na^+ channels, are subjected to very large electric fields. A gating segment of the channel protein is visualized as blocking the channel. A positive amino acid side chain on the surface of the "gate" will create an electric dipole moment. This dipole moment will affect the conformation of the segment of protein because it will tend to orient itself in the electric field to assume a conformation of minimal free energy. In the case of the Na^+ channel this could be, as in (a), the closed state. (b) If however, the membrane is depolarized the gating molecule moves *(arrow)* to realign its dipole moment with the electric field. (c) Within a fraction of a millisecond the channel is open and Na^+ ions can pass through. The minute charge movements involved in opening or closing a channel generate "gating currents" that can be detected experimentally. Within a millisecond after opening, the channel closes spontaneously and is inactive. (d) The channel can reopen only if the membrane potential is reversed to the inside-negative state (a). Studies on certain mammalian Na^+ channels suggest that, in contrast to the rapid opening and slow closing depicted here, voltage-dependent activation (opening) can be slow (0.5 to 1.0 ms) and closing more rapid. Thus, different voltage-gated Na^+ channels may have different properties.

milliseconds thus limits the number of action potentials per second that a neuron can conduct.

The Opening of Delayed Potassium Channels Repolarizes the Membrane

Within a millisecond after the increase in Na^+ permeability the voltage-dependent Na^+ channels close and are inactivated and P_{Na} returns to its original value. During the time that the Na^+ channels are closing there is a transient increase in P_K owing to the opening of K^+ channels, and the value of P_K becomes much greater than in the resting state (see Figure 17-14). The membrane potential repolarizes, overshoots the original resting potential and, for a brief instant, becomes very close to that of E_K, the potassium equilibrium potential—which is, as we have noted, *more* inside negative than is the resting potential (see page 725). The opening of the K^+ channels is induced by the membrane depolarization of the action potential. These channels are voltage-dependent; in contrast to the voltage-dependent sodium channels, the K^+ channels will open and remain open for as long as the membrane is depolarized. Because the K^+ channels open a fraction of a millisecond or so after the initial depolarization, they are called *delayed K^+ channels*. The increase in P_K accounts for the transient hyperpolarization of this region of the membrane. Eventually the K^+ and Na^+ channels close, P_K and P_{Na} return to the values characteristic of the resting state, and the membrane potential returns to its resting value.

The Na^+ and K^+ channels that determine the resting potential of the cell—the "resting" or "leakage" channels—are not voltage-dependent; they do not open or close during the action potential and, as judged by the effects of inhibitors, are different proteins from the voltage-dependent K^+ and Na^+ channels used in action potentials.

The Action Potential Is Induced in an All-or-Nothing Fashion

When the plasma membrane of a neuron is depolarized, the permeability to Na^+ ions increases slightly because of the opening of a few Na^+ channels. Sodium then moves inward, driven by large concentration and potential gradients. The entry of these positive ions tends to depolarize the membrane further, thus opening more Na^+ channels and increasing the Na^+ permeability even more (see Figure 17-16). At the same time, however, there is an outward movement of K^+ ions, which tends to *repolarize* the membrane. In fact, many neurons contain voltage-dependent K^+ channels that transiently open immediately upon depolarization, increasing the K^+ permeability. They are called *immediate K^+ channels*. Their opening tends to repolarize the membrane and hinder the generation of an action potential. Whether or not an action potential is induced depends on the ratios of the permeability to Na^+ and K^+ ions—the balance between the inward movement of Na^+ ions that depolarizes the membrane further and the outward movement of K^+ ions that repolarizes the membrane.

If the number of open Na^+ channels and inrushing Na^+ are insufficient to depolarize the cell to the threshold value, no action potential will be generated. If the threshold is exceeded, the resulting action potential will always have the same magnitude in any particular neuron. The generation of an action potential is thus said to be *all-or-nothing*. Depolarization above the threshold always leads to an action potential, depolarization below it never does. The exact value of the threshold potential can be measured. It depends on the relative numbers of the resting K^+ and Na^+ channels and the voltage-dependent Na^+ and K^+ channels in the membrane. If, for instance, the permeability of the plasma membrane to K^+ ions is high, then the neuron will require a greater increase in Na^+ permeability to induce an action potential.

Voltage Clamps Permit Measurement of Sodium and Potassium Movement across Membranes

The movement of ions across the plasma membrane during an action potential can be demonstrated with the technique of *voltage clamping*. It measures the movement of ions across the membrane when the membrane is electrically depolarized or hyperpolarized and then experimentally maintained (clamped) at this potential. Two electrodes are inserted into the nerve axon. One is used to measure the voltage across the nerve membrane, and the second passes electric current into or out of the axon to keep the membrane potential at a constant value set by the experimenter. How much current the second electrode passes is determined by measurements made by the first electrode (Figure 17-19).

Suppose the experimenter depolarizes the membrane to a particular value. The depolarization causes an opening of Na^+ channels and an influx of Na^+ ions, which would normally lead to a further depolarization of the membrane and an entry of more Na^+ ions. But when the membrane potential is clamped at this depolarized value no

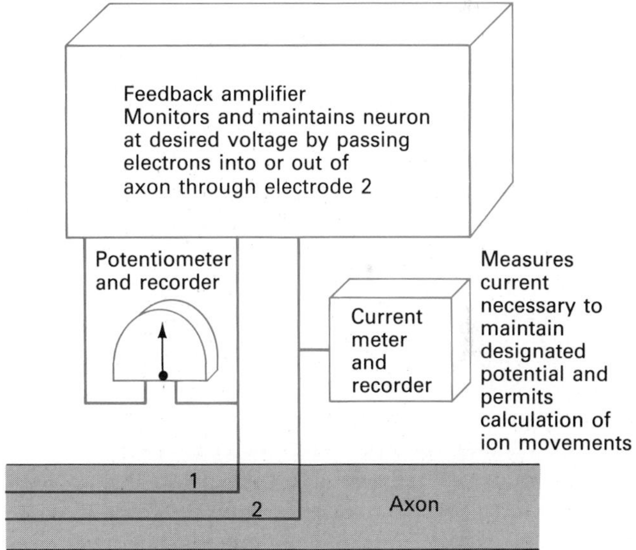

Figure 17-19 Principle of the voltage clamp technique. Two electrodes, 1 and 2, are inserted into a large axon. The electrodes are thin wires that extend the length of the cell in order to ensure that the membrane voltage is the same at all regions of the membrane. Electrode 1 is used to monitor the potential across the plasma membrane. Electrode 2 is used to pass electric current (electrons) into or out of the cell. To depolarize or hyperpolarize the cell to a predetermined value, electrons are withdrawn or fed into the cell respectively, via electrode 2. The feedback amplifier adjusts the current flow into or out of the cell to maintain the membrane potential at this prearranged value. The electrons passed into or out of the cell by electrode 2 exactly neutralize positive ions, Na^+ or K^+, that pass out of or into the cell across the plasma membrane. Thus, the flow of current through elecrode 2 is a measure of ion movements into or out of the cell.

further depolarization can occur. The inward or outward movement of ions can then be quantified from the amount of electric current needed to maintain the membrane potential at the designated value (Figure 17-20). To preserve electroneutrality the entry of each positive ion into the cell across the plasma membrane is balanced by the entry of an electron into the cytoplasm from the second electrode.

With voltage clamping controlling the membrane potential, a one-step depolarization from the resting potential of −65 mV to −9 mV (clamped at −9 mV) leads to a transient inward movement of positive ions followed by a slower outward movement of positive ions (Figure 17-20b). The positive ions moving inward are Na^+, for if the extracellular fluid is made Na^+-free by substitution with the impermeable cation choline $HOCH_2CH_2N^+[CH_3]_3$ there is no *inward* movement of ions. The outward movement of K^+ ions is unaffected. Thus the initial inward movement of Na^+ ions is indeed triggered by depolarization. The delayed outward movement of K^+ ions is likewise triggered by depolarization—and not by the initial inward movement of Na^+ ions that normally occurs.

The compound *tetraethylammonium chloride*, or *TEA* $[(CH_3CH_2)_4N^+Cl^-]$, selectively blocks the voltage-dependent K^+ channel, possibly by binding to the K^+ channel protein.* As is illustrated in Figure 17-20, it blocks the slow outward movement of K^+ ions from the cell. Using TEA as an inhibitor of K^+ movement, one can then follow the movements of Na^+ ions directly; that is, all changes in current flow will be due only to the movement of Na^+. In such a study a neuronal membrane is depolarized to a certain value in the presence of TEA and clamped at that value. The transient inward movement of Na^+ ions is then followed. The greater the initial depolarization, the greater the inward movement of Na^+ ions, owing to the opening of increased numbers of Na^+ channels. One may then conclude that the transient opening of the Na^+ channels is indeed dependent on the membrane potential.

Patch Clamps Permit Measurement of Ionic Movements through Single Sodium Channels

The studies discussed above measure the movement of ions across a large region of the nerve membrane. This movement must represent the combined action of thousands of individual Na^+ and K^+ channels. The study of the Na^+ channels themselves, and other ion channels, has been advanced by the *patch clamping*, or *single channel recording*, technique for measuring the movements of ions through small patches of plasma membrane. Patch clamping involves the use of a glass micropipette, of about 0.5-μm diameter, into which a small region of the plasma membrane is sucked (Figure 17-21). For ease of handling these studies are usually done on muscle cells, on cultured nerve tumor (neuroblastoma) cells, or on large neurons whose surfaces can be cleaned of debris from other cells. The patch membrane is depolarized about 10 mV and clamped at that voltage, and the flow of ionic current across the patch of membrane is monitored. Under these circumstances transient pulses of current cross the membrane (Figure 17-22); they result from the opening and closing of individual Na^+ channels. Each channel is either open or closed. There are no graded permeability changes for individual channels. From the recording depicted in Figure 17-22 it is determined that each channel is open for an average of 0.7 ms and the average current through each channel is 1.6 picoamperes (1.6×10^{-12} amperes; 1 ampere = 1 coulomb of charge per second). This is equivalent to the movement of about 9900 Na^+ ions per channel per millisecond:

$$\frac{(1.6 \times 10^{-12}\ \text{C/s})(10^{-3}\ \text{s})(6 \times 10^{23}\ \text{molecules/mol})}{96{,}500\ \text{C/mol}}$$

Assuming that all the Na^+ channels in these cells are identical and function independently, one expects the sum of many recordings of individual channels after depolarization to show the same properties as the Na^+ current measured in a conventional voltage clamp from thousands of channels. This was observed experimentally (Figure 17-22c).

Another conclusion emerged from the patch-clamping studies: the density of these channels is very low. Depending on the type of unmyelinated axon, there are only 5 to 500 Na^+ channels per square micrometer: roughly, one plasma membrane protein in a million is a Na^+ channel. How can we characterize and purify such rare molecules? Here neuroscientists have turned to toxins from exotic animals and plants that bind specifically to the Na^+ channel.

The Sodium Channel Protein Has Been Quantified, Purified, and Characterized

Many plants and animals contain deadly neurotoxins. Some work by blocking conduction of a nerve impulse, others by blocking synaptic transmission. Neurotoxins that bind tightly and specifically to the Na^+ channel are extremely useful in elucidating its properties. *Tetrodotoxin* (Figure 17-23) is a powerful poison concentrated in the ovaries, liver, skin, and intestines of the puffer fish, known in Japan as fugu. Fugu is a delicacy in Japan, and chefs who prepare it must be properly certified as knowledgeable of the procedures for discarding the toxic parts

*TEA does not affect the resting potential of the cell. This is evidence that the voltage-dependent K^+ channel is different from the resting K^+ channel.

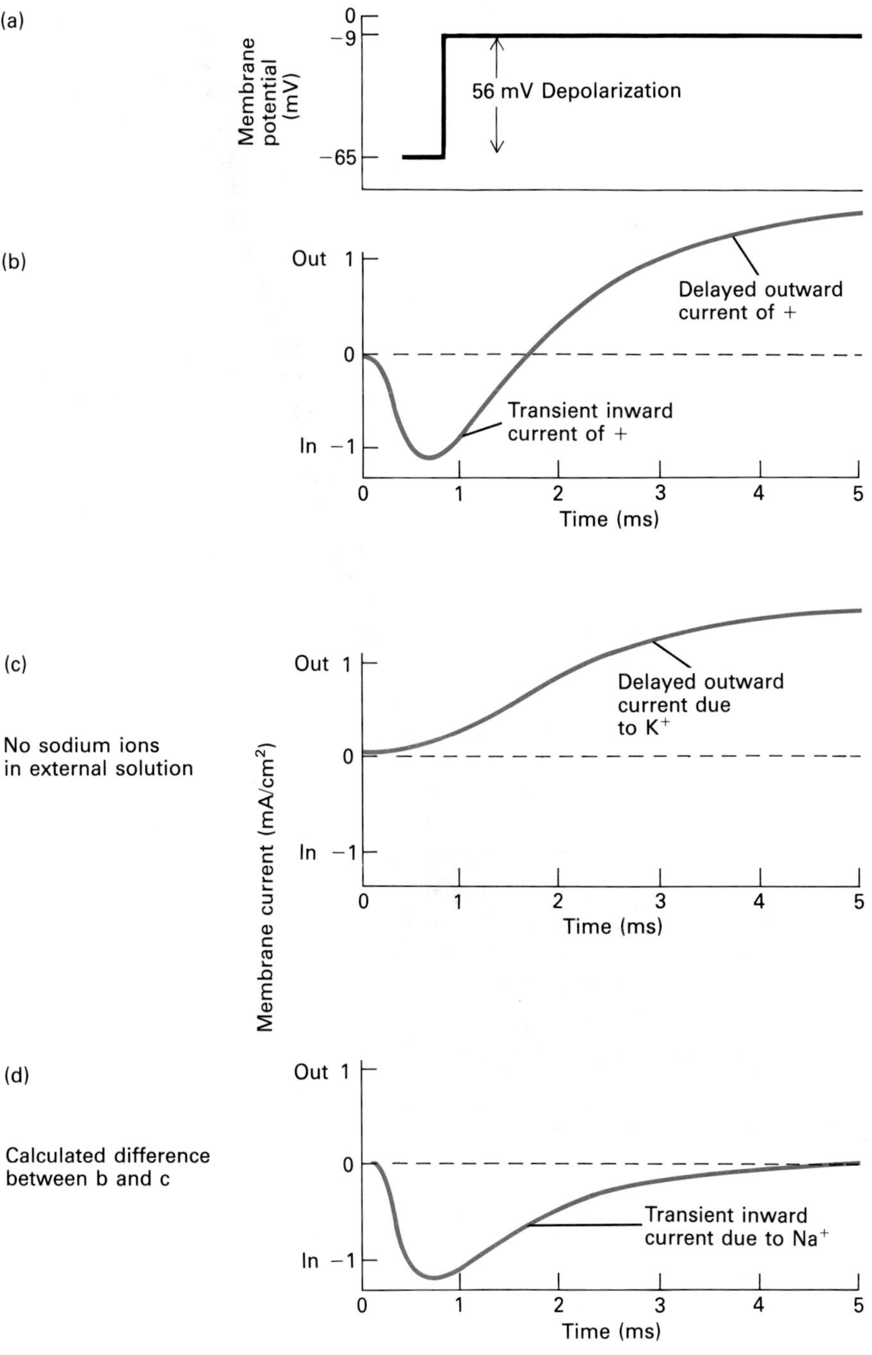

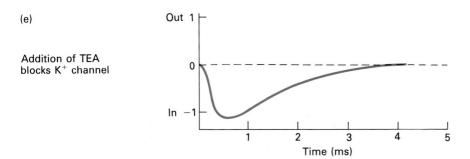

Figure 17-20 Current and ion movements across the membrane of the squid giant axon. (a) The membrane potential is suddenly changed from the resting potential of −65 mV to −9V and clamped at that value. (b) This depolarization induces a transient inward movement of positive charges, followed by a delayed outward movement of positive charges. (c) If the experiment is repeated with the axon bathed in a solution that contains no Na^+ ions, there is no transient inward movement of positive ions but the delayed outward movement of positive charges does occur normally. This result establishes that the transient inward movement of positive charges is due to Na^+ ions. (d) By calculating the difference in the current due to the ionic movements shown in (b) and (c), the ionic movements due to sodium ions can be deduced; upon depolarization there is a transient increase in membrane permeability to Na^+ ions. (e) That the late outward movement of current is due to outward movement of K^+ ions is confirmed by specifically blocking the K^+ channel with tetraethylammonium chloride (TEA).

When TEA is added to the extracellular fluid, it does not alter the inward movement of current but abolishes the later outward movement of current. The delayed outward movement of K^+ ions continues as long as the membrane is depolarized, indicating that the K^+ channels remain open as long as the membrane is depolarized. In contrast, depolarization induces only a transient opening of the voltage-dependent Na^+ channels. After it has opened and closed, the Na^+ channel becomes refractory. [See A. L. Hodgkin, and A. F. Huxley, 1952, *J. Physiol.* **117**:500.]

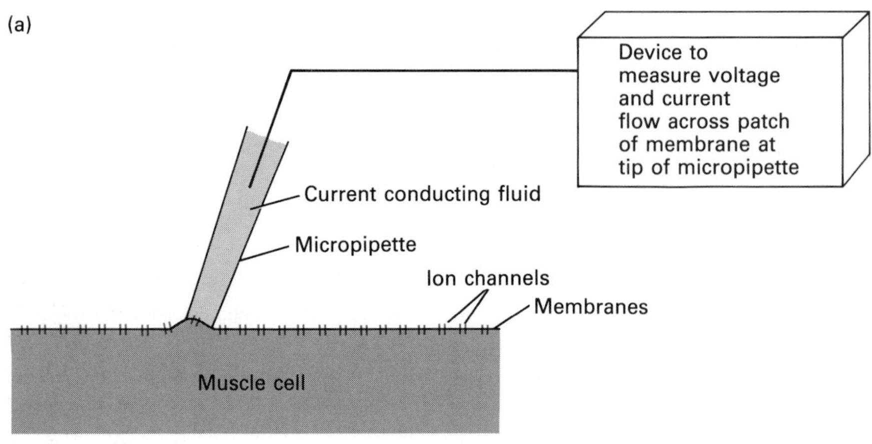

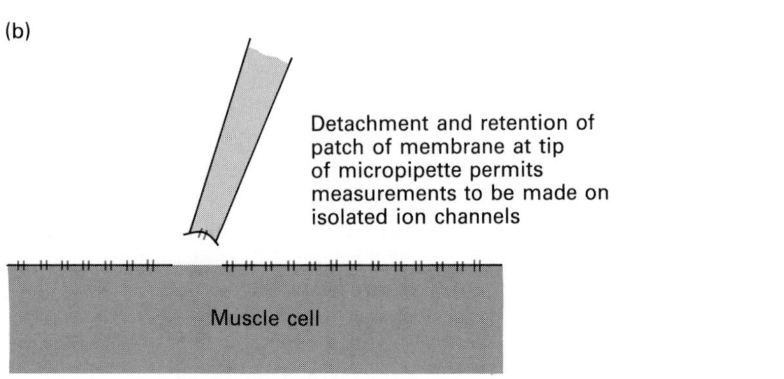

Figure 17-21 Schematic outline of the patch clamp technique. A micropipette with an opening of about 0.5-μm diameter is filled with a current-conducting saline solution. An electrode in the pipette is connected to an amplifier, which measures current flow across the membrane. (a) Recording of current flow through these channels can be made with the patch in place. When gentle suction is applied, one can remove the patch from the rest of the cell (b) and study the properties of the individual channels. One can also clamp the potential across the isolated patch to any value and observe the flow of ionic current through the single ion channel. The effects of the composition of the solution on either side of the membrane can be examined.

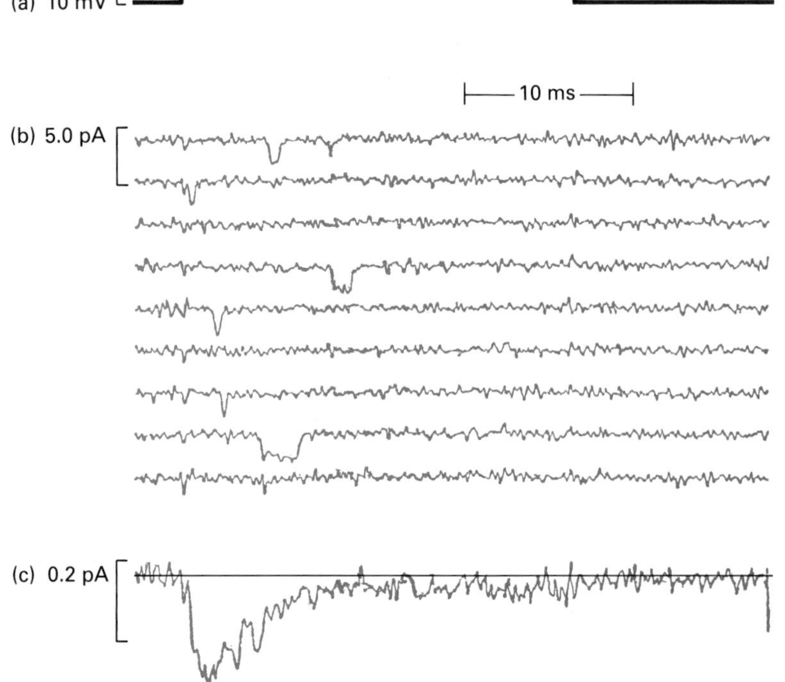

Figure 17-22 Current flux across single sodium channels. (a) The membrane of a muscle cell was depolarized by 10 mV and clamped at that value. (b) With the patch clamp technique (Figure 17-21), current flux was monitored across many small regions of the membrane. The transient pulses of current recorded as large downward deviations are due to the opening of Na^+ channels. The mean current was 1.6×10^{-12} amperes (1.6 picoamperes, pA) per open channel, and the average lifetime of a channel was 0.7 ms. The smaller deviations in current represent "background noise" in the electronic amplifier. (c) The average of 300 individual current records was computed. This record is similar to the sodium flux exhibited in the experiment depicted in Figure 17-20. [See F. J. Sigworth and E. Neher, 1980, *Nature* 287:447.]

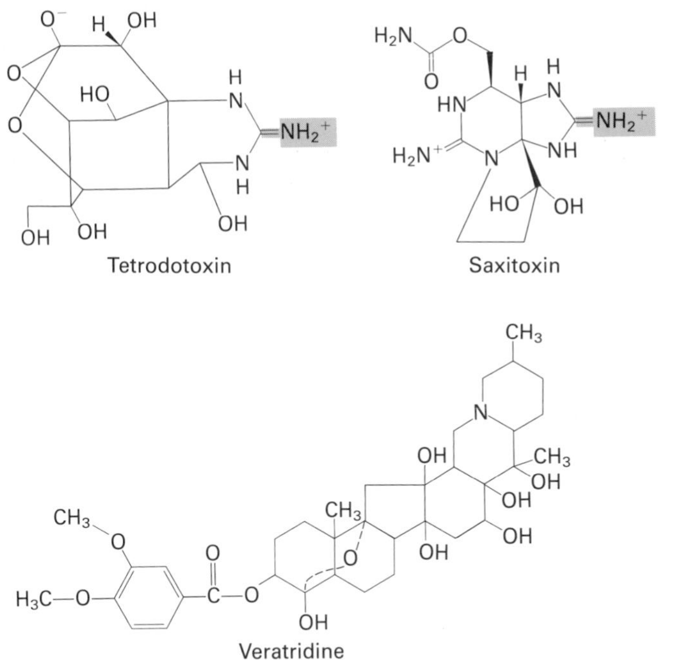

Tetrodotoxin

Saxitoxin

Veratridine

Figure 17-23 Structures of three poisons of the sodium channel. Tetrodotoxin and saxitoxin bind to the Na^+ channel in such a way as to block passage of any Na^+ ions. Veratridine binds to a different site on the channel molecule, opening the channel and keeping it in an open state. The positively charged groups on tetrodotoxin and saxitoxin (colored) may bind to negatively charged carboxylate (COO^-) groups in the Na^+ channel protein. The size of these toxins prevents their passage through the channel and leads to blockage of Na^+ transport.

of the fish. A toxin of related properties, *saxitoxin,* is produced by certain red marine dinoflagellates. When such organisms are abundant in the ocean, they create a "red tide." Filter feeding shellfish in red-tide waters concentrate the dinoflagellate toxin in their tissues and, if eaten, are highly toxic to humans. These toxins specifically bind to and block the voltage-dependent Na^+ channels in neurons, preventing all Na^+ ion movement through them. This prevents action potentials from forming, and impulse conduction is therefore blocked.

The alkaloid toxin *veratridine,* isolated from seeds of the sabadilla (*Schoenocaulon officinalis*), a Mexican plant of the lily family, binds to a different site on the Na^+ channel and causes the channel to open and remain open in the permanently activated state. When added to nerve or muscle cells, it causes a permanent depolarization of the membranes, with the membrane potential approaching that of E_{Na}.

Tetrodotoxin and saxitoxin have been particularly useful in establishing the number of Na^+ channels in a membrane. One molecule of either toxin binds to each Na^+ channel with exquisite affinity and selectivity. Measuring the amount of radioactive tetrodotoxin or saxitoxin that binds to a typical unmyelinated neuron shows that the axon contains only 5 to 500 such channels per square micrometer of area. This agrees with estimates of the numbers of channels obtained from patch clamp studies. The Na^+ channels in these membranes are thus spaced, on average, about 200 nm apart.

The ability of saxitoxin and other neurotoxins to bind Na^+ channels from detergent-solubilized membranes has made it possible to purify the Na^+ channel protein by affinity chromatography (see Figure 16-6) of detergent-extracted membrane proteins on columns of immobilized toxin. Rat brain and the electroplax of the electric eel are rich and convenient sources of this protein. The major component of the Na^+ channels from these sources is a single polypeptide of molecular weight 250,000 to 270,000. Often two smaller polypeptides copurify with this protein; they may be components of the channel itself or regulatory molecules. Purified Na^+ channel protein can be incorporated into artificial lipid membranes. The reconstituted membrane protein exhibits many of the predicted properties of the voltage-gated Na^+ channel, such as voltage dependence (Figure 17-24). In addition, the Na^+ permeability of the phospholipid bilayer is increased by veratridine, the toxin that opens Na^+ channels, and is blocked by tetrodotoxin or saxitoxin.

A major challenge in cellular neurobiology is to determine the three-dimensional structure of the voltage-dependent Na^+ and K^+ channel proteins and learn precisely how a change in voltage causes the channels to open and close. From the cloned cDNA the sequence of the 1820 amino acids of the major polypeptide of the sodium channel has been determined. It contains a fourfold internal homology: There are four homologous membrane-spanning domains of approximately 300 amino acids each, connected and flanked by shorter stretches of nonhomologous residues. The determination of the conformation of this protein in the phospholipid bilayer will be an important step in understanding how it functions.

Some Behavioral Abnormalities in *Drosophila melanogaster* Are Due to Defects in Ion-Channel Proteins

Many mutations of the fruit fly *Drosophila melanogaster* affect specific physiological functions by causing abnormalities in the nervous system. The most definitively studied *Drosophila* mutation, the X-linked *shaker,* causes a defect in a K^+ ion channel. Shaker mutants shake vigorously under ether anesthesia. This behavioral phenotype was originally a curiosity observed in laboratories when

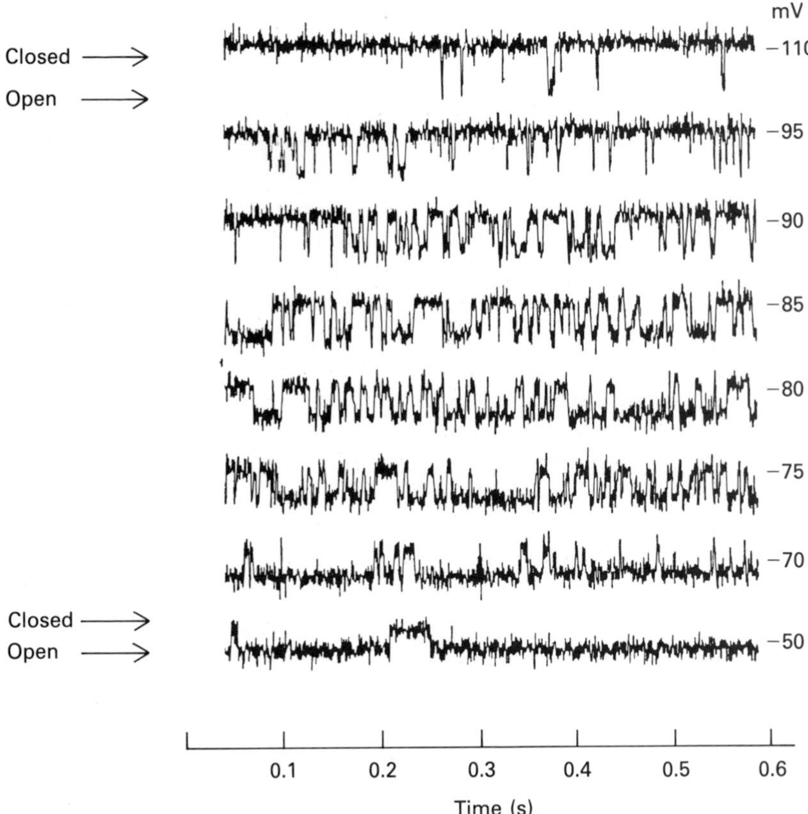

Figure 17-24 Functional reconstitution of the purified brain sodium channel into planar lipid bilayers. Purified saxitoxin binding protein, containing three polypeptides (m.w. 260,000, 39,000, and 37,000) was incorporated into an artificial phospholipid bilayer that fills an approximately 1-mm-diameter pore separating two aqueous solutions. The voltage across the membrane is clamped to a preset value, noted at the right, and the current flow across the membrane is monitored. All of the channels are closed when the voltage is clamped at −110 mV; the channels are open about half the time when the voltage is clamped at −85 mV and are open constantly when the voltage is clamped at −50 mV. This study was done in the presence of the toxin batrachotoxin. Batrachotoxin prevents closing and inactivation of the channel and shifts the voltage-dependence of the channel so that it opens at potentials that are lower than usual. Thus the results do not reflect the quantitative behavior of normal channels, but they do show that the purified channel exhibits many of the key properties of voltage-dependence, as do the native channels in nerve membranes. [See R. P. Hartshorne et al., 1985, *Proc. Nat'l Acad. Sci. USA* **82**:240.]

ether was used to anesthetize fruit flies for counting. The shaking behavior reflects a loss of motor control and a defect in excitable cells. Indeed, recordings from the axons of giant nerves in *Drosophila* showed an abnormally prolonged action potential (Figure 17-25). Further studies showed that this prolonged action potential resulted from a faulty repolarization of the membrane caused by a defect in a particular K^+ channel, the channel that normally opens immediately upon depolarization. Presumably the shaker gene encodes a defective K^+ channel protein or a defective protein necessary for the operation of the channel. Electrophysiological studies indicate that various nerve cells contain several types of K^+ channels. Genetic studies on shaker and similar mutants can establish precisely which channels function in which particular neurons, and what roles the channels play in conducting and modulating an action potential.

Impulse Transmission

We know that axons conduct nerve impulses over long distances without attenuation. The propagation of an impulse along an axon depends on the ability of the axonal membrane to amplify relatively small depolarizations and generate all-or-nothing action potentials. To understand in detail how this happens let us investigate some of the passive electrical properties of neurons.

Voltage Changes Spread Passively Along a Nerve Membrane

In its electrical properties a nerve cell resembles a long underwater telephone cable. It consists of a poorly conducting outer barrier, the cell membrane, separating two media that have a high conductivity for ions: the cell cytosol and the extracellular fluid. Suppose that the voltage-dependent Na^+ and K^+ channels in the nerve membrane are inactivated and that a single point along the nerve membrane is suddenly depolarized. The inside of the membrane at this site will have a relative excess of positive ions, principally K^+. These ions will tend to move away from the site (Figure 17-26) and thus depolarize adjacent sections of the membrane. This is called the *passive spread* of depolarization. The magnitude of depolarization, however, will diminish with distance from the

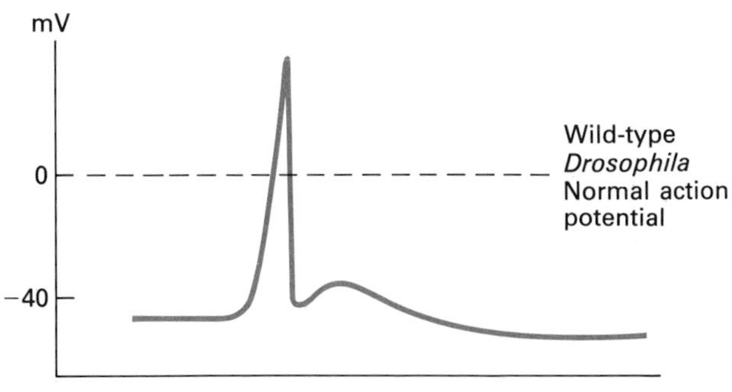

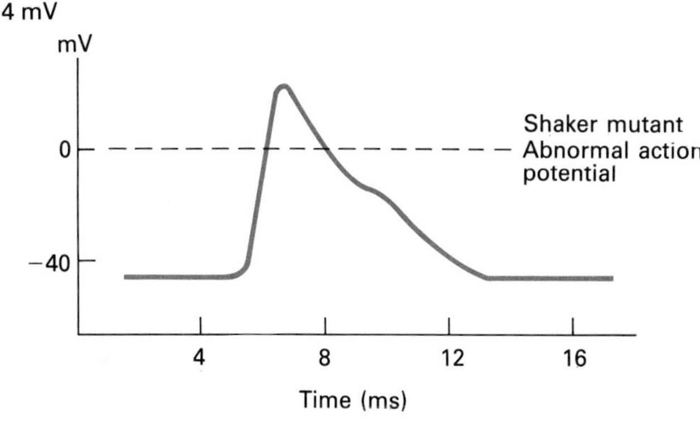

Figure 17-25 The duration of the action potential is abnormally long in axons of the shaker mutant of *Drosophila*. This is due to a defect in a K^+ channel that normally repolarizes the membrane. As a consequence of this defect, K^+ enters more slowly, increasing the time required for repolarization. [See L. A. Salkoff and R. Wyman, 1983, *Trends Neurosci.* **6**:128.]

site of initial depolarization; the depolarization becomes attenuated as some of the excess cations leak back across the membrane, which is not a perfect insulator. Only a small portion of the excess cations are carried longitudinally along the axon for long distances. The extent of this passive spread of depolarization is a function of two properties of the nerve cells, the *permeability* (or conductivity) *of the membrane* to ions, and the *conductivity of the cytosol.*

The conductivity of the cytosol is proportional to the concentration of ions dissolved in it, principally K^+ and Cl^-. In any particular organism the composition of the cytosol is believed to be very similar in all the neurons. The conductivity of the cytosol of a nerve cell will be proportional to its cross-sectional area: the larger the area, the greater the number of ions there will be (per unit length of neuron) to conduct current. Thus the passive spread of a depolarization is greater for neurons of large

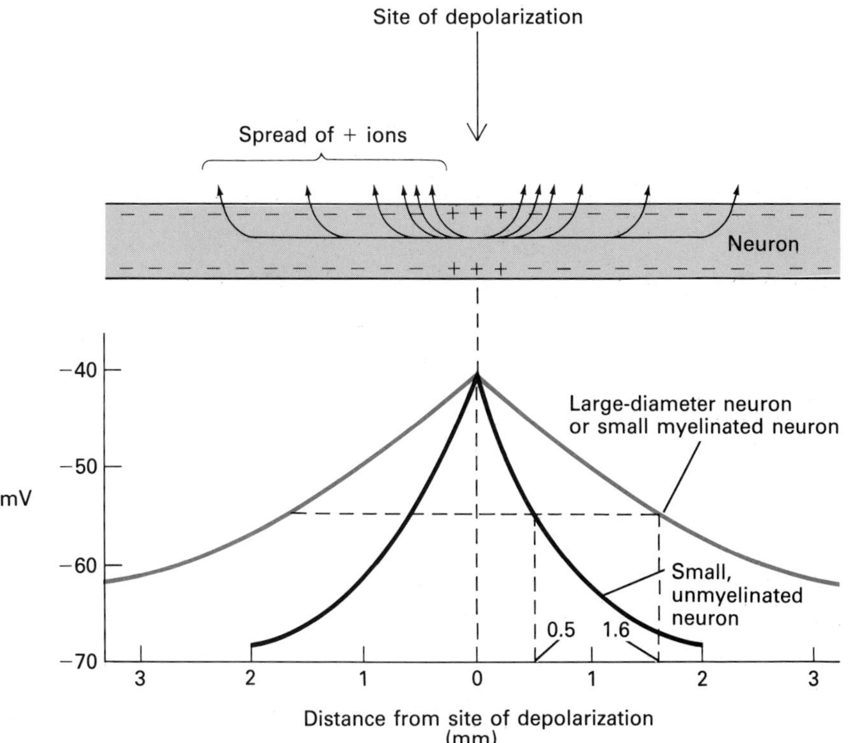

Figure 17-26 The cable properties of nerve cell membranes. In this example the voltage-dependent Na^+ and K^+ channels are inactivated and the nerve membrane is depolarized from −70 to −40 mV at a single point and clamped at this value. The voltage is then measured at various distances from this site. The extent of depolarization falls off with distance from the initial depolarization. The length constant is the distance over which the magnitude of the depolarization falls off to a value of $1/e$ (e = 2.718) of the initial depolarization. The length constant for a small neuron with a relatively leaky membrane *(black curve)* can be as small as 0.1 mm; in this example it is about 0.5 mm. For a large axon, or one with a nonleaky myelinated membrane *(colored curve),* the length constant can be as large as 5 mm. For the large neuron depicted here it is about 1.6 mm.

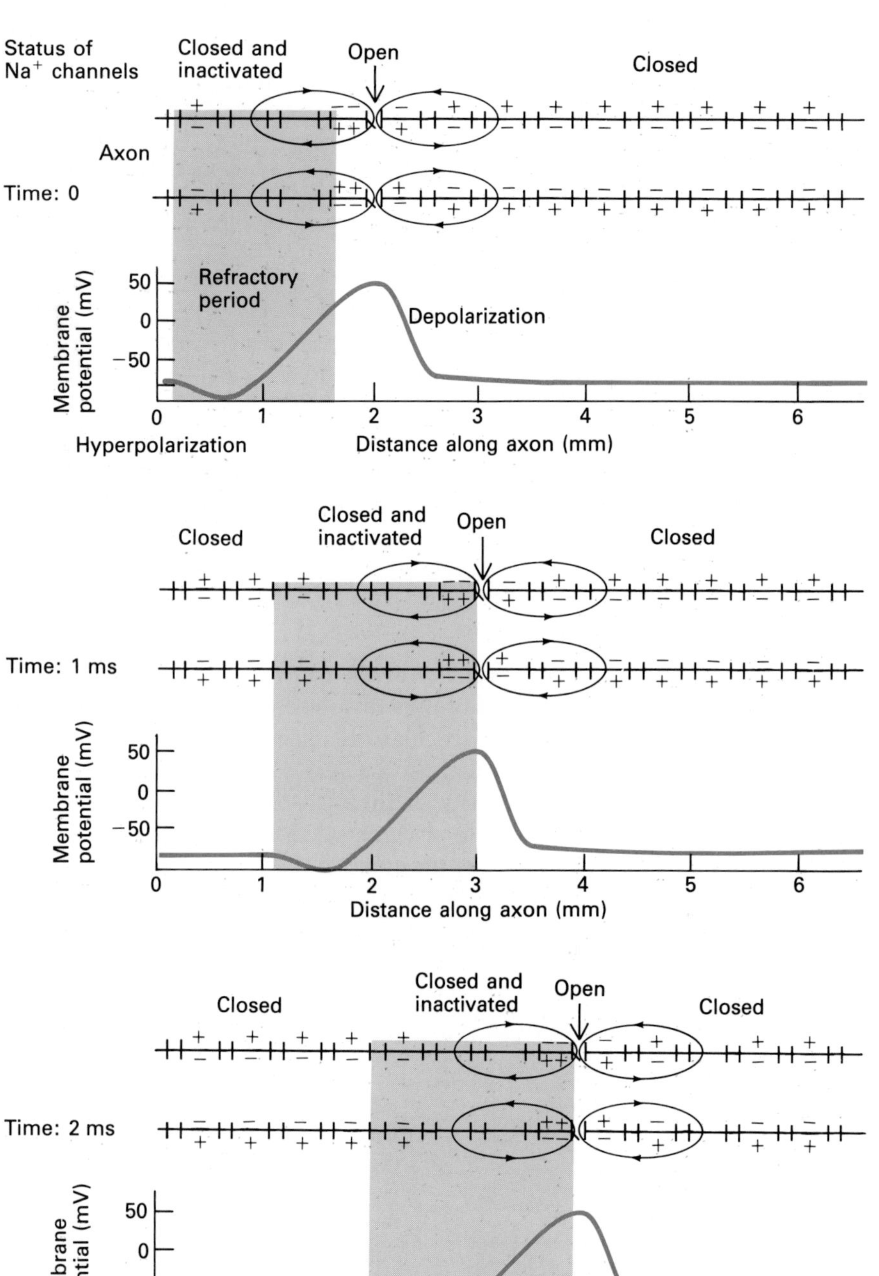

Figure 17-27 Voltage-dependent Na^+ channels allow conduction of an action potential. Multiple electrodes are inserted along the squid's giant axon to record the instantaneous membrane potentials at different positions along the axon during conduction of an action potential. At time 0, an action potential is occurring at the 2-mm position along the axon. The depolarization spreads passively down the axon, triggering action potentials downstream—at 1 ms an action potential is occurring at the 3-mm position. After the action potential has occurred, the membrane at that site becomes transiently hyperpolarized owing to an increase in K^+ permeability (see the 2-mm position at 1 ms) and then returns to the resting potential (the 2-mm position at 2 ms). Each region of the membrane becomes refractory (gray tint) for a few milliseconds after the action potential has passed; thus the action potential can be propagated only in one direction.

diameter, for the K^+ ions are able to move, on the average, farther along the axon before they leak back across the membrane. As a consequence large-diameter neurons conduct a depolarization faster and farther than thin ones. Such passive properties of a nerve membrane allow a membrane depolarization to spread only a short distance, 0.1 to about 5 mm. Depolarizations in dendrites and the cell body generally spread in this manner, though some dendrites can conduct an action potential. Similarly, neurons with very short axons conduct membrane depolarizations by passive spread.

Voltage-Dependent Sodium Channels Allow an Active Potential to Be Propagated over Long Distances

The presence of voltage-dependent Na^+ channels allows the action potential to be propagated long distances without loss of intensity. An action potential is initiated when the membrane in the axon hillock region is depolarized beyond the threshold value. As this region of membrane undergoes an action potential, its transient depolarization spreads passively to the adjacent resting region of the membrane (Figures 17-15 and 17-27). The depolarization of this region reaches threshold and generates an all-or-nothing action potential, which in turn depolarizes the next segment and opens its voltage-dependent Na^+ channels. Recall that after they have opened, the Na^+ channels become inactivated for a few milliseconds. This means that the action potential can only be propagated unidirectionally, down the axon. Since the Na^+ channels just behind the action potential are momentarily inactive, they cannot reopen in response to the depolarization and so cannot spread the action potential upstream.

The velocity of conduction varies among neurons. In the giant axon of the squid (0.6-mm diameter) it is 12 m/s. Thinner axons generally conduct at a slower rate. The conduction velocities for various vertebrate neurons vary from 1 to 100 m/s. Myelinated axons conduct action potentials faster than nonmyelinated ones of similar diameter.

Myelination Increases the Rate of Impulse Conduction

The axons of many vertebrate nerve cells are covered with a myelin sheath. We saw in Chapter 14 that myelin is a stack of specialized plasma membrane sheets produced by a glial cell that wraps itself around the axon (Figures 17-28 and 17-29; see also Figure 14-4). In the peripheral nervous system these glial cells are called *Schwann* cells; in the central nervous system they are called *oligodendrocytes*. Often several axons are surrounded by the cytoplasm of a single glial cell (Figure 17-29a). Since many nerves contain both myelinated and unmyelinated neurons, a major problem is to explain why some axons become myelinated and others do not.

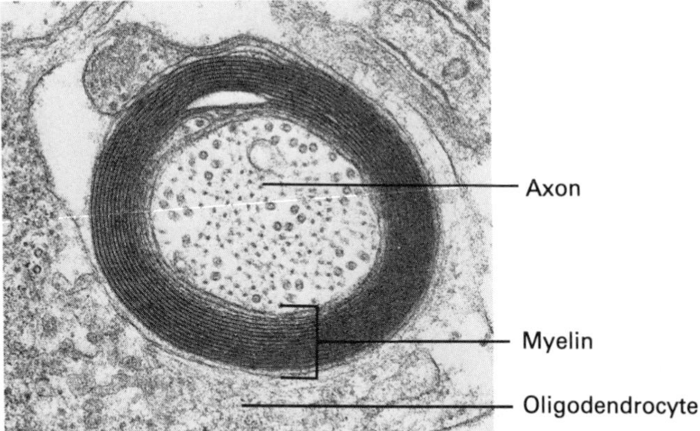

Figure 17-28 A myelinated axon in the spinal cord of an adult dog. The axon is surrounded by the oligodendrocyte that produced the axon myelin. *From C. Raine, 1981, in* Basic Neurochemistry, *3d ed., G. J. Siegal et al., eds., Little, Brown, p. 39. Copyright 1981, Little, Brown and Company.*

Myelin, like all membranes, is composed mostly of phospholipids, but it is specialized because it contains only a few types of proteins. *Myelin basic protein* and a myelin proteolipid found only in myelin in the central nervous system allow the plasma membranes to stack tightly together (Figure 17-29b). Myelin in the peripheral nervous system is constructed of other unique membrane proteins. The myelin surrounding each myelinated axon is formed from many oligodendrocytes. Each region of myelin formed by an individual oligodendrocyte is separated from the next region by an unmyelinated area called the *node of Ranvier* (or simply, *node*). It is the only site where the axonal membrane is in direct contact with the extracellular fluid (Figure 17-30).

The myelin sheath, which can be as many as 50 or 100 membranes thick, acts as an electrical insulator of the axon. It prevents the transfer of ions between the axonal cytoplasm and the extracellular fluids.

All electrical activity is confined to the nodes of Ranvier, the only sites where ions can flow across the axon membrane. Tetrodotoxin-binding studies have shown that the node regions contain a high density of voltage-dependent Na^+ channels, about 10,000 per μm^2, whereas the regions of axonal membrane between the nodes have few if any channels. An action potential at one node results in the initiation of an action potential at the next node—or even at several nodes at once (Figure 17-31). This happens because the membrane depolarization associated with an action potential spreads passively through the axonal cytoplasm to the next node with very little loss or attenuation; ions are capable of moving across the axonal membrane only at the myelin-free nodes. If the action potential were not regenerated at each node, the initial depolarization induced by an action potential at one node would decrease with distance and the signal would be lost.

The conduction velocity of myelinated nerves is much greater than that of unmyelinated nerves of the same diameter. The greater longitudinal spread of current, due to the myelin insulation, increases the velocity of conduction. A 12-μm-diameter myelinated axon conducts at 12 m/s, the same velocity as 600-μm-diameter unmyelinated squid axon. Not surprisingly, myelinated nerves are used for signaling in circuits where speed is important.

One of the leading causes of serious neurological disease among adults is *multiple sclerosis* (MS). Its characteristic feature is loss of myelin in areas of the brain and spinal cord, and it is the prototype of a *demyelinating disease*. In MS, conduction of action potentials by demyelinated neurons is slowed and the Na^+ channels spread outward from the nodes. The composition of the myelin itself appears to be normal in MS patients. The cause of the disease is not known but appears to involve either the production by the body of antibodies that react with myelin basic protein or the secretion of proteases that destroy myelin proteins.

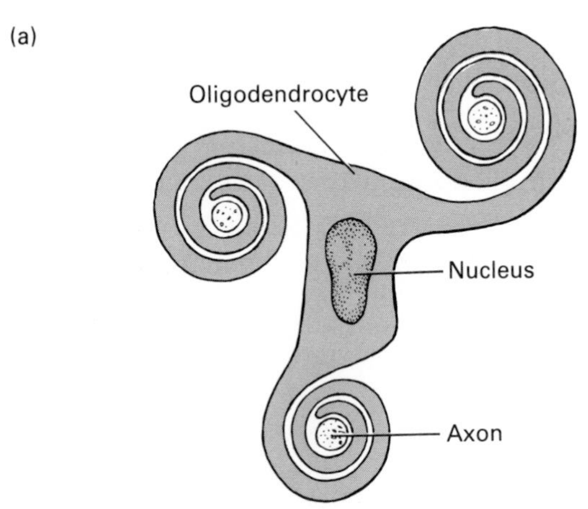

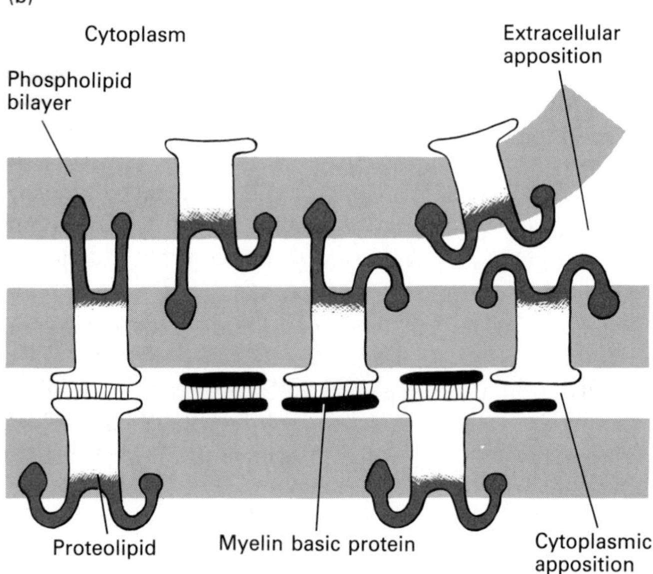

Figure 17-29 The structure and formation of myelin. (a) By wrapping itself around several axons simultaneously, a single oligodendrocyte can form a myelin sheath around multiple axons. As the oligodendrocyte continues to wrap around the axon, all the spaces between the plasma membranes, both cytoplasmic and extracytoplasmic, are reduced. The myelin sheath is characterized by a relatively high proportion of lipid (70–80 percent), but this compaction of the plasma membranes is generated by proteins unique to myelin that are synthesized only in the myelinating cells. (b) In the central nervous system the two predominant proteins are the polar *myelin basic protein,* found attached to the cytoplasmic surface of the myelin membranes, and a hydrophobic integral membrane *proteolipid* that spans the phospholipid bilayer. During maturation of myelin, there is close apposition of the cytoplasmic and extracytoplasmic faces of the membrane. As determined from sequence and structural studies, the close apposition of the cytoplasmic faces of the membrane can be due to interactions between proteolipid and proteolipid, basic protein and proteolipid, or basic protein and basic protein. Apposition of extracellular faces could be due to movement of a hydrophobic domain *(colored)* of the proteolipid into the opposite bilayer. [See R. A. Laursen, M. Samiullah, and M. B. Lees, 1984, *Proc. Nat'l Acad. Sci.* **81**:2912.]

Synapses and Impulse Transmission

Chemical Synapses Use Neurotransmitters That Affect Ion Channels

Synapses are the junctions where neurons pass signals to other neurons, to muscle cells, or to gland cells. Most neuron-neuron signaling and all known neuron-muscle and neuron-gland signaling involve the release at the synapse of chemical neurotransmitters from membrane vesicles called *synaptic vesicles.* The arrival of an action potential at the neuron terminal triggers exocytosis of the synaptic vesicles and release of transmitter. Each synaptic vesicle is believed to contain a single type of neurotransmitter, but a nerve terminal can contain two or more types of vesicles, each with a different transmitter. To effect a signal, the transmitter binds to a specific receptor on the plasma membrane of the postsynaptic cell and causes a change in its permeability to certain ions. At an *excitatory synapse,* this change in permeability leads to a depolarization of the membrane of the postsynaptic cell—a change that promotes the generation of an action potential. At an *inhibitory synapse,* binding of the neurotransmitter results in a change in ion permeability that tends to block the generation of an action potential.*

Some neurotransmitter receptors are ligand-dependent ion channels. A channel for a certain ion is part of the receptor protein, and binding of the neurotransmitter causes a conformational change in the protein that opens the channel and allows the ion to cross the membrane. More often, however, the relationship between ligand binding and the opening or closing of an ion channel is indirect. Binding of a neurotransmitter to its receptor on the postsynaptic cell triggers elevation of a chemical signal that in turn modifies the state of a separate ion channel protein. In certain receptors, as in some of those for

*In most cases the binding of an inhibitory neurotransmitter causes a *hyperpolarization* of the postsynaptic membrane. Often, though, its binding causes the permeabilities to one or several ions to be increased so that the membrane is actually *depolarized* a few millivolts. But in this case the additional depolarization to threshold for an action potential is made more difficult, since many more Na^+ channels have to open to counteract the increased ionic permeability.

(a)

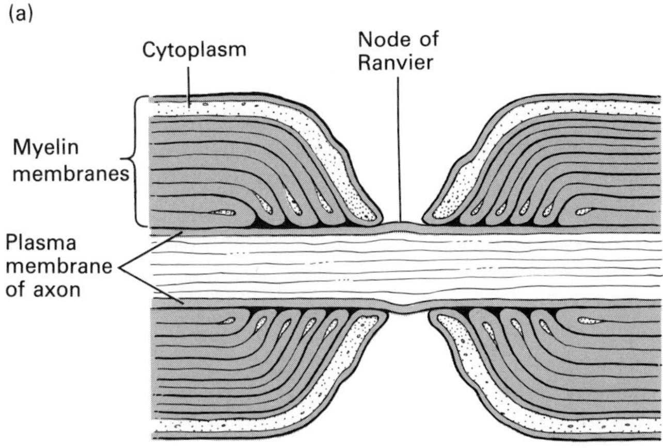

(b)

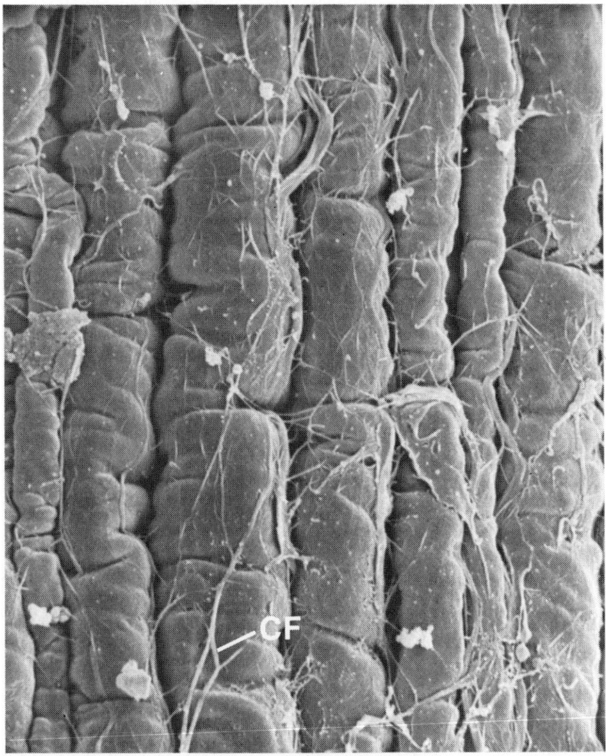

Figure 17-30 (a) The node of Ranvier is the gap that separates the portions of myelin sheath formed by two adjacent oligodendrocytes. These nodes are the only regions along the axon where the axon membrane is in direct contact with the extracellular fluid, and they are the sites where axon potentials are propagated. (b) A scanning electron micrograph of a peripheral nerve. The deep folds along the myelinated neurons are the nodes of Ranvier. Numerous strands of collagen surround individual axons and bind them together. *Part (b) from R. G. Kessel and R. H. Kardon, 1979,* Tissues and Organs: A Text-Atlas of Scanning Electron Microscopy, *W. H. Freeman and Company, p. 80.*

serotonin and norepinephrine, the chemical signal is believed to be cAMP; its action on ion channels would be mediated by the cAMP-dependent protein kinase. Many neurons, especially interneurons in the central nervous system, can receive signals at multiple excitatory and inhibitory synapses. Some of the resultant local changes in ion permeability are short-lived—a millisecond or less—while others may last several seconds.

Electrical Synapses Allow Rapid Cell-to-Cell Transmission

In an electrical synapse ions move directly from one neuron to another by means of gap junctions (Figure 17-32). The depolarization of an action potential in the presynaptic cell causes a movement of positive ions to the postsynaptic cell through gap junctions, leading to a depolarization, and thus an action potential, in the postsynaptic cell. Such cells are said to be *electrically coupled.* All electrical synapses are excitatory, as one might expect from the properties of gap junctions.

Electrical synapses have the advantage of speed; the direct transmission of impulses avoids the delay of about 0.5 ms that is characteristic of the secretory process involved in chemical signaling (Figure 17-33). In certain circumstances such as the escape of a fish from a predator, a fraction of a millisecond can mean the difference between life and death. Electrical synapses in the goldfish brain mediate a reflex action involved in strong flapping of the tail. There are also examples of electrical coupling between groups of cell bodies and dendrites, ensuring simultaneous depolarization of an entire group of coupled cells. The large number of electrical synapses in many cold-blooded fishes suggests that they may be an adaptation to low temperatures, as the lower rate of cellular metabolism in the cold reduces the rate of chemical signaling.

The efficiency with which an electrical signal is transmitted across an electrical synapse is proportional to the number of gap junctions that connect the cells. Also, as we noted in Chapter 14, the permeability of the gap junction is regulated by the level of cellular H^+ ions and possibly of Ca^{2+} ions; changes in the concentration of these ions might modulate the efficiency of transmission at electrical synapses.

A Large Number of Chemicals Are Used as Neurotransmitters

The list of substances known to function as neurotransmitters is now long (Figure 17-34). Except for acetylcholine they are either amino acids or derivatives of amino acids. Many common cellular metabolites, such as adenine and even ATP, are suspected of functioning as neurotransmitters in certain cells. Many small peptides

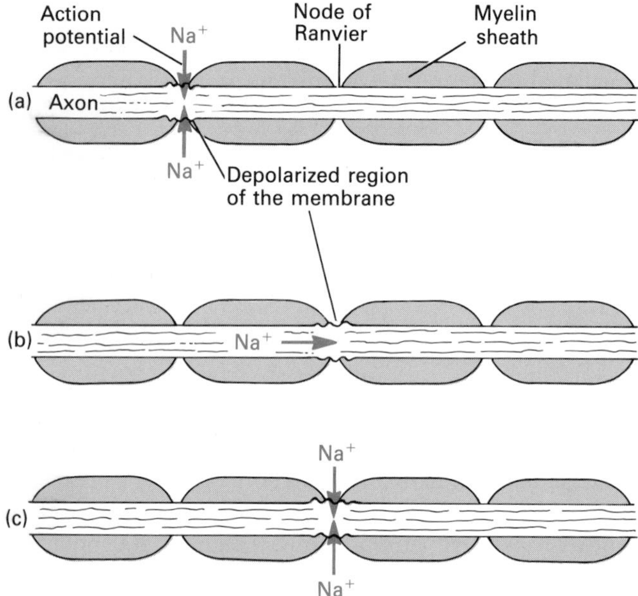

Figure 17-31 Action potentials are regenerated at the nodes of Ranvier. (a) The influx of Na^+ ions associated with the action potential at one node results in depolarization of that region of the axonal membrane. (b) The depolarization spreads rapidly down the axon to the next node, since the excess positive ions cannot diffuse across the parts of the axon membrane that are myelinated. (c) This depolarization induces an action potential at that node, thus the action potential is propagated node-to-node along the axon.

may also function as neurotransmitters. But in only a few cases do we know how the release of the chemical from the presynaptic cells is induced and how it interacts with the postsynaptic cell to cause an alteration in the membrane potential. The best characterized of these neurotransmitters is acetylcholine.

Neuron-neuron or neuron-muscle synapses using acetylcholine are termed *cholinergic*. The family of acetylcholine receptors, that is, receptors that are activated by acetylcholine, often can be distinguished from each other by postsynaptic cell responses to agonists of acetylcholine, such as nicotine and muscarine. Like many hormone receptors, acetylcholine receptors on different neurons mediate different cellular responses. Nicotine causes excitatory responses lasting only milliseconds when it binds to certain acetylcholine receptors. Such responses are called *nicotinic responses* and the receptors are called *nicotinic acetylcholine receptors*. When muscarine, a mushroom alkaloid, binds to other acetylcholine receptors in different cells, it causes responses lasting seconds. Such responses are called *muscarinic responses,* mediated by *muscarinic acetylcholine receptors;* they can be either excitatory or inhibitory. We focus here on the well-characterized nicotinic muscle acetylcholine receptor, a ligand-dependent ion channel found at many nerve-muscle synapses (Figure 17-35). Let us consider in turn the four key steps occurring at a cholinergic synapse: (1) synthesis of acetylcholine, (2) release of acetylcholine by the presynaptic neuron, (3) binding of acetylcholine by the acetylcholine receptor and depolarization of the membrane, (4) enzymatic destruction of acetylcholine. These four steps are analogous to those for many hormonal systems.

Acetylcholine Is Synthesized in the Cytosol and Stored in Synaptic Vesicles

Acetylcholine is stored within synaptic vesicles. These membrane-limited organelles are about 40 nm in diameter and accumulate in the presynaptic axon terminal,

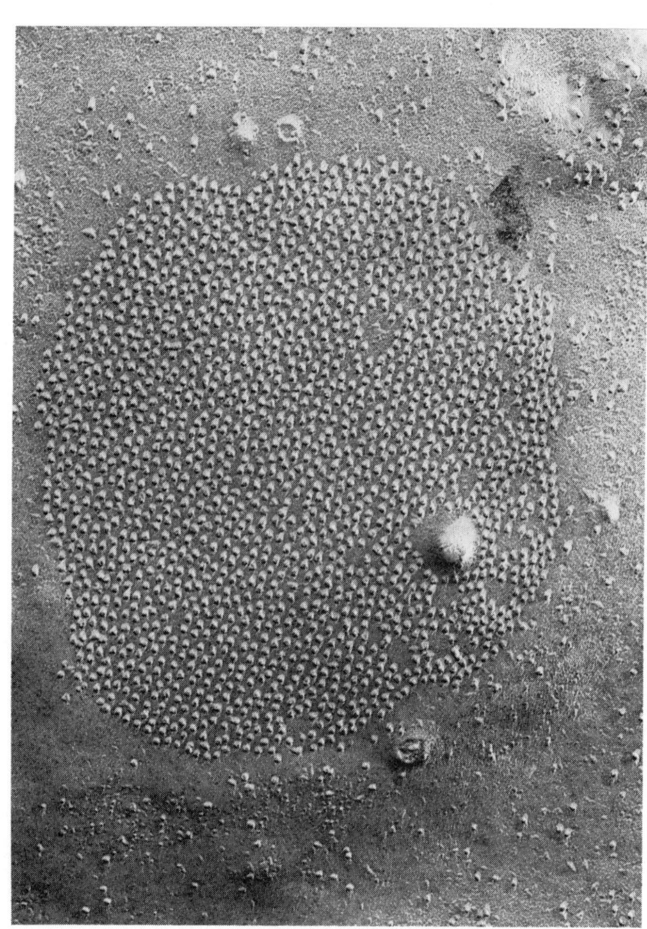

Figure 17-32 Electrical synapse between two axons. A freeze-fracture preparation from the crayfish *Procambarus.* The intramembranous particles packed in a regular array are believed to be the gap junction particles that enable ions to flow directly from the presynaptic to the postsynaptic cell. [See R. B. Hanna, J. S. Keeter, and G. D. Pappas, 1978, *J. Cell Biol.* 79:764.] *Courtesy of R. B. Hanna. Reproduced from the* Journal of Cell Biology, *1978, by copyright permission of the Rockefeller University Press.*

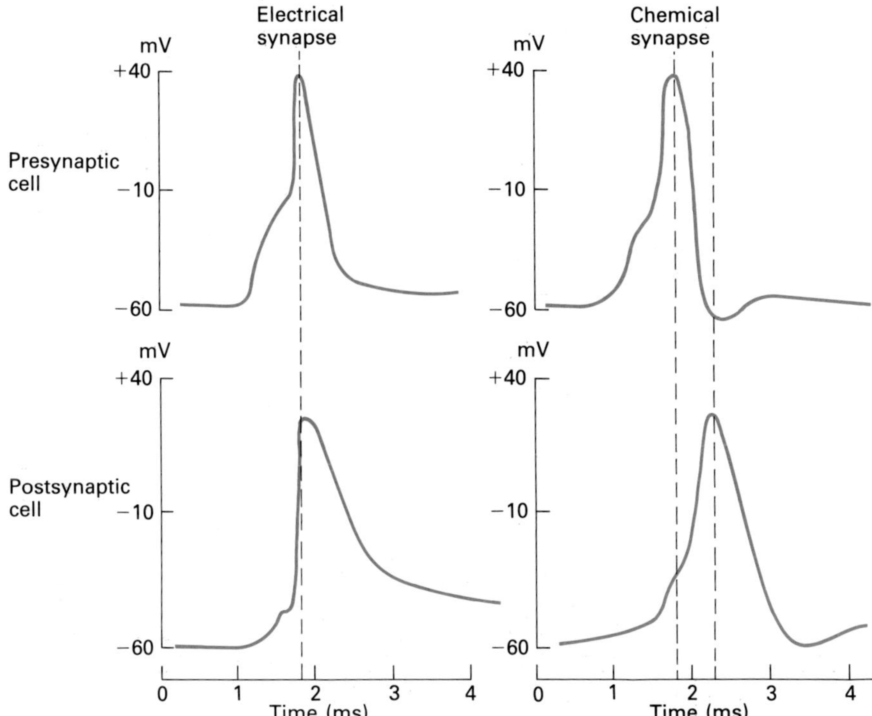

Figure 17-33 Synaptic transmission in electrical and chemical synapses. In both cases, the presynaptic neuron was stimulated and the membrane potential was followed in both the presynaptic and postsynaptic cells. In the electrical synapse, ions flow directly from the pre- to the postsynaptic cell and there is no delay in the signal transmission. In a chemical synapse, there is a delay of about 0.5 ms owing to secretion of transmitter, its diffusion, and the postsynaptic response to the signal.

often in rows just under the plasma membrane (Figures 17-35 and 17-36). The terminal of a single frog motor neuron might contain a million or more synaptic vesicles, each possibly containing between 1000 and 10,000 molecules of acetylcholine, and it might contact the surface of a skeletal muscle cell at several hundred points. Acetylcholine is synthesized in the cytosol from acetyl CoA and choline by the enzyme choline acetyltransferase (Figure 17-37). The synaptic vesicles take up and concentrate acetylcholine from the cytosol against a steep concentration gradient. How this active transport is brought about is not known.

Exocytosis of Synaptic Vesicles Is Triggered by a Rise in Cytosolic Calcium Ion Concentration

Acetylcholine is released into the *synaptic cleft,* the narrow space between the nerve and muscle cell, by fusion of the vesicle membranes with the plasma membrane. Thus neurotransmitter is released as discrete packets, or quanta. An increase in the intracellular Ca^{2+} ion concentration triggers the simultaneous exocytosis of the vesicles. The arrival of an action potential at a presynaptic terminal stimulates an influx of Ca^{2+} ions. The influx can occur even if the Na^+ channels are specifically blocked by drugs and the terminal membrane is otherwise depolarized, thus a voltage-sensitive Ca^{2+} channel is involved and the permeability to Ca^{2+} increases upon depolarization of the membrane. Patch clamp experiments show that voltage-dependent Ca^{2+} channels, like voltage-dependent Na^+ channels, open transiently upon depolarization of the membrane. As in most systems using Ca^{2+} as the intracellular messenger, the rise in cytosolic Ca^{2+} is short-lived. Ca^{2+} is pumped by Ca^{2+}-ATPases out of the cytosol into mitochondria, a major store of Ca^{2+}, or across the plasma membrane into the extracellular fluid.

Several lines of evidence indicate that the rise in Ca^{2+} ion causes the exocytosis of vesicles. First, many nerves bathed in a solution free of Ca^{2+} ion can conduct impulses but will not release neurotransmitters at their terminals. If Ca^{2+} is focally applied to a region of a nerve terminal by a micropipette, release of transmitter will occur and be confined to the region in the vicinity of the applied Ca^{2+}. Second, a microelectrode containing Ca^{2+} can be inserted into the squid giant axon. When a pulse of electric current causes *iontophoresis,* the movement of an ionized substance by an electric potential, and Ca^{2+} moves out of the micropipette into the cell, the axon terminal releases acetylcholine. Precisely how an increase of Ca^{2+} ion induces exocytosis is not known but might involve the nerve-terminal–specific phosphoprotein *synapsin I.* Synapsin I is a substrate for cAMP-dependent and calcium-calmodulin–dependent protein kinases. It is found localized in the cytoplasmic surface of all synaptic vesicle membranes, and is related in structure to erythrocyte membrane protein band 4.1, a protein that binds both actin and spectrin (Figure 14-42). Synapsin I comprises 6 percent of the protein in synaptic vesicles and might play a key role in causing adhesion or fusion of the vesicle membrane with the plasma membrane.

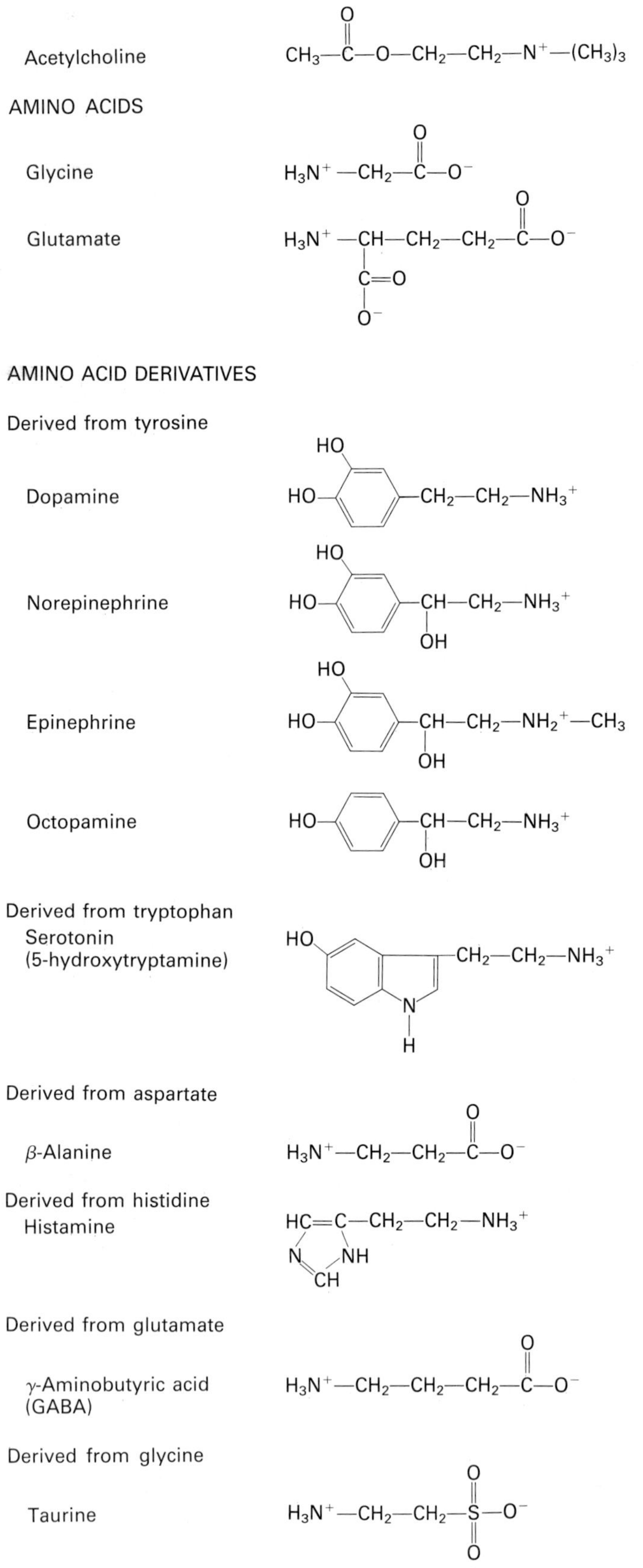

Acetylcholine	$CH_3—C(=O)—O—CH_2—CH_2—N^+—(CH_3)_3$
AMINO ACIDS	
Glycine	$H_3N^+—CH_2—C(=O)—O^-$
Glutamate	$H_3N^+—CH(C(=O)O^-)—CH_2—CH_2—C(=O)—O^-$
AMINO ACID DERIVATIVES	
Derived from tyrosine	
Dopamine	HO, HO (ring) $—CH_2—CH_2—NH_3^+$
Norepinephrine	HO, HO (ring) $—CH(OH)—CH_2—NH_3^+$
Epinephrine	HO, HO (ring) $—CH(OH)—CH_2—NH_2^+—CH_3$
Octopamine	HO (ring) $—CH(OH)—CH_2—NH_3^+$
Derived from tryptophan	
Serotonin (5-hydroxytryptamine)	HO (indole ring, N—H) $—CH_2—CH_2—NH_3^+$
Derived from aspartate	
β-Alanine	$H_3N^+—CH_2—CH_2—C(=O)—O^-$
Derived from histidine	
Histamine	$HC{=}C—CH_2—CH_2—NH_3^+$ (ring: N, NH, CH)
Derived from glutamate	
γ-Aminobutyric acid (GABA)	$H_3N^+—CH_2—CH_2—CH_2—C(=O)—O^-$
Derived from glycine	
Taurine	$H_3N^+—CH_2—CH_2—S(=O)_2—O^-$

Figure 17-34 Some small molecules that have been identified as neurotransmitters.

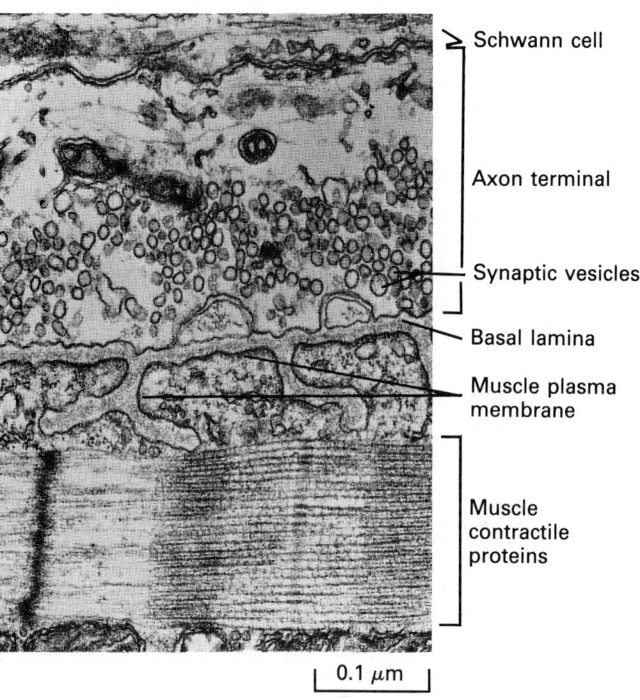

Figure 17-35 Longitudinal section through a frog nerve-muscle synapse. The plasma membrane of the muscle cell is extensively folded. Regions of the muscle plasma membrane adjacent to the axon possess acetylcholine receptors and appear thicker than the rest of the membrane. The axon terminal is filled with secretory vesicles lying just inside the plasma membrane. The basal lamina lies in the synaptic cleft separating the neuron from the muscle membrane. The axon terminal is surrounded by a Schwann cell that periodically interdigitates between the terminal and the muscle. *From J. E. Heuser and T. Reese, 1977, in* Cellular Biology of Neurons, *E. R. Kandel, ed.,* The Nervous System, *vol. 1.* Handbook of Physiology, *Williams and Wilkins, p. 266.*

Vesicle Membranes Are Recycled

The fusion of the vesicles with the plasma membrane results in an expansion of the area of the surface membrane. As in other secretory systems, there must be recycling of the membrane components of the synaptic vesicles both to preserve the vesicle membrane and to limit the surface area of the cell. Such recycling must be rapid, as nerves are capable of firing many times a second. It is also quite specific (Figure 17-38) in that several proteins unique to the synaptic vesicles are specifically internalized by endocytosis. The proteins, normally absent from the plasma membrane, may mediate exocytosis or the uptake of acetylcholine. Fluorescent antibodies to these proteins reveal that synaptic vesicle membrane proteins indeed accumulate on the plasma membrane of axon terminals when exocytosis of all acetylcholine-

(a)

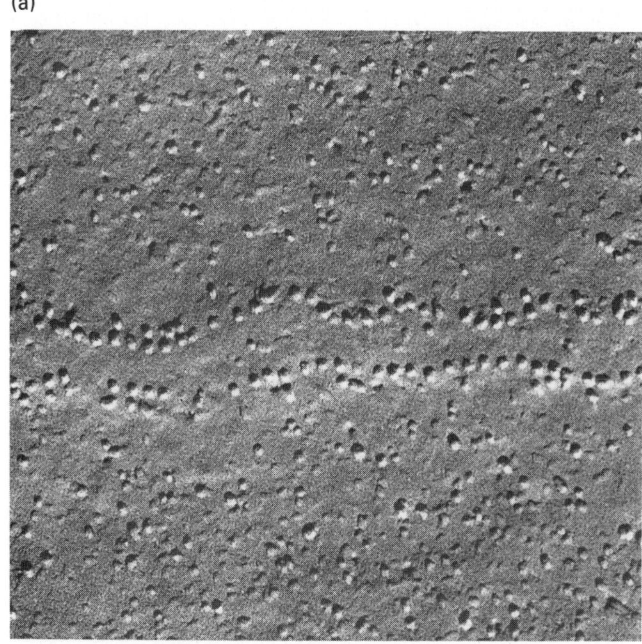

(b)

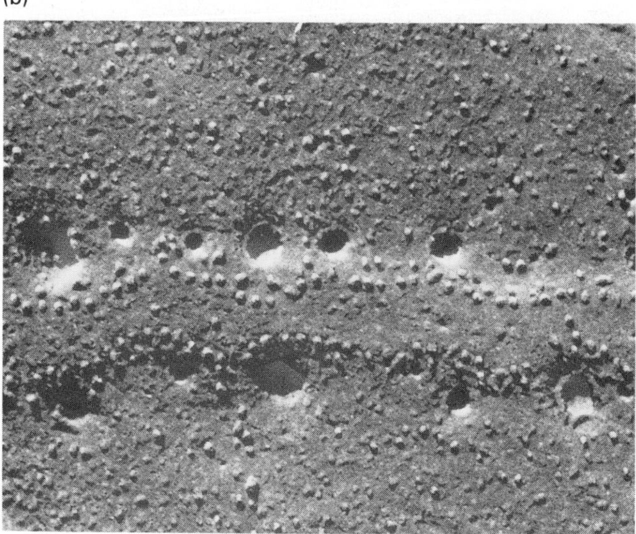

Figure 17-36 Freeze-fracture image of the axonal plasma membrane at a neuron-muscle synapse (a) in the resting state and (b) during stimulation. In the resting state, the membrane contains rows of particles that are aligned with rows of synaptic vesicles. The identity of these particles is not known, but it is suspected that they are the voltage-dependent Ca^{2+} channels. During stimulation (b) one can see, near the rows of particles, the large pits in the membrane that result from exocytosis of synaptic vesicles. *From J. E. Heuser and T. Reese, 1977, in* Cellular Biology of Neurons, *E. R. Kandel, ed.,* The Nervous System, *vol. 1.* Handbook of Physiology. *Williams and Wilkins, p. 268.*

$$\underset{\text{Acetyl CoA}}{CH_3{-}\overset{\overset{\displaystyle O}{\|}}{C}{-}S{-}CoA} + \underset{\text{Choline}}{HO{-}CH_2{-}CH_2{-}N^+{-}(CH_3)_3} \longrightarrow$$

$$\underset{\text{Acetylcholine}}{CH_3{-}\overset{\overset{\displaystyle O}{\|}}{C}{-}O{-}CH_2{-}CH_2{-}N^+{-}(CH_3)_3} + CoA{-}SH$$

Figure 17-37 Synthesis of acetylcholine by acetyltransferase. The structure for coenzyme A (CoA SH) is given in Figure 14-19.

containing vesicles is induced (Figure 17-39). Since none of the antigen is found on the surface of unstimulated neurons, the proteins presumably are recycled into vesicles.

Freeze-fracture studies on the presynaptic surface membrane reveal rows of 10-nm intramembrane particles, presumably proteins, adjacent to the rows of synaptic vesicles (see Figure 17-36). These particles possibly are involved in "guiding" the exocytosis or endocytosis of membrane constituents. Another popular idea is that the particles are the voltage-dependent Ca^{2+} channels. During stimulation one also sees deformations in the presynaptic membrane alongside the particles; they are caused by the exocytosis of a vesicle with the membrane (Figure 17-36b).

The Acetylcholine Receptor Is a Ligand-Triggered Cation Channel Protein

Once acetylcholine molecules have been released they diffuse across the synaptic cleft and combine with specific receptor molecules in the membrane of the postsynaptic neuron or muscle cell. Many studies on this receptor have been facilitated by the snake venom toxin *α-bungarotoxin,* which binds specifically and irreversibly to the receptor and prevents its function. The site of action of α-bungarotoxin was first elucidated by electrophysiologists, who showed that it blocked stimulation of the postsynaptic cell even when acetylcholine was added by micropipette to the synaptic cleft. Acetylcholine receptors are localized in the membrane of the postsynaptic cell immediately adjacent to the terminal of the presynaptic neuron, as is shown by the specific binding of α-bungarotoxin (Figure 17-40).

The interaction of acetylcholine with the receptor produces within 0.1 ms a large transient increase in the permeability of the membrane to both Na^+ and K^+ ions. In contrast to the voltage-sensitive Na^+ channel, permeability to both Na^+ and K^+ ions is increased, Na^+ slightly less than K^+. Assuming for simplicity that in this open channel $P_K = 1.5\ P_{Na} \gg P_{Cl}$, we can use Equation (17-4) to show that the resultant membrane potential will be about −15 mV. Binding of acetylcholine, there-

(a) Resting state

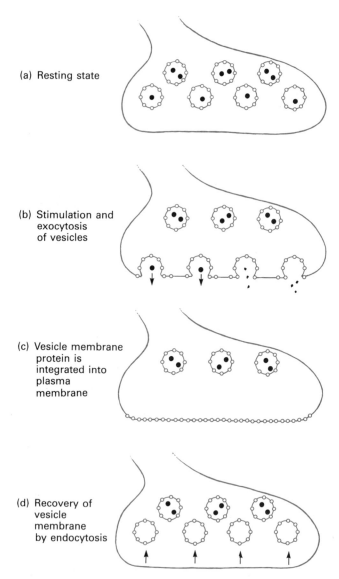

(b) Stimulation and exocytosis of vesicles

(c) Vesicle membrane protein is integrated into plasma membrane

(d) Recovery of vesicle membrane by endocytosis

Figure 17-38 Current concepts of the re-formation of secretory vesicles by a recycling of their membranes. (a) The vesicles, containing unique proteins, in their resting state. (b) and (c) During stimulation, the membrane of the vesicle fuses with the plasma membrane, releasing the neurotransmitter into the synaptic cleft. (d) The vesicle membrane proteins are specifically recovered by endocytosis during the recovery period. It is not yet clear whether noncoated vesicles or clathrin-coated vessels are used in this endocytic process nor whether endocytosis occurs near the sites of exocytosis or elsewhere.

fore, leads to a depolarization of the membrane and the generation of an action potential that will spread along the surface of the postsynaptic cell membrane (Figure 17-41). When this depolarization occurs in muscle, Ca^{2+} ions rush into the muscle cell and cause contraction. During neuron-muscle stimulation, acetylcholine is released simultaneously from all the endings of the neuron, and all the acetylcholine receptors are triggered simultaneously.

In experimentally denervated muscle, acetylcholine receptors appear at lower density over the entire surface membrane of the muscle cell. Consequently a patch of membrane 1 μm in diameter may contain only a few acetylcholine receptors, permitting their study by the patch clamp technique. Studies showed that in response to acetylcholine a single channel will remain open for several

(a)

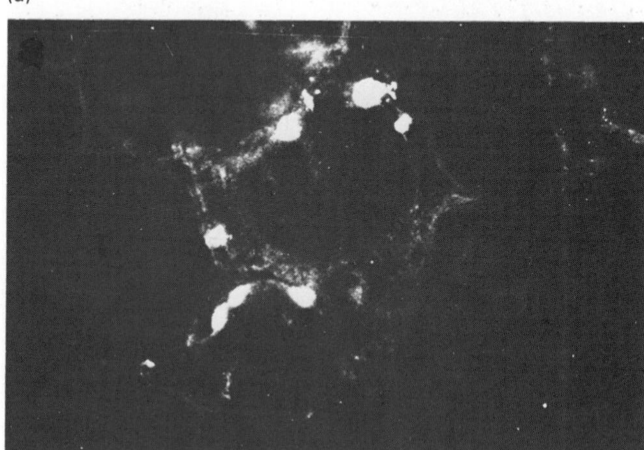

(b)

Figure 17-39 Direct evidence that vesicle membranes fuse with the plasma membrane during exocytosis. Antibodies were prepared against membrane proteins of purified synaptic vesicles and were conjugated with the fluorescent dye fluorescein. These were added to frozen sectioned preparations of frog muscle in which nerves were either unstimulated or stimulated to exocytose their secretory granules by treatment with La^{3+} (lanthanum ion). Lanthanum ion causes total exocytosis of all cholinergic synaptic vesicles. In the stimulated (a) preparations all of the end plates are stained by the antivesicle antibody. In the control preparation (b) the end plates are stained very poorly. Thus, exocytosis indeed causes fusion of the vesicle and plasma membranes. During normal electrical stimulation of the nerve, vesicle antigens do not accumulate on the cell surface; they are rapidly endocytosed to re-form secretory vesicles. [See R. von Wedel et al., 1982, *Proc. Nat'l Acad. Sci. USA* 78:1014.] *Photograph courtesy of R. von Wedel.*

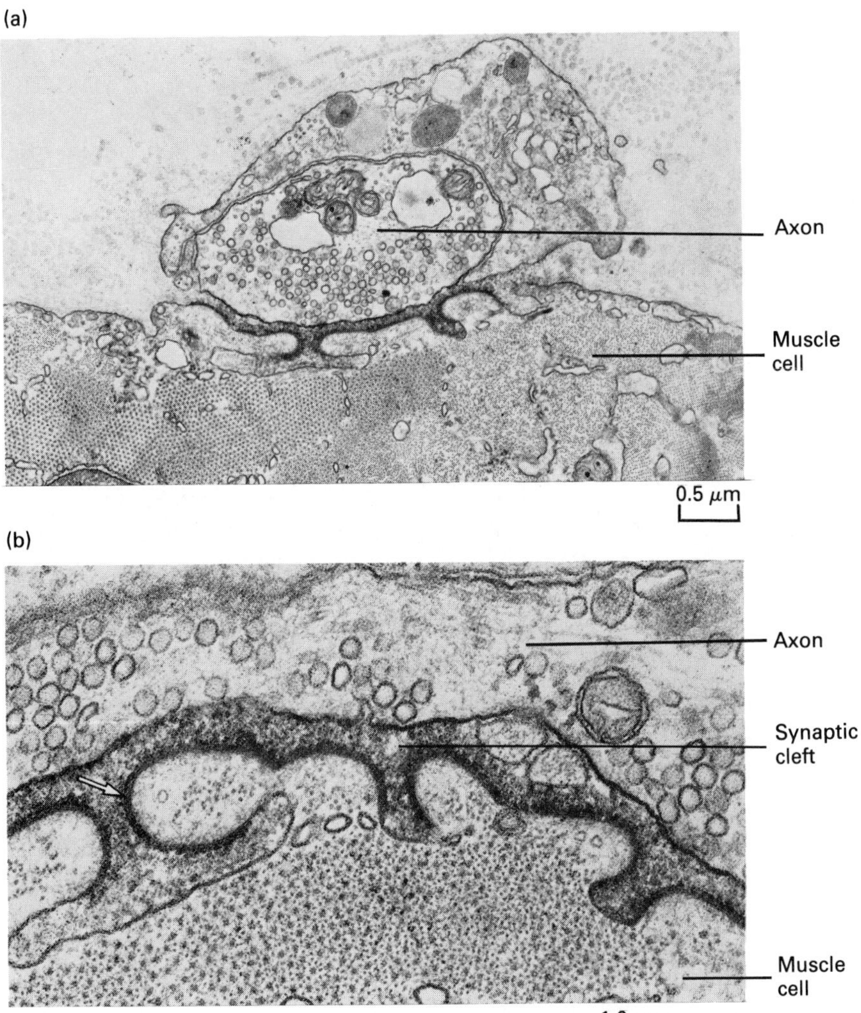

Figure 17-40 Acetylcholine receptors are highly concentrated in that part of the muscle cell membrane that is at the neuron-muscle synapse. The toxin α-bungarotoxin binds specifically and irreversibly to the acetylcholine receptor. Peroxidase-conjugated α-bungarotoxin also binds to acetylcholine receptors, and the enzyme catalyzes a reaction that produces an electron-dense product confined to the synaptic cleft. (a) Peroxidase-conjugated α-bungarotoxin is found only at the synapse. Thus the receptor is confined to the synapse; it is not on the rest of the muscle cell membrane. (b) Higher magnification shows that the receptor is concentrated at the external surface of the postsynaptic membrane at the tops and partway down the sides of the folds in the postsynaptic membrane. [See S. J. Burden, P. B. Sargent, and U. J. MacMahan, 1979, *J. Cell Biol.* 82:412.] *Courtesy of S. J. Burden. Reproduced from the* Journal of Cell Biology, *1979, by copyright permission of the Rockefeller University Press.*

milliseconds before closing spontaneously. When open, the channel is capable of transmitting 15,000 to 30,000 Na^+ or K^+ ions a millisecond. The time required for an acetylcholine molecule to provoke the opening of a channel is too small to be measured directly but is probably a few microseconds. The acetylcholine receptor is thus a ligand-dependent cation channel. By allowing both Na^+ and K^+ ions to cross the membrane, opening of the channel causes depolarization of the membrane.

Spontaneous Exocytosis of Synaptic Vesicles Produces Miniature End Plate Potentials

Careful monitoring of the membrane potential of the muscle membrane at the end plate of a synapse demonstrated spontaneous intermittent depolarizations of about 0.5 to 1.0 mV in amplitude; they occur in the absence of stimulation of the motor neuron. Each of these *miniature end plate potentials* is caused by the spontaneous release of acetylcholine from a single synaptic vesicle (Figure 17-42). Indeed, demonstration of the spontaneous small depolarizations led to the notion of *quantal release* of acetylcholine (later applied to other neurotransmitters) and thereby led to the hypothesis of vesicle exocytosis at synapses. The release of one vesicle of acetylcholine results in the opening of about 3000 ionic channels—sufficient to depolarize the region of membrane by about 1 mV, but insufficient to reach the threshold for an action potential.

Upon neuron stimulation the increased permeability of the muscle membrane, generated by the simultaneous release of many packets of acetylcholine, is generally sufficient to depolarize the postsynaptic membrane to the threshold voltage. The muscle cell will then produce and propagate an action potential along its length (Figure 17-42).

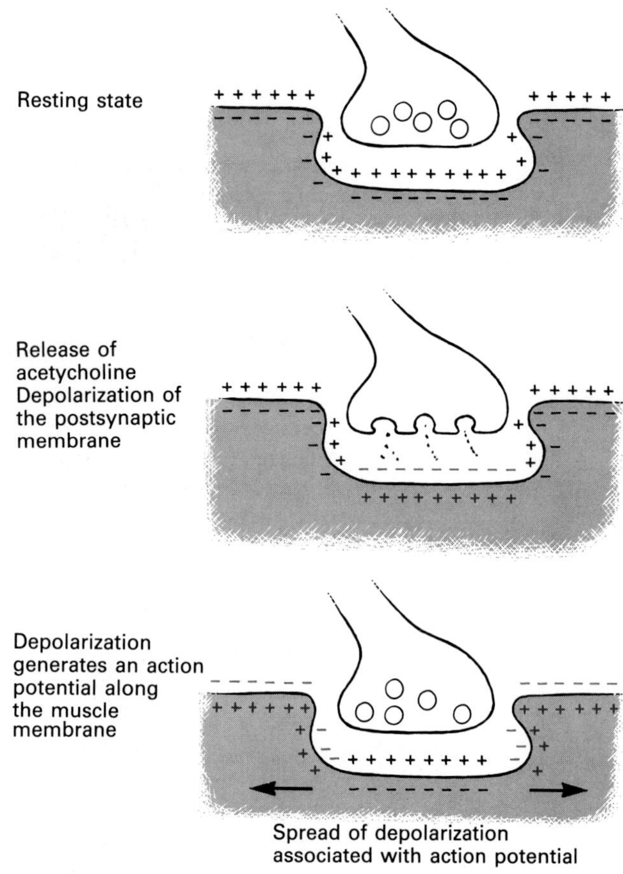

Figure 17-41 Depolarization of a muscle membrane at a synapse leads to generation of an action potential along the membrane.

The Nicotinic Acetylcholine Receptor Has Been Isolated, Cloned, and Sequenced

In most nerve and muscle tissues the acetylcholine receptor comprises a minute fraction of the total membrane protein. Electric eel and sting ray (*Torpedo*) electric organs are particularly rich in these receptors and are the sources used for its purification.

The sting ray electric organ is particularly remarkable as it can generate electrical discharges of more than 500 V, enough to kill a human being. It is made up of a column of 5000 cells called *electroplaxes*. During early development the electroplaxes are normal muscle cells. Later they lose all of the contractile proteins and become specialized for generating "stunning" voltages. One side of each cell is innervated, the other is not. Release of acetylcholine from the stimulatory neuron binds to acetylcholine receptors on the innervated side of each cell and depolarizes that membrane to about +40 mV. The depolarization does not spread to the noninnervated side, which remains at −90 mV. Upon stimulation the voltage potential between the outsides of the two faces of a cell transiently becomes 130 mV (Figure 17-43). The potential across the stack of cells is additive, like DC batteries in series, and amounts to 0.13 V per cell times 5000 cells, or 650 V.

The acetylcholine receptor can be solubilized from these membranes by nonionic detergents. The crucial step in its purification is affinity chromatography on columns of immobilized cobra toxin. Cobra toxin is similar to α-bungarotoxin but has lower affinity, and the receptor can be subsequently eluted and recovered. The monomeric m.w. of the receptor protein is 250,000 to 270,000. It consists of four different polypeptides: α (m.w. about 40,000), β (49,000), γ (57,000), and δ (65,000), with the composition $\alpha_2\beta\gamma\delta$ (Figure 17-44). Each α subunit binds one or two molecules of acetylcholine or one molecule of the acetylcholine antagonist α-bungarotoxin. Opening of

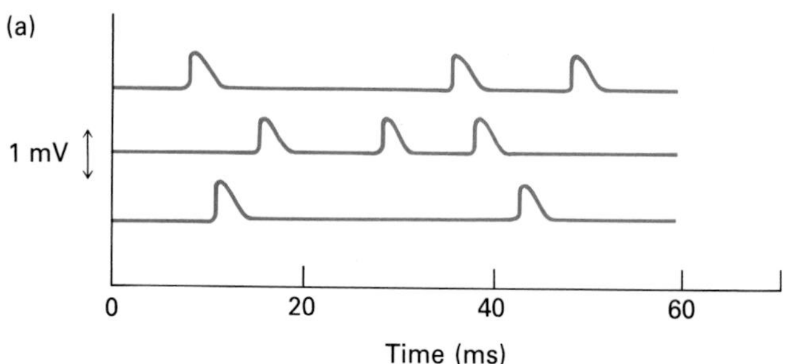

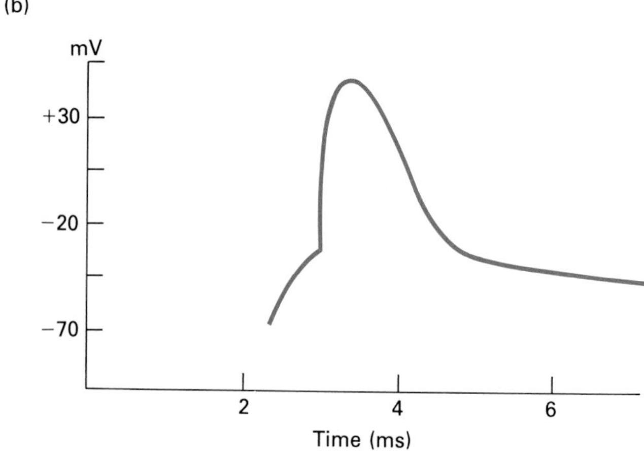

Figure 17-42 Miniature end plate potentials. (a) Recordings are taken from an intracellular electrode in an unstimulated frog muscle cell near the neuron-muscle synapse. The potentials, less than 1 mV, are generated by the spontaneous release of acetylcholine from a single synaptic vesicle. The occurrence of these small potentials in time seems to be entirely random. (b) For comparison, the action potential generated by a nerve impulse on the same muscle; the membrane change is approximately 100 mV.

the ion channel is facilitated by cooperative binding of two acetylcholine molecules.

Electron microscopy of receptor-rich membranes has revealed doughnutlike ring structures believed to be the receptor molecules (Figures 17-44 and 17-45). Each molecule has a diameter of about 9 nm with a central "pit" of 2 nm. The receptor protrudes about 7 nm from the membrane lipid matrix into the extracellular space and about 3 nm into the cytoplasm. Studies measuring the permeability of different small cations have suggested that the ion channel is about 0.65 to 0.80 nm in diameter, sufficient to allow passage of both hydrated Na^+ and K^+ ions. Neither the precise location of the cation channel nor the mechanism of its opening by acetylcholine are known; possibly there is only one channel that lies in the hole of the doughnut-shaped receptor.

In several studies the purified detergent-solubilized receptor was incorporated into liposomes. Most of the receptors were oriented right side out; α-bungaratoxin binding sites, normally extracytoplasmic, were on the outside of the liposomes. Addition of acetylcholine induced the release of radioactive Na^+ or K^+ ions trapped in the lumenal space, thus demonstrating directly that the

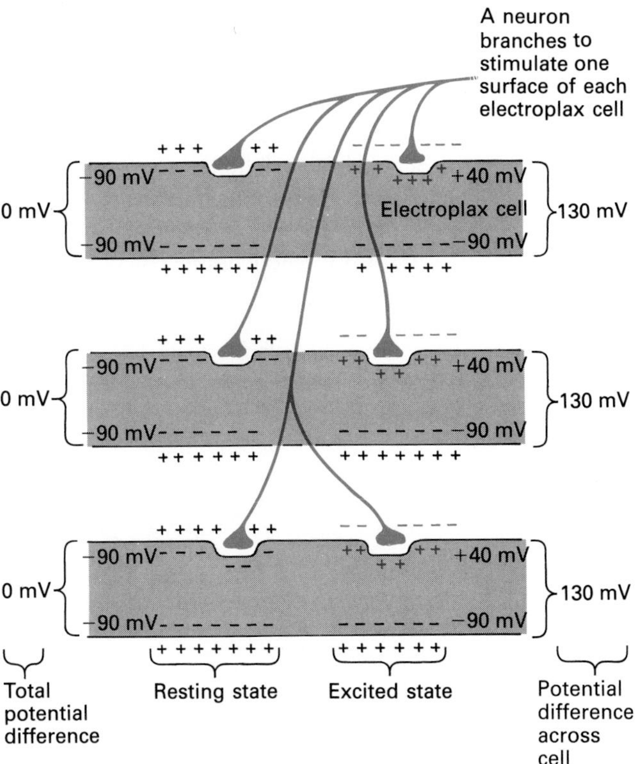

Figure 17-43 Generation of voltage by a stack of electroplax cells. Only one surface of the cell is innervated and contains acetylcholine receptors. Nerve stimulation depolarizes the innervated side to +40 mV, while the noninnervated side remains at −90 mV. The total potential difference across each cell becomes, transiently, 130 mV.

(a)

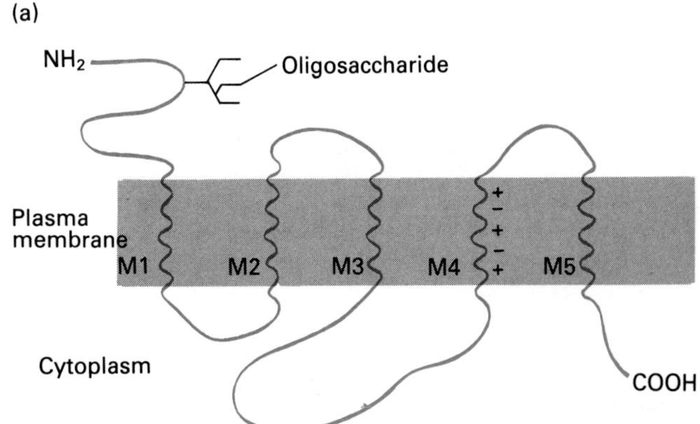

(b)

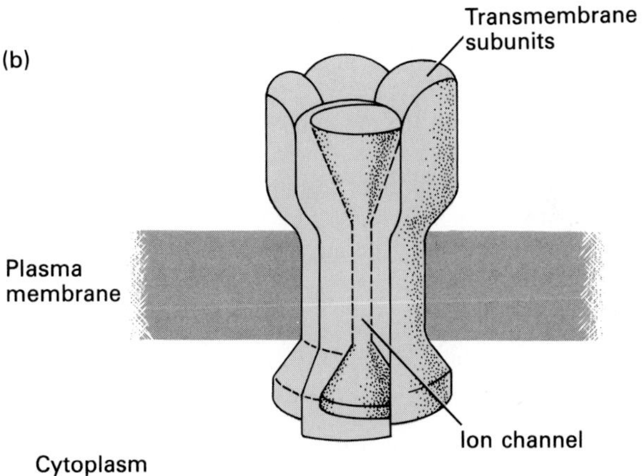

Figure 17-44 Model of the structure of the acetylcholine receptor. Messenger RNAs encoding all four distinct subunits have been cloned, and this has yielded the sequence of every polypeptide. All four subunits share regions of strong homology. On the average, about 35 to 40 percent of the residues are homologous; where substitutions do occur, one residue is replaced by another with a similar side chain. All four chains undoubtedly evolved from a common progenitor. (a) Each subunit spans the membrane five times with α helixes M1 to M5. The principal difference between the subunits is the size of the amino-terminal segment (210 to 224 amino acids) and the loop on the cytoplasmic surface between helixes M3 and M4 (63 to 89 amino acids). Helix M4 is amphipathic, one face containing negative and positive side chains and the other having only hydrophobic residues. The charged faces of the M4 helixes from the five subunits $\alpha_2\beta\gamma\delta$ probably line the ion channel itself. (b) Schematic model of the five-part, or pentameric, receptor. The distribution of protein masses is based on the amino acid sequence data and on neutron-scattering and electron-diffraction experiments. [See J.-P. Changeux, A. Devillers-Thiery, and P. Chemouilli, 1984, *Science* **225**:1335; E. F. Young et al., 1985, *Proc. Nat'l Acad. Sci. USA* **82**:626; and A. Brisson and P. N. T. Unwin, 1985, *Nature* **315**:474.]

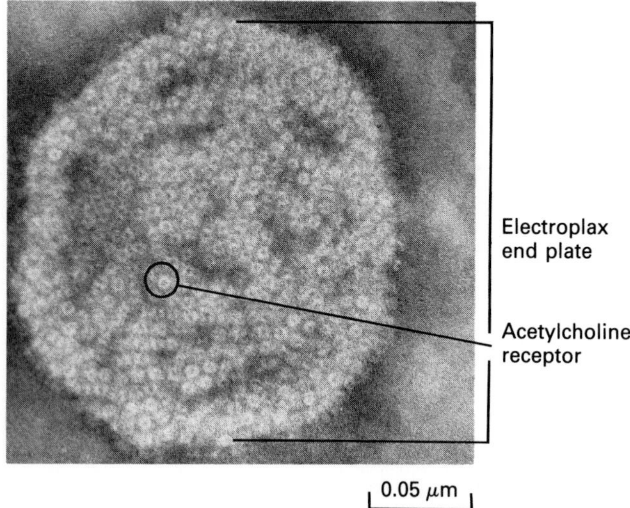

Figure 17-45 Electron micrograph of a negatively stained preparation of a membrane of an electroplax cell. The end plate region contains many doughnut-shaped objects, 9 to 12 nm in diameter, that are the acetylcholine receptors. [See F. Hucho, 1981, *Trends Biochem. Sci.* 6:242.] *Courtesy of F. Hucho.*

increase in membrane permeability is generated by acetylcholine binding and that only the acetylcholine receptor protein complex is involved in this process. Further, *tubocurarine,* an antagonist of acetylcholine, blocked the effects induced by acetylcholine. These important studies showed that all the effects of binding of acetylcholine to the postsynaptic cell occur to the purified acetylcholine receptor incorporated into liposomes. No proteins other than the four subunits of the receptor are involved.

Hydrolysis of Acetylcholine Terminates the Depolarization Signal

To restore a depolarized membrane to its excitable state it is necessary to remove or destroy the depolarizing signal. In general, there are three ways to end the signaling: (1) the transmitter diffuses away from the synaptic cleft; (2) the transmitter is taken up by the presynaptic neuron; (3) the transmitter is enzymatically degraded. Signaling by acetylcholine is terminated by enzymatic degradation of the transmitter, but different methods are used to terminate signaling by other neurotransmitters.

Acetylcholine is hydrolyzed to acetate and choline by the enzyme *acetylcholinesterase,* which is localized in the synaptic cleft between the neuron and muscle cell membranes. It is bound to a network of collagen forming the basal lamina that fills this space (see Figure 17-35). Part of the acetylcholinesterase molecule is made up of a collagen-like (Gly-Pro-X) sequence (see Figure 5-54) and forms a collagen-like triple helix, which may facilitate its binding to collagen. Another form of acetylcholinesterase is bound to the cell membrane, apparently by a phospholipid covalently attached to the C-terminus of the polypeptide.

A large number of nerve gases and other neurotoxins inhibit the function of acetylcholinesterase. Specifically, they prolong the action of acetylcholine, thus prolonging the period of membrane depolarization (Figure 17-46). They can kill by preventing relaxation of the muscles necessary for breathing. During enzymatic hydrolysis, the substrate acetylcholine reacts with a specific serine residue at the active site of acetylcholinesterase. This forms free choline and a covalent acetyl-enzyme intermediate (Figure 17-47) similar to the one formed by chymotrypsin during hydrolysis of a peptide bond (see Figure 3-24). The intermediate is subsequently hydrolyzed to yield acetate and the native enzyme. The active serine residue is the target of many potent toxins. Organic fluorophosphates such as diisopropylfluorophosphate (DIPF), or nerve gases such as sarin, or agricultural insecticides such as parathion all form stable covalent complexes with this serine residue and inhibit its function.

The Defect in Myasthenia Gravis Is a Reduction of Acetylcholine Receptors at Neuromuscular Junctions

Myasthenia gravis is a disease characterized by fatigue and weakness of skeletal muscles owing to a reduction in the number of functional acetylcholine receptors at the neuromuscular junctions. Circulating antibodies (autoantibodies) to the receptors cause their degradation. Normally only a fraction of the acetylcholine receptors in the neuromuscular junction are activated in response to a

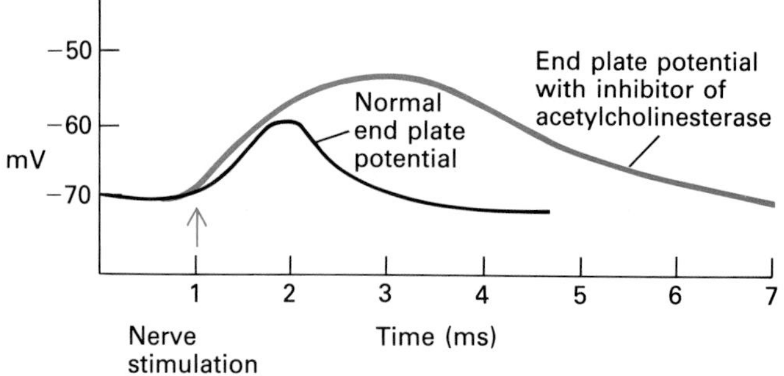

Figure 17-46 The normal depolarization, or end plate potential, induced by a subthreshold nerve stimulation of a muscle membrane is about 10 mV and lasts less than 1 ms. In the presence of an inhibitor of acetylcholinesterase, the depolarization triggered by the same stimulation is both greater and longer lasting.

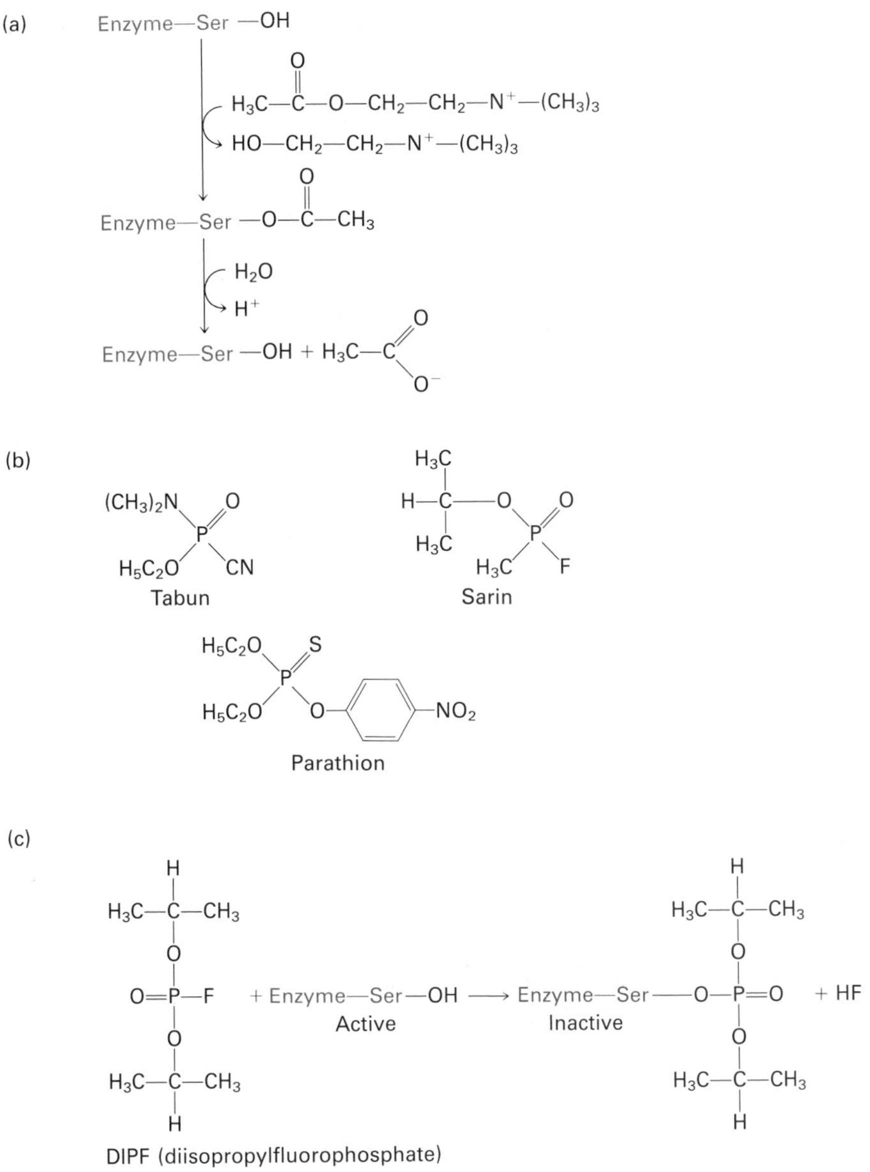

(a)

Enzyme—Ser—OH

$H_3C—C(=O)—O—CH_2—CH_2—N^+—(CH_3)_3$

$HO—CH_2—CH_2—N^+—(CH_3)_3$

Enzyme—Ser—O—C(=O)—CH_3

H_2O

H^+

Enzyme—Ser—OH + $H_3C—C(=O)O^-$

(b)

Tabun: $(CH_3)_2N$, O, P, H_5C_2O, CN

Sarin: H_3C, H—C—O, H_3C, P, O, H_3C, F

Parathion: H_5C_2O, S, P, H_5C_2O, O—C_6H_4—NO_2

(c)

DIPF (diisopropylfluorophosphate) + Enzyme—Ser—OH (Active) ⟶ Enzyme—Ser—O—P(=O)[O—CH(CH_3)$_2$]$_2$ (Inactive) + HF

Figure 17-47 (a) A mechanism of action of acetylcholinesterase, showing that acetylserine is an essential enzyme-bound intermediate. (b) Structures of several organic phosphate inhibitors of acetylcholinesterase. (c) The reaction of DIPF with acetylcholinesterase to inhibit enzyme activity. The other inhibitors also react with the active serine residue on acetylcholinesterase and inactivate the enzyme. Such inhibitors can be lethal as they prolong the action of the neurotransmitter at nerve-muscle synapses and prevent muscle relaxation.

nerve impulse. With repetitive firing of a motor neuron fewer and fewer acetylcholine molecules are released, but there is still sufficient acetylcholine transmitted to induce a muscle action potential and muscle contraction in normal muscle. Because of the reduced numbers of receptors in myasthenia gravis, the lower levels of acetylcholine released during repetitive firings are insufficient to depolarize the membrane to the threshold for triggering muscle action potentials. Persons with a loss of receptors are unable to sustain or repeat muscular contractions, resulting in neuromuscular fatigue. They are often helped by small doses of inhibitors of acetylcholinesterase. Inhibition of the enzyme permits the acetylcholine released at the nerve terminals to act over a longer period of time (Figure 17-46). The increase may be sufficient to depolarize the muscle membrane to threshold for an action potential during repetitive firing of the nerve.

Neurotransmitters

Because it causes a rapid, short-lived, and dramatic response in skeletal muscle cells, acetylcholine was one of the first neurotransmitters to be characterized definitively. Unlike the nicotinic acetylcholine receptor, a ligand-dependent ion channel, the binding of most neurotransmitters to their receptors induces permeability changes that are relatively long lasting—seconds to minutes in duration—and the changes may be either excitatory or inhibitory. Before looking at some of these transmitters in detail, let us review some of the general properties of neurotransmitters.

A Number of Criteria Must Be Met for a Compound to Be Designated a Neurotransmitter

To establish that a given compound is the neurotransmitter at a particular synapse, certain criteria must be met: (1) Synaptic vesicles in terminals of the presynaptic neuron must contain the substance, they must release it at the appropriate time in response to stimulation, and they must release a sufficient quantity to induce the appropriate response in the postsynaptic cell. (2) Iontophoresis of the substance from a micropipette into the synaptic cleft must induce the same response as does stimulation of the presynaptic nerve. (3) The substance must be removed or degraded rapidly, resulting in restoration of the resting membrane potential. Individual neurons may contain only 10^{-14} mol of a particular transmitter, and sophisticated microchemical techniques must be used to detect it.

Substances besides acetylcholine that have been identified by these criteria as neurotransmitters include dopamine, epinephrine, norepinephrine, serotonin, γ-aminobutyric acid (GABA), glycine, and glutamate (see Figure 17-34). But direct evidence for their role as neurotransmitters has been obtained at relatively few synapses. By the less restrictive criteria of presence in a synapse, and appropriate postsynaptic response of microinjected material, a number of other small chemicals, such as histamine and octopamine, are also considered to be neurotransmitters (Figure 17-34). In addition, many small peptides (Figure 17-48) are believed to be neurotransmitters as well as neurohormones. Purine nucleotides such as adenosine and ATP are found in synaptic vesicles, and they induce responses in heart muscle. Brain homogenates contain high-affinity receptors for adenosine and other purines. Thus, these nucleotides, too, may be neurotransmitters.

Catecholamines Are Widespread Neurotransmitters

The neurotransmitters *epinephrine, norepinephrine,* and *dopamine* all contain the catechol moiety

—OH
—OH

and are synthesized from the amino acid tryosine (Figure 17-49). They are referred to as the catecholamines. Nerves that synthesize and use epinephrine or norepinephrine are termed *adrenergic.*

Epinephrine and norepinephrine function as both systemic hormones and neurotransmitters. Norepinephrine is the transmitter at synapses with smooth muscles that are innervated by nerve fibers of the sympathetic nervous system, the division of the peripheral nervous system that increases the activity of the heart and internal organs in "fight or flight" reactions. Norepinephrine is also found at synapses in the central nervous system. Epinephrine is synthesized and released into the blood by the adrenal medulla, an endocrine organ that appears to be a branch of the sympathetic nervous system. During development of the embryo, cells in the region known as the neural crest that become neurons of the sympathetic nervous system also take up a position in the central part of the growing adrenal gland and form the adrenal medulla. Unlike neurons, the medulla cells do not develop axon or dendrite projections.

β-endorphin
Tyr-Gly-Gly-Phe-Met-Thr-Ser-Glu-Lys-Ser-Gln-Thr-Pro-Leu-Val-Thr-Leu-Phe-Lys-Asn-Ala-Ile-Ile-Lys-Asn-Ala-Tyr-Lys-Lys-Gly-Glu
Met-enkephalin
Tyr-Gly-Gly-Phe-Met
Leu-enkephalin
Tyr-Gly-Gly-Phe-Leu
Somatostatin
Ala-Gly-Cys-Lys-Asn-Phe-Phe-Trp-Lys-Thr-Phe-Thr-Ser-Cys
Luteinizing hormone–releasing hormone
pGlu-His-Trp-Ser-Tyr-Gly-Leu-Arg-Pro-Gly-NH_2
Thyrotropin-releasing hormone
pGlu-His-Pro-NH_2
Substance P
Arg-Pro-Lys-Pro-Glu-Glu-Phe-Phe-Gly-Leu-Met-NH_2
Neurotensin
pGlu-Leu-Tyr-Glu-Asn-Lys-Pro-Arg-Arg-Pro-Tyr-Ile-Leu
Angiotensin I
Asp-Arg-Val-Tyr-Ile-His-Pro-Phe-His-Leu
Angiotensin II
Asp-Arg-Val-Tyr-Ile-His-Pro-Phe
Vasoactive intestinal peptide
His-Ser-Asp-Ala-Val-Phe-Thr-Asp-Asn-Tyr-Thr-Arg-Leu-Arg-Lys-Glu-Met-Ala-Val-Lys-Lys-Tyr-Leu-Asn-Ser-Ile-Leu-Asn-NH_2

Figure 17-48 Some of the neuropeptides that are thought to function as neurotransmitters. An $—NH_2$ after the carboxyl-terminal residue indicates it is modified to an amide; a "p" at the beginning indicates that the glutamate has been cyclized to the "pyro" form. All of these peptides are synthesized as larger protein precursors. Like all neuronal proteins, they are synthesized in the cell body. There they are packaged into vesicles and transported down the axon to the terminals. Once exocytosed, the peptide neurotransmitters are degraded by proteases; they are not recycled or reincorporated into neurons. The opioid peptides, Met-enkephalin, Leu-enkephalin, and β-endorphin contain a common tetrapeptide sequence Tyr-Gly-Gly-Phe- that is important in the pain-modulating function of these neuropeptides. [See H. Gainer and M. J. Brownstein, 1981, in *Basic Neurochemistry,* 3d ed., G. J. Siegel et al., eds., Little, Brown, chap. 14.]

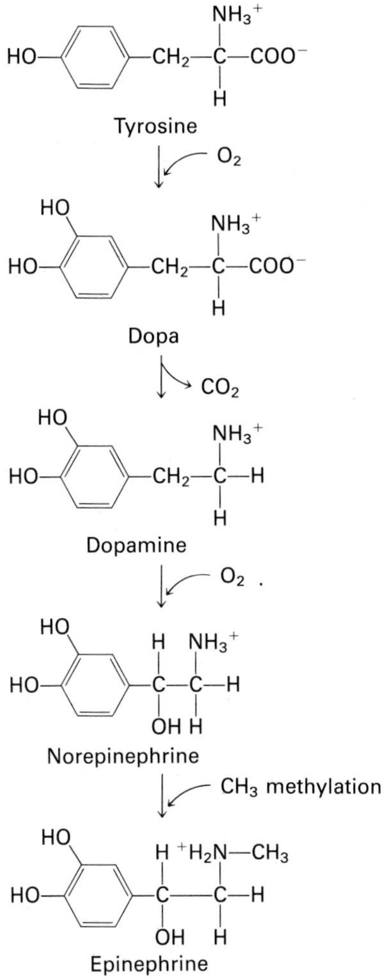

Figure 17-49 Pathway for the synthesis of the catecholamine neurotransmitters from tyrosine.

Catecholamines are synthesized in the cytosol of both medulla cells and adrenergic neurons and are then packaged into vesicles. The mechanism of uptake has been studied with purified catecholamine-containing vesicles from adrenal medulla cells. These vesicles have an ATP-dependent proton pump similar to the lysosomal H^+-ATPase (Chapter 15). The enzyme acidifies the interior of the vesicle to pH 5.5. Uptake of catecholamines into the vesicle is driven by the 50-fold concentration gradient of protons across the membrane (pH 5.5 versus pH 7.2 of the cytosol) using a proton-catecholamine antiport system. This system is believed to operate in adrenergic neurons as well. Electron micrographs of the terminals of adrenergic neurons show characteristic vesicles 40 to 150 nm in diameter that are believed to contain transmitter.

Details of the mode of action of catecholamines on postsynaptic cells are still unclear. Different cells may possess different receptors for the same neurotransmitter, and these different receptors may cause different excitatory or inhibitory responses. Recall that at least two types of β-adrenergic receptors and at least two α-adrenergic receptors have been identified (Table 17-1). All of these receptors have been found in brain tissue and heart muscle and some are enriched in the peripheral nervous system. Two receptors for dopamine can be identified by their different localization and by their responses to dopamine agonists and antagonists. The existence of multiple receptors for the same neurotransmitter allows for great flexibility in nerve-nerve signaling. A neuron generally releases only a single neurotransmitter. Two postsynaptic cells may respond very differently if they bear different receptors. One may undergo an excitatory response, the other inhibitory. In the sea snail *Aplysia*, at least six types of serotonin receptors have been identified in various synapses, some inducing an excitatory response and some an inhibitory one.

Binding of Neurotransmitters to Certain Receptors Evokes Slow Postsynaptic Potentials

The response of skeletal muscle cells to the release of acetylcholine at the neuron-muscle junction is very rapid—the permeability changes are completed within a few milliseconds. But many synapses do not work as rapidly, and many functions of the nervous system operate with time courses of seconds or minutes and are not well served by fast synaptic responses lasting milliseconds. Regulation of the heart rate, for instance, requires that action of neurotransmitters extend over several beating cycles measured in seconds.

Slow postsynaptic potentials (SPNPs) are defined as responses that begin with a lag period of milliseconds after the transmitter has been added. SPNPs may be either excitatory or inhibitory depending on whether the permeability changes in the postsynaptic membrane favor or disfavor depolarization beyond the threshold for an

Table 17-1 Effect on adenylate cyclase linkage of the same neurotransmitter with different receptors

Neurotransmitter	Receptor	Linked to adenylate cyclase?
Dopamine	D1	Yes
	D2	No
Norephinephrine	$\beta 1$	Yes
	$\beta 2$	Yes
	$\alpha 1$	No
	$\alpha 2$	No
Serotonin	5-HT_1	Yes
	5-HT_2	No?

SOURCE: S. H. Snyder and R. R. Goodman, 1980, *J. Neurochem.* 35:5.

action potential. Depending on the receptor present on the postsynaptic cell, the same neurotransmitter—acetylcholine in the example in Figure 17-50—can evoke a rapid or slow response. By binding to nicotinic receptors in skeletal muscle, acetylcholine induces a rapid (< 20 ms) depolarization of the membrane. But by binding to muscarinic acetylcholine receptors in frog cardiac muscle, it induces a long-lived (several seconds) hyperpolarization of the membrane. Thus, stimulation of these cholinergic nerves slows the rate of heart muscle contraction. Fast or slow responses, in other words, are not a consequence of the transmitter per se but of the nature of the neurotransmitter receptor.

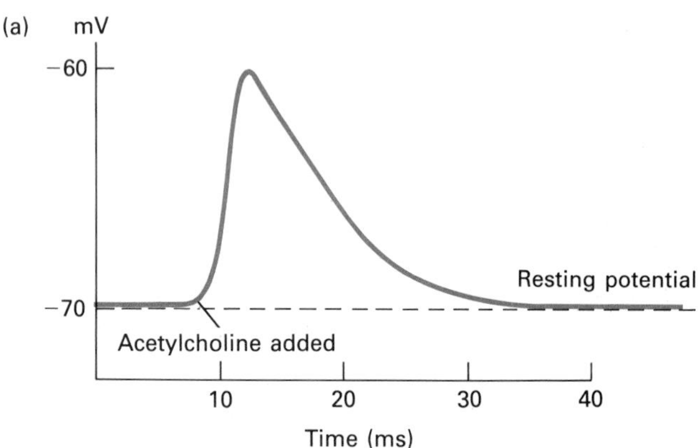

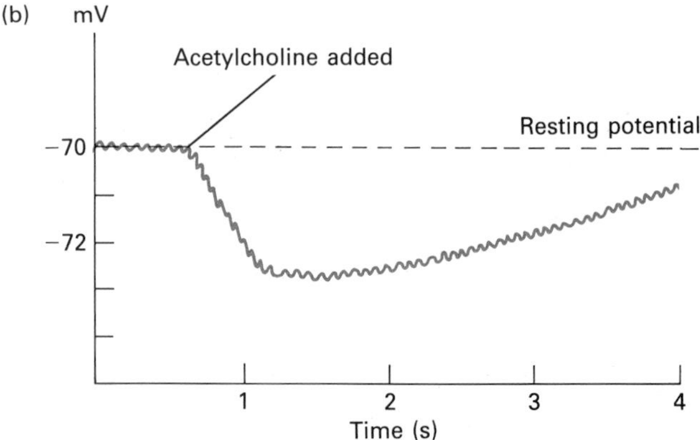

Figure 17-50 Fast and slow postsynaptic action potentials. (a) The application of acetylcholine to frog skeletal muscle produces a rapid postsynaptic depolarization of about 10 mV, and the response is completed in 20 ms. (b) Frog heart muscle responds quite differently to the application of acetylcholine. After a lag period of about 40 ms, the cell responds with a hyperpolarization of 2–3 mV, which lasts several seconds. Thus, stimulation of the parasympathetic nerve that innervates the heart and releases acetylcholine as neurotransmitter causes a reduction in the rate of cardiac contraction. Note the scale differences in the two graphs. [See H. C. Hartzell, 1981, *Nature* **291**:539.]

Nerve Cells Integrate the Excitatory and Inhibitory Responses Received at Many Synapses

Whether an action potential is generated in a neuron at any one instant depends on the balance between ionic movements generated in the dendrites and cell body by excitatory and inhibitory synapses. The electric currents that result from neurotransmitter activity are due to movement of K^+, Na^+, Cl^-, and to a lesser extent Ca^{2+} ions. The membrane depolarizations and hyperpolarizations induced at localized synapses spread passively, with diminishing intensity, along the dendrite or cell body membrane (Figure 17-51). Action potentials are generated in the axon hillock, the part of the axon that emerges from the cell body. Ionic movement that results from all of the excitatory and inhibitory synapses will, summed together, affect the membrane permeability in this region of the cell. If the *net* change in permeability in the hillock region is sufficient to depolarize the membrane potential to the threshold voltage, an action potential will be generated in the axon. Whether or not an action potential is generated depends on the spatial relationships of the stimulated synapses in the dendrites and cell body, on the timing of the incoming impulses, on the duration of the membrane effects that they induce, and on the ability of the dendrite and cell body plasma membrane to conduct a hyperpolarization or depolarization (Figure 17-51). Some neurons in the central nervous system must integrate hundreds, perhaps thousands of excitatory and inhibitory synaptic inputs. The computation is complex and depends on the microanatomy of the cell. Experimental analysis of the responses of any single neuron is not a simple task, particularly since the action of a neurotransmitter may be modified by other neurotransmitters or by neurohormones or pharmacological agents.

Some Receptors for Neurotransmitters Affect Adenylate Cyclase

In only a few cases is it known how the binding of neurotransmitters triggers a slow, long-lasting change in membrane permeability. The binding of agonists to norephinephrine β-receptors on nerve cells causes the activation of adenylate cyclase and an increase in cAMP. This is the same mechanism by which the hormonal response to this compound is mediated (see Figures 16-7 and 16-9). Certain receptors for serotonin are also believed to activate adenylate cyclase (see Table 17-1).

In one well-studied synapse of the sea slug *Aplysia*, the action of serotonin on the axon terminal of a sensory

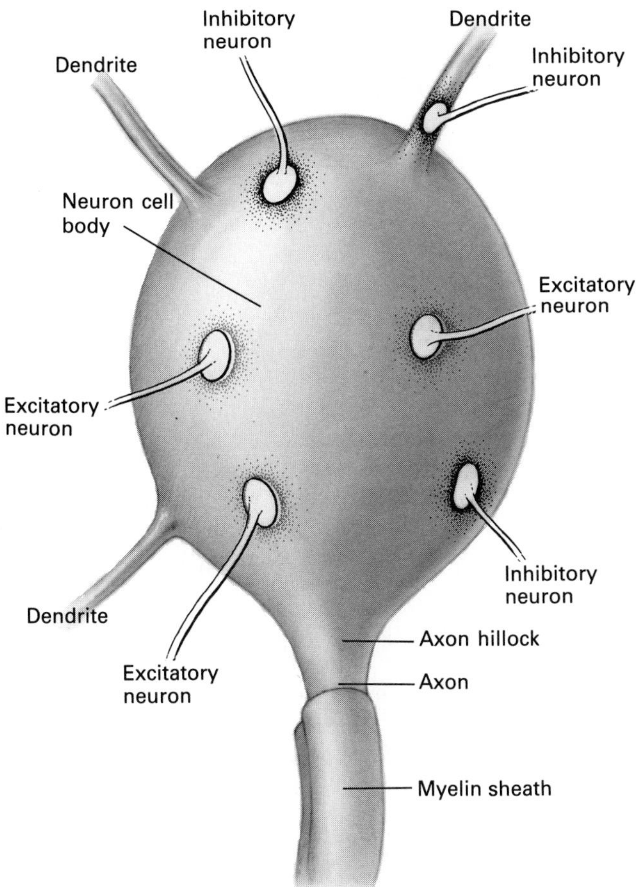

Figure 17-51 A single motor neuron in the spinal cord may receive signals from thousands of neurons that synapse with it. Only a few synaptic inputs are shown here. Some synapses may be excitatory, depolarizing regions of the motor neuron's membrane *(red)*, and some inhibitory, hyperpolarizing regions of its membrane *(black)*. The neuron may contain multiple receptors for different neurotransmitters. Which receptors are stimulated depends, of course, on their proximity to the nerve terminals that release specific neurotransmitters. An action potential is initiated at the axon hillock; firing is controlled by the summation of all the signals sensed at this point on the membrane. Glial cells *(not shown)* cover most of the surface of the motor neuron that is not covered with synaptic terminals.

neuron activates adenylate cyclase and increases the intracellular concentration of cAMP. In these cells cAMP is believed to activate a cAMP-dependent protein kinase, which may phosphorylate a K^+ channel protein or an associated protein, leading to closure of the channel (Figure 17-52). Following the administration of serotonin to the neuron there is a decrease in the outward K^+ ion movements that normally repolarize the membrane of the sensory neuron after an action potential. This prolongs the membrane depolarization during the action potential and results in an increased influx of Ca^{2+} ions through voltage-sensitive Ca^{2+} channels. Because this sensory neuron is sufficiently large, it is possible to inject into the cell the active subunit of the cAMP-dependent protein kinase. This mimics the effect of applying the natural transmitter serotonin to the nerve. Additional supporting evidence that serotonin acts by means of cAMP and a protein kinase comes from studies using patch clamping to record single channel movements in isolated sensory neuron membrane. These membranes exhibited voltage-dependent K^+ channels. When the purified active catalytic subunit of cAMP-dependent protein kinase was added to the cytoplasmic surface of the patches, the K^+ channels closed. Thus the protein kinase indeed acts on the cytoplasmic surface of the membrane to phosphorylate the channel protein itself or a membrane protein that regulates channel activity.

GABA Is the Neurotransmitter at Many Inhibitory Synapses

Gamma-aminobutryric acid (GABA) is formed from the amino acid glutamate by loss of a carboxyl group (see Figure 17-34). It is a neurotransmitter used exclusively in inhibitory synapses, and its action is well understood. In almost all cases GABA increases the permeability of the postsynaptic membrane to Cl^- ion. This tends to drive the membrane potential toward the equilibrium potential for Cl^-, which in general is *more negative* than the resting potential of the membrane (see page 725). In other words, the membrane becomes hyperpolarized. If many Cl^- channels are opened, the resultant chloride permeability is large and the membrane potential will be held near the chloride potential. As can be deduced from Equation (17-4), a much larger than normal increase in the sodium permeability will then be required to depolarize the membrane. The effect of GABA on Cl^- permeability can last for a second or more, a long time compared with the millisecond required to generate an action potential. Thus GABA induces a slow, inhibitory postsynaptic response.

A well-studied example of the role of GABA is a sensory stretch receptor cell in a lobster that responds to stretch with a steady discharge of impulses. Certain inhibitory neurons synapse with this sensory neuron, and when they are stimulated the stretch receptor neuron is hyperpolarized, an action potential is not generated, and the response of the organism to the stretch stimulus is reduced. Direct application of GABA to the stretch neuron also inhibited its firing, indicating that GABA induces the inhibitory response. Subsequent work in this and other systems showed that GABA concentrations in some axons of inhibitory neurons can reach 0.1 *M*, about 5 percent of the wet weight. The human brain content of GABA is 200 to 1000 times that of such neurotransmit-

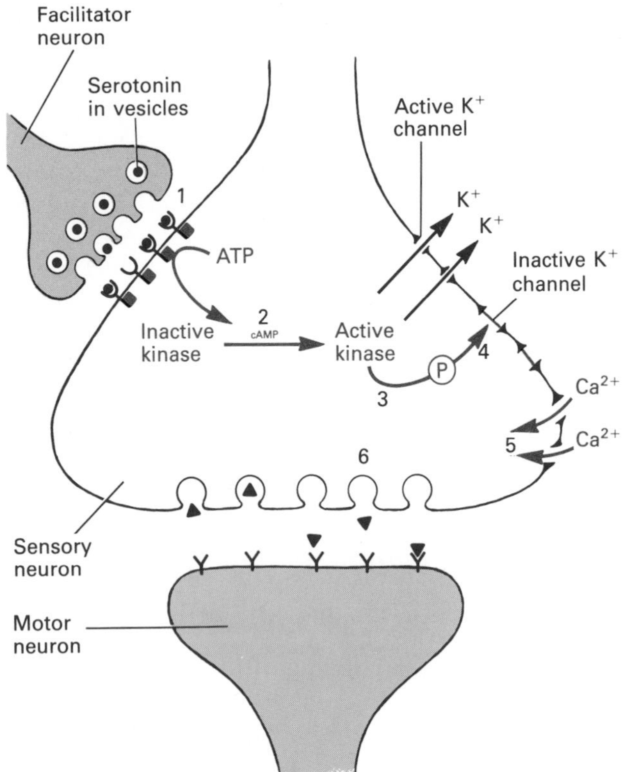

Figure 17-52 Serotonin receptors are coupled to adenylate cyclase in an *Aplysia* sensory neuron. A facilitator neuron forms an axon-axon synapse with the terminal of a sensory neuron. The sensory neuron stimulates a motor neuron, using an unknown transmitter. Stimulation of the facilitator neuron causes an increase in the ability of the sensory neuron to stimulate the motor neuron, by causing the sensory neuron to release more packets of neurotransmitter each time an action potential reaches the terminus. (1) The facilitator neuron secretes serotonin, which binds to serotonin receptors on the sensory neuron. (2) Binding triggers synthesis of cAMP, which activates a cAMP-dependent protein kinase. (3) This, in turn, phosphorylates a protein that leads to closure of certain K^+ channels (4) that usually repolarize the membrane following arrival of an action potential. (5) The prolonged depolarization results in a greater influx of Ca^{2+} ions. (6) In turn, this leads to greater exocytosis of synaptic vesicles.

Since serotonin causes some increase in intracellular Ca^{2+} concentration even in voltage-clamped sensory neurons, eliminating involvement of any membrane depolarization, cAMP-dependent phosphorylation of a Ca^{2+} channel protein might also be involved. Figure 17-55 shows other interconnections of these neurons. [See E. R. Kandel and J. Schwartz, 1982, *Science* **218**:433; M. B. Boyle et al., 1984, *Proc. Nat'l Acad. Sci.* **81**:7642; and M. J. Schuster et al., 1985, *Nature* **313**:392.]

ters as dopamine, norepinephrine, and acetylcholine. It is widely distributed in all areas of the brain and is believed to be the transmitter used in most inhibitory synapses.

Some Peptides Function as Both Neurotransmitters and Neurohormones

Nervous tissue contains an enormous number of small peptides that can affect the activity of specific neurons. A major focus of current research is to define what these peptides do and how they work. Many peptides probably function synaptically as neurotransmitters. Others act in a paracrine fashion (see Figure 16-1) as a "diffusible hormone" and seem to affect many neurons over great distances. Some peptides appear to function in both types of processes. Norepinephrine, we recall, is both a neurotransmitter and a circulating hormone, so it should not be surprising if peptides, long thought to function as hormones, are also neurotransmitters.

Capillaries in the brain are much less permeable to ions and peptides than are capillaries in other parts of the body and as a consequence most substances in the blood are excluded from the brain (this constitutes the *blood-brain barrier*); hormones in the blood, therefore, do not "confuse" central nervous system functioning.

Neurons that secrete hormones, called *neurosecretory cells,* were first discovered in connection with regulation of the function of pituitary cells by the hypothalmus (see Figure 16-29). The anterior pituitary comprises several types of cells that synthesize a number of circulating peptide hormones, among them growth hormone, FSH, LH, TSH, and ACTH. Secretion of hormones by the anterior pituitary cells is controlled by the hypothalamus, which in turn is regulated by other regions of the brain. Recall that the anterior pituitary is connected to the hypothalamus by a special closed system of blood vessels (Figure 16-29). Hypothalamic neurons secrete *hypothalamic peptide hormones* into these vessels, and the hormones then bind to receptors on the anterior pituitary cells. One hormone, TRH, stimulates secretion by the anterior pituitary of prolactin and thyrotropin. Another, LHRH, causes the secretion by the anterior pituitary of FSH and LH, which are important in regulating the growth and maturation of oocytes in the ovary (see Figure 16-29).

The application of sensitive immunohistochemical techniques has led to the discovery of these and other peptides in different areas of the brain and in the spinal cord and peripheral neurons. Increasingly they are being found in nerve terminals (Figure 17-53), where they appear to function as neurotransmitters. Indeed, there are physiological experiments indicating that in the sympathetic ganglion of the frog depicted in Figure 17-53, LHRH is the neurotransmitter that induces a long-lived excitatory postsynaptic potential. Another peptide, substance P (Figure 17-48), appears to be a neurotransmitter

(a)

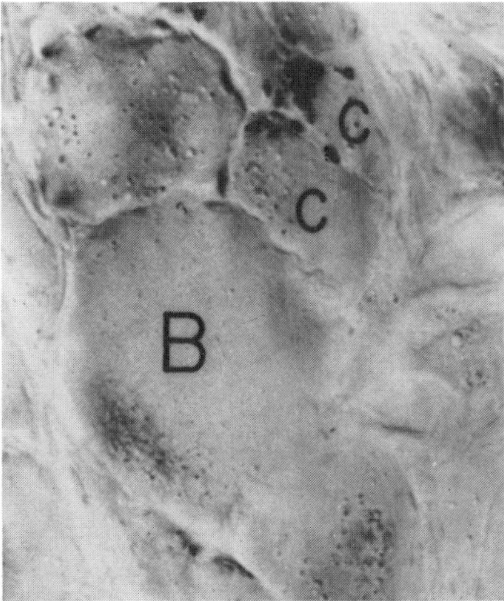

(b)

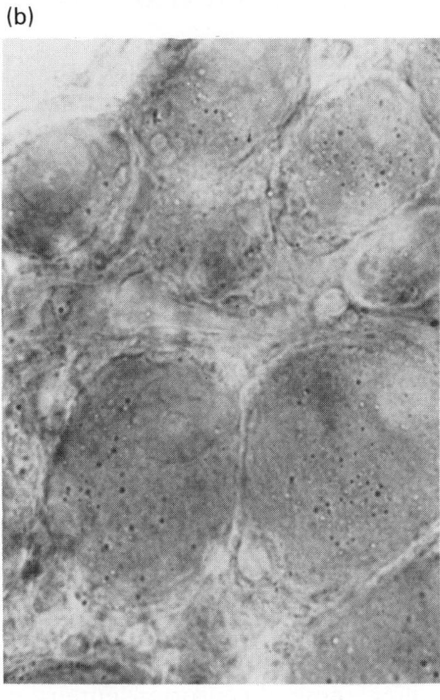

Figure 17-53 Luteinizing hormone–releasing hormone (LHRH) is present in the synaptic junctions of neurons in a frog sympathetic ganglion. (a) Sectioned ganglia were treated with anti-LHRH serum. Bound antibodies were detected with a peroxidase staining procedure that results in a black precipitate. Staining, representing the presence of LHRH, is observed in the cells labeled *C* but not in those labeled *B*. Thus, the *C* cells but not the *B* cells contain LHRH in secretory vesicles. (b) A control for this experiment. The anti-LHRH serum was pretreated with LHRH before being added to the section, removing all LHRH antibodies. Since no staining was observed, the response in (a) is specific and due to LHRH in the *C* cells. [See L. Y. Jan, Y. N. Jan, and M. S. Brownfield, 1980, *Nature* **288**:380.] *Courtesy of L. Y. Jan. Reprinted by permission from* Nature. *Copyright 1980 by Macmillan Journals Limited.*

used by sensory neurons that convey responses of pain or other noxious stimuli to the central nervous system. But the list of neuropeptides is very long, and new ones are being discovered constantly. In only a very few cases have any definitive physiological studies been done.

Endorphins and Enkephalins Are Neurohormones

The activity of neurons in both the central and peripheral nervous systems is affected by a large number of neurohormones. They are secreted by various neurons and, like other hormones, act on cells quite distant from their site of release. Neurohormones can modify the ability of nerve cells to respond to synaptic neurotransmitters. Several small polypeptides with profound effects on the nervous system have been discovered recently; examples are *Met-enkephalin, Leu-enkephalin,* and *β-endorphin.* These and other peptide hormones probably act as neurotransmitters in selected cell types but also have profound effects on general life events like mood, sleep, and body growth. Enkephalins and endorphins, for instance, function as natural pain killers or opiates and decrease the pain responses in the central nervous system.

Enkephalins were discovered during studies in the early 1970s focusing on the mechanism of opium addiction. Several groups of researchers discovered that brain membranes contain high-affinity binding sites for purified opiates such as the alkaloid morphine. The sites were presumed to be the receptors that mediated the effects of these narcotic, analgesic drugs. Since such receptors exist in the brains of all vertebrates from sharks to man, the question was raised why vertebrates should have highly specific receptors for alkaloids produced by opium poppies and why these should have survived eons of evolution. Since none of the neurotransmitters and peptides then known could serve as agonists or antagonists for the binding of opiates to brain receptors, a search was begun for natural compounds that could. This led to the discovery of two pentapeptides, Met-enkephalin and Leu-enkephalin (Figure 17-48). They both bound to the "opiate" receptors in the brain and had the same effect as morphine when injected into the ventricles (cavities) of brains of experimental animals—a profound analgesia. Enkephalins and endorphins appear to act by inhibiting neurons that transmit pain impulses to the spinal cord; presumably these neurons contain abundant endorphin or enkephalin receptors.

During periods of extreme stress, both humans and other animals can exhibit a remarkable insensitivity to pain; soldiers who lose a limb in battle sometimes do not feel the pain for many hours. The anterior pituitary is known to secrete β-endorphin and possibly other analgesic peptides during stressful periods. By diffusing through specific regions of the central nervous system, these peptides may inhibit neurons associated with pain impulses. Endorphins, then, are examples of neurohormones affecting large numbers of neurons that have a related function in the body.

The Neurotransmitter Released by a Neuron May Change during Development

For many years it was thought that each neuron released only one neurotransmitter. There is now increasing evidence that many bioactive peptides are found in the same neurons that contain a "classical" transmitter such as norephinephrine or acetylcholine. If both compounds are indeed released at a synapse, then multiple excitatory and inhibitory responses can be generated in different postsynaptic cells with different receptors.

We do know that the transmitters secreted by cells may change during the course of differentiation. The cell is affected by many developmental signals, such as hormones and various proteins released from nonneuronal cells. This is true of vertebrate sympathetic neurons that innervate cells and use acetylcholine as a transmitter. Sympathetic neurons arise from the embryonic neural crest. Cells migrate from the neural crest to different places in the embryo and give rise to a variety of neuronal cell types. Initially, sympathetic cells synthesize only catecholamines, but after migration a very small percentage of these cells receive a signal that causes them to begin acetylcholine synthesis and reduce catecholamine production. This process can also be observed in tissue culture. Cultures of dissociated sympathetic neurons from the newborn rat, free of nonneuronal cells, synthesize only norepinephrine. When certain nonneuronal cells such as heart muscle cells are added, some of these neurons are induced to begin acetylcholine synthesis. Similar changes can be evoked when proteins secreted by the heart muscle are added to the cultured neuron. This suggests a hormone-like effect.

Selectivity in Synapse Formation Is a Formidable Problem in Cell-Cell Recognition

A major focus of current research concerns the mechanism by which specific synapses are formed. For example, the cell bodies of motor neurons that innervate skeletal muscles are located in the spinal cord, and during early development their axons grow toward the exact muscles they will innervate. How do the axons "know" how to migrate toward a muscle that might be meters away from the cell body? And how do the axon terminals of a sensory neuron know with which interneurons they are meant to synapse?

Several mechanisms have been proposed. One possibility is chemoaffinity: nerves or muscles might release chemicals that attract the desired nerve cells. Another possibility is that the growth of cells and the elongation of axons or dendrites are controlled by local concentrations of trophic or growth factors. A third possibility involves construction of pathways for axonal growth. Axonal motor neurons, for instance, grow outward from the spinal cord at different times, and once a few "pioneer" neurons establish proper connections with the appropriate muscle, other axons grow alongside them toward adjacent muscles. A fourth proposed mechanism is programmed cell death. Many more neurons may grow toward a target muscle, for instance, than will ultimately form synapses with it, and those that fail to establish proper synapses will die.

Memory and Neurotransmitters

How do we learn? How do we remember? In its most general sense, learning is an aspect of behavior—it is the process by which humans and animals modify their behavior as a result of experience or as a result of acquisition of information about the environment. Memory is the process by which this information is stored and retrieved.

Psychologists have defined two types of memory, depending on how long it persists: short-term (minutes to hours) and long-term (days to years). It is generally accepted that memory results from changes in the structure or function of particular synapses, but until recently learning and memory could not be studied with the tools of cell biology or genetics. Many researchers believe that long-term memory might involve the formation or elimination of specific synapses in the brain. Short-term memory occurs too rapidly to be attributed to such gross alterations, and attention has been focused on changes in the release and functioning of neurotransmitters at particular synapses as explanations of simple learning processes. The fruit fly *Drosophila* and sea slug *Aplysia* exhibit elemental forms of learning, and some insights into the molecular events of memory have been obtained from these organisms.

Mutations in *Drosophila* Affect Learning and Memory

Remarkable as it sounds, *Drosophila* can be trained to avoid certain noxious stimuli. During the training period, a population of fruit flies is exposed to two different stimuli, either two odoriferous chemicals or two colors of light. One of the two is associated with an electric shock. The flies are then removed and placed in a new apparatus, and the two stimuli are repeated but without the electric shock. The flies are tested for their avoidance of the stimulus associated with the shock. About half the flies learn to avoid the stimulus associated with the shock, and this memory persists for at least 24 h. Painstaking observation of mutagenized flies has led to the identification of several mutations that cause defects in this learning process. Mutants homozygous at the *dunce* allele cannot learn; otherwise they appear to be normal. Mutants homozygous at the *amnesiac* allele learn normally but forget four times as quickly as wild-type flies.

The dunce mutation is understood at the molecular level: all dunce mutants are defective in one of two cAMP phosphodiesterase enzymes. Mutants previously identified as being defective in this phosphodiesterase gene also exhibit the learning defects of dunce. Thus the *dunce* gene appears to encode a cAMP phosphodiesterase. But it is not yet known which synapses are involved in the learning response nor how the defect in phosphodiesterase might affect the function of these synapses. It seems that cAMP enhances the calcium-mediated release of neurotransmitters in some neurons and, indeed, other learning mutants have a defective calmodulin-dependent adenylate cyclase enzyme.

A Simple Reflex in the Sea Snail *Aplysia* Exhibits Both Habituation and Sensitization

Aplysia, the marine snail, exhibits two of the most elementary forms of learning familiar in vertebrates: *habituation* and *sensitization. Habituation* is the decrease in a behavioral response to a stimulus when the stimulus is repeated and has no adverse effect. An animal might be startled by a loud noise. Continued repetition of the noise will cause less and less of a response. The animal has become habituated to the stimulus. Sensitization, in contrast, is an increase in behavioral response to an intense or noxious stimulus. For instance, an odor or small shock will elicit a withdrawal response, but the response is enhanced if the stimulus follows another, especially painful, stimulus such as a sharp pinch.

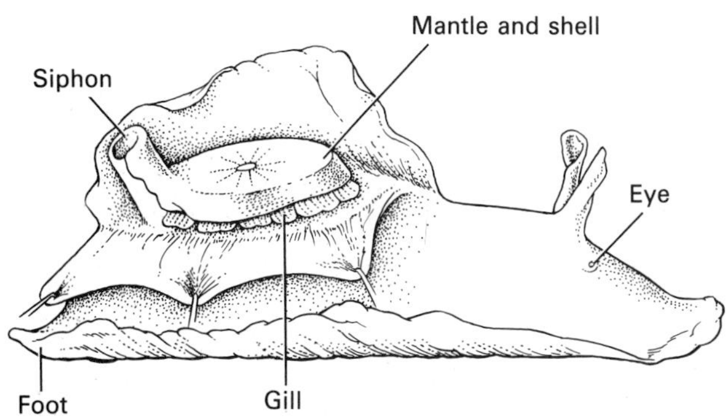

Figure 17-54 The marine snail *Aplysia punctata.* The gill is under the protective mantle; it can be seen if the overlying tissue is pulled aside. *After E. R. Kandel, 1976,* Cellular Basis of Behavior, *W. H. Freeman and Company, p. 76.*

Habituation When *Aplysia* (Figure 17-54) is touched gently on the siphon, the gill muscles contract vigorously and the gill retracts into the mantle cavity. The reflex is analogous to defensive escape responses in vertebrates, and it is mediated by a simple reflex arc (Figure 17-55). Sensory neurons in the siphon synapse with motor neurons that innervate the gill muscles.

When repeatedly touched, the animal learns to "ignore" the stimulus. After 10 or 15 stimulations on the siphon in rapid sequence the gill response is only about one-third its initial extent. By recording the electrical changes in the motor neuron, researchers discovered that this habituative response is due to changes in the amount of neurotransmitter released at the synapse between the sensory and motor neurons. With repeated stimulation of the siphon the magnitude of the excitatory postsynaptic potential is decreased because fewer quanta of neurotransmitter are released by the sensory neurons. We have noted that release of neurotransmitters is triggered by a rise in intracellular Ca^{2+} ion which, in turn, is controlled by a voltage-dependent Ca^{2+} channel. Measurement of Ca^{2+} movements in the sensory neuron showed that habituation results in a decrease in the number of voltage-dependent Ca^{2+} channels that open as a response to the arrival of the action potential at the terminal. Habituation does not affect the generation of action potentials in the sensory neuron nor the response of the receptors in the postsynaptic cells. While vertebrates exhibit a number of habituative responses similar to that of *Aplysia,* it is not yet known whether they can be explained by modifi-

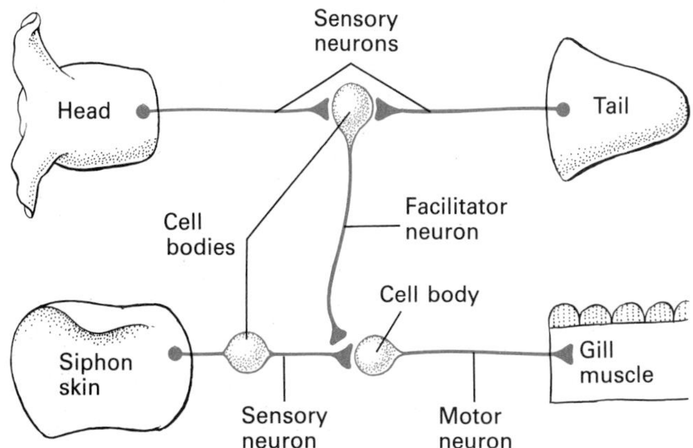

Figure 17-55 Neural circuits involved in the gill-withdrawal reflex of the sea snail *Aplysia.* For simplicity, certain of the interneurons are omitted. Figure 17-52 shows the details of the synapses between the sensory, facilitator, and motor neurons. [See E. R. Kandel and J. H. Schwartz, 1982, *Science* **218**:433.]

cations in the properties of particular channel proteins in specific synapses.

Sensitization If a habituated *Aplysia* is given a strong, noxious stimulus, such as a blow on the head or tail, it will respond to the next weak stimulus of the siphon by a rapid withdrawal of the gill. This noxious stimulation is said to *sensitize* the animal so that it responds quickly to its normal stimulus, touching of the siphon. Electrophysiological studies have shown that *Aplysia* sensitization is mediated by a set of neurons, called *facilitator neurons,* that are activated by shocks to the head or tail. Electron microscopy shows that the axon of the facilitator neuron synapses with the terminal of the sensory neuron near the site where the sensory neuron synapses with the motor neuron (Figure 17-55). Stimulation of the facilitator neuron causes the sensory neuron to release more transmitter in its synapse with the motor neuron, thus increasing the magnitude of the gill withdrawal reaction. These events provide a partial explanation for the phenomenon of sensitization in *Aplysia.*

But how does stimulation of the facilitator neuron cause the sensory neuron to release more transmitter? The facilitator neurons apparently release serotonin. Experimental application of transmitters to the sensory neuron showed that only serotonin causes the sensitization response. Since it was known that these serotonin receptors activate adenylate cyclase (see Figure 17-52), cAMP was microinjected into adapted sensory neurons. Injected cAMP had the same sensitizing effect as did a direct blow to the tail. Microinjection of the catalytic subunit of cAMP-dependent protein kinase also had the same effect. It appears that facilitating neurons mediate sensitization by secreting serotonin, which acts via receptors on the sensory neuron to increase intracellular cAMP, and an increase in cAMP results in an increase in the amount of transmitter released by the sensory neuron.

To complete this chain of events it is necessary to understand how elevated cAMP levels enhance the release of neurotransmitters. We noted (page 755) that a direct effect of the cAMP-dependent protein kinase appears to be a decrease in the ability of one of several voltage-gated K^+ channels in the membrane to be opened as a result of membrane depolarization. This effect is due to phosphorylation of the channel protein or associated protein. This K^+ channel, whose activity is thus decreased after serotonin treatment, normally contributes to the repolarization of the membrane during an action potential. By blocking this channel, action potentials reaching the nerve terminal decay more slowly. This causes a longer and larger than usual influx of Ca^{2+} ions, via the voltage-dependent Ca^{2+} channels. More internal Ca^{2+} leads to more extensive exocytosis of neurotransmitter by the sensory cell at its synapse with the motor neuron.

Sensitization in *Aplysia* is one of the few cases where short-term changes in synaptic function are understood in molecular detail. Possibly it will serve as a model for understanding more complex forms of behavior, such as long-term memory in vertebrates.

Sensory Transduction: The Visual System

The nervous system receives input from a large number of sensory receptors. Photoreceptors in the eye, taste receptors on the tongue, and touch receptors on the skin monitor various aspects of the outside environment. Stretch receptors surround many muscles and fire when the muscle is stretched. Internal receptors monitor the levels of glucose, salt, and water in body fluids. The nervous system, the brain in particular, processes and integrates this vast barrage of information and coordinates the response of the organism.

The "language" of the nervous system is electrical signals. Each of the many types of receptor cells must convert, or *transduce,* its sensory input into an electrical signal. A few sensory receptors are themselves neurons that generate action potentials in response to stimulation. Most are specialized epithelial cells that do not generate action potentials but stimulate adjacent neurons at synapses and then generate action potentials. The key question that concerns us here is: How does a sensory cell transduce its input into an electrical signal?

The system that is understood in the greatest detail is light reception by rod cells in the mammalian retina. A cascade of well-understood cellular reactions converts the light signal into changes in membrane potential. The system exhibits analogies to the cAMP cascade that mediates the adrenergic response in muscle and liver cells (see Figure 16-16).

A Retinal Rod Cell Can Be Excited by a Single Photon

The human retina contains two types of photoreceptors, *rods* and *cones.* The cones are involved in color vision and function in bright light. The rods (Figures 17-56 and 17-57) are stimulated by weak light over a range of wavelengths. These cells form synapses with *bipolar cells* that also interact with *ganglion cells* in the retina. In the outer segment of the rod cell are disks that contain photoreceptor pigments. In the region of the rod cell that forms synapses with bipolar cells are the synaptic vesicles.

The resting potential of a typical rod cell, about −30 mV, is less than that of other neurons, which are typically −60 to −90 mV. Careful recording with intracellular electrodes has shown that a pulse of light causes the membrane in the outer segment of the rod cell to become slightly hyperpolarized (Figure 17-58). In the absence of light rod cells are depolarized and release neurotransmitters. The light-induced hyperpolarization causes a decrease in transmitter release.

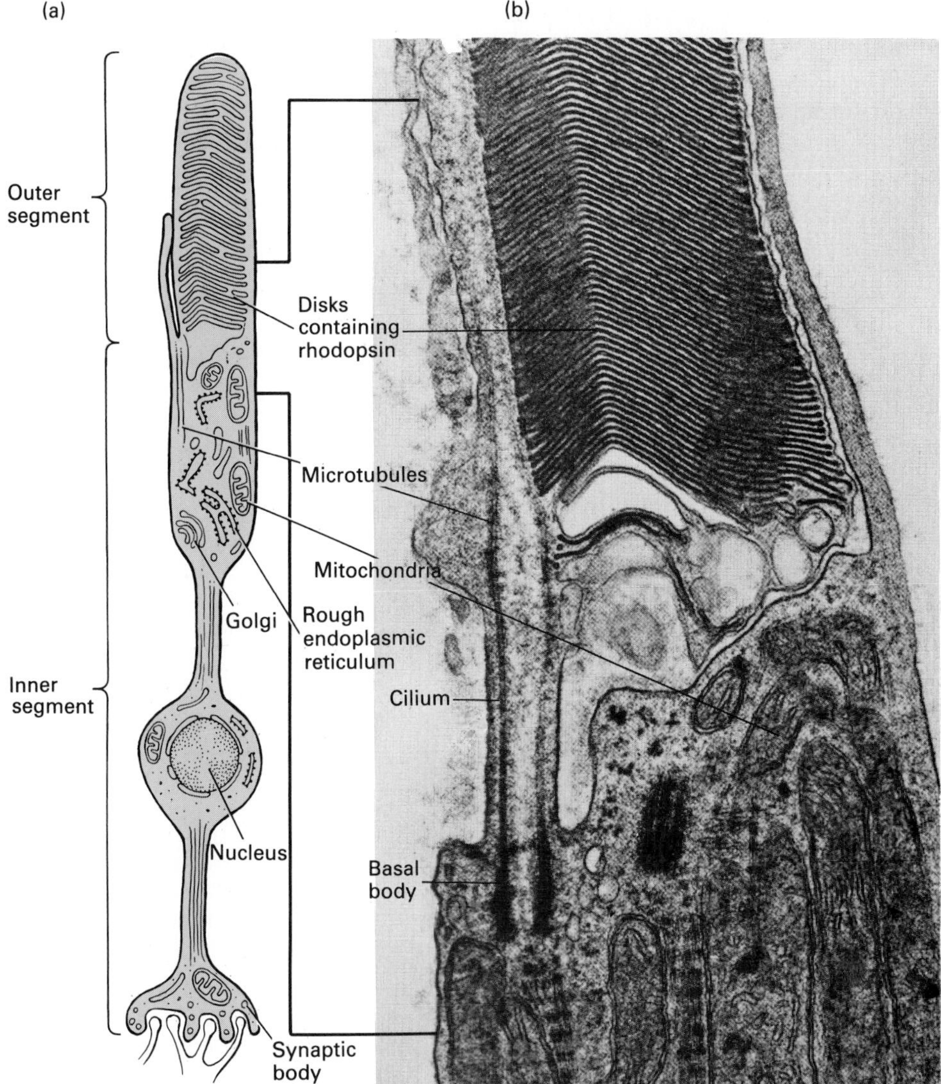

Figure 17-56 (a) Diagrammatic structure of a human rod cell and (b) an electron micrograph of a region of the rod cell at the junction of the inner and outer segments. At the synaptic body, the rod cell forms a synapse with one or more bipolar neurons, as depicted in Figure 17-57. *Part (b) from R. G. Kessel and R. H. Kardon, 1979,* Tissues and Organs: A Text-Atlas of Scanning Electron Microscopy, *W. H. Freeman and Company, p. 91.*

Remarkably, a single photon absorbed by these cells produces a measurable response, a hyperpolarization of about 1 mV that last a second or two. Only about 30 to 50 photons are necessary for half-maximal hyperpolarization of a single rod cell. This receptor, like many others, exhibits the phenomenon of *habituation.* More photons are required to cause a hyperpolarization if the rod cell is continuously exposed to light than if it is kept in the dark.

Hyperpolarization of Rod Cells Is Caused by a Closing of Sodium Channels

The plasma membrane of the rod cell is unusual because in its resting state in the dark it is highly permeable to Na^+ ions resulting from the presence of a large number of open Na^+ channels. We recall that any decrease in the permeability to Na^+ ions will cause a hyperpolarization. The effect of light on these cells is to close Na^+ channels—the more that are closed, the more negative the inside of the cell becomes. A single photon blocks the inflow of more than a million Na^+ ions.

Let us now turn to the two key questions: How is light absorbed, and how is the signal transduced into the closing of Na^+ channels?

Absorption of a Photon Triggers Isomerization of Retinal and Activation of Opsin

The photoreceptor pigment in rod cells is called *rhodopsin.* It consists of a transmembrane protein, *opsin,* to which is bound the prosthetic group 11-*cis*-retinal

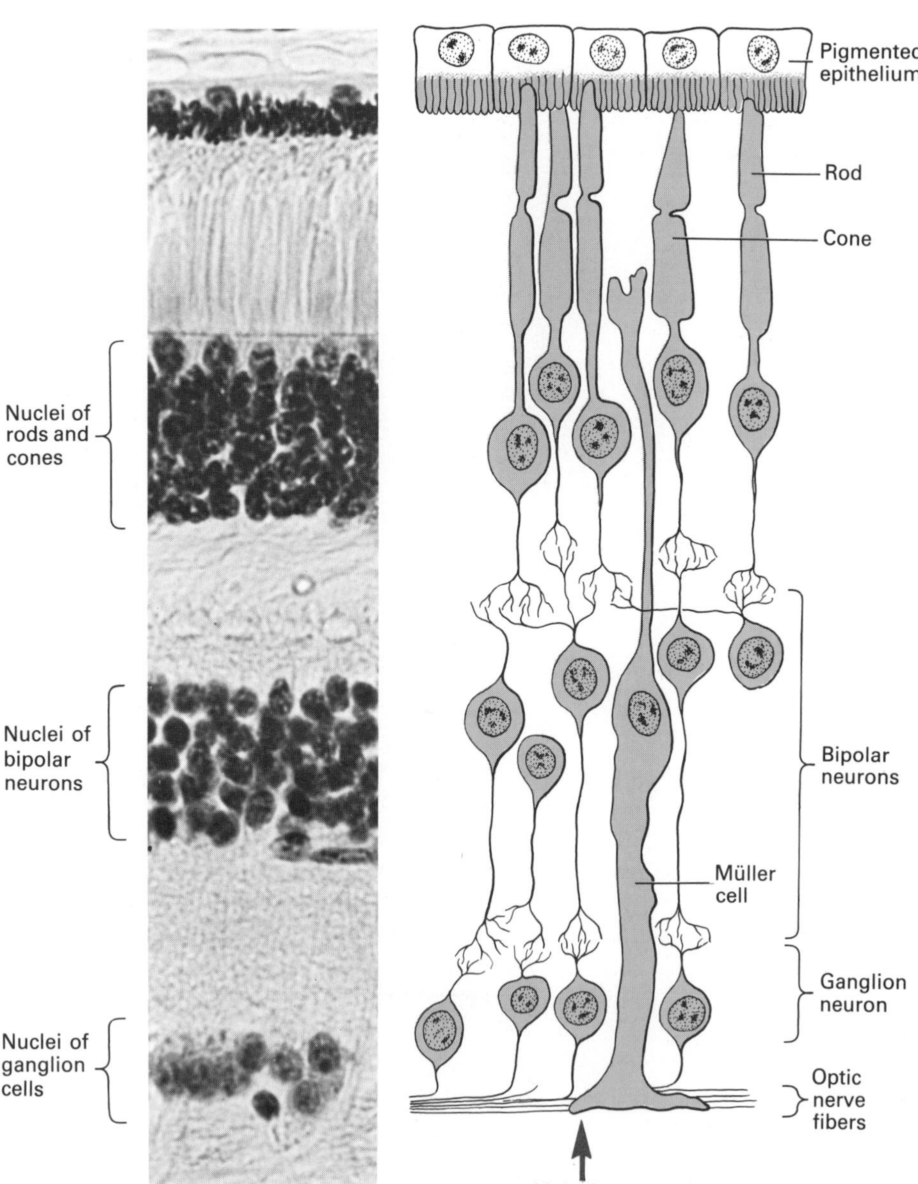

Figure 17-57 The cells of the human retina. The outermost layer of cells forms a pigmented epithelium in which the tips of the rod and cone cells are buried. The axons of the rods and cones synapse with many bipolar neurons. These, in turn, synapse with cells in the ganglion layer that send axons—optic nerve fibers—through the optic nerve to the brain. By synapsing with multiple rod cells certain bipolar cells integrate the responses of many cells. They are involved in recognizing patterns of light that fall on the retina—for instance, a band of light that excites a set of rod cells in a straight line. The layered arrangement of the retina is evident from the localization of the nuclei. Those of rods and cones are clustered in one layer, those of bipolar neurons in another, and those of the ganglion nerve cells in a third. Müller cells are supportive nonneuronal cells that fill much of the retinal spaces. *From R. G. Kessel, and R. H. Kardon, 1979,* Tissues and Organs: A Text-Atlas of Scanning Electron Microscopy, *W. H. Freeman and Company, p. 87.*

(Figure 17-59). Rhodopsin is localized to the thousand or so flattened membrane disks that make up the rod's outer segment.

11-*cis*-Retinal is the pigment that absorbs light in the visible range (400–600 nm). The primary photochemical event is the isomerization of 11-*cis*-retinal to all-*trans*-retinal. This markedly changes the conformation of the retinal; thus, the energy of light has been converted into atomic motion. The all-*trans*-retinal, unlike the 11-*cis* form, cannot form a stable complex with opsin and dissociates from it. This converts the opsin into an altered and activated form:

$$\text{O—R}_{cis} \xrightarrow{\text{light}} \text{O—R}_{trans} \longrightarrow \text{O}^* + \text{R}_{trans}$$

where O* represents the active form of opsin free of retinal (R). Eventually the cell converts the all-*trans*-retinal back to the 11-*cis* form, which then rebinds to opsin.

These events provide a good answer to the first question, how light is absorbed. Now, how does an activated opsin convey its signal to the Na^+ channels? Calcium ion and cyclic GMP (cGMP) are the most attractive candidates for the transmitter molecules that carry the signal from activated opsin to the Na^+ channels in the cell membrane.

Cyclic GMP Is a Key Transducing Molecule

Much recent work has indicated that cGMP is a key transducer molecule. Rod outer segments contain it in an

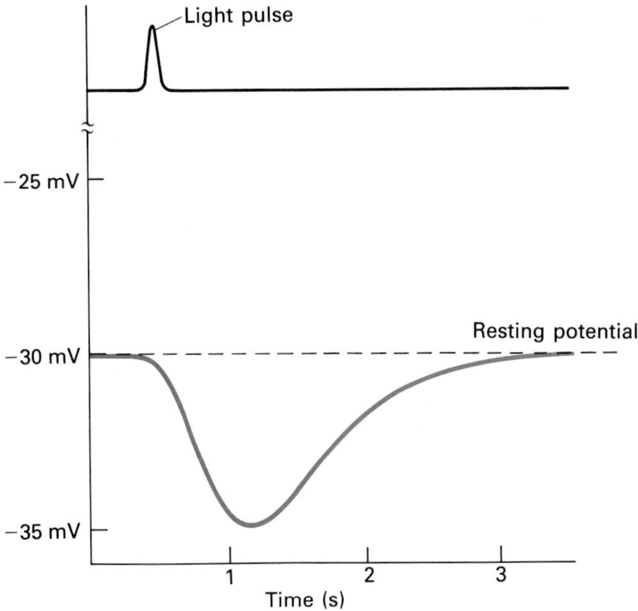

Figure 17-58 A brief pulse of light causes a transient hyperpolarization of the rod cell membrane.

unusually high concentration, about 0.07 m*M*, and its concentration *drops* upon illumination. Light appears to have no effect on the synthesis of cGMP from GTP:

$$\text{GTP} \xrightarrow{\text{guanylate cyclase}} 3',5'\text{-cGMP} + \text{PP}_i$$

However, rod outer segments contain a GMP phosphodiesterase that is activated by light:

$$3',5'\text{-cGMP} + \text{H}_2\text{O} \xrightarrow{\text{cGMP phosphodiesterase}} 5'\text{-GMP}$$

Finally, injection of more cGMP into the cell depolarizes the cell membrane. It appears that in the dark the high level of cGMP acts to keep the Na^+ channels open. Direct support for this conclusion has been obtained by the patch clamp technique. Isolated patches of rod outer segment plasma membrane contain abundant Na^+ channels. When cGMP is added to the cytoplasmic surface of these patches, there is a rapid increase in the number of open Na^+ channels. The effect occurs in the absence of protein kinases or phosphatases, and it appears that cGMP acts directly on the channels to keep them open, probably by cooperative allosteric binding. Light closes the channels by activating the cGMP phosphodiesterase and lowering the level of cGMP. Several hundred molecules of cGMP phosphodiesterase must be activated by a single photon. How does this happen?

A recently discovered rod cell protein called *transducin* (T) has been found to mediate the activation of the cGMP phosphodiesterase (Figure 17-60). In resting cells, transducin contains a tightly bound molecule of GDP and is incapable of affecting the cGMP phosphodiesterase. Activated opsin, O*, catalyzes the exchange of a free GTP for a transducin GDP, converting the transducin to a form that can activate cGMP phosphodiesterase. A single molecule of photolyzed O* still in the disk membrane can apparently activate up to 500 transducin molecules. Biochemical studies have shown that T·GTP (Figure 17-61) can indeed activate cGMP phosphodiesterase. Transducin also contains a GTPase; this activity slowly converts T·GTP to T·GDP, resulting in inactivation of cGMP phosphodiesterase.

11-*cis*-Retinal + H_2N—$(CH_2)_4$—Opsin $\overset{H^+}{\rightleftharpoons}$ R—C(H)=N$^+$(H)—$(CH_2)_4$—Opsin + H_2O

11-*cis*-Retinal   Opsin   Protonated Schiff base

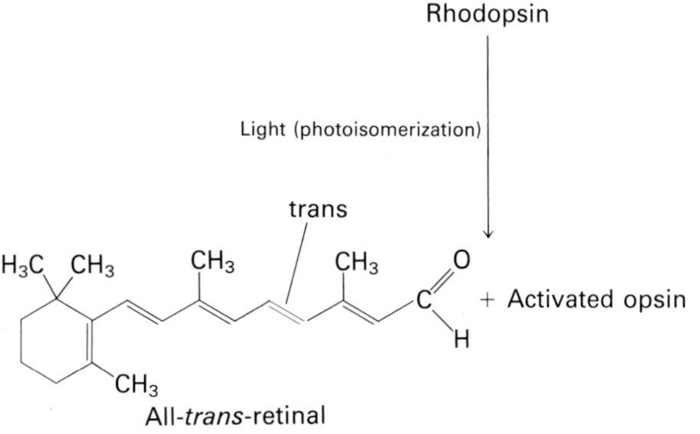

Figure 17-59 The structures of 11-*cis*-retinal, all-*trans*-retinal, and the Schiff base linkage that binds 11-*cis*-retinal to a lysine of opsin. Light causes isomerization of *cis*-retinal to the trans form; because the trans form is unable to bind to opsin, it dissociates immediately.

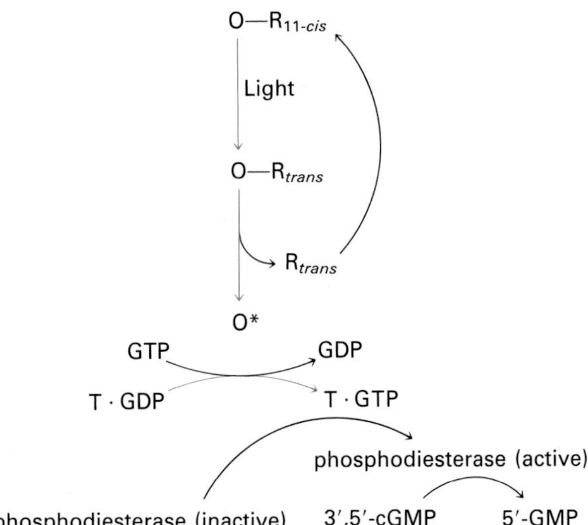

Figure 17-60 Photolyzed activated opsin, O*, catalyzes the activation of transducin, T, by catalyzing the exchange of GTP for GDP. The T · GTP, in turn, activates cGMP phosphodiesterase.

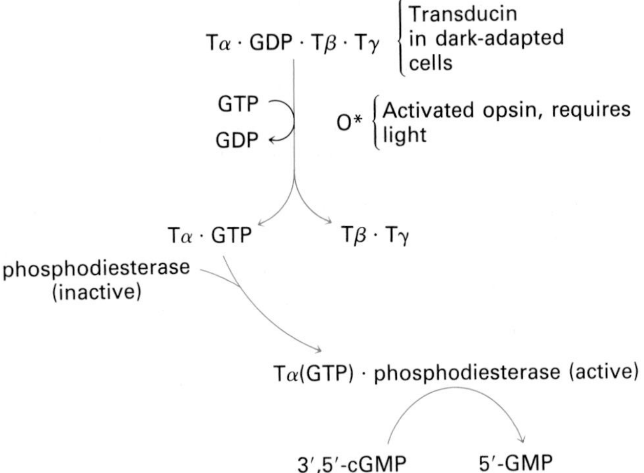

Figure 17-61 Details of the activation of transducin. Transducin consists of three subunits, Tα (m.w. 39,000), Tβ (m.w. 36,000), and Tγ (m.w. 8400). The α subunit binds GDP or GTP. In dark-adapted cells T exists as a complex of Tα · GDP · Tβ · Tγ. Exchange of GTP for GDP, catalyzed by activated opsin (O*), causes dissociation of Tα(GTP). The Tα(GTP) binds to and activates the cGMP phosphodiesterase.

The careful reader will have noted obvious similarities between transducin and the G protein of the adenylate cyclase system (see Figures 16-9 and 16-11). Activated opsin triggers GTP exchange for GDP on T, just as binding of agonists to the β-adrenergic receptor triggers exchange of GTP for GDP on G. The G·GTP complex activates adenylate cyclase, and T·GTP activates cGMP phosphodiesterase. Both G and T have hydrolizing activities that slowly convert G·GTP to G·GDP and T·GTP to T·GDP. Both G·GTP and T·GTP serve to amplify the initial signal. There may even be a strong homology between the γ and β subunits of the G and T proteins. Apparently, a single-signal amplification system evolved and was modified to mediate different kinds of responses in very different types of cells.

Summary

An electric potential exists across the plasma membrane of all the body's cells. It is caused by the different ionic compositions of the cytosol and extracellular fluid and by the different permeabilities of the plasma membrane to the principal cellular ions, sodium, potassium, chloride, and calcium ions. In most nerve and muscle cells, where the potential can be measured by a microelectrode inserted into the cell, the resting potential is about 60 mV, negative on the inside.

Impulses are conducted along the nerve axon by means of action potentials. An action potential is a sudden (less than a millisecond) depolarization of the membrane followed by a rapid hyperpolarization and a gradual return to the resting potential. These changes in membrane potential are caused by a sudden transient increase in the permeability of the membrane to Na^+ followed by a slower transient increase in its permeability to K^+. Changes in Na^+ permeability are caused by transmembrane proteins that act as specific Na^+ channels, which respond to depolarization of the membrane by transiently opening and admitting Na^+ into the cytosol. Voltage-dependent K^+ channels open in response to membrane depolarization; their opening repolarizes the membrane.

Action potentials are generated only when the nerve membrane at the region of the axon hillock is depolarized to the threshold value. An action potential generated at one point along an axon will lead to depolarization of the adjacent segment and thus to propagation of the action potential along its length. The speed of conduction depends on the diameter of the axon and conductivity of the neuronal cytosol. Thick nerves conduct faster than thin ones, and myelinated nerves conduct faster than unmyelinated nerves of similar diameter because of insulation of the neuron by the myelin sheath.

Impulses are transmitted from neurons to other cells at specialized junctions called synapses. In electrical synapses, ions pass from the presynaptic cell to the postsynaptic cell through gap junctions, and an action potential is generated in the postsynaptic cell with no time delay. In the more common chemical synapses, the arrival of an action potential in the presynpatic axon triggers the release of neurotransmitters into the synaptic cleft; from there they bind to receptors on the postsynaptic cell. Transmitters are stored in membrane-limited vesicles, and exocytosis of these vesicles is triggered by a rise in cytosolic Ca^{2+} induced by the opening of voltage-sensitive Ca^{2+} channels.

Many synapses are excitatory. Neurotransmitter acts to depolarize the postsynaptic cell and generate an action potential. In the synapse of a motor neuron and muscle cell, binding of agonist to the well-studied acetylcholine receptor triggers an increase in permeability of the membrane to both Na^+ and K^+ ions, leading to depolarization. In other postsynaptic cells the depolarization of the postsynaptic membrane is less extensive but longer-lived, on the order of seconds. Some synapses are not excitatory but inhibitory—the release of neurotransmitter triggers a hyperpolarization of the postsynaptic membrane, making it more difficult for the cell to generate an action potential. The action of GABA is mediated by an increase in permeability of the membrane Cl^- ion. In some postsynaptic cells, receptors for epinephrine and serotonin modulate the activity of adenylate cyclase. The electrical response of these cells is believed to be caused by phosphorylation of Na^+ or K^+ channel proteins by the cAMP-dependent protein kinase. Depending on the specific receptor in the postsynaptic cell, the same neurotransmitter can induce either an excitatory or inhibitory response.

Many compounds released by neurons are systemic hormones as well as neurotransmitters, affecting both distant secretory cells and adjacent neurons. Recent work suggests that small peptides such as endorphins, enkephalins, and hypothalamic releasing factors function as neurotransmitters in particular synapses in the brain and also act as hormones.

Removal of neurotransmitter from the synapse is essential for ensuring its repeated functioning. The action of acetylcholine is terminated by the enzyme acetylcholinesterase. Other neurotransmitters are removed by diffusion or by re-uptake into the presynaptic cell.

Especially in the central nervous system, many neurons must integrate excitatory and inhibitory stimuli from dozens or hundreds of other neurons, if not a thousand or more. Whether or not the threshold potential is reached at the axon hillock of the cell membrane depends on the timing and magnitude of these stimuli, the localization and duration of the resultant local hyperpolarizations and depolarizations, and the ability of the localized changes in potential to be conducted along the membrane surface.

Modifications in the activity of certain synapses are associated with short-term memory, at least in some invertebrate systems. Certain *Drosophila* mutants that cannot learn are defective in cAMP metabolism. In the sea slug *Aplysia,* a simple form of memory is linked to the opening and closing of K^+ channels in the synaptic terminal of a sensory neuron, which alters the flux of Ca^{2+} and the amount of transmitter released. A facilitator neuron forms a synapse with a sensory neuron. Serotonin is released by the facilitator neuron and causes a modification of the cAMP level in the sensory neuron, which in turn blocks the K^+ channels. This prolongs depolarization and increases exocytosis of neurotransmitter.

Many sensory transduction systems convert signals from the environment—light, taste, sound, touch—into electrical signals in certain neurons. These signals are collected, integrated, and processed by the central nervous system. The sensory system understood in most molecular detail is that of the photoreceptor rod cells. Absorption of even a single photon results in hyperpolarization of the plasma membrane and affects the release of chemical transmitters to adjacent nerve cells. The photon causes isomerization of 11-*cis* retinal bound to rhodopsin. The modified opsin causes the activation of a transducer protein called transducin (T) by catalyzing exchange of free GTP for bound GDP. Activated T·GTP, in turn, activates a cGMP phosphodiesterase. The resultant lowered level of cGMP causes closing of the membrane Na^+ channels.

References

General

The Brain. 1979. *Sci. Am.* **241**(3).

COOKE, I., and M. LIPKIN, JR., eds. 1972. *Cellular Neurophysiology: A Source Book.* Holt, Rinehart and Winston. An anthology of important papers, 1921–1967.

*HILLE, B. 1984. *Ionic Channels of Excitable Membranes,* 2d ed. Sunderland, Mass.: Sinauer Associates.

*KANDEL, E. R., and J. H. SCHWARZ. 1981. *Principles of Neural Science.* Elsevier.

*KATZ, B. 1966. *Nerve, Muscle and Synapse,* 2d ed. McGraw-Hill.

*KEYNES R. D., and D. J. AIDLEY. 1981. *Nerves and Muscle.* Cambridge University Press.

*KUFFLER, S. W., and J. G. NICHOLLS. 1974. *From Neuron to Brain,* 2d ed. Sunderland, Mass.: Sinauer Associates.

**Molecular Neurobiology.* 1983. *Cold Spring Harbor Symp. Quant. Biol.,* vol. 48.

PATTERSON, P. H., and D. PURVES, eds. 1982. *Readings in Developmental Neurobiology.* Cold Spring Harbor Laboratory. An anthology of recent papers.

*Book or review article that provides a survey of the topic.

*Science 225 (4668). September 21, 1984. An entire issue devoted to reviews in the neurosciences.

SIEGEL, G. J., R. W. ALBERS, B. W. AGRANOFF, and R. KATZMAN, eds. 1981. *Basic Neurochemistry,* 3d ed. Little, Brown.

The Origin of the Electric Potential

HILLE, B. 1981. Excitability and ionic channels. In *Basic Neurochemistry,* 3d ed., G. J. Siegel, R. W. Albers, B. W. Agranoff, and R. Katzman, eds. Little, Brown.

KUFFLER, S. W., and J. G. NICHOLLS. 1976. *From Neuron to Brain.* Sunderland, Mass.: Sinauer Associates, chaps. 5 and 6.

The Action Potential

Ionic Movements

ADAMS, P. 1982. Voltage-dependent conductances of vertebrate neurons. *Trends Neurosci.* **5**:116–119.

HODGKIN, A. L., and A. F. HUXLEY. 1952. Currents carried by sodium and potassium ions through the membrane of the giant axon of *Loligo. J. Physiol.* **116**:449–472.

HODGKIN, A. L., and A. F. HUXLEY. 1952. The dual effect of membrane potential on sodium conductance in the giant axon of *Loligo. J. Physiol.* **116**:497–506.

*HODGKIN, A. L., and A. F. HUXLEY. 1952. A quantitative description of membrane current and its application to conduction and excitation in nerve. *J. Physiol.* **117**:500–544.

HODGKIN, A. L., A. F. HUXLEY, and B. KATZ. 1952. Measurement of current-voltage relations in the membrane of the giant axon of *Loligo. J. Physiol.* **108**:37–77.

*KEYNES, R. D. 1979. Ion channels in the nerve-cell membrane. *Sci. Am.* **240**(3):126–135.

*SAKMAN, B., and E. NEHER, eds. 1983. *Single-Channel Recording.* Plenum.

STEVENS, C. F. 1984. Biophysical studies of ion channels. *Science* **225**:1346–1350.

The Voltage-Dependent Sodium Channel

ALDRICH, R. W., D. P. COREY, and C. F. STEVENS. 1983. A reinterpretation of mammalian sodium channel gating based on single channel recording. *Nature* **305**:436–441.

ARMSTRONG, C. M. 1981. Sodium channels and gating current. *Physiol. Rev.* **61**:644–683.

BARCHI, R. L., J. C. TANAKA, and R. E. FURMAN. 1984. Molecular characteristics and functional reconstitution of muscle voltage-sensitive sodium channels. *J. Cell Biochem.* **26**:135–146.

CATERALL, W. A. 1986. Voltage-dependent gating of sodium channels: correlating structure and function. *Trends Neurosci.* 9:7–10.

CATTERALL, W. A. 1985. The electroplax sodium channel revealed. *Trends Neurosci.* **8**:1–3.

KRUEGER, B. K., J. F. WORLEY III, and R. J. FRENCH. 1983. Single sodium channels from rat brain incorporated into planar lipid bilayer membranes. *Nature* **303**:172–175.

NODA, M., S. SHIMIZU, T. TANABE, T. TAKAI, T. KAYANO, T. IKEDA, H. TAKAHASHI, H. NAKAYAMA, Y. KANAOKA, N. MIKNAMINO, K. KANGAWA, H. MATSUO, M. A. RAFTERY, T. HIROSE, S. INAYAMA, H. HAYASHIDA, T. MIYATA, and S. NUMA. 1984. Primary structure of *Electrophorus electricus* sodium channel deduced from cDNA sequence. *Nature* **312**:121–127.

*SIGWORTH, F. J., and E. NEHER. 1980. Single Na^+ channel currents observed in cultured rat muscle cells. *Nature* **287**:447–449.

The Voltage-Dependent Potassium Channel

JAN, L. Y., D. PAPAZIAN, L. TIMPE, P. O'FARRELL, and Y. N. JAN. 1985. Application of *Drosophila* molecular genetics in the study of neural function studies of the *Shaker* locus for a potassium channel. *Trends Neurosci.* **8**:234–238.

*SALKOFF, L., and R. WYMAN. 1983. Ion channels in *Drosophila* muscle. *Trends Neurosci.* **6**:128.

WU, C-F., B. GANEZKY, F. N. HAUGLAND, and A-X LIU. 1983. Potassium currents in *Drosophila:* different components affected by mutations of two genes. *Science* **220**:1076–1078.

Impulse Transmission and Myelin

BRAY, G. M., M. RASMINSKY, and A. J. AGUAVO. 1981. Interactions between axons and their sheath cells. *Annu. Rev. Neurosci.* **4**:127–162.

*LEES, M. B., and S. W. BROSTOFF. 1984. Proteins of myelin. In *Myelin,* 2d ed., P. Morell, ed. Plenum.

LEMKE, G., and R. AXEL. 1985. Isolation and sequence of a cDNA encoding the major structural protein of peripheral myelin. *Cell* **40**:501–508.

MCKHANN, G. M. 1982. Multiple sclerosis. *Annu. Rev. Neurosci.* **5**:219–239.

MORELL, P., and W. T. NORTON. 1980. Myelin. *Sci. Am.* **242**(5):88–118.

*RAINE, C. D. 1984. Morphology of myelin and myelination. In *Myelin,* 2d ed., P. Morell, ed. Plenum.

*RITCHIE, J. M. 1983. On the relationship between fibre diameter and impulse transmission in myelinated nerve fibers. *Proc. R. Soc. Lond., Ser. B.* **217**:29–39.

SHINE, H. D., and R. L. SIDMAN. 1984. Immunoreactive myelin basic proteins are not detected when shiverer mutant Schwann cells and fibroblasts are co-cultured with normal neurons. *J. Cell Biol.* **98**:1291–1295.

TAKAHASHI, N., A. ROACH, D. B. TEPLOW, S. B. PRUSINER, and L. HOOD. 1985. Cloning and characterization of the myelin basic protein gene from mouse: one gene can encode both 14 kd and 18.5 kd MBPs by alternate use of exons. *Cell* **42**:138–148.

WAXMAN, S. G., and J. M. RITCHIE. 1985. Organization of ion channels in the myelinated nerve fiber. *Science* **228**:1502–1507.

Synapses and Impulse Transmission

Structure and Function of Synapses

HANNA, R. B., J. S. KEETER, and G. P. PAPPAS. 1978. The fine structure of a rectifying electrotonic synapse. *J. Cell Biol.* 79:764–773.

*HEUSER, J. E., and T. REESE. 1977. Structure of the synapse. In *Cellular Biology of Neurons,* E. R. Kandel, ed., *The Nervous System,* vol. 1. *Handbook of Physiology.* Williams and Wilkins.

McBURNEY, R. N. 1983. New approaches to the study of rapid events underlying neurotransmitter action. *Trends Neurosci.* 6:297–302.

POISNER, A. M., and J. M. TRIFARO, eds. 1982. *The Secretory Granule. The Secretory Process,* vol. 1. Elsevier Biomedical.

**The Synapse.* 1976. *Cold Spring Harbor Symp. Quant. Biol.,* vol. 40.

Synaptic Vesicles

BAINES, A. J., and V. BENNETT. 1985. Synapsin I is a spectrin-binding protein immunologically related to erythrocyte protein 4.1. *Nature* **315**:410–413.

DE CAMILLI, P., S. M. HARRIS JR., W. B. HUTTNER, and P. GREENGARD. 1983. Synapsin I (protein I), a nerve terminal-specific phosphoprotein: II. Its specific association with synaptic vesicles demonstrated by immunocytochemistry in agarose-embedded synaptosomes. *J. Cell Biol.* **96**:1355–1373.

MATTHEW, W. D., L. TSAVALER, and L. F. REICHARDT. 1981. Identification of a synaptic vesicle–specific membrane protein with a wide distribution in neuronal and neurosecretory tissue. *J. Cell Biol.* **91**:257–269.

MILJANICH, G. P., A. R. BRASIER, and R. B. KELLY. 1982. Partial purification of presynaptic plasma membrane by immunoadsorption. *J. Cell Biol.* **94**:88–96.

Exocytosis

*HAGIWARA, S., and L. BYERLY. 1983. The calcium channel. *Trends Neurosci.* **6**:189–193.

*HEUSER, J. E., T. S. REESE, M. J. DENNIS, Y. JAN, L. JAN, and L. EVANS. 1979. Synaptic vesicle exocytosis captured by quick freezing and correlated with quantal transmitter release. *J. Cell Biol.* **81**:275–300.

KATZ, B. 1969. *The Release of Neural Transmitter Substances.* Liverpool University Press.

*LENTZ, T. L. 1983. Cellular membrane reutilization and synaptic vesicle recycling. *Trends Neurosci.* **6**:48–53.

MILEDI, R. 1973. Transmitter release induced by injection of calcium ions into nerve terminals. *Proc. R. Soc. Lond. (Biol.)* **183**:421–425.

NELSON, M. T., R. J. FRENCH, and B. K. KRUEGER. 1984. Voltage-dependent calcium channels from brain incorporated into planar lipid bilayers. *Nature* **308**:77–80.

The Nicotinic Acetylcholine Receptors

ANGLISTER, L., U. J. McMAHAN and R. M. MARSHALL. 1985. Basal lamina directs acetylcholinesterase accumulation at synaptic sites in regenerating muscle. *J. Cell Biol.* **101**:735–743.

ANHOLT, R., D. R. FREDKIN, T. DEERINCK, M. ELLISMAN, M. MONTAL, and J. LINDSTROM. 1982. Incorporation of acetylcholine receptors into liposomes: vesicle structure and acetylcholine receptor function. *J. Biol. Chem.* 257:7122–7134.

BRISSON, A., and P. N. T. UNWIN. 1985. Quarternary structure of the acetylcholine receptor. *Nature* **315**:474–477.

BURDEN, S. 1982. Identification of an intracellular postsynaptic antigen at the frog neuromuscular junction. *J. Cell Biol.* **94**:521–530.

*CHANGEUX, J.-P., A. DEVILLERS-THIERY, and P. CHEMOUILLI. 1984. Acetylcholine receptor: an allosteric protein. *Science* **225**:1335–1345.

DRACHMAN, D. B. 1978. The biology of *myasthenia gravis. Annu. Rev. Neurosci.* **4**:195–225.

FATT, P., and B. KATZ. An analysis of the end-plate potential recorded with an intracellular electrode. *J. Physiol.* **115**:320–370.

MASSOULIE, J., and S. BON. 1982. The molecular forms of cholinesterase and acetylcholinesterase in vertebrates. *Annu. Rev. Neurosci.* **5**:57–106.

MERLIE, J. P., and J. R. SANES. 1985. Concentration of acetylcholine receptor mRNA in synaptic regions of adult muscle fibres. *Nature* **317**:66–68.

MISHINA, M., T. TOBIMATSU, K. IMOTO, K-I. TANAKA, Y. FUJITA, K. FUKUDA, M. KURASAKI, H. TAKAHASHI, Y. MORIMOTO, T. HIROSE, S. INAYAMA, T. TAKAHASHI, M. KUNO, and S. NUMA. 1985. Location of functional regions of acetylcholine receptor α-subunit by site-directed mutagenesis. *Nature* **313**:364–369.

NEHER, E., and B. SAKMANN. 1976. Single-channel currents recorded from membrane of denervated frog muscle fibres. *Nature* **260**:799–802.

*NODA, M., H. TAKAHASHI, T. TANABE, M. TOYOSATO, S. KIKYOTANI, Y. FURUTANI, T. HIROSE, H. TAKASHIMA, S. INAYAMA, T. MIYATA, and S. NUMA. 1983. Structural homology of *Torpedo californica* acetylcholine receptor subunits. *Nature* **302**:528–532.

SAKMANN, B., C. METHFESSEL, M. MISHINA, T. TAKAHASHI, T. TAKAI, M. KURASAKI, K. FUKUDA, and S. NUMA. 1985. Role of acetylcholine receptor subunits in gating of the channel. *Nature* **318**:538–543.

YOUNG, E. F., E. RALSTON, J. BLAKE, J. RAMACHANDRAN, Z. W. HALL, and R. M. STROUD. 1985. Topological mapping of acetylcholine receptor: evidence for a model with five transmembrane segments and a cytoplasmic COOH-terminal peptide. *Proc. Nat'l Acad. Sci. USA* **82**:626–630.

Neurotransmitters

Multiple Transmitters and Receptors

BROWN, D. A. 1983. Slow cholinergic excitation: a mechanism for increasing neuronal excitability. *Trends Neurosci.* **6**:302–307.

HARTZELL, H. C. 1981. Mechanisms of slow postsynaptic potentials. *Nature* **291**:539–544.

SIMMONDS, M. A. 1983. Multiple GABA receptors and associated regulatory sites. *Trends Neurosci.* **6**:279–281.

SNYDER, S. H., and R. R. GOODMAN. Multiple neurotransmitter receptors. *J. Neurochem.* **35**:5–15.

THAMPY, K. G., and E. M. BARNES JR. 1980. γ-Aminobutyric acid-gated chloride channels in cultured cerebral neurons. *J. Biol. Chem.* **259**:1753–1757.

Neurotransmitter Receptors and Cyclic AMP

BOYLE, M. B., M. KLEIN, S. J. SMITH, and E. R KANDEL. 1984. Serotonin increases intracellular Ca^{2+} transients in voltage-clamped sensory neurons of *Aplysia californica. Proc. Nat'l Acad. Sci. USA* **81**:7642–7646.

CASTELLUCCI, V. F., E. R. KANDEL, J. H. SCHWARTZ, F. D. WILSON, A. C. NAIRN, and P. GREENGARD. 1980. Intracellular injection of the catalytic subunit of cyclic AMP-dependent protein kinase simulates facilitation of transmitter release underlying behavioral sensitization in *Aplysia. Proc. Nat'l Acad. Sci. USA* 77:7492–7296.

GREENGARD, P. 1979. Cyclic nucleotides, phosphorylated proteins, and the nervous system. *Fed. Proc.* **38**:2208–2217.

KACZMAREK, K., K. R. JENNINGS, F. STRUMWASSER, A. C NAIRN, U. WALTER, F. D. WILSON, and P. GREENGARD. 1980. Microinjection of catalytic subunit of cyclic AMP-dependent protein kinase enhances calcium action potentials of bag cell neurons in cell culture. *Proc. Nat'l Acad. Sci. USA* 77:7487–7491.

*KUPFERMANN, I. 1980. Role of cyclic nucleotides in excitable cells. *Annu. Rev. Physiol.* **42**:629–641.

LEMOS, J. R., I. NOVAK-HOFER, and I. B. LEVITAN. 1982. Serotonin alters the phosphorylation of specific proteins inside a living nerve cell. *Nature* **298**:64–65.

*LEVITAN, I. B., J. R. LEMOS, and I. NOVAK-HOFER. 1983. Protein phosphorylation and the regulation of ion channels. *Trends Neurosci.* **6**:496–499.

SHUSTER, M. J., J. S. CAMARDO, S. A. SIEGELBAUM, and E. R. KANDEL. 1985. Cyclic AMP-dependent protein kinase closes the serotonin-sensitive K^+ channels of *Aplysia* sensory neurons in cell-free membrane patches. *Nature* **313**:392–395.

TSIEN, R. W. 1983. Modulation of gated ion channels as a mode of transmitter action. *Trends Neurosci.* **6**:307–313.

Neuropeptides

BLOCH, B., P. BRAZEAU, N. LING, P. BOHLEN, F. ESCH, W. B. WENRENBERG, R. BENOIT, F. BLOOM, and R. GUILLEMIN. 1983. Immunohistochemical detection of growth hormone-releasing factor in brain. *Nature* **301**:607–608.

*BLOOM, F. E. 1981. Neuropeptides. *Sci. Am.* **245**(4):148–168.

*GAINER, H., and M. J. BROWNSTEIN. 1981. Neuropeptides. In *Basic Neurochemistry,* 3d ed., G. J. Siegel, R. W. Albers, B. W. Agranoff, and R. Katzman, eds. Little, Brown.

JAN, L. Y., Y. N. JAN, and M. S. BROWNFIELD. 1980. Peptidergic transmitters in synaptic boutons of sympathetic ganglia. *Nature* **288**:380–382.

JAN, Y. N., and L. Y. JAN. 1983. A LHRH-like peptidergic neurotransmitter capable of "action at a distance" in autonomic ganglia. *Trends Neurosci.* **6**:320–325.

*JESSELL, T. M., and M. D. WOMACK. 1985. Substance P and the novel mammalian tachykinins: a diversity of receptors and cellular actions. *Trends Neurosci.* **8**:43–45.

*KRIEGER, D. T. 1983. Brain peptides. *Science* **222**:975–985.

NORTH, R. A., and J. T. WILLIAMS. 1983. How do opiates inhibit neurotransmitter release? *Trends Neurosci.* **6**:337–339.

*OTSUKA, M., and S. KONISHI. 1983. Substance P: the first peptide neurotransmitter? *Trends Neurosci.* **6**:317–320.

*SCHELLER, R. H., and R. AXEL. 1984. How genes control an innate behavior. *Sci. Am.* **250**(1):54–62.

SCHELLER, R. H., B. S. ROTHMAN, and E. MAYERI. 1983. A single gene encodes multiple peptide-transmitter candidates invovled in a stereotyped behavior. *Trends Neurosci.* **6**:340–345.

SIMON, E. J., and J. M. HILLER. 1981. Opioid peptides and opioid receptors. In *Basic Neurochemistry,* 3d ed., G. J. Siegel, R. W. Albers, B. W. Agranoff, and R. Katzman. eds. Little, Brown.

WEBER, E., C. J. EVANS, and J. D. BARCHAS. 1983. Multiple endogenous ligands for opioid receptors. *Trends Neurosci.* **6**:333–336.

Neurotransmitter Plasticity

*BLACK, I. B., J. E. ADLER, C. F. DREYFUS, G. M. JONAKAIT, D. M. KATZ, E. F. LAGAMMA, and K. M. MARKEY. 1984. Neurotransmitter plasticity at the molecular level. *Science* **225**:1266–1270.

PATTERSON, P. H. 1982. Cellular and hormonal interactions in the development of sympathetic neurons. In *Molecular Genetic Neurosciences,* F. O. Schmitt, S. J. Bird, and F. E. Bloom, eds. Raven.

PATTERSON, P. H., and L. L. Y. CHUN. 1977. The induction of acetylcholine synthesis in primary cultures of dissociated rat sympathetic neurons. *Dev. Biol.* **56**:263–280.

Synapse Formation and Cell-to-Cell Recognition

*GOODMAN, C. S., and M. J. BASTIANI. 1984. How embryonic cells and nerve cells recognize one another. *Sci. Am.* **251**(6):58–66.

GOODMAN, C. S., M. J. BASTIANI, C. Q. DOE, S. DU LAC, S. L. HELFAND, J. Y. KUWADA, and J. B. THOMAS. 1984. Cell recognition during neuronal development. *Science* **225**:1271–1279.

SUMMERBELL, D., and R. STIRLING. 1985. What guides growing axons? *Nature* **315**:368–369.

Memory and Neurotransmitters

Drosophila Mutations Affecting Learning

BYERS, D., R. L. DAVIS, and J. A. KIGER JR. 1981. Defect in cyclic AMP phosphodiesterase due to the *dunce* mutation of learning in *Drosophila melanogaster. Nature* **289**:79–81.

*DUDAI, Y. 1985. Genes, enzymes, and learning in *Drosophila. Trends Neurosci.* **8**:18–21.

HOTTA, Y., and S. BENZER. 1972. Mapping of behavior in *Drosophila* mosaics. *Nature* **240**:527–535.

QUINN, W. G., W. A. HARRIS, and S. BENZER. 1974. Conditioned behavior in *Drosophila melanogaster. Proc. Nat'l Acad. Sci. USA* **71**:708–712.

QUINN, W. G., P. P. SZIBER, and R. BOOKER. 1979. The *Drosophila* memory mutant *amnesiac. Nature* **277**:212–214.

Habituation and Sensitization in the Sea Snail *Aplysia*

ABRAMS, T. W., V. F. CASTELLUCCI, J. S. CAMARDO, E. R. KANDEL, and P. E. LLOYD. 1984. Two endogenous neuropeptides modulate the gill and siphon withdrawal reflex in *Aplysia* by presynaptic facilitation involving cAMP-dependent closure of a serotonin-sensitive potassium channel. *Proc. Nat'l Acad. Sci. USA* **81**:7956–7960.

KANDEL, E. R. 1981. Calcium and the control of synaptic strength by learning. *Nature* **293**:697–700.

*KANDEL, E. R., and J. H. SCHWARTZ 1982. Molecular biology of learning: modulation of transmitter release. *Science* **218**:433–443.

SIEGELBAUM, S. A., J. S. CAMARDO, and E. R. KANDEL. 1982. Serotonin and cyclic AMP close single K^+ channels in *Aplysia* sensory neurons. *Nature* **299**:413–417.

Sensory Transduction: The Visual System

*ALTMAN, J. 1985. New visions in photoreception. *Nature* 313:264–265.

FESENKO, E. E., S. S. KOLESNIKOV, and A. L. LYUBARSKY. 1985. Induction by cyclic GMP of cationic conductance in plasma membrane of retinal rod outer segment. *Nature* **313**:310–313.

FUNG, B. K.-K. 1983. Characterization of transducin from bovine retinal rod outer segments: I. Separation and reconstitution of the subunits. *J. Biol. Chem.* **258**:10495–10502.

*FUNG, B. K.-K., J. B. HURLEY, and L. STRYER. 1981. Flow of information into the light-triggered cyclic nucleotide cascade of vision. *Proc. Nat'l Acad. Sci. USA* 78:152–156.

HURLEY, J. B., H. K. W. FONG, D. B. TEPLOW, W. J. DREYER, and M. I. SIMON. 1984. Isolation and characterizations of a cDNA clone of the γ subunit of bovine retinal transducin. *Proc. Nat'l Acad. Sci. USA* 81:6948–6952.

MANNING, D. R., and A. G. GILMAN. 1983. The regulatory components of adenylate cyclase and transducin. *J. Biol. Chem.* **258**:7059–7063.

MATTHEWS, H. R., V. TORRE, and T. D. LAMB. 1985. Effects on the photoresponse of calcium buffers and cyclic GMP incorporated into the cytoplasm of retinal rods. *Nature* **313**:582–585.

MILLER, W. H., ed. 1981. Molecular mechanism of photoreceptor transduction. *Curr. Topics Membrane Transport.* **15**.

MONTEL, C., K. JONES, E. HAFEN, and G. RUBIN. 1985. Rescue of the *Drosophila* phototransduction mutation *trp* by germline transformation. *Science* 230:1040–1043.

YATSUNAMI, K., and H. G. KHORANA. 1985. GTPase of bovine rod outer segments: the amino acid sequence of the α subunit as derived from the cDNA sequence. *Prod. Natl Acad. Sci.* 82:4316–4320.

18

The Cytoskeleton and Cellular Movements: Microtubules

MOVEMENTS of cells and subcellular structures—the beating of cilia and flagella, the contraction of muscle, the movement of chromosomes, and the migration of cells along a substratum—have long fascinated cell biologists. The generation of cell shape is an equally intriguing phenomenon, especially in highly specialized cell types such as nerve cells, with their long axons, and the intestinal brush border cells, with their pencil-like microvilli. Although detailed molecular explanations are not yet available, both movement and shape characteristic of a specific cell type or structure are known to involve the complex set of protein fibers found in the cytoplasm—the *cytoskeleton*. Electron microscopy has shown that all eukaryotic cells contain three major classes of cytoskeletal fibers: 7-nm-diameter *actin microfilaments,* 24-nm-diameter *microtubules,* and 10-nm-diameter *intermediate filaments.* Both actin and microtubule filaments are formed by polymerization of protein subunits called G actin and tubulin, respectively. As we shall see, the polymerization and depolymerization of these filaments are closely regulated by the cell. Most eukaryotic cells contain one or more types of intermediate filaments, each of which is built from specific kinds of protein.

This chapter focuses on microtubules; the following chapter, on actin and intermediate filaments. Another structure important to the shape, movement, and function of cells is the *extracellular matrix,* the network of protein and polysaccharide fibers that occupies the interstitial spaces between the cells and also binds to parts of the cell surface. The extracellular matrix is discussed in Chapters 19 and 21.

Some actin and microtubule systems are more or less permanent features of a cell—for example, actin and myosin filaments in the contractile apparatus of muscle, actin filaments in the brush border of the intestinal cell, and microtubules in the core of cilia and flagella. These fiber systems lend themselves readily to biochemical and ultrastructural analyses because their highly ordered filament structures are stable, abundant, and easily prepared in pure form. Such studies have revealed a great deal about molecular structure and function in these systems.

Each type of filament occurs in various structures and participates in a wide array of movements. Particular filament types may occur only in specialized cells or may exist only transiently—for example, microtubules in the mitotic spindle and actin filaments in the lamellipodia of moving cells. The unique functions of actin and microtubule filaments are made possible by many different kinds of proteins that bind to the fibers. Some proteins play a part in determination of cell shape by anchoring the filaments to membranes or to other fibers. Other proteins enable the filament to participate in a specialized cell process or movement: the waving motions of cilia and the contraction of muscle, for instance, are the consequence of making and breaking specific protein-protein associations as two filaments slide past each other. As we shall see, the energy for these movements is generated by the hydrolysis of ATP, catalyzed by specific enzymes that bind to actin or tubulin filaments.

The fiber systems in muscle and flagella are thought to represent evolutionary specializations of microfilament and microtubule proteins that originally served functions that were common to early cells, such as cell motility or chromosome movement. Strong evidence for this theory is that many of the cytoskeletal proteins, such as tubulin, actin, and intermediate-filament proteins, are encoded in multiple gene families probably derived from an original gene in a primordial cell. Over eons of evolution, this gene was duplicated, and each copy then evolved to serve a related but often different function.

Microtubules: Structure, Function, and Assembly

With rare exceptions, such as the human erythrocyte, microtubules are found in the cytoplasm of all eukaryotic cells, from amebas to higher plants and animals, but are absent in all prokaryotes. Microtubules are long fibers about 24 nm in diameter; in cross section each microtubule appears to have a hollow center. Frequently microtubules pack into *bundles,* and cross-bridges seen between adjacent microtubules are thought to stabilize the bundles and to give them strength and rigidity (Figure 18-1).

Microtubules Fulfill Different Functions in Different Cells

Bundles of microtubules form a large and diverse collection of subcellular structures. In all nonmammalian vertebrate erythrocytes a ring of microtubules lies just beneath the cell surface (Figure 18-2). These apparently help in maintaining or restoring the shape of erythrocytes when they are exposed to deforming forces in the circulation.

In nonmitotic cultured fibroblast cells, microtubules form a complex cytoplasmic network that crisscrosses the cell (Figure 18-3). The microtubules are concentrated around the nucleus, and their "free" ends radiate outward toward the plasma membrane from one or more sites called *microtubule-organizing centers (MTOCs).* At the center of the MTOC is a *centriole.* During mitosis the centriole and MTOC form the poles of the mitotic spindle (see Figure 5-14).

During mitosis, this microtubular network disappears, and the spindle apparatus forms (Figure 18-4). The tubulin used to form the spindle apparatus is in all likelihood derived from the interphase microtubules. The spindle is important for equal partitioning of chromosomes to the daughter cells, as discussed later.

Microtubules and intermediate filaments extend along the axons and dendrites of neurons from the cell body to the terminals. It was once thought that these fibers were continuous throughout the entire axon, but observations on serial sections suggest that individual microtubules are generally 10 to 25 μm in length (Figure 18-5). The thinner intermediate filaments are associated with the axonal microtubules. As we shall see, microtubules may act as guides or tracks along which protein particles and organelles move up and down the axon during *axoplasmic transport.*

Microtubules are essential for the generation of motion by cilia and flagella (see Figure 5-48). Later in this chapter we shall examine the evidence that microtubules also generate the force that moves the chromosomes during mitosis and meiosis.

Some microtubule-containing structures, such as cilia and flagella, are permanent features of a cell and persist throughout growth and division. (There are certain exceptions: many protozoans resorb flagella before dividing, and the flagella reappear after cytokinesis is complete.) Other systems of microtubules—particularly those of the mitotic spindle—are formed by polymerization from a pool of subunits during a discrete phase of the cell cycle and are then depolymerized during another phase.

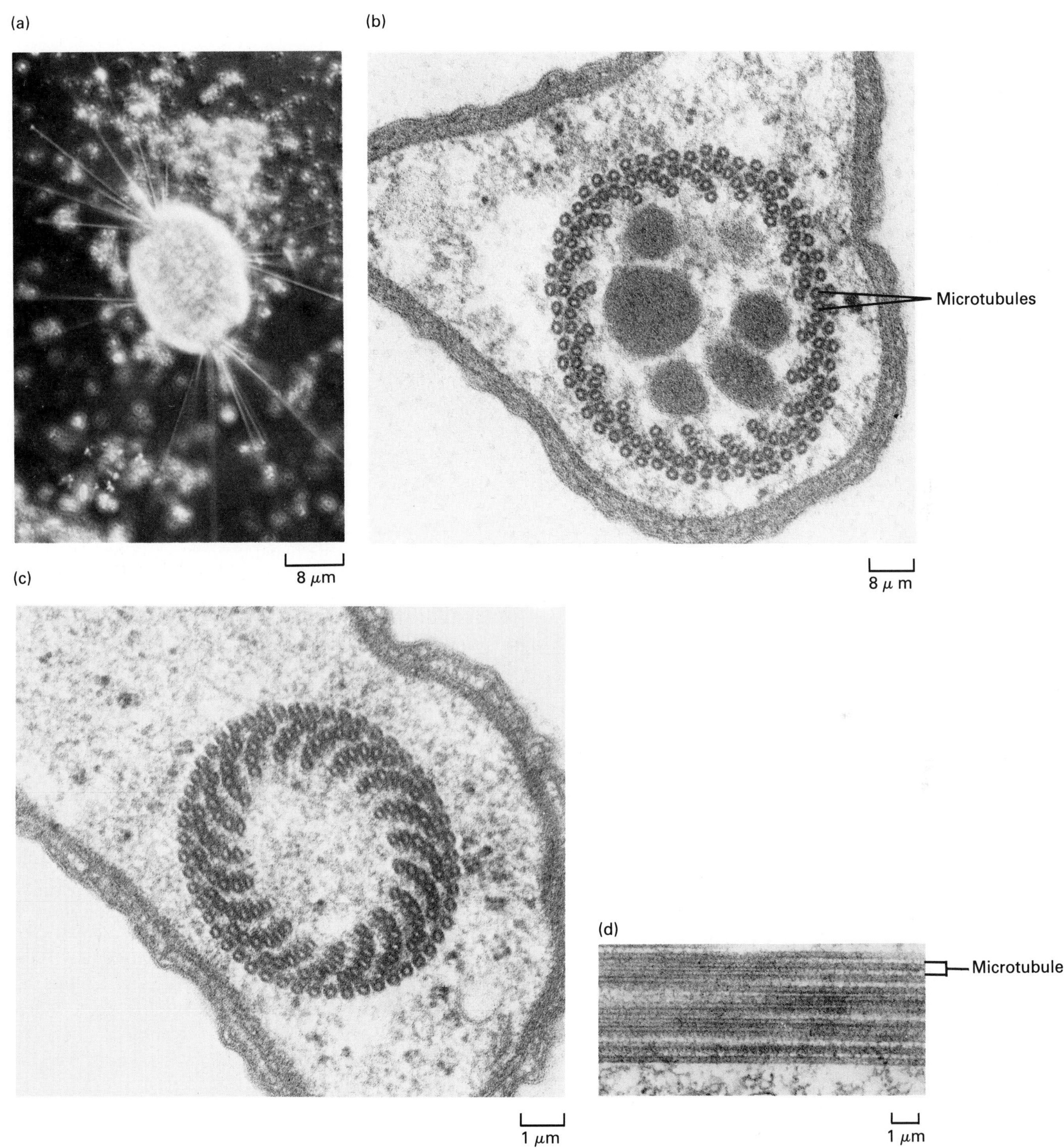

Figure 18-1 Microtubules in the tentacles of a suctorian protozoan, *Heliophrya erhardi*. (a) Overview of *H. erhardi* with mostly expanded tentacles, arranged in four bundles. The tentacles trap prey, such as other microorganisms. As the tentacles contract, the prey is moved toward the cell body, where it is ingested. (b) Cross section through the midregion of an expanded tentacle. The large number of microtubules gives the tentacles strength and rigidity. (c) Cross section through the base of a contracted tentacle. As the tentacles contract, the rows of microtubules seen in (b) spiral inward, and the microtubule network becomes thicker with a smaller overall diameter. (d) Longitudinal section of a cylinder of microtubules in the cytoplasm. Note the occasional cross-bridges between microtubules. [See M. Hauser and H. Van Eys, 1976, *J. Cell Sci.* **20**:589.] *Photographs courtesy of M. Hauser and H. Van Eys.*

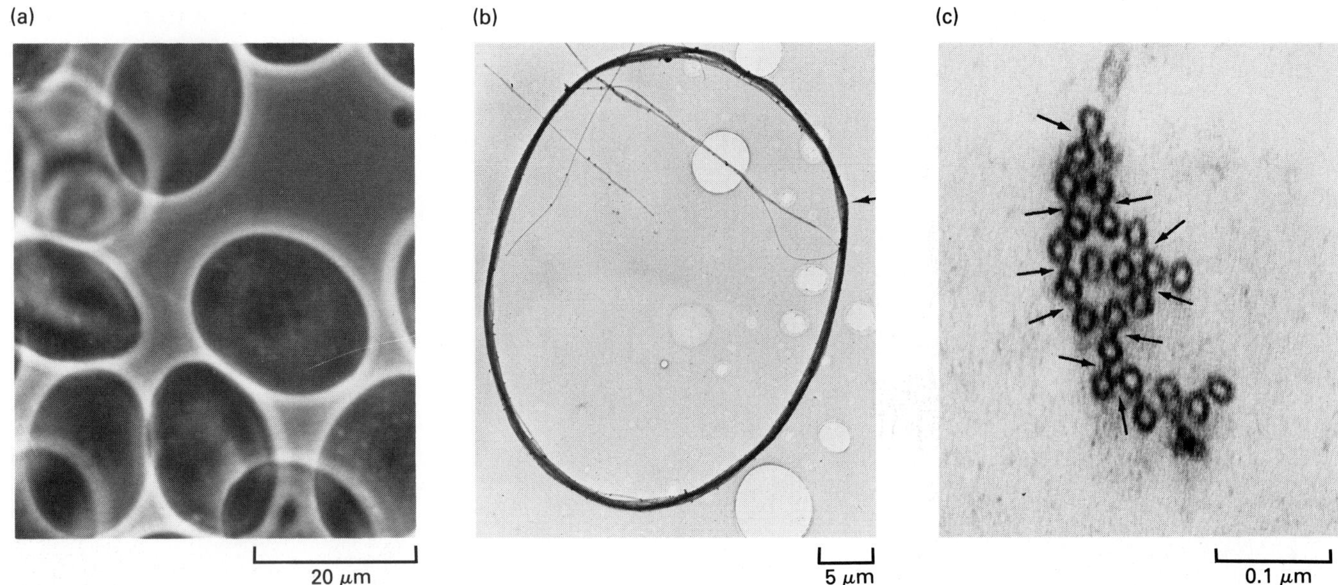

Figure 18-2 The microtubular cytoskeleton of dogfish erythrocytes. (a) Living cells in fresh blood, showing the flattened elliptical morphology of all nonmammalian vertebrate erythrocytes. (b) After lysis in detergents and additional fractionation, the marginal band of microtubules can be visualized in the electron microscope. This ringlike structure is found just inside the plasma membrane. (c) In cross section each microtubular bundle is seen to consist of a number of individual microtubules. Note the occasional cross-bridges (*arrows*) between adjacent microtubules. Similar microtubular bundles are found in erythrocytes of almost all vertebrates, including birds and amphibians; mammals are an exception. [See W. D. Cohen et al., 1982, *J. Cell Biol.* 93:828.] *Courtesy of W. D. Cohen. Reproduced from the* Journal of Cell Biology, *1982, by copyright permission of the Rockefeller University Press.*

Similarly, labile microtubular structures are also found in the axons and dendrites of nerves and in the cytoplasm of fibroblasts. The reversibility of the assembly-disassembly process for these structures has been shown experimentally. The microtubules can be made to disappear, or to depolymerize into constituent subunits, by exposure to low temperatures or high pressures, or to antimitotic drugs such as colchicine (Figure 18-6). Upon removal of colchicine by washing, or upon raising the temperature or lowering the pressure, the microtubules reappear. Such treatments do not affect the more stable microtubules in centrioles, flagella, or cilia.

All Microtubules Are Composed of α and β Tubulin Subunits

All microtubules are constructed on the same principle from similar protein subunits. In cross section, the wall of a microtubule is made up of 13 globular subunits about 4 to 5 nm in diameter, and the center appears hollow (see Figures 18-1 and 18-2c). Keith Porter (Chapter 1), using one of the earliest electron microscopes, was among the first to describe this important structural feature, which has since been confirmed for all microtubules.

Microtubules are composed of two kinds of protein

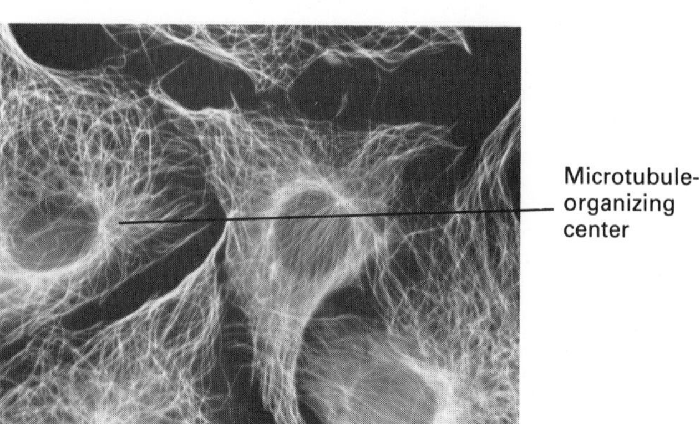

Figure 18-3 Microtubules in growing, nonmitotic mouse fibroblast cells as visualized in the fluorescence microscope. Cells were fixed in glutaraldehyde before being treated with fluorescent tubulin antibody. Note the uniform thickness of the fluorescent microtubule fibers, most of which probably correspond to individual microtubules, and the predominantly radial orientation. The microtubules radiate from the microtubule-organizing center (MTOC), lying just outside the nucleus, which contains the centriole at its center. *Courtesy of M. Osborn.*

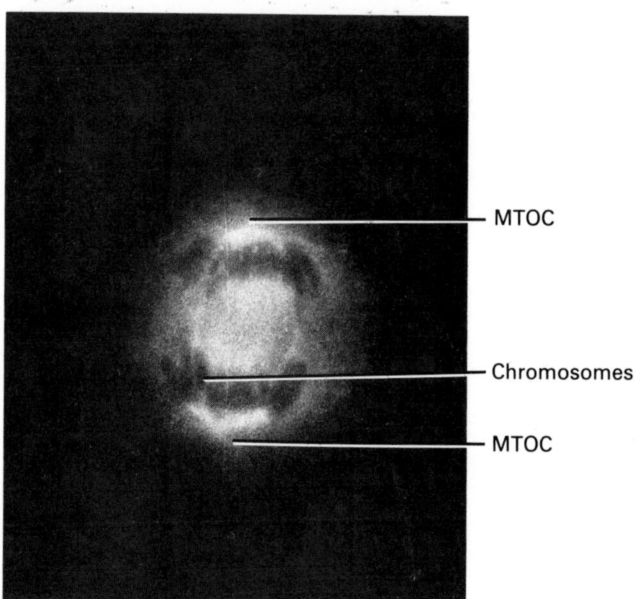

Figure 18-4 Staining of a metaphase mitotic cell with a fluorescent antitubulin antibody. Note that all of the tubulin is in the mitotic spindle apparatus. As is detailed later in the chapter, some of the microtubules radiate in all directions from the two MTOCs; others lead from the chromosomes to the MTOCs. *Photograph courtesy of M. Osborn and K. Weber.*

subunits: α *tubulin* and β *tubulin,* each of m.w. 55,000. The wall of a microtubule is made up of a helical array of repeating α and β tubulin subunits (Figure 18-7). Assembly studies (discussed in the following section) indicate that the structural unit is an $\alpha\beta$ dimer 8 nm in length. Figure 18-7 shows a model of the three-dimensional structure of a microtubule: there are 13 *protofilaments,* each composed of $\alpha\beta$ dimers that run parallel to the long axis of the tubule. Note that the repeating unit is an $\alpha\beta$ heterodimer, and that these are arranged "head to tail" within the microtubule—that is, $\alpha\beta \longrightarrow \alpha\beta \longrightarrow \alpha\beta$, as shown in the figure. Thus all microtubules have a defined polarity: the two ends are *not* structurally equivalent. This is a crucial point to which we shall return several times in the chapter.

Microtubules Are Polymerized from $\alpha\beta$ Dimers in Vitro and in the Living Cell

When purified preparations of microtubules from sperm, brain, or certain protozoan flagella, in which they are present in abundance, are subjected to chilling, the microtubules depolymerize into stable $\alpha\beta$ dimers of m.w. about 110,000. Depolymerized microtubules are called *tubulin.* The $\alpha\beta$ dimers do not dissociate into α and β monomers unless denaturing agents are added. Conversely, microtubules of normal structure can be reconstituted in vitro from a solution of tubulin under physiological conditions. These form microtubules when warmed to 37°C in the presence of GTP, provided free Ca^{2+} is not present (Figure 18-8).

By numerous cycles of polymerization (achieved by warming followed by centrifugation) and depolymerization (achieved by chilling), tubulins can be extracted in nearly pure form. In such preparations, however, a few other proteins of various molecular weights are found as well. These proteins are present in lower amounts than are α and β tubulins, but in successive cycles of polym-

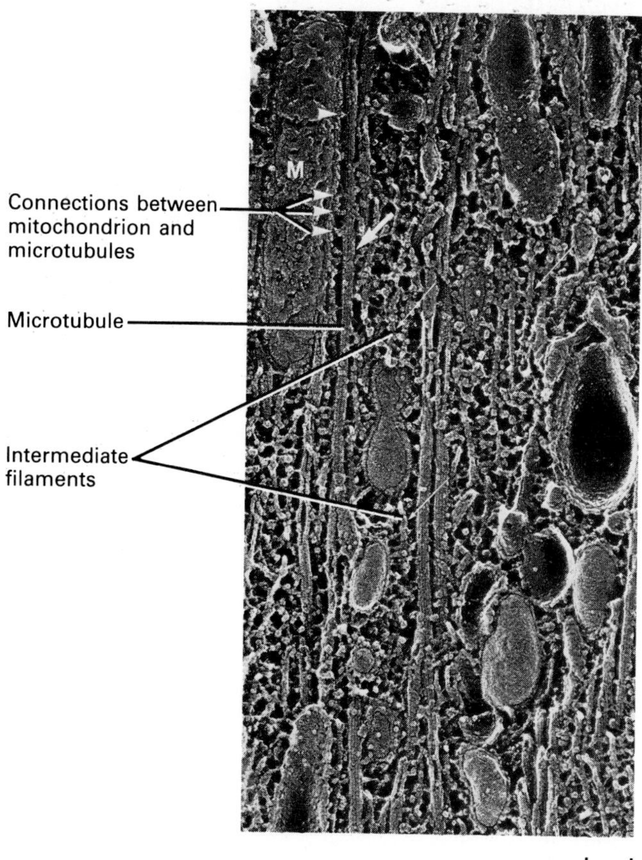

Figure 18-5 Microtubules and other filaments in a quick-frozen frog axon, visualized by the deep-etch technique. There are a number of 24-nm-diameter microtubules running longitudinally. Thinner 10-nm-diameter intermediate filaments also run longitudinally; they appear to form occasional connections with microtubules, but the composition of these connections is not known. Embedded in the cytoskeletal lattice are several mitochondria (*M*) that appear to be connected with the microtubules. Other membrane vesicles also are dispersed in the lattice. [See N. Hirokawa, 1982, *J. Cell Biol.* **94**:129.] *Courtesy of N. Hirokawa. Reproduced from the* Journal of Cell Biology, *1982, by copyright permission of the Rockefeller University Press.*

(a)

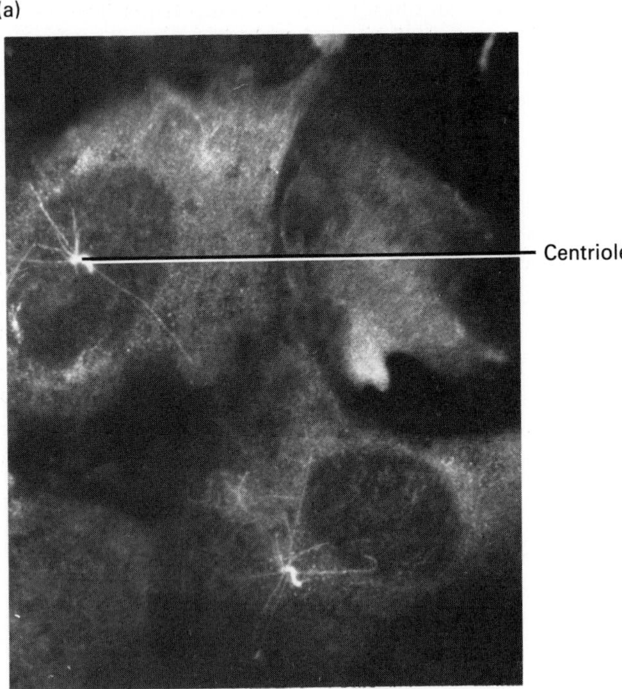

(b)

$H_3C—O$, $H_3C—O$, $NH—\overset{O}{\overset{\|}{C}}—CH_3$, $O—CH_3$, O, $O—CH_3$

Figure 18-6 Colchicine and its effects on microtubules. (a) Effect of colchicine on the organization of microtubules in a mouse fibroblast, as determined with a fluorescent antibody specific for tubulin. Colchicine blocks the addition of tubulin subunits to the ends of existing microtubules; since all microtubules are in a steady state of adding and removing tubulin subunits from their ends, the net result of colchicine addition is depolymerization of most microtubules. Comparison with Figure 18-3 shows that most cytoplasmic microtubules indeed are lost. Microtubules in cilia and centrioles are resistant to depolymerization by colchicine; these colchicine-treated cells contain a centriole. (b) Structure of colchicine, a water-soluble drug that blocks polymerization of tubulin subunits into microtubules. [See M. Osborn and K. Weber, 1975, *Proc. Nat'l Acad. Sci. USA* 73:867.] *Part (a) courtesy of M. Osborn and K. Weber.*

erization-depolymerization (assembly-disassembly), the quantitative ratio of these proteins to α and β tubulins is constant. This finding indicates that such proteins, called *microtubule-associated proteins (MAPs),* are not nonspecific contaminants but rather have a specific association with the α and β tubulins. These quantitatively minor microtubule proteins appear to function in assembly and maintenance of stability of microtubular structure.

Microtubule-Associated Proteins (MAPs) May Regulate $\alpha\beta$ Dimer Polymerization

The rate of in vitro polymerization of tubulin can be accelerated by adding certain polymeric forms of tubulin such as spirals or rings; these serve as initiators for the polymerization of $\alpha\beta$ dimers (Figure 18-9). One of several MAPs, *Tau,* may facilitate the generation of such spirals from $\alpha\beta$ dimers.

Different microtubular structures—mitotic spindles or axonal fibers—appear to contain different MAPs. Polymerization-depolymerization of mitotic spindle microtubules, for instance, yields at least one MAP that is unique to the mitotic apparatus, as well as several that are also found in interphase microtubules. These MAPs may control the polymerization of tubulin into different types of microtubules. Many MAPs are phosphorylated, suggesting that a phosphorylation-dephosphorylation cycle is an important step in MAP regulation of microtubule assembly.

In microtubule research, a working hypothesis is that each type of microtubule structure contains functionally equivalent α and β subunits, and that the unique properties of each type of microtubule are determined by unique types of MAPs that copolymerize with the $\alpha\beta$ subunits.

Microtubules Grow Preferentially at One End During in Vitro Polymerization Reactions

The addition of fragments of flagellar microtubules to a solution of soluble $\alpha\beta$ dimers greatly enhances the rate of polymerization of the dimers to microtubules. The flagellar fragments serve as "seeds" onto which dimers can add. The rate of addition of dimers at one end of the microtubule, called the *A end* (net assembly end), is several times greater than at the other end, the *D end* (net disassembly end). Depolymerization of dimers (that is, removal of subunits from the two ends of a microtubule) also occurs at different rates. Under defined conditions of ionic composition and soluble subunit concentration, a *steady state* condition is reached, where the rate of addition of new dimers to both ends of the microtubule is just balanced by the rate of loss of other dimers. Under these conditions, $\alpha\beta$ dimers add preferentially to the A end and are lost preferentially from the D end (Figure 18-10). Thus, at steady state the microtubules function as "treadmills": a tubulin dimer incorporated at the A end mi-

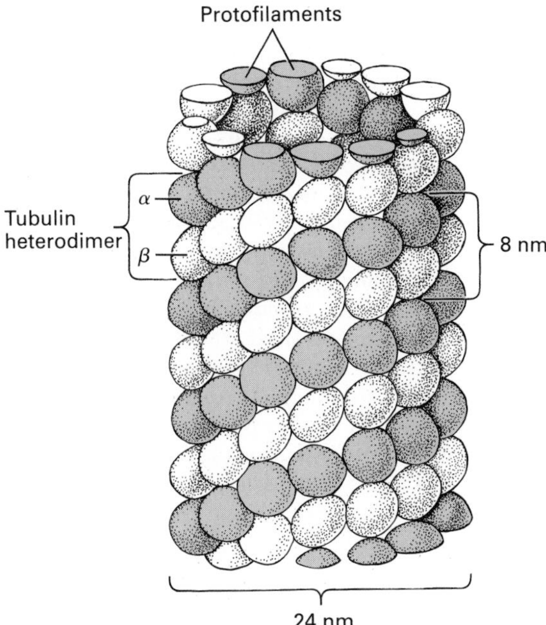

Figure 18-7 Model of the three-dimensional structure of a single microtubule. The tubulin polypeptides are aligned in 13 parallel rows, called protofilaments. Each protofilament is composed of a series of $\alpha\beta$ tubulin dimers, aligned head to tail. [See L. Amos, R. Linck, and A. Klug, 1976, *Cell Motility*, Cold Spring Harbor Laboratory, p. 853.]

grates along the microtubule and eventually is depolymerized at the D end.

A pulse-chase experiment using radiolabeled GTP ([^{3}H)GTP) to tag $\alpha\beta$ dimers, as depicted in Figure 18-10, has demonstrated this "treadmilling" function of microtubules. A free $\alpha\beta$ dimer binds two molecules of GTP. One GTP is bound so tightly that it cannot exchange with GTP in the medium. The other is bound less tightly, so it can exchange, slowly, with [^{3}H]GTP added to the medium. One GTP may be hydrolyzed to GDP during or just after the incorporation of the dimer into the microtubule. GDP and the second GTP remain bound to that dimer after it is polymerized into a microtubule. Thus, if [^{3}H]GTP is initially bound, ^{3}H remains bound to the dimer and can be used to tag that dimer in a polymerization reaction. When [^{3}H]GTP-tagged $\alpha\beta$ dimers are added for a brief period (the pulse) to a medium containing unlabeled microtubules, autoradiography shows that most of the ^{3}H incorporated into a microtubule is at one end. When an end-labeled microtubule continues the polymerization reaction in a medium containing only $\alpha\beta$ dimers with unlabeled GTP (the chase), the labeled dimers appear to "move" to the D end and are eventually lost. In the steady state, therefore, there is a net addition of unlabeled tubulin to the A end as well as a net loss from the D end.

Such experiments have established the polarity of microtubules. Accordingly, all structures that contain microtubules, such as flagella or mitotic spindles, must also be polar in nature.

The Drug Colchicine Blocks Polymerization of Tubulin

In vitro studies of microtubule assembly have suggested that a number of microtubule functions result from their capacity to assemble and disassemble. An example is seen in mitosis. Colchicine, a plant alkaloid, blocks plant and animal cells at metaphase but does not affect chromosome condensation. However, in the presence of colchicine no spindle forms, and there is no movement of chromosomes toward the poles of the cell. Thus colchicine can

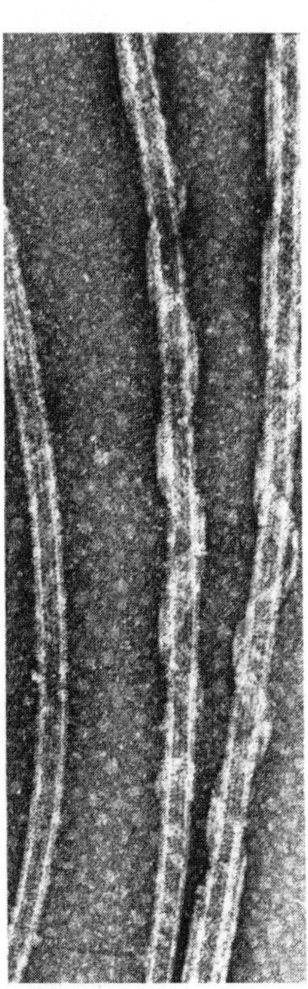

Figure 18-8 Electron micrograph of microtubules that were assembled in vitro from purified tubulin (a soluble preparation of $\alpha\beta$ dimers). A microtubule of normal appearance is shown on the far left. The two structures on the right are microtubules in which the outer wall is incomplete; they appear to be sheets of protofilaments ($\alpha\beta$ dimers linked head to tail) that have not closed to form a microtubule. Sheets of protofilaments may be normal intermediates in the assembly of microtubules: presumably, when the sheet becomes wide enough, containing 13 protofilaments, it curls up to form a tube. [See H. P. Erickson, 1976, *Cell Motility*, Cold Spring Harbor Laboratory, p. 1072.] *Photograph courtesy of H. P. Erickson.*

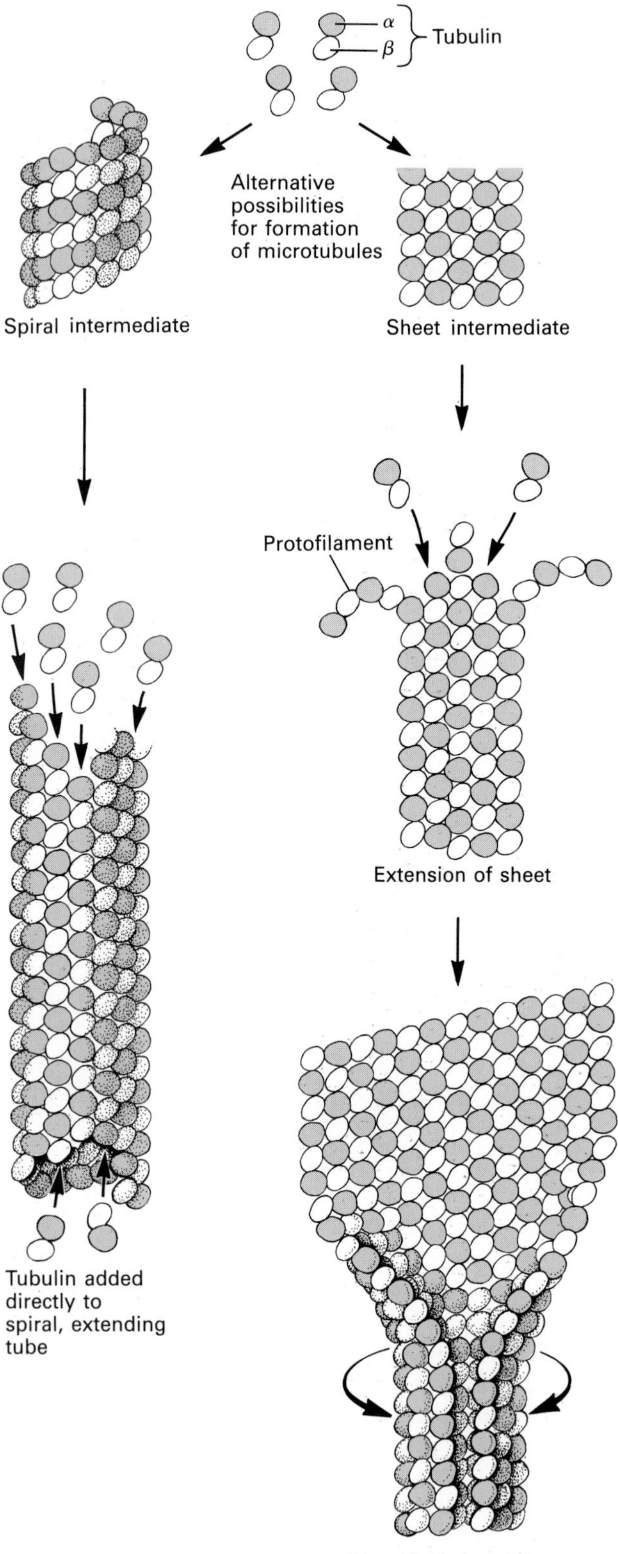

Figure 18-9 Assembly of microtubules. Microtubules can be generated in the laboratory from a concentrated solution of $\alpha\beta$ tubulin dimers. The initial step involves formation of intermediates such as spirals of tubulin. These intermediates are unstable, and their slow formation accounts for the lag period before long microtubules are formed. Certain basic proteins can accelerate the polymerization of tubulin, possibly by acting as foci for assembly of tubulin into spirals. These intermediates act as primers for polymerization of $\alpha\beta$ dimers, but the steps involved are not well established. Tubulin may add directly to the spiral, forming a tubular microtubule. Alternatively, sheets of protofilaments may be the intermediate structures. When the protofilaments become wide enough, they fold into a tube. *After P. Dustin, 1980,* Sci. Am. *243(2):66–76.*

be used to produce accumulations of metaphase chromosomes for cytogenetic studies. The drugs vinblastine and vincristine, which also inhibit microtubule polymerization, have been widely used as anticancer agents since blockage of the mitotic spindle will preferentially kill rapidly dividing cells.

The basis for the action of colchicine is its ability to bind to $\alpha\beta$ tubulin dimers. Each dimer contains one high-affinity binding site for colchicine. An $\alpha\beta$ dimer with a bound colchicine molecule can add to the end of a microtubule. However, the presence of as few as one or two such colchicine-$\alpha\beta$ dimer groups on a microtubule prevents further addition of any other $\alpha\beta$ dimers, whether or not they contain bound colchicine. Clearly, this blocking ability of colchicine explains its inhibitory action on formation of the mitotic spindle; however, it does not directly cause disassembly of microtubules.

Many microtubule-containing structures, such as the mitotic spindle, represent a "steady state" of polymerization-depolymerization of subunits (see previous section). In such structures, colchicine blockage of the assembly of $\alpha\beta$ dimers into microtubules results in a net loss of microtubules and an accumulation of $\alpha\beta$ dimers. The disappearance of cytoplasmic microtubules in nonmitotic fibroblasts treated with colchicine (Figure 18-6) suggests that these microtubules also are normally in a steady state of assembly-disassembly.

Within an Organism Both α and β Tubulins Are Heterogeneous

Although all microtubules have similar structural features, there is growing evidence that different microtubule-containing structures in each organism contain different α and β tubulin subtypes. In some cases, the α and β tubulins undergo different posttranslational modifica-

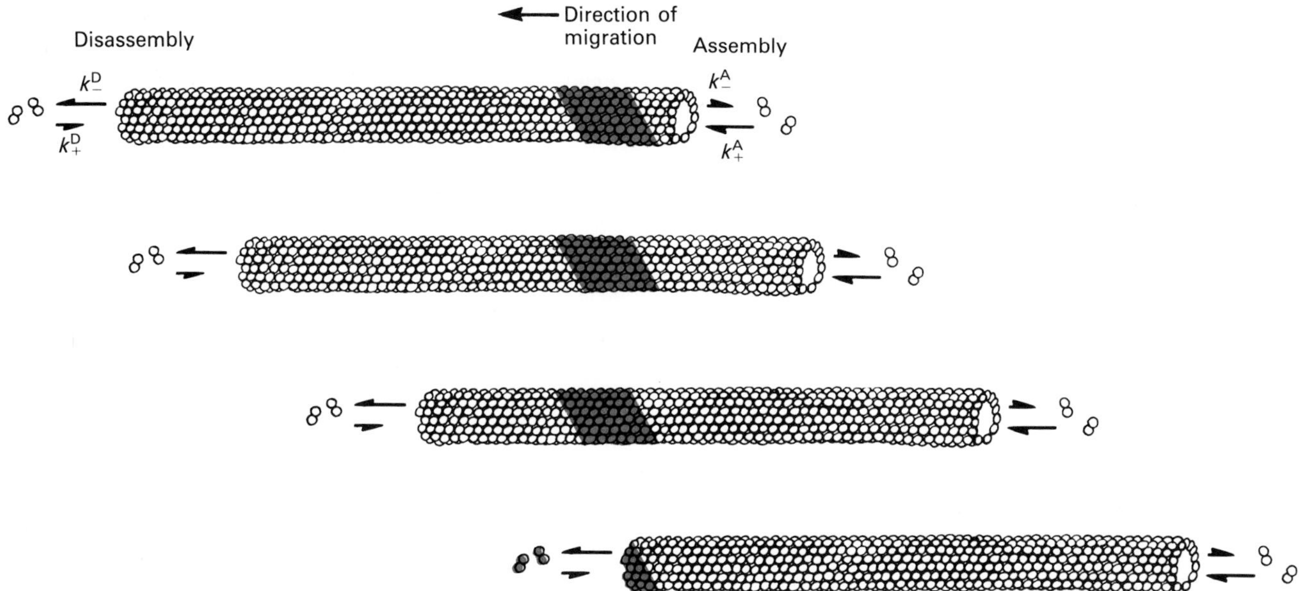

Figure 18-10 "Treadmilling" of tubulin subunits in a microtubule. In the presence of GTP, tubulin $\alpha\beta$ dimers are continually being added to, and lost from, both ends of a microtubule at different rates. The labels k_+^D and k_-^D are the rate constants for addition and removal, respectively, of an $\alpha\beta$ dimer from the D (net disassembly) end; k_+^A and k_-^A, those for the A (net assembly) end. One of the two molecules of radiolabeled GTP ([^{3}H]GTP) bound to each $\alpha\beta$ tubulin dimer does not dissociate when the dimer is incorporated into a microtubule. Thus it serves as a "tag" for newly added tubulin and can be detected by autoradiography of a microtubule preparation. After a "pulse" of labeled tubulin is added to a preparation of microtubules, the tubulin can be seen to add at one end—the A end—and following addition of unlabeled tubulin migrate to the D end, while additional (unlabeled) subunits are added to the A end. [See R. L. Margolis and L. Wilson, 1981, *Nature* **293**:705.]

tions, such as covalent addition of an acetyl group, that determine their ability to be incorporated into different microtubular structures. Alternatively, the subtypes may be products of different α and β tubulin genes. There are in fact multiple copies of both α and β genes—9 to 13 in the sea urchin, 4 in *Drosophila* and chickens—as demonstrated by studies in which cloned α and β tubulin cDNAs were hybridized to genomic DNAs. In general, it is not clear whether all gene copies are actually expressed (some could be pseudogenes), nor is it clear whether certain genes are transcribed only in certain cell types, or whether certain α or β subtypes are found in unique subcellular structures. In chickens each of the four β tubulin genes is functional. Each gene directs synthesis of a unique β tubulin mRNA; the amount of each β mRNA varies among differentiated cell types. Whether all four β (and four α) genes are necessary for microtubule function in vertebrates is not known.

Yeast cells, by contrast, contain only a single β tubulin gene, which is essential for viability (Figure 18-11). The two β tubulin genes of the unicellular flagellated alga *Chlamydomonas reinhardtii* code for identical proteins, even though they have different introns. Thus a single type of β tubulin is sufficient for all microtubular structures found in this organism—such as mitotic and meiotic spindles, cytoplasmic microtubules, MTOCs, and flagella. Similarly, the two *Chlamydomonas* α tubulins differ by only two amino acids. The α tubulin in flagella is post-translationally modified by acetylation on one lysine residue. This modification, which is not observed in cytoplasmic microtubules, occurs during or just after the incorporation of α tubulin into flagellar microtubules, so it is thought to serve a regulatory function in this process. Some α tubulins in cultured mammalian fibroblasts are posttranslationally modified by covalent addition of a tyrosine residue to the C-terminus (an unusual example of nonribosomal addition of amino acids to protein). These tyrosinylated tubulins are found preferentially in mitotic spindles, suggesting that tyrosinylation is involved in establishment of separate, functionally distinct microtubule populations.

In other cases, multiple tubulin subsets are necessary for the construction of microtubules with the same or different functions. At least in sea urchins and *Drosophila,* the tubulin subtypes used for formation of sperm flagella are different from those used for cilia or flagella in other tissues. Indeed, in *Drosophila* the gene controlling the synthesis of a tubulin species called $\beta 2$ is ex-

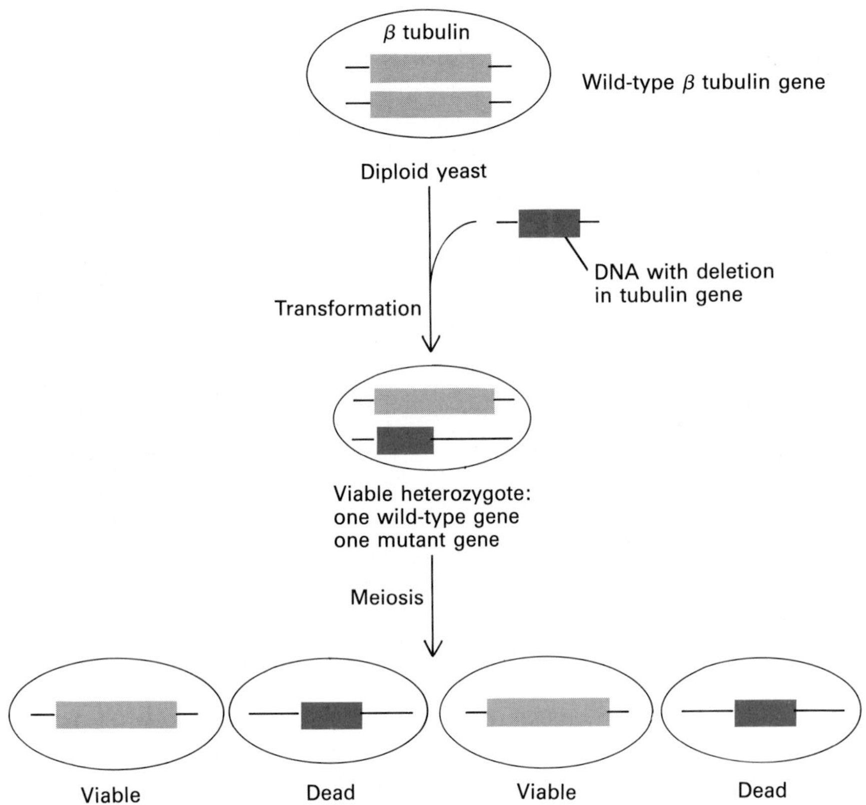

Figure 18-11 Genetic study demonstrating that the single yeast β tubulin gene is essential for viability. A diploid yeast strain was transformed with a DNA that contains a deletion in the β tubulin gene. This DNA integrates in one of the two β tubulin loci in place of the wild-type β tubulin gene. This generates a viable heterozygote with one wild-type gene and one mutant β tubulin gene. Upon sporulation, only two of the four haploid progeny are viable; these contain the wild-type tubulin genes. Thus at least one copy of the β tubulin gene is essential for viability. In the diploid containing one wild-type and one mutant gene, the wild-type gene evidently produces enough functional tubulin for the organism to grow and divide. [See N. F. Neff et al., 1983, *Cell* **33**:211.]

pressed only in developing sperm cells, and only this $\beta2$ tubulin is found in the sperm flagellum. Nevertheless, this $\beta2$ tubulin apparently interacts with other tubulins in the formation of microtubules elsewhere, as shown by the effects of a specific mutation preventing normal $\beta2$ tubulin synthesis. In *Drosophila* males homozygous for this mutation, the early meiotic divisions—which are completed *before* the time of $\beta2$ tubulin synthesis—are normal. However, all microtubule-associated events that occur *after* expression of the $\beta2$ tubulin gene are abnormal: the last meiotic division, the formation of the sperm flagellum, and the shaping of the sperm nucleus. Thus the $\beta2$ tubulin does not merely contribute to sperm structure but participates with other tubulins in multiple cellular functions.

Experiments in unicellular organisms also indicate the existence of multiple tubulin subsets. *Naegleria gruberi* is a protozoan that grows in ameboid form. In response to certain environmental stimuli such as suspension in a non-nutrient medium, it differentiates into a flagellated form with a defined body shape and two flagella. The synthesis of the new flagella takes about 1 h (Figure 18-12). Both ameboid and flagellated forms contain an abundant supply of tubulin: about 4 percent of the total protein is tubulin. However, the assembly of the two new flagella uses only a small fraction of this tubulin. Indeed, labeling studies have shown that essentially all the tubulin in the newly formed flagella is made de novo during the transformation. Furthermore, an antibody can distinguish the flagellar tubulin from tubulin in the cytoplasm. These findings indicate that a unique species of tubulin that generates the flagella is synthesized during the one-hour interval.

Understanding the differential control of the multigene tubulin family and the various posttranslational modifications presents a clear challenge for future cell biologists.

Microtubules and Intracellular Transport

Within cells, vesicles and protein particles are frequently transported long distances—many micrometers—and are targeted to a particular cellular region. Since diffusion alone cannot account for either the rate or the directionality of these processes, cell biologists have long suspected that cytoskeletal fibers play a key role. Two well-studied experimental systems—nerve cells and fish erythrophores—have provided evidence implicating microtubules in the intracellular transport of vesicles and protein particles.

Axonal Transport Requires Microtubules

Ribosomes are present only in the cell body of nerve cells, so no protein synthesis occurs in the axons and synaptic

(a)

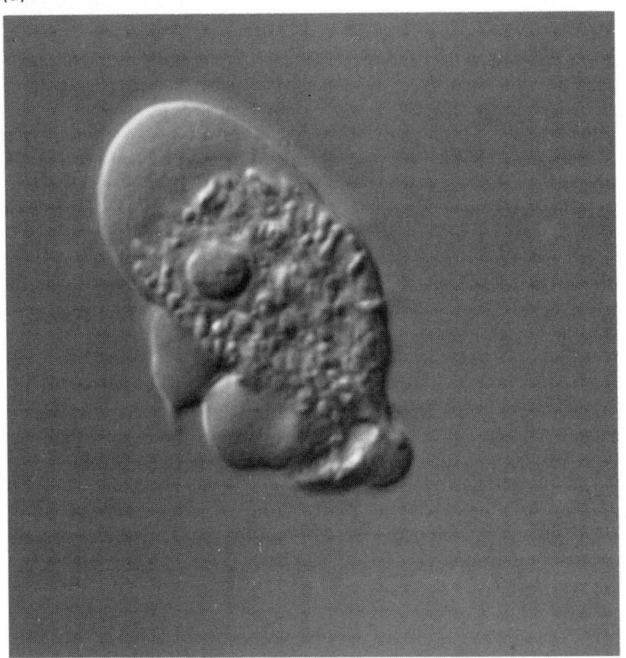

Ameboid

(b)

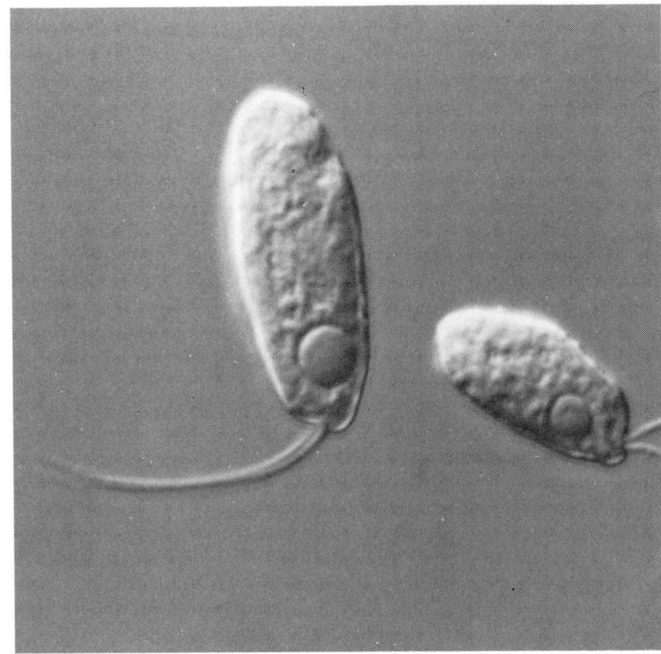

Flagellated

(c)

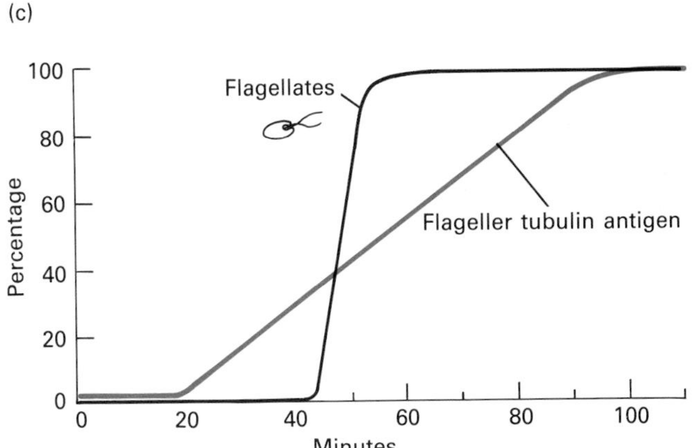

Figure 18-12 Appearance of flagellar tubulin antigen during differentiation of the ameboid form to the flagellated form of *Naegleria gruberi.* (a) The ameboid form. (b) Flagellated forms. (c) Graph correlating timing of events. Flagellated cells appear 40 to 60 min after the start of differentiation. The appearance of the flagellar tubulin antigen begins at 20 min and continues until 100 min, at which time the flagella have reached their mature length. [See C. Fulton, 1977, *J. Supramol. Struct.* 6:13.] *Photographs courtesy of C. Fulton.*

terminals. Therefore, proteins and membranous organelles must be synthesized in the cell body and transported down the axon to the synaptic regions. This process is called *axonal transport* (Chapter 17). Neurons in the sciatic nerve of mammals are particularly suitable for study of this process, because their cell bodies are conveniently located in the dorsal root ganglion near the spinal cord, and because the bundles of nerve processes are very long, extending as much as 1 m from the ganglion to innervate many leg muscles. Radiolabeled amino acids injected into the ganglion of the sciatic nerve become incorporated into proteins made in the cell body. Subsequent examination by autoradiography of nerve segments from different experimental animals at specified intervals reveals the movements of the various labeled proteins, which are identified by gel electrophoresis (Figure 18-13). Such studies have established that proteins do move down the axon, but that all proteins do not move at the same rate. The fastest group, consisting of small vesicles and particles, has a velocity of about 250 mm/day or about 3 μm/s. The slowest group, containing mostly cytoskeletal proteins, moves only a few millimeters per day (Table 18-1). Larger particles, microtubules, and intermediate filaments move down the axon at a very slow rate (Table 18-1). Indeed, tubulin is assembled into microtubules in the cell body, and the microtubules themselves are moved slowly along the axon.

In these and other neurons, fast axonal transport is in-

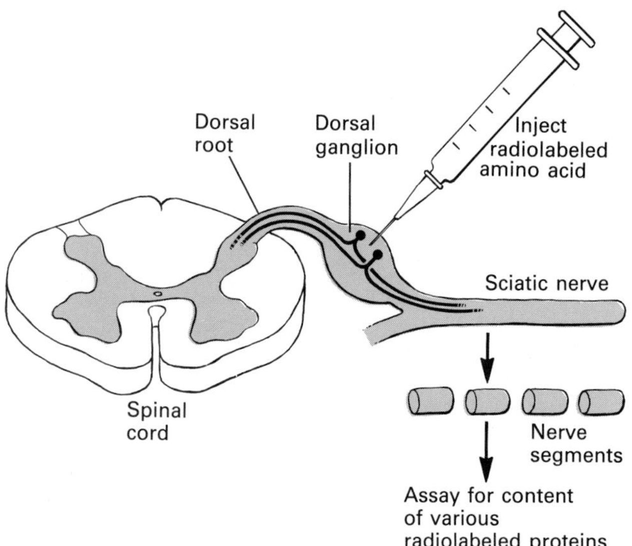

Figure 18-13 Identification of proteins and the measurement of their rate of transport down the axon of a nerve. Radiolabeled amino acids are injected into the ganglion that contains the cell bodies of a group of neurons, which incorporate them into proteins. The sciatic nerve of mammals is used in this study because cell bodies located in the dorsal root ganglion adjacent to the spinal cord can be easily labeled. The sciatic axons extend into the back of the leg, where they innervate the major striated muscles. Experimental animals are killed at different times after injection, and the sciatic nerve is dissected and cut into 5-mm segments. The amount of radiolabeled protein in each fragment is measured, and then the various proteins are resolved by gel electrophoresis. In this way, it is possible to determine how far along the axon a single protein species, such as actin, has moved in the period of time since the animal was injected. [See O. Ochs, 1981, in *Basic Neurochemistry*, 3d ed., G. J. Sigel et al., eds. Little, Brown, p. 425.]

hibited by colchicine and other agents that inhibit the function of microtubules, even though most of the microtubules in the axon are not depolymerized by such treatment. The rapidly transported proteins appear to move down the axon in organized structures, such as small membranous vesicles and mitochondria, associated with microtubules (see Figure 18-5). In the giant axon of the squid, small vesicles 30 to 50 nm in diameter move along microtubules at rates of 1.6 μm/s. Actin-myosin systems of movement (Chapter 19) do not appear to be involved, as injection of specific inhibitors of these proteins has no effect on granule movement.

Recently investigators have shown that cytoplasmic filaments separated from the cytoplasm of the squid giant axon conduct the directed movement of organelles in the presence of ATP. The organelles and filaments are visualized by phase-contrast or differential interference microscopy, in which the weak image is amplified by a video camera and television screen. Subsequent electron microscopy of the same region of the cytoplasm demonstrates that the transport filaments are indeed single microtubules. All organelles move at a rate of about 2 μm/s, comparable to the axonal transport rate in intact cells. In some cases two organelles move along the same filament in opposite directions and pass each other without colliding (Figure 18-14), indicating that each transport microtubule has several tracks for organelle movement. Clearly the interaction of organelles and microtubules is the basis for fast axonal transport. Some type of ATP-utilizing molecular "motor" powers organelle movement. Further studies showed that cytosolic proteins and ATP were required for vesicles to move along pure microtubules. Purification of axonal proteins showed that two different proteins generate movement of organelles in opposite directions along microtubules in vitro.

Microtubules Provide Tracks for Movement of Pigment Granules

The skin of many amphibians and the scales of many fish contain specialized cells called *melanophores* that contain granules of pigment. By processes under nervous and hormonal control, these pigment granules can be transported to the cell periphery to darken the color of the skin (Figure 18-15). Transport of these granules inward, toward the center of the cell, lightens the color. In this manner, the animal can adjust its color to blend better with its surroundings in order to escape predators.

A short time after a scale is removed from a fish and placed in culture, the melanophores begin spontaneously to move their pigment inward and outward. In favorable cases it has been possible to follow individual granules: after dispersal, each granule always returns to the same location in the center of the cell. During this movement, the microtubules apparently give radial direction to the pigment, acting as guides along which granules and associated material can move. As in axonal transport, these microtubules may also move or help to move the pigment.

The Motion of Cilia and Flagella

Cilia and flagella are found on many eukaryotic cells. These organelles have a very similar structure; they differ

Table 18-1 The components of axonal transport in mammalian neurons and their relationship to cytologic structures

Components	Class*	Transport rate (mm/day)	Materials transported	Cytologic structures transported
Fast	I	200–400	Glycoproteins, glycolipids, acetylcholinesterase	Vesicles, smooth endoplasmic reticulum; small granules
Intermediate	II	50	Mitochondrial proteins	Mitochondria
	III	15	Myosin-like proteins	Filaments
Slow	IV	2–4	Actin, clathrin, enolase, CPK calmodulin	Actin microfilaments
	V	0.2–1	Intermediate-filament proteins, tubulin	Microtubule–intermediate-filament network

*Roman numerals refer to the nomenclature developed by M. Willard, W. M. Cowan, and P. R. Vagelos, 1974, *Proc. Nat'l Acad. Sci. USA* 71:2183–2187 for the components of transport. [See also B. Grafstein and D. S. Forman, 1980, *Physiol. Rev.* 60:1167.]

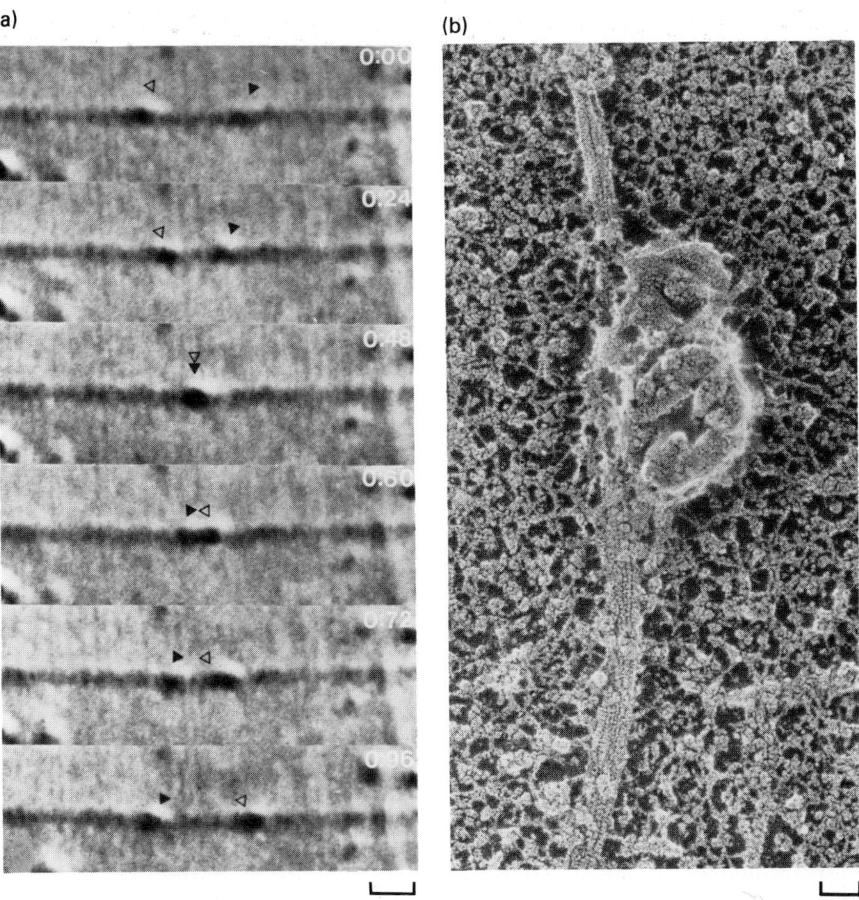

Figure 18-14 Micrographs showing bidirectional movement of two vesicular organelles on a single transport microtubule filament. (a) A piece of giant squid axon was dissected, the cytoplasm was extruded, and a buffer containing ATP was added. The preparation was then viewed in a differential interference microscope, and the images were recorded directly by a television camera mounted in place of the eyepiece and stored on video tape. The two organelles (*open versus solid triangle*), which are of different sizes, move in opposite directions along the same filament, pass each other, and continue in their original directions. Elapsed time (in seconds) appears at the top right corner of each video frame. (b) The region of cytoplasm that was studied in a similar video experiment was freeze-dried, rotary shadowed with platinum, and viewed in the electron microscope. Two large structures, presumably small vesicles, are attached to one microtubule, and were moving along it when the preparation was frozen. [See B. J. Schnapp et al., 1985, *Cell* 40:455.] *Photographs courtesy of B. J. Schnapp, R. D. Valle, M. P. Scheetz, and T. S. Reese.*

(a)

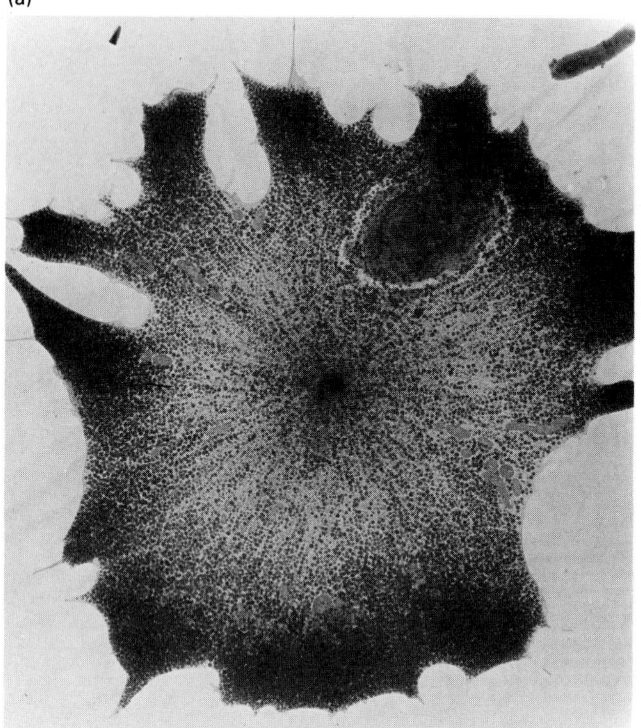

(b)

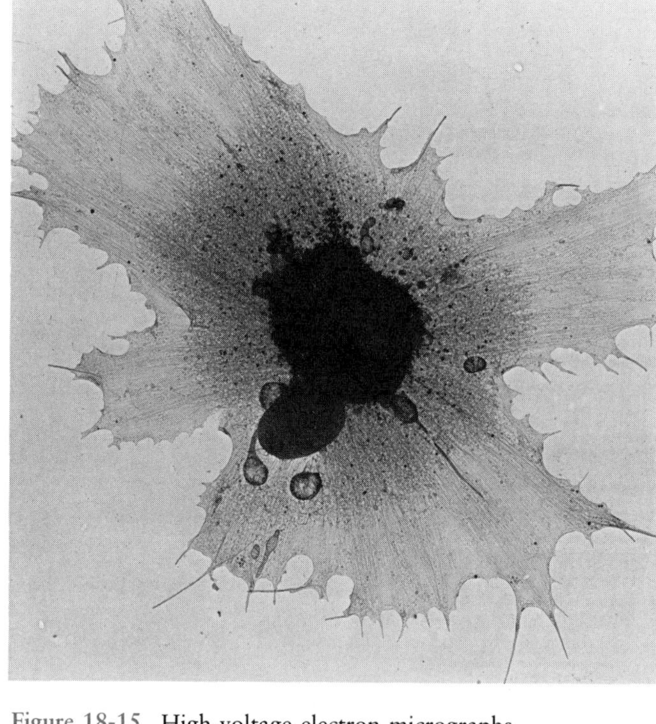

(c)

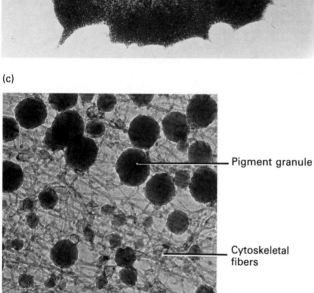

Pigment granule

Cytoskeletal fibers

Figure 18-15 High-voltage electron micrographs showing movement of pigment granules in the melanophore, or red-pigment cell, of the squirrelfish, *Holocentrus ascensionis*. (a) The pigment granules are dispersed to the cell periphery. (b) The granules are condensed around the nucleus. (c) A portion of a dispersing erythrophore, showing the pigment granules associated with tracks of cytoskeletal fibers including abundant microtubules. *Photographs courtesy of K. Porter.*

mainly in length (flagella are longer) and in pattern of beating. Sperm cells and certain protozoans use flagella for propulsion. Rows of cilia cover the surfaces of mammalian respiratory passages, such as the nose, pharynx, and trachea, where they sweep out particulate matter that collects in the mucous secretions of these tissues (see Figure 5-48).

Propulsion by both cilia and flagella is caused by bending at their base (Figure 18-16). Cilia move by a whiplike *power stroke* fueled (as we shall see) by hydrolysis of ATP, followed by a *recovery stroke*. Flagellar movement is also powered by ATP hydrolysis. In contrast to cilia, they generally move by waves that emanate from the base and spread outward toward the tip.

All Eukaryotic Cilia and Flagella Have a Similar Structure

All cilia and flagella are built on a common fundamental plan (Figure 18-17; see also Figure 5-48).

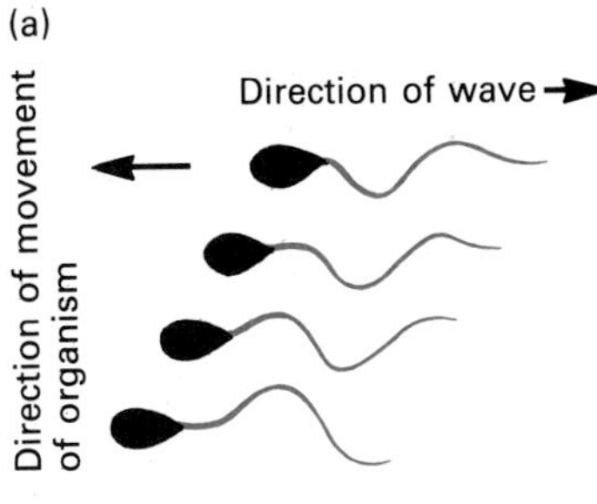

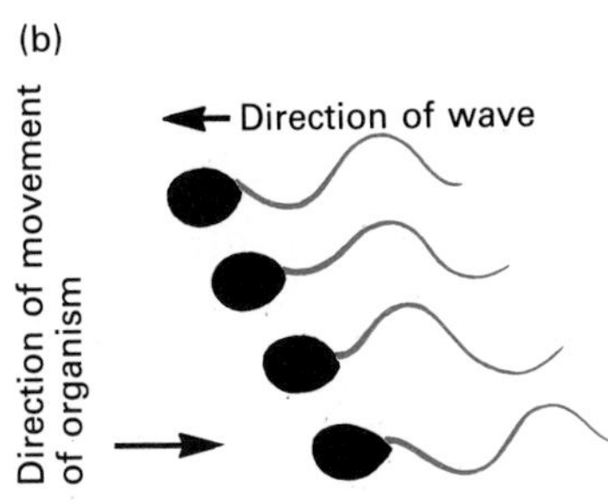

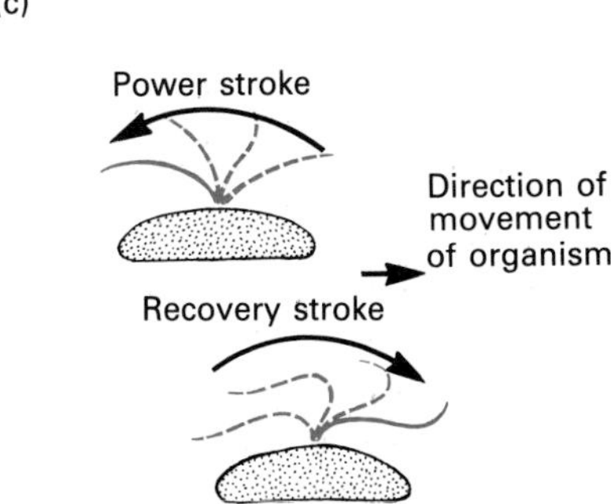

Figure 18-16 Characteristic motions of cilia and flagella. (a) In this typical sperm flagellum, successive waves of bending originate at the base and move toward the tip; these push against the water and propel the cell forward. Sperm flagella generate waves at a frequency of 30 to 40 per second. (b) A reverse type of motion is observed in certain flagellated parasitic protozoans such as trypanosomes. The wave is directed toward the cell and pulls it in the direction opposite to the wave. (c) Movement of a cilium occurs in two stages called the power stroke and the recovery stroke. In the power stroke, the cilium is extended straight outward and then is swept backward by bending at the base. This propels the organism in the opposite direction. This stroke is powered by hydrolysis of ATP. During the recovery stroke, no ATP is hydrolyzed. A wave of bending moves along the cilium from its base, pushing the cilium forward.

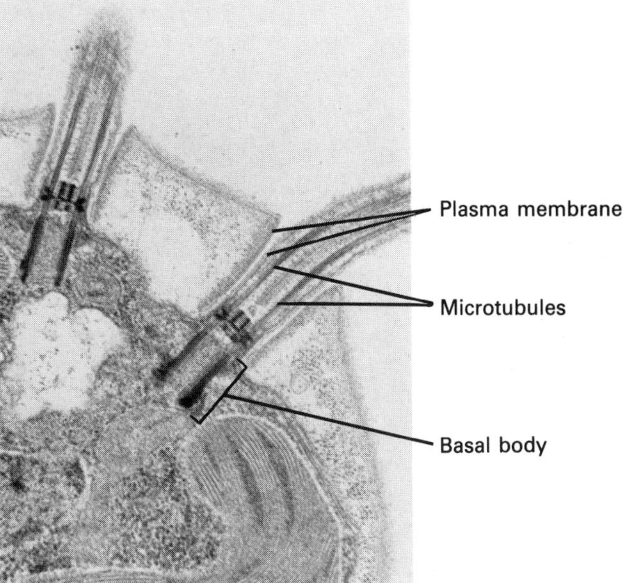

Figure 18-17 Low-magnification electron micrograph of *Chlamydomonas reinhardtii* flagella, showing the plasma membrane, the bundle of microtubules, and the basal body. *Photograph courtesy of R. Wright and J. Jarvik.*

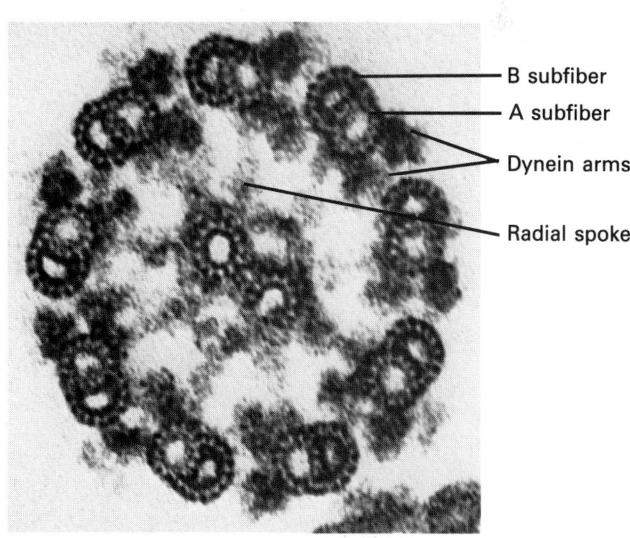

Figure 18-18 Transverse section through an isolated demembranated cilium. There are nine outer doublets, each with two dynein arms attached at regular intervals to each A subfiber, and a radial spoke connecting the A subfibers at regular intervals to the central sheath. The individual protofilaments in each of the microtubules are clearly resolved. *Photograph courtesy of L. Tilney.*

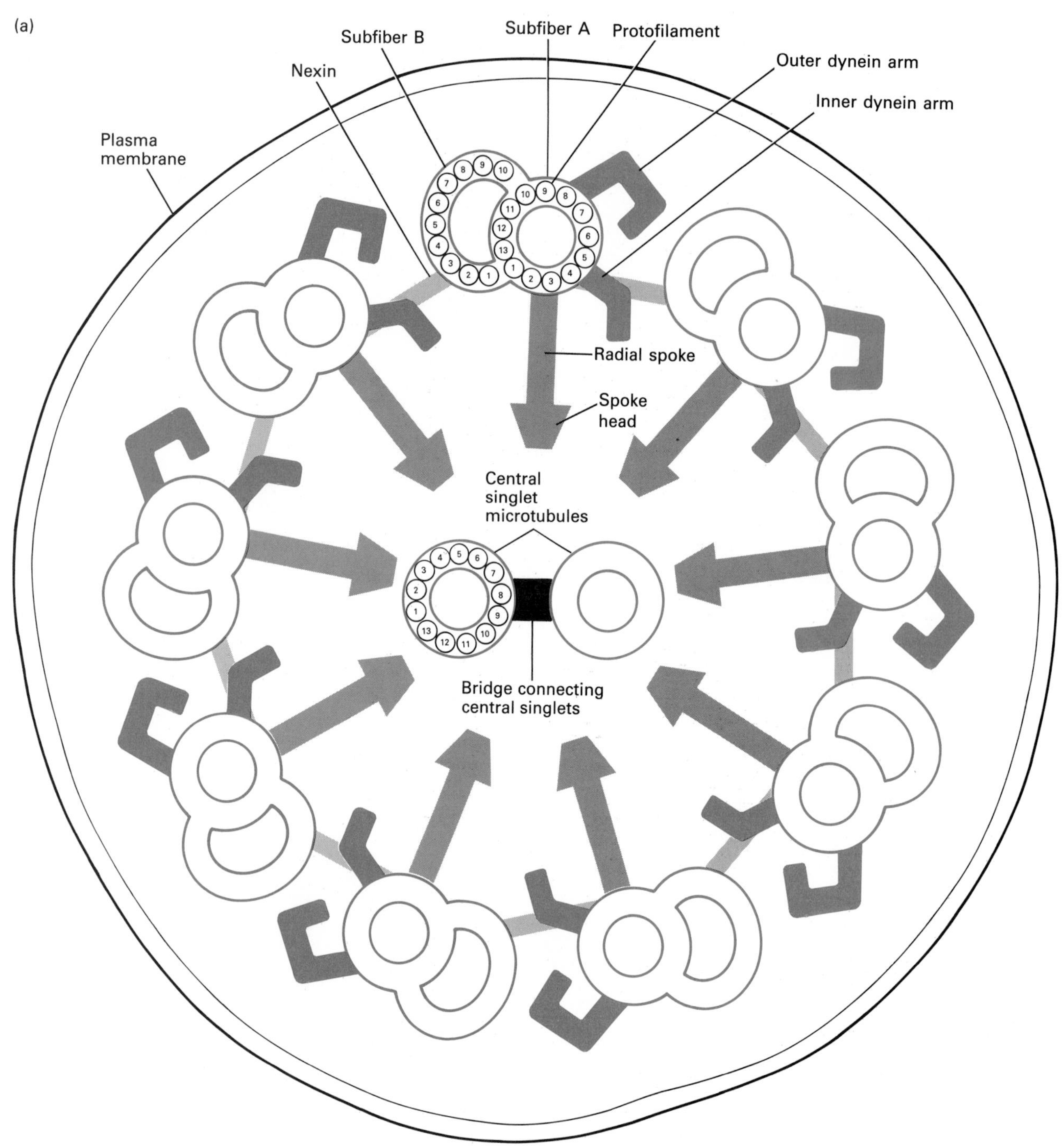

Unless otherwise noted, the terms cilia and flagella can be used interchangeably for the purpose of the following discussion. A bundle of microtubules called the *axoneme* is surrounded by a membrane that is part of the plasma membrane. Just under the main outer boundary of the cell, the axoneme connects with the *basal body*. As we shall see later, this structure also is composed of microtubules and plays an important role in initiation of growth of the axoneme.

Each axoneme contains two central singlet microtubules, each with the usual 13 protofilaments, and nine outer pairs of microtubules called *doublets*. This recurring motif is known as the 9 + 2 array (Figure 18-18). Each outer pair has a defined organization. There is one complete microtubule, called the *A subfiber,* with 13 protofilaments (Figures 18-18 and 18-19). Attached to each A subfiber is a *B subfiber,* made up of 10 (in some cases, 11) protofilaments. The organization of the A and B sub-

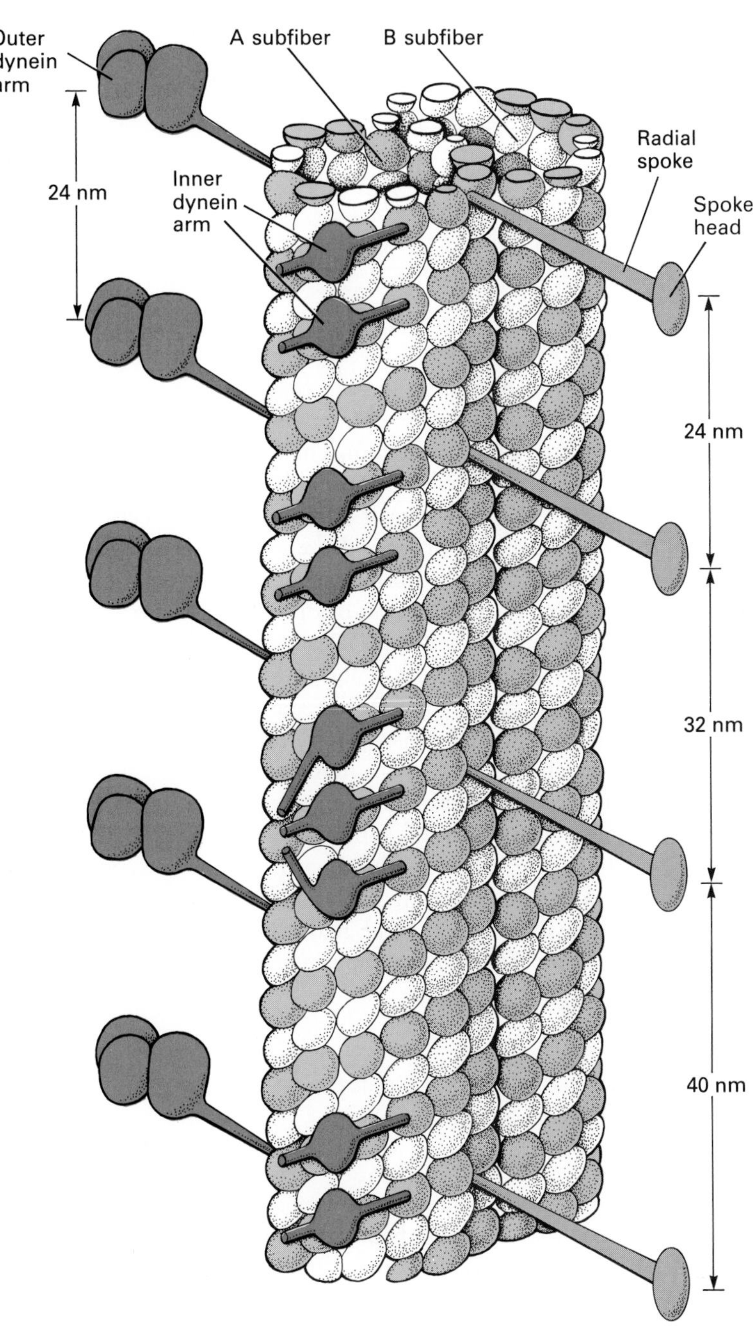

Figure 18-19 The structure of an idealized flagellum. (a) ***(page opposite)*** Cross-sectional view showing all major structures. For clarity, the individual protofilaments are indicated in only one central microtubule and one outer A subfiber–B subfiber doublet microtubule. (b) A model of an axonemal doublet with its attached dynein arms and radial spokes. There are two distinct species of inner dynein arms—one with two heads, one with three. Outer dynein arms bind with a regular period of 24 nm; inner dynein arms and radial spokes bind with periods of 24, 32, and 40 nm. A complicated and unknown mechanism must control the assembly of these tubulin-associated proteins. [See U. W. Goodenough and J. E. Heuser. 1985. *J. Cell Biol.* **100**:2008.]

fibers is so uniform that each protofilament can be assigned a number, as shown in Figure 18-19. Each A subfiber is connected to the central microtubules by *radial spokes* and structures termed *spoke heads*. Detailed analyses of electron micrographs have shown that a radial spoke is attached at every third or fourth or fifth $\alpha\beta$ dimer (or every 24 or 32 or 40 nm) to one or two specific protofilaments of the A fiber. The radial spokes are arranged asymmetrically in groups of three along the outer doublets; this asymmetry allows distinction of the "proximal" end of the flagellum (that closer to the cell) from the "distal" end (that farthest from the cell) in electron microscopic images.

Two "arms," termed the *inner dynein arm* and the *outer dynein arm,* protrude from each A subfiber at regular intervals (Figures 18-19 and 18-20). All the arms in a flagellum point in the same direction: clockwise when viewed from the base to the tip. At every third tubulin

Figure 18-20 Dynein arms on a microtubule isolated from *C. reinhardtii.* The outer dynein arms (the regularly spaced "lollipops") are seen to extend at 24-nm intervals along each microtubule. The "heads" of the arms are about 10 nm in diameter; the "stalks," less than 3 nm wide. As noted in Figure 18-19, the spacing and structure of the inner arms are less regular. For this electron micrograph, the flagellum was subjected to the deep-etch procedure, and then a platinum replica was obtained. *Photograph courtesy of J. Heuser and U. Goodenough.*

dimer an outer dynein arm is attached to two specific A protofilaments; inner arms are located at less regular intervals along the A protofilament (Figure 18-19).

Paired bridges connect the two central microtubules, like rungs on a ladder. In addition to the radial spokes, the axoneme is held together by a set of circumferential linkers that join adjacent outer doublets. These linkers, composed in part of the protein *nexin,* are highly elastic: their normal length is 30 nm, but they can be stretched to 250 nm without breaking. Nexin links are found with a periodicity of about 86 nm along the axoneme. Thus, at the level of resolution afforded by the electron microscope, flagella are quite complex structures and contain many protein components in addition to microtubules. What roles do these structures play in accomplishing the characteristic motion of flagella and cilia?

Dynein ATPases Are Essential to the Movement of Flagella

Flagella from which the membrane has been removed by nonionic detergents are called *isolated axonemes.* Such axonemes will beat when provided with ATP. Thus the "motor" resides in the organelle, not elsewhere within the cell body.

Biochemical studies have shown that all flagella, from sources as diverse as human sperm and the protozoan *Tetrahymena,* have a potent ATP-hydrolyzing activity. This ATPase activity is a property of the inner and outer dynein arms. Treatment of isolated axonemes with solutions containing high concentrations of salt removes the dynein arms, solubilizes the ATPase activity, and abolishes ATP-stimulated beating of the flagellar axonemes. The addition of a purified preparation of dynein arms restores both the ATPase activity and the beating, and electron microscopy shows that the dyneins have reattached to their proper place on the A subfibers.

Driven by ATP Hydrolysis, Dyneins "Walk" Along the Adjacent B Subfiber

In isolated axonemes, addition of ATP causes each outer doublet to slide past an adjacent one (Figure 18-21). The direction of sliding is such that the dynein arms of one doublet push the B subfiber on the adjacent doublet toward the tip of the cilium. The force producing the sliding of adjacent doublets is probably caused by successive formation and breakage of cross-bridges between the dynein arms of one doublet and the B subfiber of the adjacent doublet as the two doublets move relative to each other. Formation of such cross-bridges has in fact been observed and is known to be ATP-dependent. Removal of ATP causes the flagella axonemes to become extremely stiff and rigid. The dynein forms stable cross-bridges between adjacent outer doublets, so that they cannot move relative to each other, and the microtubules become frozen in place. Subsequent addition of ATP causes breakage of these cross-bridges, followed by hydrolysis of ATP to ADP. The successive binding of ATP to a dynein bridge, and its subsequent hydrolysis, causes the dynein arms to successively break and make bonds with the adjacent doublet. The dynein appears to "walk" along the adjacent doublet (Figure 18-21).

Radial Linkers Convert the Sliding of Microtubules into Bending of the Axoneme

In the intact flagellum, however, all of the outer doublets are constrained by their radial links to the central sheath and by their nexin connectors. The *sliding* motion of the outer doublets is converted into a *bending* of the axoneme. The theory that flagella *bend* by the sliding of outer doublets past each other was elegantly proved by careful electron microscopic studies. In the straight flagellum, all the outer doublets are of the same length and terminate at the same point. In fixed, bent flagella, all the outer doublets are also of the same length, but the ones at the inside of the bend extend farther than those on the outside—as would be expected if the tubules slide relative to one another (Figure 18-21).

To explain the generation of waves during the beating of flagella (Figure 18-16), more complex processes must be involved. The beat of a flagellum is confined to a plane; a possible molecular explanation for this is that two of the outer doublets—by convention, No. 5 and No.

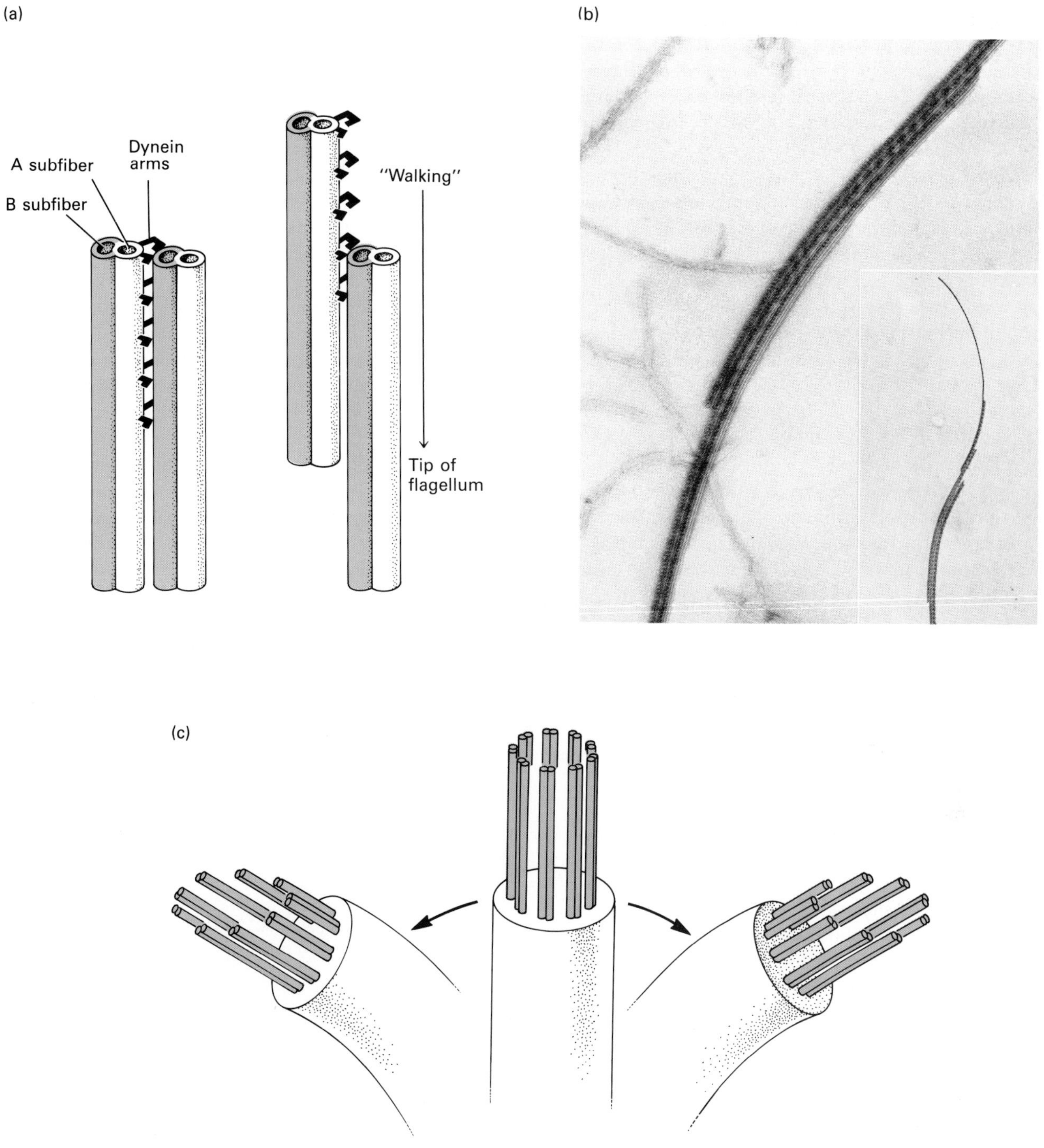

(a)

Dynein arms
A subfiber
B subfiber
"Walking"
Tip of flagellum

(b)

(c)

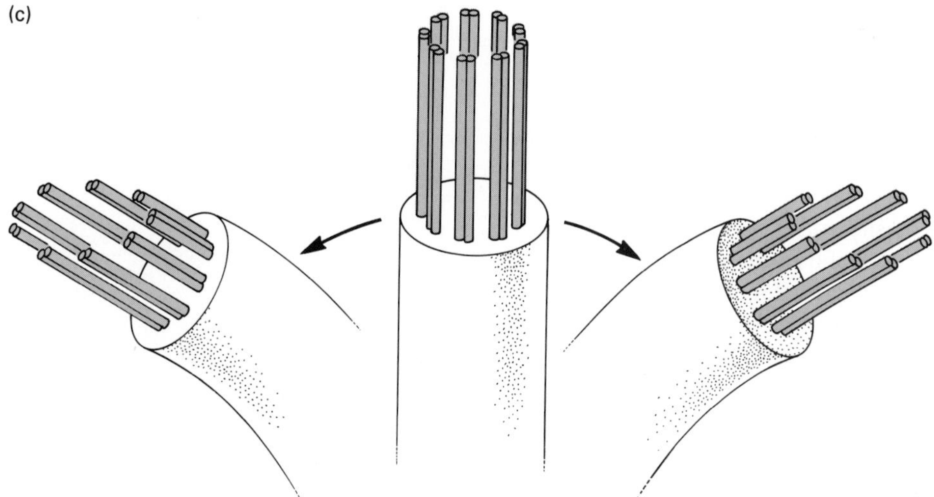

Figure 18-21 The "dynein-walking" model. (a) Diagram of two adjacent outer doublets. *Left*, Before ATP addition; *right*, after ATP addition. The dynein arms of one axonemal microtubule, powered by ATP hydrolysis, somehow "walk" along the adjacent doublet, through a series of events not yet understood at the molecular level. In an intact cilium, structures attached to the microtubules—in particular, the radial spokes—resist the sliding and produce local bending. (b) Electron micrograph was taken after incubation of a disrupted isolated axoneme with ATP. ATP hydrolysis has caused one microtubule to move relative to its neighbor. *Inset*, At lower magnification, the extent of displacement of four adjacent microtubule doublets can be seen. (c) Diagram of sliding microtubules during beating of a cilium. The membrane of the distal portion of the cilium is removed, exposing the nine outer doublets. In the straight cilium, all of the tubules end at the same point (*center*). During beating, the outer doublets slide past one another, causing a bending of the cilium and a displacement of the ends of the tubules (*left, right*). *Part (b) courtesy of F. D. Warner and D. R. Mitchell.*

6, as shown in Figure 18-19—have permanent rigid links to each other. The bond thus formed and the two central fibers define the plane along which the beat of the flagellum occurs (Figure 18-16). Because doublets No. 5 and No. 6 cannot slide past each other, any beating of the flagellum is confined to this plane.

For a flagellum to beat, the radial spokes must somehow be able to detach and reattach their connections with the central microtubules. (The specific factors regulating this process, in particular the role of ATP, have not yet been determined, however.) The elastic nexin links, by contrast, appear to be permanent, and they may prevent excessive sliding displacement between the outer doublets. Little is known about the details by which motions initiated at the base of a cilium or flagellum are transmitted to the end.

As in many other systems, the concentration of Ca^{2+} has a regulatory role in the beating of cilia and flagella. However, the detailed effects vary considerably from organism to organism: elevations in the concentration of Ca^{2+} can increase the rate of beating or can cause a reversal in the beating pattern. Whether changes in Ca^{2+} concentration are induced by membrane depolarizations, as in nerve and muscle cells, is not known. Calmodulin, a Ca^{2+}-binding protein that mediates most of the known effects of Ca^{2+} (see Chapter 16), is found in flagella.

Genetic Studies Provide an Estimate of the Number of Proteins Required for Axoneme Assembly

The biflagellate, unicellular algae *C. reinhardtii* has proved especially amenable to studies on the biochemical, genetic, and structural aspects of flagella and their assembly (Figure 18-22). By shearing a population of cells, flagella can be obtained in good purity and high yield, and the deflagellated cells will regenerate new flagella.

It is possible to isolate large numbers of viable *C. reinhardtii* mutants that are nonmotile and defective in flagellar function. The study of such mutants has allowed an estimate of the numbers of genes required to assemble a flagellum. Microscopic analysis of flagella from many of these mutants reveals they are missing an entire substructure, such as the radial spokes or inner dynein arms.

Two-dimensional gel electrophoresis (Chapter 7) of axonemal proteins resolves approximately 200 discrete polypeptides, in addition to the α and β tubulins (Figure 18-23). Frequently, nonmotile mutants that are missing a flagellar substructure are also found to lack a number of these proteins. Thus individual proteins can be correlated with specific structures and associated with a specific gene. For instance, nonmotile mutants missing only the radial spokes and spoke heads (Figure 18-23) lack 17 of these polypeptides. Therefore, all 17 of these proteins can be assumed to be located in the spoke or spoke head. Mutants missing only the spoke head lack 6 specific proteins; these 6 are a subset of the 17 lacking in the mutants missing both spokes and spoke heads. Presumably, then, these 6 proteins are localized to the spoke head, and the other 11 to the spokes. Mutants defective in spoke or spoke-head structures have been detected in at least 5 complementation groups (or genes), and each of these mutations has been correlated with 1 of the 17 proteins. Thus, at least 5 and perhaps as many as 17 discrete gene products are essential in construction of just the radial spokes and spoke heads.

By a similar mutational analysis, the inner and outer dynein arms have been shown to be composed of different proteins. A number of nonmotile mutants appear to lack only the inner dynein arms, which consist of two large proteins (m.w. ~ 300,000) as the principal structural components, as well as several smaller protein subunits. How the different structures such as the radial spokes or dynein arms are assembled from their constituent subunits is not known, but presumably specific protein-protein interactions are involved. As with MAPs, functional control of assembly may be altered by posttranslational modifications such as phosphorylation; many flagellar proteins, including some in the spoke, are phosphorylated.

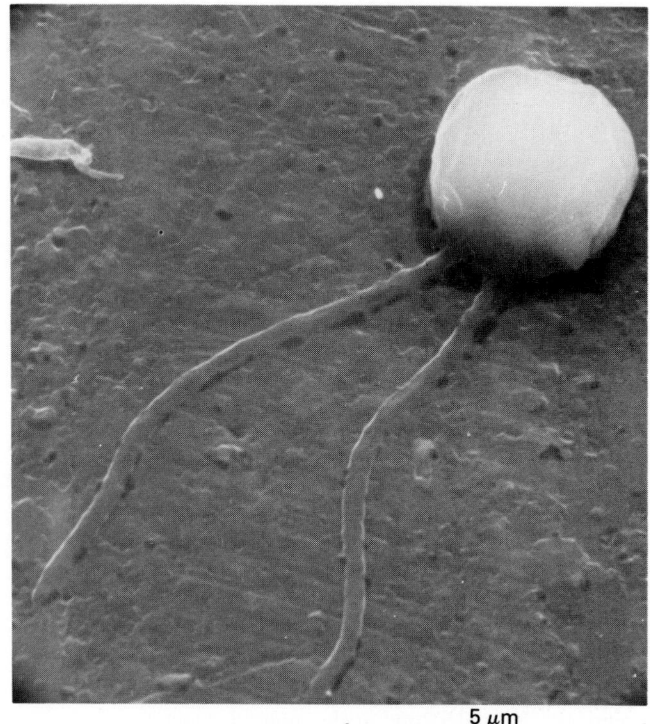

Figure 18-22 Scanning electron micrograph of the gamete stage of *C. reinhardtii*, showing the two flagella. Each flagellum is about 8 μm in length. *Photograph courtesy of B. Bean.*

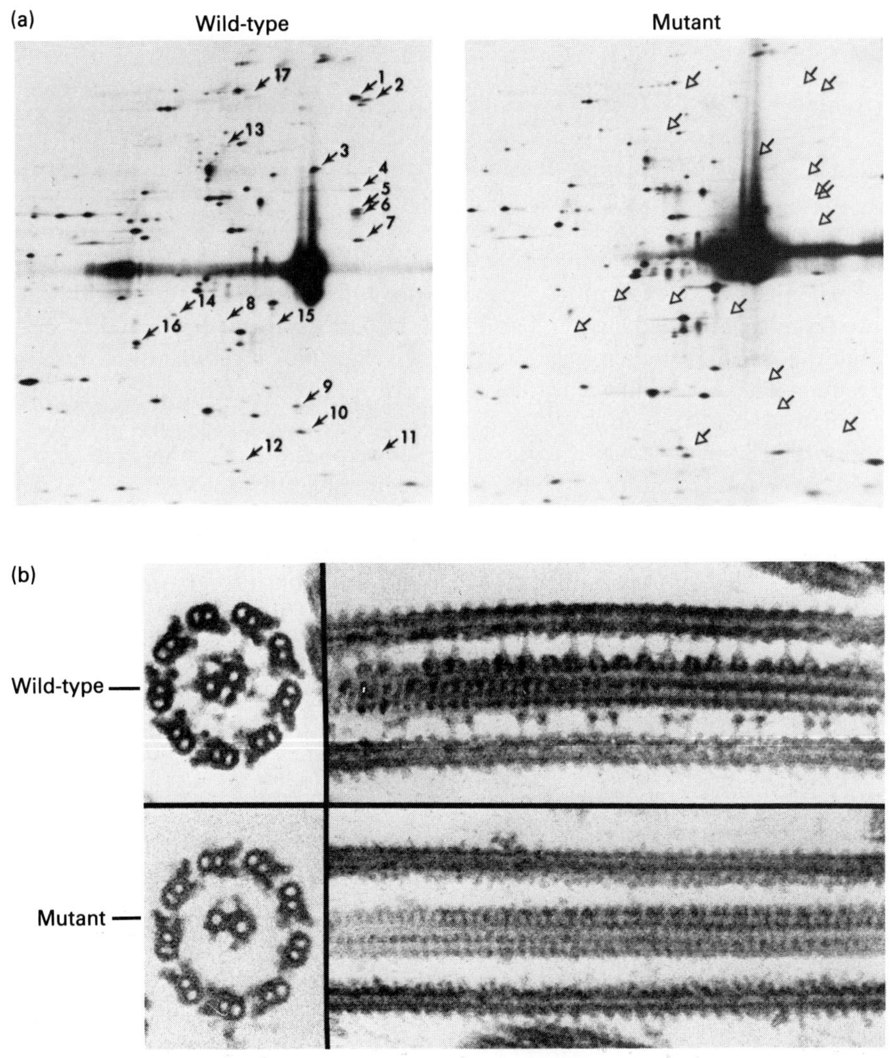

Figure 18-23 Analysis of axonemes from a wild-type *C. reinhardtii* flagellum and from a paralyzed mutant that is missing the radial spokes. (a) Two-dimensional gels of axonemal proteins. In the gel prepared from wild-type flagella, solid arrows point to a set of 17 polypeptides; in the mutant gel, these polypeptides are lacking, as indicated by open arrows. Both mutant and wild-type cells were grown in medium containing [^{35}S]sulfate for several generations, so that even small amounts of flagellar proteins are detectable by autoradiography. Tubulin subunits not resolved in the gels form the central dark streaks. (b) Electron micrographs of isolated wild-type and mutant flagellar axonemes, cut in transverse (*left*) and longitudinal (*right*) sections. Note that the mutant axonemes lack radial spokes. *Photograph (a) courtesy of G. Piperno, B. Huang, Z. Ramanis, and D. Luck; photograph (b) courtesy of B. Huang and D. Luck. Reproduced from the* Journal of Cell Biology, *1981, vol 88, p. 73 by copyright permission of the Rockefeller University Press.*

Specific defects in ciliary and flagellar substructures have been related to disease in humans. In Kartagener's syndrome, a condition characterized by severe respiratory difficulty and also male sterility, the underlying genetic defect is the absence of inner and outer dynein arms on the outer doublets of both cilia and flagella. The symptoms result from the immobility of cilia in the respiratory tract and of the flagella in the sperm. The exact nature of this human genetic defect is not known but appears analogous to that of some *Chlamydomonas* flagellar mutations.

Basal Bodies, Centrioles, and Microtubules

For microtubules to participate in movement, or to act as a structural framework, they must be anchored by at least one end. Indeed, at its cytoplasmic end each cilium or flagellum terminates and is anchored by a microtubule-containing structure called the basal body. Likewise, the microtubules in the mitotic spindle as well as those in interphase cells radiate from a region of the cell near the nucleus called the *centrosome,* which functions as a microtubule-organizing center (MTOC). In animal cells, the centrosome is an amorphous region (sometimes called the pericentriolar region) that surrounds the centriole. It is believed that microtubules in the centrosome, or possibly in the centriole itself, nucleate both the growth of microtubules in interphase cells and the growth of many of the microtubules in the mitotic spindle. Higher plants lack centrioles, and the centrosomal material in their cells acts as an MTOC.

As will become clear in the following discussion, basal bodies and centrioles are similar, if not identical, in structure and function. Both basal bodies and centrioles act as nucleating centers from which microtubules grow—either naturally, as during mitosis, or experimentally, as after microtubules have been depolymerized by exposure to cold or drugs.

Centrioles and Basal Bodies Are Built of Microtubules

Centrioles are cylinders of nine *triplet* microtubules, as is apparent in a cross-sectional view (Figure 18-24). Similarly, in most protozoans as well as in mammalian cells, each basal body consists of a cylinder of nine triplet microtubules (Figure 18-25). Each triplet contains one complete microtubule, the A subfiber, fused to the incomplete B subfiber, which in turn is fused to the incomplete C subfiber. The A and B subfibers continue into the flagellar shaft, while the C subfiber terminates within the transition zone between the basal body and flagellum. The A and B tubules of the basal body appear to initiate assembly of the nine outer doublets of the cilium or flagellum during flagellum assembly, and the subfibers grow outward from the basal body. The two central tubules in the flagellum or cilium terminate in the transition zone above the basal body. Basal bodies are, in an undetermined fashion, replicated during growth and division of a cell.

Centrioles and Basal Bodies May Be Self-Replicating Organelles

Centrioles replicate at defined points during the cell cycle, and daughter centrioles form at right angles from the mature centriole (Figure 18-24; see also Figure 5-16). Normally only a single pair of centrioles is produced; it is not known how this is accomplished. As with centriole replication, the daughter basal body is formed characteristically at right angles to the "parent." Eventually the daughter basal body separates from the "parent" and then generates a new flagellum or cilium. Basal bodies and centrioles have many of the properties of a self-replicating organelle such as a chloroplast or mitochondrion. Over the years, there have been many claims that centrioles and basal bodies contain DNA or RNA, but even the best evidence is not very convincing.

The similarity between basal bodies and centrioles extends to their mode of formation. Precursors of the ciliated epithelial cells that line the human trachea and

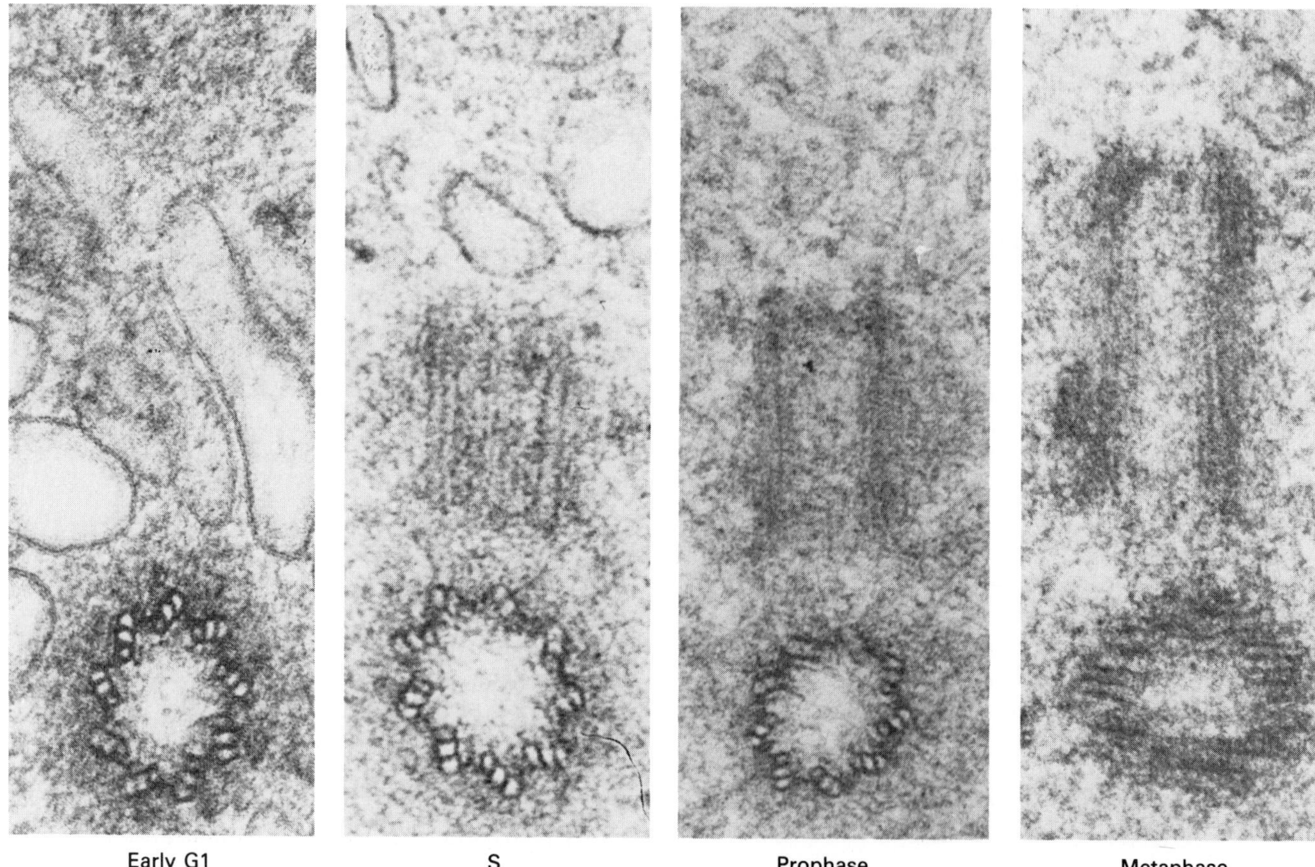

Figure 18-24 Centriole elongation during interphase and mitosis in a cultured mammalian fibroblast. *From left*: Early G1, S, prophase, metaphase. Note that centrioles are cylinders of nine triplet microtubules and that the daughter centriole grows at right angles. [See J. Rattner and S. G. Phillips, 1973, *J. Cell Biol.* 57:359.] *Photograph courtesy of J. Rattner. Reproduced from the* Journal of Cell Biology, *1983, by copyright permission of the Rockefeller University Press.*

(a)

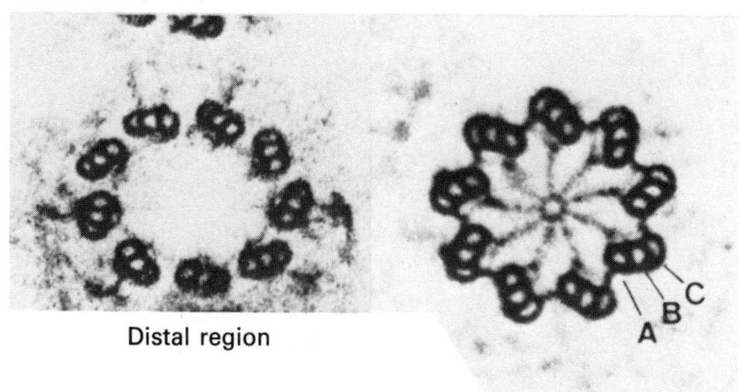

(b)

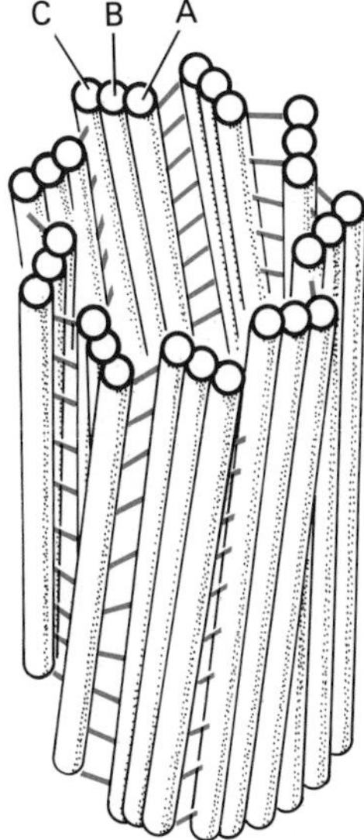

Figure 18-25 Structure of basal bodies and centrioles. (a) Cross sections of flagellar basal bodies in the protozoan *Trichonympha,* showing distal and proximal regions. The proximal region contains a central tubule and radial spokes. Note that each of the nine outer fibers are built of three tubules; by analogy with the outer doublets of flagella, these are labeled A, B, and C. (b) A schematic view of a centriole or basal body. The nine sets of triplet microtubules are connected by a set of proteinaceous fibers that are also visible in part (a). *Part (a) courtesy of I. Gibbons. Reproduced from the* Journal of Cell Biology, *1981, by copyright permission of the Rockefeller University Press.*

esophagus contain only centrioles. As these cells differentiate, the centrioles migrate from their normal position, near the nucleus, to the luminal plasma membrane. There the centriole forms numerous centriole-like structures, each of which becomes a basal body for a cilium.

In certain ciliates such as *Paramecium,* the pattern of cilia in the cell surface (Figure 18-26a) remains constant from generation to generation, but different strains have different patterns. In classic studies, rows of cilia from the cortex, or outer region, were removed from one part of a paramecium and reimplanted in another part. The new pattern of cilia thus created was inherited and remained stable for many generations in the progeny of the organism that received the transplant. Figure 18-26b shows a typical experiment.

As the cell grows and divides, new cilia are generated from newly formed basal bodies that lie in the cortical cytoplasm just under the plasma membrane. In all likelihood, the pattern of arrangement of the basal bodies determines the pattern and orientation of the cilia, although it is not known how the pattern of basal bodies in the cortical cytoplasm is inherited. Neither the presence of hereditary material—DNA or RNA—in basal bodies nor an ability to direct synthesis of their proteins is necessary to explain the stable inheritance of the pattern of cilia and basal bodies. Rather, the polarity of the rows of cilia may be due to a polarity in the way the basal bodies are connected to each other in a line from the front of the cell to the back.

Microtubules Grow from Basal Bodies and Centrioles

In interphase cells, most microtubules radiate from an amorphous area—the centrosome—around the centrioles (see Figure 18-3). The importance of this region in organizing the microtubular cytoskeleton can be seen in cells that have been treated with colchicine; all of the cytoplasmic microtubules except the centriole are depolymerized (see Figure 18-6a). When colchicine is removed by washing, polmerization occurs. Importantly, after removal of colchicine all the new microtubules radiate from the MTOC, and tubulin begins to polymerize on the distal ends of these fibers (Figure 18-27). Similarly, in mitotic cells, when the interphase microtubules depolymerize, the new mitotic spindle microtubules grow outward from the centrosome.

The role of the centriole in organizing microtubules has been substantiated experimentally. Centrioles free of most pericentriolar material cause an outgrowth of microtubules when injected into frog eggs. Also, purified centrosomes (centrioles plus attached material) organize polymerization of tubulin into microtubules (Figure 18-28). However, such findings do not rigorously prove whether the centrioles or the amorphous pericentriolar material surrounding them plays the crucial role.

Flagellar Microtubules Grow at Their Distal Ends

Studies on the *Chlamydomonas* flagellar system demonstrated a very general property of microtubular struc-

(a)

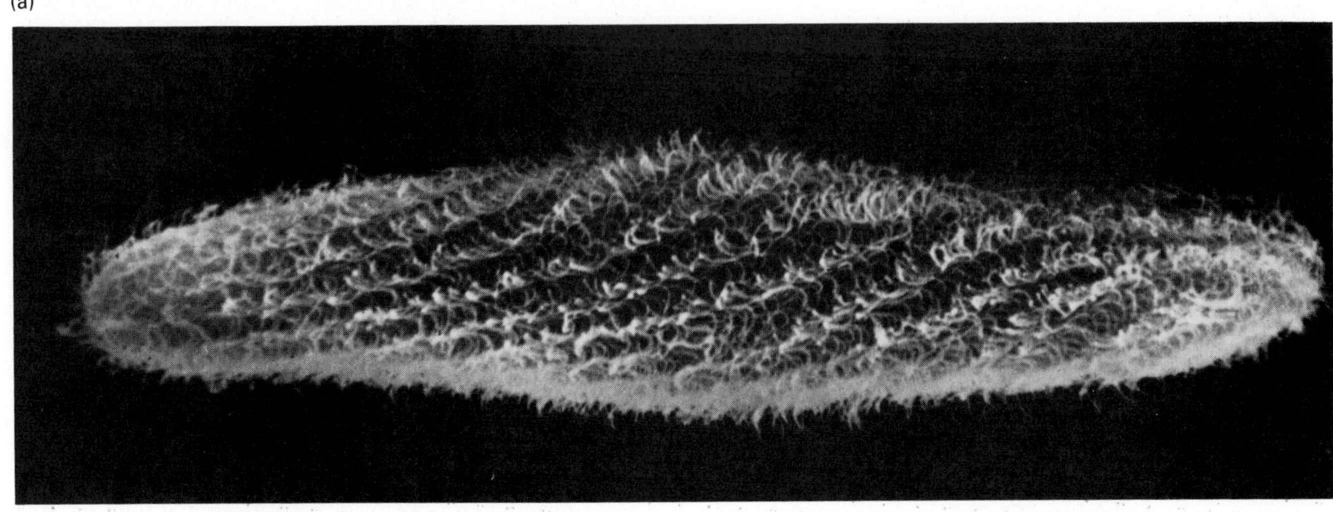

(b)

Basal bodies

Cilia

All basal bodies have same orientation

Rows of cilia have same orientation and beat in same direction

Microsurgery of cortical cytoplasm, including basal bodies

Basal bodies whose orientation was inverted by surgery

Cilia growing from inverted basal bodies are inverted and beat in opposite direction

Cell division

Abnormal pattern of basal bodies and cilia is inherited through many generations

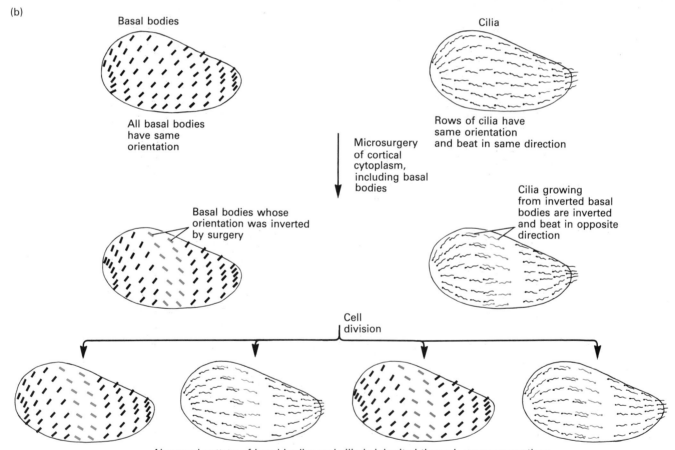

Figure 18-26 Inheritance of the pattern of cilia on the surface of *Paramecium*. (a) A scanning electron micrograph of *P. multimicronucleatum*, showing the parallel rows of cilia. Normally all of the cilia beat in the same direction, allowing a coordinated movement of the cell. (b) A row of cortical cytoplasm containing basal bodies and cilia can be removed and then reimplanted in the opposite direction. The inverted rows of cilia beat in an opposite direction, presumably because of an inversion in the orientation of the basal bodies. When the organism grows and divides, all successive generations retain the altered pattern of cilia. Thus, the progeny cells inherit a pattern of cilia rows in their cortical cytoplasm that has nothing to do with nuclear DNA, since the original and modified organisms have the same nuclear DNA. *Part (a) courtesy of G. Antipa.*

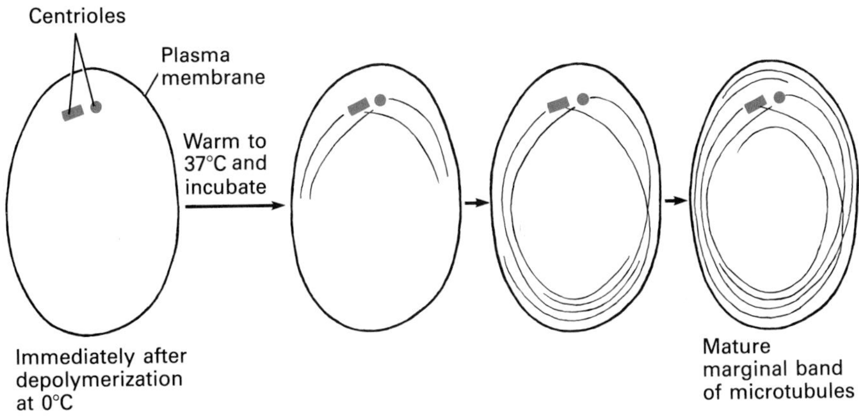

Figure 18-27 Reassembly of the microtubules in the marginal band of nucleated erythrocytes (see Figure 18-2). When the cells are kept at 0°C for a few hours, all of the microtubules in the marginal band depolymerize; only the centrioles (*color*) remain. Soon after the cells are warmed to 37°C, short microtubules are visible, focused on a "pole" of centrioles. With time, the microtubules grow into a continuous elliptical band that girds the cell. This figure is simplified for diagrammatic purposes; actually, many microtubules would grow simultaneously from each centriole. [See I. Nemhauser, J. Joseph-Silverstein, and W. D. Cohen, 1983, *J. Cell Biol.* **96**:979.]

tures: the microtubules grow in one direction, and tubulin is added at the distal end—the end opposite the basal body. When flagella are amputated by mechanical agitation, the cells regenerate a new flagellum, of precisely the same length as the old, within 1 h. Deflagellation of *Chlamydomonas* gametes triggers the synthesis of many axonemal proteins, including tubulin, dynein, and flagellar membrane protein, and their corresponding mRNAs. Clearly, the lengths of cilia and flagella and the synthesis of their component polypeptides are carefully controlled, but the mechanism involved has not been established.

A simple pulse-chase experiment showed that during flagellar growth, most of the new subunits are added to the *distal* tip of the growing flagellum. For the pulse, radiolabeled amino acids were added to a culture of *Chlamydomonas* cells. When the flagella were examined by autoradiography, most of the newly added protein was in the distal tips of the flagella (Figure 18-29). In agreement with this in vivo result, studies in which brain tubulin was added to partially disrupted *Chlamydomonas* flagella indicated that tubulin addition to the outer doublets also occurred preferentially at their distal ends. The microtubules in the outer doublets are continuous with those of the basal body. This experiment showed that microtubules grow outward from their "free" end. Curiously, polymerization to the two central microtubules occurred at their *proximal* ends, possibly indicating a reversed polarity in the central region. In most cases, however, the growth end of microtubules in vivo is the same end which tubulin adds onto preferentially in vitro.

The Length of Flagella Is Regulated

Studies on the *Chlamydomonas* flagellar system demonstrated another very general property of microtubular structures: the length of the microtubules is carefully controlled by the cell. After deflagellation, but in the absence of any protein synthesis, a *Chlamydomonas* cell can re-

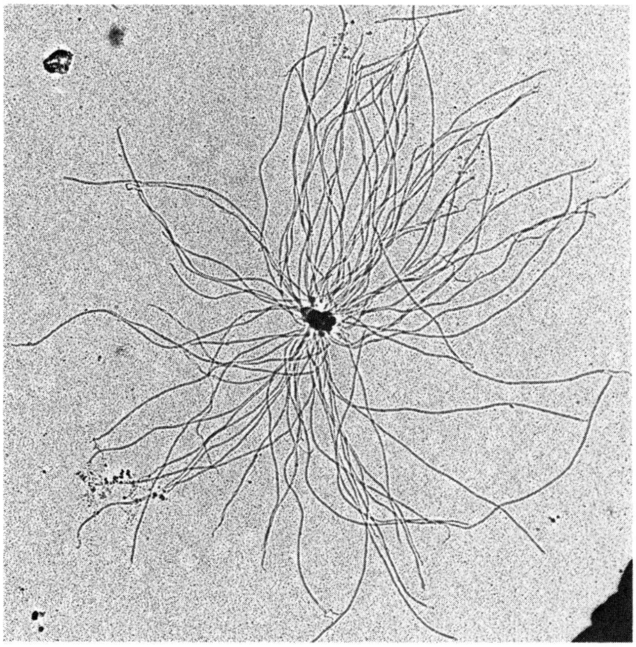

Figure 18-28 Microtubule nucleation by centrosomes. A purified preparation of centrosomes was obtained from lysed neuroblastoma cells by equilibrium density-gradient centrifugation and added to a solution of purified tubulin. After incubation at 37°C for 20 min, the material was fixed, shadowed with platinum, and viewed in the electron microscope. Several dozen microtubules are seen to radiate from this centrosome. [See T. Mitchison and M. Kirschner, 1984, *Nature* **313**:232.] *Courtesy of M. Kirschner. Reprinted by permission from* Nature. *Copyright 1984 by Macmillan Journals Limited.*

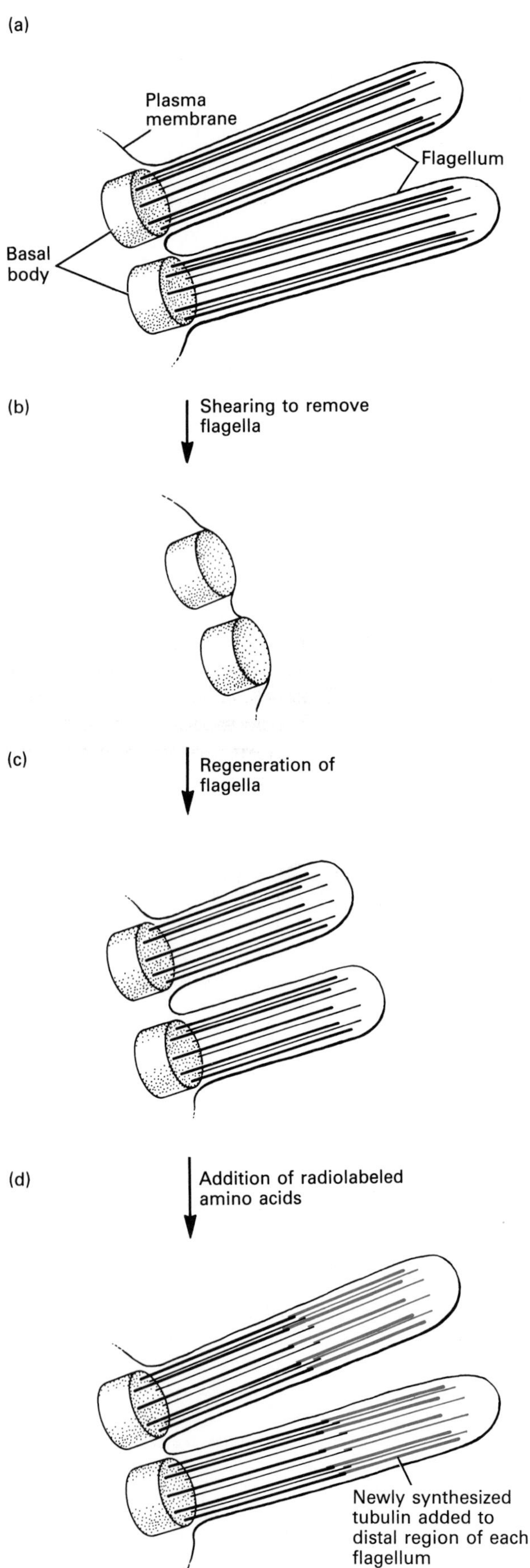

Figure 18-29 During regeneration of *Chlamydomonas* flagella, tubulin is added to the distal region of the axoneme. (a) Mature flagella. (b) Shearing of growing cells results in removal of flagella but not of basal bodies. (c) The flagellum regrows. Midway during regeneration, radiolabeled amino acids can be added to the cell. (d) By autoradiography, the newly synthesized radiolabeled tubulin (*color*) is seen only at the distal tip of the flagellar axonemes.

generate a flagellum of up to one-half its original length. Similarly, after deflagellation sea urchin embryos can regenerate full-length flagella. These results show there are large pre-existing pools of precursor proteins for flagella and cilia present during interphase but that these are not normally used to produce more or longer flagella or cilia. Obviously, then, flagellar assembly is highly regulated. In *Chlamydomonas* cells in which one of the two flagella has been amputated, the remaining old flagellum is initially resorbed at the same rate as the new one is regenerated; eventually both grow to normal length. This observation establishes that flagella microtubule assembly and disassembly can occur simultaneously in a single cell. Apparently, the processes of synthesis, assembly, and disassembly are under separate controls, but the nature of these is not yet known.

Microtubules Exhibit a Dynamic Instability

There are other situations in which one set of microtubules in the cell grows while another depolymerizes. During prophase of mitosis, for instance, spindle microtubules are growing while cytoplasmic microtubules are shrinking. Such situations are also observed in vitro in microtubule assembly reactions. We saw in Figure 18-28 that isolated centrosomes nucleate polymerization of microtubules from purified tubulin. When such preparations of nucleated microtubules are diluted, thus reducing the concentration of free tubulin, a rather surprising result occurs: some of the microtubules depolymerize (as might be expected)—but others actually grow to much longer lengths!

How can simultaneous growth and depolymerization of microtubules be explained? Figure 18-30 depicts one current theory. Recall that each $\alpha\beta$ tubulin dimer contains two bound GTP molecules. One of these is hydrolyzed to GDP during or just after incorporation of the tubulin into a microtubule. If as a result all (or most) of the tubulins at the growing end of a microtubule have a bound GDP (a GDP "cap"), the microtubule is unstable and depolymerizes rapidly. If, by contrast, additional GTP-liganded tubulin adds to the end before hydrolysis of the GTP bound to the microtubule, the microtubule not only is stable but continues to grow. Thus, one pa-

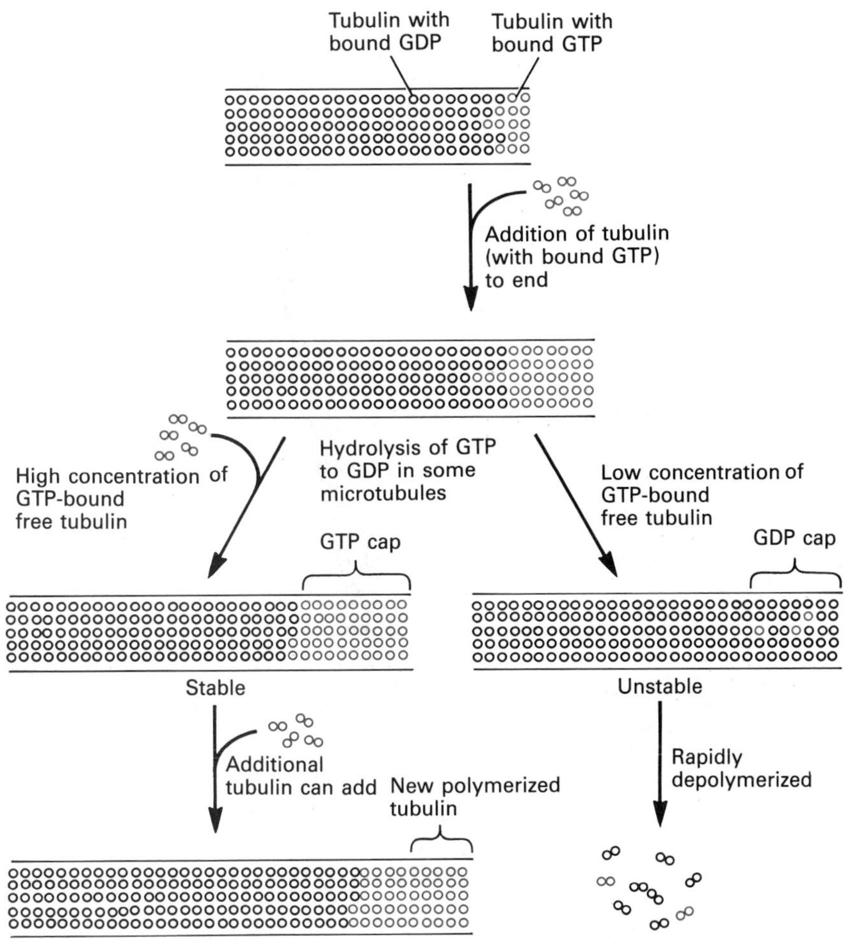

Figure 18-30 Model for the dynamic properties of microtubules. As noted in Figure 18-10, each $\alpha\beta$ tubulin dimer binds two molecules of GTP, and adds preferentially to one end of a microtubule. After incorporation of tubulin, one bound GTP is hydrolyzed to GDP. This hydrolysis is apparently catalyzed by the microtubule itself but may be facilitated by cytosolic proteins. Only microtubules with terminal tubulin associated with GTP—those with a GTP cap—are stable and can serve as primers for polymerization of additional tubulin. Microtubules with tubulin liganded to GDP at the end—with a GDP cap—are unstable, and the entire microtubule is depolymerized within 1 min. The formation of a GTP cap and elongation of the microtubule is facilitated by a high concentration of unpolymerized, GTP-liganded tubulin: addition of GTP-liganded tubulin to the end occurs at a rate faster than that of hydrolysis of the GTP bound to the tubulin in the microtubule. Conversely, at low concentrations of tubulin, hydrolysis of GTP to GDP in the terminal microtubular units tends to occur at a rate faster than that of addition of new GTP-liganded subunits; the microtubule thus acquires a GDP-capped end, and the microtubule depolymerizes. [See T. Mitchison and M. Kirschner, 1984, *Nature* **312**:237.]

rameter that determines the stability of a microtubule is the concentration of free tubulin: a high concentration favors continued growth, and a low concentration allows a GDP cap to form at the end and thus causes the microtubule to depolymerize. Another parameter is the rate of hydrolysis of GTP to GDP—a slow rate favors continued microtubule growth. But the specific factors that catalyze and regulate GTP hydrolysis in microtubules are not known and are an important subject of current research.

Mitotic Spindles, Microtubules, and Cytokinesis

Mitosis involves a number of complex mechanical processes: each of the daughter chromosomes replicated during S phase must be moved to opposite ends of the cell, and the cytoplasm must be divided in such a way that each daughter receives the appropriate amount of membranous organelles and cytoskeletal proteins. Microtubules play a major role in each of these processes. They form the mitotic spindle, which organizes the chromosomes and cytoplasm, and they may function in many of the subcellular movements that occur during mitosis. Our discussion of mitosis will focus on three key areas: first, the structure and construction of the mitotic spindle and its interaction with the chromosomes; second, the movements that line up, separate, and move the chromosomes; and, finally, the process of cytokinesis, or cell separation. Refer back to Chapter 5 for an overview of the key events in each step of mitosis (Figure 5-14).

Light-Microscopic Techniques Reveal the Mitotic Spindle in Living Cells

Very few techniques are available for microscopic visualization of live, moving cells. Most microscopic techniques involve fixation of cells or tissues, followed by selective staining to emphasize the material of greatest interest.

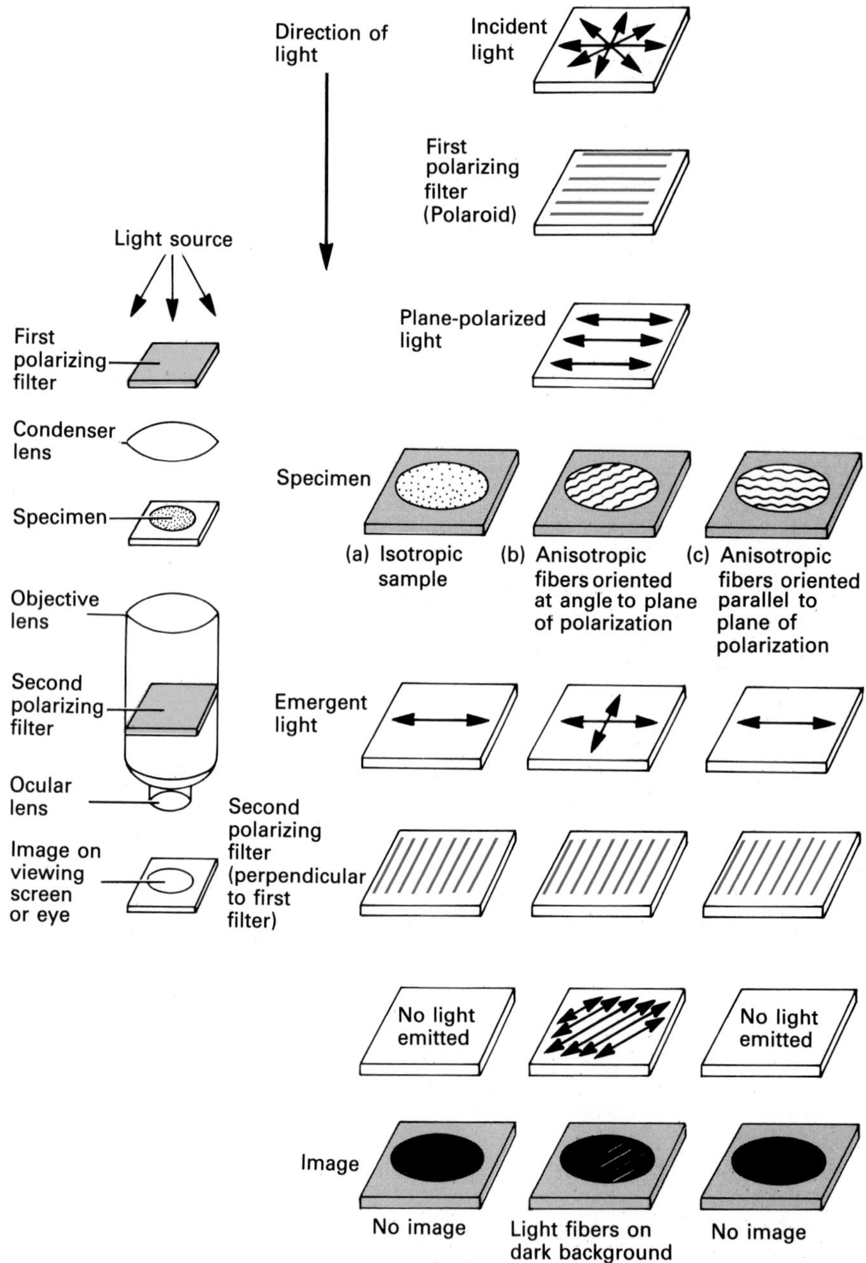

Views of killed cells, however great the resolution, provide only a static picture of cellular structure, because the cell is "frozen" in an instant of time. The very process of killing and fixing may create artifacts by changing the structures of cellular components. Generally it is quite difficult to deduce the molecular structures that are involved in movements or in other changes in architecture of cells. Which elements actually move, and how is the force generated? The static image provided by microscopy of a cell cytoskeleton is dependent on the nature of the optical system used and the kind of fixation and staining material is subjected to. Many microtubules, such as those that make up a major part of the mitotic spindle, are among the most labile cell structures, requiring special fixatives and optical systems for their visualization.

Individual microtubules or other cytoskeletal fibers, though often quite long, are far too thin to be detected by ordinary light microscopy. However, bundles of fibers can be detected by polarization microscopy (Figure 18-31). A specimen of highly ordered, parallel filaments such as microtubules is birefringent and thus becomes visible when placed between two crossed polarizing filters.

Figure 18-31 ***(opposite)*** Principles of polarization microscopy. Light waves normally have components that vibrate in all directions when viewed perpendicular to the beam, as indicated by the crossed arrows. A single polarizing filter allows only light waves that vibrate in one direction to pass through, and the emergent light is said to be *plane polarized.* If a second polarizing filter is placed at right angles to the first, no light can pass through because the plane-polarized light emergent from the first filter is effectively blocked by the second.

In a polarization microscope (*left*), one polarizing filter is placed between the light source and the specimen; a second, at right angles to the first, is placed between the specimen and the eyepiece, or viewer.

(a) When viewed in the polarizing microscope, *isotropic* specimens, such as a homogeneous solution of salts or proteins, appear black. The light that emerges from the sample is plane polarized and therefore unable to penetrate the second polarizing filter. (b) In other specimens called *anisotropic* (birefringent), the speed with which the light passes through the object depends on the direction of polarization of the incident light. Birefringence means "double refraction." Anisotropic specimens thus have two refractive indices, one for each direction of polarization. Bundles of fibers oriented in the same direction are typical anisotropic specimens; the refractive index and thus the speed of propagation of light polarized in the same direction as that of the fibers will be slightly different from that of light polarized perpendicular to the direction of the fibers. When an anisotropic specimen is situated at an angle with respect to the plane of polarization of the incident light, as shown, it effectively rotates the direction of vibration of some of the incident polarized light waves. Thus the light that emerges from the object will have some components that vibrate in a direction perpendicular to that of the initial plane-polarized light; these light waves will pass through the second polarizing filter. More precisely, the phase of the incoming polarizing light is shifted by the anisotropic object, and the emergent light rays of the same phase will combine constructively. Conversely, emergent light waves of opposite phase will destructively interfere. The image recorded on the eyepiece will be of a series of fibers similar in direction and size to those of the specimen—either a light image on a dark background (if the emergent light is in phase) or a dark image on a light background (if the emergent light is out of phase). (c) If the birefringent object is parallel to the plane of polarized light, little of the incident light will be able to pass through filters, and the specimen will appear dark to the viewer.

As shown in Figure 18-32, the bundles of microtubules in the mitotic spindle can be visualized in this way. Fibrous elements, now known to be bundles of microtubules, radiate from the two polar regions toward the chromosomes at the center of the spindle (Figure 18-33). Other microtubules (astral fibers) radiate outward in other directions from the polar region.

Several experiments have established that the birefringent fibers visualized in Figure 18-32 indeed are constructed of microtubules. When assembly of tubulin into microtubules is blocked by exposure to agents such as colchicine or conditions such as low temperature or high hydrostatic pressure, loss of the birefringent fibers of the mitotic spindle results. As might be expected, the progression of cells through mitosis is also blocked. These results also suggest that the microtubules in the mitotic spindle are in a steady state between assembly and disassembly into tubulin.

Like phase-contrast microscopy (see Figure 5-6), Nomarski interference microscopy uses a refractive index to measure differences in density among the parts of a living cell. With the Nomarski interference technique, the mitotic chromosomes, (Figure 18-32) which are the densest objects in a mitotic spindle are clearly visualized, but the much less dense microtubular spindle is seen only faintly. Such observations reinforce the conclusion that at metaphase chromosomes are indeed aligned midway between the spindle poles.

Bundles of Microtubules Form the Mitotic Spindle

Electron microscopy has shown that the mitotic spindle contains several types of microtubules: *polar fibers,* which extend from the two poles of the spindle toward the equator; *kinetochore fibers,* which attach to the centromere of each mitotic chromosome and extend toward the poles; and *astral fibers,* which radiate outward from the poles toward the periphery of the cell (Figure 18-33). The microtubules and associated structures that radiate from each pole are collectively termed a *half-spindle.*

Mitotic spindles can be isolated by gentle extraction of cells with detergents, in the presence of certain concentrated salts that allow microtubules to remain polymerized. Electron microscopy has established that these indeed are formed of individual microtubules (Figure 18-34). Besides α and β tubulin, isolated mitotic spindles contain many different microtubule-associated proteins (MAPs). These MAPs probably regulate the polymerization of tubulin into spindle fibers rather than polymerization into other microtubular fibers.

The polar microtubules often appear to run from one pole through the equator to the other pole. Although individual microtubules may indeed extend the whole dis-

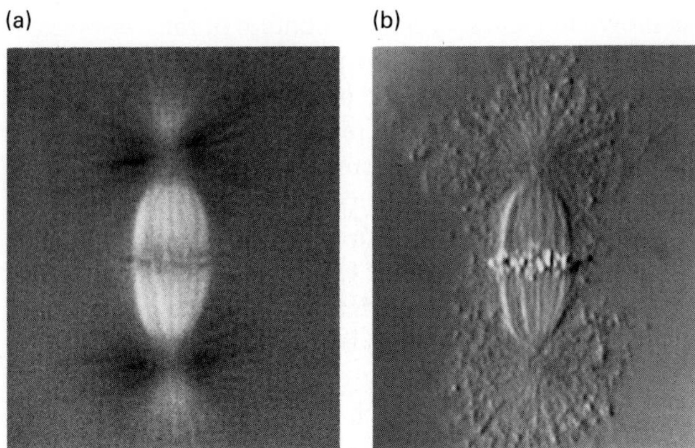

Figure 18-32 Micrographs of an isolated metaphase mitotic spindle from a sea urchin egg obtained by (a) polarization microscopy and (b) Nomarski interference microscopy. In polarized light, the pole-to-pole spindle fibers appear bright, whereas the spindle fibers perpendicular to the pole-to-pole fiber—the so-called astral rays—appear dark. The Nomarski optics technique detects differences in refractive index among different parts of the sample. It allows visualization of the dense chromosomes at the center of the mitotic spindle and also of the spindle fibers. [See E. Salmon and J. Segal, 1980, *J. Cell Biol.* **86**:355.] *Courtesy of S. Inoue. Reproduced by permission of the* Journal of Cell Biology, *1980, by copyright permission of the Rockefeller University Press.*

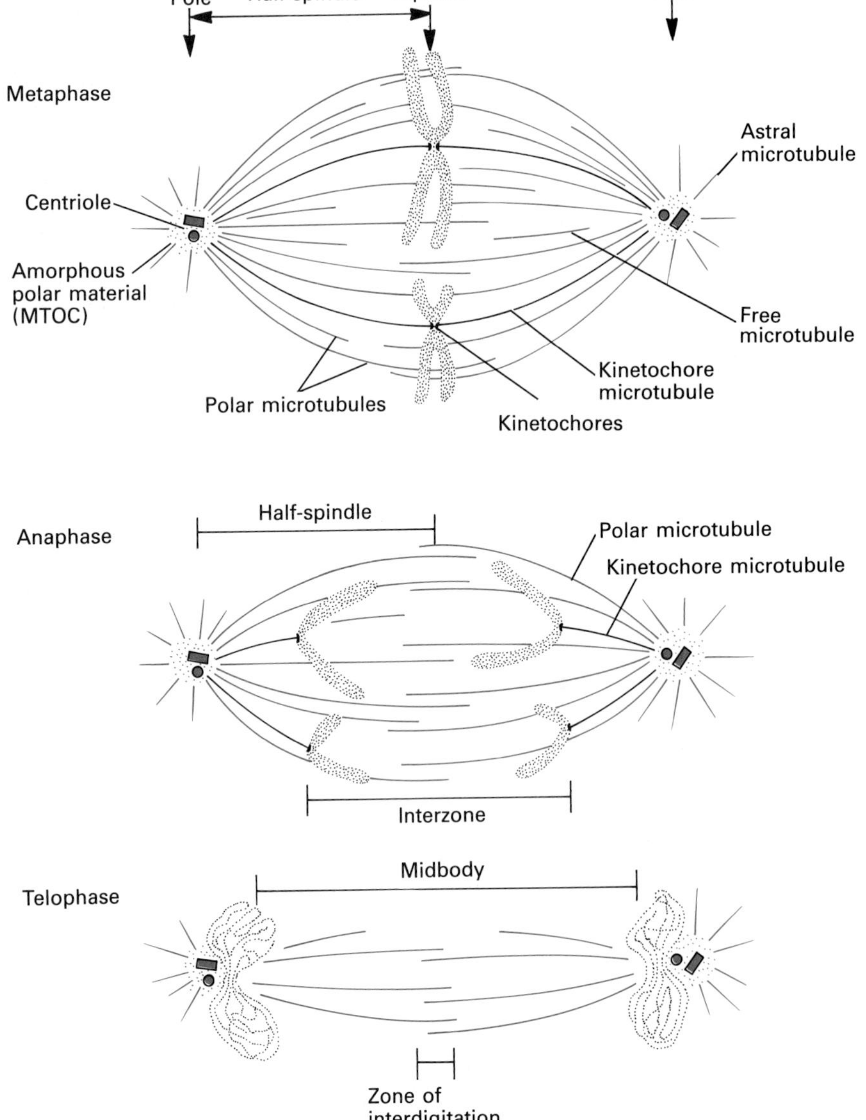

Figure 18-33 Schematic diagrams of animal spindle architecture during phases of mitosis.

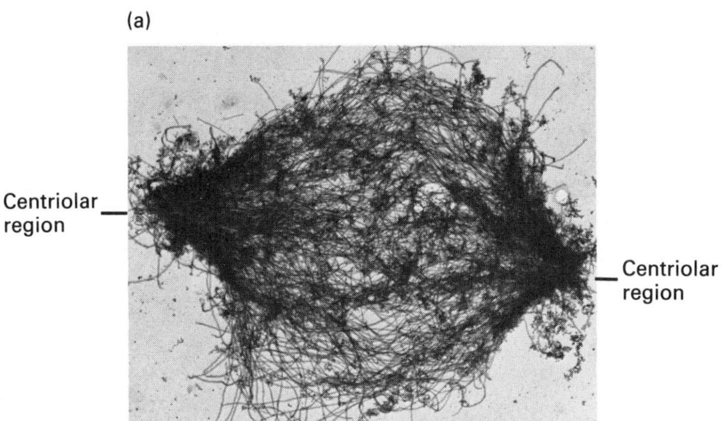

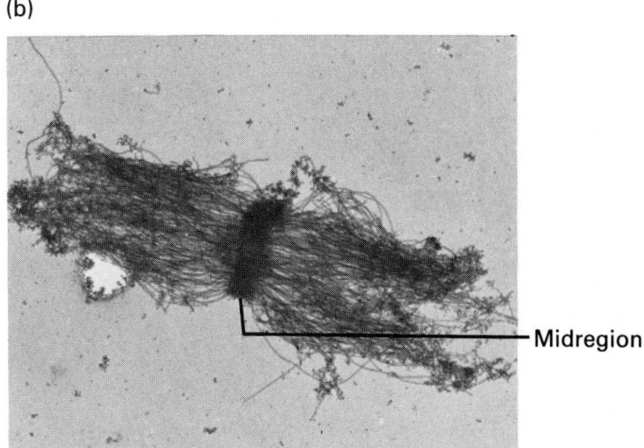

Figure 18-34 Microtubules in a mitotic cell. (a) During prophase, fibroblasts were extracted by a series of solutions of detergents and concentrated salts so that only the microtubules remain. The individual microtubules are seen radiating from the centriolar region. (b) In anaphase, all the microtubules are parallel to the spindle axis. The midregion of the spindle contains overlapping spindle microtubules. *Photographs courtesy of F. Solomon and G. Zieve.*

tance, in most micrographs the center of the metaphase spindle is seen to be constructed from two interdigitating families of microtubules, one from each spindle pole. This is clearly evident in a projection of the central spindle of the diatom *Diatoma,* reconstructed from serial sections (Figure 18-35). These polar fibers form a spindle framework that apparently provides a mechanical foundation against which chromosome movement can be generated for separation of the daughter genomes at anaphase and telophase.

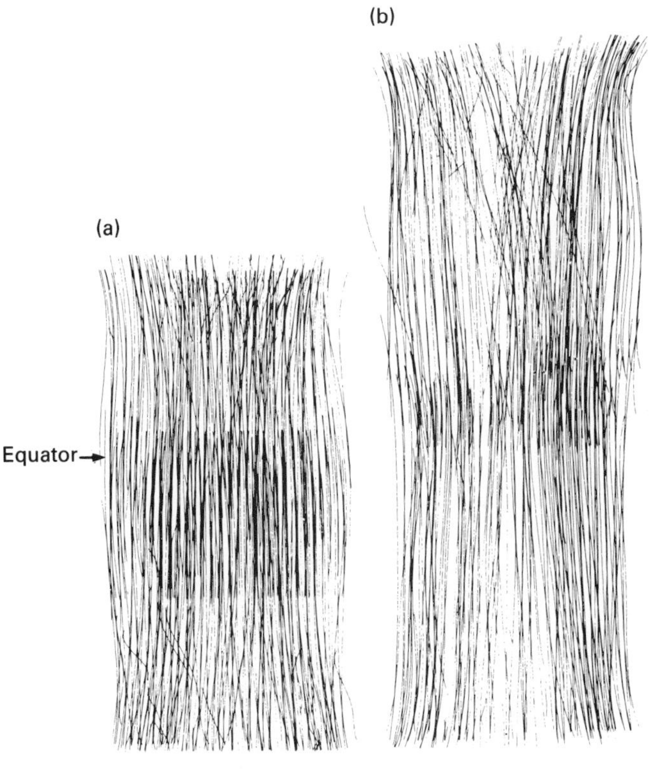

In animal cells, centrioles are at the center of the mitotic poles. However, the polar fibers radiate not from the centrioles themselves but rather from the diffusely staining material around the centriole—the microtubule-organizing center (MTOC). The centriole in animal cells is built of nine triplet microtubules (see Figure 18-25). In certain protozoans, structures containing only nine singlet microtubules are found at the center of the MTOC, whereas in others a plaque or sphere located on or near the nuclear envelope appears to function as MTOC. Higher plants, in particular, lack centrioles but nevertheless form fully functional spindles. Usually these spindles are *anastral*—that is, they have no astral fibers (Figure 18-36). It is not known what structure "substitutes" for a centriole, nor what entity organizes the spindle fibers.

Kinetochore Microtubules Connect the Chromosomes to the Poles

The metaphase chromosome is a highly condensed nucleoprotein particle. It contains two coiled daughter chroma-

Figure 18-35 Computer-generated projections of the central spindle of *Diatoma* as seen after reconstruction in three dimensions from serial transverse sections. (a) The metaphase spindle is constructed from two half-spindles. (b) By telophase, the half-spindles have moved apart, so only the longest of the microtubules can still interdigitate. There is also a slight increase in average microtubule length. In this projection, the spindle has been shortened, by a factor of 2, relative to its width and depth by increasing the average distance between the lines used represent the microtubules, in order to improve the viewing. *Courtesy of J. F. McIntosh. Reproduced by permission of the* Journal of Cell Biology, *1979, vol. 83, p. 428 by copyright permission of the Rockefeller University Press.*

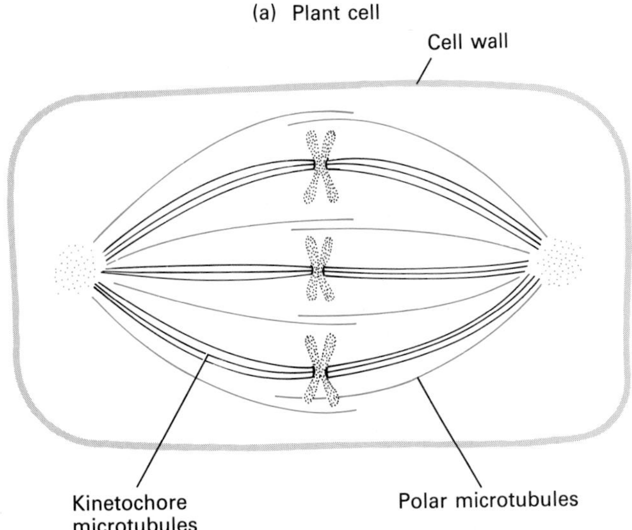

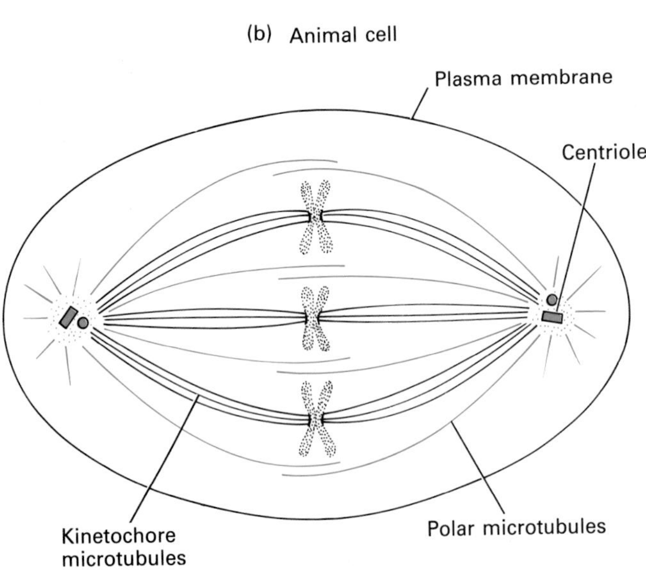

Figure 18-36 Metaphase spindle structures in (a) a plant cell and (b) an animal cell. Higher plants lack centrioles and astral fibers.

tids, each of which contains a replicated DNA molecule bound to histones and other proteins. The goal of mitosis is to distribute one chromatid to each daughter cell; this is the function of the kinetochore microtubule fibers.

The centromere and kinetochore are two closely related parts of the metaphase chromosome. The *centromere* is the area of primary constriction in the metaphase chromosome, where the chromatids are held together. The *kinetochore,* usually located at or near the centromere, is the structure to which the pole-to-chromosome microtubules are attached (Figure 18-37, see also Figures 10-12 and 10-13). The kinetochores always face the spindle poles and are essential for the proper movement of chromosomes during mitosis. During anaphase, the kinetochore appears to be the site at which force is exerted on the chromosome to pull it toward the poles.

Electron microscopy has revealed that the structure of the kinetochore is not identical in all organisms. There are two main types: in most animals and lower plants, the kinetochore is a *trilaminar,* or stratified, structure (Figures 18-37 and 18-38), whereas most higher plants have a different structure. In some organisms such as yeasts (Fig-

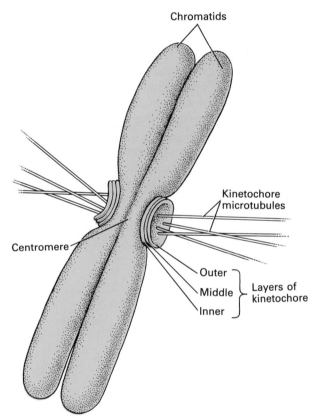

Figure 18-37 The structure of a metaphase chromosome showing the three layers of the kinetochore and the kinetochore microtubules. In metaphase, the sister chromatids have not yet separated. Each chromatid contains a kinetochore at its centromere. Microtubules insert into the outer layer of each kinetochore and run from there toward one of the two poles of the cell. At anaphase, the sister chromatids separate, and the chromosomes are pulled to opposite poles of the cell, directed by the kinetochore microtubules. [See C. J. Bostock and A. T. Sumner, 1978, *The Eukaryotic Chromosome,* North Holland Press.]

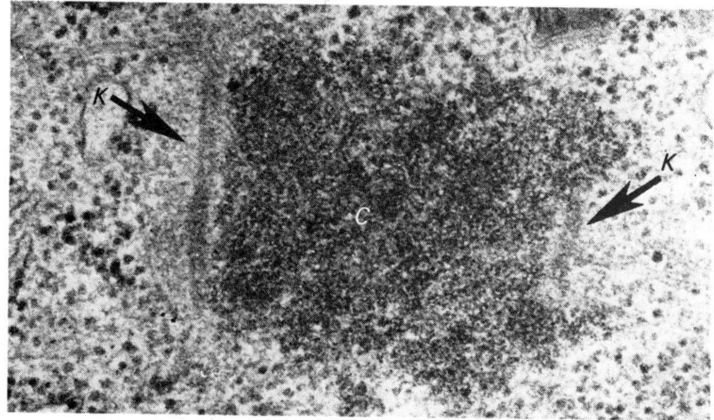

(b)

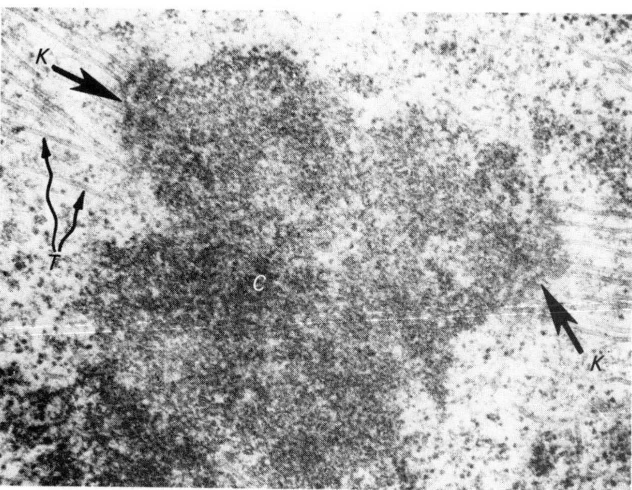

Figure 18-38 Electron micrographs of thin sections containing kinetochores and chromosomes during mitosis. (a) Portion of a chromosome (C) from a cultured Chinese hamster fibroblast. Kinetochore regions (*K*) are seen as special bands of dense material pointing toward the spindle poles. (b) Portion of a chromosome in a cultured cell from the kangaroo rat, showing kinetochore microtubules, indicated by *T*, associated with the kinetochore region. *Courtesy of B. Brinkley and E. Stubblefield.*

ure 10-13) there is no obvious kinetochore structure: the microtubules appear to insert directly into the chromosomes and extend toward the poles. What kind(s) of proteins or other molecules make up the kinetochores, or how the microtubules insert in them, is not known.

The centromere is more than just a constriction of a metaphase chromosome; it determines the attachment of the kinetochore and thus the proper movement of chromosomes during mitosis. A fragment of a chromosome without a centromere is occasionally generated by drugs or x-irradiation. Such a fragment cannot be incorporated into a metaphase spindle and is lost during mitosis.

The generation of a kinetochore is directly controlled by a unique segment of DNA termed *centromeric DNA*. As discussed in Chapter 10, several yeast DNA segments corresponding to centromeres have been cloned. These centromeric DNAs have the ability to cause any DNA segment linked to them to be replicated and distributed equally to daughter cells, exactly as in full-sized chromosomes. In some way, centromeric DNA segments bind proteins that cause a kinetochore to form.

Virtually All Microtubules in the Same Half-Spindle Have the Same Polarity

Each microtubule has a specific polarity defined by a net assembly, or A, end and a net disassembly, or D, end. In a half-spindle, the region between a pole and the equator, the polar fibers and the kinetochore microtubule fibers have a parallel orientation. The A end is at the equator, or kinetochore, and the D end faces the poles (Figure 18-39). As we shall see, this finding provides important clues to how movements occur during mitosis.

To establish this point of spindle structure, tubulin was added to a preparation of isolated mitotic spindles with an abnormally high concentration of salts. Under these conditions, tubulin polymerizes in hook-shaped sheets onto the walls of pre-existing microtubules rather than onto the ends of microtubules. The isolated spindle was then fixed, sectioned perpendicular to the pole-to-pole axis, and examined in the electron microscope. From the direction in which the tubulin sheets point, the polarity of the spindle fibers can be deduced. As can be seen in Figure 18-39, nearly all of the hooks of the tubulin sheets point in the same direction on all of the microtubules in a half-spindle; thus all of the microtubules appear to have the same polarity.

The A end of the microtubules is the end distal to the centrioles and near the equator. This feature suggests that, as with basal bodies (see Figure 18-29), all microtubules in the half-spindle are *nucleated* by the centriole and its associated MTOC material, and that tubulin is added to microtubules at the end distal to the centriole (see Figure 18-28).

Other experiments indicate, however, that isolated kinetochores can act as nucleating sites for in vitro assembly of microtubules. To resolve this seeming paradox, recall that in vitro, microtubules can grow by addition of subunits at both ends (see Figure 18-10), but that, depending on the concentration of tubulin, the rate of addition of tubulin to one end—the A end—can be severalfold that of the other end. The evidence just discussed indicates that the A end of all pole-to-kinetochore tubules is that closest to the kinetochore (Figure 18-40). If the A end of the microtubule is actually buried in the kinetochore, then no tubulin can add at this end. But there can

(a)

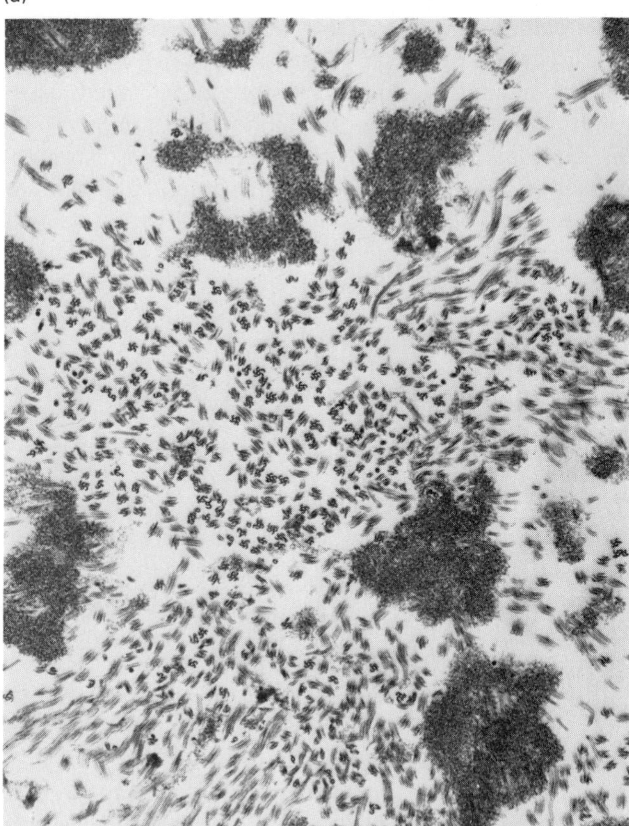

(b)

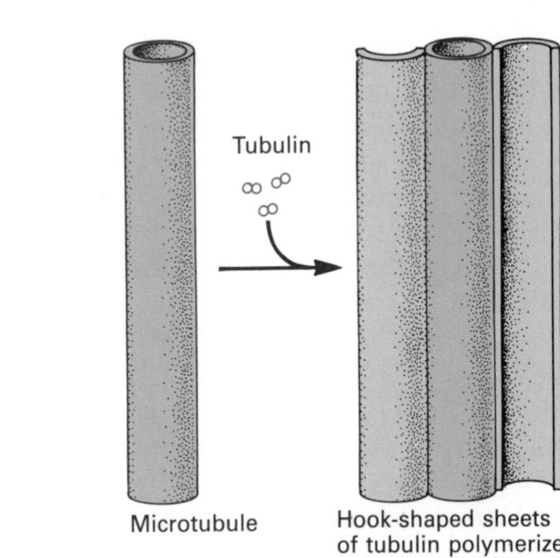

Figure 18-39 The same polarity for virtually all microtubules in a half-spindle can be shown experimentally. (a) First, anaphase mitotic fibroblasts are lysed by detergents in a solution containing a high concentration of soluble nerve tubulin. (b) Because of the abnormally high salt concentration, the nerve tubulin polymerizes in hook-shaped sheets onto the walls of pre-existing cellular microtubules. The spindle is then sectioned transversely. The section shown is midway between the polar region and the kinetochore. Most of the microtubules are "decorated" with hooks from the sheets of polymerized tubulin. Over 90 percent of all hooks "curve" clockwise; thus over 90 percent of the microtubules have the same polarity. The few microtubules of opposite polarity are scattered throughout the section. [See U. Euteneuer and J. R. McIntosh, 1981, *J. Cell Biol.* **89**:338.] *Part (a) courtesy of U. Euteneuer and J. R. McIntosh. Reproduced from the* Journal of Cell Biology, *1981, by copyright permission of the Rockefeller University Press.*

still be addition of tubulin to the opposite end—the D end—of the microtubule. Thus, kinetochores may act as nucleating sites for the pole-to-kinetochore microtubules, which then grow outward from the kinetochore. Alternatively, kinetochores may bind to the A end of unattached microtubules.

Many Events in Mitosis Do Not Depend on the Mitotic Spindle

Although the mitotic spindle is formed during prophase, it is not essential for all events in mitosis. Recall that colchicine blocks assembly of microtubules. In most animal cells, however, colchicine affects neither the breakdown of the nuclear membrane during prophase nor the condensation of chromatin into sister chromatids. In sea urchin embryos, even the separation of daughter chromatids and the re-formation of the nuclear membrane occurs in the presence of colchicine; thus, in all probability these steps do not involve the mitotic spindle. The spindle *is* required for alignment of chromosomes at the equator, for movement of daughter chromosomes toward the poles, and for separation of the daughter nuclei.

There Is a Great Diversity in the Structure of the Mitotic Spindle among Eukaryotic Cells

In higher plants and animals, the nuclear membrane breaks down during mitotic prophase, and re-forms during anaphase. Movement of chromosomes is a function of the mitotic spindle; membranes are not involved in any obvious way. In many protozoans and fungi, by contrast, the nuclear membrane remains more or less intact during mitosis, and the polar regions are seen to lie outside the nucleus. In these lower organisms, polar microtubules can form a single large pole-to-pole bundle lying in the cytoplasm in a groove of the nucleus, or several large bundles can traverse the nucleus in tunnels or channels (Figure 18-41). Chromosomes are anchored to the inner nuclear membrane, possibly by kinetochore microtu-

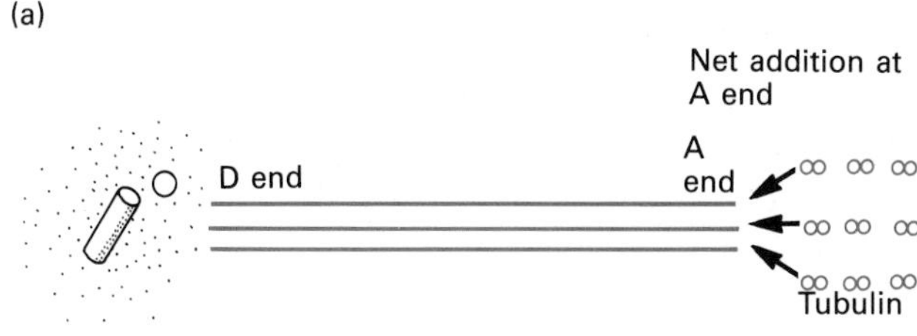

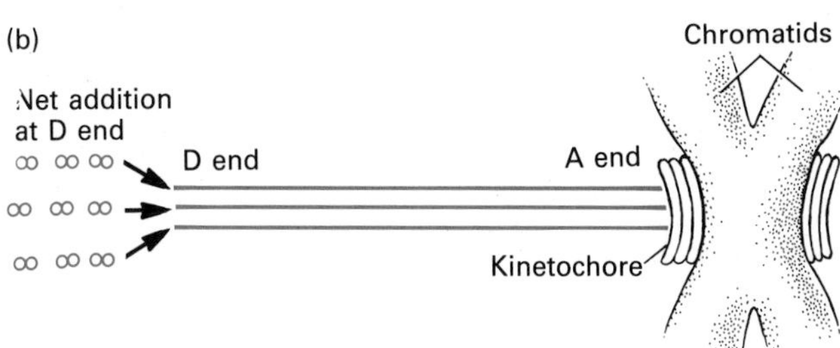

Figure 18-40 A possible model for assembly of pole-to-kinetochore spindle microtubules. The D end of all microtubules in the half-spindle faces the MTOC and centriole, and the A end faces the kinetochore. (a) Normally, addition of tubulin occurs preferentially at the A end (see Figure 18-10). (b) If the A end of pole-to-kinetochore microtubules is buried in the kinetochore and therefore unavailable, the tubule can still elongate by addition at the D end.

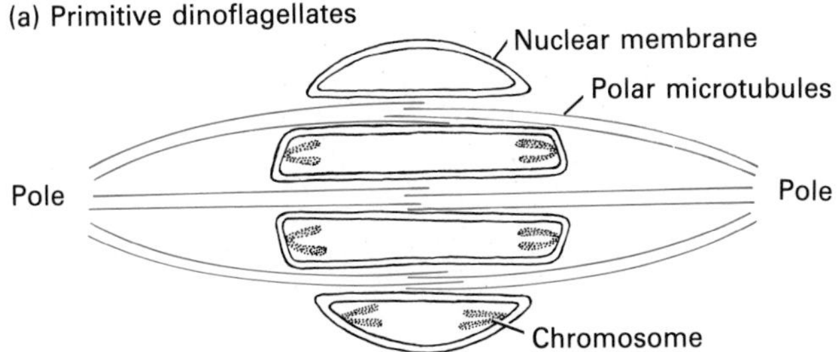

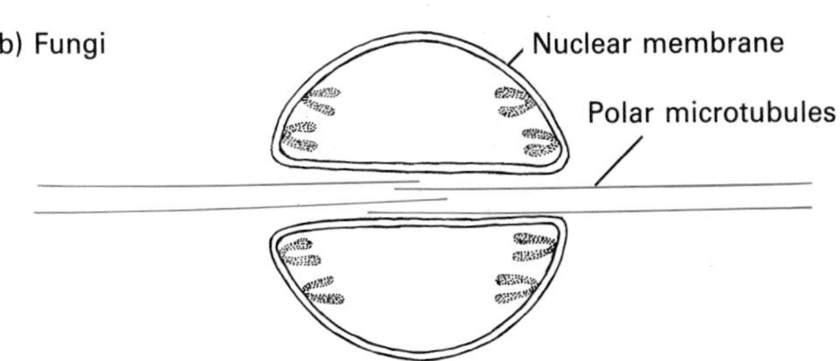

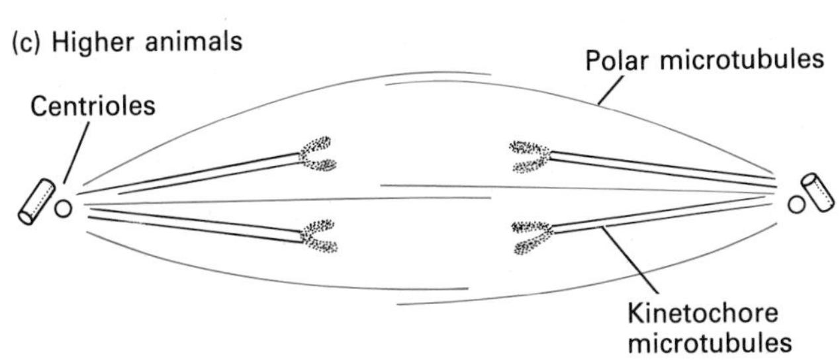

Figure 18-41 The diversity of organization of the polar (nonkinetochore) microtubules among eukaryotes. (a) In primitive dinoflagellates, one or more bundles of polar microtubules traverse the nucleus in membrane-lined tunnels. The nuclear membrane does not disappear during mitosis. The chromosomes become associated with the inner nuclear membrane by their kinetochores and move apart without obvious attachment to microtubular bundles. The pole-to-pole spindle acts as a framework to order the movements of the nuclear membrane. It is not known whether kinetochore microtubules are involved in anchoring the chromosomes to the nuclear membrane. (b) In some protozoans such as *Barbulanympha* there is a single pole-to-pole bundle of filaments that passes through a tunnel in the intact nuclear membrane. Again, the nuclear membrane remains intact during mitosis, and chromosomes separate anchored to the inner nuclear membrane. In other protozoan and in the yeast *Saccharomyces,* (Figure 6-7) the spindle is entirely within the nucleus; in different organisms the polar microtubules form into one or more bundles. At the end of yeast anaphase, there is only one polar microtubule (Fig. 18-45). (c) In higher animals, the nuclear membrane is dispersed during mitosis, and chromosomes attach directly to microtubules at their kinetochores. There is an array of variously overlapping and closely spaced microtubules of various lengths. (See I. B. Heath, 1980, *Int. Rev. Cytol.* **80**:1.]

bules, and move to the daughter nuclei in association with the nuclear membrane. This process is reminiscent of the association of bacterial DNA with the plasma membrane (see Figure 5-12): as the bacterial cell divides, the daughter DNA molecules are anchored to the membrane at opposite sides of the equator. In this way, equal partitioning of the DNA to daughter cells is assured.

Balanced Forces Align Metaphase Chromosomes at the Equator of the Spindle

Now that we have examined the structure of the mitotic spindle in some detail, let us turn to the movements that occur during mitosis and the forces that generate them.

During premetaphase the newly condensed chromosomes can be observed to move randomly between the two poles. Eventually, one of the kinetochores becomes associated with tubulin fibers from one pole; the other kinetochore quickly becomes associated with fibers from the other. This association with polar fibers is not permanent. In certain large cells, micromanipulation with fine glass needles can rotate premetaphase chromosomes 180° with respect to the axis of the spindle. This procedure breaks the attachment of chromatids and their associated kinetochore microtubules with the poles. But the chromatids reattach to fibers from the opposite pole; subsequently, the chromatid that normally is pulled into one daughter cell during anaphase is instead pulled into the other. These and other studies indicate that some force pulls the two kinetochores on sister chromatids toward opposite poles. Micromanipulation experiments suggest that the strength of these forces is proportional to the distance from the chromosome to the pole. Thus these opposing forces cause a chromosome to align at the equator of the spindle and to remain stationary there. Here the forces pulling the chromosome toward each pole are balanced. For instance, if a metaphase chromosome is displaced a bit toward one pole by micromanipulation, it quickly returns to the equator.

Anaphase Movement of Chromosomes Involves Pole-to-Kinetochore Microtubules

The anaphase stage of mitosis consists of two distinct motile events that occur at the same time. The sister chromatids break apart at their point of connection in the centromeric region and then move toward poles at opposite ends of the cell. This process, often termed anaphase A, involves shortening of the pole-to-kinetochore microtubules as the chromosomes move toward the poles. Simultaneously, there is a separation of the two poles into what will become the two daughter cells. These changes, often called anaphase B, involve elongation of the polar microtubules as the poles move apart. Microtubules appear essential for both types of anaphase movements, because both are disrupted by colchicine. Despite many years of work, however, the nature of the forces that move the chromosomes to the poles, or that separate the two poles, remains largely unknown.

The amount of force necessary to move a single chromosome to the pole, at the typical anaphase speeds of 1 μm/min, is surprisingly small—the free energy of hydrolysis of only 20 ATP molecules to ADP. One flagellar doublet microtubule, therefore, has the potential to generate many times the required power.

Three models for chromosome-to-pole movement (anaphase A) have received widespread attention. It is useful to summarize each of them briefly and to note the experimental results supporting or contradicting each. No model is totally satisfactory in explaining the mechanism of chromosome movement.

Sliding-Microtubule Model In one model, the pole-to-kinetochore microtubules slide past the polar microtubules by the directed movement of dynein cross-bridges (Figure 18-42). The mechanisms involved presumably are similar to those in the sliding of doublets past each other during beating of flagella. Some investigators claim that dynein-like proteins exist in the spindle and that ATP hydrolysis is involved in chromosome movement. Recall, however, that the framework and pole-to-kinetochore microtubules in the half-spindle are of the same polarity. A dynein-mediated movement of these tubules, similar to that in flagellar axonemes, would propel the chromosomes *away* from the poles, not toward them! Biochemical studies on isolated mitotic spindles provide additional evidence against the involvement of ATP hydrolysis. For example, detergent treatment of mitotic cells causes ATP to leak out, but does not affect poleward chromosome movement. Moreover, in preparations of permeabilized cells, inhibitors of dynein ATPase do not affect chromosome movement. This finding also indicates that ATP is not involved, although interpretation of results remains controversial.

An important criticism of the sliding-microtubule model comes from experiments in which laser microbeam irradiation was used to destroy a single kinetochore in a mitotic cell. Immediately, the paired chromatid was seen to move toward the pole to which the other kinetochore was attached. This observation suggests that metaphase chromosomes are under a constant tension, and that chromosome-to-pole movement is restrained by the microtubules connecting them (Figure 18-43).

Depolymerization Model Another model for chromosome-to-pole movement is based on depolymerization (or shortening) of the microtubules at the polar end. This process pulls the chromosomes toward the centriolar region (Figure 18-44). Spindle microtubules do not con-

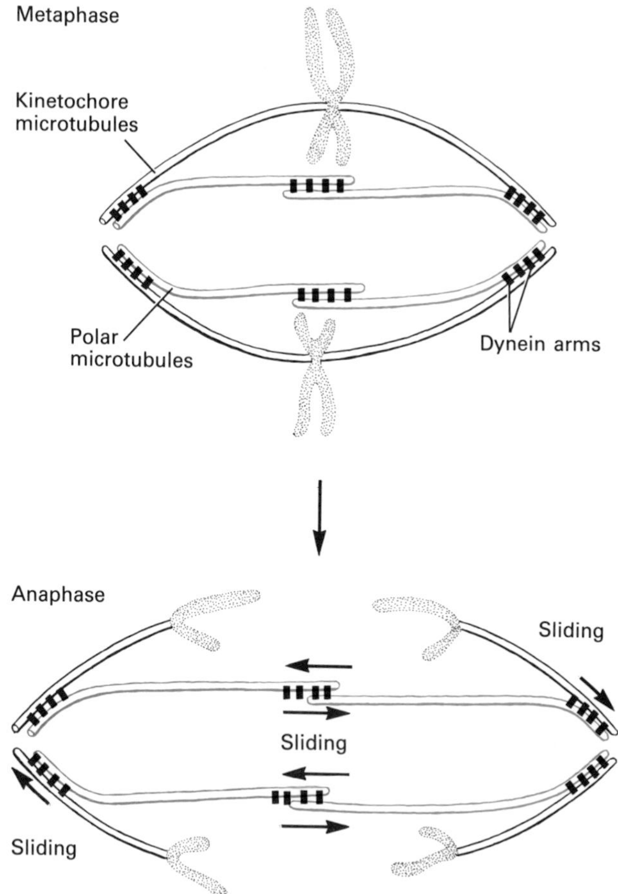

Figure 18-42 Sliding-microtubule model of chromosomal movement. Adjacent polar microtubules slide past each other to effect spindle elongation. Kinetochore microtubules slide along polar microtubules to move chromosomes toward the poles. The generator of force for sliding is presumed to be dynein arms that bridge the two sets of microtubules.

tract; they can become shorter only as a result of depolymerization at one (or both) ends. This model was suggested by the "treadmilling" of tubulin subunits along the length of a microtubule observed during in vitro polymerization (see Figure 18-10). Experiments in which the birefringence of the spindle is followed during mitosis indicate that the pole-to-kinetochore microtubules indeed decrease in length, probably by loss of subunits at the pole end—the D (net disassembly) end. But is this disassembly a cause or an effect of chromosome movement? Experimental support for the model has come from studies in which mitotic cells are treated with microtubular depolymerizing agents. For instance, slow cooling of mitotic cells, a condition that leads to depolymerization of microtubules, results in continued poleward movement of chromosomes.

Contractile-Protein Model A third model for chromosome-to-pole movement proposes that the microtubules (or other filaments in the spindle) act simply as a track along which the chromosomes move, with the actual force being generated by interactions of actin and myosin (or other contractile proteins) in the spindle. One variant of this model proposes that the kinetochore somehow "walks" along the kinetochore microtubules toward the poles, in much the same way organelles move along

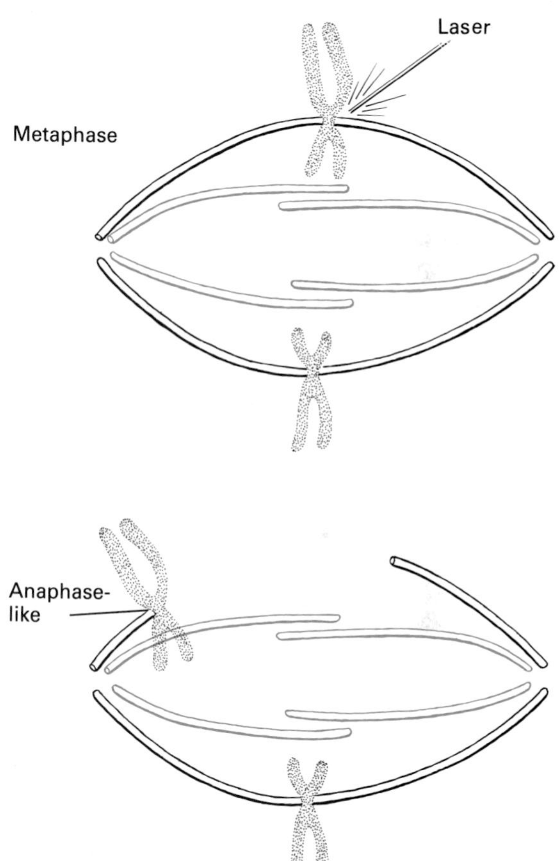

Figure 18-43 Diagrammatic representation of the laser microbeam experiments. At metaphase one of a pair of kinetochores is inactivated by laser irradiation. Immediately, the paired chromatids move to the pole that the other kinetochore faces (is attached to), and shortening of the pole-to-kinetochore microtubule occurs. Thus anaphase-type movement, including the shortening of a kinetochore microtubule, can occur during metaphase. It is as if the metaphase chromosome were connected to each pole by a "spring" attached to the kinetochores; cutting one spring causes the chromosome to move toward the opposite pole. During normal anaphase, when the two daughter chromosomes separate, the spring attached to each kinetochore pulls the chromosome to the respective pole. [See P. A. McNeill and M. W. Berns, 1981, *J. Cell Biol.* **88**:543.]

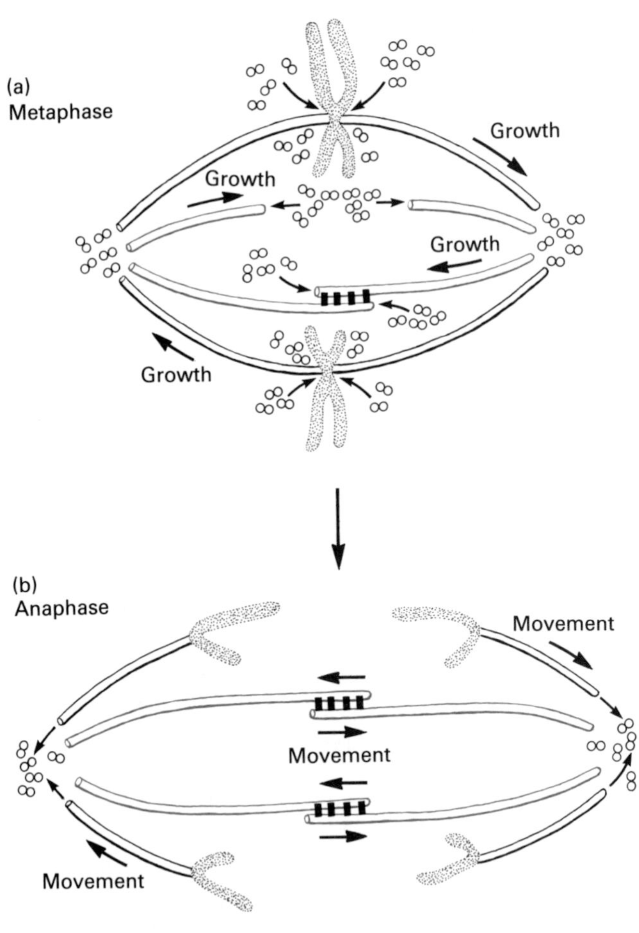

Figure 18-44 Depolymerization model of chromosomal movement. (a) Microtubules grow by polymerization of tubulin until the kinetochore microtubules are full length and the polar microtubules overlap. (b) At anaphase, the kinetochore microtubules shorten by depolymerization at the poles, and the polar microtubules move apart by a force generated at their points of overlap. According to this model, depolymerization occurs at the poles, and polymerization, at ends away from the poles.

microtubules in axonal transport (Figure 18-14). Although actin and myosin have also been localized to the spindle by immunofluorescence, there is no direct evidence that they are involved in chromosome movement. Microinjection of sea urchin eggs with antibodies to myosin does not affect anaphase chromosome movements but does inhibit the separation of daughter cells (Chapter 19). Indeed, proteins similar to those that power movement of vesicles along microtubules have been identified in mitotic spindles.

Anaphase Separation of the Poles Involves Unidentified Forces

Similar controversies exist over the forces that cause the spindle framework (the polar microtubules) to elongate and the mitotic poles to separate during anaphase B. According to one model, the elongation of the spindle (anaphase B) is driven by a dynein ATPase: polar microtubules slide by each other, moving the poles apart as depicted in Figures 18-42 and 18-44. According to another model, some unknown force pulls the two poles apart; the pole-to-pole spindle framework then acts as a track for chromosome movement and may restrain or otherwise regulate the rate of outward movement of the poles.

Some support for the latter model came from laser microbeam experiments on the central anaphase spindle of the fungus *Fusarium solani*. The spindle was studied in postanaphase cells, in which the chromosomes had already migrated to the poles and daughter nuclei had begun to re-form. In control cells the incipient daughter nuclei separated at a rate of 7 μm/min. When the central pole-to-pole spindle was destroyed by microbeam irradiation, the poles separated at more than three times this rate, or 22 μm/min. These results suggested that some unknown force pulls on the incipient daughter nuclei and that the central spindle limits the separation rate.

Although ATP hydrolysis is not required for chromosome-to-pole movements, several experiments indicate that it *is* required for outward movements of the poles. Mitotic fibroblasts permeabilized by low concentrations of detergents continue to undergo spindle elongation movements if ATP is added. Mitotic spindles have been isolated from diatoms; they, too, will undergo anaphase B movements in vitro when provided with ATP. Hydrolysis of ATP is thought to be required for movement because it is blocked by specific inhibitors—such as vanadate—of dynein ATPases. However, other ATPases may also be affected by vanadate. Thus the conclusions that dynein-like ATPases are involved in spindle elongation are not firm, and there is as yet no direct demonstration that dynein-like proteins actually exist in the spindle.

Note that a dynein-ATPase model cannot be the sole explanation for spindle elongation in *all* eukaryotic cells. For example, in yeast, as the spindle elongates the number of polar microtubules decreases until, at a length of 4 μm, only *one* from each pole remains (Figure 18-45). The spindle continues to elongate until the maximum nuclear diameter of 4 μm is reached. These microscopic observations suggest that, at least in yeast, the spindle framework is not only a force-generating system but also acts to regulate the rate of separation of the poles.

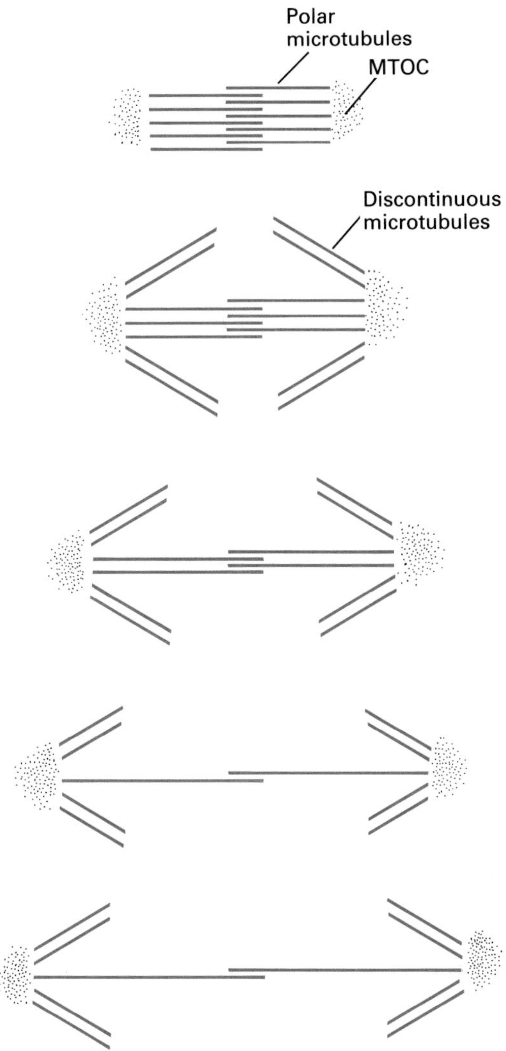

Figure 18-45 Model for spindle formation and elongation in *Saccharomyces cerevisiae.* The spindle is formed by the interdigitation of two sets of microtubules, one from each MTOC at each pole, to form a bundle. As spindle elongation occurs, some of the microtubules separate from the pole-to-pole axis. As the discontinuous microtubules shorten, the number of polar microtubules decreases to one from each pole. This suggests that growth of the polar microtubules, rather than sliding of the tubules past each other, is important for separation of the poles possibly in addition to ATP-driven sliding of the polar microtubules past each other. [See S. M. King, J. Hyams, and A. Luba, 1982, *J. Cell Biol.* **94**:341.]

Cytokinesis Is the Final Separation of the Daughter Cells

The final stage of mitosis is the division of the cytoplasm into two compartments—the process called *cytokinesis.* Cytokinesis is a contractile process and apparently does not involve microtubules; we examine it here to complete the picture of subcellular movements during mitosis.

During mitosis animal cells typically change shape, becoming more spherical. This phenomenon is especially marked in the case of cultured fibroblasts, which normally are flat and elongate when growing attached to a surface. During mitosis they "round up" and lose most of their contacts with the substrate. The process of cytokinesis in diverse animal cells appears similar to that in the much larger spherical sea urchin egg, which has been subject to intensive study.

During late anaphase in the sea urchin egg, an indentation forms around the egg surface. With time, it deepens and forms a cleavage furrow. This furrow completely encircles the egg in a plane perpendicular to the long axis of the mitotic spindle, and it continues to deepen until the opposing edges make contact in the center of the cell. Then the membranes fuse, and the original cytoplasm and the two sets of chromosomes become separated into two daughter cells. In the next chapter we shall consider the evidence that actin generates the contractile force of the furrow.

Because plant cells are bound by a relatively inextensible cell wall, cytokinesis in plants involves quite different processes. Plant cells construct a cell membrane and a cell wall from membrane vesicles. These vesicles appear first near the center of the dividing cell and then extend out to the lateral walls (Figure 18-46); they are thought to arise from the endoplasmic reticulum or Golgi complex. Such vesicles are first observed during metaphase, when they extend into the mitotic apparatus and, in some cases, even appear to contact the kinetochores. Because microtubules are also associated with these vesicles, possibly the vesicles and microtubules function in segregating the daughter chromosomes into the daughter cells, but this notion remains speculative.

The membrane vesicles also contain material for the future cell wall, such as polysaccharide precursors of cellulose and pectin. During late anaphase these vesicles fuse with one another to form large sheets near the equator of the spindle. The membranes of the vesicles become the plasma membranes of the daughter cells, and their contents form the intervening immature cell wall.

Cytokinesis in plant cells never completely separates adjacent cell cytoplasm, as occurs in animal cells. Almost all adjacent living plant cells are interconnected by a set of small cytoplasmic channels, 20 to 40 nm in diameter, called *plasmodesmata.* The plasma membranes of adjacent cells are continuous through plasmodesmata, so molecules can pass from cell to cell. Plasmodesmata are formed at the time of laying down of the cell plate; regions of the vesicular membrane remain as tunnels in the forming cell wall.

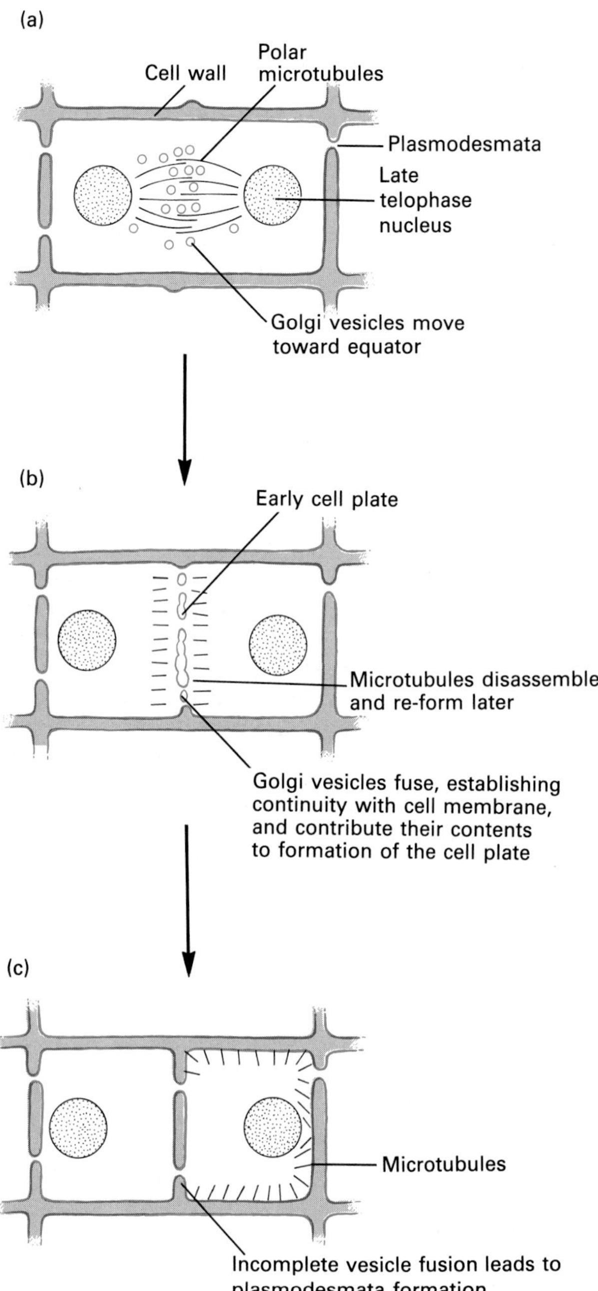

Figure 18-46 Cytokinesis is a higher plant cell. (a) In late telophase, the nuclear membrane has re-formed, and the polar microtubules have not yet dispersed. A set of small vesicles probably derived from the Golgi complex, which contain cellulose precursors of the wall, accumulate at the equatorial plate. (b) These vesicles fuse with each other to form an early cell plate—a large membrane-limited vesicle that contains cellulose. Additional vesicles fuse with the cell plate, extending it outward. (c) Eventually the cell plate fuses with the plasma membrane, and the two cells separate. The daughter cells remain connected by thin membrane-lined passages, called plasmodesmata, that penetrate the separating cell wall.

Summary

Three classes of filaments comprise the cytoskeleton of eukaryotic cells: actin-containing microfilaments, intermediate filaments, and microtubules.

Microtubules and microfilaments share two properties: they undergo reversible assembly-disassembly, depending on the need of the cell or organelle, and their assembly involves preferential addition of subunits to one end of a filament. In some cases these fibers can participate in the generation of force or movement, using the energy of hydrolysis of the terminal phosphodiester bonds in ATP. Other roles of these filaments in cells may be primarily structural.

Microtubules are heteropolymers of α and β tubulins plus microtubule-associated proteins. Together with a large number of associated proteins, they form the axoneme of cilia and flagella. Axonemes contain two singlet microtubules surrounded by a ring of nine doublet microtubules, all interconnected by several types of fibrous protein. Arms composed of the protein dynein, attached to the outer-doublet microtubules, are ATPases, which are crucial for ciliary or flagellar movement. Dyneins make and then break bonds with adjacent microtubules, causing each doublet to slide relative to its neighbor; because of the way the doublets are interconnected circumferentially, this motion is converted into the beating of the cilium or flagellum.

The length of the flagellar axoneme is closely controlled. Upon experimental deflagellation, polymerization of tubulin onto the ends of microtubules is seen to occur until the proper length is achieved. Tubulin is always added to the end distal to the basal body or centriole, but how this polymerization is regulated is not known.

Microtubules are the principal structural elements of the mitotic spindle. Bundles of microtubules radiate from the two poles, or microtubule-organizing centers (MTOCs), of the spindle. In animal cells a centriole is at the center of the polar region. Some fibers run from the pole to the kinetochore attached near the centromere of each chromosome; some bundles run from pole to the equator of the spindle. Although the evidence is contradictory, dynein ATPases may not play a role in either of the two types of anaphase movements—separation of the poles and movement of chromosomes toward the poles. Rather, the microtubule-containing spindle may act as a scaffold or framework against which other (unknown) forces may move.

Axonal transport also involves microtubules. Small vesicles or particles move rapidly—2 μm/min—in either direction along a singlet microtubule track. This movement also requires ATP hydrolysis and certain cystolic proteins, but apparently not dynein.

References

General References on Microtubules

* DUSTIN, P. 1978. *Microtubules,* 2d ed. Springer-Verlag.

* DUSTIN, P. 1980. Microtubules. *Sci. Am.* **243**(2):66–76.

* ROBERTS, K., and J. S. HYAMS, eds. 1979. *Microtubules.* Academic Press.

Structure of Microtubules

* AMOS, L. R., R. W. LINCK, and A. KLUG. 1976. Molecular structure of flagella microtubules. In *Cell Motility,* R. Goldman, T. Pollard, and J. Rosenbaum, eds. Cold Spring Harbor Laboratory, pp. 847–868.

COHEN, W. D., D. BARTLET, R. JAEGER, G. LANGFORD, and I. NEMHAUSER. 1982. The cytoskeletal system of nucleated erythrocytes. I. Composition and function of major elements. *J. Cell Biol.* **93**:828–838.

IZANT, J. G., J. A. WEATHERBEE, and J. R. MCINTOSH. 1983. A microtubule-associated protein antigen unique to mitotic spindle microtubules in PtK_1 cells. *J. Cell Biol.* **96**:424–434.

MANDELKOW, E., E.-M. MANDELKOW, and J. BORDAS. 1983. Structure of tubulin rings studied by x-ray scattering using synchrotron radiation. *J. Mol. Biol.* **167**:179–186.

* SOLOMON, F., M. MAGENDANTZ, and A. SALZMAN. 1979. Identification with cellular microtubules of one of the co-assembling microtubule-associated proteins. *Cell* **18**:431–438.

* WEBER, K., and M. OSBORN. 1979. Intracellular display of microtubular structures revealed by indirect immunofluorescent microscopy. In *Microtubules,* K. Robert and J. S. Hyams, eds. Academic Press, pp. 279–313.

ZIEVE G., and F. SOLOMON. 1982. Proteins specifically associated with the microtubules of the mammalian mitotic spindle. *Cell* **28**:233–242.

In Vitro Polymerization and Polarity of Microtubules

BERGEN, L. G., and G. G. BORISY. 1980. Head-to-tail polymerization of microtubules *in vitro. J. Cell Biol.* **84**:141–150.

BINDER, L. I., W. L. DENTLER, and J. L. ROSENBAUM. 1975. Assembly of chick brain tubulin onto flagellar microtubules from *Chlamydomonas* and sea urchin sperm. *Proc. Nat'l Acad. Sci. USA* **72**:1122–1126.

* CLEVELAND, D. W. 1982. Treadmilling of tubulin and actin. *Cell* **28**:689–691.

KIRSCHNER, M. W. 1978. Microtubule assembly and nucleation. *Int. Rev. Cytol.* **54**:1–71.

KIRSCHNER, M. W. 1980. Implication of treadmilling for the stability and polarity of actin and tubulin polymers *in vivo. J. Cell Biol.* **86**:330–334.

* MARGOLIS, R. L., and L. WILSON. 1981. Microtubule treadmills—possible molecular machinery. *Nature* **293**:705–711.

ROTHWELL, S. W., W. A. GRASSER, and D. B. MURPHY. 1985. Direct observation of microtubule treadmilling by electron microscopy. *J. Cell Biol.* **101**:1637–1642.

Multiplicity and Modification of Tubulins

GUNDERSEN, G. G., M. H. KALNOSKI and J. C. BULINSKI. 1984. Distinct populations of microtubules: tyrosinated and nontyrosinated alpha tubulin are distributed differently in vivo. *Cell* **38**:779–789.

HALL, J. L., L. DUDLEY, P. R. DOBNER, S. A. LEWIS, and N. J. COWAN. 1983. Identification of two human β-tubulin isotypes. *Mol. Cell. Biol.* **3**:854–862.

HAVERCROFT, J. C., and D. W. CLEVELAND. 1984. Programmed expression of beta-tubulin genes during development and differentiation of the chicken. *J. Cell Biol.* **99**:1927–1935.

HIRAOKA, Y., T. TODA, and M. YANAGIDA. 1984. The NDA3 gene of fission yeast encodes beta-tubulin: a cold-sensitive nda3 mutation reversibly blocks spindle formation and chromosome movement in mitosis. *Cell* **39**:349–358.

KEMPHUES, K. J., T. C. KAUFMAN, R. A. RAFF, and E. C. RAFF. 1982. The testis-specific β-tubulin subunit in *Drosophila melanogaster* has multiple functions in spermatogenesis. *Cell* **31**:655–670.

L'HERNAULT, S. W., and J. L. ROSENBAUM. 1985. Reversal of the posttranslational modification on *Chlamydomonas* flagellar alpha-tubulin occurs during flagellar resorption. *J. Cell Biol.* **100**:457–462.

LOPATA, M. A., J. C. HAVERCROFT, L. T. CHOW, and D. W. CLEVELAND. 1983. Four unique genes required for β tubulin expression in vertebrates. *Cell* **32**:713–724.

RAFF, E. C., M. T. FULLER, T. C. KAUFMAN, K. J. KEMPHUES, J. E. RUDOLPH, and R. A. RAFF. 1982. Regulation of tubulin gene expression during embryogenesis in *Drosophila melanogaster. Cell* **28**:33–40.

SILFLOW, C. D., R. L. CHISHOLM, T. W. CONNER, and L. P. W. RANUM. 1985. The two alpha-tubulin genes of *Chlamydomonas reinhardtii* code for slightly different proteins. *Mol. Cell Biol.* **5**:2389–2398.

Microtubules and Intracellular Transport

ALLEN, R. D., D. G. WEISS, J. H. HAYDEN, D. T. BROWN, H. FUJIWAKE, and M. SIMPSON. 1985. Gliding movement of and bidirection transport along single native microtubules from squid axoplasm: evidence for an active role of microtubules in cytoplasmic transport. *J. Cell Biol.* **100**:1736–1752.

BECKERLE, M. C., and K. R. PORTER. 1983. Analysis of the role of microtubules and actin in erythrophore intracellular motility. *J. Cell Biol.* **96**:354–362.

BRADY, S. T., R. J. LASEK, and R. D. ALLEN. 1982. Fast axonal transport in extruded axoplasm from squid giant axon. *Science* **218**:1129–1131.

GILBERT, S. P., R. ALLEN, and R. SLOBODA. 1985. Translocation of vesicles from squid axoplasm on flagellar microtubules. *Nature* **315**:245–248.

*Book or review article that provides a survey of the topic.

GOLDBERG, D. J., D. A. HARRIS, and J. H. SCHWARTZ. 1982. Studies on the mechanism of fast axonal transport using microinjection into single giant neurons. *Cold Spring Harbor Symp. Quant. Biol.* **46**:135–140.

* GRAFSTEIN, B., and D. S. FORMAN. 1980. Intracellular transport in neurons. *Physiol. Rev.* **60**:1167–1283.

HAYDEN, J. H., and R. D. ALLEN. 1984. Detection of single microtubules in living cells: particle transport can occur in both directions along the same microtubule. *J. Cell Biol.* **99**:1785–1793.

KOONCE, M. P. and M. SCHLIWA. 1985. Bidirectional organelle transport can occur in cell processes that contain single microtubules. *J. Cell Biol.* **100**:322–326.

PORTER, K. R., and M. A. MCNIVEN. 1982. The cytoplast: a unit structure in chromatophores. *Cell* **29**:23–32.

* SCHLIWA, M. 1984. Mechanisms of intracellular transport. In *Cell and Tissue Motility,* J. Shay, ed. Plenum.

* SCHNAPP, B. J., R. D. VALE, M. P. SHEETZ, and T. S. REESE. 1985. Single microtubules from squid axoplasm support bidirectional movement of organelles. *Cell* **40**:455–462.

VALE, R. D., B. J. SCHNAPP, T. MITCHISON, E. STEUER, T. S. REESE, and M. P. SHEETZ. 1985. Different axoplasmic proteins generate movement in opposite directions along microtubules in vitro. *Cell* **43**:623–632.

VALE, R. D., B. J. SCHNAPP, T. S. REESE, and M. P. SHEETZ. 1985. Movement of organelles along filaments dissociated from the axoplasm of the squid giant axon. *Cell* **40**:449–454.

ZIEVE, G., and F. SOLOMON. 1984. Direct isolation of neuronal microtubule skeletons. *Mol. Cell Biol.* **4**:371–374.

Structure and Function of Cilia and Flagella

* AFZELIUS, B. A. 1985. The immotile-cilia syndrome: a microtubule-associated defect. *CRC Critical Reviews in Biochemistry,* pp. 63–121.

BROKAW, C. J., D. J. L. LUCK, and B. HUANG. 1982. Analysis of the movement of *Chlamydomonas* flagella: the function of the radial-spoke system is revealed by comparison of wild-type and mutant flagella. *J.Cell Biol.* **92**:722–732.

* GIBBONS, I. R. 1981. Cilia and flagella of eukaryotes. *J. Cell Biol.* **91**:107s–124s.

GOODENOUGH, U. and J. E. HEUSER. 1985. Substructure of inner dynein arms, radial spokes, and the central pair/projection complex of cilia and flagella. *J. Cell Biol.* **100**:2008–2018.

* HAIMO, L. T., and J. L. ROSENBAUM. 1981. Cilia, flagella and microtubules. *J. Cell Biol.* **91**:125s–130s.

HOOPS, H. J., R. L. WRIGHT, J. W. JARVIK, and G. B. WITMAN. 1984. Flagellar waveform and rotational orientation in a *Chlamydomonas* mutant lacking normal striated fibers. *J. Cell Biol.* **98**:818–824.

HUANG, B., G. PIPERNO, Z. RAMANIS, and D. J. L. LUCK. 1981. Radial spokes of *Chlamydomonas* flagella: Genetic analysis of assembly and function. *J. Cell Biol.* **88**:80–88.

HYAMS, J. 1983. Dynein structure: A delicate molecular bouquet. *Nature* **302**:291.

* LUCK, D. J. L. 1984. Genetic and biochemical dissection of the eucaryotic flagellum. *J. Cell Biol.* **98**:789–794.

PIPERNO, G., and D. J. L. LUCK. 1981. Inner arm dyneins from flagella of *Chlamydomonas reinhardtii. Cell* **27**:331–340.

SILFLOW, C. D., and J. L. ROSENBAUM. 1981. Multiple α- and β-tubulin genes in *Chlamydomonas* and regulation of tubulin mRNA levels after deflagellation. *Cell* **24**:81–88.

* WARNER, F. D., and D. R. MITCHELL. 1980. Dynein, the mechanochemical coupling adenosine triphosphate of microtubule-based sliding filament mechanisms. *Int. Rev. Cytol.* **66**:1–43.

WARNER, F. D., and D. R. MITCHELL. 1981. Polarity of dynein-microtubule interactions *in vitro:* cross-bridging between parallel and antiparallel microtubules. *J. Cell Biol.* **89**:35–44.

Basal Bodies, Centrioles, and Microtubules

Structure and Function of Centrioles

AUFDERHEIDE, K. J., J. FRANKEL, and N. E. WILLIAMS. 1980. Formation and positioning of surface-related structures in protozoa. *Microbiol. Rev.* **44**:252–302.

* BRINKLEY, B. R. 1985. Microtubule organizing centers. *Ann. Rev. Cell Biol.* **1**:145–172.

FULTON, C. 1977. Intracellular regulation of cell shape and motility in *Naegleria.* First insights and a working hypothesis. *J. Supramol. Struct.* **6**:13–43.

HEIDEMANN, S. R., and M. W. KIRSCHNER. 1975. Aster formation in eggs of *Xenopus laevis:* induction by isolated basal bodies. *J. Cell Biol.* **67**:105–117.

HUANG, B., Z. RAMANIS, S. K. DUTCHER, and D. J. L. LUCK. 1982. Uniflagellar mutants of *Chlamydomonas:* evidence for the role of basal bodies in transmission of positional information. *Cell* **29**:745–753.

KURIYAMA, R., and G. G. BORISY. 1975. Centriole cycle in Chinese hamster ovary cells as determined by whole-mount electron microscopy *J. Cell Biol.* **91**:814–821.

MCINTOSH, J. R. 1983. The centrosome as an organizer of the cytoskeleton. *Mod. Cell Biol.* **2**:115–142.

NEMHAUSER, I., J. JOSPEH-SILVERSTEIN, and W. D. COHEN. 1983. Centrioles as microtubule-organizing centers for marginal bands of molluscan erythrocytes. *J. Cell Biol.* **96**:979–989.

WHEATLEY, D. N. 1982. *The Centriole: A Central Enigma of Cell Biology.* Elsevier/North Holland.

Dynamic Instability of Microtubules

CHEN, Y-D., and T. L. HILL. 1985. Monte Carlo study of the GTP cap in a five-start helix model of a microtubule. *Proc. Nat'l Acad. Sci. USA* **82**:1131–1135.

JOB, D., M. PABION, and R. L. MARGOLIS. 1985. Generation of microtubule stability subclasses by microtubule-associated proteins: implications for the microtubule "dynamic instability" model. *J. Cell Biol.* **101**:1680–1689.

* MITCHISON, T., and M. KIRSCHNER. 1984. Dynamic instability of microtubule growth. *Nature* **312**:237–242.

* MITCHISON, T., and M. KIRSCHNER. 1984. Microtubule assembly nucleated by isolated centrosomes. *Nature.* **312**:232–237.

SCHERSON, T., T. E. KREIS, J. SCHLESSINGER, U. Z. LITTAUER, G. G. BORISY, and B. GEIGER. 1984. Dynamic interactions of fluorescently labeled microtubule-associated proteins in living cells. *J. Cell Biol.* **99**:425–534.

Structure and Function of the Mitotic Spindle

AIST, J. R., and M. W. BERNS. 1981. Mechanics of chromosome separation during mitosis in *Fusarium* (Fungi Imperfecti): new evidence from ultrastructural and laser microbeam experiments. *J. Cell Biol.* **91**:446–458.

BAJER, A. S., and J. MOLE-BAJER. 1982. Asters, poles and transport properties within spindlelike microtubule arrays. *Cold Spring Harbor Symp. Quant. Biol.* **46**:263–284.

BROACH, J. R. 1986. New approaches to a genetic analysis of mitosis. *Cell* **44**:3–4.

CANDE, W. Z., and K. MCDONALD. 1985. In vitro reactivation of anaphase spindle elongation using isolated diatom spindles. *Nature* **316**:168–170.

EUTENEUER, U., and J. R. MCINTOSH. 1981. Structural polarity of kinetochore microtubules in PtK_1 cells. *J. Cell Biol.* **89**:338–345.

* HEATH, B. 1980. Variant mitoses in lower eukaryotes: indicators of the evolution of mitosis? *Int. Rev. Cytol.* **64**:1–80.

HEPLER, P. K., S. M. WICK, and S. M. WOLNIAK. 1981. The structure and role of membranes in the mitotic apparatus. In *International Cell Biology, 1980–1981,* H. G. Schweiger, ed. Springer-Verlag, pp. 673–686.

HIROKAWA, N., R. TAKEMURA, and S.-I. HISANAGA. 1985. Cytoskeletal architecture of isolated mitotic spindle with special reference to microtubule-associated proteins and cytoplasmic dynein. *J. Cell Biol.* **101**:1858–1870.

* INOUE, S. 1981. Cell division and the mitotic spindle. *J. Cell Biol.* **91**:132s–147s.

KILMARTIN, J. V., and A. E. M. ADAMS. 1984. Structural rearrangements of tubulin and actin during the cell cycle of the yeast *Saccharomyces*. *J. Cell Biol.* **98**:922–933.

KING, S. M., J. S. HYAMS, and A. LUBA. 1982. Absence of microtubule sliding and an analysis of spindle formation and elongation in isolated mitotic spindle from the yeast *Saccharomyces cerevisiae*. *J. Cell Biol.* **94**:341–349.

KUBAI, D. F. 1975. The evolution of the mitotic spindle. *Int. Rev. Cytol.* **43**:167–227.

MCNEILL, P. A., and M. W. BERNS. 1981. Chromosome behavior after laser microirradiation of a single kinetochore in mitotic PtK_2 cells. *J. Cell Biol.* **88**:543–553.

MITCHISON, T. J., and M. W. KIRSCHNER. 1985. Properties of the kinetochore in vitro. II. Microtubule capture and ATP-dependent translocation. *J. Cell Biol.* **101**:766–777.

NICKLAS, R. B., and G. W. GORDON. 1985. The total length of spindle microtubules depends on the number of chromosomes present. *J. Cell. Biol.* **100**:1–7.

* PICKETT-HEAPS, J. D., D. H. TIPPIT, and K. R. PORTER. 1982. Rethinking mitosis. *Cell* **29**:729–744.

ROOS, U-P. 1976. Light and electron microscopy of rat kangaroo cells in mitosis. III. Patterns of chromosome behavior during prometaphase. *Chromosoma* **54**:363–385.

SALMON, E. D., M. MCKEEL, and T. HAYS. 1984. Rapid rate of tubulin dissociation from microtubules in the mitotic spindle *in vivo* measured by blocking polymerization with colchicine. *J. Cell Biol.* **99**:1066–1075.

SCHOLEY, J. M., M. E. PORTER, P. M. GRISSOM, and J. R. MCINTOSH. 1985. Identification of kinesin in sea urchin eggs, and evidence for its localization in the mitotic spindle. *Nature* **318**:483–486.

TELZER, B. R., and L. T. HAIMO. 1983. Decoration of spindle microtubules with dynein: Evidence for uniform polarity. *J. Cell Biol.* **89**:373–378.

* ZIMMERMAN, A. M., and A. FORER, eds. 1981. *Mitosis/Cytokinesis.* Academic Press.

19

The Cytoskeleton and Cellular Movements: Actin-Myosin

MICROTUBULES and their associated proteins (Chapter 18) represent only one type of filamentous structure involved in cellular movements and in the determination of cell shape. This chapter focuses on the 10-nm-wide *intermediate filaments* and the 7-nm-wide *actin microfilaments* and their many and varied roles in cellular structure and motion. In addition, there are fibrous molecules outside cells; most cells are surrounded by an *extracellular matrix* composed of polysaccharides, glycoproteins, and fibrous proteins. Many aspects of cell shape and movement are determined by interactions of specific plasma membrane proteins with these matrix components.

Intermediate filaments are a group of cytoplasmic filaments comprising several distinct types of proteins. The proteins composing each class are specific to particular types of differentiated cells. For example, hair and skin contain the class called keratins, and neurons contain neurofilaments.

The protein *actin* is the major constituent of microfilaments, although different microfilaments also contain different actin-binding proteins that enable them to form unique structures or to carry out specific cellular functions. All eukaryotic cells contain actin, and in most cells actin is the single most abundant cytoplasmic protein (Table 19-1). The filamentous protein *myosin* is intimately

Table 19-1 Contractile protein content

Contractile protein	Percentage of total cellular protein	Concentration (μmol/kg)	Actin/myosin ratio
Actin			
Rabbit muscle	19	900	6
Human platelet	10	240	110
Acanthamoeba	14	150	70
Myosin			
Rabbit muscle	35	144	
Human platelet	1	2.2	
*Acanthamoeba**			
I	0.3	1.3	
II	1.2	2.3	

*The ameba *Acanthamoeba* contains two types of myosin termed I and II.

SOURCE: T. Pollard, 1981, *J. Cell Biol.* **91**:156s.

associated with actin in all muscle cells and probably in nonmuscle cells as well.

Actin and Myosin Filaments

As we shall see, a major role of actin and myosin filaments is the contraction of striated and smooth muscle cells. The mechanism of this movement—fueled by the hydrolysis of ATP and regulated by calcium ions and cAMP—is understood in considerable detail. We shall also explore the functions of microfilaments in nonmuscle cells. For example, microfilaments in the center of microvilli give them rigidity, and actin-containing *stress fibers* are involved in attachment of cells to the substrate; moreover, actin and myosin have been implicated in the contraction of the cleavage furrow in telophase cells and in the motion of cells along a substrate. Less is known of the molecular details of these processes, but we shall review the current knowledge and controversies.

Actin in muscle and actin in nonmuscle cells are the products of different genes and therefore differ slightly in their properties. However, all actins that have been studied—from sources as diverse as slime molds, fruit flies, mammalian platelets, vertebrate muscle, and plants—are similar in size, have very similar amino acid sequences, and share many other properties, suggesting that they evolved from a single ancestral gene. The apparently minor differences in amino acid sequence among the various muscle and nonmuscle actins may well be responsible for their significant differences in function. Myosin also exists in slightly different molecular forms in nonmuscle cells.

Many actin- and myosin-containing structures are permanent features of cell architecture—in muscle fibers and in cores of microvilli, for example. Other actin fibers, such as those in platelets or in the acrosomal structure of invertebrate sperm, are polymerized from subunits only when needed. In still other cases, pre-existing actin filaments can undergo rearrangements such as cross-linking into bundles of parallel fibers or into complex three-dimensional networks. Formation and subsequent rearrangements of these three-dimensional networks may be involved in movement of cells along substrates. Although evidence is incomplete, each such structure is thought to contain a specific set of actin-binding proteins. Actin polymerization-depolymerization cycles and cross-linking of filaments, as well as the regulation of these phenomena by the environment of the cell and by internal signals, are important areas of current research.

Actin Monomers Can Polymerize into Long Helical Filaments

Microfilaments are polymers of a globular, 42,000-m.w. protein subunit called *G actin*. Whether isolated from muscle or nonmuscle sources, actin filaments are constructed of an identical string of monomers (Figure 19-1; see also Figure 3-41). Two different models have been proposed for the molecular structure of actin monomers and filaments. In the more conventional model, used in the subsequent discussion, an actin monomer is roughly a sphere, and actin filaments are made up of two strings of monomers wound around each other in a right-handed helix 7 nm in diameter (Figure 19-1b). One repeat of the helix occurs every 36 nm (Figure 19-2). In the other model, actin monomers are dumbbell-shaped: in the fiber, a single chain of monomers forms a right-handed helix (Figure 19-1e; see also Figure 3-41). Which of these views is correct will be settled only when the three-dimensional molecular structure of crystals of G actin is determined. In any case, each monomer binds to its neighbors in exactly the same way. Furthermore, each actin subunit has a defined polarity, and the subunits polymerize head to tail. As a consequence, actin filaments also have a defined polarity, and all subunits "point" in the same direction (Figure 19-2).

Actin filaments depolymerize into G subunits in a solution that contains ATP and Ca^{2+} but very low concentrations of monovalent cations such as Na^{+} and K^{+}. Polym-

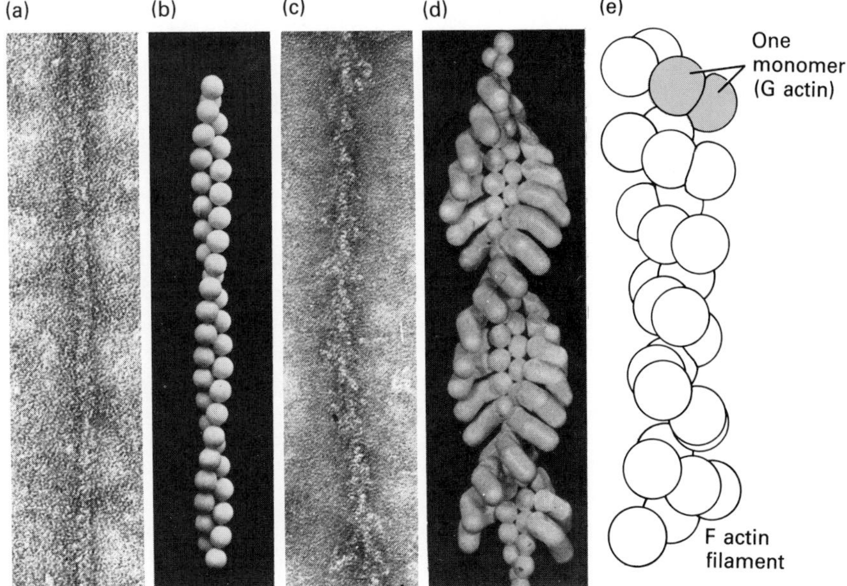

Figure 19-1 Actin filaments. (a) Electron micrograph of an *Acanthamoeba* actin filament. (b) Molecular model of the actin filament in which the fiber is constructed of two strings of spherical monomers coiled about each other in a right-handed helix. (c) Electron micrograph of an actin filament that has been reacted with a solution containing S1 fragments of myosin. The actin filament is said to be "decorated" by the bound S1. (d) Molecular model of an actin filament with attached myosin S1 heads. Note the arrowhead appearance of the decorated actin fiber. (e) Molecular model of an alternative view of the actin filament. Monomers are dumbbell-shaped, and a filament is built of a single chain of monomers. *Parts (a), (b), (c), and (d) courtesy of T. Pollard. Reproduced from the* Journal of Cell Biology, *1981, vol. 91, p. 156 by copyright permission of the Rockefeller University Press.*

erization into filaments is induced by Mg^{2+} and by solutions of K^+ or Na^+ at high concentration. As is the case with tubulin, actin can be obtained in pure form from cells by repeated cycles of depolymerization and polymerization. In vitro, actin polymerizes into filaments of structure identical to those isolated directly from cells, and actins from cells or organisms as diverse as mammals and slime molds can copolymerize. All actins, therefore, share many properties that have been highly conserved during evolution.

Actin filaments grow by addition of G actin subunits to the ends. Under appropriate conditions, addition of subunits can occur at both ends. However, the rate of addition at one end, called the "growing end," is several times that at the other. The reason for this is that G actin monomers contain a tightly bound ATP or ADP. The rate of binding to microfilaments is much faster for G actin with bound ATP than for G actin with bound ADP. Hydrolysis of the bound ATP to ADP occurs shortly after actin polymerization but is not essential for polymerization to occur. The process of ATP hydrolysis can allow actin filaments to "treadmill"—that is, to lengthen at one end by incorporation of actin-ATP monomers and at the same time to shorten at the other end by release of actin-ADP subunits. Treadmilling of actin filaments indeed has been demonstrated experimentally by means of an experimental protocol similar to that used to show treadmilling of tubulin fibers (see Figure 18-10).

Myosin Is a Bipolar, Fibrous Molecule with "Headpieces" That Bind Actin

Myosin is an unusual protein: it is both a globular enzyme and a fibrous structural protein. Myosin from smooth muscle and myosin from striated muscle have slightly different molecular properties, which account for their different contractile regulatory mechanisms. In any case, a single monomeric myosin molecule, when isolated by salt extraction from muscle, always contains two identical heavy chains (m.w. 200,000) and two pairs of light chains of two different types (m.w. ~20,000). Each heavy

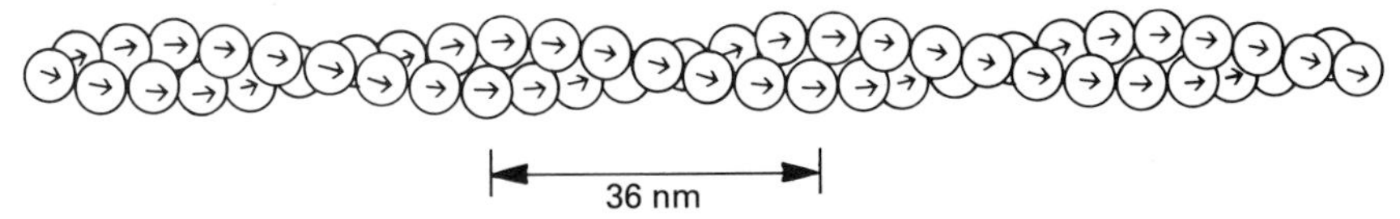

Figure 19-2 Polarity of an actin helix as depicted for the double-helical fiber. Each actin monomer is bound to its five neighbors by exactly the same set of noncovalent bonds, and each monomer "points" in the same direction. A half-rotation of the helix occurs every 36 nm; this distance is also the helix repeat length since the helix repeats itself every 36 nm.

chain consists of a globular "head," or headpiece, and a long rodlike α-helical section, or "tail." Bound to each myosin head are the two molecules of light chains, one of each type. In native myosin, the α-helical rod segments of the two heavy chains bind to each other, forming a rod from which the two heads protrude (Figure 19-3). The joints between the head and the rod of the molecule are flexible. Movements of the head are important in generating the force of contraction in muscle, as we shall see.

In muscle, myosin monomers form a specific bipolar aggregate termed the *thick filament,* containing 300 to 400 myosin molecules. The central zone of the thick filament is devoid of heads and is composed of an antiparallel overlapping array of myosin tails that is slightly longer than a single myosin tail. The terminal regions are of variable length, and myosin heads protrude from the surface in a helical array at 14-nm intervals. The thick filaments are symmetric about the center of the bare central zone, and it is important to stress that the polarity of the myosin filaments is reversed on either side of the midline (Figure 19-3). Some structural and immunochemical evidence indicates that the myosin polypeptides of the central zone are slightly different from those of the terminal regions of the aggregate.

Treatment of myosin with low concentrations of protease, such as papain, generates two fragments: the globular headpiece, termed *heavy meromyosin (HMM)* or *S1,* depending on the protease used, and the rod-shaped segment, designated *light meromyosin (LMM).* All of the myosin light chains are in the HMM or S1 fragment (Figure 19-4). Such enzymatic dissection of the molecule has been used in numerous biochemical studies to identify and localize many functional properties.

Myosin Is an Actin-Activated ATPase

Early biochemical studies on actin and myosin established several of their crucial properties. Myosin has an ATPase activity. In the absence of actin, this activity is almost undetectable. But when pure actin filaments are added, the rate of ATP hydrolysis is increased 200-fold, so that each myosin molecule hydrolyzes 5 to 10 ATP molecules per second—a rate similar to that in contracting muscle. Likewise, the S1 proteolytic fragment has this ATPase activity, which also is stimulated by actin. Presumably, then, the S1 portion of myosin binds to actin. Indeed, in the absence of ATP, myosin and actin fibers have been shown to aggregate, forming a large, complex network of fibers in which a single actin microfilament may have several myosin molecules bound to it along its length. Binding of actin to myosin must take place on the headpiece because S1 fragments also bind specifically to actin. Binding of either myosin or the S1 fragment to actin is disrupted by addition of ATP. This finding is strong evidence for the role of ATP hydrolysis in the interaction of actin and myosin—which has been amply demonstrated, as we shall see later in the discussion of muscle contraction.

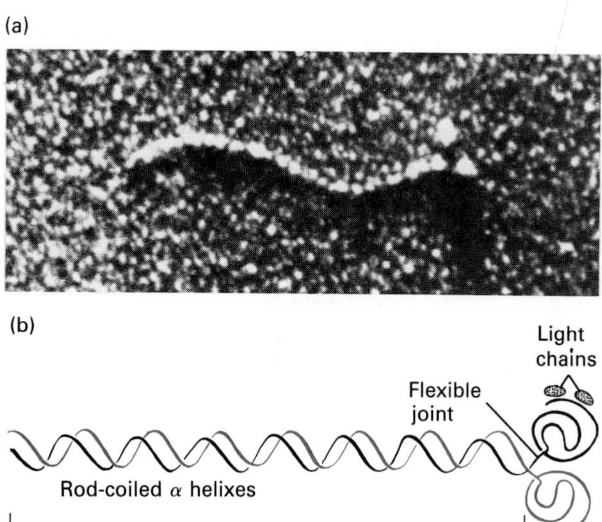

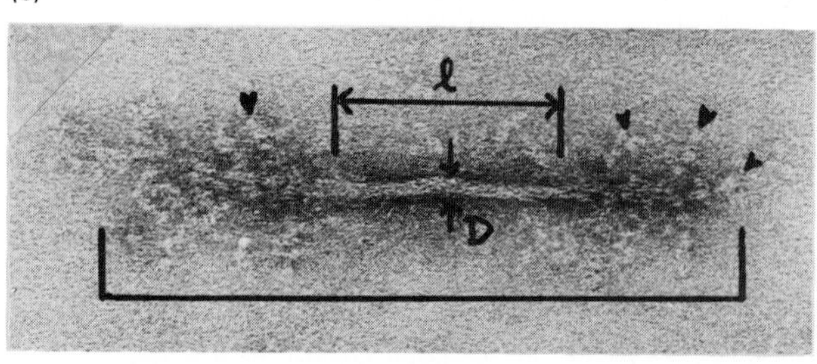

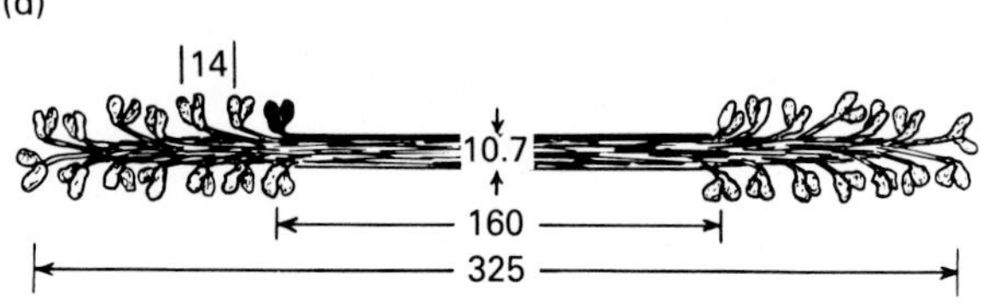

Figure 19-3 Views of myosin purified from platelets. (a) Electron micrograph of a single myosin molecule. The two heads are visible at the right. (b) Drawing of a myosin molecule showing the two constituent heavy and four light polypeptide chains. (c) Electron micrograph of a negatively stained bipolar myosin filament. The length of the bare zone comprising the rodlike sections of the molecule is designated *l,* and *D* is the diameter of this zone. Arrowheads indicate some of the myosin heads. (d) Two-dimensional model of a myosin filament, showing the dimensions of the filament parts in nanometers. *Parts (a), (c), and (d) courtesy of T. Pollard. Reproduced from the* Journal of Cell Biology, *1981, vol. 91, p. 156 by copyright permission of the Rockefeller University Press.*

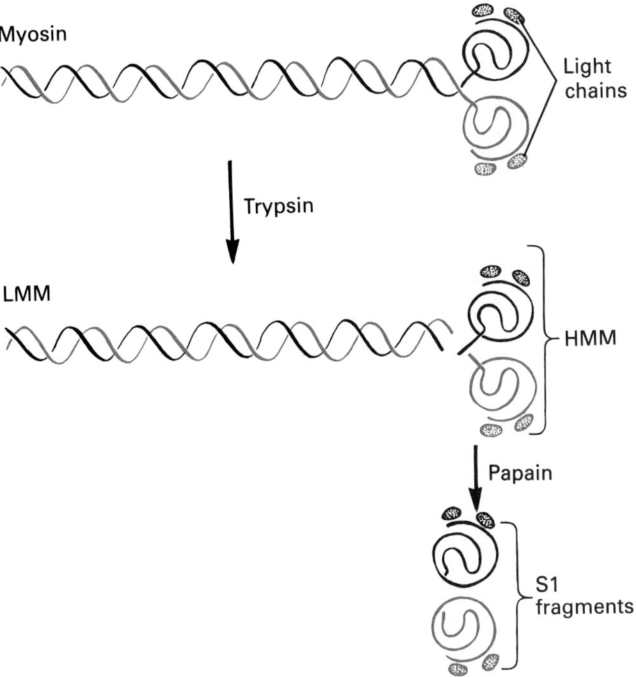

Figure 19-4 Cleavage of myosin by the protease trypsin generates two identical subfragments called heavy meromyosin, HMM. These fragments are derived from the headpieces and contain the myosin light chains. The third product of trypsin digestion is the myosin rod, called light meromyosin, or LMM. Further digestion of HMM with papain removes the remaining α-helical segments to form the globular fragments termed S1.

The Binding of Myosin to Actin Can Be Demonstrated in the Electron Microscope

The binding of the S1 fragment of myosin to actin fibers can be demonstrated by electron microscopy (see Figure 19-1). Numerous S1 fragments bind along the entire length of an actin filament, forming an array that reflects the underlying directionality and helicity of the actin filaments (Figure 19-2). When actin-S1 fragment filaments are viewed in a plane parallel to the fiber, the bound S1 fragments display a characteristic "arrowhead" pattern in the electron microscope. Throughout the length of a fiber, all the arrowheads point in the same direction. This observation establishes that actin filaments have an intrinsic polarity (see Figure 19-2). The binding of the S1 fragment to actin fibers is highly specific. S1 will not bind to filaments of spectrin, tubulin, or 10-nm filaments, for instance. In fact, the binding of muscle S1 to 7-nm-diameter filaments in detergent-extracted fibroblast cells was one of the first conclusive pieces of evidence that nonmuscle cells contain actin filaments.

Having considered the basic structural properties of actin and myosin, we can now turn to the role of these filaments in muscle contraction.

Muscle Structure and Function

In vertebrates, there are three classes of muscles: *smooth, cardiac,* and *striated* (Figure 19-5). Typically, smooth muscles are under involuntary (unconscious) control of the central nervous system. They surround internal organs such as the large and small intestines, the gallbladder, and large blood vessels. Contraction and relaxation of smooth muscles control the diameter of blood vessels and also propel food along the gastrointestinal tract. Smooth muscle cells can create and maintain tension for long periods of time.

Muscles under voluntary control have a striated appearance in the light microscope. Striated muscles, which connect the bones in the arms, legs, and spine, are used in more complex coordinated activities, such as walking or positioning of the head, and can generate rapid movements by sudden bursts of contraction. Cardiac (heart) muscle resembles striated muscle in many respects, but it is specialized for the continuous, involuntary contractions needed in pumping of the blood. Study of striated muscle cells, with their very regular organization of the actin and myosin contractile filaments, has provided important evidence about the mode of contraction in all three types of muscle.

The Contractile Apparatus in Striated Muscle Consists of a Regular Array of Actin and Myosin Filaments

Even at the level of the light microscope, the regular structure of striated muscle is evident: the dark bands, called *A bands,* alternate with light bands, called *I bands;* a narrow line, the *Z disk,* bisects each I band. The light and dark bands are perpendicular to the long axis of the muscle cell along which the muscle contracts. The segment from one Z disk to the next is termed a *sarcomere.* The sarcomere is the unit of the contraction; movements of actin and myosin within each sarcomere lead to shortening of the sarcomeres, and thus to shortening—contraction—of the muscle as a whole.

A typical striated muscle cell is cylindrical and is very large, measuring 1 to 40 mm in length and 10 to 50 μm in width. Each cell, called a myofiber, contains up to a 100 nuclei and contains many bundles of filaments termed *myofibrils* (Figure 19-6). Each myofibril is constructed of a repeating array of sarcomeres, each of which measures about 2 μm in length in resting muscle. Electron microscopy has shown that each sarcomere contains two types of filaments: thick filaments, now known to be myosin, and the thin microfilaments, containing actin

(a)

I bands

Z disks

Sarcomere

A bands

10 μm

Direction of contraction

(b)

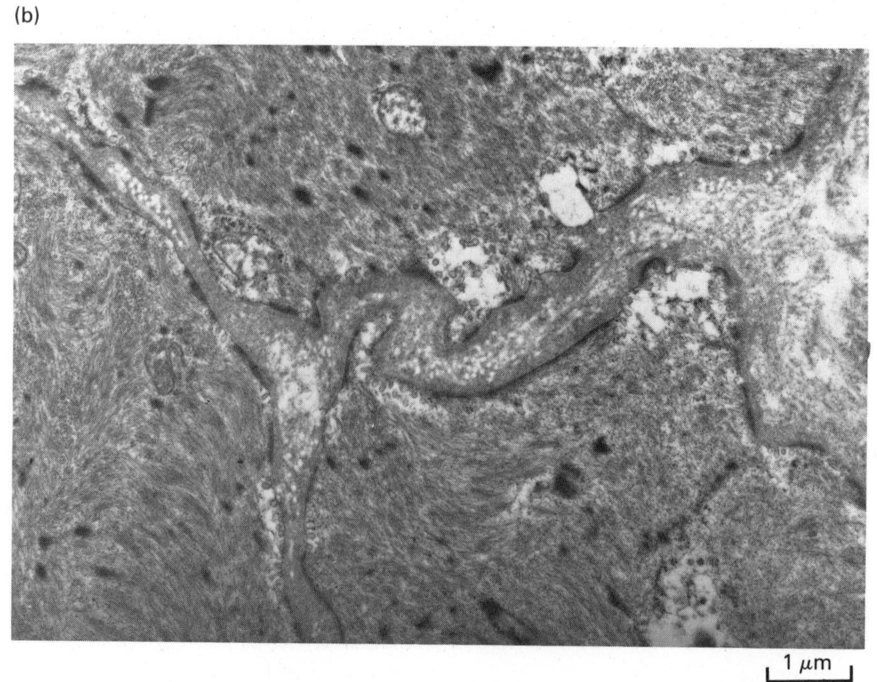

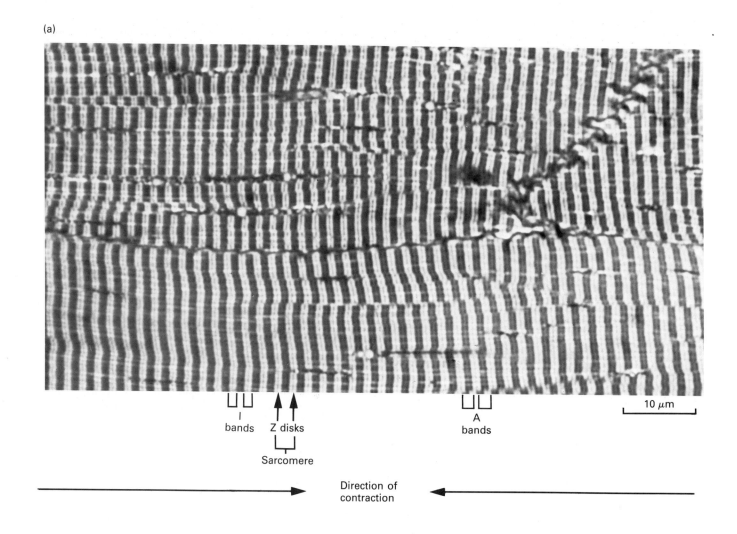

Figure 19-5 *(opposite)* Micrographs of vertebrate smooth and striated muscle. (a) A low-magnification light micrograph of a longitudinal section of a stained striated muscle fiber from a rattlesnake. The thick dark vertical stripes are composed of the A band of juxtaposed myofibrils; the light stripes are the I bands (centered by the Z disks). A sarcomere is the segment from one Z disk to the next. The tiny faint lines running horizontally across the dark A bands are areas of sarcoplasm that slightly separate the myofibrils. The bands of most myofibrils are in exact register with one another to produce continuous striations across the fiber. (b) Electron micrograph of a section of three typical smooth muscle cells from the human colon. The fibers are not organized in sarcomeres and are not all parallel to each other. Thus in any sample the cut filaments are seen in longitudinal, transverse, and oblique views. *Part (a) courtesy of D. Schulz and A. W. Clark; part (b) courtesy of A. R. Perez-Atayde.*

(Figures 19-7, 19-8, and 19-9). One end of the thin filaments is attached to the rigid Z disk. The other end of the filaments extends part way toward the center of the sarcomere.

The thick filaments, 300 to 500 nm in length, are the major constituents of the A band and have the molecular structure of myosin aggregates, as depicted in Figure 19-3. The A band is so labeled because it is anisotropic, or birefringent, when viewed in polarized light (see Figure 18-31). The property of birefringence is caused by the organized bundles of thick filaments. The I band, by contrast, is composed of thin filaments and is less anisotropic than the A band, rather than purely isotropic as erroneously suggested by the label I. The region in the center of the A band consists of the rod-shaped LMM tails of the myosin heavy chains. On both sides of this central region are protrusions from the filament. These are the heads of the myosin molecule—the HMM or S1 fragments (Figures 19-9 and 19-10). In the sarcomere region in which thick and thin filaments overlap, the so-called AI band, each thick myosin filament is surrounded by six thin actin filaments, and each actin filament is surrounded by two myosin filaments (see Figure 19-8).

Thick filaments can be selectively removed from permeabilized cells by treatment with certain salt or detergent solutions. Subsequent biochemical analysis shows that myosin is the principal protein extracted and thus is the primary structural component of the thick filaments. Added S1 myosin binds to the remaining thin filaments in the characteristic arrowhead pattern, thereby establishing that the thin filaments are indeed built of actin. Examination of the thin filaments bound with S1 shows that the myosin arrowheads always point *away from* the Z disk. Thus all of the filaments attached to one side of a Z disk have the same polarity, and filaments on opposite sides of the Z disk point in the opposite direction (see Figure 19-9).

Thick and Thin Filaments Move Relative to Each Other During Contraction

Several key experiments established that muscle contraction is caused by the sliding of thick and thin filaments past each other. As a muscle is stretched passively to different sarcomere lengths, the width of the A band—the myosin fibers—has been shown to remain constant (Figure 19-11). Electron microscopy has confirmed that the lengths of individual myosin fibers as well as actin filaments do not change as the muscle contracts or relaxes. What *does* change is the width of the I band—the part of the actin microfilaments not covered by myosin. Presumably, then, the only way movement can occur is by the sliding of the two filaments past each other.

Since the heads of the myosin fibers have the actin-

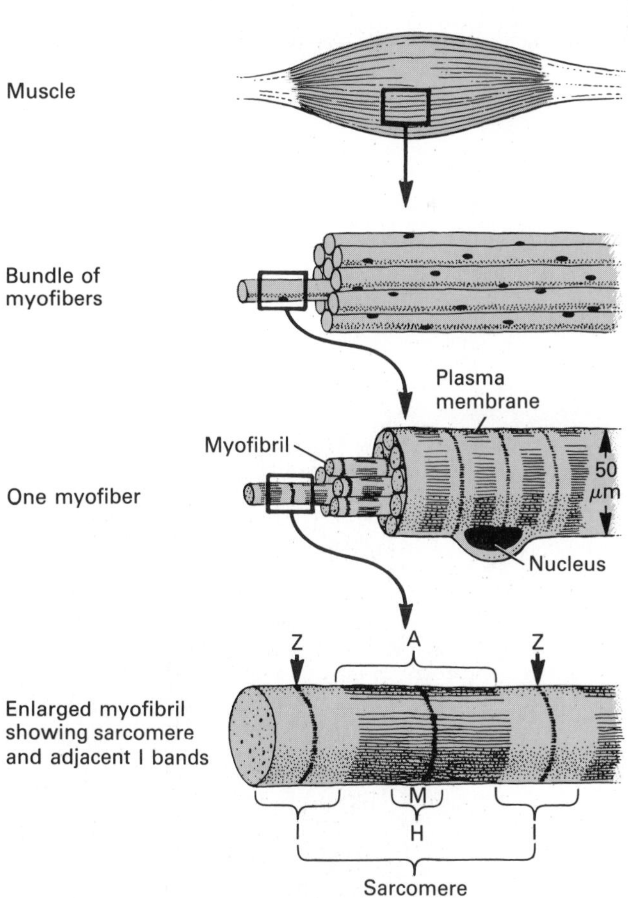

Figure 19-6 The levels of organization in striated muscle. Each muscle cell, or myofiber, is multinucleate and contains many myofibrils. The sarcomere is the functional unit of contraction and is the space between two Z disks.

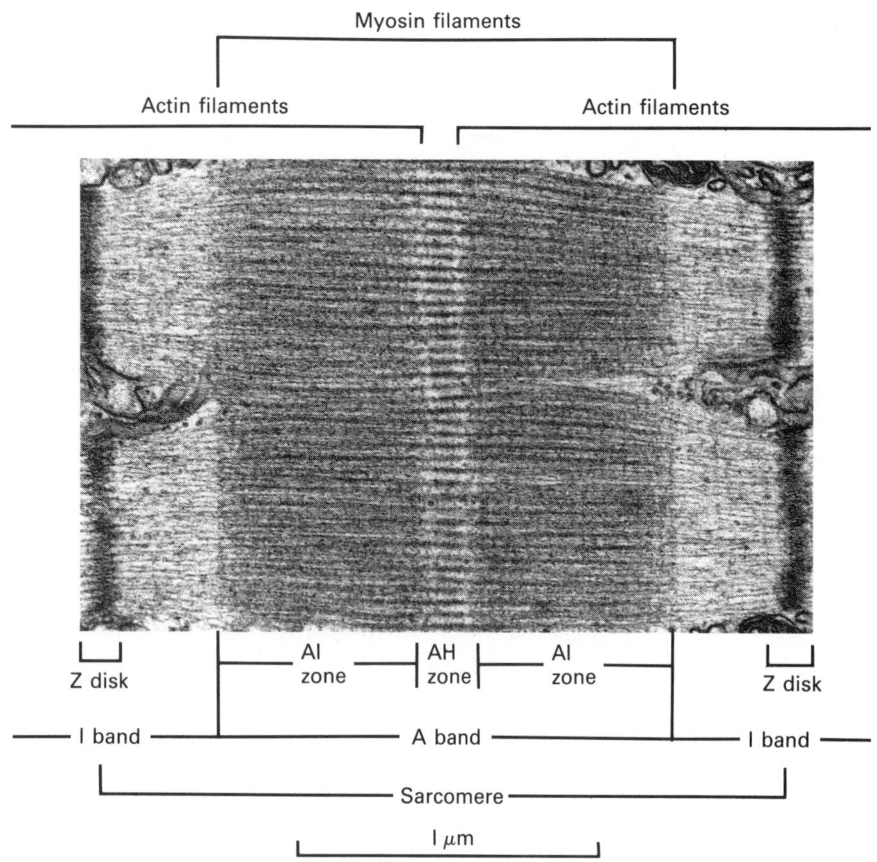

Figure 19-7 Electron micrograph of mouse striated muscle in longitudinal section. A sarcomere is seen in longitudinal aspect. On either side of the Z disks are the light-stained I bands, composed entirely of thin actin filaments. The thin filaments extend from both sides of the Z disk to interdigitate with the dark-stained myosin thick filaments that make up the A band. The region containing both thick and thin filaments (the AI zone) is darker than the area containing only thick filaments (the AH zone). *Photograph courtesy of S. P. Dadoune.*

stimulated ATPase activity and are known to bind actin, it was suggested and subsequently confirmed that the force-generating elements are the individual myosin-actin bridges. Accordingly, the force of contraction should be proportional to the overlap between the actin filaments and the head-bearing regions of the thick filaments. When the muscle is stretched to a fixed length, the number of myosin headpieces capable of interacting with actin is reduced. The greater the stretch, the less overlap, and the less tension generated when the stretched muscle is stimulated to contract. Experimental observations of such isometric contractions have borne out this predicted relationship nicely (Figure 19-12).

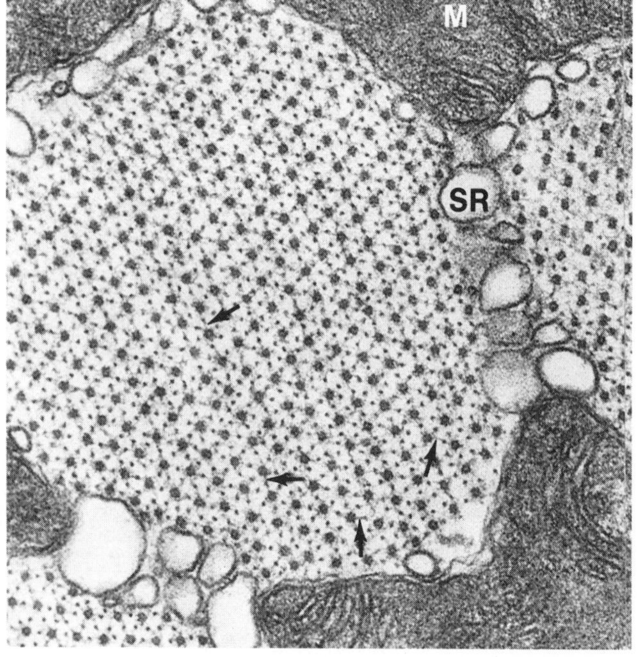

Figure 19-8 Cross section of a sarcomere at the level of the AI zone. Each thick myosin filament is surrounded by six thin actin microfilaments; each microfilament is midway between two myosin filaments. Extremely thin filaments (*arrows*) in some areas appear to interconnect the actin and myosin filaments; these represent the globular heads of the myosin molecules. SR = sarcoplasmic reticulum. *From R. G. Kessel and R. H. Kardon, 1979,* Tissues and Organs: A Text-Atlas of Scanning Electron Microscopy, *W. H. Freeman and Company.*

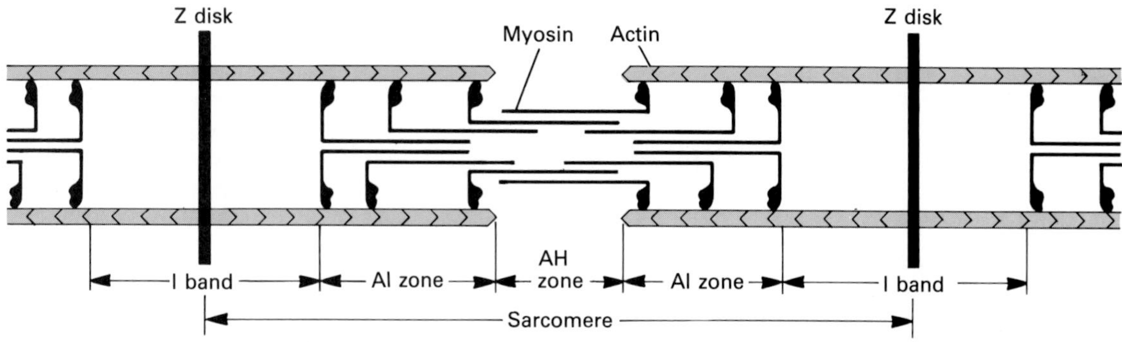

Figure 19-9 Diagram showing arrangements of thick and thin filaments in striated muscle.

ATP Hydrolysis Powers the Contraction of Muscle

Each myofiber is surrounded by mitochondria and granules of glycogen, used for generation of ATP. The contraction of muscle is fueled by the hydrolysis of ATP to ADP and P_i. If a muscle is depleted of its store of compounds with high-energy phosphate bonds, such as ATP or creatine phosphate (Chapter 2), it becomes stiff and can no longer be extended passively with small forces; this condition is known as *rigor*. (As is well known, muscles go into a state of rigor a short time after death.) In rigor, the actin and myosin filaments are tightly cross-linked together, so that the thin and thick filaments cannot easily slide past each other (see Figure 19-10). Myosin headpieces bind ATP or ADP with high affinity. Ultrastructural evidence indicates that myosin headpieces having no bound nucleotide bind at a 45° angle to an actin filament. Binding of ATP to myosin weakens the binding of the headpiece to actin (*Step 1,* Figure 19-13). In the absence of actin, ATP bound to myosin is hydrolyzed to ADP and P_i, but the two hydrolysis products remain bound to the myosin headpiece. Myosin with bound ADP and P_i binds weakly to actin, but, importantly, the myosin heads are bound at a 90° angle to the actin. Thus the hydrolysis of ATP to ADP and P_i is accompanied by a conformational change in the myosin headpiece—apparently, a 45° rotation about a hinge region of the myosin *(Step 2)*.

Contraction is accomplished by the cyclic formation and disassociation of bridges between the myosin globular heads and the sides of the adjacent actin filaments. Bridge formation pulls the thin filaments toward the center of the A band, contraction results (Figure 19-13). ATP hydrolysis is essential to the cyclic formation and dissociation of these actin-myosin bridges. As noted previously, the linkage between the myosin headpieces and the tail, or body, of the molecules is flexible. During contraction (*Step 4,* Figure 19-13), the myosin head, bound to an actin molecule, tilts toward the center of the thick filament; this conformational change is the step fueled by the energy of hydrolysis of the terminal phosphodiester bond of ATP.

Unlike the strongly exergonic hydrolysis of ATP to ADP and P_i in solution ($\Delta G^{\circ\prime} \sim -7$ kcal/mol), hydrolysis of ATP bound to myosin occurs with little change in free energy. The *release* of the bound ADP and P_i is the strongly exergonic step, and the free energy released is

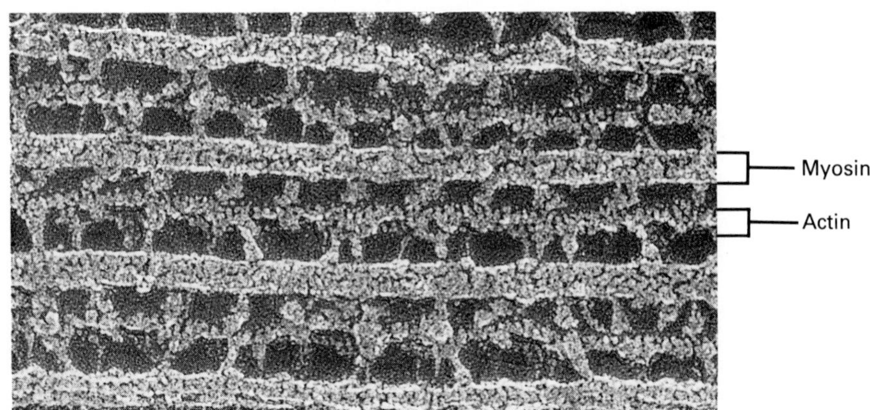

Figure 19-10 Micrograph showing actin-myosin cross-bridges in a striated insect flight muscle. This image, obtained by the quick-freeze deep-etch technique (described later in the chapter), shows a nearly crystalline array of thick myosin and thin actin filaments. The muscle was in rigor at preparation. Note the myosin headpieces connecting with the actin filaments at regular intervals. *Photograph courtesy of J. Heuser.*

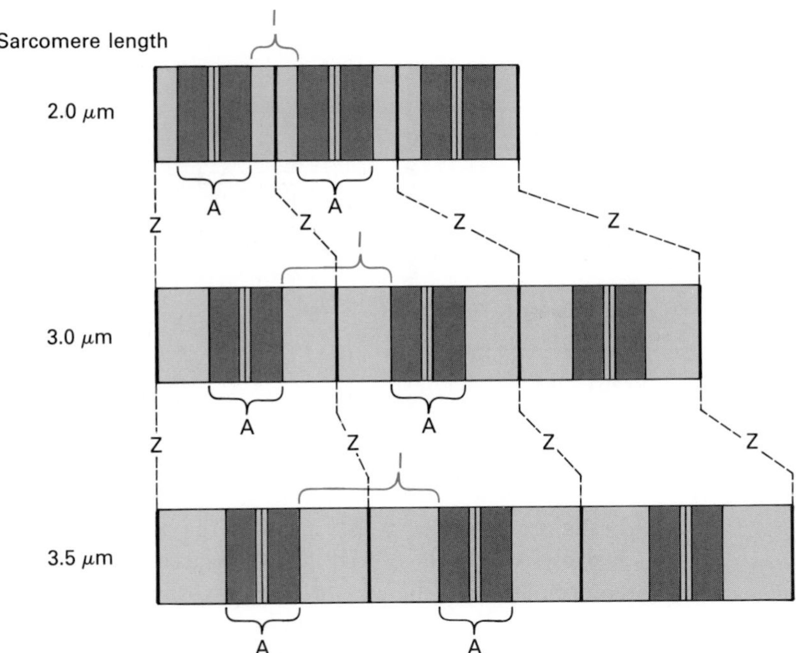

Figure 19-11 Diagrams showing a striated muscle fiber stretched to different sarcomere lengths. The A band is dark. The constant length of the A band observed in studies on muscle contraction first suggested the sliding-filament model of muscle contraction. [See A. F. Huxley and R. Niedergerke, 1954, *Nature* 173:973.]

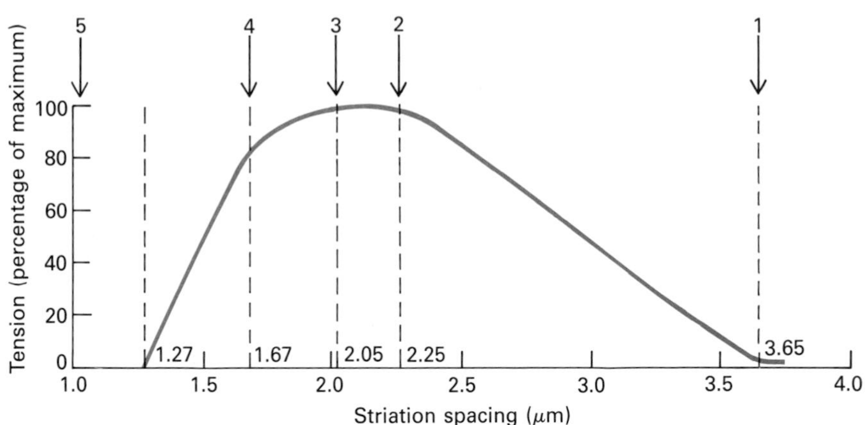

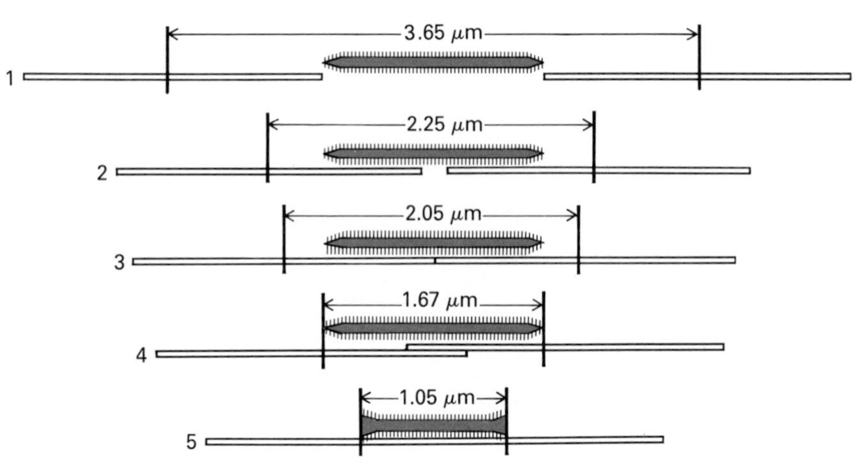

Figure 19-12 Relationship of sarcomere length and tension generated during isometric contraction of striated muscle. The numbers 1 to 5 on the graph indicate the relative positions of the thick and thin filaments in the accompanying diagram. Maximum tension is generated at sarcomere lengths that allow maximum interaction of myosin heads and actin filaments—as at positions 2 and 3. If the sarcomere length is too short—as at positions 4 and 5—actin filaments overlap each other and prevent optimum interaction with myosin heads. [See A. M. Gordon, 1966, *J. Physiol.* 184:170.]

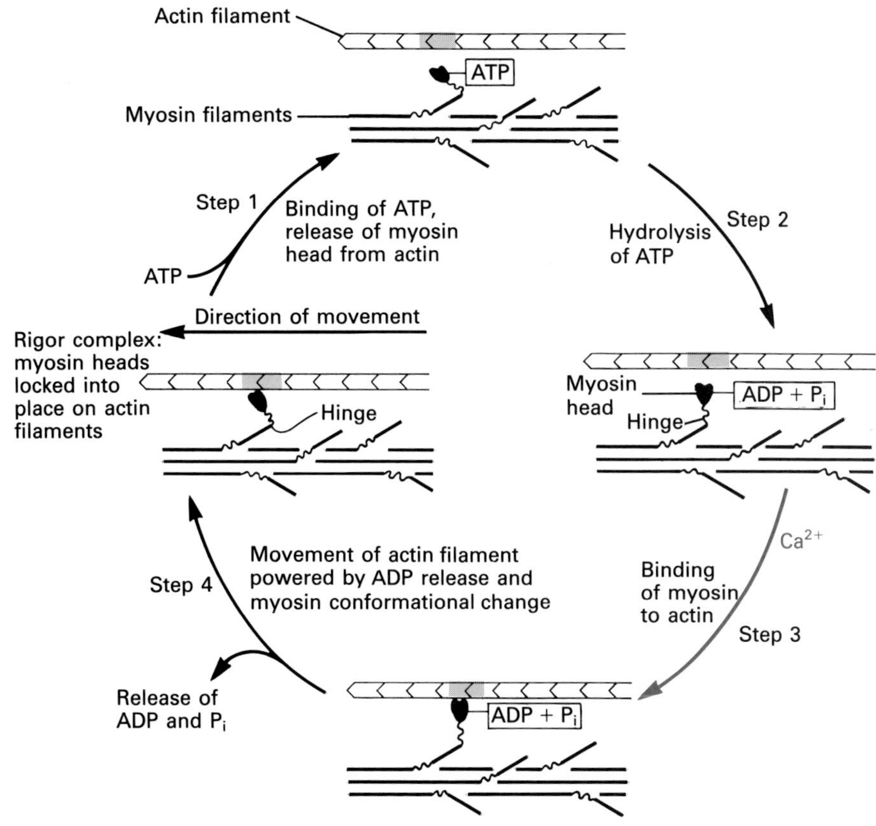

Figure 19-13 Diagram of the myosin-ATPase cycle during muscle contraction. *Step 1:* A myosin molecule binds ATP, causing weakening of the actin-myosin bonds and possibly release of the myosin head from actin. *Step 2:* ATP is hydrolyzed to ADP + P_i, but the hydrolysis products remain bound to the myosin. This generates an "energized" myosin headpiece, which has rotated so it is perpendicular to the actin filament; this is facilitated by flexible regions, or "hinges," on the myosin molecule. *Step 3:* The myosin head binds to an adjacent filament; this is the step that is dependent on the presence of Ca^{2+}. *Step 4:* The myosin head undergoes a conformational change while it is bound to the actin, so that it moves to a 45° angle with the actin filament. This causes the myosin to pull the actin filament, and the movement of the actin filament relative to the myosin then occurs. During this step, first P_i is released and then ADP. The product of this step is the so-called rigor complex, in which the myosin head is locked to the thin filament: the actin-myosin linkage is inflexible, and the thin and thick filaments cannot move past each other. Subsequent binding of ATP to the myosin head (*Step 1*) releases the myosin head from the actin. Controversy does exist as to whether or not myosin actually dissociates from actin, as depicted here, or simply binds less tightly. Also, detailed kinetic studies indicate that each of the steps depicted here can be divided into two or more substeps. [See E. Eisenberg and T. Hill, 1985, *Science* **227**:999; and M. Irving, 1985, *Nature* **316**:292.]

used to power the conformational change in the myosin. Precisely how the change is brought about is uncertain, but it generates the relative movement of the actin and myosin chains. With each cycle, which requires the hydrolysis of one ATP molecule per myosin head, the actin filament is pulled a distance of about 7 nm.

In the resting state, myosin heavy chains contain tightly bound ADP and P_i. The myosin head groups protrude outward at a 90° angle from the thick filaments but are not bound to actin or are bound weakly. Stimulation of the muscle (*Step 3,* Figure 19-13) causes each headpiece to bind to an adjacent actin filament. Subsequently, both ADP and P_i are released, and the myosin conformation changes, pulling the actin filament past the myosin filament. In the absence of additional ATP, the muscle is now in rigor (recall that myosin heads will bind, in the absence of ATP, to actin filaments in the characteristic 45° arrowhead pattern). When ATP is present it binds to the myosin headpiece and weakens the myosin-actin bond (Figure 19-14a). Hydrolysis (*Step 2,* Figure 19-13) completes the cycle.

Release of Ca^{2+} from the Sarcoplasmic Reticulum Triggers Contraction

A rise in internal Ca^{2+} concentration triggers the binding to actin of the myosin ADP · P_i complex and thus stimulates muscle contraction. So long as the Ca^{2+} concentration is sufficiently high, the myosin-actin bridges will cycle continuously, and the muscle will contract or generate tension. All the myofibrils in a myofiber contract simultaneously. But how is this accomplished in very long muscles, in which presumably the Ca^{2+} trigger must diffuse great distances (up to 100 μm)? The answer is that, as observed in the electron microscope, each myofiber is surrounded by a network of smooth membranes, collectively termed the *sarcoplasmic reticulum (SR)* (Figures 19-8, 19-15, and 19-16). The SR forms an extensive lacelike network of membrane vesicles and cisternae that surrounds the outer regions of the A band of each myofibril; this is the location of the myosin headpieces. The SR

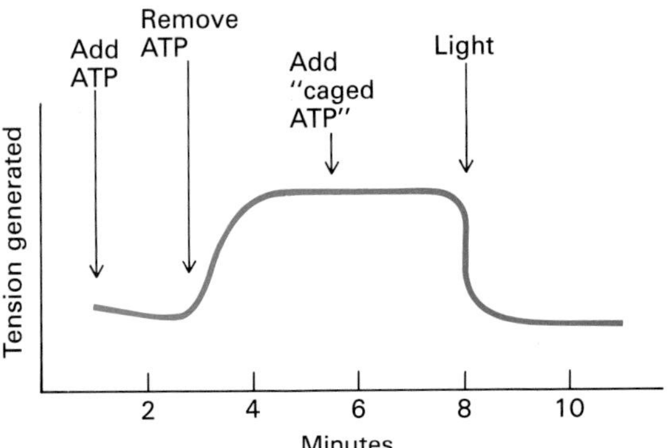

(b)

"Caged ATP"

Laser light

+ ATP

Intermediate

Figure 19-14 Demonstration that ATP relaxes muscle fibers. (a) Graph showing muscle tension at intervals corresponding to experimental manipulations. A small striated muscle fiber is first extracted with glycerol, to render it permeable to ATP and macromolecules. When ATP but no Ca^{2+} is added, the tension generated is low; the myosin heads contain bound ADP and P_i, but the myosin does not bind to actin in the absence of Ca^{2+} (*Step 3*, Figure 19-13). When ATP is removed by dilution, the muscles go into rigor; myosin headpieces, with no bound nucleotide, remain tightly bound to actin filaments (rigor complex, Figure 19-13). This causes an increase in tension. ATP cannot be simply added back to the preparation, since it will take several minutes for the compound to diffuse throughout the fiber, during which time much of the ATP will be hydrolyzed. ATP can be generated instantly and uniformly in the preparation by adding the compound called "caged ATP." This molecule does not serve as substrate for the myosin ATPase, nor can it bind to the myosin headpieces. After addition of "caged ATP" there is no effect on the tension. But when "caged ATP" is converted to ATP by a short pulse of intense laser light, immediate relaxation of the tension occurs as ATP binds to the myosin heads and dissociates the myosin from actin (*Step 1*, Figure 19-13). Since there is no Ca^{2+} in the solution, myosin ADP + P_i complexes accumulate, unattached to thin filaments. (b) Chemical equation for conversion of "caged ATP" to ATP. [See Y. E. Goldman et al., 1982, *Nature* 300:701–705.]

also forms a continuous membrane channel, called the *terminal cisterna,* that surrounds the Z disk of each myofibril (Figure 19-16). Over the Z disks the SR connects with a set of membranes termed the *transverse tubules,* or *T tubules,* but the lumen of the SR probably is not continuous with the lumen of the T tubules. The T tubules are delicate invaginations of the plasma membrane.

The SR serves as a reservoir of Ca^{2+} sequestered from the cell cytosol and myofibrils. The membrane of the SR has a potent Ca^{2+} ATPase activity, which pumps Ca^{2+} from the cytoplasm into the lumen of the SR (Chapter 15, page 628. The lumen of the SR contains two Ca^{2+}-binding proteins (Chapter 15, page 631), which store large amounts of Ca^{2+}.

Depolarization of the plasma membrane of the muscle cell results in release of Ca^{2+} from the SR into the cytosol and thus triggers contraction of the muscle. The depolarization is propagated along the membrane of the T tubules to the SR in much the way an action potential is propagated, and apparently causes formation of triphosphoinositol that, in turn, results in a flux of Ca^{2+} from the SR lumen into the cytosol (Chapter 16). Because the depolarization is conducted along the T tubule and SR membrane within milliseconds, every sarcomere in the cell contracts simultaneously. Continued stimulation of the muscle keeps the cytoplasmic Ca^{2+} level high. When stimulation ceases, the Ca^{2+} ATPase lowers the cytosolic Ca^{2+} concentration by pumping Ca^{2+} back into the

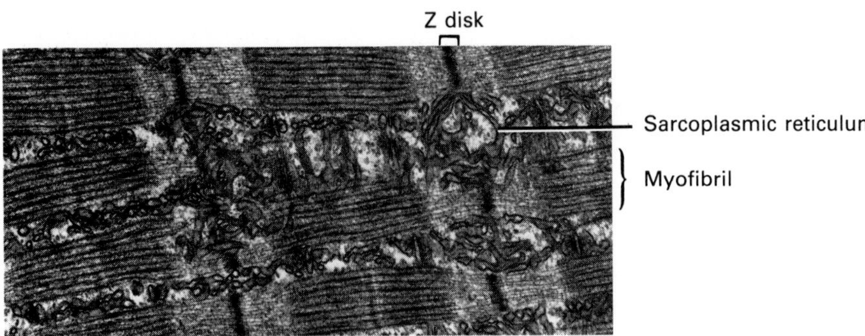

Figure 19-15 Electron micrograph showing abundant sarcoplasmic reticulum and vesicles surrounding rather small myofibrils. This is a longitudinal section through a column of myofibrils in rattlesnake muscle. *Photograph courtesy of E. Schultz and A. W. Clark.*

lumen of the SR, thus inhibiting contraction. Striated muscle can undergo very rapid increases and decreases in the level of cytosolic Ca^{2+}, thereby permitting precise control of muscular movements.

Smooth muscle does not contain a developed SR membrane. Changes in the level of cytosolic Ca^{2+} (and also of cAMP) are much slower than in striated muscle—on the order of seconds to minutes—thereby allowing a slow, steady response in contractile tension.

In Striated Muscle Troponin and Tropomyosin Mediate Ca^{2+} Stimulation of Contraction

Because of the Ca^{2+} ATPase in the SR membrane, the cytosol of resting muscle has a free Ca^{2+} concentration of about 10^{-7} *M*. Stimulation that causes an increase in Ca^{2+} concentration to 10^{-5} *M* initiates contraction. The manner in which Ca^{2+} activates contraction varies for different types of muscle; at least three types of regulation

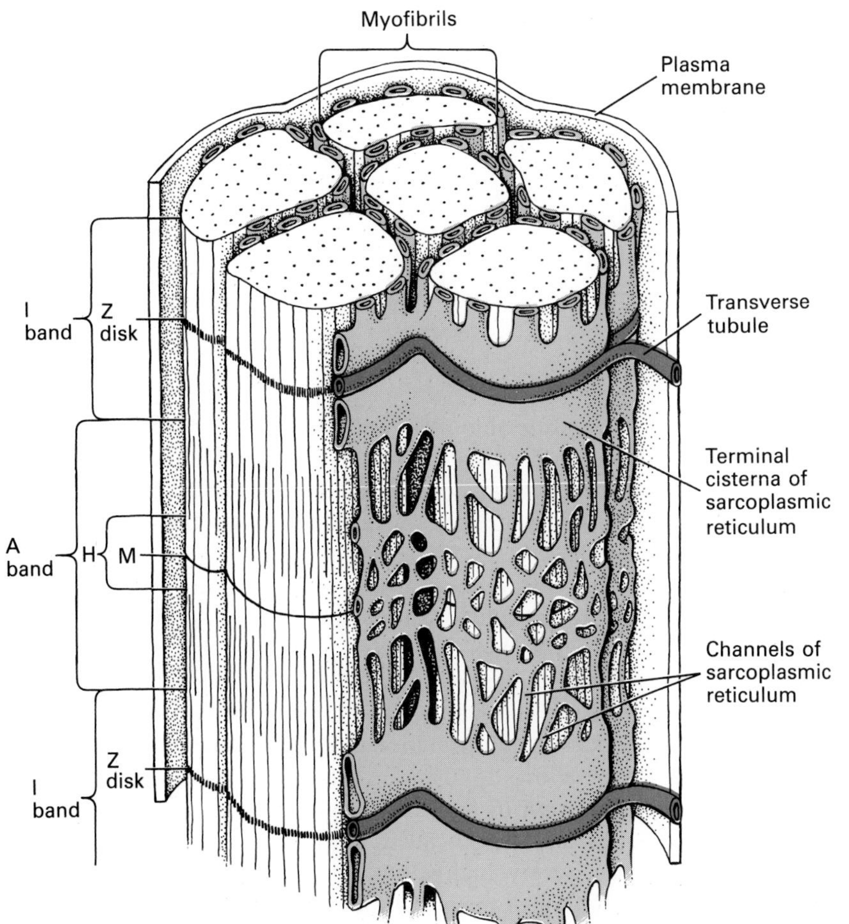

Figure 19-16 Three-dimensional structure of five myofibrils. The transverse tubules are delicate tubular invaginations of the plasma membrane, and the lumen of these tubules may connect with the extracellular medium. These tubules enter the fiber at the level of the Z disk; here they form close contact with the sarcoplasmic reticulum, which consists of cisternae and channels of smooth membrane that lie between and surround myofibrils. In the region of the Z disk they are called terminal cisternae. These connect with the lacelike network of membranous tubules, which are abundant over the H, or middle, zone of the A band.

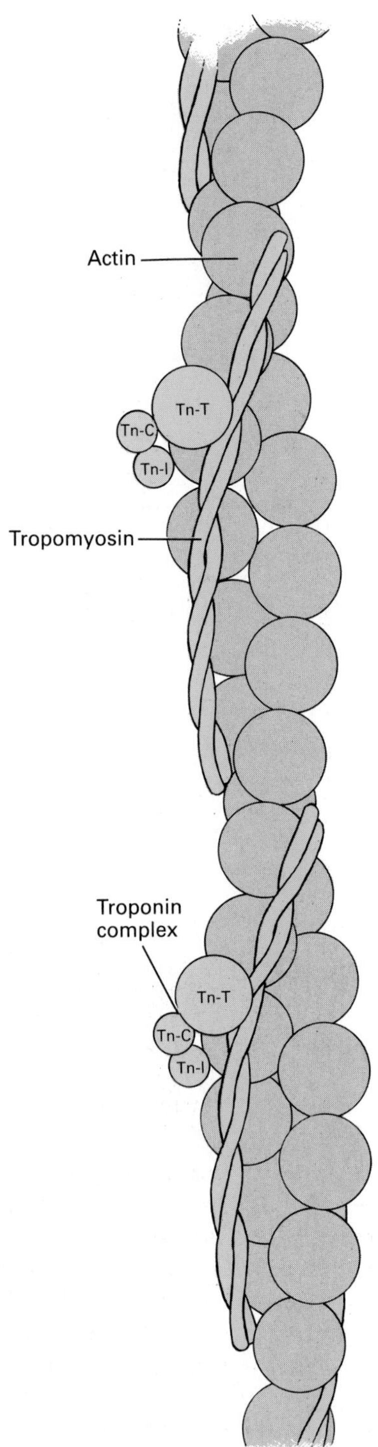

Figure 19-17 Diagram of thin filaments in vertebrate muscle, showing location of tropomyosin and the three troponin peptides, TN-T, TN-I, and TN-C.

are known. In striated vertebrate muscle the Ca^{2+} regulation affects the actin thin filaments. In smooth muscle and invertebrate muscle the regulation affects the myosin headpieces.

The thin filaments in striated muscle contain four other proteins besides actin. These are involved in the Ca^{2+} regulation of contraction (Figure 19-17). *Tropomyosin* is an elongated protein (a dimer of 35,000-m.w. subunits) that lies in each of the two grooves of the actin helix. Bound to a specific site of each tropomyosin molecule are three *troponin* (TN) peptides called troponins T, I, and C (TN-T, TN-I, TN-C, respectively). The molar ratio of actin, TN-T, TN-I, TN-C, and tropomyosin is 7:1:1:1:1.

A mixture of myosin filaments and purified thin filaments containing only actin hydrolyzes ATP at the maximum rate. The presence of troponin and tropomyosin on the thin filaments inhibits this ATPase activity by blocking the interaction of myosin heads with actin. The TN-T subunit is responsible for binding the TN complex to tropomyosin, and TN-I in concert with tropomyosin inhibits the binding of actin to myosin and thus the actin-stimulated myosin ATPase activity. TN-C is the Ca^{2+}-binding subunit with a structure and function very similar to that of calmodulin (Chapter 16). Occupation by Ca^{2+} of all of the Ca^{2+}-binding sites of TN-C results in a release of the tropomyosin–TN-I inhibition of the actin-myosin ATPase activity and in *activation* of contraction. How this function of TN-C is mediated is not clear. One popular model suggests that binding of Ca^{2+} to TN-C induces a conformational change in the tropomyosin molecule so that the myosin heads can then interact with actin. This may involve a movement of tropomyosin closer toward the center of the helical groove of the thin filament (Figure 19-18).

Contraction in Smooth Muscle and Invertebrate Muscle Is Regulated Differently by Ca^{2+}

Vertebrate smooth muscle and invertebrate muscle contain tropomyosin but not the troponin complex. Ca^{2+} regulation operates differently in these muscles and involves regulatory interactions with the myosin light chains.

In typical invertebrates, such as mollusks, at low Ca^{2+} concentrations the interaction of myosin headpieces and actin filaments that allows muscle contraction is inhibited by one of the myosin light chain pairs located in the head region. Binding of Ca^{2+} to one of the myosin light chains induces a conformational change in the myosin headpiece that allows it to bind to actin; this, in turn, causes an activation of the myosin ATPase and contraction of the muscle. The so-called "regulatory light chain protein" has a very high affinity for Ca^{2+}—similar to that of TN-C or of the ubiquitous Ca^{2+}-binding protein calmodulin.

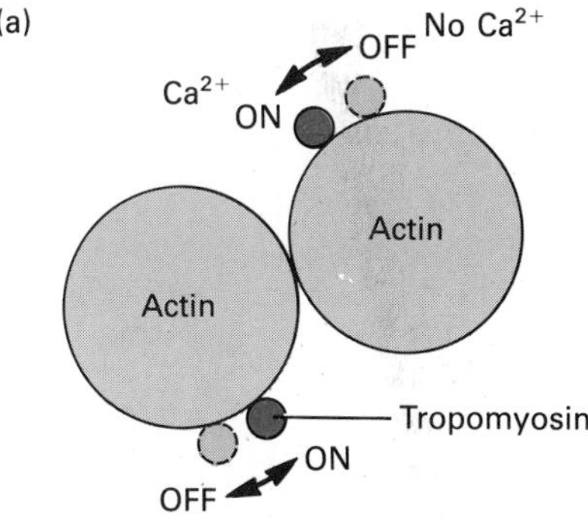

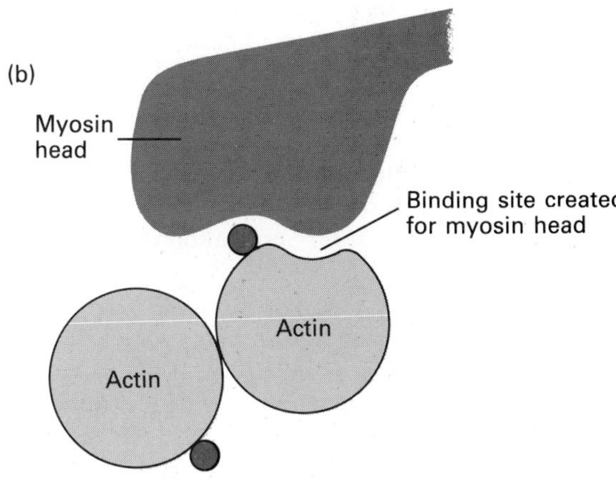

Figure 19-18 Regulation-associated movements of tropomyosin in the grooves of the actin helix. The actin thin filament is viewed down the long axis in a direction toward the Z disk. (a) Binding of Ca^{2+} to troponin causes a change in position of the tropomyosin fiber, moving it closer to the center of the groove (*ON*). (b) This movement creates a binding site on the actin helix for the myosin head.

Regulation in vertebrate smooth muscle is similar but more complex (Figure 19-19). As in mollusks, a myosin light chain attached to the myosin headpiece inhibits the actin-stimulated ATPase activity. In this system, phosphorylation of one of the myosin light chains relieves this inhibition and thus stimulates contraction. The level of Ca^{2+} regulates the extent of phosphorylation of this light chain and thus regulates muscle contraction. Ca^{2+} is required for activation of the myosin light chain kinase, which catalyzes phosphorylation of this regulatory myosin light chain: Ca^{2+} first binds to calmodulin, and the Ca^{2+}-calmodulin complex in turn binds to and activates the kinase. As we saw in Chapter 16, calmodulin is found in virtually all eukaryotic cells and is believed to mediate most intracellular effects of Ca^{2+}; as noted, its structure and primary sequence are very similar to those of troponin C.

cAMP Affects the Contractility of Smooth Muscle

The contractility of smooth muscle is affected not only by nervous stimulation but also by the levels of several hormones. Recall that epinephrine often induces *relaxation* of smooth muscles (Chapter 16). As in other cells, epinephrine causes a rise in the cellular level of cAMP in muscle cells and activates the cAMP-mediated protein kinase. In smooth muscle cells, one of the substrates of this enzyme is myosin light chain kinase. Phosphorylation *inactivates* the myosin light chain kinase by lowering its binding affinity for the Ca^{2+}-calmodulin complex (see Figure 19-19). Thus, phosphorylation of myosin light chain cannot occur, so the muscle remains in its relaxed state. This process serves as another example of how the regulatory effects of two intracellular "second messengers," Ca^{2+} and cAMP, interact in a complex manner to modulate a critical cellular process (Chapter 16).

Proteins Anchor Actin Filaments to Membranes or to the Z Disk

For the sliding filament mechanism to generate contraction of a muscle cell, one end of each actin thin filament must be immobilized. In striated muscle, actin filaments end at the Z disk; in smooth muscle, similar but smaller structures, termed *cytoplasmic plaques,* apparently serve the same function. In many smooth and striated muscle cells, actin filaments at the end of the myofibers are attached to the cytoplasmic membrane. Cardiac muscle contains structures termed *intercalated disks* adjacent to the plasma membrane to which the actin filaments are attached (Figure 19-20).

Two actin-binding proteins may play key roles in the attachment of actin filaments. One protein, *vinculin* (m.w. 130,000), is found in intercalated disks of cardiac muscle and the membrane-associated plaques in smooth muscle. Vinculin binds tightly to actin in solution, but how it interacts in vivo either with actin or with the membrane is unknown. Perhaps it binds directly to the lipid or to an integral membrane protein.

The second protein with a role in the attachment of actin filaments is *α-actinin* (m.w. 190,000), a major protein in the Z disks of striated muscle, cytoplasmic dense bodies in smooth muscle, and intercalated disks of cardiac muscle (Figures 19-21 and 19-22). An α actinin molecule is a dimer of two identical polypeptides. Electron microscopy indicates that, in attachment regions, vinculin is closer to the membrane than is α-actinin; thus the pattern of binding is thought to be membrane–vinculin–α actinin–actin (Figure 19-22). Some evidence suggests that α actinin binds tightly to the ends of actin filaments and possibly binds the ends of many such filaments together. In cardiac and striated muscle, α actinin may bundle together actin filaments into the characteristic hexagonal

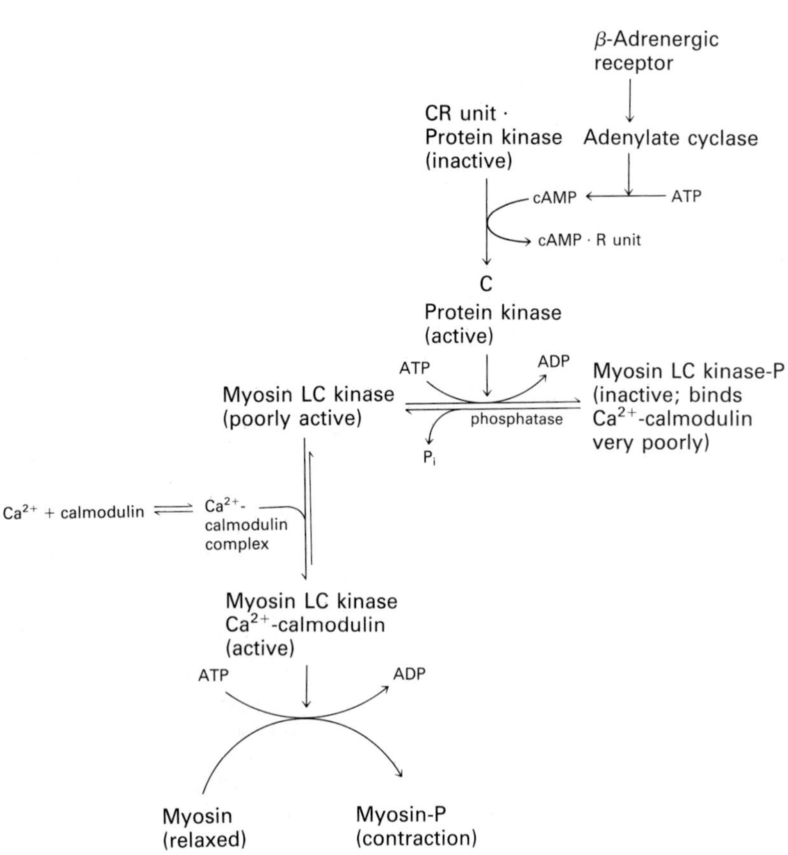

Figure 19-19 Regulation of vertebrate smooth muscle contraction by Ca^{2+} and cAMP.

Ca^{2+}: Calcium, at a concentration of 10^{-5} *M*, binds to calmodulin, and the Ca^{2+}-calmodulin complex binds to the poorly active form of myosin light chain (LC) kinase, converting it to a more active form. The active kinase catalyzes phosphorylation of a myosin light chain, which permits myosin to interact with actin, causing smooth muscle contraction. A decrease in the Ca^{2+} concentration to 10^{-7} *M* leads to a dissociation of calcium and calmodulin from the kinase, thereby inactivating the kinase. Under these conditions myosin phosphatase, which is not dependent on Ca^{2+} for activity, dephosphorylates myosin, causing smooth muscle relaxation.

cAMP: Stimulation of the β-adrenergic receptor by epinephrine results in an active form of cAMP-dependent kinase (C). Phosphorylation of myosin LC kinase by protein kinase weakens the binding of Ca^{2+}-calmodulin to the myosin LC kinase and hence favors formation of the inactive form of the myosin LC kinase. Dephosphorylation of the myosin LC kinase by phosphatase restores the kinase to the form that binds Ca^{2+}-calmodulin strongly. Phosphorylation of myosin LC kinase thus results in dephosphorylation of the myosin LC and results in smooth muscle relaxation. This may explain how epinephrine acts to relax certain smooth muscles. R is the dissociable regulatory subunit of cAMP-dependent protein kinase.

[See R. S. Adelstein and E. Eisenberg, 1980, *Annu. Rev. Biochem.* 49:92.]

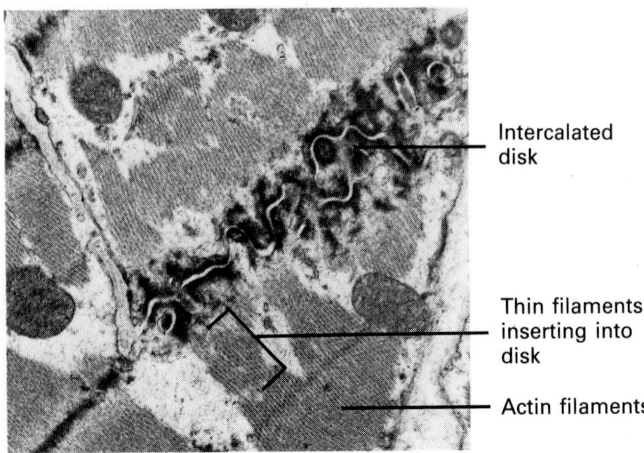

Figure 19-20 Electron micrograph of the ends of two cardiac muscle cells, showing the thin filaments inserting in the intercalated disk at the boundary between two cells. *Photograph courtesy of M. Seiler and A. Perez-Atayde.*

(a)

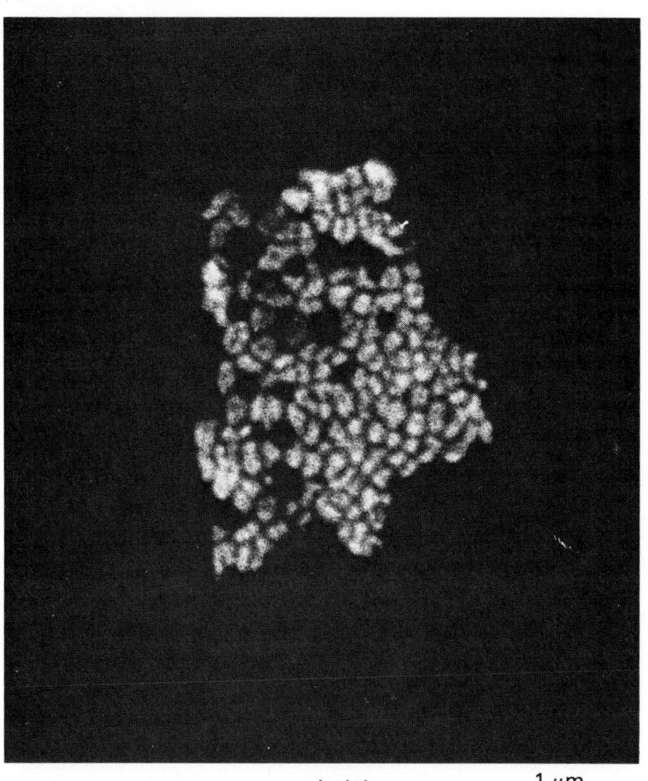

(b)

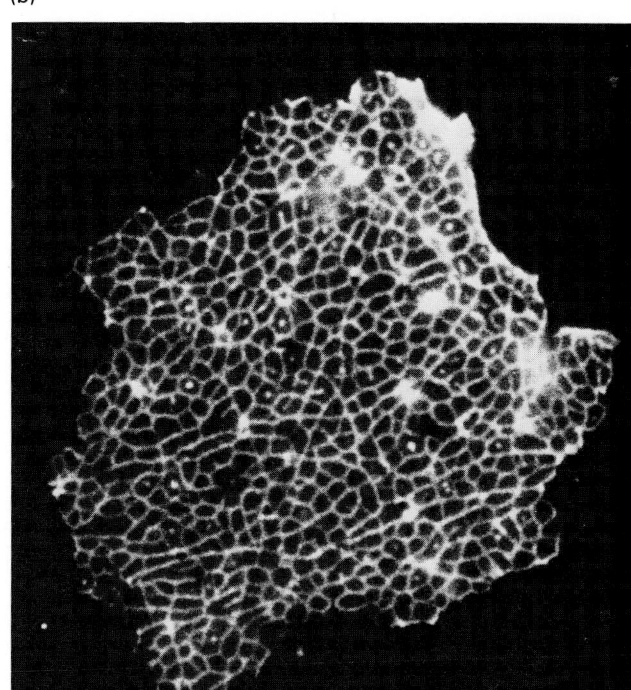

Figure 19-21 Immunofluorescence localization of (a) α actinin and (b) the intermediate filament protein desmin in sheets of Z disks from chicken skeletal muscle. α Actinin is localized in the interior of the Z disks, in which it functions to anchor the ends of actin microfilaments. Desmin is localized at the periphery of each Z disk. Fibers of desmin may interconnect adjacent Z disks to align them into a regular array of myofibrils. Desmin and other proteins may also anchor the Z disks to the plasma membrane. *Photographs courtesy of E. Lazarides and B. L. Granger. Reprinted by permission from* Nature, *vol. 283, p. 251. Copyright 1980 Macmillan Journals Limited.*

lattice (see Figure 19-8). Yet other proteins cross-link thick filaments in the M region (see Figure 19-16) keeping them in register with each other in a sarcomere.

Phosphorylated Compounds in Muscle Act as a Reservoir for ATP Needed for Contraction

Resting muscle, like other tissues, requires a constant generation of ATP for its basic metabolic functions, such as transport of ions against a concentration gradient to maintain a membrane potential and a constant intracellular environment. Muscle, especially striated muscle, is unusual in that during contraction the demand for ATP can increase 20- to 200-fold above that in the resting state. Yet mammalian skeletal muscle contains adenine nucleotides, mostly as ATP—enough for about 0.5 s of intense activity, or about 10 contractions. What mechanisms generate the extra ATP for extended muscle contraction?

The two principal types of striated muscle use different mechanisms. "White" muscle fibers—the white meat in a chicken—are called "fast-twitch" fibers because they contract and fatigue quickly; they generate ATP by anaerobic glycolysis of glucose and glycogen. The "red" fibers—the dark meat—contain abundant mitochondria and myoglobin, the red oxygen-binding pigment (see Figure 3-18). Red fibers are "slow-twitch" fibers: they contract and fatigue more slowly than fast-twitch fibers. They generate ATP by aerobic catabolism of glucose and fats, using the O_2 bound to myoglobin (see Figure 3-36). Fast-twitch muscle, especially, cannot generate sufficient ATP from glycogen and glucose to fuel rapid, intense contractions, in part because glycogen is a relatively large particle and because glycogen phosphorylase can remove only one glucose residue at a time (see Figure 16-15). Also, glycolysis yields only 2 ATPs per glucose molecule or 3 ATPs per glucose residue in glycogen (Chapter 16).

The major phosphorylated compound in vertebrate striated muscle is *creatine phosphate*. The phosphate of

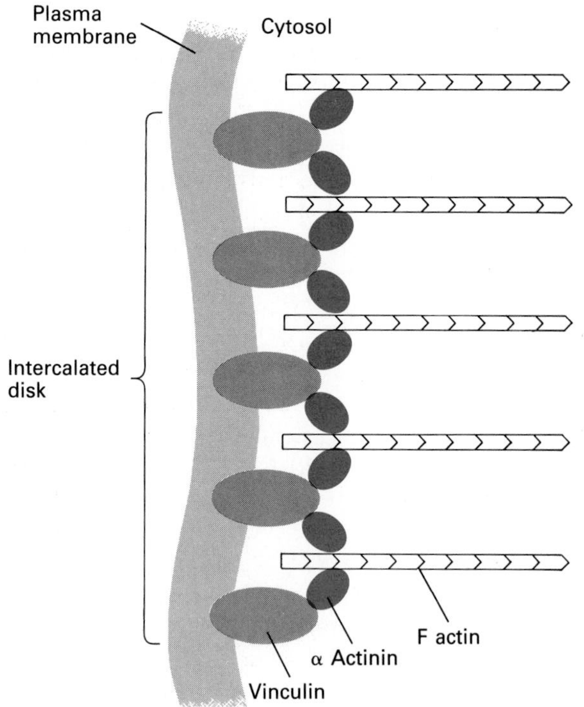

Figure 19-22 Schematic diagram of the binding of actin filaments to the plasma membrane at the intercalated disk. Vinculin is an integral membrane protein that binds actin; α-actinin is also found in the intercalated disk, which may assist in the binding of actin to vinculin and thus to the membrane. α Actinin may also organize the actin thin filaments in a regular geometric array.

this compound can be transferred to ADP in a reaction catalyzed by creatine kinase (Figure 19-23). Creatine phosphate is present in muscle at about five times the concentration of ATP. The free energy of hydrolysis of creatine phosphate ($\Delta G^{\circ\prime} = -10.3$ kcal/mol) is greater than that of the terminal phosphate of ATP ($\Delta G^{\circ\prime} = -7.3$ kcal/mol). Therefore at neutral pH the creatine kinase reaction favors formation of ATP.

Among invertebrates, other phosphorylated compounds are used as storage forms of high-energy phosphate—for example, guanidine phosphate. The phosphate residue of such compounds can be transferred to ADP by a reaction analogous to that for creatine phosphate.

Striated muscle cells also possess an enzyme, *adenylate kinase*, that catalyzes the release of the energy of both phosphoanhydride bonds in ATP:

$$2\mathrm{ADP} \xrightleftharpoons[\text{kinase}]{\text{adenylate}} \mathrm{AMP} + \mathrm{ATP}$$

In this manner, ADP, formed during muscle contraction, can generate ATP for use in additional contractions. Thus fast-twitch muscles have two highly efficient mechanisms for generating, from ADP, the large amounts of ATP to fuel rapid contraction: transfer of a phosphate from creatine phosphate (or similar molecule) to ADP and conversion of two ADPs to one ATP and one AMP. But these mechanisms suffice for only a few minutes: eventually the level of ATP must be regenerated by glycolysis of glucose that is generated from glycogen or imported from the blood.

Muscle Cells Are Formed by Fusion of Uninucleate Precursor Cells

Being multinucleate, muscle cells are unusual in that, contrary to expectation, they do not arise from a single cell by repeated nuclear division without cytokinesis. Instead, muscle cells arise by fusion of mononucleate cells called *myoblasts* (Chapter 6). (An exception is cardiac muscle, which is composed of single cells interconnected by abundant gap junctions that allow simultaneous contraction of all cells.) This fusion occurs synchronously over a period of a few hours and can be observed in cultured myoblast cells (see Figure 6-11). Differentiated muscle cells contain a number of enzymes and other proteins that are lacking in myoblasts or in nonmuscle cells. Creatine kinase, troponin, and adenylate kinase are unique to fused muscle cells, for instance. Multinucleate muscle cells synthesize a form of actin that is not made by prefusion cells. This actin, termed *α* actin, is thought to be specific for the contractile apparatus, but how it differs physiologically from actins made by myoblasts and other nonmuscle cells is not known.

All muscle-specific proteins begin to be made at the time of cell fusion. However, the regulatory mechanisms controlling synthesis of these proteins apparently do not depend on the fusion process: even when cell fusion is

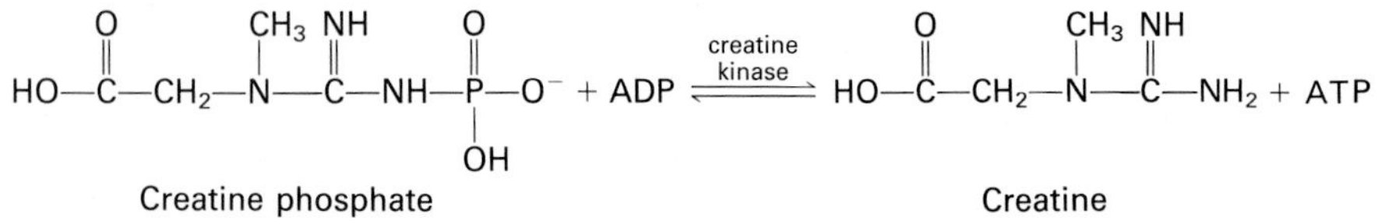

$$\mathrm{HO{-}C({=}O){-}CH_2{-}N(CH_3){-}C({=}NH){-}NH{-}P({=}O)(OH){-}O^-} + \mathrm{ADP} \xrightleftharpoons[\text{kinase}]{\text{creatine}} \mathrm{HO{-}C({=}O){-}CH_2{-}N(CH_3){-}C({=}NH){-}NH_2} + \mathrm{ATP}$$

Creatine phosphate → Creatine

Figure 19-23 Creatine phosphate is the major reservoir of high-energy phosphate in mammalian muscle. In the reaction catalyzed by creatine kinase, the phosphate group is transferred to ADP, thus forming ATP. Because the free energy of hydrolysis for creatine phosphate is greater than that for the terminal phosphate residue of ATP (Chapter 2), the reaction catalyzed by creatine kinase favors formation of ATP ($\Delta G^{\circ\prime} = -3.0$ kcal/mol).

prevented by depleting Ca^{2+} from the medium, the synthesis of creatine kinase, troponin, and α actinin occurs as usual.

Muscle Gene Expression Is Regulated at Transcriptional and Posttranscriptional Levels

Many multigene families, such as actin and myosin light chains, contain members whose expression is specific to muscle. In many cases, regulation is at the transcriptional level. During differentiation of muscle tissue the synthesis of mRNA coding for the cytoskeletal β and γ actins present in all nonmuscle cells (see following section) is reduced, whereas the sarcomeric α actins are synthesized in large amounts. The major actins present in adult heart and skeletal muscle—the cardiac and skeletal α actins—are synthesized by mRNAs made predominantly, if not uniquely, by those tissues.

Isoforms of muscle-specific proteins can confer different regulatory or contractile properties to different types of muscle cells (Figure 12-16). In some cases, these isoforms are generated by alternative splicing of an RNA transcript of a single gene. For instance, two myosin light chains, MLC_1 and MLC_3, found in striated muscle, contain identical C-terminal but different N-terminal sequences and are produced from a single gene by differential splicing. Similarly, two troponin (Tn) T isoforms differ by only 14 internal amino acids. These amino acid segments are encoded by two distinct and adjacent small exons of the Tn-T gene; alternative splicing results in the incorporation of one or the other exon into the mature Tn-T mRNA. Possible functional differences among these isoforms are now being explored. The power of recombinant DNA techniques to elucidate previously unsuspected variation in muscle proteins has opened up an entire new area of research in muscle biology.

Actin and Myosin in Nonmuscle Cells

In contrast to the orderly filamentous arrangement in striated muscle cells, the spatial organization of actin in nonmuscle cells is highly variable and complex, so that studies of its function and structure are much more difficult. Myosin is also present in nonmuscle cells, but at a much lower ratio to actin than in muscle (Table 19-1). In nonmuscle cells these proteins may serve both structural and contractile functions.

All Vertebrates Have Multiple Actin Genes and Actin Proteins

On the basis of amino acid sequence, at least six different actins have been identified in adult birds and mammals. Four are called α actins: one unique to striated skeletal muscle; one, to cardiac muscle; one, to smooth vascular muscles; and one, to smooth enteric muscles such as those that line the intestine. Additionally, two actins, termed β actin and γ actin, are found in the cytoplasm of all muscle and nonmuscle cells. The various muscle-type actins differ in only four to six amino acids and cytoplasmic β or γ actin and striated muscle actin differ in only 25 (out of about 400). Clearly genes for these actins have evolved from a common precursor. Yeast contains only a single actin gene, and most simple eukaryotic microorganisms, such as slime molds and yeasts, synthesize only a single species of actin. However, many multicellular eukaryotic organisms contain several actin genes—11 in the sea urchin and 17 in *Dictyostelium*. Since the actual number of different actins in these organisms has not yet been determined, how many of the genes are actually expressed is not known. Actins are also modified posttranslationally, by acetylation of the N-terminus and methylation of a histidine. This process may also generate multiple functional species well in excess of the number of genes.

In cell-free reactions, all known actins polymerize into the same type of double helix and bind S1 myosin in precisely the same way to form the polarized arrowhead structures. Regardless of source, all purified actins are capable of stimulating the ATPase activity of muscle actin. Recently, the gene for cardiac muscle α actin was transferred into fibroblast cells; subsequently this muscle actin was seen to be incorporated into the same actin filaments as those in which the normal fibroblast β and γ actins are found. What *unique* functions are filled by the various actin proteins is not known.

Myosin is found in small amounts in all nonmuscle cells, but short thick filaments characteristic of striated muscle cells have only very recently been reported to occur in slime molds. Tropomyosin, myosin light chain kinase, and calmodulin, the regulatory proteins in smooth muscle, have been isolated from nonmuscle cells such as amebas and platelets. Troponins, the Ca^{2+}-binding regulatory proteins in striated muscle, thus far have not been found in any other cell type. Thus, myosin and actins in nonmuscle cells have the potential to function in a contractile process. As we shall see later, such a role for these proteins has been suggested in the *contractile ring* of telophase cells, which tightens during the separation of the daughter cells.

To study the function of the single actin gene product in yeast, a temperature-sensitive mutation was created in the cloned actin gene, by in vitro mutagenesis, and used to replace the wild-type actin gene (for the experimental protocol used, see Figure 18-11). At permissive temperatures, the mutant actin functioned normally, and the cells were able to grow and divide. At nonpermissive temperatures, by contrast, the actin filaments were disrupted, secretion of proteins and synthesis of cell membrane proteins were stopped, and, importantly, the cells were

(a)

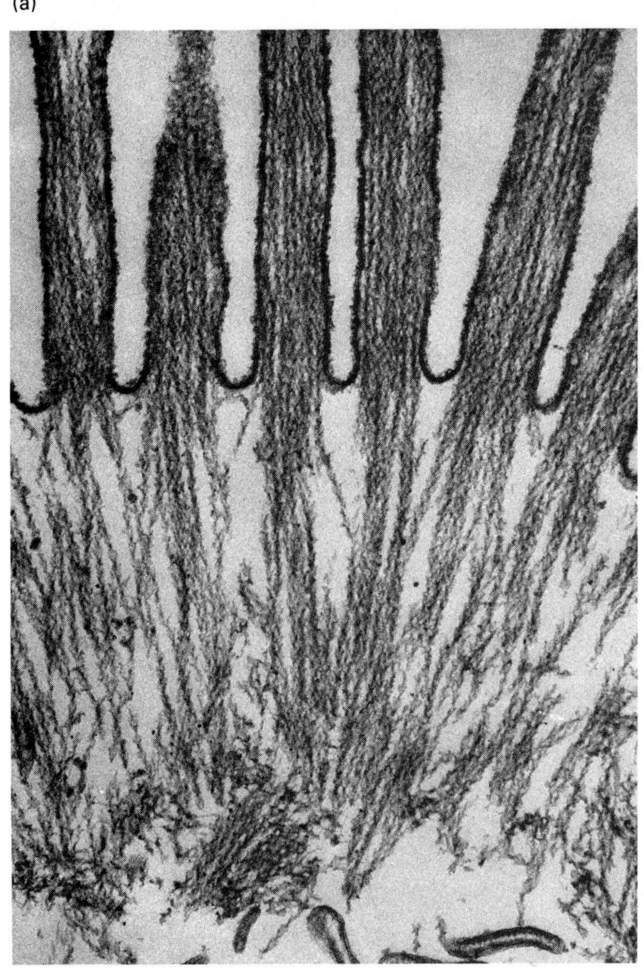

(b)

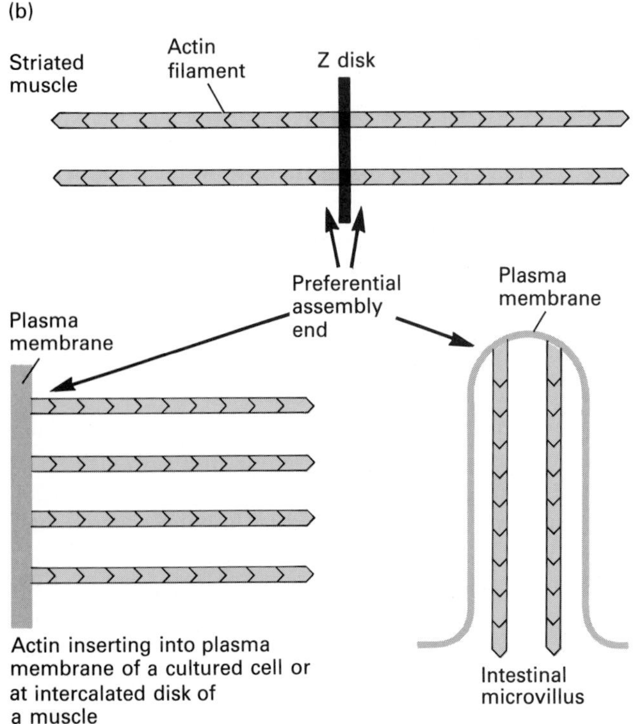

Figure 19-24 Polarized attachment of actin filaments to membranes or other anchoring structures. (a) Electron micrograph of isolated intestinal microvilli following incubation with the S1 myosin fragment, showing the arrowheads pointing away from the tip. (b) In all cases where actin filaments insert into a membrane or cytoskeletal structure, they do so with the same polarity—with the S1 myosin arrowheads pointing outward. The preferential end for polymerization of actin monomers (the A or assembly end) is the same as the "barbed" end visualized with S1 myosin staining and is always at the attachment site. In striated or cardiac muscle the A end of actin filaments inserts into the Z disk; actin filaments on either side thus have the opposite polarity. *Part (a) courtesy of D. Begg.*

unable to divide. These findings confirm that actin is required for many aspects of cellular growth and division in yeast and presumably in other organisms as well.

Many Actin-Binding Proteins Are Found in Both Muscle and Nonmuscle Cells

Vinculin and α actinin, two of the attachment proteins that interact with the ends of actin filaments in muscle cells (see Figure 19-22), are also found in nonmuscle cells. Vinculin and α actinin are found in the belt desmosomes of epithelial cells (see Figures 14-49 and 14-57) and in specialized regions of the cultured fibroblast plasma membrane called *focal adhesion plaques*. At these sites, cells adhere to the external substrate, and actin filaments attach to the cell membrane.

In nonmuscle actin filaments to which S1 myosin has been bound, the arrowheads always point *away* from the membrane or other attachment site. The opposite ends of the filaments, called the "barbed" ends, point *toward* the membrane to which they may be attached (Figure 19-24). As a consequence, any tension generated on these filaments by interaction with myosin exerts a pull on the plasma membrane, exactly as in the case of the Z disks or the intercalated disks in smooth muscle cells.

Organized Bundles of Actin Filaments Maintain Microvilli Structure

Microvilli from intestinal epithelial cells represent one of the few nonmuscle systems in which actin microfilaments occur in a well-ordered pattern (Figures 19-24 and 19-25; see also Figure 14-48). Shearing of intestinal cells produces an abundance of microvilli, which are then easily prepared in pure form for use in research work. Consequently a great deal is known about the structure of these organelles. In the center of each microvillus is a bundle of specialized actin microfilaments lacking myosin, tropomyosin, or α actinin. These unusual microfilaments pre-

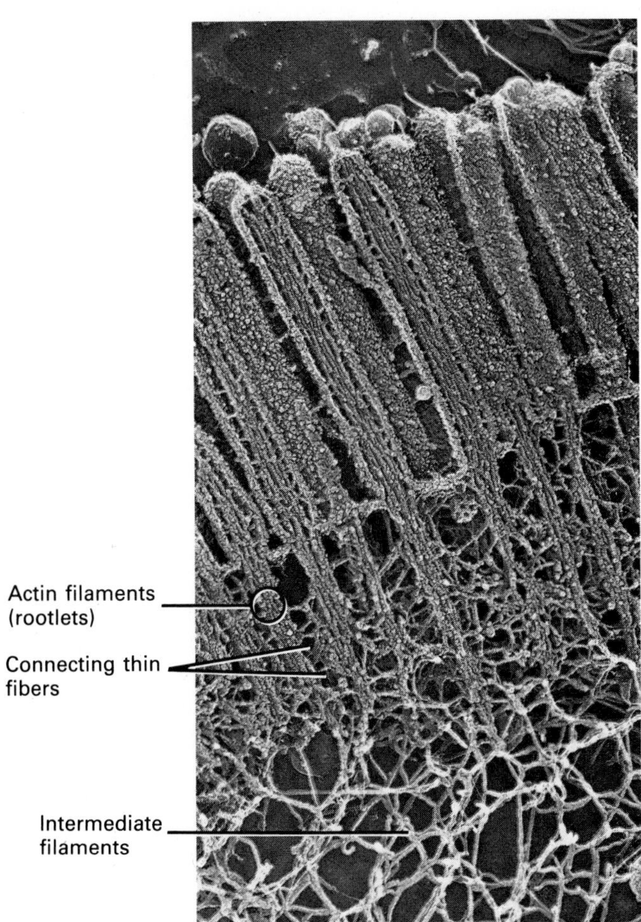

Figure 19-25 The terminal web of a preparation of fixed and quick-frozen intestinal microvilli. Tight bundles of actin filaments extend out of the microvilli to form straight "rootlets." Connecting the rootlets are a number of thin fibers that form complicated networks. Staining with specific antibodies shows that these are composed mainly of myosin and the spectrin-like protein fodrin (see Figure 19-27). These rest upon a tangled network of thicker intermediate filaments located on the bottom of the field shown here. *Photograph courtesy of N. Hirokawa. Reproduced from the* Journal of Cell Biology, *1982, vol. 94, p. 425 by copyright permission of the Rockefeller University Press.*

sumably play a structural role, maintaining the shape of the microvilli. As with all microfilaments, when these are stained with S1 myosin their arrowheads point away from the site of membrane insertion at the tip of the microvillus (see Figure 19-24).

In addition to actin, these microfilaments contain several major proteins that are important in generating their rigid structure. The protein *fimbrin* (m.w. 68,000) binds actin filaments in a ratio of 1 fimbrin molecule per 2 to 3 actin monomers, thus forming tightly packed actin bundles. Another protein, *villin* (m.w. 95,000), also cross-links the actin filaments to form bundles, but only at low Ca^{2+} concentrations. Curiously, at high (micromolar) Ca^{2+} concentrations villin severs actin filaments into short fragments; Ca^{2+} therefore may influence the length of the microvillar cores. Yet a third actin-binding protein (unnamed, m.w. 110,000) is believed to form, together with calmodulin, the cross-bridges connecting the sides of actin bundles with the inner surface of the plasma membrane (Figures 19-26 and 19-27). This 110,000-m.w. protein is a myosin-like actin-stimulated ATPase; thus fibrous cross-bridges may act to generate tension that either keeps the actin filaments positioned in the center of the microvillus or causes the microvillus to bend.

The core bundles of actin filaments end in the apical "terminal web" regions of these cells. Such regions contain many narrow myosin filaments that interconnect adjacent actin bundles with one another and with the underlying base of intermediate filaments (see Figure 14-57). Another fibrous actin-binding protein, fodrin, has also been isolated from intestinal brush borders. Fodrin,

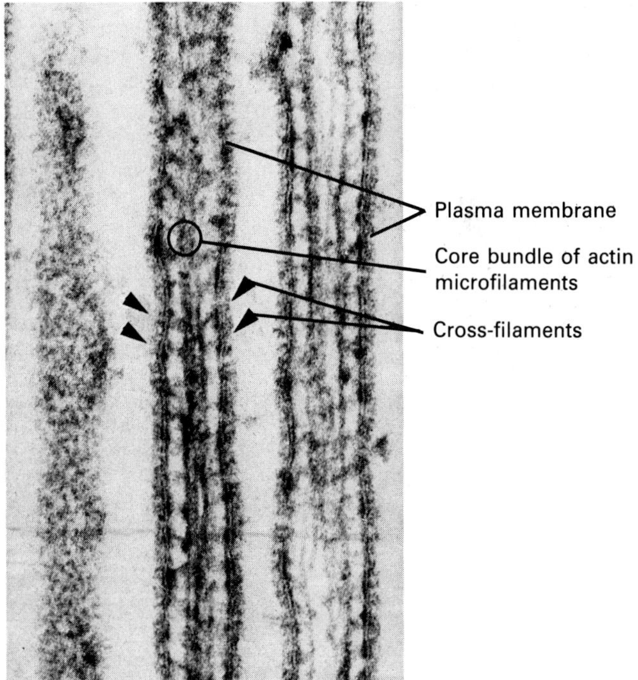

Figure 19-26 Thin section through isolated intestinal microvilli. The core bundle of actin microfilaments is connected laterally to the membrane by cross-filaments (*arrowheads*). These are composed of calmodulin and an unnamed 110,000-m.w. protein. [See P. T. Matsudaira and D. R. Burgess, 1982, *J. Cell Biol.* **92**:657.] *Photograph courtesy of P. T. Matsudaira and D. R. Burgess. Reproduced from the* Journal of Cell Biology, *1982, by copyright permission of the Rockefeller University Press.*

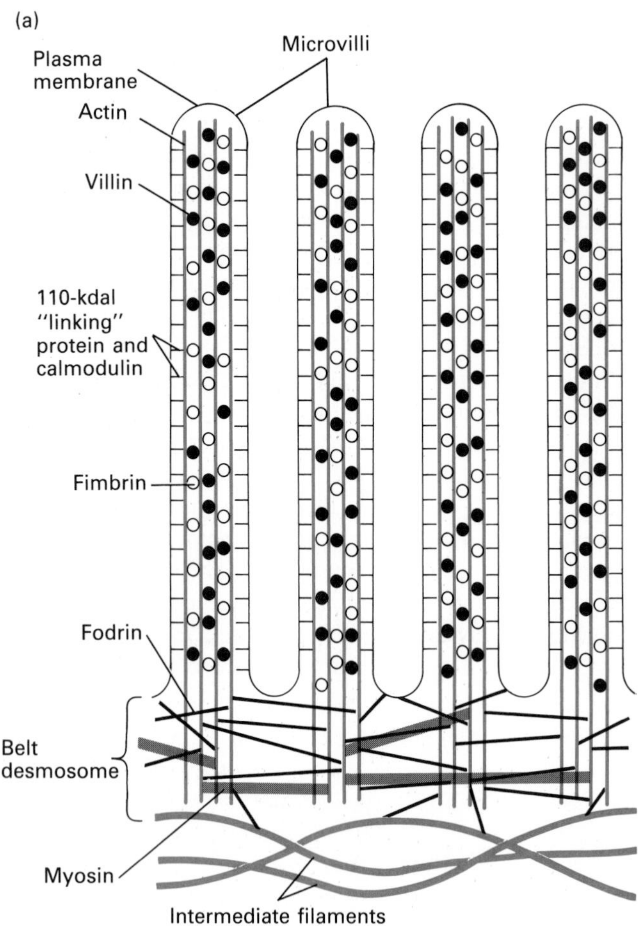

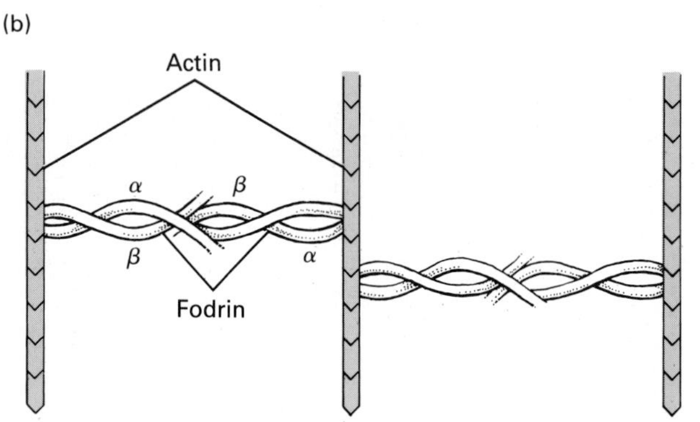

Figure 19-27 Actin filament linkage by several types of protein. (a) Architecture of the filaments in mouse intestinal brush border. Actin microfilaments are bundled together by fimbrin and villin. The 20- to 30-nm-long cross-bridges connecting actin microfilaments to the membrane are composed of the unnamed 110,000-m.w. (110-kdal) protein and calmodulin. Myosin and fodrin bind together actin filaments in adjacent bundles. Fodrin, like spectrin, has the composition $\alpha_2\beta_2$ and is 5 nm wide and 200 nm long (see part b). The α fodrin chain appears identical to erythrocyte spectrin, but the β subunits are different. Fodrin binds to the sides of actin microfilaments and cross-links them together, as does spectrin. Fodrin fibers also connect actin fibers to the cell membrane, especially in the region of the belt desmosome, and also to the layer of underlying IFs. The belt desmosome also contains vinculin and α actinin (*not shown*); many actin filaments insert into the membrane at the belt desmosome, probably anchored there by α-actinin and vinculin. (b) Hypothetical structure of fodrin-actin linkage, showing a configuration very similar to that for spectrin-actin linkage. [See N. Hirokawa, R. Cheney, and M. Willard, 1983, *Cell* **32**:953; and M. Moosekar, 1983, *Cell* **35**:11.]

which resembles spectrin in its structure and function, links actin bundles to adjacent actin bundles (Figure 19-27) and also to the plasma membrane in the region of the belt desmosome.

Actin Filaments Are Abundant in Other Nonmuscle Cells

Thus far we have considered examples of actin microfilaments that are more or less permanent features of a cell and that are found in regular geometric arrays. In many nonmuscle cells, such as cultured fibroblasts or the endothelial cells that line arteries, much of the actin occurs in long bundles of microfilaments termed *stress fibers,* which often extend the length of the cell. These fibers can be visualized in the light microscope following staining of fixed cells with fluorescent antibodies to actin (Figure 19-28). However, only about half of the actin in these cells appears to be in microfilaments; the rest is soluble G actin. Thus there may be interconversion of microfilaments with G actin monomers. But how can we be sure that actin filaments seen in nonmuscle cells are not merely artifacts created during fixation or staining?

A recently developed technique allows the detection of certain proteins inside *living* cells. First, the protein of interest is obtained in pure form and chemically linked to a highly fluorescent compound. Next, the fluorescent derivative is microinjected into cells through tiny micropipettes of the sort used to impale nerve cells. Provided the fluorescent compound has not altered functional properties of the protein, the fluorescent protein then assembles within cells into substructures identical to those formed by the unlabeled cellular equivalent. The fluorescent image shows those structures containing the protein. In a recent study using purified fluorescein-tagged actin injected into living fibroblasts, fluorescent actin was seen to become incorporated into stress fibers identical to those observed in fixed cells (Figure 19-29). This finding is conclusive evidence of the existence of stress fibers in living fibroblasts and other types of cells. It is important to real-

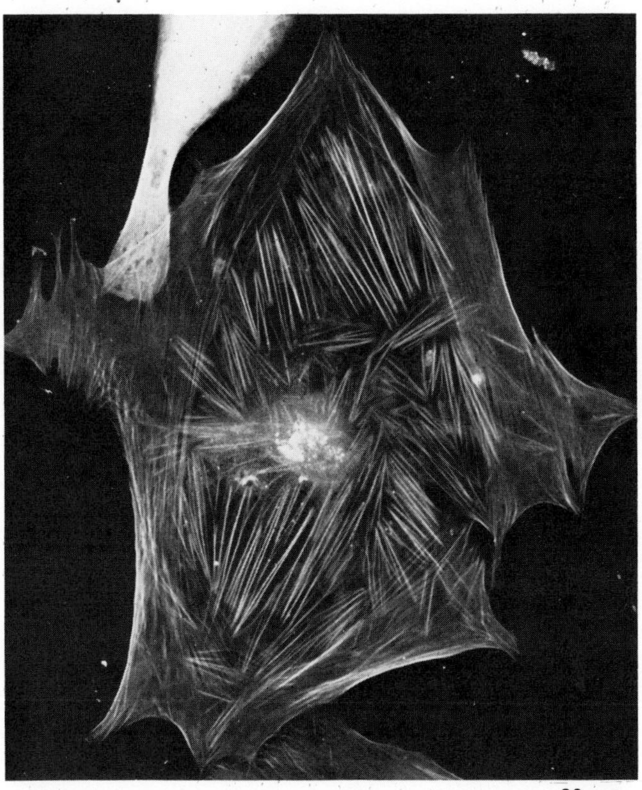

Figure 19-28 Distribution of actin in cultured fibroblast cells. This fluorescence light micrograph of a fixed human skin fibroblast stained with a fluorescent anti-actin antibody illustrates the long stress fibers. The distribution of tropomyosin is similar to that of the actin filament bundles. In cells double-stained for actin and tropomyosin, the fluorescence patterns overlap, indicating that the two molecules are found in the same fiber bundles. *Photograph courtesy of E. Lazarides.*

ize that each fluorescent fibrous structure visible with this technique is actually a bundle of parallel actin helixes—not the single 7-nm-diameter helix resolved in the electron microscope.

Stress Fibers Permit Cells to Attach to Surfaces

Stress fibers are believed to be involved in attachment to a substratum and also to generate the stress or tension that determines the flattened shape of fibroblasts. Electron microscopy has shown that these fibers contain parallel bundles of actin microfilaments—apparently with different polarities. Stress fibers are generally located in the region of cytoplasm just inside the plasma membrane and many appear to insert an end into the membrane at points of cell-substratum contact. In studies using fluorescent antibody techniques, stress fibers have been shown to contain myosin, α actinin, and tropomyosin intermittently along their length and also the regulatory protein myosin light chain kinase.

The actin in stress fibers appears to be continuous so a basic structure of alternating bands of actin and myosin, as in muscle, is not likely. However, the details of the arrangement of stress fiber proteins are not yet known. A large, rod-shaped protein called filamin (m.w. 250,000), similar in structure and function to fodrin and spectrin (see Figure 19-27), apparently cross-links the actin microfilaments into bundles.

The composition of stress fibers resembles that of smooth muscle myofibrils. Indeed, several studies suggest that stress fibers are potentially contractile. For instance, addition of ATP to detergent-permeabilized cells causes the stress fibers to shorten. However, most workers believe that stress fibers function in adhesion of cells to the substratum or, in the case of endothelial cells lining arteries, in providing the cell layer with strength and resiliency, rather than in movement along a substratum.

Rapidly moving cells, such as macrophages and amebas, lack stress fibers. Fibroblasts that are migrating out of tissue explants have few stress fibers during the time of rapid movement. However, when these cells settle on the substratum and migration ceases, the stress fibers increase in number. This interpretation of stress fiber function is supported by the observation that when fibroblasts are removed from the substratum, the cell "rounds up," and the stress fibers disappear. Subsequently some of the fibers in the randomly organized meshwork of microfila-

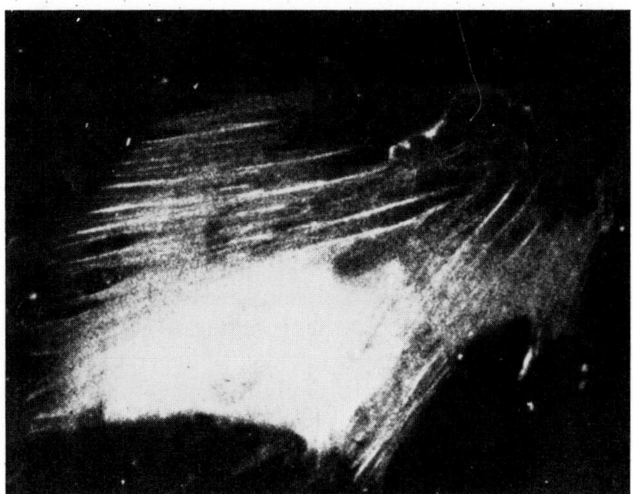

Figure 19-29 Micrograph of a living fibroblast into which actin covalently linked to a fluorescent dye has been injected. The tagged actin assembles into long stress fibers of normal morphology (compare with Figure 19-28) by copolymerizing with cellular actin. [See T. E. Kreis and W. Birchmeier, 1980, *Cell* **22**:555.] *Photograph courtesy of T. E. Kreis. Copyright, 1980, M.I.T.*

(a)

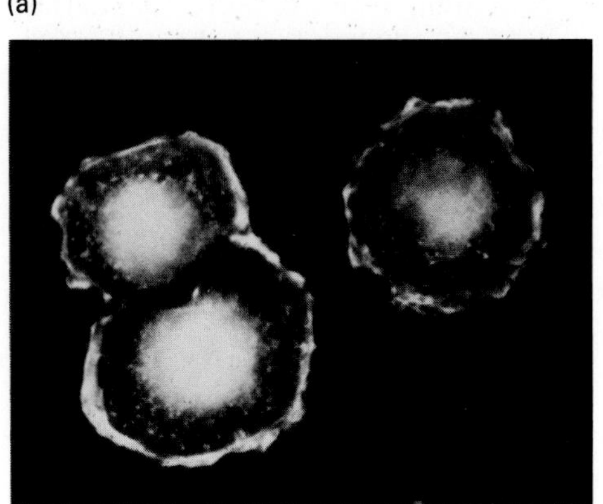

(b)

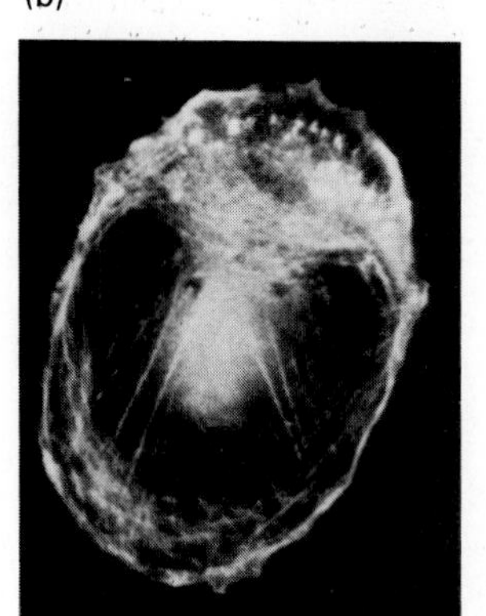

(c)

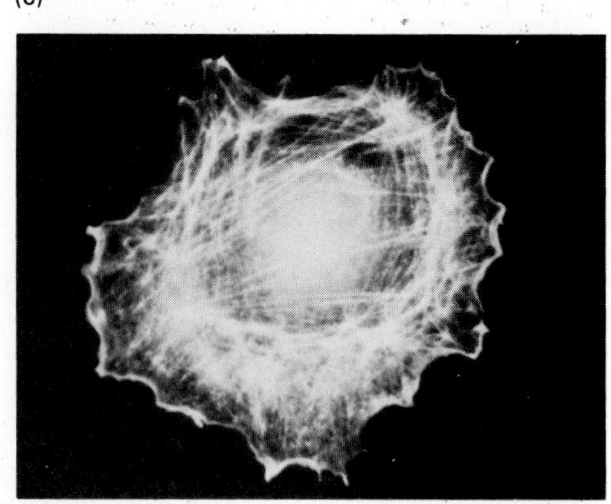

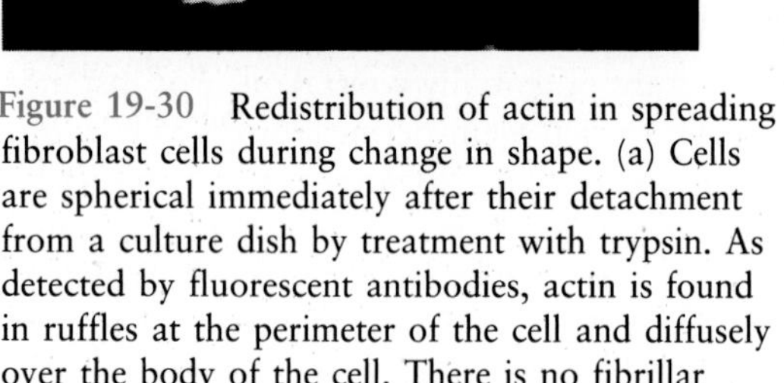

Figure 19-30 Redistribution of actin in spreading fibroblast cells during change in shape. (a) Cells are spherical immediately after their detachment from a culture dish by treatment with trypsin. As detected by fluorescent antibodies, actin is found in ruffles at the perimeter of the cell and diffusely over the body of the cell. There is no fibrillar actin at this stage. (b) After the cell has been allowed to spread on a culture dish for 3 h, much of the actin is still in circumferential ruffles, but some is found in thin microfilaments. (c) After 5 h of spreading, there are polygonal arrays of actin at the cell periphery, as well as stress fibers crossing the cell. [See R. Hynes and A. T. Destree, 1978, *Cell* **15**:875.] *Photographs courtesy of R. Hynes and A. T. Destree. Copyright, 1978, M.I.T.*

ments seen just under the cell surface appear to insert into the plasma membrane. Within a few hours after the cell settles back on its substrate, the stress fibers reappear (Figure 19-30). Apparently, adhesion of cells to a substratum somehow induces conversion of the random assortment of actin microfilaments into microfilament bundles.

Actin Microfilaments Are Found in the Cortex of Many Nonmuscle Cells

In many cells an abundance of actin microfilaments is found just under the plasma membrane in the region termed the *cortex*. The filaments often are parallel to the plasma membranes and may bind to it at several points along its length. Presumably, these fibers confer strength and rigidity to regions of the plasma membrane.

Although spectrin was for many years thought to be unique to erythrocytes, spectrin-like proteins such as fodrin (Figure 19-27) and filamin have been isolated from many cells. As in erythrocytes (Chapter 14), spectrin-like molecules are generally located just under the plasma membrane; they are particularly evident in nerve axons sectioned transversely. In all likelihood, these spectrin-like molecules participate in the binding of actin fibers to the cell membrane, but the nature of the binding is not known. Ankyrin (see Figures 14-41 and 14-42) is a spectrin-binding protein that has been identified in many nonerythroid cells. Possibly, as in erythrocytes, ankyrin links spectrin to an integral plasma membrane protein. The submembrane actin-spectrin-ankyrin cytoskeleton may also play a key role in immobilizing certain integral proteins into localized patches in the membrane.

Polymerization of Actin Monomers Is Controlled by Specific Actin-Binding Proteins

Changes in the extent of actin polymerization appear to be an important and regulated aspect of the function of certain cells. F actin (polymerized) exists in equilibrium with the globular G form, but the equilibrium constant for the polymerization-depolymerization reaction can be affected by proteins that bind selectively to G or F actin. At the concentration of ions and ATP found in cells, the formation of F actin is strongly favored. However, in platelets and sperm from several invertebrates, much of the actin is found as unpolymerized G actin even though conditions favor polymerization. In these cells a low-molecular-weight protein, *profilin,* is bound to G actin and prevents its spontaneous polymerization (Figure 19-31). Many cells contain short filaments of actin that can act as nucleating sites for polarized addition of G actin, but profilin does not prevent such chain growth. Rather, profilin inhibits the nucleation step that begins actin polymerization. Obviously, the polymeric state of cellular actin is regulated by a wide variety of factors. Among these are the hydrolysis to ADP of ATP bound to actin monomers at the ends of the filaments—recall that this tends to cause the filaments to depolymerize—and a high concentration of G actin monomer—which generally favors the growth of actin filaments. How the many factors are controlled and how the whole system is integrated are topics under active study.

A dramatic example of control of conversion of G to F actin is seen in the sperm of many species during fertilization of the egg. In sea cucumbers (see Figure 19-32), as in many species, the egg surface is covered with a vitelline layer as well as with a polysaccharide-rich jelly coat. Binding of the sperm to the jelly coat triggers the *acrosome reaction*—the extension of the long sperm acrosomal process that penetrates both the jelly coat and the vitelline layer (Figure 19-32). Fusion of the sperm and egg plasma membranes is initiated at the very tip of the acrosomal process.

The acrosome in sperm is a vesicle lying just under the tip of the sperm plasma membrane. It is separated from the nucleus by the region of cytoplasm termed the *periacrosomal region* (Figure 19-32b), which contains G actin. The acrosome reaction is triggered when receptors on the sperm membrane encounter the jelly of the egg. First the membrane surrounding the acrosome fuses with the sperm membrane, exocytosing the contents of the acrosome. Within seconds, long (90-μm) processes containing actin filaments form from the G actin and extend outward from the periacrosomal region at the tip of the sperm toward the egg plasma membrane. In the unstimulated sperm, G actin is bound to a profilin-like protein that prevents its polymerization; the complex thus formed is called *profilactin.* Contact of the sperm with the egg jelly causes an increase in the intracellular pH, which triggers the release of profilin from the G actin. The actin explosively polymerizes into filaments, all with the same polarity, providing the force for the erection of the acrosomal process. Penetration of the jelly coat and vitelline layer by the acrosomal process is aided by digestive enzymes exocytosed from the acrosomal vesicle. Thus polymerization of actin microfilaments is essential for bringing the sperm and egg plasma membrane together so that membrane fusion and fertilization can occur.

Actin polymerization also occurs during activation of blood platelets. *Platelets* are ovoid fragments of cytoplasm, 2 to 5 μm in diameter, derived from a very large cell called a megakaryocyte found in the bone marrow. They contain neither a nucleus nor internal membranes. When a blood vessel is cut, platelets from the flowing blood bind to the vessel wall at the site of the rupture. These platelets then become "activated" and change shape, producing many long filopodia that extend outward, and also develop the capacity to agglutinate, or to adhere to each other. These remarkable shape changes allow them to form a plug over the cut in the vessel. Unactivated platelets contain no microfilaments; actin appears to occur mainly as monomers associated with profilin. After activation, profilin dissociates from G actin, and microfilaments arranged as nets or bundles form throughout the platelet. The formation of this microfilament network causes the shape change of the platelets.

Several other types of protein regulate the polymerization of actin filaments. So-called *capping proteins* bind to one end of actin filaments and prevent polymerization or depolymerization at that end. *Severing,* or *fragmenting, proteins* such as villin (page 835) break up long actin filaments and prevent reassembly at high Ca^{2+} concentrations; they also appear to "cap," or bind to, one end of the fragment. At low Ca^{2+} concentrations proteins such as villin act as nucleating centers that accelerate the polymerization of G to F actin.

It is indeed remarkable that the same actin microfilament can be formed into so many different subcellular structures—such as stress fibers, muscle thin filaments, cores of intestinal microvilli, and fibers in the acrosomal process. The type of structure formed is controlled, to a large extent, by a diverse set of actin-binding proteins, summarized in Table 19-2. Myosin and possibly other actin-binding proteins are contractile and allow movements of muscle cells and, as we shall see, of a variety of nonmuscle cells as well.

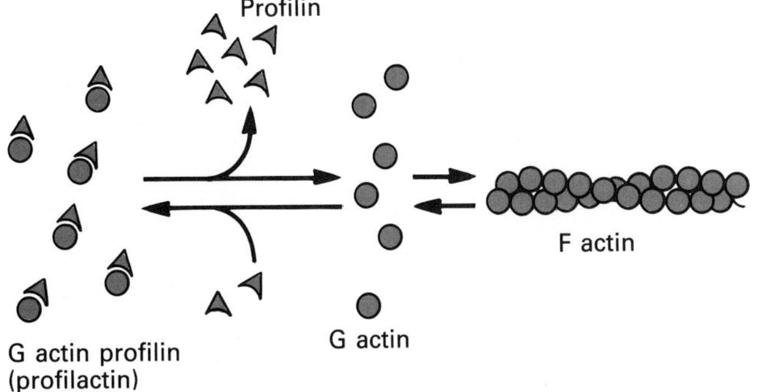

Figure 19-31 Interactions between profilin, G actin, and F actin. Profilin binds to G actin and prevents its polymerization.

Actin and Myosin Are Important for Cell Locomotion

Many free-living cells, such as amebas, are in constant motion, searching for prey. Whereas in adult animals most cells remain in a fixed place, certain cells, such as the tissue macrophages and their blood-borne precursors, the monocytes, are highly motile. They move rapidly toward points of inflammation, where their phagocytic properties constitute an initial line of defense against infection. During differentiation of tissue and organ systems of animals, many cells migrate, individually or as groups, into other areas of the embryo.

Fibroblast Motility Cell motility has been studied extensively with vertebrate fibroblasts in tissue culture. As we saw in Chapter 5, fibroblasts move jerkily along a

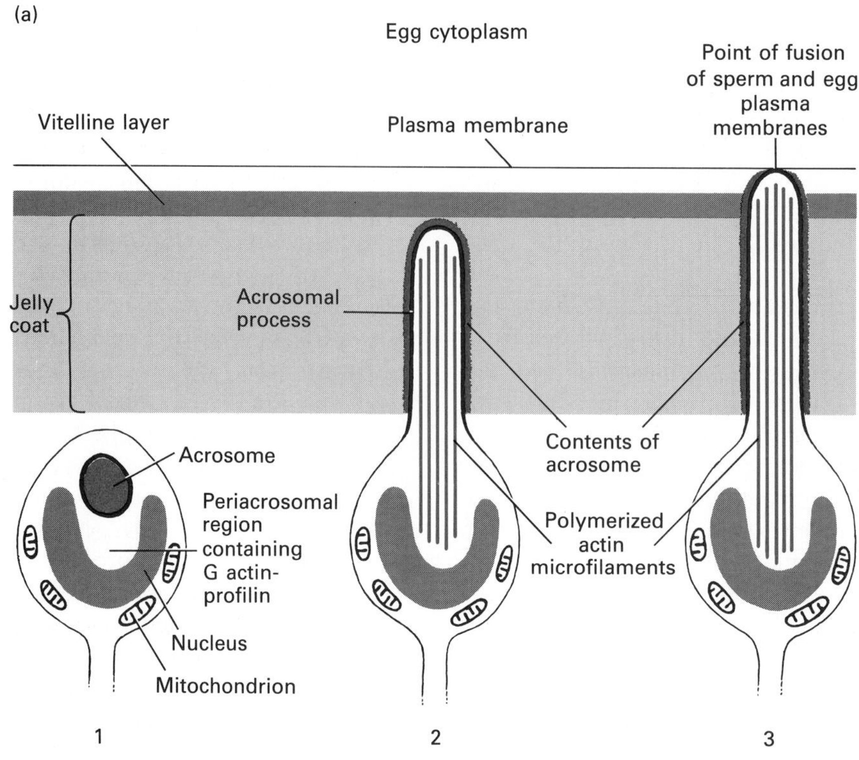
(a)
Egg cytoplasm
Point of fusion of sperm and egg plasma membranes
Vitelline layer
Plasma membrane
Jelly coat
Acrosomal process
Contents of acrosome
Acrosome
Periacrosomal region containing G actin-profilin
Polymerized actin microfilaments
Nucleus
Mitochondrion
1
2
3

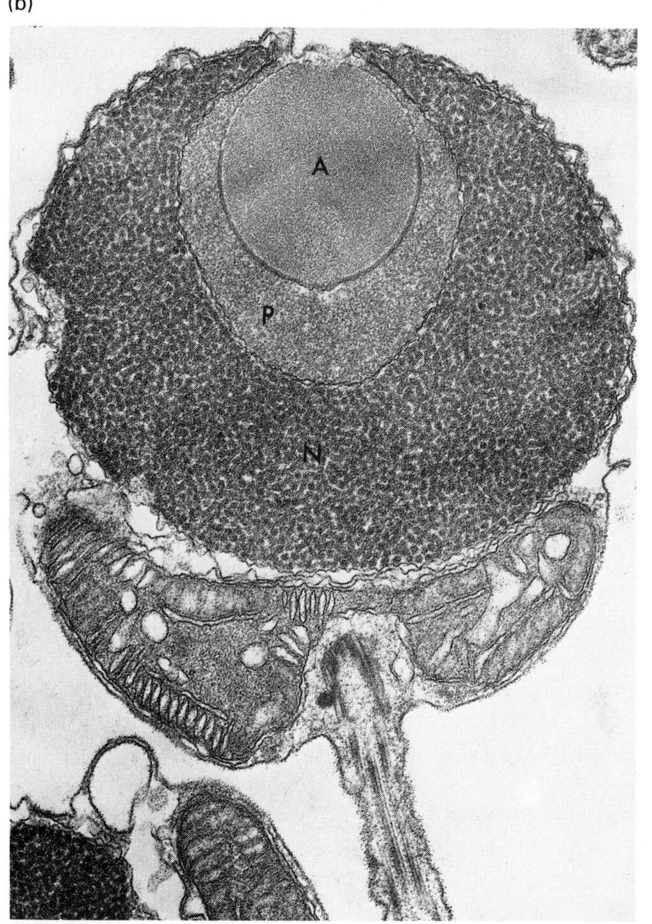
(b)
A
P
N

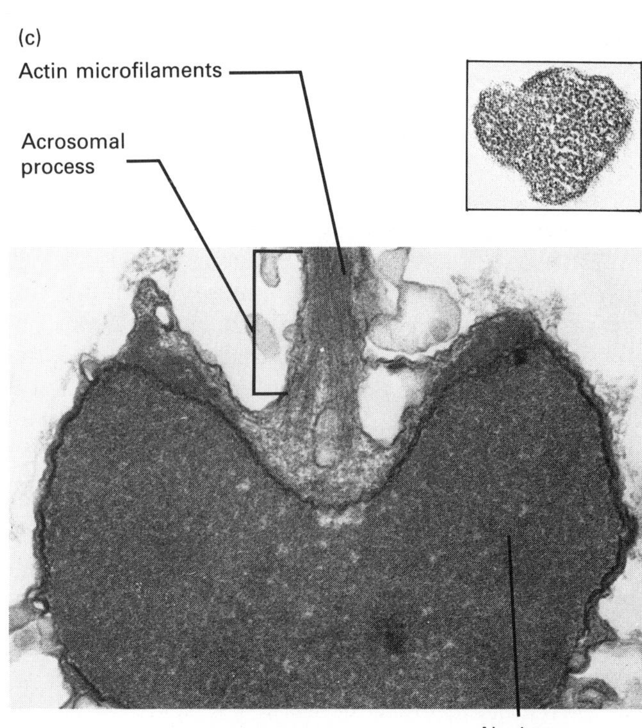
(c)
Actin microfilaments
Acrosomal process
Nucleus

Figure 19-32 *(opposite)* The acrosomal reaction in sea urchin and sea cucumber sperm. (a) Sea urchin sperm bind to the egg jelly coat by means of specific receptor proteins on their heads, as shown in 1. Binding (2) triggers exocytosis of the acrosomal vesicle that lies under the sperm plasma membrane. This releases enzymes from the acrosomal vesicle that digest both the jelly coat and also the vitelline layer covering the egg plasma membrane. Binding also triggers dissociation of profilin from G actin in the periacrosomal region of the sperm cytoplasm and results in explosive polymerization of actin microfilaments. This causes erection of the acrosomal process, a protrusion of the sperm whose plasma membrane (gray) is derived from the membrane of the acrosomal vesicle. When the acrosomal process reaches the egg plasma membrane, the sperm and egg membranes fuse (3). Fusion is triggered by surface proteins on the acrosomal process that in unactivated sperm are part of the membrane of the acrosomal vesicle. The sperm nucleus then enters the egg. (b) Thin section of an unactivated sea urchin spermatozoon. Lying within an indentation of the nucleus (*N*) is the spherical membrane-lined acrosomal vesicle (*A*). Beneath and lateral to it is the area of cytoplasm called the periacrosomal region (*P*) that contains unpolymerized profilin-actin (profilactin). Note that there are no filaments visible in this region. (c) Electron micrograph of a sea cucumber spermatozoon after the activation that occurs upon encountering an egg. Actin in the periacrosomal region (just above the nucleus) has suddenly polymerized into long fibers. *Boxed inset:* Cross section of the long, actin-filled protrusion from the sperm, showing the many actin fibers cut in transverse section. *Parts (b) and (c) courtesy of L. Tilney.*

substratum by making flattened extensions of the cell called lamellipodia. Lamellipodia can be broad but are very thin—only 0.1 to 0.4 μm thick—and contain no organelles other than a meshwork of actin-containing microfilaments (see Figure 5-51). These lamellipodia adhere to the substratum at discrete sites on their undersurface termed *adhesion plaques* (Figure 19-33). As the cells move they extend lamellipodia and microspikes in the forward direction; some adhere to the substratum, while others move backward over the top of the cell in a process called "ruffling" (Figure 19-33). Precisely how the lamellipodia are extended is not known, but the process may involve either the controlled polymerization of actin at the leading edges of the cell or rearrangements of the cross-linked F actin network in the lamellipodia that "push" the leading edge of the plasma membrane outward.

Adhesion of lamellipodia to the substratum may occur at the points of contact between actin filaments and the membrane. Actin-binding proteins such as vinculin and α actinin are found in these adhesive lamellipodia junctions (see Figure 19-22). As the fibroblast moves forward, the trailing edge remains attached to the substratum. As the cell moves forward, the tail—called the *retraction fiber*—becomes greatly elongated under the resulting tension. Eventually, the tail ruptures, leaving a bit of itself attached to the substratum, while the major part of the tail retracts into the cell body (see Figure 19-34). Ultrastructural studies indicate that the tail portion of migrating fibroblasts contains a bundle of actin-containing microfilaments with its axis parallel to the long axis of the tail. The retraction of the trailing end of the fibroblast is associated with the loss of actin microfilament bundles and their subsequent replacement by a meshwork of microfilaments. Retraction is dependent on a supply of ATP and appears to be an active contraction process involving the actin microfilament meshwork.

Ameboid Movement In unicellular eukaryotes such as amebas, and in tissue macrophages, movement occurs by a mechanism different from that for fibroblasts. Amebas can measure up to 0.5 mm across—many times larger than vertebrate fibroblasts (Figure 19-35)—and thus are

Table 19-2 Actin-binding proteins found in the cytoplasm of vertebrate cells

Protein	Function
Vinculin	Mediates binding of ends of actin filaments to cell membranes
110,000-m.w. protein (microvilli)	Links the sides of actin filaments to microvillar membranes
Filamin, fodrin	Spectrin-like proteins that cross-link adjacent actin filaments
Myosin	May cause movements of actin filaments as in muscle, may generate tension in arrays of microfilaments
Tropomysin	In striated muscle, binds in a groove along the length of an actin helix; regulates binding of actin to myosin heads
α Actinin	Participates in binding of actin microfilaments to membranes; may cross-link actin filaments into a regular array
Fimbrin	Cross-links adjacent filaments tightly to form parallel actin fibers
Villin, gelsolin	Low [Ca^{2+}]: nucleate polymerization of F actin filaments Micromolar [Ca^{2+}]: cleave actin filaments into fragments
Capping proteins	Bind to one end of a filament, preventing either addition or loss of actin monomers
Profilin	Binds to G actin monomers, blocking polymerization

(a)

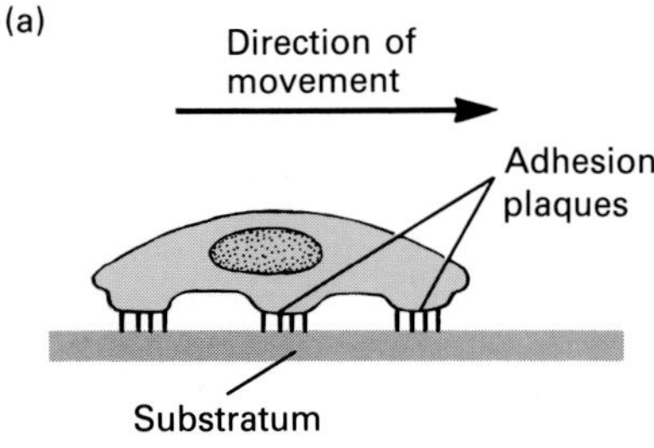

(b)

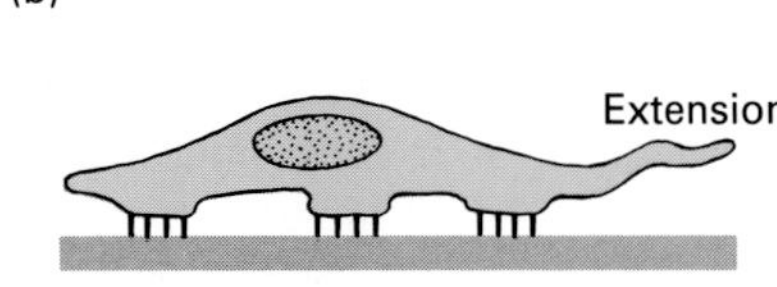

(c)

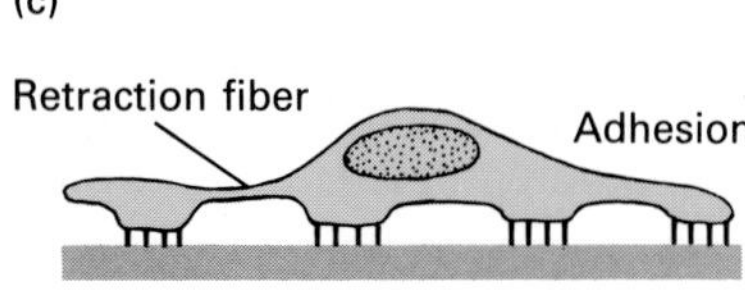

(d)

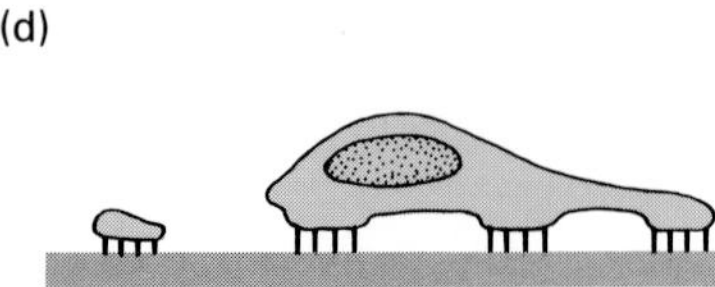

Figure 19-33 Fibroblasts jerkily move along a substratum at rates of up to 40 μm/h. (a) The cell is anchored to the substratum at adhesion plaques. (b) Lamellipodia rich in actin fibers are extended from the leading edge of the cell. (c) At several points along the ventral surface the lamellipodia adhere to the substratum, and the body of the cell containing the nucleus moves forward. At the same time ruffled extensions on the dorsal surface of the cell move backward and collapse. The tail end of the cell remains attached to the substratum by a thin strand—the retraction fiber—that gets thinner and thinner as the cell moves forward. (d) Eventually the retraction fiber contracts into the body of the cell, completing the cycle of forward movement.

frequently used to study cell movements. They move by continually extending and retracting long pseudopods along the substratum. How this elongation-retraction process occurs is not known with certainty, but it appears to involve transitions of regions of the cytoplasm from a fluidlike *sol* state to a semisolid *gel.*

The ordinary light microscope reveals two regions of the ameba cytoplasm. The central region—the *endoplasm*—contains abundant particles and membranous organelles (Figure 19-36). As can be visualized in the phase-contrast microscope, these particles are in constant, random motion; this is indicative of their freedom of movement in this sol region of the cytoplasm. The *ectoplasm*—the region of the cytoplasm just under the plasma membrane—is gel-like. Ectoplasm contains a three-dimensional network of cross-linked actin fibers; all other organelles are excluded from this region. This gel region apparently determines the shape of the pseudopod and may transmit tension from regions of cellular contraction to the sites of contact with the substratum.

It is believed that the endoplasm contains un-cross-linked actin filaments. As a pseudopod elongates and the sol-like endoplasm streams into it, the region of endoplasm near the tip of the pseudopod apparently transforms into gel-like ectoplasm. Simultaneously, the ectoplasm elsewhere in the cell transforms into sol-like endoplasm, probably by un-cross-linking of actin fibers. Proteins known to cross-link actin filaments and bundles to each other, such as α actinin, fimbrin, and fodrin, may be involved in the sol-to-gel transition. Cross-linking of actin filaments produces a network confining the movement of individual actin molecules and results in the semisolid gel state.

How movement is controlled in any of these cells is not known. Amebas cannot extend pseudopods in all directions, or they would be ripped apart! It has been shown, however, that the ability of many actin-binding proteins to cross-link actin fibers is strongly dependent on both Ca^{2+} concentration and pH. Thus Ca^{2+} and H^+ may regulate the sol-to-gel transition. Indeed, isolated ameba cytoplasm gels when the pH is lowered to 6.8 in the presence of submicromolar concentrations of Ca^{2+}; conversely, solution of the gel is induced by raising the pH or Ca^{2+} concentration. These findings implicate the involvement of a gelsolin or villin protein in the gel-to-sol transition, since these proteins fragment actin filaments in the presence of micromolar Ca^{2+}. To explain directed growth of pseudopods, differences in Ca^{2+} or H^+ concentration among various regions of the cytoplasm have been suggested; whether this is the case, however, remains to be determined.

Cytochalasins The evidence that actin (and myosin) filaments are involved in various types of nonmuscle cell movements is strong but largely circumstantial in nature, as obtained in numerous studies using a series of fungal products called *cytochalasins.* These compounds inhibit movement of fibroblasts. They cause fibroblasts to "round up" and to contract their actin microfilament bundles; they also inhibit the projections of the cellular surface membrane ("ruffling") associated with movement on the substrate. In addition, cytochalasins block the extensions of processes containing microfilaments from cells, such as axonal processes from differentiating neuronal cells.

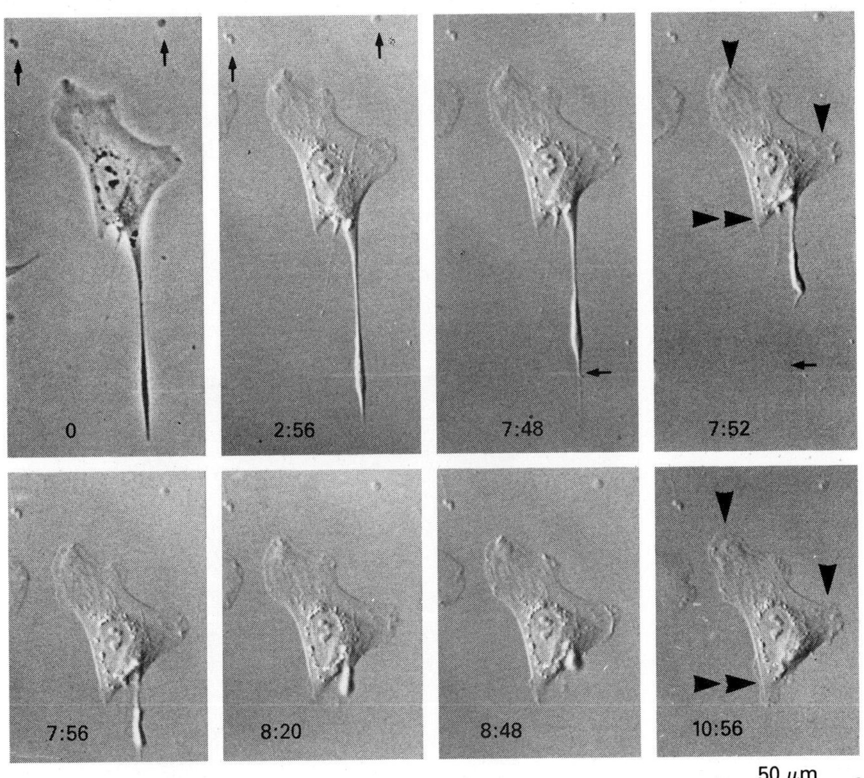

Figure 19-34 Time-lapse micrographs showing retraction of the trailing edge during fibroblast movement. Debris on the substratum serve as reference ponts (*arrowheads*). Numbers refer to elapsed minutes and seconds. Initially, at 0, the phase-contrast cellular image shows a dark tail. At 2:56, and subsequently, the same cell is viewed with Nomarski differential interference optics. Upon detachment, the trailing edge retracts very fast at first (7:48 to 7:56) and then moves more slowly (7:56 to 8:48). Adsorption of the tail remnant into the cell body occurs between 8:48 and 10:56. The retraction is followed by a surge ahead at the two existing spreading edges (10:56, *arrowheads*). Note that when it detaches, the trailing edge leaves a part of itself behind (7:48 and 7:52, *arrows*). [See W.-T. Chen, 1981, *J. Cell Biol.* 80:187.] *Photographs courtesy of W.-T. Chen. Reproduced from the* Journal of Cell Biology, *1981, by copyright permission of the Rockefeller University Press.*

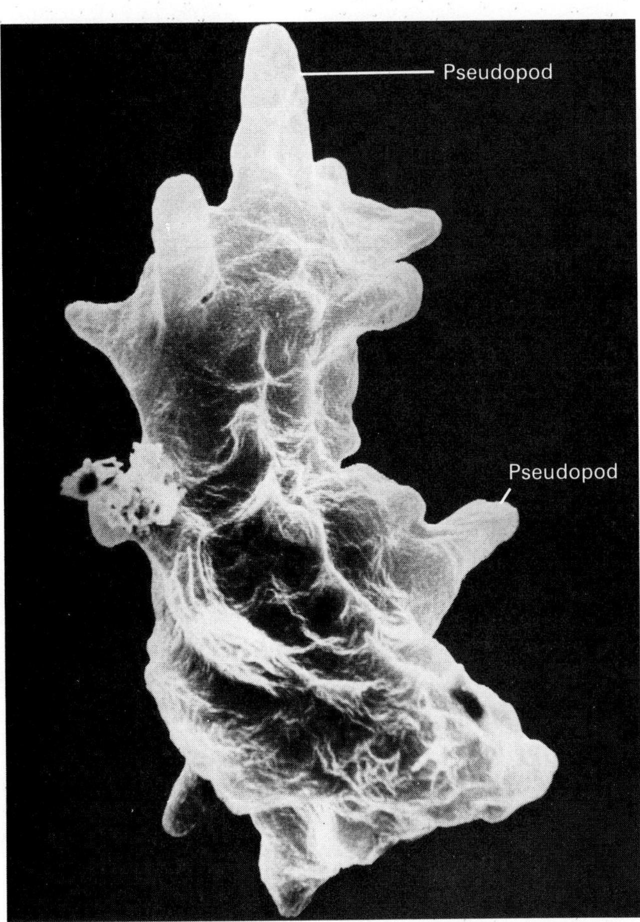

Figure 19-35 Scanning electron micrograph of the ameba *Proteus,* showing the numerous pseudopods. *Photograph courtesy of G. Antipa.*

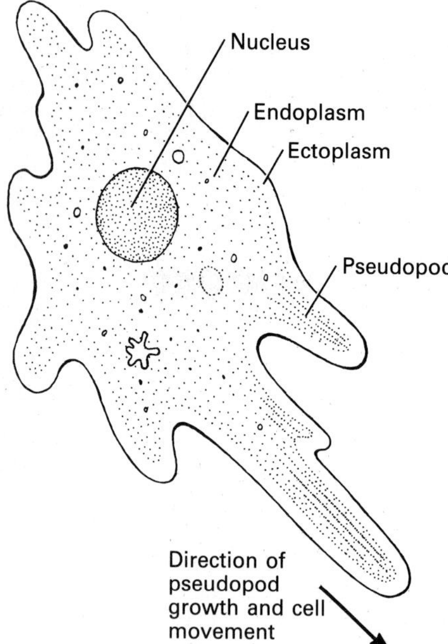

Figure 19-36 Schematic diagram of a moving *ameba* cell, showing the sol-like endoplasm and the cortical gel-like ectoplasm.

In cell-free extracts, cytochalasins specifically block the polymerization of actin filaments by binding to an end of a growing filament. Initially, G actin and F actin exist in equilibrium. But depolymerization is not affected by cytochalasins, so eventually a preponderance of G actin results. There is a good correlation between the potency of different cytochalasin analogs on actin polymerization in vitro and their in vivo effects on fibroblasts. These findings thus provide circumstantial evidence for the role of actin polymerization in fibroblast movement.

Cytokinesis Strong evidence indicates that actin-myosin interactions power cytokinesis in the separation of daughter cells during the last stage of mitosis (see Figure 5-14). Circling the cleavage furrow, the cleft between the two separating cells, is a narrow ring of actin filaments. Both myosin and α actinin have been identified in the ring, at greater concentrations than elsewhere in the cytoplasm (Figure 19-37).

As the ring of actin filaments contracts, the neck of the cleavage furrow is narrowed, and eventually the daughter cells are separated. However, as the actin ring tightens, the thickness of the ring remains constant. Thus disassembly of the ring occurs together with contraction. The ring disperses altogether as cleavage ends.

Much evidence indicates the contractile nature of the ring. For instance, antibodies to myosin injected into living sea urchin eggs inhibit cytokinesis, probably by inactivating myosin (Figure 19-38). Injection of the antimyosin antibody clearly does not inhibit the chromosome movements associated with mitosis: normal daughter nuclei form in synchrony with the control (uninjected) half of the embryo. Despite the fact that the presence of actin, myosin, and myosin light chain kinase in the mitotic spindle apparatus has been confirmed, their role in the early stages of mitosis is controversial. At least one study, depicted in Figure 19-38, suggests that they are not involved in chromosome movements.

(a)

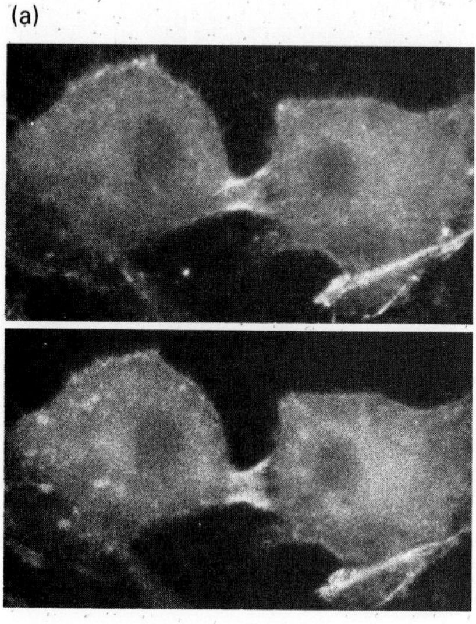

(b)

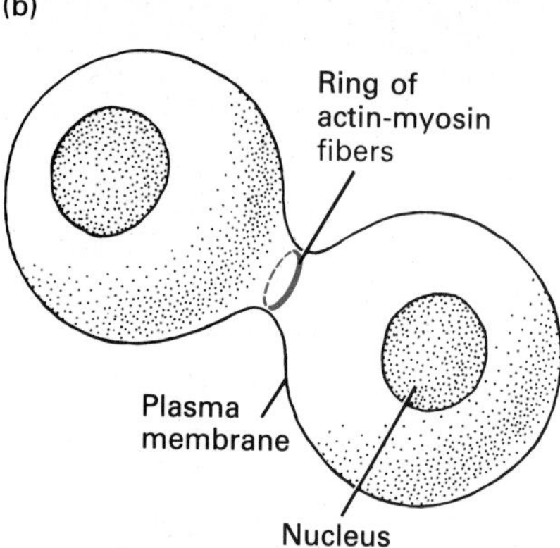

Figure 19-37 Distribution of α actinin and myosin during cytokinesis. (a) Mitotic chick embryo fibroblast cells double-stained with fluorescein-labeled α actinin-specific antibody (*top*) and rhodomine-labeled myosin-specific antibody (*bottom*). The two dyes fluoresce at different wavelengths (fluorescein generates a green color and rhodomine, red), so they can be detected sequentially by changing the colored filters in the microscope. Both proteins are concentrated in a ring around the cleavage furrow. (b) Diagram of a cleavage furrow. A ring of actin-myosin fibers runs just under the plasma membrane in the region of the cell constriction. As the ring contracts and grows smaller, the two daughter cells are gradually pinched apart. *Part (a) courtesy of T. Pollard.*

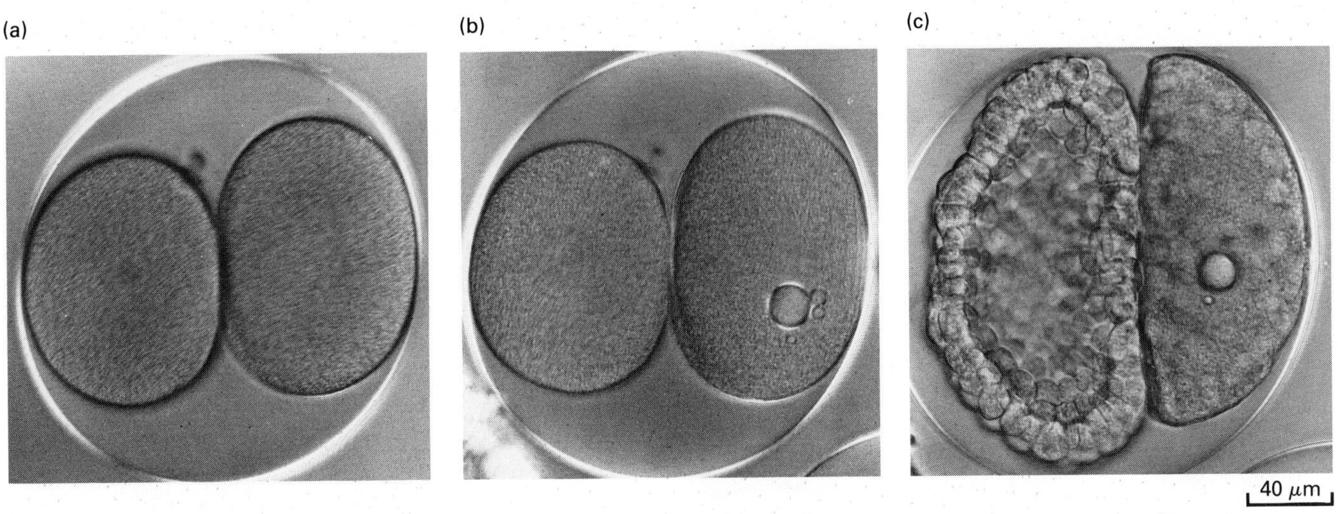

Figure 19-38 Microinjection of antimyosin antibody blocks cytokinesis but not chromosomal movement in sea urchin eggs. For these micrographs, one cell of a two-celled sea urchin embryo was injected with a myosin-specific antibody; the other cell served as an uninjected control. (a) Cells before injection in early prophase of second mitosis. (b) One cell (*right*) has been injected. The oil droplet marks the site of injection. (c) After 10 h, the control half of the embryo (*left*) is normal. In the injected cell (*right*), cytokinesis is blocked, but nuclear division proceeds normally, in approximate synchrony with the control. Nuclei are visible as rounded structures on the periphery of the cell (*right*). Cilia form on both injected and control cells. [See D. P. Kiehart, I. Mabuchi, and S. Inoue, 1982, *J. Cell Biol.* **94**:165.] *Photographs courtesy of D. P. Kiehart, I. Mabuchi, and S. Inoue. Reproduced from the* Journal of Cell Biology, *1982, by copyright permission of the Rockefeller University Press.*

Intermediate Filaments

The class of cytoplasmic filaments called intermediate filaments (IFs) is so named because the diameter—7 to 11 nm—is intermediate between those of the actin microfilaments and microtubules. Initially, IFs were regarded as degradation products of microtubules or myosin filaments. With the advent of new techniques to solubilize and characterize their protein components, IFs are now recognized as discrete entities and are thought to be the principal structural determinants in many cells and tissues. For instance, IFs form the tonofibrils that connect the spot desmosomes of epithelial cells (Chapter 14); they form the major structural proteins of skin and hair; and in muscle cells they appear to form the scaffold that holds the Z disks and the myofibrils in place. However, many basic aspects of IF function and architecture, such as the control of IF assembly within cells and the regulation of gene expression for many different cell-type-specific IF proteins, have yet to be elucidated.

Different IF Proteins Are Found in Different Cell Types

Five major classes of IFs have been distinguished on the basis of their protein composition and cell type distribution. *Keratins* form the tonofibers in epithelial cells and also the specialized structural proteins of skin and hair. *Desmin* filaments are found predominantly in muscle cells of all types, specifically in the periphery of the Z disks (see Figure 19-21). Desmin forms an interconnecting network across each muscle cell, perpendicular to the long axis of the cell. Desmin-containing fibers anchor and orient all of the neighboring Z disks so that all myofibrils in a muscle cell are in register, generating the striated pattern of these large cells. Desmin fibers may also link Z disks to the plasma membrane or to other cytoplasmic organelles.

Neurofilaments, present in axons of both central and peripheral neurons in vertebrates, are built of three discrete polypeptides (of m.w. 200,000, 150,000, and 68,000); frequently they appear to be in close association with the axonal microtubules (see Figure 18-5). The strength and rigidity of the axon may be due to these neurofilament-microtubule complexes. Microtubules and neurofilaments appear to grow by addition of new subunits in the cell body. They move together down the axon at a rate of about 1 mm/day; these filaments are thus the slowest-moving components of axonal transport.

IFs in glial cells (cells that surround neurons) are composed of still another species of IF protein, called *glial fibrillary acidic protein (GFAP);* insofar as is known, this class of proteins is expressed only in these cells.

Vimentin-containing fibers are prominent in many cultured fibroblast cells (Figure 19-39). The filaments are

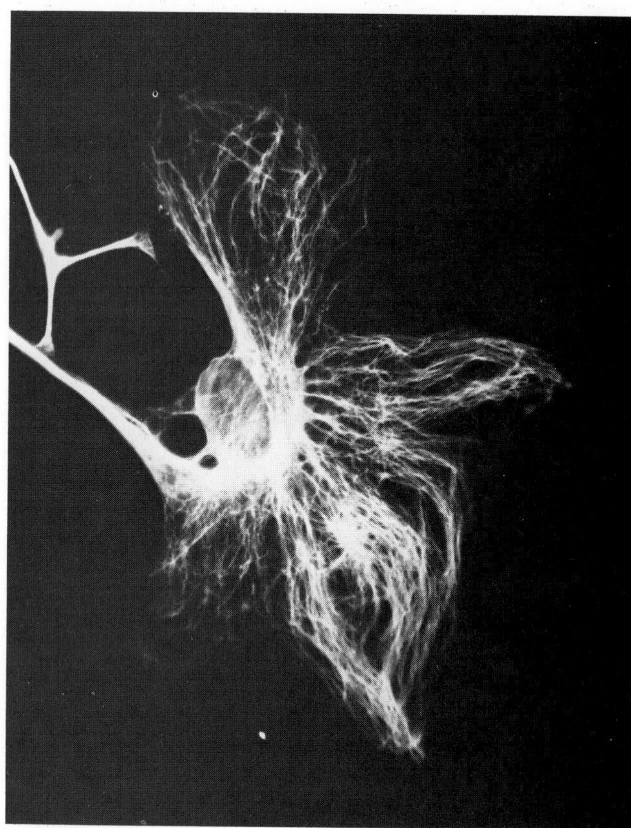

Figure 19-39 Distribution of the IF protein vimentin in a chick embryo fibroblast, as detected by immunofluorescence using a vimentin-specific antibody. The long fluorescent vimentin fibers form a network that radiates from the center into all corners of the cell. In many regions of the cell the vimentin fibers form a network lying under the plasma membrane; they may define the shape of these parts of the cell and give them rigidity. *Photograph courtesy of E. Lazarides and B. C. Granger.*

extremely insoluble—a property suggesting that they play a structural role. Fibers often terminate at the nuclear membrane and at desmosomes or adhesion plaques on the plasma membrane. They may function to keep the nucleus or other organelles in a defined place within the cell.

Rapidly growing cells, such as those of early mammalian embryos, do not contain IFs. Presumably, then, IF protein function is related to specific functions of differentiated cells rather than to cell growth and division. In this sense, IFs have a more subtle role in the regulation of cell processes than that of the ubiquitous proteins actin and tubulins. In general, the majority of cell types in the adult animal contain only a single class of IFs. A few, however, contain vimentin as well as another IF protein. Certain smooth muscle cells, for instance, contain desmin and vimentin, and some maturing neuroblasts contain vimentin and neurofilament proteins.

IF proteins are among the most abundant of the cell-type-specific proteins. Fluorescent antibodies to particular IF proteins have been very useful in typing of cells, especially for tumor classification in cases in which diagnosis cannot be made by normal histological procedures. Tumor cells retain many of the differentiated properties of the cells from which they are derived, so their tagging by IF antibodies allows identification of cell type. For example, the most common carcinomas of the breast and gastrointestinal tract contain cytokeratins and lack vimentin; thus, they are derived from epithelial cells (which contain keratins), rather than from the underlying stromal mesenchyme cells (which contain vimentin but not keratins). Such determinations are often important in the selection of treatment.

Unlike actin and tubulins, whose molecular size has remained quite constant throughout evolution, IF proteins vary greatly in molecular weight from class to class and species to species. However, they appear to share certain structural characteristics.

Amino acid and cDNA sequencing of desmin, vimentin, GFAP, and neurofilament protein fragments shows that they are at least 70 percent homologous. The keratins comprise a group of 10 to 20 different polypeptides, each of which is differently expressed in different tissues. As judged from analyses of keratin cDNAs and mRNAs, the α keratins can be grouped into two classes, called type I and type II. There is a high degree of sequence homology within each keratin class, but the keratins share only about 25 percent of their amino acid sequences with other IF proteins.

Despite the variability in size and sequence, the structure of all IFs is very similar (Figure 19-40). All IF proteins contain a central region with four α-helical domains that are separated by three conserved sites of α helix interruption. All IFs contain nonhelical regions at their N- and C-termini. Homologies between different IFs are mainly in the α-helical segments; the nonhelical segments are totally different among IF proteins and are believed to be important in the assembly of the IF subunits into filaments.

The macromolecular structures of all IF proteins are similar and resemble that of α keratin in hair and wool. The keratins are a class of fibrous, insoluble proteins. One of the two subgroups of keratins, *β keratins,* form β pleated sheets and are the major structural protein of silk and of the claws and beaks of birds. Filaments of *α keratins* are more widespread and are the principal structural protein of skin (leather is almost pure keratin), hair, wool, feathers, and nails. An α keratin molecule is a coil of three α helixes (Figure 19-41). It is unclear whether subunits of IFs are built of two or three polypeptide chains, but it is thought that they contain four segments

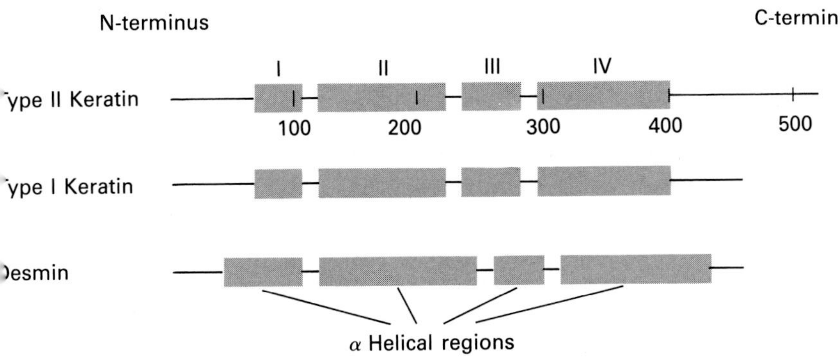

Figure 19-40 Structural homology of IF proteins. Each polypeptide contains four regions that form α helixes *(boxes)* separated by three nonhelical segments. Homologies in the sequence of amino acids are confined to the helical domains; there is no homology between the nonhelical segments of different IF proteins. *After I. Hanukoglu and E. Fuchs.*

of coiled α helixes separated by three nonhelical segments (Figure 19-41c). The coiling of the α helixes generates the rigidity and strength of the IF.

IF Proteins Are Insoluble and Are Not in Equilibrium with Soluble Monomers

IF proteins are insoluble in most aqueous buffers, unless strong detergents or acids are added. In contrast to actin and tubulin, there do not appear to be soluble pools of IF proteins. Nor are they known to undergo reversible cycles of polymerization and depolymerization, as do actin and tubulin, although they are polymers of monomer units.

When exposed to urea, most IFs depolymerize and become soluble; upon removal of the urea by dialysis, the pure proteins spontaneously polymerize into IFs of normal structure. In contrast to microfilaments and microtubules, IF proteins assemble without aid of other proteins.

When native IFs are treated with protease, the nonhelical segments at the N- and C-termini are selectively di-

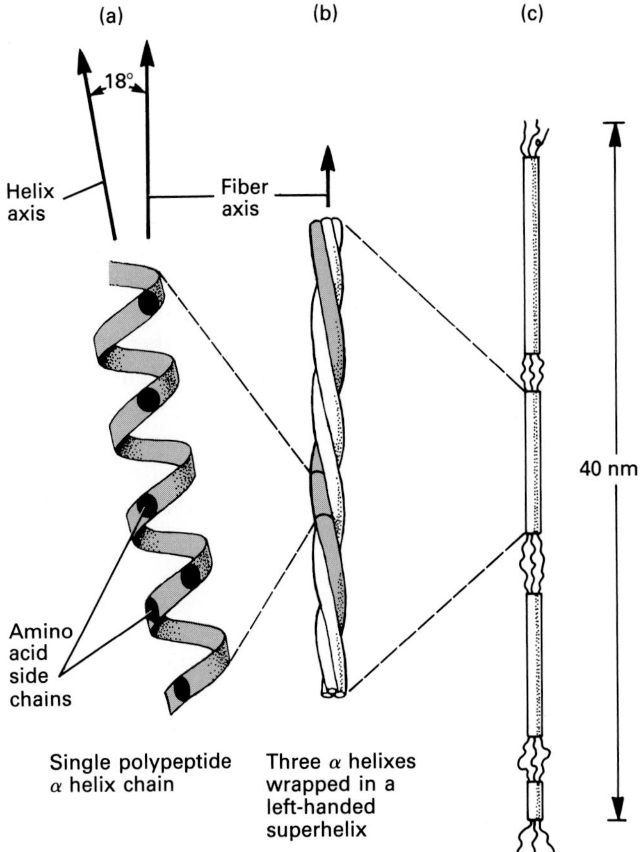

Figure 19-41 Model of the structure of α keratin in hair and wool and of cytoskeletal keratin IFs. (a) The individual polypeptide chains in hair and wool keratin are almost totally α-helical; in fact, x-ray crystallography on α keratins in wool was used to establish the structure of the α helix. (b) The fibers of α keratin are generally constructed of three α helixes wrapped around each other side by side like a rope. The α helixes are twisted so that the resulting three-stranded cable has a left-handed twist. The side chains of an individual α helix (black in part a) lie in a vertical line, along the fiber axis, when the axis of the α helix is 18° from the fiber axis, as it is in keratin; this facilitates stabilizing interactions among the side chains of the three intertwined α helixes. The chains contain a high proportion of cysteine residues that form covalent interchain disulfide cross-links. (c) A molecule of a typical IF protein, such as cytokeratin in epithelial cells, contains three coiled polypeptide chains, as depicted here, or possibly two. Each contains four segments of coiled α helixes separated by three nonhelical segments. These intertwined coils of α helixes give the molecule its rigid shape. The segments at the ends of the polypeptide are not rigid or α-helical and are believed to allow the molecules to aggregate end to end as well as side by side to form long IFs.

gested. These molecules can also be solubilized in urea, but they do not reassemble into filaments following its removal by dialysis. These findings indicate that the nonhelical regions are important for end-to-end or side-by-side adherence of IF subunits to form long filaments.

Keratins Are Unique to Epithelial Cells

The keratins represent a family of distinct but related gene products made in different epithelial cells. In intestinal epithelial cells, for instance, keratin-like proteins form the tonofilaments that traverse the cell and connect the "spot" desmosomes and hemidesmosomes (see Figure 14-56). They appear to resist shearing forces to provide strength and rigidity to the cell layer.

As many as 10 different α keratin polypeptides have been identified in human epidermis and hair follicles, and most of these have been shown to be translation products of distinct mRNAs. In all, 19 different human α keratins are known; probably, each is the product of a different gene.

The differentiation of epidermal skin cells provides an important example of the multiplicity of keratins. Cell division occurs in the germinal layer at the base of the stratified layer of cells. As the cells differentiate, they move toward the surface of the skin, until they are eventually sloughed off (see Figure 5-52). Cells in the germinal layer contain many tonofilaments of a type of keratin termed "prekeratin." As the cell undergoes terminal differentiation, it synthesizes progressively different keratin mRNAs, with the corresponding keratins. The keratins are successively cross-linked by disulfide bonds, thereby adding to their insolubility. At maturity the cell dies and becomes, literally, a bag of keratin fibers.

IFs Are often Associated with Microtubules

In both mitotic spindles and in nerve axons, IFs often surround bundles of microtubules, and there appear to be fibrous bridges (of unknown composition) between the two types of filaments (see Figure 18-5). Treatment of fibroblasts with colchicine, an inhibitor of polymerization of tubulin, causes the complete dissolution of microtubules after a period of several hours. In colchicine-treated cells vimentin remains in fibers, but these become clumped in bundles near the nucleus (Figure 19-42). These and other results suggest that the organization of the vimentin filaments is determined in some way by the lattice of microtubules, possibly by the protein connectors between the two types of filaments. Other studies indicate that there is no soluble pool of vimentin existing in equilibrium with the fibers: in the presence of colchicine the vimentin fibers form bundles but do not depolymerize.

Because IFs are so insoluble and are permanent features of the cytoplasmic architecture, it is often assumed that they play a crucial role in the regulation of cell shape. However, this may not be the case, at least in cultured fibroblasts. Microinjection of a monoclonal antibody specific to the IF protein causes a rapid collapse of IFs to a perinuclear region. This disruption occurs without any apparent alteration in the microfilament or microtubule systems and without any obvious effect on cell morphology, mitosis, or motility.

Organization of the Cytoplasm

The Cytosol Contains a High Concentration of Protein

Despite decades of research by microscopists and biochemists, the major structural features of the cytoplasm are still in dispute. One reason is that the cytosol—the part of the cytoplasm remaining after membranes and ribosomes are removed by centrifugation (see Figure 5-27)—contains about 20 to 30 percent protein. An important consequence of the high protein concentration is that water in the cytosol is found in two separate phases. Much is tightly bound to polar residues on the surfaces of proteins and other macromolecules, as the so-called *water of hydration*. A portion of the water molecules are found in the bulk phase, free to diffuse and to dissolve other molecules. Diffusion of proteins and particles is thus restricted, and formation of many weak protein-protein interactions is favored. Many proteins bind to each other with affinity constants so weak that the interactions are not detected in dilute protein solutions typically used for experiments by biochemists.

The mobility of proteins within the cytosol can be measured directly. In one method, fluorescence-tagged proteins such as albumin (a serum protein) are microinjected into the cytosol. These proteins are then studied by the technique of fluorescence recovery after photobleaching, described in Chapter 14. By this method, all microinjected protein is observed to be mobile, but the diffusion constants are only about one-fifth that of the protein in water. This reduced diffusion is probably due to weak binding of the protein to cytoplasmic fibers or membranes.

Many workers believe that most if not all enzymes in the Embden-Meyerhof pathway interact with each other to form a multienzyme complex that is stable at the high protein concentration of the cytosol. Each of the intermediates in the conversion of glucose to pyruvate, therefore, diffuses over only a tiny distance—a few nanometers—before encountering the next enzyme in the pathway. Similar complexes may function for other metabolic pathways.

(a)

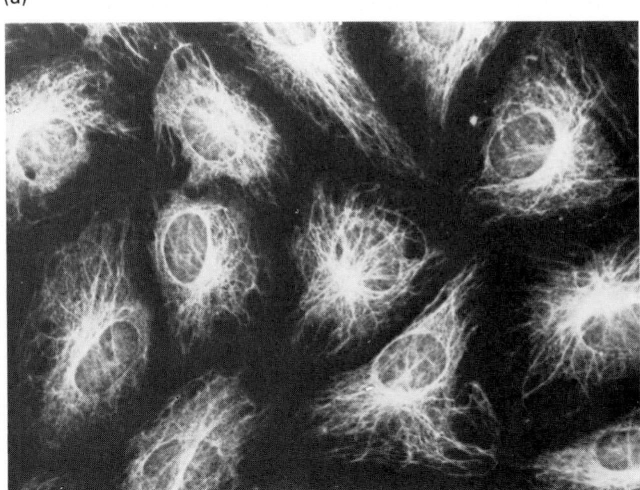

Cells stained for vimentin

(b)

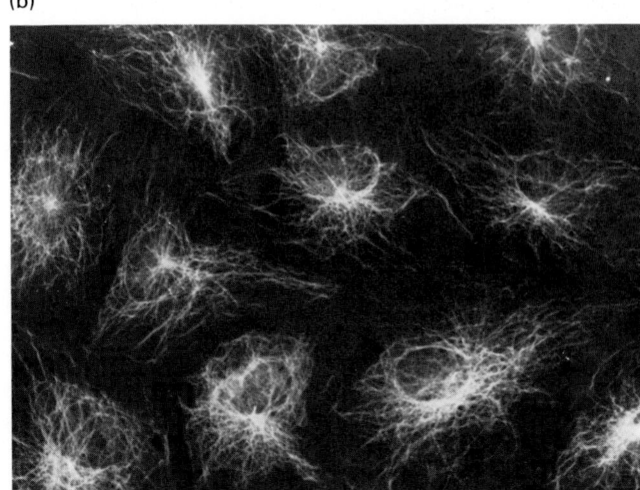

Cells stained for tubulin

(c)

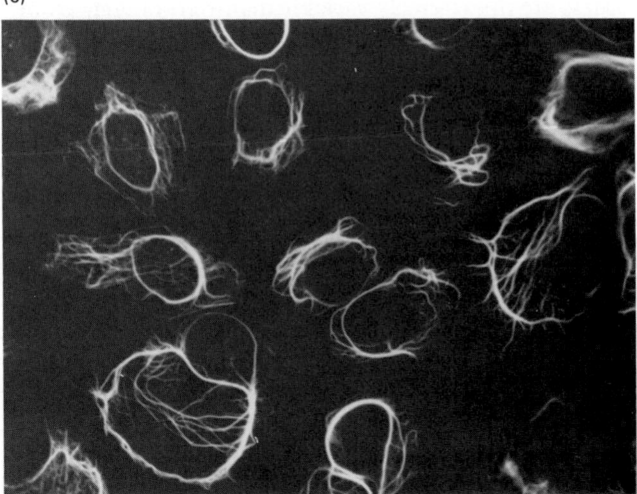

Vimentin-stained cells 10 h after colchicine treatment

(d)

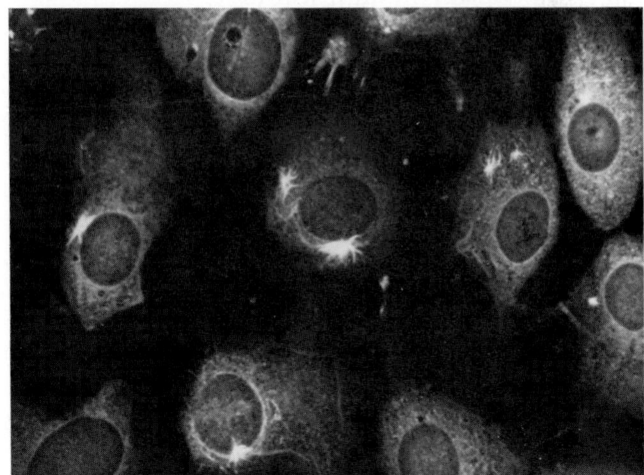

Tubulin-stained cells 10 h after colchicine treatment

Figure 19-42 The presence of vimentin filaments in fibroblast cells is dependent on the presence of microtubules. Colchicine is used to disrupt microtubules; its effect on IFs and microtubules is detected by reaction of fluorescent antibodies to tubulin and vimentin with detergent-permeabilized cells. In (a) and (b), cells have been stained for vimentin and tubulin, respectively, to illustrate the normal pattern of filaments radiating from the centriolar region near the nucleus. The patterns of fibers are similar but not identical. (c) Colchicine treatment causes the vimentin fibers to coil and clump in large thick fibers. (d) At 10 h after addition of colchicine to the tubulin-stained cells, the microtubules are depolymerized; all that remains are centrioles and a few cilia. *Photographs courtesy of R. O. Hynes and A. Destree. Copyright, 1978, M.I.T.*

Different Microscopic Techniques Generate Different Images of the Cytoplasm

Another consequence of the high concentration of protein in the cytosol is that proteins can precipitate or aggregate during the fixation or preparation procedures used for microsopy. Except for microinjection of fluorescent proteins, all light-microscopic procedures for analysis of cytoplasmic proteins—cytoskeletons in particular—involve prior disruption, extraction, or fixation of the cell. Because fluorescence-tagged antibody molecules will not penetrate live cells, the plasma membrane must be disrupted or removed with organic solvents or detergents in order to visualize the cytoskeleton by this technique (see Figure 19-28). Such procedures may significantly alter the observed characteristics of the cytoskeleton. Another possible consequence of fixation is the aggregation of proteins, which can give the erroneous impression that they are located in fibrils or particles.

Sections of cells or tissues used for standard electron microscopy can be no thicker than 0.2 μm and are usually much thinner. Electrons in conventional electron microscopes are accelerated by a potential gradient of 100,000 volts and can penetrate no more than 0.2 μm into most biological material. As a consequence, only a small piece of a cell can be observed in any one section. However, most fibrous systems are several micrometers in length, so they can be visualized as long elements only in sections that, by chance, happen to be in the plane of the fibers (Figure 19-43). Sections across the plane of the fiber bundles contain only a short bit of the fibers. Single sections contain a few or many profiles of fibers but give no visual evidence for fiber length.

These shortcomings can be compensated for by the use of serial sectioning of the tissue sample. Such serial sections are then viewed in sequence. By tracing an image from one section into the next it is often possible to reconstruct the three-dimensional architecture of a fiber. Serial sectioning, however, is rather tedious—100 sections are needed for the examination of one 20-μm-thick cell—and is used only in special cases.

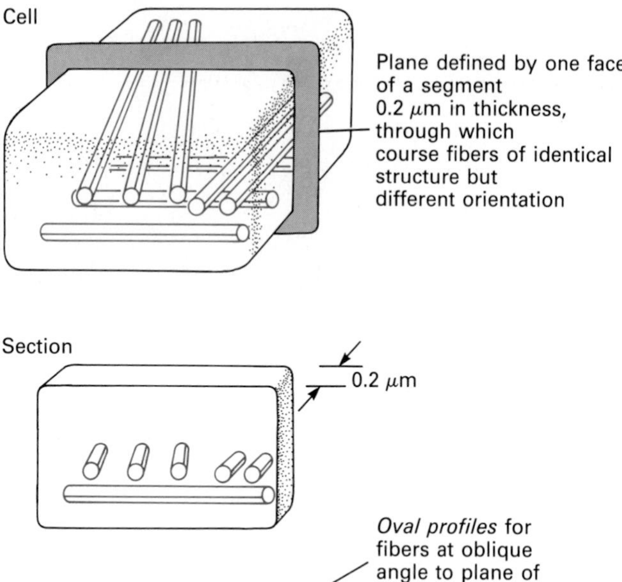

Figure 19-43 A section through a bundle of fibers can generate very different images, depending on the angle of the cut to the plane of the fibers.

Several newer techniques of electron microscopy, which involve quite different procedures for cell fixation and treatment, allow thicker sections of cells—and even entire cells—to be studied. One technique is high-voltage electron microscopy. Electrons subjected to high-voltage (typically, a million volts is used) gradients achieve tremendous velocity and can penetrate cells 10 to 20 μm thick. In fixed cells, this technique generates a unique image of the cytoplasm: it appears to be filled with a three-dimensional network of interlinked filaments of various lengths and diameters. Such a network of cytoskeletal fibers, call *microtrabeculae,* is pictured in Figure 19-44. A few fibers may be actin filaments, microtubules, or IFs, as suggested by the observed diameter, but the others are of extremely variable diameter, ranging from 3 to 15 nm. Despite differences in size, microtrabeculae appear to branch with each other at smooth, continuous junctions, rather than at single points of crossover as seen with other techniques. Cellular ribosomes and organelles appear entrapped in the microtrabecular lattice—suggesting that these structures are not freely mobile but are instead anchored to cytoskeletal fibers. The fibrils of highly variable diameter seen within the microtrabecular lattice may be fibrous entities distinct from the three principal types of cytoskeletal fibers (actin microfilaments, microtubules, and IFs). Alternatively, many workers think that they are artifacts due to precipitation of soluble proteins.

A very different view of the cytoskeleton is obtained by first treating the cells with nonionic detergents (Figure 19-45) to dissolve the membranous elements. The cytosol and organelles are extracted, leaving the insoluble elements of the cytoskeleton. Following extraction, the cell is subjected to the *quick-freeze deep-etch* technique. In this method, the cell is cooled very quickly (within milliseconds) to the temperature of liquid helium (−269°C, or 4° above absolute zero), so that ice crystals do not form and distortions in cytoskeletal architecture are avoided. While the cell is still frozen, the water is removed in a vacuum—a procedure called "freeze drying." To visualize the fibrous material, the preparations are coated with a thin molecular layer of platinum.

By this technique, the cytoplasm of cultured cells is resolved almost exclusively into three discrete filament types: actin microfilaments, microtubules, and IFs, each with a defined diameter. The filaments crisscross each other in complex patterns; at many points different types of cytoskeletal fibers contact each other. Actin microfilaments often occur in bundles of long stress fibers that contain many microfilaments; small fibrous proteins appear to connect the adjacent actin microfilaments. Al-

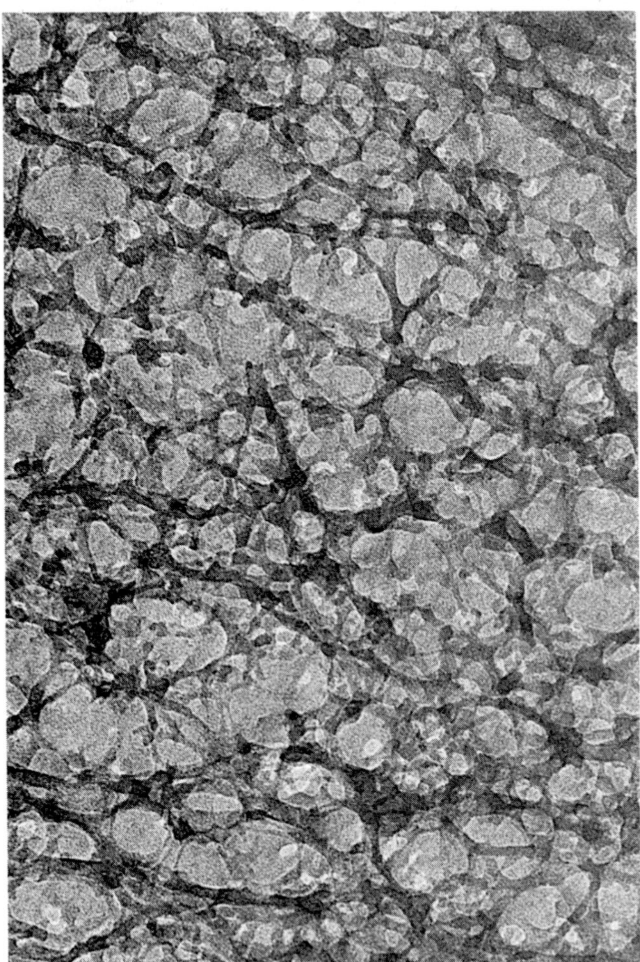

Figure 19-44 High-voltage electron micrograph showing the microtrabecular lattice of a portion of the cytoplasm of a rat fibroblast growing in tissue culture. Cells were grown directly on gold microscope grids and then fixed with glutaraldehyde. The whole cell was then frozen and dried by a method that avoids the distorting effect of surface tension ("critical-point drying"). The cytoplasm appears filled with a lacelike network of fibrils of various diameters. The thicker filaments may be microtubules; the thinner ones may be actin microfilaments or intermediate filaments or artifacts due to aggregation of cytosolic proteins. *Photograph courtesy of K. Porter.*

though the image of the cytoplasm generated with this technique contrasts sharply with that produced by high-voltage electron microscopy, in both cases ribosomes and other organelles appear anchored to the cytoskeletal network. Indeed, fluorescence microscopic studies confirm that lysosomes and perhaps mitochondria are closely associated with, if not bound to, microtubules.

All microscopic techniques indicate that the cytoskeletal network extends into all parts of the cytoplasm and serves as an anchor for ribosomes, lysosomes, and possibly other organelles. But the exact composition, structure, and arrangement of these fibers remain a source of controversy.

Attachment to Substrate

An important function of cytoskeletal fibers is in attachment of cells to the substratum. Most normal (nontumor) cells require attachment to a surface or substratum to grow and divide. Recall that collagens are the major proteins of the matrix surrounding animal cells. In tissue culture, collagen enhances growth or differentiation of numerous mammalian cell types, such as hepatocytes, myoblasts, and epithelial cells.

One of the best-documented roles for collagen in differentiation is in the fusion of myoblasts into multinucleate

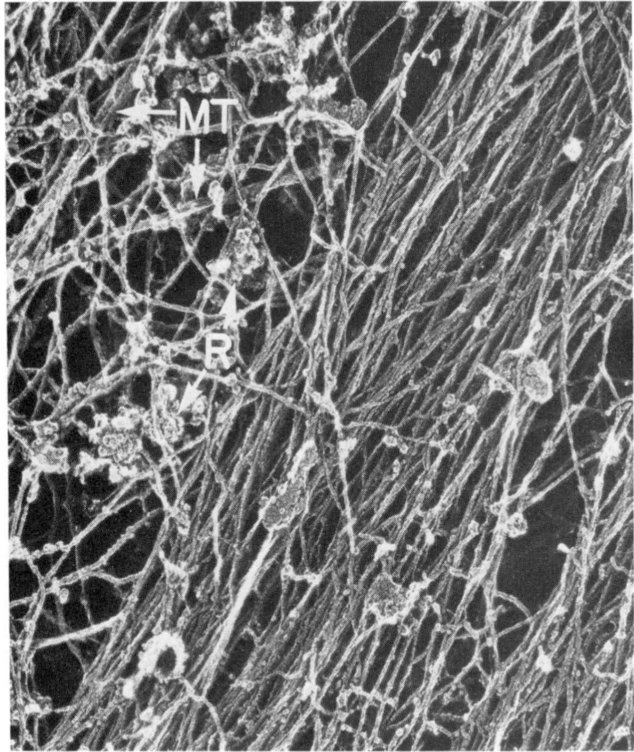

Figure 19-45 Platinum replica of a cytoskeleton prepared by the quick-freeze deep-etch technique from a fibroblast that was extracted with the detergent Triton X-100 to remove soluble cytoplasmic proteins and integral membrane proteins. The bundles of fibers in the lower right are composed of 7-nm-diameter actin filaments and are part of the cell's stress fibers. In the upper left are two thicker microtubules *(MT)* and a more diffuse network of other filaments. There are grapelike clusters of ribosomes (*R*) studding some of these filaments. *Photograph courtesy of J. E. Heuser and M. Kirschner. Reproduced from the* Journal of Cell Biology, *1980, vol. 86, p. 212 by copyright permission of the Rockefeller University Press.*

Figure 19-46 A confluent monolayer of cultured rat fibroblasts stained with fluorescent antibody to fibronectin. Note the fibrillar arrays that crisscross the cell surface and also the concentration of fibronectin at areas of cell-cell contact. Fibronectin appears to anchor cells to the substratum and to each other. *Photograph courtesy of R. Hynes.*

myofibers. When myoblasts are removed from an animal and plated in culture without the addition of collagen or of collagen-secreting fibroblasts, they are generally unable to fuse. But when plated in collagen-coated dishes, they fuse and form normal myofibers.

Fibronectins Bind Many Cells to Collagen and Other Matrix Components

Most cultured cells do not bind directly to the collagen matrix or to the plastic surface of tissue culture dishes. Rather, extracellular glycoproteins mediate this attachment. One such protein family that has been studied extensively is *fibronectin.* Fibronectin proteins are large proteins that are secreted by many cultured cells such as fibroblasts and also are normally found in the serum. At least three species of fibronectin are produced from a single gene by a pattern of alternative splicing: the "normal" protein, and two variants lacking 95 and 120 amino acids, respectively. However, no functional differences among these species have yet been elucidated.

Fibronectin forms immobilized fibrillar arrays across the surface of many cells, especially cultured fibroblasts, and interconnects cells with each other (Figure 19-46). It is not an integral membrane protein. Analysis of proteolytic fragments of fibronectin shows that the molecule consists of a number of discrete domains, or regions (Figure 19-47). Fibronectin possesses specific high-affinity binding sites for the cell surface, collagen, fibrin, sulfated proteoglycans, and other extracellular proteins and polysaccharides. Many of these binding sites are found only in one or more specific domains (see Figure 19-47). By further digestion of the cell-binding domain with proteases, it has been shown that a tetrapeptide sequence—Arg-Gly-Asp-Ser—within this domain is the minimal structure required for recognition by cells. When linked by covalent binding to an intact protein (such as albumin) and drying on a culture dish, this tetrapeptide is seen to be similar to intact fibronectin in ability to promote adhesion of cells. The fibroblast cell surface receptor for fibronectin has recently been identified and purified by means of affinity chromatography of detergent-solubilized radiolabeled plasma membrane proteins on a column to which fibronectin is bound. This receptor is a 140,000-m.w. glycoprotein that binds tightly and specifically to the cell-binding tetrapeptide of fibronectin.

As depicted in Figure 19-47, fibronectin is a V-shaped fibrous protein constructed of two large polypeptides

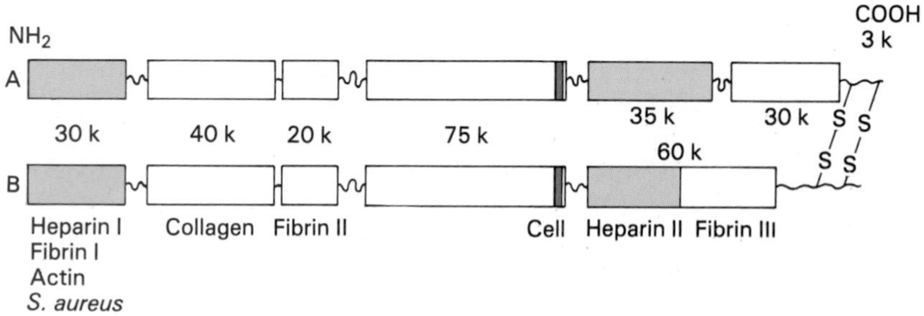

NH2
COOH
3 k
A
B
30 k
40 k
20 k
75 k
35 k
30 k
60 k
S S
S S
Heparin I
Fibrin I
Actin
S. aureus
Collagen
Fibrin II
Cell
Heparin II
Fibrin III

Figure 19-47 Model of the domain structure of fibronectin. The two nonidentical polypeptides are linked near the C-terminus by disulfide bonds. S—S indicates the relative position of the disulfide group. Proteolytic digestion of native fibronectin yields discrete fragments or domains that possess some of the binding properties of the native molecule. The sites of cleavage are denoted by the wavy lines. The sizes of the domains in kilodaltons are estimated from the sizes of proteolytic fragments. Note that there are two domains for binding to heparin and three to fibrin, which differ in affinity. Each binding domain is composed of multiple copies of short repeating sequences, each of which may contribute to the binding. [See K. Yamada et al., 1985, *J. Cell Biochem.* **28**:79.]

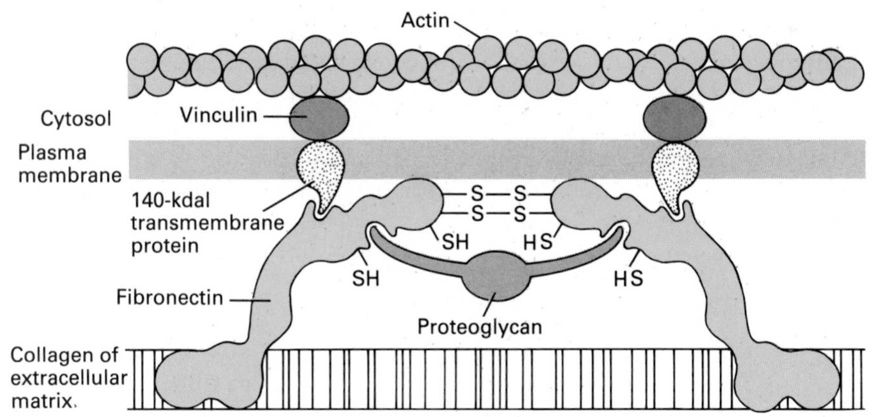

Figure 19-48 Model of the arrangement of actin and fibronectin in a region of focal contact between fibroblasts and the substratum. A key molecule is the 140,000-m.w. (140-kdal) transmembrane protein, which binds to fibronectin on the extracytoplasmic face and possibly to vinculin on the cytoplasmic face of the membrane.

(m.w. 250,000) cross-linked near one end by disulfide bonds. Not only does it mediate attachment of fibroblasts to a collagen matrix, or to the plastic surface of a culture dish, but it participates in the organization of the stress fibers and probably in other aspects of cellular organization and metabolism. In cultured cells, fibronectin is found at or close to the focal contacts, where cells stick to their substrate and at points of cell-to-cell adhesion—the points where the cytoplasmic actin microfilaments insert into the plasma membrane. Vinculin and α actinin are also found in these regions (Figure 19-48). Electron micrographs occasionally reveal exterior fibronectin fiber bundles that appear continuous with bundles of actin fibers within the cell (Figure 19-49). The membrane junction for these fibers is termed the *fibronexus.* As depicted in Figure 19-48, the 140,000-m.w. fibronectin receptor may anchor both fibers to the opposite sides of the plasma membrane.

The apparent continuity of these fibers is probably not coincidental. Many transformed cells grow in tissue culture but do not stick to plastic culture dishes. Such cells contain little fibronectin and few stress fibers. Adsorption of purified fibronectin to the culture dish frequently results in tight adhesion of the cells to the dishes and, concomitantly, to the development of stress fibers. Conversely, disrupting the microfilaments of a normal cell

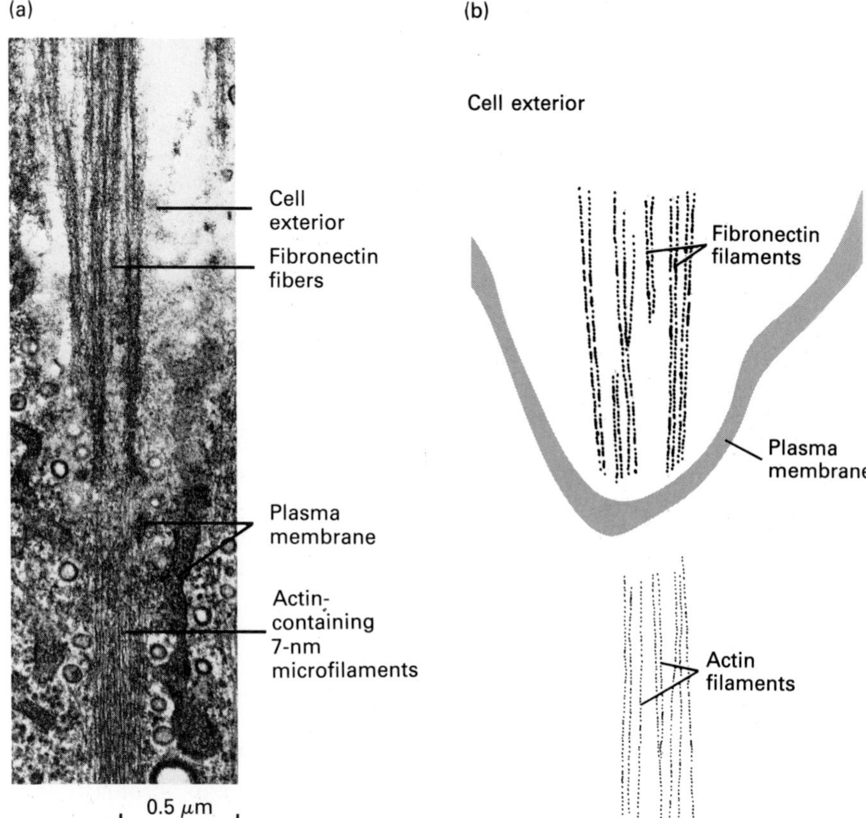

Figure 19-49 A fibronexus of a cultured fibroblast. (a) Micrograph of the fibronexus in oblique section. Individual actin-containing 7-nm microfilaments, components of a stress fiber, are seen traversing the obliquely sectioned cell membrane. The microfilaments appear in close proximity to the thicker, densely stained fibronectin fibers on the outside of the cell. (b) Interpretive drawing of the micrograph. *Part (a) courtesy of I. J. Singer. Copyright, 1978, M.I.T.*

with cytochalasin causes release of fibronectin from the cell. These results suggest that binding of the fibronectin requires actin microfilaments. It is not known how tight adhesion of a cell induces polymerization or reorganization of actin into stress fibers, nor how these fibers influence the shape or attachment of the cell. Clearly, however, the state of organization of actin and the shape of the cell are responsive to the exterior environment of the cell.

Other Cells Use Other Proteins for Binding to the Substratum

Epithelial cells, such as those lining the lumen of the intestine, attach preferentially to type IV collagen (Chapter 5), which is the major constituent of basement membranes. *Laminin,* a 400,000-m.w. fibrous glycoprotein found in basement membranes, is an attachment factor for epithelial cells (Figure 19-50). Whereas fibronectin binds to all types of collagen, laminin binds preferentially to the type IV collagen (see Table 5-6). Apart from promoting the attachment of epithelial cells, laminin may also (by an unknown mechanism) promote their differentiation.

Emerging from studies of the cytoskeleton, the membrane proteins, and the extracellular coat proteins is a picture of a continuum of interactions between the elements that allow for coordinated activities (see Figure 19-48). Indeed, the existence of such interactions is predicted by the needs of the cell: it must be able to integrate intracellular, membranal, and extracellular events if it is to be a responsible member of a cellular community.

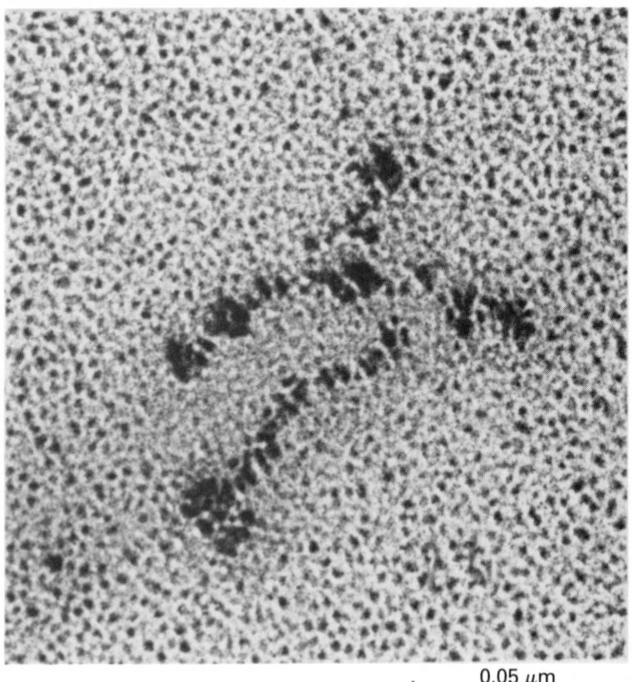

Figure 19-50 Electron micrographs of laminin, dried on a grid and shadowed with platinum. Like fibronectin, it is a long two-chain fibrous molecule. [See J. Engel et al., 1981, *J. Mol. Biol.* 150:97.] *Photograph courtesy of J. Engel, E. Odermatt, A. Engel, J. Madri, H. Furthmayer, H. Rohde, and R. Timpl.*

Summary

The role of actin microfilaments is best understood in muscle cells. In striated muscle, the ends of actin filaments are anchored to both sides of the Z disks. These interdigitate with bipolar thick filaments, constructed of myosin. Flexible headpieces of the myosin chains are energized by hydrolysis of ATP and pull the adjacent actin filaments toward the center of the myosin fiber. Hydrolysis of ATP, and hence contraction of muscle, is triggered by a rise in cytoplasmic Ca^{2+} concentration. The release of Ca^{2+} stored in the lumen of sarcoplasmic reticulum vesicles is triggered by depolarization of the surface membrane of the muscle cell. In striated muscle, Ca^{2+} induces contraction by binding to troponins, a set of proteins attached to the fibrous protein tropomyosin that lies in the groove in the thin filaments; Ca^{2+} binding, in turn, alters the conformation by which tropomyosin is bound to thin filaments so that the myosin headpieces can interact with the actin. In vertebrate smooth muscle and invertebrate muscle. Ca^{2+} acts by affecting myosin light chains located in the headpieces of myosin.

Actin-myosin interactions appear to be important for certain types of movement in nonmuscle cells—for cytokinesis and constriction of the cleavage furrow and for movement of cells along a substratum. In other cases, actin filaments may serve a structural role. Microfilaments form the major structures in the core of microvilli. Several types of actin-binding proteins cross-link the microfilaments to each other and to the microvillar membrane. Myosin, tropomyosin, and other elements of the contractile apparatus are lacking in the microvilli. Stress fibers in fibroblasts and other cells are composed primarily of actin and contain α actinin, myosin, and tropomyosin, three other proteins also found in myofibers. Stress fibers, however, are apparently involved in determination of cell shape and in attachment of the cell to the substratum, rather than in cell locomotion.

Intermediate filaments (IFs) are, in many cases, key determinants of cellular structure and comprise at least five subclasses. Desmin filaments anchor the Z disks in muscle cells. Keratin filaments form the tonofilaments of epithelial cells, and related proteins are the major structural proteins of skin, hair, and wool. Axons of nerve cells contain long neurofilaments that appear to be a major determinant of their elongated shape; still other classes of IFs are found in other cell types—vimentin in fibroblasts and glial filaments in glial cells. All IFs share common structural features; among these are intertwined coils of α helixes that give the molecules rigidity and strength.

Extracellular fibrous proteins such as fibronectin and laminin help anchor certain types of cells to the substratum, to collagens in particular. They are required for growth and differentiation of many cells. Fibronectin attaches to fibroblasts in regions in which actin fibers intersect the cell membrane; this finding suggests a functional continuity between systems of fibers inside and outside the cell.

References

Structure and Assembly of Actin and Myosin Filaments

*CLARKE, M., and J. A. SPUDICH. 1977. Nonmuscle contractile proteins: The role of actin and myosin in cell motility and shape determination. *Annu. Rev. Biochem.* **46**:797–822.

CRAIG, R. 1985. Muscle structure: First sight of crossbridge crystals. *Nature* **316**:16–17.

*DEROSIER, D. J., and L. G. TILNEY. 1982. How actin filaments pack into bundles. *Cold Spring Harbor Symp. Quant. Biol.* **46**:525–540.

MCLACHLAN, A. D., and J. KARN. 1982. Periodic charge distributions in the myosin rod amino acid sequence match crossbridge spacings in muscle. *Nature* **299**:226–229.

POLLARD, T. D. 1984. Polymerization of ADP-actin. *J. Cell Biol.* **99**:769–777.

*POLLARD, T. D., and S. W. CRAIG. 1982. Mechanism of actin polymerization. *Trends Biochem. Sci.* **7**:55–58.

*WEGNER, A. 1985. Subtleties of actin assembly. *Nature* **313**:97–98.

Structure and Contraction of Striated Muscle Cells

BESSMAN, S. P., and C. L. CARPENTER. 1985. The creatine-creatine posphate energy shuttle. *Annu. Rev. Biochem.* **54**:831–862.

CROWTHER, R. A., and P. K. LUTHER. 1984. Three-dimensional reconstruction from a single oblique section of fish muscle M-band. *Nature* **307**:569–570.

*EISENBERG, E., and T. L. HILL. 1985. Muscle contraction and free energy transduction in biological systems. *Science* **227**:999–1006.

*FRANZINI-ARMSTRONG, C., and L. D. PEACHEY. 1981. Striated muscle—contractile and control mechanisms. *J. Cell Biol.* **91**:166s–188s.

*GOODY, R. S., and K. C. HOLMS. 1983. Cross-bridges and the mechanism of muscle contraction. *Biochim. Biophys. Acta* **726**:13–39.

HUXLEY, A. F., and R. NIEDERGERKE. 1954. Structural changes in muscle during contraction. *Nature* **173**:971–973.

*HUXLEY, H. E. 1965. The mechanism of muscular contraction. *Sci. Am.* **213**(6):18–27.

MARUYAMA, K., T. YOSHIOKA, H. HIGUCHI, K. OHASHI, S. KIMURA, and R. NATORI. 1985. Connectin filaments link thick filaments and Z lines in frog skeletal muscle as revealed by immunoelectron microscopy. *J. Cell Biol.* **101**:2167–2172.

MILLER, D. M. III, I. ORTIZ, G. C. BERLINER, and H. F. EPSTEIN. 1983. Differential localization of two myosins within nematode thick filaments. *Cell* **34**:477–490.

*SQUIRE, J. 1981. *The Structural Basis of Muscle Contraction.* Plenum.

TOKUYASU, K. T., A. H. DUTTON, B. GEIGER, and S. J. SINGER. 1981. Ultrastructure of chicken cardiac muscle as studied by double immunolabeling in electron microscopy. *Proc. Nat'l Acad. Sci. USA* **78**:7619–7623.

WANG, K. 1983. Membrane skeleton of skeletal muscle. *Nature* **304**:485–486.

YATES, L. D., and M. L. GREASER. 1983. Troponin subunit stoichiometry and content in rabbit skeletal muscle and myofibrils. *J. Biol. Chem.* **258**:5770–5774.

Ca^{2+}, cAMP, and Regulation of Muscle Contraction

ADELSTEIN, R. S., J. R. SELLERS, M. A. CONTI, M. D. PATO, and P. DE LANEROLLE. 1982. Regulation of smooth muscle contractile proteins by calmodulin and cyclic AMP. *Fed. Proc.* **41**:2873–2878.

*ADELSTEIN, R. S., and E. EISENBERG. 1980. Regulation and kinetics of the actin-myosin-ATP interaction. *Annu. Rev. Biochem.* **49**:921–956.

*CHEUNG, W. Y. 1982. Calmodulin. *Sci. Am.* **246**(6):62–70.

DE LANEROLLE, P., M. NISHIKAWA, D. A. YOST, and R. S. ADELSTEIN. 1984. Increased phosphorylation of myosin light chain kinase after an increase in cyclic AMP in intact smooth muscle. *Science* **223**:1415–1417.

HARTSHORNE, D. J. 1982. Phosphorylation of myosin and the regulation of smooth-muscle actomyosin. In *Cell and Muscle Motility II*, R. M. Dowben and J. W. Shay, eds. Plenum, pp. 188–220.

MENDELSON, R. 1982. X-ray scattering by myosin S-1: Implications for the steric blocking model of muscle control. *Nature* **298**:665–667.

*PERRY, S. V., H. A. COLE, O. HUDLICKA, V. B. PATCHELL, and S. A. WESTWOOD. 1984. Role of myosin light chain kinase in muscle contraction. *Fed. Proc.* **43**:3015–3020.

*SZENT-GYORGYI, A. G. 1981. Role of regulatory light chains in myosin-linked regulation. In *Muscle Contraction: Its Regulatory Mechanisms,* S. Ebashi et al., eds. Springer-Verlag, pp. 375–389.

Actin, Myosin, and Other Actin-Binding Proteins in Nonmuscle Cells

Structure of Intestinal Microvilli

BONDER, E. M., and M. S. MOOSEKER. 1983. Direct electron microscopic visualization of barbed end capping and filament cutting by intestinal microvillar 95-dalton protein (villin): a

*Book or review article that provides a survey of the topic.

new actin assembly assay using the *Limulus* acrosomal process. *J. Cell Biol.* **96**:1097–1107.

BRETSCHER, A. 1981. Fimbrin is a cytoskeletal protein that crosslinks F-actin *in vitro*. *Proc. Nat'l Acad. Sci. USA* **78**:6849–6853.

BRETSCHER, A., and K. WEBER. 1980. Villin is a major protein of the microvillus cytoskeleton which binds both G and F actin in a calcium-dependent manner. *Cell* **20**:839–847.

BURRIDGE, K., T. KELLY, and P. MANGEAT. 1982. Nonerythrocyte spectrins: actin-membrane attachment proteins occurring in many cell types. *J. Cell Biol.* **95**:478–486.

GLENNEY, J. R. JR., and P. GLENNEY. 1985. Comparison of Ca^{++}-regulated events in the intestinal brush border. *J. Cell Biol.* **100**:754–763.

* HIROKAWA, N., R. E. CHENEY, and M. WILLARD. 1983. Location of a protein of the fodrin-spectrin-TW260/240 family in the mouse intestinal brush border. *Cell* **32**:953–965.

HIROKAWA, N., L. G. TILNEY, K. FUJIWARA, and J. E. HEUSER. 1982. Organization of actin, myosin, and intermediate filaments in the brush border of intestinal epithelial cells. *J. Cell Biol.* **94**:425–443.

MATSUDAIRA, P. T., and D. R. BURGESS. 1982. Organization of the cross-filaments in intestinal microvilli. *J. Cell Biol.* **92**:657–664.

* MOOSEKAR, M. S. 1983. Actin binding proteins of the brush border. *Cell* **35**:11–13.

* MOOSEKAR, M. S. 1985. Organization, chemistry, and assembly of the cytoskeletal apparatus of the intestinal brush border. *Annu. Rev. Cell Biol.* **1**:209–242.

MOOSEKAR, M. S., T. D. POLLARD, and K. A. WHARTON. 1982. Nucleated polymerization of actin from the membrane-associated ends of microvillar filaments in the intestinal brush border. *J. Cell Biol.* **95**:223–233.

STIDWILL, R. P., T. WYSOLMERSKI, and D. R. BURGESS. 1984. The brush border cytoskeleton is not static: *in vivo* turnover of proteins. *J. Cell Biol.* **98**:641–645.

Multiple-Actin Genes and Proteins

BAINS, W., P. PONTE, H. BLAU, and L. KEDES. 1984. Cardiac actin is the major actin gene product in skeletal muscle cell differentiation *in vitro*. *Mol. Cell Biol.* **4**:1449–1453.

GUNNING, P., P. PONTE, L. KEDES, R. J. HICKEY, and A. I. SKOULTCHI. 1984. Expression of human cardiac actin in mouse L cells: A sarcomeric actin associated with a nonmuscle cytoskeleton. *Cell* **36**:709–715.

* NOVICK, P., and D. BOTSTEIN. 1985. Phenotypic analysis of temperature-sensitive yeast actin mutants. *Cell* **40**:405–416.

Control of Actin Polymerization

CARLSSON, L., L-E. NYSTROM, I. SUNDKVIST, F. MARKEY, and U. LINDBERG. 1977. Actin polymerizability is influenced by profilin, a low molecular weight protein in non-muscle cells. *J. Mol. Biol.* **115**:465–483.

* KORN, E. D. 1982. Actin polymerization and its regulation by proteins from nonmuscle cells. *Physiol. Rev.* **62**:672–737.

* TILNEY, L. G., E. M. BONDER, L. M. COLUCCIO, and M. S. MOOSEKAR. 1983. Actin from *Thyone* sperm assembles on only one end of an actin filament: a behavior regulated by profilin. *J. Cell Biol.* **97**:112–124.

TILNEY, L. G., and S. INOUE. 1982. Acrosomal reaction of *Thyone* sperm. II. The kinetics and possible mechanism of acrosomal process elongation. *J. Cell Biol.* **93**:820–827.

Stress Fibers

BYERS, H. R., and K. FUJIWARA. 1982. Stress fibers *in situ:* immunofluorescence visualization with antiactin, antimyosin and anti-alpha-actinin. *J. Cell Biol.* **93**:804–811.

KRIES, T. E., B. GEIGER, and J. SCHLESSINGER. 1982. Mobility of microinjected rhodamine actin within living chicken gizzard cells determined by fluorescence photobleaching recovery. *Cell* **29**:835–845.

LANGANGER, B., M. MOEREMANS, G. DANEELS, A. SOBIESZEK, M. DE BRABANDER, J. DE MEY. 1986. The molecular organization of myosin in stress fibers of cultured cells. *J. Cell Biol.* **102**:200–209.

SANGER, J. W., J. M. SANGER, and B. M. JOCKUSCH. 1983. Differences in the stress fibers between fibroblasts and epithelial cells. *J. Cell Biol.* **96**:961–969.

WANG, Y.-L., F. LANNI, P. L. MCNEIL, B. R. WARE, and D. L. TAYLOR. 1982. Mobility of cytoplasmic and membrane-associated actin in living cells. *Proc. Nat'l Acad. Sci. USA* **79**:4660–4664.

Actin-Binding Proteins

BAINES, A. J. 1984. A spectrum of spectrins. *Nature* **312**:310–311.

COHEN, C. M., S. F. FOLEY, and C. KORSGREN. 1982. A protein immunologically related to erythrocyte band 4.1 is found on stress fibers of non-erythroid cells. *Nature* **299**:648–650.

GEIGER, B., K. T. TOKUYASU, A. H. DUTTON, and S. J. SINGER. 1980. Vinculin, an intracellular protein localized at specialized sites where microfilament bundles terminate at cell membranes. *Proc. Nat'l Acad. Sci. USA* **77**:4127–4131.

GLENNEY, J. R. JR., P. GLENNEY, and K. WEBER. 1982. Erythroid spectrin, brain fodrin, and intestinal brush border proteins (TW-260/240) are related molecules containing a common calmodulin-binding subunit bound to a variant cell type–specific subunit. *Proc. Nat'l Acad. Sci. USA* **28**:4002–4005.

* LAZARIDES, E., and K. WEBER. 1974. Actin antibody: the specific visualization of actin filaments in non-muscle cells. *Proc. Nat'l Acad. Sci. USA* **71**:2268–2272.

MOON, R. T., J. NGAI, B. J. WOLD, and E. LAZARIDES. 1985. Tissue-specific expression of distinct spectrin and ankyrin transcripts in erythroid and nonerythroid cells. *J. Cell Biol.* **100**:152–160.

PAYNE, M. R., and S. E. RUDNICK. 1984. Tropomyosin as a modulator of microfilaments. *Trends Biochem. Sci.* **7**:361–363.

* POLLARD, T. D. 1981. Cytoplasmic contractile proteins. *J. Cell Biol.* **91**:156s–165s.

POLLARD, T. D. 1984. Actin-binding protein evolution. *Nature* **312**:403.

ROSENFELD, G. C., D. C. HOU, J. DINGUS, I. MEZA, and J. BRYAN. 1985. Isolation and partial characterization of human platelet vinculin. *J. Cell Biol.* **100**:669–676.

* WEEDS, A. 1982. Actin-binding proteins—regulators of cell architecture and motility. *Nature* **296**:811–816.

WILKINS, J. A., and S. LIN. 1982. High-affinity interaction of vinculin with actin filaments *in vitro*. *Cell* **28**:83–90.

Variants of Actin-Binding Proteins

BASI, G. S., M. BOARDMAN, and R. V. STORTI. 1984. Alternative splicing of a *Drosophila* tropomyosin gene generates muscle tropomyosin isoforms with different carboxy-terminal ends. *Mol. Cell Biol.* **4**:2828–2836.

BREITBART, R. E., H. T. NGUYEN, R. M. MEDFORD, A. DESTREE, V. MAHDAVI, and B. NADAL-GINARD. 1985. Intricate combinatorial patterns of exon splicing generate multiple regulated troponin T isoforms from a single gene. *Cell* **41**: 67–82.

FALKENTHAL, S., V. P. PARKER, and N. DAVIDSON. 1985. Developmental variations in the splicing pattern of transcripts from the *Drosophila* gene encoding myosin alkali light chain result in different carboxyl-terminal amino sequences. *Proc. Nat'l Acad. Sci. USA* **82**:449–453.

PERIASAMY, M., E. E. STREHLER, L. I. GARFINKEL, R. M. GUBITS, N. RUIZ-OPAZO, and B. NADAL-GINARD. 1984. Fast skeletal muscle myosin light chains 1 and 3 are produced from a single gene by a combined process of differential RNA transcription and splicing. *J. Biol. Chem.* **259**:13595–13604.

Cell Motility

ALBRECHT-BUEHLER, G. 1976. The function of filopodia in spreading 3T3 mouse fibroblasts. In *Cell Motility,* R. Goldman, T. Pollard, and J. Rosenbaum, eds. Cold Spring Harbor Laboratory, pp. 247–264.

BURRIDGE, K. 1981. Are stress fibers contractile? *Nature* **294**:691–692.

CHEN, W-T. 1981. Mechanism of retraction of the trailing edge during fibroblast movement. *J. Cell Biol.* **90**:187–200.

CHEN, W-T. 1981. Surface changes during retraction-induced spreading of fibroblasts. *J. Cell Sci.* **49**:1–13.

GOLDMAN, R. D., J. A. SCHLOSS, and J. M. STARGER. 1976. Organizational changes of actin-like microfilaments during animal cell movement. In *Cell Motility,* R. Goldman, T. Pollard, and J. Rosenbaum, eds. Cold Spring Harbor Laboratory, pp. 217–245.

SHEETZ, M. P., and J. A. SPUDICH. 1983. Movement of myosin-coated fluorescent beads on actin cables *in vitro*. *Nature* **303**:31–35.

SPUDICH, J. A., S. J. KRON, and M. P. SHEETZ. 1985. Movement of myosin-coated beads on oriented filaments reconstituted from purified actin. *Nature* **315**:584–586.

STOSSEL, T. P., J. H. HARTWIG, H. L. YIN, K. S. ZANER, and O. I. STENDAHL. 1982. Actin gelation and the structure of corticol cytoplasm. *Cold Spring Harbor Symp. Quant. Biol.* **46**:569–578.

*TAYLOR, D. L., and D. S. CONDEELIS. 1979. Cytoplasmic structure and contractility in amoeboid cells. *Int. Rev. Cytol.* **56**:57–144.

TAYLOR, D. L., J. HEIPLE, Y-L. WANG, E. J. LUNA, L. TANASUGARN, J. BRIER, J. SWANSON, M. FECHHEIMER, P. AMATO, M. ROCKWELL, and G. DALEY. 1982. Cellular and molecular aspects of amoebal movement. *Cold Spring Harbor Symp. Quant. Biol.* **46**:101–112.

*TRINKAUS, J. P. 1980. Formation of protrusions of the cell surface during tissue cell movement. In *Tumor Cell Surfaces and Malignancy,* R. Hynes and C. Fred Fox, eds., Alan R. Liss, pp. 887–906.

YIN, H. L., and T. P. STOSSEL. 1979. Control of cytoplasmic actin gel-sol transformation by gelsolin—a calcium-dependent regulatory protein. *Nature* **281**:583–586.

Intermediate Filaments

*FUCHS, E., and I. HANUKOGLU. 1983. Unraveling the structure of the intermediate filaments. *Cell* **34**:332–334.

GEISLER, N., E. KAUFMAN, and K. WEBER. 1982. Proteinchemical characterization of three structurally distinct domains along the protofilament unit of desmin 10 nm filaments. *Cell* **30**:277–286.

HANUKOGLU, I., and E. FUCHS. 1983. The cDNA sequence of a type II cytoskeletal keratin reveals constant and variable structural domains among keratins. *Cell* **33**:915–924.

HYNES, R. O., and A. T. DESTREE. 1978. 10 nm filaments in normal and transformed cells. *Cell* **13**:151–163.

KLYMKOYSKY, M. W., R. H. MILLER, and E. B. LANE. 1983. Morphology, behavior, and interaction of cultured epithelial cells after the antibody-induced disruption of keratin filament organization. *J. Cell Biol.* **96**:494–509.

LAZARIDES, E. 1980. Intermediate filaments as mechanical integrators of cellular space. *Nature* **283**:249–256.

LIN, J. J.-C., and J. R. FERAMISCO. 1981. Disruption of the *in vivo* distribution of the intermediate filaments in fibroblasts through the microinjection of a specific monoclonal antibody. *Cell* **24**:185–193.

*MARCHUK, D., S. MCCROHON, and E. FUCHS. 1984. Remarkable conservation of structure among intermediate filament genes. *Cell* **39**:491–498.

*OSBORN, M., N. GEISLER, G. SHAW, G. SHARP, and K. WEBER. 1982. Intermediate filaments. *Cold Spring Harbor Symp. Quant. Biol.* **46**:413–430.

OSBORN, M., and K. WEBER. 1982. Intermediate filaments: Cell-type-specific markers in differentiation and pathology. *Cell* **31**:303–306.

QUINLAN, R. A., and W. W. FRANKE. 1982. Heteropolymer filaments of vimentin and desmin in vascular smooth muscle tissue and cultured baby hamster kidney cells demonstrated by chemical crosslinking. *Proc. Nat'l Acad. Sci. USA* **79**:3452–3456.

RAMAEKERS, F. C. S., J. J. G. PUTS, A. KANT, O. MOESKER, P. H. K. JAP, and G. P. VOOIJS. 1982. Use of antibodies to intermediate filaments in the characterization of human tumors. *Cold Spring Harbor Symp. Quant. Biol.* **46**:331–340.

SAUK, J. J., M. KRUMWEIDE, D. COCKING-JOHNSON, and J. G. WHITE. 1984. Reconstitution of cytokeratin filaments *in vitro:* further evidence for the role of nonhelical peptides in filament assembly. *J. Cell Biol.* **99**:1590–1597.

STEINERT, P. M. 1978. Structure of the three-chain unit of the bovine epidermal keratin filament. *J. Mol. Biol.* **123**:49–70.

*STEINERT, P. M., and D. A. D. PARRY. 1985. Intermediate filaments. *Annu. Rev. Cell Biol.* **1**:41.

*TRAUB, P. 1985. Intermediate filaments. Springer-Verlag.

Viewing the Cytoskeleton in Nonmuscle Cells

COLLOT, M., D. LOUVARD, and S. J. SINGER. 1984. Lysosomes are associated with microtubules and not with intermediate filaments in cultured fibroblasts. *Proc. Nat'l Acad. Sci. USA* **81**:788–792.

* FULTON, A. B. 1982. How crowded is the cytoplasm? *Cell* **30**:345–347.

* HEUSER, J. 1981. Quick-freeze, deep-etch preparation of samples for 3-D electron microscopy. *Trends Biochem. Sci.* **6**:64–68.

HEUSER, J., and M. W. KIRSCHNER. 1980. Filament organization revealed in platinum replicas of freeze-dried cytoskeletons. *J. Cell Biol.* **86**:212–234.

HIROKAWA, N. 1982. Cross-linker system between neurofilaments, microtubules, and membranous organelles in frog axons revealed by the quick-freeze, deep-etching method. *J. Cell Biol.* **94**:129–142.

JACOBSON, K., and J. WOJCIESZYN. 1984. The translational mobility of substances within the cytoplasmic matrix. *Proc. Nat'l Acad. Sci. USA* **81**:6747–6751.

KREIS, T. E., and W. BIRCHMEIER. 1980. Stress fiber sarcomeres of fibroblasts are contractile. *Cell* **22**:555–561.

MAUPIN, P., and T. D. POLLARD. 1983. Improved preservation and staining of HeLa cell actin filaments, clathrin-coated membranes, and other cytoplasmic structures by tannic acid-glutaraldehyde-saponin fixation. *J. Cell Biol.* **96**:51–62.

RIS, H. 1985. The cytoplasmic filament system in critical point-dried whole mounts and plastic-embedded sections. *J. Cell Biol.* **100**:1474–1487.

* TAYLOR, D. L., and Y-L. WANG. 1980. Fluorescently labeled molecules as probes of the structure and function of living cells. *Nature* **284**:405–410.

WOLOSEWICK, J. J., and K. R. PORTER. 1979. Microtrabecular lattice of the cytoplasmic ground substance: artifact or reality. *J. Cell Biol.* **82**:114–139.

Interactions of Cells with the Extracellular Matrix

EDELMAN, G. M. 1984. Cell-adhesion molecules: a molecular basis for animal form. *Sci. Am.* **250**(4):118–129.

* HAY, E. D. 1981. Extracellular matrix. *J. Cell Biol.* **91**:205s–223s.

* HAY, E. D., ed. 1982. *Cell Biology of Extracellular Matrix.* Plenum.

HOGAN, B. 1981. Laminin and epithelial cell attachment. *Nature* **290**:737–738.

* HYNES, R. 1985. Molecular Biology of fibronectin. *Annu. Rev. Cell Biol.* **1**:67–91.

* HYNES, R. O., and K. M. YAMADA. 1982. Fibronectins: Multifunctional modular glycoproteins. *J. Cell Biol.* **95**:369–377.

* KEFALIDES, N. A., R. ALPER, and C. C. CLARK. 1979. Biochemistry and metabolism of basement membranes. *Int. Rev. Cytol.* **61**:167–228.

KLEINMAN, H. K., R. J. KLEBE, and G. R. MARTIN. 1981. Role of collagenous matrices in the adhesion and growth of cells. *J. Cell Biol.* **88**:473–485.

OESCH, B., and W. BIRCHMEIER. 1982. New surface component of fibroblast's focal contacts identified by a monoclonal antibody. *Cell* **31**:671–679.

PIERSCHBACHER, M. D., E. G. HAYMAN, and E. RUOSLAHTI. 1985. The cell attachment determinant in fibronectin. *J. Cell Biochem.* **28**:115–126.

* PYTELA, R., M. D. PIERSCHBACHER, and E. RUOSLAHTI. 1985. Identification and isolation of a 140 kd cell surface glycoprotein with properties expected of a fibronectin receptor. *Cell* **40**:191–198.

SCHWARZBAUER, J. E., J. I. PAUL, and R. O. HYNES. 1985. On the origin of species of fibronectin. *Proc. Nat'l Acad. Sci. USA* **82**:1424–1428.

SCHWARZBAUER, J. E., J. W. TAMKUN, I. R. LEMISCHKA, and R. O. HYNES. 1983. Three different fibronectin mRNAs arise by alternative splicing within the coding region. *Cell* **35**:421–431.

SINGER, I. I. 1979. The fibronexus: a transmembrane association of fibronectin-containing fibers and bundles of 5 nm microfilaments in hamster and human fibroblasts. *Cell* **16**:675–685.

TIMPL, R. 1979. Laminin—a glycoprotein from basement membranes. *J. Biol. Chem.* **254**:9933–9937.

* TIMPL, R., J. ENGEL, and G. R. MARTIN. 1983. Laminin—a multifunctional protein of basement membranes. *Trends Neural Sci.* **8**:207–209.

* YAMADA, K., S. AKIYAMA, T. HASEGAWA, E. HASEGAWA, M. HUMPHRIES, D. KENNEDY, K. NAGATA, H. URISHIHARA, K. OLDEN, and W. CHEN. 1985. Recent advances in the research on fibronectin and other cell attachment proteins. *J. Cell Biochem.* **28**:79–97.

20

Energy Conversion: The Formation of ATP in Chloroplasts, Mitochondria, and Bacteria

All the processes involved in growth and metabolism of cells require an input of energy. In most cases this energy is supplied by hydrolysis of one (or both) of the high-energy phosphate bonds in adenosine triphosphate (see pages 42–46):

$$ATP^{4-} + H_2O \longrightarrow ADP^{3-} + P_i^{2-} + H^+$$
$$\Delta G^{\circ\prime} = -7.3 \text{ kcal/mol}$$

$$ATP^{4-} + H_2O \longrightarrow AMP^{2-} + PP_i^{3-} + H^+$$
$$\Delta G^{\circ\prime} = -7.3 \text{ kcal/mol}$$

$$ATP^{4-} + 2H_2O \longrightarrow AMP^{2-} + 2P_i^{2-} + 2H^+$$
$$\Delta G^{\circ\prime} = -14.6 \text{ kcal/mol}$$

We have seen many examples of the ways in which the energy released by the hydrolysis of these phosphoanhydride bonds is used to power reactions that are otherwise energetically unfavorable: transport of molecules against a concentration gradient; beating of cilia; contraction of muscle; and synthesis of nucleic acids and proteins from amino acids and nucleotides. ATP is the universal "currency" of chemical energy, found in all types of organisms: archaebacteria, eubacteria, animals, and plants. Thus ATP must have evolved in the earliest life forms.

Now we turn to the ways in which cells generate the high-energy phosphate bond of ATP. These processes involve an *endergonic* reaction (one requiring an input of energy):

$$H^+ + ADP^{3-} + P_i^{2-} \longrightarrow ATP^{4-} + H_2O$$
$$\Delta G^{\circ\prime} = +7.3 \text{ kcal/mol}$$

We shall focus on two of the most important processes—photosynthesis by blue-green algae and in the chloroplast of higher plants, and oxidation of metabolic products of carbohydrates, fatty acids, and amino acids in the mitochondrion and by aerobic bacteria.

At first glance, the processes of photosynthesis and aerobic oxidation appear to have little similarity. However, one of the revolutionary recent findings in cell biology is that bacteria, mitochondria, and chloroplasts actually use the same—or very nearly the same—process for generation of ATP from ADP and P_i. Proton concentration gradients and (electrical) potential gradients across these membranes are the immediate sources of energy for the synthesis of ATP from ADP and P_i (Figure 20-1). Movement of protons through specific transmembrane proteins, down their electrochemical gradient, is coupled to the synthesis of ATP from ADP and P_i. These proton gradients are also the energy sources for the transport of small molecules across these membranes against a concentration gradient. For example, the uptake of lactose by certain bacteria uses a proton-sugar symport (Figure 15-21). Rotation of bacterial flagella is likewise powered by the transmembrane proton gradient (in contrast to beating of eukaryotic flagella, which is powered by ATP hydrolysis).

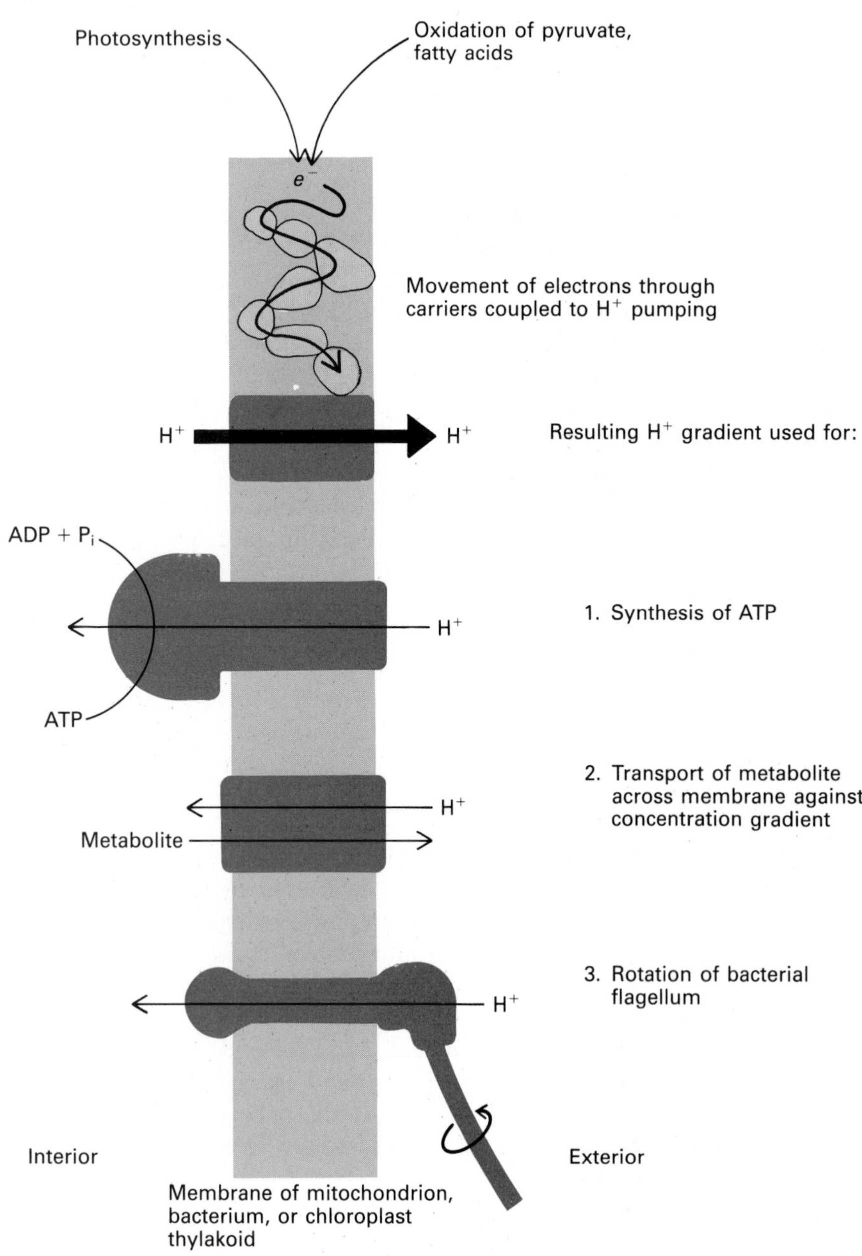

Figure 20-1 The chemiosmotic theory. In the process of photosynthesis, the energy absorbed from light is used to power the movements of electrons through a series of carriers in the chloroplast membrane (or the plasma membrane of photosynthetic bacteria). In mitochondria and aerobic bacteria, the energy liberated by oxidation of fatty acids, pyruvate, and other intermediate metabolic products is likewise used to power the movement of electrons through a series of carriers in the inner mitochondrial membrane or bacterial membrane. All electron movements are coupled to the pumping of protons across the membrane, generating a transmembrane electrochemical proton gradient, that is, a gradient of proton concentration and/or potential across the membrane.

The energy stored in this gradient can be used in several ways. (1) Movement of protons across the membrane, down their concentration gradient, is coupled to the synthesis of ATP from ADP and P_i. (2) The proton gradient is used to pump metabolites across the membrane against their concentration gradient. (3) In bacteria, the proton gradient is also used to power rotation of the flagella.

The processes of photosynthesis and mitochondrial or bacterial oxidation are also similar in the way in which the proton gradient is generated. In photosynthesis, absorbed light energy is used to power movements of electrons through a series of electron carriers located in the chloroplast membrane. In mitochondria and aerobic bacteria, electron transport across the inner mitochondrial or bacterial plasma membrane is powered by energy generated by the oxidation of metabolic products of sugars and fatty acids. In both processes, electron movements are coupled to the pumping of protons across a membrane, resulting in a proton concentration gradient and/or a potential across the membrane. We shall refer to these two collectively as the *electrochemical proton gradient.*

Many of the reactions we shall consider involve the transfer of electrons from molecule to molecule—the processes of oxidation and reduction—in which specialized electron carriers apparently operate as intermediates. Figures 20-2 and 20-3 show the structures of two important electron carriers: NAD^+ and FAD. These are essential for many of the reactions of photosynthesis and oxidative phosphorylation. Both NAD^+ and the related $NADP^+$ accept two electrons at a time to yield their respective reduced forms, NADH and NADPH. (We have already encountered NADH and NAD in the discussion of glycolysis in Chapter 5.) The carrier FAD can accept either one or two electrons, as Figure 20-3 shows. NAD is water-soluble and occurs primarily as free molecules within the cell. On the other hand, FAD is mostly bound to proteins, many of which are attached to membranes.

These processes for generation of ATP are known as *chemiosmosis.* The chemiosmotic theory is based on a principle introduced previously in the discussion of active transport (Chapter 15)—that concentration gradients of protons (and other ions) across membranes and the phosphoanhydride bonds in ATP are equivalent interconvertible storage forms of chemical potential energy.

We shall begin with a general overview of the process of energy generation in photosynthesis—together termed *photophosphorylation*—and in the oxidation reactions in the mitochondrion—referred to collectively as *oxidative phosphorylation.* We shall examine how and where the proton concentration and potential gradients are created, and how they are used to generate ATP. We shall then

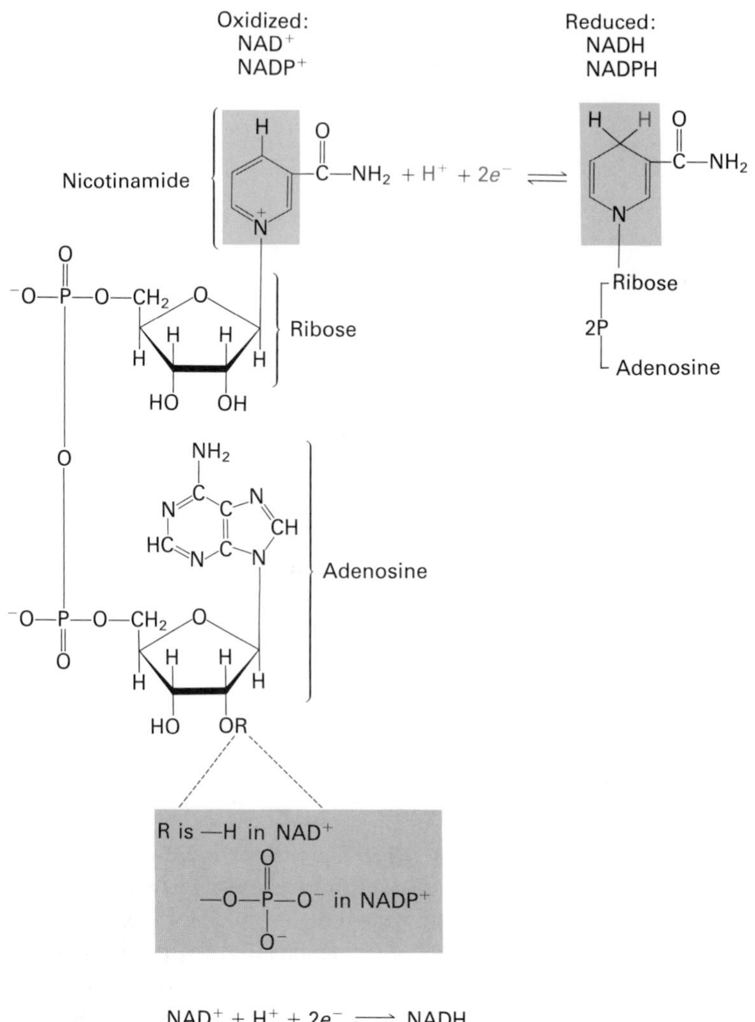

Figure 20-2 Structure of NAD^+ and NADH. Nicotinamide adenine dinucleotide (NAD^+) and the related nicotinamide adenine dinucleotide phosphate ($NADP^+$) accept only *pairs* of electrons; reduction to NADH or NADPH involves the transfer of two electrons simultaneously. In most oxidation-reduction reactions in biological systems, a pair of hydrogen atoms (two protons and two electrons) are removed from a molecule. One of the protons and both of the electrons are transferred to NAD^+, and the other proton is released into solution. Thus the overall reaction is sometimes written as $NAD^+ + 2H^+ + 2e^- \rightleftharpoons NADH + H^+$.

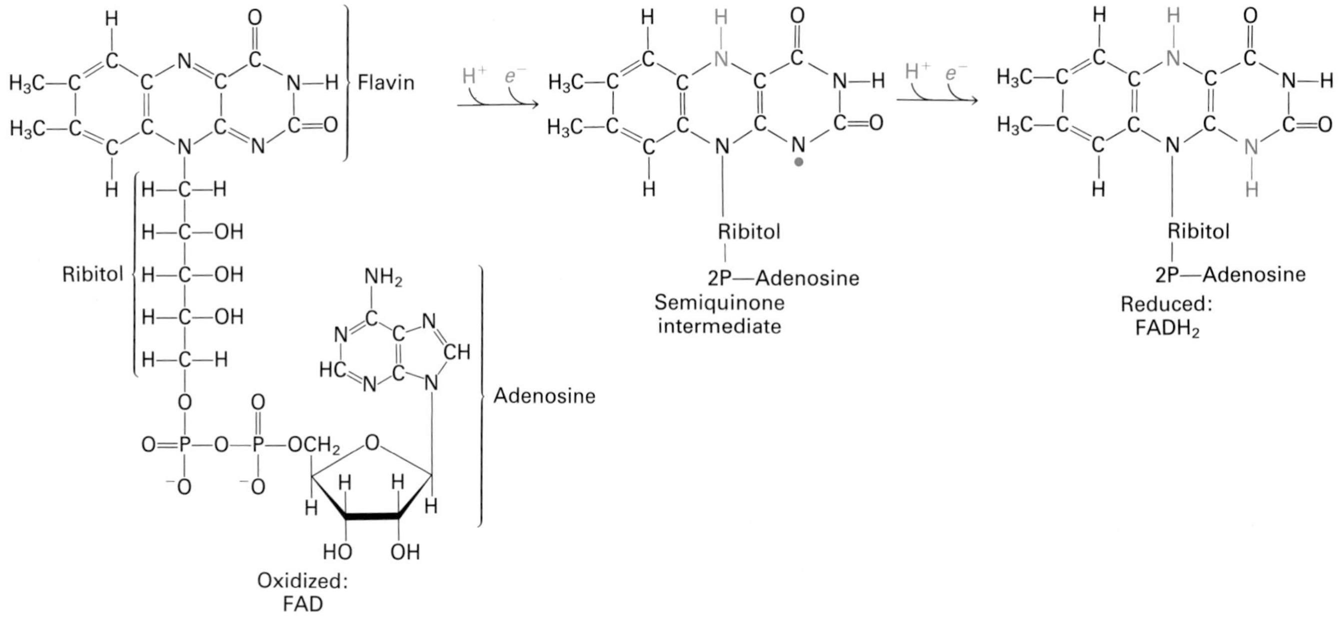

Figure 20-3 Flavin adenine dinucleotide, FAD, contains a three-ring flavin component that can accept either one or two electrons. Addition of one electron, together with a proton, to the oxidized form, FAD, generates a semiquinone intermediate. The semiquinone is a free radical because it contains an unpaired electron, denoted here by a dot. This electron is delocalized by resonance to all of the flavin ring atoms. Addition of a second electron together with a proton generates the reduced form, $FADH_2$. FMN is a related cofactor that contains only the flavin–ribitol phosphate part of FAD.

turn to a more detailed treatment of these processes, with emphasis on the role of the various membrane components and on the metabolic pathways involved in synthesis and degradation of the more important biological fuels.

Overview of Photosynthesis

Recall from Chapter 5 that photosynthesis in green plants occurs in membranes called *thylakoid membranes* that lie inside chloroplasts (Figure 20-4). As do mitochondria, chloroplasts contain inner and outer membranes. These membranes contain permeases that allow regulated movement of metabolites into and out of the organelle; they are not sites of photosynthesis. Thylakoid membrane vesicles frequently form stacks called *grana* (singular, *granum*).

Thylakoid membranes contain the pigment chlorophyll, which traps light energy. Chlorophyll is a ringed compound similar in structure to heme (see Figure 3-9) except that Mg^{2+}, rather than Fe, is in the center (see Figure 20-36). The thylakoid membrane contains several proteins involved in the trapping of light, the transport of electrons, the pumping of protons, and other enzymatic reactions. Each of these proteins and protein complexes is oriented in the thylakoid membrane in a specific fashion: some are located only on one side or the other of the membrane; some span the membrane in a unique orientation. This asymmetry allows proton gradients to be established for use in the generation of ATP.

In photosynthesis, the energy of light is trapped by chlorophyll and is used to generate a proton gradient, which is then used to synthesize ATP from ADP and P_i:

$$H^+ + ADP^{3-} + P_i^{2-} \xrightarrow{\text{Light}} ATP^{4-} + H_2O$$

Light energy is used also to split water molecules to form oxygen and to reduce the oxidized form of the electron carrier NADP:

$$2H_2O + 2NADP^+ \xrightarrow{\text{Light}} 2H^+ + 2NADPH + O_2$$

The ATP^{4-} and NADPH thus generated are used in the synthesis of glucose from CO_2 and H_2O:

$$6CO_2 + 18ATP^{4-} + 12NADPH + 12H_2O \longrightarrow C_6H_{12}O_6 + 18ADP^{3-} + 18P_i^{2-} + 12NADP^+ + 6H^+$$

All the reactions involving the formation of NADPH and ATP, and the formation of glucose from CO_2, occur within the chloroplast. The enzymes that incorporate CO_2 into chemical intermediates—by the process referred to as CO_2 *fixation*—and then convert it to glucose are soluble constituents of the chloroplast stroma (Figure 20-5).

Generation of ATP^{4-} and NADPH is associated with the specialized thylakoid membrane. These membranes contain chlorophyll and enzymes and form *closed* vesicles, which permit the generation of the transmembrane proton gradient used in the synthesis of ATP. Light energy absorbed by the chlorophylls is used to remove electrons from water. These electrons have a very high free energy content and move through a series of electron carriers located within the thylakoid membrane.

Light

Membrane-associated electron transporters

$$H_2O \rightarrow 2H^+ + \tfrac{1}{2}O_2;\quad 2e^- \longrightarrow 2e^- \longrightarrow 2e^- \longrightarrow 2e^-;\quad NADP^+ + H^+ \rightarrow NADPH$$

High-energy electrons … Low-energy electrons

At each step, these electrons lose a bit of their free energy. The final acceptor of two electrons at the end of this pathway is $NADP^+$, which is thus reduced to NADPH. Movement of electrons through certain of the carriers is coupled to the pumping of protons across the thylakoid membrane. This results in an accumulation of protons within the lumen of the thylakoid—creating an electrochemical gradient across the thylakoid membrane (Figure 20-6).

This proton gradient is the immediate source of energy for ATP synthesis. Movement of protons through specialized membrane proteins, *down* their electrochemical gradient, is coupled to the synthesis of ATP from ADP and P_i (Figure 20-6). This process can be viewed as the reversal of an ATP-dependent proton pump (Chapter 15). The membrane protein complex that catalyzes ATP synthesis is called the CF_0CF_1 complex; these complexes constitute the knobs seen protruding from the surface of the thylakoid membrane. They are similar in the structure and function to the F_0F_1 particles in mitochondria that also use transmembrane proton gradients for synthesis of ATP (see Figure 5-44).

Thus, the process of photosynthesis can be subdivided into three stages, each of which occurs in a defined area of the chloroplast:

Stage 1. The trapping of quanta of light by chlorophyll and the use of the absorbed light energy to remove electrons and protons from water (forming O_2) and the subsequent transfer of the electrons through the thylakoid membrane to the ultimate electron acceptor $NADP^+$, forming the reduced compound, NADPH, in the stroma. The movement of electrons is coupled to transport of protons across the membrane from the

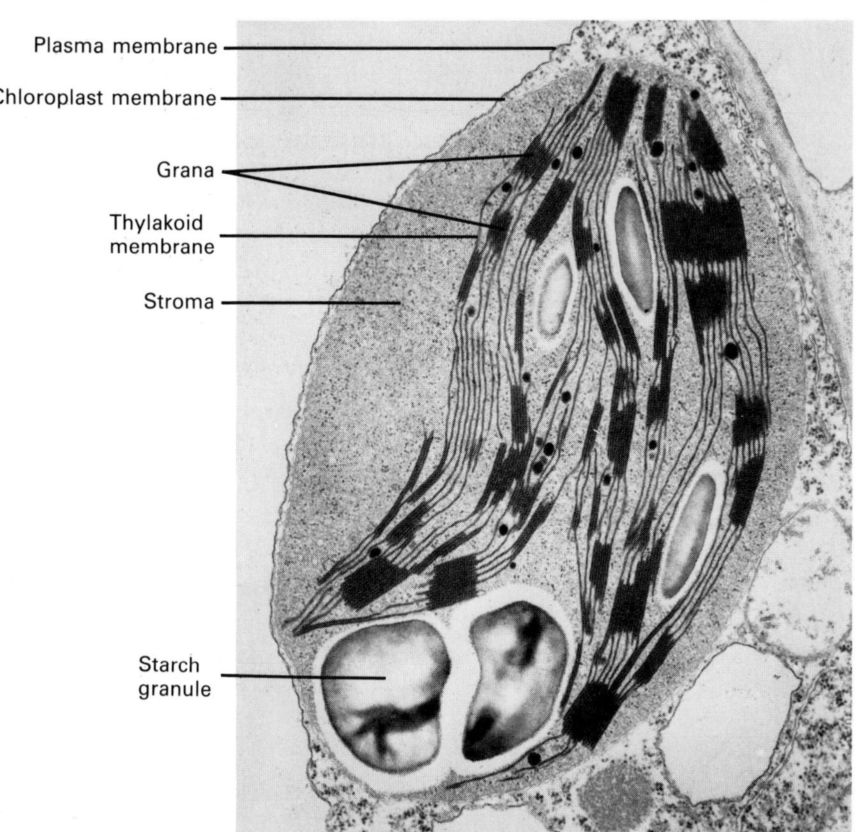

Figure 20-4 Electron micrograph of part of a *Phleum pratanese* cell showing a chloroplast. Note the well-developed grana, or stacks of thylakoid vesicles in the chloroplast, the large starch granules, and the two outer chloroplast membranes. *Photograph courtesy of Biophoto Associates/M. C. Ledbetter/Brookhaven National Laboratory.*

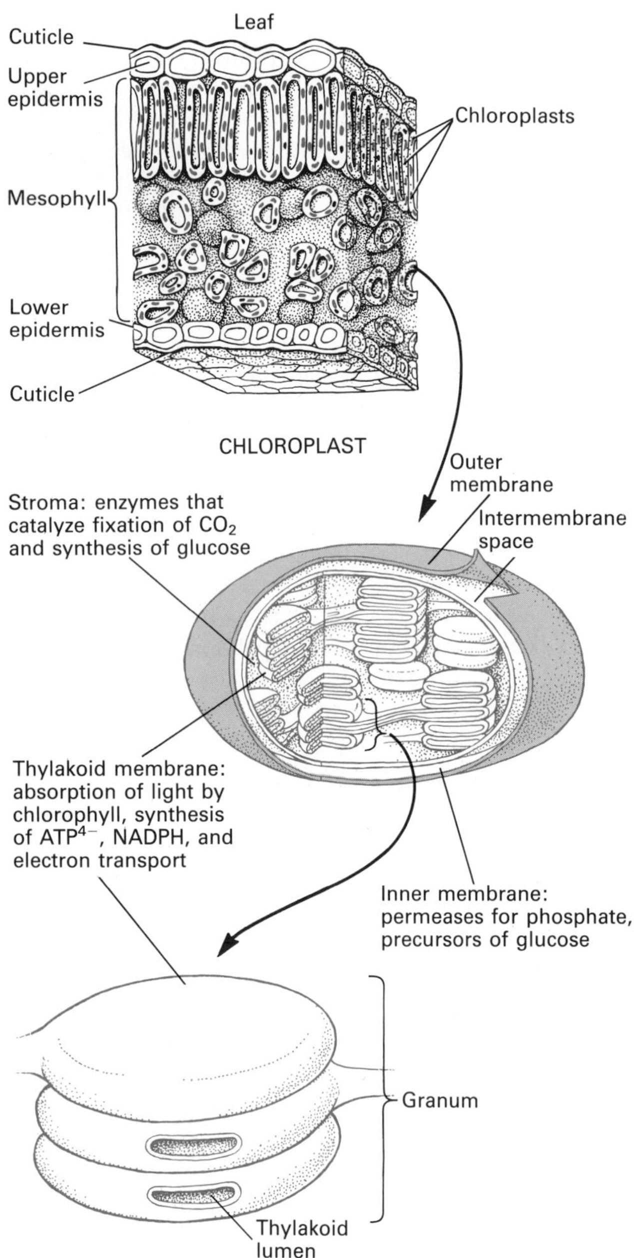

Figure 20-5 Different reactions of photosynthesis occur in different regions of the chloroplast. As shown, the chloroplast is bounded by a double membrane. The outer membrane contains porins that render it permeable to small molecules (m.w. <10,000); the inner membrane forms the permeability barrier of the organelle. Photosynthesis occurs on thylakoid membranes, a series of flat vesicles with interconnecting lumina. Stacks of thylakoid vesicles are termed grana. The stroma is the space surrounding the thylakoid vesicles.

stroma to the lumen of the thylakoid. All of these reactions are catalyzed by proteins within the thylakoid membrane.

Stage 2. The movement of protons down their concentration gradient from the lumen of the thylakoid to the stroma, through a specialized protein in the thylakoid membrane that couples this movement to the synthesis of ATP from ADP and P_i.

Stage 3. The utilization of the ATP and NADPH to "fix" CO_2 and, ultimately, to generate glucose:

$$6CO_2 + 18ATP^{4-} + 12NADPH + 12H_2O \longrightarrow C_6H_{12}O_6 + 18ADP^{3-} + 12NADP^+ + 18P_i^{2-} + 6H^+$$

The reactions of stage 3 are called the *dark reactions* of photosynthesis, because formation of glucose can occur in the dark, utilizing the supplies of ATP and NADPH generated by light energy. The stage 3 reactions are indirectly dependent on light energy, whereas the generation of NADPH and ATP are directly dependent on light energy and are called the *light reactions*.

Later in the chapter we shall examine each of these stages in some detail. First, however, let us consider the analogous processes involved in the generation of ATP by oxidative phosphorylation in the mitochondrion. We can begin by reviewing some of the key metabolic reactions used for oxidation of glucose, fatty acids, and other compounds.

Metabolism for Energy: An Overview

In animal cells, the most important fuels for generation of ATP are sugars, such as glucose, and fatty acids. Glycogen is the principal storage form of glucose, and it is converted to glucose 6-phosphate when needed by the cell (Chapter 16).

$$\text{Glucose}_n + P_i \xrightarrow{\text{glycogen phosphorylase}} \text{glucose 1-phosphate} + \text{glucose}_{n-1}$$

$$\text{Glucose 1-phosphate} \longrightarrow \text{glucose 6-phosphate}$$

Lipids are stored as triacylglycerols, primarily in adipose cells. In response to hormone, the triacylglycerols are hydrolyzed to free fatty acids and glycerol:

$$\begin{array}{l} H_3C(CH_2)_nC(=O)-O-CH_2 \\ H_3C(CH_2)_nC(=O)-O-CH + 3H_2O \longrightarrow \\ H_3C(CH_2)_nC(=O)-O-CH_2 \end{array}$$

$$\underset{\text{Fatty acid}}{3H_3C(CH_2)_nCOOH} + \underset{\text{Glycerol}}{HOCH_2-CH(OH)-CH_2OH}$$

The fatty acids are released into the blood, from which

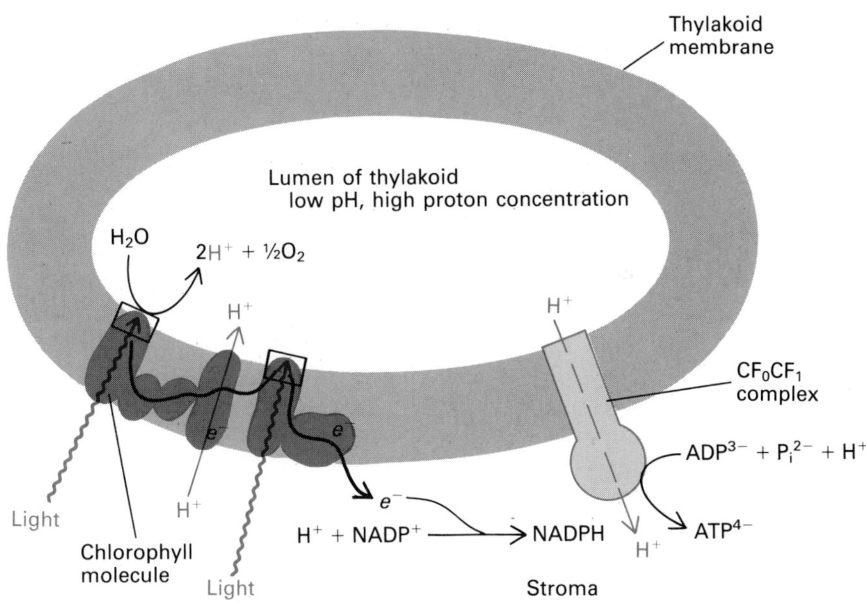

Figure 20-6 Outline of photosynthesis. Quanta of light are absorbed by a chlorophyll molecule in the thylakoid membrane. This causes the splitting of water into molecular oxygen, protons, and electrons. These released electrons have a high energy; they move through a series of membrane proteins and other carriers to the ultimate acceptor, $NADP^+$, to form NADPH. Movement of electrons is coupled to pumping of protons from the exterior to the interior of the thylakoid membranes, thus generating a transmembrane electrochemical proton gradient.
The movement of protons down their concentration gradient through another set of membrane proteins, the CF_0CF_1 complex, is coupled to the synthesis of ATP from ADP and P_i. The NADPH and ATP are generated in the stromal space of the chloroplast, where they are used to fix CO_2 and produce glucose.

they are taken up and oxidized by other cells. In humans, oxidation of fats is quantitatively more important than that of glucose as a source of ATP. In part, this is due to the fact that oxidation of 1 g of triacylglycerol to CO_2 results in the formation of about six times as much ATP as that generated by oxidation of 1 g of hydrated glycogen.

In Chapter 5, we saw that the complete oxidation of a molecule of glucose can generate as many as 36 ATP phosphoanhydride bonds:

$$6O_2 + C_6H_{12}O_6 + 36ADP^{3-} + 36P_i^{-} \longrightarrow 6CO_2 + 6H_2O + 36ATP^{4-}$$

The initial stages of glucose metabolism—glycolysis, or the Embden-Meyerhof pathway—occur in the cytoplasm. One molecule of glucose is converted into two of pyruvate. Concomitantly, two molecules of NAD^+ are reduced to NADH, and there is a net generation of two ATP phosphoanhydride bonds:

$$\underset{\text{Glucose}}{C_6H_{12}O_6} + 2ADP^{3-} + 2P_i^{2-} + 2NAD^+ \longrightarrow \underset{\text{Pyruvate}}{2C_3H_3O_3^-} + 2H^+ + 2NADH + 2ATP^{4-}$$

The balance of the ATPs are generated within the mitochondrion by complete oxidation of pyruvate to CO_2. This is coupled to reduction of molecular oxygen, the ultimate electron acceptor, to water. It is important to note that mitochondria in plant cells are essential for production of ATP during dark periods, when photosynthesis is not possible. Stored carbohydrates, mostly starch, are hydrolyzed to glucose and then metabolized to pyruvate; as in animal mitochondria, pyruvate is oxidized to CO_2, with concomitant generation of ATP.

In the glycolytic conversion of glucose to pyruvate, two phosphoanhydride bonds are consumed in phosphorylation of free glucose to produce, eventually, fructose 1,6-bisphosphate:

$$\text{Glucose} + ATP^{4-} \longrightarrow \text{glucose 6-phosphate}^{2-} + ADP^{3-} + H^+$$
$$\text{Glucose 6-phosphate}^{2-} \longrightarrow \text{fructose 6-phosphate}^{2-}$$
$$\text{Fructose 6-phosphate}^{2-} + ATP^{4-} \longrightarrow \text{fructose 1,6-bisphosphate}^{4-} + ADP^{3-} + H^+$$

Four such bonds are formed during the conversion of each fructose 1,6-bisphosphate to pyruvate, resulting in a net gain of two ATPs. In the conversion of each glucose unit in glucogen to pyruvate, there is a net gain of three ATP phosphoanhydride bonds. Only one ATP is consumed per glucose unit in forming fructose 1,6-bisphosphate, because formation of glucose 1-phosphate occurs by phosphorolysis of glycogen, and no ATP is consumed in this step.

$$\text{Glycogen} + P_i^{2-} \longrightarrow \text{glucose 1-phosphate}^{2-} \longrightarrow \text{glucose 6-phosphate}^{2-} \longrightarrow \text{fructose 6-phosphate}^{2-} \xrightarrow{ATP^{4-} \quad ADP^{3-}} \text{fructose 1,6-bisphosphate}^{4-} + H^+$$

In Glycolysis ATP Is Generated by Substrate-Level Phosphorylation

Cells utilize two basically different types of processes to synthesize ATP. Chloroplasts and mitochondria utilize ionic gradients across a membrane as the immediate energy source for ATP synthesis. The other process is used in two steps of glycolysis and involves chemical transformations of soluble substances in the cytosol by aqueous enzymes; membranes and gradients are not involved. Such processes are called *substrate-level phosphorylation.*

The first reaction in glycolysis involving substrate-level phosphorylation is the set of processes catalyzed by (1) glyceraldehyde 3-phosphate dehydrogenase and (2) phosphoglycerate kinase.

(1) Glyceraldehyde 3-phosphate^{2-} + NAD^+ + P_i^{2-} $\rightleftharpoons$ 1,3-bisphosphoglycerate^{4-} + NADH + H^+

$\Delta G^{\circ\prime} = +1.5$ kcal/mol

(2) 1,3-Bisphosphoglycerate^{4-} + ADP^{3-} $\longrightarrow$ 3-phosphoglycerate^{3-} + ATP^{4-}

$\Delta G^{\circ\prime} = -4.5$ kcal/mol

The net change in standard free energy, however, is negative ($\Delta G^{\circ\prime} = -3.0$ kcal/mol), so overall the reaction is strongly *exergonic* (accompanied by the release of free energy). Since each molecule of fructose 1,6-bisphosphate generates two molecules of glyceraldehyde 3-phosphate (see Figure 5-42), two ATPs are generated at this step during catabolism of one molecule of glucose. All the enzymes and reactants are soluble in the cytosol. In the first reaction (Figures 20-7 and 20-8), the oxidation of the aldehyde ($\text{—}\overset{\overset{\displaystyle O}{\|}}{C}\text{—H}$) group on glyceraldehyde 3-phosphate by NAD^+ is coupled to the addition of a phosphate group, forming 1,3-bisphosphoglycerate with a single high-energy phosphate bond. The $\Delta G^{\circ\prime}$ value for hydrolysis of the $\overset{\overset{\displaystyle O}{\|}}{C}\text{—O—}PO_3^{2-}$ bond in 1,3-bis-phosphoglycerate is more negative than that for the terminal phosphodiester bond in ATP (~-12 kcal/mol versus -7.3 kcal/mol). This high-energy phosphate is transferred to ADP in a strongly exergonic reaction catalyzed by phosphoglycerate kinase (Figure 20-7).

The next stages in glycolysis also generate a high-energy phosphate bond in the molecule phosphoenolpyruvate. This too is transferred to ADP in a strongly exergonic reaction, which is catalyzed by pyruvate kinase (Figure 20-9). In this set of reactions, loss of water converts the phosphate-carbon bond in 2-phosphoglycerate from a low-energy bond ($\Delta G^{\circ\prime}$ of hydrolysis -2 kcal/mol) to a high-energy bond ($\Delta G^{\circ\prime}$ of hydrolysis -15 kcal/mol) in phosphoenolpyruvate.

These two examples illustrate how interconversions of soluble chemicals by soluble enzymes are coupled to generate ATP. Recall, however, that only two of the 36 ATP molecules generated during complete oxidation of glycose to CO_2 are made during the conversion of glucose to pyruvate (Chapter 5); the rest are synthesized in the mitochondrion by a fundamentally different type of process, involving generation and utilization of proton gradients across the inner mitochondrial membrane. Both electron transfer and generation of ATP occur on the inner mitochondrial membrane, and key metabolic reactions occur within specific submitochondrial compartments.

The Inner and Outer Mitochondrial Membranes Are Structurally and Functionally Different

The inner and outer mitochondrial membranes define the two submitochondrial spaces: the *intermembrane space* between the two membranes, and the central compartment, called the matrix (Figure 20-10). Because the different membranes and compartments of the mitochondrion can be purified and studied separately, it has been possi-

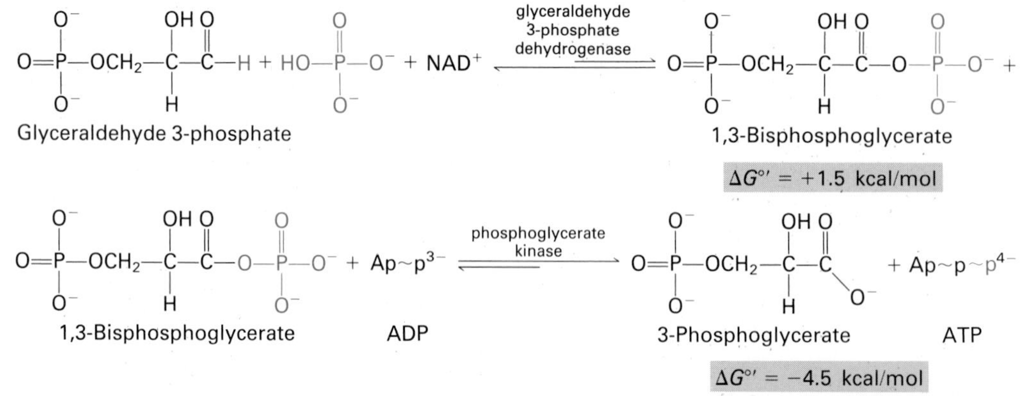

Glyceraldehyde 3-phosphate + $HO\text{—}PO_3^{2-}$ + NAD^+ $\rightleftharpoons$ (glyceraldehyde 3-phosphate dehydrogenase) 1,3-Bisphosphoglycerate + NADH + H^+

$\Delta G^{\circ\prime} = +1.5$ kcal/mol

1,3-Bisphosphoglycerate + ADP ($Ap{\sim}p^{3-}$) $\rightleftharpoons$ (phosphoglycerate kinase) 3-Phosphoglycerate + ATP ($Ap{\sim}p{\sim}p^{4-}$)

$\Delta G^{\circ\prime} = -4.5$ kcal/mol

Figure 20-7 Substrate-level phosphorylation. Glyceraldehyde 3-phosphate dehydrogenase and phosphoglycerate kinase are water-soluble enzymes that catalyze synthesis of ATP in the cytoplasm. This process of generating ATP high-energy bonds is very different from that employed in oxidative phosphorylation or photophosphorylation.

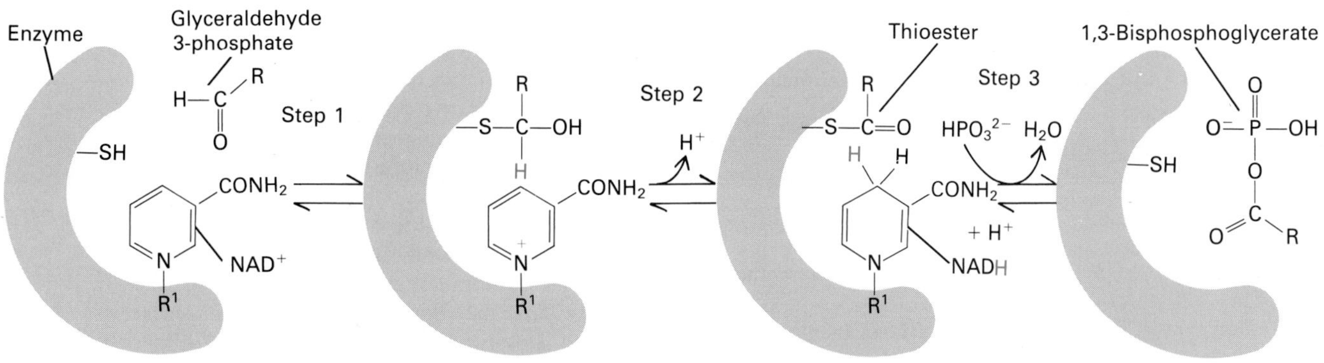

Figure 20-8 A thioester, R—C(=O)—S—R′, is an energy-rich intermediate in the glyceraldehyde 3-phosphate dehydrogenase reaction. The sulfhydryl group (—SH) is the side chain of cysteine at the active site; R symbolizes the rest of the glyceraldehyde 3-phosphate molecule.

In step 1 of the reaction, the sulfhydryl group on the enzyme reacts with glyceraldehyde 3-phosphate to form a thiohemiacetal, —S—C(R)—OH, which is the species oxidized. In step 2, a hydrogen atom *(color)*, together with two electrons, is transferred to NAD^+, thus forming the reduced species NADH. Simultaneously, a proton bound to the O atom is lost to the medium. The products of step 2 are a thioester, —S—C(R)=O, and NADH + H^+. In step 3, the thioester reacts with phosphate to produce 1,3-bisphosphoglycerate, and then regenerates the free enzyme with its sulfhydryl group.

ble to localize each reaction to a specific membrane or space (Figure 20-10).

The inner membrane and the matrix are the sites of most of the reactions involving the oxidation of pyruvate and fatty acids to CO_2 and also the coupled synthesis of ATP from ADP and P_i. These very complex processes involve many discrete steps, but they can conveniently be subdivided into three groups of reactions, each of which occurs in a discrete membrane or space within the mitochondrion (Figure 20-11):

1. Oxidation of pyruvate or fatty acids to CO_2, coupled to the reduction of the electron carriers NAD^+ and FAD to NADH and $FADH_2$. These reactions occur in the matrix, or on inner-membrane proteins that face the matrix.
2. Transfer of electrons from NADH and $FADH_2$ to O_2. These reactions occur in the inner membrane and are coupled to the generation of an electrochemical proton gradient across the inner membrane.
3. Utilization of the energy stored in the transmembrane proton gradient for synthesis of ATP. This step is analogous to the step in photosynthetic generation of ATP and utilizes extremely similar protein complexes in the inner mitochondrial membrane called the *F_0F_1-ATPase complex.*

Each of the latter two groups of interrelated operations involves sets of multisubunit proteins that are in a highly asymmetrical orientation within the inner mitochondrial membrane.

The outer mitochondrial membrane defines the smooth

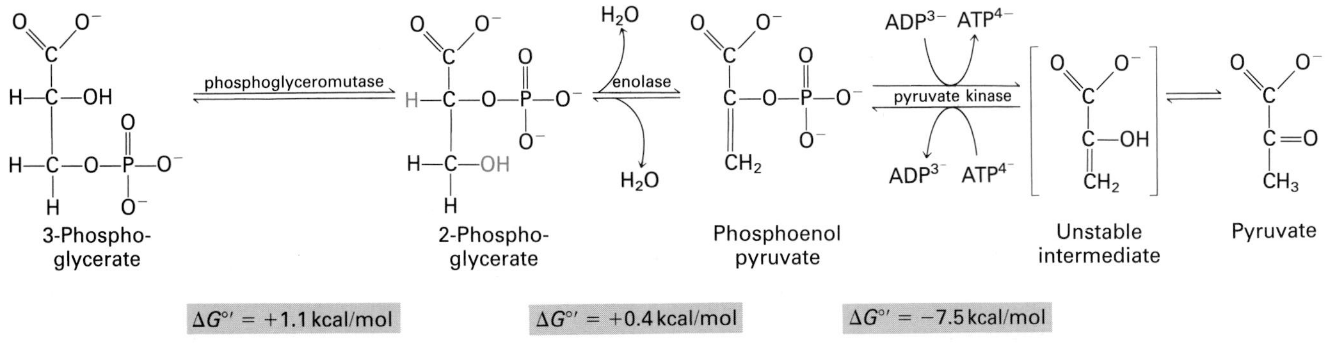

Figure 20-9 Formation of the second pair of ATP molecules during glycolysis. This reaction, catalyzed by pyruvate kinase, is another example of substrate-level phosphorylation.

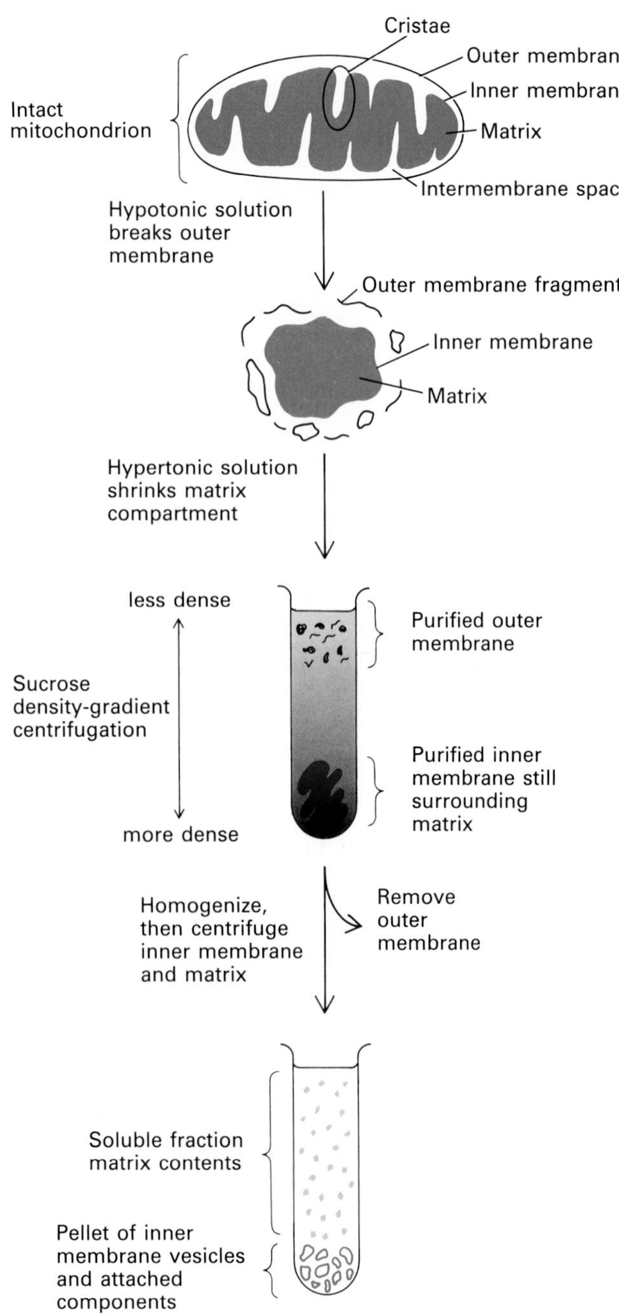

Figure 20-10 Fractionation of purified mitochondria into separate components. Studies on these purified components have elucidated each of the reactions in oxidative phosphorylation. This scheme starts with mitochondria that have been purified by a combination of velocity and density-gradient centrifugation (Chapter 5). When the mitochondria are placed in a hypotonic solution, water flows osmotically inward across the outer and inner membranes, causing the matrix to swell. But the outer membrane cannot expand its surface area, so it ruptures, releasing the contents of the intermembrane space. By centrifugation of the resultant material through a density gradient, the less dense outer membrane is separated from the denser inner membrane and the matrix. Homogenization of the latter material causes the inner membrane to fracture, thereby producing smaller vesicles and releasing the soluble components of the matrix. A simple velocity centrifugation step then separates the membrane-bound components from the soluble matrix components.

outer perimeter of the organelle. It is an unusual membrane in that it is freely permeable to most small molecules (m.w. $<10,000$); in particular, it is freely permeable to protons. It does not act as a barrier to diffusion of metabolites. The permeability is due to an outer membrane protein, *porin,* that forms a transmembrane channel. Porin is similar in structure and function to the protein lining the pores in the outer membrane of gram-negative bacteria (see Figure 5-23), which also confers permeability to most small molecules. Thus, the inner membrane is the only effective permeability barrier between the cytoplasm and the matrix of the mitochondrion.

The inner mitochondrial membrane contains a higher fraction of protein—76 percent of the total weight of the membrane—than that in any other cellular membrane. Cardiolipin (diphosphatidyl glycerol, see Figure 3-64), the lipid concentrated in this inner mitochondrial membrane, is believed to reduce the permeability of the lipid bilayer to protons, thus making it less "leaky." Freeze-fracture studies on the inner mitochondrial membrane indicate that it contains many protein-rich intramembrane particles that are laterally mobile in the plane of the membrane (see Figure 14-31a). As we shall see later in the chapter, among these are four particles that function in electron transport from NADH or $FADH_2$ to O_2 and in the synthesis of ATP. Others are permeases that allow otherwise impermeable molecules such as ADP and other phosphorylated compounds to pass from the cytosol to the matrix.

The highly convoluted foldings of the inner membrane—the *cristae*—greatly expand its surface area, thus enhancing its ability to generate ATP. In typical hepatocyte mitochondria, for example, the area of the inner membrane is about five times that of the outer membrane. In fact, in liver cells the total area of all mitochondrial inner membranes is about 15-fold that of the plasma membrane. In heart and skeletal muscle, mitochondria contain three times as many cristae as those found in typical hepatocyte mitochondria—presumably reflecting the greater demand for ATP in muscle cells. Mitochondria in certain tissues have cristae folded in different ways, but the significance of such morphological differences is not known.

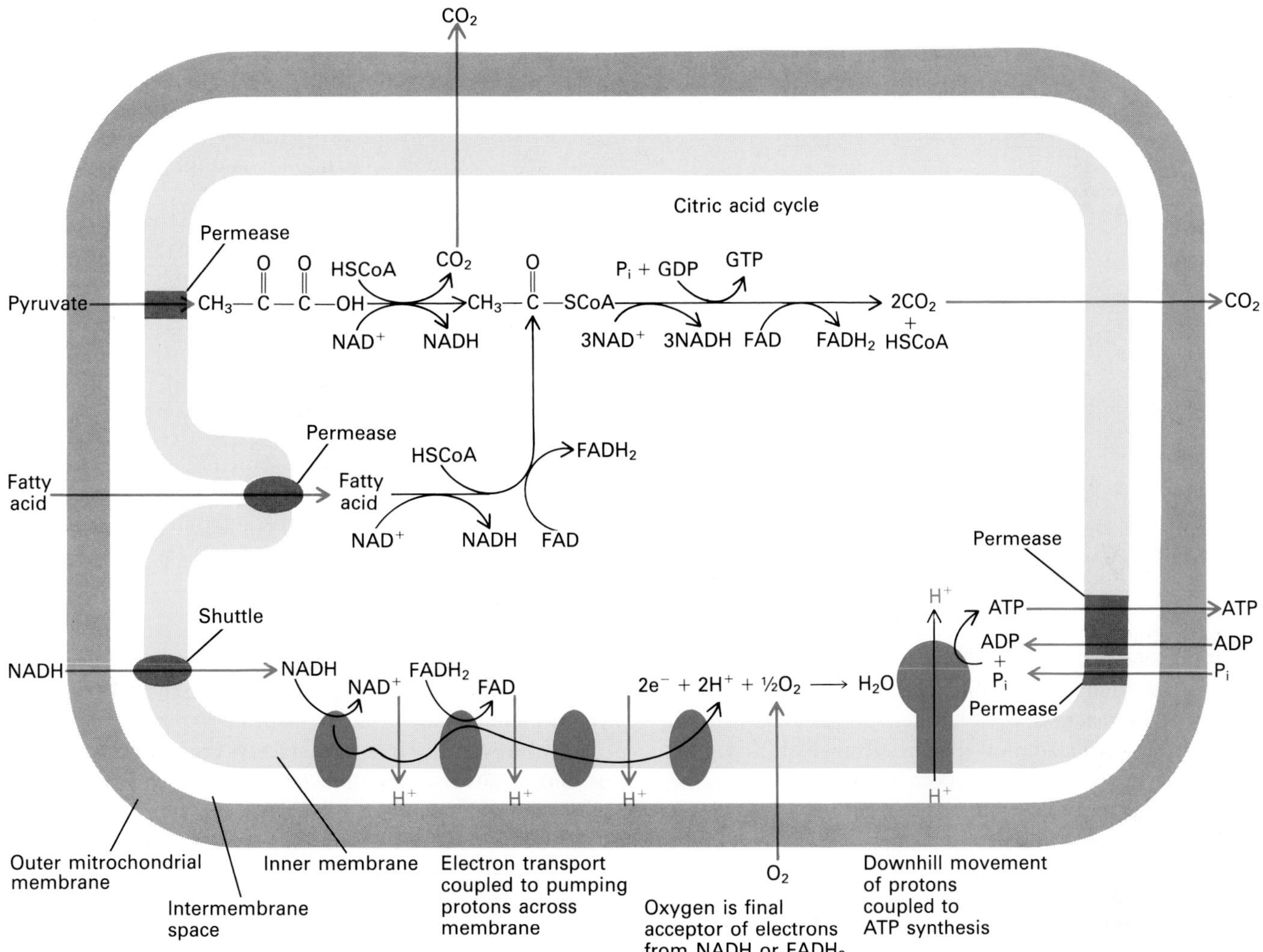

Figure 20-11 Outline of the major metabolic reactions in mitochondria. The substrates of oxidative phosphorylation—pyruvate, fatty acids, ADP, and P_i—are transported into the matrix from the cytosol by permeases; O_2 diffuses into the matrix. NADH, generated in the cytoplasm during glycolysis, is not transported directly into the matrix, because the inner membrane is impermeable to NAD^+ or NADH. Rather, a "shuttle" system (Figure 20-22) is used to transport the electrons on cytosolic NADH into the electron transport chain. The products of metabolism, such as ATP, are transported into the cytoplasm, and CO_2 diffuses into the cytoplasm across the mitochondrial membranes.

Acetyl CoA Is a Key Intermediate in the Mitochondrial Metabolism of Pyruvate and Fatty Acids

Pyruvate, generated in the cytoplasm during glycolysis, is transported across the mitochondrial membranes into the matrix of the mitochondrion (Figure 20-11). Complete oxidation of pyruvate to form CO_2 and water occurs in the mitochondrion and utilizes O_2 as the final electron acceptor (oxidizer).

Figure 20-11 traces the metabolism of pyruvate in the mitochondrion. Immediately upon entry into the matrix it is metabolized to the important intermediate acetyl CoA (Figure 20-12) by the enzyme pyruvate dehydrogenase, a soluble component of the matrix:

$$NAD^+ + \underset{\text{Pyruvate}}{CH_3{-}\overset{O}{\overset{\|}{C}}{-}\overset{O}{\overset{\|}{C}}{-}O^-} + \underset{\text{Coenzyme A}}{HSCoA} \longrightarrow \underset{\text{Acetyl CoA}}{CH_3{-}\overset{O}{\overset{\|}{C}}{-}SCoA} + CO_2 + NADH + H^+$$

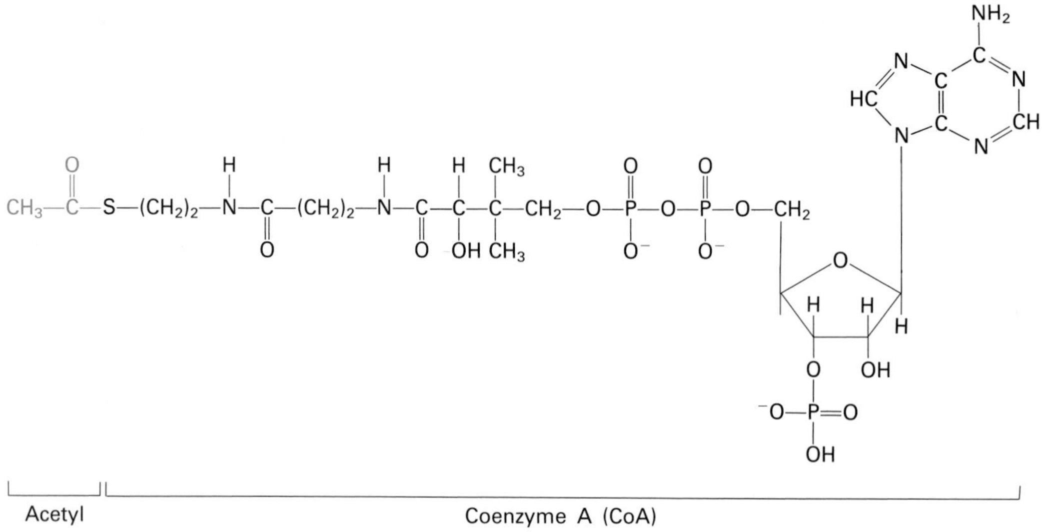

Acetyl — Coenzyme A (CoA)

Figure 20-12 Diagrammed here is the structure of acetyl CoA, an important intermediate in the metabolism of pyruvate, fatty acids, and many amino acids.

This reaction is highly exergonic ($\Delta G^{\circ\prime} = -8.0$ kcal/mol) and is essentially irreversible. Despite the apparent simplicity of the reaction, pyruvate dehydrogenase is actually one of the most complex enzymes known. It is a giant molecule (m.w. >7 million), even larger than a ribosome, and contains multiple subunits of three different enzymes, several regulatory polypeptides, and five different coenzymes, all carefully ordered in the complex.

Acetyl CoA is also a key intermediate in oxidation of fatty acids. Free fatty acids are transported into the mitochondrial matrix by a special carrier molecule. Within the matrix they react with coenzyme A to form the acyl CoA (Figure 20-13). By the four reactions depicted in Figure 20-13, each molecule of acyl CoA is oxidized to form one molecule of acetyl CoA and an acyl CoA shortened by two carbon atoms. Concomitantly one NAD^+ and one FAD are reduced to NADH and $FADH_2$, respectively. The set of reactions is repeated on the shortened acyl CoA until all of the C atoms are converted to acetyl CoA. For stearoyl CoA $\left[CH_3—(CH_2)_{16}—\overset{\overset{\displaystyle O}{\|}}{C}—SCoA\right]$, for instance, eight repetitions of these four reactions are required, generating nine acetyl CoA molecules:

$$CH_3—(CH_2)_{16}—\overset{\overset{\displaystyle O}{\|}}{C}—SCoA + 8HSCoA +$$

$$8FAD + 8NAD^+ + 8H_2O \longrightarrow$$

$$9CH_3—\overset{\overset{\displaystyle O}{\|}}{C}—SCoA + 8FADH_2 + 8NADH + 8H^+$$

Acetyl CoA occupies a central position in the oxidation of not only fatty acids and carbohydrates but also many amino acids. It is also an intermediate in many biosynthetic reactions, such as the transfer of an acetyl group to lysine residues in histone proteins and to the N-termini of many mammalian proteins. Within respiring mitochondria, however, the fate of the acetyl group of acetyl CoA is almost always oxidation to CO_2.

The Citric Acid Cycle Oxidizes the Acetyl Group of Acetyl CoA to CO_2 and Generates NADH and $FADH_2$

The final set of reactions involved in oxidation of carbohydrates and lipids is named the *citric acid cycle,* or the *Krebs cycle.* This complex set of nine reactions is depicted in Figure 20-14, but first let us consider the net reaction:

$$CH_3—\overset{\overset{\displaystyle O}{\|}}{C}—SCoA + 3NAD^+ + FAD +$$

$$GDP^{3-} + P_i^{2-} + 2H_2O \longrightarrow$$

$$2CO_2 + 3NADH + FADH_2 + 2H^+ + GTP^{4-} + HSCoA$$

Note there is no involvement of molecular oxygen at this step, and there is synthesis of only one high-energy phosphate bond (in GTP). The two carbon atoms in acetyl CoA are oxidized to two molecules of CO_2. Concomitantly, the electrons released are transferred to the electron carriers NAD and FAD, to form the reduced molecules NADH and $FADH_2$.

As shown in Figure 20-14, the cycle begins with condensation of the two-carbon acetyl group, from acetyl CoA, with the four-carbon molecule oxaloacetate. The product of reaction *1* is the six-carbon citric acid, for which the cycle is named. In reactions *2* and *3*, citrate

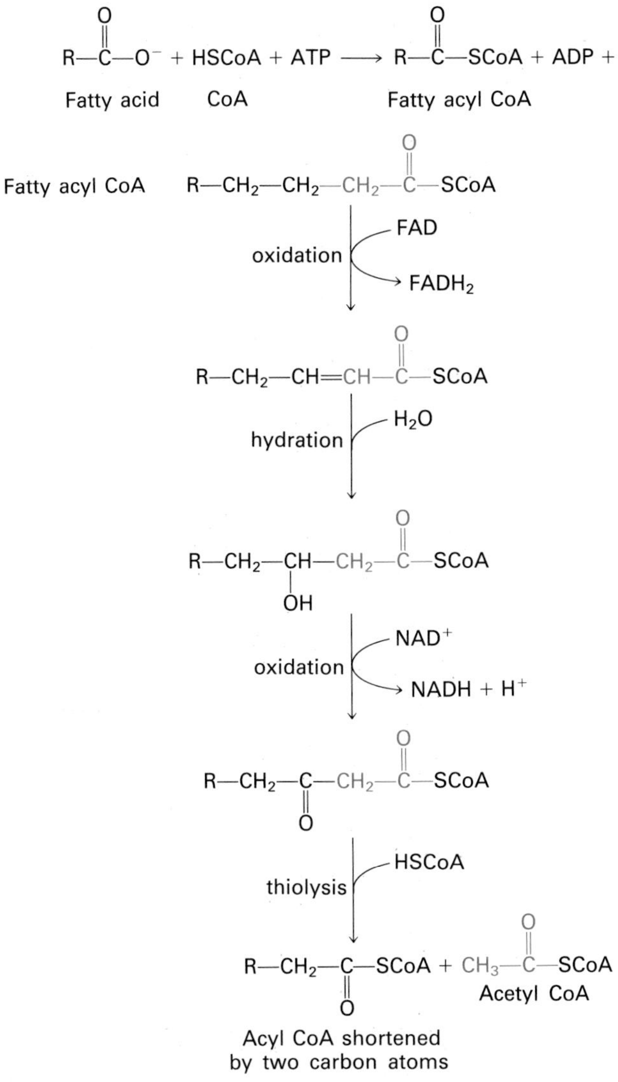

$$R{-}\overset{O}{\overset{\|}{C}}{-}O^- + HSCoA + ATP \longrightarrow R{-}\overset{O}{\overset{\|}{C}}{-}SCoA + ADP + P_i$$

Fatty acid CoA Fatty acyl CoA

Fatty acyl CoA $R{-}CH_2{-}CH_2{-}CH_2{-}\overset{O}{\overset{\|}{C}}{-}SCoA$

oxidation ($FAD \rightarrow FADH_2$)

$R{-}CH_2{-}CH{=}CH{-}\overset{O}{\overset{\|}{C}}{-}SCoA$

hydration (H_2O)

$R{-}CH_2{-}\underset{OH}{\underset{|}{CH}}{-}CH_2{-}\overset{O}{\overset{\|}{C}}{-}SCoA$

oxidation ($NAD^+ \rightarrow NADH + H^+$)

$R{-}CH_2{-}\underset{O}{\underset{\|}{C}}{-}CH_2{-}\overset{O}{\overset{\|}{C}}{-}SCoA$

thiolysis ($HSCoA$)

$R{-}CH_2{-}\underset{O}{\underset{\|}{C}}{-}SCoA + CH_3{-}\overset{O}{\overset{\|}{C}}{-}SCoA$

Acyl CoA shortened by two carbon atoms — Acetyl CoA

Figure 20-13 Oxidation of fatty acids. Four enzyme-catalyzed reactions convert a fatty acyl CoA molecule to acetyl CoA and a fatty acyl CoA shortened by two carbon atoms. Concomitantly, there is reduction of one NAD^+ to NADH, and of one FAD to $FADH_2$. The cycle is repeated on the shortened acyl CoA, until fatty acids with an even number of carbon atoms are completely converted to acetyl CoA. Fatty acids with odd numbers of C atoms are rare; they are metabolized to one molecule of propionyl CoA $\left(CH_3{-}CH_2{-}\overset{O}{\overset{\|}{C}}{-}SCoA\right)$ and multiple acetyl CoAs.

is isomerized to the six-carbon molecule isocitrate. In reaction 4, isocitrate is oxidized to the five-carbon α-ketoglutarate, generating one molecule of CO_2 and reducing one molecule of NAD^+ to NADH. In reaction 5 the α-ketoglutarate is oxidized to the four-carbon molecule succinyl CoA, generating the second CO_2 formed during each turn of the cycle, and reducing another NAD^+ to NADH. In reactions 6 to 9, succinyl CoA is oxidized to oxaloacetate, regenerating the molecule that was used initially to condense with acetyl CoA. Concomitantly, there is a reduction of one FAD (to $FADH_2$) and one NAD^+ (to NADH). The conversion of succinyl CoA to succinate (reaction 6) is coupled to the synthesis of one molecule of GTP (from GDP and P_i); this reaction is slightly exergonic ($\Delta G^{\circ\prime} = -0.8$ kcal/mol).

Most of the enzymes and small molecules involved in the citric acid cycle are soluble in aqueous solution and are localized to the matrix of the mitochondrion (Figure 20-11). The exceptions are coenzyme A (CoA), acetyl CoA, and succinyl CoA, in which the hydrophobic CoA segment is bound to the inner membrane, with the acetyl or succinyl part facing the matrix. Succinate dehydrogenase (reaction 7) is also localized to the inner membrane.

Molecular Oxygen Oxidizes NADH and $FADH_2$

The reactions detailed thus far result in the conversion of one molecule of glucose into six molecules of CO_2. Concomitantly, 10 NAD^+ molecules are reduced to 10 of NADH, and 2 of FAD to 2 of $FADH_2$. Two NADH are formed in conversion of glucose to two pyruvate molecules. Oxidation of each pyruvate to acetyl CoA generates one NADH (or a total of two per glucose). Oxidation of each acetyl CoA to CO_2 generates three molecules of NADH and one molecule of $FADH_2$, or six of NADH and two of $FADH_2$ per glucose.

Molecular oxygen is used to reoxidize these reduced coenzymes, but this does not occur in a single step. Rather, NADH and $FADH_2$ transfer their pairs of electrons to acceptor molecules in the inner mitochondrial membrane. The loss of electrons regenerates the oxidized forms of NAD^+ and FAD as well as the reduced form of the acceptor. The electrons thus released are transferred along a chain of electron carriers, all of which are integral components of the inner mitochondrial membrane. Eventually they are transferred to oxygen, the ultimate electron acceptor, thus forming water:

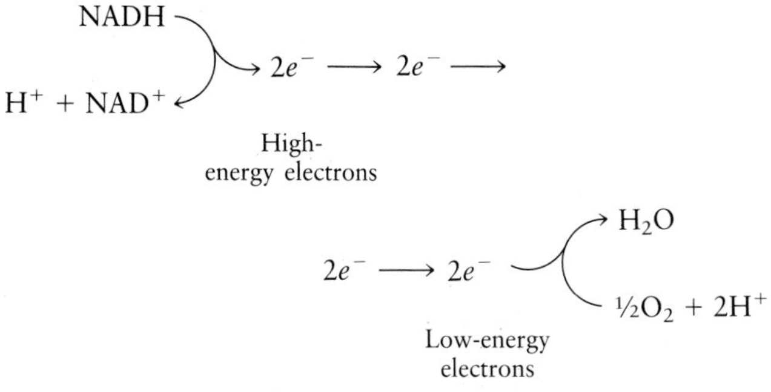

$$NADH \rightarrow H^+ + NAD^+ \;\; + \;\; 2e^- \longrightarrow 2e^- \longrightarrow$$

High-energy electrons

$$2e^- \longrightarrow 2e^- \;\; (\tfrac{1}{2}O_2 + 2H^+ \rightarrow H_2O)$$

Low-energy electrons

The following overall reactions summarize these steps:

$$NADH + H^+ + \tfrac{1}{2}O_2 \longrightarrow NAD^+ + H_2O \qquad \Delta G^{\circ\prime} = -52.6 \text{ kcal/mol}$$

$$FADH_2 + \tfrac{1}{2}O_2 \longrightarrow FAD + H_2O \qquad \Delta G^{\circ\prime} = -43.4 \text{ kcal/mol}$$

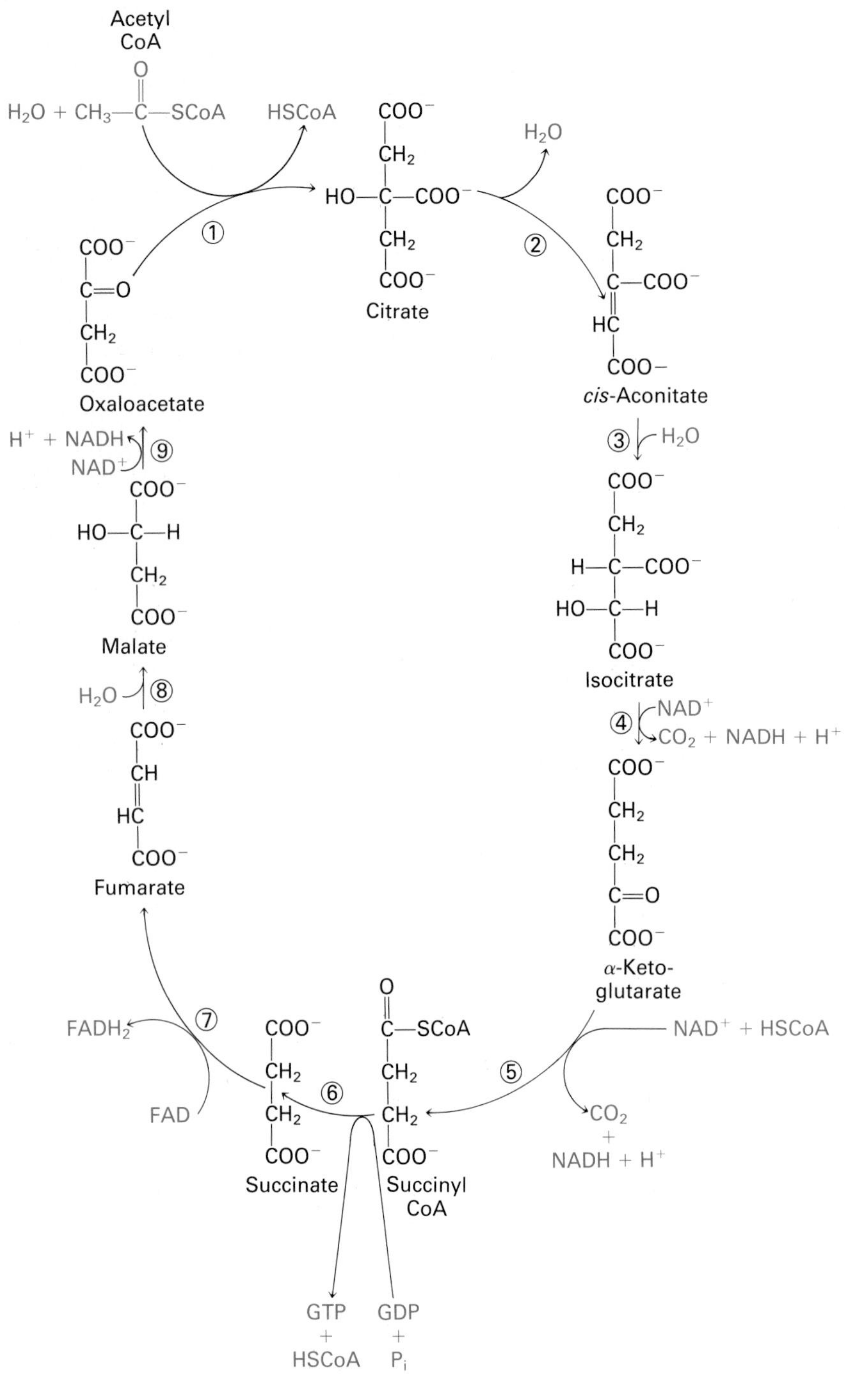

Figure 20-14 The citric acid cycle. First, a two-carbon acetyl residue ($CH_3{-}\overset{O}{\overset{\|}{C}}{-}$) condenses with a four-carbon molecule, oxaloacetate, to form the six-carbon molecule citrate. Through a sequence of enzyme-catalyzed reactions *(circled numbers 2–9)*, each molecule of citrate is converted to oxaloacetate, losing two CO_2 molecules. In four reactions, four pairs of electrons are removed from the carbon atoms: three pairs are transferred to three molecules of NAD^+, to form $3NADH + 3H^+$, and one pair is transferred to the acceptor FAD, to form $FADH_2$.

As is indicated by the negative $\Delta G^{\circ\prime}$ value, oxidation of these reduced coenzymes by O_2 is a strongly exergonic reaction. More important, most of the free energy released during oxidation of glucose to CO_2 is retained in the reduced coenzymes that are generated during glycolysis and the citric acid cycle. To see this, recall the large total change in standard free energy for the oxidation of glucose to CO_2 ($\Delta G^{\circ\prime} \sim -680$ kcal/mol); oxidation of the 10 NADH and 2 $FADH_2$ molecules yields an almost equivalent change in standard free energy [$\Delta G^{\circ\prime} = 10(-52.6) + 2(-43.4) = -613$ kcal/mol of glucose]. Thus over 90 percent of the potential free energy present in the oxidizable bonds of glucose is conserved in the reduced coenzymes. The reoxidation of these by O_2 generates the vast majority of ATP phosphoanhydride bonds.

The free energy released during oxidation of a single NADH or $FADH_2$ molecule by O_2 is sufficient to drive the synthesis of several molecules of ATP from ADP and P_i (for the reaction $ADP + P_i \rightarrow ATP$, $\Delta G^{\circ\prime} \sim -7.3$ kcal/mol). Thus it is not surprising that oxidation of NADH and synthesis of ATP do not occur in a single reaction. Rather, step-by-step transfer of electrons from NADH to O_2, via a series of proteins that constitute the *electron transport chain,* allows the free energy to be released in small increments. At several sites in electron transport from NADH to O_2, protons are pumped across the mitochondrial membrane, and a high concentration of protons is achieved in the space between the inner and outer mitochondrial membranes (Figure 20-11). A membrane potential across the inner mitochondrial membrane is also created as a result of the pumping of positively charged protons from the matrix space outward. The matrix of the mitochondrion is negative with respect to the space between the inner and outer membranes. Free energy released during oxidation of NADH or $FADH_2$ is stored in the electrochemical proton gradient across the mitochondrial membrane. As in photosynthesis, the movement of protons back across the inner mitochondrial membrane *down* their concentration gradient is coupled to the synthesis of ATP from ADP and P_i.

The Electrochemical Proton Gradient Is Used for Generation of ATP from ADP and P_i

The membrane protein complex involved in ATP synthesis is called the F_0F_1 complex. These complexes form the tadpole-shaped particles that protrude from the inner membrane (Figure 20-15). Exactly how ATP synthesis is coupled to proton movement is not clear. Later in this chapter we shall examine the structure and function of this F_0F_1 complex in some detail.

Bacteria lack mitochondria or any other internal membrane system; yet aerobic bacteria carry out the same processes of oxidative phosphorylation that occur in eukaryotic mitochondria. Enzymes catalyzing the reactions of both the Embden-Meyerhof pathway and the citric acid cycle are all localized to the bacterial cytosol. The enzymes that oxidize NADH to NAD^+ and transfer the electrons to the ultimate acceptor, O_2, are localized to the bacterial plasma membrane. Movement of electrons through these membrane carriers is coupled to the pumping of protons out of the cell (Figure 20-16). Movement of protons back into the cell down their concentration gradient is coupled to the synthesis of ATP; the membrane proteins involved in ATP synthesis have a structure and function very similar to those of the mitochondrial F_0F_1 complex.

It is important to note that the direction of proton movements is actually the same in chloroplasts, mitochondria, and bacteria (Figure 20-16). In Chapter 14 we noted that every cellular membrane has a cytoplasmic face and an extracytoplasmic face. The F_0F_1 complex is always positioned so that the tadpole-shaped F_1 segment faces the cytoplasmic face of the membrane (Figure 20-16b). In particular, thylakoid membranes are formed as invaginations of the inner chloroplast membrane (Figure 20-16b). Thus, the *outer* surface of the thylakoid membrane is the cytoplasmic surface; it is equivalent to the stromal (or inner) surface of the inner chloroplast membrane. Since the F_1 segment catalyzes ATP synthesis, ATP is formed on the cytoplasmic face of the membrane, while protons flow through the F_0F_1 complex across the membrane from the extracytoplasmic to the cytoplasmic surface. The direction of the proton gradient, established by electron transport, is such that the proton concentration is greater on the extracytoplasmic surface than on the cytoplasmic. Thus, during mitochondrial electron transport, protons are pumped out of the matrix into the intermembrane space. In photosynthesis, electrons are pumped from the stroma to the lumen of the thylakoid, from the cytoplasmic to the extracytoplasmic face of the thylakoid membrane. As discussed in Chapter 25, chloroplasts and mitochondria have probably evolved from cyanobacteria and purple sulfur bacteria, respectively.

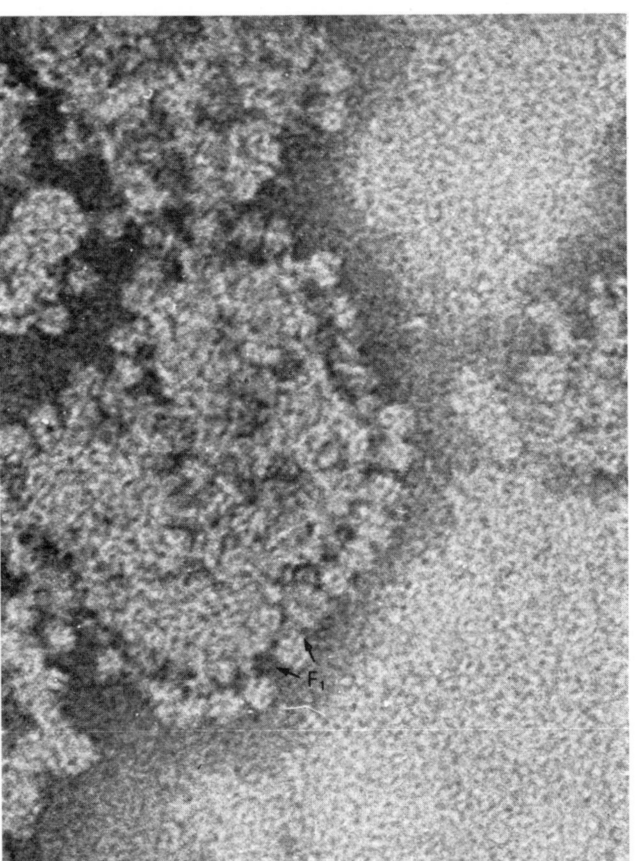

Figure 20-15 Electron micrograph of a submitochondrial particle, showing F_1 projections on its surface. These are protein complexes that synthesize ATP from ADP and P_i. *Courtesy of E. Racker.*

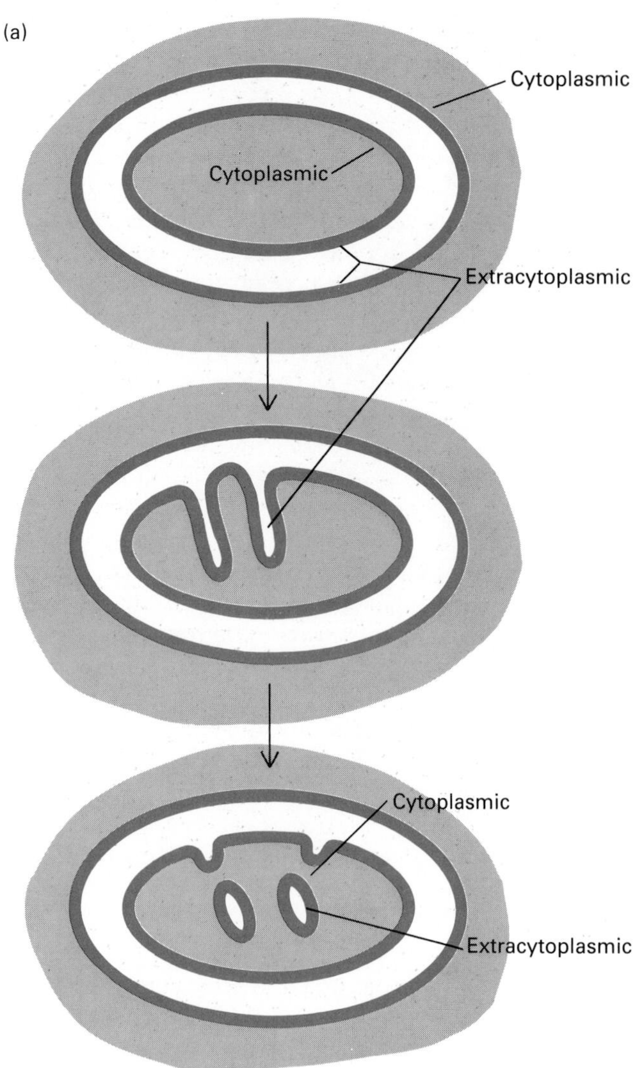

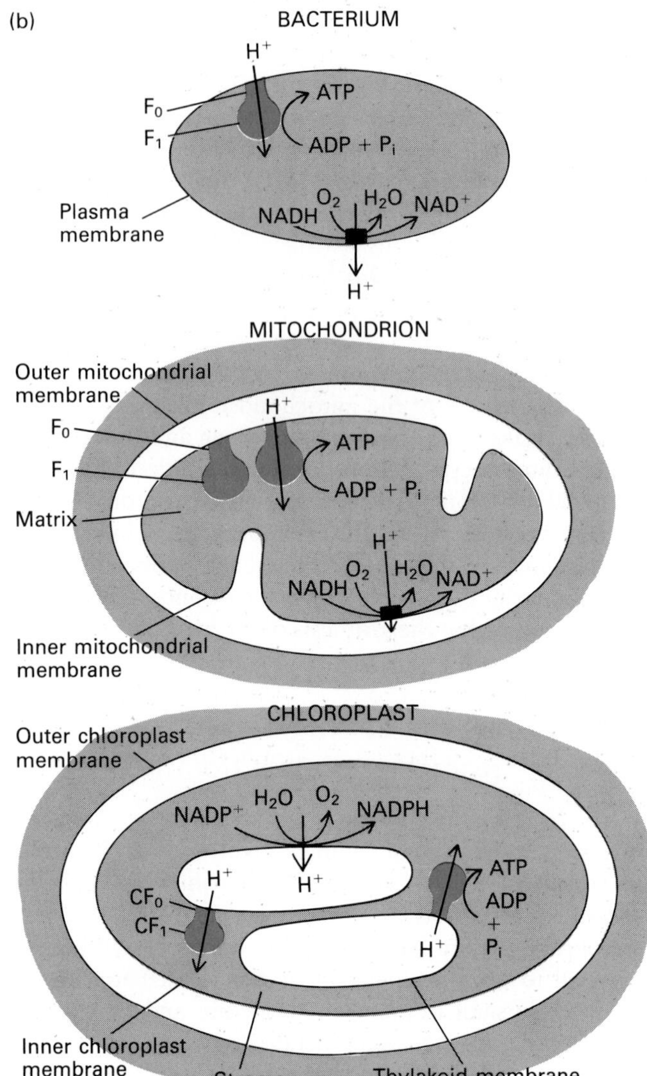

Figure 20-16 Membrane orientation and the direction of proton movement in mitochondria, bacteria, and chloroplasts. (a) The surface of a membrane facing a shaded area is designated a cytoplasmic surface; that facing an unshaded area is termed the extracytoplasmic surface. Thylakoid vesicles are formed by invagination of the inner chloroplast membrane. Thus the inner surface of the thylakoid membrane is the extracytoplasmic one; it is equivalent to the surface of the inner membrane that faces the intermembranous space. (b) In bacteria, mitochondria, and chloroplasts the F_0F_1 complexes *always* face the cytoplasmic surface of the membrane. During electron transport, protons are pumped *from* the cytoplasmic *to* the extracytoplasmic surface of the membrane. During generation of ATP, protons flow in the reverse direction, down their electrochemical gradient, through the F_0F_1 complexes.

In both photosynthesis and oxidative phosphorylation, the movement of electrons through the membrane is coupled to the generation of an electrochemical proton gradient. As we shall see, the processes of electron transport, proton pumping, and generation of ATP are interdependent. The generation of the electrochemical proton gradient—also called the *proton-motive force*—and its subsequent dissipation in the formation of ATP normally occur simultaneously, and the processes are closely coupled.

The processes of photosynthesis and oxidative phosphorylation are quite complex, and the details of many of the steps of each process remain subject to heated controversy. Still, through a combined effort of biochemists, biophysicists, microscopists, and geneticists, much has been learned. The following section describes the use of a proton gradient for the generation of ATP in oxidative phosphorylation and in photosynthesis, and the ways in which electron transport is coupled to the generation of a "proton-motive force."

The Proton-Motive Force, Generation of ATP, and Transport of Metabolites

Closed Vesicles Are Required for the Generation of ATP

Much evidence shows that, in mitochondria, the coupling between oxidation of NADH and $FADH_2$ by oxygen and the synthesis of ATP from ADP and P_i occurs *only* via the electrochemical proton gradient across the inner mitochondrial membrane. Addition of oxygen and compounds such as NADH or pyruvate to intact isolated mitochondria results in net synthesis of ATP. However, production of ATP is absolutely dependent on the integrity of the inner mitochondrial membrane. In the presence of minute amounts of detergents that make the membrane "leaky," oxidation of NADH or acetyl CoA by oxygen still occurs, but no ATP is made. Under these conditions no transmembrane proton gradient can be maintained.

Similarly, in intact, isolated chloroplasts, bright illumination generates ATP from added ADP and P_i and also results in the formation of molecular oxygen and reduction of $NADP^+$ to NADPH. If the thylakoid membranes become ruptured or leaky, light still causes the movement of electrons that results in evolution of oxygen and in reduction of $NADP^+$ to NADPH. However, no ATP is made. Thus, closed, or sealed, vesicles are essential for synthesis of ATP but not for other reactions of photosynthesis or oxidative phosphorylation.

The Proton-Motive Force Is Composed of a Proton Concentration Gradient and a Membrane Potential Gradient

In chloroplasts, bacteria, and mitochondria, the movement of protons across a membrane, down the electrochemical gradient, is coupled to the synthesis of ATP. The force that propels these protons—the proton-motive force—is a combination, or sum, of two separate forces: (1) a membrane potential gradient and (2) a concentration gradient of protons. In much the same way, movement of Na^+ into a neuron during an action potential (Chapter 17) is also due to a combination of two forces: (1) a concentration gradient of Na^+ across the nerve membrane and (2) potential gradient across the membrane.

Since the protons are positively charged, pumping protons across the membrane generates a membrane potential if the membrane is poorly permeable to anions. A proton concentration gradient can develop *only if the membrane is permeable to a major anion.* The thylakoid membrane is permeable to Cl^- and other anions. No membrane potential develops, since there is always an equal concentration of positive and negative ions on each side of the membrane. However, a large H^+Cl^- concentration gradient becomes established across the thylakoid membrane. In chloroplasts, therefore, the main component of the proton-motive force is the concentration gradient of protons. The concentration in the lumen of the thylakoid vesicles is about 100- to 1000-fold (pH 4.0–5.0) that in the stroma of the chloroplast (pH 7.0).

The mitochondrial inner membrane, by contrast, is relatively impermeable to anions. A greater proportion of the energy is stored as a membrane potential, and the actual pH (proton) gradient is smaller. In mitochondria the potential gradient across the inner mitochondrial membrane is about 200 mV, the matrix being *negative* with respect to the space between the inner and outer membrane. In mitochondria the potential gradient is the more significant component of the proton-motive force.

Since a difference of one pH unit represents a 10-fold difference in H^+ concentration, a pH gradient of one unit across a membrane is equivalent to a membrane potential of 59 mV (at 22°C). Thus we can write

$$\text{pmf (proton-motive force)} = \psi - 59(\Delta\text{pH})$$

where ψ is the transmembrane potential, and pmf is measured in millivolts. In respiring mitochondria, the pmf value is approximately 220 mV: the transmembrane potential is about 160 mV, and a ΔpH of one unit, equivalent to about 60 mV, accounts for the rest.

Mitochondria and chloroplasts are far too small to be impaled with electrodes. How then can the potential or pH gradient across the mitochondrial membrane be determined? One approach takes advantage of the lipid-solubility of the potassium ionophore valinomycin (see Figure 15-5). The mitochondrial inner membrane is normally impermeable to potassium ions (K^+). Valinomycin allows K^+ to diffuse freely across the membrane. Valinomycin is a lipid-soluble peptide that selectively binds one K^+ in its hydrophilic interior. It transports K^+ across otherwise impermeable phospholipid membranes. As in nerve cells (see Chapter 17), at equilibrium the concentration of K^+ on the two sides of the membrane is determined by the membrane potential, according to the Nernst equation:

$$E = \frac{-RT}{\mathcal{F}} \ln \frac{K_{\text{inside}}}{K_{\text{outside}}}$$

When valinomycin and radioactive potassium (^{42}K) ions are added to a suspension of respiring mitochondria or inner-membrane vesicles, oxidative phosphorylation proceeds largely unaffected. K^+ accumulates inside the mitochondria (or vesicles), in a ratio of K_{matrix} to $K_{cytoplasm}$ of about 2500. By the Nernst equation,

$$E = -59 \log \frac{K_{in}}{K_{out}} = -59 \log 2500 = -200 \text{ mV}$$

Thus the potential across the inner membrane is about 200 mV, inside (matrix) negative.

We saw in Figure 15-15 how the pH inside a lysosome can be determined from the distribution of a lipid-soluble weak base across the lysosomal membrane. Similarly, the pH across a thylakoid or mitochondrial membrane can be measured by adding a radiolabeled weak acid or base to the vesicles and then determining the concentrations inside and outside the vesicles.

The F_0F_1 Complex Couples ATP Synthesis to Proton Diffusion

One of the predominant proteins in the mitochondrial membrane is the multisubunit *coupling factor,* the enzyme that actually synthesizes ATP. An extremely similar enzyme complex is located in the thylakoid membranes of chloroplasts and in the plasma membrane of bacterial cells. The coupling factor has two principal components (Figure 20-17). F_0 is an integral membrane complex, composed of three or four distinct polypeptides and one proteolipid (a protein with covalently bound lipid), which together span the mitochondrial membrane. The F_0 complex can be extracted only with strong detergents. Attached to it is F_1, a complex of five distinct polypeptides—α, β, γ, δ, and ϵ—with the probable composition $\alpha_3\beta_3\gamma\delta\epsilon$. F_1 forms the knobs, or "tadpoles," that protrude on the matrix side of the inner membrane (Figure 20-17 and Table 20-1; see also Figure 5-44). F_1 can be detached from the membrane by mechanical agitation and is water-soluble.

When physically separated from the membrane, F_1 is capable only of catalyzing the hydrolysis of ATP. Hence, it is often called the F_1-ATPase. Its natural function, however, is in the synthesis of ATP. Submitochondrial particles from which F_1 is removed cannot catalyze synthesis of ATP. Subsequently, however, F_1 can reassociate with these particles, so that they become once again fully active in the synthesis of ATP (Figure 20-18). This observation lends further evidence in support of the role of F_1 in ATP synthesis. F_1 can be dissociated into its component polypeptides, and the functions of several of the subunits are known, at least in outline. Subunits α and β appear to bind ATP or other nucleotides, and γ and δ can bind to F_0 (Table 20-1). Possible sites of ATP or ADP binding are those at which ADP is converted to ATP or regulatory or allosteric sites at which the F_1 particle "senses" the level of ATP or ADP in the matrix and modifies the rate of ATP synthesis accordingly. The precise orientation of these

(a)

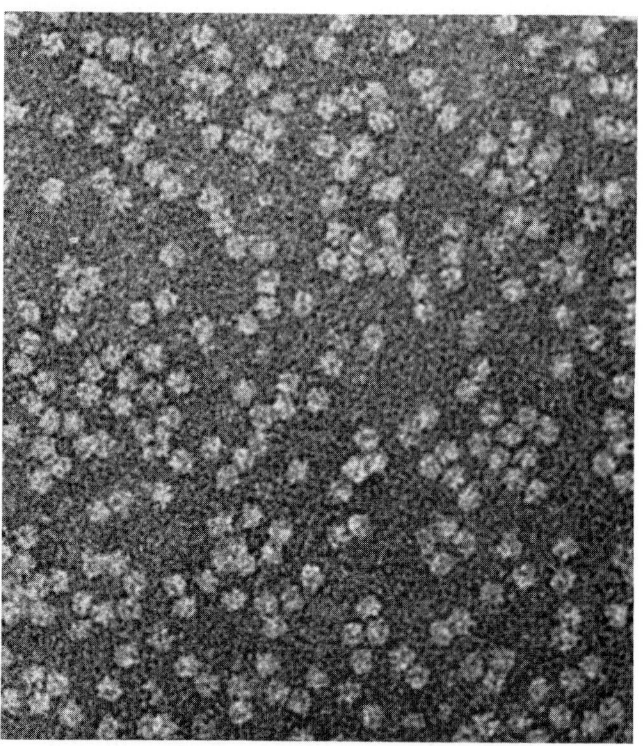

(b)

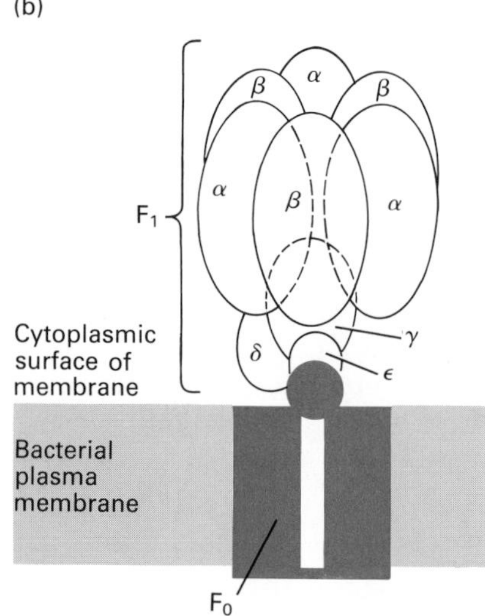

Figure 20-17 F_0 and F_1 particles, the principal components of the coupling factor in mitochondrial and other membranes. (a) Electron micrograph of purified F_1 particles from mitochondria. (b) A diagram showing the probable structure of the F_0 and F_1 components in the bacterial plasma membrane. The core of the F_1 particles is an $\alpha_3\beta_3$ complex. The δ polypeptide of F_1 and probably the γ are involved in binding F_1 to the membrane-embedded F_0 particle. The role of the F_1 ϵ subunit is not known. [See J. E. Walker, M. Saraste, and M. J. Gay, 1982, *Nature* **298**:867.] *Photograph courtesy of E. Racker.*

Table 20-1 Structure and function of subunits of mitochondrial F_1 particles

Structural or functional feature	Subunit				
	α	β	γ	δ	ϵ
Molecular weight	54,600	51,000	30,200	21,000	16,000
Required in reconstitution for net ATP synthesis and H^+ transport	+[a]	+	+	+	+
Required in reconstitution for P_i-ATP exchange	+	+	+	+	+
Required in reconstitution for ATP hydrolysis	±	+	±	+	+
ATP and ADP binding to isolated pure subunits	+	+	−	−	−

[a] + = yes; − = no; ± = results are equivocal.

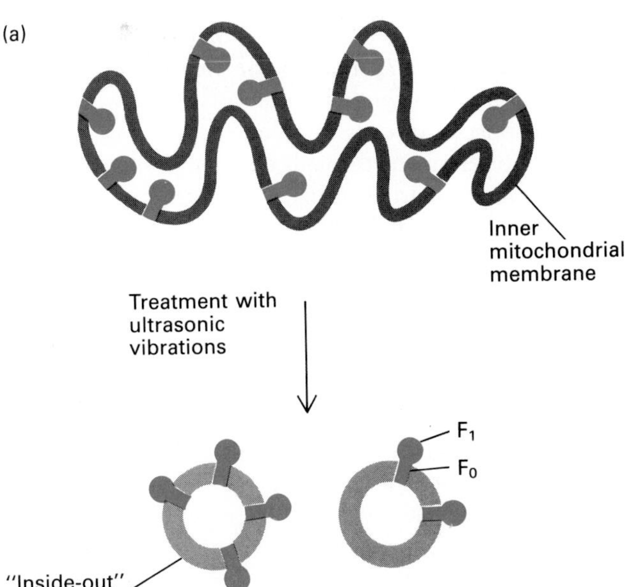

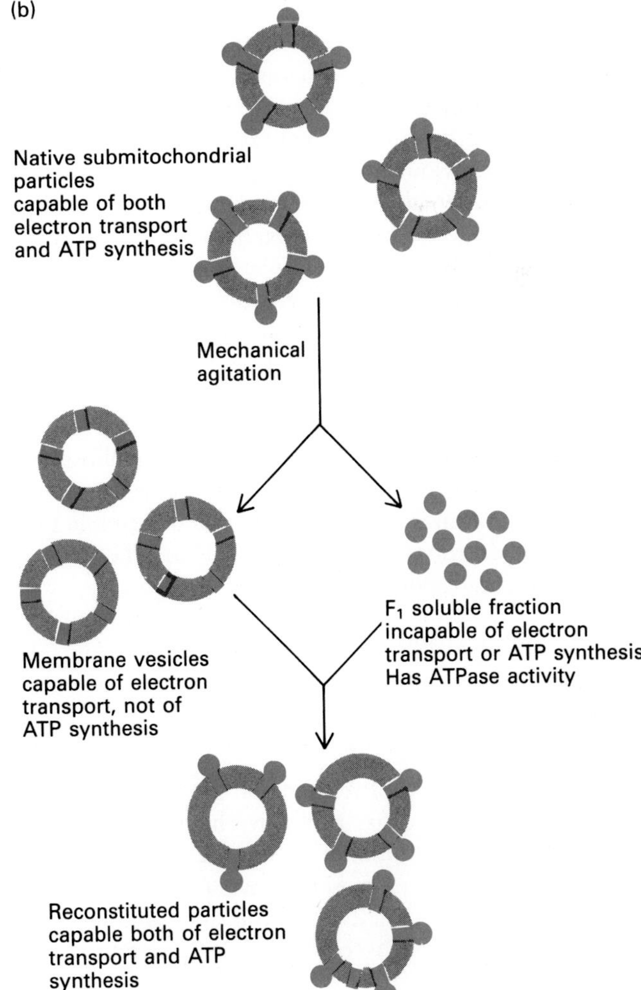

Figure 20-18 F_1 particles are required for ATP synthesis, not for electron transport. (a) "Inside-out" submitochondrial particles can be obtained from inner membranes following treatment with ultrasonic vibration. The vesicles are disrupted and resealed with the F_1 coupling factor facing outside. These vesicles can transfer electrons from added NADH to O_2 with the concomitant synthesis of ATP. (b) Exposure to mechanical agitation causes dissociation of the F_1 particles from the membranes. The membranes remain fully functional for electron transport from NADH to O_2 but cannot synthesize ATP. Subsequently, addition of F_1 particles reconstitutes the native membrane structure and restores the capacity for ATP synthesis.

subunits and their exact interactions are subjects of intense research.

It is believed that F_0 contains a transmembrane channel through which the protons flow to F_1. When the proteolipid or one of the purified protein subunits of F_0 is experimentally incorporated into phospholipid vesicles (liposomes; Chapter 14), the permeability of the vesicle to H^+ is greatly stimulated, suggesting that one or more F_0 proteins form a proton channel through the membrane. Although not an integral membrane protein, F_1 is believed to provide the permeability barrier to protons: removal of F_1 from the inner membrane makes it highly permeable to protons.

The proton permeability of this F_0 transmembrane channel can be regulated. One subunit of F_0 can bind oligomycin, an antibiotic that inhibits ATP generation in mitochondria by interfering with the utilization of the proton gradient for production of ATP. Some experiments support the role of the oligomycin-binding protein in regulating H^+ flow. As noted previously, membrane vesicles containing only F_0 are leaky to protons. The oligomycin-binding protein can be selectively removed from these vesicles by small amounts of detergent; the permeability to H^+ then increases several-fold. Readdition of this protein restores membrane permeability to its original level. Further research is required to verify these roles for the oligomycin-binding subunit. It has been suggested that this protein somehow regulates the flow of protons through F_0: when the proton gradient is very small, it prevents the reverse reaction—generation of a proton gradient by hydrolysis of ATP. Such a reaction would "waste" ATP, and it seems reasonable for the cell to protect itself against this circumstance by regulation.

The precise number of protons that must move through F_0F_1 to synthesize one molecule of ATP is uncertain. Some experiments indicate that movement of two protons is coupled to synthesis of one high-energy phosphate bond; others suggest that three are necessary. A simple calculation indicates that passage of more than one proton is required to synthesize a molecule of ATP. For example, for the reaction

$$H^+ + ADP^{3-} + P_i^{2-} \longrightarrow ATP^{4-} + H_2O$$

at standard concentrations of ATP, ADP, and P_i,

$$\Delta G^{\circ\prime} = +7.3 \text{ kcal/mol}$$

At the concentrations of reactants in the mitochondrion, however, the value is probably higher ($\sim$+10 to +12 kcal/mol). The amount of free energy released by passage of one mole of protons down an electrochemical gradient of 220 mV (0.22 V) can be calculated by using the equation from Chapter 2 (page 41), setting $n = 1$:

$$\begin{aligned} \Delta G = -n\mathcal{F}\,\Delta E &= -2306\Delta E \text{ (V)} \\ &= (-2306)(0.22) \\ &= -5080 \text{ cal/mol} \end{aligned}$$

Since just over 5 kcal/mol of free energy is made available, passage of at least two, and possibly three, protons is essential for the synthesis of each molecule of ATP from ADP and P_i.

How movement of protons through F_0F_1 is coupled to the generation of ATP is currently the subject of intense debate and experimentation. Some recent evidence suggests that ADP and P_i, after binding to the F_1 complex, spontaneously form ATP that remains tightly bound to the F_1. The step requiring proton movement is the dissociation of the completed ATP from the protein. Presumably, dissociation of ATP is induced by a conformational change in the F_1 complex, which in turn is caused by the flow of protons through it.

Much Evidence Supports the Role of the Proton-Motive Force in ATP Synthesis

The notion that a proton-motive force is the immediate source of energy for ATP synthesis, introduced in 1961, was initially opposed by virtually all workers in photophosphorylation and oxidative phosphorylation. They favored a mechanism, akin to the well-elucidated substrate-level phosphorylation in glycolysis, in which electron transport is directly coupled to synthesis of ATP. Transport of electrons through the membrane of chloroplasts or mitochondria was believed to generate directly a high-energy chemical bond (a phosphoenzyme, for instance), which was then used to convert ADP to ATP. Despite intense efforts by a large number of investigators, no such intermediate could ever be identified. Other evidence against this direct coupling model was the observation that only sealed, intact membrane vesicles can carry out oxidative phosphorylation.

But clear-cut evidence in support of the role of the proton-motive force in ATP synthesis was not easy to obtain, until the development of techniques for purification and characterization of organelle membranes and membrane proteins. Then this notion became the generally accepted hypothesis, although, as noted, many of the details are still very unclear. It is appropriate, then, to summarize here some of the early key experiments that established the basic features of the proton-motive force.

One of the most important experiments was a direct demonstration that an artificially imposed proton gradient (i.e., pH gradient) can result in the synthesis of ATP. In one study (Figure 20-19), chloroplast thylakoid vesicles were equilibrated, in the dark, with a buffered solution at pH 4.0, until the pH in the thylakoid lumen was also 4.0. These vesicles were rapidly mixed with a solution at pH 8.0 containing ADP and P_i. A burst of ATP synthesis accompanied the transmembrane movement of protons driven by a 10,000-fold (10^{-4} *M* versus 10^{-8} *M*) concentration gradient. Reciprocal experiments were done on "inside-out" preparations of submitochondrial

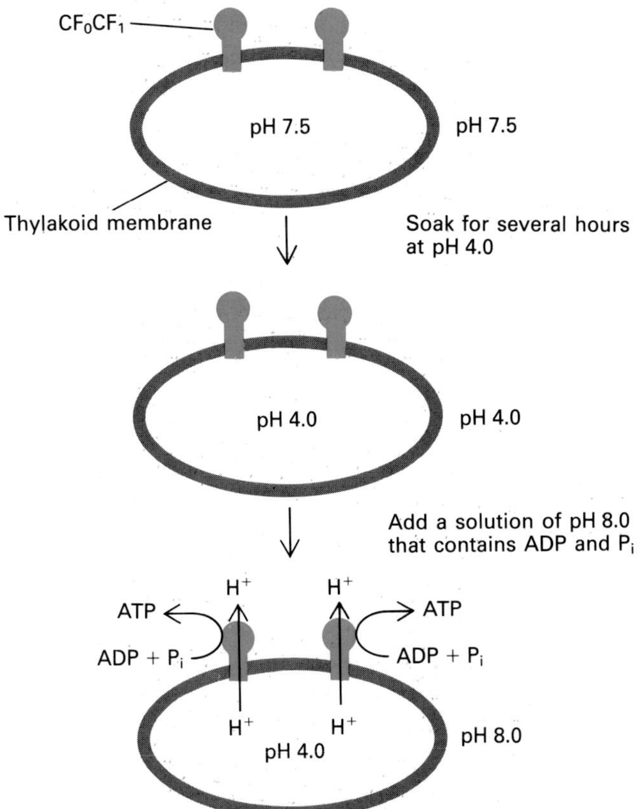

Figure 20-19 Synthesis of ATP from ADP by thylakoid membranes as a result of artificially imposing a pH gradient.

particles in which an artificially generated membrane potential gradient also resulted in synthesis of ATP.

Particularly dramatic results come from an experiment that employed a protein found in the plasma membrane of a photosynthetic bacteria, bacteriorhodopsin (Figure 14-14). When illuminated, this protein pumps protons from the inside of the bacterial cell to the outside, in a fashion similar to light-generated pumping of protons in chloroplasts of higher plants (see Figure 20-6). Bacteriorhodopsin was asymmetrically incorporated into artificial phospholipid vesicles that also contained purified mitochondrial F_0F_1 complexes. Illumination of these vesicles resulted in the accumulation of protons within the vesicles and in the generation of ATP from ADP and P_i. These results leave no doubt that the F_0F_1 complex is the ATP-generating enzyme and that the generation of ATP is dependent on an electrochemical proton gradient (Figure 20-20).

Synthesis of ATP by the F_0F_1 complex is also reversible. Addition of ATP to inside-out submitochondrial vesicles results in hydrolysis of ATP and in the *accumulation* of protons within the vesicles. Although this reaction—pumping of protons coupled to hydrolysis of ATP—is formally similar to the process of acidification of lysosomes (see Figure 15-15) or CURLs, the lysosomal and gastric ATPase are very different in structure from the mitochondrial or chloroplast F_0F_1 complex, in that they are single polypeptide chains.

Such studies with the purified F_0F_1 complex also establish directly that the phosphoanhydride bond in ATP and the proton-motive force across a membrane are interconvertible forms of energy: a proton gradient can be used to synthesize ATP from ADP and P_i, or ATP hydrolysis can be coupled to proton pumping.

Many Transporters in the Inner Mitochondrial Membrane Are Powered by the Proton-Motive Force

The inner mitochondrial membrane contains a number of permease proteins that permit various needed metabolites to enter and leave, such as pyruvate, malate, and the amino acids aspartate and glutamate. Two important permeases are those that transport phosphate and ADP from the cytosol into the mitochondrion in exchange for ATP formed by oxidative phosphorylation inside the mitochondrion.

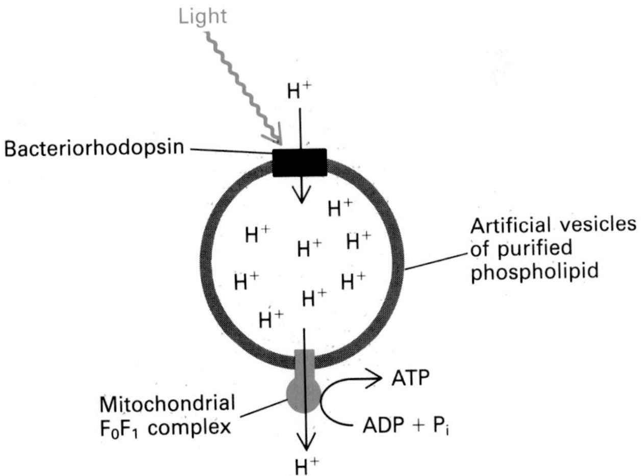

Figure 20-20 Demonstration that the F_0F_1 complex is the site of ATP synthesis. The bacterial protein bacteriorhodopsin is known to pump protons into the lumen of the vesicle upon illumination. Purified F_0F_1 complexes and the bacteriorhodopsin are incorporated into the same phospholipid vesicles (liposomes). Illumination results in synthesis of ATP from ADP and P_i. Bacteriorhodopsin pumps protons into the lumen; the resultant proton-motive force is used to power the synthesis of ATP. If no F_0F_1 is used, no ATP is made. Thus, the F_0F_1 complex alone uses a proton gradient for synthesis of ATP.

$$ADP^{3-}_{out} + H_2PO^-_{4_{out}} + ATP^{4-}_{in} \rightleftharpoons ADP^{3-}_{in} + H_2PO^-_{4_{in}} + ATP^{4-}_{out}$$

In this way, ATP can be formed in the mitochondrion and transported to the cytoplasm. Energy stored in the proton gradient is also used to power this exchange of ATP for ADP and P_i. To see this, let us examine the two components of the system: the phosphate permease and the ATP-ADP exchange protein.

The phosphate carrier is an antiport that catalyzes an exchange of one $H_2PO_4^-$ for one OH^- (Figure 20-21). Outward transport of OH^- is driven both by the proton gradient (outside pH low) and by the membrane potential (outside positive) across the inner membrane. The pH and potential gradients are generated during electron transport. Each OH^- transported outward combines with a proton, found in excess on the cytoplasmic side of the inner membrane, to form water. Thus, respiring mitochondria that are pumping protons outward can accumulate phosphate from the surrounding medium against a concentration gradient.

Entry of ADP and exit of ATP are coupled: the *ATP-ADP translocase* is an antiport that allows a molecule of ADP to enter only if one of ATP exits simultaneously. The translocase, a dimer of two 30,000-m.w. subunits, makes up 6 percent of the protein in the inner membrane, so it is one of the more abundant mitochondrial proteins.

Transport of ADP into the mitochondrion, in exchange for ATP, is driven by the transmembrane proton gradient. To understand this, recall that at neutral pH, each ATP molecule contains four negative charges, and each ADP bears three. Thus, exchange of ADP (out to in) for ATP (in to out) results in the outward movement of one additional negative charge (electron):

$$ADP^{3-}_{out} + ATP^{4-}_{in} \rightleftharpoons ADP^{3-}_{in} + ATP^{4-}_{out}$$

Net outward movement of this negative charge is powered by the membrane potential—outside positive—across the inner membrane. Thus, net uptake of ADP and net export of ATP are powered by the proton gradient. Expenditure of energy (from the proton gradient) for export of ATP from the mitochondrion, in exchange for ADP and P_i, ensures a high ratio of ATP to ADP in the cytosol, where utilization of the phosphate-bond energy of ATP takes place. Since some of the protons pumped out of the mitochondrion are utilized to power the ATP-ADP exchange, fewer are available for synthesis of ATP. For every four protons pumped out, three are used for synthesis of one molecule of ATP, and one is used to power the export of the ATP out of the mitochondrion.

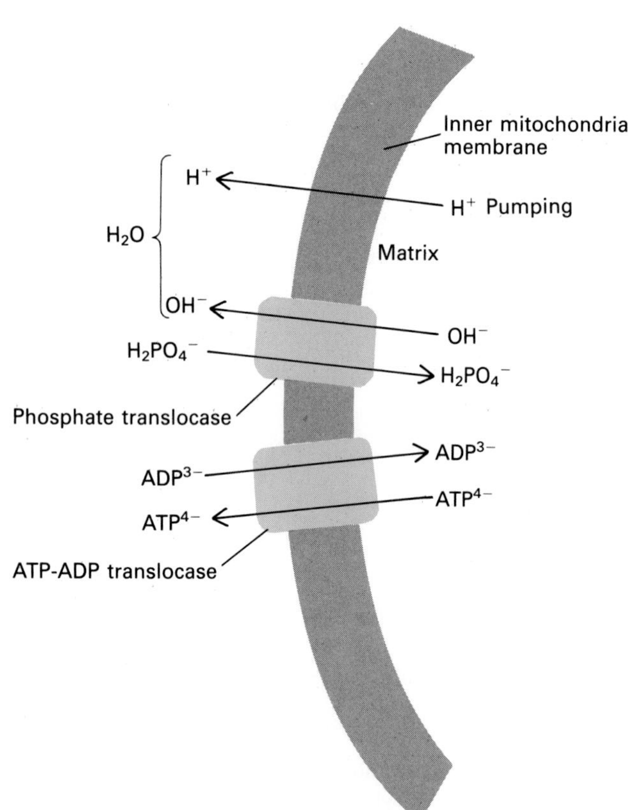

Figure 20-21 The phosphate and ATP-ADP transport system in the inner mitochondrial membrane. Uptake of $H_2PO_4^-$ is powered by the outward movement of OH^-. The transmembrane potential also powers the uptake of ADP^{3-} in exchange for export of ATP^{4-} from the mitochondrial matrix. Because the outer mitochondrial membrane is freely permeable to ATP, ADP, and phosphate, cytoplasmic supplies of these can be exchanged for intramitochondrial ATP by the transport systems of the inner membrane.

Inner-Membrane Proteins Allow Uptake of Electrons from Cytosolic NADH

During conversion of glucose to pyruvate in the cytoplasm, two molecules of NAD^+ are reduced to NADH. These electrons are ultimately transferred to oxygen, concomitant with generation of ATP. However, the inner mitochondrial membrane is impermeable to NADH. How can the electrons from cytosolic NADH enter the mitochondrial electron transport system?

Several "electron shuttles" are employed; one of these is depicted in Figure 20-22. Cytosolic NADH reduces dihydroxyacetone phosphate to glycerol 3-phosphate. The latter diffuses to the inner mitochondrial membrane; there it is reoxidized to dihydroxyacetone phosphate by an FAD-linked glycerol 3-phosphate dehydrogenase that is an integral protein of the inner mitochondrial membrane. The $FADH_2$ thus formed transfers two electrons directly into the electron transport system. The electrons from cytosolic NADH enter the electron transport system at the level of $FADH_2$, and only two ATPs are generated per pair of electrons transferred to O_2, rather than three,

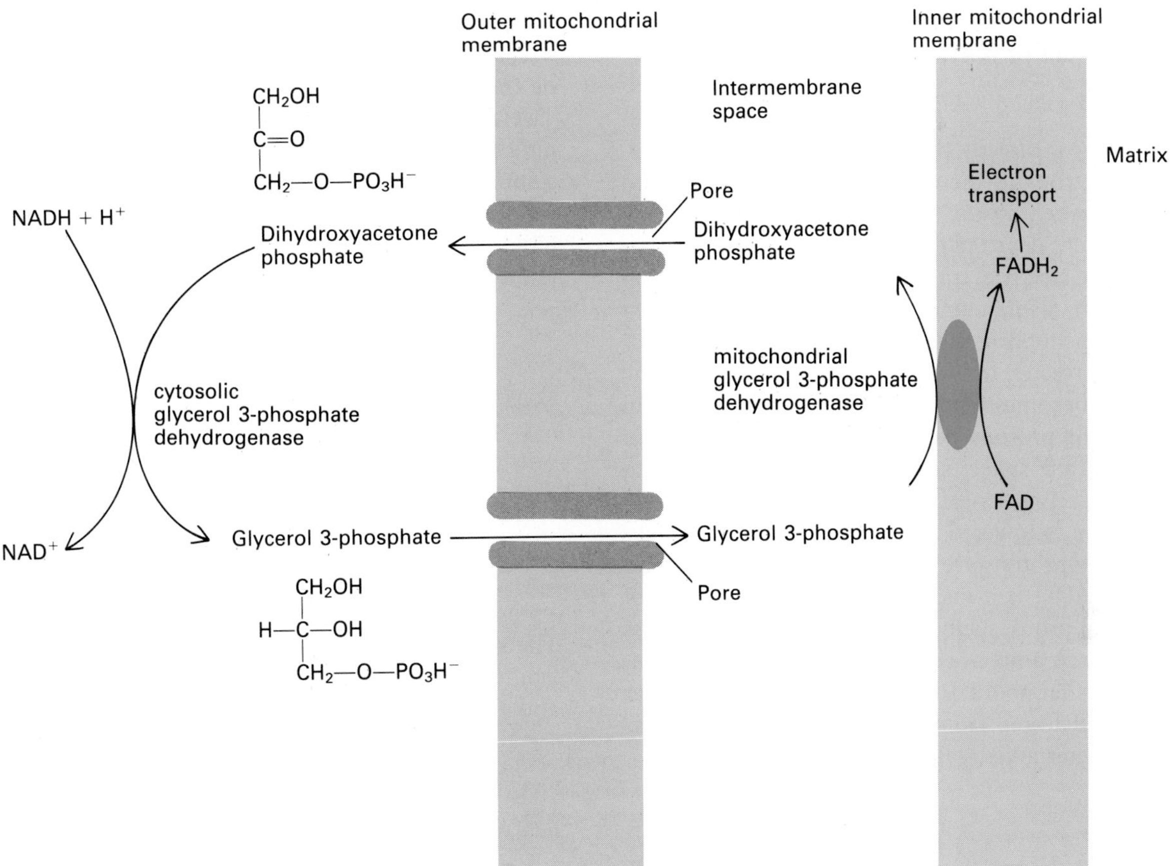

Figure 20-22 The glycerol 3-phosphate–dihydroxyacetone phosphate shuttle transports electrons from cytoplasmic NADH into the electron transport chain of the inner mitochondrial membrane. Cytoplasmic NADH is used to reduce cytoplasmic dihydroxyacetone phosphate, a glycolytic intermediate, to glycerol 3-phosphate. The latter diffuses through a pore in the outer mitochondrial membrane to the outer surface of the inner mitochondrial membrane, where it is reoxidized to dihydroxyacetone phosphate by a membrane-bound FAD-linked dehydrogenase. The electrons are transferred from the $FADH_2$ directly into the electron transport chain; the dihydroxyacetone diffuses back into the cytosol.

$$NADH_{cytosol} + H^+ + FAD_{mitochondria} \longrightarrow NAD^+_{cytosol} + FADH_{2\ mitochondria}$$

as in oxidation of mitochondrial NADH. But the energy not used to form a third ATP is required for transport of the electrons into the mitochondrion. Thus not all of the energy generated by oxidation of pyruvate or NADH is available for synthesis of ATP; as much as 25 to 30 percent is used for import of metabolites into the mitochondrion, against concentration gradients, and for export of ATP.

NADH, Electron Transport, and Proton Pumping

The studies just described were of considerable importance, since they showed that a proton gradient can be used to generate ATP. Equally important were a set of experiments establishing that movements of electrons from NADH (or $FADH_2$) to oxygen are catalyzed by a series of electron carriers in four multiprotein complexes in the inner mitochondrial membrane and that electron transport results in the generation of a proton gradient.

Electron Transport in Mitochondria Is Coupled to Proton Pumping

Isolated mitochondria maintained under conditions without oxygen do not oxidize NADH or other compounds that can donate electrons to the respiratory chain, since there is no ultimate acceptor (usually O_2) of the electrons. Nor do they generate ATP. Addition of a small amount of oxygen to such mitochondria results in oxidation of an equivalent amount of NADH:

$$NADH + H^+ + \tfrac{1}{2}O_2 \longrightarrow NAD^+ + H_2O$$

Such studies show directly that each NADH molecule releases two electrons to the electron transport chain and reduces one oxygen atom.

The experiment depicted in Figure 20-23 provides positive proof that electron transport is coupled to proton transport. When a small amount of O_2 is added to a suspension of mitochondria that were kept in the absence of O_2 and in the presence of NADH, the medium outside the mitochondria becomes acidic. The acidity is due to an accumulation of protons pumped from the matrix into the intermembrane space; since the outer mitochondrial membrane is freely permeable to protons, the pH of the outside medium is lowered briefly. As noted earlier, exactly how many protons are transported per electron pair released by NADH is a matter of controversy. Some of the transported protons subsequently diffuse back across the membrane, so a precise estimate is difficult. It is generally believed that about 10 protons are transported per 2 electrons transferred from NADH to oxygen, but experimentally determined values range from 6 to 12.

The Electron Transport System Is Composed of Many Types of Carriers

When NADH is oxidized to NAD^+, two electrons and one proton are released. Recall that protons are soluble in aqueous solution as hydronium ions (H_3O^+). Free electrons, by contrast, cannot exist in aqueous solutions. Electrons are passed from NADH or $FADH_2$ to O_2 along a chain of electron carriers, all of which are components of the inner mitochondrial membrane (Figure 20-24). Twelve or more electron carriers are grouped into four multiprotein intramembranous particles or complexes, summarized in Table 20-2. One complex, NADH–CoQ reductase, carries electrons from NADH to the small lipophilic compound ubiquinone, also termed coenzyme Q (CoQ), as shown in Figure 20-25.

$$NADH \rightarrow NAD^+ + H^+ \;:\; 2e^- \longrightarrow 2e^- \longrightarrow 2e^- \longrightarrow 2e^- + 2H^+ + CoQ \rightarrow CoQH_2$$

High-energy electrons ... Low-energy electrons

The second, succinate CoQ reductase, transfers electrons from succinate (released during its oxidation to fumarate) and $FADH_2$ to CoQ.

$$\text{Succinate} \rightarrow \text{Fumarate} + 2H^+ \;:\; 2e^- \longrightarrow 2e^- \longrightarrow 2e^- \longrightarrow 2e^- + 2H^+ + CoQ \rightarrow CoQH_2$$

$$FADH_2 \rightarrow FAD + 2H^+ \;:\; 2e^- \longrightarrow 2e^- \longrightarrow 2e^- \longrightarrow 2e^- + 2H^+ + CoQ \rightarrow CoQH_2$$

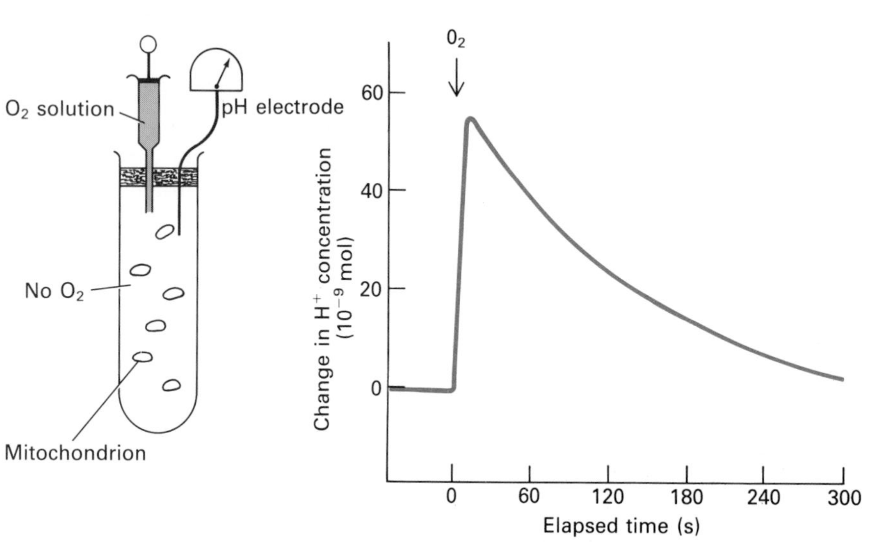

Figure 20-23 Transport of electrons from NADH or $FADH_2$ to O_2 is coupled to proton pumping. In the experiment depicted, a source of electrons for respiration, such as NADH, is added to a suspension of mitochondria that is depleted of oxygen. No NADH is oxidized. When a small amount of oxygen is added to the system *(arrow)*, the pH of the surrounding medium rises sharply, a change that corresponds to an increase in protons outside the mitochondria. Thus oxidation of NADH by oxygen is coupled to pumping of protons out of the matrix. When O_2 becomes depleted, the excess protons slowly leak back into the mitochondria, and the pH of the extracellular medium returns to its initial value.

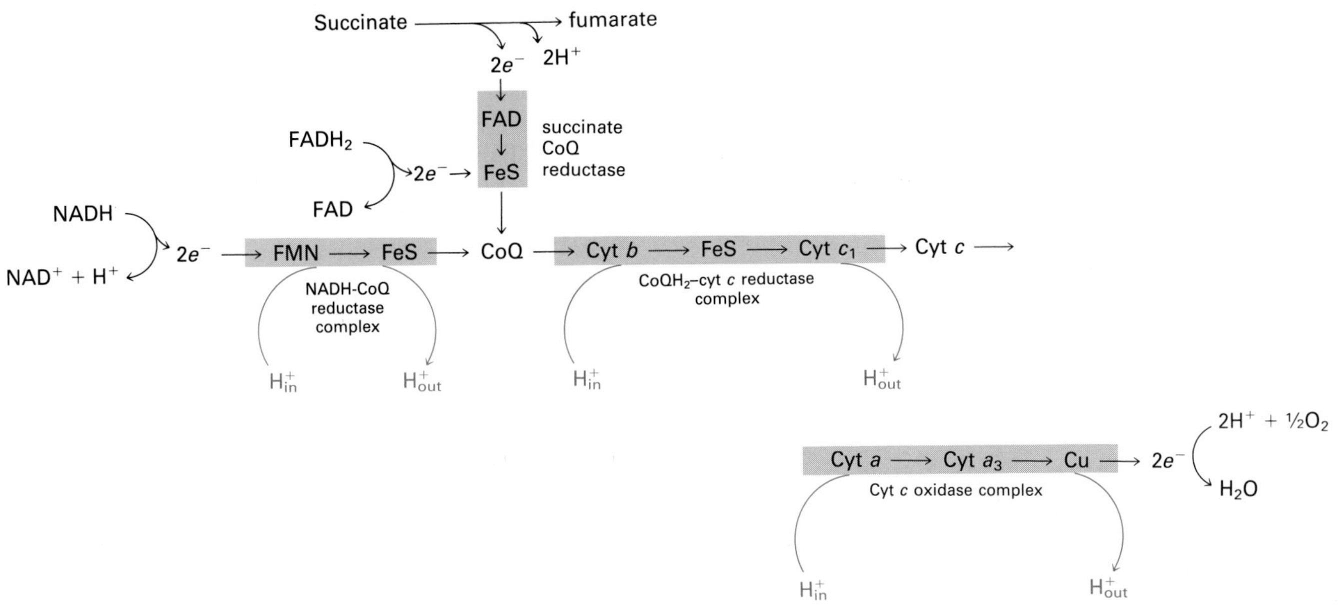

Figure 20-24 Flow of electrons from NADH to $FADH_2$ to O_2, and the three sites for pumping protons across the inner mitochondrial membrane. The arrows indicate the direction of electron transport from NADH or succinate or $FADH_2$ to O_2. The four electron transport complexes *(shaded in color)* are integral parts of the inner membrane and contain the specific electron carriers indicated. Coenzyme Q (ubiquinone) and cytochrome *c* are also associated with the membrane. Unbound $FADH_2$ can contribute electrons to the bound succinate dehydrogenase complex.

The third complex, $CoQH_2$–cytochrome *c* reductase, transfers electrons from reduced $CoQH_2$ to the water-soluble protein cytochrome *c*, to yield its reduced form.

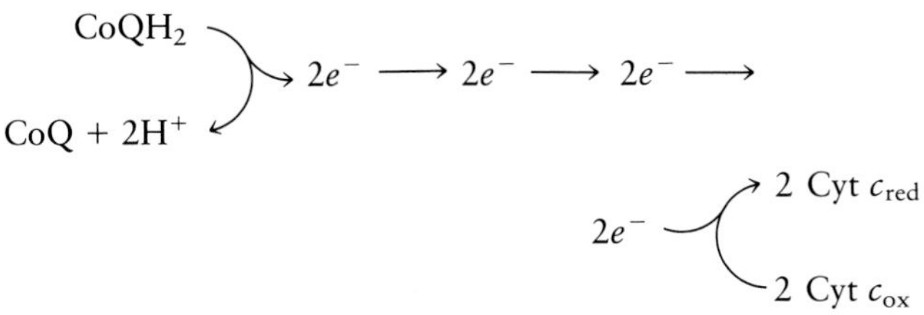

$$CoQH_2 \rightarrow CoQ + 2H^+ \quad 2e^- \longrightarrow 2e^- \longrightarrow 2e^- \longrightarrow$$

$$2e^- + 2\ \text{Cyt}\ c_{ox} \rightarrow 2\ \text{Cyt}\ c_{red}$$

The fourth, cytochrome *c* oxidase, transfers electrons from reduced cytochrome *c* to O_2, the ultimate electron acceptor.

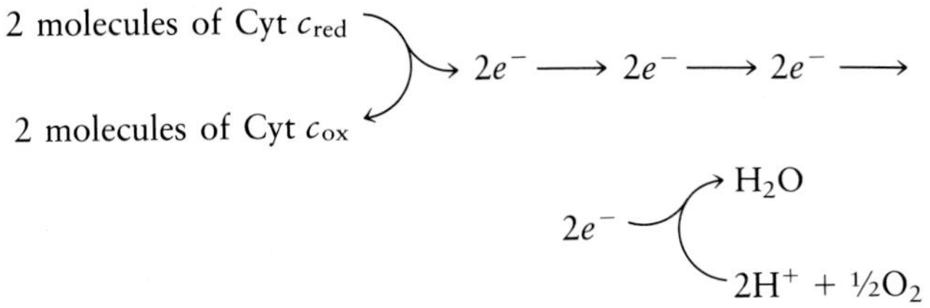

$$2\ \text{molecules of Cyt}\ c_{red} \rightarrow 2\ \text{molecules of Cyt}\ c_{ox} \quad 2e^- \longrightarrow 2e^- \longrightarrow 2e^- \longrightarrow$$

$$2e^- + 2H^+ + \tfrac{1}{2}O_2 \rightarrow H_2O$$

Three of these complexes—NADH-CoQ reductase, $CoQH_2$–cytochrome *c* reductase, and cytochrome *c* oxidase—are sites for pumping of protons across the membrane. Except for cytochrome *c*, all these enzyme complexes contain several proteins and are integral membrane proteins. (Cytochrome *c* is a water-soluble peripheral protein found on the outer face of the inner mitochondrial membrane; the role of cytochromes in electron transport is discussed in the following section.)

The large size (see Table 20-2) of these electron carrier

Table 20-2 Components of the mitochondrial electron transport chain

Enzyme complex	Mass (daltons)	Prosthetic groups
NADH–CoQ reductase	85,000	FMN FeS
Succinate–CoQ reductase	97,000	FAD FeS
$CoQH_2$–cytochrome *c* reductase	280,000	Heme b Heme c_1 FeS
Cytochrome *c* oxidase	200,000	Heme a Heme a_3 Cu
Cytochrome *c*	13,000	Heme c

SOURCE: J. W. De Pierre and L. Ernster, 1977, *Annu. Rev. Biochem.* 46:201.

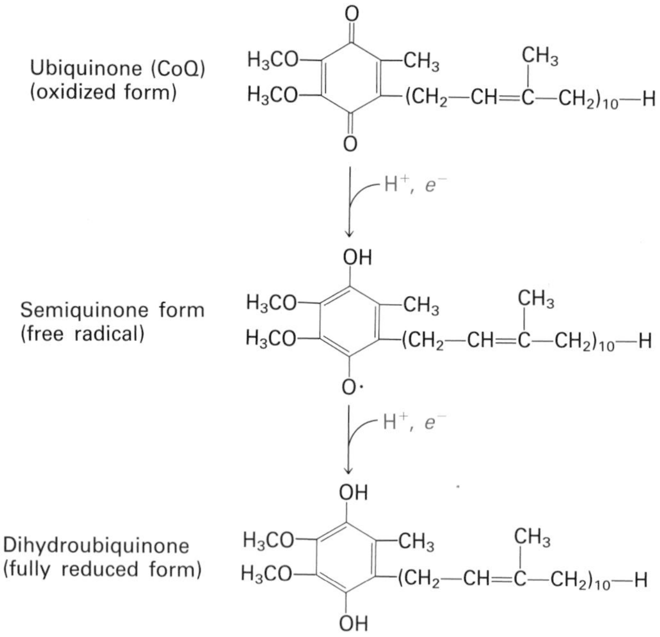

Ubiquinone (CoQ) (oxidized form)

H_3CO, H_3CO, CH_3, $(CH_2—CH{=}C(CH_3)—CH_2)_{10}—H$

H^+, e^-

Semiquinone form (free radical)

H^+, e^-

Dihydroubiquinone (fully reduced form)

Figure 20-25 Structure of ubiquinone (coenzyme Q), illustrating its ability to carry protons and electrons. Found in bacterial and mitochondrial membranes, ubiquinone is a carrier for both electrons and hydrogen ions. It is the only electron carrier in the electron transport system that is not tightly bound or covalently bonded to a protein. Ubiquinone is very mobile in the membrane.

Plastoquinone, an electron carrier in chloroplasts, has a similar structure and generally fewer isoprene units. Because of its long hydrocarbon "tail" of isoprene units, ubiquinone is soluble in the hydrophobic core of phospholipid bilayers. In the fully oxidized state the two oxygen atoms on the benzene ring are linked to carbon atoms by double bonds. In the fully reduced dihydroubiquinone, both oxygen atoms have hydrogen atoms attached to them. Addition of one proton and one electron to the oxidized ubiquinone results in a half-reduced or semiquinone form. The semiquinone is a free radical, and the unpaired electron *(dot)* is delocalized by resonance over the benzene ring and attached O atoms.

complexes and their multisubunit structure is undoubtedly related to their complicated function as both proton pumps and electron carriers. Considerably more is known about the latter than the former. A number of different prosthetic groups capable of oxidation and reduction are essential to the function of the electron carrier proteins (Table 20-2). Some of these are groups containing metal atoms; others are the flavins FAD and the related molecule FMN. The specific location of the electron carrier groups relative to one another can be determined, as we shall see, by their reduction potentials.

Cytochromes Electron carriers termed *cytochromes* are membrane proteins that contain a heme prosthetic group similar to that in hemoglobin or myoglobin. The iron in the center of the heme is the transporter of the electrons. Electron transport occurs by the interconversion of the two ionic states of iron: Fe^{3+} is the oxidized and Fe^{2+} is the reduced state. Transfer of an electron to an Fe^{3+} cytochrome causes reduction to Fe^{2+}. Because the heme ring consists of alternating double- and single-bonded atoms, a large number of resonance forms exist, and the extra electron is delocalized to the heme carbon and nitrogen atoms as well as to the Fe ion. The Fe^{2+} is reoxidized to Fe^{3+} as it passes along the electron to the next carrier on the chain (see Figure 20-24). This can be seen in the steps of the last segment of the electron transport sequence—the oxidation of cytochrome *c* at the expense of oxygen. In addition to its heme prosthetic group, cytochrome a_3, another of the polypeptides in the complex, contains a copper ion that, by interconversion of the Cu^{2+} (oxidized) and Cu^+ (reduced) forms, also participates in electron transfer.

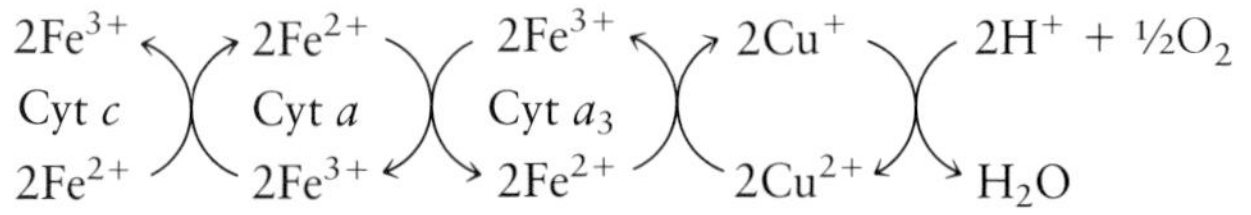

$2Fe^{3+}$ / Cyt *c* / $2Fe^{2+}$ ⇄ $2Fe^{2+}$ / Cyt *a* / $2Fe^{3+}$ ⇄ $2Fe^{3+}$ / Cyt a_3 / $2Fe^{2+}$ ⇄ $2Cu^+$ / $2Cu^{2+}$ ⇄ $2H^+ + \frac{1}{2}O_2$ / H_2O

The different cytochromes in the chain—*b*, *c*, c_1, *a*, and a_3—have slightly different heme structures and axial ligands of the Fe atom, which generate different environments for the iron. Therefore, each of the cytochromes has a different reduction potential, or tendency to accept an electron. This is important, since it means that electron flow will be unidirectional along the chain.

Other Carriers Other electron carriers used include a set of proteins that contain iron-sulfur clusters, Fe_2S_2 and Fe_4S_4, which are nonheme prosthetic groups. Specifically, these clusters consist of Fe atoms coordinated to inorganic S atoms as well as to four S atoms on cysteine residues on the protein (Figure 20-26). Some of the Fe atoms in the cluster bear a +2 charge and others a +3 charge, but because electrons in the outermost orbits move rapidly from one Fe atom to another Fe atom and are dispersed among them, each Fe atom actually bears a net charge of between +2 and +3. Additional electrons that are transferred to these carriers are also dispersed over all the Fe atoms in the cluster.

Coenzyme Q, or ubiquinone, is a lipid-soluble carrier of hydrogen atoms (i.e., protons plus electrons). As seen in Figure 20-25, the oxidized quinone can accept first a single electron and proton, to form a semiquinone, and then a second electron and proton to form the fully reduced hydroquinone.

Other electron carriers in the chain include flavins covalently bound to a protein in the NADH-CoQ reductase complex, and also Cu atoms ($Cu^{2+} + e^- \rightleftharpoons Cu^+$) bound to cytochrome a_3 in the cytochrome oxidase complex.

Cytochromes, Cu^{2+}, and FeS proteins are one-electron carriers: they accept and release a single electron at a time. Ubiquinone can accept either one or two electrons. NAD^+ is exclusively a two-electron carrier: it accepts ($H^+ + NAD^+ + 2e^- \rightarrow NADH$) or releases ($NADH \rightarrow H^+ + 2e^- + NAD^+$) only a pair of electrons at a time. Flavins can also accept two electrons (see Figure 20-3), but they do so one electron at a time:

$$\underset{\text{Fully oxidized}}{\text{FAD}} \xrightarrow{H^+e^-} \text{flavin semiquinone} \xrightarrow{H^+e^-} \underset{\text{Fully reduced}}{FADH_2}$$

In practice, most flavoprotein enzymes use only one of these steps, and cycle either between FAD and the semiquinone form or between the semiquinone and the fully reduced $FADH_2$.

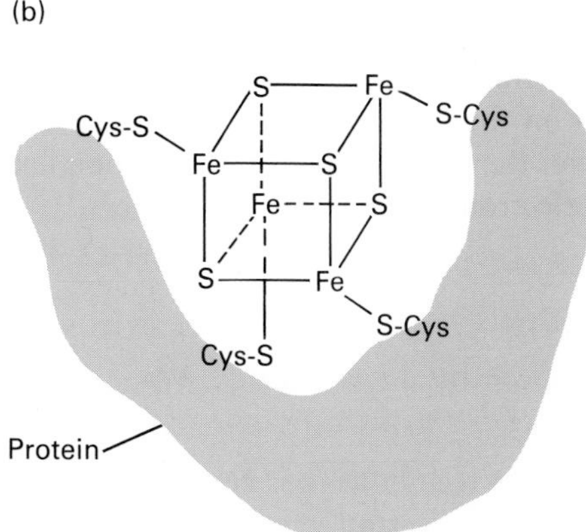

Figure 20-26 Three-dimensional structure of some FeS clusters in electron-transporting proteins. (a) Dimeric iron-sulfur (Fe_2S_2) cluster; (b) tetrameric iron-sulfur (Fe_4S_4) cluster. Each Fe atom is bonded to four S atoms. Some S atoms are molecular sulfur. Other S atoms are in cysteine side chains of a protein. [See W. H. Orme-Johnson, 1973, *Annu. Rev. Biochem.* **42**:159.]

Electron Carriers Are Oriented in the Transport Chain in Order of Their Reduction Potentials

We saw in Chapter 2 that the reduction potential for a partial reaction of reduction (oxidized + $e^- \leftrightarrows$ reduced) is a measure of the equilibrium constant of the reaction. For instance, for the reaction

$$NAD^+ + H^+ + 2e^- \rightleftharpoons NADH$$

the value of the standard reduction potential is negative:

$$E_0' = -0.32 \text{ V}$$

This means that this partial reaction tends to proceed toward the left—toward the oxidation of NADH to NAD^+. By contrast, the potential for the reaction

$$\text{Cyt } c_{ox}(Fe^{3+}) + e^- \rightleftharpoons \text{Cyt } c_{red}(Fe^{2+})$$

is positive:

$$E_0' = +0.26 \text{ V}$$

so the reaction tends to proceed to the right—to the reduction of cytochrome *c* (Fe^{3+}) to *c* (Fe^{2+}). The final step of the chain, the reduction of oxygen to water, has the most positive reduction potential:

$$2H^+ + 2e^- + \tfrac{1}{2}O_2 \rightleftharpoons H_2O \qquad E_0' = +0.816 \text{ V}$$

The reduction potential of the electron carriers in mitochondria increases steadily from NADH to O_2: thus electron transport is thermodynamically favored at each step. We can think of the electrons released by NADH as having a high energy; a fraction of this energy is lost at each step as the electrons move from NADH to O_2, the ultimate acceptor. To emphasize this point, let us calculate the $\Delta G°'$ value for cytochrome *c* oxidase, the reaction of the last complex in the electron transport chain (Figure 20-27):

$$2 \text{ Cyt } c\ (Fe^{2+}) + 2H^+ + \tfrac{1}{2}O_2 \rightleftharpoons 2 \text{ Cyt } c\ (Fe^{3+}) + H_2O$$

The half-reactions are

$$\text{Cyt } c\ (Fe^{3+}) + e^- \rightleftharpoons \text{Cyt } c\ (Fe^{2+}) \qquad E_0' = +0.26 \text{ V}$$

$$2H^+ + \tfrac{1}{2}O_2 + 2e^- \rightleftharpoons H_2O \qquad E_0' = +0.82 \text{ V}$$

The change in voltage for the total reaction is

$$\Delta E_0' = +0.82 - 0.26 = +0.56 \text{ V}$$

From Chapter 2,

$$\Delta G°' = -n\mathcal{F}\,\Delta E_0'$$

Recall that $\mathcal{F}$ is the faraday constant, 23,062, and that *n*

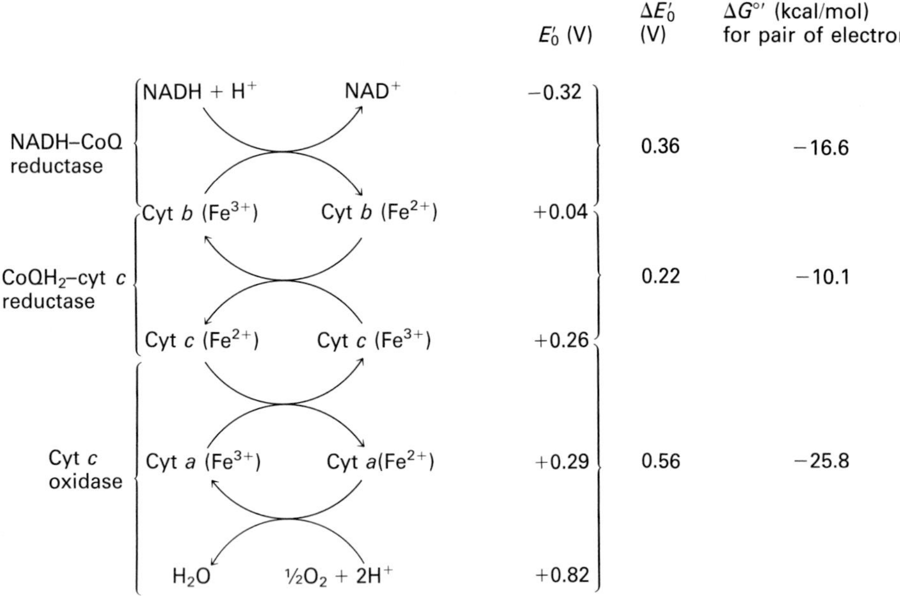

Figure 20-27 Energetics of the mitochondrial oxidation chain. Shown is the change in free energy released by transport of a pair of electrons (from NADH) through each of the complexes in which H^+ pumping occurs.

is the number of electrons involved—here, 2. Thus

$$\begin{aligned}\Delta G^{\circ\prime} &= (-2)(23{,}062)(0.56)\\ &= -25{,}829 \text{ cal/mol}\\ &= -25.8 \text{ kcal/mol}\end{aligned}$$

The reaction is strongly exergonic: transfer of a pair of electrons from cytochrome c to O_2 releases a significant amount of energy that can be made to do useful work—in this case, the pumping of protons across the membrane. An understanding of how this coupling occurs, however, requires a somewhat detailed look at the structure of the electron transport chain.

The Electron Transport Complexes Span the Inner Mitochondrial Membrane

Each of the four multiprotein complexes of the electron transport chain spans the inner mitochondrial membrane. Cytochrome c oxidase is the best characterized of these complexes. It contains seven polypeptides, including two cytochromes (the a and a_3 cytochromes) and protein-bound copper. Experimentally, the complex can be purified and then incorporated into liposomes so that all the cytochrome c oxidase molecules have the same orientation with respect to the lipid bilayer. This complex remains capable of catalyzing the oxidation of added cytochrome c (Fe^{2+}) by O_2, so only these seven polypeptides are required for electron transport from cytochrome c to O_2. As evidence that the complex spans the membrane, only three polypeptides—II, III, and V in Figure 20-28—react with membrane-impermeable reagents (see Figure 14-15) or specific antibodies added to the outside of the mitochondrion; thus these contain regions exposed to the intermembrane space. Two other polypeptides, IV and VII, react with membrane-impermeable reagents added to the matrix. A molecular model of cytochrome c oxidase (Figure 20-28) has been constructed from electron microscopic images of the purified complex incorporated in liposomes. Cytochrome c, a peripheral membrane protein, binds to a specific site on the cytoplasmic side of the complex. Addition of electrons to oxygen, by contrast, occurs on the matrix side, and cytochrome a_3, which catalyzes this step, projects from the membrane into the matrix.

Each of the electron transport complexes is laterally mobile in the plane of the membrane, as documented in Chapter 14. There do not appear to be stable contacts between two or more complexes. Transport of electrons from one complex to another may involve random collisions of the two. The lipid-soluble coenzyme Q shuttles electrons from the NADH–CoQ reductase and succinate dehydrogenase complexes to the $CoQH_2$–cytochrome c reductase complex. Cytochrome c interacts with specific sites both on the $CoQH_2$–cytochrome c reductase complex and on the cytochrome c oxidase complex and thus transfers electrons from one to the other.

The In Vivo Order of the Electron Carriers Can Be Determined with Certainty

A wide range of techniques has been employed to confirm the pathway of electron transport depicted in Figure 20-24. As stated earlier, the reduction potential of each of the electron carriers has been determined; these values increase steadily from NADH or $FADH_2$ to O_2. Thus the transport of electrons at each step is favored thermodynamically. Because oxidized and reduced forms of each cytochrome absorb light at different wavelengths, spectroscopic techniques can be used to determine the fraction of each of the cytochromes in the oxidized or reduced state. In respiring mitochondria, cytochromes that

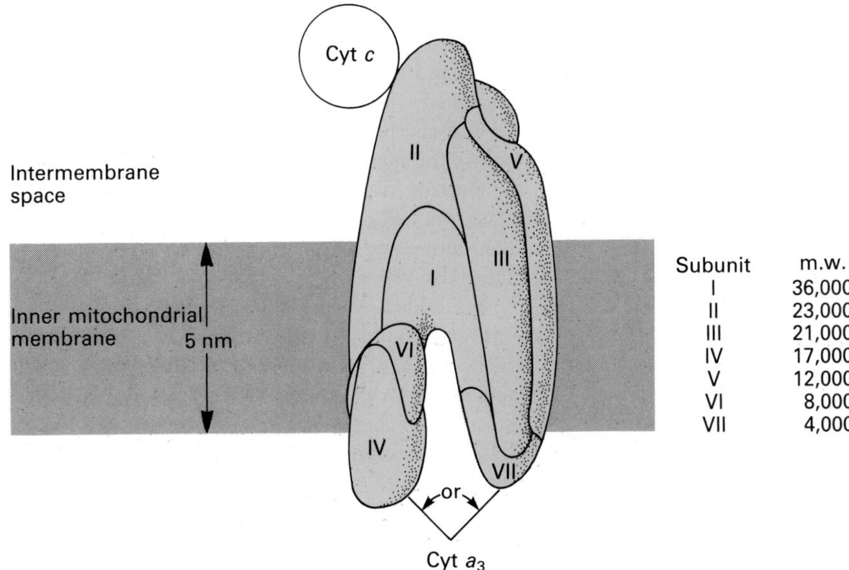

Figure 20-28 The composition and orientation of the cytochrome oxidase complex in the inner mitochondrial membrane. The complex exists as a dimer; only one of the two monomer units is shown here. The overall shape of the molecule and its relationship with the bilayer have been obtained from image reconstitution of electron micrographs. The relationship between the various subunits is based on visualization of the binding of antibodies to specific subunits and on chemical cross-linking of certain subunits. The model is rather crude at this stage, and the shapes of the individual peptides are only approximations. The binding site of cytochrome c involves mainly subunit II. Because of the ambiguities in the structure, investigators are unsure whether cytochrome a_3 is the region IV or VII of the complex. *After M. Brunori and M. T. Wilson, 1982,* Trends Biochem. Sci. *7:295.*

are closer to the oxidizing (O_2) end of the chain are more fully oxidized than are those toward the reducing (NADH) end (Figure 20-29), because they are closer to the end of the chain at which electrons are permanently removed. Similarly, in mitochondria incubated without O_2, all the cytochromes are in the reduced state. Their order in the chain is the order in which they become oxidized after the addition of oxygen.

Many inhibitors of mitochondrial function are known to block electron transport by reacting with specific complexes. As Figure 20-29 shows, these specific inhibitors have been most useful in confirming the carrier sequence. Carriers downstream of the inhibited site accumulate in the oxidized form; those upstream accumulate in the reduced state. Spectrophotometric methods permit determination of which carriers are in the oxidized or reduced state in the presence of inhibitor.

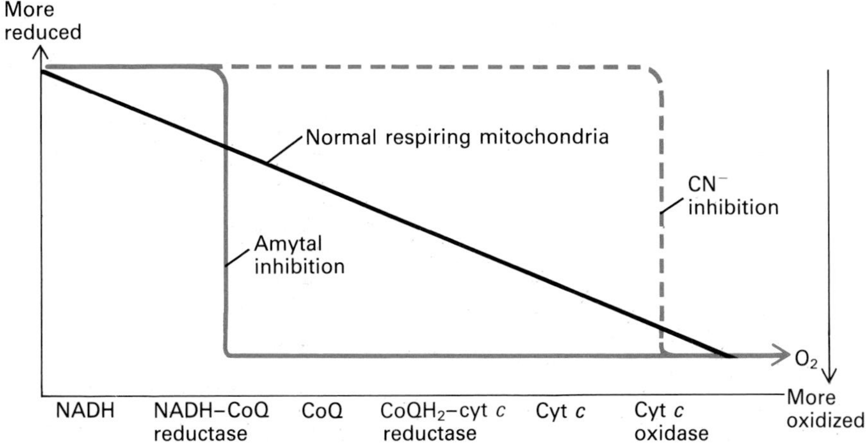

Figure 20-29 Inhibitors can be used to order the steps in electron transport from NADH to oxygen. In normal respiring mitochondria, more of the electron carriers closer to the NADH end of the chain tend to be in the reduced form, and those toward the O_2 end tend to be in the oxidized state. Each inhibitor blocks transport of electrons at a specific site. When an inhibitor such as amytal is added to a mitochondrial preparation that is oxidizing NADH, all the carriers before the blocking site become fully reduced—that is, fully loaded with electrons. All the carriers "downstream" of the block, by contrast, become fully oxidized, or depleted of electrons; any electrons carried at the time of inhibitor addition are passed onto oxygen, and there is no accumulation of additional electrons. Amytal acts at the beginning of the electron transport chain, whereas cyanide (CN^-) acts on cytochrome c oxidase at the very end of the chain.

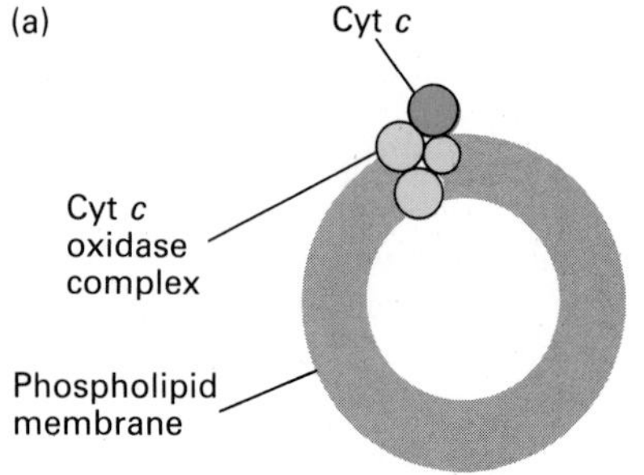

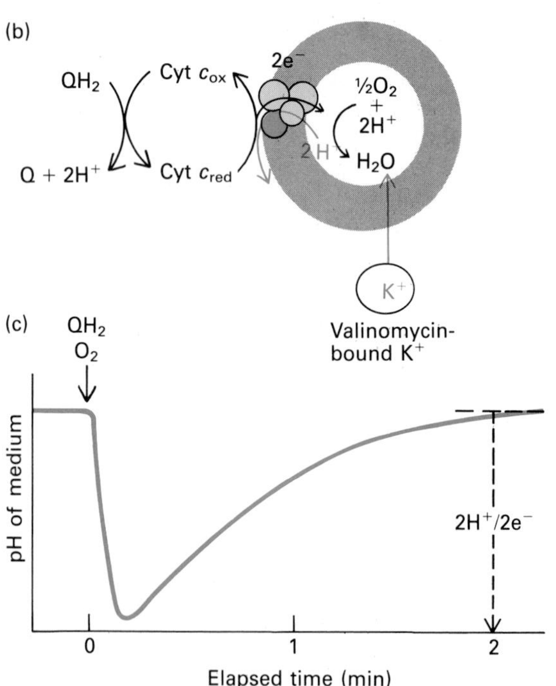

Figure 20-30 Segments of the mitochondrial electron transport system can be isolated and assembled into phospholipid membrane vesicles. When the system is supplied with either a natural or an artificial donor and acceptor of electrons, protons are pumped across the membrane. (a) The third complex in the electron transport chain—cytochrome *c* oxidase—is incorporated into vesicles; note that cytochrome *c* and its binding site are on the outer surface. (b) When O_2 and dihydroquinone (QH_2) are added, electrons are transferred from the dihydroquinone to cytochrome *c* and thence to oxygen, forming water. Concomitantly, protons are pumped from the inside to the outside of the vesicles. (c) This change is measured by a drop in the pH of the medium following oxygen addition. Valinomycin and K^+ are added to the system to dissipate the membrane potential generated by pumping of H^+, which would then reduce the number of protons pumped. As the dihydroquinone becomes fully oxidized, protons leak back into the vesicles, and the pH of the medium returns to its initial value. Q and QH_2 signify the oxidized and reduced forms, respectively, of the quinone used.

Cyanide, a potent poison, kills by blocking the final step of electron transport to oxygen. Cyanide binds irreversibly to the heme Fe atoms in cytochrome *c* oxidase and blocks the binding of oxygen. Rotenone, a plant-derived toxin that has found commercial use as an insecticide and fish poison, blocks NADH–CoQ reductase near the reduced end of the chain.

Three of the Four Electron Transport Complexes Are Sites of Proton Pumping

Except for succinate–CoQ reductase, each of the multiprotein complexes is believed to be a site for proton transport across the membrane during electron movement. By selective extraction of mitochondrial membranes with detergents, each of the complexes can be isolated in a high degree of purity and then incorporated into phospholipid vesicles. Addition of an appropriate source of electrons and an appropriate acceptor of electrons leads to pumping of protons across the membrane bilayer. For example, the cytochrome *c* oxidase complex can be incorporated into phospholipid vesicles so that the peripheral cytochrome *c* polypeptide is on the outside (Figure 20-30). Addition of the reduced dihydroquinone leads to reduction of cytochrome *c*, and the electrons then move down the chain to oxygen. Direct measurements indicate that two to four protons are transported out of the vesicles per electron pair transported (or, equivalently, hydroquinone oxidized); however, because some of the pumped protons leak back across the membrane, it is difficult to determine precisely how many are pumped.

The free energy change for transfer of a pair of electrons from NADH to cytochrome *b*—the reaction catalyzed by the NADH–CoQ reductase complex—is significant ($\Delta G^{\circ\prime} \sim -17$ kcal/mol or, equivalently, $\Delta E_0' \sim 0.36$ V). The reaction is sufficiently exergonic to drive the synthesis of one mole of ATP under standard conditions ($\Delta G^{\circ\prime} = +7.3$ kcal/mol). Similarly, the free energy changes in each of the two other complexes in the electron transport chain are sufficient to generate one mole of ATP per electron pair. Under standard conditions, oxidation of one mole of NADH can yield as much as three moles of ATP, or one ATP for each of the three complexes in the chain.

Precisely how the protons are pumped across the membrane during electron transport is not clear. In all likelihood, coenzyme Q participates in one stage of this transport process: on the matrix side of the membrane the quinone form can receive one electron (from cytochrome *b*) and one proton (from the solution), forming the partially reduced semiquinone (QH). Addition of a second electron (from an FeS protein) and a proton forms the fully reduced hydroquinone QH_2. Both the hydroquinone and the semiquinone can diffuse to the outer surface of the membrane; there they can transfer the electrons to cytochrome c_1 or another cytochrome *b*. Simultaneously the protons are lost to the medium. The oxidized quinone thus formed can diffuse back to the matrix surface, completing the cycle (Figure 20-31). The net result is transport of one or two protons from the matrix across the inner mitochondrial membrane, concomitant with electron transport.

Figure 20-31 summarizes the current understanding of the pathway of electron movements in the inner mitochondrial membrane and the sites of proton pumping. Coenzyme Q carries an electron from the first complex, NADH–CoQ reductase, to the second, $CoQH_2$–cytochrome *c* reductase, as it transports a proton across the phospholipid bilayer. Cytochrome *c*, a water-soluble protein that binds both to the $CoQH_2$–cytochrome *c* reductase and to the cytochrome *c* oxidase complex, carries electrons between these two complexes.

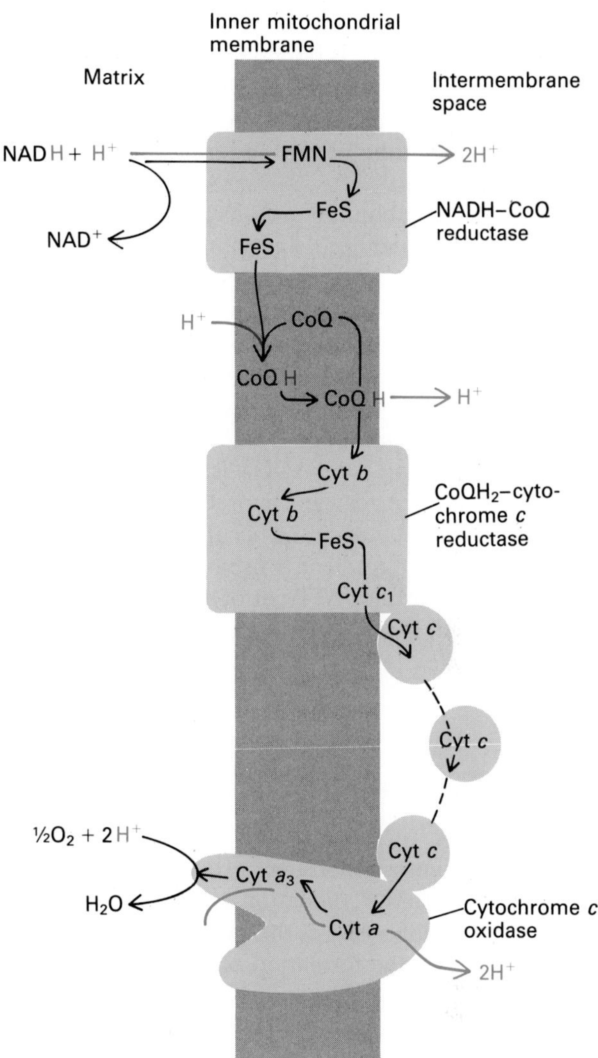

Figure 20-31 A diagrammatic representation of the pathway of electron movement *(black)* and proton pumping *(color)* in the inner mitochondrial membrane. Electrons are transferred from NADH–CoQ reductase to $CoQH_2$–cytochrome *c* reductase by coenzyme Q (ubiquinone). As depicted, coenzyme Q may move protons from the matrix to the intermembrane space; two protons are probably transported at this step (one with each electron), but neither the nature of the mechanisms involved nor the role of proteins in the $CoQH_2$–cytochrome *c* reductase complex, although under active study, is known. The peripheral protein cytochrome *c* diffuses in the intermembrane space and shuttles electrons from the $CoQH_2$–cytochrome *c* reductase complex to cytochrome *c* oxidase.

Metabolic Regulation

All enzyme-catalyzed reactions and all metabolic pathways are tightly regulated. Mitochondria synthesize ATP only as needed by the cell. The import of metabolites into the mitochondrion and the export of ATP are also coordinated with ATP synthesis and, as we shall see, the generation of heat. Similarly, the conversion of glucose, fatty acids, and other metabolites to acetyl CoA is tightly regulated so that only the needed amount of this substrate for the citric acid cycle is produced. In this section we consider how this regulation is achieved. We also examine the process of *gluconeogenesis*—the conversion of pyruvate and other small molecules to glucose—which is the apparent reverse of the Embden-Meyerhof pathway. Both the conversion of glucose to pyruvate and the conversion of pyruvate to glucose occur in liver hepatocyte cells, and much is understood of how these two competing processes are controlled. To begin with, let us focus on the overall energetics of mitochondrial oxidation.

The Ratio of ATP Production to O_2 Consumed Is a Measure of the Efficiency of Oxidative Phosphorylation

For each molecule of NADH or $FADH_2$ that is oxidized, one atom of molecular oxygen is reduced to water. A way of expressing the efficiency of oxidative phosphorylation is the ratio of phosphorus to oxygen—that is, the number of ATPs produced per oxygen atom ($\frac{1}{2}O_2$) used. Direct measurements on isolated mitochondria indicate that, for every NADH oxidized to NAD^+, two to three ATP molecules are synthesized from ADP. Between one and a half and two molecules of ATP are generated per molecule of $FADH_2$ oxidized to FAD. The principal mitochondrial reaction that generates $FADH_2$ is the oxidation of succinate to fumarate, a key citric acid cycle reaction (reaction

7, Figure 20-14). The enzyme that catalyzes this reaction, succinate dehydrogenase, is bound to the inner mitochondrial membrane and contains a tightly bound FAD. Thus the oxidation of one molecule of succinate to fumarate results in the generation of one and a half to two molecules of ATP.

If we assume that about 10 protons are transported per molecule of NADH oxidized, and that 2 to 3 protons are used to generate each ATP, then oxidation of each molecule of NADH can account for synthesis of three ATPs. The free energy released in oxidation of NADH by oxygen ($\Delta G^{\circ\prime} = -52.5$ kcal/mol) is sufficient to power the synthesis of at least three molecules of ATP from ADP and P_i. The ratio of phosphorus to oxygen observed for oxidation of succinate or $FADH_2$ is, at maximum, 2. These maxima are not achieved in practice. Some leakage of protons across the membrane occurs, and some of the energy of the proton gradient is used to transport molecules such as ADP into or ATP out of the mitochondrion. As we have noted, generally oxidation of NADH yields a phosphorus-oxygen ratio of about 2.

In Respiratory Control, Oxidation of NADH and Production of ATP Are Closely Coupled

If intact isolated mitochondria are provided with NADH and oxygen but not ADP, oxidation of NADH and reduction of oxygen rapidly cease as the amount of endogenous ADP is depleted by the formation of ATP. Oxidation of NADH is rapidly restored upon readdition of ADP. Thus mitochondria can oxidize $FADH_2$ and NADH only so long as there is a source of ADP and P_i to generate ATP. This phenomenon, termed *respiratory control,* illustrates how one key reactant can limit the rate of a complex set of interrelated reactions.

Intact cells and tissues also employ respiratory control. Cells oxidize only enough glucose to synthesize the amount of ATP required for their metabolic activities. Stimulation of a metabolic activity that utilizes ATP, such as muscle contraction, results in an increased level of cellular ADP; this, in turn, causes an increased rate of oxidation of metabolic products in the mitochondrion.

The nature of respiratory control is now well understood. Recall that oxidation of NADH or succinate or $FADH_2$ is *obligatorily* coupled to transport of protons across the inner mitochondrial membrane. If the resulting electrochemical gradient is not dissipated by utilizing the protons for synthesis of ATP, or for some other purpose, the transmembrane proton concentration and potential gradients will increase to very high levels. Oxidation of NADH will eventually cease, because it will require too much energy to pump additional protons across the membrane against the existing proton and potential gradients.

Uncoupling of Respiratory Control Dissipates Proton Gradients across the Mitochondrial Membrane

Certain poisons such as 2,4-dinitrophenol (Figure 20-32) uncouple oxidation of NADH from synthesis of ATP from ADP and P_i. Uncouplers allow oxidation of NADH and reduction of oxygen to continue at high levels, but there is no ATP synthesis. Dinitrophenol (DNP) acts as a membrane transporter for H^+, bypassing the ATP synthesis system normally associated with the H^+ movement (Figure 20-32). Both the neutral and negatively charged forms of DNP are soluble in phospholipid membranes and in aqueous solution, so DNP can act as a shuttle for protons. By making the membrane leaky to protons, DNP dissipates any transmembrane proton concentration or potential gradient. Uncouplers such as DNP abolish synthesis of ATP—and dispense with any requirement for ADP in oxidation of NADH or in transport of electrons. The energy released by the oxidation of NADH in the presence of DNP is converted to heat.

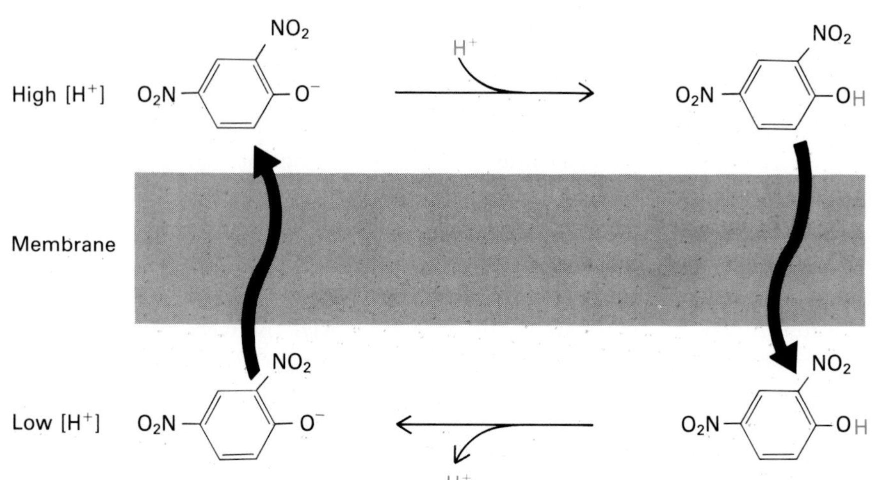

Figure 20-32 The poison 2,4-dinitrophenol functions as a membrane transporter for H^+ and acts to uncouple oxidative phosphorylation. Dinitrophenol (DNP) dissipates proton gradients generated by electron transport without causing synthesis of ATP. Both the neutral and ionic forms of DNP are soluble in the lipid bilayer; thus it can function as a shuttle for protons.

Mitochondria in Brown Fat Contain a Natural Uncoupler

Brown fat is a tissue specialized for the generation of heat. In contrast to adipocytes, which are specialized for the storage of fat (white fat tissue), brown adipose tissue contains abundant mitochondria, which impart their dark brown color to the tissue.

In brown fat mitochondria, an inner-membrane protein acts as a natural uncoupler of oxidative phosphorylation. This protein acts as a transmembrane H^+ transporter. Like other uncouplers, it acts by short-circuiting the membrane proton gradient across the inner mitochondrial membrane and then converting energy released by oxidation of NADH into heat. The action of this protein is regulated to generate heat and maintain body temperature under different environmental conditions. For instance, during adaptation of rats to cold, the ability of tissues to generate heat (thermogenesis) is increased. The increased thermogenesis is accomplished by inducing the synthesis of this inner-membrane uncoupler protein; in cold-adapted animals this H^+ pore protein may make up 15 percent of the mitochondrial membrane protein.

Adult humans have little brown fat, but human infants have a great deal, and in the newborn, thermogenesis by brown fat mitochondria is vital to survival. Brown fat mitochondria are also essential to thermogenesis in hibernating mammals. In fur seals and other animals that are naturally acclimated to the cold, mitochondria in muscle cells contain a natural uncoupler protein permitting conversion of a great deal of the energy in the proton gradient into heat used to maintain body temperature.

The Steps of Glycolysis Are Controlled by Multiple Allosteric Effectors

The activity of the Embden-Meyerhof pathway is continuously regulated, so that the production of ATP and pyruvate is adjusted to meet the needs of the cell. Phosphofructokinase, catalyzing the third reaction in the pathway of conversion of glucose to pyruvate, is the principal rate-limiting enzyme of the entire pathway (Figure 20-33). Its activity is allosterically controlled by multiple ligands in such a way that glycolysis is coordinated with other metabolic processes in the cell.

Allosteric inhibition of phosphofructokinase by NADH and citrate enables the activities of the Embden-Meyerhof pathway to be coordinated with those of the citric acid cycle (Figure 20-14). If the cytosolic concentration of NADH builds up too high (owing to a slowdown in oxidative phosphorylation), it inhibits phosphofructokinase and thus reduces the formation of pyruvate and acetyl CoA. Similarly, if citrate—the product of the first step of the citric acid cycle—accumulates, feedback inhibition of phosphofructokinase reduces the generation of pyruvate and acetyl CoA, so that less citrate is formed.

Note that phosphofructokinase is allosterically activated by ADP and allosterically inhibited by ATP. This arrangement makes the rate of glycolysis very sensitive to intracellular levels of ATP and ADP: too much ATP causes a reduction in the rate of conversion of glucose to pyruvate. Allosteric inhibition of phosphofructokinase by ATP may seem unusual, since ATP is also a substrate of this enzyme. But the affinity of the substrate-binding site for ATP is much higher (i.e., lower K_m) than is that of the allosteric site. Thus, at low concentrations of ATP, binding of ATP to the catalytic but not the inhibitory allosteric site occurs, and enzyme catalysis proceeds at near maximal rates. At high ATP concentrations, ATP binds to the allosteric site. This binding induces a conformational change that results in inactivation of phosphofructokinase and thus in reduction in the overall rate of glycolysis.

Other glycolytic enzymes are also subjected to allosteric control. Because pyruvate kinase is allosterically inhibited by ATP, buildup of ATP levels in the cell causes an inhibition of glycolysis at this step as well as at the reaction catalyzed by phosphofructokinase. The three glycolytic enzymes that are regulated by allosteric molecules are those that catalyze reactions with the most negative $\Delta G^{\circ\prime}$ values—those that are essentially irreversible under ordinary conditions. These enzymes thus are particularly suitable for regulating the entire Embden-Meyerhof pathway. Additional significance of these regulatory enzymes will become apparent when we consider the reverse pathway: conversion of pyruvate to glucose.

Glucose Can Be Synthesized from Pyruvate by Reversal of Many of the Steps of Glycolysis

Frequently cells synthesize glucose from three-carbon compounds such as pyruvate:

$$\text{2 molecules of pyruvate} + 2\text{NADH} \longrightarrow \text{glucose} + 2\text{NAD}^+$$

This process of gluconeogenesis is the apparent reverse of glycolysis. Lactate, for instance, is released into the blood during muscle contraction. In the liver it is converted first to pyruvate and then to glucose; much of the glucose is returned to the circulation to power additional contractions:

$$\underset{\text{Lactate}}{CH_3CH(OH)COO^-} + NAD^+ \xrightleftharpoons{\text{lactate dehydrogenase}} \underset{\text{Pyruvate}}{CH_3{-}\overset{\overset{\displaystyle O}{\|}}{C}{-}COO^-} + NADH + H^+$$

Many of the reactions and enzymes used in glycolysis are also used in gluconeogenesis; this is an obvious economy for the cell since it allows the same enzyme to function in two pathways. However, the Embden-Meyerhof pathway

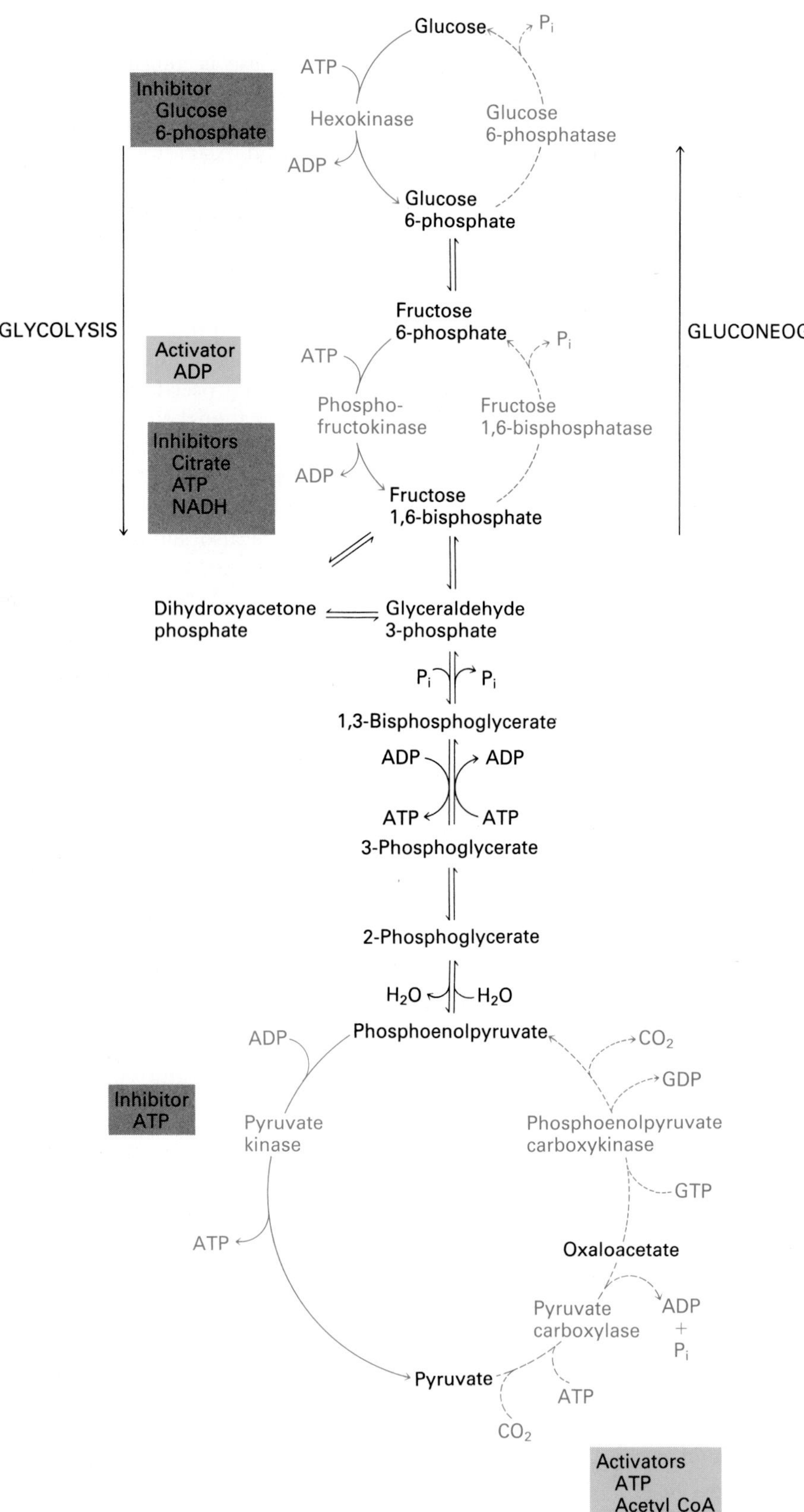

Figure 20-33 The glycolytic (Embden-Meyerhof) and gluconeogenic pathways for interconversion of glucose and pyruvate. Enzyme reactions unique to the glycolytic pathway are shown in solid color and those unique to the gluconeogenic are in dotted color. Most of these reactions are controlled by small molecules that bind at allosteric sites; allosteric activators and inhibitors appear in the shaded boxes.

is designed for the essentially irreversible conversion of glucose to pyruvate and for the generation of ATP from ADP and P_i. Glycolysis includes three steps that have a sufficiently negative $\Delta G^{\circ\prime}$ value that they are essentially irreversible under normal conditions: those catalyzed by hexokinase, phosphofructokinase, and pyruvate kinase. To reverse the pathway to convert pyruvate to glucose, the cell must bypass these steps. Moreover, the conversion of glucose to pyruvate proceeds with a net synthesis of ATP, even though there is a large negative $\Delta G^{\circ\prime}$ and the reaction is favored thermodynamically. The overall reaction is

$$\text{Glucose} + 2\text{NAD}^+ + 2\text{ADP} + 2\text{P}_i \longrightarrow 2 \text{ molecules of pyruvate} + 2\text{NADH} + 2\text{H}^+ + 2\text{ATP}$$

To reverse the pathway requires input of additional energy from hydrolysis of additional phosphoanhydride bonds. Thus the *overall* reaction for gluconeogenesis is not the reverse of glycolysis; rather, it is

$$2 \text{ pyruvate} + 2\text{NADH} + 2\text{H}^+ + 2\text{ATP} + 2\text{GTP} \longrightarrow \text{glucose} + 2\text{NAD}^+ + 2\text{ADP} + 2\text{GDP} + 4\text{P}_i$$

Four phosphoanhydride bonds, rather than two, are hydrolyzed, and the reaction proceeds thermodynamically in the direction of formation of glucose.

The first two essentially irreversible steps of glycolysis are catalyzed by hexokinase and phosphofructokinase, kinases that transfer the terminal phosphate from ATP to a sugar. Reversal of these reactions is thermodynamically unfavorable:

$$\text{Glucose} \xrightarrow[\text{hexokinase}]{\text{ATP} \;\; \text{ADP}} \text{glucose 6-phosphate} \xrightarrow[\text{isomerase}]{}$$

$$\text{fructose 6-phosphate} \xrightarrow[\text{phosphofructokinase}]{\text{ATP} \qquad \text{ADP}} \text{fructose 1,6-bisphosphate}$$

$$\Delta G^{\circ\prime} = -7.3 \text{ kcal/mol}$$

In the gluconeogenic direction, phosphatases are used at these steps, resulting in the hydrolysis of a sugar-phosphate bond rather than the synthesis of an ATP phosphoanhydride bond.

$$\text{Fructose 1,6-bisphosphate} \xrightarrow[\text{fructose 1,6-bisphosphate phosphatase}]{\text{P}_i}$$

$$\text{fructose 6-phosphate} \xrightarrow[\text{isomerase}]{}$$

$$\text{glucose 6-phosphate} \xrightarrow[\text{glucose 6-phosphatase}]{\text{P}_i} \text{glucose}$$

$$\Delta G^{\circ\prime} = -7.3 \text{ kcal/mol}$$

These reactions are favored thermodynamically and permit the conversion of fructose 1,6-bisphosphate to glucose. Thus, interconversion of glucose and fructose 1,6-bisphosphate can proceed in either direction, depending on whether kinases or phosphatases are utilized (Figure 20-33). The third step in glycolysis that is essentially irreversible is catalyzed by pyruvate kinase:

$$\text{Phosphoenolpyruvate} + \text{ADP} \xrightarrow[\text{pyruvate kinase}]{} \text{pyruvate} + \text{ATP}$$

$$\Delta G^{\circ\prime} = -7.5 \text{ kcal/mol}$$

To bypass this step in gluconeogenesis, two reactions are used, each requiring the hydrolysis of a phosphoanhydride bond. In the first reaction, CO_2 is added to pyruvate, forming the four-carbon oxaloacetate; in the second, CO_2 is lost, generating phosphoenolpyruvate:

$$\underset{\text{Pyruvate}}{CH_3{-}\overset{\overset{\displaystyle O}{\|}}{C}{-}COO^-} \xrightarrow[\text{pyruvate carboxylase}]{\text{ATP} \;\; CO_2 \;\; \text{ADP} + \text{P}_i}$$

$$\underset{\text{Oxaloacetate}}{{}^-OOC{-}\overset{\overset{\displaystyle O}{\|}}{C}{-}CH_2{-}COO^-} \xrightarrow[\text{phosphoenolpyruvate carboxykinase}]{\text{GTP} \quad \text{GDP} \quad CO_2}$$

$$\underset{\text{Phosphoenolpyruvate}}{CH_2{=}C(O{-}P(O^-)({=}O){-}OH){-}COO^-}$$

$$\Delta G^{\circ\prime} = +0.2 \text{ kcal/mol}$$

Hydrolysis of two phosphoanhydride bonds, one from ATP and one from GTP, is necessary to drive these reactions in a thermodynamically possible direction.

It is not surprising that the enzymes unique to the glycolytic (Embden-Meyerhof) or gluconeogenic pathway are the closely regulated ones. Phosphofructokinase, as we have seen, is the major rate-limiting enzyme in the reactions of glycolysis. We saw in Chapter 16 that fructose 1,6-bisphosphate phosphatase is also subject to regulation. This enzyme controls the flow of metabolites to glucose in the liver; like many other regulatory enzymes, it is modified by cAMP-dependent protein kinases.

Thus the reactions of the Embden-Meyerhof pathway and the citric acid cycle, oxidative phosphorylation, and thermogenesis are tightly controlled to produce the appropriate amount of ATP needed by the cell. Depending on the type of cell and its function in the organism, these processes can be used to produce glucose for release in the blood as nutrient for other cells, or for body heat. A common feature of these pathways is that the amounts and activities of critical enzymes are regulated by the level of small-molecule metabolites, generally by allosteric interactions or by phosphorylation.

Energy Metabolism in Bacteria

In certain species of bacteria, energy metabolism is similar to the processes just described for eukaryotic cells. Anaerobic bacteria such as *Lactobacillus* species (which cause spoilage of milk) ferment glucose to lactic acid by a glycolytic pathway identical to that used in muscle cells. Many aerobic bacteria, such as the spore-forming soil bacterium *Bacillus subtilis,* oxidize glucose to CO_2 by a combination of the Embden-Meyerhof pathway and the citric acid cycle. But in two key respects, bacterial energy metabolism is different from that in eukaryotes: (1) bacteria lack mitochondria and possess only a plasma membrane; and (2) many bacteria are able to grow in unusual environments and do not use glucose or other sugars as a source of energy. Although the details of energy metabolism in many bacteria are not well understood, it appears that most, if not all, bacteria utilize a proton-motive force across their plasma membranes. Electron transport chains are often used to generate this proton-motive force, even in certain anaerobic organisms.

There is a Proton Gradient across the Plasma Membrane of All Prokaryotes

Aerobic bacteria such as *Escherichia coli* have a respiratory chain for the transfer of electrons from NADH or $FADH_2$ to oxygen. This chain consists of fewer cytochromes and FeS proteins than in mitochondria and is localized to the plasma membrane. Electron transport is coupled to transport of protons out of the cell, and respiring *E. coli* cells generate a proton motive force across the cytoplasmic membrane equivalent to two to three pH units (see Figure 20-16).

As in mitochondria, this pH gradient can be used for many purposes. The bacterial membrane contains an F_0F_1-ATPase very similar in structure to that of mitochondria. Flow of protons inward through this multiprotein complex results in synthesis of ATP (see Figure 20-16). The proton gradient can be used to power transport of molecules into and out of the bacterial cell. We have seen that lactose uptake utilizes an H^+-lactose symport (Figure 15-21). Because bacterial cells do not have an Na^+K^+-ATPase, H^+ antiport systems are used to transport Na^+ and Ca^{2+} out of the cell, against a concentration gradient, in order to maintain the proper intracellular ionic balance:

$$H^+_{out} + Na^+_{in} \longrightarrow H^+_{in} + Na^+_{out}$$
$$2H^+_{out} + Ca^{2+}_{in} \longrightarrow 2H^+_{in} + Ca^{2+}_{out}$$

The proton-motive force is also used to power rotation of the bacterial flagellum. As Figure 20-34 shows, the basal end of the flagellum is a complex structure, consisting of four disks, or rings, that insert into the outer membrane, the peptidoglycan layer, and the bacterial plasma membrane. Although the major part of the flagellum is built of a single protein species, flagellin, the hook and rings contain at least 18 different proteins and probably more. The L, P, and S rings apparently function as bearings to allow free rotation of the flagellum. The M ring is made up of 16 identical subunits and lies in close apposition to a ring of proteins in the plasma membrane. Rotation of the flagellum requires a proton-motive force across the plasma membrane and is not dependent on the level of ATP inside the cell. Protons flowing inward through the M ring cause rotation; it is estimated that 256 protons flow inward during each revolution. As evidence for the role of the proton-motive force, uncouplers immediately depolarize the plasma membrane and immediately stop rotation, but they have slower effects on the level of intracellular ATP.

Many bacteria utilize energy sources other than carbohydrates; yet an electron transport chain is also employed to generate a proton-motive force across the cytoplasmic membrane. For instance, *Vibrio succinogenes* grows anaerobically on H_2 gas and fumarate as sole sources of energy, for which the overall equation is

$$\text{Fumarate}^{2-} + H_2 \longrightarrow \text{succinate}^{2-}$$
$$\Delta G^{\circ\prime} = -20.6 \text{ kcal/mol}$$

Electrons generated by the oxidation of H_2 gas

$$H_2 \longrightarrow 2H^+ + 2e^-$$

are transferred along a chain of electron carriers to fumarate, the terminal electron acceptor:

$$\underset{\text{Fumarate}}{(H)(COO^-)C{=}C(COO^-)(H)} + 2H^+ + 2e^- \longrightarrow \underset{\text{Succinate}}{COO^-{-}CH_2{-}CH_2{-}COO^-}$$

Again, electron transport along this chain is coupled to export of H^+ from the cell.

Other anaerobic organisms use other molecules, such as nitrate, NO_3^-, or sulfate, SO_4^{2-}, as terminal electron acceptors. Nitrate can be reduced to nitrite, NO_2^-, or to ammonia, NH_3. Certain bacteria of the genus *Clostridia,* for instance, generate ATP by the following overall reaction:

$$4H_2 + NO_3^- + H^+ \longrightarrow NH_3 + 3H_2O$$

Hydrogen gas is oxidized to protons; the electrons released are carried by a transport chain to nitrate, which serves as ultimate acceptor; each NO_3^- molecule consumes a total of eight electrons and is reduced to NH_3.

The habitat of bacteria of the genus *Desulfovibrio* is anaerobic sediments that contain organic matter and sulfate. This organism can generate ATP by converting lactate, a typical fermentation product of other organisms, into acetate and CO_2, while it reduces sulfate (SO_4^{2-}) to sulfide (S^{2-}):

$$\underset{\text{Lactate}}{2CH_3CH(OH)COO^-} + 2ADP + 2P_i + SO_4^{2-} \longrightarrow$$
$$\underset{\text{Acetate}}{2CH_3COO^-} + 2CO_2 + 2H_2O + 2ATP + S^{2-}$$

Electrons released when lactate is oxidized first to pyruvate, and then to acetate, are transferred along a chain of electron carriers to SO_4^{2-}, the ultimate electron acceptor.

(a)

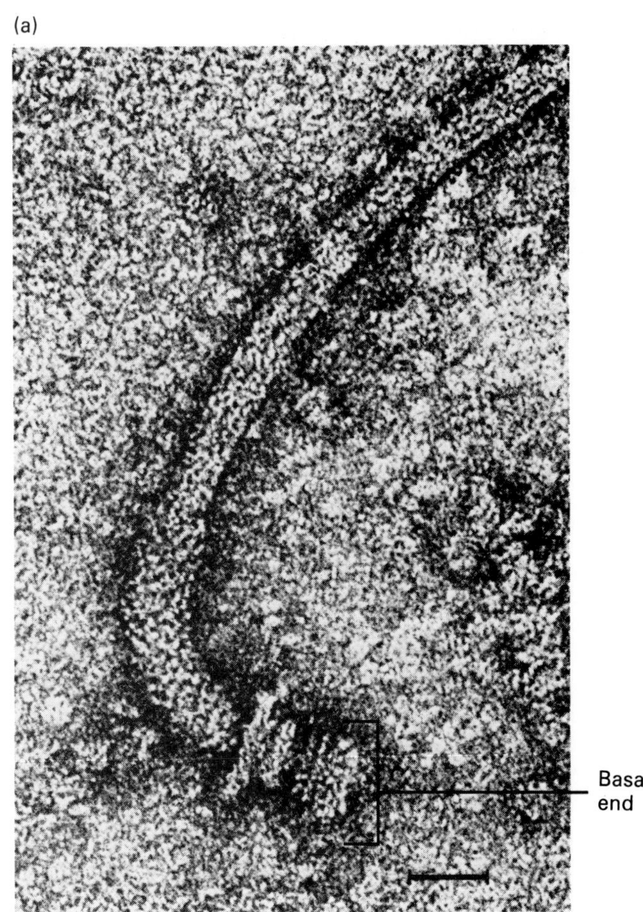

(b)

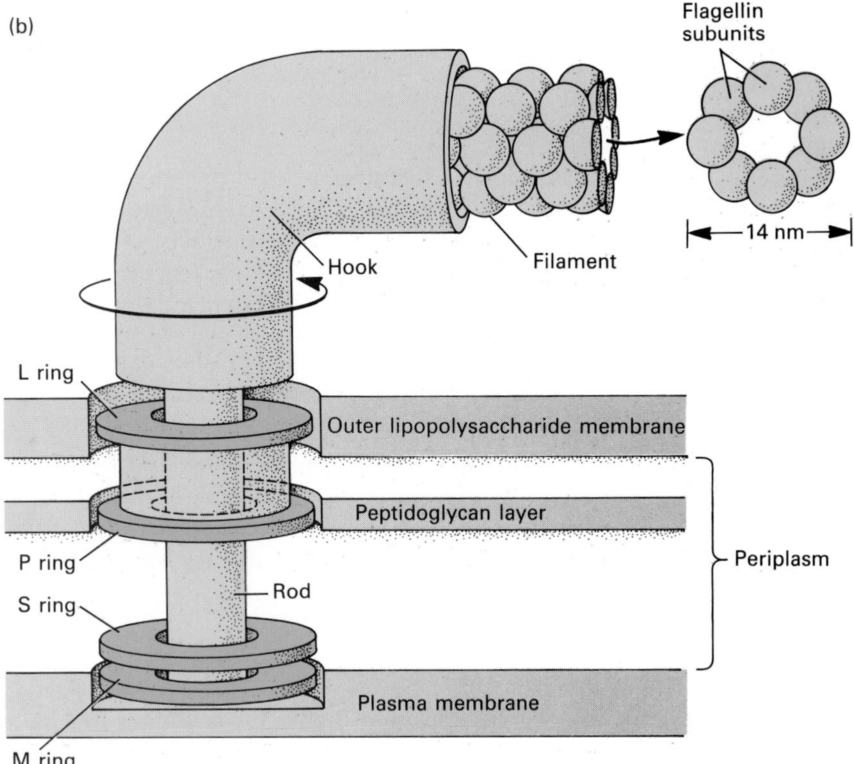

Figure 20-34 The end of a bacterial flagellum consists of disks, or rings, that function in its rotation. (a) A negatively stained micrograph of a flagellum from the gram-negative bacterium *Caulobacter crescentus*. Some of the flagellin subunits are visible on the long part of the flagellum. At the basal end of the flagellum can be seen, in side view, the complex ring-shaped structures that insert into the inner and outer bacterial membranes. (b) Model of the basal end of the flagellum of *Escherichia coli*, based on electron micrographs of the isolated organelle. Included is a suggested model for the packing of the 40,000-m.w. subunits of flagellin into a helical flagellum filament having a diameter of 14 nm. Also shown are the possible topological relations between the basal structure of the flagellum and the outer cell layers of *E. coli*. *Photograph (a) courtesy of L. Shapiro. Part (b) after M. L. de Pamphilis and J. Adler, 1971,* J. Bacteriol. ***105****:396; and M. L. de Pamphilis and J. Adler, 1971,* J. Bacteriol. ***105****:384.*

Some Anaerobic Bacteria Use the F_0F_1-ATPase in Reverse to Generate a Proton Gradient

For many years, anaerobic bacteria such as *Streptococcus faecalis* presented a puzzle. *Streptococcus faecalis* ferments glucose by the Embden-Meyerhof pathway to produce lactic acid. These reactions are the only source of ATP in this organism. Then why does *S. faecalis* contain an F_0F_1-ATPase in its membrane? Recent findings suggest that protons run through the F_0F_1 complex in reverse: hydrolysis of cytoplasmic ATP, generated by glycolysis, to ADP is coupled to export of protons from the cell. Thus the F_0F_1-ATPase catalyzes hydrolysis of ATP to generate a proton-motive force across the plasma membrane and functions to maintain intracellular pH at the appropriate value. As in other bacteria, the proton-motive force is used to power the uptake of sugars and other metabolites, as well as the export of Na^+ and Ca^{2+} (Figure 20-35). In certain organisms, then, the F_0F_1-ATPase can be used to synthesize ATP, by employing the energy of the proton-motive force, or can operate in reverse—by hydrolyzing ATP to pump protons out of the cell while accumulating other metabolites.

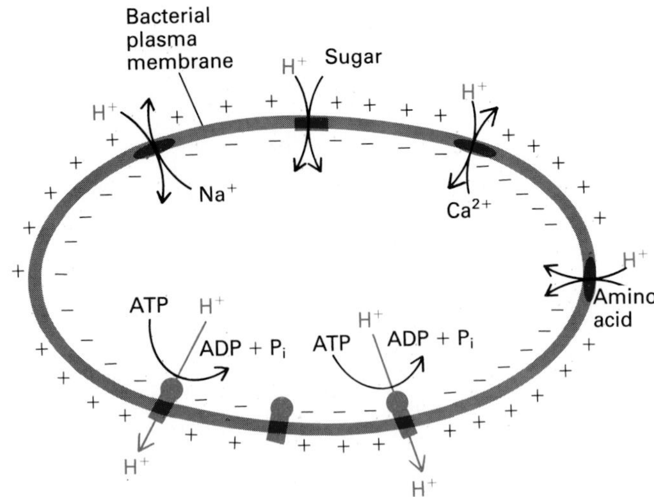

Figure 20-35 Proton-motive force across the plasma membrane of anaerobic fermentative bacteria, such as *Streptococcus faecalis.* The F_0F_1-ATPase functions as an electrogenic pump and uses ATP, generated during fermentation, to form a proton concentration gradient across the cell membrane. As in other bacteria, the proton gradient is used to power the export of Ca^{2+} and Na^+ by an antiport mechanism and also the uptake of sugars and amino acids into the cell.

Photosynthesis

Recall that the process of photosynthesis can be separated into light and dark reactions. Light is absorbed by chlorophyll molecules, thus raising them to a higher energy state. The absorbed energy is used to strip electrons from water, which also yields protons and oxygen. The electrons are transferred, by a series of carriers, to $NADP^+$, forming the reduced molecule NADPH. Concomitantly a proton gradient is formed across the thylakoid membrane. The proton-motive force is used to synthesize ATP from ADP and P_i. In reactions not directly dependent on light (the dark reactions), CO_2 is fixed and converted to glucose; these reactions are fueled by ATP and NADPH produced by the light reactions.

Synthesis of ATP from ADP and P_i utilizes a multiprotein particle, termed CF_0CF_1, that (as we have seen) is similar in structure and function to the mitochondrial F_0F_1 complex. The electron transport chain in chloroplast membranes also has some components similar to those in mitochondria. Thylakoid membranes contain, for instance, plastoquinone, a compound similar in structure and function to ubiquinone, and cytochrome *f*, which is similar to cytochrome *c*. As in mitochondria, the movement of electrons is coupled to transport of protons across the membrane. In chloroplasts, however, the flow of electrons is opposite in direction to that in mitochondria: electrons flow *from* water *to* $NADP^+$. This flow of electrons is endergonic, and the required energy is provided by absorbtion of light. Let us now consider the stages of photosynthesis in some detail, beginning with absorption of light by chlorophyll.

Light Energy Captured by Chlorophyll *a* Is Used to Transfer Electrons to Specific Acceptors

Light is a form of electromagnetic radiation; quantum mechanics established that light has properties both of waves and of particles. When it interacts with matter, light can be thought of as discrete packets of energy, or *quanta,* that are called *photons*. The energy ϵ of a photon is proportional to the frequency of the electromagnetic field:

$$\epsilon = h\nu$$

where h is Planck's constant (1.58×10^{-34} cal/s or 6.63×10^{-34} J/s), and ν is the frequency of the light. It is customary in biology to refer to the wavelength of the light wave, λ, rather than the frequency ν; the two are related by the simple equation

$$\nu = \frac{c}{\lambda}$$

where c is the velocity of light (3×10^{10} cm/s in a vacuum). Also, we can consider the energy in a mole of photons, $E = N\epsilon$, where N is Avogadro's number [6.02 ×

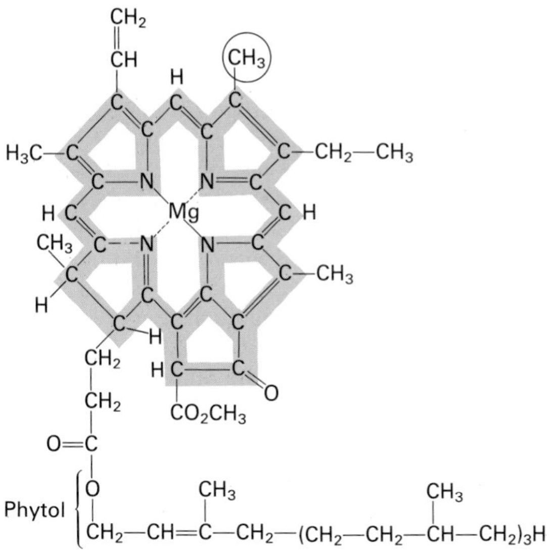

Figure 20-36 The structure of chlorophyll *a*, the principal pigment that traps the energy of light. Chlorophyll *b* differs from chlorophyll *a* in having a $-\overset{\overset{\displaystyle O}{\|}}{C}H$ group in place of the circled $-CH_3$ group. The heme group, depicted here in color, is a highly conjugated system in which electrons are delocalized. Its structure is similar to that of Fe heme, found in molecules such as hemoglobin (see Figure 3-9) or cytochromes. The hydrophobic phytol "tail" facilitates the binding of chlorophyll to membrane phospholipids and to hydrophobic regions of chlorophyll-binding proteins.

10^{23} molecules (or photons) per mole]. Thus $E = Nh\nu = Nhc/\lambda$.

Photosynthesis, as we shall see, occurs with red light of wavelength 680 nm; absorption of a mole of photons at this wavelength is indeed accompanied by absorption of a considerable amount of energy, −42 kcal. The shorter the wavelength, the greater the energy; for example, absorption of a mole of photons of blue light (400 nm) is equivalent to about 71 kcal of energy.

Chlorophyll *a* is the principal pigment involved in photosynthesis and is found in all photosynthetic eukaryotes and also in the prokaryotic blue-green algae (Figure 20-36). Vascular plants also contain chlorophyll *b*, a pigment that absorbs at wavelengths slightly different from those of chlorophyll *a* (see Figure 20-38). Carotenoids, such as *β*-carotene (Figure 20-37), absorb at still other wavelengths. These and other pigments, localized to the thylakoid membranes, funnel energy to chlorophyll *a* and greatly extend the range of light that can be absorbed and used for photosynthesis. One of the strongest pieces of evidence for the involvement of chlorophyll in photosynthesis is that the absorption spectrum of these pigments is similar to the action spectrum of photosynthesis—the ability of light of different wavelengths to cause generation of ATP (Figure 20-38) and formation of glucose.

When purified chlorophyll is dissolved in aqueous solution with a detergent and then irradiated with light of wavelength 650 to 700 nm, it fluoresces, or gives off red photons. As chlorophyll molecules absorb light energy, some electrons in chlorophyll are momentarily raised to an orbital of higher energy level. The difference in energy between the two orbitals is precisely that of the energy of the absorbed photon. The chlorophyll is said to be *excited;* the molecule can decay to the original state by releasing energy. In some cases, the excess energy is released as heat. More usually, excited chlorophyll releases a photon that is of somewhat longer wavelength—or, equivalently, of lower energy—than the exciting photon; such released light constitutes the fluorescence. But when localized to the thylakoid membrane, the excited chlorophyll does not fluoresce; rather, the energy is transferred to a neighboring molecule (the process of resonance electron transfer), or an electron can be transferred to a neighboring molecule. A particularly important process in photosynthesis is transfer of an electron from an excited chlorophyll molecule, cpl*, to an acceptor a, thus forming a negatively charged a and a positive cpl:

$$\text{cpl} \xrightarrow{\text{Light}} \text{cpl}*$$

$$\text{cpl}* + \text{a} \longrightarrow \text{cpl}^{+} + \text{a}^{-}$$

The acceptor a^-, with its extra electron of high energy, is a powerful reducing agent: it is capable of transferring the electron to still other molecules. cpl^+ is a powerful oxidant: it can remove electrons from other molecules to re-form the original cpl. In green plants, four molecules of

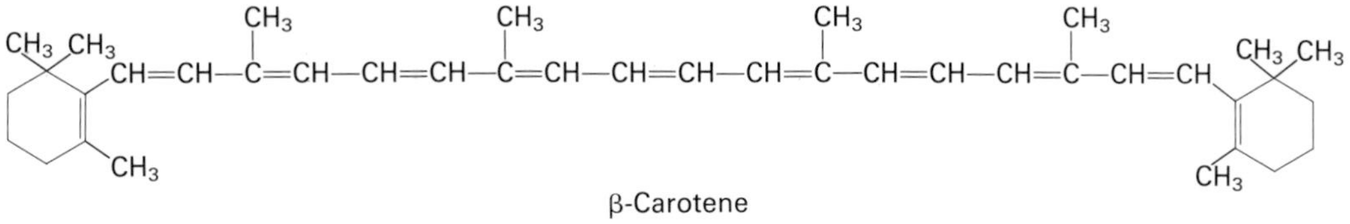

β-Carotene

Figure 20-37 The structure of *β*-carotene, a pigment related to retinal (see Figure 17-59), that assists in light absorption by chloroplasts. The pigment *β*-carotene is one of a family of carotenoids containing long hydrocarbon chains of alternating single and double bonds.

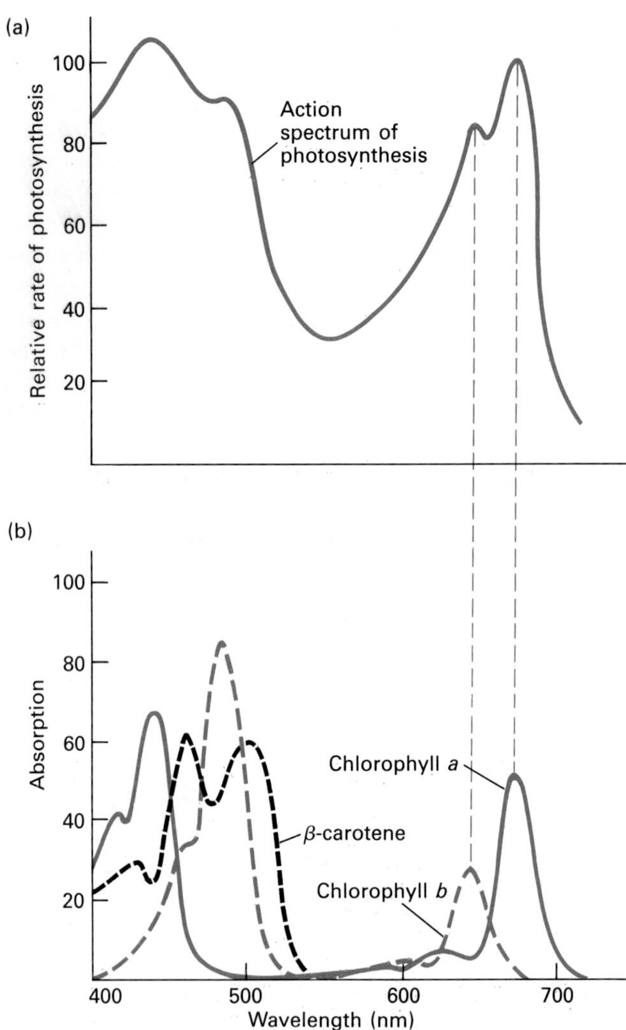

Figure 20-38 Photosynthesis at different wavelengths. (a) The action spectrum of photosynthesis in green plants. This spectrum is a measure of the ability of light of different wavelengths to support photosynthesis. (b) The absorption spectrum for three photosynthetic pigments: chlorophyll *a*, chlorophyll *b*, and β-carotene. This spectrum is the measure of the ability of light of different wavelengths to be absorbed by the pigment. These findings suggest that photosynthesis at 650 nm is due primarily to light absorbed by chlorophyll *b*; at 680 nm, to that absorbed by chlorophyll *a*; and at shorter wavelengths, β-carotene and other carotenoids are more important.

cpl^+ can remove four electrons from H_2O to form O_2:

$$2H_2O + 4cpl^+ \longrightarrow 4H^+ + O_2 + 4cpl$$

Important evidence that light causes the formation of strong oxidizing and reducing agents came from studies of photosynthetic bacteria. Blue-green bacteria, for example, oxidize water to oxygen by the same reaction that occurs in higher plants:

$$12H_2O + 6CO_2 \xrightarrow{\text{Light}} C_6H_{12}O_6 + 6O_2 + 6H_2O$$

Others, such as the purple and green bacteria, do not produce O_2. Instead, light drives the oxidation of hydrogen sulfide to produce sulfur:

$$12H_2S + 6CO_2 \xrightarrow{\text{Light}} C_6H_{12}O_6 + 12S + 6H_2O$$

In such organisms, excited chlorophyll removes electrons from hydrogen sulfide rather than water:

$$a + cpl* \longrightarrow cpl^+ + a^-$$

$$H_2S + 2cpl^+ \longrightarrow 2H^+ + S + 2cpl$$

Hydrogen gas can also be used as an electron donor by these organisms:

$$12H_2 + 6CO_2 \xrightarrow{\text{Light}} C_6H_{12}O_6 + 6H_2O$$

The oxidized chlorophyll, cpl^+, removes electrons from H_2, thus forming protons and regenerating the original chlorophyll:

$$H_2 + 2cpl^+ \longrightarrow 2H^+ + 2cpl$$

Chloroplasts Use Two Photosystems to Absorb Photons

The efficiency of photosynthesis drops sharply at wavelengths longer than 680 nm, even though chlorophyll in thylakoid membranes still absorbs light at 700 nm (Figure 20-38). The rate of photosynthesis generated by light of wavelength 700 nm can be greatly enhanced by adding light of shorter wavelength, such as 600 nm; the rate of photosynthesis supported by a combination of light at 600 and 700 nm is greater than the sum of the rates for the two wavelengths when each is used separately. Such observations led to the notion that photosynthesis in green plants involves the interaction of two separate systems of light-driven reactions. One system, called *photosystem I,* can be driven by light of wavelength 700 nm or less; the other system, *photosystem II,* requires light of wavelength shorter than 680 nm.

Each of the two photosystems is structurally distinct and is a complex of many chlorophyll molecules, carotenoid pigments, cytochromes, and other electron-transporting proteins with a defined orientation in the thylakoid membrane. By selective extraction of chloroplast membranes with certain detergents, particles containing only photosystem I complexes can be isolated. If thylakoid membranes are sheared into small membrane fragments and vesicles, membranes that are enriched in either photosystem I or II complexes can be obtained in pure form. Thus the properties of each photosystem can be studied separately in some detail.

Photosystem II is located preferentially in the grana, whereas photosystem I is located in the nonopposed thylakoid membranes (Figure 20-39). Obviously, then, the two complexes are spatially separated—a critical point, because electrons are transferred from photosystem II to photosystem I during photosynthesis. A third multiprotein particle, the *cytochrome b/f complex,* is used to transfer electrons between the two photosystems. This complex is found in both stacked and unstacked regions of the thylakoid.

Each photosystem contains several hundred chlorophyll molecules; yet only one chlorophyll per complex is capable of undergoing light-driven electron transfer. These specialized *reaction-center* chlorophyll *a* molecules are termed P_{680} (for *p*igment absorbing light at *680* nm) in photosystem II and P_{700} in photosystem I. Each of these specialized chlorophyll molecules in the reaction centers is bound to an integral membrane protein: a 110,000-m.w. protein contains the reaction center for photosystem I, and a 47,000-m.w. protein-chlorophyll complex is the reaction center for photosystem II. Associated with each photosystem is a structure called the *light-harvesting complex* (LHC), or *antennae complex.* The LHC itself has no catalytic activity; its primary function is to capture light energy and to transfer that energy to photosystem I or II. This complex contains at least two main polypeptides, each of m.w. 32,000, which bind up to half of the total chlorophyll *a* and nearly all of the chlorophyll *b*. Most of the LHCs are associated with photosystem II, but photosystem I also has a specific LHC. LHCs are located on the inner surface of the thylakoid membrane. Photons are absorbed by any of the dozens of chlorophyll molecules in each LHC. The absorbed energy is then rapidly transferred (in $<10^{-9}$ s) to the specialized chlorophyll molecule in the reaction centers of the two photosystems (Figure 20-40). In this way, the LHCs act to absorb light energy and to funnel it to a single reaction center. Thylakoid membranes contain additional protein-chlorophyll complexes that may mediate energy transfer between LHC and reaction-center chlorophylls. The role of the proteins may be to maintain an optimal orientation for energy transfer between chlorophylls, although the details of their orientation within the membrane are not known.

Photosystem II Splits Water

Absorption of a photon of light of wavelength less than 680 nm by photosystem II triggers the separation of an electron from the P_{680} chlorophyll molecule. The result

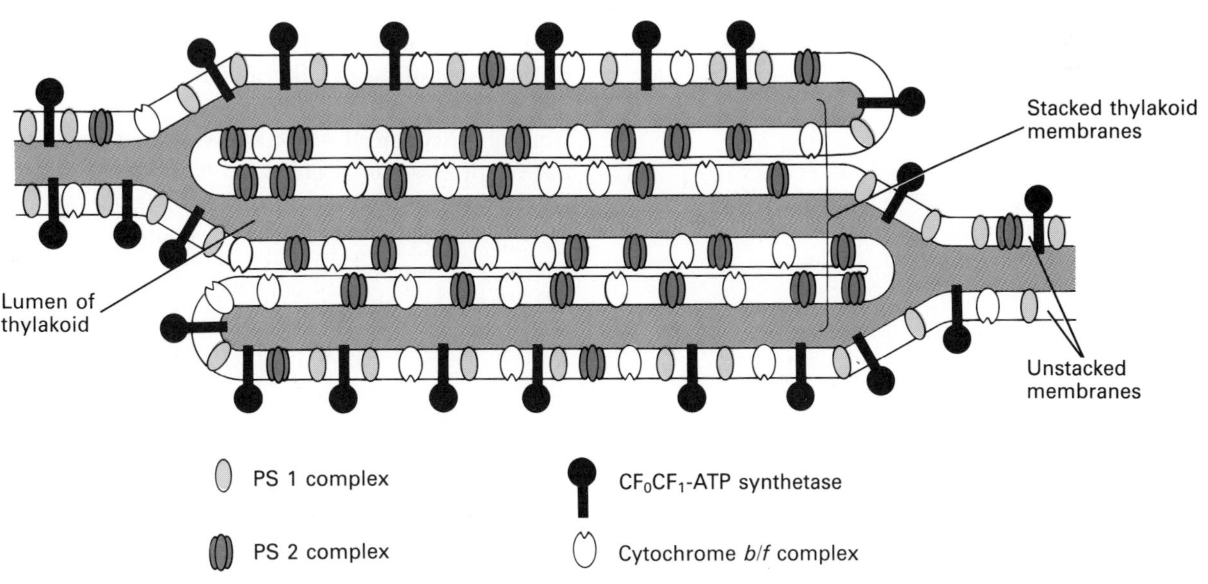

Figure 20-39 Distribution of multiprotein complexes in the thylakoid membrane. Photosystem II (PS 2) complexes are abundant in the stacked regions of the thylakoid membrane, whereas photosystem I (PS 1) complexes and the CF_0CF_1 complexes are more abundant in the nonstacked regions. A complex of cytochromes *b* and *f* transports electrons from photosystem II to photosystem I and is found in both regions. Stacking of thylakoid membranes may be due to the binding properties of the proteins in photosystem II. Evidence for this model came from studies in which thylakoids were gently fragmented by ultrasound and then fractionated by density-gradient centrifugation. Stacked and unstacked vesicles were separated and their protein and chlorophyll composition determined. *After J. M. Anderson and B. Andersson, 1982,* Trends Biochem. Sci. *7:288.*

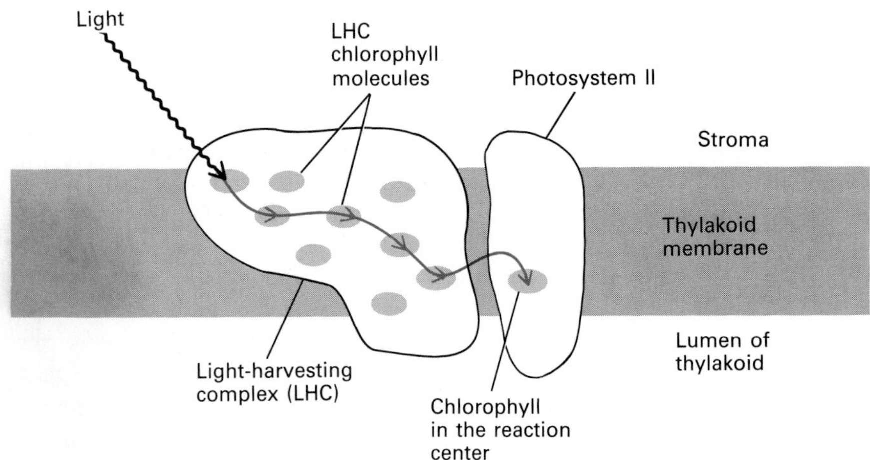

Figure 20-40 Light-harvesting complexes (LHCs) are associated with each of the two photosystem complexes. LHCs contain dozens of chlorophyll molecules. Any of these can absorb a photon; the energy is transferred to the reaction-center chlorophyll in the photosystem complex.

is an oxidized form of P_{680}; the excess positive charge is delocalized to all of the heme atoms as well as the central Mg^{2+}. Reduction of the oxidized P_{680} requires an input of electrons. These electrons are pulled from water molecules tightly bound to a manganese-containing protein found on the luminal surface of the P_{680} complex (Figures 20-41 and 20-42). The reduction of each molecule of H_2O to O_2 requires removal of four electrons. Since absorption of each photon by photosystem II results in the transfer of one electron, either several such photosystems must cooperate to reduce one H_2O molecule, or each photosystem II is capable of losing a total of four electrons. The latter explanation is believed to be correct; spectroscopic studies suggest that photosystem II can cycle through five different oxidation states, S_0 through S_4:

$$S_0 \xrightarrow{h\nu \nearrow e^-} S_1 \xrightarrow{h\nu \nearrow e^-} S_2 \xrightarrow{h\nu \nearrow e^-} S_3 \xrightarrow{h\nu \nearrow e^-} S_4 \quad (S_4 \rightarrow S_0:\ 2H_2O \rightarrow O_2 + 4H^+)$$

Bound to each photosystem II reaction center are two manganese ions (Mn^{2+}); this conformation represents one of the very few cases in which manganese has a role in a biological system. The Mn^{2+}-containing protein (Figure 20-41) appears to be involved in the water-cleavage reaction: two water molecules are split into four protons, four electrons, and one oxygen molecule (O_2). Electrons from water are transferred to the P_{680} reaction center, where they regenerate the reduced form of the chlorophyll. The protons released from water remain in the lumen of the thylakoid; these represent two of the four protons that are pumped into the lumen by transport of each pair of electrons.

Electrons split from P_{680} are immediately transported to the outer surface of the thylakoid membrane, but the nature of the initial acceptor of the electrons released from photosystem II is not known. The reduction potential of the electron is increased by absorption of the energy from the light quanta, from +0.8 V—that of the H_2O–O_2 couple—to about 0 V, equivalent to a change

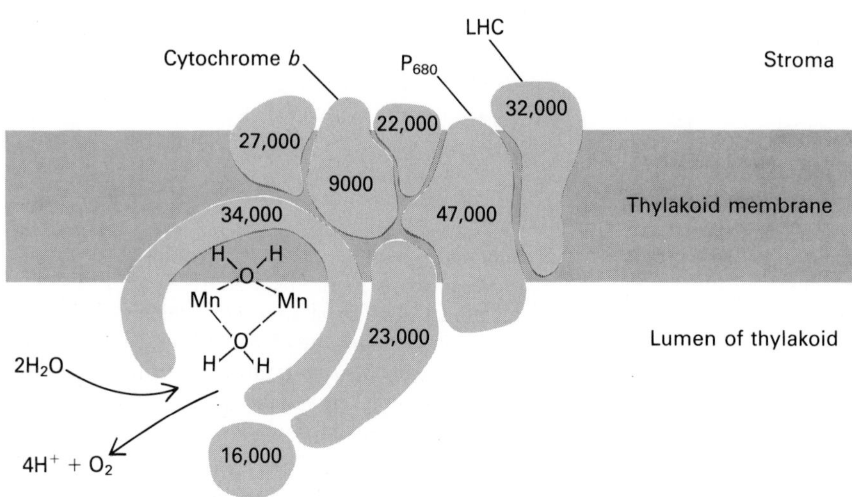

Figure 20-41 Molecular model of photosystem II. The complex includes at least eight proteins, including one (m.w. 47,000) that binds the P_{680} reaction-center chlorophyll, one (m.w. 9000) that is a cytochrome *b*, one (m.w. 32,000) from the light-harvesting complex (LHC), and one (m.w. 34,000) that contains the two manganese atoms (Mn) that play an essential role in the splitting of water. The arrangement depicted here is schematic; it is based on chemical cross-linking studies and on studies of various protein complexes isolated by detergent and salt extraction and then labeled with membrane-impermeable reagents. *After J. Barber, 1984,* Trends Biochem. Sci. *9:99.*

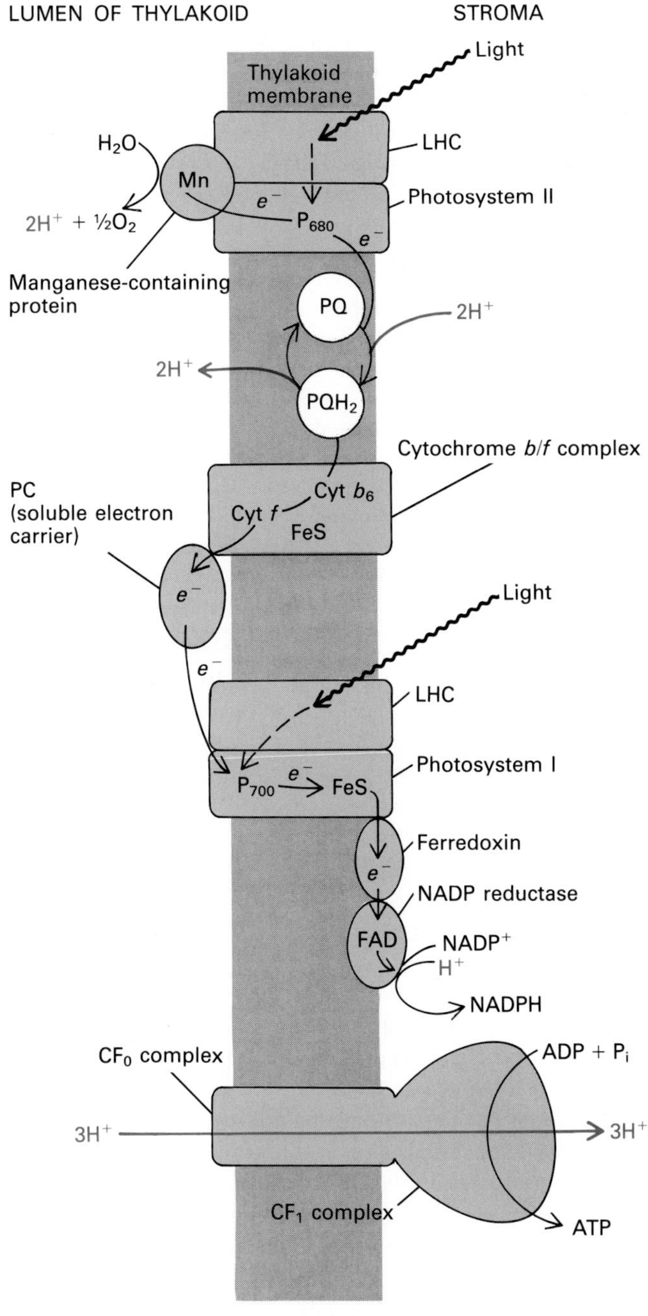

Figure 20-42 The orientation of the principal membrane components involved in photosynthesis and the pathway of electrons from water to NADP. Photons must be absorbed by two photosystems, photosystem II and photosystem I, for each electron transported. The first photon is absorbed by an array of chlorophyll and other pigments in the LHC associated with the specialized chlorophyll termed P_{680} in photosystem II. Electrons freed from P_{680} are moved to the stromal surface of the membrane and are replaced in the P_{680} by electrons transported from the manganese-containing protein that removed them from a water molecule. The protons from water remain in the lumen. Electrons from P_{680}, together with two protons from the stroma, reduce plastoquinone (PQ) to PQH_2. PQH_2 diffuses in the membrane and transfers its electrons to a second complex, containing the b_6 and f cytochromes and also FeS proteins; protons are released on the luminal side of the membrane when PQH_2 is reoxidized to PQ. These protons, together with those released from water, generate the proton-motive force across the thylakoid membrane. The electrons pass from cytochrome f to plastocyanin (PC), a soluble electron carrier. Electrons from PC are transferred to the second photosystem, photosystem I. Absorption of additional photons by the LHC associated with the P_{700} chlorophyll causes the electron to pass again across the membrane. It is transferred, via ferredoxin (another FeS protein) and FAD, to $NADP^+$, forming NADPH.

The synthesis of ATP by the CF_0CF_1 complex is the major chloroplast function known to require a proton gradient; two to three protons cross the membrane for each ATP formed.

in free energy of about 18 kcal/mol of electrons (Figure 20-43).

Electrons Are Transported from Photosystem II to Photosystem I

Electrons released from photosystem II rapidly combine with two protons from the stromal space and a molecule of plastoquinone (PQ), to generate reduced hydroquinone, PQH_2. As in mitochondria, this reduced hydroquinone diffuses to the inner surface of the thylakoid membrane; here the protons are released to the thylakoid lumen (this is the second pair of protons that are pumped).

The electrons released by PQH_2 are transferred to the cytochrome *b*/*f* complex, which is analogous in structure to the cytochrome *b*/*c* complex of mitochondria. From the cytochrome *b*/*f* complex, the electrons move to the carrier plastocyanin, which passes them on to photosystem I. Plastocyanin (PC) is a small protein with a single Cu atom coordinated to one cysteine—SH group, one methionine—CH_2—S—CH_3 group, and two histidines. The Cu alternates between the +1 and +2 states:

$$e^- + Cu^{2+} \rightleftharpoons Cu^+$$

Plastocyanin is a peripheral protein loosely bound to the luminal surface of the thylakoid membrane. It may rapidly diffuse, with its electron, in the plane of the membrane from the cytochrome *b*/*f* complex to photosystem I, or it may dissociate from the membrane and diffuse in the luminal space of the thylakoid (see Figure 20-42).

Since the photosystem I and photosystem II complexes are not uniformly distributed between stacked and un-

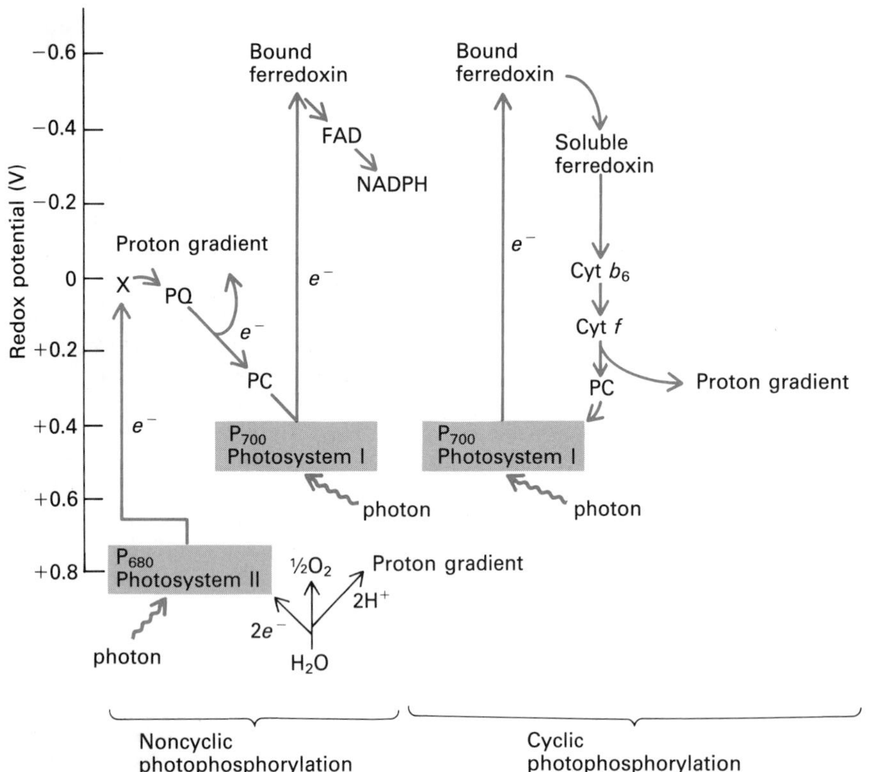

Figure 20-43 Energetics of electrons as they flow through the photosynthetic transport system. Photons absorbed by photosystem II excite electrons from a potential of +0.8 V, that of the H_2O–O_2 couple, to a potential of about 0 V. This is a change in potential equal to a gain of free energy of 18 kcal/mol of electrons. The nature of the primary electron acceptor, X, is unknown. As electrons move from plastoquinone (PQ) through the cytochrome *b*/*f* complex to plastocyanin (PC), some of this acquired energy is employed to transport protons into the thylakoid vesicles. Absorption of photons by photosystem I causes an additional increase of 1.0 V in the potential of electrons. Electrons excited by photosystem I can either be transferred directly to $NADP^+$, to form NADPH (noncyclic photophosphorylation) or be transferred via a series of ferredoxins and cytochromes back to photosystem I (cyclic photophosphorylation). Electron transport during cyclic photophosphorylation is coupled to the transport of protons into the thylakoid lumen.

stacked membranes, apparently a single structural electron transport system containing both photosystem I and photosystem II does not exist. Rather, mobile electron carriers such as PC and PQ are used to transfer electrons between these systems.

Photosystem I Is Used for Both Cyclic and Noncyclic Photophosphorylation

Photosystem I, like photosystem II, contains a single reaction-center chlorophyll termed P_{700}. Absorption of a photon leads to removal of an electron from the photosystem I chlorophyll; the oxidized chlorophyll is reduced by an electron passed from photosystem II. The activated electron is again moved to the outer surface of the membrane; there it is accepted by ferredoxin, an iron-sulfur protein. The net gain in reduction potential of the electron by photosystem I is about 1.0 V, equivalent to a gain in free energy of 23 kcal/mol of electrons.

Two fates are possible for electrons excited by photosystem I. Together with a proton, they can be transferred, via the electron carrier FAD, to $NADP^+$, thus forming the reduced molecule NADPH (see Figure 20-43). The process is called *noncyclic photophosphorylation.*

Alternatively, the activated electrons can flow through a series of cytochromes back across the membrane, until they reach photosystem I on the interior surface. A series of bound and free ferredoxins and cytochromes mediates this electron transport. Concomitant with this movement of electrons is the pumping of additional protons from the stromal space to the thylakoid lumen. In this process, termed *cyclic photophosphorylation,* additional ATP is synthesized, via the proton gradient generated during electron transport, but no NADPH is produced. Photosystem II is not involved, and no oxygen is evolved. Photosystem I is used solely for the generation of a proton gradient and thus for production of ATP. Cyclic photophosphorylation occurs when the ratio of NADPH to $NADP^+$ in the cell is high, so that there is no need to generate additional reduced nucleotides.

Considerable Evidence Supports the Two-Photosystem Scheme of Noncyclic Photosynthesis

According to the currently accepted scheme for noncyclic photosynthesis (see Figure 20-43), the two photosystems have different functions: photosystem I transfers electrons to $NADP^+$, thus forming NADPH, and photosystem II removes electrons from water. Evidence supporting this model comes from spectroscopic studies measuring the oxidation state of cytochromes such as b_6 and f that transfer electrons from photosystem II to photosystem I under different conditions of illumination. Red light of wavelength 700 nm excites only photosystem I; the wavelength of this light is too long for it to be absorbed by photosystem II. Shining only red light on chloroplasts causes all the b_6 and f cytochromes to become oxidized. (This can be determined spectroscopically since the oxidized and reduced states have different absorption maxima.) Under these conditions, electrons are withdrawn into photosystem I; none are provided from photosystem II. Addition of green light, however, activates photosystem II; immediately the b_6 and f cytochromes become partly reduced as electrons flow to them.

Many commercially important herbicides inhibit photosynthesis, and studies on their effects have proved useful in dissecting the pathway of photoelectron movements. One such class of herbicides—the *S*-triazines, such as atrazine—binds specifically to the 32,000-m.w. protein in the LHC of photosystem II. These compounds block electron transfer at the level of the second e^- acceptor in photosystem II. When added to illuminated chloroplasts, these inhibitors cause all the electron carriers downstream to accumulate in the oxidized form, since no electrons can be provided to the electron transport system from photosystem II. Triazine-resistant weeds are prevalent and present a major agricultural problem. The LHC protein to which atrazine binds is encoded in the chloroplast DNA; atrazine-resistant mutants have a single amino acid change in the protein that renders it unable to bind the herbicide.

Since in noncyclic photophosphorylation photosystems I and II act in sequence, the amount of light energy that is delivered to the two reaction centers must be controlled so that each activates the same number of electrons. The mechanism of control involves regulation of the rates of phosphorylation and dephosphorylation of the LHC associated with photosystem II. A membrane-bound protein kinase somehow senses these relative activities of the two photosystems via the oxidation-reduction state of the PQ pool that transfers electrons from photosystem II to photosystem I. If too much PQ is reduced (indicating a higher activity of photosystem II relative to photosystem I), the kinase is activated, and LHCs are phosphorylated. Phosphorylation causes the LHCs to dissociate from the photosystem II reaction center, thus reducing the efficiency of photosystem II function.

Phosphorylation of LHCs has another consequence. Recall that photosystem II is found only in the grana (see Figure 20-39). Recent work suggests that the associated LHCs are the "glue" that stacks the grana together. Phosphorylation of LHCs also causes the grana to lose cohesion and to become unstuck. But the effect of thylakoid stacking or unstacking on photosynthesis is not known.

Green plants need two photosystems. Light absorbed by chlorophyll does not provide enough energy to boost electrons from water to a reduction potential sufficient to reduce $NADP^+$ to NADPH. Thus two steps are used, each of which boosts the electrons part way (Figure 20-43). Blue-green algae, which also use water as an electron donor, likewise have two photosystems. The purple and green photosynthetic bacteria, in contrast, have only one photosystem, resembling that of photosystem I in plants. These organisms extract electrons from molecules that have a lower (less negative) reduction potential than that of water, so a single photon can boost these electrons to the reduction potential of the $NADP^+$-NADPH reaction. For instance, for the reaction

$$H_2S \longrightarrow S + 2H^+ + 2e-$$

that occurs in photosynthetic bacteria, $E_0' = -0.25$ V, compared with $E_0' = +0.86$ V for

$$H_2O \longrightarrow \tfrac{1}{2}O_2 + 2H^+ + 2e^-$$

and less energy is required to boost electrons removed from H_2S to a level sufficient to reduce $NADP^+$ to NADH. The three-dimensional structure of a bacterial reaction center has recently been determined by x-ray crystallography. It contains four chlorophylls, two molecules of the pigment bacteriophytin, two quinone molecules, and four cytochrome *c*–like molecules. Soon it will be possible to trace, in molecular detail, the pathway of the light-activated electrons within the membrane.

Fixation of Carbon Dioxide Is Catalyzed by Ribulose 1,5-Bisphosphate Carboxylase

Let us turn now to the dark reactions of photosynthesis—the fixation of carbon dioxide and its conversion to glucose, which occur in a series of enzyme-catalyzed reactions known as the *Calvin cycle.* These reactions require energy released by hydrolysis of ATP and also a reducing agent, NADPH. Formation of glucose can occur in the dark until the supply of ATP and NADPH is exhausted.

It has been established in elegant studies that the actual reaction that fixes CO_2 into carbohydrates is catalyzed by the enzyme *ribulose 1,5-bisphosphate carboxylase,* which adds CO_2 to the five-carbon sugar ribulose 1,5-bisphosphate, thus forming two molecules of 3-phosphoglycerate (Figure 20-44). This reaction occurs in the stroma of the chloroplast. Ribulose 1,5-bisphosphate carboxylase makes up 16 percent of the chloroplast protein and is believed to be the single most abundant protein on earth. It is composed of two types of subunits; of these, one is encoded in chloroplast DNA, and the other in nuclear DNA. (Chapter 21 discusses this in greater detail.)

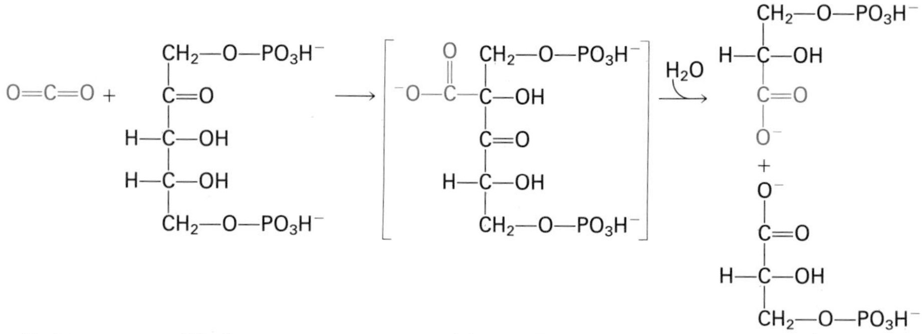

$$O{=}C{=}O + \begin{array}{c} CH_2{-}O{-}PO_3H^- \\ | \\ C{=}O \\ | \\ H{-}C{-}OH \\ | \\ H{-}C{-}OH \\ | \\ CH_2{-}O{-}PO_3H^- \end{array} \longrightarrow \left[\begin{array}{c} O \quad CH_2{-}O{-}PO_3H^- \\ \| \quad | \\ ^-O{-}C{-}C{-}OH \\ | \\ C{=}O \\ | \\ H{-}C{-}OH \\ | \\ CH_2{-}O{-}PO_3H^- \end{array} \right] \xrightarrow{H_2O} \begin{array}{c} CH_2{-}O{-}PO_3H^- \\ | \\ H{-}C{-}OH \\ | \\ C{=}O \\ | \\ O^- \\ + \\ O^- \\ | \\ C{=}O \\ | \\ H{-}C{-}OH \\ | \\ CH_2{-}O{-}PO_3H^- \end{array}$$

Carbon dioxide — Ribulose 1,5-bisphosphate — Intermediate — Two molecules of 3-phosphoglycerate

Figure 20-44 The initial reaction that fixes carbon dioxide into organic compounds involves a condensation with the five-carbon sugar ribulose 1,5-bisphosphate. The products of the reaction, catalyzed by ribulose 1,5-bisphosphate carboxylase, are two molecules of 3-phosphoglycerate. If ^{14}C-labeled CO_2 is used *(left, color)*, all the ^{14}C radioactivity is in the carboxyl carbon atom of 3-phosphoglycerate *(right, color)*.

In the key experiment implicating ribulose 1,5-bisphosphate carboxylase, photosynthetic algae were exposed to a brief pulse of ^{14}C-labeled CO_2, and then the cells were quickly disrupted. The material that was radiolabeled most rapidly was 3-phosphoglycerate, and all the ^{14}C radioactivity was in the carboxyl group, as indicated in Figure 20-44.

The fate of the 3-phosphoglycerate formed by this reaction is complex: some is converted to glucose, but some is used to regenerate ribulose 1,5-bisphosphate. Quantitatively, for every 12 molecules of 3-phosphoglycerate generated by ribulose bisphosphate carboxylase (a total of 36 C atoms), 2 (6 C atoms) are converted to 2 molecules of glyceraldehyde 3-phosphate, and 10 (30 C atoms) are converted to 6 molecules of ribulose 1,5-bisphosphate. As summarized in Figure 20-45, fixation of 6 CO_2 molecules and formation of 2 3-phosphoglycerate molecules require consumption of 18 ATPs and 12 NADPHs, generated during the light phases of photosynthesis. Glyceraldehyde 3-phosphate is transported from the chloroplast into the cytosol, wherein the final steps of glucose synthesis occur. One molecule of glyceraldehyde 3-phosphate is isomerized to dihydroxyacetone phosphate. This compound condenses with a second molecule of glyceraldehyde 3-phosphate to form fructose 1,6-bisphosphate, a normal glycolytic intermediate. However, as noted previously, the enzymes that convert fructose 1,6-bisphosphate to glucose are not those used in glycolysis. Rather, phosphatases are utilized to hydrolyze the two PO_4 groups. Formation of ATP—as would be the case if the reactions of glycolysis were simply reversed—does not occur. Hydrolysis of the phosphate residues allows the reaction to proceed in the direction of synthesis of glucose.

The regeneration of ribulose 1,5-bisphosphate from glyceraldehyde 3-phosphate is a complex process. At least seven enzymes are required in this part of the Calvin cycle. These are outlined in Figure 20-45, but their details are beyond the scope of this book.

Photorespiration Liberates CO_2 and Consumes O_2

Photosynthesis is always accompanied by *photorespiration,* a process that takes place in light and consumes O_2 and converts ribulose 1,5-bisphosphate to CO_2. As is shown in Figure 20-46, ribulose 1,5-bisphosphate carboxylase catalyzes two competing reactions: addition of CO_2 to ribulose 1,5-bisphosphate, to form two molecules of 3-phosphoglycerate, and addition of O_2, to form one molecule of 3-phosphoglycerate and one of the two-carbon compound phosphoglycolic acid. Phosphoglycolate is hydrolyzed to glycolate. This compound is transported to peroxisomes, small organelles containing a number of enzymes that generate and consume H_2O_2 (Chapter 5). In peroxisomes glycolate is oxidized to CO_2, in a process that generates neither ATP nor NADPH. Thus photorespiration uses up oxygen and generates carbon dioxide—wasteful processes for the economy of the plant. The benefits, if any, of photorespiration are not obvious. Apparently plants have not evolved a ribulose 1,5-bisphosphate carboxylase that has low activity as an oxygenase.

The C_4 Pathway for CO_2 Fixation Is Used by Several Tropical Plants

Plants such as corn, sugar cane, crabgrass, and others that can grow in a hot, dry environment must keep their stomata—the gas-exchange pores in the leaves—closed much of the time to prevent excessive loss of moisture. This causes the CO_2 level within the leaf to fall below the K_m of ribulose 1,5-bisphosphate carboxylase; under these conditions, photorespiration is greatly favored over photosynthesis. To avoid this problem, certain plants, called C_4 plants, have evolved a two-step pathway of CO_2 fixation that involves two types of cells: *mesophyll* cells, adjacent to the surface of the leaf, and interior *bundle sheath* cells (Figure 20-47a).

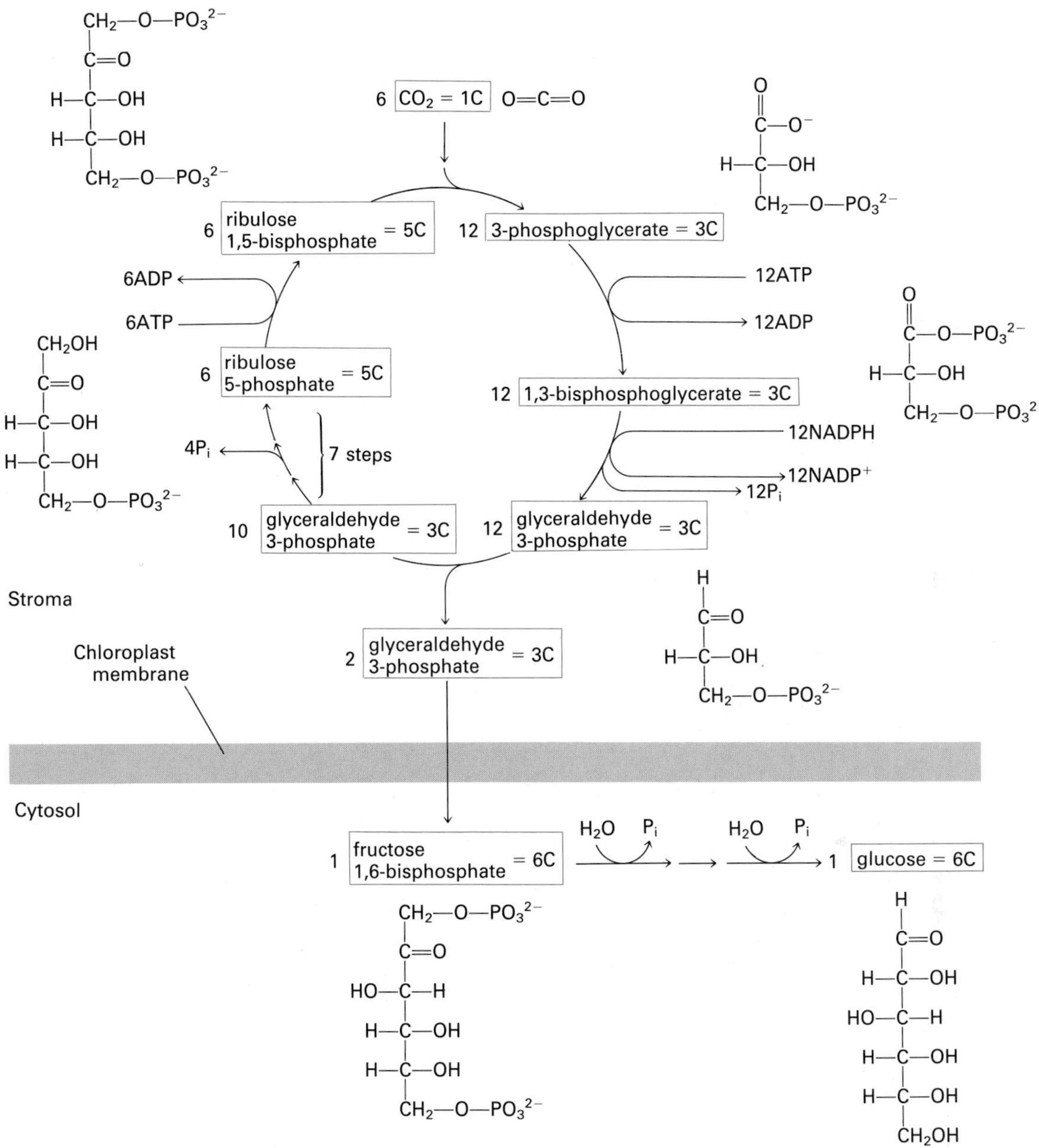

Figure 20-45 The pathway of carbon during photosynthesis. As is detailed in the text, six molecules of CO_2 are converted into two of glyceraldehyde 3-phosphate. These reactions occur in the stroma of the chloroplast. Some of the glyceraldehyde 3-phosphate is transported to the cytosol; there it is converted in an exergonic series of reactions to glucose and, ultimately, to the disaccharide storage form sucrose. Although not shown here, some glyceraldehyde 3-phosphate is also converted to amino acids and fats, compounds essential for the growth of plants.

In mesophyll cells, CO_2 from the air is assimilated by reaction with the glycolytic intermediate phosphoenolpyruvate to generate the four-carbon compound oxaloacetate, which is in turn reduced to malate (Figure 20-47b). The enzyme catalyzing this reaction, phosphoenolpyruvate carboxylase, is found only in C_4 plants. Some of the malate is converted to the amino acid aspartate. Since the first compounds to be labeled by $[^{14}C]CO_2$ are the four-carbon compounds oxaloacetate and malate, the pathway for CO_2 fixation was named the C_4 pathway.

Both malate and aspartate are transferred by special permeases to the bundle sheath cells, in which CO_2 is

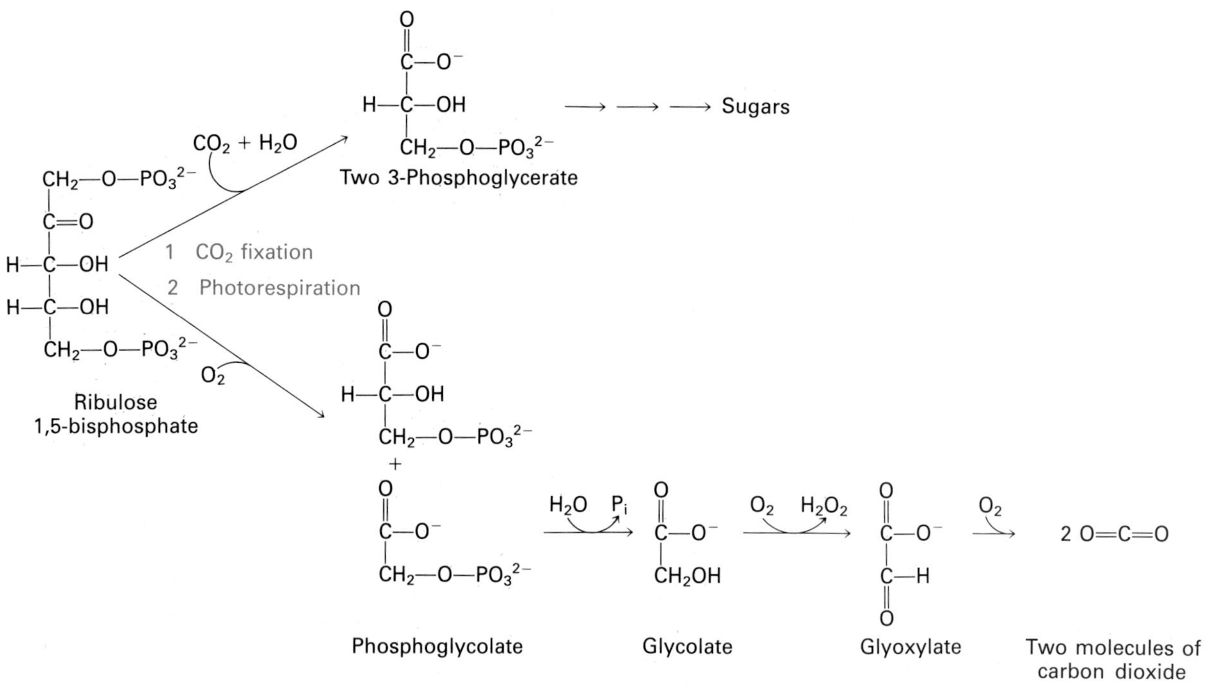

Ribulose 1,5-bisphosphate

1 CO_2 fixation

2 Photorespiration

$CO_2 + H_2O$

Two 3-Phosphoglycerate → → → Sugars

O_2

Phosphoglycolate —H_2O, P_i→ Glycolate —O_2, H_2O_2→ Glyoxylate —O_2→ $2\ O{=}C{=}O$ Two molecules of carbon dioxide

Figure 20-46 CO_2 fixation and photorespiration are competing reactions that act on ribulose 1,5-bisphosphate. Reaction 1 is favored by high CO_2 and low O_2 pressures. Reaction 2 occurs at low CO_2 and high O_2 pressures, that is, under normal atmospheric conditions.

released by decarboxylation. The CO_2 enters the Calvin cycle; the C_3 compound, pyruvate or alanine generated by decarboxylation, diffuses back to the mesophyll cell, where it is reutilized in the C_4 pathway. Because of transport of CO_2 from mesophyll cells, the concentration of CO_2 in bundle sheath cells of C_4 plants is much higher than in the normal atmosphere; this favors fixation of CO_2 to form 3-phosphoglycerate and inhibits utilization of ribulose 1,5-bisphosphate by photorespiration. Since two phosphodiester bonds of ATP are consumed in the cyclic C_4 process (to generate phosphoenolpyruvate from pyruvate), the overall efficiency of photosynthetic production of glucose from NADPH and ATP is lower than in C_3 plants, which use only the Calvin cycle for CO_2 fixation.

Oxygenation of ribulose 1,5-bisphosphate (reaction 2 in Figure 20-46) is favored by the high O_2 concentration in the atmosphere, and in C_3 plants as much as 50 percent of the photosynthetically fixed carbon may be reoxidized to CO_2 during photorespiration. Compared with C_3 plants, C_4 plants are superior utilizers of available CO_2, since the enzyme phosphoenolpyruvate carboxylase (Figure 20-47b) has a lower K_m for CO_2 than does the ribulose diphosphate carboxylase of the Calvin cycle. As a result, the net rates of photosynthesis for C_4 grasses (such as corn or sugar cane) can be two to three times those for otherwise similar C_3 grasses (such as wheat, rice, and oats).

Summary

A combination of a proton concentration (pH) gradient and a potential gradient across the inner mitochondrial membrane, the chloroplast thylakoid membrane, or the bacterial outer plasma membrane is the immediate source of energy for synthesis of ATP. Together, these two gradients are known as the proton-motive force. A multiprotein complex termed F_0F_1 catalyzes ATP synthesis as protons flow back through the membrane down their electrochemical gradient. This complex has a very similar structure in all three systems: F_0 is a transmembrane complex that generates a regulated H^+ channel; and F_1 contains the site for ATP synthesis from ADP and P_i, and is tightly bound to F_0.

In mitochondria and aerobic bacteria, NADH and $FADH_2$ are formed by the oxidation of acetyl CoA to CO_2. Acetyl CoA is a key intermediate in oxidation of carbohydrates, fats, and amino acids. Oxidation of acetyl CoA occurs by the citric acid cycle, catalyzed by a set of enzymes localized in the matrix. Acetyl CoA condenses with the four-carbon molecule oxaloacetate to form the six-carbon citrate. In a series of reactions, citrate is converted to oxaloacetate and to CO_2, concomitant with the reduction of NAD^+ to NADH and of FAD to $FADH_2$. NADH is also generated in the cytosol during the reactions of the Embden-Meyerhof pathway of glucose

(a)

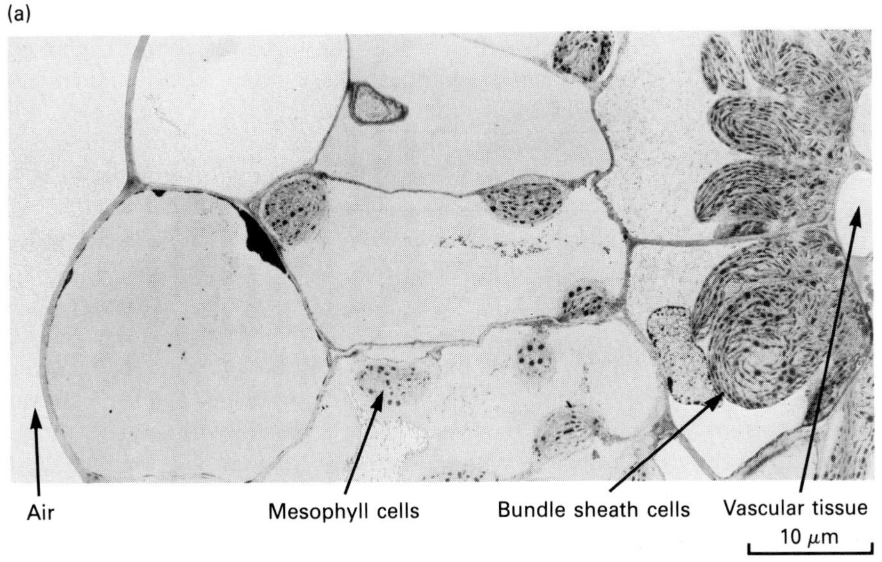

(b)

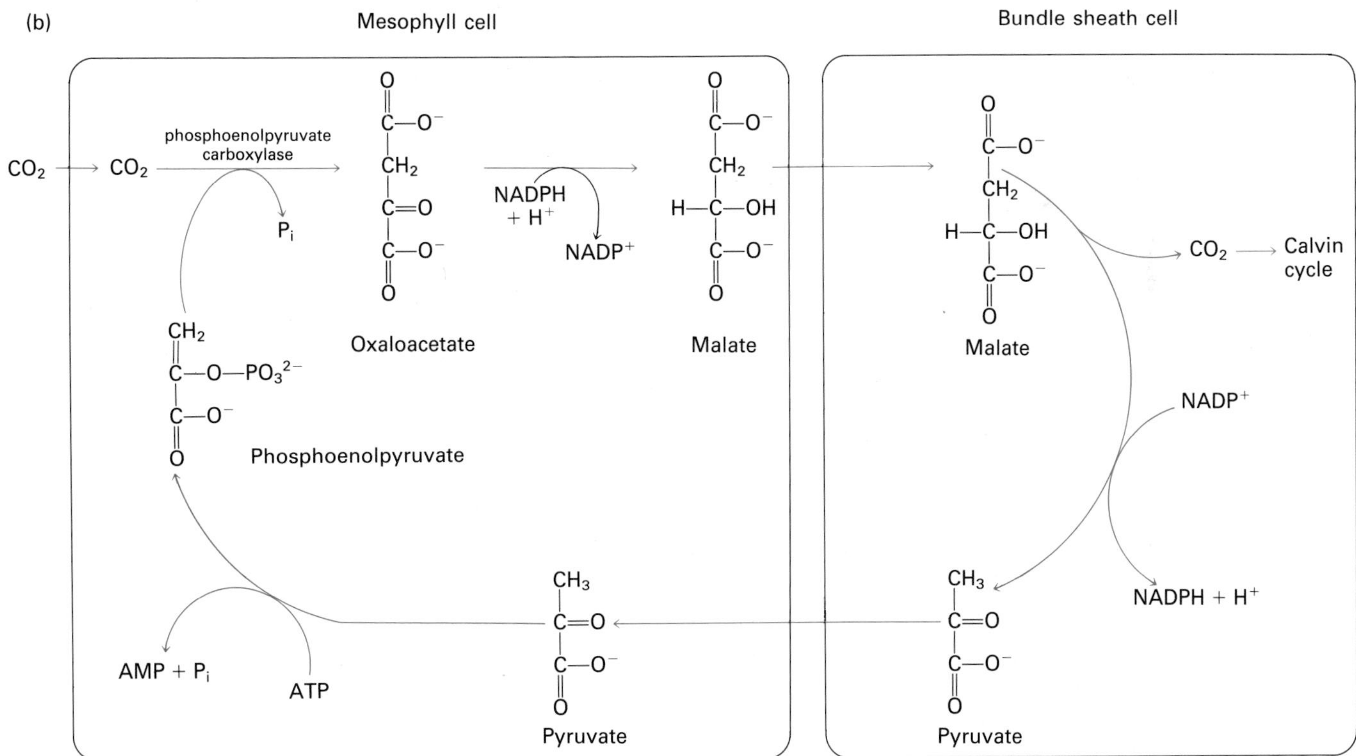

Figure 20-47 The C_4 pathway for CO_2 fixation. (a) Electron micrograph of a cross section of a leaf from a typical C_4 plant. The mesophyll cells line the surface of the leaf and thus are adjacent to the air. In these cells, CO_2 is assimilated into four-carbon molecules that are pumped to the interior bundle sheath cells. The latter cells contain abundant chloroplasts and are the sites of photophosphorylation and glucose synthesis. (b) Diagram showing that CO_2 is assimilated by phosphoenolpyruvate carboxylase in mesophyll cells. C_4 molecules such as malate are transferred from the mesophyll to the bundle sheath cell. CO_2 is then released for use in the standard Calvin cycle, and the C_3 compound pyruvate is transferred back to mesophyll cells. *Photograph (a), by S. Craig, courtesy of M. D. Hatch.*

metabolism. By means of the glycerol phosphate shuttle, cytosolic NADH is oxidized to NAD^+, concomitant with reduction of intramitochondrial FAD to $FADH_2$. This provides another source of reduced nucleotide for oxidative phosphorylation.

In mitochondria, flow of electrons from NADH or $FADH_2$ to oxygen is coupled to transport of protons across the inner mitochondrial membrane out of the mitochondrion, generating the proton-motive force. The major components of the electron transport chain are four multiprotein complexes with defined orientation in the inner membrane: succinate–CoQ reductase, NADH–CoQ reductase, $CoQH_2$–cytochrome *c* reductase, and cytochrome *c* oxidase. Cytochrome *c* oxidase transfers electrons to O_2 to form H_2O. Movement of electrons through each of the latter three complexes is coupled to proton movement. Coenzyme Q functions as a lipid-soluble reversible transporter of electrons and protons through the membrane. Because mitochondrial membranes are impermeable to anions, the predominant component of the proton-motive force is the membrane potential—about 200 mV matrix negative.

The proton-motive force across mitochondrial membranes is used to power the uptake of P_i and ADP from the cytosol, in exchange for mitochondrial ATP. The proton-motive force generated across the bacterial plasma membrane is also formed by electron transport from NADH to O_2. It is used to power uptake of nutrients, such as sugars and the export of ions such as Na^+ and Ca^{2+}, all against their concentration gradients. The proton-motive force is also used to power the rotation of bacterial flagella. In certain bacteria, a proton-motive force can be generated during oxidation of other reduced molecules, such as H_2. Either O_2 or another oxidized molecule such as SO_4^{2-} can be the ultimate electron acceptor.

During photosynthesis in chloroplasts, light energy is absorbed by chlorophyll and other pigments in light-harvesting complexes (LHCs) associated with two photosystems, photosystem I and photosystem II. The absorbed energy is transferred to a specialized chlorophyll contained in the reaction centers, called P_{680} in photosystem II and P_{700} in photosystem I. Excitation of P_{680} results in removal of electrons from H_2O on the inner surface of the thylakoid, forming O_2 and protons. The protons remain in the interior and generate part of the proton-motive force. The electrons are pumped across the membrane. They are transferred by a series of carriers to photosystem I, coupled to the transfer of additional protons into the thylakoid space. Electrons excited by photosystem I can undergo two possible fates: they can be transferred via a series of carriers to $NADP^+$, forming NADPH; this process is called noncyclic photophosphorylation. Alternatively, the electrons can be transferred back to photosystem I, during which additional protons are transported inward across the membrane. This process, termed cyclic photophosphorylation, does not involve photosystem II; neither NADPH nor O_2 is formed. Because thylakoid membranes are permeable to anions, a pH gradient (inside pH 5.0 versus matrix pH 7.0) rather than a potential gradient is the principal component of the proton-motive force. The major use of the thylakoid proton gradient is for synthesis of ATP. ATP and NADPH generated by photophosphorylation are used in a series of enzymatic reactions called the Calvin cycle for conversion of CO_2 to glucose. In C_3 plants, CO_2 is fixed in the reaction catalyzed by ribulose 1,5-bisphosphate carboxylase. In C_4 plants, CO_2 is fixed initially in the outer mesophyll cells by reaction with phosphoenolpyruvate. The four-carbon molecules so generated are shuttled into the interior bundle sheath cells, where the CO_2 is released and then used in the Calvin cycle; these reactions form part of the C_4 pathway.

References

Energy Metabolism and the Chemiosmotic Theory

* FERGUSON, S. J., and M. C. SORGATO. 1982. Proton electrochemical gradients and energy transduction processes. *Annu. Rev. Biochem.* **51**:185–218.

HATEFI, Y. 1985. The mitochondrial electron transport and oxidative phosphorylation system. *Annu. Rev. Biochem.* **45**:1015–1070.

* HINKLE, P. C., and R. E. MCCARTY. 1978. How cells make ATP. *Sci. Am.* **238**(3):104.

* MITCHELL, P. 1979. Keilin's respiratory chain concept and its chemiosmotic consequences. *Science* **206**:1148–1159. Mitchell's Nobel Prize lecture.

* NICHOLLS, D. G. 1982. *Bioenergetics: An Introduction to the Chemiosmotic Theory.* Academic Press.

RACKER, E. 1976. *A New Look at Mechanisms in Bioenergetics.* Academic Press.

RACKER, E. 1980. From Pasteur to Mitchell: a hundred years of bioenergetics. *Fed. Proc.* **39**:210–215.

SKULACHEV, V. P. 1984. Membrane bioenergetics—should we build the bridge across the river or alongside of it? *Trends Biochem. Sci.* **9**:182–185.

SKULACHEV, V. P., and P. C. HINKLE, eds. 1981. *Chemiosmotic Proton Circuits in Biological Membranes.* Addison-Wesley.

TZAGOLOFF, A. 1982. *Mitochondria.* Plenum Press.

* ZUBAY, G. 1983. *Biochemistry.* Addison-Wesley, Chapters 10 and 11.

The Citric Acid Cycle

* KREBS, H. A. 1970. The history of the tricarboxylic acid cycle. *Perspect. Biol. Med.* **14**:154–170.

MEHLMAN, M., and R. W. HANSON. 1972. *Energy Metabolism and the Regulation of Metabolic Processes in Mitochondria.* Academic Press.

* Book or review article that provides a survey of the topic.

ROBINSON, J. B., JR., and P. A. SRERE. 1985. Organization of Krebs tricarboxylic acid cycle enzymes in mitochondria. *J. Biol. Chem.* **260**:10800–10805.

* WILLIAMSON, J. R., and R. V. COOPER. 1980. Regulation of the citric acid cycle in mammalian systems. *FEBS Lett.* **117** (Suppl.):K73.

Synthesis of ATP and the F_0F_1-ATPase

AMZEL, L. M., M. MCKINNEY, P. NARAYANAN, and P. L. PEDERSEN. 1982. Structure of the mitochondrial F_1 ATPase at 9-Å resolution. *Proc. Nat'l Acad. Sci. USA* **79**:5852–5856.

* BERRY, E. A., and P. C. HINKLE. 1983. Measurement of the electrochemical proton gradient in submitochondrial particles. *J. Biol. Chem.* **258**:1474–1486.

* CROSS, R. L. 1981. The mechanism and regulation of ATP synthesis by F_1-ATPases. *Annu. Rev. Biochem.* **50**:681–714.

* FUTAI, M., and H. KANAZAWA. 1983. Structure and function of proton-translocating adenosine triphosphatase (F_0F_1): biochemical and molecular biological approaches. *Microbiol. Rev.* **47**:285–312.

* HAMMES, G. G. 1983. Mechanism of ATP synthesis and coupled proton transport: studies with purified chloroplast coupling factor. *Trends Biochem. Sci.* **8**:131–134.

* OGAWA, S., and T. M. LEE. 1984. The relation between the internal phosphorylation potential and the proton motive force in mitochondria during ATP synthesis and hydrolysis. *J. Biol. Chem.* **259**:10004–10011.

THAYER, W., and P. C. HINKLE. 1975. Synthesis of adenosine triphosphate by an artificially imposed electrochemical proton gradient in bovine heart submitochondrial particles. *J. Biol. Chem.* **250**:5330–5335.

TIEGDE, H., H. LUNSDORF, G. SCHAFER, and H. U. SCHAIRER. 1985. Subunit stoichiometry and juxtaposition of the photosynthetic coupling factor 1: Immunoelectron microscopy using monoclonal antibodies. *Proc. Nat'l. Acad. Sci. USA.* **82**:7874–7878.

WALKER, J. E., M. SARASTE, and N. J. GAY. 1982. *E. coli* F_1-ATPase interacts with a membrane protein component of a proton channel. *Nature* **298**:867–869.

Transport of Metabolites into and out of the Mitochondrion

DURAND, R., Y. BIAND, S. TOURAILLE, and S. ALZIARI. 1981. Molecular approaches to phosphate transport in mitochondria. *Trends Biochem. Sci.* **6**:211–214.

* KLINGENBERG, M. 1979. The ADP, ATP shuttle of the mitochondrion. *Trends Biochem. Sci.* **4**:249–252.

* LANOUE, K. F., and A. C. SCHOOLWERTH. 1979. Metabolite transport in mitochondria. *Annu. Rev. Biochem.* **48**:871–922.

The Electron Transport Chain in Mitochondria

BEATTIE, D. S., and A. VILLALOBO. 1982. Energy transduction by the reconstituted b-c_1 complex from yeast mitochondria: inhibitory effect of dicyclohexylcarbodimide. *J. Biol. Chem.* **257**:14745–14752.

* BRUNORI, M., and M. T. WILSON. 1982. Cytochrome oxidase. *Trends Biochem. Sci.* **7**:295–299.

* ERNSTER, L., and G. SCHATZ. 1981. Mitochondria: a historical review *J. Cell Biol.* **91**:227s–255s.

* FILLINGAM, R. H. 1980. The proton-translocating pumps of oxidative phosphorylation. *Annu. Rev. Biochem.* **49**:1079–1113.

FULLER, S. D., R. A. CAPALDI, and R. HENDERSON. 1979. Structure of cytochrome c oxidase in deoxycholate-derived two-dimensional crystals. *J. Mol. Biol.* **134**:305–327.

HACKENBROCK. C. R. 1981. Lateral diffusion and electron transfer in the mitochondrial inner membrane. *Trends Biochem. Sci.* **6**:151–154.

JOHNSON, L. V., M. L. WALSH, B. J. BOCKUS, and L. B. CHEN. 1981. Monitoring of relative mitochondrial membrane potential in living cells by fluorescence microscopy. *J. Cell Biol.* **88**:526–535.

* SARASTE, M. 1983. How complex is a respiratory complex? *Trends Biochem. Sci.* **8**:139–142.

SCHNEIDER, H., J. J. LEMASTERS, and C. R. HACKENBROCK. 1982. Lateral diffusion of ubiquinone during electron transfer in phospholipid- and ubiquinone-enriched mitochondrial membranes. *J. Biol. Chem.* **257**:10789–10793.

* SRERE, P. A. 1982. The structure of the mitochondrial inner membrane-matrix compartment. *Trends Biochem. Sci.* **7**:375–378.

* WIKSTROM, M., K. KRAB, and M. SARASTE. 1981. Proton-translocating cytochrome complexes. *Annu. Rev. Biochem.* **50**:623–655.

Thermogenesis

LIN, C. S., and E. M. KLINGENBERG. 1980. Isolation of the uncoupling protein from brown adipose tissue mitochondria. *FEBS Lett.* **113**:299–303.

* NICHOLLS, D. G., and E. RIAL. 1984. Brown fat mitochondria. *Trends Biochem. Sci.* **9**:489–491.

Metabolic Regulation

* ATKINSON, D. E. 1977. *Cellular Metabolism and Its Regulation.* Academic Press.

* ERECINSKA, A. and D. F. WILSON. 1982. Regulation of cellular energy metabolism. *J. Membr. Biol.* **70**:1–14.

KEMP, R. G., and L. G. FOE. 1983. Allosteric regulatory properties of muscle phosphofructokinase. *Mol. Cell. Biochem.* **57**:147–154.

NEWSHOLME, E. A., and C. START. 1973. *Regulation of Metabolism,* Wiley.

* UYEDA, K. 1979. Phosphofructokinase. *Adv. Enzymol.* **48**:193–244.

Energy Metabolism in Bacteria

BERG, H. C., M. D. MANSON, and M. P. CONLEY. 1982. Dynamics and energetics of flagellar rotation in bacteria. *Symp. Soc. Exp. Biol.* **35**:1–31.

GLAGOLEV, A. N. 1984. Bacterial $\Delta\mu H^+$-sensing. *Trends Biochem. Sci.* **9**:397–400.

HAMAMOTO, T., N. CARRASCO, K. MATSUSHITA, H. R. KABAK, and M. MONTAL. 1985. Direct measurement of the electrogenic activity of O-type cytochrome oxidase from *E. coli* reconstituted into planar lipid bilayers. *Proc. Nat'l Acad. Sci.* 82:2570–2573.

KOBAYASHI, H., T. SUZUKI, and T. UNEMOTO. 1986. Streptococcal cytoplasmic pH is regulated by changes in amount and activity of proton-translocating ATPase. *J. Biol. Chem.* **261**:627–630.

* MACNAB, R. M. 1984. The bacterial flagellar motor. *Trends Biochem. Sci.* **9**:185–189.

MATSHUSHITA, K., L. PATEL, R. B. GENNIS, and H. R. KABACK. 1983. Reconstitution of active transport in proteoliposomes containing cytochrome *c* oxidase and *lac* carrier protein purified from *Escherichia coli. Proc. Nat'l Acad. Sci. USA* **80**:4889–4893.

* SCHNEIDER, E., and K. ALTENDORD. 1984. The proton-translocating portion (F_0) of the *E. coli* ATP synthase. *Trends Biochem. Sci.* **9**:51–53.

SKULACHEV, V. P. 1984. Sodium bioenergetics. *Trends Biochem. Sci.* 9:483–485.

SOKATCH, J. A. 1969. *Energy Metabolism in Bacteria.* Academic Press.

* THAUER, R., K. JUNGERMANN, and K. DECKER. 1977. Energy conservation in chemotrophic anaerobic bacteria. *Bacteriol. Rev.* **41**:100–180.

Photosynthesis

Structure and Function of Chloroplasts

* BOGORAD, L. 1981. Chloroplasts. *J. Cell Biol.* **91**:256s–270s.

FOYER, C., R. LEEGOOD, and D. WALKER. 1982. What limits photosynthesis? *Nature* **22**:326.

* GOVINDJEE, ed. 1982. *Photosynthesis: Energy Conversion by Plants and Bacteria.* Academic Press. A collection of excellent review articles.

HALIWELL, B. 1981. *Chloroplast Metabolism—The Structure and Function of Chloroplasts in Green Leaf Cells.* Clarendon.

JOYARD, J., A. BILLECOCQ, S. G. BARTLETT, M. A. BLOCK, N-H. CHUA, and R. DOUCE. 1983. Localization of polypeptides to the cytosolic side of the outer envelope membrane of spinach chloroplasts. *J. Biol. Chem.* **258**:10000–10006.

* STAEHELIN, L. A., and C. J. ARNTZEN. 1983. Regulation of chloroplast membrane function: protein phosphorylation changes the spatial organization of membrane components. *J. Cell Biol.* **97**:1327–1337.

STEINBACH, K. E., S. BONITZ, C. J. ARNTZEN, and L. VOGORAD, eds. 1985. *Molecular Biology of the Photosynthetic Apparatus.* Cold Spring Harbor Laboratory.

Light Reactions and Photosystems I and II

ALLEN, J. F. 1983. Protein phosphorylation—carburetor of photosynthesis? *Trends Biochem. Sci.* 8:369–373.

* ANDERSON, J. M., and B. ANDERSSON. 1982. The architecture of photosynthetic membranes: lateral and transverse organization. *Trends Biochem. Sci.* 7:288–292.

* BARBER, J. 1984. Has the mangano-protein of the water splitting reaction of photosynthesis been isolated? *Trends Biochem. Sci.* 9:99–101.

* BENNETT, J. 1979. The protein that harvests sunlight. *Trends Biochem. Sci.* **4**:268–271.

BLANKENSHIP, R. E., and W. W. PARSON. 1978. The photochemical electron transfer reactions of photosynthetic bacteria and plants. *Annu. Rev. Biochem.* **47**:635–653.

CARRILLO, N., and R. H. VALLEJOS. 1983. The light-dependent modulation of photosynthetic electron transport. *Trends Biochem. Sci.* **8**:52–56.

* CLAYTON, R. K. 1980. *Photosynthesis: Physical Mechanisms and Chemical Patterns.* Cambridge University Press.

* DEISENHOFER, J., H. MICHEL, and R. HUBER. 1985. The structural basis of photosynthetic light reactions in bacteria. *Trends Biochem. Sci.* **10**:243–248.

HIRSCHBERG, J. A., A. BLEECKER, D. J. KYLE, L. McINTOSH, and C. J. ARNTZEN. 1984. The molecular basis of triazine-herbicide resistance in higher-plant chloroplasts. *Z. Naturforsch* 39:412–420.

LI, J. 1985. Light-harvesting chlorophyll *a*/*b*-protein: three-dimensional structure of a reconstituted membrane lattice in negative strain. *Proc. Nat'l Acad. Sci. USA* **82**:386–390.

STOECKENIUS, W., and R. A. BOGOMOLNI. 1982. Bacteriorhodopsin and related pigments of *Halobacteria. Annu. Rev. Biochem.* **51**:587–616.

STOECKENIUS, W. 1985. The rhodopsin-like pigments of halobacteria: Light-energy and signal transducers in an archaebacterium. *Trends Biochem. Sci.* **10**:483–486.

WIDGER, W. R., W. A. CRAMER, R. G. HERRMANN, and A. TREBST. 1984. Sequence homology and structural similarity between cytochrome b of mitochondrial complex III and the chloroplast b_6/f complex: Position of the cytochrome b hemes in the membrane. *Proc. Nat'l Acad. Sci. USA* **81**:674–678.

YOUVAN, D. C., and B. L. MARRS. 1984. Molecular genetics and the light reactions of photosynthesis. *Cell* **39**:1–3.

CO_2 Fixation

AKAZAWA, T., T. TAKABE, and H. KOBAYASHI. 1984. Molecular evolution of ribulose-1,5-bisphosphate carboxylase/oxygenase (RuBisCO). *Trends Biochem. Sci.* **9**:380–383.

* BASSHAM, J. A. 1962. The path of carbon in photosynthesis. *Sci. Am.* **206**(6):88–100.

* BJORKMAN, O., and J. BERRY. 1973. High-efficiency photosynthesis. *Sci. Am.* **229**(4):80–93. (A discussion of C_4 plants.)

* EDWARDS, G., and D. WALKER. 1983. *C_3, C_4 Mechanisms, and Cellular and Environmental Regulation of Photosynthesis.* University of California Press.

FLUGGE, U. I., and H. W. HELDT. 1984. The phosphate-triose phosphate-phosphoglycerate translocator of the chloroplast. *Trends Biochem. Sci.* **9**:530–533.

* HEBER, U., and G. H. KRAUSE. 1980. What is the physiological role of photorespiration? *Trends Biochem. Sci.* **5**:32–34.

LORIMER, G. H. 1981. The carboxylation and oxygenation of ribulose-1,5-bisphosphate: The primary events in photosynthesis and photorespiration. *Annu. Rev. Plant Physiol.* 32:349.

21

Assembly of Organelles

PREVIOUS chapters have focused on the complex set of macromolecular elements that combine to make a living cell. The nucleus, cytoplasm, subcellular particles, and other organelles interact with each other in order to execute the multitude of sophisticated tasks essential to the growth and metabolism of cells and organisms. A point that has been stressed throughout is the *primacy of proteins:* each type of membrane, organelle, or particle contains a discrete set of proteins that enable it to carry out its unique functions. Let us turn now to the mechanism by which individual proteins are assembled into the proper multiprotein particle or membrane-limited organelle.

Recall some of the many subcellular structures that contain proteins as an essential component. Filaments such as actin (see Figure 19-1) and microtubules (see Figure 18-7) are composed of multiple copies of one or two types of protein subunits arranged in a regular array. Multichain globular proteins such as hemoglobin (see Figure 3-8) contain two copies of each of two peptides; others, such as RNA polymerases, contain a number of different types of protein species. In addition to protein, many particles contain nucleic acid as an essential component. Nucleosomes contain four histone species complexed with 140 base pairs of DNA (see Figure 10-17), whereas ribosomal subunits are built of one or more RNA molecules and 20 to 50 discrete protein species (see Figure 4-18). Various other proteins are found in the membranous elements of the cell. The plasma membrane contains a complex set of proteins that allow communication with the extracellular environment, enabling the cell to attach to other cells or to the basal lamina, and permitting the selective uptake and extrusion of ions and other small molecules. In other organelles, certain protein constituents enable them to carry out unique functions. Lysosomes, for instance, contain a set of degradative enzymes that function at low pH, and specific ion-pumping proteins in the lysosomal membrane maintain the low intralysosomal pH. The process of protein secretion (see Figure 5-34) involves a number of membrane-limited organelles—the rough endoplasmic reticulum (ER), the Golgi complex, and secretory vesicles, each of which contains a unique set of enzymes that make specific modifications on the secreted proteins.

The process of assembly of mitochondria and chloroplasts is even more complex. Some of the mitochondrial and chloroplast proteins are encoded by the organelle DNA and are synthesized on a unique class of ribosomes found in the mitochondrion or chloroplast. However, most of the proteins in these organelles are encoded by nuclear DNA and are synthesized on cytoplasmic ribosomes. These proteins must be subsequently imported into the specific organelle and then inserted into the correct membrane or space.

The assembly of these diverse fibers, particles, and organelles, each with their individual proteins, lipids, carbohydrates, and nucleic acids, appears to be extremely complex (Figure 21-1). In particular, how does the cell direct proteins to specific membranes and organelles? Before we examine the current understanding of the assembly of subcellular particles and organelles, let us consider some of the key concepts that have guided recent work in this area:

1. Individual polypeptide chains may spontaneously assemble into multiprotein aggregates.
2. Membranes grow by expansion of existing membranes. Lipids and proteins are inserted into existing membranous elements.
3. Membrane-containing organelles such as mitochondria and chloroplasts grow by expansion of existing organelles. Proteins and lipids are added to existing organelles, and eventually the organelle divides into two or more "daughters."
4. Pure phospholipid membranes are impermeable to proteins. Certain subcellular membranes contain permeases that permit specific proteins to cross. The transport processes involved often require the expenditure of energy, which can come from ATP, from an ion gradient across the membrane, or from changes in the conformation of the protein during its passage across the membrane.
5. Specific proteins are destined to enter specific membranes or organelles. This "targeting" is due to signals on the protein itself; such signals can be amino acid sequences, carbohydrates, or phosphate residues added to the protein after its synthesis. These signals are recognized by receptors on the surface of subcellular membranes and organelles. Often the signaling or targeting group on the protein is removed after the protein reaches its final destination.
6. Many membrane proteins, such as cell surface receptors and synaptic vesicle proteins, shuttle repeatedly between two or more compartments. Specific mechanisms exist to regulate "traffic" of membranes. Both the fusion of two membranes and the budding off of a membrane region are tightly controlled processes.

As we shall see later in the chapter, these principles are applicable to synthesis of chloroplasts, mitochondria, and plasma and intracellular membranes, as well as to the process of protein secretion and to the entry of proteins into vesicles such as lysosomes and peroxisomes. Let us begin with a look at the assembly of some of the relatively simple subcellular particles, such as multisubunit enzymes, viruses, and ribosomes.

Assembly of Proteins, Viruses, and Ribosomes

Polypeptide chains can often spontaneously self-assemble into multiprotein aggregates. Such polypeptides contain specific, high-affinity binding sites that allow them to form specific complexes with other proteins. In many cases, multiprotein structures, including numerous viruses, are assembled from identical copies of one or only a few kinds of proteins. Repeated occurrence of similar protein-protein interactions leads to a symmetrical arrangement of the subunits.

Particles Containing Several Polypeptides Are Assembled in a Stepwise Process

Appropriate chemical treatments cause many types of multisubunit proteins to dissociate into individual subunits. When the dissociating agent is removed, these proteins often can reassemble spontaneously to form the original complex with all its properties. Generally a stepwise process is followed in reassembly. These steps are believed to be analogous to those followed in the cell for assembly of the protein from its newly made polypeptides, although assembly inside a cell generally occurs so quickly that it is difficult to detect and study the intermediates.

A well-studied oligomeric enzyme that illustrates these points is the *Escherichia coli* aspartate transcarbamoylase, or ATCase. Recall that ATCase consists of six copies of the C (catalytic) subunit, which binds the substrates

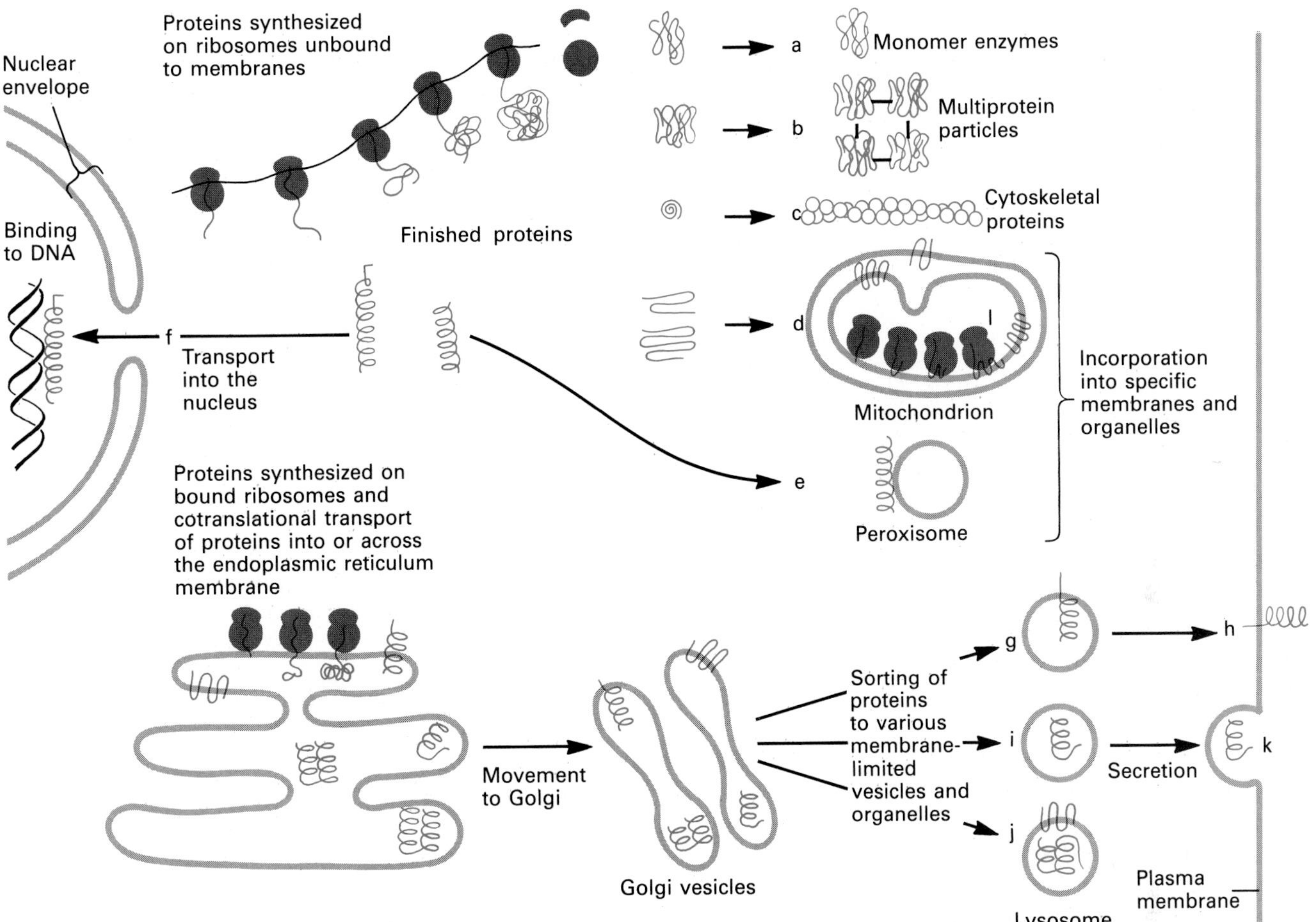

Figure 21-1 Proteins become localized in an array of particles and membranes during or after their synthesis. This schematic diagram summarizes just a few of these. Proteins synthesized on ribosomes unbound to membranes (*top left*) can assemble into multiprotein cytoplasmic particles (*a, b*) or cytoskeletal proteins (*c*). They can be incorporated into specific organelles or membranes, such as the mitochondrion (*d*) or peroxisome (*e*), or into nuclear structures (*f*). Many membrane proteins are synthesized on ribosomes bound to the endoplasmic reticulum (ER) (*lower left*). They are generally first transported to the Golgi complex and then, by means of vesicles (*g*), to the plasma membrane (*h*), or they can be sorted to various membrane-limited organelles such as the lysosome (*j*). Proteins transported to the lumen of the ER also move to the Golgi vesicles; subsequently they can be sequestered into vesicles (*i*) and exocytosed (*k*), or they can be sorted to various organelles (*j*). Organelles such as mitochondria and chloroplasts use organelle ribosomes to synthesize a discrete set of proteins, encoded by organelle DNA, that are localized to the organelle membrane (*l*).

aspartate and carbamoyl phosphate, and six copies of the R subunit, which binds the allosteric effector CTP (see Figures 3-33 and 3-34). The C subunits form two C_3 trimers that are positioned on each side of an equatorial plane. The R subunits form three R_2 dimers, each of which is bound to one C subunit above and one below the equatorial plane (Figure 21-2; see also Figure 3-35).

Upon exposure to mercury-containing compounds such as *p*-hydroxymercuribenzoate (*p*HMB), which react with sulfhydryl groups on the C subunit, ATCase dissociates into two C_3 trimers and three R_2 dimers. The C_3 trimers remain catalytically active, but as might be expected, the reaction is no longer inhibited by the allosteric effector CTP. When the *p*HMB is removed by dialysis, the C_3 and R_2 subunits spontaneously recombine, generating a fully active enzyme showing allosteric inhibition by CTP. The C_3 and R_2 oligomers can be disaggregated into totally denatured subunits that, upon removal of the denaturing agent, spontaneously re-form the native C_3 and R_2 oligomers. Such in vitro reconstitution studies defined the following sequence of events in assembly of ATCase: (1) folding of the polypeptides into α helixes and other secondary structures, (2) a conformational change that allows formation of R_2 and C_3 multipeptide aggregates,

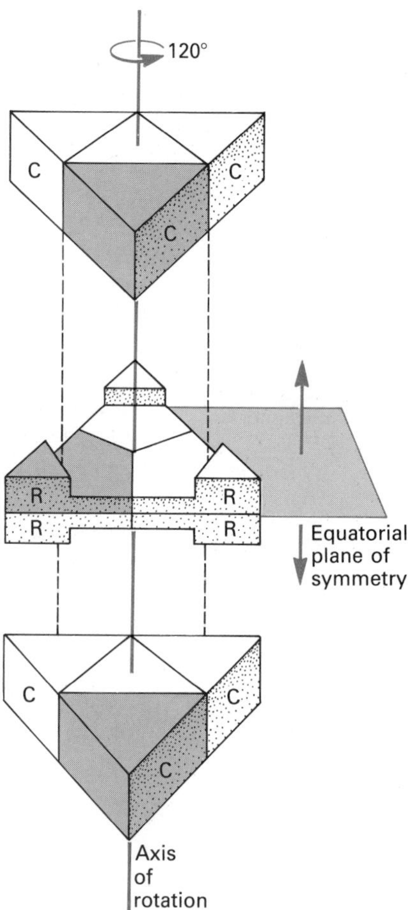

Figure 21-2 Schematic view of the ATCase molecule expanded along the threefold axis, showing the symmetry of the molecule. The central region is composed mostly of the three R_2 dimers. The two C_3 catalytic trimers are arranged above and below the plane defined by the R subunits.

A symmetry element is a transformational process—such as a rotation about an axis or a reflection in a plane of symmetry—that converts a particle into a duplicate copy of itself. ATCase has two symmetry elements. The relative positions of materials are identical on either side of the equatorial plane; thus ATCase is said to have a plane of symmetry. A rotation of 120° about the vertical axis transforms the molecule into a copy of itself, so the molecule is said to have a threefold rotational axis of symmetry. Because of the symmetry of the molecule, only three types of interactions are required to assemble all 12 polypeptides into ATCase: C-C interactions, to form C_3 trimers; R-R interactions, to form R_2 dimers; and R-C interactions, to bind R_2 dimers to C_3 trimers. [See W. M. Lipscomb et al., 1974, *J. Supramol. Struct.* 2:82.] *After W. M. Lipscomb et al., 1974.*

and (3) aggregation of R_2 and C_3 to form the mature R_6C_6 enzyme, by binding of an R to a C polypeptide. A number of intermediates are detected during in vitro assembly of the mature ATCase (Figure 21-3). Because of the symmetry of ATCase (see Figure 21-2), only three types of intrachain bonds are required to assemble 12 polypeptides into an ATCase molecule.

In the assembly of R_2 dimers or C_3 trimers, each R and C polypeptide has a binding site that is complementary to a region on its own surface. There are many other examples of a single type of polypeptide that aggregates with itself to form geometrically regular arrays. Recall that actin microfilaments and microtubules are symmetrical helical arrays of identical subunits—of G actin or of an $\alpha\beta$ tubulin heterodimer whose properties of association are dependent on compounds such as GTP and Ca^{2+}. Other proteins bind to such filaments in regular arrays, such as in binding of dynein and radial spokes to flagellar microtubules (Chapter 18). Still other proteins may interact with one or the other end of a filament, thereby blocking either polymerization or depolymerization. These filaments also can bind to specific membrane proteins; for example, in the erythrocyte, actin filaments cross-link the fibrous spectrin molecules (see Figure 14-42) to form the two-dimensional submembrane cytoskeleton.

Viral Nucleoproteins Can Assemble Spontaneously from Proteins and Nucleic Acid

Perhaps the most striking examples of symmetrical biological structures are viruses that contain only nucleic acid and one or a few *coat proteins* that cover the viral genome. Viral nucleic acid is packaged into virions, which can be either rodlike helical structures or roughly spherical ones (Chapter 6). The coats of such viruses are symmetrical objects: each coat subunit interacts with its neighbors and with the nucleic acid in the same way (or in very nearly the same way).

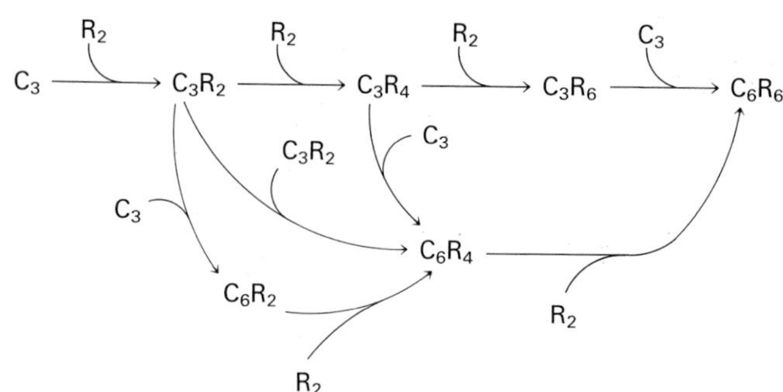

Figure 21-3 Steps in the assembly of ATCase observed during in vitro reconstitution reactions. The arrows indicate the various sequences of additions that can take place to yield the complete ATCase molecule. [See M. Bothwell and H. K. Schachman, 1974, *Proc. Nat'l Acad. Sci. USA* 71:3221.]

By means of x-ray crystallographic studies on virus crystals, the structure of several of these simpler viruses has been determined in great detail. Many of the steps in virus assembly can be reproduced experimentally by mixing the separated components in vitro. In particular, infectious TMV can be reassembled under physiological conditions from separated coat protein and RNA (see Figure 6-29). The native virus structure thus represents a state of minimum free energy: assembly can proceed without any additional template. Viral systems have provided a wealth of molecular detail on structure and assembly that is available in no other self-assembly systems. We shall review the salient features of two such viruses.

Tobacco Mosaic Virus Tobacco mosaic virus (TMV) is a rodlike particle 300 nm long and 18 nm in diameter (see Figure 6-23). It contains a single-stranded RNA of about 6400 bases and also about 2130 copies of a single protein of 158 amino acids. The proteins are arranged in a helix of 16⅓ units per turn, and the RNA winds through the protein subunits at a radius of 4 nm (Figure 21-4). Three nucleotides are bound to each subunit. Except for the initiation step in virus assembly, the actual sequence of bases is unimportant for interaction with the protein subunits. The structure is *helically symmetrical:* a rotation of 22° (=360° ÷ 16⅓) along the helix axis transforms the helix into a copy of itself. Each subunit (except at the ends) interacts with neighboring subunits and RNA in precisely the same way.

This helical symmetry suggests that the virus assembles by binding the protein subunits, one at a time, along the length of the RNA at the edge of a helix. However, the actual assembly process turns out to be more complicated. Isolated protein (coat protein) can polymerize into a virus-like helix in the absence of RNA (Figure 21-5). Thus protein-protein rather than protein-RNA interactions must be the primary driving force for assembly of virus particles. The key stable intermediate for in vitro assembly is a two-layered disk composed of 34 monomers of coat protein (Figure 21-5). Disks can be isolated from virus-infected plants; they also form spontaneously in solutions of coat protein monomer at neutral pH. Under specified conditions, such as lowered pH, several disks spontaneously polymerize into virus-like helixes. Disks always polymerize into helixes faster than do separated coat subunits—a finding that is consistent with the

(a)

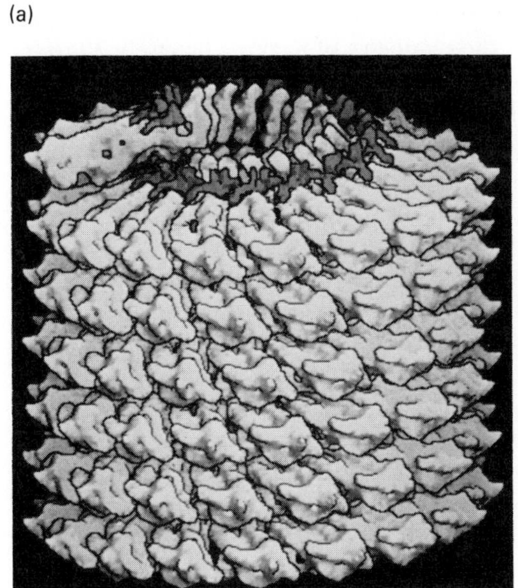

(b)

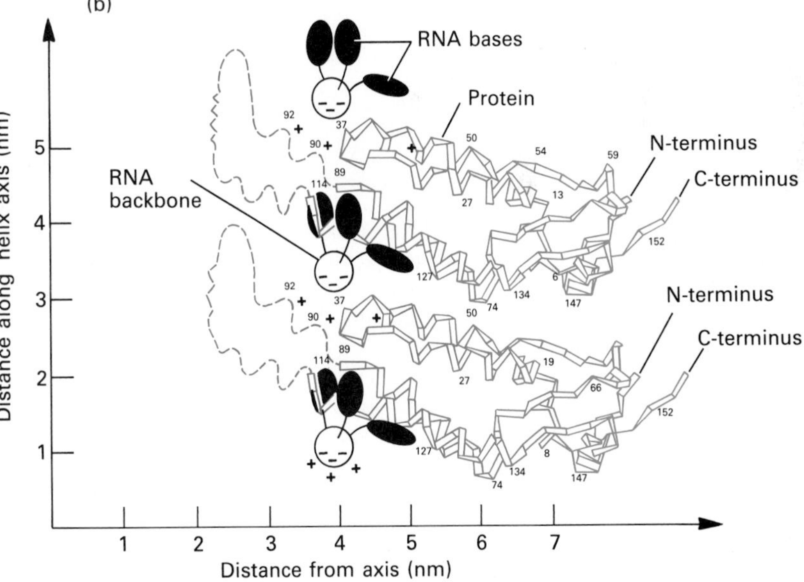

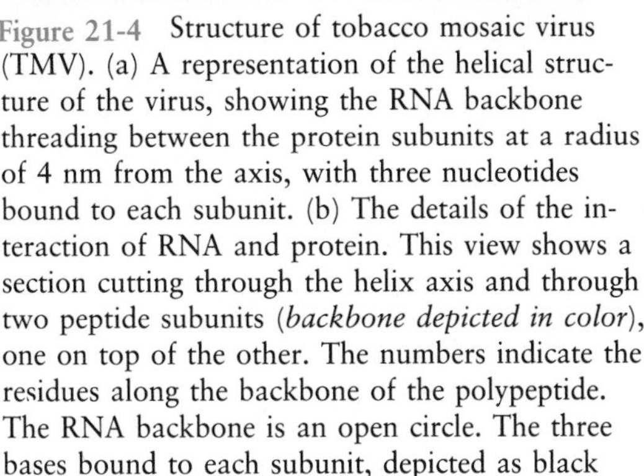

Figure 21-4 Structure of tobacco mosaic virus (TMV). (a) A representation of the helical structure of the virus, showing the RNA backbone threading between the protein subunits at a radius of 4 nm from the axis, with three nucleotides bound to each subunit. (b) The details of the interaction of RNA and protein. This view shows a section cutting through the helix axis and through two peptide subunits (*backbone depicted in color*), one on top of the other. The numbers indicate the residues along the backbone of the polypeptide. The RNA backbone is an open circle. The three bases bound to each subunit, depicted as black disks, are 4 nm from the center of the helix. The negative phosphates in the RNA chain are held in place by salt bridges to the positive side chains of Arg-90 and Arg-92. The nature of the interactions between the RNA bases and the polypeptide is not known since the x-ray crystallographic images from which this model is derived do not provide sufficient resolution. [See G. Stubb, S. Warren, and K. C. Holmes, 1977, *Nature* **267**:216; and A. Klug, 1979, *Harvey Lect.* **74**:141.] *Part (a) from K. Namba, D. Caspar, and G. Stubbs, 1985,* Science *227:773. Copyright 1985 by the AAAS.*

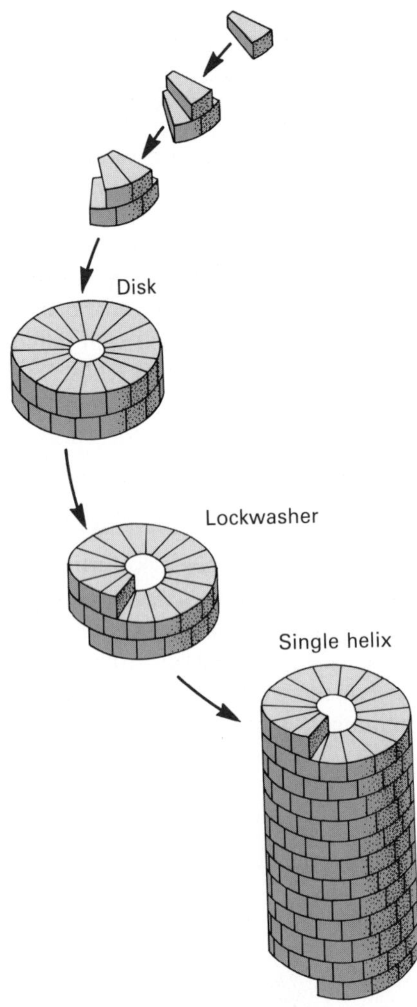

Figure 21-5 Intermediates in the cell-free assembly of a TMV protein helix, without RNA. Monomers of coat protein assemble readily into a stable two-layered disk, each layer with 17 subunits. To effect the transition from disk to helix, the disk apparently dislocates to form a "lockwasher," which tightens to 16⅓ subunits per turn—the same as in the virus. The lockwasher then spontaneously polymerizes into a helix. In cell-free systems, the transition from disk to lockwasher can be induced by lowering the pH from 7.0 to 5.0, resulting in protonation of key histidine residues. [See A. Klug, 1979, *Harvey Lect.* **74**:141.]

notion that disks are intermediates in the assembly process. The "lockwasher," depicted in Figure 21-5, is a later intermediate in assembly; it is formed by dislocation and tightening of a disk to a spiral with 16⅓ subunits per turn from the original cylindrical 17 subunits per turn. Formation of lockwashers is the rate-limiting step in assembly; lockwashers spontaneously polymerize into helixes, as shown in the figure.

In infected cells, virus particles are formed that contain only TMV RNA, not any of the cellular RNAs present in the tobacco cell. TMV RNA, but not cellular RNAs, contains a specific base sequence that initiates polymerization of disks into virus particles. A disk binds preferentially to a region of RNA about 1000 bases from the 3′ end (Figure 21-6). Since it has been found that both the 3′ and the 5′ ends of the TMV RNA protrude from the same end of partially assembled virus rods, it appears that a specific loop of the RNA inserts into the central hole of a disk, as shown in Figure 21-6a. Binding of the disk to the RNA loop causes the conformation of the disk to convert to a lockwasher, as depicted in Figure 21-6b. Addition of more disks occurs primarily by elongation toward the 5′ end, drawing a loop of RNA up through the hole and then binding a disk to the loop, as shown in Figure 21-6c, d, and e. As each additional disk binds to a loop of the RNA, it undergoes a conformational change to a lockwasher that enables it (without additional conformational change) to insert into the regular helical array of subunits. This disk-by-disk elongation process occurs until all of the TMV RNA is covered with coat protein.

Assembly of the TMV helix thus involves sequential steps: formation of the 34-subunit disk and then polymerization of the disks into the helix, concomitant with threading the TMV RNA into the protein subunits.

Spherical Viruses Spherical viruses, also called icosahedral viruses, have structures based on *icosahedral symmetry;* this is the symmetry of packing precisely 20 triangles in a regular array on the surface of a sphere. The simplest such structure is based on the so-called pentagonal dodecahedron, a structure with 12 pentagonal faces. Such a virus contains exactly 60 identical subunits, each radiating from 1 of the 12 fivefold axes (Figure 21-7a). An equivalent view is that the 60 subunits pack into 20 groups of triangles, each built of 3 subunits, that form a regular array—an icosahedron—on the surface of the sphere. Satellite tobacco necrosis virus has such a structure and contains just 60 identical coat protein subunits. Poliovirus (see Figure 6-31h) also contains 60 subunits, but each subunit consists of four different polypeptides.

Most viruses have genomes too large to be contained within a shell of 60 subunits of reasonable molecular weight, so their shells are made up of 180, 240, or 420 subunits. By the principles of solid geometry, however, no more than 60 rigid subunits can be packed on the surface of a sphere in such a way that each subunit has *exactly* the same environment and interacts with its neighbors in *exactly* the same way. To explain the packing of so many identical subunits into spherical structures, the concept of *quasiequivalence* was introduced. Quasiequivalence proposes that structural subunits have a certain flexibility in how they interact with other subunits. The coat subunits can fold into slightly different conformations and thus can have more than one type of interaction with other subunits. This allows the formation of larger polyhedral assemblies. Solid geometry also establishes that all such assemblies must be built of a multiple of 60 subunits,

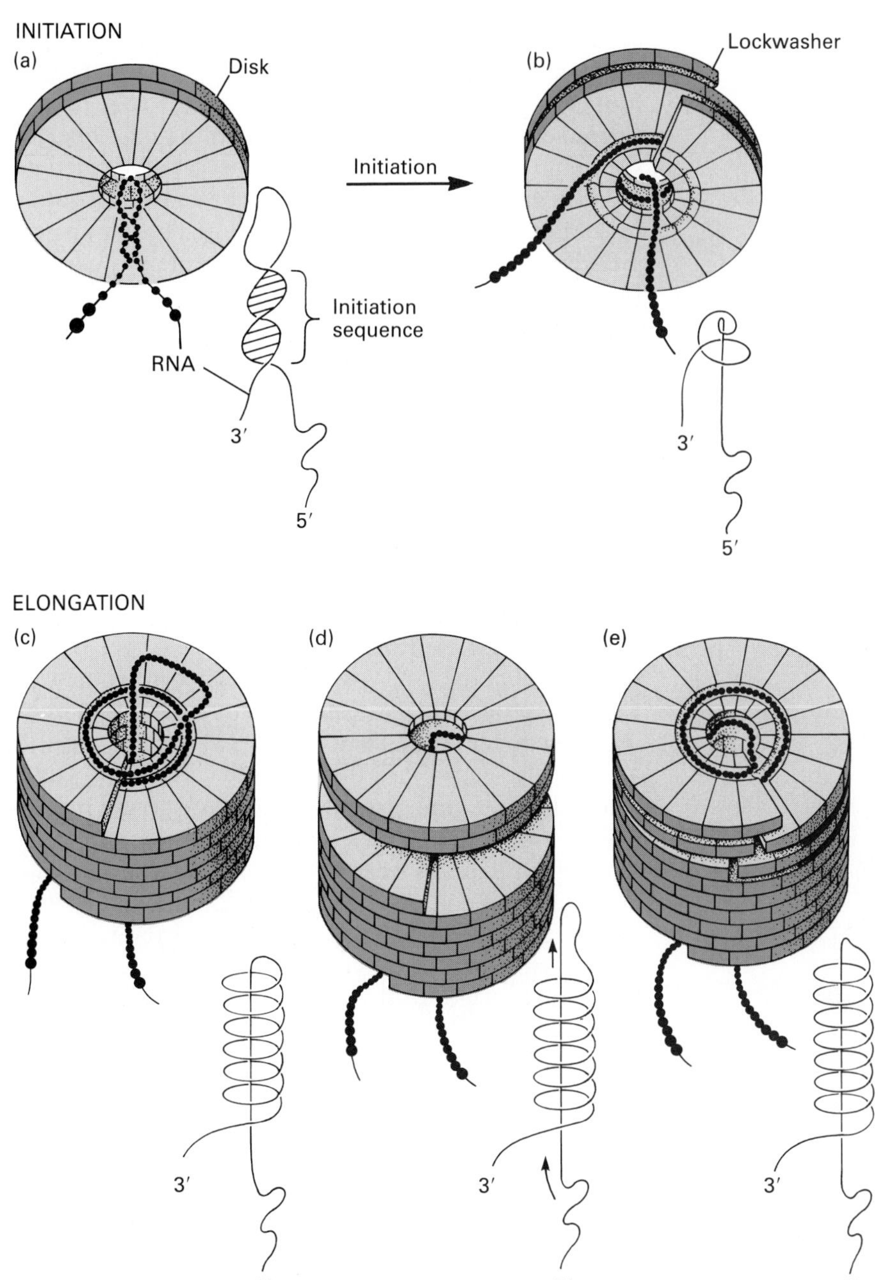

Figure 21-6 Assembly of complete tobacco mosaic virus (TMV). In each diagram, the RNA is depicted twice—once associated with the protein, and a second time alone, to show the conformation of the RNA backbone in more detail as virus formation continues.

To initiate the process, a disk binds to an "initiator" region of the RNA about 1000 bases from the 3′ end. (a) A loop of RNA threads through the central hole of the disk. (b) As it binds to the coat protein, the RNA loop induces the dislocation of a disk into a lockwasher. (c), (d), (e) The growth of a helix from six layers of subunits to eight layers by addition of another disk. (c) About 102 bases of RNA are threaded, from the 5′ end, through the center of the helix. (d) This allows another disk to bind. (e) Subsequently, the conformation of the protein in that disk changes to a lockwasher, while the loop of RNA forms a helix that binds to the newly added protein subunits. After the lockwasher locks into place in the helix, another 102 bases thread through, and the elongation process continues. [See A. Klug, 1979, *Harvey Lect.* **74**:141.] *From P. J. G. Butler and A. Klug, 1978,* The Assembly of a Virus, *Sci. Amer. 239(5)62–69. Copyright 1978 by Scientific American, Inc.*

such as 60, 120, or 180; in a structure with 60n units, there are n subunit conformations and n different packing environments.

The structure of tomato bushy stunt virus (TBSV) is known almost to atomic detail. As shown in Figure 21-7b, there are three distinct ways in which the 180 subunits are folded and interact with each other; these are labeled A, B and C. The important structural principle that emerges from the study of TBSV is that protein subunits can adopt more than one stable conformation and can interact with adjacent subunits in more than one way. Such flexibility permits the assembly of larger quasisymmetrical objects than would be possible from only 60 perfectly symmetrical subunits.

Not All Multiprotein Aggregates Self-Assemble

Many particles cannot self-assemble directly from individual subunits in cell-free reactions. Assembly of such particles may be assisted by other proteins not found in the finished structure. A striking example is seen in the assembly of the heads of icosahedral DNA bacteriophages, such as phage T4 or P22 (see Figure 6-27). The icosahedral head of phage P22 contains a single type of coat protein surrounding the phage DNA (Figure 21-8). The P22 coat protein, unlike the capsid proteins of viruses such as TBSV, is unable to polymerize by itself into a headlike shell, whether or not DNA is added to the

(a)

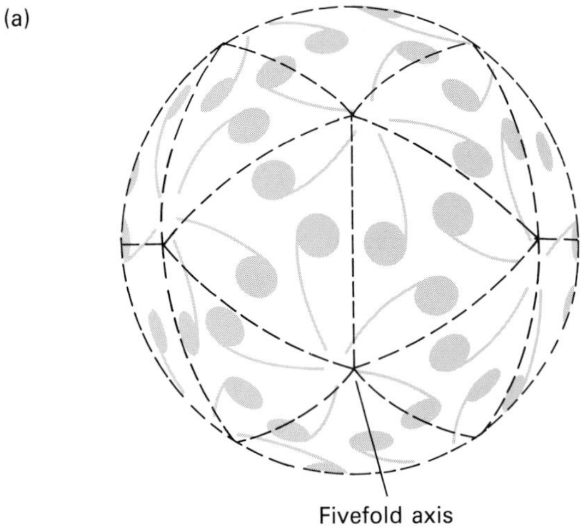

(b)

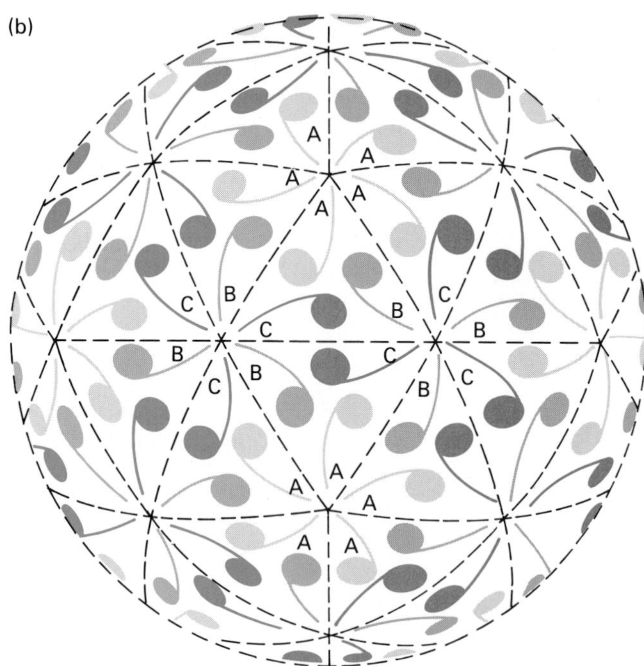

Figure 21-7 Diagram of the surface structure of icosahedral viruses, using commas as the structural subunit. (a) A symmetrical structure with 60 identical subunits, each with exactly the same environment. Satellite tobacco necrosis virus has this structure. (b) A structure with 180 subunits, such as in tomato bushy stunt virus (TBSV). However, there are no perfect symmetries for a closed shell having more than 60 subunits. Note that the *local* packing of structural units is the same in both types of structure, but, as determined from x-ray crystallographic studies, the 180 identical polypeptides in the TBSV coat pack into the virus particle in one of three configurations, denoted A, B, and C. Five A subunits radiate from the fivefold axes; three B and three C subunits radiate from each sixfold axis. *Courtesy of S. Harrison.*

cell-free reactions. The key intermediate in assembly of the capsid is the spherical DNA-free *pro-head.* It is formed of an outer layer of 420 units of coat protein surrounding approximately 200 units of an elongated *scaffolding protein.* Experiments with cell-free reactions indicate that the pro-head is formed by copolymerization of subunits of the coat and scaffolding proteins, as depicted in Figure 21-8; the scaffolding protein is essential for coat polymerization. The pro-head contains a few subunits of another protein that forms a pore, or channel, at one location. In the final stage of head assembly, phage DNA is threaded into the head through the pore, while the scaffolding protein is released and is recycled. Simultaneously the coat proteins undergo a conformational change so that the head converts from a sphere to the final icosahedral structure. Thus the mature head lacks one of the key components required for its own assembly; the virus head cannot assemble spontaneously from its own subunits.

Much of the evidence for this model comes from a study of mutants of P22 that are defective in synthesis of a specific protein. In particular, mutations in the scaffolding protein cause the coat protein to assemble into an abnormally shaped shell that cannot be filled with DNA.

In still other viral systems, a series of specific proteolytic cleavages is an irreversible and essential step in the assembly process. As we have seen, this is also the case for many multisubunit proteins, such as insulin and chymotrypsin. Such modifications serve to make irreversible certain key steps in assembly and also permit the cell to regulate various stages of the assembly process.

Assembly of Ribosomal Subunits Involves a Sequential Addition of Proteins

Thus far we have dealt with assembly of symmetrical multiprotein complexes. Assembly of ribosomal subunits involves a much more complex set of events. Recall that the small ribosomal subunit is composed of a single molecule of rRNA and, depending on the organism, 21 to 30 proteins. The large subunit contains a large rRNA and one (in prokaryotes) or two (in eukaryotes) smaller RNAs, and also 34 to 50 proteins (Chapter 4). Each RNA or protein occupies a unique position within the subunit. Unlike complexes such as ATCase (Figure 21-2), ribosomes have no axis or plane of symmetry.

Let us examine assembly of the *E. coli* 30S ribosomal subunit, because this particle has been studied in the most detail. Much less is known about the assembly of the more complex bacterial 50S subunit, or of eukaryotic ribosomes, but the principles involved are believed to be the same. Most if not all ribosomal proteins are essential for ribosomal subunit assembly or for its many functions in protein synthesis.

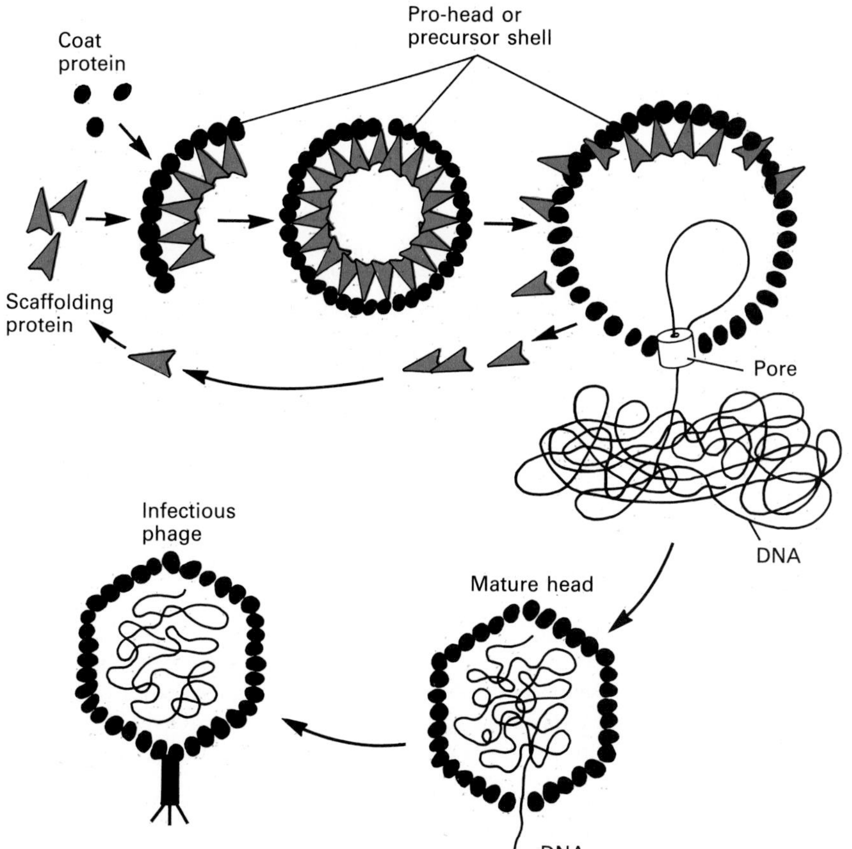

Figure 21-8 Schematic outline of the assembly of the bacteriophage P22 head. The precursor shell is formed by polymerization of two proteins. The outer layer consists of 420 units of the coat protein, which remains in the mature virus, surrounding about 200 units of a scaffolding protein. Prior to or during insertion of the DNA genome into the head through a special "pore" protein, the scaffolding proteins exit intact from the particle and take part in further rounds of pro-head assembly. To the mature head are added, in sequence, tail fibers and other components of the mature, infectious bacteriophage. The tail joins the head at the site of DNA entry; the tail may plug the pore through which the DNA entered the head, thereby keeping the DNA inside. *After J. King, 1980, in* Biological Regulation and Development, *vol. 2, R. F. Goldberg, ed., Plenum, p. 101.*

The *E. coli* 30S subunit contains the 16S rRNA (1542 nucleotides) and a single copy of each of 21 different proteins called S1 through S21. The rRNA has been completely sequenced; when isolated free of protein, it contains many long, internal, double-stranded segments (Figure 21-9; see also Figure 4-19). Each of the 21 proteins has been obtained in pure form, and all have been sequenced; in general they are small, rather basic proteins, as might be expected for proteins that bind to the negatively charged RNA. How do these combine together to generate a functional ribosome?

Recent studies demonstrated that a functional 30S subunit can be reconstituted in vitro from a mixture of the 16S rRNA and all 21 proteins. Thus all the information needed for assembly is contained in the structure of its components. For such reconstitution reactions, a solution of 16S rRNA and the 21 purified proteins is heated to 40°C for a few minutes. When all 21 proteins and 16S rRNA are included, the composition and activity in protein synthesis of the subunit formed is the same as those of control ribosomes. Ribosome self-assembly apparently involves an ordered, sequential addition of single proteins or groups of proteins to the 16S RNA.

When any one of certain proteins, such as S1, S2, S3, or S21, is omitted from the reconstitution reaction, the particles formed are inactive but resemble the structure of normal 30S subunits. Therefore, these proteins are not essential for assembly of a 30S subunit, and they are probably added at a late stage in biogenesis. Indeed, these particular proteins as well as two others can be removed from intact 30S particles by treatment with solutions containing high concentrations of salt without destroying the functional integrity of the particle. Readdition of the proteins to the depleted subunit restores normal protein synthesis activity.

Very different results are obtained when other proteins, such as S4, S17, S8, or S7, are omitted. Lack of any one of these results in the absence of any type of subunit particle containing more than a few proteins; clearly these proteins, as well as 16S rRNA, are essential for ribosome assembly. Omission of still other ribosomal proteins, such as S9, S5, or S16, results in particles that are missing multiple proteins, in addition to the one left out of the reconstitution mixture.

These studies established that ribosome assembly proceeds by an ordered addition of proteins to 16S RNA and yielded an *assembly map,* as shown in Figure 21-10. For instance, only 8 of the 21 proteins in the 30S subunit appear to bind directly to isolated 16S rRNA (see Figure 21-9). Each appears to bind to a specific region of the RNA. The binding of such "core proteins" creates binding sites for other ribosomal proteins, as seen in Figure

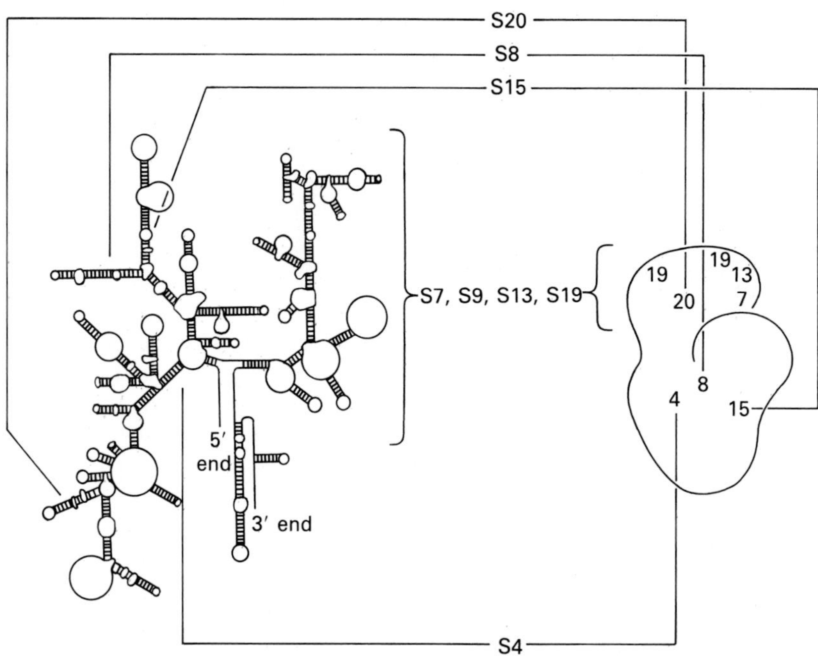

Figure 21-9 The secondary structure of 16S *Escherichia coli* ribosomal RNA, and the binding sites of eight of the ribosomal proteins that interact directly with the RNA. On the right are shown the positions of these ribosomal proteins in the complete subunit. (Figures 4-18 and 4-19 show these in more detail.)

The secondary structure of 16S RNA was deduced from results of experiments in which rRNA in solution was exposed to various reagents that modify nucleotides only when they are not in a double-helix structure. Bisulfite, for instance, modifies C residues in single-stranded regions. Ribonucleases such as RNase A (cleaves 3′ to C and U residues) and T1 (cleaves 3′ to G residues) preferentially attack nucleotides that are in single-stranded rather than base-paired regions.

The pattern of binding of ribosomal proteins to rRNA was deduced from studies in which individual proteins were added to defined fragments of rRNA, or in which rRNA and a specific protein were mixed and then degraded with ribonuclease. In these latter experiments, the RNase-resistant fragment is equated with the region of rRNA that directly binds to the protein. The location of the binding sites for S8, S15, and S20 and the multiple binding sites for S4 are known with the greatest precision. [See R. Brimacome, G. Stoffler, and H. G. Wittmann, 1978, *Annu. Rev. Biochem.* **47**:217; H. F. Noller and C. R. Woese, 1981, *Science* **212**:403; and H. F. Noller, 1984, *Annu. Rev. Biochem.* **53**:119.] *After H. F. Noller, 1984.*

21-10. The exact nature of the other protein-binding sites is not known, but they may be dependent on conformational changes either in the 16S rRNA or in other ribosomal proteins. Thus the assembly of a ribosomal subunit involves the cooperative, ordered interaction of 16S rRNA and all 21 proteins.

The pathway of ribosome biogenesis in the *E. coli* cell differs in some details from that depicted in Figure 21-10, since proteins appear to bind to the precursor rRNA in the process of its synthesis. However, several kinds of newly synthesized ribosomal subparticles can be isolated from *E. coli* cells. Analysis of their protein composition suggests that the order of addition of the proteins to 16S RNA or its precursor is similar to that depicted in Figure 21-10. For instance, all subparticles contain the eight proteins that bind in vitro to 16S RNA.

Most workers believe that no additional proteins are required for ribosome assembly; only those actually found in the finished particles are required. However, proteins such as scaffolding proteins—which accelerate one or more stages of assembly—are, as we have seen, required for the assembly of certain bacteriophages and may also be required for ribosome assembly, especially in eukaryotes. Here the ribosomal precursor RNA is formed into ribonucleoprotein particles in the nucleolus shortly after its synthesis (Chapter 9).

Synthesis of Membrane Lipids

We have already seen that most subcellular structures are built of phospholipid bilayer membranes. Even though structures such as the mitochondrial inner membrane, the plasma membrane, and the membrane of the rough endoplasmic reticulum (ER) carry out very different functions, they all are phospholipid bilayers in which are embedded

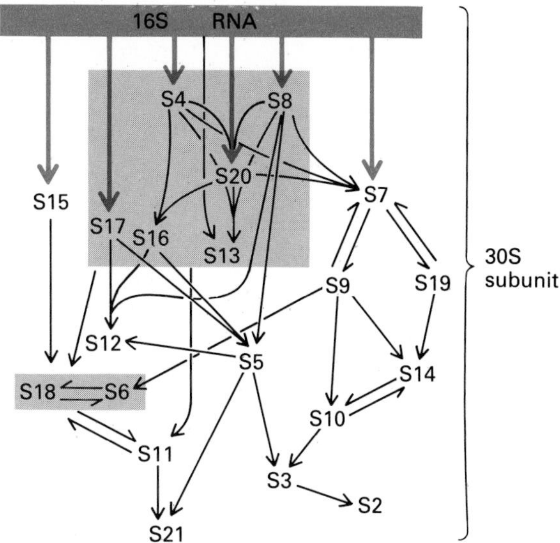

Figure 21-10 The *E. coli* 30S ribosome assembles by sequential addition of proteins to 16S RNA. In this assembly map, arrows between proteins (designated S) indicate the facilitating effect on binding of one protein on another; a thick arrow indicates a major facilitating effect. As depicted by the colored arrows, six proteins bind directly to the ribosomal RNA in the absence of other proteins. Three other proteins—S9, S13, and S19—also bind to specific regions of 16S RNA, but, as depicted here, their binding is greatly facilitated by certain ribosomal proteins. Thin arrows pointing toward S7 from S4, S8, S20, S9, and S19 indicate that the latter proteins all enhance the binding of S7 to RNA. The arrow to S11 from the large shaded box indicates that S11 binding depends on some of the proteins enclosed in the box (it is not known exactly which proteins). As one example of how to interpret this map, the binding of S5—one protein added at an intermediate stage—requires prior addition of S8, S16, and S17, as indicated by the arrows. *After W. A. Held et al., 1974,* J. Biol. Chem. *249:3103.*

different classes of proteins. Accordingly, our study of organelle assembly begins with the biosynthesis of phospholipids and their incorporation into membranes.

Phospholipids Are Synthesized in Association with Membranes

All phospholipids and other membrane lipids such as sphingomyelin and glycolipids are amphipathic molecules with an extremely hydrophobic region and dissolve very little in aqueous solutions. In fact, depending on their concentration and the ionic composition, they spontaneously form either micelles or sheets of bilayers. These physical-chemical properties of phospholipids have profound implications for the biosynthesis of membrane-containing organelles. The extremely low solubility of phospholipid in aqueous solution makes the assembly, from soluble components, of a new phospholipid bilayer energetically difficult. Phospholipids either are synthesized in association with cellular membranes or are incorporated into already existing membranes immediately after synthesis. When cells are briefly exposed to $[^{32}P]PO_4^{3-}$, or to radiolabeled fatty acids or sugars, all phospholipids and glycolipids incorporating these substances are seen to be associated with intracellular membranes. With few exceptions, membranes grow in this manner—by expansion of existing membranes.

In bacterial cells, synthesis of phospholipids occurs in association with the plasma membrane. In animal and plant cells, synthesis occurs in association with the ER membrane, usually smooth ER. The pathway for synthesis of phosphatidylethanolamine, a typical phospholipid, in animal cells is depicted in Figure 21-11. One of the substrates, a CoA ester of a fatty acid ("activated fatty acid"), is an amphipathic molecule (see Figure 14-19). The fatty acid side chain is embedded in the ER membrane, with the CoA portion protruding into the cytosol. All of the other substrates and reaction products, such as ATP, ethanolamine, and CTP, are soluble constituents of the cytoplasm. The biosynthesis of phospholipids thus occurs at the interface of the ER membrane and the cytoplasm. Most of the enzymes that catalyze these reactions are amphipathic, with one segment inserted into the ER membrane and another protruding into the cytoplasm. For example, stearoyl-CoA desaturase, which catalyzes the introduction of a double bond in stearoyl-CoA to form the CoA derivative of oleic acid, is an amphipathic enzyme having part of its active site facing the cytosol and part associated with the ER.

Special Membrane Proteins Allow Phospholipid to Equilibrate in Both Leaflets of the Membrane

All newly made phospholipid is localized to the cytoplasmic leaflet of the ER. But in the ER, as in most cellular membranes, all phospholipids are found in both membrane faces. However, phospholipids do not spontaneously flip-flop across a pure phospholipid bilayer. In many already formed membranes, such as the erythrocyte membrane or the plasma membrane of nucleated cells, an average of several days is required for a phospholipid to move from one face of the membrane to the other. How then, are newly made phospholipids moved from the cytoplasmic to the inner leaflet of the ER? It appears that ER membranes (and the plasma membrane of bacteria) may contain one or more proteins that can catalyze such a flip-flop process. The half-time for movement of a phos-

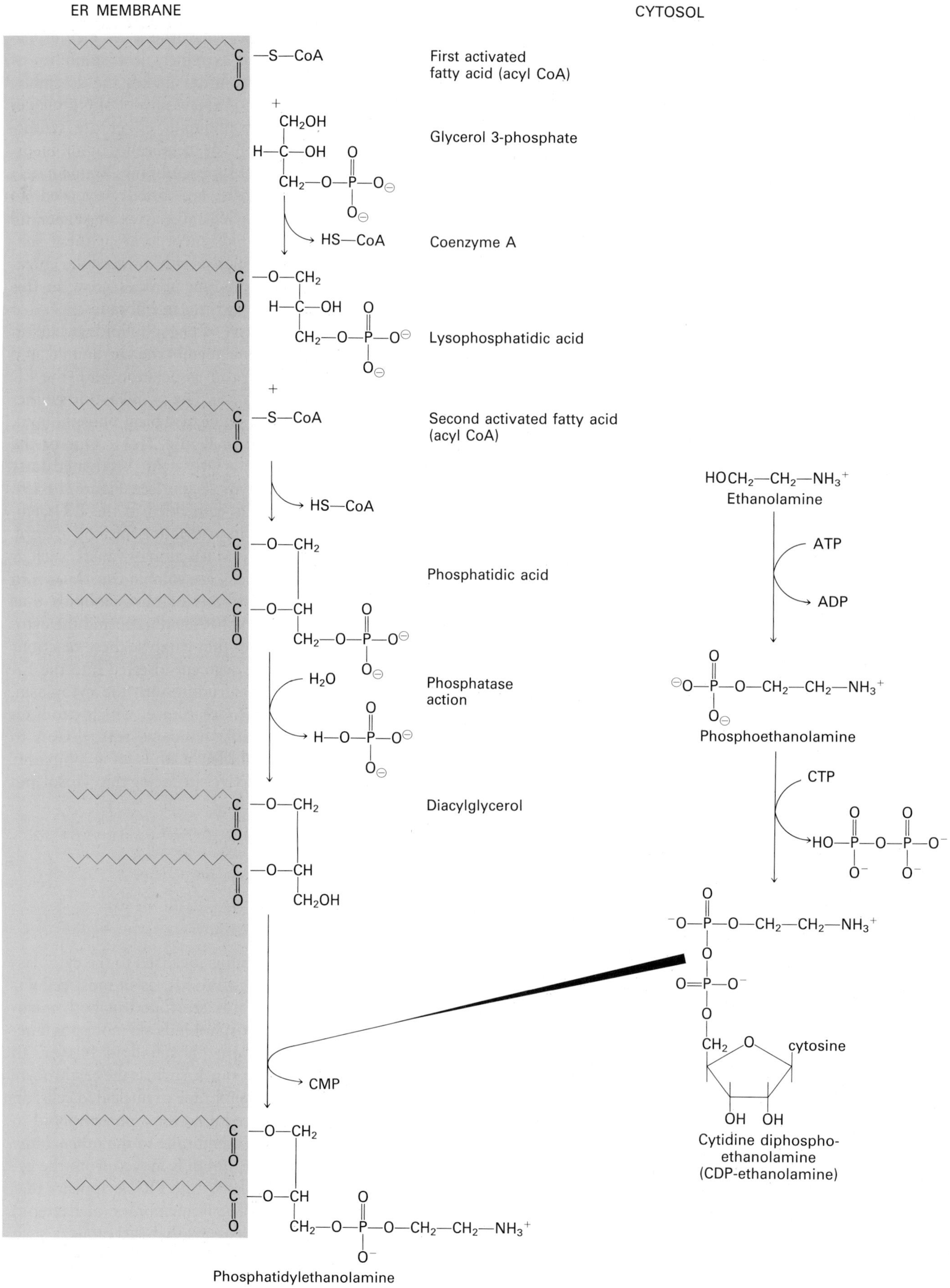

ER MEMBRANE

CYTOSOL

First activated fatty acid (acyl CoA)

+

Glycerol 3-phosphate

$HS-CoA$ Coenzyme A

Lysophosphatidic acid

+

Second activated fatty acid (acyl CoA)

$HS-CoA$

Phosphatidic acid

H_2O

Phosphatase action

$H-O-PO_3^{2-}$

Diacylglycerol

$HOCH_2-CH_2-NH_3^+$
Ethanolamine

ATP

ADP

$^{-2}O_3P-O-CH_2-CH_2-NH_3^+$
Phosphoethanolamine

CTP

$HO-P(O)(O^-)-O-P(O)(O^-)-O^-$

cytosine

OH OH

Cytidine diphospho-ethanolamine (CDP-ethanolamine)

CMP

$CH_2-O-P(O)(O^-)-O-CH_2-CH_2-NH_3^+$

Phosphatidylethanolamine

Figure 21-11 *(opposite)* Biosynthesis of phosphatidyl ethanolamine in animal cells. The precursors are fatty acyl CoA, an amphipathic molecule embedded in the ER membrane, and glycerol 3-phosphate and CDP-ethanolamine, both soluble molecules in the cytosol. Other phospholipids are assembled in the ER membrane by analogous pathways from fatty acyl CoA and soluble small molecules. Bacterial cells use a different pathway for synthesis of phosphatidyl ethanolamine, but the pathway also starts with fatty acyl CoA and occurs in the (bacterial) plasma membrane.

pholipid to the *opposite* face of the ER membrane is only a few minutes.

As we saw in Chapter 14, the two leaflets of a membrane often have different phospholipid compositions. How this lipid asymmetry is achieved is not known, but the mechanism may involve, among other factors, different affinities of phospholipids for different regions of integral membrane proteins that are localized in the two faces.

Phospholipids Move from the ER to Other Cellular Membranes

Although phospholipids are synthesized in the ER, they are found in all organelles. How do they move there? One popular model is that of *membrane budding:* a membrane vesicle, containing phospholipids, buds off the ER membrane and then fuses with the membrane of an organelle. In the formation of this vesicle, certain phospholipids may be selectively incorporated; this would explain the different phospholipid composition of different organelle membranes (Chapter 14). Such a mechanism is used for certain lipids.

A second model for phospholipid movement involves *phospholipid exchange proteins,* water-soluble proteins that can remove phospholipids from one membrane (such as the ER) and then release them into another membrane or organelle. Such proteins have been identified in liver and other types of cells. Each exchange protein binds only a single type of phospholipid. In cell-free reactions, exchange proteins function to equilibrate phospholipids among all membranes present. Whether, in the cell, such exchange proteins can move certain lipids to specific organelles is not known.

Mitochondria appear to synthesize certain lipids. Cardiolipin, in particular, is found only in the mitochondrial inner membrane (Chapter 20), and preparations of mitochondrial membranes can synthesize in vitro this phospholipid, as well as phosphatidylglycerol. Other mitochondrial lipids, such as phosphatidylethanolamine and phosphatidylcholine, are synthesized in the ER and subsequently imported into the mitochondrion.

Sites of Synthesis of Organelle and Membrane Proteins

Although in a few cases a particular lipid is concentrated in a single organelle—such as cholesterol and glycolipids in the plasma membrane and cardiolipin in the mitochondrial inner membrane—each subcellular organelle and membrane contains a unique constellation of proteins. As examples, the ATP-ADP transporter is unique to the inner mitochondrial membrane, and certain hydrolytic enzymes are greatly enriched within lysosomes. Several possibilities exist for directing proteins to the appropriate place in the cell. There could be functionally different classes of ribosomes, each occupying a specific intracellular site, that translate only certain classes of mRNAs. Mitochondria and chloroplasts do contain unique populations of ribosomes, and all proteins encoded by the mitochondrial DNA or chloroplast DNA are translated on ribosomes within the respective organelle (Table 21-1), but specialized cytoplasmic ribosomes for the synthesis of specific proteins have not been found.

Table 21-1 Proteins synthesized by different classes of cellular ribosomes

Location of ribosome	Classes of protein synthesized
Mitochondrion	All proteins encoded in mitochondrial DNA, mainly certain integral proteins of the inner membrane (see Figure 21-14)
Chloroplast	All proteins encoded by chloroplast DNA (see Table 21-5)
Cytoplasm	
Ribosomes unbound to membranes	Soluble cytosolic proteins
	Extrinsic membrane proteins localized to the cytoplasmic surface (e.g., actin, spectrin)
	Mitochondrial proteins encoded by nuclear DNA
	Chloroplast proteins encoded by nuclear DNA
	Peroxisome proteins
	Glyoxisome proteins
Ribosomes bound to ER membranes	Secreted proteins
	Integral plasma membrane glycoproteins
	Lysosomal enzymes
	Rough ER enzymes
	Golgi complex enzymes
	Extrinsic membrane proteins localized to the extracytoplasmic surface (e.g., fibronectin)

All Cytoplasmic Ribosomes Are Functionally Equivalent

Abundant evidence indicates that all ribosomes in the cytoplasm of eukaryotic cells are functionally equivalent. As noted in Chapter 5, some cytoplasmic ribosomes are tightly bound to the ER, whereas certain others are not bound to membrane-containing organelles (Figure 21-12). (Recall from Chapter 18 that many of these are bound to cytoskeletal fibers.) But these two classes of ribosomes have the same protein and rRNA composition. In cell-free protein-synthesizing systems containing a variety of added mRNAs, both classes of ribosomes function identically. Apparently, then, there is no functionally unique class of cytoplasmic ribosome that is found only in a certain intracellular site that translates only a specific class of mRNAs. Thus most if not all of the information for intracellular protein distribution is located in the amino acid sequence of the newly synthesized protein itself. These groups of amino acids are often called *signal sequences*.

Although when isolated the two classes of cytosolic ribosomes—those bound to membranes and those unbound—are equivalent, within the cell they are engaged in translating different classes of nuclear DNA–encoded mRNAs (Table 21-1). Proteins that are synthesized on membrane-bound ribosomes, such as secretory proteins, membrane proteins, glycoproteins, and lysosomal proteins, contain specific amino acid "signals" that direct the ribosome to bind to the ER membrane. As a protein chain is elongated by a ribosome, the most recently added 25 to 35 amino acids remain shielded by the large ribosomal subunit, and the N-terminal is not exposed to other cytoplasmic proteins until the peptide is at least 25 to 35 residues long. Apparently, *initiation* of synthesis of all proteins translated by cytosolic ribosomes occurs on unbound ribosomes, and consequently ribosome attachment to the membrane can only occur when the chain has grown to a length such that the *signal sequence* for attachment of the nascent chain to the membrane protrudes from the ribosome.

Proteins that are synthesized by membrane-attached ribosomes begin to cross the ER membrane before their synthesis is complete. Further sorting of these newly made proteins to their proper location requires 20 minutes to several hours. Secretory proteins are transported first to the Golgi complex and are then secreted. Lysosomal enzymes are directed to lysosomes, and so forth. These *sorting signals* either are amino acid sequences in the polypeptide or are substituents that are added post-translationally.

A large variety of proteins are synthesized by cytoplasmic ribosomes unbound to membranes. Among these are soluble cytosolic proteins, such as glycolytic enzymes, and most extrinsic membrane proteins, such as spectrin, which bind to other proteins on the cytoplasmic surface of membranes. Such proteins are released into the cytosol and either remain there or bind to other proteins (e.g., polymerization of actin or tubulin) or to membrane proteins (e.g., spectrin binding to ankyrin and band 4.1; see Chapter 14).

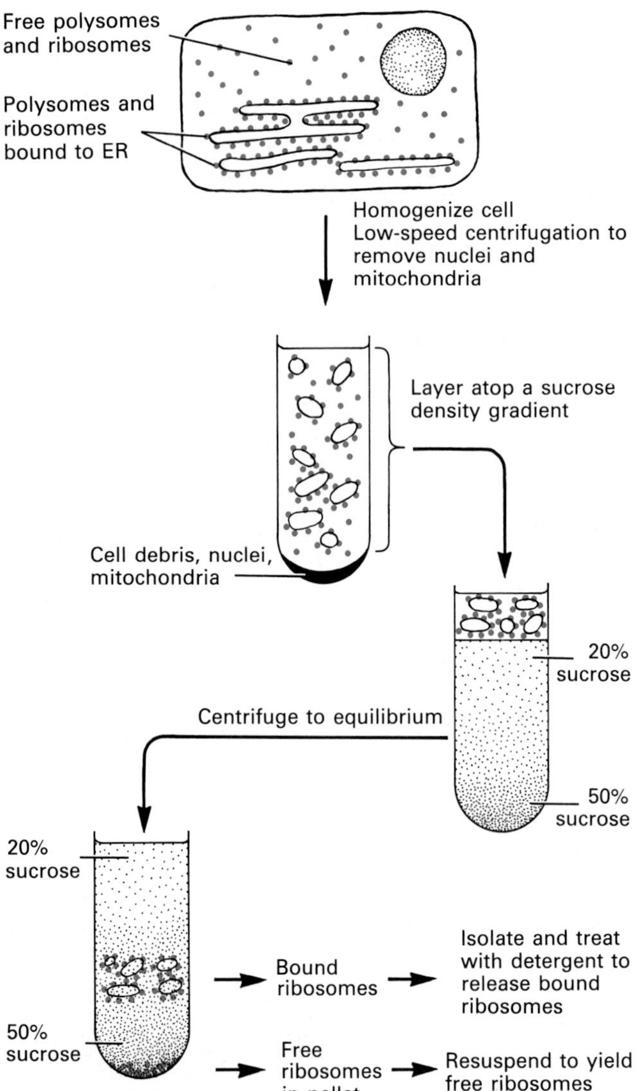

Figure 21-12 Purification of membrane-bound and membrane-free ribosomes and polysomes. A cell homogenate, freed of nuclei and mitochondria by low-speed centrifugation, is layered on a sucrose density gradient. After centrifugation, unbound ribosomes and polysomes pellet to the bottom of the tube, because they are more dense than the 50 percent sucrose solution. Membranes with bound ribosomes form a band within the gradient, owing to the low buoyant density of the phospholipids. Membrane-bound ribosomes can be freed of membranes by treatment with nonionic detergents.

The vast majority of both mitochondrial and chloroplast proteins are encoded by the nuclear DNA, not the organelle DNA (Tables 21-2 and 21-3). It was a surprise to many investigators to find that all these proteins, including some extremely hydrophobic ones, are synthesized on cytoplasmic ribosomes unbound to membranes. These proteins are released into the cytosol and are incorporated into the organelle posttranslationally. Likewise, proteins found in the lumen of the peroxisome or glyoxisome are synthesized in the cytosol by unbound ribosomes and cross into the organelle posttranslationally. Each of these organelles contains an uptake mechanism that recognizes only the appropriate proteins. The targeting of proteins that are posttranslationally incorporated into organelles is also determined by sequences of amino acids in the protein itself. Often these signal sequences are removed by specific proteolysis once the protein has reached its final destination.

Now let us consider in more detail some special problems of organelle assembly and protein targeting. The following sections describe the assembly and replication of mitochondria, a phenomenon that has fascinated cell biologists for decades. Some of the major results on chloroplast and peroxisome biogenesis are also presented, although less is known of these systems.

Synthesis and Assembly of Mitochondria

Individual mitochondria are large enough to be seen with light microscopy and can be followed by time-lapse cinematography. They increase in size, and then one or more daughter mitochondria "pinch off," in a manner reminiscent of the way bacterial cells grow and divide. The growth and division of mitochondria is not coupled to nuclear division.

Table 21-2 An inventory of genes in yeast mitochondrial DNA*

Mitochondrial component	Mitochondrial gene product
Large ribosomal subunit (21S rRNA, 38 proteins)	21S rRNA
Small ribosomal subunit (15S rRNA, 33 proteins)	15S rRNA var-1, a ribosome-associated protein
tRNAs (~30)	All of them
Cytochrome *c* oxidase (9 subunits)	Cytochrome *c* oxidase subunits I, II, III
Cytochrome b/c_1 complex (7 subunits)	Apocytochrome *b*
ATPase complex (10 subunits)	Subunit 9†, subunit 8
	Subunit 6

*Several unidentified reading frames (URFs) have been detected by DNA sequencing, but since proteins corresponding to these URFs have not yet been found, they are not included here.

†Encoded in nuclear DNA and made in the cytoplasm in *Neurospora, Aspergillus,* and humans.

SOURCE: Adapted from P. Borst, in *International Cell Biology 1980–1981,* H. G. Schweiger, ed. Springer-Verlag.

Cytoplasmic Inheritance and DNA Sequencing Established the Existence of Mitochondrial Genes

Prior to the isolation and sequencing of mitochondrial DNA (mtDNA), studies of mutants in yeast and other single-celled organisms indicated that mitochondria contain their own genetic system and exhibit cytoplasmic inheritance. "Petite" mutants in yeast are incapable of oxidative phosphorylation. They can grow by fermentation of glucose to ethanol, but the size of the yeast colonies is generally smaller (hence the term petite) since less

Table 21-3 Some mitochondrial proteins synthesized in the cytoplasm

Intramitochondrial location	Polypeptide*
Matrix	F_1-ATPase α subunit β subunit γ subunit
	Carbamoyl phosphate synthetase
	Mn^{2+}-superoxide dismutase
	RNA polymerase
	Ribosomal proteins
	Citrate synthetase and other citric acid cycle enzymes
	Ornithine transcarbamoylase
Inner membrane	Cytochrome c_1
	Subunit V of cytochrome b/c_1 complex
	ADP-ATP carrier
	Cytochrome *c* oxidase Subunit IV Subunit V Subunit VI Subunit VII
	Proteolipid of F_0-ATPase complex
Intermembrane space	Cytochrome *c*
	Cytochrome *c* peroxidase
	Cytochrome b_2
Outer membrane	Porin

*Most proteins except the ADP-ATP carrier, cytochrome *c*, and porin are fabricated as longer precursors.

ATP can be made per mole of glucose fermented. In genetic crosses between different (haploid) yeast strains, the petite mutation does not segregate with any known nuclear gene or chromosome (Figure 21-13). Because the inheritance of the petite or wild-type mitochondrial phenotype is clearly nonchromosomal, some mutable element in the *cytoplasm* must be a determinant in synthesis of a mitochondrion. (Studies much later showed that petites contain deletions of some or all of the mtDNA.)

Mitochondrial inheritance in yeast is biparental: during fusion of haploid cells, parents contribute equally to the cytoplasm of the diploid. In mammals and most other animals, the sperm contributes little if any cytoplasm to the zygote, and most if not all of the mitochondria in the embryo must be derived from those in the egg, not the sperm. Indeed, different strains of cows and rats have been found to have mtDNAs that differ slightly in DNA sequence; when two such strains are mated, the mtDNA in the offspring is invariably the maternal type.

The entire mitochondrial genome from a number of different organisms has now been cloned and sequenced. It is clear that mitochondria encode ribosomal RNAs that form mitochondrial ribosomes, although all but one or two of the ribosomal proteins are imported from the cytosol. Mitochondrial DNA encodes all of the tRNAs used for protein synthesis in the mitochondrion, and all mitochondria encode a unique class of proteins that are synthesized on mitochondrial ribosomes. Mitochondrial ribosomes differ from those in the cytoplasm not only in their RNA and protein composition but also in their sensitivity to certain antibiotics. For instance, cycloheximide inhibits protein synthesis by all eukaryotic cytoplasmic ribosomes but does not affect protein synthesis by mitochondrial ribosomes. Characterization of proteins synthesized in the presence of cycloheximide has shown that all proteins synthesized on mitochondrial ribosomes are encoded by mitochondrial DNA; most proteins localized in mitochondria are synthesized on cytosolic ribosomes and must be imported into mitochondria. All mitochondrially made polypeptides identified so far (with one possible exception) are not complete enzymes but instead are subunits of multimeric complexes. All but one of these are hydrophobic subunits of enzymes used in oxidative phosphorylation (see Table 21-2).

The Size, Coding Capacity, and Transcription of Mitochondrial DNA Vary in Different Organisms

Mitochondrial DNA is located in the matrix compartment and is sometimes found attached to the inner mitochondrial membrane. Human mtDNA is a closed circular molecule of 16,569 base pairs and has been completely sequenced (Figure 21-14). Invertebrate mtDNA is about the same size, but quite unexpectedly, yeast mtDNA contains about 78,000 base pairs—about five times the number in humans. Although very similar proteins are encoded by yeast and human mtDNA, the arrangement of the genes, the mode of transcription, and even the genetic code itself are different for the two types of mitochondria (Figure 21-14).

Human mtDNA encodes two rRNAs that are found in mitochondrial ribosomes and also 22 tRNAs that are used to translate the mitochondrial mRNAs. It has 13 sequences that begin with methionine codon, end with a chain termination codon, and are long enough to encode a polypeptide of more than 50 amino acids. Of these possible proteins, 11 have been identified: 3 subunits of cytochrome *c* oxidase, 1 or 2 of the F_1-ATPase, 7 of NADH–CoQ reductase, and cytochrome *b*. These are the major proteins known to be synthesized on mitochondrial ribosomes. The two other coding sequences are termed *URFs*—*u*nidentified (open) *r*eading *f*rames.

The Mitochondrial Genetic Code in Humans Is Unique

Remarkably, the genetic code used in human mitochondria has some differences from those of all prokaryotic and eukaryotic nuclear genes (and also from fungal mtDNA) (Table 21-4). The codon assignments are unambiguous: they are derived from a direct comparison of the sequence of mtDNA with the sequences of proteins known to be encoded by mtDNA, such as cytochrome *b* and the three subunits of cytochrome *c* oxidase. UGA, normally a stop codon, is read by the mitochondrial translation system as tryptophan, whereas AGA and AGG, normally arginine codons, are stop signals in human mitochondrial mRNAs. This similarity but nonuniversality of the genetic code has profound implications for the evolution of eukaryotic cells and their organelles. We shall take up this point in Chapter 25.

Human Mitochondrial RNAs Undergo Extensive Processing

So far as is known, all transcripts of mtDNA and their translation products remain within the organelles. There is no export of RNAs or proteins. Hybridization studies show that all RNAs found in the mitochondrion are synthesized in the mitochondrion on templates of mtDNA, so there is no import of RNA from the cytoplasm.

Like their nuclear counterparts, mitochondrial rRNAs and mRNAs are generated by enzymatic cleavage of longer precursors (Figure 21-15). However, processing of human mtDNA transcripts reveals a number of novel features. First, initiation of transcription occurs at only two points on one of the two strands, the H strand (so called because it bands at a *h*eavier density in CsCl). Transcript I initiates just upstream of the $tRNA^{phe}$ gene and terminates just after the gene for 16S rRNA; as depicted in Figure 21-15, it is cleaved to $tRNA^{phe}$, $tRNA^{val}$, and 12S

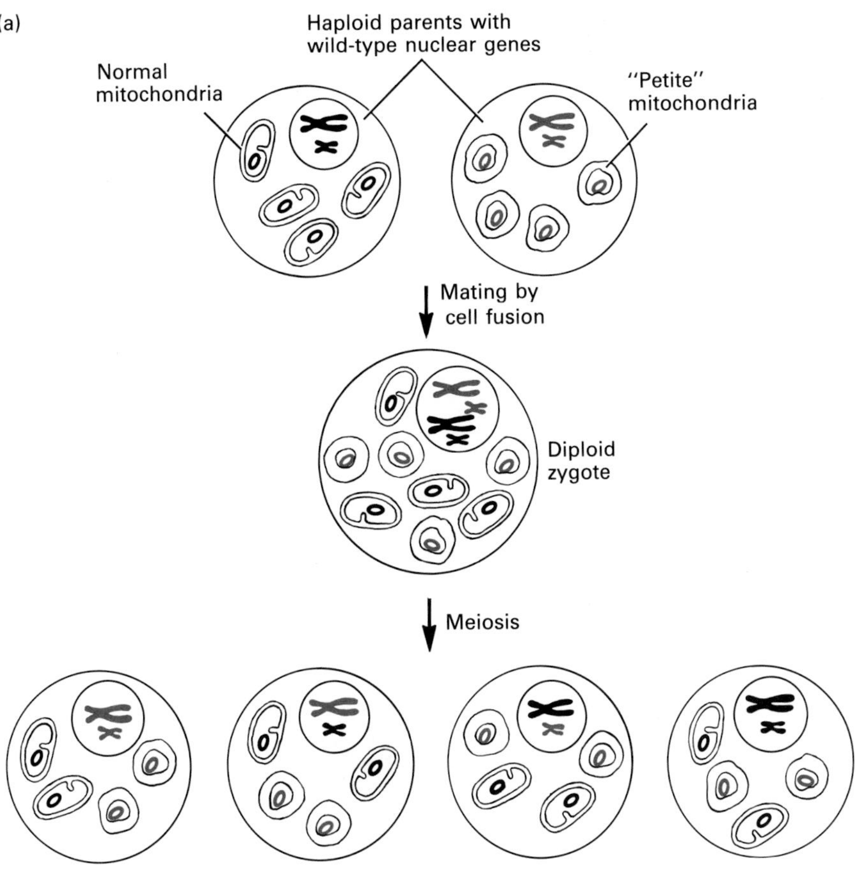

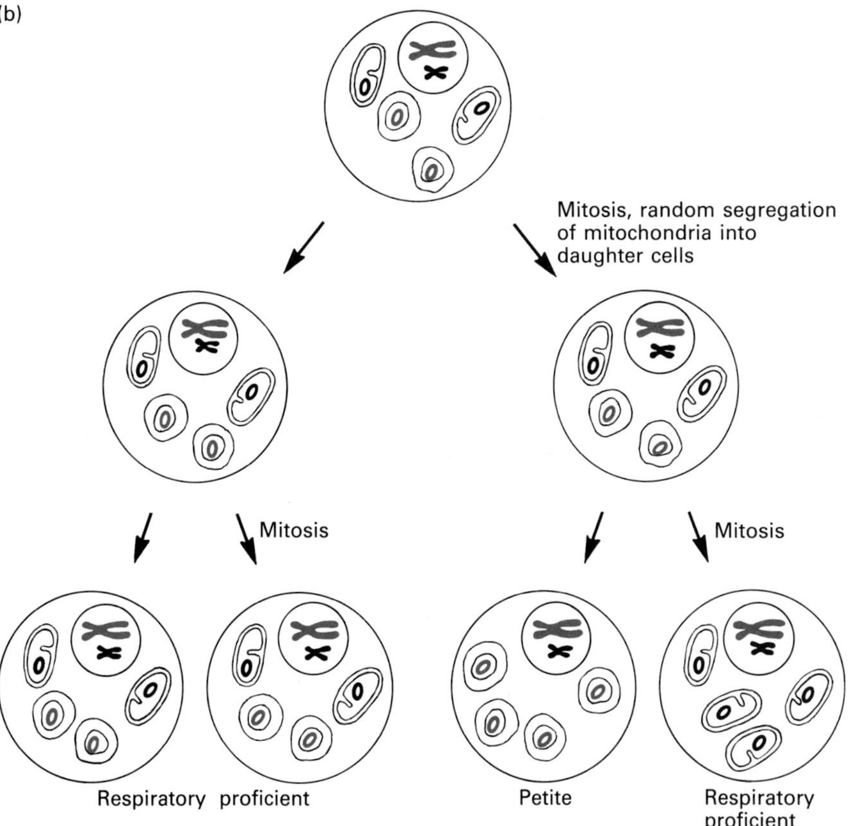

Figure 21-13 Cytoplasmic inheritance of the petite mutation in yeast. Petite-strain mitochondria are defective in oxidative phosphorylation owing to a deletion in mtDNA. (a) Haploid cells fuse to produce a diploid cell (Chapter 6) that undergoes meiosis. During meiosis there is random segregation of parental chromosomes and mitochondria containing mitochondrial DNA (mtDNA). Since yeast normally contain about 50 mtDNA molecules per cell, all products of meiosis usually contain both normal and petite mtDNAs, and are all respiratory proficient. (b) As these cells grow and divide mitotically, the cytoplasm, including mitochondria, is randomly distributed to the daughter cells. Occasionally, a cell is generated that contains only defective (petite) mtRNA and that yields a petite colony. Thus formation of this petite cell is independent of any nuclear genetic marker but is associated with an element in the cytoplasm now known to be mtDNA.

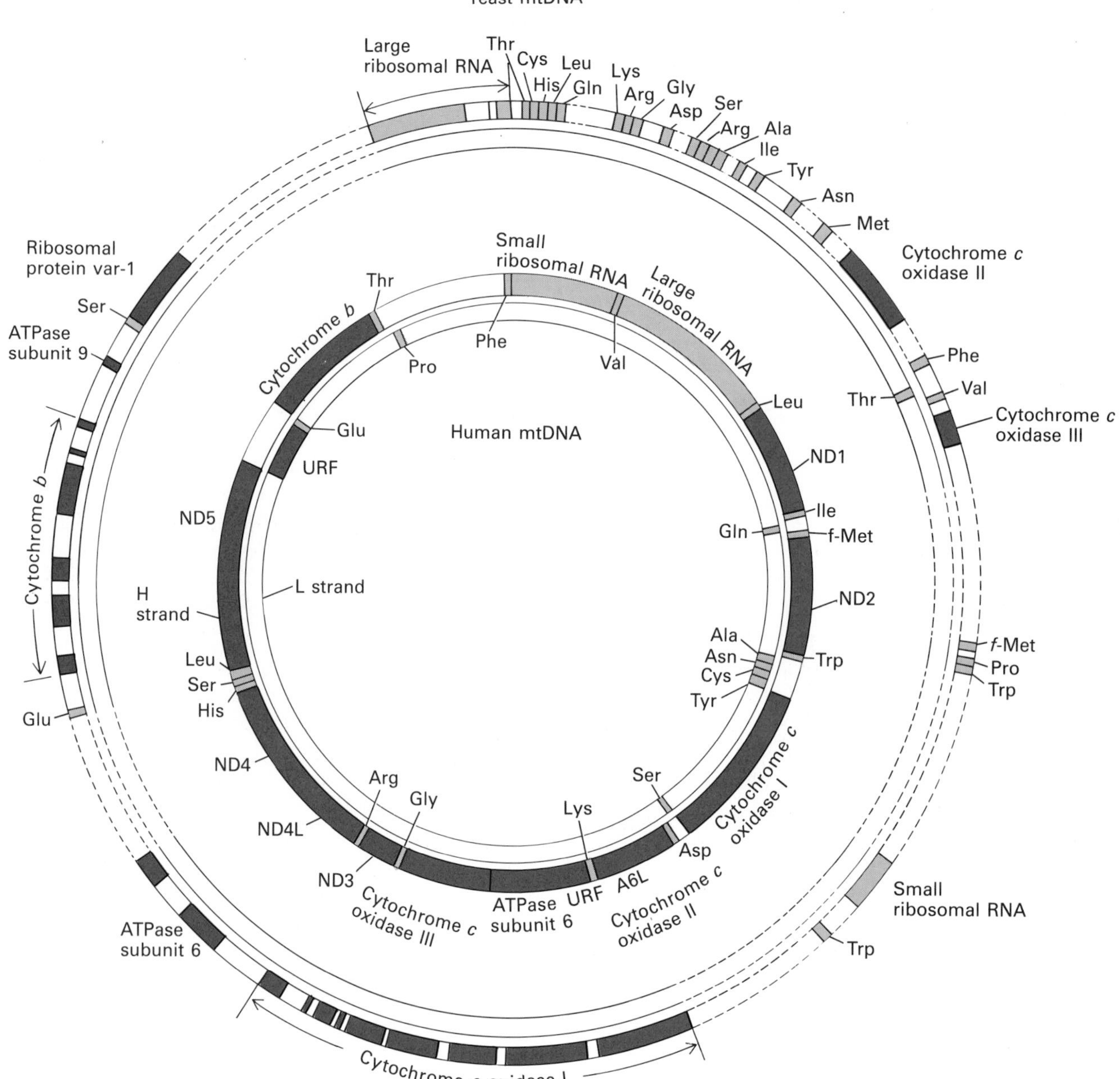

Figure 21-14 The organization of the human (*inner*) and yeast (*outer*) mtDNA. Although yeast mtDNA is five times larger than human mtDNA, here it has been drawn only two and one-half times as large. Proteins and RNAs encoded by each of the two strands are shown separately; transcription off the outer (or H) strand of each mtDNA is clockwise; off the inner (or L) strand, it is counterclockwise.

The entire human mtDNA has been sequenced. For yeast, the diagram is based on partial DNA sequence data from several laboratories. The inventory of genes is thus a provisional one, with dashed lines representing unsequenced regions. Exons of structural genes and URFs (unidentified or open reading frames) are in color; rRNA and tRNA genes are gray. The abbreviations for amino acids denote the tRNA genes. Yeast genes for cytochrome *c* oxidase subunit I, cytochrome *b*, and 21S rRNA contain introns; these are denoted by open boxes. No mammalian mtDNA genes contain introns.

Note that yeast and mammalian mtDNAs encode some different proteins. Human mtDNA encodes seven subunits of NADH–CoQ reductase—labeled ND1, ND2, ND3, ND4, ND4L, ND5, and URF—that have no counterpart in yeast mtDNA; also, F_1-ATPase subunit 9 is encoded by *mitochondrial* DNA in yeast but by *nuclear* DNA in humans. Most of the yeast mtDNA does not encode proteins, rRNA, or tRNA and is not transcribed into RNA. These nontranscribed regions contain A and T almost exclusively, and their function, if any, is not known. [See P. Borst and L. A. Grivell, 1981, *Nature* **290**:443; L. A. Grivell, 1983, *Sci. Am.* **248**(3):78–89; and A. Chomyn et al., 1985, *Nature* 314:592.] *After L. A. Grivell.*

Table 21-4 Unusual codon use in mitochondria

Codon	Normal assignment in eukaryotic and prokaryotic cells	Assignment in mitochondria		
		Humans	Yeast	Neurospora
UGA	Stop	Trp	Trp	Trp
CUU CUC CUA CUG	Leu	Leu	Thr	Leu
AUA	Ile	Met	Met	Ile
AGA AGG	Arg	Stop	Arg	Arg

SOURCES: S. Anderson et al., 1981, *Nature* **290**:457; P. Borst, in *International Cell Biology 1980–1981,* H. G. Schweiger, ed. Springer-Verlag, p. 239; and C. Breitenberger and U. L. Rajbhandary, 1985, *Trends in Biochem. Sci.* **10**:478–483.

and 16S rRNA. The second transcript, II, initiates just downstream of the first, near the 5′ end of the 12S rRNA gene. But it continues past the termination site of the first transcript, around the circular DNA to the start point. It is processed by enzymatic hydrolysis to yield the remaining tRNAs and all the polyadenylated mRNAs (but apparently not any rRNAs). Very often some of these cleavages occur on nascent chains. The sites of cleavage must be very precise to separate the mRNAs from tRNAs, since the 5′ end of each mRNA is immediately adjacent to the 3′ end of a tRNA. Transcript I is initiated about 10 times as frequently as is transcript II, accounting for the synthesis of rRNAs (and curiously, of two tRNAs) in excess of mRNAs and other tRNAs.

In contrast to cytoplasmic mRNAs, human mitochondrial mRNAs contain very few untranslated sequences. The first three bases at the 5′ end are generally the AUG (or AUA) initiator codon. A UAA terminator codon is generally at or very near the 3′ end of the mRNA. In some cases, only the U of this terminator codon is encoded in the mtDNA. The final As are part of the polyadenate sequence, which, as in cytoplasmic mRNAs, is added posttranslationally (Chapter 9).

Thus human and other mammalian mtDNAs have evolved to contain as few nonused sequences as possible. The genome is the absolute minimum required to generate the requisite mt mRNAs, tRNAs, and rRNAs, and there is extensive processing of all transcripts.

During Evolution Genes Have Moved from Mitochondrial DNA to Nuclear DNA

Yeast mtDNA, and mtDNA from other lower eukaryotes, are five times larger than mammalian mtDNAs; yet they encode essentially the same gene products: three subunits of cytochrome *c* oxidase, cytochrome *b* at least one subunit of the F_1-ATPase, one ribosomal protein (termed var-1), a 15*S* and a 21S rRNA, and multiple tRNAs. Remarkably, one mitochondrial protein, subunit 9 of the F_1-ATPase, is encoded by *mitochondrial* DNA in certain organisms such as yeast but is encoded by *nuclear* DNA in others such as *Neurospora crassa* (another fungus) and mammals. Seven subunits of NADH–CoQ reductase are encoded in mammalian mtDNA, but no homologous sequences are found in yeast mtDNA. These striking findings suggest that genes moved, during evolution, from the mitochondrion to the nucleus, or vice versa.

Translocations of short segments of mtDNA to nuclear DNA may be occurring still; DNA hybridization and sequence studies have identified short segments (about 50 base pairs long) of mtDNA interspersed randomly in the nuclear DNA of all animals and plants studied; these short segments do not appear to encode any proteins.

Yeast Mitochondrial RNAs Are Transcribed from Multiple Promoters and Spliced

In yeast mtDNA, the genes are separated by long stretches of AT-rich noncoding sequences, and at least three of the genes contain intervening sequences—those encoding cytochrome *b* and the subunits I and II of cytochrome *c* oxidase. Unlike human mtDNA, each gene in yeast mtDNA appears to use a different promoter.

In one strain of yeast ("long gene"), the gene for cytochrome *b* is split into six exons (Figure 21-16a). In another strain ("short gene"), the first three introns are missing, and the first four exons are contiguous in the DNA. Both genes encode the same protein, and in both strains the mRNA precursor for cytochrome *b* is spliced to remove the transcripts of the intervening DNA sequences. But the manner of splicing of the "long gene" transcript into cytochome *b* mRNA is apparently unique to fungal DNA. In the "long gene" strain, the entire cytochrome *b* gene is transcribed into a precursor RNA that contains tran-

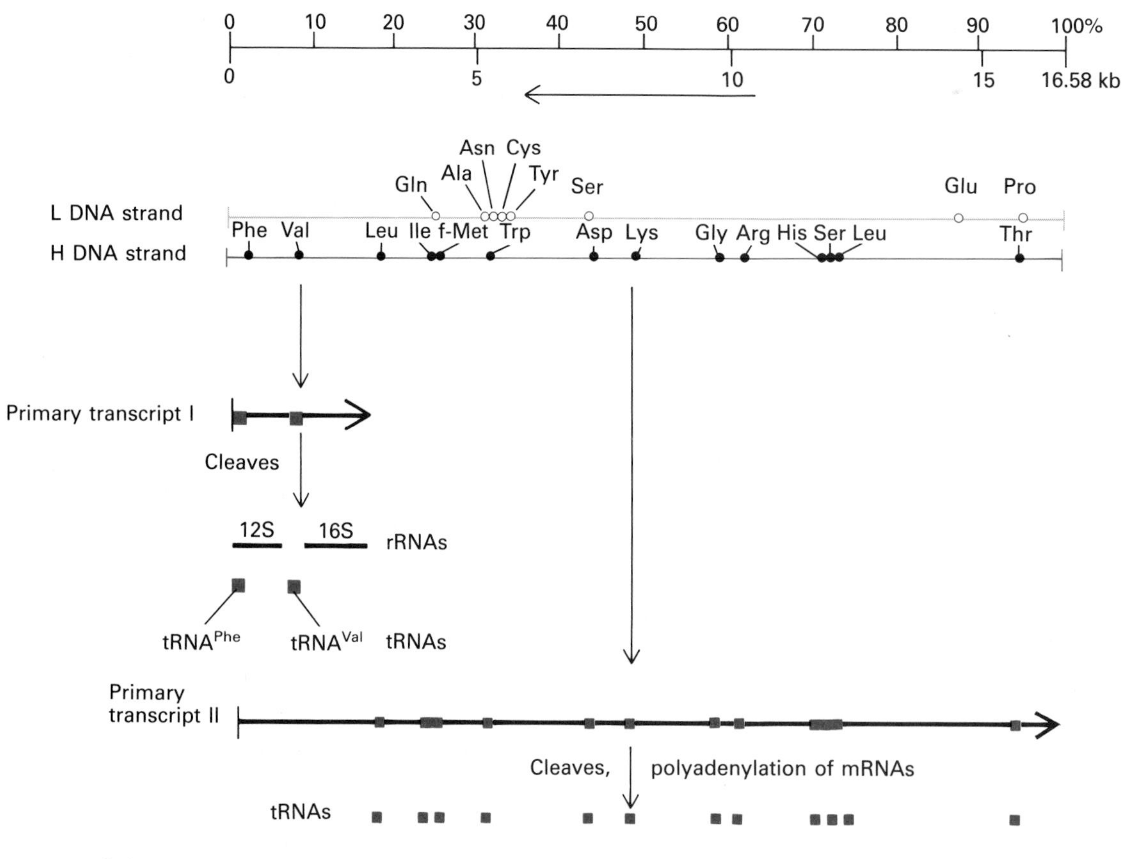

Figure 21-15 Transcription map of human mtDNA. As deduced from the DNA sequence and from RNA-DNA hybridization studies, the light (L) DNA strand encodes only eight tRNAs (*open circles, amino acid abbreviations*). This strand is transcribed right to left. The heavy (H) DNA strand encodes the 12S and 16S rRNAs, 14 tRNAs (*closed circles*), and 11 predominant species of polyadenylated RNAs. These mRNAs encode known mitochondrial proteins: cytochrome *c* oxidase subunits I, II, and III, F_1-ATPase subunit 6, seven subunits of NADH–CoQ reductase, and cytochrome *b*. There are two sites for initiation of transcription of the heavy strand. Transcript I initiates just upstream of the $tRNA^{Phe}$ gene and terminates just after the 16S rRNA; it is processed by endoribonucleases to yield one molecule each of $tRNA^{Phe}$, $tRNA^{Val}$, and 12S and 16S rRNA. The second transcript (II) initiates near the 5′ end of the 12S rRNA, and apparently continues to the other end of the mtDNA; it is processed to yield the other tRNAs and mRNAs, which are subsequently polyadenylated (A_n). Some of the cleavages of this transcript begin while the chain is nascent. [See D. Ojala et al., 1981, *Nature* **290**:470; and J. Montoya, G. L. Gaines, and G. Attardi, 1983, *Cell* **34**:151.]

scripts of all six exons. The first splice fuses exons E_1 and E_2, removing a 1000-base circular RNA that corresponds to intron 1 (Figure 21-16b). This splice is self-catalyzed by RNA, but is facilitated by one or more nucleus-encoded proteins; certain nuclear mutations result in inhibition of this splicing reaction.

Within intron 2 is an open reading frame (a URF) that is continuous with the preceding exon, E_2. The effect of removing intron 1 is to create a continuous open reading frame, beginning with the 143 N-terminal amino acids of exons E_1 and E_2 of cytochrome *b* and containing at its C-terminus about 250 amino acids encoded by intron 2. The intermediate RNA is translated by mitochondrial ribosomes to yield a chimeric protein. This type of protein, called a *maturase,* is required for the second splice, although whether it is the actual splicing enzyme is not known. This second splice—the one that requires the maturase—removes all of intron 2, including the URF that encodes the maturase part of the chimeric protein. Thus, the maturase destroys, by splicing, its own mRNA—a most novel way of regulating the extent of translation of an mRNA! This intron-encoded maturase was first identified by genetic studies. The key was the discovery of mutants in intron 2 (a supposedly noncoding segment of the cytochrome *b* gene) that abolished production of the mature cytochrome *b* polypeptide.

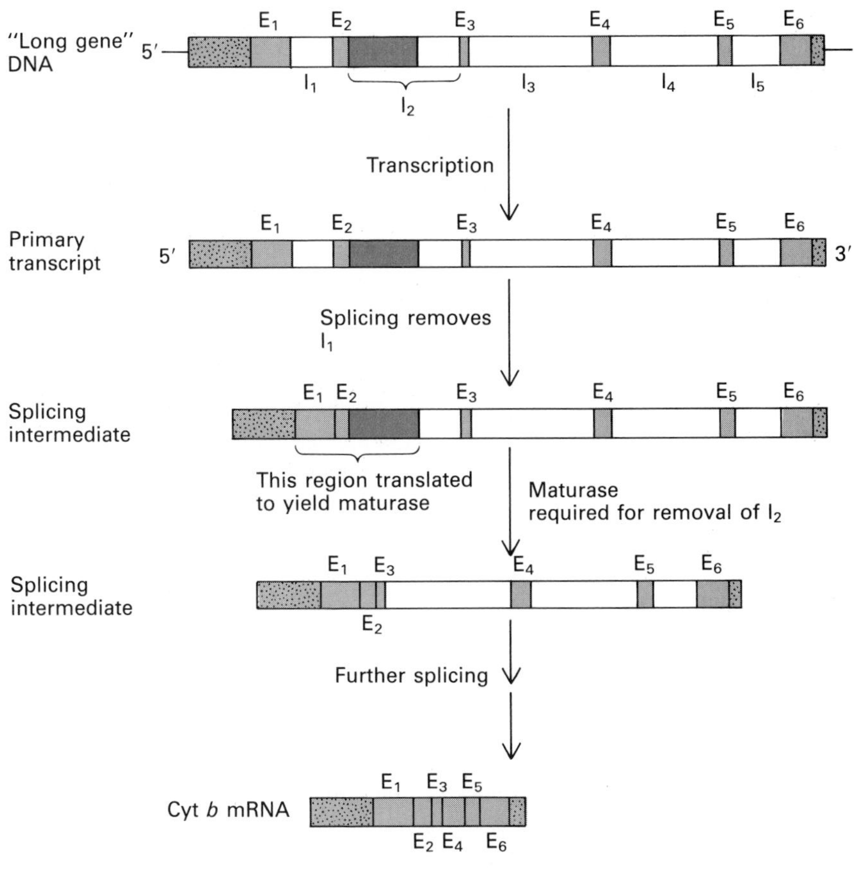

Figure 21-16 Cytochrome *b* gene of yeast. (a) The mitochondrial DNA of "long gene" yeast strains contains six exons for cytochrome *b*. The DNA from the closely related "short gene" yeast strains contains only three exons. Two of these correspond to exons 5 and 6 of the "long gene" strain; the other represents a fusion of exons 1, 2, 3, and 4 with precise deletion of the introns present in the long strain. (b) Production of the cytochrome *b* mRNA in the "long gene" strain involves several novel splicing stages. From the initial complete RNA transcript, intron 1 is removed by splicing; a protein facilitating this reaction is encoded by nuclear DNA. The resulting RNA is translated into a protein that has, at its N-terminus, sequences encoded by exons 1 and 2 of cytochrome *b*. At its C-terminus is a sequence encoded by intron 2 of cytochrome *b* (*color*). This chimeric protein has a splicing or "maturase" activity that is essential for removal of intron 2. Three other splicing reactions generate mature cytochrome *b* mRNA. Most splicing reactions of mt RNAs are self-catalyzed, similar to the self-splicing of *Tetrahymena* rRNA (see Figure 9-61). Mitochondrial RNA splicing is in some way facilitated by protein. [See T. R. Cech, 1986, *Cell* **44**:207–210.]

Any advantages of such a baroque process of mRNA splicing are unknown and mysterious—especially since the "short gene" for cytochrome *b* lacks the first three introns, including the segment that encodes the maturase. It is remarkable that, in the "long gene" for cytochrome *b*, the intron of one gene (for cytochrome *b*) is the exon of another (for the maturase).

Most Mitochondrial Proteins Are Synthesized in the Cytosol as Precursors

Most of the proteins required for oxidative phosphorylation and other mitochondrial processes are synthesized on cytoplasmic ribosomes and are imported into the mitochondrion. Even the mitochondrial DNA and RNA polymerases are synthesized in the cytosol, as are all but one of the mitochondrial ribosomal proteins. How proteins made on cytoplasmic ribosomes are transported into mitochondria has been a baffling problem. Some cytoplasmic proteins are transported to the intermembrane space (the *c* and b_2 cytochromes and cytochrome peroxidase). Some are transported to the inner membrane (cytochrome *c* oxidase subunits), and some to the matrix space [F_1-ATPase subunits, ribosomal proteins, RNA polymerase, and so on (see Table 21-3)]. How is a protein targeted to its final destination, and what drives these specific and unidirectional transport processes?

Current research suggests that cytoplasmically synthesized mitochondrial proteins are imported after their synthesis. Pulse-chase studies on yeast and *Neurospora* cells demonstrated that newly made mitochondrial proteins accumulate in the cytosol, outside the mitochondria. During the chase period, they are seen to accumulate gradually at their proper destination in the mitochondrion.

Furthermore, most cytoplasmically synthesized mitochondrial proteins of these organisms have been found to be fabricated as precursors containing an additional 20 to 60 amino acids at the N-terminus that are not found in the mature protein (Figure 21-17). The proteins include those destined for the matrix or intermembrane space or the inner mitochondrial membrane. Such precursor proteins, including those of integral membrane proteins, are soluble in the cytosol. During uptake into the mitochondrion the precursors of integral membrane proteins must undergo a major conformational change to bind to the hydrophobic core of the phospholipid membrane. One model suggests that the precursors fold so that hydrophobic amino acid side chains are on the inside and hydrophilic ones on the surface; during uptake into the mitochondrion the protein turns inside-out, placing hydrophobic residues on the outside in order to bind to the membrane.

When radiolabeled precursor proteins—either synthesized by growing cells or by cell-free protein synthesis reactions—are incubated with yeast mitochondria, the N-terminal signal sequence is removed (Figures 21-17 and 21-18). When the location of the mature protein within the mitochondrion has been determined, it is invariably in its proper place. The enzyme that removes these N-terminal signal sequences has been purified. It is a metalloprotease localized to the matrix space of the mitochondrion. During in vitro reactions, it cleaves the N-terminal extensions from several different precursor proteins, such as the pre-β subunit of F_1-ATPase, to form the mature protein.

The processing of at least two mitochondrial proteins—the b_2 and c_1 cytochromes, localized to the intermembrane space and the outer face of the inner membrane, respectively—is more complex. In the first step, a portion

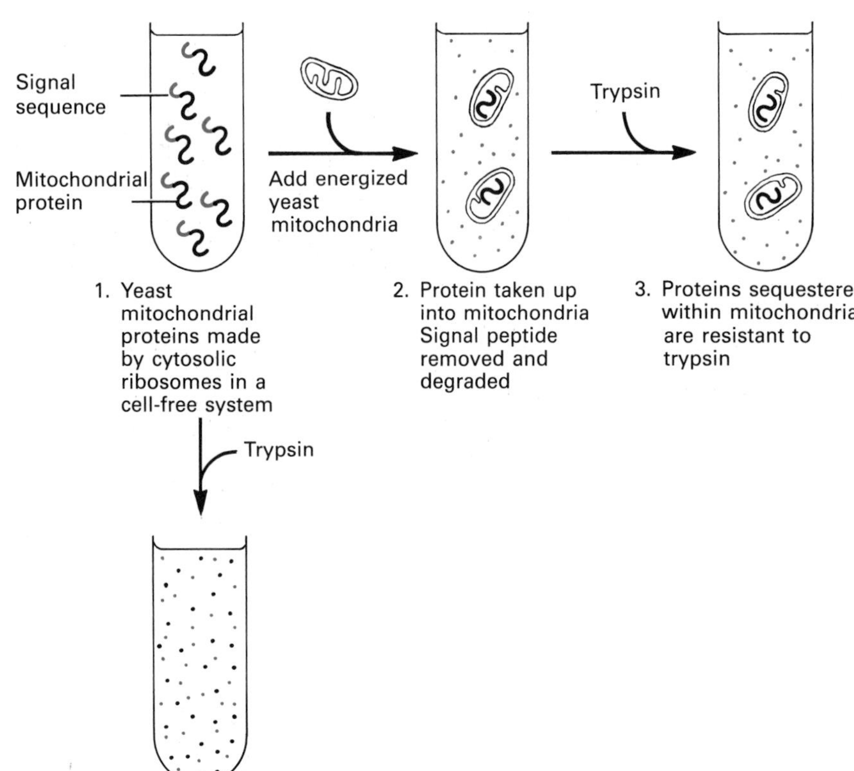
Signal sequence
Mitochondrial protein
Add energized yeast mitochondria
Trypsin
1. Yeast mitochondrial proteins made by cytosolic ribosomes in a cell-free system
2. Protein taken up into mitochondria Signal peptide removed and degraded
3. Proteins sequestered within mitochondria are resistant to trypsin
Trypsin
4. Signal sequence and mitochondrial protein degraded

Figure 21-17 Demonstration that mitochondrial proteins are imported into the organelle posttranslationally, concomitant with cleavage of the N-terminal signal sequence. Most mitochondrial proteins synthesized on cytosolic ribosomes have at their N-termini a signal sequence that is not found on the mature protein. Such precursors with signal sequences can be identified in cells following a brief pulse of radioactive amino acid or can be synthesized in a cell-free system programmed with cytosolic mRNA (*step 1*). When energized mitochondria are added, the signal sequence is cleaved from the precursor, and the protein is taken up into the organelle. Apparently the signal peptide is degraded (*step 2*). This can be demonstrated by adding trypsin or another protease to the reaction. Proteins sequestered in the mitochondrion are resistant to the protease (*step 3*) because the protease cannot penetrate the mitochondrial membranes. By contrast, the precursor protein in the cytosol is totally destroyed by protease (*step 4*).

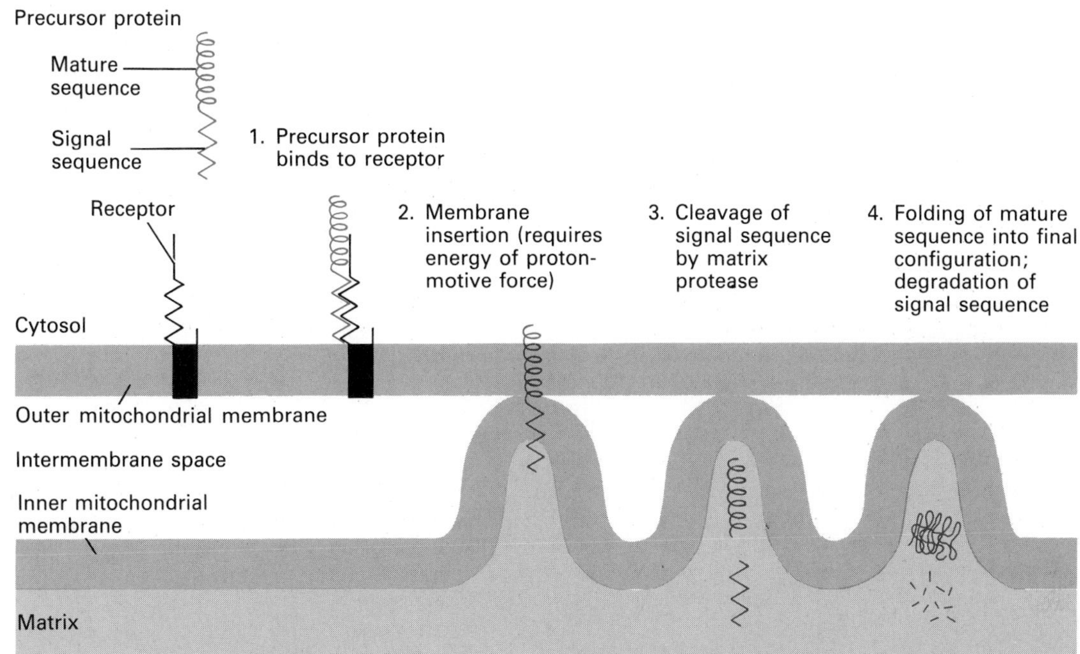

Figure 21-18 The import of a polypeptide into the matrix space of a mitochondrion. *Step 1.* The precursor protein, with its N-terminal signal peptide, is generated in the cytosol and binds first to a receptor on the outer membrane. A soluble cytosolic protein may bind to the precursor and facilitate its binding to the receptor. *Step 2.* The protein is then translocated across both the outer and inner membranes in a process that requires an electric potential (proton-motive force) across the inner membrane. Possibly, as depicted here, this occurs at rare sites at which the inner and outer membranes touch. *Step 3.* The N-terminal signal, having served its function, is removed by a matrix protease and is ultimately degraded. *Step 4.* Either concomitant with step 3 or just after, the mature protein folds into its final, active configuration. [See G. Schatz and R. A. Butow, 1983, *Cell* 32:316.]

of the protein penetrates to the matrix, where part of the N-terminal signal sequence is removed by the metalloprotease to form an intermediate (Figure 21-19). In a second step, these intermediates are converted into the mature form by removal of the rest of the signal sequence and the addition of heme, presumably concomitant with final localization. Thus import of these proteins involves a "detour" across the inner membrane.

Certain cytoplasmically synthesized mitochondrial proteins—namely, cytochrome *c* and the ADP-ATP transporter—are not synthesized as longer precursors. The cytoplasmic form of cytochrome *c* lacks a heme group and is called *apocytochrome c*. Addition of the heme group occurs concomitant with uptake. Only the apoform of cytochrome *c* is taken up by mitochondria; the mature cytochrome *c* is not. In all likelihood, addition of heme causes a conformational change in the protein so that it cannot "escape" from the mitochondrion.

Uptake of Mitochondrial Proteins Is Specific and Requires Energy

The findings described previously suggest that the N-terminal signal sequences or some other (unknown) feature of the cytosolic precursors of mitochondrial proteins is involved in directing these proteins to their final destination in the mitochondrion. But two key questions remain: How is this recognition and direction of these proteins achieved, and how are the proteins actually imported into the organelle?

To begin with, expenditure of energy is required for uptake of proteins, and high levels of ATP within the mitochondrion are essential. If mitochondria are "poisoned" with inhibitors or uncouplers of oxidative phosphorylation, such as cyanide or dinitrophenol, the level of ATP drops precipitously. Neither in the intact cell nor in cell-free reactions can such poisoned mitochondria take up added precursor proteins. The process by which the presumed hydrolysis of ATP is coupled to the transport of proteins into the mitochondrion is not known, but an electric potential, or proton-motive force, across the inner membrane presumably plays a key role. When a protein inserts part-way into the inner membrane, it becomes subjected to a transmembrane potential of 200 mV, equivalent to an electrical gradient of about 4×10^5 V/cm. Possibly, the conformation of these proteins is then altered by the electric potential, in much the same way that the conformation of voltage-dependent ion channels are

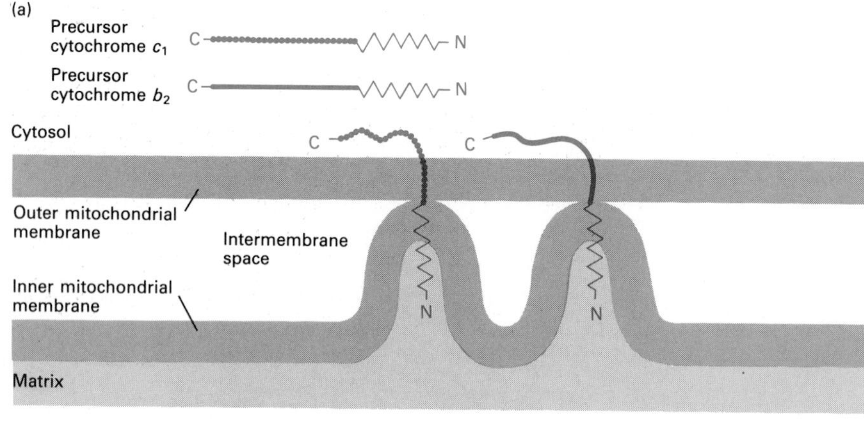

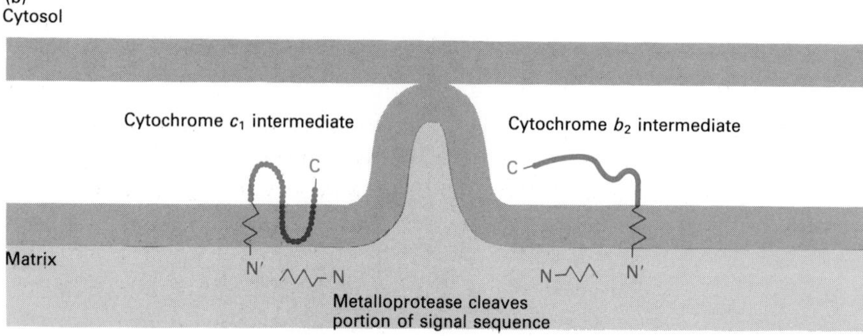

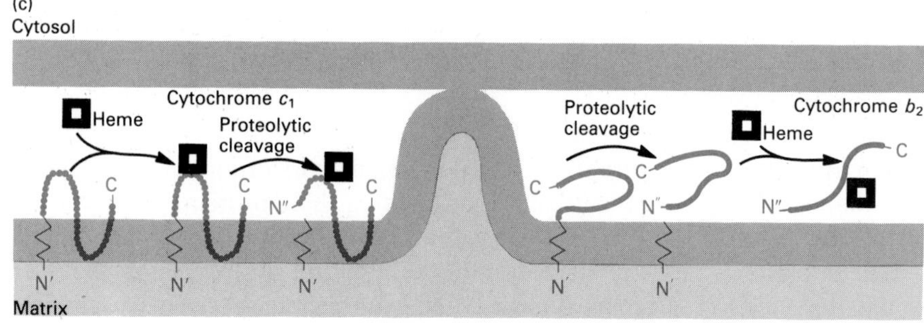

Figure 21-19 Schematic outline of membrane insertions and proteolytic cleavages during the import of two proteins into the mitochondrion. (a) The inner membrane protein cytochrome c_1 is fabricated as a precursor with a long N-terminal signal sequence. (b) Part of this sequence is cleaved by a metalloprotease in the matrix space as the polypeptide inserts into the inner membrane. (c) Subsequently heme is added, and a second proteolytic cleavage removes the rest of the signal sequence and yields cytochrome c_1 with the mature N-terminus, N″, in its final orientation in the membrane.

The processing of cytochrome b_2, finally localized in the intermembrane space, is also complex. The precursor of cytochrome b_2 contains an N-terminal signal peptide, part of which is removed by the matrix metalloprotease. Thus at least part of the N-terminal signal sequence must be exposed to the matrix, as indicated in (a). This cleavage results in an intermediate that is inserted into the inner membrane, with N-terminal N′, as indicated in (b). Subsequently, another proteolytic cleavage at the N-terminus releases mature cytochrome b_2 into the intermembrane space, wherein the heme is added, as in (c). [See S. M. Gasser et al., 1982, *Proc. Nat'l Acad. Sci. USA* 79:267.]

affected by the membrane potential (Chapter 17). Such changes in protein folding then pull the protein across the energized inner membrane.

Since purified (processed) mitochondrial proteins are not transported inside when added to mitochondria, the cleaved peptide appears to comprise a signal sequence for uptake. Supporting this function are results of recent experiments involving chimeric proteins, formed by gene fusions, that contain the cleaved signal sequence of an imported mitochondrial protein (such as cytochrome *c* oxidase) attached to a cytosolic protein. The fusion protein is imported into the matrix of isolated, energized mitochondria, and the signal sequence is cleaved. Thus the cytoplasmic protein is delivered to the matrix. Cytoplasmic proteins without such attached signal sequences are, of course, not imported into mitochondria or into any other organelle.

The outer mitochondrial membrane may contain one or more receptors for imported proteins. Mitochondria poisoned with inhibitors of energy generation still bind specifically and tightly the precursor forms of imported proteins. However, the bound proteins are not translocated into the organelle; they remain on the outer membrane (*step 1,* Figure 21-18). If the energy block is removed, these bound precursors are rapidly imported and processed.

As noted previously (Chapter 20), the outer mitochondrial membrane contains an abundant protein, porin, that apparently forms a channel through the bilayer and accounts for the unusual permeability of this membrane to small proteins. This protein is also synthesized in the cytoplasm. Its insertion into the outer mitochondrial membrane does not involve any proteolytic processing, nor is a proton-motive force across the inner membrane required.

Mitochondria appear to utilize multiple systems for binding and uptake of proteins. The outer membrane may contain multiple receptors that bind different classes of precursor proteins and serve to target them to their proper destination within the organelle. In most cases, uptake of proteins requires an expenditure of energy; proteins often undergo major changes in conformation that allow them to pass across or insert into the inner mitochondrial membrane.

Expression of Mitochondrial and Nuclear Genomes Is Coordinated

The assembly of a mitochondrion requires the close coordination of the nuclear and mitochondrial genomes. Especially in the cases of the multienzyme complexes such as cytochrome *c* oxidase and the F_0F_1-ATPase, it is essential that all the components be fabricated in the appropriate ratios.

An example of this coordination is observed in yeasts and other facultative aerobes. When grown anaerobically, yeasts lack a complete respiratory chain—cytochromes such as a, a_3, b, and c_1 are absent—and no typical mitochondria are visible in the electron microscope. The cells also lack enzymes involved in the citric acid cycle. Mitochondrial DNA, however, is replicated normally, and "promitochondria" organelles are present. Addition of oxygen to the yeast induces the synthesis of all mitochondrial components encoded by the nuclear and mitochondrial genomes, as well as the expansion of the promitochondrial membranes into mature mitochondria. Recent evidence suggests that heme synthesis is low in anaerobic yeast and that oxygen addition activates heme synthesis. Heme in turn is required for transcription of the nuclear gene for cytochrome *c*, and possibly for those of other mitochondrial proteins. Heme may serve to coordinate the transcription of yeast nuclear and mitochondrial genes and may also regulate the analogous transcription processes in mammalian cells.

Assembly of Chloroplasts and Peroxisomes

The synthesis of a chloroplast appears to be similar in many respects to that of a mitochondrion, although many key steps are not understood and chloroplast DNA has not been as intensively studied. Some chloroplast proteins are encoded by the chloroplast DNA, which is much larger than mtDNA, and are translated by chloroplast ribosomes within the organelle (Table 21-5). Others are fabricated on cytoplasmic ribosomes and are incorporated into the organelle after translation. These imported proteins are synthesized with N-terminal extension signal or recognition sequences that are cleaved as the protein is transported into the chloroplast. The development of chloroplasts from small membrane-limited pre-organelles called proplastids is triggered by the presence of light. Chloroplasts, like mitochondria, grow by expansion and then fission. This process can be observed easily in unicellular algae such as *Chlamydomonas,* which contain only a single large chloroplast (Figure 21-20).

More Proteins Are Encoded by Chloroplast DNA than by Mitochondrial DNA

A number of experiments on unicellular algae such as *Euglena* and *Chlamydomonas* suggested that some of the genetic information required for synthesis of chloroplasts resides outside the nucleus. Dark-grown *Euglena* organisms, for instance, have colorless *proplastids* that contain outer and inner chloroplast membranes but no thyla-

Table 21-5 RNAs and some proteins encoded by spinach chloroplast DNA

RNAs	
23S rRNA	
16S rRNA	
5S rRNA	
4.5S rRNA	
27 tRNAs	
Proteins	**Location**
Ribulose 1,5-bisphosphate carboxylase large subunit	Stroma
Ribosomal proteins	
50S protein 2	
30S proteins 12,19	
CF_1CF_0-ATPase subunits	Thylakoid membrane
CF_1 subunits α, β, ϵ	
CF_0 subunits I, III	
Photosystem I protein	
P_{700}	
Photosystem II proteins	
32-kdal protein	
51-kdal reaction-center protein	
44-kdal reaction-center protein	
D2 protein	
Cytochromes f, b_6, b_{559}, b/f	

SOURCE: E. J. Crouse, J. M. Schmitt, and H.-J. Bohnert, 1985, *Plant Mol. Biol. Rep.* 3:43.

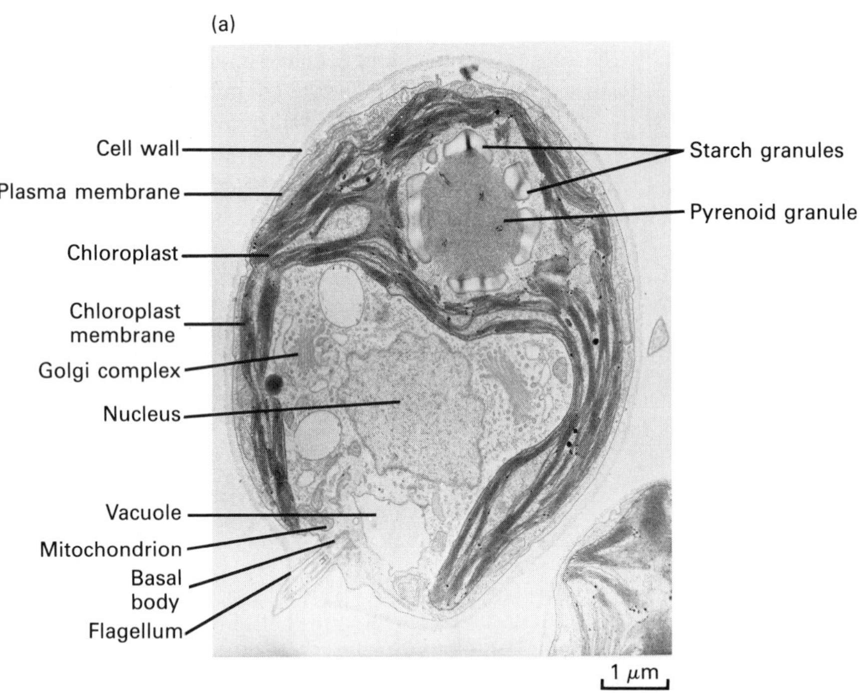

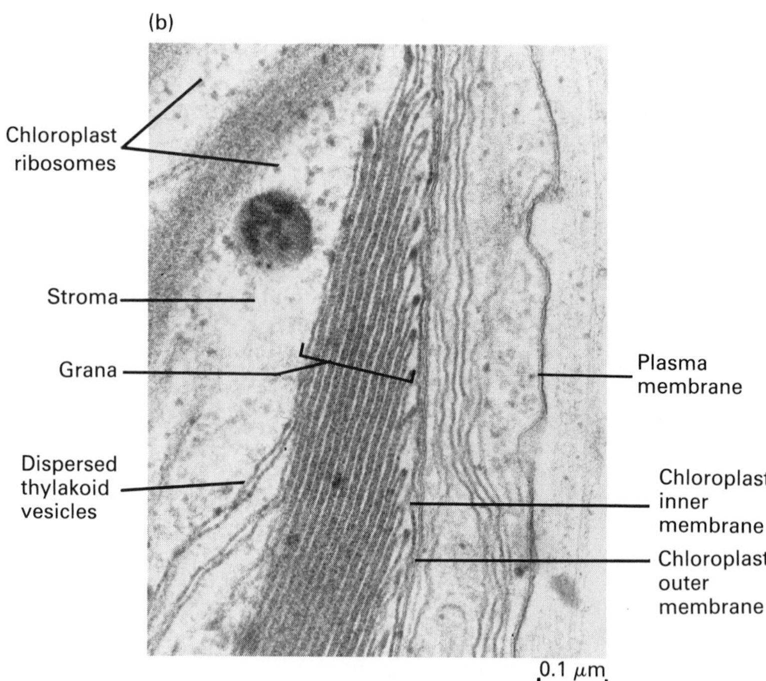

Figure 21-20 Electron micrographs of *Chlamydomonas reinhardii.* (a) The single large cup-shaped chloroplast. The stacked thylakoid vesicles are major components of the chloroplast, as is a large pyrenoid granule with its associated starch granules. (b) The chloroplast is surrounded by a double-membrane envelope, seen to advantage at this higher magnification. Here can be seen the chloroplast ribosomes and also thylakoid vesicles both dispersed and packed into grana, as well as thylakoids sectioned obliquely. *Courtesy of I. Chad, P. Siekevitz, and G. E. Palade.*

koid vesicles. When exposed to light the proplastids develop into mature chloroplasts. When the nucleus of a dark-grown cell is irradiated with a microbeam of ultraviolet light and thus partly inactivated, the cell still generates normal chloroplasts. However, when the nucleus is shielded and only the cytoplasm is irradiated, so that chloroplast DNA is damaged, many of the descendants of the cell are seen to be colorless and to lack chloroplasts. Thus chloroplast DNA is essential for development of chloroplasts.

Chloroplast DNA is much larger than mitochondrial DNA and is present in multiple copies per chloroplast. The genomes of most higher plant chloroplasts consist of a circular molecule of 145,000 base pairs encoding about 27 tRNAs (the exact number is unknown) and rRNAs, as well as mRNA for specific proteins (see Table 21-5). Many chloroplast DNAs, such as that of spinach, contain an inverted repeat that encodes the chloroplast rRNAs; consequently there are two copies of rDNA per genome (Figure 21-21). The genomes of some higher plant chloro-

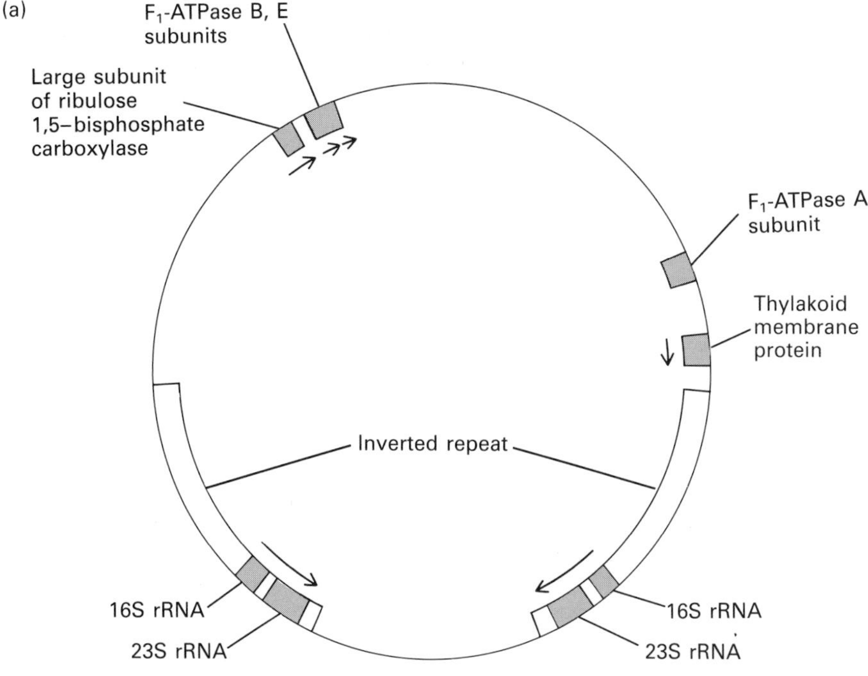

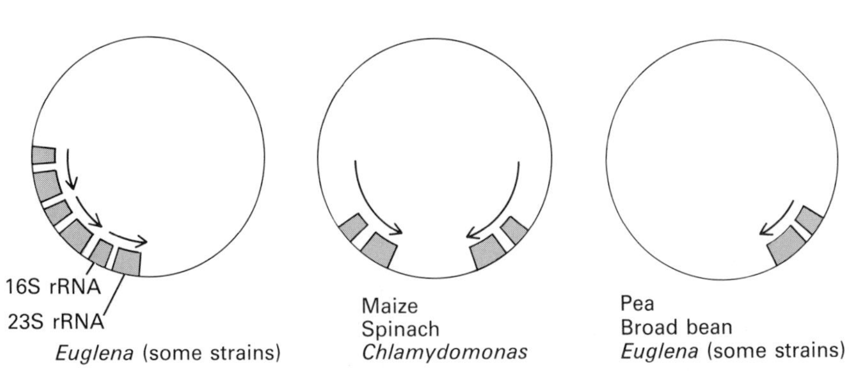

Figure 21-21 Chloroplast DNA organization. (a) Physical map of spinach chloroplast DNA. The bars represent the inverted repeat regions that contain the genes for 16S and 23S chloroplast rRNAs, shown in gray. Mapped genes encoding other chloroplast proteins are also depicted; the arrows correspond to the direction and extent of gene transcription. (b) Schematic presentation of different types of chloroplast DNA organization. The arrows denote the extent of the repeat sequences and the direction of transcription of the rRNA genes. [See J. L. Eron et al., 1981, *Proc. Nat'l Acad. Sci. USA* 78:3459; and H. J. Bohnert, E. J. Crouse, and J. M. Schmitt, 1982, in *Encyclopedia of Plant Physiology, New Series,* vol. 14B, p. 475.]

plasts, such as the pea, are only 120,000 base pairs long and contain only one set of ribosomal genes with no detectable repeat sequences. Apparently, loss or gain of one copy of the inverted repeat sequence that contains the rRNA genes has occurred during evolution of the pea family. As depicted in Figure 21-21, even closely related strains of certain organisms, such as *Euglena,* also differ in the number and organization of the genes encoding chloroplast rRNAS.

Many Cytoplasm-Synthesized Proteins Are Imported into Chloroplasts

Although the exact number is unknown, it appears that the vast majority of chloroplast proteins are made in the cytoplasm and then imported into the organelle. Three cytoplasmically synthesized chloroplast proteins have been studied with regard to biosynthesis and transport into the chloroplast. The details of these processes provide some important contrasts since one—ribulose 1,5-bisphosphate carboxylase—is an enzyme located in the stroma and only loosely bound to the thylakoid membrane, whereas the others—proteins in the light-harvesting complex, or LHC (see Figures 20-41 and 20-42)—are integral thylakoid membrane proteins.

The chlorophyll *a*/*b* binding proteins in the LHC consist of one or more related species of polypeptides that are integral membrane proteins of the thylakoid membranes. They are encoded by nuclear DNA and synthesized on cytoplasmic ribosomes. These membrane-bound LHC proteins form a regular association with 6 to 12 chlorophyll *a* or *b* molecules per polypeptide and serve to channel light energy to the reaction centers of the photosystems (Chapter 20). During import into the chloroplast, these two proteins not only must traverse both the outer and inner chloroplast envelope membranes but then must travel through the stroma and finally become inserted into a third membrane, the thylakoid membrane. The investigation of LHC protein biosynthesis and transport has been limited, but they are apparently synthesized as precursors with N-terminal signal sequences of about 40 residues that are cleaved as the proteins are incorporated into the organelle. Presumably the precursors bind to a receptor on the chloroplast's outer membrane.

Ribulose 1,5-bisphosphate carboxylase or RBPase (see Figure 20-44), is the most abundant protein in chloroplasts. The holoenzyme (m.w. 550,000) is made up of 16 subunits. Eight are identical large (m.w. ~55,000) subunits that contain the catalytic sites; the other eight are identical small (m.w. ~12,000) subunits for which the function is unknown. The enzyme is located in the stromal compartment of the chloroplast, along with the other enzymes of the Calvin cycle (Chapter 20).

The large subunit (LS) is encoded by the chloroplast DNA. The small subunit (SS) of RBPase is synthesized on free cytoplasmic polysomes and must traverse both the outer and inner chloroplast envelope membranes to reach its final destination. It is synthesized in a precursor form with an N-terminal signal sequence of about 44 amino acids. By means of an experimental protocol similar to that illustrated in Figure 21-17, it was shown that this polypeptide can be taken up by isolated chloroplasts, after which the N-terminal signal is cleaved and it is correctly assembled into the mature RBPase holoenzyme. The fact that *Chlamydomonas* chloroplasts import and process only the precursor of the *Chlamydomonas* enzyme—*not* the pea counterpart—suggests specificity in the receptor site. This is not surprising, since the sequence of the cleaved N-terminal signal sequence of the pea and *Chlamydomonas* small subunits are different. There are thought to be multiple receptors on the chloroplast membrane that bind different precursors for chloroplast proteins, perhaps to direct them to different locations within the organelle.

Import of proteins into the chloroplast requires energy. Chloroplasts isolated from cells incubated in the dark are depleted of ATP. In cell-free systems, these ATP-depleted chloroplasts do not import the precursor to the small subunit of RBPase or that of other proteins, nor do they remove signal peptides. Exposure to light of the proper wavelength restores the level of ATP and permits the organelles to take up specific chloroplast proteins.

Immature Chloroplasts Contain Few Thylakoid Proteins and Membranes

As with mitochondria, the assembly of a chloroplast shows close coordination between the nuclear and organelle components of the genome. Immature chloroplasts—proplastids—are composed of only the two closed outer and inner membranes and a small stromal space that contains the chloroplast DNA. Proplastids are found in embryonic tissue and in light-deprived eukaryotic microorganisms or leaves of higher plants (Figures 21-22 and 21-23). Proplastids do replicate but do not contain LHC proteins, chlorophyll, or electron transport systems. The presence of light stimulates synthesis of chloroplast proteins and the expansion of the inner membrane. The inner membrane buds off membrane vesicles that arrange themselves into stacks. These vesicles incorporate essential proteins and chlorophyll and become transformed into mature thylakoid vesicles (Figures 21-22 and 21-23).

The small subunit of RBPase is a well-studied example of a nuclear gene whose transcription is induced by light. When a cloned pea gene is introduced by DNA transfection into cultured petunia cells, the transferred gene is transcribed in a light-dependent fashion similar to that observed in pea leaves. More recent studies have localized a sequence of base pairs just upstream of the transcription initiation site that is essential for light induction; presumably this sequence binds a regulatory protein that responds, directly or indirectly, to the intensity of light.

Peroxisomes Contain No DNA or Ribosomes

Significantly less information is available on the mechanism of synthesis of the proteins of peroxisomes. These membrane-bound organelles are specialized vesicles that carry out reactions generating and then destroying hydrogen peroxide (H_2O_2). *Catalase* is the predominant enzyme in the lumen of the peroxisome (Chapter 5). Peroxisomes, unlike mitochondria and chloroplasts, are surrounded by a single bilayer membrane only and do not contain DNA or ribosomes. All peroxisomal proteins must therefore be imported into the organelle.

The membranes of peroxisomes are thought to arise by either pinching or budding from ER vesicles. Some intraperoxisomal enzymes may be carried along in the process. Like secretory proteins, they may be cotranslationally inserted into the lumen of the ER. However, most peroxisomal enzymes and even most peroxisomal membrane proteins are known to be synthesized on free cytoplasmic ribosomes and then to be posttranslationally transported into peroxisomes. Some but not all of these enzymes are synthesized with extra "transit" signals. Presumably, these proteins enter the peroxisomes after binding to a recognition site on the organelle membrane.

Catalase has been studied in considerable detail (Figure 21-24). Each catalase molecule contains four identical polypeptide chains, each with one attached heme group. Catalase is not synthesized as a longer precursor. Rather, newly made catalase is released into the cytosol as *apocatalase*, a single-chain polypeptide without a heme group. Apocatalase is taken up by peroxisomes some 20 min after its synthesis. Concomitant with this transport, the chains are assembled into tetramers, and heme is

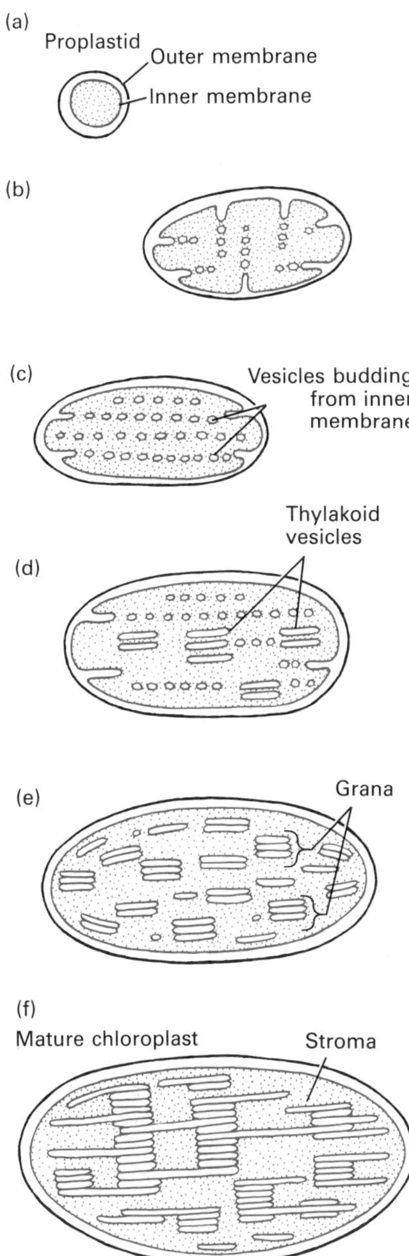

Figure 21-22 Steps in the light-induced differentiation of a proplastid into a chloroplast. (a) The proplastid in dark-adapted cells contains only the outer and inner membranes. (b), (c) Light triggers synthesis of phospholipid and of chloroplast stroma and thylakoid proteins, and a budding of small vesicles from the inner membrane. (d) As the proplastid enlarges, some of the spherical vesicles fuse, forming flattened thylakoid vesicles. Some thylakoid vesicles stack into grana. As noted in Chapter 20, adhesiveness of the vesicles is due to a protein found in the light-harvesting complex (LHC) that is synthesized in abundance during this stage. (e), (f) The proplastid enlarges and matures into a chloroplast as more thylakoid vesicles and grana are formed. [See D. von Wettstein, 1959, *J. Ultrastruc. Res.* 3:235.]

(a)

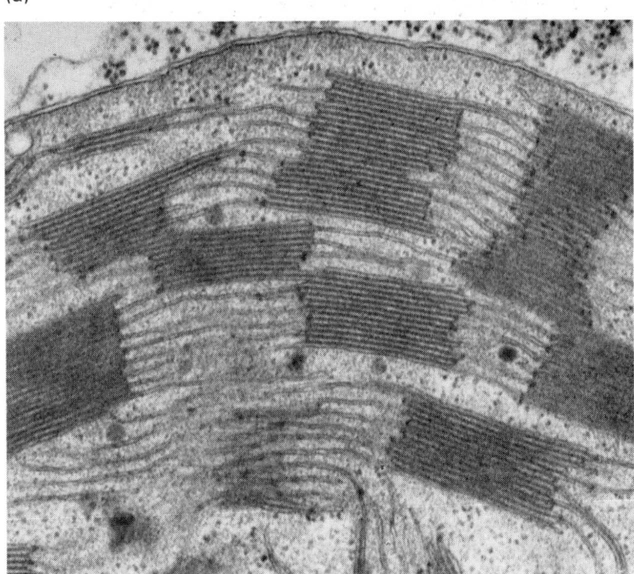

(b)

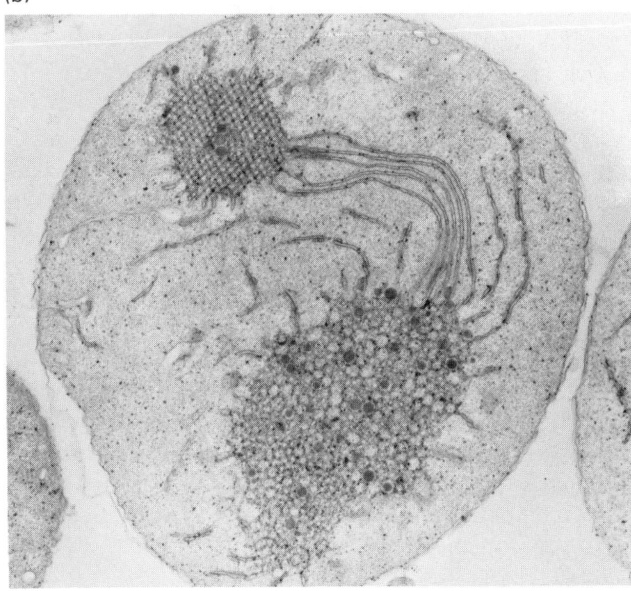

Figure 21-23 Electron micrographs of thin sections of plant tissue. (a) A chloroplast from a normal barley seedling leaf. (b) A plastid from a dark-grown barley seedling. In the absence of light, major polypeptides such as the proteins in the LHC and the chlorophyll-binding proteins are not synthesized. Plastids from such cells are termed etioplasts and contain membranes that take the form of primary lamella and interconnected vesicles containing some chloroplast pigments. As depicted in Figure 21-22, light triggers the synthesis of additional chloroplast membranes and membrane proteins and also the fusion of the small spherical vesicles to form the flattened and stacked thylakoid vesicles seen here in (a). *Courtesy of D. von Wettstein. From H. G. Schweiger, ed.* International Cell Biology 1980–1981. *Springer-Verlag.*

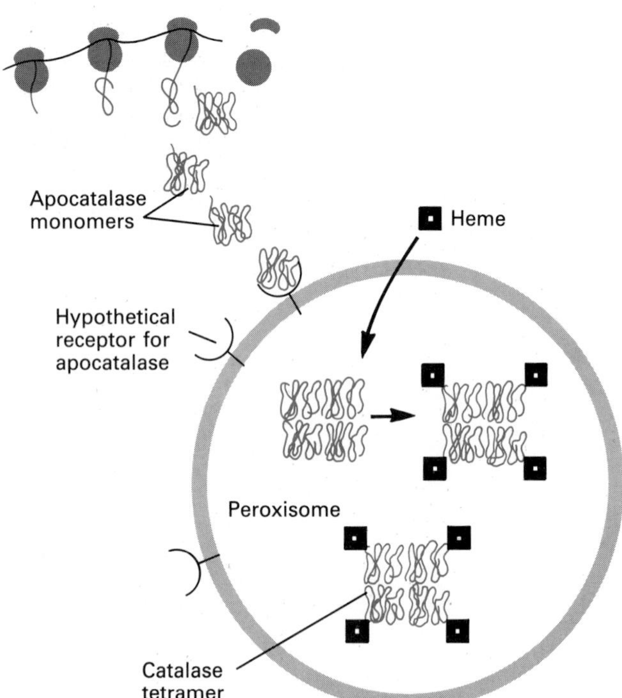

Figure 21-24 Synthesis of catalase, a peroxisomal enzyme. Catalase monomers without heme (apocatalase) are imported into the peroxisome, in a process probably involving a specific receptor on the peroxisomal membrane. Inside the organelle, the chains aggregate into tetramers, and the heme prosthetic groups are added.

added. This specific aggregation of catalase subunits may prevent the enzyme from leaking back across the peroxisomal membrane into the cytoplasm. Whether energy is required for protein import into the peroxisome is not known, nor is it yet clear whether there is a specific receptor on the organelle membrane.

Protein Secretion

We turn our attention now to the large and important class of proteins that are synthesized on ribosomes bound to the ER (Figure 21-25 and Table 21-1). These proteins entirely or partially cross the ER membrane during their synthesis. Because secretory proteins have been investigated in the greatest detail, we shall begin our discussion with these.

A vast array of different proteins are secreted by different vertebrate cells (Table 21-6), and many cells are specialized for the secretion of specific proteins. The principal function of the pancreatic acinar cells, for instance, is the secretion of digestive enzymes such as chymotrypsinogen, ribonuclease, and amylase into the intestine. Specialized cells in mammary glands synthesize and secrete milk proteins, and liver hepatocytes secrete albumin, transferrin, and lipoproteins into the serum. The B lymphocytes are specialized for synthesis and secretion of immunoglobulins. Other cells such as fibroblasts secrete proteins such as collagens and fibronectins, which form part of their extracellular matrix.

In certain cells, proteins are continuously synthesized and secreted from cells—for example, serum proteins and collagens (Figure 21-26). In other cell types, synthesis is not continuous, and proteins are stored in intracellular membrane-limited vesicles awaiting a signal for exocytosis (Figure 21-26). Examples include the exocrine cells of the pancreas, which secrete precursors of digestive enzymes, and B cells in pancreatic islets, which synthesize insulin and store it in vesicles. Release of each of these proteins is triggered by different neural and hormonal stimuli (Figure 21-27). Ciliated protozoans such as *Tetrahymena* and *Paramecium* store secretory proteins in vesicles that lie in rows just under the cell plasma membrane. Adverse nutritional conditions trigger the release en masse of these vesicles. Although these secretory processes are under different types of control, in each case secretion consists of the fusion of the membrane of the secretory vesicle with the plasma membrane.

After synthesis and translocation of the protein into the lumen of the ER, the protein moves in small transport vesicles to the membrane stacks on the cis face of the Golgi complex. The protein moves through the Golgi complex to the trans face, then to distant vesicles, and eventually to the cell exterior. Much research during the past decade has identified a large number of enzymes that act sequentially to modify secretory proteins during their maturation; each enzyme is localized to a specific organelle—the rough ER, Golgi, or secretory vesicles—and modifies proteins as they pass through. Amino acid side chains can be modified. Saccharide residues can be added and modified. Specific proteolytic cleavages may take place. Disulfide bonds can form. Sulfate and phosphate groups may be added to saccharide side chains, and polypeptide chains may assemble into multiprotein complexes. Of course, not all of these reactions occur in every cell nor for every secreted protein. Collagen is an example of a secreted protein that is modified in all of these ways.

Secretory Proteins Move from Rough ER to Golgi Membranes

Protein transport was initially established by studies on newly synthesized proteins in pancreatic acinar cells. Because most of the newly made proteins are secreted, they can be followed by electron microscopic autoradiography (Chapter 7). When the radiolabeled amino acid leucine ([^{3}H]leucine) is added to thin slices of pancreas in culture, essentially all the incorporated radioactivity in the acini is seen to be located in the rough ER (Figure 21-28). In a subsequent chase period, the tissue is incubated in abun-

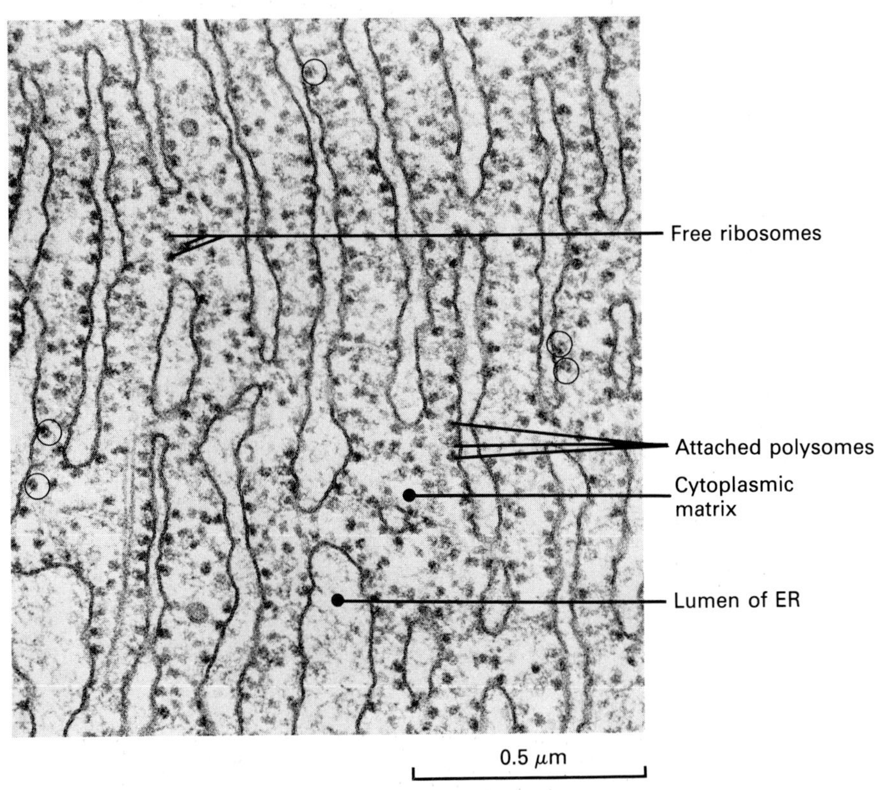

Figure 21-25 Polysomes attached to the ER in a pancreatic exocrine cell. In a few cases, the large and small ribosomal subunits are resolved (*circles*); the large subunit is attached to the membrane. *Courtesy of G. Palade.*

Table 21-6 Classes of secretory protein in vertebrates

Type	Examples	Sites of synthesis
Serum proteins	Albumin	Liver (hepatocyte)
	Transferrin (Fe transporter)	Liver
	Immunoglobins	Lymphocytes
	Lipoproteins	Liver
Structural proteins	Collagen	Fibroblasts
	Fibronectin	Fibroblasts
Peptide hormones	Insulin	Pancreatic β islet cells
	Glucagon	Pancreatic α islet cells
	Endorphins	Neurosecretory cells
	Enkephalins	Neurosecretory cells
	ACTH	Pituitary anterior lobe
Digestive enzymes	Trypsin	Pancreatic acini
	Chymotrypsin	Pancreatic acini
	Amylase	Pancreatic acini, liver, salivary glands
	Ribonuclease	Pancreatic acini
	Deoxyribonuclease	Pancreatic acini
Milk proteins	Casein	Mammary gland
	Lactalbumin	Mammary gland
Egg-white proteins	Ovalbumin	Tubular gland cells of the avian oviduct
	Conalbumin	Tubular gland cells of the avian oviduct
	Ovomucoid	Tubular gland cells of the avian oviduct
	Lysozyme	Tubular gland cells of the avian oviduct

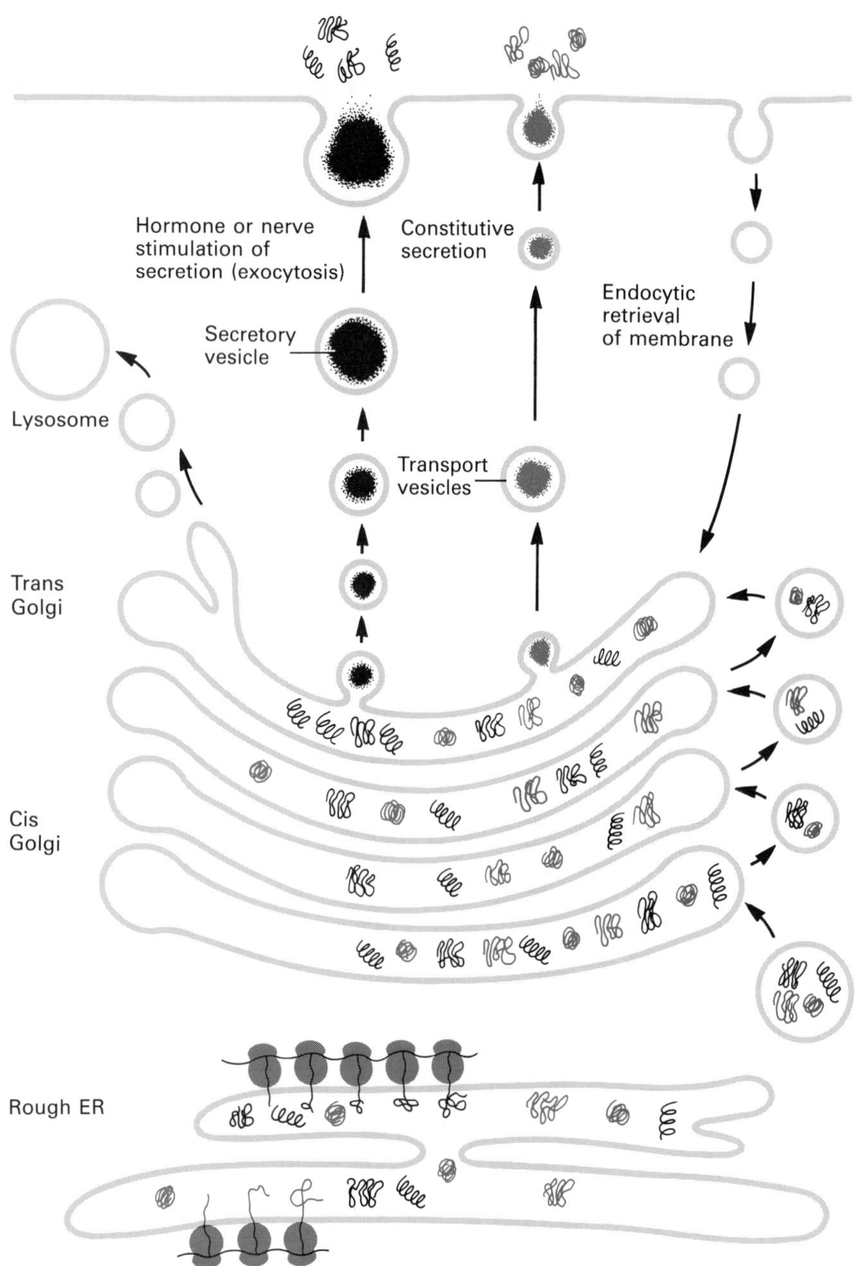

Figure 21-26 Maturation of secretory proteins. After synthesis, secretory proteins are localized to the rough ER. Always surrounded by membrane-limited vesicles, they migrate first to the vesicles on the cis face of the Golgi complex of membranes, where proteolytic cleavages and other modifications occur. Proteins migrate through the Golgi vesicles to the trans face, during which time other modifications occur. Probably, proteins are shuttled between Golgi vesicles by small transport vesicles. In some cells, such as liver cells, the proteins are secreted constitutively. In other cells, such as those of the exocrine pancreas, they are stored in secretory vesicles, there awaiting a signal for secretion. Maturation of plasma membrane glycoproteins follows the same pathway as that for constitutively secreted proteins. Sorting of proteins to lysosomes also takes place at the level of the Golgi vesicles.

dant unlabeled leucine, and there is little additional incorporation of radioactivity. After 7 min of chase, most of the labeled proteins are in the Golgi vesicles. At later times, the radioactivity is located in immature secretory vesicles, often called *condensing vesicles,* adjacent to the Golgi vesicles. Still later, the labeled protein is localized to the mature secretory vesicles. This sequence of steps is called the *maturation pathway.* It appears that the immature secretory vesicles are converted to mature secretory vesicles (zymogen granules) by progressive filling and concentration of their contents. Such electron microscopic studies have been confirmed by fractionating cells after various pulse and chase times. These results clearly indicate that secretory proteins never are found as free soluble proteins in the cytoplasm but are always sequestered within membrane vesicles.

Yeasts secrete few proteins into the growth medium, but they do secrete a number of enzymes that remain localized in the narrow space between the plasma membrane and the cell wall. The most well studied of these is invertase, an enzyme that hydrolyzes the disaccharide sucrose to glucose and fructose.

A genetic analysis of protein secretion in yeast has confirmed the maturation pathway for secretory proteins. A set of temperature-sensitive mutant yeast strains was selected in which secretion of all proteins, including invertase, is blocked at the higher temperature, but is normal at the lower temperature. At the high temperature, different

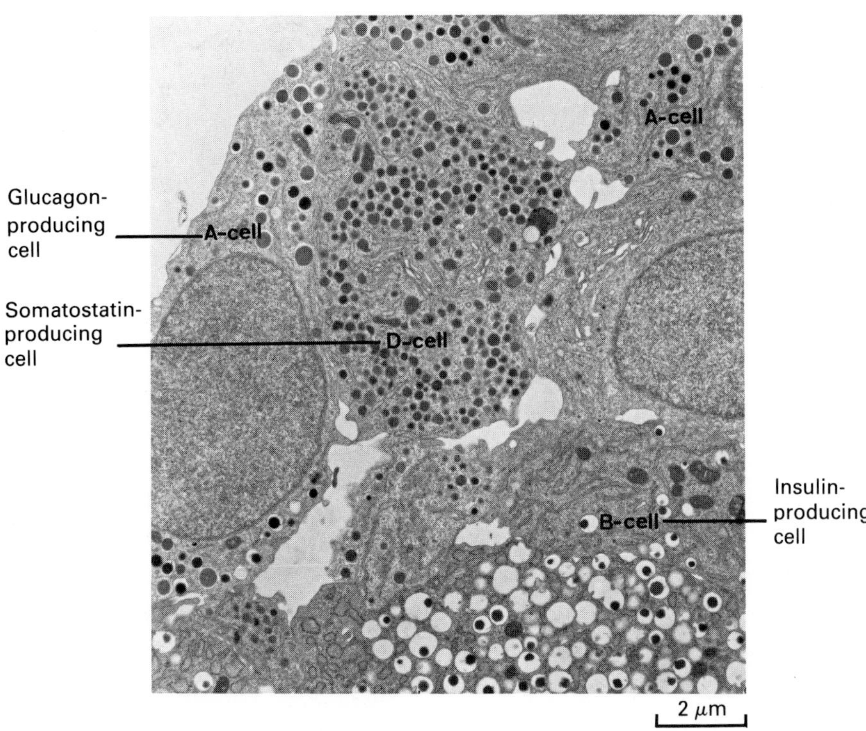

Figure 21-27 Peripheral portion of a rat pancreatic islet showing, at the electron microscopic level, the differences between the secretory vesicles of an insulin-producing cell (B cell), a glucagon-producing cell (A cell), and a somatostatin-producing cell (D cell). *Courtesy of L. Orci.*

mutants accumulate proteins either in the rough ER, in the Golgi, or in secretory vesicles. At least 25 gene products are required for the entire pathway. Studies using double mutants showed that the order of the pathway must be as follows: rough ER → Golgi → secretory vesicles → exocytosis (Figure 21-29). This maturation pathway is believed to apply to all secretory proteins in all eukaryotic organisms.

Secretory Proteins Are Localized to the Lumen of the ER

Recall from Chapter 5 that the rough ER is an extensive interconnected series of flattened sacs, generally lying in layers. The lumen, or cisterna, of the rough ER defines a space topologically distinct from the cytoplasm. When cells are homogenized, the rough ER breaks up into small closed vesicles, termed rough *microsomes,* with the same orientation—ribosomes on the outside—as that in the cell (Figure 21-30). (Similarly, smooth microsomes are formed from the network of vesicles and channels that make up the smooth ER.) Pulse-labeled secretory proteins are associated with these ER vesicles. That they are actually inside the vesicles is proved by the experiment shown in Figure 21-31. When microsomes are exposed to trypsin or other proteases, the newly made proteins remain undigested. However, when small amounts of detergents are added to the reaction, allowing proteases access to the proteins sequestered inside, the new proteins become degraded.

Secretory Proteins Are Vectorially Transported across the ER

Ribosomes that are synthesizing secretory proteins are tightly bound to the ER. Two forces bind the ribosome to the membrane: an ionic linkage and the nascent polypeptide chain itself. The evidence supporting this stems from the use of an antibiotic, *puromycin,* an analog of the 3′ end of an aminoacyl-tRNA. Puromycin binds to ribosomes in place of an incoming aminoacyl-tRNA (aa-tRNA), and the growing peptide chain is transferred to puromycin (Figure 21-32). When peptidylpuromycin is released from the ribosomes, the ribosomal subunits are released from the mRNA and from the polypeptide. The peptidylpuromycin is released into the lumen of the ER. Thus part of the growing peptide chain must have crossed the ER membrane by the time the puromycin was added.

When Mg^{2+} is removed from microsomes, usually by organic compounds called *chelators,* ribosomes dissociate from mRNA into large and small subunits. Under these conditions it is the 60S subunit, not the 40S subunit or mRNAs, that remains specifically bound to the rough ER membrane. High concentrations of Mg^{2+} chelators remove the 60S subunits from the membrane. Thus the growing protein chain passes from the large ribosomal subunit into and through the ER membrane. As the polypeptide grows, it assumes its native three-dimensional conformation within the lumen of the ER.

Precisely how the nascent chain traverses the membrane is not known. Some workers think that it passes

(a)

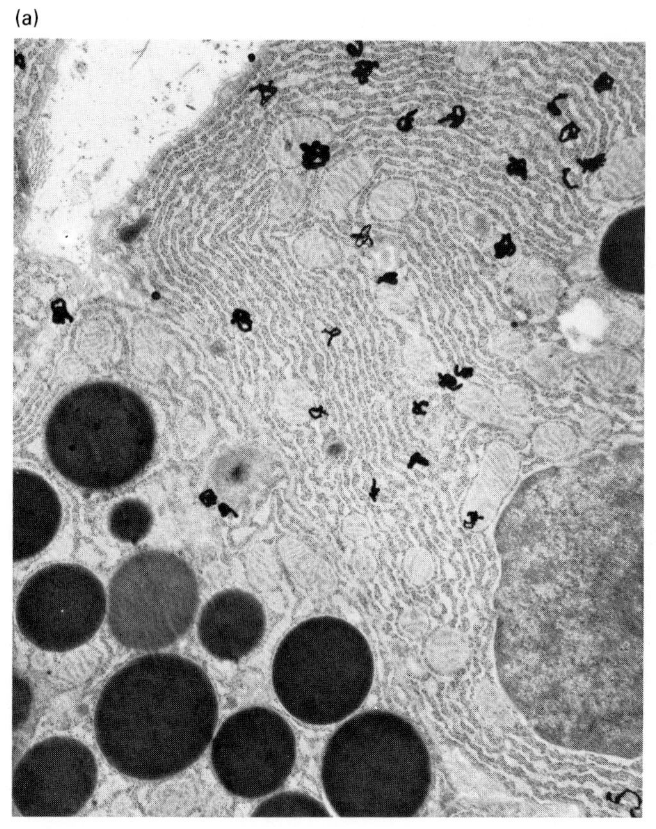

(b)

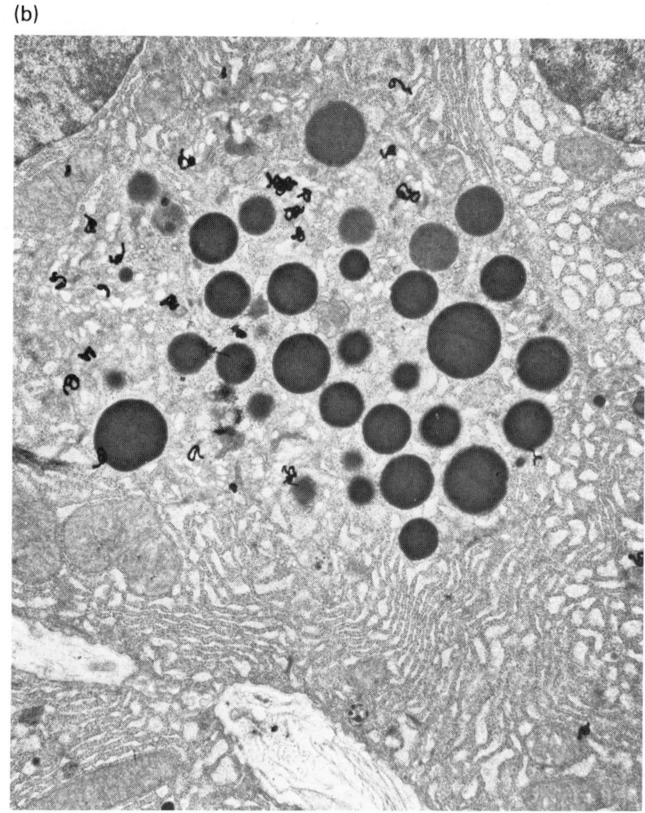

(c)

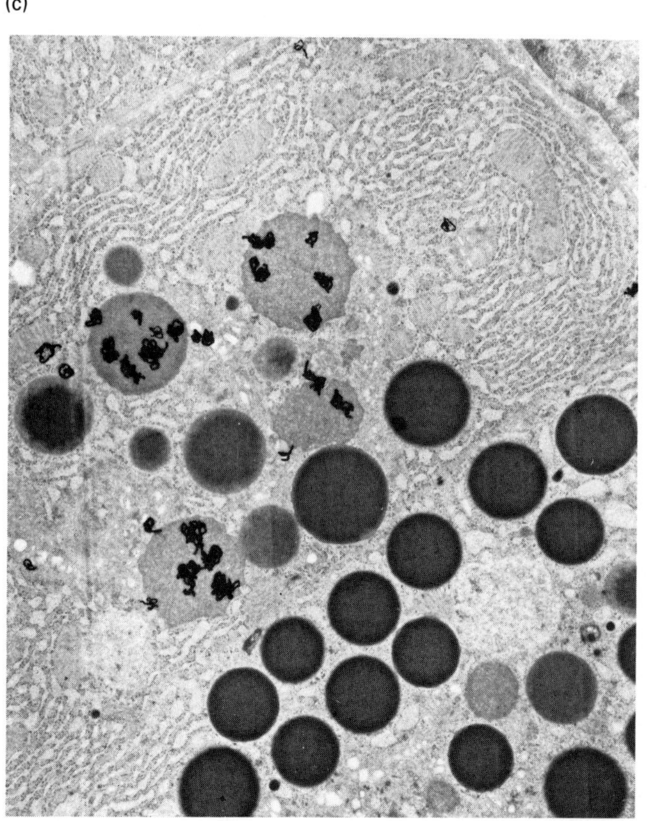

(d)

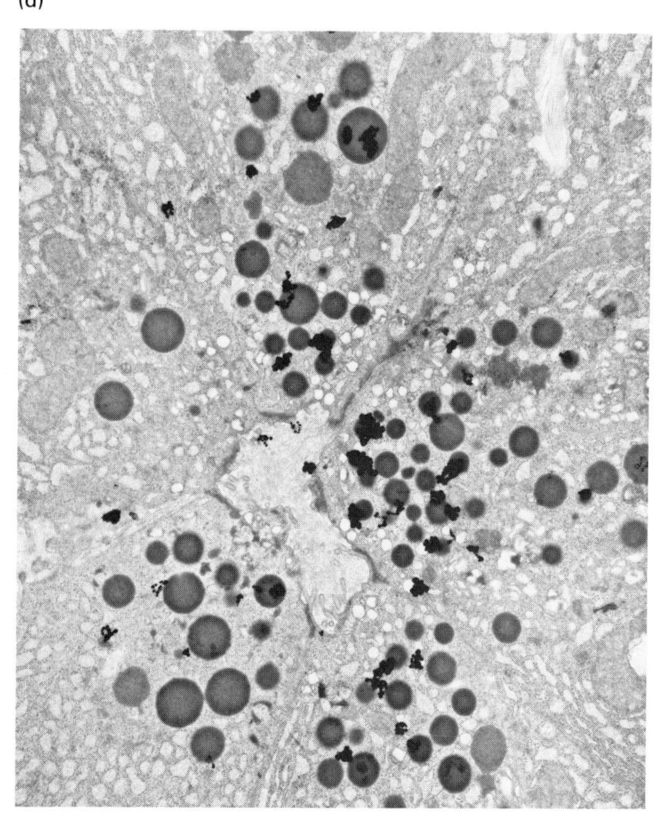

Figure 21-28 *(opposite)* Synthesis and maturation movements of guinea pig pancreatic secretory protein as revealed by electron microscopic autoradiography. (a) At the end of a 3-min labeling period with [^{3}H]leucine, the tissue was fixed, sectioned for electron microscopy, and subjected to autoradiography. Most of the labeled protein (i.e., autoradiographic grains) are over the rough ER. (b) Following a 7-min chase period with unlabeled leucine, most of the labeled proteins have moved to the Golgi. (c) After a 37-min chase, most of the grains are over immature secretory vesicles. (d) After a 117-min chase, the majority of the grains are over mature zymogen granules. *Courtesy J. Jamieson and G. Palade.*

through a protein-lined channel in the membrane (Figure 21-33b), in much the same way that small molecules pass through permease proteins. Such a transport channel may also participate in binding the ribosome to the ER membrane. Other workers suggest that the nascent protein actually penetrates the lipid bilayer directly (Figure 21-33a). Recent work has shown that energy hydrolysis of phosphoanhydride bonds in addition to that needed to polymerize aminoacyl-tRNAs into proteins (two GTPs hydrolyzed per amino acid polymerized) is required to transport the nascent chain across the membrane.

A Signal Sequence on Nascent Secretory Proteins Targets Them to the ER

Nascent secretory proteins are vectorially transported across the rough ER membrane. Most secretory proteins possess an N-terminal sequence of 16 to 30 amino acid

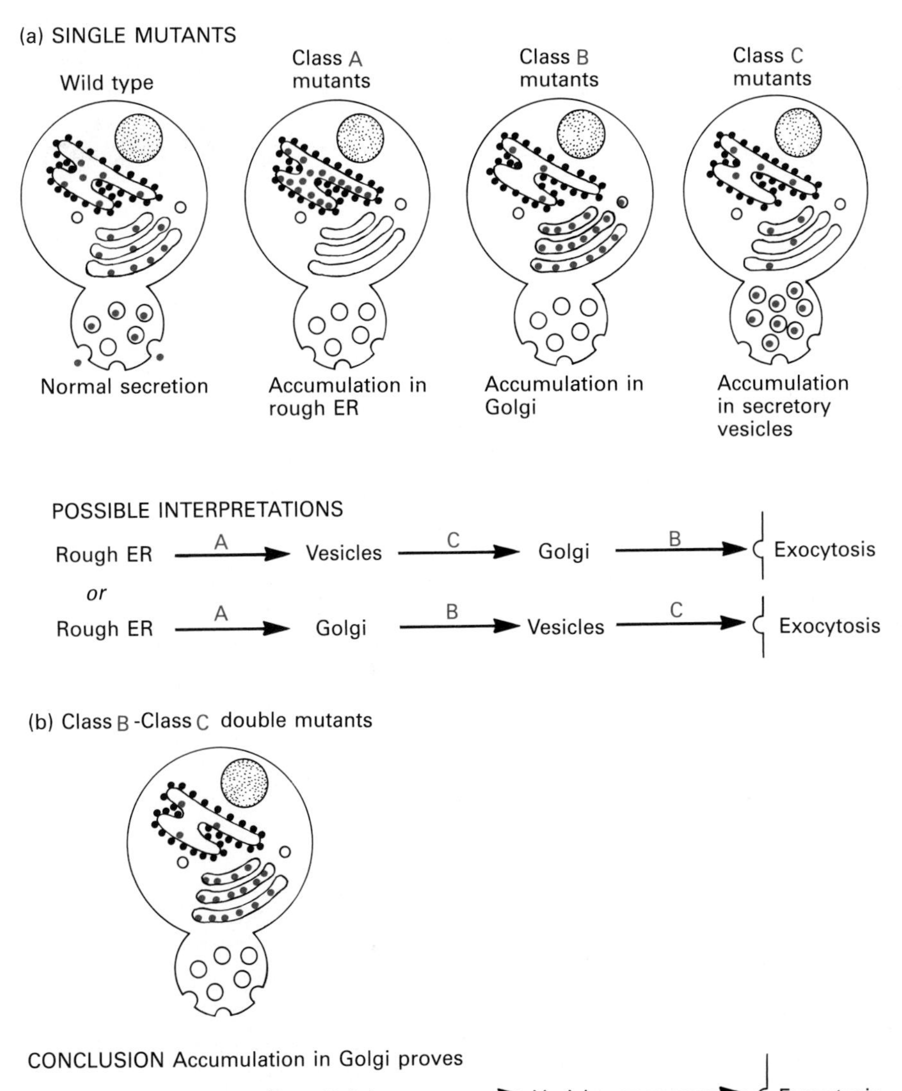

Figure 21-29 Use of double mutants in the yeast secretory pathway to show that the sequence is rough ER → Golgi → secretory vesicles → exocytosis. (a) In class A mutants, the proteins remain in the rough ER (when grown at the higher temperature). Class B mutants accumulate protein in the Golgi; class C mutants accumulate it in the vesicles. These findings do not prove the sequence ER → Golgi → vesicles → exocytosis. The two sequences depicted yield the same experimental results. (b) In these double mutants—with defects in both class B and class C secretory processes—the proteins accumulate in the Golgi, not in the vesicles. This proves that the class B mutations act at an earlier step in the maturation pathway than do class C mutations, and that the correct sequence is the lower one. [See P. Novick et al., 1981, *Cell* 25:461.]

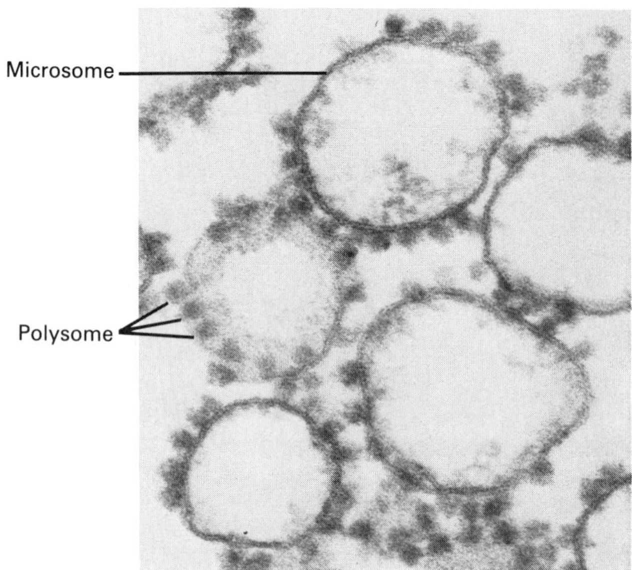

Figure 21-30 Rough microsomes isolated from the rat parotid gland. Polysomes are attached to the outer (cytoplasmic) surface of the vesicles. *Courtesy of P. Arvan and D. Castle.*

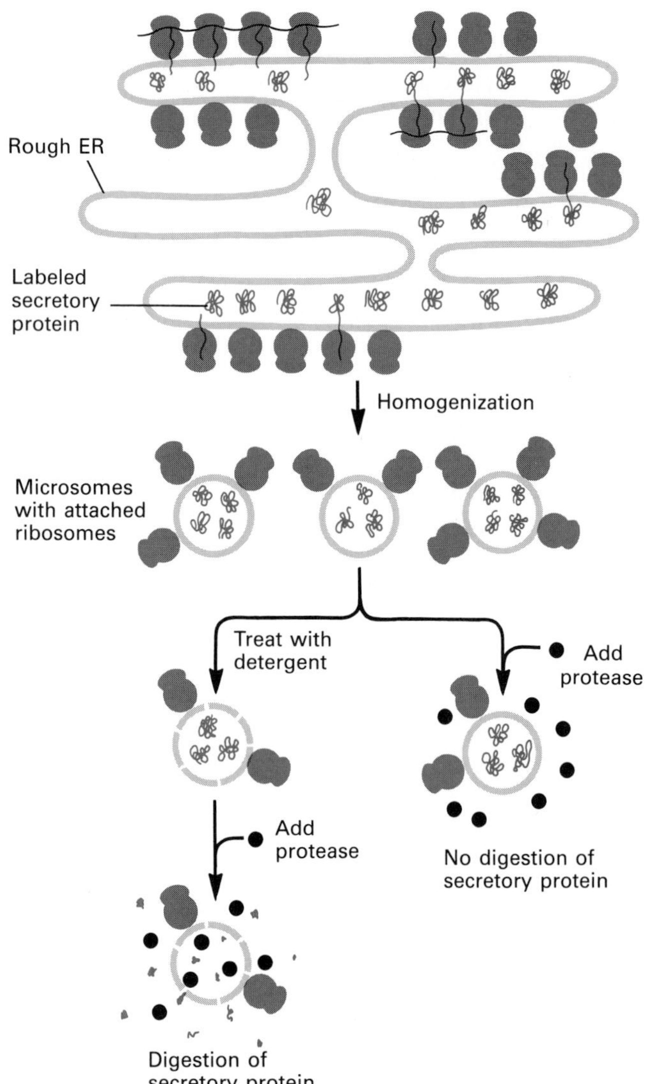

Figure 21-31 Secretory proteins are sequestered within the lumen of the rough ER, as proved in experiments similar to that depicted here. Newly synthesized protein associated with microsomes is not digested by added proteases (which remain outside the microsome). A detergent that makes the microsomal membrane permeable allows some lumenal proteins to leak out and the protease to enter; newly made proteins are destroyed by the protease.

residues (Table 21-7) that is thought to initiate transport across the ER membrane. Characteristically, these signal sequences contain 4 to 12 hydrophobic residues, but otherwise the signal sequences of various secretory proteins have little homology. Signal sequences are not normally found on completed polypeptides made in cells—implying that the signal sequence is cleaved from the protein while it is still growing on the ribosome.

In order to observe the signal sequence, the mRNA often must be translated in a cell-free system without microsomes. In these systems the role of ER membranes in removing the signal has been established; such studies make use of purified preparations of microsomal membranes that are first stripped of their own ribosomes. If these membranes are present during cell-free synthesis of a secretory protein, the protein is found in the lumen of the vesicles, with the signal sequence removed (Figure 21-34). If, however, the membranes are added to the reaction mixture *after* synthesis of the secretory protein has been completed, the protein is not incorporated into membrane vesicles, and its signal sequence remains.

To allow ER insertion of most secretory proteins, membranes must be added before the first 70 or so amino acids are polymerized. At this point about 40 amino acids, including the cleaved signal, protrude from the ribosomes, while about 30 amino acids are buried inside. Thus transport of most secretory proteins into ER vesicles is *obligatorily cotranslational*, although some yeast secretory proteins can be transported into the ER posttranslationally in an ATP-requiring process.

Recent experiments suggest that the cleaved signal peptide, or sequence, is sufficient to direct a protein into the ER. By gene fusion, the signal sequence of a secretory protein can be attached to the N-terminus of a cytosolic protein such as globin. This chimeric protein is transported into the ER lumen, and the signal peptide is cleaved off, exactly as for a "normal" secretory protein. As noted previously, all signal sequences on secretory proteins have a "core" of hydrophobic amino acids (see Table 21-7). These are indeed essential for signal se-

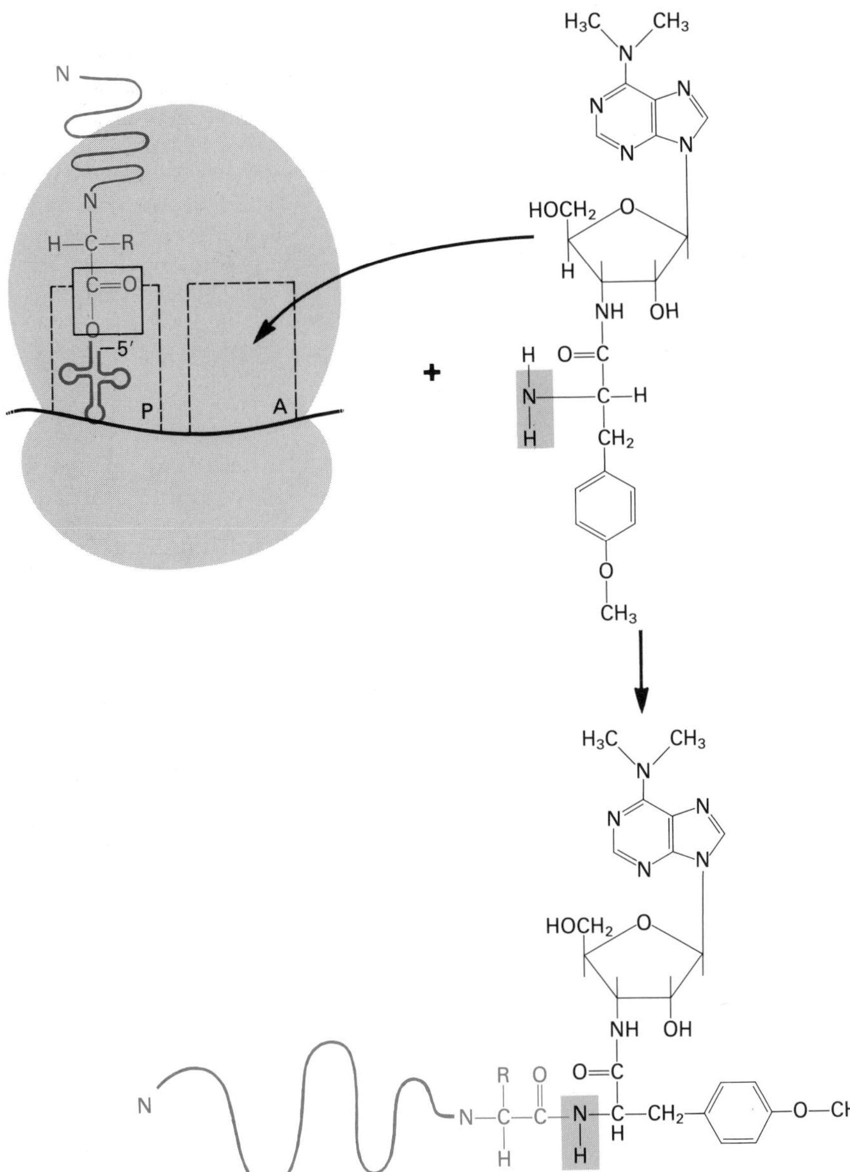

Figure 21-32 Puromycin inhibits protein synthesis by causing the premature termination of a nascent polypeptide chain and its release from the ribosome. Puromycin is an analog of the 3′ end of an aminoacyl-tRNA, containing an amino acid (methyltyrosine) linked in an amide bond to the 3′ carbon of adenosine. As discussed in the text, experiments with puromycin showed that nascent secretory proteins cross the ER membrane. [See D. Sabatini and G. Blobel, 1970, *J. Cell Biol.* **45**:146.]

quence function, since specific deletion of these from the signal sequence by in vitro mutagenesis abolishes the ability of the protein to cross the ER membrane into the lumen. Possibly these hydrophobic residues form a binding site for the "signal recognition particle" (discussed in the following section).

Not all secretory proteins have a signal sequence that is cleaved. Ovalbumin, the major protein synthesized and secreted by the hen oviduct, has no cleaved signal sequence. Experiments similar to those just described show that the sequence required for membrane transport of ovalbumin is localized to the 100 N-terminal residues. Thus secretory proteins, like other classes of proteins, bear signal sequences that serve to direct them to a particular organelle membrane.

Receptor Proteins Mediate the Interaction of Signal Peptides with the ER Membrane

Since secretory proteins are synthesized in association with the ER membrane, but not with any other cellular membrane, some specific recognition system must target them there. The identification of a *signal recognition particle,* a key element in this process, was a result of the following simple experiment. A preparation of rough ER,

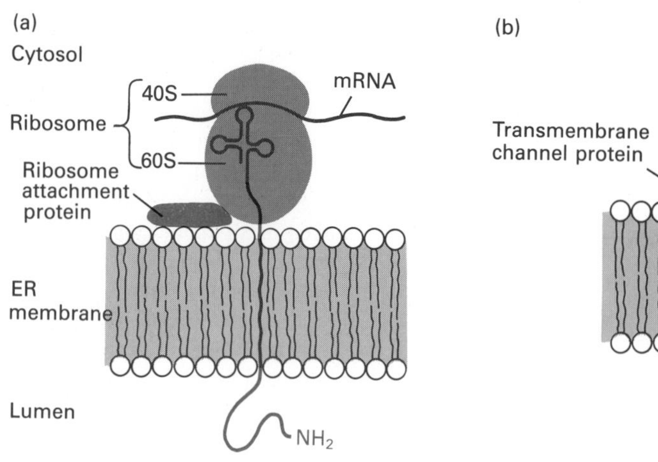

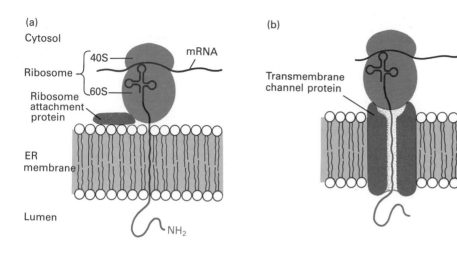

Figure 21-33 Two possible models for transfer of a nascent chain across the ER. (a) The nascent chain passes directly through the lipid bilayer, and the 60S ribosomal subunit is anchored by a hypothetical attachment protein. (b) A hypothetical pore forms a transmembrane channel through which the nascent chain passes and also anchors the 60S ribosomal subunit to the membrane. In either case, the nascent chain is cotranslationally transferred across the membrane.

freed of ribosomes, was exposed to a solution of 0.5 *M* NaCl, thereby removing several proteins from the membranes. When these "stripped" microsomal vesicles were recovered by centrifugation and added to a cell-free protein synthesis reaction, they were unable to support the insertion of nascent secretory protein. However, when the proteins removed by the salt treatment were re-added, nascent secretory proteins were cotranslationally inserted into the vesicles. A single active component—the signal recognition particle (SRP)—was purified from the mixture of stripped proteins. SRP is a particle containing 6 discrete polypeptides and a 300-nucleotide RNA and is an essential component of protein translocation across the ER (Figure 21-35).

Some aspects of the function of SRP have been clarified by studies in which mRNAs for secretory proteins, such as the pituitary hormone prolactin, are translated in a cell-free protein synthesis system (Figure 21-36).

When cell-free translation is carried out in the presence of SRP but without microsomes, synthesis is arrested at about 70 amino acids. When microsomal membranes are added to the arrested reaction, elongation of prolactin chains resumes, the ribosomes bind to the microsomal membrane, and the full-length protein is made and sequestered in the lumen of the microsomes. Thus SRP prevents synthesis of a complete secretory protein in the absence of sufficient rough ER membranes.

The rough ER membrane contains a receptor for the SRP, which can anchor the complex of ribosome and nascent chain to the membrane. This receptor, an integral membrane protein, has been found to be a single polypeptide of about 650 residues. As depicted in Figure 21-36, this solubilized SRP receptor relieves the block in elongation of nascent secretory proteins imposed by SRP in cell-free systems. It thus must bind to the SRP and possibly also to ribosomes.

Table 21-7 Amino acid sequence of signal peptides of several secretory and membrane proteins*

Protein	Sequence
Preproalbumin	Met Lys Trp Val Thr Phe Leu Leu Leu Leu Phe Ile Ser Gly Ser Ala Phe
Pre-IgG light chain	Met Asp Met Arg Ala Pro Ala Gln Ile Phe Gly Phe Leu Leu Leu Leu Phe
Prelysozyme	Met Arg Ser Leu Leu Ile Leu Val Leu Cys Phe Leu Pro Leu Ala Ala Leu
Preprolactin	Met Asn Ser Gln Val Ser Ala Arg Lys Ala Gly Thr Leu Leu Leu Leu Met
Prepenicillinase *(Escherichia coli)*	Met Ser Ile Gln His Phe Arg Val Ala Leu Ile Pro Phe Phe Ala Phe Cys
Prevesicular stomatitis virus glycoprotein	Met Lys Cys Leu Leu Tyr Leu Ala Phe Leu Phe Ile His Val Asn Cys ↓
Prelipoprotein *(E. coli)*	Met Lys Ala Thr Lys Leu Val Leu Gly Ala Val Ile Leu Gly Ser Thr Leu

*The continuous hydrophobic residues are in color. The arrows denote the sites of cleavage of these residues; this generally occurs during elongation of the growing polypeptide chain.

SOURCE: D. P. Leader, 1979, *Trends Biochem. Sci.* **4**:205.

SRP and its receptor are involved only in the initiation of transfer of the nascent chain across the membrane. Having accomplished this, they dissociate from the ribosome–nascent chain complex and recycle to direct insertion of additional protein. Thus the evidence is strong that the N-terminal signal sequence, SRP, and the SRP receptor are essential for initiation of cotranslational transport of a secretory protein across the ER membrane.

Many Integral Proteins and Glycoproteins Are Cotranslationally Inserted into the Rough ER Membrane

In Chapter 14 several of the vast array of integral proteins that occur in the plasma membrane and other cellular membranes were introduced. Because of their importance, one large class of these proteins, plasma membrane glycoproteins, have been studied in some detail. Since their synthesis and maturation resemble that of secretory proteins, it is appropriate to consider certain aspects of their biosynthesis at this point.

Several integral membrane proteins of quite diverse origin have in common a number of key structural characteristics. The surface glycoproteins of vesicular stomatitis virus and Sindbis virus; the influenza virus hemagglutinin (HA) glycoprotein; glycophorin, a major erythrocyte surface protein; and the heavy chain of the HLA-A and H2 major histocompatibility antigens all span the phospholipid bilayer only once (Chapter 14). In each, a region of 10 to 30 amino acids at the C-terminus faces the cytoplasmic surface. This region contains a number of positively charged amino acids. Adjacent to this region in all of these proteins, there is a sequence of 20 to 25 hydrophobic amino acids that is believed to span the lipid bilayer as an α helix. The remainder of the polypeptide chain (which varies considerably in size among these polypeptides) is on the extracytoplasmic surface, as are all of the attached carbohydrate chains.

All these proteins are synthesized on the ER and move to the cell surface via the Golgi complex along the same constitutive maturation pathway as that for secretory proteins (see Figure 21-26). The vesicular stomatitis virus glycoprotein (the VSV G protein; see Figure 14-20) that forms the surface spikes of the virus has been studied in the most detail; the others, however, appear to follow precisely the same maturation pathway.

Immediately after its synthesis, the G protein spans the ER membrane; thus it is a transmembrane protein (Figure 21-37). About 30 amino acids at the extreme C-terminus

(a)
Add microsomal membranes
N-terminal signal sequence
Cell-free protein synthesis
No microsomes present
Completed proteins with signal sequences
No removal of signal sequence
No transport of protein into microsome

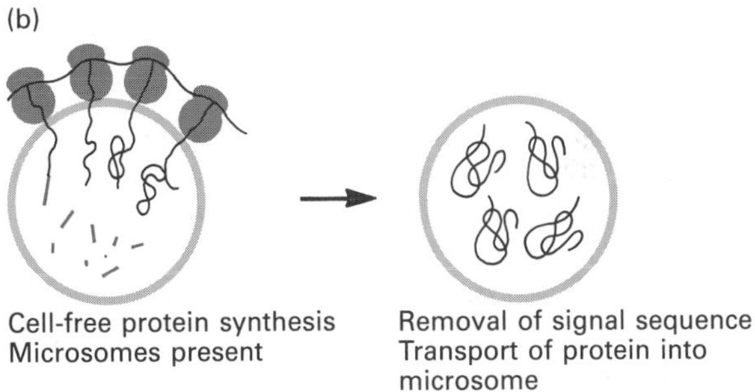

Figure 21-34 Cotranslational insertion of secretory proteins into microsomal vesicles. (a) A secretory protein produced in a cell-free reaction without microsomes retains the N-terminal signal sequence; if microsomes are added, the protein is not transported across the vesicle membrane, nor is the signal sequence removed. (b) By contrast, if microsomes are present during protein synthesis, the signal sequence is removed from the nascent chain, and the protein is transported into the lumen of the ER vesicles.

Ser ↓ Arg . . .

Pro Gly Thr Arg Cys ↓ Asp . . .

Gly ↓ Lys . . .

Met Ser Asn Leu ↓ Leu Phe Cys Gln Asn Val Gln Thr Leu . . .

Leu Pro Val Phe Ala ↓ His . . .

Lys . . .

Leu Ala Gly ↓ Cys . . .

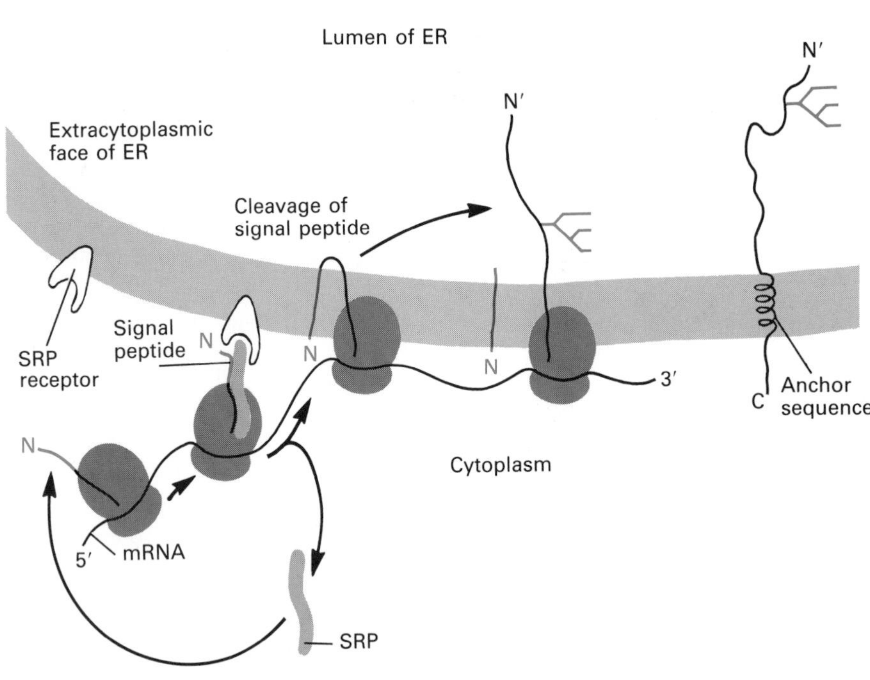

Figure 21-35 Model for synthesis of secretory proteins on the ER. The N-terminal signal sequence emerges from the ribosome only when the polypeptide is about 70 amino acids long, because about 30 amino acids remain buried in the ribosome. An elongated signal recognition particle (SRP) then binds to the signal sequence, and the SRP-nascent polypeptide plus ribosome binds to the ER membrane through the SRP receptor. This initiates cotranslational movement of the protein through the ER membrane. Following initiation, SRP dissociates from the ribosome-peptide complex and is released into the cytosol. The SRP receptor is also freed to initiate insertion of another secretory protein. Within the lumen of the ER, the signal sequence is cleaved, and carbohydrates are added to asparagine residues, both apparently on the lumenal surface. After synthesis is completed and the ribosomes are released, the remaining C-terminus of the secreted protein is transferred to the lumen of the ER. [See P. Walter, R. Gilmore, and G. Blobel, 1984, *Cell* **38**:5–8.]

remain exposed to the cytoplasm. The balance of the polypeptide, including the N-terminus (and the two asparagine-linked carbohydrate chains), is on the lumenal side of the ER membrane and is protected from protease digestion by the permeability barrier of the ER membrane. Figure 21-37 shows that the orientation of the ER G protein with respect to the membrane is the same as that of G protein on the surface of infected cells and in virions (see Figure 14-20). The same 30 C-terminal amino acids are exposed on the cytoplasmic surface, while the carbohydrate chains and the bulk of the polypeptide, including the N-terminus, remain extracytoplasmic. Thus the overall membrane topology of the G protein is preserved as it matures from the ER to the cell surface.

Much like secretory proteins, nascent G protein is extruded across the ER membrane, N-terminus first, and the 16-amino-acid signal sequence (see Table 21-7) is removed while the chain is still growing. Yet, unlike secretory proteins, G protein remains anchored into the membrane by its C-terminal sequence. The hydrophobic 23-amino-acid-long membrane-spanning region functions also as a "stop transfer" sequence: it blocks the continued extrusion of the nascent chain through the ER membrane.

Support for such a model has come from studies in which cloned cDNAs encoding the VSV G protein are expressed in tissue culture cells. Cloned intact cDNA, in an appropriate expression vector, directs synthesis of full-length G protein, which matures normally to the plasma membrane. Cloned mutant G genes were constructed in which most or all of the segment of DNA encoding the membrane-spanning region was deleted. In this case, the N-terminal fragment was secreted from the cell, implying that the hydrophobic segment near the C-terminus is essential for anchoring the protein in the membrane. Moreover, when by in vitro gene fusion manipulations the anchor sequence of VSV G protein was fused at the C-terminus of a secretory protein, the chimeric protein became anchored in the ER membrane by the G anchor sequence. Additional support for an anchoring and stop-transfer role of these hydrophobic amino acids has come from a comparison of the membrane-bound and the secreted forms of the immunoglobulin *mu* heavy chain (Chapter 24).

These findings pose an obvious problem in explaining how nascent secretory proteins become completely extruded across the ER membrane. Recall that when synthesis of a protein is complete, the ribosome detaches, leaving the 30 or so amino acids that had been buried in the large ribosomal subunit now exposed to the cytoplasm (see Figure 21-35). The C-terminus of secretory proteins, however, must be pulled through the membrane into the lumen. It is postulated that a combination of forces, including the energy of ATP hydrolysis and the pull provided by folding of the protein within the lumen, is responsible for this movement of secretory proteins.

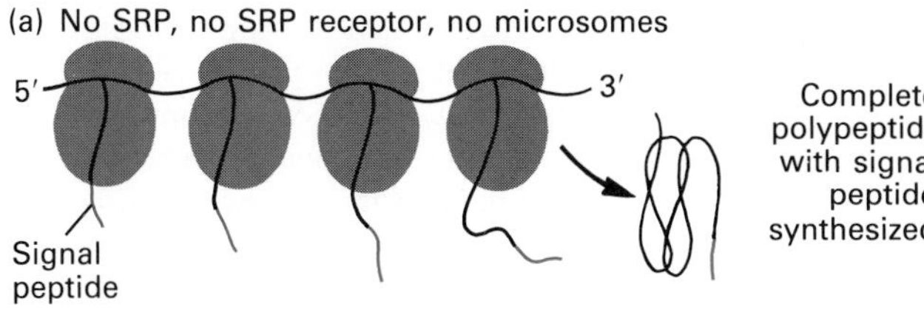

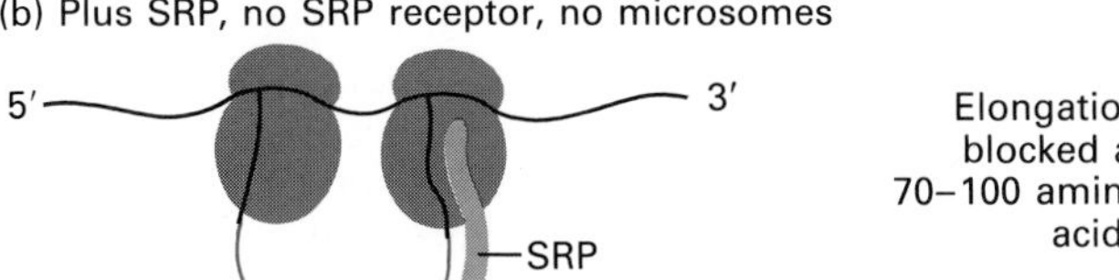

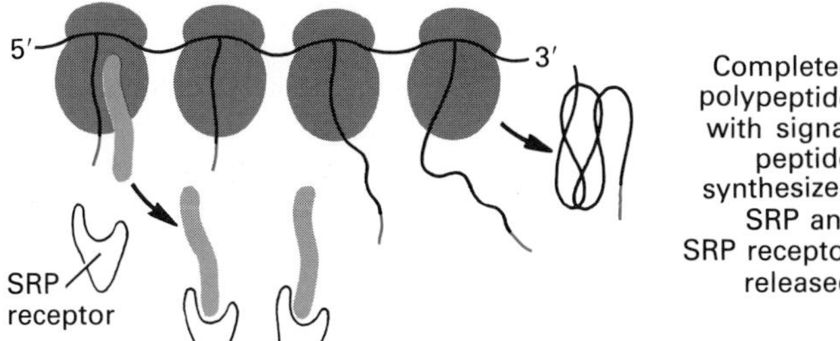

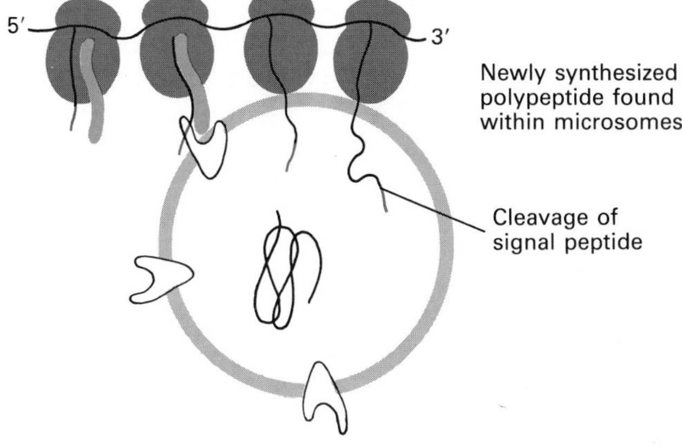

Figure 21-36 Properties of SRP and SRP receptor studied in cell-free protein synthesis reactions. Messenger RNA for preprolactin, a typical secretory protein made by the pituitary, is readily translated in a wheat germ cell-free system in the absence of any microsomes. This system contains all necessary factors for in vitro protein synthesis.

In one series of experiments, SRP, SRP plus receptor, and SRP, receptor, plus microsomes were added to the protein synthesis system to examine the effects on protein synthesis. (a) In the absence of microsomes, SRP, and SRP receptor, the complete protein with its signal sequence is synthesized. (b) Addition of only SRP causes elongation to be blocked at 70 to 100 amino acids; this is direct evidence that SRP binds to the N-terminal residues on the growing chain. (c) Addition of soluble SRP receptor removes this block in polypeptide chain elongation. The SRP receptor apparently binds to SRP and releases SRP from the growing chain. This is evidence that the SRP receptor binds the SRP. (d) If SRP, together with vesicles containing the SRP receptor, is added, the newly made protein is translocated across the vesicle membrane. [See P. Walter and G. Blobel, 1981, *J. Cell Biol.* **91**:557.]

α-Helical Sequences May Be Used for Insertion of Certain Integral Membrane Proteins

Not all integral membrane glycoproteins insert into the membrane by the route used by the VSV G protein. For instance, the erythrocyte anion transport protein—band III—has an orientation in the membrane very different from that of VSV G protein or the others noted previously (see Figure 14-39): the N-terminal 400 amino acids face the cytoplasm, while the C-terminal 450 amino acids are embedded in the erythrocyte membrane, which they span multiple times. Yet band III also is cotranslationally inserted into the ER membrane and utilizes the maturation pathway rough ER → Golgi complex → plasma membrane. Exactly how nascent band III inserts into the ER membrane is not known. Recent evidence indicates that the signal sequence for insertion occurs in the middle of the protein, in contrast to the N-terminal localization of other membrane glycoproteins and secretory proteins. As depicted in Figure 21-38, loops of two α helixes may be intermediates in insertion of the nascent chain. The two helixes are thought to pair with or bind to each other, forming a helical hairpin. The helical hairpin can insert into the ER, in a process probably requiring ATP hydrolysis, causing the ribosome to bind to the membrane.

The SRP and its receptor are known to be required for insertion of many integral proteins that span the membrane multiple times, such as the four polypeptides of the acetylcholine receptor (see Figure 17-44), and are required for insertion of band III as well. Presumably, as with other proteins, SRP directs the nascent protein to the ER membrane.

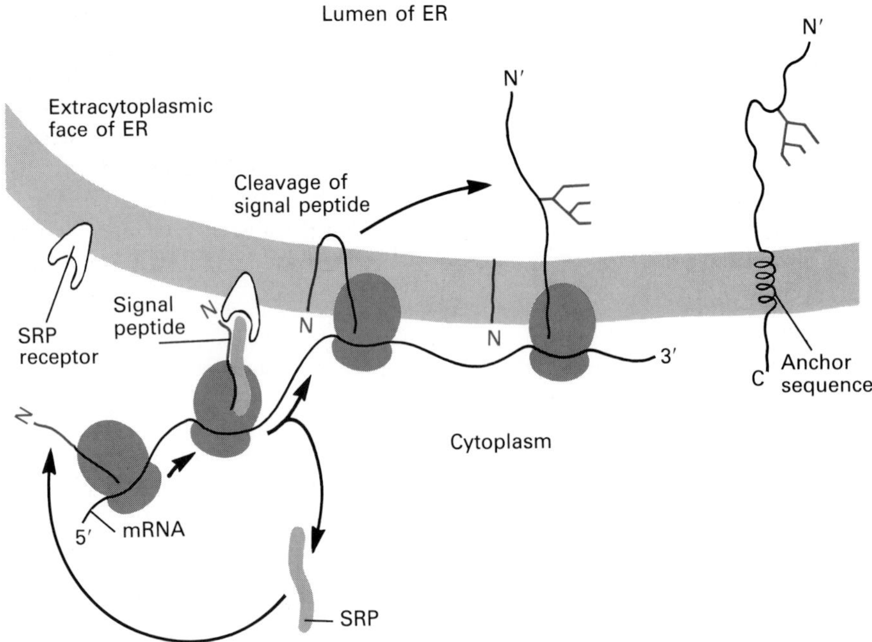

Figure 21-37 Model for synthesis and membrane insertion of the vesicular stomatitis virus (VSV) glycoprotein (G). Up until the point at which the C-terminus of the protein is synthesized, this model is the same as for secretory proteins (see Figure 21-35). When synthesis is completed, and the ribosomes are released, VSV G protein remains in the membrane anchored by a hydrophobic anchor sequence (*color*) near the C-terminus. Other membrane proteins with similar structure, such as the HLA-A polypeptides of the major histocompatibility complex, are inserted in a similar fashion.

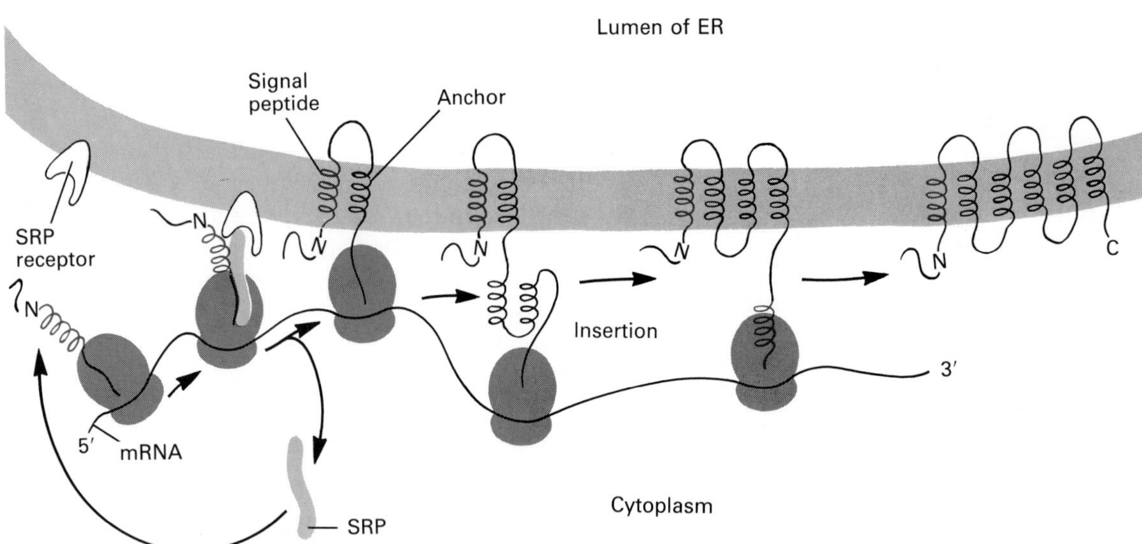

Figure 21-38 Model for transmembrane insertion of nascent band III. Band III, like many integral proteins, is believed to traverse the membrane as a series of α-helical loops, or hairpins. An internal signal sequence is known to direct binding of the nascent chains to the rough ER membrane and to initiate cotranslational insertion. In the model depicted here the signal sequence and the sequence of nascent chain following it insert into the ER membrane as a helical hairpin; SRP and the SRP receptor are involved in this step. The nascent chain then continues to grow on the cytoplasmic surface. Subsequent α-helical hairpins could insert similarly in an SRP-dependent fashion. Alternatively, as depicted here, a hairpin of two α helixes could form on the cytoplasmic surface of the ER and then insert into the ER membrane. Although only 6 membrane-spanning helixes are depicted here, band III and proteins of similar structure have 12 or more. [See W. Braell and H. F. Lodish, 1982, *Cell* **20**:23; and W. Wickner and H. F. Lodish, 1985, *Science,* **230**:400–407.]

Posttranslational Modifications of Secretory, Membrane, and Glycoproteins

Once synthesis is complete, polypeptides newly introduced into the membrane and lumen of the ER must be matured, sorted, and transported. These steps occur predominantly in a complex organelle composed of both flattened and spherical vesicles: the Golgi complex (Figure 21-39). The Golgi complex serves as a liaison between the ER and both the plasma membrane and internal organelles such as lysosomes. The elongated vesicles nearest the ER make up the *cis face* of the Golgi; those in the midportion, the *medial face;* and those nearest the periphery of the cell, the *trans face*. Secretory proteins and plasma membrane proteins are thought to travel from the ER to the Golgi, and from the Golgi to other organelles, in small vesicles that pinch off from one vesicle and fuse with other membranes of another (Figure 21-39; see also Figure 21-26). Although biologists generally refer to the Golgi as one organelle, there is growing evidence of functional divisions within the different vesicles that introduce different modifications into secretory and membrane proteins. Lysosomal proteins also are sequestered early in their maturation into elements of the Golgi and subsequently sorted to lysosomes.

Three principal types of modifications occur to many secretory or membrane proteins as they mature to the cell surface: formation of disulfide bonds, specific proteolytic cleavages, and addition and modification of carbohydrates (see Figure 4-3). These modifications are specific for the various organelles through which these proteins pass; they assist in achieving the functional form of the protein and, in some cases, act as signals to direct the protein to its ultimate destination in the cell.

Disulfide Bonds Are Formed during or Soon after Synthesis

Disulfide bonding between two cysteine residues is one of the most important stabilizing forces in the secondary structure of polypeptides. In proteins that contain more than two cysteine residues, formation of the proper arrangement of disulfide bonds (Cys—S—S—Cys) is essential for normal structure and enzymatic or hormonal action (Chapter 3). Disulfide bonds are generally confined to secretory proteins and certain membrane proteins. Cytoplasmic proteins in higher cells generally do not utilize the relatively more oxidized disulfide bond as a

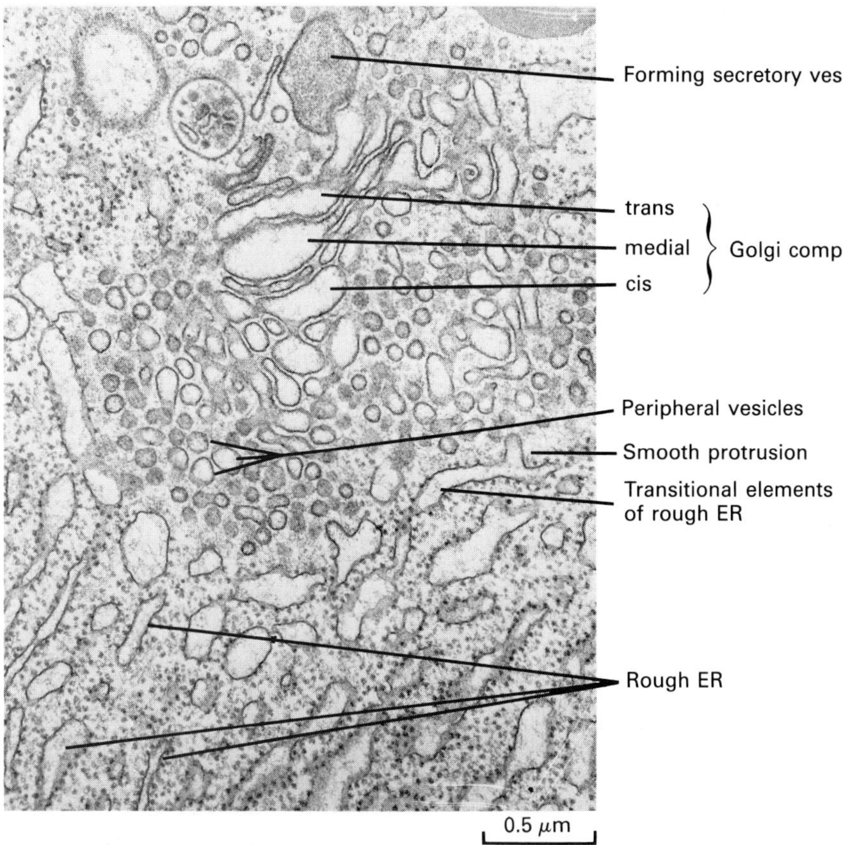

Figure 21-39 The Golgi complex and ER in an exocrine pancreatic cell. The stacked vesicles of the Golgi complex are evident, as is a forming secretory vesicle. The rough ER contains transitional elements from which smooth protrusions appear to be budding. These are believed to generate the peripheral vesicles that transport membrane and secretory proteins from the rough ER to the Golgi. Other vesicles seen at the periphery of the Golgi complex may transport proteins from one Golgi vesicle to another, or from the Golgi to the condensing vesicles. *Courtesy of G. Palade.*

stabilizing force, possibly because of a greater reducing potential in the cytoplasm.

In animal cells, formation of many intramolecular disulfide bonds occurs while the polypeptide is still growing on the ribosome (Figure 21-40). In the case of the secreted immunoglobulin light-chain polypeptide that is portrayed in Figure 21-40, the disulfide bonds are formed on the lumenal surface of the ER, and bond formation occurs sequentially: the first cysteine pairs with the second, and the third with the fourth. The disulfide bonds stabilize the two separate domains in this protein; without formation of the proper bond, the protein cannot achieve a functional conformation.

The sequential formation does not occur in all disulfide-bonded proteins. For example, proinsulin (see Figure 3-13) has three disulfide bonds that are not sequential. Presumably the folding of the nascent proinsulin in the lumen of the ER allows just the proper disulfide bonds to form. Formation of disulfide bonds between two polypeptide chains, such as between immunoglobulin H and L chains (see Figure 3-38) occurs in the lumen of the ER as the completed chains fold into their proper multimeric configuration.

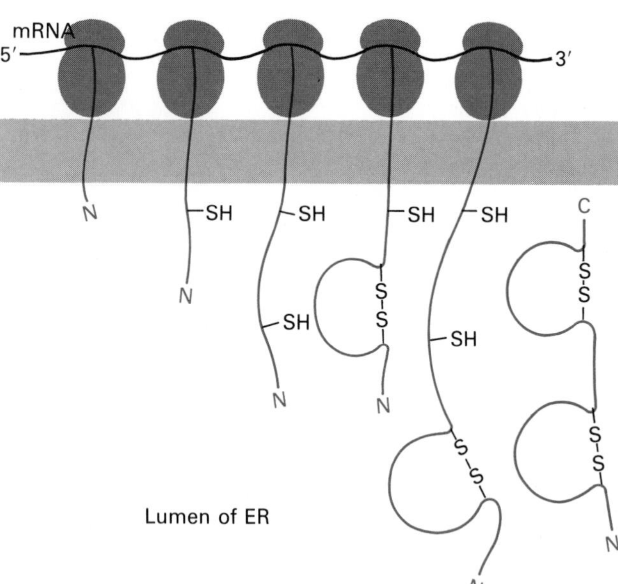

Figure 21-40 Formation of disulfide bonds on a nascent secretory protein. This diagram depicts an immunoglobulin light-chain polypeptide that contains two disulfide bonds (see Figure 3-38). Note in particular that the disulfide bonds form on the lumenal side of the ER and that in this molecule, they form sequentially while the polypeptide is growing. [See L. W. Bergman and W. M. Kuehl, 1979, *J. Biol. Chem.* **254**:8868.]

Several Proteolytic Cleavages Occur during Maturation of Secretory and Membrane Proteins

We have seen that an N-terminal signal sequence of amino acids is removed from many nascent secretory proteins and membrane proteins while the chain is still growing on the ribosome. In the case of some secretory proteins, such as growth hormone, placental lactogen, lysozyme, and ovomucoid, and certain viral membrane proteins, such as the VSV glycoprotein, removal of the signal sequence is the only known proteolytic cleavage, and removal converts the mature form directly into the mature active protein. In most cases, however, there is an additional, relatively long-lived intracellular "pro" form termed the proprotein, or prohormone. Serum proteins such as albumin, hormones such as insulin and parathyroid hormone (Figure 21-41), and membrane proteins such as the influenza virus hemagglutinin (see Figure 15-54) are synthesized as longer precursors, and one or more proteolytic cleavages are essential to generate the active, mature molecule. In general, proteolytic conversion of the proprotein to the mature molecule occurs at a late stage in intracellular maturation. The conversion of proinsulin to insulin is known to occur within secretory vesicles. The additional amino acids in the pro form can be either at the N-terminus or at both ends of the proprotein (Figure 21-42). In the case of proinsulin, the extra amino acids, collectively termed the C peptide, are located internally in the polypeptide. In proinsulin the N-terminal B chain and the C-terminal A chain of mature insulin are linked by disulfide bonds and remain attached when the C peptide is removed (Chapter 3).

In many proproteins, such as the ones depicted in Figure 21-42, the initial endoprotease cleavage occurs at the carboxyl side of two or three adjacent basic amino acid residues. An enzyme that catalyzes the proinsulin to insulin cleavage has been purified from crude secretory vesicles of islet cells, and similar (or identical) enzymes are believed to function in other secretory cells.

Why have cells evolved such complex processing pathways for secreted proteins, peptide hormones in particular? One reason is that, by delaying generation of the active hormone until it is packaged in secretory vesicles, the cell is prevented from being stimulated by the hormone it has made. Were functional insulin formed in the rough ER of the pancreatic islet cell, for example, it could bind to insulin receptors in the ER membrane and trigger an unwanted response. The secretory vesicles in which mature insulin is finally formed have a pH of about 5.0. Since insulin, like most peptide hormones, cannot bind to its receptor at this acidic pH (Chapter 15), any insulin receptors in the secretory vesicle membrane cannot be triggered by insulin.

A second rationale is that multiple copies of the same hormone can be generated by processing of a single pre-

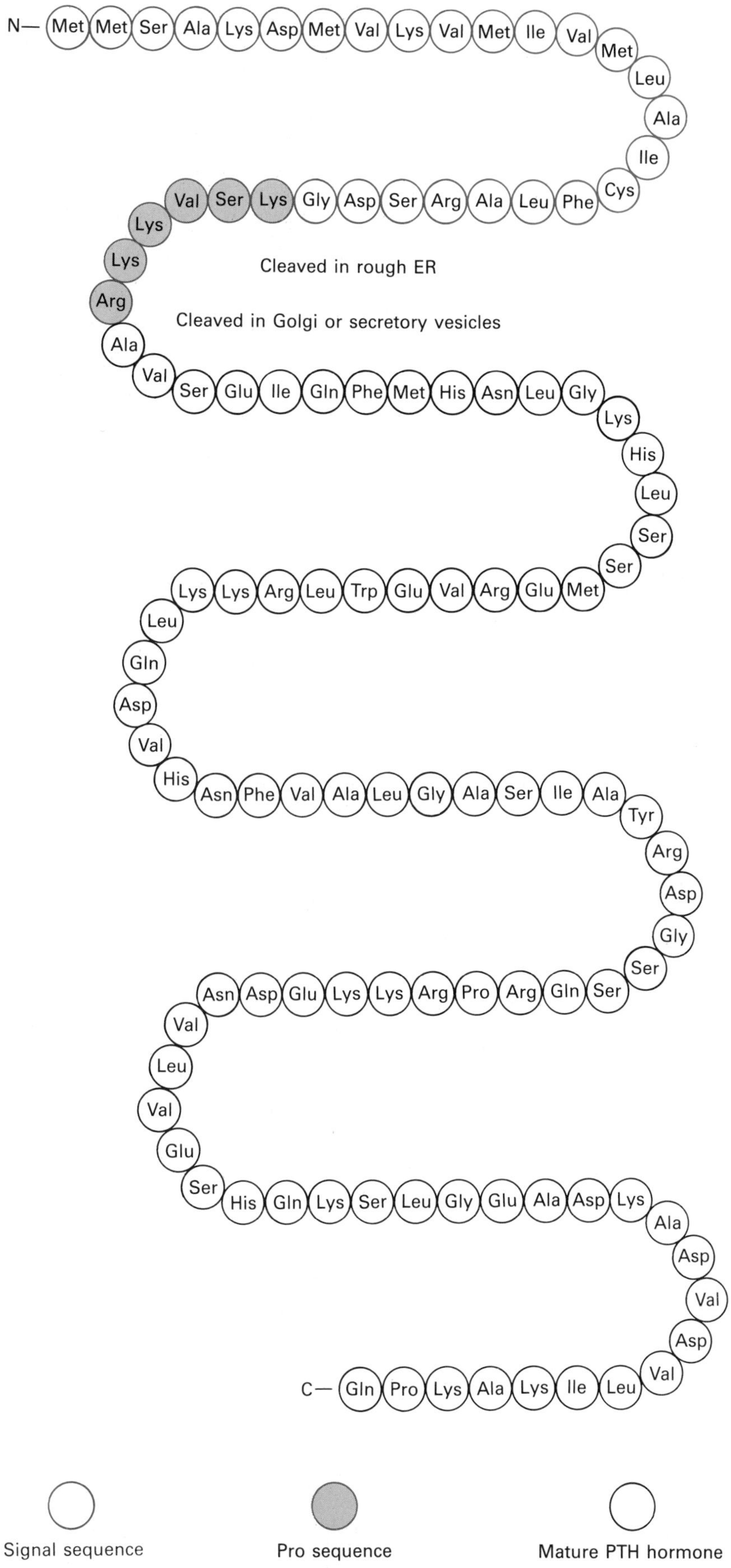

N— Met Met Ser Ala Lys Asp Met Val Lys Val Met Ile Val Met Leu Ala Ile Cys
Val Ser Lys Gly Asp Ser Arg Ala Leu Phe
Lys Lys Arg Ala Val
Cleaved in rough ER
Cleaved in Golgi or secretory vesicles
Ser Glu Ile Gln Phe Met His Asn Leu Gly Lys His Leu Ser Ser
Lys Lys Arg Leu Trp Glu Val Arg Glu Met
Leu Gln Asp Val His
Asn Phe Val Ala Leu Gly Ala Ser Ile Ala Tyr Arg Asp Gly Ser
Asn Asp Glu Lys Lys Arg Pro Arg Gln Ser
Val Leu Val Glu Ser
His Gln Lys Ser Leu Gly Glu Ala Asp Lys Ala Asp Val Asp Val
C— Gln Pro Lys Ala Lys Ile Leu

Signal sequence

Pro sequence

Mature PTH hormone

Figure 21-41 Production of parathyroid hormone (PTH) requires two proteolytic cleavages. The signal sequence, composed of 25 N-terminal residues, is cleaved from the nascent protein in the rough ER. Just before the protein is secreted—while it is in the Golgi complex or secretory vesicles—the six amino acids of the pro sequence are cleaved, to generate the mature hormone. The site of cleavage of the pro sequence is at the C-terminus of a sequence of three basic amino acids; two or three adjacent basic amino acids form the cleavage site of many other prohormones. [See J. T. Potts et al., 1980, *Ann. NY Acad. Sci.* **343**:38.]

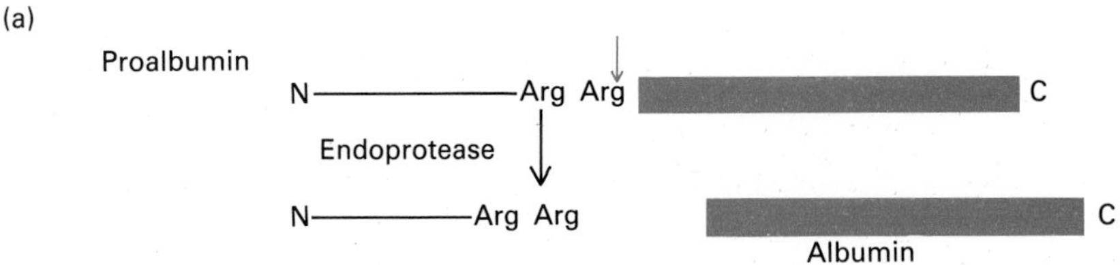

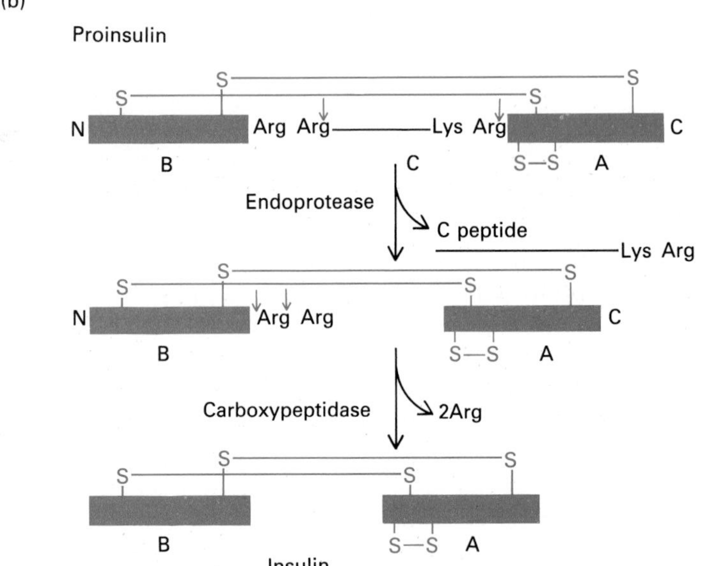

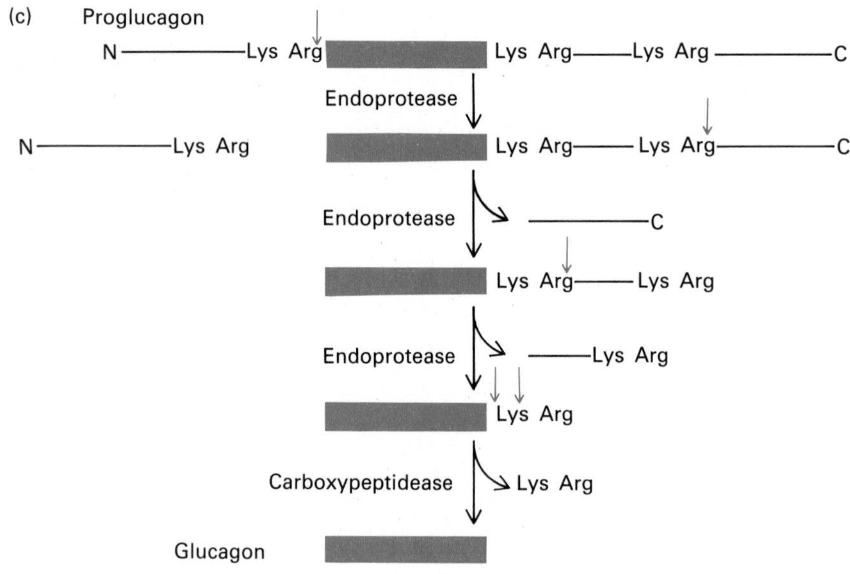

Figure 21-42 Schema for maturation of three typical proproteins. (a) Proalbumin is converted to albumin with a single cleavage by an endoprotease that cuts adjacent to a sequence of dibasic amino acids. (b) Two cleavages of proinsulin by a similar endoprotease releases the C peptide. The B chain is processed further by action of a carboxypeptidase that removes, sequentially, the two basic arginines at the C-terminus. (c) Processing of proglucagon to glucagon involves three successive cleavages by an endoprotease that cleaves after two basic amino acids. Finally, a carboxypeptidase removes the two C-terminal basic amino acids from the last intermediate to form mature glucagon. The colored boxes represent the sequences conserved in the mature protein; the black lines denote the sequences that are lost. [See J. T. Potts et al., 1980, *Ann. NY Acad. Sci.* **343**:38; and L. C. Lopez et al., 1983, *Proc. Nat'l Acad. Sci. USA* **80**:5485.]

cursor molecule, thus increasing the amount of hormone that can be packaged in a secretory vesicle. For instance, the related pentapeptides methionine-enkephalin (Tyr-Gly-Gly-Phe-Met) and leucine enkephalin (Tyr-Gly-Gly-Phe-Leu) are among the smallest of the peptide hormones (Chapter 17). Enkephalins are synthesized as prohormones containing at least five methionine enkephalins and one leucine enkephalin. Other peptide hormones, such as the yeast mating factor α, are also synthesized as a precursor that contains multiple copies of the same peptide sequence.

In still other cases a single precursor contains the sequences of several different protein hormones. Depending on the proteolytic cleavages that occur, different hormones can be generated from the same precursor. A particularly interesting example is the biogenesis of the peptide hormone adrenocorticotropic hormone, or ACTH (see Figure 10-30). In both anterior and intermediate

Figure 21-43 Melletin, a 27-amino-acid lytic peptide in bee stings, is secreted as an inactive precursor, promelletin. The processing of the precursor is unusual, since it has an N-terminal amino acid sequence composed of 11 dipeptides of the structure XY, where X is alanine, glutamine, or aspartic acid, and Y is either proline or alanine. A specific extracellular aminopeptidase that cuts dipeptides XY (*arrows*) sequentially from the N-terminus catalyzes the conversion of promelletin to melletin. This dipeptidase cuts at the C-terminal side of a proline or alanine residue only and stops when it encounters the sequence Gly-Ile-Gly, which it cannot cleave. [See G. Kreil et al., 1980, *Ann. NY Acad. Sci.* **343**:338; and C. Mollay, U. Vilas, and G. Kreil, 1982, *Proc. Nat'l Acad. Sci. USA* **79**:2260.]

lobes of the pituitary, ACTH is synthesized as a proprotein that also is a precursor of β-lipotropin (βLPH), γ-lipotropin (γLPH), β-endorphin, and other hormones. ACTH as well as the other hormones are generated by proteolytic cleavages in both types of pituitary cells. However, in intermediate lobe cells, ACTH is cleaved further into α-melanocyte-stimulating hormone (αMSH), a polypeptide with a different physiological function. By making two or more peptide hormones from a common precursor, the cell can coordinate the release of multiple hormones simultaneously, and different cells process the same protein differently.

Finally, many cells secrete toxins or potentially hazardous enzymes. Activation of precursors just before secretion, or after secretion, prevents the product from destroying the cell that made it. Consider for example, the formation of *melletin,* a 27-amino-acid peptide that is the main secretion product of the venom gland of honey bees. This peptide lyses cells (the major reason that bee stings hurt!). Proteolytic processing removes sequentially 11 dipeptides from the N-terminus of promelletin (Figure 21-43). These cleavages are necessary for activation of the lytic activity of melletin and apparently occur only after the protein has been secreted from the cells—thereby preventing active melletin from lysing the cells that made it.

Protein Glycosylation

As noted earlier, many important cell surface proteins and secretory proteins contain one or several carbohydrate groups. Such proteins are said to be *glycosylated.* These carbohydrates serve many functions. They often affect the conformation of the polypeptide, making it more resistant to proteolytic digestion. Because of their many hydroxyl groups, bound carbohydrates often increase the solubility of proteins. As we shall see, some carbohydrate residues are important in sorting of proteins to their correct organelles within a cell. All protein-linked (and lipid-linked) sugars are localized to the extracytoplasmic face of cellular membranes. Cytoplasmic proteins are never glycosylated.*

Although glycoproteins occur in all eukaryotic cells that have been studied, the eubacteria contain few if any. In eukaryotes, sugar residues are commonly linked to four different amino acid residues. These are classified as *O-linked* (serine, threonine, and, in collagen, hydroxylysine) and *N-linked* (asparagine). The structures of *N*- and *O*-linked oligosaccharides are very different, and different sugar residues are usually found in each type (Chapter 3). All *N*-linked oligosaccharides, for example, contain the sequence

$$\text{Man} \xrightarrow{1,4} \text{GlcNac} \xrightarrow{1,4} \text{GlcNAc} \longrightarrow \text{Asn}$$

linked to the asparagine, whereas in *O*-linked sugars, galactose or *N*-acetylgalactosamine is linked to the serine or threonine residue. Oligosaccharides that are *O*-linked are generally shorter, often only one to three sugar residues long (Figure 21-44: see also Figures 3-76 and 3-77). However, some *O*-linked oligosaccharides, such as those bearing the ABO blood-group antigens, can be very long (see Figure 3-79).

O-Linked Sugars Are Synthesized in the Golgi or Rough ER from Nucleotide Sugars

Incorporation of sugars into polymers such as glycoproteins requires an input of energy. The high-energy inter-

*In pathology, there is one interesting exception. A chemical reaction, non-enzyme-catalyzed, can occur between the C-1 carbon of glucose and the N-terminus of hemoglobin polypeptides, forming a stable covalent Schiff base adduct. The level of modified globin is high in diabetics, who have high levels of blood glucose and thus of cytoplasmic glucose. Measurement of this globin is often used in following the course and severity of diabetes.

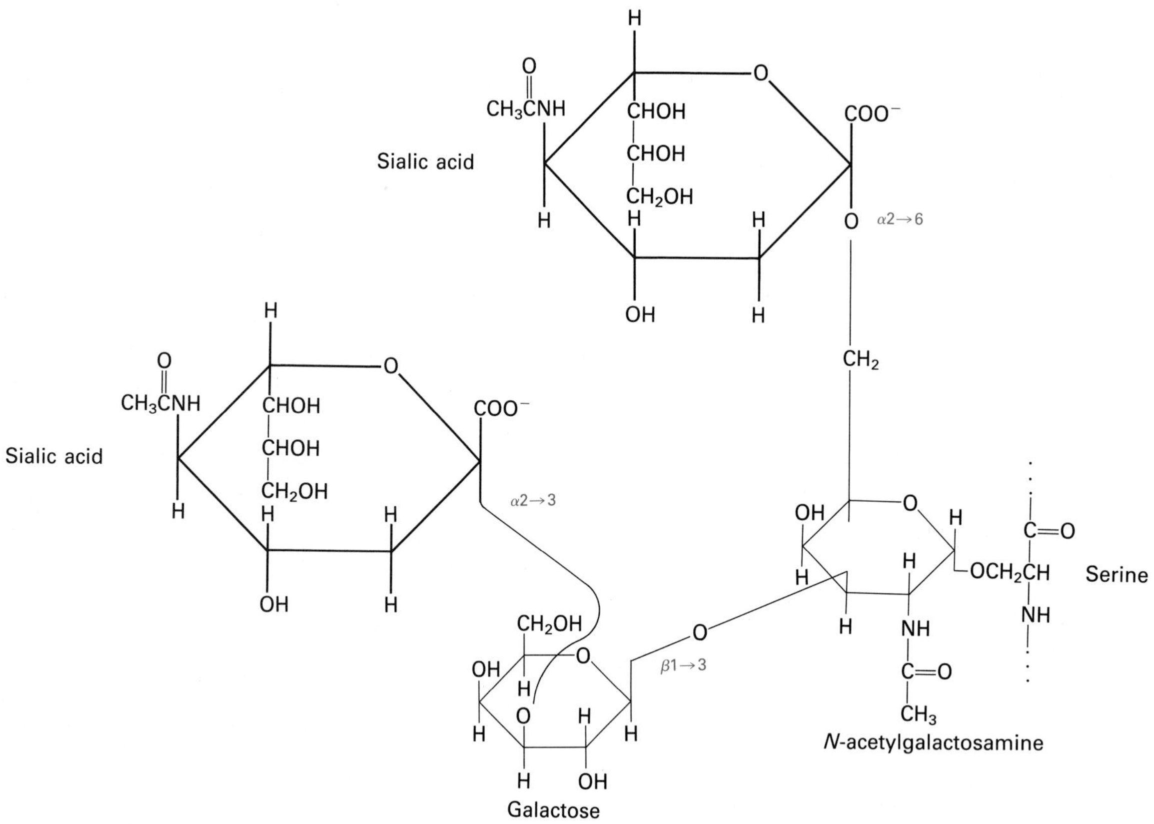

Figure 21-44 Structure of the oligosaccharide linked to serine and threonine hydroxyl groups in proteins such as glycophorin and the LDL receptor. As depicted here, two negatively charged *N*-acetylneuraminic acids (sialic acids) are attached, one to the galactose and one to the *N*-acetylgalactosamine, although in some cases only one or the other of the sialic acids is present.

mediates used for biosynthesis of oligosaccharides are nucleoside diphosphate sugars or nucleoside monophosphate sugars (Figures 21-45 and 21-46) and are brought into the Golgi lumen by special transport antiports.

In the formation of O-linked glycoproteins, sugars are added one at a time, and each sugar transfer is catalyzed by a different type of *glycosyl transferase* (Figure 21-47). For instance, in the biogenesis of the disaccharide side chains of collagen, glucose-galactose-hydroxylysine, first a galactose residue is transferred from UDP-galactose to a hydroxylysine side chain of the polypeptide. A second enzyme then catalyzes the addition of a glucose residue to the galactose. Glycosyl transferase enzymes have exquisite specificity, both for the sugar nucleotide substrate and for the specific carbon atom of the sugar or amino acid acceptor. One enzyme, for instance, adds *N*-acetylneuraminic acid (sialic acid) from CMP–sialic acid only to the 3 carbon atom of galactose; another, only to the 6 carbon. Although some O-linked glycosylation may begin on nascent proteins, most of the known glycosyl transferase enzymes face the lumen of the Golgi, and the majority of O-linked sugars are added to secretory proteins only a few minutes before secretion. Addition of sialic acid is the last step in biosynthesis of the O-linked (as well as *N*-linked) sugars and occurs in the trans-most segment of the Golgi. Similar to O-linked glycoproteins, saccharide residues in glycolipids are added to lipids in the Golgi a few minutes before the glycolipids appear on the plasma membrane. These reactions also utilize glycosyl transferases whose active sites face the lumen. Since several oligosaccharides, such as the ABO antigens, are found attached to both glycoproteins and glycolipids, the same glycosyl transferase may be utilized for synthesis of both glycoproteins and glycolipids.

The Golgi Membrane Contains Permeases for Nucleotide Sugars

Addition of sugars such as sialic acid and galactose to proteins occurs in the lumen of Golgi vesicles. The nucleoside sugar substrates, however, are fabricated in the cytosol from nucleoside triphosphates and sugar phosphates (see Figure 21-46). Presumably, the Golgi membranes contain a specific transporter or permease for these sugar nucleotides, and indeed, such specific uptake

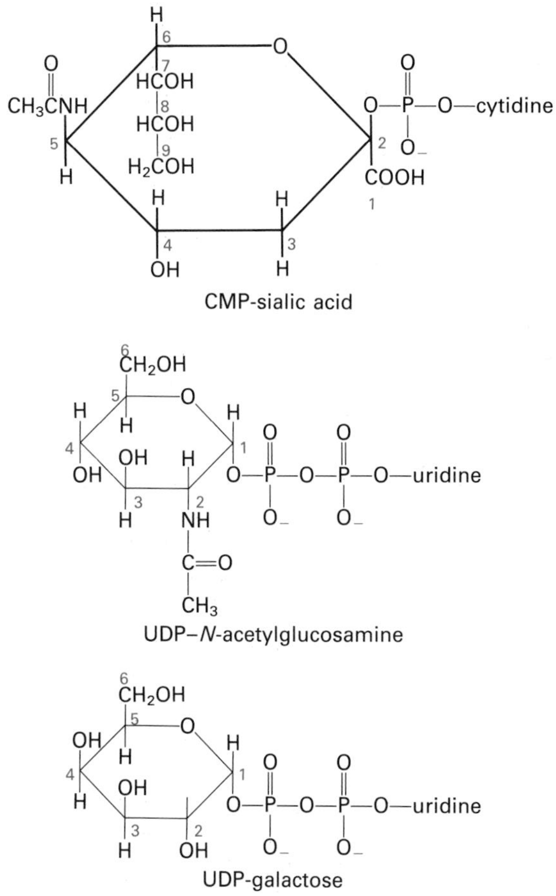

CMP-sialic acid

UDP–*N*-acetylglucosamine

UDP-galactose

Figure 21-45 Structure of sugar nucleotides that are precursors of the saccharide residues in glycoproteins. As noted in Chapter 16, UDP-glucose (see Figure 16-14) is also the substrate for synthesis of glycogen and other polysaccharides.

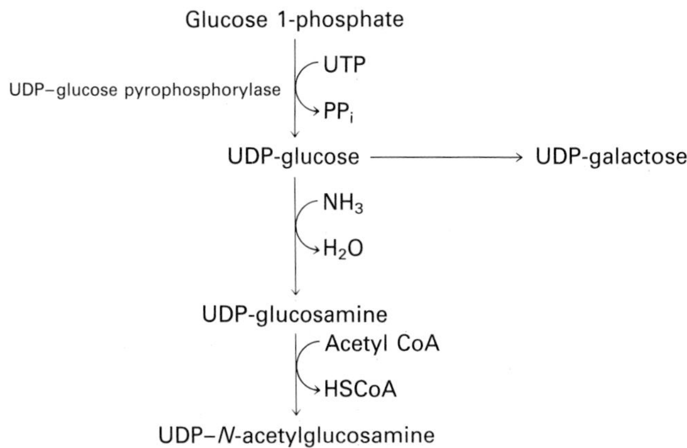

Figure 21-46 Schematic outline of the synthesis of some common sugar nucleotides.

systems for CMP–sialic acid and UDP-galactose have been identified (Figure 21-48). These permeases are antiports (Chapter 15). Nucleotides such as UDP and CMP, formed by the glycosyl transferase reaction, must be exported from the Golgi. UDP is first hydrolyzed to UMP and P_i, and UMP is transported out of the Golgi in a one-for-one exchange for UDP-galactose. Other antiports catalyze a one-for-one exchange of CMP–sialic acid for CMP and of GDP-fucose for GMP. In catalyzing this exchange, these antiports allow the concentration of nucleotide sugars within the Golgi to be maintained at a constant level, commensurate with the requirements for oligosaccharide synthesis.

The Diverse *N*-Linked Oligosaccharides Share Certain Structural Features

The *N*-linked oligosaccharides on different proteins have seemingly very different structures (Figure 21-49). One class of *N*-linked oligosaccharides—such as those found on many serum proteins and on viral glycoproteins, depicted in Figure 21-49a—are termed *complex* because they contain *N*-acetylglucosamine, mannose, fucose, galactose, and sialic acid. The different structures among complex glycoproteins result from the number of branches (ranging from two to four), and the number of sialic acid residues (from zero to four). There is additional variability, since in some cases, the 2 carbon atom of sialic acid is linked to the 3 carbon of galactose, while in others it is linked to the 6 carbon.

The second class of *N*-linked glycoproteins is termed *high-mannose*, since these proteins contain only *N*-acetylglucosamine and mannose. For example, the principal difference among members of this family of related oligosaccharides is in the number of mannose residues attached to *N*-acetylglucosamine (Figure 21-49b). In some cases a complex oligosaccharide and a high-mannose oligosaccharide are attached to different asparagine residues in the same polypeptide.

All oligosaccharides from a wide variety of proteins contain three mannose and two *N*-acetylglucosamine residues in exactly the same configuration. This observation suggested that all *N*-linked oligosaccharides, high-mannose as well as complex, are made from a common precursor that is modified differently. In fact, all *N*-linked oligosaccharides are formed from the identical precursor in a multistep process involving the sequential removal and addition of specific saccharide residues after the precursor oligosaccharide has been transferred to the polypeptide. Numerous enzymes localized to several organelles are involved in this process.

N-Linked Oligosaccharides Are Synthesized from a Common Precursor and Subsequently Processed

The structure of the precursor of all *N*-linked oligosaccharides is a branched oligosaccharide containing three glucose, nine mannose, and two *N*-acetylglucosamine molecules. It is linked by a pyrophosphoryl residue to

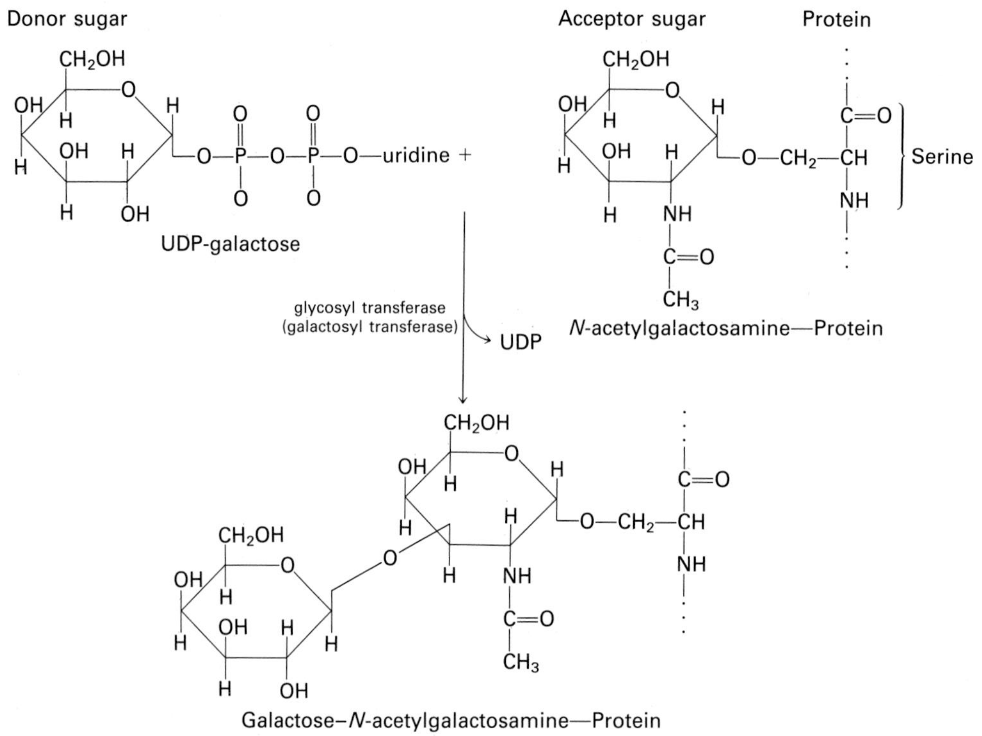

Figure 21-47 Addition of a galactose residue to the 3 carbon atom of *N*-acetylgalactosamine attached to a protein: a step in elongation of typical O-linked oligosaccharides. Where the donor is a nucleoside diphosphate sugar, a nucleoside diphosphate is one product. Where the donor is a nucleoside monophosphate sugar, such as CMP–sialic acid, a nucleoside monophosphate is a product. Glycosyl transferases are specific both for the nucleoside sugar donor and for the carbon atom of the acceptor sugar to which it is transferred.

dolichol, a long-chain (75 to 95 carbon atoms) unsaturated lipid (Figure 21-50). The dolichol pyrophosphoryl oligosaccharide is formed in the ER and is oriented in the membrane such that the oligosaccharide faces the lumen.

The biogenesis of this precursor utilizes membrane-bound enzymes of the rough ER and involves sequential addition of monosaccharides to dolichol phosphate (Figure 21-51). The dolichol portion is firmly embedded in the ER membrane, and the oligosaccharide portion extends into the lumen.

Nucleoside diphosphate sugar precursors are soluble constituents of the cytosol. Thus the sugars must cross the ER membrane at some stage of the biosynthesis of the final oligosaccharide-lipid, which projects into the lumen

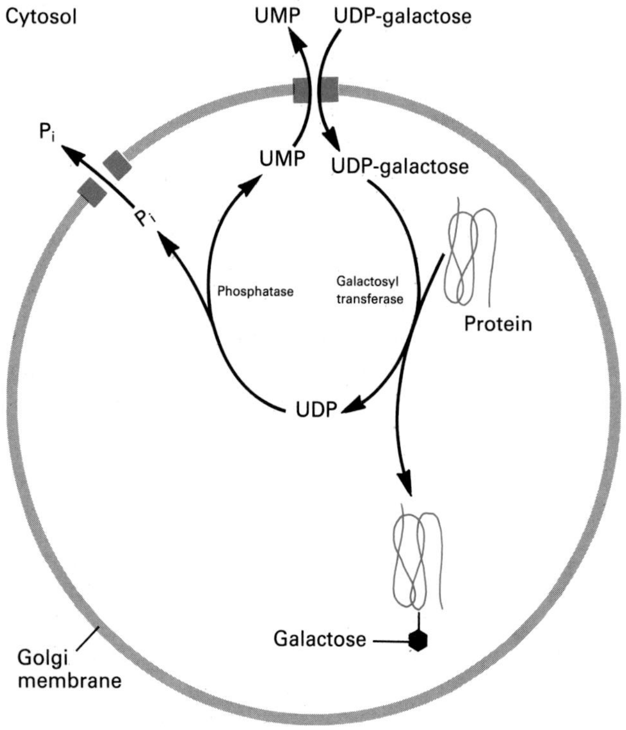

Figure 21-48 Uptake of nucleoside sugars into Golgi vesicles. UDP-galactose enters from the cytosol in exchange for UMP using an antiport located in Golgi membranes. UMP is produced by phosphatase action on UDP, a product of the galactosyl transferase reaction. A permease allows the inorganic phosphate, formed from UDP, to exit the Golgi. Other antiports allow CMP–sialic acid to enter in exchange for CMP and GDP-fucose to enter in exchange for GMP. [See S. L. Deutscher and C. B. Hirschberg, 1986, *J. Biol. Chem.* **261**:96–100.]

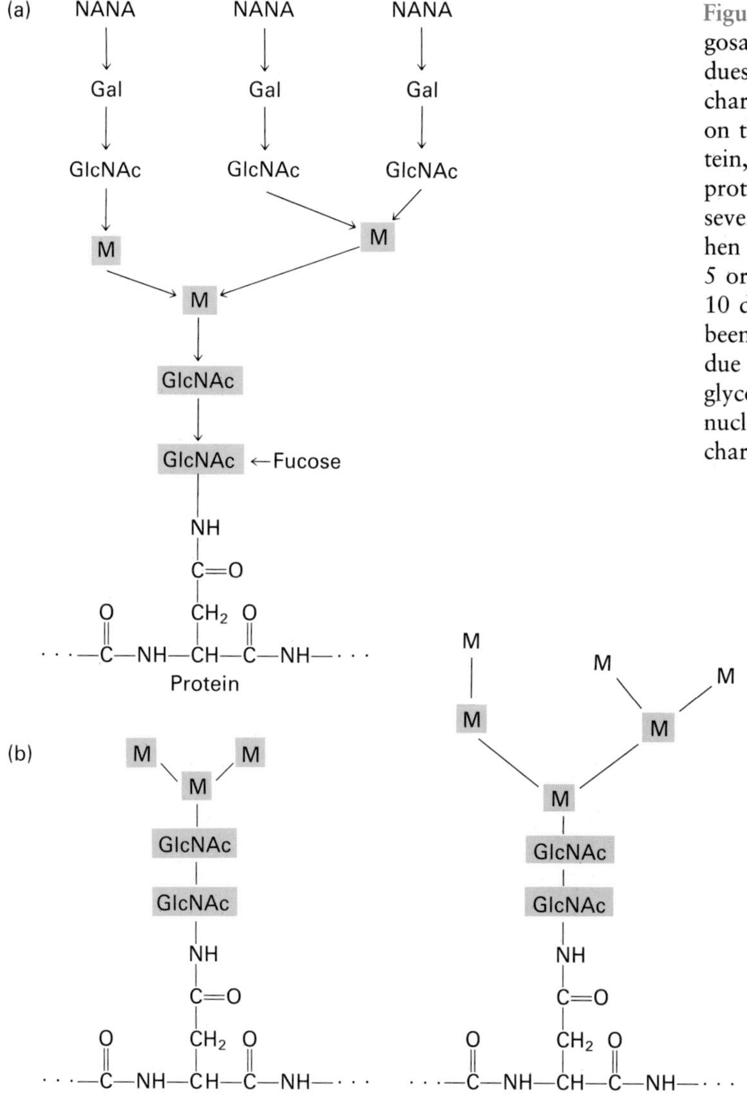

Figure 21-49 Structures of asparagine-linked oligosaccharides. Indicated in color are the five residues that are always found in *N*-linked oligosaccharides. (a) The complex oligosaccharides found on the vesicular stomatitis virus (VSV) glycoprotein, as well as on a variety of mammalian serum proteins, such as immunoglobulins. (b) Two of several related high-mannose oligosaccharides on hen ovalbumin. Other chains may have as few as 5 or as many as 9 mannose residues, and at least 10 different high-mannose oligosaccharides have been found attached to the single asparagine residue of ovalbumin that is glycosylated. Many virus glycoproteins and secretory proteins, such as ribonuclease, contain similar high-mannose oligosaccharides.

of the ER. As Figure 21-51 depicts, the first seven sugars are added on the cytoplasmic face of the ER, while the final seven are added on the lumenal surface. How translocation of the intermediates to this surface is accomplished is under active study.

The completed oligosaccharide chain is transferred en bloc by *oligosaccharide transferase* to the nascent polypeptide at tripeptide sequences Asn-X-Ser or Asn-X-Thr (where X is any amino acid). Experiments with tripeptides as substrates suggest that this tripeptide is the recognition sequence for the transferase, which is located on the lumenal surface of the ER (Figure 21-52).

Immediately after transfer to the polypeptide, while the protein is still in the rough ER, all three glucose residues are removed by two different enzymes (Figure 21-53). The glucose residues—the last to be added to the oligosaccharide as it is being formed—appear to act as a signal that the oligosaccharide is complete and ready to be transferred to a protein.

Modifications to *N*-Linked Oligosaccharides Are Completed in the Golgi

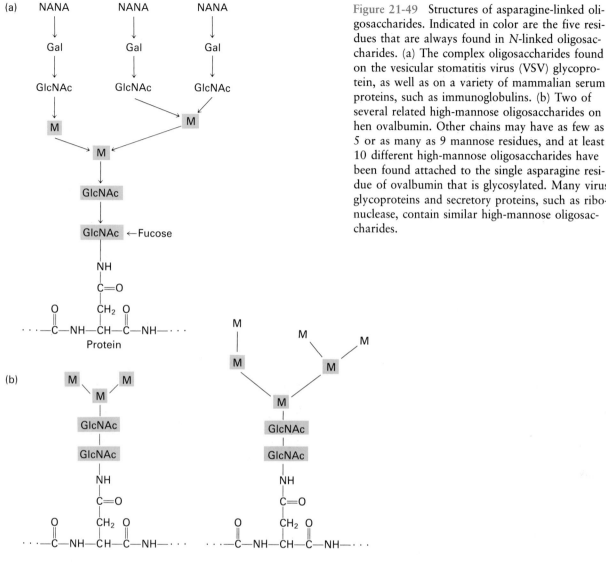

Further processing of the oligosaccharide to the complex form begins only 10 to 20 min after synthesis of the protein, at the time the protein is transferred to the Golgi

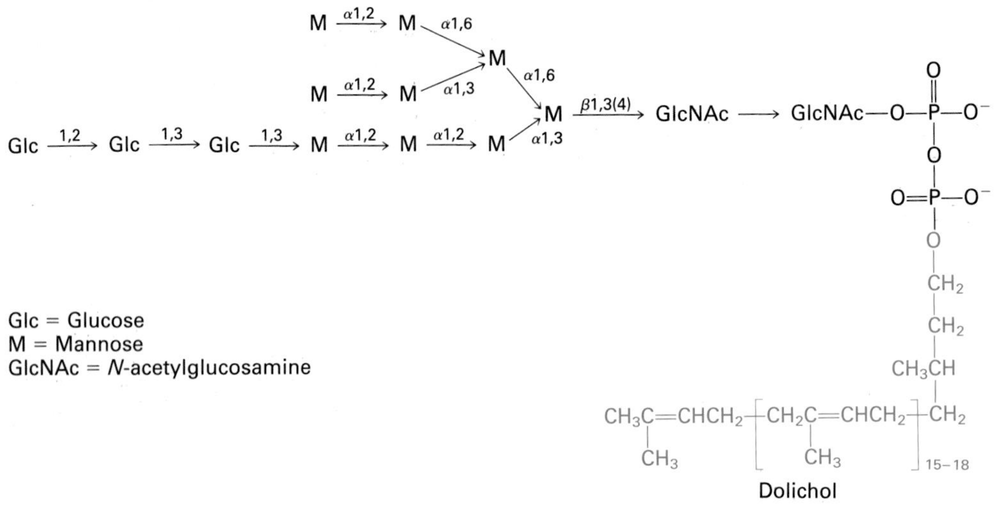

Glc = Glucose
M = Mannose
GlcNAc = *N*-acetylglucosamine

Dolichol

Figure 21-50 Structure of the dolichol pyrophosphoryl oligosaccharide precursor of *N*-linked oligosaccharides. Dolichol is strongly hydrophobic and is long enough to span a phospholipid bilayer membrane four or five times. This oligosaccharide lipid is located on the rough ER, and the sugar residues face the lumen of the organelle.

complex. In a stepwise, concerted set of reactions, six of the nine mannose residues are removed and *N*-acetylglucosamine (three residues), galactose (three residues), sialic acid (one to three residues per chain), and fucose (one residue) are added one at a time to each oligosaccharide chain (Figure 21-53). Oligosaccharide processing is completed about 10 min before the protein reaches the cell surface or is secreted.

Note that the high-mannose oligosaccharides that are found on the mature form of certain glycoproteins (see Figure 21-49) resemble intermediates ($Man_9GlcNAc_2$ to $Man_5GlcNAc_2$) in processing of the complex class of oligosaccharides (Figure 21-53). It is believed that both classes of *N*-linked oligosaccharides derive from this same intermediate, the difference being the sensitivity of the particular oligosaccharide to the *α-mannosidase* (mannose-cleaving enzyme) that catalyzes reaction 4 in Figure 21-53.

The conformation of the protein, and ultimately its primary amino acid sequence, thus determines whether or not a particular *N*-linked oligosaccharide becomes complex or remains high-mannose. Different types of cells in an organism may contain different complements of processing enzymes, and the same protein produced by various cell types may have differently processed carbohydrates. Certain oligosaccharides on the influenza virus, glycoproteins, for instance, are of the complex type when the virus is grown in one type of cell, but they are of the high-mannose form when the virus is grown in another.

Electron microscopic autoradiographic experiments show very clearly that different sugars are added to protein in different organelles. When cells are exposed briefly to [^{3}H]mannose, virtually all the incorporated radioactivity (observed as grains) is within the ER. Radioactivity from newly incorporated [^{3}H]fucose or [^{3}H]sialic acid, by contrast, is enriched over the trans Golgi region (Figure 21-54). Furthermore, galactosyl transferase is localized to the trans-most vesicles of the Golgi (Figure 21-55).

Thus the *N*-linked oligosaccharides are modified in much the same way that an automobile is built on an assembly line. The protein is transported from organelle to organelle, in each of which it is acted upon sequentially by a large set of enzymes. The reaction product of one enzyme is the substrate of the next. The segregation of the different modifying enzymes to different organelles ensures that only the properly modified substrates are presented to each enzyme.

N-Linked Oligosaccharides May Stabilize Maturing Secretory and Membrane Proteins

What is the function of *N*-linked oligosaccharides? In many cases, they appear to be required for secretion of proteins or for movement of membrane glycoproteins to the cell surface. In the presence of tunicamycin, an antibiotic that blocks the first stage in formation of the oligosaccharide-lipid donor (see Figure 21-51), the polypeptide is synthesized, but it contains no *N*-linked sugar chains. Secretion of some but not all proteins is observed to take place. For instance, the rate and extent of secretion of glycosylated and unglycosylated fibronectin by fibroblasts or transferrin by a cultured line of rat hepatoma cells are the same. However, the unglycosylated fibronectin is more susceptible to degradation by tissue proteases than is normal fibronectin—a result that implicates the carbohydrates in conferring stability to this extracellular matrix protein. Because the impact of the ab-

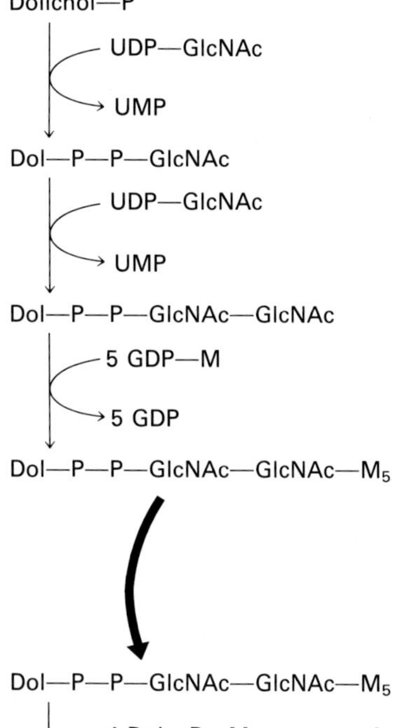

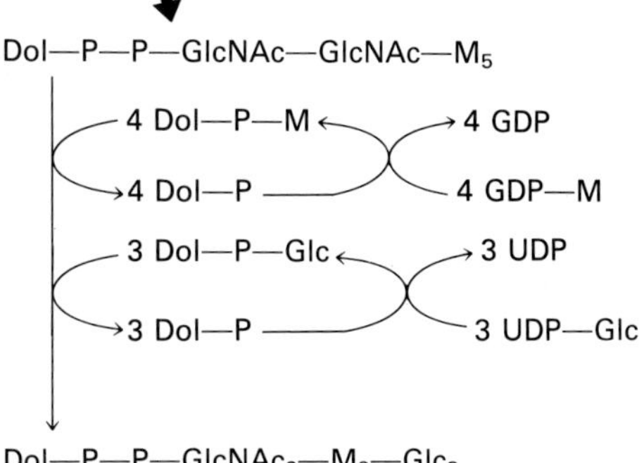

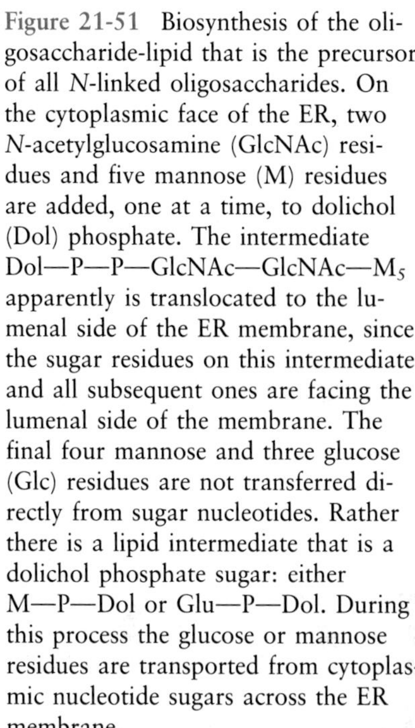
Figure 21-51 Biosynthesis of the oligosaccharide-lipid that is the precursor of all *N*-linked oligosaccharides. On the cytoplasmic face of the ER, two *N*-acetylglucosamine (GlcNAc) residues and five mannose (M) residues are added, one at a time, to dolichol (Dol) phosphate. The intermediate Dol—P—P—GlcNAc—GlcNAc—M_5 apparently is translocated to the lumenal side of the ER membrane, since the sugar residues on this intermediate and all subsequent ones are facing the lumenal side of the membrane. The final four mannose and three glucose (Glc) residues are not transferred directly from sugar nucleotides. Rather there is a lipid intermediate that is a dolichol phosphate sugar: either M—P—Dol or Glu—P—Dol. During this process the glucose or mannose residues are transported from cytoplasmic nucleotide sugars across the ER membrane.

Evidence for the "sidedness" of oligosaccharide addition came from studies in which a mannose-binding lectin was added to preparations of rough ER. The lectin was found to bind to intermediates up to M_5—$GlcNAc_2$—PP—Dol but not to the later intermediates unless the membrane was made permeable by detergents. This finding indicates that these later intermediates face the lumenal surface. [See M. Snider and O. C. Rogers, 1984, *Cell* **36**:753.]

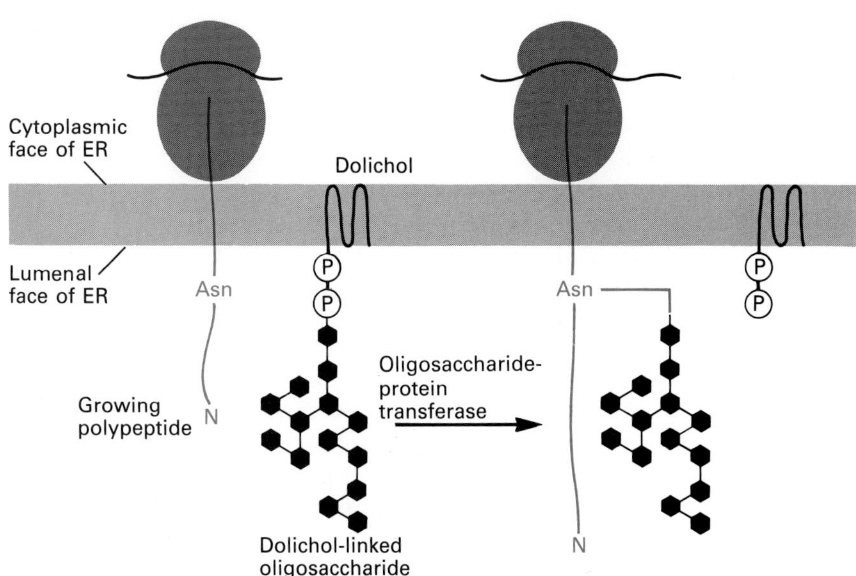

Figure 21-52 Transfer of an oligosaccharide from the dolichol carrier to a susceptible asparagine residue on a nascent protein. As indicated, the oligosaccharide is transferred en bloc as soon as the asparagine crosses to the lumenal side. The enzyme is termed an oligosaccharide-protein transferase.

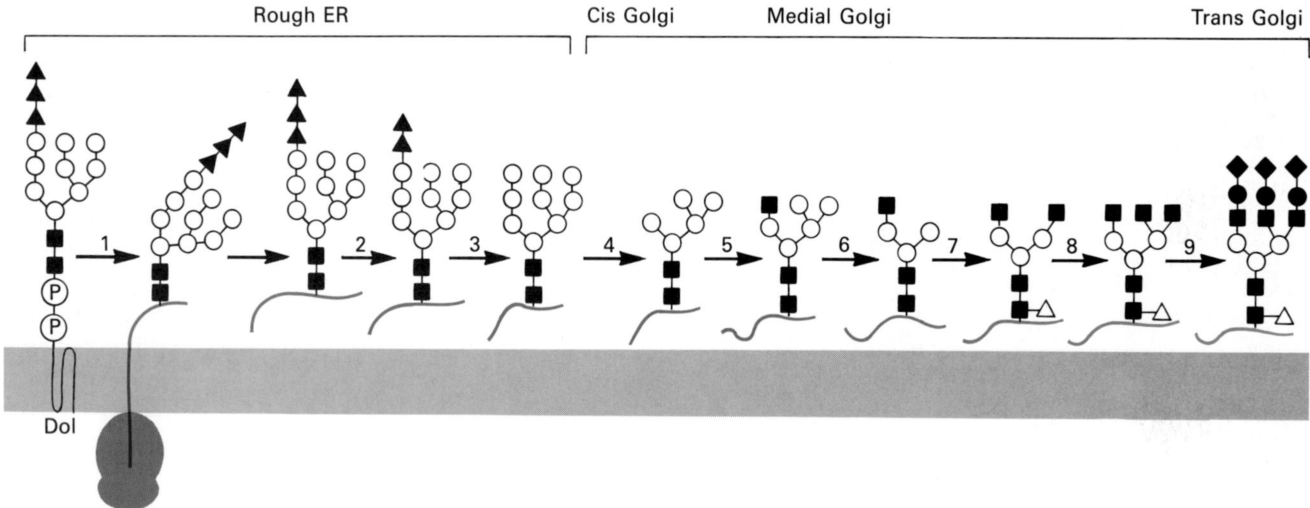

Dol dolichol

■ *N*-acetylglucosamine

○ Mannose

▲ Glucose

● Galactose

△ Fucose

◆ Sialic acid

Figure 21-53 Formation of complex *N*-linked oligosaccharides. At least 11 discrete enzymes in three organelles act sequentially to modify the common precursor of *N*-linked oligosaccharides. The numbered steps in the diagram are as follows: *2, 3,* glucose removed; *4,* four mannose removed; *5,* one *N*-acetylglucosamine (GlcNAc) added; *6,* two mannose removed; *7,* GlcNAc + fucose added; *8,* GlcNAc added; *9,* three galactose and three sialic acid added. Most likely, reaction 4 is carried out by at least two enzymes; mannose removal begins in the rough ER by one α-mannosidase and is continued in the cis Golgi by a second enzyme. [See R. Kornfield and S. Kornfield, 1985, *Annu. Rev. Biochem.* **45**:631–664.]

sence of glycosylation is so variable and by no means absolute, researchers have concluded that sugar residues play no mandatory role: they are not a "ticket" needed to move through the transport organelles. In all likelihood, carbohydrates play a role in ensuring correct charge, conformation, and stability of maturing proteins. For some proteins, this function of the carbohydrates is apparently superfluous; for others, it is clearly necessary. Certain membrane proteins lacking the appropriate *N*-linked oligosaccharides undergo extensive aggregation in the rough ER membrane, which inhibits normal flow through the various membranes and vesicles.

Membrane Dynamics and the Sorting and Maturation of Proteins

Earlier sections of this chapter discussed the localization of cytoplasmically synthesized proteins into three cytoplasmic organelles: the mitochondrion, peroxisome, and chloroplast. A parallel problem exists for the sorting of proteins that are synthesized in the rough ER and that must end up in the lysosome, ER, Golgi, or plasma membrane (see Table 21-1). The entire basis for the selective movement and localization of proteins is by no means clear, but a number of solid clues about protein sorting have been accumulated.

Phosphorylated Mannose Residues Target Proteins to Lysosomes

Lysosomal enzymes, like secretory proteins, are cotranslationally inserted into the lumen of the ER, where they receive an *N*-linked oligosaccharide identical to that of secretory proteins. One or more mannose residues of these oligosaccharides become phosphorylated, and the phosphorylated mannose apparently serves as the chemical signal that causes entry into lysosomes.

Phosphorylation of mannose residues is a two-step procedure (Figure 21-56). Both enzymes copurify with Golgi

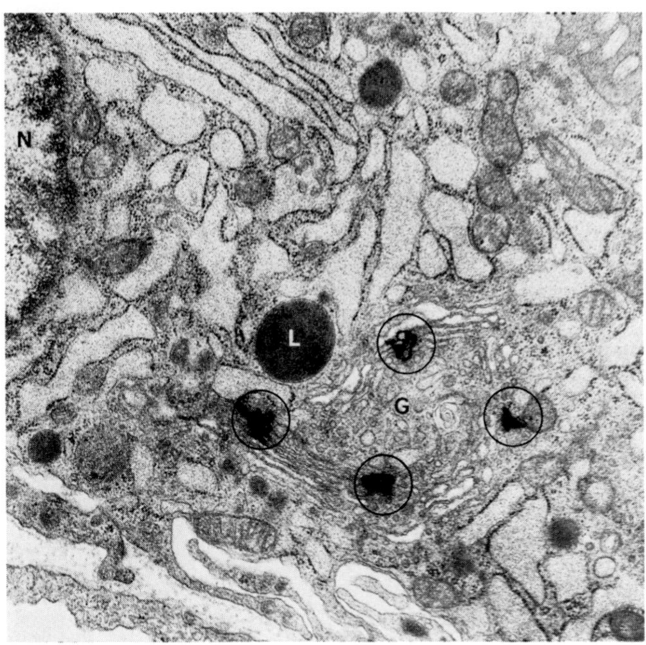

Figure 21-54 Electron microscopic autoradiography of a rat thyroid cell 5 min after injection of [^{3}H]fucose. Fucose radioactivity, observed as grains (*inside circles*), is clustered over the Golgi (*G*) region. The nucleus (*N*), lysosome (*L*), and rough ER are not radiolabeled. [See A. Haddad et al., 1971, *J. Cell Biol.* 49:856.] *Courtesy of C. P. Leblond.*

vesicles, and it is thus thought that phosphorylation of mannose residues occurs in the cis Golgi region. The first enzyme in the sequence has been purified; it utilizes only lysosomal proteins as substrate and does not catalyze reactions with other glycoproteins.

A *mannose 6-phosphate receptor* located on the lumenal face of the Golgi membrane binds the mannose 6-phosphate residue of the lysosomal protein and directs the protein to lysosomes. Regions of membrane containing the receptor and its bound lysosomal enzymes probably pinch off to become specialized vesicles that transport the enzymes to acidic (pH 5.0) sorting vesicles. The sorting vesicle is similar, if not identical, to the CURL vesicle (see Figure 15-40) in which receptors and ligands are dissociated after receptor-mediated endocytosis (Figure 21-57).

The mannose 6-phosphate receptor binds its ligand at the neutral pH of the Golgi, but not at the lower pH of the sorting vesicle. Thus, once the enzymes reach the sorting vesicle they are released from the receptor and become soluble within it. Furthermore, a phosphatase generally removes the phosphate, preventing any rebinding of the enzyme to the receptor, since nonphosphorylated oligosaccharides cannot bind. Vesicles containing the lysosomal enzyme—but not the mannose 6-phosphate receptor—bud from the sorting vesicle and fuse with lysosomes, thus delivering the enzyme to its destination. As do most cellular receptors, the mannose 6-phosphate receptor recycles; presumably specialized vesicles containing the receptor bud from the sorting vesicle and transport the receptor back to the Golgi (Figure 21-57).

Activation of Lysosomal Enzymes At a late stage in their maturation, inactive precursors of lysosomal enzymes (proenzymes) undergo a proteolytic cleavage to form an enzymatically active but smaller peptide. This cleavage may result in immediate activation of enzymatic function as the protein reaches the acidic environment of the lysosome. Use of enzymatically inactive precursor forms of lysosomal proteins may protect the cell from dangerous hydrolytic activity in acidic (pH 5.0) sorting vesicles.

A Genetic Defect in Mannose Phosphorylation The discovery of the mannose 6-phosphate pathway began with a study of patients with I cell disease. This disease results from a severe genetic abnormality causing a deficiency of multiple lysosomal enzymes in fibroblasts and several other cell types, so that large amounts of toxic intracellular wastes accumulate. Fibroblasts from these patients contain large intracellular vesicles filled with glycolipids and extracellular components that would normally be degraded by lysosomal enzymes. Cells from affected persons synthesize all lysosomal enzymes normally and have normal mannose 6-phosphate receptor on Golgi

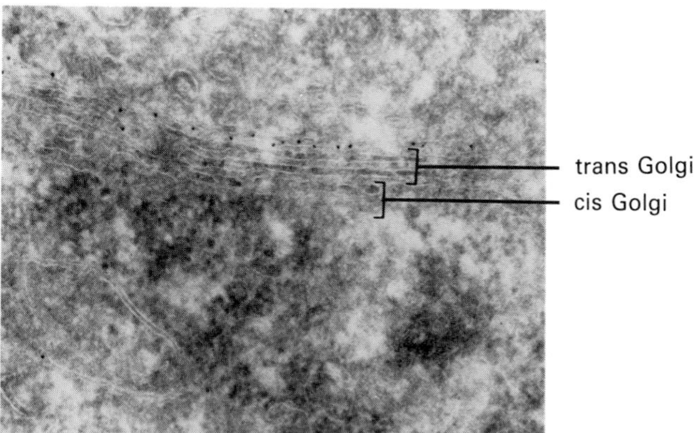

Figure 21-55 Localization of galactosyl tranferase enzyme to the trans Golgi. A frozen section of a HeLa cell was treated with an antibody to human galactosyl transferase. Bound antibody was detected by reaction with protein A attached to 8-nm gold particles. The electron-dense gold permits observation in the electron microscope. Only the two trans-most Golgi vesicles are reactive. Because the sections are only weakly stained with osmium, the cytoplasmic membranes appear as white lines on a dark background. [See J. Roth and E. Berger, 1982, *J. Cell Biol.* 93:233.] *Courtesy of G. J. A. M. Strous, E. G. Berger, and H. J. Geuze.*

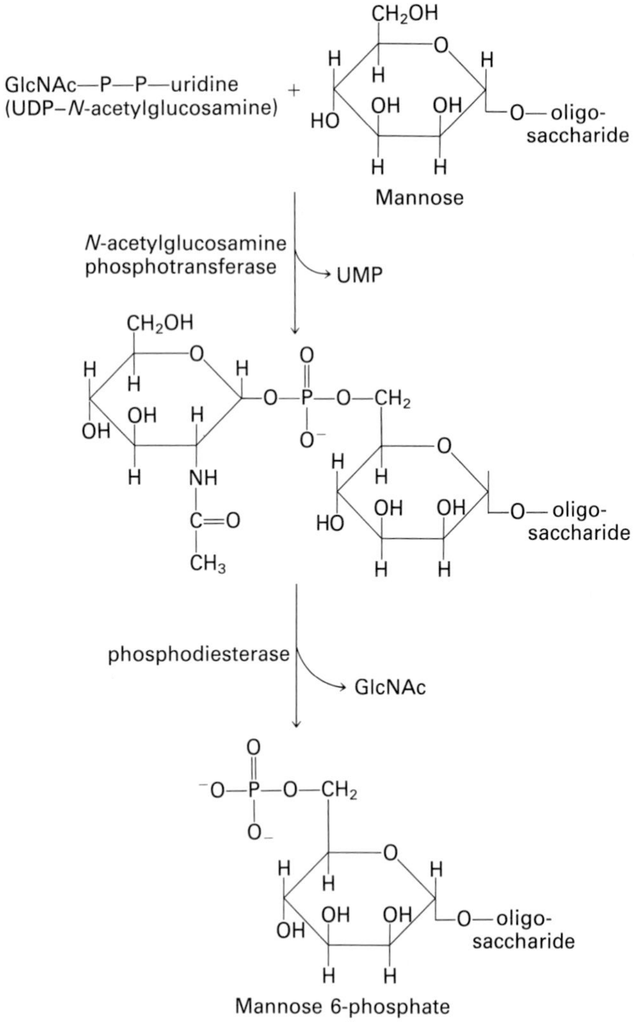

Figure 21-56 Phosphorylation of mannose residues on lysosomal enzymes occurs in two stages. The first enzyme transfers an *N*-acetylglucosamine phosphate group to one or more mannose residues. This enzyme is specific for lysosomal enzymes and does not utilize other glycoproteins as substrate. Lysosomal enzymes do not have even short stretches of homologous amino acids; thus the enzyme is likely to recognize a specific conformation rather than a primary amino acid sequence on lysosomal enzymes. The second enzyme removes the *N*-acetylglucosamine group, leaving the phosphate attached to the 6 carbon of the mannose. [See I. Tabas and S. Kornfeld, 1980, *J. Biol. Chem.* **255**:6633; and A. Hasilik and E. F. Neufeld, 1980, *J. Biol. Chem.* **255**:4946.]

membranes and on the plasma membrane. Because one of the enzymes required to phosphorylate mannose residues is lacking, the degradative enzymes are secreted, rather than sequestered in lysosomes. When fibroblasts are grown in medium containing lysosomal enzymes bearing phosphorylated mannose (such as are normally secreted by fibroblasts in small amounts), the content of lysosomal enzymes becomes normal, since lysosomal enzymes are internalized by receptor-mediated endocytosis. Lysosomes from liver cells (and other cells) from patients with I cell disease do contain a normal complement of lysosomal enzymes, even though such liver cells are also defective in mannose phosphorylation. This finding suggests that liver cells may use an alternative scheme for targeting newly made enzymes to lysosomes; the nature of the signal is unknown and is under active study.

Many genetic diseases are due to a defect in a specific lysosomal enzyme. *Hunter's syndrome* and *Hurler's syndrome* result from defects in lysosomal enzymes required for catabolism of sulfated mucopolysaccharides (see Figure 3-75). *Tay-Sachs disease* is caused by an absence of the lysosomal hydrolase β-*N*-hexosaminidase A (see Figure 5-37). Cultured fibroblasts from patients with these diseases are seen to take up the missing enzyme from the extracellular medium, via the surface mannose 6-phosphate receptor. This observation raises the possibility of treating such diseases by administration of the missing phosphorylated lysosomal enzyme to slow or reverse the clinical course.

Proteins Are Targeted to the Golgi or ER

Much less is known of the mechanism by which proteins are directed to organelles other than lysosomes. Synthesis of only one Golgi membrane–bound enzyme, galactosyl transferase (see Figure 21-55), has thus far been studied. This enzyme follows a maturation pathway similar to that of plasma membrane glycoproteins: it is synthesized on the rough ER, in which it receives an *N*-linked high-mannose oligosaccharide; it then moves to the Golgi, in which the oligosaccharide is modified. Galactosyl transferase remains in the trans Golgi.

Smooth ER of liver cells contains both cytochrome b_5, an integral membrane protein required for synthesis of unsaturated fatty acids, and a set of cytochromes known collectively as the P_{450} cytochromes, integral membrane proteins involved in modification of toxic chemicals such as carcinogens. Cytochrome P_{450} is anchored to the membrane by a hydrophobic sequence of amino acids at its N-terminus. Cytochrome b_5, by contrast, is anchored to the membrane by a set of hydrophobic residues at its C-terminus. It is synthesized on free polysomes and inserts into the smooth ER membrane only after synthesis is complete (Figure 21-58).

Vesicles Transport Proteins from Organelle to Organelle

Rough ER, smooth ER, cis and trans Golgi membranes, lysosomes, and secretory vesicles are all discrete organelles, each with its own unique constellation of enzymes. Proteins move from one to the other in membranous vesi-

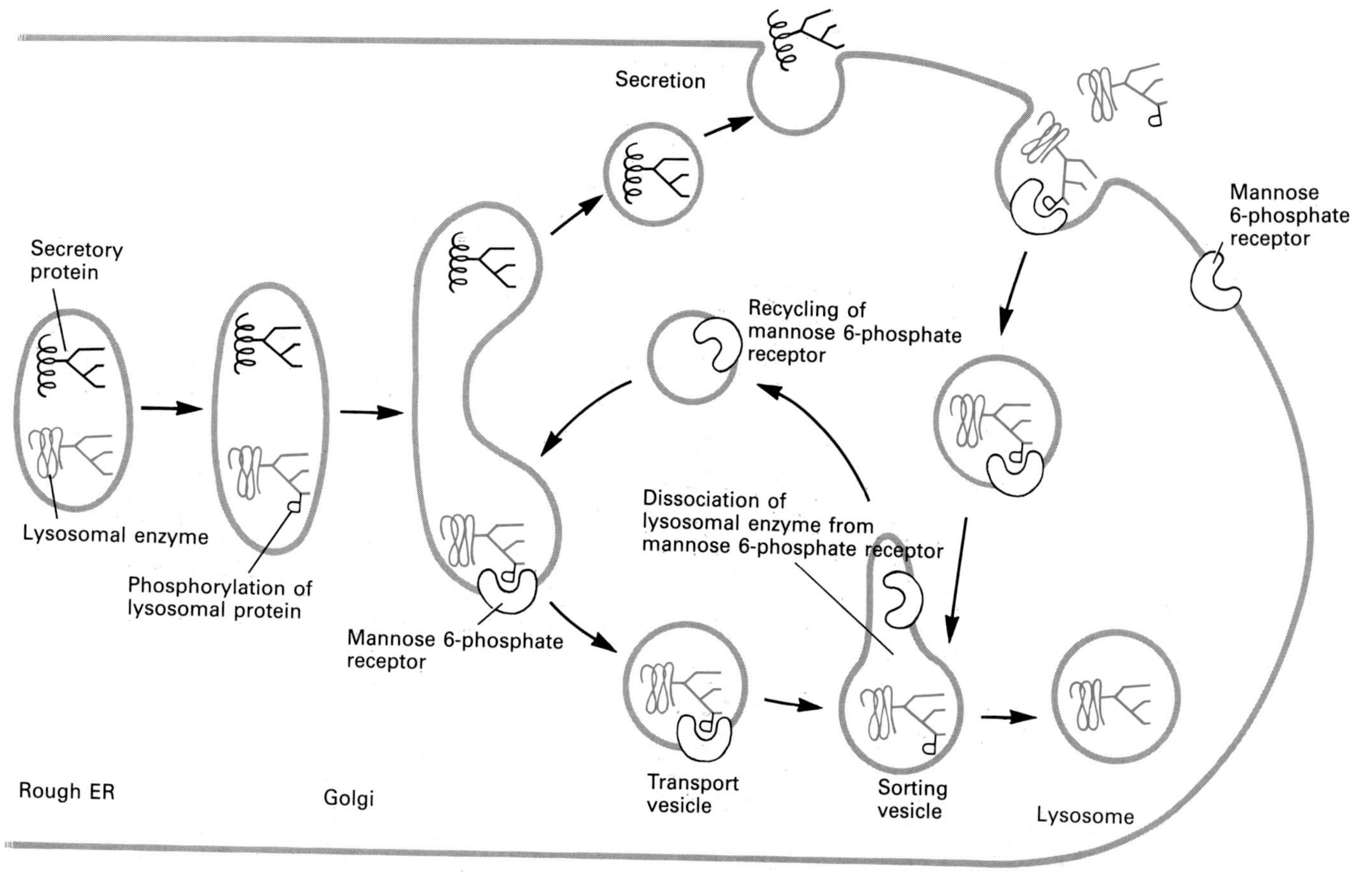

Figure 21-57 Targeting of lysosomal enzymes to lysosomes. During biosynthesis, lysosomal enzymes migrate to the cis Golgi, in which one or more mannose residues are phosphorylated. A membrane-bound mannose 6-phosphate receptor ensures the association of these proteins with vesicles that are directed to lysosomes. The low pH of the sorting vesicle causes the phosphorylated enzyme to dissociate from its receptor, and the receptor recycles back to the Golgi. The enzyme then loses its phosphate group and is transported to a lysosome. The sorting of lysosomal enzymes from secretory proteins thus occurs in the Golgi; the two classes of proteins are found in different vesicles that bud from the Golgi. The mannose 6-phosphate receptor is also found on the cell surface and mediates the endocytosis of extracellular phosphorylated lysosomal enzymes and their transport to a sorting vesicle and thence to a lysosome.

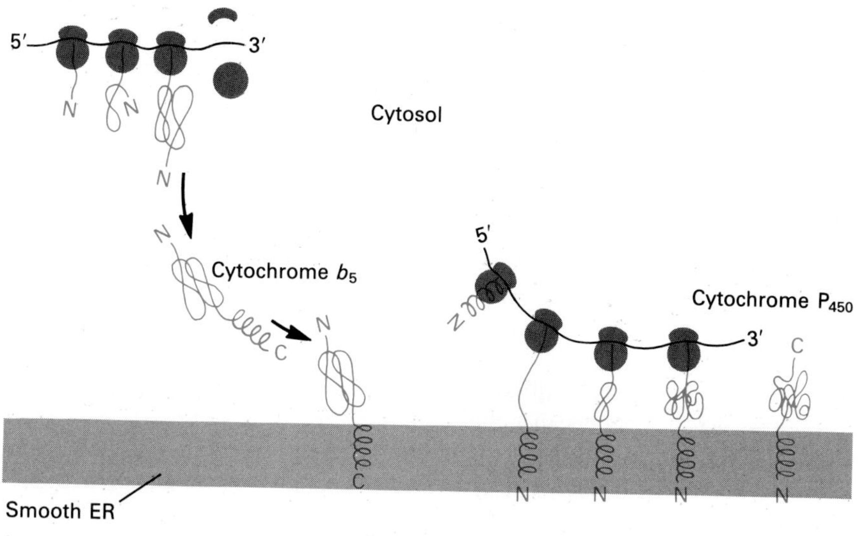

Figure 21-58 Synthesis and membrane insertion of two smooth ER proteins: cytochrome b_5 and cytochrome P_{450}. Cytochrome b_5 is anchored to the membrane by a sequence of hydrophobic amino acids at its C-terminus; cytochrome P_{450} is anchored by its N-terminus. Cytochrome b_5 inserts into the ER only after its synthesis; it is not known what factors cause it to localize in smooth ER.

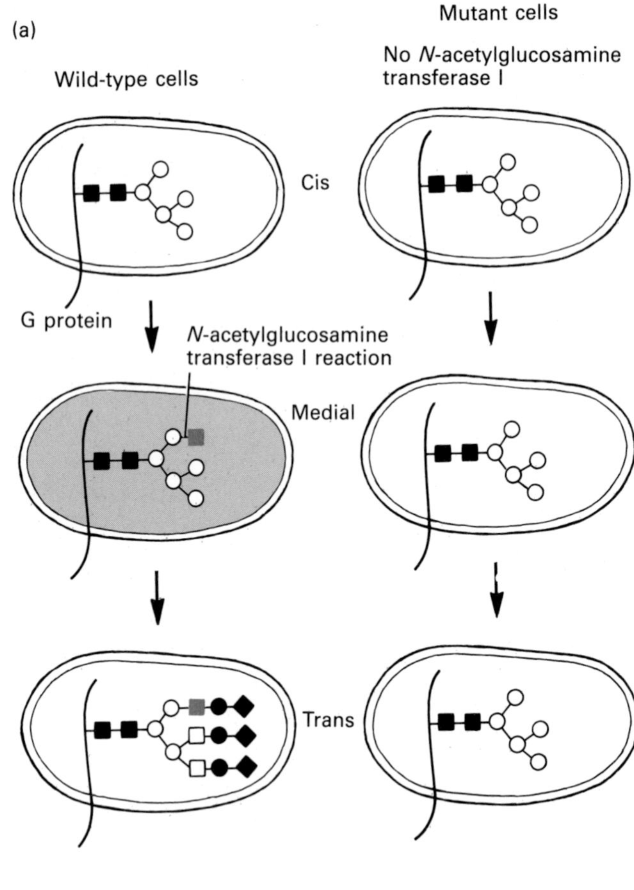

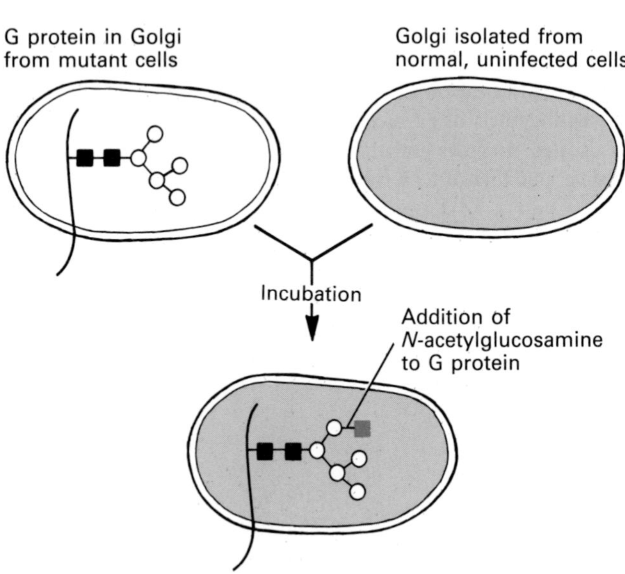

Figure 21-59 A cell-free system exhibiting movement of G protein from cis to medial Golgi vesicles. Such experiments make use of a mutant line of cultured fibroblasts that lack an enzyme (*N*-acetylglucosamine transferase I; see *step 5*, Figure 21-53) that is localized to the medial Golgi (*colored shading*) and that modifies asparagine-linked oligosaccharides by addition of one *N*-acetylglucosamine (*color*). (a) After VSV infection of wild-type cells, the oligosaccharide on the viral G protein is modified to a typical "complex" form, while in the mutant cells, G protein reaches the surface with a simpler oligosaccharide containing two *N*-acetylglucosamine (*squares*) and five mannose (*circles*) residues. (b) To detect movement of G protein from one Golgi vesicle to another in cell-free systems, the mutant cell is infected with VSV and the G protein made is radiolabeled by exposure of the cells to a radiolabeled amino acid. When Golgi vesicles are isolated from such cells, only the G protein is labeled, and all the G protein contains the mannose-rich oligosaccharide noted. Following incubation of this preparation with Golgi vesicles from normal, uninfected cells, some of the G protein is observed to contain *N*-acetylglucosamine. Electron microscopic autoradiography indicates that all such G protein is contained within sealed medial Golgi vesicles. Thus G protein has traveled from the mutant Golgi to wild-type Golgi vesicles and there received the *N*-acetylglucosamine residue. Other studies suggest that the intermediate in transport of G protein is smaller 60-nm-diameter uncoated vesicles. [See W. E. Balch et al., 1984, *Cell* 39:405; W. A. Braell et al., 1984, *Cell* 39:511; and W. E. Balch et al., 1984, *Cell* 39:525–536.]

cles. But precisely how does such a process occur, and more importantly, how are vesicles targeted from one organelle to another? How, for instance, does a vesicle containing secretory proteins from rough ER "know" to move to and fuse with cis Golgi membranes?

Electron microscopy of virtually all eukaryotic cells reveals a plethora of small membrane-limited vesicles, some coated with clathrin (see Figure 21-39). At one time it was thought that clathrin-coated vesicles are intermediates in transport of plasma membrane glycoproteins and secretory proteins from rough ER to Golgi to the cell surface. An obligatory role for coated vesicles has been made much less likely by the identification of a yeast mutant that contains a deletion of the single clathrin gene and that nevertheless secretes invertase and other proteins quite normally.

An alternative view is that uncoated vesicles (or vesicles coated with a protein other than clathrin) shuttle proteins and other components from one organelle to another. As the vesicles seal off from the "mother" organelle, they may selectively incorporate certain integral membrane proteins exposed on its surface that direct the vesicle to its destination. The pH and/or ionic content of the vesicles may be different from that of the "mother" vesicle, possibly because they contain different sets of permeases. Such changes presumably cause any coats on the vesicles to depolymerize, thereby exposing the membrane proteins that are the actual targeting signals.

Specific movement of plasma membrane glycoproteins from a cis Golgi vesicle to a medial Golgi vesicle, by means of small "transport" vesicles, can be demonstrated in a cell-free system (Figure 21-59). Such experiments should allow the eventual elucidation of all the components required for vesicle fusion, targeting, and fission.

Proteins Destined for Different Domains of the Plasma Membrane Are Sorted in the Golgi Complex

The plasma membrane of many cells is divided into two domains, separated by tight junctions, that contain different species of proteins. Recall, for example, that in the epithelial cells lining the small intestine, the apical and basolateral domains contain different enzymes and permeases that facilitate movement of products of digestion from the intestine into the blood (see Figures 14-47 and 15-18). Although it is unclear how plasma membrane proteins are targeted to the appropriate domain, the mechanism appears to involve sorting in the Golgi vesicles. A variety of microscopic and cell fractionation studies indicate that proteins destined for the apical and basolateral cellular surfaces are found together in the same Golgi vesicles. Apparently proteins destined for the different regions bud from the Golgi into separate transport vesicles that fuse with the appropriate domain of the plasma membrane.

An experimental system that has been useful in investigating this aspect of protein transport is cultured epithelial cells that maintain distinct apical and basolateral surfaces (Figure 21-60). When these cells are infected with enveloped viruses such as influenza virus, the viruses are observed to bud only from the apical surface of these cells. VSV (vesicular stomatitis virus), however, buds only from the basolateral surface. The location of virus budding is determined by the location of virus glycoproteins in the plasma membrane: influenza virus (HA) glycoproteins are transported from the Golgi exclusively to the apical region of the plasma membrane, whereas the VSV glycoprotein is transported to the basolateral region of the plasma membrane. Furthermore, in cells expressing a cloned transfected HA cDNA, all the HA accumulates

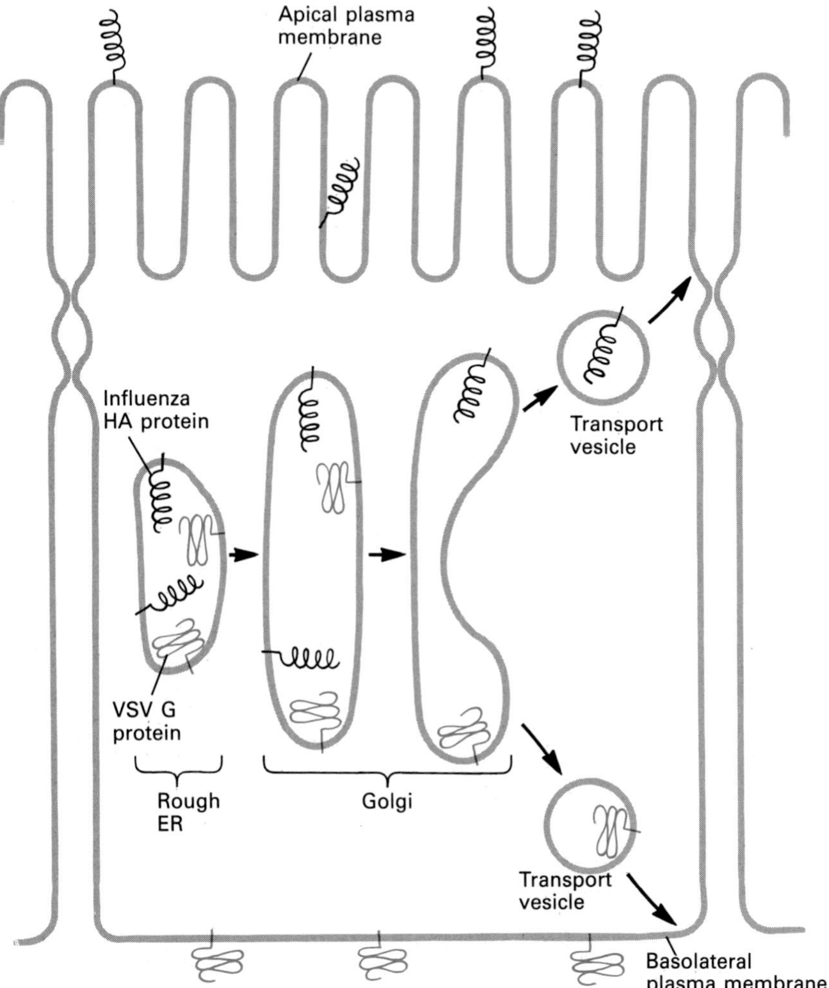

Figure 21-60 Sorting of proteins destined for the apical and basolateral domains of the plasma membrane of epithelial cells. When a cultured line of kidney epithelial cells is infected simultaneously with VSV and influenza virus, the VSV (G) glycoprotein is found only on the basolateral domain of the plasma membrane, whereas the influenza virus (HA) glycoprotein is found only on the apical. There are many other proteins besides these that are localized to a specific plasma membrane domain. By immunoelectron microscopy both proteins are observed in the same rough ER and Golgi vesicles; thus sorting into specific transport vesicles occurs when or just after the proteins have moved through the Golgi vesicles.

only in the apical membrane; this finding indicates that the targeting sequence resides in the HA protein itself and not in other viral proteins. By directed mutagenesis of regions of the glycoprotein, it should be possible to determine the regions of the HA protein that target this protein to the apical region of the plasma membrane.

There Are at Least Two Types of Post-Golgi Secretory Vesicles

The possibility of having different transport vesicles occurs again in cells that both store secretory proteins in vesicles and also constantly secrete other proteins and insert glycoproteins in the plasma membrane. In pituitary tumor cells and probably in other types of cells, two pathways of secretion are utilized (see Figure 21-26). ACTH and other hormones that undergo final proteolytic cleavage in secretory vesicles are stored in these cells and are exocytosed only upon neural or hormonal stimulation. However, in the same cells, membrane glycoproteins and certain secretory proteins are exocytosed to the cell surface constantly, independently of any known stimulus for exocytosis. Thus at least two classes of secretory proteins are sequestered into different vesicles. It has been established that all newly made secretory proteins and lysosomal enzymes are found together in the same ER and Golgi vesicles, and that segregation occurs during or just after the proteins mature through the Golgi.

Exocytosis Can Be Triggered by Neuronal or Hormonal Stimulation

Many hormones, neurotransmitters, and other agents trigger the exocytosis of stored secretory proteins in vesicles. The process is very similar to that used in exocytosis of synaptic vesicles during synaptic transmission of a nerve impulse (see Chapter 17).

The exocrine pancreatic acinar cells have been studied in considerable detail. There appear to be two fundamentally different mechanisms by which hormones or other neurotransmitters can increase enzyme secretion. One mechanism involves binding of the hormone to its surface receptor, resulting in an increase in the intracellular level of Ca^{2+} and, after a series of ill-defined steps, in stimulation of secretion. (The action of these hormones can be mimicked by Ca^{2+} ionophores, which increase the cellular levels of Ca^{2+}.) Acetylcholine, which acts in this fashion, is released in the vicinity of the acinar cells by discharge of cholinergic nerves. The peptide hormones gastrin and cholecystokinin (CCK; Chapter 16) also act to induce enzyme secretion in this manner. The former hormone is released into the circulation by the stomach when it is filling with food; the latter, by the small intestine. Thus secretion of digestive enzymes by the pancreas is induced by a specific stimulus: the presence of food in the stomach or intestine.

Other hormones act by increasing the level of cAMP in the acinar cells. They have no effect on cellular Ca^{2+} concentrations. By a poorly understood series of steps, which may involve protein phosphorylation, the elevated cAMP level triggers exocytosis. Peptide hormones called secretin and VIP, both of which are released by intestinal cells into the circulation, act in this fashion. Specific surface receptors for these similar peptides have been discovered on the basal surface of the acinar cells, where they can come into contact with fluid derived from the blood.

Membranes Recycle in Secretory Cells

During secretion of neurotransmitters and enzymes, exocytosis results in a considerable expansion of the plasma

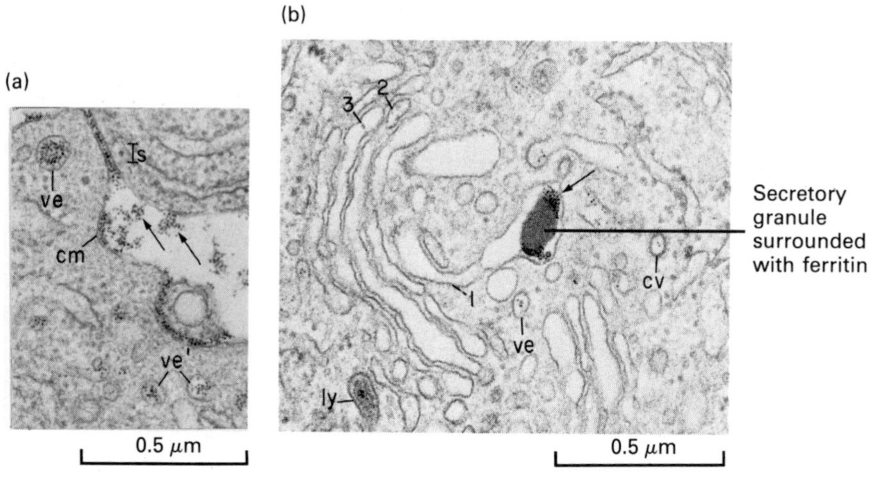

Figure 21-61 The Golgi mediates the recycling of the cell membrane in secretory cells, as shown by an experiment in which rat anterior pituitary cells were incubated with cationized ferritin, a protein that binds irreversibly to membrane components. (a) After 15 min of incubation at 37°C, the ferritin binds to the cell (plasma) membrane (*cm*) and is taken up by endocytosis. Numerous endocytic vesicles (*ve*) containing ferritin are seen in the cytoplasm near the cell membrane. (b) After incubations of 60 min or longer, the ferritin is localized in the Golgi region. The tracer is particularly concentrated around a secretory vesicle forming within the trans-most Golgi vesicle (*arrow*), suggesting that the incoming vesicles carrying ferritin fuse preferentially with trans Golgi elements. [See M. Farquhar, 1978, *J. Cell. Biol.* 77:R35.] *Courtesy of M. Farquhar.*

membrane, and the cell either must degrade the extra surface phospholipids and proteins or must recycle them by endocytosis. The bulk of cell surface material apparently is recycled, as is the case with most surface receptor proteins (Chapter 15).

One convenient tag that can be used to follow the fate of cell surface proteins inside cells is a derivative of ferritin that is positively charged and binds to negatively charged surface glycolipids and glycoproteins. When added to a suspension of anterior pituitary cells (which secrete a number of peptide hormones), it binds tightly to the cell surface (Figure 21-61a). After incubation the ferritin is taken up in endocytic vesicles; it is localized initially to both Golgi cisternae and lysosomes (Figure 21-61b)—a result suggesting that the endocytic vesicles have fused with both types of organelles. Both the cis and trans vesicles of the Golgi are enriched in the endocytosed ferritin early in this process. Later, more ferritin accumulates in trans vesicles and their associated secretory vesicles. Apparently, then, in secretory cells the Golgi vesicles are the principal site for recycling of surface membrane constituents.

The Golgi, in summary, functions to direct vesicles and proteins to a variety of specific regions of the cell. In this organelle, lysosomal enzymes, secretory proteins, and apical and basolateral plasma membrane proteins are sorted into different vesicles, each destined for a different cellular site. The Golgi also is a major site for recycling of plasma membrane components that are recovered by endocytosis. The Golgi amply earns its nickname as the "traffic policeman" of vesicular traffic within the cell.

The Nuclear Membrane Is Reassembled from Its Components during Mitosis

The nuclear membrane in animal cells presents a special problem in investigation of membrane assembly since it completely disappears during late mitotic prophase mitosis, and re-forms around the daughter chromosomes during telophase. In interphase cells two membranes surround the nucleus (Chapter 5): an outer membrane that is continuous with the rough ER and an inner membrane (Figure 21-62; see also Figures 5-11, 9-67, and 9-68). Connecting the two membranes are pores (see Figure 5-11). Lining the inner side of the nuclear membrane is a protein-rich layer termed the *nuclear lamina* (interphase in Figure 21-62). Upon purification, the lamina was found to be composed of three principal extrinsic membrane proteins, termed lamins A, B, and C (m.w. ~60,000 to 70,000) that together form a fibrous network. Lamins A and C are derived by alternative splicing of the same mRNA precursor. Lamins are highly homologous in sequence, and probably in structure to IF proteins (see Chapter 19). These proteins appear to bind regions of chromatin and thus to anchor parts of the interphase chromosomes to the nuclear membrane. In inter-

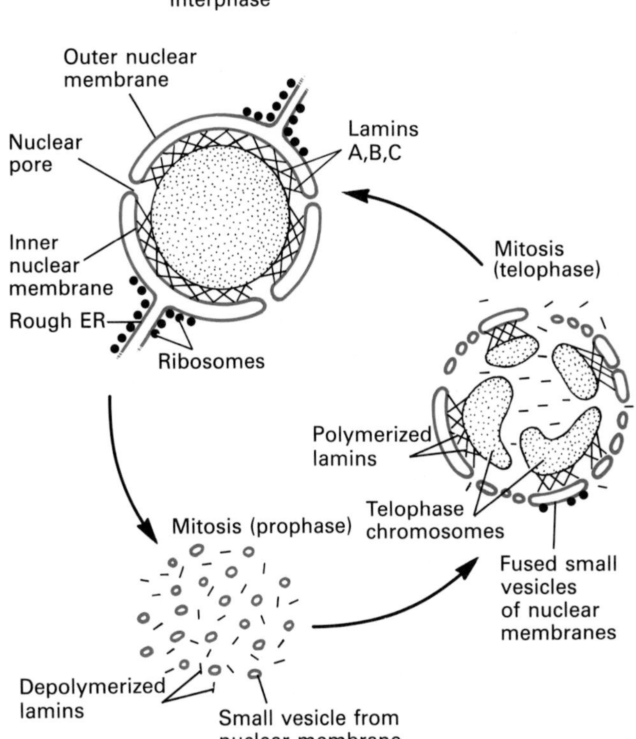

Figure 21-62 The inner and outer nuclear membranes present in interphase cells reversibly depolymerize during mitosis. In prophase, the nuclear membranes fragment into small vesicles that are dispersed in the cytoplasm. Simultaneously the nuclear lamina depolymerizes into its three component polypeptides that are also dispersed throughout the cell. In telophase, the condensed mitotic chromosomes disperse into chromatin. This apparently induces the polymerization of lamin into a fibrous network and the subsequent fusion of membrane vesicles to form the characteristic interphase nuclear membranes. Phosphorylation and dephosphorylation of lamins may regulate their depolymerization and repolymerization. How and when the pore complexes form is not known. [See L. Gerace and G. Blobel, 1981, *Cold Spring Harbor Symp.* **46**:962.]

cells all the lamins are observed to be localized to the inner surface of the nuclear membrane. Like other cellular membranes, the nuclear membranes grow continuously during the G1, S, and G2 periods of the cell cycle, doubling their size by addition of new phospholipids and proteins such as the lamins.

Observations with the light microscope suggested that, in the prophase stage of mitosis, the nuclear membrane simply disappeared as the chromatin condensed into mitotic chromosomes. Examination with the electron microscope, however, has shown that the nuclear membrane is fragmented into a number of small vesicles, indistinguishable from the ER (prophase in Figure 21-62). The lamins do not remain associated with these vesicles, however; they are depolymerized to monomeric polypeptides and dispersed throughout the cell. During telophase, after the daughter chromosomes have separated, this process is reversed. The small vesicles fuse with each other around the dispersing chromatin. Simultaneously the lamins repolymerize on the inner nuclear membrane and apparently bridge the chromatin to the membrane (telophase in Figure 21-62).

A slightly different pattern of events is found in frog embryos and in certain other cells. In these, the nuclear structure reassembles in a two-stage process: first, nuclear membranes form around individual chromosomes; these then fuse together to form a single interphase nucleus. Presumably, at the end of mitosis, a cellular signal initiates reassembly of the nucleus, and at the end of interphase, another signal initiates the breakdown of the nuclear membranes.

Recent studies using frog eggs suggest that the signal for assembly of the nuclear envelope is DNA assembled into chromatin. After fertilization of the frog egg, a nuclear division occurs every 30 min. These divisions do not require any new protein synthesis by the embryo. The (unfertilized) egg contains pools of histones and lamins sufficient for 20,000 cells. If any DNA is microinjected into an unfertilized egg, the pool of stored histones and other nuclear proteins is used to assemble the DNA into native chromatin. Soon after, nuclear envelopes assemble around the injected DNA; these envelopes have a normal appearance and contain lamins on their inner surface. It appears that chromatin induces the polymerization of lamins into a fibrous network that, in turn, causes fusion of small vesicles to form a normal interphase nuclear membrane.

During mitosis all vesicular traffic in the cell ceases. There is no endocytosis; proteins do not move from the rough ER to the Golgi; and secretory vesicles cannot be induced to fuse with the plasma membrane. Possibly the same cellular signals that induce breakdown of the nuclear membrane also block fission and fusion of other cellular membranes.

Table 21-8 Enzymatic steps and sequence of events in procollagen biosynthesis

Location	Major events	Enzyme	Function
Rough ER	Signal sequence cleaved	Signal peptidase	Removal of hydrophobic signal peptide
	Hydoroxylation of prolyl residues to 3-hydroxyprolyl	Prolyl 3-hydroxylase	?
	Hydoroxylation of prolyl residues to 4-hydroxyprolyl	Prolyl 4-hydroxylase	Stability of triple helix
	Hydoroxylation of lysyl residues to hydroxylysyl	Lysyl hydroxylase	Site for O-glycosylation
	Addition of *N*-linked high-mannose oligosaccharide	Lipid-linked oligosaccharide transferase	Secretion of protein
Golgi complex	Formation of triple helix	Self-assembly	Forms rigid collagen structure
	Addition of galactose to hydroxylysyl residues	Galactosyl hydroxylysyl transferase	?
	Addition of glucose to galactose	Glucosyl transferase	?
	Formation of inter- and intrachain disulfide bonds	?Disulfide isomerase	Stabilization of procollagen structure
Extracellular	Cleavage of N-terminal pro sequence	N-terminal procollagen peptidase	Permits alignment of tropocollagen into fibrils
	Cleavage of C-terminal pro sequence	C-terminal procollagen peptidase	Permits alignment of tropocollagen into fibrils
	Oxidation of lysyl residues to aldehyde derivates	Lysyl oxidase	Site of formation of cross-links
	Formation of intrachain cross-links	?Spontaneous	Stabilization of fibers

SOURCE: J. M. Davidson and R. A. Berg, 1980, *Methods Cell Biol.* **23**:119.

Synthesis and Assembly of Collagen

Thus far in this chapter we have been concerned with construction of structures inside cells. In multicellular organisms some of the most important structural elements are in the extracellular matrix. The most abundant protein components of this matrix are the collagens; these are in fact the most abundant proteins in the animal kingdom. Therefore let us conclude this chapter with a discussion of collagen biosynthesis. Collagens are the major fibrous structural elements of cartilage, tendon, skin, bone, lung, and blood vessels and serve to hold cells together in discrete tissues. Collagens are synthesized by fibroblasts, a type of cell localized in the interstitial spaces that surround other cells and tissues (see Figure 5-53). After synthesis, collagen polypeptides undergo extensive covalent modifications that take place in a defined sequence in the rough ER, Golgi organelles, and the extracellular space after secretion. These modifications, summarized in Table 21-8 and Figure 21-63, include the hydroxylation of prolyl and lysyl residues to form the novel amino acids 3-hydroxyproline, 4-hydroxyproline, and hydroxylysine (see Figure 3-14). Glycosylation, disulfide bonding, proteolytic processing, and formation of covalent cross-links between lysyl residues in adjacent collagen helixes also take place. These modifications allow formation of the typical collagen triple helix (see Figure 5-54) and also allow the helixes to aggregate and form the tough, flexible collagen fibrils.

Denatured Tropocollagen Peptides Cannot Renature to Form the Triple Helix

The structural unit of most collagen fibers is *tropocollagen*, a helix of three polypepide chains, each containing 1050 amino acids (see Figure 5-54). The amino acid sequence of collagen is extremely regular: every third amino acid in each chain is glycine. This feature is essential for the three chains to fold into the triple helix. Glycine is the only amino acid small enough to occupy the interior of the triple helix. Proline is also abundant, and the repeating sequences glycine-proline-X and glycine-proline-hydroxyproline occur very frequently.

The three-dimensional structure of tropocollagen is stabilized by cooperative interactions of many hydrogen bonds between the three chains. When a solution of tropocollagen is heated above 40°C, the rodlike helix becomes denatured, and the three chains separate from each other (Figure 21-64). Denatured tropocollagen cannot spontaneously renature to form completely the tropocollagen triple helix.

N-Terminal and C-Terminal Propeptides Aid in the Formation of the Triple Helix and Are Then Cleaved Off

The three chains of all collagens are synthesized as longer precursors called *procollagens*. The additional peptide segments of procollagen enable the collagen regions to form the triple helix but are subsequently removed. The synthesis of type I collagen, the fibrous form predominant in skin, bone, and tendons, has been studied in most detail.

Type I collagen, like most other secreted polypeptides, is synthesized with an N-terminal signal sequence that is cleaved from the nascent chain. The predominant cellular precursor of collagen, procollagen, is a triple-helix molecule that contains additional amino acids at the N- and C-termini of all three collagen chains. These additional amino acids constitute the pro segments, and there are about 150 additional residues at the N-terminus (Figure 21-65) and about 250 at the C-terminus. Except for a short collagen-like helical region in the N-propeptide, the additional peptide segments have a noncollagen amino acid composition and form globular domains.

There are no disulfide bonds within collagen triple helixes themselves. However, there are *intrachain* disulfide bonds within both the N- and C-terminal pro segments, and, importantly, there are five *interchain* disulfide bonds connecting the C-terminal pro segments of the three procollagen chains (Figure 21-65). The Golgi is the organelle in which interchain disulfide bonds of procollagen are formed and in which the collagen triple helixes form. The disulfide bonds in the C-terminal pro segment are believed to form first; then the three chains can zip together in the C $\rightarrow$ N direction to form the triple helix (Figure 21-65). Disulfide bonds appear to be important in the stability of the procollagen molecule and may be crucial in stabilizing the association of the three chains prior to formation of the triple collagen helix. In vitro renaturation studies provide support for this notion: the renaturation of thermally denatured procollagen is substantially accelerated if the proper interchain disulfide bonds are first formed. Electron microscopic studies of fibroblasts have identified fibrous molecules (procollagen helixes) only in the Golgi and secretory vesicles. Thus assembly of three procollagen polypeptides into a triple helix and formation of interchain disulfide bonds occur in the Golgi (see Figure 21-63).

Modifications Occur Sequentially in the Rough ER and Golgi

Besides cleavage of the signal sequence, several other modifications of collagen in the rough ER probably occur on the nascent collagen chains (see Table 21-8). Some

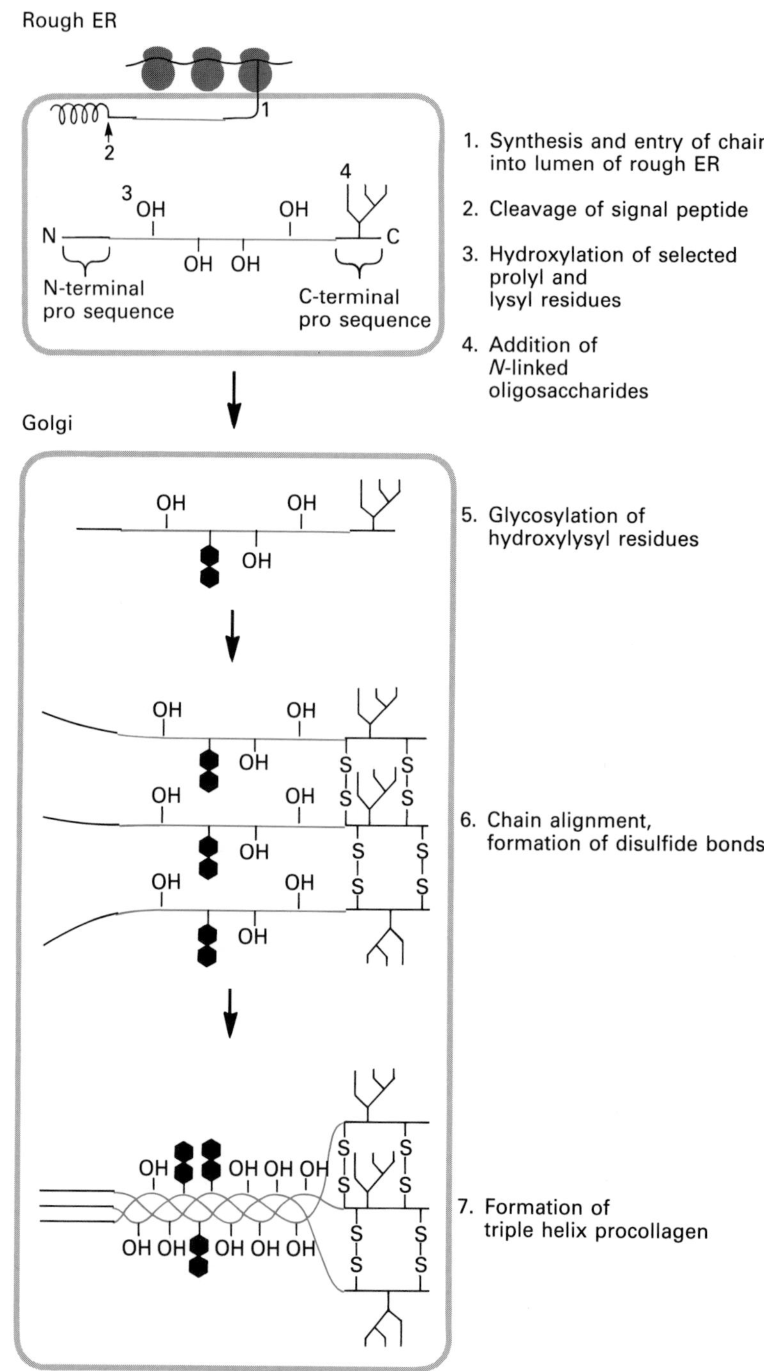

Figure 21-63 *(right and opposite)* Major events in the biosynthesis of collagen. Covalent modifications of the collagen polypeptide include hydroxylation, glycosylation, disulfide bond formation, and proteolytic cleavages; these occur in a precisely defined sequence in the rough ER, Golgi vesicles, and the extracellular space. The modifications allow formation of a stable triple-stranded collagen helix and also the lateral alignment and covalent cross-linking of tropocollagen helixes into 50-nm-diameter collagen fibrils. These in turn aggregate to form the collagen fibers, which can be as thick as 0.5 μm. Table 21-8 lists these modifications and their function in more detail.

intrachain disulfide bonds form, and specific prolyl residues in the domains destined to form the triple helix are subjected to hydroxylation by the membrane-bound enzymes *prolyl 4-hydroxylase* and *prolyl 3-hydroxylase* (Figure 21-66). Certain lysyl residues in this domain are also subjected to hydroxylation by *lysyl hydroxylase.* Ascorbic acid (vitamin C) is an essential cofactor for the enzyme prolyl hydroxylase. In the absence of ascorbate, collagen is insufficiently hydroxylated. The chains cannot form a stable triple helix (see Figure 21-64), nor can they form normal fibers. Nonhydroxylated procollagen chains are degraded within the cell. Consequently, fragility of blood vessels, tendons, and skin is characteristic of the disease *scurvy,* which results from a deficiency of vitamin C in the diet.

Procollagen Is Assembled into Fibers after Secretion

The final steps in the conversion of procollagen to tropocollagen, and in the assembly of tropocollagen into col-

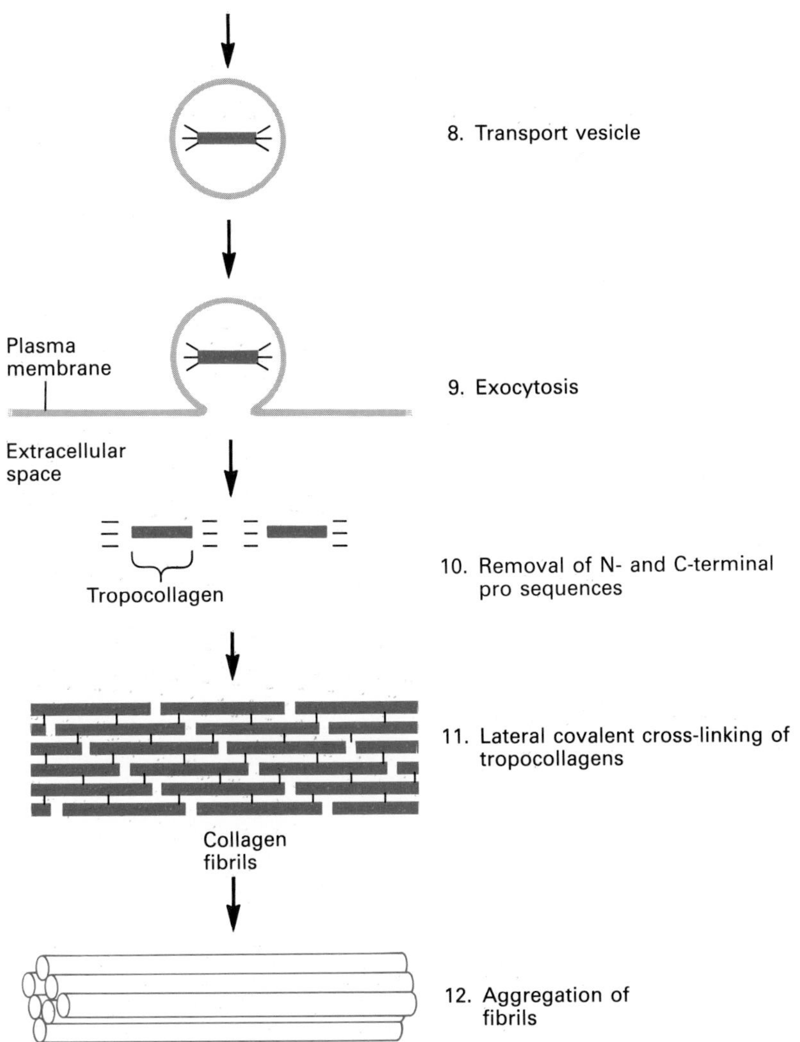

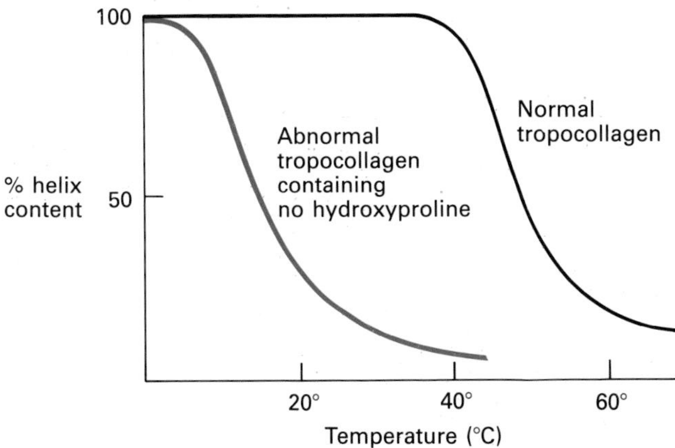

Figure 21-64 Denaturation of tropocollagen and of abnormal tropocollagen containing no hydroxyproline. Without hydrogen bonds involving hydroxyproline residues, the tropocollagen helix is unstable and is unfolded at temperatures above 20°C. Such tropocollagens are formed in the absence of ascorbic acid (vitamin C). Normal tropocollagen is more stable and resists thermal denaturation until a temperature of 40°C is reached.

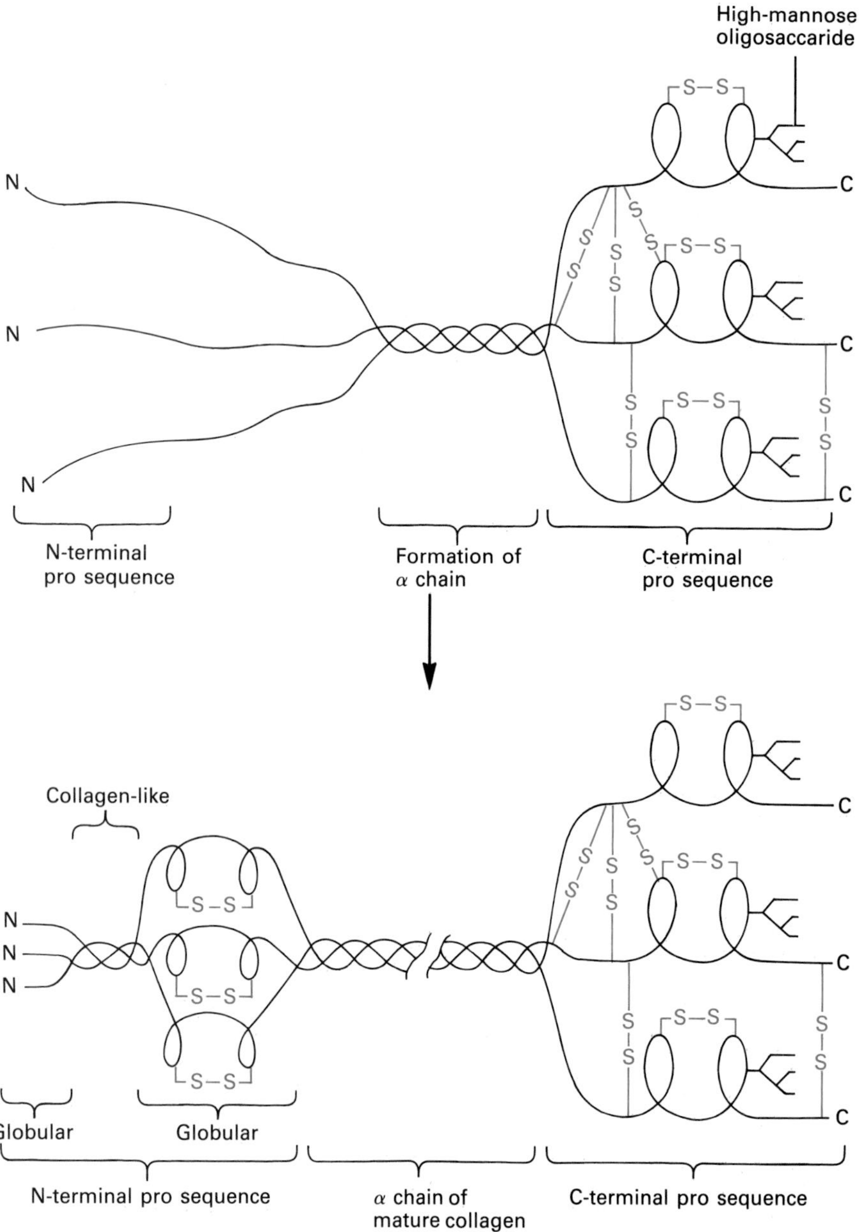

Figure 21-65 Assembly of the type I procollagen triple-stranded helix. Following removal of the N-terminal signal peptide, three procollagens spontaneously generate a collagen triple helix. This is initiated by formation of interchain disulfide bonds between the C-terminal pro sequences. Intrachain disulfide bonds form within the N-terminal pro sequences; these may act to stabilize the long procollagen triple helix, as may the short collagen-like sequence in the N-terminal pro sequence. Although not shown here, the collagen segments of procollagen contain hydroxyproline and glycosylated hydroxylysine residues. [See J. M. Davidson and R. A. Berg, 1980, *Methods Cell Biol.* **23**:119.]

lagen fibers, occur after secretion in reactions catalyzed by extracellular enzymes. Either concomitant with or shortly after exocytosis, procollagen molecules undergo proteolytic cleavage to remove the N-terminal and C-terminal pro segments (see Figure 21-63). These reactions are catalyzed by separate enzymes termed procollagen peptidases and are independent events. In some cases, removal of the N-terminal segment occurs as procollagen exits the cell, and the C-terminal segment is removed later. Once in the extracellular space, tropocollagen molecules undergo polymerization to form collagen fibrils. Eventual excision of both pro segments is essential for formation of normal fibers. In diseases such as the *Ehlers-Danlos syndrome* in humans or *dermatosparaxis* in cattle, there is a deficiency in one of the procollagen peptidases. The skin and tendons become very deformable, because collagen with pro extensions forms mostly disorganized bundles rather than the normal highly organized strong fibers.

Collagen fibrils are formed by spontaneous aggregation of the 300-nm-long tropocollagen helixes that are staggered by about 70 nm, or just less than one quarter of the length of the molecule (see Figure 5-54). Fibrils are about 50 μm in diameter and can be up to several micrometers in length. Accompanying the process of fibril formation is the oxidation of certain lysyl and hydroxylysyl residues by lysyl oxidase to form specific covalent cross-links between two chains (Figure 21-67). Such

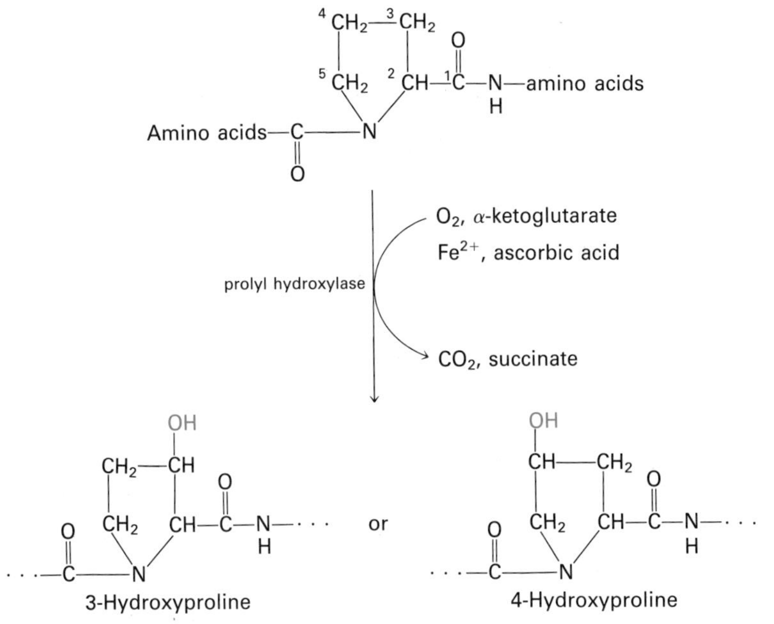

Figure 21-66 Hydroxylation of selected prolyl residues occurs in the rough ER before the triple helixes are formed. Oxygen, Fe^{2+}, ascorbic acid, and α-ketoglutarate are all required in this complex reaction. Proline can be hydroxylated in the 4 position only if it occurs in the amino acid sequence glycine-X-proline, where X is any amino acid. Hydroxylated prolines, especially the predominant 4-hydroxyproline, are important for stability of the triple helix.

bonding generates the staggered array characteristic of collagen fibers and contributes to their strength.

Collagen fibrils pack into the long fibers that occupy much of the interstitial space and form the basal lamina to which many cells are attached. In some tissues the collagen fibers are highly ordered; in others, the array appears to be nearly random. In tissues in which tensile strength is crucial, such as the Achilles tendon, collagen cross-linking is increased. Fibroblasts, therefore, can modify the properties of their extracellular matrix both by synthesizing different types of collagen peptides (see Table 5-6) and by differentially modifying the chains to influence the type and strength of packing.

Because of the large number of enzymes that must act on the collagen, mutations in many genes can affect collagen. Indeed, there are a large number of human genetic defects in collagen synthesis; blocks at almost every one of the steps outlined in Table 21-8 are known.

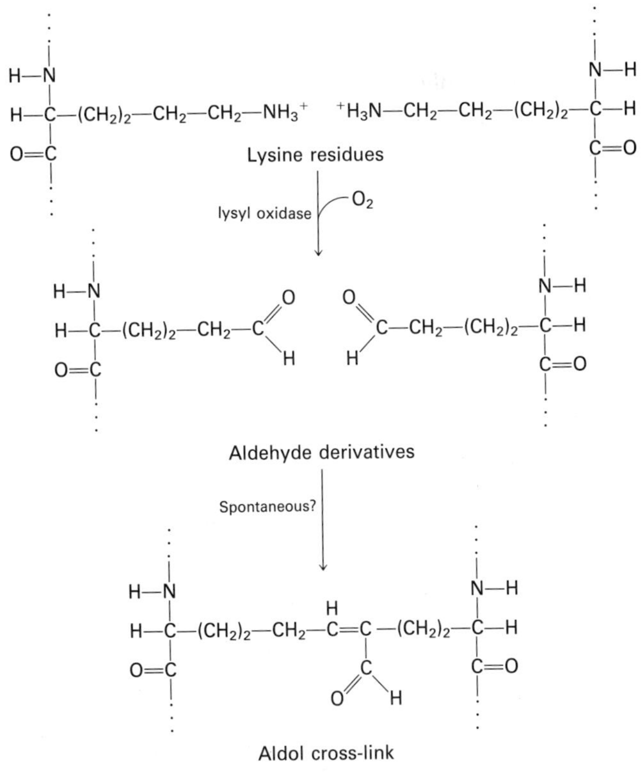

Figure 21-67 The side-by-side interactions of tropocollagen helixes are stabilized by formation of an aldol cross-link between two lysine side chains. This is accomplished by the extracellular enzyme lysyl oxidase.

Summary

To summarize the most important topics and results discussed in this chapter, let us return to the key concepts that have guided recent work on the assembly of subcellular particles, membranes, and organelles.

1. *Polypeptide chains self-assemble into complex aggregates.*

Specific protein-protein interactions allow multisubunit proteins such as aspartate transcarbamoylase (ATCase) to assemble. In molecules with one or more axes or planes of symmetry, such as ATCase, only a few kinds of interactions are required to assemble multiple subunits into place. Icosahedral (spherical) viruses are assembled by the repetitive formation of the same, or nearly the same, protein-protein interactions of the coat proteins. Helical viruses form in a multistep process; disks of coat subunits bind sequentially to the RNA to produce a conformational change that generates a helical array of proteins.

Specific interactions between ribosomal proteins and regions of rRNA, and between different ribosomal proteins, are employed during the ordered assembly of a ribosomal subunit. Tropocollagen chains self-assemble into three-stranded collagen helixes, assisted by formation of intra- and interchain disulfide bonds between pro segments of the three chains. After the triple helix is formed, the pro segments are proteolytically cleaved. Fibrous proteins such as actin and tubulin also self-assemble into rod-shaped fibers regulated by specific actin- or tubulin-binding proteins.

2. *Membranes grow by expansion of existing membranes.*

Phospholipids are synthesized on the cytoplasmic face of the smooth ER membrane or the bacterial plasma membrane. In a reaction catalyzed by certain membrane proteins, newly made phospholipids equilibrate with the extracytoplasmic membrane face. Transport of phospholipids and membrane proteins between organelles is mediated by small vesicles that bud from one organelle and fuse with another: during the processing of plasma membrane proteins and secretory proteins, vesicles migrate from the rough ER to the Golgi, from the Golgi to secretory vesicles, and from those to the plasma membrane. During endocytosis and recycling of the plasma membrane, vesicles move from the plasma membrane to sorting organelles, lysosomes, or the Golgi, or from the Golgi to lysosomes. Some of these vesicles have coats of clathrin.

3. *Membrane-containing organelles such as chloroplasts and mitochondria grow by expansion of existing organelles.*

Proteins and lipids are added to existing organelles, and eventually the organelle divides into two. Continuity of the mitochondrion or chloroplast requires the organelle DNA, which encodes the organelle rRNAs and tRNAs, but only a few of the organelle proteins. Mitochondrial gene products thus far identified include several inner membrane components of the electron transport chain: cytochrome *b*, subunits of cytochrome *c* oxidase, subunits of the F_0F_1-ATPase, subunits of NADH–CoQ reductase, and (in certain fungal mitochondria) enzymes that splice mitochondrial mRNA precursors. Besides encoding chloroplast rRNA and tRNAs, chloroplast DNA encodes one of the subunits of ribulose 1,5-bisphosphate carboxylase, the key enzyme that "fixes" CO_2, and other thylakoid proteins.

4. *Subcellular membranes contain permeases that allow certain proteins to cross.*

Uptake of cytosolic proteins into the mitochondrion or thylakoid vesicle of the chloroplast generally is dependent on energy provided from an electric potential or ionic gradient across the membrane.

5. *Proteins are targeted to specific organelles during or after synthesis.*

Most mitochondrial, chloroplast, and peroxisomal proteins are synthesized as soluble precursors in the cytoplasm. Specific receptors on the organelle surface ensure that these soluble proteins are incorporated into their proper membrane after synthesis. As noted, energy in the form of ATP or an energized membrane is essential for uptake of most chloroplast and mitochondrial enzymes. Often during uptake into the organelle a signal sequence of amino acids is proteolytically removed, or heme or other groups are added to yield an active protein. These modifications may serve to target the protein to its final destination. All of the citric acid cycle enzymes, most of the subunits of the electron transport chain and of the F_0F_1-ATPase, and mitochondrial DNA and RNA polymerases are examples of proteins imported into the mitochondrion. Chloroplasts import one of the two subunits of ribulose 1,5-bisphosphate carboxylase, the proteins in the light-harvesting complex, and many others.

An important class of proteins is synthesized on ribosomes bound to the rough ER. These include secretory proteins, lysosomal and Golgi enzymes, cell surface proteins, and glycoproteins. Synthesis of this class of proteins is initiated on ribosomes unbound to membranes. A sequence of hydrophobic amino acids, the signal sequence, is recognized by a signal receptor particle (SRP); this, in turn, binds to a receptor on the rough ER membrane. The SRP directs the insertion of the nascent protein into or across the membrane and the binding of the ribosome to the membrane. Generally, such signal sequences are at the N-terminus and are cleaved from the protein in the rough ER. Addition of mannose-containing oligosaccharides to specific asparagine residues of this class of proteins also occurs in the lumen of rough ER, as do disulfide bond formation and hydroxylation of lysyl and prolyl residues in collagen.

Proteins move from the rough ER, via membrane-limited vesicles, to the Golgi complex. There, a number of additional modifications occur, some of which are important in targeting the protein to its final destination. Processing of *N*-linked oligosaccharides to the sialic acid–containing complex form occurs, as does elongation of oligosaccharides attached to serine and threonine residues on some secretory and membrane proteins. Mannose residues in *N*-linked oligosaccharides of lysosomal enzymes are phosphorylated. A mannose 6-phosphate receptor binds these proteins and directs their transfer to lysosomes. A similar receptor on the cell surface binds extracellular phosphorylated lysosomal enzymes and, by receptor-mediated endocytosis, delivers them to lysosomes. Some secretory proteins are directed to secretory vesicles, wherein they are concentrated and stored, awaiting a neural, hormonal, or other signal for exocytosis. In the Golgi of polarized epithelial cells, plasma membrane proteins are sorted into vesicles destined for different regions of the plasma membrane. Often additional proteolytic cleavages occur in the Golgi or secretory vesicles. Many classes of hormones, such as ACTH, endorphins, and enkephalins, are synthesized together as one continuous inactive precursor protein. Proteolytic cleavage is essential for generation of the active hormone.

Other secretory proteins and most plasma membrane proteins are targeted to vesicles that constitutively fuse with the plasma membrane, thus exocytosing their contents.

Some modifications to proteins occur after secretion. Tropocollagen helixes lose both N-terminal and C-terminal pro sequences and are then incorporated into highly cross-linked collagen fibers. Activation of enzymes such as trypsinogen and chymotrypsin also occurs by specific proteolytic cleavages after secretion.

6. *Many membrane proteins shuttle repeatedly between two or more cellular compartments.*

The Golgi complex plays a key role in targeting many cellular proteins to their destinations. Current work suggests that the Golgi complex consists of at least three distinct classes of vesicles—the cis, medial, and trans Golgi, each of which has distinct enzymatic activities. Secretory proteins appear to move in the direction rough ER → cis Golgi → medial Golgi → trans Golgi → secretory vesicles. Especially in secretory cells, the Golgi also functions in recycling the plasma membrane; regions of the surface membrane are endocytosed in small vesicles, which then fuse with Golgi or lysosomal vesicles.

References

Assembly of Multiprotein Particles

BURNS, D. L., and H. K. SCHACHMAN. 1982. Assembly of the catalytic trimers of aspartate transcarbamoylase from unfolded polypeptide chains. *J. Biol. Chem.* **257**:8648–8654.

*CANTOR, C. R., and P. R. SCHIMMEL. 1980. *Biophysical Chemistry,* Part 1. W. H. Freeman and Company. Includes several chapters on the principles of protein folding and conformation.

CHOTHIA, C. 1984. Principles that determine the structure of proteins. *Annu. Rev. Biochem.* **53**:537–572.

*Book or review article that provides a survey of the topic.

*KIM, P. S., and R. L. BALDWIN. 1982. Specific intermediates in the folding reactions of small proteins and the mechanism of protein folding. *Annu. Rev. Biochem.* **51**:459–489.

THOMAS, K. A., and A. N. SCHECHTER. 1980. Protein folding: evolutionary, structural, and chemical aspects. In *Biological Regulation and Development,* vol. 2, R. F. Goldberger, ed. Plenum.

Assembly of Viruses

*BUTLER, P. J. G., and A. KLUG. 1978. The assembly of a virus. *Sci. Am.* **239**(5):62–69.

CASPAR, D. L. D., and A. KLUG. 1962. Physical principles in the construction of regular viruses. *Cold Spring Harbor Symp. Quant. Biol.* **27**:1–24.

*HARRISON, S. C. 1983. Virus structure: high resolution perspectives. *Adv. Virus Res.* **28**:175–240.

HARRISON, S. C. 1985. First comparison of two animal viruses in three dimensions. *Nature* **317**:382–384.

HARRISON, S. C. 1984. Multiple modes of subunit association in the structures of simple spherical viruses. *Trends Biochem. Sci.* **9**:345–351.

HOLMES, K. C. 1980. Protein-RNA interactions during the assembly of tobacco mosaic virus. *Trends Biochem. Sci.* **5**:4–7.

KING, J. 1980. Regulation of structural protein interactions as revealed in phage morphogenesis. In *Biological Regulation and Development,* vol. 2, R. F. Goldberger, ed. Plenum.

KLUG, A. 1979. The assembly of tobacco mosaic virus: structure and specificity. *Harvey Lect.* **74**:141–172.

*NAMBA, K., D. L. D. CASPAR, and G. J. STUBBS. 1985. Computer graphics representation of levels of organization in tobacco mosaic virus structure. *Science* **277**:773–776.

Assembly of Ribosomes

HELD, W. A., B. BALLOU, S. MIZUSHIMA, and M. NOMURA. 1974. Assembly mapping of 30S ribosomal proteins from *Escherichia coli. J. Biol. Chem.* **249**:3103–3111.

*NOLLER, H. F. 1984. Structure of ribosomal RNA. *Annu. Rev. Biochem.* **53**:119–162.

NOLLER, H. F., and C. R. WOESE. 1981. Secondary structure of 16S ribosomal RNA. *Science* **212**:403–409.

*NOMURA, M. 1973. Assembly of bacterial ribosomes. *Science* **179**:864–873.

WITTMAN, H. G. 1983. Components of bacterial ribosomes. *Annu. Rev. Biochem.* **51**:155–184.

Synthesis of Phospholipids

BISHOP, W. R., and R. M. BELL. 1985. Assembly of the endoplasmic reticulum bilayer: the phosphatidylcholine transporter. *Cell* **42**:51–60.

COLEMAN, R., and R. M. BELL, 1978. Evidence that biosynthesis of phosphatidylethanolamine, phosphatidylcholine and triacylglycerol occurs on the cytoplasmic side of microsomal vesicles. *J. Cell Biol.* **76**:245–253.

PAGANO, R., and R. SLEIGHT. 1985. Defining lipid transport pathways in animal cells. *Science* **229**:1051–1057.

* ROTHMAN, J. E., and J. LENARD. Membrane asymmetry. *Science* 195:743–753.

Assembly of Cell Membranes and Organelles: General Reviews

LODISH, H. F., W. A. BRAELL, A. L. SCHWARTZ, G. J. A. M. STROUS, and A. ZILBERSTEIN. 1981. Synthesis and assembly of membrane and organelle proteins. *Int. Rev. Cytol.* (suppl.) 12:247–307.

LODISH, H. F., and J. E. ROTHMAN. 1979. The assembly of cell membranes. *Sci. Am.* **240**(1):48–63.

SABATINI, D., G. KREIBICH, T. MORIMOTO, and M. ADESNICK. 1982. Mechanisms for the incorporation of proteins in membranes and organelles. *J. Cell Biol.* **92**:1–22.

Mitochondrial DNA and RNA

ANDERSON, S., A. T. BANKIER, B. G. BARRELL, M. H. L. DE BRUIJN, A. R. COULSON, J. DROUIN, I. C. EPERON, D. P. NIERLICH, B. A. ROE, F. SANGER, P. H. SCHREIER, A. J. H. SMITH, R. STADEN, and I. G. YOUNG. 1981. Sequence and organization of the human mitochondrial genome. *Nature* **290**:457–465.

* ATTARDI, G., P. CONTATORE, E. CHING, S. CREWS, R. GELFAND, C. MERKEL, J. MONTOYA, and D. OJALA. 1981. Organization and expression of genetic information in human mitochondrial DNA. In *International Cell Biology 1980–1981,* H. G. Schweiger, ed. Springer-Verlag, pp. 225–238.

BREITENBERGER, C. A., and U. L. RAJBHANDARY. 1985. Some highlights of mitochondrial research based on analyses of *Neurospora crassa* mitochondrial DNA. *Trends in Biochem. Sci.* **10**:478–483.

CARIGNANI, G., O. GROUDINSKY, D. FREZZA, E. SCHIAVON, E. BERGANTINO, and P. P. SLONIMSKI. 1984. An mRNA maturase is encoded by the first intron of the mitochondrial gene for the subunit I of cytochrome oxidase in *S. cerevisiae. Cell* **35**:733–742.

* CHOMYN, A., P. MARIOTTINI, M. W. J. CLEETER, C. I. RAGAN, A. MATSUNO-YAGI, Y. HATEFI, R. F. DOOLITTLE, and G. ATTARDI. 1985. Six unidentified reading frames of human mitochondrial DNA encode components of the respiratory-chain NADH dehydrogenase. *Nature* **314**:592–597.

* CLAYTON, D. A. 1984. Transcription of the mammalian mitochondrial genome. *Annu. Rev. Biochem.* **53**:573–594.

PINKHAM, J. L., and L. GUARENTE. 1985. Cloning and molecular analysis of the *HAP2* locus: a global regulator of respiratory genes in *Saccharomyces cerevisiae. Mol. Cell Biol.* 5:3410–3416.

VAN DER HORST, G., and H. F. TABAK. 1985. Self-splicing of yeast mitochondrial ribosomal and messenger RNA precursors. *Cell* **40**:759–766.

Import of Proteins into Mitochondria

DAUM, G., S. M. GASSER, and G. SCHATZ. 1982. Import of proteins into mitochondria. Energy-dependent, two-step processing of the intermembrane space enzyme cytochrome b_2 by isolated yeast mitochondria. *J. Biol. Chem.* **257**:13075–13080.

DOUGLAS, M. G., B. L. GELLER, and S. D. EMR. 1984. Intracellular targeting and import of an F_1-ATPase β-subunit–β-galactosidase hybrid protein into yeast mitochondria. *Proc. Nat'l Acad. Sci. USA* **81**:3983–3987.

HURT, E. C., B. PESOLD-HURT, and G. SCHATZ. 1984. The cleavable prepiece of an imported mitochondrial protein is sufficient to direct cytosolic dihydrofolate reductase into the mitochondrial matrix. *FEBS Lett.* **178**:306–310.

KAPUT, J., S. GOLTZ, and G. BLOBEL. 1982. Nucleotide sequence of the yeast nuclear gene for cytochrome *c* peroxidase precursor. Functional implications of the pre sequence for protein transport into mitochondria. *J. Biol. Chem.* **257**:15054–15058.

OHASHI, A., J. GIBSON, I. GREGOR, and G. SCHATZ. 1982. Import of proteins into mitochondria. The precursor of cytochrome c_1 is processed in two steps, one of them heme-dependent. *J. Biol. Chem.* **257**:13042–13047.

PFALLER, R., H. FREITAG, M. HARMEY, R. BENZ, and W. NEUPERT. 1985. A water-soluble form of porin from the mitochondrial outer membrane of *Neurospora crassa. J. Biol. Chem.* **260**:8188–8193.

* SCHATZ, G., and R. A. BUTOW. 1983. How are proteins imported into mitochondria? *Cell* **32**:316–318.

SCHLEYER, M., and W. NEUPERT. 1984. Transport of ADP/ATP carrier into mitochondria. Precursor imported in vitro acquires functional properties of the mature protein. *J. Biol. Chem.* **259**:3487–3491.

YAFFE, M. P., S. OHTA, and G. SCHATZ. 1985. A yeast mutant temperature-sensitive for mitochondrial assembly is deficient in a mitochondrial protease activity that cleaves imported precursor polypeptides. *EMBO J.* **4**:2069–2074.

ZWIZINSKI, C., M. SCHLEYER, and W. NEUPERT. 1984. Proteinaceous receptors for the import of mitochondrial precursor proteins. *J. Biol. Chem.* **259**:7850–7856.

Biogenesis of Chloroplasts

BEDBROOK, J. R., S. M. SMITH, and R. J. ELLIS. 1980. Molecular cloning and sequencing of cDNA encoding the precursor to the small subunit of chloroplast ribulose-1,5-bisphosphate carboxylase. *Nature* **287**:692–697.

* BOGORAD, L., E. J. GUBBINS, S. O. JOLLY, E. KREBBERS, I. LARRINUA, K. MUSKAVITCH, S. RODERMAL, A. SUBRAMANIAN, and A. STEINMETZ. 1983. Maize plastid genes: structure and expression. In *Gene Structure and Regulation in Development.* S. Subtelny and F. C. Kafatos, eds. Liss, pp. 13–32.

BOHNERT, J. M., E. J. BROUSE, and J. M. SCHMITT. 1982. Organization and expression of plastid genomes. In *Encyclopedia of Plant Physiology, New Series,* vol. 14B, B. Parthier and D. Boulter, eds. Springer-Verlag.

GROSSMAN, A., S. BARTLETT, and N-H. CHUA. 1980. Energy-dependent uptake of cytoplasmically synthesized polypeptides by chloroplasts. *Nature* **285**:625–627.

MORELLI, G., F. NAGY, R. T. FRALEY, S. G. ROGERS, and N.-H. CHUA. 1985. A short conserved sequence is involved in the light-inducibility of gene encoding ribulose 1,5-bisphosphate carboxylase small subunit of pea. *Nature* **315**:200–204.

VAN DEN BROECK, G., M. P. TIMKO, A. P. KAUSCH, A. R. CASHMORE, M. VAN MONTAGU, and L. HERRERA-ESTRELLA. 1985. Targeting of a foreign protein to chloroplasts by fusion to the transit peptide from the small subunit of ribulose 1,5-bisphosphate carboxylase. *Nature* **313**:358–363.

* VON WETTSTEIN, D. Chloroplast and nucleus: concerted interplay between genomes of different cell organelles. In *International Cell Biology 1980-1981,* H. G. Schweiger, ed. Springer-Verlag, pp. 250–274.

Synthesis of Peroxisomes

GOLDMAN, B. M., and G. BLOBEL. 1978. Biogenesis of peroxisomes: intracellular site of synthesis of catalase and uricase. *Proc. Nat'l Acad. Sci. USA* **75**:5066–5070.

LAZAROW, P. B. 1984. The peroxisomal membrane. In *Membrane Structure and Function,* vol. 5, E. E. Bittar, ed. Wiley, pp. 2–31.

LAZAROW, P. B., and Y. FUJIKI. 1985. Biogenesis of peroxisomes. *Annu. Rev. Cell Biol.* **1**:489–530.

* TOLBERT, N. E., and E. ESSNER. 1981. Microbodies: peroxisomes and glyoxysomes. *J. Cell Biol.* **91**:271s–283s.

Signal Recognition Particle and Protein Synthesis on the Rough ER

GILMORE, R., and G. BLOBEL. 1983. Transient involvement of signal recognition particle and its receptor in the microsomal membrane prior to protein translocation. *Cell* **35**:677–685.

GILMORE, R., G. BLOBEL, and P. WALTER. 1982. Protein translocation across the endoplasmic reticulum. I. Detection in the microsomal membrane of a receptor for the signal recognition particle. *J. Cell Biol.* **95**:463–469.

LINGAPPA, V. R., J. CHAIDEZ, C. S. YOST, and J. HEDGPETH. 1984. Determinants for protein localization: β-lactamase signal sequence directs globin across microsomal membranes. *Proc. Nat'l Acad. Sci. USA* **81**:456–460.

MEYER, D. I., E. KRAUSE, and B. DOBBERSTEIN. 1982. Secretory protein translocation across membranes—the role of the "docking protein." *Nature* **297**:647–650.

WALTER, P., and G. BLOBEL. 1981. Translocation of proteins across the endoplasmic reticulum. III. Signal recognition protein causes signal sequence–dependent and site-specific arrest of chain elongation that is released by microsomal membranes. *J. Cell Biol.* **91**:557–561.

* WALTER, P., R. GILMORE, and G. BLOBEL. 1984. Protein translocation across the endoplasmic reticulum. *Cell* **38**:5–8.

Synthesis of Membrane Proteins on the ER

ANDERSON, D. M., K. E. MOSTOV, and G. BLOBEL. 1983. Mechanisms of integration of *de novo*–synthesized polypeptides into membranes: signal-recognition particle is required for integration into microsomal membranes of calcium ATPase and of lens MP26 but not of cytochrome b_5. *Proc. Nat'l Acad. Sci. USA* **80**:7249–7253.

BRAELL, W. A., and H. F. LODISH. 1982. The erythrocyte anion transport protein is cotranslationally inserted into microsomes. *Cell* **29**:23–32.

MUECKLER, M., and H. F. LODISH. 1986. The human transporter can insert posttranslationally into microsomes. *Cell* **44**:629–637.

RAPOPORT, T. A., and M. WIEDMANN. 1985. Application of the signal hypothesis to the incorporation of integral membrane proteins. *Curr. Top. Membranes and Transport.* **24**:1–63.

ROSE, J. K., W. J. WELCH, B. M. SEFTON, F. S. ESCH, and N. C. LING. 1980. Vesicular stomatitis virus glycoprotein is anchored in the viral membrane by a hydrophobic domain near the COOH terminus. *Proc. Nat'l Acad. Sci. USA* **77**:3884–3888.

ROTHMAN, J. E., and H. F. LODISH. 1977. Synchronized transmembrane insertion and glycosylation of a nascent membrane protein. *Nature* **269**:755–778.

* WICKNER, W., and H. F. LODISH. 1985. Multiple mechanisms of insertion of proteins into and across membranes. *Science,* **230**:400–407.

YOST, C. S., J. HEDGPETH, and V. R. LINGAPPA. 1983. A stop transfer sequence confers predictable transmembrane orientation to a previously secreted protein in cell-free systems. *Cell* **34**:759–766.

The Golgi Complex and the Sorting of Secreted and Membrane Proteins

ANDERSON, R. G. W., and R. K. PATHAK. 1985. Vesicles and cisternae in the *trans* Golgi apparatus of human fibroblasts are acidic compartments. *Cell* **40**:635–643.

BALCH, W. E., W. G. DUNPHY, W. A. BRAELL, and J. E. ROTHMAN. 1984. Reconstitution of the transport of protein between successive compartments of the Golgi measured by the coupled incorporation of *N*-acetylglucosamine. *Cell* **39**:405–416.

BERGMANN, J. E., and S. J. SINGER. 1983. Immunoelectron microscopic studies of the intracellular transport of the membrane glycoprotein (G) of vesicular stomatitis virus in infected Chinese hamster ovary cells. *J. Cell Biol.* **97**:1777–1787.

BRANDS, R., M. D. SNIDER, Y. HINO, S. S. PARK, H. V. GELBOIN, and J. E. ROTHMAN. 1985. Retention of membrane proteins by the endoplasmic reticulum. *J. Cell Biol.* **101**:1724–1732.

DUNPHY, W. G., R. BRANDS, and J. E. ROTHMAN. 1985. Attachment of terminal *N*-acetylglucosamine to asparagine-linked oligosaccharides occurs in central cisternae of the Golgi stack. *Cell* **40**:463–472.

* DUNPHY, W., and J. E. ROTHMAN. 1985. Compartmental organization of the Golgi stack. *Cell* **42**:13–21.

* FARQUHAR, M. G. 1985. Progress in unraveling pathways of Golgi traffic. *Annu. Rev. Cell Biol.* **1**:447–488.

* FARQUAHR, M. G., and G. E. PALADE. 1981. The Golgi apparatus (complex)—1954–1981—from artifact to center stage. *J. Cell Biol.* **91**:77s–106s.

KELLY, R. B. 1985. Pathways of protein secretion in eukaryotes. *Science* **230**:25–32.

PAYNE, G. S., and R. SCHECKMAN. 1985. A test of clathrin function in protein secretion and cell growth. *Science* **230**:1009–1014.

ROTHMAN, J. E., R. L. MILLER, and L. J. URBANI. 1984. Intercompartmental transport in the Golgi complex is a dissociative process: facile transfer of membrane protein between two Golgi populations. *J. Cell Biol.* **99**:260–271.

SCHEKMAN, R. 1985. Protein localization and membrane traffic in yeast. *Annu. Rev. Cell Biol.* **1**:115–144.

STROUS, G. J. A. M., and E. G. BERGER. 1982. Biosynthesis, intracellular transport, and release of the Golgi enzyme galactosyltransferase (lactose synthetase A protein) in HeLa cells. *J. Biol. Chem.* **257**:7623–7628.

WARREN, G. 1985. Membrane traffic and organelle division. *Trends in Biochem. Sci.* **10**:439–443.

Synthesis of Membrane Proteins in Polarized Cells

* MISEK, D. E., E. BARD, and E. RODRIGUEZ-BOULAN. 1984. Biogenesis of epithelial cell polarity: intracellular sorting and vectorial exocytosis of an apical plasma membrane glycoprotein. *Cell* **39**:537–546.

RINDLER, M. J., I. E. IVANOV, H. PLESKEN, E. RODRIGUEZ-BOLAN, and D. D. SABATINI. 1984. Viral glycoproteins destined for apical or basolateral plasma membrane domains traverse the same Golgi apparatus during their intracellular transport in doubly infected Madin-Darby canine kidney cells. *J. Cell Biol.* **98**:1304–1319.

ROTH, M. G., R. W. COMPANS, L. GIUSTI, A. R. DAVIS, D. P. NAYAK, M.-J. GETHING, and J. SAMBROOK. 1983. Influenza virus hemagglutinin expression is polarized in cells infected with recombinant SV40 viruses carrying cloned hemagglutinin DNA. *Cell* **33**:435–443.

SIMONS, K., and S. D. FULLER. 1985. Cell surface polarity in epithelia. *Annu. Rev. Cell Biol.*

Proteolytic Processing of Secretory Proteins

BRAKE, A. J., D. J. JULIUS, and J. THORNER. 1983. A functional α-factor gene in *Saccharomyces* yeasts can contain three, four, or five repeats of the mature pheromone sequence. *Mol. Cell Biol.* **3**:1440–1450.

DOCHERTY, K., R. J. CARROLL, and D. F. STEINER. 1982. Conversion of proinsulin to insulin: involvement of a 31,500 molecular weight thiol protease. *Proc. Nat'l Acad. Sci. USA* **79**:4613–4617.

* DOUGLASS, J., O. CIVELLI, and E. HERBERT. 1984. Polyprotein gene expression: generation of diversity of neuroendocrine peptides. *Annu. Rev. Biochem.* **53**:665–715.

ORCI, L., M. RAVAZZOLA, M. AMHERDT, O. MADSEN, J.-D. VASSALLI, and A. PERRELET. 1985. Direct identification of prohormone conversion site in insulin-secreting cells. *Cell* **42**:671–681.

* STEINER, D. F., P. S. QUINN, S. J. CHAN, J. MARSH, and H. S. TAGER. 1980. Processing mechanisms in the biosynthesis of proteins. *Ann. NY Acad. Sci.* **343**:1–16.

Protein Glycosylation

CAPASSO, J. M., and C. B. HIRSCHBERG. 1984. Mechanisms of glycosylation and sulfation in the Golgi apparatus: evidence for nucleotide sugar/nucleoside monophosphate and nucleotide sulfate/nucleoside monophosphate antiports in the Golgi apparatus membrane. *Proc. Nat'l Acad. Sci. USA* **81**:7051–7055.

CUMMINGS, R. D., S. KORNFELD, W. J. SCHNEIDER, K. K. HOBGOOD, H. TOLLESHAUG, M. S. BROWN, and J. L. GOLDSTEIN. 1983. Biosynthesis of *N*- and *O*-linked oligosaccharides of the low density lipoprotein receptor. *J. Biol. Chem.* **258**:15261–15273.

DUNPHY, W. G., R. BRANDS, and J. E. ROTHMAN. 1985. Attachment of terminal *N*-acetylglucosamine to asparagine-linked oligosaccharides occurs in central cisternae of the Golgi stack. *Cell* **40**:463–472.

ELBEIN, A. D. 1981. The tunicamycins—useful tools for studies on glycoproteins. *Trends Biochem. Sci.* **6**:219–223.

ESMON, B., P. NOVICK, and R. SCHEKMAN. 1981. Compartmentalized assembly of oligosaccharides on exported glycoproteins in yeast. *Cell* **25**:451–460.

KORNFELD, R., and S. KORNFELD. 1985. Assembly of asparagine-linked oligosaccharides. *Annu. Rev. Biochem.* **45**:631–664.

* LENNARZ, W. J., ed. 1980. *The Biochemistry of Glycoproteins and Proteoglycans.* Plenum. Contains several definitive reviews on the structure and synthesis of glycoproteins.

ROTH, J. 1984. Cytochemical localization of terminal *N*-acetyl-D-galactosamine residues in cellular compartments of intestinal globlet cells: implications for the topology of *O*-glycosylation. *J. Biol. Chem.* **98**:399–406.

ROTH, J. and E. G. BERGER. 1982. Immunocytochemical localization of galactosyltransferase in HeLa cells: codistribution with thiamine pyrophosphatase in *trans*-Golgi cisternae. *J. Cell Biol.* **92**:223–229.

SNIDER, M. D., and O. C. ROGERS. 1984. Transmembrane movement of oligosaccharide-lipids during glycoprotein synthesis. *Cell* **36**:753–761.

Synthesis of Lysosomal Enzymes

CAMPBELL, C. H., and L. H. ROME. 1983. Coated vesicles from rat liver and calf brain contain lysosomal enzymes bound to mannose 6-phosphate receptors. *J. Biol. Chem.* **258**:13347–13352.

DEUTSCHER, S. L., K. E. CREEK, M. MERION, and C. B. HIRSCHBERG. 1983. Subfractionation of rat liver Golgi apparatus: separation of enzyme activities involved in the biosynthesis of the phosphomannosyl recognition marker in lysosomal enzymes. *Proc. Nat'l Acad. Sci. USA* **80**:3938–3942.

ERICKSON, A. H., G. E. CONNER, and G. BLOBEL. 1981. Biosynthesis of a lysosomal enzyme. Partial structure of two transient and functionally distinct NH_2-terminal sequences in cathepsin D. *J. Biol. Chem.* **256**:11224–11231.

GEUZE, H. J., J. W. SLOT, G. J. A. M. STROUS, A. HASILIK, and K. VON FIGURA. 1985. Possible pathways for lysosomal enzyme delivery. *J. Cell Biol.* **101**:2253–2262.

LANG, L., M. REITMAN, J. TANG, R. M. ROBERTS, and S. KORNFELD. 1984. Lysosomal enzyme phosphorylation. *J. Biol. Chem.* **259**:14663–14671.

HOFLACK, B., and S. KORNFELD. 1985. A cation-dependent mannose-6-phosphate receptor. *J. Biol. Chem.* **260**:12008–12041.

NEUFELD, E. F., T. W. LIM, and L. J. SHAPIRO. 1975. Inherited disorders of lysosomal metabolism. *Annu. Rev. Biochem.* **44**:357–376.

POHLMANN, R., A. WAHEED, A. HASILIK, and K. VON FIGURA. 1982. Synthesis of phosphorylated recognition marker in lysosomal enzymes is located in the *cis* part of Golgi apparatus. *J. Biol. Chem.* **257**:5323–5325.

REITMAN, M., A. VARKI, and S. KORNFELD. 1981. Fibroblasts from patients with I-cell disease and pseudo-Hurler polydystrophy are deficient in uridine 5′-diphosphate-*N*-acetylglucosamine: glycoprotein *N*-acetylglucosaminylphosphotransferase activity. *J. Clin. Invest.* **67**:1574–1579.

SAHAGIAN, G. G., and E. F. NEUFELD. 1983. Biosynthesis and turnover of the mannose 6-phosphate receptor in cultured Chinese hamster ovary cells. *J. Biol. Chem.* **258**:7121–7128.

Exocytosis

BAKER, P. F. 1984. Multiple controls for secretion? *Nature* **310**:629–630.

CHANDLER, D. E., and J. E. HEUSER. 1980. Arrest of membrane fusion events in mast cells by quick-freezing. *J. Cell Biol.* **86**:666–674.

ORCI, L., R. MONTESANO, and A. PERRELET. 1981. Exocytosis—endocytosis as seen with morphological probes of membrane organization. *Methods Cell Biol.* **23**:283–300.

SCHULZ, I., and H. H. STOLZE. 1980. The exocrine pancreas: the role of secretagogues, cyclic nucleotides, and calcium in enzyme secretion. *Annu. Rev. Physiol.* **42**:127–156.

Assembly of the Nuclear Membrane

* FORBES, D. J., M. W. KIRSCHNER, and J. W. NEWPORT. 1983. Spontaneous formation of nucleus-like structures around bacteriophage DNA microinjected into *Xenopus* eggs. *Cell* **34**:13–23.

* GERACE, L., and L. BLOBEL. 1981. Nuclear lamina and the structural organization of the nuclear envelope. *Cold Spring Harbor Symp.* **46**:967–978.

HESKETH, T. R., M. A. BEAVEN, J. ROGERS, B. BURKE, and G. B. WARREN. 1984. Stimulated release of histamine by a rat mast cell line is inhibited during mitosis. *J. Cell Biol.* **98**:2250–2254.

OTTAVIANO, Y., and L. GERACE. 1985. Phosphorylation of the nuclear lamina during interphase and mitosis. *J. Biol. Chem.* **260**:624–632.

WARREN, G., C. FEATHERSTONE, G. GRIFFITHS, and B. BURKE. 1983. Newly synthesized G protein of vesicular stomatitis virus is not transported to the cell surface during mitosis. *J. Cell Biol.* **97**:1623–1628.

Synthesis of Collagen

BORNSTEIN, P. 1974. The biosynthesis of collagen. *Annu. Rev. Biochem.* **43**:567–603.

* DAVIDSON, J. M., and R. A. BERG. 1981. Posttranslational events in collagen synthesis. *Methods Cell Biol.* **23**:119–136.

EYRE, D. R. 1980. Collagen: molecular diversity in the body's protein scaffold. *Science* **207**:1315–1322.

* EYRE, D. R., M. P. PAZ, and P. M. GALLOP. 1984. Cross-linking in collagen and elastin. *Annu. Rev. Biochem.* **53**:717–748.

MARTIN, G. R., R. TIMPLE, P. K. MULLER, and K. KUHN. 1985. The genetically distinct collagens. *Trends in Biochem. Sci.* **10**:285–287.

* PRUCKOP, D. J., K. I. KIVIRIKKO, L. TUDERMAN, and N. GUZMAN. 1979. The biosynthesis of collagen and its disorders. *N. Engl. J. Med.* **301**:13–23.

* PRUCKOP, D. J., K. I. KIVIRIKKO, L. TUDERMAN, and N. GUZMAN. 1979. The biosynthesis of collagen and its disorders. *N. Engl. J. Med.* **301**:77–85.

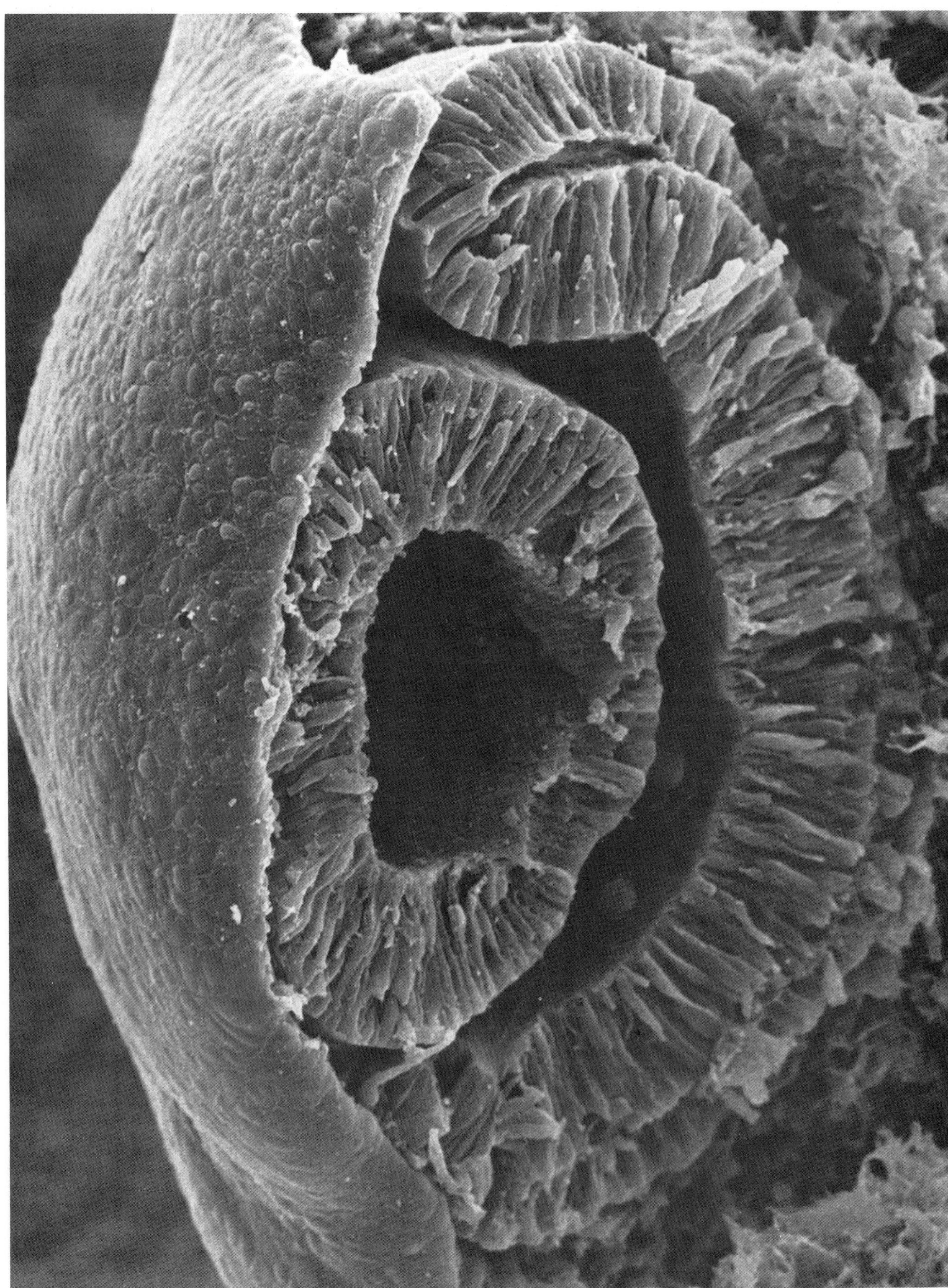

IV

Normal and Abnormal Variations in Cells

Scanning electron micrograph showing the fractured edge of a developing chick eye at about 5 to 6 days. *Courtesy of K. W. Tosney.*

IN Chapters 8 to 13 we considered the molecular aspects of gene expression and differential gene control, particularly at the level of differential transcription. Chapters 14 to 21 described the molecular properties of the structural elements of the cell and of the proteins in cell membranes. The roles of structural elements in cell division, cell movement, and changes in cell shape were emphasized, as were the functions of membrane proteins in the communication of cells with one another and with their environments. These topics and emphases reflect the approach of molecular biologists to the study of eukaryotic cells: *the cells are often viewed as individual entities.*

Much current research focuses on eukaryotic cells as a part of a developing multicellular organism. For a fertilized egg—a single cell—to grow, divide, and develop into a multicellular organism composed of perhaps thousands of different cell types, an elaborate program must operate to control gene expression in individual cells. Such a program must finally be responsible not only for cell specialization but for cell movement and cell division, which are so crucial in shaping the embryo: to give rise to the characteristic organs and structures of an embryo, the correct cells must move or divide while other, sometimes adjacent, cells do not. To the molecular biologist, a key problem in development is understanding how control of gene expression in particular cell types is connected with the dramatic events of cell growth, division, and movement.

In this concluding part of the book we shall first present a brief introduction to the materials that have traditionally been studied by developmental biologists, and we shall review some recent experiments in the field. The two chapters following treat two topics that are often considered in connection with development: cancer cells, which ignore normal developmental clues and boundaries, and immunology, a field in which the molecular bases for differentiation into different cell types from a common precursor cell have been especially well studied. The closing chapter returns to the time-honored biological theme that ultimately depends on cell variations: evolution. In that chapter we discuss how the new ideas in molecular cell biology have completely revamped our thinking about early cellular evolution.

22

Development of Cell Specificity

THE central problems of developmental biology were stated by embryologists long before all the new molecular techniques came into being. We shall review these long-standing ideas and questions because they help to define the developmental phenomena to which molecular approaches are being applied.

A central issue has been the cellular potential of embryonic cells. A fertilized egg is said to be *totipotent* because it gives rise to a whole organism by division and cell specialization. As development proceeds, the number and kinds of cells that can arise from one parental cell become restricted. Such cells are said to be *pluripotent*. Finally, cells that are destined to become a single cell type (but have not yet done so) are termed *unipotent*.

When do individual cells of the developing embryo lose totipotency? How early in development do the cells that will form, say, the heart, the skin, or the nervous system become distinct from the rest of the cells? Are the specialized cells that make up the organs descended from one single cell or from a group of cells? Are "decisions" that direct the early embryonic cells into different pathways similar to "decisions" that lead precursor cells (stem cells) to regenerate specialized adult cell populations such as red and white blood cells? Finally, do developmental decisions occur because of an irrevocable change in the cell nucleus?

One term that is widely used in discussing all of these questions about cellular potential is *determination.* A cell is said to be *determined* once it has been instructed—or once it has somehow decided for itself—to become a specific cell type *at some future point.* That is, a determined cell and its descendants have acquired the developmental potential—but not yet the ability—to produce the specialized products or to carry out the specialized function by which it will eventually be recognized.

The *differentiation* of a determined cell is the process by which genes are selectively expressed and gene products act to produce a cell with a specialized phenotype. By this definition, differentiation is often an end-stage event for cells. For example, mammalian red blood cells are said to be fully differentiated when they have lost their nuclei, at which point, of course, they can no longer grow.

The decision to enter the differentiation pathway is called *commitment.* Although there are clear conceptual and chronological differences between commitment and the earlier decisions resulting in determination, there may be no molecular difference between these two types of decisions. Combined, they may simply represent a continuous set of events acting on a series of different genes: the recognizable differentiated end product would be the outcome.

Determination has been almost impossible to study. It is not easy to obtain a pure population of cells (or pure interacting populations of cells) that will carry out the phenomenon of determination; therefore no genes or gene products that are directly associated with determination in multicellular organisms are known at present. In contrast, many experimental models for differentiation are available, and the induction of differentiation by hormones and cell-to-cell interactions has been widely studied.

Besides attempting to identify the mechanisms of differential gene expression in cells, researchers have long been trying to find out how any set of gene products can cause the exact (or nearly exact) architectural formation of an animal or plant. How are limits set on the growth and shaping of body parts? Such problems are often conceptualized in terms of cells having *positional information.* That is, seemingly similar cells in different parts of the developing body (e.g., precursors to cells making up the tissues in arms and legs) are endowed somehow with the information to make different structures. Perhaps our best hope of discovering how such positional decisions are made lies in the study of organisms such as fruit flies and worms, for which genetic defects resulting in incorrect body part formation are known.

In this chapter we shall introduce students of molecular cell biology to some of the major ideas and experimental systems of developmental biology. We shall first discuss the advantages of studying simple organisms (yeasts and slime molds) that carry out limited developmental programs but that nevertheless exhibit important general principles of development. Because most developmental studies focus on animals, we shall then describe in some detail the major steps of animal development. The usefulness of genetics in elucidating the key steps in development will be illustrated by studies of insects and worms (nematodes). Finally, we shall consider some of the major unsolved questions in development and assess the prospects for their solution.

Differentiation in Unicellular and Simple Multicellular Organisms

Differentiation is not a unique property of multicellular organisms. Like the cells of higher organisms, many unicellular microorganisms can undergo remarkable physical changes that are accompanied by changes in physiological and metabolic capacities. In microorganisms such cell differentiation is often induced by changes in the environment.

For example, the absence of adequate nutrients triggers sporulation in a number of bacteria and single-celled eukaryotes (e.g., fungi and slime molds). Spores are specialized, metabolically dormant cells that are capable of surviving desiccation, heat, exposure to toxic chemicals, and starvation for months (in some cases, years). The relatively simple developmental program resulting in sporulation illustrates some of the features of differentiation in higher organisms and provides insight into the approaches that can be used in studying differentiation. Different organisms produce spores in different ways, and of course each pathway reflects the unique evolution, metabolic properties, and physical requirements of that organism. We shall begin our discussion of cellular differentiation with a summary of sporulation in three diverse microorganisms: a bacterium, a yeast, and a slime mold.

Sporulation in Bacteria Provides Protection against Environmental Extremes

Many different types of bacteria can form spores, but we shall concentrate on *Bacillus subtilis,* which has been studied very thoroughly. A *B. subtilis* spore develops within a *vegetative* (growing) cell as an *endospore,* a cell-within-a-cell having a unique structure and chemical and enzymatic composition. The endospore is covered by a spore coat, a multilayered envelope of proteins and glycoproteins that surrounds the spore cytoplasm. The water content of spores is extremely low in comparison to that of growing cells: about 15 to 25 percent versus 75 to 85 percent. The spore coat has different chemical linkages than the cell wall of vegetative cells, and contains amino acids that are absent from the polymeric structure (the

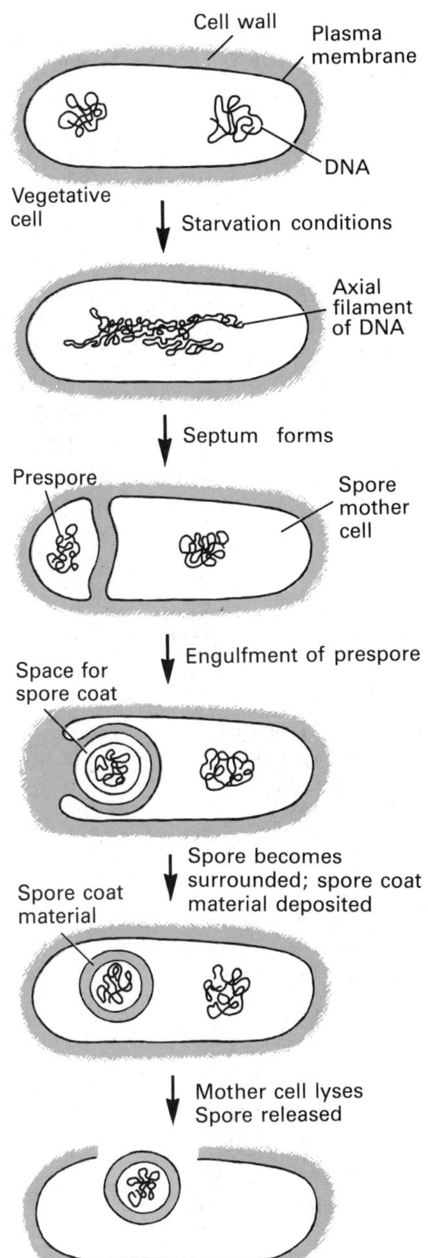

Figure 22-1 Stages in the development of a *Bacillus subtilis* spore. When a growing bacterium (the vegetative cell) is faced with starvation, the process of sporulation begins. The first morphological change is the transformation of the two identical chromosomes (the parent chromosome and the newly replicated daughter chromosome) into an axial filament of DNA. (One chromosome would have been delivered to a vegetative daughter cell if growth had continued.) A septum then divides the cell, and the smaller prespore cell is engulfed by the spore mother cell. In this cell-within-a-cell, the plasma membranes separate the cytoplasm of one cell from the cytoplasm of the other and leave a space between. In this space the spore coat material is deposited. When the internal cell, the endospore, is fully formed, it is released from the spore mother cell.

peptidoglycan) of the vegetative cell wall. About 10 to 15 percent of the dry weight of spores is made up of *dipicolinic acid,* a substance not present in growing cells:

4
3 5
2 6
HOOC N COOH
1

Mutant *B. subtilis* strains that are unable to synthesize dipicolinic acid cannot make heat-resistant spores.

The transfer of exponentially growing bacteria from a rich medium to a nutritionally poor one triggers the process of endospore development. The morphological changes associated with sporulation require about 8 h to complete. First, newly replicated DNA condenses into an axial filament—that is, a filamentous structure parallel to the long axis of the cell. Next, a septum forms toward one end of the cell and divides it into a larger cell, the *spore mother,* and a smaller cell, the *prespore,* which each contain a chromosome. The prespore then proceeds to mature into the endospore. The distal location of the septum in sporulation contrasts to the medial (central) location during vegetative cell division. Eventually the prespore, surrounded by two membranes, is totally enveloped within the cytoplasm of the spore mother cell. Next, spore coat material is deposited around the prespore, between the two membranes. Both peptidoglycan and dipicolinic acid are synthesized by the larger spore mother cell and transported across its plasma membrane to the space surrounding the prespore to become part of the spore coat. Eventually the mature endospore is released by lysis of the spore mother cell (Figures 22-1 and 22-2). When placed in a suitable environment the spore can germinate, forming a vegetative cell and completing the developmental cycle.

Sporulation involves the synthesis of a number of enzymes and other proteins that are not made by growing cells. Genetic analyses of *B. subtilis* mutants that grow normally but that are unable to sporulate have suggested that perhaps 25 to 50 genes are essential for sporulation. Different sporulation-defective mutants are blocked at different stages of the cycle depicted in Figures 22-1 and 22-2, so the sporulation genes must be expressed sequentially.

The Irreversibility of Commitment to Sporulation

For 2 to 3 h after a growing bacillus is transferred to a nutritionally defective medium, resupply of nutrients will allow resumption of normal growth. After 3 h commit-

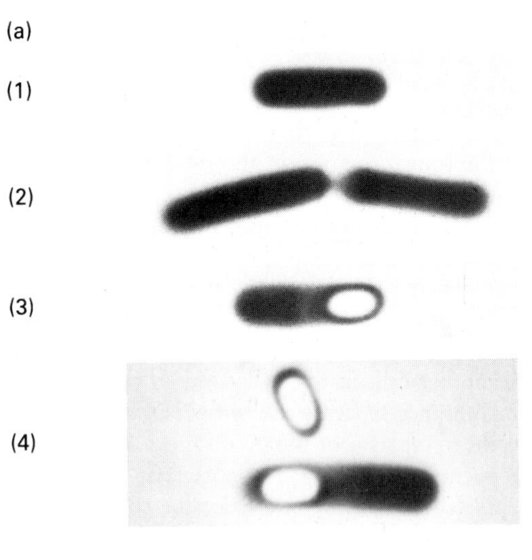

Figure 22-2 Light and electron micrographs showing spore formation in *B. subtilis.* (a) In these light micrographs, a highly refractile (clear) area appears where condensation into a spore occurs. The single bacterium (1) divides once (2) in a starvation medium before forming the spore (3) and releasing it (4). (b) Electron micrographs of sections during spore formation. (1) The axial filament of DNA is formed. (2) A septum appears on the left side. (3) The prespore region is engulfed by the plasma membrane of the spore mother cell. (4) The prespore is complete. (5) The spore coat *(white area)* is forming. (6) The spore coat is complete. [See R. H. Doi and L. Y. Santo, 1974, *Cell* **6**:154.] *Courtesy of R. H. Doi. Copyright 1974 M.I.T.*

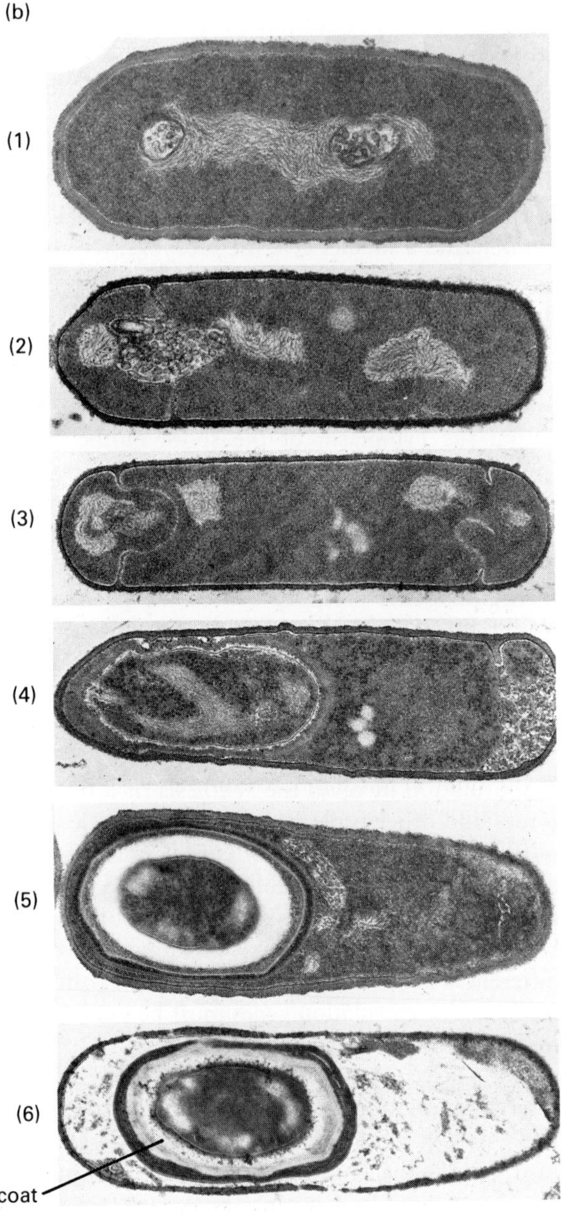

ment occurs, and the re-addition of nutrients cannot reverse the sporulation process. Some sporulation-specific gene product(s) apparently start the cell irreversibly along the sporulation pathway. Although the gene product(s) that serve as such a "switch" remain unknown, a number of molecules that confirm and enforce the change have been identified.

Certain subunits of RNA polymerase are only synthesized early in sporulation. In growing bacterial cells, RNA polymerase consists of four peptide chains—two α's, a β, and a β'—plus an additional σ (sigma) polypeptide (Chapter 8). The σ factor permits the core polymerase ($\alpha_2\beta\beta'$) to bind to DNA and initiate RNA synthesis at specific promoter sites on the DNA template. In sporulating cells, the pre-existing σ subunit is destroyed and replaced by a novel σ-like polypeptide termed σ'. In cell-free reactions, RNA polymerase containing σ' ($\alpha_2\beta\beta'\sigma'$) transcribes cloned genes known to be necessary for sporulation, but it does not transcribe genes characteristic of growing cells. The reverse is true for the RNA polymerase and the σ factor from vegetative cells. Thus the initiation of sporulation depends on—and may even be caused by—the synthesis of one or more new σ-like subunits and the loss of the vegetative σ subunit.

Sporulation in bacteria is an attractive example of a differentiated function because it offers the great advantage of being easily analyzed by microbial genetic procedures. The molecular basis for the controlled activity of the related set of sporulation genes is being successfully explored. The lessons learned from this simple example of cell differentiation may be applicable to more complicated developmental events in eukaryotic cells.

Sporulation in Yeast Involves Meiotic Division and Requires Many New Gene Products

Most yeast cells grow in nature as diploid organisms, but yeast cells, like those of other fungi, form haploid spores in response to appropriate changes in the environment. When haploid spores germinate, they form cells of two mating types (called *a* and *α* in *Saccharomyces cerevisiae*); cells opposite in mating type regularly fuse to re-form diploids (Chapters 6, 11, and 12). Because haploid cell fusion can occur between cells of different genetic backgrounds, yeasts share a large, varied gene pool that enables them to adapt to diverse ecological and environmental situations.

The entire sporulation process takes about 24 h, in contrast to vegetative cell division, which is completed in only 2 h. The meiotic process that forms sperm and egg cells in higher organisms (Chapter 5) has some similarities to sporulation in diploid yeasts. After DNA duplication, each premeiotic cell is $4n$ (that is, it has four copies of each chromosome, two maternal and two paternal). Meiotic division then reduces the $4n$ number to the $1n$ found in sperm, eggs, and yeast spores. Both chromosomal divisions in yeast occur inside a single nuclear membrane. After chromosomal segregation is complete, the original nuclear membrane reforms and surrounds each of the four new haploid spore nuclei. Simultaneously, cell wall material is being laid down around each nucleus; finally, four mature haploid spores are formed. Yeast sporulation has several investigative advantages over sperm and egg production in multicellular organisms: mutants blocked in various steps of meiosis are available in yeast; also, large numbers of diploid yeast cells can be induced to undergo meiosis synchronously.

Meiosis and spore formation in yeast are triggered by starvation conditions. During sporulation there is extensive degradation of pre-existing RNA and protein (about 50 percent of these molecules are broken down), and new mRNA and protein are made. The new multiprotein complexes generated before and during meiosis include the synaptonemal complex, which participates in chromosomal recombination (Chapter 13).

A large group of mutations that block the progress of yeast cells through the vegetative cell division cycle have been detected. The genes identified by these mutants are called CDC (cell division cycle) genes. Most of these genes provide required functions in both mitotic and meiotic cell division, and they are thus necessary in vegetative growth as well as in sporulation. In addition, some 50 to 100 genes encode proteins that are essential for sporulation and meiosis, but that are nonessential for vegetative cell growth and mitosis. Such genes include those encoding enzymes that make spore wall material and those encoding proteases that destroy pre-existing proteins during spore formation. Mutations in some of these genes interrupt the initiation of sporulation; others block discrete steps in the sporulation pathway.

Many sporulation-specific proteins have not yet been isolated, but genetic analyses have suggested that these proteins are formed sequentially during the sporulation process. The continuing molecular analysis of the events in yeast sporulation will provide information that will be important for our understanding of comparable events in higher cells.

Dictyostelium discoideum, a Slime Mold, Constructs an Elaborate Multicellular Structure for Spore Formation and Release

In a culture of yeast cells, every cell can form spores. In the slime mold *Dictyostelium discoideum,* sporulation requires an act of differentiation. Growing cells aggregate into a multinuclear mound of cells; some cells then differentiate and construct a *fruiting body* (the *sorocarp*) which holds the cells that form spores. The cells in this species remain haploid (see Figure 16-32). Liberated spores resist drying and provide a means by which the organism can be widely disseminated. This slime mold is interesting to developmental biologists because it combines the features of synchronously growing single cells with those of multicellular systems showing cell movement, cell interactions, and differentiation.

Dictyostelium cells are induced to enter the sporulation cycle by starvation (Chapter 16). As in yeast, some pre-existing RNA and protein molecules are destroyed, and new mRNA and protein are formed during differentiation. The first step in sporulation is the *chemotaxis*—the chemically induced movement—of single ameboid cells toward aggregation centers. This movement, which is governed by the pulsatile secretion of 3′,5′-cyclic AMP by some cells, leads to the aggregation of as many as 100,000 cells in a mound that is called a *slug*. About 8 h later, the cells synthesize a number of cell surface proteins that allow them to adhere tightly together in the multicellular aggregates.

Soon after the formation of the aggregates, a morphologically distinct "tip" appears on each slug. At this point the slug achieves mobility and the cells in the aggregate become determined: each is now a precursor to one of the two types of cells in the mature fruiting body—spore cells and stalk cells.

The final steps leading to fruiting body formation and sporulation are termed *culmination*. The prestalk cells in the tip elongate and large vacuoles form within them; the cells then move downward through the mass of developing spore cells. In the process, the prespore cells are forced to the surface and then up into the air.

Cellular Determination Can Be Demonstrated in the *Dictyostelium* Slug

The commitment of prespore and prestalk cells can be demonstrated by a simple experiment (Figure 22-3). The slug is cut into halves and the front and rear are induced to undergo culmination by placing them under a light.

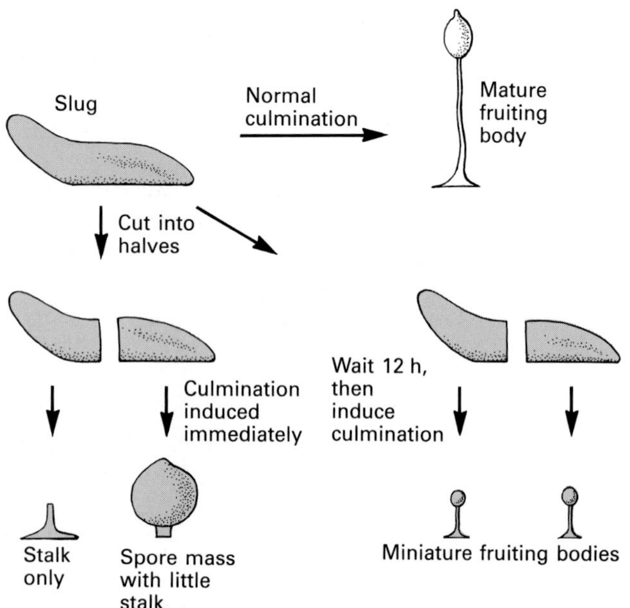

Figure 22-3 The commitment of *Dictyostelium discoideum* cells to differentiation into spore or stalk cells is reversible. If a slug is cut into two approximately equal parts and culmination is induced immediately by overhead illumination, the front (the tip) differentiates into a stalk, whereas the back half forms only spores. This result establishes that at least some cells in the aggregate are committed to the spore or the stalk pathway. However, if the slug is cut into two equal parts but culmination is delayed for 12 h, each half differentiates into a perfectly proportioned miniature fruiting body. Thus some of the prestalk cells in the front of the aggregate can redifferentiate into spore cells, and some of the prespore cells in the rear of the slug can redifferentiate into stalk cells; commitment was not complete at 4 h.

The cells from the tip of the slug form a stalky mass with few spores, whereas cells from the rear differentiate into a spore mass that has little supporting stalk. When prespore and prestalk cells are physically separated, they can be observed to transcribe different sets of mRNAs and to synthesize different proteins. However, a large number of genes that are activated during aggregation are deactivated if aggregation is interrupted. This differential gene expression can be detected by the synthesis of different sets of mRNAs in different cells at different times.

What causes a cell to differentiate along the spore or the stalk pathway? Two different models can help us think about the *Dictyostelium* results. In the first model, the cells are already determined to one or the other pathway before they enter the multicellular aggregate. After they enter the aggregate, the cells sort themselves: prestalk cells to the front, prespore cells to the rear. In the second model, all cells in the aggregate are initially the same, but a gradient of one or more diffusible substances is established from one end of the slug to the other. For example, the concentration of extracellular cAMP is several times higher in the front of the slug than in the rear. Cells in a high concentration of cAMP (or of some other substance) would become prestalk cells, while the remaining cells would become prespore cells.

This question of whether the cell environment determines cell fate or whether the features of the environment are produced by already determined cells arises again and again in developmental biology in the study of different developing systems. Only the identification of the key genes that define the activities of two different cell states and an analysis of what controls the earliest expression of these genes will shed light on such problems. No single problem of this type has been resolved yet.

No gene rearrangements such as deletion, inversion, or transposition have been identified as the basis for gene control during sporulation in *Bacillus,* yeast, or *Dictyostelium.* Rather, differential transcription, differential mRNA stabilization, and differential translation seem the likely molecular mechanisms of differentiation. In all three systems, a sequential induction of gene products appears to lead the organism through the sequential stages of development and differentiation.

Animals Are the Subjects of the Classical Developmental Studies

The ability of simple organisms to perform specialized tasks through differentiation into specialized cells clearly illustrates how cells can cooperate to vary the structure and function of an organism. However, the development of higher organisms, particularly animals, is what has always most interested developmental biologists. In the remainder of this chapter we shall concentrate on animal development.

The development of an embryo from a single cell is much more complex than the simple systems we have just discussed. The development of most animals can be divided into several stages, and we shall treat each of them in turn:

1. Fertilization—the fusion of the sperm and egg
2. Cleavage of the fertilized egg to form the blastula, a group of several thousand largely undifferentiated cells
3. Gastrulation—the organized movement of sheets of cells to generate the three principal cell layers that are the progenitors of all of the differentiated cells in the organism, including the germ cells
4. Differentiation of specialized cells into tissues and organs, and growth of the embryo

5. Growth of the newborn animal, and maintenance and regeneration of body cells after birth

A final category of events might be added: a programmed aging of the animal. Although we shall not deal with this fascinating subject, the average, disease-free life span of a species is a property of that species. For example, mice are destined to live for 2 yr; humans, for 80 to 100 yr; and Galapagos turtles for perhaps 1000 yr. The genetic makeup of each species would appear to contain this information programmed within it: as evidence, consider that mutations can cause the human disease *progeria,* which can accelerate aging so that individuals 10 yr of age appear to be very old. The afflicted person's life expectancy is reduced to 15 yr or less, often due to circulatory ailments. Which proteins are affected by the progeria mutations are not known at present.

In our treatment of embryologic development, we shall not attempt to present detailed comparisons between the developmental programs of different organisms. Readers wishing to explore these topics further are referred to the excellent textbooks in developmental biology listed at the end of the chapter. Our aim here is to illustrate some common principles of each stage of development and to highlight the aspects of development that are under scrutiny by molecular cell biologists.

Fertilization Provides an Example of the Fusion of Specialized Cells

Almost all animal development begins with a single sperm uniting with a single egg—a fusion of two highly specialized cells. The resulting diploid cell, the *zygote,* is the progenitor of all future cells in the animal body.

The Structure of Sperm and Eggs Many animal sperm are very similar in design. They consist of a head, which contains condensed packages of chromosomes, and a flexible tail, which confers mobility. The tail varies considerably in length among different species, but it is universally composed of the 9 + 2 arrays of microtubles characteristic of eukaryotic flagella and cilia (Chapter 18). The energy for movement is generated by a few mitochondria at the base of the tail. The sperm has a very limited amount of cytoplasm; besides generating movement, the only other cytoplasmic function of the sperm appears to be attachment to and penetration of the egg. This task is accomplished by the acrosome, an organelle on the head of the sperm (see Figure 19-32).

In contrast to the rather simple and constant structure of sperm, the structure of eggs varies greatly. The eggs of many invertebrates and some vertebrates such as amphibians are quite large and filled with *yolk,* a collection of stored nutrients especially rich in lipids and proteins. The yolk provides energy and building blocks for embryonic cells. As we shall see, some eggs also contain large stores of macromolecules such as ribosomes, plus supplies of "maternal" mRNA that is either never translated or translated at a much higher rate after fertilization. These stored materials in many eggs are deposited in an organized internal structure. The fixed egg structure plus the regularity of the plane of cleavage during the early divisions results in a fixed pattern of distribution of the stored cytoplasmic materials into the earliest embryonic cells. This pattern of differential inheritance of egg cytoplasm has an important impact on development in many species.

Changes in the Sperm and the Egg at Fertilization
The eggs and sperm of simple marine echinoderms such as the sea urchin and the sand dollar are popular materials for the study of fertilization because of the ease with which each cell type can be prepared and the high efficiency with which fertilization can be carried out in the laboratory. For the sperm to enter the egg, it has to be "activated" in a series of biochemical events that occur when the sperm comes into contact with the egg.

First, the contents of the *acrosome*—a membrane-limited vesicle just under the tip of the sperm—are released by exocytosis. The released proteins facilitate the eventual attachment of the sperm to the *vitelline layer,* which is located beneath the jelly coat and just outside the egg plasma membrane. Second, there is an explosive polymerization of the G actin that lies just below the acrosomal vesicle (see Figure 19-32). This actin filament enables the tip of the sperm to penetrate the jelly coat and approach the egg plasma membrane. When the plasma membrane covering the acrosomal filament reaches the egg plasma membrane, the two plasma membranes fuse, the actin filaments disassemble, and the sperm nucleus enters the egg.

In mammals, no long acrosomal filament is formed during sperm penetration. However, both invertebrate and mammalian sperm release enzymes upon contact between the sperm and the egg. The mammalian enzymes include proteases and hyaluronidase, which are essential for sperm penetration of the *zona pellucida,* the outer coat of the mammalian egg.

How More than One Sperm Is Prevented from Entering the Egg Throughout the animal kingdom fertilization of the egg is limited to a single sperm, which ensures that only a *diploid* zygote is formed. The means by which fertilization is limited appear to be similar in most animals, but they have been most thoroughly studied in marine animals. When sea urchin eggs are mixed with sperm, within 1 or 2 min after fertilization many sperm attach to the egg vitelline layer, which simultaneously lifts away from the underlying plasma membrane. Exclusion of more than one sperm per egg is probably not accomplished simply by the lifting of the vitelline layer, because this event occurs fairly slowly.

Depolarization of the egg plasma membrane occurs within a few seconds after the attachment of a sperm; the depolarization is apparently what excludes the entry of additional sperm. An initial influx of sodium ions changes the membrane potential from −60 to +25 mV. This change is followed by the rapid intracellular release of bound calcium and then by a further influx of sodium ions and an efflux of hydrogen ions (Figure 22-4). These ionic changes may be triggered by the membrane fusion between sperm and egg.

The release of the calcium ions in particular is important in activating the egg for further development. Once the acrosome of a single sperm passes through the vitelline layer, the egg and the sperm plasma membranes fuse and the sperm head enters the egg; two haploid nuclei now reside within the egg cytoplasm. The two nuclei (or *pronuclei,* as they are called after the sperm enters the egg) approach each other and fuse (Figure 22-5), a round of DNA synthesis occurs, and the process of cleavage is ready to begin.

In most mammals, after the sperm and the egg plasma membranes fuse there is a lengthy period before nuclear fusion. In fact, DNA synthesis and the first cellular cleavage begin before true nuclear fusion occurs. The time between sperm entry and the first cleavage varies greatly in different species. In mice, for example, this stage can last 10 to 12 h, whereas in the sea urchin it occurs in about 90 min.

Cleavage Is a Period of Rapid Cell or Nuclear Division

Cleavage, a period in which cells or nuclei divide quickly and repeatedly, results in the accumulation of about 1000 cells in a blastula or a blastoderm. The *blastula* is a round ball of cells formed by the cleavage of zygotes in, for example, amphibians and enchinoderms. The *blastoderm* is the flat disk of cells that results from the early cleavages of avian and mammalian zygotes. The cells of both structures are called *blastomeres.* This sequence is illustrated for the sand dollar in Figure 22-6.

In most invertebrates and in some vertebrates (e.g., frogs), cleavage does not involve growth of the embryo. The blastula, which is formed by successive divisions of material that originally constituted the egg cytoplasm, is similar in size to the fertilized egg (Figure 22-6). There are important differences in the pattern of cleavage from animal to animal. These reflect variations in the structure of the egg and in the mechanism that provides nutrition for the early embryo. In fruit flies, early cleavage is defined not as cell division but as nuclear division. After about 12 nuclear divisions (which yield more than 1000 nuclei), all nuclei still inhabit the same cytoplasm, and they form a structure called a *syncytial blastoderm.* Cellularization then occurs by the formation of membranes that divide the egg into cellular compartments, each with one nucleus (Figure 22-7).

During the initial nuclear divisions of fertilized *Drosophila melanogaster* eggs there is an incredibly rapid synthesis of DNA, perhaps the fastest known in eukaryotes. Each nuclear cleavage division occurs in only 15 min, and each complete genome duplication requires even less time, perhaps about 5 min. In contrast, DNA replication in cultured *D. melanogaster* cells takes 5 to 10 h. The speed of the bidirectional DNA replication in *D. melanogaster* embryos has been estimated to be only about 2.5 kb per minute per replication fork, a rate similar to that of fork movement in cultured cells. Thus each origin with its pair of forks would be responsible for replicating 25 kb in 5 min; 25 kb is about the average size of a replicon in eukaryotic cells. It seems likely that all possi-

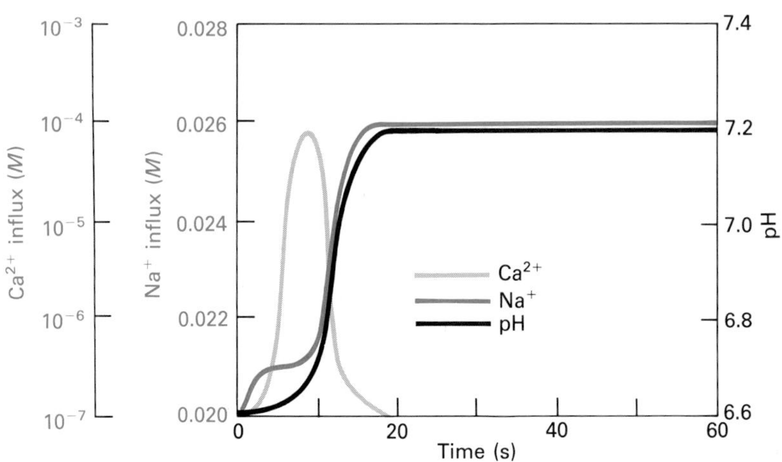

Figure 22-4 Upon fertilization, changes in the concentrations of calcium, sodium, and hydrogen ions occur within sea urchin eggs. A few seconds after fertilization there is a transient depolarization of the cell membrane associated with a small influx of sodium ions. About 10 s after fertilization calcium ions are released from intracellular stores; the result is an increase in the concentration of free calcium ions in the cytosol. Between 10 and 20 s after fertilization the sodium content of the egg begins to increase again; this change is accompanied by an efflux of protons that raises intracellular pH, probably because of the activation of a sodium/proton antiport. [See D. Epel, 1977, *Sci. Am.* 237(5):128.]

(a)

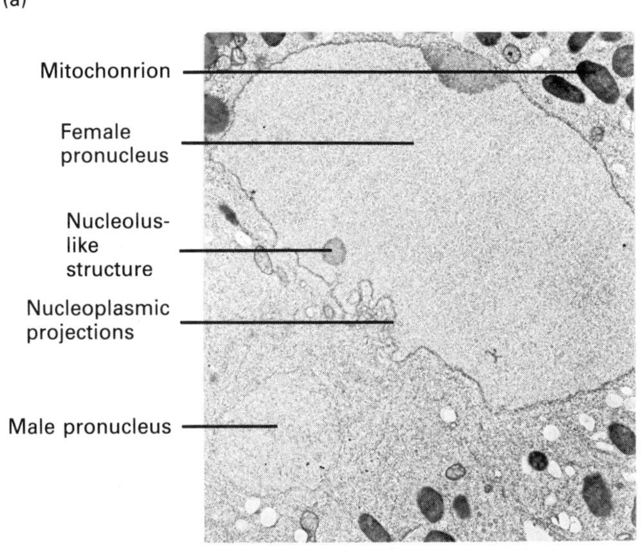

(b)

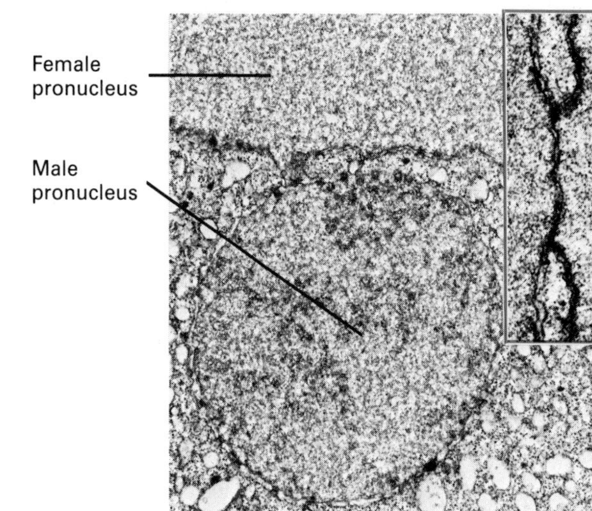

(c)

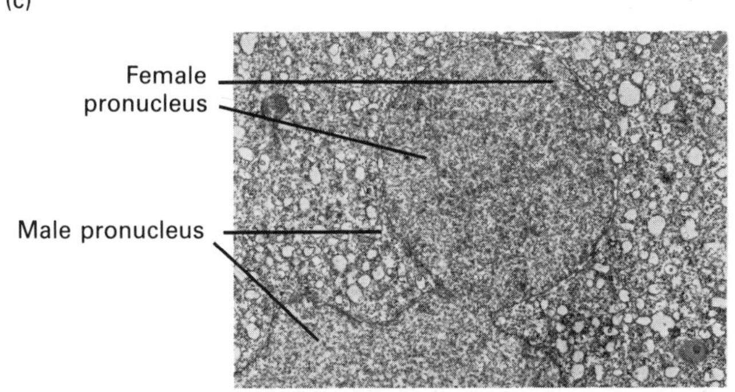

(d)

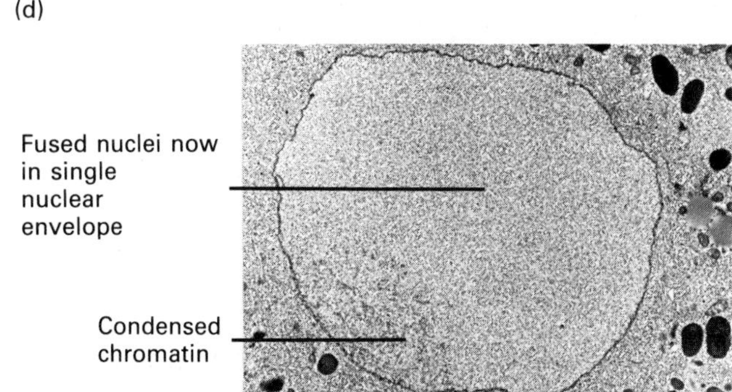

Figure 22-5 Pronuclear fusion in the sea urchin zygote. (a) The female pronucleus and the male pronucleus lie unfused within the same cytoplasm. Nucleoplasmic projections appear on the side of the female pronucleus where fusion will occur with the male pronucleus. (b) The two pronuclei are very close; in fact, the layers of the double-layered nuclear envelopes have joined at the closest point *(inset)*. (c) The two nuclei are connected. (d) A single nucleus now resides within a single nuclear envelope. In the nuclear region at the lower left, lightly condensed chromatin that was formerly in the male pronucleus is visible. [See F. J. Longo and E. Anderson, 1968, *J. Cell Biol.* 39:339.] *Courtesy of F. J. Longo. Reproduced from the* Journal of Cell Biology, *1968, by copyright permission of the Rockefeller University Press.*

ble replicons (about 10^4 totally) must be activated almost simultaneously in the embryo in order for the DNA to be so quickly duplicated.

In most animals the blastula is formed by successive cleavages of the fertilized egg, but the cleavage periods vary considerably in length. In echinoderms such as sea urchins and sand dollars, division is rapid, and a blastula of thousands of cells is achieved in a few hours (see Figure 22-6). In frogs, about 1 h is required for each of the first three or four divisions; then cleavage speeds up so that a blastula of many thousands of cells is formed within the first day after fertilization. In mammals (e.g., mice), the first three or four cleavages take 3 days or longer.

The first dozen or so cleavages in frog eggs are synchronous. These cleavages illustrate the importance of the egg structure in cleavage: the synchronous early divisions, about 12 in all, can occur *even in the absence of nuclei.* A small capillary tube can be inserted into a frog egg and the nucleus can be aspirated (removed by suction). The resulting physical damage to the egg membrane induces cleavage divisions, which continue even without the cell nucleus. Thus the machinery for cleavage of these large eggs is entirely preformed; somehow it is stored in the cytoplasm. In any case, nuclear division with the mitotic apparatus for separating chromosomes is not required for cleavage of the cytoplasm.

The embryos of echinoderms, insects, and amphibians complete the early stages of development within the preexisting egg cytoplasm and are nourished by the stored nutrients of the egg.

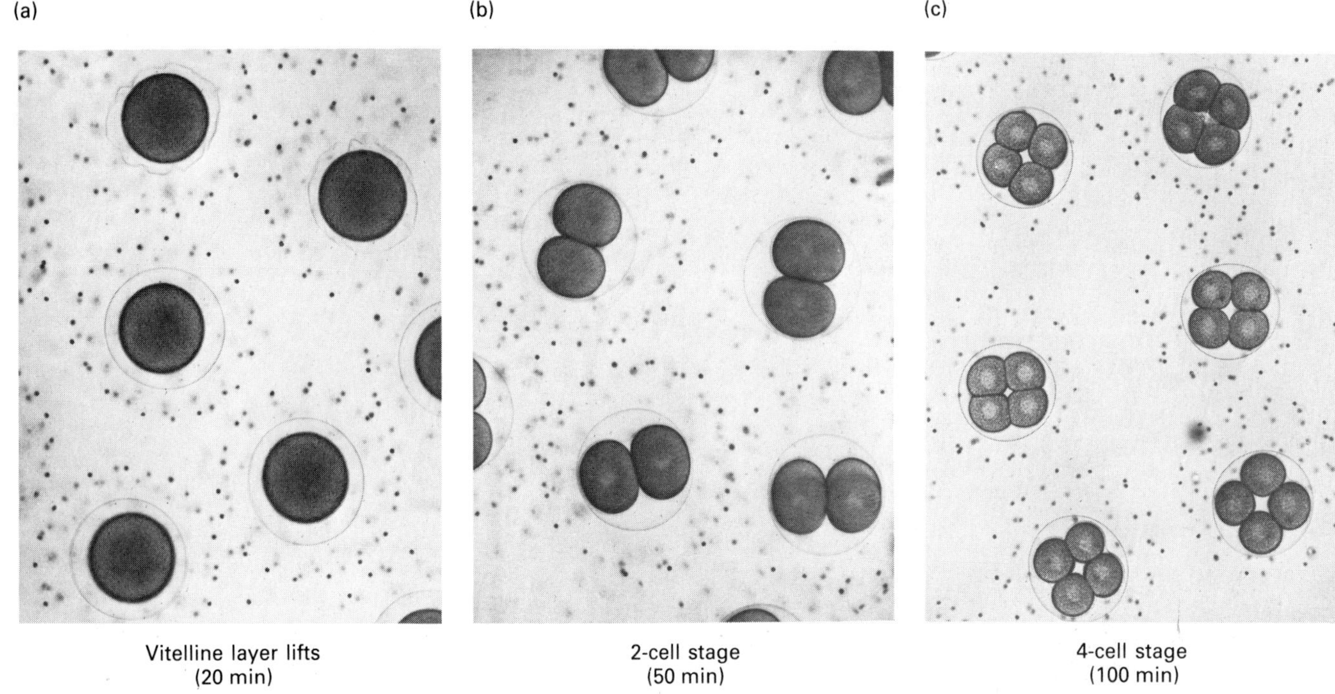

Vitelline layer lifts (20 min) | 2-cell stage (50 min) | 4-cell stage (100 min)

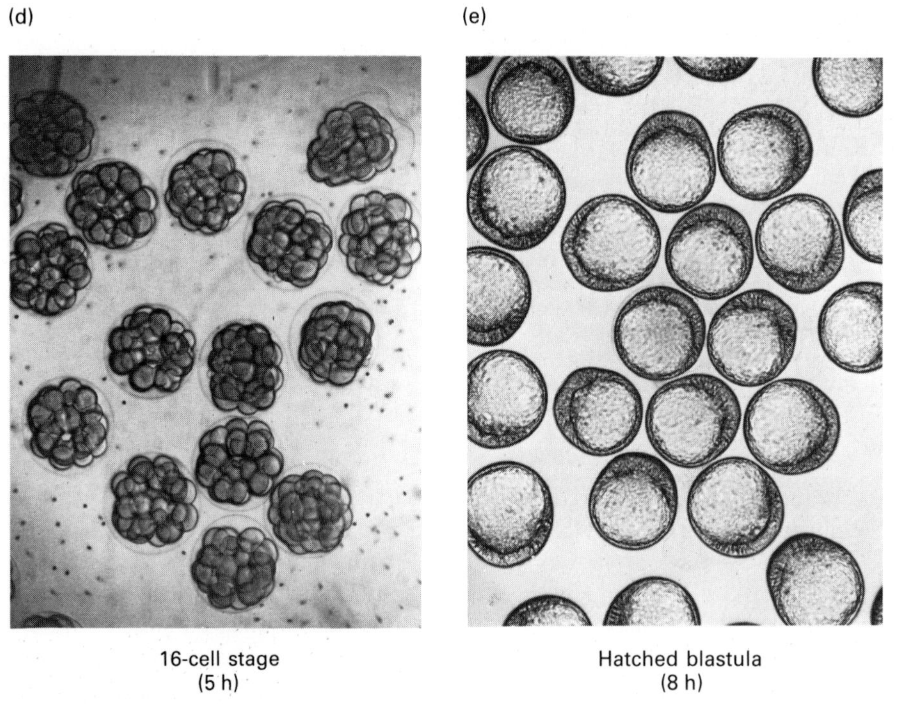

16-cell stage (5 h) | Hatched blastula (8 h)

Figure 22-6 Fertilization and cleavage in eggs of the sand dollar *(Dendraster excentricus)*. (a) The vitelline layer has not completely lifted in all eggs (see the two eggs at the top of the photo). The black dots surrounding the eggs are sperm heads. (b),(c) At the two- and four-cell stages, the nuclei are large central structures; the smaller, lighter areas in the centers of the nuclei are the nucleoli. (d) Development has proceeded within the vitelline layer for the first 5 h. (e) Together the several thousand cells of the blastula are the same size as the egg. *Courtesy of V. D. Vacquier.*

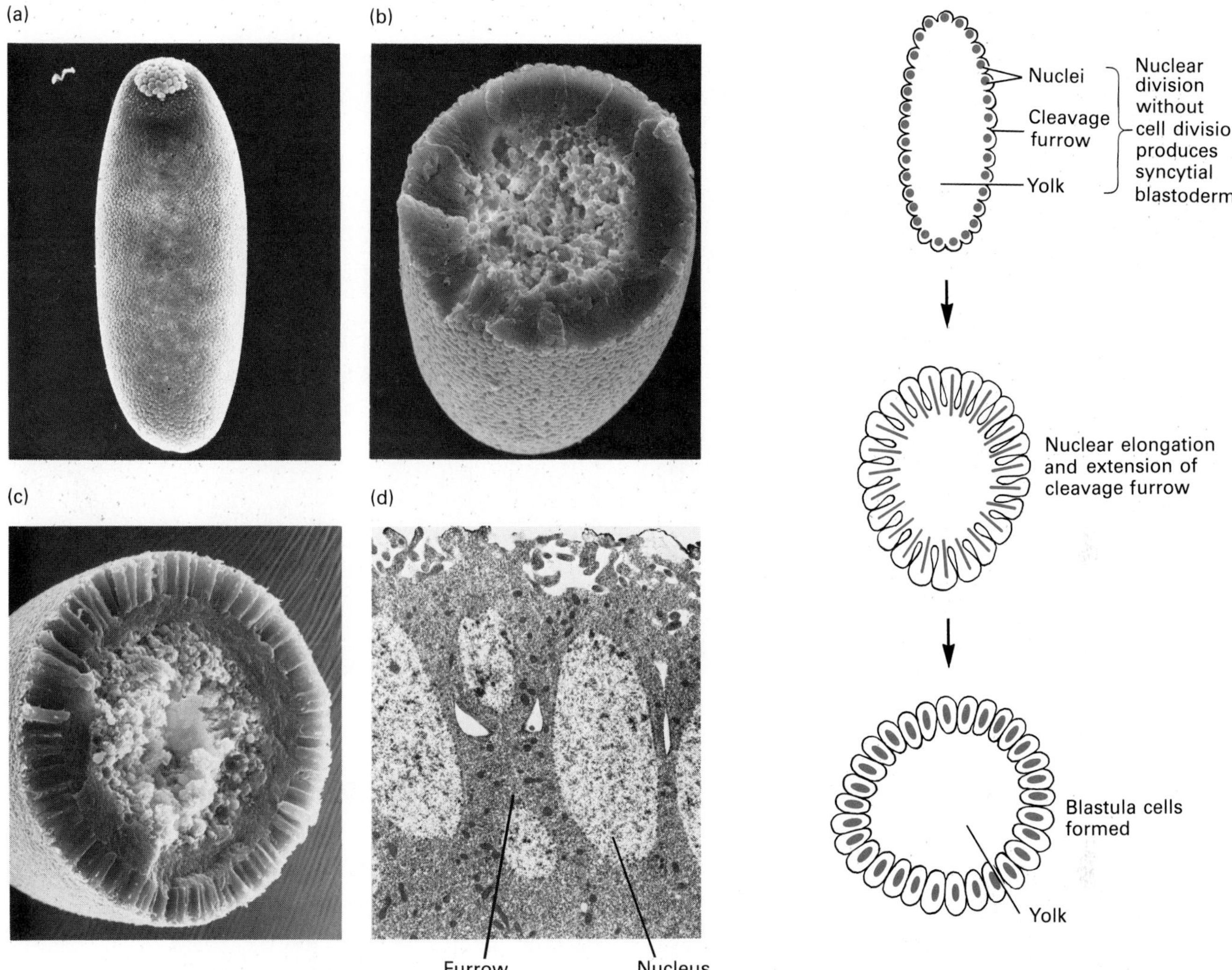

Figure 22-7 Cleavage and formation of blastula cells in a *Drosophila melanogaster* embryo. Nuclear division is not accompanied by cell division until about 1000 nuclei have been formed. (a) Prior to cell division, the surface of the embryo shows bulges that lie over individual nuclei. (b) A scanning electron micrograph of a fractured embryo at this stage shows that it is a syncytial blastoderm containing no cells. (c) When about 1000 nuclei exist, cell membranes form to divide the syncytial blastoderm into individual cells. A fractured embryo at this stage shows the partitions between cells. (d) A transmission electron micrograph of a section through an embryo shows elongated nuclei and furrows that have formed between nuclei. These furrows progress toward the center of the blastula until cell cores are divided from one another. [See F. R. Turner and A. P. Mahowald, 1976, *Devel. Biol.* 50:95.] *Photographs courtesy of A. P. Mahowald.*

Avian and Mammalian Blastulas Are Not Balls of Cells

Cleavage and early embryonic development of the fertilized egg in higher vertebrates (e.g., birds and mammals) introduce another problem. During embryonic growth a series of membranes forms around the growing embryo to provide a constant environment and to deliver nutrients. In birds the embryo develops within the *amnion,* a layer of epithelial cells plus a fibrous extracellular membrane, and nutrients are extracted from the large yolk by the amniotic cells. In mammals the nutrients come directly from the capillary circulation of the mother, mainly by diffusion through the complicated structure called the *placenta,* which actually invades the walls of the uterus. The developing fetus is buffered from physical damage by flotation in fluid encased within the amnion. In birds and mammals the earliest cleavages produce not only cells

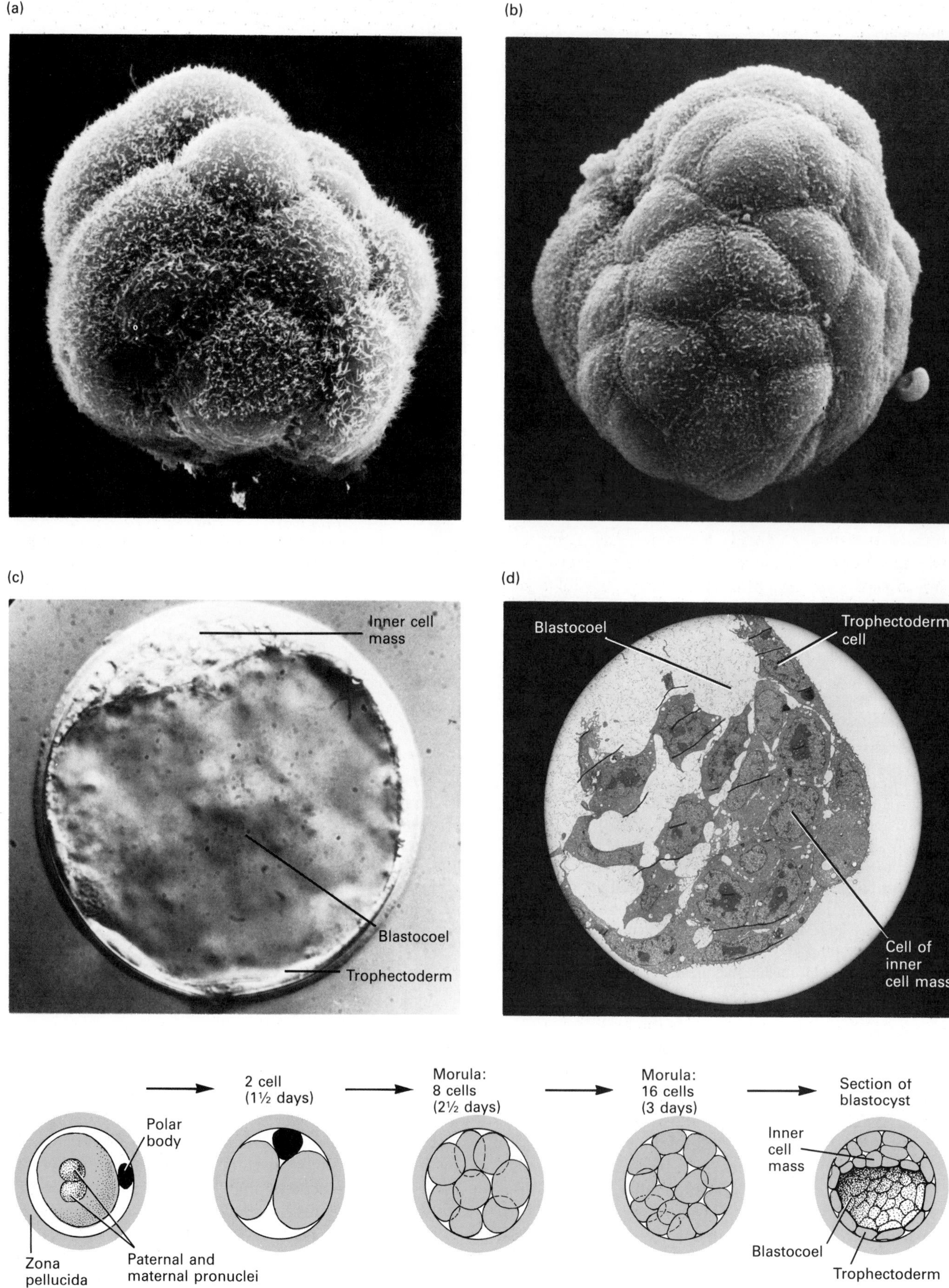
(a)
(b)
(c)
Inner cell mass
Blastocoel
Trophectoderm
(d)
Blastocoel
Trophectoderm cell
Cell of inner cell mass
2 cell (1½ days)
Morula: 8 cells (2½ days)
Morula: 16 cells (3 days)
Section of blastocyst
Polar body
Zona pellucida
Paternal and maternal pronuclei
Inner cell mass
Blastocoel
Trophectoderm

Figure 22-8 ***(opposite)*** Early development in the mouse embryo. The schematic diagram gives the times of various events after the fusion of the pronuclei. The zona pellucida, a noncellular secreted layer surrounding the egg and the early embryo, was removed to make the scanning electron micrographs in (a) and (b). These show the tight adherence of (a) morula and (b) early blastocyst cells to one another; also, many villi are visible on the free cell surfaces. (c) A blastocyst viewed by Nomarski optics is ready to implant in the uterine wall. (d) This electron micrograph of a section through a blastocyst shows the tight adherence of cells within the inner cell mass. *Courtesy of P. Calarco.*

that will participate directly in embryo formation but also cells that are the forerunners of the extraembryonic structures. The cells that will form the extraembryonic tissues surround the pre-embryonic cells, first in a clump and then in the flattened disk termed the blastoderm.

In mammals, approximately the first six to eight cleavages occur within the zona pellucida, the thick lining that is derived from the covering of the egg. The 8- to 50-cell structure is called a *morula* (from the Latin *morum,* meaning "mulberry"; Figure 22-8). As cell division proceeds toward formation of the blastoderm, an inner cavity, the *blastocoel,* is developed. At this stage the mammalian embryo becomes a *blastocyst,* a structure that has differentiated into an *inner cell mass,* which will form the embryo proper, and a surrounding epithelial lining called the *trophectoderm,* which will form the extraembryonic tissues (Figure 22-8). The blastocyst is ready to emerge from the zona pellucida.

Development up to this point can require more than 100 h, and additional divisions may occur before the egg moves out of the tube that leads from the ovary to the uterus (in humans, the fallopian tube) and attaches to the lining of the uterus, the *endometrium.* After attachment, *implantation* occurs and the blastocyst eventually becomes completely engulfed by the endometrial cells. The *trophoblasts,* the cells making up the trophectoderm, grow and differentiate into two cell types, *cytotrophoblasts,* which encircle the inner cell mass, and the *syncytiotrophoblast,* a large multinucleated cell that invades the uterine lining and attaches the blastocyst to the uterine wall (Figure 22-9). The syncytiotrophoblast is particularly interesting because of its invasive properties. Even though the cells often enter the mother's bloodstream they rarely cause any pathology. Their invasive capacity is somehow controlled and limited during normal development.

The abundant capillaries in the uterine wall then furnish nutrients to the embryo. The cytotrophoblasts and the syncytiotrophoblast eventually form the placenta, the organ of nutrient transfer from mother to fetus. Only the inner cell mass, which contributes cells to the embryo proper, is comparable to the blastula of frogs or sea urchins.

Gastrulation Involves Massive Cell Movements and Generates Three Cell Layers

Following blastula or blastoderm formation the embryo undergoes *gastrulation,* a dramatic reshaping with very little additional cell growth. This reshaping varies in its geometry from animal to animal because the blastulas and blastoderms themselves vary in shape. However, in all embryos gastrulation serves the same purpose: to create three layers of cells termed the *ectoderm,* the *endoderm,* and the *mesoderm.* These three layers represent three tissue types called the *primordial germ layers.* From the germ layers and the interactions between them the body plan of the early embryo emerges. Each organ is eventually formed from one or more of the germ layers (Table 22-1).

In blastulas that are hollow balls of cells—for example, blastulas of echinoderms and amphibians—gastrulation begins by the progressive invagination of the sphere to create a two-layered saclike structure with an opening at one end, the *vegetal pole* (Figure 22-10). Gastrulation leads to the formation of an outer epithelial cell layer—the ectoderm—and an inner layer—the endoderm—both of which surround a large central cavity. The mesoderm is derived from cells at the *blastopore,* the mouth of the invaginated cavity. In echinoderms, cells detach from this region and migrate between the endoderm and the ectoderm; in amphibians, most of the rim of the blastopore is drawn between the endoderm and the ectoderm to become mesoderm. Furthermore, in the invaginated cavity of the amphibian embryo, clusters of cells pinch off from the endoderm and lie between the first two layers. These collections of cells, termed *somites,* are also mesodermal cells (Table 22-1).

The reshaping that occurs during the gastrulation of bird and mammalian embryos is not so easy to follow visually. First, there is no hollow ball of cells with migration of cells into a central cavity. Second, as we have mentioned, many of the early embryonic cells in birds and mammals are destined to give rise to the extraembryonic tissues rather than to the embryo proper.

In birds, a group of associated cells form a *primitive streak* in the center of the blastodermal disk. The primitive streak is made up of two parallel ridges with a groove between them (Figure 22-11). The events comparable to gastrulation in frogs and echinoderms occur in and around the primitive streak. Cells stream over one another, passing down through the primitive streak to form a second layer of cells underneath the original primitive streak. These two layers of cells are the forerunners of the endoderm and the ectoderm. Cells continue to migrate through the region of the primitive streak to form a third

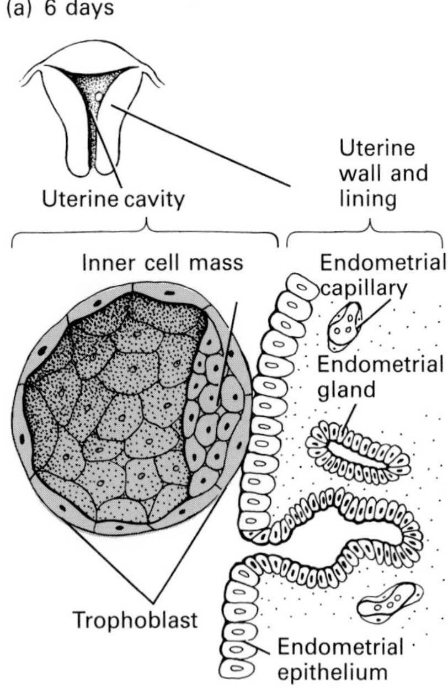

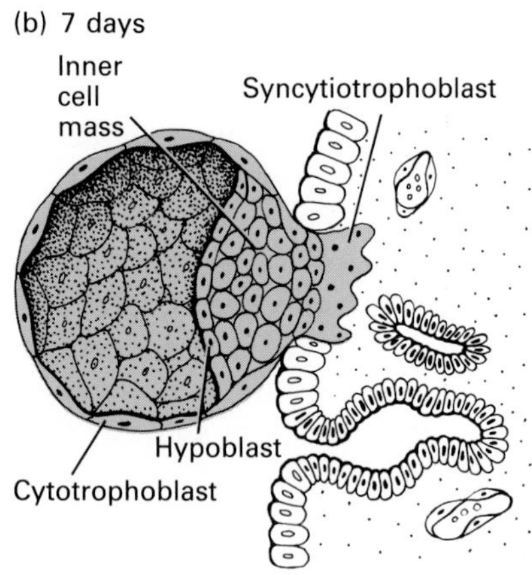

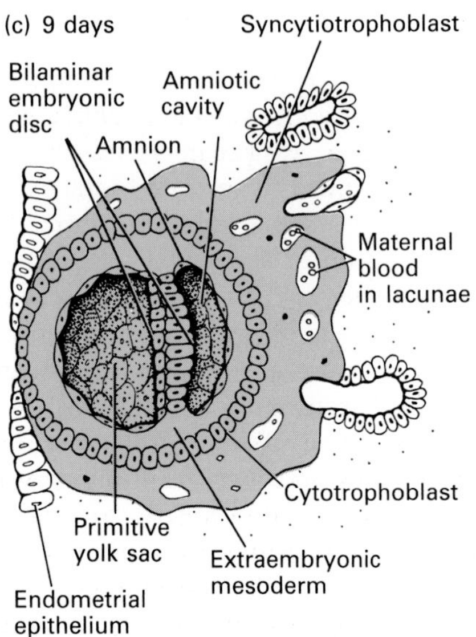

Figure 22-9 The implantation of a human embryo. (a) At 6 days, trophoblast cells attach to endometrial cells. The inner cell mass will become the embryo proper. (b) At 7 days the syncytiotrophoblast forms and penetrates the endometrial epithelium, which lines the uterine cavity. The inner cell mass begins to differentiate into hypoblasts, cells that will differentiate further into tissues of the fetus. (c) At 9 days, the embryo is implanted. The amniotic cavity is present; the cytotrophoblasts and the syncytiotrophoblast have begun to form the placenta. Soon the endometrial epithelium will grow back, sealing the embryo within the endometrium and completing implantation.

cell layer that will be the mesoderm. As the sheets of cells expand and rearrange, the long axis of the forming embryo becomes evident.

The Three Germ Layers Contribute to Specific Organ Rudiments

As the three cell layers become established and the movements of gastrulation are accomplished, the appearance of certain important landmarks signals the beginning of all later embryonic and adult structures. For example, the surface of the blastula is the part of the ectoderm that will form the epidermis (the outer layer of skin). Very early, a portion of the ectoderm opposite to the site of invagination is set aside to form the *neural ectoderm,* which is the forerunner of the nervous system. Neural ectodermal cells form a *neural plate,* a flattened structure at the anterior end of the developing embryo; the neural plate contributes cells to the brain (Figure 22-11b). A large strip of cells down the length of the embryo rolls up into the *neural tube,* the forerunner of the spinal cord (Figure 22-11c). In addition, cells of the *neural crest* migrate throughout the body to form *ganglia,* peripheral collections of neurons (see Table 22-1).

The primary endodermal cells become a tube in the center of the embryo. This tube is the forerunner of the gut as well as of associated structures such as the lungs and liver, which originate as outpouchings of the primitive gut.

Mesodermal cells participate in the widest array of differentiated functions. The central cavities of the somites—which, recall, are clumps of mesodermal cells—expand to become the chest and abdominal cavities. The somites also provide cells that eventually form the muscles and the skeletal system. Other mobile mesodermal cells and their descendants contribute to the formation of the vascular system and connective tissue. Some mesodermal cells, termed *mesenchymal* cells, participate in critical interactions with endodermal and ectodermal cells to induce characteristic cell types during organ formation.

Table 22-1
Differentiation of the three primordial germ layers in vertebrates

Embryonic components	Comments	Final differentiated cells and tissues
ECTODERM		
Outer covering of embryo proper		Skin—epidermis and associated skin organs (hair follicles, sweat and oil glands, lens of eye)
Neural plate and neural tube		Central nervous system (brain and spinal cord)
Neural crest cells	These cells migrate throughout entire embryo	Sensory and autonomic ganglia, adrenal medulla, pigment cells (including retina)
MESODERM		
	Subepidermal layer over entire embryo	Skin—dermis
Notochord	This primitive axial column remains in chordates; it is resorbed in vertebrates, and cells that will form vertebra aggregate around it	
Somites	These "pinched off" independent cell masses develop central cavities that are enlarged to form abdominal and thoracic cavities; cells migrate out to form muscles	Musculoskeletal system and connective tissue (tendons, elastic tissue, and bone), heart and blood vessels, blood cells, urogenital tract, fat cells
Mesenchyme	These migrating cells participate in forming mesodermal tissues and interact with endoderm and ectoderm in organ formation	
ENDODERM		
Primitive gut	Outpouchings from the gut interact with the mesenchyme to form associated structures	Gastrointestinal system, gut and associated organs: salivary glands, pancreas, liver, lungs

In all vertebrate embryos, early mesodermal cells from the dorsal lip of the blastopore give rise to an important elongated structure called the *notochord*. This structure, which marks the pathway along which the vertebral column and associated muscles are formed as the embryo develops, is resorbed during development. In primitive chordates—for example, sea squirts and lancelets—the notochord remains and no vertebral column is formed (see Table 22-1).

The Cells That Will Be Gametes Are Determined Early

The origin of germ cells (gametes) is of special interest because the differentiation of these cells is responsible for continuing the life cycle. The initial determination of cells as primordial germ cells occurs very early in all animals. In some animals the primordial germ cell is determined by the *germ plasm,* a special region within the egg cytoplasm.

The importance of the germ plasm has been illustrated by microinjection experiments in *D. melanogaster*. Pregametic cells normally arise from the posterior end of the syncytial blastoderm (Figure 22-12a). Irradiation of this end of an egg yields a sterile fly (Figure 22-12b). If cytoplasmic contents of the posterior end of the developing egg are aspirated (removed by suction) with a micropipette and injected into the *anterior* end of another developing egg, germ cells are formed at this abnormal site (Figure 22-12c) as well as in the posterior end of the

(a)

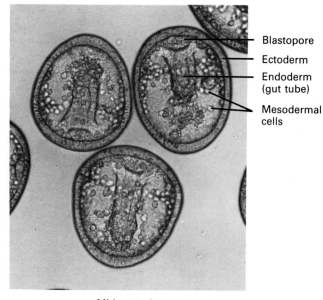

Mesenchymal cells

Vegetal pole

Mesenchyme blastula (11 h)

(b)

Blastopore

Ectoderm

Endoderm (gut tube)

Mesodermal cells

Mid-gastrula (15 h)

(c)

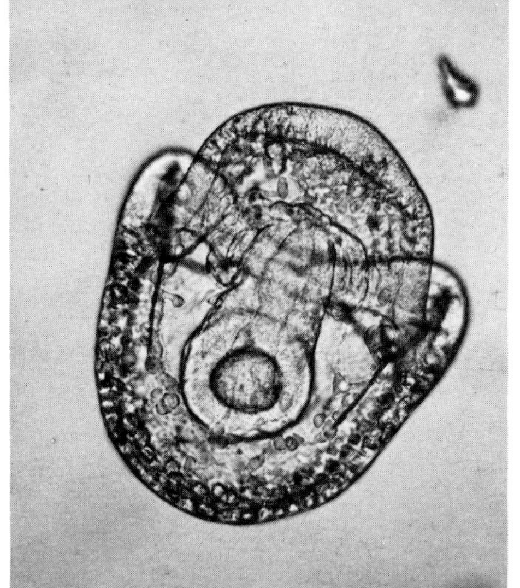

Early pluteus (20 h)

(d)

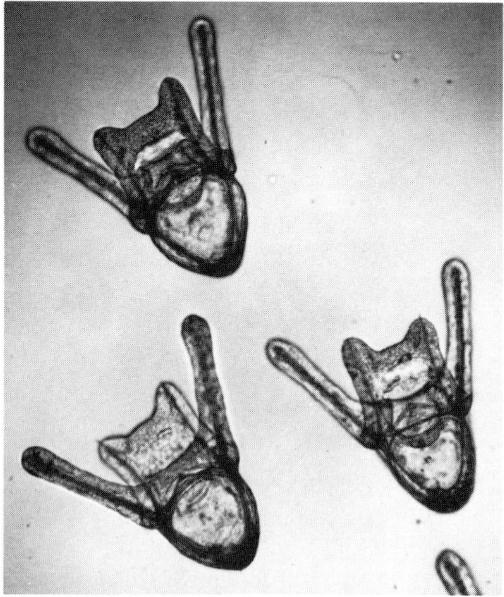

Swimming pluteus (39 h)

Figure 22-10 Gastrulation in the sand dollar *(D. excentricus)*. (a) This light micrograph of a blastula 11 h after fertilization shows the beginning of invagination at the vegetal pole. The plane of focus, which is possible because the embryo is transparent, is within the blastocoel. Primary mesenchymal cells (mesoderm precursors) have entered the blastocoel cavity. (b) Mid-gastrula invagination has progressed so that all three layers have formed. The gut tube lined with endoderm is visible in the center of the animal, and the peripheral ectoderm is also well defined. The mesodermal cells lie between. (c) Gastrulation has concluded, and the early pluteus or larval form has been reached. (d) Swimming pluteus organisms or larva; nutrition up to this point is entirely from nutrients stored in the egg. *Courtesy of V. D. Vacquier.*

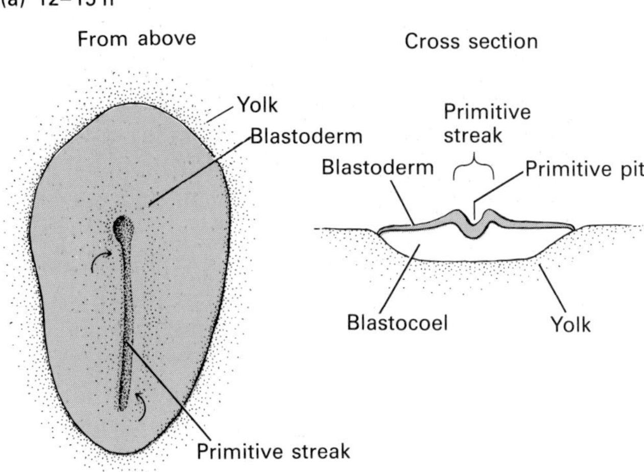

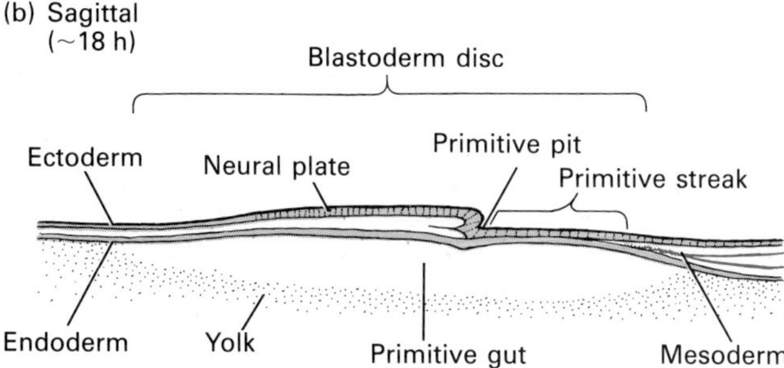

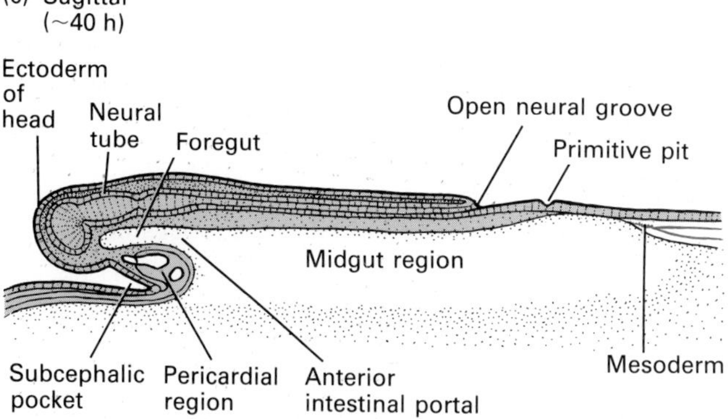

Figure 22-11 Chick embryo gastrulation and the establishment of major tissue precursors. (a) *Left:* A drawing of the primitive streak at approximately 12 to 15 h after fertilization. Primitive streak formation is complete before the egg is enclosed in a shell. The arrows show the pathway of cell migration through the primitive streak. *Right:* A cross section of the primitive streak. Cells can be observed to migrate down into the primitive gut. (b),(c) Sagittal (i.e., longitudinal) sections through a chick embryo at 18 h (when it is still in the primitive streak stage) and at 40 h (when the body plan is emerging). Note that the blastoderm has become bilayered. *Parts (b) and (c) after B. M. Patten, 1971,* Early Embryology of the Chick, *5 ed., McGraw-Hill.*

blastoderm. Because no nuclei are transferred in this experiment, some stored material in the germ plasm of the egg is the presumed inducer of germ cell formation. Although the active factors for the induction of germ cells have not been identified, the germ plasm of *D. melanogaster* contains prominent granules of RNA and protein. Similar experiments in frogs have located the germ plasm of the zygote in the pole opposite to that in which the blastopore will form.

In most species germ cells can be identified in early developmental stages by special histological stains. The pregametic cells in different animals undergo different degrees of differentiation before adulthood. For example, in the frog *Rana pipiens,* oocytes begin to mature soon after the metamorphosis of the tadpole, but the final stages of egg maturation do not occur until just before the female frog lays a batch of eggs—an event that takes place once a year. The complete cycle of mature oocyte formation, which involves the storage of yolk, extensive ribosome synthesis, and other processes, requires 3 yr. In mammals, all of the meiotic divisions and the differentiation into oocytes occur before or just after birth, but ovulation does not take place until much later. In all animals the final production and delivery of the fully competent eggs or sperm require complex hormonal stimulation that occurs in adults after the reproductive organs are fully mature (Chapter 16).

Molecular Events during Early Embryogenesis Are Now Being Explored

The various cell movements that occur during gastrulation to help establish the three cell layers and the subsequent folding and rearrangement of cell sheets are very specific, but these events are poorly understood in molecular terms. The classical studies, which were carried out with the use of vital stains and light microscopes, have been updated by electron microscopy. The latter work has demonstrated interactions at the cell surfaces and internal rearrangements of elements of the cytoskeleton. The rearrangements of cells during gastrulation require differential cell adhesiveness, differential elongation of specific cells, and the movement of whole sheets of cells, presumably mediated by contractile proteins.

The molecular bases of these events are just beginning to be explored. Proteins that promote the reaggregation of disaggregated embryonic cells have been isolated. These proteins are called *cell adhesion molecules,* or CAMs; several specific molecules of this type can be distinguished by antibodies. For example, antibodies reactive with a CAM from chicken liver cells bind to distinct groups of cells in early chick embryos (Figure 22-13). The CAMs appear to be a limited family of proteins that are modified by different polysaccharide groups. These molecules almost certainly play an important role in directing

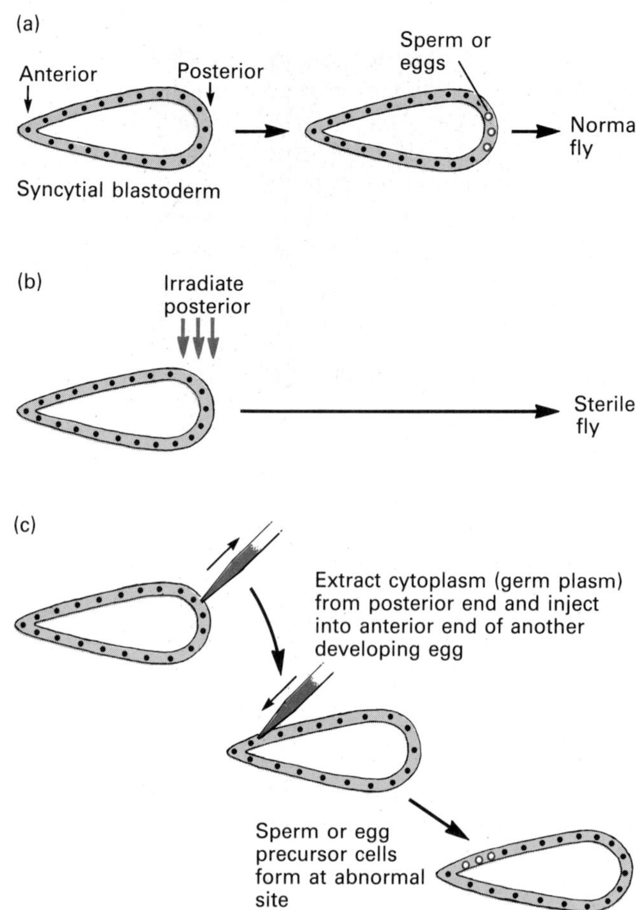

Figure 22-12 The role of polar cytoplasm in *D. melanogaster* gamete formation. The posterior end of the egg is known to develop into sperm or egg cells. This was shown by three different kinds of experiments. (a) Staining properties can be used to identify the pregametic cells. (b) Irradiation of the posterior end of the syncytial blastoderm causes sterility in the fly that develops. (c) The contents of the posterior end of a cellular preblastoderm can be removed and injected into an irradiated egg. Pregametic cells can be found in the anterior end, and the fly that develops is fertile. [See M. Okada, A. Kleinman, and H. A. Schneiderman, 1974, *Devel. Biol.* 37:43.]

cell adhesiveness, which is so important in shaping the developing form of the early embryo. However, why certain groups of cells stick specifically to other groups of cells is not known.

Because of the obvious importance of the nucleus in cell function, many studies of the synthesis of DNA, protein, and various RNAs during cleavage and gastrulation have been carried out. As we have mentioned, DNA synthesis increases along with cell number (or nuclear number) in all embryos. Protein synthesis is low in some unfertilized eggs. It is activated immediately to high levels by fertilization, and "maternal" mRNA stored in the egg is translated. (Translational control in fertilized marine eggs was discussed and illustrated in Chapter 12.) Precursors to rRNA and tRNA are not synthesized in sea urchin and frog embryos until near the end of gastrulation, but ribosomal synthesis may begin in mammalian embryos in the two- or four-cell stage.

The small RNAs produced by RNA polymerase III (Chapter 9) begin to be synthesized before pre-rRNA. U1 RNA, the small nuclear RNA that participates in RNA splicing, is among the earliest small RNAs formed during the cleavage of frog eggs. The hnRNA, which includes potential mRNA precursors, is probably synthesized from the beginning of fertilization in most embryos. It is likely that unprocessed hnRNA is stored in the cytoplasm of certain eggs—for example, in sea urchin eggs. The early synthesis of the U1 RNA might allow the stored hnRNA to become processed.

Some new mRNAs pass into the cytoplasm and enter polyribosomes during cleavage stages. This influx from the nucleus probably increases in frogs and sea urchins during gastrulation. In mammals new mRNA formation occurs soon after fertilization.

The identification of stage-specific embryonic mRNAs and proteins is obviously of great importance to researchers trying to establish how the events of early embryogenesis are mediated. Only in animals for which genetic knowledge is detailed will it be practical to first identify genes that affect early development and then isolate the molecules encoded by these genes. However, in any animal, mRNAs and proteins that are synthesized only at early developmental stages can be detected by applying the new techniques—particularly the use of monoclonal antibodies (Figure 22-13) and labeled probes made from cloned cDNA.

As an example of the latter, specific probes made from cloned early mRNAs have been hybridized in situ to cells of whole sea urchin embryos. One embryonic mRNA (known to be in the troponin family) is found only in ectodermal cells in a part of the embryo (Figure 22-14). Such methods will lead to the identification of genes encoding specific mRNAs; ultimately both the control mechanisms of these genes and the functions of their products will be described. Experiments of this type will become increasingly important in developmental biology.

Cell Determination: Progression from Totipotency to Limited Developmental Potential

The study of early embryogenesis has traditionally focused not only on the observed structural events of cleav-

(a)

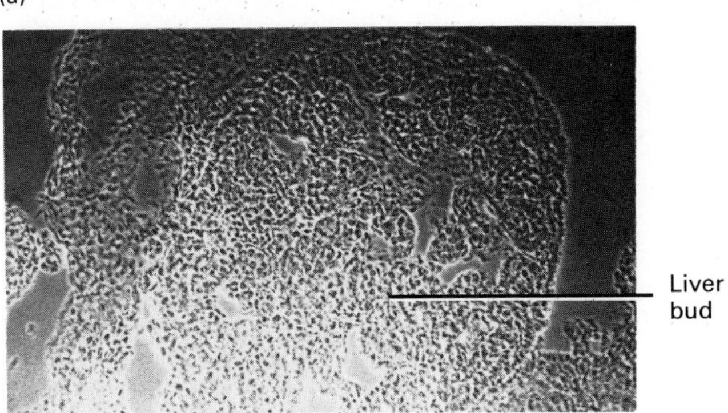

(b)

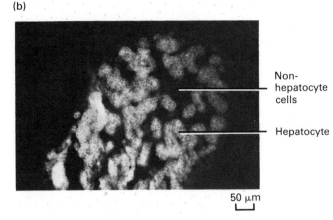

Non-
hepatocyte
cells

Hepatocyte

50 μm

Figure 22-13 The identification of embryonic cells containing specific antigens on their surfaces. Surface proteins called cell adhesion molecules (CAMs) have been identified and purified. One such protein (called L-CAM because it was originally purified from liver) was used as an antigen in mice to produce antibodies. (a) A light micrograph of a section through a liver bud in a 6-day chick embryo. The section has been exposed to fluorescent L-CAM antibodies. (b) The embryonic hepatocytes fluoresce, which indicates that they contain L-CAM at their surfaces; the surrounding cells (mesodermal and undifferentiated endodermal cells) do not. [See G. M. Edelman et al., 1983, *Proc. Nat'l Acad. Sci. USA* 80:4384.] *Courtesy of G. M. Edelman and W. J. Gallin.*

age and gastrulation but also on changes in the functional potential of cells—on *cell determination.* In all animals, different cell fates are established by the time a few hundred to a few thousand cells have formed. Two central questions have always interested embryologists: When do the descendants of the totipotent fertilized egg acquire restricted developmental potential? Is this determination an irreversible event?

The Egg Cytoplasm Is Important in Early Determination for Many Animals

Loss of totipotency occurs at different stages of development in different organisms. Substances stored in the egg cytoplasm may play a role in the early determination of the cells of some organisms.

For example, one half of the sea urchin egg is more refractile than the other half when the egg is viewed in the light microscope. The less refractile half is called the animal pole; the more refractile half is the vegetal pole. The first two cleavages are longitudinal; i.e., they divide the egg so that both daughters receive equal contributions from the two poles of the egg. Four blastomeres (cells) are thereby produced, each of which can give rise to a normal animal. The third cleavage, however, is equatorial, and produces two sets of four blastomeres. One set contains material present at the animal pole of the egg, and the other contains material from the vegetal pole.

These blastomeres are now restricted in developmental potential. The vegetal half can continue to grow when it is separated from the animal half, but it forms an abnormal organism that possesses mainly endodermal and mesodermal structures. The animal half can continue to grow also, but it develops mainly ectodermal structures (Figure 22-15). Apparently, some substances in the cytoplasm of the animal half of the egg cause the cells that form from it to become mainly ectoderm, whereas the vegetal cytoplasm contains mainly mesodermal and endodermal determinants. At the third division these unknown substances are partitioned to the two different sets of cells.

Distinct constituents of egg cytoplasm have been identified in many different invertebrates and in some vertebrates. In the cytoplasm of amphibian eggs, a clearly demarcated zone, the *gray crescent,* is caused by the presence of a pigmented material that is distributed around the equator of the egg. If the first cleavage is perpendicular to the gray crescent and the two resulting blastomeres are physically separated after such a division, a normal animal will develop from each blastomere. However, if the first cleavage is parallel to the gray crescent and the two cells are separated, only the one with the gray crescent develops normally. Thus determination in amphibians occurs very early and is related to the partitioning of substances in the egg cytoplasm.

In mammals, on the other hand, the cells resulting from the initial cleavage divisions are totipotent and are capable of giving rise to normal organisms. To show this, the 8- or 16-cell stage of a mouse or rabbit embryo can be disaggregated and each individual blastomere can be cultured to form a normal blastocyst. Such a blastocyst can be implanted in the uterus of a foster mother appropriately treated with hormones, and a normal animal will be born. The retention of totipotency for several cell divisions in mammals is probably due to the relative homoge-

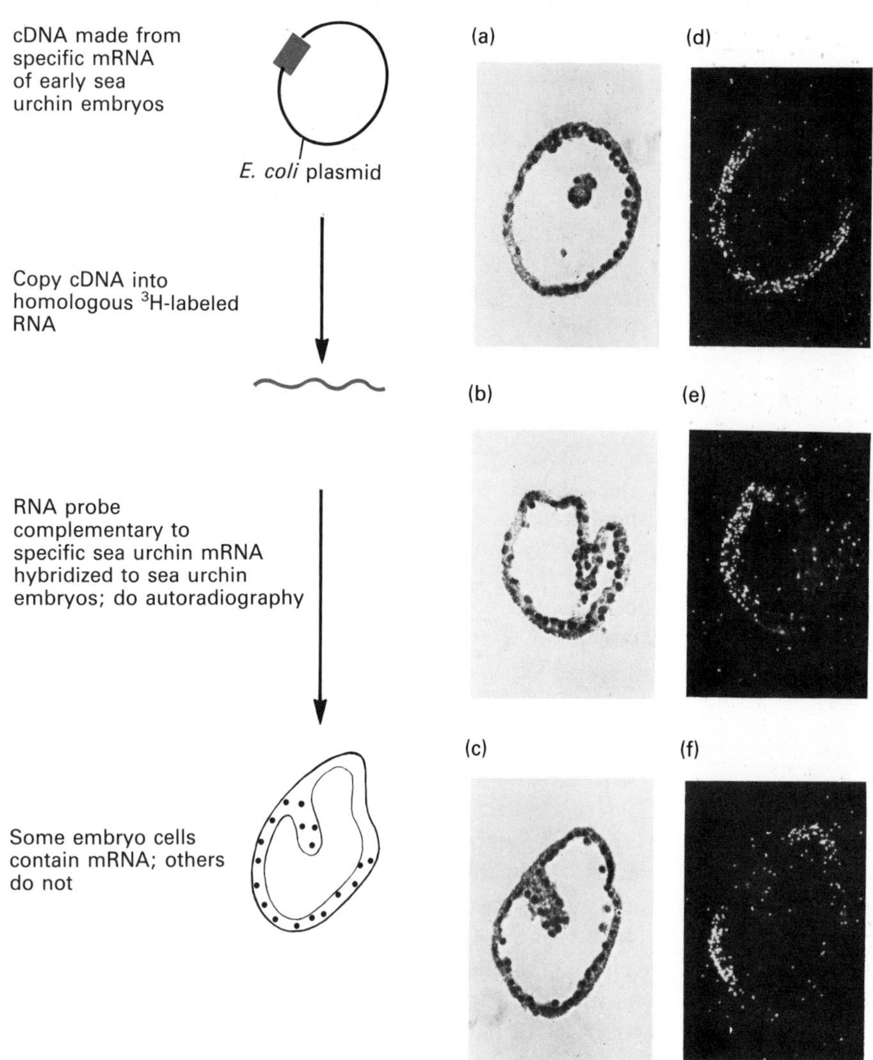

Figure 22-14 Hybridization of a cloned nucleic acid probe to sea urchin gastrulas. A cloned cDNA complementary to mRNA for an unknown protein of the early sea urchin embryo was prepared. (The mRNA was later found to be related to the troponin gene family.) A ^{3}H-labeled RNA probe was made from this cloned DNA; it specifically hybridizes to the many copies of mRNA in the fixed and mounted sea urchin gastrulas. (a),(b),(c) Three gastrulas in different orientations are shown by light microscopy after fixing and staining. (d),(e),(f) In autoradiographs *(photoprinted in reverse)* of the three gastrulas, the white dots represent the silver grains. The mRNA tested appears mainly in the ectodermal cells on the dorsal side of the gastrula. [See D. A. Lynn et al., 1983, *Proc. Nat'l Acad. Sci. USA* 80:2656.] *Photographs courtesy of Robert C. Angerer.*

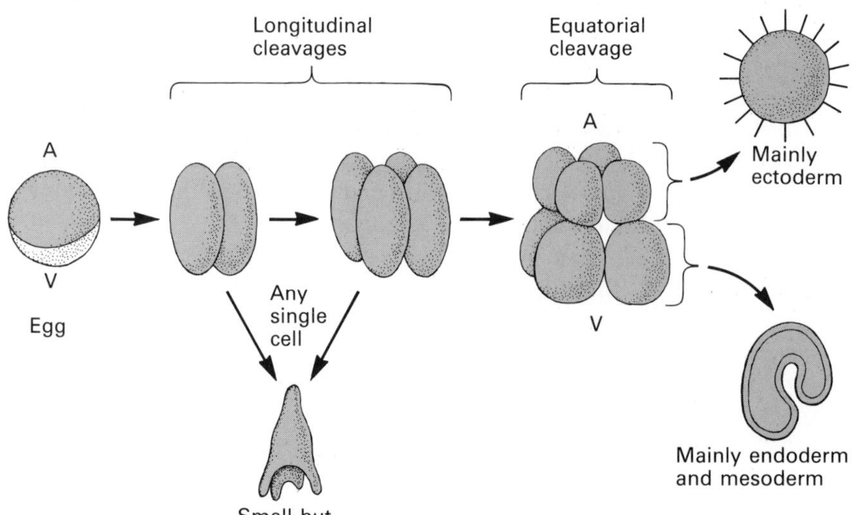

Figure 22-15 Sea urchin determination depends on the distribution of material from animal (A) and vegetal (V) poles of the egg. The third division divides the "top" and "bottom" halves of the original egg cytoplasm. If the four cells in the top half are separated from the four larger cells in the bottom half, each group will follow a different and incomplete developmental path. This divergence contrasts to the developmental capacity of the first four cells: growth and division of any one of them yields a smaller but complete normal animal.

neity of the egg cytoplasm and the consequent uniform distribution of cytoplasm to the initial cells.

A dramatic demonstration of the totipotency of individual cells may be found in the plant kingdom. Single cells from a mature organ (e.g., a leaf) can be grown in culture, where they form an undifferentiated tissue called a *callus*. If a callus is treated with appropriate plant hormones it will form a whole plant. This phenomenon has been demonstrated with a number of plants, including carrots, tobacco, and petunias (Chapter 6).

The Nuclei of Differentiated Frog Cells Can Retain Complete Developmental Potential

Unlike plant cells, no single animal cells that are present after the first several cleavage divisions will form entire animals if they are excised and grown in culture. Are the nuclei of all later cells irreversibly programmed so that they are unable to generate all of the gene products necessary for formation of an animal? This seems unlikely because germ cells, which presumably contain all necessary genes, are not set aside until many divisions have occurred. Therefore early nuclei must retain all necessary genetic potential, at least until the time when germ cells become determined.

Nuclear transplantation experiments with frog eggs have clearly demonstrated the totipotency of nuclei from blastulas. The nucleus of a frog egg can be irradiated to render it inactive, or it can be physically removed from the egg. A nucleus from another frog cell can then be implanted into the egg to test whether that nucleus can support normal frog development. When the diploid nucleus of any one of the thousands of blastomeres in a blastula is implanted into an enucleated egg, normal development occurs in a high proportion of cases. This establishes that the blastomere nuclei are, indeed, totipotent.

Although only a very small proportion of nuclei from more differentiated cells are able to support normal differentiation, it is nevertheless possible to take nuclei from intestinal cells of hatched tadpoles, inject them into enucleated eggs, and obtain normal development to the larval (tadpole) stage perhaps 1 or 2 percent of the time (Figure 22-16). Normal development to adult frogs occurs more rarely. Nuclei from cultured skin cells of tadpoles have been transplanted just one generation before they would

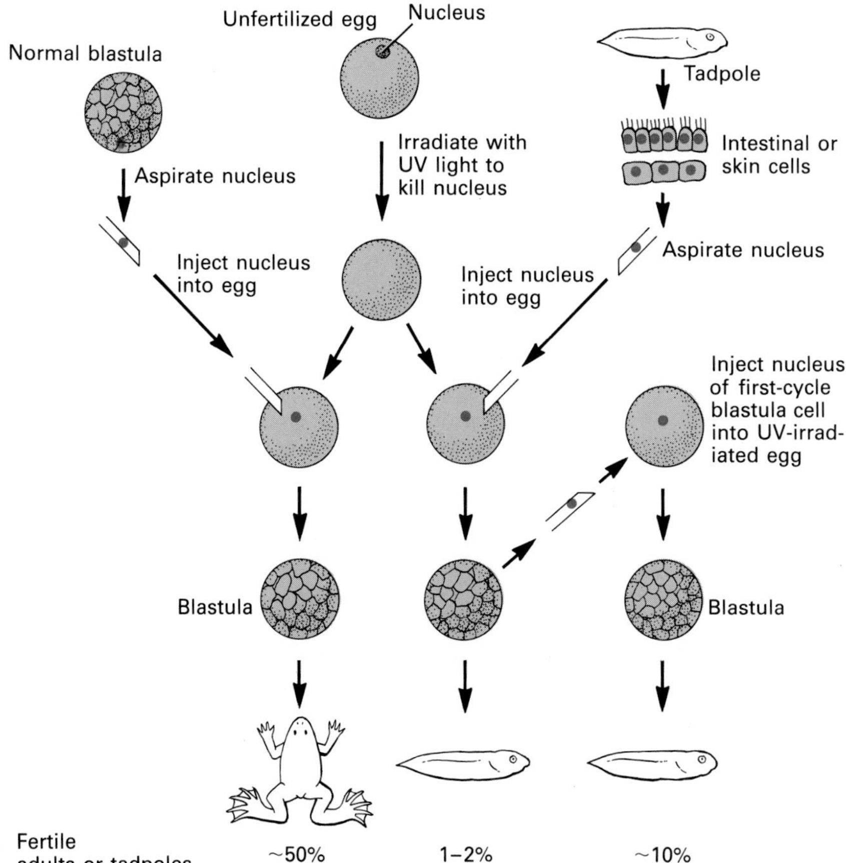

Figure 22-16 Nuclear transplantation in frog eggs. The nucleus of a frog egg can be irradiated to make it inactive, or it can be physically removed by aspiration with a micropipette. A nucleus from another frog cell can then be inserted into the enucleated egg and development can proceed. If the nucleus comes from a normal blastula cell, the chances for normal development are good (50 percent or higher). Differentiated cell nuclei can give normal development, but at a much lower rate. If differentiated cell nuclei are injected in a blastula and the nuclei of the resulting blastomeres are used in a second injection experiment, the frequency of normal development among the transplants is greatly increased. [See J. B. Gurdon, 1973, *Gene Expression during Cell Differentiation*, Oxford University Press.]

have produced large amounts of keratin, a sure sign of differentiation. These transplanted skin nuclei have yielded swimming tadpoles with eyes, muscles, blood cells, etc. Thus some differentiated cell nuclei retain their full genetic potential.

Why is it that all nuclei from specialized cells do not support the development of a complete, normal animal? A deletion or an irreversible scrambling of genes may occur during some differentiation steps. It is also possible that nuclei from some differentiated cells are capable of being reprogrammed, but that they cannot be reprogrammed fast enough to adapt to the rapid DNA replication and chromosomal segregation characteristic of early cleavage.

One way of increasing the yield of normal animals from transplanted nuclei of apparently differentiated cells is to first implant the nuclei into enucleated eggs and allow a blastula to form. The resulting blastula nuclei are then transplanted again. The proportion of frog eggs that develop to the tadpole stage can be raised to about 10 percent with this technique (Figure 22-16). Thus, at least in frogs, early cell differentiation that occurs in the first 10 to 20 cell divisions is not necessarily accompanied by irreversible nuclear changes.

However, attempts to create fertile adults by the transplantation of nuclei from the differentiated cells of adult animals have only rarely been successful. Even with nuclei from spermatogonia—the diploid sperm cell precursors that are only one step away from becoming haploid cells that can fertilize an egg to create new totipotent cells—the results have been poor. It remains possible, therefore, that in some adult vertebrate tissues gene rearrangements occur, and that these changes prevent nuclei from participating in all stages of embryo formation.

Groups of Cells—Not Single Cells—Become the "Founders" of Developing Structures

When cell determination occurs in the early embryo, how many cells become determined to a particular developmental outcome? In other words, how many cells generate the liver or the skin? An important approach to solving such problems has come from studying experimentally produced animals that have been constructed of cells from different genetic sources. Such animals, called *chimeras,* can be produced with mice by taking two genetically and morphologically distinct morula-stage embryos, disaggregating the cells, mixing them, allowing them to reaggregate, and then implanting the reaggregated embryos into a foster mother. The genetic markers permit the cells from the different sources to be followed. The progeny animals contain a mixture of genetically distinct cells from the two original embryos (Figure 22-17), each of which had two parents. Such chimeras are sometimes called *tetraparental* or *allophenic* mice.

Another type of experimentally produced animal composed of genetically distinct cells is called a *mosaic.* A genetically mosaic animal is the product of a conventional fusion of a sperm and an egg. But at some stage early in differentiation, a genetic change (a recombination or a gene inactivation; see Figure 22-18) occurs in one cell and therefore in the progeny of that cell. If the genetic differences between "normal" and "mutant" cells can be distinguished in a particular tissue, the contribution of each genetically distinct cell type to that tissue can be determined.

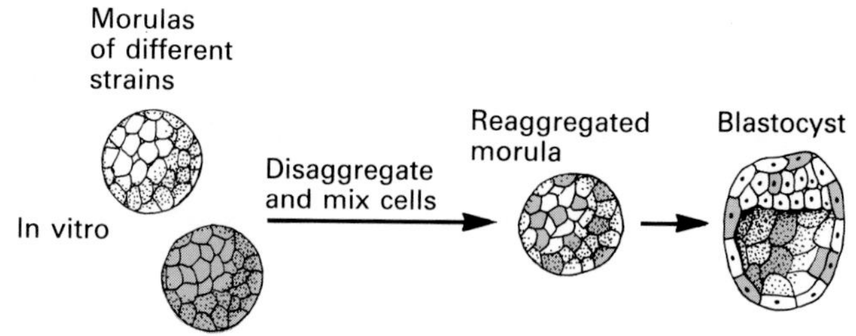

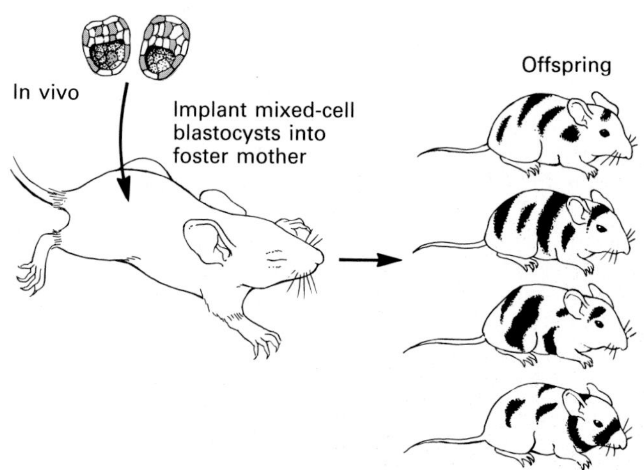

Figure 22-17 The production of chimeric mice. Cells from morulas of genetically distinct strains (e.g., strains with black and white coat color, respectively) are aggregated. The offspring have black and white hair in bands. The coats of individual animals may have different designs because individual cells from the skin overlying separate body segments and each precursor cell will be either from the black- or white-coated parent. The observation of many animals reveals this basic underlying pattern. [See B. Mintz, 1967, *Proc. Nat'l Acad. Sci. USA* **58**:344.]

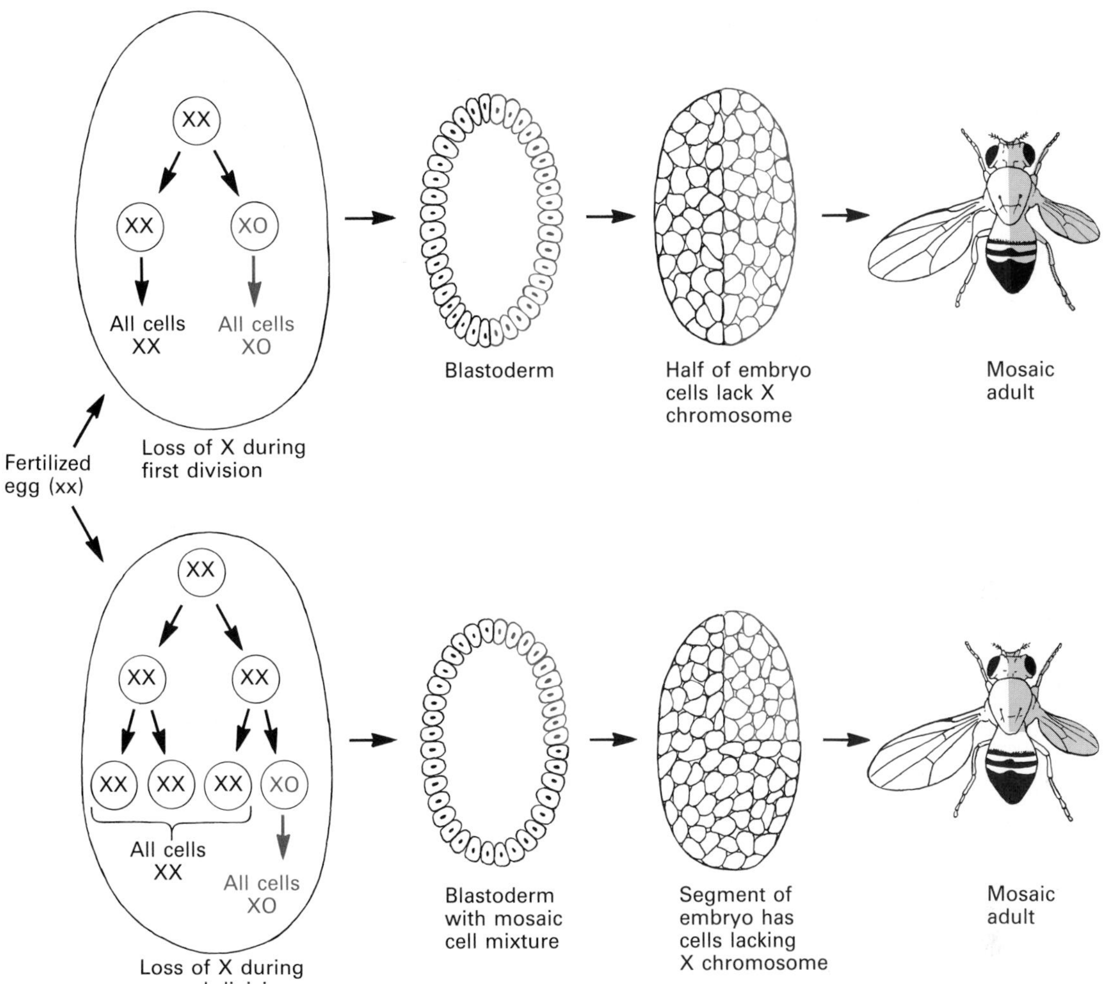

Figure 22-18 Genetic mosaics caused by X chromosome inactivation in *D. melanogaster*. The female has two X chromosomes. Individuals lacking a second X are phenotypically male; however, only the XY males are fertile. Occasionally during the early embryogenesis of a female, one of the two X chromosomes in a cell is randomly inactivated. This produces a gynandromorph, a fly whose body is made up of two genetically distinct cell types, one XX and one X0. If the two X chromosomes of the developing fly carry distinguishably different genes, such as different genes for pigment production, then the contribution of the differently marked cells to different tissues can be assessed. Examples in which X inactivation occurs during the first division or during the second are shown.

Mosaic animals are produced in two ways. A common method is based on the random X chromosome inactivation that occurs in one of the two X chromosomes in female cells (Chapter 10). Such inactivation produces descendant cells in which clonal distributions of X-linked characteristics are inherited either from the mother or the father (Figure 22-18). Mosaics are also produced by mitotic recombination in somatic cells (Figure 22-19). This event is rare, so x-ray induction of increased somatic recombination is often employed.

By analyzing a number of chimeras or genetic mosaics, it is possible to determine how many *founder* cells participate in the formation of a tissue. Suppose, for example, that only one cell were responsible for giving rise to all of the pigment cells of the skin. If cells from animals with genes to produce black coats were mixed in the early embryo with cells from animals with genes for white coats, all resulting animals would be either black or white. Suppose instead that eight founder cells were responsible and that the progeny of each of the eight founder cells were segregated in different parts of the animal. In the ideal case, an animal with eight sectors of alternating black and white color would result. Because each embryo might not get exactly four cells of each ge-

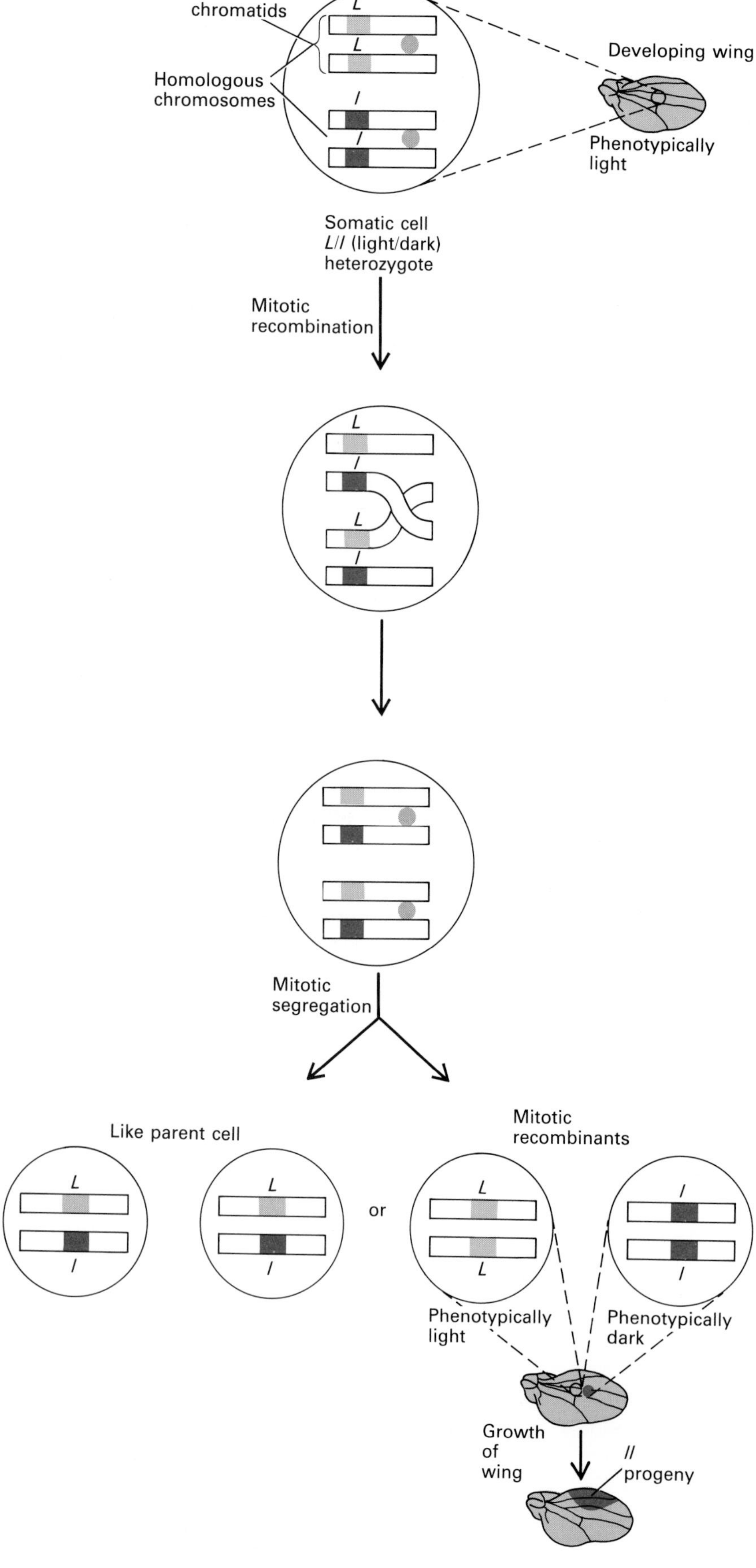

Figure 22-19 A genetic mosaic caused by mitotic recombination in somatic cells during growth of *D. melanogaster*. Although this event is observed infrequently in normal flies, it is greatly increased by exposing developing embryos to x-rays. Recombination can occur between genetically different homologous chromosomes in a heterozygote to reveal a recessive character. When such a cross-over occurs in a developing wing, a sector of marked tissue can be seen in the full-grown wing *(bottom diagram)*.

netic lineage, or because adjacent sectors might be the same color, many animals would have to be observed to identify the eight independent sectors. This outcome would support the conclusion that eight pigment-producing cells descended from the first cell division that yielded determined pigment-producing cells.

The contributions of founder cells responsible for the coat color of mice have been analyzed by producing chimeras and by studying X chromosome inactivations. In both cases mice with patches of different-colored hair have been observed. Thus more than one original determined cell is responsible for the development of hair follicles. The observation of a large number of such variegated animals has revealed that the differences in coat color fall into a predictable pattern of stripes and half-stripes (stripes that stop at the midline). Thus skin cells that will form hair must originate from groups of cells. The number of determined embryonic cells that participate in establishing the coat-color clones is probably between 15 and 30. The half-stripes on some of the animals indicate that each founder cell must give rise to progeny that reside on *either* the left or the right side of the animal.

Drosophila melanogaster embryos have been studied extensively for the purpose of determining the number of cells contributing to the *primordium* (the most rudimentary form) of each adult organ. An elegant and straightforward technique that has been used is the observation of cell lineage markers in *gynandromorphs,* female embryos that lose one X chromosome during blastoderm formation (see Figure 22-18). The gynandromorph blastoderm is a mixture of X0 and XX cells. When the parents are genetically distinct, the adult gynandomorph will have visual markers—for example, mosaic patches of cells on the wings of the fly. By estimating the area of the wing covered by the smallest mosaic patch, it is possible to estimate the number of cells that must have given rise to the wing primordium. This number has been estimated to be between 10 and 15 cells.

The studies of gynandromorphs have offered convincing proof that groups of cells, not single cells, differentiate into the individual tissues and organs that form an adult organism. It is likely that in particular sections of an early embryo, a number of cells are available to become determined (as, say, skin cell precursors). The actual number that enter each lineage may vary in individual animals, but it is always a group of cells that engages in any major determinative choice.

The Basic Body Plan Is Very Similar in All Vertebrate Embryos

The earliest stages of development—fertilization, cleavage, gastrulation, primordial germ layer formation, and cell determination—result in the production of a structure with a defined general body plan. The later stages of development constitute *organogenesis,* the period during which growth and further development are guided by the body plan. These stages of embryogenesis are by far the lengthiest ones. For example, in a mouse the organ rudiments such as the heart and liver buds can be distinguished by 7 to 8 days, but 19 or 20 days are required to produce the newborn mouse.

The basic body plan of different vertebrate embryos is remarkably similar, a fact that was pointed out a hundred years ago by Ernst Haeckel, and one that can be easily confirmed by visual comparisons alone (Figure 22-20). At early developmental stages many structures occur in all vertebrate species—even some structures that are not used in forming the adult animals. For example, human embryos develop gill pouches and tails, and then lose them. In fish, of course, gill pouches ultimately develop to function in adult respiration. The human embryonic gill pouches are not the forerunners of lungs: the embryologic program of the human retains the information to make gill pouches, but it has acquired additional information that directs it to discard this tissue and to construct lungs during development. The evolutionary rule in developing animals is modification of pre-existing features. Only in rare cases have developmental pathways been completely eliminated when their product was no longer used in the adult.

The Growth of the Embryo and Organ Formation

Two aspects of the growth and development of the embryo during organogenesis have been of particular interest to embryologists:

1. *Cell differentiation:* the development of specialized tissue that can be recognized morphologically, through biochemical or staining methods
2. *Pattern formation:* the control of cell growth so that the correct sizes and shapes of body structures are achieved

We shall discuss each of these aspects in turn.

Cell-to-Cell Interactions Trigger the Determination and Differentiation of Cells in Tissues and Organs

Tissue-specific proteins are synthesized in differentiated adult tissues such as muscle, liver, the lens of the eye, the skin, and the pancreas. During the course of organogenesis these tissue-specific proteins progressively increase in amount and number. Eventually the pattern of proteins synthesized by a developing tissue comes to resemble that of adult tissues. The mechanisms underlying the differen-

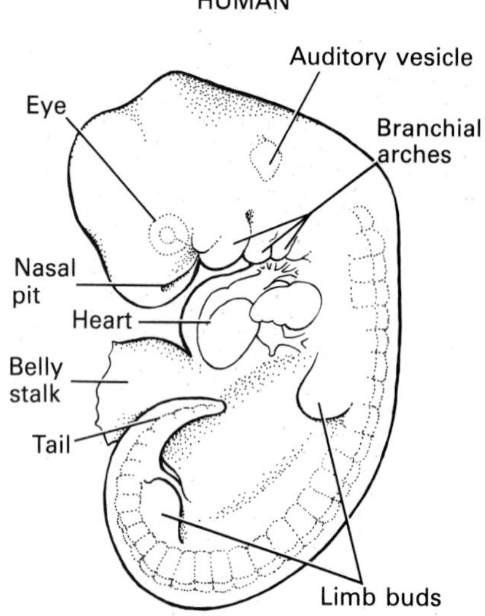

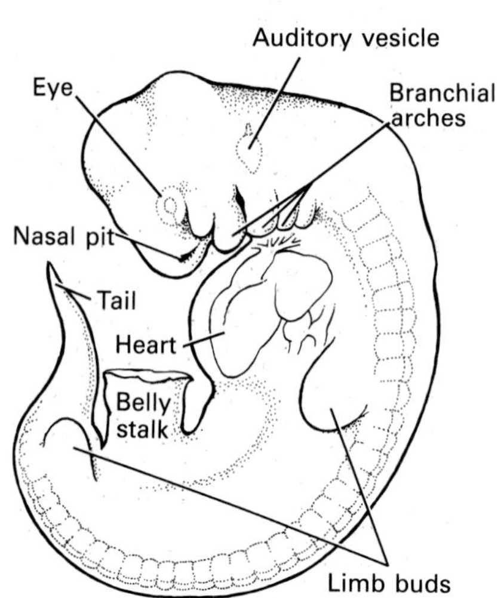

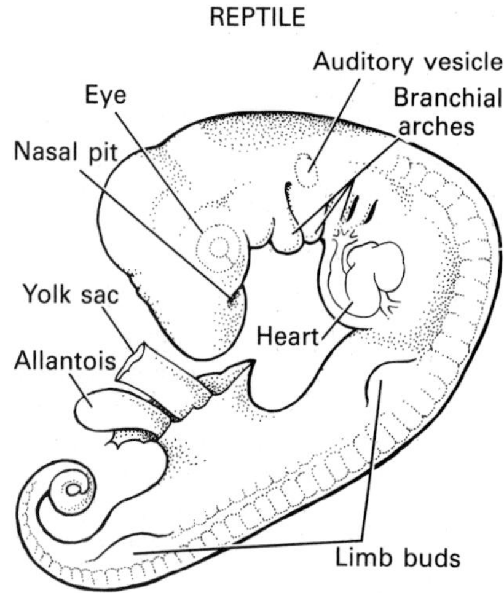

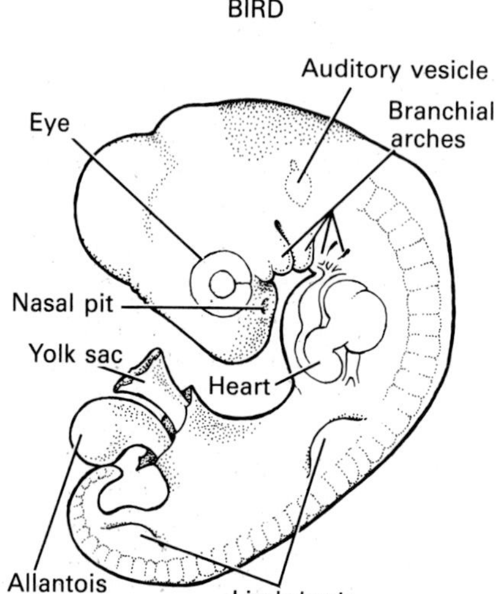

Figure 22-20 These drawings of four vertebrate embryos at comparable stages highlight their many similarities.

The following quotation is from a popular, influential book published in 1929. The authors were defending darwinism: "Note that the early stages are very like each other, and that the animals diverge as they develop. In the first stage the nerve-folds are closing in to form the brain. Then the gill-clefts appear. In the land animals these close later; in the newt and dogfish feathery gills appear. The human tail and its gradual shortening are clearly seen." *Quotation from H. G. Wells, J. Huxley, and G. P. Wells, 1929,* The Science of Life, *Doubleday. Drawings after B. M. Carlson, 1981,* Patten's Foundations of Embryology, *4th ed., McGraw-Hill.*

tiation of tissues must produce changing patterns of individual gene activations and deactivations.

The development of each of the specific cell types found in organs and the initiation of tissue-specific protein synthesis appear to involve the specific interaction of at least two types of cells that originate in different embryonic germ layers. This interaction, which calls forth the already determined possibility in one cell type, is called *induction*. Most of these interactions involve endoderm and mesenchymal cells, or ectoderm and mesenchymal cells. The interactions result in the synthesis of the tissue-specific protein(s) in the endodermal or ectodermal partner of a pair. The synthesis of the skin proteins provides an excellent example.

Mesodermal Induction of the Ectoderm in the Skin

The skin is composed of the *epidermis,* an outer layer derived from the ectoderm, and the *dermis,* an inner layer derived from the mesoderm. The usual identifiable markers of a particular region of skin (feathers, scales, hairs, or simply the thickness of the outer layers of dry cornified cells) result from the controlled synthesis of specific epidermal proteins (e.g., the family of keratins) and the controlled division of the epidermal cells (Figure 22-21).

The type of surface structure that is formed by the epidermis is governed by the underlying dermis. For example, in chickens the feathers of the legs and wings are different, and in some areas (e.g., the feet) a scaly, nonfeathered skin is produced. If presumptive epidermal or dermal cells are taken from one area of an early chick embryo and transplanted to another surface area, it becomes clear that the dermis and not the epidermis controls the outcome. For instance, when dermal cells from the leg area are transplanted underneath the epidermis overlying the wing bud, the epidermis is induced to form feathers of the type normally found on the leg (Figure 22-22, experiment 1).

On the other hand, transplantation of leg epidermis to the foot region does not give rise to feathers on the foot. Rather, the transplanted epidermis produces scales that are characteristic of the foot region (Figure 22-22, experiment 2). Thus the same area of epidermis is capable of developing several different features (e.g., feathers and scales), and the type of structure produced is determined by the underlying dermis.

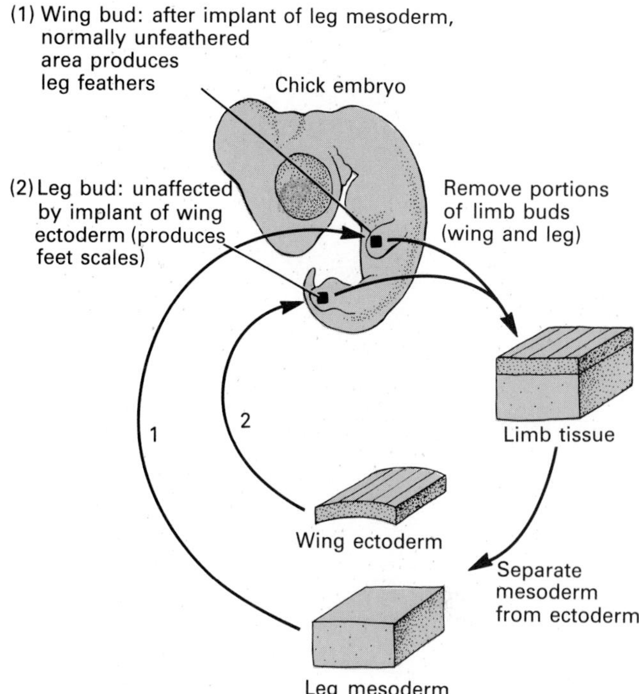

Figure 22-22 The dermis determines the type of structures produced by the epidermis. (1) Mesoderm from a leg bud induces leg feathers when it is transplanted under wing ectoderm at a site that normally does not produce feathers. (2) Ectoderm from a wing bud site that normally would produce wing feathers produces scales characteristic of the feet when it is transplanted to the leg bud. [See J. M. Cairns and J. W. Saunders, 1954, *J. Exp. Zool.* **127**:221.] *After N. K. Wessels, 1977,* Tissue Interactions and Development, *Benjamin-Cummings, p. 56.*

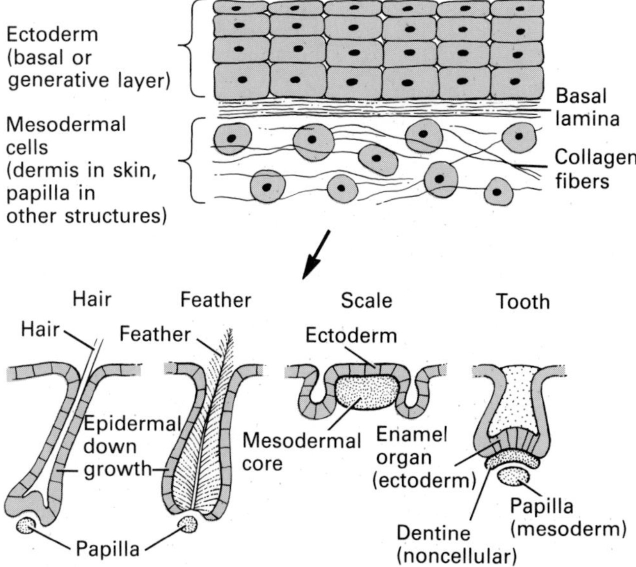

Figure 22-21 The structures formed by the vertebrate epidermis, and the nature of its interactions with dermal (mesodermal) tissue. *Drawings at bottom after E. M. Deuchar, 1975,* Cellular Interactions in Animal Development, *Chapman & Hall (London), p. 149.*

At the stage of development at which such transplant experiments are conducted, the dermis is already determined—that is, it is restricted in the type of epidermal structures it can induce—but the epidermis is not. However, dermal cells can call forth only a capacity already possessed by the epidermis they induce. For example, mouse dermis will induce cultured chick epidermal cells to form feather buds; the chick cells cannot form the hairs characteristic of the mouse skin.

In a given animal, many different structures are formed by induction of ectodermal or epidermal cells. One interesting example is the induction of ectodermal cells to produce the crystalline lens of the eye. The inducing cells are neuro-ectodermal cells that grow out from the primitive brain to form an *optic cup* (Figure 22-23). The ectodermal cells overlying the head region interact with the optic cup cells to produce the cells that form the lens, a trans-

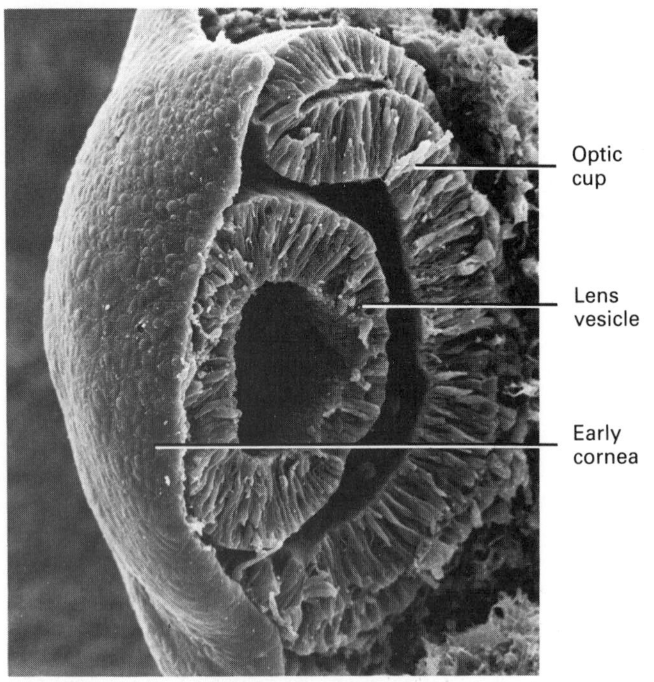

Figure 22-23 The eye of a developing chick. This scanning electron micrograph shows the fractured edge of an eye at about 5 to 6 days. *From N. K. Wessels, 1977,* Tissue Interactions in Development, *Benjamin-Cummings, p. 44. Courtesy of K. W. Tosney.*

parent tissue through which all vertebrates see. When the optic cup is transplanted underneath ectodermal cells that normally would become skin, a lens forms (Figure 22-24).

Many other structures and organs of the adult animal are constructed from ectodermally derived cells that have received and interpreted developmental signals from an inducing cell partner, most often a cell of mesodermal origin.

Cellular Interactions in the Formation of Internal Organs Interactions between the endoderm and the mesoderm are required for the formation of internal organs. The organs associated with the gut begin as endodermal outgrowths from the primitive gut into areas of mesenchymal cells. The interaction of the two cell types induces differentiation in the endodermal cells into the characteristic epithelial cells of the salivary glands, pancreas, lungs, and liver.

Mesenchymal cells from almost any part of the developing embryo will support the development in culture of the endodermal precursors of pancreatic cells into normal pancreatic acinar cells containing pancreatic digestive enzymes. In contrast, cultures of the endodermal precursor cells of salivary glands or liver will differentiate only in the presence of the mesenchymal cells that normally interact with these endodermal cells (Figure 22-25).

All response to induction does not take place in the ectodermal or endodermal tissue. The induction of muscle cells (derived from the mesoderm) by nerve cells (derived from the ectoderm) has been characterized at the molecular level. Although the experiments were performed with young rats and cats and not with embryos, nerve precursors probably instruct muscle precursors similarly during development. Thus specificity in differentiation can be realized in cells from all three germ layers.

If the nerve fiber that innervates (triggers contraction of) a muscle is cut, the muscle stops contracting. If the cut nerve is left undisturbed, it will regenerate an axon that

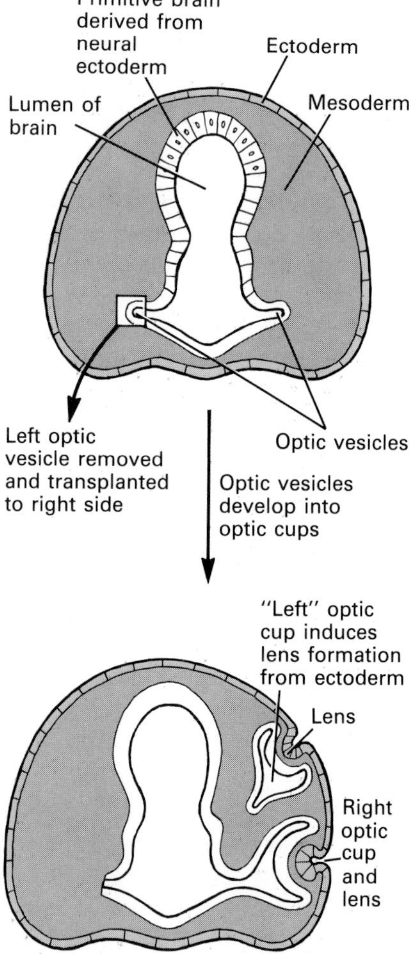

Figure 22-24 A cross section through the head of a vertebrate embryo to illustrate the interaction of the optic cup, an outgrowth of the developing central nervous system, with the overlying ectodermal cells that become the optic lens. When the left optic vesicle is transplanted to another site in the head, it induces a lens in that inappropriate site. *After N. K. Wessels, 1977,* Tissue Interaction in Development, *Benjamin-Cummings, p. 46.*

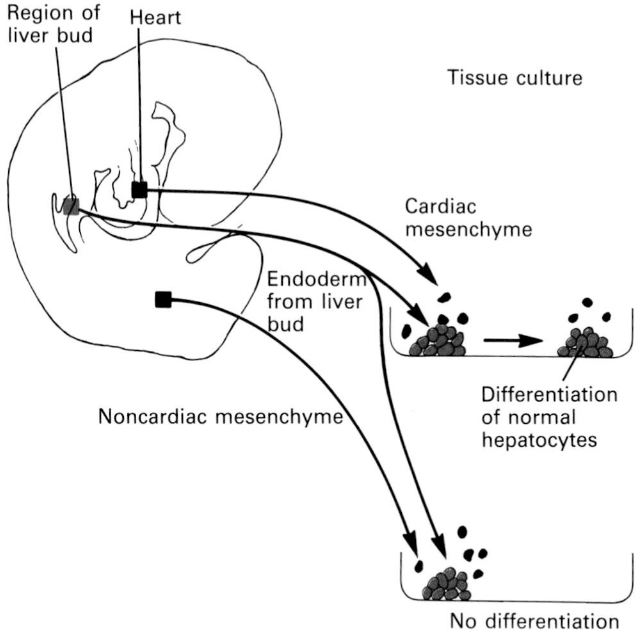

Figure 22-25 Proper differentiation of liver cells requires that presumptive hepatocytes interact with mesenchymal cells from the mid-section (cardiac region) of the embryo. This interaction can occur in culture, and it has been shown that only mesenchymal cells taken from the region near the heart are effective. (Mesenchymal cells from the head or tail will not produce differentiation.) However, there is no strict species specificity: chick cardiac mesenchyme will induce mouse endodermal cells to differentiate into hepatocytes. [See E. Houssaint, 1980, *Cell Differentiation* 9:269.]

will grow toward the muscle fiber. Eventually the nerve will reestablish a synapse with the muscle, and normal muscle function will be restored.

There are two principal types of striated muscle—fast and slow—which differ in their contractile rate and in the relaxation time of the muscle fiber. Muscle proteins such as myosin light chains are different in the two types of fibers (Chapter 19). The nerves serving fast and slow muscle fibers can be cut and their regeneration can be redirected so that the one normally innervating a fast fiber forms a synapse with the slow fiber, and vice versa. In such cases, the fast fiber differentiates into a slow fiber: the myosin light chains characteristic of the fast fiber are replaced by those characteristic of the slow. The reverse occurs in the slow fiber. It is also possible to change the behavior of a muscle fiber from fast to slow by chronic stimulation; again the myosin light chains change. Apparently a signal from the nerve cell is responsible for affecting specific protein synthesis in the muscle cell and thus helps determine the type of muscle.

The Signals between Interacting Cells in Differentiation May Function via the Extracellular Matrix

What is the signal that passes between interacting cells? In spite of a great deal of study, researchers have not derived any general conclusions about the nature of the signal that is responsible for induction. In some cases the plasma membranes of the two interacting cells do not actually have to touch. In other cases electron microscopy has shown that the two plasma membranes do come into contact. For example, the membranes of neuroectodermal cells must touch those of mesenchymal cells to induce the differentiation of the mesenchymal cells into specialized kidney cells. The same is true for the interactions of mesenchymal cells that induce endodermal cells to form salivary glands (Figure 22-26). In spite of the demon-

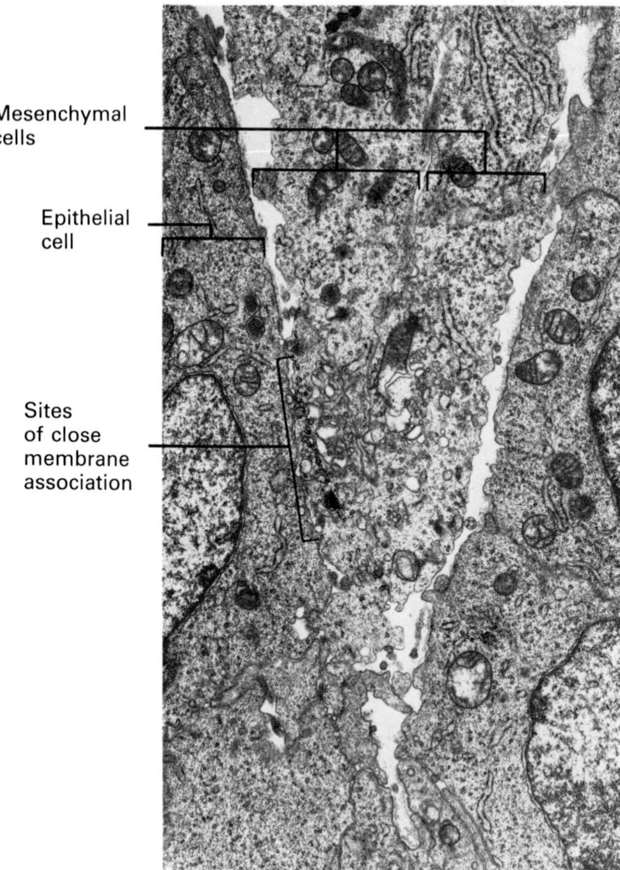

Figure 22-26 An electron micrograph showing the close association of plasma membranes between mesenchymal and epithelial cells in the salivary gland. On the left and right sides are two epithelial cells of the developing salivary gland of a mouse; the cleft between them is filled by two mesenchymal cells that appear wedge-shaped in this view. *From N. K. Wessels, 1977,* Tissue Interactions in Development, *Benjamin-Cummings, p. 203.*

strated necessity for close proximity—and sometimes outright contact—between two cell types, no protein or nucleic acid that will bring about correct induction in a responding partner has been isolated from an inducing partner. The nature of the signal passing between two such cell types remains unknown.

Communication between two types of interacting cells may occur by way of the extracellular matrix, which consists of material that is secreted by one or both of the cells into the space between the cells. This matrix frequently contains a large concentration of complex polysaccharides, such as hyaluronic acid and the glycosaminoglycans. These molecules, which are synthesized and secreted during organogenesis and which persist throughout the life of the animal, are long chains of repeating disaccharides that differ in composition in different tissues (Table 22-2). They are often linked to proteins to form *proteoglycans* (a typical structure is shown in Figure 3-75). Many other complex glycoproteins, such as *laminin, fibronectin* (Chapter 19), and various types of collagen fibers, are found in the spaces between interacting cell types. The extracellular matrix between two cell layers of different origins is often referred to as a *basal lamina.*

The involvement of the extracellular matrix in the development of the epithelium of the salivary glands has been demonstrated, although the function of the matrix is not well understood. In the developing gland, groups of epithelial cells in contact with a basal lamina grow and form clusters of cells that will be the secreting cells of the gland. Underneath the basal lamina mesodermal cells are attached. By microdissection the groups of epithelial cells plus the basal lamina can be separated from the mesodermal cells. When the epithelial component is cultured separately no further growth occurs, but when it is mixed with mesodermal cells growth and differentiation of salivary gland tissue continue. Also, if the epithelial cells are treated with hyaluronidase to destroy the basal lamina and are mixed with mesenchymal cells before the basal lamina has had time to regenerate, growth and differentiation fail to proceed (Figure 22-27).

Thus there are three elements to the interacting system: the two cell types and the basal lamina are all necessary for the epithelial cells to differentiate correctly. The molecule involved in this type of communication, the intracellular signals signifying correct cell-cell interaction, and the events of gene control within a responding cell are not yet known, but these details are currently being researched.

The Controlled Growth of Cells Shapes an Animal: Positional Information

We have already discussed tissue-specific protein synthesis as one aspect of cell differentiation. Cells are also subject to developmental instructions that govern the sizes and shapes of the many parts of the animal. These instructions presumably depend on the regulation of cell growth in addition to cell specialization. For example, proper limb formation requires the development of muscle, nerve, bone, and skin in quantities and arrangements that differ for the formation of, say, arms as opposed to legs. The limitation or stimulation of cell growth within specified regions, called *compartments,* is crucial to correct morphogenesis.

The instructions that groups of cells are given to accurately carry out morphogenetic tasks are referred to as *positional information.* Such information allows similar types of cells to form different patterns. Correct cell behavior in response to positional information has been

Table 22-2 The nature of extracellular polysaccharides

Sulfated polysaccharides	Disaccharide structure*	Chain length†	Present in proteoglycans	Location in body tissues
Chondroitin	G-GalNAc	200–2000	+	Cartilage, bone, skin
Dermatan	G- or I-GalNAc	600–2000	+	Skin, arteries, heart valves
Heparin	G- or I-GlcNAc	600–2000	+	Lung, liver, mast cells
Keratan	Gal-GlcNAc	100–500	+	Skin, cornea, cartilage, ligaments
Hyaluronic acid	G-GlcNAc	100–10,000	–	All connective tissue, joint fluid, vitreous humor of eye

*G = glucuronic acid, GalNAc = *N*-acetyl-D-galactosamine, GlcNAc = *N*-acetyl-D-glucosamine, I = iduronic acid, and Gal = D-galactose. (See Figure 3-75 for structures of complex polysaccharides.)

†Given in monosaccharide units.

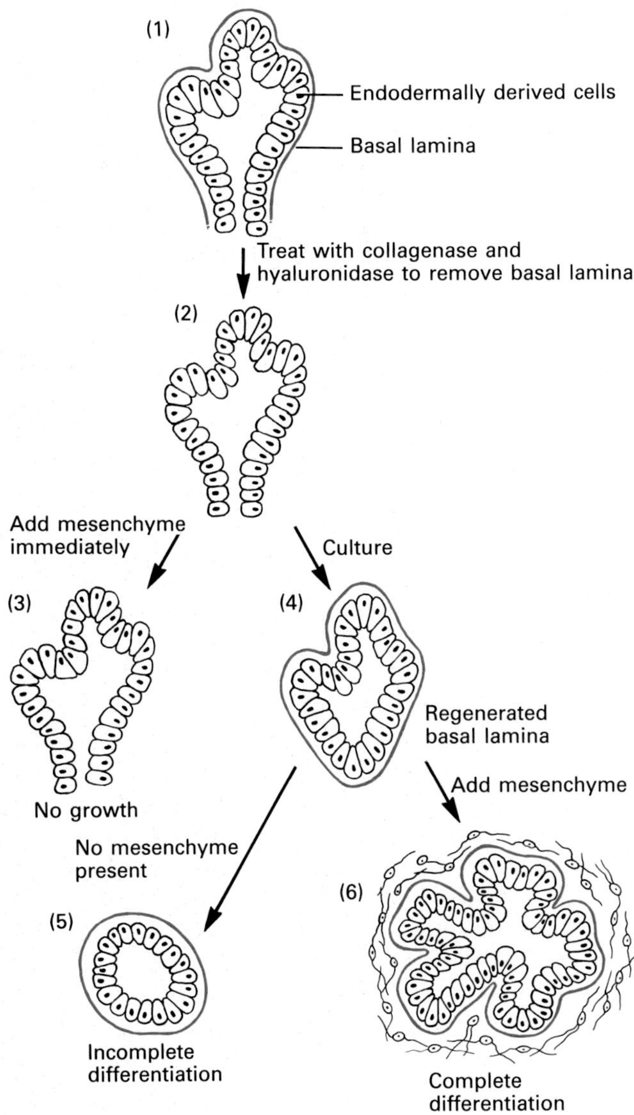

Figure 22-27 Elements necessary for salivary gland differentiation. Salivary gland explants can be dissected away from other cells (1) and treated with enzymes to remove the basal lamina (2). If the remaining epithelial cells are mixed immediately with mesenchymal cells, no growth or differentiation will result (3). If, on the other hand, the epithelial cells are cultured for a while they will regenerate the basal lamina (4), but they will not differentiate further (5) unless they are allowed to interact with mesenchymal cells (6). *After N. K. Wessels, 1977,* Tissue Interactions and Development, *Benjamin-Cummings, p. 225.*

studied in both vertebrates and invertebrates; here we shall consider vertebrate limb development as an example.

Limb Development in Chicks and Amphibians

Dramatic evidence illustrating the existence and the importance of positional information has been obtained from studies of limb formation in chick embryos. Three to four days after fertilization, groups of cells on either side of the long axis of the embryo form mounds called limb buds; these will become the legs or wings. The cell types in all limb buds appear to be identical: an outer ectodermal layer of cells envelopes an inner group of mesenchymal cells. The mesenchyme will differentiate into the muscles, cartilage, and bone forming the different structures of the legs and wings.

If before cartilage and muscle cells are first identifiable, the entire tip of a developing wing bud containing both epidermal and mesodermal cells is excised and replaced with the tip of the leg bud, the cells in the transplanted bud will form cartilage, bones, muscles, and skin. However, the limb structure will not be the normal wing, but will be rather a leg, complete with toes, claws, and scales. Thus, although the cells of 4-day leg and wing buds are histologically similar, they are not equivalent; apparently they have already received positional information. Determination includes not only the information needed for cells to develop into a particular type but also the information that instructs them to grow into a certain form.

In birds and mammals the formation of limbs is a one-time event during embryogenesis, but in the urodeles, an order of amphibians, adult limbs can be regenerated. After amputation and wound healing, mesodermal cells form a limb bud that accurately reconstructs embryologic events to produce a new and appropriate limb. Forelimbs form at anterior sites, hind limbs at posterior sites. Thus both the specialized information for differentiation and the positional information exist in the cells that remain at the stump when an adult limb is removed. In these animals that information can be activated to direct proper limb formation.

Genetic Analysis of the Formation of Adult Structures

The use of positional information is characteristic of all developmental programs, but the design of the programs varies. Most vertebrates grow from a small embryo through successively larger embryos to the adult animal with relatively little discontinuity. On the other hand, frog embryos grow into tadpoles before the animals shed their larval appearance along with their tails and their strictly aquatic life to become small frogs, amphibious animals that subsequently grow in size without much change in shape. Invertebrates display a wide variety of growth styles, but a very common feature is larval growth in several steps, followed by the gradual replacement of larval structures by adult structures.

The developmental programs of two invertebrates, *Caenorhabditis elegans* (a nematode) and *Drosophila melanogaster,* have been thoroughly studied, not only to characterize the larval and adult structures in detail, but also to identify the genes responsible for carrying out the construction of the adult. The solution of the difficult but intriguing problems of pattern formation in these and in other animals will depend on the identification of such genes and of the proteins they encode. This information will be needed to answer the question of how gene products guide cells to construct the correct body parts. We shall briefly outline special features of the development of *C. elegans* and *D. melanogaster* and then summarize some of the progress made with these animals by studying developmental mutants.

Each and Every Cell Is in the Right Place: *C. elegans*

Perhaps the ultimate example of the correct execution of a positional program is provided by the small nematode *C. elegans*. Genetically identical adults of this organism have a precisely determined number of cells (about 1000 plus a collection of gametes numbering about 2000). Each of the corresponding cells in any two adult organisms appears to be surrounded by cells of precisely the same type(s), in precisely the same configurations, in the two individuals. (Precision in body construction down to the level of individual cells has been noted in other genetically identical invertebrate animals. In the water flea *Daphnia magna,* even the locations of intracellular organelles such as mitochondria appear to be identical within the corresponding cells of different individuals.)

The embryo of *C. elegans* hatches into a larva that resembles the adult worm but that contains only about 500 cells. During three subsequent larval stages requiring about 40 h, the postembryonic cells divide in a controlled manner. Two types of adults can be formed, a male with 970 somatic cells and a hermaphrodite (a self-fertilizing "male-female") with 810 somatic cells. Each post-embryonic cell can be followed individually in the microscope during the formation of the adult worm (Figure 22-28). Genetically pure strains of worms, which are easy to maintain as self-fertilizing stocks, have been obtained. Many useful mutations affecting development—including the development of adult functions such as swimming—have been discovered and mapped; some of these mutations will be discussed later.

Controlled Divisions and Programmed Cell Death

Most of the early larval cells of *C. elegans* and the progeny of these cells divide according to a rigidly fixed program. For example, Figure 22-28c shows a larval cell labeled H1 (a hypodermal, or skin, cell at the head of the animal); this cell undergoes four divisions, producing at each division one cell that is maintained in the adult and one *stem* cell that continues to divide (Figure 22-29). The result is five cells in the adult. In contrast, the male V6 (a ventral skin cell) and its descendants undergo 30 divisions, but give rise to only 26 cells in the male adult. Along the way five cells die and are removed. This is an example of *programmed cell death,* an important aspect of embryonic development in all organisms, but one that is hard to trace except in cases where individual cells can be followed (as in *C. elegans*). The V6 cell and its descendants do not undergo the same pathway in hermaphrodites; there, another regular pathway with fewer divisions is followed.

Equivalence Groups When an individual cell in a *C. elegans* larva is experimentally destroyed by a highly focused laser beam, usually the descendants that would have resulted from the killed cell are absent in the adult. Apparently no other cell can compensate for the loss of the destroyed cell. For example, cells P8 and P12 are ventral nerve cell precursors that have seven and four descendants, respectively. These descendants are simply missing if P8 or P12 is destroyed. Thus for many cells of the adult, the overall program for the animal is set by the time the larva hatches.

In a few cases, the destruction of one larval cell causes another to pick up the task of its dead neighbor (Table 22-3). Cells P9, P10, and P11 in the ventral nerve cell lineage lie between P8 and P12, but they behave very differently in laser destruction experiments than their peripheral neighbors. If P9 is destroyed, then P10 and P11 develop normally, producing their usual nerve cell descendants. If P10 is destroyed, P9 generates P10 progeny, and P9 and P11 together form the correct descendants of the P10 and P11 type. If P11 is destroyed, P10 behaves as P11 would have, and P9 takes over the function of P10. These three larval cells are said to form an *equivalence group,* which means that each of the three is capable of doing what the primary member of the group—here, P11—normally does. In this group P10 plays the secondary role, and P9, the tertiary. Presumably P11 descendants have a more important function in the life of the organism than the other two lineages do, but the nature of this function is not known at present.

C. elegans Mutants Have Altered Patterns of Cell Division

The cell destruction experiments with *C. elegans* have uncovered some of the rules of nematode cell divisions. Through other experiments, several genes that control the pattern of regulated divisions have been recognized. Researchers following the lineage of the larval cells found that some divisions produce two end-stage cells. In other

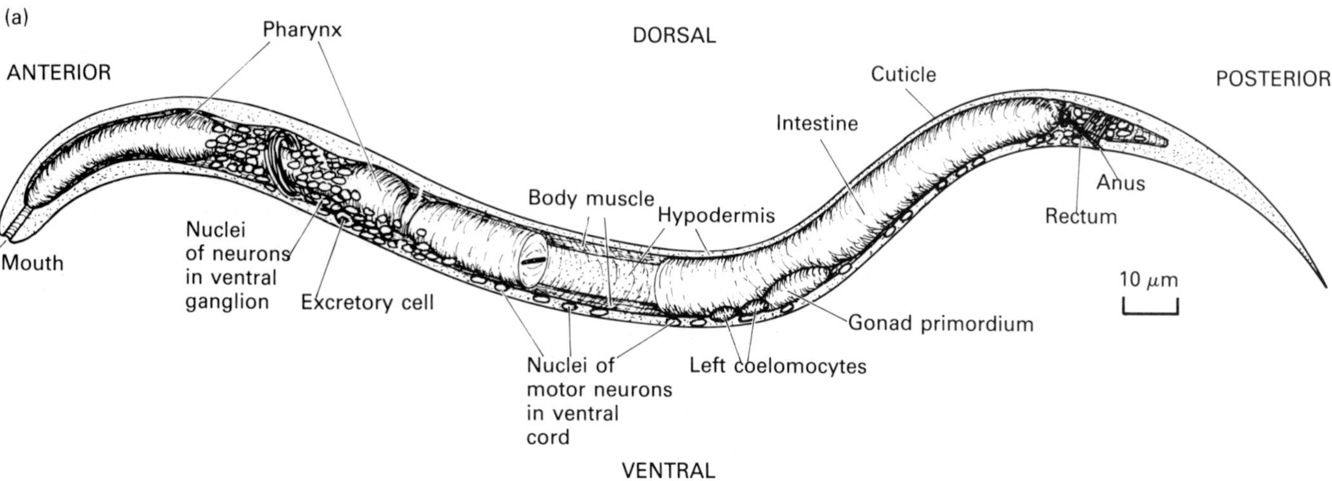

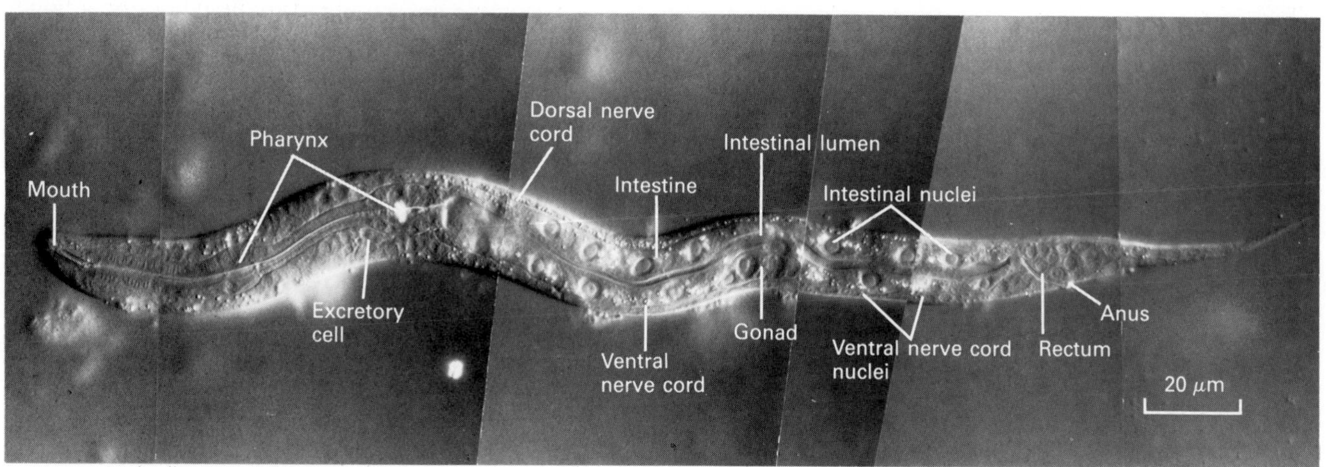

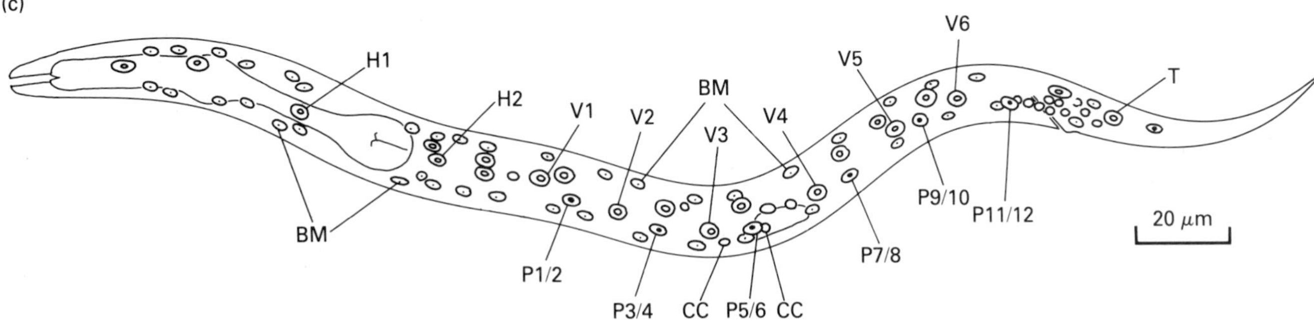

Figure 22-28 *C. elegans.* (a) The drawing shows an adult hermaphrodite with part of the midsection cut away. Body parts are labeled and nerve cell nuclei are shown in several locations. An adult animal has about 1000 cells. (b) A newly hatched final-stage larva containing about 500 cells is shown in this light micrograph taken with Nomarski optics. The individual nuclei of many cells are evident in this single plane of focus. By examination of several planes of focus, almost every nucleus within the larva can be followed as the larva develops to adulthood. (c) The diagram shows the positions of some of the postembryonic (final-larval) cells whose behavior during cell division has been charted. For example, V and H signify the cells that become the hypodermal (surface) cells of the animal on the ventral side and the head, respectively; P cells are precursors to the ventral nervous system; BM cells become body muscle; and CC cells become coelomocytes, cells lining the body cavity. *From J. E. Sulston and H. R. Horvitz, 1977,* Devel. Biol. **56***:110.*

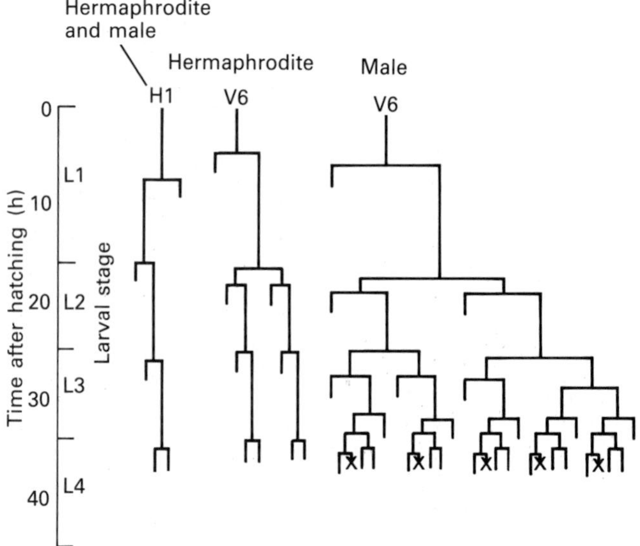

Figure 22-29 The lineages of H1 and V6 cells of *C. elegans*. The H and V cells become hypodermal (surface) cells on the head and on the ventral side, respectively, of the adult. The lineage is shown by the vertical and horizontal lines; each horizontal line represents a cell division, and the vertical lines between it are the daughter cells. The H1 cell follows the same simple pattern in the male and in the hermaphrodite. Four divisions occur and five final cells result. The divisions of V6 and its descendants vary between the male and hermaphrodite. The programmed death of particular cells in the male V6 lineage is indicated by an X. [See J. E. Sulston and H. R. Horvitz, 1977, *Devel. Biol.* **56**:110.]

Table 22-3 Cell lineage development after laser killing of particular cells in *Caenorhabditis elegans* larvae*

	Cell P*n* and its fate				
Cell ablated	**P8**	**P9**	**P10**	**P11**	**P12**
—	P8	P9	P10	P11	P12
P8	—	P9	P10	P11	P12
P9	P8	—	P10	P11	P12
P10	P8	P10	—	P11	P12
P11	P8	P10	P11	—	P12
P12	P8	P9	P10	P11	—
P9, P10	P8	—	—	P11	P12
P10, P11	P8	P11	—	—	P12
P9, P10, P11	P8	—	—	—	P12

*The first row shows the fate of each cell in the intact animal. Subsequent rows show the cell fates after ablation of the specific cells listed in the first column. For example, when P10 is ablated, P8, P11, and P12 are unaffected, but P9 takes up the position and presumably the function of P10.

SOURCE: J. Kimble, J. Sulstan, and J. White, 1979, *Cell Lineage, Stem Cells and Cell Determination*, N. LeDouarin, ed., Elsevier/North-Holland, p. 59.

cases, additional divisions are required to reach the end stage (e.g., a cell may divide once and then its two daughters may each divide again to produce a total of four end-stage cells). Individual mutations can change these patterns. For example, in some mutant strains, instead of dividing and assuming a different function and location, cells reiterate the division patterns of their parents. Such mutations turn an end-stage cell into a stem cell which gives rise to extra descendants (Figure 22-30). From the fact that mutations can cause reiterative divisions, it appears that a wild-type gene product must function in *C. elegans* to control precisely the number of cell divisions in each cell lineage. Isolation of such gene products and characterization of their mode of action should illuminate the general developmental problem of how the sizes and shapes of anatomical features are determined.

How Do Specific Genes Affect the Body Plan of *Drosophila*?

The important role of genetics in the study of pattern formation during development has also been strikingly illustrated by experiments with *D. melanogaster*. Following embryonic development, three larval stages and a pupation stage precede the appearance of *Drosophila* adults (Figure 22-31a). The larva is a segmented organism and the development of these segments has been actively studied recently. Within the internal body cavity of the larvae, there are groups of determined cells called *imaginal disks*. The adult arises from the growth and differentiation of these groups of cells during the final stages of larval development and pupation.

There are nine pairs of disks for body parts that appear in pairs in the adult (eyes, legs, etc.), plus one genital disk (Figure 22-31b). Other larval cells do not contribute to the adult. The cell grouping in imaginal disks is established early in larval development, and cells in the disks grow as undifferentiated cell clusters in the abdominal cavity of the larva. During pupation, when the larva is transformed into a fly, the disks differentiate into the specific body parts. The orderly set of molecular events that must accompany the development of a fly is now being successfully studied by examining mutations that upset the orderly sequence.

Homeotic and Segment Mutations Mutations that cause abnormal development in *D. melanogaster* body parts or that are lethal early in embryogenesis were first detected many years ago, and were subsequently mapped to individual genetic loci. One group of mutations (at least 15 in number) affects proper segmentation of the larva (there are 12 larval segments). Many of the segmentation mutations are lethal in the larval stages. A number of other mutations cause abnormal development that is expressed when imaginal disk cells differentiate.

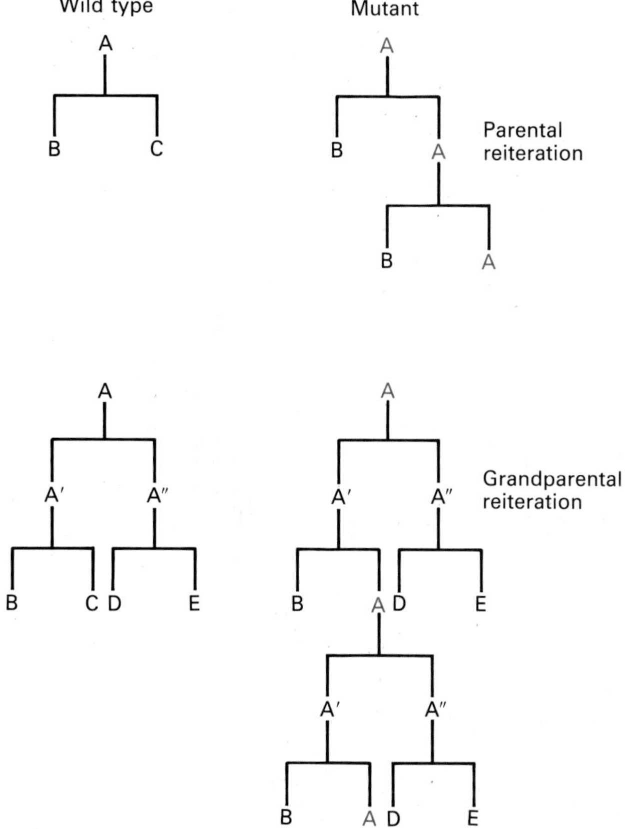

Figure 22-30 How mutations can affect two types of cell lineages in *C. elegans*. *Parental reiteration:* Cell type A normally produces two different cells, B and C. A cell from a mutant animal *(color)* can produce one new cell type B and one original cell type A that continues this cycle. *Grandparental reiteration:* Cell type A normally divides once (to give A′ and A″) and then differentiates after one more division to four cell types (B, C, D, and E). A mutation can cause one of the progeny of the second division to behave like its grandparent. This mutation requires two cell divisions to become apparent.

Homeotic mutations alter the developmental program of the cells of particular imaginal disks, usually by causing a disk to exchange one developmental pathway for another. One group of such mutations is located in the *Antennapedia* complex of genes. In the absence of one or more of the normal gene products of this complex, leg parts—instead of an antenna—grow out of the cells near the eye during the development of the eye disk. The leg is normal in form; only its location is abnormal.

Another well-studied group of homeotic mutations lies in a chromosome III region called the *bithorax* region. Flies normally have three thoracic segments—a pro-, a meso-, and a metathorax. One pair of wings is present on the second (mesothoracic) segment (abbreviated T2). Smaller rudimentary winglike structures called *halteres* are found on the metathorax, the third thoracic segment (T3); these structures balance the fly in flight. Mutations in the *bithorax* genetic locus cause T3 to develop like T2. The many different known mutations in this region affect the development of T3 with varying degrees of severity. Flies bred to have three of the mutations that are individually less severe (*bx*, *abx*, and *pbx*) can actually develop a complete extra pair of wings where halteres would normally be (Figure 22-32). The *bithorax* mutation with the greatest effect is called *Ubx* (for ultrabithorax); this mutation is lethal when the fly is homozygous for it.

Antennapedia and *bithorax* mutations apparently block normal restraining functions that are necessary to prevent certain developmental events. For example, the cells near the eye in the eye disk can form either leg parts or antennae, but these cells are normally prevented from forming legs. Similarly, the cells that normally develop into the third thoracic segment can develop into a structure like the second thoracic segment with wings, but are normally prevented from doing so by the action of the wild-type *bithorax* genes. These genes and others like them are important because their products evidently direct cells into particular developmental pathways. Such genes and gene products somehow control the final shape of the animal.

Molecular Analysis of Homeotic Loci The nature of the proteins responsible for the developmental choices controlled by *bithorax* and *Antennapedia* in *D. melanogaster* remains unknown, but regions of normal and mutant chromosomes have been cloned. To pinpoint protein-coding regions, mRNAs extracted from cells have been hybridized to the cloned genomic fragments. The mRNAs complementary to two genetic loci—*bithorax* and a locus called *notch*, which participates in controlling the growth of neuroblasts—have been found to derive sequences from regions of the genome as far apart as 90 kb. For *bithorax* and *notch*, these findings probably indicate that the mRNAs are spliced together from very long transcription units.

If the homeotic genes direct early development, their mRNAs should be found in early embryos. The synthesis of mRNA from at least one homeotic locus, the *Antennapedia* locus, occurs early in embryogenesis in a limited number of cells. Another gene called *fushi tarazu*, abbreviated *ftz*, plays a role in the proper segmentation of *Drosophila* larvae. The mRNA complementary to this gene is localized in a narrow band of cells that borders each larval segment. (This is also true for the *engrailed* locus mRNA, which is discussed in the next section.) The gene products that determine how developmental programs are established may well act only briefly in certain cell types: the *ftz* product, for example, may be present only during the time that larval tissues become segmented.

Although the different homeotic genes perform diverse functions in helping to shape the body, a striking molecu-

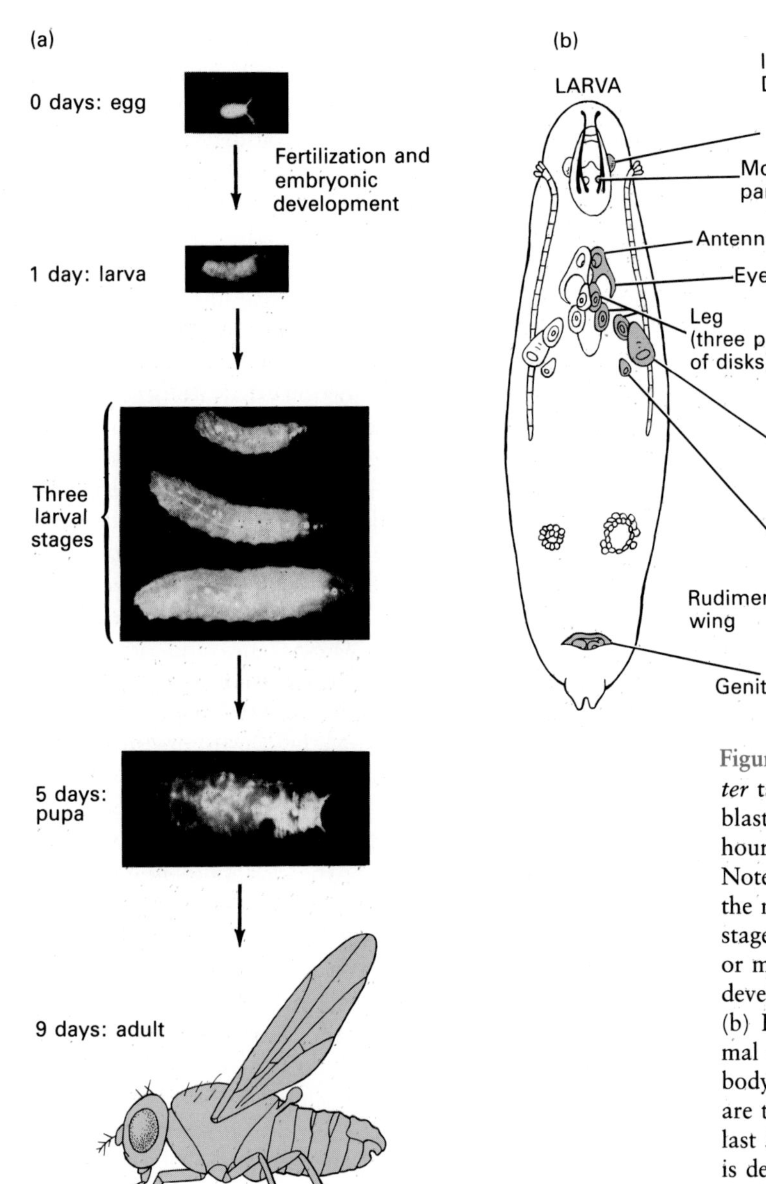

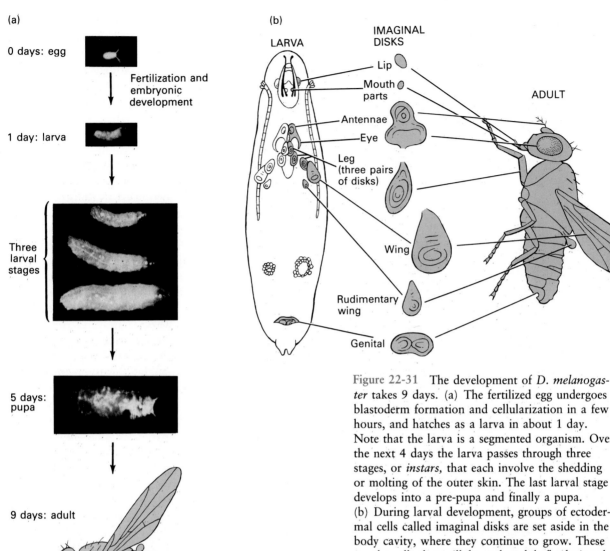

Figure 22-31 The development of *D. melanogaster* takes 9 days. (a) The fertilized egg undergoes blastoderm formation and cellularization in a few hours, and hatches as a larva in about 1 day. Note that the larva is a segmented organism. Over the next 4 days the larva passes through three stages, or *instars*, that each involve the shedding or molting of the outer skin. The last larval stage develops into a pre-pupa and finally a pupa. (b) During larval development, groups of ectodermal cells called imaginal disks are set aside in the body cavity, where they continue to grow. These are the cells that will form the adult fly during the last 3 to 4 days of pupation. The larval body itself is destroyed in the process. Imaginal disks in particular locations in the larvae give rise to specific adult structures. *Photographs from M. W. Strickberger, 1985,* Genetics, *3d ed., Macmillan, p. 38. Reprinted with permission of Macmillan Publishing Company. Part (b) larva and imaginal disks after same source and J. W. Fristrom, R. Raikow, W. Petri, and D. Stewart, 1969, in* Park City Symposium on Problems in Biology, *E. W. Hanly, ed., University of Utah Press, p. 381.*

lar link between these genes has recently been uncovered. A consensus sequence of 180 nucleotides encoding 60 amino acids appears in the *Antennapedia, bithorax,* and *ftz* loci. This sequence, which has been called the *homeobox,* is very highly conserved: 80 to 90 percent of the corresponding amino acids are identical in the three loci.

Not only is the *homeobox* conserved in fly homeotic genes, but it is also found in frogs and in mammals. A frog *homeobox* that has been cloned and sequenced shows a 75 percent amino acid homology with the fly *homeobox* (Figure 22-33). This striking degree of conservation is even higher than that found among most protein-coding genes. The mRNA containing the *homeobox* sequence has been detected in early frog embryos, which points to its possible role in early development.

Finally, a further sequence similarity between the *homeobox* product and a yeast cell protein encoded at the MAT locus (which, recall, controls mating type) has dramatic implications about the mode of action of proteins containing the amino acids encoded by the *homeobox.* The yeast protein is thought to bind to DNA as a regulatory protein; perhaps the proteins encoded by the homeotic genes are also DNA-binding regulatory pro-

(a)

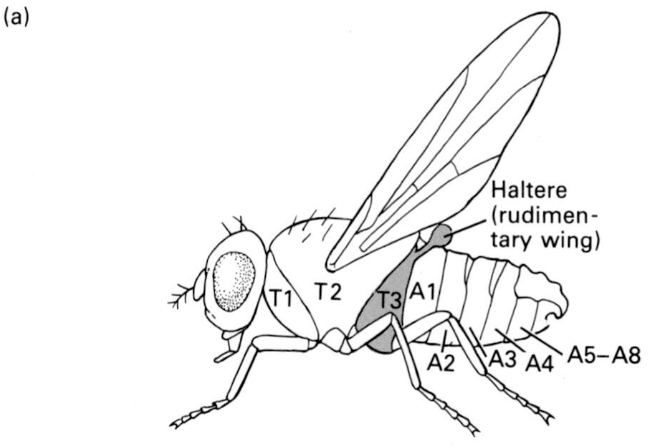

(b)

(c)

Figure 22-32 *Bithorax* mutations in *D. melanogaster.* (a) A drawing of a normal fly. Notice the appearance of the third thoracic segment (T3) and its rudimentary wing. A1, A2, etc. identify the abdominal segments. (b) A normal fly with a single set of wings. (c) A fly that is homozygous for three mutant alleles (*bx, abx,* and *pbx*) that collectively transform T3 into a structure similar to T2. A second, fully developed set of wings emerges from this transformed segment. [See E. B. Lewis, 1978, *Nature* **276**:565.] *Photographs courtesy of E. B. Lewis. Reprinted by permission from* Nature. *Copyright 1978 Macmillan Journals Limited.*

teins. If all of these related sequences in such diverse organisms turn out to play a similar crucial role in development (i.e., in regulating genes directly by DNA binding), a major advance in developmental biology will have been achieved.

The Growth of Groups of Related Cells Is under Genetic Control

Genetic studies of *D. melanogaster* wing development have established that the growth of groups of related cells that form a single recognizable structure in the adult animal is under genetic control. As we have mentioned, such a group of related cells is referred to as a *compartment.* For example, although the cells of the wing originate from a single imaginal disk, the wing is divided into an anterior and a posterior half. Normally the two portions of the wing develop from separate pools of cells within the wing disk—that is, from separate compartments that derive from distinct cell lineages.

Cells in the two halves can be distinguished after somatic recombination induced by x-rays. If a fly is heterozygous for pigment production, mitotic recombination induced early in development by x-rays (as in Figure 22-19) can create clones of cells that are homozygous for a recessive pigmentation trait. When this occurs, the clonal distribution of the individual recombinant cells can be followed in the wings of adults. Patches of cells expressing either a maternal or a paternal allele can be observed.

Such patches exist in all conceivable configurations in both the anterior and the posterior halves of *D. melanogaster* wings (Figure 22-34). This distribution shows that the wing cells of *D. melanogaster* do not follow strict lineages like the larval cells of *C. elegans.* However, patches that lie near the midline of the wing do not cross the line. Even in the case of recombinants in which a faster-growing cell type makes up a very large portion of the total wing, the patches formed by these cells do not cross the midline (Figure 22-34b). Thus, there are two cell compartments in the wing.

Genetic control of the growth in these cell compartments is further illustrated by flies bearing the *engrailed* mutation. Although in normal animals cells never grow across the anterior-posterior boundary of the wing, animals with the *engrailed* mutation exhibit clones of cells that arise in the posterior portion and invade the anterior portion. Because recombinant cell clones arising in the anterior half of the wing respect the boundary, the wild-type product of the *engrailed* gene must stop posterior cells from dividing when they reach the anterior boundary.

The same anterior/posterior compartments exist not only for wings but also for legs, thoracic segments, and other body parts: animals bearing the *engrailed* mutation show cell invasion across the anterior-posterior line in

		1	2	3	4	5	6	7	8	9	10	11	12	13	14	15	16	17	18	19	20
Fly	*Antp*	Arg	Lys	Arg	Gly	Arg	Gln	Thr	Tyr	Thr	Arg	Tyr	Gln	Thr	Leu	Glu	Leu	Glu	Lys	Glu	Phe
	ftz	Ser	Lys	Arg	Thr	Arg	Gln	Thr	Tyr	Thr	Arg	Tyr	Gln	Thr	Leu	Glu	Leu	Glu	Lys	Glu	Phe
	Ubx	Arg	Arg	Arg	Gly	Arg	Gln	Thr	Tyr	Thr	Arg	Tyr	Gln	Thr	Leu	Glu	Leu	Glu	Lys	Glu	Phe
Frog		Arg	Arg	Arg	Gly	Arg	Gln	Ile	Tyr	Ser	Arg	Tyr	Gln	Thr	Leu	Glu	Leu	Glu	Lys	Glu	Phe

		21	22	23	24	25	26	27	28	29	30	31	32	33	34	35	36	37	38	39	40
Fly	*Antp*	His	Phe	Asn	Arg	Tyr	Leu	Thr	Arg	Arg	Arg	Arg	Ile	Glu	Ile	Ala	His	Ala	Leu	Cys	Leu
	ftz	His	Phe	Asn	Arg	Tyr	Ile	Thr	Arg	Arg	Arg	Arg	Ile	Asp	Ile	Ala	Asn	Ala	Leu	Ser	Leu
	Ubx	His	Thr	Asn	His	Tyr	Leu	Thr	Arg	Arg	Arg	Arg	Ile	Glu	Met	Ala	Tyr	Ala	Leu	Cys	Leu
Frog		Arg	Phe	Asn	Arg	Tyr	Leu	Thr	Arg	Arg	Arg	Arg	Ile	Glu	Ile	Ala	Asn	Ala	Leu	Cys	Leu

		41	42	43	44	45	46	47	48	49	50	51	52	53	54	55	56	57	58	59	60
Fly	*Antp*	Thr	Glu	Arg	Gln	Ile	Lys	Ile	Trp	Phe	Gln	Asn	Arg	Arg	Met	Lys	Trp	Lys	Lys	Glu	Asn
	ftz	Ser	Glu	Arg	Gln	Ile	Lys	Ile	Trp	Phe	Gln	Asn	Arg	Arg	Met	Lys	Ser	Lys	Lys	Asp	Arg
	Ubx	Thr	Glu	Arg	Gln	Ile	Glu	Ile	Trp	Phe	Gln	Asn	Arg	Arg	Met	Lys	Leu	Lys	Lys	Glu	Ile
Frog		Thr	Glu	Arg	Gln	Ile	Lys	Ile	Trp	Phe	Gln	Asn	Arg	Arg	Met	Lys	Trp	Lys	Lys	Glu	Arg

Figure 22-33 The amino acid sequence encoded by the *homeobox* of *D. melanogaster* compared with the *homeobox* sequence from a frog gene. The DNA sequences of three *Drosophila* homeotic genes—*Antennapedia (Antp)*, *fushi tarazu (ftz)*, and *Ultrabithorax* (*Ubx*, from the *bithorax* locus)—were determined, and the amino acids they encode were deduced from the genetic code. The *homeobox* encodes 60 amino acids that are very similar for the three genes. (The amino acids that are identical in all three genes are shaded in color.) A frog DNA clone that cross-hybridized to the *homeobox* sequence was selected and sequenced, and its deduced amino acid sequence is also highly conserved. (The amino acids that are identical in all four gene loci are in color type.) [See W. McGinnis et al., 1984, *Cell* 37:403; and A. E. Carrasco et al., 1984, *Cell* 37:409.]

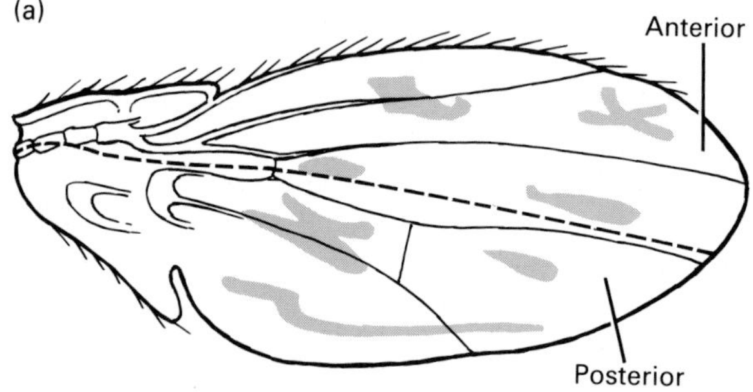

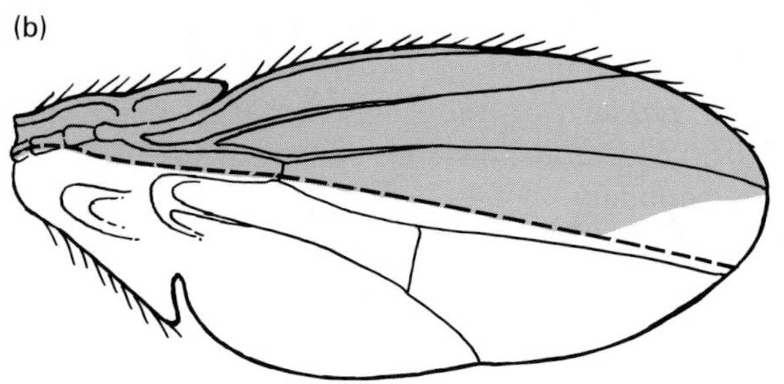

Figure 22-34 Cell lineages in the wings of *D. melanogaster*. Animals were irradiated as larvae to induce somatic recombinations in imaginal disk cells. The recombinations led to changes in wing tissue pigmentation. The wings of two flies are diagrammed here. (a) In this wing, patches of pigmented cells occur in both halves (anterior and posterior) of the wing. The patches can be of any shape and size, and several of them appear to be joined. Each patch probably originated from one recombinant cell. Here there is no restriction on the locations of pigmented versus nonpigmented cells during wing development, but none of the patches crosses the anterior-posterior dividing line. (b) This wing illustrates a case in which pigmented cells grow faster than nonpigmented cells. The pigmented cells have almost filled the anterior half of the wing, but again they have not crossed into the posterior half. The anterior and posterior portions of the wing therefore are separate compartments. [See A. Garcia-Bellido, P. A. Lawrence, and G. Morata, 1979, *Sci. Am.* **241**(1):106.] *From Garcia-Bellido. Copyright 1979 by Scientific American Inc.*

these body parts as well. Therefore, cell compartment sizes and boundaries in many parts of the animal can be affected by a single gene. At present nothing is known about the gene product(s) that can stop cells from growing beyond a defined limit. However, the *engrailed* gene has been cloned, and the cloned DNA has been used to establish that mRNA from this gene is present in each segment of early larvae when segmentation first begins (Figure 22-35). How a single gene product limits cell growth so that a compartment is filled but not overfilled is a most interesting problem.

Genetic and Molecular Studies of Invertebrates Will Illuminate Central Developmental Questions

The enormous benefit of studying the simpler animals like *D. melanogaster* and *C. elegans* is that both genetically pure normal strains and mutants of these strains exist. The existence of both strains makes it possible to chart the effects of individual genes on morphology. At least two kinds of problems that are common to all developing systems should be resolved by further studies with these simple organisms.

First, how many genes contribute to a function such as construction of an antenna, and how is the network of such genes coordinated? Molecular analysis of the mRNAs deriving from the gene complexes of *Antennapedia* and *bithorax* is already underway, and an explanation of how transcription of these products is controlled in various cell types during development should be available soon.

A second and probably a more difficult problem is to ascertain how the protein products of such important developmental genes exert their effects. If these products are regulatory elements that affect other single genes or a limited number of other genes, then we need to know what the affected genes are and what they do. If the affected genes are involved in differentiation—that is, if they encode known proteins (e.g., cuticle or bristle proteins)—then a greater depth of understanding will have been achieved. However, if, say, the homeotic mutations are found to affect only more general functions such as cell growth and division, which we do not yet understand well, then the solution of development will await a deeper understanding of cell growth itself.

Even if basic developmental issues are not resolved soon with simple organisms, these organisms have at least presented us with the opportunity to identify impor-

(a)

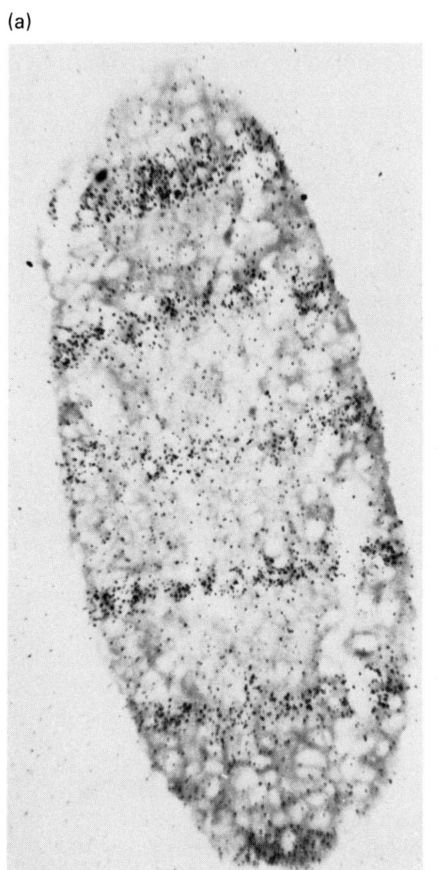

(b)

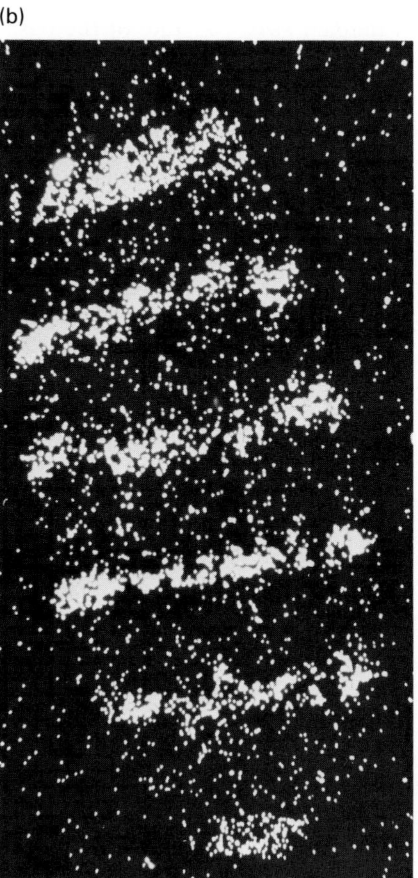

Figure 22-35 The location of RNA transcripts complementary to DNA from the *engrailed* locus in *D. melanogaster* embryos. ^{3}H-labeled cloned DNA is prepared and hybridized to the mRNA of fixed, sectioned tissue of a 6-h embryo. Exposure to x-ray film produces autoradiographs that can be printed in either of two ways: (a) with silver grains appearing as black dots or (b) with the image enhanced by reversing it via dark-field optics. The *engrailed* locus specifies the anterior-posterior boundaries of compartments in all structures of the larva and of the adult fly; note that the transcripts appear in the posterior portion of each larval segment. This accounts for the banded appearance of these photographs. [See A. Fjose, W. J. McGinnis, and W. J. Gehring, 1985, *Nature* **313**:284.] *Courtesy of W. J. Gehring. Reprinted by permission from* Nature. *Copyright 1978 Macmillan Journals Limited.*

tant developmental genes. We have not yet reached this point in our study of mammals.

Maintenance of Adult Organisms: Stem Cells

Most structures and tissues in newborn animals simply enlarge without undergoing many other types of developmental changes. The important exceptions to this rule include the making of extensive neural connections after birth in vertebrate nervous systems. However, in most animals one important lifelong event—cell regeneration—demands that some cells continue to function in adults as most cells do in early embryos. For example, certain tissues in adult vertebrates—such as skin, blood, and intestinal epithelium—are composed of short-lived cells. Many of these cells do not last longer than 2 days and must be continually replaced. Adult animals therefore share with embryos the need to control the population of the regenerative or *stem* cells in these tissues.

Stem cells can either continue to grow, in which case they divide *symmetrically* to produce two identical stem cells, or they can divide *asymmetrically* to produce one determined or fully differentiated cell and one new stem cell. The size of the stem cell population must be carefully maintained so that the correct number of differentiated cells are generated to provide the necessary differentiated functions within a total cell population of defined size. The determined cell from an asymmetric division may undergo several divisions on the way to reaching its final differentiated form. Therefore the final number of differentiated cells produced in a tissue is an outcome of the control of the stem cell population and the control of the number of divisions that determined cells undergo before differentiating fully (Figure 22-36).

Both Unipotent and Pluripotent Stem Cells Exist in the Adult

Stem cells such as those in the basal layer of the epidermis are *unipotent*. They give rise to one end-stage differentiated cell, a cornified epidermal cell. The cornified cell is a thin, nonliving plate composed mainly of a tough fibrous protein termed keratin; keratin serves as a barrier to the outside world. On the way to becoming a keratinized cell, the differentiating skin cell goes through several stages as it migrates through different epidermal layers (Figure 22-37). The keratins generated by the cell during these stages are encoded by a large multigene family (Chapter 19).

Various members of this class of proteins appear at different times in different types of skin. The number of divisions in the basal cell layer can determine the thickness of the keratin layer that is produced in different regions of the body. Compare the sole of the foot with the eyelid, for example: the cornified layer on the foot is obviously many times thicker than that on the eyelid. The division rate of the basal layer must be kept in balance with the rate of maturation and loss of the differentiated skin cells

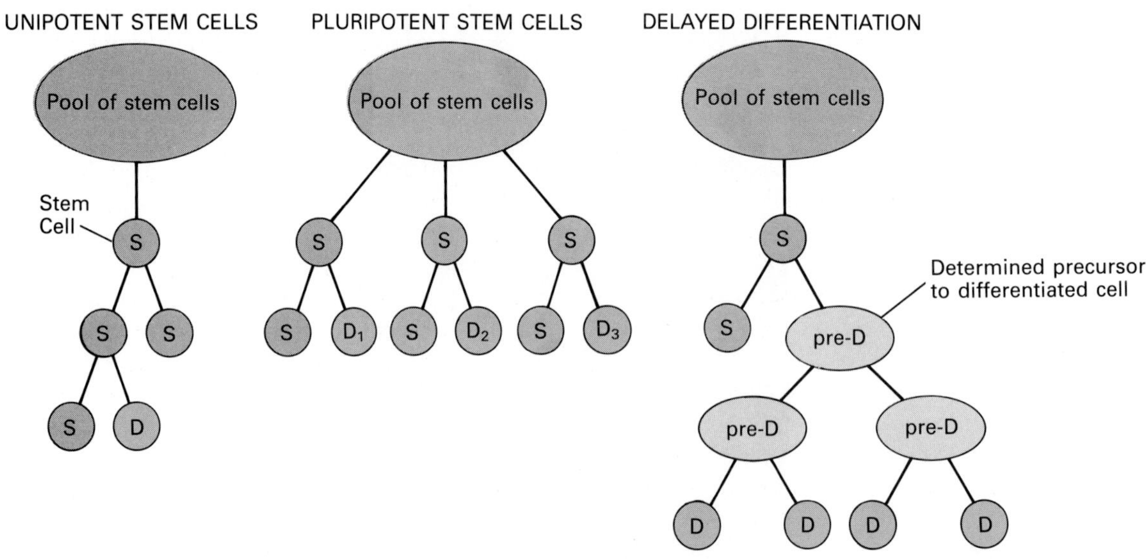

Figure 22-36 The production of differentiated cells from stem cells. Unipotent stem cells produce a single type of differentiated cell, whereas pluripotent stem cells may produce two or more types of differentiated cells. Delayed differentiation is a mechanism by which the production of differentiated cells is controlled. All stem cells remain part of the stem cell pool.

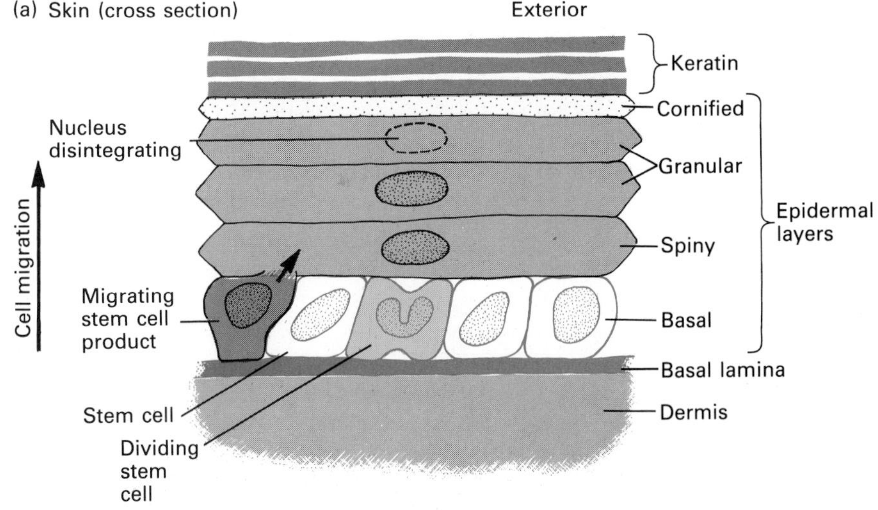

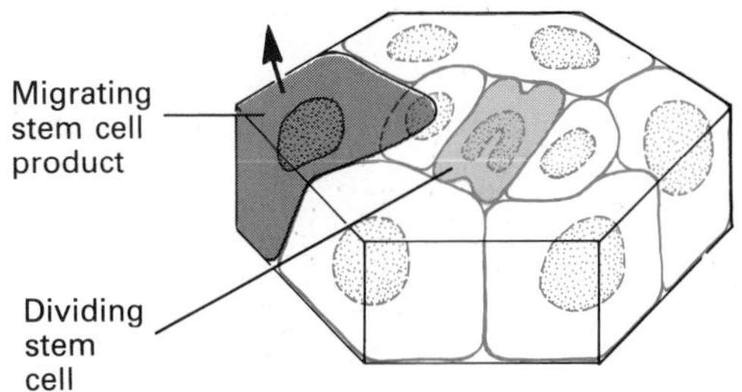

Figure 22-37 Mammalian epidermis. (a) The stem cells *(color)*, which constitute the basal layer, resupply the column of cells above them. One new cell *(gray)* is beginning to migrate upward and enter the pathway of differentiation through the spiny and the granular layers. The cell will finally emerge on the skin surface as a dried plate of keratin that will eventually be lost. The nucleus is degraded in the final stages of keratinization. Note that one basal cell is dividing. (b) In mammalian epidermis, groups of cells (including stem cells and their daughters) that replace lost cells are called epidermal proliferative units. The group diagrammed here contains nine basal cells. The dividing cell will replace the migrating product and thereby maintain the same number of cells in the unit. *After C. S. Potten, 1978, in* Stem Cells and Tissue Homeostasis, *B. I. Lord and R. J. Cole, eds., Cambridge University Press, p. 328.*

in different parts of the body. In *psoriasis,* a dermatologic disorder, the basal cell is not properly regulated and proliferation is greatly stimulated. This condition forces such a rapid turnover of the epidermis at the surface that complete keratinization is not achieved.

The hematopoietic (blood-forming) stem cells of the bone marrow are a prominent example of cells that are *pluripotent*. These cells can give rise to two or more independent types of differentiated cells (Figure 22-38). The developmental potential of bone marrow cells can be demonstrated by injecting cells of one animal into a second animal whose own blood cell production has been obliterated by heavy irradiation. To permit cytologic identification, the donor cells can be taken from an animal with a genetically distinctive cell marker such as an unusual chromosome.

The newly introduced bone marrow cells grow into colonies in the spleen of the host animal. The number of colonies is directly proportional to the number of injected cells, a strong indication that each colony arises from a single injected cell. The colonies found in the spleen contain a mixture of differentiating red blood cells and white blood cells derived from the donor animal. Because each colony is the product of a single stem cell, the stem cells must be pluripotent cells that give rise to a variety of descendants upon growth and differentiation (Figure 22-39).

Differentiation of the committed products of a stem cell division can occur within a few generations or can be suspended for considerable periods of time. For the differentiation of bone marrow cells, specific protein factors trigger the growth, and perhaps also control the growth rate, of precursors to the final differentiated cells. A sizable and growing number of these factors (e.g., erythropoeitin, T cell growth factors, granulocyte-stimulating factors, etc.) have been purified and the protein sequence determined from cloning the DNA encoding the factors. It is clear that many different factors exist; these are detected by stimulation of growth and differentiation of specific blood cells in culture.

The interaction of the protein factors with the local environment must play a role in determining the relative numbers of differentiated blood cells of each type. For example, the usual ratio of red blood cells to white blood cells produced by the spleen is about 3 : 1, but the ratio is closer to 1 : 1 in the bone marrow. The growth of red blood cell precursors is therefore favored in the spleen. The overall balance of red blood cells and white blood cells in the circulation, where red cells outnumber white cells 1000 to 1, is achieved by differential cell production plus the widely divergent turnover time of the cells. Red blood cells last an average of 120 days, whereas white blood cells (of the myelocytic series) last only a day or so.

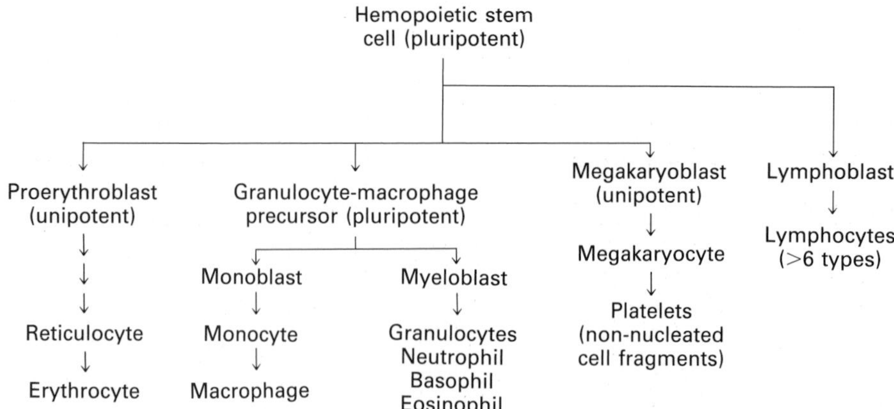

Figure 22-38 More than a dozen blood cell types arise from the pluripotent hematopoietic stem cells of bone marrow. Precursors to the lymphocytic series represent the earliest step of differentiation. These cells are discussed in detail in Chapter 24. Cultured single cells can give rise to colonies containing erythrocytic, granulocytic, and megakaryocytic cells. Proerythroblasts and megakaryoblasts are determined cells yielding only one final product. Platelets, which come from megakaryocytes, are not whole cells, but rather membrane-bound portions of cytoplasm shed when a megakaryocyte disrupts. Platelets play a critical role in blood clotting; they stop the bleeding from the cut edges of capillaries. Note that when the granulocyte-macrophage pathway is chosen, the first division of the hematopoietic stem cell yields a stem cell that is more limited than its parent but one that is still pluripotent. Monoblasts can probably migrate to other tissues and may differentiate into macrophages, mast cells, and other cell types. Myeloblasts differentiate into several types of cells that circulate in the blood but that can enter tissues in times of inflammation (infections, burns, etc.). [See D. Metcalf, 1984, *The Hemopoietic Colony-Stimulating Factors,* Elsevier/North-Holland.]

Reflections on Some Unsolved Problems in Development

In any organism, the amazing events of animal development derive from the interplay between the behavior of individual cells (cell contact, movement, and division) and the regulated expression of the genetic programs within the cells. An understanding of the molecular mechanisms of gene control, particularly transcriptional control, seems close at hand for a number of different types of genes, and this must be considered one of the most important goals in the current study of development. The molecular bases for cellular architecture and for cell movement and division are being described in ever-increasing detail. These advances represent great progress. However, much further study of the activities of the structural proteins of the cell will be required before we will be able to understand how regulated patterns of cell movement and growth can be coordinated with differential gene control to assure the precise construction of the animal body. In this section we wish to formulate some of the outstanding unsolved problems in developmental biology in terms of molecular cell biology.

Is There a Molecular Difference between Gene Control during Determination and Gene Control during Differentiation?

The traditional view of embryologists has been that the earlier events in development—those associated with determination—differ in kind from later events—those associated with differentiation. Certainly the events that determine the course of a cell's developmental potential occur *before* the events that cause it to enter a final differentiated pathway. Is there also a qualitative molecular difference in the activation of early versus late sets of genes, or is there simply a sequential relationship between regulatory decisions that are all alike at the molecular level?

If the gene control events of determination are different

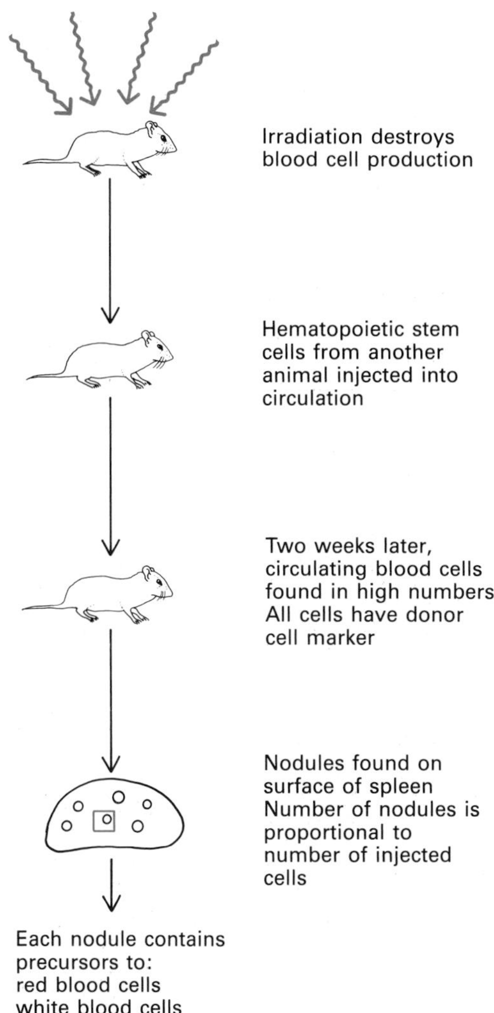
Irradiation destroys blood cell production
Hematopoietic stem cells from another animal injected into circulation
Two weeks later, circulating blood cells found in high numbers
All cells have donor cell marker
Nodules found on surface of spleen
Number of nodules is proportional to number of injected cells
Each nodule contains precursors to:
red blood cells
white blood cells
megakaryocytes

Figure 22-39 The experiment that established the pluripotent nature of the hematopoietic stem cell. The nodules (cell colonies) in the spleen are each derived from a single cell because the number of nodules is directly proportional to the number of cells injected. This proportionality would not be true if two or more cell types had to come together to produce the different cell types found in the nodules. Thus the one founder cell for each nodule is a pluripotent stem cell.

in kind from those of differentiation, what differences would be found at the molecular level? One possibility is that determination could involve DNA rearrangements that are inherited in determined cells, whereas differentiation could occur through gene activation without rearrangement. Inversion, gene conversion, specific deletion, and amplification are all known to occur in the DNA of different organisms at different times. Do such events generally underlie determination in the embryogenesis of animals?

The nuclear transplantation experiments resulting in normal tadpole development (see Figure 22-16) argue against large-scale deletions or irreversible DNA rearrangements. The high success rate for the regeneration of whole plants from single leaf cells also rules out irreversible chromosomal changes during plant development. Furthermore, the DNA for many characteristic protein products of differentiated tissue is not rearranged. In spite of this evidence, the possibility of DNA rearrangements in key genes in specific tissues will be a live issue until the idea can be safely discarded when it is proved false for enough developmentally important genes.

If determination and differentiation do not occur through structural rearrangements in the DNA, they may depend on a cascade of genetic regulatory proteins. Regulatory proteins in the egg may trigger the synthesis of additional rounds of regulatory proteins in the zygote. All such proteins might have similar molecular activities. Those acting first in the program would produce determination, and those acting later would result in differentiation. In eukaryotes, regulatory proteins that bind to DNA might increase or decrease transcription; or they might change the superhelicity of the DNA, its nucleosome packing, or its state of methylation (Chapter 12).

It seems likely that some changes in chromatin structure—for example, an untwisting of the DNA at a local site—occur to prepare a gene for transcription some time before it is actively transcribed. If such a period between untwisting and transcription were lengthy, the chromatin change itself could constitute a "determinative" event. Effectors of transcription such as hormones, cyclic AMP, or cell-to-cell contacts might act on determined cells *only after a delay;* transcriptional activation would thus cause differentiation a certain interval of time after determination. Of course, if the cycle of regulatory events were compressed into one or a few cell generations, then the distinction between determination and differentiation would become difficult if not impossible to show in experimental and molecular terms.

It is not easy to obtain pure populations of normal precursor or stem cells that might be observed to undergo determination in a controlled manner. If such populations were available, investigators could detect any stepwise effects on the structure of the chromatin (i.e., first the effects caused by determination and then those due to differentiation). At present, we simply do not know which changes, if any, in chromatin structure should be assigned to determination and which to differentiation.

Aside from the difference in the timing of regulatory events, the only other proven difference between the events of determination and those of differentiation is a quantitative difference in the production of various proteins. Very few changes in the types of proteins synthesized have yet been identified early in embryogenesis, as cells go through determination. Late in the course of their development they make large amounts of a number of known cell-specific products (e.g., muscle proteins). It therefore may seem that the molecular changes of determination must be qualitatively different from those of differentiation, when in fact the differences are really only in the chronology of the molecular events and in the amounts of products synthesized.

The Mechanisms for Cell-to-Cell Signaling Must Be Solved

Another pressing molecular question remains unanswered. The "choice" by a cell of a developmental pathway or the "decision" by a cell to undergo final differentiation often seems to involve cell-to-cell contact. At the moment there is no clear understanding of how cell-cell interactions effect or alter gene control in the nucleus. But two general explanations have been proposed.

One explanation is that the formation of gap junctions could allow the passage (either regulated or free) of material (small peptides, oligonucleotides, and small charged molecules) from one cell to another; eventually the material would enter the nucleus of the receiving cell. Such a mechanism would allow cell-cell interaction to affect gene activity.

The second explanation for how cell-cell contacts could alter gene function is based on the possibility that specific protein receptors on two neighboring cells can interact directly or that a secreted substance from one cell, such as a hormone or a glycosaminoglycan, can occupy receptors on another interacting cell. If information were passed in this manner, the occupancy of a cell surface receptor would constitute a signal that would result in differential gene control.

We do not yet have a model to explain how protein contact at the cell surface could cause the control of *specific* genes. For example, it seems highly unlikely that genes are physically connected to the cell surface in such a way that transcription could be controlled directly by the occupancy of a receptor at the cell surface. However, the occupation of a specific cell surface receptor and the subsequent intracellular biochemical events are just beginning to be studied (Chapter 16). The phosphorylation of intracellular compounds, particularly the intracellular domains of receptors themselves, is known to be a consequence of ligand binding to cell surface receptors. Further studies may reveal how this phosphorylation affects specific intranuclear events.

Is There a Cytoskeletal Counterpart to the Differentiated State?

Up to now, our attention has been focused on molecular events at the level of DNA to explain how cell potentiality might be limited and how differentiation might be triggered. This general pathway of decisions is followed for each of the several thousand differentiated cells in the vertebrate body. Do cells that are determined but as yet undifferentiated have some structural or organizational feature that is shared among different cell lineages?

For example, common structures in the nucleus or in the nucleus plus the cytoskeleton might exist in all determined cells. Suppose that during determination all genes necessary for the production of any particular differentiated state become attached to the nuclear pore as a preliminary step in activation. The activation of all such attached genes might be brought about by one general activation signal that affects the nuclear structure around the pore. Different determined precursor cells might then be able to respond in *specific* ways to a *common* signal, such as an increase in cyclic AMP, a change in pH, or an influx of calcium ions—each of which is known to cause a different differentiated response in various different cell types. Of course, such hypotheses raise new questions. They simply push the problem of specificity (i.e., the "decision" by which specific genes get associated with the proper nuclear structure) back to an earlier developmental step.

There is some evidence for a regularity in nuclear structure. The large polytene chromosomes in the nuclei of the *Drosophila* salivary gland cells (Chapter 10) appear to be placed similarly in different cells. Three-dimensional reconstructions made from serial sections of nuclei indicate that the pathways of the chromosomes through space in each salivary gland nucleus may be very similar. This could mean that each gene that is active in the differentiated salivary gland cell is in the same position with respect to the available transcriptional machinery.

Even if the active genes are proved to be in identical orientations, it will still not be clear whether this intracellular structure is formed so that the genes can be activated, or whether the structure is the *result* of the activation of the genes. If the chromosomes in all nuclei have a fixed relationship to one another, then the activation of a specific set of genes could well be the basis for the consistent orientation of these genes and not the other way around.

Cell Growth and Differentiation Are Sometimes Reciprocally Related

Cell growth and differentiation have traditionally been considered reciprocal possibilities for cells. Early in embryogenesis when determination is occurring, cell growth is prolific, but later, when differentiation takes place, cells do not divide as often. In fact, final differentiation leads in many cases to cells that cannot grow, as is true of skin cells and the nonnucleated red blood cells of mammals. This reciprocal connection between growth and differentiation can also be observed in cultured cells. For example, myoblasts continue to grow in tissue culture, but when they are placed under proper conditions growth ceases and differentiation ensues. Determined red blood cell precursors also can be stimulated to start on the final pathway of differentiation to becoming red blood cells.

Two questions arise from these considerations. Are there key proteins whose synthesis must be interrupted to stop growth in determined cells so that they will differentiate? If so, must these same proteins be stimulated to keep other cells growing? Finally, if such growth-related proteins do exist, are they the same proteins that are responsible for growth "decisions" that result in adequately filled compartments during normal development? The relationship of these hypothetical growth-related proteins and proper development is a critical one in developmental biology. We shall return to it again in our consideration of abnormal growth in Chapter 23.

Summary

Diversity in biology is ultimately rooted in the differences between individual cell types and in the different behavior of the same cell type under different conditions. Biologists have used many models to study these changes in cell capacity, which are collectively referred to as cellular differentiation. Single-celled organisms undergo regular changes of state, as exemplified by sporulation in both bacteria and yeast. Many simple multicellular organisms (e.g., slime molds) also sporulate but have the added ability to form several different cell types and to produce macroscopic differentiated structures such as spore containers and stalks. These organisms offer a number of experimental advantages to researchers trying to discover the network of interacting genes responsible for microbial differentiation. However, most biologists who are interested in development study animals.

For centuries the following phenomenon has held center stage in the biological arena: fertilization of an egg by a single sperm results in a developmental program that leads to a new animal. This program has certain similarities for all animals, but it also has features that are distinctive for each kind of animal. The major stages of development are cleavage of the fertilized egg; gastrulation, a series of movements of large cell masses to establish the ectoderm, endoderm, and mesoderm—the primordial germ layers; and organogenesis, the maturation of specific tissues and organs. The original fertilized egg and the first few cleavage products are often each capable of forming whole animals. Soon, however, this totipotency is lost. Pluripotency, the capacity to give rise to more than one cell type, persists for a while longer, but eventually each cell undergoes determination, which imposes a particular fate on the cell's progeny (i.e., they must become a specific cell type). Finally, each determined cell carries out the steps of differentiation, and its progeny acquire the characteristics of the final defined state.

Recently tissue-specific transcriptional control has been shown to underlie many decisions of differentiation. Genes that play key decision-making roles in insect and nematode development have been discovered. For example, known *D. melanogaster* genes control the formation of segments in larvae; other genes cause a group of cells to form thoracic or head structures as opposed to abdominal or leg segments. These latter genes are called homeotic genes. The products of some homeotic genes may be DNA-binding proteins; several are known to include a 60 amino acid sequence encoded by a highly conserved gene region called the *homeobox*. The conserved polypeptide region may function as the DNA-binding site.

How signals are transmitted among embryonic cells and how differentiation is triggered by cell-to-cell contacts remain largely unknown. Even though we have achieved some understanding of tissue-specific gene control, the mechanisms by which gene products construct forms of defined sizes and shapes are still a great mystery. The major questions about development have been formulated for many years; optimists now believe that in a few more decades (at most) these questions will be answered.

References

General and Historical References

BROWDER, L. W. 1984. *Developmental Biology,* 2d ed. Saunders.

DAVIDSON, E. H. 1977. *Gene Activity in Early Development,* 2d ed. Academic Press.

GEHRING, W. J., ed. 1979. *Genetic Mosaics and Cell Differentiation.* Springer-Verlag.

GURDON, J. B. 1974. *The Control of Gene Expression in Animal Development.* Harvard University Press.

HAECKEL, E. 1879. *The Evolution of Man.* London

LE DOUARIN, N., ed. 1979. *Cell Lineage, Stem Cells and Cell Determination.* Elsevier/North-Holland.

Molecular Biology of Development. 1985. Cold Spring Harbor Symp. on Quant. Biol., vol. 50.

POSTE, G., and G. L. NICHOLSON, eds. 1977. *The Cell Surface in Animal Embryogenesis and Development.* Elsevier/North-Holland.

RAFF, R. A., and T. C. KAUFMAN, eds. 1983. *Embryos, Genes, and Evolution.* Macmillan.

SPEMAN, H. 1938. *Embryonic Development and Induction.* Yale University Press.

URSPRUNG, H., and R. NOTHINGER. 1972. *The Biology of Imaginal Disks.* Springer-Verlag.

WESSELLS, N. K. 1977. *Tissue Interactions and Development.* Benjamin-Cummings.

Differentiation in Unicellular and Simple Multicellular Organisms

BARKLIS, E., and H. F. LODISH. 1983. Regulation of *Dictyostelium discoideum* mRNAs specific for prespore or prestalk cells. *Cell* 32:1139–1148.

CHUNG, S., S. M. LANDFEAT, D. D. BLUMBERG, N. S. COHEN, and H. F. LODISH. 1981. Synthesis and stability of developmentally regulated *Dictyostelium* mRNAs are affected by cell-cell contact and cAMP. *Cell* **24**:785–797.

ESPOSITO, R. E., and S. KLAPHOLZ. 1981. Meiosis and ascospore development. In *The Molecular Biology of the Yeast Saccharomyces*, J. N. Strathern, E. W. Jones, and J. R. Broach, eds. Cold Spring Harbor Laboratory.

JOHNSON, W. C., C. P. MORAN JR., and R. LOSICK. 1983. Two RNA polymerase sigma factors from *Bacillus subtilis* discriminate between over-lapping promoters for a developmentally regulated gene. *Nature* **302**:800–804.

KHESSIN, R. 1981. Conservatism in slime mold development. *Cell* **27**:241–243.

LOOMIS, W. F., ed. 1982. *The Development of Dictyostelium discoideum.* Academic Press.

LOSICK, R., and L. SHAPIR, eds. 1984. *Microbial Development.* Cold Spring Harbor Laboratory, Monograph 16.

LOSICK, R., and J. PERO. 1981. Cascades of sigma factors. *Cell* **25**:582–584.

PARISH, J. H., ed. 1979. *Developmental Biology of Prokaryotes: Studies in Microbiology,* vol. 1. Berkeley: University of California Press.

SHAPIRO, L. 1976. Differentiation in the *Caulobacter* cell cycle. *Annu. Rev. Microbiol.* **30**:377–407.

Classical Developmental Studies: Fertilization, Cleavage, Gastrulation, and Determination of Gametes

BEDFORD, J. M. 1972. An electron microscopic study of sperm penetration into the rabbit egg after natural mating. *Am. J. Anat.* **133**:213–254.

DAVIDSON, E. H., B. R. HOUGH-EVANS, and R. J. BRITTEN. 1982. Molecular biology of the sea urchin embryo. *Science* **217**:17–26.

EDDY, E. M. 1975. Germ plasm and the differentiation of the germ cell line. *Int. Rev. Cytol.* **43**:229–280.

EPEL, D. 1977. The program of fertilization. *Sci. Am.* **237**(5):128–138.

ILLMENSEE, K., A. P. MAHOWALD, and M. R. LOOMIS. 1976. The ontogeny of germ plasm during oogenesis in *Drosophila. Devel. Biol.* **49**:40–65.

LE DOUARIN, N., and C. LE LIEVRE. 1978. Cell migrations during embryogenesis. In *Birth Defects: Proceedings of the Fifth International Conference on Birth Defects,* J. W. Littlefield and J. De Grouchy, eds. Excerpta Medica.

LEVEY, I. L., G. B. STULL, and R. L. BRINSTER. 1978. Poly(A) and synthesis of polyadenylated RNA in the preimplantation mouse embryo. *Devel. Biol.* **64**:140–148.

LONGO, F. J., and E. ANDERSON. 1969. Cytologic events leading to the formation of the two-cell stage in the rabbit: association of the maternally and paternally derived genomes. *J. Ultrastruct. Res.* **29**:86–118.

LYNN, D. A., L. M. ANGERER, A. M. BRUSKIN, W. H. KLEIN, and R. C. ANGERER. 1983. Localization of a family of mRNAs in a single cell type and its precursors in sea urchin embryos. *Proc. Nat'l Acad. Sci. USA* **80**:2656–2660.

NEWPORT, J., and M. KIRSCHNER. 1982. A major developmental transition in early *Xenopus* embryos, I: Characterization and timing of cellular changes at the midblastula stage. *Cell* **30**:675–686.

TUFARO, F., and B. P. BRANDHORST. 1979. Similarity of proteins synthesized by isolated blastomeres of early sea urchin embryos. *Devel. Biol.* **72**:390–397.

TURNER, F. R., and A. P. MAHOWALD. 1976. Scanning electron microscopy of *Drosophila* embryogenesis, I: The structure of the egg envelopes and the formation of the cellular blastoderm. *Devel. Biol.* **50**:95–108.

VACQUIER, V. D. 1975. The isolation of intact cortical granules from sea urchin eggs: calcium ions trigger granule discharge. *Devel. Biol.* **43**:62–74.

WARING, G. I., C. D. ALLIS, and A. P. MAHOWALD. 1978. Isolation of polar granules and the identification of polar granule–specific protein. *Devel. Biol.* **66**:197–206.

Cell Determination: Developmental Potential

BRUN, R. B. 1978. Developmental capacities of *Xenopus* eggs, provided with erythrocyte or erythroblast nuclei from adults. *Devel. Biol.* **65**:271–284.

DI BERARDINO, M. A. 1980. Genetic stability and modulation of metazoan nuclei transplanted into eggs and oocytes. *Differentiation* **17**:17–30.

GARDNER, R. L. 1979. The relationship between cell lineage and differentiation in the early mouse embryo. In *Genetic Mosaics and Cell Differentiation,* W. J. Gehring, ed. Springer-Verlag.

GURDON, J. B., R. A. LASKEY, E. M. DE ROBERTIS, and G. A. PARTINGTON. 1979. Reprogramming of transplanted nuclei in amphibians. In *International Review of Cytology, Suppl. 9,* J. F. Danielli and M. A. Di Berardino, eds. Academic Press.

GURDON, J. B., R. A. LASKEY, and O. R. REEVES. 1975. The developmental capacity of nuclei transplanted from keratinized skin cells of adult frogs. *J. Embryol. Exp. Morph.* **34**:93–112.

MINTZ, B. 1967. Gene control of mammalian pigmentary differentiation, I: Clonal origin of melanocytes. *Proc. Nat'l Acad. Sci. USA* **58**:344–351.

NESBITT, M. N., and S. M. GARTLER. 1971. The applications of genetic mosaicism to developmental problems. *Annu. Rev. Genet.* **5**:143–162.

Cell-to-Cell Interactions in Development

BERNFIELD, M. R. 1978. The cell periphery in morphogenesis. In *Birth Defects: Proceedings of the Fifth International Conference on Birth Defects,* J. W. Littlefield and J. De Grouchy, eds. Excerpta Medica.

BULLER, A., J. ECCLES, and R. ECCLES. 1960. Differentiation of fast and slow muscles in the cat hind limb. *J. Physiol.* (London) **150**:399–416.

BULLER, A., J. ECCLES, and R. ECCLES. 1960. Interaction between motor neurons and muscles in respect of the characteristic speeds of their responses. *J. Physiol.* (London) **150**:417–439.

DEUCHAR, E. M. 1975. *Cellular Interactions in Animal Development.* London: Chapman & Hall.

HAY, E. D. 1978. Embryonic induction and tissue interaction during morphogenesis. In *Birth Defects: Proceedings of the Fifth International Conference on Birth Defects,* J. W. Littlefield and J. De Grouchy, eds. Excerpta Medica.

RUBINSTEIN, N., F. SREFER, K. MABUCHIN, F. PEPE, and J. GERGELY. 1977. Use of type-specific antimyosins to demonstrate the transformation of individual fibers in chronically stimulated fast muscle. *Fed. Proc.* **36**:584–590.

RUTISHAUSER, U. 1984. Developmental biology of neural adhesion molecule. *Nature* **310**:549–554.

Positional Information

BROWER, D. L. 1985. The sequential compartmentalization of *Drosophila* segments revisited. *Cell* **41**:361–364.

GARCIA-BELLIDO, A., P. A. LAWRENCE, and G. MORATA. 1979. Compartments in animal development. *Sci. Am.* **241**(1):102–111.

GARCIA-BELLIDO, A., R. RIPOLL, and G. MORATA. 1976. Developmental compartmentalization in the dorsal mesothoracic disk of *Drosophila. Devel. Biol.* **48**:132–147.

MALIMSKI, G., and S. BRYANT, eds. 1984. *Pattern Formation: A Primer in Developmental Biology.* Macmillan.

WOLPERT, L. 1978. Cell position and cell lineage in pattern formation and regulation. In *Stem Cells and Tissue Homeostasis,* B. I. Lord, C. S. Potten, and R. J. Cole, eds. Cambridge University Press.

Genetic Analysis of Differentiation

Caenorhabditis elegans

CHALFIE, M., H. R. HORVITZ, and J. E. SULSTON. 1982. Mutations that lead to reiterations in the cell lineages of *C. elegans. Cell* 24:59–69.

GREENWALD, I. S., P. W. STERNBERG, and H. R. HORVITZ. 1983. The *lin*-12 locus specifies cell fate in *C. elegans. Cell* 34:435–444.

SULSTON, J. E., and H. R. HORWITZ. 1977. Post-embryonic cell lineages of the nematode, *Caenorhabditis elegans. Devel. Biol.* **56**:110–156.

Drosophila melanogaster

AGARD, D. A., and J. W. SEDAT. 1983. Three-dimensional architecture of a polytene nucleus. *Nature* **302**:676–681.

BENDER, W., M. AKAM, F. KARDI, P. A. BLEACHY, M. PEIFER, P. SPIERER, E. B. LEWIS, and D. S. HOGNESS. 1983. Molecular genetics of the bithorax complex. *Science* **221**:23–29.

BENDER, W., P. SPIERER, and D. S. HOGNESS. 1983. Chromosomal walking and jumping to isolate DNA from the *Ace* and *rosy* loci and the bithorax complex in *Drosophila melanogaster. J. Mol. Biol.* **168**:17–33.

CARRASCO, A. E., W. McGINNIS, W. J. GEHRING, and E. M. DE ROBERTIS. 1984. Cloning of an *X. laevis* gene expressed during early embryogenesis that codes for a peptide region homologous to *Drosophila* homeotic genes. *Cell* **37**:409–414.

FJOSE, A., W. J. McGINNIS, and W. J. GEHRING. 1985. Isolation of a homeo box containing gene from the *engrailed* region of *Drosophila* and the spatial distribution of its transcripts. *Nature* **313**:284–289.

GARCIA-BELLIDO, A. 1977. Homeotic and atavic mutations in insects. *Am. Zool.* **17**:613–629.

GEHRING, W. J. 1985. Molecular basis of development. *Sci. Am.* **253**(4):153–162.

HAFEN, E., A. KUROIWA, and W. J. GEHRING. 1984. Spatial distribution of transcripts from the segmentation gene *fushi tarazu* during *Drosophila* embryonic development. *Cell* 37:833–841.

HAFEN, E., M. LEVINE, R. L. GARBER, and W. J. GEHRING. 1983. An improved *in situ* hybridization method for the detection of cellular RNAs in *Drosophila* tissue sections and its application for localizing transcripts of the homeotic *Antennapedia* gene complex. *EMBO J.* **2**:617–624.

LEWIS, E. B. 1978. A gene complex controlling segmentation in *Drosophila. Nature* **276**:565–570.

McGINNIS, W., R. L. GARBER, J. WIRZ, A. KUROIWA, and W. J. GEHRING. 1984. A homologous protein-coding sequence in *Drosophila* homeotic genes and its conservation in other metazoans. *Cell* **37**:403–408.

Maintenance of Adult Organisms: Stem Cells and Colony-Stimulating Factors

BROWN, D. D. 1984. The role of stable complexes that repress and activate eukaryotic genes. *Cell* **37**:359–365.

GOLUB, E. S. 1982. *In vitro* approaches to hemopoiesis. *Cell* 28:687–688.

HALL, A. K. 1983. Stem cell is a stem cell is a stem cell. *Cell* 33:11–12.

LEVENSON, R., and D. HOUSMAN. 1982. Commitment: How do cells make the decision to differentiate? *Cell* **25**:5–6.

LORD, B. I., C. S. POTTEN, and R. J. COLE, eds. 1978. *Stem Cells and Tissue Homeostasis: The Second Symposium of the British Society for Cell Biology.* Cambridge University Press.

METCALF, D. 1984. *The Hemopoietic Colony Stimulating Factors.* Elsevier/North-Holland.

METCALF, D. 1985. The granulocyte-macrophage colony-stimulating factors. *Science* **229**:16–22.

23

Cancer

CELL growth is a carefully regulated process that responds to specific needs of the body. In a young animal, cell multiplication exceeds cell death and the animal increases in size; in an adult, the processes of cell birth and death are balanced to produce a steady state. For some adult cell types, renewal is rapid: intestinal cells have a half-life of a few days before they die and are replaced; certain white blood cells are replaced as rapidly. In contrast, human red blood cells have approximately a 100-day half-life, healthy liver cells rarely die, and, in adults, there is a slow loss of brain cells with little or no replacement.

Very occasionally, the exquisite controls that regulate cell multiplication break down and a cell begins to grow and divide even though the body has no need for further cells of its type. When the descendants of such a cell inherit the propensity to grow and divide without responding to regulation, the result is a clone of cells that can expand to a considerable size. Ultimately, a mass called a *tumor* may be formed by this clone of unwanted cells. Because tumors may have devastating effects on the animals that harbor them, much research has gone into understanding how they form.

In this chapter we describe the present-day knowledge in this rapidly moving field of research, emphasizing the genetic events that transform a normally regulated cell into one that grows without responding to controls.

These genetic events are not ones inherited through the gametes. Rather, they are changes in the DNA of somatic cells. The principal type of change is the alteration of preexisting genes to *oncogenes,* whose products cause the inappropriate cell growth. Thus DNA alteration is apparently at the heart of cancer induction.

Characteristics of Tumor Cells

Although most research into the molecular basis of cancer utilizes cells growing in culture, it is important to look first at tumors as they occur in experimental animals and in humans. In this way we can see the gross properties of the disease—the properties that ultimately must be explained by analysis of cells and molecules.

Tumors arise with great frequency, especially in older animals and humans, but most pose little risk to their host because they are localized. We call such tumors *benign;* an example is warts. Rarely, tumors become life-threatening because they spread throughout the body. Such tumors are called *malignant* and are the cause of cancer.

It is usually apparent when a tumor is benign. Benign tumors contain cells that closely resemble normal cells and that may function like normal cells. Ill-understood forces keep benign tumor cells (and normal cells) localized to appropriate tissues. Benign liver tumors stay in the liver, and benign intestinal tumors stay in the intestine. A fibrous capsule usually delineates the extent of a benign tumor and makes it an easy target for a surgeon. Benign tumors become serious medical problems only if their sheer bulk interferes with normal functions or if they secrete excess amounts of biologically active substances like hormones.

The major characteristics that differentiate malignant tumors from benign ones are their properties of *invasiveness* and *spread.* Malignant tumors do not remain localized and encapsulated: they invade surrounding tissues, get into the body's circulatory system, and set up areas of proliferation away from the site of their original appearance. When tumor cells spread and engender secondary areas of growth, the process is called *metastasis;* malignant cells have the ability to *metastasize* (Figure 23-1).

Malignant cells are usually less well differentiated than benign tumor cells. Furthermore, their properties often vary over time. For example, liver cancer cells may lack certain enzymes characteristic of liver cells and may ultimately evolve to a state where they lack most liver-specific function. This variability of phenotype is often correlated to a variability of genotype: cancer cells have abnormal and unstable numbers of chromosomes, as well as many chromosomal abnormalities. Apparently, benign tumors may progress to malignancy (Figure 23-1), although the earliest stages of malignant tumors are hard to identify and pathologists are rarely sure how a malignancy began. In any case, the cells of malignant tumors have a tendency to lose differentiated traits, to acquire an altered chromosomal composition, and to become invasive and metastatic.

Cancer cells can be distinguished from normal cells by examining them in a microscope. In a specific tissue, cancer cells are usually recognized by the characteristics of rapidly growing cells: a high nucleus-to-cytoplasm ratio, prominent nucleoli, many mitoses, and relatively little specialized structure. The presence of invading cells in an otherwise normal tissue section is the most diagnostic indication of a malignancy (Figure 23-2).

Malignant cells usually have enough of the hallmarks of the normal cell type from which they were derived that it is possible to classify them by their relationship to normal tissue. Normal animal cells are often subdivided according to their embryonic tissue of origin, and the naming of tumors has followed suit. Recall that normal cells arise from one of three embryonic cell layers—endoderm, ectoderm, or mesoderm (Chapter 22). Malignant tumors are therefore classified as either *carcinomas* (those deriving from endoderm or ectoderm) or *sarcomas* (those deriving from mesoderm). The *leukemias,* a subdivision of the sarcomas, grow as individual cells in the blood, whereas most other tumors are solid masses. (The name leukemia is derived from the Latin for "white blood": the massive proliferation of leukemic cells can cause a patient's blood to appear milky.)

Over the past two decades enormous progress has been made toward understanding how some external agents initiate cancer, and this achievement is evident in much of the information in this chapter. Investigators have only scratched the surface, however, in their attempts to explain how the initiating events affect the mechanisms that regulate cell growth and why abnormal cells fail to obey the rules of normal tissue organization.

Metastasis Is Not Understood at the Molecular Level

Having emphasized that invasiveness and metastasis are the hallmarks of malignancy, it may surprise the reader that throughout most of this chapter we shall ignore these two characteristics. Our lack of attention to these important matters is a consequence of where progress in cancer research has been successful and where it has been stymied by complexity. Enormous progress has been made in areas that allow for genetic analysis because of the extraordinary power that molecular genetics has developed in recent years. Invasiveness and metastasis must be studied largely as problems of cell-to-cell interaction, and progress in this area has depended on the rate of progress in cell surface biology. Knowledge of cell surfaces is coming slowly; perhaps soon the parameters of invasiveness and spread will become better defined. For now, only a few general principles are clear.

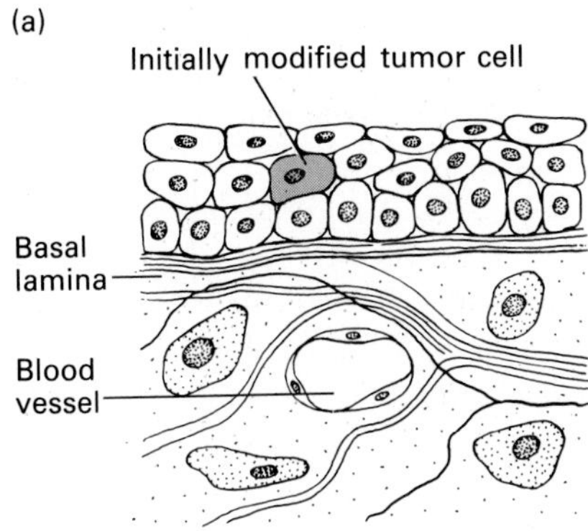

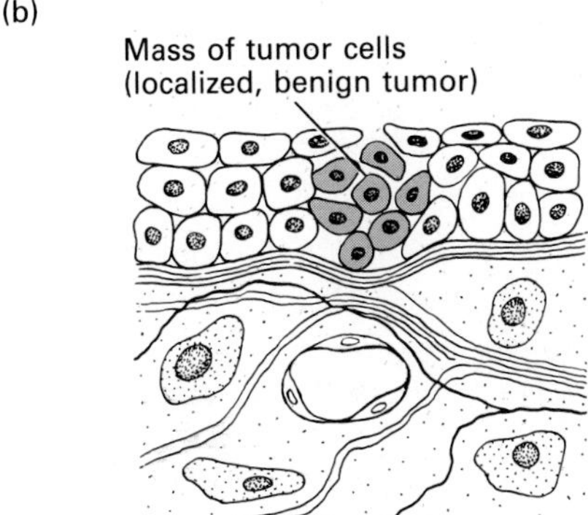

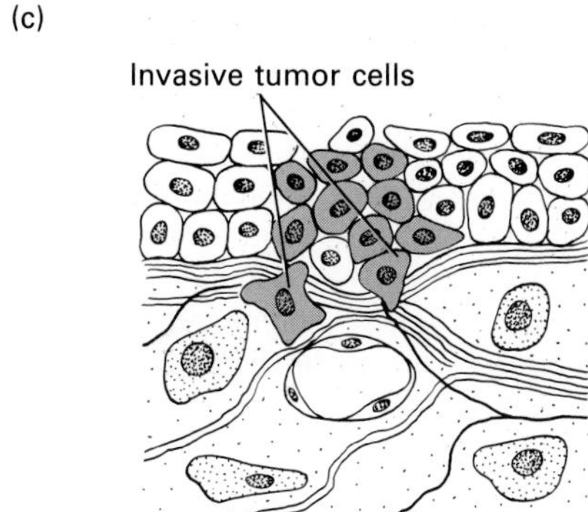

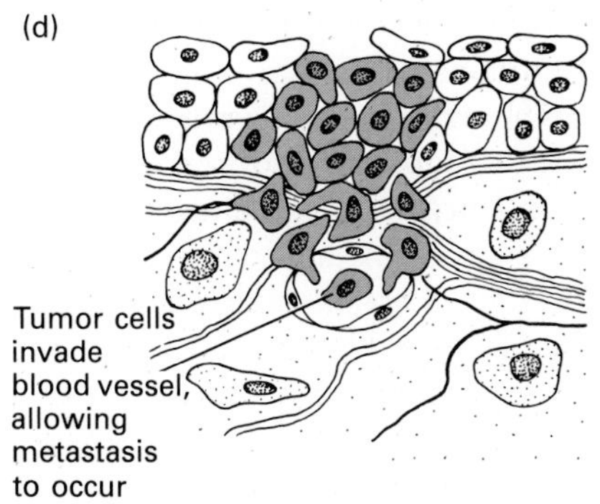

Figure 23-1 Stages in tumor growth and metastasis. (a) A single modified cell appears in a tissue. (b) The modified cell begins to grow although surrounding cells do not, and a mass of localized tumor cells form. This tumor is still benign. (c) As it progresses to malignancy the tumor invades the basal lamina that surrounds the tissue. (d) The tumor cells spread into blood vessels that will distribute them to other sites in the body. If the tumor cells can exit from the blood vessels and grow at distant sites they are considered malignant, and a patient with such a tumor is said to have cancer.

The restriction of a normal cell type to a given organ and/or tissue is maintained by cell-to-cell recognition and by physical barriers. Cells recognize one another by ill-understood surface processes, which dictate that like cells bind together and that certain cells interact. Primary among the physical barriers that keep tissues separated is the *basal lamina,* which underlies layers of epithelial cells as well as the endothelial cells of blood vessels. Basal laminas define the surfaces of external and internal epithelia and the structure of blood vessels. Metastatic tumor cells have the ability to digest their way through basal laminas (see Figure 23-1). Thus, the processes of invasion and metastasis depend in part on the ability of tumor cells to pass through basal laminas. Metastasis also depends on the ability of malignant cells to grow in new locations, surrounded by cells that they do not ordinarily contact. The ability to start growth without a mass of surrounding identical cells and the ability to ignore "foreign" cell contacts are thus qualities that tumor cells must develop. The wide range of altered behaviors that underlie malignancy must have a basis in the new proteins made by malignant cells. For example, enzymes secreted by some malignant cells degrade the collagen and other basal lamina proteins, such as proteoglycans and glycosaminoglycans.

Another aspect of malignancy that we recognize but do not understand is the ability of tumor cells to elude the immune system. It might be expected that the alterations in cell structure associated with malignancy would lead the immune system to recognize tumor cells as foreign. The blood even contains a "natural killer" cell type that can apparently recognize and kill many types of tumor cells while sparing normal cells. In spite of this, tumor cells find ways to elude immune detection. They may cover up antigens that would otherwise mark them for destruction, or they may rid themselves of the cell surface molecules that lymphocytes use to recognize foreign cells (major histocompatibility antigens; Chapter 24). As a result, immune recognition appears to play a minor role in protecting the body against tumors. When confronted with the defenses evolved by tumor cells, the immune system is largely ineffective.

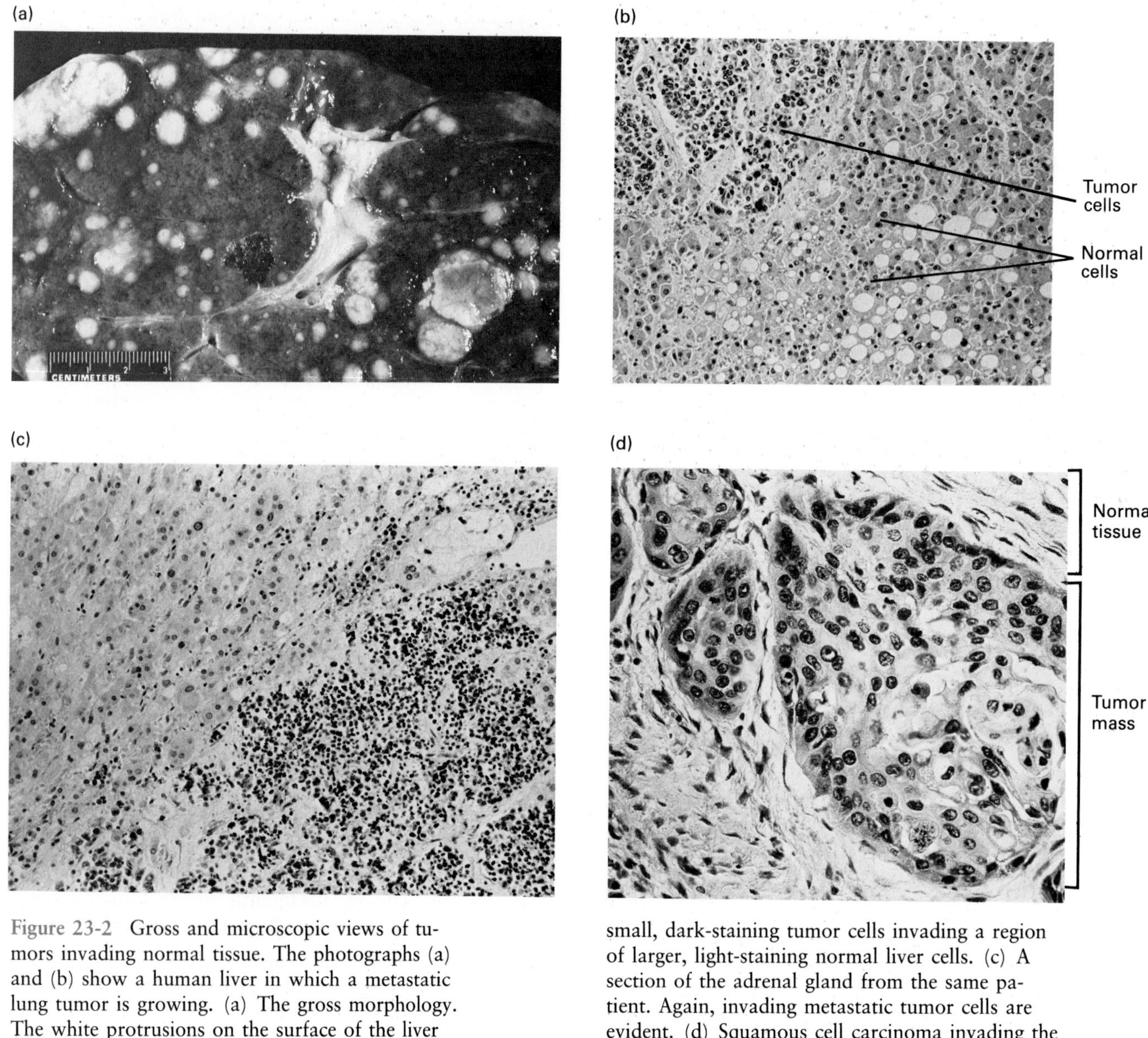

Figure 23-2 Gross and microscopic views of tumors invading normal tissue. The photographs (a) and (b) show a human liver in which a metastatic lung tumor is growing. (a) The gross morphology. The white protrusions on the surface of the liver represent the tumor masses. (b) A light micrograph of a section of the tumor showing areas of small, dark-staining tumor cells invading a region of larger, light-staining normal liver cells. (c) A section of the adrenal gland from the same patient. Again, invading metastatic tumor cells are evident. (d) Squamous cell carcinoma invading the normal connective tissue of the dermis. *Courtesy of J. Braun.*

Cells May Either Grow or Remain Quiescent

In an animal or in culture, cells are either growing or quiescent. Cell growth involves an increase in mass that leads a cell to divide; cell growth and cell division are thus interrelated concepts. A *quiescent* cell is one that is not increasing its mass or passing through the cell cycle. Quiescent cells carry out the characteristic functions of tissues, such as the synthesis of export proteins in the liver or the transmission of impulses in a nerve. Cell growth is a regulated process, so the fraction of growing cells in a given tissue is a function of both the age of the organism and the properties of the tissue. In adults, certain tissues, such as the intestine, maintain a constant size by the continued growth of new cells and the death of older cells. Such tissues may have many dividing cells. Embryos and expanding tissues of young animals also contain a large fraction of growing cells.

Cell growth involves two easily recognized, coordinated events: the duplication of cell DNA and the physical division of the cell into two daughter cells (Chapters 5 and 11). We assay DNA duplication either by measuring the incorporation of a radioactive DNA precursor or by measuring the DNA content of individual cells in a cytofluorometer; we can recognize division microscopically. If DNA synthesis and cell division are considered the key events, then cell growth cycle can be divided into four periods: G_1, the gap between the previous nuclear division and the beginning of DNA synthesis; S, the period of DNA synthesis; G_2, the gap between DNA duplication and nuclear division; and M, the period of mitosis, during which both of the chromosomes separate into the two daughter cells (see Figures 5-13 and 13-3). Cell division is usually coordinated with nuclear division, but it may be delayed. For example, in early insect embryos many nuclear divisions occur without cytoplasmic divisions; the result is highly multinucleated cells (see Figure 22-7).

In different cells, the cell cycle varies in total length and in the relative lengths of the different subdivisions. Embryonic cells can divide as frequently as rapidly growing bacteria, once every 15 to 20 min. More typically, the mammalian cell cycle is 10 to 30 h long; S, G_2, and M together require a fixed time of about 10 h, whereas G_1 is quite variable.

Quiescent cells usually have unduplicated DNA, although some epithelial cells may rest in G_2. Logically, then, stimulation of the growth of most quiescent cells must cause them to enter S (i.e., to make a new complement of DNA) before they can divide. Some workers consider quiescent cells to be in a physiological state they call G_0, which is different from G_1. Others believe that there is no physiologically distinct G_0 phase: in this view, all cells are passing through the cell cycle, but the probability that a cell will exit from G_1 to enter S is extremely variable. Rapidly growing cells would thus have a short G_1, slowly growing cells would have a long G_1, and nerve cells, which never divide, would have an infinite G_1. By either hypothesis, whether a cell divides or becomes quiescent depends on whether the cell "decides" to continue (or initiate) progression through the cell cycle. Control of cell growth would hinge on that "decision."

Control of cell growth is one of the most important aspects of an animal's physiology. Cells of an adult must divide frequently enough to allow tissues to remain in a steady state, and division must be stimulated at wounds or when special requirements are placed on a tissue. There must be many circulating cell-specific factors that signal individual cell types whether to divide or not. A few such factors are known—notably factors controlling the growth of blood cells—but many more await discovery. Uninhibited growth of cells results in tumors. How the controls of cell growth are overridden is a major area of cancer research.

Cell Culture and Transformation

Although questions about cell growth and the induction of cancer are ultimately questions about the behavior of individual cells in a living organism, the study of cells in an animal is impractical because of the difficulty of identifying the relevant cells, manipulating their behavior in a controlled manner, and separating the effects due to the intrinsic properties of the cells from the effects due to the interactions among the many cell types present in the organism. When cells growing in culture are exposed to a carcinogen—a cancer-causing chemical—these problems can be controlled. The environment of the cell can be manipulated by the investigator, the target cell can be well defined, the changes in the cell following treatment can be examined, and the fate of the carcinogen can be determined. Furthermore, in culture the cells can be quiescent or growing; they can have, in fact, precisely defined growth parameters. They can also be manipulated genetically. For these reasons, studies of normal cell growth as well as of cancer induction depend heavily on the use of cultured cells.

Fibroblastic, Epithelial, and Nonadherent Cells Can Be Cultured

To make cell cultures, cells are removed from an organism and grown at 37°C in a medium with nutrients plus 5 to 20% blood serum, as described in Chapter 6. For many years, most cell types resisted attempts to grow them in culture. Recent modifications of culture methods have allowed experimenters to grow many specialized cells in culture, and the utility of cell culture as a method of examining cell behavior has increased enormously. Most published studies, however, describe work with the few cell types that grow readily. These are not cells of a defined type; rather, they represent whatever grows when a tissue or an embryo is placed in culture.

The cell type that usually predominates in such cultures is called a *fibroblast* because it secretes the types of proteins associated with the fibroblasts that form fibrous connective tissue in animals. Cultured fibroblasts have the morphology of tissue fibroblasts, although they are not as differentiated as true fibroblasts. Fibroblasts are a cell type derived from embryonic mesoderm, and the cells that grow in culture appear to be mesodermal stem cells (see Table 22-1). With appropriate stimulation these can differentiate into many cell types: fat cells, connective tissue cells, muscle cells, and others. In most studies, however, cultured "fibroblasts" have been treated simply as convenient prototypical cells for study.

Another cell type that has been studied is the cultured *epithelial* cell. Like the cultured fibroblast, it does not necessarily correspond to a normal tissue cell; rather, it is representative of the type of cell that comes from the ectodermal or endodermal embryonic cell layers.

Cultured fibroblasts and epithelial cells are grown on glass or plastic dishes to which they adhere tightly due to their secretion of sticky proteins such as laminin, fibronectin, and collagen (see Chapter 19). Neither cell type will ordinarily grow if it is not adhering to a substratum. To prepare tissue cells for culture or to remove adherent cells from a culture dish for biochemical studies, trypsin or another protease is used. The process of putting cells into culture or of transferring cells to a new culture is often called *plating*.

Cells cultured from blood, spleen, or bone marrow cells adhere poorly, if at all, to the culture dish. In the body, such cells are held in suspension (in the blood) or they are loosely adherent (in the marrow and spleen). Because these cells often come from immature stages in the blood cell lineages, they are very useful for studying the development of leukemias.

Immortal Cell Lines Can Be Established in Culture

When cells are removed from an embryo or an adult animal, most of the adherent ones grow continuously in culture for only a limited time before they spontaneously cease growing. Such a culture dies out even if it is provided with fresh supplies of all of the known nutrients that cells need to grow, including blood serum. Human cells have been studied in some detail. After explanting, fetal cells take some time to become established in culture, during which period the majority of cells die and the "fibroblasts" become the predominant cell type. The fibroblasts then double about 50 times before they cease growth. Starting with 10^6 cells, 50 doublings can produce $10^6 \times 2^{50}$ or more than 10^{20} cells, which are equivalent to the weight of about 10^5 people. Thus, even though its lifetime is limited, a single culture, if carefully maintained, can be studied for a long time. Such a lineage of cells originating from one initial culture is called a *cell strain* (Figure 23-3a).

To be able to clone individual cells, modify cell behavior, or select mutants, it is often necessary to maintain cell cultures for many more than 100 doublings. This is possible with cells from some animal species because these cells undergo a change that endows them with the ability to grow indefinitely. A culture of cells with an indefinite life span is considered immortal; such a culture is called a *cell line* to distinguish it from an impermanent cell strain. The ability to establish a culture that will grow indefinitely varies depending on the animal species from which the cells originate. For human cells, only tumor cells grow indefinitely, and therefore the HeLa tumor cell has been invaluable for research on human cells (Chapter 6).

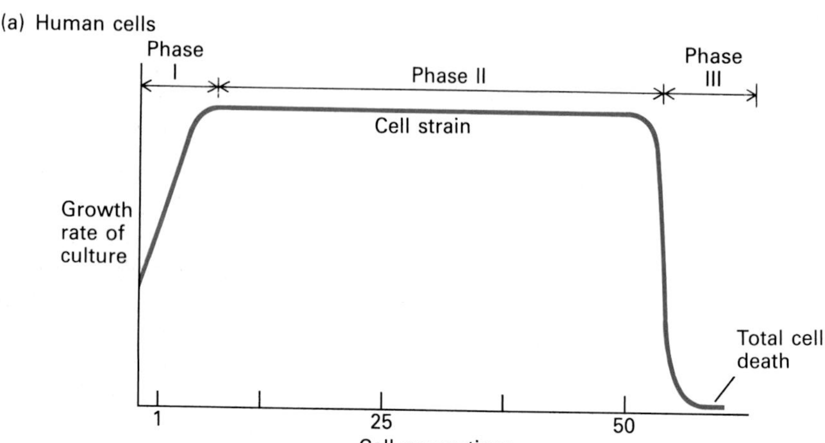

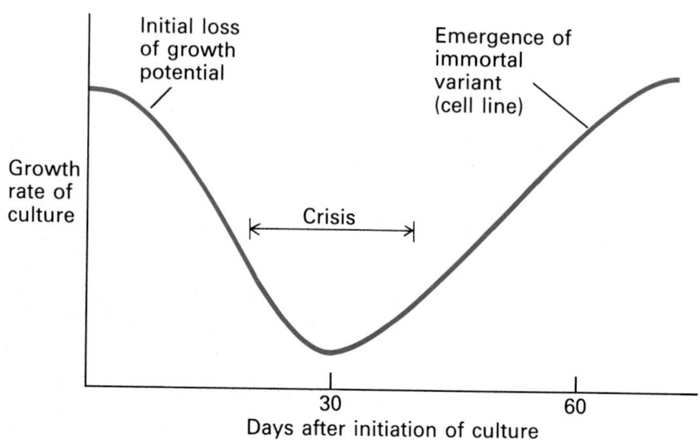

Figure 23-3 Stages in the establishment of a cell culture. (a) Human cells. When an initial explant is made (e.g., from foreskin), some cells die and others (mainly fibroblasts) start to grow; overall there is a slow increase in growth rate *(phase I)*. If the remaining cells are continually diluted, they grow as a cell strain at a constant rate for about 50 cell generations *(phase II)*, after which growth begins to slow. The ensuing period of increasing cell death *(phase III)* ultimately leads to the complete death of all of the cells in the culture. (b) Mouse or other rodent cells. When a culture is prepared from mouse embryo cells, there is initial cell death coupled with the emergence of healthy growing cells. As these are diluted and allowed to continue growth, they soon begin to lose growth potential and most cells die (the culture goes into crisis). Very rare cells do not die but continue growing until their progeny overgrow the culture. These cells constitute a cell line, which will grow forever if it is appropriately diluted and fed with nutrients: the cells are immortal.

Chicken cells die out after only a few doublings, and even tumor cells from chickens almost never become immortal. With rodent cells, however, cultures of adherent cells from embryos routinely give rise to cell lines.

When a culture of adherent rodent cells is first explanted, it grows well, but after a number of serial replatings it loses growth potential and goes into *crisis*. During crisis most of the cells die, but sometimes a rapidly growing cell variant arises spontaneously and takes over the culture. Such a variant will grow forever if it is provided with the necessary nutrients (Figure 23-3b). Cells in an established line usually have more chromosomes than the normal cell from which they arose, and their chromosome complement undergoes continual expansion and contraction in culture. The culture is said to be *aneuploid* (i.e., having an inappropriate number of chromosomes), and the cells of such a culture are obviously mutants.

If rodent cell cultures are maintained at a low cell density until a cell line emerges, the line will consist of flat cells that adhere tightly to the dish in which they are grown. A number of mouse cell lines derived in this fashion have been used extensively in cancer research; they are called 3T3 cells (because they were derived according to a schedule where $\underline{3}\times10^5$ cells were $\underline{t}$ransferred every $\underline{3}$ days into petri dishes with a 50-mm diameter to maintain the appropriate cell density). As is true for other cultured fibroblasts, the exact cell type that gives rise to 3T3 cells is uncertain, but they can differentiate into a range of mesodermally derived cell types, especially endothelial cells (those that line blood vessels). The ability to derive lines of flat cells like 3T3 set the stage for studying the transition to malignancy in cell culture because cancer cells differ dramatically from 3T3 cells in their growth properties. Before we describe the use of 3T3 cells in cancer research, however, we shall consider the control of 3T3 growth.

What Stimulates the Growth of Quiescent Cells?

If a culture of 3T3 cells is plated at 3×10^5 cells per dish in a medium with 10% blood serum, the cells will grow for a few days and then cease growth at about 10^6 cells per dish. The culture is said to have reached saturation density. The cells are not dead, however, because refeeding them with fresh medium can reinitiate their growth. The cells in the saturated culture are quiescent: they have stopped growing but they can maintain themselves in a viable state for a long time.

Among the treatments that will reinitiate growth in a quiescent 3T3 cell culture is the addition of extra serum to the medium. In fact, the density at which the cells stop growing is in direct proportion to the amount of serum with which they are initially provided (Figure 23-4). From this result, it would appear that factors in serum are the primary determinants of whether cells remain quiescent or whether they initiate growth. Serum factors may indeed be essential, but they are not the whole story.

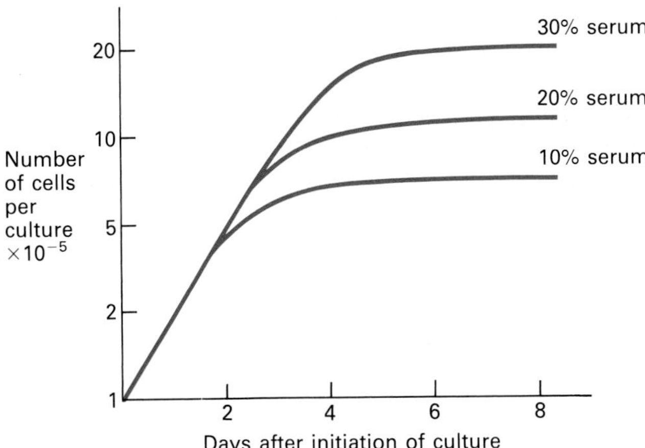

Figure 23-4 The dependence of cell growth on serum concentration. A constant number of 3T3 cells was used to initiate multiple cultures, each of which was fed with a medium containing the indicated percentages of fetal calf serum. The number of cells per culture was determined daily. The initial growth rates were indistinguishable, but the final number of cells was proportional to the amount of added serum. The experiment shows that serum factors rather than cell contacts control cell growth, because cells are already touching one another in 10% serum. [See R. W. Holley and J. A. Kiernan, 1968, *Proc. Nat'l Acad. Sci. USA* 60:300.]

For example, if a strip of cells is removed from a quiescent cell culture (the culture is said to be wounded), the cells at the border of the wound will begin growing, and they will divide a few times to fill the gap. Because the cell medium is not altered in such an experiment and because the cells that are not adjacent to the wound do not initiate growth, it is clear that local effects—cell-to-cell contacts—also control cell growth. In all probability, the degree to which crucial growth factors in serum are available to cells is limited by close contacts.

In recent years, mixtures of purified proteins have been successfully substituted for serum to allow the long-term growth of certain cell lines. A *defined medium* is one with known components and no serum (for a description of serum-free medium see Table 6-3). Defined media that will support growth differ from one type of cell to another. The nature of the components of a defined medium gives a good idea of the role of serum: the important constituents in serum are protein hormones and growth factors.

In Chapter 16 we explained that growth factors are hormones that certain cultured cells require for growth. One such hormone, *epidermal growth factor* (EGF), is needed by almost all cells. Many cells also need insulin or an insulin-like factor for growth. With these two substances plus transferrin (which makes iron available to cells), the serum requirement for most cells is largely satisfied. Individual cell types, however, often have exotic requirements, and no universal defined serum substitute has been concocted.

Growth factors and hormones facilitate cell growth by acting as signaling agents. They somehow direct the cell to carry out whatever steps are needed for growth. The factors and hormones themselves neither provide nutrient value to the cell nor play any known role in metabolic pathways. Presumably, only their ability to bind to specific cell surface receptors enables them to control cellular events (Chapter 16).

A recent clue to the mechanism of signaling by growth factors has emerged from the discovery that the receptors for EGF, platelet-derived growth factor, and insulin all have as part of their structure a tyrosine-specific protein kinase whose activity is stimulated when the receptor binds its cognate ligand. Thus, phosphorylation of tyrosine residues in proteins may be a very important device by which factors signal cells to grow; as will be evident from the following discussion, this may be a crucial clue to how the growth of certain cancer cells is stimulated.

These characteristics of cell growth in vitro provide the background for a consideration of how cancer cells differ from normal cells. Quiescent normal cells are ones that have not been stimulated to grow by factors or hormones. Because unrestricted growth is a characteristic of cancer cells, overcoming the growth inhibition in quiescent cells is a likely mechanism for inducing cancer.

Cells Growing in Culture Can Be Transformed to Malignancy

Several different treatments of adherent cells—virus infection, exposure to chemicals, irradiation—can dramatically change their growth properties in culture. Furthermore, these treatments can cause the cells to form tumors after they are injected into susceptible animals. Such changes in the growth properties of cells and their subsequent development of tumor-forming capacity are collectively referred to as *malignant transformation,* or just *transformation*. Because transformation can be carried out entirely in culture, it is widely studied as an analog of cancer induction in animals, although the exact correspondence of the two processes has not been established. Before explaining how transformation is induced, we shall first describe its consequences.

Transformation is usually recognized in adherent cells by a changed cellular morphology and growth habit. If, for instance, a growing culture of 3T3 fibroblasts is exposed to SV40 virus, a small proportion of the cells will be infected, and these will adopt a more rounded configuration. The virus-infected cells will be less adherent to one another and to the dish than are the normal surrounding cells, and they will continue to grow when the normal cells have become quiescent. A group or *focus* of these loosely adherent transformed cells can be recognized

(a)

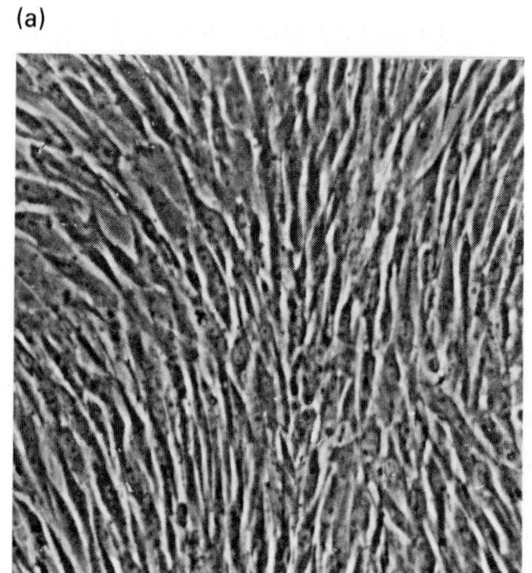

(b)

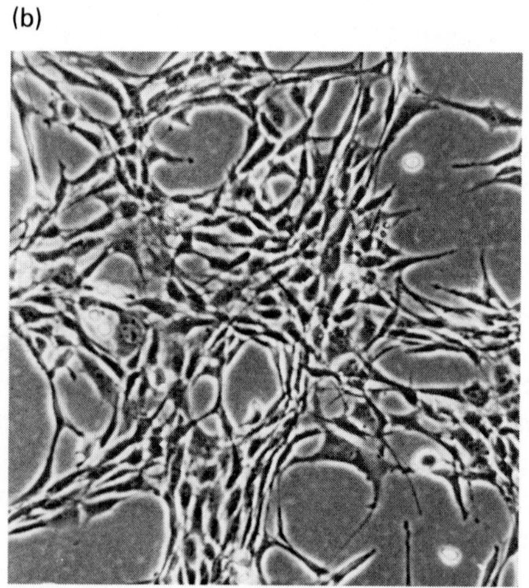

(c)

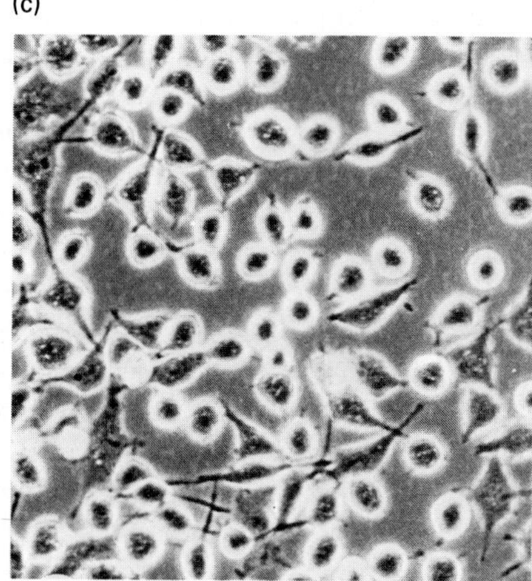

Figure 23-5 Normal and transformed rat embryo fibroblasts as viewed in the phase-contrast light microscope. (a) Cultured normal rat embryo fibroblasts. Note that the cells are aligned and closely packed in an orderly fashion. (b) Rat embryo fibroblasts that have been transformed by integration of the gene encoding the polyoma virus mid-T antigen. The cells are crisscrossed and chaotic in their growth. (c) Rat embryo fibroblasts transformed by the Abelson murine leukemia virus. The cells have lost adherence to the dish so completely that they appear almost round. Each has a white halo because of the light refraction of rounded cells. *Courtesy of L. -B. Chen.*

(a)

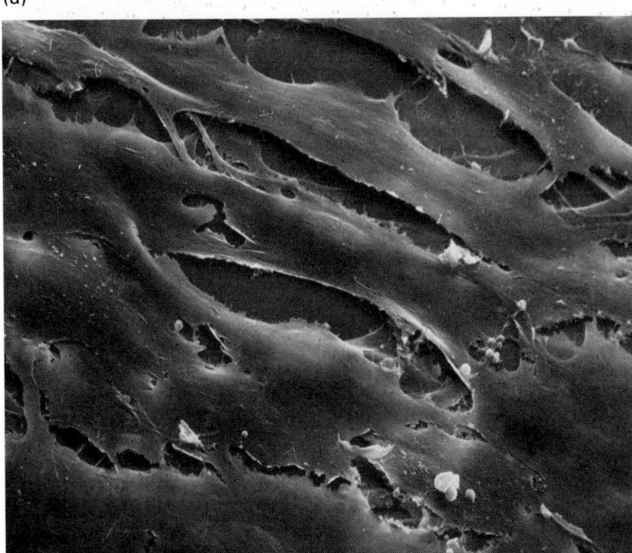

(b)

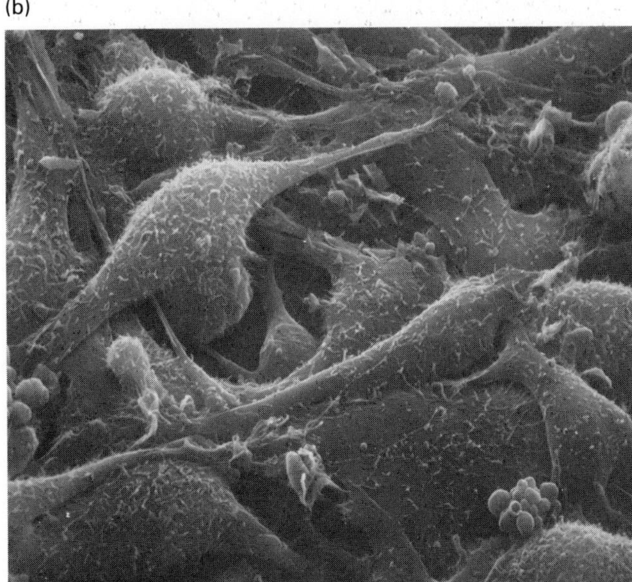

Figure 23-6 Scanning electron micrographs of normal and transformed 3T3 cells. (a) Normal 3T3 cells. Each straplike image is a cell. As in the rat embryo fibroblasts of Figure 23-5a, the cells are lined up in one direction and are spread out into thin lamellae with bulges representing the nuclei. (b) 3T3 cells transformed by Rous sarcoma virus. The cells are much more rounded, and they are covered with small hairlike processes and bulbous projections. The cells grow one atop the other and they have lost the side-by-side organization of the normal cells. These transformed cells truly appear malignant. *Courtesy of L. -B. Chen.*

under the microscope (Figures 23-5 and 23-6). If a focus of transformed cells is recovered and a culture is grown from it, the result will be a line of transformed cells (called, e.g., SV-3T3 cells). By comparing the properties of the transformed cell line and the parental line, the consequences of transformation can be assessed.

The transformation of adherent cells involves changes in a constellation of cellular properties. These include aspects of growth control, morphology, cell-to-cell interactions, membrane properties, cytoskeletal structure, protein secretion, and gene expression. Some of the changes caused by transformation are probably interrelated, but some seem independent of one another. Not all transformed cells show all of the changes that can be induced by transformation, but the following is a list of 15 changes that are usual concomitants of transformation. We know little about the molecular events that underlie the phenomena and therefore we can only describe changes in gross parameters that may someday be shown to have unsuspected relationships.

Increased Saturation Density A key characteristic of transformed cells is that they continue to grow when normal cells cease growth (Figure 23-7). This characteristic is considered an important analog of malignant cell growth in animals. There are numerous reasons why transformed cells continue to grow when normal cells have stopped: some will be described below.

Decreased Growth Factor Requirements Because transformed cells have apparently lost some of the hormone and growth factor requirements of normal cells (Figure 23-7), they grow in initial serum concentrations that are much lower than those required by normal cells. Some transformed cells produce growth factor analogs, so they may be providing their own growth factor requirement.

Loss of Capacity for Growth Arrest Normal cells suppress their own growth when the concentration of any of the many critical nutrients or factors falls below a threshold value (as is shown for EGF in Figure 23-7). When the concentration of isoleucine, phosphate, EGF, or another substance that regulates growth falls below the necessary level, normal cells go into quiescence.

Transformed cells are deficient in their ability to respond by growth arrest to lowered nutrient or factor concentrations. Cells may even kill themselves trying to continue growth in an impossible environment.

Loss of Dependence on Anchorage for Growth Normal adherent cells require firm contact with the sub-

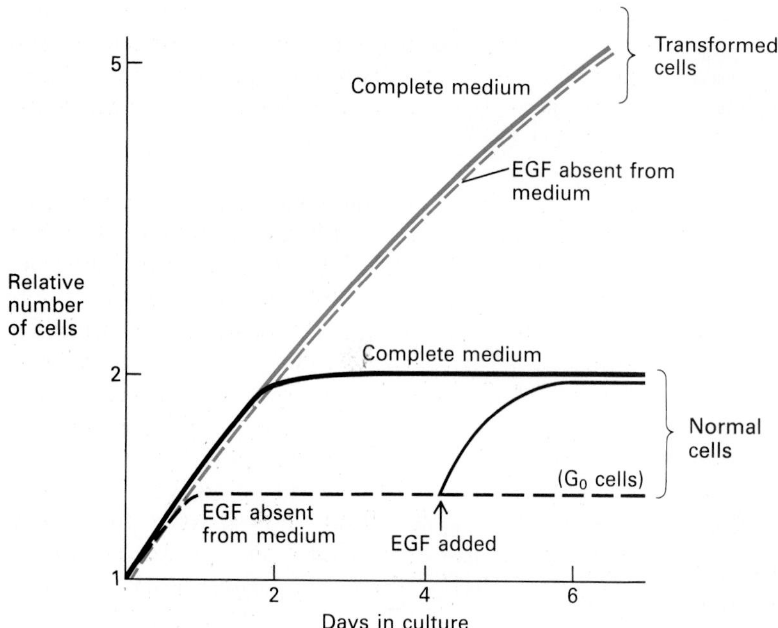

Figure 23-7 The dependence of normal and transformed cells on epidermal growth factor (EGF). Equal numbers of normal and transformed cells were plated into a defined medium with or without EGF at day 0. Cell numbers were determined daily. Transformed cells in a complete medium grew when normal cells had already ceased growth. The transformed cells had also lost the EGF requirement of normal cells.

At day 4, EGF was added to some dishes of normal cells that were initially lacking it. Because the cells responded to the addition of EGF by growth, it is evident that without EGF normal cells remain viable in a G_0 state.

stratum for growth. If they are plated onto a surface to which they cannot adhere they will not grow, although they can remain viable for long periods of time. Similarly, if normal cells are suspended in a semisolid medium such as agar, they will metabolize but they will not grow. Transformed cells have generally lost the requirement for adherence; they grow without attachment to a substratum, as indicated by their ability to form colonies when suspended as single cells in agar. This characteristic correlates extremely well with the ability of transformed cells to form tumors: cells that have lost anchorage dependence generally form tumors with high efficiency when they are injected into animals that cannot immunologically reject the cells. How the cell senses shape and why transformation abrogates the need for anchorage are both unanswered questions.

Changed Cell Morphology and Growth Habits The individual transformed cell is very different in shape and appearance from its normal counterpart (see Figures 23-5 and 23-6). It adheres much less firmly to the substratum and therefore is more rounded with fewer processes. Furthermore, transformed cells adhere poorly to each other and do not appear to sense their neighbors. As normal cells in a culture dish become more crowded they form ordered patterns; transformed cells form chaotic masses (see Figures 23-5 and 23-6). The low mutual adherence of transformed cells coupled with their loss of anchorage to the substratum allows them to grow in multiple layers, whereas normal cells grow in monolayers with some overlap at the cell borders.

Loss of Contact Inhibition of Movement In culture, normal fibroblastic cells are motile. When two normal cells moving around in a culture dish come into contact, one or both of them will stop and take off in another direction. This ensures that the cells do not overlap. When a normal cell is surrounded by others in such a way that it has nowhere to go, it ceases movement and makes gap junctions with the surrounding cells (Chapter 14). This phenomenon is referred to as *contact inhibition of movement*. Transformed cells lack contact inhibition of movement: they pass over or under one another, they grow on top of one another, and they infrequently form gap junctions.

Cell Surface Alterations Many of the foregoing properties of transformed cells relating to growth and behavior are probably consequences of cell surface events. Thus, a number of investigators have concentrated their efforts on comparing the surfaces of normal and transformed cells. Some of the known surface changes are described here and under the headings that follow.

The glycolipids and glycoproteins that occur abundantly on the surfaces of normal cells (Chapter 14) are modified in transformed cells: protein-linked *N*-acetylneuraminic (sialic) acid is decreased, as is the ganglioside content of all lipids. The general structure of the plasma membrane is not altered, however, and the lateral mobility of surface lipids remains the same. Probably the most important difference is that the proteins of the cell surface are much more mobile in transformed cells than in normal cells; for example, antibodies can more easily agglutinate a given surface protein on a transformed cell than on a normal cell. Because the lipids are not intrinsically more mobile, it may be that links between surface proteins and the underlying cytoskeletal elements are modified by transformation. Such a modified linkage may also be the basis for the altered morphology of transformed cells.

Easier Agglutination by Lectins Lectins, such as concanavalin A and wheat germ agglutinin, are plant proteins that have multiple binding sites for specific sugars. Transformed cells are agglutinated by lectins at a much lower concentration than that required to agglutinate normal cells (Figure 23-8). The number of available sugar residues in surface glycoproteins and glycolipids is not increased and may even be lowered by transformation, so the increased agglutinability is not a consequence of an increased density of binding sites. Rather, the explanation appears again to be the increased mobility of cell surface glycoproteins, which allows low concentrations of lectins to make patches of receptor-lectin complexes on the cell surface. Patches on one cell can be cross-linked by the lectin to patches on other cells, causing the cells to agglutinate.

Increased Glucose Transport Transformed cells transport glucose more rapidly than normal cells. Kinetic studies have shown that the effective K_m of sugar binding to the receptor is not altered but that the V_{max} is increased (see Chapter 15). This implies that more transporters (permeases) are available on cell surfaces, perhaps due to the increased glycolytic activity of transformed cells, which gives them a higher requirement for sugar transport.

Reduced or Absent Surface Fibronectin Normal quiescent cells in monolayer culture become covered with a dense fibrillar network containing fibronectin as a major protein component (Chapter 19). Even growing cells have a diffuse covering of fibronectin. Transformed cells either totally lack fibronectin or have greatly reduced amounts. They have difficulty binding it to their surfaces; many transformed cells also make less fibronectin, but some secrete copious amounts. The addition of high concentrations of pure fibronectin derived from normal cells can cause many tumor cells to flatten out and take on a fairly normal appearance; thus the loss of fibronectin may be an important determinant of the transformed state.

Loss of Actin Microfilaments Not only are the surfaces of transformed cells different from those of normal cells, but there are also differences in the cytoskeleton. The actin microfilaments that extend the length of normal cells (see Figures 3-40, 19-28, and 19-29) are either diffusely distributed or concentrated beneath the cell surface. The loss of cytoskeletal elements has been considered a possible cause of the increased mobility of cell surface proteins.

Release of Transforming Growth Factors *Transforming growth factors* (TGFs) are proteins secreted by transformed cells that can stimulate growth of normal cells. Specific receptors for TGFs have been found and TGFs have been identified in embryos, which implies that they have a role in normal cell physiology as well as in transformed cells.

How much of the transformed phenotype is due to autostimulation by secreted growth factors is an unanswered question, but it seems likely that such factors play a role in the growth of some, if not many, transformed cells.

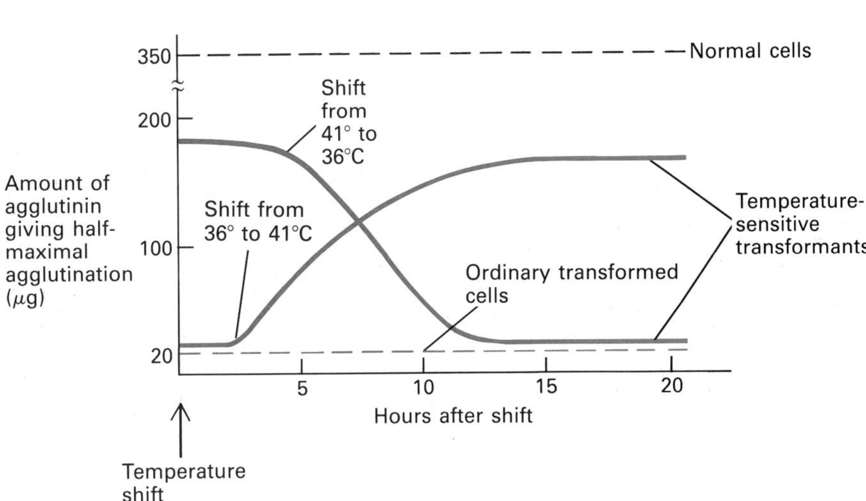

Figure 23-8 The agglutination of normal, transformed, and temperature-sensitive transformed cells by wheat germ agglutinin. The chicken embryo cells used for this experiment were of three types: normal fibroblasts, fibroblasts transformed by Rous sarcoma virus, and fibroblasts transformed by a viral mutant that produces transformed cells at 36°C and nearly normal cells at 41°C. The temperature of cells was raised or lowered, and at various times after the temperature shift the amount of wheat germ agglutinin that would give half of the maximal agglutination was determined. For normal cells, 350 μg was required; for the ordinary transformed cells, only 20 μg was required. The temperature-sensitive transformants began to change their properties a few hours after the shift: at the higher temperature they became quasinormal, whereas at the lower temperture they took on the characteristics of transformed cells. [See M. M. Burger and G. S. Martin, 1972, *Nature New Biol.* 237:356.]

Protease Secretion Transformed cells often secrete a protease called *plasminogen activator,* which cleaves a peptide bond in the serum protein *plasminogen,* converting it to the protease *plasmin.* Thus the secretion of a small amount of plasminogen activator causes a large increase in protease concentration by catalytically activating the abundant plasminogen in normal serum. External protease treatment of normal cells can cause cells to mimic some aspects of transformation (loss of actin microfilaments, growth stimulation, etc.), so it has been suggested that plasminogen activator secretion may help maintain the transformed state of certain cell lines.

Plasminogen activator secretion may be important for the invasiveness of transformed and tumor cells, because the resulting increase in plasmin may help the cells penetrate the basal lamina. The normally invasive extraembryonic cells of the fetus secrete plasminogen activator when they are implanting in the uterine wall; this provides a compelling analogy to tumor cell invasiveness. Whether plasminogen activator acts only by cleaving circulating plasminogen or whether it can attack other proteins directly is an open question.

Altered Gene Transcription All of the foregoing characteristics of transformed cells are cytoplasmic activities; yet it might be expected that the extraordinary range of differences between normal and transformed cells would be at least partly due to alterations in the transcription of specific genes and in the relative stability of the transcripts. Surprisingly, however, the mRNA populations of normal cells and transformants derived from them are quite similar. Some mRNAs have increased concentrations, some are decreased and about 3 percent of the mRNA population is specific to transformed cells. This small amount of RNA probably represents many different low-abundance mRNA molecules. The proteins encoded by these transformation-sensitive genes, although low in concentration, have profound effects on cell growth and morphology. Some of the transformation-specific mRNAs also appear in embryonic cells, and tumor cells have many proteins that are characteristic of embryonic cells; this would suggest that transformation may alter protein composition toward that characteristic of embryos.

Immortalization When cell strains, with their limited growth potential in culture, are used as targets for cell transformation, then one measurable characteristic of transformation is the induction of unlimited growth potential—the transformation of a cell strain into an immortal cell line. The ease with which transforming stimuli can generate such immortal cell lines depends on the underlying propensity of the cells to spontaneously acquire immortality. Nonadherent (blood) cells from many animals are routinely immortalized by transformation. Adherent human cells are rarely immortalized, adherent chicken cells are almost never immortalized, but adherent rodent cells are easily transformed to immortality. The other criteria of transformation described above are as relevant to adherent human and chicken cells as they are to mouse cells, which implies that transformation to immortality is a parameter that is quite distinct from the others.

Cell lines have acquired immortality before transformation by whatever spontaneous processes bring occasional cells through the crisis stage. Thus, in transformation experiments using cell lines, immortalization is not a relevant parameter.

Summary of Transformation Parameters No obvious chain of cause and effect links all of the disparate consequences of transformation. Some of the consequences listed above are obviously interrelated; others are not. For example, no linkage of immortalization to any of the other parameters is evident, and it is difficult to relate anchorage independence of growth to a changed surface mobility of proteins.

Changed patterns of gene transcription could explain all of the changes associated with transformation, but, as will become evident in the next section, the action of most intracellular transforming proteins is probably not directly related to transcriptional control. The notion of autostimulation by transforming growth factors is attractive, but not all transformed cells appear to secrete transforming factors, so this explanation lacks the generality necessary to provide an understanding of all transformation. Whatever the causal linkages, transformation appears to result from a small number of independent events with widely ramifying consequences. As we shall discuss below, in some cell types transformation is known to result from the synthesis of one or two new proteins, so there may be only a few keys to understanding the transformation of any given cell. But the labor that has brought us this far in our understanding of transformation has not yet uncovered the key events for any cell type.

Viruses as Agents of Transformation: Oncogenes

Having seen the enormous range of effects that transformation has on cells, one might expect that transforming agents would *initiate* transformation by exerting multiple new influences. Surprisingly, however, the events triggering transformation can be comparatively simple, often resulting from the transcription of one or two genes. These genes may be part of a virus, or they may be altered cellular genes. Three different types of transforming agents are known: viruses, chemicals, and radiation. Each type of influence was recognized as a *carcinogen* (a cancer-causing agent) in animals before it was studied as a transforming agent for cell cultures. We shall consider all three of these forms of carcinogens, starting with viruses.

In Chapter 6 we discussed the various types of animal viruses, emphasizing that some viruses have RNA as their genetic material and some have DNA (see Table 6-4). One group of RNA viruses, the *retroviruses,* and many types of DNA viruses can be transforming agents; such viruses are often called *tumor viruses.* Tumor viruses cause transformation as a consequence of their ability to integrate their genetic information into the host cell's DNA; most often they cause the chronic production of one or more proteins called *transforming proteins,* which are responsible for maintaining the transformed state of the infected cells. Transforming proteins are synthesized under the direction of *transforming genes* in an integrated viral genome. These intracellular products are distinct from the transforming growth factors. For DNA viruses, the known transforming genes are integral parts of the virus genome. For retroviruses, the transforming genes are normal or slightly modified cellular genes that are either appropriated from or hyperactivated in the host cell. Because the retroviruses and the DNA viruses have evolved their transforming genes in different ways, and because the ability of an RNA virus to stably integrate its genetic information into a cell is a fundamental puzzle, we shall consider these two classes of tumor viruses separately.

DNA Viruses Transform as a Consequence of Their Replication Strategy

Many different types of animal DNA viruses exist: their genome sizes vary from 5 to 200 kb (kilobases). The viruses with the smallest genomes encode very few proteins and rely mainly on host-cell functions for their replication. The viruses with the larger genomes encode many enzymes and provide many of their own replication functions. All types of DNA viruses can be tumor viruses. To simplify this discussion we shall focus our attention on one group of very small DNA tumor viruses, the *papovaviruses,* of which the best-known representatives are SV40 and polyoma.

The strategy of papovavirus multiplication is to divide its infection course into two phases: an early phase during which it replicates its DNA and primes the cell to make more virus DNA, and a late phase during which it makes coat protein and matures new virions. In the early phase, a quiescent infected cell is activated to synthesize cellular DNA. The papovaviruses have limited their size by encoding a few key proteins that induce the cell to make the enzymes necessary for DNA replication. The virus accomplishes this by causing the cell to progress from the G_0 or G_1 phase into the S phase. Thus both cellular and viral DNA are replicated during the early phase of infection. It is believed that the induction of the S phase underlies the transforming ability of papovaviruses.

The cells of certain animal species are said to be *permissive* for papovaviruses; in such cells the late phase follows the early phase, and the massive synthesis of virus is coupled to cell death (Figure 23-9). Approximately half of the genetic information in the 5-kb circular genomes of SV40 and polyoma (see Figure 9-43) specifies their coat proteins; the other half encodes the set of two or three early proteins. Each type of papovavirus has a narrow range of permissive cells in which it expresses both early and late genes to allow a productive, or *lytic,* infection. For SV40, monkey cells are permissive; for polyoma, mouse cells are permissive.

Aside from the few types of cells that are permissive for SV40 and polyoma, and the cases of slight permissivity (e.g., human cells for SV40), most cells are nonpermissive for infections by these viruses. In *nonpermissive* cells, the induction of the S phase does not ordinarily lead to viral DNA replication, probably because of some incompatibility between the cellular enzymes and a crucial viral sequence or protein. There is no switch to the late phase and no cell death. As long as the expression of the early viral functions continues, the cells cannot rest in G_0: they are continually induced to proceed through the cell cycle. However, the viral genome is usually degraded or lost by dilution during cell proliferation. Once the viral genome is lost, the cells revert to normal. This reversion results in an *abortive transformation* (Figure 23-9). If a papovavirus is to permanently transform cells, the viral genome must continually act on the cell. A small percentage of infected, nonpermissive cells do become stably and permanently transformed (Figure 23-9). Most of the criteria for transformation described in the preceding section are fulfilled by these cells; they have been transformed into cancer cells by the continual expression of the early papovavirus proteins.

A permissive cell population also can be transformed because certain viral mutants that are unable to replicate their DNA express only early proteins in permissive cells. Such mutants appear frequently enough that, say, mouse cells can be transformed by polyoma virus, and the virus can cause cancer after it is injected into baby mice. Although in wild animal populations neither polyoma nor SV40 is known to induce cancer, very distant relatives of these viruses, known as papilloma viruses, do cause cancer, even in humans.

In permanently transformed nonpermissive cells, the viral genome has integrated itself into the host-cell genome in such a way that the early viral genes are constitutively expressed (Figure 23-10). The continual activity of two or three viral gene products maintains the transformed state. This has been shown by using various temperature-sensitive viral mutants for transformation: with these mutants, the cells' morphology and growth properties can be varied from normal to transformed by

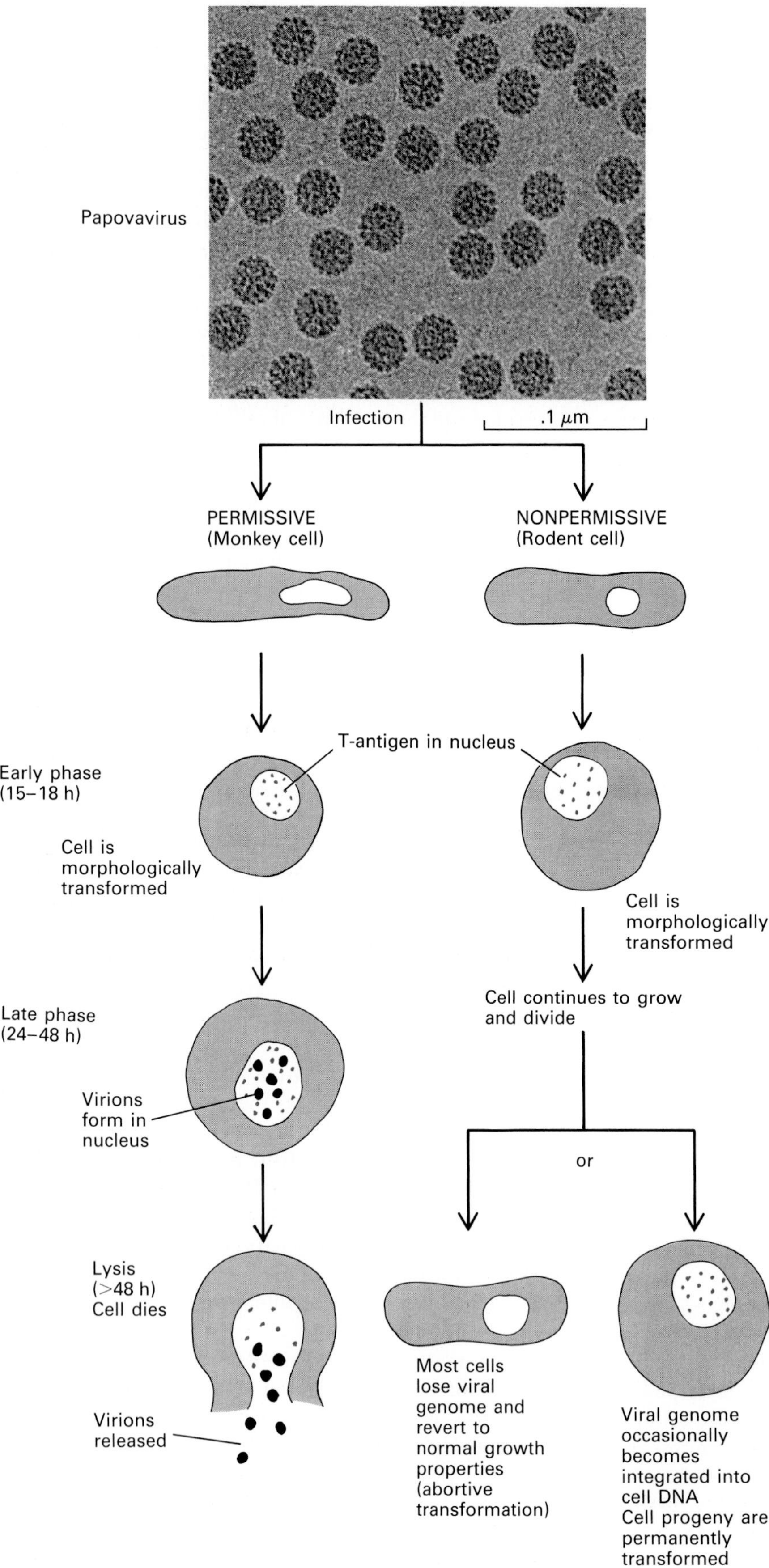

Figure 23-9 Responses of permissive and nonpermissive cells to infection by SV40, a papovavirus. In permissive cells, virus infection leads to cell death and the production of progeny virions. In nonpermissive cells, the synthesis of the early viral proteins causes cells to become transformed. If the viral DNA becomes permanently integrated into the cell DNA, permanent transformation of the cell results.

The electron micrograph shows an unstained virus preparation embedded in a thin layer of ice. *Photograph courtesy of T. Baker and M. Bina.*

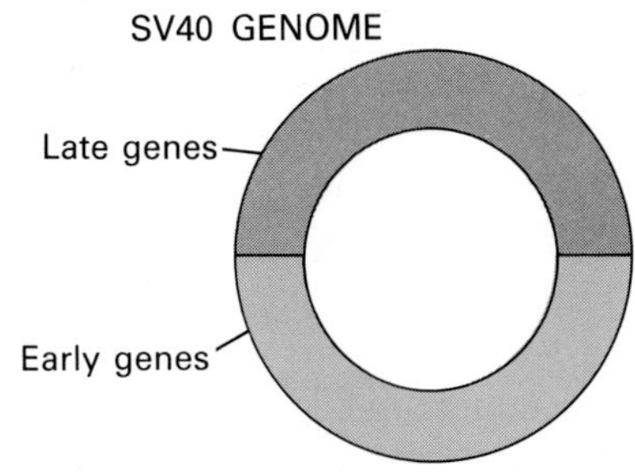

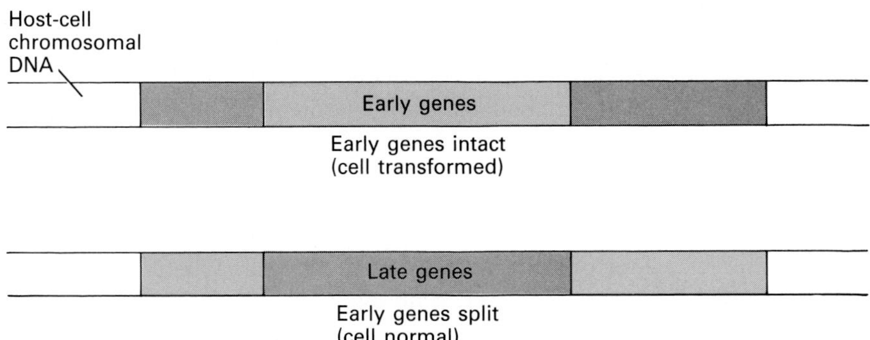

Figure 23-10 The integration of the papovavirus genome. When SV40 infects a nonpermissive cell, its DNA may integrate into a cellular chromosome. For integration, the circular viral genome must be opened to produce a linear form; the site of opening is random. If the integrated genome is opened in such a way that the early genes are intact, they will be expressed in the cell and it will be transformed. If the early genes are broken, they cannot make a functional transforming protein. The genomes shown here are schematic; in reality the integrated forms have partial duplications and deletions.

shifting the temperature. We shall use the model provided by papovaviruses in later discussions when we consider what new genes are active in transformed cells to maintain the transformed state. First, however, we shall explore the nature of viral DNA integration.

Papovavirus DNA Integrates Randomly and Causes Cancer

There are two fundamentally different mechanisms by which viral DNA could integrate into cellular DNA: site specifically and randomly. *Site-specific* integration has a defined target sequence in the cellular DNA and/or a defined integration site in the viral DNA. One or more proteins (probably viral) would bind to these sequences and direct integration. Papovaviruses do not integrate site specifically; they apparently use *random* sites both in the viral and in the host-cell DNA. The viral proteins do not appear to play an active role in the integration process. Papovavirus DNA integration is probably no different from the integration of any DNA that has been incorporated into cells (Chapter 6). Integration of viral DNA into the cellular DNA of nonpermissive cells can have no role in propagating the virus because no viral progeny are made in such cells.

These considerations make it unlikely that papovaviruses have evolved mechanisms *designed* to induce cancer. Rather, cancer induction appears to be a consequence of three particular circumstances: the occurrence in papovavirus genomes of transforming genes used by the virus for the early phase of its lytic cycle, the accidental integration of viral DNA into the DNA of nonpermissive cells, and the lack of cell death after infection of nonpermissive cells. These three circumstances allow the virus to permanently transform nonpermissive cells when viral DNA integrates in such a way that early gene expression is not impeded (Figure 23-10). This event happens frequently enough that some cell transformation occurs whenever a nonpermissive cell population is infected by SV40 or polyoma virus.

Papovaviruses Encode Two or Three Proteins That Together Transform Cells

The discovery that papovaviruses cause cancer through the agency of early viral proteins has focused attention on the types of early proteins encoded by the viruses. These proteins are really the cause of transformation; if we knew their mode of action we would have a better understanding of what causes cancer. SV40 makes two early proteins, called T (large T) and t (small t); polyoma makes three early proteins: T, mid-T, and t. T protein was originally called *T antigen* because it was first demonstrated by immunofluorescence using serum from animals bearing virus-induced tumors. The T antigens are 90,000-molecular-weight proteins found in the cell nucleus. They bind tightly to DNA and play an important role in viral DNA transcription and replication during the lytic cycle (see Figures 18-28 and 13-28). The mid-T of polyoma is a 45,000-molecular-weight plasma membrane–bound protein that has no SV40 counterpart. (Some of the SV40 T antigen, however, is bound to the plasma membrane and may serve the same function as the mid-T protein of polyoma.) The 20,000-molecular-weight t proteins are cytoplasmic. The three polyoma proteins appear to have independent roles in the transformation process.

The early region of the polyoma genome gives rise to one transcript that can be spliced in three different ways to produce the three early proteins (Figure 23-11). By deletion of specific nucleotides, it has been possible to construct three different DNA molecules that, after transfection into cells, each encode only one of the three polyoma early proteins. The DNA molecule encoding only mid-T protein can transform a rat 3T3 cell line by many of the criteria described earlier but such cells will not grow in a low-serum medium. Mid-T apparently cannot obliterate the requirement for all growth factors, but it can do almost everything else associated with transformation. In contrast, a cell that expresses both the T and the mid-T proteins can grow in low serum or even in no serum. T protein can even cause growth in low serum in the absence of mid-T, but under this condition the cells are not morphologically transformed and they do not grow to high saturation densities. This dramatic separation of transformation parameters illustrates an important general principle of cell transformation: transformation often involves the interaction of a small number of independently acting proteins. We shall see this principle at work later when we consider nonviral cancers.

For the experiments described in the previous paragraph, rat 3T3 cells were used as the targets of transformation. When plasmids encoding the T and mid-T proteins were introduced into recently explanted rat embryo fibroblasts (cells with a limited life span in culture), the division of labor among polyoma early proteins became more pronounced. A plasmid encoding mid-T produced no transformed colonies in these cells, which suggested that perhaps the mid-T protein cannot immortalize cells although it can transform 3T3 cells, which are already immortal.

Consistent with this explanation was the finding that a plasmid encoding T protein conferred immortality on rat embryo fibroblasts without changing them morphologically. The subsequent incorporation into the cell of a plasmid encoding mid-T morphologically transformed the cells immortalized by the action of T protein, and the resulting transformants behaved almost like wild-type polyoma-transformed cells. In other experiments, polyoma t protein was found to have a role in producing complete transformation.

For SV40, recombinant DNA methods have shown that different parts of T protein have different effects. Thus this single protein is responsible for several different aspects of transformation. Papovaviruses have apparently evolved a number of distinct ways of changing cell behavior. Where these changes occur independently, they do not produce a complete tumor cell. Where they act together, however, a full-blown tumor cell results.

In our discussion of transformation by DNA tumor viruses a number of important concepts have emerged. These viruses, like all of the other cancer-inducing agents to be discussed in this chapter, cause cell transformation by altering the types of proteins made in the cell. The virus adds new, active genes whose products are responsible for the transformation. In cases we shall discuss in later sections, cancer is caused by the aberrant expression of altered or normal proteins originating from the cellular genome rather than from a virus, but the principle is the same. DNA tumor viruses have also demonstrated that multiple proteins can work together to cause transformation. This cooperation and other evidence to be presented below suggest that a tumor cell is caused by a few independent activities—not many, but not just one. These activities determine the particular constellation of trans-

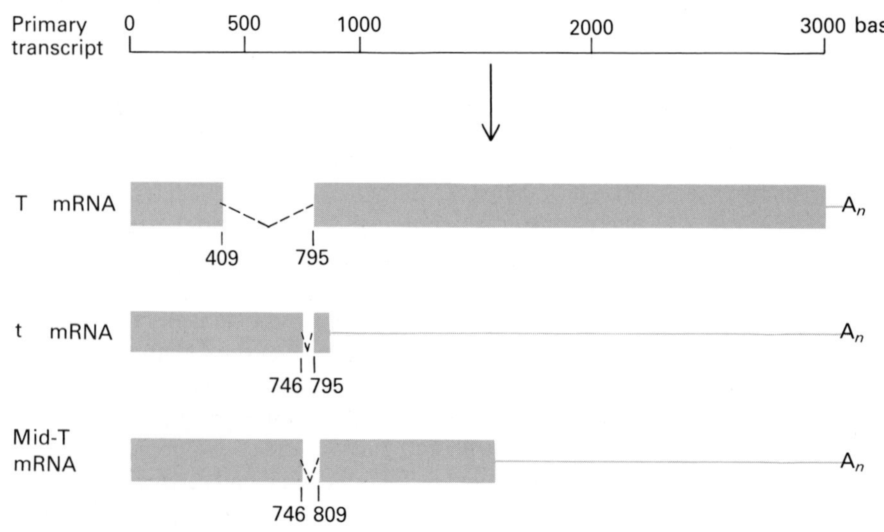

Figure 23-11 The three early proteins of polyoma virus. The early region of polyoma virus is expressed in one 3000-base long nuclear RNA that can be spliced in three ways to yield three different mRNAs and therefore three early proteins. (The thick color bars denote protein-coding regions; the thin color lines, noncoding regions.) [See M. Rassoulzadegan et al., 1982, *Nature* 300:713.]

formation parameters present, with immortalization standing out as a characteristic that is regulated separately from the others (e.g., serum factor dependence).

RNA-Containing Retroviruses Use Reverse Transcription for Infection

Transformation by retroviruses is more complicated than transformation by DNA viruses, because the basic retrovirus life cycle does not involve proteins with transforming activity. To understand the transforming ability of retroviruses we must first examine in some detail their life cycle, which was outlined in Figure 6-32.

Retroviruses have an RNA genome consisting of two apparently identical, approximately 8500-nucleotide RNA molecules noncovalently bonded to one another; each has a cellular tRNA molecule bound to it. Each virion also contains about 50 molecules of a DNA polymerase that is capable of copying the genomic RNA into DNA: this enzyme is commonly known as the *reverse transcriptase* because it reverses the normal flow of information in biological systems. The infection process brings both the RNAs and the reverse transcriptase into the cell's cytoplasm.

Once in the cytoplasm, the reverse transcriptase uses the tRNA as a primer to initiate DNA synthesis (Figure 23-12). In a complicated series of events, a complete complementary (or minus) strand of DNA is formed; this minus strand is used as a template for the formation of a plus strand of DNA. The ultimate product of reverse transcription is thus a double-stranded DNA copy of the information in the viral RNA genome (Figure 23-12).

The double-stranded DNA reverse transcript is actually longer than its template: because of the "jump" made by the initial minus-strand DNA and a second "jump" made by the initial plus-strand DNA, the small repeat at either end of the viral RNA is extended to form a *long terminal repeat* (LTR) at either end of the reverse transcript (Figure 23-12). The LTR is a very important component of the reverse transcript because, as we shall see, it contains many of the signals that allow retroviruses to function.

The Reverse Transcript Can Become Integrated into Cellular DNA as a Provirus

After its formation, the double-stranded DNA reverse transcript migrates into the cell's nucleus, where a fraction of it becomes circular. The circular DNA then integrates into an apparently random site in one of the cell's chromosomes (Figure 23-13). This integration process is directed by sequences at the ends of the LTR. The enzymes that catalyze the process have yet to be identified, but at least one is viral because viral mutations can block integration.

In contrast to the integration of DNA tumor virus genomes, the integration site within the circular retrovirus DNA is specific; that site is the unique sequence that was formed where the ends of the original linear retrovirus molecule joined to produce the circle. During integration, four base pairs at the junction are lost, two from each side of the original linear molecule. There is also a duplication of a small number of bases (typically four to six) at the chromosomal target site. By the site specificity of the integration site within the viral genome, the apparent randomness of the chromosomal target site, and the duplication of the target site, reverse transcript integration is like transposon movement (Chapter 11).

Once the reverse transcript is integrated into the host-cell DNA, the transcript—which is called *proviral DNA* in its integrated form—becomes a template for RNA synthesis (Figure 23-14). Provirus-directed transcription, like most other eukaryotic transcription, involves a *promoter*—a sequence that directs the RNA polymerase to a specific initiation site—and an *enhancer*—a sequence that facilitates transcription although it need not be located near the initiation site. The promoter and enhancer sequences, as well as the polyadenylation signals, are located in the LTR. The two LTRs at either end of the provirus are identical in sequence, but the upstream LTR acts to promote transcription whereas the downstream LTR specifies the poly A site. How these two identical sequences manage to carry out different functions is not fully understood. The RNA made from the proviral DNA has two functions: it serves as mRNA in the synthesis of viral proteins and as genomic RNA for the next generation of virus particles (Figure 23-14).

Because they lack most metabolic machinery, viruses are obligate intracellular parasites. Retroviruses are perhaps the ultimate intracellular parasites because they can link their genome stably to that of their host cell without seriously damaging the cell. Their genetic information is replicated as part of the cellular DNA and distributed to all daughter cells, which actively transcribe it so that new virus particles are made perpetually. Thus the retrovirus life cycle differs from that of DNA viruses in two fundamental respects: not only is integration site-specific within the viral genome, but once integration has occurred, the proviral DNA can continually direct the synthesis of new viral progeny without killing the infected cells.

Retroviruses can sometimes become integrated into an animal's germ-line cells; they are then inherited by all offspring of that animal, and they can become part of the inheritance of a species. Most such inherited retroviruses, as opposed to the ones that are integrated during somatic-cell growth, are transcriptionally inactive so that they do not burden the animal's transcriptional machinery. Retroviral proviruses have become integrated into the genomes of mice frequently enough that typically 0.1 to 1 percent of a mouse genome consists of proviral DNA.

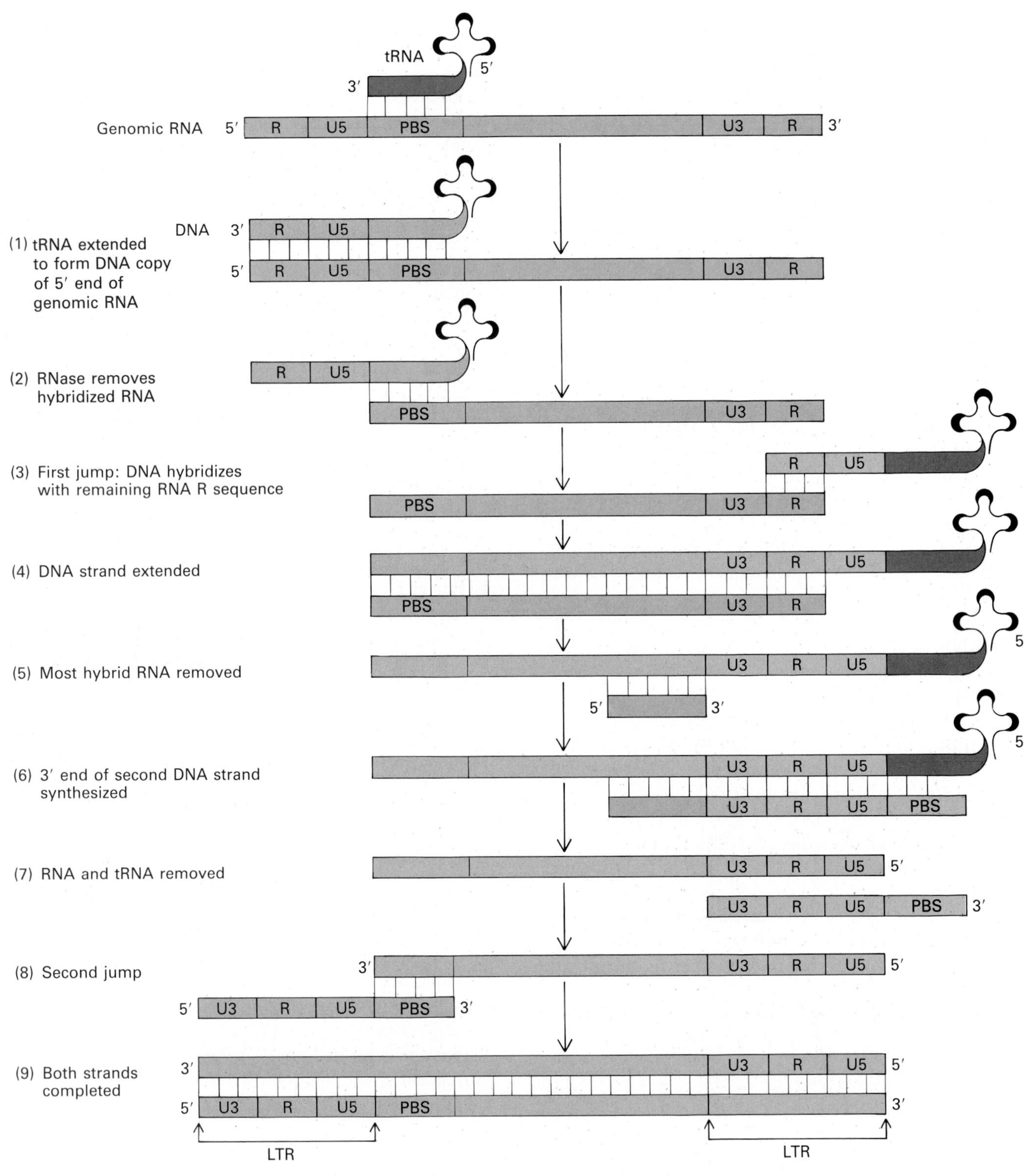
tRNA
3′
5′
Genomic RNA 5′ R U5 PBS U3 R 3′
DNA 3′ R U5
(1) tRNA extended to form DNA copy of 5′ end of genomic RNA
5′ R U5 PBS U3 R
(2) RNase removes hybridized RNA
R U5
PBS U3 R
(3) First jump: DNA hybridizes with remaining RNA R sequence
R U5
PBS U3 R
(4) DNA strand extended
U3 R U5
PBS U3 R
(5) Most hybrid RNA removed
U3 R U5 5′
5′ 3′
(6) 3′ end of second DNA strand synthesized
U3 R U5 5′
U3 R U5 PBS
(7) RNA and tRNA removed
U3 R U5 5′
U3 R U5 PBS 3′
(8) Second jump
3′ U3 R U5 5′
5′ U3 R U5 PBS 3′
(9) Both strands completed
3′ U3 R U5 5′
5′ U3 R U5 PBS 3′
LTR
LTR

Figure 23-12 ***(opposite)*** The mechanism of reverse transcription. A complicated series of nine events generates a double-stranded DNA copy of the single-stranded RNA genome of a retrovirus. The genomic RNA is packaged in the virus with a tRNA hybridized to a complementary sequence near its 5′ end at the primer binding site (PBS). The RNA has a short terminal sequence redundance (denoted R). Only one of two RNAs comprising the genome of a virion is shown; reverse transcription is not known to require the two strands. Also, the indicated terminal sequences span only a few hundred nucleotides; between the sequences denoted U5 and U3 there are actually about 7500 nucleotides. *Stage 1:* The tRNA is used as a primer by reverse transcriptase (Chapter 13) to make a short DNA copy of the 5′ end. *Stage 2:* The reverse transcriptase has a ribonuclease H activity that removes the RNA portion of the RNA-DNA hybrid. *Stage 3:* The R sequence in the "freed" DNA is complementary to the R sequence at the other end of the RNA and "jumps" to it. *Stage 4:* The DNA is extended to make a copy all the way to the 5′ end of the RNA. *Stage 5:* Again, hybrid RNA is removed, except for a short piece that is not digested. *Stage 6:* The undigested RNA segment serves as a primer for the synthesis of a second short strand of DNA. *Stage 7:* Once this second DNA strand copies the 5′ end of the first and continues through the tRNA, the tRNA and the primer for the second strand are removed by ribonuclease H. *Stage 8:* A second jump occurs, this time using the homology of the PBS. *Stage 9:* The two strands of DNA are each extended to the 5′ end of the opposing strand. These events lead to a DNA that is longer than the original RNA at both ends. A sequence from the 5′ end (U5) and a sequence from the 3′ end (U3) join with R to form a long terminal repeat (LTR). [See E. Gilboa et al., 1979, *Cell* **18**:93.]

Retroviruses Have Genes for Virion Proteins, but They Can Also Transduce Other Genes

The retrovirus genome has three regions that encode proteins, and all of these proteins are found in virions. The three regions are *gag*, which encodes a polyprotein that is cleaved to form four internal virion structural proteins; *pol*, which encodes the reverse transcriptase; and *env*, which encodes a glycoprotein that covers the virion surface (Figure 23-14). None of these proteins changes the growth properties of fibroblasts, so the basic retrovirus is not a transforming virus: in fact, infected cells are difficult to distinguish from uninfected cells by any parameter of cellular life.

Although the basic retrovirus life cycle does not include a transforming event, many retroviruses can cause cell transformation and cancers in animals and even in humans. In fact, the first tumor virus to be recognized was the Rous sarcoma virus, a chicken retrovirus described by Peyton Rous in 1911. Many other tumor-inducing retroviruses have been found, including the mouse mammary tumor virus (MMTV); leukemia viruses of chickens, mice, cats, apes, and, most recently, humans; sarcoma viruses of many species; and a few carcinoma-inducing viruses. The mechanisms by which retroviruses cause tumors have been intensively studied over the past decade, and our knowledge of the genes involved in changing cells from normal to malignant has become increasingly detailed.

The key to our present understanding of the tumor-inducing ability of transforming retroviruses came from the realization that such viruses contain genetic information other than *gag*, *pol*, and *env*. The additional genetic material acquired by these viruses is appropriated from normal cellular DNA. Because phages that have acquired cellular genes are said to *transduce* the genes they acquire, retroviruses that have acquired cellular sequences are called *transducing retroviruses*.

Transforming Retroviruses Transduce Oncogenes

The first indication of the existence of transducing retroviruses came with the discovery that Rous sarcoma virus contains nucleotide sequences that are not found in closely-related but nontransforming retroviruses. The extra nucleic acid sequence, called *src*, was then identified in the DNA of normal chickens as well as in the DNA of all vertebrate and even invertebrate species. This observation was a landmark because it showed that a gene that is able to transform cells can emerge from a cellular gene. Thus it was clear that cancer induction may involve the action of normal, or nearly normal, genes. Many other transduced cellular genes have been found in other transforming retroviruses, which implies that the normal vertebrate genome contains many potentially cancer-causing genes.

Transduced, cancer-inducing genes are called *transforming genes* or *oncogenes*, a word that derives from the Greek *onkos*, meaning a bulk or mass. (*Oncology* is the scientific study of tumors.) Oncogenes are named with three-letter italic designations (e.g., *src*); the gene in a retrovirus is denoted with the prefix v- (e.g., v-*src*), and the equivalent normal cellular gene or *protooncogene* is given the prefix c- (e.g., c-*src*). The protooncogenes that have been identified thus far mostly appear to be genes that are basic to animal life because they have been highly conserved over eons of evolutionary time. Many are evident in the DNA of arthropods (like *Drosophila*), and some are even found in yeast. Protooncogenes are thus impor-

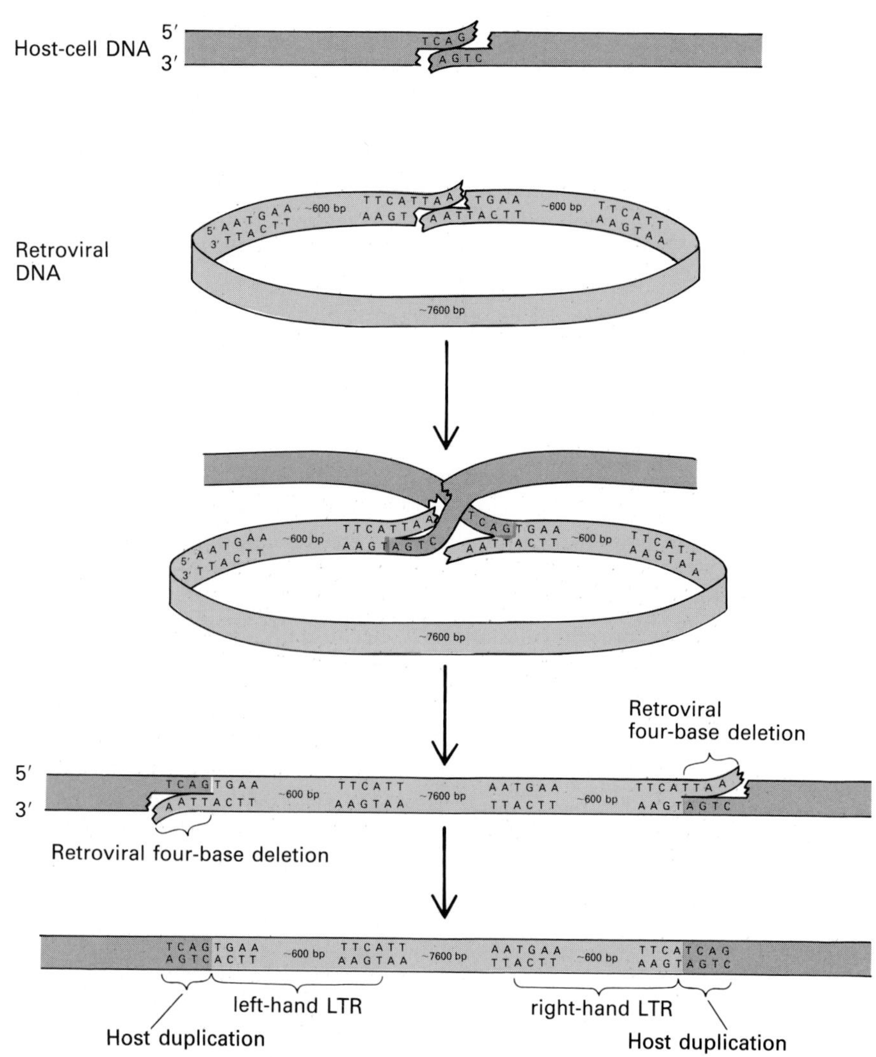

Figure 23-13 The integration of a circular retrovirus DNA molecule with two tandem LTRs. The circular DNA is shown integrating into host-cell DNA at a site at which the sequence TCAG happens to occur. The integration occurs by staggered cleavages of both the cellular DNA and the viral DNA. The illustrated case is a mouse retrovirus that makes a four-base repeat of cellular DNA at the site of integration; for other retroviruses, the repeat is four, five, or six bases long. [See C. Shoemaker and D. Baltimore, 1980, *Proc. Nat'l Acad. Sci. USA* 77:3932.]

tant normal cellular genes that have the apparently accidental property that they are easily modified into oncogenes.

Transducing retroviruses arise because of complex rearrangements following the integration of a retrovirus near a cellular protooncogene (Figure 23-15). Most commonly, the newly acquired genetic information in a transducing retrovirus replaces part or all of *gag, pol,* and *env,* making the new retrovirus defective. However, the products of *gag, pol,* and *env* can be provided by a *helper virus,* a coinfecting wild-type retrovirus, to allow the propagation of the defective virus. Examples of transducing retroviruses include Rous sarcoma virus, Abelson murine leukemia virus, and Harvey sarcoma virus (Figure 23-16).

There Are More than 20 Different Oncogenes

In many independent occurrences, cellular genetic information acquired by retroviruses has changed them into transforming viruses. At least 20 different genes have given rise to viral transforming genes, and the total possible number is not known. The transforming viruses have individual characteristics determined in part by which gene they acquire: for instance, some transform only adherent cells, whereas others transform both adherent and nonadherent cells. In animals, the ones that transform nonadherent cells cause leukemia; the remainder cause sarcomas. Abelson virus shows a remarkable specificity: in cultures of bone marrow cells it transforms only cells of the B-lymphocyte lineage (Chapter 24). In animals, however, Abelson virus occasionally causes other types of tumors, and it can affect the differentiation of fetal red blood cells. Besides causing sarcomas, Harvey virus produces tumors of the red blood cell lineage in mice. The reasons for these specificities have been hotly debated but they are not yet understood.

It should be evident that the transforming activity of a transducing retrovirus is actually a consequence of two sets of activities. The transforming activity is provided by

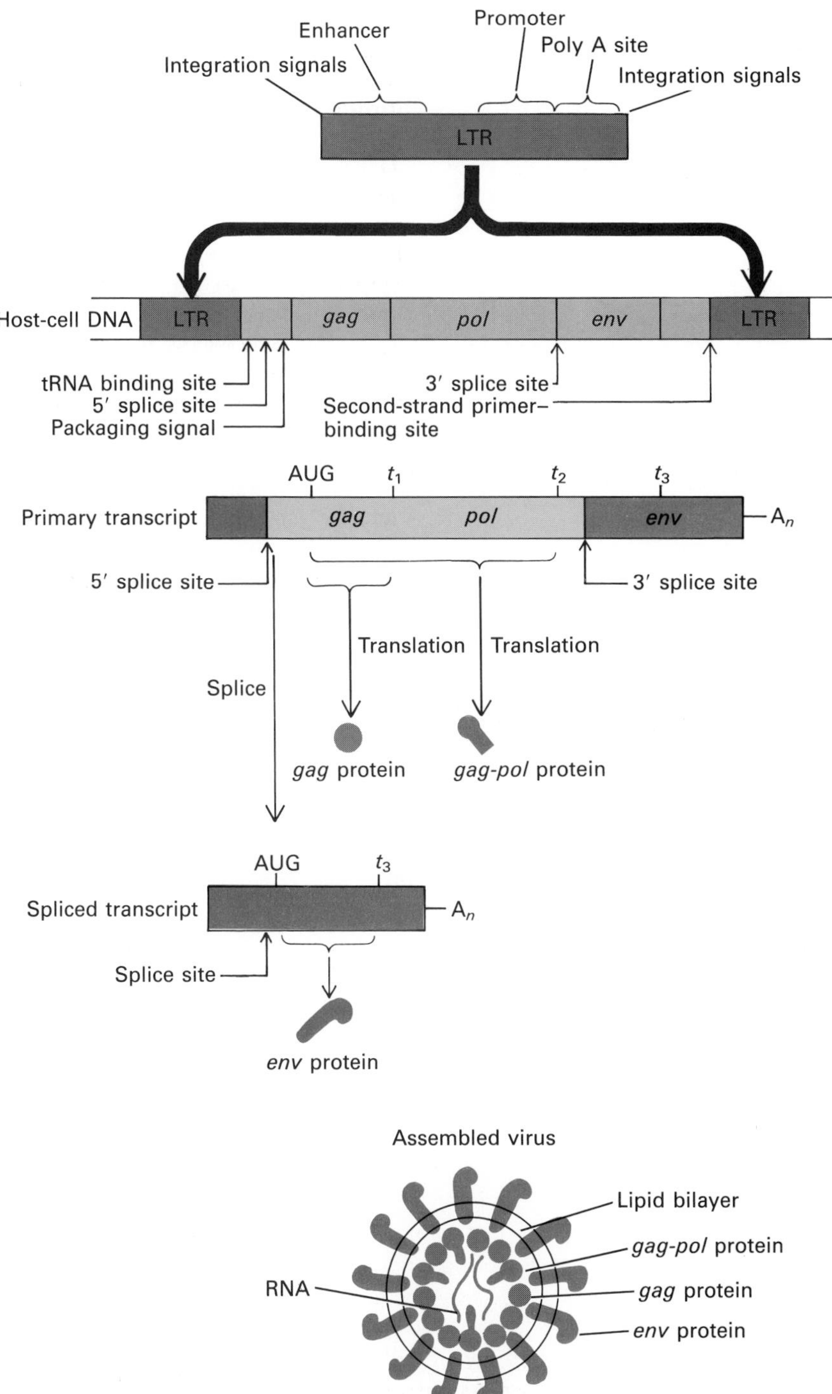

Figure 23-14 Genetic elements of the provirus and the viral gene products. A standard retroviral provirus is shown integrated into cellular DNA. The genetic structure of its flanking LTRs is indicated. Although the two LTRs are identical, their positions somehow allow the left-hand LTR to be a promoter for transcription and the right-hand LTR to specify the poly A site for the transcript.

The RNA transcript has three functional roles. It is an mRNA for the synthesis of both the *gag* and the *gag-pol* proteins because its first translation termination signal (t_1, the termination signal for *gag*) is suppressed occasionally (5 to 10 percent of the time) to give rise to some *gag-pol* protein. By virtue of its packaging signal, the transcript is also the genome of the virus. About 50 percent of the transcripts are spliced; why the 5′ and 3′ splice sites are only partially utilized remains an unanswered question. The spliced transcript, which is not packaged because the packaging signal has been spliced out, serves as the mRNA for the *env* protein.

All three of the translation products are virion proteins: the *env* protein is inserted into the lipid bilayer that surrounds the virus, the *gag* protein forms an inner shell, and the *gag-pol* protein (which contains the reverse transcriptase) is also inside the particle. After particles are formed, both the *gag* and the *gag-pol* products are extensively processed by proteolysis.

the acquired cellular genetic information, but the genetic elements that allow the oncogenes to be expressed and transferred from cell to cell are all inherited from the retroviral parent (see Figures 23-14 and 23-15). Most of the viral control elements are encoded in the LTR: these include the promoter and the enhancer, the poly A signal, and the integration signals. There are also a sequence that binds the tRNA just at the border of the left-hand LTR, a sequence for initiating plus-strand DNA synthesis at the border of the right-hand LTR, and a sequence that allows for packaging into virus particles (located between the left-hand LTR and the AUG that initiates *gag* protein synthesis). Thus the transducing retroviruses are hybrids in which both cellular and viral components play fundamental roles.

The conclusion that cellular genetic information can be the agent of transformation is remarkable because of its implication that our own genomes harbor genes that can,

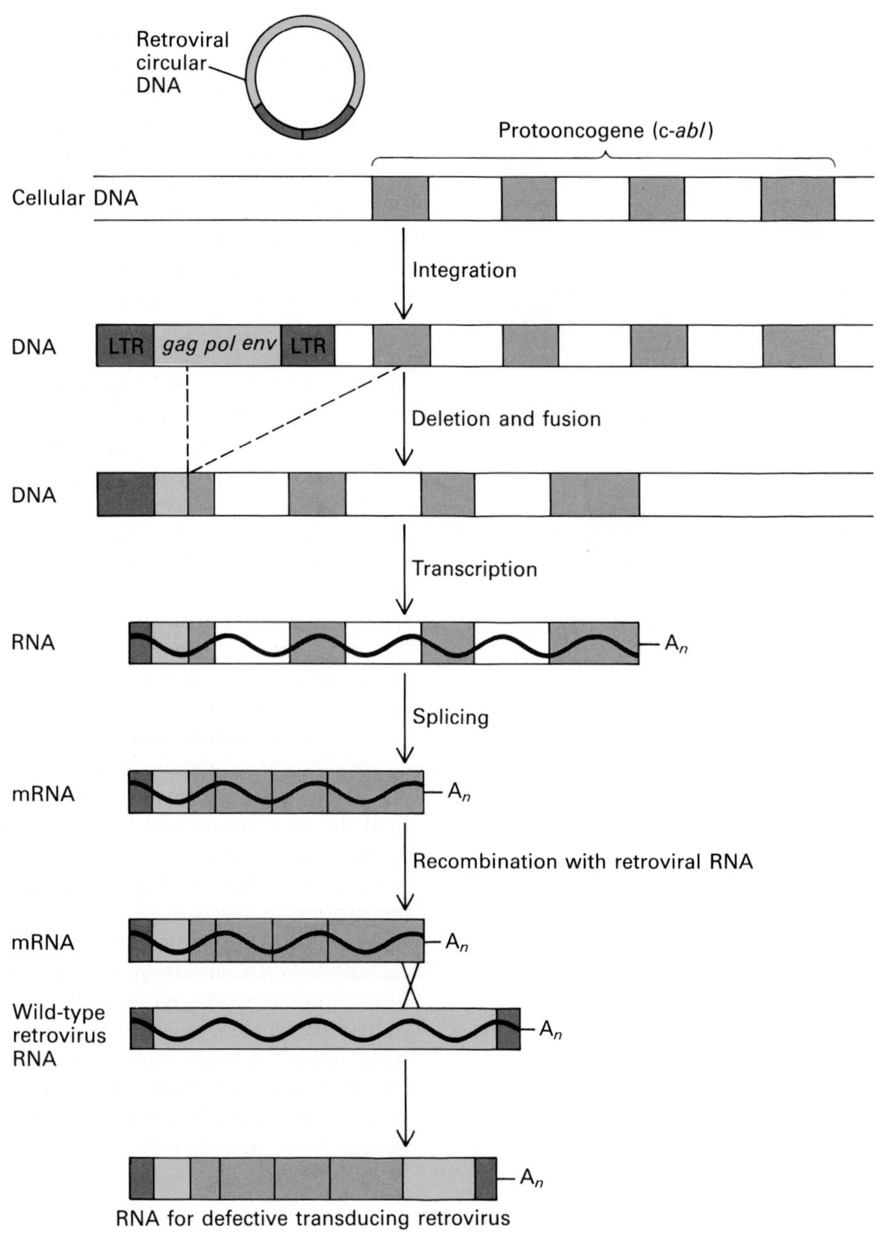

Figure 23-15 The formation of a transducing retrovirus. Although the events that generate a transducing retrovirus occur at such a low frequency that they have not been studied in the laboratory, the structure of the transducing viruses strongly suggests that they are formed as depicted here. We have chosen to illustrate Abelson virus formation because it has many features found in other retroviruses. The c-*abl* gene is shown as a series of exons in cellular DNA. A wild-type retrovirus integrates upstream of c-*abl*. This should occur randomly once in about 10^6 integrations. Then a deletion-fusion event fuses a c-*abl* exon into the *gag* region of the retrovirus. The LTR in this retrovirus–c-*abl* fusion can promote the synthesis of a transcript that will be spliced to produce an mRNA for the *gag-abl* fusion protein. It is assumed that the cell is also infected by a wild-type retrovirus that will package the *gag-abl* mRNA along with wild-type RNA. In the next generation, the two RNAs can recombine (presumably by a switch of templates during reverse transcription) to generate the RNA of the final transducing retrovirus (now called Abelson murine leukemia virus).

in the right circumstances, cause cancer. This notion has had a profound impact on cancer research, because it suggests that other agents, like chemicals and radiation, might be able to change normal genes into cancer-inducing genes by causing relatively simple mutations.

How Does a Protooncogene Become an Oncogene?

When a normal cellular gene becomes a retrovirus oncogene, it is significantly altered. Three kinds of alterations, affecting either the coding region of the gene or the expression of the gene, can occur. First, a retroviral oncogene is usually transcribed at much higher levels, and in a different range of cells, than its normal protooncogene counterpart. As part of a retrovirus, the oncogene's transcription will be determined by the viral control elements, which can promote high transcription rates in many different cell types. This quantitative difference alone appears to explain why certain normal genes transform when they are captured by a virus. Indeed, protooncogenes cloned from cellular DNA have become transforming genes when they have been inserted into a retrovirus or otherwise placed in a high-transcription environment. A second kind of alteration is the loss of parts of the cellular genes; the removal of segments apparently leaves the protein with an unregulated activity that, expressed at high levels, will transform cells. Third,

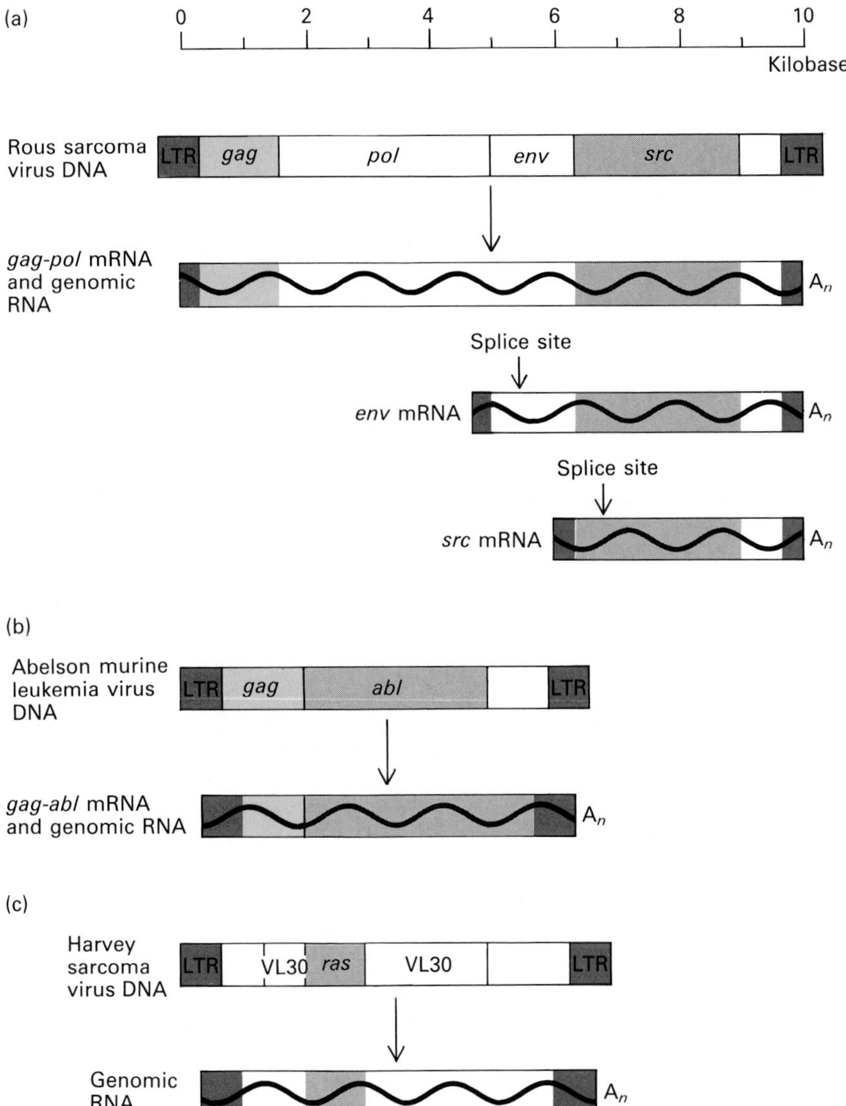

Figure 23-16 The structure and expression of the genes in three transducing retroviruses. The colored genes were derived from host-cell genomes. (a) Rous sarcoma virus—the only known nondefective transducing retrovirus. The *gag, pol, env,* and *src* products are expressed from three different mRNAs: one unspliced mRNA (for *gag* and *gag-pol*), which is also the genomic RNA, and two spliced mRNAs (one for *env,* the other for *src*). (b) Abelson murine leukemia virus. This highly defective virus yields one RNA molecule that serves as the mRNA for the *gag-abl* fusion protein and acts as the viral genome. (c) Harvey sarcoma virus. This virus has a complicated origin because it is a mouse retrovirus that acquired its oncogene during passage in a rat. Two recombination events must have occurred in the rat. One was a recombination of the mouse retrovirus with an endogenous retrovirus of the rat, called VL30. The second recombination was a fusion of a mutated form of the rat c-Ha-*ras* protooncogene into the mouse/rat recombinant retrovirus within the VL30 sequences. The only gene expressed by the Harvey virus is the v-Ha-*ras* gene, which encodes a 21,000-molecular-weight protein that is probably synthesized from the genomic RNA. The v-Ha-*ras* gene differs from the c-Ha-*ras* gene by one crucial point mutation in the codon for amino acid 12: this mutation converts the c-Ha-*ras* protooncogene into the v-Ha-*ras* oncogene.

crucial sequence differences between transforming genes and their cellular counterparts can be as subtle as single point mutations.

These alterations in the protooncogenes result in two separate consequences: a quantitative change in the level of the encoded protein and a qualitative change in the protein itself. Apparently both types of events play a role in changing protooncogenes to oncogenes.

Before we consider the proteins encoded by oncogenes, we should note that transducing retroviruses are able to acquire genes other than oncogenes. Such events probably occur naturally, but their products have not been isolated because their presence is not evident: tumor viruses make their presence known because they produce an obvious disease; other transducing retroviruses would not have such dramatic consequences, and they would therefore be very difficult to recognize. Although only oncogenic transducing viruses have been found naturally, other types of transducing retroviruses have been constructed in the laboratory by using the techniques of in vitro genetic manipulation. An example of an artificially constructed virus that can transduce both a globin cDNA and a neomycin-resistance gene is shown in Figure 23-17. Such artificial retroviruses have enabled laboratory workers to insert new or altered genes into cells and animals. It is hoped that such constructed genetic elements may be used in the future to treat human genetic diseases, such as those that affect red blood cells.

What Are the Proteins Encoded by Oncogenes?

Now that so many transforming genes and even their protein products have been identified, one might expect rapid progress in the understanding of cancer. It has been very difficult, however, to ascertain just what cellular bio-

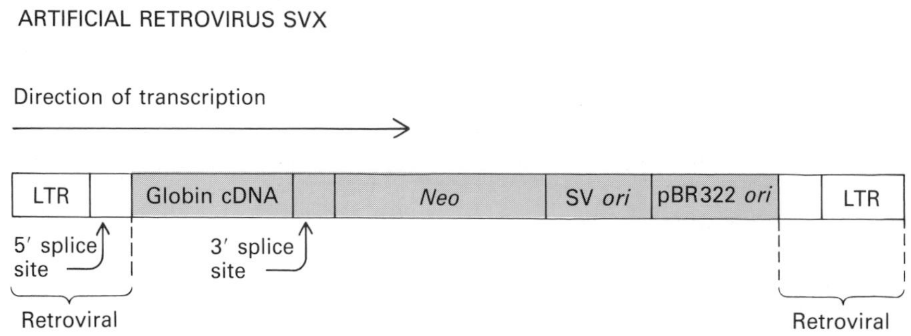

ARTIFICIAL RETROVIRUS SVX

Direction of transcription

LTR		Globin cDNA		*Neo*	SV *ori*	pBR322 *ori*		LTR

5′ splice site — 3′ splice site

Retroviral — Retroviral

Figure 23-17 The artificial retrovirus SVX. It was built as a shuttle vector, a recombinant molecule that is able to carry genes back and forth between mammalian cells and bacteria. In this vector the crucial sequence elements from the left and the right ends of a mouse retrovirus DNA molecule are coupled to the following elements: a globin cDNA that will direct globin synthesis when SVX is present in a mammalian cell, a 3′ splice site that can be joined to the 5′ splice site near the left-hand LTR to form an mRNA for the expression of the next gene (downstream), a gene (*Neo*) that encodes a protein conferring resistance to an analog of the antibiotic neomycin, an SV40 virus origin of replication, and a pBR322 bacterial plasmid origin of replication. This shuttle vector is used to introduce the globin gene into cells (other cDNAs can be substituted for globin). Cells integrating the retrovirus can be selected because of the presence of the *Neo* gene, which is expressed both in bacteria and in mammalian cells. To rescue the provirus from cells, the cells can be fused to other cells that make SV40 T antigen. This fusion activates replication from the SV40 origin, and the resulting circular DNA molecules can be purified and reinserted into bacteria. Their multiplication in bacteria is caused by the pBR322 origin of replication. Bacteria containing the recombinant retrovirus can be selected through the *Neo* marker. [See C. Cepko et al., 1984, *Cell* 37:1053.]

chemical systems are affected by the transforming proteins. There are three intriguing clues. Some transforming proteins phosphorylate tyrosine residues in other proteins; the *ras* family of oncogenes encode a guanine nucleotide–binding protein; and many oncogene products that are located in the nucleus are normally expressed when quiescent cells are stimulated to grow.

Enzymes That Phosphorylate Tyrosine Residues

About one-third of the known oncogenes encode proteins that phosphorylate tyrosine residues on other proteins; these enzymes are called *tyrosine-specific protein kinases* (Table 23-1). The reaction they catalyze is the transfer of the terminal phosphoryl group of ATP to the hydroxyl group of a tyrosine residue in a protein. Until the discovery that retrovirus oncogene products could phosphorylate tyrosine residues, the known sites of protein phosphorylation were largely limited to serine and threonine. After the tyrosine-specific phosphorylation by oncogene products was found, it was recognized that about 0.1 percent of protein-linked phosphate in normal cells is found on tyrosine residues. Cells transformed by the viruses that encode tyrosine-specific kinases were discovered to have about 10-fold higher levels of phosphotyrosine than normal cells or cells transformed by other means. The excess phosphotyrosine is distributed over many proteins, which suggests that either the viral kinases phosphorylate a number of different protein substrates or that the phosphorylation of certain key proteins sets off a spate of further phosphorylations.

Direct evidence that the viral kinases have a very low substrate specificity has come from experiments in which oncogene products have been expressed in bacteria. To accomplish this feat, oncogenes were removed from cloned retroviral DNA and placed into vectors that allowed their expression in bacteria. The kinase encoded by Abelson virus, for example, has been expressed in *Escherichia coli* with this procedure. When a bacterial cell contains the Abelson kinase, many tyrosine residues in bacterial proteins become phosphorylated. Because ordinarily no phosphotyrosine exists in *E. coli* proteins, the kinase encoded by v-*abl* must phosphorylate many substrate proteins.

These experiments and others that have been used to examine protein-bound phosphotyrosine levels in cells transformed by kinase-encoding oncogenes have shown that the kinases recognize and phosphorylate a very broad range of target proteins. The phosphorylation of one or more crucial target proteins may be the mechanism by which the kinase induces the transformed state, but if such crucial proteins exist they have thus far eluded detection. Possibly the effects on many proteins conspire to cause transformation.

Vinculin, a protein involved in linking the actin cytoskeleton to the cell membrane (Chapter 19), is one of the proteins known to have an elevated phosphotyrosine level in transformed cells. The alteration of such a protein could easily be imagined to cause some of the hallmarks of the transformed state: changed cell morphology, reduced concentrations of actin microfilaments, loss of the anchorage dependence of growth, and changed mobility of cell surface proteins. In this regard it may be pertinent that the *src* oncogene of Rous sarcoma virus encodes a tyrosine-specific protein kinase localized to *adhesion plaques,* sites of bonding between cells and their underlying support where vinculin is also concentrated.

Tyrosine phosphorylation is now known to occur in a number of normal cell systems. One example is the phosphorylation of tyrosine residues in the epidermal growth factor receptor when it interacts with its ligand. The interaction of the factor with its receptor also leads to the phosphorylation of other proteins in cells. This growth-factor receptor, like numerous others, has a tyrosine-specific protein kinase domain as part of its structure. The receptors for the platelet-derived growth factor (PDGF) and for insulin both contain tyrosine kinases, and both become phosphorylated on tyrosine after their interaction with appropriate ligands. Perhaps tyrosine-specific phosphorylation plays a role in signaling cell responses to the presence of these ligands; however, no definitive evidence for such a role has been found.

Guanine Nucleotide–Binding Proteins Encoded by the *ras* Family The Harvey and Kirsten sarcoma viruses suggest a second mechanism by which transforming proteins may affect cellular biochemical systems: these viruses encode proteins that are very similar to normal p21 proteins encoded by the three c-*ras* genes. The p21 proteins have sites for the very tight binding of guanine nucleotides, especially GDP and GTP. This feature seems highly significant, because GTP/GDP binding is a characteristic of a key regulatory protein in the cells' pathway of response to hormones that activate cAMP synthesis (Chapter 16). Also, tubulin binds GTP tightly as part of the regulation of its polymerization (Chapter 18).

Ras-related genes have been identified and specifically mutated in the yeast genome. When the two genes most like *ras* genes were eliminated by mutation, the yeast could not grow; thus *ras* genes play a critical metabolic role in yeast. Yeast with dual *ras* mutations can be rescued by a third mutation that deregulates the enzyme adenylate cyclase; this implies that the role of the *ras* genes of yeast is linked to cAMP formation. Whether mammalian *ras* genes and yeast *ras* genes have analogous functions remains to be determined.

A number of the transforming proteins—including those encoded by v-*src,* v-*abl,* and v-*ras,* as well as the polyoma mid-T—are at least partly bound to the plasma membrane of the cell (Table 23-1). This binding is not a consequence of their being processed through the endoplasmic reticulum and the Golgi complex; they are synthesized on free polyribosomes and appear to bind subsequently to the inner face of the plasma membrane. Many of these proteins contain covalently bound fatty acids that may aid the membrane binding. From their location in the cell, these proteins appear to cause transformation by direct interaction with cytoplasmic proteins.

Nuclear-Localized Oncogene Products The third clue to oncogene action comes from the nuclear oncogenes (Table 23-1). These oncogenes have certain common properties. First, their products are found in the cell nucleus while other oncogene products are cytoplasmic. Second, they are generally unmutated from their protooncogene form; they become oncogenes by virtue of the altered control of their expression, and not because of any alteration in their structure. Third, they are genes whose products are often synthesized at greatly increased rates just after quiescent cells are stimulated to grow. For example, PDGF treatment of quiescent 3T3 cells will induce a transient increase (up to 50-fold) in the production of *fos, myc,* and p53 protooncogene products. The products of these genes may be able to alter the transcriptional rates of other genes and may play some crucial role in reprogramming the transcriptional events of the cell in preparation for cell growth. In their structure and their potential ability to alter transcription, the nuclear oncogene products are similar to the polyoma and SV40 T antigens and to an adenovirus early protein (E1A).

Oncogenes May Act through Pathways Involved in Control of Normal Growth by Growth Factors

The location of certain transforming proteins in the plasma membrane and the similarities of these proteins to the epidermal growth factor receptor or the GTP-binding cAMP regulatory factor suggest that the mode of action of many transforming proteins may be to trigger responses analogous to the cell's responses to growth factors and hormones. In this model of transformation the transforming protein generates an unregulatable growth signal that is recognized by certain pre-existing cellular pathways whose function is to respond to the normal, highly regulated growth signals from factors and hormones. Increasingly so, this model seems to be the key to understanding at least some cancers, but direct evidence in support of it has yet to be accumulated.

Strong support for the idea that unregulated signaling of growth may underlie oncogenesis comes from the observation that two oncogene products are proteins that are normally involved in factor-regulated growth control. One is the *sis* gene product, which is closely related to, or even identical to, a chain of platelet-derived growth factor. The other is the *erb*B gene product, which is a portion of the epidermal growth factor receptor that lacks the hormone-binding domain and may give rise to a constitutive growth-stimulating signal. Both identifications have been made by observing that the translated nucleotide sequences derived from DNA clones of oncogenes are identical or closely related to the amino acid sequences of

Table 23-1 The oncogenes

Acronym	Oncogene found in: Animal retrovirus	Nonviral tumor	Subcellular location of protein	Nature of encoded protein
CLASS I: PROTEIN KINASES				
src	Rous avian sarcoma		Plasma membrane	Tyrosine-specific protein kinases (phosphorylate tyrosine residues)
yes	Yamaguchi avian sarcoma			
fps (fes)	Fujinami avian sarcoma (and feline sarcoma)		Cytoplasm	
abl	Abelson murine leukemia	Chronic myelogenous leukemia	Plasma membrane	
ros	UR II avian sarcoma		Cytoplasmic membranes	
fgr	Gardner-Rasheed feline sarcoma			
*erb*B	Avian erythroblastosis		Plasma membrane	Tyrosine-specific protein kinase derived from EGF receptor
neu		Neuroblastoma	Plasma membrane	Receptorlike tyrosine-specific protein kinase
fms	McDonough feline sarcoma		Cytoplasmic membranes	Receptorlike protein
mos	Moloney murine sarcoma		Cytoplasm	Protein kinases specific for serine or threonine
raf (mil)	3611 murine sarcoma			

these cellular proteins or protein fragments. These are landmark discoveries because they provide the first link between normal growth-controlling pathways and the unregulated growth of cancer cells.

Slow-Acting Carcinogenic Retroviruses Transform Cells by Insertion near a Protooncogene

The transducing retroviruses are not the only class of tumor-inducing retroviruses. Another class of these viruses also cause cancer, but over a period of months or years compared with the days or weeks required for cells to respond to a transducing virus. The genomes of the slow-acting retroviruses differ from those of the transducing viruses in one crucial respect: they lack an oncogene. The slow-acting retroviruses not only have no affect on growth of cells in culture, but they are also completely proficient to multiply themselves because they have suffered no debilitating deletions or insertions.

A mechanism that allows the avian leukemia viruses to cause cancer appears to be quite generally applicable for the other slow-acting retroviruses. Like all retroviruses, the avian leukemia viruses generally integrate into cellular chromosomes at random; thus it was a significant discovery that the proviral DNA has a highly preferred location in the cells from tumors caused by these viruses. Because it seemed possible that the site of integration in the tumor cells was near one of the established protooncogenes, all possible genes were screened; one, c-*myc*, was found to be located near the proviral DNA. This finding suggested that the slow-acting avian leukemia viruses cause disease by activating c-*myc* to become an oncogene. The viruses act slowly both because integration near c-*myc* is a random, rare event and because secondary events probably have to occur before a full-fledged tumor becomes evident.

An integrated viral genome can act in a number of ways to cause a cellular gene to become an oncogene. In some tumors, the 5′ end of the *myc* gene transcript includes a sequence from a retrovirus LTR. In such cases the right-hand LTR of an integrated retrovirus—which

Table 23-1 *Continued*

Acronym	Oncogene found in: Animal retrovirus	Oncogene found in: Nonviral tumor	Subcellular location of protein	Nature of encoded protein
CLASS II: p21 PROTEINS				
Ha-*ras*	Harvey murine sarcoma	Bladder, mammary skin carcinomas	Plasma membrane	Guanine nucleotide–binding proteins with GTPase activity
Ki-*ras*	Kirsten murine sarcoma	Lung, colon carcinomas	Plasma membrane	
N-*ras*		Neuroblastoma leukemias	Plasma membrane	
CLASS III: NUCLEAR PROTEINS				
myc	Avian MC29 myelocytomatosis		Nuclear matrix	Proteins possibly involved in regulating transcription
N-*myc*		Neuroblastoma	Nuclear matrix	
myb	Avian myeloblastosis	Leukemia	Nuclear matrix	
ski	Avian SKV770		Nucleus	
fos	FBJ osteosarcoma		Nuclear matrix	
p53		(Demonstrated by cell transformation)	Nucleus	
CLASS IV: GROWTH FACTORS				
sis	Simian sarcoma		Secreted	Derived from PDGF gene
MISCELLANEOUS				
Blym		Avian bursal lymphomas		
rel	Avian reticuloendotheliosis			
erb A	Avian erythroblastosis			
ets	Avian E26 myeloblastosis			
met		Murine osteosarcoma		

usually serves as a terminator—is believed to become a promoter and initiate the synthesis of RNA transcripts from the c-*myc* gene (Figure 23-18). Such c-*myc* transcripts apparently encode a perfectly normal c-*myc* product. The enhanced level of c-*myc* RNA due to the very strong promoting activity of the LTR appears to be the explanation of the oncogene activation: the carcinogenic effect is thus entirely the result of a change in the amount of product from a normal gene. This mechanism, which is called *promoter insertion,* has been implicated in many avian leukemia virus–induced tumors.

In a few tumors induced by avian leukemia viruses, the virus has been found integrated near the c-*myc* gene but at the 3′ end of the gene, or it has been found at the 5′ end but oriented in the transcriptional direction opposite to that of the gene. In such cases the promoter insertion model cannot be applicable, and the virus must be indirectly affecting the activity of the gene. An enhancer activity (Chapters 9 and 12) is apparently responsible for the activation of the *myc* gene by increasing its level of transcription, changing the cell type in which it is expressed, and/or altering its time of expression in the cell cycle. This mechanism is called *enhancer insertion.*

Either promoter insertion or enhancer insertion is the probable explanation for many tumors caused by retroviruses that do not carry oncogenes. This class of retroviruses includes the mouse mammary tumor virus and a number of mouse leukemia viruses. In some cases activation of c-*myc* is involved, but in others we know only that the virus has a common site of integration in many independently induced tumors. An oncogene is probably nearby, but no such oncogene has been proved to exist.

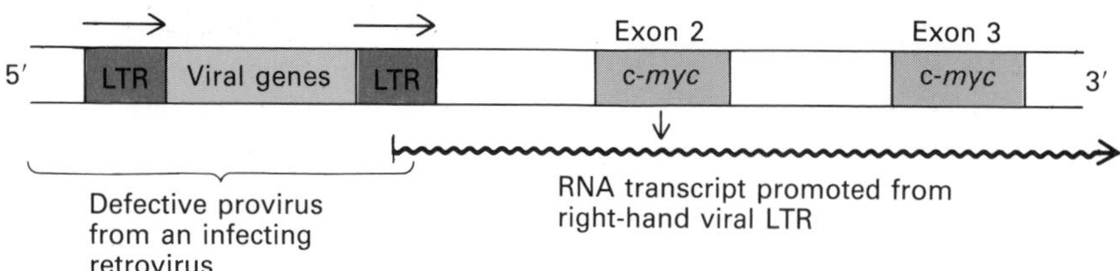

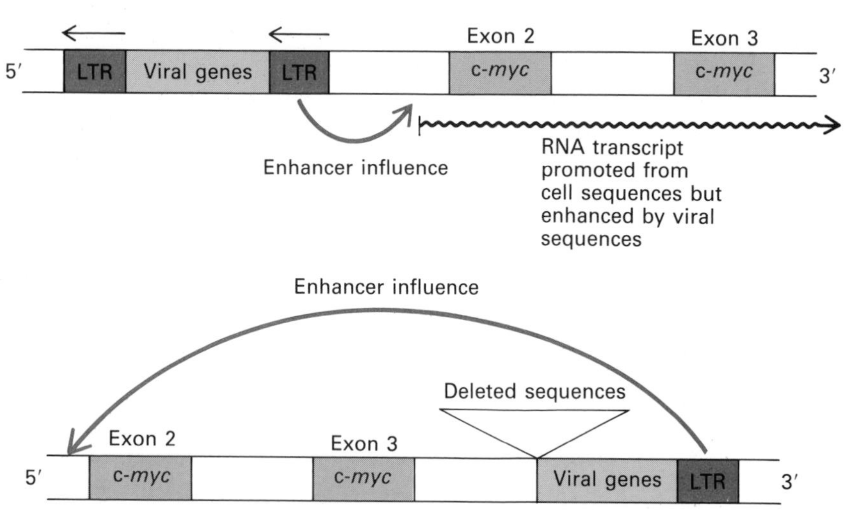

Figure 23-18 Retroviral promoter and enhancer insertions activating the c-*myc* gene. (a) The promoter effect arises when the retrovirus inserts upstream (5′) of the c-*myc* exons. The right-hand LTR can then act as a promoter. Ordinarily, right-hand LTRs act as terminators; only if the provirus has a defect preventing transcription through to the right-hand LTR does that LTR function as a promoter. The c-*myc* gene is shown as two exons; there is a further upstream exon but it has no coding sequences. (b) The enhancer effect arises when a retrovirus inserts either upstream of the c-*myc* gene but in the opposite transcriptional direction, or downstream of the gene. In both cases, a viral LTR acts on a c-*myc* promoter sequence. In the downstream case, the LTR also terminates the transcript. [Modified from actual cases of retroviral insertion described in G. G. Payne et al., 1982, *Nature* **295**:209.]

The activation of an oncogene by promoter or enhancer integration should be clearly distinguished from the acquisition of an oncogene by a transmissible virus. As is evident from Figure 23-15, the first step in the integration of an oncogene into a retrovirus may well be a promoter insertion, but several additional steps must occur before a defective, oncogene-containing retrovirus emerges. Those steps are all low-probability events, so very few promoter insertions progress to the formation of a transmissible virus. In fact, because defective transducing retroviruses depend on a helper virus for their cell-to-cell transmission, they are not maintained in natural animal populations. Most oncogene-containing viruses have arisen in laboratories or in domesticated animals and have been maintained for experimental purposes; the defective viruses are not readily spread from animal to animal. In natural populations, insertional oncogene activation is probably the major cause of retrovirus-induced cancer.

What Viruses Can Cause Human Tumors?

The occurrence of many RNA and DNA tumor viruses in animals raises the question of whether humans also have tumor viruses, and if so, what fraction of human cancer might be caused by them. In certain specific situations human tumor viruses have been found, but minimal evidence exists for a viral involvement in the types of human tumors that are prevalent in developed countries. A lack of evidence does not constitute proof of noninvolvement, however, and it remains an open question whether viruses might have a wider role in human cancer than is now suspected. In the following subsections we describe the various viruses that are presently implicated in human cancers.

Epstein-Barr Virus This herpes-like DNA virus probably plays a role in at least two human tumors: Burkitt's lymphoma and nasopharyngeal carcinoma. These tumors are rare in America and Europe, but in certain hot, wet sections of Africa Burkitt's lymphoma is the primary cancer in children, and among Southern Chinese nasopharyngeal carcinoma is a major cancer. In contrast, Epstein-Barr virus causes mainly mononucleosis in adolescents and young adults of the developed world; it leads to cancer very rarely, and then usually in people with a malfunctioning immune system. The cancer-inducing ability of the virus in Africa may be due to the presence of a second endemic factor, thought to be malaria, which works in concert with the virus.

Human Retroviruses In 1980 after many years of fruitless search for a human retrovirus, oncogenic or nononcogenic, the first plausible candidate was reported: human T-cell leukemia virus (HTLV). This virus is distinct from all known animal tumor viruses; infection with it is clearly associated with a type of T-lymphocytic human tumor that is especially prevalent in southern Japan, the Carribean, and parts of Africa. How the virus causes the tumor is being actively studied; oddly, the process appears to involve neither an oncogene within the virus nor a reproducible site of integration in tumor cell DNA. Rather, the basis of its carcinogenicity may be a unique viral genetic region totally distinct from *gag, pol,* and *env.*

The other known major human disease caused by a retrovirus is acquired immune deficiency syndrome (AIDS). The virus that causes this fatal disease is different from HTLV, although it is often called HTLV-III. The other HTLVs transform T lymphocytes into continuously growing, immortal cells. The AIDS-related retrovirus kills T lymphocytes, thus destroying a critical aspect of the body's immune defenses and leaving the victim open to a wide range of infections. One result of AIDS is a propensity of patients to develop Kaposi's sarcoma, an otherwise very rare tumor.

Hepatitis B Virus Infection with this very small DNA virus—which is mainly responsible for human "serum hepatitis"—is correlated with liver cancers, especially in underdeveloped areas of the world. Although it has a DNA genome, it multiplies through an RNA intermediate like a retrovirus. Its genome has been found integrated into the DNA of hepatomas, but no mechanism whereby the genome could be responsible for the malignant state of cells has been discovered. Whether hepatitis B virus can cause cancer by itself or only in synergy with other cancer-inducing agents is a question that is being actively investigated.

Papilloma Viruses These small DNA viruses, members of the papovavirus group, are responsible for warts and other benign human tumors and also for some malignancies. Currently over 15 distinct human papilloma viruses are recognized. Two types of papilloma virus are associated with human cervical carcinoma, and may be the cause of this sexually transmitted disease.

BK and JC Viruses These are the human counterparts of SV40 and polyoma. BK can transform cells in culture, but neither virus has been associated with any specific human cancer.

Chemical Carcinogenesis

Whereas viruses probably cause a small fraction of human cancer, chemicals are thought to be culpable in a larger number of cases. Chemicals were originally associated with cancer through experimental studies in intact animals: the classic experiment was to repeatedly paint a substance such as benzo(*a*)pyrene on the back of a mouse, causing both local and systemic tumors in the animal. In this way, many substances have been shown to be *chemical carcinogens.*

Most Chemical Carcinogens Must Be Activated by Metabolism

Chemical carcinogens have a very broad range of structures with no obvious unifying chemistry (Figure 23-19). A major class of such chemicals, the polycyclic aromatic hydrocarbons [e.g., benzo(*a*)pyrene], would seem from their structures to be highly unreactive. Early workers were puzzled why such unreactive and water-insoluble compounds should be so potent as cancer inducers. The situation was clarified when it was realized that there are two broad categories of carcinogens, *direct-acting* and *indirect-acting* (Figure 23-19), with the latter requiring metabolic activation to become carcinogens. The direct-acting carcinogens, of which there are only a few, are reactive electrophiles (compounds that seek out and react with negatively charged centers in other compounds). The metabolic activation process of indirect carcinogens makes *ultimate carcinogens* from the precursors by giving them electrophilic centers (Figure 23-20).

The metabolic activation of carcinogens is carried out by enzymes that are normally resident in the body. Animals have such enzymes, especially in the liver, because they are part of a system that detoxifies noxious chemicals that make their way into the body. Therapeutic drugs, insecticides, polycyclic hydrocarbons, and some natural products are often so fat-soluble and water-insoluble that they would accumulate continually in fat cells if there were no way for the animal to excrete them. The detoxification system works by solubilization: it adds hydrophilic groups to water-insoluble compounds, thus allowing the body to rid itself of noxious or simply insoluble materials.

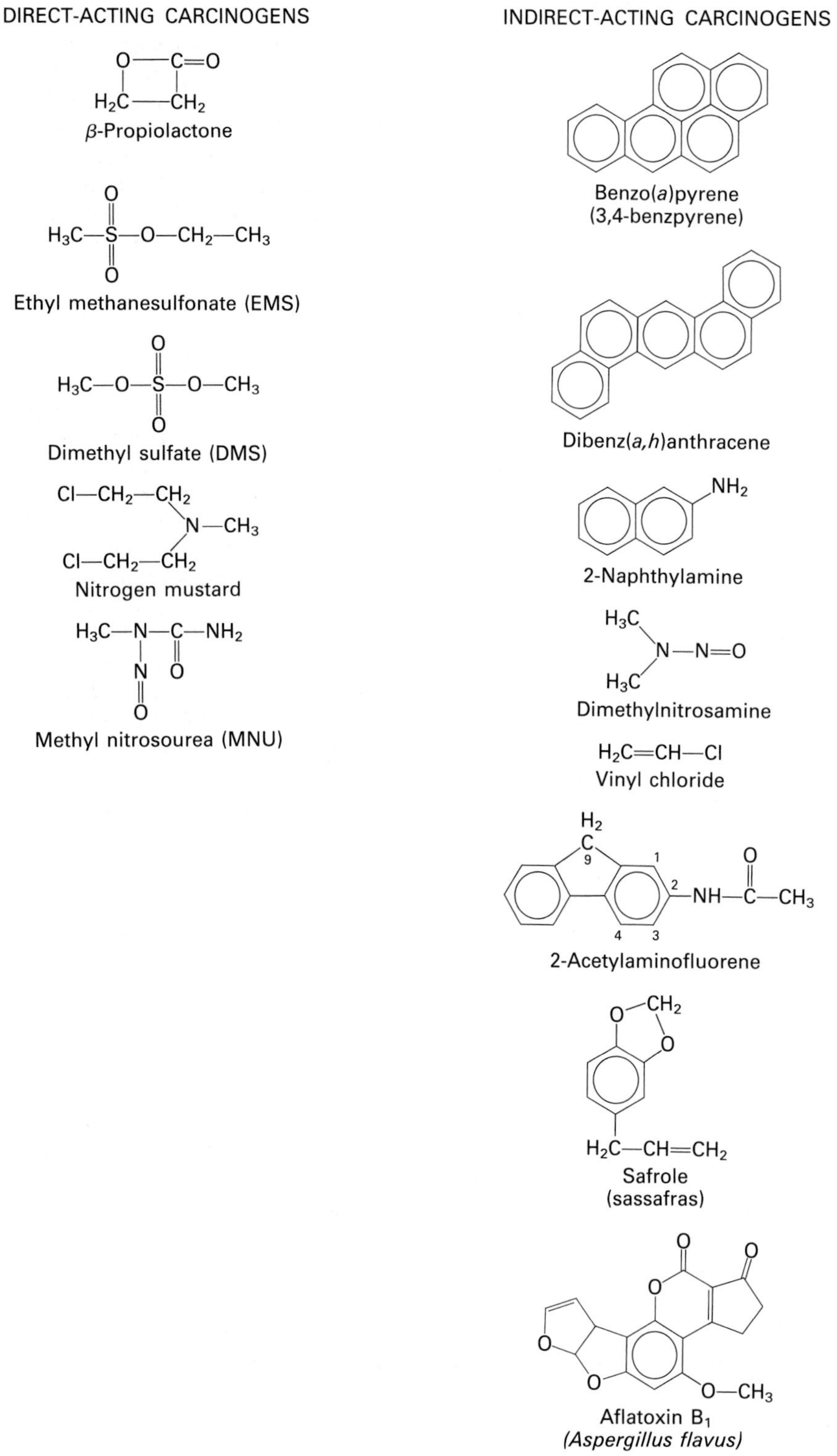

DIRECT-ACTING CARCINOGENS

β-Propiolactone

Ethyl methanesulfonate (EMS)

Dimethyl sulfate (DMS)

Nitrogen mustard

Methyl nitrosourea (MNU)

INDIRECT-ACTING CARCINOGENS

Benzo(*a*)pyrene (3,4-benzpyrene)

Dibenz(*a*,*h*)anthracene

2-Naphthylamine

Dimethylnitrosamine

$H_2C{=}CH{-}Cl$
Vinyl chloride

2-Acetylaminofluorene

Safrole (sassafras)

Aflatoxin B_1 (*Aspergillus flavus*)

Figure 23-19 Structures of selected chemical carcinogens.

The detoxification process begins with a powerful series of oxidation reactions catalyzed by a set of proteins called cytochrome P-450s (Figure 23-20). These enzymes, which are bound to endoplasmic reticulum membranes, can oxidize even highly unreactive compounds such as polycyclic aromatic hydrocarbons. The result of the oxidation of polycyclic aromatics is an epoxide, a very reactive electrophilic group. Usually the epoxides are rapidly hydrolyzed into hydroxyl groups that are then coupled to glucuronic acid or other groups to make the compounds more water-soluble. Occasional compounds, however, can be expoxidated at sites that are only slowly hydrolyzed to hydroxyl groups, probably because the relevant enzyme (epoxide hydratase) cannot get to the epoxide to

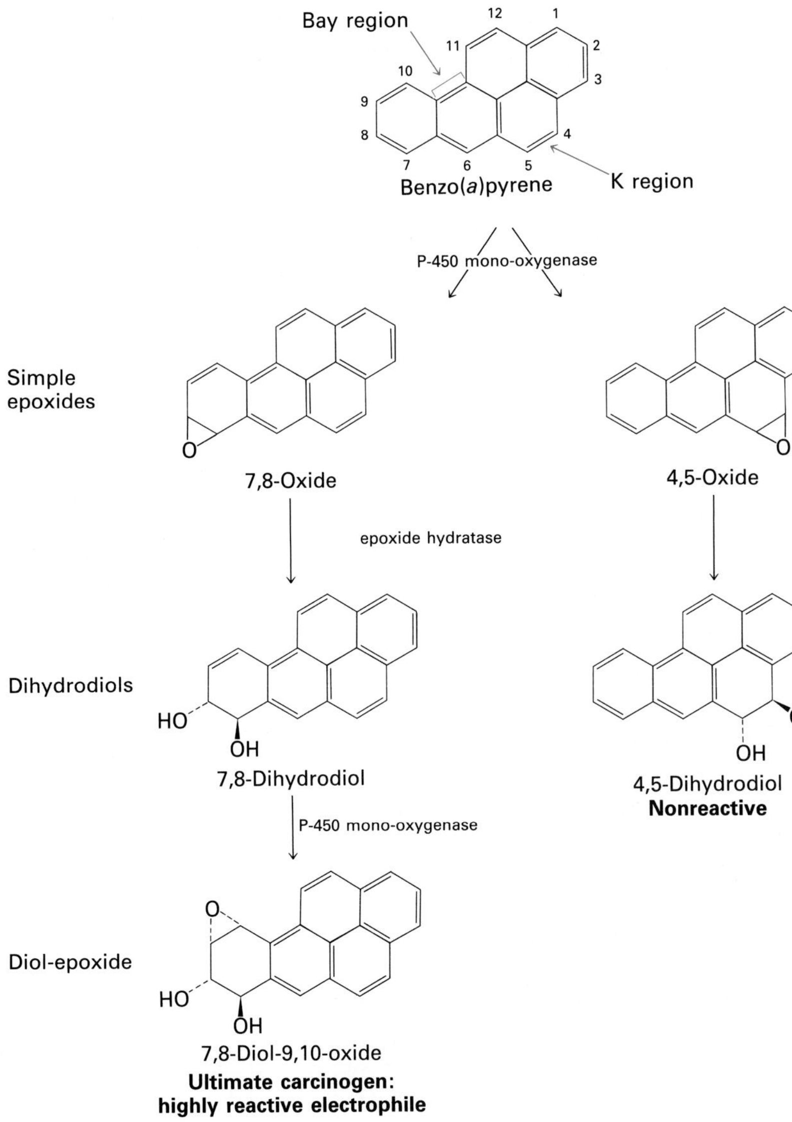

Bay region
Benzo(*a*)pyrene
K region
P-450 mono-oxygenase
Simple epoxides
7,8-Oxide
4,5-Oxide
epoxide hydratase
Dihydrodiols
7,8-Dihydrodiol
4,5-Dihydrodiol
Nonreactive
P-450 mono-oxygenase
Diol-epoxide
7,8-Diol-9,10-oxide
Ultimate carcinogen: highly reactive electrophile

Figure 23-20 The metabolic activation of a polycyclic aromatic hydrocarbon carcinogen. Benzo(*a*)pyrene is a powerful carcinogen, which although it is chemically almost inert, becomes a highly reactive electrophile due to metabolic conversions. Two metabolic pathways are shown. The right-hand pathway involves an intermediate epoxide (4,5-oxide) formed by the attack of the cytochrome P-450 system on what is called the "K region" of the molecule. An epoxide in this region is rapidly hydrolyzed to a nonreactive dihydrodiol. The left-hand pathway involves an initial oxidation at the 7,8 double bond, leading to a 7,8-oxide that is rapidly converted to a 7,8-dihydrodiol. This compound is still a good substrate for the P-450 system and it is again epoxidated, now near the "bay region" at the 9,10 double bond. The 7,8-diol-9,10-oxide (or diol-epoxide) is not a good substrate for epoxide hydratase, so it is released into the cell as a highly reactive electrophile. This form is carcinogenic because it can readily react with negatively-charged centers in DNA.

act on it. Such compounds become highly reactive electrophiles, and their precursors are the molecules we know as carcinogens (Figure 23-20). Other modes of activation occur with types of carcinogens other than polycyclic aromatic hydrocarbons, but all are catalyzed by P-450–related enzymes.

DNA Appears to Be the Relevant Target for Carcinogens

Once inside cells, electrophiles can react with negatively charged centers on many different molecules: protein, RNA, and DNA, to name the most obvious. Which reaction is important for carcinogenesis? This question has long been debated and the answer is still incomplete, but for a number of reasons the focus has turned to DNA as the most important site of reactivity for ultimate carcinogens. It is worth considering systematically the arguments that have turned the spotlight onto DNA.

1. *Ultimate carcinogens do modify DNA.* Ultimate carcinogens will react with free DNA, but, more importantly, when they are applied to cells they will react with cellular DNA. Depending on their size and structure, these compounds will react with different positions on different DNA bases.

2. *Ultimate carcinogens cause permanent changes in DNA.* Ultimate carcinogens are mutagens: the changes they cause in the base sequence of DNA are expressed as permanent changes in the phenotype of the treated cell. This characteristic is most easily assayed in bacteria, where in the presence of a liver activating system, carcinogens often show themselves to be very powerful mutagens. Because most compounds that have been identified as carcinogens for experimental animals are mutagens for bacteria, mutagenesis has become a test for carcinogens. The first and most popular of these tests is the Ames test, named for its developer Bruce Ames, a bacterial geneticist.

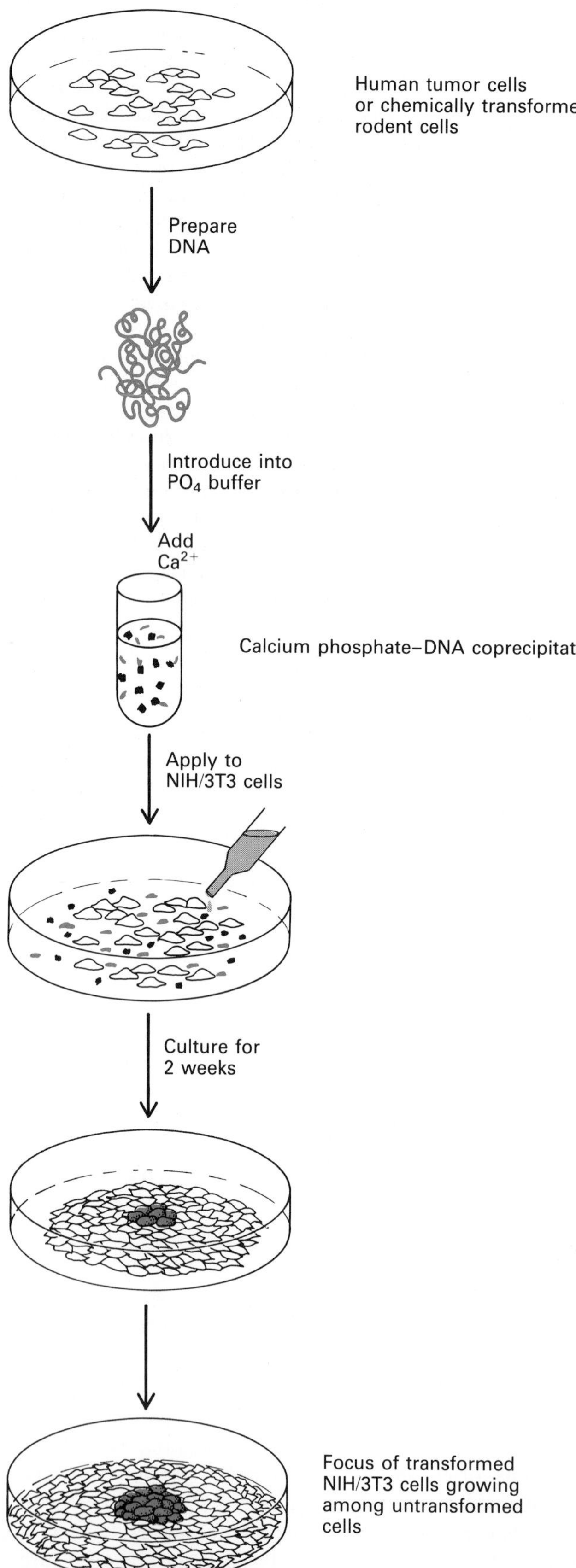

Figure 23-21 The original procedure used to assay for a cellular oncogene.

3. *Cells transformed by treatment with chemical carcinogens sometimes have oncogenes.* This very important result was first obtained by extracting DNA from chemically transformed mouse cells and treating 3T3 cells with the DNA. A small fraction of the treated cells became transformed, and their DNA acquired a new gene that could be passed on to other cells (Figure 23-21). The extension of this type of experiment to animal and human tumor cell DNA will be described below.

The transformation of cells by exposure to a chemical carcinogen often induces a permanently altered state in cellular DNA. In other words, a carcinogen causes cancer by acting as a mutagen. In theory the carcinogen could do this by binding to DNA and causing a mistake during the replication of the DNA, but the evidence suggests a very different mechanism. The usual determinant of carcinogenesis by chemicals is probably DNA sequence changes induced by the repair processes cells use to rid themselves of DNA damage (Chapter 13). We shall discuss this paradoxical situation after considering a few other aspects of carcinogenesis.

Cellular Oncogenes

In our discussion of viral carcinogenesis we emphasized the role of oncogenes that are expressed as part of proviral DNAs. Some oncogenes are not associated with proviral DNAs. *Cellular oncogenes* are mutated genes that are resident in cellular chromosomes. When cellular oncogenes are recovered from transformed or tumor cells, they can transmit the transformation characteristics to recipient cells. The prototypical experiment for detecting a cellular oncogene is that described in Figure 23-21: DNA is extracted from a transformed cell and transfected into a recipient cell population, whereupon it yields transformed cell foci in the monolayer of recipient cells.

In the first experiments cells transformed by chemicals were used as the donors of DNA; the high frequency of cellular oncogenes was realized later when DNA was extracted from many human tumor cell lines and fresh human tumors. About 20 percent of such DNA samples yield transformation in 3T3 cells, whereas comparable normal cell DNA rarely, if ever, gives transformants. Thus oncogenes are commonly present in human tumor cells. Actually, the 3T3 cell assay almost certainly underestimates their frequency.

A few of the cellular oncogenes from human tumor cells have been cloned by using recombinant DNA techniques (Figure 23-22). The sequences of these genes indicate that most are related to the c-*ras* family. The first cellular oncogene to be studied was, in fact, c-Ha-*ras*, the same gene that gave rise to the Harvey virus (see Figure 23-16). A comparison of the normal c-Ha-*ras* and the oncogene—which came from a human bladder carcinoma—has shown that the only difference is in a single

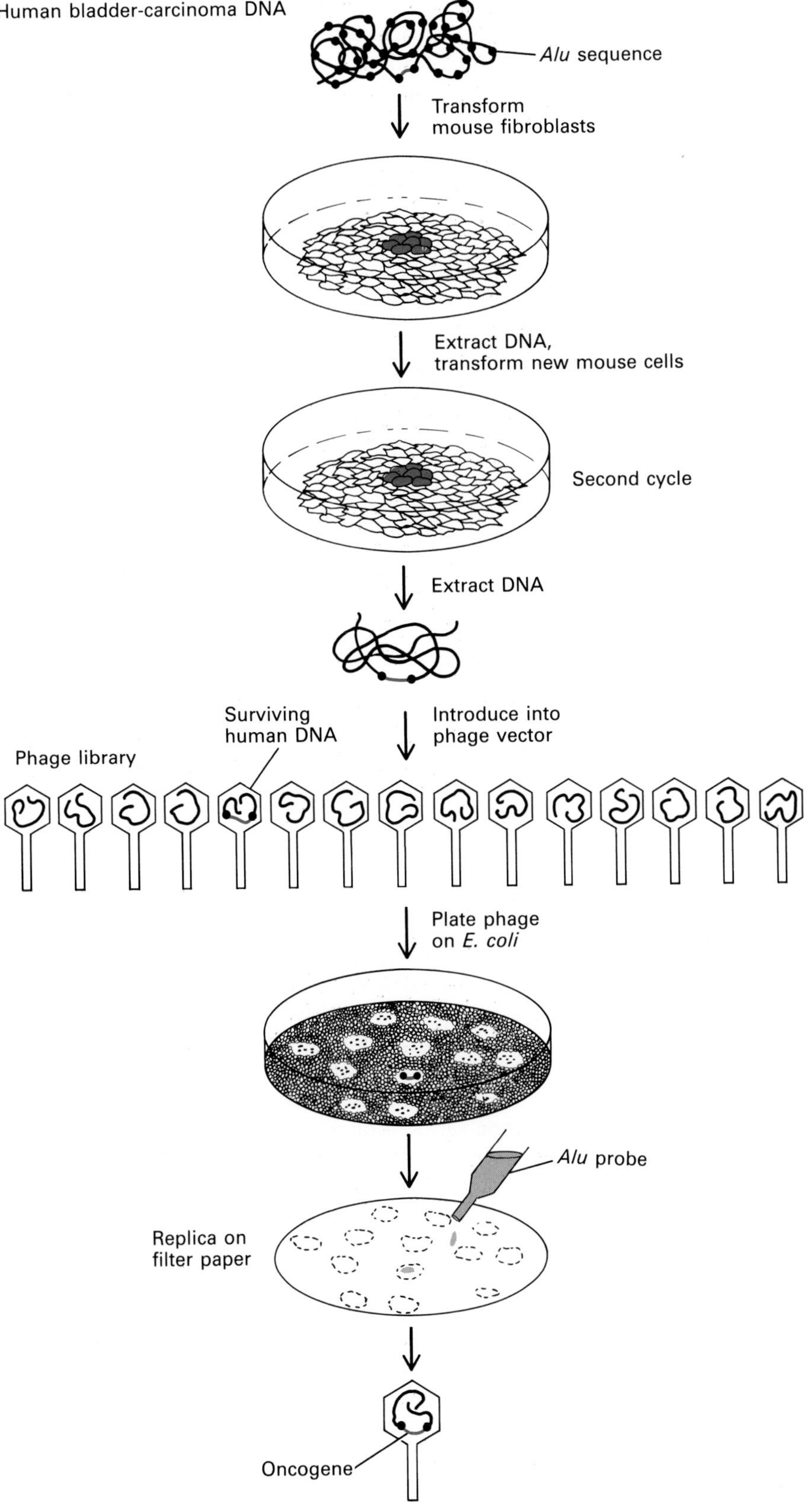

Figure 23-22 The isolation of a cellular oncogene by molecular cloning. The gene to be isolated derives from a human bladder carcinoma. Like almost all human genes, it has a nearby repetitive DNA sequence called an *Alu* sequence. The human DNA is first purified by repetitive transfer as shown in Figure 23-21. The total DNA from a secondary transfected mouse cell is cloned into bacteriophage λ, and the phage receiving human DNA is identified by hybridization to an *Alu* probe. The hybridized DNA should contain part or all of the oncogene. The expected result can be proved by showing either that the phage DNA can transform cells (if the oncogene has been completely cloned) or that the cloned piece of DNA is always present in cells transformed by DNA transfer from the original donor cell.

nucleotide, which must cause a single amino acid change. This miniscule change, which is reminiscent of the single amino acid replacement that causes sickle cell anemia (see Figure 7-20), is sufficient to change a protooncogene into a cellular oncogene. Apparently no quantitative change in gene expression need be involved in this alteration—only the qualitative one.

Thus far three c-*ras*-related genes and a number of unrelated genes have been positively identified as cellular oncogenes. The 3T3 cell assay apparently selects for c-*ras*-related genes, although it can reveal others as well. Cellular oncogenes have been found in tumors of the bladder, lung, breast, large intestine, and neural tissue. Leukemias also have oncogenes. Cellular oncogenes are not generally tissue-specific in their action: the same gene can be active in tumors of various tissues.

As was true for the transforming proteins of polyoma virus (discussed on page 1050), the use of primary cell cultures in lieu of 3T3 cells has greatly amplified our understanding of cellular oncogenes. For polyoma virus, two broad categories of transforming events were shown to involve different proteins: one protein to immortalize and one to change growth properties. For cellular oncogenes the same division of labor is evident: some gene products immortalize whereas others change cell morphology.

The initial observations were made with primary rat embryo cells. When transfected with a *ras* oncogene they showed the morphological changes associated with transformation but they were not immortalized. Transfection with *ras* plus *myc,* however, led to fully transformed, immortal, tumorigenic cell lines. Polyoma mid-T can replace *ras;* polyoma T or the various other nuclear oncogenes can replace *myc*. Thus it appears that two classes of transforming function can be defined: *ras*-like and *myc*-like. Because the *ras* proteins are cytoplasmic and the *myc* protein is nuclear, transformation seems to be a two-event process, a growth factor–like alteration of cell behavior plus a possible change in gene transcription.

This neat picture of transformation as a two-step process is clouded by observations that if *ras* or *myc* are provided with the strong promoters/enhancers of retroviral LTRs, *either* gene will morphologically transform cell lines, and *ras* will also transform primary cells to immortality. Thus high levels of the oncogene products may have different effects from lower levels, and the transformation of a given cell may be affected both by the types of oncogenes expressed in the cell and by the amount of the oncogene product in the cell.

The two-step perspective on transformation sheds some light on a phenomenon described early in this chapter. We noted that primary rodent cells have a low but reproducible probability of spontaneously becoming cell lines. We can now hypothesize that the event responsible is the activation of an immortalizing oncogene. Thus cell lines—as opposed to primary cells—may provide such a good substrate for assaying *ras* oncogenes because they already have an active *myc*-like oncogene.

The ability of an oncogene to immortalize cells is consistent with our earlier portrayal of immortalization as a parameter of transformation quite separate from the others. Cell "mortality" is probably a consequence of differentiation events that change a stem cell into an end-stage cell. Thus immortalization may actually be the outcome of a blockade of differentiation caused by the action of the oncogenes.

In most work on cellular oncogenes the genes have been recovered from human tumors, where the cause of the malignancy is difficult to ascertain. We assume that most human cancer results from the exposure of DNA to chemical carcinogens, but the exact agents involved, their metabolism, and the route of the exposure remain obscure. Thus it is significant that the same set of oncogenes as those found in human tumors—i.e., mainly *ras* and *myc*-related ones—are recovered from experimental animals or cells treated with chemical carcinogens. This correlation represents one of the primary arguments that apparently spontaneous human cancers and the tumors in carcinogen-exposed laboratory animals have a common basis.

Oncogenes Can Also Be Activated by Chromosomal Translocations and Duplications

For many years it has been known that chromosomal abnormalities abound in tumor cells. Human cells ordinarily have 23 pairs of chromosomes, each with a well-defined substructure (see Figures 10-1 and 10-5). Tumor cells are usually *aneuploid* (i.e., they have an abnormal number of chromosomes—generally too many), and they often contain *translocations*—fused elements from different chromosomes. As a rule, these unusual characteristics are not reproducible from tumor to tumor: each tumor has its own set of anomalies. Certain events recur, however: the first such regularity to be discovered, the *Philadelphia chromosome,* was found in the cells of virtually all patients with the disease chronic myelogenous leukemia. This chromosome is a fusion of most of chromosome 9 to a piece of chromosome 22. The reciprocal translocation—in which a tiny piece of chromosome 9 is fused to the broken end of chromosome 22—is also present, although it is not easily recognized.

Where such marker chromosomes are frequently present, the cause has long been thought to be an alteration of a specific gene at one of the break points. The discovery of oncogenes suggested that an oncogene might be involved, and two such cases have indeed been discovered. In the Philadelphia chromosome, the c-*abl* gene lies at the break point on chromosome 9, and the translocation causes the formation of an mRNA derived partly from chromosome 22 and partly from c-*abl*. The other case was found in the tumor known as Burkitt's lymphoma: the c-*myc* protooncogene has been located on chromosome 8 at just the place where a reproducible translocation occurs. The analogous site in the mouse genome is also involved in mouse myelomas. In both c-*myc* examples the tumor cells are antibody-producing cells, and the c-*myc* gene is translocated to regions that are involved in DNA rearrangements leading to the construction and expression of antibody-forming genes, as we shall discuss in the next chapter. The translocations alter the transcriptional activity of the c-*myc* gene and/or the stability of c-*myc* mRNA. To what extent the modification and activation of protooncogenes by chromosomal translocations causes the progressive steps giving rise to human tumors is not yet clear, but it seems increasingly likely that the chromosomal instabilities associated with malignant cells contribute importantly to malignancy.

Another common chromosomal anomaly in tumor cells is the localized reduplication of DNA to produce as many as 100 copies of a given region (usually a region spanning hundreds of kilobases). This anomaly may take either of two forms: the duplicated DNA may be tandemly organized at a single site on a chromosome, or it may exist as small, independent chromosomelike structures. The former case leads to a *homogeneously staining region* (HSR) that is visible in the light microscope at the site of the duplication; the latter case causes *double minute chromosomes* to pepper a stained chromosomal preparation (Figure 23-23). Again, oncogenes have been found in the duplicated regions. Most strikingly, a *myc*-related gene called N-*myc* has been identified in both HSRs and double minutes of a variety of human nervous system tumors.

The two-oncogene transformation of cell cultures has an interesting parallel in human and experimental tumors. Many experimental tumors resulting from expo-

(a)

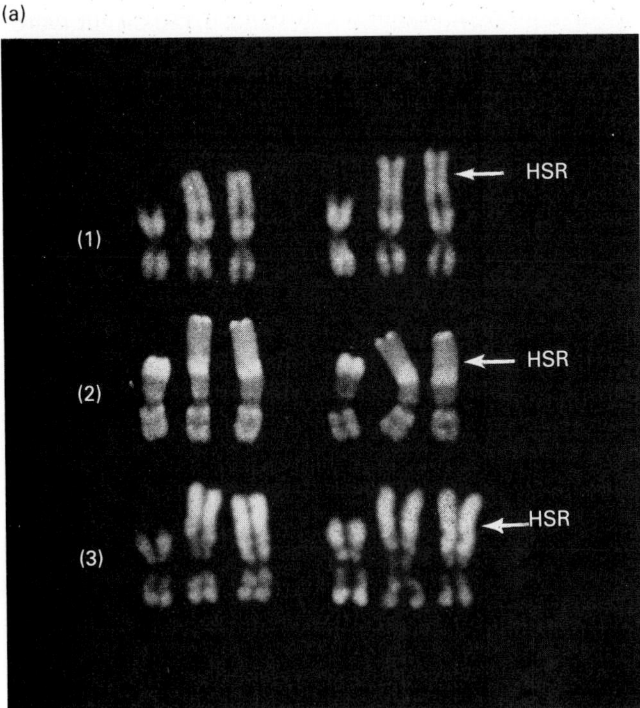

(b)

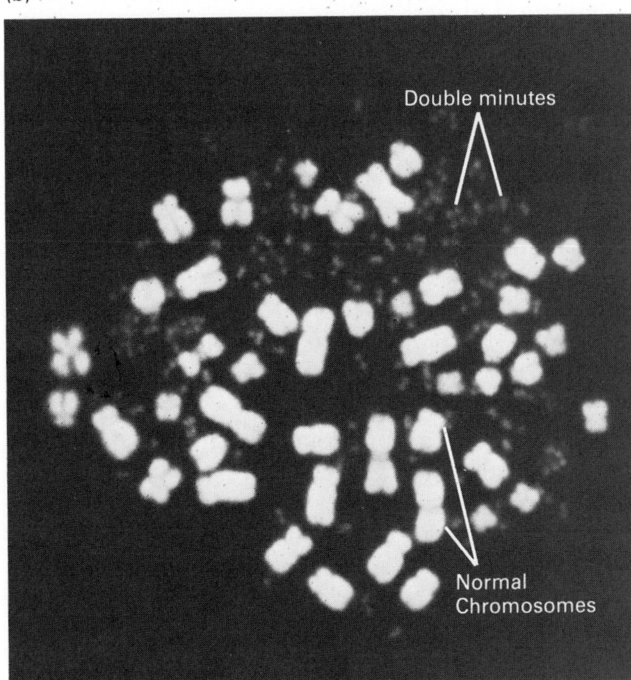

Figure 23-23 Visible forms of DNA amplification. (a) Homogeneously staining regions (HSRs) in chromosomes from two neuroblastoma cells. In each set of three chromosomes, the leftmost one is a normal chromosome 1 and the other two are HSR-containing chromosomes. The three lines (1, 2, and 3) represent three different methods of staining the chromosomes. Method 1 is quinacrine staining, which highlights AT-rich regions; method 2 is staining with chromomycin A3 plus methyl green, which highlights GC-rich areas; and method 3 is 33258 Hoechst staining after a pulse of bromodeoxyuridine late during the S phase, which highlights the early replicating regions. In all three cases the HSR shows homogeneous staining characteristics whereas the rest of the chromosome has highlights. [See S. Latt et al., 1975, *Biopolymers* 24:77.] (b) Quinacrine-stained double minute chromosomes in a human neuroblastoma cell. The normal chromosomes are the large white structures; the double minutes are the many small dots that are paired. [See N. Kohl et al., 1983, *Cell* 35:359.]

Both the HSRs and the double minute chromosomes shown here contain the N-*myc* oncogene. *Photographs courtesy of S. Latt.*

sure to a carcinogen have *ras* oncogenes, which implies that c-*ras* genes are common targets for the initial events in carcinogenesis. On the other hand, only highly malignant human tumors—and not their less malignant counterparts—often have duplicated or translocated *myc*-related genes. This observation implies that the progression from an initiated tumor to a highly malignant one may depend on *myc* activation. Thus a malignancy could often result from an initial point mutation in a *ras*-related oncogene followed by a chromosomal rearrangement that activates a *myc*-related oncogene.

Summary: Many Mechanisms Produce Oncogenes

We have encountered many mechanisms by which protooncogenes can be converted into oncogenes, and it is worth reviewing them. Oncogenes were first found in transducing retroviruses. These viruses both transform cells and cause rapidly growing tumors in animals through the action of their oncogenes. Later, oncogenes were found to be activated by the integration of retroviruses. In such cases the viruses cause tumors slowly, partly because retroviruses do not integrate at specific sites and therefore integration near a potential oncogene is a very rare event.

Another reason why these tumors arise slowly is because the de novo conversion of a protooncogene to an oncogene is probably not sufficient to cause a fully malignant tumor. Secondary effects, often associated with chromosomal instabilities, must follow the activation of the oncogene. In transducing retroviruses, the oncogenes are more nearly self-sufficient because they have undergone many generations of selection since their initial acquisition by a retrovirus. Also, some transducing retroviruses carry two oncogenes. However, even for transducing retroviruses the induction of malignancy may require further alterations in the cell beyond the initial infection.

We have discussed three mechanisms of nonviral oncogene activation: point mutation, translocation, and duplication. All three have been implicated in human tumors. Fully malignant cells appear to have more than one active oncogene: already a number of tumors with multiple active oncogenes have been found. The fragmentary evidence available at present suggests that tumors may be initiated by the activation of one oncogene, but that tumor progression involves further oncogene acquisition. There are probably some genes whose activation contributes to invasion and metastasis, and one candidate has been found.

Do Oncogenes Cause All Cancer?

Although the discovery of oncogenes was a landmark in cancer research, we do not yet know how many human cancers are caused by oncogene activation and how many of the steps in carcinogenesis are caused by oncogene activity. Human tumors usually develop in progressive stages resulting finally in the invasive, metastasizing, malignant cancer. As we have seen, multiple oncogenes are probably required to generate the malignant cell, and oncogene activation may be coupled with other changes—even nongenetic ones—to produce the life-threatening malignancy. One of the biggest questions in cancer research today is how large is the window opened by the oncogene assays. An army of investigators is seeking to answer that question.

In this chapter we have stressed the role of oncogenes in the development of cancer and ignored most other potential mechanisms. The rationale for this approach is simple: we know quite a bit about oncogenes and their role in cancer, but we have no similar highly developed body of knowledge about any other aspect of cancer. We must, however, maintain some perspective on the problem. Our knowledge of the role that oncogene activation plays in human cancer is indirect. Oncogenes (mainly of the *ras* and *myc* families) can be recovered from many human tumors, but only 10 to 20 percent of tumors yield such oncogenes. Perhaps the other tumors also have oncogenes that have not yet been assayed, but we must consider whether independent mechanisms could be at work.

The largest conceivable class of such cancer-inducing mechanisms—for which, again, there is little evidence—would depend on epigenetic inheritance. An *epigenetic* condition or process is one that is passed from a cell to its progeny without any alteration in the coding sequence of the DNA. The differentiated state of cells, which is passed on to progeny, is generally determined epigenetically. How epigenetic states are maintained is still a matter for speculation; roles for protein regulatory molecules and for DNA methylation are considered likely. Whatever its basis, a process determined epigenetically can be inherited with the same fidelity as one determined by a genetic alteration.

The *teratocarcinomas* are probably determined by a nongenetic mechanism. These are tumors of very early embryonic cells that maintain the ability to differentiate into the full range of body cell types. In fact, if teratocarcinoma cells are incorporated into a preimplantation embryo, they can contribute as normal cells to all of the tissues of the adult, including the germ-line cells, but no tumors develop. Conversely, if teratocarcinoma cells are injected into adult mice, they form lethal tumors. Thus their malignant state is conditional on their environment. Mouse teratocarcinomas can be formed by the simple procedure of transferring early embryos into older mice at sites such as the testis. These tumors can also be formed by explanting very early embryos into cell culture. No mutagenic treatment is involved and no mutation is evident: in the right environment the cells are seen to be absolutely normal. We presently believe that the teratocarcinomas are a unique case, but their occurrence certainly alerts us to the possible importance of nongenetic events in carcinogenesis.

Radiation and DNA Repair in Carcinogenesis

The living world is constantly being bombarded by radiation of many different kinds. Two types of radiation are especially dangerous because they can modify DNA: ultraviolet radiation and the ionizing radiations (x-rays and atomic particles). UV radiation of the appropriate wavelength can be absorbed by the DNA bases and can produce chemical changes in them. The most common damage is the production of dimers between adjacent pyrimidine residues in one strand of DNA (see Table 13-4 and Figure 13-37). These dimers interfere with both transcription and replication of DNA. Ionizing radiation mainly causes breaks in DNA chains. Both types of radiation can cause cancer in animals and can transform cells in culture. The ability of ionizing radiation to cause human cancer, especially leukemia, was dramatically shown in the increased rates of leukemia among survivors of the atomic bombs dropped in World War II.

UV radiation has been widely used in research on mutations and cancer because it is an easily manipulable and measurable agent that directly damages DNA. However, probably none of the primary lesions induced in DNA, including the dimers, is mutagenic. This has been shown directly in bacteria through the inactivation of *rec*A, a cellular gene involved in DNA repair: *rec*A inactivation prevents mutagenesis by UV light. When this gene's product is lost, cells are killed by UV light because DNA repair is prevented. This result indicates that the mutational effect of UV radiation is caused during repair and not as a primary consequence of the radiation.

Mutagenesis Can Be Caused by DNA Repair or by a Lack of Repair

Because cells are continually being exposed to chemicals that can react with DNA and to radiation that can damage DNA, they have evolved complex and very effective methods of repairing DNA (Chapter 13). As more complex organisms with larger genomes and longer generation times have evolved, the effectiveness of DNA repair processes has increased.

If repair processes were 100 percent effective, chemicals and radiation would pose no threat to our DNA. Unfortunately, repair is less efficient for certain types of DNA damage than for others. At least two types of DNA lesions are difficult to repair properly and can become progenitors of mutations. One type is a double-strand break of the DNA backbone. Such breaks, caused either by ionizing radiation or by chemicals, can be correctly repaired only if the free ends of the DNA rejoin exactly; however, without overlapping single-stranded regions there is no base-pair homology to catalyze the joining. Because cells will not tolerate free DNA ends, broken ends of molecules are generally joined to other broken ends. In a cell that has suffered a particular double-strand break there are often other breaks, so a broken end has a number of possible segments to which it can join. Thus the joining of broken ends can cause the translocation of pieces of DNA from one chromosome to another. As we have described, the result of a translocation can be the activation of an oncogene. Consequently, one mechanism by which chemicals or radiation can be carcinogenic is by producing double-strand DNA breaks resulting in translocations that can occasionally activate oncogenes.

A second circumstance in which proper repair of DNA is difficult occurs when cells attempt to replicate damaged DNA before repair processes have had a chance to act on the lesion. The undamaged strand can replicate, but the damaged one can neither be replicated nor properly repaired because it is no longer paired to the undamaged strand. Cells may resort to *postreplication repair* to handle such damage: undamaged double-stranded DNA denatures and one strand pairs to the damaged strand. The intact DNA duplex can then direct incorporation of appropriate nucleotides, allowing the repair of the damaged DNA.

The difficulty arises when a cell suffers so much damage over a short time that its repair systems are saturated. It then runs the danger of having repair fall so far behind the damage that DNA replicates extensively with unrepaired lesions. In such situations, both bacteria and animal cells use inducible reserve systems for repair. Such systems are not expressed in undamaged cells, but some aspect of the accumulated damage causes their derepression (induction) and expression.

One such inducible system is the SOS repair system of bacteria (see Figure 13-46). In contrast to other repair systems, this one makes errors in the DNA as it repairs lesions, so it is referred to as *error-prone*. The SOS system is responsible for UV radiation–induced mutations, and its activity is dependent on the *rec*A gene product. The errors induced by the SOS system occur at the site of lesions, which suggests that the mechanism of repair is insertion of random nucleotides in place of the damaged ones in the DNA. Apparently this system is inducible rather than constitutive because cells would be expected to accept randomization of DNA sequence only as a last resort when error-free mechanisms of repair cannot cope with damage. SOS repair may also be a way of accelerating evolution at a time when the organism is under stress. Bacteria lacking an SOS system have greatly reduced levels of mutation, which strongly implies that most of the mutations produced by treating bacteria with radiation or chemicals are caused by the error-prone SOS repair system.

Whether animal cells have an error-prone repair system is not known, but they certainly have inducible repair systems. If one or more of these systems are error-prone, inducible repair is likely to play a role in mutagenesis and therefore in carcinogenesis. In any case, many investigators believe that in animal cells, as in bacteria, most mutation is an indirect consequence of DNA lesions and not a direct result.

Defects in Human Repair Systems May Cause High Cancer Rates

The link between repair systems and carcinogenesis has been inferred from other information in addition to that about bacteria. Certain inherited genetic defects of humans make specific repair systems nonfunctional, which enormously increases the probability that carriers of the defects will get certain cancers (see Table 13-5). Such a disease is xeroderma pigmentosum, an autosomal recessive disease in which cells of patients are unable to repair UV damage or to remove bulky chemical substituents on DNA bases. The patients get skin cancers very easily if their skin is exposed to the UV rays in sunlight. The complexity of mammalian repair systems is shown by the fact that there are five different xeroderma pigmentosum lesions, all having the same phenotype and the same consequences. Hybridization of cells having one type with cells having another produces complementation: the hybrid cells can repair UV damage normally.

Other human diseases affect DNA repair and increase cancer risks. An important one is ataxia telangiectasia, a disease that sensitizes cells to x-rays. X-rays and UV light cause different types of lesions, which are handled by different repair systems: ataxia telangiectasia results in the

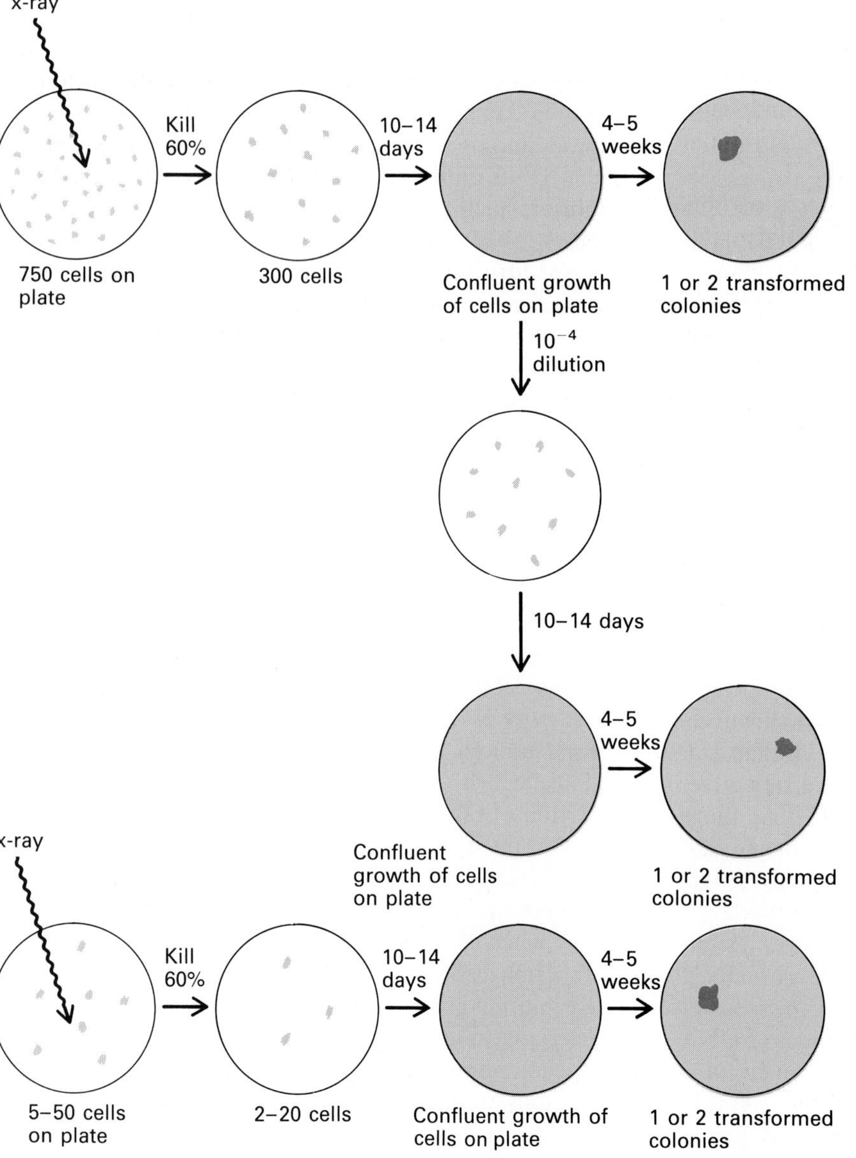

Figure 23-24 X-radiation causes cultured cells to spawn a constant fraction of transformed colonies. These experiments contradict the simpler notion that radiation causes transformation by direct damage to DNA; they suggest rather that radiation induces a propensity for cells to develop mutations long after the initial radiation effects have been "diluted out." In the experiments, low numbers of cells are irradiated, allowed to grow to confluence (saturation density), and then maintained as confluent cultures for weeks. Transformed foci arise at a frequency that is independent of the number of irradiated cells. Furthermore, dilution of the cells just after they reach confluence still gives a confluent culture that develops the same number of transformants. No transformants arise in the absence of radiation treatment. The cause of the increase in the frequency of transformation remains obscure but it seems apparent that the original radiation exposure increased the probability of a rare random event. *Adapted from the work of Dr. J. Little.*

loss of the x-ray repair pathway which separates cross-links in the DNA duplex; xeroderma pigmentosum results in the loss of the UV repair pathway, which rids the DNA of bulky chemical groups appended to it. Most chemicals mimic the effects of either x-rays or UV irradiation, so cells from patients with one of these two diseases will be hypersensitive to treatment with one or another set of chemicals. The reason why patients with DNA repair defects have greatly increased cancer rates may be that the loss of an error-free pathway in their cells causes a greater dependence on an error-prone pathway of repair.

One final observation shows just how indirect cancer induction after irradiation can be. If cultured cells are irradiated and then allowed to grow, one might expect to find a fraction of transformed cells and a fraction of normal cells, each of which would breed true. (Viral transformation produces this result.) Instead, it is found that irradiation can cause all of the cells to have a low but significant probability of giving rise to transformed progeny as they grow; no such propensity would occur in the absence of irradiation. Transformed progeny may appear many generations after irradiation, which implies that irradiation can switch on some cell process that slowly generates transforming lesions (Figure 23-24). What the process might be is obscure, but a likely guess is that an error-prone repair system similar to the SOS system in bacteria has been permanently activated by the irradiation. Treatment with chemical carcinogens can cause a similar mutation- and transformation-prone state in cells.

Promoters of Carcinogenesis

Although up to now we have implied that only lesions in DNA are responsible for cancer induction, another aspect of the process exists that has not been directly linked to a genetic lesion. This phenomenon, called *promotion,* was discovered in the early 1940s when it was found that if the skin of an animal is treated with certain chemicals (now recognized as DNA-damaging chemicals), the odds of a tumor appearing may be greatly augmented by further treatment with a second compound that has a very different structure and chemistry. The first compound is called an *initiator,* and the second is called a *promoter.* A promoter differs from an initiator because it does not need to be metabolized to a derivative; furthermore, a promoter has no tendency to react as an electrophile. Therefore the promoter must act on cells through a route quite different from that of the initiator.

Promoters are distinguished from initiators according to the schedules by which animals must be exposed to them if they are to be effective. The application of a promoter must follow the application of an initiator, and the promoter must be applied many times over many weeks or months, whereas the initiator need be applied only once. Another difference between a promoter and an initiator is that continual treatment with an initiator will often produce a tumor (i.e., the initiator may act both to initiate and to promote), but if skin that has not been initiated is treated continually with a promoter, a tumor virtually never results. Thus, promoter action depends on previous initiation.

One class of promoters, the phorbol esters (Figure 23-25), has been intensively studied and provides a model for the action of all promoters. Through the use of labeled compounds, specific cellular receptors for phorbol esters have been identified; the existence of the receptors implies that phorbol esters are analogs of some physiological signaling substance. The treatment of many types of cells with phorbol esters causes the cells to take on a transformed morphology and growth habit, but the cells revert to normal when the compound is removed. Thus phorbol esters can produce a *phenocopy* of the transformed state (i.e., they alter the phenotype but not the genotype). If cells that have the potential to differentiate are treated with phorbol esters, differentiation will often be either blocked or facilitated, which again attests to the powerful physiological effects generated by the occupation of the phorbol ester receptors.

Although it had been expected that phorbol esters would turn out to be hormone (or growth factor) analogs and that their receptor would be a cell surface hormone-binding protein, the analogy to a receptor-hormone system proved unfruitful in research. Rather, the site of action of phorbol esters became known from studies of a unique cellular enzyme called protein kinase C. This protein phosphorylates serine or threonine residues on other proteins. The activation of protein kinase C requires the

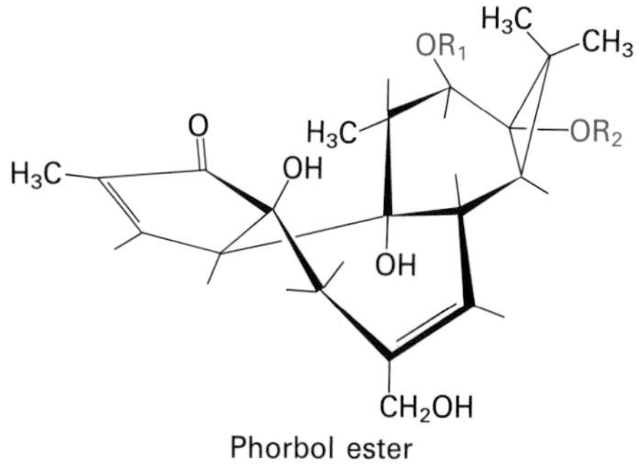

H_3C CH_3 OR_1 OR_2 O H_3C H_3C OH OH CH_2OH

Phorbol ester

Figure 23-25 The structure of phorbol esters, the prototypical tumor promoters. The molecule has four rings, one each containing seven, six, five, and three carbons. Crucial components are the two long-chain fatty acids esterified to the phorbol backbone. They are denoted OR_1 and OR_2. The molecule apparently acts as an analog to diacylglycerol, a molecule with two esterified fatty acid chains and a hydroxyl group (see Figure 16-23).

presence of a number of cofactors: Ca^{2+}, phospholipids, and diacylglycerol. This latter compound is produced when phospholipids are degraded. A specific class of progenitors of diacylglycerol is the inositol-containing phospholipids, which are degraded by a complex pathway involving phosphorylation of the inositol moiety before its cleavage (see Figure 16-23).

Phorbol esters act by substituting for diacylglycerol; the phorbol ester receptor is actually protein kinase C. This can be shown very directly: phorbol esters in association with Ca^{2+} and phospholipids strongly activate protein kinase C for phosphorylation of substrate proteins. The mechanism of action of phorbol esters appears to be a facilitation of Ca^{2+} binding. The stereochemistry of phorbol esters makes them somewhat analogous in shape to diacylglycerol; this analogy is presumably the basis for their ability to substitute for diacylglycerol and activate protein kinase C. (The linkage of promoter activity to protein kinase C does not yet explain the action of promoters, because there is no clear-cut role for protein kinase C in cell growth control. It is, however, a widely distributed enzyme and it probably has profound effects.)

A crucial aspect of promoter action is the lack of permanent effects on cell metabolism from promoter treatment. Because a cancer cell is *permanently* altered, promoters must set in motion events that lead to an irreversible alteration in the cell. The end result could be a DNA alteration, or it could be a changed state of cellular differentiation. Whatever the mechanism, a promoter must produce a second permanent alteration in a cell that has already undergone an initiator-induced mutational change.

Evidence of synergism between various cancer-inducing agents in human carcinogenesis has been discovered, and the initiator-promoter model may be just as applicable to humans as it is to animal model systems. If so, control of cancer could be achieved through control of either initiators or promoters. Short-term mutagenesis tests can only identify initiators; no reliable tests for promoters have yet been devised. Some investigators believe that a major aspect of the cancer-inducing capacity of cigarettes may lie in a promoterlike activity released from the burning of the tobacco.

The initiator-promoter model of carcinogenesis describes just one of a variety of ways in which chemicals may interact to cause tumors. For example, low concentrations of potentially carcinogenic compounds can interact when the compounds are applied together *(cocarcinogenesis)*. Some compounds, notably antioxidants, can counteract the effects of others *(anticarcinogenesis)*. In natural situations, cancer induction is probably often the result of many interacting effects, with hormonal stimulation playing an important role. The prevalence of cancer among older animals is often attributed to the long latency period required for the multiple events that cause a tumor to have a chance to occur.

Human Cancer

Our discussion thus far has focused largely on the experimental questions posed by the phenomenon of cancer. A central motivation of basic cancer research, however, is to understand human cancer well enough to be able to alter its course. We assume that all we have learned in the laboratory is relevant to the human problem, but putting it together to provide the understanding we seek has proved to be beyond the capability of today's science, for all but a few types of cancer.

In assessing the role of a substance in cancer induction, it is not useful to think in terms of a single cause. As we emphasized previously, naturally occurring cancer is probably most often a consequence of multiple factors that interact over long periods of time. Because each of these factors increases the possibility of a cancer occurring, they are called *risk factors*. For example, Epstein-Barr virus is probably a risk factor for Burkitt's lymphoma, although it is clearly not the sole cause.

Identifying risk factors for human cancer has been a slow and frustrating activity. One successful endeavor stands out: the identification of cigarette smoking as a crucial risk factor in lung cancer. A risk factor of this potency gives a clear indication of how to act to avoid lung cancer: avoid cigarettes.

Unfortunately, lung cancer is the only major human cancer for which a clear-cut risk factor has been identified. Animal fat is thought to be a risk factor for colon cancer, and many viruses and chemicals have been correlated with minor cancers; however, hard evidence that would help us avoid breast cancer, colon cancer, prostate cancer, leukemias, and so forth is simply lacking.

Is Susceptibility to Cancer Inherited?

A corollary to the belief that multiple interacting events in our environment are the major risk factors for cancer is the belief that genetic inheritance plays only a small role in the process. This proposition has been directly demonstrated by showing that people who migrate to a new environment take on the profile of cancers in their new environment within a generation. For instance, when Japanese citizens move to California they rapidly lose the oriental propensity toward stomach cancer, but they soon show the occidental propensity toward breast cancer.

Genetic inheritance does, however, play some role in human cancer. Certain inherited genes increase the probability that an individual will get a specific tumor to almost 100 percent. A classic case is retinoblastoma, which is, like most inherited tumors, a disease of childhood. Children who inherit a single copy of a specific gene defect, often seen as a small deletion on chromosome 13, will acquire an average of three retinoblastoma tumors, each derived from a single transformed cell. Because the developing retina comprises about 4×10^6 cells, only about 1 in 10^6 cells actually becomes a tumor cell. This finding suggests that even with its highly dominant inheritance, the retinoblastoma gene is acting recessively at the cell level, and a second event is needed to bring on the transformed state. Recent evidence indicates that the oncogene model is inapplicable to this tumor—as might be expected from its recessive genetics at the cellular level—and that the deletion of a gene is the cause. The evidence that deletion is causative comes from the observation that retinoblastoma cells have lost their normal chromosome 13 and have gained an extra copy (so that they now have two) of the deleted chromosome.

Cancer induction by deletion of a genetic region, rather than by activation of an oncogene, has been discovered in a variety of childhood tumors where inherited defects play an important role. These are rare tumors; whether deletional carcinogenesis might play a wider role than is now known is an important question for the future. Another unanswered question is what is the nature of genes whose deletion from both chromosomes can lead a cell to uncontrolled growth.

Thus studies of human tumors have indicated that inheritance is only rarely a crucial risk factor; however, whether inheritance might be a minor risk factor in many cancers is still being debated. Genetic propensity might, for example, explain why some smokers get lung cancer and others do not.

Summary

In focusing on a medical problem, cancer, this chapter may seem to have departed from the subject matter of the rest of the book. Cancer represents such a fundamental problem in cellular behavior, however, that many aspects of molecular biology are relevant to understanding the cancer cell. To become a cancer cell, a normal cell must undergo many fundamental changes. It must continue to multiply when normal cells would be quiescent; it must invade surrounding tissues, often breaking through the basal laminas that define the boundaries of tissues; and it must spread through the body and set up secondary areas of growth in a process called metastasis.

All of the various cell types of the body can give rise to cancer cells. Cancer cells are usually closer in their properties to immature normal cells than to more mature cell types. The retention of malignant properties by cancer cells grown in culture shows that the alteration from a normal cell to a cancer cell is caused by events within the cancer cell itself.

By growing cells in culture, experimenters can freely manipulate them to learn about the properties that distinguish normal cells from cancer cells. Carcinogens, cancer-causing agents, can also be used to alter cultured cells to produce full-fledged cancer cells in a process called transformation. Transformed cells differ from normal cells in many ways. Aspects of transformation include effects on cell growth control, cell morphology, cell-to-cell interactions, membrane properties, cytoskeletal structure, protein secretion, gene expression, and mortality (transformed cells can grow indefinitely). These alterations in cell behavior are not wholly independent, but their interrelationships remain obscure.

The three general types of carcinogens are viruses, chemicals, and radiation. Two types of carcinogenic viruses are distinguished: DNA viruses and the RNA-containing retroviruses. These viruses cause certain rare human cancers.

Both DNA and RNA viruses transform cells by permanently integrating new genes into the DNA of infected cells. The DNA-containing papovaviruses carry genes that can cause cell transformation by inducing in cells a growing state that facilitates virus multiplication. Permanent cell transformation results when virus multiplication is not possible.

Retroviruses cause cancer in two ways, both of which depend on cellular genes. Although these viruses use RNA as their genetic material, they employ reverse transcription to make an intracellular DNA copy that can integrate into cell DNA. Such DNA proviruses, by recombination with cellular genes, can acquire cellular information that turns them from relatively benign viruses to cancer-inducing agents. The cellular genes that can confer such an ability on a virus are called oncogenes. Oncogenes are initially normal genes (called protooncogenes) that by mutation or an altered context of expression become transforming genes. The second way that retroviruses can cause cancer is by integration near a protooncogene; the expression of the gene can be changed so that it becomes an oncogene. The difference between these two mechanisms of retroviral-induced cancer is that in the former case the oncogene becomes part of a transmissible virus (i.e., the gene is transduced), whereas in the latter case the protooncogene is activated in situ to an oncogene form that cannot be transmitted to other cells or animals.

More than 20 different normal cellular genes can become oncogenes by a genetic alteration. These alterations include point mutations, deletions, and altered levels or times of expression. The proteins encoded by oncogenes are related to proteins involved in normal cell growth control. Oncogenes play a critical role not only in viral cancer but also in cancer induced by chemicals.

Chemical carcinogens have a variety of structures with one unifying characteristic: electrophilic reactivity (either they are electrophiles or they are metabolized in the body to become electrophiles). Metabolic activation occurs via the cytochrome P-450 system, a pathway generally used by cells to rid themselves of noxious chemicals. The reactive electrophiles combine with many parts of the cell, but their reaction with DNA is the primary carcinogenic event. The reaction of a carcinogen and DNA leads to a

DNA alteration that produces an oncogene from a protooncogene. Oncogenes can be produced by local alterations in DNA structure, by translocations of DNA regions among chromosomes, or by extensive duplication of DNA regions including the oncogenes.

The presence of cellular oncogenes can often be demonstrated by transferring the DNA from a cancer cell into a normal cell, because the oncogene can transform the cell to malignancy. This experimental manipulation has allowed the isolation and molecular characterization of oncogenes from many human cancers.

Tumor cells may harbor more than one oncogene, which suggests that multiple genetic changes may contribute to the malignancy of a given cell. The accumulation of multiple alterations may explain the long latency period between the initial exposure to a carcinogen and the development of a full-fledged malignancy. The transformation of primary cell cultures often requires two oncogenes—further evidence that multiple influences may contribute to malignancy.

Ultraviolet, x-ray, and atomic particle radiations can all cause cancer. All affect DNA and presumably lead to oncogene activation. Cellular DNA repair processes have been implicated both in protecting against radiation-induced carcinogenesis and in causing it. The protective effect is most vividly shown by the occurrence of human diseases in which repair deficiencies lead to high rates of cancer: such deficiencies inactivate error-free repair systems. The inductive effect is seen most clearly in bacterial mutagenesis in which the loss of certain repair systems can prevent the occurrence of mutation: such losses prevent the operation of secondary, error-prone repair systems.

Although most of our understanding of cancer hinges on the role of genetic changes, there may be nongenetic changes as well. One apparent source is chemical promoters, compounds that potentiate the activity of electrophilic carcinogens. The best-understood compounds, the phorbol esters, cause nongenetic cellular changes that often mimic transformation. These substances activate a cellular protein kinase. Long-term treatment with phorbol esters leads to permanent cellular alterations that may or may not be genetic. An apparently clear-cut case of a nongenetic (or epigenetic) change that causes cancer is the alteration leading to a teratocarcinoma. These tumor cells revert to normal when they are implanted into early embryos.

The data available today have shown us the broad outlines of how a cancer can develop. Exposure of a cell to a mutagen leads to a DNA alteration that activates an oncogene. This event subtly alters the cell, giving it a growth advantage over its neighbors. As it grows, perhaps as a visible benign tumor or polyp but more likely as an inapparent clone of cells, one or more further alterations activate other oncogenes. Together, the multiple genetic modifications allow the clone to escape from all of the influences that ordinarily keep the growth of cells appropriate to the needs of the body. The clone not only grows without control, but its cells develop invasiveness and begin to spread to new sites. At this point, we recognize it as a malignant tumor. This composite picture is certainly simplistic—for one thing, it does not include nongenetic influences—but it represents a framework for future research, and, we hope, for the development of new methods of prevention and therapy for this dread disease.

References

General References*

Advances in Cancer Research (yearly volumes). Academic Press.

BECKER, F. F., ed. 1982. *Cancer: A Comprehensive Treatise.* Plenum. (Multiple volumes are released under this title.)

BOICE, J. D. JR., and J. F. FRAUMENI, eds. 1984. Radiation carcinogenesis: epidemiology and biological significance. In *Progress in Cancer Research and Therapy,* vol. 26 (same editors), Raven.

BUSCH, H., ed. 1967. *Methods in Cancer Research,* vol. 1. Academic Press.

BUYSE, M. E., M. J. STAQUET, and R. J. SYLVESTER, eds. 1984. *Cancer Clinical Trials: Methods and Practice.* Oxford University Press.

CAIRNS, J. ed. 1974. *Cancer: Science and Society.* W. H. Freeman and Company.

CAIRNS, J. 1975. The cancer problem. *Sci. Am.* 233(5):64–72, 77–78.

DEVITA, V. T. JR., S. HELLMAN, and S. A. ROSENBERG, eds. 1982. *Cancer: Principles and Practice of Oncology.* Lippincott.

GIRALDO, G., and E. BETH, eds. 1984. *Role of Viruses in Human Cancer.* Elsevier.

HIATT, H. H., J. D. WATSON, and J. A. WINSTEN, eds. 1977. *Origins of Human Cancer,* vols. 1 to 3. Cold Spring Harbor Laboratory.

HOLLAND, J. F., ed. 1982. *Cancer Medicine.* Philadelphia: Lea & Febiger.

KAISER, H. E. 1981. *Neoplasms: Comparative Pathology of Growth in Animals, Plants, and Man.* Baltimore: Williams & Wilkins.

LEVINE, A. J., W. C. TOPP, and J. D. WATSON, eds. 1984. *Cancer Cells,* vol. 1: *The Transformed Phenotype.* Cold Spring Harbor Laboratory.

MELNICK, J. L., ed. 1985. *Viruses, Oncogenes and Cancer,* vol. 2. New York: S. Karger.

REIN, R., ed. 1985. *Molecular Basis of Cancer,* part A: *Macro-Molecular Structure, Carcinogens, and Oncogenes.* Alan R. Liss.

REIN, R. ed. 1985. *Molecular Basis of Cancer,* part B: *Macro-Molecular Recognition, Chemotherapy, and Immunology.* Alan R. Liss.

*Book or review article that provides a survey of the topic.

RUDDON, R. W., ed. 1981. *Cancer Biology.* Oxford University Press.

SANDBERG, A. A., ed. 1980. *The Chromosomes in Human Cancer and Leukemia.* Elsevier.

VAN DE WOUDE, G. F., A. J. LEVINE, W. C. TOPP, and J. D. WATSON, eds. 1984. *Cancer Cells,* vol. 2: *Oncogenes and Viral Genes.* Cold Spring Harbor Laboratory.

Characteristics of Tumor Cells

* CAIRNS, J. 1975. Mutational selection and the natural history of cancer. *Nature* **255**:197–200.

CIFONE, M. A., and I. J. FIDLER. 1981. Increasing metastatic potential is associated with increasing instability of clones isolated from murine neoplasms. *Proc. Nat'l Acad. Sci. USA* **78**:6949–6952.

FIDLER, I. J., and I. R. HART. 1982. Biological diversity in metastatic neoplasms: origins and implications. *Science* **217**:998–1003.

ILLMENSEE, K., and B. MINTZ. 1976. Totipotency and normal differentiation of single teratocarcinoma cells cloned by injection into blastocysts. *Proc. Nat'l Acad. Sci. USA* **73**:549–553.

* NICHOLSON, G. L. 1979. Cancer metastasis. *Sci. Am.* **240**(3):66–76.

OSSOWSKI, L., and E. REICH. 1983. Antibodies to plasminogen activator inhibit human tumor metastasis. *Cell* **35**:611–619.

Cell Culture and Transformation

* ABERCROMBIE, M. 1970. Contact inhibition in tissue culture. *In Vitro* **6**:128–142.

ABERCROMBIE, M., and J. E. M. HEAYSMAN. 1954. Observations on the social behaviour of cells in tissue culture. *Exp. Cell Res.* **6**:293–306.

BURGER, M. M., and G. S. MARTIN. 1972. Agglutination of cells transformed by Rous sarcoma virus by wheat germ agglutinin and concanavalin A. *Nature New Biol.* **237**:356–359.

CARPENTER, G., and S. COHEN. 1975. Human epidermal growth factor and the proliferation of human fibroblasts. *J. Cell Physiol.* **88**:227–238.

* CARREL, A. 1912. On the permanent life of tissues outside of the organism. *J. Exp. Med.* **15**:516–528.

CHERINGTON, P. V., B. L. SMITH, and A. B. PARDEE. 1979. Loss of epidermal growth factor requirement and malignant transformation. *Proc. Nat'l Acad. Sci. USA* **76**:3937–3941.

DULAK, N. D., and H. M. TEMIN. 1973. A partially purified polypeptide fraction from rat liver cell conditioned medium with multiplication-stimulating activity for embryo fibroblasts. *J. Cell. Physiol.* **81**:153–160.

FOLKMAN, J., and A. MOSCONA. 1978. Role of cell shape in growth control. *Nature* **273**:345–349.

GAFFNEY, B. J. 1975. Fatty acid chain flexibility in the membranes of normal and transformed fibroblasts. *Proc. Nat'l Acad. Sci. USA* **72**:510–516.

* GEY, G. O., W. D. COFFMAN, and M. T. KUBICEK. 1952. Tissue culture studies of the proliferative capacity of cervical carcinoma and normal epithelium. *Cancer Res.* **12**:264–265.

* GOSPODAROWICZ, D., and J. S. MORAN. 1976. Growth factors in mammalian cell culture. *Annu. Rev. Biochem.* **45**:531–558.

HAKOMORI, S. 1975. Structures and organization of cell surface glycolipid: dependency on cell growth and malignant transformation. *Biochim. Biophys. Acta* **417**:58–80.

HATANAKA, M. 1974. Transport of sugars in tumor cell membranes. *Biochim. Biophys. Acta* **355**:77–104.

* HAYFLICK, L., and P. S. MOREHEAD. 1961. The serial cultivation of human diploid cell strains. *Exp. Cell Res.* **25**:585–621.

* HOLLEY, R. W. 1975. Factors that control the growth of 3T3 cells and transformed 3T3 cells. In *Proteases and Biological Control,* D. B. Rifkin and E. Shaw, eds. Cold Spring Harbor Laboratory.

* HOLLEY, R. W., and J. A. KIERNAN. 1968. "Contact inhibition" of cell division in 3T3 cells. *Proc. Nat'l Acad. Sci. USA* **60**:300–304.

HYNES, R. O. 1973. Alteration of cell-surface proteins by viral transformation and by proteolysis. *Proc. Nat'l Acad. Sci. USA* **70**:3170–3174.

* HYNES, R. O. 1976. Cell surface proteins and malignant transformation. *Biochim. Biophys. Acta* **458**:73–107.

* HYNES, R. O., ed. 1979. *Surfaces of Normal and Malignant Cells.* Wiley.

MACPHERSON, I., and L. MONTAGNIER. 1964. Agar suspension culture for the selective assay of cells transformed by polyoma virus. *Virology* **23**:291–294.

OSSOWSKI, L., J. C. UNKELESS, A. TOBIA, J. P. QUIGLEY, D. B. RIFKIN, and E. REICH. 1973. An enzymatic function associated with transformation of fibroblasts by oncogenic viruses, II: Mammalian fibroblast cultures transformed by DNA and RNA tumor viruses. *J. Exp. Med.* **137**:112–126.

SANFORD, K. K., G. D. LIKELY, and W. R. EARLE. 1954. The development of variations in transplantability and morphology within a clone of mouse fibroblasts transformed to sarcoma-producing cells *in vitro. J. Nat'l Cancer Inst.* **15**:215–237.

* SATO, G., ed. 1979. *Hormones and Cell Culture.* Cold Spring Harbor Laboratory.

SHIN, S. I., U. H. FREEDMAN, R. RISSER, and R. POLLACK. 1975. Tumorigenicity of virus-transformed cells in nude mice is correlated specifically with anchorage independent growth in vitro. *Proc. Nat'l Acad. Sci. USA* **72**:4435–4439.

TODARO, G. J. 1963. Quantitative studies of the growth of mouse embryo cells in culture and their development into established lines. *J. Cell Biol.* **17**:299–313.

TODARO, G. J., and H. GREEN. 1964. An assay for cellular transformation by SV40. *Virology* **23**:117–119.

TODARO, G. J., J. E. DELARCO, and S. COHEN. 1976. Transformation by murine and feline sarcoma viruses specifically blocks binding of epidermal growth factor to cells. *Nature* **264**:26–31.

UNKELESS, J. C., A. TOBIA, L. OSSOWSKI, J. P. QUIGLEY, D. B. RIFKIN, and E. REICH. 1973. An enzymatic function associated with transformation of fibroblasts by oncogenic viruses, I: Chick embryo fibroblast cultures transformed by avian RNA tumor viruses. *J. Exp. Med.* **137**:85–111.

WEBER, K., E. LAZARIDES, R. D. GOLDMAN, A. VOGEL, and R. POLLACK. 1974. Localization and distribution of actin fibers in normal, transformed and revertant cells. *Cold Spring Harbor Symp. Quant. Biol.* 39:363–369.

Viruses as Agents of Transformation: Oncogenes

ABRAMS, H. D., L. R. ROHRSCHNEIDER, and R. N. EISENMAN. 1982. Nuclear location of the putative transforming protein of avian myelocytomatosis virus. *Cell* 29:427–439.

BALTIMORE, D. 1970. RNA-dependent DNA polymerase in virions of RNA tumour viruses. *Nature* **226**:1209–1211.

BISHOP, J. M., and H. E. VARMUS. 1984. Functions and origins of retroviral transforming genes. In *Molecular Biology of Tumor Viruses: RNA Tumor Viruses,* R. Weiss et al., eds. Cold Spring Harbor Laboratory.

COLLETT, M. S., and R. L. ERIKSON. 1978. Protein kinase activity associated with the avian sarcoma virus *src* gene product. *Proc. Nat'l Acad. Sci. USA* 75:2021–2024.

*COOK, P. J., and D. P. BURKITT. 1971. Cancer in Africa. *Brit. Med. Bull.* 27:14–20.

DESGROSEILLERS, L., E. RASSART, and P. JOLICOEUR. 1983. Thymotropism of murine leukemia virus is conferred by its long terminal repeat. *Proc. Nat'l Acad. Sci. USA* **80**:4203–4207.

DEUEL, T. F., J. S. HUANG, S. S. HUANG, P. STROOBANT, and M. D. WATERFIELD. 1983. Expression of a platelet-derived growth factor–like protein in simian sarcoma virus transformed cells. *Science* **221**:1348–1350.

DOOLITTLE, R. F., M. W. HUNKAPILLER, L. E. HOOD, S. G. DEVARE, K. C. ROBBINS, S. A. AARONSON, and H. N. ANTONIADES. 1983. Simian sarcoma virus *onc* gene, v-*sis*, is derived from the gene (or genes) encoding a platelet-derived growth factor. *Science* **221**:275–276.

DOWNWARD, J., Y. YARDEN, E. MAYES, G. SCRACE, N. TOTTY, P. STOCKWELL, A. ULLRICH, J. SCHLESSINGER, and M. WATERFIELD. 1984. Close similarity of epidermal growth factor receptor and v-*erb*-B oncogene protein sequences. *Nature* 307:521–527.

DUESBERG, P. H., and P. K. VOGT. 1970. Differences between the ribonucleic acids of transforming and non-transforming avian tumor viruses. *Proc. Nat'l Acad. Sci. USA* **67**:1673–1680.

EPSTEIN, M. A., B. G. ACHONG, and Y. M. BARR. 1964. Virus particles in cultured lymphoblasts from Burkitt's lymphoma. Lancet **1**:702–703.

*GROSS, L., ed. 1970. *Oncogenic Viruses,* 2d ed. Pergamon.

*GROSS, L. 1983. *Oncogenic Viruses,* 3d ed. Pergamon.

HUEBNER, R. J., and G. J. TODARO. 1969. Oncogenes of RNA tumor viruses as determinants of cancer. *Proc. Nat'l Acad. Sci. USA* **64**:1087–1094.

HUNTER, T., and B. M. SEFTON. 1980. Transforming gene product of Rous sarcoma virus phosphorylates tyrosine. *Proc. Nat'l Acad. Sci. USA* 77:1311–1315.

LEVINSON, A. D., H. OPPERMANN, L. LEVINTOW, H. E. VARMUS, and J. M. BISHOP. 1978. Evidence that the transforming gene of avian sarcoma virus encodes a protein kinase associated with a phosphoprotein. *Cell* **15**:561–572.

*LURIA, S. E., J. E. DARNELL JR., D. BALTIMORE, and A. CAMPBELL, eds. 1978. *General Virology,* 3d ed. Wiley.

NUSSE, R., and H. E. VARMUS. 1982. Many tumors induced by the mouse mammary tumor virus contain a provirus integrated in the same region of the host genome. *Cell* **31**:99–109.

OPPERMANN, H., A. D. LEVINSON, H. E. VARMUS, L. LEVINTOW, and J. M. BISHOP. 1979. Uninfected vertebrate cells contain a protein that is closely related to the product of the avian sarcoma virus transforming gene *(src)*. *Proc. Nat'l Acad. Sci. USA* **76**:1804–1808.

PURCHIO, A. F., E. ERIKSON, J. S. BRUGGE, and R. L. ERIKSON. 1978. Identification of a polypeptide encoded by the avian sarcoma virus *src* gene. *Proc. Nat'l Acad. Sci. USA* 75:1567–1571.

RASSOULZADEGAN, M ., A. COWIE, A. CARR, N. GLAICHENHAUS, R. KAMEN, and F. CUZIN. 1982. The roles of individual polyoma virus early proteins in oncogenic transformation. *Nature* **300**:713–718.

RASSOULZADEGAN, M., Z. NAGHASHFAR, A. COWIE, A. CARR, M. GRISONI, R. KAMEN, and F. CUZIN. 1983. Expression of the large T protein of polyoma virus promotes the establishment in culture of "normal" rodent fibroblast cell lines. *Proc. Nat'l Acad. Sci. USA* **80**:4354–4358.

*ROUS, P. 1911. A sarcoma of the fowl transmissible by an agent separable from the tumor cells. *J. Exp. Med.* **13**:397–411.

RULEY, H. E. 1983. Adenovirus early region 1A enables viral and cellular transforming genes to transform primary cells in culture. *Nature* **304**:602–606.

SCHRIER, P. I., R. BERNARDS, R. T. M. J. VAESSEN, A. HOUWELING, and A. J. VAN DER EB. 1983. Expression of class I major histocompatibility antigens switched off by highly oncogenic adenovirus 12 in transformed rat cells. *Nature* **305**:771–775.

STEFFEN, D. 1984. Proviruses are adjacent to c-*myc* in some murine leukemia virus–induced lymphomas. *Proc. Nat'l Acad. Sci. USA* **81**:2097–2101.

STEHELIN, D., H. E. VARMUS, J. M. BISHOP, and P. K. VOGT. 1976. DNA related to the transforming gene(s) of avian sarcoma viruses is present in normal avian DNA. *Nature* **260**:170–173.

TAKEYA, T., and H. HANAFUSA. 1983. Structure and sequence of the cellular gene homologous to the RSV *src* gene and the mechanism for generating the transforming virus. *Cell* **32**:881–890.

*TEMIN, H. 1964. Nature of the provirus of Rous sarcoma. *Nat'l Cancer Inst. Monogr.* **17**:557–570.

TEMIN, H., and S. MIZUTANI. 1970. RNA-dependent DNA polymerase in virions of Rous sarcoma virus. *Nature* **226**:1211–1213.

*TOOZE, J., ed. 1980. *DNA Tumor Viruses,* 2d ed. Cold Spring Harbor Laboratory.

USHIRO, H., and S. COHEN. 1980. Identification of phosphotyrosine as a product of epidermal growth factor–activated protein kinase in A431 cell membranes. *J. Biol. Chem.* **255**:8363–8365.

*VARMUS, H., and A. J. LEVINE, eds. 1983. *Readings in Tumor Virology.* Cold Spring Harbor Laboratory.

WANG, J. Y. J. 1983. From c-*abl* to v-*abl*. *Nature* **304**:400.

WANG, J. Y. J., C. QUEEN, and D. BALTIMORE. 1982. Expression of an Abelson murine leukemia virus–encoded protein in *Escherichia coli* causes extensive phosphorylation of tyrosine residues. *J. Biol. Chem.* **257**:13181–13184.

WITTE, O. N., A. DASGUPTA, and D. BALTIMORE. 1980. The Abelson murine leukemia virus protein is phosphorylated in vitro to form phosphotyrosine. *Nature* **283**:826–831.

Chemical Carcinogenesis

* AMES, B. N. 1979. Identifying environmental chemicals causing mutations and cancer. *Science* **204**:587–593.

BROWN, P. C., T. D. TLSTY, and R. T. SCHIMKE. 1983. Enhancement of methotrexate resistance and dihydrofolate reductase gene amplification by treatment of mouse 3T6 cells with hydroxyurea. *Mol. Cell. Biol.* **3**:1097–1107.

CHEN, T. T., and C. HEIDELBERGER. 1969. Quantitative studies on the malignant transformation of mouse prostate cells by carcinogenic hydrocarbons *in vitro*. *Int. J. Cancer* **4**:166–178.

* MCCANN, J., and B. N. AMES. 1977. The Salmonella/microsome mutagenicity test: predictive value for animal carcinogenicity. In *Origins of Human Cancer,* H. H. Hiatt, J. D. Watson, and J. A. Winsten, eds. Cold Spring Harbor Laboratory.

* MILLER, J. H. 1982. Carcinogens induce targeted mutations in *Escherichia coli. Cell* **31**:5–7.

OLSSON, M., and T. LINDAHL. 1980. Repair of alkylated DNA in *Escherichia coli. J. Biol. Chem.* **255**:10569–10571.

POLAND, A., and A. KENDE. 1977. The genetic expression of aryl hydrocarbon hydroxylase activity: evidence for a receptor mutation in nonresponsive mice. In *Origins of Human Cancer,* H. H. Hiatt, J. D. Watson, and J. A. Winsten, eds. Cold Spring Harbor Laboratory.

* SIMS, P. 1980. The metabolic activation of chemical carcinogens. *Brit. Med. Bull.* **36**:11–18.

WALDSTEIN, E. A., E.-H. CAO, and R. B. SETLOW. 1982. Adaptive resynthesis of O^6-methylguanine-accepting protein can explain the differences between mammalian cells proficient and deficient in methyl excision repair. *Proc. Nat'l Acad. Sci. USA* **79**:5117–5121.

Cellular Oncogenes

ALITALO, K., M. SCHWAB, C. C. LIN, H. E. VARMUS, and J. M. BISHOP. 1983. Homogeneously staining chromosomal regions contain amplified copies of an abundantly expressed cellular oncogene (c-*myc*) in malignant neuroendocrine cells from a human colon carcinoma. *Proc. Nat'l Acad. Sci. USA* **80**:1707–1711.

BALMAIN, A., and I. B. PRAGNELL. 1983. Mouse skin carcinomas induced *in vivo* by chemical carcinogens have a transforming Harvey-*ras* oncogene. *Nature* **303**:72–74.

* BISHOP, J. M. 1983. Cellular oncogenes and retroviruses. *Annu. Rev. Biochem.* **52**:301–354.

BLAIR, D. G., M. OSKARSSON, T. G. WOOD, W. L. MCCLEMENTS, P. J. FISCHINGER, and G. G. VANDE WOUDE. 1981. Activation of the transforming potential of a normal cell sequence: a molecular model for oncogenesis. *Science* **212**:941–943.

BRODEUR, G., C. SEEGER, M. SCHWAB, H. E. VARMUS, and J. M. BISHOP. 1984. Amplification of N-*myc* in untreated human neuroblastomas correlates with advanced disease stage. *Science* **224**:1121–1124.

CANAANI, E., O. DREAZEN, A. KLAR, G. RECHAVI, D. RAM, J. B. COHEN, and D. GIVOL. 1983. Activation of the c-*mos* oncogene in a mouse plasmacytoma by insertion of an endogenous intracisternal A-particle genome. *Proc. Nat'l Acad. Sci. USA* **80**:7118–7122.

* COOPER, G. M. 1982. Cellular transforming genes. *Science* **217**:801–806.

* COWELL, J. K. 1982. Double minutes and homogeneously staining regions: gene amplification in mammalian cells. *Annu. Rev. Genet.* **16**:21–59.

CROCE, C. M., J. ERIKSON, A. AR-RUSHDI, D. ADEN, and K. NISHIKURA. 1984. Translocated c-*myc* oncogene of Burkitt lymphoma is transcribed in plasma cells and repressed in lymphoblastoid cells. *Proc. Nat'l Acad. Sci. USA* **81**:3170–3174.

DEKLEIN, A., A. GUERTS VAN KESSEL, G. GROSVELD, C. R. BARTRAM, A. HAGEMEIJER, D. BOOTSMA, N. K. SPURR, N. HEISTERKAMP, J. GROFFEN, and J. R. STEPHENSON. 1982. A cellular oncogene is translocated to the Philadelphia chromosome in chronic myelocytic leukaemia. *Nature* **300**:765–767.

ERIKSON, J., J. FINAN, P. C. NOWELL, and C. M. CROCE. 1982. Translocation of immunoglobulin V_H genes in Burkitt lymphoma. *Proc. Nat'l Acad. Sci. USA* **79**:5611–5615.

GERONDAKIS, S., S. CORY, and J. M. ADAMS. 1984. Translocation of the *myc* cellular oncogene to the immunoglobulin heavy chain locus in murine plasmacytomas is an imprecise reciprocal exchange. *Cell* **36**:973–982.

GONDA, T. J., and D. METCALF. 1984. Expression of *myb, myc* and *fos* proto-oncogenes during the differentiation of a murine myeloid leukaemia. *Nature* **310**:249–251.

HAYWARD, W. S., B. G. NEEL, and S. M. ASTRIN. 1981. Activation of a cellular *onc* gene by promoter insertion in ALV-induced lymphoid leukosis. *Nature* **209**:475–479.

HEISTERKAMP, N., J. R. STEPHENSON, J. GROFFEN, P. F. HANSEN, A. DEKLEIN, C. R. BARTRAM, and G. GROSVELD. 1983. Localization of the c-*abl* oncogene adjacent to a translocation breakpoint in chronic myelocytic leukaemia. *Nature* **306**:239–242.

* HELDIN, C.-H., and B. WESTERMARK. 1984. Growth factors: mechanisms of action and relation to oncogenes. *Cell* **37**:9–20.

HOFFMAN-FALK, H., P. EINAT, B. Z. SHILO, and F. M. HOFFMAN. 1983. *Drosophila melanogaster* DNA clones homologous to vertebrate oncogenes: evidence for a common ancestor to *src* and *abl* cellular genes. *Cell* **32**:589–598.

* KRONTIRIS, T., and G. M. COOPER. 1981. Transforming activity of human tumor DNAs. *Proc. Nat'l Acad. Sci. USA* **78**:1181–1184.

LAND, H., L. F. PARADA, and R. A. WEINBERG. 1983. Tumorigenic conversion of primary embryo fibroblasts requires at least two cooperating oncogenes. *Nature* **304**:596–601.

PARADA, L. F., C. J. TABIN, C. SHIH, and R. A. WEINBERG. 1982. Human Ej bladder carcinoma oncogene is a homologue of Harvey sarcoma virus *ras* gene. Nature **297**:474–478.

POWERS, S., T. KATAOKA, O. FASANO, M. GOLDFARB, J. STRATHERN, J. BROACH, and M. WIGLER. 1984. Genes in *S. cerevisiae* encoding proteins with domains homologous to the mammalian *ras* proteins. *Cell* **36**:607–612.

* ROWLEY, J. D. 1983. Human oncogene locations and chromosome aberrations. *Nature* **301**:290–291.

SCHWAB, M., J. ELLISON, M. BUSCH, W. ROSENAU, H. E. VARMUS, and J. M. BISHOP. 1984. Enhanced expression of the human gene N-*myc* consequent to amplification of DNA may contribute to malignant progression of neuroblastoma. *Proc. Nat'l Acad. Sci. USA* **81**:4940–4944.

SHIH, C., B.-Z. SHILO, M. P. GOLDFARB, A. DANNENBERG, and R. A. WEINBERG. 1979. Passage of phenotypes of chemically transformed cells via transfection of DNA and chromatin. *Proc. Nat'l Acad. Sci. USA* **76**:5714–5718.

STEWART, T. A., P. K. PATTENGALE, and P. LEDER. 1984. Spontaneous mammary adenocarcinomas in transgenic mice that carry and express MTV/*myc* fusion genes. *Cell* **38**:627–637.

SUKUMAR, S., V. NOTARIO, D. MARTIN-ZANCA, and M. BARBACID. 1983. Induction of mammary carcinomas in rats by nitroso-methylurea involves malignant activation of H-*ras*-1 locus by single point mutations. *Nature* **306**:658–661.

* TABIN, C. J., S. M. BRADLEY, C. I. BARGMANN, R. A. WEINBERG, A. G. PAPAGEORGE, E. M. SCOLNICK, R. DHAR, D. R. LOWY, and E. H. CHANG. 1982. Mechanism of activation of a human oncogene. *Nature* **300**:143–149.

TATCHELL, K., D. T. CHALEFF, D. DEFEO-JONES, and E. M. SCOLNICK. 1984. Requirement of either of a pair of *ras*-related genes of *Saccharomyces cerevisiae* for spore viability. *Nature* **309**:523–527.

TODA, T., I. UNO, T. ISHIKAWA, S. POWERS, T. KATAOKA, D. BROEK, S. CAMERON, J. BROACH, K. MATSUMOTO, and M. WIGLER. 1985. In yeast, *ras* proteins are controlling elements of adenylate cyclase. *Cell* **40**:27–36.

* VANDE WOUDE, G. F., A. J. LEVINE, W. C. TOPP, and J. D. WATSON, eds. 1984. *Cancer Cells*, vol. 2: *Oncogenes and Viral Genes.* Cold Spring Harbor Laboratory.

* VARMUS, H. E. 1984. The molecular genetics of cellular oncogenes. *Annu. Rev. Genet.* **18**:553–612.

WATERFIELD, M. D., G. T. SCRACE, N. WHITTLE, P. STROOBANT, A. JOHNSSON, A. WATESON, B. WESTERMARK, C. HELDIN, J. S. HUANG, and T. F. DEUEL. 1983. Platelet-derived growth factor is structurally related to the putative transforming protein $p28^{sis}$ of simian sarcoma virus. *Nature* **304**:35–39.

* WEINBERG, R. A. 1982. Oncogenes of spontaneous and chemically induced tumors. *Adv. Cancer Res.* **36**:149–164.

* WEINBERG, R. A. 1983. A molecular basis of cancer. *Sci. Am.* **249**(5):126–143.

* YUNIS, J. J. 1983. The chromosomal basis of human neoplasia. *Science* **221**:227–235.

Radiation and DNA Repair in Carcinogenesis

KENNEDY, A. R., J. CAIRNS, and J. B. LITTLE. 1984. Timing of the steps in transformation of C3H 10T1/2 cells by X-irradiation. *Nature* **307**:85–86.

KENNEDY, A. R., M. FOX, G. MURPHY, and J. B. LITTLE. 1980. Relationship between x-ray exposure and malignant transformation in C3H 10T1/2 cells. *Proc. Nat'l Acad. Sci. USA* **77**:7262–7266.

OLSSON, M., and T. LINDAHL. 1980. Repair of alkylated DNA in *Escherichia coli. J. Biol. Chem.* **255**:10569–10571.

Promoters of Carcinogenesis

* BERENBLUM, I. 1974. *Carcinogenesis as a Biological Problem.* Elsevier/North-Holland.

DAVIS, R. J., and M. P. CZECH. 1985. Tumor-promoting phorbol diesters cause the phosphorylation of epidermal growth factor receptors in normal human fibroblasts at threonine-654. *Proc. Nat'l Acad. Sci. USA* **82**:1974–1978.

DELCLOS, K. B., D. S. NAGLE, and P. M. BLUMBERG. 1980. Specific binding of phorbol ester tumor promoters to mouse skin. *Cell* **19**:1025–1032.

EBELING, J. G., G. R. VANDENBARK, L. J. KUHN, B. R. GANONG, R. M. BELL, and J. E. NIEDEL. 1985. Diacylglycerols mimic phorbol diester induction of leukemic cell differentiation. *Proc. Nat'l Acad. Sci. USA* **82**:815–819.

JETTEN, A. M., B. R. GANONG, G. R. VANDENBARK, J. E. SHIRLEY, and R. M. BELL. 1985. Role of protein kinase c in diacylglycerol-mediated induction of ornithine decarboxylase and reduction of epidermal growth factor binding. *Proc. Nat'l Acad. Sci. USA* **82**:1941–1945.

KIKKAWA, U., Y. TAKAI, Y. TANAKA, R. MIYAKE, and Y. NISHIZUKA. 1983. Protein kinase C as a possible receptor protein of tumor-promoting phorbol esters. *J. Biol. Chem.* **258**:11442–11445.

SHOYAB, M., J. E. DELARCO, and G. J. TODARO. 1979. Biologically active phorbol esters specifically alter affinity of epidermal growth factor membrane receptors. *Nature* **279**:387–391.

WEINSTEIN, I. B. 1983. Protein kinase, phospholipid and control of growth. *Nature* **302**:750.

Human Cancer

BAKHSHI, A., J. MINOWADA, A. ARNOLD, J. COSSMAN, J. P. JENSEN, J. WHANG-PENG, T. A. WALDMANN, and S. J. KORSMEYER. 1983. Lymphoid blast crises of chronic myelogenous leukemia represent stages in the development of B-cell precursors. *N. Engl. J. Med.* **309**:826–831.

CAVENEE, W. K., T. P. DRYJA, R. A. PHILLIPS, W. F. BENEDICT, R. GODBOUT, B. L. GALLIE, A. L. MURPHREE, L. C. STRONG, and R. L. WHITE. 1983. Expression of recessive alleles by chromosomal mechanisms in retinoblastoma. *Nature* **305**:779–784.

* DOLL, R., and R. PETO, eds. 1981. *The Causes of Cancer.* Oxford University Press.

KNUDSON, A. G. JR. 1977. Genetic predisposition to cancer. In *Origins of Human Cancer*, H. H. Hiatt, J. D. Watson, and J. A. Winsten, eds. Cold Spring Harbor Laboratory.

KRONTIRIS, T. G., N. A. DIMARTINO, M. COLB, and D. R. PARKINSON. 1985. Unique allelic restriction fragments of the human Ha-*ras* locus in leukocyte and tumor DNAs of cancer patients. *Nature* **313**:369–374.

YOSHIDA, M., I. MIYOSHI, and Y. HINUMA. 1982. Isolation and characterization of retrovirus from cell lines of human adult T-cell leukemia and its implication in the disease. *Proc. Nat'l Acad. Sci. USA* **79**:2031–2035.

* *Diet, Nutrition and Cancer.* 1982. Committee on Diet, Nutrition and Cancer, Assembly of Life Sciences, National Research Council, National Academy Press.

24

Immunity

THE life of every organism is constantly threatened by other organisms—this is the nature of the living world. To defend itself, each species has evolved a variety of protective mechanisms from camouflage colors, to poison, to very effective running muscles. For their continual battle with microorganisms, vertebrates have evolved a very elaborate set of protective measures called, collectively, the *immune system*. The word immune (Latin *immunis,* meaning "exempt") implies freedom from a burden: an animal that is immune to a specific infecting agent will remain free of infection by that agent. The study of the immune system constitutes the discipline of *immunology*.

The immune system works by a learning process. Our first encounter with a pathogenic bacterial, fungal, or viral infection leads to disease. But after recovery, we usually remain free of that disease forever. Our immune system has learned to recognize that specific bacterium or fungus or virus as a foreign infecting agent; should it attack us again, it will be rapidly killed.

The key process carried out by the immune system is *recognition*. The system must recognize the presence of an invader. It must also be able to discriminate between foreign invaders and the natural constituents of the body: we call this *discrimination between self and nonself*. The importance of such discrimination is underscored by disorders in which the immune system reacts against self-components of the body. Such disorders are known as the *autoimmune diseases,* and some of them are fatal. Recognition of a foreign invader is only the first step of the immune system's attack; it must be followed by steps that will kill and eliminate the invader. Thus the immune system involves two types of activities: recognition processes directed against individual, discrete aspects of a target, and generalized responses that follow from the recognition and allow the system to mount an attack against the invader, such as a killing response to an invading bacterium. The generalized responses elicited by a recognition event are called *effector* functions.

We believe that the immune system evolved to handle *pathogens,* invading microorganisms that can cause disease, but it has other properties as well. It can kill cancer cells and, in experimental animals at least, it can protect the body against certain tumors. How active the immune system is against human tumors is a debated question. The immune system also prevents tissue transplantation between individuals. Every vertebrate individual has a unique set of molecules on the surface of its cells. Its immune system recognizes those cells as self, but all cells from other individuals, even of the same species, are seen as foreign and are killed, or *rejected.* Inhibition of this rejection reaction is necessary whenever a surgeon attempts to graft a patch of skin or transplant a kidney from one person to another. Without special treatment, the only transplants that are accepted without a rejection reaction are those from an identical twin.

If we are to understand the immune system, then, we must understand some very fundamental processes. How does the system recognize the individuality of a specific pathogen? How does it discriminate foreign matter from endogenous material, nonself from self? How does it translate recognition of a foreign invader into a killing reaction? How does it learn, so that the second attack by an invader is repulsed so much faster than the first? The answers to these questions are not fully known. While we understand, for instance, a great deal about how recognition develops, we know very little about how the system avoids reacting with self-components.

This chapter concentrates on aspects of the immune

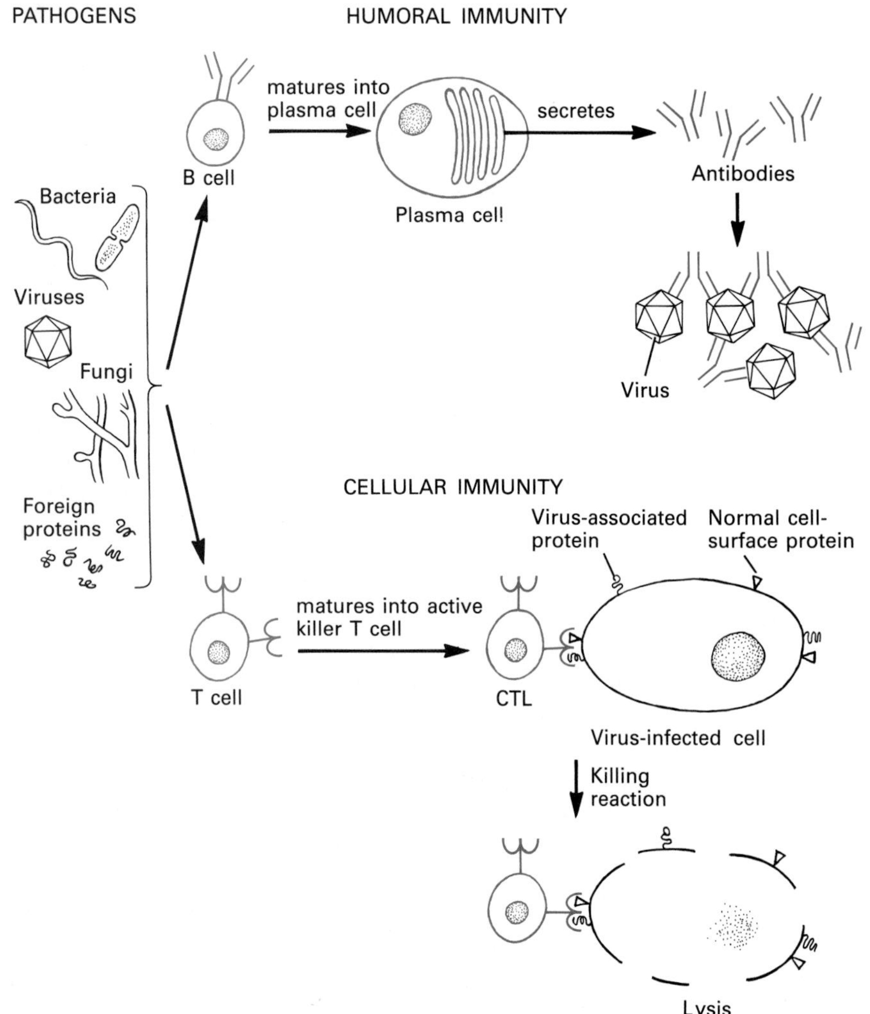

Figure 24-1 Humoral immunity and cellular immunity. When a pathogen invades the body, the immune system responds with two types of reaction. Cells of the humoral system secrete antibodies that can bind to the pathogen. Cells of the cellular system have antibody-like molecules bound to their surface. Some of the cell types that make up the cellular system support the humoral response; others kill cells that have been infected by pathogens.

system that are reasonably well known. Especially highlighted are the molecular biological mechanisms that underlie the recognition process because they have a unique aspect—the reorganization of DNA as a mechanism in cellular differentiation. An overview of the system is followed by an analysis of antibodies, the best-understood molecules of the immune system, then a consideration of the cells that make antibodies, and finally a review of cells of the immune system that act independently of antibodies.

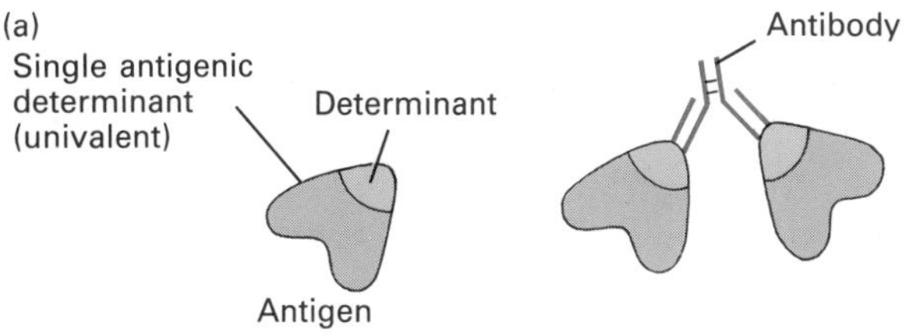

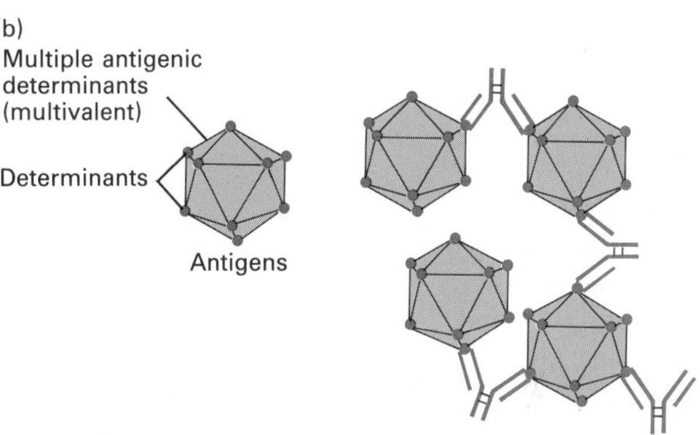

Figure 24-2 Antibodies bind to single or multiple determinants on an antigen. (a) Binding of antibody to a protein with a single determinant. The antibody's two sites can both be filled but the complex cannot grow further. Natural antibodies are mixtures of molecules that can bind to many determinants on an antigen, potentially forming a large aggregate. (b) Some antigens, notably viruses, have multiple identical determinants on a single particle. Even a pure antibody population can form large aggregates when reacting with such antigens.

Overview

The immune system works in two fundamentally different ways, by *humoral immunity* and by *cellular immunity* (Figure 24-1). The term humor refers to a fluid, and humoral immunity is immunity caused by molecules in solution in the body. The molecules of humoral immunity are proteins collectively called *immunoglobulin* and abbreviated *Ig*. They constitute 20 percent of the proteins in the blood. A single such molecule is called an *antibody,* but the term also refers to all the immunoglobulins that react with a given material. Thus, we may refer to "the antibody that binds to poliovirus" with the understanding that there are many individually different molecules together constituting such an antibody.

In the second type of immune reaction, cellular immunity, intact cells are responsible for the recognition reaction. No freely soluble molecules participate. Instead, molecules held tightly to the surfaces of the reacting immune cells act like immunoglobulins but never break their association with the plasma membrane.

The injection of almost any foreign substance into an animal elicits the formation of both antibodies and immune cells that can bind to the substance. A given substance that provokes antibody or immune cell formation, or is recognized by an antibody or immune cell, is called an *antigen.*

Antibodies Bind to Determinants and Have Two Functional Domains

The unique function performed by an antibody molecule is high-affinity binding to a limited region of an antigen. This binding involves hydrophobic forces, ionic forces, and van der Waals forces, but no covalent bond is formed between antibody and antigen. The site on an antigen at which a given antibody molecule binds is called a *determinant* (Figure 24-2; see also Figure 3-37). The surface of a protein molecule presents many possible determinants to which antibodies might bind. Some antigens have a repeating structure, so that the same determinant may occur on it many times. Such an antigen is said to be *multivalent.* A good example of a multivalent antigen is a virus particle (Figure 24-2b). By attaching a small molecule onto the surface of a protein, a new determinant can be created. If, for instance, bovine serum albumin, a fairly big protein, is coupled to the simple molecule, dinitrophenol, and the conjugate is injected into a mouse, the mouse will make antibody molecules that bind to the dinitrophenol group. This mechanism can be demonstrated by coupling dinitrophenol to another protein, and showing that this second protein now reacts with antibodies made to the original dinitrophenol-conjugated protein. In this context, dinitrophenol, or any small substituent on a protein that can elicit an antibody response, is called a *hapten* (Figure 24-3).

A key property of antibodies, which is best probed by haptens, is *specificity.* An antibody that binds well to a given hapten may bind poorly to a closely related hapten. For instance, many antidinitrophenol antibodies bind poorly to trinitrophenol. Simple changes in polypeptides also show the specificity of antibodies: a polypeptide of seven to fifteen residues, linked to a protein, can elicit antibodies that bind well to the polypeptide but poorly

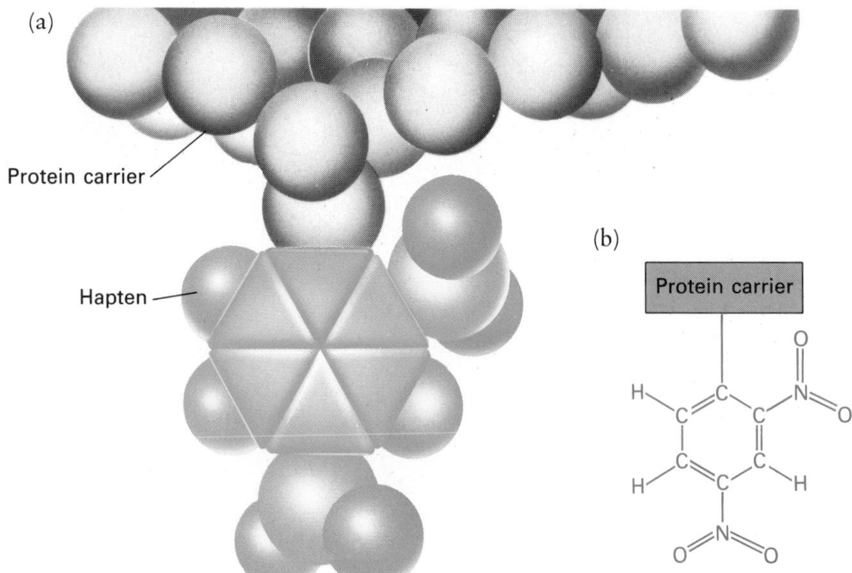

Figure 24-3 Hapten linked to a protein. The hapten shown is dinitrophenol, one commonly used in experimental immunology (see Figure 3-37). (a) The space-filling representation shows that it has a constantly varying surface that presents many different determinants to which antibodies can bind. (b) Chemical formula of dinitrophenol. *From* The Structure and Function of Antibodies, *by G. M. Edelman. Copyright 1970* vol. 223(2) *by Scientific American, Inc.*

to derivatives of the polypeptide with even one amino acid changed.

We noted before that the immune system carries out two processes: recognition and effector functions. The antibody molecule itself is divided into two analogous regions (Figure 24-4): *binding domains* that interact with the antigen and *effector domains* that signal the initiation of processes (such as phagocytosis) that are able to rid the body of the antigen bound to the antibody. The effector region also provides signals that distribute antibodies to various body fluids. Some are directed to secretions, such as saliva, mucus, or milk; others go across the placenta to protect the fetus.

A specific set of effector domains is common to many antibodies, but a given binding domain is found on only a very small set of antibodies. This dichotomy of a highly specific region and a common region in antibody molecules has its counterpart in the genes that encode antibodies. These genes are constructed from multiple variable regions which can be affixed to a single constant region.

Each antibody molecule consists of two size-classes of polypeptide chains, *light (L) chains* and *heavy (H) chains.* A single antibody molecule has two identical copies of the L chain and two of the H chain. Therefore the basic antibody structure is a four-chain molecule (Figure 24-4). The N-terminal ends of one H chain and L chain together form one binding domain. The antibody is thus bivalent; it has two binding sites for antigen.

Antibody Reaction with Antigen Is Reversible

The binding of a site on an antibody to its antigen is a simple bimolecular, reversible reaction capable of analysis by standard kinetic theory. Looking only at a single binding site we write

$$\text{Ag} + \text{Ab} \underset{k_2}{\overset{k_1}{\rightleftharpoons}} \text{Ag-Ab}$$

where Ag is the antigen, Ab is the uncomplexed antibody, and Ag-Ab the bound complex. The forward and reverse binding reactions are characterized by k_1 and k_2 rate constants, respectively. The *affinity* of the antibody binding site is measured by the ratio of complexed to free reactants at equilibrium. An *affinity constant, K,* is defined by

$$K = \frac{[\text{Ag-Ab}]}{[\text{Ag}][\text{Ab}]} = \frac{k_1}{k_2}$$

where the brackets denote concentration. Typical values of K are 10^5 to 10^{11} liters/mol. K can be determined by measuring the concentration of free antigen needed to half-fill the binding sites of an antibody (see Figure 24-5). At this point [Ag-Ab] must equal [Ab], and the above formula resolves to

$$K = \frac{1}{[\text{Ag}]}$$

Many proteins may have weak affinities for other proteins, consequently we need to define an affinity below which binding is considered to be nonspecific and above which binding is considered to be a specific antibody-antigen reaction. From experience, any K less than 10^4 liters/mol is considered nonspecific.

But antibody-antigen reactions are not necessarily the simple bimolecular events treated in the above equation. Such a treatment is appropriate for an antigen with one potential site for reaction and for a homogeneous antibody population, even though there are two binding sites per antibody molecule. If, however, binding to one site blocks a reaction with the second site, or if the antigen has two appropriately spaced identical sites so that both antibody sites can react with the same antigen molecule, then ideality is lost and more complicated equations are needed. For instance, if the two antibody sites can react with the same antigen molecule, the second site will react much faster than the first because the antigen molecule will already be very close by and the apparent affinity will be greatly increased, usually by at least 10^4. A further complication is that natural antibody populations can react with a multitude of sites on antigens and with a variety of affinities. Thus the term *avidity* is used to describe the apparent affinity of an antibody population reacting with an antigen. Avidity is measured by half-saturation of antibody, but we recognize that this measure has no simple mathematical meaning.

Antibodies Come in Many Classes

Because antibodies have to be distributed to various parts of the body and serve multiple effector functions, animals make multiple classes of antibodies. Each class has a dif-

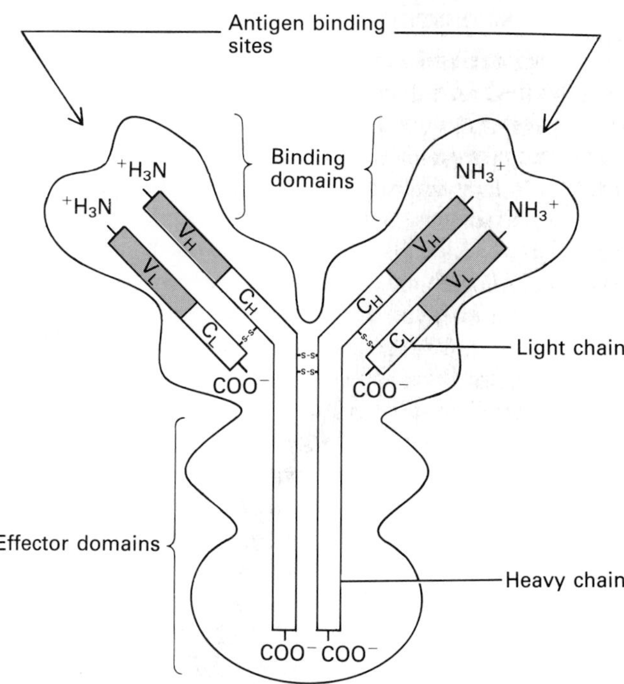

Figure 24-4 Schematic representation of an antibody molecule. Antigen binding occurs at the two upper ends of the Y-shaped molecule, where the heavy and light chains are in intimate contact. The lower portion of the Y contains the effector region that is made only from heavy chain domains. The N-terminal (NH_3^+) and C-terminal (COO^-) ends of the heavy and light chains are indicated. Disulfide bridges (S—S) link the two sets of heavy and light chains (see also Figure 3-38).

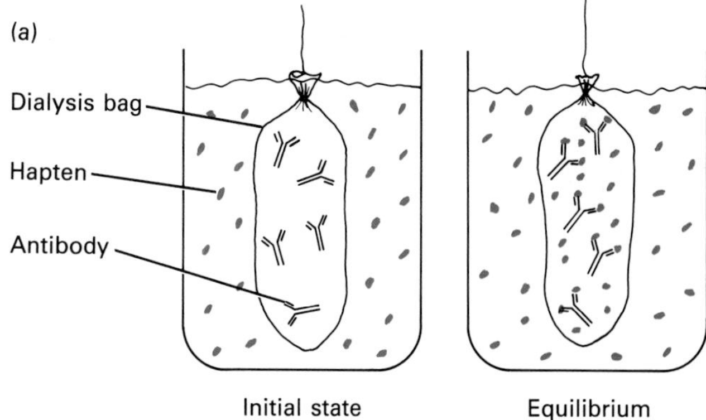

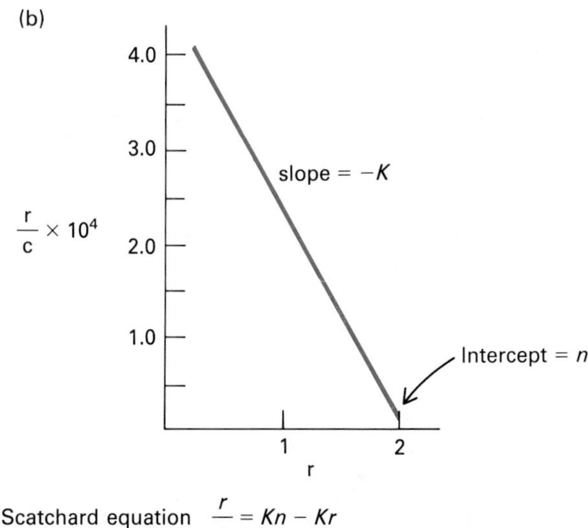

Scatchard equation $\frac{r}{c} = Kn - Kr$

where $r = \frac{\text{Moles of hapten bound to antibody}}{\text{Moles of antibody in bag}}$

c = Concentration of hapten in bulk solution

n = Number of binding sites on antibody molecule

K = Equilibrium association constant of antibody

Figure 24-5 Determining the affinity of an antibody for a hapten. (a) Equilibrium dialysis. A large volume of a solution of hapten is equilibrated through a semipermeable dialysis membrane with a solution of antibody that can bind to the hapten. As the concentration of hapten increases, the amount bound increases. (b) Typical data are plotted in a fashion that allows application of the Scatchard binding equation shown below. In the experiment shown, the dinitrophenol hapten bound to lysine was used, and a monoclonal antidinitrophenol antibody from a myeloma was placed inside the dialysis bag. The Scatchard equation gives two values, a number of binding sites per antibody molecule (n) and the equilibrium binding constant (K). The number of binding sites per molecule is given by the intercept on the x axis. It is 2, as expected, because of the two binding arms on an antibody molecule. [See H. N. Eisen, 1980, *Immunology,* 2d ed., Harper & Row, pp. 299–301.]

Table 24-1 Major antibody classes

Properties	IgM	IgD	IgG	IgE	IgA
Heavy chain	μ	δ	γ	ϵ	α
Mean human adult serum level (mg/ml)	1	0.03	12	0.0003	2
Half-life in serum (days)	5	3	25	2	6
Number of four-chain monomers	5	1	1	1	1, 2, or 3
Special properties	Early appearance; fixes complement; activates macrophages	Found mainly on cell surfaces; traces in serum	Activates complement; crosses placenta; binds to macrophages and granulocytes	Stimulates mast cells to release histamines	Found mainly in secretions

ferent set of effector domains with its own specific set of properties. There are five major classes and a number of subclasses. The important classes are *IgM, IgD, IgG, IgE,* and *IgA* (Table 24-1; Figures 24-4 and 24-6). The differences among them derive from the different H chains of which they are constituted.

When an animal begins its response to antigenic challenge (following inoculation of an antigen), the first antibody class it makes is IgM. The individual IgM molecule is a pentamer of the basic 4-chain antibody structural unit; the units are held together by disulfide bonds and by a single copy of a polypeptide known as *J chain* (Figure 24-6). Because of its multiple, closely spaced binding sites, IgM has a very high binding avidity for microorganisms, like viruses, that are covered with identical subunits. One major effector function of IgM is activation of the *complement system,* a group of proteins able to kill cells to which antibody is bound. IgM also activates *macrophages,* cells that are specialized for ingesting and killing bacteria.

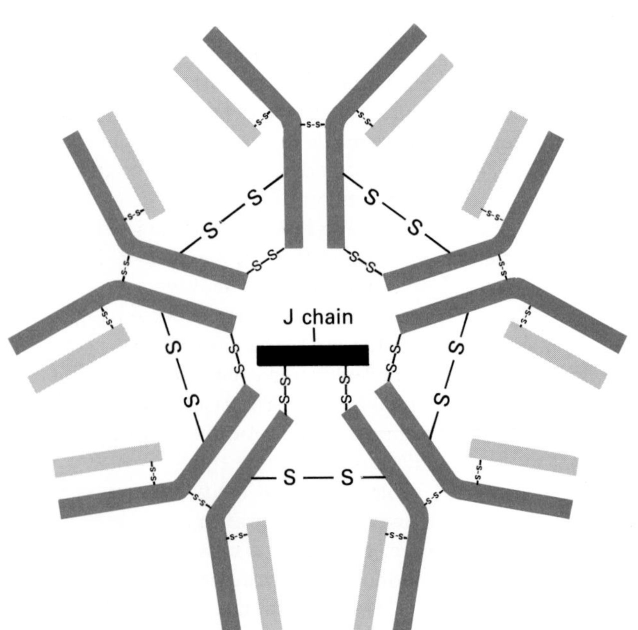

Figure 24-6 Schematic representation of an IgM molecule. The antibodies of most classes are formed of only four chains as depicted in Figure 24-4. The IgM class is a pentamer whose polymerization is initiated by an ancillary protein called the J chain. Holding the various polypeptides together are disulfide (S—S) bonds.

The IgD molecule, a monomer, is an enigma. Many cells have IgD on their surface but very little is found in bodily fluids and no special function is yet known for it.

IgG is the main serum antibody. A four-chain monomer, it is made in copious amounts in response to an antigenic challenge, especially after multiple stimulations with antigen. Like IgM, it stimulates both complement and macrophages. IgG antibodies can pass through the placenta to the fetus. Their effector regions are specialized for binding to a receptor on the maternal side of placental cells; the receptor then carries the IgG through the cell into the fetal circulation.

IgE is one of the most bothersome types of antibody. It is found mainly in tissues, where, in complex with an antigen, it activates the release of histamines from specialized, blood-derived cells called *mast cells.* Histamine release is the cause of such allergic reactions as hives, asthma, and hay fever. What positive role IgE may have is not clear; it is thought to be a defense against parasitic infections.

The final major class of antibody, IgA, is perhaps the most important. It exists as a monomer and also as a

J-chain-linked dimer or trimer. Polymerized IgA binds to a receptor on the blood-facing surface of many epithelia and is transported across the epithelial cells into whatever compartment the epithelial layer surrounds. For instance, because of a high concentration of IgA receptor molecules on the inner (basolateral) surface of intestinal cells, IgA is copiously secreted into the intestinal lumen. The IgA receptor has the unique property that during the passage of the receptor-IgA complex through the cell, most of the receptor is cut away from its membrane-binding region and remains bound to the antibody (Figure 24-7). When IgA emerges from the epithelial cell, it has a piece of the receptor bound to itself. This piece is called *secretory component* because it is bound to all secreted Ig molecules. The IgA receptor will also bind to IgM; persons unable to make IgA will secrete IgM, which also protects them. Secreted IgA (or IgM) is present in saliva, tears, the lungs, and the intestines as the body's first line of defense against infection by microorganisms.

Antibodies Are Made by B Lymphocytes

Antibodies are made by many cell types, all of which fall into a single cell lineage: the *B lymphocytes,* or *B cells.* (They are designated "B" because in birds, where they were first studied, most of the cells mature in an outpocketing at the end of the cloaca called the *bursa of Fabricius.* This organ has no counterpart in mammals, but most mammalian B lymphocytes originate from the bone marrow so the "B" remains appropriate.) B lymphocytes arise as cells having IgM or IgM plus IgD on their surfaces. Reaction of the surface-bound antibody with an antigen will activate the B lymphocyte to mature into a cell type known as a *plasma cell* which is highly specialized for antibody production and secretion (see Figure 24-1).

The Immune System Has Extraordinary Plasticity

Many haptens, like the dinitrophenol unit described earlier (Figure 24-3), are the products of synthetic chemists. During the long evolution of vertebrates, no animal outside of a laboratory ever encountered a dinitrophenylated protein. Yet at the first encounter with such a protein, most vertebrates produce dinitrophenol-specific antibodies. This response is the most remarkable aspect of immune function—the ability to recognize the whole universe of potential determinants whether or not the species has encountered them at any time in evolution. The plasticity of the immune system—its ability to extend itself to meet unprecedented challenges—makes the system's molecular events especially intriguing. All the other body systems, except the nervous system, have evolved their

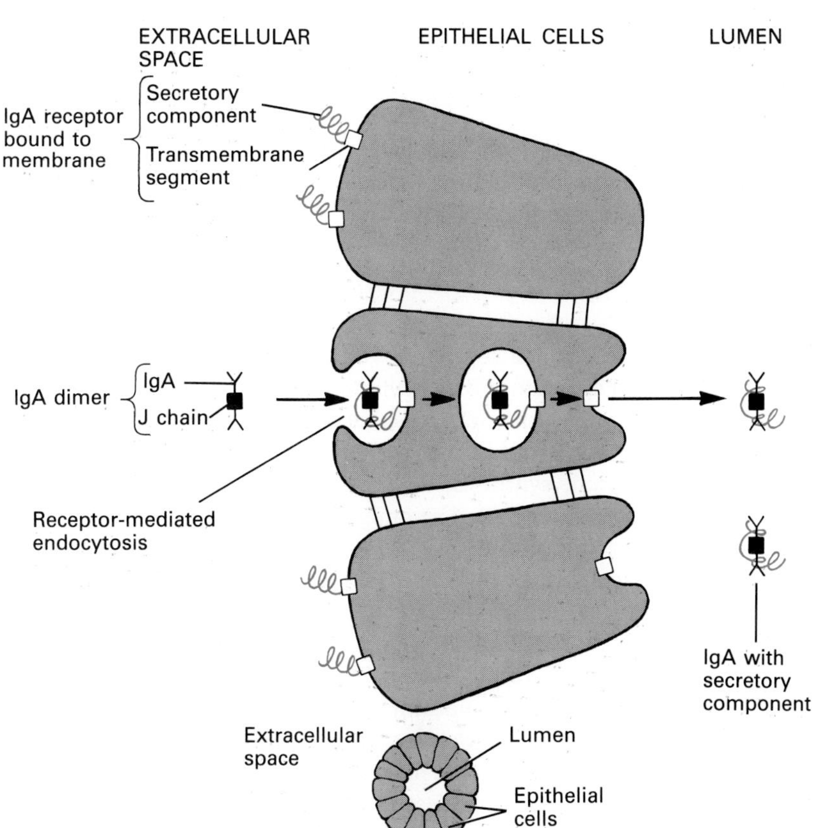

Figure 24-7 Transfer of IgA across epithelial cells by the IgA receptor. IgA molecules are shown being transported across epithelial cells from the blood-facing surface into a lumen (for instance, from the intestinal lining into the intestinal space). The receptor that carries out the transport is found on the blood-facing surface of the intestinal cell. There it binds to an IgA dimer and is carried into the cell by endocytosis. Rather than staying in the cell, the fate of most transported molecules, the IgA receptor complex with IgA is carried through the cell to the lumenal face of the cell, where the IgA separates from the cell. This separation is achieved by a cleavage of the IgA receptor, leaving "secretory component" bound to the IgA that is released into the lumen. [See K. E. Mostov, M. Friedlander, and G. Blobel, 1984, *Nature,* 308, 37–43.]

properties in response to specific environmental pressures encountered by their ancestors, that is, by natural selection. Although the immune system itself is certainly a product of evolution, not all of the genes encoding individual antibody proteins could have arisen during evolution because not all of the determinants recognized by antibodies could have been previously encountered. The genetic basis for this extraordinary plasticity of protein structure is now fairly well understood and will be a major concern later in this chapter. At this point, however, we must ask: by what cellular strategy is this plasticity achieved?

When the plasticity of the immune system was first appreciated, many suggestions were proposed concerning how the system might operate. Most of the proposals were based on an approach called *instructive theory:* it was imagined that all antibodies had one polypeptide structure that could be induced to fold in different ways by combining with antigen as a structural template; the binding specificity of the antibody would then arise from a previous encounter with the antigen (Figure 24-8). This notion has now been disproved by the observation that antibodies can be denatured and renatured in the absence of antigen and still retain their specificity. Thus the one-dimensional sequence of the polypeptides must already contain the information for the binding site. Furthermore, it was shown that antigen-specific antibodies can be detected on lymphocyte surfaces before the specific antigenic stimulus is initially presented to the animal.

Even before instructive theory had been disproved, a competing approach based on preexisting antigen-specific antibodies was being developed. It provided a consistent explanation for the cellular events that produce specificity and plasticity. This latter theory is now the basis of all thinking about immunology.

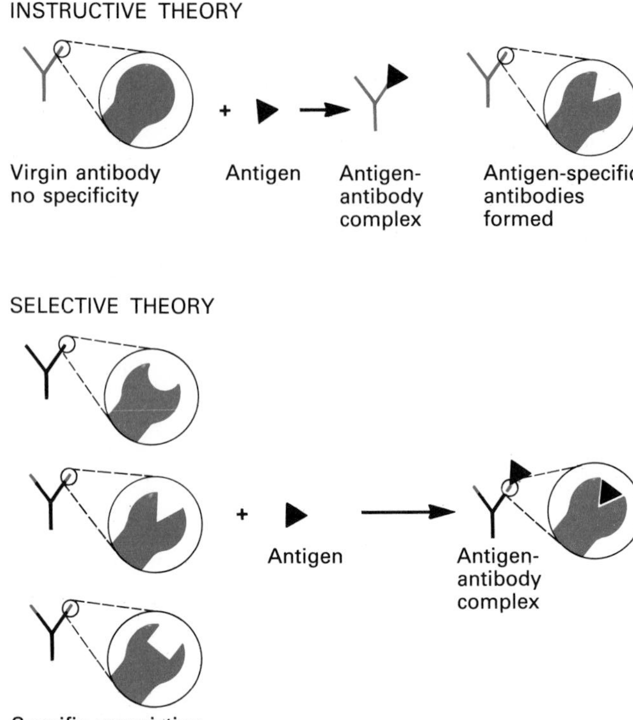

Figure 24-8 Two theories of the origin of antibody specificity. The instructive theory, now discarded, suggested that the antigen is a template for determining the structure of the antibody. With today's knowledge of genetic encoding of protein structure, such a notion is hard to imagine, but before the 1970s so much less was known about the determination of protein structure that template formation of antibodies was considered possible. Selective theory, now considered the only likely approach, suggests that the immune system generates a very large number of specific molecules and that antigens "select," or bind to, the ones that fit.

Clonal Selection Theory Underlies All Modern Immunology

The class of theories that replaced the instructive models is called *selective theory* because antigen is imagined to select the appropriate antibody from a preexisting pool of antibody molecules (see Figure 24-8). The notion of selection was brought into the field by Niels Jerne, a brilliant theorist and experimentalist of immunology, and was then incorporated by Sir MacFarlane Burnet into one of the great intellectual constructs of modern biology, the *clonal selection theory* of antibody production (Figure 24-9). This theory supposes that (1) the body is continually elaborating B lymphocytes that have on their surface immunoglobulin molecules; (2) all of the surface immunoglobulin molecules on any one cell have the same binding specificity; (3) for any one antigenic determinant, only a very small subset of the entire pool of B cells will have surface antibody with which it can bind. When a vertebrate encounters a foreign antigen in its circulation, this antigen combines with preexisting immunoglobulins on only those B cells that have molecules with the appropriate specificity. The interaction of antigen and antibody on the B-cell surface then triggers that B cell into multiplication, synthesis, and secretion of its specific antibody (Figure 24-10). The antigen has *selected* from a preexisting pool those cells with appropriate specificity and has caused their proliferation into a large population of cells all making the same antibody. Such a population of identical cells all coming from the same ancestral cell is called a *clone,* thus the term *clonal selection.* Clonal selection theory is, like natural selection, a darwinian theory because the antigen selects the cells that will multiply from a set of variants that have arisen independently of the selective force. Clonal selection has received such complete verification that it no longer is considered a theory but

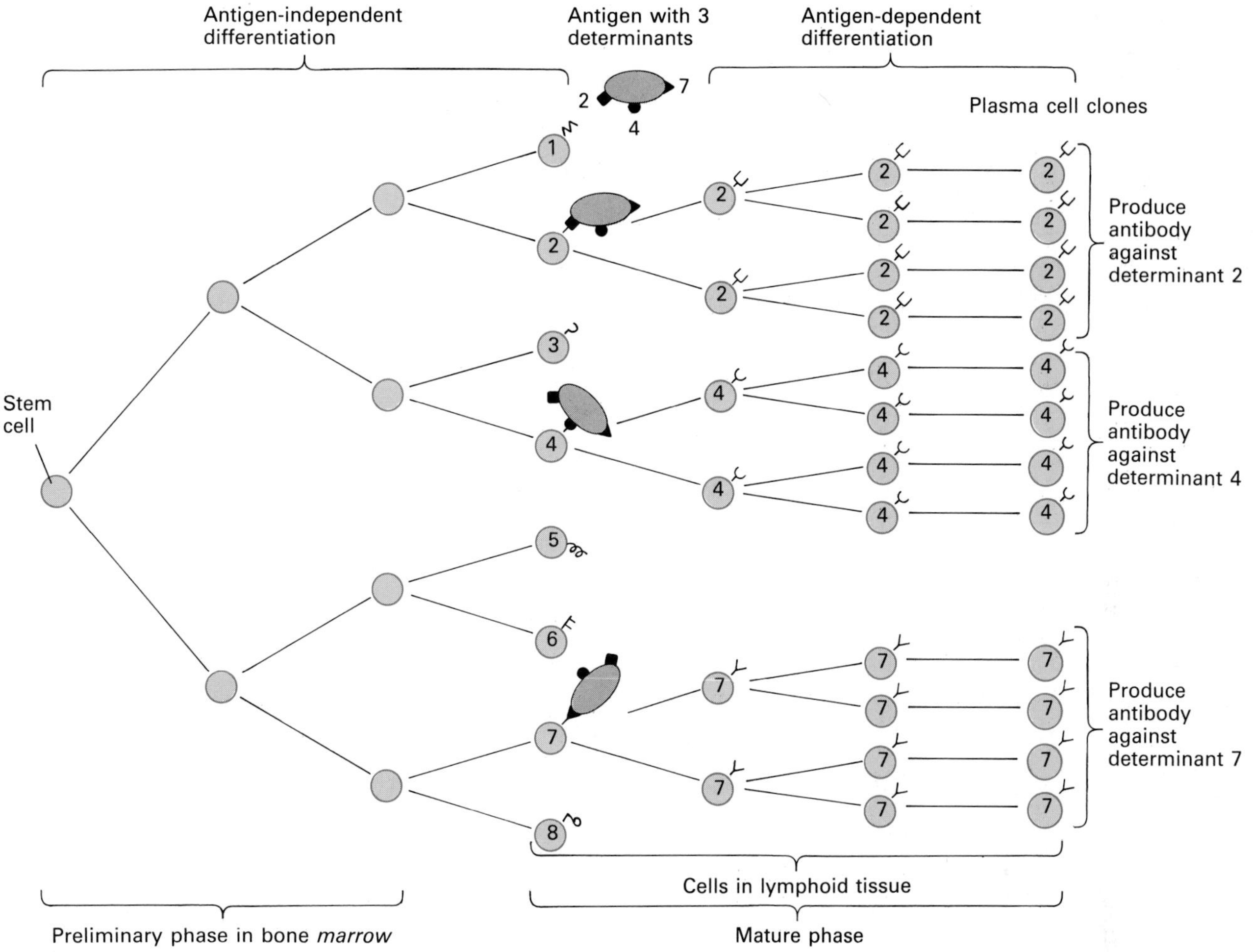

Figure 24-9 The immune system develops specificity by a process of clonal selection. From a large number of antibody-carrying cells the antigen selects the cells carrying antibodies that bind to its determinants. Then clones of these cells are produced, all carrying the same surface antibody.

rather is accorded the status of an accepted explanation of the functioning of the immune system.

Clonal selection, as it actually occurs, involves two distinct phases in the life of the B cell. In the *preliminary phase,* before any encounter with an antigen, a large number of B cells are produced, each with a different antigen-binding specificity (see Figure 24-9). B cells are continually elaborated by the bone marrow throughout the life of the animal, each newly arising cell having a lifetime of only a few days as a circulating cell of the blood or lymph. But if during its circulation a B cell encounters an antigen to which its surface antibody can bind, then the cell enters a *mature phase* of life in which it is activated to growth, division, and the secretion of antibody. Long-lived progeny called *memory B cells,* or simply *memory cells,* result from this activation so that what was a transient cell becomes permanent (Figure 24-10).

If clonal selection is to operate, there must be a mechanism for generating an enormous number of structurally different immunoglobulin molecules in the absence of any selective force (i.e., before encountering antigen). This is known as the problem of the *generation of antibody diversity* (GOD). Diversity is now known to arise from two distinct levels of variation: modifications of germ line DNA and events in somatic tissue. The somatic events are called *somatic variation,* some of which may be *somatic mutation.* Both multiple germ line genes and somatic variations of a variety of sorts play roles in generating the variability of the immune response.

The Immune System Has a Memory

Among the remarkable capabilities of the immune system is its ability to learn. A first encounter of a B cell with an antigen leads to a slowly rising synthesis of antibody, dominated by IgM *(primary response).* A second encoun-

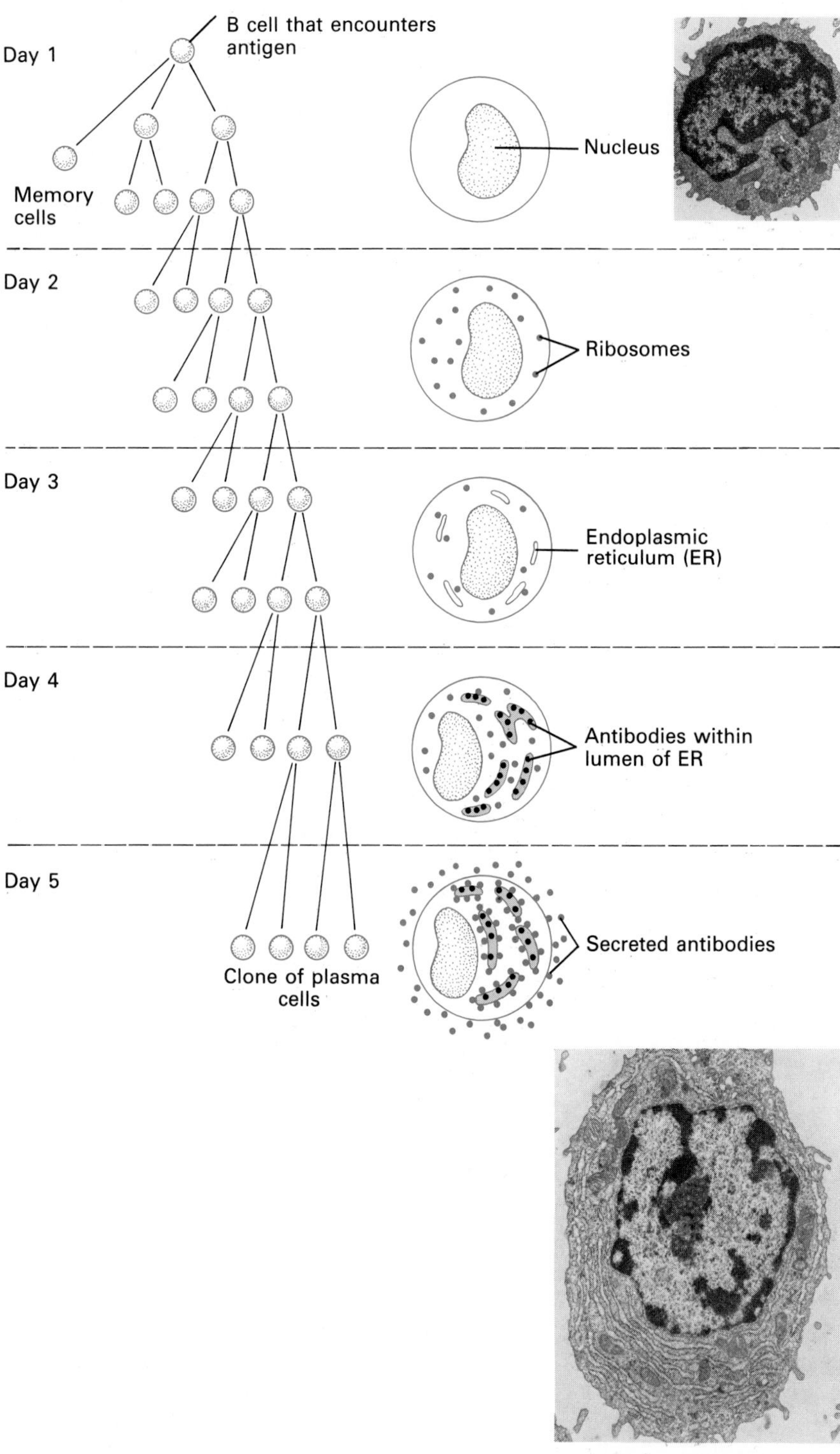

Figure 24-10 Activation of B lymphocytes leads to both growth and secretion. The diagram indicates the events that follow lymphocyte activation. The upper photograph shows a resting B cell. After an encounter with antigen and stimulation by factors produced by other immune system cells, the B cell initiates a series of divisions during which it develops the apparatus for secretion of copious amounts of antibody. The end result is a large clone of antibody-secreting plasma cells with extensive endoplasmic reticulum as seen in the lower photograph. *After L. Hood et al, 1984,* Immunology, *2d ed., Benjamin p. 11. Photographs courtesy of D. Zucker-Franklin; from* Atlas of Blood Cells: Function and Pathology, *1981, Lea and Febiger.*

ter with the same antigen leads to a more rapid and greater response, dominated by IgG *(secondary response)*. Only a previously encountered antigen provokes the secondary response: the system has learned to recognize the antigen to which it was previously exposed (Figure 24-11). The basis of learning in the immune system is the formation of long-lived memory cells (Figure 24-10). After an encounter with an antigen, memory cells with surface antibody directed to the antigen persist for the life of the organism in the total absence of further antigenic stimulation.

These circulating memory cells carry on their surfaces the particular immunoglobulins that avidly bind to reinvading antigens. The memory B cells mainly make IgG or IgA and respond so rapidly that a second encounter with an antigen leads to much faster and more effective response than the first. This is why once a person has been infected by a given virus, that virus can never catch the

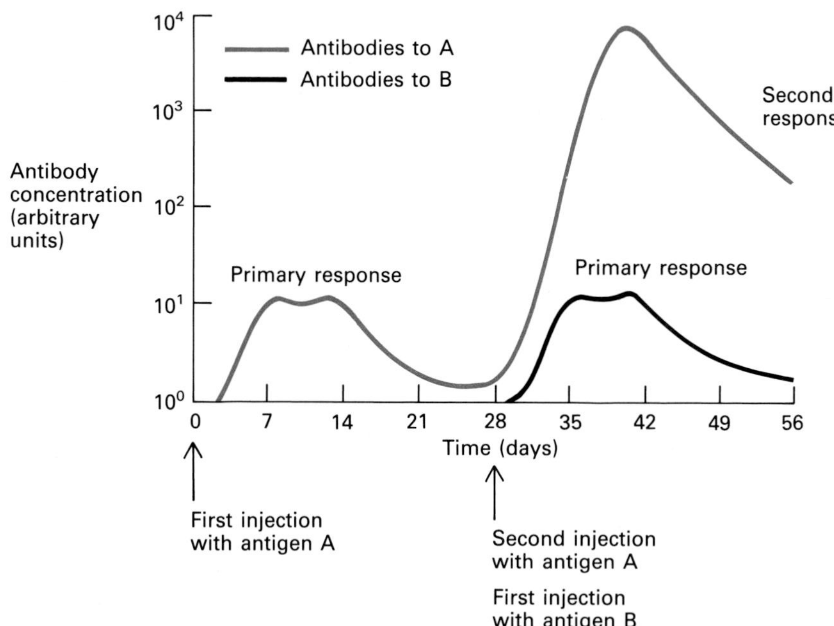

Figure 24-11 Responses to initial and later injections of an antigen. At day 0, antigen was injected into a mouse and the response was followed by measuring serum antibody levels to the antigen. At day 28, when the initial response had subsided, a second immunization was done with both antigen A and a new antigen B. The secondary response to antigen A was faster and greater than the initial response to antigen A. The response to antigen B followed the course of an initial encounter with antigen. The secondary response to antigen A is therefore a specific, not a general, response of the system.

body unprepared again. It also explains why children are so much more susceptible to infectious diseases than are adults.

The Other Cell Type in the Immune System Is the T Lymphocyte

The B lymphocyte carries surface immunoglobulin and can be activated to secrete immunoglobulin by an encounter with antigen. The immune system contains another cell population that also carries surface antigen-binding molecules, but these cells never secrete antibody molecules (see Figure 24-1). They are known collectively as the *T lymphocytes,* or simply *T cells,* because their maturation involves transit through the thymus gland. The antigen-binding molecules on the surface of T cells are structurally related to but distinct from antibody; they are called *T-cell receptors.* These receptors usually do not react with soluble antigens but rather with antigens on the surfaces of other cells. The complex reaction of surface molecules on one cell with surface molecules on another cell is at the heart of both T-cell biology and of the many cell-to-cell interactions that accompany development (see Chapter 22).

Three fundamentally different types of T lymphocytes are recognized: cytotoxic, helper, and suppressor (each has many subdivisions). The *cytotoxic T lymphocyte (CTL)* recognizes other cells that display foreign antigens on their surfaces and kills the cells (see Figure 24-1). Less genteelly, this type is also called the *killer T cell.* Precursors to CTLs display antibody-like receptors that recognize foreign proteins on cells. The recognition process apparently immortalizes the precursor CTL, providing the animal with a clone of killer T cells able to destroy cells that display the foreign protein. This killing process can be demonstrated by incubating CTLs with target cells that have previously taken up the isotope chromium 51 (^{51}Cr). If the CTLs come from an animal that has been immunized with the target cells, CTLs will kill the cells by putting holes into them. The ^{51}Cr can leak out through the holes. Thus cell destruction by the CTLs can be measured by ^{51}Cr release (Figure 24-12). The CTL is formally equivalent to an antibody in that it recognizes and eliminates unwanted substances. Antibody deals with soluble foreign materials, CTLs with cell-bound foreign materials. A classic CTL target is a virus-infected cell that displays a viral glycoprotein on its surface (Figure 24-1).

The second type of T cell is a *helper T lymphocyte,* or *T_H cell.* These cells assist B cells as well as other types of T cells in their responses to an antigen. The helper T cells are needed for B-cell stimulation, possibly because the interaction of most antigens with antibodies on the surface of B cells is insufficient to stimulate B-cell growth and secretion of soluble antibody. Only *multivalent antigen* molecules—those that contain repeated copies of a determinant (like a high-molecular-weight carbohydrate)—can stimulate B cells by themselves; other antigens require the collaboration of T_H cells with the B cells. Multivalent antigens are apparently effective alone because one antigen molecule can simultaneously react with multiple antibody molecules and cause aggregation of the surface antibody (Figure 24-2), a situation that is very stimulatory to cell growth.

Part of the function of the T_H cell is recognition of antigens followed by secretion of factors that stimulate the B cell. These factors also stimulate other cells that

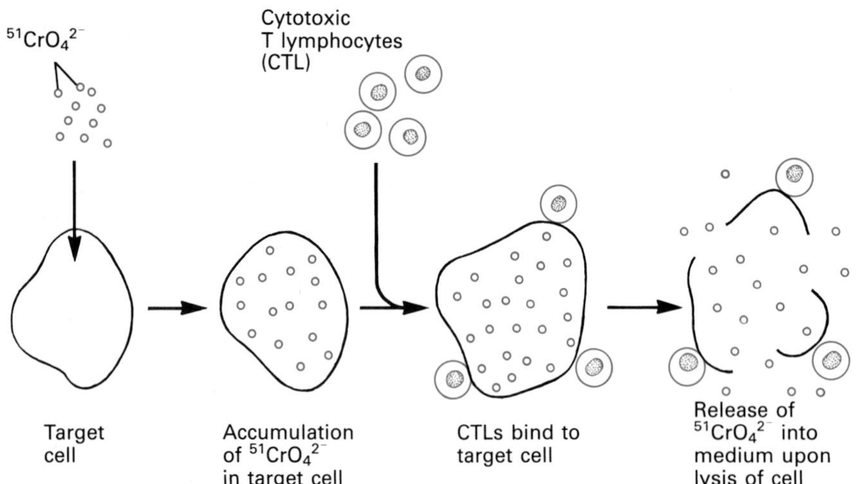

Figure 24-12 Measuring cytotoxic T-cell activity by $^{51}CrO_4$ release. First cells are loaded with $^{51}CrO_4^{2-}$. This ion, taken up by the cells from the medium, binds to cellular proteins and remains in the cell. If CTLs recognize a cell (for instance, if it is virus-infected and displays viral protein on its surface) they can bind to the cell and kill it, releasing the cell-bound $^{51}CrO_4^{2-}$ into the medium. By determining cell-bound and free $^{51}CrO_4^{2-}$, the investigator can measure the amount of cell lysis.

participate in immune reactions, such as CTLs and macrophages. The T_H cells, like CTLs, cannot recognize free antigen; they recognize only cell-bound antigens.

The third type of T cell, the *suppressor T cell* (T_S), suppresses B-cell activity. It helps to dampen responses to antigens and may also play a key role in suppressing responses to self-molecules.

The four types of lymphocyte we have described, the B lymphocyte and the three T lymphocytes, are indistinguishable in size and general morphology. To distinguish among them, immunologists use differences in the surface proteins found on the various cell types. B cells have surface antibody as well as other molecules that are not found on T cells. T cells have the antibody-like T-cell receptors and a protein called Thy-1 as well as proteins specific to T-cell subtypes (see Figure 24-13a). Operationally, the cell types are distinguished with the use of antibodies that recognize one or another of these characteristic surface proteins.

Pure populations of cells can be made by using these antibodies in one of two ways. In the first method, *negative selection,* cells with a given surface antigen are killed with specific antibody and the remaining cells are retained. A convenient way to make pure T cells or pure B cells is to use anti-immunoglobulin or anti-Thy-1, respectively. In the second method, *positive selection,* a dish is coated with the antibody and cells that bind to the dish are retained—the procedure is called *panning* (Figure 24-13b). More elegantly, a fluorescein-tagged antibody is bound to cells and a fluorescence-activated cell sorter (FACS) can examine one cell at a time to select the cells to which the antibody has bound (see Figure 5-10).

Cells Involved in the Immune Response Circulate Throughout the Body

We have briefly described some of the cells that together provide an immune response. How are these cells organized and coordinated in the body? How does antigen reach these cells? These are questions about the organization of vertebrate lymphoid tissues.

The immune system in the body has two compartments (Figure 24-14). Lymphocytes arise and go through their preliminary phase of establishing surface receptors mainly in the bone marrow and thymus, which are known as the *primary organs* of the immune system. The lymphocytes then leave these organs and are distributed to many sites in the body which together make up the *secondary organs,* or *peripheral lymphoid tissue.*

The lymphoid system guards all portals of entry into the body. The tonsils in the throat and the adenoids in the nose are lymph nodes. There are nodes in the armpit that filter fluid from the arms and nodes in the groin that filter fluid from the legs. A special series of lymphoid cells, the *Peyer's patches,* are found in the intestinal wall where they filter out antigens that enter in food or come from the bacteria growing in the intestines. The activated lymphocytes in Peyer's patches migrate out of the node into the blood. They are finally captured back in the tissue spaces just inside the intestinal lining where they secrete IgA. The IgA is then rapidly transported across the epithelial cells into the gut lumen. IgA receptor is a major component of the blood-facing surface of intestinal epithelial cells.

The peripheral lymphoid tissue is organized around the two fluid systems of the body, the blood and the lymph. These two systems are in contact. Lymph is formed by fluid transported from the blood to the spaces within and around tissues. From these extracellular spaces, lymph flows into thin-walled *lymphatic vessels,* where it is slowly moved to larger central collecting vessels. Ultimately the lymph is returned to veins, where it re-enters the blood. In blood, lymphocytes constitute 20 to 30 percent of the nucleated cells; in lymph they constitute 99 percent. The lymphocytes travel in the lymph, entering and leaving the blood circulation by squeezing between the endothelial cells that line the blood vessels. The lymph circulation is filtered through lymph nodes in which all types of lymphocytes take up temporary residence. Antigens that enter the body find their way into the blood or lymph and are filtered out by lymph nodes or by the spleen, which is the blood's filter. In the lymph nodes, phagocytes engulf and degrade antigens and eventually display pieces of antigen on their surface. The rare lymphocyte in the lymph node that carries an antibody or T-cell receptor with the appropriate specificity will bind either to free antigen or to the fragments of phagocyte-bound antigen and become activated to multiply and differentiate into secreting B lymphocytes or mature T lymphocytes. This binding reaction is the moment of clonal selection, when an antigen binds with, or selects, a specific previously established B-cell surface antibody or T-cell receptor. The collaboration between helper T lymphocytes and B lymphocytes occurs in the lymph nodes or spleen where the concentration of cells facilitate the necessary interaction and magnifies the clonal selection by antigen. The antibody made by the secreting B lymphocytes and their mature progeny, the plasma cells, leaves the node in the lymph and is transported to the blood.

Tolerance Is the Frontier of Immunology

A key attribute of the immune system is the discrimination between self and nonself. Surface recognition molecules on lymphocytes are sufficiently diverse for any antigen to find some cells with surface antibody or T-cell receptor whose binding constant for the antigen is high enough to lead to activation. But what prevents self-antigens from activating any of the body's own cells that may carry antibody able to react with them?

Immunologists call the inability to react with self *tolerance,* and they say that the immune system is tolerant to self-antigens. The mechanisms that make for tolerance are poorly understood but involve rendering unreactive the clones of both B and T cells that would otherwise carry out anti-self reactions. Most relevant evidence comes from the experimental production of a tolerant state.

The immune system in mice becomes responsive to foreign antigens in the days just after birth. All antigens present in the body at the time the system matures are considered by the immune system to be self. Thus if foreign antigens are incorporated into a newborn mouse, later on the mouse will not be able to mount an immune response when presented with those antigens. Because new clones of lymphocytes appear in an animal throughout its life,

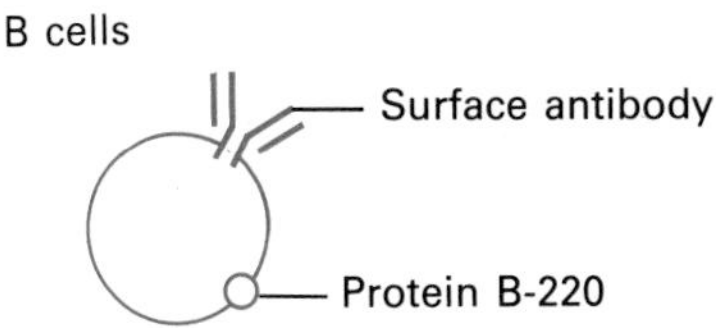

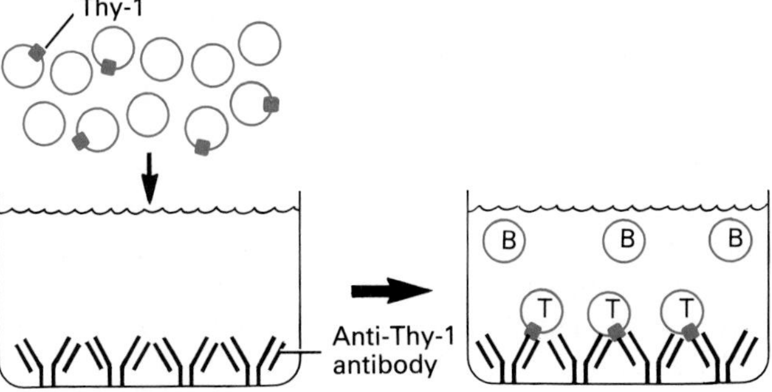

Figure 24-13 Surface structures on lymphocytes. Each class of lymphocyte has a distinct set of surface proteins. (a) Some of these are indicated for various lymphocyte subsets. The T cells fall into two categories. All have surface protein Thy-1, but CTLs and T_S cells have a common surface protein while T_H cells have a different protein. The T-cell receptor is a complex of several nonvariable proteins plus two proteins (T_α and T_β) that contain variable regions for binding to specific antigens. T_S cells may not have the standard T-cell receptor. B cells have none of the T-cell surface proteins but have two distinctive surface markers: surface antibody and a protein called B-220. All of the cell surface proteins can be specifically recognized with the use of antibodies that bind selectively to these proteins. (b) An anti-Thy-1 antibody is used to select T cells from a mixture with B cells in the process called panning. The anti-Thy-1 antibody is bound to the surface of a dish. A mixture of T and B cells is poured over the bound antibody. The T cells are bound to the antibody while the B cells remain free. The B cells are poured off, leaving a pure population of T cells on the dish. [See L. J. Wysocki and V. L. Sato, 1978, *Proc. Nat'l Acad. Sci. USA* 75:2844–2848.]

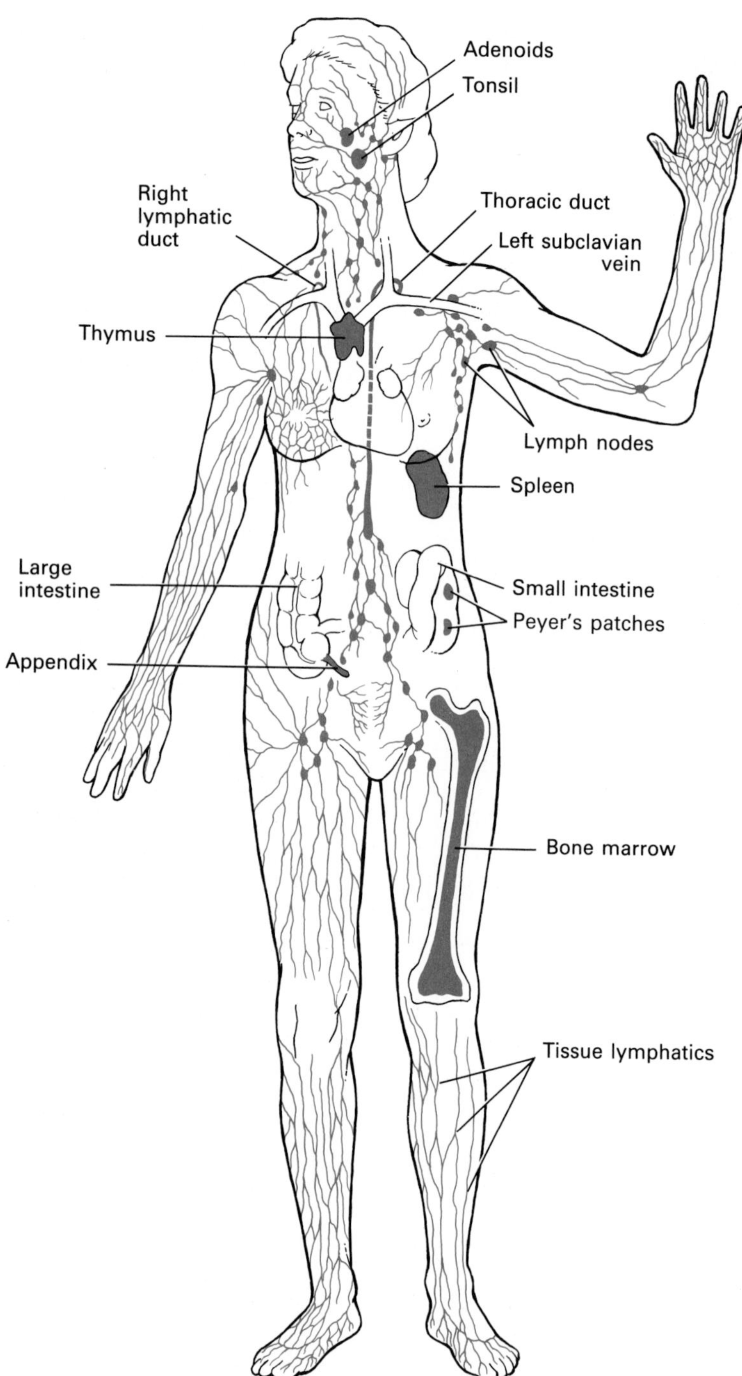

Figure 24-14 The human lymphoid system. The primary organs (thymus and bone marrow), secondary organs, and lymph vessels are shown in red. The whole system consists of the circulating lymphocytes, the vessels that carry the cells (both blood vessels and special lymphatic vessels), and the lymphoid organs. Only one bone is shown but all major bones contain marrow.

the tolerance to self must be an active, ongoing process. Suppressor T lymphocytes are thought to play an important role in maintaining tolerance by specifically suppressing lymphocyte clones that react against self-antigens.

Tolerance to a specific antigen can be induced in mature animals. Inoculation with large amounts of a protein, for instance, can make a mouse nonresponsive to a later challenge with a dose and form of antigen that would otherwise elicit a high concentration of antibody. That such tolerance is caused by T_S cells can be shown by passing the tolerant state to a second mouse by transferring the tolerant mouse's T_S cells.

The consequences of a failure of tolerance can be severe. Especially in their older years, people may begin to lose self-tolerance and make antibodies to their own proteins. This can lead to many kinds of disorders including autoimmune hemolysis (an attack on red blood cells), arthritis, and systemic lupus erythematosis. When a reaction to a major self-protein of the serum occurs, abnormally high levels of antigen-antibody complexes appear in the blood. A general problem accompanying autoimmune diseases is the deposition of these antigen-antibody complexes in the kidney, resulting in glomerulonephritis, inflammation of the filtering apparatus of the kidney. The deposits can be severe enough to cause kidney failure.

Immunopathology Is Disease Caused by the Immune System

Autoimmunity is only one type of disease caused by a faulty immune system. So many immune system problems exist that the term *immunopathology* has developed. Some disorders are due to inherited deficiencies of important components of the immune system. A person may even lack all B cells or T cells, a condition manifested in severe combined immunodeficiency disease (SCID). One form of this disease is caused by inactivation of the gene that encodes the enzyme adenosine deaminase. Excess adenosine is specifically toxic to lymphocytes. Patients with SCID totally lack lymphocytes and die at a young age of viral or bacterial infection. They can only be kept alive in a sterile environment. Many less severe deficiencies exist. Some persons lack components of the complement system. In certain cases, medical therapies for diseases not related to the immune system can affect immune function; this is especially true of treatment with steroids, which depress immune function.

Several types of cancer result from a loss of growth control in lymphocytes and macrophages. Hodgkin's disease is one example; others are acute lymphocytic leukemia, chronic lymphocytic leukemia, and many mature T-cell leukemias. Human T-lymphotropic virus (HTLV) is a retrovirus able to transform human helper T cells into tumor cells, causing adult T-cell leukemia. Another form of human retrovirus kills helper cells and appears to be responsible for the disease AIDS (acquired immune deficiency syndrome).

Along with the deficiency states and malignancies of the immune system there are hyperreactivity problems. Autoimmunity is one, but more common are the allergies, hyperreactivity to foreign antigens. Allergies result from hyperproduction of IgE in reaction to environmental substances like pollen and dust. Other types of IgE-induced hyperreactivity include asthma, food sensitivities, and reactions to toxins injected by stinging insects. In all cases, it is the local release of mast cell products induced by binding of IgE-antigen complexes that causes the often violent symptoms. These released products include histamines, heparin, and substances that cause constriction of smooth blood vessels (see Figure 3-25). Such reactions are treated with antihistamines and with epinephrine (adrenalin), which acts by raising the cyclic AMP levels of the mast cell and preventing the release of its components.

Antibodies

The previous section introduced antibodies as protein molecules with an antigen-binding region and an effector region. It also described the different classes of antibody: IgM, IgG, IgD, IgE, and IgA (see Table 24-1). We will now examine the structure of antibodies in much greater detail, raising questions of how a population of molecules with so many common properties can also have such extensive variability. The answers to this question lies in understanding the origins of antibody diversity.

Because a population of antibodies involves molecules with variable structures, it is difficult to chemically characterize antibody that comes from immunized animals. Tumors, however, initially arise as single transformed cells, therefore the antibody made by B-lymphoid tumors is homogeneous. One specific type of B-lymphoid tumor, the *plasmacytoma,* or *myeloma,* is especially useful because such tumors are analogs of plasma cells, the cells that secrete large amounts of antibody. Up to 10 percent of the protein made by a myeloma cell population can be a single homogeneous type of antibody with one light chain and one heavy chain. Myelomas occur spontaneously in humans and can be induced at will in mice of certain strains by injecting mineral oil into their peritoneum. Most of what we know about specific protein sequences of antibody chains comes from the analysis of myeloma proteins.

Heavy-Chain Structure Differentiates the Classes of Antibodies

Examination of the antibodies made by myelomas, and also the study of normal immunoglobulins, has shown that all five classes of antibody have the same general organization. Each antibody molecule has two chains, a *light (L) chain* with a molecular weight of about 23,000 and a *heavy (H) chain* of m.w. 53,000. All classes of immunoglobulin have one of two types of an L chain protein, called the kappa (κ) and lambda (λ) chains. In mice, 95 percent of the L chain is of the κ class; in humans it is close to 50 percent.

The differences between the classes arise from the heavy chain. Each class has a different type of H chain, the name of which is the Greek equivalent of the class name. Thus IgM has a μ chain, IgG has a γ chain, and the others have δ, ϵ, or α chains (see Table 24-1). There are actually four subclasses of IgG proteins with somewhat different properties, each with a different H chain—

making a total of eight types of immunoglobulin. Attached to all the H chains are asparagine-linked carbohydrate chains.

Antibodies Have a Domain Structure

Every individual antibody molecule has one type of L chain and one type of H chain. The chains are held together by disulfide bonds to form a monomer, and two monomers are linked by disulfide bonds to form the basic dimeric structure of the molecule (Figure 24-15). Within each chain, units made of about 110 amino acids fold up to form compact *domains*. Each domain has a single internal disulfide bond which holds it together. An L chain has two domains, H chains have four or five domains. The first two N-terminal domains of the H chains interact with the two L-chain domains, producing a compact unit that acts as the binding region of the antibody. In most H chains, a *hinge* region consisting of a small number of amino acids is found after the first two domains. The hinge is flexible and allows the binding regions to move freely relative to the rest of the molecule. At the hinge region are located the cysteine residues whose SH groups are linked to form the —S—S— bridges between the two monomer units of the antibody dimer. The hinge regions are the places most susceptible on the molecule to the action of protease; light protease treatment can split an antibody into two pieces, called F_{ab} and F_c fragments. The F_{ab} portion has the antigen-binding site, the F_c portion has the effector regions (see Figure 24-15).

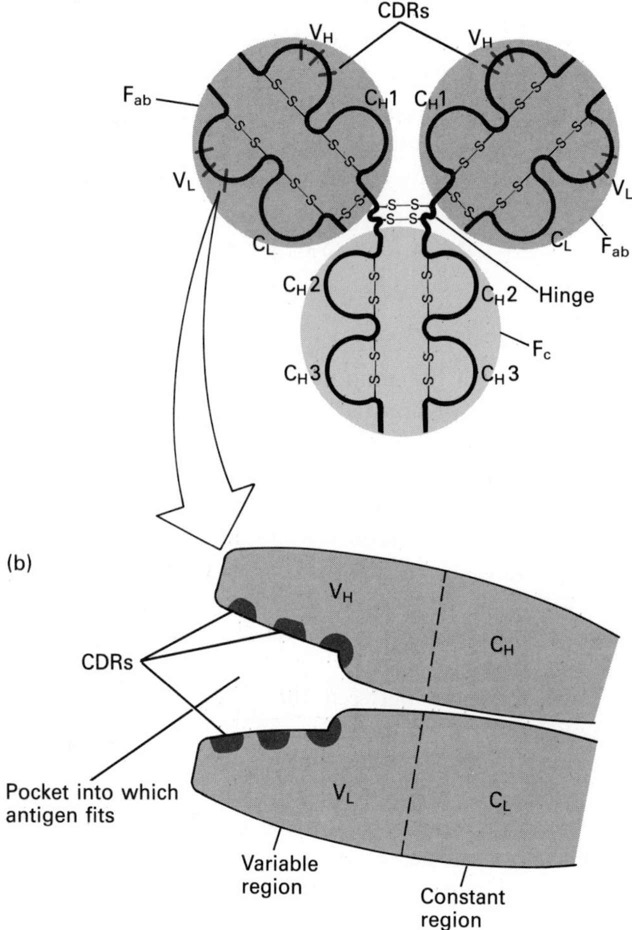

Figure 24-15 Domain structure and the complementarity-determining regions (CDRs) of antibody. (a) The molecule is organized in disulfide-bonded 110 amino acid domains, four in the heavy chain and two in the light chain. The farthest N-terminal domain of each chain is variable in sequence (V region): the other domains are constant in sequence (regions C_H1, C_H2, and C_H3 in the heavy chain and C_L in the light chain). When the molecule is protease-digested, cutting of the hinge region connecting C_H1 and C_H2, the most sensitive spot on the molecule, splits the molecule into two parts, F_{ab} (the antigen-binding domain) and F_c (the effector region). Within the V regions are segments of highly variable sequence constituting the CDRs, the amino acids that actually contact the antigen. (b) Antibody binding site. By the coordinate effort of the CDRs from light and heavy chains, a binding pocket is formed into which a specific antigen fits.

The N-Terminal Domains of H and L Chains Have Highly Variable Structures that Constitute the Antibody-Binding Site

The very first amino acid sequences determined on L chains from human myelomas made it clear that the N-terminal domain has a very variable structure while the C-terminal domain has a quite constant structure. The N-terminal domain is called the *variable region* and the C-terminal domain is called the *constant region* (Figure 24-15). H-chain sequences show the same division; within each class the N-terminal domain is highly variable and the C-terminal domains have a constant sequence. The variable domains of L and H chains are bound to one another. In fact, they interact closely to form a single compact unit (Figure 24-15). This unit is the antibody-binding site, the region of the antibody molecule that binds to antigen. This can be demonstrated with the use of an antigen that contains a reactive chemical group: the reactive group on the antigen will form a covalent bond to the variable domain of the H or L chain, showing that the variable domains form the antigen-binding pocket.

A strong correlation exists between a specific amino acid sequence and a binding specificity for certain antibodies. Antibodies of the same specificity will often have very similar variable-region sequences. Conversely, antibodies that have different specificities have different sequences. This observation illustrates a central principle of immunology, that binding specificity is determined solely by the amino acid sequence of the variable regions. The rule can be shown quite directly. If the L and H chains of a myeloma protein that binds the dinitrophenol group are separated, denatured, and then renatured together, the specificity for binding dinitrophenol, but not other haptens, returns. The experiment can be varied by substituting either the L or H chains of an antidinitrophenol antibody with other L or H chains. Such an experiment shows that both the correct L chain and the correct H chain are needed to get dinitrophenol-binding specificity. Thus it is the joint structure formed by two variable regions that produces a binding site and not either variable region alone.

X-ray crystallographers have determined the high-resolution structure of a few antibodies and have directly demonstrated the formation of binding sites by contributions from each chain (Figures 24-15, 24-16, and 24-17). They see a pocket, or cleft, on the surface of the antibody formed by three short polypeptide segments from each chain. The rest of the variable regions interact to produce the stable domain structure by forming planes of protein known as β-pleated sheets (see Chapter 3). In each chain the sheet structure of the variable domains, approximately repeated in the constant domains, is so characteristic that it has been called the *immunoglobulin fold*. The three amino acid segments of the variable region that contribute to antigen binding are loops that extend from the immunoglobulin fold (Figure 24-18). Because the binding site is complementary in structure to the antigen, these three segments are called *complementarity-determining regions*, or *CDRs*. As might be expected, the CDR structure is much more variable than that of the rest of the variable region, so that the CDRs are often called *hypervariable regions*. The variable regions therefore consist of two types of sequence, highly variable CDRs embedded in less variable *framework regions (FR)* that form the immunoglobulin fold. The overall variable region has four framework regions and three CDRs (Figure 24-19).

The Generation of Antibody Diversity Involves Several Mechanisms

Only the variable regions of an antibody contribute to the diversity of antigen-binding specificities. The problem of the generation of diversity is the problem of how one portion of a molecule can vary enormously while the other portion remains essentially invariant. The different constant regions provide structural variability to Ig molecules but this relates to effector function or to the distinction between light and heavy chains, not to antigen-binding specificity. For many years the puzzle of diversity remained unsolved because the tools did not exist with which to probe for answers. The problem so intrigued immunologists, however, that an extensive theoretical literature developed.

There are two distinct ways to imagine generating a system that would produce genes able to encode many different protein variable regions all attached to a common constant region. One way is to have a single germline gene that could undergo extensive mutation in a particular lymphoid cell, only at its variable-region end. Each cell in which this happened would make a protein with a

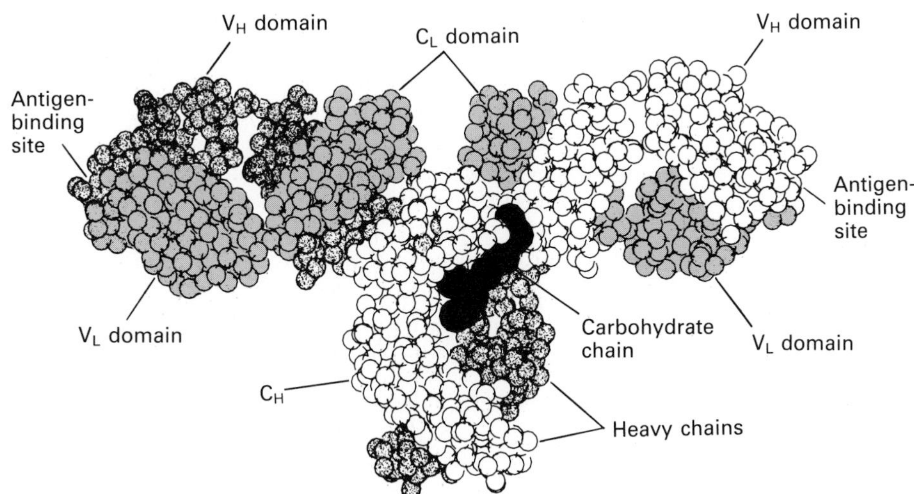

Figure 24-16 Model of antibody molecule derived from x-ray crystallographic analysis. The model shows all of the atoms as solid balls and thus describes the outer contours of the molecule. Its Y shape is evident. The individual chains are color-coded so that they can be distinguished: the two light chains are colored; one heavy chain is light gray and the other is white. *From E. W. Silverton, M. A. Navia, and D. R. Davies, 1977,* Proc. Nat'l Acad. Sci. USA *74:5140.*

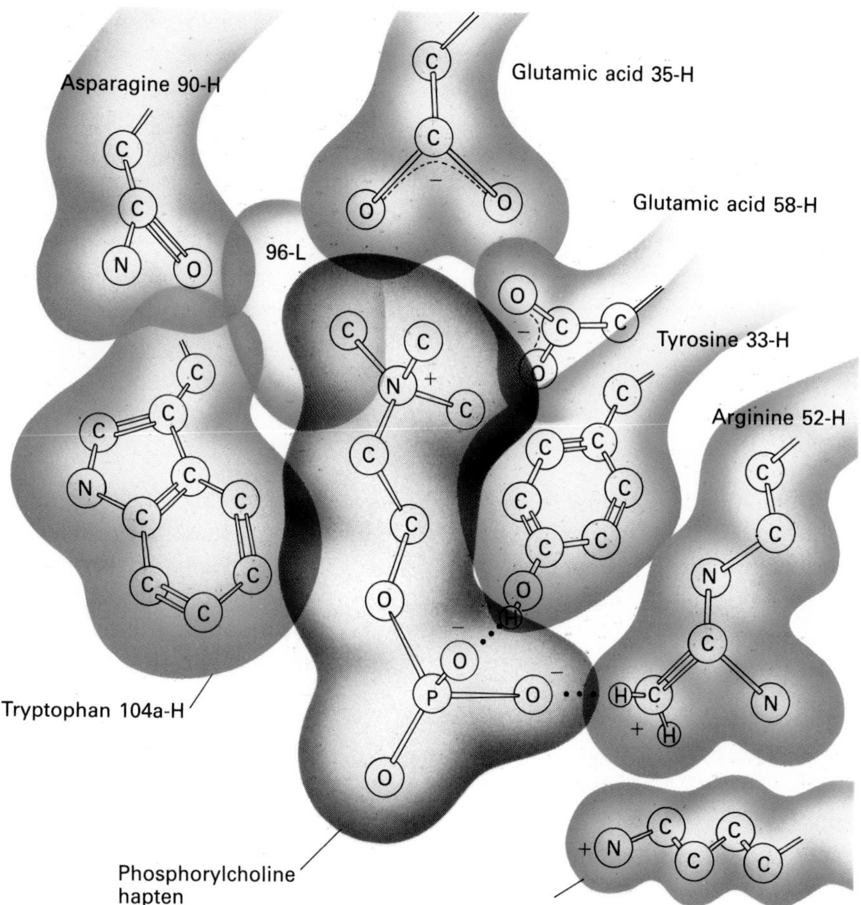

Figure 24-17 Amino acids in the CDR regions forming the binding site of an antibody molecule called McPC 603. Shown are those amino acids that contact the hapten phosphorylcholine. Each of these amino acids is part of one of the three CDRs of either the light (L) or heavy (H) chain (in this case, mainly heavy chain amino acids are involved). The CDR amino acids form a cleft or pocket into which the phosphorylcholine hapten fits. A combination of electrostatic, hydrogen-bonding and van der Waals forces hold the hapten in the cleft. *From* The Antibody Combining Site, *by J. D. Capra and A. B. Edmundson. Copyright 1977* vol. 236(1) *by Scientific American, Inc.*

unique variable-region sequence but all proteins would have the same constant region. The other way to solve the problem is to have many variable-region gene segments in the DNA, any one of them able to append to a single constant-region gene segment. These two different approaches to the diversity problem are generically described as the *somatic mutation hypothesis* and the *somatic recombination hypothesis,* respectively. Either is possible and both can be elaborated. For instance, one variant of the somatic recombination hypothesis suggests that there are only a few variable regions but that recombination among them increases variability. An important possibility—one that was ignored in the polemical phase of theoretical immunology—is that both notions are correct. This, we now know, is the answer: the system uses both mechanisms. In fact, there are really many independent mechanisms, all of which contribute to diversity.

The recombination hypothesis was the first to gain experimental support. There are, in principle, three ways that recombination could occur: in proteins, in RNA, or in DNA. Although the joining of protein segments or RNA segments are possible explanations, only recombination at the DNA level contributes to diversity.

Antibody Diversity Is Generated by DNA Rearrangement

The key experiment that convinced immunologists that DNA rearrangement is central to the immune response relied—as we have seen so often—on myeloma cells. The κ light chain mRNAs were extracted from a myeloma and some were broken in half. The 3′ half of the broken molecules was purified and used as a molecular probe for the constant region; the whole molecule served as a probe for both variable and constant regions. Mouse myeloma-cell DNA (as control) and mouse embryo DNA were cut with a restriction enzyme and the DNA fragments separated by size using electrophoresis. Then the two fragmented DNA samples were probed with the two mRNAs. The sizes of the DNA fragments that hybridized to the probes differed in the two samples (Figure 24-20). This result could only mean that in its κ chain coding region, the myeloma-cell DNA had a different organization from the DNA of the rest of the animal. Thus, DNA rearrangement was shown to be a central aspect of B cell differentiation.

Later work has used more sophisticated techniques to

(a)

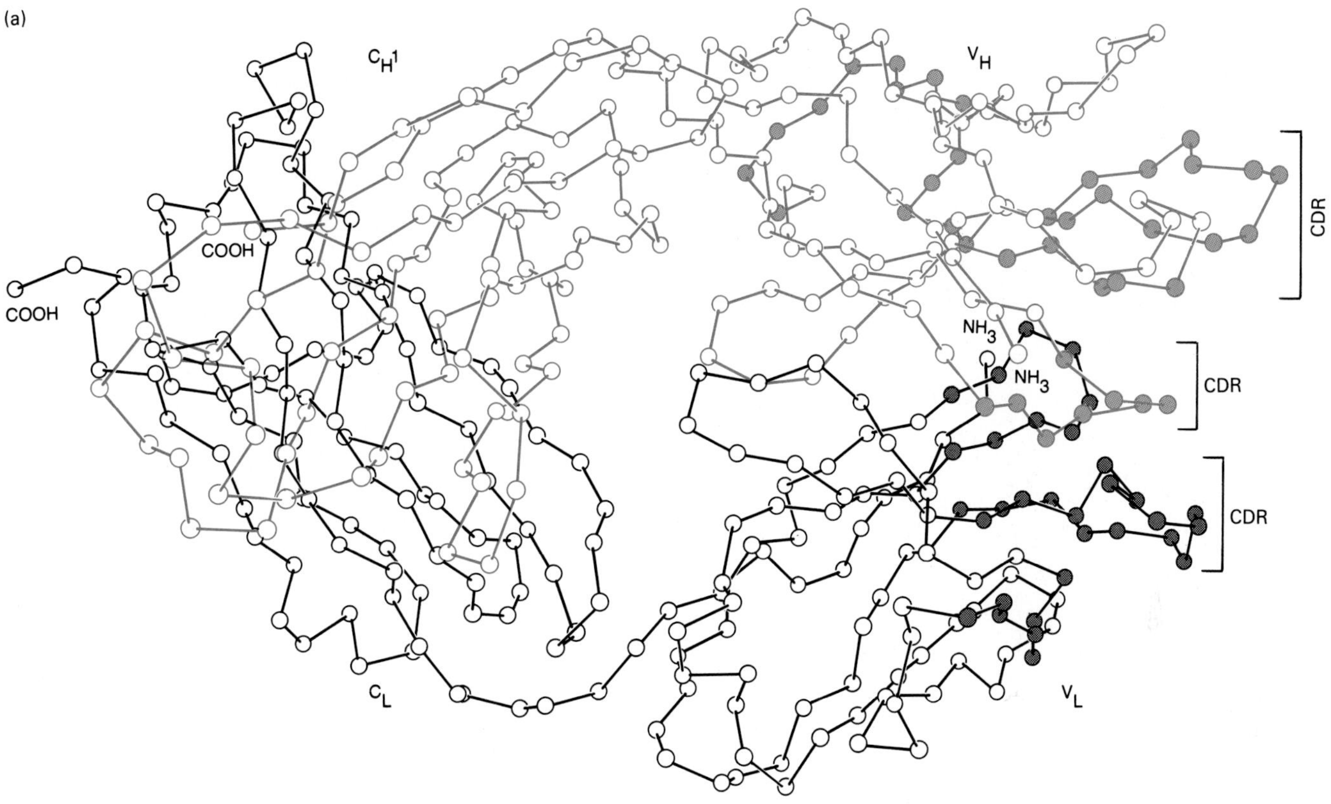

C_H1 V_H COOH COOH CDR CDR CDR NH_3 NH_3 C_L V_L

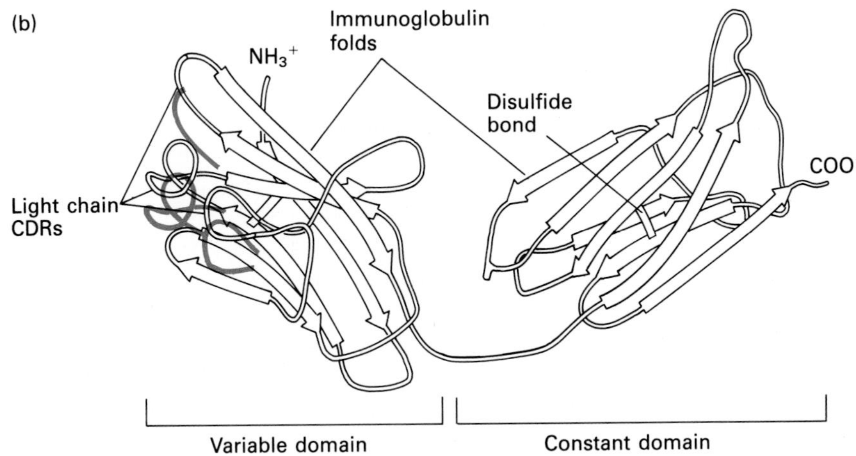

Figure 24-18 α-Carbon and sheet models of the F_{ab} segment of an antibody. The variable and first constant domains are shown. In (a), the F_{ab} molecule is represented only by the α-carbon (*balls*) of each amino acid. The loops that contain the solid balls are the CDRs; they come from both the heavy and light chain. In (b), the light chain is represented by arrows that show the strands of the β-pleated sheet that make a structure called the immunoglobulin fold. Both the variable and constant regions consist of domains organized as immunoglobulin folds. Extending from the variable domain are the hypervariable loops, the CDRs. *Part (a) from* The Antibody Combining Site, *by J. D. Capra and A. B. Edmundson, Copyright 1977 vol. 236(1) by Scientific American, Inc. p. 56; part (b) after M. Schiffer et al., 1973,* Biochemistry ***12**:4620.*

(a)

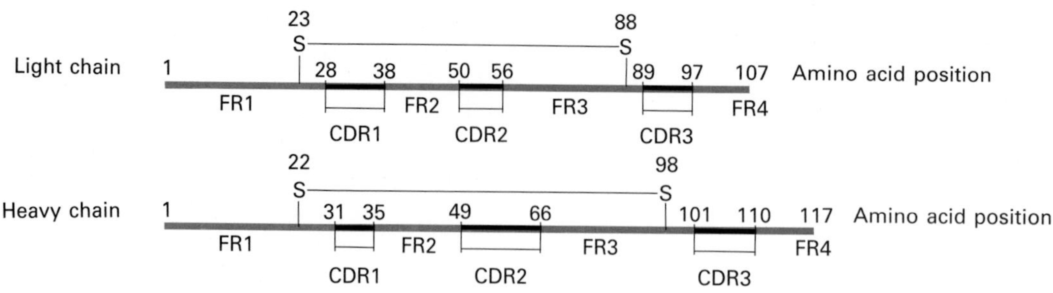

23
88
S
S
Light chain
1
28
38
50
56
89
97
107
Amino acid position
FR1
FR2
FR3
FR4
CDR1
CDR2
CDR3
22
98
Heavy chain
1
31
35
49
66
101
110
117
Amino acid position
FR1
FR2
FR3
FR4
CDR1
CDR2
CDR3

(b)

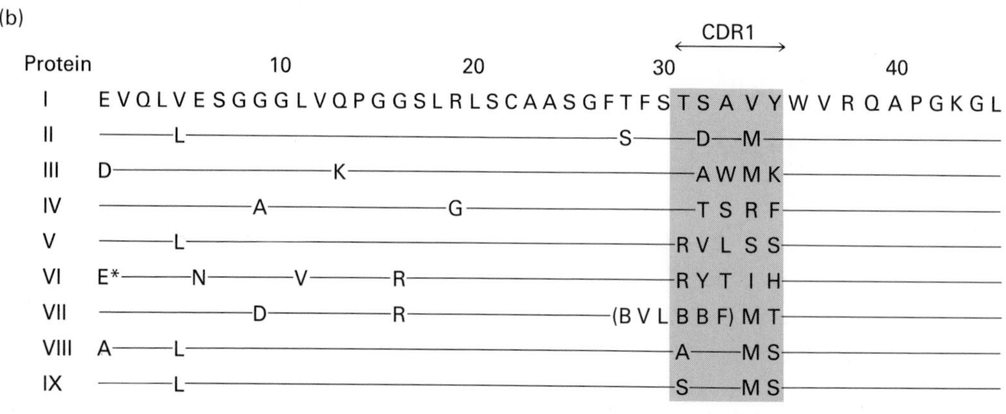

CDR1
Protein
10
20
30
40
I EVQLVESGGGLVQPGGSLRLSCAASGFTFSTSAVYWVRQAPGKGL
II L S D M
III D K AWMK
IV A G TSRF
V L RVLSS
VI E* N V R RYTIH
VII D R (BVLBBF) MT
VIII A L A MS
IX L S MS

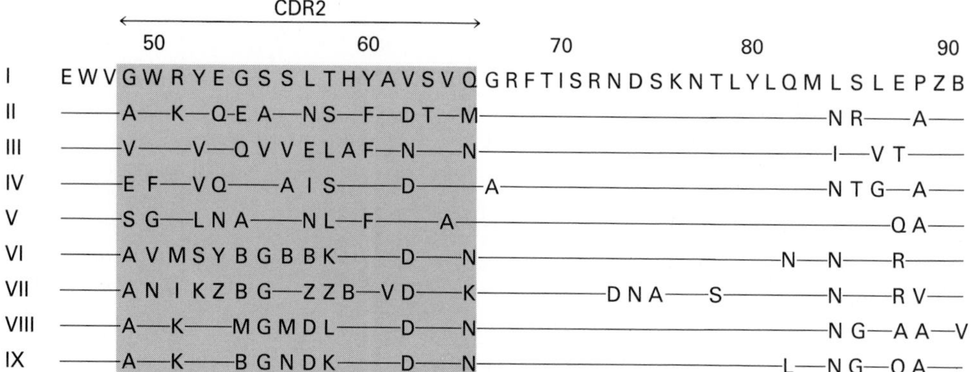

CDR2
50
60
70
80
90
I EWVGWRYEGSSLTHYAVSVQGRFTISRNDSKNTLYLQMLSLEPZB
II A K Q E A N S F D T M N R A
III V V Q V V E L A F N N I V T
IV E F V Q A I S D A N T G A
V S G L N A N L F A Q A
VI A V M S Y B G B B K D N N N R
VII A N I K Z B G Z Z B V D K D N A S N R V
VIII A K M G M D L D N N G A A V
IX A K B G N D K D N L N G Q A

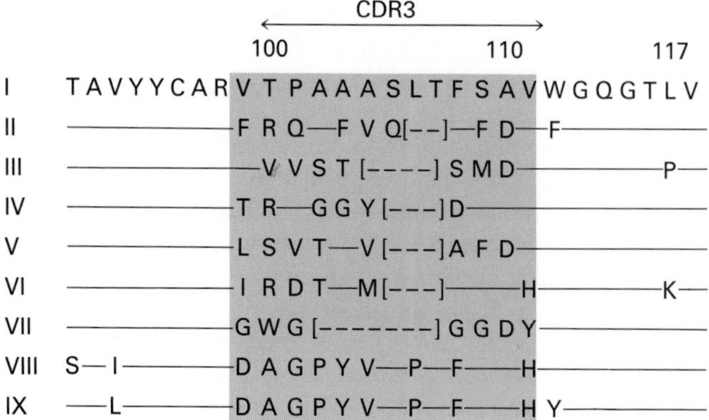

CDR3
100
110
117
I TAVYYCARVTPAAASLTFSAVWGQGTLV
II F R Q F V Q[--] F D F
III V V S T [----] S M D P
IV T R G G Y [---] D
V L S V T V [---] A F D
VI I R D T M [---] H K
VII G W G [-------] G G D Y
VIII S I D A G P Y V P F H
IX L D A G P Y V P F H Y

Figure 24-19 *(opposite)* Location of the CDRs on the variable regions of the light and heavy chain. (a) Schematic representation of the CDRs interspersed among framework (FR) segments. The FR segments in the antibody form the β-pleated sheets while the CDRs form the antigen-binding loops (see Figure 24-18). The numbers represent amino acid positions starting with amino acids 1 at the N-terminus. (b) A collection of actual sequences of human heavy chain variable regions demonstrates the extent of variation found in the CDR sequences. In the top line, the sequence of protein I is shown using the single letter amino acid code (see Figure 3-2). For proteins II-IX, the sequences are shown using the convention that a line indicates identity with protein I, a letter signifies a position where an amino acid different from that in protein I is located, parentheses indicate an uncertainty, and dashes represent deletion of a sequence relative to protein I. One region of variability between CDR2 and CDR3, at positions 84 through 88, is not in the antigen-binding site. *Part (b), see J. D. Capra and J. M. Kehoe, 1974,* Proc. Nat'l Acad. Sci. USA *71:4032.*

demonstrate the same phenomenon for κ and λ light chains and heavy chains. In myeloma-cell DNA, there are variable regions found near the constant regions. In DNA from nonlymphoid tissues there are independent variable regions and free constant regions but no variable region is found near a constant region. DNA rearrangements have been discovered to play central roles in a number of aspects of immune system behavior.

Complete analysis of immunoglobulin coding regions has shown that the heavy chains, κ light chains, and λ light chains are encoded in three genetic loci on three different chromosomes, each locus having multiple variable regions and one or a few constant regions (Figure 24-21). We call each group of variable regions a *library*. In a cell that makes an antibody consisting of a κ light chain and a heavy chain, DNA rearrangement occurs at both the κ locus and the heavy-chain locus to form the two genes that encode the two chains.

Diversity Is Generated in Light Chains by a Single Recombination Event

To appreciate how light-chain genes are constructed we need to recall that mRNAs are formed by the splicing out of regions from a nuclear precursor RNA. In the recombining of cellular DNA to form an immunoglobulin gene, the recombination has to bring together the relevant DNA into one transcriptional unit, but splicing can then eliminate any parts of the RNA from that transcriptional unit that have no role in the ultimate mRNA (introns are eliminated and exons are maintained).

κ mRNA is made from three exons. At the position closest to the 5′ end is the *L_κ exon;* it encodes a *leader* or *signal peptide* that directs newly made κ protein into the endoplasmic reticulum. The second exon is the variable region proper and the third is the constant region. In the germ line, the leader peptide and most of the variable region are encoded in one library consisting of a few hundred units (Figure 24-22). Each unit consists of one L_κ exon and one *V_κ region.* The V_κ region makes up most, but not all, of the variable region; we denote it V_κ to distinguish it from the complete variable region. The $L_\kappa + V_\kappa$ units are tandemly arrayed along one long stretch of DNA. Although each is about 400 nucleotides long, they are separated by about 7 kilobases (kb), thus 100 $L_\kappa + V_\kappa$ units would cover about 740 kb of DNA.

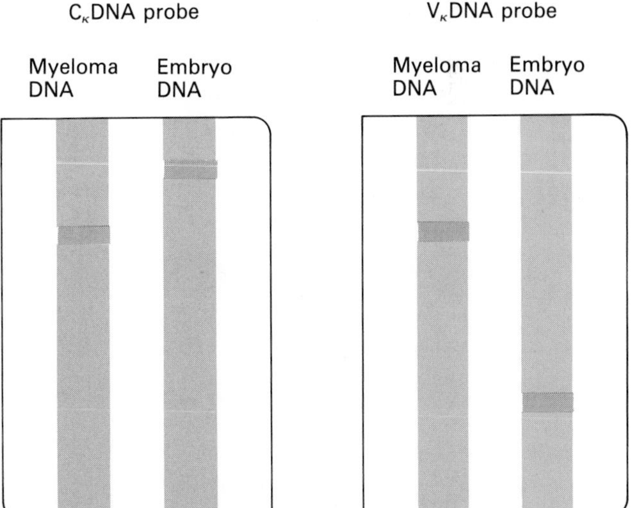

Figure 24-20 Demonstration of κ gene rearrangement during B-cell maturation. In the experiment, DNA was extracted from two sources: a clonal mouse myeloma cell and a total mouse embryo. The two DNA samples were cleaved with restriction enzymes, the DNA was fractionated by electrophoresis, and the separate DNA fragments visualized by hybridization with specific radioactive DNA probes. A κ constant-region probe was used for the left panel, a V_κ variable-region probe for the right. In both panels, the hybridizing bands in the myeloma DNA are different from those in embryo DNA. The embryo DNA bands are germline in configuration, and all adult cells (except B-lymphoid cells) have this pattern. The myeloma cell has one band that hybridizes with both the C_κ and V_κ probes indicating that the V_κ and C_κ regions, which are on separate fragments in germline DNA, have been so closely juxtaposed in myeloma DNA that they are on the same DNA fragment (within 10 kb of each other). [See N. Hozumi and S. Tonegawa, 1976, *Proc. Nat'l Acad. Sci. USA* 73:3628.]

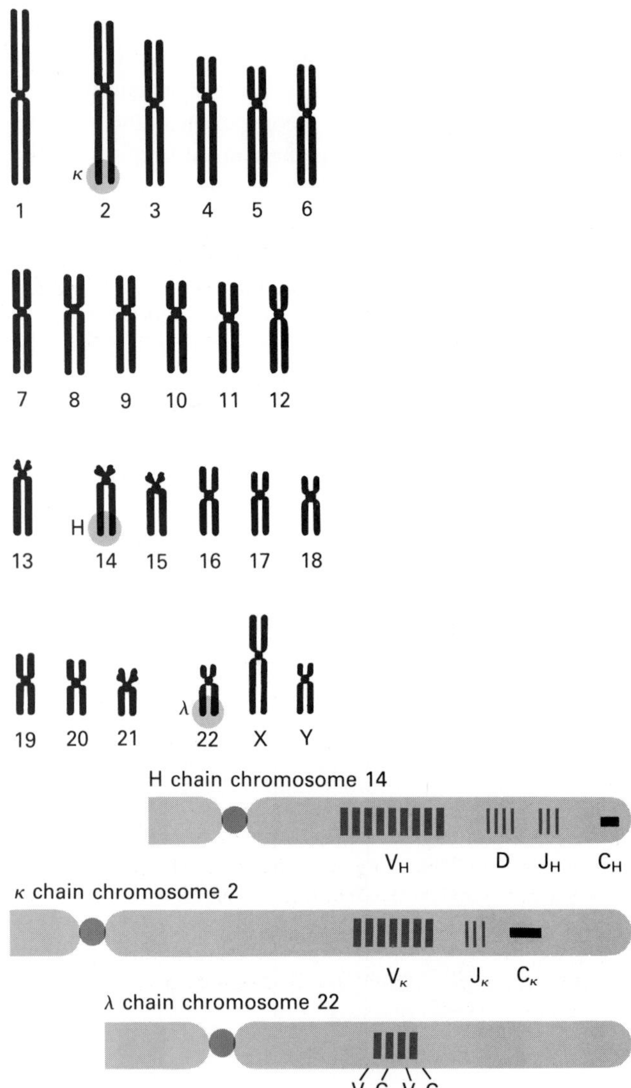

Figure 24-21 Location of immunoglobulin genes on specific human chromosomes. The chromosomal regions in which the heavy chains, κ light chains, and λ light chains are encoded are indicated on chromosomes 14, 2, and 22, respectively. Each region differs from the others in detailed organization, but each has multiple V regions. For λ, the organization shown is for the mouse; the human locus is more complicated.

This region is about a quarter of the length of an *Escherichia coli* chromosome but is less than 0.1 percent of a mammalian cell's DNA.

The formation of a complete variable region involves the joining of one $L_\kappa + V_\kappa$ unit to any one of five *joining regions*, or J_κ (to be distinguished from the J chain) located near the single constant-region exons in the cell's DNA. The five J_κ units are tandemly organized but are separated by an unknown distance from the V_κ regions. They are about 30 nucleotides long and are spread over 1.4 kb of DNA. Between the J_κ units and the C_κ region lies an intron of 2.4 kb of DNA (Figure 24-22).

The reorganization of DNA to make a functional κ gene involves one V_κ region joining to one J_κ region with the deletion or inversion of the intervening sequence (Figure 24-23). So far as is known, any V_κ can join to any J_κ, and the choice is random. Once V_κ and J_κ are joined, the whole region is transcribed into nuclear RNA and the intervening sequences between L_κ and V_κ and between J_κ and C_κ are removed by RNA splicing to produce the mature mRNA for κ protein (Figure 24-23).

Two contributions to diversity are produced by the V_κ–J_κ joining process: V_κ region diversity and J_κ region diversity. The combination of 300 V_κ regions and 4 J_κ regions can obviously produce 1200 different possible chains. But V_κ–J_κ joining generates more sequence variability than the simple combinatorial calculation would suggest because in the vicinity of the V_κ–J_κ joint the joining process is imprecise and can generate many combinations. To see the consequences of this imprecision, it is necessary to examine the joining reaction in more detail.

Imprecision of Joining Is an Important Contribution to Diversity

At the 3′ end of the coding sequence of the V_κ segments and at the 5′ edge of the J_κ segments lie DNA sequences that, because of their conservation and structure, are thought to be the signals for the joining process. They are called *recognition sequences*. The recognition sequences are organized as follows: abutting V_κ or J_κ is a seven-base palindrome, a space of about 11 or 23 nucleotides follows (one or two turns of the DNA helix), and then an AT-rich nine-base sequence is found (Figure 24-24). These sequences allow an as-yet-uncharacterized enzymatic system to carry out an orderly but imprecise joining reaction that contributes greatly to diversity.

When two pieces of DNA join, there are two products. In the joining of V_κ to J_κ, one product is a $V_\kappa J_\kappa$ unit and the second is a back-to-back joining of the recognition sequences (Figure 24-25). When many such joining events of one V_κ and J_κ were studied, it became clear that the recognition elements are joined identically in all cases, with the heptamers linked to each other. The $V_\kappa J_\kappa$ unit, however, is not precisely joined: a few nucleotides from V_κ and a few from J_κ are lost from the DNA at the joining point (see Figure 24-25). Thus, the joining process has the unique property that a small number of nucleotides at the joint are lost from DNA entirely; they appear in neither product. The random loss of nucleotides at the joining site generates significant diversity at that point but the system pays for its diversity. The cost is evident if we remember the constraints on a coding sequence.

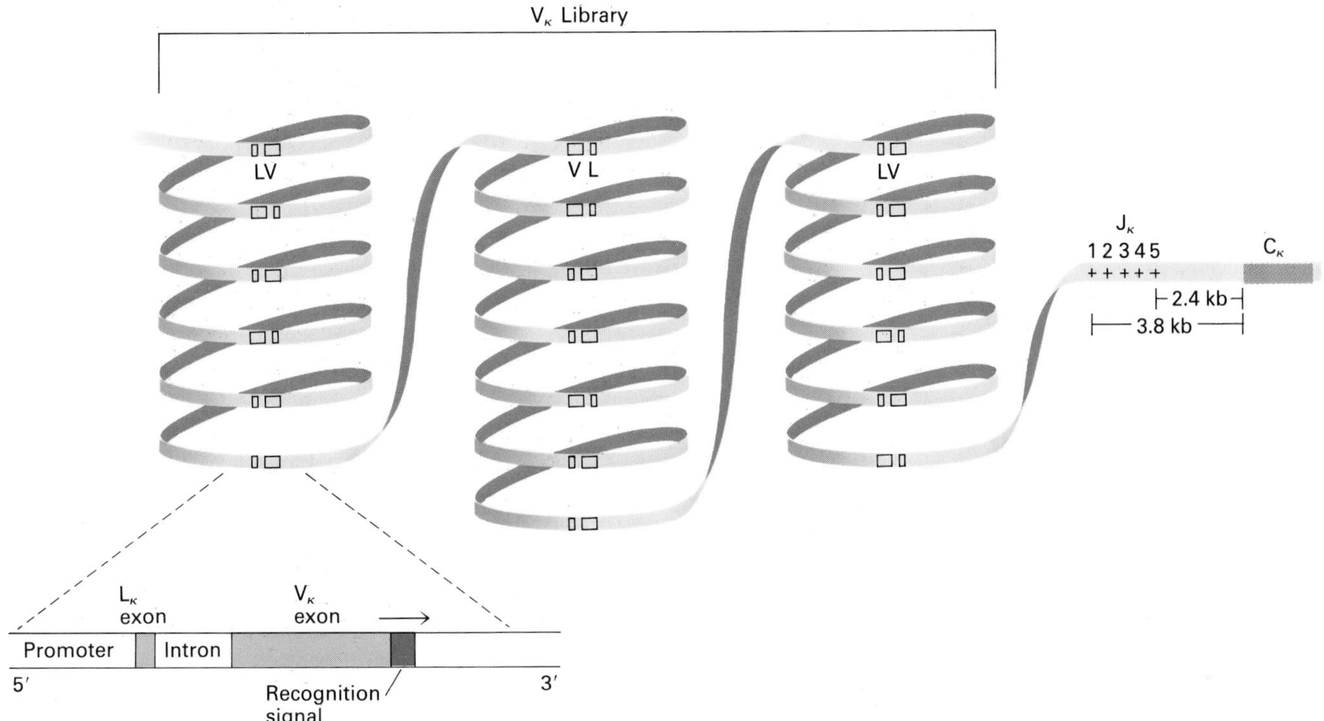

Figure 24-22 Organization of the κ locus. In a library spread over thousands of kilobases of DNA, there are hundreds of V_κ regions. They are organized, as far as is presently known, in either transcriptional orientation relative to the J_κ and C_κ segments. Both J_κ and C_κ are in the 5′-to-3′ orientation (left to right in the illustration). One L_κ–V_κ unit is shown expanded. It consists of a promoter at the 5′ end (beginning of transcription), an exon encoding a leader peptide (L_κ), an intron, an exon encoding the V_κ region, and a recognition signal that specifies the site at which the V_κ region is to join to J_κ region. The recognition signal is actually made of two pieces, one seven nucleotides long (7-mer) and the other nine nucleotides long (9-mer). The arrow shows the orientation of the 7-mer followed by the 9-mer. *After Paul D. Gottleib,* Molecular Immunology, *17:1423.*

A coding sequence in DNA must be read in threes starting with the first AUG (methionine). We say that the AUG defines a *reading frame* in which the rest of the coding region can be read. If DNA were read in one of the two other reading frames, it would encode a meaningless string of amino acids until an adventitious terminator codon were reached. Thus, when two pieces of coding DNA join, as in the joining of V_κ to J_κ, it can be an *in phase* joint that maintains a sensible reading frame or it can be an *out of phase* joint that encodes a nonsense protein (see Figure 24-25). The V_κ-to-J_κ joining process is a random one, and two out of three random joints make nonsense. Thus the system pays for its diversity in making two *nonproductive* joints for each *productive* joint.

The diversity gained by the imprecise joining process is significant. The lower part of Figure 24-25 shows how four different in-phase joinings can be made between one V_κ sequence and one J_κ sequence. All of these joints have actually been found in sequenced κ proteins.

We now have three sources of diversity: variability in the structure of the many V_κ regions in that library of sequences, variability in the structure of the four J_κ regions, and variability in the number of nucleotides deleted at V_κ/J_κ junctions. Antibody chains have three CDRs and the diversity within the V_κ library contributes to all three; the diversity at the recombination joint only contributes to CDR 3 because the site of joining is within CDR 3. The diversity in CDRs 1 and 2 is determined over evolutionary time as the individual V_κ regions have evolved. The diversity in CDR 3 is a somatic process: it happens in cells within the body of the individual animal, not over evolutionary time. Our discussion of these events has focused, for convenience, on the κ genes but similar processes occur in both the λ and heavy chain genes.

Lambda Proteins Derive from Multiple Constant Regions

We recall that antibodies have one of two light chains, κ or λ. The ratio of κ to λ is radically different from species

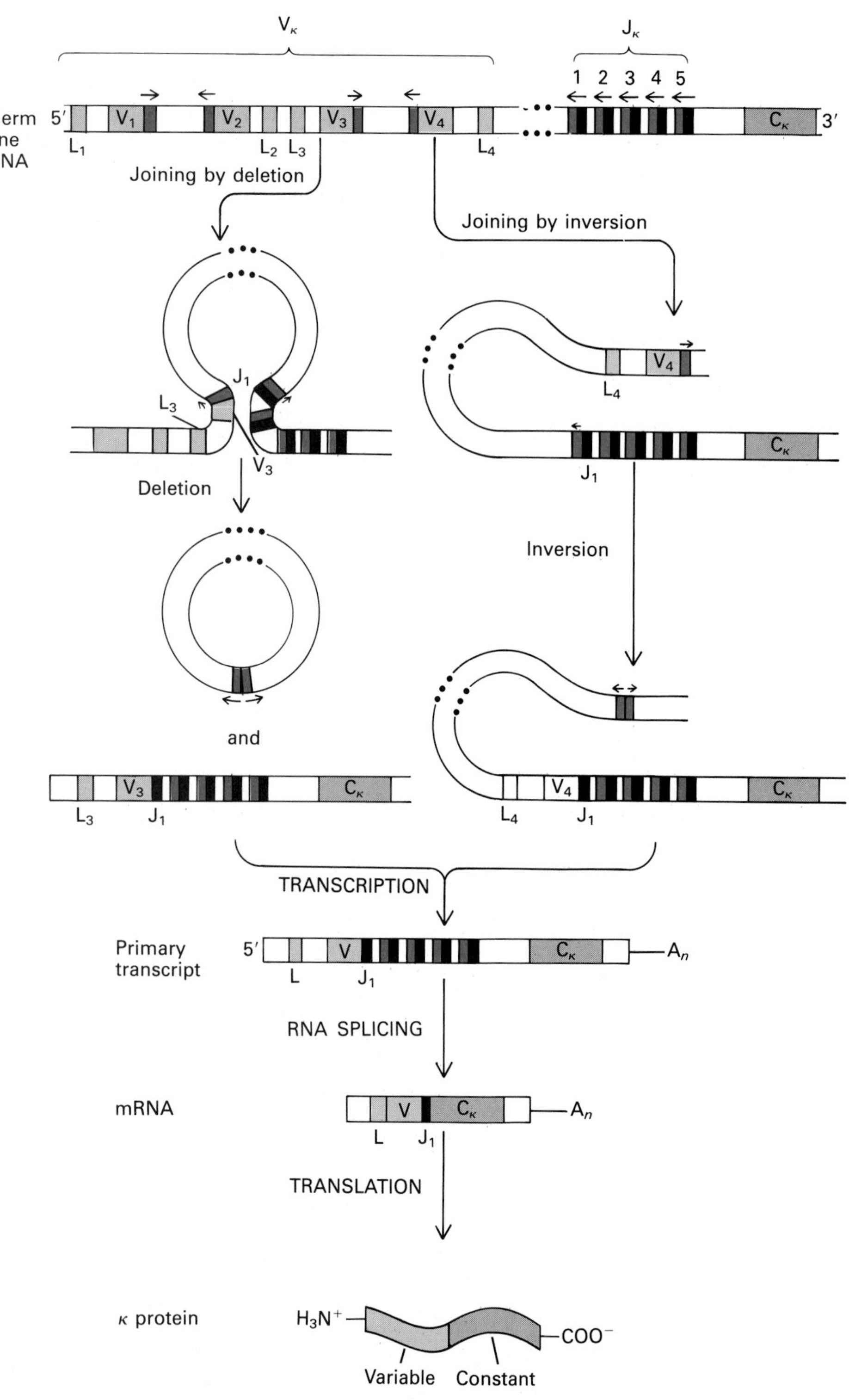

Figure 24-23 Joining of V_κ to J_κ and formation of a κ light chain. The L_κ and V_κ segments are thought to be oriented in either transcriptional direction relative to C_κ in germ-line DNA. Each V_κ and J_κ region has a recognition signal *(dark red)* to specify the points at which joining should take place. The joining process either deletes the intervening DNA or inverts it, depending on the relative orientations of V_κ and J_κ. In deletional joinings, the deleted DNA is lost from the cell. In inversional joinings, all of the DNA is conserved. Once the V_κ-to-J_κ joining has occurred, the now complete gene can be transcribed to produce a nuclear RNA that is spliced to remove all unwanted segments (including the J_κ regions that were not joined to V_κ). The spliced mRNA then encodes a complete κ light chain. The arrows indicate the same structure as that defined in Figure 24-22.

to species. Mice have 95 percent κ, humans about 50 percent, and cows about 5 percent. In all species examined, even mice, λ proteins have not one but multiple C_λ regions, each with its own J_λ region (Figure 24-21). In mice, probably because of the paucity of λ protein, there are only two V_λ regions, each associated with a J_λ–C_λ cluster. Humans have even more C_λ regions and presumably many V_λ. Mouse λ light chains have much less diversity than mouse κ light chains because V_λ and J_λ diversity are so minimal. Moreover, very little recombinational diversity has been found at the V_λ–J_λ joining site. The situation in chickens is remarkable: most chicken Ig has λ light chains, but there are few and perhaps only one functional V_λ segment. Thus, combinational diversity using libraries

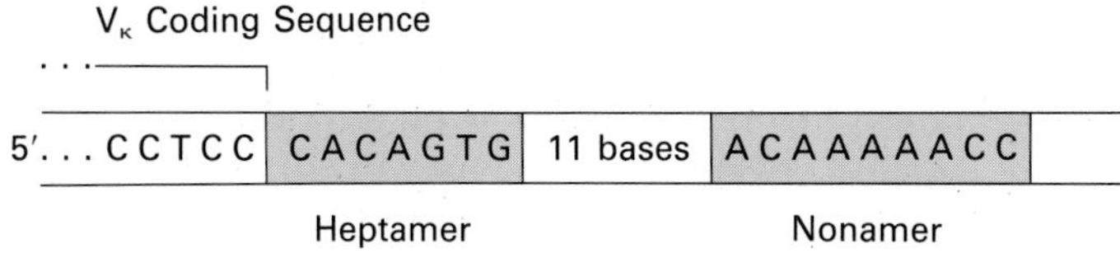

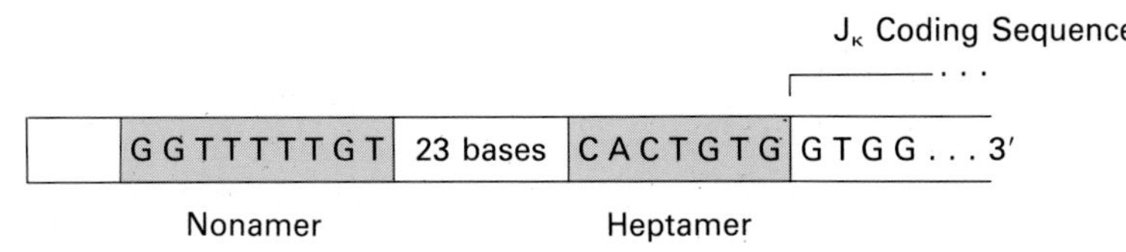

Figure 24-24 V_κ and J_κ recognition signals. Each J_κ and V_κ region has a short characteristic DNA sequence (recognition signal) preceding or following its coding sequence (see organization in Figure 24-23). This signal is recognized by DNA joining enzymes and directs those enzymes to join a V_κ to a J_κ. The recognition signals consist of a heptamer followed by a nonamer sequence. The two sequence elements are separated by either about 11 nucleotides or 23 nucleotides. The joining system always joins DNA with an 11-base spacer to DNA with a 23-base spacer, so that V_κ (11-base spacer) joins to J_κ (23-base spacer) but not to itself.

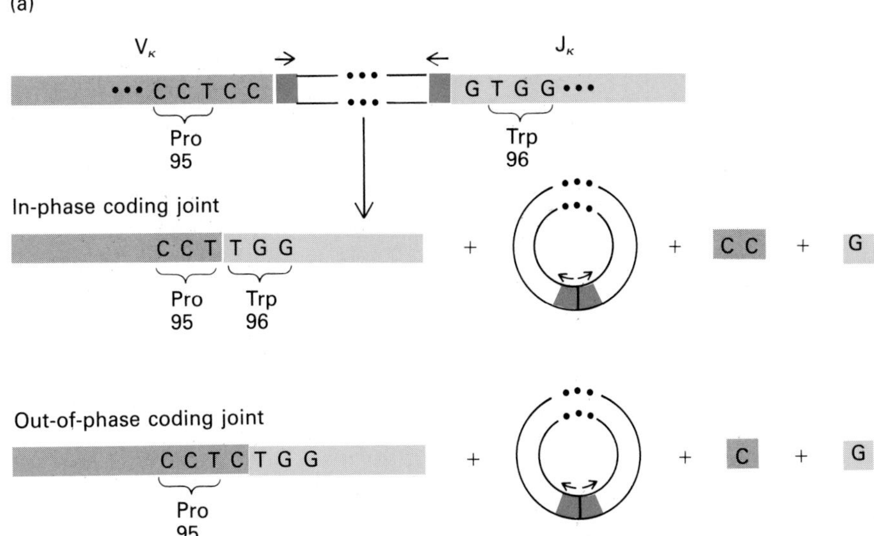

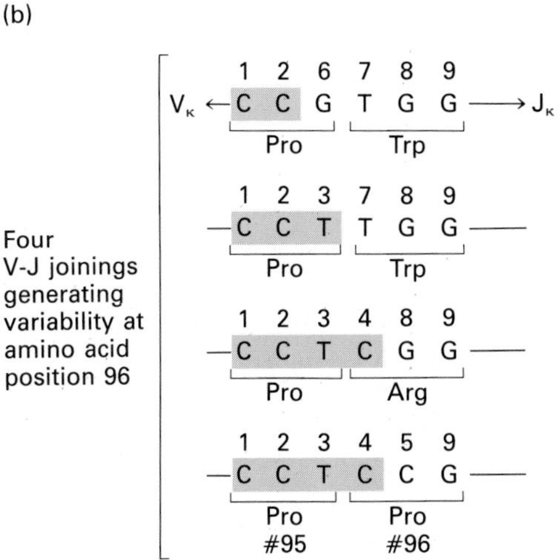

Figure 24-25 The joining of V_κ to J_κ. (a) The joining process can give rise to in-phase or out-of-phase joints. Joining involves the removal of a small, random number of bases from the ends of both V_κ and J_κ. This can leave J_κ either joined in the appropriate reading frame (Trp at position 96 encoded by TGG just following the CCT codon at position 95) or out of frame (for instance, if one more C is left on V_κ then the CCT at 95 will be followed by CTG and all of J_κ will be out of frame with V_κ). (b) The joining process can give rise to many in-phase joints with variable amino acids at the joining point. In the example, if variable ends of V_κ are joined to variable ends of J_κ, the joinings formed will encode Pro-Trp, Pro-Arg, or Pro-Pro in four different in-phase joinings.

of V regions may make a minimal contribution to the diversity of chicken Ig.

Heavy-Chain Variable Regions Derive from Three Libraries

Analysis of antibody-binding sites suggests that the heavy-chain contribution is even more important than the light-chain contribution. Consistent with a need for greater diversity, heavy-chain structure is determined by three libraries of sequence elements instead of the two that make up light chains. The third library that contributes to heavy chain is called the *D region* library (D for diversity). The D region of heavy chains constitutes the bulk of the third CDR, and the D segments are found between the other two libraries, whose segments are called V_H and J_H (Figure 24-26). Thus, variable-region formation in the heavy chain involves two joining reactions, V_H to D and D to J_H. Having three segments rather than two greatly increases the combinatorial diversity produced by drawing segments from libraries containing elements with varied sequences. There are apparently hundreds of V_H segments, perhaps 20 D segments, and 4 J_H segments. Recombinational diversity at the V_HD and DJ_H joints, like that at the $V_\kappa J_\kappa$ joint, is also created by the removal of small numbers of nucleotides.

The third CDR of heavy chains is diversified by yet another mechanism. When a D joins to a J_H, or when a V_H joins to a D, not only are nucleotides removed but a few nucleotides not found in either parental sequence are added at the joint. The enzyme responsible for the addition of nucleotides is probably *terminal deoxynucleotidyl transferase,* a template-independent DNA polymerase known to be present in cells making heavy-chain joints (but probably absent at the time cells are making light-chain joints). We call the extra nucleotides an *N region.* Thus, a complete heavy-chain variable region is a V_HNDNJ_H unit. Maintaining the reading frame presents a problem similar to that posed for light chains. The nucleotides of the NDN region between V_H and J_H must keep the reading frame correct, thus two out of three joinings will be out of phase. A danger not usually present in the pure VJ joint of light chain is that the formation of the NDN sequence could put a termination codon in the

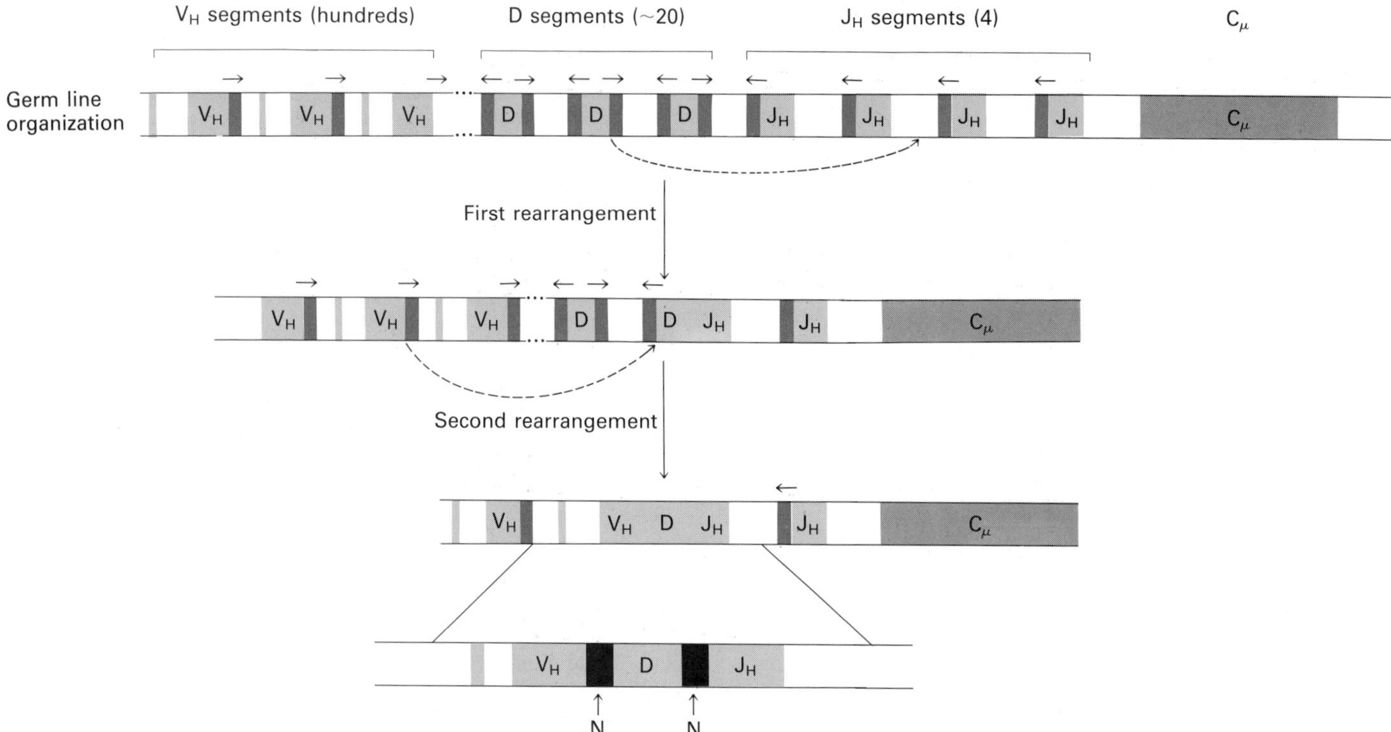

Figure 24-26 Organization and rearrangement of heavy-chain genes. In the germ line, the heavy chain-related DNA is organized with a large library of V_H regions all in the same 5′-to-3′ orientation, followed by a library of D regions, a library of J_H regions, and the constant region for the μ chain (C_μ). The V_H, D, and J_H gene segments have recognition signals like those found at the ends of V_κ and J_κ (overlined with arrows). In the first stage of rearrangement, a D segment joins to a J_H segment, deleting the intervening DNA. In the second stage a V_H joins to the preformed DJ_H unit, forming a V_HDJ_H heavy-chain variable region. At each joint, a few random nucleotides may be inserted (N regions).

reading frame. The possibility of this is minimized by the high G + C content of many N regions (terminators are UAG, UGA, and UAA) and the lack of terminators in most D regions in all three frames. The diversity generated by the NDN unit is almost incalculably vast because the N regions have wholly random sequences (with a possible bias towards G + C), and the D regions can be read in any frame depending on the length of the 5′ N region. (A strong bias towards a particular reading frame for D is evident, however, when sequences of real antibodies are examined.) NDN regions up to 30 nucleotides long have been found, and this highly variable segment can encode 0 to 10 amino acids.

Recognition Sequences for Joining Reactions Are Highly Conserved

We have described here a number of different joining events, V_κ to J_κ, V_λ to J_λ, V_H to D, and D to J_H. All of these reactions follow one inviolable rule that relates to the structure of their recognition sequences. Each of the sequence elements that participate in joining reactions have an adjacent recognition sequence. As described for κ gene segments, these recognition sequences are of two types: one has a characteristic heptamer followed by an 11- or 12-base spacer of random sequence and a high AT nonamer; the other has a similar heptamer followed by a 21- to 23-base spacer and a nonamer (see Figure 24-24). The spacers make the difference: remembering that one turn of the double-stranded DNA helix is 10.5 base pairs, one recognition sequence will have a spacer representing approximately one turn of a helix, the other spacer will be about two turns. The inviolable rule is that all unions involve a 1-turn recognition sequence combining with a 2-turn recognition sequence. Thus V_κ segments have 1-turn elements, J_κ segments have 2-turn; D segments are flanked by 1-turn elements, V_H and J_H segments have 2-turn elements. Why the spacers fall into this pattern is not totally clear, but very likely there are proteins that bind to these recognition sequences and they may be designed to recognize signals arrayed together on one side of a DNA double helix.

The enzymology of joining remains to be worked out. We imagine that recognition proteins bind to the recognition elements and specify that 1-turn and 2-turn elements join to each other. Endonucleases must cut the strands, exonucleases might degrade free ends, terminal deoxynucleotidyl transferase may add nucleotides, polymerases must provide flush ends, and then ligases must seal the gap. These events must occur in a multienzyme complex, and the activity of the complex must be highly regulated so that joining occurs only when needed and then on the correct sequence elements.

All the events considered thus far involve joining processes creating diversity by combining members from different libraries carried in the germ line. Insertion of N regions, because it involves de novo synthesis of DNA, is a form of somatic mutation that is coupled to joining. Another form of somatic mutation, separate from joining, involves the change of individual base pairs throughout a joined variable region to generate diversity across the whole region. This mutational process will be described later.

The Preliminary Phase of B-Lymphocyte Maturation

The earlier discussion of clonal selection touched upon the stages of B-lymphocyte development. In the *preliminary phase,* diversity is generated. During the *mature phase,* surface-immunoglobulin-containing B lymphocytes develop into either immunoglobulin-secreting plasma cells or memory cells. Having introduced the joining reactions, we will examine how they fit into the early events in B-lymphoid development.

B-Lymphoid Cells Go through an Orderly Process of Gene Rearrangement

B lymphocytes arise continually throughout life by differentiation from bone marrow stem cells. The bone marrow is considered a primary organ of the immune system, donating cells to the peripheral organs. Many of the intermediate stages have been defined by examining the structure of immunoglobulin-related DNA in the cells. Tumors of early-stage cells have been very helpful in defining the intermediate stages, just as myelomas have helped our understanding of end-stage cells.

The earliest recognizable B-lymphoid cell has cell surface markers (defined by antibody binding) that mark it as a B-lymphoid type, but its DNA is still germ-line in organization (Figure 24-27). The next recognizable cell has begun heavy-chain gene rearrangement but its light-chain genes are not actively rearranging. It first joins a D to a J_H and then joins a V_H to the preformed DNJ_H to form the complete V_HNDNJ_H. The cell can then make a heavy chain. Because the nearest heavy-chain constant region to the V_HNDNJ_H unit encodes a μ heavy chain, the cell makes μ and is recognized by immunofluorescence as a μ-positive, light-chain-negative bone marrow cell called a *pre-B lymphocyte.*

The next stage of B-lymphoid differentiation is rearrangement of κ light-chain genes (Figure 24-27). Most cells stop there and make a κ protein but a few, which rearrange κ genes nonproductively, go on to rearrange λ genes and become λ producers. Once a cell has constructed a complete in-phase κ or λ gene, it makes a light-chain protein. The light-chain protein can bind to heavy chain and the unit can then be processed to the cell sur-

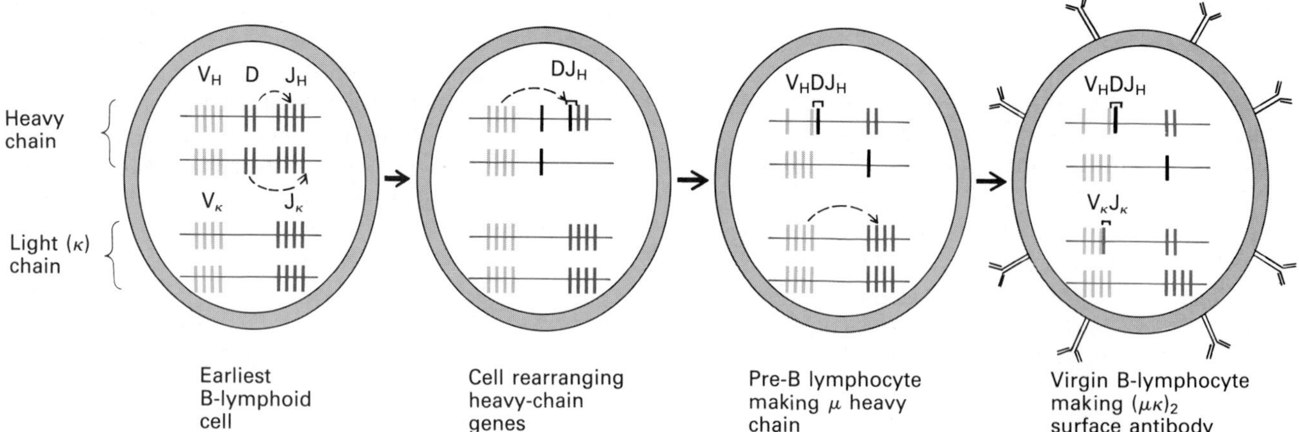

Figure 24-27 Ordered rearrangement of variable regions of antibody genes during the differentiation of B lymphocytes. The earliest cell has B lymphoid surface markers but has not started rearrangements. It begins maturation by first making D-to-J_H rearrangements at both heavy chain loci. It then begins V_H-to-DJ_H rearrangements, probably sequentially on the chromosomes. Production of an in-phase V_HNDNJ_H unit leads to heavy-chain (μ) synthesis, and that (or possibly some other signal) prevents further rearrangement. The example shown here would be an in-phase join preventing V_H-to-DJ_H rearrangement on the second chromosome. After μ chain synthesis begins, the cell is designated a pre-B lymphocyte. After an unknown number of further divisions, κ gene rearrangement begins and continues only long enough to make one κ gene that can encode a complete κ protein able to bind to μ. The $(\mu\kappa)_2$ dimers then appear on the cell surface, the cell stops division, and it becomes a mature B lymphocyte. If no in-phase κ join is produced, the cell can still rearrange its λ light chain genes. About 5 percent of mouse antibody and almost 50 percent of human antibody have λ light chains.

face, where it remains bound. A newly arisen cell with surface antibody is a *virgin B lymphocyte.*

The Preliminary Phase Can Generate 10^{11} Different Cell Types

The preliminary phase of B-lymphocyte development generates cells containing surface antibodies with a wide range of specificities. It is not possible to calculate the number of such specificities, nor is such a number very meaningful because a given antibody might bind to a range of antigens, each with different affinity. It is, however, possible to estimate the number of different antibody molecules that can be produced by the multiple diversification mechanisms available. The calculation depends upon the randomness of events. If some V or D regions are favored for rearrangement, then a disproportionate number of newly made cells may carry related antibodies. No evidence of nonrandomness has yet been uncovered, but it is likely to exist. There is some evidence that the V_H regions are used sequentially for attachment to preformed DJ_H units, but it is not clear what the effects of such a process might be. In addition, one reading frame of each D region seems to be highly favored. Nevertheless, we can assume total randomness in calculating the number of different antibodies generated because the number is so high that errors of a factor of a hundred are not important.

Assuming that in the mouse there are 300 V_H regions, 20 D regions, and 4 J_H regions, $300 \times 20 \times 4 = 2.4 \times 10^4$ different combinations of these units exist. Each DJ_H joining and V_HD joining, however, involves losses of nucleotides, so that each D can occur in each of its reading frames, and all three constituents can be truncated to various extents. We can estimate that this will increase diversity by at least 10-fold. In addition, the DJ_H and V_HD joinings have N regions that are random in length and sequence, generating at least 100 different possibilities. Thus about $2.4 \times 10^4 \times 10 \times 100 = 2.4 \times 10^7$ different heavy chains are possible. Light chains are less diverse. Ignoring λ chains because they are rare in mice, if there are 100 V_κ regions, 4 functional J_κ regions, and 10 ways of putting them together, $100 \times 4 \times 10 = 4 \times 10^3$ possibilities exist. Assuming that any heavy chain could combine with any light chain, $2.4 \times 10^7 \times 4 \times 10^3 \simeq 10^{11}$ physically different antibodies can be produced by the diversification system during the preliminary phase of B-lymphoid development. And 10^{11} is a huge number. A mouse only makes about 10^8 lymphocytes per day, so that all possible combinations might not even be expressed over the lifetime of the animal. The secret of the ability of the immune system to react to whatever pathogens nature can devise lies in the enormous diversity resulting from the mechanisms of generating antibody variable regions.

Another characteristic of the preliminary phase is that it is organized hierarchically to produce a highly diverse population of single cells. Through the early stages of the lineage, cells are continually growing and dividing about once every 12 h. How long the early stages last is not clear, but a cell with only DJ_H rearrangements can probably give rise to many progeny, each of which will have its own V_HDJ_H rearrangement. Similarly, a cell with a V_HDJ_H rearrangement can give rise to many cells with independent $V_\kappa J_\kappa$ rearrangements. But when a virgin B lymphocyte appears in the marrow it rapidly ceases growth and moves out of the marrow to the peripheral blood-lymph circulation system. Because a cell appears to cease growth very soon after it has acquired surface antibody, each cell produced by the bone marrow is virtually unique. For the cell to grow and mature any further, the surface antibody must bind strongly enough to an antigen to induce activation of further growth. T cells may assist this activation process.

The Immune System Requires Allelic Exclusion

A third aspect of the B-lymphoid maturation process to consider in examining diversification mechanisms is that cells have two chromosomes and therefore could, in principle, make two heavy chains if both chromosomes rearranged productively, that is, if they made an in-phase V_HNDNJ_H join. In addition, two κ light chains plus two or more λ light chains can conceivably be made. One cell would therefore be able to make multiple types of antibody. Such a situation would violate the precondition for clonal selection to operate. Clonal selection depends on the manufacture of only one antibody type per cell so that when the cell is activated it will go on to secrete only one type of antibody. If multiple chains were to be made, antibodies could have two different binding specificities on their dimeric molecules and would lose the advantage of high avidity gained from multiple interactions with a single antigen. Presumably as a way of avoiding these consequences, mechanisms have been evolved to ensure that B lymphocytes make one, and only one, antibody. Because this means that only one of two alleles carried by a cell will be expressed on any one cell, the process has been called *allelic exclusion*.

Allelic exclusion apparently works by shutting down rearrangement processes after one productive rearrangement has occurred. The mechanisms involved are not known, but one type of experiment clearly indicates that the process is at work. An already rearranged heavy-chain or light-chain gene can be introduced into a mouse's germ line by microinjection of a plasmid into a fertilized mouse egg (see Figure 6-22). The animals are called *transgenic*, denoting their acquisition of a foreign gene. When a mouse is transgenic for a rearranged heavy-chain gene, rearrangement of endogenous heavy-chain genes is suppressed. Transgenics that have acquired a κ chain suppress rearrangement of endogenous κ chains. In both cases, the expression of a rearranged gene suppresses rearrangement of other genes—the result expected if allelic exclusion is a consequence of suppression of secondary rearrangements.

Immunoglobulin Synthesis and B-Lymphocyte Activation

It might seem that we have already introduced sufficient complexities of the immune system but a moment's reflection will reveal numerous problems that have been left hanging. Many involve the rate of antibody synthesis and the types of antibody chains made at various stages in the differentiation of B-lymphoid cells.

All immunoglobulin synthesis takes place on the rough endoplasmic reticulum membranes of the B cell. This would be expected because immunoglobulins are either cell surface molecules or secreted molecules, and the pathways to both end points begin by the entry of newly made protein into the cisternae of the endoplasmic reticulum (see Chapter 21). Newly made polypeptides are usually targeted to the endoplasmic reticulum by a hydrophobic signal sequence. In immunoglobulins, such a sequence is found at the N-terminal of both newly made heavy chains and light chains. It is encoded by the L exon, one of which, as previously noted, is located on the 5′ side of each V region. The signal sequence is cut away from the chains shortly after its synthesis so that mature antibodies do not contain it.

Activation of B Lymphocytes Alters Cellular Behavior and Initiates Proliferation

A key event of the immune response is the encounter of an antigen with a B lymphocyte that bears surface antibody with binding specificity for that antigen. This encounter causes the virgin B cell to mature further: it carries the B cell from its transient preliminary stage to a permanent mature stage. Many molecular events transpire from an encounter with antigen. They include the activation of proliferation, secretion of antibody, production of IgG and other secondary antibody classes, somatic mutation, and the production of memory cells.

After the B lymphocyte has matured to a virgin B cell and has been expelled from the bone marrow into the periphery, it has ahead of it only a few days of life as a circulating cell unless it meets an antigen that can interact with its surface antibody and trigger further growth and maturation. That antigen will either be soluble or will be held on the surface of a macrophage, a cell that nonspecifically engulfs particles and then displays fragments of them on its surface. The requirement for growth stimulation is that the antigen be multivalent—have multiple determinants on one structure—so that many antibody molecules on the surface of one B cell can be bound together to form a tight patch on the cell surface (Figure 24-28). It is generally said that a competent antigen *cross-links* surface antibody; this cross-linking, or patching, appears to be the key to activation of a B cell. A polymeric antigen, like a carbohydrate with repeating sugars, can be a very good soluble activator by itself, but most antigens can only trigger B cells with assistance from other cells.

Surface antibody can be aggregated by multivalent antigens because of the fluidity of the lipid bilayer plasma membrane in which antibody is embedded. After island patches form on the cell surface, they coalesce into a cap on the surface (Figure 24-28) that is either shed into the surrounding fluid or internalized by endocytosis and degraded.

The macrophage plays an important role in the activation of B-lymphocyte growth. It secretes one or more factors that are absolute requirements for B-cell activation. The only characterized factor—and perhaps the only one—is called *interleukin-1,* or *IL-1.* In the absence of this factor, mature B cells will not grow even if high concentrations of multivalent antigen are present.

The achievement of B-cell activation is also assisted by helper T cells. These T_H cells have a specific receptor very much like antibody, but they usually bind to an area of the antigen different from that to which the B cell binds. For instance, when the hapten dinitrophenol is coupled to bovine serum albumin and injected into a mouse, B cells will bind to the hapten but T_H cells will be activated by the protein. In such a situation, the hapten is said to be bound to a *carrier.* The carrier does not react with T cells directly. It is first bound and internalized by macrophages, which degrade it to peptides. These peptides are then displayed on the macrophage surface, where T_H cells react with them. T_H cells that carry receptors able to interact with the macrophage-bound fragments are then stimulated to proliferate, producing a pool of carrier-specific T_H cells.

The collaboration of T_H cells and B cells that induces the B cell to secrete specific antibody is not completely understood, but a very likely mechanism based on recent experiments is shown in Figure 24-29. The key to this mechanism is that a B cell can process a protein to peptides and display them on its surface exactly the way a macrophage can. The macrophage processes all foreign proteins indiscriminately; the B cell can process specifically those proteins that bind to its surface receptor. A dinitrophenol-specific B cell will bind the dinitrophenol-carrier complex, internalize it by endocytosis, and express the carrier peptides on its surface. The carrier-stimulated T_H cells will then recognize the peptides and become further stimulated. In this interaction, the T_H cell produces a factor called *B-cell growth factor,* which activates the B cell to proliferate.

This mechanism explains why haptens must be bound to carriers to be antigenic. Free hapten can bind to B cells but cannot stimulate T_H cells and therefore cannot activate the B cells. We believe that in such a situation hapten is a surrogate for determinants constructed solely of amino acids that are present all over proteins, so that natural proteins act both as haptens and carriers and are therefore antigenic.

Activation of T_H cells induces them to secrete another important protein, *interleukin-2,* or *IL-2.* This protein autostimulates T_H cells causing them to proliferate (see Figure 24-29). It acts as an autocrine stimulator in a manner similar to the transforming growth factors described earlier (see Chapter 23). Any autocrine system of self-stimulation runs the apparent risk of producing a tumor because it is autocatalytic and has no brake. For T_H cells, regulation is achieved by control of the IL-2 receptor. That receptor is only present on T_H cells if they have bound to their macrophage-processed antigen. Thus, in the presence of a T_H antigen, the T_H cell makes IL-2 plus the IL-2 receptor and proliferates. Removal of antigen causes a loss of receptor and proliferation ceases. All of these events can be demonstrated with specific clones of T_H cells.

We have emphasized how crucial the B-cell activation step is for a successful immune response. Looking only at the proliferation response, it is evident that its importance has led to the evolution of a complicated series of cells and factors that assist activation. They include macrophages that display antigens and secrete factors and T_H cells that recognize carrier determinants and also secrete factors. But the ultimate control is the recognition of antigen by the antibody on the B cell; without that, proliferation does not occur.

As is usual with neat generalizations, there is an exception: a class of compounds that can react with all B cells and activate them independently of antigen (only IL-1 is required for activation). Called *polyclonal activators,* the compounds are typified by bacterial *lipopolysaccharide (LPS).* They are chemicals that bind to some receptor other than surface antibody. Occupancy and probably patching of the LPS receptor apparently give the B cell the same signal for growth as does patching of surface antibody (see Figure 24-28). The nature of that signal remains obscure.

The Second Step of Activation Is Secretion

Factors that induce B cells to proliferate do not necessarily induce them to secrete antibody. B-cell growth factor,

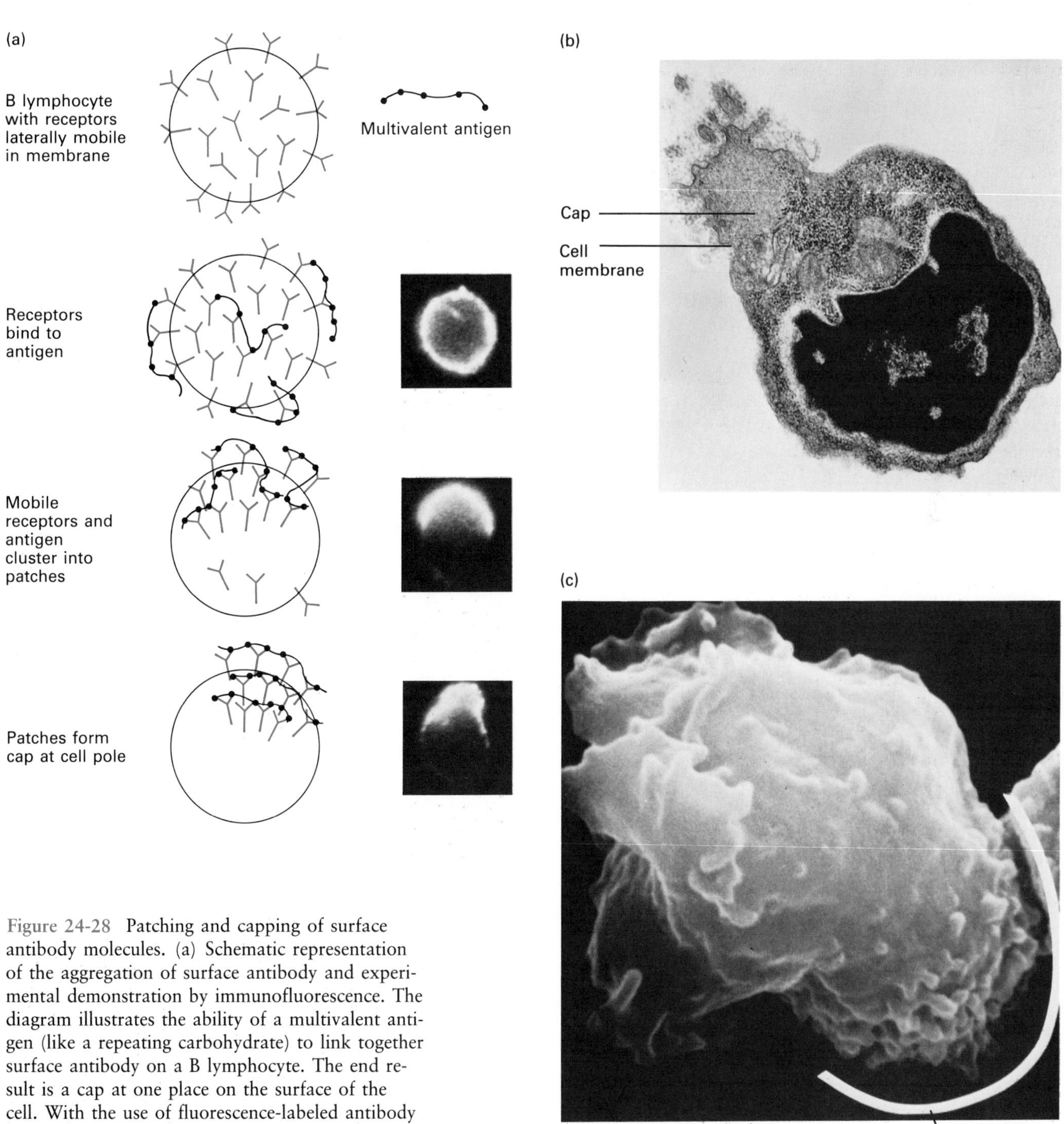

Figure 24-28 Patching and capping of surface antibody molecules. (a) Schematic representation of the aggregation of surface antibody and experimental demonstration by immunofluorescence. The diagram illustrates the ability of a multivalent antigen (like a repeating carbohydrate) to link together surface antibody on a B lymphocyte. The end result is a cap at one place on the surface of the cell. With the use of fluorescence-labeled antibody that can react with the surface antibody, the process of patch and cap formation can be visually demonstrated. (b) Transmission micrograph of a sectioned B lymphocyte with its cap region evident at the left. Within the cell, the cap region is clear of organelles because it is filled with actin and myosin that move to the cap region along with the surface antibody. The fate of the cap is either to be shed from the cell or to be taken into the cell where the antibody-antigen complexes are degraded. (c) The cap is shown in a scanning electron micrograph of a B lymphocyte. *Photographs courtesy of J. Braun.*

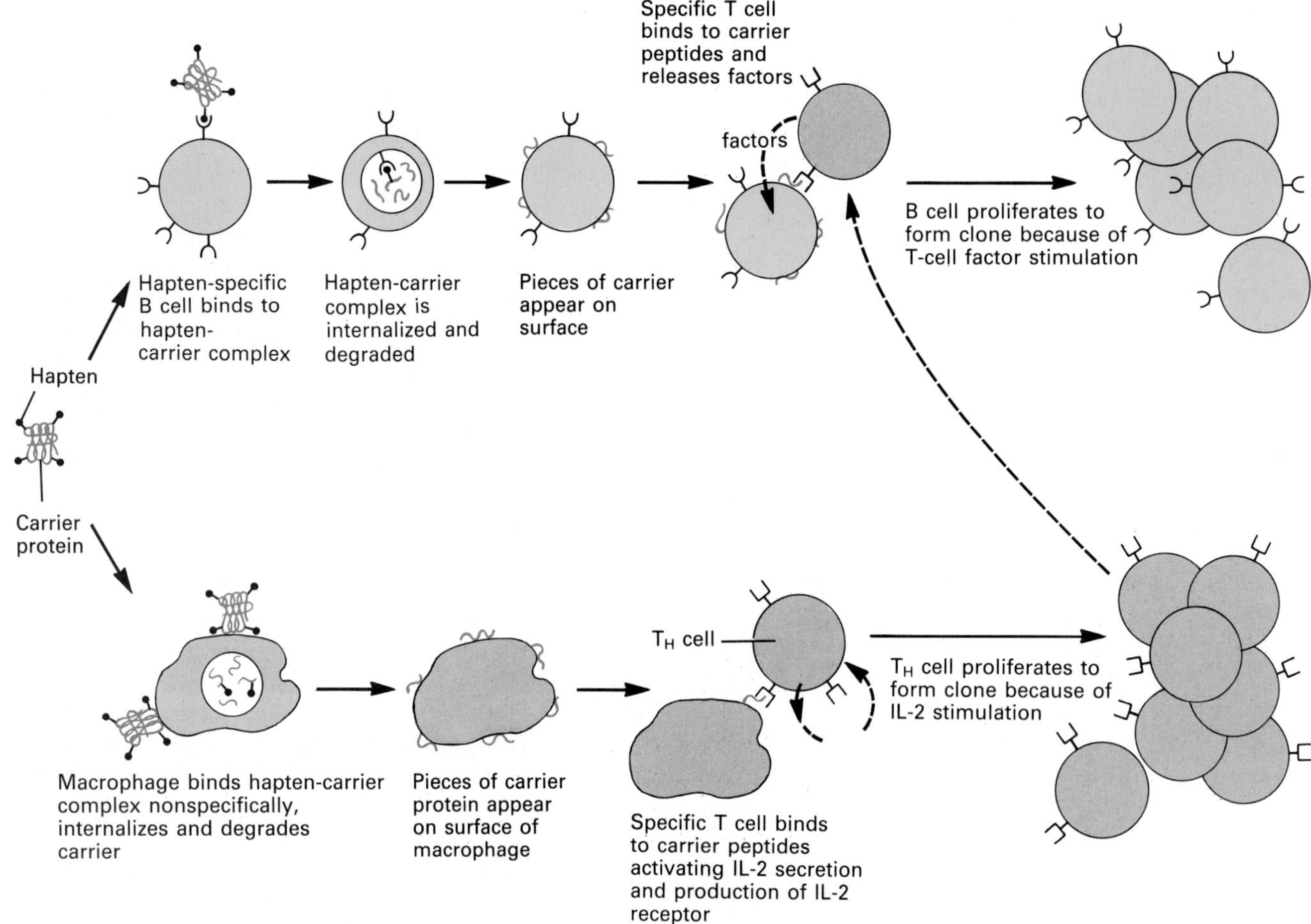

Figure 24-29 The hapten-carrier complex activates B cells by first stimulating T_H cells. The hapten-carrier complex is a protein to which several hapten molecules have been covalently coupled. The hapten portion binds to a B cell but is not at high enough concentration to initiate B-cell proliferation. The B cell internalizes and digests the hapten-carrier complex, and portions of the carrier are displayed on the B-cell surface. A macrophage cell also internalizes carrier protein and displays peptide fragments on its surface. Then T_H cells with appropriate receptors bind to the peptide displayed on the macrophage surface. Binding stimulates the T_H cells which then proliferate, recognize the identical peptide on the B cell, and secrete factors that stimulate the B cell to grow.

for instance, induces multiplication of cells but the cells retain their surface antibody and produce little secreted antibody. To activate secretion, T_H cells produce another class of molecules called *B-cell differentiation factors.* The best defined is *interferon gamma,* a molecule that gained its name from its ability to interfere with viral growth. Interferon gamma induces secretion in resting B cells without inducing proliferation. Secretion of antibody involves both a change in the type of heavy chain and a 10-fold or greater increase in the rate of antibody synthesis. Together, the B-cell growth and differentiation factors cause a massive increase in the number and nature of B-lymphoid cells after an encounter with antigen (see Figure 24-10).

To accommodate the increased rate of synthesis, the activated B cell develops a larger cytoplasm with more endoplasmic reticulum. The end stage of this maturation is a plasma cell that secretes 10 percent or more of the protein it makes as antibody. The cell is highly specialized for secretion, with a large Golgi region as well as many disks of endoplasmic reticulum (see Figure 24-10). Such a cell is the final stage of B-lymphoid development. It loses its proliferative ability entirely and dies after many days of antibody production. Interferon gamma induces both secretion and a loss of responsiveness to proliferation factors.

Secretion Involves Synthesis of an Altered Heavy Chain

One of the necessities of clonal selection is that a virgin B lymphocyte activated by an encounter with antigen must secrete antibody with exactly the binding specificity of the antibody previously carried on its surface. This means that the light- and heavy-chain variable regions must be the same in the virgin cell and the secreting cell.

Accomplishing the switch from the membrane-bound form of antibody to the secreted form without changing specificity must involve utilizing the same variable-region DNA organization for both forms of immunoglobulin. Because every variable region is individually constructed, this constraint implies that the membrane-binding portion of the constant region is altered to make the secreted form but without alteration of the variable region. Constant-region transformation is not a problem for the light chain because it does not bind directly to the membrane but rather to the heavy chain. Light chain is exactly the same in membrane-bound immunoglobulin and in secreted immunoglobulin. The heavy chain, which has a membrane-binding segment at its C-terminal, must have its molecule altered at the C-terminal end. This alteration is accomplished by a change in the type of RNA transcript made by the cell's DNA. An understanding of exactly how this occurs requires a closer look at the heavy chain's constant region.

The universal form of immunoglobulin on the surface of virgin B lymphocytes is *membrane-bound IgM*. The IgM previously described as a pentamer of the basic four-chain immunoglobulin molecule (see Figure 24-6) is *secreted IgM*. The membrane-bound form is a monomer, a single four-chain unit held to the membrane by a hydrophobic sequence at the —COOH end of each of its heavy chains. Eight exons contribute to specifying the structure of the μ heavy chain of membrane-bound IgM: a leader exon, a variable region exon, four exons that encode the four domains of the constant region, and two exons that encode the membrane-binding segment of the molecule (Figure 24-30). The mRNA for the membrane-bound form of the chain (μ_m) is polyadenylated just beyond the exon specifying its C-terminal. Splicing brings together the exons to generate the mature mRNA for μ_m.

The C-terminal of the secreted form of μ protein (μ_s) is encoded entirely by a segment of DNA contiguous with the sixth exon (Figure 24-30). The mRNA for μ_s is one that is polyadenylated just after the coding region for μ_s, excluding the two terminal exons specifying the μ_m C-terminal. The result is that intron sequence of μ_m becomes exon sequence of μ_s: the two mRNAs have different 3′ structures generating different C-terminals on the proteins.

The detailed structure of the μ_s and μ_m C-terminals are just what would be predicted for the behavior of the two proteins. The μ_m terminal includes a string of 26 un-

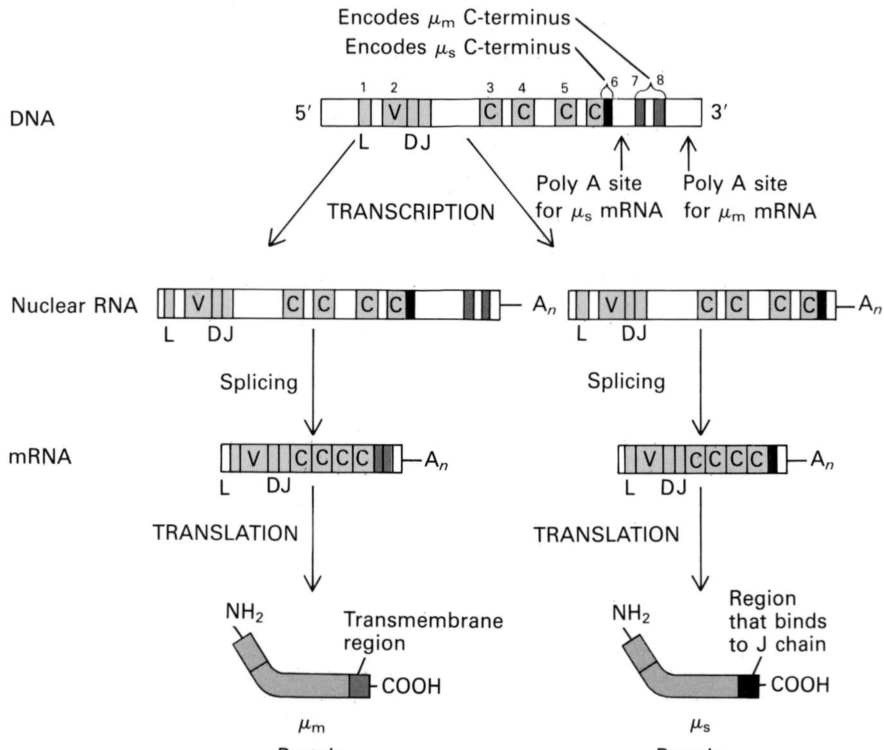

Figure 24-30 Production of μ_m and μ_s proteins by alternative polyadenylation of mRNAs. The genome organization in the C_μ region includes two polyadenylation sites. When one is used, the mRNA formed by splicing the nuclear RNA precursor has the coding region for the μ_s C-terminal. The other polyadenylation site is farther toward the 3′ end and includes two additional exons that encode the μ_m C-terminal. When this 3′ site of polyadenylation is used, the μ_s coding region is spliced out but the μ_m region is included. Thus, the C_μ region can encode both surface antibodies (with the μ_m C-terminal) and secreted antibodies (with the μ_s C-terminal).

charged amino acids preceded and followed by clusters of charged residues, a classic membrane-spanning region (see Figure 14-13). The secreted form lacks a membrane-spanning region but includes a cysteine residue that becomes the disulfide bridge linking the four-chain monomers into the pentamer of the secreted molecule. It also has a site for carbohydrate addition, presumably as a way of making the molecule more soluble.

Because the mRNAs for μ_m and μ_s differ in site of polyadenylation, it is thought that control over polyadenylation is the basis for the switch from μ_m mRNA to μ_s mRNA when a B lymphocyte is activated (see Figures 12-14, 12-15, and 24-30). This is the most obvious hypothesis but not necessarily the correct one. Some evidence exists for premature termination of transcription as a mechanism of control, and some control may be a consequence of differential protein stability.

Synthesis of μ_m and μ_s by alternative utilization of the same DNA sequence assures the identity of the variable region on the two proteins and thus is a central mechanism for allowing the clonal selection process to operate in the immune system.

The use of alternative transcription products from one region of DNA is more widespread in the immune system than this one example. For instance, there are actually two types of virgin B lymphocytes, one with solely IgM on its surface and one with both IgM and IgD on its surface. The IgM and IgD have identical variable regions: the light chains of IgM and IgD are the same and the heavy chains are formed by differential polyadenylation within transcription units that are initiated just before the VDJ complex. There are secreted forms of IgD, but the majority of IgD is membrane-bound. Why some B lymphocytes have IgD and others do not is extensively debated, but there is as yet not even agreement on the function of the molecule.

Two Cell Types Emerge from the Activation Process: Plasma Cells and Memory Cells

We have examined thus far only two consequences of activation, the proliferative response and the maturation response. Maturation, involving increased synthesis of immunoglobulin and a switch to secretion, requires changes in the amount and type of mRNA as well as changes in cell architecture. It probably involves the synthesis of many new proteins and the inhibition of synthesis of others. For instance, some experiments show that J-chain synthesis (for initiating polymerization of secreted IgM) is turned on by activation. The surface antigens also change after activation. The result is a plasma cell.

Not all progeny of an activated B lymphocyte are plasma cells. The other important product is *memory B cell*. These are cells that retain for the life of the animal a record of the antigens that it has previously encountered. Memory cells preserve variable regions that have previously proved useful so that a second encounter with an antigen can elicit a rapid, highly avid response. Like virgin B lymphocytes, they carry on their surfaces the antibody they are programmed to make. Unlike virgin B cells, they are immortal, circulating continuously as quiescent cells. An encounter with antigen activates them just as it does virgin cells. Memory cells can be helped by specific T_H cells, which also persist after a primary immune response. What is unclear is how the activated virgin cell can segregate off some of its progeny as memory B cells and others as plasma cells.

Activation Leads to Synthesis of Secondary Antibody Classes

The antibody classes IgG, IgE, and IgA play no role in the earliest stages of an immune response because they are not found on the surface of virgin B cells. For this reason they are referred to as the *secondary antibody classes*. Activated cells *switch* from IgM and IgD synthesis to the synthesis of these secondary classes. Because the various classes differ in their heavy chains and not their light chains, it is a *heavy-chain switch* that underlies the ability to change antibody classes. Analogous events described earlier, where cells switch from μ_m to μ_s synthesis and from IgM-only synthesis to IgM-plus-IgD synthesis, are known as antibody *transcript processing switches*. For synthesis of the secondary antibody classes we must consider both transcriptional switches and DNA rearrangements called *switch recombination*.

The mechanisms of switching are inherent in the organization of the heavy-chain locus. Previously we focused on the variable region components and the C_μ region (the region encoding the eight C_μ exons). Downstream of C_μ we find the segments that encode the other constant regions: C_δ, the C_γs, C_ϵ, and C_α in that order (Figure 24-31). Because there are four types of IgG and thus four C_γs, there are in total eight different C regions. Each C region is made of multiple exons that encode the domains of the individual heavy chains.

The events involved in the heavy-chain switch are still a matter of active investigation but it seems likely that they occur in two stages. The first stage—the most controversial one—would be a presumably reversible processing event whereby a new mRNA is made from unaltered DNA: a transcript processing switch. Whether the primary event is a change of termination, polyadenylation, or splicing is not known, but the end result is synthesis of a variety of mRNAs all with identical variable regions and unique constant regions. Because the switch involves aspects of transcriptional processing, cells can and do simultaneously make two or more mRNAs and therefore can simultaneously make multiple classes of antibody.

The second stage of the heavy-chain switching process appears to lock in cells irreversibly so that they can only make a single secondary antibody class (see Figure 24-31). It involves switch recombination, the deletion of

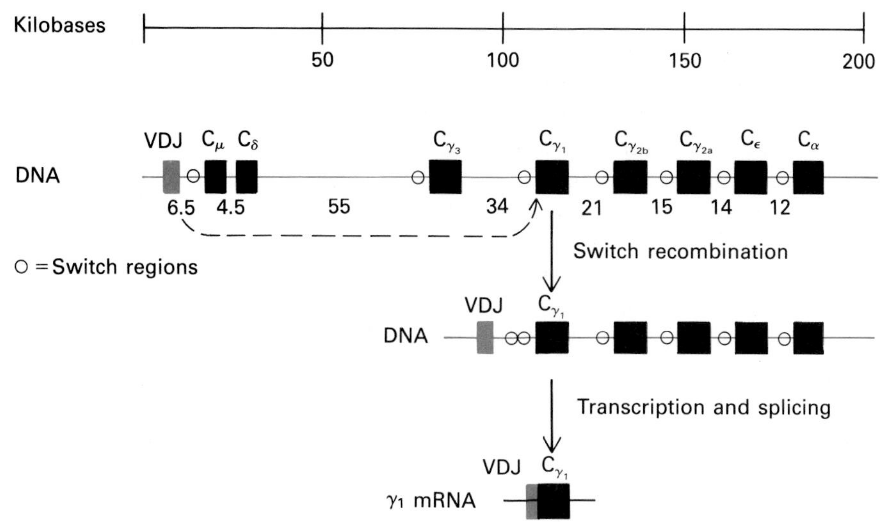

Figure 24-31 Organization of the heavy-chain region of DNA and switch recombination. The positions of the various heavy-chain constant regions are indicated. A permanent change in heavy-chain synthesis from μ to a downstream constant region is produced by recombination between switch regions located upstream of each of the constant regions. The constant regions are composed of multiple exons, and each region has alternative secreted and membrane-binding C-termini, but these details have been omitted. The numbers between units indicate the distances in kilobases.

C regions between the VDJ unit and the C region to be used. In a switch recombination event, *switch regions,* or *S regions,* recombine with each other to delete the intervening DNA. Every C region (except C_δ) has an associated S region toward the 5′ direction from its coding exons. The S regions are constructed from internally reiterated short sequences that can recombine with each other. Following switch recombination, the VDJ unit is brought close to a new C region, and polyadenylation occurs just following that C region so that the cell makes only a single class of antibody.

Switching is an event that follows activation of a B lymphocyte. Thus the progeny of an activated cell can be variable in the classes of antibody they make. When a B cell encounters an antigen it initially makes IgM but then secondary antibody classes, such as IgG, begin to predominate (Figure 24-32). It seems probable that special T_H cells help to stabilize the synthesis of specific classes (or possibly they help to direct one or both stages of switching) so that one or another class of antibody may predominate. It is important that switching be directed because different antigens require different responses. A pathogen in the intestine must elicit IgA synthesis, a parasite must elicit IgE synthesis.

The ultimate progeny of activated B lymphocytes are the plasma cells and the memory cells. The switching process increases the number of progeny types. The plasma cells can make any of seven classes of immunoglobulin, but because they undergo switch recombination during their maturation each cell makes only a single an-

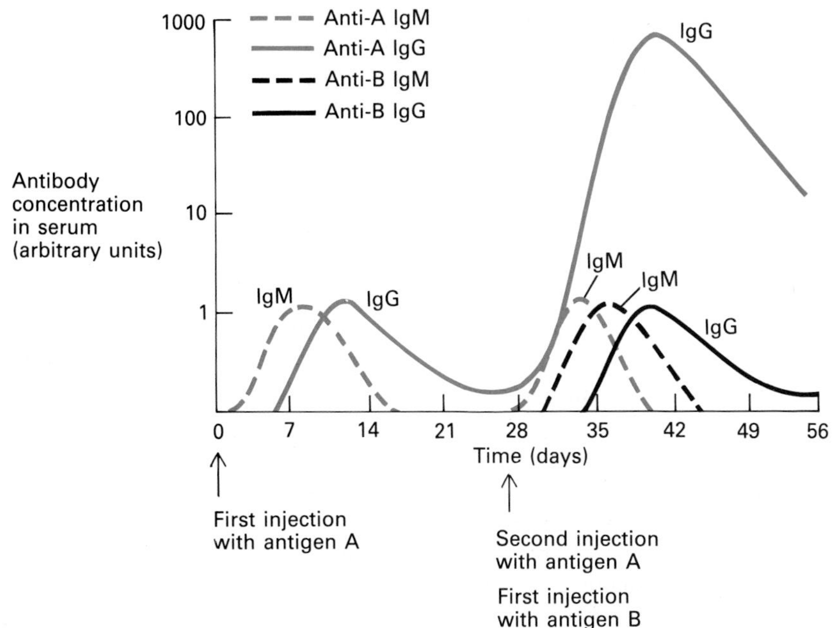

Figure 24-32 Kinetics of IgM and IgG responses after primary and secondary immunization. The diagram expands upon that in Figure 24-11 where only total antibody concentrations were indicated. Here the initial immune response to an antigen injection is seen to change with time, going from an IgM-dominated response to an IgG-dominated response. Once the initial response has subsided, a second injection with the same antigen leads to a prompt and massive IgG response because of the large number of memory cells prepared to make IgG. The secondary response also includes an IgM response, probably emanating from virgin B cells that are always present and prepared to be activated. Antigen B, a control introduced during the second injection, gives the IgM-to-IgG pattern of a primary response.

tibody class. The memory cells are also of many types: they may even go through only the first stage of switching and therefore may be able to make multiple classes of antibody. Because each C region has two alternative C terminals, one for membrane binding and one for secretion, each class of antibody can occur as either a secreted or a membrane-bound molecule. The secreted forms are made by plasma cells; the membrane-bound forms are found on memory cells.

When memory cells respond to antigen, they make mainly antibodies of the secondary classes. During a second encounter with antigen there is an immediate IgG response, and it is of much greater magnitude than the initial IgG response (see Figure 24-32) because of the large number of memory cells ready to respond. If during a second immunization with an antigen a new antigen is introduced, the animal undergoes a memory-dominated response to the previously encountered antigen and an IgM-dominated naive response to the new antigen (see Figure 24-32). This experiment very clearly highlights the specificity of immunological memory for only those antigens previously encountered.

Somatic Mutation of Variable Regions Follows from Activation

We have described the Ig variable regions as arising from combinatorial joining events among V, D, and J regions. We have followed the fate of the joined VJ and VDJ regions through transcriptional alterations and switching events with the implicit assumption that the variable regions are not altered once they are formed. This assumption, however, is false. There is a second type of mechanism that increases the variety of variable regions. Called *somatic mutation,* it involves the replacement of individual bases in a joined VDJ segment with alternative bases, causing apparently random variation throughout the VDJ segment (and probably also in flanking DNA for some distance on either side of the VDJ). Its occurrence has been documented in both L-chain and H-chain variable regions by comparing germ-line DNA sequences with expressed DNA gene sequences (Figure 24-33).

A striking aspect of somatic mutation is that it has mainly been found in VDJ segments from cells expressing the secondary antibody classes IgG or IgA. This distribution of mutational events suggests that somatic mutation is an event of immunodifferentiation following activation of the virgin B lymphocyte by antigen. The changes in amino acid sequence caused by somatic mutation have the consequence of varying the fit between the antibody and the antigen, changing the affinity of the antibody for the antigen.

It is thought that higher affinity surface antibody leads to easier activation of B lymphocytes by antigen and that therefore there is a continual selection among the somatically mutated variable regions for cells bearing higher affinity antibody. As antigen concentrations fall, this should be an especially prominent effect. Somatic mutation occurring throughout the variable region should appear to have its greatest effect in the CDRs, where mutation can lead to higher affinity of binding. In fact, somatic mutations that cause amino acid replacements are especially prevalent in the regions encoding the three CDRs.

The rate of somatic mutation has been estimated to be as high as 10^{-3} per nucleotide per cell generation, a rate 10^{6}-fold higher than the spontaneous rate of mutation in other genes. Such a high rate of mutation over the whole genome would be insupportable. There must exist mechanisms that direct mutational activity to variable-region sequences (see Figure 24-33). How this might occur is not known; possibly some sequence in the area of the variable region directs a special enzyme system to carry out point replacements of nucleotides independently of template specification. The consequence of somatic mutation is an increase in the kinds of potential antibodies an animal can make from the 10^{11} we calculated earlier to a number many orders of magnitude higher. The greatly increased diversity will generate better fitting antibodies but also, and more frequently, poorly designed antibodies. Somatic mutation occurs during the proliferative response following an encounter with antigen, and we assume that if a cell making a poor antibody is produced, its division ceases and it dies, so that only cells with appropriately designed surface antibody are continually stimulated to grow and mature. Darwinian processes continue from the first moment of antigen encounter until the mature plasma cell is made, at which time somatic mutation ceases.

Somatic mutation is presumably integrated into the many events that follow activation, but the interrelation of these events is not understood. Does somatic mutation only follow switching? Does it ever precede switching? Can a memory cell be further mutated after its activation? What role might T_H cells or their factors play in signaling cells to begin somatic mutation? These questions and others will require answers before we understand how somatic mutation fits into the overall process of immune cell differentiation.

T Lymphocytes

The T lymphocytes are key players in the immune response. They show very high specificity of recognition, recognizing one particular antigen but not its close relatives. They have on their surface a recognition molecule able to make discriminations as fine as those made by antibody. That molecule has been called the *T-cell receptor* to distinguish it from antibody. It is now evident that the T-cell receptor is closely related to antibody, but its distinctive name is retained because it has numerous

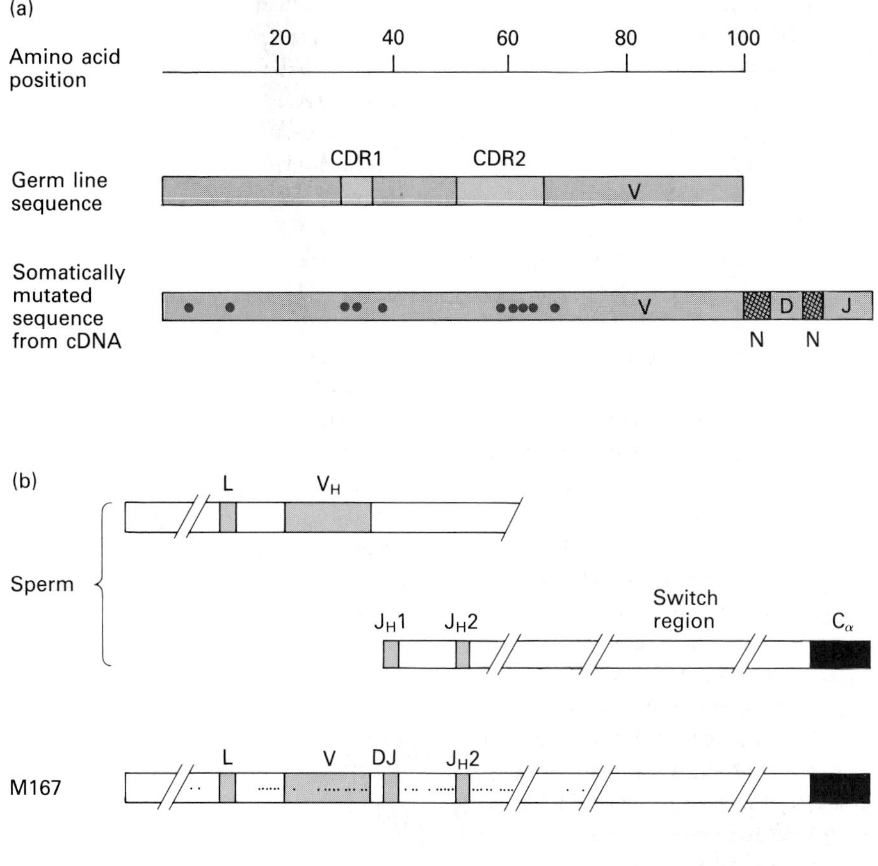

Figure 24-33 Somatic mutation of rearranged V_H regions. (a) Comparison of the nucleotide sequence of a germ-line V_H region with the sequence of the cDNA for a heavy-chain variable region that had undergone somatic mutation. The scale is given as amino acid positions but 10 nucleotide changes are indicated. [See Bothwell et al., 1981, *Cell* **24**:625–637.] (b) A comparison of germ line (sperm) DNA sequences and somatic mutations found in heavy-chain genomic DNA from a myeloma (M_{167}). The consequences of somatic mutation are evident as alterations in nucleotide sequence, indicated as dots. Somatic mutation is localized to the variable region and a small region around it. *After Kim et al., 1981,* Cell *27:573.*

properties that distinguish it from antibody. A key difference is that antibodies are produced in two forms, either cell surface molecules or soluble secreted molecules, whereas the T-cell receptor exists only at the cell surface. All T-cell functions, therefore, involve reactions on the surface of the cell. The molecules that T-cell receptors bind are designated antigens although they have radically different properties from the antigens to which antibodies bind.

T cells, like B cells, have a preliminary phase and a mature phase and go through an activation process triggered by interaction with antigen. Antigens for T cells are always cell-bound molecules, so that T-cell receptor–antigen interactions are actually cell-to-cell interactions. Given the size and complexity of cells, it is evident why we know very little about binding affinities or the kinetic behavior of T-cell receptors: it is difficult to do chemistry using whole cells as the reagents. And there are very few receptors on each cell.

The T-Cell Receptor Is Just Becoming Known

After many years of frustrating searches, immunologists isolated the T-cell receptor in the early 1980s by using as the tools of isolation antibodies that recognize it. The receptor is a two-chain molecule residing on the cell surface. The two chains, called α and β, are nearly identical in size (about 40,000 m.w.). They have asparagine-linked carbohydrate residues and are made up of two domains of approximately equal size: one variable, one constant.

The genes for both the α and β chains have been isolated. Both have organizations reminiscent of antibody genes: there are libraries of V regions, D regions, and J regions. Their recognition sequences specifying the joining reactions follow the same rules as those for antibody genes. The α and β gene clusters reside on separate chromosomes and are distinct from the antibody genes. There is also a third set of rearranging gene segments that can encode a γ chain in T cells, but its role in T-cell biology remains to be determined.

It seems certain that the α and β genes contribute to the receptor on the T helper cells (T_H) and on the cytotoxic lymphocytes (CTL), or killer T cells. For the third T-cell type, the T suppressor (T_S), the data leaves uncertainty. In all probability, the same basic T-cell receptor structure is found on all T cells and what differentiates T-cell function is the behavior of the cell rather than its receptor.

The T-cell α, β, and γ gene segments are not only organized in a pattern similar to that of antibody genes, their amino acid sequences are obviously related to those of antibody proteins. Evidently all of these recognition molecules evolved from a single primordial protein domain and all belong to a single superfamily of proteins (Figure 24-34). Any one protein in this superfamily is constructed from a gene consisting of multiple exons encoding multiple related domains. Each domain is similar in structure and is a variant of the immunoglobulin fold (see Figure 24-18b). Not only antibody and T-cell receptor genes belong to this superfamily so also do other cell surface proteins (Figure 24-34).

T-Cell Receptors Recognize Foreign Antigens as Compound Units with a Self-Molecule

Although the T-cell receptor can be highly specific for recognition of a specific foreign determinant—and it displays the specific tolerance to self-antigens also seen in antibodies—the antigen must be presented to the T-cell receptor as part of a complex with a specific self-molecule. This behavior became clear in a classic experiment done in 1974 to examine how CTLs kill infected target cells. The investigators injected a virus into an inbred mouse, producing in the mouse CTLs that when studied in culture recognized and killed cells from that mouse infected with that virus. Just as is seen with antibodies, the CTLs reacted only with cells infected by that particular virus. The experimenters then asked the apparently innocuous question: Would the CTLs kill cells from a different inbred strain of mice infected by the same virus? To almost everyone's surprise they would not kill. Nor were uninfected foreign cells killed by the CTLs. Apparently the CTL receptor recognized both the identity of the infected cell and the type of virus; it recognized self and foreign determinants together, but neither one separately. What surface self-molecule might the T-cell receptor care

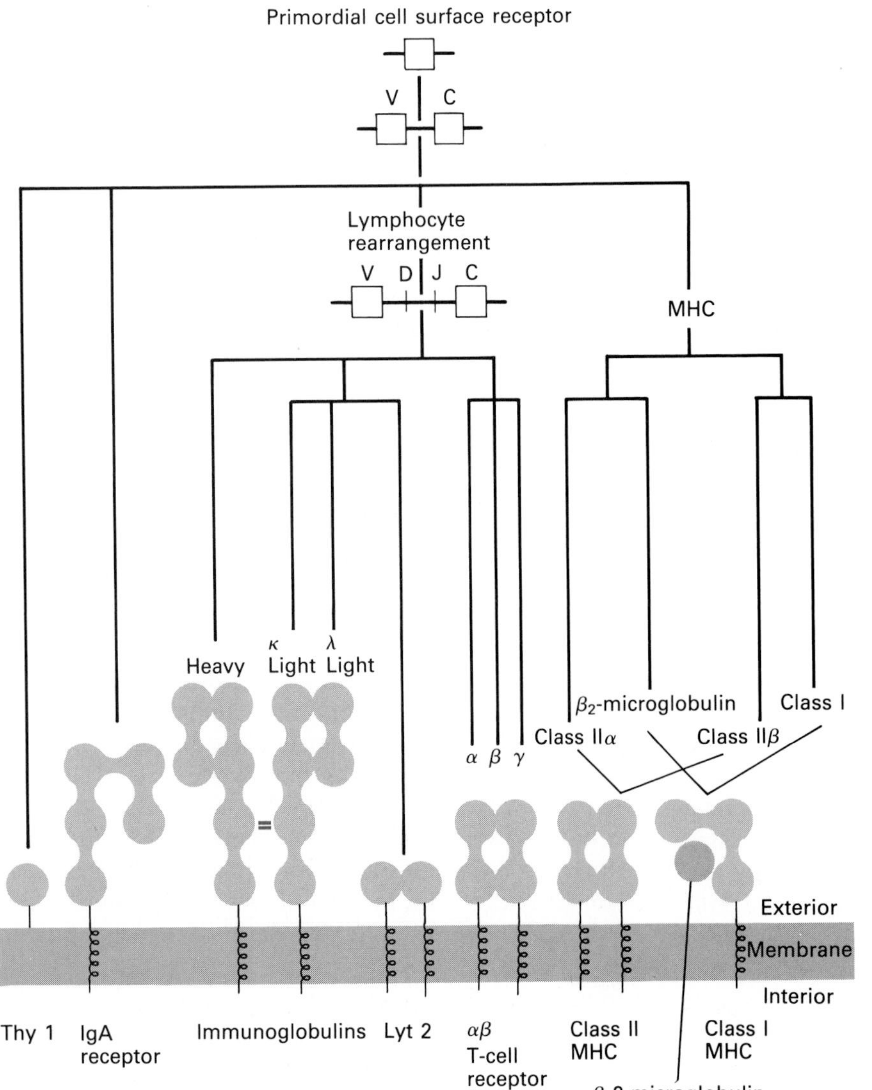

Figure 24-34 The superfamily of immunoglobulin-related proteins. For each protein the individual balls represent domains of protein structure, many derivatives of the immunoglobulin fold described in Figure 24-18b. The genes encoding these proteins all have homologies and are considered a superfamily of related genes that must have evolved over hundreds of millions of years from a common precursor. *After L. Hood, M. Kronenberg, and T. Hunkapiller, 1985,* Cell ***40**:225.*

about so much? As described below, the self-molecules recognized by CTLs are encoded in a region of the genome called the MHC genes so that the CTL killing is restricted to self-MHC (Figure 24-35).

The MHC Genes Were First Recognized in Tissue Transplantation Experiments

If skin from one inbred strain of mice is grafted to the back of another strain, the graft at first seems to settle in but soon thereafter dies. We say that the graft from a donor to a recipient strain has been *rejected*. Grafts between animals coming from the same inbred strain are not rejected. To find out which genes encode the proteins causing the rejection, experimenters minimized the genetic differences between the donor and recipient animals by inbreeding until a region containing a small number of genes could be identified as causing the rejection. In this way it was shown that a fair number of cell surface proteins can cause rejection. But one gene complex stood out in such experiments. Differences among donor and host genes in this complex caused very fast rejection. Because genes that lead to rejection are *histocompatibility genes,* this complex of genes was called the *major histocompatibility complex* or *MHC*. Many individual genes have been found within the MHC and most have been molecularly cloned. A complete map of the MHC includes some genes that elicit fast graft rejection and others that do not (Figure 24-36).

One type of MHC gene product turned out to be the target for self-recognition: the product of the *class I MHC genes,* typified by genes called *H-2D* and *H-2K* in the mouse and HLA in humans. It is a two-chain molecule with one highly polymorphic component, a polypeptide whose structure varies greatly from one inbred strain to another, and one constant chain called *β-2 microglobulin* (see Figure 14-23).

These MHC proteins are part of the immunoglobulin gene superfamily (see Figure 24-34). Most cells in a mouse's body have H-2 proteins on their surface. For a CTL taken from mouse strain A to kill a cell, the cell must have either the H2-D or the H2-K molecule from strain A on its surface. Of course the cell must also have the foreign antigen for which the CTL is specific (Figure 24-35). We call this a *joint recognition* process: both self H-2 and a foreign molecule must be jointly recognized on the same cell at the same time. Assuming that one T-cell receptor molecule recognizes both components—an assumption that could be wrong—it must simultaneously recognize both H-2 and the foreign molecule but not either separately.

For T_H cells, joint recognition of self-molecules and

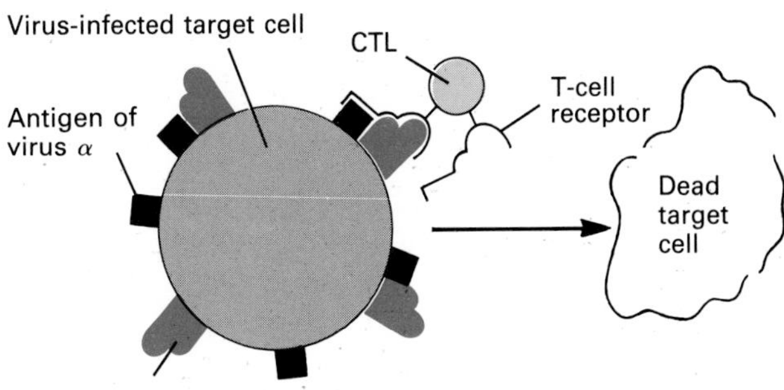

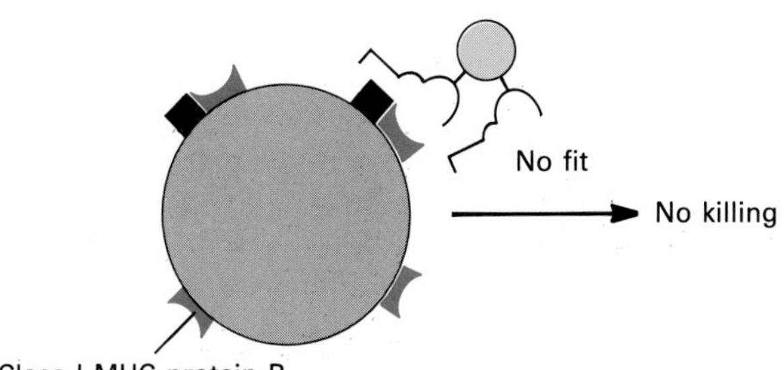

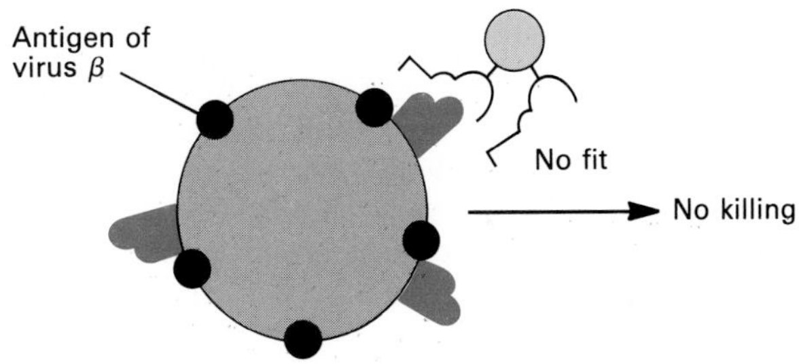

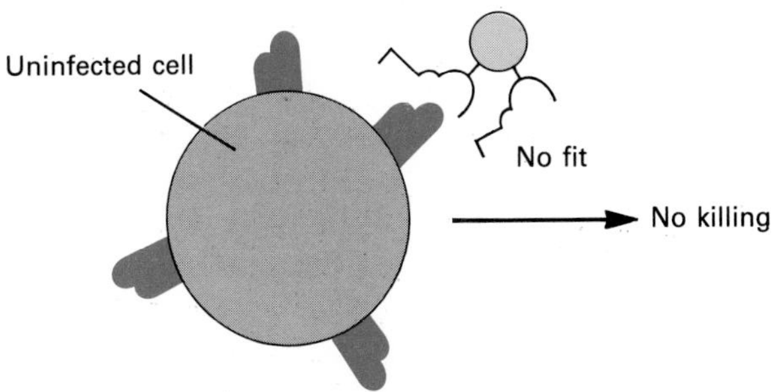

Figure 24-35 Restriction of CTL killing by class I MHC gene products. For this experiment, CTLs were prepared from animals of MHC type A inoculated with virus *α*. The CTLs killed cultured cells from A-type mice infected with virus *α*. They did not kill uninfected A-type cells, B-type cells infected with virus *α*, or A-type cells infected with virus *β*. The viruses can differ by very little; closely related strains of influenza virus show the specificity of *α* and *β*. The inbred strains of mice can also differ by very little: a few amino acid changes in an MHC class I protein can be the only difference between A and B.

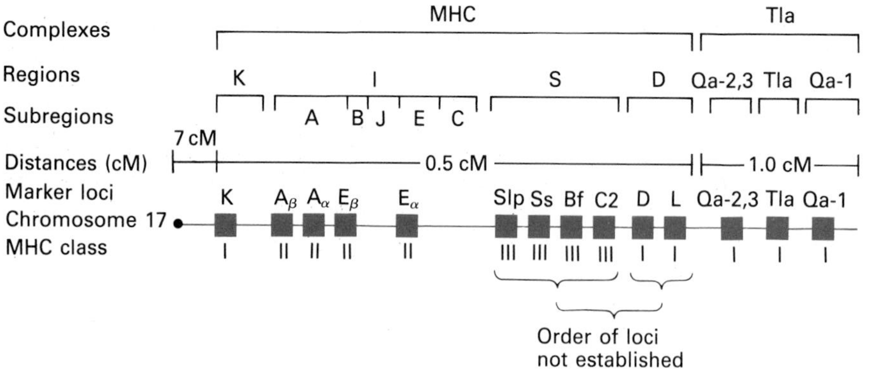

Figure 24-36 The MHC gene complex of the mouse. At first this complex of genes, called H-2 in the mouse, was identified by graft rejection. The nearby Tla region was found to encode cell surface differentiation antigens. Fine genetic mapping uncovered the K and D loci. Later work on T_H cell reactions located the I region. The S region was found to encode blood proteins (mainly of the complement system). Fine structure analysis of the I region produced subregions A, B, J, E, and C. Classical genetics defined the distances these complexes covered in units related to recombination frequency called centi-Morgans (cM). More recently, with the advent of molecular cloning, the individual protein-coding regions were identified, as shown on the bottom line of the diagram. Actually, there are many Qa and Tla loci. [See L. Hood, I. Weissman, W. Wood, and J. Wilson, 1984, *Immunology,* 2d ed. Benjamin.]

foreign cell surface molecules is also the rule. The self-molecule, however, is not a class I MHC molecule but a class II. Class II molecules have two polymorphic chains, both smaller than the polymorphic chain of class I molecules but both coming from the immunoglobulin gene superfamily (see Figure 24-34). As might be expected from our knowledge of T_H cell function (Figure 24-29), class II molecules are found mainly on two kinds of cells, macrophages and B lymphocytes. Remembering that T_H cells initially recognize fragments of antigens displayed by macrophages and later help B lymphocytes, we believe that in all probability it is to provide the T_H cell with self-recognition that class II molecules are found on these two cell types. Figure 24-29 is a simplification; in reality the peptides expressed on the surface of macrophages and B cells are associated with class II MHC proteins just as CTL targets express their antigens in association with class I molecules (see Figure 24-35). Furthermore, T_H cell is specifically restricted by MHC structure just as CTL specificity is restricted by MHC structure.

The T suppressor cells are less well understood than the T_H cells and CTLs. Their activation apparently involves joint recognition of the foreign molecule and a class II self-molecule.

Summary

The immune system evolved as the body's protective mechanism against invasion by pathogens such as viruses, bacteria, and fungi. There are two arms to the system: humoral immunity mediated by soluble protein antibodies in bodily fluids, and cellular immunity carried by surface receptors on circulating cells. B lymphocytes make soluble antibodies; T lymphocytes carry out cellular immune reactions.

Each antibody molecule has two identical binding sites that can specifically bind to an antigen. The binding reaction is a simple, reversible binding characterized by an affinity constant. Antibody molecules also have effector domains that allow the body to rid itself of the antigen. There are five major antibody classes with different effector activities: IgM, IgD, IgG, IgE, and IgA.

The mechanism by which antibodies are formed is called clonal selection. The system works by producing an enormous variety of B lymphocytes, each with a homogeneous population of cell surface antibody molecules. The antigen selects from the population of antibody-bearing cells those that carry molecules able to bind to it. Such cells are induced to multiply and mature into antibody-secreting cells. Some activated cells mature into plasma cells, which are specialized for immediate antibody production, and others mature into long-lived memory B cells, which respond rapidly to all further encounters with the antigen.

There are three major types of T cells: cytotoxic T lymphocytes (CTLs), helper T lymphocytes (T_H cells), and suppressor T lymphocytes (T_S cells). The CTLs (also known as killer T cells) directly kill target cells that they recognize with their surface antibody-like receptors. The T_H cells assist B cells in their reaction to antigens. The T_S cells dampen responses and may help prevent reaction to self-molecules.

The ability of an animal's immune system to avoid reacting to the molecules in its own body is called tolerance and is an active process maintained, at least in part, by the T cells. A failure of tolerance leads to autoimmune disease, one of a number of disease types that can be caused by the immune system. Others are failures of immune function, tumors of the immune system, and hyperreactive conditions such as allergies.

Our knowledge of the details of the structure and synthesis of antibodies is heavily dependent on studies of myelomas, tumors that secrete antibodies. By studying myeloma products, antibody proteins were found to be two-chain molecules. The heavy and light chains that constitute the molecules are each folded into a number of compact domains that form both the binding sites and the effector regions. The N-terminal portions of each chain are highly variable in amino acid sequence, producing the binding variability of the molecules. Within the variable regions, three regions directly interact with antigen and are called complementarity-determining regions (CDRs).

The structures of the genes that encode antibodies are bipartite: constant segments are attached to any of a library of variable segments. Antibody genes are not carried per se in the genome but rather are carried as gene segments that come together during lymphocyte differentiation to form the variable regions. The joining process is signaled by a DNA recognition signal consisting of a heptamer and a nonamer. They are separated by either about 11 or 22 nucleotides, that is, by one or two turns of the DNA helix. The joining rule is that a one-turn element always joins to a two-turn element. Diversity of antibody structure is partly a consequence of the large size of the variable-segment libraries and partly a consequence of combinatorial diversity in the joining of members of the libraries. Imprecision at the joints between segments further increases diversity. During heavy-chain gene formation random sequence elements are inserted into the third CDR.

The process of DNA rearrangement to make antibody genes is an orderly one. First heavy chain genes are rearranged, then light chain genes. The variability inherent in the system can make as many as 10^{11} different molecules. Thus, the immunodifferentiation process produces a vast array of cells from which antigens can choose those that fit best.

The activation of a B cell by an encounter with an antigen is the key process in successful clonal selection. This activation can be accomplished by an antigen that has multiple identical sites (is multivalent) because the clustering of surface antibody can cause a lymphocyte to begin growth and maturation. Most antigens, however, are unable to activate by themselves and require help from T_H cells. Such antigens are degraded by macrophages, which then present on their surface fragments of the antigen to T_H cells in a highly stimulatory form. Because surface antibody and secreted antibody must differ in whether or not their heavy chains have membrane-binding domains, there are alternative exons encoding the C-terminals of the heavy chain. Part of the activation of antibody involves increased utilization of the secretion exon by a process of differential polyadenylation and splicing.

The B-cell activation process can also lead to a switching process by which progeny cells may make antibodies with different effector domains than did the parental cells. Furthermore, activation involves an extensive process of specific somatic mutation of the heavy- and light-chain variable regions. Thus the clone of cells resulting from activation actually contains extensive variability. Continual darwinian selection acts to maintain those cells that make antibodies with the highest affinity for antigen.

The T-cell classes carry out their function using an antibody-like T-cell receptor encoded by a set of genes quite separate from those that encode antibodies. The receptor is a two-chain molecule, each chain having a variable and constant region. Because T-cell receptors remain membrane-bound and T cells recognize antigens on the surface of other cells, T-cell recognition is a cell-to-cell recognition process. The structure on a target cell recognized by the T-cell receptor is not solely the foreign antigen: the T-cell also specifically recognizes a self-molecule, a major histocompatibility (MHC) protein. Thus T-cell recognition is a joint process of binding to both self and foreign protein molecules. It is thought that one receptor carries out the joint recognition process.

References

General References

JERNE, N. K. 1973. The immune system. *Sci. Am.* **229**(1):52.

KABAT, E. A. 1976. *Structural Concepts in Immunology and Immunochemistry*, 2d ed., Holt.

LANDSTEINER, K. 1945. *The Specificity of Serologic Reactions.* Cambridge: Harvard University Press.

Clonal Selection

BURNET, F. M. 1957. A modification of Jerne's theory of antibody production using the concept of clonal selection. *Austral. J. Sci.* **20**:67.

MANSER, T., S. Y. HUANG, and M. L. GEFTER. 1985. The influence of clonal selection on the expression of immunoglobulin variable region genes. *Science* **226**:1283–1288.

Antibodies

AMZEL, L., R. POLJAK, F. SAUL, J. VARGA, and F. RICHARDS. 1974. The three-dimensional structure of a combining region-ligand complex of immunoglobulin NEW at 3.5 Å resolution. *Proc. Nat'l. Acad. Sci. USA* **71**:1427.

DAVIES, D. R., and H. METZGER. 1983. Structural basis of antibody function. *Ann. Rev. Immunol.* **1**:87.

DAVIES, D. R., E. A. PADLAN, and D. M. SEGEL. 1975. Three-dimensional structure of immunoglobulins. *Ann. Rev. Biochem.* **44**:639.

EDELMAN, G. M. 1970. The structure and function of antibodies. *Sci. Am.* **223**(2):34.

LOH, D. Y., A. L. M. BOTHWELL, M. WHITE-SCHARF, T. IMANISHI-KARI, and D. BALTIMORE. 1983. Molecular basis of a mouse strain-specific anti-hapten response. *Cell* **33**:153.

SIEKEVITZ, M., S. Y. HUANG, and M. L. GEFTER. 1983. The genetic basis of antibody production: a single heavy chain variable region gene encodes all molecules bearing the dominant anti-arsonate idiotype in the strain A mouse. *Eur. J. Immunol.* **13**:123.

Diversity Generated from Joining Immunoglobulin Gene Segments

BLOMBERG, B., A. TRAUNECKER, H. EISEN, and S. TONEGAWA. 1981. Organization of four mouse λ light-chain immunoglobulin genes. *Proc. Nat'l. Acad. Sci. USA* **78**:3765.

CAPRA, J. D., and A. B. EDMUNDSON. 1977. The antibody combining site. *Sci. Am.* **236**:50.

DAVIS, M. M., K. CALAME, P. W. EARLY, D. L. LIVANT, R. JOHO, I. L. WEISSMAN, and L. HOOD. 1980. An immunoglobulin heavy-chain gene is formed by two recombinational events. *Nature* **283**:733.

EARLY, P., H. HUANG, M. DAVIS, K. CALAME, and L. HOOD. 1980. An immunoglobulin heavy-chain variable region gene is generated from three segments of DNA: V_H, D, and J_H. *Cell* **19**:981.

GEARHART, P., N. JOHNSON, R. DOUGLAS, and L. HOOD. 1981. IgG antibodies to phosphorylcholine exhibit more diversity than their IgM counterparts. *Nature* **291**:29.

HILSCHMANN, H., and L. C. CRAIG. 1965. Amino acid sequence studies with Bence-Jones proteins. *Proc. Nat'l. Acad. Sci. USA* **53**:1403.

HONJO, T. 1983. Immunoglobulin genes. *Ann. Rev. Immunol.* **1**:499.

HOZUMI, N., and S. TONEGAWA. 1976. Evidence for somatic rearrangement of immunoglobulin genes coding for variable and constant regions. *Proc. Nat'l. Acad. Sci. USA* **73**:3628.

KUROSAWA, Y., and S. TONEGAWA. 1982. Organization, structure, and assembly of immunoglobulin heavy-chain diversity DNA segments. *J. Exp. Med.* **155**:201.

LEDER, P. 1982. The genetics of antibody diversity. *Sci. Am.* **246**:102.

LEWIS, S., N. ROSENBERG, F. ALT, and D. BALTIMORE. 1982. Continuing κ gene rearrangement in an Abelson murine leukemia virus transformed cell line. *Cell* **30**:807.

MAX, E. E., J. G. SEIDMAN, and P. LEDER. 1974. Sequences of five potential recombination sites encoded close to an immunoglobulin κ constant region gene. *Proc. Nat'l. Acad. Sci. USA* **76**:3450.

SAKANO, H., R. MAKI, Y. KUROSAWA, W. ROEDER, and S. TONEGAWA. 1980. Two types of somatic recombination are necessary for the generation of complete immunoglobulin heavy-chain genes. *Nature* **286**:676.

TONEGAWA, S. 1983. Somatic generation of antibody diversity. *Nature* **302**:575.

WEIGERT, M., R. PERRY, D. KELLEY, T. HUNKAPILLER, J. SCHILLING, and L. HOOD. 1980. The joining of V and J gene segments creates antibody diversity. *Nature* **283**:497.

Allelic Exclusion

ALT, F., N. ROSENBERG, S. LEWIS, E. THOMAS, and D. BALTIMORE. 1981. Organization and reorganization of immunoglobulin genes in Abelson murine leukemia virus-transformed cells: rearrangement of heavy but not light chain genes. *Cell* **27**:381.

Immunoglobulin Synthesis

CORY, S., J. JACKSON, and J. M. ADAMS. 1980. Deletions in the constant region locus can account for switches in immunoglobulin heavy-chain expression. *Nature* **285**:450.

DAVIS, M. M., S. K. KIM, and L. HOOD. 1980. DNA sequences mediating class switching in α immunoglobulins. *Science* **209**:1360.

DUTTON, R. W., and R. I. MISHELL. 1967. Cellular events in the immune response. The in vitro response of normal spleen cells to erythrocyte antigens. *Cold Spring Harbor Symp. Quant. Biol.* **32**:407.

EARLY, P., J. ROGERS, M. DAVIS, K. CALAME, M. BOND, R. WALL, and L. HOOD. 1980. Two mRNAs can be produced from a single immunoglobulin μ gene by alternative RNA processing pathways. *Cell* **20**:3131.

KOHLER, G., and C. MILSTEIN. 1975. Continuous cultures of fused cells secreting antibody of predefined specificity. *Nature* **256**:495.

MCINTYRE, B., and J. ALLISON. 1983. The mouse T cell receptor: structural heterogeneity of molecules of normal T cells defined by Xenoantiserum. *Cell* **34**:739.

MILSTEIN, C. 1980. Monoclonal antibodies. *Sci. Am.* **243**(4):66.

MOORE, K. W., J. ROGERS, I. HUNKAPILLER, P. EARLY, C. NOTTENBURG, I. WEISSMAN, H. BAZIN, R. WALL, and L. E. HOOD. 1981. Expression of IgD may use both DNA rearrangement and RNA splicing mechanisms. *Proc. Nat'l. Acad. Sci. USA* **78**:1800.

NIKAIDO, T., S. NAKAI, and T. HONJO. 1981. The switch region of the immunoglobulin C_μ gene is composed of simple tandem repetitive sequences. *Nature* **292**:845.

POTTER, M. 1972. Immunoglobulin-producing tumors and myeloma proteins of mice. *Physiol. Rev.* **62**:631.

ROGERS, J., P. EARLY, C. CARTER, K. CALAME, M. BOND, L. HOOD, and R. WALL. 1980. Two mRNAs with different 3′ ends encode membrane-bound and secreted forms of immunoglobulin μ chain. *Cell* **20**:303.

RUSCONI, S., AND G. KOHLER. 1985. Transmission and expression of a specific pair of rearranged immunoglobulin μ and κ genes in a transgenic mouse line. *Nature* **314**:330.

SHIMIZU, T., N. TAKAHASHI, Y. YAMAMAKI-KATAOKA, Y. NISHIDA, T. KATAOKA, and T. HONJO. 1981. Ordering of mouse immunoglobulin heavy-chain genes by molecular cloning. *Nature* **289**:149.

WEISSMAN, I. L. 1975. Development and distribution of immunoglobulin-bearing cells in mice. *Transplant. Rev.* **24**:159.

WHITLOCK, C. A., and O. N. WITTE. 1982. Long-term culture of B lymphocytes and their precursors from murine bone marrow. *Proc. Nat'l. Acad. Sci. USA* **79**:3608.

Somatic Mutation

ALT, F., and D. BALTIMORE. 1982. Joining of immunoglobulin heavy-chain gene segments: implications from a chromosome with evidence of three D-J_H fusions. *Proc. Nat'l. Acad. Sci. USA* **79**:4118.

BALTIMORE, D. 1981. Somatic mutation gains its place among the generators of diversity. *Cell* **26**:295.

CREWS, S., J. GRIFFIN, H. HUANG, K. CALAME, and L. HOOD. 1981. A single V_H gene segment encodes the immune response to phosphorylcholine: somatic mutation is correlated with the class of antibody. *Cell* **25**:59.

GEARHART, P., and D. BOGENHAGEN. 1983. Clusters of point mutation are found exclusively around rearranged antibody variable region genes. *Proc. Nat'l. Acad. Sci. USA* 80:3439.

KIM, S., M. M. DAVIS, E. SINN, P. PATTEN, and L. HOOD. 1981. Antibody diversity: somatic hypermutation of rearranged V_H genes. *Cell* **27**:573.

T-Lymphocytes

BARTH, R. K., B. S. KIM, N. C. LAN, T. HUNKAPILLER, N. SOBIECK, A. WINOTO, H. GERSHENFELD, C. OKADA, D. HANSBURG, I. L. WEISSMAN, and L. HOOD. 1985. The murine T-cell receptor uses a limited repertoire of expressed V_β gene segments. *Nature* **316**:517.

CANTOR, H., and E. A. BOYSE. 1975. Functional subclasses of T lymphocytes bearing different Ly antigens I: the generation of functionally distinct T cell subclasses is a differentiative process independent of antigen. *J. Exp. Med.* **141**:1375.

FORD, C. E., H. S. MICKLEM, E. P. EVANS, J. G. GRAY, and D. A. OGDEN. 1966. The inflow of bone-marrow cells to the thymus. Studies with part body-irradiated mice injected with chromosome-marked bone marrow and subjected to antigenic stimulation. *Ann. N.Y. Acad. Sci.* **129**:238.

LEDOUARIN, N. M., and F. JOTEREAU. 1975. Tracing of cells of the avian thymus through embryonic life in interspecific chimeras. *J. Exp. Med.* **142**:17.

MILLER, J. F. A. P. 1961. Immunological function of the thymus. *The Lancet* 2:748.

MOORE, M. A. S., and J. J. T. OWEN. 1967. Experimental studies on the development of the thymus. *J. Exp. Med.* **126**:715.

WEISSMAN, I. L. 1967. Thymus cell migration. *J. Exp. Med.* **126**:291.

YAGUE, J., J. WHITE, C. COLECLOUGH, J. KAPPLER, E. PALMER, and P. MARRACK. 1985. The T cell receptor: the α and β chains define idiotype, and antigen and MHC specificity. *Cell* **42**:81.

MHC Genes

BILLINGHAM, R., and W. SILVERS. 1971. *The Immunobiology of Transplantation.* Prentice-Hall.

DOHERTY, P. C., and R. M. ZINKERNAGEL. 1975. H-2 compatibility is required for T-cell-mediated lysis of target cells infected with lymphocytic choriomeningitis virus. *J. Exp. Med.* **141**:502.

HOOD, L., M. STEINMETZ, and B. MALISSNE. 1983. Genes of the major histocompatibility complex of the mouse. *Ann. Rev. Immunol.* **1**:529.

KLEIN, J. *Biology of the Mouse Histocompatibility-2 Complex,* Springer-Verlag, 1975.

KLEIN, J. 1979. The major histocompatibility complex of the mouse. *Science* **203**:516.

SHACKELFORD, D. A., J. F. KAUFMAN, A. J. KORMAN, and J. L. STOMINGER, 1982. HLA-DR antigens: structure, separation of subpopulations, gene cloning and function. *Immunol. Rev.* **66**:133.

SNELL, G. D., J. DAUSSET, and S. NATHENSON. 1976. *Histocompatibility,* Academic Press.

SNELL, G. D. 1981. Studies in histocompatibility. *Science* **213**:172.

STEINMETZ, M., K. W. MOORE, J. G. FRELINGER, B. T. SHER, F. W. SHEN, E. A. BOYSE, and L. HOOD. 1981. A pseudogene homologous to mouse transplantation antigens: transplantation antigens are encoded by eight exons that correlate with protein domains. *Cell* **25**:683.

STEINMETZ, M., A. WINOTO, K. MINARD, and L. HOOD. 1982a. Clusters of genes encoding mouse transplantation antigens. *Cell* **28**:489.

STEINMETZ, M., K. MINARD, S. HORVATH, J. MCNICHOLAS, J. FRELINGER, C. WAKE, E. LONG, B. MACH, and L. HOOD. 1982b. A molecular map of the immune response region from the major histocompatibility complex of the mouse. *Nature* **300**:35.

YANAGI, Y., Y. YOSHIKAI, K. LEGGETT, S. CLARK, I. ALEKSANDER, and T. MAK. 1984. A human T cell-specific cDNA clone encodes proteins having extensive homology to immunoglobulin chains. *Nature* **308**:145.

25

Evolution of Cells

BIOLOGICAL explorations have two major themes. One is contemporary: how the molecules of cells contribute to cell function, how cells are organized into organisms, and how organisms grow and thrive in natural populations. The other theme is historical: how life first arose and then evolved into its present forms. Until now this book has been concerned with contemporary problems in molecular cell biology. In this chapter we shall consider the impact of the newer findings in molecular cell biology on evolutionary biology, particularly on ideas dealing with early cell evolution.

Evolution traditionally has been studied through the disciplines of paleontology and comparative anatomy—that is, through examining structural relationships between present-day organisms and comparing the distinctive characteristics of these organisms to those of ancient organisms. From this research we have a comprehensive and systematic view of the evolution of plant and animal life, beginning with the fossil record of animals and plants that lived some 600 million years ago. The central axiom derived from the classic work on evolution is that complex multicellular organisms must have evolved from simple organisms (Figure 25-1). The evolution of invertebrates to vertebrates, or of fishes to amphibians to reptiles to mammals, does seem to have followed a pathway in which simple organisms evolved to more complex ones. But when we consider molecular evolution and how

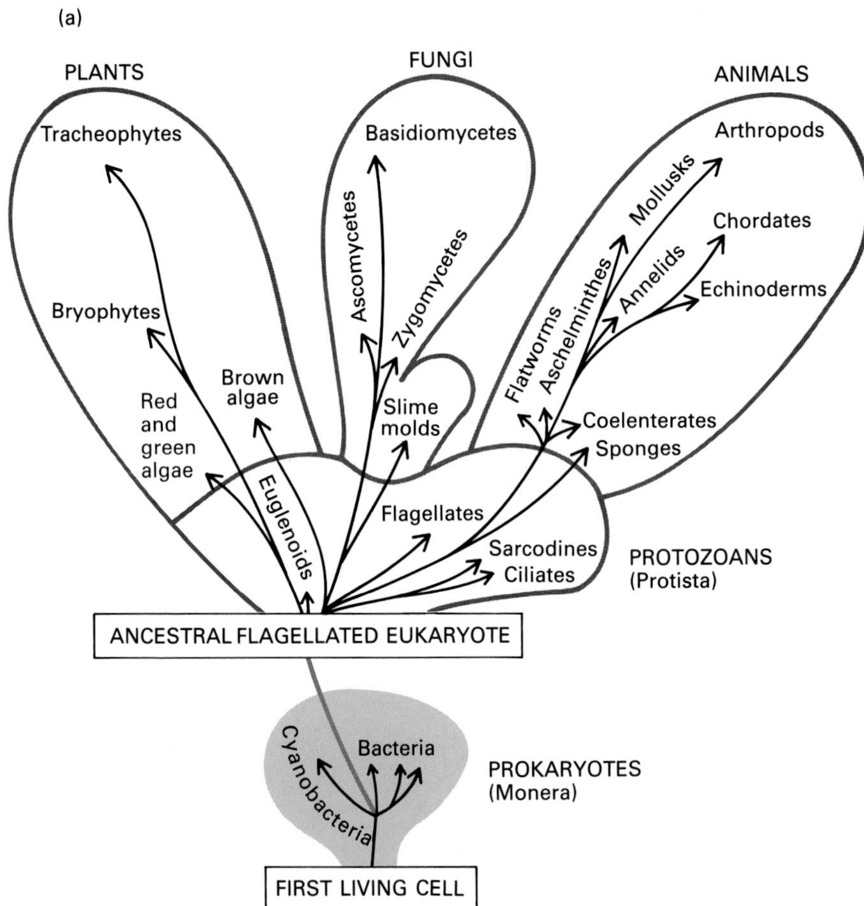

Figure 25-1 *(right and opposite)* (a) A phylogenetic chart showing the widely accepted five kingdoms: prokaryotes, protists (single-celled eukaryotes), and three multicellular kingdoms: plants, fungi, and animals. Note that at the single-cell level prokaryotes are characterized as precursors to eukaryotes. (See R. H. Whitaker, 1969, *Science* **163**:150.) (b) The same type of chart, but with emphasis on the "endosymbiotic hypothesis." According to this idea, cyanobacteria (blue-green algae) and aerobic bacteria (or some part of the bacterial genome) fused with the eukaryotic precursor cell to give rise to chloroplasts in plants and mitochondria in both plants and animals. In some form this hypothesis is widely accepted. Also, spirochete-like organisms are suggested as donors of information that resulted in cilia formation, and a heterotrophic prokaryotic anaerobe is suggested as the forerunner of the eukaryotic nucleus. These latter two proposals are controversial. [See L. Margulis, 1971, *Am. Scientist* **59**:231.]

cells arose in the first place, there are no generally accepted axioms.

When the realization came in the 1920s and the 1930s that all life forms have similar basic building blocks, J. B. S. Haldane in England and A. I. Oparin in Russia proposed that the common building blocks may have arisen by *prebiotic synthesis.* They suggested that the atmospheric and geologic conditions on primitive earth might have made possible the development of a sufficient variety and quantity of biologically important molecules to support "spontaneous" prebiotic evolution. Since the time of Haldane and Oparin, several experimenters have demonstrated how amino acids, nucleotides, and primitive polymers might have been spontaneously produced.

However, the area of greatest ignorance in evolution remains the origin of cells. The key reactions of molecular cell biology—those conferring the coding capacity of the nucleic acids and those involved in the translation of the code into protein and the replication of nucleic acids—must have arisen before the first true cell could exist. It seems unlikely, though, that *any* present-day cell is identical to the earliest cells or even very much like them. Although ancient remnants of early molecular evolution probably do exist in present-day cells, no ancient cells are available for study. The oldest forms that resemble cells in geologic samples are about the size and shape of cyanobacteria (blue-green algae), but the relationship of these microfossils (i.e., fossils observable by light- and electron-microscopic analysis of rock surfaces) to present-day single-celled organisms, either prokaryotic or eukaryotic, may never be settled conclusively.

The identity of the eukaryotic precursor cell seems to have been taken for granted by many biologists over the last several decades. Although no one has produced a good supporting argument, it has been widely assumed that present-day eukaryotes evolved from some type of prokaryote—probably because the present-day eukaryotes seem more complicated than prokaryotes. This assumption is in part almost certainly true, because eukaryotic chloroplasts and mitochondria do seem to have arisen from an ancient fusion between a prokaryote-like early organism (or at least some of the DNA from such an organism) and the precursor to nucleated eukaryotic cells. The fusion between a primitive prokaryote and the precursor to eukaryotic cells is proposed as the *endosymbiotic hypothesis;* we shall discuss this theory later in the chapter.

Despite the likelihood of some kind of prokaryotic-eukaryotic fusion, there is an enormous difference between today's prokaryotes and the nucleus of eukaryotes

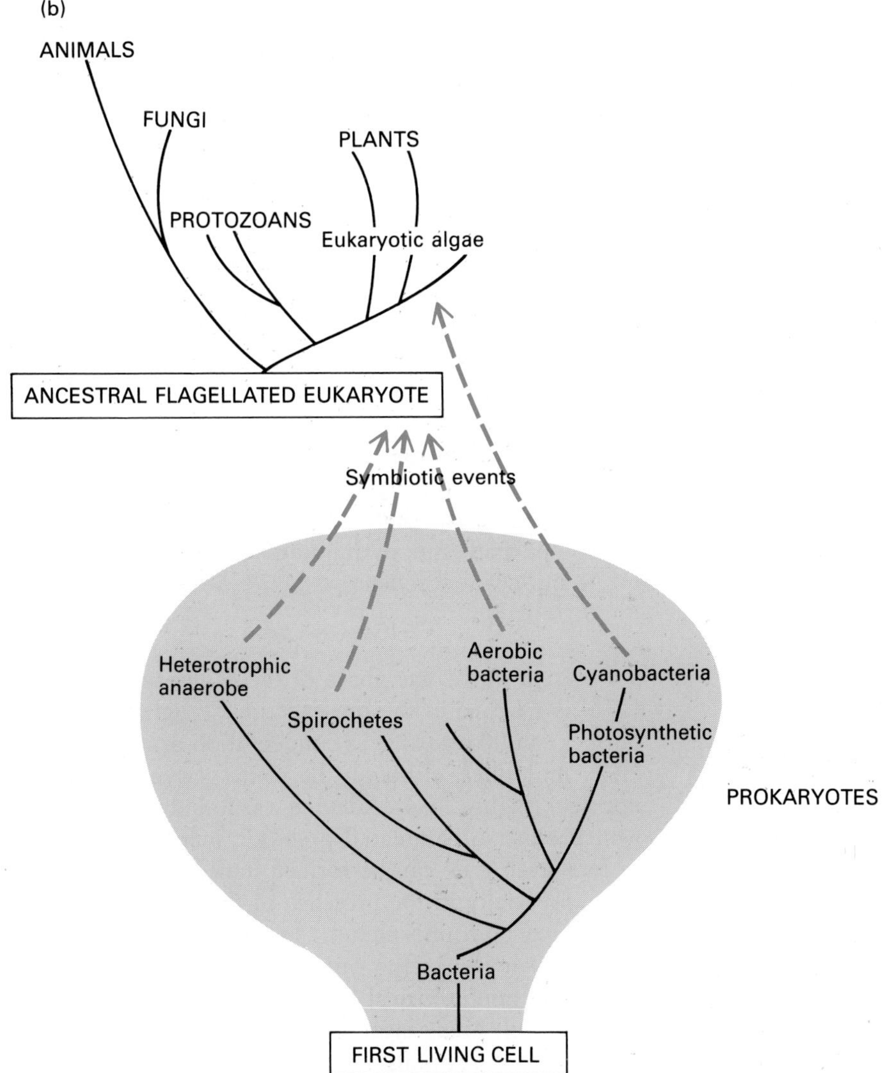

in the structure of their genes and chromosomes and in gene expression, as we have seen in Chapters 8 to 12. Among the important eukaryotic features that are not shared by prokaryotes are the large amount of "extra" (noncoding) DNA and the presence of intervening sequences—often very long ones—between protein-coding regions of genes. At present we do not know which of these very different genomic designs is most like that of primitive cells. The present-day single-celled organisms with the fewest genes and the smallest amount of DNA may have lost much of the DNA their ancestors possessed. No proof exists of a linear evolutionary connection between a primordial prokaryotic cell (one designed like today's prokaryotes) and eukaryotic cells, either those existing as unicellular organisms or those comprising multicellular organisms. As we shall see, the sequence data presently available argues against any such direct connection. One of the new twists contributed by modern molecular cell biology is the possibility that today's prokaryotes are very *highly evolved* cells whose evolution has made them the most flexible and efficient cells on the planet. In contrast, although no present-day representatives of the earliest cells remain, there may be vestiges of the earliest molecular processes (e.g., RNA splicing) in the cells of the most highly evolved organisms.

In this chapter, we shall first briefly discuss the current ideas and experiments dealing with prebiotic synthesis and then deal with the origin of cells. Some points concerning the evolution of DNA sequences and chromosome structure will also be discussed, but a general treatment of animal and plant evolution is beyond the scope of this book.

Prebiotic Synthesis

As the earth cooled and crusted over, an atmosphere was created from gases that formed beneath the surface and escaped through surface cracks. Quite likely water was delivered to the earth's surface as a gas and the oceans formed as a result of millions of years of torrential rains. The early atmosphere probably contained little or no oxygen; the present-day atmospheric oxygen on which present life depends was apparently produced from water by photosynthetic organisms that evolved later.

The nature of the atmosphere on primitive earth is a matter of controversy and conjecture. Hydrogen, ammonia, and methane were undoubtedly present, but they may not have existed free in the atmosphere in large quantities for extended times. Carbon dioxide and hydrogen sulfide were probably more permanent constituents of the early atmosphere. The sources of the energy that could produce chemical changes in the atmospheric gases (or in these gases dissolved in water) were mainly radiation from the sun and electric discharges; volcanic action discharged gases into the atmosphere and superheated hot springs also formed along rifts in the sea floor (Table 25-1).

Amino Acids and Nucleic Acid Bases Are Prominent Products of Prebiotic Conditions

In an attempt to synthesize organic compounds under possible prebiotic conditions, Stanley Miller, a student of Harold Urey, mixed hydrogen, methane, and ammonia (in the ratio 1:2:2) with water in a closed, evacuated reflux vessel (Figure 25-2). The gaseous mixture, which simulated the early atmosphere, was continuously exposed to electric discharges. When the water phase (the "ocean") was examined, more than 10 percent of the carbon from the methane was included in organic molecules such as amino acids. Glycine, alanine, aspartic acid, valine, and leucine (in both D- and L-forms) were clearly identified among the products. Hydrogen cyanide (HCN), aldehydes, and cyano compounds (—C≡N) were also present, and these could have been important intermediates in the formation of amino acids and nucleic acid bases (Figures 25-3 and 25-4).

Table 25-1 Present sources of energy (averaged over the earth's surface)

Source	Energy in cal/(cm^2·yr)
Total radiation from sun	2.6×10^5
Ultraviolet light with wavelengths of:	
300–400 nm	3.4×10^3
250–300 nm	5.6×10^2
200–250 nm	4.1×10^1
<150 nm	1.7
Electric discharges	4
Shock waves	1.1
Radioactivity (to 1.0 km depth)	8×10^{-1}
Volcanoes	1.3×10^{-1}
Cosmic rays	1.5×10^{-3}

SOURCE: L. E. Orgel, 1973, *The Origins of Life,* Wiley, p. 115.

Many different reaction conditions that mimic plausible prebiotic conditions can lead to the synthesis of amino acids and nucleic acid bases. Of particular interest are experiments showing that the simple prebiotic compounds (the carbon- and nitrogen-containing compounds mentioned above and H_2S) plus ultraviolet light as an energy source yields glycine, alanine, serine, glutamic acid, and asparagine. UV light is a much more abundant source of energy than lightning, which was simulated in the Miller experiments. Also, the success with mixtures of N_2, CO_2 or CO, and H_2 suggests that synthesis was possible even if methane and ammonia were scarce.

Both purines and pyrimidines can be formed when HCN and cyanoacetylene

$$HC{\equiv}C{-}C{\equiv}N$$

are present in the gas phase of the Miller-type reaction. HCN is reactive in the presence of UV light alone, and purines can be regarded as condensation products of five HCN molecules, although the route of synthesis is still not clear (Figure 25-4). Adenine was found to be a prominent product of simulated prebiotic synthesis reactions. In fact, even ATP can be formed under prebiotic conditions, especially in the presence of the common mineral apatite (calcium phosphate).

The earth is not the only place in the solar system where the nonbiological synthesis of organic molecules has occurred. Analyses of meteorites such as the large one that landed in Murchison, Australia in 1969 have shown these objects to contain both the free amino acids that are used in biological systems and other, nonbiological amino acids. Because amino acids have been found deep within meteorites they are not regarded as surface contaminants. The carbon- and nitrogen-containing compounds that are prominent reactants in proposed prebiotic synthesis reactions have also been identified in interstellar gas clouds: examples include H_2, HCN, cyanoacetylene, CO, H_2S, and NH_3, as well as larger molecules such as acetonitrile and acetaldehyde.

The presence of these organic substances in extraterrestrial objects suggests that the necessary building blocks for biologically important macromolecules arise readily in nature, and could have been formed on earth some 4 billion years ago through a nonbiological route. Because no free oxygen was present in the early atmosphere and cellular metabolism did not occur, these organic molecules would have been very stable. Consequently, the oceans may have accumulated relatively high concentrations of organic reactants sufficient to support further chemical activity. The primordial "organic soup" of Haldane and Oparin almost certainly existed.

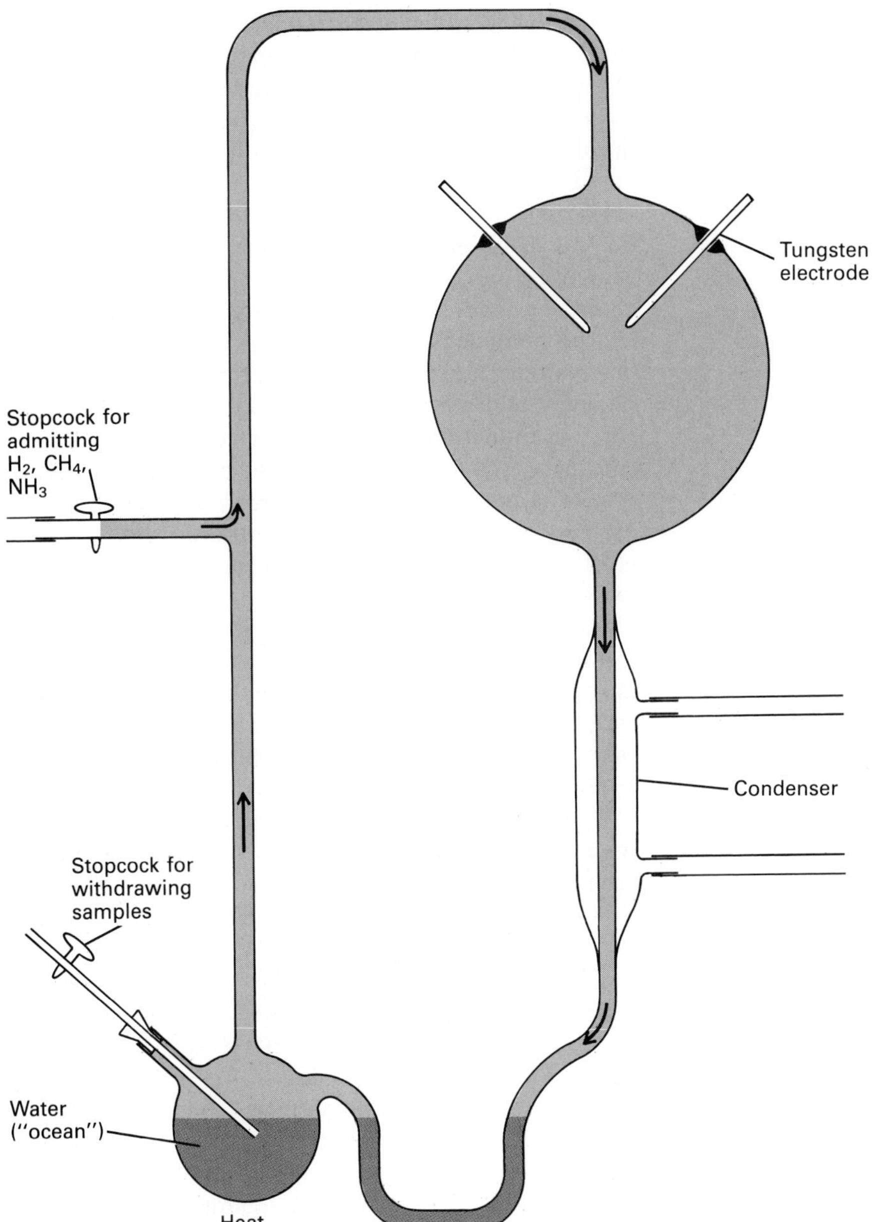

Figure 25-2 The apparatus used by Stanley Miller to simulate prebiotic organic synthesis.

Nonenzymatic Polymer Formation Can Occur

The linkage of monomeric units into polymers, which is essentially a dehydration reaction, can be demonstrated to occur at a slow rate under a number of different conditions that might have existed on primitive earth. Heating amino acid mixtures destroys some amino acids, but it also yields some random short peptide chains. Thus peptides might have been formed at temperatures that could have existed in evaporating pools of liquid or during volcanic action. If activated amino acids (aminoacyl derivatives; Chapter 4) are mixed in solution, slow spontaneous polymerization occurs. Adsorption of activated amino acids to claylike minerals greatly increases this reaction. Chains of polyalanine as long as 50 residues have been produced from alanyl adenylate. Of course, these hypothesized polymerizations raise the issue of how the amino acids may have been activated. At present we know of no efficient nonbiological mechanism that could have generated large amounts of such aminoacyl compounds.

Although all of the proposed conditions for nonbiological peptide formation raise some questions, the many successful reactions suggest convincingly that short peptides did arise during the prebiotic era. The sequences of such short peptides would presumably have reflected the

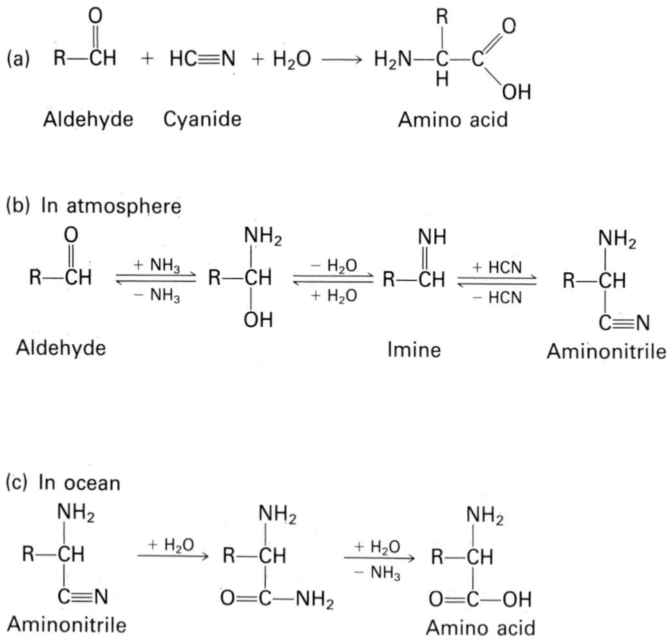

(a) Aldehyde + Cyanide + H_2O ⟶ Amino acid

$R{-}CHO + HC{\equiv}N + H_2O \longrightarrow H_2N{-}CHR{-}COOH$

(b) In atmosphere

Aldehyde $\underset{-NH_3}{\overset{+NH_3}{\rightleftharpoons}}$ $R{-}CH(NH_2)(OH)$ $\underset{+H_2O}{\overset{-H_2O}{\rightleftharpoons}}$ Imine $\underset{-HCN}{\overset{+HCN}{\rightleftharpoons}}$ Aminonitrile

(c) In ocean

Aminonitrile $\xrightarrow{+H_2O}$ $R{-}CH(NH_2){-}C({=}O){-}NH_2$ $\xrightarrow[-NH_3]{+H_2O}$ Amino acid

Figure 25-3 The overall reaction that forms amino acids from precursors in prebiotic experiments is shown in (a). The reaction probably follows the steps outlined in (b) and (c); these are collectively known as the Strecker synthesis. The first three steps in (b), which include the removal of water, could have taken place in the primitive atmosphere. If the aminonitrile then entered the ocean, hydrolysis could proceed. [See R. E. Dickerson, 1978, in *Evolution,* M. Kimura, ed., a Scientific American Book, p. 36.]

most abundant prebiotic amino acids, but no regular ordering of mixed polymers would have been possible.

Among the many unsolved problems pertaining to how random peptide assembly could have developed into ordered synthesis is the evolutionary choice of L- compared to D-amino acids. Random polymers would have contained either, of course, but only L-amino acids are found in proteins. Advantages in the more regular amino acid folding and packing may occur when all of the amino acids in proteins are either L- or D-. Also, some recent calculations show that peptides containing only L-amino acids have a slight advantage in stability over those containing only D-amino acids. Whether these slight differences are the evolutionary basis for L-amino acid choice is unknown.

In experiments mimicking prebiotic conditions, nucleotide synthesis does occur, although it proceeds more slowly than the formation of nucleic acid bases. Heating purines, pyrimidines, and ribonucleosides with ammonium chloride, urea, and inorganic phosphates or hydroxyapatite yields ribonucleotide monomers. In the presence of $MgNH_4PO_4{\cdot}6H_2O$, the mineral struvite, and urea, mononucleotides become di- and even triphosphates. Further, heating triphosphates with imidazole yields a stable activated nucleotide structure, adenosine 5′-phosphorimidazole. (Imidazole, a cyclic organic intermediate in the purine synthesis pathway, is a precursor to the histidine side chain and a possible condensation product of HCN; Figure 25-4.)

With imidazolide derivatives, the nonenzymatic formation of nucleotide polymers is readily demonstrable. Moreover, the most effective nonenzymatic polymer formation proceeds by primer extension of one chain that is hydrogen-bonded to a second chain according to the base-pairing rules of Watson and Crick (Table 25-2). This finding establishes that the presence of one polymer favors the formation of the complementary polymer. Specific metal ions may play an important role in the nonenzymatic synthesis of ribopolymers (Figure 25-5). In the presence of only Mg^{2+} the polymerization of poly G on a poly C template is slow, but when Zn^{2+} is added the formation of poly G on the poly C template is considerably stimulated. When Zn^{2+} is present the product is mostly 3′,5′-linked, as in RNA, whereas without Zn^{2+} a mixture of 3′,5′ and 2′,5′ linkages are found. Most DNA and RNA polymerases that have been studied require Mg^{2+} in solution during polymerization, and these enzymes have Zn^{2+} bound in the active enzyme.

We have no way of knowing whether these laboratory reactions were the reactions that actually took place on primitive earth to establish life in cellular form. These

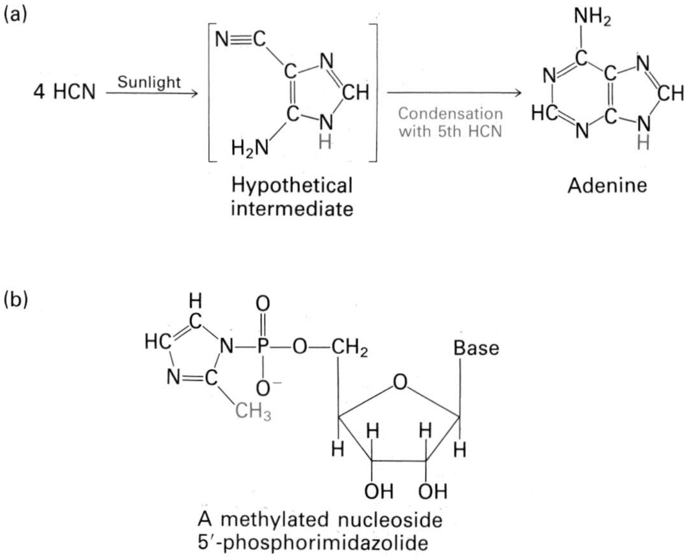

(a) 4 HCN $\xrightarrow{\text{Sunlight}}$ [Hypothetical intermediate] $\xrightarrow{\text{Condensation with 5th HCN}}$ Adenine

(b) A methylated nucleoside 5′-phosphorimidazolide

Figure 25-4 (a) The formation of adenine from hydrogen cyanide (HCN) can occur in sunlight, although the route of synthesis has not been completely worked out. When heated, HCN and ammonia yield adenine. [See S. L. Miller and L. E. Orgel, 1973, *The Origins of Life on Earth,* Prentice-Hall, p. 105.] (b) The structure of a methylated nucleoside 5′-phosphorimidazolide. Any of the four bases found in RNA can be attached to the sugar.

Table 25-2 Template-directed nonenzymatic synthesis of simple RNAs

	2-methylimidazolide*			
Template	G	C	A	U
None	<0.1	<0.1	<0.1	<0.1
Poly C	19 (96)	0.2	0.1	0.1
Poly CG (PC:G = 5:1)	13 (80)	2.3 (70)	0.1	0.1

*An activated nucleotide derivative (a methylated imidazolide) of each of the four bases (N = any base) was mixed with a poly C or a poly CG template or incubated alone. The relative amounts of each nucleotide incorporated are shown. (The numbers in parentheses indicate the percentages of added monomers incorporated.)

SOURCE: I. Inoue and L. E. Orgel, 1983, *Science* **219**:859.

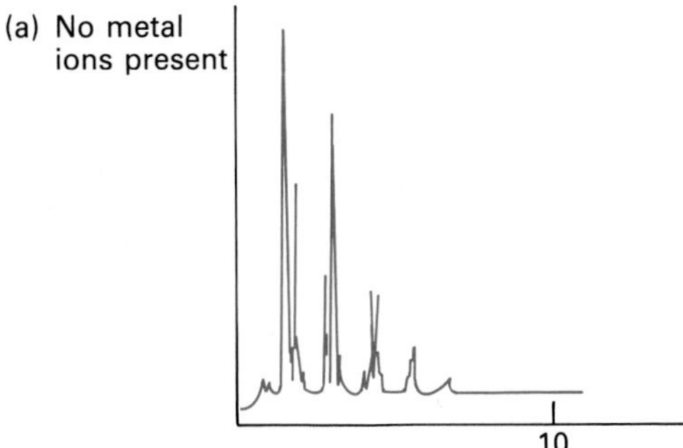

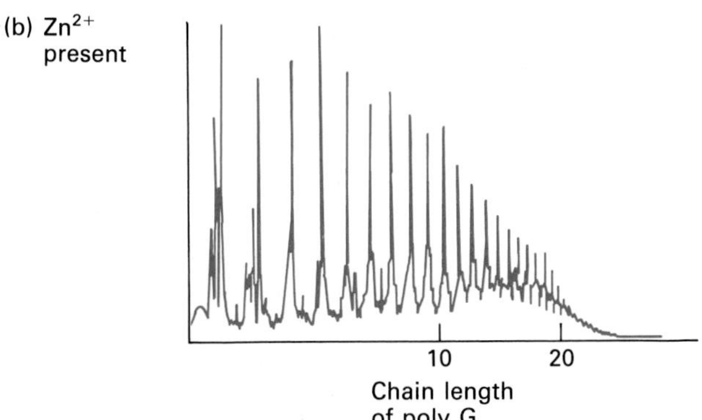

Figure 25-5 Chromatographic analyses of the products of nonenzymatic synthesis of poly G (detected by UV absorption after high-pressure liquid chromatography to separate oligonucleotides). (a) The products of guanosine 5′-phosphorimidazolide plus a poly C template in the absence of metal ions. (b) The products of the same reactants but with Zn^{2+} at 0.01 *M*. [See J. H. G. van Roode and L. E. Orgel, 1980, *J. Mol. Biol.* **144**:579.]

experiments do indicate that the chemical possibilities and sources of energy extant in the prebiotic era would almost certainly have formed amino acids and nucleotides as well as short polypeptides and oligonucleotides. The recent demonstration of RNA-RNA copying without enzymes has provided evidence that RNA could have arisen very early. In addition, as we mentioned in Chapter 9 and as we shall discuss more fully later in this chapter, protein-free site-specific RNA chain cleavage and RNA-RNA splicing have both been demonstrated. Thus quite a bit of RNA chemistry may have occurred prior to the development of proteins. However, the unsolved question, and the crux of the problem of precellular evolution, is how an *ordered* relationship between nucleic acids (RNA) and proteins began.

The Origins of the Genetic Code and the Translation Apparatus

During precellular evolution two different but coordinated problems had to be solved to enable nucleic acids to store information that could specify proteins. First, a correspondence had to be established between a linear order in one polymer and a linear order in the other—that is, a code had to develop; second, a means of translating the one linear order into the other had to be found. We know that in all cells the present-day three-letter nucleotide code in mRNA fulfills the first of these requirements and that the translation function is carried out by tRNA bound to the ribosome. However, the mechanism by which the nucleotide code "words" were chosen may always remain speculative, because there is no known chemical complementarity between the three nucleotides of a codon and its cognate amino acid.

Why Did a Code Arise?

The early systems for storing and recalling information were probably inexact; if so, they must have become more specialized and accurate as time went on. An additional assumption regarding precellular evolution is that the code developed coordinately with a translation mechanism. That is, as the nucleotide code words became recognized by the primitive translation mechanisms the utility of a particular codon increased, as presumably did the functionality of the translation apparatus. After biochemists discovered the code, the early authors on the origin of the code (Francis Crick, Carl Woese, and Leslie Orgel; see the references at the end of the chapter) discussed many of the demands that had to be met by the early evolving code and the translation apparatus. They all suggested that the code arose gradually in stages and not all at once. We shall review their ideas about how a code arose and add several thoughts about *why* a code arose.

The evolutionary driving force for the development of the code must have come from the utility of the translated product—the primitive protein. Random polypeptides would have consisted mostly of glycine and alanine because these were the most abundant early amino acids. Due to their simple composition such peptides may have presented regular structures fairly often, so that "useful" oligopeptide stretches could have existed. "Useful" for what? Perhaps random polypeptides carried out primitive functions such as pH buffering or binding a particular metal ion or binding another amino acid or nucleotide. Without such possibly useful chemical properties of polypeptides it is hard to imagine why a particular nucleic acid code specifying protein products would have become established.

The darwinian principle of natural selection has been applied directly to the earliest nucleic acid–peptide interactions. According to this view, if a particular peptide aided, however slightly, in the synthesis of an oligonucleotide with a specific nonrandom sequence, then the cycle of interdependence would have begun: the accumulation of the specific peptide would have led to the accumulation of more of the specific oligonucleotide. If a translation system (i.e., the tRNA function or something like it) arose to assure the constancy of specific sequences, the cycle of interdependence would have been established.

First position (5′ end)	Second position U	C	A	G	Third position (3′ end)
U	Phe	Ser	Tyr	Cys	U
	Phe	Ser	Tyr	Cys	C
	Leu	Ser	TERM	TERM	A
	Leu	Ser	TERM	Trp	G
C	Leu	Pro	His	Arg	U
	Leu	Pro	His	Arg	C
	Leu	Pro	Gln	Arg	A
	Leu	Pro	Gln	Arg	G
A	Ile	Thr	Asn	Ser	U
	Ile	Thr	Asn	Ser	C
	Ile	Thr	Lys	Arg	A
	Met	Thr	Lys	Arg	G
G	Val	Ala	Asp	Gly	U
	Val	Ala	Asp	Gly	C
	Val	Ala	Glu	Gly	A
	Val (Met)	Ala	Glu	Gly	G

Figure 25-6 The genetic code, with some codons of possible evolutionary significance highlighted. The frequency with which the first two positions (e.g., CU or CC) encode a single amino acid suggests that only two of the three nucleotides may have been used early in evolution. In addition, the codons for the long-chain aliphatic amino acids (Leu, Ile, and Val) have obvious similarities, as do those for the charged basic (Arg and Lys) and acidic (Asp and Glu) amino acids.

Did The Code Use Three Bases from the Outset?

Could the first code have been simpler than the present-day code—that is, could it have used either one or two bases for one amino acid? Such a combination seems unlikely because the switch to the three-base code at a later time would have invalidated the meaning of all previously encoded information. Possibly, however, only every third nucleotide in the primitive polymers was used for coding, or that two of every three were used. The use of all three nucleotides could have developed later.

Does the present-day assignment of code words have any implications for the origin of the code? With 64 code words and 20 amino acids eventually to be chosen, a random assignment would have resulted in amino acids being coded by multiple *unrelated* code words. A random codon assignment simply could not have occurred. For example, the codon XYN (where X and Y are fixed bases but N is any of the four bases) frequently codes for a single amino acid. This "works" for glycine, alanine, valine, proline, and threonine (i.e., the codons for each of these amino acids have the same base in each of the first two positions). Four of the six codons each for serine and leucine also fit the formula (Figure 25-6). When N is restricted to U or C, the codon XYN encodes a single amino acid in every single case. Thus it appears that the initial two "letters" of the code may have been decided first.

Another way in which the code is nonrandom is that chemically similar amino acids have related codons (Figure 25-6). This suggests that the recognition of classes of amino acids may have preceded the recognition of individual amino acids. A final argument for the nonrandom assignment of the code is that the charged amino acids, both the basic ones (lysine, arginine) and the acidic ones (glutamic acid and aspartic acid) use codons rich in A's and G's.

Very likely the system of codon use developed gradually with increasingly greater shades of meaning. For example, over time the coding system must have become able to distinguish and use not just a single hydrophobic amino acid (e.g. valine) but any one of several (e.g. valine, leucine, or isoleucine). As a result, small but important differences in the properties of proteins could be introduced.

Some interesting speculation concerns how the early codons might have been chosen and how these codons might have been recognized by a primitive adapter (primitive "tRNA") and translated in the absence of ribosomes (Figure 25-7). A major difficulty of preribosomal translation would have been the instability of the primitive adapter-mRNA complex because of the short region of base pairing. To overcome this problem, the first adapter

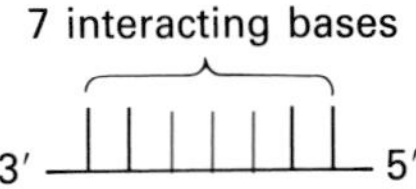

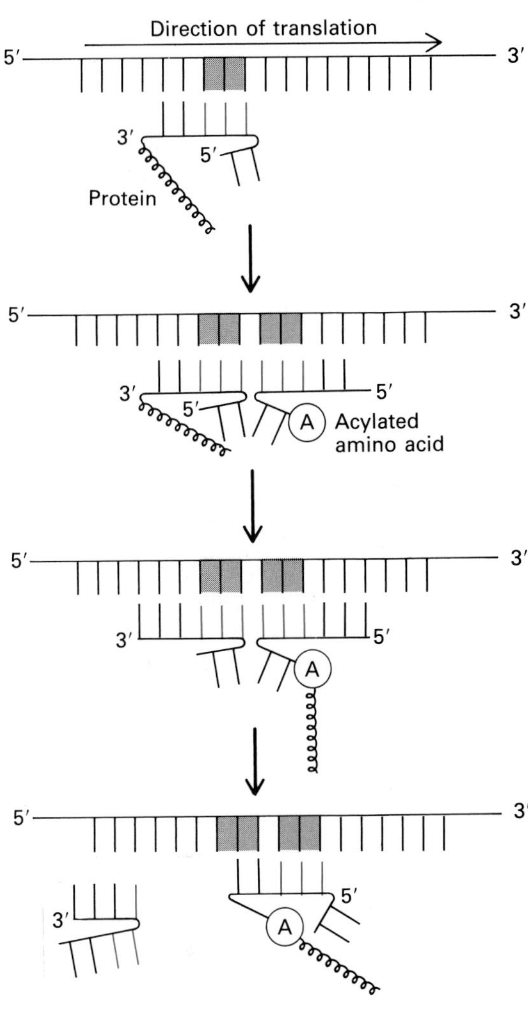

Figure 25-7 A hypothetical mechanism of protein synthesis carried out by the binding of a primitive seven-nucleotide adapter or "tRNA." (a) The adapter has three central bases *(color)* that provide the anticodon function. (b) Two different folding patterns are possible for an adapter bound to an mRNA. The adapter with the protein chain has two of its 5′ bases unpaired when a neighboring amino acyl adapter is bound. The two 3′ bases of the aminoacylated adapter are unbound. The transfer of the peptidyl chain to the acylated amino acid on the adapter on the right causes changes in the folding of the adapters. Consequent changes in base pairing and release of the adapter without an amino acid promotes movement down the mRNA chain. [See F. H. C. Crick et al., 1976, *Origins of Life* 7:389.]

may have been seven bases long; this would have enabled it to assume either of two shapes in the binding of the central "anticodon" portion to the mRNA. Three bases of the adapter would have served as the anticodon; the four remaining bases would have flanked the anticodon, two on each side. If the flanking bases, on one side or the other, paired with the primitive mRNA, an adapter might have remained bound to the mRNA long enough for a growing peptide attached to it to be linked to the amino acid on an incoming adapter. As the original adapter delivered its peptide chain the shape of the adapter would have changed, freeing its associated mRNA codon for base-pairing interaction with the incoming tRNA (Figure 25-7). An even simpler original sequence can be suggested for the original adapter. First, in order that the primordial adapter and mRNA pair for five bases instead of three, all stable pairing would be allowed, including G-U (a wobble configuration: Chapter 4). The first general anticodon structure might have been

(3′)UGNNNUU(5′)

(where N = any base; R = purine; Y = pyrimidine) to maximize the participation of G and U. A simple repetitive mRNA could have accommodated such five-base pairing: an mRNA of the sequence

$\underline{\text{RRY}}\ \underline{\text{RRY}}\ \underline{\text{RRY}}\ \underline{\text{RRY}}$

would have allowed the requisite purine-pyrimidine pairs

if the anticodon of the UGNNNUU sequence were (3′)YYR(5′):

mRNA	(5′)RRY RRY RRY(3′)
Adapter or tRNA	(3′) UG YYR UU (5′)

The mRNA would have been read 5′ → 3′; the two U's at the 5′ side of the tRNA would have become unpaired after the donation of the amino acid because of the change in tRNA conformation. Such a mechanism would have allowed the translation of three bases at a time, after which the next three bases (here, the 3′RRY) would have become available; that is, the code would have been a commaless triplet (see Figure 4-4).

Is the development of such a complementarity entirely hypothetical? The anticodon of present-day tRNAs has the general structure

(3′)NR$\underline{\text{NNN}}$UY(5′)

so the proposal is logical.

In the absence of any evidence for a direct chemical relationship between codons and particular amino acids, we can only assume that a mechanism such as that discussed above and illustrated in Figure 25-7 was responsible both for early codon choices and for sufficient translation to allow at least elementary functions. The necessary elements in this early choice must have been: (1) a simple sequence polynucleotide that could be read three nucleotides at a time by an adapter simple enough to have arisen spontaneously, (2) the adapter, which must have had two of the key elements of present day tRNAs—the base-pairing "translation" function and the ability to become attached to an amino acid, and (3) product polypeptides that somehow facilitated the overall reaction.

Perfecting the Translation System Required More Complicated Structures in rRNA

A detailed theory of evolution that would explain how the primitive oligonucleotide-oligopeptide interactions developed into a working translation system is entirely beyond the limits of present knowledge. We can only summarize and comment on the features that were acquired.

First, tRNAs came to be recognized in two different ways: (1) the anticodon with its encoded information was recognized by base pairing, and (2) each tRNA was recognized by activating enzymes that loaded the tRNA with the correct amino acid. Another essential acquisition was the ability of each tRNA to bind to ribosomes. Because all present-day cells contain ribosomes of a consistent design and function, it is tempting to assume that the precellular translation machinery included ribosomes with two subunits plus a 5S rRNA. Of course, present-day ribosomes contain more than 50 individual proteins, and the beginning translation system could hardly have made 50 precise long polypeptides. However, the earliest surfaces that allowed an interaction of the mRNA and the activated tRNA molecules may well have been other RNA surfaces. Two aspects of present-day ribosomal structure support this idea.

First, the 5′ ends, protein initiation regions, of the mRNAs of some bacteriophages and some mammalian viruses (e.g., adenovirus and reovirus) have a region of base homology with the 3′ end of the smaller rRNA (Figure 25-8). Therefore some mRNAs and rRNA do form a regular base-paired association.

Second, the nucleotide sequences of the 16-18S rRNAs from a variety of cells and the 23-28S rRNAs from a few cells have been analyzed. Although the base sequence of rRNA has clearly undergone many changes (as we shall discuss later in this chapter), certain of its overall properties have remained remarkably constant over billions of years. The precise two- and three-dimensional structures formed by the folding of the approximately 1500 to 1800 nucleotides in the small rRNA are not known with certainty. But if the molecule is folded into the two-dimensional stem-loop structure dictated by the strongest possible base-paired interactions, a structure with characteristic features emerges. When the same folding procedure is carried out with the primary sequences of small rRNAs from a number of different sources, a remarkably constant structure results (Figure 25-9).

The conservation of such a complicated two-dimensional structure *without the conservation of the base sequence* argues strongly for the importance of the structure in all living cells and implies that the structure was already present in the earliest cell. The preservation of secondary (and presumably tertiary) sequence conservation in the rRNA may be related to the binding of specific ribosomal proteins and the completion of certain steps in peptide synthesis. However, the interaction of mRNA with an rRNA forerunner that was devoid of proteins but that possessed a specific, constant secondary structure might well have been an early evolutionary development in the translation machinery.

A Reconstructive Analysis of Cell Lineages

Before presenting some results derived from studies of molecular "comparative anatomy," we shall briefly review why a more direct determination of the nature of the earlier cells is not possible. Fossils discovered in sedimentary rocks in Australia and South Africa have been identified as single spherical cells about 3 billion years old (Figures 25-10 and 25-11). The basis for identifying the imprints in these fossils as those of microorganisms is

Viral product		Viral mRNA sequence	
QB A protein	5′	——A G U A U A A G A G G A C A U A U G——	3′
R17 A protein		——U A G G A G G U U U G A C C U A U G——	
QB coat protein		——U G G G U C A A U U U G A U C A U G——	
R17 coat protein		——A C C G G A G U U U G A A G C A U G——	
QB replicase		——C U A A G G A U G A A A U G C A U G——	
R17 replicase		——A C A U G A G G A U U A C C C A U G——	

Figure 25-8 The complementary interaction between the 3′ end of the *Escherichia coli* 16S rRNA and the untranslated sequences adjacent to the 5′ ends of bacteriophage mRNA coding regions. (The mRNAs are translated by *E. coli* ribosomes.) The mRNA regions that are complementary to the 3′ segment of 16S rRNA are shown in color type. [See J. Shine and L. Dalgarno, 1974, *Proc. Nat'l Acad. Sci. USA* 71:1342.]

morphological. The sizes and shapes of the structures in microfossils are similar to those of many present-day cyanobacteria (Figure 25-11). Many of the structures are doublets, which apparently indicates that the fossilized cells were in the process of division. In addition, the ancient rock formations are very much like present-day forms consisting of sediments and precipitates around colonies of cyanobacteria and other bacterial species.

These ancient fossilized forms are called *stromatolites*. The current estimate of the earth's age is about 4.5 billion years; if the stromatolites were in fact once living cells, they must have been among the oldest cells. Unfortunately, there is little we can learn from a microfossil once we have measured its size and the thickness of its cell wall. It is unlikely that we will ever be able to compare the protein structures or gene structures of these presumed early cells with their counterparts in present-day cells. Comparison of the nucleic acid sequences of different present-day organisms is the most promising approach to discovering relationships that might have existed at the beginning of evolution.

Comparative Studies of tRNA and rRNA Structures Suggest Three Cell Lineages

In prokaryotic and eukaryotic tRNAs certain landmark features are similar (e.g., the TΨCG loop), as are the overall stem-loop structures of the tRNA molecules. In addition, many prokaryotic tRNAs (e.g., those of *Escherichia coli* and *Bacillus subtilis*) that are specific to the same amino acid show great sequence similarity across different species. Likewise, the sequences of many eukaryotic tRNAs for the same amino acid are similar in yeast, invertebrates, and mammals. However, there is considerable sequence variation between tRNAs for the same amino acid in prokaryotes and eukaryotes. Even in the case of the initiator methionyl-tRNA, which has a similar key role in all cell types, the prokaryotic and eukaryotic sequences are distinct (Table 25-3).

Because the rRNAs are longer molecules than the tRNAs, ancient genealogical similarities or separations can be more effectively traced through comparisons of rRNA sequences. Complete rRNA sequences are available in too few cases to allow wide-ranging primary sequence comparisons, but sequence relatedness has been extensively analyzed by comparing the sequences of large

Table 25-3 Comparisons of initiator methionyl-tRNA sequences of prokaryotes and eukaryotes*

Organism	(1)	(2)	(3)	(4)	(5)	(6)
	Number of base differences					
(1) *Bacillus subtilis*		6	5	9	25	25
(2) *Escherichia coli*	8		3	8	23	24
(3) *Thermus thermophilus*	7	4		8	26	24
(4) *Anacystis nidulans*	12	11	11		27	27
(5) Baker's yeast	34	31	35	36		18
(6) Mouse, sheep, rabbit	34	32	32	36	25	
	Percentage of bases differing					

*Sequences from prokaryotic species (1 to 4) and eukaryotic species (5 and 6) are compared to show both the number of base differences and the percentage of the total represented by that number. For example, *B. subtilis* and *E. coli* differ in six positions or in 8 percent of their methionyl-tRNA sequences.

SOURCE: M. Dayhoff, ed., 1978, *Atlas of Protein Sequence and Structure*, National Biomedical Research Foundation (Washington, D.C.), p. 314.

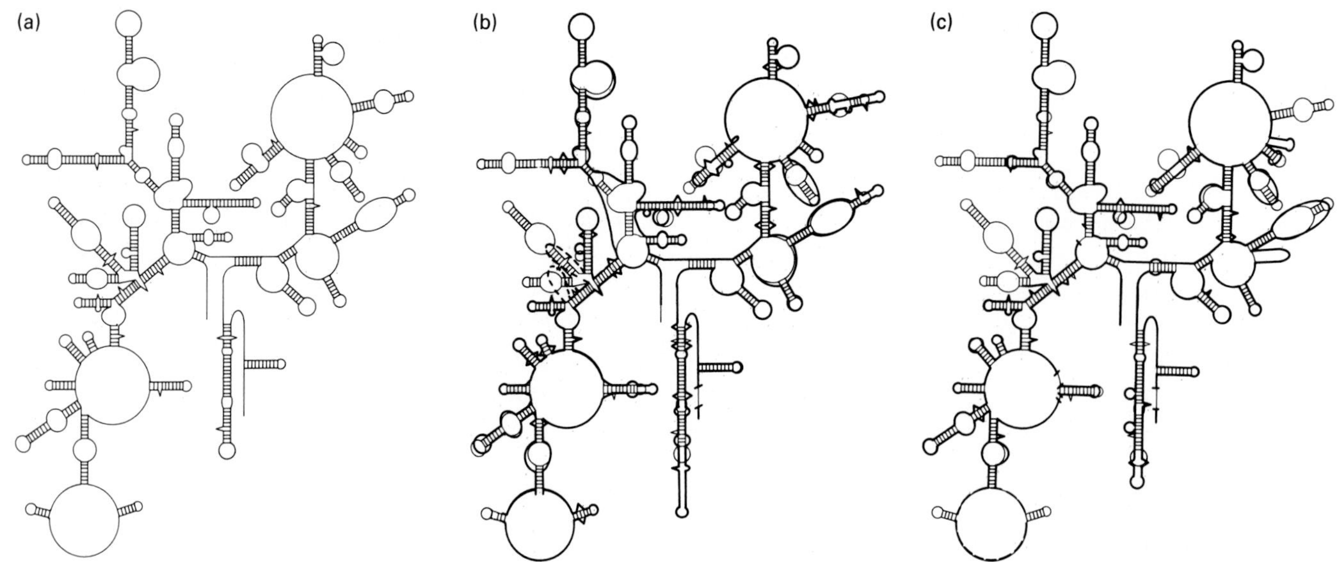

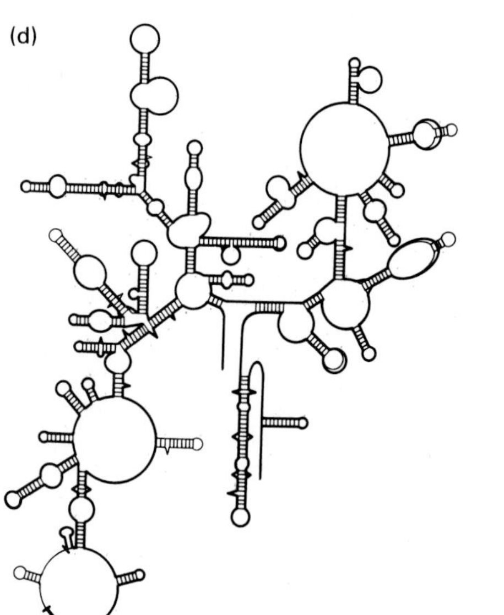

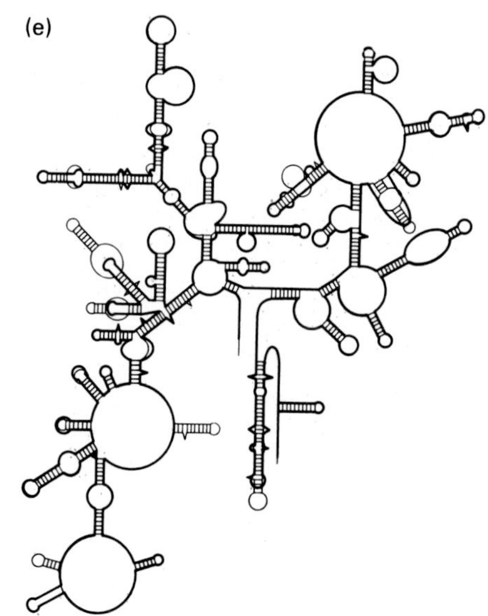

Figure 25-9 Structure conservation in the small rRNA. (a) This model for the secondary structure (i.e., the stems and loops) of the *E. coli* 16S rRNA was generated from the primary sequence data. The rules for forming the base-paired structure were to start with the 5′ end (the first to be synthesized), to search for the nearest partners that could form a stem, and then to make corrections where alternative arrangements would contribute greater stability. The folded structures of other small rRNA sequences are shown for comparison. They include (b) frog *(Xenopus laevis)* cytoplasmic 18S rRNA (encoded in the nucleus) and similar sequences from (c) a mitochondrion, (d) a chloroplast, and (e) an archaebacterium (a member of a group of bacteria thought to be evolutionarily distinct from the group that includes *E. coli*). In each case the structure being compared with *E. coli* 16S rRNA is shown as a dark line overlying the lighter line that represents the *E. coli* structure. [The dashed region in (b) represents loops that are not seen in *E. coli*.] Strong conservation of the stem-loop pattern in *all* of the small rRNAs is observed. [See H. F. Noller and C. R. Woese, 1981, *Science* **212**:403.] *Courtesy of H. F. Noller; from C. R. Woese, R. R. Gutell, R. Gupta, and H. F. Noller, 1983,* Microbiol. Rev. *47:621.*

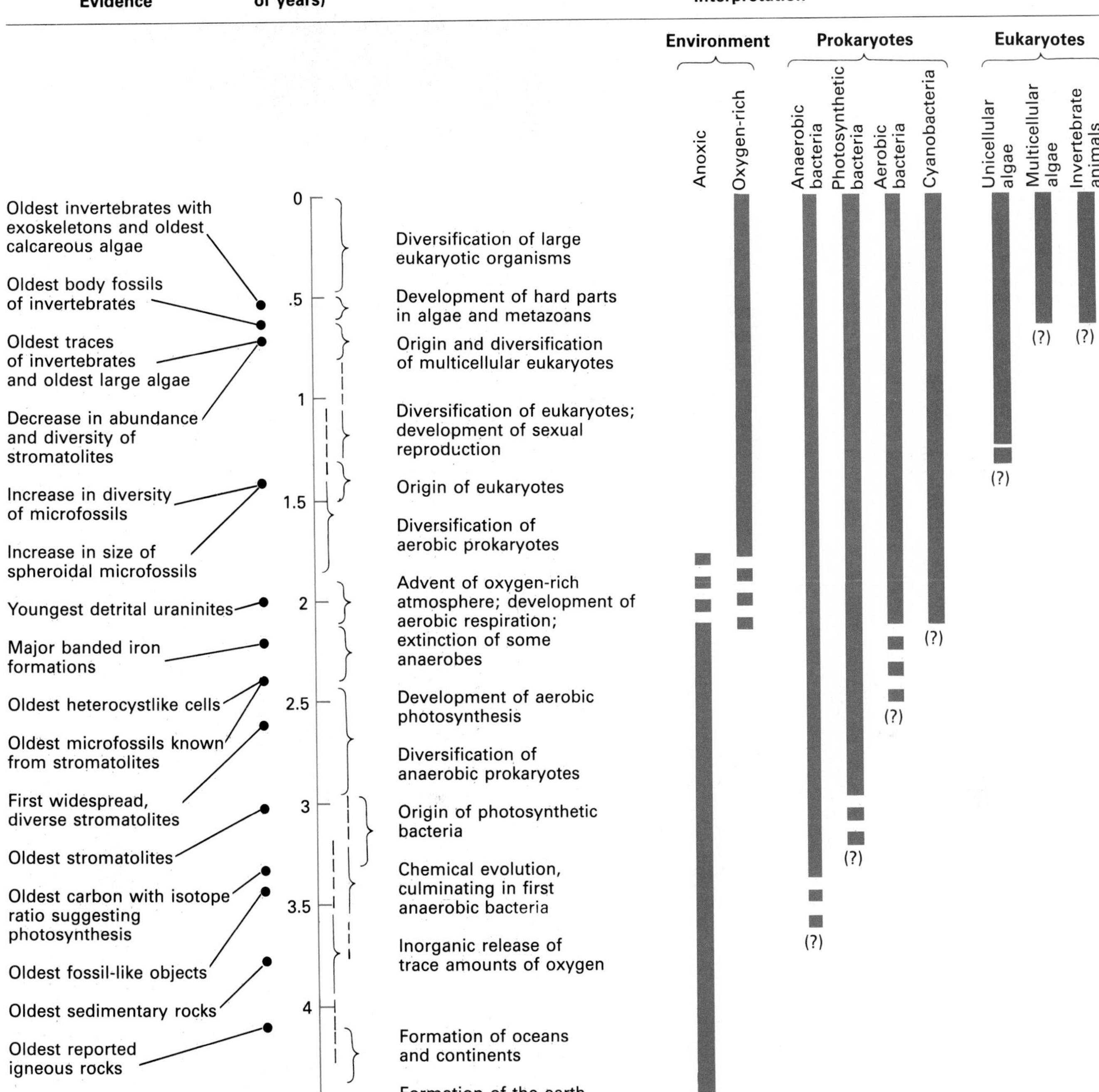

Figure 25-10 A possible time schedule for the evolution of living organisms. The fossil record of higher plants and animals dates from approximately 600 to 700 million years ago. Note that the fossil record covers only about 20 to 25 percent of evolutionary time. Much of the remainder of the chart reports conclusions drawn from microfossils preserved in ancient sedimentary rock. The chart shows important geologic events as well as proposed biological events. As in Figure 25-1, an evolutionary progression from prokaryotes to eukaryotes is strongly implied. The dashes (or square "dots") in some of the time lines and braces indicate the extent of doubt about when the indicated events occurred; the question marks at the beginnings of the prokaryotic and eukaryotic time lines reflect the uncertainty about the nature of the earliest events in cellular evolution. [See J. W. Schopf, 1978, in *Evolution*, M. Kimura, ed., a Scientific American Book, p. 48.]

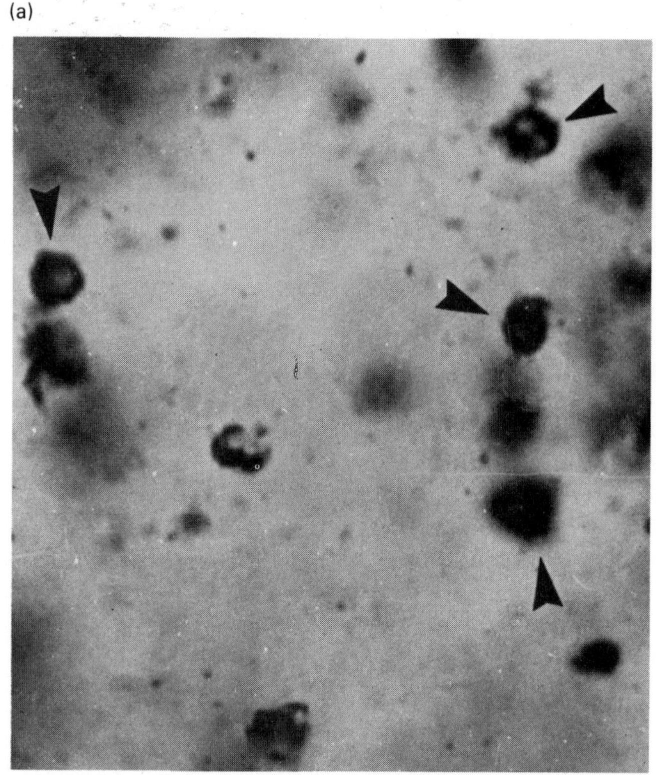

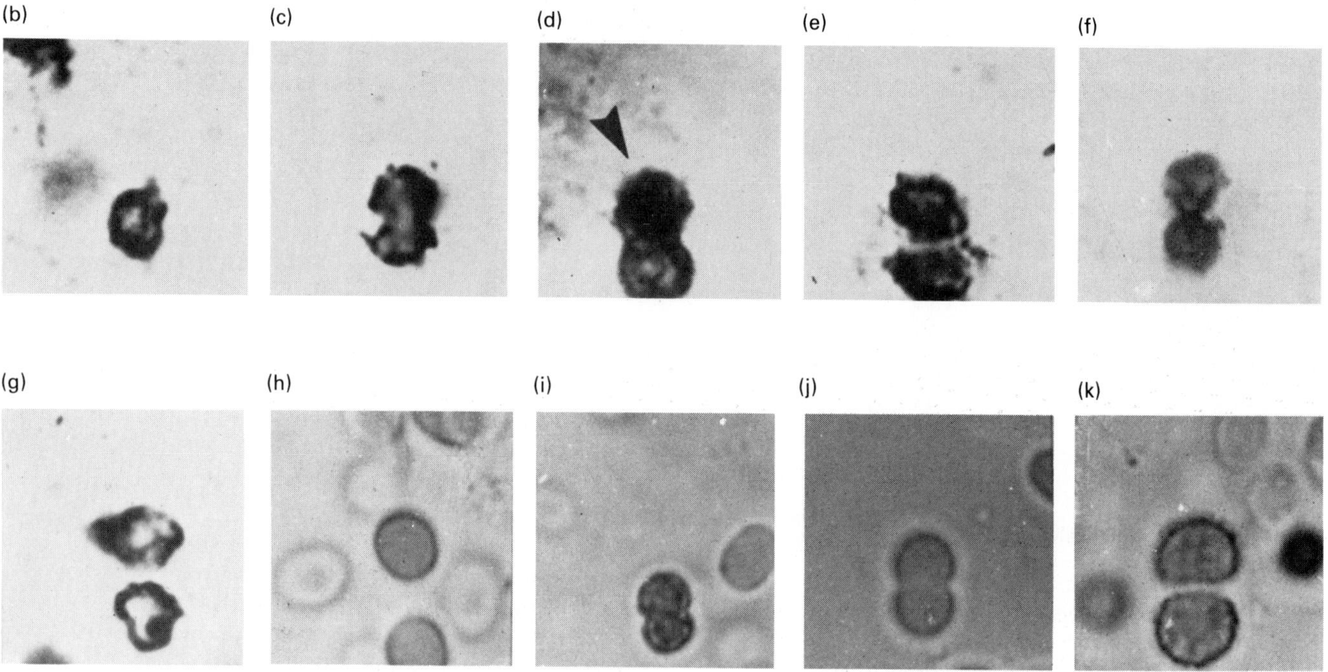

Figure 25-11 Photographs of microfossils. Carbon-containing layered rock (3 billion years old) from outcroppings in Swaziland, South Africa contains structures that resemble cells. Thin sections were cut and viewed microscopically by transmitted light (as here) or examined by surface replica techniques in the electron microscope. Spheroidal structures with a regular size distribution similar to that of cyanobacteria (about 2 μm in diameter) were readily observed. Photographs (a) and (b) show single individuals; (d) to (g) show doublets, cells that appear to be in the process of division. For comparison, (h) to (k) are micrographs of a dividing culture of the alga *Aphanocapsa spheroides* at the same magnification. [See A. H. Knoll and E. S. Barghoorn, 1977, *Science* **198**:396.] *Courtesy of E. S. Barghoorn.*

oligonucleotides within the small (16S or 18S) rRNA. The first step in these comparisons is to digest the rRNA with the endoribonuclease T1, which cuts on the 3′ side of each G residue. The resulting large oligonucleotides can then be separated and sequenced, and the sequences of oligonucleotides from different species can be compared (Table 25-4).

From this type of comparison in more than 200 species, two distinctly nonoverlapping classes of bacterial organisms have been discerned, both of which are separate from eukaryotes. The vast majority of the commonly encountered bacteria that have been studied for the last 100 years—popular laboratory organisms, soil organisms, and disease-causing organisms—have similar rRNA sequences, and are now referred to as *eubacteria* (*eu-* is from the Greek for "normal"). Although all of the eubacterial species are related by rRNA sequence analysis, the group can be divided into 8 to 10 eubacterial subgroups that have minimal intragroup variation. By comparing the extent of similarity in the rRNA structures, a phylogenetic tree called a *dendrogram* can be established for these organisms (Figure 25-12).

The division of the eubacteria on the basis of rRNA relatedness agrees well with the traditional method of organizing the bacterial species on the basis of metabolic capabilities. One important aspect of the general scheme of evolution is nicely corroborated by this arrangement. Because the first organisms were anaerobic, the sequence variation within the anaerobic group of eubacteria and between the anaerobes and other eubacteria should be the greatest; this is indeed the case. For example, the *Clostridia* are a group of clearly related anaerobes found mostly in soil. They are responsible for a number of human diseases, including tetanus and gas gangrene. The *Clostridia* have the widest intragroup sequence variation, which indicates that they represent an ancient group.

Another interesting observation about the eubacteria is that photosynthetic capacity is found in several of the eubacterial groups. This finding is consistent with the possibility that this critically important function may

Table 25-4 Comparisons of rRNA oligonucleotide sequences of 16 species

Organism	S_{AB}*															
	(1)	(2)	(3)	(4)	(5)	(6)	(7)	(8)	(9)	(10)	(11)	(12)	(13)	(14)	(15)	(16)
EUKARYOTES																
(1) *Saccharomyces cerevisiae* (yeast)	—	.29	.33	.05	.06	.08	.09	.11	.08	.11	.11	.08	.08	.10	.07	.08
(2) *Lemna minor* (duckweed)	.29	—	.36	.10	.05	.06	.10	.09	.11	.10	.10	.13	.07	.09	.07	.09
(3) *L* cell (mouse)	.33	.36	—	.06	.06	.07	.07	.09	.06	.10	.10	.09	.07	.11	.06	.07
EUBACTERIA																
(4) *Escherichia coli*	.05	.10	.06	—	.24	.25	.28	.26	.21	.11	.12	.07	.12	.07	.07	.09
(5) *Chlorobium vibrioforme*	.06	.05	.06	.24	—	.22	.22	.20	.19	.06	.07	.06	.09	.07	.05	.07
(6) *Bacillus firmus*	.08	.06	.07	.25	.22	—	.34	.26	.20	.11	.13	.06	.12	.10	.07	.09
(7) *Corynebacterium diphtheriae*	.09	.10	.07	.28	.22	.34	—	.23	.21	.12	.12	.09	.10	.10	.06	.09
(8) *Aphanocapsa*	.11	.09	.09	.26	.20	.26	.23	—	.31	.11	.11	.10	.10	.13	.10	.10
(9) Chloroplast *(Lemna)*	.08	.11	.06	.21	.19	.20	.21	.31	—	.14	.12	.10	.12	.12	.06	.07
ARCHAEBACTERIA																
(10) *Methanobacterium thermoautotrophicum*	.11	.10	.10	.11	.06	.11	.12	.11	.14	—	.51	.25	.30	.34	.17	.19
(11) *Methanobrevibacter ruminantium*	.11	.10	.10	.12	.07	.13	.12	.11	.12	.51	—	.25	.24	.31	.15	.20
(12) *Methanogenium cariaci*	.08	.13	.09	.07	.06	.06	.09	.10	.10	.25	.25	—	.32	.29	.13	.21
(13) *Methanosarcina barkeri*	.08	.07	.07	.12	.09	.12	.10	.10	.12	.30	.24	.32	—	.28	.16	.23
(14) *Halobacterium halobium*	.10	.09	.11	.07	.07	.10	.10	.13	.12	.34	.31	.29	.28	—	.19	.23
(15) *Sulfolobus acidocaldarius*	.07	.07	.06	.07	.05	.07	.06	.10	.06	.17	.15	.13	.16	.19	—	.13
(16) *Thermoplasma acidophilum*	.08	.09	.07	.09	.07	.09	.09	.10	.07	.19	.20	.21	.23	.23	.13	—

*The relatedness or similarity of two rRNAs is calculated as the S_{AB}. To obtain this number, samples of rRNA from two organisms, A and B, were digested into oligonucleotides ending in G. All oligonucleotides six units or longer were identified and sequenced, and the resulting lists of oligonucleotide sequences were compared. The total number of nucleotides in the common oligonucleotides of the two species times 2 was divided by the sum of the total number of nucleotides in all of the oligonucleotides of both species. If the two species were identical, an S_{AB} of 1.0 would result; an S_{AB} of less than 0.1 means little similarity, but an S_{AB} of 0.2 to 0.3 shows considerable similarity. Entries 15 and 16 represent the thermoacidophiles and are a subdivision of the archaebacteria. Note that their S_{AB}s are closer to the major group of archaebacteria than any other group. The color areas highlight the high S_{AB} scores obtained within groups (in comparison to the low scores obtained between groups).

SOURCE: C. R. Woese, 1981, *Sci. Am.* **244**(6):98.

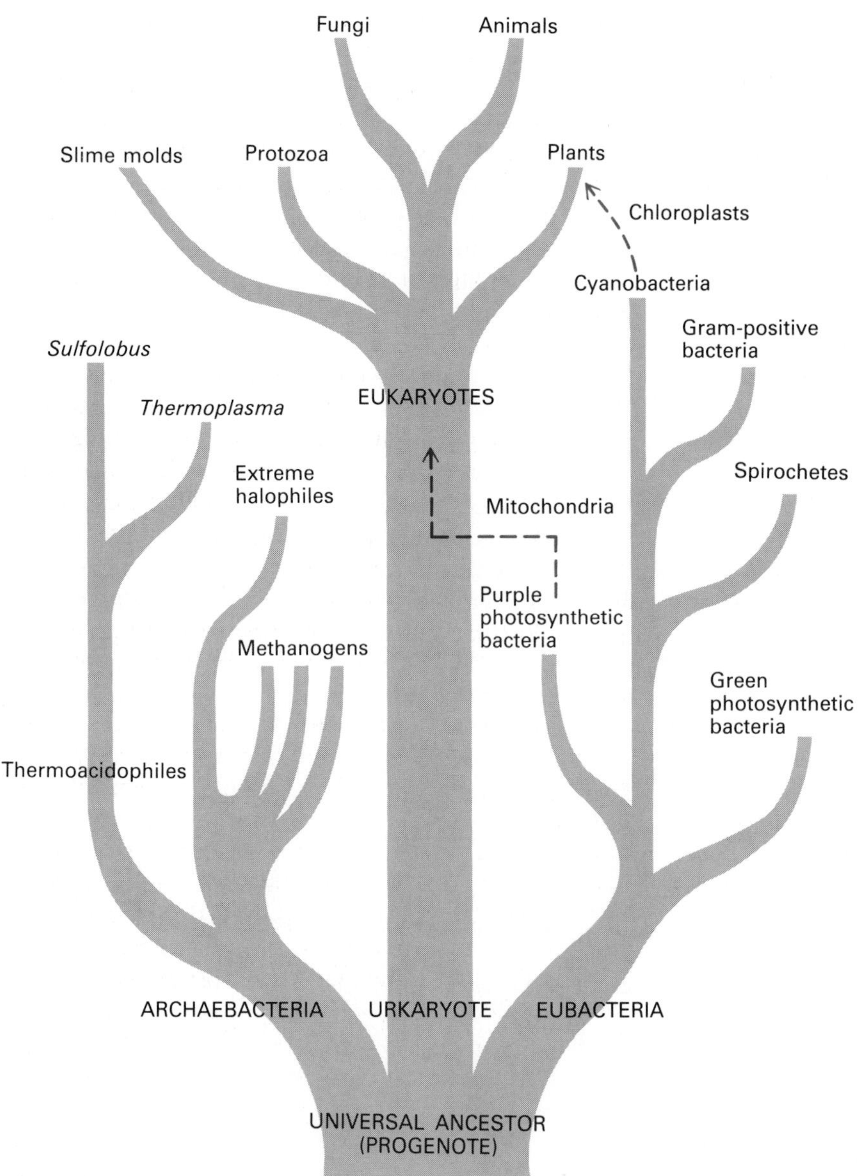

Figure 25-12 On the basis of sequence comparisons of rRNA, C. R. Woese proposed the following "cell lineage" tree of evolution. He has suggested that the universal ancestor be called a progenote, and that the ancestor to the present eukaryotic nucleus be called the urkaryote. *After C. R. Woese, 1981,* Sci. Am. *244(6):98.*

have arisen independently on several occasions. As we mentioned earlier, prokaryotic organisms that could carry out photosynthesis are thought to have fused with the progenitor of eukaryotic cells to form cells with chloroplasts. Likewise, a fusion of the eukaryotic progenitor with prokaryotic cells capable of oxidative phosphorylation (or at least with DNA fragments from such prokaryotic cells) is thought to be the origin of mitochondria. These fusions form the substance of the *endosymbiotic hypothesis*. The sequences of chloroplast and mitochondrial rRNAs do closely resemble those of specific eubacterial species, lending strong support to this hypothesis. The analysis of eubacterial rRNAs alone provides interesting and important new information, but perhaps the most striking result from the analysis of bacterial rRNAs is that not one but two evolutionarily distinct groups of bacteria-like organisms exist. The second group is made up of organisms with unusual habitats. This group of organisms has not been studied nearly as extensively as the eubacteria. About 20 species of halophilic (salt-loving) and methanogenic (methane-producing) organisms have been examined; within this group the ribosomal sequences (based on oligonucleotide comparisons) are similar, but quite distinct from those of the eubacteria. To signify the ancient origins of this second large subgroup of bacteria, they are called the *archaebacteria* (from the Greek for "ancient"). This group can also be divided, on the basis of comparative rRNA structure, into various subgroups differing in phylogenetic age.

Oligonucleotide comparisons for the rRNA of all present-day eukaryotes so far examined have produced different results than such comparisons for both the eubacteria and the archaebacteria. Although sequence comparisons have not been as extensive as those made with prokaryotes, rRNAs encoded in the nucleolus of yeast, plants, and a number of different animals have been found to be very similar (see Table 25-4). (This similarity in sequence is within the 28S and 18S regions of the pre-rRNA and *not* in the transcribed spacer sequences between the 18S and 28S coding regions.) From these comparisons it can be argued that eukaryotes represent a third independent line of cells (Figure 25-12) that is as distinct from archaebacteria and eubacteria as those two lineages are from each other. Because of the differences in rRNA sequences, it seems highly unlikely that the nuclei of eukaryotes arose from eubacteria or archaebacteria.

Despite the rRNA sequence differences that mark the three lines of cell descent, rRNA molecules have a general stem-loop structure (formed by self-complementary regions) that is highly conserved, as we have noted (see Figure 25-9). This conserved secondary structure of rRNA together with the near universality of the genetic code strongly imply that all three cell lineages emerged from a common precursor. Whether this common precursor was a functional cell is at present unknown. Perhaps the three cell lineages each arose independently from a common precellular state. This precellular stage of organization has been termed the *progenote* (Figure 25-12). The nature of the progenote is uncertain. It might have been a functional, slowly replicating, independently evolving organism; or it might have been an amorphous group of primitive transcription units plus a primitive transcription and translation apparatus from which many different cell lineages evolved, three of which have survived to the present day.

The Organization of the First Genome

If no present-day cells were the evolutionary forerunners of other lines of present-day cells, what was the progenote, the common precursor cell, like? There are two contrasting points of view on the question of how the first genome arose and how it must have been designed.

One approach—a frankly speculative one—extends our previous discussions about protein-free RNA formation in prebiotic times and the evolution of the code and the translation machinery (Figure 25-13). Because of the three central roles played by RNA in translation (i.e., mRNA, rRNA, and tRNA), RNA is thought to have preceded DNA in evolution. It is conjectured that the early generation of mRNA resulted in useful coding information—probably for domains or subdomains of proteins—interspersed with meaningless oligonucleotide stretches. RNA-RNA splicing in the absence of protein is accorded a key role in the function of the earliest organisms and even of the earliest RNA-based precellular transcription-translation systems. Finally, because reverse transcriptase–like activity is widely present in all eukaryotic cells and because the amino acid sequences of this enzyme are similar in all eukaryotes, this enzymatic activity is thought to be an ancient one: the evolutionary conversion of RNA to DNA may in fact have been achieved by the earliest form of such an enzyme.

Thus the progenote and the first slow-growing cells to arise may have shared many properties of today's eukaryotic cells, in which introns are common, extra DNA abounds, multiple chromosomes exist, and heterotrophic growth (growth requiring the supply of exogenous nutrients) is the rule. Loss of introns—the "streamlining of the genome" in Ford Doolittle's phrase—could have resulted where the environment placed a premium on fast growth, as it does for today's prokaryotes. The organisms that originally chose the fast-growth track would have been the progenitors to the autotrophic cells that survived best as the nutrients of the primordial soup were exhausted, and these latter cells would have been the forerunners of today's bacteria.

Needless to say, this view contrasts with the widely accepted notion that the first cells were similar to today's prokaryotes (see Figures 25-1 and 25-10); according to this model, the eukaryotic nuclear genome was an evolutionary product of an earlier prokaryotic genome. As we have mentioned, these traditional ideas about the nature of the earliest cells seemed logical because prokaryotic cells are simpler in design than eukaryotic cells, and because evolution is presumed to have proceeded from the simple to the complex. However, as we have seen, comparisons of rRNAs have yielded definite evidence of an ancient separation between present-day eukaryotes and present-day prokaryotes as well as two separate lines of descent among prokaryotes. Any single progenitor to all three lines of descent must have given rise to the three separate lineages a very long time ago. The common ancestor cell could have had properties of present-day eukaryotes as easily as those of present-day prokaryotes. In the next several sections we examine the evidence on which the outline of Figure 25-13 is based.

Did the First Genome Have Divided Genes?

Perhaps the greatest surprise of modern molecular genetics was the discovery that genes in the nucleus and in the organelles of eukaryotic cells contain intervening sequences (introns) that separate the protein-coding portions of the gene. To address these findings in evolutionary terms, the most critical question is: Was there a time in cellular or precellular evolution when all gene sequences were contiguous and no introns existed, or were introns present in the earliest genes? Although no conclu-

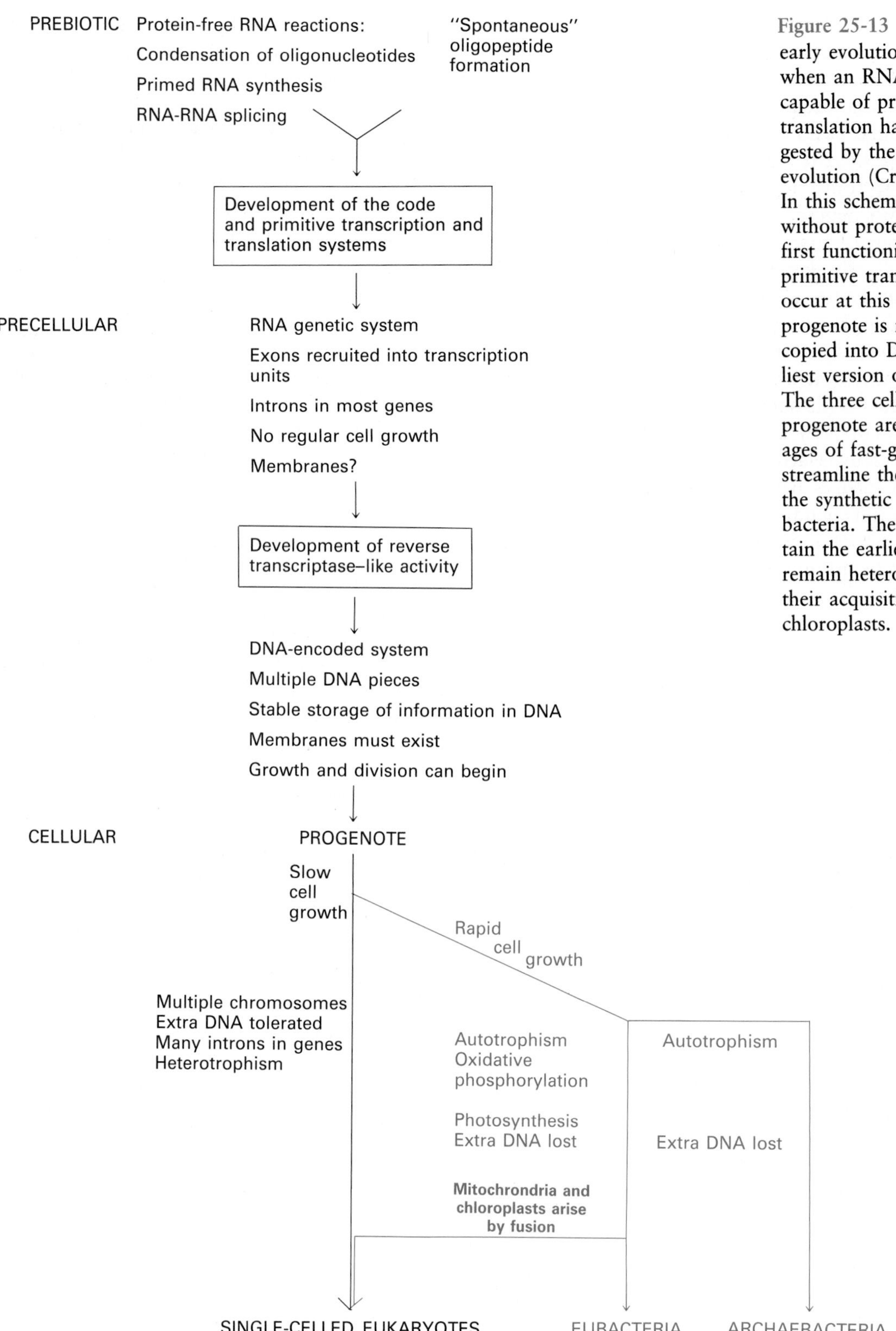

Figure 25-13 A possible course of early evolution. The prebiotic era ends when an RNA-encoded genetic system capable of primitive transcription-translation has evolved, as was suggested by the early writers on chemical evolution (Crick, Woese, and Orgel). In this scheme RNA-RNA chemistry without protein is emphasized. The first functioning genome is RNA and primitive transcription and translation occur at this stage. The stage of the progenote is reached when RNA is copied into DNA, perhaps by the earliest version of reverse transcriptase. The three cell lineages arising from the progenote are shown: The two lineages of fast-growing cells gradually streamline their genomes and acquire the synthetic capacities of modern-day bacteria. The slow-growing cells maintain the earliest genomic design and remain heterotrophic, benefiting from their acquisition of mitochondria and chloroplasts.

sive answer to this question is available now (and none may ever be), at least two general arguments leave open the possibility that introns were the norm in the first genes. We will first summarize the present evidence about RNA chemistry that occurs without proteins. Since protein-free reactions that allow synthesis and processing of RNA are well-known, it is clearly possible that RNA with intervening sequences was present at the beginning of molecular evolution. Second, we will review the distribution of introns within genes. Some genes have maintained introns in constant position for perhaps over a billion years, indicating that they are an ancient, not a recent, feature of eukaryotic genes. In addition, introns divide genes into protein domains that are the basic building blocks of proteins. Finally, while some eukaryotic genes have lost introns (documented by comparing vertebrate species that are evolutionarily relatively recent), no comparable introduction of introns has been documented. Thus, the constant position of introns for long periods and separation of protein-coding domains by introns are compatible with introns being present in the first genes and in fact for the first genes being assembled from simple protein-coding units.

RNA-Directed RNA Synthesis, Site-Specific Cleavage, and Splicing without Proteins Are All Possible

We described earlier how protein-free mixtures of activated nucleotide monomers can yield polymers, and how even template-directed RNA synthesis can be carried out without proteins (see Figure 25-5). In Chapter 9 we discussed the transesterification reaction that removes an intron from *Tetrahymena* pre-rRNA and ligates the remaining pieces (see Figure 9-61). Other recent work on RNA processing in bacteria has revealed that protein-free site-specific site cleavage can occur.

Site-Specific RNA Cleavage The simplest of the protein-free RNA reactions is site-specific RNA cleavage. This reaction occurs at two sites in a bacteriophage T4 mRNA precursor to produce two different molecules. Although the cleavage can occur at either site without a protein, an enzyme from infected cells can facilitate the cleavage at the same sites. In a second T4 mRNA, that for thymidylate synthase, an intervening sequence must be removed and the molecule re-ligated to make the functional mRNA (Figure 25-14). The excision in this case can occur in a protein-free environment. The splicing may have also occurred originally without protein, but this step is now assisted by cell protein.

The Catalytic Activity of a Small RNA in RNA Cleavage A second type of reaction involving RNA-directed RNA cleavage uses a separate facilitating RNA molecule that is distinct from the cleaved molecule. The pre-tRNAs of *E. coli* are synthesized as larger precursor molecules that must be trimmed down into finished, usable tRNAs. Perhaps the most well studied example of these is an *E. coli* tyrosyl-tRNA. RNase P, the enzyme that cleaves $tRNA^{Tyr}$, has been highly purified from *E. coli*. Part of the purified enzyme is an RNA chain 375 nucleotides long. The protein portion of the enzyme is by itself inactive. When the protein-free RNA moiety of RNase P is added to the pre-$tRNA^{Tyr}$, cleavage proceeds to produce the 5′ end of the mature tRNA (Figure 25-15). Thus the cleavage can be carried out by an RNA-directed mechanism, but in the cell this RNA has a protein associated with it.

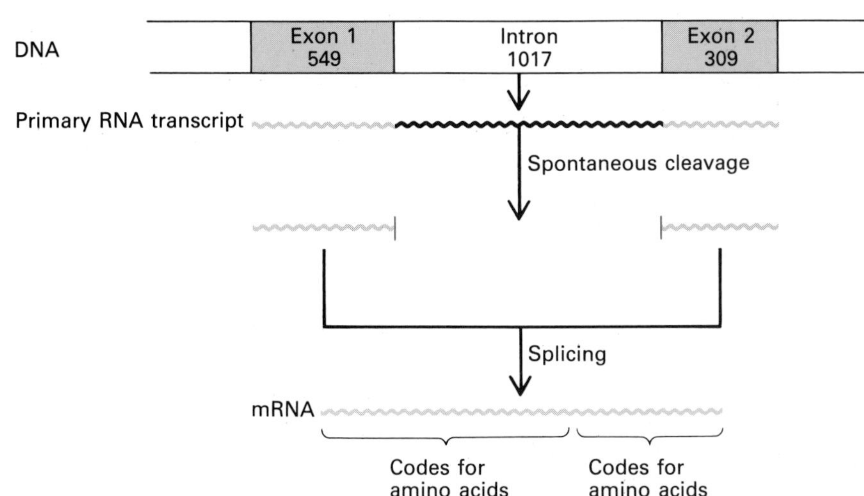

Figure 25-14 An intron in a bacteriophage gene from *E. coli*. The bacteriophage T4, which has *E. coli* as its host, encodes the enzyme thymidylate synthase. The coding sequence for thymidine kinase is in two parts (549 base pairs plus 309 base pairs) separated by an intron (1017 base pairs). The RNA transcribed from this region can undergo spontaneous cleavage to release the intron. [See M. Belfort et al., 1985, *Cell* 35:375.]

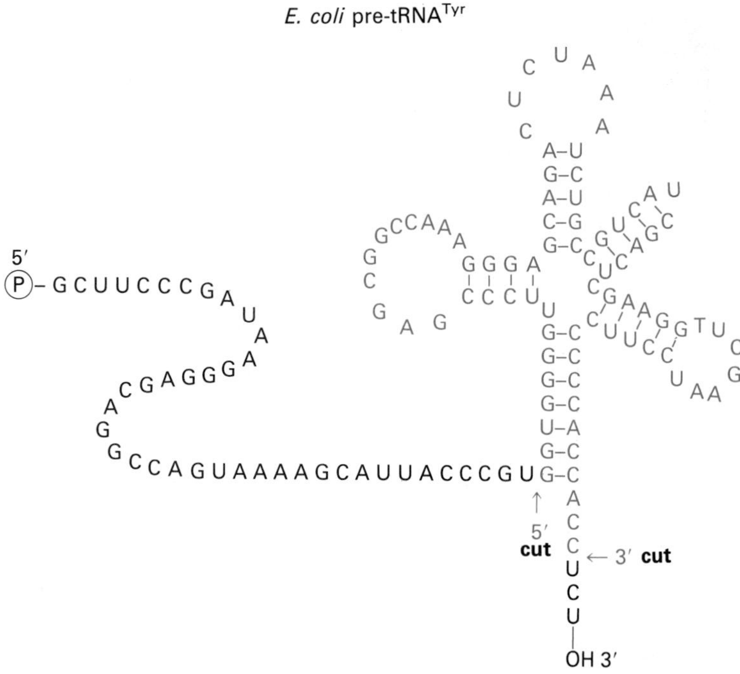

Figure 25-15 The sequence of the *E. coli* pre-tRNA^{Tyr} is cleaved to expose the 5′ and 3′ ends of the tRNA^{Tyr} *(color)*. The 5′ cleavage is catalyzed by RNase P or by M1 RNA, a 375-nucleotide RNA that is part of RNase P. The 3′ cleavage is carried out by another enzyme. [See C. Guerrier-Takada et al., 1983, *Cell* 35:849.]

Three Types of RNA Splicing Are Recognized from Biochemical Studies

As we discussed in Chapter 9, three major types of biochemical reactions result in RNA-RNA splicing, and three distinct intron structures are removed by the three different mechanisms. The distribution in nature of these three types of introns and splicing mechanisms has implications for cellular evolution, and we wish to explore these possibilities here. We shall refer to splicing mechanisms of types I, II, and III for convenience (the order of numbering is entirely arbitrary). The distribution of these three types of splices and their properties are summarized in Table 25-5.

Type I Splicing: tRNA Splicing The shortest introns within a precursor RNA are those found in the anticodon arm of pre-tRNA. They do not have recognizable consensus sequences at their borders. In yeast cells these introns are removed by two separate enzymatic reactions, an endonucleolytic cleavage and an ATP-dependent ligation (Chapter 9). The two reactions are carried out by two separate enzymes thought to consist entirely of protein. Extracts of mammalian cells also carry out a two-step tRNA splicing reaction but differ slightly in their biochemistry.

The splicing of tRNA precursors is not confined to eukaryotes. At least half a dozen different species of pre-tRNAs from *Sulfolobus solfataricus,* an archaebacterial species that is sulfur-dependent, also require splicing (Figure 25-16). The presence of an intron in exactly the same site in both archaebacterial and eukaryotic species suggests that the splicing event for tRNAs is of ancient ori-

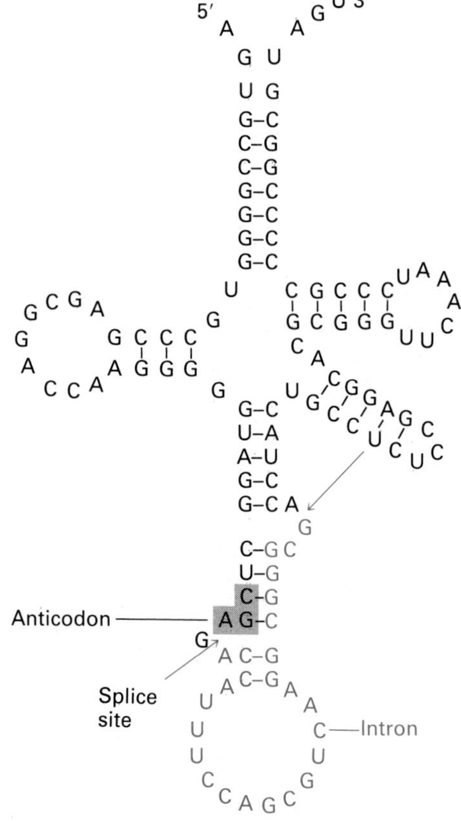

Figure 25-16 The sequence of a pre-tRNA^{Ser} gene from the archaebacterium *Sulfolobus solfataricus.* The intervening sequence in the anticodon arm is shown in color. [See B. P. Kane, R. Gupta, and C. R. Woese, 1983, *Proc. Nat'l Acad. Sci. USA* 80:3309.]

Table 25-5 Three types of splices and their distributions

Type of splice	Molecule and organism	Properties
I	tRNAs	No consensus sequences present Site: anticodon arm Occurs in two steps: cleavage and ATP-dependent ligation
II	rRNAs Nuclear *Tetrahymena* *Physarum* Chloroplast *Chlamydomonas* Mitochondria *Saccharomyces* *Neurospora* mRNAs Mitochondria *Saccharomyces* *Aspergillus*	Intron has conserved secondary structure Spontaneous splicing occurs in *Tetrahymena, Neurospora,* and *Saccharomyces* rRNAs and in some *Saccharomyces* mRNAs All other cases may require proteins or ribonucleoproteins
III	mRNAs Eukaryotic nuclei Plants and animals *Saccharomyces* mRNAs Chloroplast *(Euglena)* Mitochondrion *(Saccharomyces)*	Consensus sequences present at borders of introns Proteins, snRNPs required

gin; it may have arisen before the very early archaebacterial-eukaryotic separation indicated by differences in rRNA sequences (see Table 25-4).

Type II Splicing: rRNAs and Mitochondrial mRNAs

The second type of splicing mechanism occurs in pre-rRNAs of the protozoan *Tetrahymena* and the slime mold *Physarum polycephalum,* in the pre-rRNAs of chloroplasts of the alga *Chlamydomonas,* and in both the rRNAs and the mRNAs encoded by the mitochondrial DNA of yeasts *(Saccharomyces)* and filamentous fungi *(Neurospora crassa).* The rRNAs from *Tetrahymena, Neurospora,* and some rRNAs and mRNAs of *Saccharomyces* have been shown to undergo self-splicing. The nucleotide sequences of the introns in all of these cases show the group to have a unified type of intron structure that is important in the removal of the intron. The 3′ end of the first exon and the 5′ end of the second exon are brought together by an *internal guide sequence* (IGS) which forms four to six base pairs on either side of the junction that will be made during splicing. Within the intron there are a number of conserved blocks of sequence about 10 nucleotides long (Figure 25-17). These sequences are remarkably conserved between, say, the yeast mitochondrial mRNA precursors and the *Tetrahymena* and *Physarum* pre-rRNAs (Figure 25-18).

In addition to the similarity in the secondary structure of introns, another feature of fungal mitochondrial and *Tetrahymena* rRNA splicing is the formation of circular intermediates when the intron RNA is excised. At least half a dozen of the yeast intron sequences, like the *Tetrahymena* sequence, form such circles.

Recall that sequence analyses of chloroplast and mitochondrial rRNAs have shown that they closely resemble the rRNAs of eubacterial species. Recall also that according to the endosymbiotic hypothesis, chloroplasts and mitochondria reside within eukaryotic cells as a result of an ancient fusion between bacterial cells and the forerun-

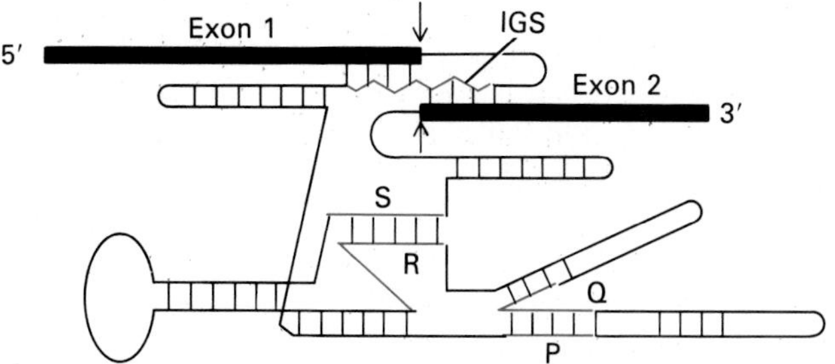

Figure 25-17 The generalized intron structure of type II RNA-RNA splices. The intron-exon junctions in such molecules are brought together by the base-pairing interaction of an internal guide sequence (IGS). Four highly conserved oligonucleotide regions (P, Q, R, and S) within the intron are indicated. [See R. W. Davis et al., 1982, *Nature* 300:719.]

	P regions	Q regions	R regions	S regions
Saccharomyces mitochondrial mRNA	7878677786 AUGCUGGAAA	7886687887 AAUCAGCAGG	888778888758 UCAGAGACUACA	867878788887 AAGAUAUAGUCC
Tetrahymena rRNA	UUGCGGGGAC	AACCAGCAGC	UCACAGACUAAA	AAGAUAUAGUCG
Physarum polycephalum rRNA	AUGCUGAGAC	UAUCAGCAAG	UCAACGACUGGA	AAGGUGCAGUCC

Figure 25-18 Base conservation in introns bordered by type II splice sites. The sequences of eight different P, Q, R, and S regions (see Figure 25-17) of yeast mitochondrial mRNA introns were compared first; the number of coinciding bases at each position are indicated above the mitochondrial sequence. The same P, Q, R, and S regions in *Tetrahymena* and *Physarum polycephalum* rRNAs were also compared, and the bases identical to the consensus bases of the yeast mRNA sequences are shown in color. Thus these four long oligonucleotide regions are highly conserved in rather different genomes. [See R. B. Waring et el., 1983, *J. Mol. Biol.* **167**:595.]

ner of the nucleated eukaryotic cell. It thus seems entirely reasonable to conclude that some primary RNA transcripts in the early eubacterial cells were subject to type II splicing, and that the descendants of these bacteria are today's chloroplasts and mitochondria. The other alternative—that all type II introns are derived from transposable elements that entered previously uninterrupted genes after the fusion of prokaryotes with eukaryotes—seems less likely because these introns do not possess the terminal repeats that are characteristic of transposons; nor is there any evidence of the usual target-site duplication. Furthermore, this latter hypothesis would require the entry of a transposon with the same internal structure into a diverse group of rRNAs and mRNAs.

Type III Splicing: Nuclear mRNA The third type of splicing is exemplified by the hundreds of nuclear genes in higher animals and plants that encode mRNA sequences containing from 2 to more than 50 exons. Comparisons of the DNA sequences from over 50 intron-exon boundaries have revealed short consensus sequences at both intron-exon junctions. At the 5′ junction the most highly conserved nucleotides are GT at the first two intron positions; a pyrimidine-rich cluster is found near the 3′ end of the intron, and there is a short conserved sequence at the branch point (see Figures 9-60 and 25-19). These sequences are also conserved in plant nuclear genes. In a number of yeast introns the general design of the intron is similar, even including the branch-point sequence. Thus it

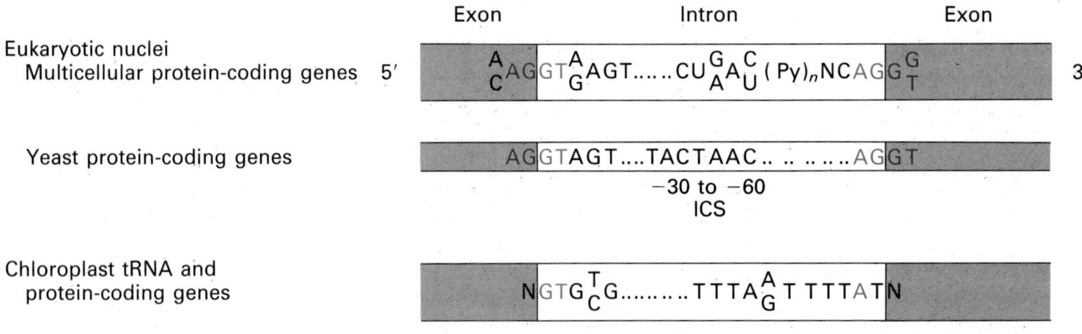

Figure 25-19 Type III splicing: consensus sequences at the intron-exon junctions of DNAs encoding primary transcripts for spliced mRNAs. The nuclear genes of multicellular plants and animals show well-preserved junctional AGGT sequences. Also, another conserved sequence followed by a pyrimidine-rich group of nucleotides occurs near the 3′ end of the intron. Yeast introns have a similar junctional sequence; in addition, they contain the internal conserved sequence (ICS) TACTAAC located about 30 to 60 bases from the 3′ splice junction; this sequence appears at the branch point of the processed primary transcript (see Figure 9-60) and probably serves the same function as the conserved sequence and the pyrimidine-rich 3′ stretch of introns from genes of multicellular organisms. Chloroplast DNAs have GT as the first pair of nucleotides in the intron and a pyrimidine-rich (T-rich) stretch at the 3′ end of the intron. These junctional sequences for different genes are clearly similar though not identical.

appears that all of these nuclear genes have introns that are removed by a mechanism that has descended from a single ancestral mechanism.

In addition to the nuclear sequences described above, the sequences that are involved in the splicing of primary transcripts from chloroplast DNA in the alga *Euglena* and in tobacco leaves are now known. One subunit of the enzyme ribulose 5′-bisphosphate carboxylase (a protein critical in the fixation of carbon dioxide in plants; see Chapter 20) is encoded by a chloroplast gene that has eight introns. A simplified version of the nuclear intron-exon consensus sequence (featuring GT in the first two intron positions) appears at each intron-exon junction. The same result has been found for the four intron-exon boundaries in psbB, an important chloroplast protein that binds herbicides. A similar sequence also exists in several tRNA genes in tobacco chloroplasts. Thus a primitive version of the generalized mRNA splice mechanism may have already been present in the eubacterial ancestor of the chloroplast.

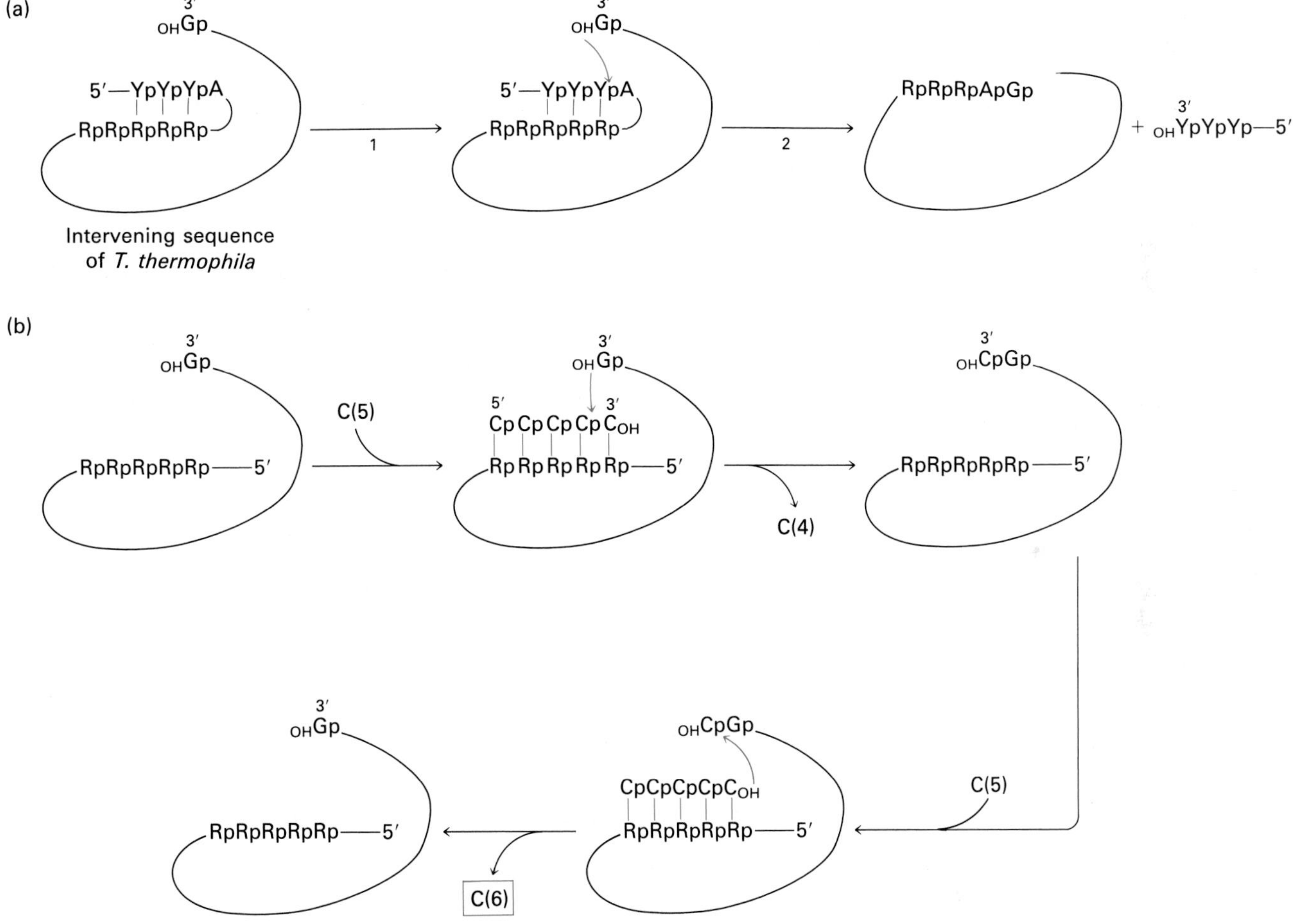

Figure 25-20 Transesterification reactions and enzyme activity of RNA. (a) This diagram illustrates the essential features of the intervening sequence (see Figure 9-61), which is spontaneously excised from the pre-rRNA of *Tetrahymena thermophila* and which then itself spontaneously cyclizes. This cyclization reaction requires a 3′ terminal guanosine (G_{OH}) that acts to break the RNA chain in a pyrimidine-rich (Ys) region that is held in place by a hydrogen bonding purine-rich (Rs) region (*reaction 1*). In the process a transesterification occurs *(reaction 2)* in which one of the phosphodiester bonds is transferred to the guanosine, releasing the previous 5′ portion of the sequence. (b) The transesterification reaction can elongate chains. Polycytidylic acid (C(5) is shown) will transiently bind to the purine-rich region of the *Tetrahymena* intervening sequence. The 3′ terminal guanosine will attack the C(5) and release the C(4). By transesterification one C residue is added to the 3′ end of the intervening sequence. This results in an active form for nucleotide transfer; when a C(5) again binds to the purine-rich region a reverse of the previous transesterification can occur yielding C(6) and the starting intervening sequence. The intervening sequence has thus acted as an enzyme by catalyzing the addition of a nucleotide to the C(5) unit. This reaction has been shown capable of making polycytidylic acid chains up to 30 units long. *After A. Zaug and T. R. Cech, 1986,* Science ***231****:470.*

Type III Self-Splicing and RNA Chain Growth through Transesterification Recent experimental results heighten the possibility of the very early existence of all type II and type III splicing. First, self-splicing has now been documented in the removal of type III introns as well as type II introns. An intron in an mRNA from yeast mitochondria is removed in a protein-free reaction. This reaction is very much like the removal of nuclear type III introns because the excised portion is removed in the form of a branched structure or "lariat" (see Figure 9-60), just as is the case with all other excised type III introns. In addition to the possibility of self-splicing there is another important similarity between type II and type III splices. In both types of splicing the phosphate that unites the two exons after splicing is present in the RNA chain to begin with; thus, both the type II and the type III mechanisms can be envisioned to proceed by a transesterification and can occur without need for an outside source of energy. Finally, recent studies with enzymatically active RNA have shown that the transesterification reaction can build up long RNA chains. Recall that the intervening sequence in *Tetrahymena* preribosomal RNA is spontaneously excised and this intervening sequence itself is still chemically active—it undergoes cyclization internally with the loss of 19 nucleotides (see Figures 9-61 and 25-20). The free 2′ and 3′ hydroxyls of the 3′ terminal guanosine residue are thought to be the active agent of this cyclization reaction acting through transesterification so that the phosphate group at the site of cyclization is transferred to the terminal guanosine (Figure 25-20a). This activity can result not only in cyclization but under appropriate conditions the transesterification can also lead to the union of two of the active intervening sequences rather than to cyclization. Another even more important reaction using transesterification has been discovered using the intervening sequence as an enzyme and poly(C) as the substrate. Poly(C) [say, 5 nucleotides long, C(5)] will bind to a purine-rich region in the part of the intervening sequence where cyclization usually occurs (Figure 25-20b). In this case a terminal C residue can be transferred to the active guanosine residue. If another C(5) binds to the purine-rich region the recently transferred 3′ terminal can be added to the 3′ OH of the C(5), elongating it by one residue to C(6). Thus transesterification can elongate chains. This reaction has been shown capable of making chains over 30 nucleotides long. If mixed oligonucleotides can be elongated this way (and there is no reason they might not be) it is possible to replicate almost any RNA. These new experiments provide very powerful evidence that protein-free RNA chemistry can be reasonably assigned a critical role in early precellular evolution. It seems highly probable that the earliest messenger RNAs would have evolved with introns.

To summarize, the protein-free RNA synthesis reactions plus those of RNA cleavage and RNA-RNA splicing illustrate clearly that RNA is capable of a variety of protein-free reactions, and that RNA could have arisen and diversified in the precellular era.

Table 25-6 Amino acid homologies between globin chains from various sources

Protein and organism	(1)	(2)	(3)	(4)	(5)	(6)	(7)	(8)	(9)
Hemoglobin α chain									
(1) Human									
(2) Dog	16								
Hemoglobin β chain									
(3) Human	57	57							
(4) Dog	59	59	10						
Myoglobin									
(5) Human	75	78	77	75					
(6) Dog	75	77	74	74	15				
Leghemoglobin									
(7) Broad bean	86	82	79	79	91	90			
(8) Kidney bean	87	85	83	83	89	89	36		
(9) Soybean	86	87	85	84	90	89	36	21	
	Percentage of amino acids differing								

SOURCE: M. Dayhoff, ed., 1979, *Atlas of Protein Sequence and Structure,* vol. 5, suppl. 3-1978, National Biomedical Research Foundation (Washington, D.C.), p. 234.

Evolutionary Implications of the Positions of Introns in Genes

From the study of gene structure in widely divergent organisms, two general points have emerged: (1) some introns have been in place for very long evolutionary periods, and (2) the positions of introns in many genes divide proteins into functional domains. Both of these conclusions are compatible with an ancient origin for introns.

The Hemoglobins, Myoglobin, and Leghemoglobin All Have Similar Intron-Exon Structures

The fossil record gives the approximate times at which the main branches of the animal kingdom diverged (see Figures 25-1 and 25-10). Comparisons of the primary amino acid sequences (and, as is now much more common, the nucleotide coding sequences) of the same protein from a variety of different animals (Table 25-6) permit an estimate of the frequency of change per unit time in evolution. Such analyses have yielded numbers ranging from 0.1 to 1.0 amino acid changes per 10^6 years. Thus the sequences of similar but nonidentical proteins can be used to derive a rough estimate of the time at which an evolutionary separation of two related genes occurred. According to sequence comparisons, the genes for the α and β chains of hemoglobin probably duplicated and separated about 500 million years ago. By similar arguments, the related oxygen-binding protein myoglobin must have separated from the hemoglobins at least 800 million years ago. Around these times, which predate the beginning of the fossil record, the simplest animal phyla may have been evolving (see Figure 25-10).

The basic design for both the α and β globin transcription units—three exons separated by two introns at similar positions—persists in all vertebrates that have been examined (frogs, chickens, rats, mice, sheep, and humans; see Figures 11-8 and 25-21). Thus the three globin exons have been present and have remained remarkably constant in position since very early in the evolution of animal life. In vertebrate muscle the single polypeptide chain of myoglobin has a three-dimensional structure similar to that of globin chains (see Chapter 3). The myoglobin gene cloned from seals and humans also has two introns located in very nearly the same places as those in globin (Figure 25-21).

A comparison of vertebrate globins and myoglobin with leghemoglobin has interesting evolutionary implications. *Leghemoglobin* is a single-chain plant protein that binds and sequesters oxygen in root nodules of legumes; thus its function is similar to that of myoglobin. The amino acid sequences of leghemoglobin, myoglobin, and the globins are clearly related although many evolutionary changes in amino acid sequences have occurred since the divergence from the presumed ancestral gene (Table 25-6). Remarkably, the genomic structure of leghemoglobin is very similar to that of the globin chains (Figure 25-21). Exons encoding the first ~30 and the last ~40 amino acids exist in all of the genes, but the central heme-binding portion of leghemoglobin is encoded by two exons, whereas this region is represented by one exon in globin

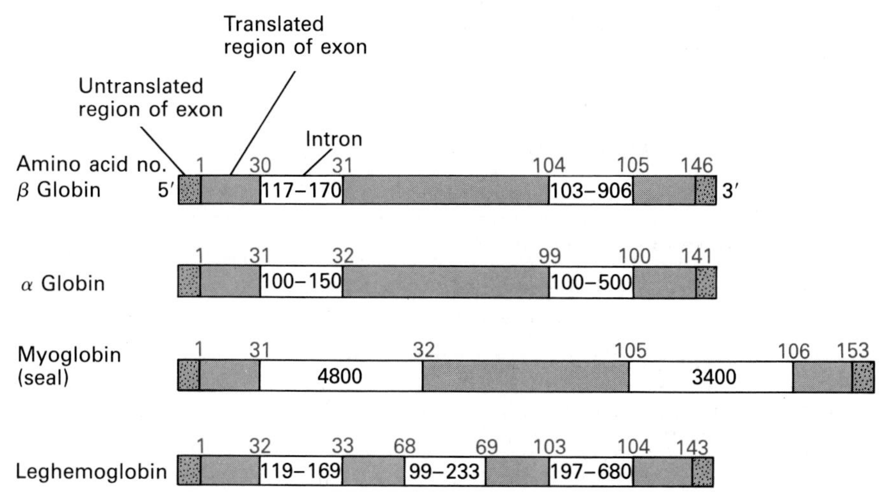

Figure 25-21 The conservation of intron positions in the globin gene family. The α and β globins of vertebrates show introns at very similar positions within the sequences coding amino acids (the numbers above the diagrams are amino acid positions in the protein chain). The lengths of the introns vary (the numbers inside the boxes indicate the ranges of lengths in nucleotides). The myoglobin gene of seals has a structure very similar to that of the globin chains but the myoglobin introns are longer. The leghemoglobin genes of plants have an extra intron dividing the central exon into two parts. [See G. Stamatoyannopoulos and A. W. Nienhuis, eds., 1981, *Organization and Expression of Globin Genes*, Alan R. Liss; E. O. Jensen et al., 1981, *Nature* **291**:677; and A. Blanchetot et al., 1983, *Nature* **301**:732.]

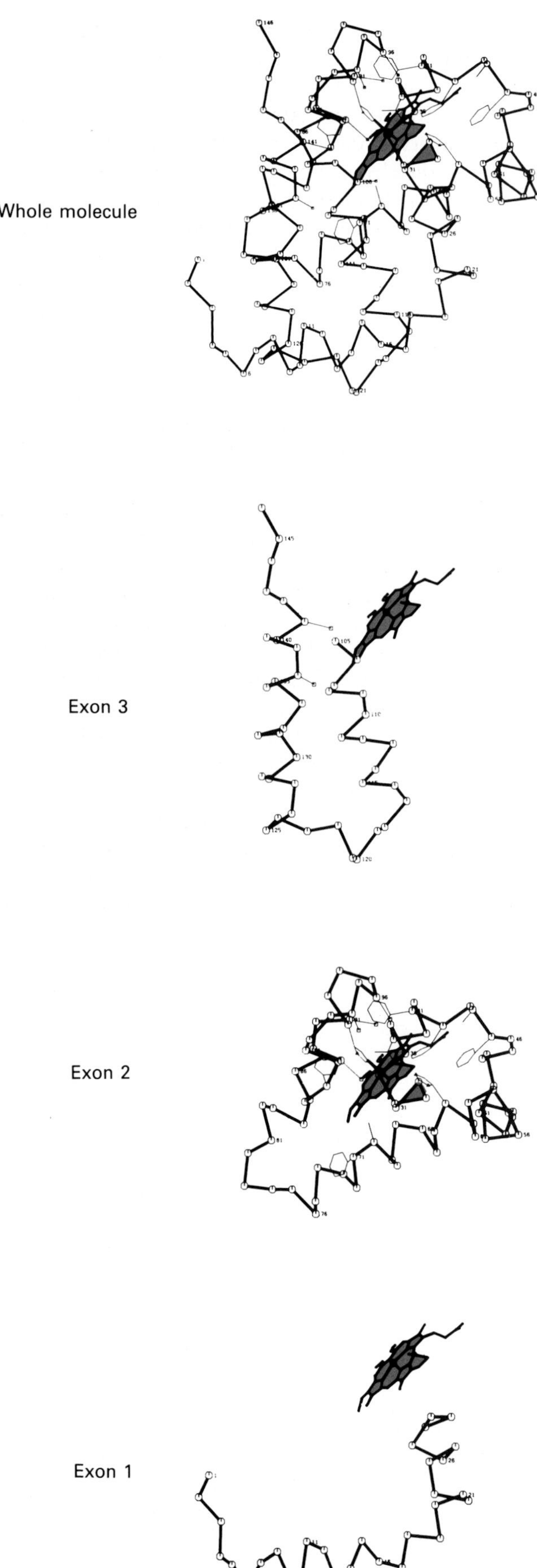

Figure 25-22 The binding of heme by the β globin chain. The diagrams show the structure of the backbone of human β globin. The heme and its contacts appear in color. The top drawing represents the whole β chain, whereas the lower three represent the parts of the chain encoded by the left (5′), center, and right (3′) exons, respectively (see Figure 25-20). Two heme contacts can be seen in the center exon (exon 2), which is the counterpart of two exons in leghemoglobin. [See M. Gō, 1981, *Nature* **291**:90.] *Courtesy of M. Gō. Reprinted by permission from* Nature. *Copyright 1981 Macmillan Journals Limited.*

and myoglobin. The polypeptide encoded by the central globin exon can by itself bind heme, and in so doing reproduces many of the physical properties (e.g., spectral shifts) of heme binding in the whole molecule. However, this central heme-binding portion of globin has two heme contacts, the sites of which could be considered separate subdomains (Figure 25-22); these would be the counterparts of the domains encoded by the two central exons in the leghemoglobin gene (see Figure 25-21). Thus it seems possible that an ancient globin precursor that predated the plant-animal division had four protein domains, that the central two became fused, and the intervening sequence was lost.

An evolutionary relationship even older than that between leghemoglobin and the other globins has been hypothesized from the similarity of the structure of the heme-binding cavity in the cytochromes and in globin chains. Although no amino acid similarity can be detected now between the two, the shape of the "pocket" in which the heme is bound is similar. This suggests that a heme-binding capacity may have been an early protein domain, and that the exon encoding this domain remains functionally intact until the present time.

Other Duplicated Genes Also Preserve Ancestral Intron-Exon Arrangements

There are many other gene families similar to the globin gene family, which has arisen from duplications and which has maintained the intron-exon arrangement after the duplication. To list several: the ovalbumin gene and the neighboring X and Y genes in chickens, the immunoglobulin genes, the genes for mouse α fetoprotein and serum albumin, and the chorion genes in silk moths (Chapter 11). In *all* of these cases each member of the family of duplicated genes maintains the same intron-exon arrangements in the DNA. The conservation of the intron-exon structure indicates that it was already in existence prior to the gene duplication.

Detailed analyses of the mouse serum albumin genes and of the ovomucoid genes in chickens have been used to carry this argument even further. Each of the genes in these sets appears to consist of three similar large sections each composed in turn of smaller domains encoded in identical sets of intron-exon arrangements (see Figure 11-11). Thus the primordial gene in each case already had a pattern of introns and exons that was preserved during the duplication that yielded the present-day genes.

Introns in the Genes for Basic Metabolic Enzymes If fully functioning genes were in fact first assembled from primitive coding elements (see Figure 25-13), it seems highly likely that these elements might exist in enzymes active in carbohydrate metabolism (such as dehydrogenases and kinases), because these enzymatic functions are necessary for all cells. A number of these enzymes are known from crystallographic studies to possess similar physical landmarks that result from common protein domains. In fact, the very idea of protein domains was advanced quite early from the physical similarities of the nicotinamide adenine dinucleotide–(NAD-) binding regions of the dehydrogenases. The intron-exon patterns in the genes for several of the enzymes of carbohydrate metabolism—glyceraldehyde 3-phosphate dehydrogenase, phosphoglycerate kinase, triosephosphate isomerase, and alcohol dehydrogenase—have been determined for various vertebrates. The intron-exon patterns for the NAD- or ATP-binding domains of the first two enzymes are shown in Figure 25-23. In each enzyme the nucleotide-binding domain is divided similarly into exons that have recognizable homology. The exon distributions in the nucleotide-binding domains of the other two enzymes mentioned above have at least a partial similarity to the distributions shown in Figure 25-23. Thus these nucleotide-binding domains appear to be very ancient, and they could certainly represent primordial protein functions that originally existed in separate exons.

Exon "Shuffling" of Already Recruited Domains Another aspect of the role of protein domains in evolution is suggested by examination of the intron-exon structures of the NAD-binding domains. Although the features of these domains are clearly related in all of the genes and enzymes mentioned above, the functional groups encoded by the enzymes' other exons are not the same. It seems very likely that once a functioning cell existed with the capacity for recombining its DNA, any exons that had been originally recruited to form genes could be duplicated and recombined wholesale into different transcription units. The arrival of the NAD-binding domain in the midst of the other coding domains appears to have resulted from such an event, which Walter Gilbert has termed "exon shuffling." Once cells began to grow and divide, exon shuffling may have been responsible for the varying arrangements of functional units in different transcription units. (Exon shuffling via the recombination of DNA segments is a different phenomenon than the recruitment of primordial exons in RNA molecules—an event that may have been instrumental in getting cells started in the first place, as we described earlier.)

Exon shuffling must certainly have occurred to compose the gene for the low-density lipoprotein receptor (Chapter 15). This gene has 18 exons; several have homology with epidermal growth factor and blood-clotting factors, and other exons are homologous to a blood protein called complement factor 9. Still other exons encode a signal sequence for the targeting of the receptor to the

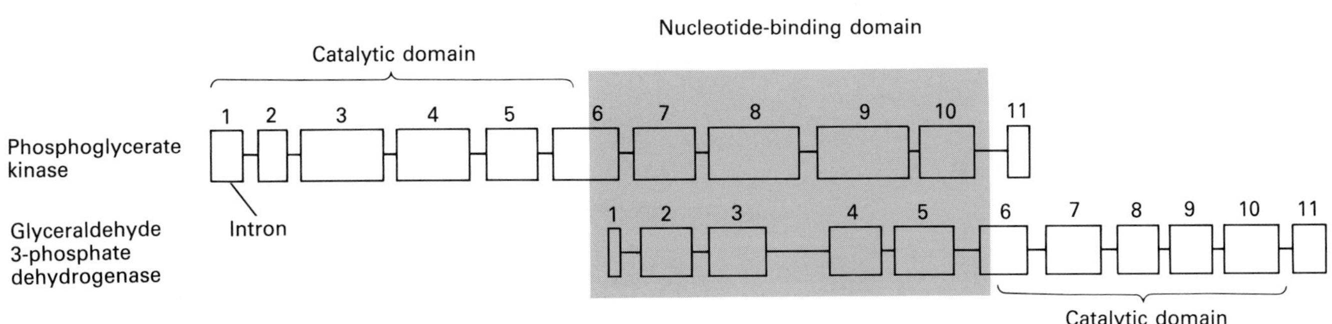

Figure 25-23 Introns in genes for mammalian glyceraldehyde 3-phosphate dehydrogenase (from a chicken) and phosphoglycerate kinase (from a horse). These two enzymes are fundamental to energy metabolism in all cells, and the amino acid sequence as well as the three-dimensional structure of each is conserved in bacteria, yeast, and mammals. This is particularly true of the structure of the nucleotide-binding domains. (Nicotinamide adenine dinucleotide binds to the first enzyme and ATP to the second.) A comparison of the intron-exon structures of these domains in the two enzymes shows a great similarity both in structure and in sequence in the regions of the introns labeled 6 to 10 in phosphoglycerate kinase and 1 to 5 in glyceraldehyde 3-phosphate dehydrogenase. The catalytic domains of the two enzyme proteins are encoded by different exons. The yeast genes (and, presumably, the bacterial genes) for these enzymes lack all of the introns shown here. [See A. M. Michelson et al., 1985, *Proc. Nat'l Acad. Sci. USA* 82:6965.]

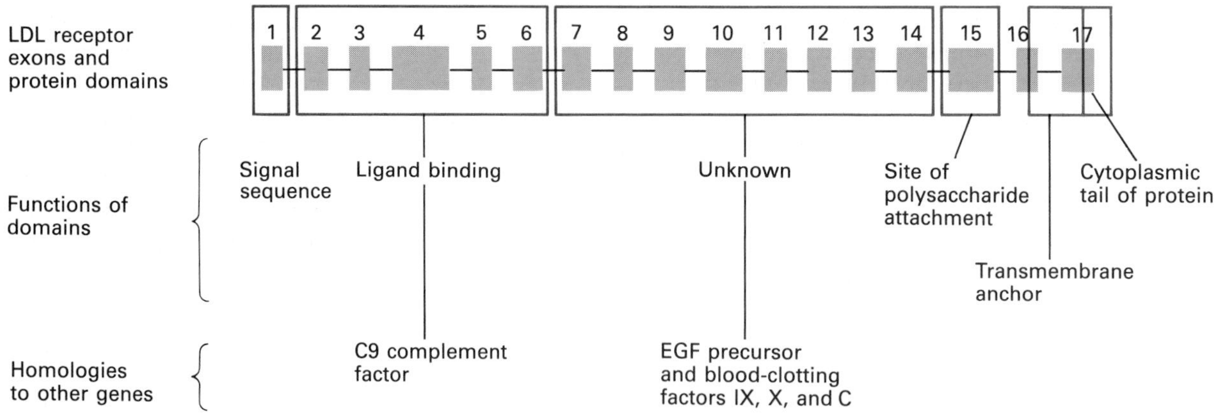

Figure 25-24 A diagram of the intron-exon structure of the gene for the low-density lipoprotein (LDL) receptor. The gene has 18 exons that encode six protein domains in the receptor. The domain that actually binds the LDLs circulating in the blood has extensive amino acid homology with the C9 complement factor (another blood protein). The large central domain, which may help to fold the receptor, has homology with the precursor to the epidermal growth factor (EGF). The LDL receptor gene seems to have been assembled by exon shuffling during evolution. [See Figure 15-44 and T. C. Sudhof et al., 1985, *Science* **228**:815.]

endoplasmic reticulum membrane, a transmembrane domain for anchoring the receptor in the membrane, and a domain to which polysaccharide side chains are attached (Figure 25-24). The various domains of this protein must have been put together after cells arose by the shuffling of exons from different transcription units.

Intron Loss Apparently Can Occur

Although the accumulated evidence that early exons represented early protein domains is impressive, it is true that many genes in yeast as well as in multicellular organisms lack intervening sequences. For example, a number of yeast genes for enzymes important in carbohydrate and amino acid metabolism have no introns; nor do mammalian α interferon genes. If during precellular times most information arose in noncontiguous bits, why are these genes free of introns? Did they lose introns they once possessed?

We do not have a completely satisfactory answer to this question, but analyses of insulin genes from several animals have suggested that the loss of introns can indeed occur. In the genomic DNA of rats there are two insulin genes encoding very similar amino acid sequences, which

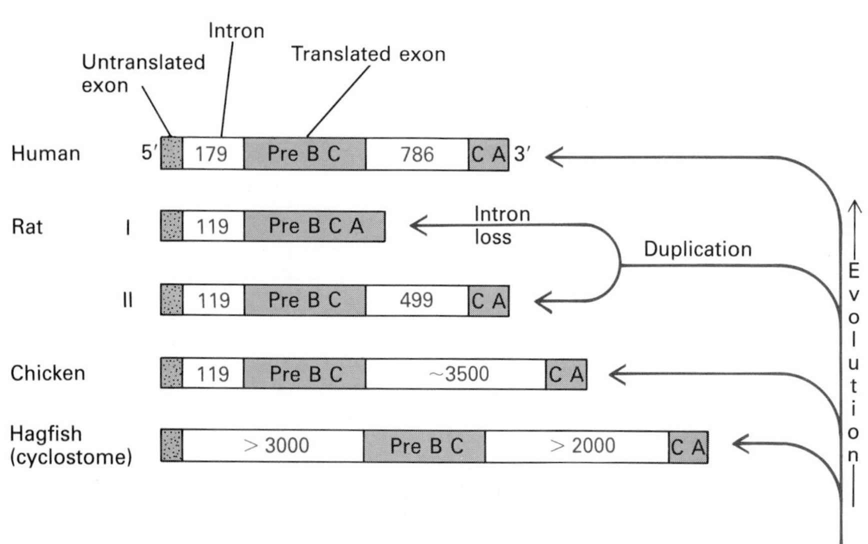

Figure 25-25 A diagram of insulin genes from humans, rats, chickens, and hagfish, which are "living fossils" of the class Agnatha. The symbols pre, B, C, and A indicate the protein-coding regions for the preproinsulin molecule. (Both the pre and the C regions are removed by proteolytic processing.) The phylogenetic relationship (hagfish most distant from humans; chickens more distant than rats) is shown at the right. The existence of two introns in the human gene as well as in the genes of animals evolutionarily older than the rat implies that intron loss was responsible for the single-intron structure of rat gene I. [See F. Perler et al., 1980, *Cell* **20**:555; and G. I. Bell et al., 1980, *Nature* **284**:26.]

indicates a recent gene duplication. One of the rat genes (rat insulin II) has two introns, whereas the other has a single intron. Does this discrepancy represent a gain or a loss of one intron?

Comparisons with other animal genomes have indicated that an intron loss resulting in rat gene I is the correct explanation. Humans, a later evolutionary arrival than rats, and chickens, whose evolutionary lineage has been traced to a time before rats, have only one insulin gene. However, in both species there are two introns in the same places as those in the rat insulin gene II. Even the hagfish, an evolutionarily ancient chordate that lacks a true vertebral column, has an insulin gene with two introns (Figure 25-25). Therefore it seems almost certain that in rats a gene with two introns was duplicated, and that one of the duplicates lost one of its introns. This situation appears to be analogous to the leghemoglobin-globin relationship: recall that the ancient precursor to all of the globin genes may have had three introns, one of which would have been lost in the precursor to the vertebrate globins. In addition, a recently studied globin gene in the water midge *Chironomus thummei* lacks introns altogether. The evolutionary divergence of this animal is clearly after the plant-animal split, which suggests that intron loss can occur in the globin family.

Following the lineage of introns in individual genes throughout evolution does not always give a clear picture, however. For example, actin genes are present in all eukaryotic species, and they have been extensively studied. Because the amino acid sequence of actins is recognizably similar in all species studied to date, it seems likely that one primordial actin gene existed; however, gene structure within this ancient family varies a great deal (Table 25-7). The single yeast actin gene has one intron, whereas all of the multiple actin genes in another lower eukaryote, the slime mold *Dictyostelium discoideum*, have no introns. One sea urchin actin gene has two introns and one has four. Each of several different *Drosophila melanogaster* actin genes has a different intron-exon arrangement. On the other hand, actin genes in mammals and chickens each have five or six introns. Do these distributions argue in favor of lost introns from a primordial gene, or do they support intron introductions during evolution?

When the exact positions of the introns in the actins are considered, some generalizations about the pattern of intron-exon arrangements can be made. There is much overlap between sea urchins, chickens, and mammals in the amino acid codons at which introns are found. Two of the three introns in plant actin genes match the positions of two introns found in some animal actin genes. Thus the existence of an actin gene with multiple introns in a precursor cell to both plants and animals is a reasonable hypothesis. However, it is clearly impossible to argue from the actin genes that introns were not introduced into some of these regions. If a splicing mechanism already existed and proper splice junctions occurred by chance, introns could have been inserted at these sites. It is equally conceivable, though, that the actin gene is among the oldest eukaryotic genes and that it consequently has had plenty of time to lose introns, particularly in rapid growing species; if so, the earliest form of actin may have contained introns at all of the locations that still exist in one species or another today.

Table 25-7 Positions of introns in actin genes

	Intron location*													
Source and type of actin gene	**5′UT**	**4**	**13**	**18**	**41**	**63**	**105**	**121**	**150**	**204**	**267**	**307**	**327**	**353**
Dictyostelium discoideum (several genes)	—	—	—	—	—	—	—	—	—	—	—	—	—	—
Saccharomyces cerevisiae	—	+	—	—	—	—	—	—	—	—	—	—	—	—
Acanthamoeba	—	—	—	—	—	—	+	—	—	—	—	—	—	—
Caenorhabditis elegans	—	—	—	+	—	+	—	—	—	—	—	—	—	—
Drosophila melanogaster Dm A-6	—	—	—	—	—	—	—	—	—	—	—	+	—	—
Drosophila melanogaster Dm A-4	—	—	+	—	—	—	—	—	—	—	—	—	—	—
Drosophila melanogaster Dm A-2	(−8)	—	—	—	—	—	—	—	—	—	—	—	—	—
Drosophila melanogaster Dm A-1	—	—	—	—	—	—	—	—	—	—	—	+	—	—
Sea urchin gene C	—	—	—	—	—	—	—	+	—	+	—	—	—	—
Sea urchin gene J	—	—	—	—	+	—	—	+	—	+	+	—	—	—
Chicken skeletal muscle	(−12)	—	—	—	+	—	—	—	+	+	+	—	+	—
Rat skeletal muscle	(−12)	—	—	—	+	—	—	—	+	+	+	—	+	
Rat β cytoplasmic	(−6)	—	—	—	+	—	—	+	—	—	+	—	+	—
Soybean	—	—	—	+	—	—	—	—	+	—	—	—	—	+
Maize	—	—	—	+	—	—	—	—	+	—	—	—	—	+

*The plusses indicate the codons at which introns are located. The positions of introns in the 5′ untranslated region (UT = untranslated) are indicated by numbers in parentheses (these give the numbers of nucleotides upstream from the initiator ATG). No introns are found at the sites indicated by dashes.

SOURCE: D. Yaffe; see R. Zokut et al., 1982, *Nature* **298**:857.

How Might Introns Be Lost?

The logical mechanism for the complete and exact loss of an intron is the copying of an mRNA (with introns removed, of course) back into DNA. The mRNA might be copied into DNA by reverse transcription followed by the reintegration of the DNA into the genome, or the mRNA might be used directly in gene conversion (Chapter 11). Deletion of an intron by exact DNA recombination at the intron-exon junctions is a less likely alternative. Some form of RNA copying or RNA-directed gene conversion is now recognized in all eukaryotic cells, so the loss of introns can be easily explained by one of these mechanisms.

Intron Insertion Presents Logical Difficulties

The possibility opposing that of intron loss during evolution is intron insertion, for example, insertion of transposition elements. Several circumstances argue against this type of genetic event. First, an unused splicing mechanism would have had to predate instances of intron introduction. In addition, intron insertion would have required precise targeting: at least at the present time, random insertion of introns is not tolerated, because the sequences for intron removal are conserved and single base changes render them unusable. Furthermore, as we have mentioned, the known sequence arrangements associated with DNA transposition (long terminal repeats and target-site duplications) are not found in or near intron sequences. Thus there is no positive evidence to suggest that introns arose by transposition.

There are other logical difficulties with intron insertion. If introns did result from insertions in contiguously coded genes, then at some point in the evolution of eukaryotes the introduction of introns would have had to be a frequent event that then ceased. Why would this have occurred? Moreover, many or most of the insertions into DNA would have required very good aim based on protein structure, because introns usually divide genes into protein domains. Why would insertion divide a collinear, continuously coding gene into such domains? Although it remains possible that some introns represent insertions, it seems unlikely that most do.

Summary: Recruitment of Noncontiguous Coding Regions Formed the First Genes

To review: The likelihood that the recruitment of exons into transcription units before DNA ever arose resulted in noncontiguous genes dating from the beginning of evolution (see Figure 25-13) is based on (1) the existence of protein-free RNA-RNA chemistry, including splicing, (2) evidence that many intron-exon arrangements are very ancient, and (3) the frequent division of DNA by introns into protein-coding domains. This conclusion traces the evolutionary history of eukaryotic cells—which still, of course, maintain many introns in protein-coding genes—back to the progenote stage. As we noted in Figure 25-13, only if a cell puts a premium on rapid growth would efficient mechanisms have developed to remove the large amounts of extra or unused DNA from every genomic region that contained a necessary transcription unit. Thus only fast-growing cells—bacteria and eukaryotes such as yeast—would have lost their ancestral extra DNA.

Loss of DNA from a *population* of organisms that only reproduce sexually must occur in gametes (either gametes being formed during meiosis or immature gametes). The shedding of unused DNA therefore might be slow in higher organisms. If any deleterious effect of the extra DNA were quite small, the evolutionary pressure to lose it would be quite low. The basic principle of genomic design for the slow-growing descendants of the earliest organisms could be a "frozen accident"—a term Crick used to describe the origin of the genetic code. This description is consistent with the fact that once a working cell arose, the code could not be changed without unacceptable damage to existing proteins. If the earliest genomic structure were a functioning set of transcription units containing introns and multiple origins of DNA replication, little advantage would have accrued from a change in the pattern, and the "accident" of the first successful arrangement might have remained "frozen" in the cells of more recently evolved organisms.

Cell Evolution: Biochemical Capacities

Once functioning, replicating cells arose, the whole marvelously diverse process of biological evolution could begin. The evolution in biological forms accompanied equally dramatic changes in the surface of the earth. The presumed primordial soup may have been fairly quickly used up; the resulting environmental changes would have exerted the first strong evolutionary pressure and created the first major niche in which evolutionary success would be rewarded. Cells that could synthesize some of the dwindling supplies of nutrients would have had a large advantage. The increased synthesis of substances previously in abundant supply would have required an increase in energy expenditure, however. The imperative to improve the efficiency and the flexibility of energy transduction probably led to several quite distinct steps in evolution.

The rRNA sequence comparisons have provided strong evidence that the oldest bacterial cells were anaerobic. One of the strongest early evolutionary pressures must have been for the development of facile energy transducing systems that could use the major available energy, sunlight.

Many different photosynthetic microorganisms are known today. Phylogenetic comparisons of their rRNAs have established several different cells in which photosynthesis arose. Some of the oldest lineages (e.g., photosynthetic anaerobes) in the eubacterial tree belong to these groups. Organisms that produced oxygen as by-products presumably changed the entire ecological system by making aerobic metabolism beneficial. These two biochemical advances—the development of photosynthesis and the development of oxygen utilization—were necessary before the construction of present-day eukaryotic cells could be achieved.

The Endosymbiotic Hypothesis Is Confirmed by rRNA Analysis

As we noted earlier, chloroplasts have rRNA sequences that are similar to the rRNA sequences of the cyanobacteria, whereas the plant cell nuclear rRNA sequences are different. Almost certainly, therefore, a primitive cyanobacterium was the progenitor of chloroplasts. The union between the eukaryotic cell precursor and the cyanobacterium occurred at least 1 to 1.5 billion years ago. Because both rRNA lineages are maintained in distinct forms today in plant cells, it seems reasonable to assume that the antecedents to the nuclear and the chloroplast ribosomal genes existed independently for some period of time before their fusion. However, the chloroplast genome is only about 10 percent (or less) the size of a present-day prokaryotic genome. The fate of the "lost" genes is unknown.

The ribosomes of plant mitochondria are similar in sequence and two-dimensional stem-loop structure (see Figure 25-9) to those of purple sulfur bacteria. It is therefore logical to propose the union of a eukaryotic precursor with a progenitor of an aerobic bacterium (or at least with some of the genes of the bacterium) as a source of genes whose products carry out oxidative phosphorylation. However, the overall sequences of mitochondrial DNAs vary considerably. More than one event of union between a prokaryotic mitochondrial precursor and the eukaryotic precursor may have occurred. For example, the DNA in mitochondria of yeast and mammals is quite different (see Figure 21-14). Alternatively, rearrangements after fusion might be sufficient to explain the variation in DNA sequences.

The origin of the nucleus in eukaryotic cells is much more speculative than the origin of the major DNA-containing organelles. A frequent textbook hypothesis is that the nucleus arose to provide better control of cell function, and perhaps this is true. The development of a separate cell compartment for the performance of difficult mechanical tasks seems plausible. For example, organizing all of the several millimeters of DNA for replication is a task that might be easier in a separate cell compartment. RNA processing, including splicing as the final step of mRNA formation, might also be easier without the interference of elements of the protein synthesis system. However, as yet we know so little about the relationship between nuclear structure and nuclear regulatory tasks (as opposed to mechanical tasks) that we cannot guess why or when in evolution the nucleus arose.

Cell Structure Is Conserved during Evolution

The development of the electron microscope led to the highly significant discovery that all eukaryotic plant and animal cells contain similar intracellular structures. Recent experiments with specific antisera have illustrated the relatedness of the proteins in many of these structures and provided additional evidence for the idea of a common evolutionary ancestry for all eukaryotic cells (Chapters 18–21); this work further emphasizes the separation between eukaryotes and prokaryotes. For example, although actins from yeast, plant, and animal cells are recognized by certain antisera, neither chloroplasts, mitochondria, nor bacteria contain any such cross-reacting proteins. Thus either the prokaryotes and the eukaryotic organelles have lost these common proteins, or the progenote had not yet developed them. The presence of the actins in both plants and animals means that they must have been present in a cell that was a common ancestor to the eukaryotes.

Study of the Evolution of Multicellular Organisms

Prebiotic chemistry and the original RNA reactions that may underlie the beginnings of cellular evolution may be buried so far in the past that a satisfying molecular description of this phase of evolution will never be achieved. However, the modern techniques of DNA sequencing and chromosome analysis will have an immense impact on the traditional evolutionary problems such as speciation and the progression of changes that allow us to distinguish between related species.

For example, consider the fact that animals such as humans and higher primates, whose last common ancestor must have existed more than 10 million years ago, have very similar chromosomal structures; differences can often be accounted for by simple changes such as inversions or translocations (Figure 25-26). The DNA sequences for the same protein-coding gene are very similar in different but related species. In addition, detailed restriction maps within large homologous regions such as the entire 60-kb globin cluster are very similar between humans and the great apes (Figure 25-27). If great tracts

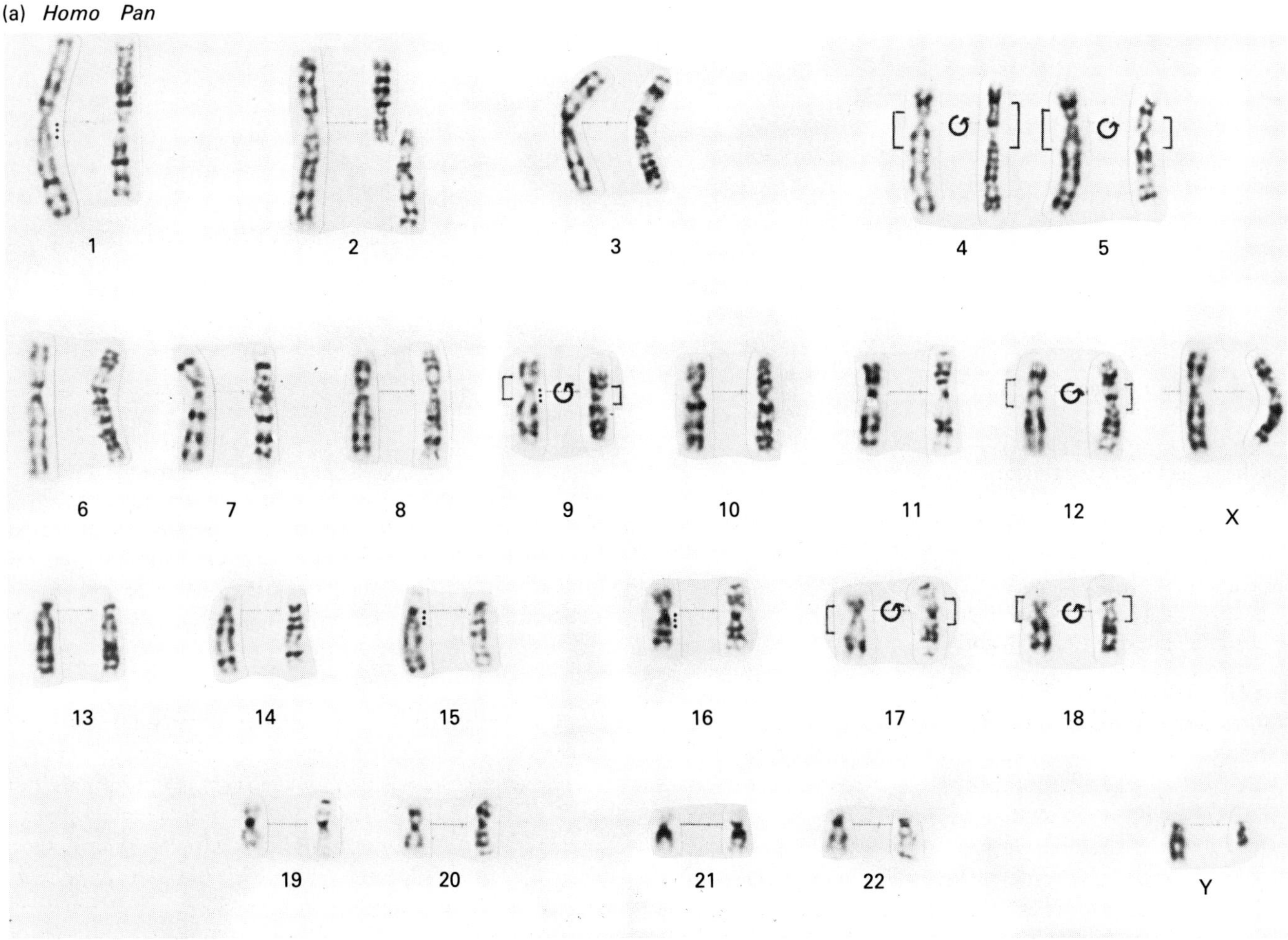

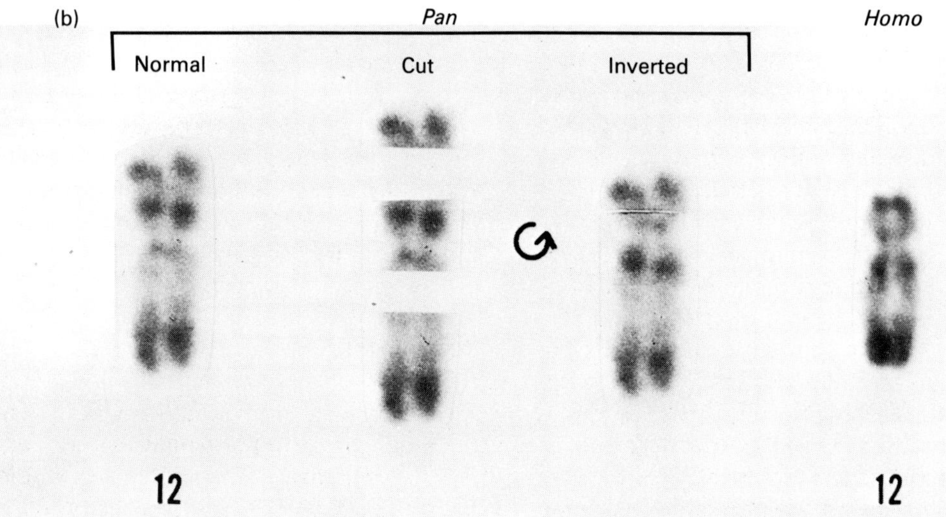

Figure 25-26 A comparison of human and chimpanzee chromosomes. (a) The photographs *(left: Homo sapiens; right: Pan paniscus)* illustrate the great similarity between the 23 human and chimpanzee chromosomes. The chromosomes were stained by the G banding technique (Chapter 10). Brackets and circular arrows indicate inverted sections; triple vertical dots indicate the acquisition of heterochromatic material. (b) A higher magnification of the number 12 chromosomes shows how cutting and inverting the chimpanzee chromosome produces a banding pattern identical to that of the human chromosome. [See J. de Grouchy, C. Turleau, and C. Finaz, 1978, *Annu. Rev. Genet.* **12**:289.] *Courtesy of J. de Grouchy. Reproduced, with permission, from the* Annual Review of Genetics. *Copyright 1978 by Annual Reviews Inc.*

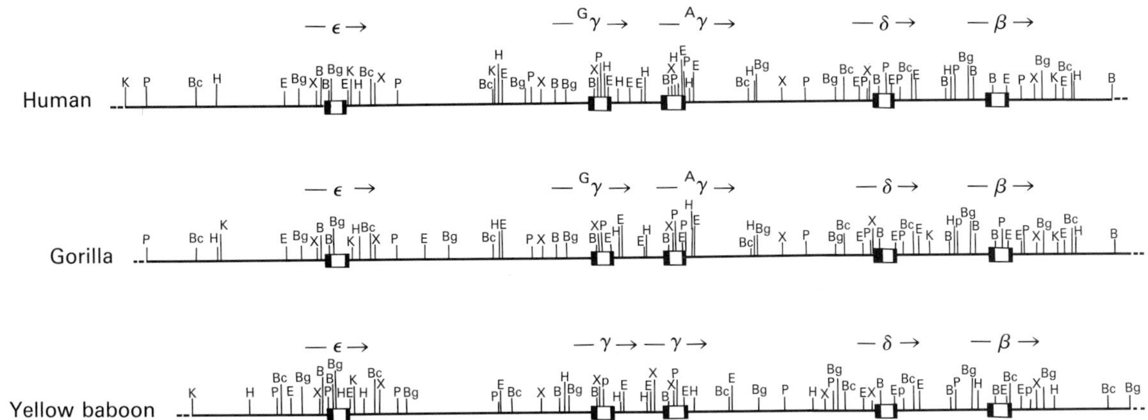

Figure 25-27 A map of the β-like globin region of humans and two primates. Included are genes expressed in embryos (ϵ and γ) as well as those of late-term fetuses and adults (δ and β). Two solid and one open box represent each gene. Actually there are two introns and three exons in each gene; see Figure 25-20. Each letter represents a restriction enzyme site. The gorilla and the yellow baboon are estimated to have diverged from humans 6 and 20 million years ago, respectively; yet the restriction sites are highly conserved. For example, 65 out of the 70 sites shown here are in the same relative positions in the humans as in the gorilla. *After P. A. Barrie, A. J. Jeffreys, and A. F. Scott, 1981,* J. Mol. Biol. **149**:*319.*

of mammalian chromosomes are similar, limited genetic changes may be responsible for what appears to be a considerable degree of evolution since the time of the common ancestor to humans and apes. With the identification of genes that are important in early embryogenesis and comparisons of such genes for two closely related species, perhaps a clearer picture of the genetic changes that are central to major evolutionary changes will emerge.

It should be remembered that speciation (defined as the evolution of organisms into groups that cannot interbreed to produce fertile offspring) does not always correlate with obvious or visible structural changes. For example, visually indistinguishable sibling species of fruit flies that do not interbreed can have very similar chromosome structures. Changes in simple sequence DNAs and in transposable DNAs between sibling species may be responsible for the reproductive incompatibility of these species. Further, the *Alu* DNA—the repetitive sequences that are present hundreds of thousands of times in the genomes of mammals—are not identical even between closely related animals. These DNA differences do correlate with evolution, but they certainly cannot yet be considered the basis for the separation of all species. Our understanding of how sequence changes are responsible for speciation will depend on investigations into how DNA changes may prevent interbreeding (on the one hand) and how genes regulate form and structure (on the other).

Although the rate of change in DNA that encodes protein may be nearly constant throughout evolution, the observable evolutionary changes do not appear to vary strictly proportionately with time. According to the fossil record, frogs have remained structurally similar—and, presumably, functionally similar—for the past 70 million years, whereas an enormous variety of mammals have evolved during the same period. However, frogs and mammals have been accumulating structural gene mutations at about the same rate. The facile explanation is that mutations in regulatory genes are responsible for evolution. Because control genes have not been adequately defined for eukaryotic cells, and specifically for vertebrate cells, this concept remains speculative.

Currently a major task for molecular cell biology is to define regulatory elements and then to pursue comparative studies of these elements for animals that are closely related phylogenetically. Thus population biology, genetics, and molecular cell biology all merge in pursuit of answers to evolution, which is perhaps the most fundamental of all biological problems.

Summary

When simple inorganic compounds are exposed to conditions designed to approximate those that probably existed on primitive earth, amino acids and nucleotides are produced. This result supports the Haldane-Oparin proposal that a "primordial soup" of organic compounds existed on earth around 3.5 billion years ago. A number of protein-free (i.e., nonenzymatic) reactions that could have been decisive in the development of the first biologically important polymers have been discovered. These reactions include RNA oligonucleotide formation from activated monomeric precursors, template-directed RNA synthesis, site-specific RNA cleavage (with and without the aid of a second RNA molecule), and finally RNA-RNA splicing. Thus RNA may have been the first polymer, and some form of reverse transcription may have given rise later to DNA. How the coordination between the RNA oligonucleotides and amino acids or peptides occurred remains unknown, but the result was a single universal three-letter code that is still used by all cells today. (Mitochondria utilize some different code words; these could be remnants of a primitive code.)

Comparisons between the rRNAs of many different cell types as well as comparison between bacterial rRNAs and the rRNAs of organelles (chloroplasts and mitochondria) have suggested two important conclusions about evolution. First, the overall two-subunit structure of ribosomes is universal, and the secondary (stem-loop) structure is extremely similar in all rRNAs. Thus the ribosome can be considered a highly preserved but fairly complicated early cellular structure. Nucleotide sequence analyses of rRNAs show three large groups of cells: eubacteria (the common soil and disease-carrying organisms), archaebacteria (methanogenic and halophilic bacteria), and eukaryotic cells. The rRNA sequence analyses have indicated that each of these lineages is separate from the other, and that none of the three is a likely progenitor of the other two. Rather, all may have descended from an earlier form. Finally, the studies show that mitochondria and chloroplast rRNAs are extremely similar to bacterial rRNAs.

The discovery of intervening sequences (mainly in eukaryotic genomes) has given rise to much speculation about the nature of the earliest genes. Because protein-free RNA reactions include cutting and splicing, and because spliced gene structures are relatively well preserved over enormous evolutionary intervals (i.e., more than a billion years), it is possible that the first genes had noncontiguous coding regions and that splicing was a prominent feature of the most primitive genomes. According to this view, cells lacking introns may have lost them due to selective pressures associated with rapid growth, whereas eukaryotic precursors were slower growers that did not need to lose their introns. The antiquity of the eukaryotic lineage is supported by the rRNA lineage studies.

Whatever the nature of the first genes, the first functioning cells were probably heterotrophic and anaerobic. Significant early biochemical changes included the development of photosynthetic metabolism and the adaptation to dwindling nutrient supplies. The earliest microfossil remains are similar in size to today's cyanobacteria (blue-green algae), which utilize sunlight and produce oxygen. Microfossil studies and geologic studies have indicated that eukaryotic cells containing a nucleus and energy-deriving organelles had arisen by 1.5 billion years ago. Analyses of proteins common to both organelles and bacteria as well as comparisons of rRNAs have left little doubt that chloroplast DNA was derived from cyanobacteria and that mitochondrial DNA is a remnant of aerobic bacteria.

Finally, the evolution of multicellular organisms from single-celled organisms occurred more than a billion years ago. The fossil record of the subsequent periods of evolutionary change begins about 600 million years ago. The structures of many genes have been preserved over this entire period, and there is still great similarity in the chromosome locations of similar blocks of genes in all primates (for example). Thus it is very likely that continued studies of chromosomes will pinpoint key structural changes that are associated with major evolutionary changes.

References

General References

BRODA, E. 1975. *The Evolution of the Bioenergetic Process.* Pergamon.

DOVER, G. A., and R. B. FLAVELL. 1983. *Genome Evolution.* Academic Press.

KIMURA, M. ed. 1978. *Evolution* (a Scientific American book). W. H. Freeman and Company.

KIMURA, M. 1984. *The Neutral Theory of Molecular Evolution.* Cambridge University Press.

MARGULIS, L., and K. V. SCHWARTZ. 1982. *Five Kingdoms.* W. H. Freeman and Company.

Prebiotic Synthesis

HALDANE, J. B. S. 1929. The origin of life. Reprinted in *On Being the Right Size,* J. M. Smith, ed. 1985, Oxford.

MILLER, S. L., and L. E. ORGEL. 1973. *The Origins of Life on Earth.* Prentice-Hall.

OPARIN, A. I. 1974. *Evolution of the Concepts on the Origin of Life: Seminar on the Origin of Life.* Moscow.

SCHOPF, J. W., ed. 1983. *Earth's Earliest Biosphere: Its Origin and Evolution.* Princeton University Press.

The Origins of the Genetic Code and the Translation Apparatus

CRICK, F. H. C. 1968. The origin of the genetic code. *J. Mol. Biol.* 38:367–379.

CRICK, F. H. C., S. BRENNER, A. KLUG, and G. PIECZNIK. 1976. A speculation on the origin of protein synthesis. *Origins of Life* 7:389–397.

EIGEN, M., and P. SCHUSTER. 1979. *The Hypercycle.* Springer-Verlag.

LAGERKVIST, U. 1981. Unorthodox codon reading and the evolution of the genetic code. *Cell* **23**:305–306.

ORGEL, L. E. 1986. Evolution of the genetic apparatus. *J. Mol. Biol.* 38:381–393.

WOESE, C. 1967. *The Origins of the Genetic Code.* Harper & Row.

A Reconstructive Analysis of Cell Lineages

Microfossils

KNOLL, R. H., and E. S. BARGHOORN. 1977. Archean microfossils showing cell division from the Swaziland system of South Africa. *Science* **198**:396–398.

VIDAL, G. 1984. The oldest eukaryotic cells. *Sci. Am.* **250**(2):48–58.

WALSH, M. M., and D. R. LOWE. 1985. Filamentous microfossils from the 3,500-Myr-old Onverwacht Group, Barberton Mountain Land, South Africa. *Nature* **314**:530–532.

Comparative Studies of rRNA Structures

FOX, G. E., E. STACKEBRANDT, R. B. HESPELL, J. GIBSON, J. MANILOFF, T. A. DYER, R. S. WOLFE, W. E. BALCH, R. S. TANNER, L. J. MAGRUM, L. B. ZABLEN, R. BLAKEMORE, R. GUPTA, L. BONEN, B. J. LEWIS, D. A. STAHL, K. R. LUEHRSEN, K. N. CHEN, and C. R. WOESE. 1980. The phylogeny of prokaryotes. *Science* **209**:457–463.

GUTELL, R. R., B. WEISER, C. R. WOESE, and H. F. NOLLER. 1985. Comparative anatomy of 16S-like ribosomal RNA. *Prog. Nucl. Acids. Res.* **32**:155–216.

NOLLER, H. F. 1984. Structure of ribosomal RNA. *Annu. Rev. Biochem.* **53**:119–162.

NOLLER, H. F., J. KOP, V. WHEATON, J. BROSIUS, R. GUTELL, A. M. KOPYLOV, F. DOHME, W. HERR, D. A. STAHL, R. GUPTA, and C. R. WOESE. 1981. Secondary structure model for 23S ribosomal RNA. *Nuc. Acids Res.* **9**:6167–6189.

NOLLER, H. F., and C. R. WOESE. 1981. Secondary structure of 16S ribosomal RNA. *Science* **212**:403–411.

WOESE, C. R. 1981. Archaebacteria. *Sci. Am.* **244**(6):98–125.

WOESE, C. R. 1983. The primary lines of descent and the universal ancestor. In *Evolution from Molecules to Man,* D. S. Bendall, ed. Cambridge University Press, pp. 209–233.

WOESE, C. R., R. R. GUTELL, R. GUPTA, and H. F. NOLLER. 1983. Detailed analysis of the higher-order structure of 16S-like ribosomal ribonucleic acids. *Microbiol. Rev.* **47**:621–669.

The Organization of the First Genome

DARNELL, J. E. 1978. Implications of RNA-RNA splicing in evolution of eukaryotic cells. *Science* **202**:1257–1260.

DARNELL, J. E. 1981. Do features of present-day eukaryotic genomes reflect ancient sequence arrangements? In *Evolution Today: Proceedings of the Second International Congress of Systematic and Evolutionary Biology,* G. G. E. Scudder and J. L. Reveal, eds. Hunt Inst. for Botanical Documentation, pp. 207–213.

DOOLITTLE, W. F. 1978. Genes in pieces: Were they ever together? *Nature* **272**:581–582.

DOOLITTLE, W. F. 1980. Revolutionary concepts in evolutionary cell biology. *Trends Biochem. Sci.* **5**:146–149.

REANNEY, D. 1974. On the origin of prokaryotes. *J. Theor. Biol.* **48**:243–251.

REANNEY, D. 1979. RNA splicing and polynucleotide evolution. *Nature* **277**:598–600.

RNA-Directed RNA Synthesis, Site-Specific Cleavage, and Splicing

BELFORT, M., J. PEDERSEN-LANE, D. WEST, K. EHRENMAR, G. MALEY, F. CHU, and F. MALEY. 1985. Processing of the intron-containing thymidylate synthease (*td*) gene of phage T4 is at the RNA level. *Cell* **41**:375–382.

BIRD, C. R., B. KOLLER, A. D. AUFFRET, A. K. HUTTLY, C. H. HOWE, T. A. DYER, and J. C. GRAY. 1985. The wheat chloroplast gene DF_0 subunit I of ATP synthase contains a large intron. *EMBO J.* **4**:1381–1388.

CECH, T. R. 1983. RNA splicing: three themes with variations. *Cell* **34**:713–716.

DAVIS, R. W., R. B. WARING, J. A. RAY, T. A. BROWN, and C. SCAZZOCCHIO. 1982. Making ends meet: a model for RNA splicing in fungal mitochondria. *Nature* **300**:719–724.

GRIVELL, L. A., L. BONEN, and P. BORST. 1983. Mosaic genes and RNA processing in mitochondria. In *Genes: Structure and Expression,* A. M. Kroon, ed. Wiley, pp. 279–306.

GUERRIER-TAKADA, C., K. DARDINER, T. MARSH, N. PACE, and S. ALTMAN. 1983. The RNA moiety of ribonuclease P is the catalytic sununit of the enzyme. *Cell* **35**:849–857.

INOUE, T., and L. E. ORGEL. 1983. A nonenzymatic RNA polymerase model. *Science* **219**:859–862.

SHARP, P. A. 1985. On the origin of RNA splicing and introns. *Cell* **42**:397–400.

ZAUG, A. J., and T. R. CECH. 1985. Oligomerization of intervening sequence RNA molecules in the absence of proteins. *Science* **229**:1060–1064.

ZAUG, A. J., and T. R. CECH. 1986. The intervening sequence RNA of *Tetrahymena* is an enzyme. *Science* **231**:470–475.

Evolutionary Implications of the Positions of Introns in Genes

ALEXANDER, F., P. R. YOUNG, and S. M. TILGHMAN. 1984. Evolution of the albumin: α-fetoprotein ancestral gene from the amplification of a 27 nucleotide sequence. *J. Mol. Biol.* **173**:159–176.

EFSTRATIADIS, A., J. W. POSAKONY, T. MANIATIS, R. M. LAWN, C. O'CONNELL, R. A. SPRITZ, J. K. DERIEL, B. G. FORGET, S. M. WEISSMAN, J. L. SLIGHTOM, A. E. BLECHL, O. SMITHIES, F. E. BARALLE, C. C. SHOULDERS, and N. J. PROUDFOOT. 1980. The structure and evolution of the human β-globin gene family. *Cell* **21**:653–668.

KAFATOS, F. C. 1983. Structure, evolution, and developmental expression of the chorion multigene families in silk moths and *Drosophila.* In *Gene Structure and Regulation in Development,* S. Subtelny and F. C. Kafatos, eds. Alan R. Liss.

Introns Separate Protein Domains and Exons "Shuffle"

BEYCHOK, S. 1984. Exons and domains in relation to protein folding. In *Protein Folding,* D. B. Wetlaufer, ed. AAAS Selected Symp. 89, pp. 145–175.

BLAKE, C. C. F. 1984. Exons—Present from the beginning? *Nature* 306:535–537.

GILBERT, W. 1978. Why genes in pieces? *Nature* **271**:501.

GILBERT, W. 1985. Genes in pieces revisited. *Science* **228**:823–824.

GŌ M. 1981. Correlation of DNA exonic regions with protein structural units in haemoglobin. *Nature* **291**:90–92.

GŌ, M. 1983. Modular structural units, exons, and function in chicken lysozyme. *Proc. Nat'l Acad. Sci. USA* **80**:1964–1968.

JUNG, A., A. E. SIPPEL, M. GREZ, and G. SCHUTZ. 1980. Exons encode functional and structural units of chicken lysozyme. *Proc. Nat'l Acad. Sci. USA* 77:5759–5763.

MICHELSON, A. M., C. C. F. BLAKE, S. T. EVANS, and S. H. ORKIN. 1985. Structure of the human phosphoglycerate kinase gene and the intron-mediated evolution and dispersal of the nucleotide-building domain. *Proc. Nat'l. Acad. Sci. USA* **82**:6965–6969.

STONE, E. M., K. N. ROTHBLUM, M. C. ALEVY, T. M. KUO, and R. J. SCHWARTZ. 1985. Complete sequence of the chicken glyceraldehyde-3-phosphate dehydrogenase gene. *Proc. Nat'l Acad. Sci. USA* **82**:1628–1632.

STRAUS, D., and W. GILBERT. 1985. Genetic engineering in the Pre-cambrian: Structure of the chicken triosephosphate isomerase gene. *Mol. Cell. Biol.* 5:3497–3506.

Intron Loss

BALTIMORE, D. 1985. Retroviruses and retrotransposons: the role of reverse transcriptase in shaping the eukaryotic genome. *Cell* **40**:481–482.

BOEKE, J. D., D. J. GARFINKEL, C. A. STYLES, and G. R. FINK. 1985. Ty elements transpose through an RNA intermediate. *Cell* **40**:491–500.

PERLER, F., A. EFSTRATIADIS, P. LOMEDICO, W. GILBERT, R. KOLODNER, and J. DODGSON. 1980. The evolution of genes: the chicken preproinsulin gene. *Cell* **20**:555–556.

Cell Evolution: Biochemical Capacities

The Endosymbiotic Hypothesis

KUNTZEL, H., and H. G. KOCHEL. 1981. Evolution of rRNA and origin of mitochondria. *Nature* **293**:751–755.

MARGULIS, L. 1981. *Symbiosis in Cell Evolution.* W. H. Freeman and Company.

MARGULIS, L. 1971. The origin of plant and animal cells. *Am. Scientist* 59:230.

SCHWARTZ, R. M., and M. O. DAYHOFF. 1978. Origins of prokaryotes, eukaryotes, mitochondria, and chloroplasts. *Science* **199**:395–403.

YANG, D., Y. OYAIZU, H. OYAIZU, G. J. OLSEN, and C. R. WOESE. 1985. Mitochondrial origins. *Proc. Nat'l Acad. Sci. USA* **82**:4443–4447.

The Evolution of Multicellular Organisms

DE GROUCHY, J., C. TURLEAU, and C. FINAZ. 1978. Chromosomal phylogeny of the primates. *Annu. Rev. Genet.* **12**:289–328.

MARTIN, S. L., K. A. VINCENT, and A. C. WILSON. 1983. Rise and fall of the delta globin gene. *J. Mol. Biol.* **164**:513–528.

OHTA, T. 1983. On the evolution of multigene families. *Theor. Population Biol.* **23**:216–240.

ROSE, M. R., and W. F. DOOLITTLE. 1983. Molecular biological mechanisms of speciation. *Science* **220**:157–162.

WILSON, A. 1985. The molecular basis of evolution. *Sci Amer.* 253(4)164–173.

Index

Page numbers in **bold face** refer to a major discussion of the entry. F after a page number indicates a figure; T after a page number indicates a table.